ASCE STANDARD

ASCE/SEI

7-22

Minimum Design Loads and Associated Criteria for Buildings and Other Structures

PROVISIONS

STRUCTURAL ENGINEERING INSTITUTE

PUBLISHED BY THE AMERICAN SOCIETY OF CIVIL ENGINEERS

Library of Congress Cataloging-in-Publication Data

Library of Congress Control Number: 2021951104

Published by American Society of Civil Engineers
1801 Alexander Bell Drive
Reston, Virginia, 20191-4382
www.asce.org/bookstore | ascelibrary.org

This standard was developed by a consensus standards development process which has been accredited by the American National Standards Institute (ANSI). Accreditation by ANSI, a voluntary accreditation body representing public and private sector standards development organizations in the United States and abroad, signifies that the standards development process used by ASCE has met the ANSI requirements for openness, balance, consensus, and due process.

While ASCE's process is designed to promote standards that reflect a fair and reasoned consensus among all interested participants, while preserving the public health, safety, and welfare that is paramount to its mission, it has not made an independent assessment of and does not warrant the accuracy, completeness, suitability, or utility of any information, apparatus, product, or process discussed herein. ASCE does not intend, nor should anyone interpret, ASCE's standards to replace the sound judgment of a competent professional, having knowledge and experience in the appropriate field(s) of practice, nor to substitute for the standard of care required of such professionals in interpreting and applying the contents of this standard.

ASCE has no authority to enforce compliance with its standards and does not undertake to certify products for compliance or to render any professional services to any person or entity.

ASCE disclaims any and all liability for any personal injury, property damage, financial loss, or other damages of any nature whatsoever, including without limitation any direct, indirect, special, exemplary, or consequential damages, resulting from any person's use of, or reliance on, this standard. Any individual who relies on this standard assumes full responsibility for such use.

Errata: Errata, if any, can be found at https://doi.org/10.1061/9780784415788.

ISBN 978-0-7844-1578-8 (soft cover)
ISBN 978-0-7844-8349-7 (PDF)
ASCE 7 Hazard Tool: https://asce7hazardtool.online/
Online platform: http://ASCE7.online
Manufactured in the United States of America.

27 26 25 24 23 22 1 2 3 4 5

ASCE STANDARDS

In 2016, the Board of Direction approved revisions to the ASCE Rules for Standards Committees to govern the writing and maintenance of standards developed by ASCE. All such standards are developed by a consensus standards process managed by the ASCE Codes and Standards Committee. The consensus process includes balloting by a balanced standards committee and reviewing during a public comment period. All standards are updated or reaffirmed by the same process every five or six years. Requests for formal interpretations shall be processed in accordance with Section 7 of ASCE Rules for Standards Committees, which are available at www.asce.org. Errata, addenda, supplements, and interpretations, if any, for this standard can also be found at www.asce.org. This standard has been prepared in accordance with recognized engineering principles and should not be used without the user's competent knowledge for a given application. The publication of this standard by ASCE is not intended to warrant that the information contained therein is suitable for any general or specific use, and ASCE takes no position respecting the validity of patent rights. The user is advised that the determination of patent rights or risk of infringement is entirely their own responsibility. A complete list of currently available standards is available in the ASCE Library (https://ascelibrary.org/).

Tips for Using This Standard

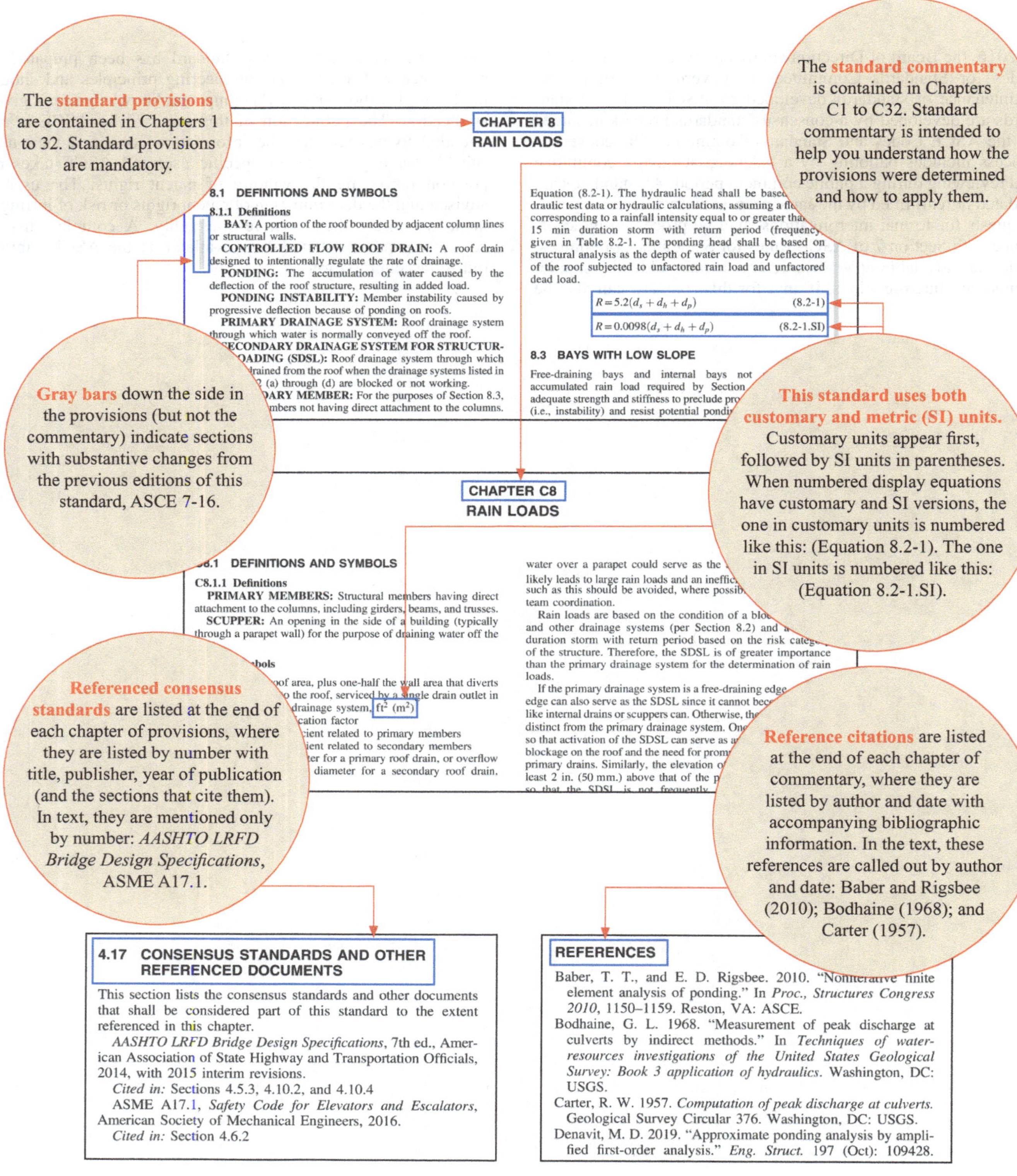

Complete hazard data in ASCE 7-22 is provided free to the user in the ASCE 7 Hazard Tool (https://asce7hazardtool.online/). See next page for more details.

Supplements, errata, and interpretations may become available in the future. Please check for important new materials at https://doi.org/10.1061/9780784415788.

Tips for Using the ASCE 7 Hazard Tool

asce7hazardtool.online

The ASCE 7 Hazard Tool provides access to the digital data defined in the hazard Geodatabases required by this standard. The digital data required for snow, seismic, and tornado are available at https://asce7hazardtool.online/, and digital data is available for flood, rain, ice, and wind, as well. Digital data required for tsunami is available at https://asce7tsunami.online/.

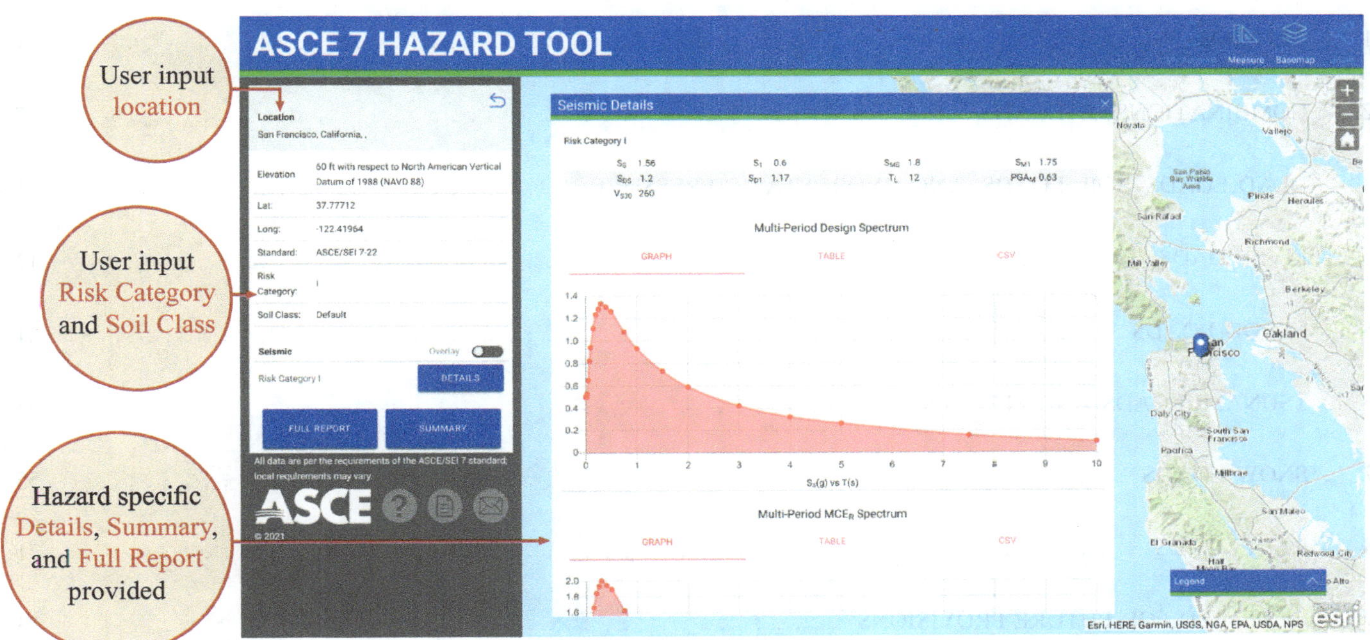

Digital Data

The ASCE 7 Hazard Tool provides digital data required by ASCE 7-22:

- Ch. 5 Flood: Flood zone and static base flood elevation, plus direct links to additional information
- Ch. 6 Tsunami: Whether the site is in a mapped tsunami design zone per the ASCE Tsunami Design Geodatabase, and link to ASCE Tsunami Design Geodatabase if required for design
- Ch. 7 Snow: Ground snow load and winter wind parameter
- Ch. 8 Rain: Median 15-minute and 60-minute duration rainfall intensities for 100-year mean recurrence interval
- Ch. 10 Ice: Radial ice thickness with concurrent 3-second gust speeds and temperature concurrent with ice thickness due to freezing rain
- Ch. 22 Seismic: Seismic coefficients S_S, S_1, S_{MS}, S_{M1}, S_{DS}, S_{D1}, T_L, PGA_M, and V_{S30}, plus the seismic design category, as well as the multi-period spectrum, the multi-period MCE_R spectrum, the two-period design spectrum, and the two-period MCE_R spectrum
- Ch. 26 Wind: Three-second gust wind speeds at 33 feet (10 meters) above ground for Exposure Category C, including identification of hurricane-prone and wind-borne debris regions
- Ch. 32 Tornado: Tornado wind speeds for 1,700-, 3,000-, 10,000-, 100,000-, 1,000,000-, and 10,000,000-year MRI, and for 1-, 2,000-, 10,000-, 40,000-, 100,000-, 250,000-, 1,000,000-, and 4,000,000-ft_2. target areas

BRIEF CONTENTS

COMMENTARY TO STANDARD ASCE/SEI 7-22

CONTENTS

COMMENTARY TO STANDARD ASCE/SEI 7-22

Commentary contents appear in second book

PREFACE

Prepared by the Minimum Design Loads and Associated Criteria for Buildings and Other Structures Standards Committee of the Codes and Standards Activity Division of the Structural Engineering Institute of ASCE

ASCE/SEI 7-22 Minimum Design Loads and Associated Criteria for Buildings and Other Structures provides the most up-to-date and coordinated loading provisions for general structural design. ASCE 7-22 prescribes design loads for all hazards including dead, live, soil, flood, tsunami, snow, rain, atmospheric ice, seismic, wind, and fire, as well as how to evaluate load combinations. The 2022 edition of ASCE 7, which supersedes ASCE 7-16, coordinates with the most current structural material standards including those from ACI, AISC, AISI, AWC, and TMS.

Significant technical changes include the following:

- General Requirements
  - New target reliability tables for tsunami and extraordinary loads
  - Removal of importance factors for snow and ice due to risk category specific maps being provided
  - Expanded provisions for in situ load testing
- Load Combinations
  - Revised load combinations to reflect changes in snow loads and new tornado loads
  - New alternative method for loads from water in soil
  - Load combinations for flood loads and atmospheric ice are now explicitly written out and numbered for improved reference and clarity
- Dead and Live Loads
  - Reformatted lateral soil loads table for improved clarity
  - New alternative method for loads from water in soil
  - Terminology change from guardrail system to guard system
  - Additions and clarifications to the live load table
  - Updated crane load vertical impact force provisions including the use of bridge crane service classes
  - New provisions for emergency vehicle loads
- Tsunami Loads and Effects
  - Clarification for inundation calculations for overwashed areas
  - Updated data for Hawaii and many populous locations in California, coordinated with the state agencies
  - New provisions for above-ground horizontal pipelines
  - Clarifications and new provisions for debris impact analysis
  - New provisions for loss of foundation strength and scour
- Snow Loads
  - Ground snow loads have been revised to reflect more recent snow load data and reliability-targeted values
  - Method for estimating drifts revised to include a wind parameter
  - A more accurate estimation of the horizontal extent of windward drifts
  - Revised thermal factors to account current trends in roof insulation and venting
- Rain Loads
  - Design rain load revised to explicitly consider a ponding head
  - New commentary for low slope roofs and drainage to existing roofs
- Atmospheric Ice Loads
  - New risk-targeted atmospheric ice load data for the continental United States and Alaska
- Seismic Design
  - Multi-period response spectrum data eliminates need for F_a and F_v coefficients
  - Increase in number of site class definitions
  - Updated provisions for two-stage analysis procedures
  - Updated provisions for calculating torsion impacts, including irregularities (new Torsional Irregularity Ratio (TIR) term] and accidental torsion
  - Updated directional loading provisions
  - Updated analysis procedure selection provisions
  - Updated displacement and drift provisions
  - Updated force equations for nonstructural components
  - New provisions for penthouses and equipment and distribution system support structures
  - New Lateral Force Resisting Systems:
    - Steel and Concrete Coupled Composite Plate Shear Walls
    - Reinforced Concrete Ductile Coupled Shear Walls
    - Cross-laminated Timber Shear Walls
    - Concrete Tabletop Structures
  - New provisions for Rigid Wall, Flexible Diaphragm buildings (big box stores/warehouse)
  - New and updated provisions for supported and interconnected (coupled) nonbuilding structures
- Wind Design
  - Updates to the wind speed maps along hurricane coastline
  - Removal of tabular methods for both the directional and the envelope procedures, and C&C
  - New provisions for MWFRS and C&C of elevated buildings
  - Updated and expanded provisions for roof and ground-mounted solar
  - Updated provisions grouped circular bins and tanks
  - Revisions to the (GC_p) graphs for external pressure coefficients on C&C
  - Updates for wind tunnel testing and adoption of new edition of ASCE 49-22
  - New chapter for tornado provisions
  - New long return period hazard maps for wind and tornado
- Digital Data Available for all Hazards
  - Required to use digital data for tsunami, snow, seismic
  - Provided for flood, rain, ice, wind, tornado

In addition to the technical changes, the 2022 edition of the ASCE 7 provisions are accompanied by a detailed commentary with explanatory and supplementary information developed to assist users of the standard, including design practitioners, building code committees, and regulatory authorities.

ASCE 7 is an integral part of building codes in the United States and around the globe and is adopted by reference into the International Building Code, International Existing Building Code, International Residential Code, and NFPA 5000 Building Construction and Safety Code. Structural engineers, architects, and those engaged in preparing and administering local building codes will find the structural load requirements essential to their practice.

ACKNOWLEDGMENTS

The American Society of Civil Engineers (ASCE) acknowledges the work of the Minimum Design Loads and Associated Criteria for Buildings and Other Structures Standards Committee of the Codes and Standards Activities Division of the Structural Engineering Institute. This group comprises individuals from many backgrounds, including consulting engineering, research, construction industry, education, government, design, and private practice.

This revision of the standard began in 2017 and incorporates information as described in the preface. This standard was prepared through the consensus standards process by balloting in compliance with procedures of ASCE's Codes and Standards Activities Committee. The individuals who serve on the Standards Committee are listed as follows.

ASCE 7-22 Minimum Design Loads and Associated Criteria for Buildings and Other Structures Standards Committees

ASCE 7-22 Main Committee

Voting Members

Ronald O. Hamburger, P.E., S.E., F.SEI *Chair*
James G. Soules, Ph.D., P.E., S.E., P.Eng, F.SEI, F.ASCE, *Vice Chair*
J. Benjamin Alper, P.E., S.E., M.ASCE
Russell A. Berkowitz, P.E., S.E., M.ASCE
Hussain Bhatia, Ph.D., P.E., S.E., M.ASCE
Joseph H. Cain, P.E., M.ASCE
Ronald J. Carrington, P.E., M.ASCE
Gary Y. K. Chock, S.E., D.CE, F.SEI, Dist.M.ASCE
Anne D. Cope, Ph.D, P.E., M.ASCE
Daniel T. Cox, Ph.D., A.M.ASCE
Bradford K. Douglas, P.E., M.ASCE
John F. Duntemann, P.E., S.E., F.SEI, M.ASCE
Gary J. Ehrlich, P.E., M.ASCE
Bruce R. Ellingwood, PhD., P.E., NAE, F.SEI, Dist.M.ASCE
David A. Fanella, P.E., S.E., F.SEI, F.ASCE
Kellie Farster, P.E., M.ASCE
Ramon E. Gilsanz, P.E., S.E., F.SEI, F.ASCE
Cole Graveen, P.E., S.E., M.ASCE
Emily Guglielmo, P.E., S.E., F.SEI, M.ASCE
James R. Harris, Ph.D., P.E., NAE, F.SEI, Dist.M.ASCE
Benchmark H. Harris, P.E.
Cherylyn Henry, P.E., F.SEI, M.ASCE
Michael Hill, P.E., S.E., M.ASCE
John D. Hooper, P.E., S.E., NAC, F.SEI, Dist.M.ASCE
Edwin T. Huston, P.E., S.E., F.SEI, M.ASCE, MIStructE
Jason J. Krohn, P.E., F.SEI, F.ASCE
Lawrence F. Kruth, P.E., M.ASCE
Marc L. Levitan, Ph.D., A.M.ASCE
Abbie B. Liel, Ph.D., P.E., F.SEI, F.ASCE
Jeffrey D. Linville, P.E., M.ASCE
Bonnie E. Manley, P.E., F.SEI, M.ASCE
Mehedy Mashnad, Ph.D., P.E.
Therese P. McAllister, Ph.D., P.E., F.SEI, M.ASCE
Kai Ki Mow, P.E., S.E., M.ASCE
Khaled Nahlawi, Ph.D., P.E., M.ASCE
Lawrence C. Novak, S.E., F.SEI, M.ASCE
G. Brent Nuttall, P.E., S.E., M.ASCE
George N. Olive, P.E., M.ASCE
Michael J. O'Rourke, P.E., F.SEI, M.ASCE
Robert G. Pekelnicky, P.E., S.E., F.SEI, M.ASCE
Ian N. Robertson, Ph.D., S.E., M.ASCE
Scott M. Rosemann, P.E., F.SEI, M.ASCE
Donald R. Scott, P.E., S.E., F.SEI, F.ASCE
Gwenyth R. Searer, P.E., S.E., M.ASCE
Randall Shackelford, P.E., M.ASCE
W. Lee Shoemaker, Ph.D., P.E., F.SEI, F.ASCE
Constadino Sirakis, P.E., M.ASCE
Thomas Sputo, Ph.D., P.E., F.SEI, F.ASCE
Seth A. Thomas, P.E., S.E., M.ASCE
Amit H. Varma, Ph.D., M.ASCE
Peter J. Vickery, Ph.D., P.E., F.SEI, F.ASCE
Andrew S. Whittaker, Ph.D., P.E., S.E., F.SEI, F.ASCE
Howard L. Zee, P.E., S.E., M.ASCE

Emeritus Members

Donald Dusenberry, P.E., F.SEI, F.ASCE
James M. Fisher, P.E.
Satyendra K. Ghosh, Ph.D., F.SEI, F.ASCE
Jonathan C. Siu, P.E., S.E., M.ASCE
John G. Tawresey, P.E., F.SEI, Dist.M.ASCE

Associate Members

Abdalsattar Alfarra, M.ASCE
Farid Alfawakhiri, Ph.D., P.E., P.Eng, M.ASCE
Mohammad AlHamaydeh, Ph.D., P.E., F.SEI, M.ASCE
Kevin P. Aswegan, P.E., S.E., M.ASCE
Samuel Baer
Nathan W. Balcirak I, EI, A.M.ASCE
James E. Beavers, Ph.D., P.E., F.ASCE
Richard M. Bennett, P.E., M.ASCE
Kevin N. Borth, P.E., S.E., M.ASCE
Cairo O. Briceno, P.E., S.E., M.ASCE
Johnathan D. Brower, P.E., M.ASCE
Ashley Cagle, P.E., S.E.
Jason E. Charalambides, Ph.D., P.E., R.A., M.ASCE, ENV SP
Francisco D. Chitty, Ph.D., A.M.ASCE
Ngai Chi Chung, P.E.
Dustin L. Cole, P.E., S.E., P.Eng, M.ASCE
Jay H. Crandell, P.E., M.ASCE
Chu Ding
Richard M. Drake, P.E., R.Eng, S.E., M.ASCE
Judian Duran, EIT, A.M.ASCE
Craig A. Durgarian, P.E., M.ASCE
Christopher Evans, M.ASCE
Quigyu Fan
Donna L.R. Friis, P.E., F.SEI, F.ASCE
Joe Gencarelli
Khaled Ghaedi, Ph.D., S.E., A.M.ASCE
John O. Grieshaber, P.E., S.E., F.SEI, F.ASCE
Jennifer Goupil, P.E., F.SEI, M.ASCE, *Secretary*
Alex Griffin, P.E., S.E., M.ASCE
Mahmoud M. Hachem, Ph.D., P.E., M.ASCE
Robert D. Hanson, P.E., F.ASCE
M.R. Hasan, P.E., M.ASCE
Drew C. Hatton, P.E., S.E., M.ASCE

Xiapin Hua, P.E., S.E., M.ASCE
Saif M. Hussain, P.E., S.E., LM.ASCE
Mohammad Iqbal, ESQ, Ph.D., P.E., F.ASCE
Kishor S. Jaiswal, Ph.D., P.E., M.ASCE
Alvaro ~~A.~~ Jaramillo-Suarez, P.E., P.Eng, F.ASCE
Hongping Jiang, Ph.D., P.E., M.ASCE
Christopher P. Jones, P.E., M.ASCE
M. R. Karim, Ph.D., P.E., S.E., M.ASCE
Logan Volkan Kebeli, Ph.D., A.M.ASCE
Ryan A. Kersting, P.E., S.E., M.ASCE
Jon P. Kiland, P.E., S.E., M.ASCE
Yoo Jae Kim, Ph.D., P.E., M.ASCE
Charles A. Kircher, Ph.D., P.E., M.ASCE
Kari Klaboe, P.E., M.ASCE
Raymond W. Kovachik, P.E., M.ASCE
James S. Lai, P.E., S.E., F.ASCE
Bryan K. Lanier, P.E., S.E., M.ASCE
Robert S. Lawson, P.E., M.ASCE
Thang H. Le, Ph.D., P.E., S.E., M.ASCE
Michael W. Lee, P.E., M.ASCE
Ji Yun Lee, Ph.D., A.M.ASCE
Aaron Lee, P.E.
Philip Line, P.E., M.ASCE
Hongchun Liu, P.E., M.ASCE
Anthony Longabard, S.E., M.ASCE
Chris Mader
Sanjeev R. Malushte, Ph.D., P.E., F.ASCE
Dion Marriott, Ph.D., S.E.
Mustafa Mashal, Ph.D., CPEng, IntPE(NZ), P.E., M.ASCE
Ferris Masri
Andrew D. Mitchell, S.E.
Kevin S. Moore, P.E., S.E., M.ASCE
Mike C. Mota, Ph.D., P.E., F.SEI, F.ASCE
Rudy Mulia, P.E., S.E., M.ASCE
Andrzej S. Nowak, F.ASCE
Frank K.H. Park, P.E., M.ASCE
Jessi Pereira
Norman F. Perkins, P.E., M.ASCE
James Kipton Ping, P.E., S.E., M.ASCE
James D. Pirnia, P.E., M.ASCE
Dylan J. Quinn, P.E., M.ASCE
Mehedi Rashid, P.E., S.E., M.ASCE
Jose G. Real Ducasa, Ing., M.ASCE
Nicholas Reid
Rossana River Guedez, P.E., M.ASCE
Eric Rutenbar, P.E., M.ASCE
Stephen P. Schneider, Ph.D., P.E., S.E., M.ASCE
Constantine Shuhaibar, Ph.D., P.E., M.ASCE
Md Sarwar Siddiqui, Ph.D., P.E.
Dan Siteman, EIT, A.M.ASCE
Harold O. Sprague, Jr., P.E., F.ASCE
Theodore Stathopoulos, Ph.D., P.E., F.SEI, F.ASCE
David A. Steele, P.E., M.ASCE
Sayed Stoman, Ph.D., P.E., S.E., M.L.S.E, SEI, M.ASCE
Reid Strain, P.E., P.Eng, M.ASCE
Thomas J. Szewczyk, P.E., S.E., M.ASCE
Mostafa Tazarv, Ph.D., P.E.
John M. Tehaney, P.E., S.E., M.ASCE
Paulos B. Tekie, Ph.D., M.D., P.E., S.E.
Christos V. Tokas, P.E., S.E., F.SEAOC, C.B.O.
Kiran Tuniki, P.E., M.ASCE
Stefka I. Vacheva, EIT, A.M.ASCE
Victoria B. Valentine, P.E., M.ASCE
Eric H. Wey, P.E., S.E., M.ASCE
Brian D. Wiese, P.E., M.ASCE
Michelle Wilkinson, P.E., S.E., M.ASCE, *Balloteer*
Peter J. G. Willse, P.E., M.ASCE
Tom C. Xia, Ph.D., P.E., M.ASCE
Riku P. Ylipelkonen, A.M.ASCE
Hyeong Jae Yoon, Ph.D., P.E., M.ASCE
Ben Yousefi
Zia Zafir, Ph.D., P.E., G.E., M.ASCE
William Zippel, P.E., M.ASCE

ASCE 7-22 Steering Committee

Ronald O. Hamburger, P.E., S.E., F.SEI, *Chair*
James G. Soules, Ph.D., P.E., S.E., P.Eng., F.SEI, F.ASCE, *Vice-Chair*
Jennifer Goupil, P.E., F.SEI, M.ASCE, *Secretary*
Cole Graveen, P.E., S.E., M.ASCE
Ronald J. Carrington, P.E., M.ASCE
Gary Y.K. Chock, S.E., D.CE, F.SEI, Dist.M.ASCE
Daniel T. Cox, Ph.D., A.M.ASCE
John F. Duntemann, P.E., S.E., F.SEI, M.ASCE
Emily Guglielmo, P.E., S.E., F.SEI, M.ASCE
James R. Harris, Ph.D., P.E., NAE, F.SEI, Dist.M.ASCE
Cherylyn Henry, P.E., F.SEI, M.ASCE
John D. Hooper, S.E., P.E., NAC, F.SEI, Dist.M.ASCE
Therese P. McAllister, Ph.D., P.E., F.SEI, M.ASCE
Robert G. Pekelnicky, P.E., S.E., F.SEI, M.ASCE
Donald R. Scott, P.E., S.E., F.SEI, F.ASCE
Michelle Wilkinson, P.E., S.E., M.ASCE, *Balloteer*

ASCE 7-22 Subcommittee on General Structural

Voting Members

Robert G. Pekelnicky, P.E., S.E., F.SEI, M.ASCE, *Chair*
James R. Harris, Ph.D., P.E., NAE, F.SEI, Dist.M.ASCE, *Vice Chair*
Farid Alfawakhiri, Ph.D., P.E., P.Eng, M.ASCE
Russell A. Berkowitz, P.E., S.E., M.ASCE
Bruce R. Ellingwood, Ph.D., P.E., NAE, F.SEI, Dist.M.ASCE
Ramon E. Gilsanz, P.E., S.E., F.SEI, F.ASCE
John D. Hooper, P.E., S.E., NAC, F.SEI, Dist.M.ASCE
Kevin J. LaMalva, P.E., F.ASCE
Philip Line, P.E., M.ASCE
Therese P. McAllister, Ph.D., P.E., F.SEI, M.ASCE
Lawrence C. Novak, S.E., F.SEI, M.ASCE
Mehrdad Sasani, Ph.D., P.E., F.SEI, F.ASCE
Donald R. Scott, P.E., S.E., F.SEI, F.ASCE
Constadino Sirakis, P.E., M.ASCE
James G. Soules, Ph.D., P.E., S.E., P.Eng, F.SEI, F.ASCE
Thomas Sputo, Ph.D., P.E., F.SEI, F.ASCE

Associate Members

Nathan W. Balcirak, EI, A.M.ASCE
James E. Beavers, Ph.D., P.E., F.ASCE
Ashley Cagle, P.E., S.E.
Francisco D. Chitty, Ph.D., A.M.ASCE
Ngai Chi Chung, P.E., *Secretary*
Bradford K. Douglas, P.E., M.ASCE
Mohammad Iqbal, ESQ, Ph.D., P.E., F.ASCE
M. R. Karim, Ph.D., P.E., S.E., M.ASCE
Eugene Kim, P.E., M.ASCE, *Balloteer*
Jason J. Krohn, P.E., F.SEI, F.ASCE
Michael W. Lee, P.E., M.ASCE
Chris Mader
Dion Marriott, Ph.D., S.E.
Andrew D. Mitchell, S.E.
Kevin S. Moore, P.E., S.E., M.ASCE

Sivakumar Munuswamy, Ph.D., P.E., M.ASCE
Chad Norvell
Frank K.H. Park, P.E., M.ASCE
Norman F. Perkins, P.E., M.ASCE
Nicholas Reid
Andrew Shuck, P.E., S.E., Aff.M.ASCE, *Historian*
David A. Steele, P.E., M.ASCE
Stefka I. Vacheva, EIT, A.M.ASCE
Amit H. Varma, Ph.D., M.ASCE
Hyeong Jae Yoon, Ph.D., P.E., M.ASCE

ASCE 7-22 Subcommittee on Load Combinations

Voting Members

Therese P. McAllister, Ph.D., P.E., F.SEI, M.ASCE, *Chair*
Philip Line, P.E., M.ASCE, *Vice-Chair*
Ruohua Zheng Guo, Ph.D., P.E., M.ASCE, *Secretary*
Emily P. Appelbaum, P.E., M.ASCE, *Balloteer*
Ji Yun Lee, Ph.D., A.M.ASCE, *Historian*
Jon-Paul Cardin, P.E., M.ASCE
Eun Jeong Cha, Ph.D., A.M.ASCE
Gary Y.K. Chock, S.E., D.CE, F.SEI, Dist.M.ASCE, *Liaison Tsunami*
Bruce R. Ellingwood, Ph.D., P.E., NAE, F.SEI, Dist.M.ASCE
James R. Harris, Ph.D., P.E., NAE, F.SEI, Dist.M.ASCE, *Liaison Snow & Rain*
Christopher P. Jones, P.E., M.ASCE
Yue Li, Ph.D., M.ASCE
Sanjeev R. Malushte, Ph.D., P.E., F.ASCE
Andrzej S. Nowak, F.ASCE
Robert G. Pekelnicky, P.E., S.E., F.SEI, M.ASCE, *Liaison General*
Vincent E. Sagan, P.E., F.ASCE
James G. Soules, Ph.D., P.E., S.E., P.Eng, F.SEI, F.ASCE, *Liaison Seismic*
Peter J. Vickery, Ph.D., P.E., F.SEI, F.ASCE, *Liaison Wind*

ASCE 7-22 Subcommittee on Dead and Live Loads

Voting Members

Cole Graveen, P.E., S.E., M.ASCE, *Chair*
Yun Jennifer Lan, P.E., S.E., M.ASCE, Vice Chair, Secretary, Balloteer, Historian
Cheryl Burwell, P.E., S.E., M.ASCE, *Past Chair*
Jason E. Charalambides, Ph.D., P.E., R.A, M.ASCE, ENV SP
Dustin L. Cole, P.E. S.E., P.Eng, M.ASCE
Manuel A. Diaz, P.E.
Kellie Farster, P.E., M.ASCE
John O. Grieshaber, P.E., S.E., F.SEI, F.ASCE
Philip Line, P.E., M.ASCE
Scott M. Rosemann, P.E., F.SEI, M.ASCE
Vincent E. Sagan, P.E., F.ASCE
Kari L. Sebern, P.E., M.ASCE
Joseph Tuttle

Associate Members

Cairo O. Briceno, P.E., S.E., M.ASCE
Brian E. Kehoe, P.E., S.E., F.ASCE
Himanshu Khurana, P.E.
Kari Klaboe, P.E., M.ASCE
Andrzej S. Nowak, F.ASCE
Constadino Sirakis, P.E., M.ASCE
Andrew P. Stam, P.E., S.E., M.ASCE
David A. Steele, P.E., M.ASCE

ASCE 7-22 Subcommittee on Flood Loads

Voting Members

Daniel T. Cox, Ph.D., A.M.ASCE, *Chair*
Christopher P. Jones, P.E., M.ASCE, *Vice Chair*
Jessica M. Mandrick, P.E., S.E., M.ASCE, *Secretary, Balloteer, Historian*
Graham R. Brasic, P.E., S.E., M.ASCE
Anthony C. Cerino, P.E., F.SEI, M.ASCE
William L. Coulbourne, P.E., F.SEI, F.ASCE
Carol Friedland, Ph.D., P.E., M.ASCE
Richard Lo, P.E., M.ASCE
Tori Tomiczek, Ph.D., EIT, A.M.ASCE
Jeffery A. Melby, Ph.D., P.E., D.CE, M.ASCE
Mehedi Rashid, P.E., S.E., M.ASCE
Seth A. Thomas, P.E., S.E., M.ASCE
Peter J. G. Willse, P.E., M.ASCE

Associate Members

Andre R. Barbosa, Ph.D., A.M. ASCE
Matthew R. Gilbertson, P.E., S.E., M. ASCE
David L. Kriebel, Ph.D., P.E., D.CE, M.ASCE
Marc L. Levitan, Ph.D., A.M.ASCE
Long T. Phan, Ph.D., P.E., M.ASCE

ASCE 7-22 Subcommittee on Tsunami Loads and Effects

Voting Members

Gary Y.K. Chock, S.E., D.CE, F.SEI, Dist.M.ASCE *Chair*
Ian N. Robertson, Ph.D., S.E., M.ASCE, *Vice Chair*
Matthew J. Francis, P.E., M.ASCE
John D. Hooper, P.E., S.E., NAC, F.SEI, Dist.M.ASCE
Patrick J. Lynett, Ph.D., A.M.ASCE
Ioan Nistor, Ph.D., P.E. M.ASCE
Hong K. Thio, Ph.D.
Seth A. Thomas, P.E., S.E., M.ASCE
Daniel J. Trisler, P.E., M.ASCE
Yong Wei, Ph.D., A.M.ASCE

Associate Members

Abbas Abdollahi, Ph.D., P.E., M.ASCE
Tatsuya Asai, Ph.D., R.A., A.M.ASCE
Jeremy D. Bricker, Ph.D., P.E.
Christina Cercone, Ph.D., A.M.ASCE
Tori Tomiczek, Ph.D., EIT, A.M.ASCE
David L. Kriebel, Ph.D., P.E., D.CE. M.ASCE
Michael R. Motley, Ph.D., P.E.
Sissy Nikolaou, Ph.D., P.E., D.GE, F.ASCE
Younes Nouri, Ph.D., P.E., P.Eng.
Catherine M. Petroff, Ph.D., P.E., M.ASCE
Tiziana Rossetto, Ph.D., FREng, FICE, A.M.ASCE
Chris Stearns, P.E., S.E., M.ASCE
Rick I. Wilson, A.M.ASCE

ASCE 7-22 Subcommittee on Snow and Rain Loads

Voting Members

John F. Duntemann, P.E., S.E, F.SEI, M.ASCE, *Chair*
Michael J. O'Rourke, P.E., F.SEI, M.ASCE, *Vice-Chair*
John Cocca, A.M.ASCE, *Secretary*
Sean M. Homem, P.E., S.E., M.ASCE, *Balloteer, Historian*
Timothy J. Allison, A.M.ASCE
Russell Benton, P.E., M.ASCE
Steven B. Brown, P.E., M.ASCE
James S. Buska, M.ASCE

Jan Dale, P.Eng. M.ASCE
Bradford K. Douglas, P.E., M.ASCE
Gary J. Ehrlich, P.E., M.ASCE
James M. Fisher, P.E.
Douglas L. Gadow, P.E., S.E., M.ASCE
James R. Harris, Ph.D., P.E., NAE, F.SEI, Dist.M.ASCE
Philip Jarrett
Aaron R. Lewis, P.E., A.M.ASCE
Abbie B. Liel, Ph.D., P.E., F.SEI, F.ASCE
Anthony Longabard, S.E., M.ASCE
Richard J. Nielsen, Ph.D, P.E., M.ASCE
George N. Olive, P.E., M.ASCE
Karl R. Pennings, P.E., S.E., M.ASCE
Scott A. Russell, P.E., P.Eng, S.E., M.ASCE
Vincent E. Sagan, P.E., F.ASCE
Joseph D. Scholze, P.E., S.E., M.ASCE
David P. Thompson, P.Eng., M.ASCE
Peter Wrenn, P.E., M.ASCE

Associate Members

Roya A. Abyaneh, P.E., M.ASCE
Samuel Baer
Melissa D. Burton, Ph.D., P.Eng, M.ASCE
Jon-Paul Cardin, P.E., M.ASCE
Mark D. Denavit, Ph.D., P.E., M.ASCE
Christopher Giffin, R.A., A.M.ASCE
Kathryn J. Jaworski, P.E., M.ASCE
Sterling Strait, P.E., S.E.
Peter J. G. Willse, P.E., M.ASCE

ASCE 7-22 Subcommittee on Atmospheric Ice

Voting Members

Ronald J. Carrington, P.E., M.ASCE, *Chair*
David Brinker
John F. Duntemann, P.E., S.E., F.SEI, M.ASCE
Kathleen Jones, A.M.ASCE
George N. Olive, P.E., M.ASCE
Alan B. Peabody, P.E., M.ASCE
Lawrence Slavin
Ronald Thorkildson

Associate Members

Bradford K. Douglas, P.E., M.ASCE
Philip Jarrett
Bryan K. Lanier, P.E., S.E., M.ASCE

ASCE 7-22 Subcommittee on Seismic Loads

Voting Members

John D. Hooper, P.E., S.E., NAC, F.SEI, Dist.M.ASCE, *Chair*
Emily Guglielmo, P.E., S.E., F.SEI, M.ASCE, *Vice Chair*
Kevin P. Aswegan, P.E., S.E., M.ASCE, *Historian*
Richard M. Bennett, P.E., M.ASCE
Russell A. Berkowitz, P.E., S.E., M.ASCE
Hussain Bhatia, Ph.D., P.E., S.E., M.ASCE
Finley Charney, Ph.D., P.E., F.ASCE, F.SEI
Kelly E. Cobeen, P.E., S.E., M.ASCE
Charles B. Crouse, Ph.D., P.E., M.ASCE
Satyendra K. Ghosh, Ph.D., F.SEI, F.ASCE
Amir S.J. Gilani, Ph.D., P.E., M.ASCE
John D. Gillengerten, P.E., S.E., M.ASCE
James R. Harris, Ph.D., P.E., NAE, F.SEI, Dist.M.ASCE
John L. Harris, III, Ph.D., S,E., P.E., F.SEI, M.ASCE
Thomas F. Heausler, P.E., S.E., F.SEI, F.ASCE
Conard A. Hohener, P.E., S.E., M.ASCE
Douglas G. Honegger, M.ASCE
Saif M. Hussain, P.E., S.E., LM.ASCE
Edwin T. Huston, P.E., S.E., F.SEI, M.ASCE, MIStructE
Dominic J. Kelly, P.E., M.ASCE
Ryan A. Kersting, P.E., S.E., M.ASCE
Charles A. Kircher, Ph.D., P.E., M.ASCE
Ronald W. LaPlante, P.E., S.E., M.ASCE
Philip Line, P.E., M.ASCE
Jeffrey D. Linville, P.E., M.ASCE
Roy F. Lobo, Ph.D., P.E., S.E., M.ASCE
Sanjeev R. Malushte, Ph.D., P.E., F.ASCE
Bonnie E. Manley, P.E., F.SEI, M.ASCE
Igor Marinovic, P.E., M.ASCE
Justin D. Marshall, Ph.D., P.E., M.ASCE
Kevin S. Moore, P.E., S.E., M.ASCE
Rudy Mulia, P.E., S.E., M.ASCE
Robert G. Pekelnicky, P.E., S.E., F.SEI, M.ASCE
Rafael Sabelli, P.E., S.E., M.ASCE
Deanna Seale, P.E.
W. Lee Shoemaker, Ph.D., P.E., F.SEI, F.ASCE
John F. Silva, P.E., S.E., F.SEI, M.ASCE
Robert E. Simmons, P.E., A.M.ASCE
James G. Soules, Ph.D., P.E., S.E., P.Eng, F.SEI, F.ASCE
Thomas J. Szewczyk, P.E., S.E., M.ASCE
Andrew W. Taylor, Ph.D., P.E., S.E., M.ASCE
John M. Tehaney, P.E., S.E., M.ASCE
Seth A. Thomas, P.E., S.E., M.ASCE
Andrew S. Whittaker, Ph.D., P.E., S.E., F.SEI, F.ASCE
Brian D. Wiese, P.E., M.ASCE
Tom C. Xia, Ph.D., P.E., M.ASCE
Zia Zafir, Ph.D., P.E., G.E., M.ASCE

Emeritus Members

Victor D. Azzi, Ph.D., P.E., M.ASCE
Robert D. Hanson, P.E., F.ASCE
M. R. Karim, Ph.D., P.E., S. E., M.ASCE
Jon P. Kiland, P.E., S.E., M.ASCE
James S. Lai, P.E., S.E., F.ASCE
Gwenyth R. Searer, P.E., S.E., M.ASCE
Christos V. Tokas, P.E., S.E., F.SEAOC, C.B.O.
Ben Yousefi

Associate Members

Mohammad AlHamaydeh, Ph.D., P.E., F.SEI, M.ASCE
Mohammad Aliaari, Ph.D., P.E., S.E., M.ASCE
J. Benjamin Alper, P.E., S.E., M.ASCE
Dennis A. Alvarez, P.E., M.ASCE
Heming Alwin, A.M.ASCE
Aswathram Balasubramanian
Adam Baxter, EIT, A.M.ASCE
James E. Beavers, Ph.D., P.E., F.ASCE
David R. Bonneville, M.ASCE
Vincent Borov, P.E., P.E., S.E., M.ASCE
Scott E. Breneman, Ph.D., P.E., S.E., M.ASCE
Cairo O. Briceno, P.E., S.E., M.ASCE
Scott Campbell, Ph.D., P.E., M.ASCE
Jennifer A. Carey, P.E., M.ASCE
Francisco D. Chitty, Ph.D., A.M.ASCE
Alex Chu, P.E., S.E., M.ASCE, *Secretary*
Andrew G. Conrad, P.E., M.ASCE
Abel Diaz Valdes, P.E., S.E., M.ASCE
Khaled Ghaedi, Ph.D., S.E., A.M.ASCE
Bradford K. Douglas, P.E., M.ASCE
Richard M. Drake, P.E., R.Eng, S.E., M.ASCE
Cynthia J. Duncan

Tarek Elkhoraibi, Ph.D., P.E., M.ASCE
Chad Fusco, P.E., M.ASCE
Michael Gannon, P.E., S.E., M.ASCE
Josh Gebelein, P.E., S.E., M.ASCE
Mahmoud M. Hachem, Ph.D., P.E., M.ASCE
Lachezar V. Handzhiyski, P.E., S.E., M.ASCE
Ronald W. Haupt, P.E., M.ASCE
Hongping Jiang, Ph.D., P.E., M.ASCE
Brian E. Kehoe, P.E., S.E., F.ASCE
Mehrshad Ketabdar, P.E., S.E., M.ASCE
Yoo Jae Kim, Ph.D., P.E., M.ASCE
Jason J. Krohn, P.E., F.SEI, F.ASCE
Yun Jennifer Lan, P.E., S.E., M.ASCE
Robert S. Lawson, P.E., M.ASCE
Thang H. Le, Ph.D., P.E., S.E., M.ASCE
Hongchun Liu, P.E., M.ASCE
Nicolas Luco, Ph.D., M.ASCE
Rafael A. Magana, P.E., M.ASCE
James Marrone
Mustafa Mashal, Ph.D., CPEng, IntPE(NZ), P.E., M.ASCE
Andrew D. Mitchell, S.E.
Mike C. Mota, Ph.D., P.E., F.SEI, F.ASCE
Joe Mugford, P.E.
Sissy Nikolaou, Ph.D., P.E., D.GE, F.ASCE
Thomas L. North, P.E., F.ASCE
Nicolas K. Oettle, Ph.D., P.E., G.E., M.ASCE
Frank K.H. Park, P.E., M.ASCE
Charles Pizzano, P.E.
Keith A. Porter, Ph.D., P.E., F.SEI, F.ASCE
Jose G. Real Ducasa, Ing., M.ASCE
Nicholas Reid
Sanaz Rezaeian, Ph.D.
Nicholas D. Robinson, A.M.ASCE
Keri L. Ryan, Ph.D, A.M.ASCE
Thomas A. Sabol, Ph.D, P.E., S.E.
Omar Sheikh, P.E., S.E., M.ASCE
Constantine Shuhaibar, Ph.D., P.E., M.ASCE
Harold O. Sprague, Jr., P.E., F.ASCE
Andrew D. Stark, P.E., S.E., M.ASCE
Casey Stevenson, P.E., M.ASCE
Sayed Stoman, Ph.D., P.E., S.E., M.L.S.E., SEI, M.ASCE
Reid Strain, P.E., P.Eng, M.ASCE
Paul O. Stuart, P.E., M.ASCE
John G. Tawresey, P.E., F.SEI, Dist.M.ASCE
Bahaar Taylor, P.E., M.ASCE
Mostafa Tazarv, Ph.D., P.E.
Paulos B. Tekie, Ph.D., M.D., P.E., S.E.
Mike Tong, Ph. D.
Nathan D. Tremblay, P.E., S.E., M.ASCE
Samuel Truthseeker, P.E.
Wang Kin Tsui
Joseph Tuttle
Chia-Ming Uang, Ph.D., M.ASCE
Victoria B. Valentine, P.E., M.ASCE
Zane B. Wells, P.E., M.ASCE
Eric H. Wey, P.E., S.E., M.ASCE
Michelle Wilkinson, P.E., S.E., M.ASCE
Jenna Wong, Ph.D., P.E., M.ASCE, *Balloteer*
Hyeong Jae Yoon, Ph.D., P.E., M.ASCE
Kent Yu, Ph.D., P.E.,S.E., M.ASCE
Wei Zheng, Ph.D., P.E., G.E., M.ASCE
Reid B. Zimmerman, P.E., S.E., M.ASCE
William Zippel, P.E., M.ASCE

TC 01: Ground Motions of ASCE 7-22 Seismic SC

Voting Members

Charles B. Crouse, Ph.D., P.E., M.ASCE, *Chair*
Zia Zafir, Ph.D., P.E., G.E., M.ASCE, *Vice Chair*
Jason Bock, P.E., M.ASCE
Ramin Golesorkhi, Ph.D., P.E., G.E., F.ASCE
James R. Harris, Ph.D., P.E., NAE, F.SEI, Dist.M.ASCE
Charles A. Kircher, Ph.D., P.E., M.ASCE
James Marrone
Sissy Nikoaou, Ph.D., P.E., D.GE, F.ASCE
Nicolas K. Oettle, Ph.D., P.E., G.E., M.ASCE
Menzer Pehlivan, Ph.D., P.E., M.ASCE
Wei Zheng, Ph.D., P.E., G.E., M.ASCE

Associate Members

Nicolas Luco, Ph.D., M.ASCE
Sanaz Rezaeian, Ph.D.

TC 02: General Provisions of ASCE 7-22 Seismic SC

Voting Members

Emily Guglielmo, P.E., S.E., F.SEI, M.ASCE, *Chair*
Ryan A. Kersting, P.E., S.E., M.ASCE, *Vice-Chair*
J. Benjamin Alper, P.E., S.E., M.ASCE
Heming Alwin, A.M.ASCE
Kevin P. Aswegan, P.E., S.E., M.ASCE
Russell A. Berkowitz, S.E., P.E., M.ASCE
Hussain Bhatia, Ph.D., P.E., S.E., M.ASCE
Scott E. Breneman, Ph.D., P.E., S.E., M.ASCE
Cairo O. Briceno, P.E., S.E., M.ASCE
Finley Charney, Ph.D., P.E., F.ASCE, F.SEI
Alex Chu, P.E., S.E., M.ASCE
Kelly E. Cobeen, P.E., M.ASCE
Bradford K. Douglas, P.E., M.ASCE
Chad Fusco, P.E., M.ASCE
Josh Gebelein, P.E., S.E., M.ASCE
Satyendra K. Ghosh, Ph.D., F.SEI, F.ASCE
James R. Harris, Ph.D., P.E., NAE, F.SEI, Dist.M.ASCE
John L. Harris, III, Ph.D., S.E., P.E., F.SEI, M.ASCE
Conard A. Hohener, P.E., M.ASCE
John D. Hooper, P.E., S.E., NAC, F.SEI, Dist.M.ASCE
Jon P. Kiland, P.E., S.E., M.ASCE
Yun Jennifer Lan, P.E., S.E., M.ASCE
Ronald W. LaPlante, P.E., S.E., M.ASCE
Philip Line, P.E., M.ASCE
Jeffrey D. Linville, P.E., M.ASCE
Hongchun Liu, P.E., M.ASCE
Roy F. Lobo, Ph.D., P.E., M.ASCE
Bonnie E. Manley, P.E., F.SEI, M.ASCE
Igor Marinovic, P.E., M.ASCE
Silvia Mazzoni, Ph.D.
Andrew D. Mitchel
Kevin S. Moore, P.E., S.E., M.ASCE
Sissy Nikolaou, Ph.D., P.E., D.GE, F.ASCE
Floriana Petrone, Ph.D., M.ASCE
Rafael Sabelli, P.E., M.ASCE
Deanna Seale, P.E.
Omar Sheikh, P.E., S.E., M.ASCE
W. Lee Shoemaker, Ph.D, P.E., F.SEI, F.ASCE
Casey Stevenson, P.E., M.ASCE
Sayed Stoman, Ph.D., P.E., S.E., M.L.S.E., SEI, M.ASCE
Thomas J. Szewczyk, P.E., M.ASCE
Seth A. Thomas, P.E., S.E., M.ASCE

Christos V. Tokas, P.E., S.E., F.SEAOC, C.B.O.
Shanshan Wang, Ph.D., P.E., M.ASCE
Zane B. Wells, P.E., M.ASCE
Tom C. Xia, Ph.D., P.E., M.ASCE

Associate Members

Huseyin Darama, Ph.D., P.E., S.E., F.ASCE
Khaled Ghaedi, Ph.D., S.E., A.M.ASCE

TC 02N: Non-Linear General Provision of ASCE 7-22 Seismic SC

Voting Members

Russell A. Berkowitz, P.E., S.E., M.ASCE, *Chair*
Kevin P. Aswegan, P.E., S.E., M.ASCE
Finley Charney Ph.D., F.ASCE
Abel Diaz M.ASCE
Amir S J Gilani Ph.D., P.E., M.ASCE
John L. Harris, III, Ph.D., S,E., P.E., F.SEI, M.ASCE
Ryan A Kersting P.E., S.E., M.ASCE
Justin Douglas Marshall Ph.D., P.E., M.ASCE
Robert George Pekelnicky P.E., S.E., F.SEI, M.ASCE
Rafael Sabelli P.E., M.ASCE
Paulos Beraki Tekie Ph.D., M.D., P.E., S.E.
Kevin Wong M.ASCE
Tom Chuan Xia Ph.D., P.E., M.ASCE
Kent Yu Ph.D., P.E., M.ASCE
Zia Zafir Ph.D., P.E., G.E., M.ASCE
Reid B. Zimmerman P.E., S.E., M.ASCE

Associate Members

Mohammad AlHamaydeh, Ph.D. P.E., F.SEI, M.ASCE
Nathan Wesley Balcirak I EI, A.M.ASCE
Cairo Obady Briceno P.E., M.ASCE
Scott Campbell
Tarek Elkhoraibi, Ph.D., P.E., M.ASCE
Khaled Ghaedi, Ph.D., S.E., A.M.ASCE
M.R. Hasan P.E., M.ASCE
Roy F Lobo Ph.D., P.E., M.ASCE
Dion Marriott Ph.D., P.E.
Guillermo Santana C.Eng, P.E., M.ASCE
Stephen Patrick Schneider Ph.D., P.E., M.ASCE
Rahul Sharma M.ASCE
Constantine Shuhaibar Ph.D., P.E., M.ASCE
Seth Alan Thomas P.E., S.E., M.ASCE
Christos V. Tokas

TC 02S: Simplified General Provision of ASCE 7-22 Seismic SC

Voting Members

Thomas F. Heausler, P.E., S.E., F.SEI, F.ASCE, *Chair*
Andrew D. Stark, P.E., S.E., M.ASCE, *Secretary*
Heming Alwin, A.M.ASCE
Bradford K. Douglas, P.E., M.ASCE
Brian E. Kehoe, P.E., S.E., F.ASCE
Ryan A. Kersting, P.E., S.E., M.ASCE
Philip Line, P.E., M.ASCE
Bonnie E. Manley, P.E., F.SEI, M.ASCE
Sissy Nikolaou, Ph.D., P.E., D.GE, F.ASCE
Deanna Seale, P.E.

TC 03: Foundations/Site Conditions of ASCE 7-22 Seismic SC

Voting Members

Ronald W. LaPlante, P.E., S.E., M.ASCE, *Chair*
Thang H. Le, Ph.D., P.E., S.E., M.ASCE, *Vice Chair*
James E. Beavers, Ph.D., P.E., F.ASCE
Jason Bock, P.E., M.ASCE
Charles B. Crouse, Ph.D., P.E., M.ASCE
Matthew J. Francis, P.E., M.ASCE
Ramin Golesorkhi, Ph.D., P.E. G.E, F. ASCE
Douglas G. Honegger, M.ASCE
Dominic J. Kelly, P.E., M.ASCE
Sissy Nikolaou, Ph.D., P.E., D.GE, F.ASCE
Nicolas K. Oettle, Ph.D., P.E., G.E., M.ASCE
Sayed Stoman, Ph.D., P.E., S.E., M.L.S.E., SEI, M.ASCE
John M. Tehaney, P.E., S.E., M.ASCE
Zia Zafir, Ph.D., P.E., G.E., M.ASCE
Wei Zheng, Ph.D., P.E., G.E., M.ASCE

Associate Members

Abdalsattar Alfarra, M.ASCE
Tarek Elkhoraibi, Ph.D., P.E., M.ASCE
Khaled Ghaedi, Ph.D., S.E., A.M.ASCE
M.R. Hasan, P.E., M.ASCE

TC 04: Concrete of ASCE 7-22 Seismic SC

Voting Members

Satyendra K. Ghosh, Ph.D., F.SEI, F.ASCE, *Chair*
Andrew W. Taylor, Ph.D., P.E., S.E., M.ASCE, *Vice-Chair*
Thomas J. Szewczyk P.E., S.E., M.ASCE
Mohammad AlHamaydeh, Ph.D., P.E., F.SEI, M.ASCE
Joe Ferzli, P.E., S.E., M.ASCE
James S. Lai, P.E., S.E., F.ASCE
Dawn Lehman, Ph.D., A.M.ASCE
Mehran Pourzanjani, P.E., M.ASCE

Associate Members

Abdalsattar Alfarra, M.ASCE
Roberto A. Anton, Ing., P.E., ENV SP, M.ASCE
Nathan W. Balcirak I, EI, A.M.ASCE
Adam Baxter, EIT, A.M.ASCE
James E. Beavers, Ph.D., P.E., F.ASCE
M.R. Hasan, P.E., M.ASCE
Yoo Jae Kim, Ph.D., P.E., M.ASCE
Jason J. Krohn, P.E., F.SEI, F.ASCE
Songtao Liao, Ph.D.,P.E., M.ASCE
Rudy Mulia, P.E., S.E., M.ASCE
Sivakumar Munuswamy, Ph.D., P.E., M.ASCE
Jose G. Real Ducasa, Ing., M.ASCE
Sanaz Saadat, P.E., M.ASCE
Guillermo Santana, Ph.D., P.E., M.ASCE
Stephen P. Schneider, Ph.D., P.E., S.E., M.ASCE
Christos V. Tokas, P.E., S.E., F.SEAOC, C.B.O.

TC 05: Masonry of ASCE 7-22 Seismic SC

Voting Members

Edwin T. Huston, P.E., S.E., F.SEI, M.ASCE, MIStructE, *Chair*
Richard M. Bennett, P.E., M.ASCE
John G. Tawresey, P.E., F.SEI, Dist.M.ASCE
John M. Tehaney, P.E., S.E., M.ASCE

TC 06: Steel of ASCE 7-22 Seismic SC

Voting Members

Bonnie E. Manley, P.E., F.SEI, M.ASCE, *Chair*
W. Lee Shoemaker, Ph.D., P.E., F.SEI, F.ASCE, *Vice Chair*
Victor D. Azzi, Ph.D., P.E., M.ASCE
Cynthia J. Duncan
John L. Harris III, Ph.D., S.E., P.E., F.SEI, M.ASCE

Justin D. Marshall, Ph.D., P.E., M.ASCE
Rudy Mulia, P.E., S.E., M.ASCE
Thomas A. Sabol, Ph.D., P.E., S.E.
Andrew D. Stark, P.E., S.E., M.ASCE

Associate Members

Adam Baxter, EIT, A.M.ASCE
Richard M. Drake, P.E., R.Eng, S.E. M.ASCE
Hongping Jiang, Ph.D., P.E., M.ASCE
Wang Kin Tsui

TC 07: Wood of ASCE 7-22 Seismic SC

Voting Members

Philip Line, P.E., M.ASCE, *Chair*
Jeffrey D. Linville, P.E., M.ASCE, *Vice Chair*
Scott Breneman, Ph.D., P.E., S.E., M.ASCE
Heming Alvin, A.M.ASCE
Kelly E. Cobeen, P.E., S.E., M.ASCE
Thang H. Le, Ph.D., P.E., S.E., M.ASCE
Deanna Seale, P.E.
Tom C. Xia, Ph.D., P.E., M.ASCE
Ben Yousefi

TC 08: Nonstructural Components of ASCE 7-22 Seismic SC

Voting Members

John D. Gillengerten, P.E., S.E., M.ASCE, *Chair*
John M. Tehaney, P.E., S.E., M.ASCE, *Vice Chair*
Dennis A. Alvarez, P.E., M.ASCE
Victor D. Azzi, Ph.D., P.E., M.ASCE
James E. Beavers, Ph.D., P.E., F.ASCE
Hussain Bhatia, Ph.D., P.E., S.E., M.ASCE
Virginia Diaz Taibo, P.E., M.ASCE
Jeffrey A. Gatscher, A.M.ASCE
Amir S.J. Gilani, Ph.D., P.E., M.ASCE
Meaghan Halligan, P.E, S.E., P.Eng, M.ASCE
Kyle D. Harris, P.E., M.ASCE
Douglas G. Honegger, M.ASCE
Hongping Jiang, Ph.D., P.E., M.ASCE
Brian E. Kehoe, P.E., S.E., F.ASCE
Ronald W. LaPlante, P.E., S.E., M.ASCE
Roy F. Lobo, Ph.D., P.E., M.ASCE
Rudy Mulia, P.E., S.E., M.ASCE
Jacob Olsen
Robert G. Pekelnicky, P.E., S.E., F.SEI, M.ASCE
Karl Peterman, P.E.
Maryann T. Phipps
Tony Shelton, A.M.ASCE
John F. Silva, P.E., S.E., F.SEI, M.ASCE
Robert E. Simmons, P.E., A.M.ASCE
James G. Soules, Ph.D., P.E., S.E., P.Eng, F.SEI, F.ASCE
Harold O. Sprague Jr., P.E., F.ASCE
Casey Stevenson, P.E., M.ASCE
Yelena K. Straight, P.E., M.ASCE
Paul O. Stuart, P.E., M.ASCE
Nathan D. Tremblay, P.E., S.E., M.ASCE
Samuel Truthseeker, P.E.
Brian D. Wiese, P.E., M.ASCE
Tom Yuschak

Associate Members

Mohammad Aliaari, Ph.D., P.E., S.E., M.ASCE
Cairo O. Briceno, P.E., S.E., M.ASCE
Joseph H. Cain, P.E., M.ASCE
Scott Campbell, Ph.D., P.E., M.ASCE
Andrew G. Conrad, P.E., M.ASCE
Andrew M. Coughlin, P.E., S.E., M.ASCE
Nate Deibler, P.E., M.ASCE
Richard M. Drake, P.E., R.Eng, S.E., M.ASCE
Ronald W. Haupt, P.E., M.ASCE
James S. Lai, P.E., S.E., F.ASCE
Ricardo A. Medina, Ph.D., P.E., M.ASCE
Andrew D. Mitchell, S.E.
Hossein Mostafaei, Ph.D., P.Eng, M.ASCE
Christos V. Tokas, P.E., S.E., F.SEAOC, C.B.O.

TC 12: Seismic Isolation and Damping Systems of ASCE 7-22 Seismic SC

Voting Members

Andrew S. Whittaker, Ph.D., P.E., S.E., F.SEI, F.ASCE, *Chair*
Reid B. Zimmerman, P.E., S.E., M.ASCE, *Vice Chair*
Ian D. Aiken, P.E., M.ASCE
Mohammad AlHamaydeh, Ph.D., P.E., F.SEI, M.ASCE
Cameron Black, Ph.D., P.E.
Abel Diaz Valdes, P.E., S.E., M.ASCE
Amir S. J. Gilani, Ph.D., P.E., M.ASCE
Saif M. Hussain, P.E., S.E., LM.ASCE
Charles A. Kircher, Ph.D., P.E., M.ASCE
Roy F. Lobo, Ph.D., S.E., M.ASCE
Aaron Malatesta, P.E.
Dion Marriott, Ph.D., S.E.
Ronald L. Mayes, Ph.D.
Robert G. Pekelnicky, P.E., S.E., F.SEI, M.ASCE
Keri L. Ryan, Ph.D.
Rafael Sabelli, P.E., S.E., M.ASCE
Constantine Shuhaibar, Ph.D., P.E., M.ASCE
Andrew W. Taylor, Ph.D., P.E., S.E., M.ASCE
Jenna Wong, Ph.D., P.E., M.ASCE

Associate Members

Mohammad Aliaari, Ph.D., P.E., S.E., M.ASCE
Huseyin Darama, Ph.D., P.E., S.E., F.ASCE
Anthony Giammona, P.E., S.E.
Amarnath Kasalanati, Ph.D., P.E.
Mohammaed S. Mohammed, Ph.D., P.E., M.ASCE Aaron Yung, P.E.
Christos V. Tokas, P.E., S.E., F.SEAOC, C.B.O.
Victor Zayas

TC 13: Nonbuilding Structures of ASCE 7-22 Seismic SC

Voting Members

James G. Soules, Ph.D., P.E., S.E., P.Eng, F.SEI, F.ASCE, *Chair*
Victor D. Azzi, Ph.D., P.E., M.ASCE
Vincent Borov, P.E., M.ASCE
Joseph H. Cain, P.E., M.ASCE
Jennifer A. Carey, P.E., M.ASCE
Richard M. Drake, P.E., R.Eng, S.E., M.ASCE
John D. Gillengerten, P.E., S.E., M.ASCE
Kyle D. Harris, P.E., M.ASCE
Ronald W. Haupt, P.E., M.ASCE
Thomas F. Heausler, P.E., S.E., F.SEI, F.ASCE
Brian E. Kehoe, P.E., S.E., F.ASCE
Mehrshad Ketabdar, P.E., S.E., M.ASCE
Hongchun Liu, P.E., M.ASCE
Roy F. Lobo, Ph.D., P.E., M.ASCE
Sanjeev R. Malushte, Ph.D., P.E., F.ASCE
Andrew D. Mitchell, S.E.
Joe Mugford, P.E.
Rudy Mulia, P.E., S.E., M.ASCE

Thomas L. North, P.E., F.ASCE
Nicholas D. Robinson, A.M.ASCE
Harold O. Sprague Jr., P.E., F.ASCE
Casey Stevenson, P.E., M.ASCE
Reid Strain, P.E., P.Eng, M.ASCE
Sayed Stoman, Ph.D., P.E., S.E., M.L.S.E., SEI, M.ASCE
John M. Tehaney, P.E., S.E., M.ASCE
Eric H. Wey, P.E., S.E., M.ASCE
Brian D. Wiese, P.E., M.ASCE

Associate Members

Douglas G. Honegger, M.ASCE
Robert E. Simmons, P.E., A.M.ASCE
Christos V. Tokas, P.E., S.E., F.SEAOC, C.B.O.

ASCE 7-22 Subcommittee on Wind Loads

Voting Members

Donald R. Scott, P.E., S.E., F.SEI, F.ASCE, *Chair*
Cherylyn Henry, P.E. F.SEI, M.ASCE, Vice Chair; Chapter 26 Task Committee, Chair
Connor J. Bruns, S.E., M.ASCE *Secretary*
Elaina J. Sutley, Ph.D., A.M.ASCE, *Balloteer*
Gregory T. Holbrook, P.E., M.ASCE, *Historian*
Melissa D. Burton, Ph.D., P.Eng, M.ASCE, Performance Based Design Task Committee, Vice Chair
William L. Coulbourne, P.E., F.SEI, F.ASCE, *Chapter 30 Task Committee, Chair*
Jay H. Crandell, P.E., M.ASCE
Richard J. Davis, P.E., M.ASCE, *Chapter 29 Task Committee, Chair*
Roy O. Denoon, Ph.D., M.ASCE
Bradford K. Douglas, P.E., M.ASCE
Gary J. Ehrlich, P.E., M.ASCE, Chapter 30 Task Committee, Secretary
Emily Guglielmo, P.E., S.E., F.SEI, M.ASCE, *Chapter 26 Task Committee, Secretary*
Michael Hill, P.E., S.E., M.ASCE
Peter A. Irwin, Ph.D., P.E., P.Eng, F.SEI, F.ASCE
Gregory A. Kopp, Ph.D., P.E., M.ASCE, *Chapter 31 Task Committee, Chair*
Bryan K. Lanier, P.E., S.E., M.ASCE
Russell Larsen, P.E., S.E., M.ASCE, Performance Based Design Task Committee, Chair
Marc L. Levitan, Ph.D., A.M.ASCE, *Tornado Task Committee, Chair*
Franklin T. Lombardo, Ph.D., EIT, A.M.ASCE, *Tornado Task Committee, Vice Chair*
Anthony Lynn Miller, P.E., F.SEI, M.ASCE
Murray J. Morrison, Ph.D., A.M.ASCE
Lawrence C. Novak, S.E., F.SEI, M.ASCE
John W. O'Brien, P.E., S.E., M.ASCE, *Chapter 30 Task Committee, Vice Chair*
Timothy A. Reinhold, Ph.D., P.E., M.ASCE, *Chapter 28 Task Committee, Chair*
Randall Shackelford, P.E., M.ASCE, Chapter 28 Task Committee, Secretary
W. Lee Shoemaker, Ph.D., P.E., F.SEI, F.ASCE, *Chapter 28 Task Committee, Vice Chair*
Thomas L. Smith, R.A., F.SEI, M.ASCE
Theodore Stathopoulos, Ph.D., P.E., F.SEI, F.ASCE
Peter J. Vickery, Ph.D., P.E., F.SEI, F.ASCE, *Chapter 26 Task Committee, Vice Chair*
Bradley Young, P.E., S.E., M.ASCE, *Chapter 27 Task Committee, Chair*

Emeritus Members

Daryl W. Boggs, P.E., M.ASCE
Lawrence G. Griffis, P.E., F.SEI, M.ASCE
Nicholas Isyumov, P.E.
Ahsan Kareem, Ph.D., NAE, F.EMI, Dist.M.ASCE

Associate Members

Timothy J. Allison, A.M.ASCE
Heather Anesta, P.E., S.E., M.ASCE
Maryam Asghari Mooneghi, Ph.D., P.E., M.ASCE
David Banks, Ph.D., M.ASCE
Michele Barbato, Ph.D., C.Eng, P.E., F.EMI, F.SEI, F.ASCE
Kevin N. Borth, P.E., S.E., M.ASCE, Performance Based Design Task Committee, Secretary
David Bott, P.E., R.A., S.E., M.ASCE
Christopher L. Bove, P.E., S.E., M.ASCE
Steven B. Brown, P.E., M.ASCE
Jason Brown, P.E., S.E., M.ASCE
Jennifer A. Carey, P.E., M.ASCE
Jason E. Charalambides, Ph.D., P.E., R.A., M.ASCE. ENV SP
Finely Charney, Ph.D., P.E., F.ASCE, F.SEI
Dustin L. Cole, P.E., S.E., P.Eng, M.ASCE
Anne D. Cope, Ph.D., P.E., M.ASCE
Mark E. Detwiler, P.E., M.ASCE
Lakshmana S. Doddipatla, Ph.D., A.M.ASCE, *Chapter 31 Task Committee, Secretary*
R. Scott Douglas, P.E., S.E., M.ASCE
John F. Duntemann, P.E., S.E., F.SEI, M.ASCE
David A. Fanella, P.E., S.E., F.SEI, F.ASCE
Thomas R. Garriott, P.E., M.ASCE
Khaled Ghaedi, Ph.D., S.E., A.M.ASCE
Mehrafarid Ghoreishi, Ph.D., P.E.
Alex Griffin, P.E., S.E., M.ASCE
Alan Hahn, P.E., S.E., M.ASCE
Benchmark H. Harris, P.E.
Drew C. Hatton, P.E., S.E., M.ASCE
Michael A. Herring, P.E., M.ASCE
Joseph R. Hetzel, P.E.
Daniel P. Hogan II, P.E., S.E., Chapter 27 Task Committee, Secretary
John D. Hooper, P.E., S.E., NAC, F.SEI, Dist.M.ASCE
Xiapin Hua, P.E., S.E., M.ASCE
Anurag Jain, Ph.D., P.E., M.ASCE
Hongping Jiang, Ph.D., P.E., M.ASCE
Johnn Judd, M.ASCE
Ramtin Kargarmoakhar, Ph.D., P.E., M.ASCE
Eugene Kim, P.E., M.ASCE
Yoo Jae Kim, Ph.D., P.E., M.ASCE
Raymond W. Kovachik, P.E., M.ASCE
James S. Lai, P.E., S.E., F.ASCE
Darin V. Lasater, P.E., M.ASCE
Robert S. Lawson, P.E., M.ASCE
Chris Mader
Bonnie E. Manley, P.E., F.SEI, M.ASCE
William H. Martin, P.E., P.Eng., S.E.
Mehedy Mashnad, Ph.D., P.E.
Therese P. McAllister, Ph.D., P.E., F.SEI, M.ASCE
Patrick W. McCarthy, P.E., M.ASCE
Rudy Mulia, P.E., S.E., M.ASCE
Joelle K. Nelson, P.E., M.ASCE, Chapter 29 Task Committee, Vice Chair
Glenn T. Overcash, P.E., M.ASCE
Long T. Phan, Ph.D., P.E., M.ASCE
James D. Pirnia, P.E., M.ASCE

Mehedi Rashid, P.E., S.E., M.ASCE
Maryam Refan, Ph.D.
Sanaz Saadat, P.E., M.ASCE
Pataya L. Scott, Ph.D., EIT, A.M.ASCE
Jason V. Smart, P.E., M.ASCE
James G. Soules, Ph.D., P.E., S.E., P.Eng, F.SEI, F.ASCE
Seymour M.J. Spence, Ph.D., A.M.ASCE
Thomas Sputo, Ph.D., P.E., F.SEI, F.ASCE
David W. Stermer, P.E., M.ASCE
Stefka I. Vacheva, EIT, A.M.ASCE
Kevin C. Warapius, P.E., M.ASCE
Dennis M. Wilson II, P.E., M.ASCE
Silky S. K. Wong, Ph.D., P.E., S.E., LEED AP, CEng MICE, M.ASCE
DongHun Yeo, Ph.D., P.E., M.ASCE
Riku P. Ylipelkonen, A.M.ASCE
Ioannis Zisis, Ph.D., M.ASCE
Ammar Motorwala, P.E., M.ASCE

DEDICATION

Jon A. Peterka, Ph.D., P.E.

May 26, 1941 – May 22, 2019

ASCE 7-22 is dedicated to Dr. Jon A Peterka, P.E., a leader in the development of codes and standards for ASCE who served on the ASCE/SEI 7 Minimum Design Loads and Associated Criteria for Buildings and Other Structures committee for several decades and was relied on for thoughtful guidance as the ASCE 7 wind load provisions evolved. Jon was a pioneer and community pillar of wind engineering. He was instrumental in the writing of the first version of ASCE 49 *Wind Tunnel Testing for Buildings and Other Structures* (and its antecedent, ASCE Manual of Practice 67). This service to our profession was only the tip of the iceberg in his passion for his work. Jon's imprint can be found throughout this standard, certainly in the knowledge and methods he contributed but also in the spirit in which we strive to provide the information necessary to improve the transparency, consistency, and quality of wind load provisions and wind tunnel testing.

CHAPTER 1
GENERAL

1.1 SCOPE

This standard provides minimum loads, hazard levels, associated criteria, and intended performance goals for buildings, other structures, and their nonstructural components that are subject to building code requirements. The loads, load combinations, and associated criteria provided herein are to be used with design strengths or allowable stress limits contained in design specifications for conventional structural materials. Used together, they are deemed capable of providing the intended performance levels for which the provisions of this standard have been developed. Procedures for applying alternative means to demonstrate acceptable performance are also described. Unless explicitly noted in this standard, specific criteria for additions, alterations, or repairs to existing buildings are beyond the scope of this standard.

1.2 DEFINITIONS AND SYMBOLS

1.2.1 Definitions The following definitions apply to all the provisions of the standard.

ALLOWABLE STRESS DESIGN: A method of proportioning structural members such that elastically computed stresses produced in the members by nominal loads do not exceed specified allowable stresses (also called "working stress design").

AUTHORITY HAVING JURISDICTION: The organization, political subdivision, office, or individual charged with the responsibility of administering and enforcing the provisions of this standard.

BUILDINGS: Structures, usually enclosed by walls and a roof, constructed to provide support or shelter for an intended occupancy.

DESIGN STRENGTH: The product of the nominal strength and a resistance factor.

DESIGNATED NONSTRUCTURAL SYSTEM: A nonstructural component or system that is essential to the intended function of a Risk Category IV structure or that is essential to Life Safety in structures assigned to other Risk Categories.

ESSENTIAL FACILITIES: Buildings and other structures that are intended to remain operational in the event of extreme environmental loading from flood, wind, tornado, snow, or earthquakes.

FACTORED LOAD: The product of the nominal load and a load factor.

HIGHLY TOXIC SUBSTANCE: As defined in 29 CFR 1910.1200, Appendix A, with Amendments as of February 1, 2000.

IMPORTANCE FACTOR: A factor that accounts for the degree of risk to human life, health, and welfare associated with damage to property or loss of use or functionality.

LIMIT STATE: A condition beyond which a structure or member becomes unfit for service and is judged either to be no longer useful for its intended function (serviceability limit state) or to be unsafe (strength limit state).

LOAD EFFECTS: Forces and deformations produced in structural members by the applied loads.

LOAD FACTOR: A factor that accounts for deviations of the actual load from the nominal load, for uncertainties in the analysis that transform the load into a load effect, and for the probability that more than one extreme load will occur simultaneously.

LOADS: Forces or other actions that result from the weight of all building materials, occupants and their possessions, environmental effects, differential movement, and restrained dimensional changes. Permanent loads are loads in which variations over time are rare or of small magnitude. All other loads are variable loads (see also *nominal loads*).

NOMINAL LOADS: The magnitudes of the loads specified in this standard for dead, live, soil, wind, tornado, snow, rain, flood, and earthquake loads.

NOMINAL STRENGTH: The capacity of a structure or member to resist the effects of loads as determined by computations using specified material strengths and dimensions and formulas derived from accepted principles of structural mechanics or by field tests or laboratory tests of scaled models, allowing for modeling effects and differences between laboratory and field conditions.

OCCUPANCY: The purpose for which a building or other structure, or part thereof, is used or intended to be used.

OTHER STRUCTURES: Structures other than buildings for which loads are specified in this standard.

P-DELTA EFFECT: The second-order effect on shears and moments of frame members induced by axial loads on a laterally displaced building frame.

PERFORMANCE-BASED PROCEDURES: An alternative to the prescriptive procedures in this standard characterized by project-specific engineering analysis, optionally supplemented by limited testing, to determine the computed reliability of an individual building or structure.

RESISTANCE FACTOR: A factor that accounts for deviations of the actual strength from the nominal strength and the manner and consequences of failure (also called "strength reduction factor").

RISK CATEGORY: A categorization of buildings and other structures for determination of flood, wind, tornado, snow, ice, and earthquake loads based on the risk associated with unacceptable performance. See Table 1.5-1.

SERVICE LOADS: Loads imparted on a building or other structure because of (1) self-weight and superimposed dead load, (2) live loads assumed to be present during normal occupancy or use of the building or other structure, (3) environmental loads that are expected to occur during the defined service life of a building or other structure, and (4) self-straining forces and effects. Service live loads and environmental loads for a particular limit state are permitted to be less than the design loads specified in the standard. Service loads shall be identified for each serviceability state being investigated.

STRENGTH DESIGN: A method of proportioning structural members such that the computed forces produced in the members by the factored loads do not exceed the member design strength (also called "load and resistance factor design").

TOXIC SUBSTANCE: As defined in 29 CFR 1910.1200, Appendix A, with Amendments as of February 1, 2000.

1.2.2 Symbols The following symbols apply only to the provisions of Chapter 1 as indicated:

D = Dead load
F_x = Minimum design lateral force applied to level x of the structure and used for purposes of evaluating structural integrity in accordance with Section 1.4.2
L = Live load
L_r = Roof live load
N = Notional load for structural integrity
R = Rain load
S = Snow load
W_x = Portion of the total dead load of the structure, D, located or assigned to level x

1.3 BASIC REQUIREMENTS

1.3.1 Strength and Stiffness Buildings and other structures, and all parts thereof, shall be designed and constructed with adequate strength and stiffness to provide structural stability, protect nonstructural components and systems, and meet the serviceability requirements of Section 1.3.2.

Acceptable strength shall be demonstrated using one or more of the following procedures:

(a) The strength procedures of Section 1.3.1.1;
(b) The allowable stress procedures of Section 1.3.1.2; or
(c) Subject to the approval of the Authority Having Jurisdiction for individual projects, the performance-based procedures of Section 1.3.1.3.

It shall be permitted to use alternative procedures for different parts of a structure and for different load combinations subject to the limitations of Chapter 2. Where resistance to extraordinary events is considered, the procedures of Section 2.5 shall be used.

1.3.1.1 Strength Procedures Structural and nonstructural components and their connections shall have adequate strength to resist the applicable load combinations of Section 2.3 of this standard without exceeding the applicable strength limit states for the materials of construction.

1.3.1.2 Allowable Stress Procedures Structural and nonstructural components and their connections shall have adequate strength to resist the applicable load combinations of Section 2.4 of this standard without exceeding the applicable allowable stresses for the materials of construction.

1.3.1.3 Performance-Based Procedures Structural and nonstructural components and their connections designed with performance-based procedures shall be demonstrated by analysis in accordance with Section 2.3.5 or by analysis procedures supplemented by testing to provide a reliability that is generally consistent with the target reliabilities stipulated in this section. Structural and nonstructural components subjected to dead, live, environmental, and other loads except earthquake, tsunami, flood, and loads from extraordinary events shall be based on the target reliabilities in Table 1.3-1. Structural systems subjected to earthquake shall be based on the target reliabilities in Tables 1.3-2 and 1.3-3. The design of structures subjected to tsunami loads shall be based on the target reliabilities in Table 1.3-4. Structures, components, and systems that are designed for extraordinary events using the requirements of Section 2.5 for scenarios approved by the Authority Having Jurisdiction shall be based on the target reliabilities in Table 1.3-5. The analysis procedures used shall account for uncertainties in loading and resistance.

Testing methods in Sections 1.3.1.3.2 shall only be applied to individual projects and shall not be applied to development of values of material resistance for general use in structural systems.

Structures and nonstructural components shall meet the serviceability and functionality requirements of Sections 1.3.2 and 1.3.3.

Performance-based design provisions for structures subjected to tsunamis shall conform to the requirements of Chapter 6.

1.3.1.3.1 Analysis Analysis shall use rational methods based on accepted principles of engineering mechanics and shall consider all significant sources of deformation and resistance. Assumptions of stiffness, strength, damping, and other properties of components and connections incorporated in the analysis shall be based on approved test data or referenced standards.

1.3.1.3.2 Project-Specific Performance Capability Testing Testing used to substantiate the project-specific performance capability of structural and nonstructural components and their connections under load shall accurately represent the materials, configuration, construction, loading intensity, and boundary conditions anticipated in the structure. Where an approved industry standard or practice that governs the testing of similar components exists, the test program and determination of design values from the test program shall be in accordance with those industry standards and practices. Where such standards or practices do not exist, specimens shall be constructed to a scale similar to that of the intended application unless it can be demonstrated that scale effects are not significant to the indicated performance. Evaluation of test results shall be made on the basis of the values obtained from not less than three tests, provided that the deviation of any value obtained from any single test does not vary from the average value for all tests by more than 15%. If such deviation from the average value for any test exceeds 15%, then additional tests shall be performed until the deviation of any test from the average value does not exceed 15% or a minimum of six tests have been performed. No test shall be eliminated unless a rationale for its exclusion is given. Test reports shall document the location, the time and date of the test, the characteristics of the tested specimen, the laboratory facilities, the test configuration, the applied loading and deformation under load, and the occurrence of any damage sustained by the specimen, together with the loading and deformation at which such damage occurred.

1.3.1.3.3 Documentation The procedures used to demonstrate compliance with this section and the results of analysis and testing shall be documented in one or more reports submitted to the Authority Having Jurisdiction and to an independent peer review.

1.3.1.3.4 Peer Review The procedures and results of analysis, testing, and calculation used to demonstrate compliance with the requirements of this section shall be subject to an independent peer review approved by the Authority Having Jurisdiction. The peer review shall comprise one or more persons having the necessary expertise and knowledge to evaluate compliance,

Table 1.3-1. Target Reliability (Annual Probability of Failure, P_F) and Associated Reliability Indices (β) for Load Conditions That Do Not Include Earthquake, Tsunami, or Extraordinary Events.

Basis	Risk Category			
	I	II	III	IV
Failure that is not sudden and does not lead to widespread progression of damage	$P_F = 1.25 \times 10^{-4}$ per year $\beta = 2.5$	$P_F = 3.0 \times 10^{-5}$ per year $\beta = 3.0$	$P_F = 1.25 \times 10^{-5}$ per year $\beta = 3.25$	$P_F = 5.0 \times 10^{-6}$ per year $\beta = 3.5$
Failure that is either sudden or leads to widespread progression of damage	$P_F = 3.0 \times 10^{-5}$ per year $\beta = 3.0$	$P_F = 5.0 \times 10^{-6}$ per year $\beta = 3.5$	$P_F = 2.0 \times 10^{-6}$ per year $\beta = 3.75$	$P_F = 7.0 \times 10^{-7}$ per year $\beta = 4.0$
Failure that is sudden and results in widespread progression of damage	$P_F = 5.0 \times 10^{-6}$ per year $\beta = 3.5$	$P_F = 7.0 \times 10^{-7}$ per year $\beta = 4.0$	$P_F = 2.5 \times 10^{-7}$ per year $\beta = 4.25$	$P_F = 1.0 \times 10^{-7}$ per year $\beta = 4.5$

Notes:
[1]The target reliability indexes are provided for a 50-year reference period, and the probabilities of failure have been annualized. The equations presented in Section 2.3.6 are based on reliability indexes for 50 years, because the load combination requirements in Section 2.3.2 are based on the maximum loads for the 50-year reference period.
[2]Commentary to Section 2.5 includes references to publications that describe the historic development of these target reliabilities for earthquake, tsunami, or extraordinary events.

Table 1.3-2. Target Reliability (Conditional Probability of Failure) for Structural Stability Caused by Earthquake.

Risk Category	Conditional Probability of Failure Caused by the MCE_R Shaking Hazard (%)
I and II	10
III	5
IV	2.5

Table 1.3-3. Target Reliability (Conditional Probability of Failure) for Ordinary Noncritical Structural Members Caused by Earthquake.

Risk Category	Conditional Probability of Component or Anchorage Failure Caused by the MCE_R Shaking Hazard (%)
I and II	25
III	15
IV	9

Table 1.3-4. Target Reliability (Conditional Probability of Failure) for Structural Elements Subject to Tsunami Inundation.

Risk Category	Conditional Probability of Failure Caused by the Maximum Considered Tsunami Hydrodynamic Pressure (%)
I	Not applicable
II	10
III	5
IV	3
Tsunami Vertical Evacuation Refuge Structure	1

including knowledge of the expected performance, the structural and component behavior, the particular loads considered, structural analysis of the type performed, the materials of construction, and laboratory testing of elements and components to determine structural resistance and performance characteristics. The review shall include assumptions, criteria, procedures, calculations, analytical models, test setup, test data, final drawings, and reports. Upon satisfactory completion, the peer reviewers shall submit a letter to the Authority Having Jurisdiction indicating the scope of their review and their findings.

1.3.2 Serviceability Structural systems, and members thereof, shall be designed under service loads to have adequate stiffness to limit deflections, lateral drift, vibration, or any other deformations that adversely affects the intended use and performance of buildings and other structures based on requirements set forth in the applicable codes and standards or as specified in the project design criteria.

1.3.3 Functionality Structural systems and members and connections thereof assigned to Risk Category IV shall be designed with reasonable probability to have adequate structural strength and stiffness to limit deflections, lateral drift, or other deformations such that their behavior would not prevent function of the facility immediately following any of the design-level environmental hazard events specified in this standard. Designated nonstructural systems and their attachment to the structure shall be designed with sufficient strength and stiffness such that their behavior would not prevent function immediately following any of the design-level environmental hazard events specified in this standard. Components of designated nonstructural systems shall be designed, qualified, or otherwise protected such

Table 1.3-5. Target Reliability (Maximum Conditional Probability of Failure) for Structural Strength and Stability Limit States Caused by Extraordinary Load Events.

Risk Category	Conditional Limit State Probability (%)
I	15
II	10
III	5
IV	2

that they shall be demonstrated capable of performing their critical function after the facility is subjected to any of the design-level environmental hazards specified in this standard.

The provisions in Sections 1.3.1.1 and 1.3.1.2 in this standard are deemed to comply with the requirements of this section.

1.3.4 Self-Straining Forces and Effects Provision shall be made for anticipated self-straining forces and effects arising from differential settlements of foundations and from restrained dimensional changes caused by temperature, moisture, shrinkage, creep, and similar effects.

1.3.5 Analysis Load effects on individual structural members shall be determined by methods of structural analysis that take into account equilibrium, general stability, geometric compatibility, and both short- and long-term material properties. Members that tend to accumulate residual deformations under repeated service loads shall have included in their analysis the added eccentricities expected to occur during their service life.

1.3.6 Counteracting Structural Actions All structural members and systems and all components and cladding in a building or other structure shall be designed to resist forces caused by earthquakes, wind, and tornadoes, with consideration of overturning, sliding, and uplift, and continuous load paths shall be provided for transmitting these forces to the foundation. Where sliding is used to isolate the elements, the effects of friction between sliding elements shall be included as a force. Where all or a portion of the resistance to these forces is provided by dead load, the dead load shall be taken as the minimum dead load likely to be in place during the event causing the considered forces. Consideration shall be given to the effects of vertical and horizontal deflections resulting from such forces.

1.3.7 Fire Resistance Structural fire resistance shall be provided in accordance with the requirements specified in the applicable building code. As an alternative, the performance-based design procedures in Appendix E are permitted, where approved.

1.4 GENERAL STRUCTURAL INTEGRITY

All structures shall be provided with a continuous load path in accordance with the requirements of Section 1.4.1 and shall have a complete lateral force-resisting system with adequate strength to resist the forces indicated in Section 1.4.2. All members of the structural system shall be connected to their supporting members in accordance with Section 1.4.3. Structural walls shall be anchored to diaphragms and supports in accordance with Section 1.4.4. The effects on the structure and its components caused by the forces stipulated in this section shall be taken as the notional load, N, and combined with the effects of other loads in accordance with the load combinations of Section 2.6. Where material resistance depends on load duration, notional loads are permitted to be taken as having a duration of 10 min. Structures designed in conformance with the requirements of this standard for Seismic Design Categories B, C, D, E, or F shall be deemed to comply with the requirements of Sections 1.4.2, 1.4.3, and 1.4.4.

1.4.1 Load Path Connections All parts of the structure between separation joints shall be interconnected to form a continuous path to the lateral force-resisting system, and the connections shall be capable of transmitting the lateral forces induced by the parts being connected. Any smaller portion of the structure shall be tied to the remainder of the structure with elements having the strength to resist a force of not less than 5% of the portion's weight.

1.4.2 Lateral Forces Each structure shall be analyzed for the effects of static lateral forces applied independently in each of two orthogonal directions. In each direction, the static lateral forces at all levels shall be applied simultaneously. For purposes of analysis, the force at each level shall be determined using Equation (1.4-1):

$$F_x = 0.01 W_x \quad (1.4\text{-}1)$$

where F_x is the design lateral force applied at story x, and W_x is the portion of the total dead load of the structure, D, located or assigned to level x.

Structures explicitly designed for stability, including second-order effects, shall be deemed to comply with the requirements of this section.

1.4.3 Connection to Supports A positive connection for resisting a horizontal force acting parallel to the member shall be provided for each beam, girder, or truss either directly to its supporting elements or to slabs designed to act as diaphragms. Where the connection is through a diaphragm, the member's supporting element shall also be connected to the diaphragm. The connection shall have the strength to resist a force of 5% of the unfactored dead load plus live load reaction imposed by the supported member on the supporting member.

1.4.4 Anchorage of Structural Walls Walls that provide vertical load bearing or lateral shear resistance for a portion of the structure shall be anchored to the roof and to all floors and members that provide lateral support for the wall or that are supported by the wall. The anchorage shall provide a direct connection between the walls and the roof or floor construction. The connections shall be capable of resisting a strength-level horizontal force perpendicular to the plane of the wall equal to 0.2 times the weight of the wall tributary to the connection, but not less than 5 lb/ft^2 (0.24 kN/m^2).

1.4.5 Extraordinary Loads and Events When considered, design for resistance to extraordinary loads and events shall be in accordance with the procedures of Section 2.5.

1.5 CLASSIFICATION OF BUILDINGS AND OTHER STRUCTURES

1.5.1 Risk Categorization Buildings and other structures shall be classified based on the risk to human life, health, and welfare associated with their damage or failure by nature of their occupancy or use, according to Table 1.5-1, for the purposes of applying flood, wind, tornado, snow, earthquake, and ice provisions. Each building or other structure shall be assigned to the highest applicable risk category or categories. Minimum design loads for structures shall incorporate the applicable Importance Factors given in Table 1.5-2, as required by other sections of this standard. Assignment of a building or other structure to multiple risk categories based on the type of load condition being evaluated (e.g., snow or seismic) shall be permitted.

When the building code or other referenced standard specifies an occupancy category, the risk category shall not be taken as lower than the occupancy category specified therein.

1.5.2 Multiple Risk Categories Where buildings or other structures are divided into portions with independent structural systems, the classification for each portion shall be permitted to be determined independently. Where building systems, such as required egress, HVAC, or electrical power, for a portion with a higher Risk Category pass through or depend on other portions of

Table 1.5-1. Risk Category of Buildings and Other Structures for Flood, Wind, Tornado, Snow, Earthquake, and Ice Loads.

Use or Occupancy of Buildings and Structures	Risk Category
Buildings and other structures that represent low risk to human life in the event of failure	I
All buildings and other structures except those listed in Risk Categories I, III, and IV	II
Buildings and other structures, the failure of which could pose a substantial risk to human life	III
Buildings and other structures not included in Risk Category IV, with potential to cause a substantial economic impact and/or mass disruption of day-to-day civilian life in the event of failure	
Buildings and other structures not included in Risk Category IV (including, but not limited to, facilities that manufacture, process, handle, store, use, or dispose of such substances as hazardous fuels, hazardous chemicals, hazardous waste, or explosives) containing toxic or explosive substances where the quantity of the material exceeds a threshold quantity established by the Authority Having Jurisdiction and is sufficient to pose a threat to the public if released*	
Buildings and other structures designated as Essential Facilities	IV
Buildings and other structures, the failure of which could pose a substantial hazard to the community	
Buildings and other structures (including, but not limited to, facilities that manufacture, process, handle, store, use, or dispose of such substances as hazardous fuels, hazardous chemicals, or hazardous waste) containing sufficient quantities of highly toxic substances where the quantity of the material exceeds a threshold quantity established by the Authority Having Jurisdiction and is sufficient to pose a threat to the public if released*	
Buildings and other structures required to maintain the functionality of other Risk Category IV structures	

*Buildings and other structures containing toxic, highly toxic, or explosive substances shall be eligible for classification to a lower risk category if it can be demonstrated to the satisfaction of the Authority Having Jurisdiction by a hazard assessment as described in Section 1.5.3 that a release of the substances is commensurate with the risk associated with that risk category.

the building or other structure having a lower risk category, those portions shall be assigned to the higher risk category.

1.5.3 Toxic, Highly Toxic, and Explosive Substances Buildings and other structures containing toxic, highly toxic, or explosive substances are permitted to be classified as Risk Category II structures if it can be demonstrated to the satisfaction of the Authority Having Jurisdiction by a hazard assessment as part of an overall risk management plan (RMP) that a release of the toxic, highly toxic, or explosive substances is not sufficient to pose a threat to the public.

To qualify for this reduced classification, the owner or operator of the buildings or other structures containing the toxic, highly toxic, or explosive substances shall have an RMP that incorporates three elements as a minimum: a hazard assessment, a prevention program, and an emergency response plan.

As a minimum, the hazard assessment shall include the preparation and reporting of worst-case release scenarios for each structure under consideration, showing the potential effect on the public for each. As a minimum, the worst-case event shall include the complete failure, for example, instantaneous release of the entire contents of a vessel, piping system, or other storage structure. A worst-case event includes, but is not limited to, a release during the design wind, design tornado, or design seismic event. In this assessment, the evaluation of the effectiveness of subsequent measures for accident mitigation shall be based on the assumption that the complete failure of the primary storage structure has occurred. The off-site impact shall be defined in terms of the population in the potentially affected area. To qualify for the reduced classification, the hazard assessment shall demonstrate that a release of the toxic, highly toxic, or explosive substances from a worst-case event does not pose a threat to the public outside the property boundary of the facility.

As a minimum, the prevention program shall consist of the comprehensive elements of process safety management, which is based on accident prevention through the application of management controls in the key areas of design, construction, operation, and maintenance. Secondary containment of the toxic, highly toxic, or explosive substances—including, but not limited to, double-wall tank, dike of sufficient size to contain a spill, or other means to contain a release of the toxic, highly toxic, or explosive substances within the property boundary of the facility and prevent release of harmful quantities of contaminants to the air, soil, groundwater, or surface water—is permitted to be used to mitigate the risk of release. Where secondary containment is provided, it shall be designed for all environmental loads and is not eligible for this reduced classification. In hurricane-prone regions, mandatory practices and procedures that effectively diminish the effects of wind on critical structural elements, or, alternatively, that protect against harmful releases during and after hurricanes, are permitted to be used to mitigate the risk of release.

As a minimum, the emergency response plan shall address public notification, emergency medical treatment for accidental exposure to humans, and procedures for emergency response to releases that have consequences beyond the property boundary of the facility. The emergency response plan shall address the potential that resources for response could be compromised by the event that has caused the emergency.

Table 1.5-2. Importance Factors by Risk Category of Buildings and Other Structures for Earthquake Loads.

Risk Category from Table 1.5-1	Seismic Importance Factor, I_e
I	1.00
II	1.00
III	1.25
IV	1.50

Notes: The component importance factor, I_p, applicable to earthquake loads is not included in this table because it depends on the importance of the individual component rather than that of the building as a whole, or its occupancy (see Section 13.1.3).

1.6 IN SITU LOAD TESTS

An in situ load test of any construction shall be conducted when required by the Authority Having Jurisdiction whenever there is reason to question its safety for the intended use. In situ load tests shall be conducted in accordance with this section. The load test procedures of this section shall not be used to evaluate seismic load effects, tsunami loads, or loads from extraordinary events.

1.6.1 Load Test Procedure Specified Elsewhere Where the applicable building code or a material standard approved by the Authority Having Jurisdiction includes an applicable load test procedure and acceptance criteria, the load test shall be conducted in accordance with that code or standard.

1.6.2 Load Test Procedure Not Specified Elsewhere In the absence of an applicable load test procedure within the building code or an applicable material standard approved by the Authority Having Jurisdiction, a load test procedure shall be developed and overseen by a registered design professional, who will also be responsible for interpreting the results of the load test. At a minimum, the applied load shall be determined using the load factors specified in the basic combinations in Section 2.3.1. The applied load shall simulate applicable loading and deformation conditions relevant to the performance of the element or structure being tested. For statically loaded components, the test load shall be left in place for a period of at least 1 h. For components that carry dynamic loads such as impact loads, the load shall be left in place for a period consistent with the component's actual function. For materials that have strengths that are dependent on load duration, the test load shall be adjusted to account for the difference in load duration of the test compared to the expected duration of the design loads being considered. The component or structure shall be considered to have satisfied the test requirements where the following criteria are met:

1. During and after the test, the component or structure remains stable;
2. Within 24 h after removal of the test load, the component or structure has recovered at least 75% of the maximum deflection;
3. Deflection that remains after the load test (if any) does not substantively affect serviceability as determined by the registered design professional.

1.7 CONSENSUS STANDARDS AND OTHER REFERENCED DOCUMENTS

This section lists the consensus standards and other documents that shall be considered part of this standard to the extent referenced in this chapter.

29 CFR (Code of Federal Regulations) Part 1910.1200. *OSHA Standards for General Industry*. Appendix A, with Amendments as of February 1, 2000. Occupational Safety and Health Administration, US Department of Labor, 2005.

Cited in: Section 1.2

CHAPTER 2
COMBINATIONS OF LOADS

2.1 GENERAL

Buildings and other structures shall be designed using the provisions of either Section 2.3 or 2.4. Where elements of a structure are designed by a particular material standard or specification, they shall be designed exclusively by either Section 2.3 or 2.4.

2.2 SYMBOLS

A_k = Load or load effect arising from extraordinary event, A
D = Dead load
D_i = Weight of ice
E = Earthquake load
F = Load caused by fluids with well-defined pressures and maximum heights other than those caused by groundwater pressure
F_a = Flood load
H = Load due to lateral earth pressure (including lateral earth pressure from fixed or moving surcharge loads), ground water pressure, or pressure of bulk materials
L = Live load
L_r = Roof live load
N = Notional load for structural integrity, Section 1.4
R = Rain load
S = Snow load
T = Cumulative effect of self-straining forces and effects arising from contraction or expansion resulting from environmental or operational temperature changes, shrinkage, moisture changes, creep in component materials, movement caused by differential settlement, or combinations thereof
W = Wind load
W_i = Wind-on-ice, determined in accordance with Chapter 10
W_T = Tornado load, determined in accordance with Chapter 32

2.3 LOAD COMBINATIONS FOR STRENGTH DESIGN

2.3.1 Basic Combinations Structures, components, and foundations shall be designed so that their design strength equals or exceeds the effects of the factored loads in the following combinations. Effects of one or more loads not acting shall be considered. Seismic load effects shall be combined loads, in accordance with Section 2.3.6. The most unfavorable effects from wind loads, tornado loads, and earthquake loads shall be considered, where appropriate, but they need not be assumed to act simultaneously. Refer to Sections 1.4, 2.3.6, 12.4, and 12.14.3 for the specific definition of the earthquake load effect, E. Each relevant strength limit state shall be investigated.

1a. $1.4D$
2a. $1.2D + 1.6L + (0.5L_r \text{ or } 0.3S \text{ or } 0.5R)$
3a. $1.2D + (1.6L_r \text{ or } 1.0S \text{ or } 1.6R) + (L \text{ or } 0.5W)$
4a. $1.2D + 1.0(W \text{ or } W_T) + L + (0.5L_r \text{ or } 0.3S \text{ or } 0.5R)$
5a. $0.9D + 1.0(W \text{ or } W_T)$

EXCEPTIONS:

1. The load factor on L in combinations 3a and 4a is permitted to equal 0.5 for all occupancies in which L_o in Chapter 4, Table 4.3-1, is less than or equal to 100 psf (4.78 kN/m^2), with the exception of garages or areas occupied as places of public assembly.
2. In combinations 2a and 4a, the companion load, S, shall be taken as either the flat roof snow load (p_f) or the sloped roof snow load (p_s).
3. Where using W_T in combination 4a, $(0.5L_r$ or $0.3S$ or $0.5R)$ is permitted to be replaced with 0.5 (L_r or R).

Where fluid loads, F, are present, they shall be included with the same load factor as dead load, D, in combinations 1a through 4a. Where loads, H, are present, they shall be included as follows:

1. Where the effect of H adds to the principal load effect, include H with a load factor of 1.6;
2. Where the effect of H resists the principal load effect, include H with a load factor of 0.9, where the load H is permanent, or a load factor of 0 for all other conditions.

Each relevant strength limit state shall be investigated.

2.3.2 Load Combinations Including Flood Load When a structure is located in a flood zone (Section 5.3.1), the following load combinations shall be considered in addition to the basic combinations in Section 2.3.1:

In V-Zones or Coastal A-Zones:

4b. $1.2D + 1.0W + 2.0F_a + L + (0.5L_r \text{ or } 0.3S \text{ or } 0.5R)$
5b. $0.9D + 1.0W + 2.0F_a$

In noncoastal A-Zones:

4b. $1.2D + 0.5W + 1.0F_a + L + (0.5L_r \text{ or } 0.3S \text{ or } 0.5R)$
5b. $0.9D + 0.5W + 1.0F_a$

2.3.3 Load Combinations Including Atmospheric Ice and Wind-on-Ice Loads When a structure is subjected to atmospheric ice and wind-on-ice loads, the following load combinations shall be considered:

2b. $1.2D + 1.6L + 0.2D_i + 0.3S$
4c. $1.2D + L + D_i + W_i + 0.3S$
4d. $1.2D + D_i$
5c. $0.9D + D_i + W_i$

2.3.4 Load Combinations Including Self-Straining Forces and Effects Where the structural effects of T are expected to adversely affect structural safety or performance, T shall be considered in combination with other loads. The load factor on T shall be established considering the uncertainty associated with the likely magnitude of the structural forces and effects, the probability that the maximum effect of T will occur simultaneously with other applied loadings, and the potential adverse consequences if the effect of T is greater than assumed. The load factor on T shall not have a value less than 1.0.

2.3.5 Load Combinations for Nonspecified Loads Where approved by the Authority Having Jurisdiction, the registered design professional is permitted to determine the combined load effect for strength design using a method that is consistent with the method on which the load combination requirements in Section 2.3.1 are based. Such a method must be probability based and must be accompanied by documentation regarding the analysis and collection of supporting data that are acceptable to the Authority Having Jurisdiction.

2.3.6 Basic Combinations with Seismic Load Effects When a structure is subject to seismic load effects, the following load combinations shall be considered in addition to the basic combinations in Section 2.3.1. The most unfavorable effects from seismic loads shall be investigated where appropriate, but they need not be considered to act simultaneously with wind or tornado loads.

Where the prescribed seismic load effect, $E = f(E_v, E_h)$, defined in Sections 12.4.2 or 12.14.3.1, is combined with the effects of other loads, the following seismic load combinations shall be used:

6. $1.2D + E_v + E_h + L + 0.15S$
7. $0.9D - E_v + E_h$

Where the seismic load effect with overstrength, $E_m = f(E_v, E_{mh})$, defined in Sections 12.4.3 or 12.14.3.2, is combined with the effects of other loads, the following seismic load combination for structures shall be used:

6. $1.2D + E_v + E_{mh} + L + 0.15S$
7. $0.9D - E_v + E_{mh}$

EXCEPTION:

1. The load factor on L in combination 6 is permitted to equal 0.5 for all occupancies in which L_o in Chapter 4, Table 4.3-1, is less than or equal to 100 psf (4.78 kN/m^2), with the exception of garages or areas occupied as places of public assembly.
2. In combination 6, the companion load, S, shall be taken as either the flat roof snow load (p_f) or the sloped roof snow load (p_s).

Where fluid loads, F, are present, they shall be included with the same load factor as dead load D in combinations 6 and 7.

Where loads, H, are present, they shall be included as follows:

1. Where the effect of H adds to the primary variable load effect, include H with a load factor of 1.6.
2. Where the effect of H resists the primary variable load effect, include H with a load factor of 0.9 where the load H is permanent, or a load factor of 0 for all other conditions.

2.3.7 Alternative Method for Loads from Water in Soil This section is permitted as an alternate for combining loads from soil and water in soil to the requirements in Section 2.3.1. The alternate is only permitted when using the loads defined in Section 3.3. For the purposes of this section, replace Symbol H from Section 2.2 with new symbols H_{eb} and H_w as follows:

H_{eb} = Load due to lateral earth pressure or pressure of bulk materials, including the effect of buoyancy from ground water pressure on the lateral pressure of earth or bulk materials but otherwise excluding ground water pressure

H_w = Load due to ground water pressure in soil

Where loads H_{eb} and H_w are present, they shall be included in the basic load combinations of Section 2.3.1 as follows:

1. Where the effect of H_{eb} adds to the principal load effect, compute H_{eb} based on the maximum ground water elevation and include H_{eb} with a load factor of 1.6.
2. Where the effect of H_{eb} resists the principal load effect, compute H_{eb} based on the minimum ground water elevation, and include H_{eb} with a load factor of 0.9 where the load H_{eb} is permanent or a load factor of 0 for all other conditions.
3. Where the effect of H_w adds to the principal load effect, include H_w based on the maximum ground water elevation with a load factor of 1.0.
4. Where the effect of H_w resists the principal load effect and the soil is permanent, compute H_w based on the minimum ground water elevation and include H_w with a load factor of 1.0, otherwise assign a load factor of 0.0 to H_w.

2.4 LOAD COMBINATIONS FOR ALLOWABLE STRESS DESIGN

2.4.1 Basic Combinations Loads listed herein shall be considered to act in the following combinations; whichever produces the most unfavorable effect on the building, foundation, or structural member shall be considered. Effects of one or more loads not acting shall be considered. Seismic load effects shall be combined with other loads, in accordance with Section 2.4.5. The most unfavorable effects from wind loads, tornado loads, and earthquake loads shall be considered, where appropriate, but they need not be assumed to act simultaneously. Refer to Sections 1.4, 2.4.5, 12.4, and 12.14.3 for the specific definition of the earthquake load effect, E.

Increases in allowable stress shall not be used with the loads or load combinations given in this standard unless it can be demonstrated that such an increase is justified by structural behavior caused by rate or duration of load.

1a. D
2a. $D + L$
3a. $D + (L_r \text{ or } 0.7S \text{ or } R)$
4a. $D + 0.75L + 0.75(L_r \text{ or } 0.7S \text{ or } R)$
5a. $D + 0.6(W \text{ or } W_T)$
6a. $D + 0.75L + 0.75(0.6(W \text{ or } W_T)) + 0.75(L_r \text{ or } 0.7S \text{ or } R)$
7a. $0.6D + 0.6(W \text{ or } W_T)$

EXCEPTIONS:

1. In combinations 4a and 6a, the companion load, S, shall be taken as either the flat roof snow load (p_f) or the sloped roof snow load (p_s).
2. For nonbuilding structures in which the wind or tornado load is determined from force coefficients, C_f, identified in Figures 29.4-1, 29.4-2, and 29.4-3 and the projected area contributing wind or tornado force to a foundation element exceeds 1,000 ft^2 (93 m^2) on either a vertical or a horizontal plane, it shall be permitted to replace (W or W_T) with 0.9 (W or W_T) in combination 7a for design of the foundation, excluding anchorage of the structure to the foundation.

3. Where using W_T in combination 6a, 0.75 (L_r or 0.7S or R) is permitted to be replaced with 0.75 (L_r or R).

Where fluid loads, F, are present, they shall be included in combinations 1a through 6a, with the same factor as that used for dead load, D.

Where loads, H, are present, they shall be included as follows:

1. Where the effect of H adds to the principal load effect, include H with a load factor of 1.0.
2. Where the effect of H resists the principal load effect, include H with a load factor of 0.6 where the load H is permanent, or a load factor of 0 for all other conditions.

2.4.2 Load Combinations Including Flood Load When a structure is located in a flood zone, the following load combinations shall be considered:

In V-Zones or Coastal A-Zones:

5b. $D + 0.6W + 1.5F_a$
6b. $D + 0.75L + 0.75(0.6W) + 0.75(L_r \text{ or } 0.7S \text{ or } R) + 1.5F_a$
7b. $0.6D + 0.6W + 1.5F_a$

In noncoastal A-Zones:

5b. $D + 0.6W + 0.75F_a$
6b. $D + 0.75L + 0.75(0.6W) + 0.75(L_r \text{ or } 0.7S \text{ or } R) + 0.75F_a$
7b. $0.6D + 0.6W + 0.75F_a$

2.4.3 Load Combinations Including Atmospheric Ice and Wind-on-Ice Loads When a structure is subjected to atmospheric ice and wind-on-ice loads, the following load combinations shall be considered:

1b. $D + 0.7D_i$
2b. $D + L + 0.7D_i$
3b. $D + 0.7D_i + 0.7W_i + 0.7S$
7c. $0.6D + 0.7D_i + 0.7W_i$

2.4.4 Load Combinations Including Self-Straining Forces and Effects Where the structural effects of T are expected to adversely affect structural safety or performance, T shall be considered in combination with other loads. Where the maximum effect of load, T, is unlikely to occur simultaneously with the maximum effects of other variable loads, it shall be permitted to reduce the magnitude of T considered in combination with these other loads. The fraction of T considered in combination with other loads shall not be less than 0.75.

2.4.5 Basic Combinations with Seismic Load Effects When a structure is subject to seismic load effects, the following load combinations shall be considered in addition to the basic combinations and associated exceptions detailed in Section 2.4.1.

Where the prescribed seismic load effect, $E = f(E_v, E_h)$, defined in Section 12.4.2 or 12.14.3.1, is combined with the effects of other loads, the following seismic load combinations shall be used:

8. $1.0D + 0.7E_v + 0.7E_h$
9. $1.0D + 0.525E_v + 0.525E_h + 0.75L + 0.1S$
10. $0.6D - 0.7E_v + 0.7E_h$

Where the seismic load effect with overstrength, $E_m = f(E_v, E_{mh})$, defined in Section 12.4.3 or 12.14.3.2, is combined with the effects of other loads, the following seismic load combinations for structures not subject to flood or atmospheric ice loads shall be used:

8. $1.0D + 0.7E_v + 0.7E_{mh}$
9. $1.0D + 0.525E_v + 0.525E_{mh} + 0.75L + 0.1S$
10. $0.6D - 0.7E_v + 0.7E_{mh}$

Where allowable stress design methodologies are used with the seismic load effect defined in Section 12.4.3 or 12.14.3.2 and applied in load combinations 8, 9, or 10, allowable stresses are permitted to be determined using an allowable stress increase factor of 1.2. This increase shall not be combined with increases in allowable stresses or load combination reductions otherwise permitted by this standard, or the material reference document, except for increases caused by adjustment factors, in accordance with AWC NDS.

EXCEPTIONS:

1. In combination 9, the companion load, S, shall be taken as either the flat roof snow load (p_f) or the sloped roof snow load (p_s).
2. It shall be permitted to replace 0.6D with 0.9D in combination 10 for the design of special reinforced masonry shear walls where the walls satisfy the requirement of Section 14.4.2.

Where fluid loads, F, are present, they shall be included in combinations 8, 9, and 10, with the same factor as that used for dead load, D.

Where loads H are present, they shall be included as follows:

1. Where the effect of H adds to the primary variable load effect, include H with a load factor of 1.0.
2. Where the effect of H resists the primary variable load effect, include H with a load factor of 0.6 where the load H is permanent or a load factor of 0 for all other conditions.

2.5 LOAD COMBINATIONS FOR EXTRAORDINARY EVENTS

2.5.1 Applicability Where required by the owner or applicable code, strength and stability shall be checked to ensure that structures are capable of withstanding the effects of extraordinary (i.e., low probability) events, such as fires, explosions, and vehicular impact without disproportionate collapse.

2.5.2 Load Combinations

2.5.2.1 Capacity For checking the capacity of a structure or structural element to withstand the effect of an extraordinary event, the following gravity load combination shall be considered:

$$(0.9 \text{ or } 1.2)D + A_k + 0.5L + 0.15S \qquad (2.5\text{-}1)$$

in which A_k is the load or load effect resulting from the extraordinary event, A.

2.5.2.2 Residual Capacity For checking the residual load-carrying capacity of a structure or structural element following the occurrence of a damaging event, selected load-bearing elements identified by the registered design professional shall be notionally removed, and the capacity of the damaged structure shall be evaluated using the following gravity load combination:

$$(0.9 \text{ or } 1.2)D + 0.5L + 0.2(L_r \text{ or } 0.7S \text{ or } R) \qquad (2.5\text{-}2)$$

2.5.3 Stability Requirements Stability shall be provided for the structure as a whole and for each of its elements. Any method that considers the influence of second-order effects is permitted.

2.6 LOAD COMBINATIONS FOR GENERAL STRUCTURAL INTEGRITY LOADS

The notional loads, N, specified in Section 1.4 for structural integrity shall be combined with other loads, in accordance with Section 2.6.1 for strength design and Section 2.6.2 for allowable stress design.

2.6.1 Strength Design Notional Load Combinations

1. $1.2D + 1.0N + L + 0.15S$
2. $0.9D + 1.0N$

2.6.2 Allowable Stress Design Notional Load Combinations

1. $D + 0.7N$
2. $D + 0.75(0.7N) + 0.75L + 0.75$ (L_r or $0.7S$ or R)
3. $0.6D + 0.7N$

2.7 CONSENSUS STANDARDS AND OTHER REFERENCED DOCUMENTS

This section lists the consensus standards and other documents that shall be considered part of this standard to the extent referenced in this chapter.

AWC NDS, *National Design Specification for Wood Construction, Including Supplements*, 2018 edition. American Wood Council, 2017.

Cited in: Section 2.4.5

CHAPTER 3
DEAD LOADS, SOIL LOADS, AND HYDROSTATIC PRESSURE

3.1 DEAD LOADS

3.1.1 Definition Dead loads consist of the weight of all materials of construction incorporated into the building including, but not limited to, walls, floors, roofs, ceilings, stairways, built-in partitions, finishes, cladding, and other similarly incorporated architectural and structural items, and the weight of fixed service equipment, including cranes and material handling systems.

3.1.2 Weights of Materials of Construction In determining dead loads for purposes of design, the actual weights of materials of construction shall be used. In the absence of definite information, values approved by the Authority Having Jurisdiction shall be used.

3.1.3 Weight of Fixed Service Equipment In determining dead loads for purposes of design, the weight of fixed service equipment, including the maximum weight of the contents of fixed service equipment, shall be included. The components of fixed service equipment that are variable, such as liquid contents and movable trays, shall not be used to counteract forces causing overturning, sliding, and uplift conditions in accordance with Section 1.3.6.

EXCEPTIONS:

1. Where force effects are the result of the presence of the variable components, the components are permitted to be used to counter those load effects. In such cases, the structure shall be designed for force effects with the variable components present and with them absent.
2. For the calculation of seismic force effects, the components of fixed service equipment that are variable, such as liquid contents and movable trays, need not exceed those expected during normal operation.

3.1.4 Vegetative and Landscaped Roofs The weight of all landscaping and hardscaping materials for vegetative and landscaped roofs shall be considered as dead load. The weight shall be computed considering both fully saturated soil and drainage layer materials, and fully dry soil and drainage layer materials, to determine the most severe load effects on the structure.

3.1.5 Solar Panels The weight of solar panels, their support system, and ballast shall be considered as dead load.

Table 3.2-1. Lateral Soil Load.

Description of Backfill Material	Unified Soil Classification	Lateral Soil Load* psf per foot of depth (kN/m^2 per meter of depth)	
		Active Pressure	At-rest Pressure
Well-graded, clean gravels, gravel-sand mixes	GW	35 (5.50)	60 (9.43)
Poorly graded, clean gravels, gravel-sand mixes	GP	35 (5.50)	60 (9.43)
Silty gravels, poorly graded gravel-sand mixes	GM	35 (5.50)	60 (9.43)
Clayey gravels, poorly graded gravel-and-clay mixes	GC	45 (7.07)	60 (9.43)
Well-graded, clean sands, gravel-sand mixes	SW	35 (5.50)	60 (9.43)
Poorly graded, clean sands, sand-gravel mixes	SP	35 (5.50)	60 (9.43)
Silty sands, poorly graded sand-silt mixes	SM	45 (7.07)	60 (9.43)
Sand-silt clay mix with plastic fines	SM-SC	85 (13.35)	100 (15.71)
Clayey sands, poorly graded sand-clay mixes	SC	85 (13.35)	100 (15.71)
Inorganic silts and clayey silts	ML	85 (13.35)	100 (15.71)
Mixture of inorganic silt and clay	ML-CL	85 (13.35)	100 (15.71)
Inorganic clays of low to medium plasticity	CL	100 (15.71)	100 (15.71)
Organic silts and silt-clays, low plasticity	OL	Unsuitable as backfill material	
Inorganic clayey silts, elastic silts	MH	Unsuitable as backfill material	
Inorganic clays of high plasticity	CH	Unsuitable as backfill material	
Organic clays and silty clays	OH	Unsuitable as backfill material	

*Lateral soil loads are given for moist conditions for the specified soils at their optimum densities. Actual field conditions shall govern. Submerged or saturated soil pressures shall include the weight of the buoyant soil plus the hydrostatic loads.

3.2 SOIL LOADS AND HYDROSTATIC PRESSURE

3.2.1 Lateral Pressures Structures below grade shall be designed to resist lateral soil loads from adjacent soil. If lateral soil loads are not given in a geotechnical report approved by the Authority Having Jurisdiction, then the lateral soil loads specified in Table 3.2-1 shall be used as the minimum design lateral soil loads. Foundation walls and other walls in which horizontal movement is restricted at the top shall be designed for at-rest pressure. Walls that are free to move and rotate at the top, such as retaining walls, shall be permitted to be designed for active pressure.

Where applicable, lateral pressure from fixed or moving surcharge loads shall be added to the lateral soil loads. When a portion or the whole of the adjacent soil is below a free-water surface, computations shall be based on the weight of the soil diminished by buoyancy, plus full hydrostatic pressure. The lateral pressure shall be increased if expansive soils are present at the site, as determined by a geotechnical investigation.

EXCEPTION: Foundation walls extending not more than 8 ft (2.44 m) below grade and laterally supported at the top by flexible diaphragms shall be permitted to be designed for active pressure.

3.2.2 Uplift Loads on Floors and Foundations Basement floors, slabs on ground, foundations, and similar approximately horizontal elements below grade shall be designed to resist uplift loads where applicable. The upward pressure of water shall be taken as the full hydrostatic pressure applied over the entire area. The hydrostatic load shall be determined based on the elevation of the underside of the element being evaluated.

Foundations, slabs on ground, and other components placed on expansive soils shall be designed to tolerate the movement or resist the upward loads caused by the expansive soils, or the expansive soil shall be removed or stabilized around and beneath the structure.

3.3 ALTERNATIVE METHOD FOR LOADS FROM WATER IN SOIL

This section is permitted as an alternative method to the requirements in Section 3.2. The alternative separates the computation of lateral loads from soil pressure from ground water pressure in soil. If this alternative is used, the load factors and combinations in Section 2.3.7 shall be used. The lateral pressure from soil shall be computed as required in Section 3.2.1, but the ground water pressure in soil shall be included in H_w, not in H. H_w for lateral and uplift loads shall be based upon the maximum ground water elevation. The maximum ground water elevation shall be established such that the annual probability of exceedance does not exceed the following values:

1. 0.0024 for Risk Category I
2. 0.0012 for Risk Category II
3. 0.0006 for Risk Category III
4. 0.0003 for Risk Category IV

For structures not located in a Design Flood Zone (see Chapter 5), the maximum ground water elevation need not be above the ground surface. Where lateral pressures are used to resist other variable loads and the soil is considered to be permanent, H_w shall be based upon the minimum ground water elevation. The minimum ground water elevation shall be established so that the annual probability of being at a lower elevation does not exceed the same values specified for the maximum ground water elevation. The minimum ground water elevation need not be taken below the lowest portion of the structure.

3.4 CONSENSUS STANDARDS AND OTHER REFERENCED DOCUMENTS

No consensus standards and other documents that shall be considered part of this standard are referenced in this chapter.

CHAPTER 4
LIVE LOADS

4.1 DEFINITIONS

The following definitions apply to the provisions of this chapter.

FIXED LADDER: A ladder that is permanently attached to a structure, building, or equipment.

GRAB BAR SYSTEM: A bar and associated anchorages and attachments to the structural system, for the support of body weight in locations such as toilets, showers, and tub enclosures.

GUARD SYSTEM: A component or assembly of components, including anchorages and attachments to the structural system, near open sides of an elevated surface for the purpose of minimizing the possibility of a fall from the elevated surface by people, equipment, or material.

HANDRAIL SYSTEM: A rail grasped by hand for guidance and support and associated anchorages and attachments to the structural system.

HELIPAD: A structural surface that is used for landing, taking off, taxiing, and parking helicopters.

LIVE LOAD: A load produced by the use and occupancy of the building or other structure that does not include construction or environmental loads, such as wind load, snow load, rain load, earthquake load, flood load, or dead load.

ROOF LIVE LOAD: A load on a roof produced (1) during maintenance by workers, equipment, and materials, and (2) during the life of the structure by movable objects, such as planters or other similar small decorative appurtenances that are not occupancy related. An occupancy-related live load on a roof such as rooftop assembly areas, rooftop decks, and vegetative or landscaped roofs with occupiable areas, is considered to be a live load rather than a roof live load.

SCREEN ENCLOSURE: A building or part thereof, in whole or in part self-supporting, having walls and a roof of insect or sun screening using fiberglass, aluminum, plastic, or similar lightweight netting material, which encloses an occupancy or use such as outdoor swimming pools, patios or decks, and horticultural and agricultural production facilities.

VEHICLE BARRIER SYSTEM: A system of components, including anchorages and attachments to the structural system near open sides or walls of garage floors or ramps, that acts as a restraint for vehicles.

4.2 LOADS NOT SPECIFIED

For occupancies or uses not designated in this chapter, the live load shall be determined in accordance with a method approved by the Authority Having Jurisdiction.

4.3 UNIFORMLY DISTRIBUTED LIVE LOADS

4.3.1 Required Live Loads The live loads used in the design of buildings and other structures shall be the maximum loads expected by the intended use or occupancy but shall not be less than the minimum uniformly distributed live loads required by Table 4.3-1.

4.3.2 Provision for Partitions In office buildings and in other buildings where partition locations are subject to change, provisions for partition weight shall be made, whether or not partitions are shown on the plans. The partition load shall not be less than 15 psf (0.72 kN/m^2) and shall not be reduced per Section 4.7.

EXCEPTION: A partition live load is not required where the minimum specified live load is 80 psf (3.83 kN/m^2) or greater.

4.3.3 Partial Loading The uniform live load applied only to a portion of a structure or member shall be accounted for if it produces a more unfavorable load effect than the uniform live load applied over the full structure or member. Uniform live loads applied to a portion of a structure or member are permitted to be reduced in accordance with Section 4.7.

4.3.3.1 Partial Loading of Roofs Where uniform roof live loads are reduced to less than 20 psf (0.96 kN/m^2) in accordance with Section 4.8.2 and are applied to the design of structural members arranged so as to create continuity, the reduced roof live load shall be applied to adjacent spans or to alternate spans, whichever produces the greatest unfavorable load effect.

4.3.4 Interior Walls and Partitions Interior walls and partitions that exceed 6 ft (1.83 m) in height, including their finish materials, shall have adequate strength and stiffness to resist the loads to which they are subjected but not less than a horizontal load of 5 psf (0.240 kN/m^2).

4.4 CONCENTRATED LIVE LOADS

Floors, roofs, and other similar surfaces shall be designed to support the uniformly distributed live loads prescribed in Section 4.3 or the concentrated live loads given in Table 4.3-1, whichever produces the greater load effects. Unless otherwise specified, the indicated concentrated load shall be assumed to be uniformly distributed over an area of 2.5 ft (762 mm) by 2.5 ft (762 mm) and shall be located so as to produce the maximum load effects in the members.

4.5 LOADS ON HANDRAIL, GUARD, GRAB BAR, AND VEHICLE BARRIER SYSTEMS, AND ON SHOWER SEATS AND FIXED LADDERS

4.5.1 Handrail and Guard Systems Handrail and guard systems shall be designed to resist a single concentrated load of 200 lb (0.89 kN) applied in any direction at any point on the

Table 4.3-1. Minimum Uniformly Distributed Live Loads, L_o, and Minimum Concentrated Live Loads.

Occupancy or Use	Uniform, L_o psf (kN/m^2)	Live Load Reduction Permitted? (Section No.)	Multiple-Story Live Load Reduction Permitted? (Section No.)	Concentrated lb (kN)	Also See Section
Apartments (See Residential)					
Access floor systems					
Office use	50 (2.40)	Yes (4.7.2)	Yes (4.7.2)	2,000 (8.90)	
Computer use	100 (4.79)	Yes (4.7.2)	Yes (4.7.2)	2,000 (8.90)	
Armories and drill rooms	150 (7.18)	No (4.7.5)	No (4.7.5)		
Assembly areas					
Fixed seats (fastened to floors)	60 (2.87)	No (4.7.5)	No (4.7.5)		
Lobbies	100 (4.79)	No (4.7.5)	No (4.7.5)		
Movable seats	100 (4.79)	No (4.7.5)	No (4.7.5)		
Platforms (assembly)	100 (4.79)	No (4.7.5)	No (4.7.5)		
Stage floors	150 (7.18)	No (4.7.5)	No (4.7.5)		
Bleachers, folding and telescopic seating, and grandstands	100 (4.79)	No (4.7.5)	No (4.7.5)		4.14
Stadiums and arenas with fixed seats (fastened to the floor)	60 (2.87)	No (4.7.5)	No (4.7.5)		4.14
Other assembly areas	100 (4.79)	No (4.7.5)	No (4.7.5)		
Balconies and decks	1.5 times the live load for the area served. Not required to exceed 100 psf (4.79 kN/m^2)	Yes (4.7.2)	Yes (4.7.2)		
Catwalks for maintenance and service access	40 (1.92)	Yes (4.7.2)	Yes (4.7.2)	300 (1.33)	
Corridors					
First floor	100 (4.79)	Yes (4.7.2)	Yes (4.7.2)		
Other floors	Same as occupancy served except as indicated				
Dining rooms and restaurants	100 (4.79)	No (4.7.5)	No (4.7.5)		
Dwellings (See Residential)					
Elevator machine room and control room grating (on area of 2 in. by 2 in. [50 mm by 50 mm])		—	—	300 (1.33)	
Finish light floor plate construction (on area of 1 in. by 1 in. [25 mm by 25 mm])		—	—	200 (0.89)	
Fire escapes	100 (4.79)	Yes (4.7.2)	Yes (4.7.2)		
On single-family dwellings only	40 (1.92)	Yes (4.7.2)	Yes (4.7.2)		
Fixed ladders		—	—	See Sec. 4.5.4	
Garages and Vehicle Floors					4.10
Passenger vehicle garages	40 (1.92)	No (4.7.4)	Yes (4.7.4)	See Sec. 4.10.1.	
Trucks and bus garages	See Section 4.10.2	—	—	See Sec. 4.10.2.	
Emergency vehicles		—	—	See Sec. 4.10.4	
Handrails and Guard systems	See Section 4.5.1	—	—	See Sec. 4.5.1.	
Grab bars		—	—	See Sec. 4.5.2	
Helipads (See Section 4.11)					
Helicopter takeoff weight 3,000 lb (13.35 kN) or less	40 (1.92)	No (4.11.1)	—	See Sec. 4.11.2.	
Helicopter takeoff weight more than 3,000 lb (13.35 kN)	60 (2.87)	No (4.11.1)	—	See Sec. 4.11.2	
Hospitals					
Operating rooms, laboratories	60 (2.87)	Yes (4.7.2)	Yes (4.7.2)	1,000 (4.45)	
Patient rooms	40 (1.92)	Yes (4.7.2)	Yes (4.7.2)	1,000 (4.45)	
Corridors above first floor	80 (3.83)	Yes (4.7.2)	Yes (4.7.2)	1,000 (4.45)	
Hotels (See Residential)					
Libraries					
Reading rooms	60 (2.87)	Yes (4.7.2)	Yes (4.7.2)	1,000 (4.45)	
Stack rooms	150 (7.18)	No (4.7.3)	Yes (4.7.3)	1,000 (4.45)	4.13
Corridors above first floor	80 (3.83)	Yes (4.7.2)	Yes (4.7.2)	1,000 (4.45)	

continues

Table 4.3-1 *(Continued)*. Minimum Uniformly Distributed Live Loads, L_o, and Minimum Concentrated Live Loads.

Occupancy or Use	Uniform, L_o psf (kN/m^2)	Live Load Reduction Permitted? (Section No.)	Multiple-Story Live Load Reduction Permitted? (Section No.)	Concentrated lb (kN)	Also See Section
Manufacturing					
Light	125 (6.00)	No (4.7.3)	Yes (4.7.3)	2,000 (8.90)	
Heavy	250 (11.97)	No (4.7.3)	Yes (4.7.3)	3,000 (13.35)	
Office buildings					
File and computer rooms shall be designed for heavier loads based on anticipated occupancy					
Lobbies and first-floor corridors	100 (4.79)	Yes (4.7.2)	Yes (4.7.2)	2,000 (8.90)	
Offices	50 (2.40)	Yes (4.7.2)	Yes (4.7.2)	2,000 (8.90)	
Corridors above first floor	80 (3.83)	Yes (4.7.2)	Yes (4.7.2)	2,000 (8.90)	
Penal institutions					
Cell blocks	40 (1.92)	Yes (4.7.2)	Yes (4.7.2)		
Corridors	100 (4.79)	Yes (4.7.2)	Yes (4.7.2)		
Public restrooms	Same as live load for area served but not required to exceed 60 (2.87)	Yes (4.7.2)	Yes (4.7.2)		
Recreational uses					
Bowling alleys, poolrooms, and similar uses	75 (3.59)	No (4.7.5)	No (4.7.5)		
Dance halls and ballrooms	100 (4.79)	No (4.7.5)	No (4.7.5)		
Gymnasiums	100 (4.79)	No (4.7.5)	No (4.7.5)		
Theater projection, control, and follow spot rooms	50 (2.40)	Yes (4.7.2)	Yes(4.7.2)		
Roller skating rinks	100 (4.79)	No (4.7.5)	No (4.7.5)		
Residential					
One- and two-family dwellings					
Uninhabitable attics without storage	10 (0.48)	Yes (4.7.2)	Yes (4.7.2)		4.12.1
Uninhabitable attics with storage	20 (0.96)	Yes (4.7.2)	Yes (4.7.2)		4.12.2
Habitable attics and sleeping areas	30 (1.44)	Yes (4.7.2)	Yes (4.7.2)		
All other areas except stairs	40 (1.92)	Yes (4.7.2)	Yes (4.7.2)		
All other residential occupancies					
Private rooms and corridors serving them	40 (1.92)	Yes (4.7.2)	Yes (4.7.2)		
Public rooms	100 (4.79)	No (4.7.5)	No (4.7.5)		
Corridors serving public rooms	100 (4.79)	Yes (4.7.2)	Yes (4.7.2)		
Roofs					
Ordinary flat, pitched, and curved roofs	20 (0.96)	Yes (4.8.2)	—		4.8.1
Roof areas used for assembly purposes	100 (4.79)	No (4.7.5)			
Roof areas used for occupancies other than assembly	Same as occupancy served	Yes (4.8.3)	—		
Vegetative and landscaped roofs					
Roof areas not intended for occupancy	20 (0.96)	Yes (4.8.2)	—		
Roof areas used for assembly purposes	100 (4.79)	No (4.7.5)	—		
Roof areas used for occupancies other than assembly	Same as occupancy served	Yes (4.8.3)	—		
Awnings and canopies					
Fabric construction supported by a skeleton structure	5 (0.24)	No (4.8.2)	—		
Screen enclosure support frame	5 (0.24) based on the tributary area of the roof supported by the frame member	No (4.8.2)	—	200 (0.89)	
All other construction	20 (0.96)	Yes (4.8.2)	—		4.8.1

continues

Table 4.3-1 *(Continued)*. Minimum Uniformly Distributed Live Loads, L_o, and Minimum Concentrated Live Loads.

Occupancy or Use	Uniform, L_o psf (kN/m²)	Live Load Reduction Permitted? (Section No.)	Multiple-Story Live Load Reduction Permitted? (Section No.)	Concentrated lb (kN)	Also See Section
Primary roof members, exposed to a work floor					
Single panel point of lower chord of roof trusses or any point along primary structural members supporting roofs over manufacturing, storage warehouses, and repair garages				2,000 (8.90)	
All other primary roof members		—	—	300 (1.33)	
All roof surfaces subject to maintenance workers		—	—	300 (1.33)	
Schools					
Classrooms	40 (1.92)	Yes (4.7.2)	Yes (4.7.2)	1,000 (4.45)	
Corridors above first floor	80 (3.83)	Yes (4.7.2)	Yes (4.7.2)	1,000 (4.45)	
First-floor corridors	100 (4.79)	Yes (4.7.2)	Yes (4.7.2)	1,000 (4.45)	
Scuttles, skylight ribs, and accessible ceilings				200 (0.89)	
Sidewalks, vehicular driveways, and yards subject to trucking	250 (11.97)	No (4.7.3)	Yes (4.7.3)	8,000 (35.60)	4.10.3 4.10.4
Stairs and exit ways	100 (4.79)	Yes (4.7.2)	Yes (4.7.2)	300 (1.33)	4.15
One- and two-family dwellings only	40 (1.92)	Yes (4.7.2)	Yes (4.7.2)	300 (1.33)	4.15
Storage areas above ceilings	20 (0.96)	Yes (4.7.2)	Yes (4.7.2)		
Storage warehouses (shall be designed for heavier loads if required for anticipated storage)					
Light	125 (6.00)	No (4.7.3)	Yes (4.7.3)		
Heavy	250 (11.97)	No (4.7.3)	Yes (4.7.3)		
Stores					
Retail					
First floor	100 (4.79)	Yes (4.7.2)	Yes (4.7.2)	1,000 (4.45)	
Upper floors	75 (3.59)	Yes (4.7.2)	Yes (4.7.2)	1,000 (4.45)	
Wholesale, all floors	125 (6.00)	No (4.7.3)	Yes (4.7.3)	1,000 (4.45)	
Vehicle barriers				See Sec. 4.5.3	
Walkways and elevated platforms (other than exit ways)	60 (2.87)	Yes (4.7.2)	Yes (4.7.2)		
Yards and terraces, pedestrian	100 (4.79)	No (4.7.5)	No (4.7.5)		

handrail or top rail to produce the maximum load effect on the element being considered and to transfer this load through the supports to the structure.

4.5.1.1 Uniform Load Handrail and guard systems shall also be designed to resist a load of 50 lb/ft (pound-force per linear foot) (0.73 kN/m) applied in any direction along the handrail or top rail and to transfer this load through the supports to the structure. This load need not be assumed to act concurrently with the concentrated load specified in Section 4.5.1.

EXCEPTIONS: The uniform load need not be considered for the following occupancies:

1. One- and two-family dwellings;
2. Factory, industrial, and storage occupancies in areas that are not accessible to the public and that serve an occupant load not greater than 50; and
3. Roofs not intended for occupancy.

4.5.1.2 Guard System Component Loads Balusters, panel fillers, and guard system infill components, including all rails except the handrail and the top rail, shall be designed to resist a horizontally applied normal load of 50 lb (0.22 kN) on an area not to exceed 12 in. × 12 in. (305 mm × 305 mm), including openings and space between rails and located so as to produce the maximum load effects. Reactions due to this loading are not required to be superimposed with the loads specified in Sections 4.5.1 and 4.5.1.1.

4.5.2 Grab Bar Systems and Shower Seats Grab bar systems and shower seats shall be designed to resist a single concentrated load of 250 lb (1.11 kN) applied in any direction at any point on the grab bar or shower seat to produce the maximum load effect.

4.5.3 Vehicle Barrier Systems Vehicle barrier systems for passenger vehicles shall be designed to resist a single concentrated load of 6,000 lb (26.70 kN) applied horizontally in any direction to the barrier system and shall have anchorages or attachments capable of transferring this load to the structure. For design of the system, the load shall be assumed to act at heights between 1 ft 6 in. (460 mm) and 2 ft 3 in. (686 mm) above the floor or ramp surface, located to produce the maximum load effects. The load shall be applied on an area not to exceed 12 in. × 12 in. (305 mm × 305 mm). This load is not required to act concurrently with any handrail or guard system loadings specified in Section 4.5.1. Vehicle barrier systems in garages accommodating trucks and buses shall be designed in accordance with *AASHTO LRFD Bridge Design Specifications*.

4.5.4 Fixed Ladders Fixed ladders with rungs shall be designed to resist a single concentrated load of 300 lb (1.33 kN) applied at any point to produce the maximum load effect on the element being considered. The number and position of additional concentrated live load units shall be a minimum of 1 unit of 300 lb (1.33 kN) for every 10 ft (3.05 m) of ladder height.

Where rails of fixed ladders extend above a floor or platform at the top of the ladder, each side rail extension shall be designed to resist a single concentrated live load of 100 lb (0.445 kN) applied in any direction at any height up to the top of the side rail extension. Ships ladders with treads instead of rungs shall be designed to resist the stair loads given in Table 4.3-1.

4.6 IMPACT LOADS

4.6.1 General The live loads specified in Sections 4.3 through 4.5 shall be assumed to include adequate allowance for ordinary impact conditions. Provision shall be made in the structural design for uses and loads that involve unusual vibration and impact forces.

4.6.2 Elevators All elements subject to dynamic loads from elevators shall be designed for impact loads and deflection limits prescribed by ASME A17.1.

4.6.3 Machinery For the purpose of design, the weight of machinery and moving loads shall be increased as follows to allow for impact: (1) light machinery, shaft- or motor-driven, 20%; and (2) reciprocating machinery or power-driven units, 50%. All percentages shall be increased where specified by the manufacturer.

4.6.4 Elements Supporting Hoists for Façade Access and Building Maintenance Equipment Structural elements that support hoists for façade access and building maintenance equipment shall be designed for a live load of 2.5 times the rated load of the hoist or the stall load of the hoist, whichever is larger.

4.6.5 Fall Arrest, Lifeline, and Rope Descent System Anchorages Fall arrest, lifeline, and rope descent system anchorages and the structural elements that support these anchorages shall be designed for a live load of 3,100 lb (13.8 kN) for each attached line in any direction that the load may be applied.

Anchorages of horizontal lifelines and the structural elements that support the anchorages shall be designed for the maximum tension that develops in the horizontal lifeline from these live loads.

4.7 REDUCTION IN UNIFORM LIVE LOADS

4.7.1 General Except for roof uniform live loads, all other minimum uniformly distributed live loads, L_o, in Table 4.3-1, are permitted to be reduced in accordance with the requirements of Sections 4.7.2 through 4.7.6.

4.7.2 Reduction in Uniform Live Loads Subject to the limitations of Sections 4.7.3 through 4.7.6, members for which the value of $K_{LL}A_T$ is 400 ft² (37.16 m²) or more are permitted to be designed for a reduced uniform live load in accordance with the following formula:

$$L = L_o\left(0.25 + \frac{15}{\sqrt{K_{LL}A_T}}\right) \quad (4.7\text{-}1)$$

$$L = L_o\left(0.25 + \frac{4.57}{\sqrt{K_{LL}A_T}}\right) \quad (4.7\text{-}1.\text{SI})$$

where

L = Reduced design live load per ft² (m²) of area supported by the member;
L_o = Unreduced design live load per ft² (m²) of area supported by the member (see Table 4.3-1);
K_{LL} = Live load element factor (see Table 4.7-1); and
A_T = Tributary area in ft² (m²).

L shall not be less than $0.50L_o$ for members supporting one floor, and L shall not be less than $0.40L_o$ for members supporting two or more floors.

4.7.3 Heavy Live Loads Live loads that exceed 100 lb/ft² (4.79 kN/m²) shall not be reduced.

EXCEPTION: Live loads for members supporting two or more floors are permitted to be reduced by a maximum of 20%, but the reduced live load shall not be less than L as calculated in Section 4.7.2.

4.7.4 Passenger Vehicle Garages The live loads shall not be reduced in passenger vehicle garages.

EXCEPTION: Live loads for members supporting two or more floors are permitted to be reduced by a maximum of 20%, but the reduced live load shall not be less than L as calculated in Section 4.7.2.

4.7.5 Assembly Area Loads Live loads shall not be reduced in assembly areas.

Table 4.7-1. Live Load Element Factor, K_{LL}.

Element	K_{LL}*
Interior columns	4
Exterior columns without cantilever slabs	4
Edge columns with cantilever slabs	3
Corner columns with cantilever slabs	2
Edge beams without cantilever slabs	2
Interior beams	2
All other members not identified, including	1
Edge beams with cantilever slabs	
Cantilever beams	
One-way slabs	
Two-way slabs	
Members without provisions for continuous shear transfer normal to their span	

*In lieu of the preceding values, K_{LL} is permitted to be calculated.

4.7.6 Limitations on One-Way Slabs The tributary area, A_T, for one-way slabs shall not exceed an area defined by the slab span times a width normal to the span of 1.5 times the slab span.

4.8 REDUCTION IN UNIFORM ROOF LIVE LOADS

4.8.1 General The minimum uniformly distributed roof live loads, L_o, in Table 4.3-1, are permitted to be reduced in accordance with the requirements of Sections 4.8.2 and 4.8.3.

4.8.2 Ordinary Roofs, Awnings, and Canopies Ordinary flat, pitched, and curved roofs, and awning and canopies other than those of fabric construction supported by a skeleton structure, are permitted to be designed for a reduced uniform roof live load, as specified in Equation (4.8-1), or other controlling combinations of loads, as specified in Chapter 2, whichever produces the greater load effect. In structures such as greenhouses, where special scaffolding is used as a work surface for workers and materials during maintenance and repair operations, a lower roof load than specified in Equation (4.8-1) shall not be used unless approved by the Authority Having Jurisdiction. On such structures, the minimum roof live load shall be 12 psf (0.58 kn/m^2).

$$L_r = L_o R_1 R_2 \quad \text{where} \quad 12 \le L_r \le 20 \tag{4.8-1}$$

$$L_r = L_o R_1 R_2 \quad \text{where} \quad 0.58 \le L_r \le 0.96 \tag{4.8-1.SI}$$

where L_r is the reduced roof live load per ft^2 (m^2) of horizontal projection supported by the member, and L_o is the unreduced design roof live load per ft^2 (m^2) of horizontal projection supported by the member (see Table 4.3-1).

The reduction factors R_1 and R_2 shall be determined as follows:

$$R_1 = \begin{cases} 1 & \text{for} \quad A_T \le 200 \text{ ft}^2 \\ 1.2 - 0.001A_T & \text{for} \quad 200 \text{ ft}^2 < A_T < 600 \text{ ft}^2 \\ 0.6 & \text{for} \quad A_T \ge 600 \text{ ft}^2 \end{cases}$$

in SI:

$$R_1 = \begin{cases} 1 & \text{for} \quad A_T \le 18.58 \text{ m}^2 \\ 1.2 - 0.011A_T & \text{for} \quad 18.58 \text{ m}^2 < A_T < 55.74 \text{ m}^2 \\ 0.6 & \text{for} \quad A_T \ge 55.74 \text{ m}^2 \end{cases}$$

where A_T is the tributary area in ft^2 (m^2) supported by the member and

$$R_2 = \begin{cases} 1 & \text{for} \quad F \le 4 \\ 1.2 - 0.05F & \text{for} \quad 4 < F < 12 \\ 0.6 & \text{for} \quad F \ge 12 \end{cases}$$

where, for a pitched roof, F is the number of inches of rise per foot (in SI: $F = 0.12 \times$ slope, with slope expressed in percentage points) and, for an arch or dome, F is the rise-to-span ratio multiplied by 32.

4.8.3 Occupiable Roofs Roofs that have an occupancy function, such as roof gardens or other special purposes, are permitted to have their uniformly distributed live load reduced in accordance with the requirements of Section 4.7.

Roofs used for other special purposes shall be designed for appropriate loads as approved by the Authority Having Jurisdiction.

4.9 CRANE LOADS

4.9.1 General The crane live load shall be the rated capacity of the crane. Design loads for the runway beams, including connections and support brackets, of moving bridge cranes and monorail cranes shall include the maximum wheel loads of the crane and the vertical impact, lateral, and longitudinal forces induced by the moving crane.

4.9.2 Maximum Wheel Load The maximum wheel loads shall be the wheel loads produced by the weight of the bridge, as applicable, plus the sum of the rated capacity and the weight of the trolley with the trolley positioned on its runway at the location where the resulting load effect is maximized.

4.9.3 Vertical Impact Force The maximum wheel loads of the crane determined in accordance with Section 4.9.2 shall be increased by the percentages shown in the following text to account for the effects of vertical impact or vibration:

Monorail cranes (powered)	25%
Bridge crane service class D, E, or F	25%
Bridge crane service class A, B, or C	10%
Cranes with hand-geared bridge, trolley, and hoist	0%

4.9.3.1 Bridge Crane Service Class For the purpose of determining the vertical impact force, one of the following bridge crane service classes shall be assigned based on the actual service conditions, including the frequency of use, variability in load lifted, and the operation speed.

Bridge Crane Service Class A (Standby or infrequent service): This service class shall include cranes used in installations such as powerhouses, public utilities, turbine rooms, motor rooms, and transformer stations where precise handling of equipment at slow speeds is required and cranes are infrequently used or are idled for long periods. Full rated loads shall be handled for initial installation of equipment and for infrequent maintenance.

Bridge Crane Service Class B (Light service): This service class shall include cranes used in repair shops, light assembly operations, service buildings, light warehousing, etc. where service requirements are light and the speed is slow. Loads are permitted to vary, but full rated loads shall occur only occasionally, with two to five lifts per hour, averaging 10 ft (3.05 m) per lift.

Bridge Crane Service Class C (Moderate service): This service class shall include cranes used in machine shops or paper mill machine rooms, etc. where service requirements are moderate. In this type of service, the crane shall handle loads which average 50% of the rated capacity with five to ten lifts per hour, averaging 15 ft (4.6 m), not over 50% of the lifts at rated capacity.

Bridge Crane Service Class D (Heavy service): This service class shall include cranes used in heavy machine shops, foundries, fabricating plants, steel warehouses, container yards, lumber mills, etc., and the standard duty bucket and magnet operations where heavy duty production is required. In this type of service, the crane shall handle loads approaching 50% of the rated capacity constantly during the working period. High speeds are used for this type of service with 10 to 20 lifts per hour averaging 15 ft (4.6 m), not over 65% of the lifts at rated capacity.

Bridge Crane Service Class E (Severe service): This service class shall include cranes capable of handling loads approaching rated capacity throughout its life. Applications include magnet, bucket, magnet/bucket combination cranes for scrap yards, cement mills, lumber mills, fertilizer plants, container handling, etc., with 20 or more lifts per hour at or near the rated capacity.

Bridge Crane Service Class F (Continuous severe service): This service class shall include cranes capable of handling loads approaching rated capacity continuously under severe service conditions throughout its life. Applications include custom designed specialty cranes essential to performing the critical work tasks affecting the total production facility. These cranes shall provide the highest reliability with special attention to ease of maintenance features.

4.9.4 Lateral Force The lateral force on crane runway beams with electrically powered trolleys shall be calculated as 20% of the sum of the rated capacity of the crane and the weight of the hoist and trolley. The lateral force shall be assumed to act horizontally at the traction surface of a runway beam, in either direction perpendicular to the beam, and shall be distributed with due regard to the lateral stiffness of the runway beam and supporting structure.

4.9.5 Longitudinal Force The longitudinal force on crane runway beams, except for bridge cranes with hand-geared bridges, shall be calculated as 10% of the maximum wheel loads of the crane. The longitudinal force shall be assumed to act horizontally at the traction surface of a runway beam in either direction parallel to the beam.

4.10 GARAGE AND VEHICULAR FLOOR LOADS

4.10.1 Passenger Vehicle Garages Floors in garages and portions of a building used for the storage of motor vehicles shall be designed for the uniformly distributed live loads of Table 4.3-1 or the following concentrated load: (1) for garages restricted to passenger vehicles accommodating not more than nine passengers, 3,000 lb (13.35 kN) acting on an area of 4.5 in. × 4.5 in. (114 mm × 114 mm); and (2) for mechanical parking structures without slab or deck that are used for storing passenger vehicles only 2,250 lb (10 kN) per wheel.

4.10.2 Truck and Bus Garages Live loads in garages and portions of a building used for the storage of trucks and buses shall be in accordance with the *AASHTO LRFD Bridge Design Specifications*; however, provisions for fatigue and dynamic load allowance therein are not required to be applied.

EXCEPTION: The vehicular live loads and load placement in garages and portions of a building used for the storage of trucks or buses are permitted to be determined using the actual vehicle weights for the vehicles allowed onto the garage floors, provided that such loads and placement are based on rational engineering principles and are approved by the Authority Having Jurisdiction. The load effects from the actual vehicle weights shall not be less than the effects from a uniform live load of 50 psf (2.40 kN/m^2). This live load shall not be reduced.

4.10.3 Sidewalks, Vehicular Driveways, and Yards Subject to Trucking Uniform loads for sidewalks, vehicular driveways, and yards subject to truck loads, other than that provided in Table 4.3-1, shall also be considered where appropriate, in accordance with an approved method which contains provisions for truck loads. The concentrated wheel load provided in Table 4.3-1 shall be applied on an area of 4.5 in. × 4.5 in. (114 mm × 114 mm).

4.10.4 Emergency Vehicle Loads Where a structure or portions of a structure are accessed by fire department vehicles and other emergency vehicles, those portions of the structure subjected to such loads shall be designed for one of the following:

(a) The actual operational loads, including the total vehicle load, the individual wheel loads, and the outrigger reactions, whichever produces the greater load effects, as stipulated and approved by the Authority Having Jurisdiction. The loads shall include the combined weight of the vehicle and equipment. The total vehicle load, individual wheel loads, and outrigger reactions do not act concurrently.
(b) The design truck and design tandem live loads in the *AASHTO LRFD Bridge Design Specifications*. The design truck and the design tandem need not be assumed to act concurrently. The AASHTO provisions for fatigue and dynamic load allowance are not required to be applied.

Emergency vehicle loads need not be assumed to act concurrently with other uniform live loads.

EXCEPTION: Emergency vehicle loads need not be considered for portions of the structure where physical barriers such as bollards, vehicle barriers, or overhead clearance bars restrict emergency vehicle access.

4.11 HELIPAD LOADS

4.11.1 General The live loads shall not be reduced. The labeling of helicopter capacity shall be as required by the Authority Having Jurisdiction.

4.11.2 Concentrated Helicopter Loads Two single concentrated loads, 8 ft (2.44 m) apart, shall be applied on the landing area (representing the helicopter's two main landing gear, whether skid type or wheeled type), each having a magnitude of 0.75 times the maximum takeoff weight of the helicopter and located to produce the maximum load effect on the structural elements under consideration. The concentrated loads shall be applied over an area of 8 in. × 8 in. (200 mm × 200 mm) and are not required to act concurrently with other uniform or concentrated live loads.

A single concentrated load of 3,000 lb (13.35 kN) shall be applied over an area of 4.5 in. × 4.5 in. (114 mm × 114 mm), located so as to produce the maximum load effects on the structural elements under consideration. The concentrated load is not required to act concurrently with other uniform or concentrated live loads.

4.12 UNINHABITABLE ATTICS

4.12.1 Uninhabitable Attics without Storage In residential occupancies, uninhabitable attic areas without storage are those where the maximum clear height between the joist and rafter is less than 42 in. (1,067 mm) or where there are not two or more adjacent trusses with web configurations capable of accommodating an assumed rectangle 42 in. (1,067 mm) in height by 24 in. (610 mm) in width, or greater, within the plane of the trusses. The live load in Table 4.3-1 need not be assumed to act concurrently with any other live load requirement.

4.12.2 Uninhabitable Attics with Storage In residential occupancies, uninhabitable attic areas with storage are those where the maximum clear height between the joist and rafter is 42 in. (1,067 mm) or greater or where there are two or more adjacent trusses with web configurations capable of accommodating an assumed rectangle 42 in. (1,067 mm) in height by 24 in. (610 mm) in width, or greater, within the

Reduced uplift and lateral loads on surfaces of enclosed spaces below the DFE shall apply only if provision is made for entry and exit of floodwater.

5.4.3 Hydrodynamic Loads Dynamic effects of moving water shall be determined by a detailed analysis utilizing basic concepts of fluid mechanics.

EXCEPTION: Where water velocities do not exceed 10 ft/s (3.05 m/s), dynamic effects of moving water shall be permitted to be converted into equivalent hydrostatic loads by increasing the DFE for design purposes by an equivalent surcharge depth, d_h, on the headwater side and above the ground level only, equal to

$$d_h = \frac{aV^2}{2g} \quad (5.4\text{-}1)$$

where

V = Average velocity of water, ft/s (m/s);
g = Acceleration due to gravity, 32.2 ft/s^2 (9.81 m/s^2); and
a = Coefficient of drag or shape factor (not less than 1.25).

The equivalent surcharge depth shall be added to the DFE design depth and the resultant hydrostatic pressures applied to, and uniformly distributed across, the vertical projected area of the building or structure that is perpendicular to the flow. Surfaces parallel to the flow or surfaces wetted by the tailwater shall be subject to the hydrostatic pressures for depths to the DFE only.

5.4.4 Wave Loads Wave loads shall be determined by one of the following three methods: (1) by using the analytical procedures outlined in this section, (2) by more advanced numerical modeling procedures, or (3) by laboratory test procedures (physical modeling).

Wave loads are those loads that result from water waves propagating over the water surface and striking a building or other structure. Design and construction of buildings and other structures subject to wave loads shall account for the following loads: waves breaking on any portion of the building or structure; uplift forces caused by shoaling waves beneath a building or structure, or portion thereof; wave runup striking any portion of the building or structure; wave-induced drag and inertia forces; and wave-induced scour at the base of a building or structure, or its foundation. Wave loads shall be included for both V-Zones and A-Zones. In V-Zones, waves are 3 ft (0.91 m) high, or higher; in coastal floodplains landward of the V-Zone, waves are less than 3 ft (0.91 m) high.

Nonbreaking and broken wave loads shall be calculated using the procedures described in Sections 5.4.2 and 5.4.3 that show how to calculate hydrostatic and hydrodynamic loads.

Breaking wave loads shall be calculated using the procedures described in Sections 5.4.4.1 through 5.4.4.4. Breaking wave heights used in the procedures described in Sections 5.4.4.1 through 5.4.4.4 shall be calculated for V-Zones and Coastal A-Zones using Equations (5.4-2) and (5.4-3):

$$H_b = 0.78 d_s \quad (5.4\text{-}2)$$

where H_b is the breaking wave height in ft (m), and d_s is the local still water depth in ft (m).

The local still water depth shall be calculated using Equation (5.4-3), unless more advanced procedures or laboratory tests permitted by this section are used:

$$d_s = 0.65(BFE - G) \quad (5.4\text{-}3)$$

where BFE is in ft (m), and G is ground elevation in ft (m).

5.4.4.1 Breaking Wave Loads on Vertical Pilings and Columns The net force resulting from a breaking wave acting on a rigid vertical pile or column shall be assumed to act at the still water elevation and shall be calculated by the following:

$$F_D = 0.5\gamma_w C_D D H_b^2 \quad (5.4\text{-}4)$$

where

F_D = Net wave force, lb (kN);
γ_w = Unit weight of water, lb/ft^3 (kN/m^3), = 62.4 lb/ft^3 (9.80 kN/m^3) for freshwater and 64.0 lb/ft^3 (10.05 kN/m^3) for saltwater;
C_D = Coefficient of drag for breaking waves, = 1.75 for round piles or columns and = 2.25 for square piles or columns;
D = Pile or column diameter, ft (m) for circular sections, or for a square pile or column, 1.4 times the width of the pile or column, ft (m); and
H_b = Breaking wave height, ft (m).

5.4.4.2 Breaking Wave Loads on Vertical Walls Maximum pressures and net forces resulting from a normally incident breaking wave (depth limited in size, with $H_b = 0.78d_s$) acting on a rigid vertical wall shall be calculated by the following:

$$P_{max} = C_p\gamma_w d_s + 1.2\gamma_w d_s \quad (5.4\text{-}5)$$

and

$$F_t = 1.1C_p\gamma_w d_s^2 + 2.4\gamma_w d_s^2 \quad (5.4\text{-}6)$$

where

P_{max} = Maximum combined dynamic ($C_p\gamma_w d_s$) and static ($1.2\gamma_w d_s$) wave pressures, also referred to as shock pressures, lb/ft^2 (kN/m^2);
F_t = Net breaking wave force per unit length of structure, also referred to as shock, impulse, or wave impact force, lb/ft (kN/m), acting near the still water elevation;
C_p = Dynamic pressure coefficient ($1.6 < C_p < 3.5$) (see Table 5.4-1);
γ_w = Unit weight of water, lb/ft^3 (kN/m^3), = 62.4 lb/ft^3 (9.80 kN/m^3) for freshwater and 64.0 lb/ft^3 (10.05 kN/m^3) for saltwater; and
d_s = Still water depth, ft (m), at base of building or other structure where the wave breaks.

This procedure assumes the vertical wall causes a reflected or standing wave against the waterward side of the wall with the crest of the wave at a height of $1.2d_s$ above the still water level. Thus, the dynamic static and total pressure distributions against the wall are as shown in Figure 5.4-1.

This procedure also assumes the space behind the vertical wall is dry, with no fluid balancing the static component of the wave force on the outside of the wall. If free water exists behind the wall, a portion of the hydrostatic component of the wave pressure and force disappears (see Figure 5.4-2) and the net force shall be computed by Equation (5.4-7) [the maximum combined wave pressure is still computed with Equation (5.4-5)]:

$$F_t = 1.1C_p\gamma_w d_s^2 + 1.9\gamma_w d_s^2 \quad (5.4\text{-}7)$$

where

F_t = Net breaking wave force per unit length of structure, also referred to as shock, impulse, or wave impact force, lb/ft (kN/m), acting near the still water elevation;

C_p = Dynamic pressure coefficient ($1.6 < C_p < 3.5$) (see Table 5.4-1);
γ_w = Unit weight of water, lb/ft^3 (kN/m^3), = 62.4 lb/ft^3 (9.80 kN/m^3) for freshwater and 64.0 lb/ft^3 (10.05 kN/m^3) for saltwater; and
d_s = Still water depth, ft (m) at base of building or other structure where the wave breaks.

5.4.4.3 Breaking Wave Loads on Nonvertical Walls Breaking wave forces given by Equations (5.4-6) and (5.4-7) shall be modified in instances where the walls or surfaces upon which the breaking waves act are nonvertical. The horizontal component of breaking wave force shall be given by

$$F_{nv} = F_t \sin^2\alpha \qquad (5.4\text{-}8)$$

where

F_{nv} = Horizontal component of breaking wave force, lb/ft (kN/m);
F_t = Net breaking wave force acting on a vertical surface, lb/ft (kN/m); and
α = Vertical angle between nonvertical surface and the horizontal.

Table 5.4-1. Value of Dynamic Pressure Coefficient, C_p.

Risk Category*	C_p
I	1.6
II	2.8
III	3.2
IV	3.5

*For risk category, see Table 1.5-1.

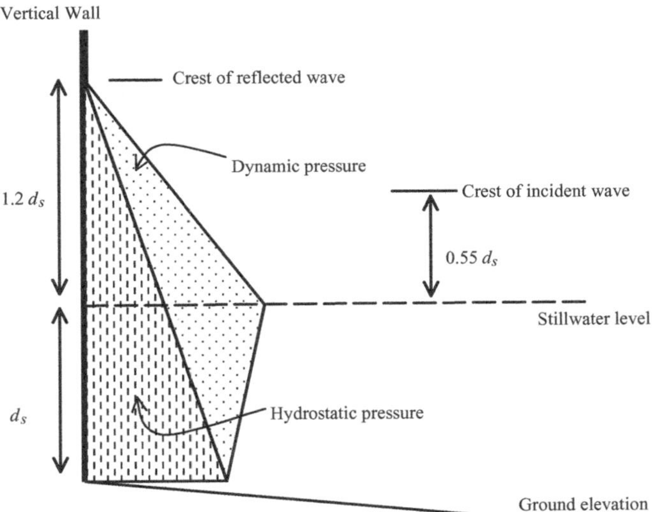

Figure 5.4-1. Normally Incident Breaking Wave Pressures against a Vertical Wall (Space behind Vertical Wall Is Dry).

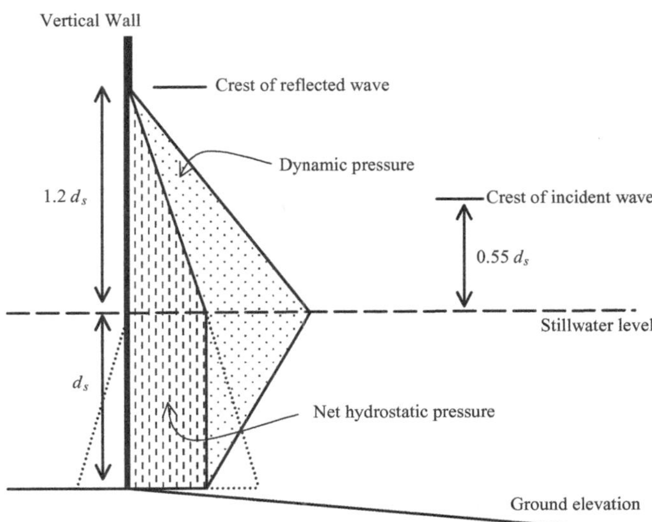

Figure 5.4-2. Normally Incident Breaking Wave Pressures against a Vertical Wall (Still Water Level Equal on Both Sides of Wall).

5.4.4.4 Breaking Wave Loads from Obliquely Incident Waves Breaking wave forces given by Equations (5.4-6) and (5.4-7) shall be modified in instances where waves are obliquely incident. Breaking wave forces from non-normally incident waves shall be given by

$$F_{oi} = F_t \sin^2\alpha \qquad (5.4\text{-}9)$$

where

F_{oi} = Horizontal component of obliquely incident breaking wave force, lb/ft (kN/m);
F_t = Net breaking wave force (normally incident waves) acting on a vertical surface, lb/ft (kN/m); and
α = Horizontal angle between the direction of wave approach and the vertical surface.

5.4.5 Impact Loads Impact loads result from debris, ice, and any object transported by floodwaters striking against buildings and structures or parts thereof. Impact loads shall be determined using a rational approach as concentrated loads acting horizontally at the most critical location at or below the DFE.

5.5 CONSENSUS STANDARDS AND OTHER REFERENCED DOCUMENTS

This section lists the consensus standards and other documents that shall be considered part of this standard to the extent referenced in this chapter.

ASCE/SEI 24 *Flood resistant design and construction,* ASCE, 2014.
Cited in: Section 5.3.3

CHAPTER 6
TSUNAMI LOADS AND EFFECTS

6.1 GENERAL REQUIREMENTS

6.1.1 Scope The following buildings and other structures located within the Tsunami Design Zone shall be designed for the effects of Maximum Considered Tsunami, including hydrostatic and hydrodynamic forces, waterborne debris accumulation and impact loads, subsidence, and scour effects in accordance with this chapter:

a Tsunami Risk Category IV buildings and structures;
b Tsunami Risk Category III buildings and structures with inundation depth greater than 3 ft (0.914 m) at any location within the intended footprint of the structure; and
c Where required by a state or locally adopted building code statute to include design for tsunami effects, Tsunami Risk Category II buildings with mean height above grade plane greater than the height designated in the statute and having inundation depth greater than 3 ft (0.914 m) at any location within the intended footprint of the structure.

EXCEPTION: Tsunami Risk Category II single-story buildings of any height without mezzanines or any occupiable roof level and not having any critical equipment or systems need not be designed for the tsunami loads and effects specified in this chapter.

For the purposes of this chapter, Tsunami Risk Category shall be as determined in accordance with Section 6.4.

Tsunami Design Zone shall be determined using the ASCE Tsunami Design Geodatabase of geocoded reference points shown in Figure 6.1-1. The ASCE Tsunami Design Geodatabase of geocoded reference points of runup and associated inundation limits of the Tsunami Design Zone is available at [http://asce7tsunami.online].

EXCEPTION: For coastal regions subject to tsunami inundation and not covered by Figure 6.1-1, Tsunami Design Zone, inundation limits, and runup elevations shall be determined using the site-specific procedures of Section 6.7, or for Tsunami Risk Category II or III structures, determined in accordance with the procedures of Section 6.5.1.1 using Figure 6.7-1.

Designated nonstructural components and systems associated with Tsunami Risk Category III Critical Facilities and Tsunami Risk Category IV structures subject to this chapter shall be located above, protected from, or otherwise designed for inundation in accordance with Section 6.15 so that they are able to provide their essential functions immediately following the Maximum Considered Tsunami event.

User Note: The ASCE Tsunami Design Geodatabase of geocoded reference points of runup and associated inundation limits of the Tsunami Design Zone is available at http://asce7tsunami.online. Sea level rise has not been incorporated into the Tsunami Design Zone maps, and any additive effect on the inundation at the site should be explicitly evaluated.

6.2 DEFINITIONS

The following definitions apply only to the tsunami requirements of this chapter. Also see Figure 6.2-1 for an illustration of some key terms.

ASCE TSUNAMI DESIGN GEODATABASE: The ASCE database (version 2022-1.0) of geocoded reference points of Offshore 328 ft (100 m) depth Tsunami Amplitude, H_T, and Predominant Period, T_{TSU}, of the Maximum Considered Tsunami, disaggregated hazard source contribution figures, probabilistic subsidence, Runup Elevation and Inundation geocoded reference points, and Tsunami Design Zone maps.

BATHYMETRIC PROFILE: A cross section showing ocean depth plotted as a function of horizontal distance from a reference point (such as a coastline).

CHANNELIZED SCOUR: Scour that results from broad flow that is diverted to a focused area, such as return flow in a preexisting stream channel or alongside a seawall.

CLOSURE RATIO (of Inundated Projected Area): Ratio of the area of enclosure, not including glazing and openings, that is inundated to the total projected vertical plane area of the inundated enclosure surface exposed to flow pressure.

COLLAPSE PREVENTION STRUCTURAL PERFORMANCE LEVEL: A postevent damage state in which a structure has damaged components and continues to support gravity loads but retains little or no margin against collapse.

CRITICAL EQUIPMENT or CRITICAL SYSTEMS: Nonstructural components designated essential for the functionality of the Critical Facility or Essential Facility or that are necessary to maintain safe containment of hazardous materials.

CRITICAL FACILITY: Buildings and structures that provide services that are designated by federal, state, local, or tribal governments to be essential for the implementation of the response and recovery management plan or for the continued functioning of a community, such as facilities for power, fuel, water, communications, public health, major transportation infrastructure, and essential government operations. Critical Facilities comprise all public and private facilities deemed by a community to be essential for the delivery of vital services, protection of special populations, and the provision of other services of importance for that community.

DEADWEIGHT TONNAGE (DWT): Deadweight Tonnage (DWT) is a vessel's Displacement Tonnage (DT) minus its Lightship Weight (LWT). DWT is a classification used for the carrying capacity of a vessel that is equal to the sum of the weights of cargo, fuel, fresh water, ballast water, provisions, passengers, and crew; it does not include the weight of the vessel itself. Displacement Tonnage is the total weight of a fully loaded vessel. Lightship Weight is the weight of the vessel without cargo, crew, fuel, fresh water, ballast water, provisions, passengers, or crew.

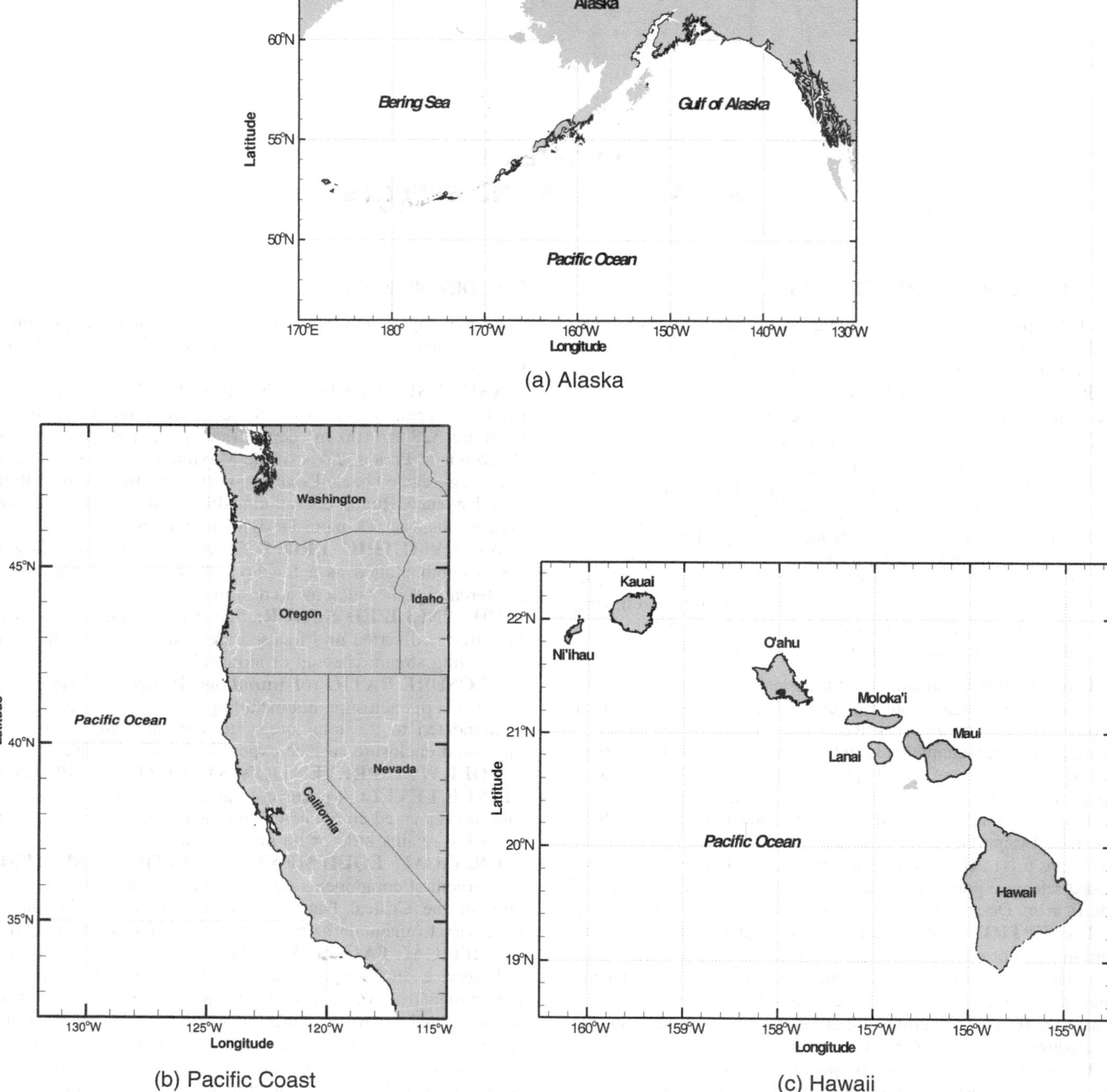

Figure 6.1-1. Extent of ASCE Tsunami Design Geodatabase[1] of geocoded reference points of runup and associated inundation limits of the Tsunami Design Zone.[2]

[1]The ASCE Tsunami Design Geodatabase of geocoded reference points of runup and associated inundation limits of the Tsunami Design Zone is available at http://asce7tsunami.online.

[2]Sea level rise has not been incorporated into the Tsunami Design Zone maps, and any additive effect on the inundation at the site should be explicitly evaluated.

DESIGN STRENGTH: Nominal strength multiplied by a resistance factor, ϕ.

DESIGN TSUNAMI PARAMETERS: The tsunami parameters used for design, consisting of the inundation depths and flow velocities at the stages of inflow and outflow most critical to the structure and momentum flux.

DESIGNATED NONSTRUCTURAL COMPONENTS AND SYSTEMS: Nonstructural components and systems that

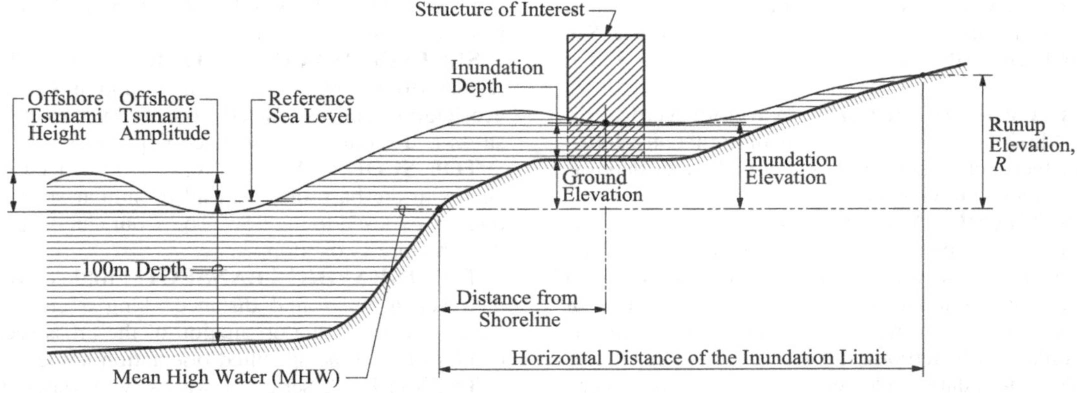

Figure 6.2-1. Illustration of key definitions along a flow transect in a tsunami design zone.

are assigned a component Importance Factor, I_p, equal to 1.5 per Section 13.1.3 of this standard.

DUCTILITY-GOVERNED ACTION: Any action on a structural component characterized by a postelastic force versus deformation curve that (1) has sufficient ductility and (2) results from an impulsive short-term force that is not sustained.

FORCE-SUSTAINED ACTIONS: Any action on a structural component characterized by a sustained force or a postelastic force versus deformation curve that is not ductility-governed due to lack of sufficient ductility.

FROUDE NUMBER, F_r: A dimensionless number defined by $u/\sqrt{(gh)}$, (where u is the flow velocity averaged over the cross section perpendicular to the flow), which is used to quantify the normalized tsunami flow velocity as a function of water depth.

GENERAL EROSION: A general wearing away and erosion of the land surface over a significant portion of the inundation area, excluding localized scour actions.

GRADE PLANE: A horizontal reference plane at the site representing the average elevation of finished ground level adjoining the structure at all exterior walls. Where the finished ground level slopes away from the exterior walls, the grade plane is established by the lowest points within the area between the structure and the property line or, where the property line is more than 6 ft (1.83 m) from the structure, between the structure and points 6 ft (1.83 m) from the structure.

HAZARD-CONSISTENT TSUNAMI SCENARIO: One or more surrogate tsunami scenarios generated from the principal disaggregated seismic source regions that replicate the Offshore Tsunami Amplitudes of the ASCE Tsunami Design Geodatabase or the offshore tsunami waveform characteristics of Figure 6.7-1 and Table 6.7-1 for the site of interest, taking into account the net effect of the probabilistic treatment of uncertainty into the offshore wave amplitude of the scenario(s).

HYDRODYNAMIC LOADS: Loads imposed on an object by water flowing against and around it.

HYDROSTATIC LOADS: Loads imposed on an object by a standing mass of water.

IMMEDIATE OCCUPANCY STRUCTURAL PERFORMANCE LEVEL: The postevent damage state in which a structure remains safe to occupy.

IMPACT LOADS: Loads that result from debris or other objects transported by the design tsunami striking a structure or portion thereof.

INUNDATION DEPTH: The depth of design tsunami water level, including relative sea level change, with respect to the grade plane at the structure.

INUNDATION ELEVATION: The elevation of the design tsunami water surface, including relative sea level change, with respect to vertical datum at Mean High Water (MHW).

INUNDATION LIMIT: The maximum horizontal inland extent of flooding for the Maximum Considered Tsunami, where the inundation depth above grade becomes zero; the horizontal distance that is flooded, relative to the shoreline defined where the elevation is at the Mean High Water (MHW).

LIFE SAFETY STRUCTURAL PERFORMANCE LEVEL: The postevent damage state is that in which a structure has damaged components but retains a margin against onset of partial or total collapse.

LIQUEFACTION SCOUR: The limiting case of pore pressure softening associated with hydrodynamic flow, where the effective stress drops to zero. In noncohesive soils, the shear stress required to initiate sediment transport also drops to zero during liquefaction scour.

LOCAL COSEISMIC TSUNAMI: A tsunami preceded by an earthquake with damaging effects felt within the subsequently inundated area.

LOCAL SCOUR: Removal of material from a localized portion of land surface, resulting from flow around, over, or under a structure or structural element.

MAXIMUM CONSIDERED TSUNAMI: A probabilistic tsunami having a 2% probability of being exceeded in a 50-year period or a 2,475-year mean recurrence interval.

MEAN HIGH WATER: The average of all the high water heights observed over the National Tidal Datum Epoch. For stations with shorter series, comparison of simultaneous observations with a control tide station is made in order to derive the equivalent datum of the National Tidal Datum Epoch.

MOMENTUM FLUX: The quantity $\rho_s hu^2$ for a unit width based on the depth-averaged flow speed, u, over the inundation depth, h, for equivalent fluid density, ρ_s, having the units of force per unit width.

NEARSHORE PROFILE: Cross-sectional bathymetric profile from the shoreline to a water depth of 328 ft (100 m).

NEARSHORE TSUNAMI AMPLITUDE: The Maximum Considered Tsunami amplitude immediately off the coastline at 33 ft (10 m) of water depth.

NONBUILDING CRITICAL FACILITY STRUCTURE: Nonbuilding structure whose Tsunami Risk Category is designated as either III or IV.

NONBUILDING STRUCTURE: A structure other than a building.

OFFSHORE TSUNAMI AMPLITUDE: Maximum Considered Tsunami amplitude relative to the Reference Sea Level, measured where the undisturbed water depth is 328 ft (100 m).

OFFSHORE TSUNAMI HEIGHT: Waveform vertical dimension of the Maximum Considered Tsunami from consecutive trough to crest, measured where the undisturbed water depth is 328 ft (100 m), after removing the tidal variation.

OPEN STRUCTURE: A structure in which the portion within the inundation depth has no greater than 20% closure ratio, and in which the closure does not include any Tsunami Breakaway Walls, and which does not have interior partitions or contents that are prevented from passing through and exiting the structure as unimpeded waterborne debris.

PILE: A deep foundation element, which includes piers, caissons, and piles.

PILE SCOUR: A special case of enhanced local scour that occurs at a pile, bridge pier, or similar slender structure.

PLUNGING SCOUR: A special case of enhanced local scour that occurs when the flow passes over a complete or nearly complete obstruction, such as a barrier wall, and drops steeply onto the ground below, scouring out a depression.

PORE PRESSURE SOFTENING: A mechanism that enhances scour through increased pore-water pressure generated within the ground during rapid tsunami loading and the release of that pressure during drawdown.

PORE PRESSURE SOFTENING PARAMETER, Λ_T: A dimensionless parameter defined by Equation (6.12-1) characterizing the fraction of the weight of soil grains supported by tsunami-induced excess pore water pressure.

PRIMARY STRUCTURAL COMPONENT: Structural components required to resist tsunami forces and actions and inundated structural components of the gravity-load-carrying system.

RECOGNIZED LITERATURE: Published research findings and technical papers that are approved by the Authority Having Jurisdiction.

REFERENCE SEA LEVEL: The sea level datum used in site-specific inundation modeling that is typically taken to be Mean High Water (MHW).

RELATIVE SEA LEVEL CHANGE: The local change in the level of the ocean relative to the land, which might be caused by ocean rise and/or subsidence of the land.

RUNUP ELEVATION: Ground elevation at the maximum tsunami inundation limit, including relative sea level change, with respect to the Mean High Water (MHW) datum.

SECONDARY STRUCTURAL COMPONENT: A structural component that is not primary.

SHOALING: The increase in wave height and wave steepness caused by the decrease in water depth as a wave travels into shallower water.

SOLITON FISSION: Short-period waves generated on the front edge of a tsunami waveform under conditions of shoaling on a long and gentle seabed slope or having abrupt seabed discontinuities, such as fringing reefs.

STRUCTURAL COMPONENT: A component of a building that provides gravity-load-carrying or lateral-force resistance as part of a continuous load path to foundation, including beams, columns, slabs, braces, walls, wall piers, coupling beams, and connections.

STRUCTURAL WALL: A wall that provides gravity-load-carrying support or one that is designed to provide lateral-force resistance.

SURGE: Rapidly rising water level resulting in horizontal flow inland.

SUSCEPTIBLE DEPTH, D_e: The depth of soil susceptible to pore pressure softening.

SUSTAINED FLOW SCOUR: Enhanced local scour that results from flow acceleration around a structure. The flow acceleration and associated vortices increase the bottom shear stress and scour out a localized depression.

TOE SCOUR: A special case of enhanced local scour that occurs at the base of a seawall or similar structure on the side directly exposed to the flow. Toe scour can occur whether or not the structure is overtopped.

TOPOGRAPHIC TRANSECT: Profile of vertical elevation data versus horizontal distance along a cross section of the terrain, in which the orientation of the cross section is perpendicular or at some specified orientation angle to the shoreline.

TSUNAMI: A series of waves with variable long periods, typically resulting from earthquake-induced uplift or subsidence of the seafloor.

TSUNAMI AMPLITUDE: The absolute value of the difference between a particular peak or trough of the tsunami and the undisturbed sea level at the time.

TSUNAMI BORE: A steep and turbulent broken wavefront generated on the front edge of a long-period tsunami waveform when shoaling over mild seabed slopes or abrupt seabed discontinuities such as fringing reefs, or in a river estuary, per Section 6.6.4. Soliton fission in the Nearshore Profile can often lead to the occurrence of tsunami bores.

TSUNAMI BORE HEIGHT: The height of a broken tsunami surge above the water level in front of the bore, or above the grade elevation if the bore arrives on nominally dry land.

TSUNAMI BREAKAWAY WALL: Any type of wall subject to flooding that is not required to provide structural support to a building or other structure and that is designed and constructed such that, before the development of the design flow conditions of inundation Load Case 1, as defined in Section 6.8.3.1, the wall will collapse or detach in such a way that (1) it allows substantially free passage of floodwaters and external or internal waterborne debris, including unattached building contents, and (2) it does not damage the structure or supporting foundation system.

TSUNAMI DESIGN ZONE: An area identified on the Tsunami Design Zone Map between the shoreline and the inundation limit, within which structures are analyzed and designed for inundation by the Maximum Considered Tsunami.

TSUNAMI DESIGN ZONE MAP: The map given in Figure 6.1-1 designating the potential horizontal inundation limit of the Maximum Considered Tsunami, or a state or local jurisdiction's probabilistic map produced in accordance with Section 6.7 of this chapter.

TSUNAMI EVACUATION MAP: An evacuation map based on a tsunami inundation map based on assumed scenarios that is developed and provided to a community by either the applicable state agency or NOAA under the National Tsunami Hazard Mitigation Program. Tsunami inundation maps for evacuation may be significantly different in extent than the Probabilistic Tsunami Design Zone, and Tsunami Evacuation Maps are not intended for design or land use purposes.

TSUNAMI-PRONE REGION: The coastal region in the United States addressed by this chapter with quantified probability in the recognized literature of tsunami inundation hazard with runup greater than 3 ft (0.914 m) caused by tsunamigenic earthquakes in accordance with the Probabilistic Tsunami Hazard Analysis method given in this chapter.

TSUNAMI RISK CATEGORY: The Risk Category from Section 1.5, as modified for specific use related to this chapter per Section 6.4.

TSUNAMI VERTICAL EVACUATION REFUGE STRUCTURE: A structure designated and designed to serve as a point of refuge to which a portion of the community's population can evacuate above a tsunami when high ground is not available.

6.3 SYMBOLS AND NOTATION

A_{beam} = Vertical projected area of an individual beam element
A_{col} = Vertical projected area of an individual column element
A_d = Vertical projected area of obstructing debris accumulated on structure
A_{wall} = Vertical projected area of an individual wall element
a_1 = Amplitude of the leading pulse (negative for a leading depression tsunami)
a_2 = Amplitude of the following, or second, pulse
B = Overall building width
b = Width subject to force
C_{bs} = Force coefficient with breakaway slab
C_{cx} = Proportion of closure coefficient
C_d = Drag coefficient based on quasi-steady forces
C_{dis} = Discharge coefficient for overtopping
C_o = Orientation coefficient (of debris)
c_v = Coefficient of consolidation
c_{2V} = Plunging scour coefficient
D = Dead load
D_a = Diameter of rock armor
D_s = Scour depth
d_d = Additional drop in grade to the base of wall on the side of a seawall or freestanding retaining wall subject to plunging scour
DT = Displacement Tonnage
DWT = Deadweight Tonnage of vessel
E = Earthquake load
E_g = Hydraulic head in the Energy Grade Line Analysis
E_{mh} = Horizontal Seismic Load Effect, including overstrength factor, defined in Section 12.4.3.1
F_d = Drag force on an element or component
F_{dx} = Drag force on the building or structure at each level
F_h = Unbalanced hydrostatic lateral force
F_i = Debris impact design force
F_{ni} = Nominal maximum instantaneous debris impact force
F_{pw} = Hydrodynamic force on a perforated wall
F_r = Froude number $= u/\sqrt{gh}$
F_{TSU} = Tsunami load or effect
F_v = Buoyancy force
F_w = Load on wall or pier
$F_{w\theta}$ = Force on a wall oriented at an angle θ to the flow
F_Λ = Pore pressure softening vertical flux index
f_{uw} = Equivalent uniform lateral force per unit width
g = Acceleration caused by gravity
H_B = Barrier height of a levee, seawall, or freestanding retaining wall
H_O = Depth to which a barrier is overtopped above the barrier height
H_T = Offshore Tsunami Amplitude determined from Figure 6.7-1
H_{TSU} = Load caused by tsunami-induced lateral earth pressure under submerged conditions
h = Tsunami inundation depth above grade plane at the structure
h_{draw} = Difference between the maximum inundation depth, h_{max}, and the residual water level after drawdown of the associated wave
h_e = Inundated height of an individual element
h_i = Inundation depth at point i
h_{max} = Maximum inundation depth above grade plane at the structure
h_o = Offshore water depth
h_r = Residual water height within a building
h_s = Height of structural floor slab above grade plane at the structure
h_{ss} = Height of the bottom of the structural floor slab, taken above grade plane at the structure
h_{sx} = Story height of story x
h_w = Height of wall
I_{tsu} = Importance Factor for tsunami forces to account for additional uncertainty in estimated parameters
I_Λ = Pore pressure softening drawdown inertia index
k = Effective stiffness of the impacting debris or the lateral stiffness of the impacted structural element
k_s = Fluid density factor to account for suspended soil and other smaller flow-embedded objects that are not considered in Section 6.11
L = Live load
L_{refuge} = Public assembly live load effect in the tsunami refuge floor area
l_w = Length of a structural wall
LWT = Lightship Weight of vessel
m = Component demand modification factor accounting for expected ductility, applied to the expected strength of a ductility-governed element action, to obtain the acceptable structural component capacity at a particular performance level when using a linear static analysis procedure
$m_{contents}$ = Mass of contents in a shipping container
m_d = Mass of debris object
MCT = Maximum Considered Tsunami
n = Manning's coefficient
P_u = Uplift pressure on slab or building horizontal element
P_{ur} = Reduced uplift pressure for slab with opening
Q_{CE} = Expected strength of the structural element
Q_{CS} = Specified strength of the structural element
Q_{UD} = Ductility-governed force caused by gravity and tsunami loading
Q_{UF} = Maximum force generated in the element caused by gravity and tsunami loading
q = Discharge per unit width over an overtopped structure
R = Mapped tsunami runup elevation
R_{max} = Dynamic response ratio
R_s = Net upward resistance from foundation elements
S = Snow load
s = Friction slope of the energy grade line
t = Time
t_d = Duration of debris impact
t_o = Offset time of the wave train
T_{draw} = Site-specific drawdown time, that is, the time from the peak of the transformed onshore waveform to the following trough
T_{TSU} = Predominant wave period, or the time from the start of the first pulse to the end of the second pulse at the 328 ft (100 m) bathymetric depth
U = Jet velocity of plunging flow
u = Tsunami flow velocity
u_{max} = Maximum tsunami flow velocity at the structure
u_v = Vertical component of tsunami flow velocity
V_w = Displaced water volume
W_d = Weight of debris
W_s = Weight of the structure

w_g = Width of opening gap in slab
x = Horizontal distance inland from the Mean High Water (MHW) shoreline
x_R = Mapped inundation limit distance inland from the Mean High Water (MHW) shoreline
z = Ground elevation above the Mean High Water (MHW) datum
α = Froude number coefficient in the Energy Grade Line Analysis
β = Effective wake angle downstream of an obstructing structure to the structure of interest
γ_b = Buoyant weight density of soil
γ_s = Minimum fluid weight density for design hydrostatic loads
γ_{sw} = Effective weight density of seawater
Δx_i = Incremental distance used in the Energy Grade Line Analysis
ξ_{100} = Surf similarity parameter using 328 ft (100 m) nearshore wave characteristics
η = Free surface elevation as a function of time, t, used to drive the offshore boundary condition at the 328 ft (100 m) depth contour
κ_d = Drainage factor
θ = Angle between the longitudinal axis of a wall and the flow direction
ϕ = Structural resistance factor
ρ_s = Minimum fluid mass density for design hydrodynamic loads
ρ_{sw} = Effective mass density of seawater
ϕ = Average slope of grade at the structure
ϕ_i = Average slope of grade at point i
$\phi_{i,\ 200}$ = Topographic slope averaged over a distance of 200 ft (60.96 m) along a transect
Λ_T = Pore pressure softening parameter
Ω = Angular frequency of the waveform, equal to $2\pi/T$, where T is the wave period
Ω_O = Overstrength factor for the lateral-force-resisting system given in Table 12.2-1
Φ = Mean slope angle of the Nearshore Profile
ψ = Angle between the plunging jet at the scour hole and the horizontal

6.4 TSUNAMI RISK CATEGORIES

For the purposes of this chapter, tsunami risk categories for buildings and other structures shall be the risk categories given in Section 1.5 with the following modifications:

1. Federal, state, local, or tribal governments shall be permitted to include Critical Facilities in Tsunami Risk Category III, such as power-generating stations, water-treatment facilities for potable water, wastewater-treatment facilities, and other public utility facilities not included in Risk Category IV.
2. The following structures need not be included in Tsunami Risk Category IV, and state, local, or tribal governments shall be permitted to designate them as Tsunami Risk Category II or III:
   (a) Fire stations, ambulance facilities, and emergency vehicle garages;
   (b) Earthquake or hurricane shelters;
   (c) Emergency aircraft hangars; and
   (d) Police stations that do not have holding cells and that are not uniquely required for postdisaster emergency response as a Critical Facility.
3. Tsunami Vertical Evacuation Refuge Structures shall be included in Tsunami Risk Category IV.

6.5 ANALYSIS OF DESIGN INUNDATION DEPTH AND FLOW VELOCITY

6.5.1 Tsunami Risk Category II and III Buildings and Other Structures The Maximum Considered Tsunami inundation depth and tsunami flow velocity characteristics at a Tsunami Risk Category II or III building or other structure shall be determined by using the Energy Grade Line Analysis of Section 6.6 and the inundation limit and runup elevation of the Maximum Considered Tsunami given in Figure 6.1-1.

The site-specific Probabilistic Tsunami Hazard Analysis (PTHA) in Section 6.7 shall be permitted as an alternate to the Energy Grade Line Analysis. Site-specific velocities determined by PTHA shall be subject to the limitation in Section 6.7.6.8.

EXCEPTION: For Tsunami-Prone Regions not covered by Figure 6.1-1, the procedures of Section 6.5.1.1 shall apply to Tsunami Risk Category II and III buildings and other structures.

6.5.1.1. Runup Evaluation for Areas Where No Map Values Are Given For Tsunami Risk Category II and III buildings and other structures where no mapped inundation limit is shown in Figure 6.1-1, the ratio of tsunami runup elevation above Mean High Water Level to Offshore Tsunami Amplitude, R/H_T, shall be permitted to be determined using the surf similarity parameter ξ_{100}, according to Equation (6.5-2a, b, c, d, or e) and Figure 6.5-1.

The surf similarity parameter, ξ_{100}, for this application to tsunami engineering shall be determined in accordance with Equation (6.5-1).

$$\xi_{100} = \frac{T_{\text{TSU}}}{\cot \Phi} \sqrt{\frac{g}{2\pi H_T}} \qquad (6.5\text{-}1)$$

where Φ is the mean slope angle of the Nearshore Profile taken from the 328 ft (100 m) water depth to the Mean High Water elevation along the axis of the topographic transect for the site. H_T is the Offshore Tsunami Amplitude, and T_{TSU} is the wave period of the tsunami at 328 ft (100 m) water depth. H_T and T_{TSU} are given in Figure 6.7-1.

$$\text{For} \quad \xi_{100} \le 0.6,\ R/H_T = 1.5 \qquad (6.5\text{-}2a)$$

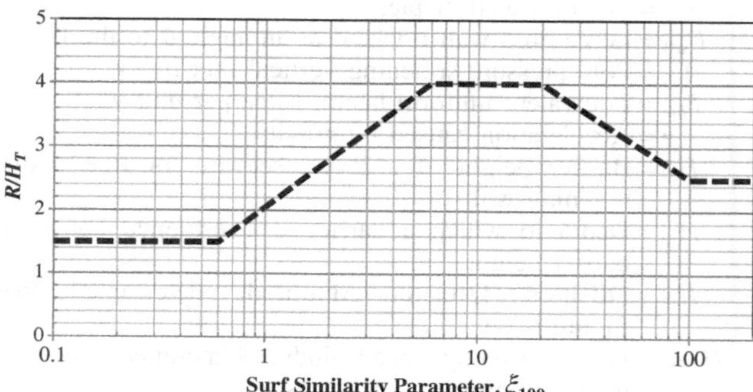

Figure 6.5-1. Runup ratio, R/H_T, as a function of the mean slope of the surf similarity parameter, ξ_{100}, where no mapped inundation limit exists.

For $\xi_{100} > 0.6$ and $\xi_{100} \leq 6$,

$$R/H_T = 2.50[\log_{10}(\xi_{100})] + 2.05 \quad (6.5\text{-}2b)$$

For $\xi_{100} > 6$ and $\xi_{100} \leq 20$, $R/H_T = 4.0$ (6.5-2c)

For $\xi_{100} > 20$ and $\xi_{100} \leq 100$,

$$R/H_T = -2.15[\log_{10}(\xi_{100})] + 6.80 \quad (6.5\text{-}2d)$$

For $\xi_{100} > 100$, $R/H_T = 2.5$ (6.5-2e)

EXCEPTION: These equations shall not be used where there is an expectation of wave focusing such as at headlands, in V-shaped bays, or where the on-land flow fields are expected to vary significantly in the direction parallel to the shoreline because of longshore variability of topography.

6.5.2 Tsunami Risk Category IV Buildings and Other Structures The Energy Grade Line Analysis of Section 6.6 shall be performed for Tsunami Risk Category IV buildings and other structures, and the site-specific Probabilistic Tsunami Hazard Analysis (PTHA) of Section 6.7 shall also be performed. Site-specific velocities determined by site-specific PTHA determined to be less than the Energy Grade Line Analysis shall be subject to the limitation in Section 6.7.6.8. Site-specific velocities determined to be greater than the Energy Grade Line Analysis shall be used.

EXCEPTION: For structures other than Tsunami Vertical Evacuation Refuge Structures, a site-specific Probabilistic Tsunami Hazard Analysis need not be performed where the inundation depth resulting from the Energy Grade Line Analysis is determined to be less than 12 ft (3.66 m) at any point within the location of the Tsunami Risk Category IV structure.

6.5.3 Sea Level Change The direct physical effects of potential relative sea level change shall be considered in determining the maximum inundation depth during the project lifecycle. A project lifecycle of not less than 50 years shall be used. The minimum rate of potential relative sea level change shall be the historically recorded sea level change rate for the site. The potential increase in relative sea level during the project lifecycle of the structure shall be added to the tsunami runup elevations.

6.6 INUNDATION DEPTHS AND FLOW VELOCITIES BASED ON RUNUP

6.6.1 Maximum Inundation Depth and Flow Velocities Based on Runup The maximum inundation depths and flow velocities associated with the stages of tsunami flooding shall be determined in accordance with Section 6.6.2. Calculated flow velocity shall not be taken as less than 10 ft/s (3.0 m/s) and need not be taken as greater than the lesser of $1.5(gh_{max})^{1/2}$ and 50 ft/s (15.2 m/s).

Where the maximum topographic elevation along the topographic transect between the shoreline and the inundation limit is greater than the runup elevation, one of the following methods shall be used.

1. The site-specific procedure of Section 6.7.6 shall be used to determine inundation depth and flow velocities at the site, subject to the range of calculated velocities defined in the first paragraph of this section.
2. For determination of the inundation depth and flow velocity at the site, the procedure of Section 6.6.2, Energy Grade Line Analysis, shall be used, assuming a runup elevation and horizontal inundation limit that has at least 100% of the maximum topographic elevation along the topographic transect.
3. Where the site lies within a completely overwashed area for which Inundation Depth Points are provided in the ASCE Tsunami Design Geodatabase, the inundation elevation profiles shall be determined using the Energy Grade Line Analysis with the following modifications.
   (a) The Energy Grade Line Analysis shall be initiated from the inland edge of the overwashed land with an inundation elevation equal to the maximum topographic elevation of the overwashed portion of the transect.
   (b) The Froude number shall be 1 at the inland edge of the overwashed land and shall vary linearly with distance to match the value of the Froude number determined at the shoreline per the coefficient α.
   (c) The Energy Grade Line Analysis flow elevation profile shall be uniformly adjusted with a vertical offset such that the computed inundation depth at the Inundation Depth Point is at least the depth specified by the ASCE Tsunami Design Geodatabase, but the flow elevation profile shall not be adjusted lower than the topographic elevations of the overwashed land transect.

6.6.2 Energy Grade Line Analysis of Maximum Inundation Depths and Flow Velocities The maximum velocity and maximum inundation depth along the ground elevation profile up to the inundation limit shall be determined using the Energy Grade Line Analysis. The orientations of the topographic transect profiles used shall be determined considering the requirements of Section 6.8.6.1. The ground elevation along the transect, z_i, shall be represented as a series of linear sloped segments each with a Manning's coefficient consistent with the equivalent terrain macroroughness friction of that terrain segment. The Energy Grade Line Analysis shall be performed incrementally in accordance with Equation (6.6-1) across the topographic transect in a stepwise procedure. Equation (6.6-1) shall be applied across the topographic transect from the runup where the hydraulic head at the inundation limit, x_R, is zero, and the water elevation is equal to the runup, R, by calculating the change in hydraulic head at each increment of terrain segment toward the shoreline until the site of interest is reached, as shown in Figure 6.6-1.

$$E_{g,i} = E_{g,i-1} + (\varphi_i + s_i)\Delta x_i \quad (6.6\text{-}1)$$

where

$E_{g,i}$ = Hydraulic head at point $i = h_i + u_i^2/2g = h_i(1 + 0.5F_{ri}^2)$;
h_i = Inundation depth at point i;
u_i = Maximum flow velocity at point i;
φ_i = Average ground slope between points i and $i-1$;
F_{ri} = Froude number = $u/(gh)^{1/2}$ at point i;
Δx_i = Increment of horizontal distance $(x_{i-1} - x_i)$ distance, which shall not be coarser than 100 ft (30.5 m) spacing;
x_i = Horizontal distance inland from the Mean High Water (MHW) shoreline at point i; and
s_i = Friction slope of the energy grade line between points i and $i-1$, is calculated per Equation (6.6-2).

$$s_i = (u_i)^2/((1.49/n)^2 h_i^{4/3}) = gF_{ri}^2/((1.49/n)^2 h_i^{1/3}) \quad (6.6\text{-}2)$$

$$s_i = (u_i)^2/((1.00/n)^2 h_i^{4/3}) = gF_{ri}^2/((1.00/n)^2 h_i^{1/3}) \quad (6.6\text{-}2.SI)$$

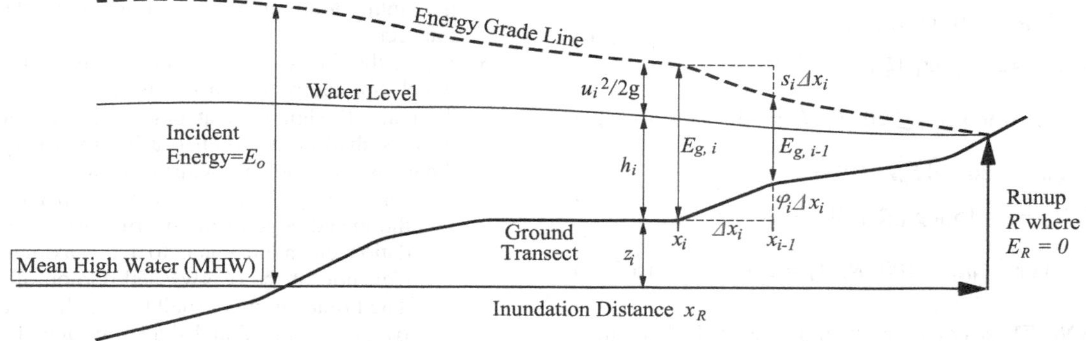

Figure 6.6-1. Energy method for overland tsunami inundation depth and velocity.

Note: R is the design tsunami runup elevation above MHW datum; x_R is the design inundation distance inland from MHW shoreline; and z_i is the ground elevation above MHW datum at point i.

Table 6.6-1. Manning's Roughness, *n*, for Energy Grade Line Analysis.

Description of Frictional Surface	n
Coastal water nearshore bottom friction	0.025 to 0.03
Open land or field	0.025
All other cases	0.03
Buildings of at least urban density	0.04

where n is Manning's coefficient of the terrain segment being analyzed, according to Table 6.6-1, and E_R is the hydraulic head of zero at the point of runup.

Velocity shall be determined as a function of inundation depth, in accordance with the prescribed value of the Froude number calculated according to Equation (6.6-3).

$$F_r = \alpha\left(1 - \frac{x}{x_R}\right)^{0.5} \tag{6.6-3}$$

where a value of 1.0 shall be used for α, the Froude number coefficient. Where tsunami bores are required to be considered per Section 6.6.4, the tsunami bore conditions specified in Sections 6.10.2.3 and 6.10.3.3 shall be applied using the values of h_e and $(h_e u^2)_{bore}$ evaluated with $\alpha = 1.3$.

At locations along the topographic transect where the average slope, $\varphi_{i,\,200}$, is negative, the velocity determined in accordance with Equation (6.6-3) shall be increased by an amount Δu_i per Equation (6.6-4), which need not be taken greater than 6 ft/s (1.83 m/s).

$$\Delta u_i = u_i |(\varphi_{i,200})|^{0.25} \tag{6.6-4}$$

where $\varphi_{i,\,200}$ is the slope averaged over a distance of 200 ft (60.96 m) centered on the transect point, x_i.

As a final step in the analysis, apply the minimum flow velocity required by Section 6.6.1 where necessary. The resulting values at the site are used as the maximum inundation depth, h_{max}, and the maximum velocity, u_{max}.

6.6.3 Terrain Roughness It shall be permitted to perform inundation analysis assuming bare-earth conditions with equivalent macroroughness. Bed roughness shall be prescribed using the Manning's coefficient n. It shall be permitted to use the values listed in Table 6.6-1 or other values based on terrain analysis in the recognized literature or as specifically validated for the inundation model used.

6.6.4 Tsunami Bores Tsunami bores shall be considered where any of the following conditions exist:

1. The prevailing nearshore bathymetric slope is 1/100 or milder,
2. Shallow fringing reefs or other similar step discontinuities in nearshore bathymetric slope occur,
3. Where historically documented,
4. As described in the recognized literature, or
5. As determined by a site-specific inundation analysis.

Where tsunami bores are deemed to occur, the tsunami bore conditions specified in Sections 6.10.2.3 and 6.10.3.3 shall be applied.

6.6.5 Amplified Flow Velocities Flow velocities determined in this section shall be adjusted for flow amplification in accordance with Section 6.8.5 as applicable. The adjusted value need not exceed the maximum limit specified in Section 6.6.1.

6.7 INUNDATION DEPTHS AND FLOW VELOCITIES BASED ON SITE-SPECIFIC PROBABILISTIC TSUNAMI HAZARD ANALYSIS

When required by Section 6.5, the inundation depths and flow velocities shall be determined by site-specific inundation studies complying with the requirements of this section. Site-specific analyses shall use the ASCE Tsunami Design Geodatabase of geocoded reference points of Offshore Tsunami Amplitude and dominant waveform period shown in Figure 6.7-1 as input to an inundation numerical model or shall use an integrated generation, propagation, and inundation model that replicates the given offshore tsunami waveform amplitude and period from the seismic sources given in Section 6.7.1. The ASCE Tsunami Design Geodatabase of geocoded reference points of offshore 328 ft (100 m) depth, Tsunami Amplitude, H_T, and predominant period, T_{TSU}, of the Maximum Considered Tsunami is available at http://asce7tsunami.online.

6.7.1 Tsunami Waveform The tsunami waveform used along the offshore boundary of 328 ft (100 m) bathymetry shall be constructed in accordance with Equation (6.7-1), as illustrated in Figure 6.7-2.

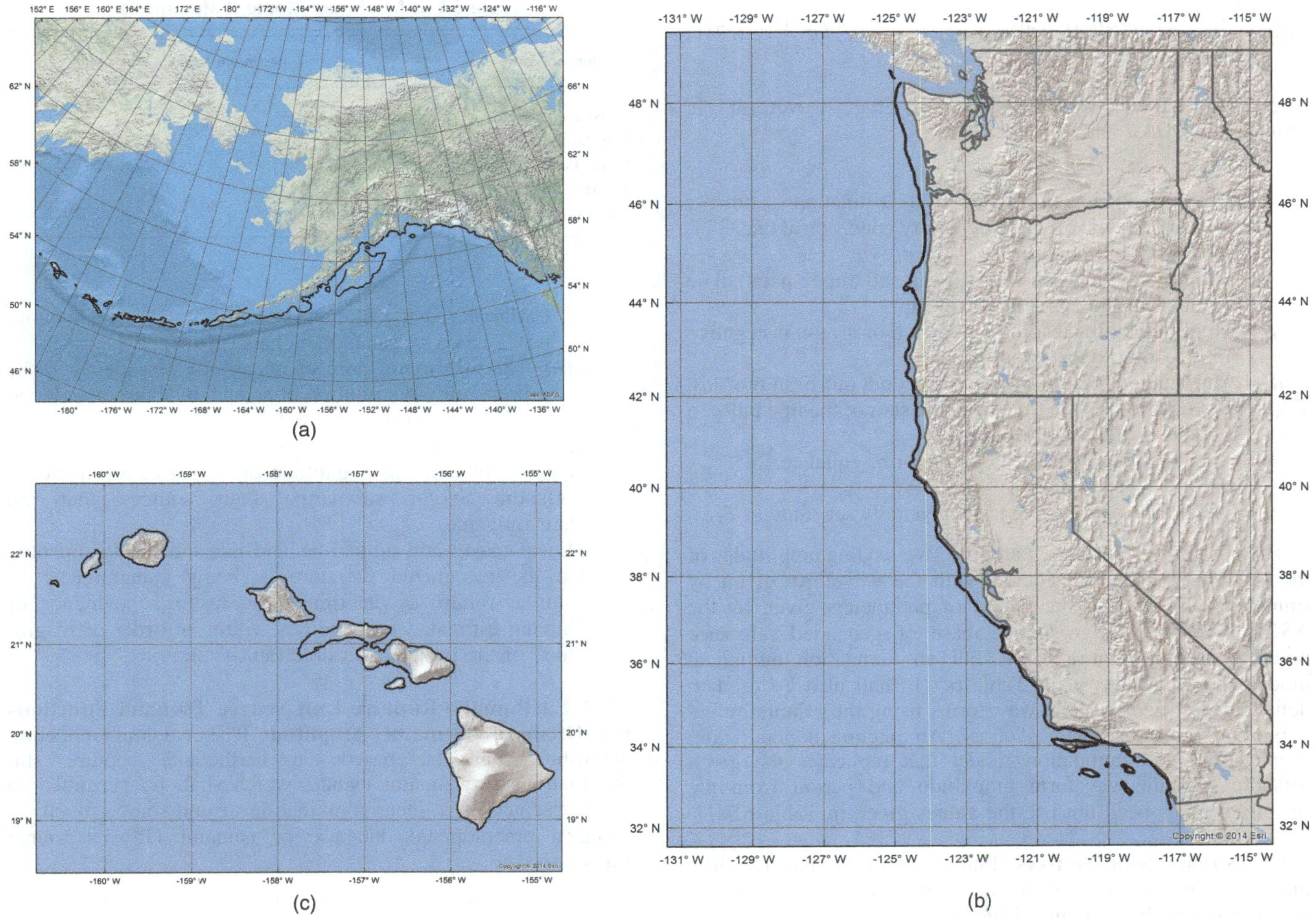

(a)

(c)

(b)

Figure 6.7-1. Extent of ASCE Tsunami Design Geodatabase of geocoded reference points of Offshore Tsunami Amplitude, H_T[at 328 ft (100 m) depth], and predominant period, T_{TSU}, of the Maximum Considered Tsunami.

Note: The ASCE Tsunami Design Geodatabase of geocoded points of Offshore Tsunami Amplitude, H_T,[at 328 ft (100 m)] depth, and predominant period, T_{TSU}, of the Maximum Considered Tsunami is available at http://asce7tsunami.online.

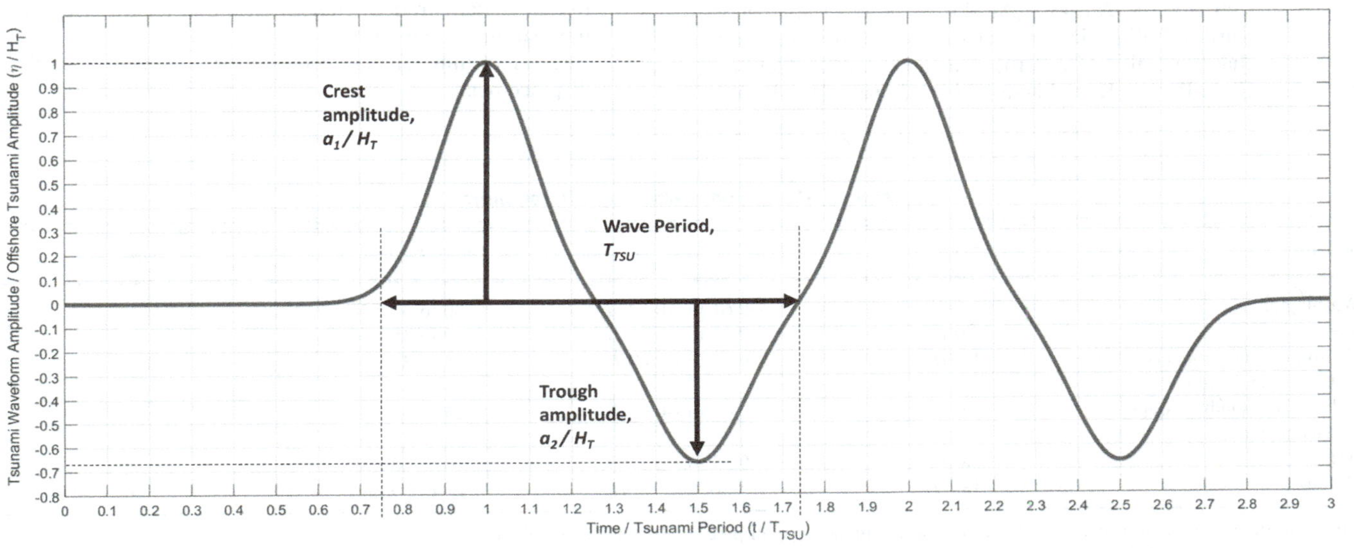

Figure 6.7-2. Illustration of tsunami offshore incident waveform parameters at 328 ft (100 m) depth with $N = 2$.

$$\eta=\sum_{i=1}^{N}[a_1 e^{-\{\omega[t-t_o-(i-1)T_{TSU}]\}^2}+a_2 e^{-\{\omega[t-t_o-(i-0.5)T_{TSU}]\}^2}] \quad (6.7\text{-}1)$$

where the total wave height of the waveform is $abs(a_1)+abs(a_2)$, and

η = Free surface elevation (ft or m) as a function of time, t, used to drive the offshore boundary condition at the 328 ft (100 m) depth contour;
N = Number of individual waves (peak and trough pairs) to be included in the tsunami waveform;
a_1 = Amplitude of the leading pulse (in ft or m)—it is negative for a leading depression tsunami;
a_2 = Amplitude of the following, or second, pulse (in ft or m);
T_{TSU} = Wave period, or the time from the start of the first pulse to the end of the second pulse;
ω = Angular frequency of the waveform, equal to $2\pi/T_{TSU}$; and
t_o = Offset time of the wave train, generally set equal to T_{TSU}.

The possibility of negative and positive leading amplitudes of the tsunami shall be considered, with the waveform given by Equation (6.7-1) using the values of parameters given by the ASCE Tsunami Design Geodatabase of geocoded reference points shown in Figure 6.7-1. For an inundation numerical model, the values given in Table 6.7-1 shall also be used to define at least two possible waveforms using the minimum and maximum prescribed values of a_2. An integrated generation, propagation, and inundation model that replicates the given offshore tsunami waveform amplitude and period from the seismic sources need not use the values given in Table 6.7-1.

6.7.2 Tsunamigenic Sources Tsunami sources shall consider the following to the extent that probabilistic hazards are documented in the recognized literature:

1. Local and distant subduction zone sources. It shall be permitted to use a system of delineated and discretized subduction zones in the Pacific basin composed of systems of rectangular subfaults and their corresponding tectonic parameters.
   (a) Principal seismic sources shall include but are not restricted to Alaska: (Alaska-Aleutian, Kamchatka-Kurile), California: (Alaska-Aleutian, Cascadia, Kamchatka-Kuril, Chile-Peru), Hawaii: (Alaska-Aleutian, Chile-Peru, Kamchatka-Kuril, Japan, Izu-Bonin-Mariana Islands), and Oregon and Washington: (Cascadia, Alaska, Kamchatka-Kuril).
   (b) The maximum moment magnitude considered in the probability distribution of seismicity shall include the values given in Table 6.7-2.
2. Local, non-subduction-zone seismic fault sources capable of moment magnitude of 7 or greater, including offshore and/or submarine fault sources that are tsunamigenic.
3. Local coastal and submarine landslide sources documented in the recognized literature as being tsunamigenic of similar runup, as determined by historical evidence or having estimated probabilities within an order of magnitude of the principal seismic fault sources.

Table 6.7-2. Maximum Moment Magnitude.

Subduction Zone	Maximum Moment Magnitude, M_w
Alaska-Aleutian	9.5
Cascadia	9.2
Chile-Peru	9.5
Izu-Bonin-Marianas	9.0
Kamchatka-Kurile and Japan Trench	9.4

6.7.3 Earthquake Rupture Unit Source Tsunami Functions for Offshore Tsunami Amplitude The tsunami modeling algorithm shall be based on earthquake rupture slip distributions for tsunami events, which shall be permitted to be represented by a linear combination of unit source functions using a precomputed database of tsunami Green's source functions.

1. Tsunami waveform generation shall be permitted to be modeled by deconstructing a tsunami that is generated by an earthquake into a sum of individual tsunami waveforms composed from a scaled set of unit source subfaults that describe the earthquake rupture in terms of location, orientation, and rupture direction and sequence.
2. The waveforms defining the time series of wave height and velocity from a unit slip on each subfault shall be weighted by the actual slip or rupture distribution for the event and then summed linearly.
3. The algorithm shall account for coseismic vertical displacement.

Table 6.7-1. Regional Waveform Parameters.

Region	a_1	a_2[a]	T_{TSU}[b]	N
Washington	1.0 H_T	−0.61 to −0.82 H_T	30 to 40 min	2
Oregon	1.0 H_T	−0.55 to −0.67 H_T	30 to 45 min	2
California – north inclusive of Cape Mendocino	1.0 H_T	−0.55 to −0.67 H_T	25 to 35 min	2
California – south of Cape Mendocino	1.0 H_T	−0.43 to −0.67 H_T	25 to 35 min	3
Alaska	1.0 H_T	−0.55 to −0.82 H_T	20 to 40 min	2
Hawaii	1.0 H_T	−0.67 to −1.0 H_T	25 to 30 min	3

[a] For a leading depression waveform, the trough of amplitude a_2 shall precede the crest of amplitude a_1.
[b] The value of T_{TSU} shall be used if no mapped value is given in Figure 6.7-1.

6.7.4 Treatment of Modeling and Natural Uncertainties A statistically weighted logic tree approach shall be used to account for epistemic uncertainties in the model parameters and shall provide a sample of tsunamigenic earthquakes and their occurrence probabilities from tectonic, geodetic, historical, and paleotsunami data, and estimated plate convergence rates, as follows:

1. Subdivide the occurrence probability systematically to account for variations in the parameters of magnitude, fault depth and geometry, and location, slip distribution, and rupture extent of events consistent with maximum magnitudes, and tidal variation considering at least the Reference Sea Level.
2. To the extent practical and quantifiable, follow a similar logic tree approach to determine samples of tsunami sources such as nonsubduction zone earthquakes, landslides, and volcanic eruptions.

Aleatory uncertainties, such as the natural variability in the source processes, modeling uncertainties, and tidal variation as they relate to nearshore processes and wave runup, shall be included in the probabilistic analysis. When accounting for long wave durations with multiple maxima in the tsunami time series, it shall be permitted to consider tidal variability by selecting a rational tidal elevation independently from a probabilistic distribution of tide stages for each wave maximum. Truncation of aleatory distributions shall be chosen at an appropriate level for the return period but shall not be less than one standard deviation based on a regression analysis of computed versus observed data of Section 6.7.6.7.2.

6.7.5 Offshore Tsunami Amplitude The probabilistic analysis shall be performed either by direct computation according to Section 6.7.5.2 or by performing a Probabilistic Tsunami Hazard Analysis for a region of interest to produce site-specific Offshore Tsunami Amplitude hazard maps and predominant wave period at 328 ft (100 m) depth in accordance with the following:

1. A Digital Elevation Model (DEM) from global, regional, and coastal data sets shall be used to cover the computational domain from the tsunami sources to the site. The bathymetry grid of the ocean shall have a DEM resolution finer than 4.35 mi (7,000 m), and the offshore model regime with depth greater than 656 ft (200 m) shall have a DEM resolution finer than 3,281 ft (1,000 m).
2. The Earth surface deformation shall be determined from the seismic source parameters using a planar fault model accounting for vertical changes to the seafloor.

6.7.5.1 Offshore Tsunami Amplitude for Distant Seismic Sources OFFSHORE TSUNAMI AMPLITUDE shall be probabilistically determined in accordance with the following:

1. A weighted combination of tsunami waveforms determined for each unit fault segment in accordance with the slip distribution shall be used to propagate tsunamis in deep water using the linear long wave equations, also termed the shallow water wave equations, where water depth is much less than the wavelength, to take into account spatial variations in seafloor depth.
2. The offshore wave amplitude distribution and associated wave parameters, including period, shall be determined for the design exceedance rate of the 2,475-year Maximum Considered Tsunami taking into consideration uncertainties per Section 6.7.4.
3. The analysis shall include the disaggregation of the seismic sources and associated moment magnitudes that together contribute at least 90% to the net offshore tsunami hazard at the site under consideration.
4. The extent of offshore tsunami amplitude points considered for the site shall include the following:
   (a) For sites within Washington, Oregon, California, and Hawaii, the extent shall include points within at least 40 mi (64.4 km) but not exceeding 50 mi (80.5 km) of projected length along the coastline, centered on the site within a tolerance of plus or minus 6 mi (9.7 km);
   (b) For sites within Alaska, the extent shall include points within at least 100 mi (161 km) but not exceeding 125 mi (201 km), centered on the site within a tolerance of plus or minus 15 mi (24.1 km);
   (c) For sites within bays, the designated center of the computed offshore tsunami amplitude points shall be taken either offshore of the mouth of the bay or centered in accordance with criteria (a) or (b) above, whichever produces the more severe flow conditions at the site.
   (d) For island locations where the projected width of the island is less than 40 mi (64.4 km), it shall be permitted to consider the extent of offshore tsunami amplitude points corresponding to the projected width of the island. Shorter extents of offshore tsunami amplitude points shall be permitted for island locations, but shall not be less than 10 mi (16.1 km). In addition, the tsunami source development and inundation modeling are subject to an independent peer review by a tsunami modeler approved by the Authority Having Jurisdiction, who shall present a written report to the Authority Having Jurisdiction as to the hazard consistency of the modeling with the requirements of Section 6.7.
5. The mean value of the computed offshore tsunami amplitudes shall be not less than 100% of the mean value for the coinciding offshore tsunami amplitude data given by the ASCE Tsunami Design Geodatabase.
6. The individual values of computed offshore tsunami wave amplitude shall be not less than 80% of the coinciding-offshore tsunami amplitude values given by the ASCE Tsunami Design Geodatabase.

6.7.5.2 Direct Computation of Probabilistic Inundation and Runup It shall be permitted to compute probabilistic inundation and runup directly from a probabilistic set of sources, source characterizations, and uncertainties consistent with Section 6.7.2, Section 6.7.4, and the computing conditions set out in Section 6.7.6. The offshore wave amplitudes computed shall comply with the requirements of Sections 6.7.5.1.4, 6.7.5.1.5, and 6.7.5.1.6.

6.7.5.3 Use of Higher-Order Tsunami Model Features Higher-order tsunami model features, including wave dispersion, elasticity of the earth, variation in gravity, water compressibility, or water density stratification, shall be permitted as analysis enhancements for use within the same model software and parameters selected to comply with Section 6.7, including but not limited to bathymetric and terrain data, grid configuration, and bottom friction, provided these features are also validated in accordance with Section 6.7.6.7 and the higher-order tsunami model satisfies the following conditions:

1. The computed offshore wave amplitudes comply with the requirements of Section 6.7.5.1, or the model with the

higher-order features complies with all of the following requirements:

(a) Sections 6.7.5.1.1 through 6.7.5.1.4,
(b) The mean value of the computed offshore tsunami amplitudes is at least 85% of the mean value for the coinciding offshore tsunami amplitude data of the ASCE Tsunami Design Geodatabase,
(c) The values of computed offshore tsunami wave amplitude are not less than 75% of the coinciding offshore tsunami amplitude values of the ASCE Tsunami Design Geodatabase.
(d) Documentation of the tsunami source development, inundation modeling features and their prior validations, computed offshore tsunami amplitudes and runups for the model, with and without the higher order features shall be subject to an independent peer review by a tsunami modeler approved by the Authority Having Jurisdiction, who shall present a written report to the Authority Having Jurisdiction as to the model's compliance with the requirements of Section 6.7.

2. Use of a tsunami model with higher-order features shall comply with the following procedure:
   (a) The tsunami model without the higher-order features shall determine a Hazard Consistent Tsunami Scenario that satisfies the requirements of Sections 6.7.5.1.5 and 6.7.5.1.6.
   (b) The Hazard Consistent Tsunami Scenario shall be used for the same tsunami model with its higher-order features engaged, and one of the following criteria shall be satisfied:
      (i) If Section 6.7.5.1 is satisfied, then the results of the higher-order model shall be deemed acceptable;
      (ii) If Section 6.7.5.3.1 is satisfied, then a report documenting the results of a peer review of the higher-order model results shall be submitted to the Authority Having Jurisdiction for approval;

If neither Section 6.7.5.1 nor Section 6.7.5.3.1 is satisfied, then the Hazard Consistent Tsunami Scenario shall be modified until one of the above criteria is satisfied using the higher-order model features.

6.7.6 Procedures for Determining Tsunami Inundation and Runup

6.7.6.1 Representative Design Inundation Parameters Each disaggregated tsunami event shall be analyzed to determine representative design parameters consisting of maxima of runup, inundation depth, flow velocity, and momentum flux.

6.7.6.2 Seismic Subsidence before Tsunami Arrival Where the seismic source is a local subduction event, the Maximum Considered Tsunami inundation shall be determined for an overall elevation subsidence value shown in Figure 6.7-3(a) and 6.7-3(b) or shall be directly computed for the seismic source mechanism. The GIS digital map layers of subsidence are available in the ASCE Tsunami Design Geodatabase at http://asce7tsunami.online.

6.7.6.3 Model Macroroughness Parameter It shall be permitted to perform inundation mapping under bare-earth conditions with macroroughness. Bed roughness shall be permitted to be prescribed using the Manning's coefficient, n. Unless otherwise determined for the site, a default value of 0.025 or 0.030 shall be used for the ocean bottom and on land. Use of other values based on terrain analysis shall be justified in the recognized literature or shall be specifically validated for the inundation model for field benchmarks of historical tsunami. Where values other than the defaults are used, the effects of degradation of roughness because of damaging flow characteristics shall be considered in the choice of Manning coefficient.

6.7.6.4 Nonlinear Modeling of Inundation Nonlinear shallow water wave equations or equivalent modeling techniques shall be used to transform the offshore wave amplitude from 328 ft (100 m) depth toward the shore to its nearshore tsunami amplitude and maximum inundation. The following effects shall be included as applicable to the bathymetry:

1. Shoaling, refraction, and diffraction to determine nearshore tsunami amplitude;
2. Dispersion effects in the case of short-wavelength sources, such as landslides and volcanic sources;
3. Reflected waves;
4. Channeling in bays;
5. Edge waves, and shelf and bay resonances;
6. Bore formation and propagation; and
7. Harbor and port breakwaters and levees.

6.7.6.5 Model Spatial Resolution A Digital Elevation Model (DEM) for the nearshore bathymetry depth of less than 656 ft (200 m) shall have a resolution not coarser than 295 ft (90 m). At bathymetric depths of less than 32.8 ft (10 m) and on land, the DEM shall have a resolution not coarser than the highest resolution available from the Tsunami Inundation Digital Elevation Models of the NOAA National Geophysical Data Center (NGDC). If a nested grid approach is used, the reduction in grid-spacing between consecutive grids shall not be more than a factor of 5. Where the NOAA NGDC models are not available, use of the best available USGS integrated Digital Elevation Model data shall be permitted when approved by the Authority Having Jurisdiction.

6.7.6.6 Built Environment If buildings and other structures are included for the purposes of more detailed flow analysis, the Digital Elevation Model resolution shall have a minimum resolution of 10 ft (3.0 m) to capture flow deceleration and acceleration in the built environment.

6.7.6.7 Inundation Model Validation

6.7.6.7.1 Historical or Paleotsunami Inundation Data Model scenario results shall be validated with available historical and/or paleotsunami records.

6.7.6.7.2 Model Validation by Benchmark Tests The inundation model shall be validated using the certification criteria of the National Tsunami Hazard Mitigation Program (NTHMP) within 10% by providing satisfactory performance in a series of benchmark tests of known data sets designated by the Tsunami Model Validation Advisory Group in NOAA Technical Memorandum OAR PMEL-135, "Standards, Criteria, and Procedures for NOAA Evaluation of Tsunami Numerical Models," as modified by the NTHMP.

6.7.6.7.3 Tsunami Bore Formation or Soliton Fission In regions where bore formation may occur, the model shall be validated with an independent scenario in the recognized literature, and acceptability shall be determined using modeled runup.

6.7.6.8 Determining Site-Specific Inundation Flow Parameters Inundation parameters for the scenarios from each disaggregated source region shall be determined. Probabilistic flow parameters shall be developed for the site

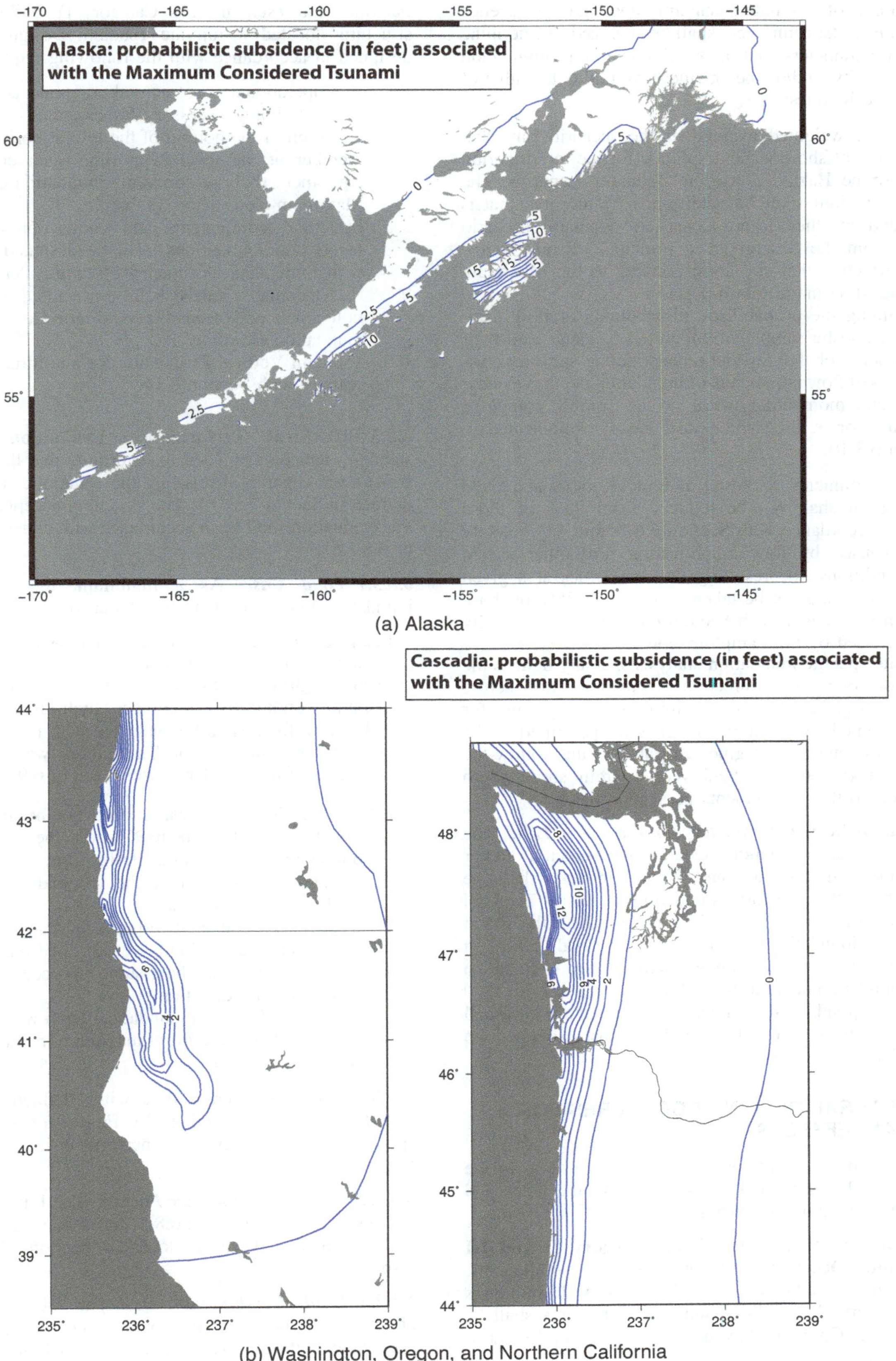

(a) Alaska

(b) Washington, Oregon, and Northern California

Figure 6.7-3. Earthquake-induced regional ground subsidence (in feet) associated with a Maximum Considered Tsunami caused by a local subduction earthquake (1 ft = 0.305 m).

Note: The GIS digital map layers of subsidence are available from the ASCE Tsunami Design Geodatabase at http://asce7tsunami.online.

from the sample of computed tsunamis and their occurrence probabilities. Each tsunami event shall be analyzed to determine representative parameters such as maximum runup, inundation depth, flow velocity, and/or specific momentum flux by either of the following techniques:

1. Taking the weighted average of the scenario runs that bracket the offshore wave amplitude for the return period to determine Hazard-Consistent Tsunami scenarios. The inundation limit shall be determined by the area that is inundated by the Hazard-Consistent Tsunami scenario waves from the disaggregated principal seismic source zones affecting that site corresponding to the Maximum Considered Tsunami return period.
2. Determining the probabilistic distributions of flow parameters from the sample of computed tsunamis and their occurrence probabilities and reconstructing statistical distributions of flow parameters of inundation depth, velocity, and specific momentum flux at the site from the computed scenarios for at least three load cases, as indicated in Section 6.8.10.

In urban environments, the resulting flow velocities at a given structure location shall not be reduced from 90% of those determined in accordance with Section 6.6 before any velocity adjustments caused by flow amplification. For other terrain roughness conditions, the resulting flow velocities at a given structure location shall not be taken as less than 75% of those determined in accordance with Section 6.6 before any velocity adjustments caused by flow amplification.

Where a site-specific inundation analysis spatially defines the topographic diversion of the direction of inundation flow across the Tsunami Design Zone, the minimum flow velocity limits for the site determined by Section 6.6 shall be permitted to be determined using angularly segmented transects that follow the site-specific inundation flow field vectors from shoreline to Runup, subject to the requirements of Section 6.8.6.2.

6.7.6.9 Tsunami Design Parameters for Flow over Land The flow parameters of inundation depth, flow velocity, and/or specific momentum flux at the site of interest shall be captured from a time history inundation analysis. Tsunami inundation depth and velocity shall be evaluated for the site at the stages of inundation defined by the load cases in Section 6.8.3.1. If the maximum momentum flux is found to occur at an inundation depth different than Load Case 2, the flow conditions corresponding to the maximum momentum flux shall be considered in addition to the load cases defined in Section 6.8.3.1.

6.8 STRUCTURAL DESIGN PROCEDURES FOR TSUNAMI EFFECTS

Structures, components, and foundations shall conform to the requirements of this section when subjected to the loads and effects of the Maximum Considered Tsunami.

6.8.1 Performance of Tsunami Risk Category II and III Buildings and Other Structures Structural components, connections, and foundations of Risk Category II buildings and Risk Category III buildings and other structures shall be designed to meet Collapse Prevention Structural Performance criteria or better.

6.8.2 Performance of Tsunami Risk Category III Critical Facilities and Tsunami Risk Category IV Buildings and Other Structures Tsunami Risk Category III Critical Facilities and Tsunami Risk Category IV buildings and other structures located within the Tsunami Design Zone shall be designed in accordance with the following requirements.

1. The operational nonstructural components and equipment of the building necessary for essential functions and the elevation of the bottom of the lowest horizontal structural member at the level supporting such components and equipment shall be above the inundation elevation of the Maximum Considered Tsunami.
2. Structural components and connections in occupiable levels and foundations shall be designed in accordance with Immediate Occupancy Structural Performance criteria. Occupiable levels shall be permitted where the elevation equals or exceeds the Maximum Considered Tsunami inundation elevation.
3. Tsunami Vertical Evacuation Refuge Structures shall also comply with Section 6.14.

6.8.3 Structural Performance Evaluation Strength and stability shall be evaluated to determine that the design of the structure is capable of resisting the tsunami at the Load Cases defined in Section 6.8.3.1. The structural acceptance criteria for this evaluation shall be in accordance with either Section 6.8.3.4 or 6.8.3.5.

6.8.3.1 Load Cases As a minimum, the following three inundation load cases shall be evaluated.

Load Case 1: At an exterior inundation depth not exceeding the maximum inundation depth nor the lesser of one story or the height of the top of the first-story windows, the minimum condition of combined hydrodynamic force with buoyant force shall be evaluated with respect to the depth of water in the interior. The interior water depth shall be evaluated in accordance with Section 6.9.1.

EXCEPTION: Load Case 1 need not be applied to Open Structures nor to structures where the soil properties or foundation and structural design prevent detrimental hydrostatic pressurization on the underside of the foundation and lowest structural slab.

Load Case 2: Depth at two-thirds of maximum inundation depth when the maximum velocity and maximum specific momentum flux shall be assumed to occur in either incoming or receding directions.

Load Case 3: Maximum inundation depth when velocity shall be assumed at one-third of maximum in either incoming or receding directions.

The inundation depths and velocities defined for Load Cases 2 and 3 shall be determined by Figure 6.8-1, unless a site-specific tsunami analysis is performed in accordance with Section 6.7.

6.8.3.2 Tsunami Importance Factors The Tsunami Importance Factors, I_{tsu}, given in Table 6.8-1 shall be applied to the tsunami hydrodynamic and impact loads in Sections 6.10 and 6.11, respectively.

6.8.3.3 Load Combinations Principal tsunami forces and effects shall be combined with other specified loads in accordance with the load combinations of Equation (6.8-1a):

$$0.9D + F_{TSU} + H_{TSU} \quad (6.8\text{-}1a)$$

$$1.2D + F_{TSU} + 0.5L + 0.2S + H_{TSU} \quad (6.8\text{-}1b)$$

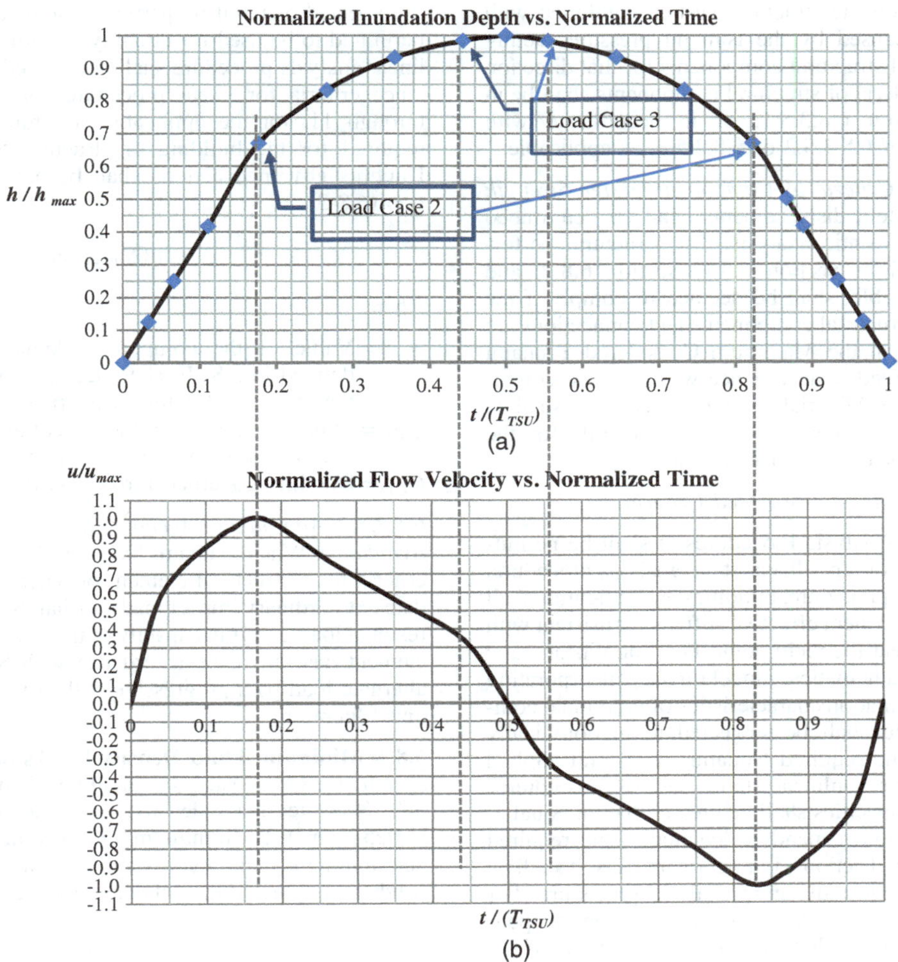

Figure 6.8-1. Inundation Load Cases 2 and 3.

Table 6.8-1. Tsunami Importance Fctors for Hydrodynamic and Impact Loads.

Tsunami Risk Category	I_{tsu}
II	1.0
III, Tsunami Risk Category IV, Vertical Evacuation Refuges, and Tsunami Risk Category III Critical Facilities	1.25
Tsunami Risk Category IV where the inundation depth is less than 12 ft (3.66 m) and a site-specific Probabilistic Tsunami Hazard Analysis is not performed, per the exception to Section 6.5.2	1.5

where

F_{TSU} = Tsunami load effect for incoming and receding directions of flow, and

H_{TSU} = Load caused by tsunami-induced lateral foundation pressures developed under submerged conditions. Where the net effect of H_{TSU} counteracts the principal load effect, the load factor for H_{TSU} shall be 0.9.

6.8.3.4 Lateral-Force-Resisting System Acceptance Criteria For structures designed to the requirements for Seismic Design Category A, B, or C, the lateral-force-resisting system shall be designed to resist the Maximum Considered Tsunami. For structures designed to the requirements for Seismic Design Category D, E, or F, the lateral-force-resisting system shall be deemed adequate if 0.75 $\Omega_0 E_h$ exceeds the required tsunami force F_{TSU}, where E_h is the required seismic resistance and Ω_0 is the system overstrength as defined in Chapter 12 of this standard. If the value of 0.75 $\Omega_0 E_h$ does not exceed F_{TSU}, the lateral-force-resisting system at all levels below the maximum inundation depth shall be designed for a minimum seismic force $Q_E \geq F_{TSU}/(0.75\ \Omega_0)$, or the lateral-force-resisting system shall be explicitly analyzed to validate its resistance to the Maximum Considered Tsunami. For Immediate Occupancy Structural Performance objectives, the lateral-force-resisting system shall be explicitly analyzed and evaluated.

6.8.3.5 Structural Component Acceptance Criteria Structural components shall be designed for the forces that result from the overall tsunami forces on the structural system combined with any resultant actions caused by the tsunami pressures acting locally on the individual structural components for that direction of flow. Acceptance criteria of structural components shall be in accordance with Section 6.8.3.5.1, or in accordance with alternative procedures of 6.8.3.5.2 or 6.8.3.5.3, as applicable.

6.8.3.5.1 Acceptability Criteria by Component Design Strength Internal forces and system displacements shall be determined using a linearly elastic, static analysis. The structural performance criteria required in Section 6.8.1, Section 6.8.2, and Section 6.8.3, as applicable, shall be deemed to comply if the design strength of the structural components and connections are shown to be greater than the Maximum Considered Tsunami loads and effects computed in accordance with the load combinations of Section 6.8.3.3. Material resistance factors, φ, shall be used as prescribed in the material-specific standards for the component and behavior under consideration.

6.8.3.5.2 Alternative Performance-Based Criteria

6.8.3.5.2.1 Alternative Analysis Procedures It shall be permitted to use either a linear or a nonlinear static analysis procedure. In a linear static analysis procedure, buildings and structures shall be modeled using an equivalent effective stiffness consistent with the secant value at or near the yield point. For a nonlinear static analysis procedure, a mathematical model directly incorporating the nonlinear load-deformation characteristics of individual components of the structure shall be subjected to monotonically increasing loads until the required tsunami forces and applied actions are reached. For nonlinear static analysis procedures, expected deformation capacities shall be greater than or equal to the maximum deformation demands calculated at the required tsunami forces and applied actions. For debris impacts, it shall be permitted to use a nonlinear dynamic analysis procedure. For Tsunami Risk Category IV buildings and structures, an independent peer review shall be conducted as part of a review of the performance-based design by the Authority Having Jurisdiction.

6.8.3.5.2.2 Alternative Structural Component Acceptability Criteria All actions shall be classified as either ductility-governed actions or force-sustained actions based on component inelastic behavior and the duration of the load effect, as follows:

1. Fluid forces in primary and secondary structural components detailed in accordance with the requirements of Seismic Design Category D, E, or F shall be evaluated as force-sustained actions.
2. Debris impacts and foundation settlement effects on primary and secondary structural components shall be evaluated as ductility-governed actions.
3. Debris impacts and foundation settlement effects on primary and secondary structural components not detailed in accordance with Seismic Design Category D, E, or F shall be evaluated as force-sustained actions.

For force-sustained actions, structural components shall have specified design strengths greater than or equal to the maximum design forces. Force-sustained actions shall be permitted to satisfy Equation (6.8-2):

$$Q_{CS} \geq Q_{UF} \tag{6.8-2}$$

where Q_{CS} is the specified strength of the structural element and Q_{UF} is the maximum force generated in the element because of gravity and tsunami loading.

Expected material properties as defined in ASCE 41 shall be permitted to be used for ductility-governed actions. Results of a linear analysis procedure shall not exceed the component acceptance criteria for linear procedures of ASCE 41, Chapters 9 through 11, for the applicable structural performance criteria required for the building or structure tsunami risk category. Ductility-governed actions shall be permitted to satisfy Equation (6.8-3):

$$mQ_{CE} \geq Q_{UD} \tag{6.8-3}$$

where

m = Value of the component demand modification factor defined in ASCE 41 to account for expected ductility at the required structural performance level;

Q_{CE} = Expected strength of the structural element determined in accordance with ASCE 41; and

Q_{UD} = Ductility-governed force caused by tsunami loading.

6.8.3.5.3 Alternative Acceptability by Progressive Collapse Avoidance Where tsunami loads or effects exceed acceptability criteria for a structural element or where required to accommodate extraordinary impact loads, it shall be permitted to check the residual load-carrying capacity of the structure, assuming that the element has failed, in accordance with Section 2.5.2.2 and an alternate load path progressive collapse procedure in the recognized literature.

6.8.4 Minimum Fluid Density for Tsunami Loads Seawater specific weight density, γ_{sw}, shall be taken as 64.0 lb/ft^3 (10 kN/m^3). Seawater mass density, ρ_{sw}, shall be taken as 2.0 sl/ft^3 (1,025 kg/m^3). The minimum fluid specific weight density, γ_s, for determining tsunami hydrostatic loads accounting for suspended solids and debris flow-embedded smaller objects shall be

$$\gamma_s = k_s \gamma_{sw} \tag{6.8-4}$$

The minimum fluid mass density, ρ_s, for determining tsunami hydrodynamic loads accounting for suspended solids and debris flow-embedded smaller objects shall be

$$\rho_s = k_s \rho_{sw} \tag{6.8-5}$$

where k_s, the fluid density factor, shall be taken as 1.1.

6.8.5 Flow Velocity Amplification The effect of upstream obstructing buildings and structures shall be permitted to be considered at a site that is exposed to the flow diffracting conditions given in Section 6.8.5.1 by any of the following:

1. A site-specific inundation analysis that includes modeling of the built environment in accordance with Section 6.7.6.6, or
2. The built environment is considered in the selection of Manning's roughness of Table 6.6-1 in accordance with the Energy Grade Line Analysis of Section 6.6.2, or
3. Site-specific physical or numerical modeling in accordance with Section 6.8.5.2 or Section 6.8.10, as applicable.

6.8.5.1 Upstream Obstructing Structures The effect of upstream obstructions on flow shall be considered where the obstructions are enclosed structures of concrete, masonry, or structural steel construction located within 500 ft (152 m) of the site, and both of the following apply:

1. Structures have plan width greater than 100 ft (30.5 m) or 50% of the width of the downstream structure, whichever is greater; and
2. The structures exist within the sector between 10 and 55 degrees to either side of the flow vector aligned with the center third of the width of the downstream structure.

6.8.5.2 Flow Velocity Amplification by Physical or Numerical Modeling The effect of upstream structures on the flow velocity at a downstream site shall be permitted to be evaluated using site-specific numerical or physical modeling, as described in Section 6.7.6.6 or 6.8.10. The velocity determined for a "bare-earth" inundation shall be amplified for the conditions of Section 6.8.5.1. This analysis is not permitted to reduce the flow velocity except for structural countermeasures designed in accordance with Section 6.13.

6.8.6 Directionality of Flow

6.8.6.1 Flow Direction Design of structures for tsunami loads and effects shall consider both incoming and outgoing flow conditions. The principal inflow direction shall be assumed to vary by ± 22.5 degrees from the transect perpendicular to the orientation of the shoreline averaged over 500 ft (152 m) to either side of the site. The center of rotation of the variation of transects shall be located at the geometric center of the structure in plan at the grade plane.

6.8.6.2 Site-Specific Directionality A site-specific inundation analysis performed in accordance with Section 6.7.6 shall be permitted to be used to determine directionality of flow, provided that the directionalities so determined shall be assumed to vary by at least ± 10 degrees.

6.8.7 Minimum Closure Ratio for Load Determination Loads on buildings shall be calculated assuming a minimum closure ratio of 70% of the inundated projected area along the perimeter of the structure, unless it is an Open Structure as defined in Section 6.2. The load effect of debris accumulation against or within the Open Structure shall be considered by using a minimum closure ratio of 50% of the Inundated Projected Area along the perimeter of the Open Structure. Open Structures need not be subject to Load Case 1 of Section 6.8.3.1.

6.8.8 Minimum Number of Tsunami Flow Cycles Design shall consider a minimum of two tsunami inflow and outflow cycles, the first of which shall be based on an inundation depth at 80% of the Maximum Considered Tsunami (MCT), and the second of which shall be assumed to occur with the MCT inundation depth at the site. Local scour effects caused by the first cycle, and determined in accordance with Section 6.12, shall be assumed to occur at 80% of the MCT inundation depth at the site and shall be considered as an initial condition of the second cycle.

6.8.9 Seismic Effects on the Foundations Preceding Local Subduction Zone Maximum Considered Tsunami Where designated in Figure 6.7-3 as a site subject to a local subduction zone tsunami from an offshore subduction earthquake, the structure shall be designed for the preceding coseismic effects. The foundation of the structure shall be designed to resist the preceding earthquake ground motion and its associated effects per Chapter 11 of this standard using the Maximum Considered Earthquake Geometric Mean (MCE_G) Peak Ground Acceleration of Figures 22-7, 22-8, and 22-9. Building foundation design shall include changes in the site surface and the in situ soil properties resulting from the design seismic event as initial conditions for the subsequent design tsunami event. The geotechnical investigation report shall include evaluation of foundation effects in reference to seismic effects preceding the tsunami, with consideration of slope instability, liquefaction, total and differential settlement, surface displacement caused by faulting, and seismically induced lateral spreading or lateral flow. The additional requirements of Section 6.12 shall also be evaluated.

6.8.10 Physical Modeling of Tsunami Flow, Loads, and Effects Physical modeling of tsunami loads and effects shall be permitted as an alternative to the prescriptive procedures in Sections 6.8.5 (flow velocity amplification), 6.10 (hydrodynamic loads), 6.11 (debris impact loads), and 6.12 (foundation design), provided that it meets all the following criteria.

1. The facility or facilities used for physical modeling shall be capable of generating appropriately scaled flows and inundation depths as specified for Load Cases in Section 6.8.3.1.
2. The test facility shall be configured so that reflections and edge effects shall not significantly affect the test section during the duration of the experiments.
3. The scale factors used in the physical modeling shall not be less than those shown in Table 6.8-2. Scale model tests not directly addressed in Table 6.8-2 shall include a justification of the model applicability and scaling procedures.
4. Debris impacts of full or partial components shall be tested at full scale unless accompanied by a justification of the appropriateness of scaled testing in terms of hydrodynamics and structural mechanics as well as material properties.
5. The report of test results shall include a discussion of the accuracy of load condition generation and scale effects caused by dynamic and kinematic considerations, including dynamic response of test structures and materials.
6. Test results shall be adjusted to account for effective density, as calculated in Section 6.8.4.
7. Test results shall be adjusted by the Importance Factor from Section 6.8.3.2.
8. Test results shall include the effects of flow directionality in accordance with Section 6.8.6. This inclusion can be accomplished either by direct testing of flow at varying angles of incidence or by a combination of numerical and physical modeling that takes into account directionality of flow.

6.9 HYDROSTATIC LOADS

6.9.1 Buoyancy Reduced net weight caused by buoyancy shall be evaluated for all inundated structural and designated nonstructural elements of the building in accordance with Equation (6.9-1).

$$F_v = \gamma_s V_w \qquad (6.9\text{-}1)$$

Table 6.8-2. Minimum Scale Factors for Physical Modeling.

Model Element	Minimum Scale Factor
Individual buildings	1:25
Flow modeling for groups of buildings	1:200
Structural components (e.g., walls, columns, piers)	1:10
Geotechnical investigations	1:5

Uplift caused by buoyancy shall include enclosed spaces without tsunami breakaway walls that have an opening area less than 25% of the inundated exterior wall area. Buoyancy shall also include the effect of air trapped below floors, including integral structural slabs, and in enclosed spaces where the walls are not designed to break away. All windows, except those designed for large missile wind-borne debris impact or blast loading, shall be permitted to be considered openings when the inundation depth reaches the top of the windows or the expected strength of the glazing, whichever is less. The volumetric displacement of foundation elements, excluding deep foundations, shall be included in this calculation of uplift.

6.9.2 Unbalanced Lateral Hydrostatic Force Inundated structural walls with openings less than 10% of the wall area and either longer than 30 ft (9.14 m) without adjacent tsunami breakaway walls, or having a two- or three-sided perimeter structural wall configuration regardless of length, shall be designed to resist an unbalanced hydrostatic lateral force given by Equation (6.9-2), occurring during the Load Case 1 and the Load Case 2 inflow cases defined in Section 6.8.3.1.

Where the flow does not overtop the wall, the lateral hydrostatic force on the wall shall be determined using Equation (6.9-2).

$$F_h = \frac{1}{2}\gamma_s b h^2 \tag{6.9-2}$$

Where the flow overtops the wall, the lateral hydrostatic force on the wall shall be determined using Equation (6.9-3).

$$F_h = 0.6\gamma_s (2h - h_w) b h_w \tag{6.9-3}$$

6.9.3 Residual Water Surcharge Load on Floors and Walls All horizontal floors below the maximum inundation depth shall be designed for dead load plus a residual water surcharge pressure, p_r, given by Equation (6.9-4).

$$p_r = \gamma_s h_r \quad \text{where } h_r = h_{\max} - h_s \tag{6.9-4}$$

where h_s is the top of floor slab elevation. However, h_r need not exceed the height of the continuous portion of any perimeter structural element at the floor.

Structural walls that have the potential to retain water during drawdown shall also be designed for residual water hydrostatic pressure.

6.9.4 Hydrostatic Surcharge Pressure on Foundation Hydrostatic surcharge pressure caused by tsunami inundation shall be calculated as

$$p_s = \gamma_s h_{\max} \tag{6.9-5}$$

6.10 HYDRODYNAMIC LOADS

Hydrodynamic loads shall be determined in accordance with this section. The structure's lateral-force-resisting system and all structural components below the inundation elevation at the site shall be designed for the hydrodynamic loads given in either Section 6.10.1 or 6.10.2. All wall and slab components shall also be designed for all applicable loads given in Section 6.10.3.

6.10.1 Simplified Equivalent Uniform Lateral Static Pressure It shall be permitted to account for the combination of any unbalanced lateral hydrostatic and hydrodynamic loads by applying an equivalent maximum uniform pressure, p_{uw}, determined in accordance with Equation (6.10-1), applied over 1.3 times the calculated maximum inundation depth, $h_{\max}$, at the site, in each direction of flow.

$$p_{uw} = 1.25 I_{\text{tsu}} \gamma_s h_{\max} \tag{6.10-1}$$

6.10.2 Detailed Hydrodynamic Lateral Forces

6.10.2.1 Overall Drag Force on Buildings and Other Structures The building's lateral-force-resisting system shall be designed to resist overall drag forces at each level caused either by incoming or outgoing flow at Load Case 2 given by Equations (6.10-2) and (6.10-3):

$$F_{dx} = \frac{1}{2}\rho_s I_{tsu} C_d C_{cx} B (h_{sx} u^2) \tag{6.10-2}$$

where C_d is the drag coefficient for the building as given in Table 6.10-1, and where C_{cx} is determined as

$$C_{cx} = \frac{\sum(A_{\text{col}} + A_{\text{wall}}) + 1.5 A_{\text{beam}}}{B h_{sx}} \tag{6.10-3}$$

and where A_{col} and A_{wall} are the vertical projected areas of all individual column and wall elements, and A_{beam} is the combined vertical projected area of the slab edge facing the flow and the deepest beam laterally exposed to the flow. The summation of these column, wall, and beam areas is divided by the overall building wall area of width B times the average of the story heights, h_{sx}, above and below each level for each story below the tsunami inundation height, for each of the three load cases specified in Section 6.8.3.1. Any structural or nonstructural wall that is not a tsunami breakaway wall shall be included in the A_{wall}. C_{cx} shall not be taken as less than the closure ratio value given in Section 6.8.7 but need not be taken as greater than 1.0.

6.10.2.2 Drag Force on Components The lateral hydrodynamic load given by Equation (6.10-4) shall be applied as a pressure resultant on the projected inundated height, h_e, of all structural components and exterior wall assemblies below the inundation depth:

$$F_d = \frac{1}{2}\rho_s I_{\text{tsu}} C_d b (h_e u^2) \tag{6.10-4}$$

Table 6.10-1. Drag Coefficients for Rectilinear Structures.

Width to Inundation Depth* Ratio, B/h	Drag Coefficient, C_d
≤12	1.25
60	1.75
≥120	2.0

* Inundation depth for each of the three Load Cases of inundation specified in Section 6.8.3.1. Interpolation shall be used for intermediate values of width to inundation depth ratio B/h. Where building setbacks occur, drag coefficients shall be determined for each portion of a constant width. For each portion along the inundated height of the building, its equivalent inundated depth is taken as its submerged vertical dimension.

Table 6.10-2. Drag Coefficients for Structural Components.

Structural Element Section	Drag Coefficient, C_d
Round column or equilateral polygon with six sides or more	1.2
Rectangular column of at least 2:1 aspect ratio with longer face oriented parallel to flow	1.6
Triangular column pointing into flow	1.6
Freestanding wall submerged in flow	1.6
Square or rectangular column with longer face oriented perpendicular to flow	2.0
Triangular column pointing away from flow	2.0
Wall or flat plate, normal to flow	2.0
Diamond-shape column, pointed into the flow (based on face width, not projected width)	2.5
Rectangular beam, normal to flow	2.0
I, L, and channel shapes	2.0

where for interior components, except vertical load-bearing components in storage warehouses, and truck and bus garages, the values of C_d given in Table 6.10-2 shall be used, and b is the component width perpendicular to the flow. For exterior components, a C_d value of 2.0 shall be used, and width dimension b shall be taken as the tributary width multiplied by the closure ratio value given in Section 6.8.7. For interior vertical load-bearing components in storage warehouses, a C_d value of 2.0 shall be used, and width dimension b shall be taken as 9 ft (2.74 m) or the length of structural wall, whichever is greater. For interior vertical load-bearing components in truck and bus garages, a C_d value of 2.0 shall be used, and width dimension b shall be taken as 40 ft (12.19 m), or the tributary width multiplied by the closure ratio value given in Section 6.8.7, whichever is smaller.

The drag force on component elements shall not be additive to the overall drag force computed in Section 6.10.2.1.

6.10.2.3 Tsunami Loads on Vertical Structural Components, F_w The force F_w on vertical structural components shall be determined as the hydrodynamic drag forces in accordance with Equation (6.10-5a). Where flow of a tsunami bore occurs with a Froude number at the site that is greater than 1.0 and where individual wall, wall pier, or column components have a width-to-inundation-depth ratio of 3 or more, F_w shall be determined by Equation (6.10-5b). The following two conditions shall be considered: (1) force F_w applied to all vertical structural components that are wider than three times the inundation depth corresponding to Load Case 2 during inflow as defined in Section 6.8.3, and (2) force F_w based on an inundation depth equal to one-third of the structural wall width, with the corresponding flow velocity determined using Figure 6.8-1.

$$F_w = \frac{1}{2}\rho_s I_{\text{tsu}} C_d b(h_e u^2) \qquad (6.10\text{-}5a)$$

$$F_w = \frac{3}{4}\rho_s I_{\text{tsu}} C_d b(h_e u^2)_{\text{bore}} \qquad (6.10\text{-}5b)$$

6.10.2.4 Hydrodynamic Load on Perforated Walls, F_{pw} For walls with openings that allow flow to pass between wall piers, the force on the elements of the perforated wall, F_{pw}, shall be permitted to be determined using Equation (6.10-6), but shall not be less than F_d per Equation (6.10-4).

$$F_{pw} = (0.4C_{cx} + 0.6)F_w \qquad (6.10\text{-}6)$$

6.10.2.5 Walls Angled to the Flow For walls oriented at an angle less than 90 degrees to the flow directions considered in Section 6.8.3, the transient lateral load per unit width, $F_{w\theta}$, shall be determined in accordance with Equation (6.10-7).

$$F_{w\theta} = F_w \sin^2\theta \qquad (6.10\text{-}7)$$

where θ is the included angle between the wall and the direction of the flow.

6.10.2.6 Loads on Above-Ground Horizontal Pipelines Above-ground horizontal pipelines necessary for the functionality of Tsunami Risk Category III and IV buildings and other structures shall be designed to resist the loads given in Sections 6.10.2.6.1 and 6.10.2.6.2.

6.10.2.6.1 Hydrodynamic Loads on Above-Ground Horizontal Pipelines The horizontal force per unit length shall be calculated as

$$F_{rp} = C_{cp} C_r \rho_s I_{\text{tsu}} D_p u^2 \qquad (6.10\text{-}8)$$

where

C_{cp} = 1.5;
C_r = Pipe resistance coefficient as given in Table 6.10-3 or Figure 6.10-1;
ρ_s = Minimum fluid mass density for design hydrodynamic loads;
I_{tsu} = Importance Factor;
D_p = Pipe diameter; and
u = Tsunami flow velocity.

Table 6.10-3. Pipe Resistance Coefficient, C_r.

Froude number, F_r	Resistance Coefficient, C_r
$0 < F_r < 0.25$	3.6
$0.25 < F_r < 1.3$	$4.22 - 2.48F_r$
$1.3 < F_r < 2.75$	1.0

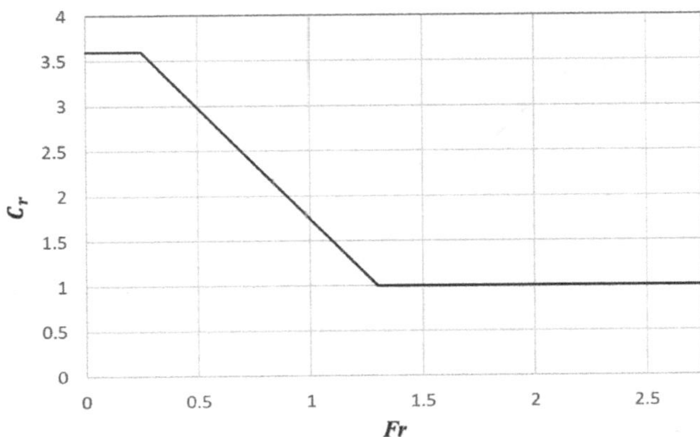

Figure 6.10-1. Pipe resistance coefficient, C_r, as a function of the Froude number, F_r.

Table 6.10-4. Upward Lift Coefficient, C_l^+, for Pipelines.

Froude number, F_r	Upward Lift Coefficient, C_l^+
$0 < F_r < 0.25$	2.8
$0.25 < F_r < 1.3$	$3.23 - 1.71F_r$
$1.3 < F_r < 2.75$	1.0

The upward vertical force per unit length shall be calculated as

$$F_{l+} = C_{cp} C_l^+ \rho_s I_{tsu} D_p u^2 \quad (6.10\text{-}9)$$

where

$C_{cp} = 1.5$,
C_l^+ = Upward lift coefficient as given in Table 6.10-4 or in Figure 6.10-2,
ρ_s = Minimum fluid mass density for design hydrodynamic loads,
I_{tsu} = Importance Factor,
D_p = Pipe diameter, and
u = Tsunami flow velocity.

The downward vertical force per unit length shall be calculated as

$$F_{l-} = C_{cp} C_l^- \rho_s I_{tsu} D_p u^2 \quad (6.10\text{-}10)$$

where

$C_{cp} = 1.5$,
C_l^- = Downward lift coefficient as given in Table 6.10-5 or in Figure 6.10-3,
ρ_s = Minimum fluid mass density for design hydrodynamic loads,
I_{tsu} = Importance Factor,
D_p = Pipe diameter, and
u = Tsunami flow velocity.

6.10.2.6.2 Debris Impacts on Above-Ground Horizontal Pipelines Debris impact loads on above-ground horizontal pipelines shall be in accordance with Section 6.11.

6.10.3 Hydrodynamic Pressures Associated with Slabs

6.10.3.1 Flow Stagnation Pressure The walls and slabs of spaces in buildings that are subject to flow stagnation

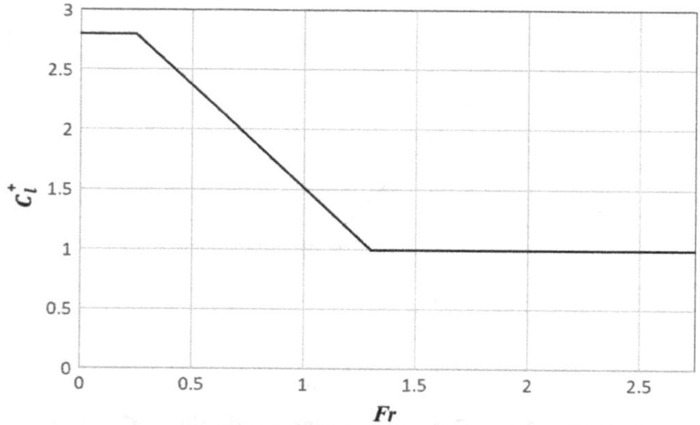

Figure 6.10-2. Upward lift coefficient, C_l^+, as a function of the Froude number, F_r.

Table 6.10-5. Downward Lift Coefficient, C_l^-, for Pipelines.

Froude number, F_r	Downward Lift Coefficient, C_l^-
$0 < F_r < 0.25$	−2.8
$0.25 < F_r < 1.3$	$2.19\ F_r - 3.35$
$1.3 < F_r < 2.75$	−0.5

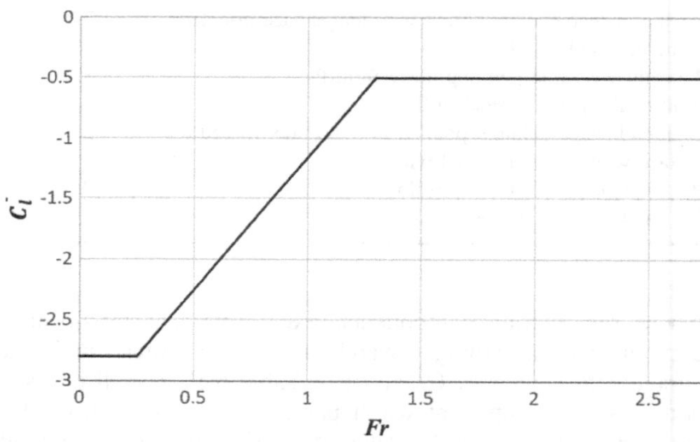

Figure 6.10-3. Downward lift coefficient, C_l^-, as a function of the Froude number, F_r.

pressurization shall be designed to resist pressure determined in accordance with Equation (6.10-11).

$$P_p = \frac{1}{2} \rho_s I_{tsu} u^2 \quad (6.10\text{-}11)$$

where u is the maximum free flow velocity at that location and load case.

6.10.3.2 Hydrodynamic Surge Uplift at Elevated Horizontal Slabs Slabs and other horizontal components shall be designed to resist the applicable uplift pressures given in this section.

6.10.3.2.1 Slabs Submerged during Tsunami Inflow Horizontal slabs that become submerged during tsunami inundation inflow shall be designed for a minimum hydrodynamic uplift pressure of 20 lb/ft^2 (0.958 kPa) applied to the soffit of the slab. This uplift is an additional Load Case to any hydrostatic buoyancy effects required by Section 6.9.1.

6.10.3.2.2 Slabs over Sloping Grade Horizontal slabs located over grade slope, φ, greater than 10 degrees shall be designed for a redirected uplift pressure applied to the soffit of the slab, given by Equation (6.10-12), but not less than 20 lb/ft^2 (0.958 kPa).

$$P_u = 1.5 I_{tsu} \rho_s u_v^2 \quad (6.10\text{-}12)$$

where

$u_v = u \tan \varphi$;
u = Horizontal flow velocity corresponding to a water depth equal to or greater than h_{ss}, the elevation of the soffit of the floor system; and
φ = Average slope of grade plane beneath the slab.

6.10.3.3 Tsunami Bore Flow Entrapped in Structural Wall-Slab Recesses Hydrodynamic loads for bore flows entrapped in structural wall-slab recesses shall be determined in accordance with this section. The reductions of load given in Sections 6.10.3.3.2 to 6.10.3.3.5 may be combined multiplicatively, but the net load reduction shall not exceed the maximum individual reduction given by any one of these sections.

6.10.3.3.1 Pressure Load in Structural Wall-Slab Recesses Where flow of a tsunami bore beneath an elevated slab is prevented by a structural wall located downstream of the upstream edge of the slab, the wall and the slab within h_s of the wall shall be designed for the outward pressure, P_u, of 350 lb/ft^2 (16.76 kPa). Beyond h_s, but within a distance of $h_s + l_w$ from the wall, the slab shall be designed for an upward pressure of half of P_u [i.e., 175 lb/ft^2 (8.38 kPa)]. The slab outside a distance of $h_s + l_w$ from the wall shall be designed for an upward pressure of 30 lb/ft^2 (1.436 kPa).

6.10.3.3.2 Reduction of Load with Inundation Depth Where the inundation depth is less than two-thirds of the clear story height, the uplift pressures specified in Section 6.10.3.3.1 shall be permitted to be reduced in accordance with Equation (6.10-13) but shall not be taken as less than 30 lb/ft^2 (1.436 kPa).

$$P_u = I_{\text{tsu}}\left(590 - 160\frac{h_s}{h}\right)[\text{lb/ft}^2] \quad (6.10\text{-}13)$$

$$P_u = I_{\text{tsu}}\left(28.25 - 7.66\frac{h_s}{h}\right)[\text{kPa}] \quad (6.10\text{-}13.\text{SI})$$

where h_s/h is the ratio of slab height to inundation depth.

6.10.3.3.3 Reduction of Load for Wall Openings Where the wall blocking the bore below the slab has openings through which the flow can pass, the reduced pressure on the wall and slab shall be determined in accordance with Equation (6.10-14).

$$P_{ur} = C_{cx} P_u \quad (6.10\text{-}14)$$

where C_{cx} is the ratio of the solid area of the wall to the total inundated area of the vertical plane of the inundated portion of the wall at that level.

6.10.3.3.4 Reduction in Load for Slab Openings Where the slab is provided with an opening gap or breakaway panel designed to create a gap of width w_g adjacent to the wall, then the uplift pressure on the remaining slab shall be determined in accordance with Equation (6.10-15).

$$P_{ur} = C_{bs} P_u \quad (6.10\text{-}15)$$

$$\text{where for } w_g < 0.5h_s,\ C_{bs} = 1 - \frac{w_g}{h_s} \quad (6.10\text{-}16)$$

$$\text{and for } w_g \geq 0.5h_s,\ C_{bs} = 0.56 - 0.12\frac{w_g}{h_s} \quad (6.10\text{-}17)$$

The value of C_{bs} shall not be taken as less than zero.

6.10.3.3.5 Reduction in Load for Tsunami Breakaway Wall If the wall restricting the flow is designed as a tsunami breakaway wall, then the uplift on the slab shall be permitted to be determined in accordance with Section 6.10.3.1, but it need not exceed the pressure equivalent to the total nominal shear force necessary to cause disengagement of the breakaway wall from the slab.

6.11 DEBRIS IMPACT LOADS

Debris impact loads shall be determined in accordance with this section. These loads need not be combined with other tsunami-related loads as determined in other sections of this chapter.

Where the minimum inundation depth is 3 ft (0.914 m) or greater, design shall include the effects of debris impact forces. The most severe effect of impact loads within the inundation depth shall be applied to the perimeter gravity-load-carrying structural components located on the principal structural axes perpendicular to the range of inflow or outflow directions defined in Section 6.8.7. Except as specified below, loads shall be applied at points critical for flexure and shear on all such members in the inundation depth being evaluated. Inundation depths and velocities corresponding to Load Cases 1, 2, and 3, defined in Section 6.8.3.1, shall be used. Impact loads need not be applied simultaneously to all affected structural components.

All buildings and other structures meeting the above requirement shall be designed for impact by floating wood poles, logs, and vehicles, and for tumbling boulders and concrete debris, per Sections 6.11.2 to 6.11.4. Where a site is proximate to a port or container yard, the potential for strikes from shipping containers and ships and barges shall be determined by the procedure in Section 6.11.5. Buildings and other structures determined by that procedure to lie in the hazard zone for strikes by shipping containers shall be designed for impact loads in accordance with Section 6.11.6. In lieu of Sections 6.11.2 to 6.11.6, it shall be permitted to alternatively evaluate the impacts by poles, logs, vehicles, tumbling boulders, concrete debris, and shipping containers by applying the alternative simplified static load of Section 6.11.1.

Where a building or other structure falls into one or more of the following categories, interior gravity-load-carrying structural components subject to inundation shall be designed for the specified debris impact:

1. Garages: vehicle debris impact per Section 6.11.3.
2. Open Structures with slab-on-grade at ground level: submerged tumbling boulder and concrete debris impact per Section 6.11.4, except where the slab-on-grade is explicitly designed per Section 6.12.4.2.
3. Buildings and other structures in which the slab-on-grade at ground level has isolation joints designed to reduce buoyancy effects: submerged tumbling boulder and concrete debris impact per Section 6.11.4.
4. Buildings with isolated breakaway concrete slab panels above grade designed to dislodge and relieve hydrodynamic uplift forces: the columns and structural walls nearest to the breakaway panels in all directions shall be designed for submerged tumbling boulder and concrete debris impact per Section 6.11.4.

Tsunami Risk Category III Critical Facilities and Tsunami Risk Category IV buildings and structures determined to be in the hazard zone for strikes by ships and barges in excess of 88,000 lb (39,916 kg) Deadweight Tonnage (DWT), as determined by the procedure of Section 6.11.5, shall be designed for impact by these vessels in accordance with Section 6.11.7.

6.11.1 Alternative Simplified Debris Impact Static Load It shall be permitted to account for debris impact by applying the force given by Equation (6.11-1) as a maximum static load, in lieu of the loads defined in Sections 6.11.2 to 6.11.6. This force shall be applied at points critical for flexure and shear on all such members in the inundation depth corresponding to Load Case 3 defined in Section 6.8.3.1.

$$F_i = 330 C_o I_{tsu} [\text{kips}] \quad (6.11\text{-}1)$$

$$F_i = 1,470 C_o I_{tsu} [\text{kN}] \quad (6.11\text{-}1.\text{SI})$$

where C_o is the orientation coefficient, equal to 0.65.

Where it is determined by the site hazard assessment procedure of Section 6.11.5 that the site is not in an impact zone for shipping containers, ships, and barges, then it shall be permitted to reduce the simplified debris impact force to 50% of the value given by Equation (6.11-1).

6.11.2 Wood Logs and Poles The nominal maximum instantaneous debris impact force, F_{ni}, shall be determined in accordance with Equation (6.11-2).

$$F_{ni} = u_{max} \sqrt{k m_d} \quad (6.11\text{-}2)$$

The design instantaneous debris impact force, F_i, shall be determined in accordance with Equation (6.11-3).

$$F_i = I_{tsu} C_o F_{ni} \quad (6.11\text{-}3)$$

where

I_{tsu} = Importance Factor (given in Table 6.8-1);
C_o = Orientation coefficient, equal to 0.65 for logs and poles;
u_{max} = Maximum flow velocity at the site occurring at depths sufficient to float the debris;
k = Effective stiffness of the impacting debris or the lateral stiffness of the impacted structural element(s) deformed by the impact, whichever is less; and
m_d = Mass (W_d/g) of the debris.

Logs and poles are assumed to strike longitudinally for calculation of debris stiffness in Equation (6.11-2). The stiffness of the log or pole shall be calculated as $k = EA/L$, in which E is the longitudinal modulus of elasticity of the log, A is its cross-sectional area, and L is its length. A minimum weight of 1,000 lb (454 kg) and minimum log stiffness of 350 kip/in. (61,300 kN/m) shall be assumed.

Table 6.11-1. Dynamic Response Ratio for Impulsive Loads, R_{max}.

Ratio of Impact Duration to Natural Period of the Impacted Structural Element	R_{max} (Response Ratio)
0.0	0.0
0.1	0.4
0.2	0.8
0.3	1.1
0.4	1.4
0.5	1.5
0.6	1.7
0.7	1.8
0.9	1.8
1.0	1.7
1.1	1.7
1.2	1.6
1.3	1.6
≥1.4	1.5

The impulse duration for elastic impact shall be calculated from Equation (6.11-4):

$$t_d = \frac{2 m_d u_{max}}{F_{ni}} \quad (6.11\text{-}4)$$

For an equivalent elastic static analysis, the impact force shall be multiplied by the dynamic response factor, R_{max}, specified in Table 6.11-1. To obtain intermediate values of R_{max}, linear interpolation shall be used. For a wall, the impact shall be assumed to act along the horizontal center of the wall, and the natural period shall be permitted to be determined based on the fundamental period of an equivalent column with width equal to one-half of the vertical span of the wall. It also shall be allowed to use an alternative method of analysis per Section 6.11.8.

6.11.3 Impact by Vehicles An impact of floating vehicles shall be applied to vertical structural element(s) at any point greater than 3 ft (0.914 m) above grade up to the maximum depth. The impact force shall be taken as 30 kip (130 kN) multiplied by I_{tsu}.

6.11.4 Impact by Submerged Tumbling Boulder and Concrete Debris Where the maximum inundation depth exceeds 6 ft (1.83 m), an impact force of 8,000 lb (36 kN) multiplied by I_{tsu} shall be applied to vertical structural element(s) at 2 ft (0.61 m) above grade.

6.11.5 Site Hazard Assessment for Shipping Containers, Ships, and Barges Shipping containers and ships or barges dispersed from container yards, ports, and harbors shall be evaluated as potential debris impact objects. In such cases, a probable dispersion region shall be identified for each source to determine if the structure is located within a debris impact hazard region, as defined by the procedure in this section. If the structure is within the debris impact hazard region, then impact by shipping containers and/or ships and barges, as appropriate, shall be evaluated per Sections 6.11.6 and 6.11.7.

6.11.5.1 Debris Impact Hazard Region The debris impact hazard region shall be determined in accordance with this section. The geographic center of the debris source shall be identified, together with the primary flow direction, as defined in Section 6.8.6.1. A line shall be drawn through the source center and perpendicular to the primary flow direction with a length equal to the projected width of the debris source. From the ends of this projected width line, lines shall be drawn at ±22.5 degrees with respect to the direction of tsunami inflow until they reach the grounding limit, as shown in Figure 6.11-1. If topography (such as hills) will bound the water from this sector, the direction of the sector shall be rotated to accommodate hill lines or the wedge shall be narrowed where it is constrained on two or more sides. Where a site-specific inundation analysis spatially defines the topographic diversion of the direction of inundation flow across the Tsunami Design Zone, the primary flow direction shall be permitted to follow the site-specific inundation flow field vectors from the site to the grounding limit, while including 22.5-degree lateral spreading from the predominant flow field vectors.

The inland extent of the debris hazard region shall be curtailed in accordance with the following:

(a) Where the maximum inundation depth is less than 3 ft (0.914 m), or in the case of ships, where the inundation depth is less than the ballasted draft plus 2 ft (0.61 m).
(b) Where structural steel, reinforced concrete, or reinforced solid grouted masonry wall structures act as effective barriers to debris transport. To be considered an effective

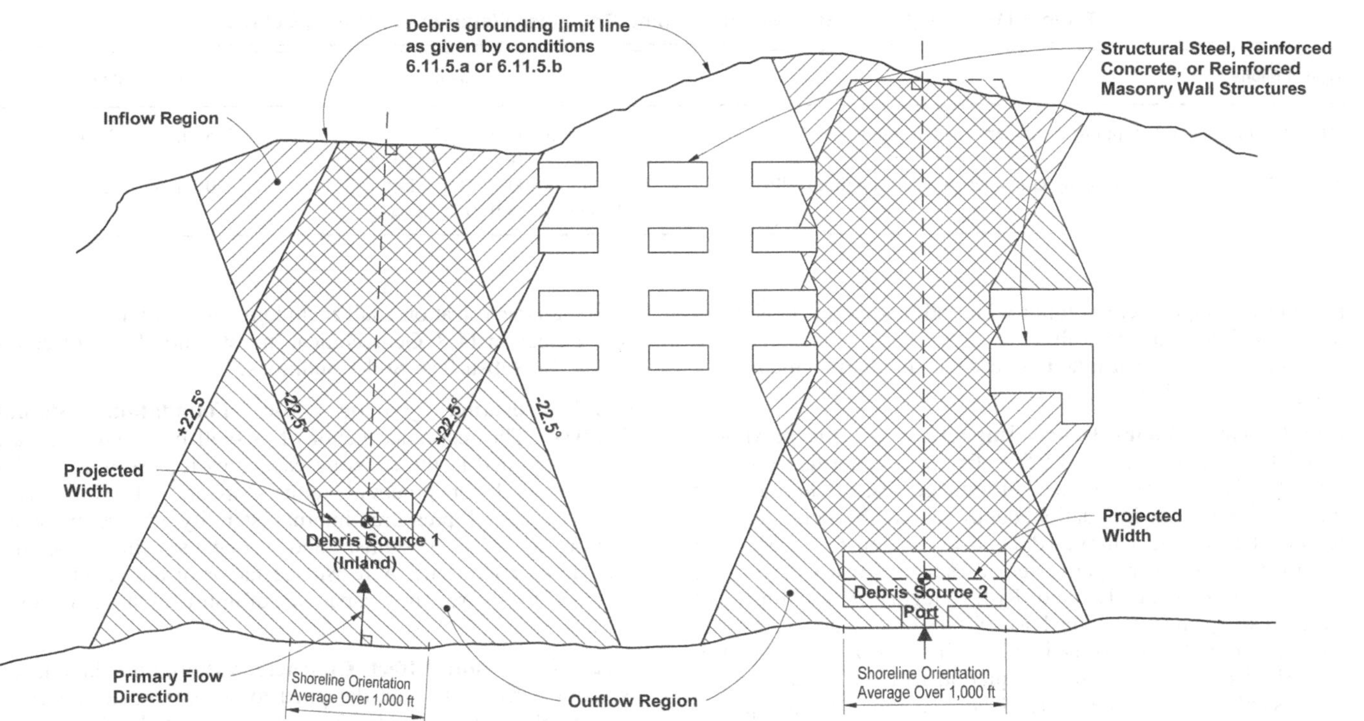

Figure 6.11-1. Illustration of determination of floating debris impact hazard region.
Note: 1 ft = 0.3048 m.

barrier, the height of these structures shall be at least equal to (1) for containers and barges, the inundation depth minus 2 ft (0.61 m), or (2) for ships, the inundation depth minus the sum of the ballasted draft and 2 ft (0.61 m).

(c) Numerical debris transport modeling shall be permitted to determine the inland extent of the debris impact hazard region, using massless tracers with grounding limit parameters set in accordance with condition 6.11.5.1(a), to represent debris transported by the fluid as provided by a site-specific inundation analysis. Debris transport modeling shall be subject to an independent peer review by a qualified tsunami modeler approved by the Authority Having Jurisdiction, who shall present a written report indicating the scope of their review and their findings to the Authority Having Jurisdiction. Debris impact speeds for design are subject to the minimum flow speeds determined using the procedures of Section 6.6 in accordance with the percentages given in Section 6.7.6.8.

The debris impact hazard region for outflow shall be determined by drawing a line perpendicular to the centerline of the inward flow at the grounding limit line. The line shall be equal in width to the projected width of the source. At each end of this projected width, lines shall be drawn outward at 22.5 degrees from the outflow direction, as shown in Figure 6.11-1. Buildings and other structures contained only in the first sector shall be designed for strikes by a container and/or other vessel carried with the inflow. Buildings and other structures contained only in the second sector shall be designed for strikes by a container and/or other vessel carried in the outflow. Buildings and other structures contained in both sectors shall be designed for strikes by a container and/or other vessel moving in either direction.

6.11.6 Shipping Containers The impact force from shipping containers shall be calculated from Equations (6.11-2) and (6.11-3). The mass m_d is the mass of the empty shipping container. It shall be assumed that the strike contact is from one bottom corner of the front (or rear) of the container. The container stiffness is $k = EA/L$, in which E is the modulus of elasticity of the bottom rail of the container, A is the cross-sectional area of the bottom rail, and L is the length of the bottom rail of the container. Minimum values are provided in Table 6.11-2. C_o, the orientation factor, shall be taken as equal to 0.65 for shipping containers.

The nominal design impact force, F_{ni}, from Equation (6.11-2) for shipping containers need not be taken as greater than 220 kips (980 kN).

For empty shipping containers, the impulse duration for elastic impact shall be calculated from Equation (6.11-4).

For loaded shipping containers the duration of the pulse is determined from Equation (6.11-5):

$$t_d = \frac{(m_d + m_{\text{contents}})u_{\max}}{F_{ni}} \tag{6.11-5}$$

in which m_{contents} shall be taken to be 50% of the maximum rated content capacity of the shipping container. Minimum values of $(m_d + m_{\text{contents}})$ are given in Table 6.11-2 for loaded shipping containers. The design shall consider both empty and loaded shipping containers.

For an equivalent static analysis, the impact force shall be multiplied by the dynamic response factor, $R_{\max}$, specified in Table 6.11-1. To obtain intermediate values of $R_{\max}$, linear interpolation shall be used. For a wall, the impact shall be assumed to act along the horizontal center of the wall, and the natural period shall be permitted to be determined based on the

Table 6.11-2. Weight and Stiffness of Shipping Container Waterborne Floating Debris.

Type of Debris	Weight	Debris Stiffness (k)
20 ft (6.1 m) standard shipping container oriented longitudinally	Empty: 5,000 lb (2,270 kg) Loaded: 29,000 lb (13,150 kg)	245 kip/in. (42,900 kN/m)
40 ft (12.19 m) standard shipping container oriented longitudinally	Empty: 8,400 lb (3,810 kg) Loaded: 38,000 lb (17,240 kg)	170 kip/in. (29,800 kN/m)

period of an equivalent column with width equal to one-half of the vertical span of the wall.

It also shall be permitted to use an alternative method of analysis per Section 6.11.8.

6.11.7 Extraordinary Debris Impacts Where the maximum inundation depth exceeds 12 ft (3.66 m), extraordinary debris impacts of the largest deadweight tonnage vessel with ballasted draft less than the inundation depth within the debris hazard region of piers and wharves defined in Section 6.11.5 shall be assumed to impact the perimeter of Tsunami Risk Category III Critical Facilities and Tsunami Risk Category IV buildings and structures anywhere from the base of the structure up to 1.3 times the inundation depth plus the height to the deck of the vessel. The load shall be calculated from Equation (6.11-3), based on the stiffness of the impacted structural element and a weight equal to the Lightship Weight (LWT) plus 30% of Deadweight Tonnage (DWT). An alternative analysis of Section 6.11.8 shall be permitted. Either as the primary approach, or where the impact loads exceed acceptability criteria for any structural element subject to impact, it is permitted to accommodate the impact through the alternative load path progressive collapse provisions of Section 6.8.3.5.3, applied to all framing levels from the base up to the story level above 1.3 times the inundation depth plus the height to the deck of the vessel as measured from the waterline.

6.11.8 Alternative Methods of Response Analysis A dynamic analysis is permitted to be used to determine the structural response to the force applied as a rectangular pulse of duration time, t_d, with the magnitude calculated in accordance with Equation (6.11-3). If the impact is large enough to cause inelastic behavior in the structure, it shall be permitted to use an equivalent single degree of freedom mass-spring system with a nonlinear stiffness that considers the ductility of the impacted structure for the dynamic analysis. Alternatively, for inelastic impact, the structural response shall be permitted to be calculated based on a work-energy method with nonlinear stiffness that incorporates the ductility of the impacted structure. The velocity applied in the work-energy method of analysis shall be u_{max} multiplied by the product of Importance Factor, I_{tsu}, and the orientation factor, C_o.

6.12 FOUNDATION DESIGN

Design of structure foundations and tsunami barriers shall provide resistance to the loads and effects of Section 6.12.2, shall provide capacity to support the structural load combinations defined in Section 6.8.3.1, and shall accommodate the displacements determined in accordance with Section 6.12.2.6. Foundation embedment depth and the capacity of the exposed piles to resist structural loads, including grade beam loads, shall both be determined taking into account the cumulative effects of general erosion and local scour. Alternatively, it shall be permitted to use the performance-based criteria of Section 6.12.3.

Site characterization shall include relevant information specified in Section 11.8, Geotechnical Investigation Report Requirements for Subsurface Soil Conditions.

6.12.1 Resistance Factors for Foundation Stability Analyses The resistance factor of φ shall be assigned a value of 0.67 applied to the resisting capacities for use with stability analyses and for potential failures associated with bearing capacity, lateral pressure, internal stability of geotextile and reinforced earth systems, and slope stability, including drawdown conditions. A resistance factor of 0.67 shall also be assigned for the resisting capacities of uplift-resisting anchorage elements.

6.12.2 Load and Effect Characterization Foundations and tsunami barriers shall be designed to accommodate the effects of lateral earth pressure in accordance with Section 3.2, hydrostatic forces computed in accordance with Section 6.9, hydrodynamic loads computed in accordance with Section 6.10, and uplift and underseepage forces computed in accordance with Section 6.12.2.1. Foundations shall provide the capacity to withstand uplift and overturning from tsunami hydrostatic, hydrodynamic, and debris loads applied to the building superstructure. In addition, the effect of soil strength loss, general erosion, and scour shall be considered in accordance with the requirements of this section. A minimum of two wave cycles shall be considered for such effects.

6.12.2.1 Uplift and Underseepage Forces Tsunami uplift and underseepage forces shall be evaluated as described in this section.

1. Uplift and underseepage forces shall include the three inundation Load Cases defined in Section 6.8.3.1.
2. Uplift and underseepage forces on the foundation shall be determined for cases where
   (a) The soil is expected to be saturated before the tsunami, or
   (b) Soil saturation is anticipated to occur over the course of the incoming series of tsunami waves, or
   (c) The area of concern is expected to remain inundated after the tsunami.
3. The effect of live load and snow load shall not be used for uplift resistance.

6.12.2.2 Loss of Strength The effects of the loss of soil shear strength on erosion and scour shall be considered in accordance with Sections 6.12.2.3 and 6.12.2.4, respectively, and foundations shall be designed to accommodate loss of soil shear strength through tsunami-induced pore pressure softening when

1. The maximum Froude number at the site is 0.5 or greater, and
2. Figure 6.12-2 indicates that the pore pressure softening parameter at the ground surface, $\Lambda_7(0)$, is 0.5 or greater.

$\Lambda_T(0)$ shall be determined in accordance with Equation 6.12-1:

$$\Lambda_T(0) = \min\left[1, \frac{2F_\Lambda}{\sqrt{\pi I_\Lambda}}\right] \quad (6.12\text{-}1)$$

where

F_Λ = Pore pressure softening vertical flux index, shall be determined in accordance with Equation (6.12-2):

$$F_\Lambda = \frac{\gamma_s}{\gamma_b}(1-\kappa_d)h_{draw} \quad (6.12\text{-}2)$$

I_Λ = Pore pressure softening drawdown inertia index, shall be determined in accordance with Equation (6.12-3):

$$I_\Lambda = c_v \cdot T_{draw} \quad (6.12\text{-}3)$$

γ_b = Buoyant weight density of the soil, assuming the pore water is clean sea water with specific weight density equal to γ_{sw};

γ_s = Fluid weight density for tsunami loads per Equation (6.8-4);

h_{draw} = Difference between the maximum inundation depth and the residual water level after drawdown of the associated wave, as shown in Figure 6.12-1. Where a site-specific time series inundation analysis is not available, h_{draw} shall be taken as the maximum inundation depth, h_{max}, determined per Section 6.6.2;

T_{draw} = Site-specific drawdown timescale as defined per Figure 6.12-1. Where a site-specific time series inundation analysis is not available, T_{draw} shall be taken as one-quarter of the predominant wave period, T_{TSU};

c_v = Coefficient of consolidation in ft^2/s (m^2/s); and

κ_d = Drainage factor, determined from Table 6.12-1. If variable soil conditions are present, it shall be permitted to use $\kappa_d = 0.2$.

The minimum depth of soil shear strength loss through tsunami-induced pore pressure softening shall be the lesser of:

1. A depth of 1.2 times h_{draw}, or
2. The susceptible depth, D_e, as determined from Figure 6.12-3.

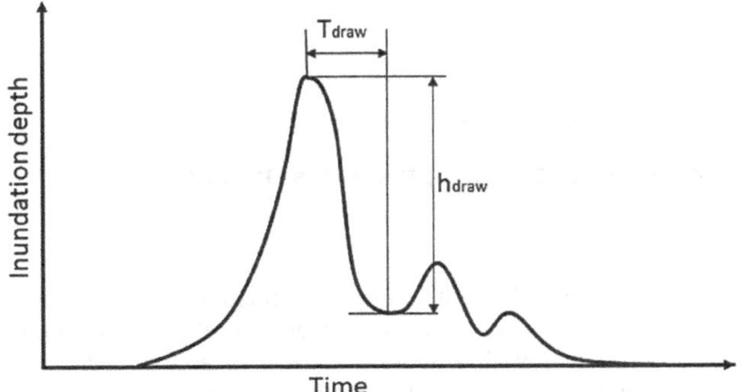

Figure 6.12-1. Site-specific tsunami inundation wave parameters used for pore pressure softening evaluation.

Table 6.12-1. Drainage Factor, κ_d, for Different Soil Types.

Soil Type	κ_d
Very soft to soft clay	0.3
Medium stiff clay	0.2
Stiff clay	0.1
Loose sand	0.3
Dense sand	0.1
Very dense sand and gravel, sandy clay, or silty sand	0.05

EXCEPTION: The depth of soil shear strength loss shall not extend below the depth where bedrock is encountered.

The softened soil shear strength shall be determined by multiplying the unsoftened soil shear strength by a factor of $(1-\Lambda_T)$, where Λ_T is determined by Equation (6.12-1). This reduction of strength shall be permitted to be applied uniformly throughout the susceptible depth, D_e, or varied linearly using $\Lambda_T = \Lambda_T(0)$ at the ground surface and $\Lambda_T = 0.5$ at D_e. Alternatively, the magnitude and distribution of the loss of strength shall be determined by physical or numerical modeling of soil–structure–fluid interactions.

6.12.2.3 General Erosion General erosion during tsunami inundation runup and drawdown conditions shall be considered. Analysis of general erosion shall account for flow amplification as described in Section 6.8.5 and soil strength loss as described in Section 6.12.2.2.

EXCEPTION: Analysis of general erosion is not required for rock or other nonerodible strata that are capable of preventing erosion due to tsunami flow velocities of 30 ft/s (9.14 m/s).

General erosion during drawdown conditions shall consider flow concentration in channels, including channels newly formed during tsunami inundation and drawdown (channelized scour). Analysis of erosion due to channelized scour need not include enhancement caused by pore pressure softening.

6.12.2.4 Scour The depth and extent of scour adjacent to foundation elements shall be evaluated using the methods of Sections 6.12.2.4.1 and 6.12.2.4.2. Analysis of scour shall take into account tsunami flow cycles as described in Section 6.8.8 and loss of strength as described in Section 6.12.2.2.

EXCEPTION: Scour evaluation is not required for rock or other nonerodible strata that are capable of preventing scour from tsunami flow velocities of 30 ft/s (9.14 m/s) nor for open structures.

6.12.2.4.1 Sustained Flow Scour Scour, including the effects of sustained flow around structures, shall be evaluated at the foundations of building corners and at the extreme ends of structural piers or walls within 22.5 degrees of perpendicular to the flow and having a projected width to maximum inundation depth ratio of 3 or more. Calculated sustained flow scour depth shall be taken to be at the flow-facing surface of the foundation element. The assumed spatial extent shall be considered to encompass the indicated building perimeter elements and to extend from those foundation elements at a 1 vertical to 1 horizontal slope for consolidated or cohesive soils and at a 1 vertical to 3 horizontal slope for unconsolidated or granular soils. Scour depth need not be taken lower than the top surface of intact rock strata or other nonerodible strata that are capable of preventing scour from tsunami flow of 30 ft/s (9.14 m/s), nor be determined for open structures. Sustained flow scour design depth and extent shall be determined by one of the following methods:

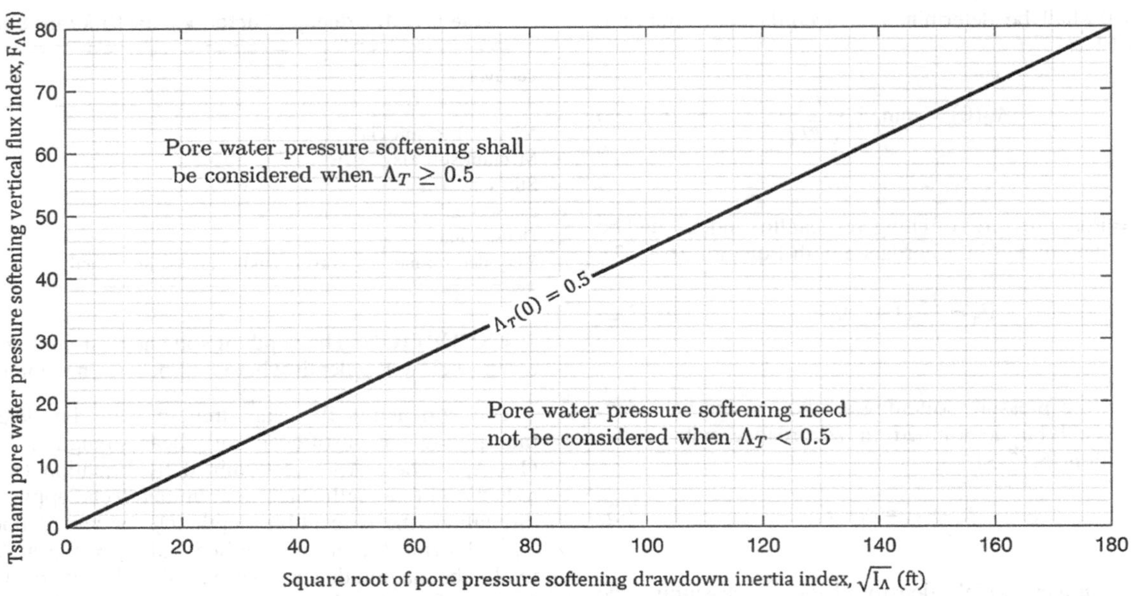

Figure 6.12-2. Susceptibility to pore pressure softening due to tsunami inundation.
Note: 1 ft = 0.305 m.

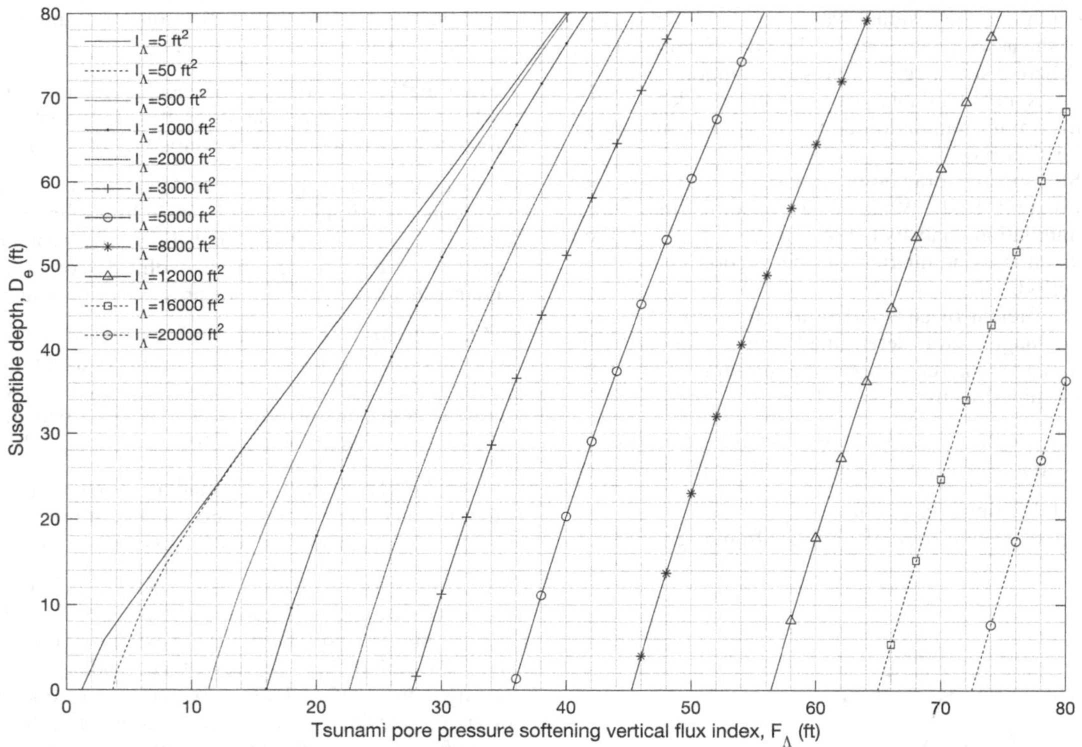

Figure 6.12-3. Susceptible depth, D_e, affected by tsunami-induced pore pressure softening
Note: 1 ft = 0.305 m; 1 ft^2 = 0.093 m^2.

(a) Dynamic numerical or physical modeling or empirical methods in the recognized literature.
(b) Sustained flow scour with associated pore pressure softening effects determined in accordance with Table 6.12-2 and Figure 6.12-4. Local scour depth caused by sustained flow given by Table 6.12-2 and Figure 6.12-4 shall be permitted to be reduced by an adjustment factor in areas where the maximum flow Froude number determined in accordance with Equation (6.6-3) is less than 0.5. The adjustment factor shall be taken as varying linearly from 0 at the horizontal inundation limit to 1.0 at the point where the Froude number is 0.5.

Table 6.12-2. Design Scour Depth Caused by Sustained Flow and Pore Pressure Softening.

Inundation Depth, h	Scour Depth, D^*
<10 ft (3.05 m)	1.2h
≥10 ft (3.05 m)	12 ft (3.66 m)

*Not applicable to scour at sites with intact rock strata.

Table 6.12-3. Vertical Extent of Pile Scour to Projected Element Width Ratio, D_s/b.

Froude Number, F_r	D_s/b
<1.0	1.3F_r
≥1.0	1.3

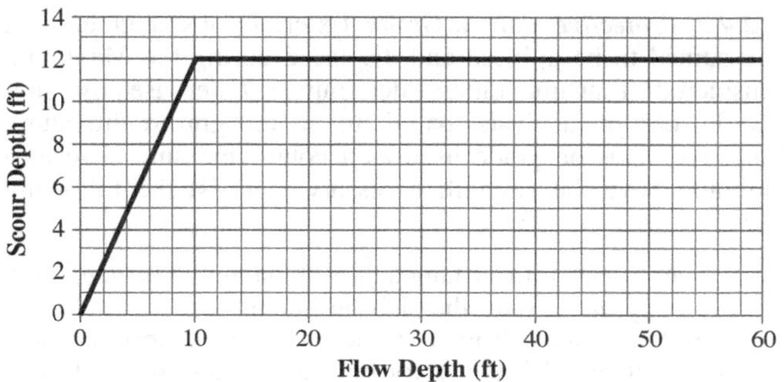

Figure 6.12-4. Scour depth caused by sustained flow and pore pressure softening.
Note: 1 ft = 0.305 m.

(c) Where vertical foundation elements, with cross sections not exceeding an aspect ratio of 2:1, become exposed to flow through local scour determined in accordance with method (a) or (b), or through general erosion as determined in accordance with Section 6.12.2.3, and where the Froude number is less than 1.7, it shall be permitted to calculate the vertical extent of pile scour along the length of the vertical foundation element using Table 6.12-3 and Figure 6.12-5. The width, b, is the projected foundation element width in the direction perpendicular to the flow. The total scour for the foundation need not exceed the depth determined by Figure 6.12-4.

6.12.2.4.2 Plunging Scour Plunging scour horizontal extent and depth shall be determined by dynamic numerical or physical modeling or by empirical methods. In the absence of site-specific dynamic modeling and analysis, the plunging scour depth, D_s shall be determined by Equation (6.12-4).

$$D_s = c_{2V}\sqrt{\frac{qU\sin\psi}{g}}\text{[U.S.standard or SI units]} \quad (6.12\text{-}4)$$

where

c_{2V} = Dimensionless scour coefficient, permitted to be taken as equal to 2.8;

ψ = Angle between the jet at the scour hole and the horizontal, taken as the lesser value of 75 degrees and the side slope of the overtopped structure on the scoured side, in the absence of other information;

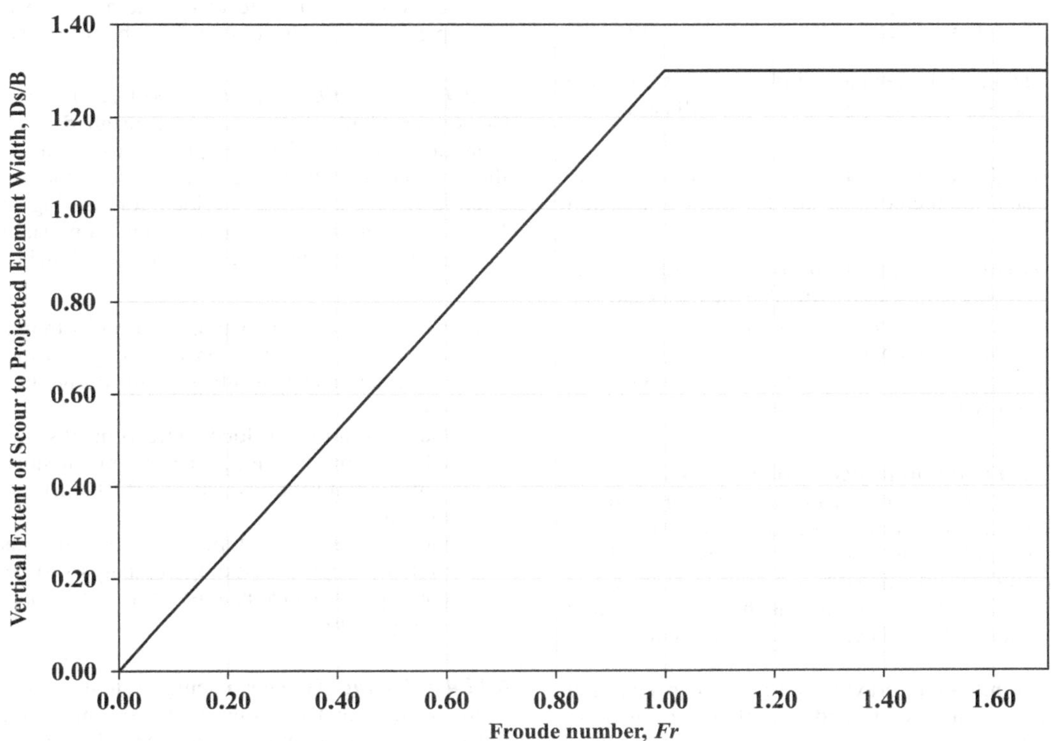

Figure 6.12-5. Vertical extent of pile scour to projected element width ratio.

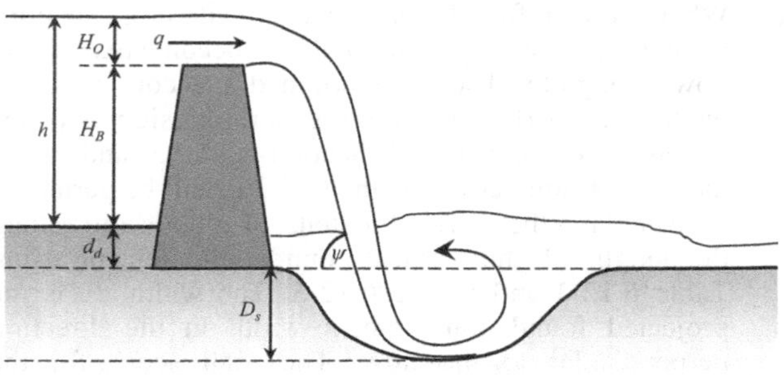

Figure 6.12-6. Plunging scour parameters.

g = Acceleration caused by gravity;
q = Discharge per unit width over the overtopped structure, as illustrated in Figure 6.12-6 and calculated in accordance with Equation (6.12-5); and
U = Jet velocity approaching the scour hole, obtained in accordance with Equation (6.12-7).

$$q = C_{dis} \frac{2}{3} \sqrt{2g} H_O^{3/2} \quad (6.12\text{-}5)$$

where C_{dis} is a dimensionless discharge coefficient obtained in accordance with Equation (6.12-3):

$$C_{dis} = 0.611 + 0.08 \frac{H_O}{H_B} \quad (6.12\text{-}6)$$

U is the jet velocity approaching the scour hole, resulting from the drop between the height h of the upstream water surface, plus any additional elevation difference d_d on the scouring side, in accordance with Equation (6.12-7):

$$U = \sqrt{2g(h + d_d)} \quad (6.12\text{-}7)$$

where d_d is the additional elevation difference between the upstream and scouring sides of the structure, as illustrated in Figure 6.12-3.

6.12.2.5 Horizontal Soil Loads Horizontal soil loads caused by unbalanced scour shall be included in the design of foundation elements.

6.12.2.6 Displacements Vertical and horizontal displacements of foundation elements and slope displacements shall be determined using empirical or elastoplastic analytical or numerical methods in the recognized literature by applying tsunami loads determined in Section 6.12.2 together with other applicable geotechnical and foundation loads required by this standard.

6.12.3 Alternative Foundation Performance-Based Design Criteria In situ soil stresses from tsunami loads and effects shall be included in the calculation of foundation pressures. For local coseismic tsunami hazards that occur as a result of a local earthquake, the in situ soil and site surface condition at the onset of tsunami loads shall be those existing at the end of seismic shaking, including liquefaction, lateral spread, and fault rupture effects.

Building foundations shall provide sufficient capacity and stability to resist structural loads and the effects of general erosion and scour in accordance with the recognized literature. For Tsunami Risk Category IV buildings and structures, it shall be permitted to evaluate the overall performance of the foundation system for potential pore pressure softening by performing a two- or three-dimensional tsunami–soil–structure interaction numerical modeling analysis. The results shall be evaluated to demonstrate consistency with the structural performance acceptance criteria in Section 6.8. For Tsunami Risk Category IV buildings and structures, an independent peer review shall be conducted as part of a review of the performance-based design by the Authority Having Jurisdiction.

6.12.4 Foundation Countermeasures Fill, protective slab on grade, geotextiles and reinforced earth systems, facing systems, and ground improvement shall be permitted to reduce the effects of tsunamis.

6.12.4.1 Fill Fill used for structural support and protection shall be placed in accordance with ASCE 24, Sections 1.5.4 and 2.4.1. Structural fill shall be designed to be stable during inundation and to resist the loads and effects specified in Section 6.12.2.

6.12.4.2 Protective Slab on Grade Exterior slabs on grade shall be assumed to be uplifted and displaced during the Maximum Considered Tsunami unless determined otherwise by site-specific design analysis based upon recognized literature. Protective slabs on grade used as a countermeasure shall at a minimum have the strength necessary to resist the following loads:

1. Shear forces from sustained flow at maximum tsunami flow velocity, u_{max}, over the slab on grade;
2. Uplift pressures from flow acceleration at upstream and downstream slab edges for both inflow and return flow;
3. Seepage flow gradients under the slab if the potential exists for soil saturation during successive tsunami waves;
4. Pressure fluctuations over slab sections and at joints;
5. Pore pressure increases from liquefaction and from the passage of several tsunami waves; and
6. Erosion of substrate at upstream, downstream, and flow parallel slab edges, as well as between slab sections.

6.12.4.3 Geotextiles and Reinforced Earth Systems Geotextiles shall be designed and installed in accordance with manufacturers' installation requirements and as recommended in the recognized literature. Resistance factors required in Section 6.12.1 shall be provided for bearing capacity, uplift, lateral pressure, internal stability, and slope stability.

The following reinforced earth systems shall be permitted to be used:

1. Geotextile tubes constructed of high-strength fabrics capable of achieving full tensile strength without constricting deformations when subject to the design tsunami loads and effects;
2. Geogrid earth and slope reinforcement systems that include adequate protection against general erosion and scour, and a maximum lift thickness of 1 ft (0.3 m) and facing protection; and
3. Geocell earth and slope reinforcement erosion protection system designs, including an analysis to determine anticipated performance against general erosion and scour if no facing is used.

6.12.4.4 Facing Systems Facing systems and their anchorage shall be sufficiently strong to resist uplift and displacement during design load inundation. The following facing methods for reinforced earth systems shall be permitted to be used:

1. Vegetative facing for general erosion and scour resistance where tsunami flow velocities are less than 12.5 ft/s (3.81 m/s). Design shall be in accordance with methods and requirements in the recognized literature.
2. Geotextile filter layers, including primary filter protection of countermeasures using a composite grid assuming high contact stresses and high-energy wave action design criteria in AASHTO M288-06, including soil retention, permeability, clogging resistance, and survivability.
3. Mattresses providing adequate flexibility and including energy dissipation characteristics. Edges shall be embedded to maintain edge stability under design inundation flows.
4. Concrete facing provided in accordance with protective slab-on-grade countermeasures in Section 6.12.4.2 and containing adequate anchorage to the reinforced earth system under design inundation flows.
5. Stone armoring and riprap provided to withstand tsunami shall be designed as follows: Stone diameter shall not be less than the size determined based on tsunami inundation depth and currents using design criteria in the recognized literature. Where the maximum Froude number, F_r, is 0.5 or greater, the high-velocity turbulent flows associated with tsunamis shall be specifically considered, using methods in the recognized literature.

Subject to independent review, it shall be permitted to base designs on physical or numerical modeling.

6.12.4.5 Ground Improvement Ground improvement countermeasures shall be designed using soil–cement mixing to provide nonerodible scour protection per Section 6.12.2.4 and at minimum provide soil–cement mass strength reinforcement of 100 psi (0.69 MPa) average unconfined compressive strength.

6.13 STRUCTURAL COUNTERMEASURES FOR TSUNAMI LOADING

The following countermeasures shall be permitted to reduce the structural effects of tsunamis.

6.13.1 Open Structures Open Structures shall not be subject to Load Case 1 of Section 6.8.3.1. The load effect of debris accumulation against or within the Open Structure shall be evaluated by assuming a minimum closure ratio of 50% of the Innundated Projected Area along the perimeter of the Open Structure.

6.13.2 Tsunami Barriers Tsunami barriers used as an external perimeter structural countermeasure shall be designed consistent with the protected structure performance objectives to jointly achieve the performance criteria. These criteria include barrier strength, stability, slope erosion protection, toe scour, and geotechnical stability requirements and barrier height and footprint to fully prevent inundation during the Maximum Considered Tsunami. Where a barrier is designed to be overtopped by the design event or intended to provide only partial impedance of the design event, the protected structure and its foundation shall be designed for the residual inundation resulting from the design event. The foundation system treatment requirements in Section 6.12 of this chapter shall also be applied.

6.13.2.1 Information on Existing Buildings and Other Structures to Be Protected As-built information on building configuration, building components, site, and foundation shall be permitted to be evaluated in accordance with ASCE 41, Chapters 9 through 11.

6.13.2.2 Site Layout The spatial limits of the layout of tsunami barriers shall include the following:

1. The tsunami barrier shall be set back from the protected structure for perimeter protection. Any alignment change shall have a minimum radius of curvature equal to at least half the maximum inundation depth.
2. For overtopping or partial impedance to inundation, at a minimum the barrier limits shall protect the structure from inundation flow based on an approach angle of ±22.5 degrees from the shoreline. The flow approach angle shall be evaluated in accordance with Sections 6.8.6.1 and 6.8.6.2.

6.14 TSUNAMI VERTICAL EVACUATION REFUGE STRUCTURES

Tsunami Vertical Evacuation Refuge Structures designated as a means of alternative evacuation by the Authority Having Jurisdiction shall be designed in accordance with the additional requirements of this section.

6.14.1 Minimum Inundation Elevation and Depth Tsunami refuge floors shall be located not less than the greater of 10 ft (3.0 m) or one-story height above 1.3 times the Maximum Considered Tsunami inundation elevation at the site as determined by a site-specific inundation analysis, as indicated in Figure 6.14-1. This same Maximum Considered Tsunami

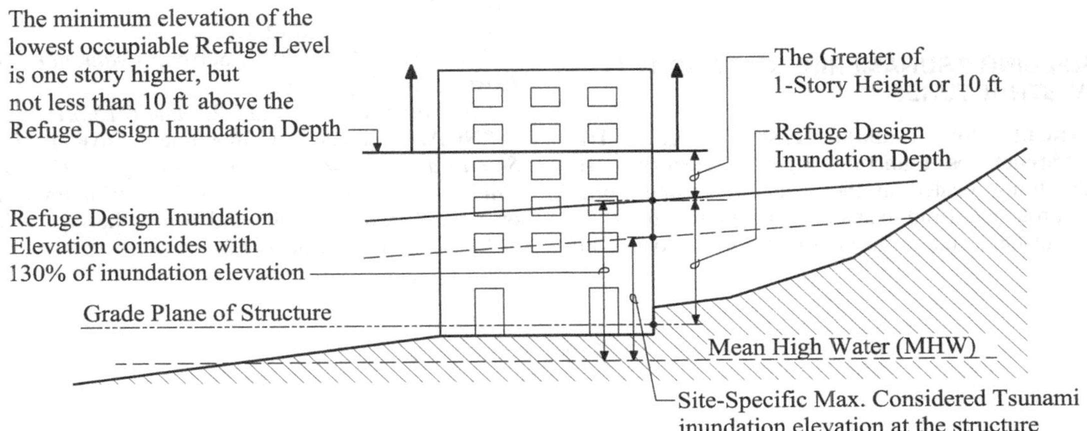

Figure 6.14-1. Minimum refuge level elevation (1 ft = 0.305 m).

site-specific inundation elevation, factored by 1.3, shall also be used for design of the Tsunami Vertical Evacuation Refuge Structure in accordance with Sections 6.8 to 6.12. The 1.3 factor is not applicable to relative sea level change and regional seismic subsidence.

6.14.2 Refuge Live Load An assembly live load, L_{refuge}, of 100 lb/ft^2 (4.8 kPa) shall be used in any designated evacuation floor area within a tsunami refuge floor level.

6.14.3 Laydown Impacts Where the maximum inundation depth exceeds 6 ft (1.83 m), the laydown impact of adjacent pole structures collapsing onto occupied portions of the building shall be considered.

6.14.4 Information on Construction Documents Construction documents shall include tsunami design criteria and the occupancy capacity of the tsunami refuge area. Floor plans shall indicate all refuge areas of the facility and exiting routes from each area. The latitude and longitude coordinates of the building shall be recorded on the construction documents.

6.14.5 Peer Review Design shall be subject to independent peer review by an appropriately licensed design professional, who shall present a written report to the Authority Having Jurisdiction as to the design's compliance with the requirements of this standard.

6.15 DESIGNATED NONSTRUCTURAL COMPONENTS AND SYSTEMS

6.15.1 Performance Requirements Designated nonstructural components and systems in structures located in the Tsunami Design Zone shall be either protected from tsunami inundation effects or positioned in the structure above the inundation elevation of the Maximum Considered Tsunami, such that the designated nonstructural components and systems will be capable of performing their critical function during and after the Maximum Considered Tsunami. Tsunami barriers used as inundation protection shall have a top-of-wall elevation that is not less than 1.3 times the maximum inundation elevation at the barrier. The 1.3 factor is not applicable to relative sea level change and regional seismic subsidence. The tsunami barrier shall also satisfy the requirements of Section 6.13. Alternatively, it shall be permitted to design the designated nonstructural components and systems directly for tsunami effects, provided that inundation would not inhibit them from performing their critical function during and after the Maximum Considered Tsunami.

6.16 NONBUILDING TSUNAMI RISK CATEGORY III AND IV STRUCTURES

6.16.1 Requirements for Tsunami Risk Category III Nonbuilding Structures Tsunami Risk Category III nonbuilding structures located in the Tsunami Design Zone shall be either protected from tsunami inundation effects or designed to withstand the effects of tsunami loads in accordance with Section 6.8 of this chapter and in accordance with the specific performance requirements of Section 6.8.3. Tsunami barriers used as inundation protection shall have a top-of-wall elevation that is not less than 1.3 times the maximum inundation elevation at the barrier. The 1.3 factor is not applicable to relative sea level change and regional seismic subsidence. The tsunami barrier shall also satisfy the requirements of Section 6.13.

6.16.2 Requirements for Tsunami Risk Category IV Nonbuilding Structures Tsunami Risk Category IV designated nonstructural systems in nonbuilding structures located in the Tsunami Design Zone shall be (1) protected from tsunami inundation effects, (2) positioned above 1.3 times the inundation elevation of the Maximum Considered Tsunami in such a manner that the Tsunami Risk Category IV nonbuilding structure will be capable of performing its critical function during and after the Maximum Considered Tsunami, or (3) designed to withstand the effects of tsunami loads in accordance with Section 6.8 of this chapter and the specific performance requirements of Section 6.8.3. Tsunami barriers used as inundation protection shall have a top-of-wall elevation that is not less than 1.3 times the maximum inundation elevation at the barrier. The 1.3 factor is not applicable to relative sea level change and regional seismic subsidence. The tsunami barrier shall also satisfy the requirements of Section 6.13.

6.17 CONSENSUS STANDARDS AND OTHER REFERENCED DOCUMENTS

This section lists the consensus standards and other documents that shall be considered part of this standard to the extent referenced in this chapter. Those referenced documents identified by an asterisk (*) are not consensus standards; rather, they are documents developed within the industry and represent acceptable procedures for design and construction to the extent referred to in the specified section.

AASHTO Guide Specifications and Commentary for Vessel Collision Design of Highway Bridges, 2nd ed., 2009, with 2010 interim revisions. American Association of State Highway and Transportation Officials.
Cited in: Section 6.11.5

AASHTO M288-06. 2006. *Standard Specification for Geotextile Specification for Highway Applications*. American Association of State Highway and Transportation Officials.
Cited in: Section 6.12.4.4

ASCE/SEI 24-14. 2015. *Flood Resistant Design and Construction*. ASCE.
Cited in: Section 6.12.4.1

ASCE/SEI 41-13. 2014. *Seismic Evaluation and Retrofit of Existing Buildings*. ASCE.
Cited in: Sections 6.8.3.5.2.2 and 6.13.2.1

*NOAA Technical Memorandum OAR PMEL-135. 2007. *Standards, Criteria, and Procedures for NOAA Evaluation of Tsunami Numerical Models*. Pacific Marine Environmental Laboratory, National Oceanic and Atmospheric Administration.
Cited in: Section 6.7.6.7.2

CHAPTER 7
SNOW LOADS

7.1 DEFINITIONS AND SYMBOLS

7.1.1 Definitions

ASCE DESIGN GROUND SNOW LOAD GEODATABASE: The ASCE database (version 2022-1.0) of geocoded values of risk-targeted design ground snow load values.

DRIFT: The accumulation of wind-driven snow that results in a local surcharge load on the roof structure at locations such as a parapet or roof step.

FLAT ROOF SNOW LOAD: Uniform load for flat roofs.

FREEZER BUILDINGS: Buildings in which the inside temperature is kept at or below freezing. Buildings with an air space between the roof insulation layer above and a ceiling of the freezer area below are not considered freezer buildings.

GROUND SNOW LOAD: The site-specific weight of the accumulated snow at the ground level used to develop roof snow loads on the structure.

MINIMUM SNOW LOAD: Snow load on low sloped roofs, including the roof snow load immediately after a single snowstorm without wind.

PONDING: Refer to definitions in Chapter 8, "Rain Loads."

PONDING INSTABILITY: Refer to definitions in Chapter 8, "Rain Loads."

R-VALUE: A measure of the resistance to heat flow through a roof component or assembly per unit area.

SLIPPERY SURFACE: Membranes with a smooth surface, for example, glass, metal, or rubber. Membranes with an embedded aggregate or mineral granule surface are not considered a slippery surface.

SLOPED ROOF SNOW LOAD: Uniform load on horizontal projection of a sloped roof, also known as the balanced load.

VENTILATED ROOF: Roof that allows exterior air to naturally circulate between the roof surface above and the insulation layer below. The exterior air commonly flows from the eave to the ridge.

7.1.2 Symbols

C_e = Exposure factor, as determined from Table 7.3-1

C_s = Slope factor, as determined from Figure 7.4-1

C_t = Thermal factor, as determined from Table 7.3-2

h = Vertical separation distance, feet (m), between the edge of a higher roof including any parapet and the edge of a lower adjacent roof excluding any parapet

h_b = Height of balanced snow load, determined by dividing p_s by γ, ft (m)

h_c = Clear height from top of balanced snow load to (1) closest point on adjacent upper roof, (2) top of parapet, or (3) top of a projection on the roof, ft (m)

h_d = Height of snow drift, ft (m)

h_{d1} or h_{d2} = Heights of snow drifts where two intersecting snow drifts can form, ft (m)

h_o = Height of obstruction above the surface of the roof, ft (m)

l_u = Length of the roof upwind of the drift, ft (m)

p_d = Maximum intensity of drift surcharge load, lb/ft^2 (kN/m^2)

p_f = Snow load on flat roofs ("flat" = roof slope $\leq 5°$), lb/ft^2 (kN/m^2)

p_g = Ground snow load, as determined from Figure 7.2-1 and Table 7.2-1; or a site-specific analysis, lb/ft^2 (kN/m^2)

p_m = Minimum snow load for low-slope roofs, lb/ft^2 (kN/m^2)

p_s = Sloped roof (balanced) snow load, lb/ft^2 (kN/m^2)

s = Horizontal separation distance between the edges of two adjacent buildings, ft (m)

S = Roof slope run for a rise of one

w = Width of snow drift, ft (m)

w_1 or w_2 = Widths of snow drifts where two intersecting snow drifts can form, ft (m)

W = Horizontal distance from eave to ridge, ft (m)

W_2 = Percent time wind speed is above 10 mph (4.6 m/s) during winter (October through April); winter wind parameter from Figure 7.6-1 and Table 7.2-1 for Alaska

γ = Snow density, as determined from Equation (7.7-1), lb/ft^3 (kN/m^3)

θ = Roof slope on the leeward side, degrees

7.2 GROUND SNOW LOADS, p_g

Ground snow loads, p_g, to be used in the determination of design snow loads shall be determined using the ASCE Design Ground Snow Load Geodatabase. A graphical representation of the data in the ASCE Design Ground Snow Load Geodatabase is shown in Figures 7.2-1A through 7.2-1D for the conterminous United States and Table 7.2-1 for Alaska. Where the results from the geodatabase indicate that a case study needs to be conducted for a specific location, the ground snow load determination for the location shall be based on an analysis of data available in the vicinity of the site, shall meet the reliability targets set forth in Table 1.3-1, and shall be approved by the Authority Having Jurisdiction.

User Note: The ASCE Design Ground Snow Load Geodatabase of geocoded design ground snow load values for all four risk categories is available at https://asce7hazardtool.online/ or approved equivalent.

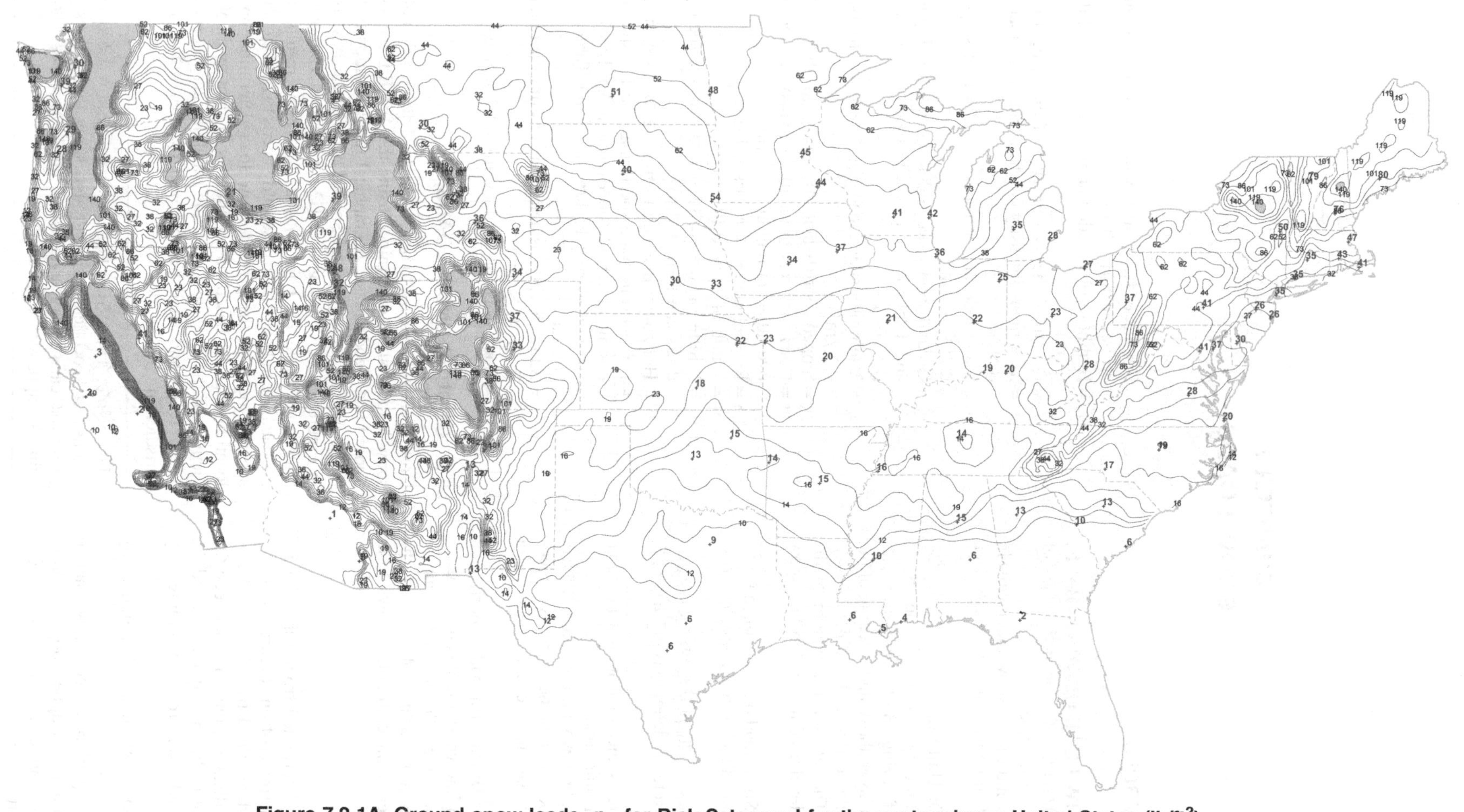

Figure 7.2-1A. Ground snow loads, p_g, for Risk Category I for the conterminous United States (lb/ft^2).

Note:

1. This figure is a representation of data from the Ground Snow Load Geodatabase of geocoded design ground snow load values, available at [https://asce7hazardtool.online/].
2. Values for specific locations can most accurately be determined by accessing the Geodatabase.
3. Lines shown on the figure are contours separated by a constant ratio of 1.18 with values of 10, 12, 14, 16, 19, 23, 27, 32, 38, 44, 52, 62, 73, 86, 101, 119 and 140 psf.
4. Values denoted with a "+" symbol indicate design ground snow loads at state capitals or other cities with large populations.
5. Areas shown in gray represent areas with ground snow loads exceeding 140 psf. Ground snow load values for these locations must be determined from the Geodatabase.

Figure 7.2-1B. Ground snow loads, p_g, for Risk Category II for the conterminous United States (lb/ft^2).

Note:

1. This figure is a representation of data from the Ground Snow Load Geodatabase of geocoded design ground snow load values, available at [https://asce7hazardtool.online/].
2. Values for specific locations can most accurately be determined by accessing the Geodatabase.
3. Lines shown on the figure are contours separated by a constant ratio of 1.18 with values of 10, 12, 14, 16, 19, 23, 27, 32, 38, 44, 52, 62, 73, 86, 101, 119 and 140 psf.
4. Values denoted with a "+" symbol indicate design ground snow loads at state capitals or other cities with large populations.
5. Areas shown in gray represent areas with ground snow loads exceeding 140 psf. Ground snow load values for these locations must be determined from the Geodatabase.

Figure 7.2-1C. Ground snow loads, p_g, for Risk Category III for the conterminous United States (lb/ft^2).

Note:

1. This figure is a representation of data from the Ground Snow Load Geodatabase of geocoded design ground snow load values, available at [https://asce7hazardtool.online/].
2. Values for specific locations can most accurately be determined by accessing the Geodatabase.
3. Lines shown on the figure are contours separated by a constant ratio of 1.18 with values of 10, 12, 14, 16, 19, 23, 27, 32, 38, 44, 52, 62, 73, 86, 101, 119 and 140 psf.
4. Values denoted with a “+” symbol indicate design ground snow loads at state capitals or other cities with large populations.
5. Areas shown in gray represent areas with ground snow loads exceeding 140 psf. Ground snow load values for these locations must be determined from the Geodatabase.

Figure 7.2-1D. Ground snow loads, p_g, for Risk Category IV for the conterminous United States (lb/ft^2).

Note:

1. This figure is a representation of data from the Ground Snow Load Geodatabase of geocoded design ground snow load values, available at [https://asce7hazardtool.online/].
2. Values for specific locations can most accurately be determined by accessing the Geodatabase.
3. Lines shown on the figure are contours separated by a constant ratio of 1.18 with values of 10, 12, 14, 16, 19, 23, 27, 32, 38, 44, 52, 62, 73, 86, 101, 119 and 140 psf.
4. Values denoted with a "+" symbol indicate design ground snow loads at state capitals or other cities with large populations.
5. Areas shown in gray represent areas with ground snow loads exceeding 140 psf. Ground snow load values for these locations must be determined from the Geodatabase.

Table 7.2-1. Ground Snow Loads, p_g, and Winter Wind Parameter, W_2, for Alaskan Locations.

City/Town	Elevation (ft)	Ground Snow Load, $p_g^{1,2,4}$ (lb/ft^2) Risk Category I	II	III	IV	Winter Wind Parameter, W_2
Adak	100	32	40	46	50	0.7
Anchorage/Eagle River[3]	500	64	80	92	100	0.2
Arctic Village	2,100	38	48	55	60	0.2
Bethel	100	51	64	74	80	0.7
Bettles	700	102	128	147	160	0.2
Cantwell	2,100	109	136	156	170	0.3
Cold Bay	100	45	56	64	70	0.8
Cordova	100	128	160	184	200	0.3
Deadhorse	100	32	40	46	50	0.6
Delta Junction	400	51	64	74	80	0.5
Dillingham	100	141	176	202	220	0.5
Emmonak	100	128	160	184	200	0.7
Fairbanks	1,200	77	96	110	120	0.1
Fort Yukon	400	64	80	92	100	0.2
Galena	200	77	96	110	120	0.3
Girdwood	200	179	224	258	280	0.2
Glennallen	1,400	58	72	83	90	0.2
Haines	100	237	296	340	370	0.7
Holy Cross	100	154	192	221	240	0.2
Homer[3]	500	58	72	83	90	0.5
Iliamna	200	102	128	147	160	0.5
Juneau	100	90	112	129	140	0.5
Kaktovik	100	58	72	83	90	0.6
Kenai/Soldotna	200	83	104	120	130	0.4
Ketchikan	100	38	48	55	60	0.5
Kobuk	200	115	144	166	180	0.6
Kodiak	100	45	56	64	70	0.6
Kotzebue	100	77	96	110	120	0.6
McGrath	400	83	104	120	130	0.2
Nenana	400	96	120	138	150	0.2
Nikiski	200	102	128	147	160	0.4
Nome	100	90	112	129	140	0.6
Palmer/Wasilla	500	64	80	92	100	0.2
Petersburg	100	122	152	175	190	0.2
Point Hope	100	58	72	83	90	0.6
Saint Lawrence Island	100	122	152	175	190	0.8
Saint Paul Island	100	51	64	74	80	0.9
Seward	100	77	96	110	120	0.5
Sitka	100	64	80	92	100	0.4
Talkeetna	400	154	192	221	240	0.2
Tok	1,700	45	56	64	70	0.2
Umiat	300	38	48	55	60	0.2
Unalakleet	100	45	56	64	70	0.7
Unalaska	100	96	120	138	150	0.6
Utqiaġvik (Barrow)	100	32	40	46	50	0.6
Valdez	100	205	256	294	320	0.3
Wainwright	100	32	40	46	50	0.6
Whittier	100	346	432	497	540	0.3
Willow	300	102	128	147	160	0.2
Yakutat	100	179	224	258	280	0.3

Notes: To convert lb/ft^2 to kN/m^2, multiply by 0.0479. To convert feet to meters, multiply by 0.3048.

1. Statutory requirements of the Authority Having Jurisdiction are not included in this state ground snow load table.
2. For locations where there is substantial change in altitude over the city/town, the load applies at and below the cited elevation within the jurisdiction and up to 100 ft above the cited elevation, unless otherwise noted.
3. For locations in Anchorage/Eagle River and Homer above the cited elevation, the ground snow load shall be increased by 15% for every 100 ft above the cited elevation.

Snow loads are zero for Hawaii, except in mountainous regions, as determined by the Authority Having Jurisdiction.

EXCEPTION: The snow load provisions of this chapter need not be considered for roofs or roof members with no potential for drift accumulation or unbalanced snow loading where the ground snow load, p_g, is less than the factored roof live load used in design. For all other roofs or roof members, the snow load provisions in this chapter do not need to be considered in either of the following situations:

- The ground snow load, p_g, is less than or equal to 10 lb/ft^2 (0.48 kN/m^2) and the length of the roof upwind from any potential drifting location, l_u, is less than or equal to 100 ft (30.48 m).
- The ground snow load, p_g, is less than or equal to 5 lb/ft^2 (0.24 kN/m^2) and the length of the roof upwind from any potential drifting location, l_u, is less than or equal to 300 ft (91.44 m).

The ground snow load, p_g, shall be used as the balanced snow load for snow accumulation surfaces, such as decks, balconies, and other near-ground level surfaces or roofs of subterranean spaces, whose height above the ground surface is less than the depth of the ground snow, h_g ($h_g = p_g/\gamma$).

7.3 FLAT ROOF SNOW LOADS, p_f

The flat roof snow load, p_f, shall be calculated in lb/ft^2 (kN/m^2) using the following formula:

$$p_f = 0.7 C_e C_t p_g \quad (7.3\text{-}1)$$

7.3.1 Exposure Factor, C_e The value for C_e shall be determined from Table 7.3-1.

7.3.2 Thermal Factor, C_t The value for C_t shall be determined from Tables 7.3-2 and 7.3-3.

For values of P_g and R_{roof} that fall between those shown in Table 7.3-3, linear interpolation may be used to determine the value of C_t.

7.3.3 Minimum Snow Load for Low-Slope Roofs, p_m A minimum roof snow load, p_m, shall only apply to monoslope, hip, and gable roofs with slopes less than 15 degrees and to curved roofs where the vertical angle from the eaves to the crown is less than 10 degrees. The minimum roof snow load for low-slope roofs shall be obtained as follows.

Where p_g is equal to or less than the value of the minimum snow load upper limit, $p_{m,\max}$, shown in Table 7.3-4:

$$p_m = p_g$$

Where p_g is greater than the value of the minimum snow load upper limit, $p_{m,\max}$, shown in Table 7.3-4:

$$p_m = p_{m,\max}$$

This minimum roof snow loads hall be a separate uniform load case. It need not be used in determining, or in combination with, drift, sliding, unbalanced, or partial loads.

7.4 SLOPED ROOF SNOW LOADS, p_s

Snow loads acting on a sloping surface shall be assumed to act on the horizontal projection of that surface. The sloped roof (balanced) snow load, p_s, shall be obtained by multiplying the flat roof snow load, p_f, by the roof slope factor, C_s:

$$p_s = C_s p_f \quad (7.4\text{-}1)$$

Table 7.3-1. Exposure Factor, C_e.

	Exposure of Roof[a]		
Surface Roughness Category	Fully Exposed[b]	Partially Exposed	Sheltered
B (see Section 26.7)	0.9	1.0	1.2
C (see Section 26.7)	0.9	1.0	1.1
D (see Section 26.7)	0.8	0.9	1.0
Above the tree line in windswept mountainous areas	0.7	0.8	N/A
In Alaska, in areas where trees do not exist within a 2 mi (3 km) radius of the site	0.7	0.8	N/A

The terrain category and roof exposure condition chosen shall be representative of the anticipated conditions during the life of the structure. An exposure factor shall be determined for each roof of a structure.

[a] Partially Exposed: All roofs not Fully Exposed or Sheltered. Fully Exposed: Roofs exposed on all sides with no shelter[b] afforded by terrain, higher structures, or trees. Roofs that contain several large pieces of mechanical equipment, parapets that extend above the height of the balanced snow load (h_b), or other obstructions are not in this category. Sheltered: Roofs located tight in among conifers that qualify as obstructions.

[b] Obstructions within a distance of $10h_o$ provide "shelter," where h_o is the height of the obstruction above the roof level. If the only obstructions are a few deciduous trees that are leafless in winter, the "fully exposed" category shall be used. Note that these are heights above the roof. Heights used to establish the Exposure Category in Section 26.7 are heights above the ground.

Table 7.3-2. Thermal Factor, C_t.

Thermal condition[a]	C_t
All structures except as indicated as follows	See Table 7.3-3
Unheated structures, open-air structures, structures kept just above freezing [40 to 50 °F (4 to 10 °C)], and other structures with cold, ventilated roofs meeting the minimum requirements of the applicable energy code	1.2
Freezer building	1.3
Continuously heated greenhouses[b] with a roof having a thermal resistance (R-value) less than 2.0 h·ft^2·°F/Btu (0.4 m^2·K/W) or a thermal transmittance (U-factor) greater than 0.5 Btu/h·ft^2·°F (2.5 W/m^2·K)	0.85

[a] These conditions shall be representative of the anticipated conditions during winters for the life of the structure.

[b] Greenhouses with a constantly maintained interior temperature of 50°F (10°C) or more, at any point 3 ft (0.9 m) above the floor level during winters and having either a maintenance attendant on duty at all times or a temperature alarm system to provide warning in the event of a heating failure.

Table 7.3-3. Thermal Factor, C_t, for Heated Structures with Unventilated Roofs.[a]

R_{roof} (h·ft²·°F/Btu [m²·K/W])	U_{roof} (Btu/h·ft²·°F [W/m²·K])	P_g (psf [kPa]) ≤10 [0.48]	20 [0.96]	30 [1.44]	40 [1.92]	50 [2.40]	60 [2.88]	≥70 [3.36]
≤20 [3.52]	≥0.050 [0.284]	1.20	1.11	1.05	1.01	1.00	1.00	1.00
30 [5.28]	0.033 [0.189]	1.20	1.17	1.14	1.13	1.12	1.11	1.10
40 [7.04]	0.025 [0.142]	1.20	1.19	1.17	1.16	1.16	1.15	1.15
50 [8.80][b]	0.020 [0.114][b]	1.20	1.20	1.19	1.19	1.19	1.18	1.18

[a] For values of P_g and R_{roof} that fall between those shown in the table, linear interpolation may be used to determine the value of C_t.
[b] For values of R_{roof} > 50 h·ft²·°F/Btu (8.80 m²·K/W) or U_{roof} < 0.020 Btu/h·ft²·°F (0.114 W/m²·K), C_t should be taken as equal to 1.2.

Table 7.3-4. Minimum Snow Loads for Low-Slope Roofs.

Risk Category	$p_{m,max}$
I	25 lb/ft² (1.20 kN/m²)
II	30 lb/ft² (1.44 kN/m²)
III	35 lb/ft² (1.68 kN/m²)
IV	40 lb/ft² (1.92 kN/m²)

Values of C_s for warm roofs, cold roofs, curved roofs, and multiple roofs are determined from Sections 7.4.1 through 7.4.4. The thermal factor, C_t, from Table 7.3-2 determines if a roof is "cold" or "warm." "Slippery surface" values shall be used only where the roof's surface is unobstructed and sufficient space is available below the eaves to accept all the sliding snow. A roof shall be considered unobstructed if no objects exist on it that prevent snow on it from sliding. Roof areas with snow retention devices shall not be considered unobstructed. Slippery surfaces shall include metal, slate, glass, and bituminous, rubber, and plastic membranes with a smooth surface. Membranes with an embedded aggregate or mineral granule surface shall not be considered smooth. Asphalt shingles, wood shingles, and shakes shall not be considered slippery.

7.4.1 Slope Factor, C_s The slope factor, C_s, shall be determined using Figure 7.4-1, as outlined in Table 7.4-1.

7.4.2 Slope Factor for Curved Roofs Portions of curved roofs that have a slope exceeding 70 degrees shall be considered free of snow load (i.e., $C_s = 0$). Balanced loads shall be determined from the balanced load diagrams in Figure 7.4-2, with C_s determined from the appropriate curve in Figure 7.4-1.

7.4.3 Slope Factor for Multiple Folded Plate, Sawtooth, and Barrel Vault Roofs Multiple folded plate, sawtooth, or barrel vault roofs shall have a $C_s = 1.0$, with no reduction in snow load because of slope (i.e., $p_s = p_f$).

7.4.4 Ice Dams and Icicles along Eaves Unventilated roofs that drain water over their eaves shall be capable of sustaining a uniformly distributed load of $2p_f$ on all overhanging portions if the structure's thermal factor, C_t, is equal to or less than 1.1 per Section 7.3.2. The load on the overhang shall be based on the flat

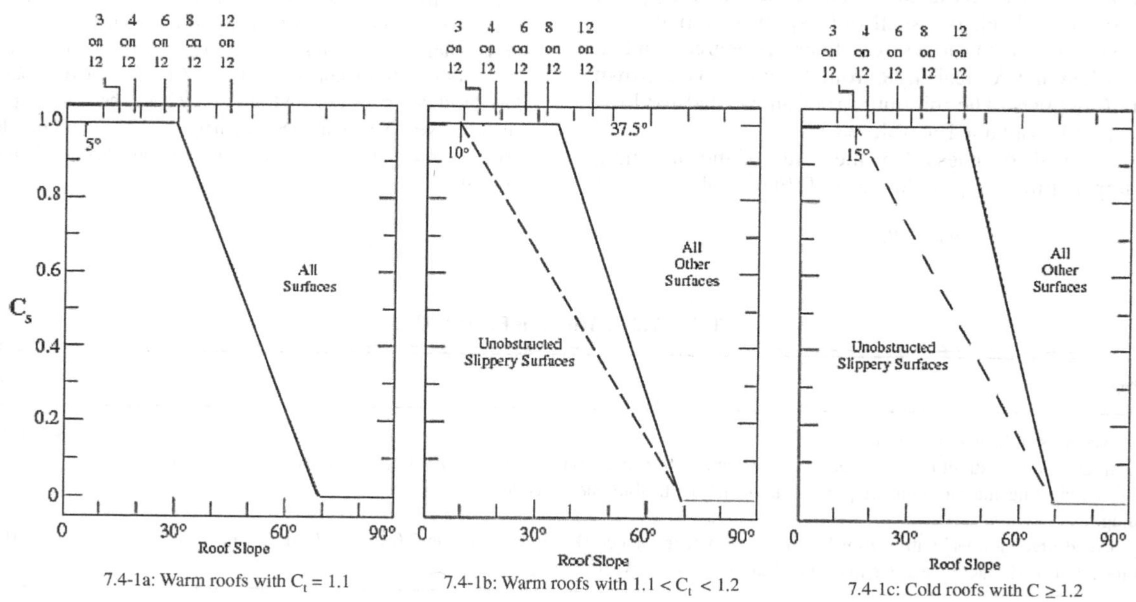

Figure 7.4-1. Graphs for determining slope factor, C_s.
Note: See Tables 7.3-2 and 7.3-3 for C_t definitions.

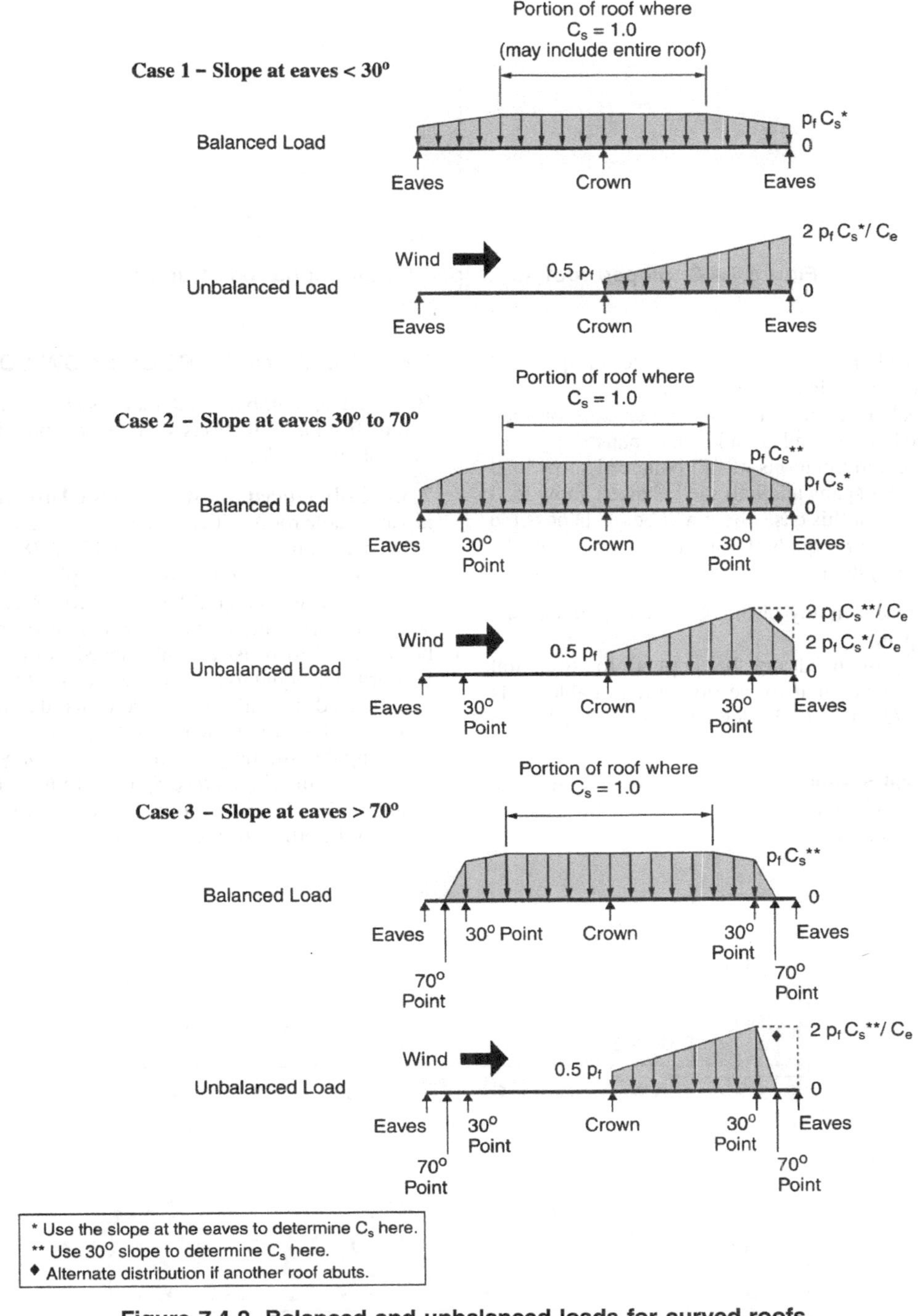

Figure 7.4-2. Balanced and unbalanced loads for curved roofs.

roof snow load for the heated portion of the roof upslope of the exterior wall. No other loads, except dead loads, shall be present on the roof when this uniformly distributed load is applied.

Where C_t is less than or equal to 1.1, the slope factor, C_s, shall be determined in accordance with Figure 7.4-1(a). Where C_t is greater than 1.1 and less than 1.2, the slope factor, C_s, shall be determined in accordance with Figure 7.4-1(b). Where C_t is equal to, or greater than, 1.2, the slope factor, C_s, shall be determined in accordance with Figure 7.4-1(c).

7.4.5 Sloped Roof Snow Loads for Air-Supported Structures Roof snow loading for air-supported structures with vinyl coated exterior fabric shall be as shown in Figure 7.4-3.

7.5 PARTIAL LOADING

The effect of having selected spans loaded with the balanced snow load, and remaining spans loaded with half the balanced snow load, shall be investigated as follows.

7.5.1 Continuous Beam Systems Continuous beam systems shall be investigated for the effects of the three loadings shown in Figure 7.5-1:

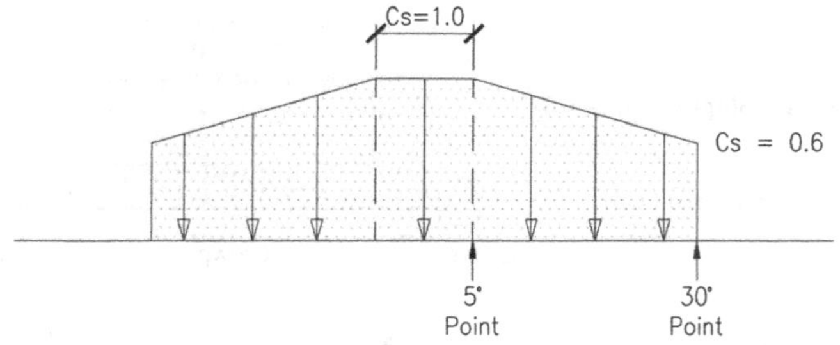

Figure 7.4-3. Sloped roof snow load for air-supported structures.

Case 1: Full balanced snow load on either exterior span and half the balanced snow load on all other spans.
Case 2: Half the balanced snow load on either exterior span and full balanced snow load on all other spans.
Case 3: All possible combinations of full balanced snow load on any two adjacent spans and half the balanced snow load on all other spans. For this case, there will be $(n–1)$ possible combinations, where n equals the number of spans in the continuous beam system.

If a cantilever is present in any of the above cases, it shall be considered to be a span.

Partial load provisions need not be applied to structural members that span perpendicular to the ridgeline in gable roofs with slopes between 1/2 on 12 (2.38 degrees) and 7 on 12 (30.3 degrees).

7.5.2 Other Structural Systems Areas sustaining only half the balanced snow load shall be chosen so as to produce the greatest effects on members being analyzed.

7.6 UNBALANCED ROOF SNOW LOADS

Balanced and unbalanced loads shall be analyzed separately. Winds from all directions shall be accounted for when establishing unbalanced loads.

7.6.1 Unbalanced Snow Loads for Hip and Gable Roofs For hip and gable roofs with a slope exceeding 7 on 12 (30.2 degrees), or with a slope less than 1/2 on 12 (2.38 degrees), unbalanced snow loads are not required to be applied. Roofs with an eave to ridge distance, W, of 20 ft (6.1 m) or less that have simply supported prismatic members spanning from ridge to eave shall be designed to resist an unbalanced uniform snow load on the leeward side equal to p_g. For these roofs, the windward side shall be unloaded. For all other gable roofs, the unbalanced load shall consist of $0.3p_s$ on the windward side, p_s on the leeward side plus a rectangular surcharge with magnitude $h_d\gamma/\sqrt{S}$ and horizontal extent from the ridge $8h_d\sqrt{S}/3$ where h_d is the drift height from Equation (7.6-1) with l_u equal to the eave to ridge distance for the windward portion of the roof, W.

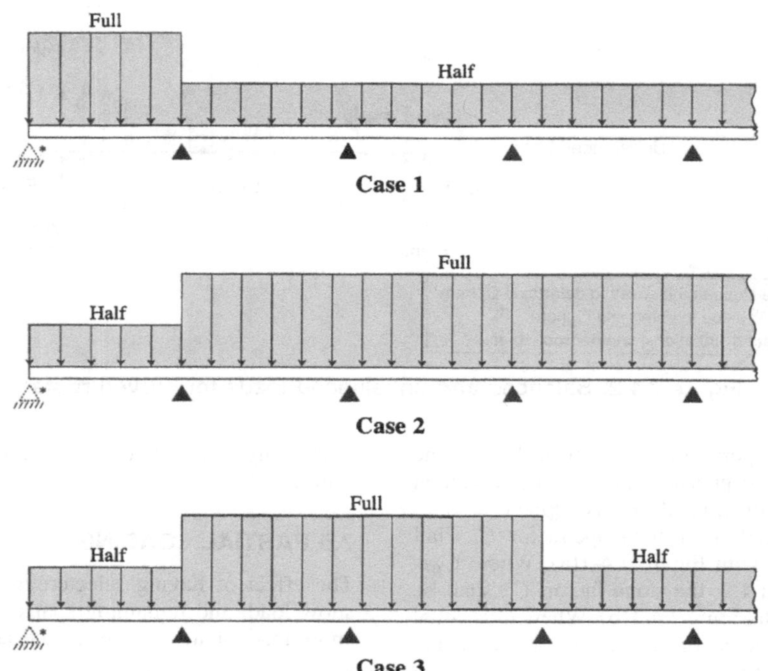

Figure 7.5-1. Partial loading diagrams for continuous beams.

The drift height h_d is given by

$$h_d = 1.5\sqrt{\frac{P_g^{0.74} l_u^{0.70} W_2^{1.7}}{\gamma}} \quad (7.6\text{-}1)$$

where the winter wind parameter W_2 for the site is given in Figure 7.6-1 and Table 7.2-1 for Alaska.

Balanced and unbalanced loading diagrams are presented in Figure 7.6-2.

> User Note: The winter wind parameter W_2 is available at https://asce7hazardtool.on-line/.

7.6.2 Unbalanced Snow Loads for Curved Roofs Portions of curved roofs that have a slope exceeding 70 degrees shall be considered free of snow load. If the slope of a straight line from the eaves (or the 70-degree point, if present) to the crown is less than 10 degrees or greater than 60 degrees, unbalanced snow loads shall not be taken into account.

Unbalanced loads shall be determined according to the loading diagrams in Figure 7.4-2. In all cases, the windward side shall be considered free of snow. If the ground or another roof abuts a Case 2 or Case 3 (see Figure 7.4-2) curved roof at or within 3 ft (0.9 m) of its eaves, the snow load shall not be decreased between the 30 degree point and the eaves but shall remain constant at the 30-degree point value. This distribution is shown as a dashed line in Figure 7.4-2.

7.6.3 Unbalanced Snow Loads for Multiple Folded Plate, Sawtooth, and Barrel Vault Roofs Unbalanced loads shall be applied to folded plate, sawtooth, and barrel-vaulted multiple roofs with a slope exceeding 3/8 in. on 12 (1.79 degrees). According to Section 7.4.4, $C_s = 1.0$ for such roofs, and the balanced snow load equals p_f. The unbalanced snow load shall increase from one-half the balanced load at the ridge or crown (i.e., $0.5p_f$) to two times the balanced load given in Section 7.4.4 divided by C_e at the valley (i.e., $2p_f/C_e$). Balanced and unbalanced loading diagrams for a sawtooth roof are presented in Figure 7.6-3. However, the snow surface above the valley shall not be at an elevation higher than the snow above the ridge. Snow depths shall be determined by dividing the snow load by the density of that snow from Equation (7.7-1).

7.6.4 Unbalanced Snow Loads for Dome Roofs Unbalanced snow loads shall be applied to domes and similar rounded structures. Snow loads, determined in the same manner as for curved roofs in Section 7.6.2, shall be applied to the downwind 90 degree sector in plan view. At both edges of this sector, the load shall decrease linearly to zero over sectors of 22.5 degrees each. There shall be no snow load on the remaining 225 degree upwind sector.

7.7 DRIFTS ON LOWER ROOFS (AERODYNAMIC SHADE)

Roofs shall be designed to sustain localized loads from snowdrifts that form in the wind shadow of (1) higher portions of the same structure, and (2) adjacent structures and terrain features.

7.7.1 Lower Roof of a Structure Snow that forms drifts comes from a higher roof or, with the wind from the opposite direction, from the roof on which the drift is located. These two kinds of drifts ("leeward" and "windward," respectively) are shown in Figure 7.7-1. The geometry of the surcharge load due to snow drifting shall be approximated by a triangle, as shown in Figure 7.7-2. Drift loads shall be superimposed on the balanced snow load. If h_c/h_b is less than 0.2, drift loads are not required to be applied.

For leeward drifts, the drift height, h_d, shall be determined directly from Equation (7.6-1) using the length of the upper roof. However, the drift height need not be taken as larger than 60% of the length of the lower level roof. For windward drifts, the drift height shall be determined by substituting the length of the lower roof for l_u into Equation (7.6-1) and using three-quarters of h_d as the drift height. The larger of these two heights shall be used in design. For leeward drifts, if h_d is equal to or less than h_c, the drift width, w, shall equal $4h_d$ and the drift height shall equal h_d. If this height exceeds h_c, the drift width, w, shall equal $4h_d^2/h_c$ and the drift height shall equal h_c.

For leeward drifts, the drift width, w, shall not be greater than $8h_c$. For windward drifts, the drift width shall be taken as eight times the windward drift height or $(8(.75h_d) = 6h_d)$. If the drift width for either windward or leeward drifts, w, exceeds the width of the lower roof, the drift shall taper linearly to zero at the far end of the lower level roof. Windward and leeward drifts shall be checked independently to determine which controls the structural design of each member. The maximum intensity of the drift surcharge load, p_d, equals $h_d\gamma$, where snow density, γ, is defined in Equation (7.7-1):

$$\gamma = 0.13p_g + 14 \text{ but not more than } 30\ \text{lb/ft}^3 \quad (7.7\text{-}1)$$

$$\gamma = 0.426p_g + 2.2, \text{ but not more than } 4.7\ \text{kN/m}^3 \quad (7.7\text{-}1.\text{SI})$$

This density shall also be used to determine h_b by dividing p_s by γ (in SI: also multiply by 102 to get the depth in m).

7.7.2 Adjacent Structures If the horizontal separation distance between adjacent structures, s, is less than 20 ft (6.1 m) and less than six times the vertical separation distance ($s < 6h$), then the requirements for the leeward drift of Section 7.7.1 shall be used to determine the drift load on the lower structure. The height of the snow drift shall be the smaller of h_d, based on the length of the adjacent higher structure, and $(6h-s)/6$. The horizontal extent of the drift shall be the smaller of $6h_d$ or $(6h-s)$.

For windward drifts, the requirements of Section 7.7.1 shall be used. The resulting drift is permitted to be truncated.

7.7.3 Intersecting Drifts Intersecting drifts shall be evaluated at reentrant corners, parapet wall corners, intersections of gable roof with the roof step wall of a taller roof and other similar geometries. Section 7.7.1 shall be used to determine the snow drift geometry for each direction. Where the two snowdrifts intersect, the larger of the two snowdrift depths shall govern, as shown in Figure 7.7-3. Intersecting snowdrift loads shall be considered to occur concurrently, except that the two drift loads need not be superimposed.

For leeward intersecting snowdrifts at reentrant corners, the length of the upper roof applicable for each snowdrift shall be used with l_u parallel to w_1 for the first drift and l_u parallel to w_2 for the second drift. For windward snowdrifts, the lengths of the lower roof shall be used for l_u.

7.8 ROOF PROJECTIONS AND PARAPETS

The method for windward drifts in Section 7.7.1 shall be used to calculate drift loads on all sides of roof projections and at parapet walls. The height of such drifts shall be taken as three-quarters the drift height from Equation 7.6-1 (i.e., $0.75h_d$). For parapet walls, l_u shall be taken equal to the length of the roof upwind of the wall. For roof projections, l_u shall be taken equal to the greater of the length of the roof upwind or downwind of the projection.

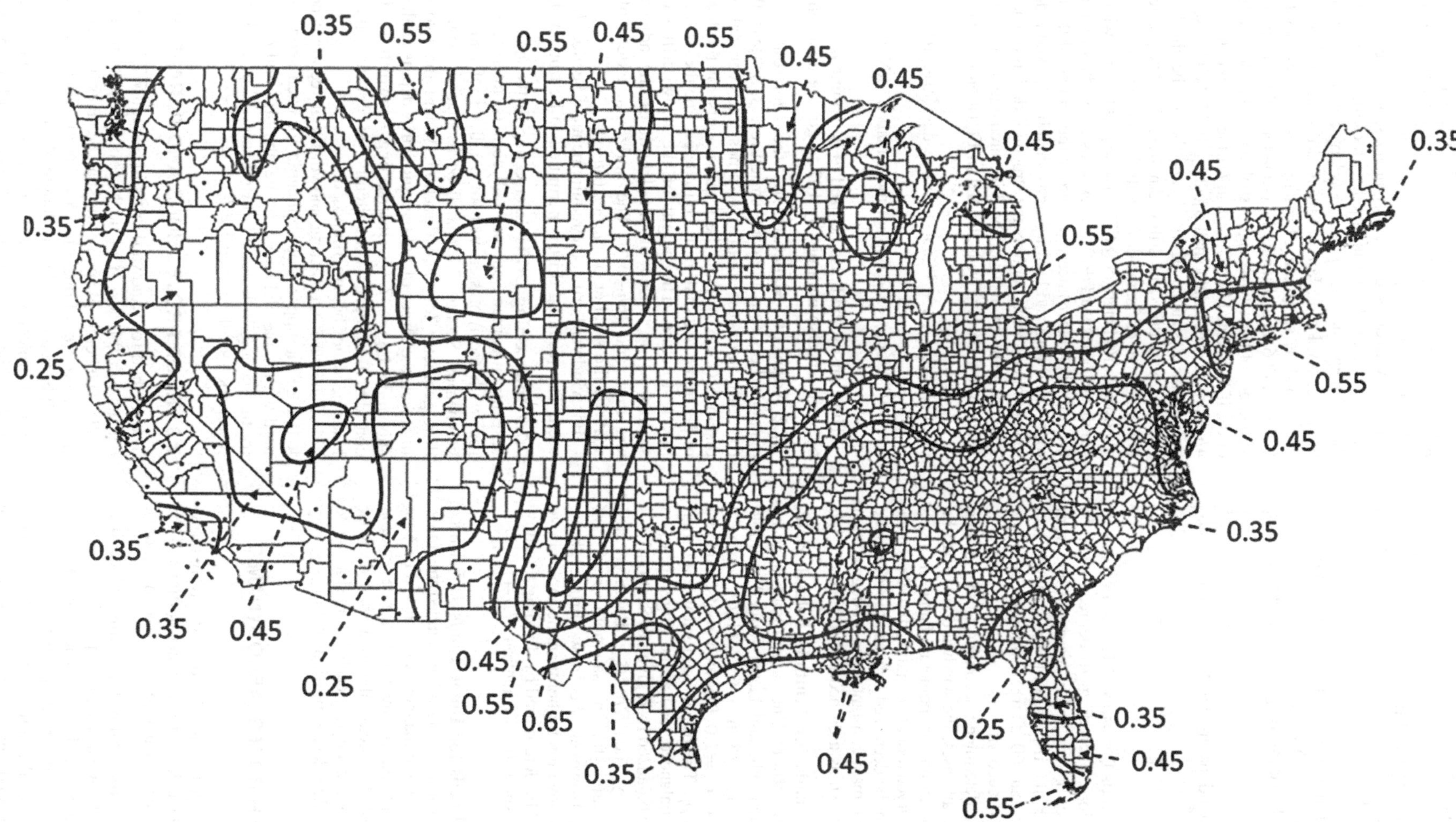

Figure 7.6-1. Map of winter wind parameter, W_2.

Note: The winter wind parameter W_2 is available at https://asce7hazardtool.on-line.

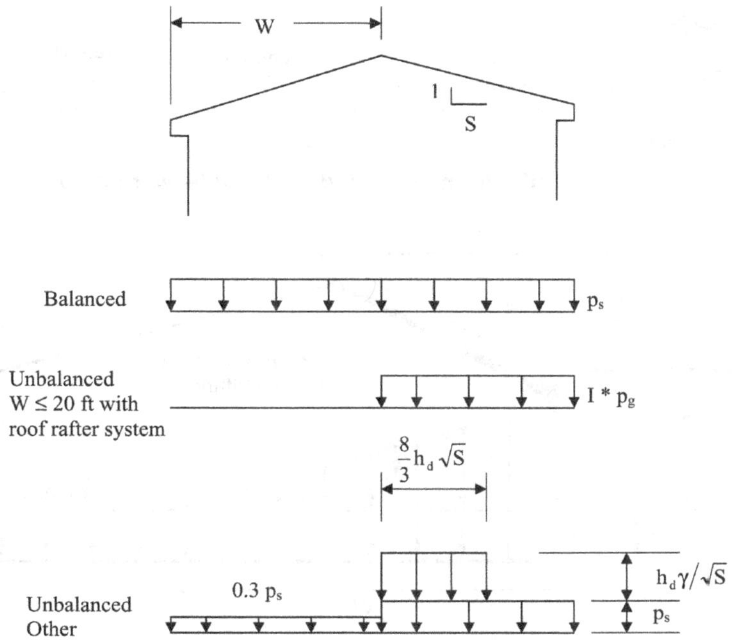

Figure 7.6-2. Balanced and unbalanced snow loads for hip and gable roofs.
Note: Unbalanced loads need not be considered for θ > 30.2° (7 on 12) or for θ < 2.38° (1/2 on 12).

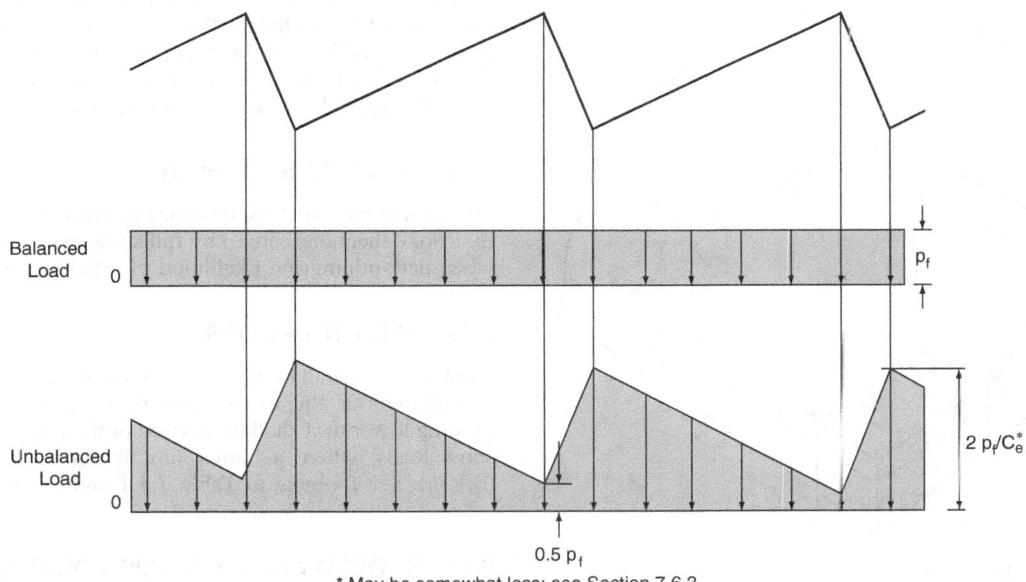

Figure 7.6-3. Balanced and unbalanced snow loads for a sawtooth roof.

EXCEPTION: Drift loads shall not be required where the side of the roof projection is less than 15 ft (4.6 m) or the clear distance between the height of the balanced snow load, h_b, and the bottom of the projection (including horizontal supports) is at least 2 ft (0.61 m).

7.9 SLIDING SNOW

The load caused by snow sliding off a sloped roof onto a lower roof shall be determined for slippery upper roofs with slopes greater than ¼ on 12, and for other (i.e., nonslippery) upper roofs with slopes greater than 2 on 12. The total sliding load per unit length of eave shall be $0.4p_f W$, where W is the horizontal distance from the eave to ridge for the sloped upper roof. The sliding load shall be distributed uniformly on the lower roof over a distance of 15 ft (4.6 m) from the upper roof eave. If the width of the lower roof is less than 15 ft (4.6 m), the sliding load shall be reduced proportionally.

The sliding snow load shall not be further reduced unless a portion of the snow on the upper roof is blocked from sliding onto the lower roof by snow already on the lower roof.

For separated structures, sliding loads shall be considered when $h/s > 1$ and $s < 15$ ft ($s < 4.6$ m). The horizontal extent of the sliding load on the lower roof shall be $15-s$ with s in feet

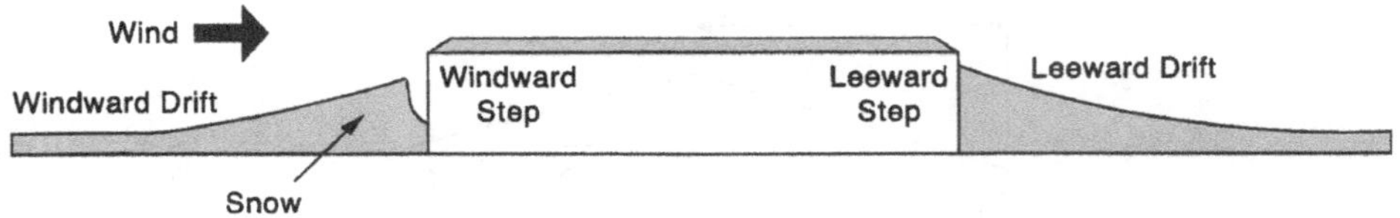

Figure 7.7-1. Drifts formed at windward and leeward steps.

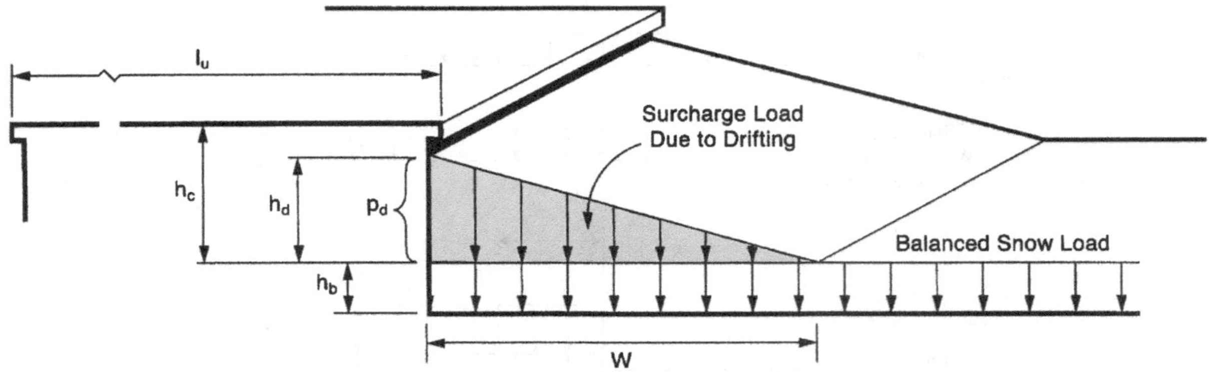

Figure 7.7-2. Configuration of snowdrifts on lower roofs.

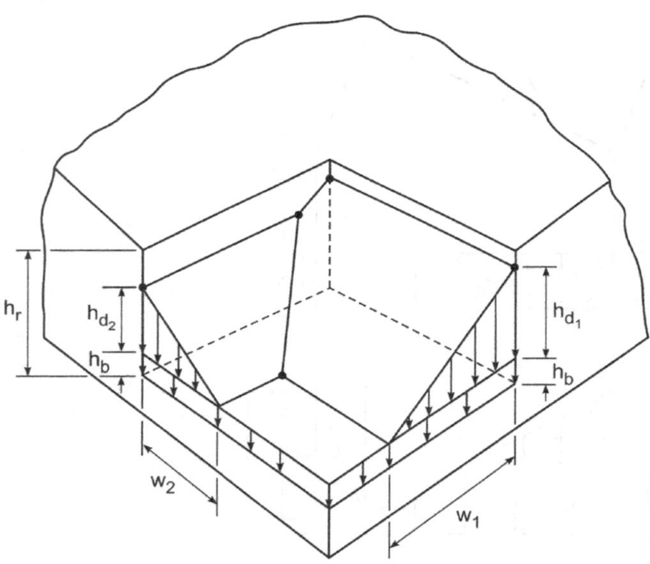

Uniform Snow Loads on Upper Roof not Shown

Figure 7.7-3. Configuration of intersecting snowdrifts at lower roof.

(4.6–s with s in meters), and the load per unit length shall be $0.4p_f W(15–s)/15$ with s in feet ($0.4p_f W(4.6–s)/4.6$ with s in meters).

Sliding loads shall be superimposed on the balanced snow load and need not be used in combination with drift, unbalanced, partial, or rain-on-snow loads.

7.10 RAIN-ON-SNOW SURCHARGE LOAD

For locations where p_g is less than, or equal to, the value of $p_{m,max}$ given in Table 7.3-4, but not zero, all roofs with slopes (in degrees) less than $W/50$ with W in ft (in SI: $W/15.2$ with W in m) shall include a 8 lb/ft^2 (0.38 kN/m^2) rain-on-snow surcharge load. This additional load applies only to the sloped roof (balanced) load case and need not be used in combination with drift, sliding, unbalanced, minimum, or partial loads.

7.11 PONDING INSTABILITY

Susceptible bays shall be designed to preclude ponding instability. Roof deflections caused by full snow loads shall be evaluated when determining the likelihood of ponding instability.

7.12 EXISTING ROOFS

Existing roofs shall be evaluated for increased snow loads caused by additions or alterations. Owners or agents for owners of an existing lower roof shall be advised of the potential for increased snow loads, where a higher roof is constructed within 20 ft (6.1 m). See footnote to Table 7.3-1 and Section 7.7.2.

7.13 SNOW ON OPEN-FRAME EQUIPMENT STRUCTURES

Open-frame equipment structures shall be designed for snow loads in accordance with Sections 7.13.1 through 7.13.4. The thermal factor, $C_t = 1.2$, shall be used in determination of snow loads for unheated open-frame equipment structures.

7.13.1 Snow at Top Level Flat roof snow loads (p_f) and drift loads shall be applied at the top level of the structure where there is flooring or elements that can retain snow. Open-frame members with a width of more than 8 in. (200 mm) shall be considered snow retaining surfaces. The top level shall be designed for snowdrifts, in accordance with Sections 7.7 and 7.9, where there are wind walls or equivalent obstructions.

7.13.2 Snow at Levels below the Top Level At all levels with flooring (grating, checkered plate, etc.) located below a level with flooring, the flat roof snow load shall be applied over a portion of that flooring level near any open edge in accordance with Figure 7.13-1. The flat roof snow load shall extend from the upwind edge of the flooring a horizontal distance equal to the vertical difference in elevation between the level in question and the next floor above.

7.13.3 Snow Loads on Pipes and Cable Trays Individual pipes and cable trays with a diameter (pipe) or width (tray) less than or equal to $0.73p_f/\gamma$ shall be designed for a triangular snow load in accordance with Figure 7.13-2(a). Individual pipes and cable trays with a diameter (pipe) or width (tray) greater than $0.73p_f/\gamma$ shall be designed for a trapezoidal snow load, in accordance with Figure 7.13-2(b). Snow loads on pipes are not required to be considered if the wintertime external surface temperature of the pipe is greater than 45 °F (7.2 °C).

Where the spacing between multiple adjacent pipes or cable trays at the same elevations is less than the height of the flat roof snow load (p_f/γ), an additional uniform cornice load of p_f shall be applied in the spaces between the pipes or cable trays, as shown in Figure 7.13-3. For $S_p \geq h$, the additional cornice loads need not be applied.

7.13.4 Snow Loads on Equipment and Equipment Platforms Snow loads on the structure shall include snow accumulation on equipment and equipment platforms that can retain snow. Snow accumulation need not be considered on equipment with a wintertime external surface temperature greater than 45°F (7.2°C).

7.14 ALTERNATE PROCEDURE

In lieu of the requirements specified in Sections 7.3 through 7.13, the design snow loads shall be permitted to be determined by thermal performance studies and scale-model studies in wind tunnels or water flumes conducted in accordance with the requirements of ASCE 49 and this section. Snow loads determined using these studies shall be based on the ground snow loads outlined in Section 7.2 and shall be derived to be consistent with the reliability targets outlined in Chapter 1.

7.14.1 Limitations on Snow Loads Derived from Scale Model Studies Snow loads derived based on scale-model studies shall not be taken as less than 80% of those specified in Sections 7.2 through 7.13 unless an independent peer review

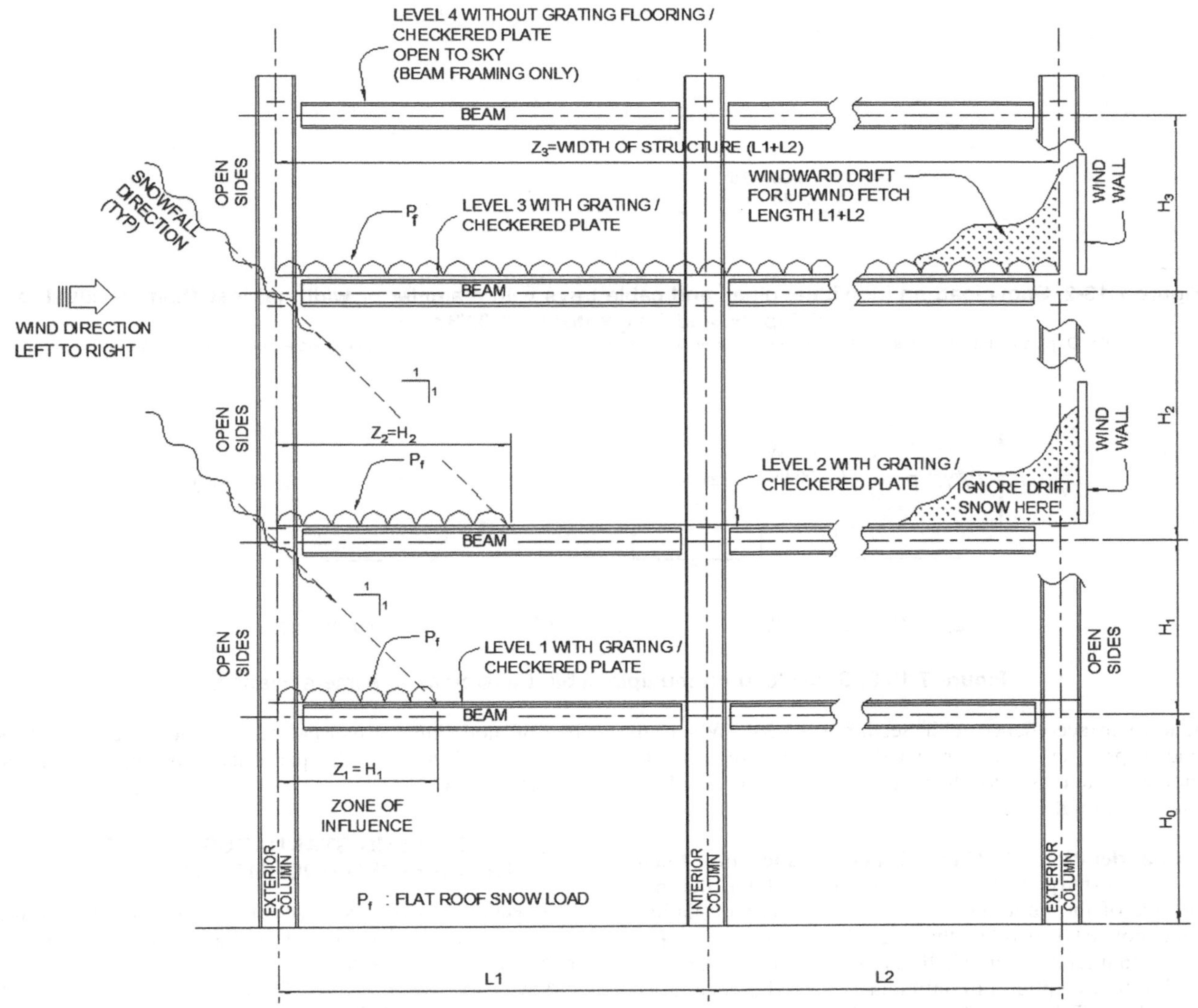

Figure 7.13-1. Open-frame equipment structures.

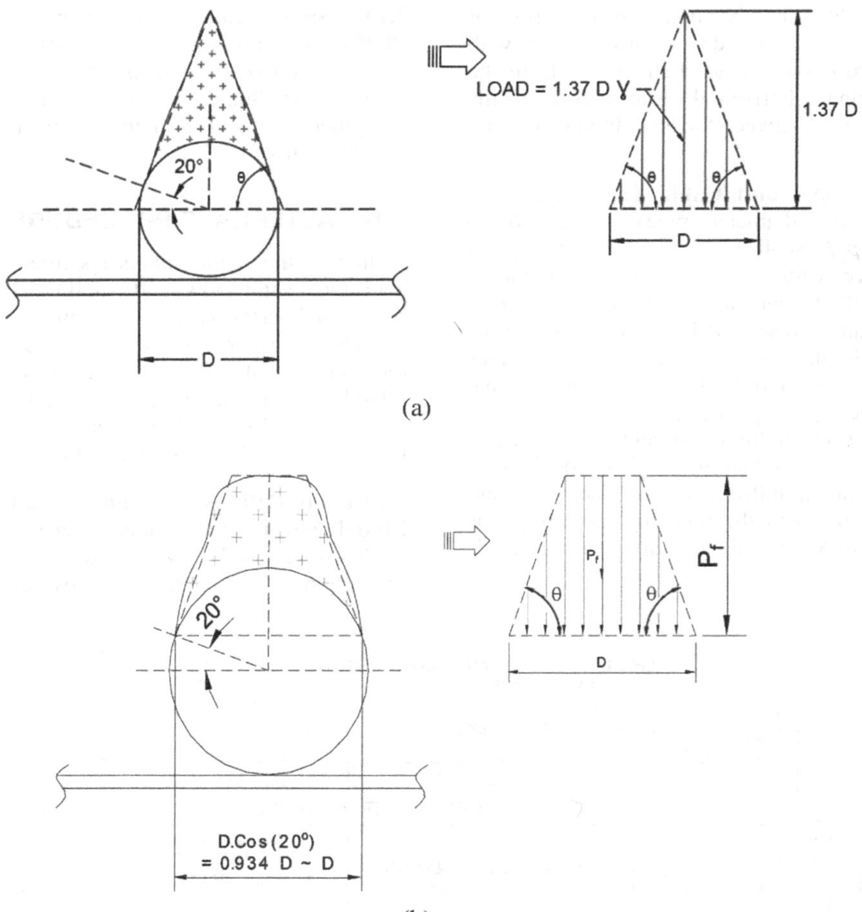

Figure 7.13-2. Snow load on individual pipes and cable trays with diameter or width (a) less than, or equal to, $0.73p_f/\gamma$, and (b) greater than $0.73p_f/\gamma$.

Note: D, pipe diameter $+2x$ insulation thickness (as applicable); P_f, flat roof snow load; θ, assumed angle of repose = 70 degrees.

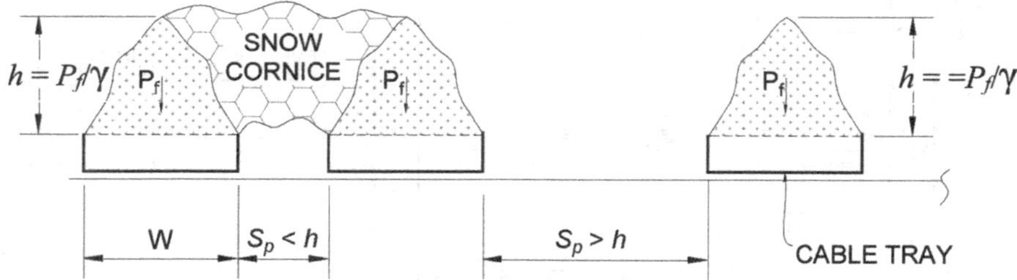

Figure 7.13-3. Snow load on multiple cable trays/pipes at same elevation.

is conducted in accordance with Section 1.3.1.3.4. When an independent peer review is conducted, the design snow loads shall not be taken as less than 60% of those specified in Sections 7.2 through 7.13.

7.14.2 Consideration of Thermal Performance in Model Studies If the study includes an assessment of the thermal performance of the building or structure, the roof insulation value shall not be taken as less than R20 [U=0.05 Btu/(h ft^2 °F)] and the internal temperature shall not be taken as greater than 70 °F, with the exception of continuously heated greenhouses as outlined in Table 7.3-2, where the expected thermal properties and internal temperatures may be considered. In lieu of modeling the thermal performance, the thermal factor outlined in Section 7.3.2 is permitted to be applied to the scale model snow load.

7.15 CONSENSUS STANDARDS AND OTHER REFERENCED DOCUMENTS

This section lists the consensus standards and other documents that shall be considered part of this standard to the extent referenced in this chapter.

ASCE 49, *Wind Tunnel Testing for Buildings and Other Structures*, ASCE, 2020.

Cited in: Section 7.14

CHAPTER 8
RAIN LOADS

8.1 DEFINITIONS AND SYMBOLS

8.1.1 Definitions

BAY: A portion of the roof bounded by adjacent column lines or structural walls.

CONTROLLED FLOW ROOF DRAIN: A roof drain designed to intentionally regulate the rate of drainage.

PONDING: The accumulation of water caused by the deflection of the roof structure, resulting in added load.

PONDING INSTABILITY: Member instability caused by progressive deflection because of ponding on roofs.

PRIMARY DRAINAGE SYSTEM: Roof drainage system through which water is normally conveyed off the roof.

SECONDARY DRAINAGE SYSTEM FOR STRUCTURAL LOADING (SDSL): Roof drainage system through which water is drained from the roof when the drainage systems listed in Section 8.2 (a) through (d) are blocked or not working.

SECONDARY MEMBER: For the purposes of Section 8.3, structural members not having direct attachment to the columns.

8.1.2 Symbols

d_h = Hydraulic head equal to the depth of water on the undeflected roof above the inlet of the secondary drainage system for structural loading (SDSL) required to achieve the design flow, in. (mm)

d_s = Static head equal to the depth of water on the undeflected roof up to the inlet of the secondary drainage system for structural loading (SDSL), in. (mm)

d_p = Ponding head equal to the depth of water due to deflections of the roof subjected to unfactored rain load and unfactored dead load in. (mm)

L_s = Span of secondary members, in. (mm)

S = Spacing of secondary members, in. (mm)

R = Rain load, lb/ft^2 (kN/m^2)

8.2 DESIGN RAIN LOADS

Each portion of a roof shall be designed to sustain the load of all rainwater that will accumulate on it, assuming all drainage systems that meet any of the following criteria are blocked:

(a) The primary drainage system,
(b) Secondary drainage systems with an inlet that is vertically separated from the inlet to the primary drainage system by less than 2 in. (51 mm),
(c) Secondary drainage systems that share drain lines with the primary drainage system, or
(d) Secondary drainage systems with controlled flow roof drains.

Rain loads shall be based on the summation of the static head, d_s, hydraulic head, d_h, and ponding head, d_p, using Equation (8.2-1). The hydraulic head shall be based on hydraulic test data or hydraulic calculations, assuming a flow rate corresponding to a rainfall intensity equal to or greater than the 15 min duration storm with return period (frequency) given in Table 8.2-1. The ponding head shall be based on structural analysis as the depth of water caused by deflections of the roof subjected to unfactored rain load and unfactored dead load.

$$R = 5.2(d_s + d_h + d_p) \qquad (8.2\text{-}1)$$

$$R = 0.0098(d_s + d_h + d_p) \qquad (8.2\text{-}1.\text{SI})$$

8.3 BAYS WITH LOW SLOPE

Free-draining bays and internal bays not subjected to accumulated rain load required by Section 8.2 shall have adequate strength and stiffness to preclude progressive deflection (i.e., instability) and resist potential ponding rain loads where either of the following conditions are met:

1. The roof slope is less than 1/4 in. per foot (1.19 degrees), or
2. The bay is adjacent to a free-draining edge with secondary members parallel to the free-draining edge and the roof slope is less than β where $\beta = (L_s/S + \pi)/20$ (in. per foot).

8.4 DRAINAGE TO EXISTING ROOFS

Drainage systems for new construction shall not discharge water onto existing roofs unless the existing roof is evaluated and either is able to support the loads determined using this chapter or is upgraded to support the loads determined using this chapter.

8.5 CONSENSUS STANDARDS AND OTHER REFERENCED DOCUMENTS

No consensus standards and other documents that shall be considered part of this standard are referenced in this chapter.

Table 8.2-1. Design Storm Return Period by Risk Category.

Risk Category	Design Storm Return Period
I and II	100 years
III	200 years
IV	500 years

CHAPTER 9
RESERVED FOR FUTURE PROVISIONS

CHAPTER 10
ICE LOADS—ATMOSPHERIC ICING

10.1 GENERAL

Atmospheric ice loads caused by freezing rain, snow, and in-cloud icing shall be considered in the design of ice-sensitive structures. In areas where records or experience indicate that snow or in-cloud icing produces larger loads than freezing rain, site-specific studies shall be used. Structural loads caused by hoarfrost are not a design consideration. Roof snow loads are covered in Chapter 7.

10.1.1 Site-Specific Studies Site-specific studies that are acceptable to the Authority Having Jurisdiction shall be used to determine the ice thickness or load for the applicable Risk Category, concurrent wind speed, and concurrent temperature in

1. Areas where records or experience indicate that snow or in-cloud icing produce larger loads than freezing rain;
2. Special icing regions, as shown in Figures 10.4-2 through 10.4-5; and
3. Complex terrain, where examination indicates unusual icing conditions exist.

10.1.2 Dynamic Loads Dynamic loads, such as those resulting from galloping, ice shedding, and aeolian vibrations, which are caused, or enhanced, by an ice accretion on a flexible structural member, component, or appurtenance, are not covered in this section.

10.1.3 Exclusions Electrical supply and communication systems, communications towers and masts, and other structures for which national standards exist are excluded from the requirements of this section. Applicable standards and documents include the National Electrical Safety Code, ASCE Manual of Practice 74, and ANSI/TIA-222.

10.2 DEFINITIONS

The following definitions apply only to the provisions of this chapter.

ATMOSPHERIC ICE GEODATABASE: The ASCE database (version 2022-1.0) of geocoded nominal ice thickness, concurrent wind, and concurrent temperature data.

COMPONENTS AND APPURTENANCES: Nonstructural elements that may be exposed to atmospheric icing. Examples include ladders, handrails, netting, antennas, waveguides, radio frequency (RF) transmission lines, pipes, electrical conduits, and cable trays.

FREEZING RAIN: Rain or drizzle that falls into a layer of subfreezing air at the Earth's surface and freezes on the ground or an object to form glaze ice.

GLAZE: Clear, high-density ice.

HOARFROST: An accumulation of ice crystals formed by the direct deposition of water vapor from the air onto an object.

ICE-SENSITIVE STRUCTURES: Structures for which the effect of an atmospheric icing load governs the design of part, or all, of the structure. These structures include, but are not limited to, lattice structures, guyed masts, overhead electric and communication lines, light suspension and cable-stayed bridges, aerial cable systems (e.g., for ski lifts and logging operations), amusement rides, open catwalks and platforms, flagpoles, and signs.

IN-CLOUD ICING: Icing that occurs when supercooled cloud or fog droplets, carried by the wind, freeze on objects. In-cloud icing usually forms rime but may also form glaze.

RIME: White or opaque ice with entrapped air.

SNOW: Snow that adheres to objects by some combination of capillary forces, freezing, and sintering.

10.3 SYMBOLS

A_i = Cross-sectional area of ice, ft^2 (m^2)
A_s = Surface area of one side of a flat plate or the projected area of complex shapes, ft^2 (m^2)
D = Diameter of a circular structure or member, as defined in Chapter 29, ft (m)
D_c = Diameter of the cylinder circumscribing an object, ft (m)
f_z = Factor to account for the increase in ice thickness with height
K_{zt} = Topographic factor, as defined in Chapter 26
q_z = Velocity pressure evaluated at height, z, as defined in Chapter 29, lb/ft^2 (N/m^2)
r = Radius of the maximum cross section of a dome or radius of a sphere, ft (m)
t = Nominal ice thickness on a cylinder caused by freezing rain at a height of 33 ft (10 m), from Figures 10.4-2 through 10.4-5, for the applicable risk category, in. (mm)
t_d = Design ice thickness from Equation (10.4-5), in. (mm)
V_c = Concurrent gust speed, from Figures 10.5-1 and 10.5-2, mph (m/s)
V_i = Volume of ice, ft^3 (m^3)
z = Height above ground, ft (m)
ε = Solidity ratio, as defined in Chapter 29

10.4 ICE LOADS CAUSED BY FREEZING RAIN

10.4.1 Ice Load The ice load shall be determined using the volume or cross-sectional area of glaze ice formed on all exposed surfaces of structural members, guys, components, appurtenances, and cable systems. On structural shapes, prismatic members, and other similar shapes, the cross-sectional area of ice shall be determined by

$$A_i = \pi \frac{t_d}{12}\left(D_c + \frac{t_d}{12}\right) \quad (10.4\text{-}1)$$

$$A_i = \pi \frac{t_d}{1000}\left(D_c + \frac{t_d}{1000}\right) \quad (10.4\text{-}1\text{SI})$$

D_c is shown for a variety of cross-sectional shapes in Figure 10.4-1.

On flat plates and large three-dimensional objects such as domes and spheres, the volume of ice shall be determined by

$$V_i = \pi \frac{t_d}{12} A_s \quad (10.4\text{-}2)$$

$$V_i = \pi \frac{t_d}{1000} A_s \quad (10.4\text{-}2\text{SI})$$

For a flat plate, A_s shall be the area of one side of the plate; for domes and spheres, A_s shall be determined by

$$A_s = \pi r^2 \quad (10.4\text{-}3)$$

It is acceptable to multiply V_i by 0.8 for vertical plates and 0.6 for horizontal plates.

The ice density shall be not less than 56 pcf (900 kg/m^3).

10.4.2 Nominal Ice Thickness The nominal ice thickness is the equivalent radial glaze ice thickness. Figures 10.4-2 through 10.4-5 show the nominal ice thicknesses, t, from freezing rain, at a height of 33 ft (10 m), over the contiguous 48 states and Alaska for Risk Categories I, II, III, and IV. Ice thicknesses for Hawaii shall be obtained from local meteorological studies.

10.4.3 Height Factor The height factor, f_z, used to increase the radial thickness of ice for height above ground, z, shall be determined by

$$f_z = \left(\frac{z}{33}\right)^{0.10} \quad \text{for } 0\text{ ft} < z \le 900\text{ ft} \quad (10.4\text{-}4)$$

where $f_z = 1.4$ for $z > 900$ ft.

$$f_z = \left(\frac{z}{10}\right)^{0.10} \quad \text{for } 0\text{ m} < z \le 275\text{ m} \quad (10.4\text{-}4.\text{SI})$$

where $f_z = 1.4$ for $z > 275$ m.

10.4.4 Topographic Factor Both the ice thickness and concurrent gust speed for structures on hills, ridges, and escarpments are higher than those on level terrain because of wind speed-up effects. The topographic factor for the concurrent wind pressure is K_{zt}, and the topographic factor for ice thickness is $(K_{zt})^{0.35}$, where K_{zt} is obtained from Equation (26.8-1).

10.4.5 Design Ice Thickness for Freezing Rain The design ice thickness t_d shall be calculated from Equation (10.4-5).

$$t_d = t f_z (K_{zt})^{0.35} \quad (10.4\text{-}5)$$

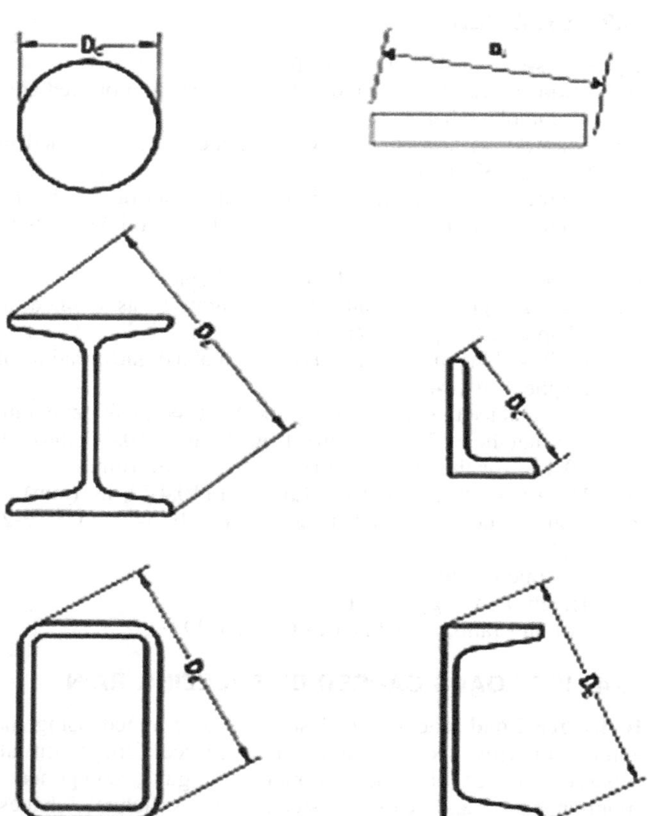

Figure 10.4-1. Characteristic dimension, D_c, for calculating the ice area for a variety of cross-sectional shapes.

10.5 WIND ON ICE-COVERED STRUCTURES

Ice accreted on structural members, components, and appurtenances increases the projected area of the structure exposed to wind. The projected area shall be increased by adding t_d to all free edges of the projected area. Wind loads on this increased projected area shall be used in the design of ice-sensitive structures. Figures 10.5-1 and 10.5-2 show 3s gust speeds at 33 ft (10 m) above grade that are concurrent with the ice loads caused by freezing rain. These gust speeds shall be used with nominal ice thicknesses for all risk categories. Wind loads shall be calculated in accordance with Chapters 26 through 31, as modified by Sections 10.5.1 through 10.5.5.

10.5.1 Wind on Ice-Covered Chimneys, Tanks, and Similar Structures Force coefficients, C_f, for structures with square, hexagonal, and octagonal cross sections shall be as given in Figure 29.4-1. Force coefficients, C_f, for structures with round cross sections shall be as given in Figure 29.4-1 for the entry round, $D\sqrt{q_z} \le 2.5$ for all ice thicknesses, wind speeds, and structure diameters.

10.5.2 Wind on Ice-Covered Solid Freestanding Walls and Solid Signs Force coefficients, C_f, shall be as given in Figure 29.3-1, based on the dimensions of the wall or sign, including ice.

10.5.3 Wind on Ice-Covered Open Signs and Lattice Frameworks The solidity ratio, ε, shall be based on the projected area, including ice. The force coefficient, C_f, for the projected area of flat members shall be as given in Figure 29.4-2. The force coefficient, C_f, for rounded members and for the additional projected area caused by ice on both flat and rounded members shall be as given in Figure 29.4-2 for rounded members with $D\sqrt{q_z} \le 2.5$ for all ice thicknesses, wind speeds, and member diameters.

10.5.4 Wind on Ice-Covered Trussed Towers The solidity ratio, ε, shall be based on the projected area, including ice. The force coefficients, C_f, shall be as given in Figure 29.4-3. It is acceptable to reduce the force coefficients, C_f, for the additional projected area caused by ice on both round and flat members by the factor for rounded members in Note 3 of Figure 29.4-3.

10.5.5 Wind on Ice-Covered Guys and Cables The force coefficient, C_f, (as defined in Chapter 29) for ice-covered guys and cables shall be 1.2.

10.6 DESIGN TEMPERATURES FOR FREEZING RAIN

The design temperatures for ice, and wind-on-ice, caused by freezing rain shall be either the concurrent temperature for the site shown in Figures 10.6-1 and 10.6-2 or 32 °F (0° C), whichever gives the maximum load effect. The temperature for Hawaii shall be 32 °F (0 °C). For temperature-sensitive structures, the load shall include the effect of temperature change from everyday conditions to the design temperature for ice and wind on ice. These temperatures shall be used with nominal ice thicknesses for all Risk Categories.

10.7 PARTIAL LOADING

The full intensity of the ice load applied to a portion of a structure or member shall be accounted for when it produces a critical load effect. It is permitted to consider this to be a static load.

10.8 DESIGN PROCEDURE

1. The nominal ice thickness, t, the concurrent wind speed, V_c, and the concurrent temperature for the site shall be determined from Figures 10.4-2 through 10.4-5, 10.5-1, 10.5-2, 10.6-1, and 10.6-2 or a site-specific study.
2. The topographic factor for the site, K_{zt}, shall be determined in accordance with Section 10.4.4.
3. The height factor, f_z, shall be determined in accordance with Section 10.4.3 for each design segment of the structure.
4. The design ice thickness, t_d, shall be determined in accordance with Section 10.4.5, Equation (10.4-5).
5. The ice load shall be calculated for the design ice thickness, t_d, in accordance with Section 10.4.1.
6. The velocity pressure, q_z, for wind speed, V_c, shall be determined in accordance with Section 29.3.
7. The wind force coefficients, C_f, shall be determined in accordance with Section 10.5.
8. The gust effect factor shall be determined in accordance with Section 26.11.
9. The design wind force shall be determined in accordance with Chapter 29.
10. The iced structure shall be analyzed for the load combinations in either Section 2.3 or 2.4.

10.9 CONSENSUS STANDARDS AND OTHER REFERENCED DOCUMENTS

This section lists the consensus standards and other documents that shall be considered part of this standard to the extent referenced in this chapter.

ASCE Manual of Practice 74, *Guidelines for Electrical Transmission Line Structural Loading*, American Society of Civil Engineers, 2020.
Cited in: Section 10.1.3

ANSI/TIA-222, *Structural Standard for Antenna Supporting Structures, Antennas, and Small Wind Turbine Support Structures*, American National Standards Institute, 2019.
Cited in: Section 10.1.3

NESC, *National Electrical Safety Code*, IEEE, 2017.
Cited in: Section 10.1.3

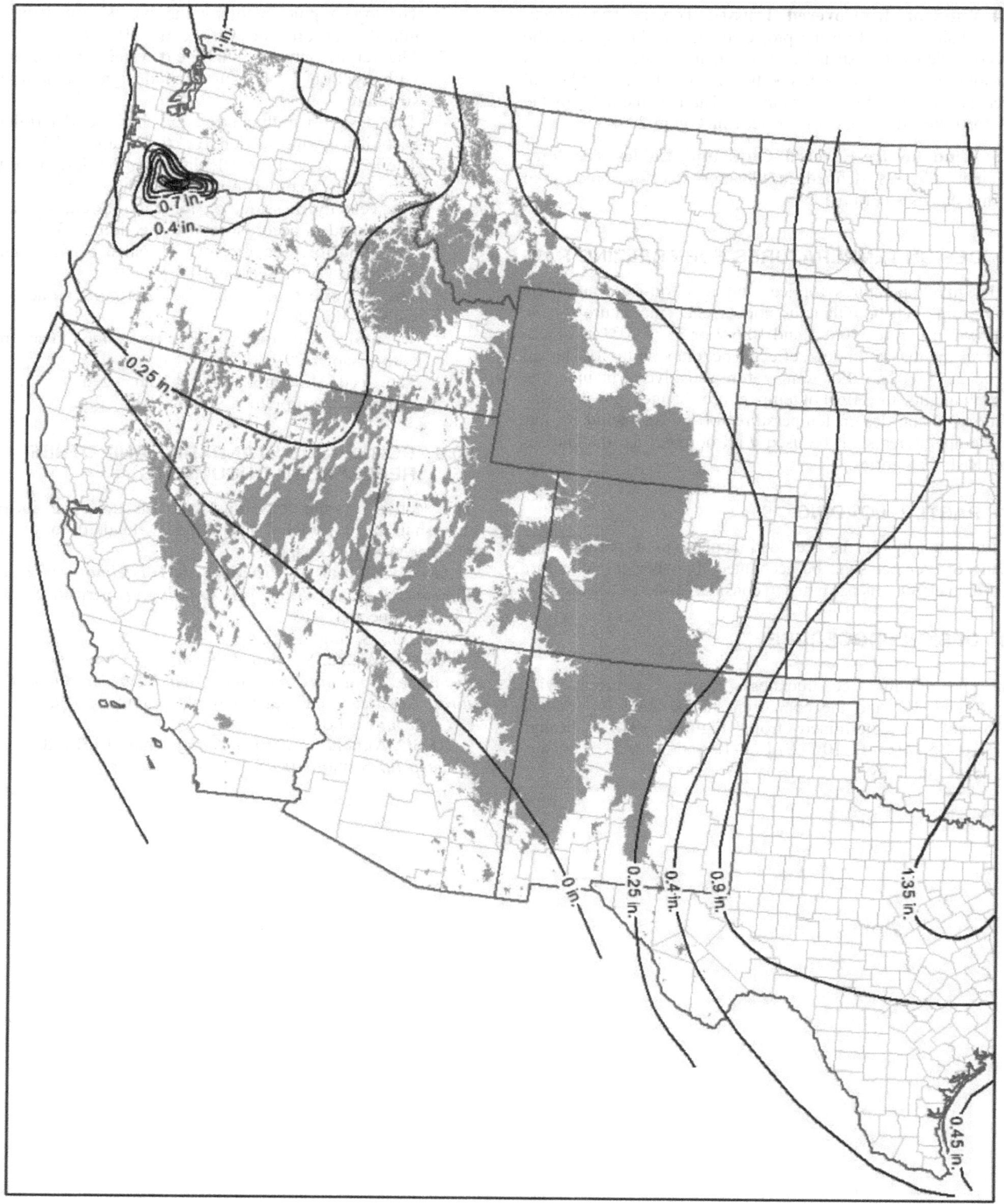

Figure 10.4-2A. Nominal ice thicknesses caused by freezing rain for the contiguous 48 states for Risk Category I (250-year mean recurrence interval) structures.

Notes:

1. Values are nominal ice thicknesses above ground, in inches at 33 ft (10 m). Metric conversion: 1 in. = 25.4 mm.
2. Linear interpolation between contours is permitted.
3. Islands, coastal areas, and land boundaries outside the last contour shall use the last contour.

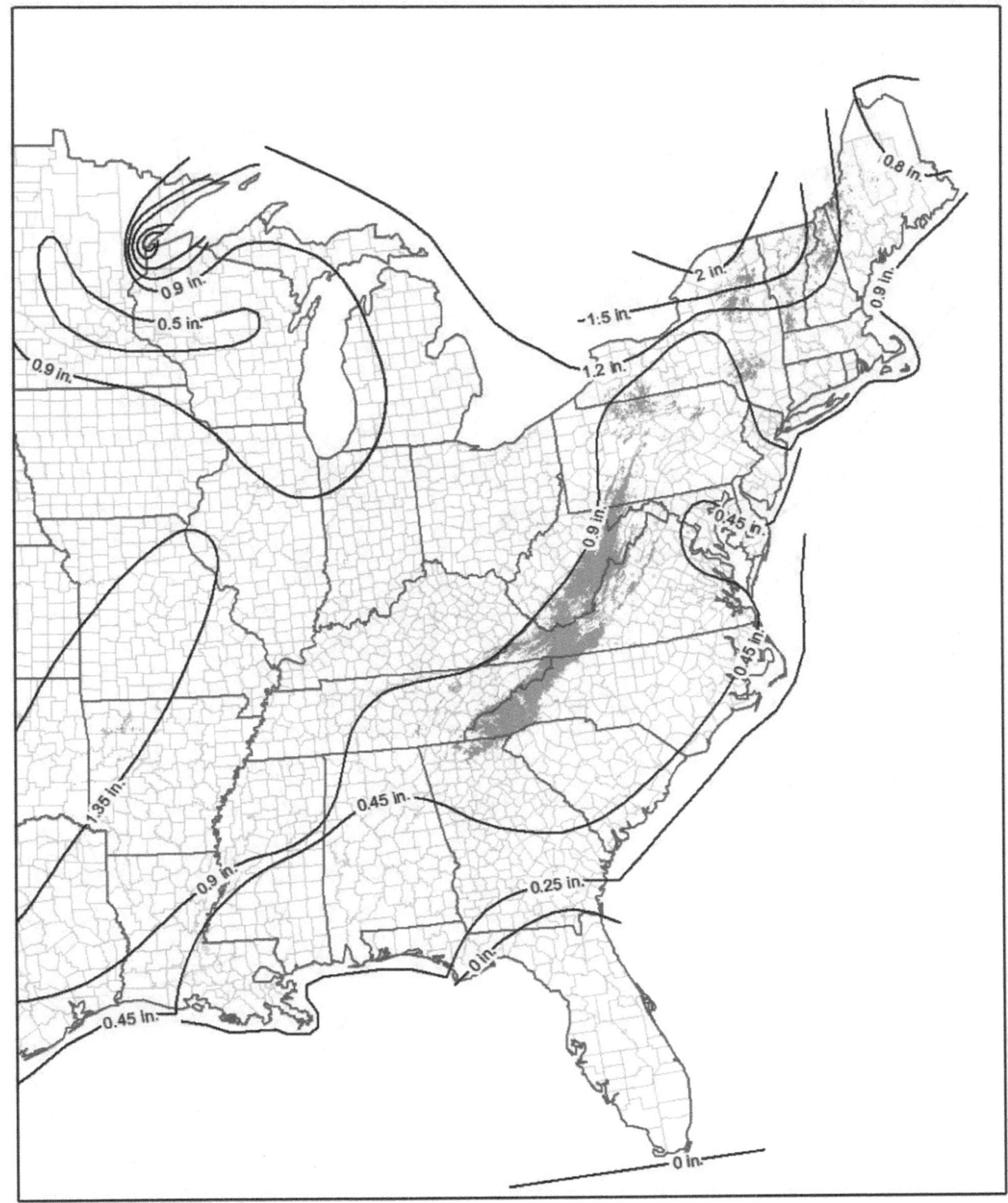

Figure 10.4-2A. ***(Continued)*** **Nominal ice thicknesses caused by freezing rain for the contiguous 48 states for Risk Category I (250-year mean recurrence interval) structures.**

4. The shading indicates special icing regions with elevations above 2,100 ft (640 m) in the east and 6,000 ft (1,829 m) in the west, with sparse weather station data for determining design ice loads.
5. For Columbia River Gorge Region, see Figure 10.4-4A. For Lake Superior Region, see Figure 10.4-5A.
6. Location-specific nominal ice thicknesses shall be permitted to be determined using the ASCE Atmospheric Ice Geodatabase, which can be accessed at the ASCE 7 Hazard Tool (https://asce7hazardtool.online).

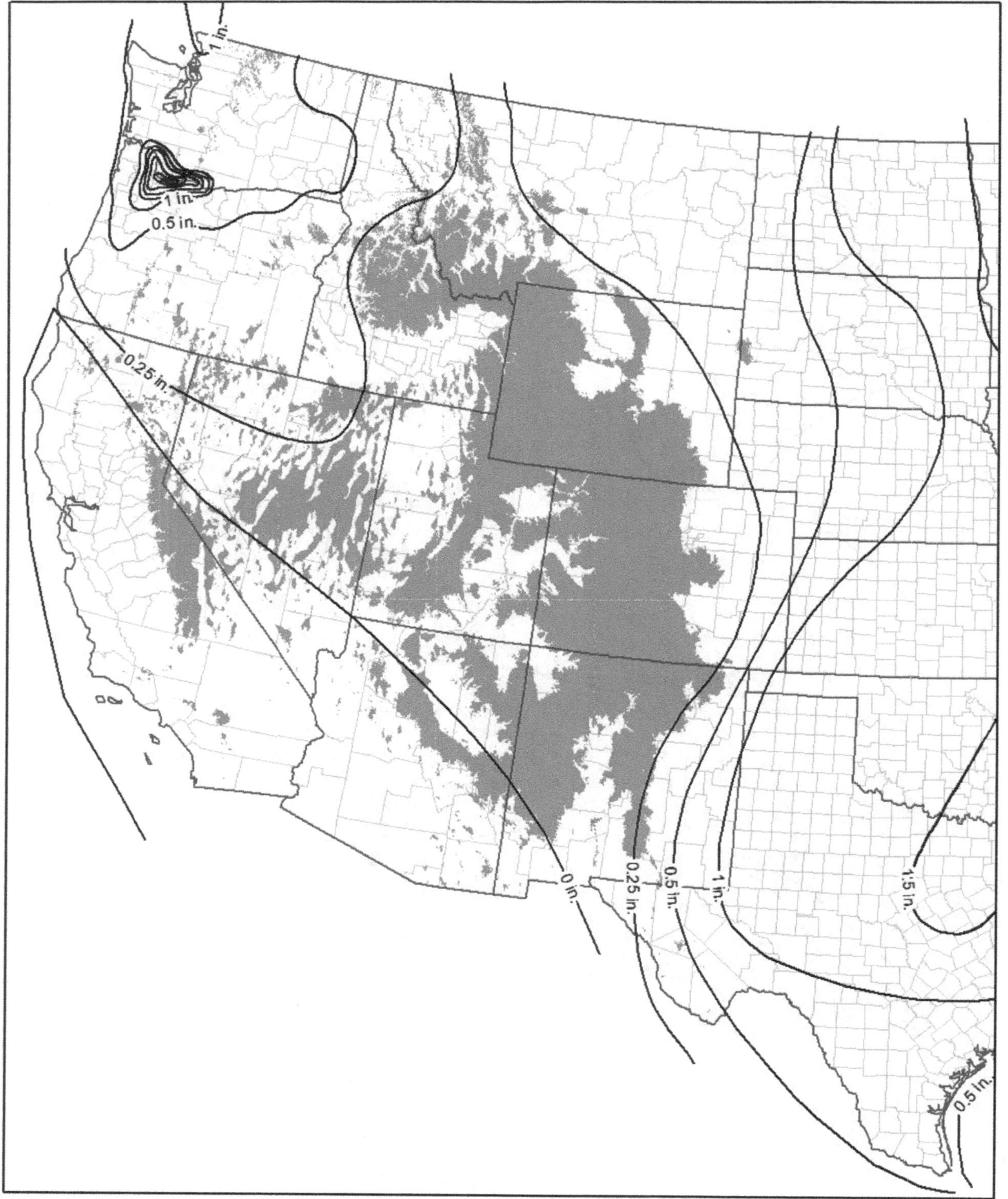

Figure 10.4-2B. Nominal ice thicknesses caused by freezing rain for the contiguous 48 states for Risk Category II (500-year mean recurrence interval) structures.

Notes:

1. Values are nominal ice thicknesses, in inches, at 33 ft (10 m) above ground. Metric conversion: 1 in. = 25.4 mm.
2. Linear interpolation between contours is permitted.

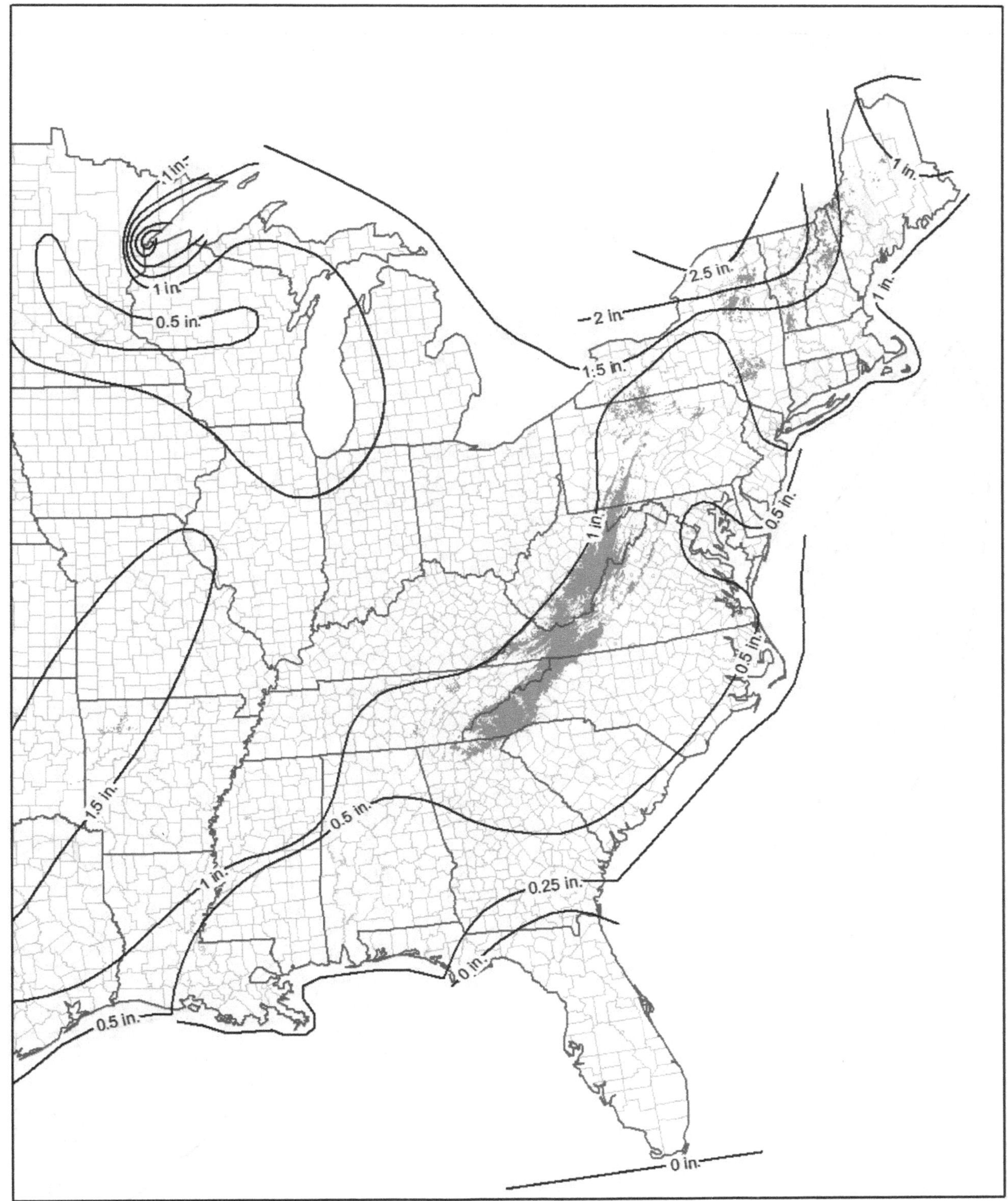

Figure 10.4-2B. ***(Continued)*** **Nominal ice thicknesses caused by freezing rain for the contiguous 48 states for Risk Category II (500-year mean recurrence interval) structures.**

3. Islands, coastal areas, and land boundaries outside the last contour shall use the last contour.
4. The shading indicates special icing regions, elevations above 2,100 ft (640 m) in the east and 6,000 ft (1,829 m) in the west, with sparse weather station data for determining design ice loads.
5. For Columbia River Gorge Region, see Figure 10.4-4B. For Lake Superior Region, see Figure 10.4-5B.
6. Location-specific nominal ice thicknesses shall be permitted to be determined using the ASCE Atmospheric Ice Geodatabase, which can be accessed at the ASCE 7 Hazard Tool (https://asce7hazardtool.online).

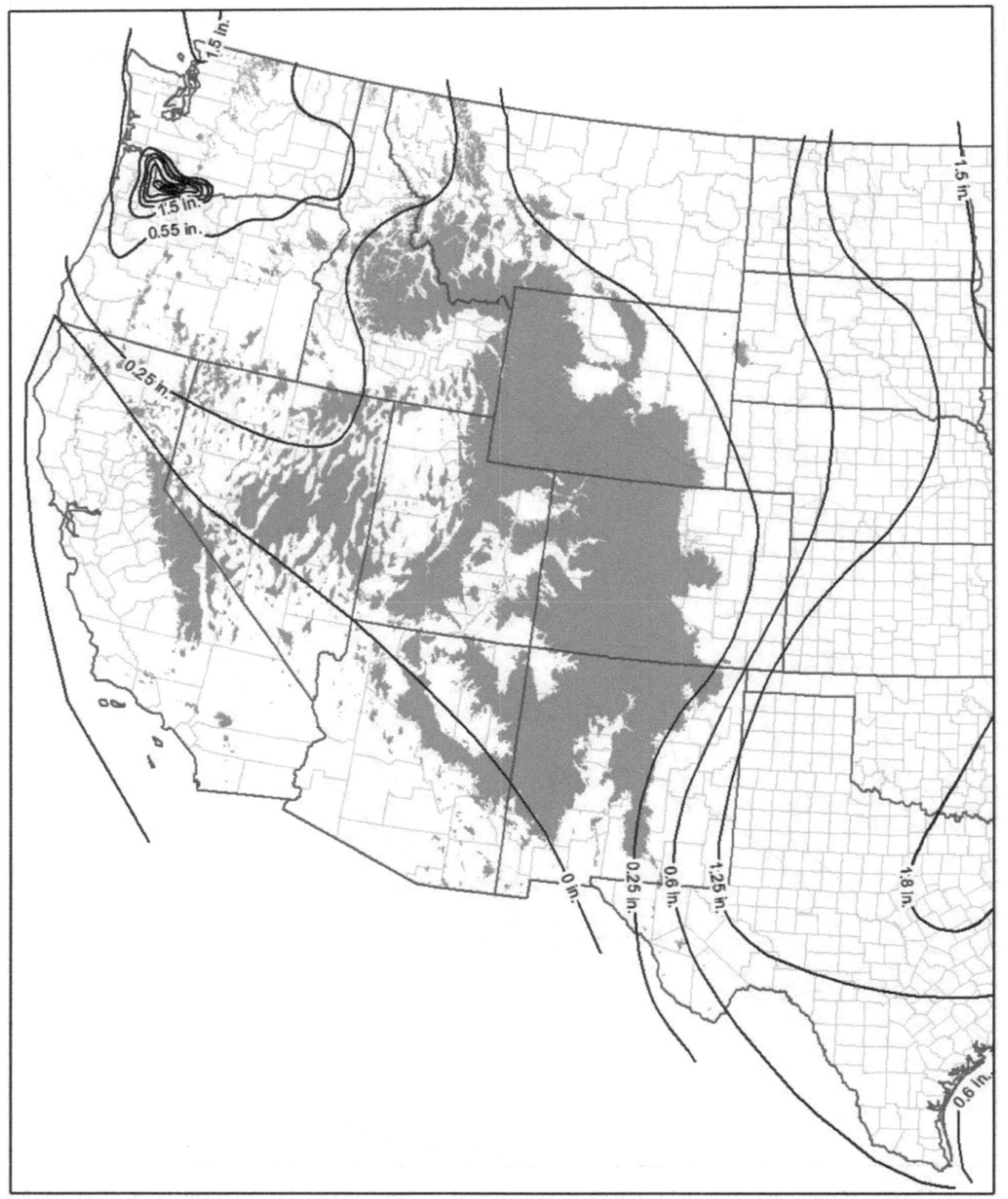

Figure 10.4-2C. Nominal ice thicknesses caused by freezing rain for the contiguous 48 states for Risk Category III (1,000-year mean recurrence interval) structures.

Notes:

1. Values are nominal ice thicknesses, in inches, at 33 ft (10 m) above ground. Metric conversion: 1 in. = 25.4 mm.
2. Linear interpolation between contours is permitted.
3. Islands, coastal areas, and land boundaries outside the last contour shall use the last contour.

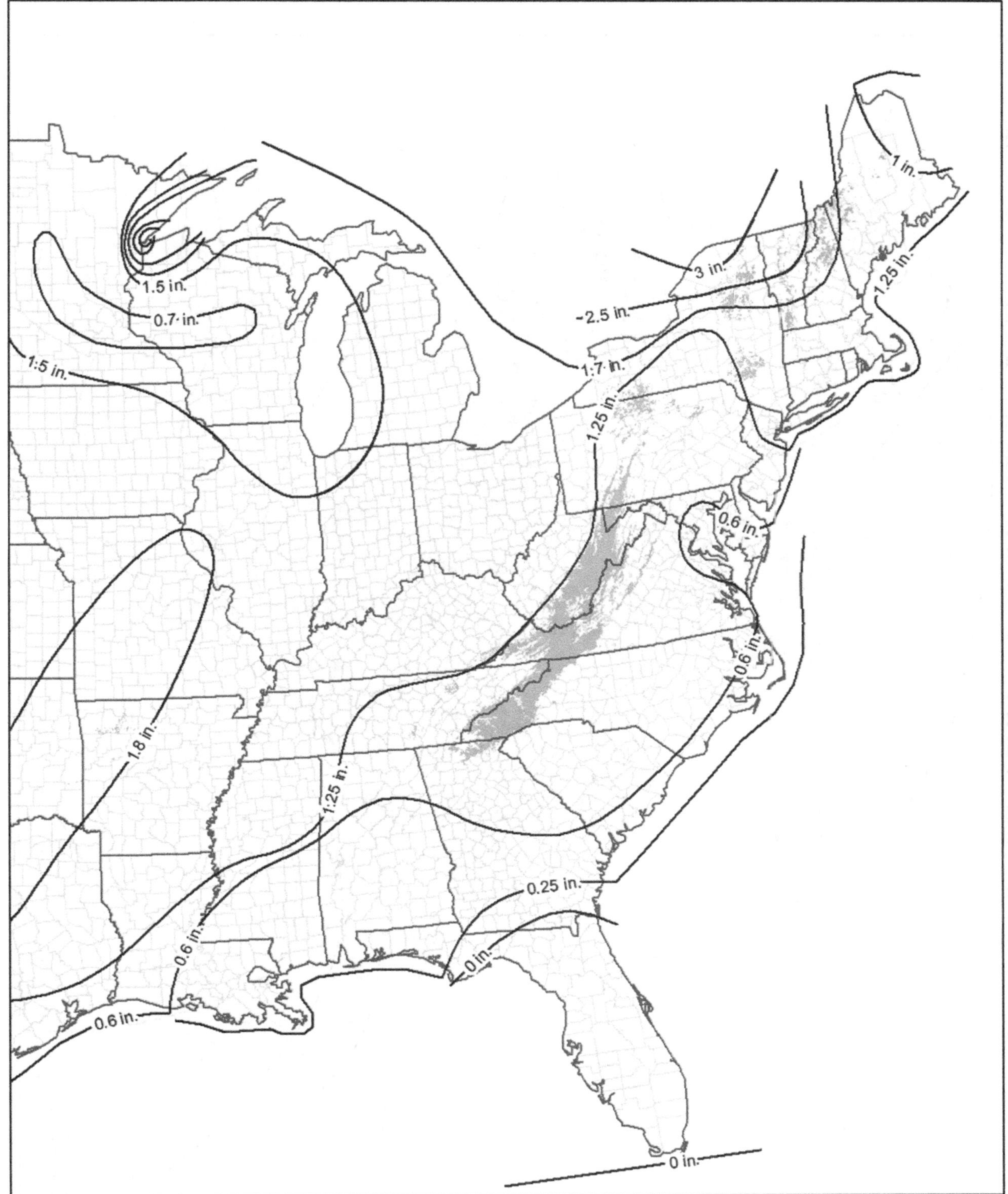

Figure 10.4-2C. ***(Continued)*** **Nominal ice thicknesses caused by freezing rain for the contiguous 48 states for Risk Category III (1,000-year mean recurrence interval) structures.**

4. The shading indicates special icing regions, elevations above 2,100 ft (640 m) in the east and 6,000 ft (1,829 m) in the west, with sparse weather station data for determining design ice loads.
5. For Columbia River Gorge Region, see Figure 10.4-4C. For Lake Superior Region, see Figure 10.4-5C.
6. Location-specific nominal ice thicknesses shall be permitted to be determined using the ASCE Atmospheric Ice Geodatabase, which can be accessed at the ASCE 7 Hazard Tool (https://asce7hazardtool.online).

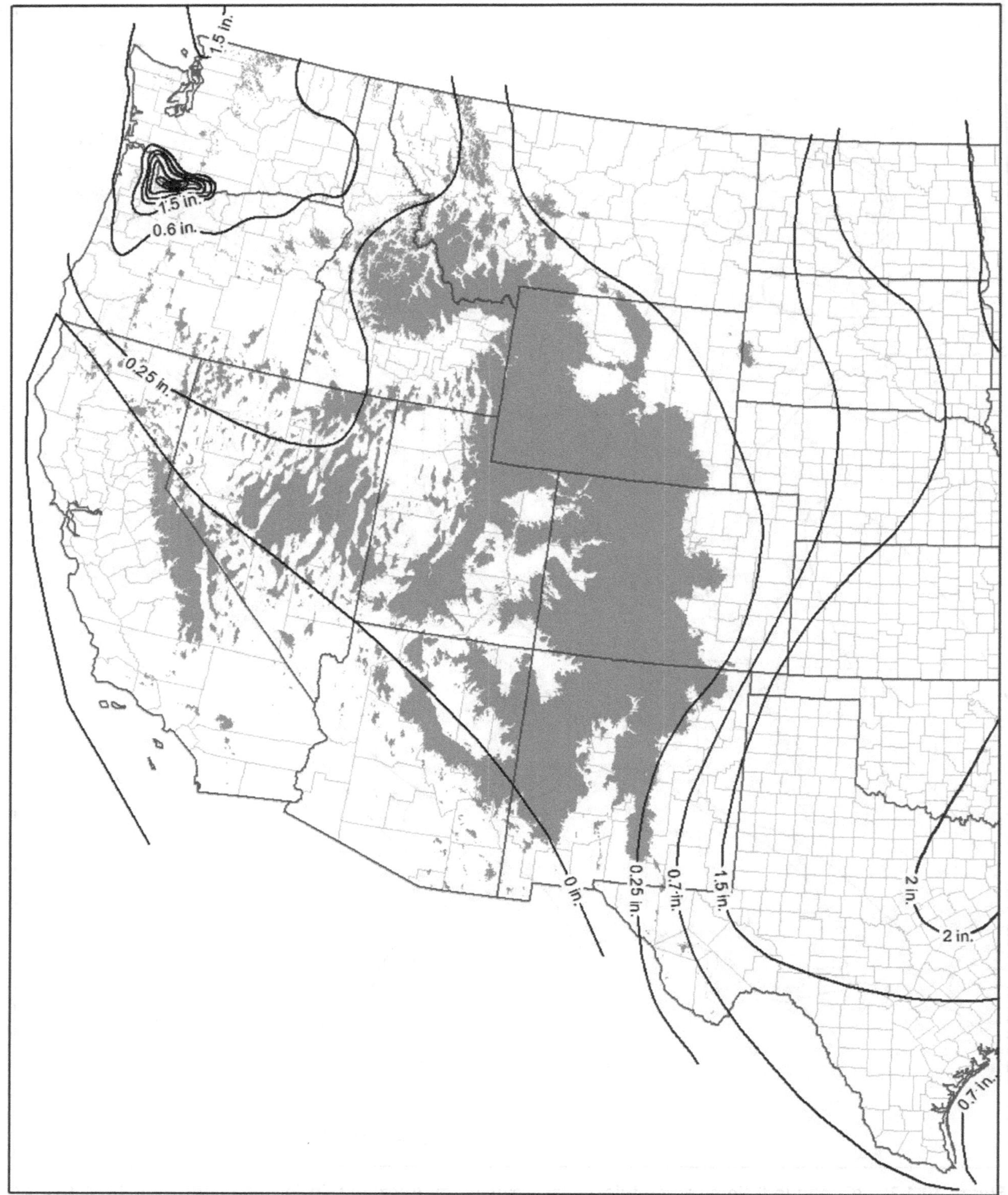

Figure 10.4-2D. Nominal ice thicknesses caused by freezing rain for the contiguous 48 states for Risk Category IV (1,400-year mean recurrence interval) structures.

Notes:

1. Values are nominal ice thicknesses, in inches, at 33 ft (10 m) above ground. Metric conversion: 1 in. = 25.4 mm.
2. Linear interpolation between contours is permitted.
3. Islands, coastal areas, and land boundaries outside the last contour shall use the last contour.

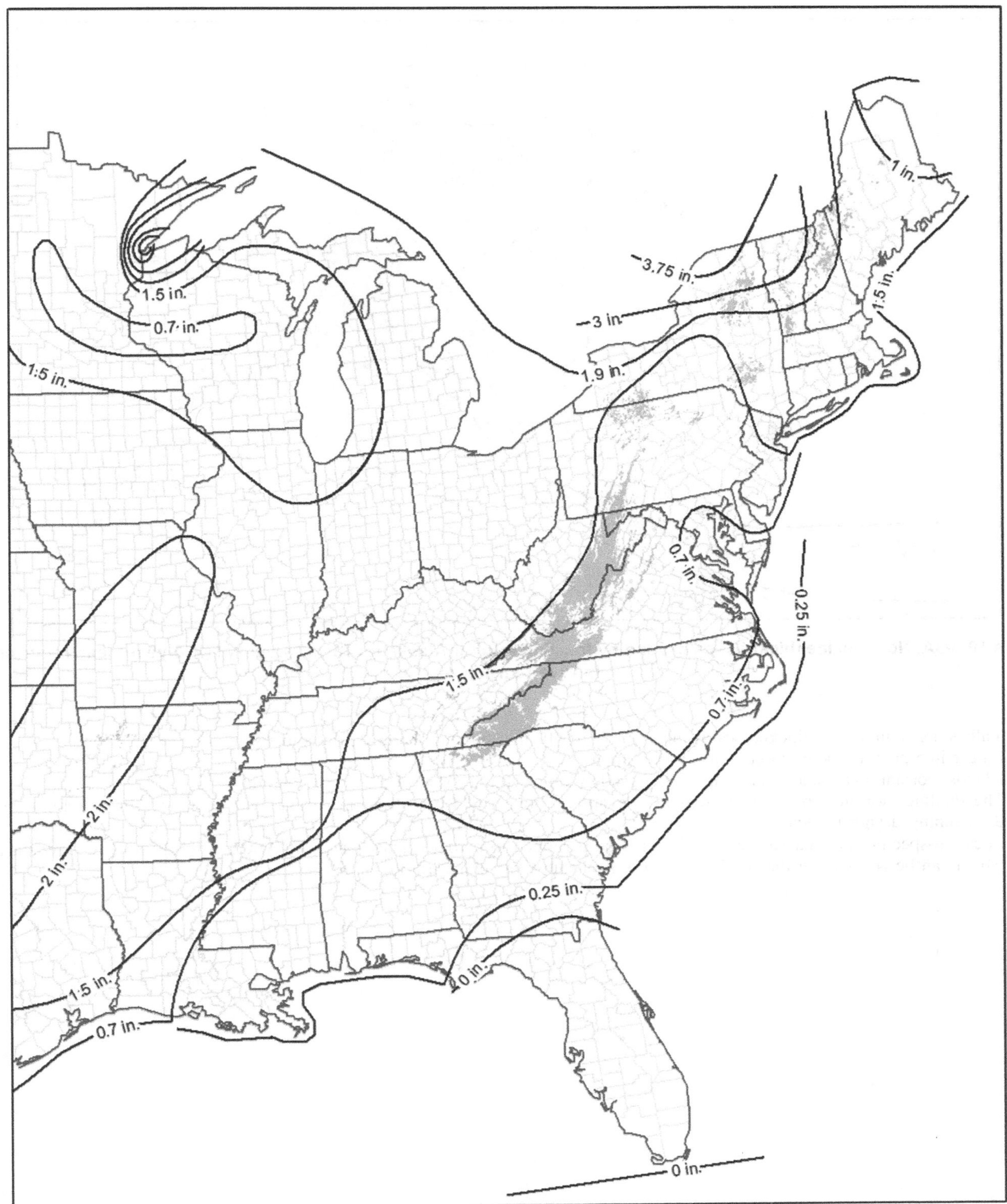

Figure 10.4-2D. ***(Continued)*** **Nominal ice thicknesses caused by freezing rain for the contiguous 48 states for Risk Category IV (1,400-year mean recurrence interval) structures.**

4. The shading indicates special icing regions, elevations above 2,100 ft (640 m) in the east and 6,000 ft (1,829 m) in the west, with sparse weather station data for determining design ice loads.
5. For Columbia River Gorge Region, see Figure 10.4-4D. For Lake Superior Region, see Figure 10.4-5D.
6. Location-specific nominal ice thicknesses shall be permitted to be determined using the ASCE Atmospheric Ice Geodatabase, which can be accessed at the ASCE 7 Hazard Tool (https://asce7hazardtool.online).

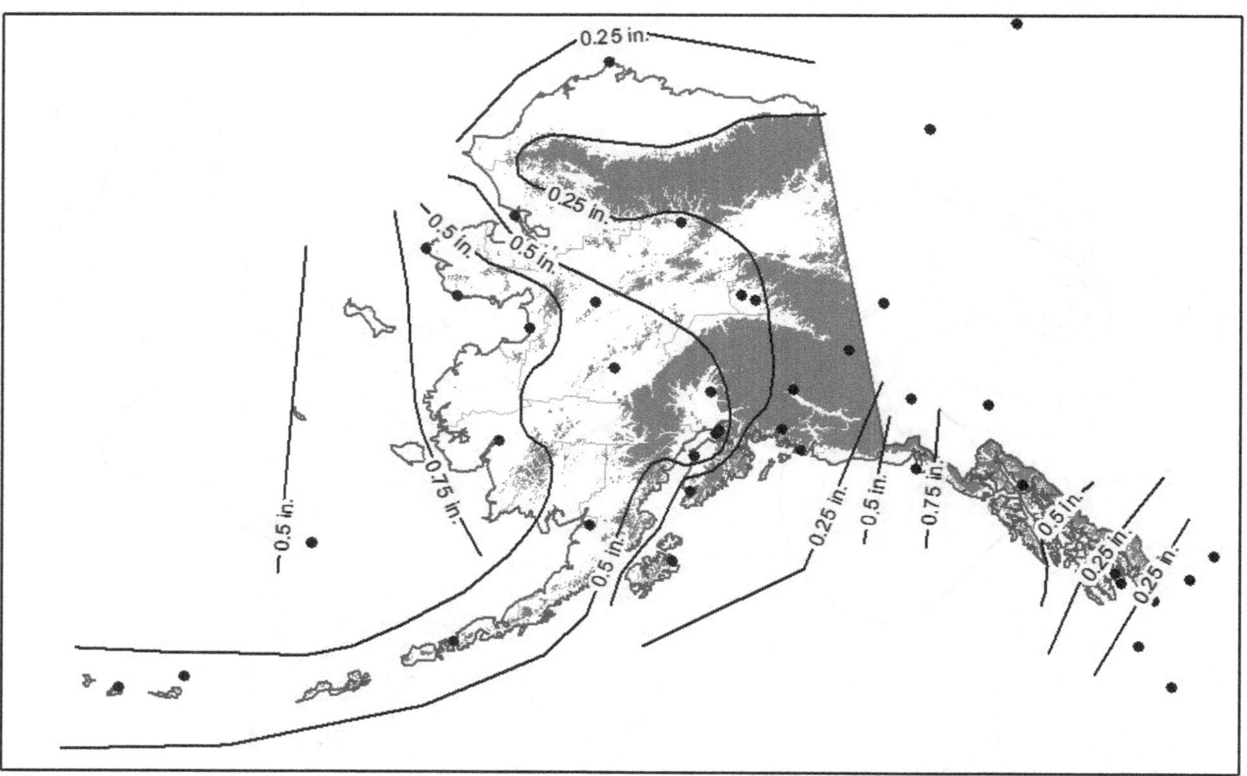

Figure 10.4-3A. Nominal ice thicknesses caused by freezing rain for Alaska for Risk Category I (250-year mean recurrence interval) structures.

Notes:

1. Values are nominal ice thicknesses in inches at 33 ft (10 m) above ground. Metric conversion: 1 in. = 25.4 mm.
2. Linear interpolation between contours is permitted.
3. Islands, coastal areas, and land boundaries outside the last contour shall use the last contour.
4. The shading indicates special icing regions, elevations above 1,600 ft (488 m) in Alaska, with sparse weather station data for determining design ice loads.
5. Location-specific nominal ice thicknesses shall be permitted to be determined using the ASCE Atmospheric Ice Geodatabase, which can be accessed at the ASCE 7 Hazard Tool (https://asce7hazardtool.online/).

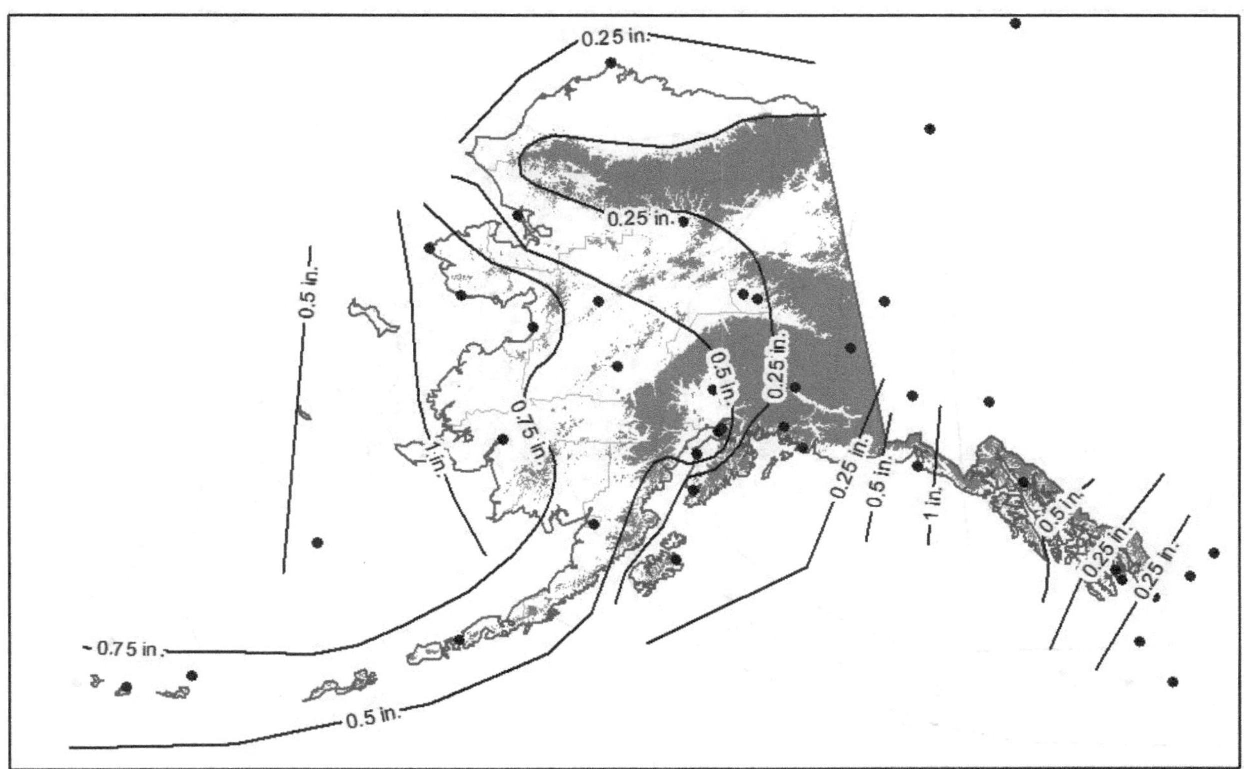

Figure 10.4-3B. Nominal ice thicknesses caused by freezing rain for Alaska for Risk Category II (500-year mean recurrence interval) structures.

Notes:

1. Values are nominal ice thicknesses, in inches, at 33 ft (10 m) above ground. Metric conversion: 1 in. = 25.4 mm.
2. Linear interpolation between contours is permitted.
3. Islands, coastal areas, and land boundaries outside the last contour shall use the last contour.
4. The shading indicates special icing regions, elevations above 1,600 ft (488 m) in Alaska, with sparse weather station data for determining design ice loads.
5. Location-specific nominal ice thicknesses shall be permitted to be determined using the ASCE Atmospheric Ice Geodatabase, which can be accessed at the ASCE 7 Hazard Tool (https://asce7hazardtool.online/).

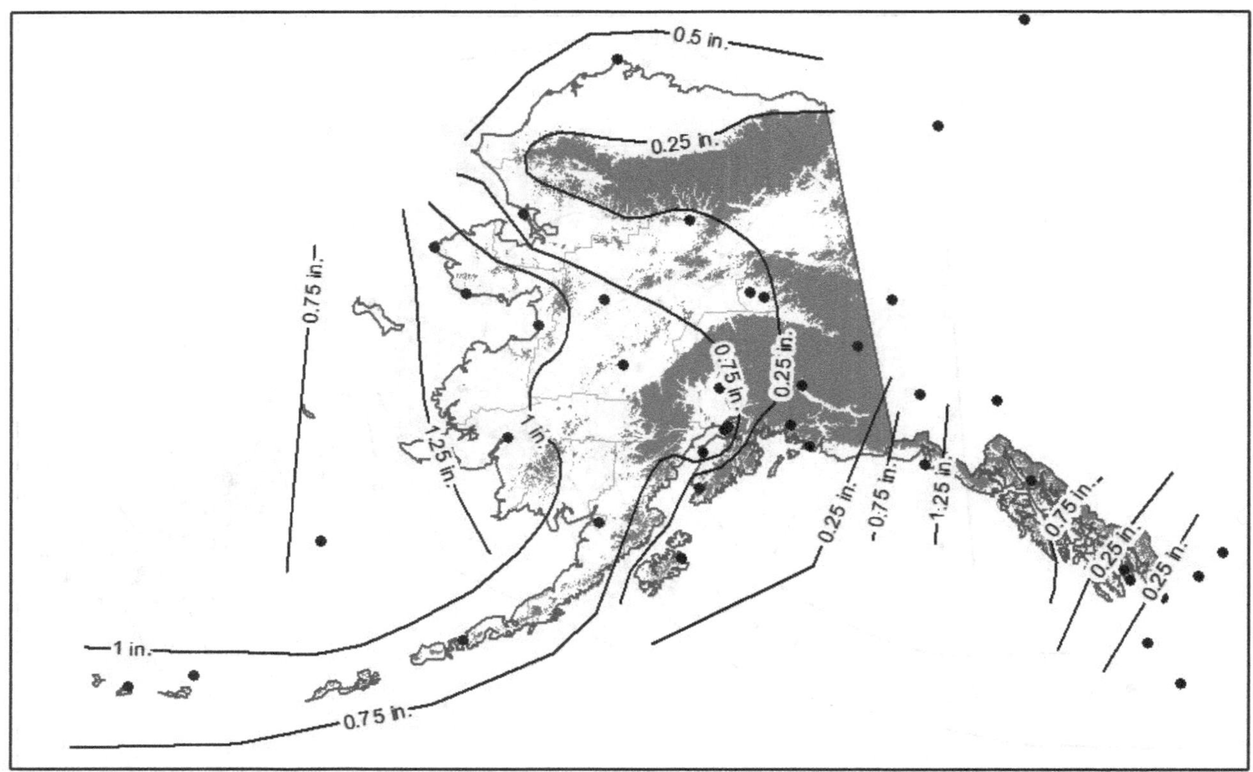

Figure 10.4-3C. Nominal ice thicknesses caused by freezing rain for Alaska for Risk Category III (1,000-year mean recurrence interval) structures.

Notes:

1. Values are nominal ice thicknesses, in inches, at 33 ft (10 m) above ground. Metric conversion: 1 in. = 25.4 mm.
2. Linear interpolation between contours is permitted.
3. Islands, coastal areas, and land boundaries outside the last contour shall use the last contour.
4. The shading indicates special icing regions, elevations above 1,600 ft (488 m) in Alaska, with sparse weather station data for determining design ice loads.
5. Location-specific nominal ice thicknesses shall be permitted to be determined using the ASCE Atmospheric Ice Geodatabase, which can be accessed at the ASCE 7 Hazard Tool (https://asce7hazardtool.online/).

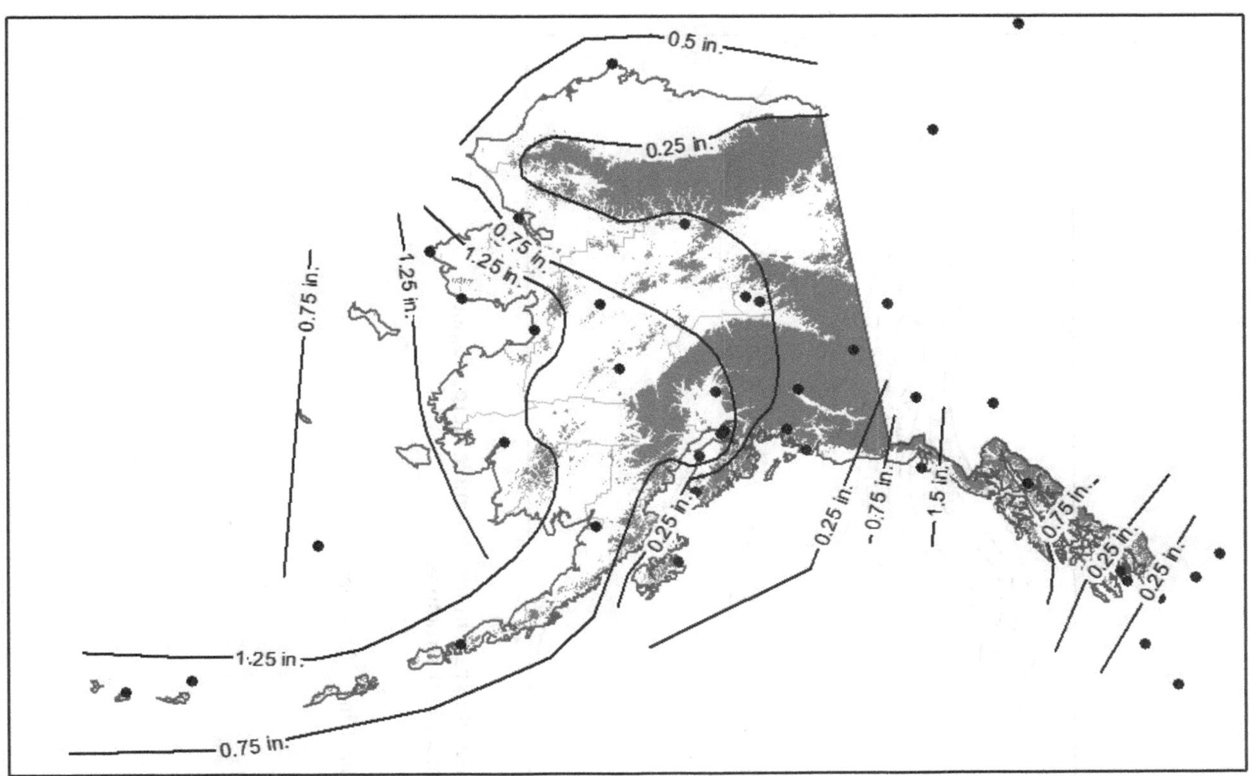

Figure 10.4-3D. Nominal ice thicknesses caused by freezing rain for Alaska for Risk Category IV (1,400-year mean recurrence interval) structures.

Notes:

1. Values are nominal ice thicknesses, in inches, at 33 ft (10 m) above ground. Metric conversion: 1 in. = 25.4 mm.
2. Linear interpolation between contours is permitted.
3. Islands, coastal areas, and land boundaries outside the last contour shall use the last contour
4. The shading indicates special icing regions, elevations above 1,600 ft (488 m) in Alaska, with sparse weather station data for determining design ice loads.
5 Location-specific nominal ice thicknesses shall be permitted to be determined using the ASCE Atmospheric Ice Geodatabase, which can be accessed at the ASCE 7 Hazard Tool (https://asce7hazardtool.online/).

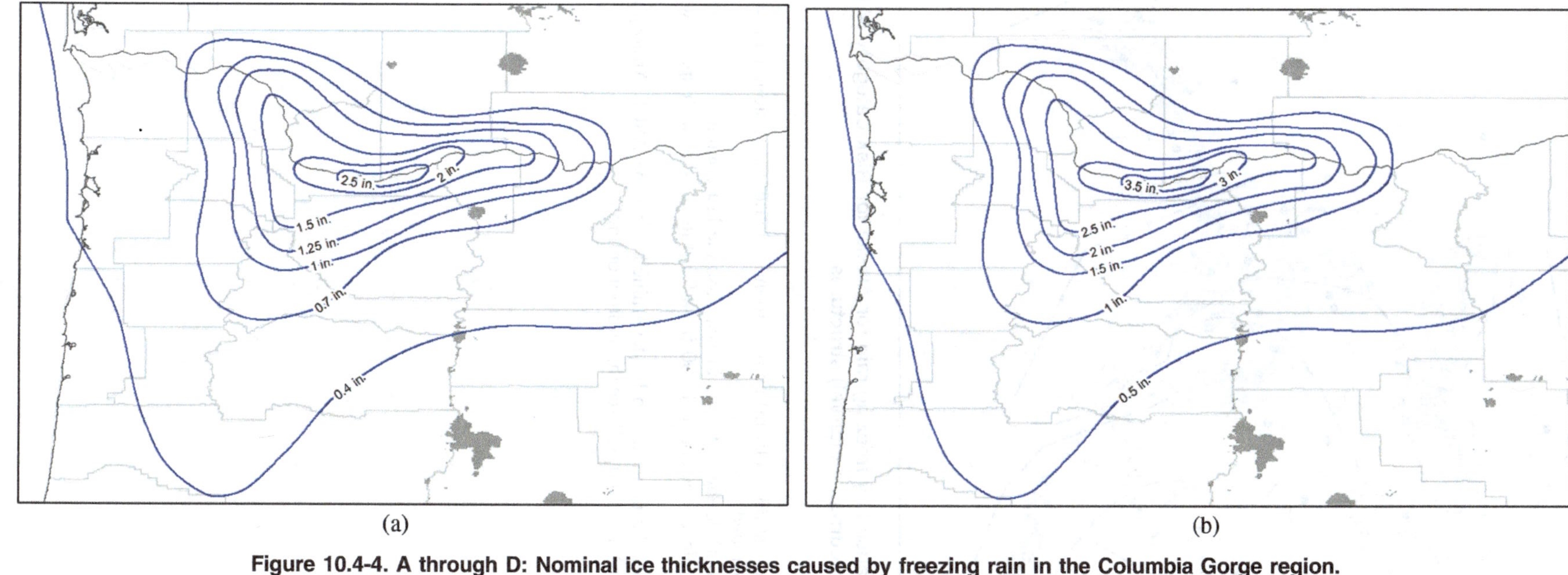

Figure 10.4-4. A through D: Nominal ice thicknesses caused by freezing rain in the Columbia Gorge region. (a) Risk Category I structures. (b) Risk Category II structures.

Notes:

1. Values are nominal ice thicknesses, in inches, at 33 ft (10 m) above ground. Metric conversion: 1 in. = 25.4 mm.
2. Linear interpolation between contours is permitted.
3. Islands, coastal areas, and land boundaries outside the last contour shall use the last contour.

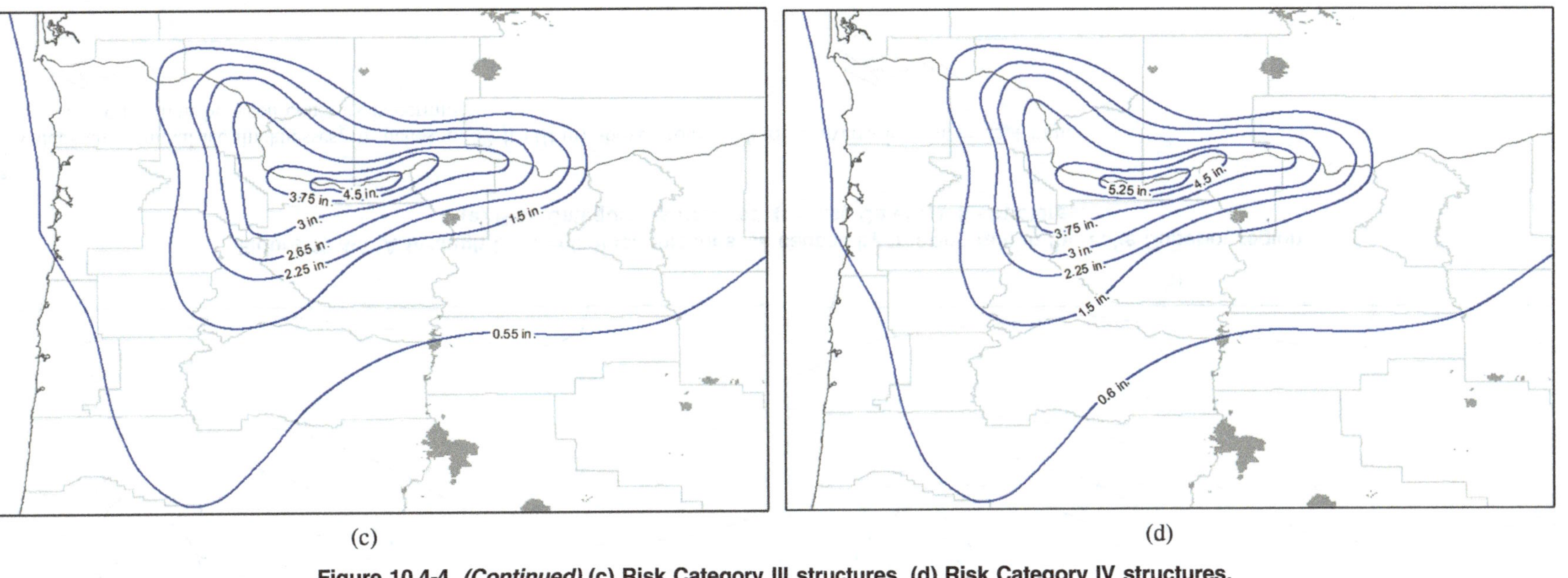

Figure 10.4-4. ***(Continued)*** **(c) Risk Category III structures. (d) Risk Category IV structures.**

4. The shading indicates special icing regions, with elevations above 6,000 ft (1,829 m), with sparse weather station data for determining design ice loads.
5. Location-specific nominal ice thicknesses shall be permitted to be determined using the ASCE Atmospheric Ice Geodatabase, which can be accessed at the ASCE 7 Hazard Tool (https://asce7hazardtool.online/).

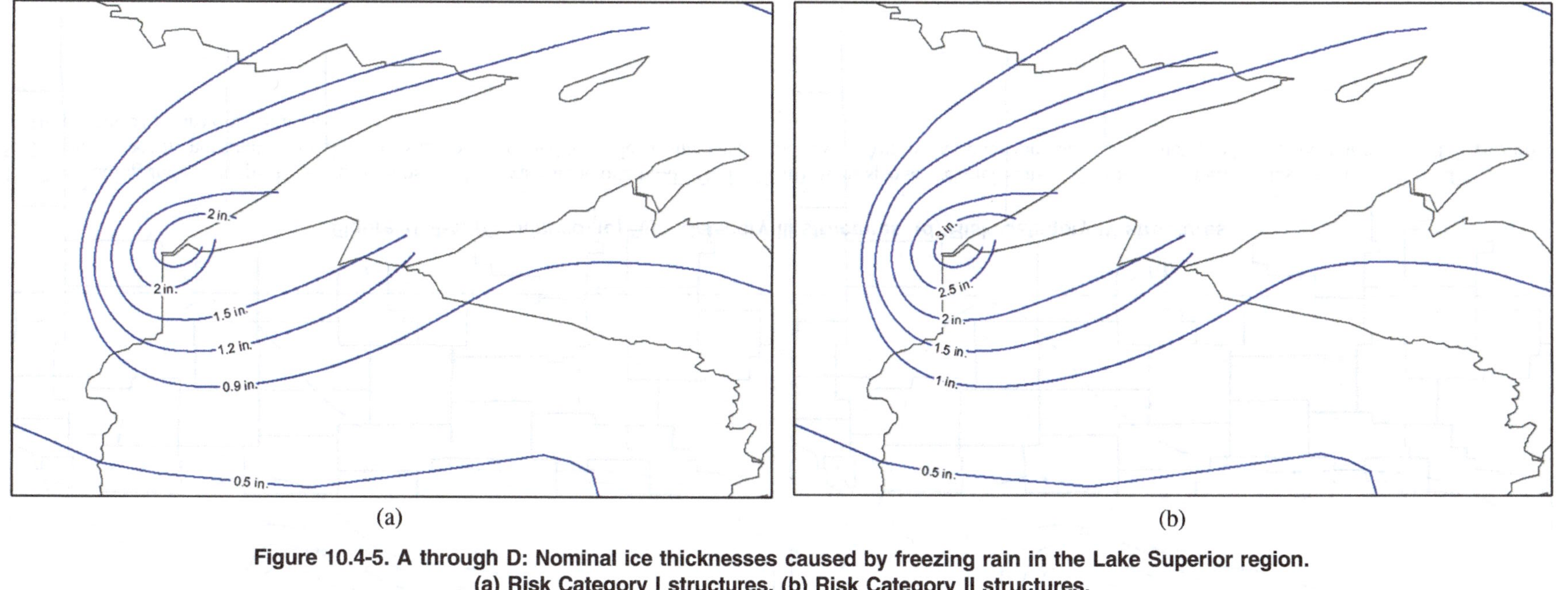

Figure 10.4-5. A through D: Nominal ice thicknesses caused by freezing rain in the Lake Superior region. (a) Risk Category I structures. (b) Risk Category II structures.

Notes:

1. Values are nominal ice thicknesses, in inches, at 33 ft (10 m) above ground. Metric conversion: 1 in. = 25.4 mm.
2. Linear interpolation between contours is permitted.

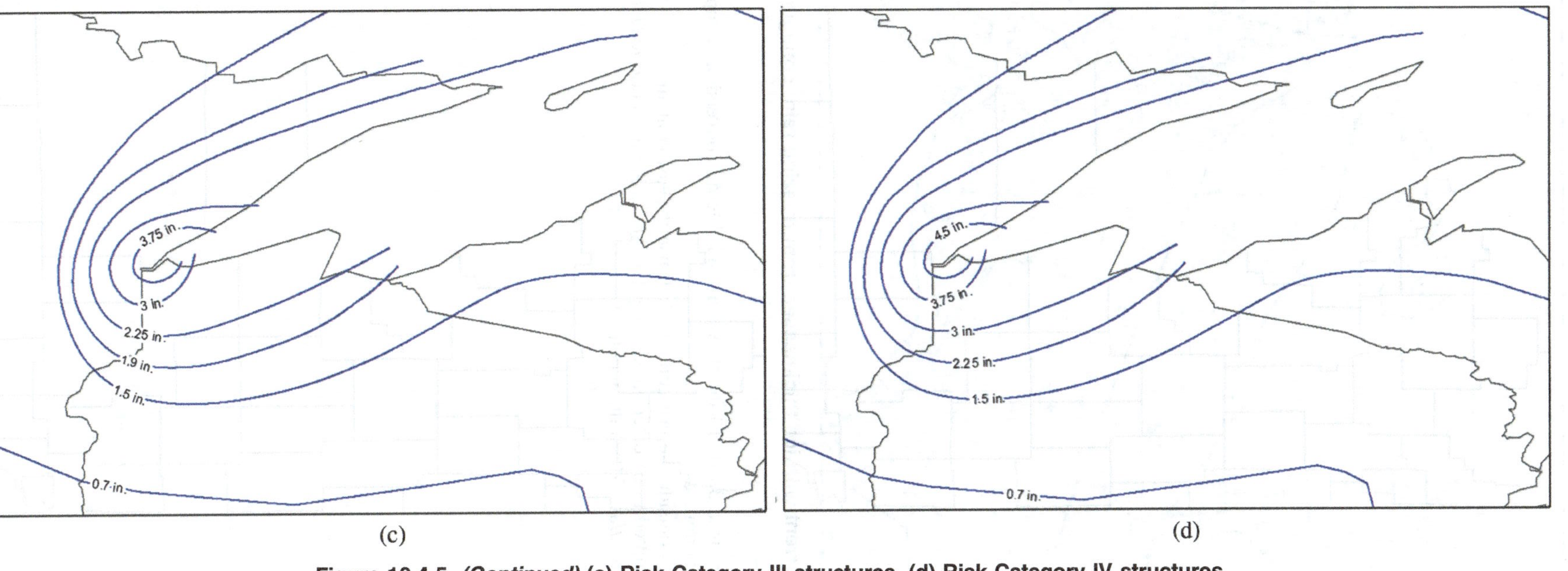

Figure 10.4-5. ***(Continued)*** **(c) Risk Category III structures. (d) Risk Category IV structures.**

3. Islands, coastal areas, and land boundaries outside the last contour shall use the last contour
4. Location-specific nominal ice thicknesses shall be permitted to be determined using the ASCE Atmospheric Ice Geodatabase, which can be accessed at the ASCE 7 Hazard Tool (https://asce7hazardtool.online/).

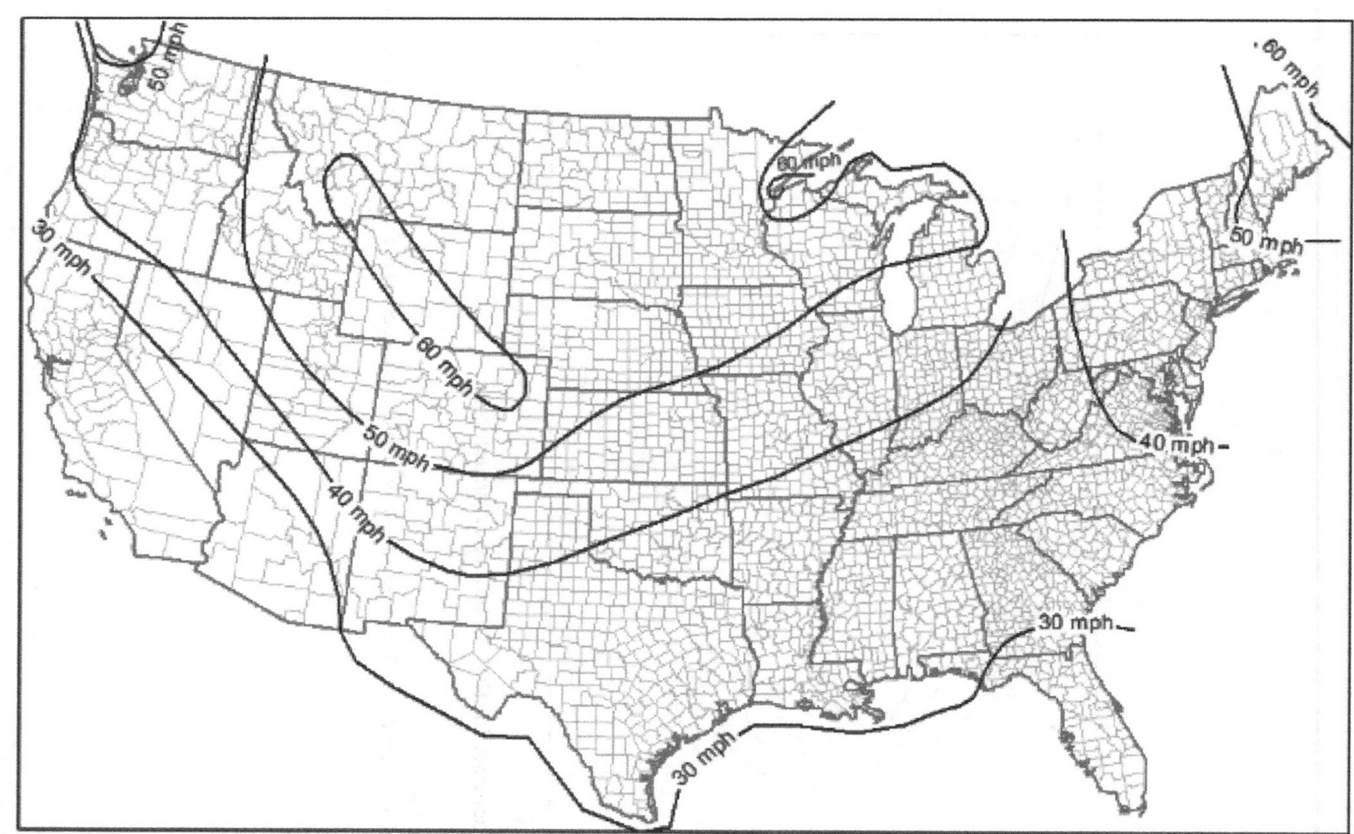

Figure 10.5-1. Gust speeds concurrent with ice thicknesses from freezing rain; contiguous 48 states.

Notes:

1. Values are 3-s gust speeds in miles per hour at 33 ft (10 m) above ground. Metric conversion: 1 mph = 0.45 m/s.
2. Linear interpolation between contours is permitted.
3. Islands, coastal areas, and land boundaries outside the last contour shall use the last contour.
4. Location-specific gust speed shall be permitted to be determined using the ASCE Atmospheric Ice Geodatabase, which can be accessed at the ASCE 7 Hazard Tool (https://asce7hazardtool.online).

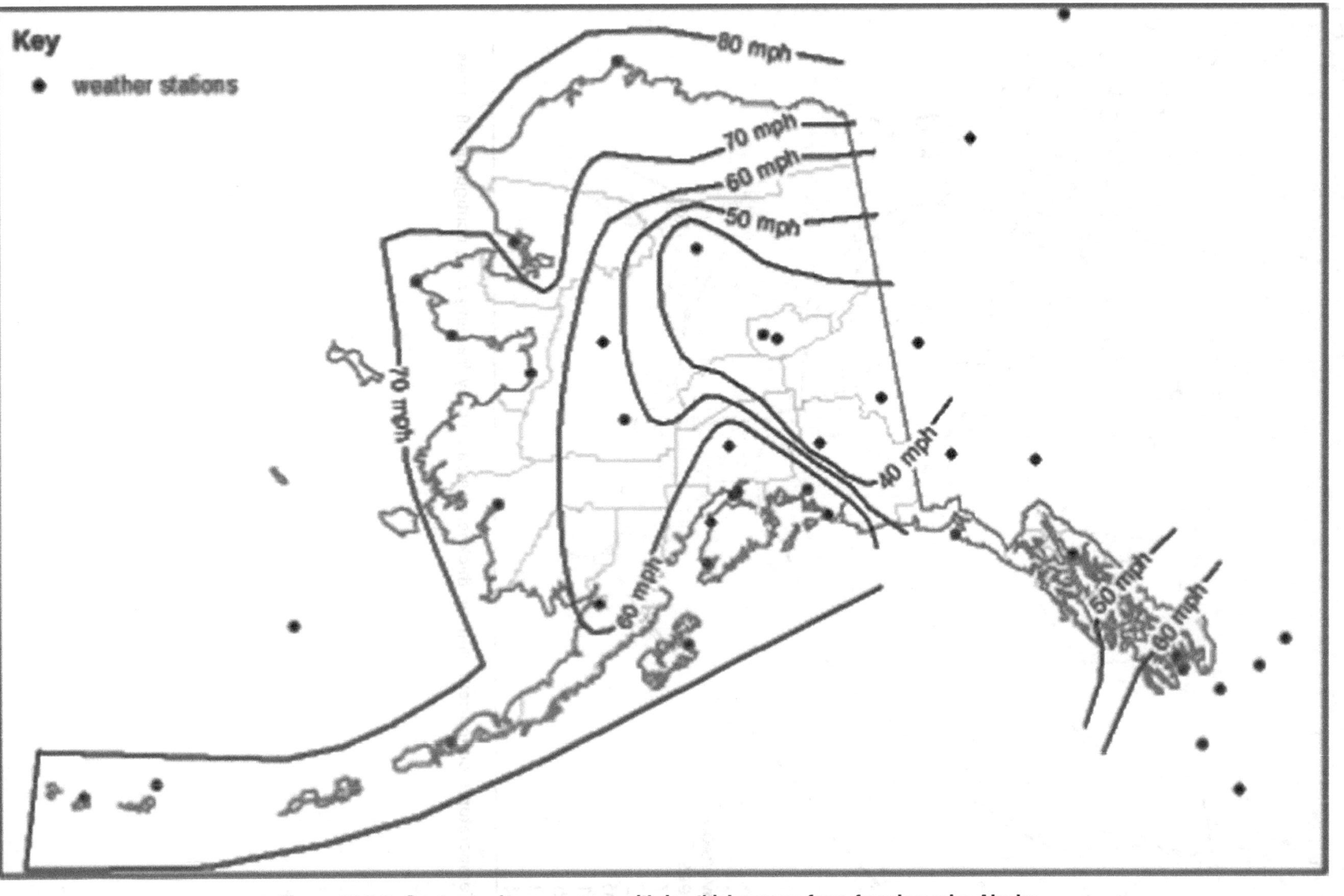

Figure 10.5-2. Gust speeds concurrent with ice thicknesses from freezing rain; Alaska.

Notes:

1. Values are 3s gust speeds in miles per hour at 33 ft (10 m) above ground. Metric conversion: 1 mph = 0.45 m/s.
2. Linear interpolation between contours is permitted.
3. Islands, coastal areas, and land boundaries outside the last contour shall use the last contour.
4. Location-specific gust speed shall be permitted to be determined using the ASCE Atmospheric Ice Geodatabase, which can be accessed at the ASCE 7 Hazard Tool (https://asce7hazardtool.online).

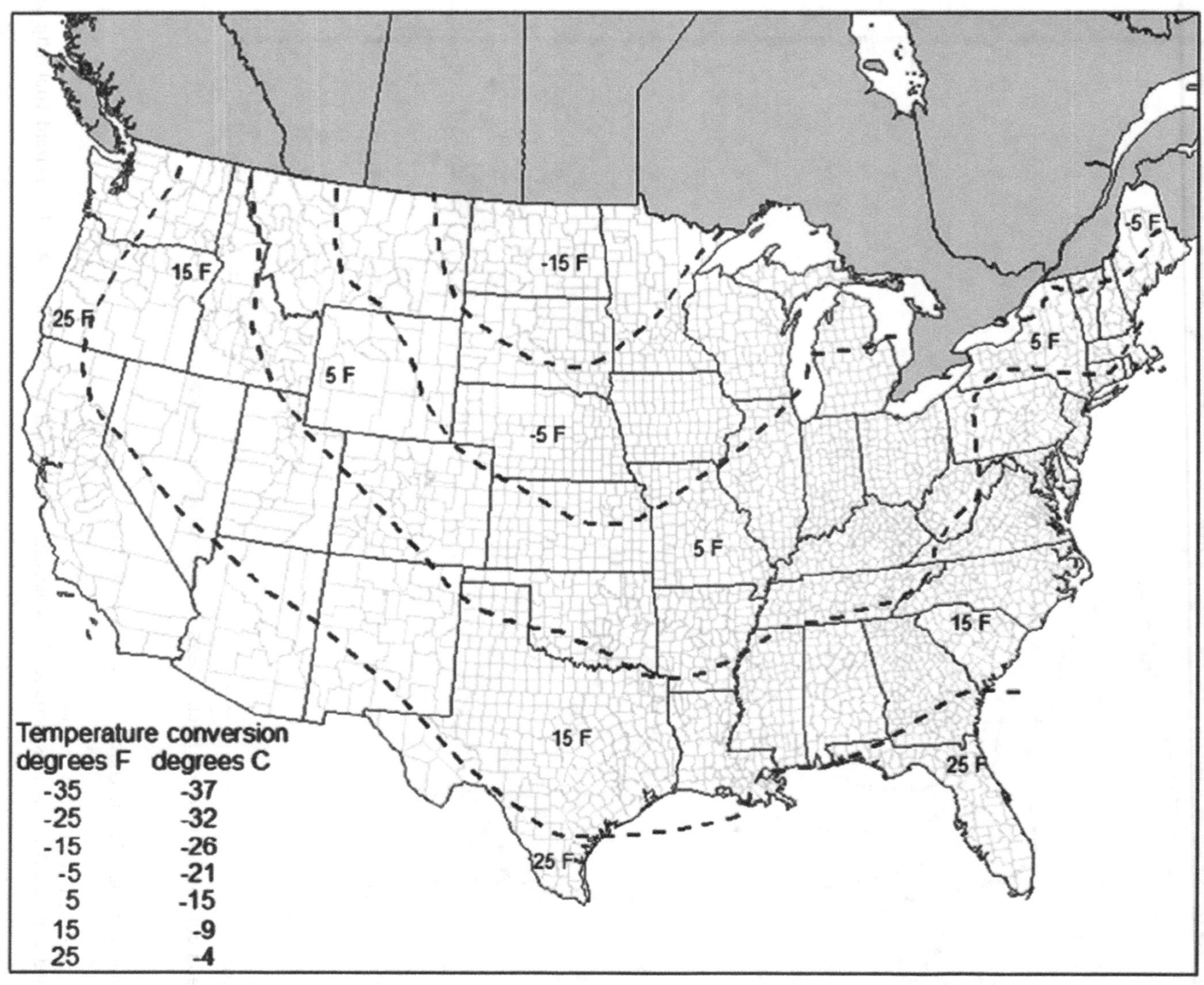

Figure 10.6-1. Temperatures concurrent with ice thicknesses caused by freezing rain; contiguous 48 states.

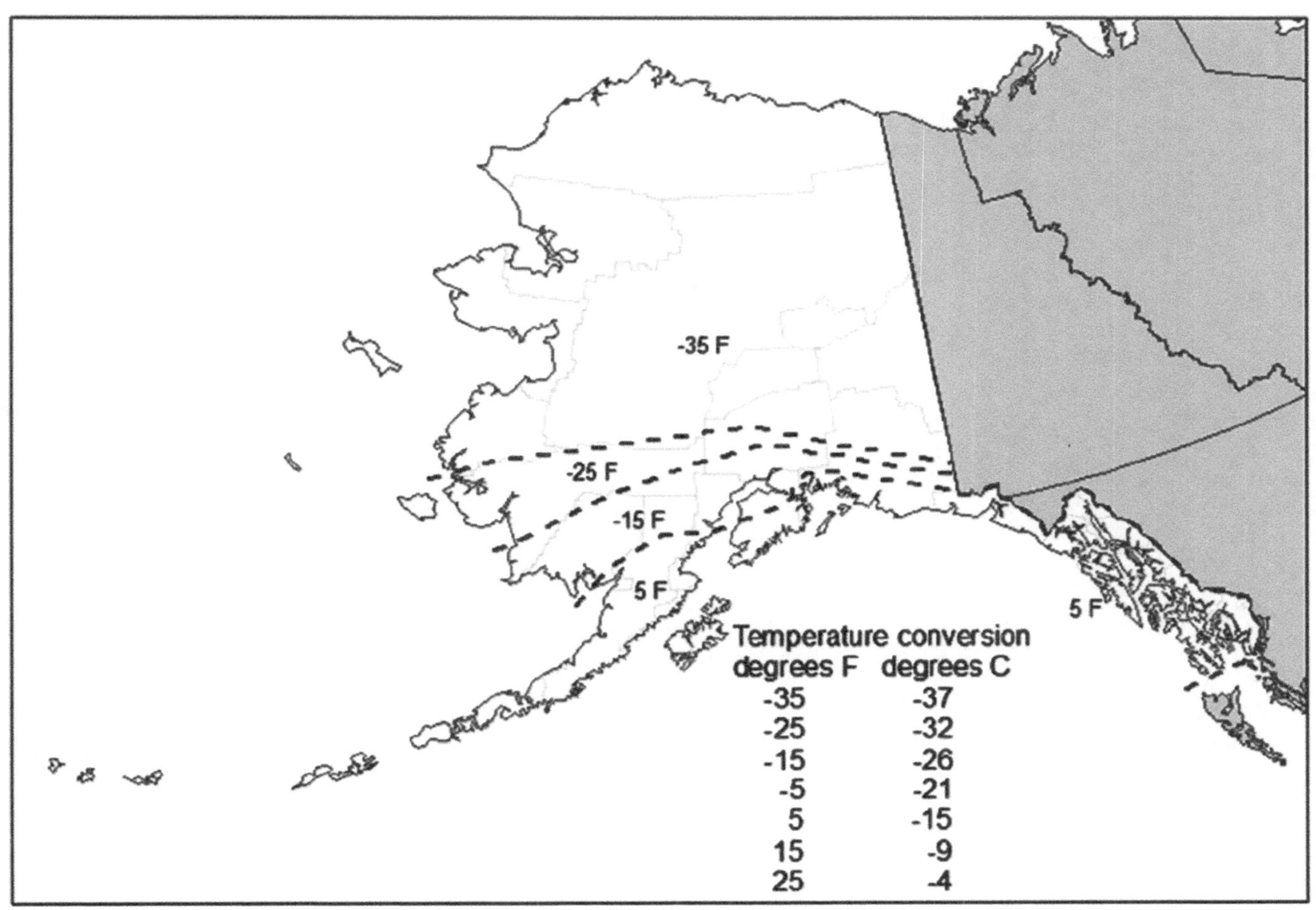

Figure 10.6-2. Temperatures concurrent with ice thicknesses caused by freezing rain; Alaska.

CHAPTER 11
SEISMIC DESIGN CRITERIA

11.1 GENERAL

11.1.1 Purpose Chapter 11 presents criteria for the design and construction of buildings and other structures subject to earthquake ground motions. The specified earthquake loads are based on postelastic energy dissipation in the structure. Because of this fact, the requirements for design, detailing, and construction shall be satisfied, even for structures and members for which load combinations that do not include earthquake loads indicate larger demands than combinations that include earthquake loads.

11.1.2 Scope Every structure and portion thereof, including nonstructural components, shall be designed and constructed to resist the effects of earthquake motions as prescribed by the seismic requirements of this standard. Certain nonbuilding structures, as described in Chapter 15, are also within the scope and shall be designed and constructed in accordance with the requirements of Chapter 15. The following structures are exempt from the seismic requirements of this standard:

1. Detached one- and two-family dwellings that are located where the mapped, short period, spectral response acceleration parameter, S_S, is less than 0.4 or where the seismic design category determined in accordance with Section 11.6 is A, B, or C.
2. Detached one- and two-family wood-frame dwellings not included in Exemption 1 with not more than two stories above grade plane, satisfying the limitations of and constructed in accordance with the IRC.
3. Agricultural storage structures that are intended only for incidental human occupancy.
4. Nonbuilding structures that require special consideration of their response characteristics and environment that are not addressed in Chapter 15 and for which other regulations provide seismic criteria, such as vehicular bridges, electrical transmission towers, hydraulic structures, buried utility lines and their appurtenances, and nuclear reactors.
5. Piers and wharves that are not accessible to the general public.

11.1.3 Applicability Structures and their nonstructural components shall be designed and constructed in accordance with the requirements of the following chapters based on the type of structure or component:

Buildings: Chapter 12
Nonbuilding structures: Chapter 15
Nonstructural components: Chapter 13
Seismically isolated structures: Chapter 17
Structures with damping systems: Chapter 18

Buildings whose purpose is to enclose equipment or machinery and whose occupants are engaged in maintenance or monitoring of that equipment, machinery, or their associated processes shall be permitted to be classified as nonbuilding structures designed and detailed in accordance with Section 15.5 of this standard.

11.1.4 Alternate Materials and Methods of Construction Alternate materials and methods of construction to those prescribed in the seismic requirements of this standard shall not be used unless approved by the Authority Having Jurisdiction. Substantiating evidence shall be submitted demonstrating that the proposed alternate will be at least equal in strength, durability, and seismic resistance for the purpose intended.

11.1.5 Quality Assurance Quality assurance for seismic force-resisting systems and other designated seismic systems defined in Section 13.2.2 shall be provided in accordance with the requirements of the Authority Having Jurisdiction.

Where the Authority Having Jurisdiction has not adopted quality assurance requirements, or where the adopted requirements are not applicable to the seismic force-resisting system or designated seismic systems as described in Section 13.2.2, the registered design professional in responsible charge of designing the seismic force-resisting system or other designated seismic systems shall submit a quality assurance plan to the Authority Having Jurisdiction for approval. The quality assurance plan shall specify the quality assurance program elements to be implemented.

11.2 DEFINITIONS

The following definitions apply only to the seismic provisions of Chapters 11 through 22 of this standard.

ACTIVE FAULT: A fault determined to be active by the Authority Having Jurisdiction from properly substantiated data (e.g., most recent mapping of active faults by the US Geological Survey).

ADDITION: An increase in building area, aggregate floor area, height, or number of stories of a structure.

ALTERATION: Any construction or renovation to an existing structure other than an addition.

APPENDAGE: An architectural component such as a canopy, marquee, ornamental balcony, or statuary.

APPROVAL: The written acceptance by the Authority Having Jurisdiction of documentation that establishes the qualification of a material, system, component, procedure, or person to fulfill the requirements of this standard for the intended use.

ATTACHMENTS: Means by which nonstructural components or supports of nonstructural components are secured or connected to the seismic force-resisting system of the structure. Such attachments include anchor bolts, welded connections, and mechanical fasteners.

BASE: The level at which the horizontal seismic ground motions are considered to be imparted to the structure.

BASE SHEAR: Total design lateral force or shear at the base.

BOUNDARY ELEMENTS: Portions along wall and diaphragm edges for transferring or resisting forces. Boundary elements include chords and collectors at diaphragm and shear wall perimeters, edges of openings, discontinuities, and reentrant corners.

BUILDING: Any structure whose intended use includes shelter of human occupants.

CANTILEVERED COLUMN SYSTEM: A seismic force-resisting system in which lateral forces are resisted entirely by columns acting as cantilevers from the base.

CHARACTERISTIC EARTHQUAKE: An earthquake assessed for an active fault having a magnitude equal to the best estimate of the maximum magnitude capable of occurring on the fault but not less than the largest magnitude that has occurred historically on the fault.

COLLECTOR (DRAG STRUT, TIE, DIAPHRAGM STRUT): A diaphragm or shear wall boundary element parallel to the applied load that collects and transfers diaphragm shear forces to the vertical elements of the seismic force-resisting system or distributes forces within the diaphragm or shear wall.

COMPONENT: A part of an architectural, electrical, or mechanical system.

Component, Flexible: Nonstructural component that has a fundamental period greater than 0.06 s.

Component, Nonstructural: A part of an architectural, mechanical, or electrical system within or without a building or nonbuilding structure.

Component, Rigid: Nonstructural component that has a fundamental period less than or equal to 0.06 s.

Component, Rugged: A nonstructural component that has been shown to consistently function after design earthquake level or greater seismic events (based on past earthquake experience data or past seismic testing) when adequately anchored or supported. The classification of a nonstructural component as rugged shall be based on a comparison of the specific component with components of similar strength and stiffness. Common examples of rugged components are AC motors, compressors, and base-mounted horizontal pumps.

CONCRETE:

Plain Concrete: Concrete that is either unreinforced or contains less reinforcement than the minimum amount specified in ACI 318 for reinforced concrete.

Reinforced Concrete: Prestressed or nonprestressed concrete reinforced with no less reinforcement than the minimum amount required by ACI 318 and designed on the assumption that the two materials act together in resisting forces.

CONSTRUCTION DOCUMENTS: The written, graphic, electronic, and pictorial documents describing the design, locations, and physical characteristics of the project required to verify compliance with this standard.

COUPLING BEAM: A beam that is used to connect adjacent concrete wall elements to make them act together as a unit to resist lateral loads.

DEFORMABILITY: The ratio of the ultimate deformation to the limit deformation.

High-Deformability Element: An element whose deformability is not less than 3.5 where subjected to four fully reversed cycles at the limit deformation.

Limited-Deformability Element: An element that is neither a low-deformability nor a high-deformability element.

Low-Deformability Element: An element whose deformability is 1.5 or less.

DEFORMATION:

Limit Deformation: Two times the initial deformation that occurs at a load equal to 40% of the maximum strength.

Ultimate Deformation: The deformation at which failure occurs and that shall be deemed to occur if the sustainable load reduces to 80% or less of the maximum strength.

DESIGN EARTHQUAKE: The earthquake effects that are two-thirds of the corresponding risk-targeted maximum considered earthquake (MCE_R).

DESIGN EARTHQUAKE DISPLACEMENT: *See* DISPLACEMENT AND DRIFT. *See* DISPLACEMENT AND DRIFT. *See* **DESIGN STORY DRIFT**.

DESIGN EARTHQUAKE GROUND MOTION: The earthquake ground motions that are two-thirds of the corresponding MCE_R ground motions.

DESIGNATED SEISMIC SYSTEMS: Those nonstructural components that require design in accordance with Chapter 13 and for which the component Importance Factor, I_p, is greater than 1.0.

DIAPHRAGM: Roof, floor, or other membrane or bracing system acting to transfer the lateral forces to the vertical resisting elements.

Flexure-Controlled Diaphragm: Diaphragm with a flexural yielding mechanism, which limits the maximum forces that develop in the diaphragm, and a design shear strength or factored nominal shear capacity greater than the shear corresponding to the nominal flexural strength.

Shear-Controlled Diaphragm: Diaphragm that does not meet the requirements of a flexure-controlled diaphragm.

Transfer Forces, Diaphragm: Forces that occur in a diaphragm caused by transfer of seismic forces from the vertical seismic force-resisting elements above the diaphragm to other vertical seismic force-resisting elements below the diaphragm because of offsets in the placement of the vertical elements or changes in relative lateral stiffnesses of the vertical elements.

Vertical Diaphragm: *See* WALL: Shear Wall.

DIAPHRAGM BOUNDARY: A location where shear is transferred into or out of the diaphragm element. Transfer is either to a boundary element or to another force-resisting element.

DIAPHRAGM CHORD: A diaphragm boundary element perpendicular to the applied load that is assumed to take axial stresses caused by the diaphragm moment.

DIAPHRAGM DEFORMATION: The relative horizontal displacement of portions of a diaphragm due to strain in the diaphragm elements and not due to deformations of the vertical elements of the seismic-force resisting system or to diaphragm-rotation effects.

DIAPHRAGM-ROTATION EFFECTS: Relative horizontal displacement of portions of a diaphragm due to unequal deformations of the vertical elements of the seismic-force resisting system.

DISPLACEMENT AND DRIFT:

Design Earthquake Displacement: The displacement at a given location of the structure corresponding to the Design Earthquake.

Design Story Drift: The story drift corresponding to the Design Earthquake, taken at a representative plan location (center of mass or building perimeter, as required by Section 12.8.6).

Maximum Considered Earthquake Displacement: The displacement at a given location of the structure corresponding to the Risk-Targeted Maximum Considered Earthquake (MCE_R).

Story Drift: The horizontal displacement at the top of the story relative to the bottom of the story at vertically aligned points corresponding to the given loading.

Story Drift Ratio: The story drift divided by the story height, h_{sx}.

DISTRIBUTION SYSTEM: An interconnected system composed primarily of linear components including piping, tubing, conduit, raceways, or ducts. Distribution systems include in-line components such as valves, in-line suspended pumps, and mixing boxes.

ELEMENT ACTION: Element axial, shear, or flexural behavior.

Critical Action: An action, failure of which would result in the collapse of multiple bays or multiple stories of the building or would result in a significant reduction in the structure's seismic resistance.

Deformation-Controlled Action: Element actions for which reliable inelastic deformation capacity is achievable without critical strength decay.

Force-Controlled Action: Any element actions modeled with linear properties and element actions not classified as deformation-controlled.

Noncritical Actions: An action, failure of which would not result in either collapse or significant loss of the structure's seismic resistance.

Ordinary Action: An action, failure of which would result in only local collapse, comprising not more than one bay in a single story, and would not result in a significant reduction of the structure's seismic resistance.

ENCLOSURE: An interior space surrounded by walls.

EQUIPMENT SUPPORT: Those structural members or assemblies of members or manufactured elements, including braces, frames, legs, lugs, snuggers, hangers, or saddles, that transmit gravity loads and operating loads between the equipment and the structure.

Equipment Support Structures and Platforms: Assemblies of members or manufactured elements other than integral supports, including, but not limited to, moment frames, braced frames, skids, legs longer than 24 in. (600 mm), or walls that support one or more nonstructural components or systems.

Distribution System Support: Members that provide vertical or lateral seismic resistance for distribution systems, including but not limited to, hangers, braces, pipe racks, and trapeze assemblies.

Equipment Support, Integral: Assemblies of members or manufactured elements and their associated attachments and base plates that provide vertical or lateral support for nonstructural components, are directly connected to both the nonstructural component and the attachment to the structure or foundation, and where the nonstructural component acts as part of the lateral force resisting system of the equipment support. Integral equipment supports include but are not limited to legs less than or equal to 24 in. (600 mm) in length, lugs, skirts, and saddles.

FLEXIBLE CONNECTIONS: Those connections between equipment components that permit rotational and/or translational movement without degradation of performance. Examples include universal joints, bellows expansion joints, and flexible metal hose.

FOUNDATION GEOTECHNICAL CAPACITY: The maximum pressure or strength design capacity of a foundation based on the supporting soil, rock, or controlled low-strength material.

FOUNDATION STRUCTURAL CAPACITY: The design strength of foundations or foundation components as provided by adopted material standards and as altered by the requirements of this standard.

FRAME:

Braced Frame: An essentially vertical truss, or its equivalent, of the concentric or eccentric type that is provided in a building frame system or dual system to resist seismic forces.

Concentrically Braced Frame (CBF): A braced frame in which the members are subjected primarily to axial forces. CBFs are categorized as ordinary concentrically braced frames (OCBFs) or special concentrically braced frames (SCBFs).

Eccentrically Braced Frame (EBF): A diagonally braced frame in which at least one end of each brace frames into a beam a short distance from a beam-column or from another diagonal brace.

Moment Frame: A frame in which members and joints resist lateral forces by flexure and along the axis of the members. Moment frames are categorized as intermediate moment frames (IMFs), ordinary moment frames (OMFs), and special moment frames (SMFs).

STRUCTURAL SYSTEM:

Building Frame System: A structural system with an essentially complete space frame providing support for vertical loads. Seismic force resistance is provided by shear walls or braced frames.

Dual System: A structural system with an essentially complete space frame providing support for vertical loads. Seismic force resistance is provided by moment-resisting frames and shear walls or braced frames as prescribed in Section 12.2.5.1.

Shear Wall–Frame Interactive System: A structural system that uses combinations of ordinary reinforced concrete shear walls and ordinary reinforced concrete moment frames designed to resist lateral forces in proportion to their rigidities considering interaction between shear walls and frames on all levels.

Space Frame System: A 3-D structural system composed of interconnected members, other than bearing walls, that is capable of supporting vertical loads and, where designed for such an application, is capable of providing resistance to seismic forces.

FRICTION CLIP: A device that relies on friction to resist applied loads in one or more directions to anchor a nonstructural component. Friction is provided mechanically and is not due to gravity loads.

GLAZED CURTAIN WALL: A nonbearing wall that extends beyond the edges of building floor slabs and includes a glazing material installed in the curtain wall framing.

GLAZED STOREFRONT: A nonbearing wall that is installed between floor slabs, typically including entrances, and includes a glazing material installed in the storefront framing.

GRADE PLANE: A horizontal reference plane representing the average of finished ground level adjoining the structure at all exterior walls. Where the finished ground level slopes away from the exterior walls, the grade plane is established by the lowest points within the area between the structure and the property line or, where the property line is more than 6 ft (1,829 mm) from the structure, between the structure and points 6 ft (1,829 mm) from the structure.

HEATING, VENTILATING, AIR-CONDITIONING, AND REFRIGERATION (HVACR): The equipment, distribution systems, and terminals, excluding interconnecting piping and ductwork that provide, either collectively or individually, the processes of heating, ventilating, air-conditioning, or refrigeration to a building or portion of a building.

INSPECTION, SPECIAL: The observation of the work by a special inspector to determine compliance with the approved construction documents and these standards in accordance with the quality assurance plan.

Continuous Special Inspection: The full-time observation of the work by a special inspector who is present in the area where work is being performed.

Periodic Special Inspection: The part-time or intermittent observation of the work by a special inspector who is present in the area where work has been or is being performed.

INSPECTOR, SPECIAL: A person approved by the Authority Having Jurisdiction to perform special inspection, and who shall be identified as the owner's inspector.

INVERTED PENDULUM-TYPE STRUCTURES: Structures in which more than 50% of the structure's mass is concentrated at the top of a slender, cantilevered structure and in which the stability of the mass at the top of the structure relies on rotational restraint to the top of the cantilevered element.

JOINT: The geometric volume common to intersecting members.

LIGHT-FRAME CONSTRUCTION: A method of construction where the structural assemblies (e.g., walls, floors, ceilings, and roofs) are primarily formed by a system of repetitive wood or cold-formed steel framing members or subassemblies of these members (e.g., trusses).

LONGITUDINAL REINFORCEMENT RATIO: Area of longitudinal reinforcement divided by the cross-sectional area of the concrete.

MAXIMUM CONSIDERED EARTHQUAKE DISPLACEMENT: *See* DISPLACEMENT AND DRIFT

MAXIMUM CONSIDERED EARTHQUAKE (MCE) GROUND MOTION: The most severe earthquake effects considered by this standard, more specifically defined in the following two terms:

Maximum Considered Earthquake Geometric Mean (MCE_G) Peak Ground Acceleration: The most severe earthquake effects considered by this standard determined for geometric mean peak ground acceleration and without adjustment for targeted risk. The MCE_G peak ground acceleration adjusted for site effects (PGA_M) is used in this standard for evaluation of liquefaction, lateral spreading, seismic settlements, and other soil-related issues. In this standard, general procedures for determining PGA_M are provided in Section 11.8.3; site-specific procedures are provided in Section 21.5.

Risk-Targeted Maximum Considered Earthquake (MCE_R) Ground Motion Response Acceleration: The most severe earthquake effects considered by this standard determined for the orientation that results in the largest maximum response to horizontal ground motions and with adjustment for targeted risk. In this standard, general procedures for determining the MCE_R ground motion values are provided in Section 11.4.4; site-specific procedures are provided in Sections 21.1 and 21.2.

MECHANICALLY ANCHORED TANKS OR VESSELS: Tanks or vessels provided with mechanical anchors to resist overturning moments.

NONBUILDING STRUCTURE: A structure, other than a building, constructed of a type included in Chapter 15 and within the limits of Section 15.1.1.

NONBUILDING STRUCTURE SIMILAR TO A BUILDING: A nonbuilding structure that is designed and constructed in a manner similar to buildings, responds to strong ground motion in a fashion similar to buildings, and has a basic lateral and vertical seismic force-resisting system conforming to one of the types indicated in Table 12.2-1 or 15.4-1.

OPEN-TOP TANK: A tank without a fixed roof or cover, floating cover, gas holder cover, or dome.

ORTHOGONAL: In two horizontal directions, at 90 degrees to each other.

OWNER: Any person, agent, firm, or corporation that has a legal or equitable interest in a property.

P-DELTA EFFECT: The secondary effect on shears and moments of structural members caused by the action of the vertical loads induced by horizontal displacement of the structure resulting from various loading conditions.

PARTITION: A nonstructural interior wall that spans horizontally or vertically from support to support. The supports may be the basic building frame, subsidiary structural members, or other portions of the partition system.

PILE: Deep foundation element, which includes piers, caissons, and piles.

PILE CAP: Foundation elements to which piles are connected, including grade beams and mats.

PREMANUFACTURED MODULAR MECHANICAL AND ELECTRICAL SYSTEM: A prebuilt, fully or partially enclosed assembly of mechanical and electrical components.

REGISTERED DESIGN PROFESSIONAL: An architect or engineer registered or licensed to practice professional architecture or engineering, as defined by the statutory requirements of the professional registration laws of the state in which the project is to be constructed.

REINFORCED CONCRETE DUCTILE COUPLED WALL: A seismic force-resisting system as defined in ACI 318 Section 2.3 and complying with ACI 318 Section 18.10.9.

SEISMIC DESIGN CATEGORY: A classification assigned to a structure based on its Risk Category and the severity of the design earthquake ground motion at the site, as defined in Section 11.4.

SEISMIC FORCE-RESISTING SYSTEM: That part of the structural system that has been considered in the design to provide the required resistance to the seismic forces prescribed herein.

SEISMIC FORCES: The assumed forces prescribed herein, related to the response of the structure to earthquake motions, to be used in the design of the structure and its components.

SELF-ANCHORED TANKS OR VESSELS: Tanks or vessels that are stable under design overturning moment without the need for mechanical anchors to resist uplift.

SHEAR PANEL: A floor, roof, or wall element sheathed to act as a shear wall or diaphragm.

SITE CLASS: A classification assigned to a site based on the types of soils present and their engineering properties, as defined in Chapter 20.

STORAGE RACKS, STEEL: A framework or assemblage, comprised of cold-formed or hot-rolled steel structural members, intended for storage of materials, including, but not limited to, pallet storage racks, selective racks, movable-shelf racks, rack-supported systems, automated storage and retrieval systems (stacker racks), push-back racks, pallet-flow racks, case-flow racks, pick modules, and rack-supported platforms. Other types of racks, such as drive-in or drive-through racks, cantilever racks, portable racks, or racks made of materials other than steel, are not considered steel storage racks for the purpose of this standard.

STORAGE RACKS, STEEL CANTILEVERED: A framework or assemblage comprised of cold-formed or hot-rolled steel structural members, primarily in the form of vertical columns, extended bases, horizontal arms projecting from the faces of the columns, and longitudinal (down-aisle) bracing between columns. There may be shelf beams between the arms, depending on the products being stored; this definition does not include other types of racks such as pallet storage racks, drive-in racks, drive-through racks, or racks made of materials other than steel.

STORY: The portion of a structure between the tops of two successive floor surfaces and, for the topmost story, from the top of the floor surface to the top of the roof surface.

STORY ABOVE GRADE PLANE: A story in which the floor or roof surface at the top of the story is more than 6 ft (1,829 mm) above grade plane or is more than 12 ft (3,658 mm) above the finished ground level at any point on the perimeter of the structure.

STORY DRIFT: *See* DISPLACEMENT AND DRIFT

STORY SHEAR: The summation of design lateral seismic forces at levels above the story under consideration.

STRENGTH:

Design Strength: Nominal strength multiplied by a strength reduction factor, ϕ.

Nominal Strength: Strength of a member or cross section calculated in accordance with the requirements and assumptions of the strength design methods of this standard (or the reference documents) before application of any strength-reduction factors.

Required Strength: Strength of a member, cross section, or connection required to resist factored loads or related internal moments and forces in such combinations as stipulated by this standard.

STRUCTURAL HEIGHT: The vertical distance from the base to the highest level of the seismic force-resisting system of the structure. For pitched or sloped roofs, the structural height is from the base to the average height of the roof.

STRUCTURAL OBSERVATIONS: The visual observations to determine that the seismic force-resisting system is constructed in general conformance with the construction documents.

STRUCTURE: That which is built or constructed and limited to buildings and nonbuilding structures as defined herein.

SUBDIAPHRAGM: A portion of a diaphragm used to transfer wall anchorage forces to diaphragm crossties.

SUPPORTS: Those members, assemblies of members, or manufactured elements, including braces, frames, legs, lugs, snubbers, hangers, saddles, or struts, and associated fasteners that transmit loads between nonstructural components and their attachments to the structure.

TESTING AGENCY: A company or corporation that provides testing and/or inspection services.

TRUSSED TOWER: A lattice-type structure, freestanding or guyed, which supports static equipment such as chimneys, stacks, lights, or other lightweight components.

USGS SEISMIC DESIGN GEODATABASE: A US Geological Survey (USGS) database of geocoded values of seismic design parameters S_S, S_1, S_{MS}, S_{M1}, and PGA_M and geocoded sets of multi-period 5%-damped risk-targeted maximum considered earthquake (MCE_R) response spectra.

> **User Note:** The USGS Seismic Design Geodatabase is intended to be accessed through a USGS Seismic Design Web Service that allows the user to specify the site location, by latitude and longitude, and the site class to obtain the seismic design data. The USGS web service spatially interpolates between the gridded data of the USGS geodatabase. Both the USGS geodatabase and the USGS web service can be accessed at https://doi.org/10.5066/F7NK3C76. The USGS Seismic Design Geodatabase is available at the ASCE 7 Hazard Tool https://asce7hazardtool.online/ or an approved equivalent.

VENEERS: Facings or ornamentation of brick, concrete, stone, tile, or similar materials attached to a backing.

WALL: A component that has a slope of 60 degrees or greater with the horizontal plane used to enclose or divide space.

Bearing Wall: Any wall meeting either of the following classifications:

1. Any metal or wood stud wall that supports more than 100 lb per linear ft (1,459 N/m) of vertical load in addition to its own weight.
2. Any concrete or masonry wall that supports more than 200 lb per linear ft (2,919 N/m) of vertical load in addition to its own weight.

Light Frame Wall: A wall with wood or steel studs.

Light Frame Wood Shear Wall: A wall constructed with wood studs and sheathed with material rated for shear resistance.

Nonbearing Wall: Any wall that is not a bearing wall.

Nonstructural Wall: A wall other than a bearing wall or shear wall.

Shear Wall (Vertical Diaphragm): A wall, bearing or nonbearing, designed to resist lateral forces acting in the plane of the wall (sometimes referred to as a "vertical diaphragm").

Structural Wall: A wall that meets the definition for bearing wall or shear wall.

WALL SYSTEM:

BEARING: A structural system with bearing walls providing support for all or major portions of the vertical loads. Shear walls or braced frames provide seismic force resistance.

WOOD STRUCTURAL PANEL: A wood-based panel product that meets the requirements of DOC PS1 or DOC PS2 and is bonded with a waterproof adhesive. Included under this designation are plywood, oriented strand board, and composite panels.

11.3 SYMBOLS

The unit dimensions used with the items covered by the symbols shall be consistent throughout except where specifically noted. Symbols presented in this section apply only to the seismic provisions of Chapters 11 through 22 in this standard.

A_0 = Area of the load-carrying foundation, ft^2 (m^2)

A_{ch} = Cross-sectional area, in.^{2} (mm^2), of a structural member measured out-to-out of transverse reinforcement

A_{sh} = Total cross-sectional area of hoop reinforcement, in.^{2} (mm^2), including supplementary crossties, having a spacing of s_h and crossing a section with a core dimension of h_c

A_{vd} = Required area of leg of diagonal reinforcement, in.^{2} (mm^2)

A_x = Torsional amplification factor (Section 12.8.4.3)

a_i = Acceleration at level i obtained from a modal analysis (Section 13.3.1)

a_i = the maximum acceleration at level i obtained from the nonlinear response history analysis at the Design Earthquake ground motion

b_p = Width of the rectangular glass panel

C_{AR} = Component resonance ductility factor that converts the peak floor or ground acceleration into the peak component acceleration as determined in Section 13.3.1.3

C_d = Deflection amplification factor as given in Table 12.2-1, 15.4-1, or 15.4-2

C_{dX} = Deflection amplification factor in the X direction (Section 12.9.2.5)

C_{dY} = Deflection amplification factor in the Y direction (Section 12.9.2.5)

$C_{s\text{-}diaph}$ = Seismic response coefficient for design of diaphragms using the alternative diaphragm design method of Section 12.10.4

$C_{d\text{-}diaph}$ = Deflection amplification factor for diaphragm deflection (12.10.4)

C_{p0} = Diaphragm design acceleration coefficient at the structure base (Section 12.10.3.2.1)

C_{pi} = Diaphragm design acceleration coefficient at 80% of the structural height above the base, h_n (Section 12.10.3.2.1)

C_{pn} = Diaphragm design acceleration coefficient at the structural height, h_n (Section 12.10.3.2.1)

C_{px} = Diaphragm design acceleration coefficient at level x (Section 12.10.3.2.1)

C_s = Seismic response coefficient determined in Section 12.8.1.1 or 19.3.1 (dimensionless)

C_{s2} = Higher mode seismic response coefficient (Section 12.10.3.2.1)

C_t = Building period coefficient (Section 12.8.2.1)

C_v = Vertical response spectral coefficient as given in Table 11.9-1

C_{vs} = Coefficient of variation of soil shear modulus, defined as the standard deviation divided by the mean (Section 12.13.3)

C_{vx} = Vertical distribution factor as determined in Section 12.8.3

c = Distance from the neutral axis of a flexural member to the fiber of maximum compressive strain, in. (mm)

D = Effect of dead load

D_{clear} = Relative horizontal (drift) displacement, measured over the height of the glass panel under consideration, which causes initial glass-to-frame contact. For rectangular glass panels within a rectangular wall frame, D_{clear} is set forth in Section 13.5.9.1

D_{pI} = Seismic relative displacement (Section 13.3.2)

D_s = Total depth of stratum in Equation (19.3-4), ft (m)

d_c = Total thickness of cohesive soil layers in the top 100 ft (30 m), ft (m) (Section 20.4.3)

d_i = Thickness of any soil or rock layer i [between 0 and 100 ft (0 and 30 m)]; see Section 20.4.1

d_S = Total thickness of cohesionless soil layers in the top 100 ft (30 m), ft (m) (Section 20.4.2)

E = Effect of horizontal and vertical earthquake-induced forces (Section 12.4)

E_{cl} = Capacity-limited horizontal seismic load effect, equal to the maximum force that can develop in the element as determined by a rational, plastic mechanism analysis

F_i, F_n, F_x = Portion of the seismic base shear, V, induced at level i, n, or x, respectively, as determined in Section 12.8.3

F_{md} = Factor to convert the geometric mean spectral ordinate to a maximum direction spectral ordinate

F_p = Seismic force acting on a component of a structure as determined in Sections 12.11.1 and 13.3.1

F_{px} = Diaphragm seismic design force at level x

f'_c = Specified compressive strength of concrete used in design

f'_s = Ultimate tensile strength of the bolt, stud, or insert leg wires, psi (MPa). For ASTM A307 bolts or ASTM A108 studs, it is permitted to be assumed to be 60,000 psi (415 MPa)

f_y = Specified yield strength of reinforcement, psi (MPa)

f_{yh} = Specified yield strength of the special lateral reinforcement, psi (kPa)

$G = \gamma v_s^2/g$ = Average shear modulus for the soils beneath the foundation at large strain levels, lb/ft² (Pa)

$G_0 = \gamma v_{s0}^2/g$ = Average shear modulus for the soils beneath the foundation at small strain levels, lb/ft² (Pa)

g = Acceleration due to gravity

H = Thickness of soil

H_f = Factor for force amplification as a function of height in the structure as determined in Section 13.3.1.1

h = Height of a shear wall measured as the maximum clear height from top of foundation to bottom of diaphragm framing above, or the maximum clear height from top of diaphragm to bottom of diaphragm framing above

h = Average roof height of structure with respect to the base (Chapter 13)

h^* = Effective height of the building, ft (m), as determined in Chapter 19

h_c = Core dimension of a component measured to the outside of the special lateral reinforcement, in. (mm)

h_i, h_x = Height above the base to level i or x, respectively

h_n = Structural height as defined in Section 11.2

h_p = Height of the rectangular glass panel

h_{sx} = Story height below level $x = (h_x - h_{x-1})$

I_e = Importance Factor as prescribed in Section 11.5.1

I_p = Component Importance Factor as prescribed in Section 13.3.1

i = Building level referred to by a subscript i; $i = 1$ designates the first level above the base

K_p = Stiffness of the component or attachment (Section 13.3.3)

K_{xx}, K_{rr} = Rotational foundation stiffness in Equations (19.3-9) and (19.3-19), ft-lb/degree (N-m/rad)

K_y, K_r = Translational foundational stiffness in Equations (19.3-8) and (19.3-18), lb/in. (N/m)

KL/r = Lateral slenderness ratio of a compression member measured in terms of its effective length, KL, and the least radius of gyration of the member cross section, r

k = Distribution exponent, given in Section 12.8.3

k_a = Coefficient for amplification factor for diaphragm flexibility defined in Sections 12.11.2.1 and 12.14.7.5

L = Overall length of the building at the base in the direction being analyzed, ft (m)

L_{diaph} = Span in feet of the horizontal diaphragm or diaphragm segment being considered, measured between vertical elements or collectors that provide support to the diaphragm or diaphragm segment (Section 12.10.4)

M_t = Torsional moment resulting from eccentricity between the locations of the center of mass and the center of rigidity (Section 12.8.4.1)

M_{ta} = Accidental torsional moment as determined in Section 12.8.4.2

m = Subscript denoting the mode of vibration under consideration; $m = 1$ for the fundamental mode

N = Standard penetration resistance per ASTM D1586

N = Number of stories above the base (Section 12.8.2.1)

$\overline{N}$ = Average field standard penetration resistance for the top 100 ft (30 m) (Sections 20.3.3 and 20.4.2)

$\overline{N}_{ch}$ = Average standard penetration resistance for cohesionless soil layers for the top 100 ft (30 m) (Sections 20.3.3 and 20.4.2)

N_i = Standard penetration resistance of any soil or rock layer i [between 0 and 100 ft (0 and 30 m)]; see Section 20.4.2

n = Designation for the level that is uppermost in the main portion of the building

PGA_G = Lower-bound limit on deterministic maximum considered earthquake geometric mean peak ground acceleration (Table 21.2-1)

PGA_M = Mapped MCE_G peak ground acceleration as defined in Section 11.8.3

PI = Plasticity index per ASTM D4318

P_x = Total unfactored vertical design load at and above level x, for use in Section 12.8.7

Q_E = Effect of horizontal seismic (earthquake-induced) forces

R = Response modification coefficient as given in Tables 12.2-1, 12.14-1, 15.4-1, and 15.4-2

R_{diaph} = Response modification coefficient for design of diaphragms using the alternative diaphragm design method of Section 12.10.4

R_{po} = Component strength factor as determined in Section 13.3.1.4

R_μ = Structure ductility reduction factor as determined in Section 13.3.1.2

R_s = Diaphragm design force reduction factor (Section 12.10.3.5)

R_X = Response modification coefficient in the X direction (Section 12.9.2.5)

R_Y = Response modification coefficient in the Y direction (Section 12.9.2.5)

S_1 = MCE_R, 5% damped, spectral response acceleration parameter at a period of 1 s for Site Class BC site conditions as determined in accordance with Section 11.4.3

S_a = 5% damped design spectral response acceleration parameter at any period as defined in Section 11.4.5

S_{aM} = Site-specific, 5% damped, MCE_R spectral response acceleration parameter at any period

S_{av} = 5% damped, design vertical response spectral acceleration parameter at any period as determined in accordance with Section 11.9.3

S_{aMv} = 5% damped, MCE_R vertical response spectral acceleration parameter at any period as determined in accordance with Section 11.9.2

S_{D1} = Design, 5% damped, spectral response acceleration parameter at a period of 1 s as defined in Section 11.4.4

S_{DpS} = Design, 5% damped, spectral response acceleration parameter at short periods as defined in Section 11.4.4

S_{M1} = MCE_R, 5% damped, spectral response acceleration parameter at a period of 1 s adjusted for site class effects as defined in determined in accordance with Section 11.4.3

S_{MS} = MCE_R, 5% damped, spectral response acceleration parameter at short periods adjusted for site class effects as determined in accordance with Section 11.4.3

S_S = MCE_R, 5% damped, spectral response acceleration parameter at a period of 0.2 s for Site Class BC site conditions as determined in accordance with Section 11.4.3

s_h = Spacing of special lateral reinforcement, in. (mm)

s_u = Undrained shear strength (Section 20.4.3)

$\bar{s}_u$ = Average undrained shear strength in top 100 ft (30 m); see Sections 20.3.3 and 20.4.3, ASTM D2166, or ASTM D2850

s_{ui} = Undrained shear strength of any cohesive soil layer i [between 0 and 100 ft (0 and 30 m)]; see Section 20.4.3

T = Fundamental period of the building

$T_0 = 0.2S_{D1}/S_{DS}$

$\tilde{T}$ = Fundamental period as determined in Chapter 19

T_a = Approximate fundamental period of the building as determined in Section 12.8.2

T_{diaph} = Period of diaphragm for design of diaphragm using the alternative diaphragm design method of Section 12.10.4

T_L = Long-period transition period(s) shown in Figures 22-14 through 22-17

T_{lower} = Period of vibration at which 90% of the actual mass has been recovered in each of the two orthogonal directions of response (Section 12.9.2). The mathematical model used to compute T_{lower} shall not include accidental torsion and shall include P-delta effects.

T_p = Fundamental period of the component and its attachment (Section 13.3.3)

$T_S = S_{D1}/S_{DS}$

T_{upper} = Larger of the two orthogonal fundamental periods of vibration (Section 12.9.2). The mathematical model used to compute T_{upper} shall not include accidental torsion and shall include P-delta effects

TIR = Torsional Irregularity Ratio defined in Section 12.3.2.1.1

T_v = Vertical period of vibration

V = Total design lateral force or shear at the base

V_{EX} = Maximum absolute value of elastic base shear computed in the X direction among all three analyses performed in that direction (Section 12.9.2.5)

V_{EY} = Maximum absolute value of elastic base shear computed in the Y direction among all three analyses performed in that direction (Section 12.9.2.5)

V_{IX} = Inelastic base shear in the X direction (Section 12.9.2.5)

V_{IY} = Inelastic base shear in the Y direction (Section 12.9.2.5)

V_t = Design value of the seismic base shear as determined in Section 12.9.1.4.1

V_X = Equivalent lateral force (ELF) base shear in the X direction (Section 12.9.2.5)

V_x = Seismic design shear in story x as determined in Section 12.8.4

V_Y = Equivalent lateral force (ELF) base shear in the Y direction (Section 12.9.2.5)

$\tilde{V}$ = Reduced base shear accounting for the effects of soil structure interaction as determined in Section 19.3.1

$\tilde{V}_1$ = Portion of the reduced base shear, $\tilde{V}_1$, contributed by the fundamental mode, kip (kN) (Section 19.3)

ΔV = Reduction in V as determined in Section 19.3.1, kip (kN)

ΔV_1 = Reduction in V_1 as determined in Section 19.3.1, kip (kN)

v_s = Shear wave velocity, in ft/s (m/s), at small shear strains (greater than 10^{-3}% strain); see Section 19.2.1

$\bar{v}_s$ = Average shear wave velocity at small shear strains in top 100 ft (30 m) (Sections 20.3.3 and 20.4.1)

v_{si} = Shear wave velocity of any soil or rock layer i [between 0 and 100 ft (0 and 30 m)] (Section 20.4.1)

v_{so} = Average shear wave velocity, ft/s (m/s), for the soils beneath the foundation at small strain levels (Section 19.2.1.1)

W = Effective seismic weight of the building as defined in Section 12.7.2. For calculation of seismic-isolated building period, W is the total effective seismic weight of the building, in kip (kN), as defined in Sections 19.2 and 19.3

W = Effective seismic weight of the building, in kip (kN), as defined in Sections 19.2 and 19.3
W_c = Gravity load of a component of the building
W_P = Component operating weight, lb (N)
w_{px} = Weight tributary to the diaphragm at level x
w = Moisture content, in percent, per ASTM D2216
w_i, w_n, w_x = Portion of W that is located at or assigned to level i, n, or x, respectively
x = Level under consideration; $x = 1$ designates the first level above the base
z = Height in structure of point of attachment of component with respect to the base (Section 13.3.1)
z_s = Mode shape factor (Section 12.10.3.2.1)
β = Ratio of shear demand to shear capacity for the story between levels x and $x - 1$
$\bar{\beta}$ = Fraction of critical damping for the coupled structure–foundation system, determined in Section 19.2.1
β_0 = Foundation damping factor as specified in Section 19.2.1.2
Γ_{m1}, Γ_{m2} = First and higher modal contribution factors, respectively (Section 12.10.3.2.1)
γ = Average unit weight of soil, lb/ft^3 (N/m^3)
Δ = Design story drift as determined in Section 12.8.6
Δ_{avg} = Average of the story drifts at the two opposing edges of the building, as defined in Section 12.3.2.1.1
Δ_{max} = Maximum story drift at the building's edge subjected to lateral forces, as defined in Section 12.3.2.1.1
$\Delta_{fallout}$ = Relative seismic displacement (drift) at which glass fallout from the curtain wall, storefront, or partition occurs
Δ_a = Allowable story drift as specified in Section 12.12.1
Δ_{ADVE} = Average drift of adjoining vertical elements of the seismic force-resisting system over the story below the diaphragm under consideration, under tributary lateral load equivalent to that used in the computation of δ_{MDD} in Figure 12.3-1, in. (mm)
δ_{DE} = Design earthquake displacement as determined in Section 12.8.6
δ_{di} = Displacement due to diaphragm deformation corresponding to the design earthquake including Section 12.10 diaphragm forces (Section 12.8.6)
δ_e = Elastic displacement computed under design earthquake forces (Section 12.8.6)
δ_{MCE} = Maximum Considered Earthquake Displacement as determined in Section 12.8.6
δ_{MDD} = Computed maximum in-plane deflection of the diaphragm under lateral load, in. (mm) (Figure 12.3-1)
δ_{max} = Maximum displacement at level x considering torsion (Section 12.8.4.3)
δ_M = Maximum inelastic response displacement considering torsion (Section 12.12.3)
δ_{MT} = Total separation distance between adjacent structures on the same property (Section 12.12.3)
δ_{avg} = Average of the displacements at the extreme points of the structure at level x (Section 12.8.4.3)
δ_{xm} = Modal deflection of level x at the center of the mass at and above level x as determined by Section 19.3.2
$\bar{\delta}_x, \bar{\delta}_{x1}$ = Deflection of level x at the center of the mass at and above level x in Equations (19.2-13) and (19.3-3), in. (mm)
θ = Stability coefficient for P-delta effects as determined in Section 12.8.7
ηx = Force scale factor in the X direction (Section 12.9.2.5)
ηy = Force scale factor in the Y direction (Section 12.9.2.5)
ρ = Redundancy factor based on the extent of structural redundancy present in a building, as defined in Section 12.3.4
λ = Time effect factor
Ω_0 = Overstrength factor as defined in Tables 12.2-1, 15.4.-1, and 15.4-2
$\Omega_{0\text{-diaph}}$ = Diaphragm overstrength factor (Section 12.10.4)
Ω_{0p} = The anchorage overstrength factor given in Tables 13.5-1 and 13.6-1

11.4 SEISMIC GROUND MOTION VALUES

11.4.1 Near-Fault Sites Sites satisfying either of the following conditions shall be classified as near fault:

- 9.5 mi (15 km) or less from the surface projection of a known active fault capable of producing M_w7 or larger events, or
- 6.25 mi (10 km) or less from the surface projection of a known active fault capable of producing events M_w6 or larger, but smaller than M_w7.

EXCEPTIONS:

1. Faults with estimated slip rate less than 0.04 in. (1 mm) per year shall not be used to determine whether a site is a near-fault site.
2. The surface projection used in the determination of near-fault site classification shall not include portions of the fault at depths of 6.25 mi (10 km) or greater.

11.4.2 Site Class The site shall be classified as Site Class A, B, BC, C, CD, D, DE, E, or F in accordance with Chapter 20.

11.4.2.1 Default Site Conditions Where the soil properties are not known in sufficient detail to determine the site class, risk-targeted maximum considered earthquake (MCE_R) spectral response accelerations shall be based on the most critical spectral response acceleration at each period of Site Class C, Site Class CD, and Site Class D, unless the Authority Having Jurisdiction determines, based on geotechnical data, that Site Class DE, E, or F soils are present at the site.

11.4.3 Risk-Targeted Maximum Considered Earthquake (MCE_R) Spectral Response Acceleration Parameters Risk-targeted maximum considered earthquake (MCE_R) spectral response acceleration parameters S_S, S_1, S_{MS}, and S_{M1} shall be obtained from the USGS Seismic Design Geodatabase for the applicable site class.

EXCEPTION: Where a site-specific ground motion analysis is performed in accordance with Section 11.4.7, risk-targeted maximum considered earthquake (MCE_R) spectral response acceleration parameters S_{MS} and S_{M1} shall be determined in accordance with Section 21.4 and risk-targeted maximum considered earthquake (MCE_R) spectral response acceleration parameters S_S and S_1 shall be either (1) determined from the site-specific MCE_R response spectrum calculated in accordance with the requirements of Section 21.2.3 assuming Site Class BC site condition or (2) obtained from the USGS Seismic Design Geodatabase.

11.4.4 Design Spectral Acceleration Parameters Design earthquake spectral response acceleration parameters at short periods, S_{DS}, and at 1-s periods, S_{D1}, shall be determined from Equations (11.4-1) and (11.4-2), respectively. Where the simplified alternative design procedure of Section 12.14 is used, the value of S_{DS} shall be determined in accordance with Section 12.14.8.1, and the value for S_{D1} need not be determined.

$$S_{DS} = \frac{2}{3} S_{MS} \quad (11.4\text{-}1)$$

$$S_{D1} = \frac{2}{3} S_{M1} \quad (11.4\text{-}2)$$

where

S_{MS} = MCE_R, 5%-damped, spectral response acceleration parameter at short periods adjusted for site effects as determined in accordance with Section 11.4.3, and

S_{M1} = MCE_R, 5%-damped, spectral response acceleration parameter at a period of 1 s adjusted for site effects as determined in accordance with Section 11.4.3.

11.4.5 Design Response Spectrum Where a design response spectrum is required by this standard, the design response spectrum shall be determined in accordance with the requirements of Section 11.4.5.1.

EXCEPTIONS:

1. Where a site-specific ground motion analysis is performed in accordance with Section 11.4.7, the design response spectrum shall be determined in accordance with Section 21.3.
2. Where values of the multi-period 5%-damped MCE_R response spectrum are not available from the USGS Seismic Design Geodatabase, the design response spectrum shall be permitted to be determined in accordance with Section 11.4.5.2.

11.4.5.1 Multi-Period Design Response Spectrum The multi-period design response spectrum shall be developed as follows:

1. At discrete values of period, T, equal to 0.0 s, 0.01 s, 0.02 s, 0.03 s, 0.05 s, 0.075 s, 0.1 s, 0.15 s, 0.2 s, 0.25 s, 0.3 s, 0.4 s, 0.5 s, 0.75 s, 1.0 s, 1.5 s, 2.0 s, 3.0 s, 4.0 s, 5.0 s, 7.5 s, and 10 s, the 5%-damped design spectral response acceleration parameter, S_a, shall be taken as 2/3 of the multi-period 5%-damped MCE_R response spectrum from the USGS Seismic Design Geodatabase for the applicable site class.
2. At each response period, T, less than 10 s and not equal to one of the discrete values of period, T, listed in Item 1 above, S_a, shall be determined by linear interpolation between values of S_a, of Item 1 above.
3. At each response period, T, greater than 10 s, S_a, shall be taken as the value of S_a at the period of 10 s of Item 1 above, factored by $10/T$, where the value of T is less than or equal to that of the long-period transition period, T_L, and shall be taken as the value of S_a at the period of 10 s factored by $10T_L/T^2$, where the value of T is greater than that of the long-period transition period, T_L.

11.4.5.2 Two-Period Design Response Spectrum The two-period design response spectrum shall be developed as indicated in Figure 11.4-1 and as follows:

1. For periods less than T_0, the design spectral response acceleration parameter, S_a, shall be taken as given in Equation (11.4-3):

$$S_a = S_{DS}\left(0.4 + 0.6\frac{T}{T_0}\right) \quad (11.4\text{-}3)$$

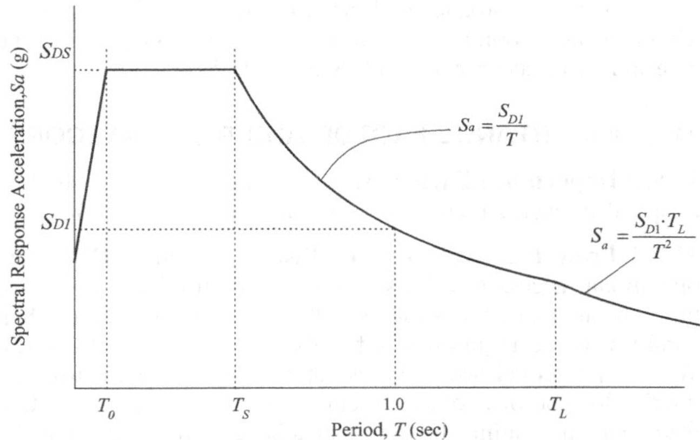

Figure 11.4-1. Two-period design response spectrum.

2. For periods greater than or equal to T_0 and less than or equal to T_S, the design spectral response acceleration parameter, S_a, shall be taken as equal to S_{DS}.
3. For periods greater than T_S and less than or equal to T_L, the design spectral response acceleration parameter, S_a, shall be taken as given in Equation (11.4-4):

$$S_a = \frac{S_{D1}}{T} \quad (11.4\text{-}4)$$

4. For periods greater than T_L, S_a shall be taken as given in Equation (11.4-5):

$$S_a = \frac{S_{D1} T_L}{T^2} \quad (11.4\text{-}5)$$

where

S_{DS} = Design spectral response acceleration parameter at short periods;

S_{D1} = Design spectral response acceleration parameter at a 1 s period;

T = Fundamental period of the structure, S;

$T_0 = 0.2(S_{D1}/S_{DS})$;

$T_S = S_{D1}/S_{DS}$; and

T_L = Long-period transition period(s) shown in Figures 22-14 through 22-17.

11.4.6 Risk-Targeted Maximum Considered Earthquake (MCE_R) Response Spectrum Where an MCE_R response spectrum is required, it shall be determined by multiplying the design response spectrum by 1.5.

11.4.7 Site-Specific Ground Motion Procedures A site response analysis shall be performed in accordance with Section 21.1 for structures on Site Class F sites, unless exempted in accordance with Section 20.3.1.

It shall be permitted to perform a site response analysis in accordance with Section 21.1 and/or a ground motion hazard analysis in accordance with Section 21.2 to determine ground motions for any structure.

When the procedures of either Section 21.1 or 21.2 are used, the design response spectrum shall be determined in accordance with Section 21.3, the design acceleration parameters shall be determined in accordance with Section 21.4, and, if required, the MCE_G peak ground acceleration parameter PGA_M shall be determined in accordance with Section 21.5.

11.5 IMPORTANCE FACTOR AND RISK CATEGORY

11.5.1 Importance Factor An Importance Factor, I_e, shall be assigned to each structure in accordance with Table 1.5-2.

11.5.2 Protected Access for Risk Category IV Where operational access to a Risk Category IV structure is required through an adjacent structure, the adjacent structure shall conform to the requirements for Risk Category IV structures. Where operational access is less than 10 ft (3.048 m) from an interior lot line or another structure on the same lot, protection from potential falling debris from adjacent structures shall be provided by the owner of the Risk Category IV structure.

11.6 SEISMIC DESIGN CATEGORY

Structures shall be assigned a seismic design category in accordance with this section.

Risk Category I, II, or III structures located where the mapped spectral response acceleration parameter at 1-s period, S_1, is greater than or equal to 0.75 shall be assigned to Seismic Design Category E. Risk Category IV structures located where the mapped spectral response acceleration parameter at 1-s period, S_1, is greater than or equal to 0.75 shall be assigned to Seismic Design Category F. All other structures shall be assigned to a seismic design category based on their risk category and the design spectral response acceleration parameters, S_{DS} and S_{D1}, determined in accordance with Section 11.4.4. Each building and structure shall be assigned to the more severe seismic design category in accordance with Table 11.6-1 or 11.6-2, irrespective of the fundamental period of vibration of the structure, T. The provisions in Chapter 19 shall not be used to modify the spectral response acceleration parameters for determining seismic design category.

Where S_1 is less than 0.75, the seismic design category is permitted to be determined from Table 11.6-1 alone where all of the following apply:

1. In each of the two orthogonal directions, the approximate fundamental period of the structure, T_a, determined in accordance with Section 12.8.2.1 is less than $0.8T_s$, where T_s is determined in accordance with Section 11.4.5.
2. In each of two orthogonal directions, the fundamental period of the structure used to calculate the story drift is less than T_s.
3. Equation (12.8-2) is used to determine the seismic response coefficient, C_s.

Table 11.6-1. Seismic Design Category Based on Short-Period Response Acceleration Parameter.

Value of S_{DS}	Risk Category I or II or III	Risk Category IV
$S_{DS} < 0.167$	A	A
$0.167 \leq S_{DS} < 0.33$	B	C
$0.33 \leq S_{DS} < 0.50$	C	D
$0.50 \leq S_{DS}$	D	D

Table 11.6-2. Seismic Design Category Based on 1 s Period Response Acceleration Parameter.

Value of S_{D1}	Risk Category I or II or III	Risk Category IV
$S_{D1} < 0.067$	A	A
$0.067 \leq S_{D1} < 0.133$	B	C
$0.133 \leq S_{D1} < 0.20$	C	D
$0.20 \leq S_{D1}$.	D	

4. The diaphragms are rigid in accordance with Section 12.3; or, for diaphragms that are not rigid, the horizontal distance between vertical elements of the seismic force-resisting system does not exceed 40 ft (12.192 m).

Where the alternate simplified design procedure of Section 12.14 is used, the seismic design category is permitted to be determined from Table 11.6-1 alone, using the value of S_{DS} determined in Section 12.14.8.1, except that where S_1 is greater than or equal to 0.75, the seismic design category shall be E.

11.7 DESIGN REQUIREMENTS FOR SEISMIC DESIGN CATEGORY A

Buildings and other structures assigned to Seismic Design Category A need only comply with the requirements of Section 1.4. Nonstructural components in SDC A are exempt from seismic design requirements. In addition, tanks assigned to Risk Category IV shall satisfy the freeboard requirement in Section 15.6.5.1.

11.8 GEOLOGIC HAZARDS AND GEOTECHNICAL INVESTIGATION

11.8.1 Site Limitation for Seismic Design Categories E and F A structure assigned to Seismic Design Category E or F shall not be located where a known potential exists for an active fault to cause rupture of the ground surface at the structure.

11.8.2 Geotechnical Investigation Report Requirements for Seismic Design Categories C through F A geotechnical investigation report shall be provided for a structure assigned to Seismic Design Category C, D, E, or F in accordance with this section. An investigation shall be conducted, and a report shall be submitted that includes an evaluation of the following potential geologic and seismic hazards:

(a) Slope instability,
(b) Liquefaction,
(c) Total and differential settlement, and
(d) Surface displacement caused by faulting or seismically induced lateral spreading or lateral flow.

The report shall contain recommendations for foundation designs or other measures to mitigate the effects of the previously mentioned hazards.

EXCEPTION: Where approved by the Authority Having Jurisdiction, a site-specific geotechnical report is not required where prior evaluations of nearby sites with similar soil conditions provide direction relative to the proposed construction.

11.8.3 Additional Geotechnical Investigation Report Requirements for Seismic Design Categories D through F The geotechnical investigation report for a structure assigned to Seismic Design Category D, E, or F shall include all of the following, as applicable:

1. The determination of dynamic seismic lateral earth pressures on basement and retaining walls caused by design earthquake ground motions.
2. The potential for liquefaction, seismically-induced permanent ground displacement, and soil strength loss evaluated for site peak ground acceleration, earthquake magnitude, and source characteristics consistent with the MCE_G peak ground acceleration. Peak ground acceleration shall be determined based on either (1) a site-specific study taking into account soil amplification effects as specified in Section 11.4.7 or (2) the value of the MCE_G peak ground acceleration parameter PGA_M from the USGS Seismic Design Geodatabase for the applicable site class.
3. Assessment of potential consequences of liquefaction, seismically-induced permanent ground displacement, and soil strength loss, including, but not limited to, estimation of total and differential settlement, lateral soil movement, lateral soil loads on foundations, reduction in foundation soil-bearing capacity and lateral soil reaction, soil downdrag and reduction in axial and lateral soil reaction for pile foundations, increases in soil lateral pressures on retaining walls, and flotation of buried structures.
4. Discussion of mitigation measures such as, but not limited to, selection of appropriate foundation type and depths, selection of appropriate structural systems to accommodate anticipated displacements and forces, ground stabilization, or any combination of these measures and how they shall be considered in the design of the structure.

11.9 VERTICAL GROUND MOTIONS FOR SEISMIC DESIGN

11.9.1 General If the option to incorporate the effects of vertical seismic ground motions is exercised in lieu of the requirements of Section 12.4.2.2, the requirements of this section are permitted to be used in the determination of the vertical design earthquake ground motions. The requirements of Section 11.9.2 shall only apply to structures in Seismic Design Categories C, D, E, and F located in the conterminous United States at or west of −105 degrees longitude, Alaska, Hawaii, Puerto Rico, US Virgin Islands, Guam and Northern Mariana Islands, and American Samoa. For structures in Seismic Design Categories C, D, E, and F located in the conterminous United States east of −105 degrees longitude, the value of S_{aMv} shall be taken as two-thirds of the value of S_{aM}. The requirements of Section 11.9.3 shall apply to all structures in the Seismic Design Categories C, D, E, and F.

11.9.2 MCE_R Vertical Response Spectrum Where a vertical response spectrum is required by this standard and site-specific procedures are not used, the MCE_R vertical response spectral acceleration, S_{aMv}, shall be developed as follows:

1. For vertical periods (T_v) less than or equal to 0.025 s, S_{aMv} shall be determined in accordance with Equation (11.9-1) as follows:

$$S_{aMv} = 0.65C_v(S_{aM}/F_{md}) \qquad (11.9\text{-}1)$$

2. For vertical periods greater than 0.025 s and less than or equal to 0.05 s, S_{aMv} shall be determined in accordance with Equation (11.9-2) as follows:

$$S_{aMv} = 16_{Cv}(S_{aM}/F_{md})(T_v - 0.025) + 0.65C_v(S_{aM}/F_{md}) \qquad (11.9\text{-}2)$$

3. For vertical periods greater than 0.05 s and less than or equal to 0.1 s, S_{aMv} shall be determined in accordance with Equation (11.9-3) as follows:

$$S_{aMv} = 1.05C_v(S_{aM}/F_{md}) \qquad (11.9\text{-}3)$$

4. For vertical periods greater than 0.1 s and less than or equal to 2.0 s, S_{aMv} shall be determined in accordance with Equation (11.9-4) as follows:

$$S_{aMv} = 1.05C_v\left(S_{aM}/F_{md}\right)\left(\frac{0.1}{T_v}\right)^{0.5} \qquad (11.9\text{-}4)$$

The value of S_{aMv} shall not be less than 0.5 (S_{aM}/F_{md}).

5. For vertical periods greater than 2.0 s, S_{aMv} shall be determined in accordance with Equation (11.9-5) as follows:

$$S_{aMv} = 0.5(S_{aM}/F_{md}) \qquad (11.9\text{-}5)$$

where

Cv = Is defined in terms of SS in Table 11.9-1;
S_{aM} = MCE_R spectral response acceleration parameter at the same period as S_{aMv};
F_{md} = Factor to convert the geometric mean spectral ordinate to a maximum direction spectral ordinate; and
T_v = Vertical period of vibration.

The maximum-component factor, F_{md}, shall be taken as follows:

$$T_v \le 0.2\,\text{s} : F_{md} = 1.2 \qquad (11.9\text{-}6)$$

$$0.2 < T_v \le 1.0\,\text{s} : F_{md} = 1.2 + 0.0625(T_v - 0.2) \qquad (11.9\text{-}7)$$

$$1 < T_v \le 1.0\,\text{s} : F_{md} = 1.25 + 0.05(T_v - 1.0)/9 \qquad (11.9\text{-}8)$$

In lieu of using the above procedure, a site-specific study is permitted to be performed to obtain S_{aMv}, but the value so

Table 11.9-1. Values of Vertical Coefficient, C_v.

MCE_R Spectral Response Parameter at Short Periods*	Site Classes A, B	Site Class BC	Site Class C	Site Class CD	Site Classes D, DE, E, F
$S_{MS} \ge 2.0$	0.9	1.1	1.3	1.4	1.5
$S_{MS} = 12.0$	0.9	1.0	1.1	1.2	1.3
$S_{MS} = 0.6$	0.9	0.95	1.0	1.05	1.1
$S_{MS} = 0.3$	0.8	0.8	0.8	0.85	0.9
$S_{MS} \le 0.2$	0.7	0.7	0.7	0.7	0.7

*Use straight-line interpolation for intermediate values of S_{MS}.

determined shall not be less than 80% of the S_{aMv} value determined from Equations (11.9-1) through (11.9-5).

11.9.3 Design Vertical Response Spectrum The design vertical response spectral acceleration, S_{av}, shall be taken as two-thirds of the value of S_{aMv} determined in Sections 11.9.1 or 11.9.2.

11.10 CONSENSUS STANDARDS AND OTHER REFERENCED DOCUMENTS

See Chapter 23 for the list of consensus standards and other documents that shall be considered part of this standard to the extent referenced in this chapter.

CHAPTER 12
SEISMIC DESIGN REQUIREMENTS FOR BUILDING STRUCTURES

12.1 STRUCTURAL DESIGN BASIS

12.1.1 Basic Requirements The seismic analysis and design procedures to be used in the design of building structures and their members shall be as prescribed in this section. The building structure shall include complete lateral and vertical force-resisting systems capable of providing adequate strength, stiffness, and energy dissipation capacity to withstand the design ground motions within the prescribed limits of deformation and strength demand. The design ground motions shall be assumed to occur along any horizontal direction of a building structure. The adequacy of the structural systems shall be demonstrated through the construction of a mathematical model and evaluation of this model for the effects of design ground motions. The design seismic forces and their distribution over the height of the building structure shall be established in accordance with one of the applicable procedures indicated in Section 12.6, and the corresponding internal forces and deformations in the members of the structure shall be determined. An approved alternative procedure shall not be used to establish the seismic forces and their distribution unless the corresponding internal forces and deformations in the members are determined using a model consistent with the procedure adopted.

EXCEPTION: As an alternative, the simplified design procedures of Section 12.14 are permitted to be used in lieu of the requirements of Sections 12.1 through 12.12, subject to all of the limitations contained in Section 12.14.

12.1.2 Member Design, Connection Design, and Deformation Limit Individual members, including those not part of the seismic force-resisting system, shall be provided with adequate strength to resist the shears, axial forces, and moments determined in accordance with this standard, and connections shall develop the strength of the connected members or the forces indicated in Section 12.1.1. The deformation of the structure shall not exceed the prescribed limits where the structure is subjected to the design seismic forces.

12.1.3 Continuous Load Path and Interconnection A continuous load path, or paths, with adequate strength and stiffness shall be provided to transfer all forces from the point of application to the final point of resistance. All parts of the structure between separation joints shall be interconnected to form a continuous path to the seismic force-resisting system, and the connections shall be capable of transmitting the seismic force, F_p, induced by the parts being connected. Any smaller portion of the structure shall be tied to the remainder of the structure with elements that have a design strength capable of transmitting a seismic force of 0.133 times the short-period design spectral response acceleration parameter, S_{DS}, times the weight of the smaller portion or 5% of the portion's weight, whichever is greater. This connection force does not apply to the overall design of the seismic force-resisting system. Connection design forces need not exceed the maximum forces that the structural system can deliver to the connection.

12.1.4 Connection to Supports A positive connection for resisting a horizontal force acting parallel to the member shall be provided for each beam, girder, or truss, either directly to its supporting elements or to slabs designed to act as diaphragms. Where the connection is through a diaphragm, then the member's supporting element must also be connected to the diaphragm. The connection shall have a minimum design strength of 5% of the dead plus live load reaction.

12.1.5 Foundation Design The foundation shall be designed to resist the forces developed and to accommodate the movements imparted to the structure and foundation by the design ground motions. The dynamic nature of the forces, the expected ground motion, the design basis for strength and energy dissipation capacity of the structure, and the dynamic properties of the soil shall be included in the determination of the foundation design criteria. The design and construction of foundations shall comply with Section 12.13.

When calculating load combinations using the load combinations specified in either Section 2.3 or Section 2.4, the weights of foundations shall be considered dead loads in accordance with Section 3.1.2. The dead loads are permitted to include overlying fill and paving materials.

12.1.6 Material Design and Detailing Requirements Structural elements, including foundation elements, shall conform to the material design and detailing requirements set forth in Chapter 14.

12.2 STRUCTURAL SYSTEM SELECTION

12.2.1 Selection and Limitations Except as noted in Section 12.2.1.1, the basic lateral and vertical seismic force-resisting system shall conform to one of the types indicated in Table 12.2-1 or a combination of systems as permitted in Sections 12.2.2, 12.2.3, and 12.2.4. Each system is subdivided by the types of vertical elements used to resist lateral seismic forces. The structural systems used shall be in accordance with the structural system limitations and the limits on structural height, h_n, contained in Table 12.2-1. The appropriate response modification coefficient, R; overstrength factor, Ω_0; and deflection amplification factor, C_d, indicated in Table 12.2-1 shall be used in determining the base shear, element design forces, and design story drift.

Each selected seismic force-resisting system shall be designed and detailed in accordance with the specific requirements for the

Table 12.2-1. Design Coefficients and Factors for Seismic Force-Resisting Systems.

Seismic Force-Resisting System	ASCE 7 Section Where Detailing Requirements Are Specified	Response Modification Coefficient, R^a	Overstrength Factor, $\Omega_0{}^b$	Deflection Amplification Factor, $C_d{}^c$	Structural System Limitations Including Structural Height, h_n, Limits (ft)[d] — Seismic Design Category B	C	D[e]	E[e]	F[f]
A. BEARING WALL SYSTEMS									
1. Special reinforced concrete shear walls[g,h]	14.2	5	2½	5	NL	NL	160	160	100
2. Reinforced concrete ductile coupled walls[q]	14.2	8	2½	8	NL	NL	160	160	100
3. Ordinary reinforced concrete shear walls[g]	14.2	4	2½	4	NL	NL	NP	NP	NP
4. Detailed plain concrete shear walls[g]	14.2	2	2½	2	NL	NP	NP	NP	NP
5. Ordinary plain concrete shear walls[g]	14.2	1½	2½	1½	NL	NP	NP	NP	NP
6. Intermediate precast shear walls[g]	14.2	4	2½	4	NL	NL	40[i]	40[i]	40[i]
7. Ordinary precast shear walls[g]	14.2	3	2½	3	NL	NP	NP	NP	NP
8. Special reinforced masonry shear walls	14.4	5	2½	3½	NL	NL	160	160	100
9. Intermediate reinforced masonry shear walls	14.4	3½	2½	2¼	NL	NL	NP	NP	NP
10. Ordinary reinforced masonry shear walls	14.4	2	2½	1¾	NL	160	NP	NP	NP
11. Detailed plain masonry shear walls	14.4	2	2½	1¾	NL	NP	NP	NP	NP
12. Ordinary plain masonry shear walls	14.4	1½	2½	1¼	NL	NP	NP	NP	NP
13. Prestressed masonry shear walls	14.4	1½	2½	1¾	NL	NP	NP	NP	NP
14. Ordinary reinforced AAC masonry shear walls	14.4	2	2½	2	NL	35	NP	NP	NP
15. Ordinary plain AAC masonry shear walls	14.4	1½	2½	1½	NL	NP	NP	NP	NP
16. Light-frame (wood) walls sheathed with wood structural panels rated for shear resistance	14.5	6½	3	4	NL	NL	65	65	65
17. Light-frame (cold-formed steel) walls sheathed with wood structural panels rated for shear resistance or steel sheets	14.1	6½	3	4	NL	NL	65	65	65
18. Light-frame walls with shear panels of all other materials	14.1 and 14.5	2	2½	2	NL	NL	35	NP	NP
19. Light-frame (cold-formed steel) wall systems using flat strap bracing	14.1	4	2	3½	NL	NL	65	65	65
20. Cross-laminated timber shear walls	14.5	3	3	3	65	65	65	65	65
21. Cross-laminated timber shear walls with shear resistance provided by high-aspect-ratio panels only	14.5	4	3	4	65	65	65	65	65
B. BUILDING FRAME SYSTEMS									
1. Steel eccentrically braced frames	14.1	8	2	4	NL	NL	160	160	100
2. Steel special concentrically braced frames	14.1	6	2	5	NL	NL	160	160	100
3. Steel ordinary concentrically braced frames	14.1	3¼	2	3¼	NL	NL	35[j]	35[j]	NP[j]
4. Special reinforced concrete shear walls[g,h]	14.2	6	2½	5	NL	NL	160	160	100
5. Reinforced concrete ductile coupled walls[q]	14.2	8	2½	8	NL	NL	160	160	100
6. Ordinary reinforced concrete shear walls[g]	14.2	5	2½	4½	NL	NL	NP	NP	NP
7. Detailed plain concrete shear walls[g]	14.2 and 14.2.2.7	2	2½	2	NL	NP	NP	NP	NP
8. Ordinary plain concrete shear walls[g]	14.2	1½	2½	1½	NL	NP	NP	NP	NP
9. Intermediate precast shear walls[g]	14.2	5	2½	4½	NL	NL	40[i]	40[i]	40[i]
10. Ordinary precast shear walls[g]	14.2	4	2½	4	NL	NP	NP	NP	NP
11. Steel and concrete composite eccentrically braced frames	14.3	8	2½	4	NL	NL	160	160	100
12. Steel and concrete composite special concentrically braced frames	14.3	5	2	4½	NL	NL	160	160	100
13. Steel and concrete composite ordinary braced frames	14.3	3	2	3	NL	NL	NP	NP	NP
14. Steel and concrete composite plate shear walls	14.3	6½	2½	5½	NL	NL	160	160	100
15. Steel and concrete composite special shear walls	14.3	6	2½	5	NL	NL	160	160	100
16. Steel and concrete composite ordinary shear walls	14.3	5	2½	4½	NL	NL	NP	NP	NP
17. Special reinforced masonry shear walls	14.4	5½	2½	4	NL	NL	160	160	100

18. Intermediate reinforced masonry shear walls	14.4	4	2½	4	NL	NL	NP	NP	NP
19. Ordinary reinforced masonry shear walls	14.4	2	2½	2	NL	160	NP	NP	NP
20. Detailed plain masonry shear walls	14.4	2	2½	2	NL	NP	NP	NP	NP
21. Ordinary plain masonry shear walls	14.4	1½	2½	1¼	NL	NP	NP	NP	NP
22. Prestressed masonry shear walls	14.4	1½	2½	1¾	NL	NP	NP	NP	NP
23. Light-frame (wood) walls sheathed with wood structural panels rated for shear resistance	14.5	7	2½	4½	NL	NL	65	65	65
24. Light-frame (cold-formed steel) walls sheathed with wood structural panels rated for shear resistance or steel sheets	14.1	7	2½	4½	NL	NL	65	65	65
25. Light-frame walls with shear panels of all other materials	14.1 and 14.5	2½	2½	2½	NL	NL	35	NP	NP
26. Steel buckling-restrained braced frames	14.1	8	2½	5	NL	NL	160	160	100
27. Steel special plate shear walls	14.1	7	2	6	NL	NL	160	160	100
28. Steel and concrete coupled composite plate shear walls	14.3	8	2½	5½	NL	NL	160	160	100
C. MOMENT-RESISTING FRAME SYSTEMS									
1. Steel special moment frames	14.1 and 12.2.5.5	8	3	5½	NL	NL	NL	NL	NL
2. Steel special truss moment frames	14.1	7	3	5½	NL	NL	160	100	NP
3. Steel intermediate moment frames	12.2.5.7 and 14.1	4½	3	4	NL	NL	35[k]	NP[k]	NP[k]
4. Steel ordinary moment frames	12.2.5.6 and 14.1	3½	3	3	NL	NL	NP[l]	NP[l]	NP[l]
5. Special reinforced concrete moment frames[m]	12.2.5.5 and 14.2	8	3	5½	NL	NL	NL	NL	NL
6. Intermediate reinforced concrete moment frames	14.2	5	3	4½	NL	NL	NP	NP	NP
7. Ordinary reinforced concrete moment frames	14.2	3	3	2½	NL	NP	NP	NP	NP
8. Steel and concrete composite special moment frames	12.2.5.5 and 14.3	8	3	5½	NL	NL	NL	NL	NL
9. Steel and concrete composite intermediate moment frames	14.3	5	3	4½	NL	NL	NP	NP	NP
10. Steel and concrete composite partially restrained moment frames	14.3	6	3	5½	160	160	100	NP	NP
11. Steel and concrete composite ordinary moment frames	14.3	3	3	2½	NL	NP	NP	NP	NP
12. Cold-formed steel—special bolted moment frame[n]	14.1	3½	3[o]	3½	35	35	35	35	35
D. DUAL SYSTEMS WITH SPECIAL MOMENT FRAMES CAPABLE OF RESISTING AT LEAST 25% OF PRESCRIBED SEISMIC FORCES	12.2.5.1								
1. Steel eccentrically braced frames	14.1	8	2½	4	NL	NL	NL	NL	NL
2. Steel special concentrically braced frames	14.1	7	2½	5½	NL	NL	NL	NL	NL
3. Special reinforced concrete shear walls[g,h]	14.2	7	2½	5½	NL	NL	NL	NL	NL
4. Reinforced concrete ductile coupled walls[q]	14.2	8	2½	8	NL	NL	NL	NL	NL
5. Ordinary reinforced concrete shear walls[g]	14.2	6	2½	5	NL	NL	NP	NP	NP
6. Steel and concrete composite eccentrically braced frames	14.3	8	2½	4	NL	NL	NL	NL	NL
7. Steel and concrete composite special concentrically braced frames	14.3	6	2½	5	NL	NL	NL	NL	NL
8. Steel and concrete composite plate shear walls	14.3	7½	2½	6	NL	NL	NL	NL	NL
9. Steel and concrete composite special shear walls	14.3	7	2½	6	NL	NL	NL	NL	NL
10. Steel and concrete composite ordinary shear walls	14.3	6	2½	5	NL	NL	NP	NP	NP
11. Special reinforced masonry shear walls	14.4	5½	3	5	NL	NL	NL	NL	NL
12. Intermediate reinforced masonry shear walls	14.4	4	3	3½	NL	NL	NP	NP	NP
13. Steel buckling-restrained braced frames	14.1	8	2½	5	NL	NL	NL	NL	NL
14. Steel special plate shear walls	14.1	8	2½	6½	NL	NL	NL	NL	NL
15. Steel and concrete coupled composite plate shear walls	14.3	8	2½	5½	NL	NL	NL	NL	NL
E. DUAL SYSTEMS WITH INTERMEDIATE MOMENT FRAMES CAPABLE OF RESISTING AT LEAST 25% OF PRESCRIBED SEISMIC FORCES	12.2.5.1								
1. Steel special concentrically braced frames[p]	14.1	6	2½	5	NL	NL	35	NP	NP
2. Special reinforced concrete shear walls[g,h]	14.2	6½	2½	5	NL	NL	160	100	100
3. Ordinary reinforced masonry shear walls	14.4	3	3	2½	NL	160	NP	NP	NP

continues

Table 12.2-1 *(Continued)*. Design Coefficients and Factors for Seismic Force-Resisting Systems.

Seismic Force-Resisting System	ASCE 7 Section Where Detailing Requirements Are Specified	Response Modification Coefficient, R^a	Overstrength Factor, $\Omega_0{}^b$	Deflection Amplification Factor, $C_d{}^c$	Structural System Limitations Including Structural Height, h_n, Limits (ft)[d] — Seismic Design Category				
					B	C	D[e]	E[e]	F[f]
4. Intermediate reinforced masonry shear walls	14.4	3½	3	3	NL	NL	NP	NP	NP
5. Steel and concrete composite special concentrically braced frames	14.3	5½	2½	4½	NL	NL	160	100	NP
6. Steel and concrete composite ordinary braced frames	14.3	3½	2½	3	NL	NL	NP	NP	NP
7. Steel and concrete composite ordinary shear walls	14.3	5	3	4½	NL	NL	NP	NP	NP
8. Ordinary reinforced concrete shear walls[g]	14.2	5½	2½	4½	NL	NL	NP	NP	NP
F. SHEAR WALL-FRAME INTERACTIVE SYSTEM WITH ORDINARY REINFORCED CONCRETE MOMENT FRAMES AND ORDINARY REINFORCED CONCRETE SHEAR WALLS[g]	12.2.5.8 and 14.2	4½	2½	4	NL	NP	NP	NP	NP
G. CANTILEVERED COLUMN SYSTEMS DETAILED TO CONFORM TO THE REQUIREMENTS FOR:	12.2.5.2								
1. Steel special cantilever column systems	14.1	2½	2½	2½	35	35	35	35	35
2. Steel ordinary cantilever column systems	14.1	1¼	1¼	1¼	35	35	NP[l]	NP[l]	NP[l]
3. Special reinforced concrete moment frames[m]	12.2.5.5 and 14.2	2½	2½	2½	35	35	35	35	35
4. Intermediate reinforced concrete moment frames	14.2	1½	1½	1½	35	35	NP	NP	NP
5. Ordinary reinforced concrete moment frames	14.2	1	1¼	1	35	NP	NP	NP	NP
6. Timber frames	14.5	1½	1½	1½	35	35	35	NP	NP
H. STEEL SYSTEMS NOT SPECIFICALLY DETAILED FOR SEISMIC RESISTANCE, EXCLUDING CANTILEVER COLUMN SYSTEMS	14.1	3	3	3	NL	NL	NP	NP	NP

[a] Response modification coefficient, R, for use throughout the standard. Note that R reduces forces to a strength level, not an allowable stress level.
[b] Where the tabulated value of the overstrength factor, Ω_0, is greater than or equal to 2½, Ω_0 is permitted to be reduced by subtracting the value of 1/2 for structures with flexible diaphragms.
[c] Deflection amplification factor, C_d, for use in Sections 12.8.6, 12.8.7, 12.9.1.2, and 12.12.
[d] NL = Not Limited, and NP = Not Permitted. For metric SI units, multiply by 0.348 m/ft and round to the nearest 0.1 m.
[e] See Section 12.2.5.4 for a description of seismic force-resisting systems limited to buildings with a structural height, h_n, of 240 ft (73.2 m) or less.
[f] See Section 12.2.5.4 for seismic force-resisting systems limited to buildings with a structural height, h_n, of 160 ft (48.8 m) or less.
[g] In Section 2.3 of ACI 318. A shear wall is defined as a structural wall.
[h] In Section 2.3 of ACI 318. The definition of "special structural wall" includes precast and cast-in-place construction.
[i] An increase in structural height, h_n, to 45 ft (13.7 m) is permitted for single-story storage warehouse facilities.
[j] Steel ordinary concentrically braced frames are permitted in single-story buildings up to a structural height, h_n, of 60 ft (18.3 m) where the dead load of the roof does not exceed 20 lb/ft^2 (0.96 kN/m^2) and in penthouse structures.
[k] See Section 12.2.5.7 for limitations in structures assigned to Seismic Design Categories D, E, or F.
[l] See Section 12.2.5.6 for limitations in structures assigned to Seismic Design Categories D, E, or F.
[m] In Section 2.3 of ACI 318, the definition of "special moment frame" includes precast and cast-in-place construction.
[n] Cold-formed steel—special bolted moment frames shall be limited to one story in height in accordance with ANSI/AISI S400.
[o] Alternately, the seismic load effect including overstrength, E_{mh}, is permitted to be based on the expected strength determined in accordance with ANSI/AISI S400.
[p] Ordinary moment frame is permitted to be used in lieu of intermediate moment frame for Seismic Design Category B or C.
[q] Structural height, h_n, shall not be less than 60 ft (18.3m).

system as set forth in the applicable reference document listed in Table 12.2-1 and the additional requirements set forth in Chapter 14.

Nothing contained in this section shall prohibit the use of alternative procedures for the design of individual structures that demonstrate acceptable performance in accordance with the requirements of Section 1.3.1.3 of this standard.

12.2.1.1 Alternative Structural Systems Use of seismic force-resisting systems not contained in Table 12.2-1 shall be permitted contingent on submittal to and approval by the Authority Having Jurisdiction and independent structural design review of an accompanying set of design criteria and substantiating analytical and test data. The design criteria shall specify any limitations on system use, including Seismic Design Category and height; required procedures for designing the system's components and connections; required detailing; and the values of the response modification coefficient, R; overstrength factor, Ω_0; and deflection amplification factor, C_d. The submitted data shall establish the system's nonlinear dynamic characteristics and demonstrate that the design criteria result in a probability of collapse conditioned on the occurrence of MCE_R shaking not greater than 10% for Risk Category II structures. The conditional probability of collapse shall be determined based on a nonlinear analytical evaluation of the system and shall account for sources of uncertainty in quality of the design criteria, modeling fidelity, laboratory test data, and ground motions. Structural design review shall conform to the criteria of Section 16.5.

12.2.1.2 Elements of Seismic Force-Resisting Systems Elements of seismic force-resisting systems, including members and their connections, shall conform to the detailing requirements specified in Table 12.2-1 for the selected structural system.

EXCEPTION: Substitute elements that do not conform to the requirements specified in Table 12.2-1 shall be permitted contingent on submittal to and approval by the Authority Having Jurisdiction of all of the following:

1. In-depth description of the methodology used to evaluate equivalency of the substitute element for the seismic force-resisting system of interest, or reference to published documentation describing the methodology in depth.
2. Justification of the applicability of the equivalency methodology, including but not limited to consideration of the similarity of the forces transferred across the connection between the substitute and conforming elements and the balance of the seismic force-resisting system, and the similarity between the substitute and conforming element on the distribution of forces and displacements in the balance of the structure.
3. A design procedure for the substitute elements, including procedures to determine design strength stiffness, detailing, connections, and limitations to applicability and use.
4. Requirements for the manufacturing, installation, and maintenance of the substitute elements.
5. Experimental evidence demonstrating that the hysteretic characteristics of the conforming and substitute elements are similar through deformation levels anticipated in response to MCE_R shaking. The evaluation of experimental evidence shall include assessment of the ratio of the measured maximum strength to design strength; the ratio of the measured initial stiffness to design stiffness; the ultimate deformation capacity; and the cyclic strength and stiffness deterioration characteristics of the conforming and substitute elements.
6. Evidence of independent structural design review, in accordance with Section 16.5, or review by a third party acceptable to the Authority Having Jurisdiction, of conformance to the requirements of this section.

12.2.2 Combinations of Framing Systems in Different Directions Different seismic force-resisting systems are permitted to be used to resist seismic forces along each of the two orthogonal axes of the structure. Where different systems are used, the respective R, C_d, and Ω_0 coefficients shall apply to each system, including the structural system limitations contained in Table 12.2-1.

12.2.3 Combinations of Framing Systems in the Same Direction Where different seismic force-resisting systems are used in combination to resist seismic forces in the same direction, other than those combinations considered as dual systems, the design shall comply with the requirements of this section. The most stringent applicable structural system limitations contained in Table 12.2-1 shall apply, except as otherwise permitted by this section.

12.2.3.1 R, C_d, and Ω_0 Values for Vertical Combinations Where a structure has a vertical combination in the same direction, the following requirements shall apply.

1. Where the lower system has a lower response modification coefficient, R, the design coefficients (R, Ω_0, and C_d) for the upper system are permitted to be used to calculate the forces and drifts of the upper system. For the design of the lower system, the design coefficients (R, Ω_0, and C_d) for the lower system shall be used. Forces transferred from the upper system to the lower system shall be increased by multiplying by the ratio of the higher response modification coefficient to the lower response modification coefficient.
2. Where the upper system has a lower response modification coefficient, the design coefficients (R, Ω_0, and C_d) for the upper system shall be used for both systems.

EXCEPTIONS:

1. Rooftop structures not exceeding two stories in height and 10% of the total structure weight.
2. Other supported structural systems with a weight equal to or less than 10% of the weight of the structure.
3. Detached one- and two-family dwellings of light-frame construction.

12.2.3.2 Two-Stage Analysis Procedure for Vertical Combinations of Systems A two-stage equivalent lateral force procedure is permitted to be used for structures that have a flexible upper portion above a rigid lower portion, provided the design of the structure complies with all of the following.

(a) The stiffness of the lower portion shall be at least 10 times the stiffness of the upper portion. For purposes of determining this ratio, the base shear shall be computed and distributed vertically according to Section 12.8. Using these forces, the stiffness for each portion shall be computed as the ratio of the base shear for that portion to the elastic displacement, δ_e, computed at the top of that portion, considering the portion fixed at its base. For the lower portion, the applied forces shall include the reactions from the upper portion, modified as required in Item (d).

(b) The period of the entire structure shall not be greater than 1.1 times the period of the upper portion considered as a separate structure supported at the transition from the upper to the lower portion.
(c) The upper portion shall be designed as a separate structure using the appropriate values of R and ρ.
(d) The lower portion shall be designed as a separate structure using the appropriate values of R and ρ while meeting the requirements of Section 12.2.3.1. The reactions from the upper portion shall be those determined from the analysis of the upper portion, where the effects of the horizontal seismic load, E_h, are amplified by the ratio of the R/ρ of the upper portion over R/ρ of the lower portion. This ratio shall not be less than 1.0.
(e) The upper portion is analyzed with the equivalent lateral force or modal response spectrum procedure, and the lower portion is analyzed with the equivalent lateral force procedure.
(f) The structural height of the upper portion shall not exceed the height limits of Table 12.2-1 for the seismic force-resisting system used, where the height is measured from the base of the upper portion.
(g) Where Horizontal Irregularity Type 4 or Vertical Irregularity Type 3 exists at the transition from the upper to the lower portion, the reactions from the upper portion shall be amplified in accordance with Sections 12.3.3.4, 12.10.1.1, and 12.10.3.3, in addition to amplification required by Item (d).

12.2.3.3 R, C_d, and Ω_0 Values for Horizontal Combinations The value of the response modification coefficient, R, used for design in the direction under consideration shall not be greater than the least value of R for any of the systems used in that direction. The deflection amplification factor, C_d, and the overstrength factor, Ω_0, shall be consistent with R required in that direction.

EXCEPTION: Resisting elements are permitted to be designed using the least value of R for the different structural systems found in each independent line of resistance if the following three conditions are met: (1) Risk Category I or II building, (2) two stories or fewer above grade plane, and (3) use of light-frame construction or flexible diaphragms. The value of R used for design of diaphragms in such structures shall not be greater than the least value of R for any of the systems used in that same direction.

12.2.3.4 Two-Stage Analysis Procedure for One-Story Structures with Flexible Diaphragms and Rigid Vertical Elements A two-stage equivalent lateral force procedure shall be permitted to be used for determination of seismic design forces in vertical elements of one-story structures having flexible diaphragms supported by rigid vertical elements, provided that the structure conforms to all of the following:

(a) The structure shall comply with the requirements of Section 12.10.4.
(b) The seismic design forces to the vertical elements in each horizontal direction shall be taken as the sum of forces contributed by the effective seismic weight tributary to the diaphragm and forces contributed by the effective seismic weight tributary to the in-plane vertical elements, as follows:
   1. Forces contributed by the effective seismic weight tributary to the diaphragm shall be the reactions from diaphragm forces determined in accordance with Section 12.10.4.2.1, amplified by the ratio of R_{diaph} divided by R/ρ of the vertical seismic force-resisting system. This ratio shall not be taken as less than 1.0.
   2. Forces contributed by the effective seismic weight tributary to the in-plane rigid vertical elements shall be determined in accordance with Section 12.8 using the period calculated in accordance with Section 12.8.

12.2.4 Combination Framing Detailing Requirements Structural members common to different framing systems used to resist seismic forces in any direction shall be designed using the detailing requirements of Chapter 12 required by the highest response modification coefficient, R, of the connected framing systems.

12.2.5 System-Specific Requirements The structural framing system shall also comply with the following system-specific requirements of this section.

12.2.5.1 Dual System For a dual system, the moment frames shall be capable of resisting at least 25% of the design seismic forces. The total seismic force resistance is to be provided by the combination of the moment frames and the shear walls or braced frames in proportion to their rigidities.

12.2.5.2 Cantilever Column Systems Cantilever column systems are permitted as indicated in Table 12.2-1 and as follows. Where not otherwise limited by the applicable material standard, the required axial strength of individual cantilever column elements, considering only the load combinations that include seismic load effects, shall not exceed 15% of the available axial strength, including slenderness effects.

Foundation and other elements used to provide overturning resistance at the base of cantilever column elements shall be designed to resist the seismic load effects, including overstrength of Section 12.4.3.

12.2.5.3 Inverted Pendulum-Type Structures Regardless of the structural system selected, inverted pendulums as defined in Section 11.2 shall comply with this section. Supporting columns or piers of inverted pendulum-type structures shall be designed for the bending moment calculated at the base determined using the procedures given in Section 12.8 and varying uniformly to a moment at the top equal to one-half the calculated bending moment at the base.

12.2.5.4 Increased Structural Height Limit for Steel Eccentrically Braced Frames, Steel Special Concentrically Braced Frames, Steel Buckling-Restrained Braced Frames, Steel Special Plate Shear Walls, Steel and Concrete Coupled Composite Plate Shear Walls, Reinforced Concrete Ductile Coupled Walls, and Special Reinforced Concrete Shear Walls The limits on structural height, h_n, in Table 12.2-1 are permitted to be increased from 160 ft (50 m) to 240 ft (75 m) for structures assigned to Seismic Design Category D or E and from 100 ft (30 m) to 160 ft (50 m) for structures assigned to Seismic Design Category F, provided that the seismic force-resisting systems are limited to steel eccentrically braced frames, steel special concentrically braced frames, steel buckling-restrained braced frames, steel special plate shear walls, steel and concrete coupled composite plate shear walls, reinforced concrete ductile coupled walls, or special reinforced concrete cast-in-place shear walls and both of the following requirements are met:

1. The structure shall not have a Type 1 horizontal irregularity with a TIR > 1.4, and
2. The steel eccentrically braced frames, steel special concentrically braced frames, steel buckling-restrained braced frames, steel special plate shear walls, steel and concrete coupled composite plate shear walls, reinforced concrete

ductile coupled walls, or special reinforced cast-in-place concrete shear walls in any one plane shall resist no more than 60% of the total seismic forces in each direction, neglecting accidental torsional effects.

12.2.5.5 Special Moment Frames in Structures Assigned to Seismic Design Categories D through F For structures assigned to Seismic Design Categorie D, E, or F, where a special moment frame is required by Table 12.2-1 because of the structural system limitations, the frame shall be continuous to the base.

A special moment frame that is used but not required by Table 12.2-1 is permitted to be discontinued above the base and supported by a more rigid system with a lower response modification coefficient, R, provided that the requirements of Sections 12.2.3.1 and 12.3.3.5 are met.

12.2.5.6 Steel Ordinary Moment Frames

12.2.5.6.1 Seismic Design Category D or E

(a) Single-story steel ordinary moment frames in structures assigned to Seismic Design Category D or E are permitted up to a structural height, h_n, of 65 ft (20 m) where the dead load supported by and tributary to the roof does not exceed 20 lb/ft^2 (0.96 kN/m^2). In addition, the dead load of the exterior walls more than 35 ft (10.6 m) above the base tributary to the moment frames shall not exceed 20 lb/ft^2 (0.96 kN/m^2).

EXCEPTION: Single-story structures with steel ordinary moment frames whose purpose is to enclose equipment or machinery and whose occupants are engaged in maintenance or monitoring of that equipment, machinery, or their associated processes shall be permitted to be of unlimited height where the sum of the dead and equipment loads supported by and tributary to the roof does not exceed 20 lb/ft^2 (0.96 kN/m^2). In addition, the dead load of the exterior wall system, including exterior columns more than 35 ft (10.6 m) above the base, shall not exceed 20 lb/ft^2 (0.96 kN/m^2). For determining compliance with the exterior wall or roof load limits, the weight of equipment or machinery, including cranes, not self-supporting for all loads shall be assumed to be fully tributary to the area of the adjacent exterior wall or roof not to exceed 600 ft^2 (55.8 m^2), regardless of its height above the base of the structure.

(b) Steel ordinary moment frames in structures assigned to Seismic Design Category D or E not meeting the limitations set forth in Section 12.2.5.6.1.a are permitted within light-frame construction up to a structural height, h_n, of 35 ft (10.6 m) where neither the roof dead load nor the dead load of any floor above the base supported by and tributary to the moment frames exceeds 35 lb/ft^2 (1.68 kN/m^2). In addition, the dead load of the exterior walls tributary to the moment frames shall not exceed 20 lb/ft^2 (0.96 kN/m^2).

12.2.5.6.2 Seismic Design Category F Single-story steel ordinary moment frames in structures assigned to Seismic Design Category F are permitted up to a structural height, h_n, of 65 ft (20 m) where the dead load supported by and tributary to the roof does not exceed 20 lb/ft^2 (0.96 kN/m^2). In addition, the dead load of the exterior walls tributary to the moment frames shall not exceed 20 lb/ft^2 (0.96 kN/m^2).

12.2.5.7 Steel Intermediate Moment Frames

12.2.5.7.1 Seismic Design Category D

(a) Single-story steel intermediate moment frames in structures assigned to Seismic Design Category D are permitted up to a structural height, h_n, of 65 ft (20 m) where the dead load supported by and tributary to the roof does not exceed 20 lb/ft^2 (0.96 kN/m^2). In addition, the dead load of the exterior walls more than 35 ft (10.6 m) above the base tributary to the moment frames shall not exceed 20 lb/ft^2 (0.96 kN/m^2).

EXCEPTION: Single-story structures with steel intermediate moment frames whose purpose is to enclose equipment or machinery and whose occupants are engaged in maintenance or monitoring of that equipment, machinery, or their associated processes shall be permitted to be of unlimited height where the sum of the dead and equipment loads supported by and tributary to the roof does not exceed 20 lb/ft^2 (0.96 kN/m^2). In addition, the dead load of the exterior wall system, including exterior columns more than 35 ft (10.6 m) above the base, shall not exceed 20 lb/ft^2 (0.96 kN/m^2). For determining compliance with the exterior wall or roof load limits, the weight of equipment or machinery, including cranes, not self-supporting for all loads shall be assumed to be fully tributary to the area of the adjacent exterior wall or roof not to exceed 600 ft^2 (55.8 m^2), regardless of its height above the base of the structure.

(b) Steel intermediate moment frames in structures assigned to Seismic Design Category D not meeting the limitations set forth in Section 12.2.5.7.1.a are permitted up to a structural height, h_n, of 35 ft (10.6 m).

12.2.5.7.2 Seismic Design Category E

(a) Single-story steel intermediate moment frames in structures assigned to Seismic Design Category E are permitted up to a structural height, h_n, of 65 ft (20 m) where the dead load supported by and tributary to the roof does not exceed 20 lb/ft^2 (0.96 kN/m^2). In addition, the dead load of the exterior walls more than 35 ft (10.6 m) above the base tributary to the moment frames shall not exceed 20 lb/ft^2 (0.96 kN/m^2).

EXCEPTION: Single-story structures with steel intermediate moment frames whose purpose is to enclose equipment or machinery and whose occupants are engaged in maintenance or monitoring of that equipment, machinery, or their associated processes shall be permitted to be of unlimited height where the sum of the dead and equipment loads supported by and tributary to the roof does not exceed 20 lb/ft^2 (0.96 kN/m^2). In addition, the dead load of the exterior wall system, including exterior columns more than 35 ft (10.6 m) above the base, shall not exceed 20 lb/ft^2 (0.96 kN/m^2). For determining compliance with the exterior wall or roof load limits, the weight of equipment or machinery, including cranes, not self-supporting for all loads shall be assumed fully tributary to the area of the adjacent exterior wall or roof not to exceed 600 ft^2 (55.8 m^2), regardless of its height above the base of the structure.

(b) Steel intermediate moment frames in structures assigned to Seismic Design Category E not meeting the limitations set forth in Section 12.2.5.7.2.a are permitted up to a structural height, h_n, of 35 ft (10.6 m) where neither the roof dead load nor the dead load of any floor above the base supported by and tributary to the moment frames exceeds 35 lb/ft^2 (1.68 kN/m^2). In addition, the dead load of the exterior walls tributary to the moment frames shall not exceed 20 lb/ft^2 (0.96 kN/m^2).

12.2.5.7.3 Seismic Design Category F

(a) Single-story steel intermediate moment frames in structures assigned to Seismic Design Category F are permitted up to a structural height, h_n, of 65 ft (20 m) where the dead load supported by and tributary to the roof does not exceed 20 lb/ft^2 (0.96 kN/m^2). In addition, the dead load of the exterior walls tributary to the moment frames shall not exceed 20 lb/ft^2 (0.96 kN/m^2).

(b) Steel intermediate moment frames in structures assigned to Seismic Design Category F not meeting the limitations set forth in Section 12.2.5.7.3.a are permitted within light-frame construction up to a structural height, h_n, of 35 ft (10.6 m) where neither the roof dead load nor the dead load of any floor above the base supported by and tributary to the moment frames exceeds 35 lb/ft^2 (1.68 kN/m^2). In addition, the dead load of the exterior walls tributary to the moment frames shall not exceed 20 lb/ft^2 (0.96 kN/m^2).

12.2.5.8 Shear Wall–Frame Interactive Systems The shear strength of the shear walls of the shear wall–frame interactive system shall be at least 75% of the design story shear at each story. The frames of the shear wall–frame interactive system shall be capable of resisting at least 25% of the design story shear in every story.

12.3 DIAPHRAGM FLEXIBILITY, CONFIGURATION IRREGULARITIES, AND REDUNDANCY

12.3.1 Diaphragm Flexibility The structural analysis shall consider the relative stiffnesses of diaphragms and the vertical elements of the seismic force-resisting system. Unless a diaphragm can be idealized as either flexible or rigid in accordance with Sections 12.3.1.1, 12.3.1.2, or 12.3.1.3, the structural analysis shall explicitly include consideration of the stiffness of the diaphragm (i.e., semirigid modeling assumption). Diaphragms in hillside structures of light-frame construction shall be modeled in accordance with Section 12.3.1.4.

12.3.1.1 Flexible Diaphragm Condition Diaphragms constructed of untopped steel decking or wood structural panels are permitted to be idealized as flexible if any of the following conditions exist:

1. In structures where the vertical elements are steel braced frames; steel and concrete composite braced frames; or concrete, masonry, steel, or steel and concrete composite shear walls.
2. In one- and two-family dwellings.
3. In structures of light-frame construction where all of the following conditions are met:
   (a) Topping of concrete or similar materials is not placed over wood structural panel diaphragms except for nonstructural topping no greater than 1.5 in. (38 mm) thick, and
   (b) Each line of vertical elements of the seismic force-resisting system complies with the allowable story drift of Table 12.12-1.

12.3.1.2 Rigid Diaphragm Condition Diaphragms of concrete slabs or concrete-filled metal deck with span-to-depth ratios of 3 or less in structures that do not have a Type 2, 3, 4, or 5 Horizontal Structural Irregularity are permitted to be idealized as rigid.

12.3.1.3 Calculated Flexible Diaphragm Condition Diaphragms not satisfying the conditions of Sections 12.3.1.1 or 12.3.1.2 are permitted to be idealized as flexible provided

$$\frac{\delta_{MDD}}{\Delta_{ADVE}} > 2 \qquad (12.3\text{-}1)$$

where δ_{MDD} and Δ_{ADVE} are as shown in Figure 12.3-1. The loading used in this calculation shall be that prescribed in Section 12.8.

12.3.1.4 Diaphragms in Hillside Light-Frame Structures Diaphragms in hillside structures of light-frame construction shall be modeled as rigid or semirigid when they are seismically braced on one or more sides directly by the foundation or foundation stem wall or by a framed wall with a clear height of 2 ft (0.61 m) or less, and are also seismically braced on other sides by framed walls with any wall having a clear height of more than 7 ft (2.13 m). Other diaphragms shall be idealized as flexible or rigid in accordance with Sections 12.3.1.1, 12.3.1.2, or 12.3.1.3, or shall be modeled as semirigid.

12.3.2 Irregular and Regular Classification Structures shall be classified as having a structural irregularity based on the criteria in this section. Such classification shall be based on their structural configurations.

12.3.2.1 Horizontal Irregularity Structures that have one or more of the irregularity types listed in Tables 12.3-1 and 12.3-1a shall be designated as having a horizontal structural irregularity. Such structures assigned to the Seismic Design Categories listed in Table 12.3-1 shall comply with the requirements in the sections referenced in the tables.

12.3.2.1.1 Torsional Irregularity Ratio A Torsional Irregularity Ratio (TIR) shall be calculated for each story and for each accidental torsion case:

$$TIR = \frac{\Delta_{\max}}{\Delta_{\text{avg}}} \qquad (12.3\text{-}2)$$

where $\Delta_{\max}$ is the maximum story drift at the building's edge subjected to lateral forces using the equivalent lateral force procedure per Section 12.8 with the application of accidental torsion per Section 12.8.4 and $A_x = 1.0$; and Δ_{avg} is the average of the story drifts at the two opposing edges of the building

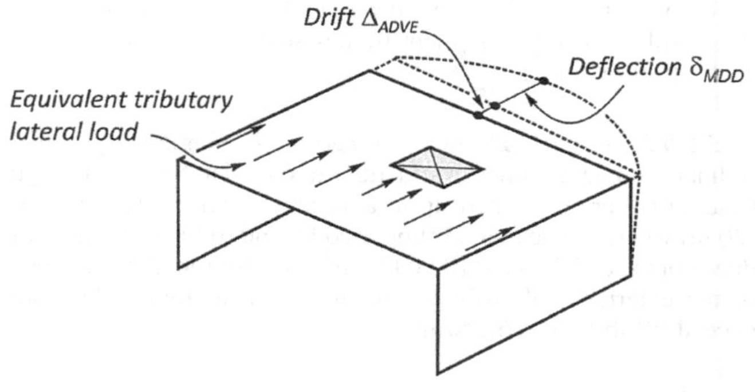

Figure 12.3-1. Flexible diaphragm.

Table 12.3-1. Horizontal Structural Irregularities.

Type	Description	Reference Section*	Seismic Design Category Application
1	**Torsional Irregularity:** Torsional irregularity, defined to exist where either: • More than 75% of any story's lateral strength below the diaphragm is provided at or on one side of the center of mass, or • The Torsional Irregularity Ratio (TIR) exceeds 1.2. The story lateral strength is the total strength of all seismic-resisting elements sharing the story shear for the direction under consideration.	Table 12.3-1a, 12.3.3.5, 12.5.1.2 12.7.3 12.8.4.3 12.8.6 16.3.4	B, C, D, E, F C, D, E, F B, C, D, E, F C, D, E, F C, D, E, F B, C, D, E, F
2.	**Reentrant Corner Irregularity:** Reentrant corner irregularity, defined to exist where both plan projections of the structure beyond a reentrant corner are greater than 20% of the plan dimension of the structure in the given direction.	12.3.3.5	D, E, F
3.	**Diaphragm Discontinuity Irregularity:** Diaphragm discontinuity irregularity, defined to exist where there is a diaphragm with an abrupt discontinuity or variation in stiffness, including one that has a cutout or open area greater than 25% of the gross enclosed diaphragm area, or a change in effective diaphragm stiffness of more than 50% from one story to the next.	12.3.3.5	D, E, F
4.	**Out-of-Plane Offset Irregularity:** Out-of-plane offset irregularity, defined to exist where there is a discontinuity in a lateral force-resistance path, such as an out-of-plane offset of at least one of the vertical elements.	12.3.3.5 12.3.3.5 12.7.3 16.3.4	B, C, D, E, F D, E, F B, C, D, E, F B, C, D, E, F
5.	**Nonparallel System Irregularity:** Nonparallel system irregularity, defined to exist where vertical lateral force-resisting elements are not parallel to the major orthogonal axes of the seismic force-resisting system.	12.5.3 12.5.4 12.7.3 16.3.4	C, D, E, F B, C, D, E, F B, C, D, E, F

*See Section 12.8.1.3 for requirement for any structure with an irregularity listed in this table.

Table 12.3-1a. Reference Sections for Type 1 Horizontal Torsional Irregularity (TIR).

Horizontal Torsional Irregularity (TIR)	Seismic Design Category	Reference Section	Description[a]
>1.2[b]	B, C, D, E, F	15.4.1	Inclusion of accidental torsion in nonbuilding structures not similar to buildings
		16.3.4	Mass offset in Nonlinear Response History analysis
		17.2.2	Irregularities in Seismically Isolated Buildings
		18.2.3.2	Applicability of Equivalent Lateral Force Procedure for structures with damping systems
	C, D, E, F	12.7.3	Use of 3D analytical model
	D, E, F	12.5.4	Use of orthogonal load combinations
		12.8.4.2 and 12.8.4.3	Inclusion of amplified accidental torsion
		12.8.6	Story drift for purpose of comparing to drift limits, computing P-Delta, and checking deformation compatibility must be computed at the edge with the largest drift.
	D, E, F	12.3.3.4	Amplification of design force for collectors, their connections, and connections of diaphragms to collectors or vertical elements of the seismic force-resisting system (SFRS)
>1.4	B, C, D, E, F	18.2.1.1	Minimum seismic base shear for structures with damping systems
	D, E, F	12.2.5.4	Increased structural height limits not allowed for certain systems
	D, E, F	12.5.4	Use of orthogonal load combinations
		12.8.4.2	Include accidental torsion
>1.4 in both directions	D, E, F	12.3.4.2.1	Criteria for setting redundancy factor = 1.3
>1.6	D, E, F	12.9.1.5	Restrictions on use mass offset for accidental torsion in Modal Response Spectrum analysis

[a]Refer to the reference section for the precise requirement.

[b]The requirements for TIR > 1.2 also apply if TIR ≤ 1.2 and if more than 75% of the story's lateral strength below the diaphragm is provided at or on one side of the center of mass per Table 12.3-1.

Notes: The requirements in this table are cumulative; thus for TIR > 1.4 or 1.6, all the requirements for TIR > 1.2 also apply.

Table 12.3-2. Vertical Structural Irregularities.

Type	Description	Reference Section*	Seismic Design Category Application
1a.	**Stiffness–Soft Story Irregularity** is defined to exist where there is a story in which the lateral stiffness is less than 70% of that in the story above or, where there are at least three stories above, less than 80% of the average stiffness of the three stories above.	-	-
1b.	**Stiffness–Extreme Soft Story Irregularity** is defined to exist where there is a story in which the lateral stiffness is less than 60% of that in the story above or, where there are at least three stories above, less than 70% of the average stiffness of the three stories above.	12.3.3.1	E, F
2.	**Vertical Geometric Irregularity** is defined to exist where the horizontal dimension of the seismic force-resisting system in any story is more than 130% of that in an adjacent story.	-	-
3.	**In-Plane Discontinuity in Vertical Lateral Force-Resisting Element Irregularity** is defined to exist where there is an in-plane offset of a vertical seismic force-resisting element resulting in overturning demands on supporting structural elements.	12.3.3.4 12.3.3.5	B, C, D, E, F D, E, F
4a.	**Discontinuity in Lateral Strength–Weak Story Irregularity** is defined to exist where the story lateral strength is less than that in the story above. The story lateral strength is the total lateral strength of all seismic force-resisting system elements resisting the story shear for the direction under consideration.	12.3.3.1	E, F
4b.	**Discontinuity in Lateral Strength–Extreme Weak Story Irregularity** is defined to exist where the story lateral strength is less than 65% of that in the story above. The story lateral strength is the total lateral strength of all seismic force-resisting system elements resisting the story shear for the direction under consideration.	12.3.3.1 12.3.3.2 12.3.3.3	E, F D B, C

*See Section 12.8.1.3 for requirement for any structure with an irregularity listed in this table.

determined using the same loading and diaphragm rigidity as applied for the determination of Δ_{max}.

For the purpose of computing Δ_{max} and Δ_{avg}, it shall be permitted to assume the diaphragm is rigid for structures with rigid or semirigid diaphragms. The TIR for the building is the maximum value from the values computed for each story and each direction. The TIR shall not apply to structures with flexible diaphragms.

12.3.2.2 Vertical Irregularity Structures that have one or more of the irregularity types listed in Table 12.3-2 shall be designated as having a vertical structural irregularity. Such structures assigned to the Seismic Design Categories listed in Table 12.3-2 shall comply with the requirements in the sections referenced in that table.

EXCEPTIONS:

1. Vertical structural irregularities of Types 1a and 1b in Table 12.3-2 do not apply where no design story drift ratio is greater than 130% of the story drift ratio of the next story above. For this exception the following need not be considered:
   (a) Torsional effects, and
   (b) The design story-drift ratio relationship of the top two stories of the structure.
2. Vertical structural irregularities of Types 1a and 1b in Table 12.3-2 are not required to be considered for one-story buildings in any Seismic Design Category or for two-story buildings assigned to Seismic Design Category B, C, or D.

12.3.3 Limitations and Additional Requirements for Systems with Structural Irregularities

12.3.3.1 Prohibited Vertical Irregularities for Seismic Design Category E or F Structures assigned to Seismic Design Category E or F that have vertical irregularities Type 1b, 4a, or 4b of Table 12.3-2 shall not be permitted.

EXCEPTION: Structures assigned to Seismic Design Category E or F that have vertical irregularity Type 4a shall be permitted where the story lateral strength is not less than 80% of that in the story above.

12.3.3.2 Prohibited Vertical Irregularities for Seismic Design Category D Structures assigned to Seismic Design Category D that have vertical irregularity Type 4b of Table 12.3-2 shall not be permitted.

12.3.3.3 Extreme Weak Stories for Seismic Design Categories B or C Structures assigned to Seismic Design Category B or C that have a vertical irregularity Type 4b, as defined in Table 12.3-2, shall not be more than two stories or 30 ft (9 m) in structural height, h_n.

EXCEPTION: The limit does not apply where the "weak" story is capable of resisting a total seismic force equal to Ω_0 times the design force prescribed in Section 12.8.

12.3.3.4 Elements Supporting Discontinuous Walls or Frames Structural elements supporting discontinuous walls or frames of structures that have horizontal irregularity Type 4 of Table 12.3-1 or vertical irregularity Type 3 of Table 12.3-2 shall be designed to resist the seismic load effects, including overstrength of Section 12.4.3. The connections of such discontinuous walls or frames to the supporting members shall be adequate to transmit the forces for which the discontinuous walls or frames were required to be designed.

12.3.3.5 Increase in Forces Caused by Irregularities for Seismic Design Categories D through F For structures assigned to Seismic Design Category D, E, or F and having a horizontal structural irregularity of Type 1, 2, 3, or 4 in Table 12.3-1 or a vertical structural irregularity of Type 3 in Table 12.3-2, the design forces determined from Section 12.10.1.1 shall be increased 25% at each diaphragm level where the irregularity occurs for the following elements of the seismic force-resisting system:

1. Connections of diaphragms to vertical elements and to collectors, and
2. Collectors and their connections, including connections to vertical elements of the seismic force-resisting system.

EXCEPTION: Forces calculated using the seismic load effects including overstrength of Section 12.4.3 need not be increased.

12.3.4 Redundancy A redundancy factor, ρ, shall be assigned to the seismic force-resisting system in each of two orthogonal directions for all structures in accordance with this section.

12.3.4.1 Conditions Where Value of ρ is 1.0 The value of ρ is permitted to equal 1.0 for the following:

1. Structures assigned to Seismic Design Category B or C;
2. Drift calculation and P-delta effects;
3. Design of nonstructural components;
4. Design of nonbuilding structures that are not similar to buildings;
5. Design of collector elements, splices, and their connections for which the seismic load effects including overstrength of Section 12.4.3 are used;
6. Design of members or connections where the seismic load effects, including overstrength of Section 12.4.3, are required for design;
7. Diaphragm seismic design forces determined using Equation (12.10-1), including the limits imposed by Equations (12.10-2) and (12.10-3);
8. Diaphragm seismic design forces determined in accordance with Section 12.10.4;
9. Structures with damping systems designed in accordance with Chapter 18; and
10. Design of structural walls for out-of-plane forces, including their anchorage.

12.3.4.2 Redundancy Factor, ρ, for Seismic Design Categories D through F For structures assigned to Seismic Design Categorie D, E, or F in the direction of interest, ρ shall be taken as 1.3 unless one of the following two conditions is met, whereby ρ is permitted to be taken as 1.0.

1. For each story where the story shear is greater than 35% of the base shear in the direction of interest, the following conditions shall be met with the notional removal of any lateral force resisting element or connection as indicated in Table 12.3-3. Structures that consist of a lateral force-resisting system (LFRS) with elements not listed in Table 12.3-3 need not consider the requirements of this provision.
   (a) There are at least two bays of seismic force-resisting framing on each side of the center of mass.
   (b) The reduction in lateral strength of the story in the direction of interest does not exceed 35%.
   (c) The resulting system with consideration of removal of the element does not have a Type 1 Horizontal Irregularity with a TIR > 1.4.
2. The structure does not have any horizontal irregularities as defined in Section 12.3.2.1 in the direction of interest at all levels, and the seismic force-resisting systems shall consist of at least two bays of seismic force-resisting perimeter framing on each side of the structure at each story resisting more than 35% of the base shear. The number of bays of shear walls shall be as defined in Table 12.3-3.

12.3.4.2.1 Redundancy Factor, ρ, for Structures with Type 1 Horizontal (Torsional) Irregularity with TIR>1.4 in Seismic Design Categories D through F For a structure assigned to Seismic Design Category D, E, or F, and having a Type 1 Horizontal irregularity with TIR > 1.4 in both orthogonal directions at any given level as defined in Table 12.3-1, ρ shall equal 1.3 for the entire structure in both orthogonal directions. For all other structures, including structures with a Type 1 irregularity, the requirements of 12.3.4.2 shall apply.

12.4 SEISMIC LOAD EFFECTS AND COMBINATIONS

12.4.1 Applicability All members of the structure, including those not part of the seismic force-resisting system, shall be designed using the seismic load effects of Section 12.4 unless otherwise exempted by this standard. Seismic load effects are the axial, shear, and flexural member forces resulting from application of horizontal and vertical seismic forces as set forth in Section 12.4.2. Where required, seismic load effects shall include overstrength, as set forth in Section 12.4.3.

12.4.2 Seismic Load Effect The seismic load effect, E, includes effects of both horizontal and vertical ground motions as shown in Eq. (12.4-1) and Eq. (12.4-2).

$$E = E_h + E_v \quad (12.4\text{-}1)$$

$$E = E_h - E_v \quad (12.4\text{-}2)$$

where E_h is the effect of horizontal seismic forces as defined in Section 12.4.2.1, and E_v is the vertical seismic effect applied in the vertical downward direction as determined in Section 12.4.2.2. E_v shall be subject to reversal to the upward direction in accordance with the applicable load combinations.

E_h, and E_v shall be used in load combinations 6 and 7 in Section 2.3.6 and in load combinations 8, 9, and 10 in Section 2.4.5.

12.4.2.1 Horizontal Seismic Load Effect The horizontal seismic load effect, E_h, shall be determined in accordance with Equation (12.4-3) as follows:

Table 12.3-3. Requirements for Each Story Resisting More than 35% of the Base Shear.

Lateral Force-Resisting Element	Condition
Braced frames, Light-frame walls with flat strap bracing	Removal of an individual brace, or connection thereto
Moment frames	Loss of moment resistance at the beam-to-column connections at both ends of a single beam
Shear walls or wall piers with a height-to-length ratio greater than 1.0	Removal of a shear wall bay, or wall pier, with a height-to-length ratio greater than 1.0, or connections thereto. The shear wall and wall pier height-to-length ratios are determined as shown in Figure 12.3-2. A shear wall bay is defined as the length of the wall divided by the story height (rounded down).
Cantilever columns	Loss of moment resistance at the base of any single cantilever column

$$E_h = \rho Q_E \quad (12.4\text{-}3)$$

where Q_E is the effects of horizontal seismic forces from V or F_p (where required by Section 12.5.3 or 12.5.4, such effects shall result from application of horizontal forces simultaneously in two directions at right angles to each other), and ρ is the redundancy factor, as defined in Section 12.3.4.

12.4.2.2 Vertical Seismic Load Effect The vertical seismic load effect, E_v, shall be determined in accordance with Equation (12.4-4a) as follows:

$$E_v = 0.2 S_{DS} D \quad (12.4\text{-}4a)$$

where S_{DS} is the design spectral response acceleration parameter at short periods obtained from Section 11.4.5, and D is the effect of dead load.

EXCEPTIONS:

1. Where the option to incorporate the effects of vertical seismic ground motions using the provisions of Section 11.9 is required elsewhere in this standard, the vertical seismic load effect, E_v, shall be determined in accordance with Equation (12.4-4b) as follows:

$$E_v = 0.3 S_{av} D \quad (12.4\text{-}4b)$$

   where S_{av} is the design vertical response spectral acceleration obtained from Section 11.9.3, and D is the effect of dead load.
2. The vertical seismic load effect, E_v, is permitted to be taken as zero for either of the following conditions:
   (a) In Equations (12.4-1), (12.4-2), (12.4-5), and (12.4-6) for structures assigned to Seismic Design Category B, and
   (b) In Equation (12.4-2) where determining demands on the soil–structure interface of foundations.

12.4.3 Seismic Load Effects Including Overstrength The seismic load effect including overstrength, E_m, includes effects of both horizontal and vertical ground motions as shown in Eq. (12.4-5) and Eq. (12.4-6). Where required, the seismic load effects including overstrength shall be determined in accordance with the following:

$$E_m = E_{mh} + E_v \quad (12.4\text{-}5)$$

$$E_m = E_{mh} - E_v \quad (12.4\text{-}6)$$

where

E_m = Seismic load effect including overstrength;
E_{mh} = Effect of horizontal seismic forces, including overstrength as defined in Section 12.4.3.1 or Section 12.4.3.2; and
E_v = Vertical seismic load effect as defined in Section 12.4.2.2. E_v is an applied load in the vertical downward direction. E_v shall be subject to reversal to the upward direction as per the associated load combinations.

E_{mh} and E_v shall be used in load combinations 6 and 7 in Section 2.3.6 and in load combinations 8, 9, and 10 in Section 2.4.5.

12.4.3.1 Horizontal Seismic Load Effect Including Overstrength The effect of horizontal seismic forces including overstrength, E_{mh}, shall be determined in accordance with Equation (12.4-7) as follows:

$$E_{mh} = \Omega_0 Q_E \quad (12.4\text{-}7)$$

where Q_E is the effects of horizontal seismic forces from V, F_{px}, or F_p as specified in Sections 12.8.1, 12.10, or 13.3.1 (where required by Section 12.5.3 or 12.5.4, such effects shall result from application of horizontal forces simultaneously in two directions at right angles to each other); and Ω_0 is the overstrength factor.

E_{mh} need not be taken as larger than E_{cl}, where E_{cl} is the capacity-limited horizontal seismic load effect as defined in Section 11.3.

12.4.3.2 Capacity-Limited Horizontal Seismic Load Effect Where capacity-limited design is required by the material reference document, the seismic load effect, including overstrength, shall be calculated with the capacity-limited horizontal seismic load effect, E_{cl}, substituted for E_{mh} in the load combinations of Section 2.3.6 and Section 2.4.5.

12.4.4 Minimum Upward Force for Horizontal Cantilevers for Seismic Design Categories D through F For a structure assigned to Seismic Design Category D, E, or F, horizontal cantilever structural members shall be designed for a supplemental basic load combination consisting of a minimum net upward force of 0.2 times the dead load for strength design, or 0.14 times the dead load for allowable stress design.

12.5 DIRECTION OF LOADING

12.5.1 Direction of Loading Criteria The directions of application of seismic forces used in the design shall be those that produce the most critical load effects. In lieu of an analysis that finds the critical direction of loading for each element of the seismic force resisting system it is permitted to satisfy this requirement using one of the two methods defined in 12.5.1.1 and 12.5.1.2 subject to the conditions specified in Section 12.5.2, 12.5.3, or 12.5.4.

12.5.1.1 Independent Directional Procedure The design seismic forces shall be applied independently in each of the two orthogonal directions.

12.5.1.2 Orthogonal Directional Combination Procedure The design seismic forces shall be applied in two orthogonal directions simultaneously using one of the two following approaches. Seismic force resisting system member and foundation strength and drift requirements shall be satisfied for all combinations.

(a) For a structure analyzed using the equivalent lateral force analysis procedure of Section 12.8 or the modal response spectrum analysis procedure of Section 12.9.1, the effects of 100% of the seismic forces for one direction shall be combined with the effects of 30% of the forces for the perpendicular direction. Combinations of force effects shall be computed such that each axis has the 100% factor in positive and negative directions.
(b) For a structure analyzed using the linear response history procedure of Section 12.9.2 or the nonlinear response history procedure of Chapter 16, orthogonal pairs of ground motion acceleration histories shall be applied simultaneously.

12.5.2 Seismic Design Category B For a structure assigned to Seismic Design Category B, any method in Section 12.5.1 shall be permitted.

12.5.3 Seismic Design Category C For a structure assigned to Seismic Design Category C, any method in Section 12.5.1 shall be permitted, unless the structure has a Type 5 horizontal irregularity, in which case the Independent Directional Procedure shall not be permitted.

12.5.4 Seismic Design Categories D through F For a structure assigned to Seismic Design Category D, E, or F, the Independent Directional Procedure method of Section 12.5.1.1 shall be permitted unless one or more of the conditions exist.

(a) The structure has a Type 1 horizontal irregularity.
(b) The structure has a Type 5 horizontal irregularity.
(c) A column that forms part of two or more intersecting seismic force-resisting systems and is subjected to axial load due to seismic forces acting along either principal plan axis exceeding 20% of the axial design strength of the column.
(d) A concrete shear wall that has connected intersecting wall planes and an axial demand exceeding 20% of the axial design strength at any location in the wall due to seismic forces acting along either principal plan axis.
(e) A masonry shear wall that has connected intersecting wall planes and an axial demand exceeding 20% of the axial design strength at any location in the wall due to seismic forces acting along either principal plan axis.
(f) A wood or cold-formed steel light-frame shear wall system that has connected intersecting wall planes and an axial load exceeding 20% of the axial design strength of that wall element at any location within the wall due to seismic forces acting along either principal plan axis.
(g) Cantilever column systems.

EXCEPTION:
Where Items (a) and (b) do not apply, the Independent Directional Procedure is permitted to be used for components of the seismic force-resisting system not listed in Items (c) through (g).

Foundation elements and connections that support elements of structures with any of the preceding conditions (c) through (g) shall be designed using the Orthogonal Direction Combination Procedure of Section 12.5.1.2. In addition, foundations and connections supporting lateral force-resisting system (LFRS) elements that are not coplanar shall be designed using the Orthogonal Direction Combination Procedure.

12.6 ANALYSIS PROCEDURE SELECTION

The structural analysis required by Chapter 12 shall be completed in accordance with the requirements of (a) Equivalent Lateral Force Procedure of Section 12. 8, (b) Modal Response Spectrum Analysis of Section 12.9.1, (c) Linear Response History Analysis of Section 12.9.2, or (d) an analysis procedure approved by the Authority Having Jurisdiction. Nonlinear Response History Analysis of Chapter 16 is allowed for any structure.

12.7 MODELING CRITERIA

12.7.1 Foundation Modeling For purposes of determining seismic loads, it is permitted to consider the structure to be fixed at the base. Alternatively, where foundation flexibility is considered, it shall be in accordance with Section 12.13.3 or Chapter 19.

12.7.2 Effective Seismic Weight The effective seismic weight, W, of a structure shall include the dead load above the base and other loads above the base as follows:

1. In areas used for storage, a minimum of 25% of the floor live load shall be included.
2. Where provision for partitions is made or where required by Section 4.3.2 in the floor load design, the actual partition weight or a minimum weight of 10 lb/ft^2 (0.48 kN/m^2) of floor area, whichever is greater.
3. Total operating weight of permanent equipment.
4. Where the flat roof snow load, P_f, exceeds 45 lb/ft^2 (2.16 kN/m^2), 15% of the uniform design snow load, regardless of actual roof slope.
5. Weight of landscaping and other materials at roof gardens and similar areas, as defined in Section 3.1.4.
6. Weight of fluids and bulk material expected to be present during normal use.

EXCEPTIONS:

(a) Where the total weight of Items 1, 3, 5, and 6 adds no more than 5% to the effective seismic weight at that level, the weight of these items need not be included.
(b) Floor live load in public garages and open parking structures need not be included.
(c) The weights of fluids included as effective seismic weight are permitted to be reduced as justified by a rational fluid-structure interaction analysis.

12.7.3 Structural Modeling A mathematical model of the structure shall be constructed for the purpose of determining member forces and structure displacements resulting from applied loads and any imposed displacements or P-delta effects. The model shall include the stiffness and strength of elements that are significant to the distribution of forces and deformations in the structure and represent the spatial distribution of mass and stiffness throughout the structure.

In addition, the model shall comply with the following:

1. Stiffness properties of concrete and masonry elements shall consider the effects of cracked sections.
2. For steel moment frame systems, the contribution of panel zone deformations to displacement and drift shall be included.

Structures that have horizontal structural irregularity Type 1, 4, or 5 of Table 12.3-1 shall be analyzed using a 3D representation. Where a 3D model is used, a minimum of three degrees of freedom consisting of translation in two orthogonal plan directions and rotation about the vertical axis shall be included at each level of the structure. Where the diaphragms have not been classified as rigid or flexible in accordance with Section 12.3.1, the model shall include representation of the diaphragm's stiffness characteristics and, when dynamic analysis is performed, sufficient degrees of freedom as are required to account for the participation of the diaphragm in the structure's dynamic response. When modal response spectrum or response history analysis is performed, a minimum of three dynamic degrees of freedom consisting of translation in two orthogonal plan directions and torsional rotation about the vertical axis at each level of the structure shall be used.

EXCEPTION: Analysis using a 3D representation is not required for structures with flexible diaphragms that have Type 4 horizontal structural irregularities.

12.7.4 Interaction Effects Moment-resisting frames that are enclosed or adjoined by elements that are more rigid and not considered to be part of the seismic force-resisting system shall be designed so that the action or failure of those elements will not impair the vertical load and seismic force-resisting capability of the frame. The design shall provide for the effect of these rigid elements on the structural system at structural deformations corresponding to the design story drift (Δ) as determined in Section 12.8.6. In addition, the effects of these elements shall be considered where determining whether a structure has one or more of the irregularities defined in Section 12.3.2.

12.8 EQUIVALENT LATERAL FORCE (ELF) PROCEDURE

12.8.1 Seismic Base Shear The seismic base shear, V, in a given direction shall be determined in accordance with the following equation:

$$V = C_s W \tag{12.8-1}$$

where

C_s = The seismic response coefficient determined in accordance with Section 12.8.1.1, and
W = The effective seismic weight per Section 12.7.2.

12.8.1.1 Calculation of Seismic Response Coefficient Where the design spectral acceleration parameter S_a determined in accordance with either Section 11.4.5.1 or Chapter 21 is available, either Method 1 or Method 2 is permitted to determine the seismic response coefficient, C_s. Where Exception 2 of Section 11.4.5 applies, Method 1 shall not be used. The lower bound for the seismic response coefficient, C_s, provided in Item 3 shall be applicable for both Method 1 and Method 2.

Method 1: The seismic response coefficient, C_s, shall be determined in accordance with Equation (12.8-2).

$$C_s = \frac{S_a}{\left(\frac{R}{I_e}\right)} \tag{12.8-2}$$

where

S_a = Design spectral response acceleration parameter defined in Section 11.4.5.1 and determined for the period T defined in Section 12.8.2;
R = Response modification factor in Table 12.2-1; and
I_e = Importance Factor determined in accordance with Section 11.5.1.

Where Equation (12.8-2) is used and the period T is less than the period at which S_a is maximum, the maximum value of S_a shall be used.

Method 2: The seismic response coefficient, C_s, shall be determined in accordance with the following:

$$C_s = \frac{S_{DS}}{\left(\frac{R}{I_e}\right)} \tag{12.8-3}$$

where

S_{DS} = Design spectral response acceleration parameter in the short period range as determined from Section 11.4.5 or 11.4.8;
R = Response modification factor in Table 12.2-1; and
I_e = Importance Factor determined in accordance with Section 11.5.1.

The value of C_s computed in accordance with Equation (12.8-3) need not exceed the following:

for $T \le T_L$,

$$C_s = \frac{S_{D1}}{T\left(\frac{R}{I_e}\right)} \tag{12.8-4}$$

and for $T > T_L$,

$$C_s = \frac{S_{D1} T_L}{T^2\left(\frac{R}{I_e}\right)} \tag{12.8-5}$$

where I_e and R are as defined in this section;

S_{D1} = Design spectral response acceleration parameter at a period of 1.0 s, as determined from Section 11.4.5 or 11.4.6;
T = Fundamental period of the structure(s) determined in Section 12.8.2; and
T_L = Long-period transition period(s) determined in Section 11.4.6.

Lower Bounds for C_s for Both Methods: C_s shall not be less than

$$C_s = 0.044 S_{DS} I_e \ge 0.01 \tag{12.8-6}$$

In addition, for structures located where S_1 is equal to or greater than 0.6, C_s shall not be less than

$$C_s = 0.5 S_1 / (R/I_e) \tag{12.8-7}$$

where S_1 is the mapped maximum considered earthquake spectral response acceleration parameter determined in accordance with Section 11.4.2 or 11.4.4.

12.8.1.2 Soil–Structure Interaction Reduction A soil–structure interaction reduction is permitted where determined using Chapter 19 or other generally accepted procedures approved by the Authority Having Jurisdiction.

12.8.1.3 Maximum S_{DS} Value in Determination of C_s and E_v The values of C_s and E_v are permitted to be calculated using a value of S_{DS} equal to 1.0, but not less than 70% of the value of S_{DS}, as defined in Section 11.4.5, provided that all of the following criteria are met:

1. The structure does not have irregularities, as defined in Section 12.3.2;
2. The structure does not exceed five stories above the lower of the base or grade plane as defined in Section 11.2. Where present, each mezzanine level shall be considered a story for the purposes of this limit;
3. The structure has a fundamental period, T, that does not exceed 0.5 s, as determined using Section 12.8.2;
4. The structure meets the requirements necessary for the redundancy factor, ρ, to be permitted to be taken as 1.0, in accordance with Section 12.3.4.2;
5. The site soil properties are not classified as Site Class E or F, as defined in Section 11.4.3;
6. The structure is classified as Risk Category I or II, as defined in Section 1.5.1; and
7. The response modification coefficient, R, as defined in Table 12.2-1, is 3 or greater.

12.8.2 Period Determination The fundamental period of the structure, T, in the direction under consideration shall be established using the structural properties and deformational characteristics of the resisting elements in a properly

Table 12.8-1. Coefficient for Upper Limit on Calculated Period.

Design Spectral Response Acceleration Parameter at 1 s, S_{D1}	Coefficient C_u
≥ 0.4	1.4
0.3	1.4
0.2	1.5
0.15	1.6
≤ 0.1	1.7

substantiated analysis. The fundamental period, T, shall not exceed the product of the coefficient for upper limit on calculated period, C_u, from Table 12.8-1 and the approximate fundamental period, T_a, determined in accordance with Section 12.8.2.1. As an alternative to performing an analysis to determine the fundamental period, T, it is permitted to use the approximate building period, T_a, calculated in accordance with Section 12.8.2.1, directly.

12.8.2.1 Approximate Fundamental Period The approximate fundamental period (T_a), in seconds, shall be determined from the following equation:

$$T_a = C_t h_n^x \tag{12.8-8}$$

where h_n is the structural height as defined in Section 11.2, and the coefficients C_t and x are determined from Table 12.8-2.

Alternatively, it is permitted to determine the approximate fundamental period, T_a, in seconds, from the following equation for structures not exceeding 12 stories above the base as defined in Section 11.2 where the seismic force-resisting system consists entirely of concrete or steel moment-resisting frames and the average story height is at least 10 ft (3 m):

$$T_a = 0.1N \tag{12.8-9}$$

where N is the number of stories above the base.

The approximate fundamental period, T_a, in seconds, for masonry or concrete shear wall structures not exceeding 120 ft (36.6 m) in height is permitted to be determined from Equation (12.8-10) as follows:

$$T_a = \frac{C_q}{\sqrt{C_w}} h_n \tag{12.8-10}$$

where

C_q = 0.0019 ft (0.00058 m), and

C_w is calculated from Equation (12.8-11) as follows:

$$C_w = \frac{100}{A_B} \sum_{x}^{i=1} \frac{A_i}{\left[1 + 0.83\left(\frac{h_n}{D_i}\right)^2\right]} \tag{12.8-11}$$

where

A_B = Area of base of structure [ft^2 (m^2)];
A_i = Web area of shear wall i [ft^2 (m^2)];
D_i = Length of shear wall i [ft (m)]; and
x = Number of shear walls in the building effective in resisting lateral forces in the direction under consideration.

12.8.3 Vertical Distribution of Seismic Forces The lateral seismic force, F_x (kip or kN), induced at any level shall be determined from the following equations:

$$F_x = C_{vx} V \tag{12.8-12}$$

$$C_{vx} = \frac{w_x h_x^k}{\sum_n^{i=1} w_i h_i^k} \tag{12.8-13}$$

where

C_{vx} = Vertical distribution factor;
V = Total design lateral force or shear at the base of the structure [kip (kN)];
w_i, w_x = Portion of the total effective seismic weight of the structure (W) located or assigned to level i or x;
h_i, h_x = Height [ft (m)] from the base to level i or x; and
k = Exponent related to the structure period as follows:

For structures that have a period of 0.5 s or less, $k = 1$;
for structures that have a period of 2.5 s or more, $k = 2$; and
for structures that have a period between 0.5 and 2.5 s, k shall be 2 or shall be determined by linear interpolation between 1 and 2.

12.8.4 Horizontal Distribution of Forces The seismic design story shear in any story, V_x [kip (kN)], shall be determined from the following equation:

$$V_x = \sum_{i=x}^{n} F_i \tag{12.8-14}$$

where F_i is the portion of the seismic base shear, V [kip (kN)], induced at level i.

Table 12.8-2. Values of Approximate Period Parameters C_t and x.

Structure Type	C_t	x
Moment-resisting frame systems in which the frames resist 100% of the required seismic force and are not enclosed or adjoined by components that are more rigid and will prevent the frames from deflecting where subjected to seismic forces:		
Steel moment-resisting frames	0.028 (0.0724)*	0.8
Concrete moment-resisting frames	0.016 (0.0466)*	0.9
Steel eccentrically braced frames in accordance with Table 12.2-1, line B1 or D1	0.03 (0.0731)*	0.75
Steel buckling-restrained braced frames	0.03 (0.0731)*	0.75
All other structural systems	0.02 (0.0488)*	0.75

*SI equivalents in parentheses.

The seismic design story shear, V_x [kip (kN)], shall be distributed to the various vertical elements of the seismic force-resisting system in the story under consideration based on the relative lateral stiffness of the vertical resisting elements and the diaphragm.

12.8.4.1 Inherent Torsion For diaphragms that are not flexible, the distribution of lateral forces at each level shall consider the effect of the inherent torsional moment, M_t, resulting from eccentricity between the locations of the center of mass and the center of rigidity. For flexible diaphragms, the distribution of forces to the vertical elements shall account for the position and distribution of the masses supported.

12.8.4.2 Accidental Torsion

12.8.4.2.1 Accidental Torsion Irregularities Where diaphragms are not flexible, accidental torsion shall be applied to all structures for determination if a horizontal irregularity exists. Accidental torsion shall be included in the analysis and design of structures with the following irregularities as specified in Table 12.3-1:

1. A structure assigned to Seismic Design Category B with Type 1 horizontal structural irregularity and TIR > 1.4.
2. A structure assigned to Seismic Design Category C, D, E, or F with Type 1 horizontal structural irregularity.

12.8.4.2.2 Accidental Torsion Applications Where the consideration of accidental torsion is required, it shall be applied as follows.

1. Accidental torsional moments, M_{ta}, shall be combined with the inherent torsional moment, M_t, resulting from differences in the locations of the center of mass and the center of rigidity.
2. Accidental torsional moments, M_{ta}, shall be determined using an assumed displacement of the center of mass each way from its actual location by a distance equal to 5% of the dimension of the structure perpendicular to the direction of the applied forces.
3. Where earthquake forces are applied concurrently in two orthogonal directions, the required 5% displacement of the center of mass need not be applied in both of the orthogonal directions simultaneously but shall be applied in the direction that produces the greater effect for each element considered.

12.8.4.3 Amplification of Accidental Torsional Moment A structure assigned to Seismic Design Category C, D, E, or F, where Type 1 horizontal structural irregularity exists as defined in Table 12.3-1 shall have the effects accounted for by multiplying M_{ta} at each level by a torsional amplification factor, A_x, as illustrated in Figure 12.8-1 and determined from the following equation:

$$A_x = \left(\frac{\delta_{max}}{1.2\delta_{avg}}\right)^2 \quad (12.8\text{-}15)$$

where δ_{max} is the maximum displacement [in. (mm)] at level x, computed assuming $A_x = 1$, and δ_{avg} is the average of the displacements [in. (mm)] at the extreme points of the structure at level x, computed assuming $A_x = 1$.

For the purpose of computing δ_{max} and δ_{avg}, it shall be permitted to assume the diaphragm is rigid. The torsional amplification factor, A_x, shall not be less than 1 and is not required to exceed 3.0. The more severe loading for each element shall be considered for design.

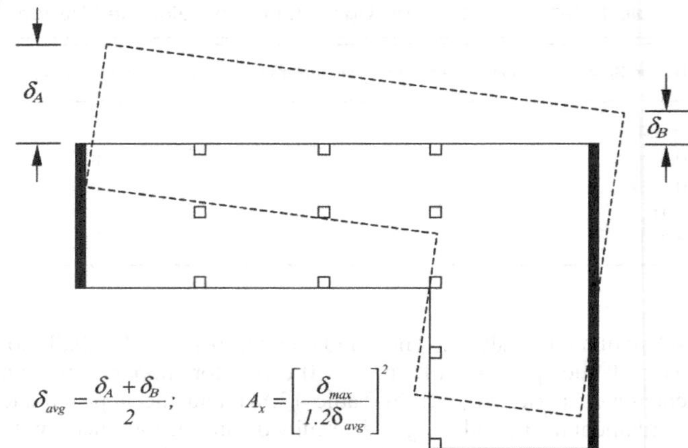

Figure 12.8-1. Torsional amplification factor, A_x.

12.8.5 Overturning The structure shall be designed to resist overturning effects caused by the seismic forces determined in Section 12.8.3.

12.8.6 Displacement and Drift Determination Displacements and drifts shall be determined as required by this section.

12.8.6.1 Minimum Base Shear and Load Combination for Computing Displacement and Drift The elastic analysis of the seismic force-resisting system for computing displacement and drift shall be made using the prescribed seismic design forces of Section 12.8 using a load factor of 1.0 on E_h, in the presence of expected gravity loads. Expected gravity loads shall be taken as no less than $1.0D + 0.5L$, L shall be taken as $0.8L_O$ for live loads that exceed 100 lb/ft^2 (4.79 kN/m^2) and $0.4L_O$ for all other live loads, and L_O is the unreduced design live load (see Table 4-1).

EXCEPTION: Equation (12.8-6) need not be considered for computing drift.

12.8.6.2 Period for Computing Displacement and Drift For determining displacements and drifts, it is permitted to determine the elastic displacements, δ_e, using seismic design forces based on the computed fundamental period of the structure without the upper limit, C_uT_a, specified in Section 12.8.2.

12.8.6.3 Design Earthquake Displacement The Design Earthquake Displacement, δ_{DE}, shall be determined at the location of an element or component using Equation (12.8-16) or as permitted in Chapter 16, 17, or 18.

$$\delta_{DE} = \frac{C_d\delta_e}{I_e} + \delta_{di} \quad (12.8\text{-}16)$$

where

C_d = Deflection amplification factor in Table 12.2-1.
I_e = Importance Factor determined in accordance with Section 11.5.1;
δ_e = Elastic displacement computed under design earthquake forces, including the effects of accidental torsion and torsional amplification as applicable; and
δ_{di} = Displacement due to diaphragm deformation corresponding to the design earthquake, including Section 12.10 diaphragm forces.

12.8.6.4 Maximum Considered Earthquake Displacement The Maximum Considered Earthquake Displacement, δ_{MCE}, shall be determined at the location under consideration using Equation (12.8-17) or as permitted in Chapter 16, 17, or 18.

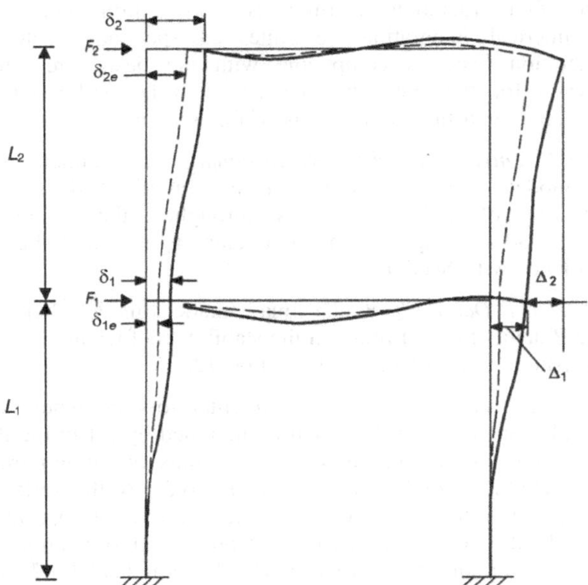

Note: Δ_i = story drift; Δ_i / L_i = story drift ratio; δ_x = total displacement; i = level under consideration.

Story Level 1: F_1 = strength-level design earthquake force; δ_{1e} = elastic displacement computed under strength-level design earthquake forces; $\delta_1 = C_d \delta_{1e} / I_E$ = amplified displacement; $\Delta_1 = \delta_1 \leq \Delta_a$ (Table 12.12-1).

Story Level 2: F_2 = strength-level design earthquake force; δ_{2e} = elastic displacement computed under strength-level design earthquake forces; $\delta_2 = C_d \delta_{2e} / I_E$ = amplified displacement; $\Delta_2 = C_d(\delta_{2e} - \delta_{1e}) / I_E \leq \Delta_a$ (Table 12.12-1).

Figure 12.8-2. Story drift determination.

$$\delta_{MCE} = 1.5\left[\frac{R\delta_e}{I_e} + \delta_{di}\right] \quad (12.8\text{-}17)$$

where R is the response modification coefficient in Table 12.2-1.

12.8.6.5 Design Story Drift Determination The Design Story Drift, Δ, shall be computed as the difference of the Design Earthquake Displacements, δ_{DE}, as determined in accordance with Section 12.8.6, at the centers of mass at the top and bottom of the story under consideration. Where centers of mass do not align vertically, it is permitted to compute the deflection at the bottom of the story based on the vertical projection of the center of mass at the top of the story.

Diaphragm deformation may be neglected in determining the design story drift. Diaphragm rotation shall be considered as follows. For structures assigned to Seismic Design Category C, D, E, or F that have horizontal irregularity Type 1 of Table 12.3-1, the design story drift, Δ, shall be computed as the largest difference of the Design Earthquake Displacements of vertically aligned points at the top and bottom of the story under consideration along any of the edges of the structure, including the effects of diaphragm rotation.

12.8.7 P-Delta Effects P-delta effects on story shears and moments, the resulting member forces and moments, and the story drifts induced by these effects are not required to be considered where the stability coefficient, θ, as determined by the following equation is equal to or less than 0.10:

$$\theta = \frac{P_x / h_{sx}}{V_x / \Delta_{xe}} \quad (12.8\text{-}18)$$

where

P_x = Total vertical design load [kip (kN)] at and above level x per Section 12.8.6.1 (where computing P_x, no individual load factor need exceed 1.0);

V_x / Δ_{xe} = Story stiffness at level x [kip/in (kN/mm)], calculated as the ratio of the seismic design shear, V_x, divided by the corresponding elastic story drift, Δ_{xe}; and

h_{sx} = Story height below level x [in. (mm)].

The story stiffness, V_x / Δ_{xe}, shall be computed at the location required for the design story drift per Section 12.8.6.5 and may be determined without consideration of diaphragm deformation.

The stability coefficient, θ, shall not exceed θ_{max}, determined as follows:

$$\theta_{max} = \frac{0.5}{\beta C_d} \leq 0.25 \quad (12.8\text{-}19)$$

where C_d is the deflection amplification factor in Table 12.2-1, and β is the ratio of shear demand to design shear capacity for the story between levels x and $x - 1$. The value of β is permitted to be conservatively taken as 1.0, and shall not be taken less than $1.25/\Omega_0$. The value of θ_{max} determined from Equation (12.8-19) need not be taken less than 0.1.

Where the stability coefficient, θ, is greater than 0.10 but less than or equal to θ_{max}, the incremental factor related to P-delta effects on displacements and member forces shall be determined by rational analysis. Alternatively, it is permitted to multiply displacements and member forces by $1.0/(1 - \theta)$. Where θ is greater than θ_{max}, the structure is potentially unstable and shall be redesigned. Where the P-delta effect is included in an automated analysis, Equation (12.8-19) shall still be satisfied; however, the value of θ computed from Equation (12.8-18) using the results of the P-delta analysis is permitted to be divided by $(1 + \theta)$ before checking Equation (12.8-19).

12.9 LINEAR DYNAMIC ANALYSIS

12.9.1 Modal Response Spectrum Analysis

12.9.1.1 Number of Modes An analysis shall be conducted to determine the natural modes of vibration for the structure. The analysis shall include a sufficient number of modes to obtain a combined modal mass participation of 100% of the structure's mass. For this purpose, it shall be permitted to represent all modes with periods less than 0.05 s in a single rigid body mode that has a period of 0.05 s.

EXCEPTION: Alternatively, the analysis shall be permitted to include a minimum number of modes to obtain a combined modal mass participation of at least 90% of the actual mass in each orthogonal horizontal direction of response considered in the model.

12.9.1.2 Modal Response Parameters The value for each force-related design parameter of interest, including story drifts, support forces, and individual member forces for each mode of response, shall be computed using the properties of each mode and the response spectra defined in either Section 11.4.5 divided by the quantity R/I_e. The value for displacement and drift quantities shall be multiplied by the quantity C_d/I_e.

12.9.1.3 Combined Response Parameters The value for each parameter of interest calculated for the various modes shall be combined using the square root of the sum of the squares (SRSS) method, the complete quadratic combination (CQC) method, the complete quadratic combination method as modified by ASCE 4 (CQC-4), or an approved equivalent approach. The CQC or the CQC-4 method shall be used for each of the modal values where closely spaced modes have significant cross-correlation of translational and torsional response.

12.9.1.4 Scaling Design Values of Combined Response A base shear, V, shall be calculated in each of the two orthogonal horizontal directions using the calculated fundamental period of the structure, T, in each direction and the procedures of Section 12.8.

12.9.1.4.1 Scaling of Forces Where the calculated fundamental period exceeds $C_u T_a$ in a given direction, $C_u T_a$ shall be used in lieu of T in that direction. Where the combined response for the modal base shear, V_t, is less than 100% of the calculated base shear, V, using the equivalent lateral force procedure, the forces shall be multiplied by V/V_t, where V is the equivalent lateral force procedure base shear, calculated in accordance with this section and Section 12.8, and V_t is the base shear from the required modal combination.

12.9.1.4.2 Scaling of Drifts Where the combined response for the modal base shear, V_t, is less than $C_s W$, and where C_s is determined in accordance with Equation (12.8-7), displacements shall be multiplied by $C_s W/V_t$.

12.9.1.5 Horizontal Shear Distribution The distribution of horizontal shear shall be in accordance with Section 12.8.4, The effects of accidental torsion shall be accounted for by applying a static accidental torsional moment, M_{ta}, determined in accordance with Sections 12.8.4.2 and 12.8.4.3, to the mathematical model, and combining the results with the scaled design values computed in accordance with Section 12.9.1.4.

EXCEPTION: For structures without a horizontal irregularity Type 1, the effects of accidental torsion may be included in the dynamic analysis model in lieu of applying M_{ta}. When the effects of accidental torsion are included in the dynamic analysis model and TIR $\leq$ 1.6, amplification of torsion in accordance with Section 12.8.4.3 is not required. If TIR > 1.6, accidental torsion shall be added as a static load case in accordance with Section 12.8.4.3.

12.9.1.6 P-Delta Effects The P-delta effects shall be determined in accordance with Section 12.8.7. The base shear used to determine the displacement and drifts shall be determined in accordance with Section 12.8.6.

12.9.1.7 Soil–Structure Interaction Reduction A soil–structure interaction reduction is permitted where determined using Chapter 19 or other generally accepted procedures approved by the Authority Having Jurisdiction.

12.9.1.8 Structural Modeling A mathematical model of the structure shall be constructed in accordance with Section 12.7.3, except that all structures designed in accordance with this section shall be analyzed using a three dimensional (3D) representation. Where the diaphragms have not been classified as rigid in accordance with Section 12.3.1, the model shall include representation of the diaphragm's stiffness characteristics and additional dynamic degrees of freedom as required to account for the participation of the diaphragm in the structure's dynamic response.

12.9.2 Linear Response History Analysis

12.9.2.1 General Requirements Linear response history analysis shall consist of an analysis of a linear mathematical model of the structure to determine its response through methods of numerical integration, to suites of spectrally matched acceleration histories compatible with the design response spectrum for the site. The analysis shall be performed in accordance with the requirements of this section.

12.9.2.2 General Modeling Requirements Three-dimensional (3D) models of the structure shall be required. Modeling the distribution of stiffness and mass throughout the structure's lateral load-resisting system and diaphragms shall be in accordance with Section 12.7.3.

12.9.2.2.1 P-Delta Effects The mathematical model shall include P-delta effects. Limits on the stability coefficient, θ, shall be satisfied in accordance with Section 12.8.7.

12.9.2.2.2 Accidental Torsion Accidental torsion, where required by Section 12.8.4.2, shall be included by offsetting the center of mass in each direction (i.e., plus or minus) from its expected location by a distance equal to 5% of the horizontal dimension of the structure at the given floor measured perpendicular to the direction of loading. Amplification of accidental torsion in accordance with Section 12.8.4.3 is not required.

12.9.2.2.3 Foundation Modeling Where foundation flexibility is included in the analysis, modeling of the foundation shall be in accordance with Section 12.13.3.

12.9.2.2.4 Number of Modes to Include in Modal Response History Analysis Where the modal response history analysis procedure is used, the number of modes to include in the analysis shall be in accordance with Section 12.9.1.1.

12.9.2.2.5 Damping Linear viscous damping shall not exceed 5% critical for any mode with a vibration period greater than or equal to T_{lower}.

12.9.2.3 Ground Motion Selection and Modification Ground acceleration histories used for analysis shall consist of a suite of no fewer than three pairs of spectrally matched orthogonal components derived from artificial or recorded ground motion events. The target response spectrum for each spectrally matched set shall be developed in accordance with Sections 11.4.6 or 21.3, as applicable.

12.9.2.3.1 Procedure for Spectrum Matching Each component of ground motion shall be spectrally matched over the period range $0.8T_{lower}$ to $1.2T_{upper}$. Over the same period range and in each direction of response, the average of the 5% damped pseudoacceleration ordinates computed using the spectrum-matched records shall not fall above or below the target spectrum by more than 10% in each direction of response.

12.9.2.4 Application of Ground Acceleration Histories Two orthogonal directions of response, designated as X and Y, shall be selected and used for all response history analysis. Ground motions shall be applied independently in the X and Y directions.

12.9.2.5 Modification of Response for Design

12.9.2.5.1 Determination of Maximum Elastic and Inelastic Base Shear For each ground motion analyzed, a maximum elastic base shear, designated as V_{EX} and V_{EY} in the X and Y directions, respectively, shall be determined. The mathematical model used for computing the maximum elastic base shear shall not include accidental torsion.

For each ground motion analyzed, a maximum inelastic base shear, designated as V_{IX} and V_{IY} in the X and Y directions, respectively, shall be determined as follows:

$$V_{IX} = \frac{V_{EX} I_e}{R_X} \qquad (12.9\text{-}1)$$

$$V_{IY} = \frac{V_{EY} I_e}{R_Y} \qquad (12.9\text{-}2)$$

where I_e is the Importance Factor and R_X and R_Y are the response modification coefficients for the X and Y directions, respectively.

12.9.2.5.2 Determination of Base Shear Scale Factor Design base shears, V_X and V_Y, shall be computed in the X and Y directions, respectively, in accordance with Section 12.8.1. For each ground motion analyzed, base shear scale factors in each direction of response shall be determined as follows:

$$\eta_X = \frac{V_X}{V_{IX}} \geq 1.0 \qquad (12.9\text{-}3)$$

$$\eta_Y = \frac{V_Y}{V_{IY}} \geq 1.0 \qquad (12.9\text{-}4)$$

12.9.2.5.3 Determination of Combined Force Response For each direction of response and for each ground motion analyzed, the combined force response shall be determined as follows:

(a) The combined force response in the X direction shall be determined as $I_e \eta_X / R_X$ times the computed elastic response in the X direction using the mathematical model with accidental torsion (where required) plus $I_e \eta_Y / R_Y$ times the computed elastic response in the Y direction using the mathematical model without accidental torsion.

(b) The combined force response in the Y direction shall be determined as $I_e \eta_Y / R_Y$ times the computed elastic response in the Y direction using the mathematical model with accidental torsion (where required), plus $I_e \eta_X / R_X$ times the computed elastic response in the X direction using the mathematical model without accidental torsion.

12.9.2.5.4 Determination of Combined Displacement Response Response modification factors C_{dX} and C_{dY} shall be assigned in the X and Y directions, respectively. For each direction of response and for each ground motion analyzed, the combined displacement responses shall be determined as follows:

(a) The combined displacement response in the X direction shall be determined as $\eta_X C_{dX} / R_X$ times the computed elastic response in the X direction using the mathematical model with accidental torsion (where required), plus $\eta_Y C_{dY} / R_Y$ times the computed elastic response in the Y direction using the mathematical model without accidental torsion.

(b) The combined displacement response in the Y direction shall be determined as $\eta_Y C_{dY} / R_Y$ times the computed elastic response in the Y direction using the mathematical model with accidental torsion (where required), plus $\eta_X C_{dX} / R_X$ times the computed elastic response in the X direction using the mathematical model without accidental torsion.

EXCEPTION: Where the design base shear in the given direction is not controlled by Equation (12.8-7), the factors η_X or η_Y, as applicable, are permitted to be taken as 1.0 for the purpose of determining combined displacements.

12.9.2.6 Enveloping of Force Response Quantities Design force response quantities shall be taken as the envelope of the combined force response quantities computed in both orthogonal directions and for all ground motions considered. Where force interaction effects are considered, demand to capacity ratios are permitted to be enveloped in lieu of individual force quantities.

12.9.2.7 Enveloping of Displacement Response Quantities Story drift quantities shall be determined for each ground motion analyzed and in each direction of response using the combined displacement responses defined in Section 12.9.2.5.4. For the purpose of complying with the drift limits specified in Section 12.12, the envelope of story drifts computed in both orthogonal directions and for all ground motions analyzed shall be used.

12.10 DIAPHRAGMS, CHORDS, AND COLLECTORS

Diaphragms, chords, and collectors shall be designed in accordance with Sections 12.10.1 and 12.10.2.

EXCEPTIONS:

(a) Precast concrete diaphragms, including chords and collectors in structures assigned to Seismic Design Category C, D, E, or F, shall be designed in accordance with Section 12.10.3.

(b) Precast concrete diaphragms in Seismic Design Category B, cast-in-place concrete diaphragms, cast-in-place concrete equivalent precast diaphragms, and wood-sheathed diaphragms supported by wood diaphragm framing, bare steel deck diaphragms, and concrete-filled steel deck diaphragms are permitted to be designed in accordance with Section 12.10.3.

(c) Diaphragms, chords, and collectors in one-story structures that conform to the limitations of Section 12.10.4.1 are permitted to be designed in accordance with Section 12.10.4.

12.10.1 Diaphragm Design Diaphragms shall be designed for both the shear and the bending stresses resulting from design forces. At diaphragm discontinuities, such as openings and reentrant corners, the design shall ensure that the dissipation or transfer of edge (chord) forces, combined with other forces in the diaphragm, is within the shear and tension capacity of the diaphragm.

12.10.1.1 Diaphragm Design Forces Floor and roof diaphragms shall be designed to resist in-plane seismic design forces from the structural analysis but shall not be less than that determined in accordance with Equation (12.10-1) as follows:

$$F_{px} = \frac{\sum_{n}^{i=x} F_i}{\sum_{n}^{i=x} w_i} w_{px} \qquad (12.10\text{-}1)$$

where

F_{px} = Diaphragm design force at level x;
F_i = Design force applied to level i;
w_i = Weight tributary to level i; and
w_{px} = Weight tributary to the diaphragm at level x.

The force determined from Equation (12.10-1) shall not be less than

$$F_{px} = 0.2 S_{DS} I_e w_{px} \qquad (12.10\text{-}2)$$

The force determined from Equation (12.10-1) need not exceed

$$F_{px} = 0.4S_{DS}I_e w_{px} \quad (12.10\text{-}3)$$

Diaphragms shall be designed for the inertial forces determined from Equations (12.10-1) through (12.10-3) and for applicable transfer forces resisted by the diaphragm between vertical seismic force-resisting elements. For structures that have a horizontal structural irregularity of Type 4 in Table 12.3-1, the transfer forces between horizontally offset vertical seismic force-resisting elements shall be increased by the overstrength factor of Section 12.4.3 before being added to the diaphragm inertial forces. For structures that have horizontal or vertical structural irregularities of the types indicated in Section 12.3.3.5, the requirements of that section shall also apply.

EXCEPTION: One- and two-family dwellings of light-frame construction shall be permitted to use $\Omega_0 = 1.0$.

12.10.2 Collector Elements Collector elements shall be provided that are capable of transferring the seismic forces originating in other portions of the structure to the element providing the resistance to those forces.

12.10.2.1 Collector Elements Requiring Load Combinations Including Overstrength for Seismic Design Categories C through F In structures assigned to Seismic Design Category C, D, E, or F, collector elements (Figure 12.10-1) and their connections, including connections to vertical elements, shall be designed to resist the maximum of the following:

(a) Forces calculated using the seismic load effects including overstrength of Section 12.4.3 with seismic forces determined by the equivalent lateral force procedure of Section 12.8 or the modal response spectrum analysis procedure of Section 12.9.1;
(b) Forces calculated using the seismic load effects including overstrength of Section 12.4.3 with seismic forces determined by Equation (12.10-1); and
(c) Forces calculated using the load combinations of Section 2.3.6 with seismic forces determined by Equation (12.10-2).

Transfer forces as described in Section 12.10.1.1 shall be considered.

EXCEPTION: In structures or portions thereof braced entirely by wood light-frame shear walls, collector elements and their connections, including connections to vertical elements, need only be designed to resist forces using the load combinations of Section 2.3.6 with seismic forces determined in accordance with Section 12.10.1.1.

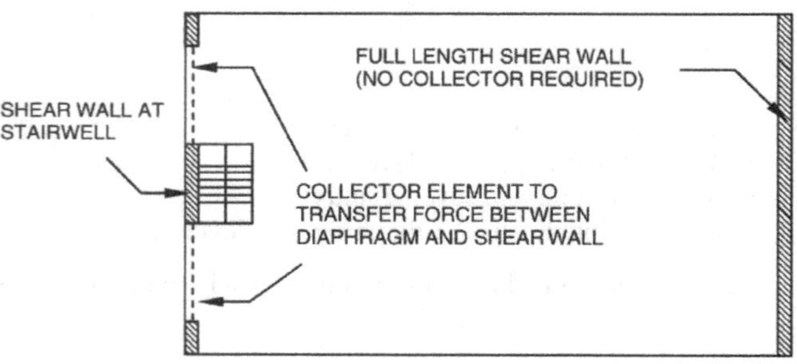

Figure 12.10-1. Collectors.

12.10.3 Alternative Design Provisions for Diaphragms, Including Chords and Collectors Where required or permitted in Section 12.10, diaphragms, including chords and collectors, shall be designed using the provisions in Sections 12.10.3.1 through 12.10.3.5 and the following:

1. Footnote *b* to Table 12.2-1 shall not apply.
2. Section 12.3.3.5 shall not apply.
3. Section 12.3.4.1, Item 5, shall be replaced with the following: "Design of diaphragms, including chords, collectors, and their connections to the vertical elements" are used.
4. Section 12.3.4.1, Item 7, shall not apply.

12.10.3.1 Design Diaphragms, including chords, collectors, and their connections to the vertical elements, shall be designed in two orthogonal directions to resist the in-plane design seismic forces determined in Section 12.10.3.2. Collectors shall be provided that are capable of transferring the seismic forces originating in other portions of the structure to the vertical elements providing the resistance to those forces. Design shall provide for transfer of forces at diaphragm discontinuities, such as openings and reentrant corners.

12.10.3.2 Seismic Design Forces for Diaphragms, Including Chords and Collectors Diaphragms, including chords, collectors, and their connections to the vertical elements, shall be designed to resist in-plane seismic design forces given by Equation (12.10-4):

$$F_{px} = \frac{C_{px}}{R_s} w_{px} \quad (12.10\text{-}4)$$

The force F_{px} determined from Equation (12.10-4) shall not be less than:

$$F_{px} = 0.2S_{DS}I_e w_{px} \quad (12.10\text{-}5)$$

C_{px} shall be determined as illustrated in Figure 12.10-2.

12.10.3.2.1 Design Acceleration Coefficients C_{p0}, C_{pi}, and C_{pn} Design acceleration coefficients C_{p0} and C_{pn} shall be calculated by Equations (12.10-6) and (12.10-7):

$$C_{p0} = 0.4S_{DS}I_e \quad (12.10\text{-}6)$$

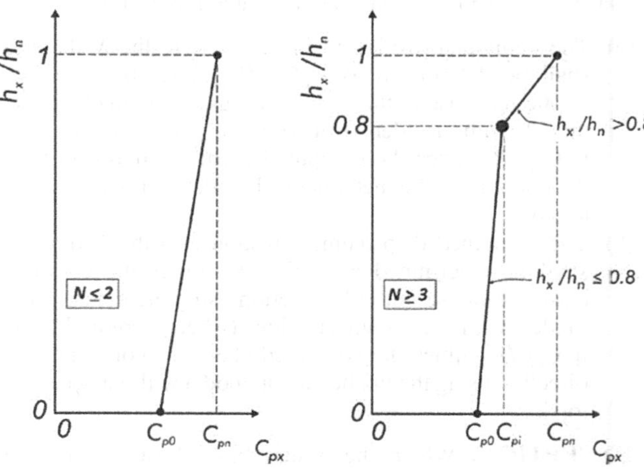

Figure 12.10-2. Calculating the design acceleration coefficient, C_{px}, in buildings with $N \leq 2$ and in buildings with $N \geq 3$.

and

$$C_{pn} = \sqrt{(\Gamma_{m1}\Omega_0 C_s)^2 + (\Gamma_{m2}C_{s2})^2} \geq C_{pi} \quad (12.10\text{-}7)$$

Design acceleration coefficient, C_{pi}, shall be the greater of the values given by Equations (12.10-8) and (12.10-9):

$$C_{pi} = 0.8C_{p0} \quad (12.10\text{-}8)$$

$$C_{pi} = 0.9\Gamma_{m1}\Omega_0 C_s \quad (12.10\text{-}9)$$

where Ω_0 is the overstrength factor given in Table 12.2-1; C_s is determined in accordance with Section 12.8 or 12.9; and C_{s2} shall be the smallest of the values calculated from Equations (12.10-10), (12.10-11), and (12.10-12).

$$C_{s2} = (0.15N + 0.25)I_e S_{DS} \quad (12.10\text{-}10)$$

$$C_{s2} = I_e S_{DS} \quad (12.10\text{-}11)$$

$$\text{For } N \geq 2 \quad C_{s2} = \frac{I_e S_{D1}}{0.03(N-1)} \quad (12.10\text{-}12a)$$

$$\text{For } N = 1 \quad C_{s2} = 0 \quad (12.10\text{-}12b)$$

The modal contribution factors Γ_{m1} and Γ_{m2} in Equation (12.10-7) shall be calculated from Equations (12.10-13) and (12.10-14).

$$\Gamma_{m1} = 1 + \frac{z_s}{2}\left(1 - \frac{1}{N}\right) \quad (12.10\text{-}13)$$

and

$$\Gamma_{m2} = 0.9z_s\left(1 - \frac{1}{N}\right)^2 \quad (12.10\text{-}14)$$

where the mode shape factor, z_s, is to be taken as 0.3 for buildings designed with buckling restrained braced frame systems defined in Table 12.2-1; 0.7 for buildings designed with moment-resisting frame systems defined in Table 12.2-1; 0.85 for buildings designed with dual systems defined in Table 12.2-1 with special or intermediate moment frames capable of resisting at least 25% of the prescribed seismic forces; and 1.0 for buildings designed with all other seismic force-resisting systems.

12.10.3.3 Transfer Forces in Diaphragms Diaphragms designed in accordance with this section shall be designed for the inertial forces determined from Equations (12.10-4) and (12.10-5) and for applicable transfer forces acting through the diaphragm between vertical seismic force-resisting elements. For structures that have a horizontal structural irregularity of Type 4 in Table 12.3-1, the transfer forces between horizontally offset vertical seismic force-resisting elements shall be increased by the overstrength factor of Section 12.4.3 before being added to the diaphragm inertial forces. For structures that have other horizontal or vertical structural irregularities of the types indicated in Section 12.3.3.5, the requirements of that section shall also apply.

EXCEPTION: One- and two-family dwellings of light-frame construction shall be permitted to use $\Omega_0 = 1.0$.

12.10.3.4 Collectors—Seismic Design Categories C through F In structures assigned to Seismic Design Category C, D, E, or F, collectors and their connections, including connections to vertical elements, shall be designed to resist 1.5 times the diaphragm inertial forces from Section 12.10.3.2 plus 1.5 times the design transfer forces.

EXCEPTIONS:

1. Any transfer force increased by the overstrength factor of Section 12.4.3 need not be further amplified by 1.5.
2. For moment frame and braced frame systems, collector forces need not exceed the lateral strength of the corresponding frame line below the collector, considering only the moment frames or braced frames. In addition, diaphragm design forces need not exceed the forces corresponding to the collector forces so determined.
3. In structures or portions thereof braced entirely by light-frame shear walls, collector elements and their connections, including connections to vertical elements, need only be designed to resist the diaphragm seismic design forces, without the 1.5 multiplier.

12.10.3.5 Diaphragm Design Force Reduction Factor The diaphragm design force reduction factor, R_s, shall be determined in accordance with Table 12.10-1.

12.10.4 Alternative Diaphragm Design Provisions for One-Story Structures with Flexible Diaphragms and Rigid Vertical Elements Where permitted by Section 12.10 and subject to the limitations of Section 12.10.4.1, diaphragm design forces, including design forces for chords, collectors, and their in-plane connections to vertical elements, shall be determined in accordance with Section 12.10.4.2.

12.10.4.1 Limitations Diaphragms in one-story structures are permitted to be designed in accordance with Section 12.10.4 provided all of the following limitations are satisfied.

1. All portions of the diaphragm shall be designed using the provisions of this section in both orthogonal directions.
2. The diaphragm shall consist of either (a) a wood structural panel diaphragm designed in accordance with AWC SDPWS and fastened to wood framing members or wood nailers with sheathing nailing in accordance with the AWC SDPWS Section 4.2 nominal shear capacity tables, or (b) a bare (untopped) steel deck diaphragm meeting the requirements of AISI S400 and AISI S310.
3. Toppings of concrete or similar materials that affect diaphragm strength or stiffness shall not be placed over the wood structural panel or bare steel deck diaphragm.
4. The diaphragm shall not contain horizontal structural irregularities, as specified in Table 12.3-1, except that Horizontal Structural Irregularity Type 2 is permitted.
5. The diaphragm shall be rectangular in shape or shall be divisible into rectangular segments for purpose of seismic design, with vertical elements of the seismic force-resisting system or collectors provided at each end of each rectangular segment span.
6. The vertical elements of the seismic force-resisting system shall be limited to one or more of the following: concrete shear walls, precast concrete shear walls, masonry shear walls, steel concentrically braced frames, steel and concrete composite braced frames, or steel and concrete composite shear walls.
7. The vertical elements of the seismic force-resisting system shall be designed in accordance with Section 12.8, except that they shall be permitted to be designed using the two-stage analysis procedure of Section 12.2.3.4, where applicable.

Table 12.10-1. Diaphragm Design Force Reduction Factor, R_s.

Diaphragm System		Shear-Controlled	Flexure-Controlled
Cast-in-place concrete designed in accordance with ACI 318	—	1.5	2
Precast concrete designed in accordance with ACI 318	Elastic design option	0.7	0.7
	Basic design option	1.0	1.0
	Reduced design option	1.4	1.4
Wood sheathed designed in accordance with Section 14.5 and AWC SDPWS	—	3.0	NA*
Bare steel deck diaphragm designed in accordance with Section 14.1.5	With special seismic detailing	2.5	NA*
	Other	1.0	NA*
Concrete-filled steel deck diaphragm designed in accordance with Section 14.1.6	—	2.0	NA*

*Not Applicable

12.10.4.2 Design Diaphragms, including chords, collectors, and their connections to vertical elements, shall be designed in two orthogonal directions to resist the in-plane design seismic forces determined in accordance with this section. Multi-span diaphragms and diaphragms that are not rectangular in shape shall be divided into rectangular segments for purposes of design in accordance with this section, with lateral support provided at each end of each diaphragm segment span by a vertical element or collector element.

12.10.4.2.1 Seismic Design Forces The diaphragm seismic design force, F_{px}, shall be determined in accordance with Equation (12.10-15).

$$F_{px} = C_{s\text{-diaph}}(w_{px}) \tag{12.10-15}$$

where

w_{px} = The effective seismic weight tributary to the diaphragm, and $C_{s\text{-diaph}}$ is calculated as

$$C_{s\text{-diaph}} = \frac{S_{DS}}{\frac{R_{\text{diaph}}}{I_e}} \tag{12.10-16a}$$

and need not be greater than

$$C_{s\text{-diaph}} = \frac{S_{D1}}{T_{\text{diaph}}\left(\frac{R_{\text{diaph}}}{I_e}\right)} \tag{12.10-16b}$$

where

S_{DS} = Design spectral response parameter in the short period range as determined from Section 11.4.5 or 11.4.8;

R_{diaph} = 4.5 for wood structural panel diaphragms, 4.5 for bare steel deck diaphragms that meet the special seismic detailing requirements of AISI S400, and 1.5 for all other bare steel deck diaphragms;

I_e = Importance Factor determined in accordance with Section 11.5.1; and

T_{diaph} = $0.002L_{\text{diaph}}$ for wood structural panel diaphragms, or $0.001L_{\text{diaph}}$ for profiled steel deck panel diaphragms, determined for each rectangular segment of the diaphragm in each orthogonal direction [seconds].

12.10.4.2.2 Diaphragm Shears Diaphragm design shears shall be computed for each diaphragm segment in accordance with Section 12.10.4.2.1. Where the diaphragm segment span, L_{diaph}, is less than 100 ft (30.5 m), the diaphragm design shear, from loading perpendicular to the span, shall be the diaphragm shear calculated using the F_{px} forces of Section 12.10.4.2.1 multiplied by 1.5. Where the diaphragm segment span, L_{diaph}, is greater than or equal to 100 ft (30.5 m), the diaphragm design shear shall be amplified to 1.5 times the shear calculated using the F_{px} forces of Section 12.10.4.2.1, over an amplified shear boundary zone having a minimum width of 10% of the diaphragm segment span, L_{diaph}. The amplified shear boundary zone shall be provided at each supporting end of the diaphragm segment span under consideration.

12.10.4.2.3 Diaphragm Chords Diaphragm chords shall be provided at each edge of each diaphragm segment to resist tension and compression forces resulting from diaphragm moments. Diaphragm chord forces shall be computed using the F_{px} forces of Section 12.10.4.2.1.

12.10.4.2.4 Collector Elements and Their Connections Collector elements shall be provided that are capable of transferring the seismic forces originating in other portions of the structure to the vertical elements of the seismic force-resisting system. Collector element forces shall be computed using the F_{px} forces of Section 12.10.4.2.1. Collectors and their connections to vertical elements of the seismic force-resisting system in structures assigned to Seismic Design Categories C through F shall be designed to resist the forces calculated using the seismic load effects including overstrength factor of Section 12.4.3, with diaphragm overstrength factor, $\Omega_{0\text{-diaph}}$, taken equal to 2; however, $W_{0\text{-diaph}}$ need not exceed R_{diaph}. This need not be combined with the shear amplification of 1.5 specified in Section 12.10.4.2.2.

12.10.4.2.5 Diaphragm Deflection Where required by this standard, the deflection amplification factor, $C_{d\text{-diaph}}$, for diaphragm deflection shall be taken as one of the following:

$C_{d\text{-diaph}}$ = 3.0 for wood structural panel diaphragms,
= 3.0 for bare steel deck diaphragms that meet the special seismic detailing requirements of AISI S400, and
= 1.5 for all other bare steel deck diaphragms.

Diaphragm deflections shall be calculated using Section 12.10.4.2.1 seismic design forces.

12.10.4.2.6 Modifications to Diaphragm Seismic Design For structures in which the diaphragm design forces are determined in accordance with Section 12.10.4, footnote *b* to Table 12.2-1 shall not apply, and Section 12.3.3.5 shall not apply.

12.11 STRUCTURAL WALLS AND THEIR ANCHORAGE

12.11.1 Design for Out-of-Plane Forces Structural walls shall be designed for a force normal to the surface equal to $F_p = 0.4S_{DS}I_e$ times the weight of the structural wall with a minimum force of 10% of the weight of the structural wall.

12.11.2 Anchorage of Structural Walls and Transfer of Design Forces into Diaphragms or Other Supporting Structural Elements

12.11.2.1 Wall Anchorage Forces The anchorage of structural walls to supporting construction shall provide a direct connection capable of resisting the following:

$$F_p = 0.4S_{DS}k_aI_eW_p \quad (12.11\text{-}1)$$

where

F_p = Design force in the individual anchors;
S_{DS} = Design spectral response acceleration parameter at short periods per Section 11.4.5;
I_e = Importance Factor determined in accordance with Section 11.5.1;
k_a = Amplification factor for diaphragm flexibility;
L_f = Span, in feet, of a flexible diaphragm that provides the lateral support for the wall (the span is measured between vertical elements that provide lateral support to the diaphragm in the direction considered; use zero for rigid diaphragms); and
W_p = Weight of the wall tributary to the anchor.

F_p shall not be taken as less than the larger of $0.2k_aI_eW_p$ and 5 lb/ft^2 (0.24 kN/m^2) times the area of the wall tributary to the anchor.

$$k_a = 1.0 + \frac{L_f}{100} \quad (12.11\text{-}2)$$

but k_a need not be taken as larger than 2.0, and need not be taken as larger than 1.0 when the connection is not at a flexible diaphragm.

Where the anchorage is not located at the roof and all diaphragms are not flexible, the value from Equation (12.11-1) is permitted to be multiplied by the factor $(1 + 2z/h)/3$, where z is the height of the anchor above the base of the structure, and h is the height of the roof above the base; however, F_p shall not be less than required by Section 12.11.1 with a minimum anchorage force of $F_p = 0.2W_p$ but not less than 5 lb/ft^2 (0.24 kN/m^2) times the area of the wall tributary to the anchor.

Structural walls shall be designed to resist bending between anchors where the anchor spacing exceeds 4 ft (1,219 mm). Interconnection of structural wall elements and connections to supporting framing systems shall have sufficient strength, rotational capacity, and ductility to resist shrinkage, thermal changes, and differential foundation settlement when combined with seismic forces.

12.11.2.2 Additional Requirements for Anchorage of Concrete or Masonry Structural Walls to Diaphragms in Structures Assigned to Seismic Design Categories C through F

12.11.2.2.1 Transfer of Anchorage Forces into Diaphragm Diaphragms shall be provided with continuous ties or struts between diaphragm chords to distribute these anchorage forces into the diaphragms. Diaphragm connections shall be positive, mechanical, or welded. Added chords are permitted to be used to form subdiaphragms to transmit the anchorage forces to the main continuous crossties. The maximum length-to-width ratio of structural subdiaphragms that serve as part of the continuous tie system shall be 2.5 to 1. Connections and anchorages capable of resisting the prescribed forces shall be provided between the diaphragm and the attached components. Connections shall extend into the diaphragm a sufficient distance to develop the force transferred into the diaphragm.

12.11.2.2.2 Steel Elements of Structural Wall Anchorage System The strength design forces for steel elements of the structural wall anchorage system, with the exception of anchor bolts and reinforcing steel, shall be increased by 1.4 times the forces otherwise required by this section.

12.11.2.2.3 Wood Diaphragms The anchorage of concrete or masonry structural walls to wood diaphragms shall be in accordance with AWC SDPWS 4.1.5.1 and this section. Continuous ties required by this section shall be in addition to the diaphragm sheathing. Anchorage shall not be accomplished by use of toenails or nails subject to withdrawal, nor shall wood ledgers or framing be used in cross-grain bending or cross-grain tension. The diaphragm sheathing shall not be considered effective for providing the ties or struts required by this section.

12.11.2.2.4 Metal Deck Diaphragms In metal deck diaphragms, the metal deck shall not be used as the continuous ties required by this section in the direction perpendicular to the deck span.

12.11.2.2.5 Embedded Straps Diaphragm to structural wall anchorage using embedded straps shall be attached to, or hooked around, the reinforcing steel or otherwise terminated so as to effectively transfer forces to the reinforcing steel.

12.11.2.2.6 Eccentrically Loaded Anchorage System Where elements of the wall anchorage system are loaded eccentrically or are not perpendicular to the wall, the system shall be designed to resist all components of the forces induced by the eccentricity.

12.11.2.2.7 Walls with Pilasters Where pilasters are present in the wall, the anchorage force at the pilasters shall be calculated considering the additional load transferred from the wall panels to the pilasters. However, the minimum anchorage force at a floor or roof shall not be reduced.

12.12 DRIFT AND DEFORMATION

12.12.1 Story Drift Limit The design story drift, Δ, as determined in Section 12.8.6, 12.9.1, or 12.9.2 shall not exceed the allowable story drift, Δ_a, as obtained from Table 12.12-1 for any story.

12.12.1.1 Moment Frames in Structures Assigned to Seismic Design Categories D through F For seismic force-resisting systems solely comprising moment frames in structures assigned to Seismic Design Category D, E, or F, the design story drift, Δ, shall not exceed Δ_a/ρ for any story. ρ shall be determined in accordance with Section 12.3.4.2.

12.12.2 Structural Separation All portions of the structure shall be designed and constructed to act as an integral unit in resisting seismic forces unless separated structurally by a distance sufficient to avoid damaging contact as set forth in this section.

Separations shall allow for the Design Earthquake Displacements, δ_{DE}, as determined in accordance with Section 12.8.6.

Table 12.12-1. Allowable Story Drift, Δ_a.

Structure	Risk Category		
	I or II	III	IV
Structures, other than masonry shear wall structures, four stories or less above the base as defined in Section 11.2, with interior walls, partitions, and ceilings that have been designed to accommodate the drifts associated with the Design Earthquake Displacements	$0.025h_{sx}$[a]	$0.020h_{sx}$	$0.015h_{sx}$
Masonry cantilever shear wall structures[b]	$0.010h_{sx}$	$0.010h_{sx}$	$0.010h_{sx}$
Other masonry shear wall structures	$0.007h_{sx}$	$0.007h_{sx}$	$0.007h_{sx}$
All other structures	$0.020h_{sx}$	$0.015h_{sx}$	$0.010h_{sx}$

[a]There shall be no drift limit for single-story structures in which the interior walls, partitions, and ceilings have been designed to accommodate story drifts associated with the Design Earthquake Displacement. The structural separation requirement of Section 12.12.3 is not waived.

[b]Structures in which the basic structural system consists of masonry shear walls designed as vertical elements cantilevered from their base or foundation support that are so constructed that moment transfer between shear walls (coupling) is negligible.

Notes: h_{sx} is the story height below level x. For seismic force-resisting systems solely comprising moment frames in Seismic Design Categories D, E, and F, the allowable story drift shall comply with the requirements of Section 12.12.1.1.

Adjacent structures on the same property shall be separated by at least δ_{SS}, determined as

$$\delta_{SS} = \sqrt{(\delta_{DE1})^2 + (\delta_{DE2})^2} \qquad (12.12\text{-}2)$$

where δ_{DE1} and δ_{DE2} are the Design Earthquake Displacements of the adjacent structures at their adjacent edges. Where a structure adjoins a property line not common to a public way, the structure shall be set back from the property line by at least the displacement δ_{DE} of that structure.

EXCEPTION: Smaller separations or property line setbacks are permitted where justified by rational analysis based on inelastic response to design ground motions.

12.12.3 Members Spanning between Structures Gravity connections or supports for members spanning between structures or seismically separate portions of structures shall be designed for the maximum anticipated relative displacements. These displacements shall be calculated using Maximum Considered Earthquake Displacement (δ_{MCE}), as determined in accordance with Section 12.8.6 and assuming that the two structures are moving in opposite directions and using the absolute sum of the displacements.

12.12.4 Deformation Compatibility for Seismic Design Categories D through F For structures assigned to Seismic Design Category D, E, or F, every structural component not included in the seismic force-resisting system in the direction under consideration shall be designed to be adequate for the gravity load effects and the seismic forces resulting from the Design Earthquake Displacements (δ_{DE}) and the associated drifts as determined in accordance with Section 12.8.6.

EXCEPTION: Reinforced concrete frame members not designed as part of the seismic force-resisting system shall comply with Section 18.14 of ACI 318. Where determining the moments and shears induced in components that are not included in the seismic force-resisting system in the direction under consideration, the stiffening effects of adjoining rigid structural and nonstructural elements shall be considered, and a rational value of member and restraint stiffness shall be used.

12.13 FOUNDATION DESIGN

12.13.1 Design Basis The design basis for foundations shall be as set forth in Section 12.1.5.

12.13.2 Materials of Construction Materials used for the design and construction of foundations shall comply with the requirements of Chapter 14 and the additional requirements of Section 12.13.9 for foundations on liquefiable sites. Design and detailing of steel piles shall comply with Section 14.1.8 and the additional requirements for Section 12.13.9 where applicable. Design and detailing of concrete piles shall comply with ACI 318 and the additional requirements for Section 12.13.9 where applicable.

12.13.3 Foundation Load-Deformation Characteristics Where foundation flexibility is included for the linear analysis procedures in Chapter 12, the load-deformation characteristics of the foundation–soil system shall be modeled in accordance with the requirements of this section. The linear load-deformation behavior of foundations shall be represented by an equivalent linear stiffness using soil properties that are compatible with the soil strain levels associated with the design earthquake motion. The strain-compatible shear modulus, G, and the associated strain-compatible shear wave velocity, v_S, needed for the evaluation of equivalent linear stiffness shall be determined using the criteria in Chapter 19 or based on a site-specific study. The strain-compatible shear modulus, G, shall be varied over a range defined with an upper bound equal to $G(1 + C_{vs})$ and a lower bound equal to $G/(1 + C_{vs})$, where C_{vs} is the coefficient of variation which accounts for uncertainties in the SSI analysis and soil properties. The factor C_{vs} shall be taken as 1.0 unless smaller variations can be justified based on field measurements of dynamic soil properties or direct measurements of dynamic foundation stiffness. In no case shall the value of C_{vs} be taken as less than 0.5. The largest values of response shall be used in design.

12.13.4 Reduction of Foundation Overturning Overturning effects at the soil–foundation interface are permitted to be reduced by 25% for foundations of structures that satisfy both of the following conditions:

(a) The structure is designed in accordance with the equivalent lateral force analysis as set forth in Section 12.8, and
(b) The structure is not an inverted pendulum or cantilevered column type structure.

Overturning effects at the soil–foundation interface are permitted to be reduced by 10% for foundations of structures designed in accordance with the modal analysis requirements of Section 12.9.

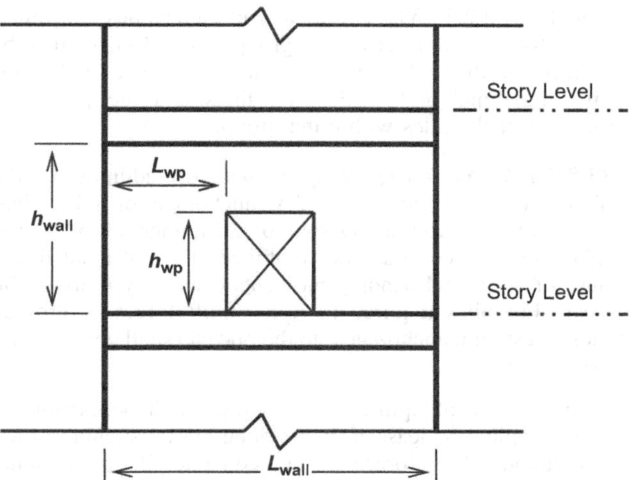

Figure 12.3-2. Shear wall and wall pier height-to-length ratio determination.

Note: h_{wall} = height of shear wall; h_{wp} = height of wall pier; L_{wall} = length of shear wall; L_{wp} = length of wall pier. Shear wall height-to-length ratio: h_{wall}/L_{wall}. Wall pier height-to-length ratio: h_{wp}/L_{wp}.

12.13.5 Strength Design for Foundation Geotechnical Capacity Where basic combinations for strength design listed in Section 2.3 are used, combinations that include earthquake loads, E, are permitted to include reduction of foundation overturning effects defined in Section 12.13.4. The following sections shall apply for determination of the applicable nominal strengths and resistance factors at the soil–foundation interface.

12.13.5.1 Nominal Strength The nominal foundation geotechnical capacity, Q_{ns}, shall be determined using any of the following methods:

1. Presumptive load-bearing values;
2. By a registered design professional based on geotechnical site investigations that include field and laboratory testing to determine soil classification and as-required active, passive, and at-rest soil strength parameters; or
3. By in situ testing of prototype foundations.

For structures that are supported on more than one foundation, the method used to determine the nominal strength of all foundations shall be the same. Nominal strength values are permitted to be based on either a limitation of maximum expected foundation deformation, or by the nominal strength that is associated with an anticipated failure mechanism.

12.13.5.1.1 Soil Strength Parameters For competent soils that do not undergo strength degradation under seismic loading, strength parameters for static loading conditions shall be used to compute nominal foundation geotechnical capacities for seismic design unless increased seismic strength values based on site conditions are provided by a registered design professional. For sensitive cohesive soils or saturated cohesionless soils, the potential for earthquake-induced strength degradation shall be considered. Nominal foundation geotechnical capacities for vertical, lateral, and rocking loading shall be determined using accepted foundation design procedures and principles of plastic analysis, and shall be best-estimate values using soil properties that are representative average values.

Total resistance to lateral loads is permitted to be determined by taking the sum of the values derived from lateral bearing pressure plus horizontal sliding resistance (from some combination of friction and cohesion).

1. Lateral sliding resistance from friction shall be limited to sand, silty sand, clayey sand, silty gravel, and clayey gravel soils (SW, SP, SM, SC, GM, and GC), and rock. Lateral sliding resistance from friction shall be calculated as the most unfavorable dead load factor multiplied by dead load, D, and multiplied by a coefficient of friction.
2. Lateral sliding resistance from cohesion shall be limited to clay, sandy clay, clayey silt, silt, and sandy silt (CL, ML, SC, and SM). Lateral sliding resistance from cohesion shall be calculated as the contact area multiplied by the cohesion.
3. Horizontal friction sliding resistance and cohesion sliding resistance shall be taken as zero for areas of foundations supported by piles.

Where presumptive load bearing values for supporting soils are used to determine nominal soil strengths, organic silt, organic clays, peat, or nonengineered fill shall not be assumed to have a presumptive load capacity.

12.13.5.2 Resistance Factors The resistance factors prescribed in this section shall be used for vertical, lateral, and rocking resistance of all foundation types. Nominal foundation geotechnical capacities, Q_{ns}, shall be multiplied by the resistance factor, φ, shown in Table 12.13-1. Alternatively, a vertical resistance factor, $\varphi = 0.80$, is permitted to be used when the nominal strength (upward or downward) is determined by in-situ testing of prototype foundations, based on a test program that is approved by the Authority Having Jurisdiction.

12.13.5.3 Acceptance Criteria For linear seismic analysis procedures in accordance with Sections 12.8 and 12.9, factored loads, including reductions permitted in Section 12.13.4, shall not exceed foundation design strengths, ϕQ_{ns}.

12.13.6 Allowable Stress Design for Foundation Geotechnical Capacity Where basic combinations for allowable stress design listed in Section 2.4 are used for design, combinations that include earthquake loads, E, are permitted to include reduction of foundation overturning effects defined in Section 12.13.4. Allowable foundation load capacities, Q_{as}, shall be determined using allowable stresses in geotechnical materials that have been determined by geotechnical investigations required by the Authority Having Jurisdiction.

12.13.7 Requirements for Structures Assigned to Seismic Design Category C In addition to the requirements of Section 11.8.2, the following foundation design requirements shall apply to structures assigned to Seismic Design Category C.

12.13.7.1 Pole-Type Structures Where construction using posts or poles as columns embedded in earth or embedded in concrete

Table 12.13-1. Resistance Factors for Strength Design of Soil–Foundation Interface.

Direction and Type of Resistance	Resistance Factor, φ
Vertical Resistance	
Compression (bearing) strength	0.45
Pile friction (either upward or downward)	0.45
Lateral Resistance	
Lateral bearing pressure	0.5
Sliding (by either friction or cohesion)	0.85

footings in the earth is used to resist lateral loads, the depth of embedment required for posts or poles to resist seismic forces shall be determined by means of the design criteria established in the foundation investigation report.

12.13.7.2 Foundation Ties Individual pile caps, drilled piers, or caissons shall be interconnected by ties. All ties shall have a design strength in tension or compression at least equal to a force equal to 10% of S_{DS} times the larger pile cap or column factored dead plus factored live load unless it is demonstrated that equivalent restraint will be provided by reinforced concrete beams within slabs on grade or reinforced concrete slabs on grade or confinement by competent rock, hard cohesive soils, very dense granular soils, or other approved means.

12.13.7.3 Pile Anchorage Requirements In addition to the requirements of ACI 318, anchorage of piles shall comply with this section. Where required for resistance to uplift forces, anchorage of steel pipe [round hollow structural steel (HSS) sections], concrete-filled steel pipe, or H piles to the pile cap shall be made by means other than concrete bond to the bare steel section.

EXCEPTION: Anchorage of concrete-filled steel pipe piles is permitted to be accomplished using deformed bars developed into the concrete portion of the pile.

12.13.8 Requirements for Structures Assigned to Seismic Design Categories D through F In addition to the requirements of Sections 11.8.2, 11.8.3 and 14.1.8, the following foundation design requirements shall apply to structures assigned to Seismic Design Category D, E, or F. Design and construction of concrete foundation elements shall conform to the requirements of ACI 318, Section 18.13, except as modified by the requirements of this section.

EXCEPTION: Detached one- and two-family dwellings of light-frame construction not exceeding two stories above grade plane need only comply with the requirements for Sections 11.8.2, 11.8.3 (items 2 through 4), 12.13.2, and 12.13.7.

12.13.8.1 Pole-Type Structures Where construction using posts or poles as columns embedded in earth or embedded in concrete footings in the earth is used to resist lateral loads, the depth of embedment required for posts or poles to resist seismic forces shall be determined by means of the design criteria established in the foundation investigation report.

12.13.8.2 Foundation Ties Individual pile caps, drilled piers, or caissons shall be interconnected by ties. In addition, individual spread footings founded on soil defined in Chapter 20 as Site Class E or F shall be interconnected by ties. All ties shall have a design strength in tension or compression at least equal to a force equal to 10% of S_{DS} times the larger pile cap or column factored dead plus factored live load unless it is demonstrated that equivalent restraint is provided by reinforced concrete beams within slabs on grade or reinforced concrete slabs on grade or confinement by competent rock, hard cohesive soils, very dense granular soils, or other approved means.

12.13.8.3 General Pile Design Requirement Piling shall be designed and constructed to withstand deformations from earthquake ground motions and structure response. Deformations shall include both free-field soil strains (without the structure) and deformations induced by lateral pile resistance to structure seismic forces, all as modified by soil–pile interaction.

12.13.8.4 Batter Piles Batter piles and their connections shall be capable of resisting forces and moments from the load combinations including overstrength from Chapter 2 or Section 12.14.3.2.3. Where vertical and batter piles act jointly to resist foundation forces as a group, these forces shall be distributed to the individual piles in accordance with their relative horizontal and vertical rigidities and the geometric distribution of the piles within the group.

12.13.8.5 Pile Anchorage Requirements In addition to the requirements of Section 12.13.7.3, anchorage of piles shall comply with this section. Design of anchorage of piles into the pile cap shall consider the combined effect of axial forces because of uplift and bending moments caused by fixity to the pile cap. For piles required to resist uplift forces or provide rotational restraint, anchorage into the pile cap shall comply with the following:

1. In the case of uplift, the anchorage shall be capable of developing the least of the nominal tensile strength of the longitudinal reinforcement in a concrete pile, the nominal tensile strength of a steel pile, and 1.3 times the pile pullout resistance, or shall be designed to resist the axial tension force resulting from the seismic load effects including overstrength of Section 12.4.3 or 12.14.3.2. The pile pullout resistance shall be taken as the ultimate frictional or adhesive force that can be developed between the soil and the pile plus the pile weight.
2. In the case of rotational restraint, the anchorage shall be designed to resist the axial and shear forces and moments resulting from the seismic load effects including overstrength of Section 12.4.3 or 12.14.3.2 or shall be capable of developing the full axial, bending, and shear nominal strength of the pile.

12.13.8.6 Splices of Pile Segments Splices of pile segments shall develop the nominal strength of the pile section.

EXCEPTION: Splices designed to resist the axial and shear forces and moments from the seismic load effects including overstrength of Section 12.4.3 or 12.14.3.2.

12.13.8.7 Pile–Soil Interaction Pile moments, shears, and lateral deflections used for design shall be established considering the interaction of the pile and soil. Where the ratio of the depth of embedment of the pile to the pile diameter or width is less than or equal to 6, the pile is permitted to be assumed to be flexurally rigid with respect to the soil.

12.13.8.8 Pile Group Effects Pile group effects from soil on lateral pile nominal strength shall be included where pile center-to-center spacing in the direction of lateral force is less than eight pile diameters or widths. Pile group effects on vertical nominal strength shall be included where pile center-to-center spacing is less than three pile diameters or widths.

12.13.9 Requirements for Foundations Subject to Seismically-Induced Soil Displacement or Strength Loss Where the geotechnical investigation report required in Section 11.8 identifies the potential for liquefaction, seismically-induced permanent ground displacement, or soil strength loss in MCE_G earthquake motions, structures shall be designed to accommodate the effects in accordance with the requirements of Sections 12.13.9.1 through 12.13.9.3. Such structures shall also be designed to resist the seismic load effects of Section 12.4, presuming these effects do not occur.

EXCEPTION: Structures on shallow foundations need not be designed for the requirements of this section where the geotechnical investigation report indicates a negligible potential for the occurrence of permanent horizontal ground displacement, no bearing capacity loss, no lateral resistance loss, and differential settlements of site soils or improved site soils do not exceed one-fourth of the differential settlement threshold specified in Table 12.3-3.

Where the geotechnical investigation report indicates the potential for flow failure or seismically-induced long runout landslides, the provisions of Section 12.13.9 are not applicable and the condition shall be mitigated.

12.13.9.1 Foundation Design Foundations shall be designed to support gravity and design earthquake loads, as indicated in the basic load combinations of Section 12.4, using the reduced soil bearing capacity, as indicated in the geotechnical investigation report, considering the effects of liquefaction, seismically-induced permanent ground displacement, and soil strength loss caused by MCE_G earthquake motions. It shall be permitted to include the mitigating effects of any planned improvements for the site.

12.13.9.2 Shallow Foundations Building structures shall be permitted to be supported on shallow foundations provided that the foundations are designed and detailed in accordance with Section 12.13.9.2.1 and the conditions provided in items (a) and (b) of Section 12.13.9.2 are met.

(a) The geotechnical investigation report indicates that seismically-induced permanent horizontal ground displacement does not exceed the value in Table 12.13-2.

(b) The geotechnical investigation report indicates that the differential settlement over a defined length, L, does not exceed the differential settlement threshold specified in Table 12.13-3.

EXCEPTION: It is permitted to exceed the differential settlement threshold specified in Table 12.13-3 for shallow foundations, provided the foundation and superstructure are designed to accommodate seismically-induced differential settlements without loss of the ability to support gravity loads in addition to the requirements of Section 12.13.9.2.1. For structures assigned to Risk Category II or III, residual strength of members and connections shall not be less than 67% of the undamaged nominal strength, considering the nonlinear behavior of the structure or, alternatively, demands on all members and connections shall not exceed the element's nominal strength when subjected to differential settlements. For structures assigned to Risk Category IV, demands on all members and connections shall not exceed the element's nominal strength when subjected to differential settlements.

12.13.9.2.1 Shallow Foundation Design Shallow foundations shall satisfy the design and detailing requirements of Sections 12.13.9.2.1.1 or 12.13.9.2.1.2 as required.

12.13.9.2.1.1 Foundation Ties Individual footings shall be interconnected by ties in accordance with Section 12.13.8.2 and the additional requirements of this section. Reinforced concrete sections shall be detailed in accordance with Sections 18.13.4 of ACI 318. Where the geotechnical investigation report indicates that seismically-induced permanent horizontal ground displacement exceeding 3 in. (76.2 mm) will occur in MCE_G earthquake motions, both of the following requirements shall be met.

1. Ties between individual footings on the same column or wall line shall, in lieu of the force requirements of Section 12.13.8.2, have a design strength in tension and compression at least equal to the design tie force, F_{tie}, as indicated in Equation (12.13-1). These effects shall be combined with the load effects from design earthquake lateral loads.

$$F_{tie} = 0.5\mu P_u \quad (12.13\text{-}1)$$

where μ is the coefficient of friction between the bottom of the footing and the soil, as indicated in the geotechnical report, or taken as 0.5 in the absence of other information; and P_u is the total of the supported gravity loads of all footings along the same column or wall line, determined in accordance with load combination 6 in Section 2.3.6.

2. Individual footings shall be integral with or connected to a reinforced concrete slab-on-ground, at least 5 in. (127.0 cm) thick and reinforced in each horizontal direction with a minimum reinforcing ratio of 0.0025. Alternately, individual footings shall be integral with or connected to a posttensioned concrete slab on ground designed according to PTI DC10.5 with a minimum effective compression after losses of 100 psi (690 kPa). For sites with expansive soils, movements from both expansive soils and seismically-induced permanent ground displacement need not be considered concurrently. For purposes of this section, concrete slab on ground need not satisfy Section 18.6.4 of ACI 318.

EXCEPTION: A system of diagonal reinforced concrete ties is permitted to be used, if the system of ties provides equivalent

Table 12.13-2. Upper Limit on Permanent Horizontal Ground Displacement for Shallow Foundations.

Risk Category	I or II	III	IV
Limit, in. (mm)	18 (455)	12 (305)	4 (100)

Table 12.13-3. Differential Settlement Threshold.

Structure Type	Risk Category		
	I or II	III	IV
Single-story structures with concrete or masonry wall systems	0.0075L	0.005L	0.002L
Other single-story structures	0.015L	0.010L	0.002L
Multistory structures with concrete or masonry wall systems	0.005L	0.003L	0.002L
Other multistory structures	0.010L	0.006L	0.002L

lateral shear strength and stiffness to a slab on ground as defined above.

12.13.9.2.1.2 Mat Foundations Mat foundations shall be designed to accommodate the expected vertical differential settlements indicated in the geotechnical investigation report, considering any increased loads induced by differential settlements of adjacent columns. The flexural demands caused by seismically-induced differential settlement need not be considered if the mat is detailed in accordance with the requirements of Section 18.13.3.1 of ACI 318. Mat foundations shall have longitudinal reinforcement in both directions at the top and bottom.

12.13.9.3 Deep Foundations Deep foundations shall be designed to support vertical loads as indicated in the basic load combinations of Section 12.4, in combination with the moments and shears caused by lateral deformation of deep foundation elements in response to lateral inertial loads. Axial capacity of the deep foundation and lateral resistance of the soil shall be reduced to account for the effects of liquefaction and reduced soil strength. Deep foundations shall satisfy the design and detailing requirements of Sections 12.13.9.3.1 through 12.13.9.3.5.

12.13.9.3.1 Downdrag Design of piles shall incorporate the effects of downdrag caused by seismically-induced settlement. For geotechnical design, the downdrag load shall be determined as the downward skin friction on the pile within and above the settling zone(s). The net geotechnical ultimate capacity of the pile shall be the ultimate geotechnical capacity of the pile below the settling zone(s) reduced by the downdrag load. For structural design, the downdrag load induced by seismic settlement shall be treated and factored as a seismic load, though it need not be considered concurrently with axial loads resulting from inertial response of the structure, determined according to Section 12.4.

EXCEPTION: In lieu of the procedure defined in this section, rational analysis is permitted to be used to determine the downdrag load and the net geotechnical ultimate capacity of the pile if the seismically-induced foundation differential settlement is indicated in the geotechnical investigation report and one of the following requirements is met:

1. The seismically-induced foundation differential settlement does not exceed one-fourth of the differential settlement threshold in Table 12.13-3.
2. The seismically-induced foundation differential settlement does not exceed the differential settlement threshold in Table 12.13-3 and does not exceed $0.0075L$; foundation ties comply with the additional detailing requirements of Section 12.13.9.2.1.1; and mat foundations comply with the requirements of Section 12.13.9.2.1.2.
3. The seismically-induced foundation differential settlement complies with the design requirements of Section 12.13.9.2(b); foundation ties comply with the additional detailing requirements of Section 12.13.9.2.1.1; and mat foundations comply with the requirements of Section 12.13.9.2.1.2.

12.13.9.3.2 Lateral Resistance Passive pressure and friction mobilized against walls, pile caps, and grade beams, when reduced for the effects of liquefaction and seismically-induced reduction in soil strength, shall be permitted to resist lateral inertial loads in combination with piles. Resistance provided by the combination of piles, passive pressure, and friction shall be determined based on compatible lateral deformations.

12.13.9.3.3 Concrete Deep Foundation Detailing Concrete piles, including cast-in-place and precast piles, shall be detailed to comply with Section 18.13.5.5 of ACI 318 from the top of the pile to a depth exceeding that of the deepest liquefiable soil by at least 7 times the member cross-sectional dimension.

12.13.9.3.4 Permanent Horizontal Ground Displacement Where the geotechnical investigation report indicates a potential for seismically-induced permanent horizontal ground displacement will occur in the event of MCE_G earthquake motions, pile design shall be based on a detailed analysis incorporating the expected lateral deformation, the depths over which the deformation is expected to occur, and the nonlinear behavior of the piles. Where nonlinear behavior of piles occurs caused by seismically-induced permanent horizontal ground displacement, the pile deformations shall not result in loss of the pile's ability to carry gravity loads, nor shall the deteriorated pile's lateral strength be less than 67% of the undamaged nominal strength. In addition, the following requirements shall be satisfied:

1. Structural steel H-piles shall satisfy the width-thickness limits for highly ductile H-pile members in ANSI/AISC 341.
2. Unfilled structural steel pipe piles shall satisfy the width-thickness limits for highly ductile round hollow structural steel (HSS) elements in ANSI/AISC 341.
3. Concrete piles shall be detailed to comply with Section 18.13.5.5 of ACI 318 from the top of the pile to a depth exceeding that of the deepest layer of soil prone to seismically-induced permanent horizontal ground displacement by at least 7 times the pile diameter. Nominal shear strength shall exceed the maximum forces that can be generated because of pile deformations determined in the detailed analysis.

12.13.9.3.5 Foundation Ties Individual pile caps shall be interconnected by ties in accordance with Section 12.13.8.2. Where the geotechnical investigation report indicates seismically-induced permanent horizontal ground displacement, the design forces for ties shall include the additional pressures applied to foundation elements because of the lateral displacement in accordance with the recommendations of the geotechnical investigation report. These effects shall be combined with the load effects from design earthquake lateral loads.

12.14 SIMPLIFIED ALTERNATIVE STRUCTURAL DESIGN CRITERIA FOR SIMPLE BEARING WALL OR BUILDING FRAME SYSTEMS

12.14.1 General

12.14.1.1 Simplified Design Procedure The procedures of this section are permitted to be used in lieu of other analytical procedures in Chapter 12 for the analysis and design of simple buildings with bearing wall or building frame systems, subject to all of the limitations listed in this section. Where these procedures are used, the Seismic Design Category shall be determined from Table 11.6-1 using the value of S_{DS} from Section 12.14.8.1, except that where S_1 is greater than or equal to 0.75, the Seismic Design Category shall be E. The simplified design procedure is permitted to be used if the following limitations are met.

1. The structure shall qualify for Risk Category I or II in accordance with Table 1.5-1.

2. The site class, defined in Chapter 20, shall not be Site Class E or F.
3. The structure shall not exceed three stories above grade plane.
4. The seismic force-resisting system shall be either a bearing wall system or a building frame system, as indicated in Table 12.14-1.
5. The structure shall have at least two lines of lateral resistance in each of two major axis directions. At least one line of resistance shall be provided on each side of the center of weight in each direction.
6. The center of weight in each story shall be located not further from the geometric centroid of the diaphragm than 10% of the length of the diaphragm parallel to the eccentricity.
7. For structures with cast-in-place concrete diaphragms, overhangs beyond the outside line of shear walls or braced frames shall satisfy the following:

$$a \leq d/3 \qquad (12.14\text{-}1)$$

where a is the distance perpendicular to the forces being considered from the extreme edge of the diaphragm to the line of vertical resistance closest to that edge, and d is the depth of the diaphragm parallel to the forces being considered at the line of vertical resistance closest to the edge.

All other diaphragm overhangs beyond the outside line of shear walls or braced frames shall satisfy

$$a \leq d/5 \qquad (12.14\text{-}2)$$

8. For buildings with a diaphragm that is not flexible, the forces shall be apportioned to the vertical elements as if the diaphragm were flexible. The following additional requirements shall be satisfied:
   (a) For structures with two lines of resistance in a given direction, the distance between the two lines is at least 50% of the length of the diaphragm perpendicular to the lines; and
   (b) For structures with more than two lines of resistance in a given direction, the distance between the two most extreme lines of resistance in that direction is at least 60% of the length of the diaphragm perpendicular to the lines.

   Where two or more lines of resistance are closer together than one-half the horizontal length of the longer of the walls or braced frames, it shall be permitted to replace those lines by a single line at the centroid of the group for the initial distribution of forces, and the resultant force to the group shall then be distributed to the members of the group based on their relative stiffnesses.
9. Lines of resistance of the seismic force-resisting system shall be oriented at angles of no more than 15 degrees from alignment with the major orthogonal horizontal axes of the building.
10. The simplified design procedure shall be used for each major orthogonal horizontal axis direction of the building.
11. System irregularities caused by in-plane or out-of-plane offsets of lateral force-resisting elements shall not be permitted.

    EXCEPTION: Out-of-plane and in-plane offsets of shear walls are permitted in two-story buildings of light-frame construction provided that the framing supporting the upper wall is designed for seismic force effects from overturning of the wall amplified by a factor of 2.5.
12. The story lateral strength of any story shall not be less than 80% of that in the story above. The story lateral strength is the total lateral strength of all seismic force-resisting system elements resisting the story shear for the direction under consideration.

12.14.1.2 Reference Documents The reference documents listed in Chapter 23 shall be used as indicated in Section 12.14.

12.14.1.3 Definitions The definitions listed in Section 11.2 shall be used in addition to the following:

PRINCIPAL ORTHOGONAL HORIZONTAL DIRECTIONS: The orthogonal directions that overlay the majority of lateral force-resisting elements.

12.14.1.4 Notation

D = Effect of dead load
E = Effect of horizontal and vertical earthquake-induced forces
F_a = Acceleration-based site coefficient, see Section 12.14.8.1
F_i = Portion of the seismic base shear, V, induced at level i
F_p = Seismic design force applicable to a particular structural component
F_x = See Section 12.14.8.2
h_i = Height above the base to level i
h_x = Height above the base to level x
Level i = Building level referred to by the subscript i; $i = 1$ designates the first level above the base
Level n = Level that is uppermost in the main portion of the building
Level x = See Level i
Q_E = Effect of horizontal seismic forces
R = Response modification coefficient as given in Table 12.14-1
S_{DS} = See Section 12.14.8.1
S_S = See Section 11.4.1
S_1 = See Section 11.4.1
V = Total design shear at the base of the structure in the direction of interest, as determined using the procedure of Section 12.14.8.1
V_x = Seismic design shear in story x; see Section 12.14.8.3
W = See Section 12.14.8.1
W_c = Weight of wall
w_i = Portion of the effective seismic weight, W, located at or assigned to level i
W_p = Weight of structural component
w_x = See Section 12.14.8.2

12.14.2 Design Basis The structure shall include complete lateral and vertical force-resisting systems with adequate strength to resist the design seismic forces, specified in this section, in combination with other loads. Design seismic forces shall be distributed to the various elements of the structure and their connections using a linear elastic analysis in accordance with the procedures of Section 12.14.8. The members of the seismic force-resisting system and their connections shall be detailed to conform with the applicable requirements for the selected structural system as indicated in Section 12.14.4.1. A continuous load path, or paths, with adequate strength and stiffness shall be provided to transfer all forces from the point of application to the final point of resistance. The foundation shall be designed to accommodate the forces developed.

12.14.3 Seismic Load Effects All members of the structure, including those not part of the seismic force-resisting system, shall be designed using the seismic load effects of

Table 12.14-1. Design Coefficients and Factors for Seismic Force-Resisting Systems for Simplified Design Procedure.

Seismic Force-Resisting System	ASCE 7 Section Where Detailing Requirements Are Specified	Response Modification Coefficient, R^a	Limitations[b]		
			Seismic Design Category		
			B	C	D, E
A. BEARING WALL SYSTEMS					
1. Special reinforced concrete shear walls	14.2	5	P	P	P
2. Ordinary reinforced concrete shear walls	14.2	4	P	P	NP
3. Detailed plain concrete shear walls	14.2	2	P	NP	NP
4. Ordinary plain concrete shear walls	14.2	1½	P	NP	NP
5. Intermediate precast shear walls	14.2	4	P	P	40[c]
6. Ordinary precast shear walls	14.2	3	P	NP	NP
7. Special reinforced masonry shear walls	14.4	5	P	P	P
8. Intermediate reinforced masonry shear walls	14.4	3½	P	P	NP
9. Ordinary reinforced masonry shear walls	14.4	2	P	NP	NP
10. Detailed plain masonry shear walls	14.4	2	P	NP	NP
11. Ordinary plain masonry shear walls	14.4	1½	P	NP	NP
12. Prestressed masonry shear walls	14.4	1½	P	NP	NP
13. Light-frame (wood) walls sheathed with wood structural panels rated for shear resistance	14.5	6½	P	P	P
14. Light-frame (cold-formed steel) walls sheathed with wood structural panels rated for shear resistance or steel sheets	14.1	6½	P	P	P
15. Light-frame walls with shear panels of all other materials	14.1 and 14.5	2	P	P	NP[d]
16. Light-frame (cold-formed steel) wall systems using flat strap bracing	14.1	4	P	P	P
17. Cross-laminated timber shear walls	14.5	3	P	P	P
18. Cross-laminated timber shear walls with shear resistance provided by high-aspect-ratio panels only	14.5	4	P	P	P
B. BUILDING FRAME SYSTEMS					
1. Steel eccentrically braced frames	14.1	8	P	P	P
2. Steel special concentrically braced frames	14.1	6	P	P	P
3. Steel ordinary concentrically braced frames	14.1	3¼	P	P	P
4. Special reinforced concrete shear walls	14.2	6	P	P	P
5. Ordinary reinforced concrete shear walls	14.2	5	P	P	NP
6. Detailed plain concrete shear walls	14.2 and 14.2.2.7	2	P	NP	NP
7. Ordinary plain concrete shear walls	14.2	1½	P	NP	NP
8. Intermediate precast shear walls	14.2	5	P	P	40[c]
9. Ordinary precast shear walls	14.2	4	P	NP	NP
10. Steel and concrete composite eccentrically braced frames	14.3	8	P	P	P
11. Steel and concrete composite special concentrically braced frames	14.3	5	P	P	P
12. Steel and concrete composite ordinary braced frames	14.3	3	P	P	NP
13. Steel and concrete composite plate shear walls	14.3	6½	P	P	P
14. Steel and concrete composite special shear walls	14.3	6	P	P	P
15. Steel and concrete composite ordinary shear walls	14.3	5	P	P	NP
16. Special reinforced masonry shear walls	14.4	5½	P	P	P
17. Intermediate reinforced masonry shear walls	14.4	4	P	P	NP
18. Ordinary reinforced masonry shear walls	14.4	2	P	NP	NP
19. Detailed plain masonry shear walls	14.4	2	P	NP	NP
20. Ordinary plain masonry shear walls	14.4	1½	P	NP	NP
21. Prestressed masonry shear walls	14.4	1½	P	NP	NP
22. Light-frame (wood) walls sheathed with wood structural panels rated for shear resistance	14.5	7	P	P	P
23. Light-frame (cold-formed steel) walls sheathed with wood structural panels rated for shear resistance or steel sheets	14.1	7	P	P	P

continues

Table 12.14-1 *(Continued).*

Seismic Force-Resisting System	ASCE 7 Section Where Detailing Requirements Are Specified	Response Modification Coefficient, R^a	Limitations[b] Seismic Design Category		
			B	C	D, E
24. Light-frame walls with shear panels of all other materials	14.1 and 14.5	2½	P	P	NP[d]
25. Steel buckling-restrained braced frames	14.1	8	P	P	P
26. Steel special plate shear walls	14.1	7	P	P	P

[a] Response modification coefficient, R, for use throughout the standard.
[b] P = permitted; NP = not permitted.
[c] Light-frame walls with shear panels of all other materials are not permitted in Seismic Design Category E.
[d] Light-frame walls with shear panels of all other materials are permitted up to 35 ft (10.6 m) in structural height, h_n, in Seismic Design Category D and are not permitted in Seismic Design Category E.

Section 12.14.3 unless otherwise exempted by this standard. Seismic load effects are the axial, shear, and flexural member forces resulting from application of horizontal and vertical seismic forces as set forth in Section 12.14.3.1. Where required, seismic load effects shall include overstrength, as set forth in Section 12.14.3.2.

12.14.3.1 Seismic Load Effect The seismic load effect, E, includes effects of both horizontal and vertical ground motions as shown in Eq. (12.14-3) and Eq. (12.14-4).

$$E = E_h + E_v \quad (12.14\text{-}3)$$

$$E = E_h - E_v \quad (12.14\text{-}4)$$

where E_h is the effect of horizontal seismic forces as defined in Section 12.14.3.1.1, and E_v is the effect of vertical seismic forces as defined in Section 12.14.3.1.2.

E_h and E_v shall be used in load combinations 6 and 7 in Section 2.3.6 and in load combinations 8, 9, and 10 in Section 2.4.5.

12.14.3.1.1 Horizontal Seismic Load Effect The horizontal seismic load effect, E_h, shall be determined in accordance with Equation (12.14-5) as follows:

$$E_h = Q_E \quad (12.14\text{-}5)$$

where Q_E is the effects of horizontal seismic forces from V or F_p as specified in Sections 12.14.7.5, 12.14.8.1, and 13.3.1.

12.14.3.1.2 Vertical Seismic Load Effect The vertical seismic load effect, E_v, shall be determined in accordance with Equation (12.14-6) as follows:

$$E_v = 0.2S_{DS}D \quad (12.14\text{-}6)$$

where S_{DS} is the design spectral response acceleration parameter at short periods obtained from Section 11.4.5, and D is the effect of dead load.

EXCEPTION: The vertical seismic load effect, E_v, is permitted to be taken as zero for either of the following conditions:

1. In Equations (12.14-3), (12.14-4), (12.14-7), and (12.14-8) where S_{DS} is equal to or less than 0.125; and
2. In Equation (12.14-4) where determining demands on the soil–structure interface of foundations.

12.14.3.2 Seismic Load Effect Including Overstrength The seismic load effect including overstrength, E_m includes effects of both horizontal and vertical ground motions as shown in Eq. (12.14-7) and Eq. (12.14-8). Where required, the seismic load effects, including overstrength, shall be determined in accordance with the following.

$$E_m = E_{mh} + E_v \quad (12.14\text{-}7)$$

$$E_m = E_{mh} - E_v \quad (12.14\text{-}8)$$

where

E_m = Seismic load effect including overstrength
E_{mh} = The effect of horizontal seismic forces, including overstrength, as defined in Section 12.14.3.2.1 or 12.14.3.2.2; and
E_v = The vertical seismic load effect as defined in Section 12.14.3.1.2.

E_{mh} and E_v shall be used in load combinations 6 and 7 in Section 2.3.6 and in load combinations 8, 9, and 10 in Section 2.4.5.

12.14.3.2.1 Horizontal Seismic Load Effect with a 2.5 Overstrength The effect of horizontal seismic forces, including overstrength, E_{mh}, shall be determined in accordance with Equation (12.14-9) as follows:

$$E_{mh} = 2.5Q_E \quad (12.14\text{-}9)$$

where Q_E is the effects of horizontal seismic forces from V or F_p as specified in Sections 12.14.7.5, 12.14.8.1, and 13.3.1.

E_{mh} need not be taken as larger than E_{cl}, where E_{cl} = The capacity-limited horizontal seismic load effect as defined in Section 11.3.

12.14.3.2.2 Capacity-Limited Horizontal Seismic Load Effect Where capacity-limited design is required by the material reference document, the seismic load effect including overstrength shall be calculated with the capacity-limited horizontal seismic load effect, E_{cl}, substituted for E_{mh} in the load combinations of Section 2.3.6 and Section 2.4.5.

12.14.4 Seismic Force-Resisting System

12.14.4.1 Selection and Limitations The basic lateral and vertical seismic force-resisting system shall conform to one of the types indicated in Table 12.14-1 and shall conform to all of the detailing requirements referenced in the table. The appropriate response modification coefficient, R, indicated in Table 12.14-1 shall be used in determining the base shear and element design forces as set forth in the seismic requirements of this standard.

Special framing and detailing requirements are indicated in Section 12.14.7 and in Sections 14.1, 14.2, 14.3, 14.4, and 14.5 for structures assigned to the various Seismic Design Categories.

12.14.4.2 Combinations of Framing Systems

12.14.4.2.1 Horizontal Combinations Different seismic force-resisting systems are permitted to be used in each of the two principal orthogonal building directions. Where a combination of different structural systems is used to resist lateral forces in the same direction, the value of R used for design in that direction shall not be greater than the least value of R for any of the systems used in that direction.

EXCEPTION: For buildings of light-frame construction or buildings that have flexible diaphragms and that are two stories or fewer above grade plane, resisting elements are permitted to be designed using the least value of R of the different seismic force-resisting systems found in each independent line of framing. The value of R used for design of diaphragms in such structures shall not be greater than the least value for any of the systems used in that same direction.

12.14.4.2.2 Vertical Combinations Different seismic force-resisting systems are permitted to be used in different stories. The value of R used in a given direction shall not be greater than the least value of any of the systems used in that direction.

12.14.4.2.3 Combination Framing Detailing Requirements The detailing requirements of Section 12.14.7 required by the higher response modification coefficient, R, shall be used for structural members common to systems that have different response modification coefficients.

12.14.5 Diaphragm Flexibility Diaphragms constructed of steel decking (untopped), wood structural panels, or similar panelized construction techniques are permitted to be considered flexible.

12.14.6 Application of Loading The effects of the combination of loads shall be considered as prescribed in Section 12.14.3. The design seismic forces are permitted to be applied separately in each orthogonal direction, and the combination of effects from the two directions need not be considered. Reversal of load shall be considered.

12.14.7 Design and Detailing Requirements The design and detailing of the members of the seismic force-resisting system shall comply with the requirements of this section. The foundation shall be designed to resist the forces developed and accommodate the movements imparted to the structure by the design ground motions. The dynamic nature of the forces, the expected ground motion, the design basis for strength and energy dissipation capacity of the structure, and the dynamic properties of the soil shall be included in the determination of the foundation design criteria. The design and construction of foundations shall comply with Section 12.13. Structural elements including foundation elements shall conform to the material design and detailing requirements set forth in Chapter 14.

12.14.7.1 Connections All parts of the structure between separation joints shall be interconnected, and the connection shall be capable of transmitting the seismic force, F_p, induced by the parts being connected. Any smaller portion of the structure shall be tied to the remainder of the structure with elements that have a strength of 0.20 times the short-period design spectral response acceleration coefficient, S_{DS}, times the weight of the smaller portion or 5% of the portion's weight, whichever is greater.

A positive connection for resisting a horizontal force acting parallel to the member shall be provided for each beam, girder, or truss, either directly to its supporting elements or to slabs designed to act as diaphragms. Where the connection is through a diaphragm, then the member's supporting element must also be connected to the diaphragm. The connection shall have minimum design strength of 5% of the dead plus live load reaction.

12.14.7.2 Openings or Reentrant Building Corners Except where otherwise specifically provided for in this standard, openings in shear walls, diaphragms, or other plate-type elements shall be provided with reinforcement at the edges of the openings or reentrant corners designed to transfer the stresses into the structure. The edge reinforcement shall extend into the body of the wall or diaphragm a distance sufficient to develop the force in the reinforcement.

EXCEPTION: Shear walls of wood structural panels are permitted where designed in accordance with AWC SDPWS for perforated shear walls or ANSI/AISI S400 for Type II shear walls.

12.14.7.3 Collector Elements Collector elements shall be provided with adequate strength to transfer the seismic forces originating in other portions of the structure to the element providing the resistance to those forces (Figure 12.10-1). Collector elements, splices, and their connections to resisting elements shall be designed to resist the forces defined in Section 12.14.3.2.

EXCEPTION: In structures, or portions thereof, braced entirely by light-frame shear walls, collector elements, splices, and connections to resisting elements are permitted to be designed to resist forces in accordance with Section 12.14.7.4.

12.14.7.4 Diaphragms Floor and roof diaphragms shall be designed to resist the design seismic forces at each level, F_x, calculated in accordance with Section 12.14.8.2. Where the diaphragm is required to transfer design seismic forces from the vertical-resisting elements above the diaphragm to other vertical-resisting elements below the diaphragm because of changes in relative lateral stiffness in the vertical elements, the transferred portion of the seismic shear force at that level, V_x, shall be added to the diaphragm design force. Diaphragms shall provide for both the shear and bending stresses resulting from these forces. Diaphragms shall have ties or struts to distribute the wall anchorage forces into the diaphragm. Diaphragm connections shall be positive, mechanical, or welded type connections.

12.14.7.5 Anchorage of Structural Walls Structural walls shall be anchored to all floors, roofs, and members that provide out-of-plane lateral support for the wall or that are supported by the wall. The anchorage shall provide a positive direct connection between the wall and floor, roof, or supporting member with the strength to resist the out-of-plane force given by Equation (12.14-10):

$$F_p = 0.4k_a S_{DS} W_p \quad (12.14\text{-}10)$$

where

$$k_a = 1.0 + \frac{L_f}{100} \quad (12.14\text{-}11)$$

and k_a need not be taken as larger than 2.0;

L_f = Span, in feet, of a flexible diaphragm that provides the lateral support for the wall (the span is measured between vertical elements that provide lateral support to the diaphragm in the direction considered—use zero for rigid diaphragms);
F_p = Design force in the individual anchors;
k_a = Amplification factor for diaphragm flexibility;
S_{DS} = Design spectral response acceleration at short periods per Section 12.14.8.1;
W_p = Weight of the wall tributary to the anchor; and
F_p = Shall not be taken as less than the larger of $0.2k_aW_p$ and 5 lb/ft^2 (0.24 kN/m^2) times the area of the wall tributary to the anchor.

12.14.7.5.1 Transfer of Anchorage Forces into Diaphragms Diaphragms shall be provided with continuous ties or struts between diaphragm chords to distribute these anchorage forces into the diaphragms. Added chords are permitted to be used to form subdiaphragms to transmit the anchorage forces to the main continuous crossties. The maximum length-to-width ratio of the structural subdiaphragm shall be 2.5 to 1. Connections and anchorages capable of resisting the prescribed forces shall be provided between the diaphragm and the attached components. Connections shall extend into the diaphragm a sufficient distance to develop the force transferred into the diaphragm.

12.14.7.5.2 Wood Diaphragms The anchorage of concrete or masonry structural walls to wood diaphragms shall be in accordance with AWC SDPWS 4.1.5.1 and this section. Continuous ties required by this section shall be in addition to the diaphragm sheathing. Anchorage shall not be accomplished by use of toenails or nails subject to withdrawal, nor shall wood ledgers or framing be used in cross-grain bending or cross-grain tension. The diaphragm sheathing shall not be considered effective as providing the ties or struts required by this section.

12.14.7.5.3 Metal Deck Diaphragms In metal deck diaphragms, the metal deck shall not be used as the continuous ties required by this section in the direction perpendicular to the deck span.

12.14.7.5.4 Embedded Straps Diaphragm to wall anchorage using embedded straps shall be attached to or hooked around the reinforcing steel or otherwise terminated so as to effectively transfer forces to the reinforcing steel.

12.14.7.6 Bearing Walls and Shear Walls Exterior and interior bearing walls and shear walls and their anchorage shall be designed for a force equal to 40% of the short-period design spectral response acceleration, S_{DS}, times the weight of wall, W_c, normal to the surface, with a minimum force of 10% of the weight of the wall. Interconnection of wall elements and connections to supporting framing systems shall have sufficient ductility, rotational capacity, or strength to resist shrinkage, thermal changes, and differential foundation settlement where combined with seismic forces.

12.14.7.7 Anchorage of Nonstructural Systems Where required by Chapter 13, all portions or components of the structure shall be anchored for the seismic force, F_p, prescribed therein.

12.14.8 Simplified Lateral Force Analysis Procedure An equivalent lateral force analysis shall consist of the application of equivalent static lateral forces to a linear mathematical model of the structure. The lateral forces applied in each direction shall sum to a total seismic base shear given by Section 12.14.8.1 and shall be distributed vertically in accordance with Section 12.14.8.2. For purposes of analysis, the structure shall be considered fixed at the base.

12.14.8.1 Seismic Base Shear The seismic base shear, V, in a given direction shall be determined in accordance with Equation (12.14-12):

$$V = \frac{FS_{DS}}{R} W \quad (12.14\text{-}12)$$

where $S_{DS} = \frac{2}{3} F_a S_s$, where F_a is permitted to be taken as 1.0 for rock sites, 1.4 for soil sites, or determined in accordance with Section 11.4.4. For the purpose of this section, sites are permitted to be considered to be rock if there is no more than 10 ft (3 m) of soil between the rock surface and the bottom of spread footing or mat foundation. In calculating S_{DS}, S_s shall be in accordance with Section 11.4.4 but need not be taken as larger than 1.5;

F = 1.0 for buildings that are one story above grade plane, 1.1 for buildings that are two stories above grade plane, and 1.2 for buildings that are three stories above grade plane;
R = Response modification factor from Table 12.14-1; and
W = Effective seismic weight of the structure that includes the dead load, as defined in Section 3.1, above grade plane and other loads above grade plane as listed in the following text:

1. In areas used for storage, a minimum of 25% of the floor live load shall be included.
   EXCEPTIONS:
   (a) Where the inclusion of storage loads adds no more than 5% to the effective seismic weight at that level, it need not be included in the effective seismic weight.
   (b) Floor live load in public garages and open parking structures need not be included.
2. Where provision for partitions is required by Section 4.3.2 in the floor load design, the actual partition weight, or a minimum weight of 10 lb/ft^2 (0.48 kN/m^2) of floor area, whichever is greater.
3. Total operating weight of permanent equipment.
4. Where the flat roof snow load, P_f, exceeds 30 lb/ft^2 (1.44 kN/m^2), 20% of the uniform design snow load, regardless of actual roof slope.
5. Weight of landscaping and other materials at roof gardens and similar areas.

12.14.8.2 Vertical Distribution The forces at each level shall be calculated using the following equation:

$$F_x = \frac{w_x}{W} V \quad (12.14\text{-}13)$$

where w_x is the portion of the effective seismic weight of the structure, W, at level x.

12.14.8.3 Horizontal Shear Distribution The seismic design story shear in any story, V_x [kip (kN)], shall be determined from the following equation:

$$V_x = \sum_{1=x}^{n} F_i \quad (12.14\text{-}14)$$

where F_i is the the portion of the seismic base shear, V [kip (kN)], induced at level i.

12.14.8.3.1 Flexible Diaphragm Structures The seismic design story shear in stories of structures with flexible diaphragms, as defined in Section 12.14.5, shall be distributed to the vertical elements of the seismic force-resisting system using tributary area rules. Two-dimensional analysis is permitted where diaphragms are flexible.

12.14.8.3.2 Structures with Diaphragms That Are Not Flexible For structures with diaphragms that are not flexible, as defined in Section 12.14.5, the seismic design story shear, V_x [kip (kN)], shall be distributed to the various vertical elements of the seismic force-resisting system in the story under consideration based on the relative lateral stiffnesses of the vertical elements and the diaphragm.

12.14.8.3.2.1 Torsion The design of structures with diaphragms that are not flexible shall include the torsional moment, M_t [kip-ft (kN-m)], resulting from eccentricity between the locations of center of mass and the center of rigidity.

12.14.8.4 Overturning The structure shall be designed to resist overturning effects caused by the seismic forces determined in Section 12.14.8.2. The foundations of structures shall be designed for not less than 75% of the foundation overturning design moment, M_f [kip-ft (kN-m)], at the foundation–soil interface.

12.14.8.5 Drift Limits and Building Separation Structural drift need not be calculated. Where a drift value is needed for use in material standards, to determine structural separations between buildings or from property lines, for design of cladding, or for other design requirements, it shall be taken as 1% of structural height, h_n, unless computed to be less. Each portion of the structure shall be designed to act as an integral unit in resisting seismic forces unless it is separated structurally by a distance sufficient to avoid damaging contact under the total deflection.

12.15 CONSENSUS STANDARDS AND OTHER REFERENCED DOCUMENTS

See Chapter 23 for the list of consensus standards and other documents that shall be considered part of this standard to the extent referenced in this chapter.

CHAPTER 13
SEISMIC DESIGN REQUIREMENTS FOR NONSTRUCTURAL COMPONENTS

13.1 GENERAL

13.1.1 Scope This chapter establishes minimum design criteria for nonstructural components, including their supports and attachments.

Nonstructural components include

1. Components that are in or supported by a structure,
2. Components that are outside of a structure (except for nonbuilding structures within the scope of Chapter 15) and are permanently connected to the mechanical or electrical systems, or
3. Components that are part of the egress system of a structure.

Where the weight of a nonstructural component is greater than or equal to 20% of the combined effective seismic weight, W, of the nonstructural component and the supporting structure as defined in Section 12.7.2, the component shall be designed in accordance with Section 13.2.9.

13.1.2 Seismic Design Category For the purposes of this chapter, nonstructural components shall be assigned to the same seismic design category as

1. The structure that they occupy or are supported by, or
2. The structure to which they are permanently connected by mechanical or electrical systems, or
3. For parts of an egress system, the structure it serves.

If more than one of these criteria is applicable, the highest seismic design category shall be used.

13.1.3 Component Importance Factor All components shall be assigned a component Importance Factor as indicated in this section. The component Importance Factor, I_p, shall be taken as 1.5 if any of the following conditions apply:

1. The component is required to function for life-safety purposes after an earthquake, including fire protection sprinkler systems and egress stairways.
2. The component conveys, supports, or otherwise contains toxic, highly toxic, or explosive substances where the quantity of the material exceeds a threshold quantity established by the Authority Having Jurisdiction and is sufficient to pose a threat to the public if released.
3. The component is in or supported by a Risk Category IV structure or permanently connected by mechanical or electrical systems to a Risk Category IV structure, and the component is required for the continued operation of a structure designated an Essential Facility, or its failure would impair the continued operation of a structure designated an Essential Facility.
4. The component conveys, supports, or otherwise contains hazardous substances and is attached to a structure or portion thereof classified by the Authority Having Jurisdiction as a hazardous occupancy.

All other components shall be assigned a component Importance Factor, I_p, equal to 1.0.

13.1.4 Exemptions The nonstructural components listed in Table 13.1-1 are exempt from the requirements of this chapter.

13.1.5 Premanufactured Modular Mechanical and Electrical Systems Premanufactured mechanical and electrical modules 6 ft (1.8 m) high and taller that are not otherwise prequalified in accordance with Chapter 13 and that contain or support mechanical and electrical components shall be designed in accordance with the provisions for nonbuilding structures similar to buildings in Chapter 15. Nonstructural components contained or supported within modular systems shall be designed in accordance with Chapter 13.

13.1.6 Application of Nonstructural Component Requirements to Nonbuilding Structures Nonbuilding structures (including storage racks and tanks) that are supported by other structures shall be designed in accordance with Chapter 15. Where Section 15.3 requires that seismic forces be determined in accordance with Chapter 13 and values for C_{AR} and R_{po} are not provided in Table 13.5-1 or 13.6-1, the term $[\frac{C_{AR}}{R_{po}}]$ in Equation (13.3-1) shall be taken as equal to $2.5/R$, where the value of R for the nonbuilding structure is obtained from Tables 15.4-1 or 15.4-2.

13.1.7 Reference Documents Where a reference document provides a basis for the earthquake-resistant design of a particular type of nonstructural component, that document is permitted to be used, subject to the approval of the Authority Having Jurisdiction and the following conditions:

1. The design earthquake forces shall not be less than those determined in accordance with Section 13.3.1.
2. Each nonstructural component's seismic interactions with all other connected components and with the supporting structure shall be accounted for in the design. The component shall accommodate drifts, deflections, and relative displacements determined in accordance with the applicable seismic requirements of this standard.
3. Nonstructural component anchorage requirements shall not be less than those specified in Section 13.4.

13.1.8 Reference Documents Using Allowable Stress Design Where a reference document provides a basis for the earthquake-resistant design of a particular type of component,

Table 13.1-1. Nonstructural Components Exempt from the Requirements of This Chapter.

Seismic Design Category (SDC)	Nonstructural Components Exempt from the Requirements of this Chapter
All Categories	• Furniture (except storage cabinets, as noted in Table 13.5-1) • Temporary components that remain in place for 180 days or less • Mobile units and equipment including components that are moved from one point in the structure to another during ordinary use
A	• All components
B	• Architectural Components, other than parapets, provided that the component Importance Factor, I_p, is equal to 1.0 • Mechanical and Electrical Components
C	• Mechanical and Electrical Components, provided that either ∘ The component Importance Factor, I_p, is equal to 1.0 and the component is positively attached to the structure; or ∘ The component weighs 20 lb (89 N) or less
D, E, F	• Mechanical and electrical components positively attached to the structure, provided that ∘ For discrete mechanical and electrical components, the component weighs 400 lb (1,779 N) or less, the center of mass is located 4 ft (1.22 m) or less above the adjacent floor level, flexible connections are provided between the component and associated ductwork, piping, and conduit, and the component Importance Factor, I_p, is equal to 1.0; or ∘ For discrete mechanical and electrical components, the component weighs 20 lb (89 N) or less; or ∘ For distribution systems, the component Importance Factor, I_p, is equal to 1.0 and the operating weight of the system is 5 lb/ft (73 N/m) or less. • Distribution systems included in the exceptions for conduit, cable tray, and raceways in Section 13.6.5, duct systems in 13.6.6, and piping and tubing systems in 13.6.7.3. Where in-line components, such as valves, in-line suspended pumps, and mixing boxes require independent support, they shall be addressed as discrete components and shall be braced considering the tributary contribution of the attached distribution system.

and the same reference document defines acceptance criteria in terms of allowable stresses rather than strengths, that reference document is permitted to be used. The allowable stress load combination shall consider dead, live, operating, and earthquake loads in addition to those in the reference document. The earthquake loads determined in accordance with Section 13.3.1 shall be multiplied by a factor of 0.7. The allowable stress design load combinations of Section 2.4 need not be used. The component shall also accommodate the relative displacements specified in Section 13.3.2.

13.2 GENERAL DESIGN REQUIREMENTS

13.2.1 Applicable Requirements for Architectural, Mechanical, and Electrical Components, Supports, and Attachments Architectural, mechanical, and electrical components, supports, and attachments shall comply with the sections referenced in Table 13.2-1. These requirements shall be satisfied by one of the following methods:

1. Project-specific design and documentation submitted for approval to the Authority Having Jurisdiction after review and acceptance by a registered design professional; or
2. Submittal of the manufacturer's certification that the component is seismically qualified by at least one of the following:
   (a) Analysis, or
   (b) Testing in accordance with the alternative set forth in Section 13.2.6, or
   (c) Experience data in accordance with the alternative set forth in Section 13.2.7.

13.2.2 Load Combinations Nonstructural components, including their supports and attachments, covered by this chapter and not otherwise exempt by Section 13.1.8 shall comply with Section 1.3, including consideration of load combinations of either Section 2.3 or 2.4, as appropriate. For the purposes of combining load effects, F_p shall be used per Section 12.4.2.1, and horizontal seismic design forces including overstrength shall be used per Section 12.4.3.1.

13.2.3 Special Certification Requirements for Designated Seismic Systems Certifications shall be provided for designated seismic systems assigned to Seismic Design Categories C through F as follows:

1. Active mechanical and electrical equipment that must remain operable following the design earthquake ground motion shall be certified by the manufacturer as operable whereby active parts or energized components shall be certified exclusively on the basis of approved shake table testing in accordance with Section 13.2.6 or experience data in accordance with Section 13.2.7, unless it can be

Table 13.2-1. Applicable Requirements for Architectural, Mechanical, and Electrical Components: Supports and Attachments.

Nonstructural Element (i.e., Component, Support, or Attachment)	General Design Requirements (Section 13.2)	Force and Displacement Requirements (Section 13.3)	Attachment Requirements (Section 13.4)	Architectural Component Requirements (Section 13.5)	Mechanical and Electrical Component Requirements (Section 13.6)
Architectural components and supports and attachments for architectural components	X	X	X	X	
Mechanical and electrical components	X	X	X		X
Supports and attachments for mechanical and electrical components	X	X	X		X

Note: X = applicable requirements.

shown that the component is inherently rugged by comparison with similar seismically qualified components. Evidence demonstrating compliance with this requirement shall be submitted for approval to the Authority Having Jurisdiction after review and acceptance by a registered design professional.

2. Components with hazardous substances and assigned a component Importance Factor, I_p, of 1.5 in accordance with Section 13.1.3 shall be certified by the manufacturer as maintaining containment following the design earthquake ground motion by (a) analysis, (b) approved shake table testing in accordance with Section 13.2.6, or (c) experience data in accordance with Section 13.2.7. Evidence demonstrating compliance with this requirement shall be submitted for approval to the Authority Having Jurisdiction after review and acceptance by a registered design professional.
3. Certification of components through analysis shall be limited to nonactive components and shall be based on seismic demand considering $\left[\frac{C_{AR}}{R_{po}}\right] = 2.5$. The value of R_μ shall be taken as 1.3 for components located above the grade plane.

EXCEPTION: If the period of the component, T_p, is less than or equal to 0.06 seconds, $\left[\frac{C_{AR}}{R_{po}}\right]$ may be taken as 1.0.

13.2.4 Consequential Damage The functional and physical interrelationship of components, their supports, and their effect on each other shall be considered so that the failure of an essential or nonessential architectural, mechanical, or electrical component shall not cause the failure of an essential architectural, mechanical, or electrical component. Where not otherwise established by analysis or test, required clearances for sprinkler system drops and sprigs shall not be less than those specified in Section 13.2.4.1.

13.2.4.1 Clearances between Equipment, Distribution Systems, Supports, and Sprinkler System Drops and Sprigs The installed clearance between any sprinkler drop or sprig and the following items shall be at least 3 in. (76 mm) in all directions:

1. Permanently attached equipment, including its structural supports and bracing; and
2. Other distribution systems, including their structural supports and bracing.

EXCEPTION: Sprinklers installed using flexible sprinkler hose need not meet the installed clearance requirement of this section.

13.2.5 Flexibility The design and evaluation of components, their supports, and their attachments shall consider their flexibility and their strength.

13.2.6 Testing Alternative for Seismic Capacity Determination As an alternative to the analytical requirements of Sections 13.2 through 13.6, testing shall be deemed an acceptable method to determine the seismic capacity of components and their supports and attachments. Seismic qualification by testing based on a nationally recognized testing standard procedure, such as ICC-ES AC 156, acceptable to the Authority Having Jurisdiction shall be deemed to satisfy the design and evaluation requirements provided that the substantiated seismic capacities equal or exceed the seismic demands determined in accordance with Sections 13.3.1 and 13.3.2. For the testing alternative, the maximum seismic demand determined in accordance with Equation (13.3-2) is not required to exceed $3.2 I_p W_p$.

13.2.7 Experience Data Alternative for Seismic Capacity Determination As an alternative to the analytical requirements of Sections 13.2 through 13.6, use of experience data shall be deemed an acceptable method to determine the seismic capacity of components and their supports and attachments. Seismic qualification by experience data based on nationally recognized procedures acceptable to the Authority Having Jurisdiction shall be deemed to satisfy the design and evaluation requirements provided that the substantiated seismic capacities equal or exceed the seismic demands determined in accordance with Sections 13.3.1 and 13.3.2.

13.2.8 Construction Documents Where design of nonstructural components or their supports and attachments is required by Table 13.2-1, such design shall be shown in construction documents prepared by a registered design professional for use by the owner, Authorities Having Jurisdiction, contractors, and inspectors.

13.2.9 Supported Nonstructural Components with Greater than or Equal to 20% Combined Weight For the condition where the weight of a nonstructural component is equal to or greater than 20% of the combined effective seismic weight, W, of the nonstructural component and the supporting structure, an analysis combining the structural characteristics of both the nonstructural component and the supporting structure shall be performed to determine the seismic design forces. The nonstructural component and the supporting structure shall be designed for forces and displacements determined in accordance with Chapter 12 or Section 15.5, as appropriate, with the R value of the combined system taken as the lesser of $0.40\left[\frac{C_{AR}}{R_{po}}\right]$ of the nonstructural component or the R value of the supporting structure. The nonstructural component and attachments shall be designed for the forces and displacements resulting from the combined analysis. Design criteria for the nonstructural component shall otherwise be in accordance with this chapter.

EXCEPTION: Where the ratio of the fundamental period of the nonstructural component and its attachment (to the structure) to the fundamental period of the supporting structure (including the lumped weight of the nonstructural component) is less than 0.5 or greater than 2.0, the supporting structure is permitted to be designed in accordance with the requirements of Chapter 12 or Section 15.5, as appropriate. The supported nonstructural component shall follow the requirements of Chapter 13 as if the weight of the nonstructural component were less than 20% of the combined effective seismic weight, W, of the nonstructural component and supporting structure.

13.3 SEISMIC DEMANDS ON NONSTRUCTURAL COMPONENTS

13.3.1 Horizontal Seismic Design Forces The horizontal seismic design force, F_p, shall be applied at the component's center of gravity and distributed relative to the component's mass distribution. The redundancy factor, ρ, is permitted to be taken as equal to 1.

The directions of F_p used shall be those that produce the most critical load effects on the component, the component supports, and attachments. Alternatively, it is permitted to use the more severe of the following two load cases:

- Case 1: A combination of 100% of F_p in any one horizontal direction and 30% of F_p in a perpendicular horizontal direction applied simultaneously.
- Case 2: The combination from Case 1 rotated 90 degrees.

The horizontal seismic design force shall be calculated as

$$F_p = 0.4 S_{DS} I_p W_p \left[\frac{H_f}{R_\mu}\right]\left[\frac{C_{AR}}{R_{po}}\right] \qquad (13.3\text{-}1)$$

F_p is not required to be taken as greater than

$$F_p = 1.6 S_{DS} I_p W_p \qquad (13.3\text{-}2)$$

and shall not be taken as less than

$$F_p = 0.3 S_{DS} I_p W_p \qquad (13.3\text{-}3)$$

where

F_p = Seismic design force;
S_{DS} = Spectral acceleration, short period, as determined in accordance with Section 11.4.5;
I_p = Component Importance Factor as determined in accordance with Section 13.1.3;
W_p = Component operating weight;
H_f = Factor for force amplification as a function of height in the structure as determined in Section 13.3.1.1;
R_μ = Structure ductility reduction factor as determined in Section 13.3.1.2;
C_{AR} = Component resonance ductility factor that converts the peak floor or ground acceleration into the peak component acceleration, as determined in Section 13.3.1.3; and
R_{po} = Component strength factor as determined in Section 13.3.1.4.

13.3.1.1 Amplification with Height, H_f For nonstructural components supported at or below grade plane, the factor for force amplification with height H_f, is 1.0. For components supported above grade plane by a building or nonbuilding structure, H_f is permitted to be determined by Equation (13.3-4) or Equation (13.3-5). Where the approximate fundamental period of the supporting building or nonbuilding structure is unknown, H_f is permitted to be determined by Equation (13.3-5).

$$H_f = 1 + a_1\left(\frac{z}{h}\right) + a_2\left(\frac{z}{h}\right)^{10} \qquad (13.3\text{-}4)$$

$$H_f = 1 + 2.5\left(\frac{z}{h}\right) \qquad (13.3\text{-}5)$$

where

$a_1 = 1/T_a \leq 2.5$;
$a_2 = [1 - (0.4/T_a)^2] \geq 0$;
z = Height above the base of the structure to the point of attachment of the component. For items at or below the base, z shall be taken as 0. The value of $\frac{z}{h}$ need not exceed 1.0;
h = Average roof height of structure with respect to the base; and
T_a = Lowest approximate fundamental period of the supporting building or nonbuilding structure in either orthogonal direction. For structures with combinations of seismic force-resisting systems (SFRSs), the SFRS that produces the lowest value of T_a shall be used.

For the purposes of computing H_f, T_a is determined using Equation (12.8-8) for buildings. Where the SFRS is unknown, T_a is permitted to be determined by Equation (12.8-8) using the approximate period parameters for "all other structural systems."

For nonbuilding structures, T_a is permitted to be taken as

(a) The period of the nonbuilding structure, T, determined using the structural properties and deformation characteristics of the resisting elements in a properly substantiated analysis as indicated in Section 12.8.2; or
(b) The period of the nonbuilding structure, T, determined using Equation (15.4-6); or
(c) The period T_a determined by Equation (12.8-8), using the approximate period parameters for "all other structural systems."

13.3.1.2 Structure Ductility Reduction Factor, R_μ For components supported by a building or nonbuilding structure, the reduction factor for ductility of the supporting structure, R_μ, is calculated as

$$R_\mu = [1.1R/(I_e\Omega_0)]^{1/2} \geq 1.3 \qquad (13.3\text{-}6)$$

where

I_e = Importance Factor as prescribed in Section 11.5.1 for the building or nonbuilding structure supporting the component;
R = Response modification factor for the building or nonbuilding structure supporting the component, from Table 12.2-1, 15.4-1, or 15.4-2; and
Ω_0 = Overstrength factor for the building or nonbuilding structure supporting the component, from Table 12.2-1, 15.4-1, or 15.4-2.

For components supported at or below grade plane, R_μ shall be taken as 1.0. When the SFRS of the building or nonbuilding structure is not specified, R_μ shall be taken as 1.3 for components above grade plane. When the SFRS of the building or nonbuilding structure is not listed in Table 12.2-1, 15.4-1, or 15.4-2, R_μ shall be taken as 1.3 for components above grade plane, unless seismic design parameters for the SFRS have been approved by the Authority Having Jurisdiction.

If the building or nonbuilding structure supporting the component contains combinations of SFRSs in different directions, or vertical combinations of SFRSs, the structure ductility reduction factor for the entire structure shall be based on the SFRS that produces the lowest value of R_μ. Where a nonbuilding structure type listed in Table 15.4-1 has multiple entries based on permissible height increases, the value of R_μ is permitted to be determined using values of R and Ω_0 for the "with permitted height increase" entry.

13.3.1.3 Component Resonance Ductility Factor, C_{AR} Components shall be assigned a component resonance ductility factor, C_{AR}, based on whether the component is supported at or below grade plane, or is supported above grade plane by a building or nonbuilding structure. Components that are in or supported by a building or nonbuilding structure and are at or below grade plane are considered supported at or below grade. All other components in or supported by a building or nonbuilding structure are considered supported above grade.

Architectural components shall be assigned a component resonance ductility factor in Table 13.5-1.

Mechanical and electrical equipment shall be assigned a component resonance ductility factor in Table 13.6-1. The component resonance ductility factor for mechanical and electrical equipment mounted on the equipment support structures or platforms shall not be less than the component resonance ductility factor used for the equipment support structure or platform itself.

The component resonance ductility factor for equipment support structures or platforms shall be determined in accordance with Section 13.6.4.6. The weight of supported mechanical and electrical components shall be included when calculating the component operating weight, W_p, of equipment support structures or platforms.

Distribution systems shall be assigned component resonance ductility factors in Table 13.6-1, to be used for the design of the distribution system itself (e.g., the piping, ducts, and raceways). The component resonance ductility factor for distribution system supports shall be determined in accordance with Section 13.6.4.7

13.3.1.4 Component Strength, R_{po} The component strength factor, R_{po}, for nonstructural components is given in Table 13.5-1 or 13.6-1.

13.3.1.5 Nonlinear Response History Analysis In lieu of the forces determined in accordance with Equation (13.3-1), the nonlinear response history analysis procedures of Chapters 16, 17, and 18 may be used to determine the seismic design force for nonstructural components. Where the dynamic properties of the nonstructural component are not explicitly modeled in the nonlinear response history analysis, the seismic design force, Fp, shall be calculated as

$$F_p = I_p W_p a_i \left[\frac{C_{AR}}{R_{po}}\right] \tag{13.3-7}$$

where a_i is the maximum acceleration at level i obtained from the nonlinear response history analysis at the Design Earthquake ground motion. When a_i is determined using nonlinear response history analysis, a suite of not less than seven ground motions shall be used. If the supporting structure is designed using nonlinear response history analysis, the entire suite of ground motions used to design the structure shall be used to determine a_i. The value of the parameter a_i shall be taken as the mean of the maximum values of acceleration at the center of mass of the support level, obtained from each analysis. The upper and lower limits of F_p determined by Equations (13.3-2) and (13.3-3) shall apply.

13.3.1.6 Vertical Seismic Force The component, including its supports and attachments, shall be designed for a concurrent vertical seismic design force equal to E_v per Section 12.4.2.2.

EXCEPTION: The concurrent vertical seismic force need not be considered for lay-in access floor panels and lay-in ceiling panels.

13.3.1.7 Nonseismic Loads Where nonseismic loads on nonstructural components exceed F_p, such loads shall govern the strength design, but the detailing requirements and limitations prescribed in this chapter shall apply.

13.3.2 Seismic Relative Displacements The effects of seismic relative displacements shall be considered in combination with displacements caused by other loads as appropriate. Seismic relative displacements, D_{pI}, shall be calculated as

$$D_{pI} = D_p I_e \tag{13.3-8}$$

where I_e is the Importance Factor in Section 11.5.1, and D_p is the displacement determined in accordance with the equations set forth in Sections 13.3.2.1 and 13.3.2.2.

13.3.2.1 Displacements within Structures For two connection points on the same structure A or the same structural system, one at a height h_x and the other at a height h_y, D_p shall be determined as

$$D_p = \delta_{xA} - \delta_{yA} \tag{13.3-9}$$

where

D_p = Relative seismic displacement the component must be designed to accommodate;

δ_{xA} = Deflection at building level x of structure A, determined in accordance with Equation (12.8-16); and

δ_{yA} = Deflection at building level y of structure A, determined in accordance with Equation (12.8-16).

Alternatively, D_p is permitted to be determined using the linear dynamic procedures described in Section 12.9. For structures in which the story drift associated with the Design Earthquake Displacement does not exceed the allowable story drift as defined in Table 12.12-1, D_p is not required to be taken as greater than

$$D_p = \frac{(h_x - h_y)\Delta_{aA}}{h_{sx}} \tag{13.3-10}$$

where Δ_{aA} is the allowable story drift for structure A as designed in Table 12.12-1, and h_{sx} is the story height used in the definition of the allowable drift, Δ_a, in Table 12.12-1.

Where single-story structures are designed in accordance with note a of Table 12.12-1, Equation (13.3-10) shall not apply.

13.3.2.2 Displacements between Structures For two connection points on separate structures A and B or separate structural systems, one at a height h_x and the other at a height h_y, D_p shall be determined as

$$D_p = |\delta_{xA}| + |\delta_{yB}| \tag{13.3-11}$$

For structures in which the story drifts associated with the Design Earthquake Displacement does not exceed the allowable story drift as defined in Table 12.12-1, the drift is not required to be taken as greater than

$$D_p = \frac{h_x \Delta_{aA}}{h_{sx}} + \frac{h_y \Delta_{aB}}{h_{sx}} \tag{13.3-12}$$

Where single-story structures are designed in accordance with note a of Table 12.12-1, Equation (13.3-12) shall not apply where

D_p = Relative seismic displacement the component must be designed to accommodate;

δ_{xA} = Deflection at building level x of structure A at the Design Earthquake Displacement; determined in accordance with Equation (12.8-16);

δ_{yA} = Deflection at building level y of structure A at the Design Earthquake Displacement; determined in accordance with Equation (12.8-16);

δ_{yB} = Deflection at building level y of structure B at the Design Earthquake Displacement; determined in accordance with Equation (12.8-16);

h_x = Height of level x to which upper connection point is attached;

h_y = Height of level y to which lower connection point is attached;

Δ_{aA} = Allowable story drift for structure A as defined in Table 12.12-1;

Δ_{aB} = Allowable story drift for structure B as defined in Table 12.12-1; and

h_{sx} = Story height used in the definition of allowable drift, Δ_a, in Table 12.12-1. Note that Δ_a/h_{sx} is the drift index.

The effects of seismic relative displacements shall be considered in combination with displacements caused by other loads as appropriate.

13.3.3 Component Period The fundamental period, T_p, of the nonstructural component, including its supports and attachment to the structure, shall be determined by the following equation, provided the component, supports, and attachment can be reasonably represented analytically by a simple single-degree-of-freedom spring-and-mass system:

$$T_p = 2\pi\sqrt{\frac{W_p}{K_p g}} \quad (13.3\text{-}13)$$

where

T_p = Component fundamental period,
W_p = Component operating weight,
g = Gravitational acceleration, and
K_p = Combined stiffness of the component, supports, and attachments, determined in terms of load per unit deflection at the center of gravity of the component.

Alternatively, the fundamental period of the component, T_p, in seconds is permitted to be determined from experimental test data or by a properly substantiated analysis.

13.4 NONSTRUCTURAL COMPONENT ANCHORAGE AND ATTACHMENT

Nonstructural components and their supports shall be attached (or anchored) to the structure in accordance with the requirements of this section, and the attachment shall satisfy the requirements for the parent material as set forth elsewhere in this standard.

Except where permitted in Section 13.6.12, component attachments shall be bolted, welded, or otherwise positively fastened without consideration of frictional resistance produced by the effects of gravity. A continuous load path of sufficient strength and stiffness between the component and the supporting structure shall be provided. Local elements of the structure, including connections, shall be designed and constructed for the component forces where they control the design of the elements or their connections. The component forces shall be those determined in Section 13.3.1. The design documents shall include sufficient information relating to the attachments to verify compliance with the requirements of this section.

13.4.1 Design Force in the Attachment The force in the attachment shall be determined based on the prescribed forces and displacements for the component as determined in Sections 13.3.1 and 13.3.2.

13.4.2 Anchors in Concrete or Masonry When it is required to apply the seismic load effects including overstrength in Section 12.4.3, Ω_0 shall be taken as the anchorage overstrength factor, Ω_{0p}, given in Tables 13.5-1 and 13.6-1.

13.4.2.1 Anchors in Concrete Anchors in concrete shall be designed in accordance with Chapter 17 of ACI 318.

13.4.2.2 Anchors in Masonry Anchors in masonry shall be designed in accordance with TMS 402. Anchors shall be designed to be governed by the tensile or shear strength of a ductile steel element.

EXCEPTION: Anchors shall be permitted to be designed so that either

1. The support or component that the anchor is connecting to the structure undergoes ductile yielding at a load level corresponding to anchor forces not greater than the design strength of the anchors, or
2. The anchors shall be designed to resist the load combinations in accordance with Section 13.2.2, including Ω_{0p} as given in Tables 13.5-1 and 13.6-1.

13.4.2.3 Post-Installed Anchors in Concrete and Masonry Post-installed anchors in concrete shall be prequalified for seismic applications in accordance with ACI 355.2 or other approved qualification procedures. Post-installed anchors in masonry shall be prequalified for seismic applications in accordance with approved qualification procedures.

13.4.3 Installation Conditions Determination of forces in attachments shall take into account the expected conditions of installation, including eccentricities and prying effects.

13.4.4 Multiple Attachments Determination of force distribution of multiple attachments at one location shall take into account the stiffness and ductility of the component, component supports, attachments, and structure, and the ability to redistribute loads to other attachments in the group. Designs of anchorage in concrete in accordance with Chapter 17 of ACI 318 shall be considered to satisfy this requirement.

13.4.5 Power-Actuated Fasteners Power-actuated fasteners in concrete or steel shall not be used for sustained tension loads or for brace applications in Seismic Design Category D, E, or F unless approved for seismic loading. Power-actuated fasteners in masonry are not permitted unless approved for seismic loading.

EXCEPTIONS:

1. Power-actuated fasteners in concrete used for support of acoustical tile or lay-in panel suspended ceiling applications and distributed systems where the service load on any individual fastener does not exceed 90 lb (400 N), and
2. Power-actuated fasteners in steel where the service load on any individual fastener does not exceed 250 lb (1,112 N).

13.4.6 Friction Clips Friction clips in Seismic Design Category D, E, or F shall not be used for supporting permanent loads in addition to resisting seismic forces. C-type beam and large flange clamps are permitted for hangers, provided they are equipped with restraining straps equivalent to those specified in NFPA 13, Section 9.3.7. Lock nuts or equivalent shall be provided to prevent loosening of threaded connections.

13.5 ARCHITECTURAL COMPONENTS

13.5.1 General. Architectural components, and their supports and attachments, shall satisfy the requirements of this section. Appropriate coefficients shall be selected from Table 13.5-1.

EXCEPTION: Components supported by chains or otherwise suspended from the structure are not required to satisfy the seismic force and relative displacement requirements, provided they meet all of the following criteria:

1. The design load for such items shall be equal to 1.4 times the operating weight acting down with a simultaneous horizontal load equal to 1.4 times the operating weight. The horizontal load shall be applied in the direction that results in the most critical loading for design.

Table 13.5-1. Coefficients for Architectural Components.

Architectural Component	C_{AR} Supported at or below grade plane	C_{AR} Supported above grade plane by a structure	R_{po}	Ω_{op}[a]
Interior nonstructural walls and partitions[b]				
Light frame ≤ 9 ft (2.74 m) in height	1	1	1.5	2
Light frame > 9 ft (2.74 m) in height	1.4	1.4	1.5	2
Reinforced masonry	1	1	1.5	2
All other walls and partitions	2.2	2.8	1.5	1.5
Cantilever elements (unbraced or braced to structural frame below its center of mass)				
Parapets and cantilever interior nonstructural walls	1.8	2.2	1.5	1.75
Chimneys where laterally braced or supported by the structural frame	1.8	2.2	1.5	1.75
Cantilever elements (braced to structural frame above its center of mass)				
Parapets	1	1	1.5	2
Chimneys	1	1	1.5	2
Exterior nonstructural walls[b]	1	1	1.5	2
Exterior nonstructural wall elements and connections[b]				
Wall element	1	1	1.5	2
Body of wall panel connections	1	1	1.5	2
Fasteners of the connecting system	2.2	2.8	1.5	1
Veneer				
Limited-deformability elements and attachments	1	1	1.5	2
Low-deformability elements and attachments	1	1	1.5	2
Penthouses (except where framed by an extension of the building frame)				
Seismic force-resisting systems with $R \geq 6$	N/A	1.4	2	2
Seismic force-resisting systems with $4 \leq R < 6$	N/A	2.2	2	1.75
Seismic force-resisting systems with $R < 4$	N/A	2.8	2	1.5
Other systems	N/A	2.8	1.5	1.5
Ceilings				
All	1	1	1.5	2
Cabinets				
Permanent floor-supported storage cabinets more than 6 ft (1.8 m) tall, including contents	1	1	1.5	2
Permanent floor-supported library shelving, book stacks, and bookshelves more than 6 ft (1.8 m) tall, including contents	1	1	1.5	2
Laboratory equipment	1	1	1.5	2
Access floors				
Special access floors (designed in accordance with Section 13.5.7.2)	1	1	2	2
All other	2.2	2.8	1.5	1.5
Appendages and ornamentations	1.8	2.2	1.5	1.75
Signs and billboards	1.8	2.2	1.5	1.75
Other rigid components	1	1	1.5	2
Other flexible components				
High-deformability elements and attachments	1.4	1.4	1.5	2
Limited-deformability elements and attachments	1.8	2.2	1.5	1.75
Low-deformability materials and attachments	2.2	2.8	1.5	1.5
Egress stairways not part of the building seismic force-resisting system	1	1	1.5	2
Egress stairs and ramp fasteners and attachments	1.8	2.2	1.5	1.75

[a] Overstrength factor, where required for nonductile anchorage to concrete and masonry (see Section 13.4.2).
[b] Where flexible diaphragms provide lateral support for concrete or masonry walls and partitions, *the design forces for anchorage to the diaphragm shall be as specified in Section 12.11.2.*
Note: N/A = not applicable.

2. Seismic interaction effects shall be considered in accordance with Section 13.2.4.
3. The connection to the structure shall allow a 360-degree range of motion in the horizontal plane.

13.5.2 Forces and Displacements All architectural components, and their supports and attachments, shall be designed for the seismic forces defined in Section 13.3.1.

Architectural components that could pose a life-safety hazard shall be designed to accommodate the seismic relative displacement requirements of Section 13.3.2. Architectural components shall be designed considering vertical deflection caused by joint rotation of cantilever structural members.

13.5.3 Exterior Nonstructural Wall Elements and Connections Exterior nonstructural wall panels or elements that are attached to or enclose the structure shall be designed to accommodate the seismic relative displacements defined in Section 13.3.2 and movements caused by temperature changes. Such elements shall be supported by means of positive and direct structural supports or by mechanical connections and fasteners in accordance with the following requirements:

1. Connections and panel joints shall allow for the story drift caused by relative seismic displacements, D_{pl}, determined in Section 13.3.2, or 0.5 in. (13 mm), whichever is greater.
2. Connections accommodating story drift through sliding mechanisms or bending of threaded steel rods shall satisfy all of the following.
   (a) Threaded rods or bolts shall be fabricated of low-carbon or stainless steel. Where cold-worked carbon steel threaded rods are used, the rods as fabricated shall meet or exceed the reduction of area, elongation, and tensile strength requirements of ASTM F1554, Grade 36. Grade 55 rods shall also be permitted provided they meet the requirements of Supplement 1; and
   (b) Where threaded rods connecting the panel to the supports are used in connections using slotted or oversize holes, the rods shall have length-to-diameter ratios of 4 or less, where the length is the clear distance between the nuts or threaded plates. The slots or oversized holes shall be proportioned to accommodate the full in-plane design story drift in each direction, the nuts shall be installed finger-tight, and a positive means to prevent the nut from backing off shall be used; and
   (c) Connections that accommodate story drift by bending of threaded rods shall satisfy

$$(L/d)/D_{pl} \geq 6.0(1/\text{in.}) \quad (13.5\text{-}1)$$

$$(L/d)/D_{pl} \geq 0.24(1/\text{mm}) \quad (13.5\text{-}1.\text{SI})$$

where

L = Clear length of rod between nuts or threaded plates [in. (mm)],
d = Rod diameter [in. (mm)], and
D_{pl} = Relative seismic displacement the connection must be designed to accommodate [in. (mm)].

3. The connecting member itself shall have sufficient ductility and rotation capacity to preclude fracture of the concrete or brittle failures at or near welds.
4. All fasteners in the connecting system, such as bolts, inserts, welds, and dowels, and the body of the connectors, shall be designed for the force, F_p, determined by Section 13.3.1 using the applicable design coefficients taken from Table 13.5-1, applied at the center of mass of the panel. The connecting system shall include both the connections between the wall panels or elements and the structure, and the interconnections between wall panels or elements.
5. Where anchorage is achieved using flat straps embedded in concrete or masonry, such straps shall be attached to or hooked around reinforcing steel or otherwise terminated so as to effectively transfer forces to the reinforcing steel or to ensure that pullout of anchorage is not the initial failure mechanism.

13.5.4 Glass Glass in glazed curtain walls and storefronts shall be designed and installed to accommodate without breakage or dislodgement the relative-displacement requirement of Section 13.5.9.

Where glass is secured to the window system framing by means of structural sealant glazing, the requirements contained in the reference standards listed in Table 13.5-2 shall also apply.

13.5.5 Out-of-Plane Bending Transverse or out-of-plane bending or deformation of a component or system that is subjected to forces as determined in Section 13.5.2 shall not exceed the deflection capability of the component or system.

13.5.6 Suspended Ceilings Suspended ceilings shall be in accordance with this section.

EXCEPTIONS:

1. Suspended ceilings with areas less than or equal to 144 ft^2 (13.4 m^2) that are surrounded by walls or soffits that are laterally braced to the structure above are exempt from the requirements of this section.
2. Suspended ceilings constructed of screw- or nail-attached gypsum board on one level that are surrounded by and connected to walls or soffits that are laterally braced to the structure above are exempt from the requirements of this section.

13.5.6.1 Seismic Forces The weight of the ceiling, W_p, shall include the ceiling grid; ceiling tiles or panels; light fixtures if attached to, clipped to, or laterally supported by the ceiling grid; and other components that are laterally supported by the ceiling. W_p shall be taken as not less than 4 lb/ft^2 (192 N/m^2).

The seismic force, F_p, shall be transmitted through the ceiling attachments to the building structural elements or the ceiling–structure boundary.

13.5.6.2 Industry Standard Construction for Acoustical Tile or Lay-In Panel Ceilings Unless designed in accordance with Section 13.5.6.3, or seismically qualified in accordance with Section 13.2.6 or 13.2.7, acoustical tile or lay-in panel ceilings shall be designed and constructed in accordance with this section.

13.5.6.2.1 Seismic Design Category C. Acoustical tile or lay-in panel ceilings in structures assigned to Seismic Design Category

Table 13.5-2. Reference Standards for Structural Sealant Glazing.

ASTM C1087, *Test Method for Determining Compatibility of Liquid-Applied Sealants with Accessories Used in Structural Glazing Systems*
ASTM C1135, *Test Method for Determining Tensile Adhesion Properties of Structural Sealants*
ASTM C1184, *Specification for Structural Silicone Sealants*
ASTM C1265, *Test Method for Determining the Tensile Properties of an Insulating Glass Edge Seal for Structural Glazing Applications*
ASTM C1294, *Test Method for Compatibility of Insulating Glass Edge Sealants with Liquid-Applied Glazing Materials*
ASTM C1369, *Specification for Secondary Edge Sealants for Structurally Glazed Insulating Glass Units*

C shall be designed and installed in accordance with ASTM C635, ASTM C636, and ASTM E580, Section 4, Seismic Design Category C.

13.5.6.2.2 Seismic Design Categories D through F. Acoustical tile or lay-in panel ceilings in structures assigned to Seismic Design Categories D, E, and F shall be designed and installed in accordance with ASTM C635, ASTM C636, and ASTM E580, Section 5, Seismic Design Categories D, E, and F, as modified by this section.

Acoustical tile or lay-in panel ceilings shall also comply with the following:

(a) The width of the perimeter supporting closure angle or channel shall be not less than 2.0 in. (50 mm), unless qualified perimeter supporting clips are used. Closure angles or channels shall be screwed or otherwise positively attached to wall studs or other supporting structures. Perimeter supporting clips shall be qualified in accordance with approved test criteria per Section 13.2.6. Perimeter supporting clips shall be attached to the supporting closure angle or channel with a minimum of two screws per clip and shall be installed around the entire ceiling perimeter. In each orthogonal horizontal direction, one end of the ceiling grid shall be attached to the closure angle, channel, or perimeter supporting clip. The other end of the ceiling grid in each horizontal direction shall have a minimum 0.75 in. (19 mm) clearance from the wall and shall rest on and be free to slide on a closure angle, channel, or perimeter supporting clip.

(b) For ceiling areas exceeding 2,500 ft^2 (232 m^2), a seismic separation joint or full-height partition that breaks the ceiling up into areas not exceeding 2,500 ft^2 (232 m^2), each with a ratio of the long to short dimension less than or equal to 4, shall be provided, unless structural analyses are performed of the ceiling bracing system for the prescribed seismic forces which demonstrate that ceiling penetrations and closure angles or channels provide sufficient clearance to accommodate the anticipated lateral displacement. Each area shall be provided with closure angles or channels in accordance with Section 13.5.6.2.2.a and horizontal restraints or bracing.

13.5.6.3 Integral Construction As an alternative to providing large clearances around sprinkler system penetrations through ceilings, the sprinkler system and ceiling grid are permitted to be designed and tied together as an integral unit. Such a design shall consider the mass and flexibility of all elements involved, including the ceiling, sprinkler system, light fixtures, and mechanical (HVACR) appurtenances. Such design shall be performed by a registered design professional.

13.5.7 Access Floors

13.5.7.1 General The weight of the access floor, W_p, shall include the weight of the floor system, 100% of the weight of all equipment fastened to the floor, and 25% of the weight of all equipment supported by but not fastened to the floor. The seismic force, F_p, shall be transmitted from the top surface of the access floor to the supporting structure.

Overturning effects of equipment fastened to the access floor panels also shall be considered. The ability of "slip on" heads for pedestals shall be evaluated for suitability to transfer overturning effects of equipment.

Where checking individual pedestals for overturning effects, the maximum concurrent axial load shall not exceed the portion of W_p assigned to the pedestal under consideration.

13.5.7.2 Special Access Floors Access floors shall be considered to be "special access floors" if they are designed to comply with the following considerations:

1. Connections transmitting seismic loads consist of mechanical fasteners, anchors satisfying the requirements of Chapter 17 of ACI 318, welding, or bearing. Design load capacities shall comply with recognized design codes and/or certified test results.
2. Seismic loads are not transmitted by friction, power-actuated fasteners, adhesives, or by friction produced solely by the effects of gravity.
3. The design analysis of the bracing system includes the destabilizing effects of individual members buckling in compression.
4. Bracing and pedestals are of structural or mechanical shapes produced to ASTM specifications that specify minimum mechanical properties. Electrical tubing shall not be used.
5. Floor stringers are used that are designed to carry axial seismic loads and that are mechanically fastened to the supporting pedestals.

13.5.8 Partitions

13.5.8.1 General Partitions that are tied to the ceiling, and all partitions greater than 6 ft (1.8 m) high, shall be laterally braced to the building structure. Such bracing shall be independent of any ceiling lateral force bracing. Bracing shall be spaced to limit horizontal deflection at the partition head, to be compatible with ceiling deflection requirements as determined in Section 13.5.6 for suspended ceilings and elsewhere in this section for other systems.

EXCEPTION: Partitions that meet all of the following conditions:

1. The partition height does not exceed 9 ft (2.7 m).
2. The linear weight of the partition does not exceed 10 lb (0.479 kN) times the height in ft (m) of the partition.
3. The partition horizontal seismic load does not exceed 5 lb/ft^2 (0.24 kN/m^2).

13.5.8.2 Glass Glass in glazed partitions shall be designed and installed in accordance with Section 13.5.9.

13.5.9 Glass in Glazed Curtain Walls, Glazed Storefronts, and Glazed Partitions

13.5.9.1 General Glass in glazed curtain walls, glazed storefronts, and glazed partitions shall meet the relative displacement requirement of Equation (13.5-2):

$$\Delta_{fallout} \geq 1.25 D_{pI} \quad (13.5\text{-}2)$$

or 0.5 in. (13 mm), whichever is greater, where $\Delta_{fallout}$ is the relative seismic displacement (drift) at which glass fallout from the curtain wall, storefront wall, or partition occurs (Section 13.5.9.2); and D_{pI} is the relative seismic displacement the component must be designed to accommodate (Section 13.3.2). D_{pI} shall be applied over the height of the glass component under consideration.

EXCEPTIONS:

1. Glass need not comply with this requirement if it has sufficient clearances from its frame such that physical contact between the glass and frame does not occur at the design drift:

$$D_{\text{clear}} \geq 1.25 D_{pl} \tag{13.5-3}$$

where D_{clear} is the relative horizontal (drift) displacement, measured over the height of the glass panel under consideration, which causes initial glass-to-frame contact. For rectangular glass panels within a rectangular wall frame,

$$D_{\text{clear}} = 2c_1\left(1 + \frac{h_p c_2}{b_p c_1}\right)$$

where

h_p = Height of the rectangular glass panel,
b_p = Width of the rectangular glass panel,
c_1 = Average of the clearances (gaps) on both sides between the vertical glass edges and the frame, and
c_2 = Average of the clearances (gaps) at the top and bottom between the horizontal glass edges and the frame.

2. Fully tempered monolithic glass in Risk Categories I, II, and III located no more than 10 ft (3 m) above a walking surface need not comply with this requirement.
3. Annealed or heat-strengthened laminated glass in single thickness with interlayer no less than 0.030 in. (0.76 mm) that is captured mechanically in a wall system glazing pocket and whose perimeter is secured to the frame by a wet-glazed, gunable, curing elastomeric sealant perimeter bead of 0.5 in. (13 mm) minimum glass contact width, or other approved anchorage system, need not comply with this requirement.

13.5.9.2 Seismic Drift Limits for Glass Components Δ_{fallout}, the drift causing glass fallout from the curtain wall, storefront, or partition, shall be determined in accordance with AAMA 501.6 or by engineering analysis.

13.5.10 Egress Stairs and Ramps Egress stairs and ramps not part of the seismic force-resisting system (SFRS) of the structure to which they are attached shall be detailed to accommodate the seismic relative displacements, D_{pI}, defined in Section 13.3.2, including diaphragm deformation. The net relative displacement shall be assumed to occur in any horizontal direction. Such elements shall be supported by means of positive and direct structural supports or by mechanical connections and fasteners in accordance with the following requirements:

(a) Sliding connections with slotted or oversize holes, sliding bearing supports with keeper assemblies or end stops, and connections that permit movement by deformation of metal attachments, shall accommodate a displacement D_{pI}, but not less than 0.5 in. (13 mm), without loss of vertical support or inducement of displacement-related compression forces in the stair.
(b) Sliding bearing supports without keeper assemblies or end stops shall be designed to accommodate a displacement $1.5D_{pI}$, but not less than 1.0 in. (25 mm), without loss of vertical support. Breakaway restraints are permitted if their failure does not lead to loss of vertical support.
(c) Metal supports shall be designed with rotation capacity to accommodate seismic relative displacements as defined in item (b). The strength of such metal supports shall not be limited by bolt shear, weld fracture, or other brittle modes.
(d) All fasteners and attachments such as bolts, inserts, welds, dowels, and anchors shall be designed for the seismic design forces determined in accordance with Section 13.3.1 using the applicable design coefficients as given in Table 13.5-1.

EXCEPTION: If sliding or ductile connections are not provided to accommodate seismic relative displacements, the stiffness and strength of the stair or ramp structure shall be included in the building structural model of Section 12.7.3, and the stair shall be designed with Ω_0 corresponding to the SFRS but not less than 2-1/2.

13.5.11 Penthouses and Rooftop Structures Penthouses and rooftop structures shall be designed in accordance with this section. The horizontal seismic design force, F_p, shall be determined in accordance with Section 13.3.1, using the design coefficients listed in Table 13.5-1.

EXCEPTION: Penthouses and rooftop structures framed by an extension of the building frame and designed in accordance with the requirements of Chapter 12.

13.5.11.1 Seismic Force-Resisting Systems for Penthouses and Rooftop Structures The seismic force-resisting system for penthouses and rooftop structures shall conform to one of the types indicated in Table 12.2-1 or Table 15.4-1. The structural systems used shall be in accordance with the structural system limitations noted in the tables and shall be designed and detailed in accordance with the specific requirements for the system as set forth in the applicable reference documents listed in Table 12.2-1 or 15.4-1 and the additional requirements set forth in Chapter 14. Height limits for penthouses and rooftop structures shall be measured from the top of the roof structure to the average height of the penthouse roof.

EXCEPTION: Penthouses and rooftop structures designed using the coefficients for Other Systems in Table 13.5-1 and which also conform to the requirements of relevant material standards need not conform to one of the types indicated in Table 12.2-1 or 15.4-1. The height limit for penthouses and rooftop structures designed using the coefficients for Other Systems shall be 28 ft (8.5 m).

13.6 MECHANICAL AND ELECTRICAL COMPONENTS

13.6.1 General Mechanical and electrical components and their supports shall satisfy the requirements of this section. The attachment of mechanical and electrical components and their supports to the structure shall meet the requirements of Section 13.4. Appropriate coefficients shall be selected from Table 13.6-1.

EXCEPTION: Light fixtures, lighted signs, and ceiling fans not connected to ducts or piping, which are supported by chains or otherwise suspended from the structure, are not required to satisfy the seismic force and relative displacement requirements provided they meet all of the following criteria:

1. The design load for such items shall be equal to 1.4 times the operating weight acting down with a simultaneous horizontal load equal to 1.4 times the operating weight.

Table 13.6-1. Seismic Coefficients for Mechanical and Electrical Components.

MECHANICAL AND ELECTRICAL COMPONENTS	C_{AR} Supported at or below grade plane	C_{AR} Supported above grade plane by a structure	R_{po}	Ω_{op}[b]
Air-side HVACR, fans, air handlers, air conditioning units, cabinet heaters, air distribution boxes, and other mechanical components constructed of sheet metal framing	1.4	1.4	2	2
Wet-side HVACR, boilers, furnaces, atmospheric tanks and bins, chillers, water heaters, heat exchangers, evaporators, air separators, manufacturing or process equipment, and other mechanical components constructed of high-deformability materials	1	1	1.5	2
Air coolers (fin fans), air-cooled heat exchangers, condensing units, dry coolers, remote radiators, and other mechanical components elevated on integral structural steel or sheet metal supports	1.8	2.2	1.5	1.75
Engines, turbines, pumps, compressors, and pressure vessels not supported on skirts and not within the scope of Chapter 15	1	1	1.5	2
Skirt-supported pressure vessels not within the scope of Chapter 15	1.8	2.2	1.5	1.75
Elevator and escalator components	1	1	1.5	2
Generators, batteries, inverters, motors, transformers, and other electrical components constructed of high-deformability materials	1	1	1.5	2
Motor control centers, panel boards, switch gear, instrumentation cabinets, and other components constructed of sheet metal framing	1.4	1.4	2	2
Communication equipment, computers, instrumentation, and controls	1	1	1.5	2
Roof-mounted stacks, cooling and electrical towers laterally braced below their center of mass	1.8	2.2	1.5	1.75
Roof-mounted stacks, cooling and electrical towers laterally braced above their center of mass	1	1	1.5	2
Lighting fixtures	1	1	1.5	2
Other mechanical or electrical components	1	1	1.5	2
Manufacturing or process conveyors (nonpersonnel)	1.8	2.2	1.5	1.75
VIBRATION-ISOLATED COMPONENTS AND SYSTEMS[a]				
Components and systems isolated using neoprene elements and neoprene isolated floors with built-in or separate elastomeric snubbing devices or resilient perimeter stops	1.8	2.2	1.3	1.75
Spring-isolated components and systems and vibration-isolated floors closely restrained using built-in or separate elastomeric snubbing devices or resilient perimeter stops	1.8	2.2	1.3	1.75
Internally isolated components and systems	1.8	2.2	1.3	1.75
Suspended vibration-isolated equipment, including in-line duct devices and suspended internally isolated components	1.8	2.2	1.3	1.75
EQUIPMENT SUPPORT STRUCTURES AND PLATFORMS				
Support structures and platforms where $T_p/T_a < 0.2$, or $T_p \leq 0.06$ s, per Section 13.6.4.6	NA	1	1.5	2
Seismic force-resisting systems with $R > 3$	1.4	1.4	1.5	2
Seismic force-resisting systems with $R \leq 3$	1.8	2.2	1.5	1.75
Other systems	2.2	2.8	1.5	1.5
DISTRIBUTION SYSTEM SUPPORTS				
Tension-only and cable bracing	1	1	1.5	2
Cold-formed steel rigid bracing	1	1	1.5	2
Hot-rolled steel bracing	1	1	1.5	2
Other rigid bracing	1	1	1.5	2
Lateral resistance provided by rods in flexure	1.8	2.2	1.5	1.75
Vertical cantilever supports such as pipe tees and moment frames above and supported by a floor or roof	1.8	2.2	1.5	1.75
DISTRIBUTION SYSTEMS				
Piping in accordance with ASME B31 (2001, 2002, 2008, 2010), including in-line components with joints made by welding or brazing	1	1	3	2
Piping in accordance with ASME B31, including in-line components, constructed of high- or limited-deformability materials, with joints made by threading, bonding, compression couplings, or grooved couplings	1	1	2	2
Piping and tubing not in accordance with ASME B31, including in-line components, constructed of high-deformability materials, with joints made by welding or brazing	1	1	2	2
Piping and tubing not in accordance with ASME B31, including in-line components, constructed of high- or limited-deformability materials, with joints made by threading, bonding, compression couplings, or grooved couplings	1.8	2.2	2	1.75
Piping and tubing constructed of low-deformability materials, such as cast iron, glass, and nonductile plastics	1.8	2.2	1.5	1.75
Duct systems, including in-line components, constructed of high-deformability materials, with joints made by welding or brazing	1	1	2	2

continues

Table 13.6-1 *(Continued).*

MECHANICAL AND ELECTRICAL COMPONENTS	C_{AR} Supported at or below grade plane	C_{AR} Supported above grade plane by a structure	R_{po}	Ω_{op}[b]
Duct systems, including in-line components, constructed of high- or limited-deformability materials, with joints made by means other than welding or brazing	1	1	1.5	2
Duct systems, including in-line components, constructed of low-deformability materials, such as cast iron, glass, and nonductile plastics	1.8	2.2	1.5	1.75
Electrical conduit, cable trays, and raceways	1	1	1.5	2
Bus ducts	1	1	1.5	2
Plumbing	1	1	1.5	2
Pneumatic tube transport systems	1	1	1.5	2

[a] Components mounted on vibration isolators shall have a bumper restraint or snubber in each horizontal direction. The design force shall be taken as $2F_p$ if the nominal clearance (air gap) between the equipment support frame and restraint is greater than 0.25 in. (6 mm). If the nominal clearance specified on the construction documents is not greater than 0.25 in. (6 mm), the design force is permitted to be taken as F_p.

[b] Overstrength factor as required for anchorage to concrete and masonry (see Section 13.4.2).

The horizontal load shall be applied in the direction that results in the most critical loading for the design.

2. Seismic interaction effects shall be considered in accordance with Section 13.2.4.
3. The connection to the structure shall allow a 360-degree range of motion in the horizontal plane.

Where design of mechanical and electrical components for seismic effects is required, consideration shall be given to the dynamic effects of the components, their contents, and where appropriate, their supports and attachments. In such cases, the interaction between the components and the supporting structures, including other mechanical and electrical components, shall also be considered.

13.6.2 Mechanical Components HVACR ductwork shall meet the requirements of Section 13.6.6. Piping systems shall meet the requirements of Section 13.6.7. Boilers and vessels shall meet the requirements of Section 13.6.10. Elevators shall meet the requirements of Section 13.6.11. All other mechanical components shall meet the requirements of Section 13.6.13. Mechanical components with I_p greater than 1.0 shall be designed for the seismic forces and relative displacements defined in Sections 13.3.1 and 13.3.2 and shall satisfy the following additional requirements:

1. Provision shall be made to eliminate seismic impact for components vulnerable to impact, for components constructed of nonductile materials, and in cases where material ductility will be reduced because of service conditions (e.g., low-temperature applications).
2. The possibility of loads imposed on components by attached utility or service lines, caused by differential movement of support points on separate structures, shall be evaluated.
3. Where piping or HVACR ductwork components are attached to structures that could displace relative to one another, and for isolated structures where such components cross the isolation interface, the components shall be designed to accommodate the seismic relative displacements defined in Section 13.3.2.

13.6.2.1 HVACR Equipment HVACR equipment that has been qualified in accordance with the requirements of Chapters 1 through 10 of ANSI/AHRI Standard 1270 (I-P) or ANSI/AHRI Standard 1271 (SI) shall be deemed to meet the seismic qualification requirements of Section 13.2.3, provided all of the following requirements are met:

(a) Active and/or energized components shall be seismically certified exclusively through shake table testing or experience data.

(b) The horizontal seismic design force, F_p, considered in the certification of non-active components through analysis shall be based on the procedures in Section 13.3.1 using $[\frac{C_{AR}}{R_{po}}] = 2.5$. The value of R_μ shall be taken as 1.3 for components located above the grade plane.

(c) Capacity of non-active components used in seismic certification by analysis shall be based on the provisions of ASCE 7.

(d) Rugged components shall conform to the definition in Chapter 11.

EXCEPTION: If the period of the component, T_p, is less than or equal to 0.06 seconds, $[\frac{C_{AR}}{R_{po}}]$ may be taken as 1.0.

13.6.3 Electrical Components Conduit, cable trays, and raceways shall meet the requirements of Section 13.6.5. Utility and service lines shall meet the requirements of Section 13.6.9. Other electrical components shall meet the requirements of Section 13.6.13. All electrical components with I_p greater than 1.0 shall be designed for the seismic forces and relative displacements defined in Sections 13.3.1 and 13.3.2 and shall satisfy the following additional requirements:

1. Provision shall be made to eliminate seismic impact between components.
2. Loads imposed on the components by attached utility or service lines that are attached to separate structures shall be evaluated.
3. Batteries on racks shall have wraparound restraints to ensure that the batteries do not fall from the racks. Spacers shall be used between restraints and cells to prevent damage to cases. Racks shall be evaluated for sufficient lateral load capacity.
4. Internal coils of dry-type transformers shall be positively attached to their supporting substructure within the transformer enclosure.

5. Electrical control panels, computer equipment, and other items with slide-out components shall have a latching mechanism to hold the components in place.
6. Electrical cabinet design shall comply with the applicable National Electrical Manufacturers Association (NEMA) standards. Cutouts in the lower shear panel that have not been made by the manufacturer and that significantly reduce the strength of the cabinet shall be specifically evaluated.
7. The attachments for additional external items weighing more than 100 lb (445 N) shall be specifically evaluated if not provided by the manufacturer.
8. Where conduit, cable trays, or similar electrical distribution components are attached to structures that could displace relative to one another, and for isolated structures where such components cross the isolation interface, the components shall be designed to accommodate the seismic relative displacements defined in Section 13.3.2.

13.6.4 Component Supports Mechanical and electrical component supports (including those with $I_p = 1.0$) and the means by which they are attached to the component shall be designed for the forces and displacements determined in Sections 13.3.1 and 13.3.2. Such supports include structural members, braces, frames, skirts, legs, saddles, pedestals, cables, guys, stays, snubbers, tethers, and elements forged or cast as a part of the mechanical or electrical component.

13.6.4.1 Design Basis If standard supports designed in accordance with nationally recognized standards such as ASME B31, NFPA 13, or MSS SP-58 are used, they shall be designed by either load rating (i.e., testing) or for the calculated seismic forces. If proprietary supports are used, load rating shall be determined by approved test standards or approved material-specific design procedures. The stiffness of the support, where appropriate, shall be designed such that the seismic load path for the component performs its intended function.

13.6.4.2 Design for Relative Displacement Component supports shall be designed to accommodate the seismic relative displacements between points of support determined in accordance with Section 13.3.2.

13.6.4.3 Support Attachment to Component The means by which supports are attached to the component, except where integral (i.e., cast or forged), shall be designed to accommodate both the forces and displacements determined in accordance with Sections 13.3.1 and 13.3.2. If $I_p = 1.5$ for the component, the local region of the support attachment point to the component shall be evaluated for the effect of the load transfer on the component wall.

13.6.4.4 Material Detailing Requirements The materials comprising supports and the means of attachment to the component shall be constructed of materials suitable for the application, including the effects of service conditions, for example low-temperature applications. Materials shall be in conformance with a nationally recognized standard.

13.6.4.5 Additional Requirements The following additional requirements shall apply to mechanical and electrical component supports:

1. Seismic supports shall be constructed so that support engagement is maintained.
2. Reinforcement (e.g., stiffeners or Belleville washers) shall be provided at bolted connections through sheet metal equipment housings as required to transfer the equipment seismic loads specified in this section from the equipment to the structure. Where equipment has been certified per Section 13.2.3, 13.2.6, or 13.2.7, anchor bolts or other fasteners and associated hardware as included in the certification shall be installed in conformance with the manufacturer's instructions. For those cases where no certification exists or where instructions for such reinforcement are not provided, reinforcement methods shall be as specified by a registered design professional or as approved by the Authority Having Jurisdiction.
3. Where weak-axis bending of cold-formed steel supports is relied on for the seismic load path, such supports shall be specifically evaluated.
4. Components mounted on vibration isolators shall have a bumper restraint or snubber in each horizontal direction, and vertical restraints shall be provided where required to resist overturning. Isolator housings and restraints shall be constructed of ductile materials. (See additional design force requirements in note *a* to Table 13.6-1.) A viscoelastic pad or similar material of appropriate thickness shall be used between the bumper and components to limit the impact load.

13.6.4.6 Equipment Support Structures and Platforms Equipment support structures and platforms shall be designed in accordance with this section. The horizontal seismic design force, F_p, shall be determined in accordance with Section 13.3.1, using the design coefficients listed in Table 13.6-1. The SFRS for equipment support structures and platforms shall conform to one of the types indicated in Table 12.2-1 or 15.4-1. The SFRS used shall be in accordance with the structural system limitations noted in the tables. The selected SFRS shall be designed and detailed in accordance with the specific requirements for the system as set forth in the applicable reference documents listed in Table 12.2-1 or 15.4-1 and the additional requirements set forth in Chapter 14.

EXCEPTION: Equipment support structures and platforms designed using the coefficients for Other Systems in Table 13.6-1, under Equipment Supports, and which also conform to the requirements of relevant material standards, need not conform to one of the types indicated in Table 12.2-1 or Table 15.4-1.

Equipment support structures or platforms that are supported by a building or nonbuilding structure are permitted to be designed using $C_{AR} = 1$, $R_{po} = 1.5$, and $\Omega_{0p} = 2.0$ if the ratio $T_p/T_a < 0.2$, or if $T_p \leq 0.06$ s. The value of T_p for the equipment support structure or platform shall include consideration of the mass and stiffness of the components being supported.

13.6.4.7 Distribution System Supports Distribution system supports are assigned a component resonance ductility factor from Table 13.6-1, based on the type of support system.

Vertical and lateral supports for distribution systems, including trapeze assemblies, shall be designed for seismic forces and seismic relative displacements as required in Section 13.3, except as noted in Sections 13.6.5, 13.6.6, and 13.6.7. Distribution systems shall be braced to resist vertical, transverse, and longitudinal seismic loads. Seismic loads for distribution system supports and trapeze assemblies shall be based on the weight of the distribution system tributary to the supports, including fittings and in-line components.

13.6.5 Distribution Systems: Conduit, Cable Tray, and Raceways Cable trays and raceways shall be designed for seismic forces and seismic relative displacements as required in Section 13.3. Conduit greater than 2.5 in. (64 mm) trade size and attached to panels, cabinets, or other equipment subject to seismic relative displacement, D_{pI}, shall be provided with flexible connections or designed for seismic forces and seismic relative displacements as required in Section 13.3.

EXCEPTIONS:

1. Design for the seismic forces and relative displacements of Section 13.3 shall not be required for raceways with $I_p = 1.0$ where flexible connections or other assemblies are provided between the cable tray or raceway and associated components to accommodate the relative displacement, where the cable tray or raceway is positively attached to the structure, and where one of the following apply:
   (a) Trapeze assemblies are used with 0.375 in. (10 mm) diameter rod hangers not exceeding 12 in. (305 mm) in length from the conduit, cable tray, or raceway support point to the connection at the supporting structure to support raceways, and the total weight supported by any single trapeze is 100 lb (445 N) or less; or
   (b) Trapeze assemblies with 0.5 in. (13 mm) diameter rod hangers not exceeding 12 in. (305 mm) in length from the conduit, cable tray, or raceway support point to the connection at the supporting structure are used to support the cable tray or raceway, and the total weight supported by any single trapeze is 200 lb (890 N) or less; or
   (c) Trapeze assemblies with 0.5 in. (13 mm) diameter rod hangers not exceeding 24 in. (610 mm) in length from the conduit, cable tray, or raceway support point to the connection at the supporting structure are used to support the cable tray or raceway, and the total weight supported by any single trapeze is 100 lb (445 N) or less; or
   (d) The conduit, cable tray, or raceway is supported by individual rod hangers 0.375 in. (10 mm) or 0.5 in. (13 mm) in diameter, and each hanger in the raceway run is 12 in. (305 mm) or less in length from the conduit, cable tray, or raceway support point connection to the supporting structure, and the total weight supported by any single rod is 50 lb (220 N) or less.
2. Design for the seismic forces and relative displacements of Section 13.3 shall not be required for conduit, regardless of the value of I_p, where the conduit is less than 2.5 in. (64 mm) trade size.

Design for the displacements across seismic joints shall be required for conduit, cable trays, and raceways with $I_p = 1.5$ without consideration of conduit size.

13.6.6 Distribution Systems: Duct Systems HVACR and other duct systems shall be designed for seismic forces and seismic relative displacements as required in Section 13.3.

EXCEPTIONS: The following exceptions pertain to ducts not designed to carry toxic, highly toxic, or flammable gases or not used for smoke control.

1. Design for the seismic forces and relative displacements of Section 13.3 shall not be required for duct systems with $I_p = 1.0$ where flexible connections or other assemblies are provided to accommodate the relative displacement between the duct system and associated components, the duct system is positively attached to the structure, and where one of the following apply:
   (a) Trapeze assemblies with 0.375 in. (10 mm) diameter rod hangers not exceeding 12 in. (305 mm) in length from the duct support point to the connection at the supporting structure are used to support the duct, and the total weight supported by any single trapeze is 100 lb (445 N) or less; or
   (b) Trapeze assemblies with 0.5 in. (13 mm) diameter rod hangers not exceeding 12 in. (305 mm) in length from the duct support point to the connection at the supporting structure are used to support the duct, and the total weight supported by any single trapeze is 200 lb (890 N) or less; or
   (c) Trapeze assemblies with 0.5 in. (13 mm) diameter rod hangers not exceeding 24 in. (610 mm) in length from the duct support point to the connection at the supporting structure are used to support the duct, and the total weight supported by any single trapeze is 100 lb (445 N) or less; or
   (d) The duct is supported by individual rod hangers 0.375 in. (10 mm) or 0.5 in. (13 mm) in diameter, and each hanger in the duct run is 12 in. (305 mm) or less in length from the duct support point to the connection at the supporting structure, and the total weight supported by any single rod is 50 lb (220 N) or less.
2. Design for the seismic forces and relative displacements of Section 13.3 shall not be required where provisions are made to avoid impact with other ducts or mechanical components or to protect the ducts in the event of such impact, the distribution system is positively attached to the structure, and HVACR ducts have a cross-sectional area of less than 6 ft^2 (0.557 m^2) and weigh 20 lb/ft (292 N/m) or less.

Components that are installed in line with the duct system and have an operating weight greater than 75 lb (334 N), such as fans, terminal units, heat exchangers, and humidifiers, shall be supported and laterally braced independent of the duct system, and such braces shall meet the force requirements of Section 13.3.1. Components that are installed in line with the duct system, have an operating weight of 75 lb (334 N) or less (such as small terminal units, valves, and dampers), and are otherwise not independently braced shall be positively attached with mechanical fasteners to rigid ducts on both sides. Appurtenances such as louvers and diffusers shall be positively attached to ducts with mechanical fasteners. Where such components are installed in a braced duct system, their weight shall be included in the design of the lateral bracing. Piping and conduit attached to in-line equipment shall be provided with adequate flexibility to accommodate the seismic relative displacements of Section 13.3.2.

13.6.7 Distribution Systems: Piping and Tubing Systems Unless otherwise noted in this section, piping and tubing systems shall be designed for the seismic forces and seismic relative displacements of Section 13.3. ASME pressure piping systems shall satisfy the requirements of Section 13.6.7.1. Fire protection sprinkler piping shall satisfy the requirements of Section 13.6.7.2. Elevator system piping shall satisfy the requirements of Section 13.6.11.

Where other applicable material standards or recognized design bases are not used, piping design including consideration of service loads shall be based on the following allowable stresses:

(a) For piping constructed with ductile materials (e.g., steel, aluminum, or copper), 90% of the minimum specified yield strength;

(b) For threaded connections in piping constructed with ductile materials, 70% of the minimum specified yield strength;
(c) For piping constructed with nonductile materials (e.g., cast iron or ceramics), 10% of the material minimum specified tensile strength; and
(d) For threaded connections in piping constructed with nonductile materials, 8% of the material minimum specified tensile strength.

Piping not detailed to accommodate the seismic relative displacements at connections to other components shall be provided with connections that have sufficient flexibility to avoid failure of the connection between the components.

Suspended components that are installed in-line and rigidly connected to and supported by the piping system, such as valves, strainers, traps, pumps, air separators, and tanks, are permitted to be considered part of the piping system for the purposes of determining the need for and sizing of lateral bracing. Where components are braced independently because of their weight but the associated piping is not braced, flexibility shall be provided as required to accommodate relative movement between the components.

13.6.7.1 ASME Pressure Piping Systems Pressure piping systems, including their supports, designed and constructed in accordance with ASME B31 shall be deemed to meet the force, displacement, and other requirements of this section. In lieu of specific force and displacement requirements in ASME B31, the force and displacement requirements of Section 13.3 shall be used. Materials meeting the toughness requirements of ASME B31 shall be considered high-deformability materials.

13.6.7.2 Fire Protection Sprinkler Piping Systems Fire protection sprinkler piping, pipe hangers, and bracing designed and constructed in accordance with NFPA 13 shall be deemed to meet the force and displacement requirements of this section. Clearances for sprinkler drops and sprigs and other equipment shall conform to 13.2.4.1. The exceptions of Section 13.6.7.3 shall not apply.

13.6.7.3 Exceptions Design for the seismic forces of Section 13.3 shall not be required for piping systems where flexible connections, expansion loops, or other assemblies are provided to accommodate the relative displacement between component and piping, where the piping system is positively attached to the structure, and where one of the following apply:

1. Trapeze assemblies are supported by 0.375 in. (10 mm) diameter rod hangers not exceeding 12 in. (305 mm) in length from the pipe support point to the connection at the supporting structure, do not support piping with I_p greater than 1.0, no single pipe exceeds the limits set forth in items 4a, 4b, or 4c below, and the total weight supported by any single trapeze is 100 lb (445 N) or less;
2. Trapeze assemblies are supported by 0.5 in. (13 mm) diameter rod hangers not exceeding 12 in. (305 mm) in length from the pipe support point to the connection at the supporting structure, do not support piping with I_p greater than 1.0, no single pipe exceeds the diameter limits set forth in items 4a, 4b, or 4c below, and the total weight supported by any single trapeze is 200 lb (890 N) or less;
3. Trapeze assemblies are supported by 0.5 in. (13 mm) diameter rod hangers not exceeding 24 in. (610 mm) in length from the pipe support point to the connection at the supporting structure, do not support piping with I_p greater than 1.0, no single pipe exceeds the diameter limits set forth in items 4a, 4b, or 4c below, and the total weight supported by any single trapeze is 100 lb (445 N) or less;
4. Piping is supported by rod hangers and provisions are made to avoid impact with other structural or nonstructural components or to protect the piping in the event of such impact; or pipes are supported by individual rod hangers 0.375 in. (10 mm) or 0.5 in. (13 mm) in diameter, where each hanger in the pipe run is 12 in. (305 mm) or less in length from the pipe support point to the connection at the supporting structure, and the total weight supported by any single hanger is 50 lb (220 N) or less. In addition, the following limitations on the size of piping shall be observed:
   (a) In structures assigned to Seismic Design Category C where I_p is greater than 1.0, the nominal pipe size shall be 2 in. (50 mm) or less.
   (b) In structures assigned to Seismic Design Category D, E, or F where I_p is greater than 1.0, the nominal pipe size shall be 1 in. (25 mm) or less.
   (c) In structures assigned to Seismic Design Category D, E, or F where $I_p = 1.0$, the nominal pipe size shall be 3 in. (80 mm) or less.
5. Pneumatic tube systems supported with trapeze assemblies using 0.375 in. (10 mm) diameter rod hangers not exceeding 12 in. (305 mm) in length from the tube support point to the connection at the supporting structure and the total weight supported by any single trapeze is 100 lb (445 N) or less.
6. Pneumatic tube systems supported by individual rod hangers 0.375 in. (10 mm) or 1.5 in. (13 mm) in diameter, and each hanger in the run is 12 in. (305 mm) or less in length from the tube support point to the connection at the supporting structure, and the total weight supported by any single rod is 50 lb (220 N) or less.

13.6.8 Distribution Systems: Trapezes with a Combination of Systems Trapezes that support a combination of distribution systems (electrical conduit, raceway, duct, piping, etc.) shall be designed using the most restrictive requirements for the supported distribution systems from Sections 13.6.5 through 13.6.7 for the aggregate weight of the supported system. If any distribution system on the trapeze is not exempted, the trapeze shall be braced.

13.6.9 Utility and Service Lines At the interface of adjacent structures or portions of the same structure that may move independently, utility lines shall be provided with adequate flexibility to accommodate the anticipated differential movement between the portions that move independently. Differential displacement shall be calculated in accordance with Section 13.3.2.

The possible interruption of utility service shall be considered in relation to designated seismic systems in Risk Category IV as defined in Table 1.5-1. Specific attention shall be given to the vulnerability of underground utilities and utility interfaces between the structure and the ground where Site Class E or F soil is present, and where the seismic coefficient S_{DS} at the underground utility or at the base of the structure is equal to or greater than 0.33.

13.6.10 Boilers and Pressure Vessels Boilers or pressure vessels designed and constructed in accordance with ASME BPVC shall be deemed to meet the force, displacement, and other requirements of this section. In lieu of the specific force and displacement requirements in ASME BPVC, the force and displacement requirements of Sections 13.3.1 and 13.3.2 shall be used. Materials that meet the toughness requirements of ASME BPVC shall be considered high-deformability materials. Other boilers and pressure vessels designated as having $I_p = 1.5$, but not designed and constructed in accordance with the requirements of ASME BPVC, shall comply with the requirements of Section 13.6.13.

13.6.11 Elevator and Escalator Design Requirements Elevators and escalators designed in accordance with the seismic requirements of ASME A17.1 shall be deemed to meet the seismic force requirements of this section, except as modified in the following text. The exceptions of Section 13.6.7.3 shall not apply to elevator piping.

13.6.11.1 Escalators, Elevators, and Hoistway Structural Systems Escalators, elevators, and hoistway structural systems shall be designed to meet the force and displacement requirements of Sections 13.3.1 and 13.3.2.

13.6.11.2 Elevator Equipment and Controller Supports and Attachments Elevator equipment and controller supports and attachments shall be designed to meet the force and displacement requirements of Sections 13.3.1 and 13.3.2.

13.6.11.3 Seismic Controls for Elevators Elevators operating with a speed of 150 ft/min (46 m/min) or greater shall be provided with seismic switches. Seismic switches shall provide an electric signal indicating that structural motions are of such a magnitude that the operation of the elevators may be impaired. Seismic switches in accordance with Section 8.4.10.1.2 of ASME A17.1 shall be deemed to meet the requirements of this section.

EXCEPTION: In cases where seismic switches cannot be located near a column in accordance with ASME A17.1, they shall have two horizontal axes of sensitivity and have a trigger level set to 20% of the acceleration of gravity where located at or near the base of the structure and 50% of the acceleration of gravity in all other locations.

On activation of the seismic switch, elevator operations shall conform to requirements of ASME A17.1, *except* as noted in the following text.

In facilities where the loss of the use of an elevator is a life-safety issue, the elevator shall only be used after the seismic switch has triggered, provided that

1. The elevator shall operate no faster than the service speed, and
2. Before the elevator is occupied, it is operated from top to bottom and back to top to verify that it is operable.

13.6.11.4 Retainer Plates Retainer plates are required at the top and bottom of the car and counterweight.

13.6.12 Rooftop Solar Panels Rooftop solar panels and their attachments shall be designed for the forces and displacements determined in Section 13.3.

EXCEPTION: Ballasted solar panels without positive direct attachment to the roof structure are permitted on Risk Categories I, II, and III structures six stories or fewer in height and having a maximum roof slope equal to or less than 1 in 20, provided that they comply with the following.

1. The height of the center of mass of any panel above the roof surface is less than half the least spacing in plan of the panel supports, but in no case greater than 3 ft (0.9 m).
2. Each panel or array of panels is designed to accommodate without impact, instability, or loss of support a seismic displacement, δ_{mpv}, of any panel relative to any roof edge or offset and any other curb or obstruction to sliding on the roof surface where δ_{mpv} is determined in accordance with Equation (13.6-1), but is not taken as less than 2 ft (0.6 m):

$$\delta_{mpv} = 5I_e(S_{DS} - 0.4)^2[\text{ft}] \quad (13.6\text{-}1)$$

$$\delta_{mpv} = 1.5I_e(S_{DS} - 0.4)^2[\text{m}] \quad (13.6\text{-}1.\text{SI})$$

   Any portion of an unattached array of panels that is not interconnected as specified in Item 3, shall be provided with a minimum separation between adjacent unattached panels of not less than $0.5\delta_{mpv}$. Signage or roof markings (e.g., yellow stripes) shall be provided to delineate the area around the panel that must be kept free of obstructions. Alternatively, δ_{mpv} may be determined by shake table testing or nonlinear response history analysis, whereby the value of δ_{mpv} shall not be taken as less than 80% of the value given by Equation (13.6-1) unless independent peer review is conducted in accordance with Section 1.3.1.3.4.
3. Each array of panels is interconnected as an integral unit to form a continuous load path such that the members and connections have the design strength to resist a horizontal force of $0.2S_{DS}W_{pi}$ in tension and compression considering any eccentricities, across any section cut by a vertical plane, where W_{pi} is the weight of the smaller of the two portions. The solar panels shall not be considered as part of the load path that resists the interconnection force unless the panels have been evaluated or tested for such loading.
4. Panel framing and supports are designed for a seismic force path from the center of mass of each component to locations of friction resistance equal to the lesser of F_p from Section 13.3.1 and $0.6W_p$, where W_p is the weight of each component.
5. All electrical cables leading from a panel to another panel or to another roof object are designed to accommodate, without rupture or distress, differential movements between cable connection points of $1.0\delta_{mpv}$, with consideration given to torsional movement of the panel and its possible impingement on the electrical cables.
6. All edges and offsets of roof surfaces on which panels are placed are bounded by a curb or parapet not less than 12 in. (305 mm) in height and designed to resist a concentrated load applied at the probable points of impact between the curb or parapet and the panel of not less than $0.2S_{DS}$ times the weight of the panel. Alternatively, a panel may be placed so that all parts of the panel are a minimum of $2.0\delta_{mpv}$, but not less than 4 ft (1.22 m), from any roof edge or offset.
7. Where justified by testing and analysis, the maximum roof slope for structures assigned to Seismic Design Categories C and D shall be permitted to be 1 in 12 provided that independent peer review is conducted in accordance with Section 1.3.1.3.4.

13.6.13 Other Mechanical and Electrical Components Mechanical and electrical components, including conveyor systems, not designed and constructed in accordance with the reference documents in Chapter 23 shall meet the following:

1. Components and their supports and attachments shall comply with the requirements of Sections 13.4, 13.6.2, 13.6.3, and 13.6.4.
2. For mechanical components with hazardous substances and assigned a component Importance Factor, I_p, of 1.5 in accordance with Section 13.1.3, and for boilers and pressure vessels not designed in accordance with ASME BPVC, the design strength for seismic loads in combination with other service loads and appropriate environmental effects shall be based on the following material properties:
   (a) For mechanical components constructed with ductile materials (e.g., steel, aluminum, or copper), 90% of the minimum specified yield strength;
   (b) For threaded connections in components constructed with ductile materials, 70% of the minimum specified yield strength;
   (c) For mechanical components constructed with nonductile materials (e.g., plastic, cast iron, or ceramics), 10% of the material minimum specified tensile strength; and
   (d) For threaded connections in components constructed with nonductile materials, 8% of the material minimum specified tensile strength.

13.7 CONSENSUS STANDARDS AND OTHER REFERENCED DOCUMENTS

See Chapter 23 for the list of consensus standards and other documents that shall be considered part of this standard to the extent referenced in this chapter.

CHAPTER 14
MATERIAL-SPECIFIC SEISMIC DESIGN AND DETAILING REQUIREMENTS

14.0 SCOPE

Structural elements, including foundation elements, shall conform to the material design and detailing requirements set forth in this chapter or as otherwise specified for nonbuilding structures in Tables 15.4-1 and 15.4-2.

14.1 STEEL

Structures, including foundations, constructed of steel to resist seismic loads shall be designed and detailed in accordance with this standard including the reference documents and additional requirements provided in this section.

14.1.1 Reference Documents The design, construction, and quality of steel members that resist seismic forces shall conform to the applicable requirements, as amended herein, of ANSI/AISC 341, ANSI/AISC 360, ANSI/AISI S100, ANSI/AISI S230, ANSI/AISI S240, ANSI/AISI S310, ANSI/AISI S400, ANSI/SDI-QA/QC, ANSI/SDI-SD, ANSI/SJI-100, ANSI/SJI-200, ASCE 8, and ASCE 19.

14.1.2 Structural Steel

14.1.2.1 General The design of structural steel for buildings and structures shall be in accordance with ANSI/AISC 360. Where required, the seismic design of structural steel structures shall be in accordance with the additional provisions of Section 14.1.2.2.

14.1.2.2 Seismic Requirements for Structural Steel Structures The design of structural steel structures to resist seismic forces shall be in accordance with the provisions of Section 14.1.2.2.1 or 14.1.2.2.2, as applicable.

14.1.2.2.1 Seismic Design Categories B and C. Structural steel structures assigned to Seismic Design Category B or C shall be of any construction permitted by the applicable reference documents in Section 14.1.1. Where a response modification coefficient, *R*, is used in accordance with Table 12.2-1 for the design of structural steel structures assigned to Seismic Design Category B or C, the structures shall be designed and detailed in accordance with the requirements of AISC 341.

EXCEPTION: The response modification coefficient, *R*, designated for "steel systems not specifically detailed for seismic resistance, excluding cantilever column systems" in Table 12.2-1 shall be permitted for systems designed and detailed in accordance with AISC 360 and need not be designed and detailed in accordance with AISC 341.

14.1.2.2.2 Seismic Design Categories D through F. Structural steel structures assigned to Seismic Design Category D, E, or F shall be designed and detailed in accordance with AISC 341, except as permitted in Table 15.4-1.

14.1.3 Cold-Formed Steel

14.1.3.1 General The design of cold-formed carbon or low-alloy steel structural members shall be in accordance with the requirements of AISI S100, and the design of cold-formed stainless steel structural members shall be in accordance with the requirements of ASCE 8. Where required, the seismic design of cold-formed steel structures shall be in accordance with the additional provisions of Section 14.1.3.2.

14.1.3.2 Seismic Requirements for Cold-Formed Steel Structures Where a response modification coefficient, *R*, in accordance with Table 12.2-1, is used for the design of cold-formed steel structures, the structures shall be designed and detailed in accordance with the requirements of AISI S100, ASCE 8, and AISI S400, as applicable.

14.1.4 Cold-Formed Steel Light-Frame Construction

14.1.4.1 General Cold-formed steel light-frame construction shall be designed in accordance with AISI S240. Where required, the seismic design of cold-formed steel light-frame construction shall be in accordance with the additional provisions of Section 14.1.4.2.

14.1.4.2 Seismic Requirements for Cold-Formed Steel Light-Frame Construction The design of cold-formed steel light-frame construction to resist seismic forces shall be in accordance with the provisions of Section 14.1.4.2.1 or 14.1.4.2.2, as applicable.

14.1.4.2.1 Seismic Design Categories B and C. Where a response modification coefficient, *R*, in accordance with Table 12.2-1 is used for the design of cold-formed steel light-frame construction assigned to Seismic Design Category B or C, the structures shall be designed and detailed in accordance with the requirements of AISI S400.

EXCEPTION: The response modification coefficient, *R*, designated for "steel systems not specifically detailed for seismic resistance, excluding cantilever column systems" in Table 12.2-1 shall be permitted for systems designed and detailed in accordance with AISI S240 and need not be designed and detailed in accordance with AISI S400.

14.1.4.2.2 Seismic Design Categories D through F. Cold-formed steel light-frame construction structures assigned to Seismic Design Category D, E, or F shall be designed and detailed in accordance with AISI S400.

14.1.4.3 Prescriptive Cold-Formed Steel Light-Frame Construction Cold-formed steel light-frame construction for one- and two-family dwellings is permitted to be designed and constructed in accordance with the requirements of AISI S230 subject to the limitations therein.

14.1.5 Cold-Formed Steel Deck Diaphragms Cold-formed steel deck diaphragms shall be designed in accordance with the requirements of AISI S100, AISI S240, SDI-SD, or ASCE 8, as applicable. Nominal strengths shall be determined in accordance with AISI S310. The required strength of diaphragms, including bracing members that form part of the diaphragm, shall be determined in accordance with Section 12.10. Where required by this standard, special seismic detailing requirements shall be in accordance with AISI S400. Special inspections and qualification of welding special inspectors for cold-formed steel floor and roof deck shall be in accordance with the quality assurance inspection requirements of SDI-QA/QC.

14.1.6 Concrete-Filled Steel Deck Diaphragms Concrete-filled steel deck diaphragms shall be of any construction permitted by the applicable reference documents in Section 14.1.1. Where a Diaphragm Design Force Reduction Factor, R_s, in accordance with Table 12.10-1, is used for design, the concrete-filled steel deck diaphragm and shear transfer to structural steel members shall be designed and detailed in accordance with the requirements of AISC 341.

14.1.7 Open Web Steel Joists and Joist Girders The design, manufacture, and use of open web steel joists and joist girders shall be in accordance with SJI-100 and SJI-200, as applicable.

14.1.8 Steel Cables The design strength of steel cables serving as main structural load carrying members shall be determined by the requirements of ASCE 19.

14.1.9 Additional Detailing Requirements for Steel Piles in Seismic Design Categories D through F In addition to the foundation requirements set forth in Sections 12.1.5 and 12.13, design and detailing of H-piles shall conform to the requirements of AISC 341, and the connection between the pile cap and steel piles or unfilled steel pipe piles in structures assigned to Seismic Design Category D, E, or F shall be designed for a tensile force not less than 10% of the pile compression capacity.

EXCEPTION: Connection tensile capacity need not exceed the strength required to resist the seismic load effects including overstrength of Section 12.4.3 or 12.14.3.2. Connections need not be provided where the foundation or supported structure does not rely on the tensile capacity of the piles for stability under the design seismic forces.

14.2 CONCRETE

Structures, including foundations, constructed of concrete to resist seismic loads shall be designed and detailed in accordance with this standard, including the reference documents and additional requirements provided in this section.

14.2.1 Reference Documents The quality and testing of concrete materials and the design and construction of structural concrete members that resist seismic forces shall conform to the requirements of ACI 318, except as modified in Section 14.2.2.

14.2.2 Modifications to ACI 318 The text of ACI 318 shall be modified as indicated in Sections 14.2.2.1 through 14.2.2.7. Italics are used for text within Sections 14.2.2.1 through 14.2.2.7 to indicate requirements that differ from ACI 318.

14.2.2.1 Definitions Add the following definitions to ACI 318, Section 2.3.

Cast-In-Place Concrete Equivalent Precast Diaphragm: A cast-in-place noncomposite topping slab diaphragm, as defined in ACI 318, Section 18.12.5, or a diaphragm constructed with precast concrete components that uses closure strips between precast components with detailing that meets the requirements of ACI 318 for the seismic design category of the structure.

Detailed Plain Concrete Structural Wall: A wall complying with the requirements of ACI 318, Chapter 14.

Ordinary Precast Structural Wall: A precast wall complying with the requirements of ACI 318 excluding Chapters 14, 18, and 27.

Precast Concrete Diaphragm: A diaphragm constructed with precast concrete components, with or without a cast-in-place topping, that includes the use of discrete connectors or joint reinforcement to transmit diaphragm forces.

14.2.2.2 ACI 318, Section 10.7.6 Modify Section 10.7.6 by revising Section 10.7.6.1.5 to read as follows:

10.7.6.1.5 *In structures assigned to Seismic Design Categories C, D, E, or F, the ties or hoops shall have a hook on each free end that complies with Section 25.3.4.*

14.2.2.3 Scope Modify ACI 318, Section 18.2.1.2, to read as follows:

18.2.1.2 *All members shall satisfy requirements of Chapters 1 to 17 and 19 to 26. Structures assigned to SDC B, C, D, E, or F also shall satisfy Sections 18.2.1.3 through 18.2.1.7, as applicable, except as modified by the requirements of Chapters 14 and 15 of ASCE 7. Where ACI 318, Chapter 18, conflicts with other ACI 318 chapters, Chapter 18 shall govern.*

14.2.2.4 Intermediate Precast Structural Walls Modify ACI 318, Section 18.5, by renumbering Sections 18.5.2.2 and 18.5.2.3 to Sections 18.5.2.3 and 18.5.2.4, respectively, and adding a new Section 18.5.2.2 to read as follows:

18.5.2.2 *Connections that are designed to yield shall be capable of maintaining 80% of their design strength at the deformation induced by design displacement or shall use type 2 mechanical splices.*

18.5.2.3 *For elements of the connection that are not designed to yield, the required strength shall be based on* $1.5S_y$ *of the yielding portion of the connection.*

18.5.2.4 *In structures assigned to SDC D, E, or F, wall piers shall be designed in accordance with 18.10.8 or 18.14.*

14.2.2.5 Special Precast Structural Walls Modify ACI 318, Section 18.11.2.1, to read as follows:

18.11.2.1 *Special structural walls constructed using precast concrete shall satisfy 18.10 in addition to 18.5.2 as modified by Section 14.2.2 of ASCE 7, except 18.10.2.4 shall not apply for precast walls where deformation demands are concentrated at the panel joints.*

14.2.2.6 Foundations Modify ACI 318, Section 18.13.1.1, to read as follows:

18.13.1.1 *Section 18.13, as modified by Sections 12.1.5 or 12.13 of ASCE 7, shall apply to foundations resisting earthquake-induced forces or transferring earthquake-induced forces between structure and ground.*

14.2.2.7 Detailed Plain Concrete Shear Walls Modify ACI 318, Section 14.6, by adding a new Section 14.6.2 to read:

14.6.2 Detailed Plain Concrete Shear Walls

14.6.2.1 *Detailed plain concrete shear walls are walls conforming to the requirements for ordinary plain concrete shear walls and Section 14.6.2.2.*

14.6.2.2 *Reinforcement shall be provided as follows:*

(a) *Vertical reinforcement of at least 0.20 in^2 (129 mm^2) in cross-sectional area shall be provided continuously from support to support at each corner, at each side of each opening, and at the ends of walls. The continuous vertical bar required beside an opening is permitted to substitute for the No. 5 bar required by Section 11.7.5.1.*
(b) *Horizontal reinforcement of at least 0.20 in^2 (129 mm^2) in cross-sectional area shall be provided:*
   1. *Continuously at structurally connected roof and floor levels and at the top of walls;*
   2. *At the bottom of load-bearing walls or in the top of foundations where doweled to the wall; and*
   3. *At a maximum spacing of 120 in. (3,048 mm).*

Reinforcement at the top and bottom of openings, where used in determining the maximum spacing specified in Item 3 in the preceding text, shall be continuous in the wall.

14.3 COMPOSITE STEEL AND CONCRETE STRUCTURES

Structures, including foundations, constructed of composite steel and concrete to resist seismic loads shall be designed and detailed in accordance with this standard, including the reference documents and additional requirements provided in this section.

14.3.1 Reference Documents The design, construction, and quality of composite steel and concrete members that resist seismic forces shall conform to the applicable requirements of AISC 341, AISC 360, and ACI 318, excluding Chapter 14.

14.3.2 General Systems of structural steel acting compositely with reinforced concrete shall be designed in accordance with AISC 360 and ACI 318, excluding Chapter 14. Where required, the seismic design of composite steel and concrete systems shall be in accordance with the additional provisions of Section 14.3.3.

14.3.3 Seismic Requirements for Composite Steel and Concrete Structures Where a response modification coefficient, *R*, in accordance with Table 12.2-1 is used for the design of systems of structural steel acting compositely with reinforced concrete, the structures shall be designed and detailed in accordance with the requirements of AISC 341.

14.3.4 Metal-Cased Concrete Piles Metal-cased concrete piles shall be designed and detailed in accordance with ACI 318.

14.4 MASONRY

Structures, including foundations, constructed of masonry to resist seismic loads shall be designed and detailed in accordance with this standard, including the references and additional requirements provided in this section.

14.4.1 Reference Documents The design, construction, and quality assurance of masonry members that resist seismic forces shall conform to the requirements of TMS 402 and TMS 602, except as modified by Section 14.4.

14.4.2 *R* Factors To qualify for the response modification coefficients, *R*, set forth in this standard, the requirements of TMS 402 and TMS 602, as amended in subsequent sections, shall be satisfied.

Special reinforced masonry shear walls designed in accordance with Section 8.3 or 9.3 of TMS 402 shall also comply with the additional requirements in Section 14.4.4 or 14.4.5.

14.4.3 Modifications to Chapter 7 of TMS 402

14.4.3.1 Separation Joints Add the following new Section 7.5.1 to TMS 402:

7.5.1 Separation Joints *Where concrete abuts structural masonry and the joint between the materials is not designed as a separation joint, the concrete shall be roughened so that the average height of aggregate exposure is 0.125 in. (3 mm) and shall be bonded to the masonry in accordance with these requirements as if it were masonry. Vertical joints not intended to act as separation joints shall be crossed by horizontal reinforcement as required by Section 5.1.1.2.*

14.4.4 Modifications to Chapter 6 of TMS 402

14.4.4.1 Reinforcement Requirements and Details

14.4.4.1.1 Reinforcing Bar Size Limitations. Modify TMS 402, Section 6.1.2, as follows.

Delete TMS 402, Section 6.1.2.1 and replace with:

6.1.2.1 Reinforcing bars used in masonry shall not be larger than No. 9 (M#29).

Delete TMS 402, Section 6.1.2.2 and replace with:

6.1.2.2 The nominal bar diameter shall not exceed one-quarter of the least clear dimension of the cell, course, or collar joint in which it is placed.

Add the following sentence to the end of TMS 402, Section 6.1.2.4:

The area of reinforcing bars placed in a cell or in a course of hollow unit construction shall not exceed 4% of the cell area.

14.4.4.1.2 Splices in Reinforcement. Add the following new Sections 6.1.6.1.1.4 and 6.1.6.1.2.1 to TMS 402:

6.1.6.1.1.4 Where $M/V_u d_v$ exceeds 1.5 and the seismic load associated with the development of the nominal shear capacity exceeds 80% of the seismic load associated with development of the nominal flexural capacity, lap splices shall not be used in plastic hinge zones of special reinforced masonry shear walls. The length of the plastic hinge zone shall be taken as at least 0.15 times the distance between the point of zero moment and the point of maximum moment.

6.1.6.1.2.1 Where $M/V_u d_v$ exceeds 1.5 and the seismic load associated with the development of the nominal shear capacity exceeds 80% of the seismic load associated with development of the nominal flexural capacity, welded splices shall not be permitted in plastic hinge zones of special reinforced walls of masonry.

14.4.5 Modifications to Chapter 9 of TMS 402

14.4.5.1 Anchoring to Masonry Add the following as the first paragraph in TMS 402, Section 9.1.6:

9.1.6 Anchor Bolts Embedded in Grout *Anchorage assemblies connecting masonry elements that are part of the seismic force-resisting system to diaphragms and chords shall be designed so that the strength of the anchor is governed by steel tensile or shear yielding. Alternatively, the anchorage assembly is permitted to be designed so that it is governed by masonry breakout or anchor pullout, provided the anchorage assembly is designed to resist not less than 2.0 times the factored forces transmitted by the assembly.*

14.4.5.2 Coupling Beams Add the following new Section 9.3.4.2.5 to TMS 402:

9.3.4.2.5 Coupling Beams. *Structural members that provide coupling between shear walls shall be designed to reach their moment or shear nominal strength before either shear wall* *reaches its moment or shear nominal strength. Analysis of coupled shear walls shall comply with accepted principles of mechanics.*

The design shear strength, ϕV_n, of the coupling beams shall satisfy the following criterion:

$$\phi V_n \geq \frac{1.25(M_1 + M_2)}{L_c} + 1.4V_g$$

where

M_1, M_2 = *Nominal moment strength at the ends of the beam,*
L_c = *Length of the beam between the shear walls, and*
V_g = *Unfactored shear force caused by gravity loads.*

The calculation of the nominal flexural moment shall include the reinforcement in reinforced concrete roof and floor systems. The width of the reinforced concrete used for calculations of reinforcement shall be six times the floor or roof slab thickness.

14.4.6 Modifications to Chapter 12 of TMS 402

14.4.6.1 Corrugated Sheet Metal Anchors Add Section 12.2.2.11.1.1 to TMS 402 as follows:

12.2.2.11.1.1 Corrugated sheet metal anchors shall not be used.

14.5 WOOD

Structures, including foundations, constructed of wood to resist seismic loads shall be designed and detailed in accordance with this standard including the references and additional requirements provided in this section.

14.5.1 Reference Documents The quality, testing, design, and construction of members and their fastenings in wood systems that resist seismic forces shall conform to the requirements of the applicable following reference documents: AWC NDS and AWC SDPWS.

14.6 CONSENSUS STANDARDS AND OTHER REFERENCED DOCUMENTS

See Chapter 23 for the list of consensus standards and other documents that shall be considered part of this standard to the extent referenced in this chapter.

CHAPTER 15
SEISMIC DESIGN REQUIREMENTS FOR NONBUILDING STRUCTURES

15.1 GENERAL

15.1.1 Nonbuilding Structures The provisions of this chapter shall apply to unoccupied nonbuilding structures and buildings whose primary purpose is to enclose equipment or machinery and whose occupants are engaged in maintenance or monitoring of that equipment, machinery, or their processes. Nonbuilding structures at or below the grade plane or supported by other structures shall be designed and detailed to resist the minimum lateral forces specified in this chapter. Design shall conform to the applicable requirements of other sections as modified by this section. Foundation design shall comply with the requirements of Sections 12.1.5 and 12.13, and Chapter 14.

15.1.2 Design The design of nonbuilding structures shall provide sufficient stiffness, strength, and ductility consistent with the requirements specified herein for buildings to resist the effects of seismic ground motions as represented by these design forces:

(a) Applicable strength and other design criteria shall be obtained from other portions of the seismic requirements of this standard or its reference documents.
(b) Where applicable strength and other design criteria are not contained in or referenced by the seismic requirements of this standard, such criteria shall be obtained from reference documents. Where reference documents define acceptance criteria in terms of allowable stresses as opposed to strength, the design seismic forces shall be obtained from this section and used in combination with other loads as specified in Section 2.4 of this standard and used directly with allowable stresses specified in the reference documents. Detailing shall be in accordance with the reference documents.

15.1.3 Structural Analysis Procedure Selection Structural analysis procedures for nonbuilding structures that are similar to buildings shall be selected in accordance with Section 12.6. Nonbuilding structures that are not similar to buildings shall be designed using the equivalent lateral force procedure in accordance with Section 12.8, the linear dynamic analysis procedures in accordance with Section 12.9, the nonlinear response history analysis procedure in accordance with Chapter 16, or the procedure prescribed in the specific reference document.

15.1.4 Nonbuilding Structures Sensitive to Vertical Ground Motions Tanks, vessels, hanging structures, and nonbuilding structures incorporating horizontal cantilevers shall use Section 11.9 to determine the vertical seismic design ground motion to be applied to the design of the nonbuilding structure. For these structures, the design forces in members and connections shall be determined by modal analysis performed in accordance with Section 12.9 or response history analysis performed in accordance with Chapter 12 or 16, except that the vertical ground motion component shall be included in the analysis using the spectra defined in Section 11.9. Alternatively, the equivalent static procedures of Section 12.8 and Section 15.4 are permitted to be used to determine the seismic design force with the vertical component motions. For tanks and vessels, vertical ground motions shall be applied as required by Section 15.7.2, Item (c). For hanging structures and nonbuilding structures incorporating horizontal cantilevers, the design vertical response spectral acceleration, S_{av}, shall be taken as the peak value from the response spectrum of Section 11.9. Alternatively, the design vertical response spectral acceleration, S_{av}, is permitted to be determined using the provisions of Section 11.9, with the vertical period of the structure determined by rational analysis. Horizontal seismic effects shall be combined with vertical seismic effects using the direction of loading criteria specified in Section 15.1.4.1 for hanging structures and structures incorporating horizontal cantilevers. The response modification factor, R, for use with vertical seismic design ground motions shall be taken as 1.0, except in the determination of hydrodynamic hoop forces in cylindrical tank walls. The determination of hydrodynamic hoop forces caused by vertical seismic design ground motions in cylindrical tank walls shall comply with the requirements of Section 15.7.2, Item (c), ii.

15.1.4.1 Direction of Loading Criteria for Nonbuilding Structures Sensitive to Vertical Ground Motions The following orthogonal load combinations of horizontal and vertical seismic load effects shall be applied to hanging structures and structures incorporating horizontal cantilevers.

15.1.4.1.1 Strength The directions of application of seismic forces used in the strength design of structure elements shall be those that produce the most critical load effects. This requirement is deemed satisfied if the structure elements are designed to the more stringent demands of the following load directions:

1. 100% of the forces for one horizontal direction plus 30% of the forces for the perpendicular horizontal direction plus 30% of the forces for the vertical direction. The combination requiring the maximum component strength shall be used.
2. 100% of the forces for the vertical direction plus 30% of the forces for a horizontal direction plus 30% of the forces for the perpendicular horizontal direction. The combination requiring the maximum component strength shall be used.

15.1.4.1.2 Overturning and Stability The directions of application of seismic forces used in the evaluation of overturning stability and sliding of the structure shall be those that produce the most critical load effects. This requirement is deemed satisfied if the structures and their foundations are evaluated for overturning stability and sliding using the following load directions:

1. 100% of the forces for one horizontal direction plus 30% of the forces for the perpendicular horizontal direction plus 30% of the forces for the vertical direction. The combination requiring the maximum component strength shall be used.

15.2 NONBUILDING STRUCTURES CONNECTED BY NONSTRUCTURAL COMPONENTS TO OTHER ADJACENT STRUCTURES

15.2.1 General Requirements For nonbuilding structures connected by nonstructural components to other adjacent structures, an analysis combining the structural characteristics of the nonbuilding structure, the adjacent structure, and the connecting nonstructural components in accordance with the requirements of Chapter 12 or Section 15.5, as appropriate, shall be performed to determine the seismic forces.

EXCEPTIONS: Regular nonbuilding structures connected to regular adjacent structures are permitted to be designed independently, with the tributary weight of the nonstructural components considered in the determination of the effective seismic weight, W, for any of the following conditions:

(a) The ratio of the fundamental period of the nonbuilding structure to that of the adjacent connected structure in the direction of motion is greater than 0.9 and less than 1.1.

(b) The ratio of the fundamental period of the nonbuilding structure to that of the adjacent structure in the direction of motion is greater than 0.8 and less than 1.2 and the ratio of the effective seismic weight of the nonbuilding structure and the adjacent structure is greater than 0.8 and less than 1.2.

(c) The ratio of the stiffness of the connecting nonstructural components to that of the nonbuilding structure in the direction of motion and the ratio of the stiffness of the connecting nonstructural components to that of the adjacent structure in the direction of motion are both less than 0.2.

15.2.2 Nonstructural Components Spanning between Nonbuilding Structures Nonstructural components spanning between nonbuilding structures shall comply with all applicable requirements of Chapter 13. Design and detailing for relative displacement between structures shall be in accordance with Section 13.3.2.2.

15.3 NONBUILDING STRUCTURES SUPPORTED BY OTHER STRUCTURES

Where nonbuilding structures identified in Table 15.4-2 are supported by other structures and nonbuilding structures are not part of the primary seismic force-resisting system, one of the following methods shall be used.

15.3.1 Supported Nonbuilding Structures with Less Than 20% of Combined Weight For the condition where the weight of the supported nonbuilding structure is less than 20% of the combined effective seismic weights of the nonbuilding structure and supporting structure, the design seismic forces of the nonbuilding structure shall be determined in accordance with Chapter 13, where the values of C_{AR} and R_{po} shall be determined in accordance with Section 13.1.6. The supporting structure shall be designed in accordance with the requirements of Chapter 12 or Section 15.5, as appropriate, with the weight of the nonbuilding structure considered in the determination of the effective seismic weight, W.

EXCEPTIONS:

(a) Where the ratio of the fundamental period of the nonbuilding structure to that of the supporting structure (including the lumped weight of the nonbuilding structure) is greater than 2.0, the supporting structure is permitted to be designed in accordance with the requirements of Chapter 12 or Section 15.5, as appropriate, with the nonbuilding structure modeled as attached to a rigid base.

(b) Where the ratio of the fundamental period of the nonbuilding structure to that of the supporting structure (including the lumped weight of the nonbuilding structure) is less than 0.5, the supporting structure is permitted to be designed in accordance with the requirements of Chapter 12 or Section 15.5, as appropriate, with the weight of the nonbuilding structure considered in the determination of the effective seismic weight, W. The supported nonbuilding structure shall follow the requirements of Section 15.3.1.

15.3.2 Supported Nonbuilding Structures with Greater Than or Equal to 20% of Combined Weight For the condition where the weight of the supported nonbuilding structure is equal to or greater than 20% of the combined effective seismic weight of the nonbuilding structure and supporting structure, an analysis combining the structural characteristics of both the nonbuilding structure and the supporting structures shall be performed to determine the seismic design forces.

The combined structure shall be designed in accordance with Section 15.5, with the R value of the combined system taken as the lesser of the R values of the nonbuilding structure or the supporting structure. The nonbuilding structure and attachments shall be designed for the forces determined for the nonbuilding structure in the combined analysis.

EXCEPTIONS:

(a) Where the ratio of the fundamental period of the nonbuilding structure to that of the supporting structure (including the lumped weight of the nonbuilding structure) is greater than 2.0, the supporting structure is permitted to be designed in accordance with the requirements of Chapter 12 or Section 15.5, as appropriate, with the nonbuilding structure modeled as attached to a rigid base.

(b) Where the ratio of the fundamental period of the nonbuilding structure to that of the supporting structure (including the lumped weight of the nonbuilding structure) is less than 0.5, the supporting structure is permitted to be designed in accordance with the requirements of Chapter 12 or Section 15.5, as appropriate, with the weight of the nonbuilding structure considered in the determination of the effective seismic weight, W. The supported nonbuilding structure shall follow the requirements of Section 15.3.1.

15.3.3 Nonstructural Components Supported by Nonbuilding Structures Nonstructural components supported by nonbuilding structures shall be designed in accordance with Chapter 13 of this standard.

15.4 STRUCTURAL DESIGN REQUIREMENTS

15.4.1 Design Basis Nonbuilding structures that have specific seismic design criteria established in reference documents shall be designed using the standards as amended herein. Where reference documents are not cited herein, nonbuilding structures shall be designed in compliance with Sections 15.5 and 15.6 to resist minimum seismic lateral forces that are not less than the requirements of Section 12.8, with the following additions and exceptions.

1. The seismic force-resisting system shall be selected as follows:
   (a) For nonbuilding structures similar to buildings, a system shall be selected from among the types indicated in Table 12.2-1 or Table 15.4-1 subject to the system limitations and limits on structural height, h_n, based on the seismic design category indicated in the table. The appropriate values of R, Ω_0, and C_d indicated in the selected table shall be used in determining the base shear, element design forces, and design story drift as indicated in this standard. Design and detailing requirements shall comply with the sections referenced in the selected table.
   (b) For nonbuilding structures not similar to buildings, a system shall be selected from among the types

Table 15.4-1. Seismic Coefficients for Nonbuilding Structures Similar to Buildings.

					Structural System and Structural Height, h_n, Limits (ft)[a] — Seismic Design Category				
Nonbuilding Structure Type	**Detailing Requirements**	R	Ω_0	C_d	B	C	D[b]	E[b]	F[c]
Steel storage racks	Section 15.5.3.1	4	2	3.5	NL	NL	NL	NL	NL
Steel cantilever storage racks, hot-rolled steel:									
Ordinary moment frame (cross-aisle)	15.5.3.2 and AISC 360	3	3	3	NL	NL	NP	NP	NP
Ordinary moment frame (cross-aisle)[d]	15.5.3.2 and AISC 341	2.5	2	2.5	NL	NL	NL	NL	NL
Ordinary braced frame (cross-aisle)	15.5.3.2 and AISC 360	3	3	3	NL	NL	NP	NP	NP
Ordinary braced frame (cross-aisle)[d]	15.5.3.2 and AISC 341	3.25	2	3.25	NL	NL	NL	NL	NL
Steel cantilever storage racks, cold-formed steel:[e]									
Ordinary moment frame (cross-aisle)	15.5.3.2 and AISI S100	3	3	3	NL	NL	NP	NP	NP
Ordinary moment frame (cross-aisle)	15.5.3.2 and AISI S100	1	1	1	NL	NL	NL	NL	NL
Ordinary braced frame (cross-aisle)	15.5.3.2 and AISI S100	3	3	3	NL	NL	NP	NP	NP
Building frame systems:									
Steel special concentrically braced frames	AISC 341	6	2	5	NL	NL	160	160	100
Steel ordinary concentrically braced frame	AISC 341	3¼	2	3¼	NL	NL	35[f]	35[f]	NP[f]
With permitted height increase	AISC 341	2½	2	2½	NL	NL	160	160	100
With unlimited height	AISC 360	1.5	1	1.5	NL	NL	NL	NL	NL
Moment-resisting frame systems:									
Steel special moment frames	AISC 341	8	3	5.5	NL	NL	NL	NL	NL
Special reinforced concrete moment frames[g]	ACI 318, including Ch. 18	8	3	5.5	NL	NL	NL	NL	NL
Steel intermediate moment frames:	AISC 341	4.5	3	4	NL	NL	35[h,i]	NP[h,i]	NP[h,i]
With permitted height increase	AISC 341	2.5	2	2.5	NL	NL	160	160	100
With unlimited height	AISC 341	1.5	1	1.5	NL	NL	NL	NL	NL
Intermediate reinforced concrete moment frames:	ACI 318, including Ch. 18	5	3	4.5	NL	NL	NP	NP	NP
With permitted height increase	ACI 318, including Ch. 18	3	2	2.5	NL	NL	50	50	50
With unlimited height	ACI 318, including Ch. 18	0.8	1	1	NL	NL	NL	NL	NL
Steel ordinary moment frames:	AISC 341	3.5	3	3	NL	NL	NP[h,i]	NP[h,i]	NP[h,i]
With permitted height increase	AISC 341	2.5	2	2.5	NL	NL	100	100	NP[h,i]
With unlimited height	AISC 360	1	1	1	NL	NL	NL	NL	NL
Ordinary reinforced concrete moment frames:	ACI 318, excluding Ch. 18	3	3	2.5	NL	NP	NP	NP	NP
With permitted height increase	ACI 318, excluding Ch. 18	0.8	1	1	NL	NL	50	50	50

[a] NL = no limit; NP = not permitted.
[b] See Section 12.2.5.4 for a description of seismic force-resisting systems limited to structures with a structural height, h_n, of 240 ft (73.2 m) or less.
[c] See Section 12.2.5.4 for seismic force-resisting systems limited to structures with a structural height, h_n, of 160 ft (48.8 m) or less.
[d] The column-to-base connection shall be designed to the lesser of M_n of the column or the factored moment at the base of the column for the seismic load case using the overstrength factor.
[e] Cold-formed sections that meet the requirements of AISC 341, Table D1.1, are permitted to be designed in accordance with AISC 341.
[f] Steel ordinary braced frames are permitted in pipe racks up to 65 ft (20 m).
[g] In Section 2.3 of ACI 318, the definition of "special moment frame" includes precast and cast-in-place construction.
[h] Steel ordinary moment frames and intermediate moment frames are permitted in pipe racks up to 65 ft (20 m) where the moment joints of field connections are constructed of bolted end plates.
[i] Steel ordinary moment frames and intermediate moment frames are permitted in pipe racks up to 35 ft (11 m).

indicated in Table 15.4-2 subject to the system limitations and limits on structural height, h_n, based on seismic design category indicated in the table. The appropriate values of R, Ω_0, and C_d indicated in Table 15.4-2 shall be used in determining the base shear, element design forces, and design story drift as indicated in this standard. Design and detailing requirements shall comply with the sections referenced in Table 15.4-2.

2. For nonbuilding systems that have an R value provided in Table 15.4-2, the minimum specified value in Equation (12.8-5) shall be replaced by

$$C_s = 0.044 S_{DS} I_e \quad (15.4\text{-}1)$$

Table 15.4-2. Seismic Coefficients for Nonbuilding Structures Not Similar to Buildings.

					Structural System and Structural Height, h_n, Limits (ft)[a,b]				
					Seismic Design Category				
Nonbuilding Structure Type	Detailing Requirements[c]	R	Ω_0	C_d	B	C	D	E	F
Elevated tanks, vessels, bins, or hoppers:									
On symmetrically braced integral legs (not similar to buildings)	Section 15.7.10	3	2^d	2.5	NL	NL	160	100	100
On unbraced integral legs or asymmetrically braced integral legs (not similar to buildings)	15.7.10	2	2^d	2.5	NL	NL	100	60	60
Horizontal, saddle-supported welded steel vessels	15.7.14	3	2^d	2.5	NL	NL	NL	NL	NL
Flat-bottom ground-supported tanks:	15.7								
Steel or fiber-reinforced plastic:									
Mechanically anchored		3	2^d	2.5	NL	NL	NL	NL	NL
Self-anchored		2.5	2^d	2	NL	NL	NL	NL	NL
Reinforced or prestressed concrete:									
Reinforced nonsliding base		2	2^d	2	NL	NL	NL	NL	NL
Anchored flexible base		3.25	2^d	2	NL	NL	NL	NL	NL
Unanchored and unconstrained flexible base		1.5	1.5^d	1.5	NL	NL	NL	NL	NL
All other		1.5	1.5^d	1.5	NL	NL	NL	NL	NL
Cast-in-place concrete silos that have walls continuous to the foundation	15.6.2	3	1.75	3	NL	NL	NL	NL	NL
Reinforced concrete tabletop structure (not similar to buildings)[g] supporting elevated tanks, vessels, bins, hoppers, engines, turbines, pumps, fans, generators, or compressors	15.6.9	1.5	1.5	1.5	NL	NL	70	70	70
	15.6.9.1	2	2	2	NL	NL	70	70	70
	15.6.9.2	2.5	2	2	NL	NL	70	70	70
	15.6.9.3	4	2	2.5	NL	NL	70	70	70
All other reinforced masonry structures not similar to buildings detailed as intermediate reinforced masonry shear walls	14.4.1[e]	3	2	2.5	NL	NL	50	50	50
All other reinforced masonry structures not similar to buildings detailed as ordinary reinforced masonry shear walls	14.4.1	2	2.5	1.75	NL	160	NP	NP	NP
All other nonreinforced masonry structures not similar to buildings	14.4.1	1.25	2	1.5	NL	NP	NP	NP	NP
Concrete chimneys and stacks	15.6.2 and ACI 307	2	1.5	2.0	NL	NL	NL	NL	NL
Steel chimneys and stacks	15.6.2 and ASME STS-1	2	2	2	NL	NL	NL	NL	NL
All steel and reinforced concrete distributed mass cantilever structures not otherwise covered herein, including stacks, chimneys, silos, skirt-supported vertical vessels; single-pedestal or skirt-supported	15.6.2								
Welded steel	15.7.10	2	2^d	2	NL	NL	NL	NL	NL
Welded steel with special detailing[f]	15.7.10, 15.7.10.5 a and b	3	2^d	2	NL	NL	NL	NL	NL
Prestressed or reinforced concrete	15.7.10	2	2^d	2	NL	NL	NL	NL	NL
Prestressed or reinforced concrete with special detailing	15.7.10 and ACI 318, Secs. 18.2, 18.10	3	2^d	2	NL	NL	NL	NL	NL
Trussed towers (freestanding or guyed), guyed stacks, and guyed chimneys	15.6.2	3	2	2.5	NL	NL	NL	NL	NL
Steel tubular support structures for onshore wind turbine generator systems	15.6.7	1.5	1.5	1.5	NL	NL	NL	NL	NL
Cooling towers:									
Concrete or steel		3.5	1.75	3	NL	NL	NL	NL	NL
Wood frames		3.5	3	3	NL	NL	NL	50	50

continues

Table 15.4-2 *(Continued)*. Seismic Coefficients for Nonbuilding Structures Not Similar to Buildings.

					Structural System and Structural Height, h_n, Limits (ft)[a,b]				
					Seismic Design Category				
Nonbuilding Structure Type	Detailing Requirements[c]	R	Ω_0	C_d	B	C	D	E	F
Telecommunication towers:	15.6.6								
Truss: Steel		3	1.5	3	NL	NL	NL	NL	NL
Pole: Steel		1.5	1.5	1.5	NL	NL	NL	NL	NL
Wood		1.5	1.5	1.5	NL	NL	NL	NL	NL
Concrete		1.5	1.5	1.5	NL	NL	NL	NL	NL
Frame: Steel		3	1.5	1.5	NL	NL	NL	NL	NL
Wood		1.5	1.5	1.5	NL	NL	NL	NL	NL
Concrete		2	1.5	1.5	NL	NL	NL	NL	NL
Amusement structures and monuments	15.6.3	2	2	2	NL	NL	NL	NL	NL
Inverted-pendulum-type structures (except elevated tanks, vessels, bins, and hoppers)	12.2.5.3	2	2	2	NL	NL	NL	NL	NL
Ground-supported cantilever walls or fences	15.6.8	1.25	2	2.5	NL	NL	NL	NL	NL
Signs and billboards		3.0	1.75	3	NL	NL	NL	NL	NL
Steel lighting system support pole structures	15.6.10	1.5	1.5	1.5	NL	NL	NL	NL	NL
All other self-supporting structures, tanks, or vessels not covered above or by reference standards that are not similar to buildings		1.25	2	2.5	NL	NL	50	50	50

[a] NL = no limit; NP = not permitted.
[b] For the purpose of height limit determination, the height of the structure shall be taken as the height to the top of the structural frame making up the primary seismic force-resisting system.
[c] If a section is not indicated in the detailing requirements column, no specific detailing requirements apply.
[d] See Section 15.7.3.a for the application of the overstrength factors, Ω_0, for tanks and vessels.
[e] Detailed with an essentially complete vertical load-carrying frame.
[f] Sections 15.7.10.5(a) and 15.7.10.5(b) shall be applied for any risk category.
[g] The concrete columns shall have slenderness ratio $L/r \leq 22$.

The value of C_s shall not be taken as less than 0.03.

In addition, for nonbuilding structures located where $S_1 \geq 0.6g$, the minimum specified value in Equation (12.8-6) shall be replaced by

$$C_s = 0.8S_1/(R/I_e) \qquad (15.4\text{-}2)$$

EXCEPTION: Tanks and vessels that are designed to AWWA D100, AWWA D103, API 650 (Appendix E), and API 620 (Appendix L), as modified by this standard, and stacks and chimneys that are designed to ACI 307 as modified by this standard, shall be subject to the larger of the minimum base shear value defined by the reference document or the value determined by replacing Equation (12.8-5) with the following:

$$C_s = 0.044S_{DS}I_e \qquad (15.4\text{-}3)$$

The value of C_s shall not be taken as less than 0.01.

In addition, for nonbuilding structures located where $S_1 \geq 0.6g$, the minimum specified value in Equation (12.8-6) shall be replaced by

$$C_s = 0.5S_1/(R/I_e) \qquad (15.4\text{-}4)$$

Minimum base shear requirements need not apply to the convective (sloshing) component of liquid in tanks.

3. The Importance Factor, I_e, shall be as set forth in Section 15.4.1.1.
4. The vertical distribution of the lateral seismic forces in nonbuilding structures covered by this section shall be determined
   (a) Using the requirements of Section 12.8.3, or
   (b) Using the procedures of Section 12.9.1, or
   (c) In accordance with the reference document applicable to the specific nonbuilding structure.
5. Provided the mass locations for the structure, any contents, and any supported structural or nonstructural elements (including but not limited to piping and stairs) that could contribute to the mass or stiffness of the structure are accounted for and quantified in the analysis, the accidental torsion requirements of Section 12.8.4.2 need not be accounted for:
   (a) Rigid nonbuilding structures, or
   (b) Nonbuilding structures not similar to buildings and designed with R values less than or equal to 3.5, or
   (c) Nonbuilding structures similar to buildings and designed with R values less than or equal to 3.5, provided one of the following conditions is met:

(i) The calculated center of rigidity at each diaphragm is greater than 5% of the plan dimension of the diaphragm in each direction from the calculated center of mass of the diaphragm, or

(ii) The structure does not have a Type 1 horizontal torsional irregularity, and the structure has at least two lines of lateral resistance in each of two major axis directions. At least one line of lateral resistance shall be provided a distance of not less than 20% of the structure's plan dimension from the center of mass on each side of the center of mass.

In addition, structures designed to this section shall be analyzed using a 3D representation in accordance with Section 12.7.3.

6. For nonbuilding structural systems containing liquids, gases, and granular solids supported at the base as defined in Section 15.7.1, the minimum seismic design force shall not be less than that required by the reference document for the specific system.
7. Where a reference document provides a basis for the earthquake-resistant design of a particular type of nonbuilding structure covered by Chapter 15, such a standard shall not be used unless the following limitations are met:
   (a) The seismic ground accelerations and seismic coefficients shall be in conformance with the requirements of Section 11.4.
   (b) The values for total lateral force and total base overturning moment used in design shall not be less than 80% of the base shear value and overturning moment, each adjusted for the effects of soil–structure interaction that is obtained using this standard.
8. The base shear is permitted to be reduced in accordance with Section 19.2 to account for the effects of foundation damping from soil–structure interaction. In no case shall the reduced base shear be less than $0.7V$.
9. Unless otherwise noted in Chapter 15, the effects on the nonbuilding structure caused by gravity loads and seismic forces shall be combined in accordance with the factored load combinations as presented in Section 2.3.
10. Where specifically required by Chapter 15, the design seismic force on nonbuilding structures shall be as defined in Section 12.4.3.

15.4.1.1 Importance Factor The Importance Factor, I_e, and risk category for nonbuilding structures are based on the relative hazard of the contents and the function. The value of I_e shall be the largest value determined by the following:

(a) Applicable reference document listed in Chapter 23,
(b) The largest value as selected from Table 1.5-2, or
(c) As specified elsewhere in Chapter 15.

15.4.2 Rigid Nonbuilding Structures Nonbuilding structures that have a fundamental period, T, less than 0.06 s, including their anchorages, shall be designed for the lateral force obtained from the following:

$$V = 0.30 S_{DS} W I_e \tag{15.4-5}$$

where

V = Total design lateral seismic base shear force applied to a nonbuilding structure,
S_{DS} = Site design response acceleration as determined from Section 11.4.5,
W = Nonbuilding structure operating weight, and
I_e = Importance Factor determined in accordance with Section 15.4.1.1.

The force shall be distributed with height in accordance with Section 12.8.3.

15.4.3 Loads The seismic effective weight, W, for nonbuilding structures shall include the dead load and other loads as defined for structures in Section 12.7.2. For purposes of calculating design seismic forces in nonbuilding structures, W also shall include all normal operating contents for items such as tanks, vessels, bins, hoppers, and the contents of piping. W shall include snow and ice loads where these loads constitute 25% or more of W or where required by the Authority Having Jurisdiction based on local environmental characteristics.

15.4.4 Fundamental Period The fundamental period, T, of the nonbuilding structure shall be determined using the structural properties and deformation characteristics of the resisting elements in a properly substantiated analysis as indicated in Section 12.8.2. Alternatively, the fundamental period is permitted to be computed from the following equation:

$$T = 2\pi \sqrt{\frac{\sum_{i=1}^{n} w_i \delta_i^2}{g \sum_{i=1}^{n} f_i \delta_i}} \tag{15.4-6}$$

where the values of f_i represent any lateral force distribution in accordance with the principles of structural mechanics. The elastic deflections, δ_i, shall be calculated using the applied lateral forces, f_i. Equations (12.8-7), (12.8-8), (12.8-9), and (12.8-10) shall not be used for determining the period of a nonbuilding structure.

15.4.5 Drift Limit The drift limit of Section 12.12.1 need not apply to nonbuilding structures if a rational analysis indicates that the limit can be exceeded without adversely affecting structural stability or attached or interconnected components and elements such as walkways and piping.

15.4.6 P-Delta For nonbuilding structures similar to buildings, P-delta effects shall be evaluated using the requirements of Section 12.8.7. For nonbuilding structures not similar to buildings, the requirements of Section 12.8.7 do not apply. For nonbuilding structures not similar to buildings, P-delta effects on design forces and drifts induced by these effects shall be explicitly considered in the design of the structure. P-delta effects for nonbuilding structures not similar to buildings shall be based on displacements determined by an elastic analysis multiplied by C_d/I_e, using the appropriate C_d value from Table 15.4-2.

15.4.7 Drift, Deflection, and Structure Separation Drift, deflection, and structure separation calculated using strength-level seismic forces shall be determined in accordance with this standard unless specifically amended in Chapter 15.

15.4.8 Site-Specific Response Spectra Where required by a reference document or the Authority Having Jurisdiction, specific types of nonbuilding structures shall be designed for site-specific criteria that account for local seismicity and geology, expected recurrence intervals, and magnitudes of events from known seismic hazards (see Section 11.4.8 of this standard). If a longer recurrence interval is defined in the reference document for the nonbuilding structure, such as liquefied natural gas (LNG) tanks (NFPA 59A), the recurrence interval required in the reference document shall be used.

15.4.9 Anchors in Concrete or Masonry

15.4.9.1 Anchors in Concrete Anchors in concrete used for nonbuilding structure anchorage shall be designed in accordance with Chapter 17 of ACI 318.

15.4.9.2 Anchors in Masonry Anchors in masonry used for nonbuilding structure anchorage shall be designed in accordance with TMS 402. Anchors shall be designed to be governed by the tensile or shear strength of a ductile steel element.

EXCEPTION: Anchors shall be permitted to be designed so that either

(a) The attachment that the anchor is connecting to the structure undergoes ductile yielding at a load level corresponding to anchor forces not greater than the design strength of the anchor, or
(b) The anchors resist the load combinations in accordance with Section 12.4.3, including Ω_0 as given in Tables 15.4-1 and 15.4-2.

15.4.9.3 Post-Installed Anchors in Concrete and Masonry Post-installed anchors in concrete shall be prequalified for seismic applications in accordance with ACI 355.2 or other approved qualification procedures. Post-installed anchors in masonry shall be prequalified for seismic applications in accordance with approved qualification procedures.

15.4.9.4 ASTM F1554 Anchors When ASTM F1554 Grade 36 anchors are specified and are designed as ductile anchors in accordance with ACI 318, Section 17.10.5.3(a), or where the design must meet the requirements of Section 15.7.5 or 15.7.11.7(b) of this standard, substitution of weldable ASTM F1554 Grade 55 (with Supplementary Requirement S1) anchors shall be prohibited.

15.4.10 Requirements for Nonbuilding Structure Foundations on Liquefiable Sites Nonbuilding structure foundations on liquefiable sites shall comply with Section 12.13.9 and the requirements of Section 15.4.10.1.

15.4.10.1 Nonbuilding Structures on Shallow Foundations Nonbuilding structures shall not be permitted to be supported on shallow foundations at liquefiable sites unless it can be demonstrated that the structure's foundation, superstructure, and connecting systems can be designed to accommodate the soil strength loss, lateral spreading, and total and differential settlements induced by MCE_G earthquake ground motions indicated in the geotechnical investigation report.

15.4.11 Material Requirements The requirements regarding specific materials in Chapter 14 shall be applicable unless specifically exempted in Chapter 15.

15.5 NONBUILDING STRUCTURES SIMILAR TO BUILDINGS

15.5.1 General Nonbuilding structures similar to buildings as defined in Section 11.2 shall be designed in accordance with this standard as modified by this section and the specific reference documents. This general category of nonbuilding structures shall be designed in accordance with the seismic requirements of this standard and the applicable portions of Section 15.4. The combination of load effects, E, shall be determined in accordance with Section 12.4.

15.5.2 Pipe Racks

15.5.2.1 Design Basis In addition to the requirements of Section 15.5.1, pipe racks supported at the base of the structure shall be designed to meet the force requirements of Section 12.8 or 12.9.1. Displacements of the pipe rack and potential for interaction effects (pounding of the piping system) shall be considered using the amplified deflections calculated as

$$\delta_x = \frac{C_d \delta_{xe}}{I_e} \tag{15.5-1}$$

where

C_d = Deflection amplification factor in Table 15.4-1,
δ_{xe} = Deflections determined using the prescribed seismic design forces of this standard, and
I_e = Importance Factor determined in accordance with Section 15.4.1.1.

See Section 13.6.2 for the design of piping systems and their attachments. Friction resulting from gravity loads shall not be considered to provide resistance to seismic forces.

15.5.3 Storage Racks Storage racks constructed from steel and supported at or below grade shall be designed in accordance with Section 15.5.3.1 or 15.5.3.2, as applicable, and the requirements in Section 15.5.3.3.

15.5.3.1 Steel Storage Racks Steel storage racks supported at or below grade shall be designed in accordance with ANSI/RMI MH 16.1, its force and displacement requirements, and the seismic design ground motion values determined according to Section 11.4, except as follows:

15.5.3.1.1 Modify Section 7.1.2 of ANSI/RMI MH 16.1 as follows:

7.1.2 Base Plate Design

Once the required bearing area has been determined from the allowable bearing stress, F'_p, the minimum thickness of the base plate is determined by rational analysis or by appropriate test using a test load 1.5 times the ASD design load or the factored LRFD load. Design forces that include seismic loads for anchorage of steel storage racks to concrete or masonry shall be determined using load combinations with overstrength provided in Section 12.4.3.1 of ASCE/SEI 7. The overstrength factor shall be taken as 2.0.

Anchorage of steel storage racks to concrete shall be in accordance with the requirements of Section 15.4.9 of ASCE/SEI 7. Upon request, information shall be given to the owner or the owner's agent on the location, size, and pressures under the column base plates of each type of upright frame in the installation. When rational analysis is used to determine base plate thickness, and other applicable standards do not apply, the base plate shall be permitted to be designed for the following loading conditions, where applicable: [balance of section unchanged]

15.5.3.1.2 Modify Section 7.1.4 of ANSI/RMI MH 16.1 as follows:

7.1.4 Shims

Shims may be used under the base plate to maintain the plumbness and/or levelness of the storage rack. The shims shall be made of a material that meets or exceeds the design bearing strength (LRFD) or allowable bearing strength (ASD) of the floor. The shim size and location under the base plate shall be equal to or greater than the required base plate size and location.

~~*In no case shall the total thickness of a shim stack under a base plate exceed six times the diameter of the largest anchor bolt used in that base.*~~

Shims stacks ~~*having a total thickness greater than two and less than or equal to six times the anchor bolt diameter under bases with only one anchor bolt*~~ *shall be interlocked or welded together in a fashion that is capable of transferring all the shear forces at the base.*

~~*Shims stacks having a total thickness of less than or equal to two times the anchor bolt diameter need not be interlocked or welded together.*~~

Bending in the anchor associated with shims or grout under the base plate shall be taken into account in the design of anchor bolts.

15.5.3.2 Steel Cantilevered Storage Racks Steel cantilevered storage racks supported at or below grade shall be designed in accordance with ANSI/RMI MH 16.3, its force and displacement requirements, and the seismic design ground motion values determined according to Section 11.4, except as follows.

15.5.3.2.1 Modify Section 8.5.1 of ANSI/RMI MH 16.3 as follows:

8.5.1 Anchor Bolt Design

Anchorage of steel cantilevered storage racks to concrete shall be in accordance with the requirements of Section 15.4.9 of ASCE/SEI 7. The redundancy factor in the load combinations in Section 2.1 and 2.2 shall be 1.0. Design forces that include seismic loads for anchorage of steel cantilevered storage racks to concrete or masonry shall be determined using load combinations with overstrength provided in Sections 2.3.6 or 2.4.5 of ASCE/SEI 7.

If shims are used under the base plate to maintain the plumbness and/or levelness of the steel cantilevered storage rack, the shims stacks shall be interlocked or welded together in a fashion that is capable of transferring all the shear forces at the base. Bending in the anchor associated with shims or grout under the base plate shall be taken into account in the design of anchor bolts.

15.5.3.3 Alternative As an alternative to ANSI/RMI MH 16.1 or 16.3, as modified, storage racks shall be permitted to be designed in accordance with the requirements of Sections 15.1, 15.2, 15.3, 15.5.1, and 15.5.3.3.1 through 15.5.3.3.4 of this standard.

15.5.3.3.1 General Requirements Storage racks shall satisfy the force requirements of this section. The Importance Factor, I_e, for storage racks in structures open to the public, such as warehouse retail stores, shall be taken equal to 1.5.

EXCEPTION: Steel storage racks supported at the base are permitted to be designed as structures with an R of 4, provided the seismic requirements of this standard are met. Higher values of R are permitted to be used where the detailing requirements of reference documents listed in Section 14.1.1 are met. The Importance Factor, I_e, for steel storage racks in structures open to the public, such as warehouse retail stores, shall be taken as equal to 1.5.

15.5.3.3.2 Operating Weight Storage racks shall be designed for each of the following conditions of operating weight, W or W_p:

(a) Weight of the rack plus every storage level loaded to 67% of its rated load capacity; and
(b) Weight of the rack plus the highest storage level only, loaded to 100% of its rated load capacity.

The design shall consider the actual height of the center of mass of each storage load component.

15.5.3.3.3 Vertical Distribution of Seismic Forces For all storage racks, the vertical distribution of seismic forces shall be as specified in Section 12.8.3 and in accordance with the following:

(a) The base shear, V, of the typical structure shall be the base shear of the steel storage rack where loaded in accordance with Section 15.5.3.2.
(b) The base of the structure shall be the floor supporting the steel storage rack. Each steel storage level of the rack shall be treated as a level of the structure with heights h_i and h_x measured from the base of the structure.
(c) The factor k is permitted to be taken as 1.0.

15.5.3.3.4 Seismic Displacements Storage rack installations shall accommodate the seismic displacement of the steel storage racks and their contents relative to all adjacent or attached components and elements. The assumed total relative displacement for steel storage racks shall be not less than 5% of the structural height above the base, h_n, unless a smaller value is justified by test data or analysis in accordance with Section 11.1.4.

15.5.4 Electrical Power-Generating Facilities

15.5.4.1 General Electrical power-generating facilities are power plants that generate electricity by steam turbines, combustion turbines, diesel generators, or similar turbomachinery.

15.5.4.2 Design Basis In addition to the requirements of Section 15.5.1, electrical power-generating facilities shall be designed using this standard and the appropriate factors contained in Section 15.4.

15.5.5 Structural Towers for Tanks and Vessels

15.5.5.1 General In addition to the requirements of Section 15.5.1, structural towers that support tanks and vessels and are not integral with the tank shall be designed to meet the requirements of Section 15.3. In addition, the following special considerations shall be included:

(a) The distribution of the lateral base shear from the tank or vessel onto the supporting structure shall consider the relative stiffness of the tank and resisting structural elements.
(b) The distribution of the vertical reactions from the tank or vessel onto the supporting structure shall consider the relative stiffness of the tank and resisting structural elements. Where the tank or vessel is supported on grillage beams, the calculated vertical reaction caused by weight and overturning shall be increased at least 20% to account for nonuniform support. The grillage beam and vessel attachment shall be designed for this increased design value.
(c) Seismic displacements of the tank and vessel shall consider the deformation of the support structure where determining P-delta effects or evaluating required clearances to prevent pounding of the tank on the structure. P-delta effects shall be based on displacements determined by an elastic analysis multiplied by C_d/I_e using the appropriate C_d value from Table 15.4-2.

Tanks and vessels supported by structural towers that are integral to the tank or vessel shall be designed according to Section 15.7.10.1.

15.5.6 Piers and Wharves

15.5.6.1 General Piers and wharves are structures located in waterfront areas that project into a body of water or that parallel the shoreline.

15.5.6.2 Design Basis In addition to the requirements of Section 15.5.1, piers and wharves that are accessible to the general public, such as cruise ship terminals and piers with retail or commercial offices or restaurants, shall be designed to comply with this standard. Piers and wharves that are not accessible to the general public are beyond the scope of this section.

The design shall account for the effects of liquefaction and soil failure collapse mechanisms and shall consider all applicable marine loading combinations, such as mooring, berthing, wave, and current on piers and wharves, as required. Structural detailing shall consider the effects of the marine environment.

15.6 GENERAL REQUIREMENTS FOR NONBUILDING STRUCTURES NOT SIMILAR TO BUILDINGS

Nonbuilding structures that do not have lateral and vertical seismic force-resisting systems that are similar to buildings shall be designed in accordance with this standard as modified by this section and the specific reference documents. Loads and load distributions shall not be less demanding than those determined in this standard. The combination of earthquake load effects, E, shall be determined in accordance with Section 12.4.2.

EXCEPTION: The redundancy factor, ρ, per Section 12.3.4 shall be taken as 1.

15.6.1 Earth-Retaining Structures This section applies to all earth-retaining structures assigned to Seismic Design Category D, E, or F. The lateral earth pressures caused by earthquake ground motions shall be determined in accordance with Section 11.8.3. The risk category shall be determined by the proximity of the earth-retaining structure to other buildings and structures. If failure of the earth-retaining structure would affect the adjacent building or structure, the risk category shall not be less than that of the adjacent building or structure.

Earth-retaining walls are permitted to be designed for seismic loads as either yielding or nonyielding walls. Cantilevered reinforced concrete or masonry retaining walls shall be assumed to be yielding walls and shall be designed as simple flexural wall elements.

15.6.2 Trussed Towers, Chimneys, and Stacks

15.6.2.1 General Trussed towers shall be designed to resist seismic lateral forces determined in accordance with Section 15.4. A system that meets the definition of braced frame and conforms to Section 15.5, "Nonbuilding Structures Similar to Buildings," and Table 15.4-1 shall not be considered a trussed tower in Table 15.4-2.

Chimneys and stacks are permitted to be either lined or unlined and shall be constructed from concrete, steel, or masonry. Steel stacks, concrete stacks, steel chimneys, concrete chimneys, and liners shall be designed to resist seismic lateral forces determined from a substantiated analysis using reference documents. Interaction of the stack or chimney with the liners shall be considered. A minimum separation shall be provided between the liner and chimney equal to C_d times the calculated differential lateral drift.

15.6.2.2 Concrete Chimneys and Stacks Concrete chimneys and stacks shall be designed in accordance with the requirements of ACI 307, except that (1) the design base shear shall be determined based on Section 15.4.1 of this standard, (2) the seismic coefficients shall be based on the values provided in Table 15.4-2, and (3) openings shall be detailed as described in the next paragraph. When modal response spectrum analysis is used for design, the procedures of Section 12.9 shall be permitted to be used.

For concrete chimneys and stacks assigned to Seismic Design Categories D, E, and F, splices for vertical rebar shall be staggered such that no more than 50% of the bars are spliced at any section and alternate lap splices are staggered by the development length. In addition, where the loss of cross-sectional area is greater than 10%, cross sections in the regions of breachings/openings shall be designed and detailed for vertical force, shear force, and bending moment demands along the vertical direction, determined for the affected cross section using an overstrength factor of 1.5. The region where the overstrength factor applies shall extend above and below the opening(s) by a distance equal to half of the width of the largest opening in the affected region. Appropriate reinforcement development lengths shall be provided beyond the required region of overstrength. The jamb regions around each opening shall be detailed using the column tie requirements in Section 10.7.6 of ACI 318. Such detailing shall extend for a jamb width of a minimum of two times the wall thickness and for a height of the opening height plus twice the wall thickness above and below the opening but no less than the development length of the longitudinal bars. Where the existence of a footing or base mat precludes the ability to achieve the extension distance below the opening and within the stack, the jamb reinforcing shall be extended and developed into the footing or base mat. The percentage of longitudinal reinforcement in jamb regions shall meet the requirements of Section 10.6.1.1 of ACI 318 for compression members.

15.6.2.3 Steel Chimneys and Stacks Steel chimneys and stacks shall be designed in accordance with the requirements of ASME STS-1, except that (1) the design base shear shall be determined based on Section 15.4.1 of this standard, (2) the seismic coefficients shall be based on the values provided in Table 15.4-2, and (3) ASME STS-1, Section 4.3.5, and Nonmandatory Appendix D Table D-1, Table D-2, Figure D-2, and Figure D-2a shall not be used. When modal response spectrum analysis is used for design, the procedures of Section 12.9 shall be permitted to be used.

15.6.3 Amusement Structures Amusement structures are permanently fixed structures constructed primarily for the conveyance and entertainment of people. Amusement structures shall be designed to resist seismic lateral forces determined in accordance with Section 15.4.

15.6.4 Special Hydraulic Structures Special hydraulic structures are structures that are contained inside liquid-containing structures. These structures are exposed to liquids on both wall surfaces at the same head elevation under normal operating conditions. Special hydraulic structures are subjected to out-of-plane forces only during an earthquake, when the structure is subjected to differential hydrodynamic fluid forces. Examples of special hydraulic structures include separation walls, baffle walls, weirs, and other similar structures.

15.6.4.1 Design Basis Special hydraulic structures shall be designed for out-of-phase movement of the fluid. Unbalanced forces from the motion of the liquid must be applied simultaneously "in front of" and "behind" these elements.

Structures subject to hydrodynamic pressures induced by earthquakes shall be designed for rigid-body and sloshing-liquid forces and their own inertia force. The height of sloshing shall be determined and compared with the freeboard height of the structure. Interior elements, such as baffles or roof supports, also shall be designed for the effects of unbalanced forces and sloshing.

15.6.5 Secondary Containment Systems Secondary containment systems, such as impoundment dikes and walls, shall meet the requirements of the applicable standards for tanks and vessels and of the Authority Having Jurisdiction.

Secondary containment systems shall be designed to withstand the effects of the maximum considered earthquake ground motion where empty and two-thirds of the maximum considered earthquake ground motion where full, including all hydrodynamic forces, as determined in accordance with the procedures of Section 11.4. Where it is determined by the risk assessment required by Section 1.5.3 or by the Authority Having Jurisdiction that the site may be subject to aftershocks of the same magnitude as the maximum considered motion, secondary containment systems shall be designed to withstand the effects of the maximum considered earthquake ground motion where full, including all hydrodynamic forces, as determined in accordance with the procedures of Section 11.4.

15.6.5.1 Freeboard Sloshing of the liquid within the secondary containment area shall be considered in determining the height of the impoundment. Where the primary containment has not been designed with a reduction in the structure category (i.e., no reduction in Importance Factor, I_e) as permitted by Section 1.5.3, no freeboard provision is required. Where the primary containment has been designed for a reduced structure category (i.e., Importance Factor I_e reduced) as permitted by Section 1.5.3, a minimum freeboard, δ_s, shall be provided where

$$\delta_s = 0.42DS_{ac} \quad (15.6\text{-}1)$$

where S_{ac} is the spectral acceleration of the convective component and is determined according to the procedures of Section 15.7.6.1 using 0.5% damping. For circular impoundment dikes, D shall be taken as the diameter of the impoundment dike. For rectangular impoundment dikes, D shall be taken as the plan dimension of the impoundment dike, L, for the direction under consideration.

15.6.6 Telecommunication Towers Self-supporting and guyed telecommunication towers shall be designed to resist seismic lateral forces determined in accordance with ANSI/TIA-222-H. The seismic design ground motion values shall be determined according to Section 11.4.

15.6.7 Steel Tubular Support Structures for Onshore Wind Turbine Generator Systems Steel tubular support structures for onshore wind turbine generator systems shall be designed to resist seismic lateral forces determined in accordance with Section 15.4.

15.6.8 Ground-Supported Cantilever Walls or Fences

15.6.8.1 General Ground-supported cantilever walls or fences 6 ft (1.83 m) or greater in height shall satisfy the requirements of this section. Earth-retaining structures shall comply with Section 15.6.1.

15.6.8.2 Design Basis Walls or fences shall be designed to resist earthquake ground motions in accordance with Section 15.4. Detailed plain and ordinary plain concrete or masonry walls or fences and ordinary plain autoclaved aerated concrete (AAC) walls or fences are not permitted in Seismic Design Category C, D, E, and F.

15.6.9 Reinforced Concrete Tabletop Structure for Rotating Equipment and Process Vessels or Drums Slab-column reinforced concrete moment frames designed per Section 15.4 qualify for the seismic design parameters for the tabletop structure entries in Table 15.4-2 if the following requirements are satisfied:

(a) The flexural capacity of each column under all load combinations is less than two-thirds the flexural capacity of the slab and less than two-thirds the flexural capacity of the foundation that supports the column,
(b) The thickness of the slab is not less than 3 ft (0.91 m), and
(c) The design of the reinforced concrete members satisfies ACI 318, with the exception of Chapter 18.

***15.6.9.1 Tabletop Structures Designed with* R = 2.0** Tabletop structures designed with $R = 2.0$ shall be supported by at least six columns, satisfy the requirements listed in Section 15.6.9, and be detailed to satisfy Section 18.3.3 of ACI 318.

***15.6.9.2 Tabletop Structures Designed with* R = 2.5** Tabletop structures designed with $R = 2.5$ shall be detailed to satisfy the requirements of ACI 318 for ordinary moment frames. Items (a), (b), and (c) in Section 15.6.9 need not be satisfied. Section 18.3.4 of ACI 318 shall be deemed to be satisfied if the total area of vertical reinforcement in the column does not exceed 2% of the gross concrete cross section of the column. If there is more reinforcement in the column and if the flexural capacity of the column is greater than the flexural capacity of the slab or beam, the joint shear force, V_u, shall be computed with forces in the slab reinforcement that are in equilibrium with the seismic load assuming the vertical bars on the tension face of the column have reached yield.

***15.6.9.3 Tabletop Structures Designed with* R = 4.0** Tabletop structures designed with $R = 4.0$ shall be detailed to satisfy the requirements of ACI 318 for intermediate moment frames. Items (a), (b), and (c) of 15.6.9 need not be satisfied.

15.6.10 Steel Lighting System Support Pole Structures Steel lighting system support pole structures shall be designed in accordance with the seismic requirements of ASCE 72, except that the seismic design ground motion values shall be determined according to Section 11.4.

15.7 TANKS AND VESSELS

15.7.1 General This section applies to all tanks, vessels, bins, silos, and similar containers storing liquids, gases, and granular solids supported at the base (hereafter referred to generically as "tanks and vessels"). Tanks and vessels covered herein include reinforced concrete, prestressed concrete, steel, aluminum, and fiber-reinforced plastic materials. Tanks supported on elevated levels in buildings shall be designed in accordance with Section 15.3.

15.7.2 Design Basis Tanks and vessels storing liquids, gases, and granular solids shall be designed in accordance with this standard and shall be designed to meet the requirements of the applicable reference documents listed in Chapter 23. Resistance to seismic forces shall be determined from a substantiated analysis based on the applicable reference documents listed in Chapter 23.

(a) Damping for the convective (sloshing) force component shall be taken as 0.5%.
(b) Impulsive and convective components shall be combined by the direct sum or the square root of the sum of the squares (SRSS) method where the modal periods are separated. If significant modal coupling may occur, the complete quadratic combination (CQC) method shall be used.

(c) Vertical earthquake forces shall be considered in accordance with the applicable reference document. If the reference document permits the user the option of including or excluding the vertical earthquake force to comply with this standard, it shall be included. For tanks and vessels not covered by a reference document, the forces caused by the vertical acceleration shall be defined as follows:
   (i) *Hydrodynamic vertical and lateral forces in noncylindrical tank walls.* The increase in hydrostatic pressures caused by the vertical excitation of the contained liquid shall correspond to an effective increase in unit weight, γ_L, of the stored liquid equal to $0.4S_{av}\gamma_L$, where S_{av} is taken as the peak of the vertical response spectrum defined in Section 11.9.
   (ii) *Hydrodynamic hoop forces in cylindrical tank walls.* In a cylindrical tank wall, the hoop force per unit height, N_h, at height y from the base, associated with the vertical excitation of the contained liquid, shall be calculated as follows:

$$N_h = \frac{S_{av}}{R}\gamma_L(H_L - y)\left(\frac{D_i}{2}\right) \qquad (15.7\text{-}1)$$

where

D_i = Inside tank diameter,
H_L = Liquid height inside the tank,
y = Distance from base of the tank to height being investigated,
γ_L = Unit weight of stored liquid, and
S_{av} = Vertical seismic parameter from Section 11.9, determined natural period of vibration of vertical liquid motion.

The hoop force associated with the vertical excitation of the liquid shall be combined with the impulsive and convective components by the direct sum or SRSS method.
   (iii) *Vertical inertia forces in cylindrical and rectangular tank walls.* Vertical inertia forces associated with the vertical acceleration of the structure itself shall be taken as equal to $0.4S_{av}W$, where S_{av} is taken as the peak of the vertical response spectrum defined in Section 11.9.

15.7.3 Strength and Ductility Structural members that are part of the seismic force-resisting system shall be designed to provide the following.

(a) Connections to seismic force-resisting elements, excluding anchors (bolts or rods) embedded in concrete, shall be designed to develop forces calculated using the seismic load effects including overstrength of Section 12.4.3. For anchors (bolts or rods) embedded in concrete, the design of the anchor embedment shall meet the requirements of Section 15.7.5. In addition, the connection of the anchors to the tank or vessel shall be designed to develop the lesser of the strength of the anchor in tension as determined by API 650 or AWWA D100 or forces calculated using the seismic load effects including overstrength of Section 12.4.3. The overstrength requirements of Section 12.4.3 and the Ω_0 values tabulated in Table 15.4-2 do not apply to the design of walls, including interior walls, of tanks or vessels.
(b) Penetrations, manholes, and openings in shell elements shall be designed to maintain the strength and stability of the shell to carry tensile and compressive membrane shell forces.
(c) Support towers for tanks and vessels, where the support tower is integral with the tank or vessel, with irregular bracing, unbraced panels, asymmetric bracing, or concentrated masses shall be designed using the requirements of Section 12.3.2 for irregular structures. Support towers using chevron or eccentrically braced framing shall comply with the seismic requirements of this standard. Support towers using tension-only bracing shall be designed such that the full cross section of the tension element can yield during overload conditions.
(d) In support towers for tanks and vessels, where the support tower is integral with the tank or vessel, compression struts that resist the reaction forces from tension braces shall be designed to resist the lesser of the yield load of the brace, A_gF_y, or forces calculated using the seismic load effects including overstrength of Section 12.4.3.
(e) The vessel stiffness relative to the support system (foundation, support tower, skirt, etc.) shall be considered in determining forces in the vessel, the resisting elements, and the connections.
(f) For concrete liquid-containing structures, system ductility and energy dissipation under allowable stress design level loads shall not be allowed to be achieved by inelastic deformations to such a degree as to jeopardize the serviceability of the structure. Stiffness degradation and energy dissipation shall be allowed to be obtained either through limited microcracking, or by means of lateral force resistance mechanisms that dissipate energy without damaging the structure.

15.7.4 Flexibility of Piping Attachments Design of piping systems connected to tanks and vessels shall consider the potential movement of the connection points during earthquakes and provide sufficient flexibility to avoid release of the product by failure of the piping system. The piping system and supports shall be designed so as not to impart significant mechanical loading on the attachment to the tank or vessel shell. Mechanical devices that add flexibility, such as bellows, expansion joints, and other flexible apparatus, are permitted to be used where they are designed for seismic displacements and defined operating pressure.

Unless otherwise calculated, the minimum displacements in Table 15.7-1 shall be assumed. For attachment points located above the support or foundation elevation, the displacements in Table 15.7-1 shall be increased to account for drift of the tank or vessel relative to the base of support. The piping system and tank connection shall also be designed to tolerate C_d times the displacements given in Table 15.7-1 without rupture, although permanent deformations and inelastic behavior in the piping supports and tank shell are permitted. For attachment points located above the support or foundation elevation, the displacements in Table 15.7-1 shall be increased to account for drift of the tank or vessel. The values given in Table 15.7-1 do not include the influence of relative movements of the foundation and piping anchorage points caused by foundation movements (e.g., settlement or seismic displacements). The effects of the foundation movements shall be included in the piping system design, including the determination of the mechanical loading on the tank or vessel, and the total displacement capacity of the mechanical devices intended to add flexibility.

Table 15.7-1. Minimum Design Displacements for Piping Attachments.

Condition	Minimum Design Displacement, in. (mm)
Mechanically Anchored Tanks and Vessels	
Upward vertical displacement relative to support or foundation	1 (25.4)
Downward vertical displacement relative to support or foundation	0.5 (12.7)
Range of horizontal displacement (radial and tangential) relative to support or foundation	0.5 (12.7)
Self-Anchored Tanks or Vessels (at Grade)	
Upward vertical displacement relative to support or foundation:	
If designed in accordance with a reference document as modified by this standard:	
Anchorage ratio less than or equal to 0.785 (indicates no uplift)	1 (25.4)
Anchorage ratio greater than 0.785 (indicates uplift)	4 (101.1)
If designed for seismic loads in accordance with this standard but not covered by a reference document:	
For tanks and vessels with diameter less than 40 ft (12.2 m)	8 (202.2)
For tanks and vessels with diameter equal to or greater than 40 ft (12.2 m)	12 (0.305)
Downward vertical displacement relative to support or foundation:	
For tanks with a ringwall/mat foundation	0.5 (12.7)
For tanks with a berm foundation	1 (25.4)
Range of horizontal displacement (radial and tangential) relative to support or foundation	2 (50.8)

Table 15.7-2. Anchorage Ratio.

Anchorage Ratio, J	Criteria
$J < 0.785$	No uplift under the design seismic overturning moment. The tank is self-anchored.
$0.785 < J < 1.54$	Tank is uplifting, but the tank is stable for the design load, provided that the shell compression requirements are satisfied. The tank is self-anchored.
$J > 1.54$	Tank is not stable and shall be mechanically anchored for the design load.

The anchorage ratio, J, for self-anchored tanks shall comply with the criteria shown in Table 15.7-2 and is defined as

$$J = \frac{M_{rw}}{D^2(w_t + w_a)} \tag{15.7-2}$$

where

$$w_t = \frac{W_s}{\pi D} + w_r \tag{15.7-3}$$

w_r = Roof load acting on the shell [lb/ft (N/m)] of shell circumference. Only permanent roof loads shall be included. Roof live load shall not be included;

w_a = Maximum weight of the tank contents that may be used to resist the shell overturning moment [lb/ft (N/m)] of shell circumference (w_a usually consists of an annulus of liquid limited by the bending strength of the tank bottom or annular plate);

M_{rw} = Overturning moment applied at the bottom of the shell caused by the seismic design loads [ft-lb (N-m)] (also known as the ringwall moment);

D = Tank diameter, ft (m), [ft (m)]; and

W_s = Total weight of tank shell [lb (N)].

15.7.5 Anchorage Tanks and vessels at grade are permitted to be designed without anchorage where they meet the requirements for self-anchored tanks in reference documents. Tanks and vessels supported above grade on structural towers or building structures shall be anchored to the supporting structure.

For the anchorage of steel tanks and vessels with a diameter or width greater than 5 ft (1.5 m) or a height greater than 10 ft (3.0 m) in Seismic Design Category B, anchorage shall be designed in accordance with Section 15.4.9.

For the anchorage of steel tanks and vessels with a diameter or width greater than 5 ft (1.5 m) or a height greater than 10 ft (3.0 m) in Seismic Design Categories C, D, E, and F, all of the following special detailing requirements shall apply:

(a) Anchorage shall be in accordance with Section 15.4.9.1, whereby the anchor embedment into the concrete shall be designed to develop the steel strength of the anchor in tension in accordance with ACI 318 Section, 17.10.5.3 (a), as given by ACI 318, Equation (17.6.1.2), or shall be designed using anchor reinforcement in accordance with ACI 318, Section 17.4.2.9, to develop the steel strength of the anchor in tension per ACI 318, Equation (17.6.1.2).

(b) The minimum gauge length of anchor rod, which is the length over which its elongation can occur, shall be at least eight times the rod diameter.

(c) Post-installed anchors are permitted to be used in accordance with Section 15.4.9.3, provided the anchor embedment into the concrete is sufficient to develop the steel strength of the anchor rod in tension.

(d) Where the special detailing requirements of this section apply, the load combinations of Section 12.4.3 that include overstrength do not apply.

For steel tanks and vessels with a diameter or width less than or equal to 5 ft (1.5 m) and a height less than or equal space to 10 ft (3.0 m) in Seismic Design Categories C, D, E, and F, anchorage shall be in accordance with Section 15.4.9.

15.7.6 Ground-Supported Storage Tanks for Liquids

15.7.6.1 General Ground-supported, flat-bottom tanks storing liquids shall be designed to resist the seismic forces as follows:

(a) For tanks or vessels storing liquids with a diameter or width less than or equal to 5 ft (1.5 m), the base shear and overturning moment shall be calculated as if the tank and the entire contents act as an impulsive mass, using Equation (15.7-5). The convective mass shall be set equal to zero in Equation (15.7-6). The lateral force distribution shall be per API 650, API 620, AWWA D100, AWWA D110, AWWA D115, or ACI 350.3. The requirements of Section 15.7, including Section 15.7.6.2, shall apply.

(b) Tanks or vessels storing liquids with a diameter or width greater than 5 ft (1.5 m) shall be designed to consider the hydrodynamic pressures of the liquid in determining the equivalent lateral forces and lateral force distribution per the applicable reference documents listed in Chapter 23. The requirements of Section 15.7 including 15.7.6.2 shall apply.

15.7.6.2 Design Basis The design of tanks storing liquids shall consider the impulsive and convective (sloshing) effects and their consequences on the tank, foundation, and attached elements. The impulsive component corresponds to the high-frequency amplified response to the lateral ground motion of the tank roof, the shell, and the portion of the contents that moves in unison with the shell. The convective component corresponds to the low-frequency amplified response of the contents in the fundamental sloshing mode. Damping for the convective component shall be 0.5% for the sloshing liquid unless otherwise defined by the reference document. The following definitions shall apply:

D_i = Inside diameter of tank or vessel;
H_L = Design liquid height inside the tank or vessel;
L = Inside length of a rectangular tank, parallel to the direction of the earthquake force being investigated;
N_h = Hydrodynamic hoop force per unit height in the wall of a cylindrical tank or vessel;
T_c = Natural period of the first (convective) mode of sloshing;
T_i = Fundamental period of the tank structure and impulsive component of the content;
V_i = Base shear caused by impulsive component from weight of tank and contents;
V_c = Base shear caused by the convective component of the effective sloshing mass;
y = Distance from base of the tank to level being investigated; and
γ_L = Unit weight of stored liquid.

The seismic base shear is the combination of the impulsive and convective components:

$$V = V_i + V_c \quad (15.7\text{-}4)$$

where

$$V_i = \frac{S_{ai} W_i}{\left(\frac{R}{I_e}\right)} \quad (15.7\text{-}5)$$

and

$$V_c = \frac{S_{ac} I_e}{1.5} W_c \quad (15.7\text{-}6)$$

where

W_i = Impulsive weight (impulsive component of liquid, roof and equipment, shell, bottom, and internal elements);
W_c = Portion of the liquid weight sloshing; and
S_{ai} = Spectral acceleration as a multiplier of gravity, including the site impulsive components at period T_i and 5% damping. For $T_i \le T_s$,

$$S_{ai} = S_{DS} \quad (15.7\text{-}7)$$

For $T_s < T_i \le T_L$,

$$S_{ai} = \frac{S_{D1}}{T_i} \quad (15.7\text{-}8)$$

For $T_i > T_L$,

$$S_{ai} = \frac{S_{D1} T_L}{T_i^2} \quad (15.7\text{-}9)$$

NOTES:

1. Where a reference document is used in which the spectral acceleration for the tank shell and the impulsive component of the liquid are independent of T_i, then $S_{ai} = S_{DS}$.
2. S_{ai} determined from Equations (15.7-8) and (15.7-9) shall not be less than the minimum values required in Section 15.4.1,
Item 2, multiplied by R/I_e.
3. Impulsive and convective seismic forces for tanks are permitted to be combined using the SRSS method in lieu of the direct sum method shown in Section 15.7.6 and its related subsections.

S_{ac} = Spectral acceleration of the sloshing liquid (convective component), based on the sloshing period T_c and 0.5% damping.

For $T_c \le T_L$,

$$S_{ac} = \frac{1.5 S_{D1}}{T_c} \le S_{DS} \quad (15.7\text{-}10)$$

For $T_c > T_L$,

$$S_{ac} = \frac{1.5 S_{D1} T_L}{T_c^2} \quad (15.7\text{-}11)$$

EXCEPTIONS:

1. Where the design spectral acceleration parameter, S_a, determined in accordance with either Section 11.4.5.1 or Chapter 21 is available, the value of S_{ai} determined by Equation (15.7-7), (15.7-8), or (15.7-9) is permitted to be taken as the value of S_a at the period T_i. Where the period T_i is less than the period at which S_a is maximum, the maximum value of S_a shall be used. The value of S_{ai} shall not be less than the minimum values required in Section 15.4.1, Item 2, multiplied by R/I_e. When Exception 1 is invoked, the value of S_{ac} must be determined using Exception 2.
2. Where the design spectral acceleration parameter, S_a, determined in accordance with either Section 11.4.5.1 or Chapter 21 is available, the value of S_{ac} determined by Equation (15.7-10) or (15.7-11) is permitted to be taken as 1.5 times the value of S_a at the period T_c. The value of 1.5 S_{ac} need not exceed the maximum value of S_a. When Exception 2 is invoked, the value of S_{ai} must be determined using Exception 1.
3. For $T_c > 4\,s$, S_{ac} is permitted to be determined by a site-specific study using one or more of the following methods: (a) the procedures in Chapter 21, provided such procedures, which rely on ground-motion attenuation equations for computing response spectra, cover the natural period band containing T_c; (b) ground-motion simulation methods that use seismological models of fault rupture and wave

propagation; and (c) analysis of representative strong-motion accelerogram data with reliable long-period content extending to periods greater than T_c. Site-specific values of S_{ac} shall be based on one-standard-deviation determinations. However, in no case shall the value of S_{ac} be taken as less than the value determined in accordance with Equation (15.7-11) using 50% of the mapped value of T_L from Chapter 22.

The 80% limit on S_a required by Sections 21.3 and 21.4 shall not apply to the determination of site-specific values of S_{ac}, which satisfy the requirements of this exception. In determining the value of S_{ac}, the value of T_L shall not be less than 4 s, where

$$T_c = 2\pi\sqrt{\frac{D}{3.68\ g\tanh\left(\frac{3.68\ H}{D}\right)}} \qquad (15.7\text{-}12)$$

where

D = Tank diameter, ft (m), [ft (m)],
H = Liquid height, ft (m), [ft (m)], and
g = Acceleration caused by gravity in consistent units.

15.7.6.2.1 Distribution of Hydrodynamic and Inertial Forces Unless otherwise required by the appropriate reference document listed in Chapter 23, the method given in ACI 350.3 is permitted to be used to determine the vertical and horizontal distribution of the hydrodynamic and inertial forces on the walls of circular and rectangular tanks.

15.7.6.2.2 Sloshing Sloshing of the stored liquid shall be taken into account in the seismic design of tanks and vessels in accordance with the following requirements:

(a) The height of the sloshing liquid, δ_s, above the product design height shall be computed using Equation (15.7-13):

$$\delta_s = 0.42 D_i I_e S_{ac} \qquad (15.7\text{-}13)$$

For cylindrical tanks, D_i shall be the inside diameter of the tank; for rectangular tanks, D_i shall be replaced by the longitudinal plan dimension of the tank, L, for the direction under consideration.

(b) For tanks in Risk Category IV, the Importance Factor, I_e, used for freeboard determination only, shall be taken as 1.0.

(c) For tanks in Risk Categories I, II, and III, the value of T_L used for freeboard determination is permitted to be set equal to 4 s. The value of the Importance Factor, I_e, used for freeboard determination for tanks in Risk Categories I, II, and III shall be determined from Table 1.5-1.

(d) The effects of sloshing shall be accommodated by means of one of the following:
1. A minimum freeboard in accordance with Table 15.7-3,
2. A roof and supporting structure designed to contain the sloshing liquid in accordance with Subsection (e),
3. Secondary containment is provided to control the product spill, or
4. For open-top tanks or vessels only, an overflow spillway around the tank or vessel perimeter.

EXCEPTION: No minimum freeboard is required for open-top tanks where the following conditions are met:
1. Contained fluid is not toxic, explosive, or highly toxic and has been approved by the Authority Having Jurisdiction as acceptable for product spill.
2. Site-specific product spill prevention, control, and countermeasure plan (SPCC) has been developed and approved by the Authority Having Jurisdiction to properly handle resulting spill. The SPCC shall account for proper site drainage, infiltration, foundation scour, and protection of adjacent facilities from sloshing spill.

(e) If the sloshing is restricted because the freeboard is less than the computed sloshing height, then the roof and supporting structure shall be designed for an equivalent hydrostatic head equal to the computed sloshing height less the freeboard. Also, the design of the tank shall use the confined portion of the convective (sloshing) mass as an additional impulsive mass.

Table 15.7-3. Minimum Required Freeboard.

Value of S_{DS}	Risk Category		
	I or II	III	IV
$S_{DS} < 0.33g$	Not required	Not required	δ_s
$S_{DS} \geq 0.33g$	Not required	$0.7\delta_s$	δ_s

15.7.6.2.3 Equipment and Attached Piping Equipment, piping, and walkways or other appurtenances attached to the structure shall be designed to accommodate the displacements imposed by seismic forces. For piping attachments, see Section 15.7.4.

15.7.6.2.4 Internal Elements The attachments of internal equipment and accessories that are attached to the primary liquid or pressure retaining shell or bottom or that provide structural support for major elements (e.g., a column supporting the roof rafters) shall be designed for the lateral loads caused by the sloshing liquid, in addition to the inertial forces, by a substantiated analysis method.

15.7.6.2.5 Sliding Resistance The transfer of the total lateral shear force between the tank or vessel and the subgrade shall be considered as follows.

(a) For flat-bottom steel tanks, the overall horizontal seismic shear force is permitted to be resisted by friction between the tank bottom and the foundation or subgrade. Storage tanks shall be designed such that sliding does not occur where the tank is full of stored product. The maximum calculated seismic base shear, V, shall not exceed

$$V < W \tan 30° \qquad (15.7\text{-}14)$$

(b) W shall be determined using the effective seismic weight of the tank, roof, and contents after reduction for coincident vertical earthquake. Lower values of the friction factor shall be used if the design of the tank bottom to supporting foundation does not justify the friction value given by Equation (15.7-14) (e.g., leak detection membrane beneath the bottom with a lower friction factor, smooth bottoms, etc.). Alternatively, the friction factor is permitted to be determined by testing in accordance with Section 11.1.4.

(c) No additional lateral anchorage is required for steel tanks designed in accordance with reference documents.

(d) The lateral shear transfer behavior for special tank configurations (e.g., shovel bottoms, highly crowned tank bottoms, or tanks on grillage) can be unique and is beyond the scope of this standard.

15.7.6.2.6 Local Shear Transfer Local transfer of the shear from the roof to the wall and the wall of the tank into the base shall be considered. For cylindrical tanks and vessels, the peak local tangential shear per unit length shall be calculated as

$$v_{max} = \frac{2\,V}{\pi D} \quad (15.7\text{-}15)$$

(a) Tangential shear in flat-bottom steel tanks shall be transferred through the welded connection to the steel bottom. This transfer mechanism is deemed acceptable for steel tanks designed in accordance with the reference documents where $S_{DS} < 1.0g$.
(b) For concrete tanks with a sliding base where the lateral shear is resisted by friction between the tank wall and the base, the friction coefficient value used for design shall not exceed tan 30°.
(c) Fixed-base or hinged-base concrete tanks transfer the horizontal seismic base shear shared by membrane (tangential) shear and radial shear into the foundation. For anchored flexible-base concrete tanks, most of the base shear is resisted by membrane (tangential) shear through the anchoring system, with only insignificant vertical bending in the wall. The connection between the wall and the floor shall be designed to resist the maximum tangential shear.

15.7.6.2.7 Pressure Stability For steel tanks, the internal pressure from the stored product stiffens thin cylindrical shell structural elements subjected to membrane compression forces. This stiffening effect is permitted to be considered in resisting seismically induced compressive forces if they are permitted by the reference document or the Authority Having Jurisdiction.

15.7.6.2.8 Shell Support Steel tanks resting on concrete ringwalls or slabs shall have a uniformly supported annulus under the shell. Uniform support shall be provided by one of the following methods:

(a) Shimming and grouting the annulus;
(b) Using fiberboard or other suitable padding;
(c) Using butt-welded bottom or annular plates resting directly on the foundation; or
(d) Using closely spaced shims (without structural grout), provided the localized bearing loads are considered in the tank wall and foundation to prevent local crippling and spalling.

Mechanically anchored tanks shall be shimmed and grouted. Local buckling of the steel shell for the peak compressive force caused by operating loads and seismic overturning shall be considered.

15.7.6.2.9 Repair, Alteration, or Reconstruction Repairs, modifications, or reconstruction (i.e., cut down and re-erect) of a tank or vessel shall conform to industry standard practice and this standard. For welded steel tanks storing liquids, see API 653 and the applicable reference document listed in Chapter 23. Tanks that are relocated shall be reevaluated for the seismic loads at the new site and the requirements of new construction in accordance with the appropriate reference document and this standard.

15.7.7 Water Storage and Water Treatment Tanks and Vessels

15.7.7.1 Welded Steel Welded steel water storage tanks and vessels shall be designed in accordance with the seismic requirements of AWWA D100, with the following exception:

(a) The seismic design ground motion values shall be determined according to Section 11.4(b). Modify AWWA D100, Section 13.5.4.4, as follows:
13.5.4.4 Freeboard. Sloshing shall be considered in determining the freeboard above the MOL. Freeboard is defined as the distance from the MOL to the lowest level of roof framing. The freeboard provided shall meet the requirements of Table 29. The sloshing wave height. . . .

15.7.7.2 Bolted Steel Bolted steel water storage structures shall be designed in accordance with the seismic requirements of AWWA D103, with the following exceptions:

(a) The seismic design ground motion values shall be determined according to Section 11.4.
(b) For Type 6 tanks, the overturning ratio, J, as determined using AWWA D103, Equation (14-32); shall not exceed 0.785.

15.7.7.3 Reinforced and Prestressed Concrete Reinforced and prestressed concrete tanks shall be designed in accordance with the seismic requirements of AWWA D110, AWWA D115, or ACI 350.3, except that the Importance Factor, I_e, shall be determined according to Section 15.4.1.1; the response modification coefficient, R, shall be taken from Table 15.4-2; the seismic design ground motion values shall be determined according to Section 11.4; and the design input forces for strength design procedures shall be determined using the procedures of ACI 350.3, except that S_{ac} shall be substituted for C_c in ACI 350.3, Section 9.4.2, using Equations (15.7-10) for $T_c \leq T_L$ and (15.7-11) for $T_c > T_L$ from Section 15.7.6.1.

15.7.7.4 Corrugated Steel Corrugated steel water storage tanks, except those assigned to Risk Category I, shall meet the requirements of Sections 15.7.1 through 15.7.6 and the following requirements:

(a) Component allowable stresses and materials shall be in accordance with AWWA D103 for the hot-rolled steel elements of the tank. Minimum thicknesses per AWWA D103 shall not apply.
(b) Component stresses and materials shall be in accordance with AISI S100 for the cold-formed steel elements and connections of the tank.
(c) Vertical elements shall be added to the tank shell to resist compressive seismic forces. The buckling capacity of the vertical elements shall be determined using AISC 360 for hot-rolled steel elements or AISI S100 for cold-formed steel elements.
(d) The base shear and overturning moment shall be calculated as if the tank and the entire contents act as an impulsive mass using Equation (15.7-5). The convective mass shall be set equal to zero in Equation (15.7-6).
(e) Tank sliding, relative to the foundation, shall be resisted by embedment of the tank shell in the foundation.

Alternatively, the shell is permitted to be confined by a concrete curb designed to transfer the base shear forces from the tank to the foundation.

(f) The tank shall be mechanically anchored to the foundation to resist seismic uplift.

15.7.8 Petrochemical and Industrial Tanks and Vessels Storing Liquids

15.7.8.1 Welded Steel Welded steel flat-bottom, ground-supported petrochemical tanks, industrial tanks, and vessels storing liquids under an internal pressure less than or equal to 2.5 psig (17.2 kPa g) shall be designed in accordance with the seismic requirements of API 650. Welded steel flat-bottom, ground-supported petrochemical tanks, industrial tanks, and vessels storing liquids under an internal pressure greater than 2.5 psig (17.2 kPa g) and less than or equal to 15 psig (104.4 kPa g) shall be designed in accordance with the seismic requirements of API 620.

15.7.8.2 Bolted Steel Bolted steel tanks are used for storage of production liquids. API 12B covers the material, design, and erection requirements for vertical, cylindrical, and above-ground bolted tanks in nominal capacities of 100 barrels (15.9 m^3) to 10,000 barrels (1590 m^3) for production service. Unless required by the Authority Having Jurisdiction, these temporary structures need not be designed for seismic loads. If design for seismic load is required, the loads are permitted to be adjusted for the temporary nature of the anticipated service life.

15.7.8.3 Reinforced and Prestressed Concrete Reinforced concrete tanks for the storage of petrochemical and industrial liquids shall be designed in accordance with the force requirements of Section 15.7.7.3.

15.7.8.4 Corrugated Steel Corrugated steel petroleum and industrial liquid storage tanks, except those assigned to Risk Category I, shall meet the requirements of Section 15.7.7.4.

15.7.8.5 Reinforced Thermoset Plastic and Fiber-Reinforced Plastic Reinforced thermoset plastic and fiber-reinforced plastic tanks shall be designed in accordance with the requirements of ASME RTP-1, with the following exceptions:

(a) The seismic design ground motion values shall be determined according to Section 11.4.
(b) The seismic design base shear and overturning moment shall be determined using the seismic provisions of API 650, Annex E.
(c) ASME RTP-1, Nonmandatory Appendix NM-3, shall not be used.
(d) Where anchors in concrete or masonry are required to resist seismic forces, anchorage shall be in accordance with Section 15.4.9.

15.7.9 Ground-Supported Storage Tanks for Granular Materials

15.7.9.1 General The intergranular behavior of the material shall be considered in determining effective mass and load paths, including the following behaviors:

(a) Increased lateral pressure (and the resulting hoop stress) caused by loss of the intergranular friction of the material during seismic shaking;
(b) Increased hoop stresses generated from temperature changes in the shell after the material has been compacted; and
(c) Intergranular friction, which can transfer seismic shear directly to the foundation.

15.7.9.2 Lateral Force Determination The lateral forces for tanks and vessels storing granular materials at grade shall be determined by the requirements and accelerations for short-period structures (i.e., S_{DS}).

15.7.9.3 Force Distribution to Shell and Foundation

15.7.9.3.1 Increased Lateral Pressure The increase in lateral pressure on the tank wall shall be added to the static design lateral pressure but shall not be used in the determination of pressure stability effects on the axial buckling strength of the tank shell.

15.7.9.3.2 Effective Mass A portion of a stored granular mass acts with the shell (the effective mass). The effective mass is related to the physical characteristics of the product, the height-to-diameter ratio (H/D) of the tank, and the intensity of the seismic event. The effective mass shall be used to determine the shear and overturning loads resisted by the tank.

15.7.9.3.3 Effective Density The effective density factor (that part of the total stored mass of product that is accelerated by the seismic event) shall be determined in accordance with ACI 313.

15.7.9.3.4 Lateral Sliding For granular storage tanks that have a steel bottom and are supported such that friction at the bottom-to-foundation interface can resist lateral shear loads, no additional anchorage to prevent sliding is required. For tanks without steel bottoms (i.e., the material rests directly on the foundation), shear anchorage shall be provided to prevent sliding.

15.7.9.3.5 Combined Anchorage Systems If separate anchorage systems are used to prevent overturning and sliding, the relative stiffness of the systems shall be considered in determining the load distribution.

15.7.9.4 Welded Steel Structures Welded steel granular storage structures shall be designed in accordance with the seismic requirements of this standard. Component allowable stresses and materials shall be per AWWA D100, except that the allowable circumferential membrane stresses and material requirements in API 650 shall apply.

15.7.9.5 Bolted Steel Structures Bolted steel granular storage structures shall be designed in accordance with the seismic requirements of this section. Component allowable stresses and materials shall be per AWWA D103.

15.7.9.6 Reinforced Concrete Structures Reinforced concrete structures for the storage of granular materials shall be designed in accordance with the seismic force requirements of this standard and the requirements of ACI 313.

15.7.9.7 Prestressed Concrete Structures Prestressed concrete structures for the storage of granular materials shall be designed in accordance with the seismic force requirements of this standard and the requirements of ACI 313.

15.7.10 Elevated Tanks and Vessels for Liquids and Granular Materials

15.7.10.1 General This section applies to tanks, vessels, bins, and hoppers that are elevated above grade and where the supporting tower is an integral part of the structure. Tanks and vessels that are supported by another structure are considered mechanical equipment and shall be designed in accordance with Section 15.3.

Elevated tanks shall be designed for the force and displacement requirements of the applicable reference document or Section 15.4.

15.7.10.2 Effective Mass The design of the supporting tower or pedestal, anchorage, and foundation for seismic overturning shall assume that the stored material is a rigid mass acting at the volumetric center of gravity. The effects of fluid–structure interaction are permitted to be considered in determining the forces, effective period, and mass centroids of the system if the following requirements are met:

(a) The sloshing period, T_c, is greater than $3T$, where T is the natural period of the tank with confined liquid (rigid mass) and supporting structure, and
(b) The sloshing mechanism (i.e., the percentage of convective mass and centroid) is determined for the specific configuration of the container by detailed fluid–structure interaction analysis or testing.

Soil–structure interaction is permitted to be included in determining T, provided the requirements of Chapter 19 are met.

15.7.10.3 P-Delta Effects The lateral drift of the elevated tank shall be considered as follows:

(a) The design drift, as determined by an elastic analysis, shall be increased by the factor C_d/I_e for evaluating the additional load in the support structure.
(b) The base of the tank shall be assumed to be fixed rotationally and laterally.
(c) Deflections caused by bending, axial tension, or compression shall be considered. For pedestal tanks with a height-to-diameter ratio less than 5, shear deformations of the pedestal shall be considered.
(d) The dead load effects of roof-mounted equipment or platforms shall be included in the analysis.
(e) Initial tilt need not be considered in the P-delta analysis if constructed within the plumbness tolerances specified by the reference document.

15.7.10.4 Transfer of Lateral Forces into Support Tower For post-supported tanks and vessels that are cross-braced:

(a) The bracing shall be installed in such a manner as to provide uniform resistance to the lateral load (e.g., pretensioning or tuning to attain equal sag).
(b) The additional load in the brace caused by the eccentricity between the post-to-tank attachment and the line of action of the bracing shall be included.
(c) Eccentricity of compression strut line of action (elements that resist the tensile pull from the bracing rods in the seismic force-resisting systems) with their attachment points shall be considered.
(d) The connection of the post or leg with the foundation shall be designed to resist both the vertical and lateral resultant from the yield load in the bracing, assuming that the direction of the lateral load is oriented to produce the maximum lateral shear at the post-to-foundation interface. Where multiple rods are connected to the same location, the anchorage shall be designed to resist the concurrent tensile loads in the braces.

15.7.10.5 Evaluation of Structures Sensitive to Buckling Failure Shell structures that support substantial loads may exhibit a primary mode of failure from localized or general buckling of the support pedestal or skirt caused by seismic loads. Such structures may include single-pedestal water towers, skirt-supported process vessels, and similar single-member towers. Where the structural assessment concludes that buckling of the support is the governing primary mode of failure, structures specified in this standard to be designed to Subsections (a) and (b) and those that are assigned as Risk Category IV shall be designed to resist the seismic forces as follows:

(a) The seismic response coefficient for this evaluation shall be in accordance with Section 12.8.1.1 of this standard, with I_e/R set equal to 1.0. Soil–structure and fluid–structure interactions are permitted to be used in determining the structural response. Vertical or orthogonal combinations need not be considered.
(b) The resistance of the structure shall be defined as the critical buckling resistance of the element, that is, a factor of safety set equal to 1.0.

15.7.10.6 Welded Steel Water Storage Structures Welded steel elevated water storage structures shall be designed and detailed in accordance with the seismic requirements of AWWA D100, with the structural height limits imposed by Table 15.4-2.

15.7.10.7 Concrete Pedestal (Composite) Tanks Concrete pedestal (composite) elevated water storage structures shall be designed in accordance with the requirements of AWWA D107, except that the seismic design ground motion values shall be determined according to Section 11.4.

15.7.11 Boilers and Pressure Vessels

15.7.11.1 General Attachments to the pressure boundary, supports, and seismic force-resisting anchorage systems for boilers and pressure vessels shall be designed to meet the force and displacement requirements of Section 15.3 or 15.4 and the additional requirements of this section. Boilers and pressure vessels categorized as Risk Category III or IV shall be designed to meet the force and displacement requirements of Section 15.3 or 15.4.

15.7.11.2 ASME Boilers and Pressure Vessels Boilers or pressure vessels designed and constructed in accordance with the ASME *Boiler and Pressure Vessel Code* (BPVC) shall be deemed to meet the requirements of this section, provided the force and displacement requirements of Section 15.3 or 15.4 are used, with appropriate scaling of the force and displacement requirements to the working stress design basis.

15.7.11.3 Attachments of Internal Equipment and Refractory Attachments to the pressure boundary for internal and external ancillary components (refractory, cyclones, trays, etc.) shall be designed to resist the seismic forces specified in this standard to safeguard against rupture of the pressure boundary. Alternatively, the element attached is permitted to be designed to fail before damaging the pressure boundary, provided the consequences of the failure do not place the pressure boundary in jeopardy. For boilers or vessels containing liquids, the effect of sloshing on the internal equipment shall be considered if the equipment can damage the integrity of the pressure boundary.

15.7.11.4 Coupling of Vessel and Support Structure Where the mass of the operating vessel or vessels supported is greater than 20% of the total mass of the combined structure, the structure and vessel designs shall consider the effects of dynamic coupling between each other. Coupling with adjacent, connected structures such as multiple towers shall be considered if the structures are interconnected with elements that transfer loads from one structure to the other.

15.7.11.5 Effective Mass Fluid–structure interaction (sloshing) shall be considered in determining the effective mass of the stored material, provided sufficient liquid surface exists for sloshing to occur and that T_c is greater than $3T$. Changes to or variations in material density with pressure and temperature shall be considered.

15.7.11.6 Other Boilers and Pressure Vessels Boilers and pressure vessels designated as being in Risk Category IV, but not designed and constructed in accordance with the requirements of the ASME BPVC, shall meet the following requirements.

The seismic loads, in combination with other service loads and appropriate environmental effects, shall not exceed the material strength shown in Table 15.7-4.

Consideration shall be made to mitigate seismic impact loads for boiler or vessel elements constructed of nonductile materials or vessels operated in such a way that material ductility is reduced (e.g., low-temperature applications).

15.7.11.7 Supports and Attachments for Boilers and Pressure Vessels Attachments to the pressure boundary and support for boilers and pressure vessels shall meet the following requirements:

(a) Attachments and supports transferring seismic loads shall be constructed of ductile materials suitable for the intended application and environmental conditions.
(b) Anchorage shall be in accordance with Section 15.4.9, whereby the anchor embedment into the concrete is designed to develop the steel strength of the anchor in tension. The steel strength of the anchor in tension shall be determined in accordance with ACI 318, Equation (17.6.1.2). The anchor shall have a minimum gauge length of eight diameters. The load combinations including overstrength of Section 12.4.3 are not to be used to size the anchor bolts for tanks and horizontal and vertical vessels.
(c) Seismic supports and attachments to structures shall be designed and constructed so that the support or attachment remains ductile throughout the range of reversing seismic lateral loads and displacements.
(d) Vessel attachments shall consider the potential effect on the vessel and the support for uneven vertical reactions based on variations in relative stiffness of the support members, dissimilar details, nonuniform shimming, or irregular supports. Uneven distribution of lateral forces shall consider the relative distribution of the resisting elements, the behavior of the connection details, and vessel shear distribution.

The requirements of Sections 15.4 and 15.7.10.5 shall also be applicable to this section.

Table 15.7-4. Maximum Material Strength.

Material	Minimum Ratio F_u/F_y	Max. Material Strength of Vessel Material (%)	Max. Material Strength of Threaded Material (%)[a]
Ductile (e.g., steel, aluminum, copper)	1.33[b]	90[c]	70[c]
Semiductile	1.2[d]	70[c]	50[c]
Nonductile (e.g., cast iron, ceramics, fiberglass)	N/A	25[e]	20[e]

[a] Threaded connection to vessel or support system.
[b] Minimum 20% elongation per the ASTM material specification.
[c] Based on material minimum specified yield strength.
[d] Minimum 15% elongation per the ASTM material specification.
[e] Based on material minimum specified tensile strength.
Note: N/A = not applicable.

15.7.12 Liquid and Gas Spheres

15.7.12.1 General Attachments to the pressure or liquid boundary, supports, and seismic force-resisting anchorage systems for liquid and gas spheres shall be designed to meet the force and displacement requirements of Section 15.3 or 15.4 and the additional requirements of this section. Spheres categorized as Risk Category III or IV shall themselves be designed to meet the force and displacement requirements of Section 15.3 or 15.4.

15.7.12.2 ASME Spheres Spheres designed and constructed in accordance with Section VIII of the ASME BPVC shall be deemed to meet the requirements of this section, provided the force and displacement requirements of Section 15.3 or 15.4 are used with appropriate scaling of the force and displacement requirements to the working stress design basis.

15.7.12.3 Attachments of Internal Equipment and Refractory Attachments to the pressure or liquid boundary for internal and external ancillary components (refractory, cyclones, trays, etc.) shall be designed to resist the seismic forces specified in this standard to safeguard against rupture of the pressure boundary. Alternatively, the element attached to the sphere could be designed to fail before damaging the pressure or liquid boundary, provided the consequences of the failure do not place the pressure boundary in jeopardy. For spheres containing liquids, the effect of sloshing on the internal equipment shall be considered if the equipment can damage the pressure boundary.

15.7.12.4 Effective Mass Fluid–structure interaction (sloshing) shall be considered in determining the effective mass of the stored material, provided sufficient liquid surface exists for sloshing to occur and T_c is greater than $3T$. Changes to or variations in fluid density shall be considered.

15.7.12.5 Post- and Rod-Supported Spheres For post-supported spheres that are cross-braced,

(a) The requirements of Section 15.7.10.4 shall also be applicable to this section.
(b) The stiffening effect (reduction in lateral drift) from pretensioning of the bracing shall be considered in determining the natural period.
(c) The slenderness and local buckling of the posts shall be considered.
(d) Local buckling of the sphere shell at the post attachment shall be considered.
(e) For spheres storing liquids, bracing connections shall be designed and constructed to develop the minimum published yield strength of the brace. For spheres storing gas vapors only, bracing connection shall be designed for Ω_0 times the maximum design load in the brace. Lateral

bracing connections directly attached to the pressure or liquid boundary are prohibited.

15.7.12.6 Skirt-Supported Spheres For skirt-supported spheres, the following requirements shall apply:

(a) The requirements of Section 15.7.10.5 shall also apply.
(b) The local buckling of the skirt under compressive membrane forces caused by axial load and bending moments shall be considered.
(c) Penetration of the skirt support (manholes, piping, etc.) shall be designed and constructed to maintain the strength of the skirt without penetrations.

15.7.13 Refrigerated Gas Liquid Storage Tanks and Vessels

15.7.13.1 General Tanks and facilities for the storage of liquefied hydrocarbons and refrigerated liquids shall meet the requirements of this standard. Low-pressure welded steel flat-bottom, ground-supported storage tanks for liquefied hydrocarbon gas (e.g., liquefied petroleum gas or butane) and refrigerated liquids (e.g., ammonia) shall be designed in accordance with the requirements of Section 15.7.8 and API 620.

15.7.14 Horizontal, Saddle-Supported Vessels for Liquid or Vapor Storage

15.7.14.1 General Horizontal vessels supported on saddles (sometimes referred to as "blimps") shall be designed to meet the force and displacement requirements of Section 15.3 or 15.4.

15.7.14.2 Effective Mass Changes to or variations in material density shall be considered. The design of the supports, saddles, anchorage, and foundation for seismic overturning shall assume that the material stored is a rigid mass acting at the volumetric center of gravity.

15.7.14.3 Vessel Design Unless a more rigorous analysis is performed,

(a) Horizontal vessels with a length-to-diameter ratio of 6 or more are permitted to be assumed to be a simply supported beam spanning between the saddles for determining the natural period of vibration and global bending moment.
(b) For horizontal vessels with a length-to-diameter ratio of less than 6, the effects of "deep beam shear" shall be considered where determining the fundamental period and stress distribution.
(c) Local bending and buckling of the vessel shell at the saddle supports caused by seismic load shall be considered. The stabilizing effects of internal pressure shall not be considered to increase the buckling resistance of the vessel shell.
(d) If the vessel is a combination of liquid and gas storage, the vessel and supports shall be designed (both with and without gas pressure acting assume that piping has ruptured and pressure does not exist).

15.8 CONSENSUS STANDARDS AND OTHER REFERENCED DOCUMENTS

See Chapter 23 for the list of consensus standards and other documents that shall be considered part of this standard to the extent referenced in this chapter.

CHAPTER 16
NONLINEAR RESPONSE HISTORY ANALYSIS

16.1 GENERAL REQUIREMENTS

16.1.1 Scope It shall be permitted to use nonlinear response history analysis, in accordance with the requirements of this chapter, to demonstrate acceptable strength, stiffness, and ductility to resist maximum considered earthquake (MCE_R) shaking with acceptable performance. When nonlinear response history analysis is performed, the design shall also satisfy the requirements of Section 16.1.2. Nonlinear response history analysis shall include the effects of horizontal motion, and where required by Section 16.1.3, vertical motion. Documentation of the design and analysis shall be prepared in accordance with Section 16.1.4. Ground motion acceleration histories shall be selected and modified in accordance with the procedures of Section 16.2. The structure shall be modeled and analyzed in accordance with the criteria in Section 16.3. Analysis results shall meet the acceptance criteria of Section 16.4. Independent structural design review shall be performed in accordance with the requirements of Section 16.5.

16.1.2 Linear Analysis In addition to nonlinear response history analysis, a linear analysis in accordance with one of the applicable procedures of Chapter 12 shall also be performed. The structure's design shall meet all applicable criteria of Chapter 12.

EXCEPTIONS:

1. For Risk Category I, II, and III structures, Sections 12.12.1 and 12.12.5 do not apply to the linear analysis. Where mean computed drifts from the nonlinear analyses exceed 150% of the permissible story drifts per Section 12.12.1, deformation-sensitive nonstructural components shall be designed for 2/3 of these mean drifts.
2. The overstrength factor, Ω_0, is permitted to be taken as 1.0 for the seismic load effects of Section 12.4.3.
3. The redundancy factor, ρ, is permitted to be taken as 1.0.
4. Where accidental torsion is explicitly modeled in the nonlinear analysis, it shall be permitted to take the value of A_x as unity in the Chapter 12 analysis.

16.1.3 Vertical Response Analysis Nonlinear response history analysis shall explicitly include the effects of vertical response where any of the following occur:

1. Vertical elements of the gravity force-resisting system are discontinuous.
2. For nonbuilding structures, when Chapter 15 requires consideration of vertical earthquake effects.

16.1.4 Documentation Before performing the nonlinear analysis, project-specific design criteria shall be approved by the independent structural design reviewer(s) and the Authority Having Jurisdiction. The project-specific criteria shall identify the following:

1. The selected seismic and gravity force-resisting systems and procedures used in the structural design.
2. Geotechnical parameters including soil characteristics, recommended foundation types, design parameters, seismic hazard evaluation, target spectra, and selection and scaling of acceleration histories.
3. Design loading, including gravity and environmental loads.
4. Analytical modeling approach and assumptions, including software to be used, definition of mass, identification of force-controlled and deformation-controlled behaviors, description of which component actions are modeled elastically and inelastically, expected material properties, basis for hysteretic component modeling, component initial stiffness assumptions, joint stiffness assumptions, diaphragm modeling, damping, and procedure for modeling foundation–soil interaction.
5. Summaries of laboratory test and other applicable data used to justify the hysteretic component modeling or used to justify acceptable structural performance.
6. Specific acceptance criteria values used for evaluating performance of elements of the seismic force-resisting system. Associated documentation shall also include identification of component failure modes deemed indicative of collapse.
7. Where drifts exceed 150% of the values permitted in Section 12.12, the criteria used to demonstrate acceptable deformation compatibility of components of the gravity force-resisting system.

Following completion of the analysis process, the following documentation shall be prepared and presented to the independent structural design reviewer(s) and the Authority Having Jurisdiction:

1. Final geotechnical report, including soil shear strength, stiffness, and damping characteristics; recommended foundation types and design parameters; and seismic hazard evaluation, including both the target spectra and selection and scaling of ground motions.
2. Overall building dynamic behavior, including natural frequencies, mode shapes, and modal mass participation.
3. Key structural system response parameter results and comparisons with the acceptance criteria of Section 16.4.
4. Detailing of critical elements.

16.2 GROUND MOTIONS

16.2.1 Target Response Spectrum A target, 5%-damped, MCE_R response spectrum shall be developed using either the procedures of Section 16.2.1.1 or Section 16.2.1.2. It shall be permitted to consider the effects of base slab averaging and foundation embedment in accordance with Chapter 19.

Where the effects of vertical earthquake shaking are included in the analysis, a target MCE_R vertical spectrum shall also be constructed.

16.2.1.1 Method 1 A single target response spectrum shall be developed, based on the requirements of either Section 11.4.6 or Section 11.4.7.

16.2.1.2 Method 2 Two or more site-specific target response spectra shall be developed. When this method is used, the following requirements shall be fulfilled, in addition to the other requirements of this chapter:

1. Two or more periods shall be selected, corresponding to those periods of vibration that significantly contribute to the inelastic dynamic response of the building in two orthogonal directions. In the selection of periods, lengthening of the elastic periods of the model shall be considered.
2. For each selected period, a target spectrum shall be created that either matches or exceeds the MCE_R value at that period. When developing the target spectrum, (1) site-specific disaggregation shall be performed to identify earthquake events that contribute most to the MCE_R ground motion at the selected period and (2) the target spectrum shall be developed to capture one or more spectral shapes for dominant magnitude and distance combinations revealed by the disaggregation.
3. The envelope of the target spectra shall not be less than 75% of the spectral values computed using Method 1 of Section 16.2.1.1, for all periods in the range specified in Section 16.2.3.1.
4. For each target response spectrum, a ground motion suite for response history analyses shall be developed and used in accordance with Sections 16.2.3 through 16.2.4. The acceptance criteria requirements of Section 16.4 shall be evaluated independently for each of the ground motion suites.

Variations on the procedures described in this section are permitted to be used when approved by the design review.

16.2.2 Ground Motion Selection A suite of not less than 11 ground motions shall be selected for each target spectrum. Ground motions shall consist of pairs of orthogonal horizontal ground motion components and, where vertical earthquake effects are considered, a single vertical ground motion component. Ground motions shall be selected from events within the same general tectonic regime and having generally consistent magnitudes and fault distances as those controlling the target spectrum and shall have a spectral shape similar to the target spectrum. For near-fault sites, as defined in Section 11.4.1, and other sites where MCE_R shaking can exhibit directionality and impulsive characteristics, the proportion of ground motions with near-fault and rupture directivity effects shall represent the probability that MCE_R shaking will exhibit these effects. Where the required number of recorded ground motions is not available, it shall be permitted to supplement the available records with simulated ground motions. Ground motion simulations shall be consistent with the magnitudes, source characteristics, fault distances, and site conditions controlling the target spectrum.

16.2.3 Ground Motion Modification Ground motions shall either be amplitude-scaled in accordance with the requirements of Section 16.2.3.2 or spectrally matched in accordance with the requirements of Section 16.2.3.3. Spectral matching shall not be used for near-fault sites unless the pulse characteristics of the ground motions are retained after the matching process has been completed.

16.2.3.1 Period Range for Scaling or Matching A period range shall be determined, corresponding to the vibration periods that contribute significantly to the building's lateral dynamic response. This period range shall have an upper bound, greater than or equal to twice the largest first-mode period in the principal horizontal directions of response, unless a lower value, not less than 1.5 times the largest first-mode period, is justified by dynamic analysis under MCE_R ground motions. The lower bound period shall be established such that the period range includes at least the number of elastic modes necessary to achieve 90% mass participation in each principal horizontal direction. The lower bound period shall not exceed 20% of the smallest first-mode period for the two principal horizontal directions of response. Where vertical response is considered in the analysis, the lower bound period used for modification of vertical components of ground motion need not be taken as less than the larger of 0.1 seconds, or the lowest period at which significant vertical mass participation occurs.

16.2.3.2 Amplitude Scaling For each horizontal ground motion pair, a maximum-direction spectrum shall be constructed from the two horizontal ground motion components. Each ground motion shall be scaled, with an identical scale factor applied to both horizontal components, such that the average of the maximum-direction spectra from all ground motions generally matches or exceeds the target response spectrum over the period range defined in Section 16.2.3.1. The average of the maximum-direction spectra from all the ground motions shall not fall below 90% of the target response spectrum for any period within the same period range. Where vertical response is considered in the analysis, the vertical component of each ground motion shall be scaled such that the average of the vertical response spectra envelops the target vertical response spectrum over the period range specified in Section 16.2.3.1.

16.2.3.3 Spectral Matching Each pair of ground motions shall be modified such that the average of the maximum-direction spectra for the suite equals or exceeds 110% of the target spectrum over the period range defined in Section 16.2.3.1. Where vertical response is considered in the analysis, the vertical component of each ground motion shall be spectrally matched to the target vertical response spectrum, such that the average of the matched spectra does not fall below the target vertical spectrum in the scaling range of Section 16.2.3.1.

16.2.4 Application of Ground Motions to the Structural Model Ground motions shall be applied to the supports of the structural model. For near-fault sites, as defined in Section 11.4.1, each pair of horizontal ground motion components representative of a nearby fault source shall be rotated to the fault-normal and fault-parallel directions of the causative fault and applied to the building in such orientation. For all other selected ground motions at near-fault sites, and for all ground motions at other sites, each pair of horizontal ground motion components shall be applied to the building at orthogonal orientations, such that the average (or mean) of the component response spectrum for the records applied in each direction is within ±10% of the mean of the component response spectra of all records applied for the period range specified in Section 16.2.3.1.

16.3 MODELING AND ANALYSIS

16.3.1 Modeling Mathematical models shall be three-dimensional and shall conform to the requirements of this section and Section 12.7. For structures that have subterranean levels, the structural model shall extend to the foundation level and ground motions shall be input at the foundation level. All elements that significantly affect seismic response when subjected to MCE_R ground motions shall be included. Modeling of element nonlinear hysteretic behavior shall be consistent with ASCE 41 or applicable laboratory test data. Test data shall not be extrapolated beyond tested deformation levels. Degradation in element strength or stiffness shall be included in the hysteretic models unless it can be demonstrated that response is not sufficient to produce these effects.

Analysis models shall be capable of representing the flexibility of floor diaphragms where this is significant to the structure's response. Diaphragms at horizontal and vertical discontinuities in lateral resistance shall be explicitly modeled in a manner that permits capturing the force transfers and resulting deformations.

16.3.2 Gravity Load The modeling of, and demands on, elements in the analysis model shall be determined considering earthquake effects acting in combination with expected gravity loads, both with and without live load. Gravity loads with live load shall be taken as $1.0D + 0.5L$, where L shall be taken as 80% of unreduced live loads that exceed $100\,\text{lb/ft}^2$ $(4.79\,\text{kN/m}^2)$ and 40% of all other unreduced live loads. Gravity loads without live load shall be taken as $1.0D$.

EXCEPTION: Where the sum, over the entire structure, of the expected live load ($0.5L$), as defined above, does not exceed 25% of the total dead load, D, and the live load intensity, L_0, over at least 75% of the structure is less than 100 psf $(4.79\,\text{kN/m}^2)$, the case without live load need not be considered.

16.3.3 P-Delta Effects P-delta effects considering the spatial distribution of gravity loads shall be included in the analysis.

16.3.4 Torsion Inherent eccentricity resulting from any offset in the centers of mass and stiffness at each level shall be accounted for in the analysis. In addition, where a Type 1 horizontal structural irregularly exists, as defined in Section 12.3.2.1, accidental eccentricity consisting of an assumed displacement of the center of mass each way from its actual location by a distance equal to 5% of the diaphragm dimension of the structure parallel to the direction of mass shift shall be considered. The required 5% displacement of the center of mass need not be applied in both orthogonal directions at the same time.

16.3.5 Damping Hysteretic energy dissipation of structural members shall be modeled directly. Additional inherent damping, not associated with inelastic behavior of elements, shall be modeled appropriate to the structure type and shall not exceed 2.5% equivalent viscous damping in the significant modes of response.

16.3.6 Explicit Foundation Modeling When soil spring and/or dashpot elements are included in the structural model, horizontal input ground motions shall be applied to the horizontal soil elements rather than being applied to the foundation directly.

16.4 ANALYSIS RESULTS AND ACCEPTANCE CRITERIA

Structures shall be demonstrated to meet the global acceptance criteria of Section 16.4.1 and the element-level acceptance criteria of Section 16.4.2.

The mean value of story drift, and element demand, Q_u, shall be used to evaluate acceptability.

EXCEPTION: Where a ground motion produces unacceptable response as permitted in Section 16.4.1.1, 120% of the median value, but not less than the mean value obtained from the suite of analyses producing acceptable response shall be used.

16.4.1 Global Acceptance Criteria

16.4.1.1 Unacceptable Response Unacceptable response to ground motion shall consist of any of the following:

1. Analytical solution fails to converge,
2. Predicted demands on deformation-controlled elements exceed the valid range of modeling,
3. Predicted demands on critical or ordinary force-controlled elements, as defined in Section 16.4.2, exceed the element capacity,
4. Predicted deformation demands on elements not explicitly modeled exceed the deformation limits at which the members are no longer able to carry their gravity loads,
5. Peak transient story drift ratio exceeds 150% of the permissible value of mean transient story drift, as per Section 16.4.1.2, or
6. For structures exceeding 240 ft (73m) in height, the residual story drift for any story exceeds a value of $0.015\,h_{sx}$.

Unacceptable response to ground motion shall not be permitted.

EXCEPTION: For Risk Category I and II structures, where spectral matching of ground motion is not used, not more than one motion shall be permitted to produce unacceptable response.

16.4.1.2 Transient Story Drift The mean transient story drift for all building heights, $\overline{\Delta}$ in ft (m), shall not exceed two times the limits of Table 12.12-1. Additionally, for structures exceeding 100 ft (30 m) in height, $\overline{\Delta}$ shall not exceed the value obtained from Equation (16.4-1):

$$\overline{\Delta} \le h_{sx}(4.71x10^{-2} - 7.14x10^{-5}h_n) \quad (16.4\text{-}1)$$

$$\overline{\Delta} \le h_{sx}(4.71x10^{-2} - 2.34x10^{-4}h_n) \quad (16.4\text{-}1.SI)$$

where h_n and h_{sx} are measured in ft (m). The value obtained from Equation (16.4-1) need not be taken less than $0.03\,h_n$.

The transient story drift ratio shall be computed as the absolute value of the largest difference of the deflections of vertically aligned points at the top and bottom of the story under consideration along any of the edges of the structure within a single response history analysis. For masonry shear wall structures, the limits of Table 12.12-1 applicable to masonry cantilever wall structures and other masonry wall structures shall not apply and these structures shall instead comply with the limits for other structures.

16.4.1.3 Residual Story Drift For structures exceeding 240 ft (73 m) in height, the mean residual story drift shall not exceed $0.01\,h_{sx}$, where residual story drift is taken as the maximum value of story drift in a structure at rest, following response to an earthquake motion.

16.4.2 Element-Level Acceptance Criteria All element actions shall be classified either as force-controlled or deformation-controlled, in accordance with ACI 318 for reinforced concrete elements or ASCE 41 for elements of other materials.

For each element action, the quantity, Q_u, shall be computed. Q_u shall be taken as the mean value of the response parameter of interest obtained from the suite of analyses.

Force-controlled actions shall be evaluated for acceptability in accordance with Section 16.4.2.1. Deformation-controlled actions shall be evaluated for acceptability in accordance with Section 16.4.2.2. Where required by Section 16.4.2.1, element actions shall be categorized as Critical, Ordinary, or Noncritical.

16.4.2.1 Force-Controlled Actions Force-controlled actions shall satisfy Equation (16.4-1) and (16.4-2):

$$(1.2 + 0.12S_{MS})D + 0.5L + 1.3I_e(Q_u - Q_{ns}) \leq \phi BR_n \quad (16.4\text{-}1)$$

$$(0.9 - 0.12S_{MS})D + 1.3I_e(Q_u - Q_{ns}) \leq \phi BR_n \quad (16.4\text{-}2)$$

where D and L are as defined in Section 16.3.2, S_{MS} is the site-adjusted Maximum Considered Earthquake Spectral Acceleration at a period of 0.2 seconds; I_e is the Importance Factor prescribed in Section 1.5.1; Q_u is the mean value of the demand computed from the suite of analyses; Q_{ns} is the portion of the demand caused by loads other than seismic; R_n is the nominal strength specified by the applicable material standard. The resistance factor ϕ for Critical elements shall be taken as the value specified by the applicable material standard. The resistance factor ϕ for Ordinary elements shall be taken as 0.9. The resistance factor ϕ for Noncritical elements shall be taken as 1.0. B is a factor to account for differences between expected strength R_{ne} and nominal resistance R_n. It is permitted to assign B a value of 1.0, or, alternatively, B can be taken as $0.9R_{ne}/R_n$, where R_{ne} is the expected strength of the element.

Where an industry standard referenced in Chapter 14 defines expected strength, that value shall be used. Where this is not defined, it shall be permitted to calculate expected strength as the nominal strength defined in industry standards, except that expected material properties as defined in ACI 318 for reinforced concrete elements and ASCE 41 for elements of other materials shall be used in lieu of specified values.

EXCEPTIONS:

1. Noncritical force-controlled actions that are modeled, including consideration of strength loss effects, need not satisfy Equations (16.4-1) or (16.4-2).
2. Force-controlled actions limited by formation of a yield mechanism, other than shear in structural walls, need only satisfy Equations (16.4-3) and (16.4-4):

$$(1.2 + 0.12S_{MS})D + 0.5L + 0.2S + E_{mc} \leq \phi BR_n \quad (16.4\text{-}3)$$

$$(0.9 - 0.12S_{MS})D + E_{mc} \leq \phi BR_n \quad (16.4\text{-}4)$$

   Where E_{mc} is the capacity-limited earthquake effect associated with developing the plastic capacity of yielding components, determined in accordance with the applicable material standard, or alternatively, determined by rational analysis considering expected material properties including strain hardening effects where applicable.
3. Where response to vertical earthquake shaking is directly included in the analysis, the first term in equations 16.4-1 through 16.4-4 can be taken as 1.2D (16.4-1 and 16.4-3) or 0.9D (16.4-2 and 16.4-4).

16.4.2.2 Deformation-Controlled Actions The valid range of modeling for deformation-controlled element actions shall be as established in the applicable material design standard. Where the material design standard does not specify the valid range of modeling, this parameter shall be established as the maximum value of the parameter at which the element model is capable of replicating the hysteretic behavior and load-carrying capability observed in laboratory testing of similar elements. Where suitable test data is not available, either the valid range of modeling of the deformation capacity specified in ACI 318 for reinforced concrete elements or the maximum deformation parameter as specified by ASCE 41 for elements of other materials shall be used. It shall be permitted to extend the valid range of modeling for an element beyond these deformations if the element strength and stiffness are degraded to negligible values once these deformations are reached.

16.4.2.3 Elements of the Gravity Force-Resisting System Elements that are not part of the seismic force-resisting system shall be demonstrated to be capable of supporting gravity loads using the mean building displacements from the suite of nonlinear response history analyses.

16.5 DESIGN REVIEW

An independent structural design review shall be performed in accordance with the requirements of this section. Upon completion of the review, the reviewer(s) shall provide the Authority Having Jurisdiction and the registered design professional with a letter attesting to:

1. Scope of review performed,
2. Whether the reviewer(s) concur with the analysis and its applicability to the design,
3. Conformance of the design to applicable requirements of the standard, and
4. Any items relating to the design or analysis that require further resolution by the Authority Having Jurisdiction.

16.5.1 Reviewer Qualifications Reviewer(s) shall consist of one or more individuals acceptable to the Authority Having Jurisdiction and possessing knowledge of the following items:

1. The requirements of this standard and the standards referenced herein, as they pertain to design of the type of structure under consideration.
2. Selection and scaling of ground motions for use in nonlinear response history analysis.
3. Analytical structural modeling for use in nonlinear response history analysis, including use of laboratory tests in the creation and calibration of the structural analysis models, and including knowledge of soil–structure interaction if used in the analysis or the treatment of ground motions.
4. Behavior of structural systems, of the type under consideration, when subjected to earthquake loading.

At least one reviewer shall be a registered design professional.

16.5.2 Review Scope The scope of review shall include the items identified in Section 16.1.4, as well as the associated project documentation that demonstrates conformance to the design criteria.

16.6 CONSENSUS STANDARDS AND OTHER REFERENCED DOCUMENTS

See Chapter 23 for the list of consensus standards and other documents that shall be considered part of this standard to the extent referenced in this chapter.

CHAPTER 17
SEISMIC DESIGN REQUIREMENTS FOR SEISMICALLY ISOLATED STRUCTURES

17.1 GENERAL

Every seismically isolated structure and every portion thereof shall be designed and constructed in accordance with the requirements of this section and the applicable requirements of this standard.

17.1.1 Definitions The following definitions only apply to the seismically isolated structure provisions of Chapter 17 and are in addition to the definitions presented in Chapter 11.

BASE LEVEL: The first level of the isolated structure above the isolation interface.

DISPLACEMENT RESTRAINT SYSTEM: A collection of structural elements that limits lateral displacement of seismically isolated structures caused by the maximum considered earthquake.

EFFECTIVE DAMPING: The value of equivalent viscous damping corresponding to energy dissipated during cyclic response of the isolation system.

EFFECTIVE STIFFNESS: The value of the lateral force in the isolation system, or an element thereof, divided by the corresponding lateral displacement.

ISOLATION INTERFACE: The boundary between the upper portion of the structure, which is isolated, and the lower portion of the structure, which moves rigidly with the ground.

ISOLATION SYSTEM: The collection of structural elements that includes all individual isolator units, all structural elements that transfer force between elements of the isolation system, and all connections to other structural elements. The isolation system also includes the wind-restraint system, energy-dissipation devices, and/or the displacement restraint system if such systems and devices are used to meet the design requirements of this chapter.

ISOLATOR UNIT: A horizontally flexible and vertically stiff structural element of the isolation system that permits large lateral deformations under design seismic load. An isolator unit is permitted to be used either as part of, or in addition to, the weight-supporting system of the structure.

MAXIMUM DISPLACEMENT: The maximum lateral displacement, excluding additional displacement caused by actual and accidental torsion, required for design of the isolation system. The maximum displacement is to be computed separately using upper bound and lower bound properties.

SCRAGGING: Cyclic loading or working of rubber products, including elastomeric isolators, to effect a reduction in stiffness properties, a portion of which is recovered over time.

TOTAL MAXIMUM DISPLACEMENT: The total maximum lateral displacement, including additional displacement caused by actual and accidental torsion, required for the verification of the stability of the isolation system or elements thereof, design of structure separations, and vertical load testing of isolator unit prototypes. The total maximum displacement is to be computed separately using upper bound and lower bound properties.

WIND-RESTRAINT SYSTEM: The collection of structural elements that provides restraint of the seismically isolated structure for wind loads. The wind-restraint system is permitted to be either an integral part of isolator units or a separate device.

17.1.2 Symbols Symbols presented in this section only apply to the seismically isolated structure provisions of Chapter 17 and are in addition to the symbols presented in Chapter 11.

b = Shortest plan dimension of the structure, ft (mm) measured perpendicular to d

B_M = Numerical coefficient as set forth in Table 17.5-1, for effective damping equal to β_M

C_{vx} = Vertical distribution factor

d = Longest plan dimension of the structure, ft (mm) measured perpendicular to b

D_M = Maximum displacement, in. (mm) at the center of rigidity of the isolation system in the direction under consideration, as prescribed by Equation (17.5-1)

D'_M = Maximum displacement, in. (mm) at the center of rigidity of the isolation system in the direction under consideration, as prescribed by Equation (17.6-1)

D_{TM} = Total maximum displacement, in. (mm) of an element of the isolation system including both translational displacement at the center of rigidity and the component of torsional displacement in the direction under consideration, as prescribed by Equation (17.5-3)

e = Actual eccentricity, ft (mm) measured in plan between the center of mass of the structure above the isolation interface and the center of rigidity of the isolation system, plus accidental eccentricity [ft (mm)] taken as 5% of the maximum building dimension perpendicular to the direction of force under consideration

E_{loop} = Energy dissipated, kip/in. (kN-mm), in an isolator unit during a full cycle of reversible load over a test displacement range from Δ^+ to Δ^-, as measured by the area enclosed by the loop of the force-deflection curve

F^+ = Maximum positive force, kips (kN) in an isolator unit during a single cycle of prototype testing at a displacement amplitude of Δ^+

F^- = Minimum negative force, kips (kN) in an isolator unit during a single cycle of prototype testing at a displacement amplitude of Δ^-

F_x = Lateral seismic force, kips (kN) at level x, as prescribed by Equation (17.5-9)

Table 17.5-1. Damping Factor, B_M.

Effective Damping, β_M (percentage of critical)[a,b]	B_M Factor
≤2	0.8
5	1.0
10	1.2
20	1.5
30	1.7
40	1.9
≥50	2.0

[a] The damping factor shall be based on the effective damping of the isolation system, determined in accordance with the requirements of Section 17.2.8.6.
[b] The damping factor shall be based on linear interpolation for effective damping values other than those given.

h_i, h_l, h_x = Height [ft (m)] above the isolation interface of level i, l, or x
h_{sx} = Height of story below level x
k_{eff} = Effective stiffness [kip/in (kN/mm)] of an isolator unit, as prescribed by Equation (17.8-1)
k_M = Effective stiffness [kip/in. (kN/mm)] of the isolation system in the horizontal direction under consideration
L = Effect of live load in Chapter 17
N = Number of isolator units
P_T = Ratio of the effective translational period of the isolation system to the effective torsional period of the isolation system, as calculated by dynamic analysis or as prescribed by Equation (17.5-4) but need not be taken as less than 1.0
r_I = Radius of gyration of the isolation system [ft (mm)]
R_I = Numerical coefficient related to the type of seismic force-resisting system above the isolation system
T_{fb} = Fundamental period of the structure above the isolation interface, determined using a modal analysis assuming fixed-base conditions, s
T_M = Effective period of the seismically isolated structure at the displacement D_M in the direction under consideration, as prescribed by Equation (17.5-2), s
V_b = Total lateral seismic design force or shear on elements of the isolation system or elements below the isolation system [kips (kN)], as prescribed by Equation (17.5-5)
V_s = Total lateral seismic design force or shear on elements above the base level [kips (kN)], as prescribed by Equation (17.5-6) and the limits of Section 17.5.4.3
V_{st} = Total unreduced lateral seismic design force or shear on elements above the base level [kips (kN)], as prescribed by Equation (17.5-7)
$\overline{V}_s$ = Sheer wave velocity
w_i, w_l, w_x = Portion of W that is located at or assigned to level i, l, or x [kips (kN)]
W = Effective seismic weight [kips (kN)] of the structure above the isolation interface, as defined by Section 12.7.2
W_s = Effective seismic weight [kips (kN)] of the structure above the isolation interface, as defined by Section 12.7.2, excluding the effective seismic weight [kips (kN)] of the base level
x_i, y_i = Horizontal distances [ft (mm)] from the center of mass to the ith isolator unit in the two horizontal axes of the isolation system
y = Distance, ft (mm) between the center of rigidity of the isolation system and the element of interest measured perpendicular to the direction of seismic loading under consideration
β_{eff} = Effective damping of the isolation system, as prescribed by Equation (17.8-2)
β_M = Effective damping of the isolation system at the displacement D_M, as prescribed by Equation (17.2-4)
Δ^+ = Maximum positive displacement, in. (mm) of an isolator unit during each cycle of prototype testing
Δ^- = Minimum negative displacement, in. (mm) of an isolator unit during each cycle of prototype testing
$\lambda_{(ae,max)}$ = Property modification factor for calculation of the maximum value of the isolator property of interest, used to account for aging effects and environmental conditions, as defined in Section 17.2.8.4
$\lambda_{(ae,min)}$ = Property modification factor for calculation of the minimum value of the isolator property of interest, used to account for aging effects and environmental conditions, as defined in Section 17.2.8.4
λ_{max} = Property modification factor for calculation of the maximum value of the isolator property of interest, used to account for all sources of isolator property variability, as defined in Section 17.2.8.4
λ_{min} = Property modification factor for calculation of the minimum value of the isolator property of interest, used to account for all sources of isolator property variability, as defined in Section 17.2.8.4
$\lambda_{(spec,max)}$ = Property modification factor for calculation of the maximum value of the isolator property of interest, used to account for permissible manufacturing variation on the average properties of a group of same-sized isolators, as defined in Section 17.2.8.4
$\lambda_{(spec,min)}$ = Property modification factor for calculation of the minimum value of the isolator property of interest, used to account for permissible manufacturing variation on the average properties of a group of same-sized isolators, as defined in Section 17.2.8.4
$\lambda_{(test,max)}$ = Property modification factor for calculation of the maximum value of the isolator property of interest, used to account for heating, rate of loading, and scragging, as defined in Section 17.2.8.4
$\lambda_{(test,min)}$ = Property modification factor for calculation of the minimum value of the isolator property of interest, used to account for heating, rate of loading, and scragging, as defined in Section 17.2.8.4
ΣE_M = Total energy dissipated [kip/in. (kN-mm)] in the isolation system during a full cycle of response at displacement D_M
$\Sigma|F_D^+|_{max}$ = Sum, for all isolator units, of the maximum absolute value of force [kips (kN)], at a positive displacement equal to D_M
$\Sigma|F_D^-|_{max}$ = Sum, for all isolator units, of the maximum absolute value of force [kips (kN)], at a negative displacement equal to D_M

17.2 GENERAL DESIGN REQUIREMENTS

17.2.1 Importance Factor All portions of the structure, including the structure above the isolation system, shall be assigned a risk category in accordance with Table 1.5-1. The Importance Factor, I_e, shall be taken as 1.0 for a seismically isolated structure, regardless of its risk category assignment.

17.2.2 Configuration Each isolated structure shall be designated as having a structural irregularity if the structural configuration above the isolation system has a Type 1 horizontal structural irregularity with a *TIR* greater than 1.4 or Type 1a, 1b, 4a, 4b vertical irregularity, as defined in Table 12.3-2.

17.2.3 Redundancy A redundancy factor, ρ, shall be assigned to the structure above the isolation system based on requirements of Section 12.3.4. The value of the redundancy factor, ρ, is permitted to be equal to 1.0 for isolated structures that do not have a structural irregularity, as defined in Section 17.2.2.

17.2.4 Isolation System

17.2.4.1 Environmental Conditions In addition to the requirements for vertical and lateral loads induced by wind and earthquake, the isolation system shall provide for other environmental conditions, including aging effects, creep, fatigue, operating temperature, and exposure to moisture or damaging substances.

17.2.4.2 Wind Forces Isolated structures shall resist design wind loads at all levels above the isolation interface. At the isolation interface, a wind-restraint system shall be provided to limit lateral displacement in the isolation system to a value equal to that required between floors of the structure above the isolation interface, in accordance with Section 17.5.6.

17.2.4.3 Fire Resistance Fire resistance for the isolation system shall provide at least the same degree of protection as the fire resistance required for the columns, walls, or other such gravity-bearing elements in the same region of the structure.

17.2.4.4 Lateral Restoring Force The isolation system shall be configured, for both upper bound and lower bound isolation system properties, to produce a restoring force such that the lateral force at the corresponding maximum displacement is at least 0.025 W greater than the lateral force at 50% of the corresponding maximum displacement.

17.2.4.5 Displacement Restraint The isolation system shall not be configured to include a displacement restraint that limits lateral displacement caused by risk-targeted maximum considered earthquake (MCE_R) ground motions to less than the total maximum displacement, D_{TM}, unless the seismically isolated structure is designed in accordance with all of the following criteria:

1. MCE_R response is calculated in accordance with the dynamic analysis requirements of Section 17.6, explicitly considering the nonlinear characteristics of the isolation system and the structure above the isolation system.
2. The ultimate capacity of the isolation system and structural elements below the isolation system shall exceed the strength and displacement demands of the MCE_R response.
3. The structure above the isolation system is checked for stability and ductility demand of the MCE_R response.
4. The displacement restraint does not become effective at a displacement less than 0.6 times the total maximum displacement.

17.2.4.6 Vertical-Load Stability Each element of the isolation system shall be designed to be stable under the design vertical load where it is subjected to a horizontal displacement equal to the total maximum displacement. The design vertical load shall be computed using load combination 2 of Section 17.2.7.1 for the maximum vertical load and load combination 3 of Section 17.2.7.1 for the minimum vertical load.

17.2.4.7 Overturning The factor of safety against global structural overturning at the isolation interface shall not be less than 1.0 for required load combinations. All gravity and seismic loading conditions shall be investigated. Seismic forces for overturning calculations shall be based on MCE_R ground motions, and W shall be used for the vertical restoring force.

Local uplift of individual elements shall not be allowed unless the resulting deflections do not cause overstress or instability of the isolator units or other structure elements.

17.2.4.8 Inspection and Replacement All the following items shall be addressed as part of the long-term inspection and replacement program:

1. Access for inspection and replacement of all components of the isolation system shall be provided.
2. Registered design professional (RDP) shall complete a final series of observations of structure separation areas and components that cross the isolation interface before the issuance of the certificate of occupancy for the seismically isolated structure. Such observations shall verify that conditions allow free and unhindered displacement of the structure up to the total maximum displacement and that components that cross the isolation interface have been constructed to accommodate the total maximum displacement.
3. Seismically isolated structures shall have a monitoring, inspection, and maintenance plan for the isolation system established by the RDP responsible for the design of the isolation system.
4. Remodeling, repair, or retrofitting at the isolation system interface, including that of components that cross the isolation interface, shall be performed under the direction of a RDP.

17.2.4.9 Quality Control A quality control testing program for isolator units shall be established by the RDP responsible for the structural design, incorporating the production testing requirements of Section 17.8.5.

17.2.5 Structural System

17.2.5.1 Horizontal Distribution of Force A horizontal diaphragm or other structural elements shall provide continuity above the isolation interface and shall have adequate strength and ductility to transmit forces from one part of the structure to another.

17.2.5.2 Minimum Building Separations Minimum separations between the isolated structure and surrounding retaining walls or other fixed obstructions shall not be less than the total maximum displacement.

17.2.5.3 Nonbuilding Structures Nonbuilding structures shall be designed and constructed in accordance with the requirements of Chapter 15 using design displacements and forces, calculated in accordance with Sections 17.5 or 17.6.

17.2.5.4 Steel Ordinary Concentrically Braced Frames Steel ordinary concentrically braced frames are permitted as the seismic force-resisting system in seismically isolated structures assigned to Seismic Design Categories D, E, and F and are permitted to a height of 160 ft (48.4 m) or less, provided all the following design requirements are satisfied:

1. The value of R_I, as defined in Section 17.5.4 is 1.0.
2. The total maximum displacement (D_{TM}), as defined in Equation (17.5-3), shall be increased by a factor of 1.2.

17.2.5.5 Isolation System Connections Moment-resisting connections of structural steel elements of the seismic isolation system below the base level are permitted to conform to the requirements for ordinary steel moment frames of AISC 341, E1.6a and E1.6b.

17.2.6 Elements of Structures and Nonstructural Components Parts or portions of an isolated structure, permanent nonstructural components and the attachments to them, and the attachments for permanent equipment supported by a structure shall be designed to resist seismic forces and displacements, as prescribed by this section and the applicable requirements of Chapter 13.

17.2.6.1 Components at or above the Isolation Interface Elements of seismically isolated structures and nonstructural components, or portions thereof, which are at or above the isolation interface, shall be designed to resist a total lateral seismic force equal to the maximum dynamic response of the element or component under consideration, determined using a response history analysis.

EXCEPTION: Elements of seismically isolated structures and nonstructural components or portions designed to resist seismic forces and displacements, as prescribed in Chapter 12 or 13 as appropriate, are not required to meet this provision.

17.2.6.2 Components Crossing the Isolation Interface Elements of seismically isolated structures and nonstructural components, or portions thereof, which cross the isolation interface, shall be designed to withstand the total maximum displacement and to accommodate, on a long-term basis, any permanent residual displacement.

17.2.6.3 Components below the Isolation Interface Elements of seismically isolated structures and nonstructural components, or portions thereof, which are below the isolation interface, shall be designed and constructed in accordance with the requirements of Section 12.1 and Chapter 13.

17.2.7 Seismic Load Effects and Load Combinations All members of the isolated structure, including those that are not part of the seismic force-resisting system, shall be designed using the seismic load effects of Section 12.4, and the additional load combinations of Section 17.2.7.1, for the design of the isolation system and for the testing of prototype isolator units.

17.2.7.1 Isolator Unit Vertical Load Combinations The average, minimum, and maximum vertical load on each isolator unit type shall be computed from application of horizontal seismic forces, Q_E, caused by MCE_R ground motions and the following applicable vertical load combinations:

1. Average vertical load: load corresponding to 1.0 dead load plus 0.5 live load.
2. Maximum vertical load: load combination 6 of Section 2.3.6, where E is given by Equation (12.4-1) and $0.2S_{DS}$ is replaced by $0.12S_{MS}$ in Equation (12.4-4a).
3. Minimum vertical load: load combination 7 of Section 2.3.6, where E is given by Equation (12.4-2) and $0.2S_{DS}$ is replaced by $0.12S_{MS}$ in Equation (12.4-4a).

17.2.8 Isolation System Properties

17.2.8.1 Isolation System Component Types All components of the isolation system shall be categorized and grouped in terms of common type and size of isolator unit and common type and size of supplementary damping device, if such devices are also components of the isolation system.

17.2.8.2 Isolator Unit Nominal Properties Isolator unit nominal design properties shall be based on average properties over the three cycles of prototype testing, specified by Item 2 of Section 17.8.2.2. Variation in isolator unit properties with vertical load are permitted to be established based on a single representative deformation cycle, by averaging the properties determined using the three vertical load combinations specified in Section 17.2.7.1, at each displacement level, where required to be considered by Section 17.8.2.2.

EXCEPTION: If the measured values of isolator unit effective stiffness and effective damping for load combination 1 of Section 17.2.7.1 differ by less than 15% from those based on the average of measured values for the three vertical load combinations of Section 17.2.7.1, then nominal design properties are permitted to be computed only for load combination 1 of Section 17.2.7.1.

17.2.8.3 Bounding Properties of Isolation System Components Bounding properties of isolation system components shall be developed for each isolation system component type. Bounding properties shall include variation in all of the following component properties:

1. Measured by prototype testing, Item 2 of Section 17.8.2.2, considering variation in prototype isolator unit properties caused by required variation in vertical test load, rate of test loading or velocity effects, effects of heating during cyclic motion, history of loading, scragging (temporary degradation of mechanical properties with repeated cycling), and other potential sources of variation measured by prototype testing;
2. Permitted by manufacturing specification tolerances used to determine acceptability of production isolator units, as required by Section 17.8.5; and
3. Because of aging and environmental effects, including creep, fatigue, contamination, operating temperature and duration of exposure to that temperature and wear over the life of the structure.

17.2.8.4 Property Modification Factors Maximum and minimum property modification (λ) factors shall be used to account for the variation of the nominal design parameters of each isolator unit type for the effects of heating caused by cyclic dynamic motion, loading rate, scragging and recovery, variability in production bearing properties, temperature, aging, environmental exposure, and contamination. When manufacturer-specific qualification test data, in accordance with Section 17.8, have been approved by the RDP these data are permitted to be used to develop the property modification factors, and the maximum and minimum limits of Equations (17.2-1) and (17.2-2) need not apply. When qualification test data, in accordance with Section 17.8, have not been approved by the RDP the maximum and minimum limits of Equations (17.2-1) and (17.2-2) shall apply.

Property modification factors (λ) shall be developed for each isolator unit type, and when applied to the nominal design parameters shall envelop the hysteretic response for the range of demands from $\pm 0.5D_M$ up to and including the maximum displacement, $\pm D_M$. Property modification factors for environmental conditions are permitted to be developed from data that need not satisfy the similarity requirements of Section 17.8.2.7.

For each isolator unit type, the maximum property modification factor, λ_{max}, and the minimum property modification factor, λ_{min}, shall be established from contributing property modification factors in accordance with Equations (17.2-1) and (17.2-2), respectively:

$$\lambda_{max} = (1 + (0.75 \times (\lambda_{(ae,max)} - 1))) \times \lambda_{(test,max)} \times \lambda_{(spec,max)} \geq 1.8 \quad (17.2\text{-}1)$$

$$\lambda_{min} = (1 - (0.75 \times (1 - \lambda_{(ae,min)}))) \times \lambda_{(test,min)} \times \lambda_{(spec,min)} \leq 0.60 \quad (17.2\text{-}2)$$

where

$\lambda_{(ae,max)}$ = Property modification factor for calculation of the maximum value of the isolator property of interest, used to account for aging effects and environmental conditions;

$\lambda_{(ae,min)}$ = Property modification factor for calculation of the minimum value of the isolator property of interest, used to account for aging effects and environmental conditions;

$\lambda_{(test,max)}$ = Property modification factor for calculation of the maximum value of the isolator property of interest, used to account for heating, rate of loading, and scragging;

$\lambda_{(test,min)}$ = Property modification factor for calculation of the minimum value of the isolator property of interest, used to account for heating, rate of loading, and scragging;

$\lambda_{(spec,max)}$ = Property modification factor for calculation of the maximum value of the isolator property of interest, used to account for permissible manufacturing variation on the average properties of a group of same-sized isolators; and

$\lambda_{(spec,min)}$ = Property modification factor for calculation of the minimum value of the isolator property of interest, used to account for permissible manufacturing variation on the average properties of a group of same-sized isolators.

EXCEPTION: If the prototype isolator testing is conducted on a full-scale specimen that satisfies the dynamic test data of Section 17.8.2.3, then the values of the property modification factors shall be based on the test data, and the upper and lower limits of Equations (17.2-1) and (17.2-2) need not apply.

17.2.8.5 Upper Bound and Lower Bound Force-Deflection Behavior of Isolation System Components A mathematical model of upper bound force-deflection (loop) behavior of each type of isolation system component shall be developed. Upper bound force-deflection behavior of isolation system components that are essentially hysteretic devices (e.g., isolator units) shall be modeled using the maximum values of isolator properties calculated using the property modification factors of Section 17.2.8.4. Upper bound force-deflection behavior of isolation system components that are essentially viscous devices (e.g., supplementary viscous dampers) shall be modeled in accordance with the requirements of Chapter 18 for such devices.

A mathematical model of lower bound force-deflection (loop) behavior of each type of isolation system component shall be developed. Lower bound force-deflection behavior of isolation system components that are essentially hysteretic devices (e.g., isolator units) shall be modeled using the minimum values of isolator properties calculated using the property modification factors of Section 17.2.8.4. Lower bound force-deflection behavior of isolation system components that are essentially viscous devices (e.g., supplementary viscous dampers) shall be modeled in accordance with the requirements of Chapter 18 for such devices.

17.2.8.6 Isolation System Properties at Maximum Displacements The effective stiffness, k_M, of the isolation system at the maximum displacement, D_M, shall be computed using both upper bound and lower bound force-deflection behavior of individual isolator units, in accordance with Equation (17.2-3):

$$k_M = \frac{\sum |F_M^+| + \sum |F_M^-|}{2D_M} \quad (17.2\text{-}3)$$

The effective damping, β_M, of the isolation system at the maximum displacement, D_M [in. (mm)], shall be computed using both upper bound and lower bound force-deflection behavior of individual isolator units, in accordance with Equation (17.2-4):

$$\beta_M = \frac{\sum E_M}{2\pi k_M D_M^2} \quad (17.2\text{-}4)$$

where

$\sum E_M$ = Total energy dissipated in the isolation system during a full cycle of response at the displacement D_M, kips-in. (kN-mm);

$\sum F_M^+$ = Sum, for all isolator units, of the absolute value of force at a positive displacement equal to D_M, kips (kN); and;

$\sum F_M^-$ = Sum, for all isolator units, of the absolute value of force at a negative displacement equal to D_M, kips (kN).

17.2.8.7 Upper Bound and Lower Bound Isolation System Properties at Maximum Displacement The analysis of the isolation system and structure shall be performed separately for upper bound and lower bound properties, and the governing case for each response parameter of interest shall be used for design. In addition, the analysis shall comply with all of the following:

1. For the equivalent lateral force procedure, and for the purposes of establishing minimum forces and displacements for dynamic analysis, the following variables shall be calculated independently for upper bound and lower bound isolation system properties: k_M and β_M per Section 17.2.8.6 [Equations (17.2-3) and (17.2-4)], D_M per Section 17.5.3.1 [Equation (17.5-1)], T_M per Section 17.5.3.2 [Equation (17.5-2)], D_{TM} per Section 17.5.3.3 [Equation (17.5-3)], V_b per Section 17.5.4.1 [Equation (17.5-5)], and V_s and V_{st} per Section 17.5.4.2 [Equations (17.5-6) and (17.5-7)];
2. Limitations on V_s established in Section 17.5.4.3 shall be evaluated independently for both upper bound and lower bound isolation system properties, and the most adverse requirement shall govern.
3. For the equivalent lateral force procedure and for the purposes of establishing minimum story shear forces for response spectrum analysis, the vertical force distribution from Section 17.5.5 shall be determined separately for upper bound and lower bound isolation system properties. This determination will require independent calculation of F_1, F_x, C_{vx}, and k, per Equations (17.5-8) through (17.5-11), respectively.

17.3 SEISMIC HAZARD

17.3.1 Spectral Response Acceleration Parameters and Response Spectrum The MCE_R spectral response acceleration parameters (S_{MS} and S_{M1}) and the MCE_R spectrum shall be determined in accordance with Section 11.4.

17.3.2 Ground Motions for Response History Analysis Where response history analysis in accordance with Section 17.6.3.4 is used to design seismically isolated structures, the provisions of Section 16.2 shall apply except that in lieu of the requirements of Section 16.2.3.1, the period range shall be determined in accordance with the following:

A period range shall be determined corresponding to the vibration periods that significantly contribute to the structure's lateral dynamic response. The upper bound of this period range shall be greater than or equal to $1.25T_M$, as determined using lower bound isolation system properties. The lower bound period of this period range shall be established such that the period range includes at least the number of modes necessary to achieve 90% mass participation in each principal horizontal direction and shall not exceed T_{fb}. Where vertical response is considered in the analysis, the lower bound period of the period range used for the modification of vertical components of ground motion need not be taken as less than the larger of 0.1 s or the lowest period at which significant vertical mass participation occurs.

17.4 ANALYSIS PROCEDURE SELECTION

Seismically isolated structures, except those defined in Section 17.4.1, shall be designed using the dynamic procedures of Section 17.6. Where supplementary viscous dampers are used, the response history analysis procedures of Section 17.4.2.2 shall be used.

17.4.1 Equivalent Lateral Force Procedure The equivalent lateral force procedure of Section 17.5 is permitted to be used for design of a seismically isolated structure provided that all of the following items are satisfied. These requirements shall be evaluated separately for upper bound and lower bound isolation system properties, and the more restrictive requirement shall govern.

1. The structure is located on a Site Class A, B, C, or D site.
2. The effective period of the isolated structure at the maximum displacement, D_M, is less than or equal to 5.0 s.
3. The structure above the isolation interface is less than or equal to four stories or 65 ft (19.8 m) in structural height measured from the base level.

   EXCEPTION: These limits are permitted to be exceeded if there is no tension or uplift on the isolators.
4. The effective damping of the isolation system at the maximum displacement, D_M, is less than or equal to 30%.
5. The effective period of the isolated structure, T_M, is greater than three times the elastic, fixed-base period of the structure above the isolation system, determined using a rational modal analysis.
6. The structure above the isolation system does not have a structural irregularity, as defined in Section 17.2.2.
7. The isolation system meets all of the following criteria:
   (a) The effective stiffness of the isolation system at the maximum displacement is greater than one-third of the effective stiffness at 20% of the maximum displacement.
   (b) The isolation system is capable of producing a restoring force, as specified in Section 17.2.4.4.
   (c) The isolation system does not limit maximum earthquake displacement to less than the total maximum displacement, D_{TM}.

17.4.2 Dynamic Procedures The dynamic procedures of Section 17.6 are permitted to be used as specified in this section.

17.4.2.1 Response Spectrum Analysis Procedure Response spectrum analysis procedure shall not be used for design of a seismically isolated structure unless the structure, site, and isolation system meet the criteria of Section 17.4.1, Items 1, 2, 3, 4, and 6.

17.4.2.2 Response History Analysis Procedure The response history analysis procedure is permitted to be used for design of any seismically isolated structure and shall be used for design of all seismically isolated structures not meeting the criteria of Section 17.4.2.1.

17.5 EQUIVALENT LATERAL FORCE PROCEDURE

17.5.1 General Where the equivalent lateral force procedure is used to design seismically isolated structures, the requirements of this section shall apply.

17.5.2 Deformation Characteristics of the Isolation System Minimum lateral earthquake design displacements and forces on seismically isolated structures shall be based on the deformation characteristics of the isolation system. The deformation characteristics of the isolation system include the effects of the wind-restraint system if such a system is used to meet the design requirements of this standard. The deformation characteristics of the isolation system shall be based on properly substantiated prototype tests performed in accordance with Section 17.8 and shall incorporate property modification factors in accordance with Section 17.2.8.4.

The analysis of the isolation system and structure shall be performed separately for upper bound and lower bound properties, and the governing case for each response parameter of interest shall be used for design.

17.5.3 Minimum Lateral Displacements Required for Design

17.5.3.1 Maximum Displacement The isolation system shall be designed and constructed to withstand, at a minimum, the maximum displacement, D_M, determined using upper bound and lower bound properties, in the most critical direction of horizontal response, calculated using Equation (17.5-1):

$$D_M = \frac{g S_{M1} T_M}{4\pi^2 B_M} \tag{17.5-1}$$

where

g = Acceleration caused by gravity [in./s^2 (mm/s^2)] if the units of the displacement D_M are in in. (mm);

S_{M1} = MCE_R 5% damped spectral acceleration parameter at 1 s period in units of g-sec, as determined in Section 11.4.4 or 11.4.8;

T_M = Effective period of the seismically isolated structure (s) at the displacement D_M in the direction under consideration, as prescribed by Equation (17.5-2); and

B_M = Numerical coefficient as set forth in Table 17.5-1 for the effective damping of the isolation system β_M at the displacement D_M.

17.5.3.2 Effective Period at the Maximum Displacement The effective period of the isolated structure, T_M, at the maximum displacement, D_M, shall be determined using upper bound and lower bound deformational characteristics of the isolation system and Equation (17.5-2):

$$T_M = 2\pi\sqrt{\frac{W}{k_M g}} \quad (17.5\text{-}2)$$

where

W = Effective seismic weight of the structure above the isolation interface as defined in Section 12.7.2;

k_M = Effective stiffness [kip/in. (kN/mm)] of the isolation system at the maximum displacement, D_M, as prescribed by Equation (17.2-3); and

g = Acceleration caused by gravity [in./s^2 (mm/s^2)] if the units of k_M are in kip/in. (kN/mm).

17.5.3.3 Total Maximum Displacement The total maximum displacement, D_{TM}, of elements of the isolation system shall include additional displacement caused by actual and accidental torsion, calculated from the spatial distribution of the lateral stiffness of the isolation system and the most disadvantageous location of eccentric mass. The total maximum displacement, D_{TM}, of elements of an isolation system shall not be taken as less than that prescribed by Equation (17.5-3):

$$D_{TM} = D_M\left[1 + \left(\frac{y}{P_T^2}\right)\frac{12e}{b^2 + d^2}\right] \quad (17.5\text{-}3)$$

where

D_M = Displacement at the center of rigidity of the isolation system in the direction under consideration, as prescribed by Equation (17.5-1);

y = Distance [in. (mm)] between the centers of rigidity of the isolation system and the element of interest measured perpendicular to the direction of seismic loading under consideration;

e = Actual eccentricity measured in plan between the center of mass of the structure above the isolation interface and the center of rigidity of the isolation system, plus accidental eccentricity [ft (mm)], taken as 5% of the longest plan dimension of the structure perpendicular to the direction of force under consideration;

b = Shortest plan dimension of the structure [ft (mm)] measured perpendicular to d;

d = Longest plan dimension of the structure [ft (mm)]; and

P_T = Ratio of the effective translational period of the isolation system to the effective torsional period of the isolation system, as calculated by dynamic analysis, or as prescribed by Equation (17.5-4) but need not be taken as less than 1.0.

$$P_T = \frac{1}{r_I}\sqrt{\frac{\sum_N^{i=1}(x_i^2 + y_i^2)}{N}} \quad (17.5\text{-}4)$$

where

x_i, y_i = Horizontal distances [ft (mm)] from the center of mass to the ith isolator unit in the two horizontal axes of the isolation system;

N = Number of isolator units; and

r_I = Radius of gyration of the isolation system [ft (mm)], which is equal to $[(b^2 + d^2)/12]^{1/2}$ for isolation systems of rectangular plan dimension, $b \times d$.

The total maximum displacement, D_{TM}, shall not be taken as less than 1.15 times D_M.

17.5.4 Minimum Lateral Forces Required for Design

17.5.4.1 Isolation System and Structural Elements below the Base Level The isolation system, the foundation, and all structural elements below the base level shall be designed and constructed to withstand a minimum lateral seismic force, V_b, using all of the applicable requirements for a nonisolated structure, as prescribed by the value of Equation (17.5-5), determined using both upper bound and lower bound isolation system properties:

$$V_b = k_M D_M \quad (17.5\text{-}5)$$

where k_M is the effective stiffness [kip/in. (kN/mm)] of the isolation system at the displacement D_M, as prescribed by Equation (17.2-3), and D_M is the maximum displacement [in. (mm)] at the center of rigidity of the isolation system in the direction under consideration, as prescribed by Equation (17.5-1).

V_b shall not be taken as less than the maximum force in the isolation system at any displacement up to and including the maximum displacement, D_M, as defined in Section 17.5.3

Overturning loads on elements of the isolation system, the foundation, and structural elements below the base level caused by lateral seismic force, V_b, shall be based on the vertical distribution of force of Section 17.5.5, except that the unreduced lateral seismic design force, V_{st}, shall be used in lieu of V_s in Equation (17.5-9).

17.5.4.2 Structural Elements above the Base Level The structure above the base level shall be designed and constructed using all of the applicable requirements for a nonisolated structure for a minimum shear force, V_s, determined using upper bound and lower bound isolation system properties, as prescribed by Equation (17.5-6):

$$V_s = \frac{V_{st}}{R_I} \quad (17.5\text{-}6)$$

where R_I is the numerical coefficient related to the type of seismic force-resisting system above the isolation system, and V_{st} is the total unreduced lateral seismic design force or shear on elements above the base level, as prescribed by Equation (17.5-7).

The R_I factor shall be based on the type of seismic force-resisting system used for the structure above the base level in the direction of interest and shall be three-eighths of the value of R given in Table 12.2-1, with a maximum value not greater than 2.0 and a minimum value not less than 1.0.

EXCEPTION: The value of R_i is permitted to be taken as greater than 2.0, provided the strength of the structure above the base level in the direction of interest, as determined by nonlinear static analysis at a roof displacement corresponding to a maximum story drift the lesser of the MCE$_R$ drift or $0.015h_{sx}$, is not less than 1.1 times V_b.

The total unreduced lateral seismic force or shear on elements above the base level shall be determined using upper bound and lower bound isolation system properties, as prescribed by Equation (17.5-7):

$$V_{st} = V_b \left(\frac{W_s}{W}\right)^{(1-2.5\beta m)} \quad (17.5\text{-}7)$$

where W is the effective seismic weight [kips (kN)] of the structure above the isolation interface, as defined in Section 12.7.2; and W_s is the effective seismic weight [kips (kN)] of the structure above the isolation interface, as defined in Section 12.7.2, excluding the effective seismic weight [kips (kN)] of the base level.

The effective seismic weight W_s in Equation (17.5-7) shall be taken as equal to W when the average distance from the top of the isolator to the underside of the base level floor framing above the isolators exceeds 3 ft (0.9 m).

EXCEPTION: For isolation systems whose hysteretic behavior is characterized by an abrupt transition from preyield to postyield or preslip to postslip behavior, the exponent term $(1 - 2.5\beta_M)$ in Equation (17.5-7) shall be replaced by $(1 - 3.5\beta_M)$.

17.5.4.3 Limits on V_s The value of V_s shall not be taken as less than each of the following:

1. The lateral seismic force required by Section 12.8 for a fixed-base structure of the same effective seismic weight, W_s, and a period equal to the period of the isolation system using the upper bound properties, T_M;
2. The base shear corresponding to the factored design wind load; and
3. The lateral seismic force, V_{st}, calculated using Equation (17.5-7), and with V_b set equal to the force required to fully activate the isolation system using the greater of the upper bound properties, or
   (a) 1.5 times the nominal properties for the yield level of a softening system,
   (b) The ultimate capacity of a sacrificial wind-restraint system,
   (c) The breakaway friction force of a sliding system, or
   (d) The force at zero displacement of a sliding system following a complete dynamic cycle of motion at D_M.

17.5.5 Vertical Distribution of Force The lateral seismic force, V_s, shall be distributed over the height of the structure above the base level, using upper bound and lower bound isolation system properties, using the following equations:

$$F_1 = \frac{(V_b - V_{st})}{R_I} \quad (17.5\text{-}8)$$

and

$$F_x = C_{vx} V_s \quad (17.5\text{-}9)$$

and

$$C_{vx} = \frac{w_x h_x^k}{\sum_{i=2}^{n} w_i h_i^k} \quad (17.5\text{-}10)$$

and

$$k = 14\beta_M T_{fb} \quad (17.5\text{-}11)$$

where

F_1 = Lateral seismic force [kips (kN)] induced at level 1, the base level;
F_x = Lateral seismic force [kips (kN)] induced at level x, $x > 1$;
C_{vx} = Vertical distribution factor;
V_s = Total lateral seismic design force or shear on elements above the base level, as prescribed by Equation (17.5-6) and the limits of Section 17.5.4.3;
$w_i w_x$ = Portion of W_s that is located at or assigned to level i or x;
$h_i h_x$ = Height above the isolation interface of level i or x; and
T_{fb} = Fundamental period, in s, of the structure above the isolation interface determined using a rational modal analysis assuming fixed-base conditions.

EXCEPTION: In lieu of Equations (17.5-6) and (17.5-9), the lateral seismic force F_x is permitted to be calculated as the average value of the force at level x in the direction of interest, using the results of a simplified stick model of the building and a lumped representation of the isolation system using response history analysis scaled to V_b/R_I at the base level.

17.5.6 Drift Limits The maximum story drift of the structure above the isolation system shall not exceed $0.015h_{sx}$. The drift shall be calculated by Equation (12.8-15) with C_d for the isolated structure equal to R_I, as defined in Section 17.5.4.2.

17.6 DYNAMIC ANALYSIS PROCEDURES

17.6.1 General Where dynamic analysis is used to design seismically isolated structures, the requirements of this section shall apply.

17.6.2 Modeling The mathematical models of the isolated structure, including the isolation system, the seismic force-resisting system, and other structural elements, shall conform to Section 12.7.3 and to the requirements of Sections 17.6.2.1 and 17.6.2.2.

17.6.2.1 Isolation System The isolation system shall be modeled using deformational characteristics developed in accordance with Section 17.2.8. The lateral displacements and forces shall be computed separately for upper bound and lower bound isolation system properties, as defined in Section 17.2.8.5. The isolation system shall be modeled with sufficient detail to capture all of the following:

1. Spatial distribution of isolator units;
2. Translation, in both horizontal directions, and torsion of the structure above the isolation interface considering the most disadvantageous location of eccentric mass;
3. Overturning and uplift forces on individual isolator units; and
4. Effects of vertical load, bilateral load, and/or the rate of loading if the force-deflection properties of the isolation system are dependent on one or more of these attributes.

The total maximum displacement, D_{TM}, across the isolation system shall be calculated using a model of the isolated structure that incorporates the force-deflection characteristics of nonlinear elements of the isolation system and the seismic force-resisting system.

17.6.2.2 Isolated Structure The maximum displacement of each floor and design forces and displacements in elements of the seismic force-resisting system are permitted to be calculated using a linear elastic model of the isolated structure, provided that all elements of the seismic force-resisting system of the structure above the isolation system remain essentially elastic.

Seismic force-resisting systems with essentially elastic elements include, but are not limited to, regular structural systems designed for a lateral force not less than 100% of V_s determined in accordance with Sections 17.5.4.2 and 17.5.4.3.

The analysis of the isolation system and structure shall be performed separately for upper bound and lower bound properties, and the governing case for each response parameter of interest shall be used for design.

17.6.3 Description of Procedures

17.6.3.1 General Response spectrum analysis shall be performed in accordance with Section 12.9 and the requirements of Section 17.6.3.3. Response history analysis shall be performed in accordance with the requirements of Section 17.6.3.4.

17.6.3.2 MCE$_R$ Ground Motions The MCE$_R$ ground motions of Section 17.3 shall be used to calculate the lateral forces and displacements in the isolated structure, the total maximum displacement of the isolation system, and the forces in the isolator units, isolator unit connections, and supporting framing immediately above and below the isolators used to resist isolator P-delta demands.

17.6.3.3 Response Spectrum Analysis Procedure Response spectrum analysis shall be performed using a modal damping value for the fundamental mode in the direction of interest not greater than the effective damping of the isolation system or 30% of critical, whichever is less. Modal damping values for higher modes shall be selected consistent with those that would be appropriate for response spectrum analysis of the structure above the isolation system assuming a fixed base.

Response spectrum analysis used to determine the total maximum displacement shall include simultaneous excitation of the model by 100% of the ground motion in the critical direction and 30% of the ground motion in the perpendicular, horizontal direction. The maximum displacement of the isolation system shall be calculated as the vector sum of the two orthogonal displacements.

17.6.3.4 Response History Analysis Procedure Response history analysis shall be performed for a set of ground motion pairs selected and scaled in accordance with Section 17.3.2. Each pair of ground motion components shall be applied simultaneously to the model, considering the most disadvantageous location of eccentric mass. The maximum displacement of the isolation system shall be calculated from the vector sum of the two orthogonal displacements at each time step.

The parameters of interest shall be calculated for each ground motion used for the response history analysis, and the average value of the response parameter of interest shall be used for design.

17.6.3.4.1 Accidental Mass Eccentricity Torsional response resulting from lack of symmetry in mass and stiffness shall be accounted for in the analysis. In addition, accidental eccentricity consisting of displacement of the center of mass from the computed location by an amount equal to 5% of the diaphragm dimension shall be considered separately in each of two orthogonal directions at the level under consideration.

The effects of accidental eccentricity are permitted to be accounted for by amplifying forces, drifts, and deformations determined from an analysis, using only the computed center of mass, provided that factors used to amplify forces, drifts, and deformations of the center-of-mass analysis are shown to produce results that bound all the mass-eccentric cases.

17.6.4 Minimum Lateral Displacements and Forces

17.6.4.1 Isolation System and Structural Elements below the Base Level The isolation system, foundation, and all structural elements below the base level shall be designed using all of the applicable requirements for a nonisolated structure and the forces obtained from the dynamic analysis without reduction, but the design lateral force shall not be taken as less than 90% of V_b determined by Equation (17.5-5).

The total maximum displacement of the isolation system shall not be taken as less than 80% of D_{TM}, as prescribed by Section 17.5.3.3, except that D'_M is permitted to be used in lieu of D_M where

$$D'_M = \frac{D_M}{\sqrt{1 + (T/T_M)^2}} \quad (17.6\text{-}1)$$

and

D_M = Maximum displacement [in. (mm)] at the center of rigidity of the isolation system in the direction under consideration, as prescribed by Equation (17.5-1);

T = Elastic, fixed-base period, in s, of the structure above the isolation system, as determined by Section 12.8.2, including the coefficient C_u, if the approximate period formulas are used to calculate the fundamental period; and

T_M = Effective period, in s, of the seismically isolated structure, at the displacement D_M in the direction under consideration, as prescribed by Equation (17.5-2).

17.6.4.2 Structural Elements above the Base Level Subject to the procedure-specific limits of this section, structural elements above the base level shall be designed using the applicable requirements for a nonisolated structure and the forces obtained from the dynamic analysis reduced by a factor of R_I, as determined in accordance with Section 17.5.4.2.

For response spectrum analysis, the design shear at any story shall not be less than the story shear resulting from application of the forces calculated using Equation (17.5-9) and a value of V_b equal to the base shear obtained from the response spectrum analysis in the direction of interest.

For response history analysis of regular structures, the value of V_b shall not be taken as less than 80% of that determined in accordance with Section 17.5.4.1, and the value V_s shall not be taken as less than 100% of the limits specified by Section 17.5.4.3.

For response history analysis of irregular structures, the value of V_b shall not be taken as less than 100% of that determined in accordance with Section 17.5.4.1, and the value V_s shall not be taken as less than 100% of the limits specified by Section 17.5.4.3.

17.6.4.3 Scaling of Results Where the factored lateral shear force on structural elements, determined using either the response spectrum or response history procedure, is less than the minimum values prescribed by Sections 17.6.4.1 and 17.6.4.2, all design parameters shall be adjusted upward proportionally.

17.6.4.4 Drift Limits Maximum story drift corresponding to the design lateral force, including displacement caused by vertical deformation of the isolation system, shall comply with either of the following limits:

1. Where response spectrum analysis is used, the maximum story drift of the structure above the isolation system shall not exceed $0.015h_{sx}$.
2. Where response history analysis based on the force-deflection characteristics of nonlinear elements of the seismic force-resisting system is used, the maximum story drift of the structure above the isolation system shall not exceed $0.020h_{sx}$.

Drift shall be calculated using Equation (12.8-15), with the C_d of the isolated structure equal to R_I as defined in Section 17.5.4.2.

The secondary effects of the maximum lateral displacement of the structure above the isolation system, combined with gravity forces, shall be investigated if the story drift ratio exceeds $0.010/R_I$.

17.7 DESIGN REVIEW

An independent design review of the isolation system and related test programs shall be performed by one or more individuals possessing knowledge of the following items, with a minimum of one reviewer being a registered design professional (RDP). Isolation system design review shall include, but not be limited to, all the following:

1. Project design criteria, including site-specific spectra and ground motion histories.
2. Preliminary design, including the selection of the devices, determination of the maximum displacement, the total maximum displacement, and the lateral force level.
3. Review of qualification data and appropriate property modification factors for the manufacturer and device selected.
4. Prototype testing program (Section 17.8.2).
5. Final design of the entire structural system and all supporting analyses, including modeling of isolators for response history analysis, if performed.
6. Isolator production testing program (Section 17.8.5).

17.8 TESTING

17.8.1 General The deformation characteristics and damping values of the isolation system used in the design and analysis of seismically isolated structures shall be based on tests of a selected sample of the components before construction, as described in this section. The isolation system components to be tested shall include the wind-restraint system if such a system is used in the design.

The tests specified in this section are for establishing and validating the isolator unit and isolation system test properties that are used to determine design properties of the isolation system, in accordance with Section 17.2.8.

17.8.1.1 Qualification Tests Isolation device manufacturers shall submit, for approval by the RDP the results of qualification tests, analysis of test data, and supporting scientific studies that are permitted to be used to quantify the effects of heating caused by cyclic dynamic motion, loading rate, scragging, variability and uncertainty in production bearing properties, temperature, aging, environmental exposure, and contamination. The qualification testing shall be applicable to the component types, models, materials, and sizes to be used in the construction. The qualification testing shall have been performed on components manufactured by the same manufacturer supplying the components to be used in the construction. When scaled specimens are used in the qualification testing, principles of scaling and similarity shall be used in the interpretation of the data.

17.8.2 Prototype Tests Prototype tests shall be performed separately on two full-size specimens (or sets of specimens, as appropriate) of each predominant type and size of isolator unit of the isolation system. The test specimens shall include the wind-restraint system if such a system is used in the design. Specimens tested shall not be used for construction unless they are accepted by the RDP responsible for the design of the structure.

17.8.2.1 Record For each cycle of each test, the force-deflection behavior of the test specimen shall be recorded.

17.8.2.2 Sequence and Cycles Each of the following sequence of tests shall be performed for the prescribed number of cycles, at a vertical load equal to the average dead load plus one-half the effects caused by live load on all isolator units of a common type and size. Before these tests, the production set of tests specified in Section 17.8.5 shall be performed on each isolator:

1. Twenty fully reversed cycles of loading at a lateral force corresponding to the wind design force.
2. The sequence of either Item (a) or Item (b) shall be performed:
   (a) Three fully reversed cycles of loading at each of the following increments of the displacement: $0.25D_M$, $0.5D_M$, $0.67D_M$, and $1.0D_M$, where D_M is determined in Section 17.5.3.1 or Section 17.6, as appropriate.
   (b) The following sequence, performed dynamically at the effective period, T_M: continuous loading of one fully reversed cycle at each of the increments of the maximum displacement, that is, $1.0D_M$, $0.67D_M$, $0.5D_M$, and $0.25D_M$ followed by continuous loading of one fully reversed cycle at $0.25D_M$, $0.5D_M$, $0.67D_M$, and $1.0D_M$. A rest interval is permitted between these two sequences.
3. Three fully reversed cycles of loading at the maximum displacement, $1.0D_M$.
4. The sequence of either Item (a) or Item (b) shall be performed:
   (a) $30S_{M1}/(S_{MS}B_M)$, but not fewer than 10, continuous fully reversed cycles of loading at 0.75 times the maximum displacement, $0.75D_M$.
   (b) The test of Item (a), performed dynamically at the effective period, T_M. This test may comprise separate sets of multiple cycles of loading, with each set consisting of not fewer than five continuous cycles.

If an isolator unit is also a vertical load-carrying element, then Item 3 of the sequence of cyclic tests specified previously shall be performed for two additional vertical load cases specified in Section 17.2.7.1. The load increment caused by earthquake overturning, Q_E, shall be equal to or greater than the peak earthquake vertical force response corresponding to the test displacement being evaluated. In these tests, the combined vertical load shall be taken as the typical or average downward force on all isolator units of a common type and size. Axial load and displacement values for each test shall be the greater of those determined by analysis using upper bound and lower bound values of isolation system properties, determined in accordance with Section 17.2.8.5. The effective period T_M shall be the lower of those determined by analysis using upper bound and lower bound values.

17.8.2.3 Dynamic Testing Tests specified in Section 17.8.2.2 shall be performed dynamically at the lower of the effective periods, T_M, determined using upper bound and lower bound properties.

Dynamic testing shall not be required if the prototype testing has been performed dynamically on similar-sized isolators that meet the requirements of Section 17.8.2.7, and the testing was conducted at similar loads and accounted for the effects of velocity, amplitude of displacement, and heating effects. The prior dynamic prototype test data shall be used to establish factors that adjust three-cycle average values of k_d and E_{loop} to account for the difference in test velocity and heating effects and to establish $\lambda_{(test,min)}$ and $\lambda_{(test,max)}$.

Only if full-scale testing is not possible, reduced-scale prototype specimens can be used to quantify rate-dependent properties of isolators. The reduced-scale prototype specimens shall be of the same type and material and shall be manufactured with the same processes and quality as full-scale prototypes and shall be tested at a frequency that represents full-scale prototype loading rates.

17.8.2.4 Units Dependent on Bilateral Load If the force-deflection properties of the isolator units exhibit bilateral load dependence, the tests specified in Sections 17.8.2.2 and 17.8.2.3 shall be augmented to include bilateral load at the following increments of the maximum displacement, D_M: 0.25 and 1.0, 0.5 and 1.0, 0.67 and 1.0, and 1.0 and 1.0.

If reduced-scale specimens are used to quantify bilateral-load-dependent properties, they shall meet the requirements of Section 17.8.2.7; the reduced-scale specimens shall be of the same type and material and manufactured with the same processes and quality as full-scale prototypes.

The force-deflection properties of an isolator unit shall be considered to be dependent on bilateral load if the effective stiffness when subjected to bilateral loading is different by more than 15% from the effective stiffness subjected to unilateral loading.

17.8.2.5 Maximum and Minimum Vertical Load Isolator units that carry vertical load shall be subjected to one fully reversed cycle of loading at the total maximum displacement, D_{TM}, and at each of the vertical loads corresponding to the maximum and minimum downward vertical loads, as specified in Section 17.2.7.1 on any one isolator of a common type and size. Axial load and displacement values for each test shall be the greater of those determined by analysis using the upper bound and lower bound values of isolation system properties, determined in accordance with Section 17.2.8.5.

EXCEPTION: In lieu of envelope values for a single test, it shall be acceptable to perform two tests, one each for the combination of vertical load and horizontal displacement obtained from analysis using the upper bound and lower bound values of isolation system properties, respectively, determined in accordance with Section 17.2.8.5.

17.8.2.6 Sacrificial Wind-Restraint Systems If a sacrificial wind-restraint system is to be used, its ultimate capacity shall be established by test.

17.8.2.7 Testing Similar Units Prototype tests need not be performed if an isolator unit, when compared to another tested unit, complies with all of the following criteria:

1. The isolator is not more than 15% larger nor more than 30% smaller than the previously tested prototype, in terms of governing device dimensions;
2. The isolator uses the same materials;
3. The isolator has an energy dissipated per cycle, E_{loop}, that is not less than 85% of the previously tested unit;
4. The isolator is fabricated by the same manufacturer using the same or more stringent documented manufacturing and quality control procedures;
5. For elastomeric type isolators, the design shall not be subject to a greater shear strain nor greater vertical stress than that of the previously tested prototype; and
6. For sliding type isolators, the design shall not be subject to a greater vertical stress or sliding velocity than that of the previously tested prototype using the same sliding material.

The prototype testing exemption above shall be approved by independent design review, as specified in Section 17.7.

When the results of tests of similar isolator units are used to establish dynamic properties in accordance with Section 17.8.2.3, in addition to Items 2 through 4, the following criteria shall be satisfied:

7. The similar unit shall be tested at a frequency that represents design full-scale loading rates, in accordance with principles of scaling and similarity; and
8. The length scale of reduced-scale specimens shall not be greater than two.

17.8.3 Determination of Force-Deflection Characteristics The force-deflection characteristics of an isolator unit shall be based on the cyclic load tests of prototype isolators specified in Section 17.8.2.

As required, the effective stiffness of an isolator unit, k_{eff}, shall be calculated for each cycle of loading as prescribed by Equation (17.8-1):

$$k_{eff} = \frac{|F^+| + |F^-|}{|\Delta^+| + |\Delta^-|} \quad (17.8\text{-}1)$$

where F^+ and F^- are the positive and negative forces, at the maximum positive and minimum negative displacements Δ^+ and Δ^-, respectively.

As required, the effective damping, β_{eff}, of an isolator unit shall be calculated for each cycle of loading by Equation (17.8-2):

$$\beta_{eff} = \frac{2}{\pi} \frac{E_{loop}}{k_{eff}(|\Delta^+| + |\Delta^-|)^2} \quad (17.8\text{-}2)$$

where the energy dissipated per cycle of loading, E_{loop}, and the effective stiffness, k_{eff}, shall be based on peak test displacements of Δ^+ and Δ^-.

The postyield stiffness, k_d, of each isolator unit shall be calculated for each cycle of loading using the following assumptions:

1. A test loop shall be assumed to have bilinear hysteretic characteristics with values of k_1, k_d, f_o, f_y, k_{eff}, and E_{loop}, as shown in Figure 17.8-1;
2. The computed loop shall have the same values of effective stiffness, k_{eff}, and energy dissipated per cycle of loading, E_{loop}, as the test loop; and
3. The assumed value of k_1 shall be a visual fit to the elastic stiffness of the isolator unit during unloading, immediately after D_M.

It is permitted to use different methods for fitting the loop, such as a straight-line fit of k_d directly to the hysteresis curve extending to D_M and then determining k_1 to match E_{loop}.

17.8.4 Test Specimen Adequacy The performance of the test specimens shall be deemed adequate if all of the following conditions are satisfied:

1. The force-deflection plots for all tests specified in Section 17.8.2 have a positive incremental force-resisting capacity.
2. The average postyield stiffness, k_d, and energy dissipated per cycle, E_{loop}, for the three cycles of test specified in Section 17.8.2.2, Item 3, for the vertical load equal to the average dead load plus one-half the effects caused by live load, including the effects of heating and rate of loading, in

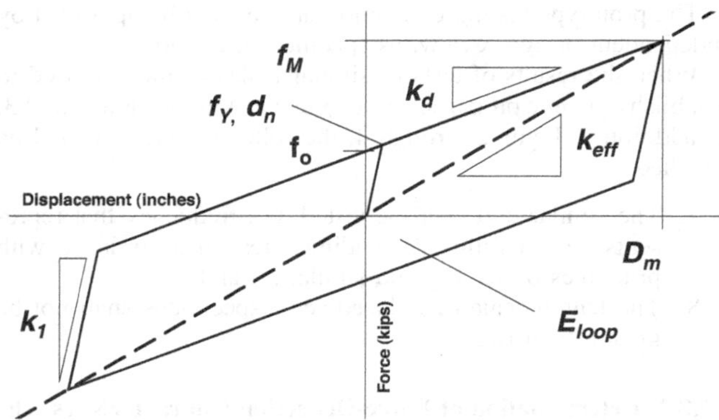

Figure 17.8-1. Nominal properties of the isolator bilinear force-deflection model.

accordance with Section 17.2.8.3, fall within the range of the nominal design values defined by the permissible individual isolator range, which are typically +/−5% greater than the $\lambda_{(spec,min)}$ and $\lambda_{(spec,max)}$ range for the average of all isolators.

3. For each increment of test displacements $0.67D_M$ and $1.0D_M$ specified in Item 2 and Item 3 of Section 17.8.2.2, and for each vertical load case specified in Section 17.8.2.2, the value of the postyield stiffness, k_d, at each of the cycles of test at a common displacement shall fall within the range defined by $\lambda_{(test,min)}$ and $\lambda_{(test,max)}$ multiplied by the nominal value of postyield stiffness.
4. For each specimen, there is no greater than a 20% change in the initial effective stiffness over the cycles of test specified in Item 4 of Section 17.8.2.2.
5. For each test specimen, the value of the postyield stiffness, k_d, and energy dissipated per cycle, E_{loop}, for any cycle of test in Section 17.8.2.2, Item 4(a) shall fall within the range of the nominal design values, defined by $\lambda_{(test,min)}$ and $\lambda_{(test,max)}$.
6. For each specimen, there is no greater than a 20% decrease in the initial effective damping over the cycles of test specified in Item 4 of Section 17.8.2.2.
7. All specimens of vertical load-carrying elements of the isolation system remain stable where tested, in accordance with Section 17.8.2.5.

EXCEPTION: The RDP is permitted to adjust the limits of Items 3, 4, and 6 to account for the property variation factors of Section 17.2.8.4 used for design of the isolation system.

17.8.5 Production Tests A test program for the isolator units used in the construction shall be established by the RDP. The test program shall evaluate the consistency of measured values of nominal isolator unit properties by testing 100% of the isolators in combined compression and shear, at not less than two-thirds of the maximum displacement, D_M, determined using lower bound properties.

The mean results of all tests shall fall within the range of values defined by the $\lambda_{(spec,max)}$ and $\lambda_{(spec,min)}$ values established in Section 17.2.8.4. A different range of values is permitted to be used for individual isolator units and for the average value across all isolators of a given unit type, provided that differences in the ranges of values are accounted for in the design of each element of the isolation system, as prescribed in Section 17.2.8.4.

17.9 CONSENSUS STANDARDS AND OTHER REFERENCED DOCUMENTS

See Chapter 23 for the list of consensus standards and other documents that shall be considered part of this standard to the extent referenced in this chapter.

CHAPTER 18
SEISMIC DESIGN REQUIREMENTS FOR STRUCTURES WITH DAMPING SYSTEMS

18.1 GENERAL

18.1.1 Scope Every structure with a damping system, and every portion thereof, shall be designed and constructed in accordance with the requirements of this standard, as modified by this chapter. Where damping devices are used across the isolation interface of a seismically isolated structure, displacements, velocities, and accelerations shall be determined in accordance with Chapter 17.

18.1.2 Definitions The following definitions apply only to the structures with damping system provisions of Chapter 18 and are in addition to the definitions presented in Chapter 11.

DAMPING DEVICE: A flexible structural element of the damping system that dissipates energy caused by relative motion of each end of the device. Damping devices include all pins, bolts, gusset plates, brace extensions, and other components required to connect damping devices to the other elements of the structure. Damping devices are classified as either displacement-dependent or velocity-dependent, or a combination thereof, and are permitted to be configured to act in either a linear or nonlinear manner.

DAMPING SYSTEM: The collection of structural elements that includes all the individual damping devices, all structural elements or bracing required to transfer forces from damping devices to the base of the structure, and the structural elements required to transfer forces from damping devices to the seismic force-resisting system.

DISPLACEMENT-DEPENDENT DAMPING DEVICE: The force response of a displacement-dependent damping device is primarily a function of the relative displacement between each end of the device. The response is substantially independent of the relative velocity between each of the devices and/or the excitation frequency.

FORCE-CONTROLLED ELEMENTS: Element actions for which reliable inelastic deformation capacity is not achievable without critical strength decay.

VELOCITY-DEPENDENT DAMPING DEVICE: The force-displacement relation for a velocity-dependent damping device is primarily a function of the relative velocity between each end of the device and could also be a function of the relative displacement between each end of the device.

18.1.3 Symbols Symbols presented in this section apply only to the structures with damping system provisions of Chapter 18 and are in addition to the symbols presented in Chapter 11.

B_{1D} = Numerical coefficient, as set forth in Table 18.7-1, for effective damping equal to $\beta_{mD} (m = 1)$ and period of structure equal to T_{1D}

B_{1E} = Numerical coefficient, as set forth in Table 18.7-1, for the effective damping equal to $\beta_I + \beta_{V1}$ and period equal to T_1

B_{1M} = Numerical coefficient, as set forth in Table 18.7-1, for effective damping equal to $\beta_{mM} (m = 1)$ and period of structure equal to T_{1M}

B_{mD} = Numerical coefficient, as set forth in Table 18.7-1, for effective damping equal to β_{ml} and period of structure equal to T_m

B_{mM} = Numerical coefficient, as set forth in Table 18.7-1, for effective damping equal to β_{mM} and period of structure equal to T_m

B_R = Numerical coefficient, as set forth in Table 18.7-1, for effective damping equal to β_R and period of structure equal to T_R

B_{V+I} = Numerical coefficient, as set forth in Table 18.7-1, for effective damping equal to the sum of viscous damping in the fundamental mode of vibration of the structure in the direction of interest, $\beta_{Vm} (m = 1)$, plus inherent damping, β_I, and period of structure equal to T_1

C_{mFD} = Force coefficient, as set forth in Table 18.7-2

C_{mFV} = Force coefficient, as set forth in Table 18.7-3

C_{S1} = Seismic response coefficient of the fundamental mode of vibration of the structure in the direction of interest, Section 18.7.1.2.4 or 18.7.2.2.4 $(m = 1)$

C_{Sm} = Seismic response coefficient of the mth mode of vibration of the structure in the direction of interest, Section 18.7.1.2.4 $(m = 1)$ or Section 18.7.1.2.6 $(m > 1)$

C_{SR} = Seismic response coefficient of the residual mode of vibration of the structure in the direction of interest, Section 18.7.2.2.8

D_{1D} = Fundamental mode design displacement at the center of rigidity of the roof level of the structure in the direction under consideration, Section 18.7.2.3.2

D_{1M} = Fundamental mode MCE_R displacement at the center of rigidity of the roof level of the structure in the direction under consideration, Section 18.7.2.3.5

D_{mD} = Design displacement at the center of rigidity of the roof level of the structure caused by the mth mode of vibration in the direction under consideration, Section 18.7.1.3.2

D_{mM} = MCE_R displacement at the center of rigidity of the roof level of the structure caused by the mth mode of vibration in the direction under consideration, Section 18.7.1.3.5

D_{RD} = Residual mode design displacement at the center of rigidity of the roof level of the structure in the direction under consideration, Section 18.7.2.3.2

D_{RM} = Residual mode MCE_R displacement at the center of rigidity of the roof level of the structure in the direction under consideration, Section 18.7.2.3.5

D_Y = Displacement at the center of rigidity of the roof level of the structure at the effective yield point of the seismic force-resisting system, Section 18.7.3.3

E_{loop} = Area of one load-displacement hysteresis loop, Section 18.6.2.5

f_i = Lateral force at level i of the structure distributed approximately in accordance with Section 12.8.3, Section 18.7.2.2.3

F_{i1} = Inertial force at level i (or mass point i) in the fundamental mode of vibration of the structure in the direction of interest, Section 18.7.2.2.9

F_{im} = Inertial force at level i (or mass point i) in the mth mode of vibration of the structure in the direction of interest, Section 18.7.1.2.7

F_{iR} = Inertial force at level i (or mass point i) in the residual mode of vibration of the structure in the direction of interest, Section 18.7.2.2.9

h_i = Height above the base to level i, Section 18.7.2.2.3

h_n = Structural height, Section 18.7.2.2.3

q_H = Hysteresis loop adjustment factor, as determined in Section 18.7.3.2.2.1

Q_{DSD} = Force in an element of the damping system required to resist design seismic forces of displacement-dependent damping devices, Section 18.7.4.5

Q_E = Seismic design force in each element of the damping system, Section 18.7.4.5

Q_{mDSV} = Force in an element of the damping system required to resist design seismic forces of velocity-dependent damping devices caused by the mth mode of vibration of the structure in the direction of interest, Section 18.7.4.5

Q_{mSFRS} = Force in an element of the damping system equal to the design seismic force of the mth mode of vibration of the structure in the direction of interest, Section 18.7.4.5

T_1 = Fundamental period of the structure in the direction under consideration

T_{1D} = Effective period, in seconds, of the fundamental mode of vibration of the structure at the design displacement in the direction under consideration, as prescribed by Section 18.7.1.2.5 or 18.7.2.2.5

T_{1M} = Effective period, in seconds, of the fundamental mode of vibration of the structure at the MCE_R displacement in the direction under consideration, as prescribed by Section 18.7.1.2.5 or 18.7.2.2.5

T_m = Period, in seconds, of the mth mode of vibration of the structure in the direction under consideration, Section 18.7.1.2.6

T_R = Period, in seconds, of the residual mode of vibration of the structure in the direction under consideration, Section 18.7.2.2.7

V = Seismic base shear in the direction of interest, Section 18.2.1.1

V_1 = Design value of the seismic base shear of the fundamental mode in a given direction of response, as determined in Sections 18.7.2.2.1 and 18.7.2.2.2,

V_m = Design value of the seismic base shear of the mth mode of vibration of the structure in the direction of interest, Section 18.7.1.2.2

V_{min} = Minimum allowable value of base shear permitted for design of the seismic force-resisting system of the structure in the direction of interest, Section 18.2.1.1

V_R = Design value of the seismic base shear of the residual mode of vibration of the structure in a given direction, as determined in Section 18.7.2.2.6

w_i = Effective seismic weight of the ith floor of the structure, Section 18.7.1.2.2

$\overline{W}_1$ = Effective fundamental mode seismic weight, determined in accordance with Equation (18.7-2b) for $m = 1$

$\overline{W}_m$ = Effective seismic weight of the mth mode of vibration of the structure, Section 18.7.1.2.2

W_m = Maximum strain energy in the mth mode of vibration of the structure in the direction of interest at modal displacements, δ_{im}, Section 18.7.3.2.2.1

W_{mj} = Work done by jth damping device in one complete cycle of dynamic response corresponding to the mth mode of vibration of the structure in the direction of interest at modal displacements, δ_{im}, Section 18.7.3.2.2.1

$\overline{W}_R$ = Effective residual mode seismic weight, determined in accordance with Equation (18.7-30)

α = Velocity exponent relating damping device force to damping device velocity

β_{HD} = Component of effective damping of the structure in the direction of interest caused by postyield hysteretic behavior of the seismic force-resisting system and elements of the damping system at effective ductility demand, μ_D, Section 18.7.3.2.2

β_{HM} = Component of effective damping of the structure in the direction of interest caused by postyield hysteretic behavior of the seismic force-resisting system and elements of the damping system at effective ductility demand, μ_M, Section 18.7.3.2.2

β_I = Component of effective damping of the structure caused by the inherent dissipation of energy by elements of the structure, at or just below the effective yield displacement of the seismic force-resisting system, Section 18.7.3.2.1

β_{mD} = Total effective damping of the mth mode of vibration of the structure in the direction of interest at the design displacement, Section 18.7.3.2

β_{mM} = Total effective damping of the mth mode of vibration of the structure in the direction of interest at the MCE_R displacement, Section 18.7.3.2

β_R = Total effective damping in the residual mode of vibration of the structure in the direction of interest, calculated in accordance with Section 18.7.3.2 (using $\mu_D = 1.0$ and $\mu_M = 1.0$)

β_{Vm} = Component of effective damping of the mth mode of vibration of the structure in the direction of interest caused by viscous dissipation of energy by the damping system, at or just below the effective yield displacement of the seismic force-resisting system, Section 18.7.3.2.3

Γ_1 = Participation factor of the fundamental mode of vibration of the structure in the direction of interest, Section 18.7.1.2.3 or 18.7.2.2.3 ($m = 1$)

Γ_m = Participation factor in the mth mode of vibration of the structure in the direction of interest, Section 18.7.1.2.3

Γ_R = Participation factor of the residual mode of vibration of the structure in the direction of interest, Section 18.7.2.2.7

δ_i = Elastic deflection of level i of the structure caused by applied lateral force, f_i, Section 18.7.2.2.3

δ_{i1D} = Fundamental mode design deflection of level i at the center of rigidity of the structure in the direction under consideration, Section 18.7.2.3.1

δ_{iD} = Total design deflection of level i at the center of rigidity of the structure in the direction under consideration, Section 18.7.2.3

δ_{iM} = Total MCE_R deflection of level i at the center of rigidity of the structure in the direction under consideration, Section 18.7.2.3

δ_{im} = Deflection of level i in the mth mode of vibration at the center of rigidity of the structure in the direction under consideration, Section 18.7.3.2.3

δ_{imD} = Design deflection of level i in the mth mode of vibration at the center of rigidity of the structure in the direction under consideration, Section 18.7.1.3.1

δ_{iRD} = Residual mode design deflection of level i at the center of rigidity of the structure in the direction under consideration, Section 18.7.2.3.1

Δ_{1D} = Design story drift caused by the fundamental mode of vibration of the structure in the direction of interest, Section 18.7.2.3.3

Δ_D = Total design story drift of the structure in the direction of interest, Section 18.7.2.3.3

Δ_M = Total MCE_R story drift of the structure in the direction of interest, Section 18.7.2.3

Δ_{mD} = Design story drift caused by the mth mode of vibration of the structure in the direction of interest, Section 18.7.1.3.3

Δ_{RD} = Design story drift caused by the residual mode of vibration of the structure in the direction of interest, Section 18.7.2.3.3

$\lambda_{(ae,\max)}$ = Factor to represent possible variation in damper properties above the tested values caused by aging and environmental effects; this is a multiple of all the individual aging and environmental effects, Section 18.2.4.5

$\lambda_{(ae,\min)}$ = Factor to represent possible variation in damper properties below the tested values caused by aging and environmental effects; this is a multiple of all the individual aging and environmental effects, Section 18.2.4.5

$\lambda_{\max}$ = Factor to represent possible total variation in damper properties above the nominal properties, Section 18.2.4.5

$\lambda_{\min}$ = Factor to represent possible total variation in damper properties below the nominal properties, Section 18.2.4.5

$\lambda_{(\text{spec},\max)}$ = Factor to represent permissible variation in production damper nominal properties above those assumed in design, Section 18.2.4.5

$\lambda_{(\text{spec},\min)}$ = Factor to represent permissible variation in production damper nominal properties below those assumed in design, Section 18.2.4.5

$\lambda_{(\text{test},\max)}$ = Factor to represent possible variations in damper properties above the nominal values obtained from the prototype tests; this is a multiple of all the testing effects, Section 18.2.4.5

$\lambda_{(\text{test},\min)}$ = Factor to represent possible variations in damper properties below the nominal values obtained from the prototype tests; this is a multiple of all the testing effects, Section 18.2.4.5

μ = Effective ductility demand on the seismic force-resisting system in the direction of interest

μ_D = Effective ductility demand on the seismic force-resisting system in the direction of interest caused by the design earthquake ground motions, Section 18.7.3.3

μ_M = Effective ductility demand on the seismic force-resisting system in the direction of interest caused by the MCE_R ground motions, Section 18.7.3.3

$\mu_{\max}$ = Maximum allowable effective ductility demand on the seismic force-resisting system caused by the design earthquake ground motions, Section 18.7.3.4

ϕ_{i1} = Displacement amplitude at level i of the fundamental mode of vibration of the structure in the direction of interest, normalized to unity at the roof level, Section 18.7.2.2.3

ϕ_{im} = Displacement amplitude at level i of the mth mode of vibration of the structure in the direction of interest, normalized to unity at the roof level, Section 18.7.1.2.2

ϕ_{iR} = Displacement amplitude at level i of the residual mode of vibration of the structure in the direction of interest normalized to unity at the roof level, Section 18.7.2.2.7

∇_{1D} = Design story velocity caused by the fundamental mode of vibration of the structure in the direction of interest, Section 18.7.2.3.4

∇_D = Total design story velocity of the structure in the direction of interest, Section 18.7.1.3.4

∇_M = Total MCE_R story velocity of the structure in the direction of interest, Section 18.7.2.3

∇_{mD} = Design story velocity caused by the mth mode of vibration of the structure in the direction of interest, Section 18.7.1.3.4

∇_{RD} = Design story velocity caused by the residual mode of vibration of the structure in the direction of interest, Section 18.7.2.3.4

18.2 GENERAL DESIGN REQUIREMENTS

18.2.1 System Requirements Design of the structure shall consider the basic requirements for the seismic force-resisting system and the damping system, as defined in the following sections. The seismic force-resisting system shall have the required strength to meet the forces defined in Section 18.2.1.1. The combination of the seismic force-resisting system and the damping system is permitted to be used to meet the drift requirement.

18.2.1.1 Seismic Force-Resisting System Structures that contain a damping system shall have a seismic force-resisting system that, in each lateral direction, conforms to one of the types indicated in Table 12.2-1.

The design of the seismic force-resisting system in each direction shall satisfy the minimum base shear requirements of this section and the requirements of Section 18.4, if the nonlinear response history procedure of Section 18.3 is used, or Section 18.7.4, if either the response spectrum procedure of Section 18.7.1 or the equivalent lateral force procedure of Section 18.7.2 is used.

The seismic base shear used for design of the seismic force-resisting system shall not be less than $V_{\min}$, where $V_{\min}$ is determined as the greater of the values computed using Equations (18.2-1) and (18.2-2):

$$V_{\min} = \frac{V}{B_{V+1}} \tag{18.2-1}$$

$$V_{\min} = 0.75V \tag{18.2-2}$$

where V is the seismic base shear in the direction of interest, determined in accordance with Section 12.8, and B_{V+I} is the numerical coefficient, as set forth in Table 18.7-1, for effective damping equal to the sum of viscous damping in the fundamental mode of vibration of the structure in the direction of interest, $\beta_{Vm}(m=1)$, plus inherent damping, β_I, and period of structure equal to T_1.

EXCEPTION: The seismic base shear used for design of the seismic force-resisting system shall not be taken as less than $1.0V$ if either of the following conditions apply:

1. In the direction of interest, the damping system has fewer than two damping devices on each floor level, configured to resist torsion.
2. The seismic force-resisting system has Torsional Irregularity Ratio $TIR > 1.4$, or vertical irregularity Type 1b (Table 12.3-2).

18.2.1.2 Damping System Damping devices and all other components required to connect damping devices to the other elements of the structure shall be designed to remain elastic for MCE_R loads. Other elements of the damping system are permitted to have inelastic response at MCE_R if it is shown by analysis or test that inelastic response of these elements would not adversely affect damping system function. If either the response spectrum procedure of Section 18.7.1 or the equivalent lateral force procedure of Section 18.7.2 is used, the inelastic response shall be limited, in accordance with the requirements of Section 18.7.4.6.

Force-controlled elements of the damping system shall be designed for seismic forces that are increased by 20% from those corresponding to average MCE_R response.

18.2.2 Seismic Hazard

18.2.2.1 Spectral Response Acceleration Parameters and Response Spectrum The design earthquake and MCE_R spectral response acceleration parameters (S_{DS}, S_{D1}, S_{MS}, and S_{M1}) and the design earthquake and MCE_R spectra shall be determined in accordance with Section 11.4.

18.2.2.2 Ground Motions for Response History Analysis Where response history analysis in accordance with Section 18.3 is used to design structures with damping systems, the provisions of Section 16.2 shall apply. For establishing the period range in Section 16.2.3.1, nominal properties of damping devices at the MCE_R per Section 18.2.4.4 shall be assumed. Design earthquake ground motions shall be taken as two-thirds times MCE_R ground motions.

18.2.3 Procedure Selection Structures with a damping system provided for seismic resistance shall be analyzed and designed using the nonlinear response history procedure of Section 18.3.

EXCEPTION: It shall be permitted to analyze and design the structure using the response spectrum procedure of Section 18.7.1, subject to the limits of Section 18.2.3.1 or the equivalent lateral force procedure of Section 18.7.2, subject to the limits of Section 18.2.3.2.

18.2.3.1 Response Spectrum Procedure The response spectrum procedure of Section 18.7.1 is permitted to be used for analysis and design provided all the following conditions apply:

1. In each principal direction, the damping system has at least two damping devices in each story, configured to resist torsion; and
2. The total effective damping of the fundamental mode, $\beta_{mD}(m=1)$, of the structure in the direction of interest is not greater than 35% of critical.

18.2.3.2 Equivalent Lateral Force Procedure The equivalent lateral force procedure of Section 18.7.2 is permitted to be used for analysis and design, provided all the following conditions apply:

1. In each principal direction, the damping system has at least two damping devices in each story, configured to resist torsion.
2. The total effective damping of the fundamental mode, $\beta_{mD}(m=1)$, of the structure in the direction of interest is not greater than 35% of critical.
3. The seismic force-resisting system does not have horizontal irregularity Type 1a or 1b (Table 12.3-1) or vertical irregularity Type 1 or 2 (Table 12.3-2).
4. Floor diaphragms are rigid, as defined in Section 12.3.1.
5. The height of the structure above the base does not exceed 100 ft (30 m).

18.2.4 Damping System

18.2.4.1 Device Design The design, construction, and installation of damping devices shall be based on response to MCE_R ground motions and consideration of all of the following:

1. Low-cycle, large-displacement degradation caused by seismic loads.
2. High-cycle, small-displacement degradation caused by wind, thermal, or other cyclic loads.
3. Forces or displacements caused by gravity loads.
4. Adhesion of device parts caused by corrosion or abrasion, biodegradation, moisture, or chemical exposure.
5. Exposure to environmental conditions, including, but not limited to, temperature, humidity, moisture, radiation (e.g., ultraviolet light), and reactive or corrosive substances (e.g., saltwater).

Devices using bimetallic interfaces subject to cold welding of the sliding interface shall be prohibited from use in a damping system.

Damping devices subject to failure by low-cycle fatigue shall resist wind forces without slip, movement, or inelastic cycling.

The design of damping devices shall incorporate the range of thermal conditions, device wear, manufacturing tolerances, and other effects that cause device properties to vary during the design life of the device, in accordance with Section 18.2.4.4. Ambient temperature shall be the normal in-service temperature of the damping device. The design temperature range shall cover the annual minimum and maximum in-service temperatures of the damping device.

18.2.4.2 Multiaxis Movement Connection points of damping devices shall provide sufficient articulation to accommodate simultaneous longitudinal, lateral, and vertical displacements of the damping system.

18.2.4.3 Inspection and Periodic Testing Means of access for the inspection and removal of all damping devices shall be provided.

The registered design professional (RDP) responsible for design of the structure shall establish an inspection, maintenance, and testing schedule for each type of damping device to ensure that the devices respond in a dependable manner throughout their design life. The degree of inspection and testing shall reflect the established in-service history of the damping devices and the likelihood of change in properties over the design life of the devices.

18.2.4.4 Nominal Design Properties Nominal design properties for energy-dissipation devices shall be established from either project-specific prototype test data or prior prototype tests on devices of a similar type and size. The nominal design properties shall be based on data from prototype tests specified in Section 18.6.2.2 (2) and determined by Section 18.6.2.4 (2). These nominal design properties shall be modified by property variation or lambda (λ) factors, as specified in Section 18.2.4.5.

18.2.4.5 Maximum and Minimum Damper Properties Maximum and minimum property modification (λ) factors shall be used to account for variation of the nominal design parameters of each damping device type for the effects of heating caused by cyclic dynamic motion, loading rate, duration of seismic and wind loading, variability and uncertainty in production device properties, operating temperature, aging, environmental exposure, and contamination. Manufacturer-specific qualification test data, in accordance with Section 18.6.1.1 and prototype test data in accordance with Section 18.6.2, shall be used to develop the property modification factors.

Maximum and minimum property modification (λ) factors shall be established, in accordance with Equations (18.2-3a) and (18.2-3b) for each device, by the RDP and used in analysis and design to account for the variation from nominal properties:

$$\lambda_{\max} = [(1 + (0.75 \times (\lambda_{(ae,\max)} - 1))) \times \lambda_{(\text{test},\max)} \times \lambda_{(\text{spec},\max)}] \geq 1.2 \quad (18.2\text{-}3a)$$

$$\lambda_{\min} = [(1 - (0.75 \times (1 - \lambda_{(ae,\min)}))) \times \lambda_{(\text{test},\min)} \times \lambda_{(\text{spec},\min)}] \leq 0.85 \quad (18.2\text{-}3b)$$

where

$\lambda_{(ae,\max)}$ = Factor to represent possible variation in damper properties above the nominal values caused by aging and environmental effects; this is a multiple of all the individual aging and environmental effects;

$\lambda_{(ae,\min)}$ = Factor to represent possible variation in damper properties below the nominal values caused by aging and environmental effects; this is a multiple of all the individual aging and environmental effects;

$\lambda_{(\text{test},\max)}$ = Factor to represent possible variations in damper properties above the nominal values obtained from the prototype tests; this is a multiple of all the testing effects;

$\lambda_{(\text{test},\min)}$ = Factor to represent possible variations in damper properties below the nominal values obtained from the prototype tests; this is a multiple of all the testing effects;

$\lambda_{(\text{spec},\max)}$ = Factor established by the RDP to represent permissible variation in production damper properties above the nominal values; and

$\lambda_{(\text{spec},\min)}$ = Factor established by the RDP to represent permissible variation in production damper properties below the nominal values.

EXCEPTION: With test data reviewed by the RDP and accepted by peer review, it is permitted to use $\lambda_{\max}$ less than 1.2 and $\lambda_{\min}$ greater than 0.85.

Maximum and minimum analysis and design properties for each device shall be determined in accordance with Equations (18.2-4a) and (18.2-4b), for each modeling parameter as follows:

$$\text{Maximum design property} = \text{nominal design property} \times \lambda_{\max} \quad (18.2\text{-}4a)$$

$$\text{Minimum design property} = \text{nominal design property} \times \lambda_{\min} \quad (18.2\text{-}4b)$$

A maximum and minimum analysis and design property shall be established for each modeling parameter as necessary for the selected method of analysis. Maximum velocity coefficients, stiffness, strength, and energy dissipation shall be considered together as the maximum analysis and design case, and minimum velocity coefficients, strength, stiffness, and energy dissipation shall be considered together as the minimum analysis and design case.

Separate maximum and minimum properties shall be established for loads and displacements corresponding to the design level conditions and the MCE_R conditions.

18.2.4.6 Damping System Redundancy If fewer than four energy-dissipation devices are provided in any story of a building in either principal direction, or fewer than two devices are located on each side of the center of stiffness of any story in either principal direction, all energy-dissipation devices shall be capable of sustaining displacements equal to 130% of the maximum calculated displacement in the device under MCE_R. A velocity-dependent device shall be capable of sustaining the force and displacement associated with a velocity equal to 130% of the maximum calculated velocity for that device under MCE_R.

18.3 NONLINEAR RESPONSE HISTORY PROCEDURE

The stiffness and damping properties of the damping devices used in the models shall be based on or verified by testing of the damping devices, as specified in Section 18.6. The nonlinear force-velocity-displacement characteristics of damping devices shall be modeled, as required, to explicitly account for device dependence on frequency, amplitude, and duration of seismic loading.

A nonlinear response history analysis shall use a mathematical model of the seismic force-resisting system and the damping system, as provided in this section. The model shall directly account for the nonlinear hysteretic behavior of all members and connections undergoing inelastic behavior, in a manner consistent with applicable laboratory test data. Test data shall not be extrapolated beyond tested deformation levels. If the analysis results indicate that degradation in element strength or stiffness can occur, the hysteretic models shall include these effects.

EXCEPTION: If the calculated force in an element of the seismic force-resisting system or the damping system does not exceed 1.5 times its expected strength using strength reduction factor $\phi = 1$, that element is permitted to be modeled as linear.

Inherent damping of the structure shall not be taken as greater than 3% of critical unless test data consistent with levels of deformation at or just below the effective yield displacement of the seismic force-resisting system support higher values.

Analysis shall be performed at both the design earthquake and at the MCE_R earthquake levels. The design earthquake analysis need not include the effects of accidental eccentricity. Results from the design earthquake analysis shall be used to design the seismic force-resisting system. Results from the MCE_R analysis shall be used to design the damping system.

18.3.1 Damping Device Modeling Mathematical models of displacement-dependent damping devices shall include the hysteretic behavior of the devices consistent with test data and accounting for all significant changes in strength, stiffness, and hysteretic loop shape. Mathematical models of velocity-dependent damping devices shall include the velocity coefficient consistent with test data. If damping device properties change with time and/or temperature, such behavior shall be modeled explicitly. The flexible elements of damping devices connecting damper units to the structure shall be included in the model.

EXCEPTION: If the properties of the damping devices are expected to change during the duration of the response history analysis, the dynamic response is permitted to be enveloped by the maximum and minimum device properties from Section 18.2.4.5. All these limit cases for variable device properties shall satisfy the same conditions as if the time-dependent behavior of the devices were explicitly modeled.

18.3.2 Accidental Mass Eccentricity Inherent eccentricity resulting from lack of symmetry in mass and stiffness shall be accounted for in the MCE_R analysis. In addition, accidental eccentricity consisting of displacement of the center of mass from the computed location by an amount equal to 5% of the diaphragm dimension separately, in each of two orthogonal directions at each diaphragm level, shall be accounted for in the analysis.

EXCEPTION: It is permitted to account for the effects of accidental eccentricity through the establishment of amplification factors on forces, drifts, and deformations that permit results determined from an analysis, using only the computed center-of-mass configuration to be scaled to bound the results of all the mass-eccentric cases.

18.3.3 Response Parameters Maximum values of each response parameter of interest shall be calculated for each ground motion used for the response history analysis. Response parameters shall include the forces, displacements, and velocities (in the case of velocity-dependent devices) in each discrete damping device. The average value of a response parameter of interest across the suite of design earthquake or MCE_R motions is permitted to be used for design.

18.4 SEISMIC LOAD CONDITIONS AND ACCEPTANCE CRITERIA FOR NONLINEAR RESPONSE HISTORY PROCEDURE

For the nonlinear response history procedure of Section 18.3, the seismic force-resisting system, damping system, loading conditions, and acceptance criteria for response parameters of interest shall conform with the requirements of the following subsections.

18.4.1 Seismic Force-Resisting System The seismic force-resisting system shall satisfy the strength requirements of Section 12.2.1 using both

1. The seismic base shear, V_{min}, as given by Section 18.2.1.1, and
2. The demands from the design earthquake nonlinear response history analysis.

The story drifts shall be determined using the MCE_R ground motions with the combined model of the seismic force-resisting system and the damping system. Accidental eccentricity shall be included.

The maximum drift at MCE_R shall neither exceed 3%, nor the drift limits specified in Table 12.12-1 times the smaller of $1.5R/C_d$ and 1.9. C_d and R shall be taken from Table 12.2-1 for the building framing under consideration.

18.4.2 Damping System The damping devices and their connections shall be sized to resist the forces, displacements, and velocities from the MCE_R ground motions. Force-controlled elements of the damping system shall be designed for seismic forces that are increased by 20% from those corresponding to the average MCE_R response.

18.4.3 Combination of Load Effects The effects on the damping system caused by gravity loads and seismic forces shall be combined, in accordance with Section 12.4, using the effect of horizontal seismic forces, Q_E, except that Q_E shall be determined in accordance with the MCE_R analysis. When load combinations are used that include live loading, it is permitted to use a load factor of 25% on live load for nonlinear response history analysis. The redundancy factor, ρ, shall be taken equal to 1.0 in all cases, and the seismic load effect, including overstrength of Section 12.4.3, need not apply to the design of the damping system.

18.4.4 Acceptance Criteria for the Response Parameters of Interest The damping system components shall be evaluated by the strength design criteria of this standard using the seismic forces and seismic loading conditions determined from the MCE_R nonlinear response history analyses and strength reduction factor $\phi = 1.0$.

18.5 DESIGN REVIEW

An independent design review of the damping system and related test programs shall be performed by one or more individuals possessing knowledge of the subsequent items; a minimum of one reviewer shall be a RDP. Damping system design review shall include, but need not be limited to, all of the following:

1. Project design criteria including site-specific spectra and ground motion histories;
2. Preliminary design of the seismic force-resisting system and the damping system, including selection of the devices and their design parameters;
3. Review of manufacturer test data and property modification factors for the manufacturer and device selected;
4. Prototype testing program (Section 18.6.2);
5. Final design of the entire structural system and supporting analyses, including modeling of the damping devices for response history analysis, if performed; and
6. Damping device production testing program (Section 18.6.3).

18.6 TESTING

18.6.1 General The force-velocity-displacement relationships and damping properties, assumed as the damping device nominal design properties in Section 18.2.4.4, shall be confirmed by the tests conducted in accordance with Section 18.6.2 or shall be based on prior tests of devices meeting the similarity requirements of Section 18.6.2.3.

The prototype tests specified in Section 18.6.2 shall be conducted to confirm the force-velocity-displacement properties of the damping devices assumed for analysis and design and to demonstrate the robustness of individual devices under seismic excitation. These tests shall be conducted prior to production of devices for construction.

The production testing requirements are specified in Section 18.6.3.

Device nominal properties determined from the prototype testing shall meet the acceptance criteria established using $\lambda_{(spec,max)}$ and $\lambda_{(spec,min)}$ from Section 18.2.4.5. These criteria shall account for likely variations in material properties.

Device nominal properties determined from the production testing of Section 18.6.3 shall meet the acceptance criteria established using $\lambda_{(spec,max)}$ and $\lambda_{(spec,min)}$ from Section 18.2.4.5.

The fabrication and quality control procedures used for all prototype and production devices shall be identical. These procedures shall be approved by the RDP prior to the fabrication of prototype devices.

18.6.1.1 Qualification Tests Damping device manufacturers shall submit, for approval by the RDP, the results of qualification tests, analysis of test data, and supporting studies used to quantify the effects of heating caused by cyclic dynamic motion, loading rate, duration of seismic and wind loading, variability and uncertainty in production device properties, operating temperature, aging, environmental exposure, and contamination. The qualification testing shall be applicable to the component types, materials, and force-velocity-displacement response to be used in the proposed construction.

18.6.2 Prototype Tests The following tests shall be performed separately on two full-size damping devices of each type and size used in the design, in the order listed as follows.

Representative sizes of each type of device are permitted to be used for prototype testing, provided that both of the following conditions are met:

1. Fabrication and quality control procedures are identical for each type and size of device used in the structure.
2. Prototype testing of representative sizes is approved by the RDP responsible for the design of the structure.

Test specimens shall not be used for construction unless they are approved by the RDP responsible for the design of the structure and meet the requirements for prototype and production tests.

18.6.2.1 Data Recording The force-deflection relationship for each cycle of each test shall be recorded electronically.

18.6.2.2 Sequence and Cycles of Testing For all the following test sequences, each damping device shall be subjected to gravity load effects and thermal environments representative of the installed condition. For seismic testing, the displacement in the devices calculated for the MCE_R ground motions, termed herein as the maximum device displacement, shall be used.

Prior to the sequence of prototype tests defined in this section, a production test, in accordance with Section 18.6.3, shall be performed, and data from this test shall be used as a baseline for comparison with subsequent tests on production dampers.

1. Each damping device shall be subjected to the number of cycles expected in the design windstorm, but not less than 2,000 continuous fully reversed cycles of wind load. Wind load shall be at amplitudes expected in the design windstorm and shall be applied at a frequency equal to the inverse of the fundamental period of the structure, $1/T_1$.

   It is permitted to use alternate loading protocols, representative of the design windstorm, that apportion the total wind displacement into its expected static, pseudostatic, and dynamic components.

   EXCEPTION: Damping devices need not be subjected to these tests if they are not subject to wind-induced forces or displacements, or if the design wind force is less than the device yield or slip force.
2. Each damping device shall be brought to ambient temperature and loaded with the following sequence of fully reversed, sinusoidal cycles at a frequency equal to $1/(1.5T_1)$:
   (a) Ten fully reversed cycles at the displacement in the energy-dissipation device, corresponding to 0.33 times the MCE_R device displacement;
   (b) Five fully reversed cycles at the displacement in the energy-dissipation device, corresponding to 0.67 times the MCE_R device displacement;
   (c) Three fully reversed cycles at the displacement in the energy-dissipation device, corresponding to 1.0 times the MCE_R device displacement; and
   (d) (d) Where test (c) produces a force in the energy-dissipation device that is less than the MCE_R force in the device from analysis, test (c) shall be repeated at a frequency that produces a force equal to, or greater than, the MCE_R force from analysis.
3. Where the damping device characteristics vary with operating temperature, the tests of Section 18.6.2.2, 2(a) through 2(d) shall be conducted on at least one device at a minimum of two additional temperatures (minimum and maximum) that bracket the design temperature range.

   EXCEPTION: Damping devices are permitted to be tested by alternative methods, provided all the following conditions are met:
   (a) Alternative methods of testing are equivalent to the cyclic testing requirements of this section.
   (b) Alternative methods capture the dependence of the damping device response on ambient temperature, frequency of loading, and temperature rise during testing.
   (c) Alternative methods are approved by the RDP responsible for the design of the structure.
4. If the force-deformation properties of the damping device at any displacement less than or equal to the maximum device displacement change by more than 15% for changes in testing frequency from $1/(1.5T_1)$ to $2.5/T_1$, then the preceding tests [2(a) through 2(c)] shall also be performed at frequencies equal to $1/T_1$ and $2.5/T_1$.

EXCEPTION: When full-scale dynamic testing is not possible because of test machine limitations, it is permitted to use reduced-scale prototypes to qualify the rate-dependent properties of damping devices, provided scaling principles and similitude are used in the design of the reduced-scale devices and the test protocol.

18.6.2.3 Testing Similar Devices Prototype tests need not be performed on a particular damping device if a previously prototype-tested unit exists that meets all the following conditions:

1. It is of similar dimensional characteristics, internal construction, and static and dynamic internal pressures (if any) to the subject damping device; and
2. It is of the same type and materials as the subject damping device; and

3. It was fabricated using identical documented manufacturing and quality control procedures that govern the subject damping device; and
4. It was tested under similar maximum strokes and forces to those required of the subject damping device.

18.6.2.4 Determination of Force-Velocity-Displacement Characteristics The force-velocity-displacement characteristics of the prototype damping device shall be based on the cyclic displacement tests specified in Section 18.6.2.2 and all the following requirements:

1. The maximum force and minimum force at zero displacement, the maximum force and minimum force at maximum device displacement, and the area of hysteresis loop (E_{loop}) shall be calculated for each cycle of deformation. Where required, the effective stiffness of a damping device shall be calculated for each cycle of deformation using Equation (17.8-1).
2. Damping device nominal test properties for analysis and design shall be based on the average value for the first three cycles of test at a given displacement. For each cycle of each test, corresponding lambda factors (λ_{test}) for cyclic effects shall be established by comparison of nominal and per-cycle properties.
3. Lambda (λ) factors for velocity and temperature shall be determined simultaneously with those for cyclic effects where full-scale prototype test data are available. Where these or similar effects are determined from separate tests, lambda factors shall be established by the comparison of properties determined under prototype test conditions with corresponding properties determined under the range of test conditions applicable to the property variation parameter.

18.6.2.5 Device Adequacy The performance of a prototype damping device shall be deemed adequate if all the conditions listed as follows are satisfied. The 15% limits specified in the following text are permitted to be increased by the RDP responsible for the design of the structure, provided that the increased limit has been demonstrated by analysis not to have a deleterious effect on the response of the structure.

18.6.2.5.1 Displacement-Dependent Damping Devices The performance of the prototype displacement-dependent damping devices shall be deemed adequate if all the following conditions, based on tests specified in Section 18.6.2.2, are satisfied:

1. For Test 1, no signs of damage including leakage, yielding, or breakage.
2. For Tests 2, 3, and 4, the maximum force and minimum force at zero displacement for a damping device for any one cycle does not differ by more than 15% from the average maximum and minimum forces at zero displacement, as calculated from all cycles in that test at a specific frequency and temperature.
3. For Tests 2, 3, and 4, the maximum force and minimum force at maximum device displacement for a damping device for any one cycle does not differ by more than 15% from the average maximum and minimum forces at the maximum device displacement, as calculated from all cycles in that test at a specific frequency and temperature.
4. For Tests 2, 3, and 4, the area of hysteresis loop (E_{loop}) of a damping device for any one cycle does not differ by more than 15% from the average area of the hysteresis loop, as calculated from all cycles in that test at a specific frequency and temperature.
5. The average maximum and minimum forces at zero displacement and maximum displacement, and the average area of the hysteresis loop (E_{loop}), as calculated for each test in the sequence of Tests 2, 3, and 4, shall not differ by more than 15% from the target values specified by the RDP responsible for the design of the structure.
6. The average maximum and minimum forces at zero displacement and the maximum displacement, and the average area of the hysteresis loop (E_{loop}), as calculated for Test 2(c), shall fall within the limits specified by the RDP, as described by the nominal properties and the lambda factor for specification tolerance ($\lambda_{(spec,max)}$ and $\lambda_{(spec,min)}$) from Section 18.2.4.5.
7. The test lambda factors for damping units, determined in accordance with Section 18.6.2.4, shall not exceed the values specified by the RDP, in accordance with Section 18.2.4.5.

18.6.2.5.2 Velocity-Dependent Damping Devices The performance of the prototype velocity-dependent damping devices shall be deemed adequate if all the following conditions, based on tests specified in Section 18.6.2.2, are satisfied:

1. For Test 1, no signs of damage including leakage, yielding, or breakage.
2. For velocity-dependent damping devices with stiffness, the effective stiffness of a damping device in any one cycle of Tests 2, 3, and 4 does not differ by more than 15% from the average effective stiffness, as calculated from all cycles in that test at a specific frequency and temperature.
3. For Tests 2, 3, and 4, the maximum force and minimum force at zero displacement for a damping device for any one cycle does not differ by more than 15% from the average maximum and minimum forces at zero displacement, as calculated from all cycles in that test at a specific frequency and temperature.
4. For Tests 2, 3, and 4, the area of hysteresis loop (E_{loop}) of a damping device for any one cycle does not differ by more than 15% from the average area of the hysteresis loop, as calculated from all cycles in that test at a specific frequency and temperature.
5. The average maximum and minimum forces at zero displacement, effective stiffness (for damping devices with stiffness only), and average area of the hysteresis loop (E_{loop}), calculated for Test 2(c), shall fall within the limits specified by the RDP, as described by the nominal properties and the lambda factor for specification tolerance ($\lambda_{(spec,max)}$ and $\lambda_{(spec,min)}$) from Section 18.2.4.5.
6. The test lambda factors for damping units determined in accordance with Section 18.6.2.4 shall not exceed the values specified by the RDP, in accordance with Section 18.2.4.5.

18.6.3 Production Tests Prior to installation in a building, damping devices shall be tested in accordance with the requirements of this section.

A test program for the production damping devices shall be established by the RDP. The test program shall validate the nominal properties by testing 100% of the devices for three cycles at 0.67 times the MCE_R stroke at a frequency equal to $1/(1.5T_1)$. The measured values of the nominal properties shall fall within the limits provided in the project specifications. These limits shall agree with the specification tolerances on nominal design properties established in Section 18.2.4.5.

EXCEPTION: Production damping devices need not be subjected to this test program if it can be shown by other means that their properties meet the requirements of the project specifications. In such cases, the RDP shall establish an alternative program to ensure the quality of the installed damping devices. This alternative program shall include production testing of at least one device of each type and size unless project-specific prototype tests have been conducted on that identical device type and size. Devices that undergo inelastic action, or are otherwise damaged during this test, shall not be used in construction.

18.7 ALTERNATE PROCEDURES AND CORRESPONDING ACCEPTANCE CRITERIA

Structures analyzed by the response spectrum procedure shall meet the requirements of Sections 18.7.1, 18.7.3, and 18.7.4. Structures analyzed by the equivalent lateral force procedure shall meet the requirements of Sections 18.7.2, 18.7.3, and 18.7.4.

Table 18.7-1. Damping Coefficient, B_{V+I}, B_{1D}, B_{1E}, B_R, B_{1M}, B_{mD}, B_{mM} (Where Period of the Structure ≥ T_0).

Effective Damping, β (percentage of critical)	B_{V+I}, B_{1D}, B_{1E}, B_R, B_{1M}, B_{mD}, B_{mM} (where period of the structure ≥ T_0)
≤2	0.8
5	1.0
10	1.2
20	1.5
30	1.8
40	2.1
50	2.4
60	2.7
70	3.0
80	3.3
90	3.6
≥100	4.0

18.7.1 Response Spectrum Procedure Where the response spectrum procedure is used to analyze a structure with a damping system, the requirements of this section shall apply.

18.7.1.1 Modeling A mathematical model of the seismic force-resisting system and damping system shall be constructed that represents the spatial distribution of mass, stiffness, and damping throughout the structure. The model and analysis shall comply with the requirements of Section 12.9 for the seismic force-resisting system and with the requirements of this section for the damping system. The stiffness and damping properties of the damping devices used in the models shall be based on or verified by testing of the damping devices, as specified in Section 18.6.

The elastic stiffness of elements of the damping system other than damping devices shall be explicitly modeled. Stiffness of damping devices shall be modeled depending on damping device type as follows:

1. For displacement-dependent damping devices: Displacement-dependent damping devices shall be modeled with an effective stiffness that represents damping device force at the response displacement of interest (e.g., design story drift). Alternatively, the stiffness of hysteretic and friction damping devices is permitted to be excluded from response spectrum analysis (RSA), provided that design forces in displacement-dependent damping devices, Q_{DSD}, are applied to the model as external loads (Section 18.7.4.5).
2. For velocity-dependent damping devices: Velocity-dependent damping devices that have a stiffness component (e.g., viscoelastic damping devices) shall be modeled with an effective stiffness corresponding to the amplitude and frequency of interest.

18.7.1.2 Seismic Force-Resisting System

18.7.1.2.1 Seismic Base Shear The seismic base shear, V, of the structure in a given direction shall be determined as the combination of modal components, V_m, subject to the limits of Equation (18.7-1):

Table 18.7-2. Force Coefficient,[a,b] C_{mFD}.

Effective Damping	$\mu \le 1.0$[c] $\alpha \le 0.25$	$\alpha = 0.5$	$\alpha = 0.75$	$\alpha \ge 1.0$	$C_{mFD} = 1.0$
≤0.05	1.00	1.00	1.00	1.00	μ ≥ 1.0
0.1	1.00	1.00	1.00	1.00	μ ≥ 1.0
0.2	1.00	0.95	0.94	0.93	μ ≥ 1.1
0.3	1.00	0.92	0.88	0.86	μ ≥ 1.2
0.4	1.00	0.88	0.81	0.78	μ ≥ 1.3
0.5	1.00	0.84	0.73	0.71	μ ≥ 1.4
0.6	1.00	0.79	0.64	0.64	μ ≥ 1.6
0.7	1.00	0.75	0.55	0.58	μ ≥ 1.7
0.8	1.00	0.70	0.50	0.53	μ ≥ 1.9
0.9	1.00	0.66	0.50	0.50	μ ≥ 2.1
≥1.0	1.00	0.62	0.50	0.50	μ ≥ 2.2

[a] Unless analysis or test data support other values, the force coefficient C_{mFD} for viscoelastic systems shall be taken as 1.0.
[b] Interpolation shall be used for intermediate values of velocity exponent, α, and ductility demand, μ.
[c] C_{mFD} shall be taken as equal to 1.0 for values of ductility demand, μ, greater than or equal to the values shown.

Table 18.7-3. Force Coefficient,[a,b] C_{mFV}.

Effective Damping	$\alpha \le 0.25$	$\alpha = 0.5$	$\alpha = 0.75$	$\alpha \ge 1.0$
≤0.05	1.00	0.35	0.20	0.10
0.1	1.00	0.44	0.31	0.20
0.2	1.00	0.56	0.46	0.37
0.3	1.00	0.64	0.58	0.51
0.4	1.00	0.70	0.69	0.62
0.5	1.00	0.75	0.77	0.71
0.6	1.00	0.80	0.84	0.77
0.7	1.00	0.83	0.90	0.81
0.8	1.00	0.90	0.94	0.90
0.9	1.00	1.00	1.00	1.00
≥1.0	1.00	1.00	1.00	1.00

[a] Unless analysis or test data support other values, the force coefficient C_{mFV} for viscoelastic systems shall be taken as 1.0.

[b] Interpolation shall be used for intermediate values of velocity exponent, α.

$$V \ge V_{\min} \quad (18.7\text{-}1)$$

The seismic base shear, V, of the structure shall be determined by the square root of the sum of the squares method (SRSS) or complete quadratic combination of modal base shear components, V_m.

18.7.1.2.2 Modal Base Shear Modal base shear of the mth mode of vibration, V_m, of the structure in the direction of interest shall be determined in accordance with Equations (18.7-2a) and (18.7-2b):

$$V_m = C_{Sm}\overline{W}_m \quad (18.7\text{-}2a)$$

$$\overline{W}_m = \frac{\left(\sum_{n}^{i=1} w_i \phi_{im}\right)^2}{\sum_{n}^{i=1} w_i \phi_{im}^2} \quad (18.7\text{-}2b)$$

where

C_{Sm} = Seismic response coefficient of the mth mode of vibration of the structure in the direction of interest, as determined from Section 18.7.1.2.4 ($m = 1$) or Section 18.7.1.2.6 ($m > 1$);

$\overline{W}_m$ = Effective seismic weight of the mth mode of vibration of the structure; and

ϕ_{im} = Displacement amplitude at the ith level of the structure in the mth mode of vibration in the direction of interest, normalized to unity at the roof level.

18.7.1.2.3 Modal Participation Factor The modal participation factor of the mth mode of vibration, Γ_m, of the structure in the direction of interest shall be determined in accordance with Equation (18.7-3):

$$\Gamma_m = \frac{\overline{W}_m}{\sum_{n}^{i=1} w_i \phi_{im}} \quad (18.7\text{-}3)$$

18.7.1.2.4 Fundamental Mode Seismic Response Coefficient The fundamental mode ($m = 1$) seismic response coefficient, C_{S1}, in the direction of interest shall be determined in accordance with Equations (18.7-4) and (18.7-5):

For $T_{1D} < T_S$,

$$C_{S1} = \left(\frac{R}{C_d}\right)\frac{S_{DS}}{\Omega_0 B_{1D}} \quad (18.7\text{-}4)$$

For $T_{1D} \ge T_S$,

$$C_{S1} = \left(\frac{R}{C_d}\right)\frac{S_{D1}}{T_{1D}(\Omega_0 B_{1D})} \quad (18.7\text{-}5)$$

18.7.1.2.5 Effective Fundamental Mode Period Determination The effective fundamental mode ($m = 1$) period at the design earthquake ground motion, T_{1D}, and at the MCE_R ground motion, T_{1M}, shall be based on either explicit consideration of the postyield force deflection characteristics of the structure or determined in accordance with Equations (18.7-6) and (18.7-7):

$$T_{1D} = T_1\sqrt{\mu_D} \quad (18.7\text{-}6)$$

$$T_{1M} = T_1\sqrt{\mu_M} \quad (18.7\text{-}7)$$

18.7.1.2.6 Higher Mode Seismic Response Coefficient Higher mode ($m > 1$) seismic response coefficient, C_{Sm}, of the mth mode of vibration ($m > 1$) of the structure in the direction of interest shall be determined in accordance with Equations (18.7-8) and (18.7-9):

For $T_m < T_S$,

$$C_{Sm} = \left(\frac{R}{C_d}\right)\frac{S_{DS}}{\Omega_0 B_{mD}} \quad (18.7\text{-}8)$$

For $T_m \ge T_S$,

$$C_{Sm} = \left(\frac{R}{C_d}\right)\frac{S_{D1}}{T_m(\Omega_0 B_{mD})} \quad (18.7\text{-}9)$$

where T_m is the period(s), of the mth mode of vibration of the structure in the direction under consideration, and B_{mD} is the numerical coefficient as set forth in Table 18.7-1 for effective damping equal to β_{mD} and period of the structure equal to T_m.

18.7.1.2.7 Design Lateral Force Design lateral force at level i caused by the mth mode of vibration, F_{im}, of the structure in the direction of interest shall be determined in accordance with Equation (18.7-10):

$$F_{im} = w_i \phi_{im} \frac{\Gamma_m}{\bar{W}_m} V_m \quad (18.7\text{-}10)$$

Design forces in elements of the seismic force-resisting system shall be determined by the SRSS or complete quadratic combination of modal design forces.

18.7.1.3 Damping System Design forces in damping devices and other elements of the damping system shall be determined on the basis of the floor deflection, story drift, and story velocity response parameters described in the subsequent sections.

Displacements and velocities used to determine maximum forces in damping devices at each story shall account for the angle of orientation of each device from the horizontal and consider the effects of increased response caused by torsion required for design of the seismic force-resisting system.

Floor deflections at level i, δ_{iD} and δ_{iM}, story drifts, Δ_D and Δ_M, and story velocities, ∇_D and ∇_M, shall be calculated for both the design earthquake ground motions and the MCE_R ground motions, respectively, in accordance with this section.

18.7.1.3.1 Design Earthquake Floor Deflection The deflection of the structure caused by the design earthquake ground motions at level i in the mth mode of vibration, δ_{imD}, of the structure in the direction of interest shall be determined in accordance with Equation (18.7-11):

$$\delta_{imD} = D_{mD} \phi_{im} \quad (18.7\text{-}11)$$

The total design deflection at each floor of the structure shall be calculated by the SRSS or complete quadratic combination of modal design earthquake deflections.

18.7.1.3.2 Design Earthquake Roof Displacement Fundamental ($m = 1$) and higher mode ($m > 1$) roof displacements caused by the design earthquake ground motions, D_{1D} and D_{mD}, of the structure in the direction of interest shall be determined in accordance with Equations (18.7-12) and (18.7-13):

For $m = 1$,

$$D_{1D} = \left(\frac{g}{4\pi^2}\right)\Gamma_1 \frac{S_{DS}T_{1D}^2}{B_{1D}} \geq \left(\frac{g}{4\pi^2}\right)\Gamma_1 \frac{S_{DS}T_1^2}{B_{1E}}, \quad T_{1D} < T_S$$

(18.7-12a)

$$D_{1D} = \left(\frac{g}{4\pi^2}\right)\Gamma_1 \frac{S_{D1}T_{1D}}{B_{1D}} \geq \left(\frac{g}{4\pi^2}\right)\Gamma_1 \frac{S_{D1}T_1}{B_{1E}}, \quad T_{1D} \geq T_S$$

(18.7-12b)

For $m > 1$,

$$D_{mD} = \left(\frac{g}{4\pi^2}\right)\Gamma_m \frac{S_{D1}T_m}{B_{mD}} \leq \left(\frac{g}{4\pi^2}\right)\Gamma_m \frac{S_{DS}T_m^2}{B_{mD}} \quad (18.7\text{-}13)$$

18.7.1.3.3 Design Earthquake Story Drift Design story drift in the fundamental mode, Δ_{1D}, and higher modes, Δ_{mD} ($m > 1$), of the structure in the direction of interest shall be calculated in accordance with Section 12.8.6, using modal roof displacements of Section 18.7.1.3.2.

Total design story drift, ∇_D, shall be determined by the SRSS or complete quadratic combination of modal design earthquake drifts.

18.7.1.3.4 Design Earthquake Story Velocity Design story velocity in the fundamental mode, ∇_{1D}, and higher modes, ∇_{mD} ($m > 1$), of the structure in the direction of interest shall be calculated in accordance with Equations (18.7-14) and (18.7-15):

$$\text{For } m = 1, \quad \nabla_{1D} = 2\pi \frac{\Delta_{1D}}{T_{1D}} \quad (18.7\text{-}14)$$

$$\text{For } m > 1, \quad \nabla_{mD} = 2\pi \frac{\Delta_{mD}}{T_m} \quad (18.7\text{-}15)$$

Total design story velocity, ∇_D, shall be determined by the SRSS or complete quadratic combination of modal design velocities.

18.7.1.3.5 MCE_R Response Total modal maximum floor deflection at level i, MCE_R story drift values, and MCE_R story velocity values shall be based on Sections 18.7.1.3.1, 18.7.1.3.3, and 18.7.1.3.4, respectively, except design roof displacement shall be replaced by MCE_R roof displacement. MCE_R roof displacement of the structure in the direction of interest shall be calculated in accordance with Equations (18.7-16) and (18.7-17):

For $m = 1$,

$$D_{1M} = \left(\frac{g}{4\pi^2}\right)\Gamma_1 \frac{S_{MS}T_{1M}^2}{B_{1M}} \geq \left(\frac{g}{4\pi^2}\right)\Gamma_1 \frac{S_{MS}T_1^2}{B_{1E}}, \quad T_{1M} < T_S$$

(18.7-16a)

$$D_{1M} = \left(\frac{g}{4\pi^2}\right)\Gamma_1 \frac{S_{M1}T_{1M}}{B_{1M}} \geq \left(\frac{g}{4\pi^2}\right)\Gamma_1 \frac{S_{M1}T_1}{B_{1E}}, \quad T_{1M} \geq T_S$$

(18.7-16b)

For $m > 1$,

$$D_{mM} = \left(\frac{g}{4\pi^2}\right)\Gamma_m \frac{S_{M1}T_m}{B_{mM}} \leq \left(\frac{g}{4\pi^2}\right)\Gamma_m \frac{S_{MS}T_m^2}{B_{mM}} \quad (18.7\text{-}17)$$

where B_{mM} is a numerical coefficient as set forth in Table 18.7-1 for effective damping equal to β_{mM} and period of the structure equal to T_m.

18.7.2 Equivalent Lateral Force Procedure Where the equivalent lateral force procedure is used to design a structure with a damping system, the requirements of this section shall apply.

18.7.2.1 Modeling Elements of the seismic force-resisting system shall be modeled in a manner consistent with the requirements of Section 12.8. For purposes of analysis, the structure shall be considered to be fixed at the base.

Elements of the damping system shall be modeled as required to determine design forces transferred from the damping devices to both the ground and the seismic force-resisting system. The effective stiffness of velocity-dependent damping devices shall be modeled.

Damping devices need not be explicitly modeled provided that effective damping is calculated in accordance with the procedures of Section 18.7.4 and used to modify response, as required in Sections 18.7.2.2 and 18.7.2.3.

The stiffness and damping properties of the damping devices used in the models shall be based on or verified by testing of the damping devices, as specified in Section 18.6.

18.7.2.2 Seismic Force-Resisting System

18.7.2.2.1 Seismic Base Shear. The seismic base shear, V, of the seismic force-resisting system in a given direction shall be determined as the combination of the two modal components, V_1 and V_R, in accordance with Equation (18.7-18):

$$V = \sqrt{V_1^2 + V_R^2} \geq V_{\min} \qquad (18.7\text{-}18)$$

where

V_1 = Design value of the seismic base shear of the fundamental mode in a given direction of response, as determined in Section 18.7.2.2.2;

V_R = Design value of the seismic base shear of the residual mode in a given direction, as determined in Section 18.7.2.2.6; and

$V_{\min}$ = Minimum allowable value of base shear permitted for design of the seismic force-resisting system of the structure in the direction of interest, as determined in Section 18.2.1.1.

18.7.2.2.2 Fundamental Mode Base Shear The fundamental mode base shear, V_1, shall be determined in accordance with Equation (18.7-19):

$$V_1 = C_{S1}\overline{W}_1 \qquad (18.7\text{-}19)$$

where C_{S1} is the fundamental mode seismic response coefficient, as determined in Section 18.7.2.2.4, and $\overline{W}_1$ is the effective fundamental mode seismic weight, including portions of the live load, as defined by Equation (18.7-2b) for $m = 1$.

18.7.2.2.3 Fundamental Mode Properties The fundamental mode shape, ϕ_{i1}, and participation factor, Γ_1, shall be determined by either dynamic analysis using the elastic structural properties and deformational characteristics of the resisting elements or using Equations (18.7-20) and (18.7-21):

$$\phi_{i1} = \frac{h_i}{h_n} \qquad (18.7\text{-}20)$$

$$\Gamma_1 = \frac{\overline{W}_1}{\sum_{i=1}^{n} w_i \phi_{i1}} \qquad (18.7\text{-}21)$$

where

h_i = The height above the base to level i;

h_n = The structural height as defined in Section 11.2; and

w_i = The portion of the total effective seismic weight, W, located at or assigned to level i.

The fundamental period, T_1, shall be determined either by dynamic analysis using the elastic structural properties and deformational characteristics of the resisting elements, or using Equation (18.7-22) as follows:

$$T_1 = 2\pi\sqrt{\frac{\sum_{i=1}^{n} w_i \delta_i^2}{g\sum_{i=1}^{n} f_i \delta_i}} \qquad (18.7\text{-}22)$$

where f_i is the lateral force at level i of the structure distributed in accordance with Section 12.8.3, and δ_i is the elastic deflection at level i of the structure caused by applied lateral forces f_i.

18.7.2.2.4 Fundamental Mode Seismic Response Coefficient The fundamental mode seismic response coefficient, C_{S1}, shall be determined using Equations (18.7-23) or (18.7-24):

For $T_{1D} < T_S$,

$$C_{S1} = \left(\frac{R}{C_d}\right)\frac{S_{DS}}{\Omega_0 B_{1D}} \qquad (18.7\text{-}23)$$

For $T_{1D} \geq T_S$,

$$C_{S1} = \left(\frac{R}{C_d}\right)\frac{S_{D1}}{T_{1D}(\Omega_0 B_{1D})} \qquad (18.7\text{-}24)$$

where

S_{DS} = The design spectral response acceleration parameter in the short period range,

S_{D1} = The design spectral response acceleration parameter at a period of 1 s, and

B_{1D} = Numerical coefficient as set forth in Table 18.7-1 for effective damping equal to $\beta_{mD}(m = 1)$ and period of the structure equal to T_{1D}.

18.7.2.2.5 Effective Fundamental Mode Period Determination The effective fundamental mode period at the design earthquake, T_{1D}, and at the MCE$_R$, T_{1M}, shall be based on explicit consideration of the postyield force deflection characteristics of the structure or shall be calculated using Equations (18.7-25) and (18.7-26):

$$T_{1D} = T_1\sqrt{\mu_D} \qquad (18.7\text{-}25)$$

$$T_{1M} = T_1\sqrt{\mu_M} \qquad (18.7\text{-}26)$$

18.7.2.2.6 Residual Mode Base Shear Residual mode base shear, V_R, shall be determined in accordance with Equation (18.7-27):

$$V_R = C_{SR}\overline{W}_R \qquad (18.7\text{-}27)$$

where C_{SR} is the residual mode seismic response coefficient as determined in Section 18.7.2.2.8, and $\overline{W}_R$ is the effective residual mode weight of the structure determined using Equation (18.7-30).

18.7.2.2.7 Residual Mode Properties Residual mode shape, ϕ_{iR}, participation factor, Γ_R, effective residual mode seismic weight of the structure, $\overline{W}_R$, and effective period, T_R, shall be determined using Equations (18.7-28) through (18.7-31):

$$\phi_{iR} = \frac{1 - \Gamma_1\phi_{i1}}{1 - \Gamma_1} \qquad (18.7\text{-}28)$$

$$\Gamma_R = 1 - \Gamma_1 \qquad (18.7\text{-}29)$$

$$\overline{W}_R = W - \overline{W}_1 \qquad (18.7\text{-}30)$$

$$T_R = 0.4T_1 \qquad (18.7\text{-}31)$$

18.7.2.2.8 Residual Mode Seismic Response Coefficient The residual mode seismic response coefficient, C_{SR}, shall be determined in accordance with Equation (18.7-32):

$$C_{SR} = \left(\frac{R}{C_d}\right)\frac{S_{DS}}{\Omega_0 B_R} \quad (18.7\text{-}32)$$

where B_R is a numerical coefficient as set forth in Table 18.7-1 for effective damping equal to β_R and period of the structure equal to T_R.

18.7.2.2.9 Design Lateral Force The design lateral force in elements of the seismic force-resisting system at level i caused by fundamental mode response, F_{i1}, and residual mode response, F_{iR}, of the structure in the direction of interest shall be determined in accordance with Equations (18.7-33) and (18.7-34):

$$F_{i1} = w_i \phi_{i1} \frac{\Gamma_1}{\overline{W}_1} V_1 \quad (18.7\text{-}33)$$

$$F_{iR} = w_i \phi_{iR} \frac{\Gamma_R}{\overline{W}_R} V_R \quad (18.7\text{-}34)$$

Design forces in elements of the seismic force-resisting system shall be determined by taking the SRSS of the forces caused by fundamental and residual modes.

18.7.2.3 Damping System Design forces in damping devices and other elements of the damping system shall be determined on the basis of the floor deflection, story drift, and story velocity response parameters described in the following sections.

Displacements and velocities used to determine maximum forces in damping devices at each story shall account for the angle of orientation of each device from the horizontal and consider the effects of increased response caused by torsion required for design of the seismic force-resisting system.

Floor deflections at level i, δ_{iD} and δ_{iM}, story drifts, Δ_D and Δ_M, and story velocities, ∇_D and ∇_M, shall be calculated for both the design earthquake ground motions and the MCE_R ground motions, respectively, in accordance with the following sections.

18.7.2.3.1 Design Earthquake Floor Deflection The total design deflection at each floor of the structure in the direction of interest shall be calculated as the SRSS of the fundamental and residual mode floor deflections. The fundamental and residual mode deflections caused by the design earthquake ground motions, δ_{i1D} and δ_{iRD}, at the center of rigidity of level i of the structure in the direction of interest shall be determined using Equations (18.7-35) and (18.7-36):

$$\delta_{i1D} = D_{1D}\phi_{i1} \quad (18.7\text{-}35)$$

$$\delta_{iRD} = D_{RD}\phi_{iR} \quad (18.7\text{-}36)$$

where D_{1D} is the fundamental mode design displacement at the center of rigidity of the roof level of the structure in the direction under consideration, Section 18.7.2.3.2, and D_{RD} is the residual mode design displacement at the center of rigidity of the roof level of the structure in the direction under consideration, Section 18.7.2.3.2.

18.7.2.3.2 Design Earthquake Roof Displacement Fundamental and residual mode displacements caused by the design earthquake ground motions, D_{1D} and D_{1R}, at the center of rigidity of the roof level of the structure in the direction of interest shall be determined using Equations (18.7-37) and (18.7-38):

$T_{1D} < T_S$

$$D_{1D} = \left(\frac{g}{4\pi^2}\right)\Gamma_1 \frac{S_{DS}T_{1D}^2}{B_{1D}} \geq \left(\frac{g}{4\pi^2}\right)\Gamma_1 \frac{S_{DS}T_1^2}{B_{1E}} \quad (18.7\text{-}37a)$$

$T_{1D} \geq T_S$

$$D_{1D} = \left(\frac{g}{4\pi^2}\right)\Gamma_1 \frac{S_{D1}T_{1D}}{B_{1D}} \geq \left(\frac{g}{4\pi^2}\right)\Gamma_1 \frac{S_{D1}T_1}{B_{1E}} \quad (18.7\text{-}37b)$$

$$D_{RD} = \left(\frac{g}{4\pi^2}\right)\Gamma_R \frac{S_{D1}T_R}{B_R} \leq \left(\frac{g}{4\pi^2}\right)\Gamma_R \frac{S_{DS}T_R^2}{B_R} \quad (18.7\text{-}38)$$

18.7.2.3.3 Design Earthquake Story Drift Design story drifts, Δ_D, in the direction of interest shall be calculated using Equation (18.7-39):

$$\Delta_D = \sqrt{\Delta_{1D}^2 + \Delta_{RD}^2} \quad (18.7\text{-}39)$$

where Δ_{1D} is the design story drift caused by the fundamental mode of vibration of the structure in the direction of interest and Δ_{RD} is the design story drift caused by the residual mode of vibration of the structure in the direction of interest.

Modal design story drifts, Δ_{1D} and Δ_{RD}, shall be determined as the difference of the deflections at the top and bottom of the story under consideration using the floor deflections of Section 18.7.2.3.1.

18.7.2.3.4 Design Earthquake Story Velocity Design story velocities, ∇_D, in the direction of interest shall be calculated in accordance with Equations (18.7-40) through (18.7-42):

$$\nabla_D = \sqrt{\nabla_{1D}^2 + \nabla_{RD}^2} \quad (18.7\text{-}40)$$

$$\nabla_{1D} = 2\pi \frac{\Delta_{1D}}{T_{1D}} \quad (18.7\text{-}41)$$

$$\nabla_{RD} = 2\pi \frac{\Delta_{RD}}{T_R} \quad (18.7\text{-}42)$$

where ∇_{1D} is the design story velocity caused by the fundamental mode of vibration of the structure in the direction of interest, and ∇_{RD} is the design story velocity caused by the residual mode of vibration of the structure in the direction of interest.

18.7.2.3.5 MCE_R Response Total modal MCE_R floor deflections at level i, maximum story drifts, and maximum story velocities shall be based on the equations in Sections 18.7.2.3.1, 18.7.2.3.3, and 18.7.2.3.4, respectively, except that design roof displacements shall be replaced by MCE_R roof displacements. MCE_R roof displacements shall be calculated in accordance with Equations (18.7-43) and (18.7-44):

$T_{1M} < T_S$

$$D_{1M} = \left(\frac{g}{4\pi^2}\right)\Gamma_1 \frac{S_{MS}T_{1M}^2}{B_{1M}} \geq \left(\frac{g}{4\pi^2}\right)\Gamma_1 \frac{S_{MS}T_1^2}{B_{1E}} \quad (18.7\text{-}43a)$$

$T_{1M} \geq T_S$

$$D_{1M} = \left(\frac{g}{4\pi^2}\right)\Gamma_1 \frac{S_{M1}T_{1M}}{B_{1M}} \geq \left(\frac{g}{4\pi^2}\right)\Gamma_1 \frac{S_{M1}T_1}{B_{1E}} \quad (18.7\text{-}43b)$$

$$D_{RM} = \left(\frac{g}{4\pi^2}\right)\Gamma_R \frac{S_{M1}T_R}{B_R} \leq \left(\frac{g}{4\pi^2}\right)\Gamma_R \frac{S_{MS}T_R^2}{B_R} \quad (18.7\text{-}44)$$

where

S_{M1} = MCE$_R$, 5% damped, spectral response acceleration parameter at a period of 1 s adjusted for site class effects, as defined in Section 11.4.4;

S_{MS} = MCE$_R$, 5% damped, spectral response acceleration parameter at short periods adjusted for site class effects, as defined in Section 11.4.4; and

B_{1M} = Numerical coefficient as set forth in Table 18.7-1 for effective damping equal to $\beta_{mM}(m=1)$ and period of structure equal to T_{1M}.

18.7.3 Damped Response Modification As required in Sections 18.7.1 and 18.7.2, response of the structure shall be modified for the effects of the damping system.

18.7.3.1 Damping Coefficient Where the period of the structure is greater than or equal to T_0, the damping coefficient shall be as prescribed in Table 18.7-1. Where the period of the structure is less than T_0, the damping coefficient shall be linearly interpolated between a value of 1.0 at a 0 s period for all values of effective damping and the value at period T_0, as indicated in Table 18.7-1.

18.7.3.2 Effective Damping The effective damping at the design displacement, β_{mD}, and at the MCE$_R$ displacement, β_{mM}, of the mth mode of vibration of the structure in the direction under consideration shall be calculated using Equations (18.7-45) and (18.7-46):

$$\beta_{mD} = \beta_I + \beta_{Vm}\sqrt{\mu_D} + \beta_{HD} \qquad (18.7\text{-}45)$$

$$\beta_{mM} = \beta_I + \beta_{Vm}\sqrt{\mu_M} + \beta_{HM} \qquad (18.7\text{-}46)$$

where

β_{HD} = Component of effective damping of the structure in the direction of interest caused by postyield hysteretic behavior of the seismic force-resisting system and elements of the damping system at effective ductility demand, μ_D;

β_{HM} = Component of effective damping of the structure in the direction of interest caused by postyield hysteretic behavior of the seismic force-resisting system and elements of the damping system at effective ductility demand, μ_M;

β_I = Component of effective damping of the structure caused by the inherent dissipation of energy by elements of the structure, at or just below the effective yield displacement of the seismic force-resisting system;

β_{Vm} = Component of effective damping of the mth mode of vibration of the structure in the direction of interest caused by viscous dissipation of energy by the damping system, at or just below the effective yield displacement of the seismic force-resisting system;

μ_D = Effective ductility demand on the seismic force-resisting system in the direction of interest caused by the design earthquake ground motions; and

μ_M = Effective ductility demand on the seismic force-resisting system in the direction of interest caused by the MCE$_R$ ground motions.

Unless analysis or test data support other values, the effective ductility demand of higher modes of vibration in the direction of interest shall be taken as 1.0.

18.7.3.2.1 Inherent Damping Inherent damping, β_I, shall be based on the material type, configuration, and behavior of the structure and nonstructural components responding dynamically at or just below yield of the seismic force-resisting system. Unless analysis or test data support other values, inherent damping shall be taken as not greater than 3% of critical for all modes of vibration.

18.7.3.2.2 Hysteretic Damping Hysteretic damping of the seismic force-resisting system and elements of the damping system shall be based either on test or analysis or shall be calculated using Equations (18.7-47) and (18.7-48):

$$\beta_{HD} = q_H(0.64 - \beta_I)\left(1 - \frac{1}{\mu_D}\right) \qquad (18.7\text{-}47)$$

$$\beta_{HM} = q_H(0.64 - \beta_I)\left(1 - \frac{1}{\mu_M}\right) \qquad (18.7\text{-}48)$$

where

q_H = Hysteresis loop adjustment factor, as defined in Section 18.7.3.2.2.1;

μ_D = Effective ductility demand on the seismic force-resisting system in the direction of interest caused by the design earthquake ground motions; and

μ_M = Effective ductility demand on the seismic force-resisting system in the direction of interest caused by the MCE$_R$ ground motions.

Unless analysis or test data support other values, the hysteretic damping of higher modes of vibration in the direction of interest shall be taken as zero.

18.7.3.2.2.1 Hysteresis Loop Adjustment Factor. The calculation of hysteretic damping of the seismic force-resisting system and elements of the damping system shall consider pinching and other effects that reduce the area of the hysteresis loop during repeated cycles of earthquake demand. Unless analysis or test data support other values, the fraction of full hysteretic loop area of the seismic force-resisting system used for design shall be taken as equal to the factor, q_H, calculated using Equation (18.7-49):

$$q_H = 0.67\frac{T_S}{T_1} \qquad (18.7\text{-}49)$$

where T_S is the period defined by the ratio, S_{D1}/S_{DS}, and T_1 is the period of the fundamental mode of vibration of the structure in the direction of interest.

The value of q_H shall not be taken as greater than 1.0 and need not be taken as less than 0.5.

18.7.3.2.3 Viscous Damping. Viscous damping of the mth mode of vibration of the structure, β_{Vm}, shall be calculated using Equations (18.7-50) and (18.7-51):

$$\beta_{Vm} = \frac{\sum_j W_{mj}}{4\pi W_m} \qquad (18.7\text{-}50)$$

$$W_m = \frac{1}{2}\sum_j F_{im}\delta_{im} \qquad (18.7\text{-}51)$$

where

W_{mj} = Work done by jth damping device in one complete cycle of dynamic response corresponding to the mth mode of vibration of the structure in the direction of interest at modal displacements, δ_{im};

W_m = Maximum strain energy in the mth mode of vibration of the structure in the direction of interest at modal displacements, δ_{im};

F_{im} = mth mode inertial force at level i; and

δ_{im} = Deflection of level i in the mth mode of vibration at the center of rigidity of the structure in the direction under consideration.

Viscous modal damping of displacement-dependent damping devices shall be based on a response amplitude equal to the effective yield displacement of the structure.

The calculation of the work done by individual damping devices shall consider orientation and participation of each device with respect to the mode of vibration of interest. The work done by individual damping devices shall be reduced as required to account for the flexibility of elements, including pins, bolts, gusset plates, brace extensions, and other components that connect damping devices to other elements of the structure.

18.7.3.3 Effective Ductility Demand The effective ductility demand on the seismic force-resisting system caused by the design earthquake ground motions, μ_D, and caused by the MCE_R ground motions, μ_M, shall be calculated using Equations (18.7-52) through (18.7-54):

$$\mu_D = \frac{D_{1D}}{D_Y} \geq 1.0 \quad (18.7\text{-}52)$$

$$\mu_M = \frac{D_{1M}}{D_Y} \geq 1.0 \quad (18.7\text{-}53)$$

$$D_Y = \left(\frac{g}{4\pi^2}\right)\left(\frac{\Omega_0 C_d}{R}\right)\Gamma_1 C_{S1} T_1^2 \quad (18.7\text{-}54)$$

where

D_{1D} = Fundamental mode design displacement at the center of rigidity of the roof level of the structure in the direction under consideration, Section 18.7.1.3.2 or 18.7.2.3.2;

D_{1M} = Fundamental mode maximum displacement at the center of rigidity of the roof level of the structure in the direction under consideration, Section 18.7.1.3.5 or 18.7.2.3.5;

D_Y = Displacement at the center of rigidity of the roof level of the structure at the effective yield point of the seismic force-resisting system;

R = Response modification coefficient from Table 12.2-1;

C_d = Deflection amplification factor from Table 12.2-1;

Ω_0 = Overstrength factor from Table 12.2-1;

Γ_1 = Participation factor of the fundamental mode of vibration of the structure in the direction of interest, Section 18.7.1.2.3 or 18.7.2.2.3 ($m \doteq 1$);

C_{S1} = Seismic response coefficient of the fundamental mode of vibration of the structure in the direction of interest, Section 18.7.1.2.4 or 18.7.2.2.4 ($m = 1$); and

T_1 = Period of the fundamental mode of vibration of the structure in the direction of interest.

The design ductility demand, μ_D, shall not exceed the maximum value of effective ductility demand, μ_{max}, given in Section 18.7.3.4.

EXCEPTION: It is permitted to use nonlinear modeling, as described in Section 18.3, to develop a force-displacement (pushover) curve of the seismic force-resisting system. It is permitted to use this curve in lieu of the effective yield displacement, D_Y, of Equation (18.7-54) to calculate the effective ductility demand caused by the design earthquake ground motions, μ_D, and caused by the MCE_R ground motions, μ_M, in Equations (18.7-52) and (18.7-53), respectively. In this case, the value of (R/C_d) shall be taken as 1.0 in Equations (18.7-4), (18.7-5), (18.7-8), and (18.7-9).

18.7.3.4 Maximum Effective Ductility Demand For determination of the hysteresis loop adjustment factor, hysteretic damping, and other parameters, the maximum value of effective ductility demand, μ_{max}, shall be calculated using Equations (18.7-55) and (18.7-56):

For $T_{1D} \leq T_S$,

$$\mu_{max} = 0.5[(R/(\Omega_0 I_e))^2 + 1] \quad (18.7\text{-}55)$$

For $T_1 \geq T_S$,

$$\mu_{max} = R/(\Omega_0 I_e) \quad (18.7\text{-}56)$$

where I_e is the Importance Factor determined in accordance with Section 11.5.1, and T_{1D} is the effective period of the fundamental mode of vibration of the structure at the design displacement in the direction under consideration.

For $T_1 < T_S < T_{1D}$, μ_{max} shall be determined by linear interpolation between the values of Equations (18.7-55) and (18.7-56).

18.7.4 Seismic Load Conditions and Acceptance Criteria for RSA and ELF Procedures Design forces and displacements determined in accordance with the response spectrum procedure of Section 18.7.1, or the equivalent lateral force (ELF) procedure of Section 18.7.2, shall be checked using the strength design criteria of this standard and the seismic loading conditions of Section 18.7.4.3.

The seismic force-resisting system, damping system, seismic loading conditions, and acceptance criteria shall conform to the following subsections.

18.7.4.1 Seismic Force-Resisting System The seismic force-resisting system shall satisfy the requirements of Section 12.2.1 using seismic base shear and design forces determined in accordance with Section 18.7.1.2 or 18.7.2.2.

The design story drift, Δ_D, as determined in either Section 18.7.1.3.3 or 18.7.2.3.3, shall not exceed (R/C_d) times the allowable story drift, as obtained from Table 12.12-1, considering the effects of torsion, as required in Section 12.12.1.

18.7.4.2 Damping System The damping system shall satisfy the requirements of Section 12.2.1 for seismic design forces and seismic loading conditions determined in accordance with Section 18.7.4.3. Force-controlled elements of the damping system shall be designed for seismic forces that are increased by 20% from those corresponding to average MCE_R response.

18.7.4.3 Combination of Load Effects The effects on the damping system and its components caused by gravity loads and seismic forces shall be combined in accordance with Section 12.4, using the effect of horizontal seismic forces, Q_E, determined in accordance with Section 18.7.4.5. The redundancy factor, ρ, shall be taken equal to 1.0 in all cases, and the seismic load effect including overstrength of Section 12.4.3 need not apply to the design of the damping system.

18.7.4.4 Modal Damping System Design Forces Modal damping system design forces shall be calculated on the basis of the type of damping devices and the modal design story displacements and velocities determined in accordance with either Section 18.7.1.3 or 18.7.2.3.

Modal design story displacements and velocities shall be increased as required to envelop the total design story displacements and velocities determined in accordance with Section 18.3, where peak response is required to be confirmed by response history analysis.

For displacement-dependent damping devices: Design seismic force in displacement-dependent damping devices shall be based on the maximum force in the device at displacements up to, and including, the design story drift, Δ_D.

For velocity-dependent damping devices: Design seismic force in each mode of vibration in velocity-dependent damping devices shall be based on the maximum force in the device at velocities up to, and including, the design story velocity for the mode of interest.

Displacements and velocities used to determine design forces in damping devices at each story shall account for the angle of orientation of the damping device from the horizontal and consider the effects of increased floor response caused by torsional motions.

18.7.4.5 Seismic Load Conditions and Combination of Modal Responses Seismic design force, Q_E, in each element of the damping system shall be taken as the maximum force of the following three loading conditions:

1. Stage of maximum displacement: Seismic design force at the stage of maximum displacement shall be calculated in accordance with Equation (18.7-57):

$$Q_E = \Omega_0 \sqrt{\sum_m (Q_{mSFRS})^2} \pm Q_{DSD} \quad (18.7\text{-}57)$$

where Q_{mSFRS} is the force in an element of the damping system equal to the design seismic force of the mth mode of vibration of the structure in the direction of interest, and Q_{DSD} is the force in an element of the damping system required to resist design seismic forces of displacement-dependent damping devices.

Seismic forces in elements of the damping system, Q_{DSD}, shall be calculated by imposing design forces of displacement-dependent damping devices on the damping system as pseudostatic forces. Design seismic forces of displacement-dependent damping devices shall be applied in both positive and negative directions at peak displacement of the structure.

2. Stage of maximum velocity: Seismic design force at the stage of maximum velocity shall be calculated in accordance with Equation (18.7-58):

$$Q_E = \sqrt{\sum_m (Q_{mDSV})^2} \quad (18.7\text{-}58)$$

where Q_{mDSV} is the force in an element of the damping system required to resist design seismic forces of velocity-dependent damping devices caused by the mth mode of vibration of the structure in the direction of interest.

Modal seismic design forces in elements of the damping system, Q_{mDSV}, shall be calculated by imposing modal design forces of velocity-dependent devices on the nondeformed damping system as pseudostatic forces. Modal seismic design forces shall be applied in directions consistent with the deformed shape of the mode of interest. Horizontal restraint forces shall be applied at each floor level, i, of the nondeformed damping system concurrent with the design forces in velocity-dependent damping devices, such that the horizontal displacement at each level of the structure is zero. At each floor level, i, restraint forces shall be proportional to and applied at the location of each mass point.

3. Stage of maximum acceleration: Seismic design force at the stage of maximum acceleration shall be calculated in accordance with Equation (18.7-59):

$$Q_E = \sqrt{\sum_m (C_{mFD}\Omega_0 Q_{mSFRS} + C_{mFV} Q_{mDSV})^2} \pm Q_{DSD} \quad (18.7\text{-}59)$$

The force coefficients, C_{mFD} and C_{mFV}, shall be determined from Tables 18.7-2 and 18.7-3, respectively, using values of effective damping determined in accordance with the following requirements:

For fundamental mode response ($m = 1$) in the direction of interest, the coefficients, C_{1FD} and C_{1FV}, shall be based on the velocity exponent, α, which relates device force to damping device velocity. The effective fundamental mode damping shall be taken as equal to the total effective damping of the fundamental mode less the hysteretic component of damping ($\beta_{1D} - \beta_{HD}$ or $\beta_{1M} - \beta_{HM}$) at the response level of interest ($\mu = \mu_D$ or $\mu = \mu_M$).

For higher mode ($m > 1$) or residual mode response in the direction of interest, the coefficients, C_{mFD} and C_{mFV}, shall be based on a value of α equal to 1.0. The effective modal damping shall be taken as equal to the total effective damping of the mode of interest (β_{mD} or β_{mM}). For determination of the coefficient C_{mFD}, the ductility demand shall be taken as equal to that of the fundamental mode ($\mu = \mu_D$ or $\mu = \mu_M$).

18.7.4.6 Inelastic Response Limits Elements of the damping system are permitted to exceed strength limits for design loads, provided it is shown by analysis or test that each of the following conditions are satisfied:

1. Inelastic response does not adversely affect damping system function.
2. Element forces, calculated in accordance with Section 18.7.4.5, using a value of Ω_0 taken as equal to 1.0, do not exceed the strength required to satisfy the load combinations of Section 12.4.

18.8 CONSENSUS STANDARDS AND OTHER REFERENCED DOCUMENTS

See Chapter 23 for the list of consensus standards and other documents that shall be considered part of this standard to the extent referenced in this chapter.

CHAPTER 19
SOIL–STRUCTURE INTERACTION FOR SEISMIC DESIGN

19.1 GENERAL

19.1.1 Scope Determination of the design earthquake forces and he corresponding displacements of the structure is permitted to consider the effects of soil–structure interaction (SSI) in accordance with this section. If soil–structure interaction effects are considered, the analytical model of the structure shall directly incorporate horizontal, vertical, and rotational foundation and soil flexibility. For the purpose of this section, both upper and lower bound estimates for the foundation and soil stiffnesses per Section 12.13.3 shall be considered. The case that results in the smaller reduction, or greater amplification, in response parameters shall be used for design. SSI may be used in conjunction with the nonlinear response history procedure of Section 19.2.3 when the structure is located on Site Class C, D, E, or F. SSI base slab averaging and embedment effects may not be used in conjunction with the equivalent lateral procedure of Section 19.2.1 nor the linear dynamic procedure of Section 19.2.2.

If the provisions of this chapter are used, then Sections 12.8.1.3 and 12.13.4 shall not apply.

19.1.2 Definitions The following definitions apply to the provisions of Chapter 19 and are in addition to the definitions presented in Chapter 11:

BASE SLAB AVERAGING: Kinematic SSI of a shallow (nonembedded) foundation caused by wave incongruence over the base area.

FOUNDATION INPUT MOTION: Motion that effectively excites the structure and its foundation.

FREE-FIELD MOTION: Motion at ground surface in absence of structure and its foundation.

INERTIAL SSI: The dynamic interaction between the structure, its foundation, and the surrounding soil caused by the foundation input motion.

KINEMATIC SSI: The modification of free-field ground motion caused by nonvertical incident seismic waves and spatial incoherence; the modification yields the foundation input motion.

RADIATION DAMPING: The damping in the soil–structure system caused by the generation and propagation of waves away from the foundation, which are caused by dynamic displacements of the foundation relative to the free-field displacements.

SOIL DAMPING: The hysteretic (material) damping of the soil.

19.1.3 Symbols The following symbols apply only to the provisions of Chapter 19 as indicated and are in addition to the symbols presented in Chapter 11:

a_o = Dimensionless frequency [Equations (19.3-11) and (19.3-21)]

B = Half the smaller dimension of the base of the structure

B_{SSI} = Factor to adjust the design response spectrum and MCE_R response spectrum in accordance with Sections 11.4.6 and 11.4.7 or a site-specific response spectrum for damping ratios other than 0.05 [Equation (19.2-4)]

b_e = Effective foundation size [Equations (19.4-4) and (19.4-4SI)]

$\tilde{C}_s$ = Seismic response coefficient determined in accordance with Section 12.8.1.1, assuming a flexible structural base at the foundation–soil interface in accordance with Section 19.1

D_s = Depth of a soft layer overlaying a stiff layer [Equation (19.3-4)]

e = Foundation embedment depth

G_{rd} = Effective shear modulus used in determining radiation damping effects derived or approximated based on $G_{0,rd}$ and Table 19.3-2

$G_{0,rd}$ = Average shear modulus for the soils used in determining radiation damping effects computed using Equation (20.4-1), over a depth of B or r_f below the base of the structure at small strain levels

h^* = Effective structure height

K_{xx}, K_{rr} = Rotational foundation stiffness [Equations (19.3-9) and (19.3-19)]

K_y, K_r = Translational foundational stiffness [Equations (19.3-8) and (19.3-18)]

L = Half the larger dimension of the base of the structure

M^* = Effective modal mass for the fundamental mode of vibration in the direction under consideration

RRS_{bsa} = Site-specific response spectral modification factor for base-slab averaging [Equation (19.4-1)]

RRS_e = Site-specific response spectral modification factor for foundation embedment [Equation (19.4-5)]

r_f = Radius of the circular foundation

$\tilde{S}_a$ = Response spectral acceleration including the effects of SSI [Equations (19.2-5) through (19.2-8)]

T = Fundamental period of the structure determined in accordance with Section 12.8.2, based on a mathematical model with a fixed base condition. The upper bound limitation of C_uT_a on the fundamental period from Section 12.8.2 shall not apply, and the approximate structural period, T_a, shall not be used

$\tilde{T}$ = Fundamental period of the structure using a model with a flexible base in accordance with Section 19.1.1. The upper bound limitation of C_uT_a on the fundamental period from Section 12.8.2 shall not apply, and the approximate structural period, T_a, shall not be used

$(\tilde{T}/T)_{eff}$ = Effective period lengthening that depends on expected ductility demand, μ [Equation (19.3-2)]

T_{xx}, T_{rr} = Fundamental translational period of SSI system [Equations (19.3-7) and (19.3-17)]

T_y, T_r = Fundamental translational period of SSI system [Equations (19.3-6) and (19.3-16)]

$\tilde{V}$ = Base shear adjusted for soil–structure interaction [Equation (19.2-1)]

$\tilde{V}_t$ = Base shear adjusted for soil–structure interaction determined through modal response spectrum analysis

$v_{s,e}$ = Average effective shear wave velocity for site soil conditions used in determining embedment effects, taken as average value of velocity over the embedment depth of the foundation determined using $v_{so,e}$ and Table 19.3-1 or a site-specific study and shall not be taken as less than 650 ft/s (200 m/s)

$v_{s,rd}$ = Average effective shear wave velocity used in determining radiation damping effects over a depth of B or r_f below the base of the structure, determined using $v_{so,rd}$ and Table 19.3-1 or a site-specific study

$v_{so,e}$ = Average low strain shear wave velocity used in determining embedment effects computed using Equation (20.4-1) over the embedment depth of the foundation

$v_{so,rd}$ = Average low strain shear wave velocity used in determining radiation damping effects computed using Equation 20.4-1 over a depth of B or r_f below the base of the structure

$\overline{W}$ = Weight caused by the modal mass in the fundamental mode, or alternatively, the effective seismic weight in accordance with Section 12.7.2

α = Coefficient that accounts for the reduction in base shear caused by foundation damping SSI

α_{xx}, α_{rr} = Dimensionless factor, function of dimensionless frequency, a_0 [Equations (19.3-14) and (19.3-24)]

β = Effective viscous damping ratio of the structure, taken as 5% unless otherwise justified by analysis

β_0 = Effective viscous damping ratio of the soil–structure system, based on Section 19.3.2 [Equation (19.3-1)]

β_f = Effective viscous damping ratio relating to foundation–soil interaction [Equation (19.3-3)]

β_{rd} = Radiation damping ratio determined in accordance with Section 19.3.3 or 19.3.4 [Equations (19.3-5) and (19.3-15)]

β_s, β_s' = Soil hysteretic damping ratio determined in accordance with Section 19.3.5

β_{xx}, β_{rr} = Rotational foundation damping coefficient [Equations (19.3-12) and (19.3-22)]

β_y, β_r = Translational foundation damping coefficient [Equations (19.3-10) and (19.3-20)]

γ = Average unit weight of the soils over a depth of B below the base of the structure

μ = Expected ductility demand

ν = Poisson's ratio; it is permitted to use 0.3 for sand and 0.45 for clay soils

Ψ = Dimensionless factor, function of Poisson's ratio [Equations (19.3-13) and (19.3-23)]

19.2 SSI ADJUSTED STRUCTURAL DEMANDS

19.2.1 Equivalent Lateral Force Procedure The inclusion of kinematic interaction effects, in accordance with Section 19.4 or other methods, is not permitted with the equivalent lateral force procedure. To account for the effects of SSI using a linear static procedure, the base shear, V, determined from Equation (12.8-1) is permitted to be modified as follows:

$$\tilde{V} = V - \Delta V \qquad (19.2\text{-}1)$$

$$\Delta V = \left(C_s - \frac{\widetilde{C_s}}{B_{SSI}}\right)\bar{W} \le 0.3\ V \qquad (19.2\text{-}2)$$

$$B_{SSI} = 4/[5.6 - \ln(100\beta_0)] \le \begin{cases} 1.4 & \text{for } R \le 3 \\ 1.7 - \frac{R}{10} & \text{for } 3 < R < 6 \\ 1.1 & \text{for } R \ge 6 \end{cases} \qquad (19.2\text{-}3)$$

where

$\tilde{V}$ = Base shear adjusted for SSI;

V = Fixed-base structure base shear computed in accordance with Section 12.8.1;

R = Response modification factor in Table 12.2-1;

Table 19.3-1. Effective Shear Wave Velocity Ratio $\left(\frac{v_{s,rd}}{v_{so,rd}} \text{ or } \frac{v_{s,e}}{v_{so,e}}\right)$.

Site Class	Effective Peak Acceleration, $S_{DS}/2.5$[a,c]			
	$S_{DS}/2.5=0$	$S_{DS}/2.5=0.1$	$S_{DS}/2.5=0.4$	$S_{DS}/2.5\ge 0.8$
A	1.00	1.00	1.00	1.00
B	1.00	1.00	0.97	0.95
BC	1.00	0.98	0.92	0.86
C	1.00	0.97	0.87	0.77
CD	1.00	0.96	0.79	0.50
D	1.00	0.95	0.71	0.32
DE	1.00	0.86	0.40	0.18
E	1.00	0.77	0.22	[b]
F	[b]	[b]	[b]	[b]

[a] Use straight-line interpolation for intermediate values of $S_{DS}/2.5$.
[b] Site-specific geotechnical investigation and dynamic site response analyses shall be performed.
[c] It is permitted to apply this table to the MCE$_R$ response spectrum, determined according to Section 11.4.7, the $S_{DS}/2.5$ headers shall be replaced by $S_{MS}/2.5$; to apply the table to the site-specific MCE$_R$ response spectrum, determined according to Chapter 21, the $S_{DS}/2.5$ headers shall be replaced by the MCE$_R$ response spectral acceleration at 0.01 s natural period.

C_s = Seismic response coefficient determined in accordance with Section 12.8.1.1, assuming a fixed structural base at the foundation–soil interface;

$\tilde{C}_s$ = Seismic response coefficient determined in accordance with Section 12.8.1.1, assuming flexibility of the structural base at the foundation–soil interface in accordance with Section 19.1.1, using $\tilde{T}$ as the fundamental period of the structure in lieu of the fundamental period of the structure, T, as determined by Section 12.8.2;

$\overline{W}$ = Weight caused by the effective modal mass in the fundamental mode, alternatively taken as the effective seismic weight in accordance with Section 12.7.2; and

β_0 = Effective viscous damping ratio of the soil–structure system, in accordance with Section 19.3.2.

19.2.2 Linear Dynamic Analysis The inclusion of kinematic interaction effects, in accordance with Section 19.4 or other methods, is not permitted with the linear dynamic procedure. To account for the effects of SSI, a linear dynamic analysis is permitted to be performed in accordance with Section 12.9, using either the SSI modified design response spectrum and MCE_R response spectrum in accordance with Sections 11.4.6 and 11.4.7 or SSI modified site-specific response spectrum, per Section 19.2.2.1 or an SSI modified site-specific response spectrum in accordance with Section 19.2.2.2 for spectral response acceleration, $\tilde{S}_a$, versus structural period, T. The resulting response spectral acceleration shall be divided by R/I_e, where I_e is prescribed in Section 11.5.1. The mathematical model used for the linear dynamic analysis shall include flexibility of the foundation and underlying soil in accordance with Section 19.1.1.

Scaling of the lateral forces from the modal response analysis shall be in accordance with Section 12.9.1.4 with calculated base shear, V, replaced with SSI adjusted base shear, $\tilde{V}$, determined in accordance with Equation (19.2-1) and the modal base shear, V_t, replaced by the modal base shear calculated with the effects of SSI, $\tilde{V}_t$.

The modal base shear calculated with the effects of SSI, $\tilde{V}_t$, shall not be less than $0.7V$.

19.2.2.1 SSI Modified General Design Response Spectrum The general design response spectrum, which includes the effects of SSI to be used with the modal analysis procedure in Section 19.2.2, shall be developed as follows:

$$\tilde{S}_a = \left[\left(\frac{5}{B_{\text{SSI}}} - 2\right) \times \frac{T}{T_S} + 0.4\right] \times S_{DS} \tag{19.2-4}$$

For $0 < T < T_0$, and

$$\tilde{S}_a = S_{DS}/B_{SSI} \quad \text{for } T_0 \le T \le T_S, \text{ and}$$

$$\tilde{S}_a = S_{D1}/(B_{SSI}T), \quad \text{for } T_S < T \le T_L, \text{ and}$$

$$\tilde{S}_a = S_{D1}T_L/(B_{SSI}T^2), \quad \text{for } T > T_L$$

where S_{DS} and S_{D1} are defined in Section 11.4.5; T_S, T_0, and T_L are as defined in Section 11.4.6; T is the period at the response spectrum ordinate; and B_{SSI} is defined in Equation (19.2-3).

19.2.2.2 SSI Site-Specific Response Spectrum A site-specific response spectrum that incorporates modifications caused by SSI is permitted to be developed in accordance with the requirements of Chapter 21. The spectrum is permitted to be adjusted for the effective viscous damping ratio of β_0, of the soil–structure system, as defined in Section 19.3.2, in the development of the site-specific spectrum.

19.2.3 Nonlinear Response History Procedure It is permitted to account for the effects of SSI using a nonlinear response history analysis performed in accordance with Chapter 16 using acceleration histories scaled to a site-specific response spectrum modified for kinematic interaction, in accordance with Section 19.4 or other approved methods. The mathematical model used for the analysis shall include foundation and soil flexibility, per Section 19.1.1, and shall explicitly incorporate the effects of foundation damping, per Section 19.3 or by other approved methods. Kinematic interaction effects, per Section 19.4, are permitted to be included in the determination of the site-specific response spectrum.

The site-specific response spectrum shall be developed, per the requirements of Chapter 21, with the following additional requirements:

1. The spectrum is permitted to be adjusted for kinematic SSI effects by multiplying the spectral acceleration ordinate at each period by the corresponding response spectrum ratios for either base slab averaging or embedment, or both base slab averaging and embedment ($RRS_{\text{bsa}} \times RRS_e$) per Section 19.4, or by directly incorporating one or both of these effects into the development of the spectrum.
2. For structures embedded in the ground, the site-specific response spectrum is permitted to be developed at the depth of the embedded base level in lieu of at grade. For this case, the response spectrum ratio for embedment effects (RRS_e) shall be taken as 1.0.
3. The site-specific response spectrum modified for kinematic interaction shall not be taken as less than 80% of S_a as determined from a site-specific response spectrum in accordance with Section 21.3.
4. The site-specific response spectrum modified for kinematic interaction shall not be taken as less than 70% of S_a, as determined from the design response spectrum and MCE_R response spectrum in accordance with Sections 11.4.6 and 11.4.7.

If the acceleration histories are scaled to a site-specific spectrum modified for kinematic soil–structure interaction it is permitted to include kinematic soil–structure interaction in the equivalent lateral force procedure or the linear dynamic procedure conducted per Section 16.1.2 using the site-specific response spectrum modified for kinematic soil–structure interaction, subject to the limitations herein. Where foundation damping is included in the nonlinear model, the equivalent lateral force procedure or the linear dynamic procedure, in conjunction with Section 16.1.2, shall be based on the provisions of 19.2.1 or 19.2.2, respectively.

19.3 FOUNDATION DAMPING EFFECTS

19.3.1 Foundation Damping Requirements Foundation damping effects are permitted to be considered through direct incorporation of soil hysteretic damping and radiation damping in the mathematical model of the structure. The use of the procedures in this section are permitted in conjunction with the equivalent lateral procedure modifications of Section 19.2.1 or the linear dynamic procedure modifications of Section 19.2.2, unless any of the following conditions occur:

1. A foundation system consisting of discrete footings that are not interconnected and that are spaced less than the larger

dimension of the supported lateral force-resisting element in the direction under consideration.

2. A foundation system consisting of, or including, deep foundations such as piles or piers.
3. A foundation system consisting of structural mats interconnected by concrete slabs that are characterized as flexible in accordance with Section 12.3.1.3 or that are not continuously connected to grade beams or other foundation elements.

19.3.2 Effective Damping Ratio The effects of foundation damping shall be represented by the effective damping ratio of the soil–structure system, β_0, determined in accordance with Equation (19.3-1):

$$\beta_0 = \beta_f + \frac{\beta}{(\tilde{T}/T)^2_{\text{eff}}} \le 0.20 \quad (19.3\text{-}1)$$

where

β_f = Effective viscous damping ratio relating to foundation–soil interaction;

β = Effective viscous damping ratio of the structure, taken as 5%, unless otherwise justified by analysis; and

$(\tilde{T}/T)_{\text{eff}}$ = Effective period lengthening ratio defined in Equation (19.3-2).

The effective period lengthening ratio shall be determined in accordance with Equation (19.3-2):

$$\left(\frac{\tilde{T}}{T}\right)_{\text{eff}} = \left\{1 + \frac{1}{\mu}\left[\left(\frac{\tilde{T}}{T}\right)^2 - 1\right]\right\}^{0.5} \quad (19.3\text{-}2)$$

where μ is the expected ductility demand. For equivalent lateral force or modal response spectrum analysis procedures, μ is the maximum base shear divided by the elastic base shear capacity; alternately, μ is permitted to be taken as R/Ω_0, where R and Ω_0 are per Table 12.2-1. For the response history analysis procedures, μ is the maximum displacement divided by the yield displacement of the structure measured at the highest point above grade.

The foundation damping ratio caused by soil hysteretic damping and radiation damping, β_f, is permitted to be determined in accordance with Equation (19.3-3) or by other approved methods.

$$\beta_f = \left[\frac{(\tilde{T}/T)^2 - 1}{(\tilde{T}/T)^2}\right]\beta_s + \beta_{rd} \quad (19.3\text{-}3)$$

where β_s is the soil hysteretic damping ratio, determined in accordance with Section 19.3.5, and β_{rd} is the radiation damping ratio, determined in accordance with Section 19.3.3 or 19.3.4.

If a site more than a depth B or R below the base of the building consists of a relatively uniform layer of depth, D_s, overlaying a very stiff layer with a shear wave velocity more than twice that of the surface layer and $4D_s/v_s\tilde{T} < 1$, then the damping values, β_r, in Equation (19.3-3) shall be replaced by β_s', per Equation (19.3-4):

$$\beta_s' = \left(\frac{4D_s}{v_s\tilde{T}}\right)^4 \beta_s \quad (19.3\text{-}4)$$

19.3.3 Radiation Damping for Rectangular Foundations The effects of radiation damping for structures with a rectangular foundation plan shall be represented by the effective damping ratio of the soil–structure system, β_{rd}, determined in accordance with Equation (19.3-15):

$$a_0 = \frac{2\pi B}{\tilde{T} v_{s,rd}} \quad (19.3\text{-}5)$$

$$\psi = \sqrt{\frac{2(1-\nu)}{(1-2\nu)}} \le 2.5 \quad (19.3\text{-}6)$$

$$\alpha_{xx} = 1.0 - \left[\frac{(0.55 + 0.01\sqrt{(L/B) - 1})a_0^2}{(2.4 - \frac{0.4}{(L/B)^3}) + a_0^2}\right] \quad (19.3\text{-}7)$$

$$G_{0,rd} = \gamma v_{so,rd}^2 / g \quad (19.3\text{-}8)$$

$$K_y = \frac{G_{rd}B}{2-\nu}\left[6.8\left(\frac{L}{B}\right)^{0.65} + 0.8\left(\frac{L}{B}\right) + 1.6\right] \quad (19.3\text{-}9)$$

$$K_{xx} = \frac{G_{rd}B^3}{1-\nu}\left[3.2\left(\frac{L}{B}\right) + 0.8\right] \quad (19.3\text{-}10)$$

$$\beta_y = \left[\frac{4(L/B)}{(K_y/G_{rd}B)}\right]\left[\frac{a_0}{2}\right] \quad (19.3\text{-}11)$$

$$\beta_{xx} = \left[\frac{(4\psi/3)(L/B)a_0^2}{(\frac{K_{xx}}{G_{rd}B^3})[(2.2 - \frac{0.4}{(L/B)^3}) + a_0^2]}\right]\left[\frac{a_0}{2\alpha_{xx}}\right] \quad (19.3\text{-}12)$$

$$T_y = 2\pi\sqrt{\frac{M^*}{K_y}} \quad (19.3\text{-}13)$$

$$T_{xx} = 2\pi\sqrt{\frac{M^*(h^*)^2}{\alpha_{xx}K_{xx}}} \quad (19.3\text{-}14)$$

$$\beta_{rd} = \frac{1}{(\tilde{T}/T_y)^2}\beta_y + \frac{1}{(\tilde{T}/T_{xx})^2}\beta_{xx} \quad (19.3\text{-}15)$$

where

M^* = Effective modal mass for the fundamental mode of vibration in the direction under consideration;

h^* = Effective structure height taken as the vertical distance from the foundation to the centroid of the first mode shape for multistory structures. Alternatively, h^* is permitted to be approximated as 70% of the total structure height for multistory structures or as the full height of the structure for one-story structures;

L = Half the larger dimension of the base of the structure;

B = Half the smaller dimension of the base of the structure;

$v_{s,rd}$ = Average effective shear wave velocity over a depth of B below the base of the structure determined using $v_{so,rd}$ and Table 19.3-1 or a site-specific study;

$v_{so,rd}$ = Average low strain shear wave velocity computed using Equation (20.4-1) over a depth of B below the base of the structure;

G_{rd} = Effective shear modulus derived or approximated based on $G_{0,rd}$ and Table 19.3-2;

$G_{0,rd}$ = Average shear modulus for the soils computed using Equation (20.4-1) over a depth B below the base of the structure at small strain levels;

Table 19.3-2. Effective Shear Modulus Ratio ($G_{rd}/G_{0,rd}$).

Site Class	Effective Peak Acceleration, $S_{DS}/2.5$[a,c]			
	$S_{DS}/2.5 = 0$	$S_{DS}/2.5 = 0.1$	$S_{DS}/2.5 = 0.4$	$S_{DS}/2.5 \geq 0.8$
A	1.00	1.00	1.00	1.00
B	1.00	1.00	0.95	0.90
BC	1.00	0.97	0.84	0.73
C	1.00	0.95	0.75	0.60
CD	1.00	0.92	0.62	0.25
D	1.00	0.90	0.50	0.10
DE	1.00	0.73	0.16	0.03
E	1.00	0.60	0.05	[b]
F	[b]	[b]	[b]	[b]

[a] Use straight-line interpolation for intermediate values of $S_{DS}/2.5$.
[b] Site-specific geotechnical investigation and dynamic site response analyses shall be performed.
[c] It is permitted to apply this table to the MCE_R response spectrum, determined according to Section 11.4.7, the $S_{DS}/2.5$ headers shall be replaced by $S_{MS}/2.5$; to apply the table to the site-specific MCE_R response spectrum determined according to Chapter 21, the $S_{DS}/2.5$ headers shall be replaced by the MCE_R response spectral acceleration at 0.01 s natural period.

γ = Average unit weight of the soils over a depth of B below the base of the structure; and

ν = Poisson's ratio; it is permitted to use 0.3 for sandy and 0.45 for clayey soils.

19.3.4 Radiation Damping for Circular Foundations The effects of radiation damping for structures with a circular foundation plan shall be represented by the effective damping ratio of the soil–structure system, β_{rd}, [Equation (19.3-25)] determined in accordance with Equation (19.3-15):

$$a_0 = \left[\frac{2\pi r_f}{\tilde{T} v_{s,rd}}\right] \quad (19.3\text{-}16)$$

$$\psi = \sqrt{\frac{2(1-\nu)}{(1-2\nu)}} \leq 2.5 \quad (19.3\text{-}17)$$

$$\alpha_{rr} = 1.0 - \left[\frac{0.35a_0^2}{1.0 + a_0^2}\right] \quad (19.3\text{-}18)$$

$$K_r = \frac{8G_{rd}r_f}{2-\nu} \quad (19.3\text{-}19)$$

$$K_{rr} = \frac{8G_{rd}r_f^3}{3(1-\nu)} \quad (19.3\text{-}20)$$

$$\beta_r = \left[\frac{\pi}{(K_r/G_{rd}r_f)}\right]\left[\frac{a_0}{2}\right] \quad (19.3\text{-}21)$$

$$\beta_{rr} = \left[\frac{(\pi\psi/4)a_0^2}{(K_{rr}/G_{rd}r_f^3)[2 + a_0^2]}\right]\left[\frac{a_0}{2\alpha_{rr}}\right] \quad (19.3\text{-}22)$$

$$T_r = 2\pi\sqrt{\frac{M^*}{K_r}} \quad (19.3\text{-}23)$$

$$T_{rr} = 2\pi\sqrt{\frac{M^*(h^*)^2}{\alpha_{rr}K_{rr}}} \quad (19.3\text{-}24)$$

$$\beta_{rd} = \frac{1}{(\tilde{T}/T_r)^2}\beta_r + \frac{1}{(\tilde{T}/T_{rr})^2}\beta_{rr} \quad (19.3\text{-}25)$$

where

r_f = Radius of the circular foundation;

$v_{s,rd}$ = The average effective shear wave velocity used in determining radiation damping effects over a depth of r_f below the base of the structure determined using $v_{so,rd}$ and Table 19.3-1 or a site-specific study;

$v_{so,rd}$ = The average low strain shear wave velocity used in determining radiation damping effects computed using Equation (20.4-1) over a depth of r_f below the base of the structure;

γ = The average unit weight of the soils over a depth of r_f below the base of the structure;

G_{rd} = Effective shear modulus used in determining radiation damping effects derived or approximated based on $G_{0,rd}$ and Table 19.3-2; and

$G_{0,rd} = \gamma v_{so,rd}^2/g$, the average shear modulus for the soils used in determining radiation damping effects computed using Equation (20.4-1) over a depth of r_f below the base of the structure at small strain levels.

19.3.5 Soil Damping The effects of soil hysteretic damping shall be represented by the effective soil hysteretic damping ratio, β_s, determined based on a site-specific study. Alternatively, it is permitted to determine β_s in accordance with Table 19.3-3.

19.4 BASE SLAB AVERAGING AND EMBEDMENT (KINEMATIC) SSI EFFECTS

Kinematic SSI effects are permitted to be represented by response spectral modification factors RRS_{bsa} for base slab averaging and RRS_e for embedment, which are multiplied by the spectral acceleration ordinates of the response spectrum at each period. The modification factors are calculated in accordance with Sections 19.4.1 and 19.4.2. Modifications of the response spectrum for kinematic SSI effects are permitted only for use with the nonlinear response history analysis provisions of Chapter 16, using the site-specific response spectrum developed

Table 19.3-3. Soil Hysteretic Damping Ratio, β_s.

Site Class	Effective Peak Acceleration, $S_{DS}/2.5$[a]			
	$S_{DS}/2.5 = 0$	$S_{DS}/2.5 = 0.1$	$S_{DS}/2.5 = 0.4$	$S_{DS}/2.5 \geq 0.8$
C	0.01	0.01	0.03	0.05
CD	0.01	0.01	0.05	0.09
D	0.01	0.02	0.07	0.15
DE	0.01	0.03	0.12	0.20
E	0.01	0.05	0.20	[b]
F	[b]	[b]	[b]	[b]

[a] Use straight-line interpolation for intermediate values of $S_{DS}/2.5$.

[b] Site-specific geotechnical investigation and dynamic site response analyses shall be performed.

in accordance with Chapter 21 and subject to the limitations in Sections 19.2.3, 19.4.1, and 19.4.2.

The product of $RRS_{\text{bsa}} \times RRS_e$ shall not be less than 0.7.

19.4.1 Base Slab Averaging Consideration of the effects of base slab averaging through the development of site-specific transfer functions that represent the kinematic SSI effects expected at the site for a given foundation configuration is permitted.

Alternatively, modifications for base slab averaging using the procedures of this section are permitted for the following cases:

1. All structures located on Site Class C, CD, D, DE, or E; and
2. Structures that have structural mats or foundation elements interconnected with concrete slabs or that are continuously connected with grade beams or other foundation elements of sufficient lateral stiffness, so as not to be characterized as flexible under the requirements of Section 12.3.1.3.

The *modification* factor for base slab averaging, RRS_{bsa}, shall be determined using Equation (19.4-1) for each period required for analysis.

$$RRS_{\text{bsa}} = 0.25 + 0.75 \times \left\{ \frac{1}{b_0^2} [1 - (\exp(-2b_0^2)) \times B_{\text{bsa}}] \right\}^{1/2} \quad (19.4\text{-}1)$$

where

$$B_{\text{bsa}} = \begin{cases} 1 + b_0^2 + b_0^4 + \frac{b_0^6}{2} + \frac{b_0^8}{4} + \frac{b_0^{10}}{12} & b_0 \leq 1 \\ [\exp(2b_0^2)] \times \left[\frac{1}{\sqrt{\pi} b_0}\left(1 - \frac{1}{16 b_0^2}\right)\right] & b_0 > 1 \end{cases} \quad (19.4\text{-}2)$$

$$b_0 = 0.00071 \times \left(\frac{b_e}{T}\right) \quad (19.4\text{-}3)$$

where b_e is the effective foundation size (ft),

$$b_0 = 0.0023 \times \left(\frac{b_e}{T}\right) \quad (19.4\text{-}3.\text{SI})$$

where b_e is the effective foundation size (m),

$$b_e = \sqrt{A_{\text{base}}} \leq 260\,\text{ft} \quad (19.4\text{-}4)$$

$$b_e = \sqrt{A_{\text{base}}} \leq 80\,\text{m} \quad (19.4\text{-}4.\text{SI})$$

T is the response spectra ordinate period, which shall not be taken as less than 0.20 s when used in Equations (19.4-3) or (19.4-3.SI); and A_{base} is the area of the base of the structure [ft^2 (m^2)].

19.4.2 Embedment The response spectrum shall be developed based on a site-specific study at the depth of the base of the structure. Alternatively, modifications for embedment are permitted using the procedures of this section.

The modification factor for embedment, RRS_e, shall be determined using Equation (19.4-5) for each period required for analysis.

$$RRS_e = 0.25 + 0.75 \times \cos\left(\frac{2\pi e}{T v_{s,e}}\right) \quad (19.4\text{-}5)$$

where

e = Foundation embedment depth [ft (m)], not greater than 20 ft (6.1 m) and not greater than the bottom of the base slab diaphragm; a minimum of 75% of the foundation footprint shall be present at the embedment depth, and the foundation embedment for structures located on sloping sites shall be the shallowest embedment;

$v_{s,e}$ = The average effective shear wave velocity for site soil conditions used in determining embedment effects, taken as average value of velocity over the embedment depth of the foundation determined using $v_{so,e}$ and Table 19.3-1 or a site-specific study and shall not be taken as less than 650 ft/s (200 m/s);

$v_{so,e}$ = The average low strain shear wave velocity used in determining embedment effects computed using Equation (20.4-1) over the embedment depth of the foundation; and

T = Response spectra ordinate period, which shall not be taken as less than 0.20 s when used in Equation (19.4-5).

19.5 CONSENSUS STANDARDS AND OTHER REFERENCED DOCUMENTS

See Chapter 23 for the list of consensus standards and other documents that shall be considered part of this standard to the extent referenced in this chapter.

CHAPTER 20
SITE CLASSIFICATION PROCEDURE FOR SEISMIC DESIGN

20.1 SITE CLASSIFICATION

The site soil shall be classified in accordance with Table 20.2-1 and Section 20.2 based on the average shear wave velocity parameter, $\bar{v}_s$, which is derived from the measured shear wave velocity profile from the ground surface to a depth of 100 ft (30 m). Where shear wave velocity is not measured, appropriate generalized correlations between shear wave velocity and standard penetration test (SPT) blow counts, Cone Penetration Test (CPT) tip resistance, shear strength, or other geotechnical parameters shall be used to obtain an estimated shear wave velocity profile, as described in Section 20.3. Where site-specific data (measured shear wave velocities or other geotechnical data that can be used to estimate shear wave velocity) are available only to a maximum depth less than 100 ft (30 m), $\bar{v}_s$ shall be estimated as described in Section 20.3. Where the soil properties are not known in sufficient detail to determine the site class, the most critical site conditions of Site Class C, Site Class CD and Site Class D, as defined in Section 11.4.2, shall be used unless the Authority Having Jurisdiction or geotechnical data determine that Site Class DE, E, or F soils are present at the site. Site Classes A and B shall not be assigned to a site if there is more than 10 ft (3.1 m) of soil between the rock surface and the bottom of the spread footing or mat foundation.

20.2 SITE CLASS DEFINITIONS

Site class types shall be assigned in accordance with the definitions provided in Table 20.2-1 and this section.

20.2.1 Site Class F Where any of the following conditions is satisfied, the site shall be classified as Site Class F and a site response analysis, in accordance with Section 21.1, shall be performed:

1. Soil profile that includes soils vulnerable to potential failure or collapse under seismic loading, such as liquefiable soils, quick and highly sensitive clays, and collapsible weakly cemented soils.

   EXCEPTION: For structures that have fundamental periods of vibration equal to, or less than, 0.5 s, site response analysis is not required to determine spectral accelerations for liquefiable soils. Rather, for the purpose of determining spectral accelerations only, a site class is permitted to be determined in accordance with Section 20.2 and the corresponding response spectrum determined from Section 11.4.
2. Soil profile includes peats and/or highly organic clays [$H > 10$ ft ($H > 3$ m)] where H = thickness of soil.
3. Soil profile includes very high plasticity clays [$H > 25$ ft ($H > 7.6$ m) with $PI > 75$] in a soil profile that would otherwise be classified as Site Class CD, D, DE, or E.

   EXCEPTION: Site response analysis is not required for this clay category for Seismic Design Category A and Seismic Design Category B, where the seismic design category is based on the values S_{DS} and S_{D1}.
4. Soil profile includes very thick soft/medium stiff clays [$H > 120$ ft ($H > 37$ m)] with $s_u < 1,000$ psf ($s_u < 50$ kPa).

   EXCEPTION: Site response analysis is not required for this clay category for Seismic Design Category A and Seismic Design Category B, where the seismic design category is based on the values S_{DS} and S_{D1}.

20.2.2 Soft Clay Site Class E Where a site does not qualify under the criteria for Site Class F and there is a total thickness of soft clay greater than 10 ft (3 m), where a soft clay layer is defined by $s_u < 500$ psf ($s_u < 25$ kPa), $w \geq 40\%$, and $PI > 20$, it shall be classified as Site Class E. This classification is made regardless of $\bar{v}_s$, as computed in Section 20.4.

20.2.3 Site Classes C, CD, D, DE, and E The assignment of Site Class C, CD, D, DE, and E soils shall be made based on the average shear wave velocity, $\bar{v}_s$, which is derived from the site shear wave velocity profile from the ground surface to a depth of 100 ft (30 m), as described in Section 20.4.

20.2.4 Site Classes B and BC (Medium Hard and Soft Rock) Site Class B can only be assigned to a site on the basis of shear wave velocity measured on site. If shear wave velocity data are not available and the site condition is estimated by a geotechnical engineer, engineering geologist, or seismologist as Site Class B or BC on the basis of site geology, consisting of competent rock with moderate fracturing and weathering, the site shall be classified as Site Class BC. Softer and more highly fractured and weathered rock shall either be measured on site for shear wave velocity or classified as Site Class C.

20.2.5 Site Class A (Hard Rock) The hard rock, Site Class A, category shall be supported by shear wave velocity measurement, either on site or on profiles of the same rock type in the same formation with an equal or greater degree of weathering and fracturing. Where hard rock conditions are known to be continuous to a depth of 100 ft (30 m), surficial shear wave velocity measurements to maximum depths less than 100 ft are permitted to be extrapolated to assess $\bar{v}_s$.

Table 20.2-1. Site Classification.

Site Class	$\bar{v}_s$ Calculated Using Measured or Estimated Shear Wave Velocity Profile (ft/s)
A. Hard rock	>5,000
B. Medium hard rock	>3,000 to 5,000
BC. Soft rock	>2,100 to 3,000
C. Very dense sand or hard clay	>1,450 to 2,100
CD. Dense sand or very stiff clay	>1,000 to 1,450
D. Medium dense sand or stiff clay	>700 to 1,000
DE. Loose sand or medium stiff clay	>500 to 700
E. Very loose sand or soft clay	≥500
F. Soils requiring site response analysis in accordance with Section 21.1	See Section 20.2.1

Note: For SI: 1 ft = 0.3048 m; 1 ft/s = 0.3048 m/s.

20.3 ESTIMATION OF SHEAR WAVE VELOCITY PROFILES

Where measured shear wave velocity data are not available, shear wave velocity shall be estimated as a function of depth using correlations with suitable geotechnical parameters, including standard penetration test (SPT) blow counts, shear strength, overburden pressure, void ratio, or cone penetration test (CPT) tip resistance, measured at the site.

Site class based on estimated values of $\bar{v}_s$ shall be derived using $\bar{v}_s$, $\bar{v}_s/1.3$, and $1.3\bar{v}_s$ when correlation models are used to derive shear wave velocities. Where correlations derived for specific local regions can be demonstrated to have greater accuracy, factors less than 1.3 can be used if approved by the Authority Having Jurisdiction. If the different average velocities result in different site classes per Table 20.2-1, the most critical of the site classes for ground motion analysis at each period shall be determined by a geotechnical engineer, as described in Section 11.4.2.

Where the available data used to establish the shear wave velocity profile extends to depths less than 100 ft (30 m) but more than 50 ft (15 m), and the site geology is such that soft layers are unlikely to be encountered between 50 and 100 ft, the shear wave velocity of the last layer in the profile shall be extended to 100 ft for the calculation of $\bar{v}_s$ in Equation (20.4-1). Where the data does not extend to depths of 50 ft (15 m), default site classes, as described in Section 20.1, shall be used unless another site class can be justified on the basis of the site geology.

20.4 DEFINITIONS OF SITE CLASS PARAMETERS

The definitions presented in this section shall apply to the upper 100 ft (30 m) of the site profile. Profiles containing distinct soil and rock layers shall be subdivided into those layers designated by a number that ranges from 1 to n at the bottom, where there is a total of n distinct layers in the upper 100 ft (30 m). The symbol i refers to any one of the layers between 1 and n.

20.4.1 $\bar{v}_s$, Average Shear Wave Velocity The average shear wave velocity, $\bar{v}_s$, shall be determined in accordance with the following formula:

$$\bar{v}_s = \frac{\sum_{i=1}^{n} d_i}{\sum_{i=1}^{n} \frac{d_i}{v_{si}}} \qquad (20.4\text{-}1)$$

where

d_i = Thickness of any layer between 0 and 100 ft (30 m),
v_{si} = Shear wave velocity in ft/s (m/s), and
$\sum_{n}^{i=1} d_i$ = 100 ft (30 m).

20.5 CONSENSUS STANDARDS AND OTHER REFERENCED DOCUMENTS

See Chapter 23 for the list of consensus standards and other documents that shall be considered part of this standard to the extent referenced in this chapter.

CHAPTER 21
SITE-SPECIFIC GROUND MOTION PROCEDURES FOR SEISMIC DESIGN

21.1 SITE RESPONSE ANALYSIS

The requirements of Section 21.1 shall be satisfied where site response analysis is performed or required by Section 11.4.7. The analysis shall be documented in a report.

21.1.1 Base Ground Motions An MCE_R response spectrum shall be developed for a base condition consisting of bedrock, or when bedrock is very deep, firm soil conditions below softer surficial layers, using the procedure of Sections 11.4.6 or 21.2. Unless a site-specific ground motion hazard analysis described in Section 21.2 is carried out, the MCE_R base response spectrum shall be developed using the procedure of Section 11.4.6, assuming a site condition representative of the geological conditions at the base (represented by a base-condition average shear wave velocity, $\bar{v}_s$). At least five recorded or simulated horizontal ground motion acceleration time histories shall be selected from events that have magnitudes and fault distances that are consistent with those that control the MCE_R ground motion. Each selected time history shall be modified so that its response spectrum is, on average, approximately at the level of the MCE_R rock response spectrum over the period range of significance to structural response.

21.1.2 Site Condition Modeling A site response model based on low strain shear wave velocities, nonlinear or equivalent linear shear stress–strain relationships, and unit weights shall be developed. Low strain shear wave velocities shall be determined from field measurements at the site or from measurements from similar soils in the site vicinity. Nonlinear or equivalent linear shear stress–strain relationships and unit weights shall be selected on the basis of laboratory tests or published relationships for similar soils. The uncertainties in soil properties shall be estimated. Where very deep soil profiles make the development of a soil model to bedrock impractical, the model is permitted to be terminated where the soil stiffness is at least as great as the values used to define Site Class C in Chapter 20.

21.1.3 Site Response Analysis and Computed Results Base ground motion time histories shall be input to the soil profile as outcropping motions. Using appropriate computational techniques that treat nonlinear soil properties in a nonlinear or equivalent-linear manner, the response of the soil profile shall be determined and surface ground motion time histories shall be calculated. Ratios of 5% damped response spectra of surface ground motions to input base ground motions shall be calculated. The recommended surface MCE_R ground motion response spectrum shall not be lower than the MCE_R response spectrum of the base motion multiplied by the average surface-to-base response spectral ratios (calculated period by period) obtained from the site response analyses. The recommended surface ground motions that result from the analysis shall reflect consideration of sensitivity of response to uncertainty in soil properties, depth of soil model, and input motions.

21.2 RISK-TARGETED MAXIMUM CONSIDERED EARTHQUAKE (MCE_R) GROUND MOTION HAZARD ANALYSIS

The requirements of Section 21.2 shall be satisfied where a ground motion hazard analysis is performed or required by Section 11.4.7. The ground motion hazard analysis shall account for the regional tectonic setting, geology, and seismicity; the expected recurrence rates and maximum magnitudes of earthquakes on known faults and source zones; the characteristics of ground motion near source effects, if any, on ground motions; and the effects of subsurface site conditions on ground motions. The characteristics of subsurface site conditions shall be considered either using ground motion models that represent regional and local geology or in accordance with Section 21.1. The analysis shall incorporate current seismic interpretations, including uncertainties for models and parameter values for seismic sources and ground motions. If the spectral response accelerations predicted by the ground motion models do not represent the maximum response in the horizontal plane, then the response spectral accelerations computed from the hazard analysis shall be scaled by factors to increase the motions to the maximum response. If the ground motion models predict the geometric mean or similar metric of the two horizontal components, then the scale factors shall be 1.2 for periods less than, or equal to, 0.2 s, 1.25 for a period of 1.0 s, and 1.3 for periods greater than or equal to 10 s, unless it can be shown that other scale factors more closely represent the maximum response, in the horizontal plane, to the geometric mean of the horizontal components. Scale factors between these periods shall be obtained by linear interpolation. The analysis shall be documented in a report.

21.2.1 Probabilistic (MCE_R) Ground Motions The probabilistic spectral response accelerations shall be taken as the spectral response accelerations in the direction of maximum horizontal response represented by a 5% damped acceleration response spectrum that is expected to achieve a 1% probability of collapse within a 50-year period.

At each spectral response period for which the acceleration is computed, ordinates of the probabilistic ground motion response spectrum shall be determined from iterative integration of a site-specific hazard curve with a lognormal probability density function representing the collapse fragility (i.e., probability of collapse as a function of spectral response acceleration). The ordinate of the probabilistic ground motion response spectrum at each period shall achieve a 1% probability of collapse within a

50-year period for a collapse fragility that has (1) a 10% probability of collapse at said ordinate of the probabilistic ground motion response spectrum and (2) a logarithmic standard deviation value of 0.6.

21.2.2 Deterministic (MCE_R) Ground Motions The deterministic spectral response acceleration at each period shall be calculated as an 84th-percentile 5% damped spectral response acceleration in the direction of maximum horizontal response computed at that period. The largest such acceleration calculated for scenario earthquakes on all known faults within the region shall be used. The scenario earthquakes shall be determined from deaggregation for the probabilistic spectral response acceleration at each period. Scenario earthquakes contributing less than 10% of the largest contributor at each period shall be ignored.

For the purpose of this standard, the deterministic response spectral acceleration at each period shall be taken as not less than the deterministic lower limit response spectrum, of Table 21.2-1 of the site class, determined in accordance with the site class requirements of Section 11.4.2.

EXCEPTION: The deterministic ground motion response spectrum need not be calculated where the probabilistic ground motion response spectrum of 21.2.1 is, at all response periods, less than the deterministic lower limit response spectrum of Table 21.2-1 for the site class, determined in accordance with the site class requirements of Section 11.4.2.

21.2.3 Site-Specific MCE_R Response Spectrum The site-specific MCE_R spectral response acceleration at any period, S_{aM}, shall be taken as the lesser of the spectral response accelerations from the probabilistic ground motions of Section 21.2.1 and the deterministic ground motions of Section 21.2.2.

EXCEPTION: The site-specific MCE_R response spectrum may be taken as equal to the MCE_R response spectrum obtained from the USGS Seismic Design Geodatabase for the applicable site class.

The site-specific MCE_R spectral response acceleration at any period shall not be taken as less than 80% of the MCE_R response spectrum obtained from the USGS Seismic Design Geodatabase for the applicable site class.

21.2.3.1 Site Class F Sites For sites classified as Site Class F requiring site-specific analysis in accordance with Section 11.4.7, the site-specific MCE_R spectral response acceleration at any period shall not be less than 80% of the MCE_R response spectrum obtained from the USGS Seismic Design Geodatabase for Site Class E.

EXCEPTION: Where a different site class can be justified using the site-specific classification procedures of Section 20.3.3, a lower limit of 80% of S_{aM} for the justified site class shall be permitted to be used.

21.3 DESIGN RESPONSE SPECTRUM

The design spectral response acceleration at any period shall be determined from Equation (21.3-1):

$$S_a = \frac{2}{3} S_{aM} \quad (21.3\text{-}1)$$

where S_{aM} is the MCE_R spectral response acceleration obtained from Section 21.1 or 21.2.

Table 21.2-1. Deterministic Lower Limit Values of MCE_R Response Spectra and PGA_G (g).

	Site Class							
Period T(s)	A	B	BC	C	CD	D	DE	E
0.00	0.50	0.57	0.66	0.73	0.74	0.69	0.61	0.55
0.01	0.50	0.57	0.66	0.73	0.75	0.70	0.62	0.55
0.02	0.52	0.58	0.68	0.74	0.75	0.70	0.62	0.55
0.03	0.60	0.66	0.75	0.79	0.78	0.70	0.62	0.55
0.05	0.81	0.89	0.95	0.96	0.89	0.76	0.62	0.55
0.075	1.04	1.14	1.21	1.19	1.08	0.90	0.71	0.62
0.10	1.12	1.25	1.37	1.37	1.24	1.04	0.82	0.72
0.15	1.12	1.29	1.53	1.61	1.50	1.27	1.00	0.87
0.20	1.01	1.19	1.50	1.71	1.66	1.44	1.15	1.01
0.25	0.90	1.07	1.40	1.71	1.77	1.58	1.30	1.15
0.30	0.81	0.98	1.30	1.66	1.83	1.71	1.44	1.30
0.40	0.69	0.83	1.14	1.53	1.82	1.80	1.61	1.48
0.50	0.60	0.72	1.01	1.38	1.73	1.80	1.68	1.60
0.75	0.46	0.54	0.76	1.07	1.41	1.57	1.60	1.59
1.0	0.37	0.42	0.60	0.86	1.17	1.39	1.51	1.58
1.5	0.26	0.29	0.41	0.60	0.84	1.09	1.35	1.54
2.0	0.21	0.23	0.31	0.45	0.64	0.88	1.19	1.46
3.0	0.15	0.17	0.21	0.31	0.45	0.63	0.89	1.11
4.0	0.12	0.13	0.16	0.24	0.34	0.47	0.66	0.81
5.0	0.10	0.11	0.13	0.19	0.26	0.36	0.49	0.61
7.5	0.063	0.068	0.080	0.11	0.15	0.19	0.26	0.31
10	0.042	0.045	0.052	0.069	0.089	0.11	0.14	0.17
PGA_G	0.37	0.43	0.50	0.55	0.56	0.53	0.46	0.42

21.4 DESIGN ACCELERATION PARAMETERS

Where the site-specific procedure is used to determine the design ground motion, in accordance with Section 21.3, the parameter S_{DS} shall be taken as 90% of the maximum spectral acceleration, S_a, obtained from the site-specific spectrum at any period within the range from 0.2 to 5 s, inclusive. The parameter S_{D1} shall be taken as 90% of the maximum value of the product, TS_a, for periods from 1 to 2 s for sites with $\bar{v}_s > 1{,}450$ ft/s ($\bar{v}_s > 442$ m/s) and for periods from 1 to 5 s for sites with $\bar{v}_s \leq 1{,}450$ ft/s ($\bar{v}_s \leq 442$ m/s), but not less than 100% of the value of S_a at 1 s. The parameters S_{MS} and S_{M1} shall be taken as 1.5 times S_{DS} and S_{D1}, respectively.

For use with the equivalent lateral force procedure, the site-specific spectral acceleration, S_a, at T shall be permitted to replace S_{D1}/T in Equation (12.8-3) and $S_{D1}T_L/T^2$ in Equation (12.8-4). The parameter S_{DS} calculated per this section shall be permitted to be used in Equations (12.8-2), (12.8-5), (15.4-1), and (15.4-3). The mapped value of S_1 shall be used in Equations (12.8-6), (15.4-2), and (15.4-4).

21.5 MAXIMUM CONSIDERED EARTHQUAKE GEOMETRIC MEAN (MCE_G) PEAK GROUND ACCELERATION

21.5.1 Probabilistic MCE_G Peak Ground Acceleration The probabilistic geometric mean peak ground acceleration shall be taken as the geometric mean peak ground acceleration with a 2% probability of exceedance within a 50-year period.

21.5.2 Deterministic MCE_G Peak Ground Acceleration The deterministic geometric mean peak ground acceleration shall be calculated as the largest 84th-percentile geometric mean peak ground acceleration for scenario earthquakes on all known active faults within the site region. The scenario earthquakes shall be determined from deaggregation for the probabilistic geometric mean peak ground acceleration. Scenario earthquakes contributing less than 10% of the largest contributor shall be ignored. The deterministic geometric mean peak ground acceleration shall not be taken as lower than the value of PGA_G of Table 21.2-1 of the site class, determined in accordance with the site class requirements of Section 11.4.2.

21.5.3 Site-Specific MCE_G Peak Ground Acceleration PGA_M The site-specific MCE_G peak ground acceleration, PGA_M, shall be taken as the lesser of the probabilistic geometric mean peak ground acceleration of Section 21.5.1 and the deterministic geometric mean peak ground acceleration of Section 21.5.2. The site-specific MCE_G peak ground acceleration shall not be taken as less than 80% of the value of the MCE_G peak ground acceleration parameter PGA_M obtained from the USGS Seismic Design Geodatabase for the applicable site class.

21.6 CONSENSUS STANDARDS AND OTHER REFERENCED DOCUMENTS

See Chapter 23 for the list of consensus standards and other documents that shall be considered part of this standard to the extent referenced in this chapter.

CHAPTER 22
SEISMIC GROUND MOTION AND LONG-PERIOD TRANSITION MAPS

Contained in this chapter are Figures 22-1 through 22-8, which map the risk-targeted maximum considered earthquake (MCE_R) spectral response acceleration parameters, S_{MS} and S_{M1}, for default site conditions, based on the most critical spectral response acceleration of Site Classes C, CD, and D; Figures 22-9 through 22-13, which map the maximum considered earthquake geometric mean (MCE_G) peak ground acceleration parameter, PGA_M, for the default site conditions; and Figures 22-14 through 22-17, which map the long-period transition period parameter, T_L. In accordance with Section 11.4.3, S_{MS} and S_{M1} values for Site Classes A, B, BC, C, CD, D, DE, and E—as well as values of the MCE_R spectral response acceleration parameters, S_S and S_1, (for Site Class BC)—are contained in the USGS Seismic Design Geodatabase defined in Section 11.2; values of PGA_M for all the site classes are also contained in this geodatabase, in accordance with Section 11.8.3. For the definitions of these ground motion parameters, see Section 11.3.

These maps and the USGS Seismic Design Geodatabase were prepared by the US Geological Survey in collaboration with the Building Seismic Safety Council (BSSC) Provisions Update Committee and the ASCE 7 Seismic Subcommittee and have been updated for this standard.

Maps of T_L for Guam and the Northern Mariana Islands and for American Samoa are not provided because this parameter has not been developed for those islands via the same deaggregation computations done for the other US regions. Therefore, as in previous editions of this standard, the value of T_L shall be 12 s for those islands.

The following is a list of the maps contained in this chapter:

User Note: The USGS Seismic Design Geodatabase is available at the ASCE 7 Hazard Tool https://asce7hazardtool.online/ or an approved equivalent.

22.1 CONSENSUS STANDARDS AND OTHER REFERENCED DOCUMENTS

See Chapter 23 for the list of consensus standards and other documents that shall be considered part of this standard to the extent referenced in this chapter.

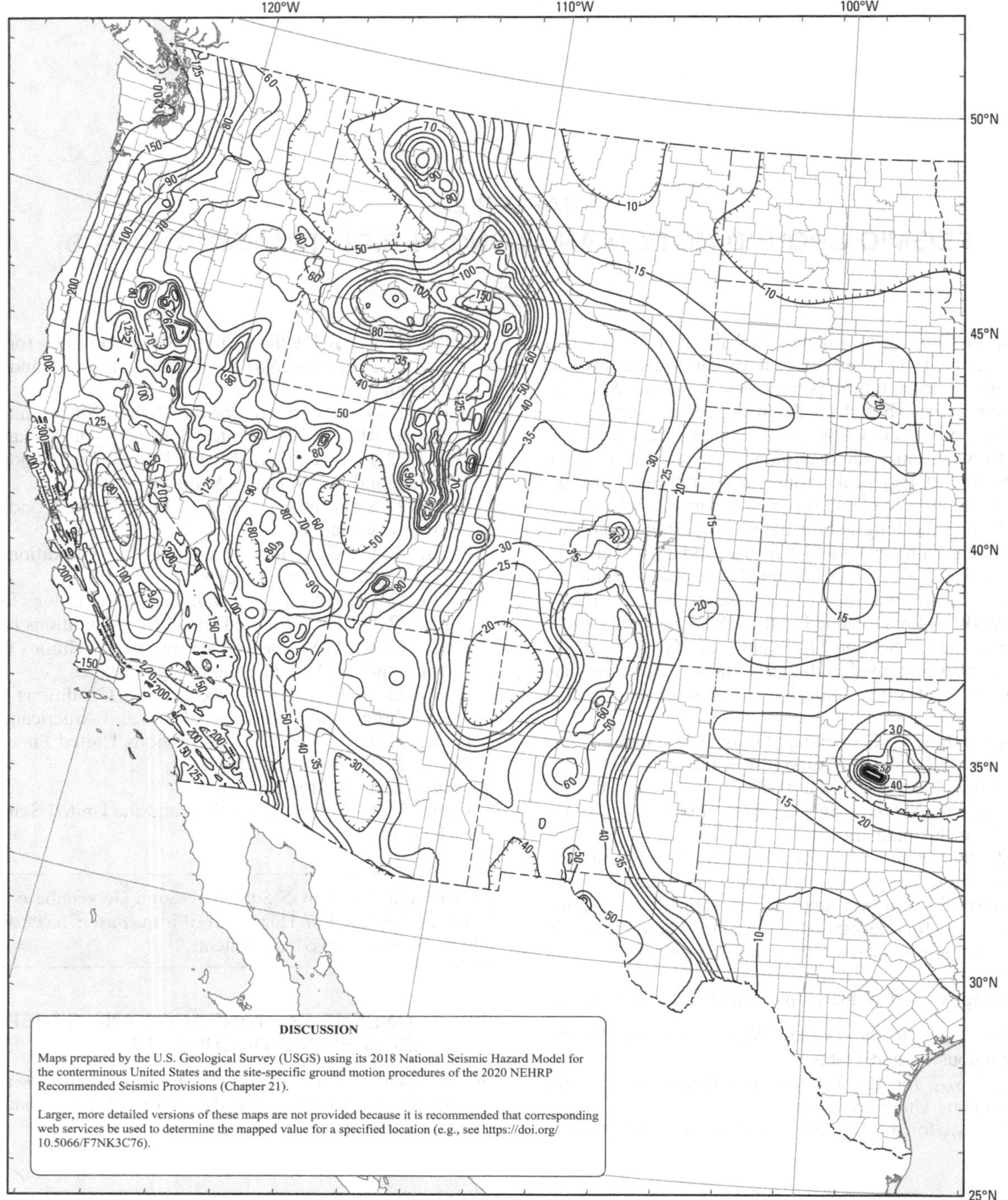

Notes:

Maps prepared by USGS in collaboration with the FEMA-funded Building Seismic Safety Council (BSSC) and ASCE. The basis is explained in commentaries prepared by BSSC and ASCE and in the references.

Ground motion values contoured on these maps incorporate

- A target risk of structural collapse equal to 1% in 50 years based upon a generic structural fragility,
- A factor of 1.1 to adjust from a geometric mean to the maximum response regardless of direction, and
- Deterministic upper limits imposed near large, active faults, which are taken as 1.8 times the estimated median response to the characteristic earthquake for the governing fault (1.8 is used to represent the 84th percentile response), but not less than 150% *g*.

As such, the values are different from those on the uniform-hazard 2014 USGS National Seismic Hazard Maps posted at https://doi.org/10.5066/F7HT2MHG.

Larger, more detailed versions of these maps are not provided because it is recommended that the corresponding USGS web tool (https://doi.org/10.5066/F7NK3C76) be used to determine the mapped value for a specified location.

User Note: The USGS Seismic Design Geodatabase is available at the ASCE 7 Hazard Tool https://asce7hazardtool.online/ or an approved equivalent.

Figure 22-1. S_{MS} for the default site conditions for the conterminous United States.

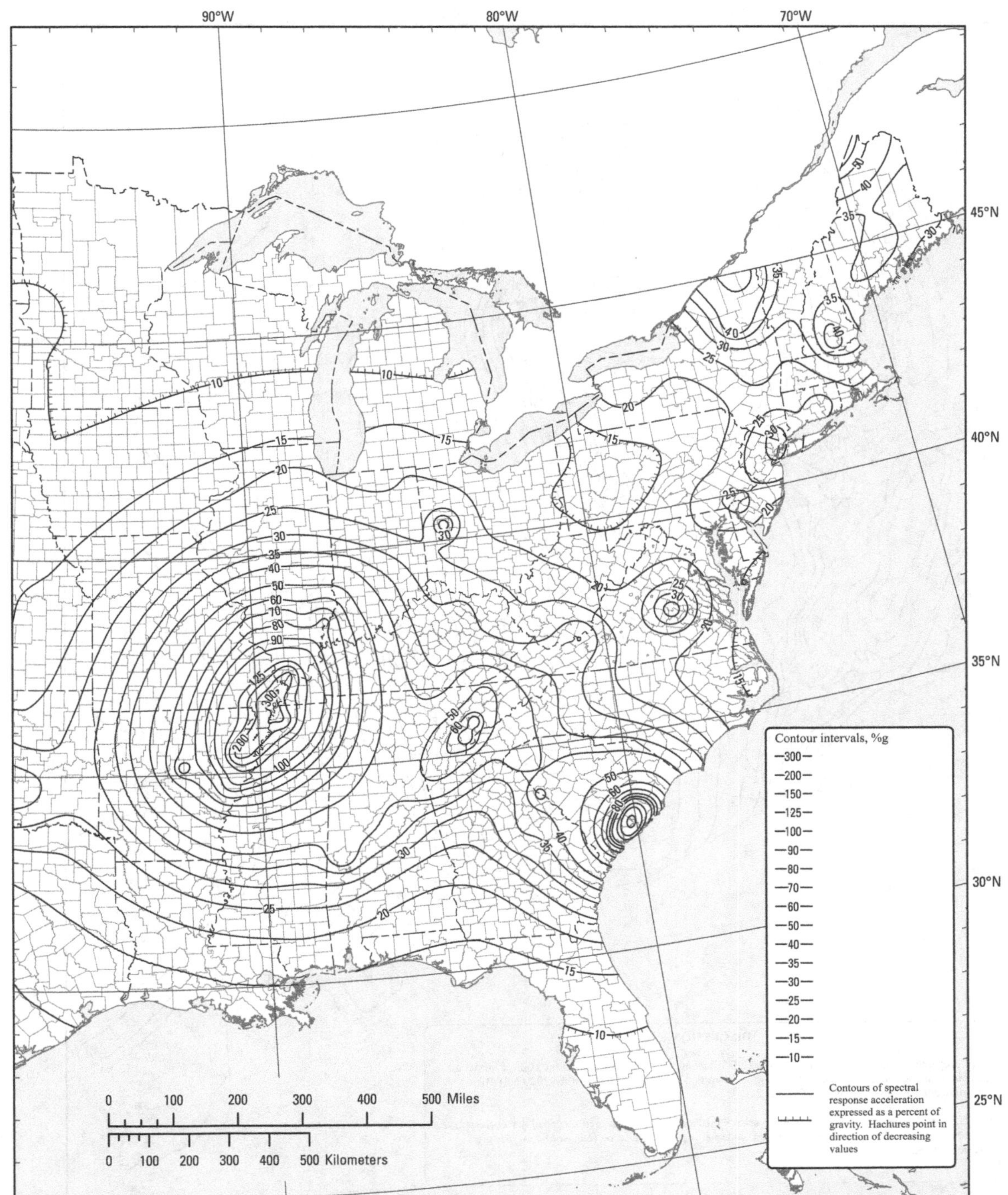

Figure 22-1 (*Continued*). S_{MS} for the default site conditions for the conterminous United States.

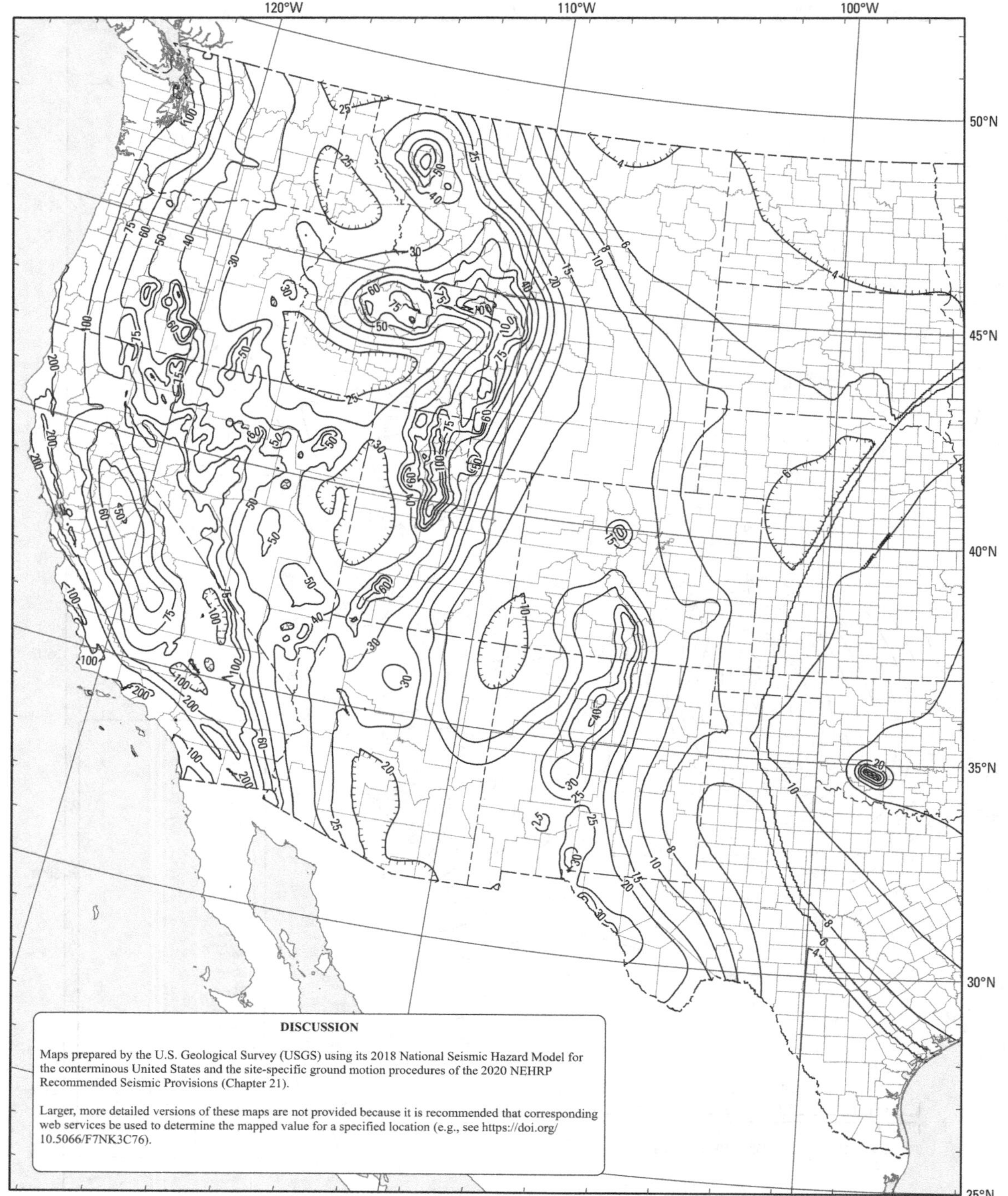

Notes:

Maps prepared by USGS in collaboration with the FEMA-funded BSSC and ASCE. The basis is explained in commentaries prepared by BSSC and ASCE and in the references.

Ground motion values contoured on these maps incorporate

- A target risk of structural collapse equal to 1% in 50 years based upon a generic structural fragility,
- A factor of 1.3 to adjust from a geometric mean to the maximum response regardless of direction, and
- Deterministic upper limits imposed near large, active faults, which are taken as 1.8 times the estimated median response to the characteristic earthquake for the governing fault (1.8 is used to represent the 84th percentile response), but not less than 60% *g*.

As such, the values are different from those on the uniform-hazard 2014 USGS National Seismic Hazard Maps posted at https://doi.org/10.5066/F7HT2MHG.

Larger, more detailed versions of these maps are not provided because it is recommended that the corresponding USGS web tool (https://doi.org/10.5066/F7NK3C76) be used to determine the mapped value for a specified location.

User Note: The USGS Seismic Design Geodatabase is available at the ASCE 7 Hazard Tool https://asce7hazardtool.online/ or an approved equivalent.

Figure 22-2. S_{M1} for the default site conditions for the conterminous United States.

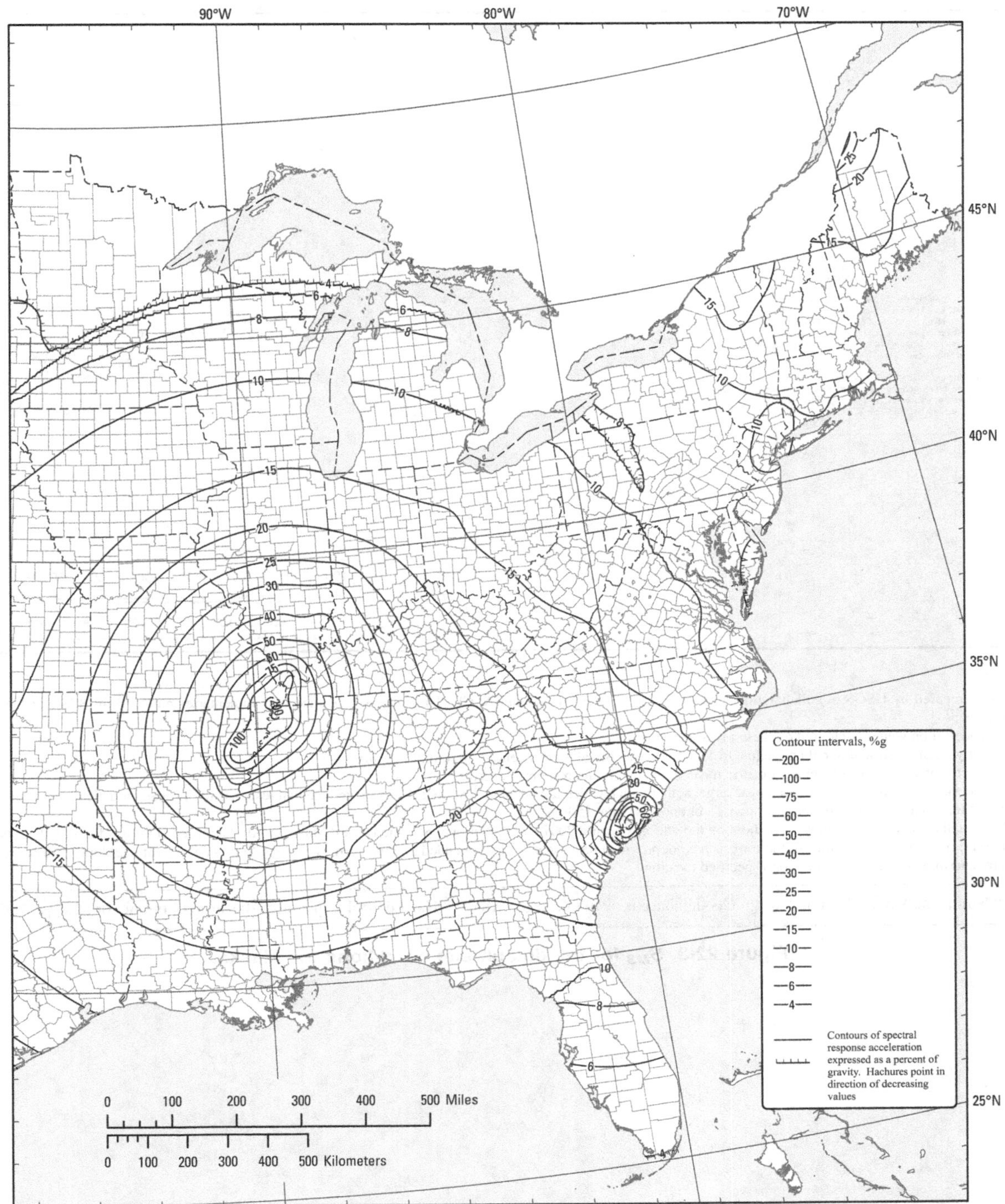

Figure 22-2 (*Continued*). S_{M1} for the default site conditions for the conterminous United States.

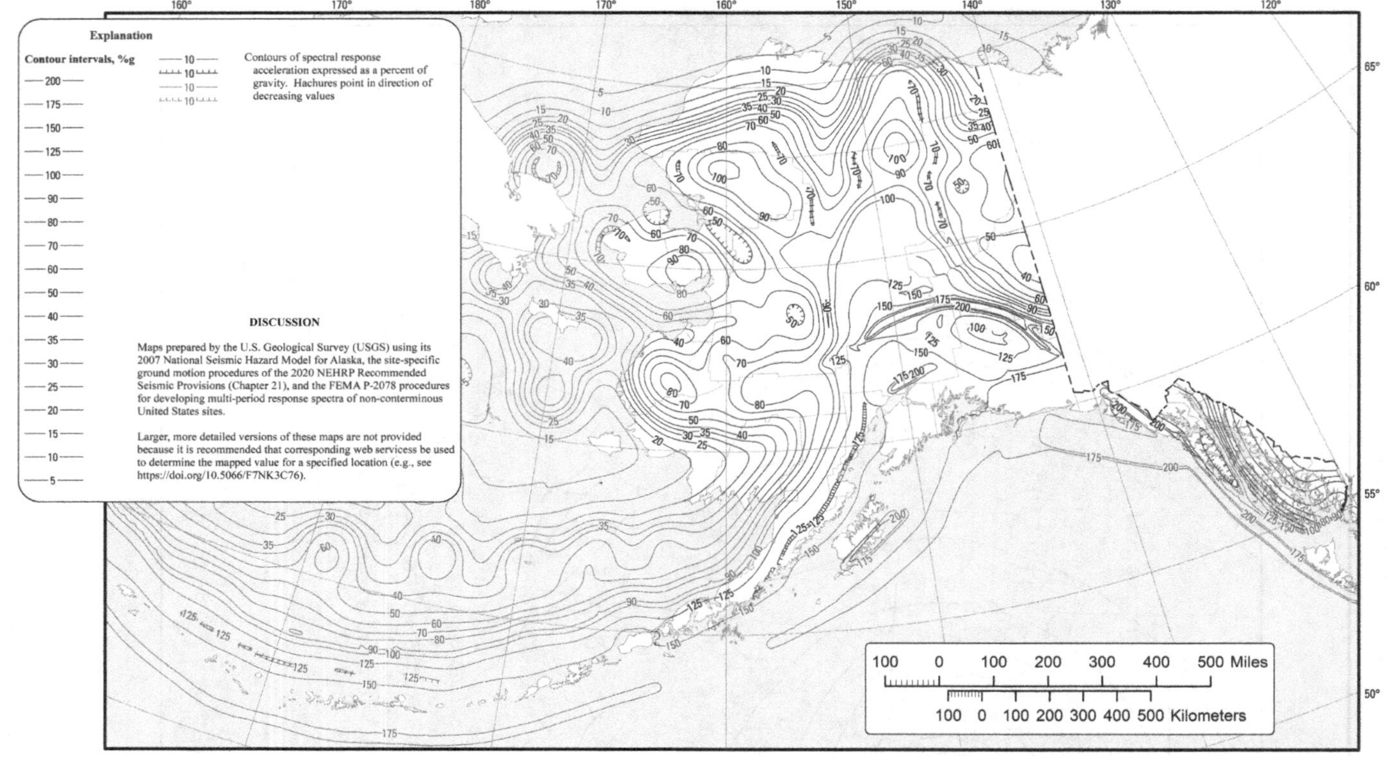

Notes:

Maps prepared by USGS in collaboration with the FEMA-funded BSSC and ASCE. The basis is explained in commentaries prepared by BSSC and ASCE and in the references.

Ground motion values contoured on these maps incorporate

- A target risk of structural collapse equal to 1% in 50 years based upon a generic structural fragility,
- A factor of 1.1 to adjust from a geometric mean to the maximum response regardless of direction, and
- Deterministic upper limits imposed near large, active faults, which are taken as 1.8 times the estimated median response to the characteristic earthquake for the fault (1.8 is used to represent the 84th percentile response), but not less than 150% *g*.

As such, the values are different from those on the uniform-hazard 2007 USGS National Seismic Hazard Maps for Alaska posted at https://doi.org/10.5066/F7HT2MHG.

Larger, more detailed versions of these maps are not provided because it is recommended that the corresponding USGS web tool (https://doi.org/10.5066/F7NK3C76) be used to determine the mapped value for a specified location.

User Note: The USGS Seismic Design Geodatabase is available at the ASCE 7 Hazard Tool https://asce7hazardtool.online/ or an approved equivalent.

Figure 22-3. S_{MS} for the default site conditions for Alaska.

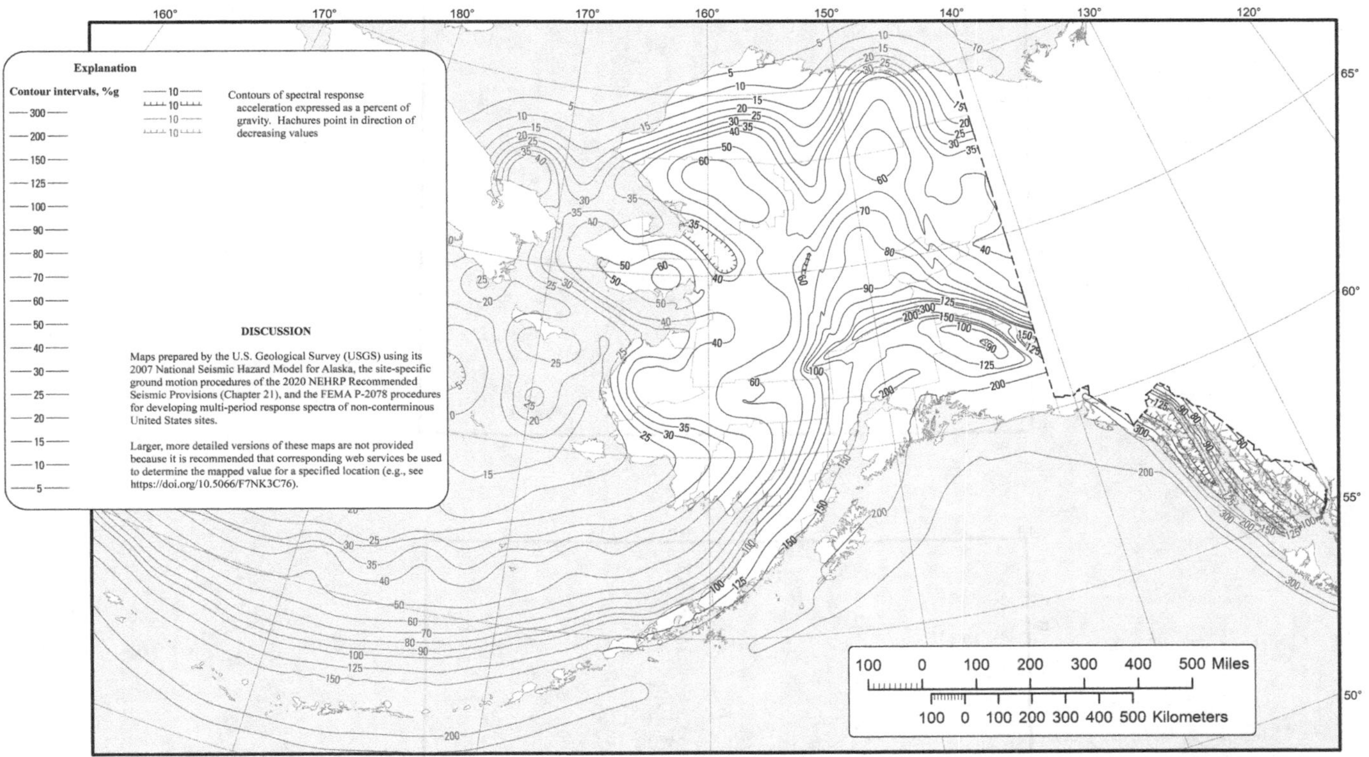

Notes:

Maps prepared by USGS in collaboration with the FEMA-funded BSSC and ASCE. The basis is explained in commentaries prepared by BSSC and ASCE and in the references.

Ground motion values contoured on these maps incorporate

- A target risk of structural collapse equal to 1% in 50 years based upon a generic structural fragility,
- A factor of 1.3 to adjust from a geometric mean to the maximum response regardless of direction, and
- Deterministic upper limits imposed near large, active faults, which are taken as 1.8 times the estimated median response to the characteristic earthquake for the fault (1.8 is used to represent the 84th percentile response), but not less than 60% g.

As such, the values are different from those on the uniform-hazard 2007 USGS National Seismic Hazard Maps for Alaska posted at https://doi.org/10.5066/F7HT2MHG.

Larger, more detailed versions of these maps are not provided because it is recommended that the corresponding USGS web tool (https://doi.org/10.5066/F7NK3C76) be used to determine the mapped value for a specified location.

User Note: The USGS Seismic Design Geodatabase is available at the ASCE 7 Hazard Tool https://asce7hazardtool.online/ or an approved equivalent.

Figure 22-4. S_{M1} for the default site conditions for Alaska.

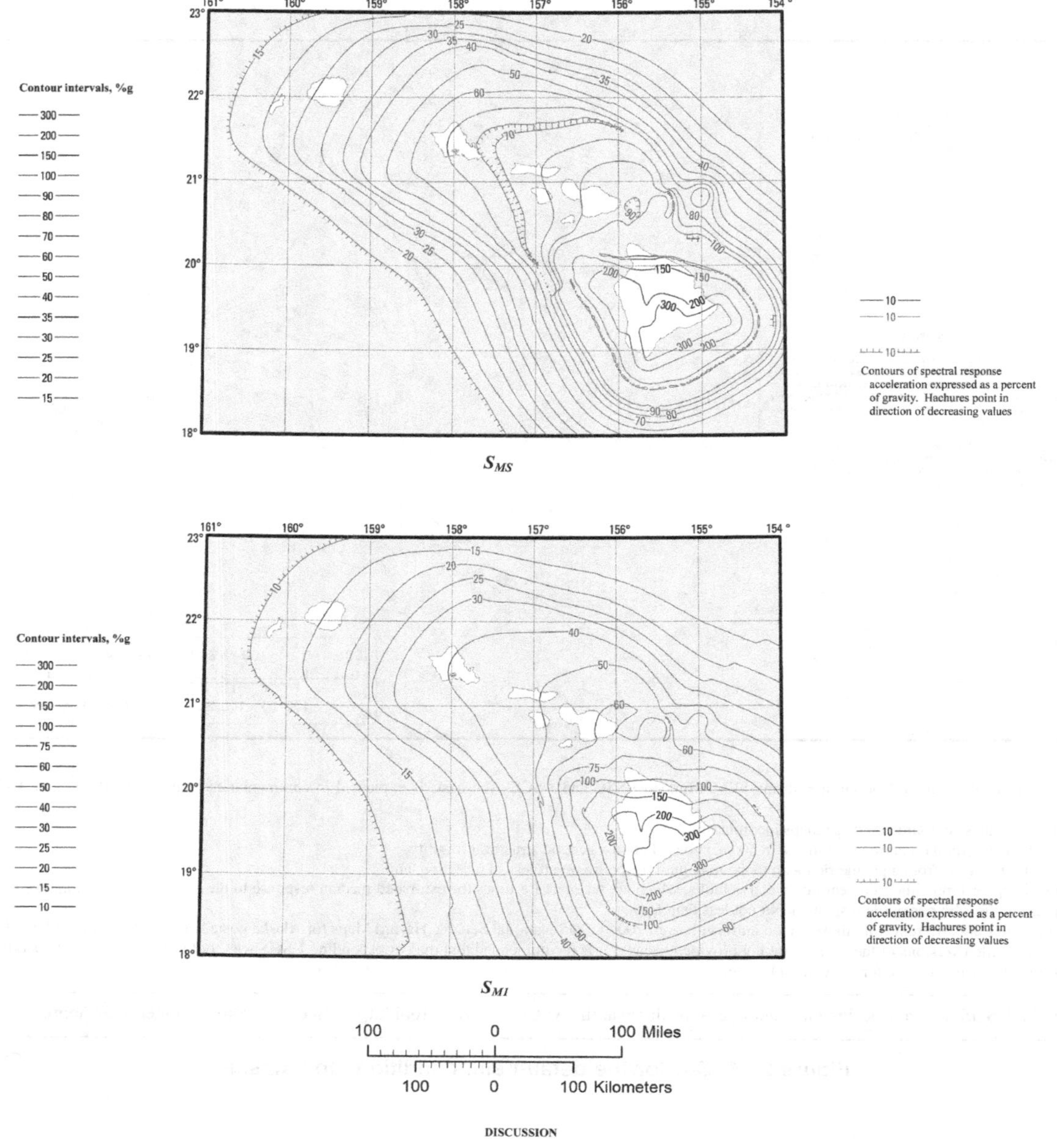

DISCUSSION

Maps prepared by the U.S. Geological Survey (USGS) using its 1998 National Seismic Hazard Model for Hawaii, the site-specific ground motion procedures of the 2020 NEHRP Recommended Seismic Provisions (Chapter 21), and the FEMA P-2078 procedures for developing multi-period response spectra of non-conterminous United States sites.

Larger, more detailed versions of these maps are not provided because it is recommended that corresponding web services be used to determine the mapped value for a specified location (e.g., see https://doi.org/10.5066/F7NK3C76).

Notes:

Maps prepared by USGS in collaboration with the FEMA-funded BSSC and ASCE. The basis is explained in commentaries prepared by BSSC and ASCE and in the references.

Ground motion values contoured on these maps incorporate

- A target risk of structural collapse equal to 1% in 50 years based upon a generic structural fragility, and
- Deterministic upper limits imposed near large, active faults, which are taken as 1.8 times the estimated median response to the characteristic earthquake for the fault (1.8 is used to represent the 84th percentile response), but not less than 150% and 60% g for 0.2 and 1.0 s, respectively.

As such, the values are different from those on the uniform-hazard 1998 USGS National Seismic Hazard Maps for Hawaii posted at https://doi.org/10.5066/F7HT2MHG.

Larger, more detailed versions of these maps are not provided because it is recommended that the corresponding USGS web tool (https://doi.org/10.5066/F7NK3C76) be used to determine the mapped value for a specified location.

User Note: The USGS Seismic Design Geodatabase is available at the ASCE 7 Hazard Tool https://asce7hazardtool.online/ or an approved equivalent.

Figure 22-5. S_{MS} and S_{M1} for the default site conditions for Hawaii.

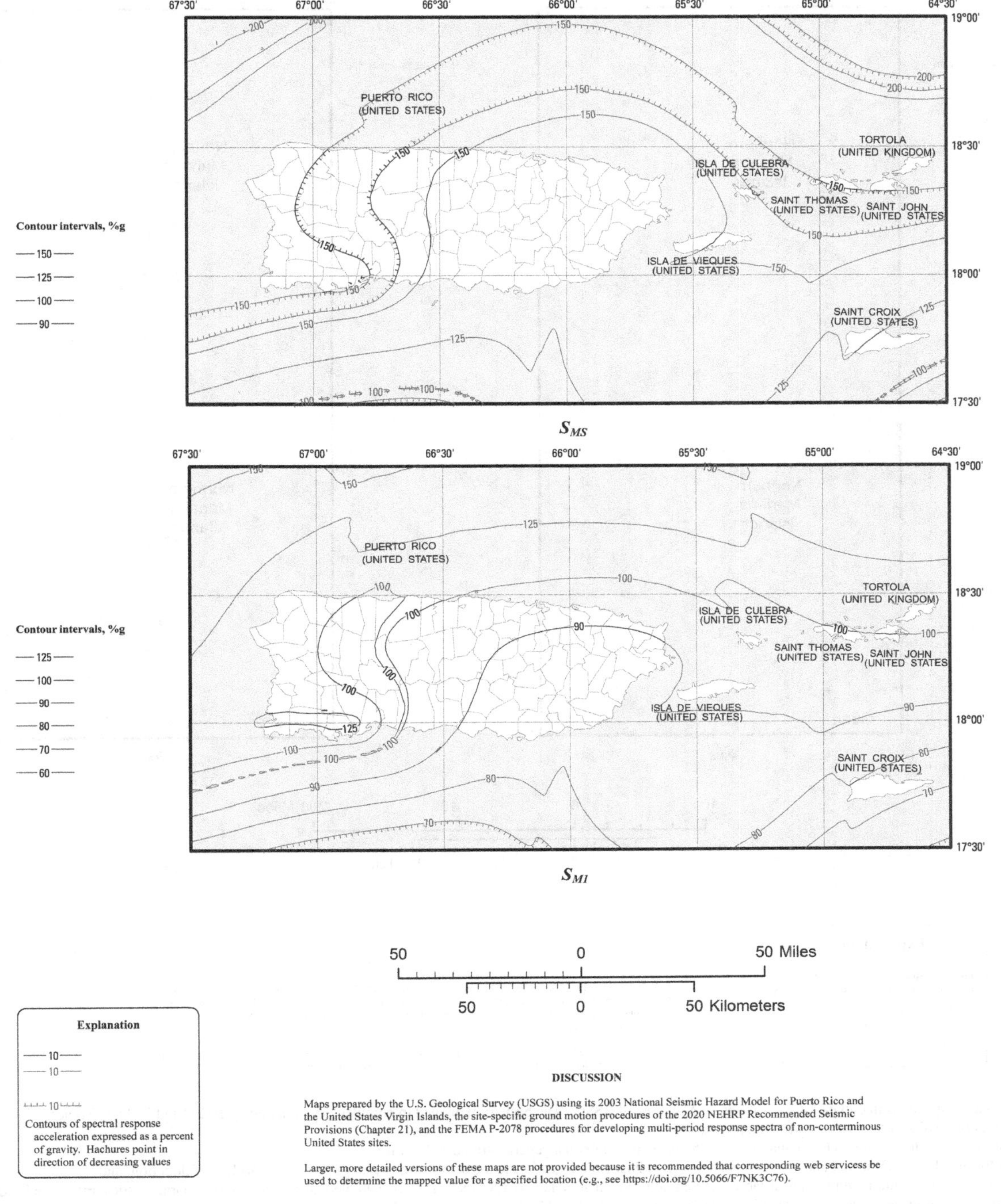

Notes:

Maps prepared by USGS in collaboration with the FEMA-funded BSSC and ASCE. The basis is explained in commentaries prepared by BSSC and ASCE and in the references.

Ground motion values contoured on these maps incorporate

- A target risk of structural collapse equal to 1% in 50 years based upon a generic structural fragility,
- A factor of 1.1 and 1.3 for 0.2 and 1.0 s, respectively, to adjust from a geometric mean to the maximum response regardless of direction, and
- Deterministic upper limits imposed near large, active faults, which are taken as 1.8 times the estimated median response to the characteristic earthquake for the fault (1.8 is used to represent the 84th percentile response), but not less than 150% and 60% g for 0.2 and 1.0 s, respectively.

As such, the values are different from those on the uniform-hazard 2003 USGS National Seismic Hazard Maps for Puerto Rico and the US. Virgin Islands posted at https://doi.org/10.5066/F7HT2MHG.

Larger, more detailed versions of these maps are not provided because it is recommended that the corresponding USGS web tool (https://doi.org/10.5066/F7NK3C76) be used to determine the mapped value for a specified location.

User Note: The USGS Seismic Design Geodatabase is available at the ASCE 7 Hazard Tool https://asce7hazardtool.online/ or an approved equivalent.

Figure 22-6. S_{MS} **and** S_{M1} **for the default site conditions for Puerto Rico and the US Virgin Islands.**

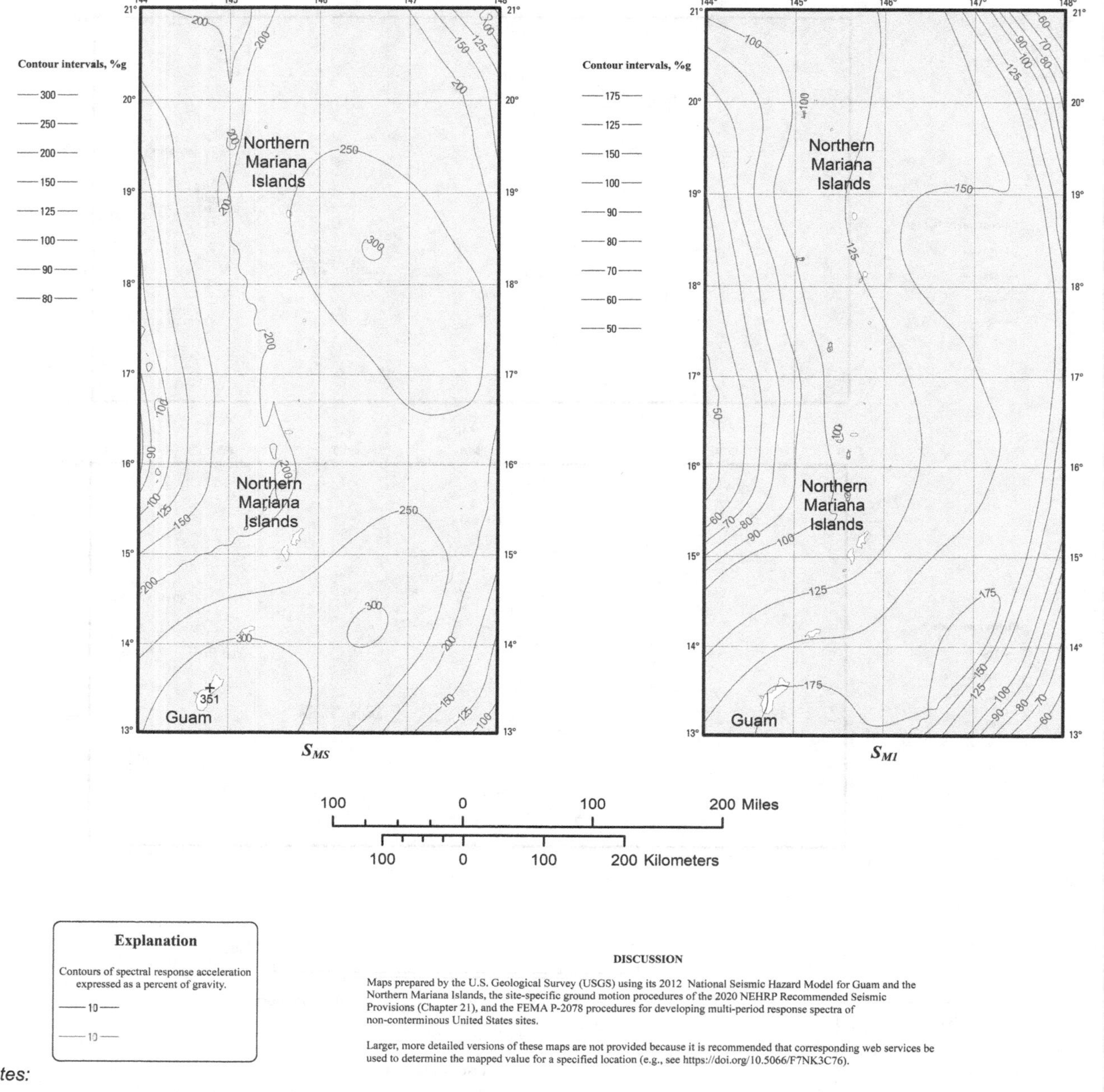

Notes:

Maps prepared by USGS in collaboration with the FEMA-funded BSSC. The basis is explained in commentary prepared by BSSC and in the references.

Ground motion values contoured on these maps incorporate

- A target risk of structural collapse equal to 1% in 50 years based upon a generic structural fragility,
- A factor of 1.1 and 1.3 for 0.2 and 1.0 s, respectively, to adjust from a geometric mean to the maximum response regardless of direction, and
- Deterministic upper limits imposed near large, active faults, which are taken as 1.8 times the estimated median response to the characteristic earthquake for the governing fault (1.8 is used to represent the 84th percentile response), but not less than 150% and 60% g for 0.2 and 1.0 s, respectively.

As such, the values are different from those on the uniform-hazard 2012 USGS National Seismic Hazard Maps for Guam and the Northern Mariana Islands posted at https://doi.org/10.5066/F7HT2MHG.

Larger, more detailed versions of these maps are not provided because it is recommended that the corresponding USGS web tool (https://doi.org/10.5066/F7NK3C76) be used to determine the mapped value for a specified location.

User Note: The USGS Seismic Design Geodatabase is available at the ASCE 7 Hazard Tool https://asce7hazardtool.online/ or an approved equivalent.

Figure 22-7. S_{MS} and S_{M1} for the default site conditions for Guam and the Northern Mariana Islands.

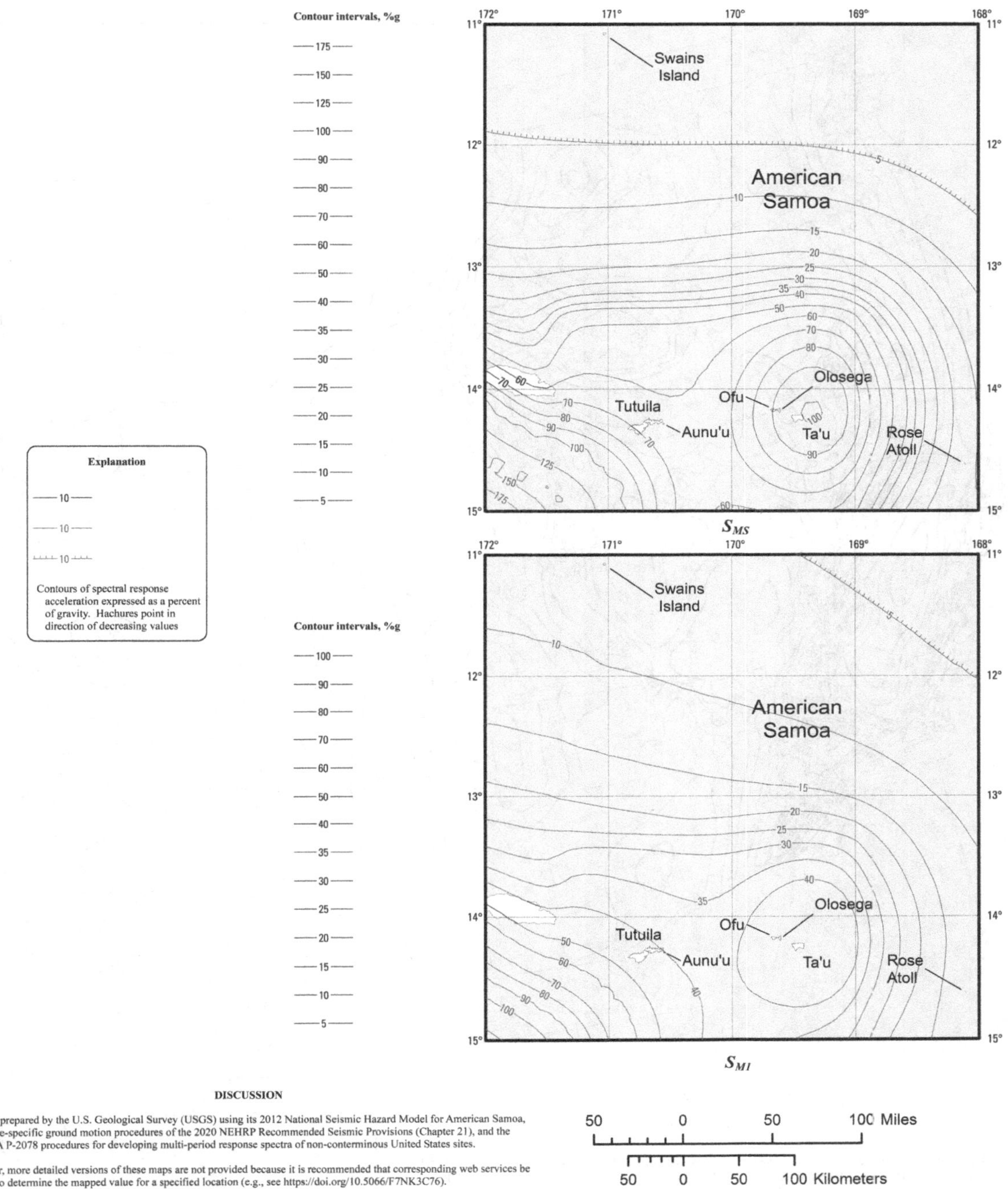

Notes:

Maps prepared by USGS in collaboration with the FEMA-funded BSSC. The basis is explained in commentary prepared by BSSC and in the references. Ground motion values contoured on these maps incorporate

- A target risk of structural collapse equal to 1% in 50 years based upon a generic structural fragility,
- A factor of 1.1 and 1.3 for 0.2 and 1.0 s, respectively, to adjust from a geometric mean to the maximum response regardless of direction, and
- Deterministic upper limits imposed near large, active faults, which are taken as 1.8 times the estimated median response to the characteristic earthquake for the fault (1.8 is used to represent the 84th percentile response), but not less than 150% and 60% g for 0.2 and 1.0 s, respectively.

As such, the values are different from those on the uniform-hazard 2012 USGS National Seismic Hazard Maps for American Samoa posted at https://doi.org/10.5066/F7HT2MHG.

Larger, more detailed versions of these maps are not provided because it is recommended that the corresponding USGS web tool (https://doi.org/10.5066/F7NK3C76) be used to determine the mapped value for a specified location.

User Note: The USGS Seismic Design Geodatabase is available at the ASCE 7 Hazard Tool https://asce7hazardtool.online/ or an approved equivalent.

Figure 22-8. S_{MS} and S_{M1} for the default site conditions for American Samoa.

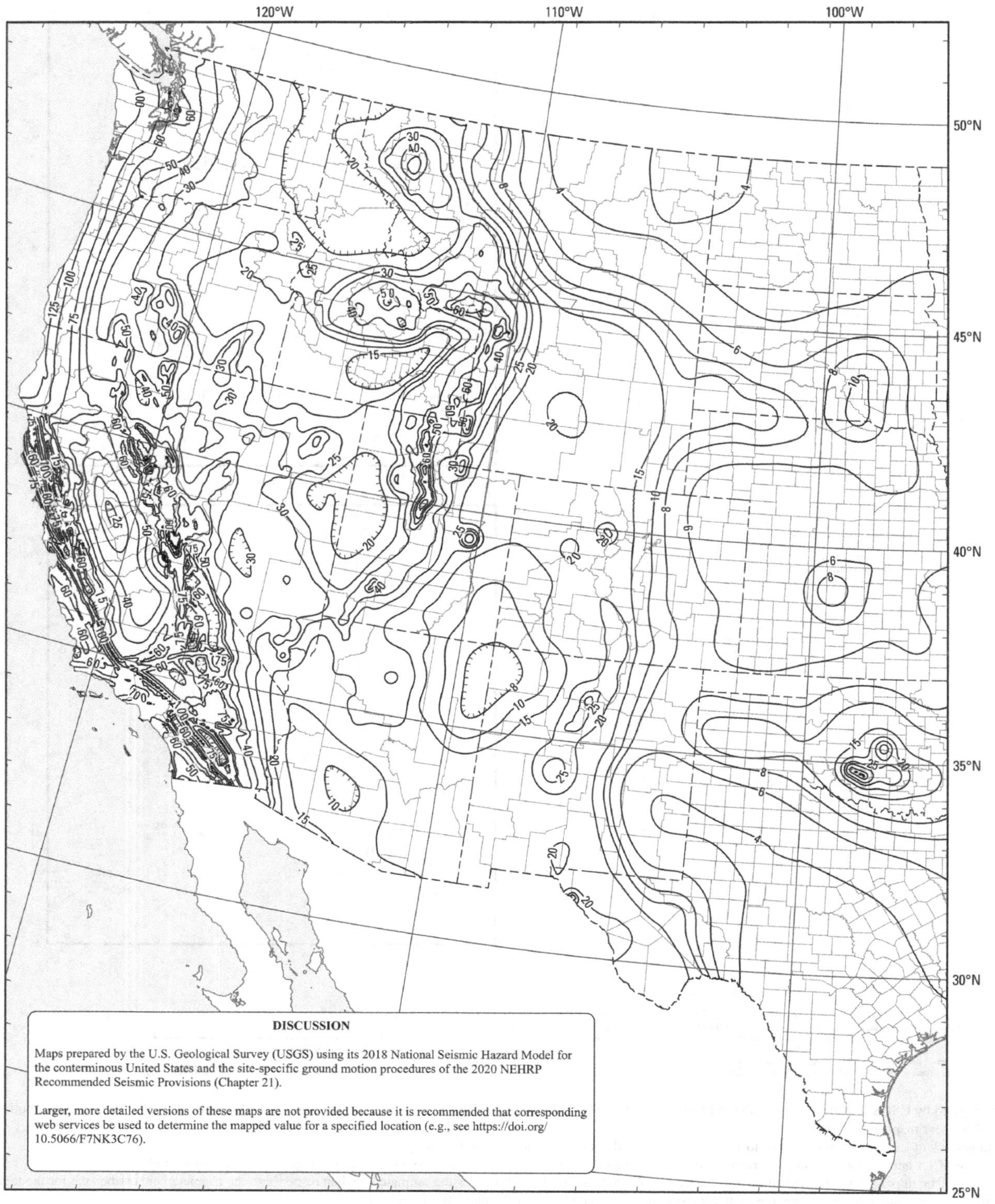

Figure 22-9. PGA_M for the default site conditions for the conterminous United States.

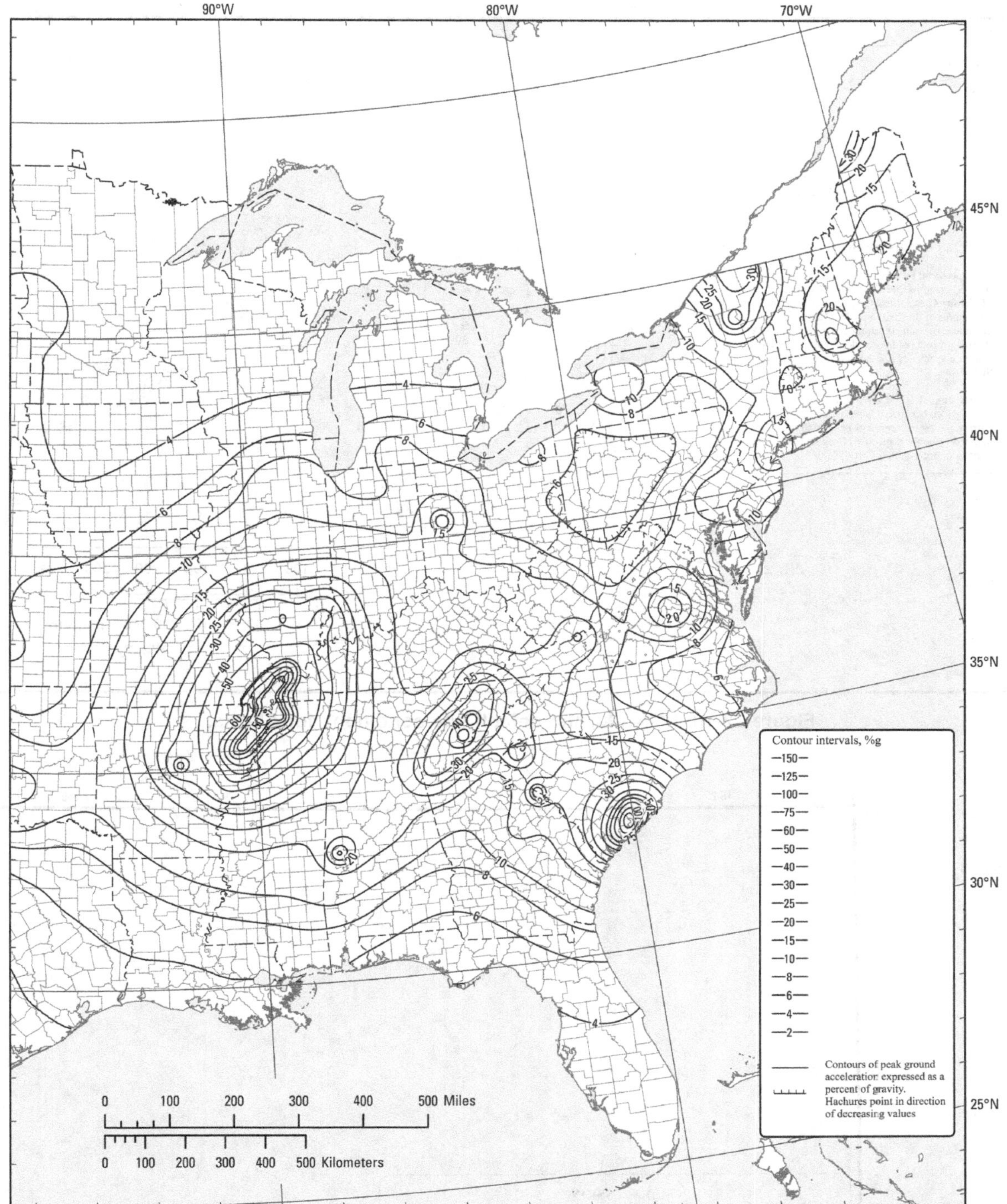

Figure 22-9 (*Continued*). PGA_M for the default site conditions for the conterminous United States.

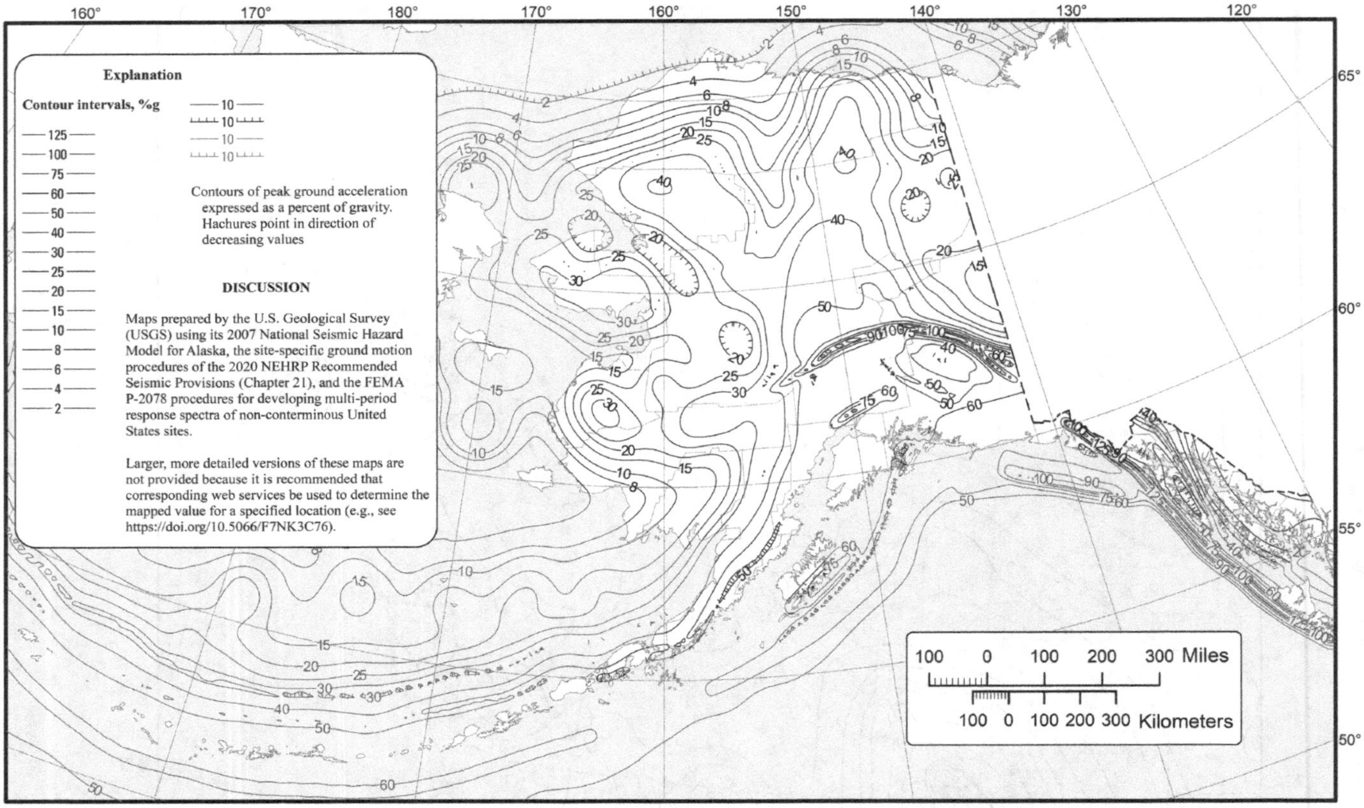
Explanation
Contour intervals, %g
125
100
75
60
50
40
30
25
20
15
10
8
6
4
2
10
10
10
10
Contours of peak ground acceleration expressed as a percent of gravity. Hachures point in direction of decreasing values
DISCUSSION
Maps prepared by the U.S. Geological Survey (USGS) using its 2007 National Seismic Hazard Model for Alaska, the site-specific ground motion procedures of the 2020 NEHRP Recommended Seismic Provisions (Chapter 21), and the FEMA P-2078 procedures for developing multi-period response spectra of non-conterminous United States sites.
Larger, more detailed versions of these maps are not provided because it is recommended that corresponding web services be used to determine the mapped value for a specified location (e.g., see https://doi.org/10.5066/F7NK3C76).
100 0 100 200 300 Miles
100 0 100 200 300 Kilometers
160° 170° 180° 170° 160° 150° 140° 130° 120°
65° 60° 55° 50°

Figure 22-10. PGA_M for the default site conditions for Alaska.

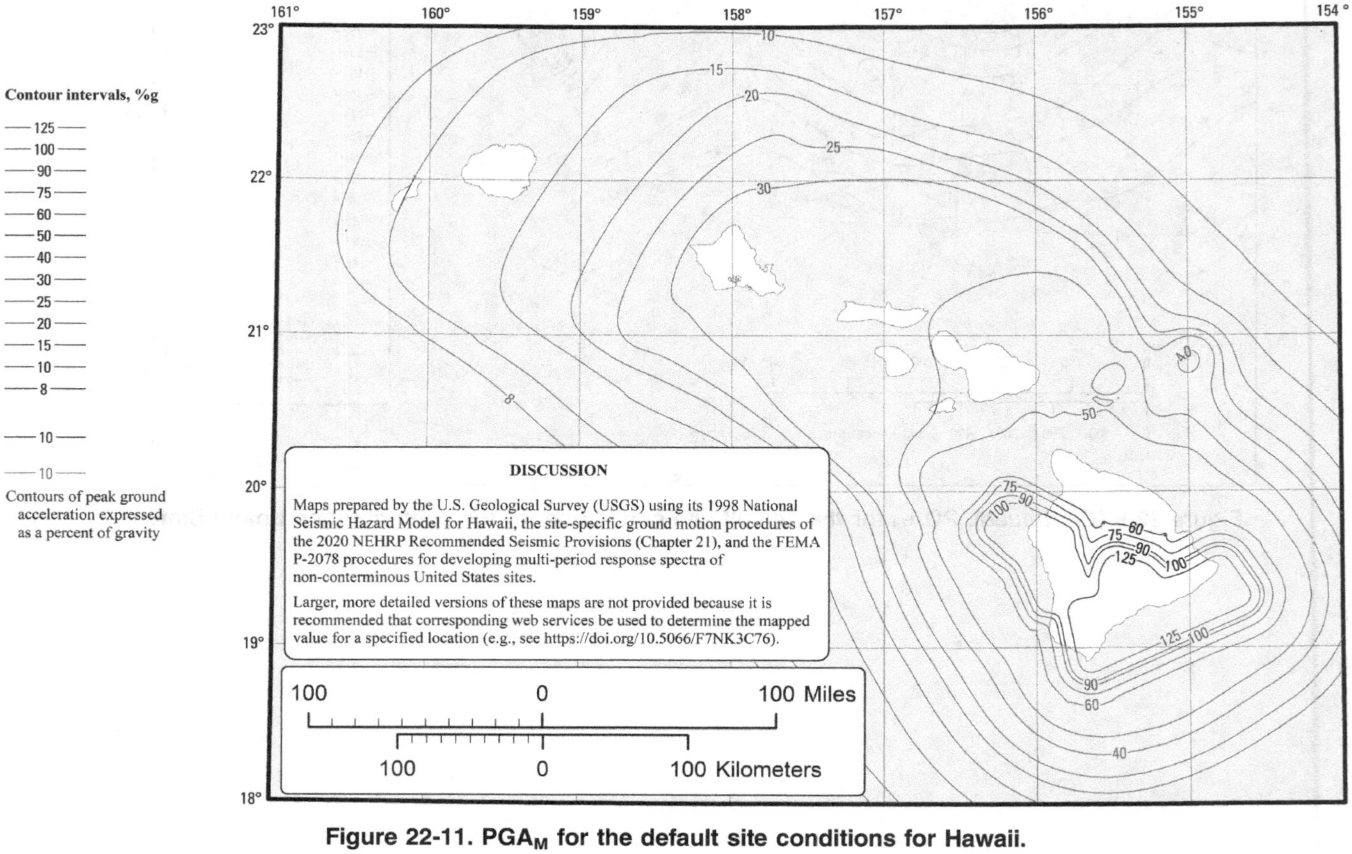
Contour intervals, %g
125
100
90
75
60
50
40
30
25
20
15
10
8
10
10
Contours of peak ground acceleration expressed as a percent of gravity
DISCUSSION
Maps prepared by the U.S. Geological Survey (USGS) using its 1998 National Seismic Hazard Model for Hawaii, the site-specific ground motion procedures of the 2020 NEHRP Recommended Seismic Provisions (Chapter 21), and the FEMA P-2078 procedures for developing multi-period response spectra of non-conterminous United States sites.
Larger, more detailed versions of these maps are not provided because it is recommended that corresponding web services be used to determine the mapped value for a specified location (e.g., see https://doi.org/10.5066/F7NK3C76).
100 0 100 Miles
100 0 100 Kilometers
161° 160° 159° 158° 157° 156° 155° 154°
23° 22° 21° 20° 19° 18°

Figure 22-11. PGA_M for the default site conditions for Hawaii.

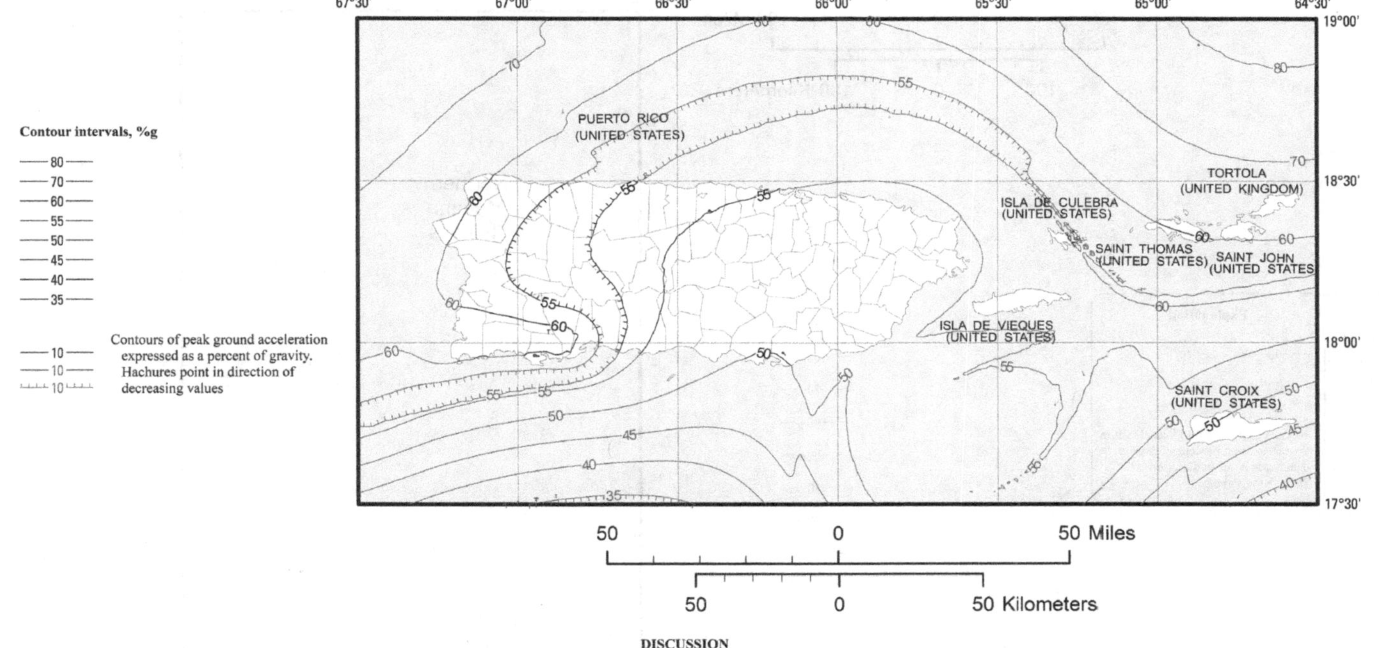

DISCUSSION

Maps prepared by the U.S. Geological Survey (USGS) using its 2003 National Seismic Hazard Model for Puerto Rico and the United States Virgin Islands, the site-specific ground motion procedures of the 2020 NEHRP Recommended Seismic Provisions (Chapter 21), and the FEMA P-2078 procedures for developing multi-period response spectra of non-conterminous United States sites.

Larger, more detailed versions of these maps are not provided because it is recommended that corresponding web servicess be used to determine the mapped value for a specified location (e.g., see https://doi.org/10.5066/F7NK3C76).

Figure 22-12. PGA_M for the default site conditions for Puerto Rico and the US Virgin Islands.

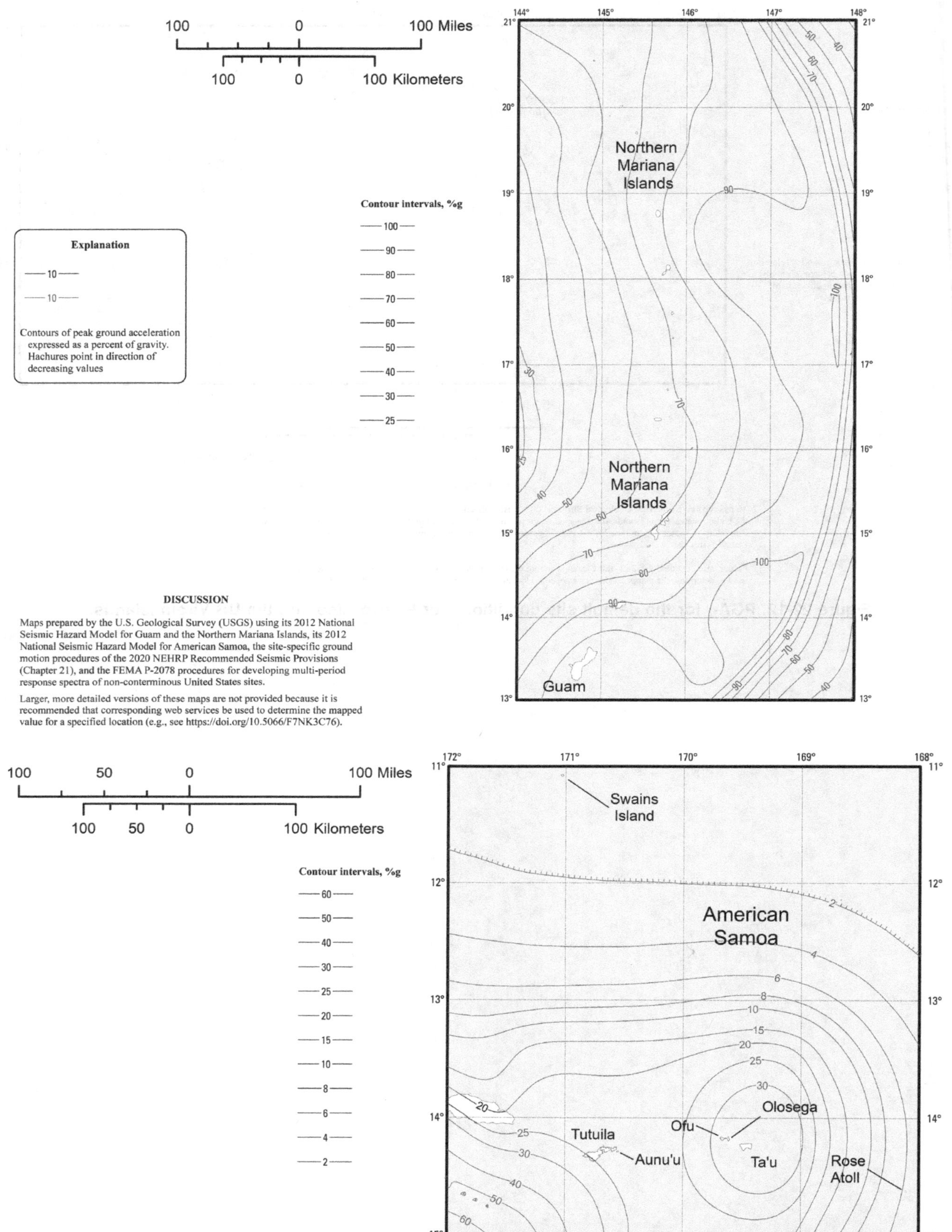

Figure 22-13. PGA_M for the default site conditions for Guam and the Northern Mariana Islands and American Samoa.

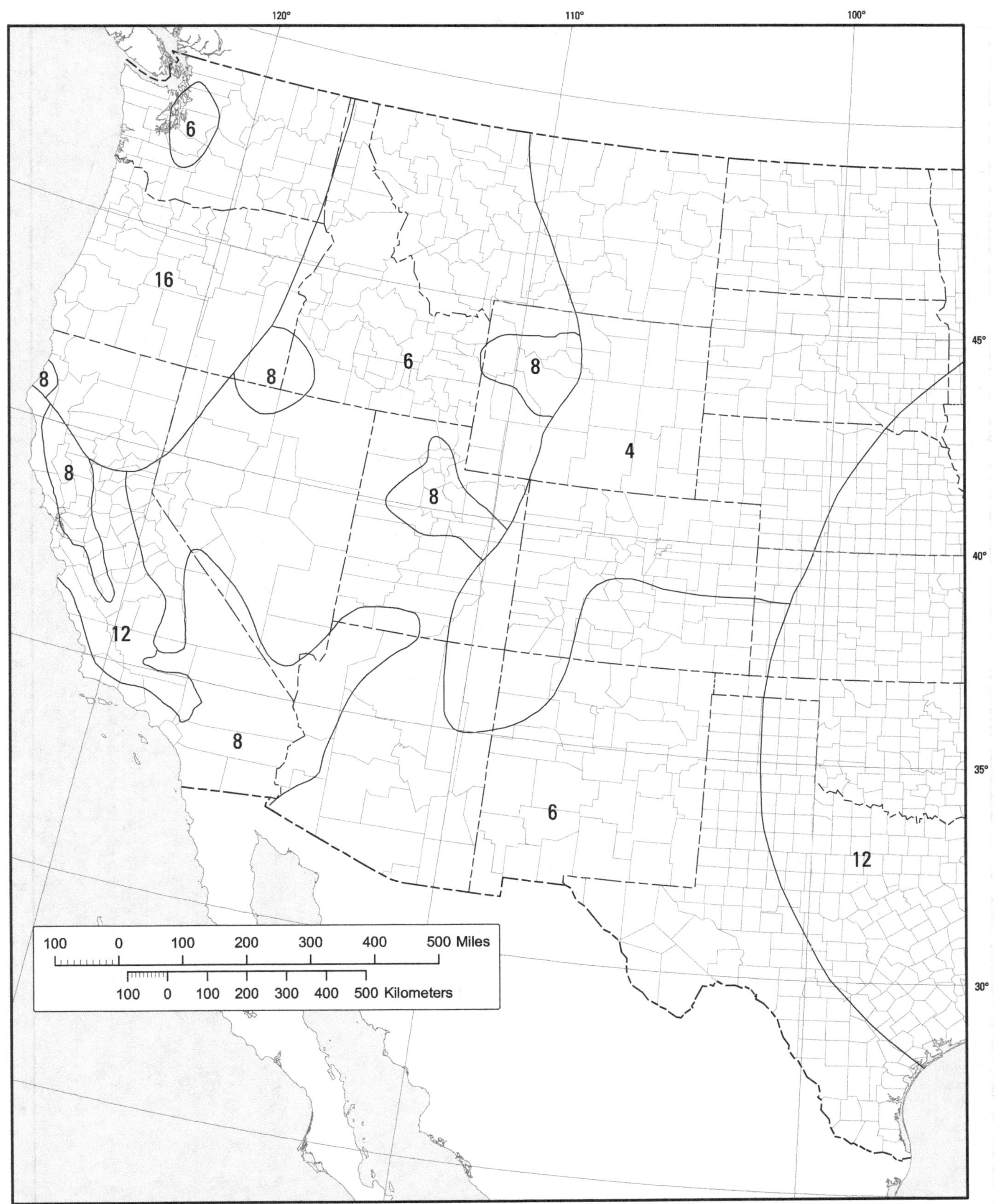

Figure 22-14. T_L for the conterminous United States.

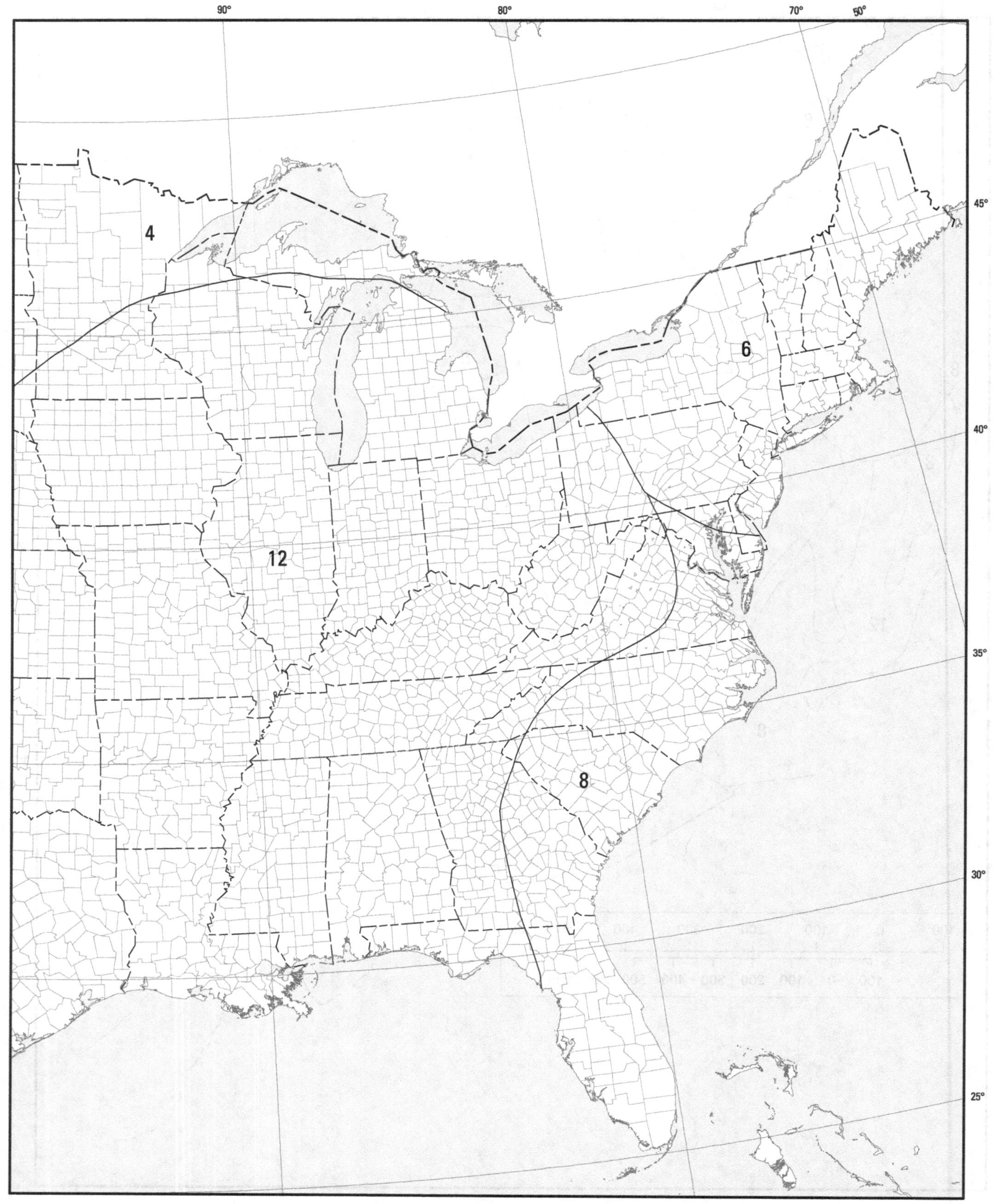

Figure 22-14 (*Continued*). T_L for the conterminous United States.

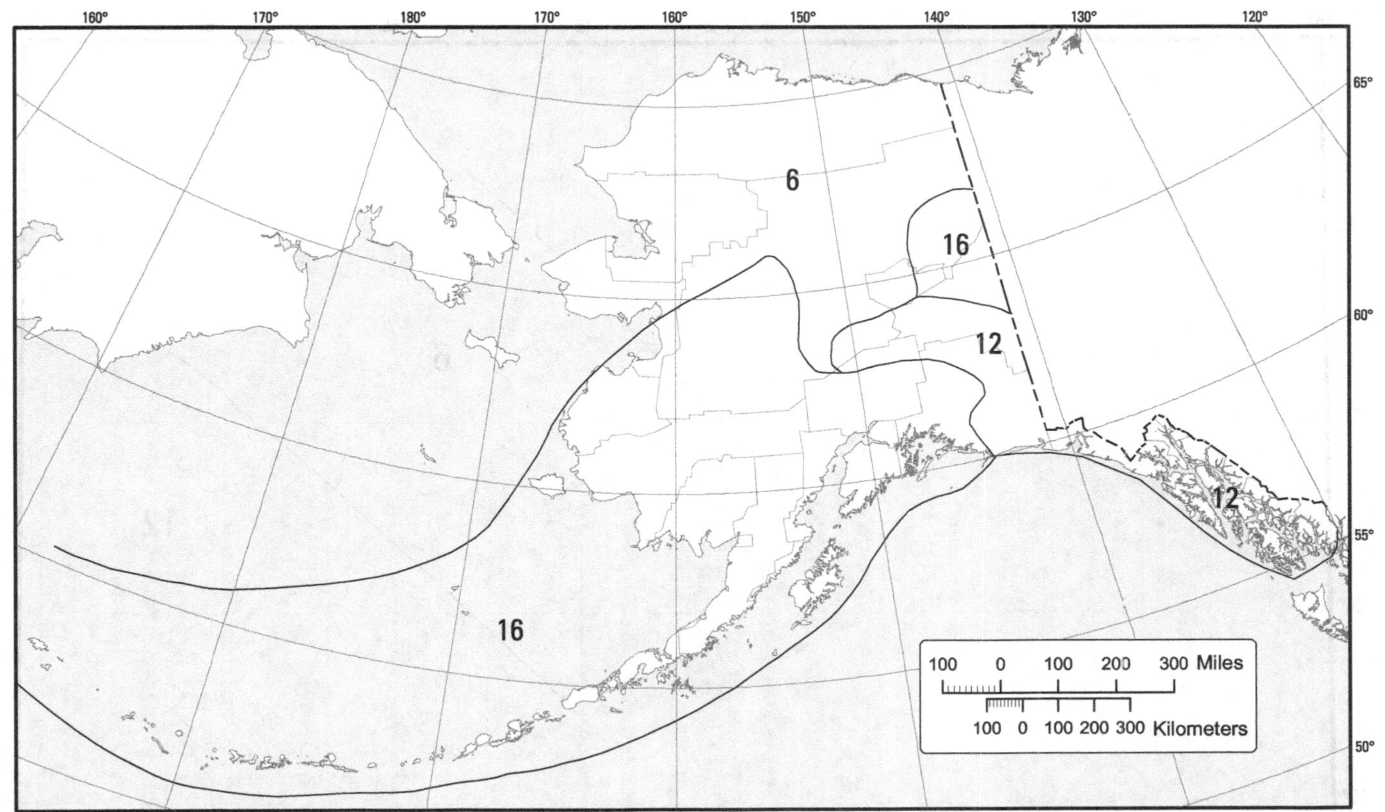

Figure 22-15. T_L **for Alaska.**

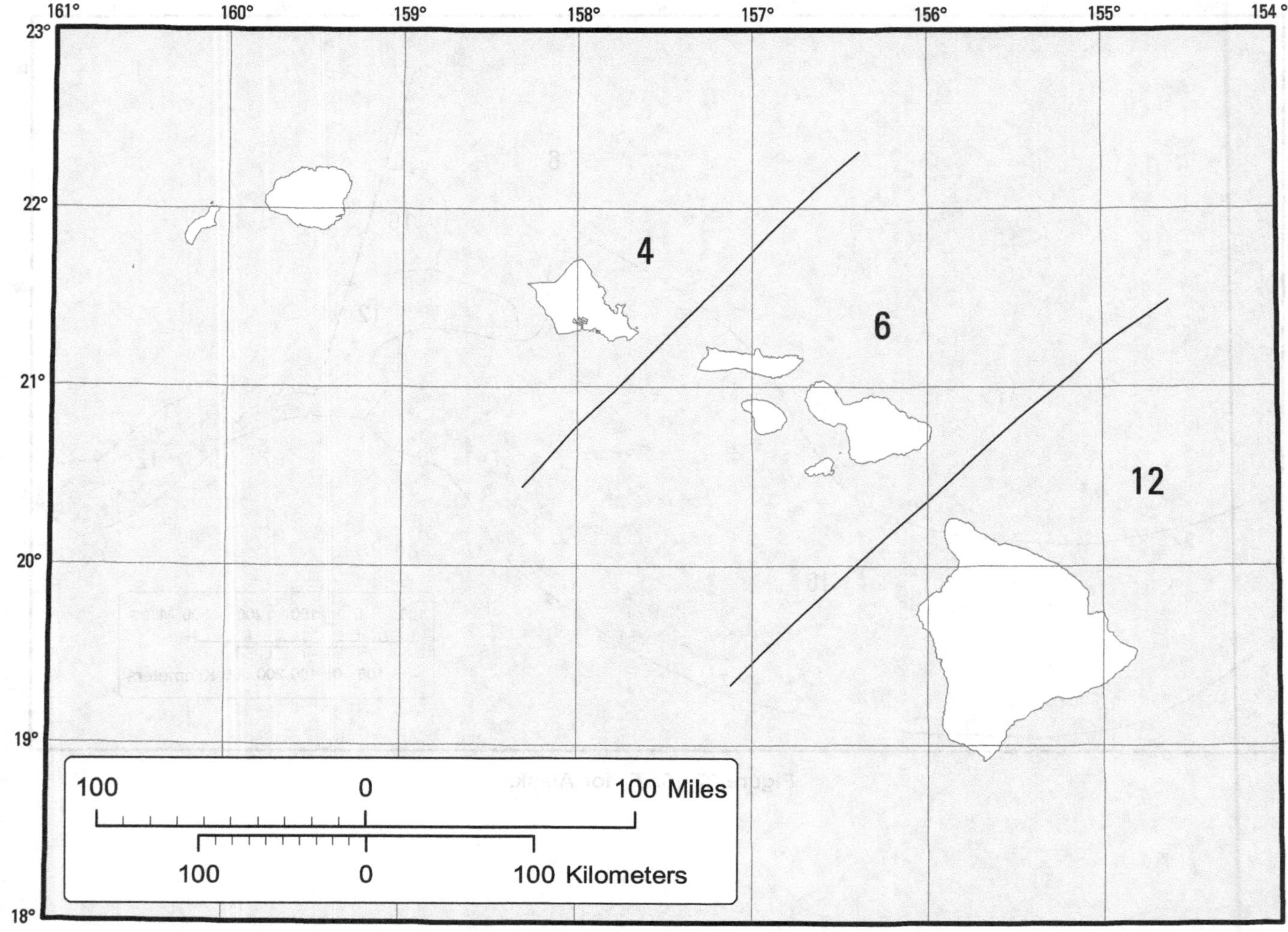

Figure 22-16. T_L for Hawaii.

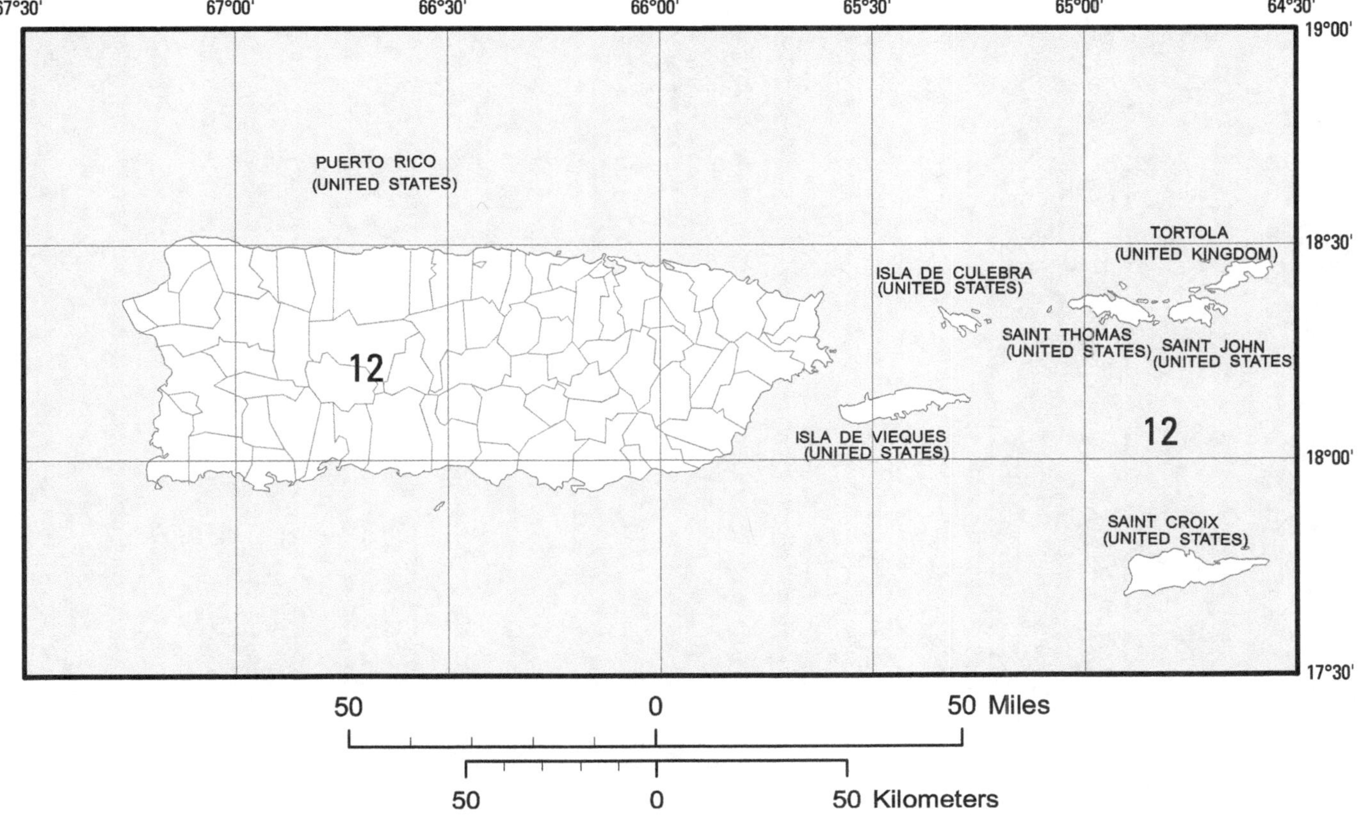

Figure 22-17. T_L **for Puerto Rico and the US Virgin Islands.**

CHAPTER 23
SEISMIC DESIGN REFERENCE DOCUMENTS

23.1 CONSENSUS STANDARDS AND OTHER REFERENCE DOCUMENTS

This section lists the consensus standards and other documents that shall be considered part of this standard to the extent referenced in Chapters 11 through 22. Those referenced documents identified by an asterisk (*) are not consensus standards; rather, they are documents developed within the industry and represent acceptable procedures for design and construction to the extent referred to in the specified section.

***AAMA 501.6**, *Recommended Dynamic Test Method for Determining the Seismic Drift Causing Glass Fallout from Window Wall, Curtain, and Storefront Systems*, American Architectural Manufacturers Association, 2018.

Cited in: Section 13.5.9.2

ACI 307, *Code Requirements for Reinforced Concrete Chimneys and Commentary*, American Concrete Institute, 2008.

Cited in: Sections 15.4.1, 15.6.2.2, Table 15.4-2

ACI 313, *Design Specification for Concrete Silos and Stacking Tubes for Storing Granular Materials and Commentary*, American Concrete Institute, 2019.

Cited in: Sections 15.7.9.3.3, 15.7.9.6, 15.7.9.7

ACI 318, *Building Code Requirements for Structural Concrete and Commentary*, American Concrete Institute, 2019.

Cited in: Sections 11.2, 12.12.5, 12.13.8, 12.13.9.2.1.1, 12.13.9.2.1.2, 12.13.9.3.3, 12.13.9.3.4, 13.4.2.1, 13.4.4, 13.5.7.2, 14.2.1, 14.2.2, 14.2.2.1, 14.2.2.2, 14.2.2.3, 14.2.2.4, 14.2.2.5, 14.2.2.6, 14.2.2.7, 14.2.4.3.5, 14.3.1, 14.3.2, 14.4.4.1.2, 15.4.9.1, 15.4.9.4, 15.6.2.2, 15.7.5, 15.7.11.7, Tables 12.2-1, 12.10-1, 15.4-1, 15.4-2

ACI 350.3, *Seismic Design of Liquid-Containing Concrete Structures and Commentary*, American Concrete Institute, 2006.

Cited in: Sections 15.7.6.1.1, 15.7.7.3

ACI 355.2-19, *Qualification of Post-Installed Mechanical Anchors in Concrete and Commentary*, American Concrete Institute, 2019.

Cited in: Sections 13.4.2.3, 15.4.9.3

ANSI/AHRI Standards 1270 (I-P) and 1271 (SI) Sections 1 through 10, excluding Appendixes, *Requirements for Seismic Qualification of HVACR Equipment*, Air-Conditioning, Heating, and Refrigeration Institute, 2015.

Cited in: Section 13.6.2.1

ANSI/AISC 341, *Seismic Provisions for Structural Steel Buildings*, American Institute of Steel Construction, 2022.

Cited in: Sections 12.13.9.3.4, 14.1.1, 14.1.2.2.1, 14.1.2.2.2, 14.1.3, 14.1.6, 14.1.8, 14.3.1, 14.3.3, 17.2.5.5, Table 15.4-1

ANSI/AISC 360, *Specification for Structural Steel Buildings*, American Institute of Steel Construction, 2022.

Cited in: Sections 14.1.1, 14.1.2.1, 14.1.2.2.1, 14.1.3, 14.3.1, 14.3.2, Table 15.4-1

ANSI/AISI S100, *North American Specification for the Design of Cold-Formed Steel Structural Members*, 2016 Edition (Reaffirmed 2020), with Supplement 2, 2020 Edition, AISI S100-16 (2020) w/S2-20, American Iron and Steel Institute.

Cited in: Sections 14.1.1, 14.1.3.1, 14.1.3.2, 14.1.5, Table 15.4-1

ANSI/AISI S230, *Standard for Cold-Formed Steel Framing —Prescriptive Method for One- and Two-Family Dwellings*, 2019 Edition, AISI S230-19, American Iron and Steel Institute.

Cited in: Sections 14.1.1, 14.1.4.3

ANSI/AISI S240, *North American Standard for Cold-Formed Steel Framing*, 2020 Edition, AISI S240-20, American Iron and Steel Institute.

Cited in: Sections 14.1.1, 14.1.4.1, 14.1.4.2.1, 14.1.5

ANSI/AISI S310, *North American Standard for the Design of Profiled Steel Diaphragm Panels*, 2020 Edition, AISI S310-20, American Iron and Steel Institute.

Cited in: Sections 14.1.1, 14.1.5

ANSI/AISI S400, *North American Standard for Seismic Design of Cold-Formed Steel Structural Systems*, 2020 Edition, AISI S400-20, American Iron and Steel Institute.

Cited in: Sections 12.14.7.2, 14.1.1, 14.1.3.2, 14.1.4.2.1, 14.1.4.2.2, 14.1.5.1, Table 12.2-1

ANSI/RMI MH 16.1, *Specification for the Design, Testing, and Utilization of Industrial Steel Storage Racks*, Rack Manufacturers Institute, 2012.

Cited in: Sections 15.5.3.1, 15.5.3.1.1, 15.5.3.1.2, 15.5.3.3

ANSI/RMI MH 16.3 *Specification for the Design, Testing, and Utilization of Industrial Steel Cantilevered Storage Racks*, Rack Manufacturers Institute, 2016.

Cited in: Section 15.5.3.2, 15.5.3.2.1, 15.5.3.3

ANSI/SDI-QA/QC, *Standard for Quality Control and Quality Assurance for Installation of Steel Deck*, Steel Deck Institute, 2022.

Cited in: Sections 14.1.1, 14.1.5

ANSI/SDI-SD, *Standard for Steel Deck*, Steel Deck Institute, 2022.

Cited in: Sections 14.1.1, 14.1.5

ANSI/SJI-100, *Standard Specification for K-Series, LH-Series and DLH-Series Open Web Steel Joists and for Joist Girders*, Steel Joist Institute, 2020.

Cited in: Sections 14.1.1, 14.1.6

ANSI/SJI-200, *Standard Specification for Composite Steel Joists*, CJ-Series, Steel Joist Institute, 2015.

Cited in: Section 14.1.1, 14.1.6

ANSI/TIA-222-H, *Structural Standard for Antenna Supporting Structures, Antennas and Small Wind Turbine Support Structures*, Addendum 1, Telecommunications Industry Association, 2019.

Cited in: Section 15.6.6

API 12B, *Specification for Bolted Tanks for Storage of Production Liquids, 12B*, 16th Edition, American Petroleum Institute, 2014.

Cited in: Section 15.7.8.2

API 620, *Design and Construction of Large, Welded, Low-Pressure Storage Tanks*, 12th Edition, Addendum 2, American Petroleum Institute, 2018.

Cited in: Sections 15.4.1, 15.7.8.1, 15.7.13.1

API 650, *Welded Tanks for Oil Storage*, 12th Edition, Addendum 3, American Petroleum Institute, 2018.

Cited in: Sections 15.4.1, 15.7.8.1, 15.7.9.4

API 653, *Tank Inspection, Repair, Alteration, and Reconstruction*, 5th Edition, Addendum 1, American Petroleum Institute, 2018.

Cited in: Section 15.7.6.1.9, Table 15.4-1

ASCE 4, *Seismic Analysis of Safety-Related Nuclear Structures and Commentary*, ASCE, 1998.

Cited in: Section 12.9.1.3

ASCE 8, *Specification for the Design of Cold-Formed Stainless Steel Structural Members*, ASCE, 2002.

Cited in: Sections 14.1.1, 14.1.3.1, 14.1.3.2, 14.1.5

ASCE 19, *Structural Applications of Steel Cables for Buildings*, ASCE, 2010.

Cited in: Sections 14.1.1, 14.1.7

ASCE 41, *Seismic Evaluation and Retrofit of Existing Buildings*, ASCE, 2017.

Cited in: Sections 16.3, 16.4.2, 16.4.2.1, 16.4.2.2

ASCE 72, *Design of Steel Lighting System Support Pole Structures*, ASCE, 2022.

Cited in: Section 15.6.10

ASME A17.1-2019/CSA B44-19, *Safety Code for Elevators and Escalators*, CSA Group, 2019.

Cited in: Sections 13.6.11, 13.6.11.3

ASME B31, Code for Pressure Piping, American Society of Mechanical Engineers (consists of the following sections):

ASME B31.1, *Power Piping*, 2018.

ASME B31.3, *Process Piping*, 2018.

ASME B31.4, *Pipeline Transportation Systems for Liquids and Slurries*, 2019.

ASME B31.5, *Refrigeration Piping and Heat Transfer Components*, 2019.

ASME B31.8, *Gas Transmission and Distribution Piping Systems*, 2018.

ASME B31.9, *Building Services Piping*, 2017.

ASME B31.12, *Hydrogen Piping and Pipelines*, 2019.

ASME B31EA-2010, Addenda to ASME B31E-2008 *Standard for the Seismic Design and Retrofit of Above-Ground Piping Systems*, Addendum A, 2010.

Cited in: Sections 13.6.4.1, 13.6.7.1, Table 13.6-1

ASME BPVC, *Boiler and Pressure Vessel Code*, American Society of Mechanical Engineers (consists of the following sections):

BPVC-I, *Rules for Construction of Power Boilers*, 2019.

BPVC-IV, *Rules for Construction of Heating Boilers*, 2019.

BPVC-VIII Division 1, *Rules for Construction of Pressure Vessels*, 2019.

BPVC-VIII Division 2, *Rules for Construction of Pressure Vessels*, Alternative Rules, 2019.

BPVC-VIII Division 3, *Rules for Construction of Pressure Vessels*, Alternative Rules for Construction of High Pressure Vessels, 2019.

Cited in: Sections 13.6.10, 13.6.13, 15.7.11.2, 15.7.11.6, 15.7.12.2

ASME RTP-1, *Reinforced Thermoset Plastic Corrosion-Resistant Equipment*, American Society of Mechanical Engineers, 2019.

Cited in: Section 15.7.8.5

ASME STS-1, *Steel Stacks*, American Society of Mechanical Engineers, 2016

Cited in: Section 15.6.2.3, Table 15.4-2

ASTM A108, *Standard Specification for Steel Bar, Carbon and Alloy, Cold-Finished*, ASTM International, 2007.

Cited in: Section 11.3

ASTM A307, *Standard Specification for Carbon Steel Bolts and Studs, 60 000 PSI Tensile Strength*, ASTM, 2007.

Cited in: Section 11.3

ASTM A500, *Standard Specification for Cold-Formed Welded and Seamless Carbon Steel Structural Tubing in Rounds and Shapes*, ASTM, 2009.

Cited in: Section 14.1.3.3.4

ASTM A615, *Standard Specification for Deformed and Plain Carbon-Steel Bars for Concrete Reinforcement*, ASTM 2016.

Cited in: Section 14.2.4.5.5

ASTM A653, *Standard Specification for Steel Sheet, Zinc-Coated (Galvanized) or Zinc-Iron Alloy-Coated (Galvannealed) by the Hot-Dip Process*, ASTM, 2009.

Cited in: Section 14.1.3.3.3

ASTM A706/A706M, *Standard Specification for Low-Alloy Steel Deformed and Plain Bars for Concrete Reinforcement*, ASTM, 2004.

Cited in: Section 14.2.4.5.5

ASTM C635/C635M-17, *Standard Specification for the Manufacture, Performance, and Testing of Metal Suspension Systems for Acoustical Tile and Lay-In Panel Ceilings*, ASTM, 2017.

Cited in: Sections 13.5.6.2.1, 13.5.6.2.2

ASTM C636/C636M-19, *Standard Practice for Installation of Metal Ceiling Suspension Systems for Acoustical Tile and Lay-in Panels*, ASTM, 2019.

Cited in: Sections 13.5.6.2.1, 13.5.6.2.2

ASTM C1087-16, *Standard Test Method for Determining Compatibility of Liquid-Applied Sealants with Accessories Used in Structural Glazing Systems*, ASTM, 2016.

Cited in: Table 13.5-2

ASTM C1135-19, *Standard Test Method for Determining Tensile Adhesion Properties of Structural Sealants*, ASTM, 2019.

Cited in: Table 13.5-2

ASTM C1184-18e1, *Standard Specification for Structural Silicone Sealants*, ASTM, 2018.

Cited in: Table 13.5-2

ASTM C1265-17, *Standard Test Method for Determining the Tensile Properties of an Insulating Glass Edge Seal for Structural Glazing Applications*, ASTM, 2017.

Cited in: Table 13.5-2

ASTM C1294-15, *Standard Test Method for Compatibility of Insulating Glass Edge Sealants with Liquid-Applied Glazing Materials*, ASTM, 2015.

Cited in: Table 13.5-2

ASTM C1369-19, *Standard Specification for Secondary Edge Sealants for Structurally Glazed Insulating Glass Units*, ASTM, 2019.

Cited in: Table 13.5-2

ASTM D1586, *Standard Test Method for Penetration Test and Split-Barrel Sampling of Soils*, ASTM, 2004.

Cited in: Sections 11.3, 20.4.2

ASTM D2166, *Standard Test Method for Unconfined Compressive Strength of Cohesive Soil*, ASTM, 2000.
Cited in: Section 20.4.3

ASTM D2216, *Standard Test Method for Laboratory Determination of Water (Moisture) Content of Soil and Rock by Mass*, ASTM, 1998.
Cited in: Section 20.4.3

ASTM D2850, *Standard Test Method for Unconsolidated-Undrained Triaxial Compression Test on Cohesive Soils*, ASTM, 2003.
Cited in: Sections 11.3, 20.4.3

ASTM D4318, *Standard Test Methods for Liquid Limit, Plastic Limit, and Plasticity Index of Soils*, ASTM, 2000.
Cited in: Sections 11.3, 20.4.3

ASTM E580/E580M-20, *Standard Practice for Installation of Ceiling Suspension Systems for Acoustical Tile and Lay-In Panels in Areas Subject to Earthquake Ground Motions*, ASTM, 2020.
Cited in: Sections 13.5.6.2.1, 13.5.6.2.2

ASTM F1554-18, *Standard Specification for Anchor Bolts, Steel, 36, 55, and 105-ksi Yield Strength*, ASTM, 2018.
Cited in: Sections 13.5.3, 15.4.9.4

AWC NDS-18, *National Design Specification for Wood Construction*, Including Supplements, American Wood Council, 2017.
Cited in: Section 14.5.1

AWC SDPWS-21, *Special Design Provisions for Wind and Seismic*, American Wood Council, 2020.
Cited in: Sections 12.11.2.2.3, 12.14.7.2, 12.14.7.5.2, 14.5.1, Table 12.10-1

AWWA D100-11, *Welded Carbon Steel Tanks for Water Storage*, American Water Works Association, 2011.
Cited in: Sections 15.4.1, 15.7.7.1, 15.7.9.4, 15.7.10.6

AWWA D103, *Factory-Coated Bolted Carbon Steel Tanks for Water Storage*, American Water Works Association, 2009 with Addendum 2014.
Cited in: Sections 15.4.1, 15.7.7.2, 15.7.9.5

AWWA D107, *Composite Elevated Tanks for Water Storage*, American Water Works Association, 2016 with Errata 2019.
Cited in: Section 15.7.10.7

AWWA D110, *Wire- and Strand-Wound, Circular, Prestressed Concrete Water Tanks*, American Water Works Association, 2013.
Cited in: Section 15.7.7.3

AWWA D115, *Tendon-Prestressed Concrete Water Tanks*, American Water Works Association, 2017.
Cited in: Section 15.7.7.3

DOC PS 1-09, *Structural Plywood*, US Department of Commerce, National Institute of Standards and Technology, 2009.
Cited in: Section 11.2

DOC PS 2-10, *Performance Standard for Wood-Based Structural-Use Panels*, US Department of Commerce, National Institute of Standards and Technology, 2011.
Cited in: Section 11.2

***FEMA P-795**, *Quantification of Building Seismic Performance Factors: Component Equivalency Methodology*. Applied Technology Council, 2011.

***ICC-ES AC 156**, editorially revised September 2019, *Acceptance Criteria for Seismic Certification by Shake-Table Testing of Nonstructural Components and Systems*, International Code Council Evaluation Service, 2019.
Cited in: Section 13.2.5

***IRC**, *2015 International Residential Code*, International Code Council, 2012.
Cited in: Section 11.1.2

***MSS SP-58-2018**, *Pipe Hangers and Supports—Materials, Design, Manufacture, Selection, Application, and Installation*, Manufacturers Standardization Society of the Valve and Fittings, Industry, 2018.
Cited in: Section 13.6.4.1

NFPA 13, *Standard for the Installation of Sprinkler Systems*, National Fire Protection Association, 2019.
Cited in: Sections 13.4.6, 13.6.4.1, 13.6.7.2

NFPA 59A, *Standard for the Production, Storage, and Handling of Liquefied Natural Gas (LNG)*, National Fire Protection Association, 2019.
Cited in: Section 15.4.8

***NIST GCR 10-917-8**, *Evaluation of the FEMA P-695 Methodology for Quantification of Building Seismic Performance Factors*, NEHRP Consultants Joint Venture, a Partnership of the Applied Technology Council and the Consortium of Universities for Research in Earthquake Engineering. US Department of Commerce, 2010.

***NIST GCR 12-917-20**, *Tentative Framework for Development of Advanced Seismic Design Criteria for New Buildings*. NEHRP Consultants Joint Venture, a Partnership of the Applied Technology Council and the Consortium of Universities for Research in Earthquake Engineering. US Department of Commerce, 2012.

PTI DC10.5, *Standard Requirements for Design and Analysis of Post-Tensioned Concrete Foundations on Expansive Soils*, Post-Tensioning Institute, 2012.
Cited in: Section 12.13.9.2.1.1

TMS 402, *Building Code Requirements for Masonry Structures*, The Masonry Society, 2016.
Cited in: Sections 13.4.2.2, 14.4.1, 14.4.2, 14.4.3, 14.4.3.1, 14.4.4.1.1, 14.4.4.1.2, 14.4.5, 14.4.5.1, 14.4.5.2, 14.4.5.3, 14.4.5.4, 14.4.5.5, 14.4.6, 14.4.6.1, 15.4.9.2

TMS 602, *Specification for Masonry Structures*, The Masonry Society, 2016.
Cited in: Sections 14.4.1, 14.4.2, 14.4.7, 14.4.7.1

CHAPTER 24
RESERVED FOR FUTURE PROVISIONS

CHAPTER 25
RESERVED FOR FUTURE PROVISIONS

CHAPTER 26
WIND LOADS: GENERAL REQUIREMENTS

26.1 PROCEDURES

26.1.1 Scope Buildings and other structures, including the main wind force resisting system (MWFRS) and all components and cladding (C&C) thereof, shall be designed and constructed to resist the wind loads determined in accordance with Chapters 26 through 31.

Risk Category III and IV buildings and other structures, including the MWFRS and all C&C thereof, shall also be designed and constructed to resist tornado loads determined in accordance with Chapter 32, as applicable.

The provisions of this chapter define basic wind parameters for use with other provisions contained in this standard.

> **User Note:** A building or other structure designed for wind loads determined exclusively in accordance with Chapter 26 cannot be designated as a storm shelter without meeting additional critical requirements provided in the applicable building code and ICC 500, the *ICC/NSSA Standard for the Design and Construction of Storm Shelters*. See Commentary Section C26.1.1 for an in-depth discussion on Storm Shelters.

26.1.2 Permitted Procedures The design wind loads for buildings and other structures, including the MWFRS and C&C elements thereof, shall be determined using one of the procedures as specified in this section. An outline of the overall process for the determination of the wind loads, including section references, is provided in Figure 26.1-1.

Additional outlines and User Notes are provided at the beginning of Chapters 27 through 31 for more detailed step-by-step procedures for determining the wind loads.

26.1.2.1 Main Wind Force Resisting System Wind loads for the MWFRS shall be determined using one of the following procedures:

1. Directional Procedure for buildings of all heights as specified in Chapter 27 for buildings meeting the requirements specified therein;
2. Envelope Procedure for low-rise buildings as specified in Chapter 28 for buildings meeting the requirements specified therein;
3. Directional Procedure for building appurtenances (rooftop structures and rooftop equipment) and other structures (such as solid freestanding walls and solid freestanding signs, chimneys, tanks, open signs, single-plane open frames, and trussed towers) as specified in Chapter 29; or
4. Wind Tunnel Procedure for all buildings and all other structures as specified in Chapter 31.

26.1.2.2 Components and Cladding Wind loads on C&C on all buildings and other structures shall be designed using one of the following procedures:

1. Analytical Procedures provided in Parts 1 through 5, as appropriate, of Chapter 30; or
2. Wind Tunnel Procedure as specified in Chapter 31.

26.1.3 Performance-Based Procedures Wind design of buildings and other structures using performance-based procedures shall be permitted subject to the approval of the Authority Having Jurisdiction. The performance-based wind design procedures used shall, at a minimum, conform to Section 1.3.1.3.

26.2 DEFINITIONS

The following definitions apply to the provisions of Chapters 26 through 31.

APPROVED: Acceptable to the Authority Having Jurisdiction.

ASCE WIND DESIGN GEODATABASE: The ASCE database (version 2022-1.0) of geocoded wind speed design data.

> **User Note:** The ASCE Wind Design Geodatabase of geocoded wind speed design data is available at https://asce7hazardtool.online/.

ATTACHED CANOPY: A horizontal (maximum slope of 2%) patio cover attached to the building wall at any height; it is different from an overhang, which is an extension of the roof surface.

BASIC WIND SPEED, V: Three-second gust speed at 33 ft (10 m) above the ground in Exposure C (see Section 26.7.3) as determined in accordance with Section 26.5.1.

BUILDING, ELEVATED: A building supported on structural elements where wind can pass beneath the building.

BUILDING, ENCLOSED: A building that has the total area of openings in each wall that receives positive external pressure less than or equal to 4 ft² (0.37 m²) or 1% of the area of that wall, whichever is smaller. This condition is expressed for each wall by the following equation:

$$A_o < 0.01A_g \text{ or } 4\ \text{ft}^2\ (0.37\ \text{m}^2), \text{ whichever is smaller,}$$

where A_o and A_g are as defined for Open Buildings.

BUILDING, LOW-RISE: An enclosed, partially enclosed, or partially open building that complies with the following conditions:

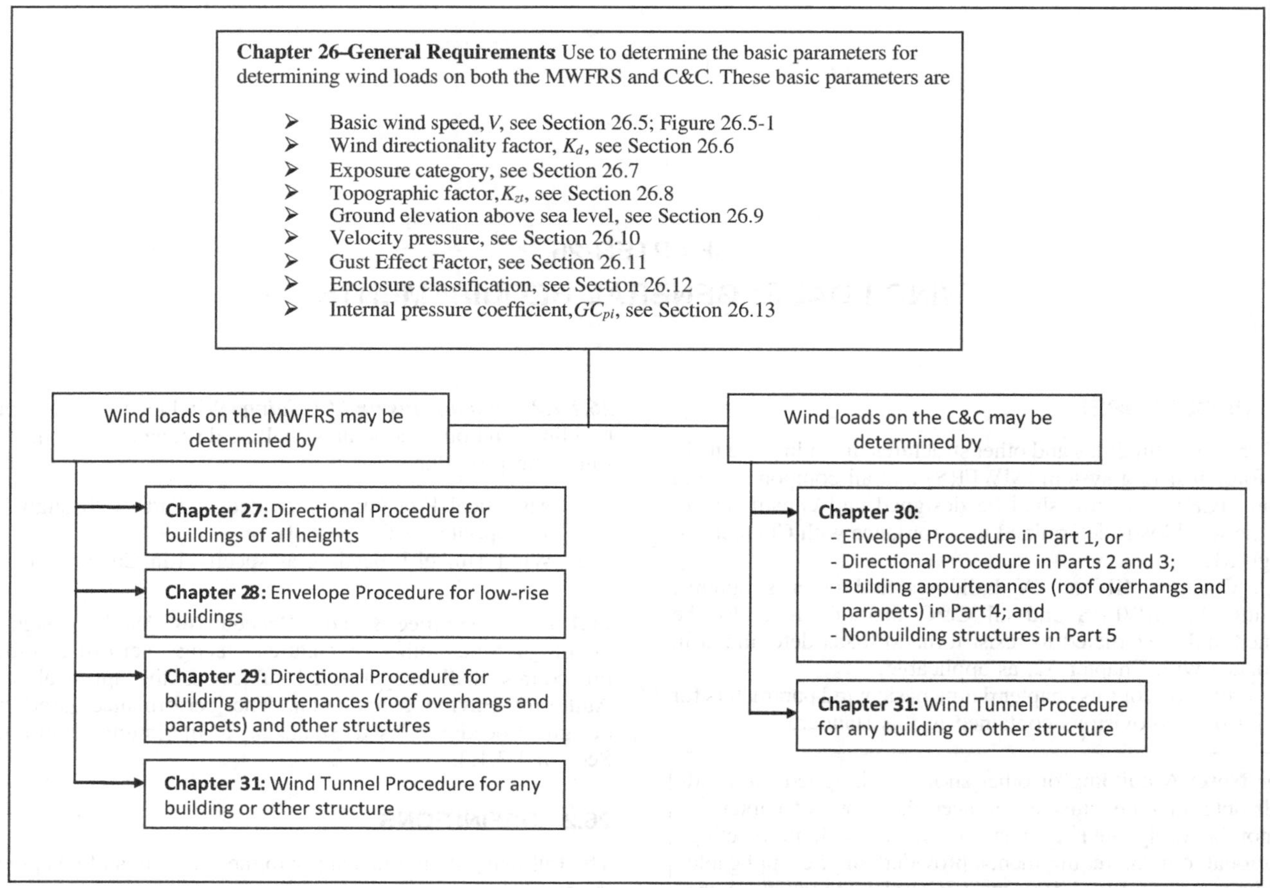

Figure 26.1-1. Outline of process for determining wind loads.

Note: Additional outlines and User Notes are provided at the beginning of each chapter for more detailed step-by-step procedures for determining the wind loads.

1. Mean roof height h less than or equal to 60 ft (18 m), and
2. Mean roof height h does not exceed the least horizontal dimension.

BUILDING, OPEN: A building that has each wall at least 80% open. This condition is expressed for each wall by the equation $A_o \geq 0.8A_g$

where A_o is the total area of openings in a wall that receives positive external pressure, ft² (m²), and A_g is the gross area of that wall in which A_o is identified, ft² (m²).

BUILDING, PARTIALLY ENCLOSED: A building that complies with both of the following conditions:

1. Total area of openings in a wall that receives positive external pressure exceeds the sum of the areas of openings in the balance of the building envelope (walls and roof) by more than 10%.
2. Total area of openings in a wall that receives positive external pressure exceeds 4 ft² (0.37 m²) or 1% of the area of that wall, whichever is smaller, and the percentage of openings in the balance of the building envelope does not exceed 20%.

These conditions are expressed by the following equations:

$$A_o > 1.10A_{oi}$$

$$A_o > 4\ \text{ft}^2 (0.37\ \text{m}^2)\ \text{or}$$

$$> 0.01A_g,\ \text{whichever is smaller, and}\ A_{oi}/A_{gi} \leq 0.20$$

where A_o and A_g are as defined for open buildings; A_{oi} is the sum of the areas of openings in the building envelope (walls and roof), not including A_o, ft² (m²); and A_{gi} is the sum of the gross surface areas of the building envelope (walls and roof), not including A_g, ft² (m²).

BUILDING, PARTIALLY OPEN: A building that does not comply with the requirements for open, partially enclosed, or enclosed buildings.

BUILDING, TORSIONALLY REGULAR UNDER WIND LOAD: A building with the MWFRS about each principal axis proportioned so that the maximum displacement at each story under Case 2, the torsional wind load case, of Figure 27.3-8 does not exceed the maximum displacement at the same location under Case 1 of Figure 27.3-8, the basic wind load case.

BUILDING ENVELOPE: Cladding, roofing, exterior walls, glazing, door assemblies, window assemblies, skylight assemblies, and other components enclosing the building.

BUILDING OR OTHER STRUCTURE, FLEXIBLE: Slender buildings and other structures that have a fundamental natural frequency less than 1 Hz.

BUILDING OR OTHER STRUCTURE, REGULAR-SHAPED: A building or other structure that has no unusual geometrical irregularity in spatial form.

BUILDING OR OTHER STRUCTURE, RIGID: A building or other structure whose fundamental frequency is greater than or equal to 1 Hz.

COMPONENTS AND CLADDING (C&C): Elements of the building envelope or elements of building appurtenances and rooftop structures and equipment that do not qualify as part of the MWFRS.

DESIGN FORCE, *F*: Equivalent static force to be used in the determination of wind loads for other structures.

DESIGN PRESSURE, *p*: Equivalent static pressure to be used in the determination of wind loads for buildings.

DIAPHRAGM: Roof, floor, or other membrane or bracing system acting to transfer lateral forces to the vertical MWFRS. For analysis under wind loads, diaphragms constructed of untopped steel decks, concrete-filled steel decks, and concrete slabs, each having a span-to-depth ratio of 2 or less, shall be permitted to be idealized as *rigid*. Diaphragms constructed of wood structural panels are permitted to be idealized as *flexible*.

DIRECTIONAL PROCEDURE: A procedure for determining wind loads on buildings and other structures for specific wind directions, in which the external pressure coefficients used are based on past wind tunnel testing of prototypical building models for the corresponding direction of wind.

EAVE HEIGHT, h_e: The distance from the ground surface adjacent to the building to the roof eave line at a particular wall. If the height of the eave varies along the wall, the average height shall be used.

EFFECTIVE WIND AREA, *A*: The area used to determine the external pressure coefficient, (GC_p) and (GC_{rn}). For C&C elements, the effective wind area in Figures 30.3-1 through 30.3-7, 30.4-1, 30.5-1, and 30.7-1 through 30.7-3 is the span length multiplied by an effective width that need not be less than one-third the span length. For rooftop solar arrays, the effective wind area in Figure 29.4-7 is equal to the tributary area for the structural element being considered, except that the width of the effective wind area need not be less than one-third its length. For cladding fasteners, the effective wind area shall not be greater than the area that is tributary to an individual fastener.

ENVELOPE PROCEDURE: A procedure for determining wind load cases on buildings in which pseudoexternal pressure coefficients are derived from past wind tunnel testing of prototypical building models successively rotated through 360 degrees, such that the pseudopressure cases produce key structural actions (e.g., uplift, horizontal shear, and bending moments) that envelop their maximum values among all possible wind directions.

ESCARPMENT: With respect to topographic effects in Section 26.8, a cliff or steep slope generally separating two levels or gently sloping areas (see Figure 26.8-1). Also known as a scarp.

FREE ROOF: A roof with a configuration generally conforming to those shown in Figures 27.3-4 through 27.3-6 (monoslope, pitched, or troughed) in an open building with no enclosing walls underneath the roof surface.

GLAZING: Glass or transparent or translucent plastic sheet used in windows, doors, skylights, or curtain walls.

GLAZING, IMPACT-RESISTANT: Glazing that has been shown by testing to withstand the impact of test missiles. See Section 26.12.3.2.

HILL: With respect to topographic effects in Section 26.8, a land surface characterized by strong relief in any horizontal direction (see Figure 26.8-1).

HURRICANE-PRONE REGIONS: Areas vulnerable to hurricanes; in the United States and its territories, these are defined as

1. The US Atlantic Ocean and Gulf of Mexico coasts where the basic wind speed for Risk Category II buildings is greater than 115 mi/h (51.4 m/s); and
2. Hawaii, Puerto Rico, Guam, US Virgin Islands, Northern Mariana Islands, and American Samoa.

IMPACT-PROTECTIVE SYSTEM: Construction that has been shown by testing to withstand the impact of test missiles and that is applied, attached, or locked over exterior glazing. See Section 26.12.3.2.

MAIN WIND FORCE RESISTING SYSTEM (MWFRS): An assemblage of structural elements assigned to provide support and stability for the overall building or other structure. The system generally receives wind loading from more than one surface.

MEAN ROOF HEIGHT, *h*: The average of the roof eave height and the height to the highest point on the roof surface, except that, for roof angles less than or equal to 10 degrees, the mean roof height is permitted to be taken as the roof eave height.

OPENINGS: Holes that allow air to flow through the building envelope during a design wind event.

RECOGNIZED LITERATURE: Published research findings and technical papers that are approved.

RIDGE: With respect to topographic effects in Section 26.8, an elongated crest of a hill characterized by strong relief in two directions (see Figure 26.8-1).

ROOFTOP SOLAR PANEL: A device to receive solar radiation and convert it into electricity or heat energy. Typically, this is a photovoltaic module or solar thermal panel.

SOLAR ARRAY: Any number of rooftop solar panels grouped closely together.

WIND-BORNE DEBRIS REGIONS: Areas within hurricane-prone regions where impact protection is required for glazed openings. See Section 26.12.3.

WIND TUNNEL PROCEDURE: A procedure for determining wind loads on buildings and other structures in which pressures and/or forces and moments are determined for each wind direction considered, from a model of the building or other structure and its surroundings, in accordance with Chapter 31.

26.3 SYMBOLS

The following symbols apply only to the provisions of Chapters 26 through 31:

A = Effective wind area, ft^2 (m^2)

A_f = Area of open buildings and other structures either normal to the wind direction or projected on a plane normal to the wind direction, ft^2 (m^2)

A_g = Gross area of that wall in which A_o is identified, ft^2 (m^2)

A_{gi} = Sum of the gross surface areas of the building envelope (walls and roof) not including A_g, ft^2 (m^2)

A_n = Normalized wind area for rooftop solar panels in Figure 29.4-7, ft^2 (m^2)

A_o = Total area of openings in a wall that receives positive external pressure, ft^2 (m^2)

A_{og} = Total area of openings in the building envelope, ft^2 (m^2)

A_{oi} = Sum of the areas of openings in the building envelope (walls and roof) not including A_o, ft^2 (m^2)

A_s = Gross area of the solid freestanding wall or solid sign, ft^2 (m^2)

A_1, A_2 = Effective wind area for use with ground-mounted solar panels in Figure 29.4-11, ft^2 (m^2)

a = Width of pressure coefficient zone, ft (m)

a_B = Width of pressure coefficient zone for elevated buildings, ft (m)

B = Horizontal dimension of a building measured normal to wind direction, ft (m)

$\bar{b}$ = Mean hourly wind speed factor in Equation (26.11-16), from Table 26.11-1

$\hat{b}$ = Three-second gust speed factor from Table 26.11-1

C_f = Force coefficient to be used in determination of wind loads for other structures

C_N = Net pressure coefficient to be used in determination of wind loads for open buildings

C_p = External pressure coefficient to be used in determination of wind loads for buildings

c = Turbulence intensity factor in Equation (26.11-7), from Table 26.11-1

D = Diameter of a circular structure or member, ft (m)

D' = Depth of protruding elements such as ribs and spoilers, ft (m)

d_p = Horizontal distance of Zone 2 for ground-mounted solar panels in Figure 29.4.9, ft (m)

d_s = Horizontal offset for ground-mounted solar panels in Figure 29.4.9, ft (m)

d_1 = For rooftop solar arrays, horizontal distance orthogonal to the panel edge to an adjacent panel or the building edge, ignoring any rooftop equipment in Figure 29.4-7, ft (m)

d_2 = For rooftop solar arrays, horizontal distance from the edge of one panel to the nearest edge in the next row of panels in Figure 29.4-7, ft (m)

F = Design wind force for other structures, lb (N)

F_n = Normal force for ground-mounted solar panels in Figure 29.4-9, lb (N)

G = Gust-effect factor

G_f = Gust-effect factor for MWFRS of flexible buildings and other structures

$(GC_{gn_{\text{dynamic}}})$ = Dynamic net pressure coefficient for ground-mounted solar panels from Figure 29.4-11

$(GC_{gm_{\text{dynamic}}})$ = Dynamic moment coefficient for ground-mounted solar panels from Figure 29.4-11

$(GC_{gn_{\text{static}}})$ = Static net pressure coefficient for ground-mounted solar panels from Figure 29.4-10

$(GC_{gm_{\text{static}}})$ = Static moment coefficient for ground-mounted solar panels from Figure 29.4-10

(GC_p) = Product of external pressure coefficient and gust-effect factor to be used in determination of wind loads for buildings

(GC_{pf}) = Product of the equivalent external pressure coefficient and gust-effect factor to be used in determination of wind loads for MWFRS of low-rise buildings

(GC_{pi}) = Product of internal pressure coefficient and gust-effect factor to be used in determination of wind loads for buildings

(GC_{pn}) = Combined net pressure coefficient for a parapet

(GC_r) = Product of external pressure coefficient and gust-effect factor to be used in determination of wind loads for rooftop structures

(GC_{rn}) = Net pressure coefficient for rooftop solar panels in Equations (29.4-4) and (29.4-5)

$(GC_{rn})_{nom}$ = Nominal net pressure coefficient for rooftop solar panels determined from Figure 29.4-7

g_Q = Peak factor for background response in Equations (26.11-6) and (26.11-10)

g_R = Peak factor for resonant response in Equation (26.11-10)

g_v = Peak factor for wind response in Equations (26.11-6) and (26.11-10)

H = Height of hill, ridge, or escarpment in Figure 26.8-1, ft (m)

h = Mean roof height of a building or height of other structure, except that eave height shall be used for roof angle θ less than or equal to 10 degrees, in ft (m)

h_B = Height above grade of the bottom surface of the elevated building, ft (m)

h_c = Height of the solid cylinder, ft (m)

h_e = Roof eave height at a particular wall, or the average height if the eave varies along the wall

h_p = Height to top of parapet in Figures 27.5-2 and 30.6-1

h_{pt} = Mean parapet height above the adjacent roof surface for use with Equation (29.4-5), ft (m)

h_1 = Height of a solar panel above the roof at the lower edge of the panel, ft (m)

h_2 = Height of a solar panel above the roof at the upper edge of the panel, ft (m)

$I_{\bar{z}}$ = Intensity of turbulence from Equation (26.11-7)

K_1, K_2, K_3 = Multipliers in Figure 26.8-1 to obtain K_{zt}

K_d = Wind directionality factor from Section 26.6 and Table 26.6-1

K_e = Ground elevation factor

K_h = Velocity pressure exposure coefficient evaluated at height $z = h$

K_z = Velocity pressure exposure coefficient evaluated at height z

K_{zt} = Topographic factor as defined in Section 26.8

L = Horizontal dimension of a building measured parallel to the wind direction, ft (m)

L_b = Normalized building length, for use with Figure 29.4-7, ft (m)

L_c = Panel chord length for ground-mounted solar panels in Figure 29.4-9, ft (m)

L_h = Distance upwind of crest of hill, ridge, or escarpment in Figure 26.8-1 to where the difference in ground elevation is half the height of the hill, ridge, or escarpment, ft (m)

L_p = Panel chord length for use with rooftop solar panels in Figure 29.4-7, ft (m)

L_r = Horizontal dimension of return corner for a solid freestanding wall or solid sign from Figure 29.3-1, ft (m)

$L_{\bar{z}}$ = Integral length scale of turbulence, ft (m)

ℓ = Integral length scale factor from Table 26.11-1, ft (m)

M_c = Moment for ground-mounted solar panels in Figure 29.4.9, lb-ft (N-m)

N_1 = Reduced frequency from Equation (26.11-14)

N_s = Reduced frequency for ground-mounted solar panels
n = Lowest natural frequency for ground-mounted solar panels, Hz
η_1 = Fundamental natural frequency, Hz
η_a = Approximate lower bound natural frequency from Section 26.11.2, Hz
p = Design pressure to be used in determination of wind loads for buildings, lb/ft^2 (N/m^2)
P_L = Wind pressure acting on leeward face in Figure 27.3-8, lb/ft^2 (N/m^2)
Q = Background response factor from Equation (26.11-8)
q = Velocity pressure, lb/ft^2 (N/m^2)
q_h = Velocity pressure evaluated at height $z = h$, lb/ft^2 (N/m^2)
q_i = Velocity pressure for internal pressure determination, lb/ft^2 (N/m^2)
q_p = Velocity pressure at top of parapet, lb/ft^2 (N/m^2)
q_z = Velocity pressure evaluated at height z above ground, lb/ft^2 (N/m^2)
R = Resonant response factor from Equation (26.11-12)
r = Rise-to-span ratio for arched roofs
R_B, R_h, R_L = Values from Equations (26.11-15a) and (26.11-15b)
R_i = Reduction factor from Equation (26.13-1)
R_n = Value from Equation (26.11-13)
S = Center-to-center row spacing for ground-mounted solar panels in Figure 29.4-9, ft (m)
S_L = Open distance in longitudinal direction for ground-mounted solar panels in Figure 29.4-9, ft (m)
S_T = Open distance in transverse direction for ground-mounted solar panels in Figure 29.4-9, ft (m)
s = Vertical dimension of the solid freestanding wall or solid sign from Figure 29.3-1, ft (m)
s_p = Gap between adjacent panels for ground-mounted solar panels in Figure 29.4-9, ft (m)
V = Basic wind speed obtained from Figures 26.5-1A through 26.5-1D, mi/h (m/s); the basic wind speed corresponds to a 3 s gust speed at 33 ft (10 m) above the ground in Exposure Category C
V_i = Unpartitioned internal volume, in ft^3 (m^3)
$\overline{V}_{\bar{z}}$ = Mean hourly wind speed at height $\bar{z}$, in ft/s (m/s)
W = Width of building in Figures 30.3-3, 30.3-5A, and 30.3-5B and width of span in Figures 30.3-4 and 30.3-6, ft (m)
W_g = Shortest row length for ground-mounted solar panels in Figure 29.4-9, ft (m)
W_L = Width of a building on its longest side in Figure 29.4-7, ft (m)
W_S = Width of a building on its shortest side in Figure 29.4-7, ft (m)
x = Distance upwind or downwind of crest in Figure 26.8-1, ft (m)
z = Height above ground level, ft (m)
$\bar{z}$ = Equivalent height of structure, ft (m)
z_e = Ground elevation above sea level, ft (m)
z_g = Nominal height of the atmospheric boundary layer used in this standard (values appear in Table 26.11-1)
z_{min} = Exposure constant from Table 26.11-1
α = 3-s gust-speed power law exponent from Table 26.11-1
$\hat{\alpha}$ = Reciprocal of α from Table 26.11-1
$\bar{\alpha}$ = Mean hourly wind-speed power law exponent in Equation (26.11-16) from Table 26.11-1
β = Damping ratio, percent critical for buildings or other structures
γ_c = Panel chord factor for use with rooftop solar panels in Equation (29.4-5)
γ_E = Array edge factor for use with rooftop solar panels in Figure 29.4-7 and Equations (29.4-4) and (29.4-5)
γ_p = Parapet height factor for use with rooftop solar panels in Equation (29.4-5)
ε = Ratio of solid area to gross area for solid freestanding wall, solid sign, open sign, face of a trussed tower, or lattice structure
$\bar{\varepsilon}$ = Integral length scale power law exponent in Equation (26.11-9) from Table 26.11-1
η = Value used in Equations (26.11-15a) and (26.11-15b); see Section 26.11.4
θ = Angle of plane of roof from horizontal, degrees
ν = Height-to-width ratio for solid sign
ω = Angle that the solar panel makes with the roof surface in Figure 29.4-7, or that a ground-mounted solar panel makes with the ground in Figure 29.4-9, degrees

26.4 GENERAL

26.4.1 Sign Convention Positive pressure acts toward the surface, and negative pressure acts away from the surface.

26.4.2 Critical Load Condition Values of external and internal pressures shall be combined algebraically to determine the most critical load.

26.4.3 Wind Pressures Acting on Opposite Faces of Each Building Surface In the calculation of design wind loads for the MWFRS and for C&C for buildings, the algebraic sum of the pressures acting on opposite faces of each building surface shall be taken into account.

26.5 WIND HAZARD MAP

26.5.1 Basic Wind Speed The basic wind speed, V, used in the determination of design wind loads on buildings and other structures shall be determined from Figure 26.5-1 as follows, except as provided in Sections 26.5.2 and 26.5.3:

For Risk Category I buildings and structures, use Figure 26.5-1A.
For Risk Category II buildings and structures, use Figure 26.5-1B.
For Risk Category III buildings and structures, use Figure 26.5-1C.
For Risk Category IV buildings and structures, use Figure 26.5-1D.

Alternatively, it shall be permitted to use the basic wind speeds from the ASCE Wind Design Geodatabase for the contiguous United States, Alaska, and selected special wind regions.

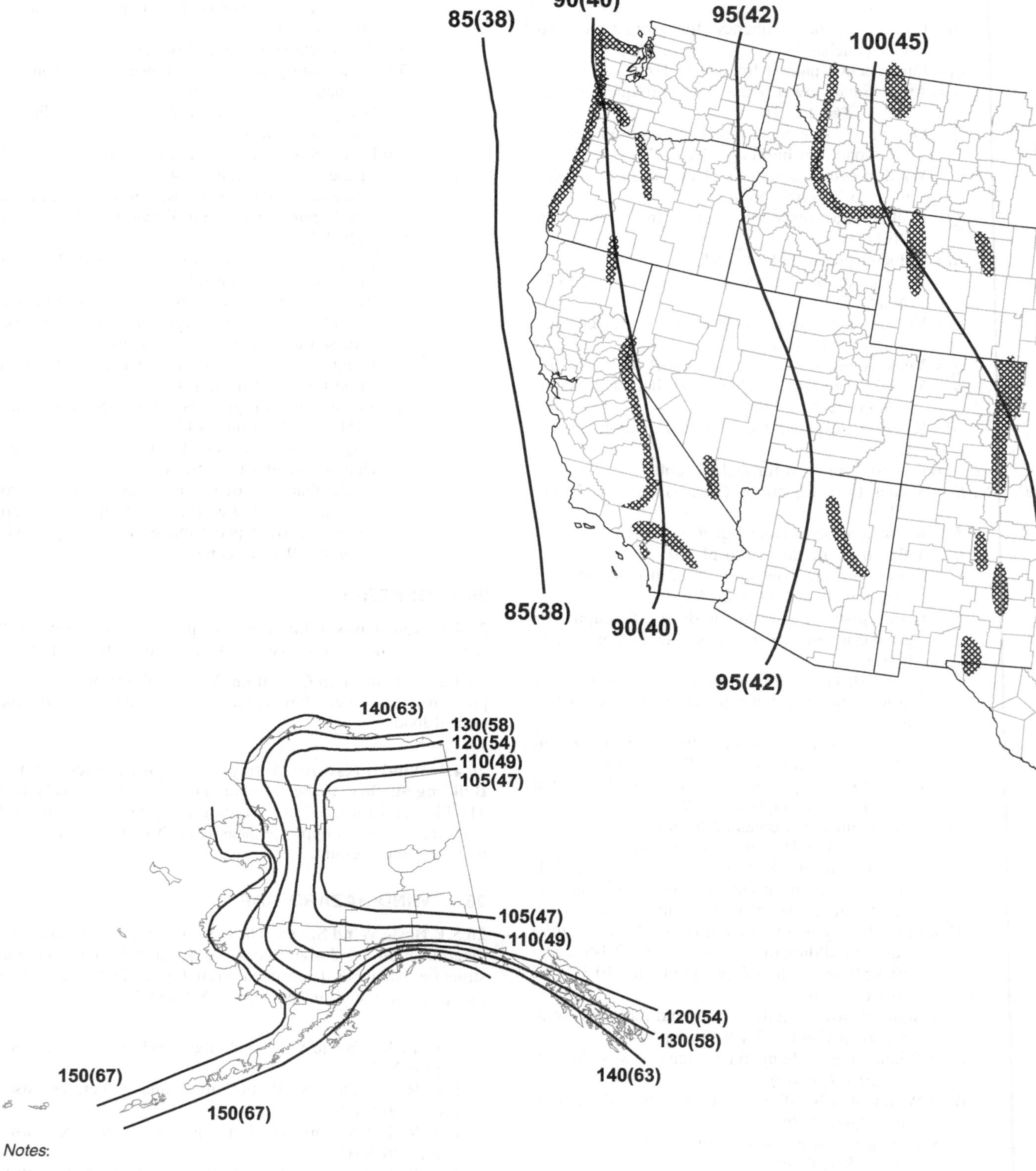

Notes:

1. Values are 3 s gust wind speeds in mi/h (m/s) at 33 ft (10 m) above ground for Exposure Category C.
2. Linear interpolation is permitted between contours. Point values are provided to aid with interpolation.
3. Islands, coastal areas, and land boundaries outside the last contour shall use the last wind speed contour.
4. Location-specific basic wind speeds shall be permitted to be determined using the ASCE Wind Design Geodatabase.

Figure 26.5-1A. Basic wind speeds for Risk Category I buildings and other structures.

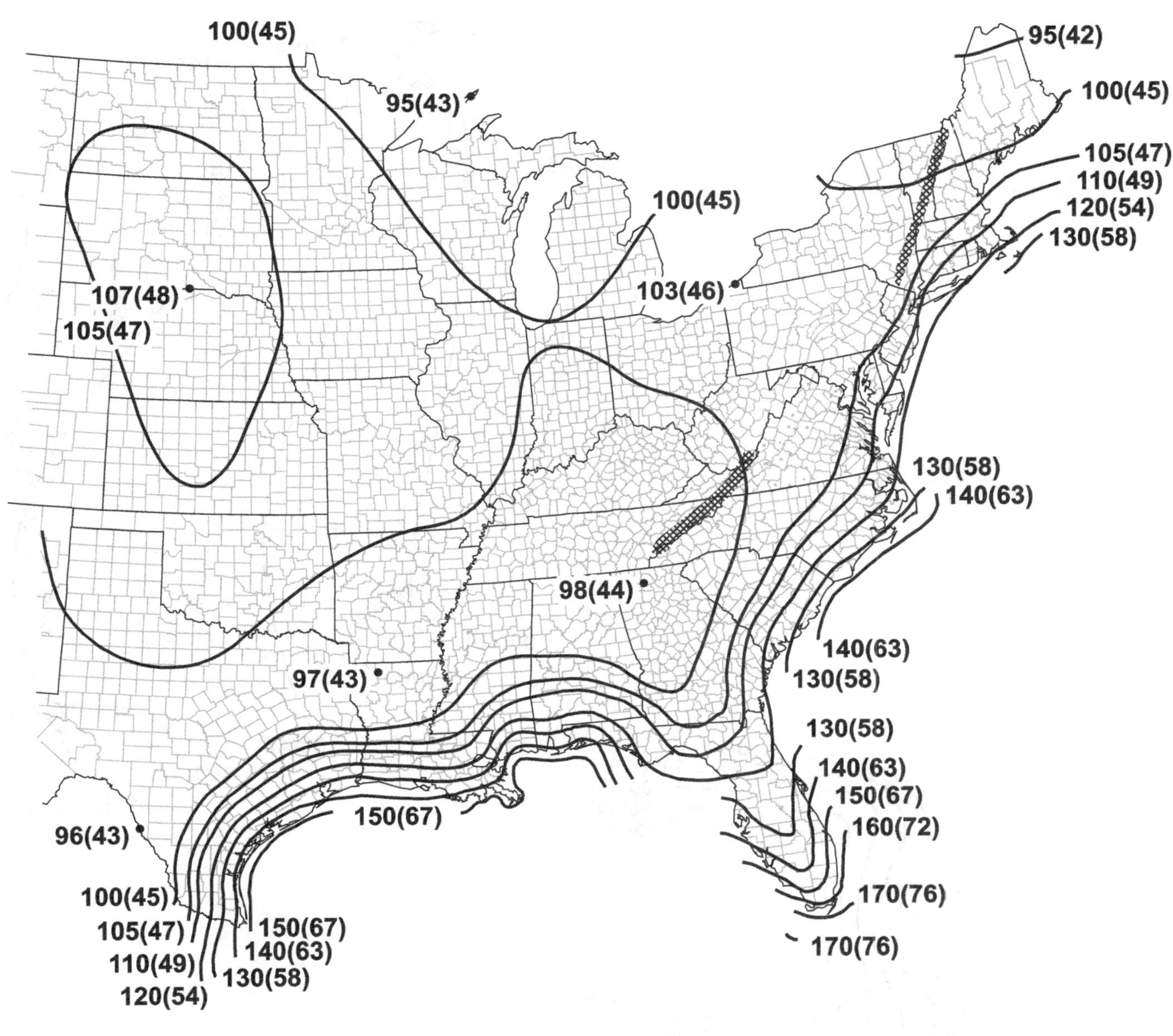

Location	V (mi/h)	V (m/s)
American Samoa	150	(67)
Guam & Northern Mariana Islands	180	(80)
Hawaii	ASCE Wind Design Geodatabase	
Puerto Rico	ASCE Wind Design Geodatabase	
U.S. Virgin Islands	ASCE Wind Design Geodatabase	

5. Wind speeds for Hawaii, US Virgin Islands, and Puerto Rico shall be determined from the ASCE Wind Design Geodatabase.
6. Mountainous terrain, gorges, ocean promontories, and special wind regions shall be examined for unusual wind conditions. Site-specific values for selected special wind regions shall be permitted to be determined using the ASCE Wind Design Geodatabase.
7. Wind speeds correspond to approximately a 15% probability of exceedance in 50 years (Annual Exceedance Probability = 0.00333, MRI = 300 years).
8. The ASCE Wind Design Geodatabase can be accessed at the ASCE 7 Hazard Tool (https://asce7hazardtool.online) or approved equivalent.

Figure 26.5-1A (*Continued*). Basic wind speeds for Risk Category I buildings and other structures.

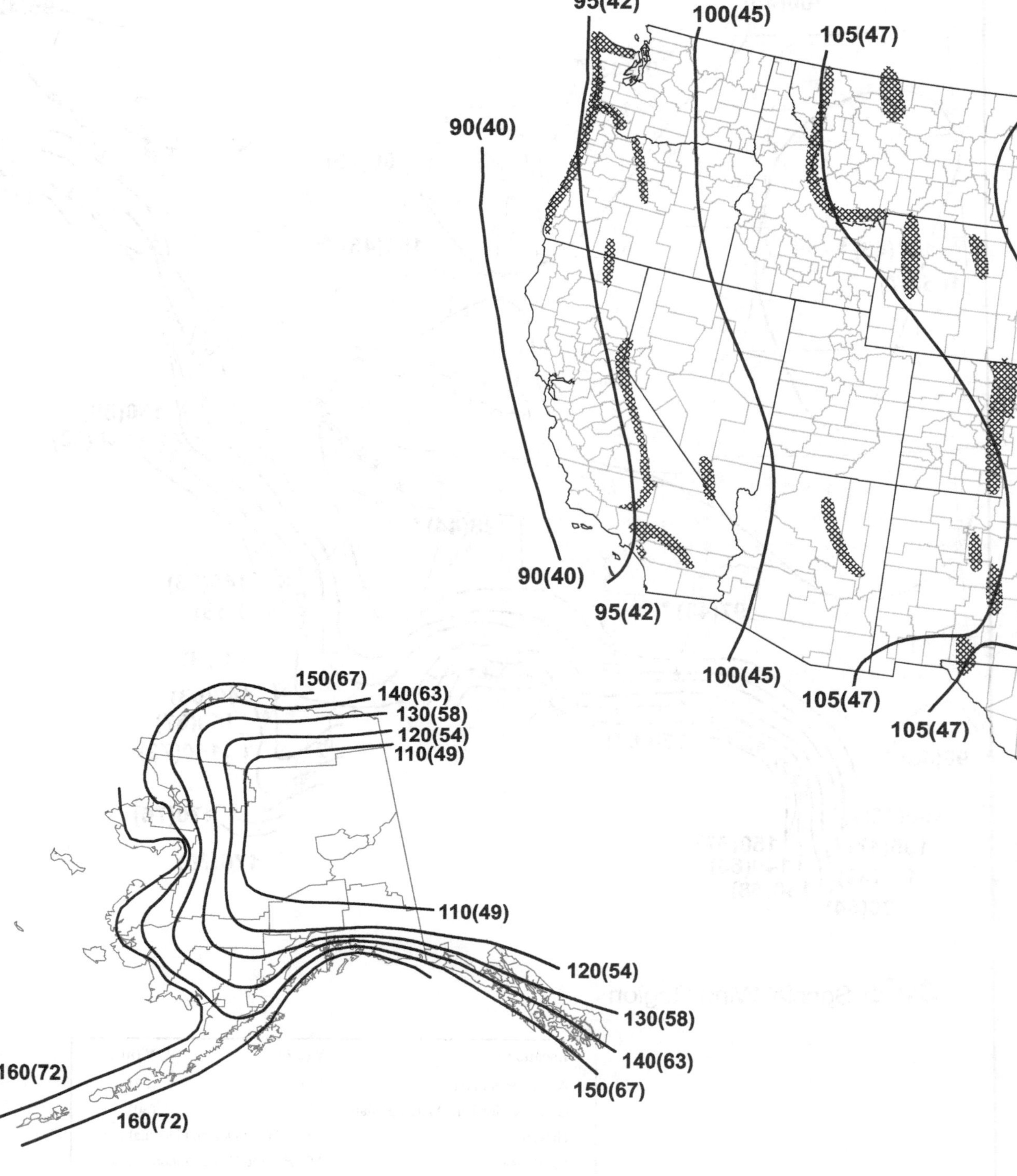

Notes:

1. Values are 3 s gust wind speeds in mi/h (m/s) at 33 ft (10 m) above ground for Exposure Category C.
2. Linear interpolation is permitted between contours. Point values are provided to aid with interpolation.
3. Islands, coastal areas, and land boundaries outside the last contour shall use the last wind speed contour.
4. Location-specific basic wind speeds shall be permitted to be determined using the ASCE Wind Design Geodatabase.

Figure 26.5-1B. Basic wind speeds for Risk Category II buildings and other structures.

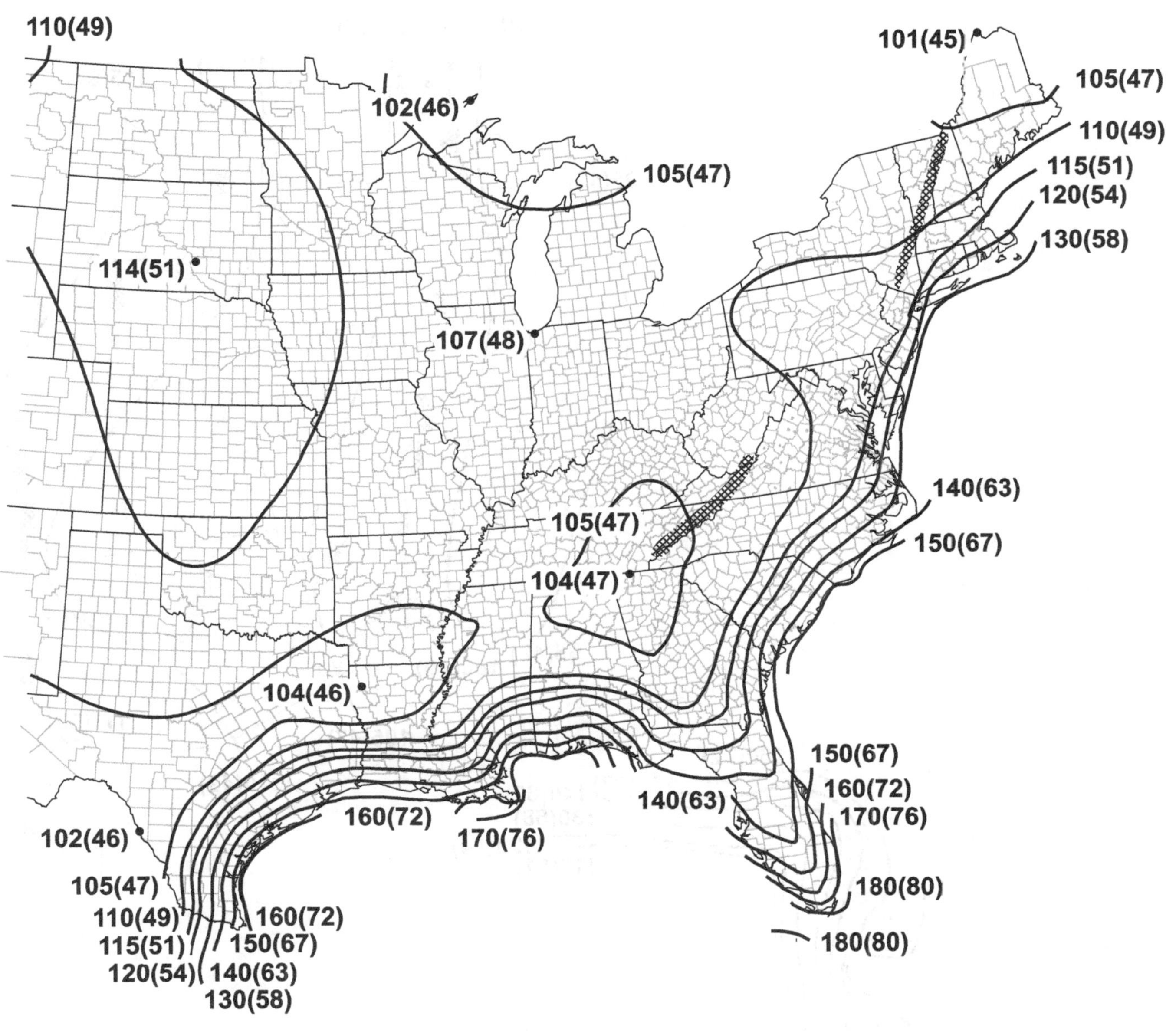

Location	V (mi/h)	V (m/s)
American Samoa	160	(72)
Guam & Northern Mariana Islands	195	(87)
Hawaii	ASCE Wind Design Geodatabase	
Puerto Rico	ASCE Wind Design Geodatabase	
U.S. Virgin Islands	ASCE Wind Design Geodatabase	

5. Wind speeds for Hawaii, US Virgin Islands, and Puerto Rico shall be determined from the ASCE Wind Design Geodatabase.
6. Mountainous terrain, gorges, ocean promontories, and special wind regions shall be examined for unusual wind conditions. Site-specific values for selected special wind regions shall be permitted to be determined using the ASCE Wind Design Geodatabase.
7. Wind speeds correspond to approximately a 15% probability of exceedance in 50 years (Annual Exceedance Probability = 0.00143, MRI = 700 years).
8. The ASCE Wind Design Geodatabase can be accessed at the ASCE 7 Hazard Tool (https://asce7hazardtool.online) or approved equivalent.

Figure 26.5-1B (*Continued*). Basic wind speeds for Risk Category II buildings and other structures.

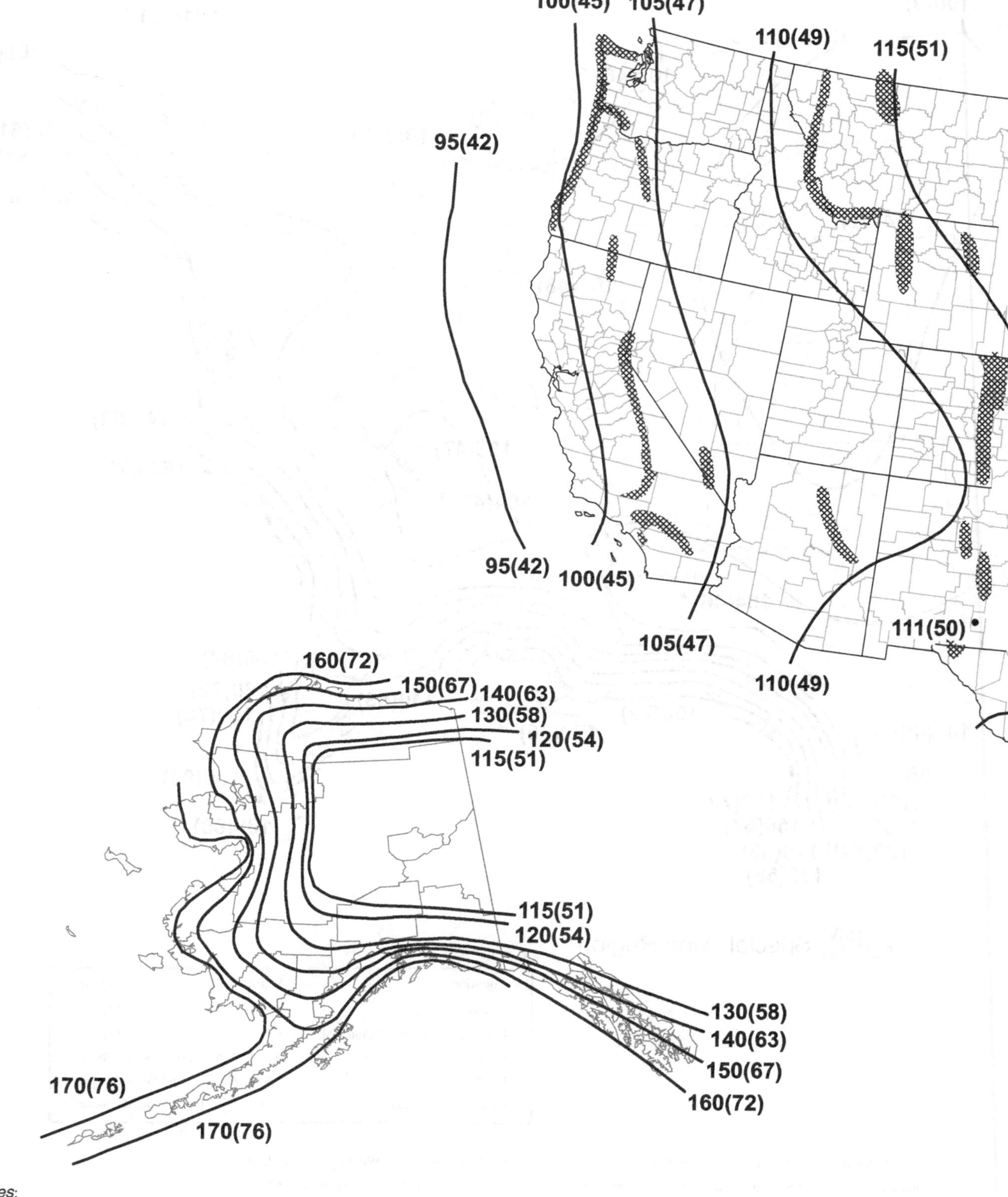

Notes:

1. Values are 3 s gust wind speeds in mi/h (m/s) at 33 ft (10 m) above ground for Exposure Category C.
2. Linear interpolation is permitted between contours. Point values are provided to aid with interpolation.
3. Islands, coastal areas, and land boundaries outside the last contour shall use the last wind speed contour.
4. Location-specific basic wind speeds shall be permitted to be determined using the ASCE Wind Design Geodatabase.

Figure 26.5-1C. Basic wind speeds for Risk Category III buildings and other structures.

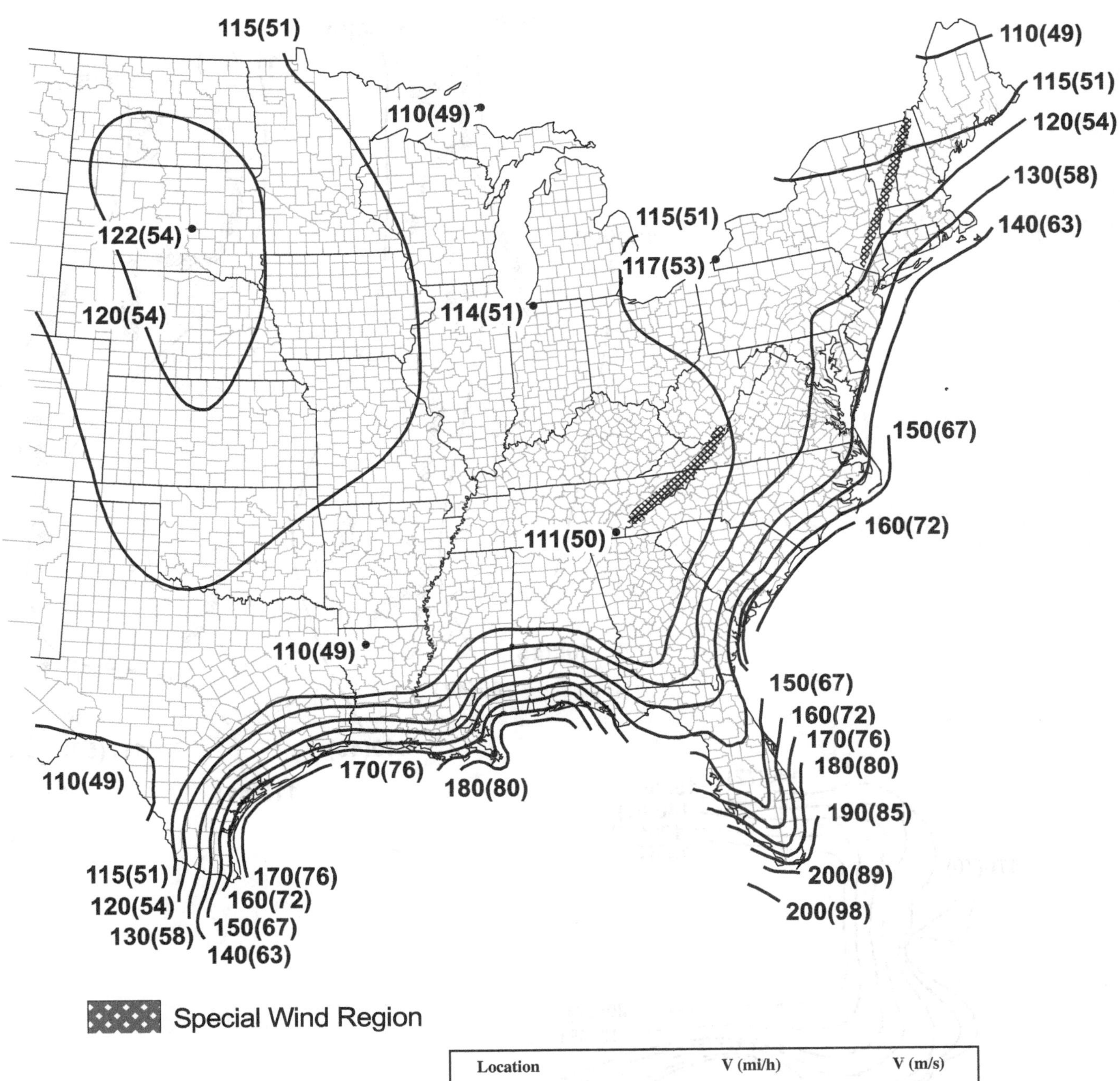

Location	V (mi/h)	V (m/s)
American Samoa	170	(76)
Guam & Northern Mariana Islands	210	(94)
Hawaii	ASCE Wind Design Geodatabase	
Puerto Rico	ASCE Wind Design Geodatabase	
U.S. Virgin Islands	ASCE Wind Design Geodatabase	

5. Wind speeds for Hawaii, US Virgin Islands, and Puerto Rico shall be determined from the ASCE Wind Design Geodatabase.
6. Mountainous terrain, gorges, ocean promontories, and special wind regions shall be examined for unusual wind conditions. Site-specific values for selected special wind regions shall be permitted to be determined using the ASCE Wind Design Geodatabase.
7. Wind speeds correspond to approximately a 15% probability of exceedance in 50 years (Annual Exceedance Probability = 0.00588, MRI = 1,700 years).
8. The ASCE Wind Design Geodatabase can be accessed at the ASCE 7 Hazard Tool (https://asce7hazardtool.online) or approved equivalent.

Figure 26.5-1C (*Continued*). Basic wind speeds for Risk Category III buildings and other structures.

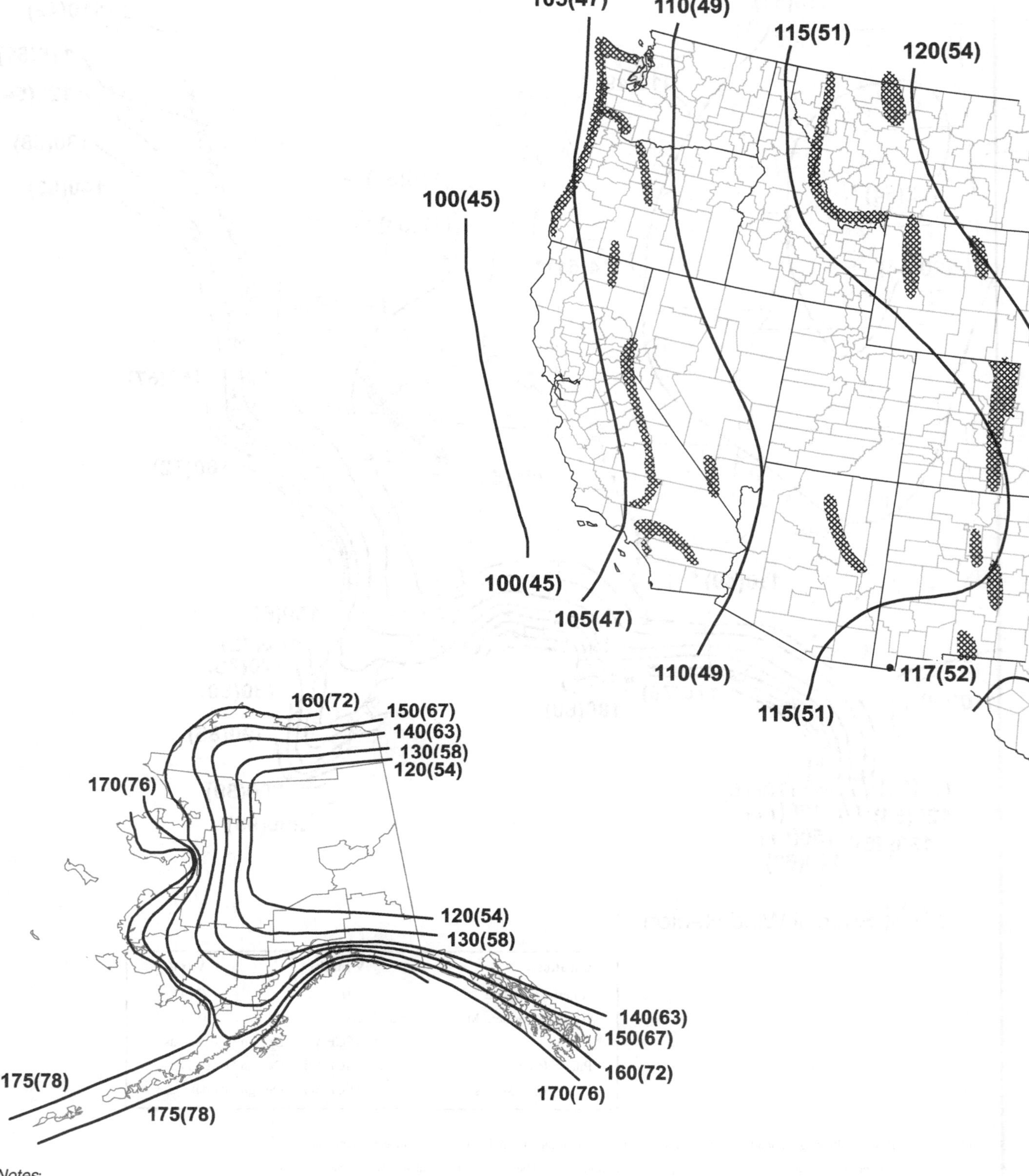

Notes:

1. Values are 3 s gust wind speeds in mi/h (m/s) at 33 ft (10 m) above ground for Exposure Category C.
2. Linear interpolation is permitted between contours. Point values are provided to aid with interpolation.
3. Islands, coastal areas, and land boundaries outside the last contour shall use the last wind speed contour.
4. Location-specific basic wind speeds shall be permitted to be determined using the ASCE Wind Design Geodatabase.

Figure 26.5-1D. Basic wind speeds for Risk Category IV buildings and other structures.

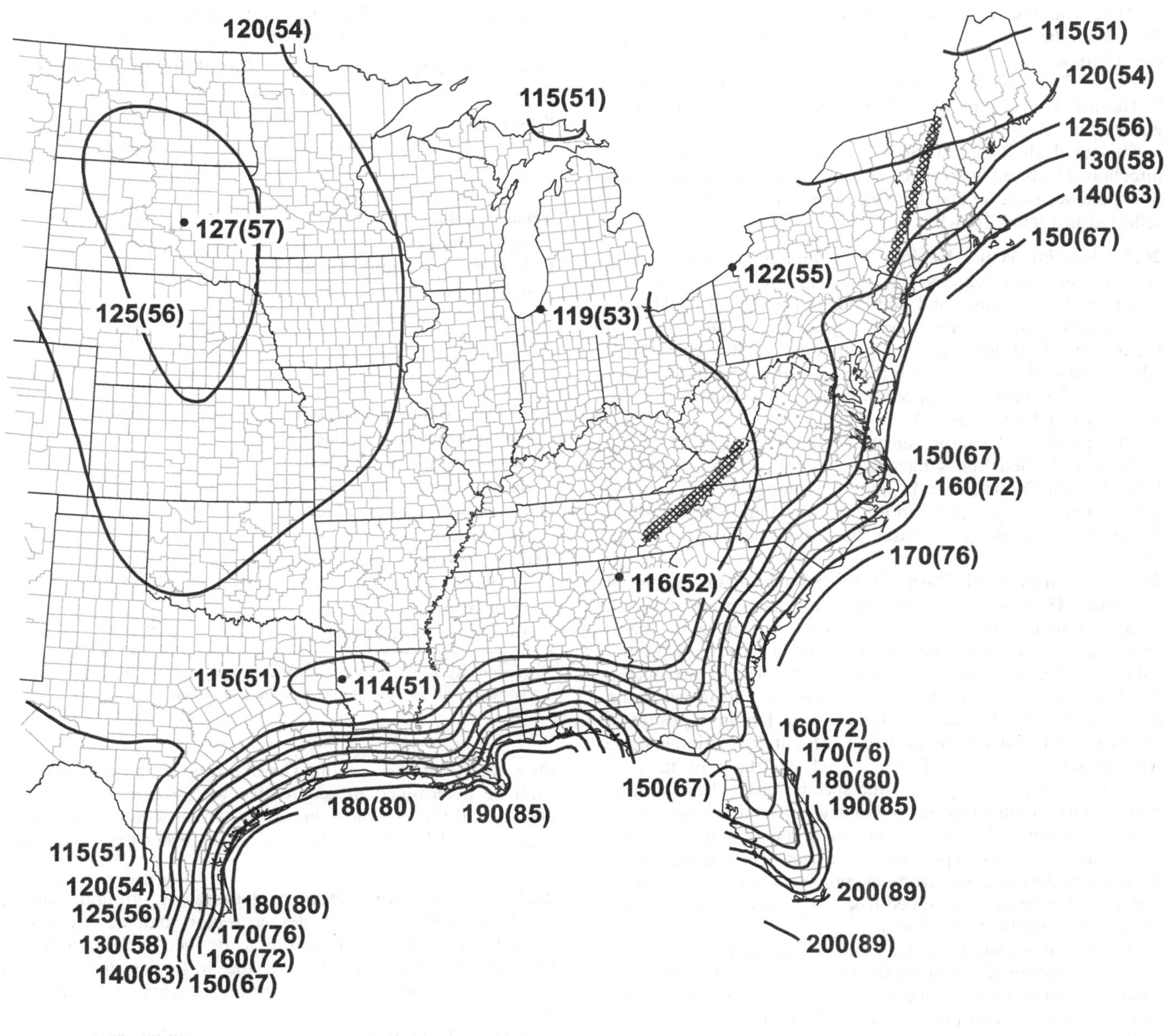

Location	V (mi/h)	V (m/s)
American Samoa	180	(80)
Guam & Northern Mariana Islands	220	(98)
Hawaii	ASCE Wind Design Geodatabase	
Puerto Rico	ASCE Wind Design Geodatabase	
U.S. Virgin Islands	ASCE Wind Design Geodatabase	

5. Wind speeds for Hawaii, US Virgin Islands, and Puerto Rico shall be determined from the ASCE Wind Design Geodatabase.
6. Mountainous terrain, gorges, ocean promontories, and special wind regions shall be examined for unusual wind conditions. Site-specific values for selected special wind regions shall be permitted to be determined using the ASCE Wind Design Geodatabase.
7. Wind speeds correspond to approximately a 15% probability of exceedance in 50 years (Annual Exceedance Probability = 0.00033, MRI = 3,000 years).
8. The ASCE Wind Design Geodatabase can be accessed at the ASCE 7 Hazard Tool (https://asce7hazardtool.online) or approved equivalent.

Figure 26.5-1D (*Continued*). Basic wind speeds for Risk Category IV buildings and other structures.

Basic wind speeds for Hawaii, US Virgin Islands, and Puerto Rico shall be determined by using the ASCE Wind Design Geodatabase.

The ASCE Wind Design Geodatabase is available at the ASCE 7 Hazard Tool (https://asce7hazardtool.online), or approved equivalent.

The wind shall be assumed to come from any horizontal direction. The basic wind speed shall be increased where records or experience indicate that the wind speeds are higher than those reflected in Figure 26.5-1.

26.5.2 Special Wind Regions Mountainous terrain, gorges, and special wind regions shown in Figure 26.5-1 shall be examined for unusual wind conditions. The Authority Having Jurisdiction shall, if necessary, adjust the values given in Figure 26.5-1 to account for higher local wind speeds. Such adjustment shall be based on meteorological information and an estimate of the basic wind speed obtained in accordance with the provisions of Section 26.5.3.

Site-specific values for select special wind regions in the contiguous United States have been included in the ASCE 7 Wind Design Geodatabase. It shall be permitted to use the wind speed values for Special Wind Regions from the ASCE 7 Wind Design Geodatabase (https://asce7hazardtool.online/).

26.5.3 Estimation of Basic Wind Speeds from Regional Climatic Data In areas outside hurricane-prone regions, regional climatic data shall only be used in lieu of the basic wind speeds given in Figure 26.5-1 when (1) approved extreme-value statistical analysis procedures have been used in reducing the data, and (2) the length of record, sampling error, averaging time, anemometer height, data quality, and terrain exposure of the anemometer have been taken into account. Reduction in basic wind speed below that of Figure 26.5-1 shall be permitted.

In hurricane-prone regions, wind speeds derived from simulation techniques shall only be used in lieu of the basic wind speeds given in Figure 26.5-1 when approved simulation and extreme-value statistical analysis procedures are used. The use of regional wind-speed data obtained from anemometers is not permitted to define the hurricane wind-speed risk along the Gulf and Atlantic coasts, the Caribbean, or Hawaii.

When the basic wind speed is estimated from regional climatic data or simulation, the estimate shall correspond to the applicable mean recurrence interval, and the estimate shall be adjusted for equivalence to a 3 s gust wind speed at 33 ft (10 m) above ground in Exposure C.

26.6 WIND DIRECTIONALITY FACTOR

The wind directionality factor, K_d, shall be determined from Table 26.6-1 and shall be included in the wind loads calculated in Chapters 27 through 30. The effect of wind directionality in determining wind loads in accordance with Chapter 31 shall be based on a rational analysis of the wind speeds conforming to the requirements of Section 26.5.3 and of Section 31.4.3.

26.7 EXPOSURE

For each wind direction considered, the upwind exposure shall be based on ground surface roughness that is determined from natural topography, vegetation, and constructed facilities.

26.7.1 Wind Directions and Sectors For each selected wind direction at which the wind loads are to be determined, the exposure of the building or structure shall be determined for the two upwind sectors extending 45 degrees on either side of the selected wind direction. The exposure in these two sectors shall be determined in accordance with Sections 26.7.2 and 26.7.3, and the exposure the use of which would result in the highest wind loads shall be used to represent the winds from that direction.

Table 26.6-1. Wind Directionality Factor, K_d.

Structure Type	Directionality Factor K_d
Buildings	
Main wind force resisting system	0.85
Components and cladding	0.85
Arched roofs	0.85
Circular domes	1.0*
Chimneys, tanks, and similar structures	
Square	0.90
Hexagonal	0.95
Octagonal	1.0*
Round	1.0*
Solid freestanding walls, roof top equipment, and solid freestanding and attached signs	0.85
Open signs and single-plane open frames	0.85
Trussed towers	
Triangular, square, or rectangular	0.85
All other cross sections	0.95

*Directionality factor $K_d = 0.95$ shall be permitted for round or octagonal structures with nonaxisymmetric structural systems.

26.7.2 Surface Roughness Categories A ground surface roughness within each 45 degree sector shall be determined for a distance upwind of the site, as defined in Section 26.7.3, from the categories defined in the following text, for the purpose of assigning an exposure category as defined in Section 26.7.3.

Surface Roughness B. Urban and suburban areas, wooded areas, or other terrain with numerous, closely spaced obstructions that have the size of single-family dwellings or larger.

Surface Roughness C. Open terrain with scattered obstructions that have heights generally less than 30 ft (9.1 m). This category includes flat, open country and grasslands.

Surface Roughness D. Flat, unobstructed areas and water surfaces. This category includes smooth mud flats, salt flats, and unbroken ice.

26.7.3 Exposure Categories

Exposure B. For buildings or other structures with a mean roof height less than or equal to 30 ft (9.1 m), Exposure B shall apply where the ground surface roughness, as defined by Surface Roughness B, prevails in the upwind direction for a distance greater than 1,500 ft (457 m). For buildings or other structures with a mean roof height greater than 30 ft (9.1 m), Exposure B shall apply where Surface Roughness B prevails in the upwind direction for a distance greater than 2,600 ft (792 m) or 20 times the height of the building or structure, whichever is greater.

Exposure C. Exposure C shall apply for all cases where Exposure B or D does not apply.

Exposure D. Exposure D shall apply where the ground surface roughness, as defined by Surface Roughness D, prevails in the upwind direction for a distance greater than 5,000 ft (1,524 m) or 20 times the building or structure height, whichever is greater. Exposure D shall also apply where the ground surface roughness immediately upwind of the site is B or C, and the building or structure is within a distance of 600 ft (183 m) or 20 times the building or structure height, whichever is greater, from an Exposure D condition as defined in the previous sentence.

For a building or structure located in the transition zone between exposure categories, the category resulting in the largest wind forces shall be used.

EXCEPTION: An intermediate exposure between the preceding categories is permitted in a transition zone, provided it is determined by a rational analysis method defined in the recognized literature.

26.7.4 Exposure Requirements

26.7.4.1 Directional Procedure (Chapter 27) For each wind direction considered, wind loads for the design of the MWFRS of enclosed and partially enclosed buildings using the Directional Procedure of Chapter 27 shall be based on the exposures as defined in Section 26.7.3. Wind loads for the design of open buildings with monoslope, pitched, or troughed free roofs shall be based on the exposures, as defined in Section 26.7.3, resulting in the highest wind loads for any wind direction at the site.

26.7.4.2 Envelope Procedure (Chapter 28) Wind loads for the design of the MWFRS for all low-rise buildings designed using the Envelope Procedure of Chapter 28 shall be based on the exposure category resulting in the highest wind loads for any wind direction at the site.

26.7.4.3 Directional Procedure for Building Appurtenances and Other Structures (Chapter 29) Wind loads for the design of building appurtenances (such as rooftop structures and equipment) and other structures (such as solid freestanding walls and freestanding signs, chimneys, tanks, open signs, single-plane open frames, and trussed towers) as specified in Chapter 29 shall be based on the appropriate exposure for each wind direction considered.

26.7.4.4 Components and Cladding (Chapter 30) Design wind pressures for C&C shall be based on the exposure category resulting in the highest wind loads for any wind direction at the site.

26.8 TOPOGRAPHIC EFFECTS

26.8.1 Wind Speed-Up over Hills, Ridges, and Escarpments Wind speed-up effects at isolated hills, ridges, and escarpments constituting abrupt changes in the general topography, located in any exposure category, shall be included in the determination of the wind loads when site conditions and locations of buildings and other structures meet all the following conditions:

1. The building or other structure is located as shown in Figure 26.8-1 in the upper one-half of a hill or ridge or near the crest of an escarpment.
2. $H/L_h \geq 0.2$.
3. H is greater than or equal to 15 ft (4.5 m) for Exposures C and D and 60 ft (18 m) for Exposures B.

26.8.2 Topographic Factor The wind speed-up effect shall be included in the calculation of design wind loads by using the factor K_{zt}:

$$K_{zt} = (1 + K_1 K_2 K_3)^2 \quad (26.8\text{-}1)$$

where K_1, K_2, and K_3 are given in Figure 26.8-1. The values for K_2 and K_3 shall not be less than 0. The equations for K_1, K_2, and K_3 may be used instead of using the tabular values when increased accuracy in determining K_{zt} is required. For $H/L_h > 0.5$, assume that $H/L_h = 0.5$ for evaluating K_1 and substitute $2H$ for L_h for evaluating K_2 and K_3.

If site conditions and locations of buildings and other structures do not meet all the conditions specified in Section 26.8.1, then $K_{zt} = 1.0$.

26.9 GROUND ELEVATION FACTOR

The ground elevation factor to adjust for air density, K_e, shall be determined in accordance with Table 26.9-1. It is permitted to take $K_e = 1$ for all elevations.

Table 26.9-1. Ground Elevation Factor, K_e.

Ground Elevation above Sea Level		
ft	m	Ground Elevation Factor, K_e
<0	<0	See note 2
0	0	1.00
1,000	305	0.96
2,000	610	0.93
3,000	914	0.90
4,000	1,219	0.86
5,000	1,524	0.83
6,000	1,829	0.80
>6,000	>1,829	See note 2

Notes:

1. Conservative approximation $K_e = 1.00$ is permitted in all cases.
2. Factor K_e shall be determined from Table 26.9-1 using interpolation or from the following formula for all elevations: $K_e = e^{-0.0000362 z_e}$ (z_e= ground elevation above sea level, ft); or $K_e = e^{-0.000119 z_e}$ (z_e= ground elevation above sea level, m).
3. K_e is permitted to be taken as 1.00 in all cases.

26.10 VELOCITY PRESSURE

26.10.1 Velocity Pressure Exposure Coefficient Based on the exposure category determined in Section 26.7.3, a velocity pressure exposure coefficient, K_z or K_h, as applicable, shall be determined from Table 26.10-1. For a site located in a transition zone between exposure categories that is near to a change in ground surface roughness, intermediate values of K_z or K_h, between those shown in Table 26.10-1 are permitted provided they are determined by a rational analysis method defined in the recognized literature.

26.10.2 Velocity Pressure Velocity pressure, q_z, evaluated at height z above ground shall be calculated by the following equation:

Diagrams

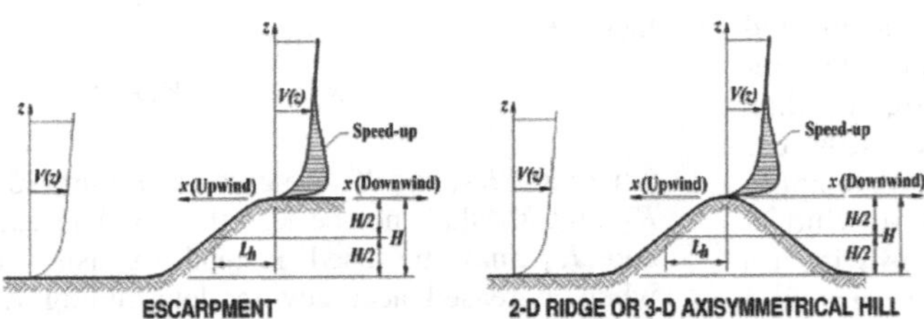

Topographic Multipliers[a,b,c,d]

	K_1 Multiplier				K_2 Multiplier			K_3 Multiplier		
H/L_h	2D Ridge	2D Escarpment	3D Axisym-metrical Hill	x/L_h	2D Escarpment	All Other Cases	z/L_h	2D Ridge	2D Escarpment	3D Axisym-metrical Hill
0.20	0.29	0.17	0.21	0.00	1.00	1.00	0.00	1.00	1.00	1.00
0.25	0.36	0.21	0.26	0.50	0.88	0.67	0.10	0.74	0.78	0.67
0.30	0.43	0.26	0.32	1.00	0.75	0.33	0.20	0.55	0.61	0.45
0.35	0.51	0.30	0.37	1.50	0.63	0.00	0.30	0.41	0.47	0.30
0.40	0.58	0.34	0.42	2.00	0.50	0.00	0.40	0.30	0.37	0.20
0.45	0.65	0.38	0.47	2.50	0.38	0.00	0.50	0.22	0.29	0.14
0.50	0.72	0.43	0.53	3.00	0.25	0.00	0.60	0.17	0.22	0.09
				3.50	0.13	0.00	0.70	0.12	0.17	0.06
				4.00	0.00	0.00	0.80	0.09	0.14	0.04
							0.90	0.07	0.11	0.03
							1.00	0.05	0.08	0.02
							0.50	0.01	0.02	0.00
							2.00	0.00	0.00	0.00

[a]For values of H/L_h, x/L_h, and z/L_h other than those shown, linear interpolation is permitted.

[b]For $H/L_h > 0.5$, assume that $H/L_h = 0.5$ for evaluating K_1 and substitute $2H$ for L_h for evaluating K_2 and K_3.

[c]Multipliers are based on the assumption that wind approaches the hill or escarpment along th direction of maximum slope.e

[d]Multipliers shall be used for any exposure.

Notation

H = Height of hill or escarpment relative to the upwind terrain, ft (m);
K_1 = Factor to account for shape of topographic feature and maximum speed-up effect;
K_2 = Factor to account for reduction in speed-up with distance upwind or downwind of crest;
K_3 = Factor to account for reduction in speed-up with height above local terrain;
L_h = Distance upwind of crest to where the difference in ground elevation is half the height of hill or escarpment, ft (m);
x = Distance (upwind or downwind) from the crest to the site of the building or other structure, ft (m);
z = Height above ground surface at the site of the building or other structure, ft (m);
μ = Horizontal attenuation factor;
γ = Height attenuation factor.

Equations

$K_{zt} = (1 + K_1 K_2 K_3)^2$

K_1 = Determined from table above

$K_2 = (1 - |x| / \mu L_h)$

$$K_3 = e^{-\gamma z/L_h}$$

Parameters for Speed-Up over Hills and Escarpments

	$K_1/(H/L_h)$				μ	
	Exposure					
Hill Shape	B	C	D	γ	Upwind of Crest	Downwind of Crest
2D ridges (or valleys with negative H in $K_1/(H/L_h)$)	1.30	1.45	1.55	3	1.5	1.5
2D escarpments	0.75	0.85	0.95	2.5	1.5	4
3D axisymmetrical hill	0.95	1.05	1.15	4	1.5	1.5

Figure 26.8-1. Topographic factor, K_{zt}.

$$q_z = 0.00256K_zK_{zt}K_eV^2(\text{lb/ft}^2);\ V, \text{mi/h} \quad (26.10\text{-}1)$$

$$q_z = 0.613K_zK_{zt}K_eV^2(\text{N/m}^2);\ V, \text{m/s} \quad (26.10\text{-}1.\text{SI})$$

where

K_z = Velocity pressure exposure coefficient, see Section 26.10.1;
K_{zt} = Topographic factor, see Section 26.8.2;
K_e = Ground elevation factor, see Section 26.9;
V = Basic wind speed, see Section 26.5; and
q_z = Velocity pressure at height z.

The velocity pressure at mean roof height is computed as $q_h = q_z$ evaluated from Equation (26.10-1) using K_z at mean roof height h.

The basic wind speed, V, used in determination of design wind loads on rooftop structures, rooftop equipment, and other building appurtenances shall consider the Risk Category equal to the greater of the following:

1. Risk category for the building on which the equipment or appurtenance is located, or
2. Risk category for any building or other structure to which the equipment or appurtenance provides a necessary service.

26.11 GUST EFFECTS

26.11.1 Gust-Effect Factor The gust-effect factor for a rigid building or other structure is permitted to be taken as 0.85.

26.11.2 Frequency Determination To determine whether a building or other structure is rigid or flexible as defined in Section 26.2, the fundamental natural frequency, n_1, shall be established using the structural properties and deformational characteristics of the resisting elements in a properly substantiated analysis. Low-rise buildings, as defined in Section 26.2, are permitted to be considered rigid.

26.11.2.1 Limitations for Approximate Natural Frequency As an alternative to performing an analysis to determine n_1, the approximate building natural frequency, n_a, shall be permitted to be calculated in accordance with Section 26.11.3 for structural steel, concrete, or masonry buildings meeting the following requirements:

1. The building height is less than or equal to 300 ft (91 m).
2. The building height is less than four times its effective length, L_{eff}.

The effective length, L_{eff}, in the direction under consideration shall be determined from the following equation:

$$L_{\text{eff}} = \frac{\sum_{i=1}^{n} h_i L_i}{\sum_{i=1}^{n} h_i} \quad (26.11\text{-}1)$$

The summations are over the height of the building, where h_i is the height above grade of level i, and L_i is the building length at level i parallel to the wind direction.

26.11.3 Approximate Natural Frequency The approximate lower bound natural frequency (n_a), in Hz, of concrete or structural steel buildings meeting the conditions of Section 26.11.2.1 is permitted to be determined from one of the following equations.

For structural steel moment-resisting frame buildings,

$$n_a = 22.2/h^{0.8} \quad (26.11\text{-}2)$$

$$n_a = 8.58/h^{0.8} \quad (26.11\text{-}2.\text{SI})$$

For concrete moment-resisting frame buildings,

$$n_a = 43.5/h^{0.9} \quad (26.11\text{-}3)$$

$$n_a = 14.93/h^{0.9} \quad (26.11\text{-}3.\text{SI})$$

Table 26.10-1. Velocity Pressure Exposure Coefficients, K_h and K_z.

Height above Ground Level, z or h		Exposure		
ft	m	B	C	D
0–15	0–4.6	0.57 (0.70)*	0.85	1.03
20	6.1	0.62 (0.70)*	0.90	1.08
25	7.6	0.66 (0.70)*	0.94	1.12
30	9.1	0.70	0.98	1.16
40	12.2	0.74	1.04	1.22
50	15.2	0.79	1.09	1.27
60	18.3	0.83	1.13	1.31
70	21.3	0.86	1.17	1.34
80	24.4	0.90	1.21	1.38
90	27.4	0.92	1.24	1.40
100	30.5	0.95	1.26	1.43
120	36.6	1.00	1.31	1.48
140	42.7	1.04	1.34	1.52
160	48.8	1.08	1.39	1.55
180	54.9	1.11	1.41	1.58
200	61.0	1.14	1.44	1.61
250	76.2	1.21	1.51	1.68
300	91.4	1.27	1.57	1.73
350	106.7	1.33	1.62	1.78
400	121.9	1.38	1.66	1.82
450	137.2	1.42	1.70	1.86
500	152.4	1.46	1.74	1.89

*Use 0.70 in Chapter 28, Exposure B, when $z < 30$ ft (9.1 m).

Notes:

1. Velocity pressure exposure coefficient K_z may be determined from the following formula:

For $z < 15$ ft	$K_z = 2.41\,(15/z_g)^{2/\alpha}$
For $z < 4.6$ m	$K_z = 2.41\,(4.6/z_g)^{2/\alpha}$
For 15 ft (4.6 m) $\leq z \leq z_g$	$K_z = 2.41\,(z/z_g)^{2/\alpha}$
For $z_g < z \leq 3{,}280$ ft (1,000 m)	$K_z = 2.41$

2. α and z_g are tabulated in Table 26.11-1.
3. Linear interpolation for intermediate values of height z is acceptable.
4. Exposure categories are defined in Section 26.7.

For structural steel and concrete buildings with other lateral-force-resisting systems,

$$n_a = 75/h \quad (26.11\text{-}4)$$

$$n_a = 22.86/h \quad (26.11\text{-}4.SI)$$

For concrete or masonry shear wall buildings, it is also permitted to use

$$n_a = 385(C_w)^{0.5}/h \quad (26.11\text{-}5)$$

$$n_a = 117.3(C_w)^{0.5}/h \quad (26.11\text{-}5.SI)$$

where

$$C_w = \frac{100}{A_B}\sum_{i=1}^{n}\left(\frac{h}{h_i}\right)^2 \frac{A_i}{\left[1+0.83\left(\frac{h_i}{D_i}\right)^2\right]}$$

where
h = Mean roof height, ft (m);
n = Number of shear walls in the building effective in resisting lateral forces in the direction under consideration;
A_B = Base area of the building, ft^2 (m^2);
A_i = Horizontal cross-sectional area of shear wall i, ft^2 (m^2);
D_i = Length of shear wall i, ft (m); and
h_i = Height of shear wall i, ft (m).

26.11.4 Rigid Buildings or Other Structures For rigid buildings or other structures as defined in Section 26.2, the gust-effect factor shall be taken as 0.85 or calculated by this formula:

$$G = 0.925\left(\frac{1+1.7g_Q I_{\bar{z}} Q}{1+1.7g_v I_{\bar{z}}}\right) \quad (26.11\text{-}6)$$

$$I_{\bar{z}} = c\left(\frac{33}{\bar{z}}\right)^{1/6} \quad (26.11\text{-}7)$$

$$I_{\bar{z}} = c\left(\frac{10}{\bar{z}}\right)^{1/6} \quad (26.11\text{-}7.SI)$$

where $I_{\bar{z}}$ is the intensity of turbulence at height $\bar{z}$, and $\bar{z}$ is the equivalent height of the building or structure defined as $0.6h$, but not less than z_{min}, for all building or structure heights h. z_{min} and c are listed for each exposure in Table 26.11-1; g_Q and g_v shall be taken as 3.4. The background response Q is given by

$$Q = \sqrt{\frac{1}{1+0.63\left(\frac{B+h}{L_{\bar{z}}}\right)^{0.63}}} \quad (26.11\text{-}8)$$

where B and h are defined in Section 26.3, and $L_{\bar{z}}$ is the integral length scale of turbulence at the equivalent height given by

$$L_{\bar{z}} = \ell\left(\frac{\bar{z}}{33}\right)^{\bar{\varepsilon}} \quad (26.11\text{-}9)$$

$$L_{\bar{z}} = \ell\left(\frac{\bar{z}}{10}\right)^{\bar{\varepsilon}} \quad (26.11\text{-}9.SI)$$

in which ℓ and $\bar{\varepsilon}$ are constants listed in Table 26.11-1.

26.11.5 Flexible Buildings or Other Structures For flexible buildings or other structures as defined in Section 26.2, the gust-effect factor shall be calculated by

$$G_f = 0.925\left(\frac{1+1.7I_{\bar{z}}\sqrt{g_Q^2Q^2+g_R^2R^2}}{1+1.7g_v I_{\bar{z}}}\right) \quad (26.11\text{-}10)$$

The background peak factors g_Q and g_v shall be taken as 3.4, and the resonant peak factor g_R is

$$g_R = \sqrt{2\ln(3{,}600 n_1)} + \frac{0.577}{\sqrt{2\ln(3{,}600 n_1)}} \quad (26.11\text{-}11)$$

The background response factor, Q, and the intensity of turbulence at height z, I_z, are defined in Section 26.11.4. The resonant response factor is

$$R = \sqrt{\frac{1}{\beta}R_n R_h R_B(0.53+0.47R_L)} \quad (26.11\text{-}12)$$

where n_1 is the fundamental natural frequency, and β is the damping ratio, fraction of critical (e.g., for 2% use 0.02 in the equation).

Table 26.11-1. Terrain Exposure Constants.

Customary Units										
Exposure	α	z_g (ft)	$\hat{a}$	$\hat{b}$	$\bar{\alpha}$	$\bar{b}$	c	l (ft)	$\bar{\varepsilon}$	z_{min} (ft)*
B	7.5	3,280	1/7.5	0.84	1/4.5	0.47	0.30	320	1/3.0	30
C	9.8	2,460	1/9.8	1.00	1/6.4	0.66	0.20	500	1/5.0	15
D	11.5	1,935	1/11.5	1.09	1/8.0	0.78	0.15	650	1/8.0	7
SI Units										
Exposure	α	z_g (m)	$\hat{a}$	$\hat{b}$	$\bar{\alpha}$	$\bar{b}$	c	l (m)	$\bar{\varepsilon}$	z_{min} (m)*
B	7.5	1,000	1/7.5	0.84	1/4.5	0.47	0.30	97.54	1/3.0	9.14
C	9.8	750	1/9.8	1.00	1/6.4	0.66	0.20	152.40	1/5.0	4.57
D	11.5	590	1/11.5	1.09	1/8.0	0.78	0.15	198.12	1/8.0	2.13

* z_{min} = Minimum height used to ensure that the equivalent height $\bar{z}$ is the greater of $0.6h$ or z_{min}. For buildings or other structures with $h \le z_{min}$, $\bar{z}$ shall be taken as z_{min}.

The power spectral density of turbulence at the equivalent height of the structure $\bar{z}$, evaluated at the structure's natural reduced frequency, N_1, is

$$R_n = \frac{7.47N_1}{(1 + 10.3N_1)^{5/3}} \tag{26.11-13}$$

$$N_1 = \frac{n_1 L_{\bar{z}}}{\overline{V}_{\bar{z}}} \tag{26.11-14}$$

where $L_{\bar{z}}$ is defined in Equation (26.11-9).

The size effect factors related to the height, breadth, and depth of the building are

$$R_h = \frac{1}{\eta_h} - \frac{1}{2\eta_h^2}(1 - e^{-2\eta_h})$$
$$R_B = \frac{1}{\eta_B} - \frac{1}{2\eta_B^2}(1 - e^{-2\eta_B}) \tag{26.11-15a}$$
$$R_L = \frac{1}{\eta_L} - \frac{1}{2\eta_L^2}(1 - e^{-2\eta_L})$$

where the turbulent coherence (correlation) factors in the corresponding directions, evaluated at the natural reduced frequency, are

$$\eta_h = 4.6n_1 h/\overline{V}_z$$
$$\eta_B = 4.6n_1 B/\overline{V}_z \tag{26.11-15b}$$
$$\eta_L = 15.4n_1 L/\overline{V}_z$$

The mean hourly wind speed (in ft/s or m/s) at the equivalent structure height, $\bar{z}$, is

$$\overline{V}_{\bar{z}} = \bar{b}\left(\frac{\bar{z}}{33}\right)^{\bar{\alpha}}\left(\frac{88}{60}\right)V \tag{26.11-16}$$

$$\overline{V}_{\bar{z}} = \bar{b}\left(\frac{\bar{z}}{10}\right)^{\bar{\alpha}}V \tag{26.11-16.SI}$$

where $\bar{b}$ and $\bar{\alpha}$ are constants listed in Table 26.11-1, $\bar{z}$ is obtained from Section 26.11.4, and V is the basic wind speed, mi/h (m/s).

26.11.6 Rational Analysis In lieu of the procedure defined in Sections 26.11.4 and 26.11.5, determination of the gust-effect factor by any rational analysis defined in the recognized literature is permitted.

26.11.7 Limitations Where combined gust-effect factors and pressure coefficients (GC_p), (GC_{pi}), and (GC_{pf}) are given in figures and tables, the gust-effect factor shall not be determined separately.

26.12 ENCLOSURE CLASSIFICATION

26.12.1 General For the purpose of determining internal pressure coefficients, buildings and other structures for which internal pressure coefficients (GC_{pi}) apply shall be classified as enclosed, partially enclosed, partially open, or open, as defined in Section 26.2. If a building or other structure satisfies both the "open" and "partially enclosed" enclosure classification definitions, it shall be classified as a "partially open" building or other structure.

26.12.2 Openings A determination shall be made of the amount of openings in the building envelope for use in determining the enclosure classification. To make this determination, each building wall shall be assumed as the windward wall for consideration of the amount of openings present with respect to the remaining building envelope.

26.12.3 Protection of Glazed Openings Glazed openings in Risk Category II, III, or IV buildings located in hurricane-prone regions shall be protected as specified in this section.

26.12.3.1 Wind-Borne Debris Regions Glazed openings shall be protected in accordance with Section 26.12.3.2 in the following locations:

1. Within 1 mi (1.6 km) of the mean high water line where an Exposure D condition exists upwind of the waterline and the basic wind speed is equal to or greater than 130 mi/h (58 m/s), or
2. In areas where the basic wind speed is equal to or greater than 140 mi/h (63 m/s).

For Risk Category II buildings and other structures and Risk Category III buildings and other structures, except health care facilities, the wind-borne debris region shall be based on Figure 26.5-1B. For Risk Category III health care facilities, the wind-borne debris region shall be based on Figure 26.5-1C. For Risk Category IV buildings and structures, the wind-borne debris region shall be based on Figure 26.5-1D. Risk categories shall be determined in accordance with Section 1.5.

EXCEPTION: Glazing located more than 60 ft (18.3 m) above the ground and more than 30 ft (9.2 m) above aggregate-surfaced roofs, including roofs with gravel or stone ballast, located within 1,500 ft (458 m) of the building shall be permitted to be unprotected.

26.12.3.2 Protection Requirements for Glazed Openings Glazing in buildings requiring protection shall be impact resistant or be protected with an impact-protective system.

Impact-protective systems and impact-resistant glazing shall be subjected to missile test and cyclic pressure differential tests in accordance with ASTM E1996 as applicable. Testing to demonstrate compliance with ASTM E1996 shall be in accordance with ASTM E1886. Impact-resistant glazing and impact-protective systems shall comply with the pass/fail criteria of Section 7 of ASTM E1996 based on the missile required by Table 3 or Table 4 of ASTM E1996. Glazing in sectional doors, rolling doors, and flexible doors shall be subjected to missile tests and cyclic pressure differential tests in accordance with ANSI/DASMA 115 as applicable.

Glazing and impact-protective systems in buildings and other structures classified as Risk Category IV in accordance with Section 1.5 shall comply with the "enhanced protection" requirements of Table 3 of ASTM E1996. Glazing and impact-protective systems in all other buildings covered by Section 26.12.3.1 shall comply with the "basic protection" requirements of Table 3 of ASTM E1996.

EXCEPTION: When approved, other testing methods and/or performance criteria shall be permitted to be used to demonstrate compliance with all provisions of this section.

26.13 INTERNAL PRESSURE COEFFICIENTS

Internal pressure coefficients (GC_{pi}) shall be determined from Table 26.13-1 based on building enclosure classifications determined from Section 26.12.

26.13.1 Reduction Factor for Large-Volume Buildings, R_i For a partially enclosed building containing a single,

Table 26.13-1. Main Wind Force Resisting System and Components and Cladding (All Heights): Internal Pressure Coefficient, (GC_{pi}), for Enclosed, Partially Enclosed, Partially Open, and Open Buildings (Walls and Roof).

Enclosure Classification	Criteria for Enclosure Classification	Internal Pressure	Internal Pressure Coefficient (GC_{pi})
Enclosed buildings	A_o is less than the smaller of $0.01A_g$ or 4 ft^2 (0.37 m^2), and $A_{oi}/A_{gi} \le 0.2$	Moderate	+0.18 −0.18
Partially enclosed buildings	$A_o > 1.1A_{oi}$, and $A_o >$ the lesser of $0.01A_g$ or 4 ft^2 (0.37 m^2), and $A_{oi}/A_{gi} \le 0.2$	High	+0.55 −0.55
Partially open buildings	A building that does not comply with Enclosed, Partially Enclosed, or Open classifications	Moderate	+0.18 −0.18
Open buildings	Each wall is at least 80% open	Negligible	0.00

Notes:

1. Plus and minus signs signify pressures acting toward and away from the internal surfaces, respectively.
2. Values of (GC_{pi}) shall be used with q_z or q_h as specified.
3. Two cases shall be considered to determine the critical load requirements for the appropriate condition:
   (a) A positive value of (GC_{pi}) applied to all internal surfaces, or
   (b) A negative value of (GC_{pi}) applied to all internal surfaces.

unpartitioned large volume, the internal pressure coefficient, (GC_{pi}), shall be multiplied by the reduction factor, R_i:

$$R_i = 1.0 \quad \text{or} \quad R_i = 0.5\left(1 + \frac{1}{\sqrt{1 + \frac{V_i}{22{,}800A_{og}}}}\right) < 1.0 \quad (26.13\text{-}1)$$

where A_{og} is the total area of openings in the building envelope [walls and roof (ft^2)], and V_i is the unpartitioned internal volume (ft^3).

26.14 CONSENSUS STANDARDS AND OTHER REFERENCED DOCUMENTS

This section lists the consensus standards and other documents that shall be considered part of this standard to the extent referenced in this chapter.

ANSI A58.1, *Minimum Design Loads for Buildings and Other Structures*, American National Standards Institute, 1982.

Cited in: Section C26.5.2.

ANSI/DASMA 115, *Standard Method for Testing Sectional Doors, Rolling Doors, and Flexible Doors: Determination of Structural Performance under Missile Impact and Cyclic Wind Pressure*, Door and Access Systems Manufacturers Association International, 2017.

Cited in: Sections 26.12.3.2, C26.12.

ASTM E330, *Standard Test Method for Structural Performance of Exterior Windows, Doors, Skylights, and Curtain Walls by Uniform Static Air Pressure Difference*, ASTM International, 2014.

Cited in: Section C26.5.1.

ASTM E1886, *Standard Test Method for Performance of Exterior Windows, Curtain Walls, Doors, and Impact Protective Systems Impacted by Missile(s) and Exposed to Cyclic Pressure Differentials*, ASTM, 2019.

Cited in: Sections 26.12.3.2, C26.12

ASTM E1996, *Standard Specification for Performance of Exterior Windows, Curtain Walls, Doors and Impact Protective Systems Impacted by Windborne Debris in Hurricanes*, ASTM International, 2020.

Cited in: Sections 26.12.3.2, C26.12

CAN/CSA A123.21, *Standard Test Method for the Dynamic Wind Uplift Resistance of Membrane-Roofing Systems*, CSA Group, 2014.

Cited in: Section C26.5.1

CHAPTER 27

WIND LOADS ON BUILDINGS: MAIN WIND FORCE RESISTING SYSTEM (DIRECTIONAL PROCEDURE)

27.1 SCOPE

27.1.1 Building Types This chapter applies to the determination of Main Wind Force Resisting System (MWFRS) wind loads on enclosed, partially enclosed, partially open, and open buildings of all heights using the Directional Procedure, in which it is necessary to separate applied wind loads onto the windward, leeward, and sidewalls of the building to properly assess the internal forces in the MWFRS members.

27.1.2 Conditions A building that has design wind loads determined in accordance with this chapter shall comply with both of the following conditions:

1. The building is a regular-shaped building as defined in Section 26.2, and
2. The building does not have response characteristics that make it subject to across-wind loading, vortex shedding, or instability caused by galloping or flutter; nor does it have a site location for which channeling effects or buffeting, in the wake of upwind obstructions, warrant special consideration.

27.1.3 Limitations The provisions of this chapter take into consideration the load magnification effect caused by gusts, in resonance with along-wind vibrations of flexible buildings. Buildings that do not meet the requirements of Section 27.1.2 or that have unusual shapes or response characteristics shall be designed using recognized literature, documenting such wind load effects, or shall use the Wind Tunnel Procedure specified in Chapter 31.

27.1.4 Shielding There shall be no reductions in velocity pressure caused by apparent shielding afforded by buildings and other structures or terrain features.

27.1.5 Minimum Design Wind Loads The wind pressure to be used in the design of the MWFRS for an enclosed, or partially enclosed, building shall not be less than 16 lb/ft² (0.77 kN/m²) multiplied by the wall area of the building, and 8 lb/ft² (0.38 kN/m²) multiplied by the roof area of the building, projected onto a vertical plane normal to the assumed wind direction. Wall and roof loads shall be applied simultaneously. The design wind force for open buildings shall be not less than 16 lb/ft² (0.77 kN/m²) multiplied by the area, A_f.

27.2 GENERAL REQUIREMENTS

User Note: Use Chapter 27 to determine wind pressures on the MWFRS of buildings with any general plan shape, building height, or roof geometry that matches the figures provided. These provisions use the traditional "all heights" method (Directional Procedure) by calculating wind pressures using *specific wind pressure equations* applicable to each building surface.

The steps to determine the wind loads on the MWFRS for enclosed, partially enclosed, partially open, and open buildings of all heights are provided in Table 27.2-1.

27.2.1 Wind Load Parameters Specified in Chapter 26 The following wind load parameters shall be determined in accordance with Chapter 26:

- Basic wind speed, V (Section 26.5);
- Wind directionality factor, K_d (Section 26.6);
- Exposure category (Section 26.7);
- Topographic factor, K_{zt} (Section 26.8);
- Ground elevation factor, K_e (Section 26.9);
- Gust-effect factor (Section 26.11);
- Enclosure classification (Section 26.12); and
- Internal pressure coefficient, (GC_{pi}) (Section 26.13).

27.3 WIND LOADS: MAIN WIND FORCE RESISTING SYSTEM

27.3.1 Enclosed, Partially Enclosed, and Partially Open Rigid and Flexible Buildings Design wind pressures for the MWFRS of buildings of all heights in lb/ft² (N/m²), shall be determined by the following equation:

$$p = qK_dGC_p - q_iK_d(GC_{pi}) \quad (27.3\text{-}1)$$

where

$q = q_z$ For windward walls evaluated at height z above the ground;

$q = q_h$ For leeward walls, sidewalls, and roofs evaluated at height h;

Table 27.2-1. Steps to Determine MWFRS Wind Loads for Enclosed, Partially Enclosed, Partially Open, and Open Buildings of All Heights.

Step 1: Determine Risk Category of building; see Table 1.5-1.

Step 2: Determine the basic wind speed, V, for the applicable risk category; see Figure 26.5-1.

Step 3: Determine wind load parameters:

- Wind directionality factor, K_d; see Section 26.6 and Table 26.6-1.
- Exposure category; see Section 26.7.
- Topographic factor, K_{zt}; see Section 26.8 and table in Figure 26.8-1.
- Ground elevation factor, K_e; see Section 26.9.
- Gust-effect factor, G or G_f; see Section 26.11.
- Enclosure classification; see Section 26.12.
- Internal pressure coefficient, (GC_{pi}); see Section 26.13 and Table 26.13-1.

Step 4: Determine velocity pressure exposure coefficient, K_z or K_h; see Table 26.10-1.

Step 5: Determine velocity pressure q_z or q_h, Equation (26.10-1).

Step 6: Determine external pressure coefficient, C_p or C_N:

- Figure 27.3-1 for walls and flat, gable, hip, monoslope, or mansard roofs;
- Figure 27.3-2 for domed roofs;
- Figure 27.3-3 for arched roofs;
- Figure 27.3-4 for monoslope roof, open building;
- Figure 27.3-5 for pitched roof, open building;
- Figure 27.3-6 for troughed roof, open building; and
- Figure 27.3-7 for along-ridge/valley wind load case for monoslope, pitched, or troughed roof, open building.

Step 7: Calculate wind pressure, p, on each building surface:

- Equation (27.3-1) for rigid and flexible buildings, and
- Equation (27.3-2) for open buildings.

Step 8: Evaluate design wind load cases according to Section 27.3.5 and Figure 27.3-8.

$q_i = q_h$ For windward walls, sidewalls, leeward walls, and roofs of enclosed buildings, partially open buildings, and for negative internal pressure evaluation in partially enclosed buildings;

$q_i = q_z$ For positive internal pressure evaluation in partially enclosed buildings, where height z is defined as the level of the highest opening in the building that could affect the positive internal pressure; For buildings sited in wind-borne debris regions, the enclosure classification with respect to glazed openings shall be in accordance with Section 26.12.3, and for positive internal pressure evaluation, q_i may conservatively be evaluated at height $h(q_i = q_h)$;

K_d = Wind directionality factor (see Section 26.6);

G = Gust-effect factor (see Section 26.11); For flexible buildings, G_f, determined in accordance with Section 26.11.5, shall be substituted for G;

C_p = External pressure coefficient from Figures 27.3-1, 27.3-2, and 27.3-3; and

(GC_{pi}) = Internal pressure coefficient from Table 26.13-1.

Both q and q_i shall be evaluated using exposure defined in Section 26.7.3. Pressure shall be applied simultaneously on windward and leeward walls and on roof surfaces, as defined in Figures 27.3-1, 27.3-2, and 27.3-3.

27.3.1.1 Elevated Buildings The MWFRS loads for rigid or flexible elevated buildings meeting both of the following geometric limitations, for any principal wind direction, shall be determined in accordance with this section for that principal direction.

1. The sum of the cross-sectional areas of the supporting structural elements and any enclosed or partially enclosed spaces beneath the building, as seen in a horizontal cross-section through the space under the building, does not exceed the indicated percentage of the plan cross-sectional area of the aforementioned building for the indicated L/B ratio (linear interpolation is allowed):

   L/B Ratio Percentage of Plan Area of Building Blocked by Elements or Enclosures Beneath the Building

L/B	Plan Area Blocked
<2.5	50%
3.0	45%
3.5	40%
4.0	36%
4.5	33%
5.0	30%

2. The sum of the areas of the supporting structural elements and any enclosed, partially enclosed or partially open spaces, as projected onto a vertical plane normal to the wind direction, does not exceed 75% of the projected area outlined by the height of these elements and the projected width of the building above.

For buildings that do not meet both of these criteria for a particular principal wind direction, wind loads for that direction shall be determined using Section 27.3.1. The wind loads on the portion of the building consisting of the structural elements, and enclosed or partially enclosed spaces beneath the elevated building, shall be calculated assuming the wind is unable to pass beneath the building and that portion of the building has no openings.

For buildings that meet both of these criteria for a particular principal direction, the overall lateral MWFRS loads shall include the combined effects of wind loads determined for the elevated building (Section 27.3.1.1.1) and for the supporting structural elements including any enclosed, partially enclosed, or partially open spaces below the elevated building (Section 27.3.1.1.2). MWFRS wind loads for the bottom surface of the elevated building shall not be used to reduce the effects of wind loads on overturning moments.

27.3.1.1.1 Elevated Building Loads Wind loads on the roof and walls of the elevated building shall be calculated using the methods and pressure coefficients defined in Section 27.3.1.

27.3.1.1.2 Lateral Loads on Elements Below the Elevated Building Lateral wind loads on elements below the elevated building shall be calculated using

1. Projected areas of all the structural elements below the elevated building perpendicular to the wind direction being considered, neglecting potential shielding;

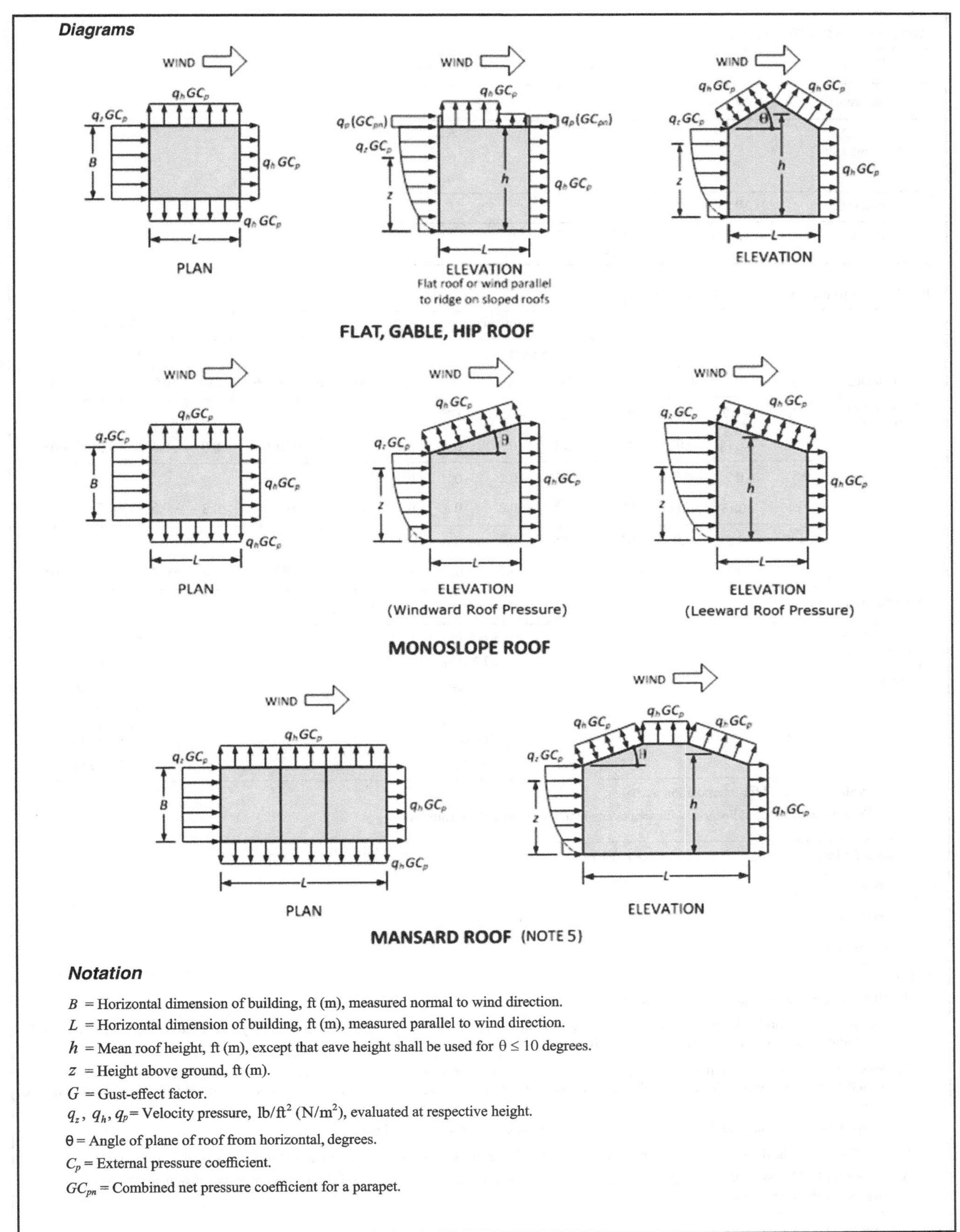

Notation

B = Horizontal dimension of building, ft (m), measured normal to wind direction.

L = Horizontal dimension of building, ft (m), measured parallel to wind direction.

h = Mean roof height, ft (m), except that eave height shall be used for $\theta \le 10$ degrees.

z = Height above ground, ft (m).

G = Gust-effect factor.

q_z, q_h, q_p = Velocity pressure, lb/ft^2 (N/m^2), evaluated at respective height.

θ = Angle of plane of roof from horizontal, degrees.

C_p = External pressure coefficient.

GC_{pn} = Combined net pressure coefficient for a parapet.

Figure 27.3-1. Main wind force resisting system: external pressure coefficients, C_p, for enclosed, partially enclosed, and partially open buildings—walls and roofs.

Wall Pressure Coefficients, C_p

Surface	L/B	C_p	Use with
Windward wall	All values	0.8	q_z
Leeward wall	0–1	−0.5	q_h
	2	−0.3	q_h
	≥ 4	−0.2	q_h
Sidewall	All values	−0.7	q_h
Parapet	All values	See Section 27.3.4 for GC_{pn}	q_p

Roof Pressure Coefficients, C_p, for use with q_h

		windward										Leeward		
		Angle, θ										Angle, θ		
Wind Direction	h/L	10°	15°	20°	25°	30°	35°	45°	60°	60° < θ ≤ 80°	> 80	10°	15°	≥ 20°
Normal to Ridge for θ ≥ 10°	≤0.25	−0.7	−0.5	−0.3	−0.2	−0.2	0.0[a]							
		−0.18	0.0[a]	0.2	0.3	0.3	0.4	0.4	0.6	0.01 θ	0.8	−0.3	−0.5	−0.6
	0.5	−0.9	−0.7	−0.4	−0.3	−0.2	−0.2	0.0[a]						
		−0.18	−0.18	0.0[a]	0.2	0.2	0.3	0.4	0.6	0.01 θ	0.8	−0.5	−0.5	−0.6
	≥1.0	−1.3[b]	−1.0	−0.7	−0.5	−0.3	−0.2	0.0[a]						
		−0.18	−0.18	−0.18	0.0[a]	0.2	0.2	0.3	0.6	0.01 θ	0.8	−0.7	−0.6	−0.6

Wind Direction	h/L	Horizontal Distance from Windward Edge	C_p
Normal to Ridge for θ < 10° and Parallel to Ridge for all θ	≤ 0.5	0 to $h/2$	−0.9, −0.18
		$h/2$ to h	−0.9, −0.18
		h to $2h$	−0.5, −0.18
		$>2h$	−0.3, −0.18
	≥ 1.0	0 to $h/2$	−1.3[b], −0.18
		$>h/2$	−0.7, −0.18

[a]Value is provided for interpolation purposes.

[b]Value can be reduced linearly, with area over which it is applicable as follows:

Area, ft^2 (m^2)	Reduction Factor
≤100 (9.3)	1.0
250 (23.2)	0.9
≥1,000 (92.9)	0.8

Notes:

1. Plus and minus signs signify pressures acting toward and away from the surfaces, respectively.
2. Linear interpolation is permitted for values of L/B, h/L, and θ, other than shown. Interpolation shall only be carried out between values of the same sign Where no value of the same sign is given, assume 0.0 for interpolation purposes.
3. Where two values of C_p are listed, this indicates that the windward roof slope is subjected to either positive or negative pressures and the roof structure shall be designed for both conditions. Interpolation for intermediate ratios of h/L shall only be carried out between C_p values of like sign.
4. Parapets are shown only on the flat roof elevation but may occur on any roof type shown.
5. For mansard roofs, the top horizontal surface and leeward inclined surface shall be treated as leeward surfaces according to the table
6. Except for MWFRSs at the roof consisting of moment-resisting frames, the total horizontal shear shall not be less than that determined by neglecting wind forces on roof surfaces

Figure 27.3-1 (*Continued*). Main wind force resisting system: external pressure coefficients, C_p, for enclosed, partially enclosed, and partially open buildings—walls and roofs.

Diagrams

External Pressure Coefficients for Domes with Circular Base

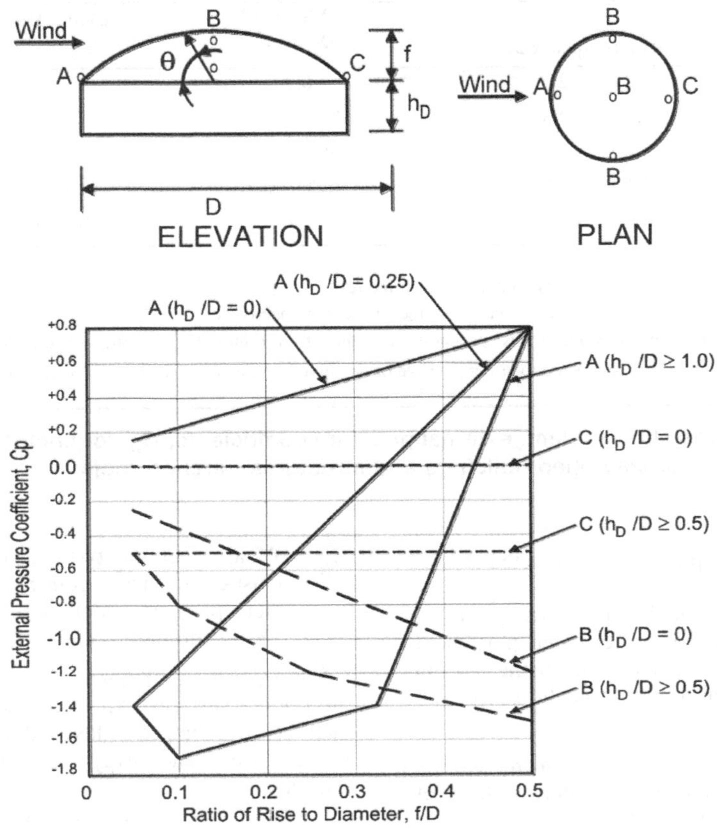

Notation

f = Dome rise, ft (m)

h_D = Height to base of dome, ft (m)

D = Diameter, ft (m)

θ = Angle of plane of roof from horizontal, degrees

Notes

1. Two load cases shall be considered:
   Case A: C_p values between A and B and between B and C shall be determined by linear interpolation along arcs on the dome parallel to the wind direction;
   Case B: C_p shall be the constant value of A for $\theta \leq 25°$ and shall be determined by linear interpolation from 25 degrees to B and from B to C.
2. Values denote C_p to be used with $q_{(h_{D+f})}$, where h_{D+f} is the height at the top of the dome.
3. Plus and minus signs signify pressures acting toward and away from the surfaces, respectively.
4. C_p is constant on the dome surface for arcs of circles perpendicular to the wind direction; for example, the arc passing through $B-B-B$ and all arcs parallel to $B-B-B$
5. For values of h_D/D between those listed on the graph curves, linear interpolation shall be permitted.
6. $\theta = 0$ degrees on dome springline $\theta = 90$ degrees at dome center top point. f is measured from springline to top. ,
7. The total horizontal shear shall not be less than that determined by neglecting wind forces on roof surfaces.
8. For f/D values less than 0.05, use Figure 27.3-1.

Figure 27.3-2. Main wind force resisting system: external pressure coefficients, C_p, for enclosed, partially enclosed, and partially open buildings and structures—domed roofs with a circular base.

External Pressure Coefficient, C_p

Conditions	Rise-to-Span Ratio, r	C_p Windward Quarter	C_p Center Half	C_p Leeward Quarter
Roof on elevated structure	$0 < r < 0.2$	-0.9	$-0.7 - r$	-0.5
	$0.2 \le r < 0.3$	$1.5r - 0.3$ $6r - 2.1$	$-0.7 - r$	-0.5
	$0.3 \le r \le 0.6$	$2.75r - 0.7$	$-0.7 - r$	-0.5
Roof springing from ground level	$0 < r \le 0.6$	$1.4r$	$-0.7 - r$	-0.5

Notes

1. Values listed are for the determination of average loads on main wind-force resisting systems.
2. Plus and minus signs signify pressures acting toward and away from the surfaces, respectively.
3. For wind directed parallel to the axis of the arched roofs, use pressure coefficients from Figure 27.3-1, with wind directed parallel to ridge.
4. Where two values of C_p are listed, the roof is subjected to either positive or negative pressures and the roof structure shall be designed for both conditions.

Figure 27.3-3. Main wind force resisting system: external pressure coefficients, C_p, for enclosed, partially enclosed, and partially open buildings and structures—arched roofs.

2. Net force coefficient of 1.3 applied to the projected area of all the elements; and,
3. q_z where z is set equal to the height above grade of the bottom horizontal surface of the elevated portion of the building plus 25% of the distance from the bottom of the elevated portion of the building to the mean roof height.

27.3.1.1.3 External Loads on Horizontal Bottom Surfaces of the Elevated Building Vertical wind loads on the bottom surfaces of the elevated building shall be calculated using

1. Roof pressure coefficients for the wind direction "normal to the ridge for $\theta < 10°$ and parallel to ridge for all θ" provided in Figure 27.3-1. Negative roof pressure coefficients denote downward loading on the bottom floor assembly, while positive roof pressure coefficients denote upward loading.
2. Value of h, used to define the distance from the windward edge, shall be taken as the height of the top of the elements below the elevated building; and,
3. Velocity pressure, used to calculate external pressures on this floor, shall be q_z, where z is set equal to the height above grade of the bottom horizontal surface of the elevated portion of the building plus 25% of the distance from the bottom of the elevated portion of the building to the mean roof height.

27.3.2 Open Buildings with Monoslope, Pitched, or Troughed Free Roofs The net design pressure for the MWFRS of open buildings with monoslope, pitched, or troughed free roofs in lb/ft^2 (N/m^2), shall be determined by Equation (27.3-2):

$$p = q_h K_d G C_N \quad (27.3\text{-}2)$$

where

q_h = Velocity pressure evaluated at mean roof height, h, using the exposure as defined in Section 26.7.3 that results in the highest wind loads for any wind direction at the site;

K_d = Wind directionality factor (see Section 26.6);

G = Gust-effect factor from Section 26.11; and

C_N = Net pressure coefficient determined from Figures 27.3-4 through 27.3-7.

Net pressure coefficients, C_N, include contributions from top and bottom surfaces. All load cases shown for each roof angle shall be investigated. Plus and minus signs signify pressure acting toward and away from the top surface of the roof, respectively.

For free roofs with an angle of plane of roof from horizontal θ less than or equal to 5 degrees and containing fascia panels, the fascia panel shall be considered to be an inverted parapet. The contribution of loads on the fascia to the MWFRS loads shall be determined using Section 27.3.5, with q_p equal to q_h. For an open or partially enclosed building with transverse frames and a pitched roof ($\theta \le 45°$), an additional horizontal force in the longitudinal direction (parallel to the ridge) that acts in combination with the roof load calculated in Section 27.3.3 shall be determined in accordance with Section 28.3.5.

27.3.3 Roof Overhangs The positive external pressure on the bottom surface of windward roof overhangs shall be determined using $C_p = 0.8$ and combined with the top surface pressures determined using Figure 27.3-1.

27.3.4 Parapets The design wind pressure for the effect of parapets on MWFRS of rigid or flexible buildings in lb/ft^2 (N/m^2) shall be determined by Equation (27.3-3):

$$p_p = q_p K_d (GC_{pn}) \quad (27.3\text{-}3)$$

where

p_p = Combined net pressure on the parapet caused by the combination of the net pressures from the front and back parapet surfaces in which plus (and minus) signs signify net pressure acting toward (and away from) the front (exterior) side of the parapet;

q_p = Velocity pressure evaluated at the top of the parapet;

K_d = Wind directionality factor (see Section 26.6); and

(GC_{pn}) = Combined net pressure coefficient, which is $+1.5$ for windward parapet or -1.0 for leeward parapet.

Diagrams

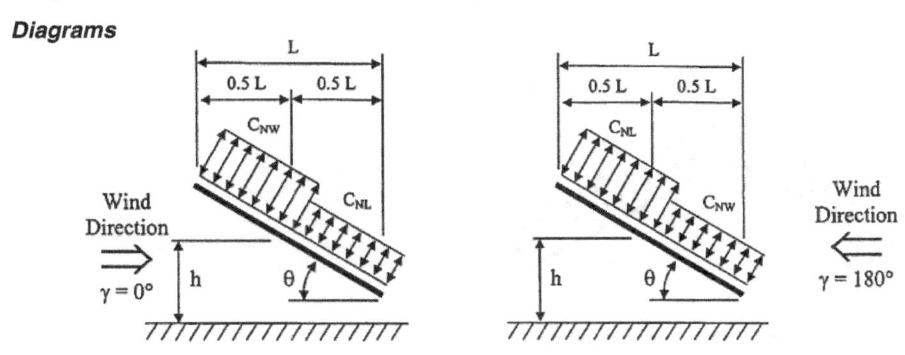

Notation

L = Horizontal dimension of roof, measured in the along-wind direction, ft (m).

h = Mean roof height, ft (m).

γ = Direction of wind, degrees.

θ = Angle of plane of roof from horizontal, degrees.

Net Pressure Coefficient, C_N

Roof Angle, θ	Load Case	Wind Direction, $\gamma = 0°$				Wind Direction, $\gamma = 180°$			
		Clear Wind Flow		Obstructed Wind Flow		Clear Wind Flow		Obstructed Wind Flow	
		C_{NW}	C_{NL}	C_{NW}	C_{NL}	C_{NW}	C_{NL}	C_{NW}	C_{NL}
<7.5°	A	1.2	0.3	–0.5	–1.2	1.2	0.3	–0.5	–1.2
	B	–1.1	–0.1	–1.1	–0.6	–1.1	–0.1	–1.1	–0.6
7.5°	A	–0.6	–1.0	–1.0	–1.5	0.9	1.5	–0.2	–1.2
	B	–1.4	0.0	–1.7	–0.8	1.6	0.3	0.8	–0.3
15°	A	–0.9	–1.3	–1.1	–1.5	1.3	1.6	0.4	–1.1
	B	–1.9	0.0	–2.1	–0.6	1.8	0.6	1.2	–0.3
22.5°	A	–1.5	–1.6	–1.5	–1.7	1.7	1.8	0.5	–1.0
	B	–2.4	–0.3	–2.3	–0.9	2.2	0.7	1.3	0.0
30°	A	–1.8	–1.8	–1.5	–1.8	2.1	2.1	0.6	–1.0
	B	–2.5	–0.5	–2.3	–1.1	2.6	1.0	1.6	0.1
37.5°	A	–1.8	–1.8	–1.5	–1.8	2.1	2.2	0.7	–0.9
	B	–2.4	–0.6	–2.2	–1.1	2.7	1.1	1.9	0.3
45°	A	–1.6	–1.8	–1.3	–1.8	2.2	2.5	0.8	–0.9
	B	–2.3	–0.7	–1.9	–1.2	2.6	1.4	2.1	0.4

Notes

1. C_{NW} and C_{NL} denote net pressures (contributions from top and bottom surfaces) for windward and leeward half of roof surfaces, respectively.
2. Clear wind flow denotes relatively unobstructed wind flow with blockage less than or equal to 50%. Obstructed wind flow denotes objects below roof inhibiting wind flow (>50% blockage).
3. For values of θ between 7.5 and 45 degrees, linear interpolation is permitted.
4. For free roofs with $0.05 \leq h/L < 0.25$ and $\theta < 5$ degrees, use Figure 27.3-7.
5. Plus and minus signs signify pressures acting toward and away from the top roof surface, respectively.
6. All load cases shown for each roof angle shall be investigated.

Figure 27.3-4. Main wind force resisting system ($0.25 \leq h/L \leq 1.0$): net pressure coefficient, C_N, for open buildings with monoslope free roofs ($\theta \leq 45°$, $\gamma = 0°$, $180°$).

Diagrams

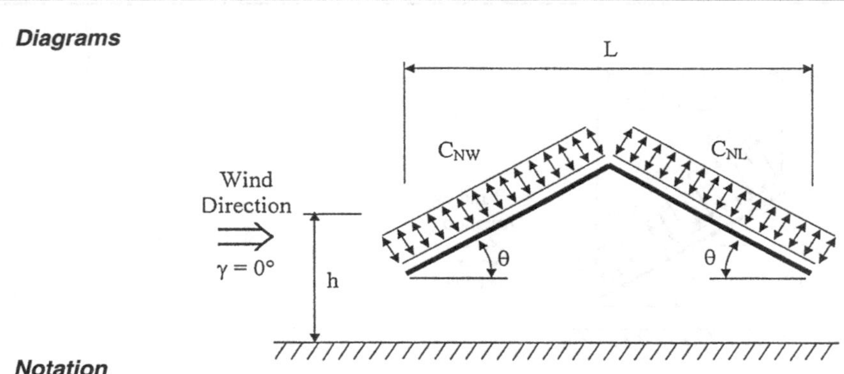

Notation

L = Horizontal dimension of roof, measured in the along-wind direction, ft (m).

h = Mean roof height, ft (m).

γ = Direction of wind, degrees.

θ = Angle of plane of roof from horizontal, degrees.

Net Pressure Coefficient, C_N

Roof Angle, θ	Load Case	Wind Direction, γ = 0°, 180°			
		Clear Wind Flow		Obstructed Wind Flow	
		C_{NW}	C_{NL}	C_{NW}	C_{NL}
7.5°	A	1.1	−0.3	−1.6	−1.0
	B	0.2	−1.2	−0.9	−1.7
15°	A	1.1	−0.4	−1.2	−1.0
	B	0.1	−1.1	−0.6	−1.6
22.5°	A	1.1	0.1	−1.2	−1.2
	B	−0.1	−0.8	−0.8	−1.7
30°	A	1.3	0.3	−0.7	−0.7
	B	−0.1	−0.9	−0.2	−1.1
37.5°	A	1.3	0.6	−0.6	−0.6
	B	−0.2	−0.6	−0.3	−0.9
45°	A	1.1	0.9	−0.5	−0.5
	B	−0.3	−0.5	−0.3	−0.7

Notes

1. C_{NW} and C_{NL} denote net pressures (contributions from top and bottom surfaces) for windward and leeward half of roof surfaces, respectively.
2. Clear wind flow denotes relatively unobstructed wind flow with blockage less than, or equal to, 50%. Obstructed wind flow denotes objects below roof inhibiting wind flow (>50% blockage).
3. For values of θ between 7.5 and 45 degrees, linear interpolation is permitted. For values of θ less than 7.5 degrees, use C_N from Figure 27.3-4.
4. Plus and minus signs signify pressures acting toward and away from the top roof surface, respectively.
5. All load cases shown for each roof angle shall be investigated.

Figure 27.3-5. Main wind force resisting system ($0.25 \le h/L \le 1.0$): net pressure coefficient, C_N, for open buildings with pitched free roofs ($\theta \le 45°$, $\gamma = 0°$, 180°).

Diagrams

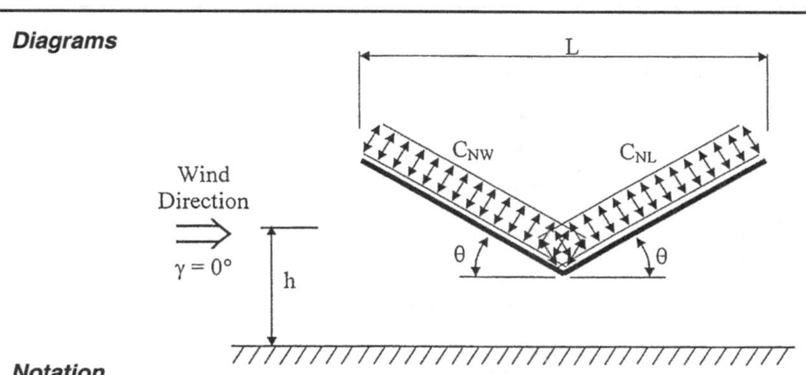

Notation

L = Horizontal dimension of roof, measured in the along-wind direction, ft (m).

h = Mean roof height, ft (m).

γ = Direction of wind, degrees.

θ = Angle of plane of roof from horizontal, degrees.

Net Pressure Coefficient, C_N

Roof Angle, θ	Load Case	Wind Direction, γ = 0°, 180°			
		Clear Wind Flow		Obstructed Wind Flow	
		C_{NW}	C_{NL}	C_{NW}	C_{NL}
7.5°	A	−1.1	0.3	−1.6	−0.5
	B	−0.2	1.2	−0.9	−0.8
15°	A	−1.1	0.4	−1.2	−0.5
	B	0.1	1.1	−0.6	−0.8
22.5°	A	−1.1	−0.1	−1.2	−0.6
	B	−0.1	0.8	−0.8	−0.8
30°	A	−1.3	−0.3	−1.4	−0.4
	B	0.1	0.9	−0.2	−0.5
37.5°	A	−1.3	−0.6	−1.4	−0.3
	B	0.2	0.6	−0.3	−0.4
45°	A	−1.1	−0.9	−1.2	−0.3
	B	0.3	0.5	−0.3	−0.4

Notes

1. C_{NW} and C_{NL} denote net pressures (contributions from top and bottom surfaces) for windward and leeward half of roof surfaces, respectively.
2. Clear wind flow denotes relatively unobstructed wind flow with blockage less than, or equal to, 50%. Obstructed wind flow denotes objects below roof inhibiting wind flow (>50% blockage).
3. For values of θ between 7.5 and 45 degrees, linear interpolation is permitted. For values of θ less than 7.5 degrees, use C_N from Figure 27.3-4.
4. Plus and minus signs signify pressures acting toward and away from the top roof surface, respectively.
5. All load cases shown for each roof angle shall be investigated.

Figure 27.3-6. Main wind force resisting system ($0.25 \leq h/L \leq 1.0$): net pressure coefficient, C_N, for open buildings with troughed free roofs ($\theta \leq 45°$, $\gamma = 0°$, 180°).

Diagrams

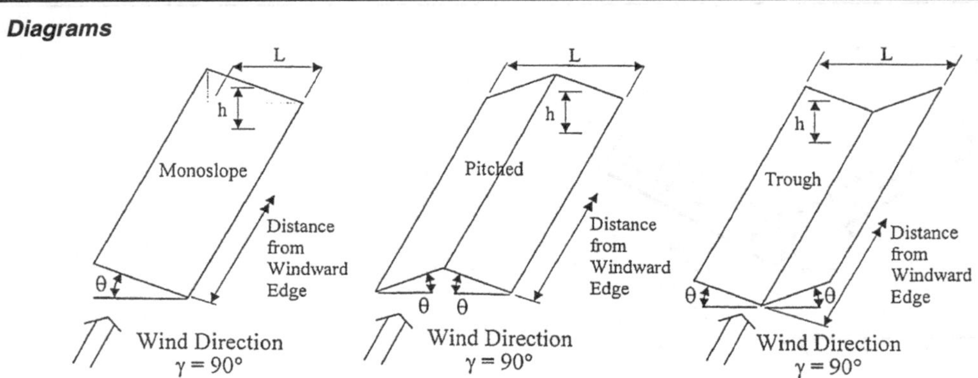

Notation

h = Mean roof height, ft (m). See Figures 27.3-4, 27.3-5, or 27.3-6, for a graphical depiction of this dimension.

γ = Direction of wind, degrees.

θ = Angle of plane of roof from horizontal, degrees.

Net Pressure Coefficient, C_N

Horizontal Distance from Windward Edge	Roof Angle θ	Load Case	Clear Wind Flow C_N	Obstructed Wind Flow C_N
$\leq h$	All shapes	A	−0.8	−1.2
	$\theta \leq 45°$	B	0.8	0.5
$>h, \leq 2h$	All shapes	A	−0.6	−0.9
	$\theta \leq 45°$	B	0.5	0.5
$>2h$	All shapes	A	−0.3	−0.6
	$\theta \leq 45°$	B	0.3	0.3

Notes

1. C_N denotes net pressures (contributions from top and bottom surfaces).
2. Clear wind flow denotes relatively unobstructed wind flow with blockage less than or equal to 50%. Obstructed wind flow denotes objects below roof inhibiting wind flow (>50% blockage).
3. Plus and minus signs signify pressures acting toward and away from the top roof surface, respectively.
4. All load cases shown for each roof angle shall be investigated.

Figure 27.3-7. Main wind force resisting system: net pressure coefficient, C_N, for open buildings with free roofs ($\theta \leq 45°$, $\gamma = 90°$, 270°).

Diagrams

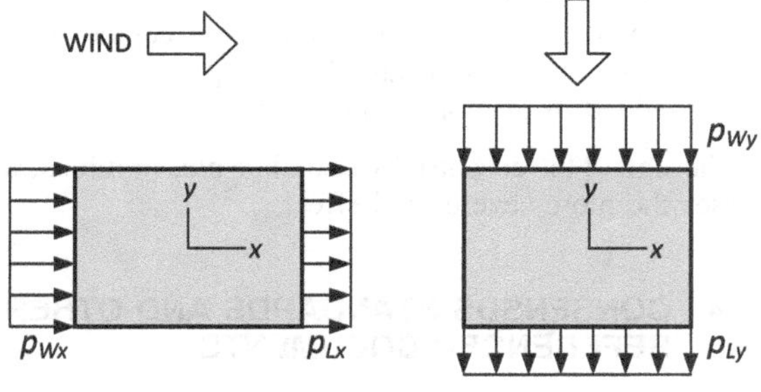

CASE 1

Full design pressure on the projected wall area perpendicular to each principal axis of the structure, considered separately along each principal axis. Full design pressure on side walls and roof areas for wind along each principal axis as specified in Figures 27.3-1 through 27.3-7. All pressures act simultaneously for each principal wind direction.

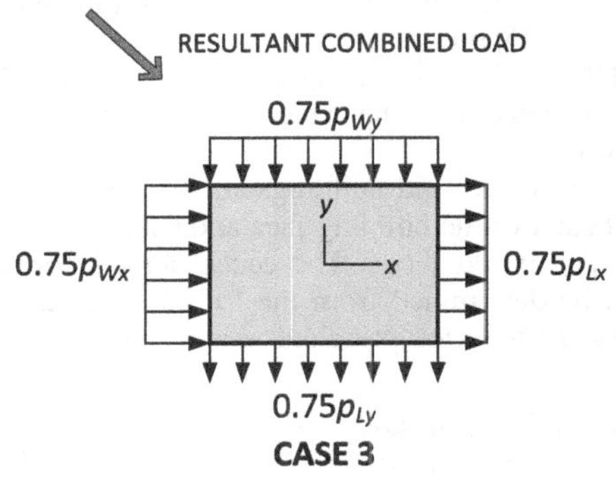

CASE 3

Wall pressures are 75% of Case 1. For roof pressures, see Note 2. All pressures act simultaneously.

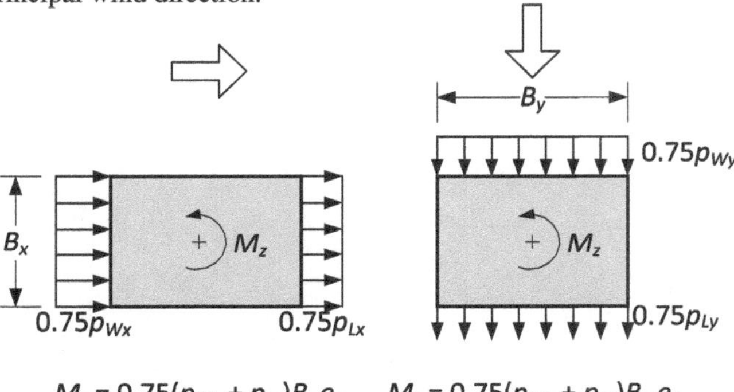

$M_z = 0.75(p_{Wx} + p_{Lx})B_x e_x$ $\quad$ $M_z = 0.75(p_{Wy} + p_{Ly})B_y e_y$

$e_x = \pm 0.15B_x$ $\quad$ $e_y = \pm 0.15B_y$

CASE 2

Three-quarters of design wind pressure on the projected wall area perpendicular to each principal axis of the structure and sidewalls in conjunction with a torsional moment, considered separately along each principal direction. Roof pressures are 75% of Case 1. All pressures and torsion act simultaneously for each principal wind direction.

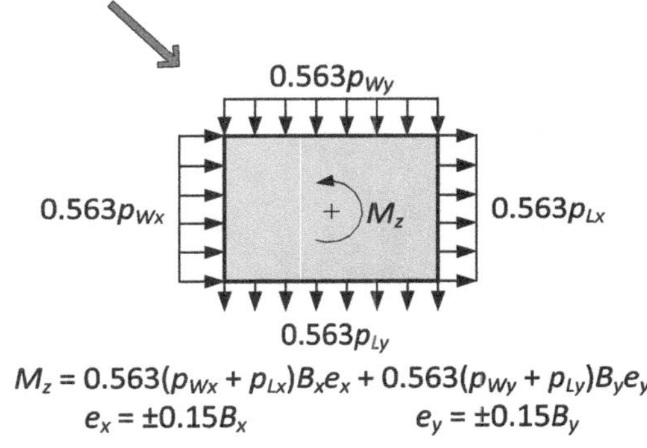

$M_z = 0.563(p_{Wx} + p_{Lx})B_x e_x + 0.563(p_{Wy} + p_{Ly})B_y e_y$

$e_x = \pm 0.15B_x$ $\quad$ $e_y = \pm 0.15B_y$

CASE 4

Wall pressures and torsional moment are 75% of Case 2 (wall pressures are 56.3% of Case 1). For roof pressures see Note 2. All pressures and torsion act simultaneously.

Notation

p_{Wx}, p_{Wy} = Windward wall design pressure acting in the x, y principal direction, respectively.
p_{Lx}, p_{Ly} = Leeward wall design pressure acting in the x, y principal direction, respectively.
B_x, B_y = Horizontal dimension of building normal to the wind in the x, y principal direction, respectively.
e_x, e_y = Eccentricity from the x, y principal axis of the structure, respectively; as shown for rigid buildings, or as defined by Equation (27.3-4) for flexible buildings. See also Note 4.
M_z = Torsional moment per unit height acting about a vertical axis of the building.

Notes

1. Diagrams show plan views of buildings.
2. Pressures on roof are not shown for clarity. For Cases 3 and 4 the resulting pressure on any roof area defined by the two principal wind directions of Cases 1 and 2 shall be 100% of the larger value of the roof pressures defined for Cases 1 and 2, respectively.
3. Pressures on sidewalls for Cases 1 and 2 are not shown for clarity and need not be considered at floors with rigid diaphragms continuous with the sidewalls.
4. M_z shall be applied on rigid diaphragms. On floors with semi-rigid or flexible diaphragms, or without diaphragms, it shall be permitted to apply M_z using an equivalent distributed pressure block on all walls receiving normal wind pressure, applied in the proportion specified for each wall in Figure 27.3-1, or using other rational methods.

Figure 27.3-8. Main wind force resisting system, design wind load cases.

27.3.5 Design Wind Load Cases The MWFRS of buildings of all heights, the wind loads of which have been determined under the provisions of this chapter, shall be designed for the wind load cases as defined in Figure 27.3-8.

EXCEPTION: Buildings meeting the requirements of Section D.1 of Appendix D need only be designed for Case 1 and Case 3 of Figure 27.3-8.

The eccentricity e for rigid buildings shall be measured from the geometric center of the building face and shall be considered for each principal axis (e_X,e_Y). The eccentricity e for flexible buildings shall be determined from the following equation and shall be considered for each principal axis (e_X,e_Y):

$$e = \frac{e_Q + 1.7 I_z \sqrt{(g_Q Q e_Q)^2 + (g_R R e_R)^2}}{1 + 1.7 I_z \sqrt{(g_Q Q)^2 + (g_R R)^2}} \quad (27.3\text{-}4)$$

where

e_Q = Eccentricity e as determined for rigid buildings in Figure 27.3-8; and

e_R = Distance between the elastic shear center and center of mass of each floor, and I_z, g_Q, Q, g_R, and R shall be as defined in Section 26.11.

The sign of the eccentricity e shall be plus or minus, whichever causes the more severe load effect.

27.4 CONSENSUS STANDARDS AND OTHER REFERENCED DOCUMENTS

No consensus standards and other documents that shall be considered part of this standard are referenced in this chapter.

CHAPTER 28

WIND LOADS ON BUILDINGS: MAIN WIND FORCE RESISTING SYSTEM (ENVELOPE PROCEDURE)

28.1 SCOPE

28.1.1 Building Types This chapter applies to the determination of Main Wind Force Resisting System (MWFRS) wind loads, using the Envelope Procedure, on low-rise buildings as defined in Section 26.2.

28.1.2 Conditions The design wind loads determined in accordance with this section shall apply to buildings complying with all the following conditions:

1. Building is a regular-shaped building as defined in Section 26.2.
2. Building does not have response characteristics that make it subject to across-wind loading, vortex shedding, or instability caused by galloping or flutter, nor does it have a site location for which channeling effects or buffeting in the wake of upwind obstructions warrant special consideration.
3. Building does not have an arched, barrel, or unusually shaped roof.

28.1.3 Limitations The provisions of this chapter take into consideration mean and fluctuating loads caused by winds acting on rigid buildings. Buildings that do not meet the conditions of Section 28.1.2 shall be designed using appropriate procedures in Chapter 27, recognized literature documenting such wind load effects, or the Wind Tunnel Procedure specified in Chapter 31.

28.1.4 Shielding There shall be no reductions in velocity pressure caused by shielding that may be afforded by other buildings and other structures or terrain features.

User Note: Use Chapter 28 to determine the wind pressure on the MWFRS of *low-rise buildings* that have a flat, gable, or hip roof. These provisions use the Envelope Procedure by calculating wind pressures from the *specific equation* applicable to each building surface. For building shapes and heights for which these provisions are applicable, this method generally yields the lowest wind pressure of all the analytical methods specified in this standard.

28.2 GENERAL REQUIREMENTS

The steps required for the determination of MWFRS wind loads on low-rise buildings are shown in Table 28.2-1.

Table 28.2-1. Steps to Determine Wind Loads on MWFRS Low-Rise Buildings.

Step 1: Determine risk category of building; see Table 1.5-1.
Step 2: Determine the basic wind speed, V, for applicable risk category; see Figure 26.5-1.
Step 3: Determine wind load parameters:

- Wind directionality factor, K_d; see Section 26.6 and Table 26.6-1.
- Exposure Category (B, C, or D); see Section 26.7.
- Topographic factor, K_{zt}; see Section 26.8 and Figure 26.8-1.
- Ground elevation factor, K_e; see Section 26.9 and Table 26.9-1.
- Enclosure classification; see Section 26.12.
- Internal pressure coefficient, (GC_{pi}); see Section 26.13 and Table 26.13-1.

Step 4: Determine velocity pressure exposure coefficient, K_z or K_h; see Table 26.10-1.
Step 5: Determine velocity pressure, q_z or q_h, from Equation (26.10-1).
Step 6: Determine external pressure coefficient, (GC_{pf}), for each load case using Section 28.3.2 for flat and gable roofs.

User Note: See Commentary Figure C28.3-2 for guidance on hip roofs.

Step 7: Calculate wind pressure, p, from Equation (28.3-1).

28.2.1 Wind Load Parameters Specified in Chapter 26 The following wind load parameters shall be determined in accordance with Chapter 26:

- Basic wind speed, V (Section 26.5);
- Wind directionality factor, K_d (Section 26.6);
- Exposure category (Section 26.7);
- Topographic factor, K_{zt} (Section 26.8);
- Ground elevation factor, K_e (Section 26.9);
- Velocity pressure exposure coefficient, K_z or K_h (Section 26.10);
- Enclosure classification (Section 26.12); and
- Internal pressure coefficient, (GC_{pi}) (Section 26.13).

28.3 WIND LOADS: MAIN WIND FORCE RESISTING SYSTEM

28.3.1 Design Wind Pressure for Low-Rise Buildings Design wind pressures for the MWFRS of low-rise buildings shall be determined by the following equation:

$$p = q_h K_d[(GC_{pf}) - (GC_{pi})](\mathrm{lb/ft^2}) \quad (28.3\text{-}1)$$

$$p = q_h K_d[(GC_{pf}) - (GC_{pi})](\mathrm{N/m^2}) \quad (28.3\text{-}1.\mathrm{SI})$$

where

q_h = Velocity pressure evaluated at mean roof height h, as defined in Section 26.3;
K_d = Wind directionality factor, see Section 26.6;
(GC_{pf}) = External pressure coefficient from Section 28.3.2; and
(GC_{pi}) = Internal pressure coefficient from Table 26.13-1.

28.3.1.1 External Pressure Coefficients, (GC_{pf}) The combined gust-effect factor and external pressure coefficients used in this chapter, (GC_{pf}), are not permitted to be separated.

28.3.2 Load Cases Buildings shall be evaluated using each of the basic load cases and the torsional load cases acting independently in accordance with this section.

EXCEPTION: Design for the torsional cases is not required for

1. One-story buildings with h less than or equal to 30 ft (9.1 m),
2. One-story and two-story light-frame buildings, or
3. One-story and two-story buildings with flexible diaphragms.

28.3.2.1 Basic Load Cases The external pressure coefficient, (GC_{pf}), for basic load cases shall be determined in accordance with Figure 28.3-1. The building shall be evaluated with each corner taken as the windward corner, with loading patterns applied as shown, and with all zones being loaded simultaneously. Each corner shall be evaluated separately for the two load cases. In each load case, Zones 2E and 3E shall be located along the roof edge perpendicular to the ridge that is nearest the corner being evaluated. Combinations of external and internal pressures (see Table 26.13-1) shall be evaluated as required to obtain the most severe loadings.

For flat roofs, the roof angle, θ, shall be taken as 0, with the Zone 2/3 and Zone 2E/3E boundary located at the mid-width of the building.

The roof pressure coefficient, (GC_{pf}), when negative in Zones 2 and 2E, shall be applied in Zone 2/2E for a distance from the edge of roof equal to 0.5 times the horizontal dimension of the building parallel to the direction of the MWFRS being designed or 2.5 times the eave height at the windward wall, whichever is less; the remainder of Zone 2/2E extending to the ridge line shall use the pressure coefficient (GC_{pf}) for Zone 3/3E.

28.3.2.2 Torsional Load Cases The external pressure coefficient, (GC_{pf}), for torsional load cases shall be determined in accordance with Figure 28.3-2. For Zones 1 to 6 and 1E to 6E, Load Case 3 shall use Load Case 1 pressure coefficients from Figure 28.3-1, and Load Case 4 shall use Load Case 2 pressure coefficients from Figure 28.3-1. For zones designated with a "T" (1T to 6T), the pressure coefficients from Figure 28.3-2 shall be used.

The building shall be evaluated with each corner taken as the windward corner with loading patterns applied as shown and with all zones being loaded simultaneously. Each corner shall be evaluated separately for the two load cases. In each load case, Zones 2E and 3E shall be located along the roof edge perpendicular to the ridge that is nearest the corner being evaluated. Combinations of external and internal pressures (see Table 26.13-1) shall be evaluated as required to obtain the most severe loadings.

For flat roofs, the roof angle, θ, shall be taken as 0 with the Zone 2/3 and Zone 2E/3E boundary at the mid-width of the building. The roof pressure coefficient, (GC_{pf}), when negative in Zones 2 and 2E, shall be applied in Zone 2/2E for a distance from the edge of roof equal to 0.5 times the horizontal dimension of the building parallel to the direction of the MWFRS being designed or 2.5 times the eave height at the windward wall, whichever is less; the remainder of Zone 2/2E extending to the ridge line shall use the pressure coefficient, (GC_{pf}), for Zone 3/3E.

28.3.3 Total Horizontal Load The total horizontal shear shall not be less than that determined by neglecting the wind forces on the roof.

EXCEPTION: This provision does not apply to buildings using moment frames for the MWFRS.

28.3.4 Parapets The design wind pressure for the effect of parapets on MWFRS of low-rise buildings with flat, gable, or hip roofs shall be determined by the following equation:

$$p_p = q_p K_d(GC_{pn})(\mathrm{lb/ft^2}) \quad (28.3\text{-}2)$$

$$p_p = q_p K_d(GC_{pn})(\mathrm{N/m^2}) \quad (28.3\text{-}2.\mathrm{SI})$$

where

p_p = Combined net pressure on the parapet caused by the combination of the net pressures from the front and back parapet surfaces; plus (and minus) signs signify net pressure acting toward (and away from) the front (exterior) side of the parapet;
q_p = Velocity pressure evaluated at the top of the parapet;
K_d = Wind directionality factor, see Section 26.6; and
(GC_{pn}) = Combined net pressure coefficient: +1.5 for windward parapet, −1.0 for leeward parapet.

28.3.5 Roof Overhangs The positive external pressure on the bottom surface of windward roof overhangs shall be determined using $GC_p = 0.7$ in combination with the top surface pressures determined using Figure 28.3-1.

28.3.6 Minimum Design Wind Loads The wind pressure to be used in the design of the MWFRS for an enclosed or partially enclosed building shall not be less than 16 lb/ft² (0.77 kN/m²) multiplied by the wall area of the building, and 8 lb/ft² (0.38 kN/m²) multiplied by the roof area of the building projected onto a vertical plane normal to the assumed wind direction.

28.3.7 Horizontal Wind Loads on Open or Partially Enclosed Buildings with Transverse Frames and Pitched Roofs A horizontal pressure in the longitudinal direction (parallel to the ridge) that acts in combination with the roof load calculated in Section 27.4.3 for an open or partially enclosed building with transverse frames and a pitched roof (θ < 45 degrees) shall be determined by the following equation:

$$p = q_h K_d[(GC_{pf})_{\mathrm{windward}} - (GC_{pf})_{\mathrm{leeward}}]K_B K_S \quad (28.3\text{-}3)$$

where

q_h = Velocity pressure evaluated at mean roof height h using the exposure as defined in Section 26.7.3;

Basic Load Cases

Diagrams

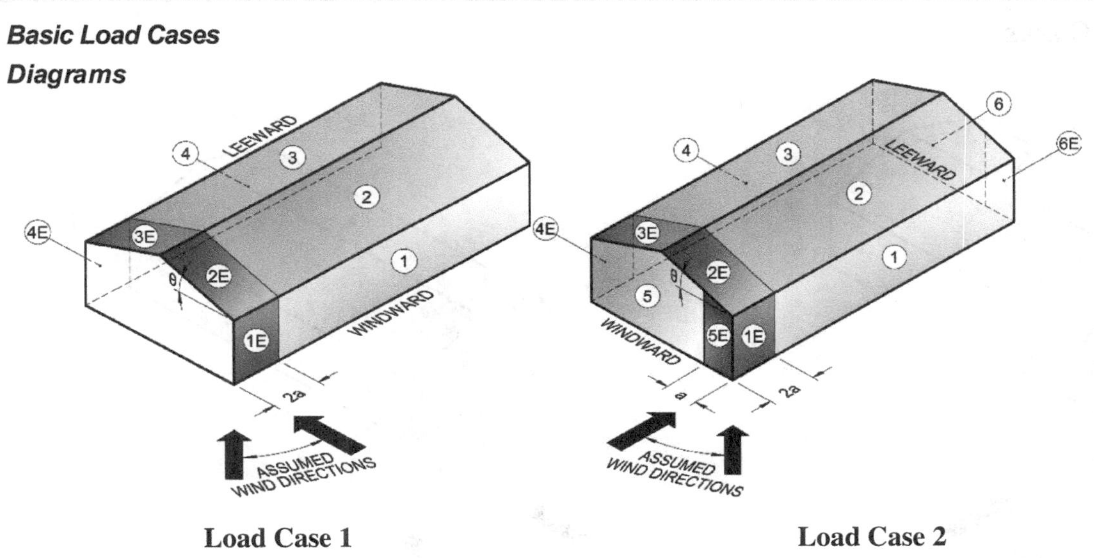

Notation

a = 10% of least horizontal dimension or 0.4 h,whichever is smaller, but not less than either 4% of least horizontal dimension or 3 ft (0.9 m).
EXCEPTION: For buildings with $\theta = 0$ to 7° and a least horizontal dimension greater than 300 ft (90 m), dimension a shall be limited to a maximum of 0.8 h.

h = Mean roof height, ft (m), except that eave height shall be used for $\theta \leq 10°$.

θ = Angle of plane of roof from horizontal, in degrees.

Load Case 1

Roof Angle θ (degrees)	Building Surface							
	1	2	3	4	1E	2E	3E	4E
0–5	0.40	−0.69	−0.37	−0.29	0.61	−1.07	−0.53	−0.43
20	0.53	−0.69	−0.48	−0.43	0.80	−1.07	−0.69	−0.64
30–45	0.56	0.21	−0.43	−0.37	0.69	0.27	−0.53	−0.48
90	0.56	0.56	−0.37	−0.37	0.69	0.69	−0.48	−0.48

Load Case 2

Roof Angle θ (degrees)	Building Surface											
	1	2	3	4	5	6	1E	2E	3E	4E	5E	6E
0–90	−0.45	−0.69	−0.37	−0.45	0.40	−0.29	−0.48	−1.07	−0.53	−0.48	0.61	−0.43

Notes

1. Plus and minus signs signify pressures acting toward and away from the surfaces, respectively.
2. For values of θ other than those shown, linear interpolation is permitted.

Figure 28.3-1. Basic load cases for main wind force resisting system [$h \leq 60$ ft ($h \leq 18.3$ m)]: external pressure coefficients, (GC_{pf}), for enclosed, partially enclosed, and partially open buildings—low-rise walls and roofs.

Torsional Load Cases

Diagrams

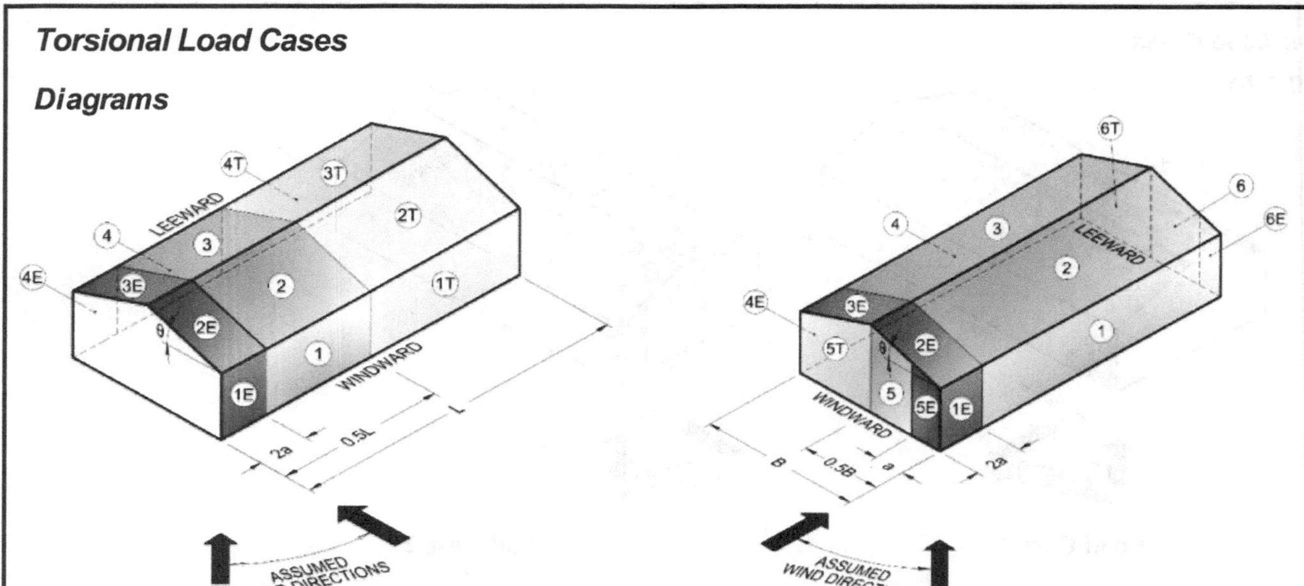

Load Case 3 **Load Case 4**

Notation

a = 10% of least horizontal dimension or 0.4 h, whichever is smaller, but not less than either 4% of least horizontal dimensionor 3ft (0.9 m).

EXCEPTION: For buildings with $\theta = 0$ to 7° and a least horizontal dimension greater than 300 ft (90 m), dimension a shall be limited to a maximum of 0.8 h.

h = Mean roof height, in feet (meters), except that eave height shall be used for $\theta \leq 10°$.

θ = Angle of plane of roof from horizontal, in degrees.

Load Case 3

	Building Surface			
Roof Angle θ (degrees)	**1T**	**2T**	**3T**	**4T**
0–5	0.10	–0.17	–0.09	–0.07
20	0.13	–0.17	–0.12	–0.11
30–45	0.14	0.05	–0.11	–0.09
90	0.14	0.14	–0.09	–0.09

Load Case 4

	Building Surface	
Roof Angle θ (degrees)	**5T**	**6T**
0–90	0.10	–0.07

Notes

1. Plus and minus signs signify pressures acting toward and away from the surfaces, respectively.
2. For values of θ other than those shown, linear interpolation is permitted.

Figure 28.3-2. Torsional load cases for main wind force resisting system [$h \leq 60$ ft ($h \leq 18.3$ m)]: external pressure coefficients, (GC_{pf}), for enclosed, partially enclosed, and partially open buildings—low-rise walls and roofs.

Diagram

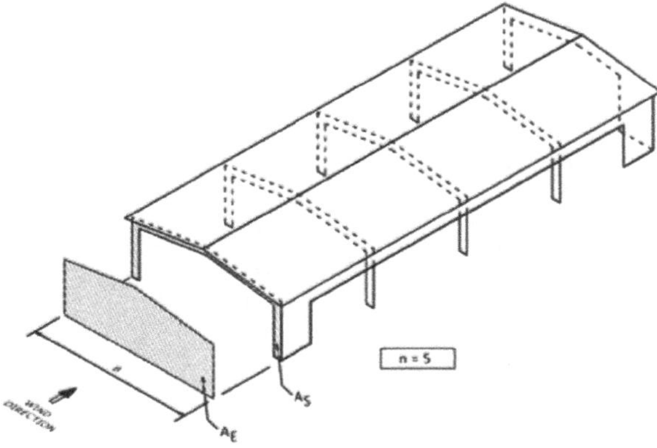

Notation

B = Width of the building perpendicular to the ridge, ft (m)

A_S = Effective solid area of the end wall (i.e., the projected area of any portion of the end wall that would be exposed to the wind)

A_E = Total end wall area for an equivalent enclosed building

n = Number of frames, but shall not be taken as less than 3, even for 2-frame building.

Figure 28.3-3. Horizontal wind loads on open or partially enclosed buildings with transverse frames and pitched roofs: definitions of geometric terminology.

K_d = Wind directionality factor, see Section 26.6;

(GC_{pf}) = External pressure coefficient given in Figure 28.3-1 for Load Case 2, where building surfaces 5 and 5E shall be used to compute the average windward end wall pressure and building surfaces 6 and 6E shall be used to compute the average leeward end wall pressure;

K_B = Frame width factor = $1.8–0.01B$, $B < 100$ ft ($B < 30.5$ m), or 0.8, $B \geq 100$ ft ($B \geq 30.5$ m);

K_S = Shielding factor = $0.60 + 0.073(n - 3) + (1.25\phi^{1.8})$;

ϕ = Solidity ratio = A_S/A_E;

B = Width of the building perpendicular to the ridge, ft (m);

n = Number of frames, but shall not be taken as less than 3, even for small 2-frame building;

A_S = Effective solid area of the end wall, that is, the projected area of any portion of the end wall that would be exposed to the wind (Figure 28.3-3); and

A_E = Total end wall area for an equivalent enclosed building (Figure 28.3-3).

The total longitudinal force F to be resisted by the MWFRS shall be determined by Equation (28.3-4):

$$F = pA_E \qquad (28.3\text{-}4)$$

Equation (28.3-3) is applicable to buildings with open end walls and with end walls fully or partially enclosed with cladding. For all cases, A_E shall be the area that is equivalent to the end wall fully enclosed. The longitudinal force, F, given by Equation (28.3-4) represents the total force for which the MWFRS longitudinal bracing shall be designed. The distribution to each sidewall shall be based on the force F applied at the centroid of the end wall area A_E.

Fascia load need not be considered separately if fascia areas are included in the A_S calculation.

28.4 CONSENSUS STANDARDS AND OTHER REFERENCED DOCUMENTS

No consensus standards and other documents that shall be considered part of this standard are referenced in this chapter.

CHAPTER 29

WIND LOADS ON BUILDING APPURTENANCES AND OTHER STRUCTURES: MAIN WIND FORCE RESISTING SYSTEM (DIRECTIONAL PROCEDURE)

29.1 SCOPE

29.1.1 Structure Types This chapter applies to the determination of wind loads on building appurtenances (such as rooftop structures and rooftop equipment) and other structures of all heights (such as solid freestanding walls and freestanding solid signs, chimneys, tanks, circular bins, silos, open signs, single-plane open frames, ground-mounted fixed-tilt solar panel systems, and trussed towers) using the Directional Procedure.

The steps required for the determination of wind loads on building appurtenances and other structures are shown in Table 29.1-1. The steps required to determine wind loads on the Main Wind Force Resisting System (MWFRS) on circular bins, silos, and tanks are shown in Table 29.1-2.

User Note: Use Chapter 29 to determine wind pressures on the MWFRS of solid freestanding walls, freestanding solid signs, chimneys, tanks, open signs, single-plane open frames, ground-mounted fixed-tilt solar panel systems, and trussed towers. Wind loads on rooftop structures and equipment may be determined from the provisions of this chapter. The wind pressures are calculated using specific equations, based upon the Directional Procedure.

29.1.2 Conditions An appurtenance or structure that has design wind loads, determined in accordance with this section, shall comply with all of the following conditions:

1. Structure is a regular-shaped structure as defined in Section 26.2; and
2. Structure does not have response characteristics making it subject to across-wind loading, vortex shedding, or instability caused by galloping or flutter; nor does it have a site location for which channeling effects or buffeting in the wake of upwind obstructions warrant special consideration.

29.1.3 Limitations The provisions of this chapter take into consideration the load magnification effect caused by gusts in resonance with along-wind vibrations of flexible structures. Structures that do not meet the requirements of Section 29.1.2, or that have unusual shapes or response characteristics, shall be designed using recognized literature documenting such wind load effects or shall use the Wind Tunnel Procedure specified in Chapter 31.

29.1.4 Shielding There shall be no reductions in velocity pressure caused by apparent shielding afforded by buildings and other structures or terrain features.

Table 29.1-1. Steps to Determine Wind Loads on MWFRS Rooftop Equipment and Other Structures.

Step 1: Determine the Risk Category of building or other structure; see Table 1.5-1.

Step 2: Determine the basic wind speed, V, for applicable Risk Category; see Figure 26.5-1.

Step 3: Determine wind load parameters:

- Wind directionality factor, K_d; see Section 26.6 and Table 26.6-1.
- Exposure Category B, C, or D; see Section 26.7.
- Topographic factor, K_{zt}; see Section 26.8 and Figure 26.8-1.
- Ground elevation factor, K_e; see Section 26.9 and Table 26.9-1.
- Gust-effect factor, G; see Section 26.11, except for rooftop equipment and ground-mounted fixed-tilt solar panel systems.
- Combined (GC_r) factor for rooftop equipment; see Section 29.4.1.
- Combined (GC_{pn} and GC_{pm}) factors for ground-mounted fixed-tilt solar panel systems; see Section 29.4.5.

Step 4: Determine velocity pressure exposure coefficient, K_z or K_h; see Table 26.10-1.

Step 5: Determine velocity pressure q_z or q_h; see Equation (26.10-1).

Step 6: Determine force coefficient, C_f, except for rooftop equipment and ground-mounted fixed-tilt solar panel systems:

- Solid freestanding signs or solid freestanding walls, Figure 29.3-1.
- Chimneys, tanks, Figure 29.4-1.
- Open signs, single-plane open frames, Figure 29.4-2.
- Trussed towers, Figure 29.4-3.
- Rooftop equipment, using combined (GC_r) factors listed in Section 29.4.1.
- Rooftop solar panels, Figure 29.4-7 and Equation (29.4-6), or Figure 29.4-8.
- Ground-mounted fixed-tilt solar panel systems, using combined (GC_{pn} and GC_{pm}) factors, see Section 29.4.5.2.

Step 7: Calculate wind force, F, or pressure, p, and moment, M:

- Equation (29.3-1) for signs and walls.
- Equations (29.4-2) and (29.4-3) for rooftop structures and equipment.
- Equation (29.4-1) for other structures.
- Equation (29.4-5) or (29.4-7) for rooftop solar panels.
- Equations (29.4-8) and (29.4-9) for ground-mounted fixed-tilt solar panel systems.

Table 29.1-2. Steps to Determine Wind Loads on MWFRS Circular Bins, Silos, and Tanks.

Step 1: Determine the Risk Category of structure; see Table 1.5-1.
Step 2: Determine the basic wind speed, V, for applicable Risk Category; see Figure 26.5-1.
Step 3: Determine wind load parameters:
- Wind directionality factor, K_d; see Section 26.6 and Table 26.6-1.
- Exposure Category B, C, or D; see Section 26.7.
- Topographic factor, K_{zt}; see Section 26.8 and Figure 26.8-1.
- Ground elevation factor, K_e; see Section 26.9 and Table 26.9-1.
- Enclosure classification, see Section 26.12.
- Internal pressure coefficient, (GC_{pi}), see Table 26.13-1.
- Gust-effect factor, G; see Section 26.11.

Step 4: Determine velocity pressure exposure coefficient, K_z or K_h; see Table 26.10-1.
Step 5: Determine velocity pressure q_h; see Equation (26.10-1).
Step 6: Determine force coefficient for walls, see Sections 29.4.2.1 and 29.4.2.4.
Step 7: Determine external pressure coefficient (GC_p) for roofs and undersides if elevated, see Sections 29.4.2.2 and 29.4.2.3.
Step 8: Calculate wind force, F, or pressure, p:
- Equation (29.4-1) for walls.
- Equation (29.4-4) for roofs.

29.2 GENERAL REQUIREMENTS

29.2.1 Wind Load Parameters Specified in Chapter 26 The following wind load parameters shall be determined in accordance with Chapter 26:

- Basic wind speed, V (Section 26.5);
- Wind directionality factor, K_d (Section 26.6);
- Exposure category (Section 26.7);
- Topographic factor, K_{zt} (Section 26.8);
- Ground elevation factor, K_e (Section 26.9); and
- Enclosure classification (Section 26.12).

29.3 DESIGN WIND LOADS: SOLID FREESTANDING WALLS AND SOLID SIGNS

29.3.1 Solid Freestanding Walls and Solid Freestanding Signs The design wind force for solid freestanding walls and solid freestanding signs shall be determined by the following formula:

$$F = q_h K_d G C_f A_s \text{ (lb)} \quad (29.3\text{-}1)$$

$$F = q_h K_d G C_f A_s \text{ (N)} \quad (29.3\text{-}1.\text{SI})$$

where

q_h = Velocity pressure evaluated at height h (defined in Figure 29.3-1) as determined in accordance with Section 26.10;
K_d = Wind directionality factor, see Section 26.6;
G = Gust-effect factor from Section 26.11;
C_f = Net force coefficient from Figure 29.3-1; and
A_s = Gross area of the solid freestanding wall or freestanding solid sign, ft^2 (m^2).

29.3.2 Solid Attached Signs The design wind pressure on a solid sign attached to the wall of a building, where the plane of the sign is parallel to and in contact with the plane of the wall, and the sign does not extend beyond the side or top edges of the wall, shall be determined using procedures for wind pressures on walls in accordance with Chapter 30 and setting the internal pressure coefficient, (GC_{pi}), equal to 0.

This procedure shall also be applicable to solid signs attached to but not in direct contact with the wall, provided that the gap between the sign and wall is no more than 3 ft (0.9 m) and the edge of the sign is at least 3 ft (0.9 m) in from free edges of the wall (i.e., side and top edges and bottom edges of elevated walls).

29.4 DESIGN WIND LOADS: OTHER STRUCTURES

The design wind force for other structures (chimneys, tanks, open signs, single-plane open frames, and trussed towers), whether ground or roof mounted, shall be determined by the following equation:

$$F = q_z K_d G C_f A_f \text{ (lb)} \quad (29.4\text{-}1)$$

$$F = q_z K_d G C_f A_f \text{ (N)} \quad (29.4\text{-}1.\text{SI})$$

where

q_z = Velocity pressure evaluated at height z, as defined in Section 26.10, of the centroid of area A_f;
K_d = Wind directionality factor, see Section 26.6;
G = Gust-effect factor from Section 26.11;
C_f = Force coefficients from Figures 29.4-1 through 29.4-4; and
A_f = Projected area normal to the wind except where C_f is specified for the actual surface area, ft^2 (m^2).

Determination of G, C_f, and A_f for structures found in petrochemical and other industrial facilities that are not otherwise addressed in ASCE 7 is permitted in accordance with *Wind Loads for Petrochemical and Other Industrial Facilities* (ASCE 2011). Determination of G, C_f, and A_f for lighting system support poles is permitted in accordance with ASCE 72.

29.4.1 Rooftop Structures and Equipment for Buildings The lateral force, F_h, and vertical force, F_v, for rooftop structures and equipment, except as otherwise specified for roof-mounted solar panels (Sections 29.4.3 and 29.4.4) and structures identified in Section 29.4, shall be determined as specified.

The resultant lateral force, F_h, shall be determined from Equation (29.4-2) and applied at a height above the roof surface equal to or greater than the centroid of the projected area, A_f

$$F_h = q_h K_d (GC_r) A_f \text{ (lb)} \quad (29.4\text{-}2)$$

$$F_h = q_h K_d (GC_r) A_f \text{ (N)} \quad (29.4\text{-}2.\text{SI})$$

where

(GC_r) = 1.9 for rooftop structures and equipment with A_f less than $0.1Bh$; (GC_r) shall be permitted to be reduced linearly from 1.9 to 1.0 as the value of A_f is increased from $0.1Bh$ to Bh;
q_h = Velocity pressure evaluated at mean roof height of the building;
K_d = Wind directionality factor, see Section 26.6; and
A_f = Vertical projected area of the rooftop structure or equipment on a plane normal to the direction of wind, ft^2 (m^2).

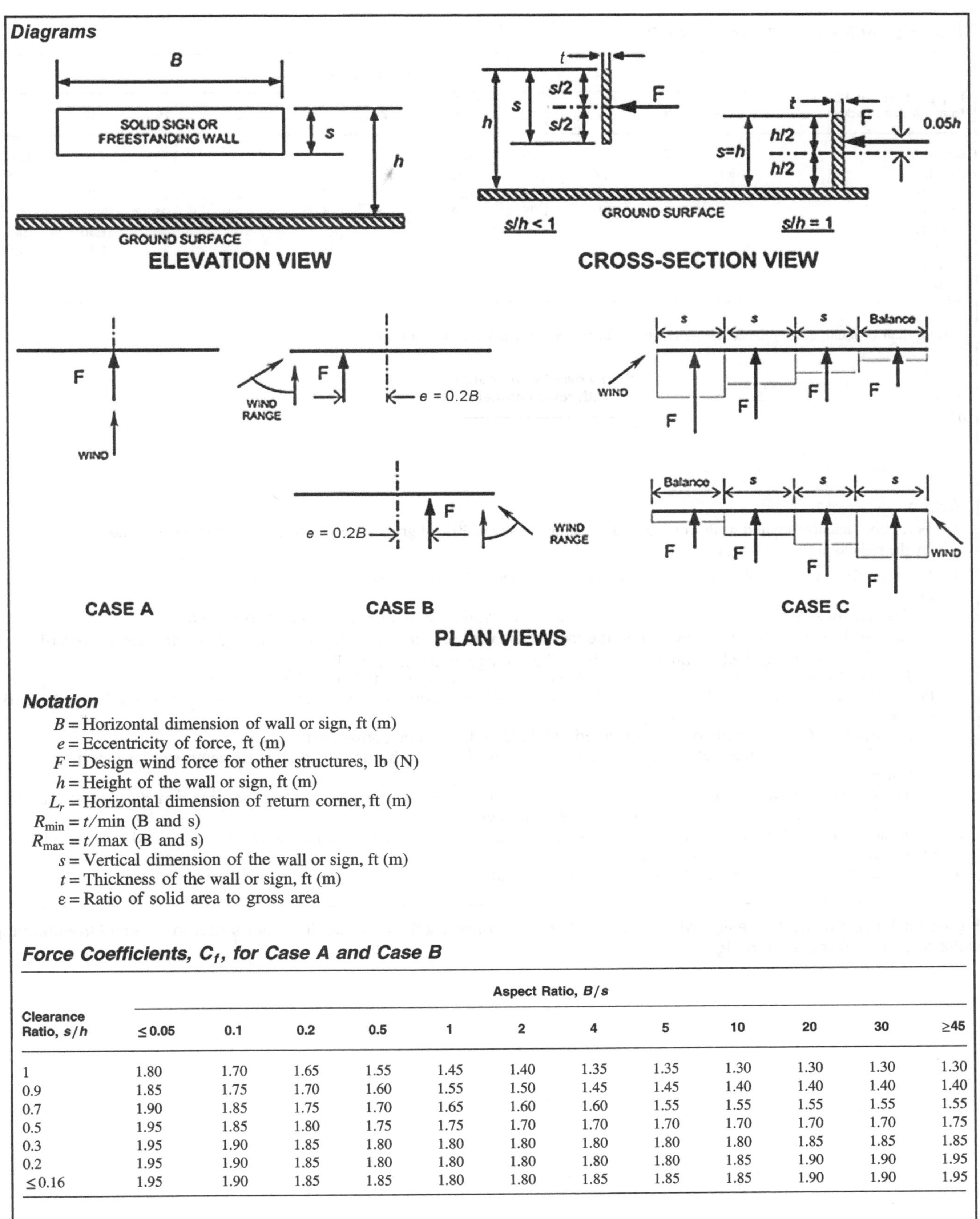

Notation

B = Horizontal dimension of wall or sign, ft (m)
e = Eccentricity of force, ft (m)
F = Design wind force for other structures, lb (N)
h = Height of the wall or sign, ft (m)
L_r = Horizontal dimension of return corner, ft (m)
R_{min} = t/min (B and s)
R_{max} = t/max (B and s)
s = Vertical dimension of the wall or sign, ft (m)
t = Thickness of the wall or sign, ft (m)
ε = Ratio of solid area to gross area

Force Coefficients, C_f, for Case A and Case B

Clearance Ratio, s/h	Aspect Ratio, B/s											
	≤0.05	0.1	0.2	0.5	1	2	4	5	10	20	30	≥45
1	1.80	1.70	1.65	1.55	1.45	1.40	1.35	1.35	1.30	1.30	1.30	1.30
0.9	1.85	1.75	1.70	1.60	1.55	1.50	1.45	1.45	1.40	1.40	1.40	1.40
0.7	1.90	1.85	1.75	1.70	1.65	1.60	1.60	1.55	1.55	1.55	1.55	1.55
0.5	1.95	1.85	1.80	1.75	1.75	1.70	1.70	1.70	1.70	1.70	1.70	1.75
0.3	1.95	1.90	1.85	1.80	1.80	1.80	1.80	1.80	1.80	1.85	1.85	1.85
0.2	1.95	1.90	1.85	1.80	1.80	1.80	1.80	1.80	1.85	1.90	1.90	1.95
≤0.16	1.95	1.90	1.85	1.85	1.80	1.80	1.85	1.85	1.85	1.90	1.90	1.95

Figure 29.3-1. Design wind loads (all heights): force coefficients, C_f, for other structures—solid freestanding walls and solid freestanding signs.

Force Coefficients, C_f, for Case C

Region (horizontal distance from windward edge)	Aspect Ratio, B/s										
	2	3	4	5	6	7	8	9	10	13	≥45
0 to s	2.25	2.60	2.90	3.10*	3.30*	3.40*	3.55*	3.65*	3.75*	4.00*	4.30*
s to $2s$	1.50	1.70	1.90	2.00	2.15	2.25	2.30	2.35	2.45	2.60	2.55
$2s$ to $3s$		1.15	1.30	1.45	1.55	1.65	1.70	1.75	1.85	2.00	1.95
$3s$ to $10s$			1.10	1.05	1.05	1.05	1.05	1.00	0.95		
$3s$ to $4s$										1.50	1.85
$4s$ to $5s$										1.35	1.85
$5s$ to $10s$										0.90	1.10
$> 10s$										0.55	0.55

*Values shall be multiplied by the following reduction factor when a return corner is present:

L_r/S	*Reduction Factor*
0.3	0.90
1.0	0.75
≥2	0.60

Plan view of wall or sign with return corner

L_r

WIND

B

Notes

1. Force coefficients for solid walls or signs with openings less than 30% of gross area shall be permitted to be multiplied by the reduction factor $(1 - (1 - \varepsilon)^{1.5})$.
2. To allow for both normal and oblique wind directions, the following cases shall be considered:
   For $s/h < 1$:
   Case A: Resultant force acts normal to the face of the wall or sign through the geometric center.
   Case B: Resultant force acts normal to the face of the wall or sign at a distance from the geometric center toward the windward edge equal to 0.2 times the average width of the wall or sign.
   For double-faced signs with all sides enclosed and $R_{max} \leq 0.4$, it is permitted to use force eccentricity, $e = (0.2 - 0.25R_{max})B$.
   For double-faced signs with all sides enclosed and $R_{min} \leq 0.75$, it is permitted to multiply tabulated C_f values in Cases A and B by the reduction factor, $(1 - 0.133R_{min})$.
   For $B/s \geq 2$, Case B need not be considered, while Case C must be considered;
   Case C: Resultant forces act normal to the face of the wall or sign through the geometric centers of each region.
   For $s/h = 1$:
   The same cases as above except that the vertical locations of the resultant forces occur at a distance above the geometric center equal to 0.05 times the average height of the wall or sign.
3. For Case C where $s/h > 0.8$, for efficients shall be multiplied by the reduction factor $(1.8 - s/h)$. It is permitted to apply this reduction with those specified in Note 2.
4. Linear interpolation is permitted for values of s/h, B/s, and L_r/s other than shown.

Figure 29.3-1 (*Continued*). Design wind loads (all heights): force coefficients, C_f, for other structures—solid freestanding walls and solid freestanding signs.

Force Coefficients, C_f

Cross Section	Type of Surface	h/D: 1	h/D: 7	h/D: 25
Square (wind normal to face)	All	1.3	1.4	2.0
Square (wind along diagonal)	All	1.0	1.1	1.5
Hexagonal or octagonal	All	1.0	1.2	1.4
Round, $D\sqrt{q_z} > 2.5$ $D\sqrt{q_z} > 5.3$ (in SI)	Moderately smooth ($D'/D<0.02$)	0.5	0.6	0.7
	Rough ($0.02 \le D'/D < 0.08$)	0.7	0.8	0.9
	Very rough ($D'/D = 0.08$)	0.8	1.0	1.2
Round, $D\sqrt{q_z} \le 2.5$ $D\sqrt{q_z} \le 5.3$ (in SI)	All	0.7	0.8	1.2

Notation

D = Diameter of circular cross section and least horizontal dimension of square, hexagonal, or octagonal cross sections at elevation under consideration, ft (m).
D' = Depth of protruding elements such as ribs and spoilers, ft (m).
h = Height of structure, ft (m).
q_z = Velocity pressure evaluated at height z above ground, lb/ft^2 (N/m^2).

Notes

1. Design wind force shall be calculated based on the area of the structure projected on a vertical plane normal to the wind direction. The force shall be assumed to act parallel to the wind direction.
2. Linear interpolation is permitted for h/D values other than shown.

Figure 29.4-1. Other structures (all heights): force coefficients, C_f, for chimneys, tanks, and similar structures.

Force Coefficients, C_f

ε	Flat-Sided Members	Rounded Members: $D\sqrt{q_z} \le 2.5$ ($D\sqrt{q_z} \le 5.3$) s.I	Rounded Members: $D\sqrt{q_z} > 2.5$ ($D\sqrt{q_z} > 5.3$) s.I
<0.1	2.0	1.2	0.8
0.1 to 0.29	1.8	1.3	0.9
0.3 to 0.7	1.6	1.5	1.1

Notation

ε = Ratio of solid area to gross area.
D = Diameter of a typical round member, ft (m).
q_z = Velocity pressure evaluated at height z above ground, lb/ft^2 (N/m^2).

Notes

1. Signs with openings making up 30% or more of the gross area are classified as open signs.
2. Calculation of the design wind forces shall be based on the area of all exposed members and elements projected on a plane normal to the wind direction. Forces shall be assumed to act parallel to the wind direction.
3. Area, A_f, consistent with these force coefficients is the solid area projected normal to the wind direction.

Figure 29.4-2. Other structures (all heights): force coefficients, C_f, for open signs and single-plane open frames.

Force Coefficients, C_f

Tower Cross Section	C_f
Square	$4.0\varepsilon^2 - 5.9\varepsilon + 4.0$
Triangle	$3.4\varepsilon^2 - 4.7\varepsilon + 3.4$

Notation

ε = Ratio of solid area to gross area of one tower face for the segment under consideration.

Notes

1. For all wind directions considered, the area, A_f, consistent with the specified force coefficients shall be the solid area of a tower face projected on the plane of that face for the tower segment under consideration.
2. Specified force coefficients are for towers with structural angles or similar flat-sided members.
3. For towers containing rounded members, it is acceptable to multiply the specified force coefficients by the following factor when determining wind forces on such members:

$$0.51\varepsilon^2 + 0.57, \text{ but not} > 1.0$$

4. Wind forces shall be applied in the directions resulting in maximum member forces and reactions. For towers with square cross sections, wind forces shall be multiplied by the following factor when the wind is directed diagonally along a tower

$$1 + 0.75\varepsilon, \text{ but not} > 1.2$$

5. Wind forces on tower appurtenances, such as ladders, conduits, lights, and elevators, shall be calculated using appropriate force coefficients for these elements.
6. Loads caused by ice accretion, as described in Chapter 10, shall be accounted for.

Figure 29.4-3. Other structures (all heights): force coefficients, C_f, for open structures—trussed towers.

The vertical uplift force, F_v, on rooftop structures and equipment shall be determined from Equation (29.4-3)

$$F_v = q_h K_d (GC_r) A_r \text{ (lb)} \quad (29.4\text{-}3)$$

$$F_v = q_h K_d (GC_r) A_r \text{ (N)} \quad (29.4\text{-}3.\text{SI})$$

where

(GC_r) = 1.5 for rooftop structures and equipment with A_r less than $0.1BL$; (GC_r) shall be permitted to be reduced linearly from 1.5 to 1.0 as the value of A_r is increased from $0.1BL$ to BL;

q_h = Velocity pressure evaluated at the mean roof height of the building;

K_d = Wind directionality factor, see Section 26.6; and

A_r = Horizontal projected area of rooftop structure or equipment, ft² (m²).

29.4.2 Design Wind Loads: Circular Bins, Silos, and Tanks with $h \le 120$ ft ($h \le 36.5$ m), $D \le 120$ ft ($D \le 36.5$ m), and $0.25 \le HD \le 4$ Grouped circular bins, silos, and tanks of similar size with center-to-center spacing greater than two diameters shall be treated as isolated structures. For spacings less than 1.25 diameters, the structures shall be treated as grouped, and the wind pressure shall be determined from Section 29.4.2.4. For intermediate spacings, linear interpolation of the C_p (or C_f) values shall be used.

29.4.2.1 External Walls of Isolated Circular Bins, Silos, and Tanks To determine the overall drag on circular bins, silos, and tanks using Equation (29.4-1), a force coefficient C_f of 0.63, based on the projected wall area Dh_c, is permitted to be used in lieu of Figure 29.4-1, provided that all of the following conditions are met:

(a) h_c/D is in the range of 0.25 to 4.0 and the cylinder (diameter D) is standing on the ground or supported by columns; and
(b) Clearance height, C, is less than or equal to the height of the solid, cylinder h_c, as shown in Figure 29.4-4; and
(c) Tank surface is moderately smooth, as defined in Figure 29.4-1.

29.4.2.2 Roofs of Isolated Circular Bins, Silos, and Tanks The net design pressures on the roofs of circular bins, silos, and tanks shall be determined from Equation (29.4-4)

$$p = q_h K_d (GC_p - (GC_{pi})) \text{ (lb/ft}^2\text{)} \quad (29.4\text{-}4)$$

$$p = q_h K_d (GC_p - (GC_{pi})) \text{(N/m}^2\text{)} \quad (29.4\text{-}4.\text{SI})$$

where

q_h = Velocity pressure for all surfaces evaluated at mean roof height h;

K_d = Wind directionality factor, see Section 26.6;

C_p = External pressure coefficient from Figure 29.4-5 for roofs;

(GC_{pi}) = Internal pressure coefficient for roofed structures from Section 26.13; and

G = Gust-effect factor from Section 26.11.

The external pressures on the conical, flat, or dome roofs (roof angle less than 10 degrees) of circular bins, silos, or tanks shall be

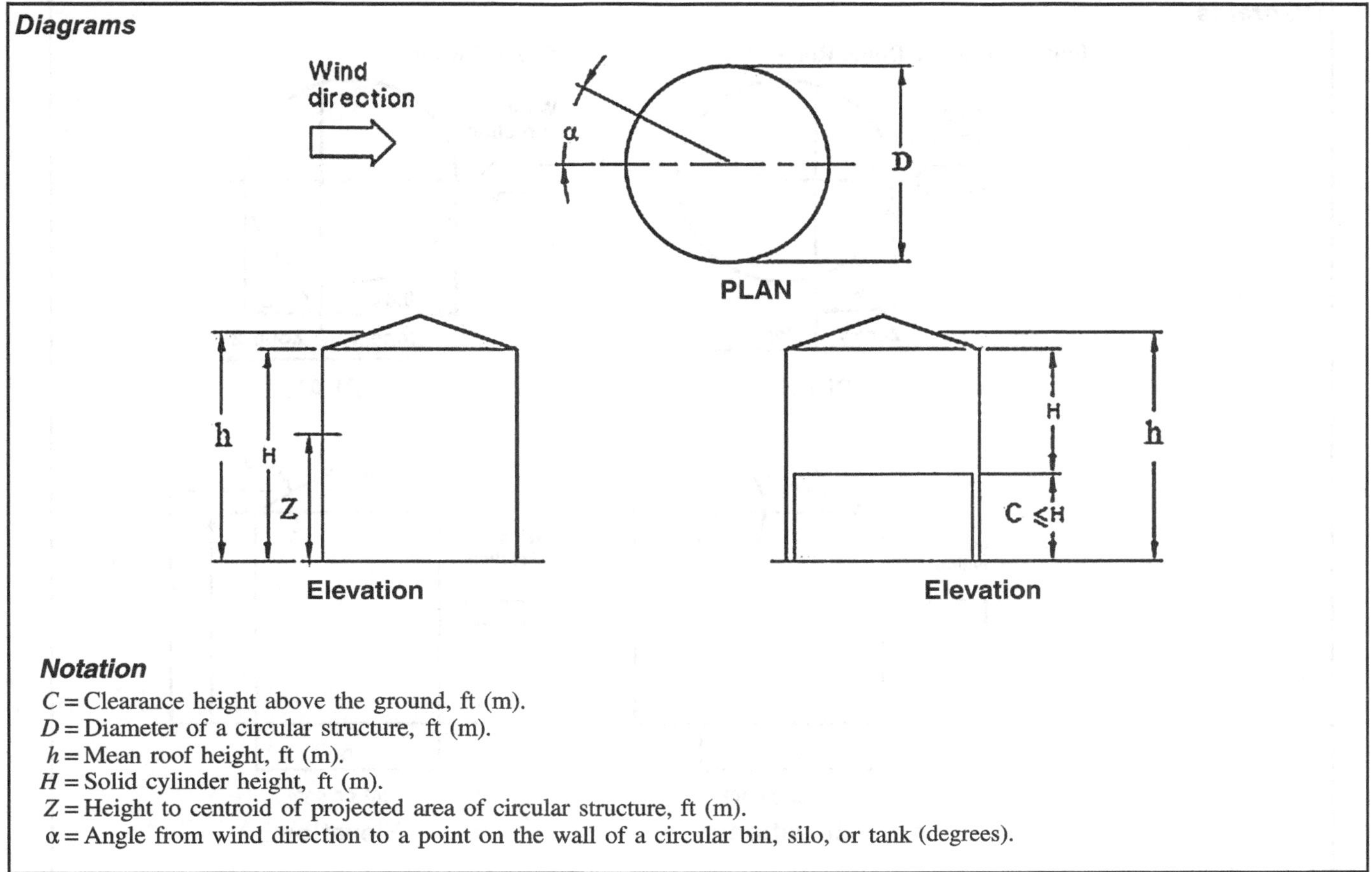

Figure 29.4-4. Other structures, design wind loads for main wind force resisting systems [$h < 120$ ft ($h < 36.6$ m)]: circular bins, silos, and tanks on the ground or supported by columns, where $D \le 120$ ft ($D \le 36.6$ m), $0.25 \le H/D < 4.0$.

equal to the external pressure coefficients, C_p, given in Figure 29.4-5 for Zones 1 and 2. The external pressures for dome roofs (roof angle greater than 10 degrees) shall be determined from Figure 27.3-2.

29.4.2.3 Undersides of Isolated Elevated Circular Bins, Silos, and Tanks External pressure coefficients, C_p, for the underside of elevated circular bins, silos, or tanks with clearance height, C, above ground less than or equal to the solid cylinder height, H, shall be taken as 0.8 and −0.6. For structures with clearance height above ground of less than or equal to one-third of the cylinder height, use linear interpolation between these values and $C_p = 0.0$, according to the ratio of C/h, where C and h are defined, as shown in Figure 29.4-4.

29.4.2.4 Roofs and Walls of Grouped Circular Bins, Silos, and Tanks For closely spaced groups of three or more circular bins, silos, or tanks with center-to-center spacing less than $1.25D$, roof pressure coefficients, C_p, and drag force coefficient, C_f, on projected walls shall be calculated using Figure 29.4-6. The net design pressures on the roofs shall be determined from Equation (29.4-4). The overall drag shall be calculated based on Equation (29.4-1).

29.4.3 Rooftop Solar Panels for Buildings of All Heights with Flat Roofs or Gable or Hip Roofs with Slopes Less Than 7 Degrees As illustrated in Figure 29.4-7, the design wind pressure for rooftop solar panels apply to those located on enclosed or partially enclosed buildings of all heights with flat roofs, or with gable or hip roof slopes with $\theta \le 7°$, with panels conforming to

$L_p \le 6.7$ ft (2.04 m),
$\omega \le 35°$,
$h_1 \le 2$ ft (0.61 m),
$h_2 \le 4$ ft (1.22 m),

and with a minimum gap of 0.25 in. (6.4 mm) provided between all panels, and the spacing of gaps between panels not exceeding 6.7 ft (2.04 m). In addition, the minimum horizontal clear distance between the panels and the edge of the roof shall be the larger of $2(h_2 - h_{pt})$ and 4 ft (1.2 m) for the design pressures in this section to apply. The design wind pressure for rooftop solar panels shall be determined by Equations (29.4-5) and (29.4-6):

$$p = q_h K_d (GC_{rn}) \text{ (lb/ft}^2\text{)} \quad (29.4\text{-}5)$$

$$p = q_h K_d (GC_{rn}) \text{ (N/m}^2\text{)} \quad (29.4\text{-}5.\text{SI})$$

where q_h is the velocity pressure for all surfaces evaluated at mean roof height, h, and K_d is the wind directionality factor; see Section 26.6.

Diagrams

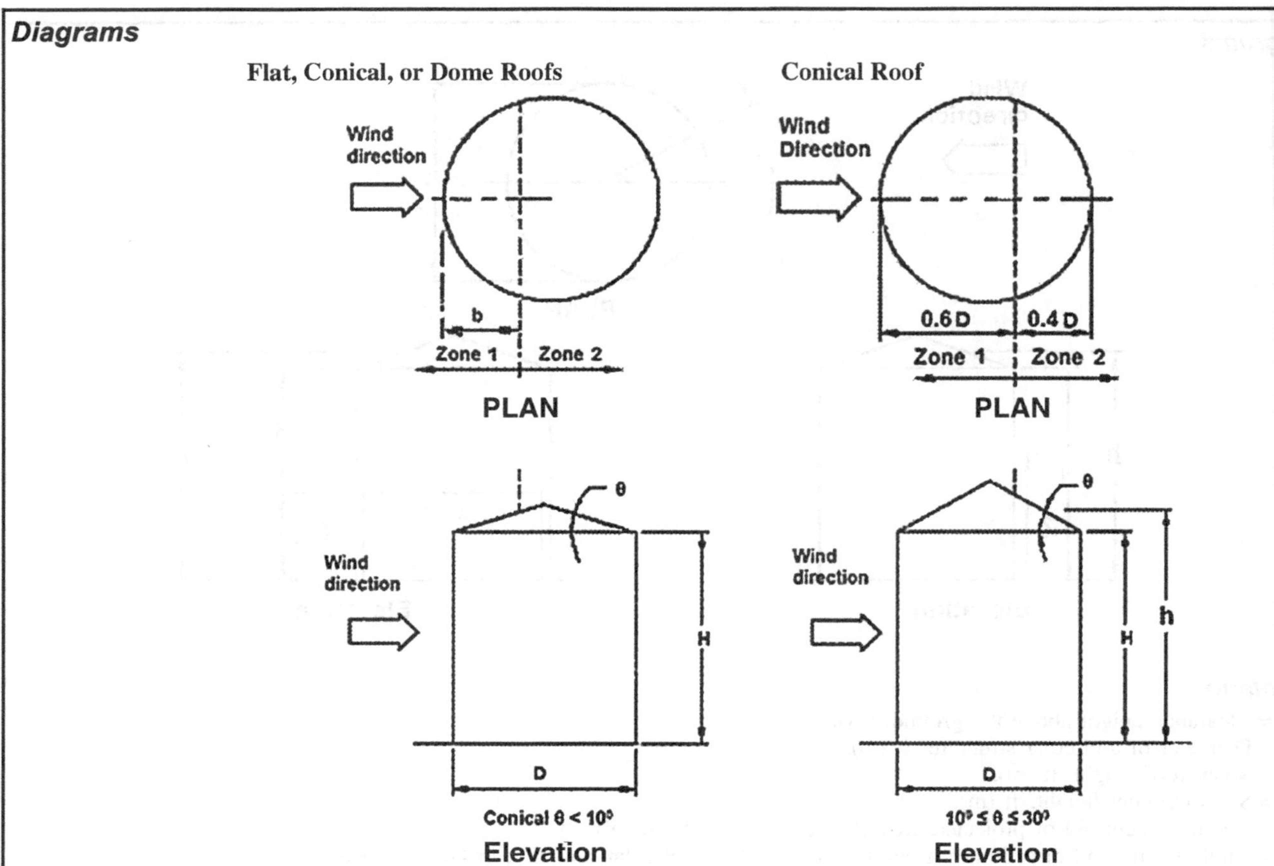

Notation

b = Determined below, ft (m), dependent on H/D for roofs with average θ less than 10 degrees.
h = Mean roof height, ft (m).
H = The solid cylinder height, ft (m).
D = Diameter of a circular structure, ft (m).
θ = Angle of plane of roof from horizontal, in degrees.

External Pressure Coefficients, C_p

Zone 1	−0.8
Zone 2	−0.5

Notes

For roofs with average θ less than 10 degrees, the dimension, b, shall be determined as follows:

H/D	b
0.25	0.2D
0.5	0.5D
≥1.0	0.1h + 0.6D

Linear interpolation shall be permitted.

Figure 29.4-5. Other structures, design wind loads for main wind force resisting systems [h<120 ft (h<36.6 m)]: external pressure coefficients, C_p, for isolated roofs of circular bins, silos, and tanks, where $D \leq 120$ ft ($D \leq 36.6$ m), $0.25 \leq H/D < 4.0$.

Diagrams

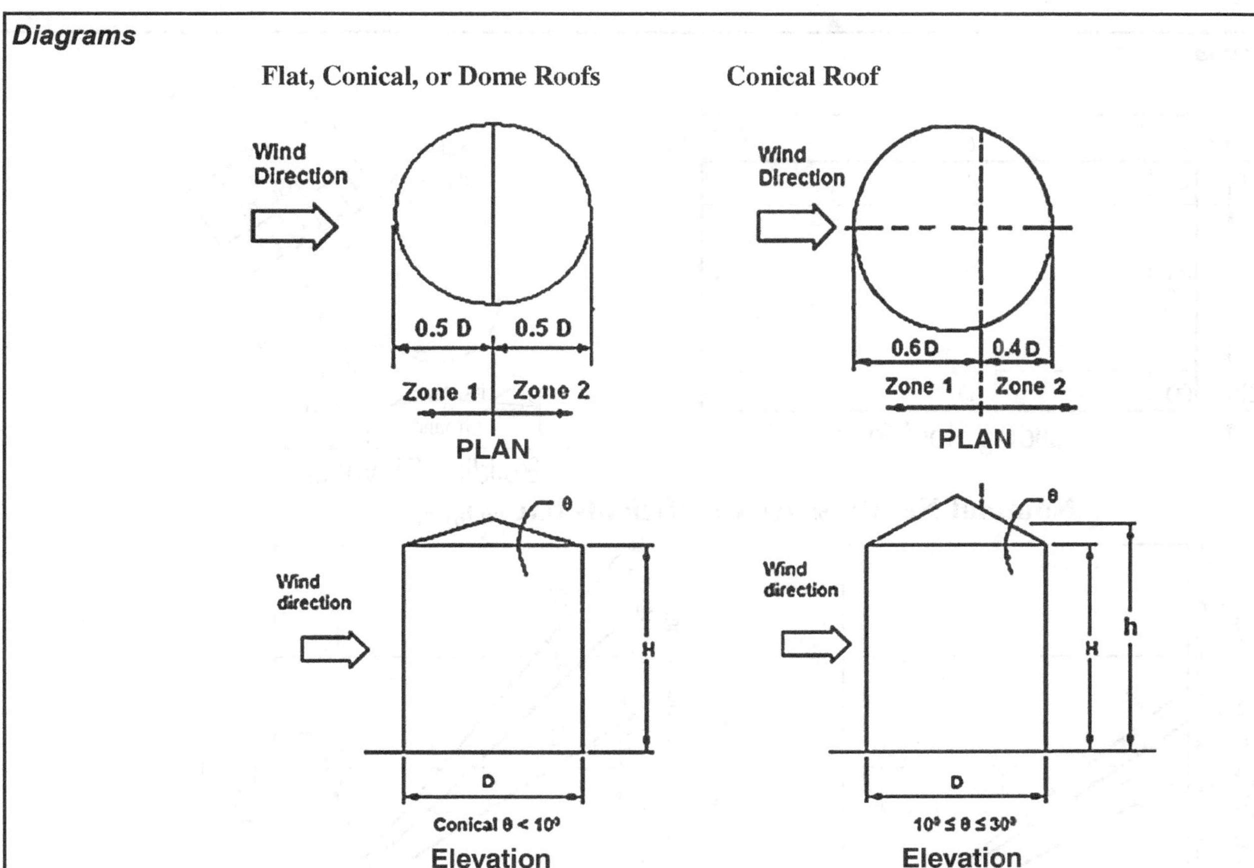

Notation

D = Diameter of a circular structure, ft (m).
h = Mean roof height, ft (m).
H = The solid cylinder height, ft (m).
θ = Angle of plane of roof from horizontal, degrees.

Drag Force Coefficient (C_f) on Projected Walls

H/D	C_f	Use with
<1	1.3	q_h
2	1.1	q_h
4	1.0	q_h

Roof Pressure Coefficients, C_p, for Use with q_h

	H/D	Zone 1	Zone 2
$\theta < 10°$	≤0.5	−0.9	−0.5
	≥1.0	−1.3	−0.7
$10° < \theta < 30°$	≤4	−1.0	−0.6

Figure 29.4-6. Other structures, design loads for main wind force resisting systems [$h < 120$ ft ($h < 36.6$ m)]: drag force coefficients, c_f, and roof pressure coefficients, c_p, for grouped circular bins, silos, and tanks on the ground or supported by columns, where $D \leq 120$ ft ($D \leq 36.6$ m), $0.25 \leq H/D < 4.0$, and center-to-center spacing ≤ 1.25.

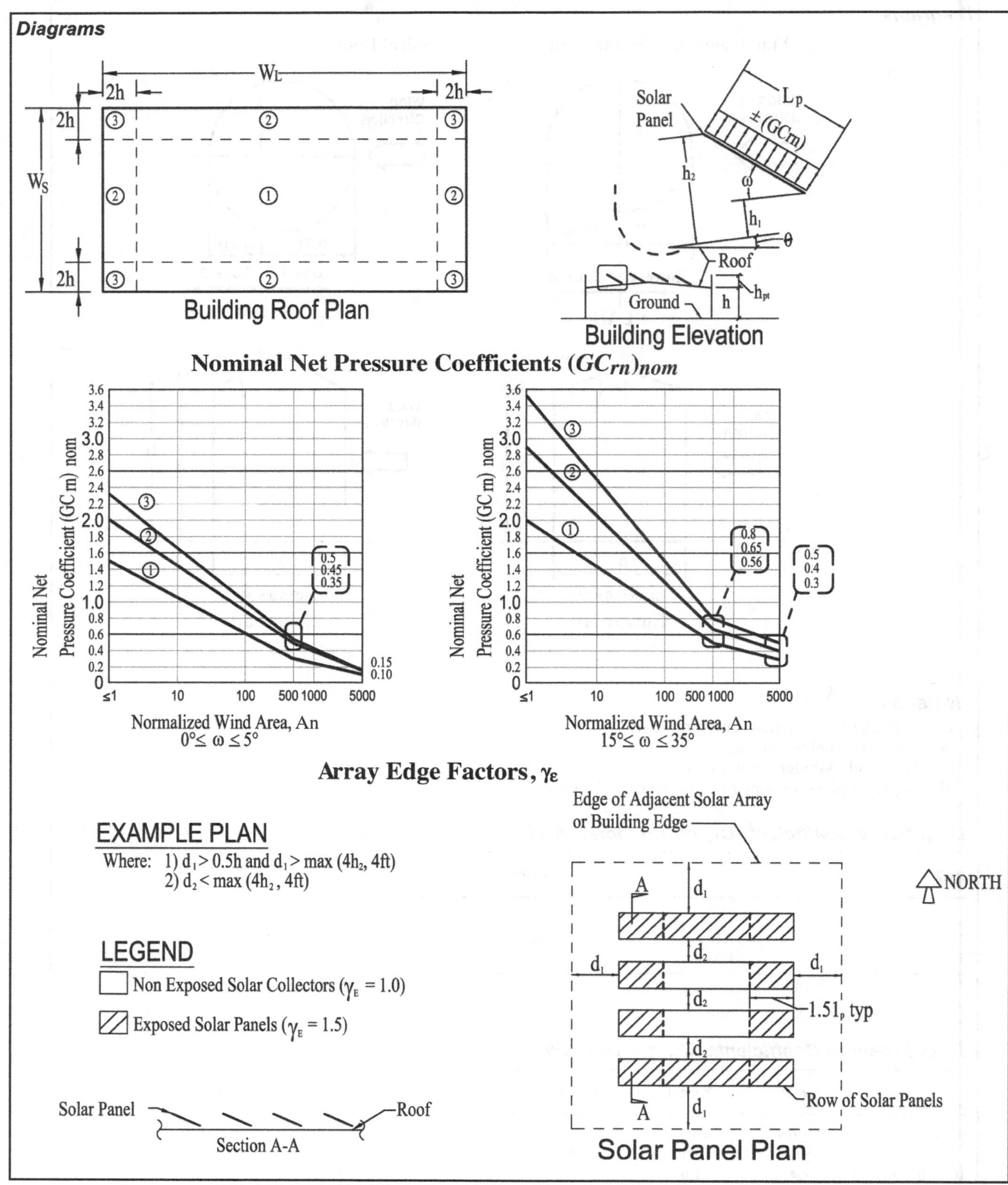

Figure 29.4-7. Design wind loads (all heights): rooftop solar panels for enclosed and partially enclosed buildings, roof $\theta \leq 7°$.

Notation

A = Effective wind area, ft^2 (m^2).
A_n = Normalized wind area, non dimensional.
d_1 = For rooftop solar array, horizontal distance orthogonal to the panel edge to an adjacent panel or the building edge, ignoring any rooftop equipment in Figure 29.4-7, ft (m).
d_2 = For rooftop solar arrays, horizontal distance from the edge of one panel to the nearest edge in the next row in Figure 29.4-7, ft (m).
h = Mean roof height of a building except that eave height shall be used for roof angle θ less than or equal to 10°, ft (m).
h_1 = Height of the gap between the panels and the roof surface, ft (m).
h_2 = Height of a solar panel above the roof at the upper edge of the panel, ft (m).
h_{pt} = Mean parapet height above the adjacent roof surface for use with Equation (29.4-5), ft (m).
L_p = Panel chord length.
W_L = Width of a building on its longest side in Figure 29.4-7, ft (m).
W_S = Width of a building on its shortest side in Figure 29.4-7, ft (m).
γ_E = Array edge factor as defined in Section 29.4.4.
θ = Angle of plane of roof from horizontal, in degrees.
ω = Angle that the solar panel makes with the roof surface in Figure 29.4-7, degrees.

Notes

1. (GC_m) acts toward (+) and away (–) from the top surface of the panels.
2. Linear interpolation is allowed for ω between 5 and 15 degrees.
3. $A_n = (1,000/[\max(L_b, 15)^2]A$, where A is the effective wind area of the structural element of the solar panel being considered, and L_b is the minimum of $0.4(hW_L)^{0.5}$ or h or W_s, ft (m).

Figure 29.4-7 (*Continued*). Design wind loads (all heights): rooftop solar panels for enclosed and partially enclosed buildings, roof $\theta \leq 7°$.

$$(GC_{rn}) = (\gamma_p)(\gamma_c)(\gamma_E)(GC_{rn})_{\text{nom}} \quad (29.4\text{-}6)$$

where

γ_p = $\min(1.2, 0.9 + h_{pt}/h)$;
γ_c = $\max(0.6 + 0.06L_p, 0.8)$; and
γ_E = 1.5 for uplift loads on panels that are exposed and within a distance $1.5(L_p)$ from the end of a row at an exposed edge of the array; $\gamma_E = 1.0$ elsewhere for uplift loads and for all downward loads, as illustrated in Figure 29.4-7; a panel is defined as exposed if d_1 to the roof edge $> 0.5h$ and one of the following applies:

1. d_1 to the adjacent array $> \max(4h_2$, 4 ft (1.2 m), or
2. d_2 to the next adjacent panel $> \max(4h_2$, 4 ft (1.2 m).

$(GC_{rn})_{nom}$ is the nominal net pressure coefficient for rooftop solar panels as determined from Figure 29.4-7.

When $\omega \leq 2°$, $h_2 \leq 0.83$ ft (0.25 m), a minimum gap of 0.25 in. (6.4 mm) is provided between all panels, and the spacing of gaps between panels does not exceed 6.7 ft (2.04 m), the procedure of Section 29.4.4 shall be permitted.

The roof shall be designed for both of the following:

1. The case where solar collectors are present. Wind loads acting on solar collectors in accordance with this section shall be applied simultaneously with roof wind loads specified in other sections acting on areas of the roof not covered by the plan projection of solar collectors. For this case, roof wind loads specified in other sections need not be applied on areas of the roof covered by the plan projection of solar collectors.
2. Cases where the solar arrays have been removed.

29.4.4 Rooftop Solar Panels Parallel to the Roof Surface on Buildings of All Heights and Roof Slopes The design wind pressures for rooftop solar panels located on enclosed or partially enclosed buildings of all heights, with panels parallel to the roof surface, with a tolerance of 2 degrees and with a maximum height above the roof surface, h_2, not exceeding 10 in. (0.25 m), shall be determined in accordance with this section. A minimum gap of 0.25 in. (6.4 mm) shall be provided between all panels, with the spacing of gaps between panels not exceeding 6.7 ft (2.04 m). In addition, the array shall be located at least $2h_2$ from the roof edge, a gable ridge, or a hip ridge. The design wind pressure for rooftop solar collectors shall be determined by Equation (29.4-7)

$$p = q_h K_d (GC_p)(\gamma_E)(\gamma_a) \text{ (lb/ft}^2) \quad (29.4\text{-}7)$$

$$p = q_h K_d (GC_p)(\gamma_E)(\gamma_a) \text{ (N/m}^2) \quad (29.4\text{-}7.\text{SI})$$

where

q_h = Velocity pressure for all surfaces evaluated at mean roof height, h;
K_d = Wind directionality factor, see Section 26.6;
(GC_p) = External pressure coefficient for C&C of roofs with respective roof zoning, determined from Figures 30.3-2A-I through 30.3-7 or 30.5-1;
γ_E = Array edge factor = 1.5 for uplift loads on panels that are exposed and within a distance of $2h_2$ from an edge of the array; $\gamma_E = 1.0$ elsewhere for uplift loads and for all downward loads, as illustrated in Figure 29.4-7; a panel is defined as exposed if d_1 to the roof edge $>0.5h$ and one of the following applies:
d_1 to the adjacent array $>2h_2$ or
d_2 to the next adjacent panel $>2h_2$; and
γ_a = Solar panel pressure equalization factor, defined in Figure 29.4-8.

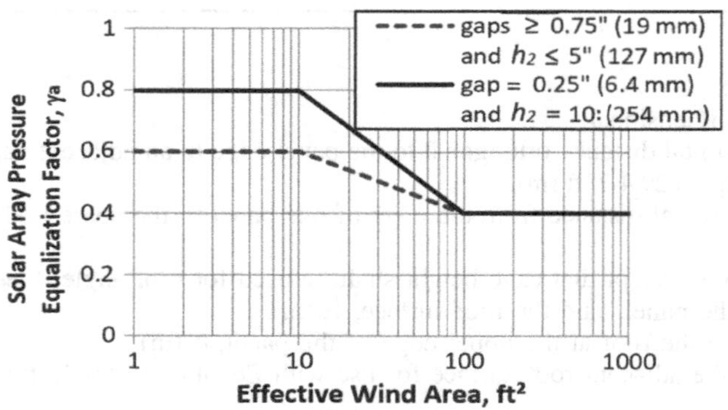

Figure 29.4-8. Solar panel pressure equalization factor, γ_a, for enclosed and partially enclosed buildings of all heights.
Note: Bilinear interpolation between the solid line and the dashed line is permitted. Change to = signs for solid line.

The roof shall be designed for both of the following:

1. In the case where solar panels are present, wind loads acting on solar collectors in accordance with this section shall be applied simultaneously with roof wind loads specified in other sections acting on areas of the roof not covered by the plan projection of solar collectors. For this case, roof wind loads specified in other sections need not be applied on areas of the roof covered by the plan projection of solar collectors.
2. Case where the solar panels have been removed.

29.4.5 Ground-Mounted Fixed-Tilt Solar Panel Systems

29.4.5.1 Scope Wind loads shall be calculated in accordance with Sections 29.4.5.2 and 29.4.5.3 for ground-mounted fixed-tilt solar photovoltaic (PV) panel systems installed in rows satisfying the following geometric limitations:

$6\text{ ft }(1.8\text{ m}) \leq L_c \leq 14\text{ ft }(4.4\text{ m})$
$(W_g/L_c) \geq 7$
$0° \leq \omega \leq 60°$
$0.5 \leq (h/L_c) \leq 0.8$
$0.20 \leq (L_c/S) \leq 0.60$
$s_p \leq 0.014L_c$
$S_L \leq 0.25L_c$
$S_T \leq 2S$

Number of rows ≥ 3
The rows have the same chord length.
The ratio of area blocked by support framing to total area below lowest edge of panels $\leq 8\%$ over any length of $4L_c$,

where

L_c = Panel chord length, ft (m).
W_g = Shortest row length in array, ft (m).
ω = Angle between the solar panels and the ground surface, degrees.
h = Mean height of panel, ft (m).
S = Center-to-center row spacing, ft (m).
s_p = Gap between adjacent panels in both directions.
S_L = Horizontal distance in longitudinal direction of open area within a single row.
S_T = Horizontal distance in transverse direction of open area between adjacent rows.

29.4.5.2 Design Wind Loads The design force, F_n, and design moment about the center-of-plane of panels, M_c, for ground-mounted solar panels shall be determined by Equation (29.4-8) and Equation (29.4-9), respectively:

$$F_n = q_h K_d[\pm(GC_{gn})]A\ (\text{lb}) \quad (29.4\text{-}8)$$

$$F_n = q_h K_d[\pm(GC_{gn})]A\ (\text{N}) \quad (29.4\text{-}8.\text{SI})$$

$$M_c = q_h K_d[\pm(GC_{gm})]AL_c\ (\text{lb/ft}) \quad (29.4\text{-}9)$$

$$M_c = q_h K_d[\pm(GC_{gm})]AL_c\ (\text{N/m}) \quad (29.4\text{-}9.\text{SI})$$

where

$$GC_{gn} = [\pm(GC_{gn_{\text{static}}}) \pm (GC_{gn_{\text{dynamic}}})] \quad (29.4\text{-}10)$$

$$GC_{gm} = [\pm(GC_{gm_{\text{static}}}) \pm (GC_{gm_{\text{dynamic}}})] \quad (29.4\text{-}11)$$

q_h = Velocity pressure for all surfaces evaluated at the mean height of the panel, h in lb/ft² (N/m²);
K_d = Directionality Factor; see Section 26.6
$(GC_{gn_{\text{static}}})$ = Static net pressure coefficient, as determined from Figure 29.4-10;
$(GC_{gm_{\text{dynamic}}})$ = Dynamic net pressure coefficient, as determined from Figure 29.4-11, and shall be applied with the sign that yields the most adverse load effect;
$(GC_{gn_{\text{static}}})$ = Static moment coefficient, as determined from Figure 29.4-10;
$(GC_{gm_{\text{dynamic}}})$ = Dynamic moment coefficient, as determined from Figure 29.4-11, and shall be applied with the sign that yields the most adverse load effect; and
A = Effective wind area of element, ft² (m²).

Interior shielded Zone 1 and exterior perimeter Zone 2 shall be designed and designated, as shown in Figure 29.4-9.

For arrays with $0.20 \leq (L_c/S) < 0.25$, all rows shall be designed using Zone 2 static and dynamic coefficients.

For all locations, K_{zt} shall be determined per Equation (26.8-1). Where K_{zt} exceeds 1.0 per Equation (26.8-1), the static and dynamic coefficients for the Zone 2 shall be used throughout the entire array.

29.4.5.3 Design Support Posts and Foundations The support posts and foundations shall be designed for the simultaneous

Diagrams

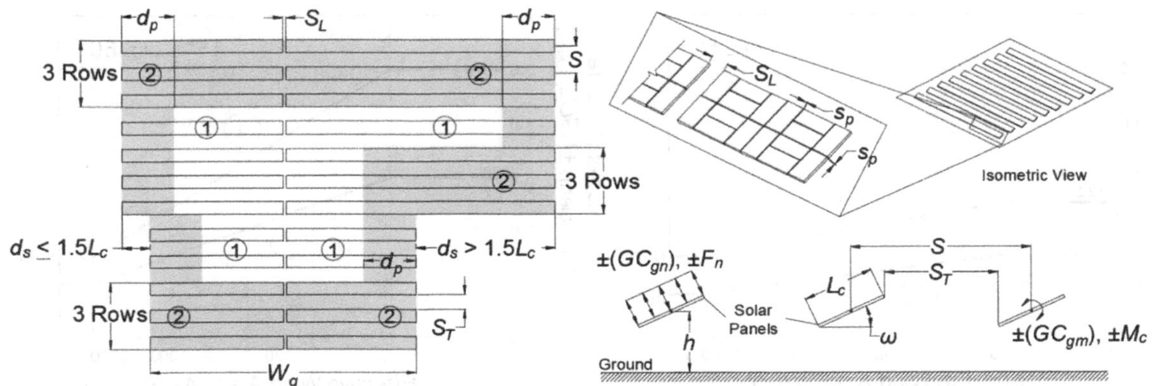

Parameters for Application of Ground-Mounted Fixed-Tilt Solar Panel System Requirements

Notation

d_p = horizontal distance of Zone 2 from row end, ft (m); $d_p = 4L_c$ or 30 ft (9.1 m), whichever is smaller.

d_s = horizontal offset between adjacent rows with staggered row ends, ft (m);

1. Where an open area for access and/or clearance purposes exceeds a distance equal to $0.25L_c$ in the longitudinal direction or $2S$ in the transverse direction, a new Zone 2 is formed where the array meets it.
2. Where $d_s > 1.5L_c$, the Zone 2 width shall be increased to d_s in the longitudinal direction.
3. The Zone 2 width shall always be three rows in the transverse direction.

Figure 29.4-9. Parameters for application of ground-mounted fixed-tilt solar panel system requirements.

application of the design force, F_n, and design moment, M_c. The design value used for the horizontal component of F_n shall not be less than 0.1 times the vertical component of F_n.

29.4.5.4 Reduced Frequency for Ground-Mounted Solar Panel Systems The reduced frequency, N_s, for ground-mounted solar panel systems shall be determined by Equation (29.4-12)

$$N_s = 0.682nL_c/V \quad (29.4\text{-}12)$$

$$N_s = nL_c/V \quad (29.4\text{-}12.\text{SI})$$

where n is the lowest natural frequency of the mode of interest (moment or bending), Hz, and V is the basic wind speed mi/h (m/s).

29.5 PARAPETS

Wind loads on parapets are specified in Section 27.3.4 for buildings of all heights designed using the Directional Procedure and in Section 28.3.4 for low-rise buildings designed using the Envelope Procedure.

29.6 ROOF OVERHANGS

Wind loads on roof overhangs are specified in Section 27.3.3 for buildings of all heights designed using the Directional Procedure and in Section 28.3.5 for low-rise buildings designed using the Envelope Procedure.

29.7 MINIMUM DESIGN WIND LOADING

The design wind force for other structures shall be not less than 16 lb/ft^2 (0.77 kN/m^2) multiplied by the area, A_f.

29.8 CONSENSUS STANDARDS AND OTHER REFERENCED DOCUMENTS

The following consensus standards and other documents shall be considered part of this standard and are referenced in this chapter.

ASCE 72, *Design of Steel Lighting System Support Pole Structures*, American Society of Civil Engineers, 2022.

Cited in: Sections 29.4 and C29.4

ASCE Task Committee on Wind-Induced Forces, *Wind Loads for Petrochemical and Other Industrial Facilities*, American Society of Civil Engineers, 2011.

Cited in: Sections 29.4 and C29.4

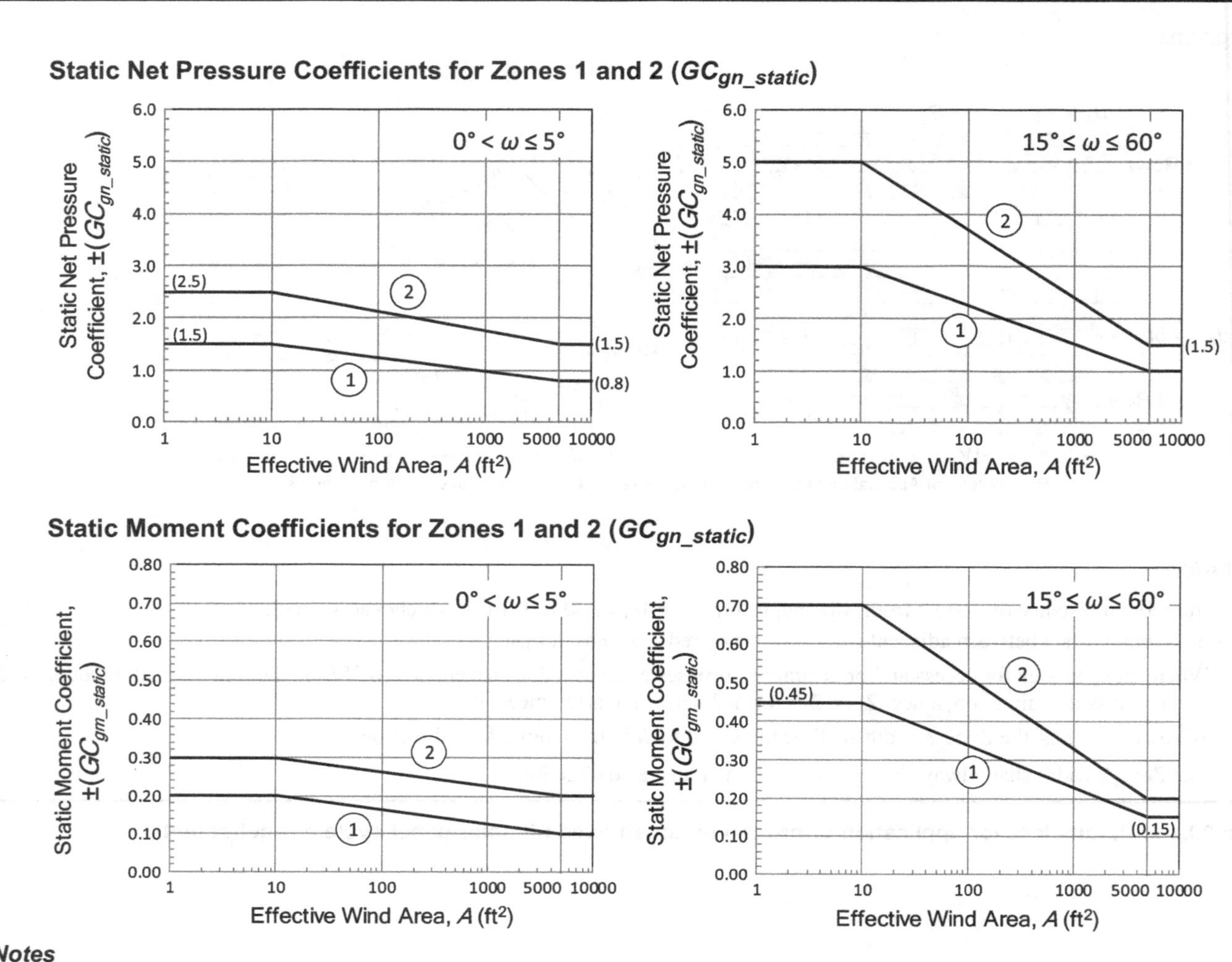

Notes

1. Linear interpolation shall be permitted for values of ω between 5 and 15 degrees.
2. (GC_{gn}) denotes net pressures (contributions from top and bottom surfaces of the solar panels); plus and minus signs signify forces acting toward and away from the panels, respectively; each structural element shall be designed to with stand the worst structural effects produced by these two load cases.
3. (GC_{gn}) denotes moments about the center-of-plane of the panels (contributions from top and bottom surfaces of the solar panels) that comprise the full chord length (L_c); positive coefficients denote moments acting in a counter clockwise direction (see Figure 29.4-9), whereas negative coefficients denote moments acting in a clockwise direction; each structural element shall be designed to with stand the worst structural effects produced by these two load cases.

Figure 29.4-10. Design wind loads: static wind load coefficients for ground-mounted fixed-tilt solar panel systems.

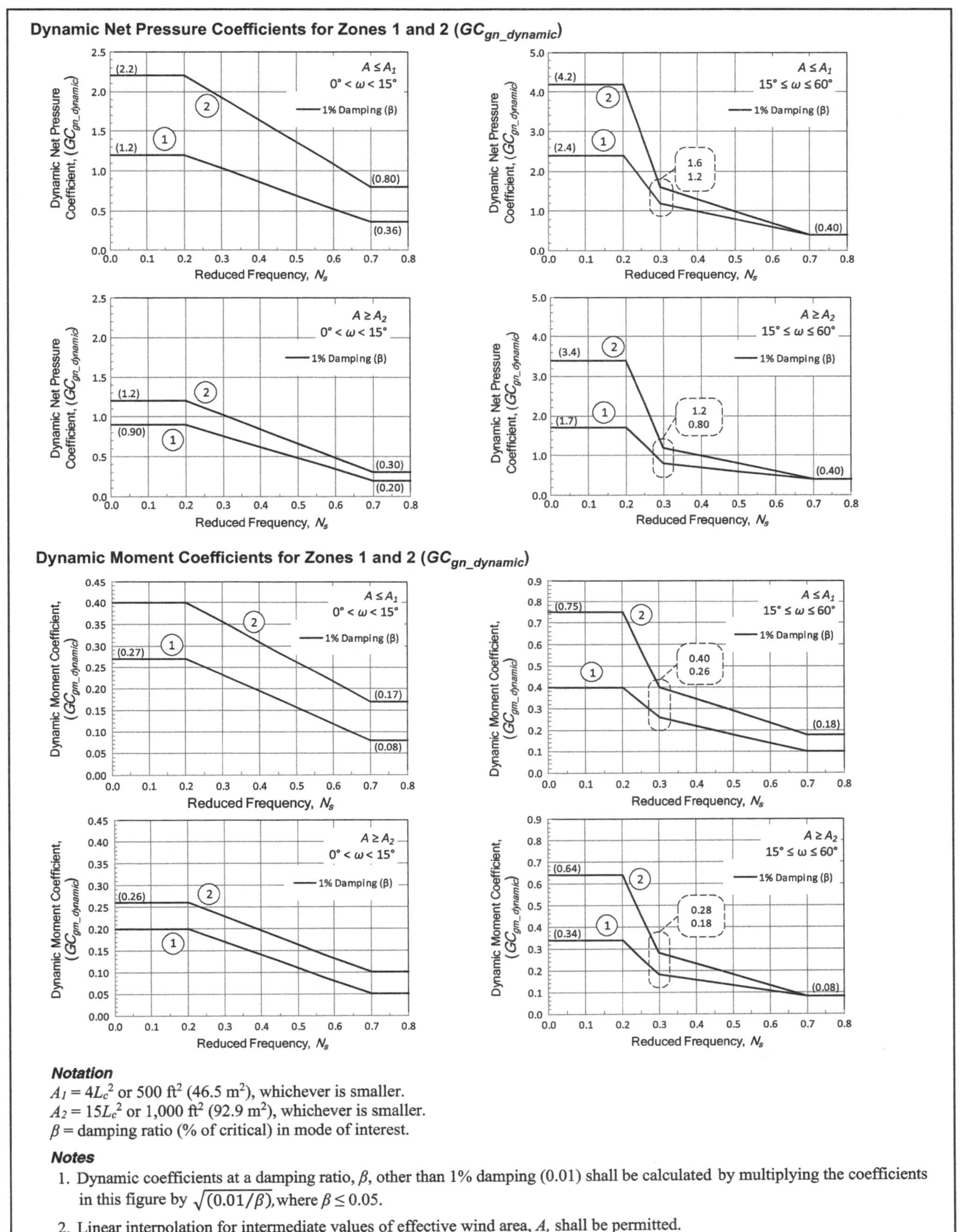

Notation

$A_1 = 4L_c^2$ or 500 ft² (46.5 m²), whichever is smaller.
$A_2 = 15L_c^2$ or 1,000 ft² (92.9 m²), whichever is smaller.
β = damping ratio (% of critical) in mode of interest.

Notes

1. Dynamic coefficients at a damping ratio, β, other than 1% damping (0.01) shall be calculated by multiplying the coefficients in this figure by $\sqrt{(0.01/\beta)}$, where $\beta \leq 0.05$.
2. Linear interpolation for intermediate values of effective wind area, A, shall be permitted.

Figure 29.4-11. Design wind loads: dynamic wind load coefficients for ground-mounted fixed-tilt solar panel systems.

CHAPTER 30
WIND LOADS: COMPONENTS AND CLADDING

30.1 SCOPE

30.1.1 Building Types This chapter applies to the determination of wind pressures on components and cladding (C&C) on buildings.

1. Part 1 is applicable to an enclosed, partially enclosed, or partially open
   - Low-rise building (see definition in Section 26.2); or
   - Building with $h \leq 60$ ft (18.3 m).

   The building has a flat roof, gable roof, multispan gable roof, hip roof, monoslope roof, stepped roof, or sawtooth roof, and the wind pressures are calculated from a wind pressure equation.
2. Part 2 is applicable to an enclosed, partially enclosed, or partially open
   - Building with $h > 60$ ft (18.3 m).

   The building has a flat roof, pitched roof, gable roof, hip roof, mansard roof, arched roof, or domed roof, and the wind pressures are calculated from a wind pressure equation.
3. Part 3 is applicable to an open building of all heights that has a pitched free roof, monoslope free roof, or troughed free roof.
4. Part 4 is applicable to building appurtenances such as roof overhangs, parapets, and rooftop equipment.
5. Part 5 is applicable to non-building structures – circular bins, silos, and tanks; rooftop solar panels and roof pavers.
   - Circular bins, silos, and tanks with $h \leq 120$ ft (38.6 m);
   - Rooftop solar panels: Buildings of all heights with flat roofs or gable or hip roofs with roof slopes less than or equal to 7 degrees; and
   - Roof pavers: Buildings of all heights with roof slopes less than or equal to 7 degrees.

30.1.2 Conditions A building that has design wind loads determined in accordance with this chapter shall comply with all of the following conditions:

1. The building is a regular-shaped building as defined in Section 26.2; and
2. The building does not have response characteristics that make it subject to across-wind loading, vortex shedding, or instability caused by galloping or flutter; nor does it have a site location for which channeling effects or buffeting in the wake of upwind obstructions warrant special consideration.

30.1.3 Limitations The provisions of this chapter take into consideration the load magnification effect caused by gusts in resonance with along-wind vibrations of flexible buildings. The loads on buildings that do not meet the requirements of Section 30.1.2 or that have unusual shapes or response characteristics shall be determined using recognized literature documenting such wind load effects or shall use the wind tunnel procedure specified in Chapter 31.

30.1.4 Shielding There shall be no reductions in velocity pressure caused by apparent shielding afforded by buildings and other structures or terrain features.

30.1.5 Air-Permeable Cladding Design wind loads determined from Chapter 30 shall be used for air-permeable claddings, including modular vegetative roof assemblies, unless approved test data or recognized literature demonstrates lower loads for the type of air-permeable cladding being considered.

30.2 GENERAL REQUIREMENTS

30.2.1 Wind Load Parameters Specified in Chapter 26 The following wind load parameters are specified in Chapter 26:

- Basic wind speed, V (Section 26.5),
- Wind directionality factor, K_d (Section 26.6),
- Exposure category (Section 26.7),
- Topographic factor, K_{zt} (Section 26.8),
- Ground elevation factor, K_e (Section 26.9),
- Velocity pressure exposure coefficient, K_z or K_h (Section 26.10.1); Velocity pressure, q_z (Section 26.10.2),
- Gust-effect factor (Section 26.11),
- Enclosure classification (Section 26.12), and
- Internal pressure coefficient, (GC_{pi}) (Section 26.13).

30.2.2 Minimum Design Wind Pressures The design wind pressure for C&C of buildings shall not be less than a net pressure of 16 lb/ft^2 (0.77 kN/m^2) acting in either direction normal to the surface.

30.2.3 Tributary Areas Greater than 700 ft^2 (65 m^2) C&C elements with tributary areas greater than 700 ft^2 (65 m^2) shall be permitted to be designed using the provisions for main wind force resisting systems.

30.2.4 External Pressure Coefficients Combined gust-effect factor and external pressure coefficients for C&C, (GC_p), are given in the figures associated with this chapter. The pressure coefficient values and gust-effect factor shall not be separated.

PART 1: LOW-RISE BUILDINGS

User Note: Use Part 1 of Chapter 30 to determine wind pressures on C&C of *enclosed, partially enclosed, or partially open low-rise buildings* that have roof shapes as specified in the applicable figures. The provisions in Part 1 are based on the Envelope Procedure, with *wind pressures calculated using the specified equation* as applicable to each building surface. For buildings for which these provisions are applicable, this method generally yields the lowest wind pressures of all analytical methods contained in this standard.

30.3 BUILDING TYPES

The provisions of Section 30.3 are applicable to an enclosed, partially enclosed, or partially open

- Low-rise building (see definition in Section 26.2); or
- Building with $h \leq 60$ ft (18.3 m).

The building has a flat roof, gable roof, multispan gable roof, hip roof, monoslope roof, stepped roof, or sawtooth roof. The steps required for the determination of wind loads on C&C for these building types are shown in Table 30.3-1.

30.3.1 Conditions For the determination of the design wind pressures on the C&C using the provisions of Section 30.3.2, the conditions indicated on the selected figure(s) shall be applicable to the building under consideration.

30.3.2 Design Wind Pressures Design wind pressures on C&C elements of low-rise buildings and buildings with $h \leq 60$ ft (18.3 m) shall be determined from the following equation:

$$p = q_h K_d[(GC_p) - (GC_{pi})](\text{lb/ft}^2) \qquad (30.3\text{-}1)$$

$$p = q_h K_d[(GC_p) - (GC_{pi})](\text{N/m}^2) \qquad (30.3\text{-}1.\text{SI})$$

where

q_h = Velocity pressure evaluated at mean roof height h as defined in Section 26.10;

K_d = Wind directionality factor, see Section 26.6; and

(GC_p) = External pressure coefficients given in

- Figure 30.3-1 (walls),
- Figures 30.3-2A–G (flat roofs, gable roofs, and hip roofs) and 30.5.2 (pitched free roofs),
- Figure 30.3-3 (stepped roofs),
- Figure 30.3-4 (multispan gable roofs),
- Figsure 30.3-5A–B (monoslope roofs),
- Figure 30.3-6 (sawtooth roofs),
- Figure 30.3-7 (domed roofs),
- Figure 30.3-8 (arched roofs),
- Figure 30.3-2A (bottom surfaces of elevated buildings).

(GC_{pi}) = Internal pressure coefficient given in Table 26.13-1.

30.3.2.1 Bottom Horizontal Surface of Elevated Buildings. Design wind pressures for C&C elements on the bottom flat horizontal surface of elevated buildings shall be determined using the roof pressure coefficients from Figure 30.3-2A with the following modifications:

1. h_B shall be the height above grade of the bottom surface of the elevated building, as depicted in Figure 30.3-1A. The value of h shall equal h_B for determining zone dimensions from Figure 30.3-2A. For elevated buildings with a flat bottom horizontal building surface and situated on a slope, h_B shall be taken as the maximum height between the slope and the bottom of the elevated building.
2. Areas of the horizontal surface above partially enclosed spaces and areas extending a_B perpendicular to walls beneath the elevated building with plan dimension greater than 4 ft (1.2 m), as shown in the shaded regions in Figure 30.3-1A, shall be designed to resist positive pressures equal to the Zone 4 wall pressures obtained using Figure 30.3-1. The value of a_B shall equal 0.4 h_B or the width of the wall, whichever is smaller for determining zone dimensions from Figure 30.3-1A.

The loading convention shall denote downward loading on the bottom surface with negative pressure coefficients and upward loading on the bottom surface with positive pressure coefficients.

EXCEPTION: The provisions of Section 30.3.2.1 do not apply to buildings with $h_B < 2$ ft (0.61 m).

Table 30.3-1. Steps to Determine C&C Wind Loads for Enclosed, Partially Enclosed, and Partially Open Low-Rise Buildings.

Step 1: Determine risk category; see Table 1.5-1.
Step 2: Determine the basic wind speed, V, for applicable risk category; see Figure. 26.5-1.
Step 3: Determine the wind load parameters:

- Wind directionality factor, K_d; see Section 26.6 and Table 26.6-1.
- Exposure Category B, C, or D; see Section 26.7.
- Topographic factor, K_{zt}; see Section 26.8 and Figure 26.8-1.
- Ground elevation factor, K_e; Section 26.9 and Table 26.9-1.
- Enclosure classification; see Section 26.12.
- Internal pressure coefficient, (GC_{pi}); see Section 26.13 and Table 26.13-1.

Step 4: Determine velocity pressure exposure coefficient, K_h; see Table 26.10-1.
Step 5: Determine velocity pressure, q_h, Equation (26.10-1).
Step 6: Determine external pressure coefficient, (GC_p):

- Walls; see Figure 30.3-1.
- Flat roofs, gable roofs, hip roofs; see Figure 30.3-2.
- Stepped roofs; see Figure 30.3-3.
- Multispan gable roofs; see Figure 30.3-4.
- Monoslope roofs; see Figure 30.3-5.
- Sawtooth roofs; see Figure 30.3-6.
- Domed roofs; see Figure 30.3-7.
- Arched roofs; see Figure 30.3-8.
- Bottom horizontal surface of elevated buildings; see Section 30.3.2.1.

Step 7: Calculate wind pressure, p; Equation (30.3-1).

PART 2: BUILDINGS WITH $h > 60$ ft [($h > 18.3$ m)]

User Note: Use Part 2 of Chapter 30 for determining wind pressures for C&C of *enclosed, partially enclosed, or partially open buildings with h* $\leq$ 60 ft (18.3 m) that have roof shapes as specified in the applicable figures. These provisions are based on the Directional Procedure with *wind pressures calculated from the specified equation* applicable to each building surface.

Diagram

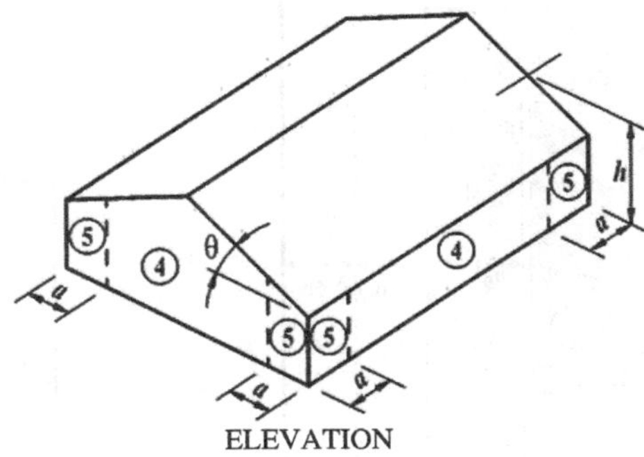

Notation

a = 10% of least horizontal dimension or $0.4h$, whichever is smaller, but not less than either 4% of least horizontal dimension or 3 ft (0.9 m).

Exception: For buildings with $\theta = 0°$ to $7°$ and a least horizontal dimension greater than 300 ft (90 m), dimension a shall be limited to a maximum of $0.8h$.

h = Mean roof height, ft (m), except that eave height shall be used for $\theta \leq 10°$.

θ = Angle of plane of roof from horizontal, degrees.

External Pressure Coefficient, (GC_p) - Walls

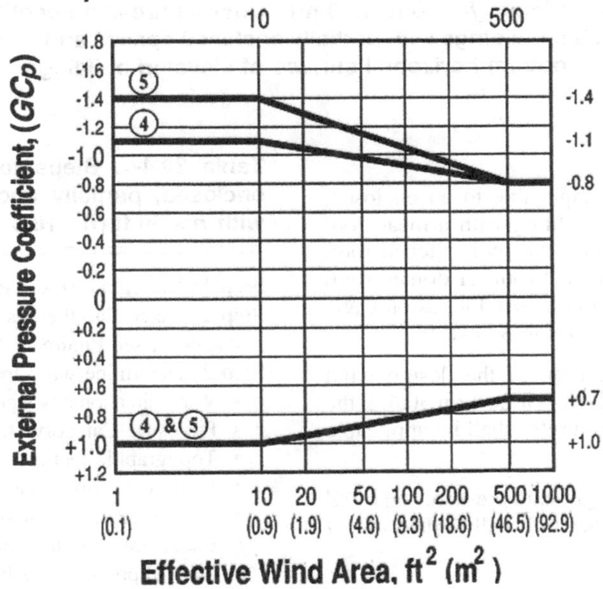

Notes

1. Vertical scale denotes (GCp) to be used with qh.
2. Horizontal scale denotes effective wind area A, ft^2 (m^2).
3. Plus and minus signs signify pressures acting toward and away from the surfaces, respectively.
4. Each component shall be designed for maximum positive and negative pressures.
5. Values of (GCp) for walls shall be reduced by 10% when $\theta \leq 10°$.

Figure 30.3-1. Components and cladding, $h \leq 60$ ft (18.3 m): external pressure coefficients (GC_p) for enclosed, partially enclosed, and partially open buildings—walls.

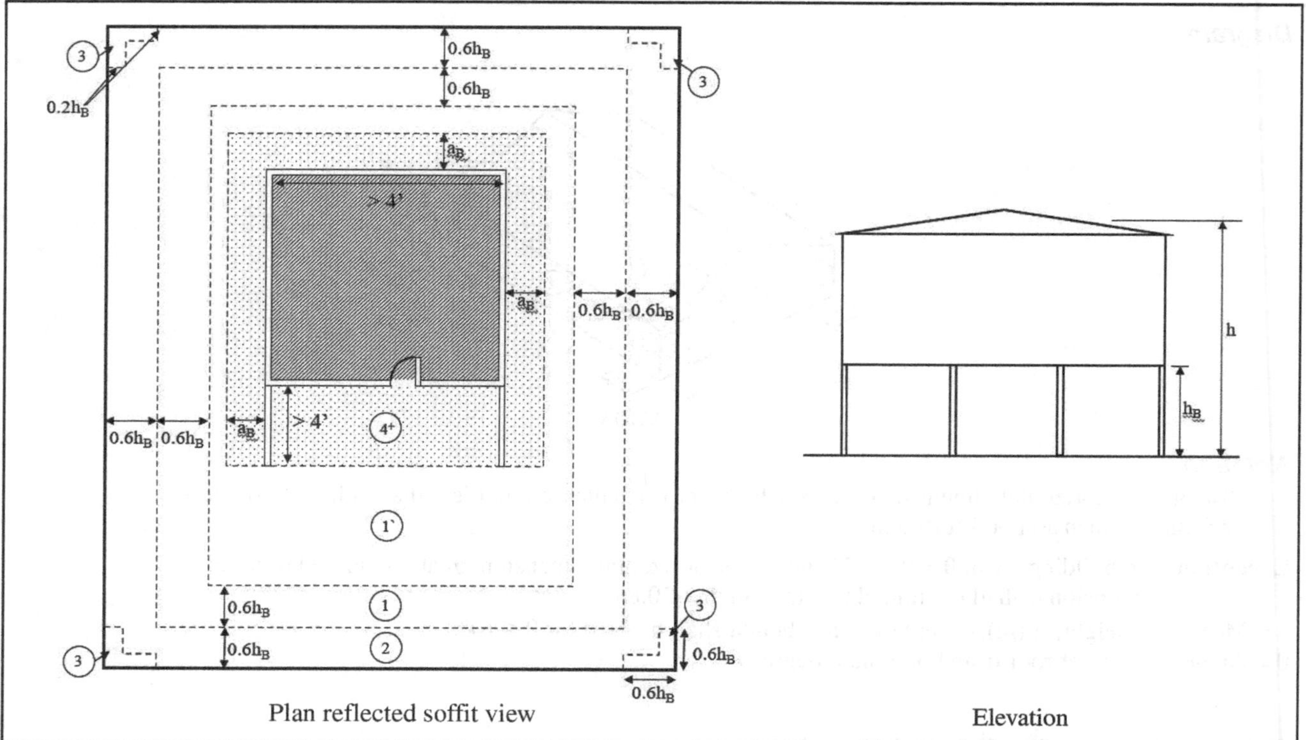

Figure 30.3-1A. Components and cladding [$h \leq 60$ ft (18.3 m)]: external pressure coefficient zones for enclosed, partially enclosed, and partially open elevated buildings with partially enclosed spaces and areas beneath the elevated building—bottom horizontal surface of elevated buildings.

30.4 BUILDING TYPES

The provisions of Section 30.4 are applicable to an enclosed, partially enclosed, or partially open building with a mean roof height [$h > 60$ ft ($h > 18.3$ m)] with a flat roof, pitched roof, gable roof, hip roof, mansard roof, arched roof, or domed roof. The steps required for the determination of wind loads on C&C for these building types are shown in Table 30.4-1.

30.4.1 Conditions For the determination of the design wind pressures on the C&C using the provisions of Section 30.4.2, the conditions indicated on the selected figure(s) shall be applicable to the building under consideration.

30.4.2 Design Wind Pressures Design wind pressures on C&C for all buildings with [$h > 60$ ft ($h < 18.3$ m)] shall be determined from the following equation:

$$p = qK_d(GC_p) - q_iK_d(GC_{pi})(\text{lb/ft}^2) \quad (30.4\text{-}1)$$

$$p = qK_d(GC_p) - q_iK_d(GC_{pi})(\text{N/m}^2) \quad (30.4\text{-}1.\text{SI})$$

where

$q = q_z$ For windward walls calculated at height z above the ground;

$q = q_h$ For leeward walls, sidewalls, and roofs evaluated at height h;

K_d = Wind directionality factor, see Section 26.6;

$q_i = q_h$ For windward walls, sidewalls, leeward walls, and roofs of enclosed and partially open buildings and for negative internal pressure evaluation in partially enclosed buildings;

Table 30.4-1. Steps to determine C&C wind loads for enclosed, partially enclosed, or partially open building with $h > 60$ ft ($h > 18.3$ m)].

Step 1: Determine risk category; see Table 1.5-1.

Step 2: Determine the basic wind speed, V, for applicable risk category; see Figure 26.5-1.

Step 3: Determine wind load parameters:
- Wind directionality factor, K_d; see Section 26.6 and Table 26.6-1.
- Exposure Category B, C, or D; see Section 26.7.
- Topographic factor.
- K_{zt}; see Section 26.8 and Figure 26.8-1.
- Ground elevation factor, K_e; see Section 26.9 and Table 26.9-1.
- Enclosure classification; see Section 26.12.
- Internal pressure coefficient, GC_{pi}; see Section 26.13 and Table 26.13-1.

Step 4: Determine velocity pressure exposure coefficient, K_z or K_h; see Table 26.10-1.

Step 5: Determine velocity pressure, q_h, Equation (26.10-1).

Step 6: Determine external pressure coefficient (GC_p):
- Walls and flat roofs ($\theta < 10°$); see Figure 30.4-1.
- Gable and hip roofs; see Figure 30.3-2 per Note 6 of Figure 30.4-1.
- Arched roofs; see Figure 30.3-8.
- Domed roofs; see Figure 30.3-7.
- Bottom horizontal surface of elevated buildings; see Section 30.4.2.1.

Step 7: Calculate wind pressure, p, Equation (30.4-1).

Diagrams

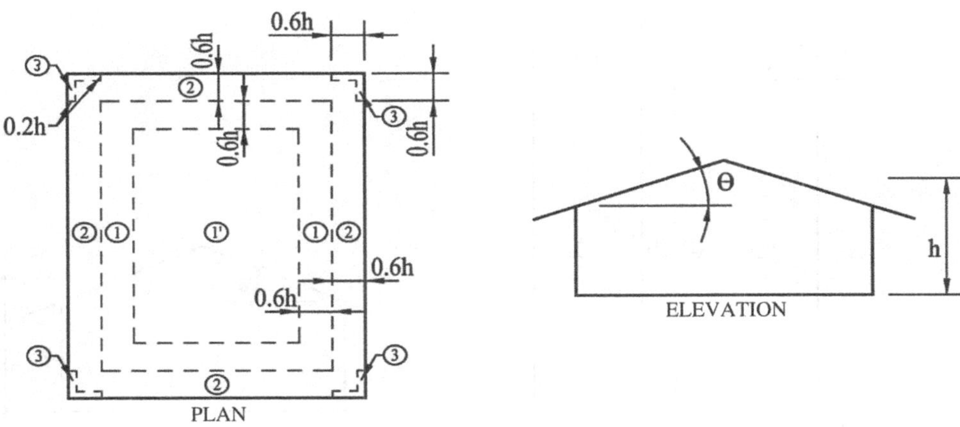

Notation

B = Horizontal dimension of building measured normal to wind direction, ft (m).
h = Eave height shall be used for $\theta = 10°$.
θ = Angle of plane of roof from horizontal, degrees.

External Pressure Coefficients

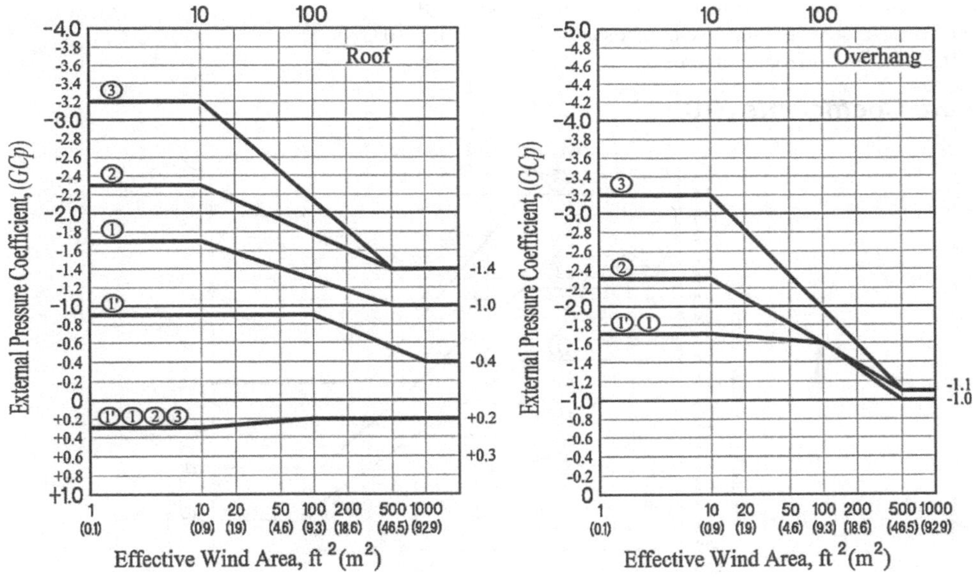

Notes

1. Vertical scale denotes (GC_p) to be used with q_h.
2. Horizontal scale denotes effective wind area A, ft² (m²).
3. Plus and minus signs signify pressures acting toward and away from the surfaces, respectively.
4. Each component shall be designed for maximum positive and negative pressures.
5. If a parapet equal to or higher than 3 ft (0.9 m) is provided around the perimeter of the roof with $\theta \leq 7°$, the negative values of (GC_p) in Zone 3 shall be equal to those for Zone 2, and positive values of (GC_p) in Zones 2 and 3 shall be set equal to those for wall Zones 4 and 5, respectively, in Figure 30.3-1.
6. Values of (GC_p) for roof overhangs include pressure contributions from both upper and lower surfaces.
7. If overhangs exist, the lesser horizontal dimension of the building shall not include any overhang dimension, but the edge distance, a, shall be measured from the outside edge of the overhang.

Figure 30.3-2A. Components and cladding [$h \leq 60$ ft ($h \leq 18.3$ m)]: external pressure coefficients (GC_p) for enclosed, partially enclosed, and partially open buildings—gable roofs, $\theta \leq 7°$.

Diagrams

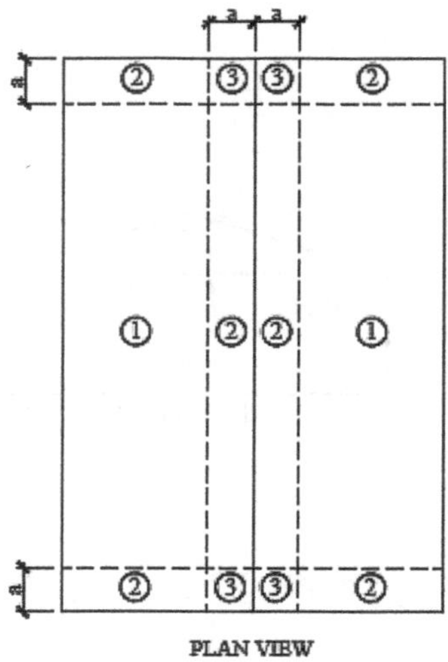

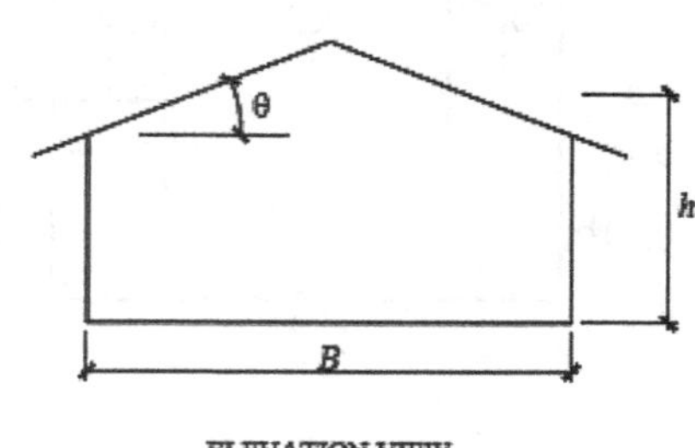

Notation

a = 10% of least horizontal dimension or $0.4h$, whichever is smaller, but not less than either 4% of least horizontal dimension or 3 ft (0.9 m). If an overhang exists, the edge distance shall be measured from the outside edge of the overhang. The horizontal dimensions used to compute the edge distance shall not include any overhang dimensions.

B = Horizontal dimension of building measured normal to wind direction, ft (m).

h = Mean roof height, in ft (m), except that eave height shall be used for $\theta \leq 10°$.

θ = Angle of plane of roof from horizontal, degrees.

External Pressure Coefficients (GC_p)

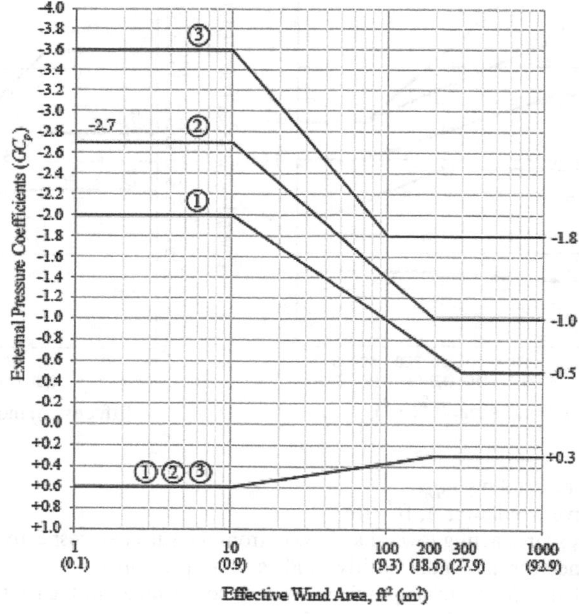

Notes

1. Vertical scale denotes (GCp) to be used with q_h.
2. Horizontal scale denotes effective wind area A, ft^2 (m^2).
3. Plus and minus signs signify pressures acting toward and away from the surfaces, respectively.
4. Each component shall be designed for maximum positive and negative pressures.
5. Values of (GCp) for roof overhangs to be determined in accordance with Section 30.7 Roof Overhangs.

Figure 30.3-2B. Components and cladding [$h \leq 60$ ft ($h \leq 18.3$ m)]: external pressure coefficients, (GC_p), for enclosed, partially enclosed, and partially open buildings—gable roofs, $7° < \theta \leq 20°$.

Diagrams

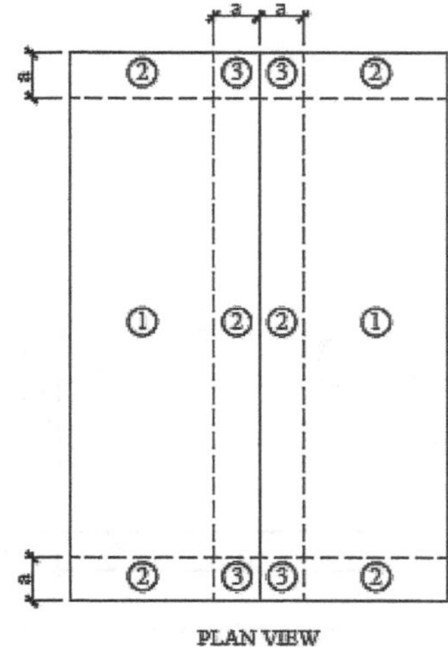

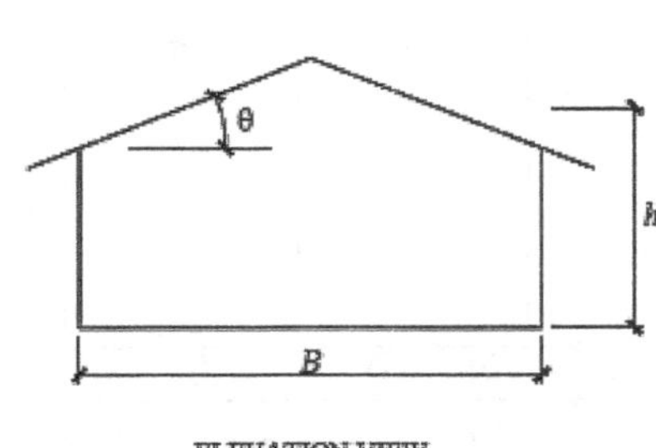

Notation

a = 10% of least horizontal dimension or $0.4h$, whichever is smaller, but not less than either 4% of least horizontal dimension or 3 ft (0.9 m). If an overhang exists, the edge distance shall be measured from the outside edge of the overhang. The horizontal dimensions used to compute the edge distance shall not include any overhang dimensions.

B = Horizontal dimension of building measured normal to wind direction, ft (m).

h = Mean roof height, in ft (m), except that eave height shall be used for $\theta \leq 10°$.

θ = Angle of plane of roof from horizontal, degrees.

External Pressure Coefficients (GC_p)

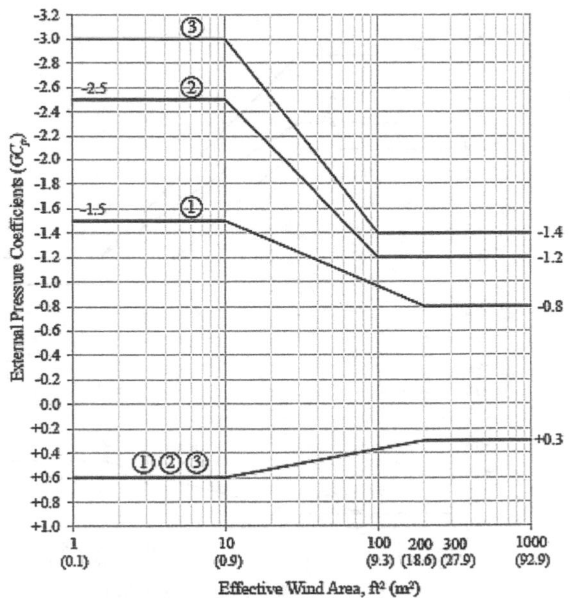

Notes

1. Vertical scale denotes (GCp) to be used with q_h.
2. Horizontal scale denotes effective wind area A, ft^2 (m^2).
3. Plus and minus signs signify pressures acting toward and away from the surfaces, respectively.
4. Each component shall be designed for maximum positive and negative pressures.
5. Values of (GCp) for roof overhangs to be determined in accordance with Section 30.7 Roof Overhangs.

Figure 30.3-2C. Components and cladding [$h \leq$ 60 ft ($h \leq$ 18.3 m)]: external pressure coefficients, (GC_p), for enclosed, partially enclosed, and partially open buildings—gable roofs, 20° < θ ≤ 27°.

Diagrams

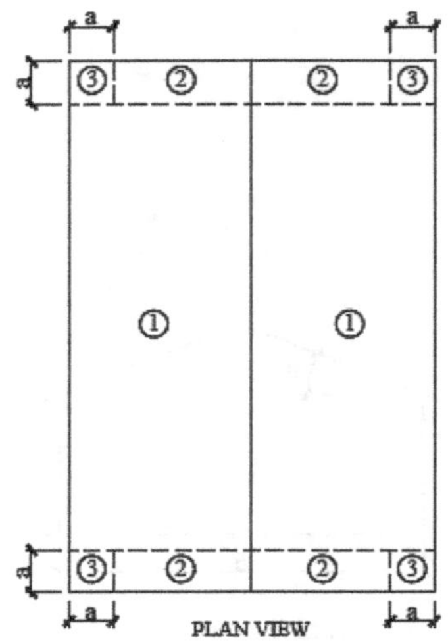

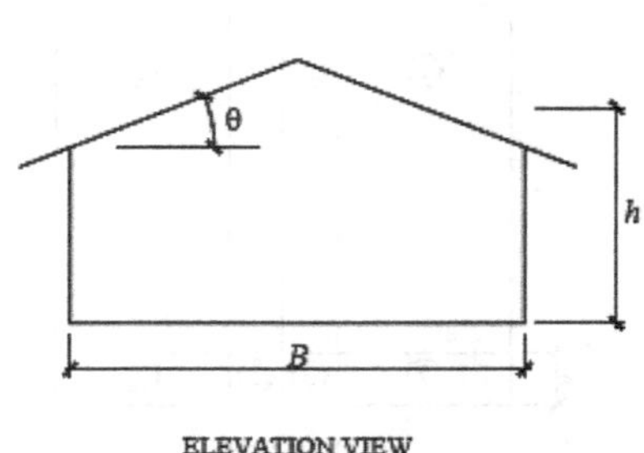

Notation

a = 10% of least horizontal dimension or $0.4h$, whichever is smaller, but not less than either 4% of least horizontal dimension or 3 ft (0.9 m). If an overhang exists, the edge distance shall be measured from the outside edge of the overhang. The horizontal dimensions used to compute the edge distance shall not include any overhang dimensions.

B = Horizontal dimension of building measured normal to wind direction, ft (m).

h = Mean roof height, in ft (m), except that eave height shall be used for $\theta \leq 10°$.

θ = Angle of plane of roof from horizontal, degrees.

External Pressure Coefficients (GC_p)

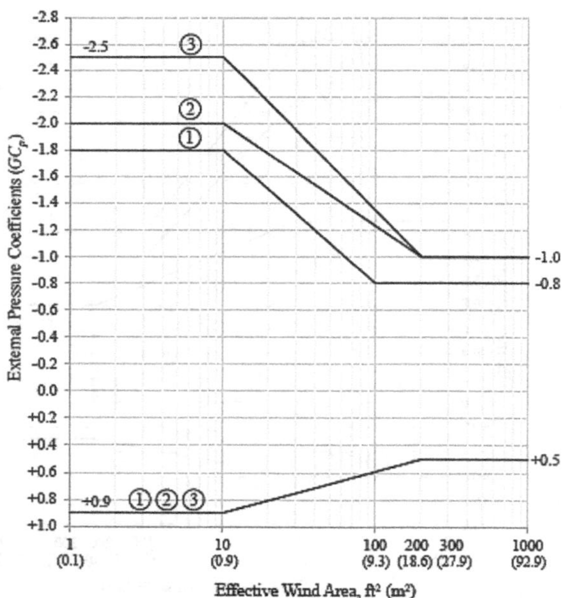

Notes

1. Vertical scale denotes (GCp) to be used with q_h.
2. Horizontal scale denotes effective wind area A, ft^2 (m^2).
3. Plus and minus signs signify pressures acting toward and away from the surfaces, respectively.
4. Each component shall be designed for maximum positive and negative pressures.
5. Values of (GCp) for roof overhangs to be determined in accordance with Section 30.7 Roof Overhangs.

Figure 30.3-2D. Components and cladding [$h \leq 60$ ft ($h \leq 18.3$ m)]: external pressure coefficients, (GC_p), for enclosed, partially enclosed, and partially open buildings—gable roofs, $27° < \theta \leq 45°$.

Diagrams

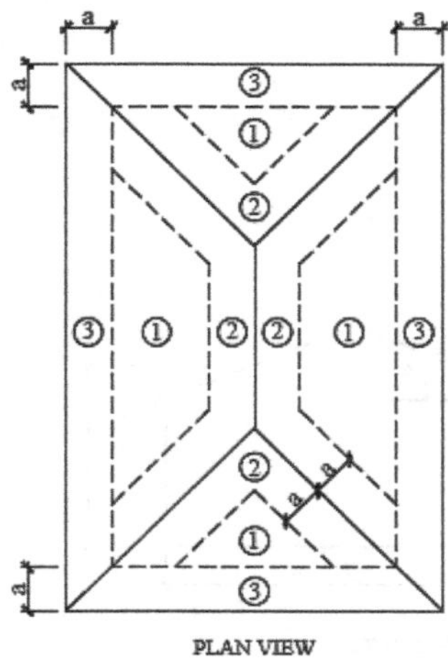

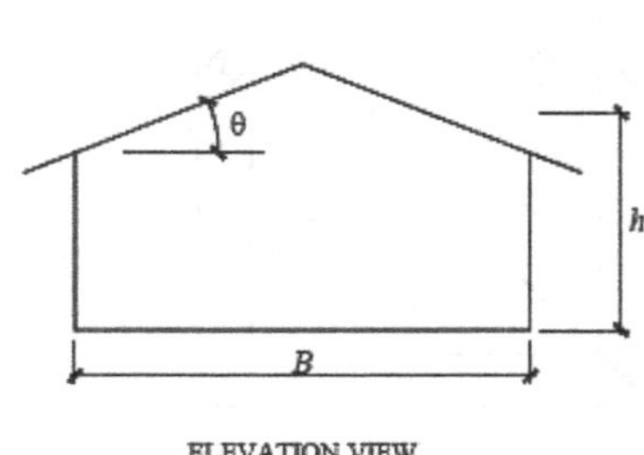

Notation

a = 10% of least horizontal dimension or $0.4h$, whichever is smaller, but not less than either 4% of least horizontal dimension or 3 ft (0.9 m). If an overhang exists, the edge distance shall be measured from the outside edge of the overhang. The horizontal dimensions used to compute the edge distance shall not include any overhang dimensions.

B = Horizontal dimension of building measured normal to wind direction, ft (m).

h = Mean roof height, in ft (m), except that eave height shall be used for $\theta \leq 10°$.

θ = Angle of plane of roof from horizontal, degrees.

External Pressure Coefficients (GC_p)

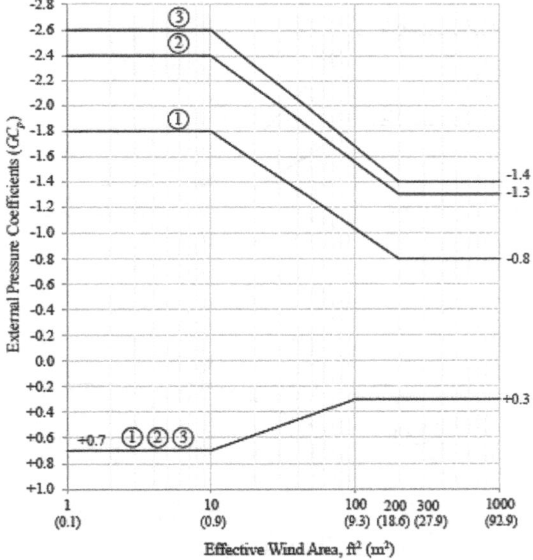

Notes

1. Vertical scale denotes (GC_p) to be used with q_h.
2. Horizontal scale denotes effective wind area A, ft^2 (m^2).
3. Plus and minus signs signify pressures acting toward and away from the surfaces, respectively.
4. Each component shall be designed for maximum positive and negative pressures.
5. Values of (GC_p) for roof overhangs to be determined in accordance with Section 30.7 Roof Overhangs.

Figure 30.3-2E. Components and cladding [$h \leq 60$ ft ($h \leq 18.3$ m)]: external pressure coefficients, (GC_p), for enclosed and partially enclosed, and partially open buildings—hip roofs, 7° < θ ≤ 20°.

Diagrams

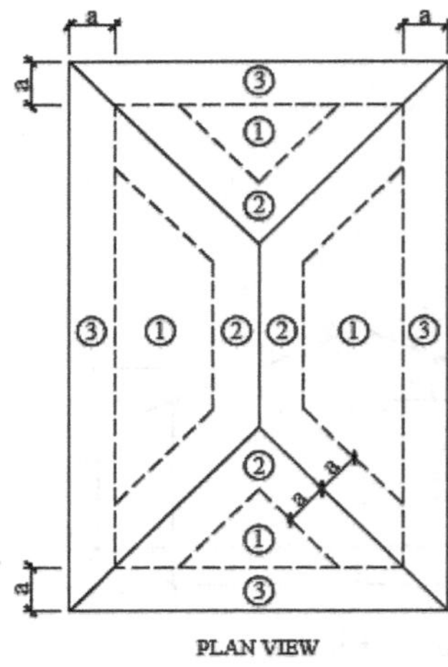

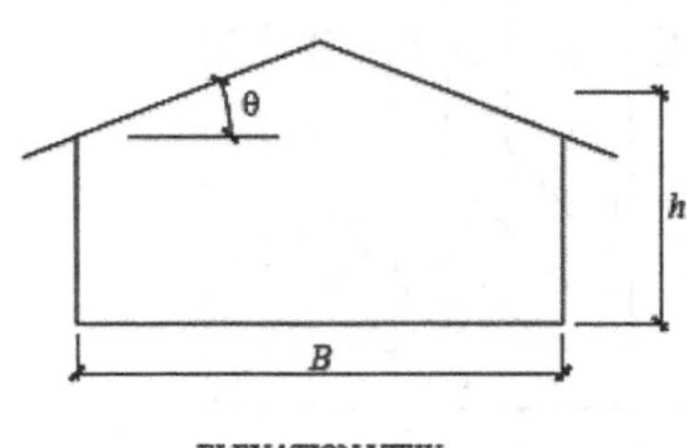

Notation

a = 10% of least horizontal dimension or $0.4h$, whichever is smaller, but not less than either 4% of least horizontal dimension or 3 ft (0.9 m). If an overhang exists, the edge distance shall be measured from the outside edge of the overhang. The horizontal dimensions used to compute the edge distance shall not include any overhang dimensions.
B = Horizontal dimension of building measured normal to wind direction, ft (m).
h = Mean roof height, in ft (m), except that eave height shall be used for $\theta \leq 10°$.
θ = Angle of plane of roof from horizontal, degrees.

External Pressure Coefficients (GC_p)

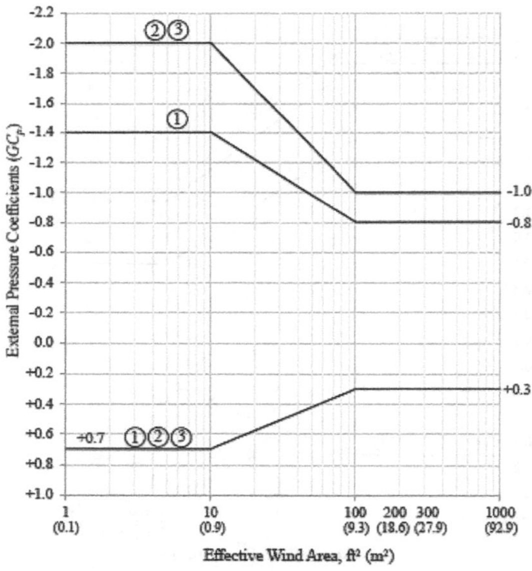

Notes

1. Vertical scale denotes (GCp) to be used with q_h.
2. Horizontal scale denotes effective wind area A, ft^2 (m^2).
3. Plus and minus signs signify pressures acting toward and away from the surfaces, respectively.
4. Each component shall be designed for maximum positive and negative pressures.
5. Values of (GCp) for roof overhangs to be determined in accordance with Section 30.7 Roof Overhangs.

Figure 30.3-2F. Components and cladding [$h \leq 60$ ft ($h \leq 18.3$ m)]: external pressure coefficients, (GC_p), for enclosed, partially enclosed, and partially open buildings—hip roofs, $20° < \theta \leq 27°$.

Diagrams

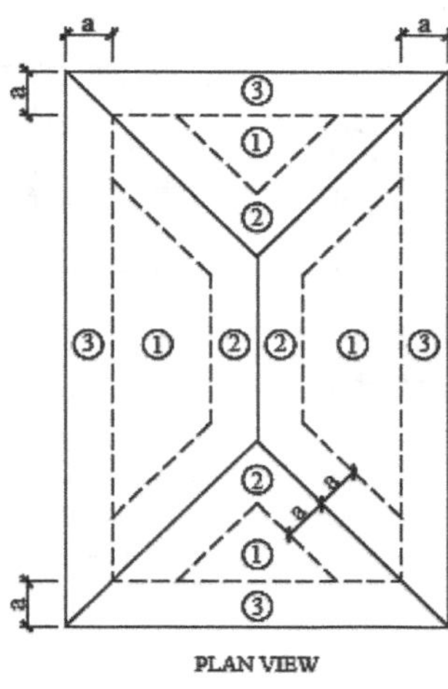

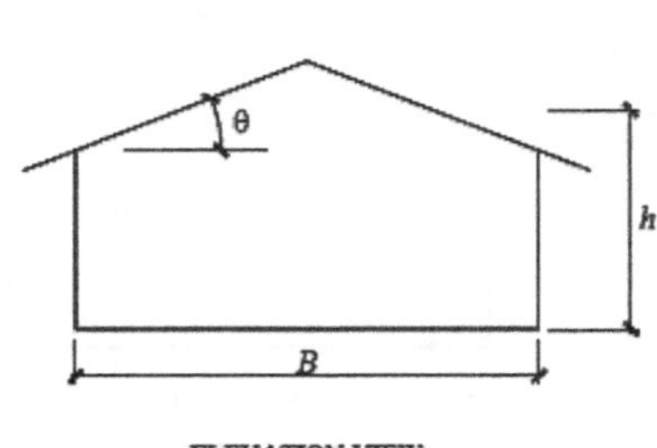

Notation

a = 10% of least horizontal dimension or $0.4h$, whichever is smaller, but not less than either 4% of least horizontal dimension or 3 ft (0.9 m). If an overhang exists, the edge distance shall be measured from the outside edge of the overhang. The horizontal dimensions used to compute the edge distance shall not include any overhang dimensions.

B = Horizontal dimension of building measured normal to wind direction, ft (m).

h = Mean roof height, in ft (m), except that eave height shall be used for $\theta \leq 10°$.

θ = Angle of plane of roof from horizontal, degrees.

External Pressure Coefficients (GC_p)

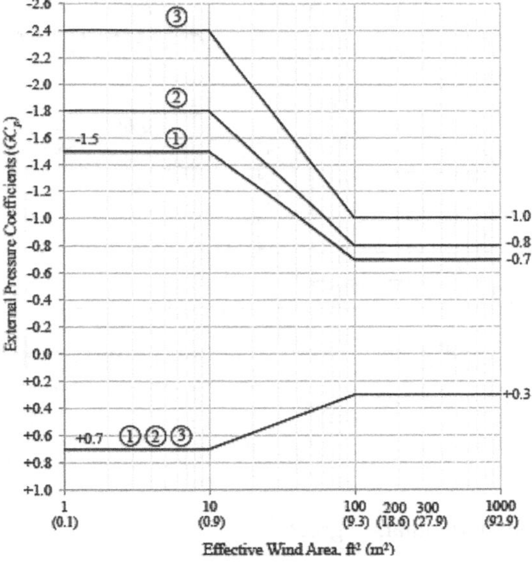

Notes

1. Vertical scale denotes (GCp) to be used with q_h.
2. Horizontal scale denotes effective wind area A, ft^2 (m^2).
3. Plus and minus signs signify pressures acting toward and away from the surfaces, respectively.
4. Each component shall be designed for maximum positive and negative pressures.
5. Values of (GCp) for roof overhangs to be determined in accordance with Section 30.7 Roof Overhangs.
6. For roof slopes $27° < \theta_1 < 45°$, interpolate the (GC_P) coefficients from Figures 30.3-2F and 30.3-2G, for each zone of interest. Use the following interpolation formula: $\frac{[GC_P(2G) - GC_P(2F)]*(\theta_1 - 27°)}{(45°-27°)} + GC_P(2F)$

Figure 30.3-2G. Components and cladding [$h \leq 60$ ft ($h \leq 18.3$ m)] external pressure coefficients, (GC_p), for enclosed, partially enclosed, and partially open buildings—hip roofs, $\theta = 45°$.

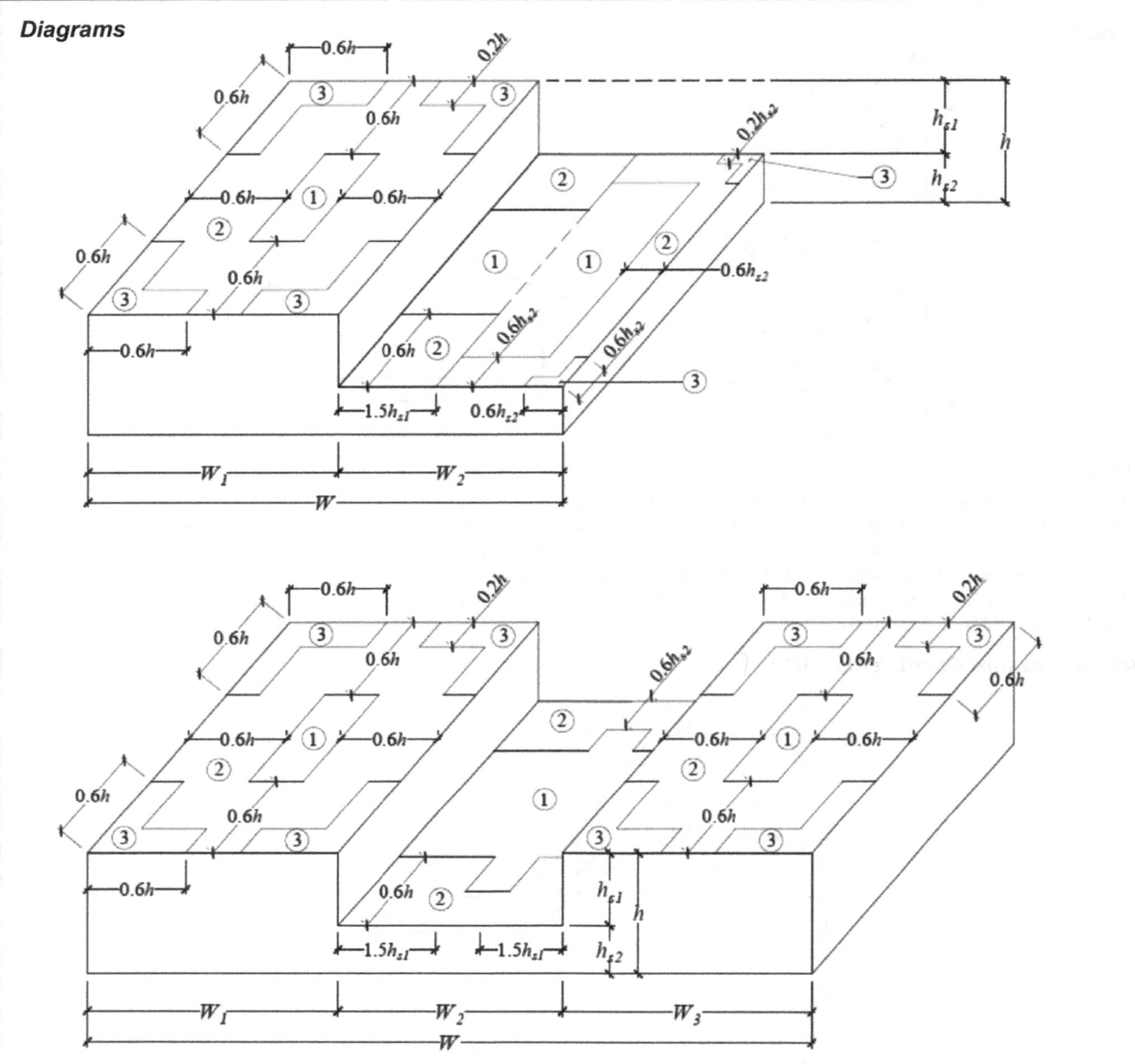

Notation

h = Mean roof height, ft (m)
W = Building width
Θ = Angle of the plane of the roof from horizontal, degrees

Notes

On the lower level of flat, stepped roofs shown here, the zone designations and pressure coefficients shown in Figure 30.3-2A shall apply. For the upper figure, the zones for the lower height roof are to be applied from the edge of the roof inward towards the taller building.

Figure 30.3-3. Components and cladding [$h \leq 60$ ft ($h \leq 18.3$ m)]: external pressure coefficients, (GC_p), for enclosed, partially enclosed, and partially open buildings, $\theta \leq 7°$—stepped roofs.

Diagrams

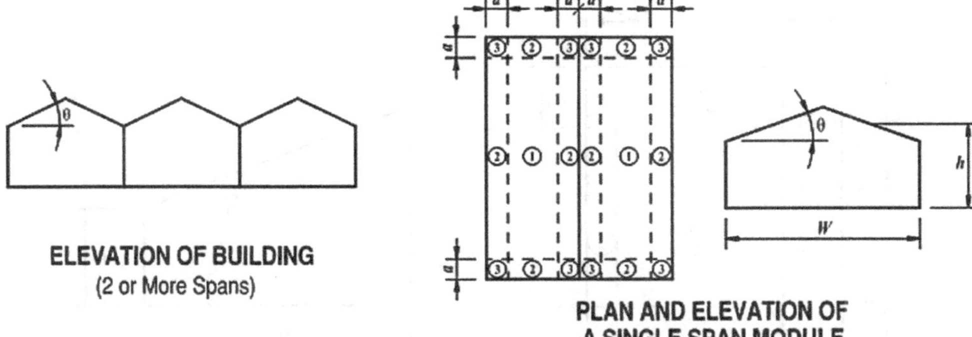

Notation

a = 10% of least horizontal dimension of a single-span module or $0.4h$, whichever is smaller, but not less than either 4% of least horizontal dimension of a single-span module or 3 ft (0.9 m).

h = Mean roof height, ft (m), except that eave height shall be used for $\theta \leq 10°$.

W = Building module width, ft (m).

θ = Angle of plane of roof from horizontal, degrees.

External Pressure Coefficients

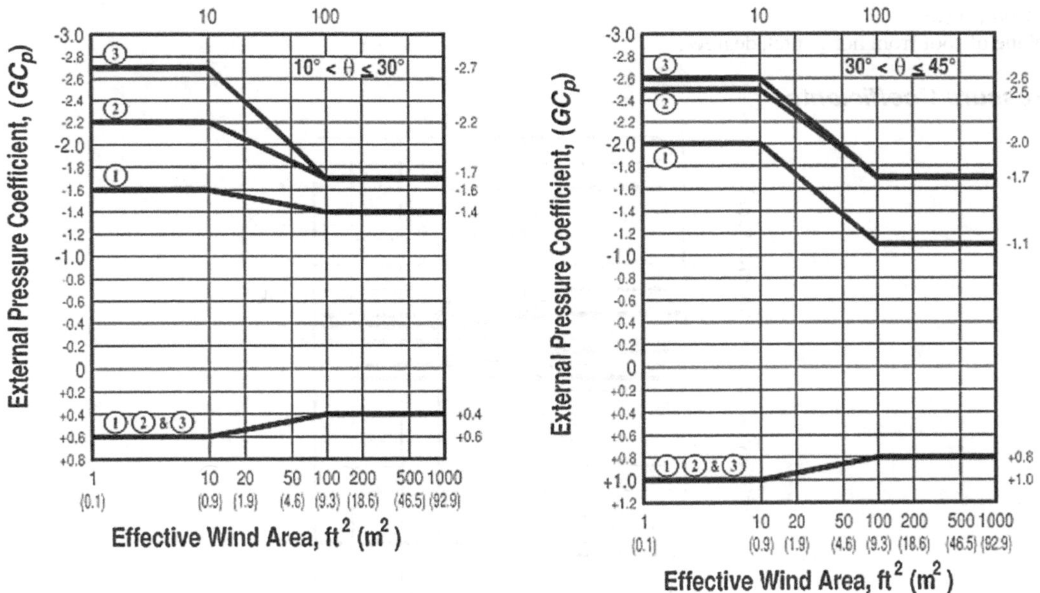

Notes

1. Vertical scale denotes (GCp) to be used with q_h.
2. Horizontal scale denotes effective wind area A, ft^2 (m^2).
3. Plus and minus signs signify pressures acting toward and away from the surfaces, respectively.
4. Each component shall be designed for maximum positive and negative pressures.
5. For $\theta \leq 10°$, values of (GCp) from Fig. 30.3-2A shall be used.

Figure 30.3-4. Components and cladding [$h \leq$ 60 ft ($h \leq$ 18.3 m)]: external pressure coefficients, (GC_p), for enclosed, partially enclosed, partially open buildings—multispan gable roofs.

Diagrams

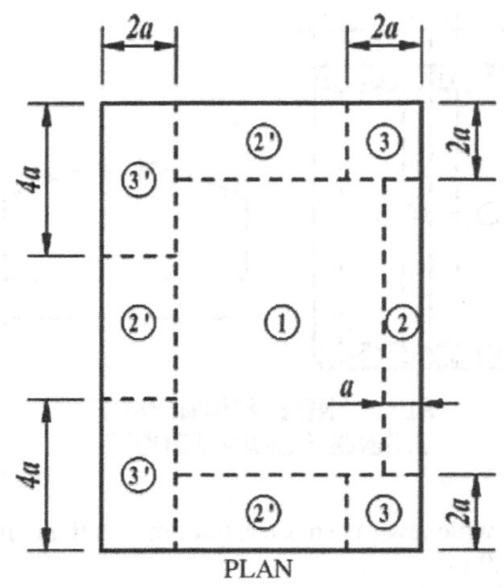

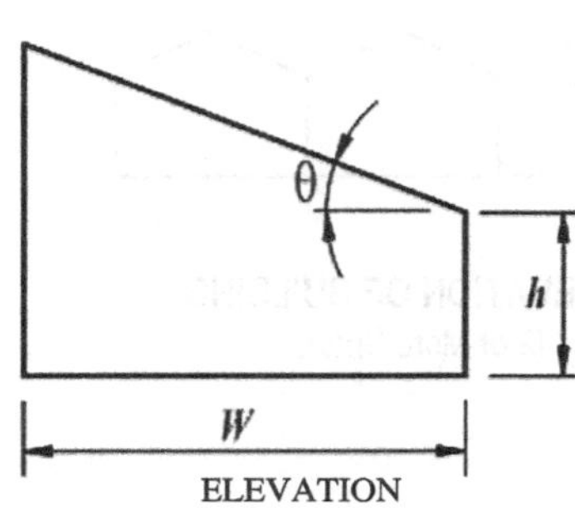

Notation

a = 10% of least horizontal dimension or $0.4h$, whichever is smaller, but not less than either 4% of least horizontal dimension or 3 ft (0.9 m).

h = Eave height shall be used for $\theta \leq 10°$.

W = Building width, ft (m).

θ = Angle of plane of roof from horizontal, degrees.

External Pressure Coefficients

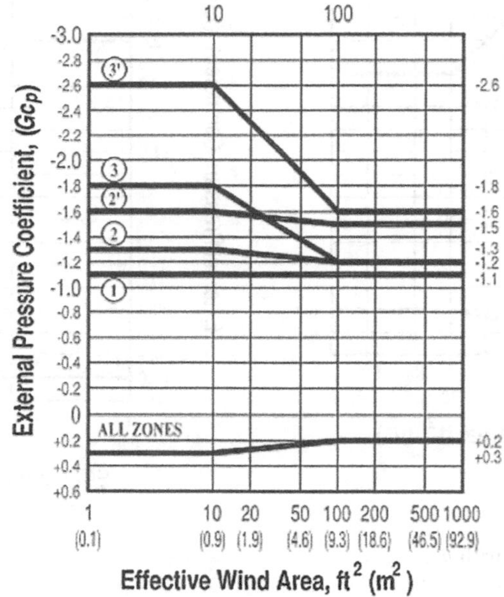

Notes

1. Vertical scale denotes (GCp) to be used with q_h.
2. Horizontal scale denotes effective wind area A, ft^2 (m^2).
3. Plus and minus signs signify pressures acting toward and away from the surfaces, respectively.
4. Each component shall be designed for maximum positive and negative pressures.
5. For $\theta \leq 3°$, values of (GCp) from Fig. 30.3-2A shall be used.

Figure 30.3-5A. Components and cladding [$h \leq 60$ ft ($h \leq 18.3$ m)]: external pressure coefficients, (GC_p), for enclosed, partially enclosed, and partially open buildings—monoslope roofs, $3° < \theta \leq 10°$.

Diagrams

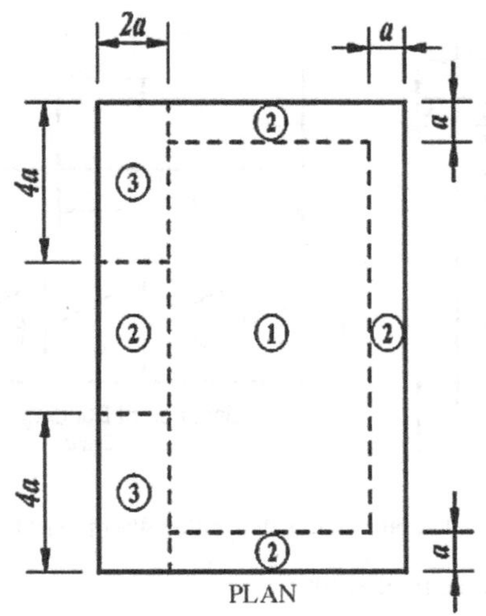

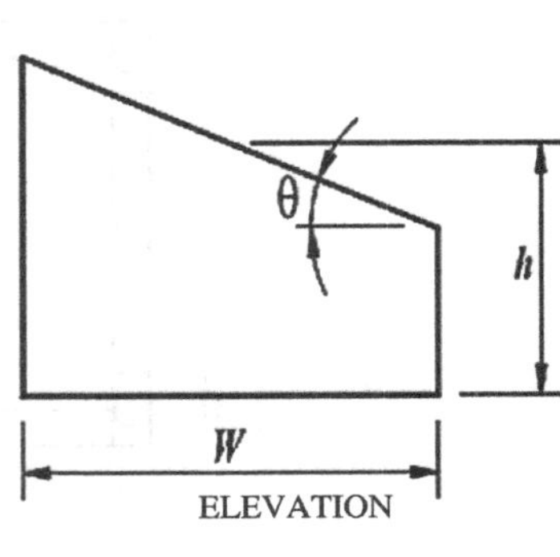

Notation

a = 10% of least horizontal dimension or $0.4h$, whichever is smaller, but not less than either 4% of least horizontal dimension or 3 ft (0.9 m).
h = Mean roof height, ft (m).
W = Building width, ft (m).
θ = Angle of plane of roof from horizontal, degrees.

External Pressure Coefficients

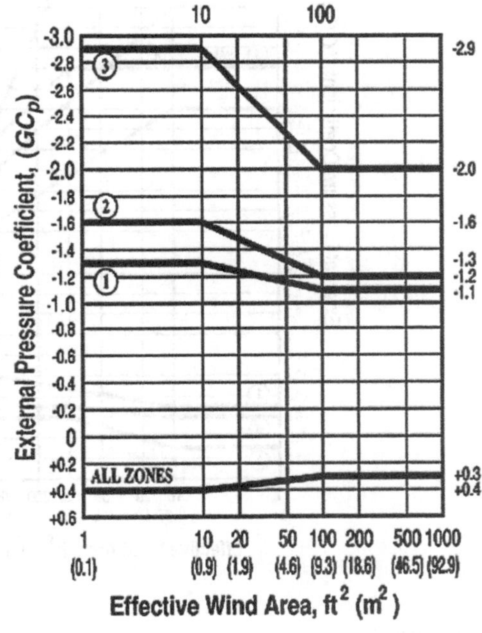

Notes

1. Vertical scale denotes (GCp) to be used with q_h.
2. Horizontal scale denotes effective wind area A, ft^2 (m^2).
3. Plus and minus signs signify pressures acting toward and away from the surfaces, respectively.
4. Each component shall be designed for maximum positive and negative pressures.

Figure 30.3-5B. Components and cladding [$h \leq 60$ ft ($h \leq 18.3$ m)]: external pressure coefficients, (GC_p), for enclosed, partially enclosed, and partially open buildings—monoslope roofs, $10° < \theta \leq 30°$.

Diagrams

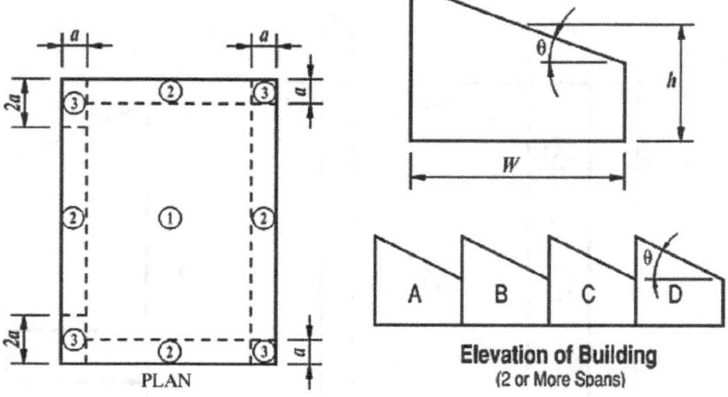

Notation

a = 10% of least horizontal dimension or $0.4h$, whichever is smaller, but not less than either 4% of least horizontal dimension or 3 ft (0.9 m).
h = Mean roof height, ft (m), except that eave height shall be used for $\theta \leq 10°$.
W = Building module width, ft (m).
θ = Angle of plane of roof from horizontal, degrees.

External Pressure Coefficients

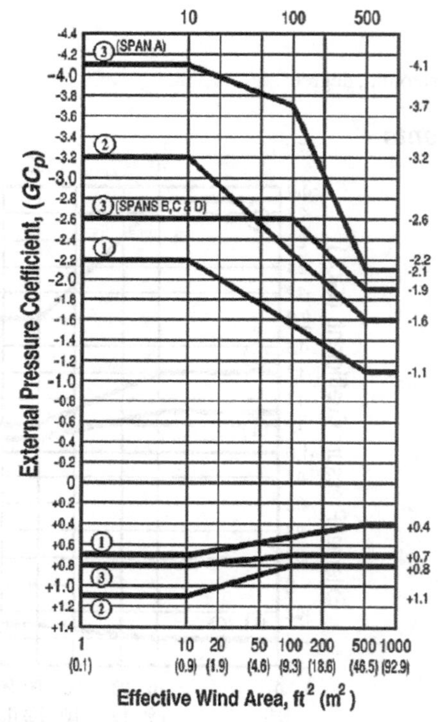

Notes

1. Vertical scale denotes (GCp) to be used with q_h.
2. Horizontal scale denotes effective wind area A, ft^2 (m^2).
3. Plus and minus signs signify pressures acting toward and away from the surfaces, respectively.
4. Each component shall be designed for maximum positive and negative pressures.
5. For $\theta \leq 10°$, values of (GCp) from Figure. 30.3-2A shall be used.

Figure 30.3-6. Components and cladding [$h \leq 60$ ft ($h \leq 18.3$ m)]: external pressure coefficients, (GC_p), for enclosed, partially enclosed, and partially open buildings—sawtooth roofs.

Diagram

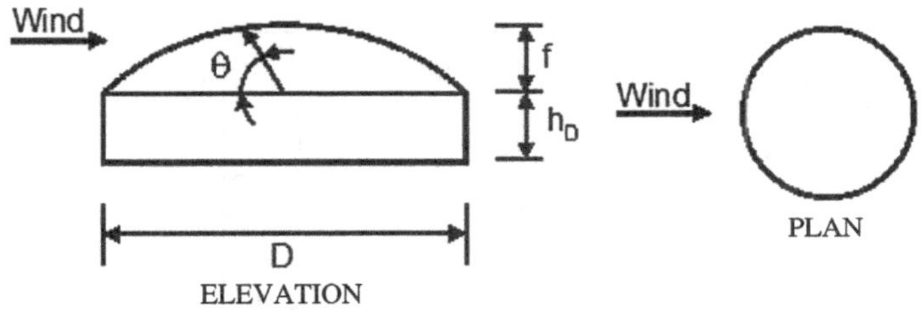

Notation

f = Dome rise, ft (m).
D = Diameter of a circular structure or member, ft (m).
h_D = Height to base of dome, ft (m).
θ = Angle of plane of roof from horizontal, degrees.

Coefficients for Domes with a Circular Base

External Pressure	Negative Pressures	Positive Pressures	Positive Pressures
θ, degrees	0–90	0–60	61–90
(GCp)	–0.9	+0.9	+0.5

Notes

1. Values denote (GCp) to be used with $q(h_D+f)$ where $h_D + f$ is the height at the top of the dome.
2. Plus and minus signs signify pressures acting toward and away from the surfaces, respectively.
3. Each component shall be designed for the maximum positive and negative pressures.
4. Values apply to $0 \le h_D/D \le 0.5$, $0.2 \le f/D \le 0.5$.
5. $\theta = 0$ degrees on dome springline, $\theta = 90$ degrees at dome center top point. f is measured from springline to top.

Figure 30.3-7. Components and cladding (all heights): external pressure coefficients, (GC_p), for enclosed, partially enclosed, and partially open buildings and structures—domed roofs.

$q_i = q_z$ For positive internal pressure evaluation in partially enclosed buildings where height z is defined as the level of the highest opening in the building that could affect the positive internal pressure. For positive internal pressure evaluation, q_i may conservatively be evaluated at height $h(q_i = q_h)$;

(GC_p) = External pressure coefficients given in

- Figure 30.4-1 for walls and flat roofs,
- Figure 30.3-8 for arched roofs,
- Figure 30.3-7 for domed roofs,
- Note 6 of Figure 30.4-1 for other roof angles and geometries; and
- Figure 30.4-1 for bottom surfaces of elevated buildings.

(GC_{pi}) = Internal pressure coefficient given in Table 26.13-1.

q and q_i shall be evaluated using exposure as defined in Section 26.7.3.

30.4.2.1 Bottom Horizontal Surface of Elevated Buildings Design wind pressures for C&C elements on the bottom flat horizontal surface of elevated buildings shall be determined using the roof pressure coefficients from Figure 30.4-1 with the following modifications:

1. The velocity pressure, q, used in Equation 30.4-1 shall be calculated at a height equal to the height above grade of the bottom horizontal surface plus 25% of the height of the elevated building above the horizontal bottom surface, calculated as $[h_B + 0.25(h - h_B)]$. For elevated buildings with a flat bottom horizontal building surface and situated on a slope, h_B shall be taken as the maximum height between the slope and the bottom of the elevated building.
2. Areas of the horizontal surface above partially enclosed spaces and areas extending a distance a_B perpendicular to walls with plan dimension greater than 4 ft (1.2 m), as shown by the shaded regions in Figure 30.4-1A, shall be designed to resist positive pressures equal to the Zone 4 wall pressures obtained using Figure 30.4-1. The value of a_B shall equal 0.4 h_B or the width of the wall, whichever is smaller, for determining zone dimensions from Figure 30.4-1A.

The loading convention shall denote downward loading on the bottom surface with negative pressure coefficients and upward loading on the bottom surface with positive pressure coefficients.

EXCEPTION: In buildings with a mean roof height h greater than 60 ft (18.3 m) and less than 90 ft (27.4 m), Figures 30.3-1 through 30.3-6 shall be permitted to be used if the mean roof height h does not exceed least horizontal dimension.

PART 3: OPEN BUILDINGS

User Note: Use Part 3 of Chapter 30 for determining wind pressures for C&C of *open buildings* that have pitched, monoslope, or troughed roofs. These provisions are based on the Directional Procedure with *wind pressures calculated from the specified equation* applicable to each roof surface.

Diagrams

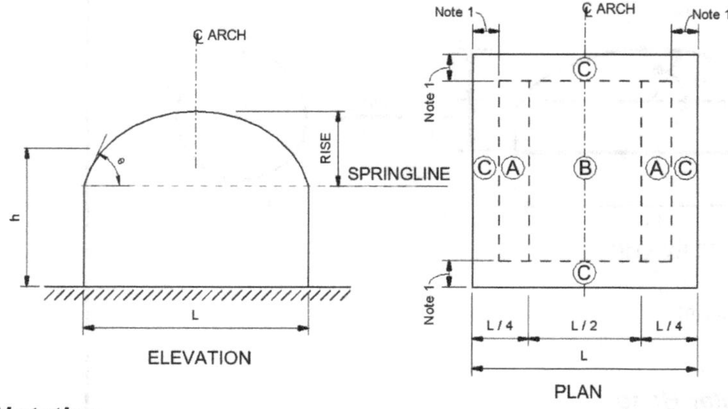

Notation

L = Horizontal dimension of building, ft (m), measured normal to ridge.

r = Rise-to-span ratio = 'rise'/L.

h = Roof height as defined in Chapter 26 for roof angle θ.

θ = Angle of plane of roof from horizontal, in degrees, measured at eave.

External Pressure Coefficient, GC_p

Conditions	Rise-to-Span Ratio, r	(GC_p)	
		Zone A	Zone B
Roof on elevated structure	$0 < r < 0.2$	–1.08	$-0.84-1.2r$
	$0.2 \le r < 0.3$	$1.8r-0.36$ $7.2r-2.52$ –0.6	$-0.84-1.2r$
	$0.3 \le r \le 0.6$	$3.3r-0.84$ –0.6	$-0.84-1.2r$
Roof springing from ground level	$0 < r \le 0.6$	$1.68r$ –0.6	$-0.84-1.2r$

Notes

1. At roof perimeter Zone C, use the roof zones (Zone 2 and 3) and external pressure coefficients in Fig. 30.3-2A, B, C, and D with "θ" based on springline slope. In the case of Fig. 30.3-2A, use 0.6h as the size of the roof perimeter Zone C. For roof Zones A and B, use external pressure coefficients shown in this table.
2. Plus and minus signs signify pressures acting toward and away from the surfaces, respectively.
3. Where multiple values of (GC_p) are listed, the roof zone is subjected to either positive or negative pressures and the structure shall be designed for all conditions.

Figure 30.3-8. Components and cladding (all heights): external pressure coefficients, (GC_p), for enclosed, partially enclosed, and partially open buildings and structures—arched roofs.

30.5 BUILDING TYPES

The provisions of Section 30.5 are applicable to an open building of all heights that has a pitched free roof, monosloped free roof, or troughed free roof. The steps required for the determination of wind loads on C&C for these building types is shown in Table 30.5-1.

30.5.1 Conditions For the determination of the design wind pressures on C&Cs using the provisions of Section 30.5.2, the conditions indicated on the selected figure(s) shall be applicable to the building under consideration.

30.5.2 Design Wind Pressures The net design wind pressure for component and cladding elements of open buildings of all heights with monoslope, pitched, and troughed roofs shall be determined by the following equation:

$$p = q_h K_d G C_N \quad (30.5\text{-}1)$$

where

q_h = Velocity pressure evaluated at mean roof height h using the exposure as defined in Section 26.7.3 that results in the highest wind loads for any wind direction at the site; and
K_d = Wind directionality factor, see Section 26.6;
G = Gust-effect factor from Section 26.11; and
C_N = Net pressure coefficient given in

- Figure 30.5-1 for monosloped roof,
- Figure 30.5-2 for pitched roof, and
- Figure 30.5-3 for troughed roof.

Net pressure coefficients, C_N, include contributions from top and bottom surfaces. All load cases shown for each roof angle shall be investigated. Plus and minus signs signify pressure acting toward and away from the top surface of the roof, respectively.

PART 4: BUILDING APPURTENANCES, ROOFTOP STRUCTURES AND EQUIPMENT

> **User Note:** Use Part 4 of Chapter 30 for determining wind pressures for C&C on roof overhangs or parapets of buildings. These provisions are based on the Directional Procedure with *wind pressures calculated from the specified equation* applicable to each roof overhang or parapet surface.

Table 30.5-1. Steps to Determine C&C Wind Loads for Open Buildings.

Step 1: Determine risk category; see Table 1.5-1.
Step 2: Determine the basic wind speed, V, for applicable risk category; see Figure 26.5-1.
Step 3: Determine wind load parameters:

- Wind directionality factor, K_d, see Section 26.6 and Table 26.6-1.
- Exposure Category B, C, or D; see Section 26.7.
- Topographic factor, K_{zt}; see Section 26.8 and Figure 26.8-1.
- Ground elevation factor, K_e; see Section 26.9 and Table 26.9-1.
- Gust-effect factor, G; see Section 26.11.

Step 4: Determine velocity pressure exposure coefficient, K_z or K_h; see Table 26.10-1.
Step 5: Determine velocity pressure, q_h, Equation (26.10-1).
Step 6: Determine net pressure coefficients, C_N:

- Monoslope roof; see Figure 30.5-1.
- Pitched roof; see Figure 30.5-2.
- Troughed roof; see Figure 30.5-3.

Step 7: Calculate wind pressure, p, Equation (30.5-1).

30.6 PARAPETS

The design wind pressure for C&C elements of parapets for all building types and heights shall be determined from the following equation:

$$p = q_p K_d((GC_p) - (GC_{pi})) \quad (30.6\text{-}1)$$

where

q_p = Velocity pressure evaluated at the top of the parapet;
K_d = Wind directionality factor, see Section 26.6;
(GC_p) = External pressure coefficient given in

- Figure 30.3-1 for walls with $h \le 60$ ft (18.3 m);
- Figure 30.3-2A–C for flat roofs, gable roofs, and hip roofs;
- Figure 30.3-3 for stepped roofs;
- Figure 30.3-4 for multispan gable roofs;
- Figure 30.3-5A–B for monoslope roofs;
- Figure 30.3-6 for sawtooth roofs;
- Figure 30.3-7 for domed roofs of all heights;
- Figure 30.4-1 for walls and flat roofs with $h > 60$ ft (18.3 m);
- Figure 30.3-8 for arched roofs.

(GC_{pi}) = Internal pressure coefficient from Table 26.13-1, based on the porosity of the parapet envelope.

Two load cases from Figure 30.6-1 shall be considered:

- *Load Case A:* Windward parapet shall consist of applying the applicable positive wall pressure from Figure 30.3-1 [$h \le 60$ft ($h \le 18.3m$)] or Figure 30.4-1 [$h > 60$ft ($h > 18.3m$)] to the

Table 30.6-1. Steps to Determine C&C Wind Loads for Parapets.

Step 1: Determine risk category of building; see Table 1.5-1.
Step 2: Determine the basic wind speed, V, for applicable risk category; see Figure. 26.5-1.
Step 3: Determine wind load parameters:

- Wind directionality factor, K_d; see Section 26.6 and Table 26.6-1.
- Exposure Category B, C, or D; see Section 26.7.
- Topographic factor, K_{zt}; see Section 26.8 and Figure 26.8-1.
- Ground elevation factor, K_e; see Section 26.9 and Table 26.9-1.
- Enclosure classification; see Section 26.12.
- Internal pressure coefficient (GC_{pi}); see Section 26.13 and Table 26.13-1.

Step 4: Determine velocity pressure exposure coefficient, K_h, at top of the parapet; see Table 26.10-1.
Step 5: Determine velocity pressure, q_p, at the top of the parapet using Equation (26.10-1).
Step 6: Determine external pressure coefficient for wall and roof surfaces adjacent to parapet, (GC_p):

- Walls with $h \le 60ft$ (18.3 m); see Figure 30.3-1.
- Flat, gable, and hip roofs; see Figure 30.3-2A–I.
- Stepped roofs; see Figure 30.3-3.
- Multispan gable roofs; see Figure 30.3-4.
- Monoslope roofs; see Figure 30.3-5A–B.
- Sawtooth roofs; see Figure 30.3-6.
- Domed roofs of all heights; see Figure 30.3-7.
- Walls and flat roofs with $h > 60ft$ ($h > 18.3m$); see Figure 30.4-1.
- Arched roofs; see Figure 30.3-8.

Step 7: Calculate wind pressure, p, using Equation (30.6-1) on windward and leeward face of parapet, considering two load cases (Case A and Case B) as shown in Figure 30.6-1.

Diagrams

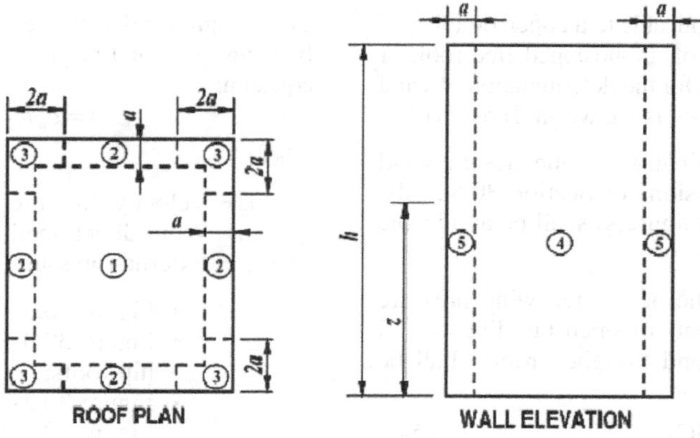

Notation

a = 10% of least horizontal dimension, but not less than 3 ft (0.9 m).
h = Mean roof height, ft (m), except that eave height shall be used for $\theta \leq 10°$.
z = Height above ground, ft (m).
θ = Angle of plane of roof from horizontal, degrees.

External Pressure Coefficients

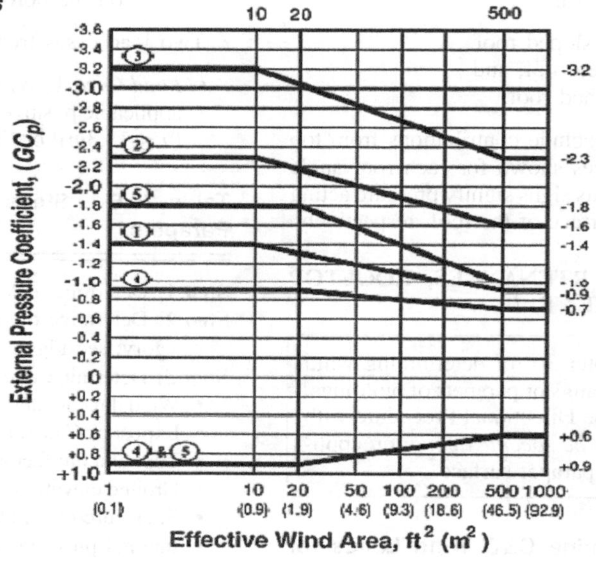

Notes

1. Vertical scale denotes (GCp) to be used with appropriate q_z or q_h.
2. Horizontal scale denotes effective wind area A, ft^2 (m^2)
3. Plus and minus signs signify pressures acting toward and away from the surfaces, respectively.
4. Use q_z with positive values of (GCp) and q_h with negative values of (GCp).
5. Each component shall be designed for maximum positive and negative pressures.
6. Coefficients are for roofs with angle $\theta \leq 7°$. For other roof angles and geometry, use (GCp) values from Fig. 30.3-2A–2I and Fig. 30.3-5A,5B and attendant q_h based on exposure defined in Section 26.7.
7. If a parapet equal to or higher than 3 ft (0.9 m) is provided around the perimeter of the roof with $\theta \leq 10°$, Zone 3 shall be treated as Zone 2.

Figure 30.4-1. Components and cladding, part 2 [h > 60ft (h > 18.3m)]: external pressure coefficients, (GC_p), for enclosed, partially enclosed, partially open buildings—walls and roofs.

Diagrams

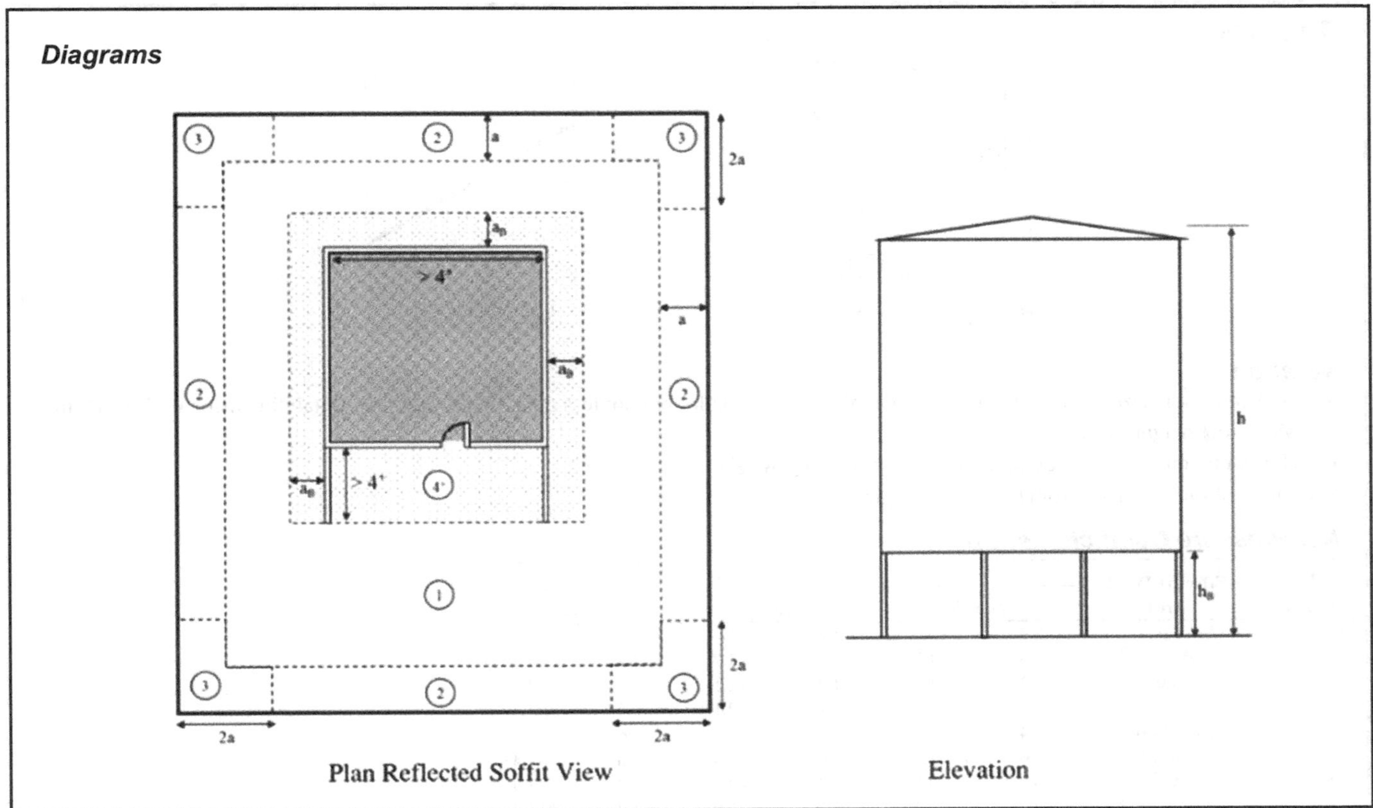

Figure 30.4-1A. Components and cladding, part 3 [$h > 60$ ft ($h > 18.3$ m)]: external pressure coefficient zones for enclosed, partially enclosed, and partially open elevated buildings with partially enclosed spaces and areas beneath the elevated building—bottom horizontal surface of elevated buildings.

windward surface of the parapet while applying the applicable negative edge or corner zone roof pressure from Figures 30.3-2A, B, or C, 30.3-3, 30.3-4, 30.3-5A or B, 30.3-6, 30.3-7, Figure 30.3-8, or Figure 30.4-1 [$h > 60$ft ($h > 18.3m$)] as applicable to the leeward surface of the parapet.

- *Load Case B:* Leeward parapet shall consist of applying the applicable positive wall pressure from Figure 30.3-1 [$h \leq 60$ft ($h \leq 18.3m$)] or Figure 30.4-1 [$h > 60$ft ($h > 18.3m$)] to the windward surface of the parapet, and applying the applicable negative wall pressure from Figure 30.3-1 [$h \leq 60$ft ($h \leq 18.3m$)] or Figure 30.4-1 [$h > 60$ft ($h > 18.3m$)] as applicable to the leeward surface. Edge and corner zones shall be arranged as shown in the applicable figures. (GC_p) shall be determined for appropriate roof angle and effective wind area from the applicable figures.

If internal pressure is present, both load cases should be evaluated under positive and negative internal pressure.

The steps required for the determination of wind loads on component and cladding of parapets are shown in Table 30.6-1.

30.7 ROOF OVERHANGS

The design wind pressure for roof overhangs of enclosed, partially enclosed, and partially open buildings of all heights shall be determined from the following equation:

$$p = q_h K_d[(GC_p) - (GC_{pi})](\text{lb/ft}^2) \quad (30.7\text{-}1)$$

$$p = q_h K_d[(GC_p) - (GC_{pi})](\text{N/m}^2) \quad (30.7\text{-}1.\text{SI})$$

where

q_h = Velocity pressure from Section 26.10 evaluated at mean roof height h using exposure defined in Section 26.7.3;

K_d = Wind directionality factor; see Section 26.6;

(GC_p) = External pressure coefficient at the overhang, (GC_p), given in Figure 30.3-2A for gable roofs $\theta \leq 7°$ [$h \leq 60$ ft (18.3 m)], or calculated as the sum of the GC_p of the overhang's top and bottom surfaces determined by the applicable roof and wall external pressure coefficient GC_p figures. The GC_p of the overhang's top surface is the same as the applicable roof surface's GC_p; the GC_p of the overhang's bottom surface is the same as the adjacent wall's GC_p, adjusted for effective wind area.

(GC_{pi}) = Internal pressure coefficient given in Table 26.13-1.

The steps required for the determination of wind loads on C&C of roof overhangs are shown in Table 30.7-1.

Diagrams

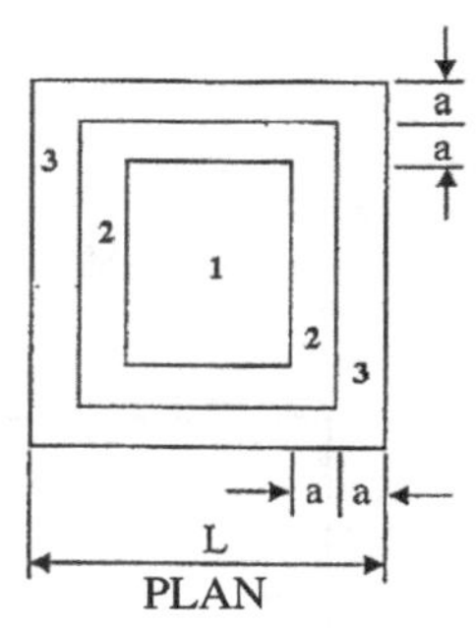

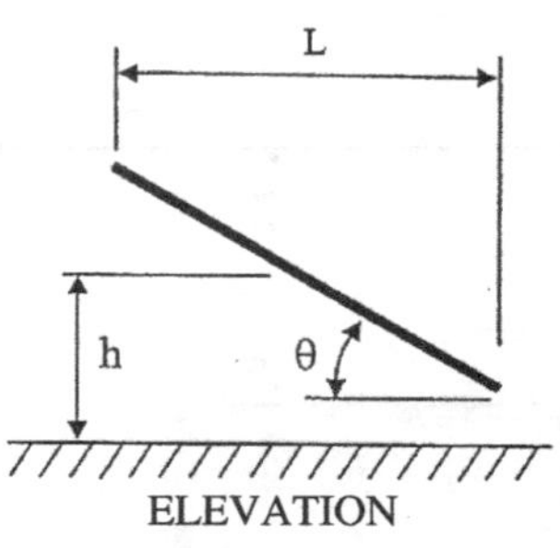

Notation

a = 10% of least horizontal dimension or $0.4h$, whichever is smaller but not less than 4% of least horizontal dimension or 3 ft (0.9 m).

h = Mean roof height, ft (m).

L = Horizontal dimension of building, measured in along-wind direction, ft (m).

θ = Angle of plane of roof from horizontal, degrees.

Net Pressure Coefficients, C_N

Roof Angle, θ	Effective Wind Area	Clear Wind Flow					
		Zone 3		Zone 2		Zone 1	
0°	$\le a^2$	2.4	−3.3	1.8	−1.7	1.2	−1.1
	$> a^2, \le 4.0a^2$	1.8	−1.7	1.8	−1.7	1.2	−1.1
	$> 4.0a^2$	1.2	−1.1	1.2	−1.1	1.2	−1.1
7.5°	$\le a^2$	3.2	−4.2	2.4	−2.1	1.6	−1.4
	$> a^2, \le 4.0a^2$	2.4	−2.1	2.4	−2.1	1.6	−1.4
	$> 4.0a^2$	1.6	−1.4	1.6	−1.4	1.6	−1.4
15°	$\le a^2$	3.6	−3.8	2.7	−2.9	1.8	−1.9
	$> a^2, \le 4.0a^2$	2.7	−2.9	2.7	−2.9	1.8	−1.9
	$> 4.0a^2$	1.8	−1.9	1.8	−1.9	1.8	−1.9
30°	$\le a^2$	5.2	−5	3.9	−3.8	2.6	−2.5
	$> a^2, \le 4.0a^2$	3.9	−3.8	3.9	−3.8	2.6	−2.5
	$> 4.0a^2$	2.6	−2.5	2.6	−2.5	2.6	−2.5
45°	$\le a^2$	5.2	−4.6	3.9	−3.5	2.6	−2.3
	$> a^2, \le 4.0a^2$	3.9	−3.5	3.9	−3.5	2.6	−2.3
	$> 4.0a^2$	2.6	−2.3	2.6	−2.3	2.6	−2.3
		Obstructed Wind Flow					
		Zone 3		Zone 2		Zone 1	
0°	$\le a^2$	1	−3.6	0.8	−1.8	0.5	−1.2
	$> a^2, \le 4.0a^2$	0.8	−1.8	0.8	−1.8	0.5	−1.2
	$> 4.0a^2$	0.5	−1.2	0.5	−1.2	0.5	−1.2
7.5°	$\le a^2$	1.6	−5.1	1.2	−2.6	0.8	−1.7
	$> a^2, \le 4.0a^2$	1.2	−2.6	1.2	−2.6	0.8	−1.7
	$> 4.0a^2$	0.8	−1.7	0.8	−1.7	0.8	−1.7
15°	$\le a^2$	2.4	−4.2	1.8	−3.2	1.2	−2.1
	$> a^2, \le 4.0a^2$	1.8	−3.2	1.8	−3.2	1.2	−2.1
	$> 4.0a^2$	1.2	−2.1	1.2	−2.1	1.2	−2.1
30°	$\le a^2$	3.2	−4.6	2.4	−3.5	1.6	−2.3
	$> a^2, \le 4.0a^2$	2.4	−3.5	2.4	−3.5	1.6	−2.3
	$> 4.0a^2$	1.6	−2.3	1.6	−2.3	1.6	−2.3
45°	$\le a^2$	4.2	−3.8	3.2	−2.9	2.1	−1.9
	$> a^2, \le 4.0a^2$	3.2	−2.9	3.2	−2.9	2.1	−1.9
	$> 4.0a^2$	2.1	−1.9	2.1	−1.9	2.1	−1.9

Notes

1. C_N denotes net pressures (contributions from top and bottom surfaces).
2. Clear wind flow denotes relatively unobstructed wind flow with blockage less than or equal to 50%. Obstructed wind flow denotes objects below roof inhibiting wind flow (>50% blockage).
3. For values of θ other than those shown, linear interpolation is permitted.
4. Plus and minus signs signify pressures acting toward and away from the top roof surface, respectively.
5. Components and cladding elements shall be designed for positive and negative pressure coefficients shown.

Figure 30.5-1. Components and cladding ($0.25 \le h/L \le 1.0$): net pressure coefficient, C_N, for open buildings—monoslope free roofs, $\theta \le 45°$.

Diagrams

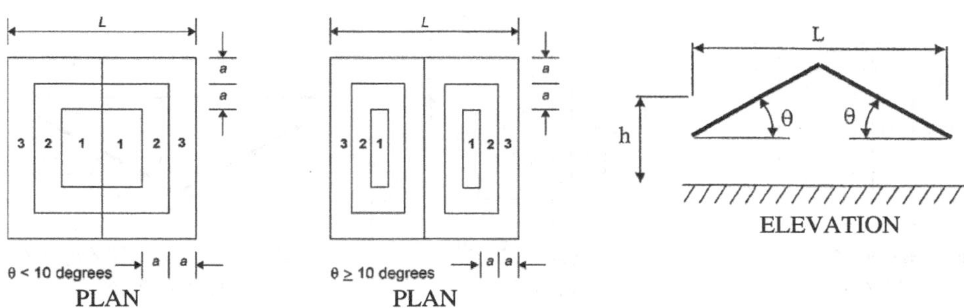

Notation

a = 10% of least horizontal dimension or $0.4h$, whichever is smaller, but not less than 4% of least horizontal dimension or 3 ft (0.9 m). Dimension a is as shown in Fig. 30.7-1.

h = Mean roof height, ft (m).

L = Horizontal dimension of building, measured in along-wind direction, ft (m).

θ = Angle of plane of roof from horizontal, degrees.

Net Pressure Coefficients, C_N

Roof Angle, θ	Effective Wind Area	Clear Wind Flow Zone 3		Zone 2		Zone 1	
0°	$\leq a^2$	2.4	−3.3	1.8	−1.7	1.2	−1.1
	$> a^2, \leq 4.0a^2$	1.8	−1.7	1.8	−1.7	1.2	−1.1
	$> 4.0a^2$	1.2	−1.1	1.2	−1.1	1.2	−1.1
7.5°	$\leq a^2$	2.2	−3.6	1.7	−1.8	1.1	−1.2
	$> a^2, \leq 4.0a^2$	1.7	−1.8	1.7	−1.8	1.1	−1.2
	$> 4.0a^2$	1.1	−1.2	1.1	−1.2	1.1	−1.2
15°	$\leq a^2$	2.2	−2.2	1.7	−1.7	1.1	−1.1
	$> a^2, \leq 4.0a^2$	1.7	−1.7	1.7	−1.7	1.1	−1.1
	$> 4.0a^2$	1.1	−1.1	1.1	−1.1	1.1	−1.1
30°	$\leq a^2$	2.6	−1.8	2	−1.4	1.3	−0.9
	$> a^2, \leq 4.0a^2$	2	−1.4	2	−1.4	1.3	−0.9
	$> 4.0a^2$	1.3	−0.9	1.3	−0.9	1.3	−0.9
45°	$\leq a^2$	2.2	−1.6	1.7	−1.2	1.1	−0.8
	$> a^2, \leq 4.0a^2$	1.7	−1.2	1.7	−1.2	1.1	−0.8
	$> 4.0a^2$	1.1	−0.8	1.1	−0.8	1.1	−0.8

Roof Angle, θ	Effective Wind Area	Obstructed Wind Flow Zone 3		Zone 2		Zone 1	
0°	$\leq a^2$	1	−3.6	0.8	−1.8	0.5	−1.2
	$> a^2, \leq 4.0a^2$	0.8	−1.8	0.8	−1.8	0.5	−1.2
	$> 4.0a^2$	0.5	−1.2	0.5	−1.2	0.5	−1.2
7.5°	$\leq a^2$	1	−5.1	0.8	−2.6	0.5	−1.7
	$> a^2, \leq 4.0a^2$	0.8	−2.6	0.8	−2.6	0.5	−1.7
	$> 4.0a^2$	0.5	−1.7	0.5	−1.7	0.5	−1.7
15°	$\leq a^2$	1	−3.2	0.8	−2.4	0.5	−1.6
	$> a^2, \leq 4.0a^2$	0.8	−2.4	0.8	−2.4	0.5	−1.6
	$> 4.0a^2$	0.5	−1.6	0.5	−1.6	0.5	−1.6
30°	$\leq a^2$	1	−2.4	0.8	−1.8	0.5	−1.2
	$> a^2, \leq 4.0a^2$	0.8	−1.8	0.8	−1.8	0.5	−1.2
	$> 4.0a^2$	0.5	−1.2	0.5	−1.2	0.5	−1.2
45°	$\leq a^2$	1	−2.4	0.8	−1.8	0.5	−1.2
	$> a^2, \leq 4.0a^2$	0.8	−1.8	0.8	−1.8	0.5	−1.2
	$> 4.0a^2$	0.5	−1.2	0.5	−1.2	0.5	−1.2

Notes

1. C_N denotes net pressures (contributions from top and bottom surfaces).
2. Clear wind flow denotes relatively unobstructed wind flow with blockage less than or equal to 50%. Obstructed wind flow denotes objects below roof inhibiting wind flow (>50% blockage).
3. For values of θ other than those shown, linear interpolation is permitted.
4. Plus and minus signs signify pressures acting toward and away from the top roof surface, respectively.
5. Components and cladding elements shall be designed for positive and negative pressure coefficients shown.

Figure 30.5-2. Components and cladding ($0.25 \leq h/L \leq 1.0$): net pressure coefficient, C_N, for open buildings—pitched free roofs, $\theta \leq 45°$.

Diagrams

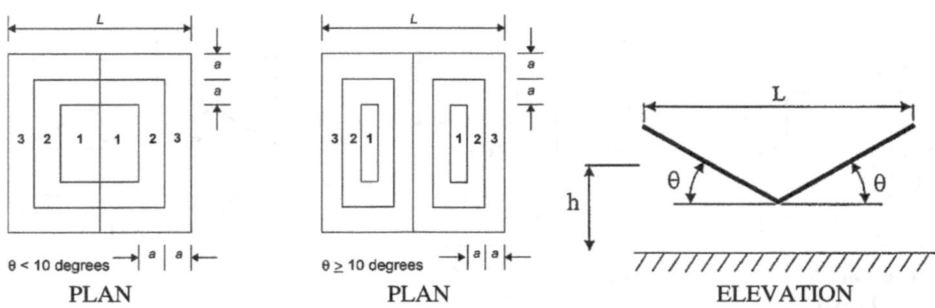

Notation

a = 10% of least horizontal dimension or $0.4h$, whichever is smaller, but not less than 4% of least horizontal dimension or 3 ft (0.9 m). Dimension a is as shown in Fig. 30.7-1.

h = Mean roof height, ft (m).

L = Horizontal dimension of building, measured in along-wind direction, ft (m).

θ = Angle of plane of roof from horizontal, degrees.

Net Pressure Coefficients, C_N

Roof Angle, θ	Effective Wind Area	Clear Wind Flow					
		Zone 3		Zone 2		Zone 1	
0°	$\leq a^2$	2.4	−3.3	1.8	−1.7	1.2	−1.1
	$> a^2, \leq 4.0a^2$	1.8	−1.7	1.8	−1.7	1.2	−1.1
	$> 4.0a^2$	1.2	−1.1	1.2	−1.1	1.2	−1.1
7.5°	$\leq a^2$	2.4	−3.3	1.8	−1.7	1.2	−1.1
	$> a^2, \leq 4.0a^2$	1.8	−1.7	1.8	−1.7	1.2	−1.1
	$> 4.0a^2$	1.2	−1.1	1.2	−1.1	1.2	−1.1
15°	$\leq a^2$	2.2	−2.2	1.7	−1.7	1.1	−1.1
	$> a^2, \leq 4.0a^2$	1.7	−1.7	1.7	−1.7	1.1	−1.1
	$> 4.0a^2$	1.1	−1.1	1.1	−1.1	1.1	−1.1
30°	$\leq a^2$	1.8	−2.6	1.4	−2	0.9	−1.3
	$> a^2, \leq 4.0a^2$	1.4	−2	1.4	−2	0.9	−1.3
	$> 4.0a^2$	0.9	−1.3	0.9	−1.3	0.9	−1.3
45°	$\leq a^2$	1.6	−2.2	1.2	−1.7	0.8	−1.1
	$> a^2, \leq 4.0a^2$	1.2	−1.7	1.2	−1.7	0.8	−1.1
	$> 4.0a^2$	0.8	−1.1	0.8	−1.1	0.8	−1.1
		Obstructed Wind Flow					
		Zone 3		Zone 2		Zone 1	
0°	$\leq a^2$	1	−3.6	0.8	−1.8	0.5	−1.2
	$> a^2, \leq 4.0a^2$	0.8	−1.8	0.8	−1.8	0.5	−1.2
	$> 4.0a^2$	0.5	−1.2	0.5	−1.2	0.5	−1.2
7.5°	$\leq a^2$	1	−4.8	0.8	−2.4	0.5	−1.6
	$> a^2, \leq 4.0a^2$	0.8	−2.4	0.8	−2.4	0.5	−1.6
	$> 4.0a^2$	0.5	−1.6	0.5	−1.6	0.5	−1.6
15°	$\leq a^2$	1	−2.4	0.8	−1.8	0.5	−1.2
	$> a^2, \leq 4.0a^2$	0.8	−1.8	0.8	−1.8	0.5	−1.2
	$> 4.0a^2$	0.5	−1.2	0.5	−1.2	0.5	−1.2
30°	$\leq a^2$	1	−2.8	0.8	−2.1	0.5	−1.4
	$> a^2, \leq 4.0a^2$	0.8	−2.1	0.8	−2.1	0.5	−1.4
	$> 4.0a^2$	0.5	−1.4	0.5	−1.4	0.5	−1.4
45°	$\leq a^2$	1	−2.4	0.8	−1.8	0.5	−1.2
	$> a^2, \leq 4.0a^2$	0.8	−1.8	0.8	−1.8	0.5	−1.2
	$> 4.0a^2$	0.5	−1.2	0.5	−1.2	0.5	−1.2

Notes

1. C_N denotes net pressures (contributions from top and bottom surfaces).
2. Clear wind flow denotes relatively unobstructed wind flow with blockage less than or equal to 50%. Obstructed wind flow denotes objects below roof inhibiting wind flow (>50% blockage).
3. For values of θ other than those shown, linear interpolation is permitted.
4. Plus and minus signs signify pressures acting toward and away from the top roof surface, respectively.
5. Components and cladding elements shall be designed for positive and negative pressure coefficients shown.

Figure 30.5-3. Components and cladding ($0.25 \leq h/L \leq 1.0$): net pressure coefficient, C_N, for open buildings—troughed free roofs, $\theta \leq 45°$.

Diagrams

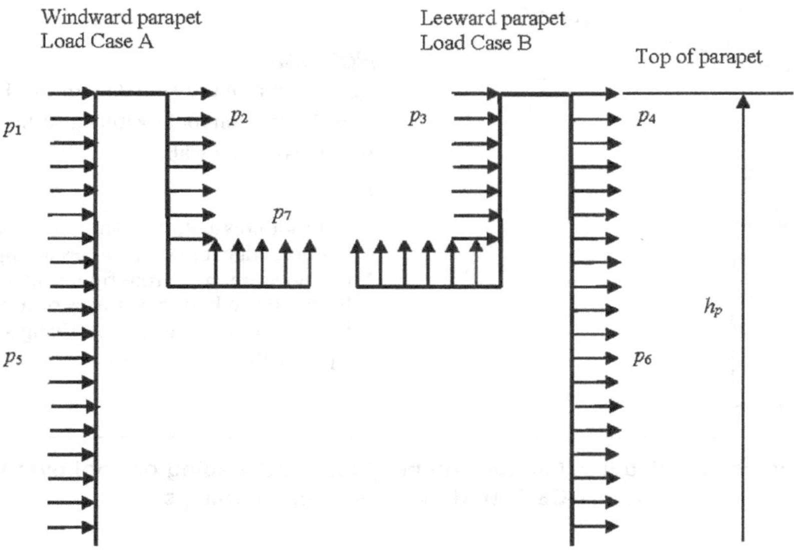

Notes

Windward Parapet: Load Case A

1. Windward parapet pressure (p_1) is determined using the positive wall pressure (p_5) Zones 4 or 5 from the applicable figure.
2. Leeward parapet pressure (p_2) is determined using the negative roof pressure (p_7) Zones 2 or 3 from the applicable figure.

Leeward Parapet: Load Case B

1. Windward parapet pressure (p_3) is determined using the positive wall pressure (p_5) Zones 4 or 5 from the applicable figure.
2. Leeward parapet pressure (p_4) is determined using the negative wall pressure (p_6) Zones 2 or 3 from the applicable figure.

User Note: See Note 5 in Figure 30.3-2A and Note 7 in Figure 30.5-1 for reductions in component and cladding roof pressures when parapets 3 ft (0.9 m) or higher are present.

Figure 30.6-1. Components and cladding, part 4 (all building heights): parapet wind loads for all building types—parapet wind loads.

Table 30.7-1. Steps to Determine C&C Wind Loads for Roof Overhangs.

Step 1: Determine risk category of building; see Table 1.5-1.

Step 2: Determine the basic wind speed, V, for applicable risk category; see Fig. 26.5-1.

Step 3: Determine wind load parameters:

- Wind directionality factor, K_d; see Section 26.6 and Table 26.6-1.
- Exposure Category B, C, or D; see Section 26.7.
- Topographic factor, K_{zt}; see Section 26.8 and Figure 26.8-1.
- Ground elevation factor, K_e; see Section 26.9 and Table 26.9-1.
- Enclosure classification; see Section 26.12.
- Internal pressure coefficient (GC_{pi}); see Section 26.13 and Table 26.13-1.

Step 4: Determine velocity pressure exposure coefficient, K_h; see Table 26.10-1.

Step 5: Determine velocity pressure, q_h, at mean roof height h using Equation (26.10-1).

continues

Table 30.7-1 *(Continued)*. Steps to Determine C&C Wind Loads for Roof Overhangs.

Step 6: Determine the external pressure coefficient (GC_p) at the overhang, using Figure 30.3-2A for gable roofs $\theta \leq 7°$ [$h \leq 60$ ft (18.3 m)], or calculated as the sum of the GC_p coefficients on the overhang's top and bottom surfaces, as determined by the applicable roof and wall GC_p given in figures:

- Gable roofs $7° < \theta \leq 45°$, hip roofs $\theta \leq 45°$ [$h \leq 60$ ft (18.3 m)]: see Figure 30.3-2B-G.
- Other roofs [$h \leq 60$ ft (18.3 m)]: see Table 30.3-1 Step 6.
- Flat roofs $\theta \leq 7°$ [$h > 60$ ft (18.3 m)]: see Figure 30.5-1.
- Gable and hip roofs [$h > 60$ ft (18.3 m)]: see Figure 30.5-1.
- Other roofs [$h > 60$ ft (18.3 m)]: see Table 30.5-1 Step 6.
- Walls [$h \leq 60$ ft (18.3 m)]: see Figure 30.3-1.
- Walls [$h > 60$ ft (18.3 m)]: see Figure 30.5-1.

Step 7: Calculate wind pressure, p, using Equation (30.7-1); refer to Figure 30.7-1.

Diagrams

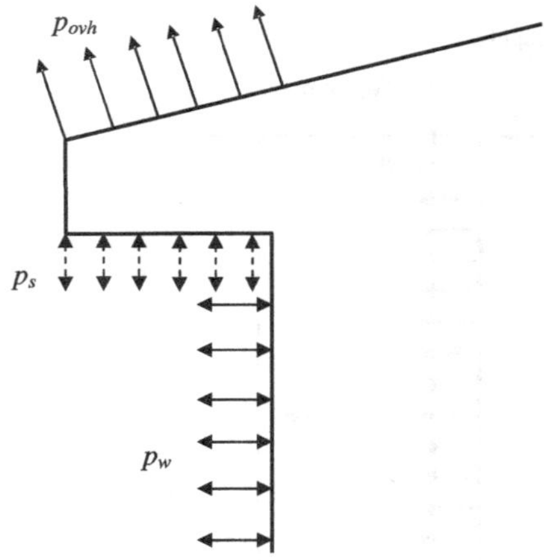

Notation

p_{ovh} = Net roof pressure on roof overhangs.
p_s = Pressure on roof overhang soffit.
p_w = Pressure on wall.

Notes

1. Net roof pressure, p_{ovh}, on roof overhangs is determined from interior, edge, or corner zones as applicable from figures.
2. Net pressure, p_{ovh}, from figures includes pressure contribution from top and bottom surfaces of roof overhang.
3. Positive pressure at roof overhang soffit p_s shall be taken as equal to the wall pressure p_w.

Figure 30.7-1. Components and cladding (all building heights): wind loading on roof overhangs for all building types—C&C wind loads on roof overhangs.

30.8 ROOFTOP STRUCTURES AND EQUIPMENT FOR BUILDINGS

The C&C pressure on each wall of the rooftop structure shall be equal to the lateral force determined in accordance with Section 29.4.1 divided by the respective wall surface area of the rooftop structure and shall be considered to act inward and outward. The C&C pressure on the roof shall be equal to the vertical uplift force determined in accordance with Section 29.4.1 divided by the horizontal projected area of the roof of the rooftop structure and shall be considered to act in the upward direction.

30.9 ATTACHED CANOPIES ON BUILDINGS

The design wind pressure for canopies attached to the walls of buildings shall be determined from the following equation:

$$p = q_h K_d (GC_p)(\text{lb/ft}^2) \quad (30.9\text{-}1)$$

$$p = q_h K_d (GC_p)(\text{N/m}^2) \quad (30.9\text{-}1.\text{SI})$$

where

q_h = Velocity pressure from Section 26.10 evaluated at mean roof height h using exposure defined in Section 26.7.3; and

K_d = Wind directionality factor; see Section 26.6; and

(GC_p) = Net pressure coefficients for attached canopies given in Figure 30.9-1A–B [buildings with $h < 60$ ft ($h < 18.3$ m)] and 30.9-2A-B [buildings with $h > 60$ ft ($h > 18.3$ m)] for contributions from both upper and lower surfaces individually and their combined (net) effect on attached canopies.

The steps required for the determination of wind loads on attached canopies are shown in Table 30.9-1.

EXCEPTIONS:

1. As an alternative to using the (GC_p) value from Figure 30.9-2A for buildings with a mean roof height between 60 ft (18.3 m) and 90 ft (27.4 m), the value of GC_p may be linearly interpolated between the value for a mean roof height of 60 ft (18.3 m) from Figure 30.9-1A and the value for a mean roof height of 90 ft (27.4 m) from Figure 30.9-2A for each specific h_c/h_e of the attached canopy.
2. As an alternative to using the (GC_{pn}) value from Figure 30.9-2B for buildings with a mean roof height between 60 ft (18.3 m) and 90 ft (27.4 m), the value of GC_{pn} may be linearly interpolated between the value for a mean roof height of 60 ft (18.3 m) from Figure 30.9-1B and the value for a mean roof height of 90 ft (27.4 m) from Figure 30.9-2B for each specific h_c/h_e of the attached canopy.

Table 30.9-1. Steps to Determine C&C Wind Loads on Attached Canopies.

Step 1: Determine risk category of building; see Table 1.5-1.
Step 2: Determine the basic wind speed, V, for applicable risk category; see Figure 26.5-1.
Step 3: Determine wind load parameters:

- Wind directionality factor, K_d; see Section 26.6 and Table 26.6-1.
- Exposure category B, C, or D; see Section 26.7.
- Topographic factor, K_{zt}; see Section 26.8 and Figure 26.8-1.
- Ground elevation factor, K_e; see Section 26.9 and Table 26.9-1.

Step 4: Determine velocity pressure exposure coefficient, K_h; see Table 26.10-1.
Step 5: Determine velocity pressure, q_h, at mean roof height h using Equation (26.10-1).
Step 6: Determine surface or net pressure coefficient (GC_p) or (GC_{pn}) using Figure 30.9-1A-B [buildings with $h \leq 60$ ft ($h \leq 18.3$ m)] or Figure 30.9-2A-B [buildings with $h > 60$ ft ($h > 18.3$ m)].
Step 7: Calculate wind pressure, p, using Equation (30.9-1).

Diagram

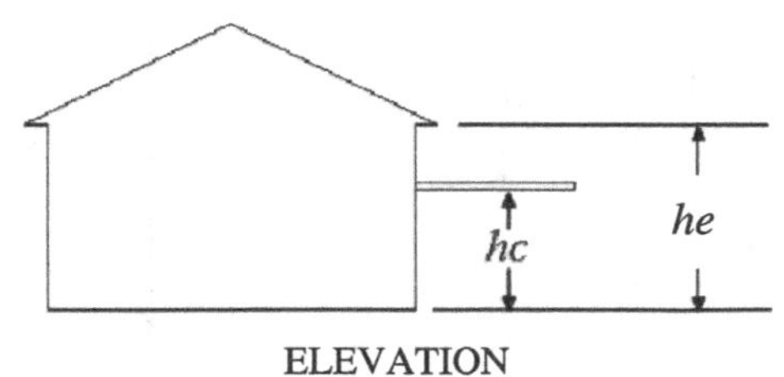

Pressure Coefficient

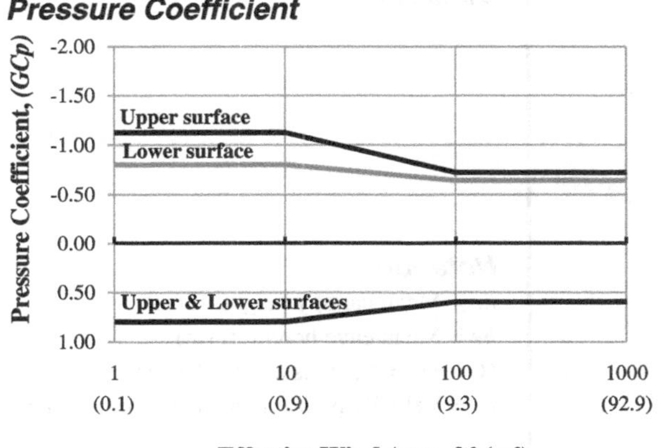

Notation

hc = Mean canopy height, ft (m).

he = Mean eave height, ft (m).

(GCp) = Pressure coefficients.

q_h = Velocity pressure evaluated at height $z = h$, lb/ft2 (N/m2).

Notes

1. Pressures are based on the most critical values for all ratios of hc/he.
2. Vertical scale denotes (GCp) to be used with q_h.
3. Horizontal scale denotes effective wind area, ft^2 (m^2).
4. Negative signs signify pressures acting away from the surface.

Figure 30.9-1A. Pressure coefficients on separate surfaces of attached canopies on buildings with $h \leq 60$ ft ($h \leq 18.3$ m).

PART 5: NONBUILDING STRUCTURES

30.10 CIRCULAR BINS, SILOS, AND TANKS WITH $h \leq 120$ft ($h \leq 36.6m$)

Wind pressures on surfaces of isolated circular bins, silos, and tanks shall be calculated from Sections 30.10.1 through 30.10.5.

Grouped circular bins, silos, and tanks of similar size with center-to-center spacing greater than 2 diameters shall be treated as isolated structures. For spacings less than 1.25 diameters, the structures shall be treated as grouped and the wind pressure shall be determined from Section 30.10.6. For intermediate spacings, linear interpolation of the C_p (or C_f) values shall be used.

The steps required for the determination of wind loads for circular bins, silos, and tanks are shown in Table 30.10-1.

30.10.1 Design Wind Pressure Design wind pressure on C&C for isolated circular bins, silos, and tanks in $(\text{lb/ft}^2)(N/m^2)$ shall be determined from the following equation:

$$p = q_h K_d[(GC_p) - (GC_{pi})] \qquad (30.10\text{-}1)$$

where

q_h = Velocity pressure for all surfaces evaluated at mean roof height h

K_d = Wind directionality factor; see Section 26.6.

(GC_p) = External pressure coefficients given in

- Section 30.10.2 for walls
- Section 30.10.5 for underneath sides
- Section 30.10.4 for roofs

(GC_{pi}) = Internal pressure coefficient given in Table 26.13-1 and Section 30.11.3.

30.10.2 External Walls of Isolated Circular Bins, Silos, and Tanks The external pressures on the walls of circular bins, silos, and tanks shall be determined from the external pressure

Table 30.10-1. Steps to Determine C&C Wind Loads for Circular Bins, Silos, and Tanks.

Step 1: Determine risk category; see Table 1.5-1.

Step 2: Determine the basic wind speed, V, for applicable risk category; see Figure. 26.5-1.

Step 3: Determine wind load parameters:

- Wind directionality factor, K_d; see Section 26.6 and Table 26.6-1.
- Exposure Category B, C, or D; see Section 26.7.
- Topographic factor, K_{zt}; see Section 26.8 and Figure 26.8-1.
- Ground elevation factor, K_e; see Section 26.9 and Table 26.9-1.
- Enclosure classification; see Section 26.12.
- Internal pressure coefficient (GC_{pi}); see Section 26.13 and Section 30.11.3.

Step 4: Determine velocity pressure exposure coefficient, K_z or K_h; see Table 26.10-1.

Step 5: Determine velocity pressure, q_h, Equation (26.10-1).

Step 6: Determine external pressure coefficient (GC_p).

- Walls; see Sections 30.10.2 and 30.10.6.
- Roofs; see Sections 30.10.4 and 30.10.6.

Step 7: Calculate wind pressure, p, using Equation (30.10-1).

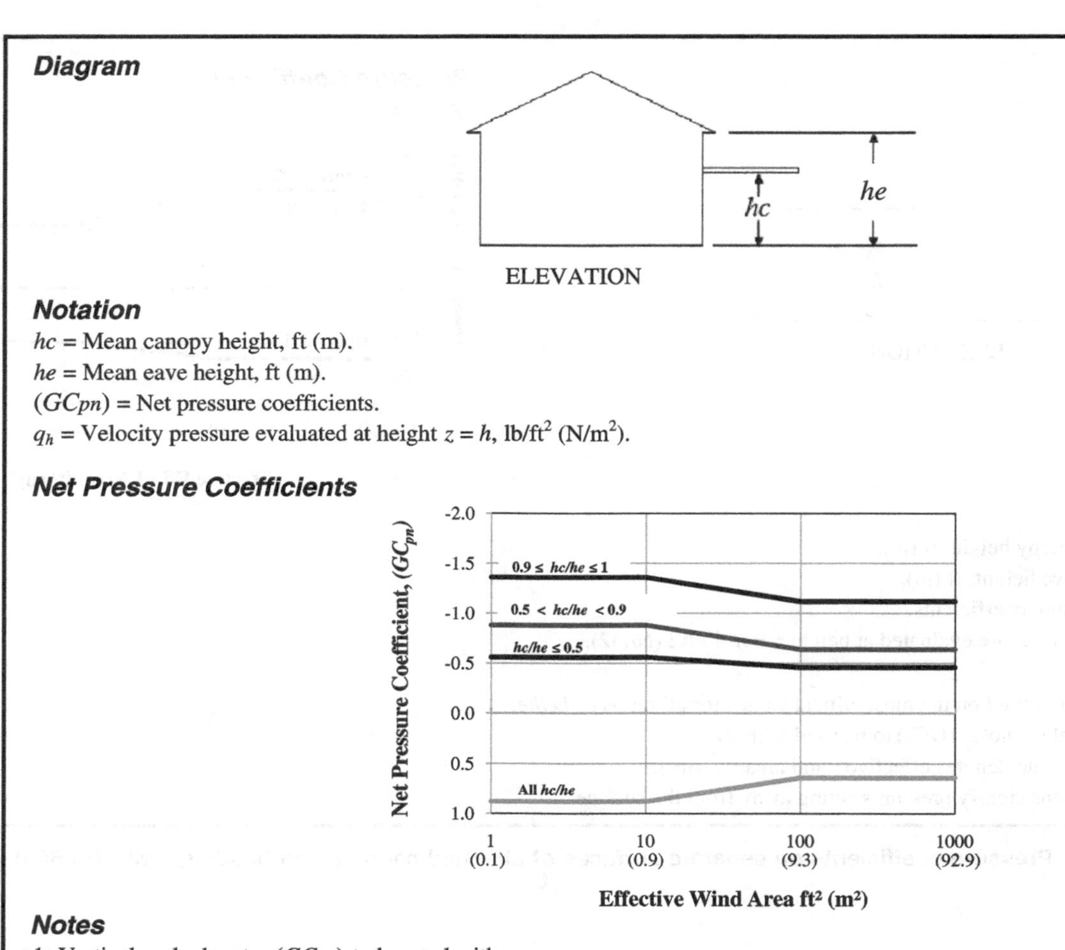

Notes

1. Vertical scale denotes (*GCpn*) to be used with q_h.
2. Horizontal scale denotes effective wind area, ft^2 (m^2).
3. Negative and positive signs signify uplifting and downward pressures, respectively.
4. Each component shall be designed for maximum positive and negative pressures.
5. Use linear interpolation for intermediate values of *hc*/*he*.

Figure 30.9-1B. Net pressure coefficients considering simultaneous contributions from upper and lower surfaces on attached canopies on buildings with $h \leq 60$ ft ($h \leq 18.3$ m).

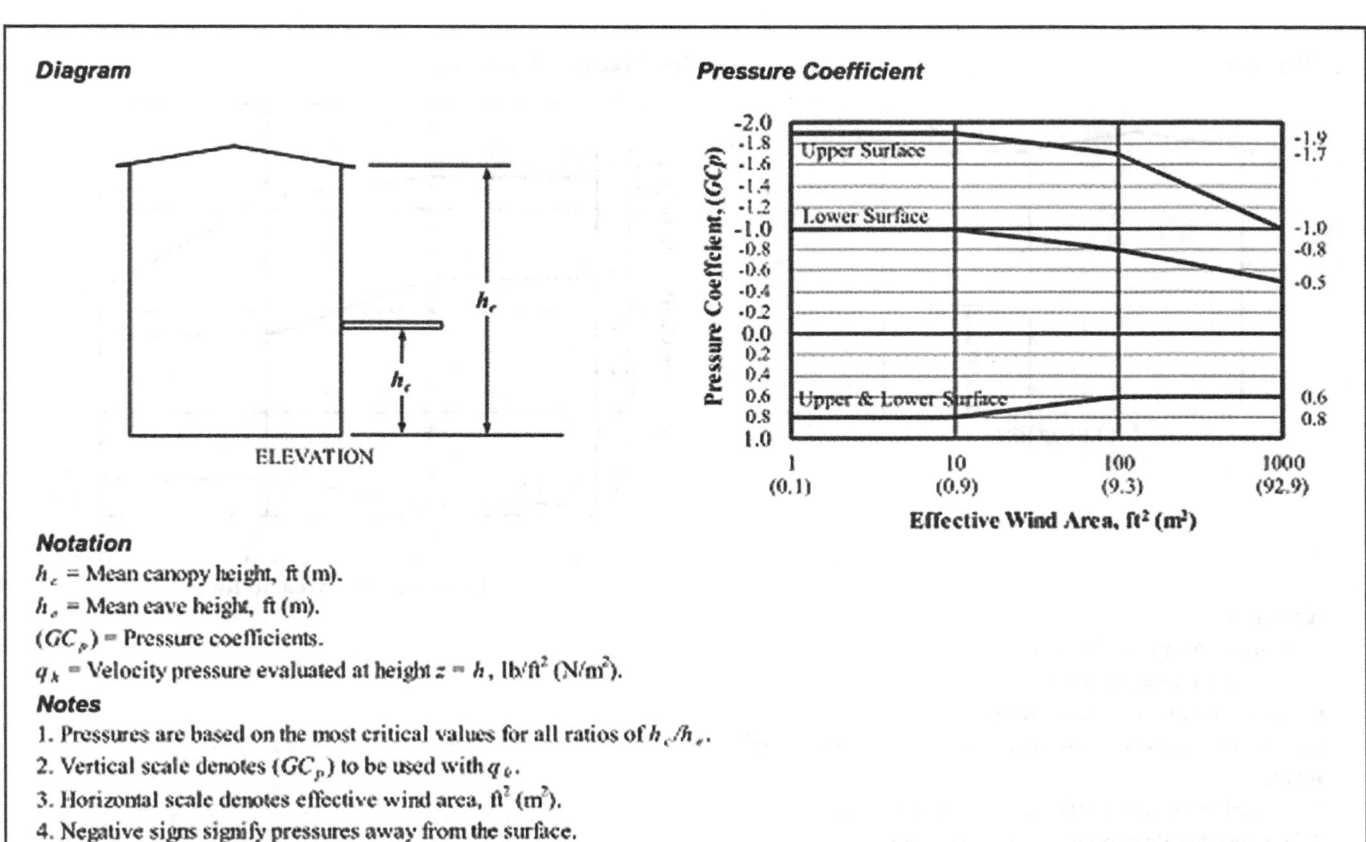

Figure 30.9-2A. Pressure coefficients on separate surfaces of attached canopies on buildings with $h > 60$ ft ($h > 18.3$ m).

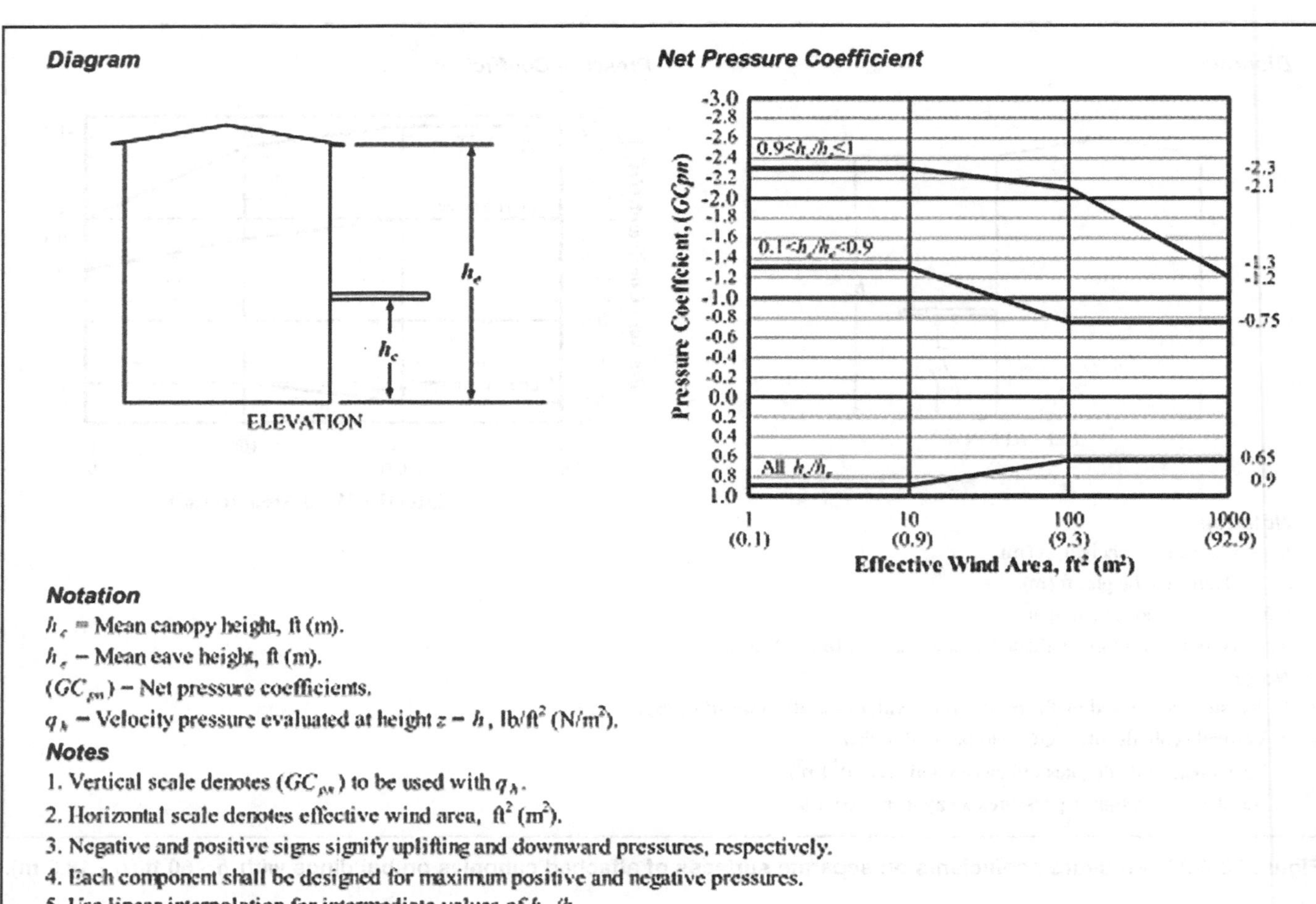

Figure 30.9-2B. Net pressure coefficients considering simultaneous contributions from upper and lower surfaces on attached canopies on buildings with $h > 60$ ft ($h > 18.3$ m).

coefficients (GC_p) as a function of the angle α, given as follows for the shape ranges indicated:

$$GC_{p(\alpha)} = k_b C_{(\alpha)} \quad (30.10\text{-}2)$$

where the cylinder (diameter D) is standing on the ground or supported by columns giving a clearance height (C) less than the height of the cylinder (H), as shown in Figure 30.10-1.

H/D is in the range 0.25 to 4.0 inclusive. α = angle from the wind direction to a point on the wall of a circular bin, silo, or tank, in degrees.

$$k_b = 1.0 \text{ for } C_{(\alpha)} \geq -0.15, \text{ or}$$

$$= 1.0 - 0.55(C(\alpha) + 0.15)\log_{10}(H/D) \quad (30.10.3)$$

$$\text{for } C_{(\alpha)} < -0.15$$

$$C_{(\alpha)} = -0.5 + 0.4\cos\alpha + 0.8\cos 2\alpha + 0.3\cos 3\alpha$$
$$1 - 0.1\cos 4\alpha - 0.05\cos 5\alpha \quad (30.10\text{-}4)$$

Figure 30.10-1 lists the external pressure coefficients for walls, which include the graphic distribution of the external pressure, $GC_{p(\alpha)}$, around the perimeter of the wall.

30.10.3 Internal Surface of Exterior Walls of Isolated Open-Topped Circular Bins, Silos, and Tanks The pressures on the internal surface of exterior walls of open-topped circular bins, silos, and tanks shall be determined from Equation (30.10-5):

$$GC_{pi} = -0.9\text{–}0.35\log_{10}(H/D) \quad (30.10\text{-}5)$$

30.10.4 Roofs of Isolated Circular Bins, Silos, and Tanks The external pressures on the roofs or lids of bins, silos, or tanks of circular cross section shall be equal to the external pressure coefficients, (GC_p), given in Figure 30.10-2 for Zones 1, 2, 3, and 4.

Zone 3 is applicable to the windward edges of roofs with slope less than or equal to 30 degrees, and Zone 4 is applicable to the region near the cone apex for roofs with slope greater than 15 degrees. The applicable areas are shown in Figure 30.10-2.

30.10.5 Undersides of Isolated Elevated Circular Bins, Silos, and Tanks (GC_p) values for the undersides of elevated circular bins, silos, and tanks shall be taken as 1.2 and −0.9 for Zone 3 and 0.8 and –0.6 for Zone 1 and Zone 2, as shown in Figure 30.10-2.

30.10.6 Roofs and Walls of Grouped Circular Bins, Silos, and Tanks Closely spaced groups with center-to-center spacing less than $1.25D$, the external pressures of grouped bins, silos, or tanks, shall be equal to the external pressure coefficients, (GC_p), given in Figure 30.10-3 for Zones 1, 2, 3a, 3b, and 4 for roofs and Figure 30.10-4 for Zones 5a, 5b, 8, and 9 for walls.

30.11 ROOFTOP SOLAR PANELS FOR BUILDINGS OF ALL HEIGHTS WITH FLAT ROOFS OR GABLE OR HIP ROOFS WITH SLOPES LESS THAN 7 DEGREES

The design wind pressures for rooftop solar modules and panels shall be determined in accordance with Section 29.4.3 for rooftop solar arrays that conform to the geometric requirements specified in Section 29.4.3.

30.12 ROOF PAVERS FOR BUILDINGS OF ALL HEIGHTS WITH ROOF SLOPES LESS THAN OR EQUAL TO 7 DEGREES

The design net uplift pressures for roof pavers, for buildings of all heights with roof slopes less than or equal to 7 degrees, shall be determined from the following equations:

$$p = q_h K_d C_{L_{net}} (\text{lb/ft}^2) \quad (30.12\text{-}1)$$

where

K_d = Wind directionality factor, see Section 26.6

q_h = Velocity pressure evaluated at mean height h, as defined in Section 26.10, of the roof on which the pavers are located;

$C_{L_{net}}$ = Design net uplift pressure coefficient which shall be taken as

(a) The value of external pressure coefficient, GC_p, from Figures 30.3-2A and 30.5-1,

or

(b) Determined by approved wind tunnel tests performed in accordance with Chapter 31,

or

(c) Determined by methods described in the recognized literature.

30.13 CONSENSUS STANDARDS AND OTHER REFERENCED DOCUMENTS

No consensus standards and other documents that shall be considered part of this standard are referenced in this chapter.

Diagrams

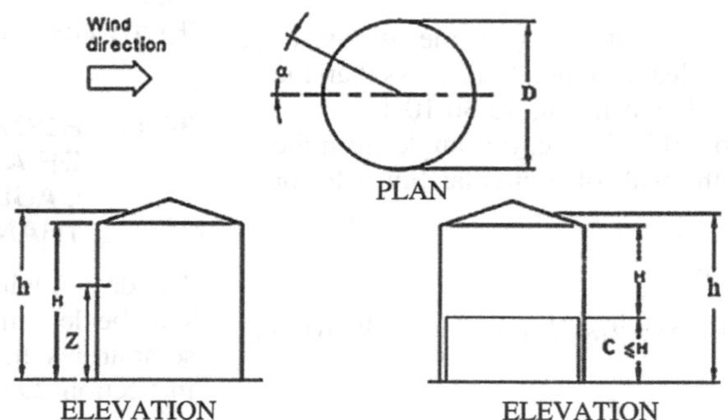

Circular Bins, Silos, and Tanks on Ground or Supported by Columns

Notation

h = Mean roof height, ft (m).
H = The solid cylinder height, ft (m).
Z = Height to centroid of projected area of solid circular structure, ft (m).
α = Angle from wind direction to a point on the wall of a circular bin, silo, or tank in degrees.
D = Diameter of a circular structure, ft (m).
C = Clearance height above the ground, ft (m).

External Pressure Coefficients, $(GC_{p(\alpha)})$, on Walls of Circular Bins, Silos, and Tanks

Angle, α	Aspect Ratio, *H*/*D*					
(degrees)	0.25	0.50	1	2	3	4
0°	1.00	1.00	1.00	1.00	1.00	1.00
15°	0.70	0.70	0.70	0.70	0.70	0.70
30°	0.30	0.30	0.30	0.30	0.30	0.30
45°	–0.30	–0.30	–0.30	–0.30	–0.30	–0.30
60°	–0.70	–0.80	–1.00	–1.10	–1.20	–1.20
75°	–0.80	–1.10	–1.40	–1.70	–1.90	–2.00
90°	–0.80	–1.10	–1.40	–1.70	–1.90	–2.00
105°	–0.70	–0.90	–1.10	–1.30	–1.40	–1.40
120°	–0.60	–0.70	–0.70	–0.80	–0.80	–0.90
135°	–0.40	–0.50	–0.50	–0.50	–0.60	–0.60
150°	–0.40	–0.40	–0.40	–0.50	–0.50	–0.50
165°	–0.40	–0.40	–0.40	–0.50	–0.50	–0.50
180°	–0.40	–0.40	–0.40	–0.50	–0.50	–0.50

Distribution of the External Pressure, $(GC_{p(\alpha)})$, around the Perimeter of the Wall

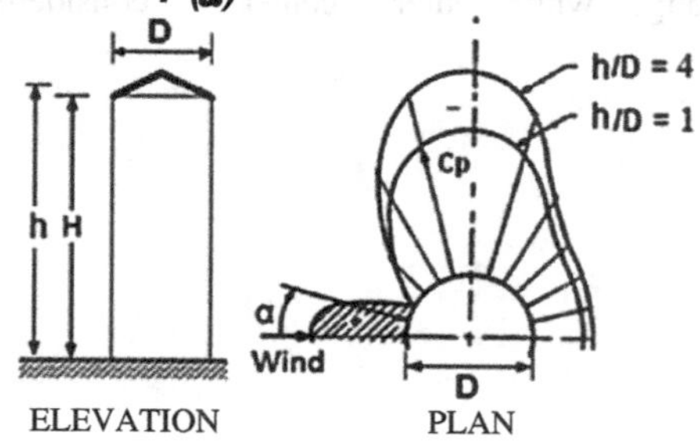

Figure 30.10-1. Components and cladding [$h \leq 120$ft ($h \leq 36.6m$)]: external pressure coefficients, GC_p, for walls of isolated circular bins, silos, and tanks with $D < 120$ ft (36.6 m) and $0.25 < H/D < 4.0$—other structures.

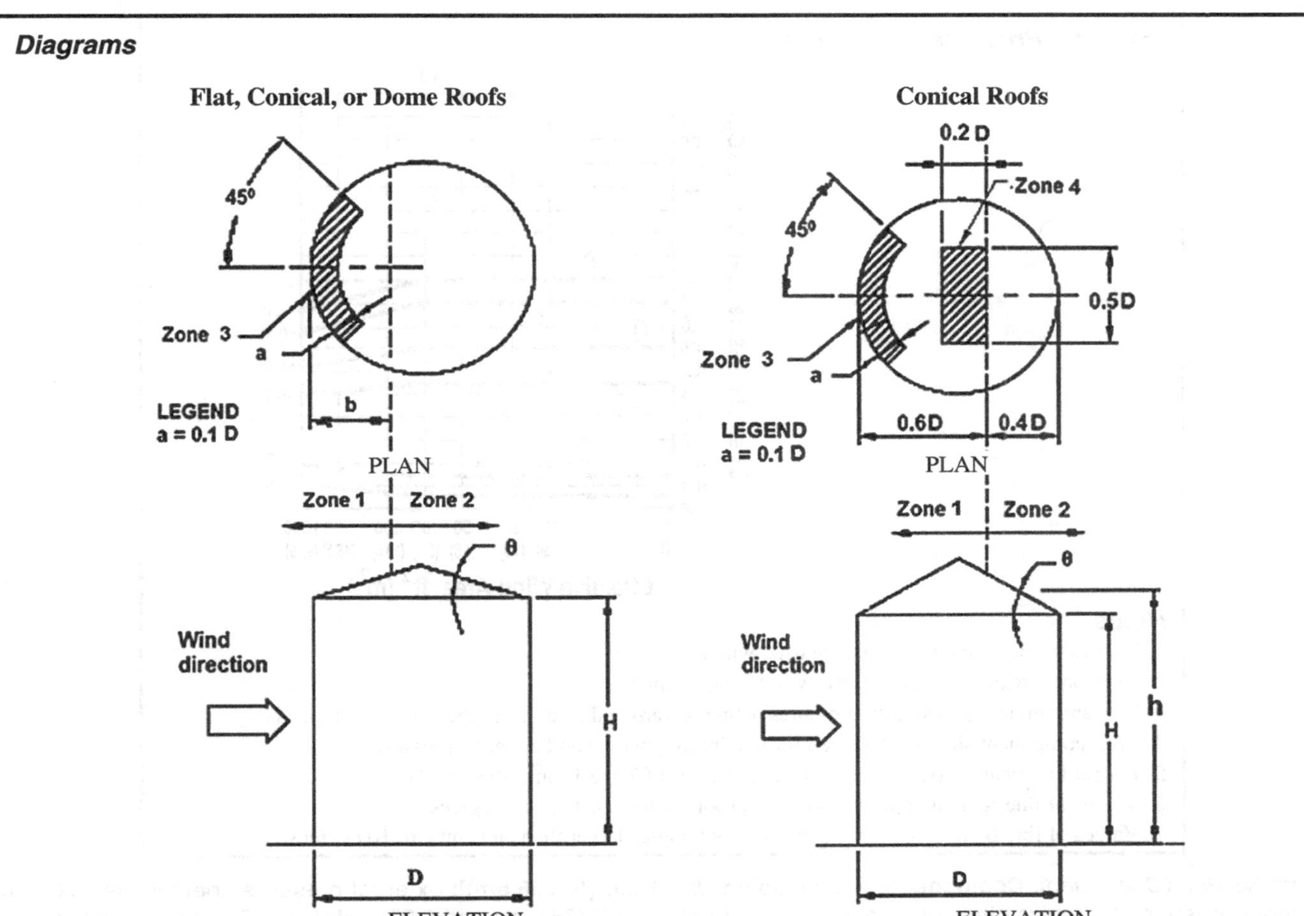

Notation

a =10% of least horizontal dimension.
b = Horizontal dimension specified for Zone 1 of a conical roof, ft (m).
D = Diameter of a circular structure, ft (m).
h = Mean roof height, ft (m).
H = Solid cylinder height, ft (m).
θ = Angle of plane of roof from horizontal, degrees.

External Pressure Coefficients, ($GC_{p(\alpha)}$), for Roofs of Isolated Circular Bins, Silos, and Tanks

H/D	0.25	0.5	≥1.0
b	$0.2D$	$0.5D$	$0.1h + 0.6D$

Notes

For roofs with average roof angles less than 10 degrees, b shall be determined from this table.
Linear interpolation shall be permitted.

Figure 30.10-2. Components and cladding [$h \leq 120$ft ($h \leq 36.6m$)]: external pressure coefficients, GC_p, for roofs of isolated circular bins, silos, and tanks with $D < 120$ ft (36.6 m) and $0.25 < H/D < 4.0$—other structures.

External Pressure Coefficient

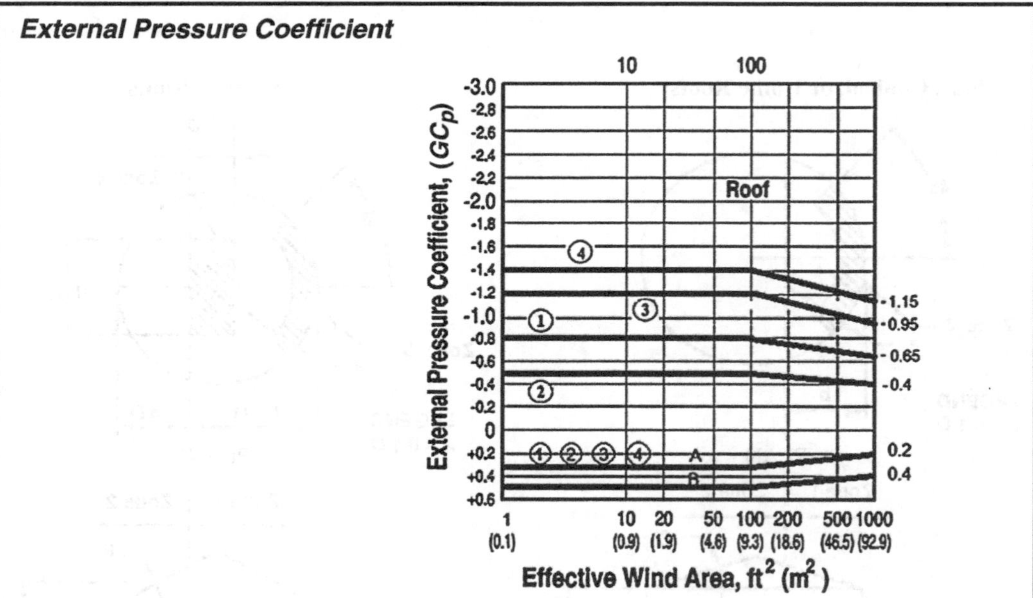

Notes

1. Vertical scale denotes (GCp) to be used with q_h.
2. Horizontal scale denotes effective wind area, ft^2 (m^2).
3. Plus and minus signs signify pressures acting toward and away from the surfaces, respectively.
4. Each component shall be designed for maximum positive and negative pressures.
5. For roof overhangs, (GCp) shall equal the values of Zone 1 multiplied by 2.0.
6. Values of line A shall apply to roofs with roof angles less than 10 degrees.
7. Values of line B shall apply to roofs with roof angles larger than and equal to 10 degrees.

Figure 30.10-2 (*Continued*). Components and cladding [$h \leq 120$ft ($h \leq 36.6m$)]: external pressure coefficients, GC_p, for roofs of isolated circular bins, silos, and tanks with $D < 120$ ft (36.6 m) and $0.25 < H/D < 4.0$—other structures.

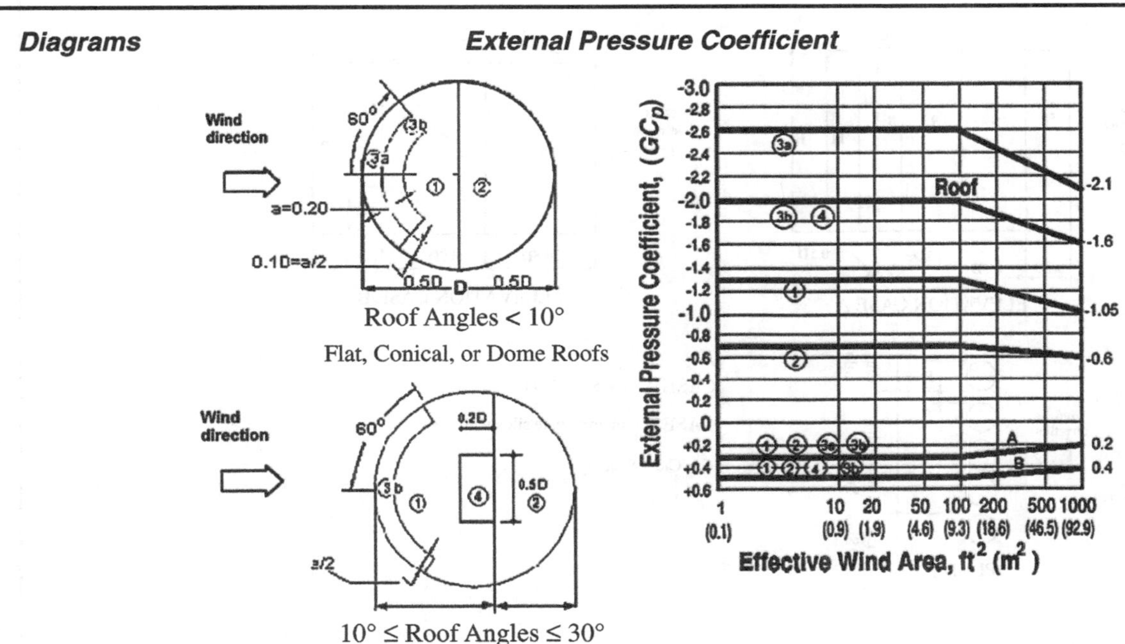

Notation

a = 20% of least horizontal dimension.
D = Diameter of a circular structure, ft (m).
h = Mean roof height, ft (m), see Fig. 30.12-4.
θ = Angle of plane of roof from horizontal, degrees.

Notes

1. Vertical scale denotes (GCp) to be used with q_h.
2. Horizontal scale denotes effective wind area, ft^2 (m^2).
3. Plus and minus signs signify pressures acting toward and away from the surfaces, respectively.
4. Each component shall be designed for maximum positive and negative pressures.
5. Values of line A shall apply to roofs with roof angles less than 10 degrees.
6. Values of line B shall apply to roofs with roof angles larger than and equal to 10 degrees.
7. Zone 4 shall apply to roofs with roof angles larger than 15 degrees.
8. For roof overhangs, (GCp) shall equal the values of Zone 1 multiplied by 2.0.

Figure 30.10-3. Components and cladding [$h \leq 120$ft ($h \leq 36.6m$)]: external pressure coefficients, GC_p, for roofs of grouped circular bins, silos, and tanks with $D < 120$ ft (36.6 m) and 0.25 < H/D < 4.0 (center-to-center spacing < 1.25D)—other structures.

Diagrams

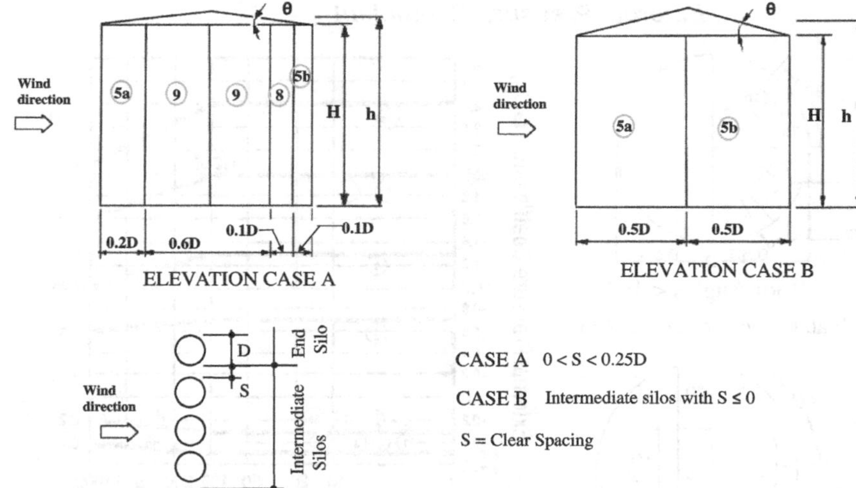

Notation

D = Diameter of a circular structure, ft (m).

h = Mean roof height, ft (m).

H = The solid cylinder height, ft (m).

θ = Angle of plane of roof from horizontal, degrees.

External Pressure Coefficient

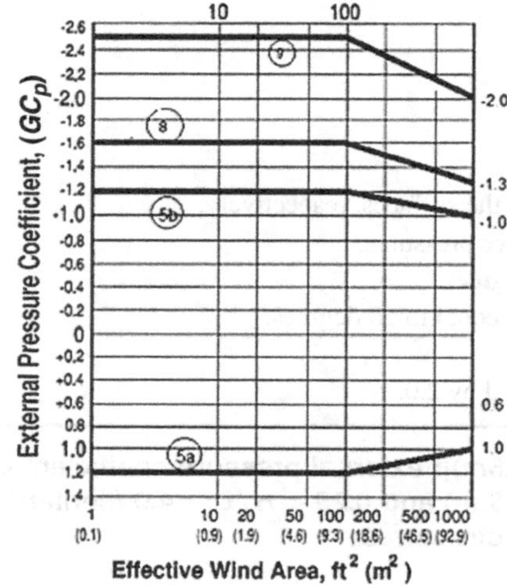

Notes

1. Vertical scale denotes (GCp) to be used with q_h.
2. Horizontal scale denotes effective wind area, ft^2 (m^2).
3. Plus and minus signs signify pressures acting toward and away from the surfaces, respectively.
4. Each component shall be designed for maximum positive and negative pressures.
5. Zone 9 shall be the region with the shortest distance between the adjacent silos and at the outside corners of the groups.
6. Case A is applicable for the silos with clear spacing larger than 0 and less than 0.25D. Case B is applicable for the intermediate silos of connected silo array - excluding end silos - with clear spacing equal to or less than 0.

Figure 30.10-4. Components and cladding [$h \leq 120$ft ($h \leq 36.6m$)]: external pressure coefficients, GC_p, for walls of grouped circular bins, silos, and tanks with $D < 120$ ft (36.6 m) and $0.25 < H/D < 4.0$ (center-to-center spacing $< 1.25D$)—other structures.

CHAPTER 31
WIND TUNNEL PROCEDURE

31.1 SCOPE

The Wind Tunnel Procedure shall be used where required by Sections 27.1.3, 28.1.3, 29.1.3, and 30.1.3. The Wind Tunnel Procedure shall be permitted for any building or other structure in lieu of the design procedures specified in Chapter 27 [main wind force resisting system (MWFRS) for buildings of all heights], Chapter 28 (MWFRS of low-rise buildings), Chapter 29 (MWFRS of all other structures), and Chapter 30 (components and cladding [C&C] for all building types and other structures).

The results of numerical Wind Tunnel Procedures (i.e., the use of computational fluid dynamics for wind engineering purposes) shall be verified, validated by a physical wind tunnel procedure in accordance with ASCE 49, and subject to an independent peer review, in accordance with ASCE 7 Section 1.3.1.3.4.

User Note: Chapter 31 may always be used for determining wind pressures for the MWFRS and/or for C&C of any building or other structure. This method is considered to produce the most accurate wind pressures and loads of any method specified in this standard.

31.2 TEST CONDITIONS

Wind tunnel tests, or similar tests using fluids other than air, used for the determination of design wind loads for any building, other structure, portion of structure, or component, shall be conducted in accordance with this section. Tests for the determination of the mean, fluctuating, and peak forces and pressures shall meet the requirements of ASCE 49.

31.3 DYNAMIC RESPONSE

Tests for the purpose of determining the dynamic response of a building, other structure, portion of structure, or component shall be in accordance with Section 31.2. The tested model and associated analysis shall account for mass distribution, stiffness, and damping.

31.4 SITE SPECIFIC LOAD EFFECTS FOR BUILDINGS, OTHER STRUCTURES, AND COMPONENTS

Load effects for buildings, other structures, and components at a single, specific site shall be permitted to be determined in accordance with this section. Wind tunnel tests shall conform with the requirements of Sections 31.2 and 31.3.

31.4.1 Mean Recurrence Intervals of Load Effects The load effect required for strength design shall be determined for the same mean recurrence interval as for the analytical method by using a rational analysis method, as defined in the recognized literature, for combining the directional wind tunnel data with the directional meteorological data or probabilistic models based thereon. The load effect required for allowable stress design shall be equal to the load effect required for strength design divided by 1.6. For buildings or other structures that are sensitive to possible variations in the values of the dynamic parameters, sensitivity studies shall be required to provide a rational basis for design recommendations.

31.4.2 Limitations on Wind Speeds The wind speeds, and probabilistic estimates based thereon, shall be subject to the limitations described in Section 26.5.3.

31.4.3 Wind Directionality The directional wind climate based on recorded or simulated directional wind speed data shall be considered in determining wind loads, and the data shall be presented as part of the wind tunnel report submitted to the Authority Having Jurisdiction. The method for combining wind tunnel model data with information on wind speed and direction at the project site shall also be clearly stated in the Wind Tunnel Report. Variation in the wind direction, based upon uncertainty in the wind climate data, shall be considered when determining the wind loading, and the design wind loads shall be based on the largest values that result from this uncertainty. Consideration of uncertainty in the wind direction is not required in the determination of serviceability-related wind effects.

31.4.4 Limitations on Loads Loads for the MWFRS, determined by wind tunnel testing, shall be limited such that the overall principal loads in the x and y directions are not less than 80% of those that would be obtained from Chapter 27 or Chapter 28 for buildings, or Chapter 29 for appurtenances or other structures. The overall principal load for buildings shall be based on the overturning moment for flexible buildings and the base shear for other buildings. The overall principal load for other structures shall be based on the overturning moment for flexible structures and the base shear for other structures.

Pressures for C&C, determined by wind tunnel testing, shall be limited to not less than 80% of those calculated for Zone 4 for walls and Zone 1 for roofs, using the procedure in Chapter 30. These zones refer to those shown in Figures. 30.3-1, 30.3-2A–C, 30.3-3, 30.3-4, 30.3-5A–B, 30.3-6, 30.3-7, and 30.5-1.

The limiting values of 80% may be reduced to 50% for the MWFRS and 65% for C&C if either of the following conditions applies:

1. There were no specific influential buildings or objects within the detailed proximity model.
2. Loads and pressures from supplemental tests for all significant wind directions in which specific influential buildings or objects are replaced by the roughness representative of

the adjacent roughness condition, but not rougher than Exposure B, are included in the test results.

31.4.5 Limitations on Wind Loads for Ground-Mounted Fixed-Tilt Solar Panel Systems For ground-mounted fixed-tilt solar panel systems that meet the limitations and geometry requirements of Section 29.4.5.1, the minimum design wind load based on a wind tunnel study shall not be less than 65% for components and cladding (C&C) and 50% for the main wind force resisting system (MWFRS) of the values resulting from Section 29.4.5, subject to the conditions of Section 31.4.3. Wind load values lower than these limits shall be permitted when an independent peer review of the wind tunnel test is performed, in accordance with Section 31.6. The minimum design wind force based on a wind tunnel study for ground-mounted fixed-tilt solar panel systems need not comply with the minimum net pressure of 16 lb/ft^2 (0.77 kN/m^2) per Section 30.2.2.

31.5 LOAD EFFECTS FOR BUILDINGS, OTHER STRUCTURES, AND COMPONENTS USED AT MULTIPLE SITES

31.5.1 Wind Loads Wind loads on buildings, other structures, portions of structures, or components used at multiple sites, shall be permitted to be determined by wind tunnel tests by determining load coefficients for use in the analysis equations of the Directional Procedure in Chapters 27 and 29 for MWFRS and in Part 4 of Chapter 30 for C&C. Alternatively, the loads are permitted to be specified with an analysis method defined in the wind tunnel test report. It is not required to include specific nearby buildings in the testing when results are to be used for multiple sites.

The wind tunnel tests shall conform with the requirements of Sections 31.2 and 31.3.

Data analysis shall consider wind loads from all wind directions. Generic load coefficients shall be calculated to be consistent with coefficients in Chapters 27, 29, and 30, or shall be defined to apply to an analysis procedure specified in the test report.

The test report shall include data collection methods, data analysis, wind field modeling, wind tunnel model details, measured wind loads, conversion of data into generic coefficients, and conditions of applicability of results. Wind tunnel results shall not be extrapolated to geometric configurations that were not anticipated by the wind tunnel study. Interpolation between two or more tests shall be permitted. The limitations of the wind tunnel study shall be clearly reported.

31.5.2 Limitations on Wind Loads for Rooftop Solar Panels For photovoltaic solar panel systems that meet the limitations and geometry requirements of Figure 29.4-7, the minimum design wind load based on a wind tunnel study shall not be less than 65% of the values resulting from 29.4-7, subject to the conditions of Section 31.4.3. The minimum design wind force based on a wind tunnel study for roof-mounted solar panel systems need not comply with the minimum net pressure of 16 lb/ft^2 (0.77 kN/m^2) as per ASCE 7, Section 30.2.2.

31.5.3 Peer Review Requirements for Wind Tunnel Tests of Buildings, Other Structures, and Components used at Multiple Sites For wind loads on buildings, other structures, and components used at multiple sites determined by wind tunnel tests, an independent peer review of the wind tunnel test shall be performed in accordance with Section 31.6. For photovoltaic solar panel systems that meet the limitations and geometrical requirements of Figure 29.4-7, wind load values lower than those indicated in Section 31.5.2 shall be permitted when an independent peer review of the wind tunnel test is performed, in accordance with Section 31.6.

31.6 PEER REVIEW REQUIREMENT FOR WIND TUNNEL TESTS

When an independent peer review is required by Section 31.5.3 or 31.4.5, it shall be an objective, technical review performed by knowledgeable reviewer(s) experienced in performing wind tunnel studies on buildings and similar systems, and in properly simulated atmospheric boundary layers or wind fields. The minimum qualifications for the peer reviewer shall include the following:

- Peer reviewer shall be independent from the wind tunnel laboratory that performed the tests and report and shall bear no conflict of interest.
- Peer reviewer shall have technical expertise in the application of wind tunnel studies on buildings, other structures, or components, similar to that being reviewed.
- Peer reviewer shall have experience in performing or evaluating boundary layer wind tunnel studies and shall be familiar with the technical issues and regulations governing the Wind Tunnel Procedure in ASCE 49, as it is applied to the building, other structure, or component under consideration.

The peer reviewer shall review the wind tunnel report, including, but not limited to, data collection methods, data analysis, wind field modeling, building, other structure, or component wind tunnel model, resulting wind loads, conversion of data into design values, conditions of applicability of results, and other relevant issues identified by the reviewer.

The peer reviewer shall submit a written report to the Authority Having Jurisdiction and the client. The report shall include, at a minimum, statements regarding the following: scope of peer review with limitations defined; the status of the wind tunnel study at time of review; conformance of the wind tunnel study with the requirements of ASCE 49 and Section 31.5.1; conclusions of the reviewer identifying areas that need further review, investigation, and/or clarification; recommendations; and whether, in the reviewer's opinion, the wind loads derived from the wind tunnel study are in conformance with ASCE 7 for the intended use(s).

31.7 WIND-BORNE DEBRIS

Glazing in buildings in wind-borne debris regions shall be in accordance with Section 26.12.3.

31.8 CONSENSUS STANDARDS AND OTHER REFERENCED DOCUMENTS

This section lists the consensus standards and other documents that shall be considered part of this standard to the extent referenced in this chapter.

ASCE 49, *Wind Tunnel Testing for Buildings and Other Structures*, American Society of Civil Engineers, 2020.

Cited in: Sections 31.2, 31.6, 31.6.1.2, C31, C31.4.2, C31.6.1

CHAPTER 32
TORNADO LOADS

32.1 PROCEDURES

32.1.1 Scope Buildings and other structures classified as Risk Category III or IV and located in the tornado-prone region as shown in Figure 32.1-1, including the main wind force resisting system (MWFRS) and all components and cladding (C&C) thereof, shall be designed and constructed to resist the greater of the tornado loads determined in accordance with the provisions of this chapter or the wind loads determined in accordance with Chapters 26 through 31, using the load combinations provided in Chapter 2.

> **User Note:** The tornado loads specified in this chapter provide reasonable consistency with the reliability delivered by the existing criteria in Chapters 26 and 27 for MWFRS, and therefore are only required for Risk Category III and IV buildings and other structures (see Return Period discussion in Section C32.5.1 for more information). The tornado loads are based on tornado speeds using 1,700- and 3,000-year return periods for Risk Category III and IV, respectively (which are the same return periods used for basic wind speeds in Chapter 26). The tornado speed at any given geographic location will range from approximately Enhanced Fujita Scale EF0 – EF2 intensity, depending on the risk category and effective plan area of the building or other structure (see Section C32.5.1). Options for protection of life and property from more intense tornadoes include construction of a storm shelter and/or design for longer-return-period tornado speeds as provided in Appendix G, including performance-based design. A building or other structure designed for tornado loads determined exclusively in accordance with Chapter 32 cannot be designated as a storm shelter without meeting additional critical requirements provided in the applicable building code and ICC 500, the ICC/NSSA *Standard for the Design and Construction of Storm Shelters*. See Commentary Section C32.1.1 for an in-depth discussion on storm shelters.

32.1.2 Permitted Procedures The design tornado loads for buildings and other structures, including the MWFRS and C&C elements thereof, shall be determined using one of the procedures as specified in this section and subject to the applicable limitations of Chapters 26 through 32, excluding Chapter 28.

An outline of the overall process for the determination of the tornado loads, including section references, is provided in Figure 32.1-3.

32.1.2.1 Tornado Loads on the Main Wind Force Resisting System Tornado loads for the MWFRS shall be determined using one or more of the following procedures, as modified by Chapter 32:

1. Directional Procedure for buildings of all heights as specified in Chapter 27 for buildings meeting the requirements specified therein;
2. Directional Procedure for Building Appurtenances (such as rooftop structures and rooftop equipment) and Other Structures (such as solid freestanding walls and solid freestanding signs, chimneys, tanks, open signs, single-plane open frames, and trussed towers) as specified in Chapter 29 for buildings meeting the requirements specified therein; or
3. Wind Tunnel Procedure for all buildings and all other structures as specified in Chapter 31 for buildings meeting the requirements specified therein.

32.1.2.2 Tornado Loads on Components and Cladding Tornado loads on the C&C of all buildings and other structures shall be determined using one or more of the following procedures, as modified by Chapter 32:

1. Analytical Procedures as specified in Parts 1 through 5, as appropriate, of Chapter 30, for buildings or other structures meeting the requirements specified therein; or
2. Wind Tunnel Procedure for all buildings and other structures as specified in Chapter 31, for buildings meeting the requirements specified therein.

32.1.3 Performance-Based Procedures Tornado design of buildings and other structures using performance-based procedures shall be permitted subject to the approval of the Authority Having Jurisdiction. The performance-based tornado design procedures used shall, at a minimum, conform to Section 1.3.1.3 and be documented and submitted to the Authority Having Jurisdiction in accordance with Section 1.3.1.3.

32.2 DEFINITIONS

The following definitions apply to the provisions of Chapter 32. Terms not defined in this chapter shall be defined in accordance with Chapters 26 through 31, as appropriate, excluding Chapter 28.

ASCE TORNADO DESIGN GEODATABASE: The ASCE database (version 2022-1.0) of geocoded tornado speed design data.

OTHER STRUCTURES, SEALED: A structure that is completely sealed or has controlled ventilation such that tornado-induced atmospheric pressure changes will not be transmitted to the inside of the structure, including but not limited to certain tanks and vessels.

TORNADO-PRONE REGION: The area of the conterminous United States most vulnerable to tornadoes, as shown in Figure 32.1-1.

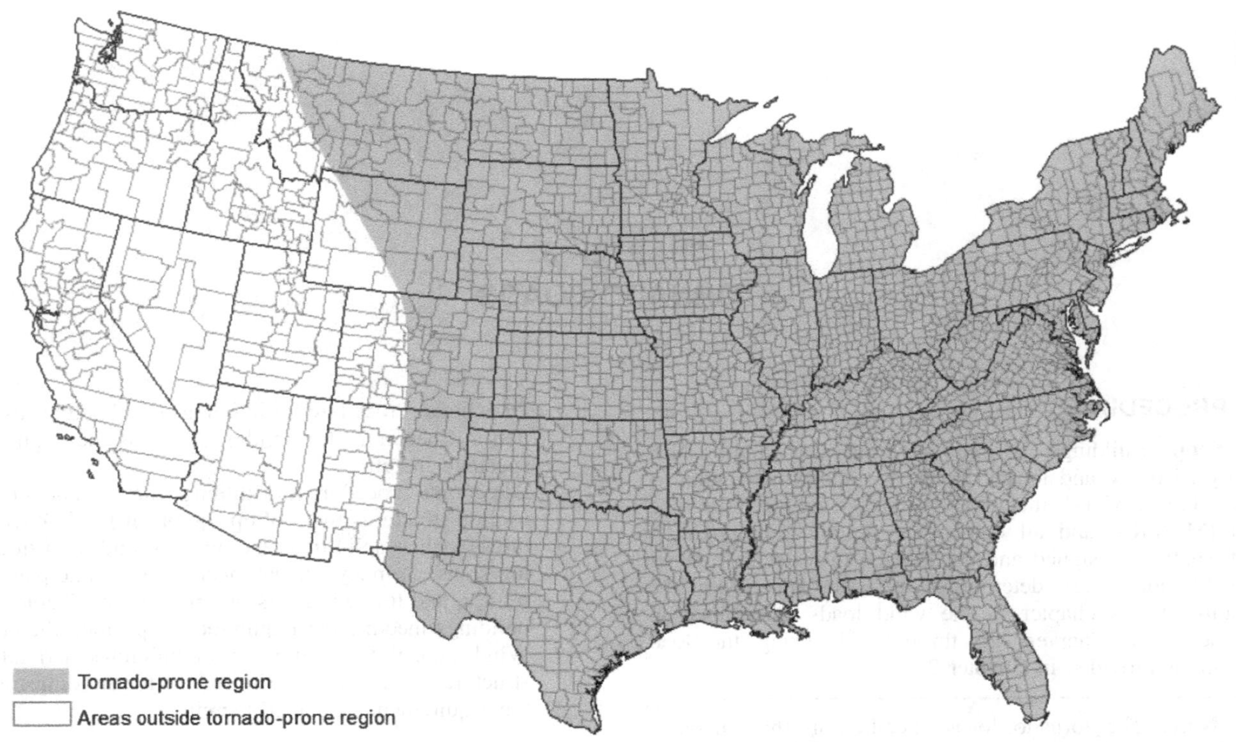

Figure 32.1-1. Tornado-prone region.

32.3 SYMBOLS AND NOTATION

The following symbols apply only to the provisions of Chapter 32. Symbols and notations not defined in this chapter shall be defined in accordance with Chapters 26 through 31, as appropriate, excluding Chapter 28.

A_e = Effective plan area of the building, other structure, or facility, ft² (m²), as defined in Section 32.5.4

F_{hT} = Lateral design tornado force for rooftop structures and equipment from Equation (32.16-3), lb (N)

F_T = Design tornado force for certain other structures from Equation (32.16-2), lb (N)

F_{vT} = Vertical design tornado force for rooftop structures and equipment from Equation (32.16-4), lb (N)

G_T = Tornado gust-effect factor as defined in Section 32.11

(GC_{piT}) = Product of internal pressure coefficient that includes the effects of atmospheric pressure change and gust-effect factor, to be used in determination of tornado loads for buildings and some other structures, as determined in Section 32.13

K_{dT} = Tornado directionality factor as defined in Section 32.6

K_{hTor} = Tornado velocity pressure exposure coefficient evaluated at height $z = h$, as determined in Section 32.10

K_{vT} = Tornado pressure coefficient adjustment factor for vertical winds as defined in Section 32.14

K_{zTor} = Tornado velocity pressure exposure coefficient evaluated at height z, as determined in Section 32.10

p_{pT} = Combined net tornado design pressure on a parapet from Equation (32.15-3), lb/ft² (N/m²)

p_T = Design tornado pressure to be used in determination of tornado loads for buildings and for certain other structures, lb/ft² (N/m²)

q_{hT} = Tornado velocity pressure evaluated at height $z = h$, lb/ft² (N/m²)

q_{pT} = Tornado velocity pressure evaluated at top of parapet from Equations (32.15-3) and (32.17-4), lb/ft² (N/m²)

q_{zop} = Tornado velocity pressure internal pressure determination for partially enclosed buildings, lb/ft² (N/m²)

q_{zT} = Tornado velocity pressure evaluated at height z above ground, lb/ft² (N/m²)

V_T = Tornado speed obtained from Figures 32.5-1 and 32.5-2, mi/h (m/s). The tornado speed corresponds to a 3 s gust speed at 33 ft (10 m) above the ground.

z_{op} = Level of the lowest opening in the building that could affect the positive internal pressure, lb/ft² (N/m²)

32.4 GENERAL

32.4.1 Sign Convention The combined effects of internal pressures and atmospheric pressure change, expressed in coefficient form by (GC_{piT}), shall follow the same sign convention as provided in Section 26.4, where positive pressure acts toward the surface and negative pressure acts away from the surface.

32.4.2 Critical Load Condition Values of external pressures shall be combined algebraically with the combined effects of internal pressures and atmospheric pressure change to determine the most critical load.

32.5 TORNADO HAZARD MAPS

32.5.1 Tornado Speed The tornado speed, V_T, used in the determination of tornado loads on buildings and other structures shall be determined from Figures 32.5-1 and 32.5-2 as follows:

1. For Risk Category III buildings and structures, use Figures 32.5-1A through 32.5-1H.

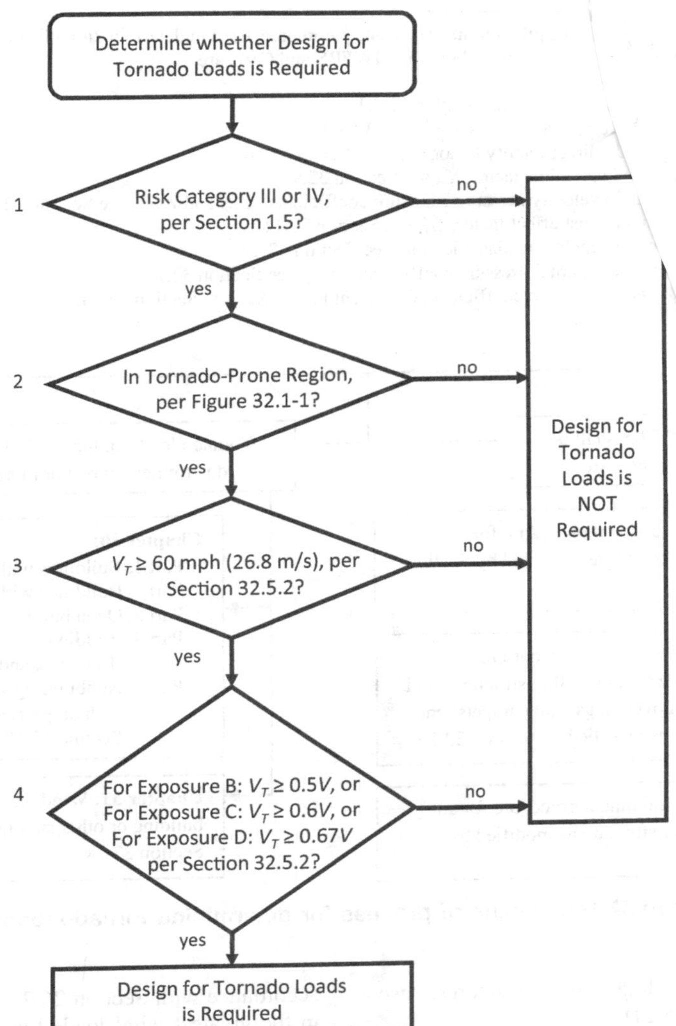

Figure 32.1-2. Flowchart of process for determining when design for tornado loads is required.

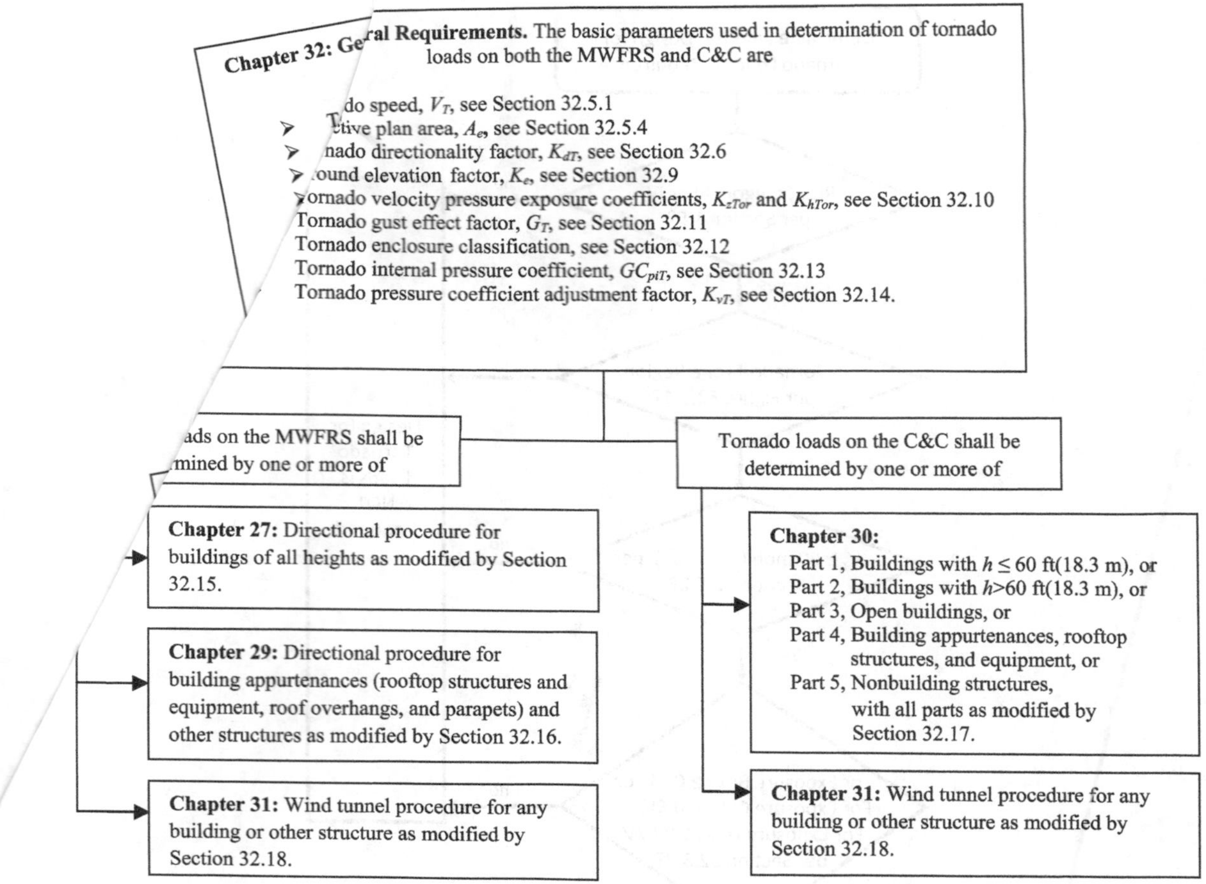

Figure 32.1-3. Outline of process for determining tornado loads.

2. For Risk Category IV buildings and structures, use Figures 32.5-2A through 32.5-2H.

To select the appropriate tornado hazard map to use for the assigned risk category, the effective plan area, A_e, of the building, other structure, or facility, shall be determined in accordance with Section 32.5.4 and shall be rounded up to next available mapped A_e, including 1; 2,000; 10,000; 40,000; 100,000; 250,000; 1,000,000; and 4,000,000 ft^2 (0.1; 186; 929; 3,716; 9,290; 23,226; 92,903; and 371,612 m^2). Alternatively, linear interpolation of tornado speed between maps using the logarithm of the effective plan area size is permitted.

Alternatively, it shall be permitted to use the tornado speeds from the ASCE Tornado Design Geodatabase. The ASCE Tornado Design Geodatabase is available at the ASCE 7 Hazard Tool (https://asce7hazardtool.online) or approved equivalent.

32.5.2 Design for Tornado Loads Not Required For Risk Category III and IV buildings and other structures determined to have tornado speeds V_T < 60 mi/h (26.8 m/s), design for tornado loads shall not be required. Where $V_T \geq 60$ mi/h (26.8 m/s) but is less than the following threshold speeds then design for tornado loads shall not be required:

1. For Exposure B: $V_T < 0.5V$, or
2. For Exposure C: $V_T < 0.6V$, or
3. For Exposure D: $V_T < 0.67V$,

where V is the basic wind speed determined in accordance with Section 26.5 and the exposure category is determined in accordance with Section 26.7.3, based on the exposure resulting in the greatest wind loads for any wind direction at the site.

32.5.3 Direction of Tornadic Wind The tornadic wind shall be assumed to come from any horizontal direction.

32.5.4 Effective Plan Area The effective plan area, A_e, of the building or other structure shall be determined in accordance with this section.

32.5.4.1 Essential Facilities For Essential Facilities and buildings and other structures required to maintain the functionality of Essential Facilities, the effective plan area shall be equal to the area of the smallest convex polygon enclosing both the Essential Facility and all of the buildings and other structures that maintain the functionality of the Essential Facility.

32.5.4.2 Other than Essential Facilities For buildings and structures that are not designated as Essential Facilities and are not required to maintain the functionality of Essential Facilities, the effective plan area shall be equal to the area of the smallest convex polygon enclosing the plan of the building, other structure, or facility. It is permitted to reduce the effective plan area to that of the effective plan area of the largest structurally independent building or other structure, which does not share structural components with adjacent buildings or other structures.

32.5.4.3 Ground-Mounted Photovoltaic Panel Systems The effective plan area, A_e, of ground-mounted photovoltaic panel systems shall be equal to the effective plan area of the largest structurally independent photovoltaic support structure that does not share structural components with other adjacent structures.

32.6 TORNADO DIRECTIONALITY FACTOR

The tornado directionality factor, K_{dT}, shall be determined from Table 32.6-1 and shall be used to determine the tornado loads in accordance with Sections 32.15 through 32.17.

32.7 TORNADO EXPOSURE

Tornado velocity pressure exposure coefficients K_{zTor} and K_{hTor} are determined in Section 32.10.1. Exposure requirements in Section 26.7 shall not apply to the determination of K_{zTor} and K_{hTor}.

32.8 TOPOGRAPHIC FACTOR

Tornado velocity pressures q_{zT} and q_{hT} are determined in Section 32.10.2. Topographic speedup effects in Section 26.8 shall not apply to the determination of q_{zT} and q_{hT}.

32.9 GROUND ELEVATION FACTOR

The ground elevation factor to adjust for air density, K_e, shall be determined in accordance with Section 26.9 and shall be used to determine the tornado velocity pressure in accordance with Section 32.10.2.

32.10 TORNADO VELOCITY PRESSURE

32.10.1 Tornado Velocity Pressure Exposure Coefficient A tornado velocity pressure exposure coefficient, K_{zTor} or K_{hTor}, as applicable, shall be determined from Table 32.10-1.

32.10.2 Tornado Velocity Pressure Tornado velocity pressure, q_{zT}, evaluated at height z above ground, shall be determined in accordance with the following equation:

$$q_{zT} = 0.00256 K_{zTor} K_e V_T^2 \text{ (lb/ft}^2\text{); } V_T \text{ in mi/h} \quad (32.10\text{-}1)$$

$$q_{zT} = 0.613 K_{zTor} K_e V_T^2 \text{ (N/m}^2\text{); } V_T \text{ in m/s} \quad (32.10\text{-}1.\text{SI})$$

where

K_{zTor} = Tornado velocity pressure exposure coefficient, see Section 32.10.1;
K_e = Ground elevation factor, see Section 32.9;
V_T = Tornado speed, see Section 32.5; and
q_{zT} = Tornado velocity pressure at height z.

The velocity pressure at mean roof height shall be computed as $q_{hT} = q_{zT}$ evaluated from Equation (32.10-1) using K_{zTor} at mean roof height h.

32.11 TORNADO GUST EFFECTS

32.11.1 Tornado Gust-Effect Factor The tornado gust-effect factor, G_T, for a building or other structure shall be 0.85 or calculated from Equation (26.11-6) using Exposure C terrain exposure constants.

32.11.2 Limitations Where the combined gust-effect factors and pressure coefficients (GC_p), (GC_{piT}), and (GC_{pf}) are given in figures and tables, the gust-effect factor shall not be determined separately.

32.12 TORNADO ENCLOSURE CLASSIFICATION

32.12.1 General For the purpose of determining internal pressure coefficients for tornadoes, buildings and other structures for which tornado internal pressure coefficients, (GC_{piT}), apply shall have an enclosure classification assigned in accordance with this section. If a building or other structure satisfies both the "open" and "partially enclosed" tornado enclosure classification definitions, it shall be classified as a "partially open" building or other structure.

32.12.2 Openings To assign the tornado enclosure classification, the amount of openings in the building envelope shall be determined by taking each wall of the building or other structure, assuming it functions as the windward wall, and summing the total area of openings present with respect to the area of the remaining building envelope. Buildings shall be classified as enclosed, partially enclosed, partially open, or open as defined in Section 26.2. Other structures shall be classified as sealed, as defined in Section 32.2, or enclosed, partially enclosed, partially open, or open as defined in Section 26.2.

Where not required by Section 32.12.3 to protect glazed openings, enclosed buildings and other structures shall either (1) be reevaluated for classification as partially enclosed, with all unprotected glazed openings on each assumed windward wall considered as openings; or (2) be protected in accordance with Section 32.12.3.1.

32.12.3 Protection of Glazed Openings Glazed openings shall be protected as specified in this section for Essential Facilities and for buildings and other structures required to maintain the functionality of Essential Facilities.

32.12.3.1 Protection Requirements for Glazed Openings Glazing in buildings requiring protection shall be protected with an impact-protective system or shall be impact-resistant glazing. Impact-protective systems shall be either (a) permanently affixed non-operable systems or (b) permanently affixed operable systems capable of being fully deployed from inside the building within five minutes and used in buildings that are staffed 24 hours per day.

Impact-protective systems and impact-resistant glazing shall be subjected to missile tests in accordance with ASTM E1996 using missile level D or E as described in Table 2 of ASTM E1996. Testing to demonstrate compliance with ASTM E1996 shall be in accordance with ASTM E1886. Impact-resistant glazing and impact-protective systems shall comply with the pass/fail criteria of Section 7 of ASTM E1996. Glazing in sectional doors, rolling doors, and flexible doors shall be subjected to missile tests in accordance with ANSI/DASMA 115 as applicable. Glazing and impact-protective systems shall comply with the "Enhanced Protection" requirements of Table 3 of ASTM E1996, with tornado speed used in place of basic wind speed for determination of wind zone.

EXCEPTION: Other testing methods and/or performance criteria are permitted to be used where approved.

32.13 TORNADO INTERNAL PRESSURE COEFFICIENTS

Tornado internal pressure coefficients, (GC_{piT}), shall be determined from Table 32.13-1 based on building and other structure enclosure classifications determined in accordance with Section 32.12.1.

Notes:

1. Values are 3 s gust speeds in mi/h at 33 ft (10 m) above ground.
2. To convert tornado speeds from mi/h to m/s, multiply mapped values by 0.447.
3. Linear interpolation is permitted between contours. Point values (where shown) are provided to aid with interpolation.

Figure 32.5-1A. Tornado speeds for Risk Category III buildings and other structures, for effective plan area of 1 ft^2 (0.1 m^2).

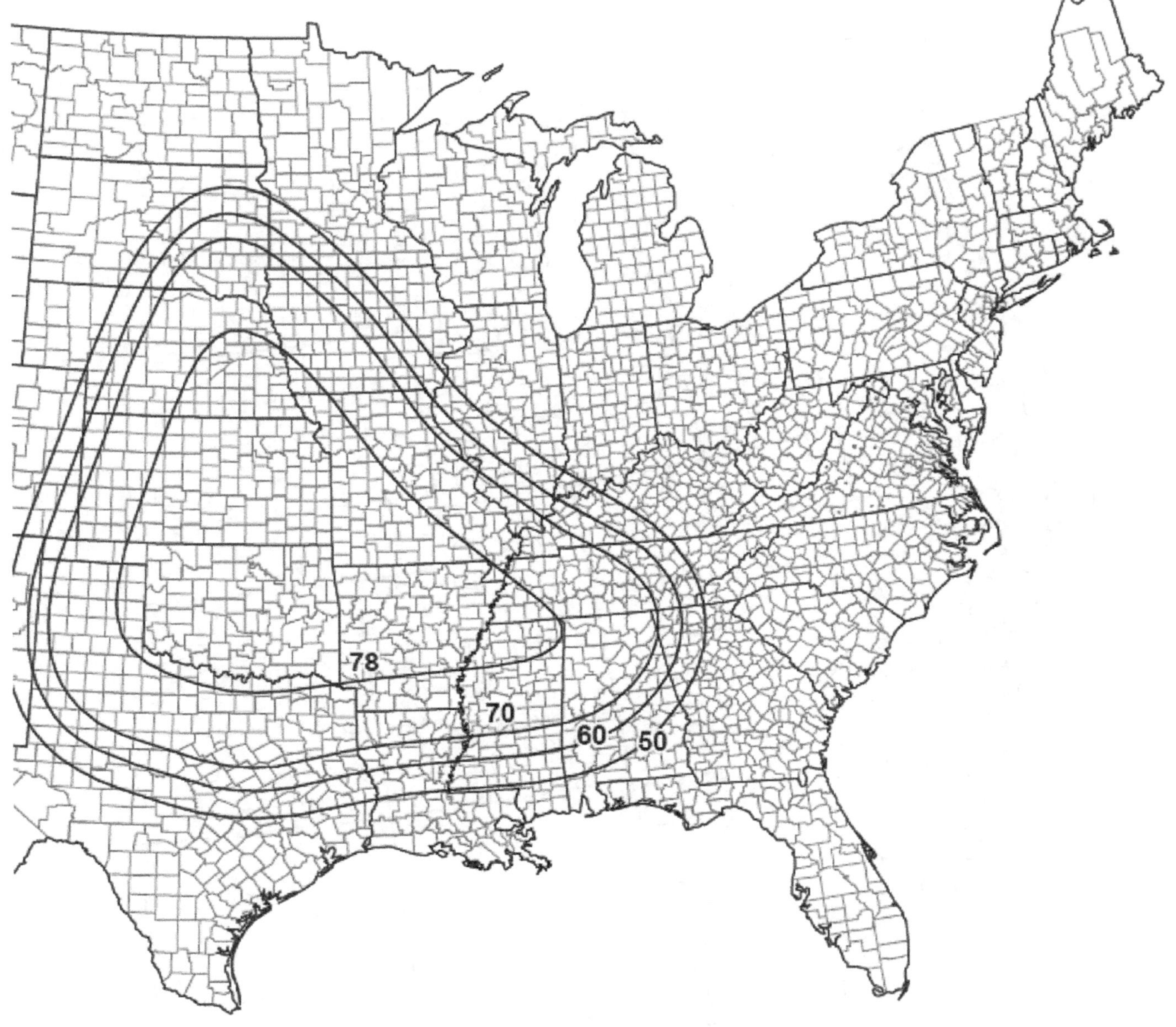

4. Islands, coastal areas, and land boundaries outside the last contour shall use the last tornado speed contour.
5. Tornado speeds correspond to approximately a 3% probability of exceedance in 50 years (annual exceedance probability =0.000588, MRI = 1,700 years).
6. Location-specific tornado speed is permitted to be determined using the ASCE Tornado Design Geodatabase, available at the ASCE 7 Hazard Tool (http://asce7hazardtool.online) or approved equivalent.

Figure 32.5-1A (*Continued*). Tornado speeds for Risk Category III buildings and other structures, for effective plan area of 1 ft^2 (0.1 m^2).

Notes:

1. Values are 3 s gust speeds in mi/h at 33 ft (10 m) above ground.
2. To convert tornado speeds from mi/h to m/s, multiply mapped values by 0.447.
3. Linear interpolation is permitted between contours. Point values (where shown) are provided to aid with interpolation.

Figure 32.5-1B. Tornado speeds for Risk Category III buildings and other structures, for effective plan area of 2,000 ft^2 (186 m^2).

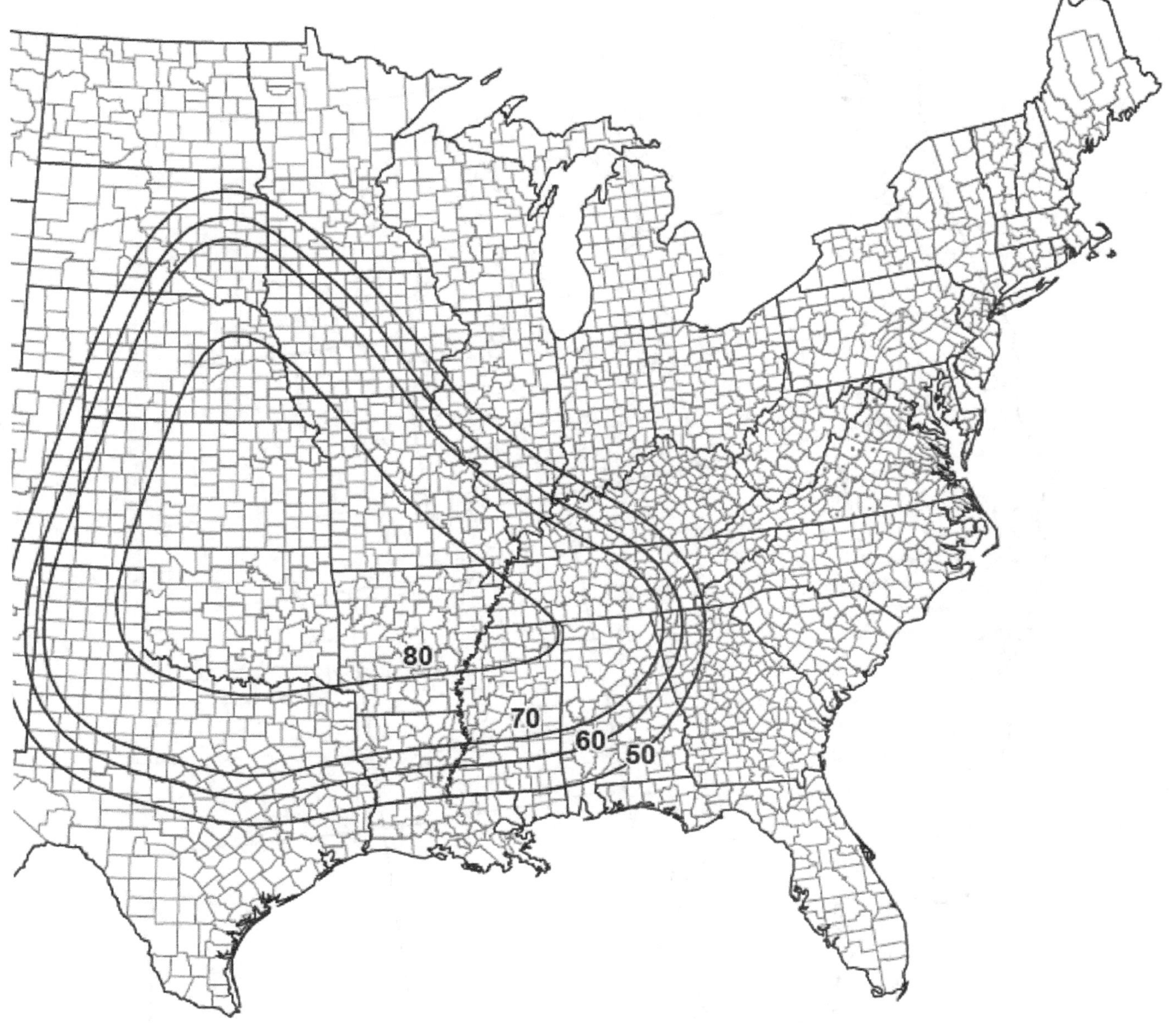

4. Islands, coastal areas, and land boundaries outside the last contour shall use the last tornado speed contour.
5. Tornado speeds correspond to approximately a 3% probability of exceedance in 50 years (annual exceedance probability = 0.000588, MRI = 1,700 years).
6. Location-specific tornado speed is permitted to be determined using the ASCE Tornado Design Geodatabase, available at the ASCE 7 Hazard Tool (http://asce7hazardtool.online) or approved equivalent.

Figure 32.5-1B (*Continued*). Tornado speeds for Risk Category III buildings and other structures, for effective plan area of 2,000 ft^2 (186 m^2).

Notes:

1. Values are 3 s gust speeds in mi/h at 33 ft (10 m) above ground.
2. To convert tornado speeds from mi/h to m/s, multiply mapped values by 0.447.
3. Linear interpolation is permitted between contours. Point values (where shown) are provided to aid with interpolation.

Figure 32.5-1C. Tornado speeds for Risk Category III buildings and other structures, for effective plan area of 10,000 ft^2 (929 m^2).

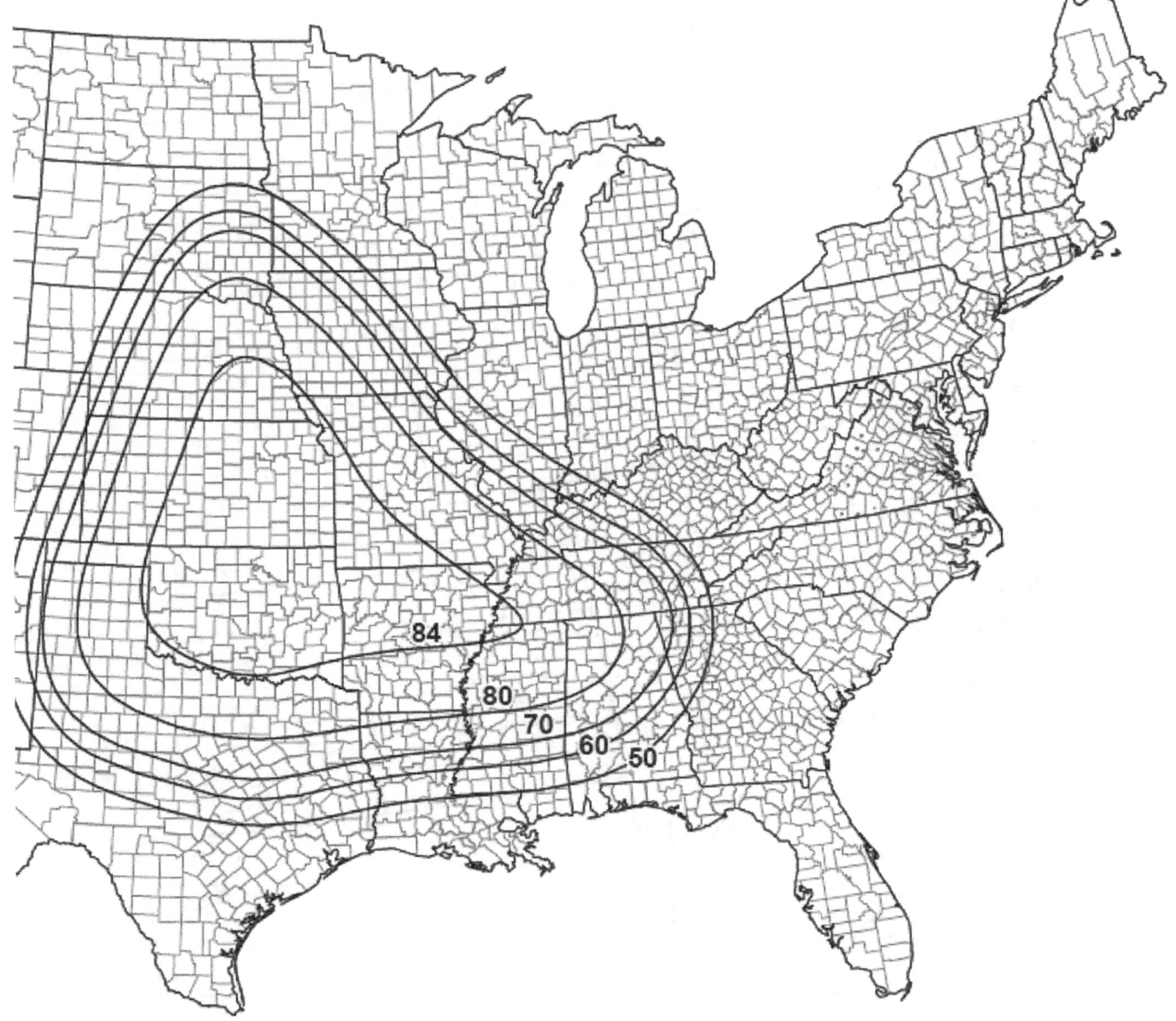

4. Islands, coastal areas, and land boundaries outside the last contour shall use the last tornado speed contour.
5. Tornado speeds correspond to approximately a 3% probability of exceedance in 50 years (annual exceedance probability = 0.000588, MRI = 1,700 years).
6. Location-specific tornado speed is permitted to be determined using the ASCE Tornado Design Geodatabase, available at the ASCE 7 Hazard Tool (http://asce7hazardtool.online) or approved equivalent.

Figure 32.5-1C (*Continued*). Tornado speeds for Risk Category III buildings and other structures, for effective plan area of 10,000 ft^2 (929 m^2).

Notes:

1. Values are 3 s gust speeds in mi/h at 33 ft (10 m) above ground.
2. To convert tornadc speeds from mi/h to m/s, multiply mapped values by 0.447.
3. Linear interpolation is permitted between contours. Point values (where shown) are provided to aid with interpolation.

Figure 32.5-1D. Tornado speeds for Risk Category III buildings and other structures, for effective plan area of 40,000 ft^2 (3,716 m^2).

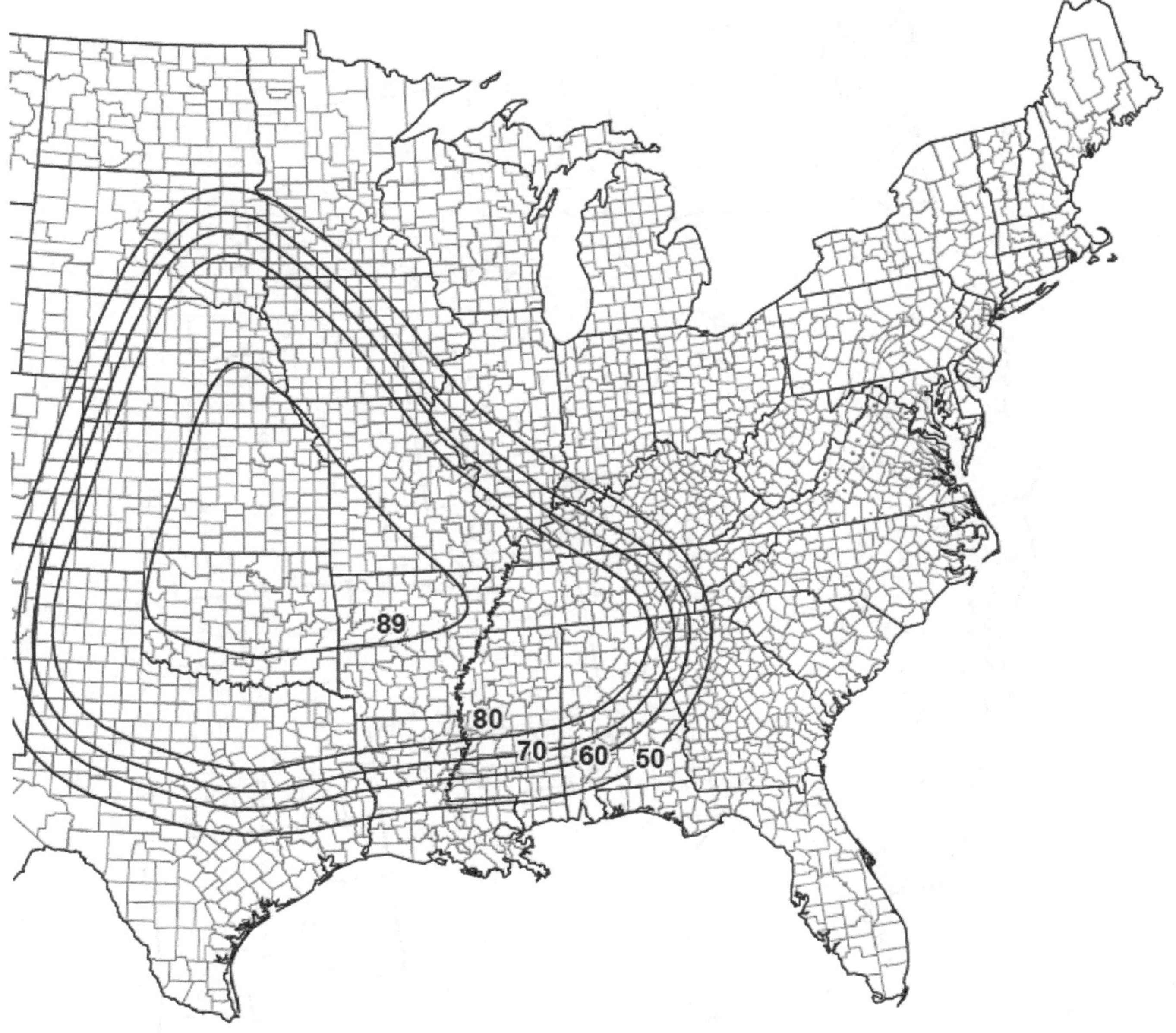

4. Islands, coastal areas, and land boundaries outside the last contour shall use the last tornado speed contour.
5. Tornado speeds correspond to approximately a 3% probability of exceedance in 50 years (annual exceedance probability = 0.000588, MRI = 1,700 years).
6. Location-specific tornado speed is permitted to be determined using the ASCE Tornado Design Geodatabase, available at the ASCE 7 Hazard Tool (http://asce7hazardtool.online) or approved equivalent.

Figure 32.5-1D (*Continued*). Tornado speeds for Risk Category III buildings and other structures, for effective plan area of 40,000 ft^2 (3,716 m^2).

Notes:

1. Values are 3 s gust speeds in mi/h at 33 ft (10 m) above ground.
2. To convert tornado speeds from mi/h to m/s, multiply mapped values by 0.447.
3. Linear interpolation is permitted between contours. Point values (where shown) are provided to aid with interpolation.

Figure 32.5-1E. Tornado speeds for Risk Category III buildings and other structures, for effective plan area of 100,000 ft^2 (9,290 m^2).

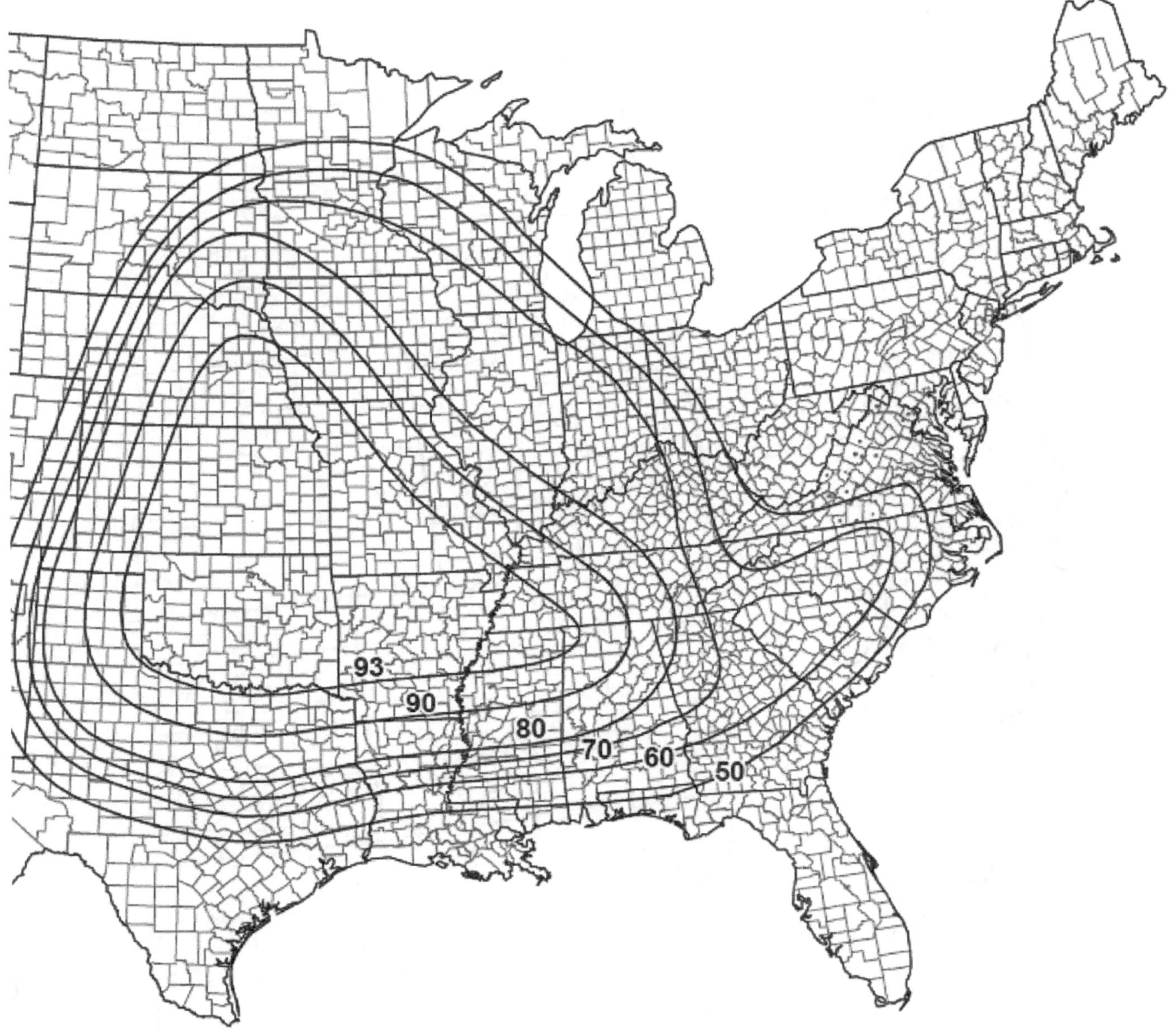

4. Islands, coastal areas, and land boundaries outside the last contour shall use the last tornado speed contour.
5. Tornado speeds correspond to approximately a 3% probability of exceedance in 50 years (annual exceedance probability = 0.000588, MRI = 1,700 years).
6. Location-specific tornado speed is permitted to be determined using the ASCE Tornado Design Geodatabase, available at the ASCE 7 Hazard Tool (http://asce7hazardtool.online) or approved equivalent.

Figure 32.5-1E (*Continued*). Tornado speeds for Risk Category III buildings and other structures, for effective plan area of 100,000 ft^2 (9,290 m^2).

Notes:

1. Values are 3 s gust speeds in mi/h at 33 ft (10 m) above ground.
2. To convert tornado speeds from mi/h to m/s, multiply mapped values by 0.447.
3. Linear interpolation is permitted between contours. Point values (where shown) are provided to aid with interpolation.

Figure 32.5-1F. Tornado speeds for Risk Category III buildings and other structures, for effective plan area of 250,000 ft² (23,226 m²).

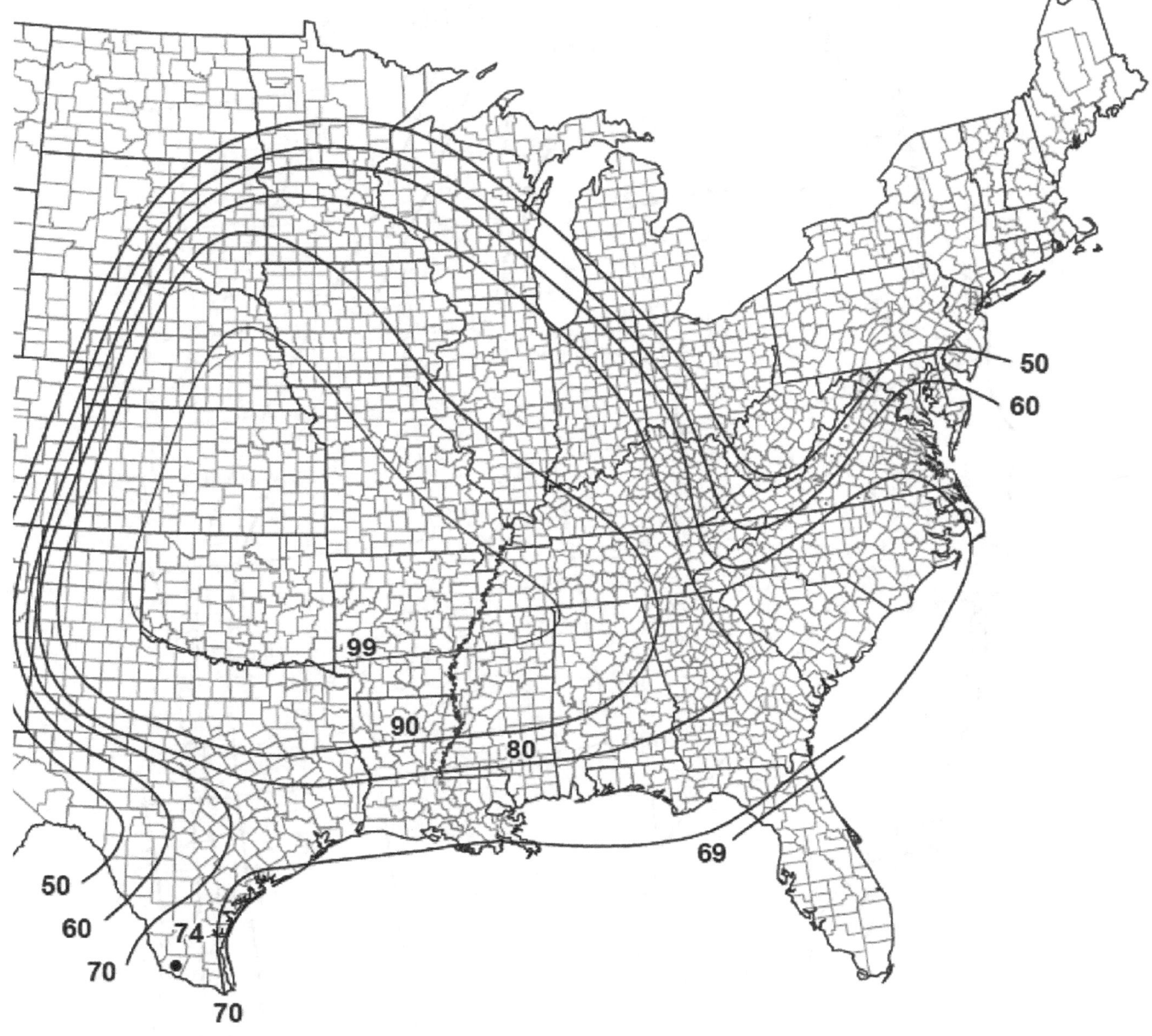

4. Islands, coastal areas, and land boundaries outside the last contour shall use the last tornado speed contour.
5. Tornado speeds correspond to approximately a 3% probability of exceedance in 50 years (annual exceedance probability = 0.000588, MRI = 1,700 years).
6. Location-specific tornado speed is permitted to be determined using the ASCE Tornado Design Geodatabase, available at the ASCE 7 Hazard Tool (http://asce7hazardtool.online) or approved equivalent.

Figure 32.5-1F (*Continued*). Tornado speeds for Risk Category III buildings and other structures, for effective plan area of 250,000 ft^2 (23,226 m^2).

Notes:

1. Values are 3 s gust speeds in mi/h at 33 ft (10 m) above ground.
2. To convert tornado speeds from mi/h to m/s, multiply mapped values by 0.447.
3. Linear interpolation is permitted between contours. Point values (where shown) are provided to aid with interpolation.

Figure 32.5-1G. Tornado speeds for Risk Category III buildings and other structures, for effective plan area of 1,000,000 ft^2 (92,903 m^2).

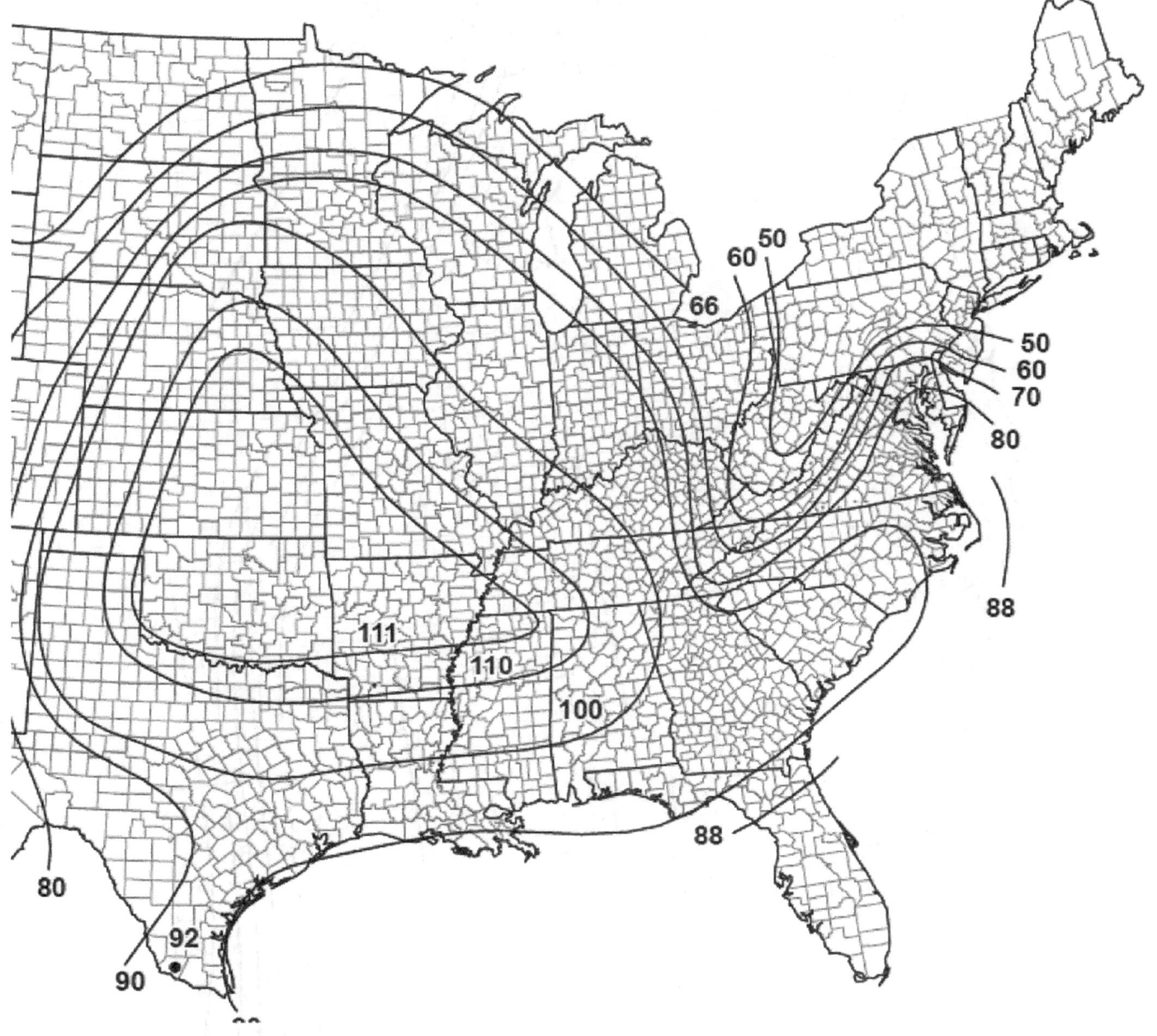

4. Islands, coastal areas, and land boundaries outside the last contour shall use the last tornado speed contour.
5. Tornado speeds correspond to approximately a 3% probability of exceedance in 50 years (annual exceedance probability = 0.000588, MRI = 1,700 years).
6. Location-specific tornado speed is permitted to be determined using the ASCE Tornado Design Geodatabase, available at the ASCE 7 Hazard Tool (http://asce7hazardtool.online) or approved equivalent.

Figure 32.5-1G (*Continued*). Tornado speeds for Risk Category III buildings and other structures, for effective plan area of 1,000,000 ft^2 (92,903 m^2).

Notes:

1. Values are 3 s gust speeds in mi/h at 33 ft (10 m) above ground.
2. To convert tornado speeds from mi/h to m/s, multiply mapped values by 0.447.
3. Linear interpolation is permitted between contours. Point values (where shown) are provided to aid with interpolation.

Figure 32.5-1H. Tornado speeds for Risk Category III buildings and other structures, for effective plan area of 4,000,000 ft^2 (371,612 m^2).

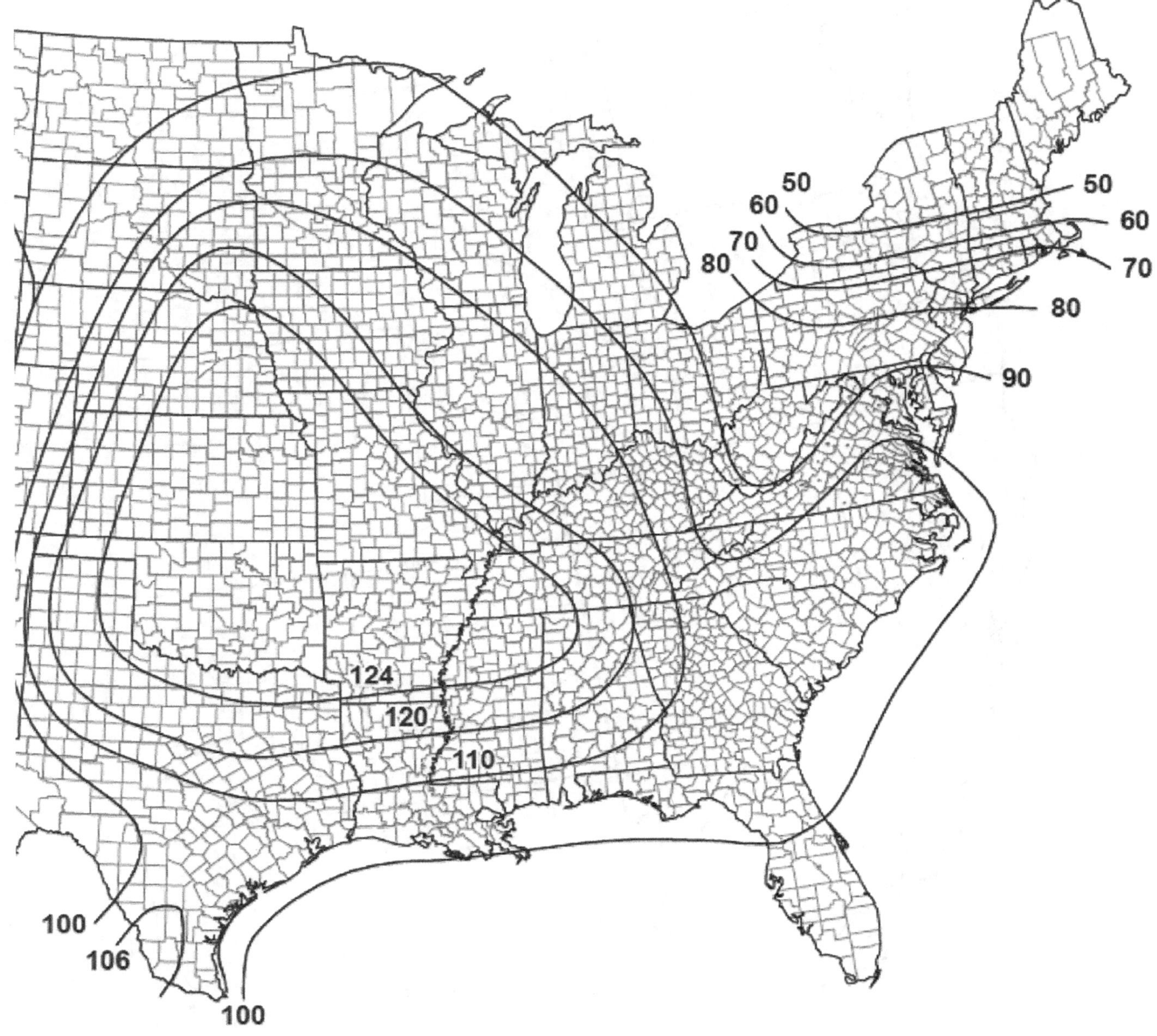

4. Islands, coastal areas, and land boundaries outside the last contour shall use the last tornado speed contour.
5. Tornado speeds correspond to approximately a 3% probability of exceedance in 50 years (annual exceedance probability = 0.000588, MRI = 1,700 years).
6. Location-specific tornado speed is permitted to be determined using the ASCE Tornado Design Geodatabase, available at the ASCE 7 Hazard Tool (http://asce7hazardtool.online) or approved equivalent.

Figure 32.5-1H (*Continued*). Tornado speeds for Risk Category III buildings and other structures, for effective plan area of 4,000,000 ft^2 (371,612 m^2).

Notes:

1. Values are 3 s gust speeds in mi/h at 33 ft (10 m) above ground.
2. To convert tornado speeds from mi/h to m/s, multiply mapped values by 0.447.
3. Linear interpolation is permitted between contours. Point values (where shown) are provided to aid with interpolation.

Figure 32.5-2A. Tornado speeds for Risk Category IV buildings and other structures, for effective plan area of 1 ft^2 (0.1 m^2).

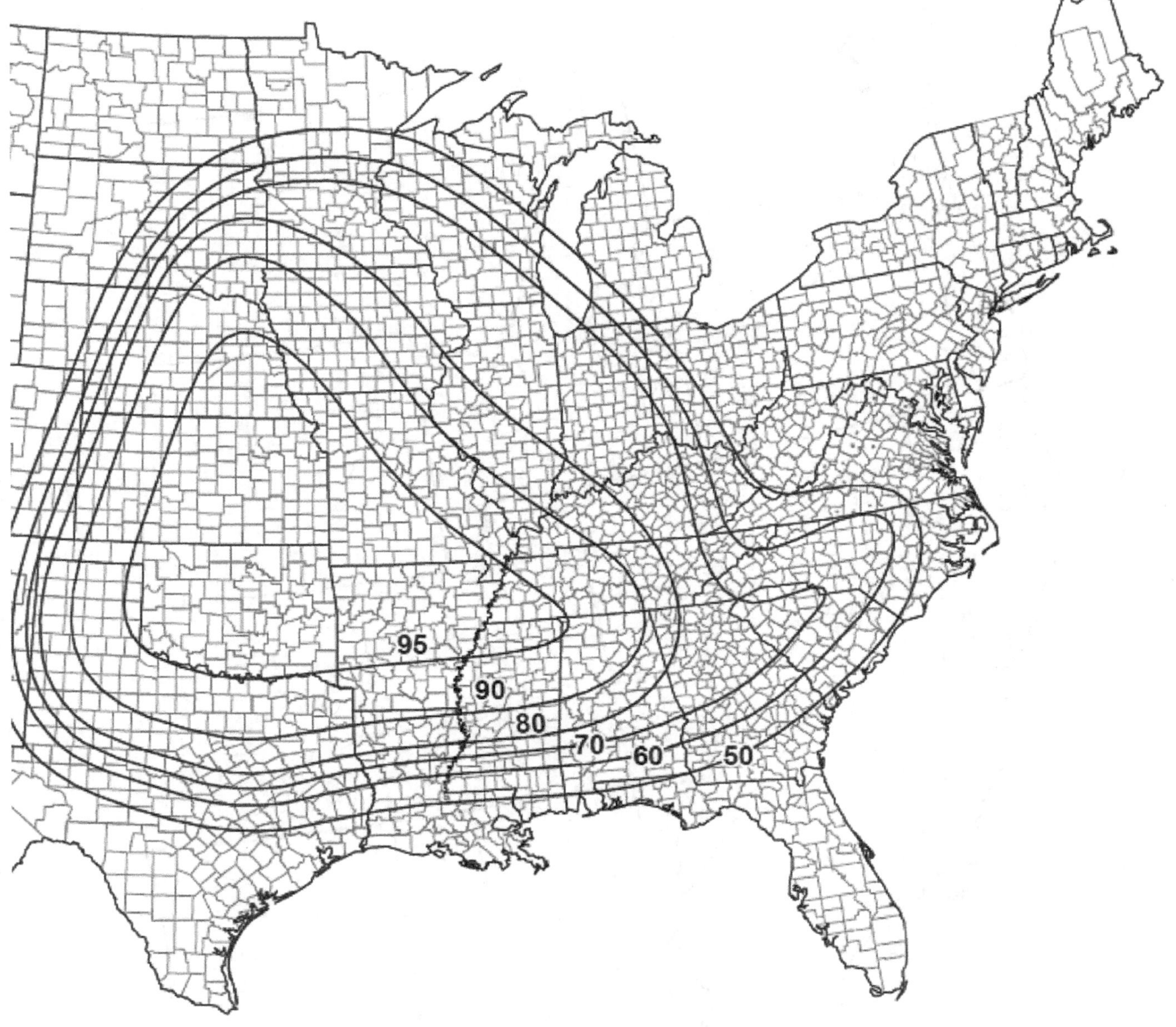

4. Islands, coastal areas, and land boundaries outside the last contour shall use the last tornado speed contour.
5. Tornado speeds correspond to approximately a 1.7% probability of exceedance in 50 years (annual exceedance probability = 0.00033, MRI = 3,000 years).
6. Location-specific tornado speed is permitted to be determined using the ASCE Tornado Design Geodatabase, available at the ASCE 7 Hazard Tool (http://asce7hazardtool.online) or approved equivalent.

Figure 32.5-2A (*Continued*). Tornado speeds for Risk Category IV buildings and other structures, for effective plan area of 1 ft^2 (0.1 m^2).

Notes:

1. Values are 3 s gust speeds in mi/h at 33 ft (10 m) above ground.
2. To convert tornado speeds from mi/h to m/s, multiply mapped values by 0.447.
3. Linear interpolation is permitted between contours. Point values (where shown) are provided to aid with interpolation.

Figure 32.5-2B. Tornado speeds for Risk Category IV buildings and other structures, for effective plan area of 2,000 ft² (186 m²).

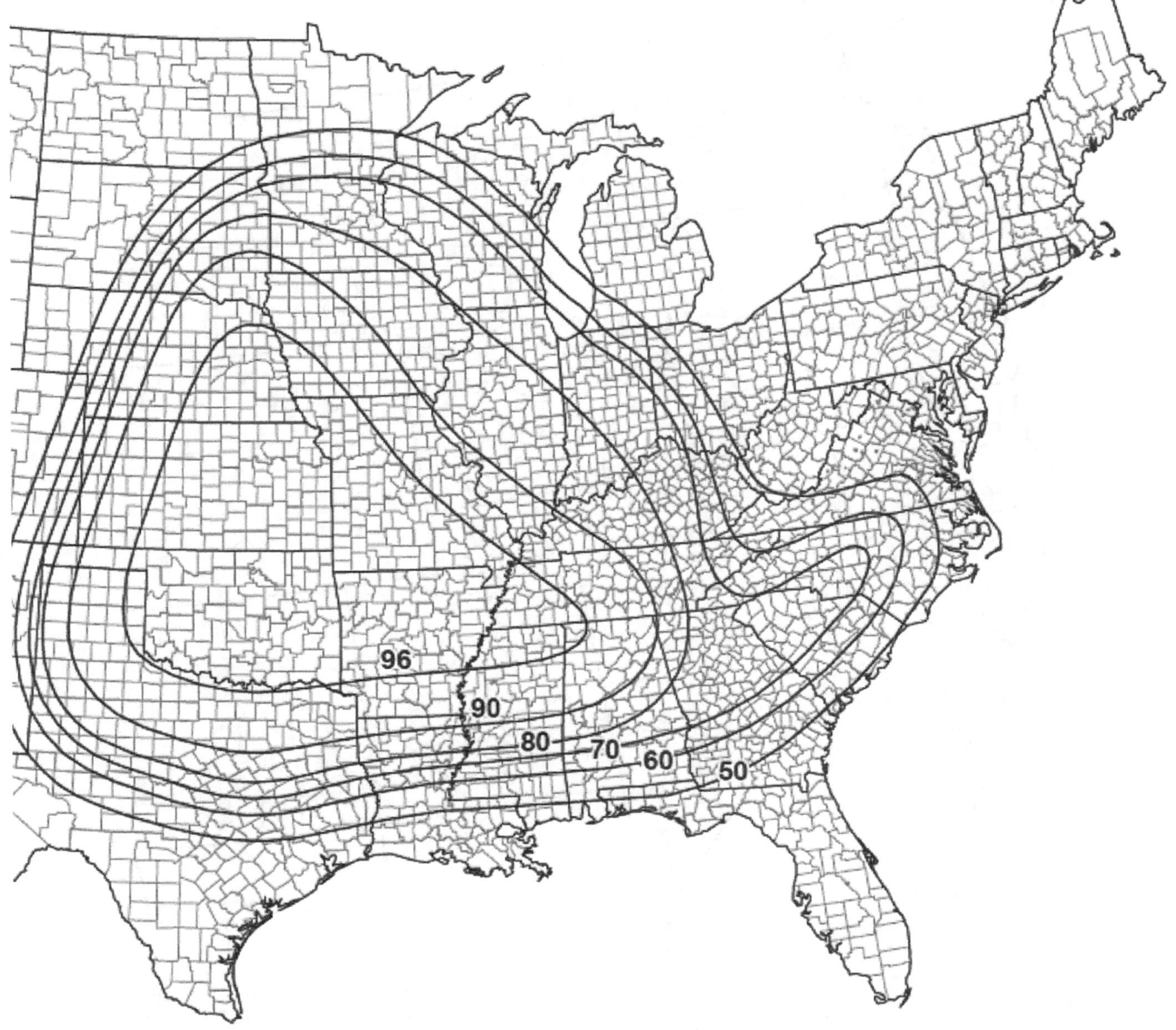

4. Islands, coastal areas, and land boundaries outside the last contour shall use the last tornado speed contour.
5. Tornado speeds correspond to approximately a 1.7% probability of exceedance in 50 years (annual exceedance probability = 0.00033, MRI = 3,000 years).
6. Location-specific tornado speed is permitted to be determined using the ASCE Tornado Design Geodatabase, available at the ASCE 7 Hazard Tool (http://asce7hazardtool.online) or approved equivalent.

Figure 32.5-2B (*Continued*). Tornado speeds for Risk Category IV buildings and other structures, for effective plan area of 2,000 ft^2 (186 m^2).

Notes:

1. Values are 3 s gust speeds in mi/h at 33 ft (10 m) above ground.
2. To convert tornado speeds from mi/h to m/s, multiply mapped values by 0.447.
3. Linear interpolation is permitted between contours. Point values (where shown) are provided to aid with interpolation.

Figure 32.5-2C. Tornado speeds for Risk Category IV buildings and other structures, for effective plan area of 10,000 ft^2 (929 m^2).

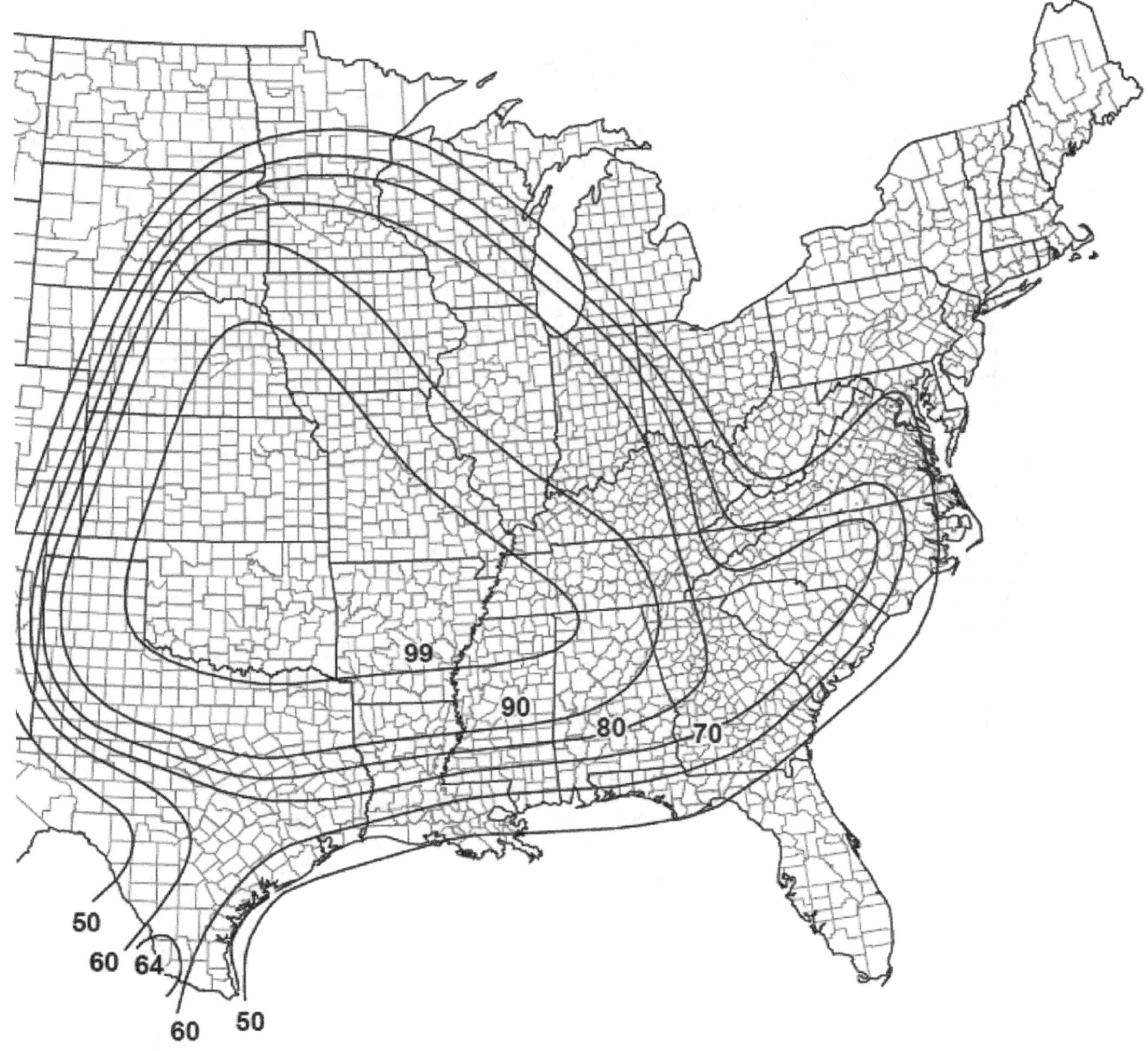

4. Islands, coastal areas, and land boundaries outside the last contour shall use the last tornado speed contour.
5. Tornado speeds correspond to approximately a 1.7% probability of exceedance in 50 years (annual exceedance probability = 0.00033, MRI = 3,000 years).
6. Location-specific tornado speed is permitted to be determined using the ASCE Tornado Design Geodatabase, available at the ASCE 7 Hazard Tool (http://asce7hazardtool.online) or approved equivalent.

Figure 32.5-2C (*Continued*). Tornado speeds for Risk Category IV buildings and other structures, for effective plan area of 10,000 ft^2 (929 m^2).

Notes:

1. Values are 3 s gust speeds in mi/h at 33 ft (10 m) above ground.
2. To convert tornado speeds from mi/h to m/s, multiply mapped values by 0.447.
3. Linear interpolation is permitted between contours. Point values (where shown) are provided to aid with interpolation.

Figure 32.5-2D. Tornado speeds for Risk Category IV buildings and other structures, for effective plan area of 40,000 ft^2 (3,716 m^2).

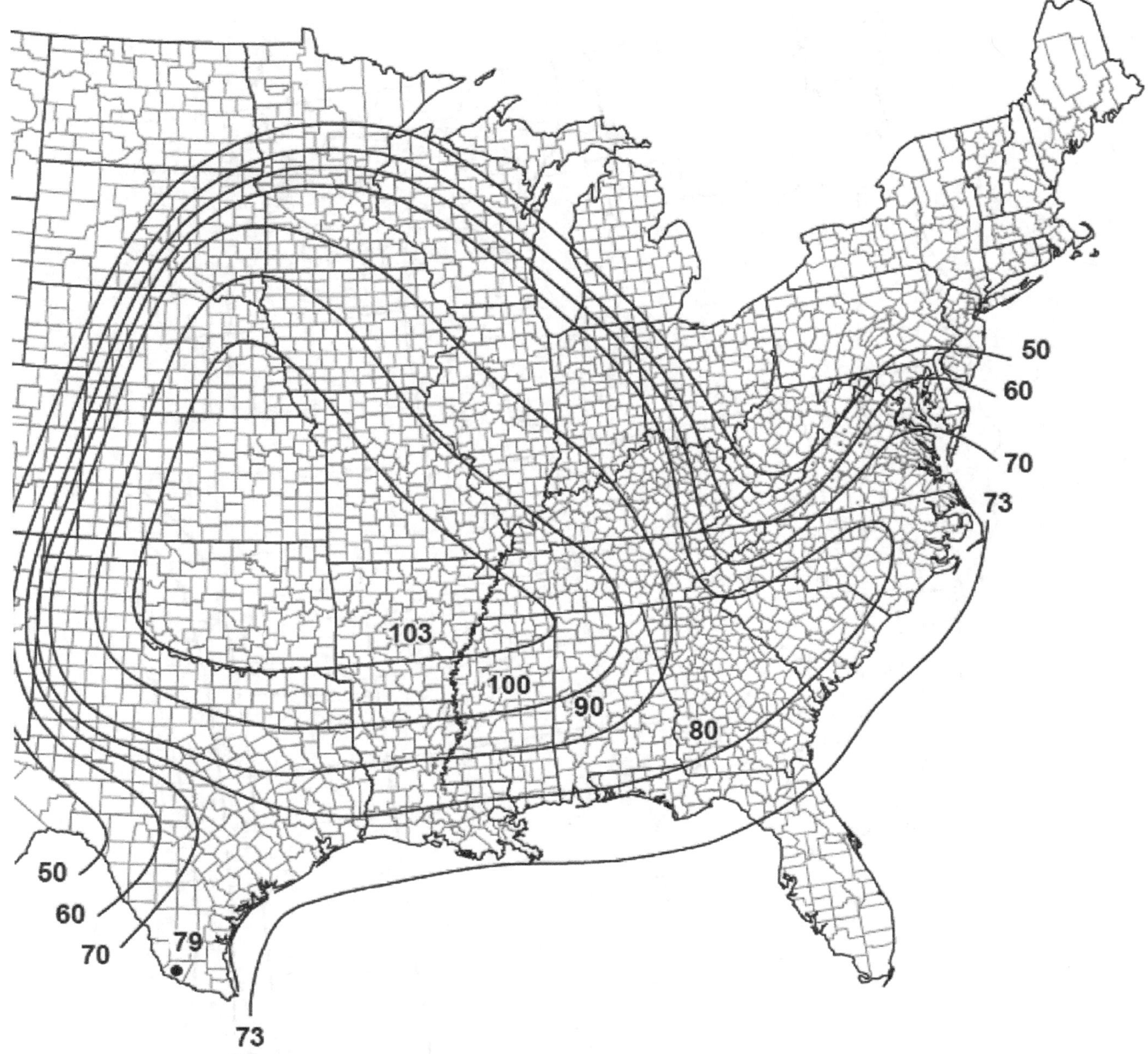

4. Islands, coastal areas, and land boundaries outside the last contour shall use the last tornado speed contour.
5. Tornado speeds correspond to approximately a 1.7% probability of exceedance in 50 years (annual exceedance probability = 0.00033, MRI = 3,000 years).
6. Location-specific tornado speed is permitted to be determined using the ASCE Tornado Design Geodatabase, available at the ASCE 7 Hazard Tool (http://asce7hazardtool.online) or approved equivalent.

Figure 32.5-2D (*Continued*). Tornado speeds for Risk Category IV buildings and other structures, for effective plan area of 40,000 ft^2 (3,716 m^2).

Notes:

1. Values are 3 s gust speeds in mi/h at 33 ft (10 m) above ground.
2. To convert tornado speeds from mi/h to m/s, multiply mapped values by 0.447.
3. Linear interpolation is permitted between contours. Point values (where shown) are provided to aid with interpolation.

Figure 32.5-2E. Tornado speeds for Risk Category IV buildings and other structures, for effective plan area of 100,000 ft^2 (9,290 m^2).

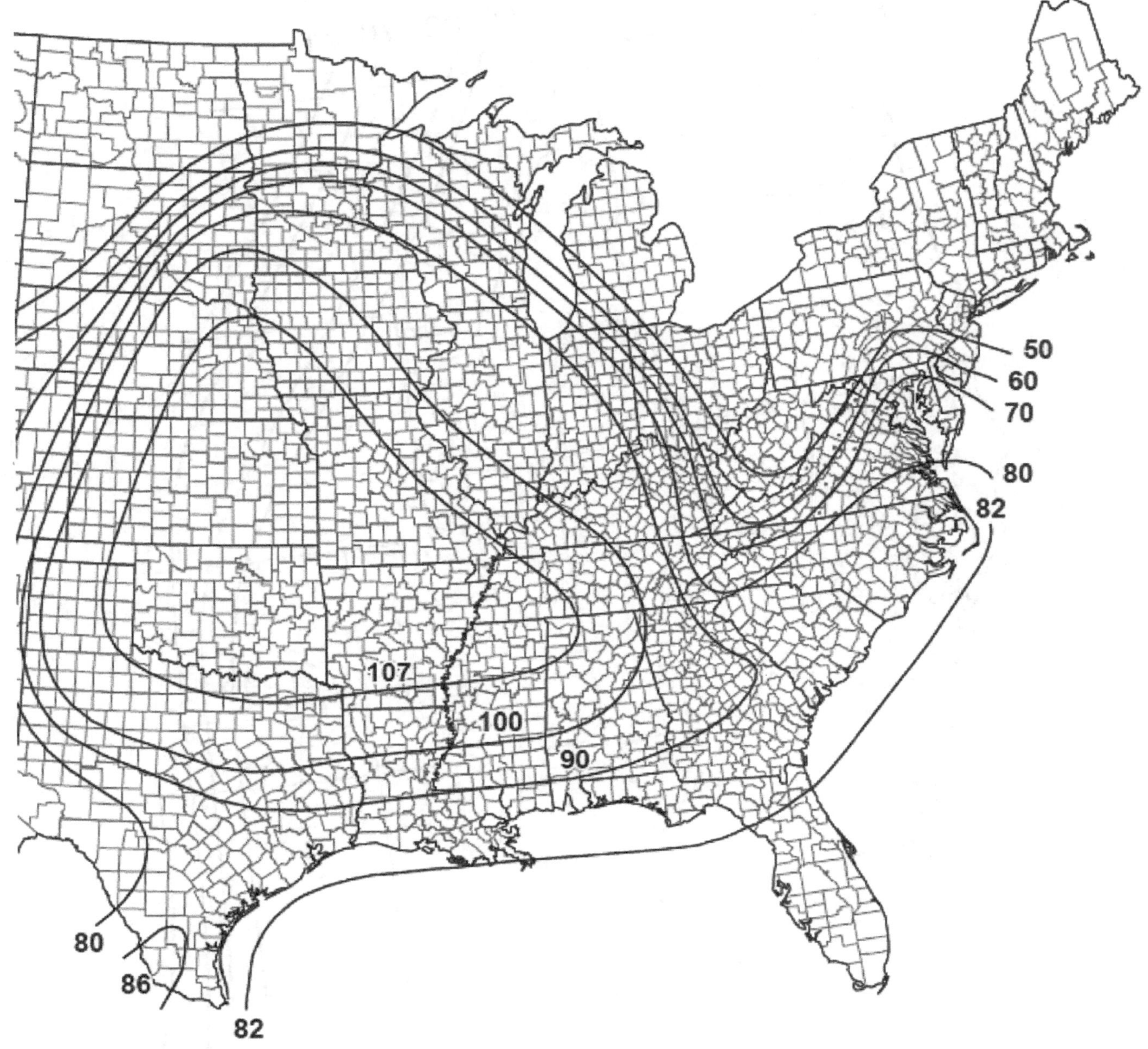

4. Islands, coastal areas, and land boundaries outside the last contour shall use the last tornado speed contour.
5. Tornado speeds correspond to approximately a 1.7% probability of exceedance in 50 years (annual exceedance probability = 0.00033, MRI = 3,000 years).
6. Location-specific tornado speed is permitted to be determined using the ASCE Tornado Design Geodatabase, available at the ASCE 7 Hazard Tool (http://asce7hazardtool.online) or approved equivalent.

Figure 32.5-2E (*Continued*). Tornado speeds for Risk Category IV buildings and other structures, for effective plan area of 100,000 ft² (9,290 m²).

Notes:

1. Values are 3 s gust speeds in mi/h at 33 ft (10 m) above ground.
2. To convert tornado speeds from mi/h to m/s, multiply mapped values by 0.447.
3. Linear interpolation is permitted between contours. Point values (where shown) are provided to aid with interpolation.

Figure 32.5-2F. Tornado speeds for Risk Category IV buildings and other structures, for effective plan area of 250,000 ft^2 (23,226 m^2).

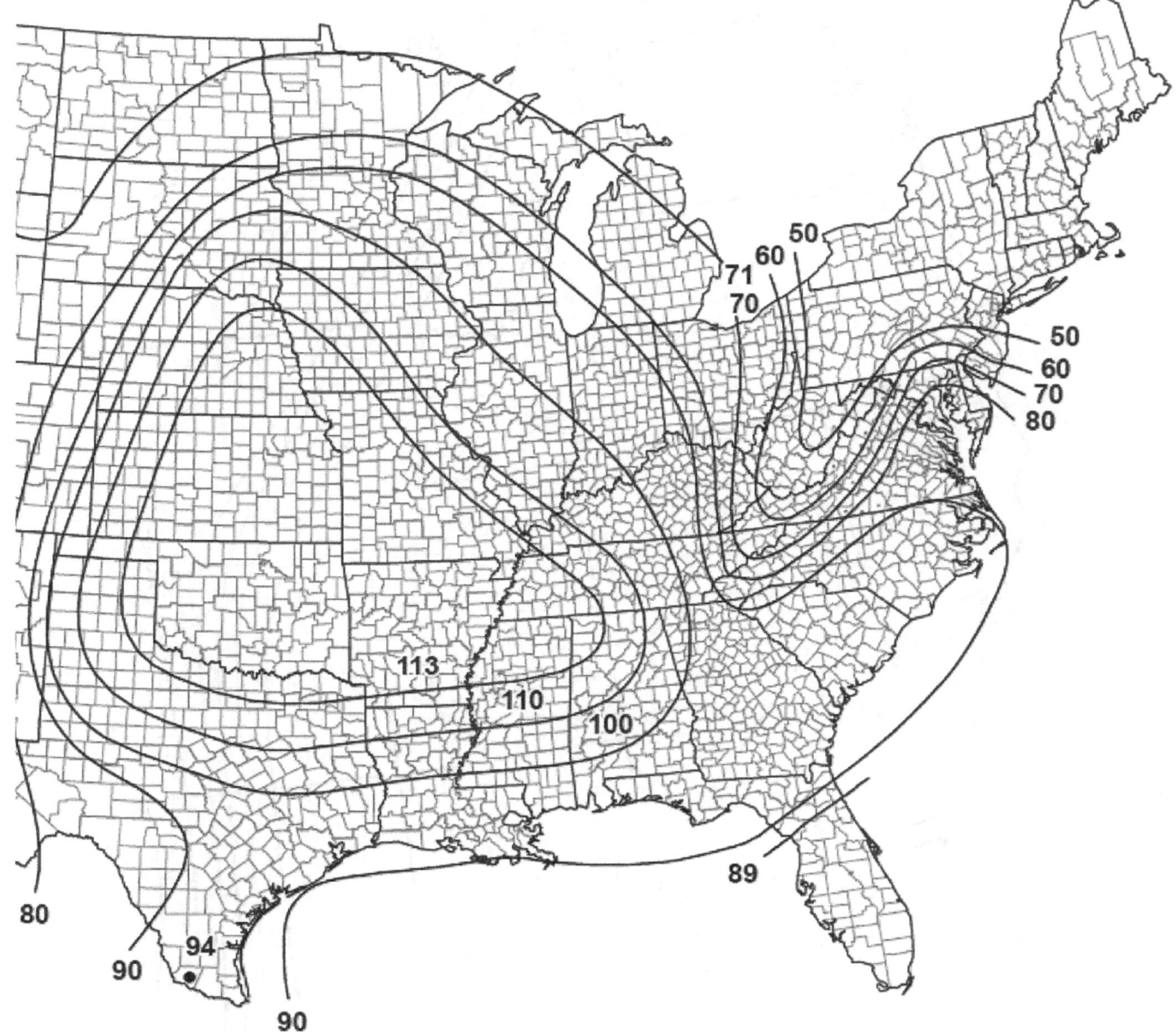

4. Islands, coastal areas, and land boundaries outside the last contour shall use the last tornado speed contour.
5. Tornado speeds correspond to approximately a 1.7% probability of exceedance in 50 years (annual exceedance probability = 0.00033, MRI = 3,000 years).
6. Location-specific tornado speed is permitted to be determined using the ASCE Tornado Design Geodatabase, available at the ASCE 7 Hazard Tool (http://asce7hazardtool.online) or approved equivalent.

Figure 32.5-2F (*Continued*). Tornado speeds for Risk Category IV buildings and other structures, for effective plan area of 250,000 ft^2 (23,226 m^2).

Notes:

1. Values are 3 s gust speeds in mi/h at 33 ft (10 m) above ground.
2. To convert tornado speeds from mi/h to m/s, multiply mapped values by 0.447.
3. Linear interpolation is permitted between contours. Point values (where shown) are provided to aid with interpolation.

Figure 32.5-2G. Tornado speeds for Risk Category IV buildings and other structures, for effective plan area of 1,000,000 ft^2 (92,903 m^2).

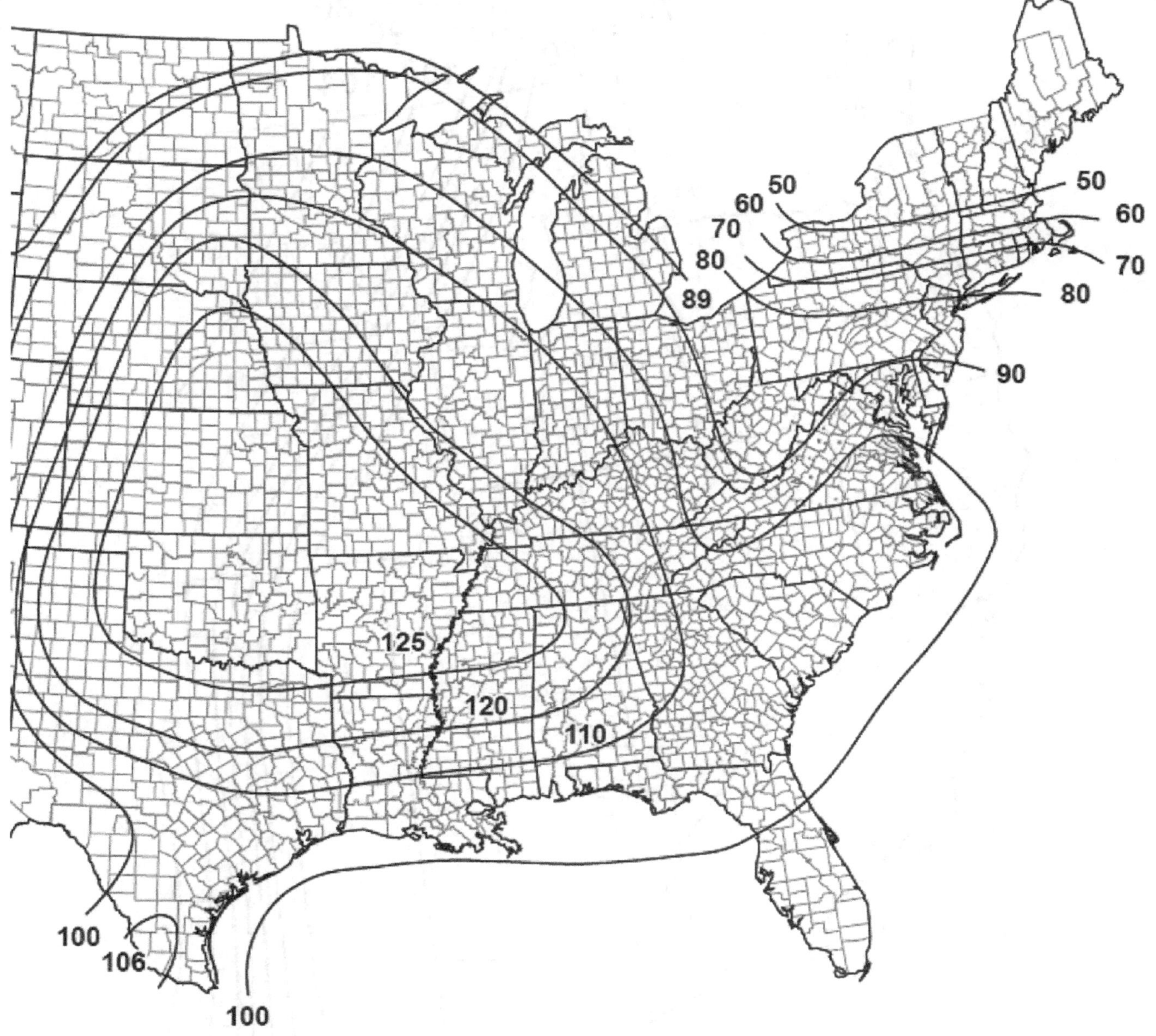

4. Islands, coastal areas, and land boundaries outside the last contour shall use the last tornado speed contour.
5. Tornado speeds correspond to approximately a 1.7% probability of exceedance in 50 years (annual exceedance probability = 0.00033, MRI = 3,000 years).
6. Location-specific tornado speed is permitted to be determined using the ASCE Tornado Design Geodatabase, available at the ASCE 7 Hazard Tool (http://asce7hazardtool.online) or approved equivalent.

Figure 32.5-2G (*Continued*). Tornado speeds for Risk Category IV buildings and other structures, for effective plan area of 1,000,000 ft^2 (92,903 m^2).

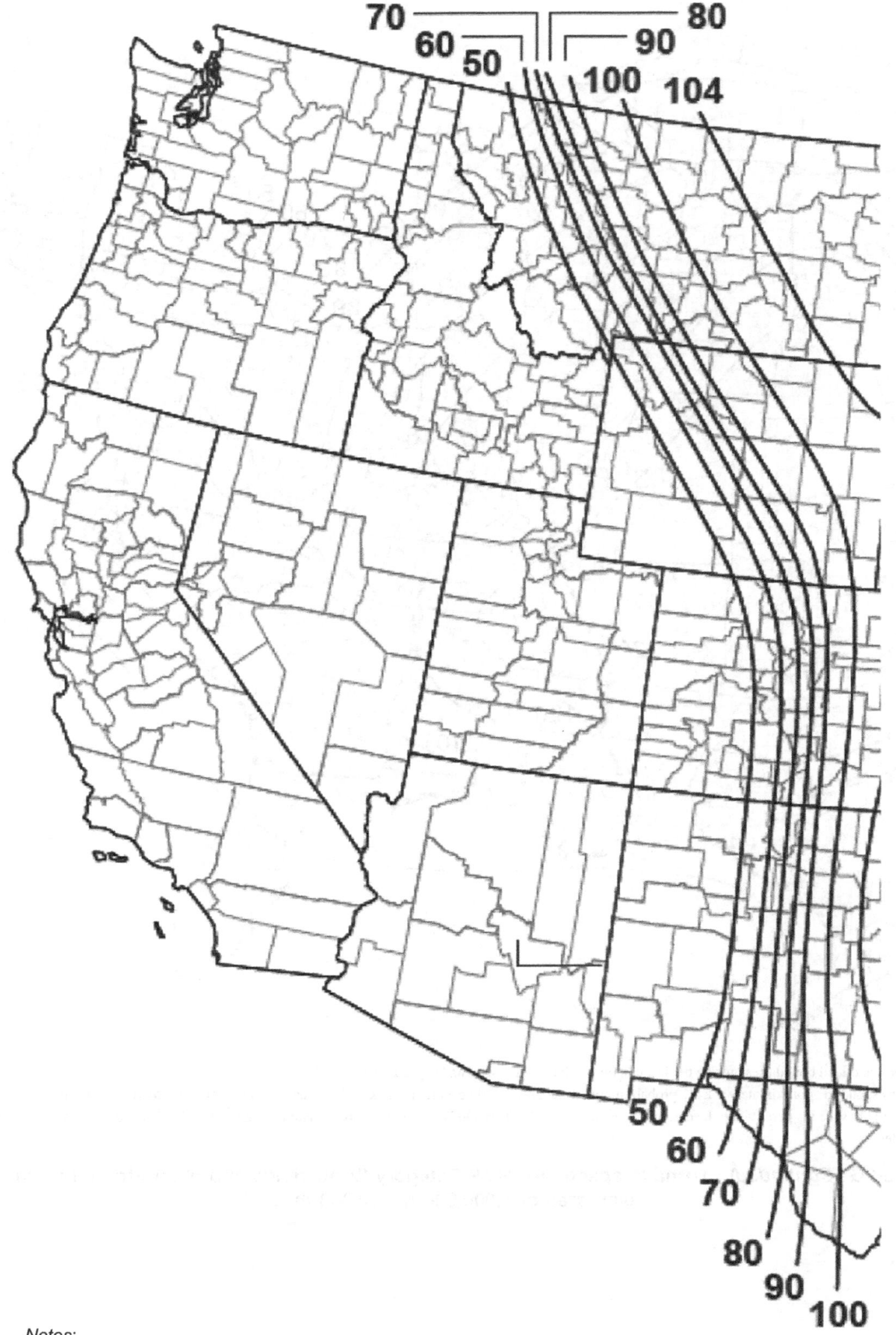

Notes:

1. Values are 3 s gust speeds in mi/h at 33 ft (10 m) above ground.
2. To convert tornado speeds from mi/h to m/s, multiply mapped values by 0.447.
3. Linear interpolation is permitted between contours. Point values (where shown) are provided to aid with interpolation.

Figure 32.5-2H. Tornado speeds for Risk Category IV buildings and other structures, for effective plan area of 4,000,000 ft^2 (371,612 m^2).

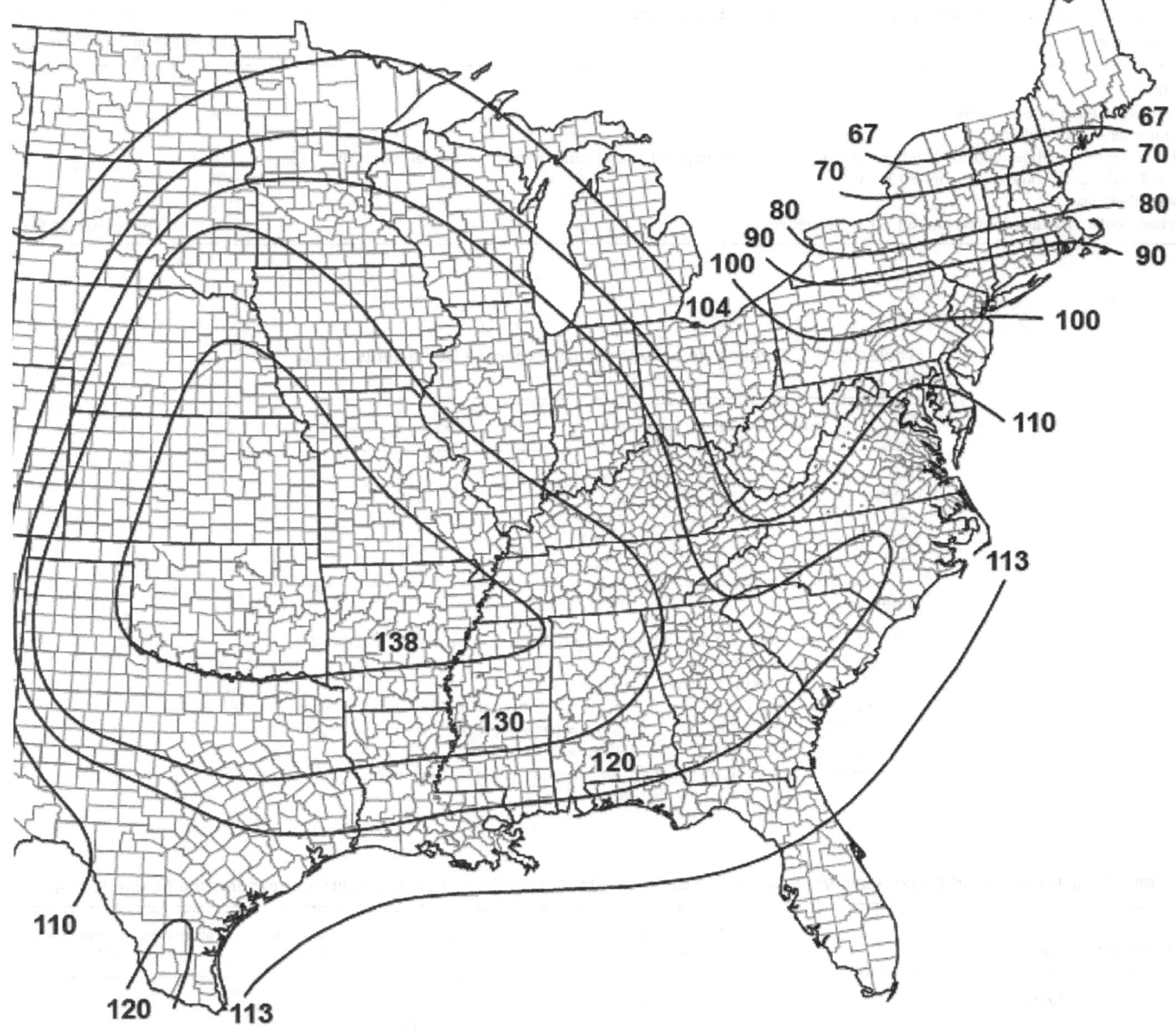

4. Islands, coastal areas, and land boundaries outside the last contour shall use the last tornado speed contour.
5. Tornado speeds correspond to approximately a 1.7% probability of exceedance in 50 years (annual exceedance probability = 0.00033, MRI = 3,000 years).
6. Location-specific tornado speed is permitted to be determined using the ASCE Tornado Design Geodatabase, available at the ASCE 7 Hazard Tool (http://asce7hazardtool.online) or approved equivalent.

Figure 32.5-2H (*Continued*). Tornado speeds for Risk Category IV buildings and other structures, for effective plan area of 4,000,000 ft^2 (371,612 m^2).

Table 32.6-1. Tornado Directionality Factor, K_{dT}.

Structure Type	Tornado Directionality Factor, K_{dT}
Buildings	
Main wind force resisting system	0.80
Components and cladding	
For Essential Facilities and for buildings and other structures required to maintain the functionality of Essential Facilities	1.0
Roof Zone 1′ as shown in Figure 30.3-2A	0.90
All other cases	0.75
Arched roofs, circular domes, and all other structures	Use value from Table 26.6-1

Table 32.10-1. Tornado Velocity Pressure Exposure Coefficients, K_{zTor} and K_{hTor}.

Height above Ground Level, z or h		K_{zTor} and K_{hTor}
ft	m	
0–200	0–61.0	1.0
250	76.2	0.96
300	91.4	0.92
>328	>100	0.90

Notes:

1. The tornado velocity pressure exposure coefficient, K_{zTor}, is permitted to be determined as follows:
   For $0 < z \leq 200$ ft (60 m) $K_{zTor} = 1.0$
   For $200 < z \leq 328$ ft $K_{zTor} = [(2820 - z)/2620]^2$, where z is in ft
   (For $61 < z \leq 100$ m) $K_{zTor} = [(861 - z)/800]^2$, where z is in m)
   For $z > 328$ ft (100 m) $K_{zTor} = 0.90$
2. Linear interpolation for intermediate values of height z is permitted.

Table 32.13-1. Main Wind Force Resisting System and Components and Cladding Tornado Internal Pressure Coefficient, (GC_{piT}).

Enclosure Classification	Criteria for Enclosure Classification	Internal Pressure Combined with Atmospheric Pressure Change	Tornado Internal Pressure Coefficient, (GC_{piT})
Sealed other structures	See Section 32.2	Extreme	+1.0
Enclosed buildings and other structures	See Table 26.13-1	High	+0.55 −0.18
Partially enclosed buildings and other structures	See Table 26.13-1	High	+0.55 −0.55
Partially open buildings and other structures	See Table 26.13-1	Moderate	+0.18 −0.18
Open buildings and other structures	See Table 26.13-1	Negligible	0.00

Notes:

1. Plus and minus signs signify pressures acting toward and away from the internal surfaces, respectively.
2. Values of (GC_{piT}) shall be used with q_z or q_h as specified.
3. Two cases shall be considered to determine the critical load requirements for the appropriate condition:
   (a) A positive value of (GC_{piT}) applied to all internal surfaces, or
   (b) A negative value of (GC_{piT}) applied to all internal surfaces.

32.13.1 Reduction Factor for Large-Volume Buildings, R_i For a partially enclosed building containing a single, unpartitioned large volume, the internal pressure coefficient, (GC_{piT}), is permitted to be multiplied by the reduction factor, R_i, as determined in Section 26.13.1.

32.14 TORNADO EXTERNAL PRESSURE COEFFICIENTS

External pressure coefficients C_p and (GC_p) shall be determined from the applicable sections of Chapters 27, 29, 30, and 31, and multiplied by the tornado pressure coefficient adjustment factor for vertical winds, K_{vT}, as specified in Table 32.14-1 and in accordance with Sections 32.15 through 32.17.

32.15 TORNADO LOADS ON BUILDINGS: MAIN WIND FORCE RESISTING SYSTEM

32.15.1 Enclosed, Partially Enclosed, and Partially Open Buildings Section 27.3.1 shall apply for determination of MWFRS loads for buildings of all heights, as modified in this section. Design tornado pressures, p_T, for the MWFRS of enclosed, partially enclosed, and partially open buildings of all heights shall be determined in accordance with the following equation, which replaces Equation (27.3-1):

$$p_T = qG_TK_{dT}K_{vT}C_p - q_i(GC_{piT})\ (\text{lb/ft}^2) \qquad (32.15\text{-}1)$$

$$p_T = qG_TK_{dT}K_{vT}C_p - q_i(GC_{piT})\ (\text{N/m}^2) \qquad (32.15\text{-}1.\text{SI})$$

where

$q = q_{zT}$ For external pressure on all walls evaluated at height z above the ground, lb/ft^2 (N/m^2),

$q = q_{hT}$ For external pressure on roofs evaluated at height h, lb/ft^2 (N/m^2),

$q_i = q_{hT}$ For internal pressure evaluation of roofs of enclosed and partially open buildings, lb/ft^2 (N/m^2),

$q_i = q_{zT}$ For internal pressure evaluation of walls of enclosed and partially open buildings, lb/ft^2 (N/m^2),

$q_i = q_{zop}$ For internal pressure evaluation of the roof and all walls in partially enclosed buildings, where height z_{op} is defined as the level of the lowest opening in the building that could affect the positive internal pressure, lb/ft^2 (N/m^2). Glazed openings not meeting the protection requirements of Section 32.12.3.1 shall be considered as openings,

G_T = Tornado gust-effect factor from Section 32.11,

K_{dT} = Tornado directionality factor from Section 32.6,

K_{vT} = Tornado pressure coefficient adjustment factor from Section 32.14,

C_p = External pressure coefficient from Section 27.3.1, and

(GC_{piT}) = Tornado internal pressure coefficient from Section 32.13.

32.15.1.1 Elevated Buildings Section 27.3.1.1 shall apply for determination of MWFRS loads on elevated buildings, as modified in this section. The design tornado pressure, p_T, for the effects of tornado pressure on the MWFRS shall be determined in accordance with Equation (32.15-1), where K_{vT} = 1.0 for lateral loads on elements below the elevated building and vertical loads on the horizontal bottom surfaces of the elevated building.

32.15.2 Open Buildings with Monoslope, Pitched, or Troughed Free Roofs Section 27.3.2 shall apply for determination of MWFRS loads on open buildings, as modified in this section. The net design tornado pressure, p_T, for the MWFRS of open buildings with monoslope, pitched, or troughed free roofs shall be determined in accordance with Section 27.3.2 and the following equation, which replaces Equation (27.3-2):

Table 32.14-1. Tornado Pressure Coefficient Adjustment Factor for Vertical Winds, K_{vT}.

STRUCTURE TYPE	K_{vT}
Buildings	
Negative (Uplift) Pressures on Roofs	
Main Wind Force Resisting System	1.1
Components and Cladding	
Roof slope ≤ 7 degrees	
Zone 1	1.2
Zone 2	1.05
Zone 3	1.05
Roof slope > 7 degrees	
Zone 1	1.2
Zone 2	1.2
Zone 3	1.3
Positive Pressures (Downward Acting) on Roofs	1.0
Wall Pressures	1.0
All Other Cases	1.0
Other Structures	
Negative (Uplift) Pressures on Rooftop Structures and Equipment and Rooftop Solar Panels Parallel to the Roof Surface	
Main Wind Force Resisting System	1.1
Components and Cladding	Use values for building C&C
Negative (Uplift) Pressures on Roofs of Bins, Silos, and Tanks	
Main Wind Force Resisting System	1.1
Components and Cladding	See Section 32.17.5
All Other Cases	1.0

$$p_T = q_{hT} G_T K_{dT} C_N \text{ (lb/ft}^2) \quad (32.15\text{-}2)$$

$$p_T = q_{hT} G_T K_{dT} C_N \text{ (N/m}^2) \quad (32.15\text{-}2.\text{SI})$$

where

q_{hT} = Tornado velocity pressure from Section 32.10.2 evaluated at mean roof height h, lb/ft^2 (N/m^2),
G_T = Tornado gust-effect factor from Section 32.11,
K_{dT} = Tornado directionality factor from Section 32.6, and
C_N = Net pressure coefficient from Section 27.3.2.

32.15.3 Roof Overhangs Section 27.3.3 shall apply for determination of MWFRS loads on roof overhangs.

32.15.4 Parapets Section 27.3.4 shall apply for determination of MWFRS loads on parapets, as modified in this section. The design tornado pressure, p_{pt}, for the effects of parapets on the MWFRS shall be determined in accordance with the following equation, which replaces Equation (27.3-3):

$$p_{pT} = q_{pT} K_{dT} (GC_{pn}) \text{ (lb/ft}^2) \quad (32.15\text{-}3)$$

$$p_{pT} = q_{pT} K_{dT} (GC_{pn}) \text{ (N/m}^2) \quad (32.15\text{-}3.\text{SI})$$

where

p_{pT} = Combined net tornado design pressure on the parapet caused by the combination of the net pressures from the front and back parapet surfaces, lb/ft^2 (N/m^2). Plus (and minus) signs signify net pressure acting toward (and away from) the front (exterior) side of the parapet;
q_{pT} = Tornado velocity pressure from Section 32.10.2 evaluated at the top of the parapet, lb/ft^2 (N/m^2);
K_{dT} = Tornado directionality factor from Section 32.6; and
(GC_{pn}) = Combined net pressure coefficient from Section 27.3.4.

32.15.5 Design Load Cases The design load cases shall be determined using Section 27.3.5. The exception for buildings meeting the requirements of Section D1 of Appendix D shall not apply.

32.16 TORNADO LOADS ON BUILDING APPURTENANCES AND OTHER STRUCTURES: MAIN WIND FORCE RESISTING SYSTEM

32.16.1 General Requirements The design tornado speed in Section 32.5 for rooftop structures and equipment, including rooftop solar panels, shall equal that used for design of the building.

The effective plan area for freestanding walls and solid signs shall be equal to the wall or sign length squared divided by 20 but shall not be taken smaller than the length multiplied by the width. The length of the wall or sign shall be calculated as the distance between expansion joints.

32.16.2 Solid Freestanding Walls and Solid Signs Section 29.3 shall apply for determination of MWFRS loads on solid freestanding walls and solid signs, as modified in this section. The design tornado force, F_T, for solid freestanding walls and solid freestanding signs shall be determined in accordance with following equation, which replaces Equation (29.3-1):

$$F_T = q_{zT} G_T K_{dT} C_f A_s \text{ (lb)} \quad (32.16\text{-}1)$$

$$F_T = q_{zT} G_T K_{dT} C_f A_s \text{ } (N) \quad (32.16\text{-}1.\text{SI})$$

where

q_{zT} = Tornado velocity pressure from Section 32.10.2 evaluated at the centroid of area, A_s, of the wall or sign, lb/ft2 (N/m^2);
G_T = Tornado gust-effect factor from Section 32.11;
K_{dT} = Tornado directionality factor from Section 32.6;
C_f = Net force coefficient from Figure 29.3-1; and
A_s = Gross area of the solid freestanding wall or freestanding solid sign, ft^2 (m^2).

32.16.3 Other Structures Section 29.4 shall apply for determination of MWFRS loads on other structures (chimneys, tanks, open signs, single-plane open frames, and trussed towers), as modified in this section. The design tornado force, F_T, for other structures, whether ground or roof mounted, shall be determined in accordance with the following equation, which replaces Equation (29.4-1):

$$F_T = q_{zT} G_T K_{dT} C_f A_f \text{ (lb)} \quad (32.16\text{-}2)$$

$$F_T = q_{zT} G_T K_{dT} C_f A_f \text{ (N)} \quad (32.16\text{-}2.\text{SI})$$

where

q_{zT} = Tornado velocity pressure from Section 32.10.2 evaluated at height z, as defined in Section 26.10, of the centroid of area A_f, lb/ft^2 (N/m^2);
G_T = Tornado gust-effect factor from Section 32.11;
K_{dT} = Tornado directionality factor from Section 32.6;
C_f = Force coefficients from Section 29.4; and
A_f = Projected area normal to the tornadic wind from Section 29.4, ft^2 (m^2).

32.16.3.1 Trussed Towers Trussed towers and other open lattice-type structures shall be designed for an additional 40 ft^2 (3.7 m^2) of projected surface area of clinging debris at midheight of the structure or 50 ft (15 m), whichever is less.

32.16.3.2 Rooftop Structures and Equipment for Buildings Section 29.4.1 shall apply for determination of MWFRS loads on rooftop structures and equipment, as modified in this section. The lateral design tornado force, F_{hT}, and vertical design tornado force, F_{vT}, for rooftop structures and equipment, except as otherwise specified for roof-mounted solar panels (Sections 32.16.3.4 and 32.16.3.5) and structures identified in Section 32.16.3, shall be determined as follows.

The resultant lateral force, F_{hT}, shall be applied at a height above the roof surface equal to or greater than the centroid of the projected area, A_f, and determined in accordance with the following equation, which replaces Equation (29.4-2):

$$F_{hT} = q_{hT} K_{dT} (GC_r) A_f \text{ (lb)} \quad (32.16\text{-}3)$$

$$F_{hT} = q_{hT} K_{dT} (GC_r) A_f \text{ (N)} \quad (32.16\text{-}3.\text{SI})$$

where

q_{hT} = Tornado velocity pressure from Section 32.10.2 evaluated at mean roof height h, lb/ft^2 (N/m^2),
K_{dT} = Tornado directionality factor from Section 32.6,
(GC_r) = Product of external pressure coefficient and gust-effect factor from Section 29.4.1, and
A_f = Vertical projected area of the rooftop structure or equipment from Section 29.4.1, ft^2 (m^2).

The vertical uplift design tornado force, F_{vT}, on rooftop structures and equipment shall be determined in accordance with the following equation, which replaces Equation (29.4-3):

$$F_{vT} = q_{hT} K_{dT} K_{vT} (GC_r) A_r \text{ (lb)} \quad (32.16\text{-}4)$$

$$F_{vT} = q_{hT} K_{dT} K_{vT} (GC_r) A_r \text{ (N)} \quad (32.16\text{-}4.\text{SI})$$

where

q_{hT} = Tornado velocity pressure from Section 32.10.2 evaluated at mean roof height h, lb/ft^2 (N/m^2),
K_{dT} = Tornado directionality factor from Section 32.6,
K_{vT} = Tornado pressure coefficient adjustment factor from Section 32.14,
(GC_r) = Product of external pressure coefficient and gust-effect factor from Section 29.4.1, and
A_r = Horizontal projected area of rooftop structure or equipment, ft^2 (m^2).

32.16.3.3 Roofs of Isolated Circular Bins, Silos, and Tanks Section 29.4.2.2 shall apply for determination of MWFRS loads on the roofs of isolated circular bins, silos, and tanks, as modified in this section. The net design tornado pressures shall be determined in accordance with the following equation, which replaces Equation (29.4-4):

$$p_T = q_{hT} [G_T K_{dT} K_{vT} C_p - (GC_{piT})] \text{ (lb/ft}^2) \quad (32.16\text{-}5)$$

$$p_T = q_{hT} [G_T K_{dT} K_{vT} C_p - (GC_{piT})] \text{ (N/m}^2) \quad (32.16\text{-}5.\text{SI})$$

where

q_{hT} = Tornado velocity pressure from Section 32.10.2 evaluated at mean roof height h, lb/ft^2 (N/m^2),
G_T = Tornado gust-effect factor from Section 32.11,
K_{dT} = Tornado directionality factor from Section 32.6,
K_{vT} = Tornado pressure coefficient adjustment factor from Section 32.14,
C_p = External pressure coefficient from Section 29.4.2.2, and
(GC_{piT}) = Tornado internal pressure coefficient from Section 32.13.

32.16.3.4 Rooftop Solar Panels for Buildings of All Heights with Flat Roofs or Gable or Hip Roofs with Slopes Less Than 7 Degrees Section 29.4.3 shall apply for determination of MWFRS loads on rooftop photovoltaic panels for buildings of all heights with flat roofs or gable or hip roofs with slopes less than 7 degrees, as modified in this section. The design tornado pressure, p_T, for rooftop photovoltaic panels shall be determined by the following equation, which replaces Equation (29.4-5):

$$p_T = q_{hT} K_{dT} (GC_{rn}) \text{ (lb/ft}^2) \quad (32.16\text{-}6)$$

$$p_T = q_{hT} K_{dT} (GC_{rn}) \text{ (N/m}^2) \quad (32.16\text{-}6.\text{SI})$$

where

q_{hT} = Tornado velocity pressure from Section 32.10.2 evaluated at mean roof height h, lb/ft^2 (N/m^2),
K_{dT} = Tornado directionality factor from Section 32.6, and
(GC_{rn}) = Net pressure coefficient from Section 29.4.3.

32.16.3.5 Rooftop Solar Panels Parallel to the Roof Surface on Buildings of All Heights and Roof Slopes Section 29.4.4 shall apply for determination of MWFRS loads on rooftop photovoltaic panels parallel to the roof surface on buildings of all heights and roof slopes as modified in this section. The design tornado pressure, p_T, for rooftop photovoltaic panels shall be determined in accordance with the following equation, which replaces Equation (29.4-7):

$$p_T = q_{hT} K_{dT} K_{vT} (GC_p)(\gamma_E)(\gamma_a) \text{ (lb/ft}^2) \quad (32.16\text{-}7)$$

$$p_T = q_{hT} K_{dT} K_{vT} (GC_p)(\gamma_E)(\gamma_a) \text{ (N/m}^2) \quad (32.16\text{-}7.\text{SI})$$

where

q_{hT} = Tornado velocity pressure from Section 32.10.2 evaluated at mean roof height h, lb/ft^2 (N/m^2);
K_{dT} = Tornado directionality factor from Section 32.6;
K_{vT} = Tornado pressure coefficient adjustment factor from Section 32.14;
(GC_p) = External pressure coefficient from Section 29.4.4;
γ_E = Array edge factor from Section 29.4.4; and
γ_a = Solar panel pressure equalization factor from Section 29.4.4.

32.17 TORNADO LOADS: COMPONENTS AND CLADDING

32.17.1 Low-Rise Buildings Section 30.3 shall apply for determination of component and cladding tornado loads on low-rise buildings, as modified in this section. The design tornado pressures, p_T, on C&C elements in low-rise buildings and buildings with $h \leq 60$ ft ($h \leq 18.3$ m) shall be determined in accordance with the following equation, which replaces Equation (30.3-1):

$$p_T = q_{hT} [K_{dT} K_{vT} (GC_p) - (GC_{piT})] \text{ (lb/ft}^2) \quad (32.17\text{-}1)$$

$$p_T = q_{hT} [K_{dT} K_{vT} (GC_p) - (GC_{piT})] \text{ (N/m}^2) \quad (32.17\text{-}1.\text{SI})$$

where

q_{hT} = Tornado velocity pressure from Section 32.10.2 evaluated at mean roof height h, lb/ft^2 (N/m^2);
K_{dT} = Tornado directionality factor from Section 32.6;
K_{vT} = Tornado pressure coefficient adjustment factor from Section 32.14;
(GC_p) = External pressure coefficient from Section 30.3; and
(GC_{piT}) = Tornado internal pressure coefficient from Section 32.13.

32.17.1.1 Bottom Horizontal Surfaces of Elevated Buildings Section 30.3.2.1 shall apply for determination of C&C loads on bottom horizontal surfaces of elevated buildings, as modified in this section. The design tornado pressure, p_T, for the effects of tornado pressure on C&C shall be determined in accordance with Equation (32.17-1), where $K_{vT} = 1.0$.

32.17.2 Buildings with $h > 60$ ft ($h > 18.3$ m) Section 30.4 shall apply for the determination of component and cladding tornado loads on buildings with $h > 60$ ft ($h > 18.3$ m), as modified in this section. The design tornado pressures, p_T, on C&C elements for all buildings with $h > 60$ ft ($h > 18.3$ m) shall be determined in accordance with the following equation, which replaces Equation (30.4-1):

$$p_T = q K_{dT} K_{vT} (GC_p) - q_i (GC_{piT}) \text{ (lb/ft}^2) \quad (32.17\text{-}2)$$

$$p_T = q K_{dT} K_{vT} (GC_p) - q_i (GC_{piT}) \text{ (N/m}^2) \quad (32.17\text{-}2.\text{SI})$$

where

$q = q_{zT}$ For external pressure on all walls evaluated at height z above the ground, lb/ft^2 (N/m^2);

q = For external pressures on roofs evaluated at height h, lb/ft² (N/m²);

$q_i = q_{hT}$ For internal pressure evaluation of roofs of enclosed and partially open buildings, lb/ft² (N/m²);

$q_i = q_{zT}$ For internal pressure evaluation of walls of enclosed and partially open buildings, lb/ft² (N/m²);

$q_i = q_{zop}$ For internal pressure evaluation of the roof and all walls in partially enclosed buildings, where height z_{op} is defined as the level of the lowest opening in the building that could affect the positive internal pressure, lb/ft² (N/m²). Glazed openings not meeting the protection requirements of Section 32.12.3.1 shall be considered as openings;

K_{dT} = Tornado directionality factor from Section 32.6;

K_{vT} = Tornado pressure coefficient adjustment factor from Section 32.14;

(GC_p) = External pressure coefficient from Section 30.4; and

(GC_{piT}) = Tornado internal pressure coefficient from Section 32.13.

32.17.2.1 Bottom Horizontal Surfaces of Elevated Buildings Section 30.5.2.1 shall apply for determination of C&C loads on bottom horizontal surfaces of elevated buildings, as modified in this section. The design tornado pressure, p_T, for the effects of tornado pressure on C&C shall be determined in accordance with Equation (32.17-1), where K_{vT} = 1.0, and the tornado velocity pressure, q, shall be calculated at the height specified in Section 30.5.2.1, subparagraph 1.

32.17.3 Open Buildings Section 30.5 shall apply for the determination of component and cladding tornado loads on open buildings, as modified in this section. The net tornado design pressures, p_T, for C&C elements of open buildings of all heights shall be determined in accordance with the following equation, which replaces Equation (30.5-1):

$$p_T = q_{hT} G_T K_{dT} C_N \text{ (lb/ft}^2\text{)} \quad (32.17\text{-}3)$$

$$p_T = q_{hT} G_T K_{dT} C_N \text{ (N/m}^2\text{)} \quad (32.17\text{-}3.\text{SI})$$

where

q_{hT} = Tornado velocity pressure from Section 32.10.2 evaluated at mean roof height h, lb/ft² (N/m²);

G_T = Tornado gust-effect factor from Section 32.11;

K_{dT} = Tornado directionality factor from Section 32.6; and

C_N = Net pressure coefficient from Section 30.5.

32.17.4 Building Appurtenances and Rooftop Structures and Equipment

32.17.4.1 Parapets Section 30.6 shall apply for determination of component and cladding tornado loads on parapets, as modified in this section. The design tornado pressures, p_T, on C&C elements of parapets for all building types and heights shall be determined in accordance with the following equation, which replaces Equation (30.6-1):

$$p_T = q_{pT}[K_{dT}(GC_p) - (GC_{piT})] \text{(lb/ft}^2\text{)} \quad (32.17\text{-}4)$$

$$p_T = q_{pT}[K_{dT}(GC_p) - (GC_{piT})] \text{ (N/m}^2\text{)} \quad (32.17\text{-}4.\text{SI})$$

where

q_{pT} = Tornado velocity pressure from Section 32.10.2 evaluated at the top of the parapet, lb/ft² (N/m²);

K_{dT} = Tornado directionality factor from Section 32.6;

(GC_p) = External pressure coefficient from Section 30.8; and

(GC_{piT}) = Tornado internal pressure coefficient from Section 32.13, based on the porosity of the parapet envelope.

32.17.4.2 Roof Overhangs Section 30.7 shall apply for determination of component and cladding tornado loads on roof overhangs, as modified in this section. The design tornado pressures, p_T, on C&C elements of roof overhangs for all building types and heights shall be determined in accordance with the following equation, which replaces Equation (30.7-1):

$$p_T = q_{hT}[K_{dT}K_{vT}(GC_p) - (GC_{piT})] \text{ (lb/ft}^2\text{)} \quad (32.17\text{-}5)$$

$$p_T = q_{hT}[K_{dT}K_{vT}(GC_p) - (GC_{piT})] \text{ (N/m}^2\text{)} \quad (32.17\text{-}5.\text{SI})$$

where

q_{hT} = Tornado velocity pressure from Section 32.10.2 evaluated at mean roof height h, lb/ft² (N/m²);

K_{dT} = Tornado directionality factor from Section 32.6;

K_{vT} = Tornado pressure coefficient adjustment factor from Section 32.14;

(GC_p) = External pressure coefficient from Section 30; and

(GC_{piT}) = Tornado internal pressure coefficient from Section 32.13, based on the porosity of the parapet envelope.

32.17.4.3 Attached Canopies on Buildings with h ≤ 60 ft (h ≤ 18.3 m) Section 30.9 shall apply for determination of component and cladding tornado loads on attached canopies on buildings with $h \leq 60$ ft ($h \leq 18.3$ m), as modified in this section. The design tornado pressures, p_T, on C&C elements of attached canopies shall be determined in accordance with the following equation, which replaces Equation (30.9-1):

$$p_T = q_{hT} K_{dT}(GC_p) \text{ (lb/ft}^2\text{)} \quad (32.17\text{-}6)$$

$$p_T = q_{hT} K_{dT}(GC_p) \text{ (N/m}^2\text{)} \quad (32.17\text{-}6.\text{SI})$$

where

q_{hT} = Tornado velocity pressure from Section 32.10.2 evaluated at mean roof height h, lb/ft² (N/m²);

K_{dT} = Tornado directionality factor from Section 32.6; and

(GC_p) = Net pressure coefficient from Section 30.9.

32.17.5 Nonbuilding Structures Section 30.10 shall apply for determination of component and cladding tornado loads on circular bins, silos, and tanks with $h \leq 120$ ft ($h \leq 36.6$ m), as modified in this section.

32.17.5.1 Isolated Circular Bins, Silos, and Tanks The design tornado pressures, p_T, on C&C elements of isolated circular bins, silos and tanks shall be determined in accordance with the following equation, which replaces Equation (30.10-1):

$$p_T = q_{hT}[K_{dT}K_{vT}(GC_p) - (GC_{piT})] \text{ (lb/ft}^2\text{)} \quad (32.17\text{-}7)$$

$$p_T = q_{hT}[K_{dT}K_{vT}(GC_p) - (GC_{piT})] \text{ (N/m}^2\text{)} \quad (32.17\text{-}7.\text{SI})$$

where

q_{hT} = Tornado velocity pressure from Section 32.10.2 evaluated at mean roof height h, lb/ft² (N/m²);

K_{dT} = Tornado directionality factor from Section 32.6;

K_{vT} = Tornado pressure coefficient adjustment factor. The roof slope for determination of K_{vT} shall be as defined in Figure 30.12-2:

(1) For roof zones 1 and 2 in Figure 30.12-2, K_{vT} shall equal the Zone 1 value for building roofs from Table 32.14-1, and

(2) For roof zones 3 and 4 in Figure 30.12-2, K_{vT} shall equal the Zone 2 value for building roofs from Table 32.14-1;

(GC_p) = External pressure coefficient from Section 30.10; and

(GC_{piT}) = Tornado internal pressure coefficient, as follows:

(1) For internal surface of exterior walls of isolated open-topped circular bins, silos, and tanks, use (GC_{piT}) from Section 30.12.3, and

(2) In all other cases, use (GC_{piT}) from Section 32.13.

32.18 TORNADO LOADS: WIND TUNNEL PROCEDURE

The wind tunnel procedure, as described in Chapter 31, is permitted for determination of external pressure coefficients and force coefficients for use with the tornado loading provisions of Sections 32.15 through 32.17. The wind tunnel test shall be performed on an isolated building model (without a proximity model) in a boundary layer wind tunnel for open (Exposure C) terrain.

32.19 CONSENSUS STANDARDS AND OTHER REFERENCED DOCUMENTS

This section lists the consensus standards and other documents that shall be considered part of this standard to the extent referenced in this chapter.

ASTM E1886. 2019. *Standard Test Method for Performance of Exterior Windows, Curtain Walls, Doors, and Impact Protective Systems Impacted by Missile(s) and Exposed to Cyclic Pressure Differentials*. ASTM International.

Cited in: Section 32.12.3.1

ASTM E1996. 2020. *Standard Specification for Performance of Exterior Windows, Curtain Walls, Doors, and Impact Protective Systems Impacted by Windborne Debris in Hurricanes.* ASTM International.

Cited in: Section 32.12.3.1, C32.12.3.1

ANSI/DASMA 115. 2017. *Standard Method for Testing Sectional Doors, Rolling Doors, and Flexible Doors: Determination of Structural Performance under Missile Impact and Cyclic Wind Pressure*. Door and Access Systems Manufacturers Association International.

Cited in: Section 32.12.3.1

APPENDIX A
RESERVED FOR FUTURE PROVISIONS

APPENDIX B
RESERVED FOR FUTURE PROVISIONS

APPENDIX C
SERVICEABILITY CONSIDERATIONS

C.1 SERVICEABILITY CONSIDERATIONS

This appendix is not a mandatory part of the standard but provides guidance for design for serviceability to maintain the function of a building and the comfort of its occupants during normal usage. Serviceability limits (e.g., maximum static deformations or accelerations) shall be chosen with due regard to the intended function of the structure.

Serviceability shall be checked using appropriate loads for the limit state being considered.

C.2 DEFLECTION, DRIFT, AND VIBRATION

C.2.1 Vertical Deflections Deformations of floor and roof members and systems caused by service loads shall not impair the serviceability of the structure.

C.2.2 Drift of Walls and Frames Lateral deflection or drift of structures and deformation of horizontal diaphragms and bracing systems caused by wind effects shall not impair the serviceability of the structure.

C.2.3 Vibrations Floor systems supporting large open areas free of partitions or other sources of damping, where vibration caused by pedestrian traffic might be objectionable, shall be designed with due regard for such vibration.

Mechanical equipment that can produce objectionable vibrations in any portion of an inhabited structure shall be isolated to minimize the transmission of such vibrations to the structure.

Building structural systems shall be designed so that wind-induced vibrations do not cause occupant discomfort or damage to the building, its appurtenances, or contents.

C.3 DESIGN FOR LONG-TERM DEFLECTION

Where required for acceptable building performance, members and systems shall be designed to accommodate long-term irreversible deflections under sustained load.

C.4 CAMBER

Special camber requirements that are necessary to bring a loaded member into proper relation with the work of other trades shall be set forth in the design documents.

Beams detailed without specified camber shall be positioned during erection so that any minor camber is upward. If camber involves the erection of any member under preload, this shall be noted in the design documents.

C.5 EXPANSION AND CONTRACTION

Dimensional changes in a structure and its elements caused by variations in temperature, relative humidity, or other effects shall not impair the serviceability of the structure.

Provision shall be made either to control crack widths or to limit cracking by providing relief joints.

C.6 DURABILITY

Buildings and other structures shall be designed to tolerate long-term environmental effects or shall be protected against such effects.

APPENDIX D
BUILDINGS EXEMPTED FROM TORSIONAL WIND LOAD CASES

D.1 SCOPE

The torsional load cases in Figure 27.3-8 (Case 2 and Case 4) need not be considered for a building meeting the conditions of Sections D.2, D.3, D.4, or D.5 if it can be shown by other means that the torsional load cases of Figure 27.3-8 do not control the design.

D.2 ONE- AND TWO-STORY BUILDINGS MEETING THE FOLLOWING REQUIREMENTS

One-story buildings with h less than or equal to 30 ft (9.2 m), buildings two stories or fewer framed with light-frame construction, and buildings two stories or fewer designed with flexible diaphragms are exempted.

D.3 BUILDINGS CONTROLLED BY SEISMIC LOADING

D.3.1 Buildings with Diaphragms at Each Level That Are Not Flexible Building structures that are regular under seismic load (as defined in Section 12.3.2) and conform to the following are exempted:

1. Eccentricity between the center of mass and the geometric centroid of the building at that level shall not exceed 15% of the overall building width along each principal axis considered at each level, and
2. Design story shear determined for earthquake load, as specified in Chapter 12, at each floor level shall be at least 1.5 times the design story shear determined for wind loads as specified herein.

The design earthquake and wind load cases considered when evaluating this exception shall be the load cases without torsion.

D.3.2 Buildings with Diaphragms at Each Level That Are Flexible Building structures that are regular under seismic load (as defined in Section 12.3.2) and conform to the following are exempted:

1. Design earthquake shear forces resolved to the vertical elements of the lateral load-resisting system shall be at least 1.5 times the corresponding design wind shear forces resisted by those elements.

The design earthquake and wind load cases considered when evaluating this exception shall be the load cases without torsion.

D.4 BUILDINGS CLASSIFIED AS TORSIONALLY REGULAR UNDER WIND LOAD

Buildings meeting the definition of torsionally regular buildings under wind load contained in Section 26.2 are exempted.

EXCEPTION: If a building does not qualify as being torsionally regular under wind load, it is permissible to base the design on the basic wind load Case 1 that is proportionally increased so that the maximum displacement at each level is not less than the maximum displacement for the torsional load Case 2.

D.5 BUILDINGS WITH DIAPHRAGMS THAT ARE FLEXIBLE AND DESIGNED FOR INCREASED WIND LOADING

The torsional wind load cases need not be considered if the design wind pressure in Cases 1 and 3 of Figure 27.3-8 is increased by a factor of 1.5.

APPENDIX E
PERFORMANCE-BASED DESIGN PROCEDURES FOR FIRE EFFECTS ON STRUCTURES

E.1 SCOPE

This appendix is not a mandatory part of the standard. It provides procedures for performance-based design and evaluation of structures for fire conditions that result in fire-induced effects on a structure's members and connections. The use of performance-based structural fire design procedures constitutes an alternative methodology to meet project design requirements, as is permitted by Section 1.3.7 and the alternative materials, design, and methods of construction provision in the building codes. This appendix does not provide for standard fire resistance design with prescriptive methods, nor does it address explosions.

E.2 DEFINITIONS

The following definitions apply to the information presented in this appendix.

FIRE: An oxidation process that results in the burning of flammable materials and produces heat.

FIRE EFFECTS: Thermal and structural response caused by fire exposure and subsequent cooling.

FIRE EXPOSURE: The extent to which materials, products, or assemblies are subjected to the conditions created by fire.

FIRE RESISTANCE: The ability of a material, product, or assembly to withstand fire or provide protection from it for a period of time.

FUEL LOAD: The total quantity of combustible contents within a building, space, or area, expressed either as total energy or as equivalent mass.

HEAT TRANSFER: The exchange of thermal energy caused by a temperature difference.

PERFORMANCE-BASED STRUCTURAL FIRE DESIGN: The explicit design of structural members and connections to satisfy performance objectives for structural design fires.

STANDARD FIRE RESISTANCE DESIGN: The selection of fire-resistive assemblies to meet code requirements for structural fire resistance (also known as prescriptive design). The rating of a fire-resistive assembly is based on its performance in standard fire testing.

STRUCTURAL DESIGN FIRE: A fire that has the potential to affect the integrity and stability of a structure that is used for the design and evaluation of the structure.

THERMAL BOUNDARY CONDITION: The temperature and/or heat flux to which an assembly or the structure is subjected during or after fire exposure based on the radiative and convective heating and/or cooling conditions at exposed surfaces.

THERMAL INSULATION: A material or medium that reduces the heat transfer between objects in thermal contact or in range of radiative or convective influence.

THERMAL RESPONSE: The temperature distribution of members and connections when exposed to thermal boundary conditions.

THERMAL RESTRAINT: A condition in which thermal expansion or contraction of structural members is resisted by forces external to the members. The level of restraint depends on the adjacent framing and connection details.

E.3 GENERAL REQUIREMENTS

Performance-based structural fire design and evaluation procedures shall comply with the requirements of Section 1.3.1.3.

Design and evaluation of structures for fire effects shall include the following tasks: identify performance objectives, quantify the fuel load, identify and evaluate structural design fires, determine temperature histories of structural members and connections, and determine the response of the structure. These tasks shall include evaluating the heating and subsequent cooling of the structure under fire exposure, as appropriate.

The structural response shall be evaluated against limit states based on the performance objectives. The analyses shall account for temperature-dependent material properties, boundary conditions, and thermally induced failure modes and shall evaluate structural stability, strength, deformation, and load path continuity.

E.4 PERFORMANCE OBJECTIVES

Performance objectives shall be expressed as quantifiable metrics for the design and evaluation of structural response to structural design fires. Performance objectives for structural integrity, including strength, stiffness, and stability, shall meet the minimum criteria specified in Section E.4.1. Additional project-specific performance objectives shall meet the requirements in Section E.4.2.

E.4.1 Structural Integrity Structural integrity shall be provided for buildings and other structures subject to structural design fires such that the structural system remains stable with a continuous load path to the extent necessary to ensure occupant life safety.

The performance of the structural system under structural design fires shall allow building occupants to travel safely to refuge areas within the building or exit the building to a public way. Structural support of building egress routes shall be maintained for a period of time necessary to ensure a safe and complete evacuation of building occupants. Structural support of building refuge areas shall be maintained throughout the heating and cooling of the structure.

E.4.2 Project-Specific Performance Objectives Buildings and other structures shall be designed to meet project-specific performance objectives required by the owner, Authority Having Jurisdiction, or applicable building code in addition to those in Section E.4.1.

E.5 THERMAL ANALYSIS OF FIRE EFFECTS

The thermal response of structural members and connections during and after structural design fires shall be determined for input to the structural analyses of fire effects.

E.5.1 Fuel Load The fuel load shall be quantified for use in the evaluation of structural design fires.

E.5.2 Structural Design Fires Structural design fires shall be identified and used to develop time-dependent thermal boundary conditions for use in heat transfer analyses.

E.5.3 Heat Transfer Analysis The temperature history of structural members and connections shall be determined using heat transfer analysis methods based on the time-dependent thermal boundary conditions for structural design fires.

Temperature-dependent thermal properties of materials comprising the structural system and thermal insulation shall be used in heat transfer analyses to determine the thermal response. It shall be permissible to use constant thermal property values if they yield conservative results.

E.6 STRUCTURAL ANALYSIS OF FIRE EFFECTS

Structural analyses shall include portions of the structural system that are subject to fire effects from the structural design fires as determined in Section E.5, with consideration of unheated portions of the structural system that provide thermal restraint. It is permissible for the analyses to consider alternate load paths that are capable of being maintained following structural damage or degradation caused by fire effects.

E.6.1 Temperature History for Structural Members and Connections Temperature histories for structural members and connections shall be determined from thermal analysis of structural design fires and shall be used to analyze fire effects on structural performance.

E.6.2 Temperature-Dependent Properties Temperature-dependent properties of structural materials shall be used to determine the performance of structural members and connections subject to structural design fires.

E.6.3 Load Combinations Load combinations in Section 2.5 for extraordinary events shall be used for analysis of fire effects and shall include time- and sequence-dependent effects. To check the residual capacity of a structure damaged by a structural design fire, the approach identified in Section 2.5.2.2 shall be used.

APPENDIX F
WIND HAZARD MAPS FOR LONG RETURN PERIODS

F.1 SCOPE

This appendix provides wind hazard maps for longer mean recurrence intervals (MRIs) than included in Chapter 26.

F.2 WIND SPEEDS

Except as provided in Sections 26.5.2 and 26.5.3, it shall be permitted to determine the wind speed used for design wind loads on buildings and other structures from Figures F.2-1 through F.2-3 as follows:

For 10,000-year MRI, use Figure F.2-1.
For 100,000-year MRI, use Figure F.2-2.
For 1,000,000-year MRI, use Figure F.2-3.

Alternatively, it shall be permitted to use the wind speed from the ASCE Wind Design Geodatabase for the contiguous United States.

User Note: The ASCE Wind Design Geodatabase is available at the ASCE 7 Hazard Tool (https://asce7hazardtool.online/), or approved equivalent.

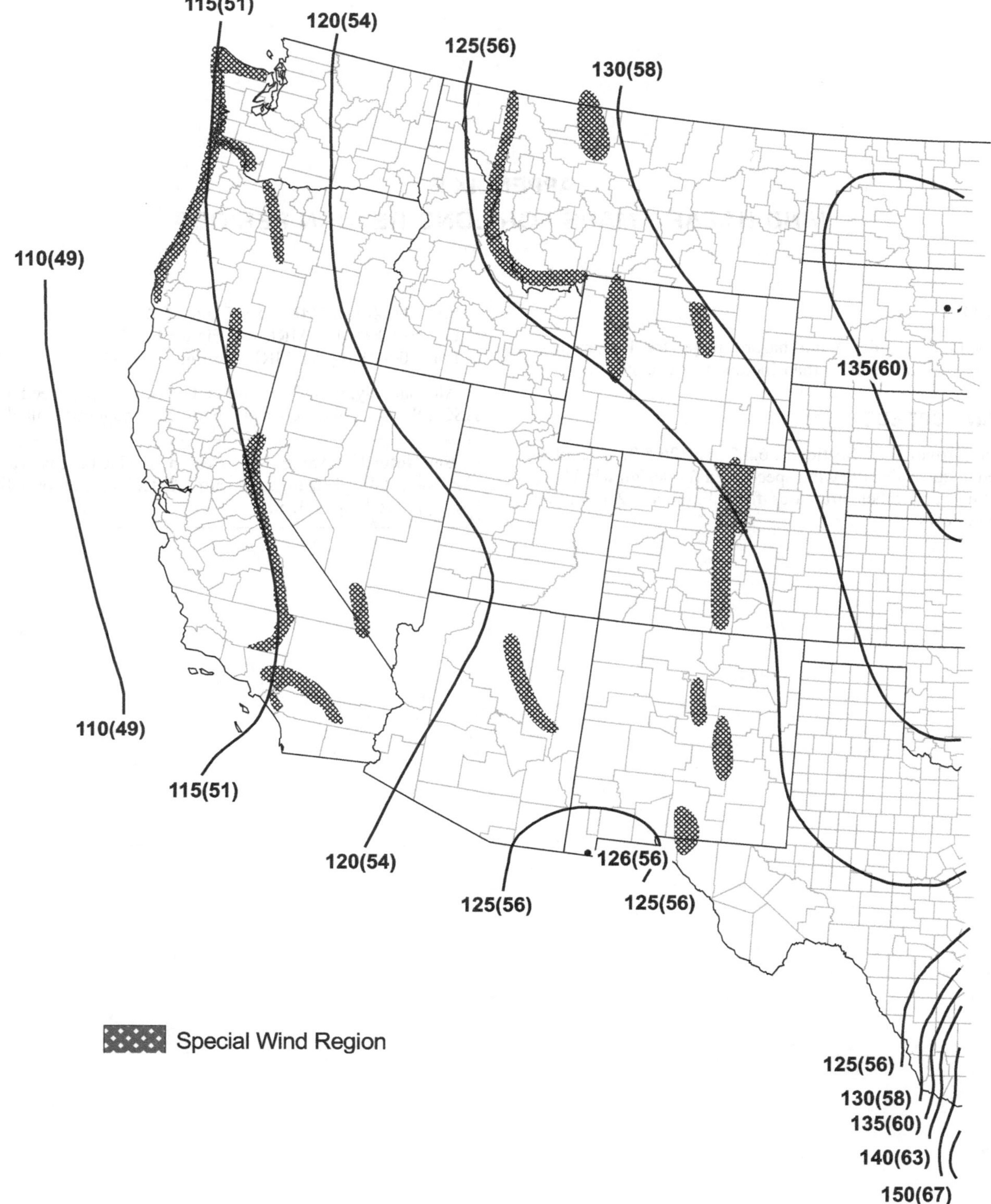

Notes:

1. Values are 3 s gust wind speeds in mi/h (m/s) at 33 ft (10 m) above ground for Exposure Category C.
2. Linear interpolation is permitted between contours. Point values are provided to aid with interpolation.
3. Islands, ccastal areas, and land boundaries outside the last contour shall use the last wind speed contour.
4. Location-specific basic wind speeds shall be permitted to be determined using the ASCE Wind Design Geodatabase.

Figure F.2-1. 10,000-Year MRI gust wind speed, mi/h (m/s).

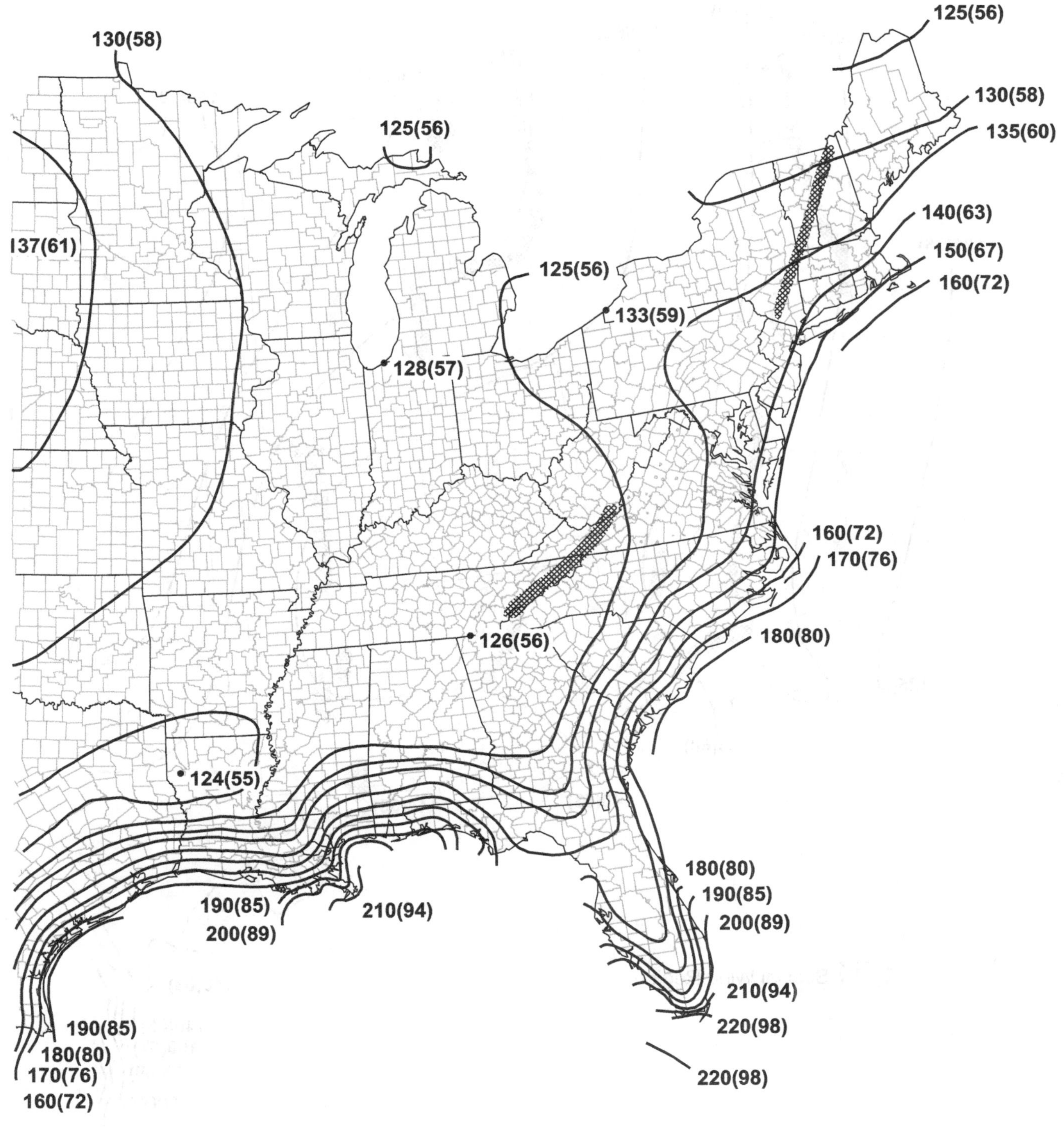

5. Mountainous terrain, gorges, ocean promontories, and special wind regions shall be examined for unusual wind conditions.

6. Wind speeds correspond to approximately a 0.5% probability of exceedance in 50 years (Annual Exceedance Probability = 0.0001, MRI = 10,000 years).

7. The ASCE Wind Design Geodatabase is available at the ASCE 7 Hazard Tool (https://asce7hazardtool.online), or approved equivalent.

Figure F.2-1. (*Continued*) 10,000-Year MRI gust wind speed, mi/h (m/s).

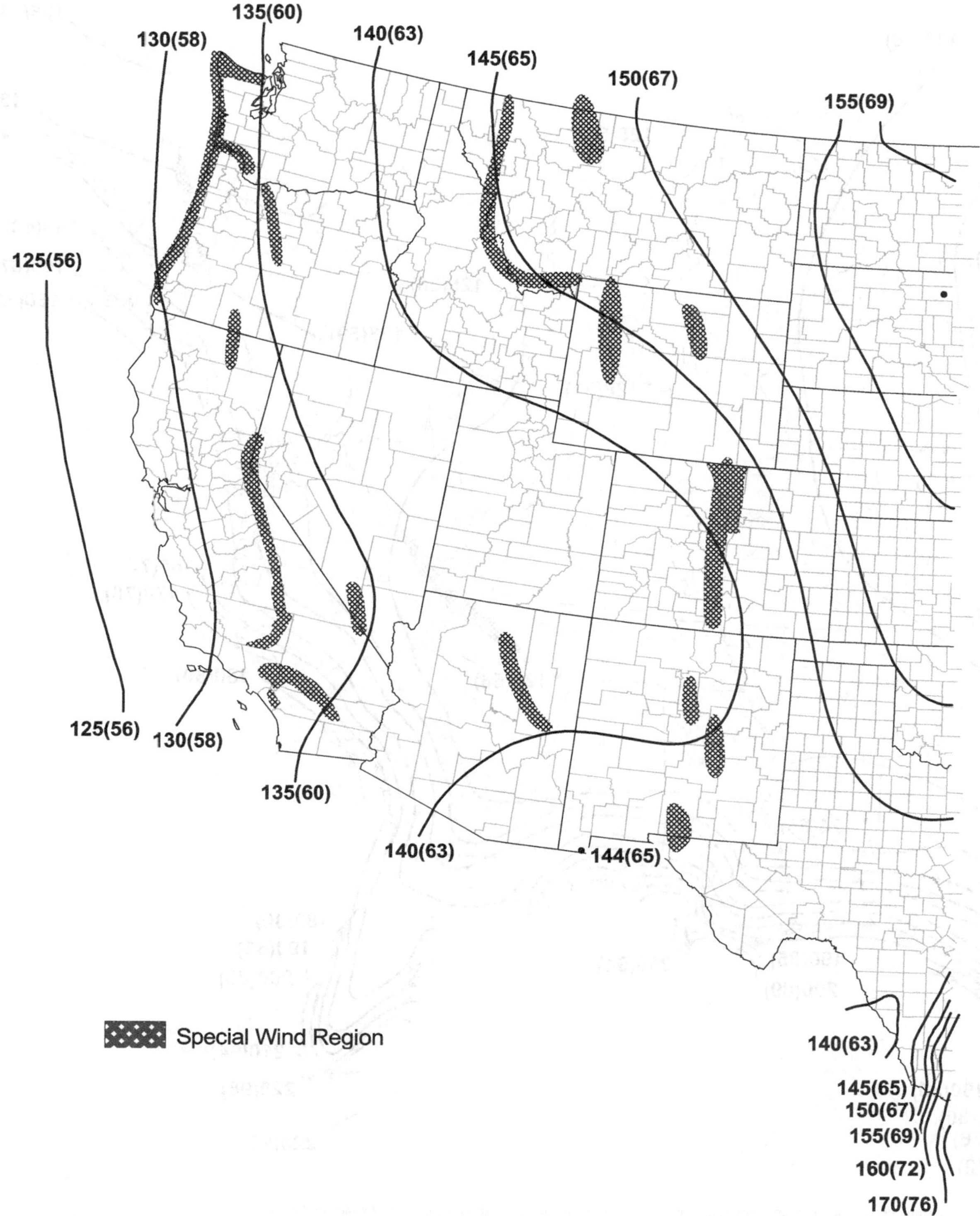

Notes:

1. Values are 3 s gust wind speeds in mi/h (m/s) at 33 ft (10 m) above ground for Exposure Category C.
2. Linear interpolation is permitted between contours. Point values are provided to aid with interpolation.
3. Islands, coastal areas, and land boundaries outside the last contour shall use the last wind speed contour.
4. Location-specific basic wind speeds shall be permitted to be determined using the ASCE Wind Design Geodatabase.

Figure F.2-2. 100,000-Year MRI gust wind speed, mi/h (m/s).

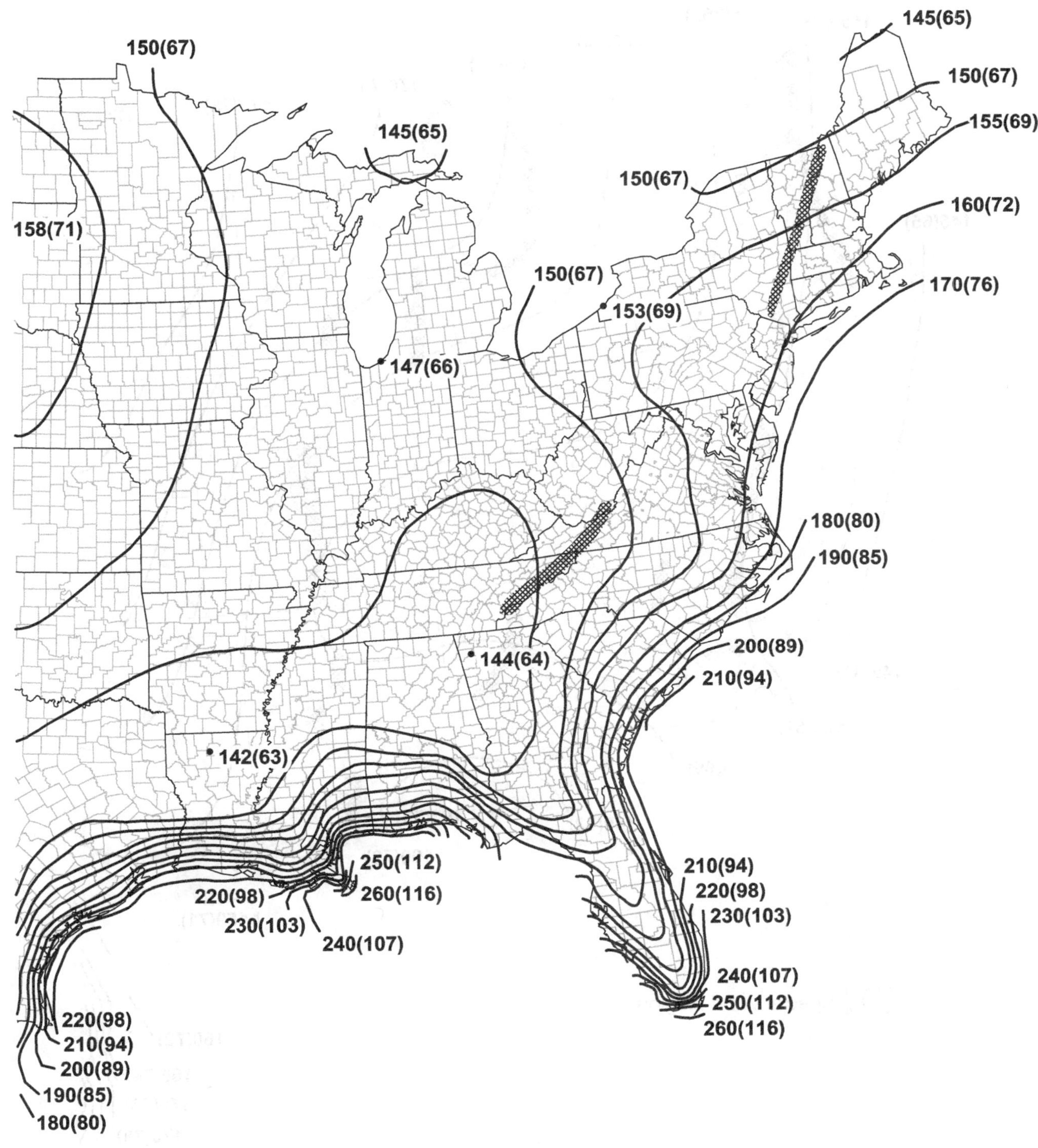

5. Mountainous terrain, gorges, ocean promontories, and special wind regions shall be examined for unusual wind conditions.

6. Wind speeds correspond to approximately a 0.5% probability of exceedance in 50 years (Annual Exceedance Probability = 0.00001, MRI = 100,000 years).

7. The ASCE Wind Design Geodatabase is available at the ASCE 7 Hazard Tool (https://asce7hazardtool.online), or approved equivalent.

Figure F.2-2. (*Continued*) 100,000-Year MRI gust wind speed, mi/h (m/s).

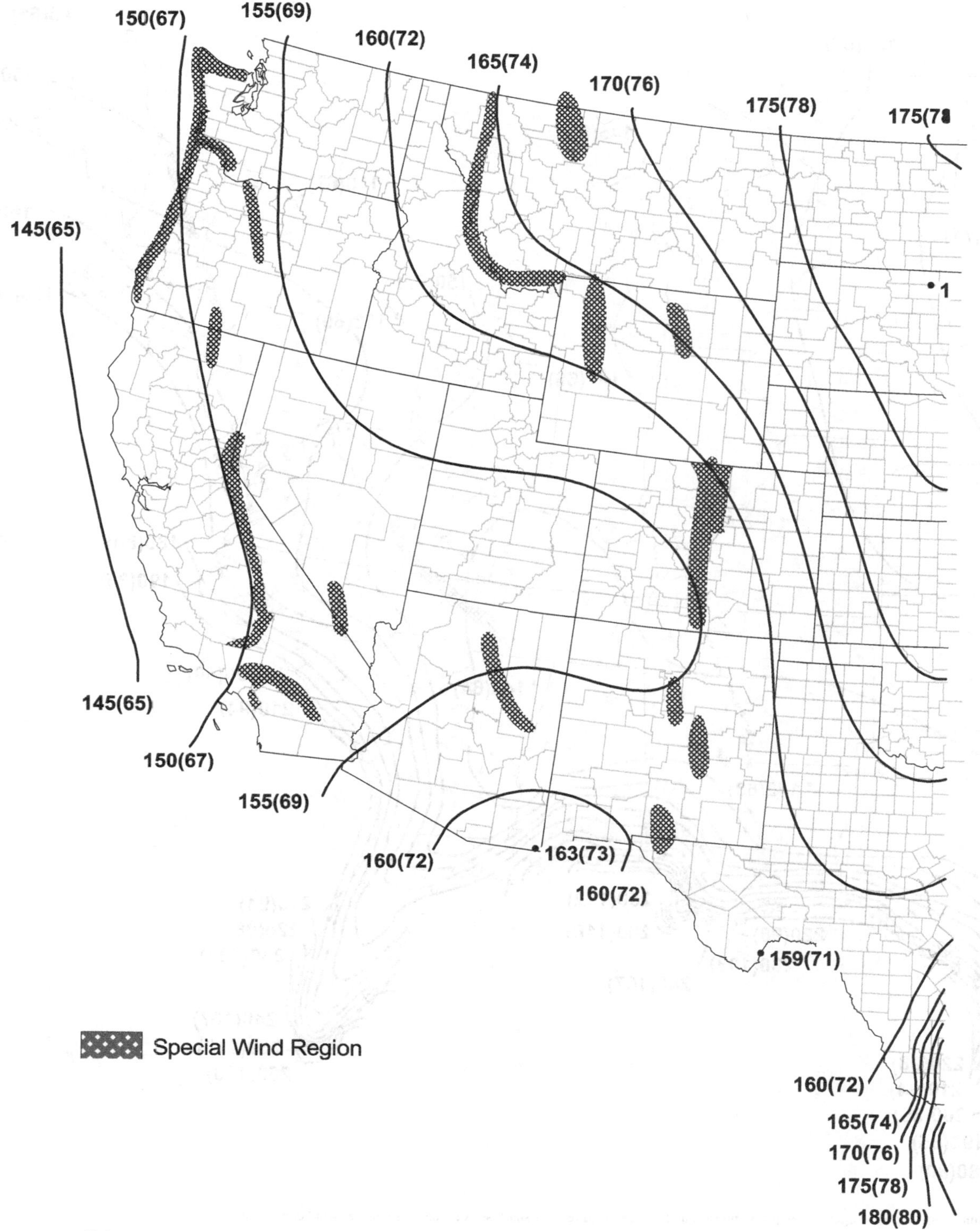

Notes:

1. Values are 3 s gust wind speeds in mi/h (m/s) at 33 ft (10 m) above ground for Exposure Category C.
2. Linear interpolation is permitted between contours. Point values are provided to aid with interpolation.
3. Islands, coastal areas, and land boundaries outside the last contour shall use the last wind speed contour.
4. Location-specific basic wind speeds shall be permitted to be determined using the ASCE Wind Design Geodatabase.

Figure F.2-3. 1,000,000-year MRI gust wind speed, mi/h (m/s).

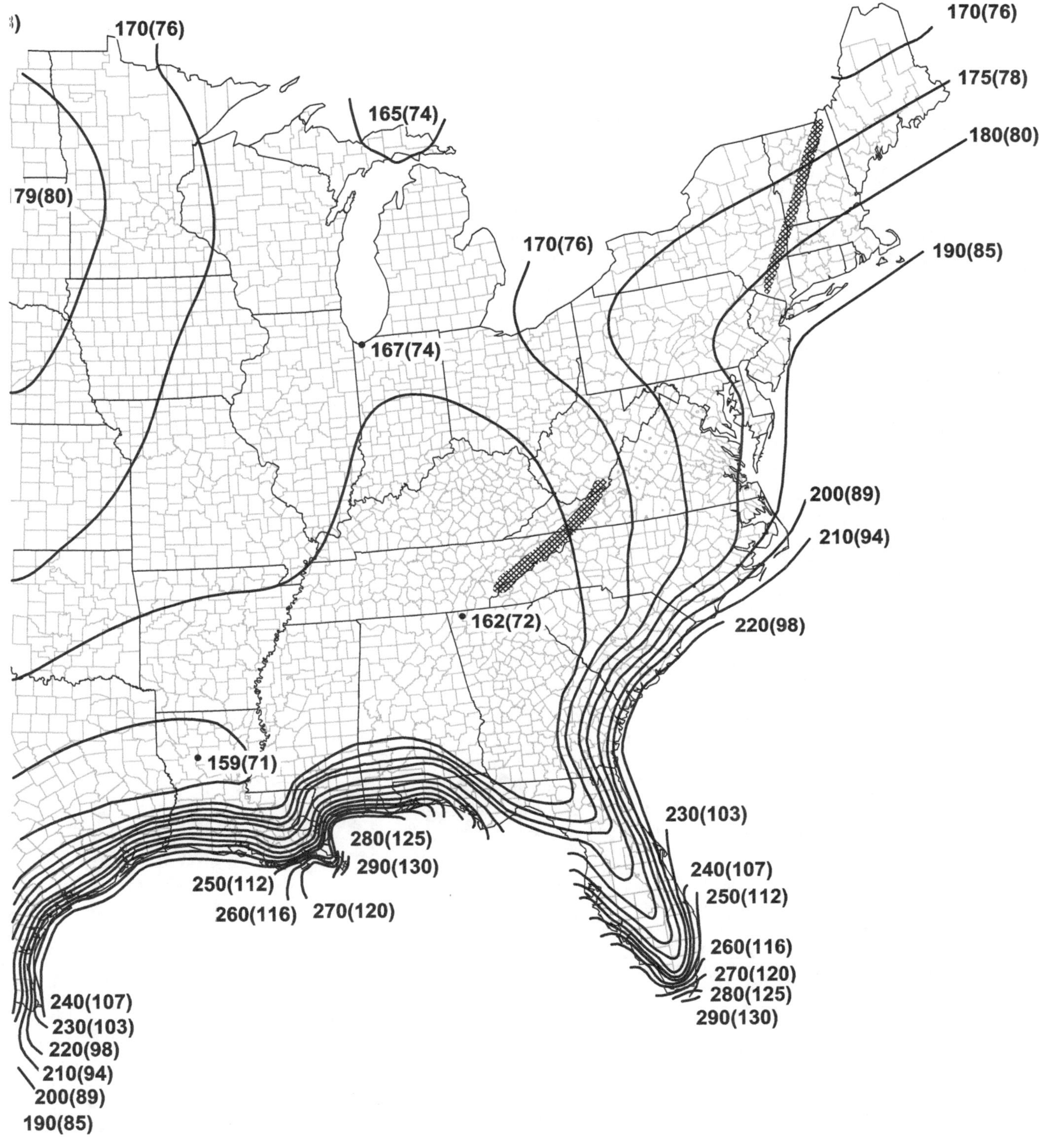

5. Mountainous terrain, gorges, ocean promontories, and special wind regions shall be examined for unusual wind conditions.

6. Wind speeds correspond to approximately a 0.005% probability of exceedance in 50 years (Annual Exceedance Probability = 0.000001, MRI = 1,000,000 years).

7. The ASCE Wind Design Geodatabase is available at the ASCE 7 Hazard Tool (https://asce7hazardtool.online), or approved equivalent.

Figure F.2-3. (*Continued*) 1,000,000-year MRI gust wind speed, mi/h (m/s).

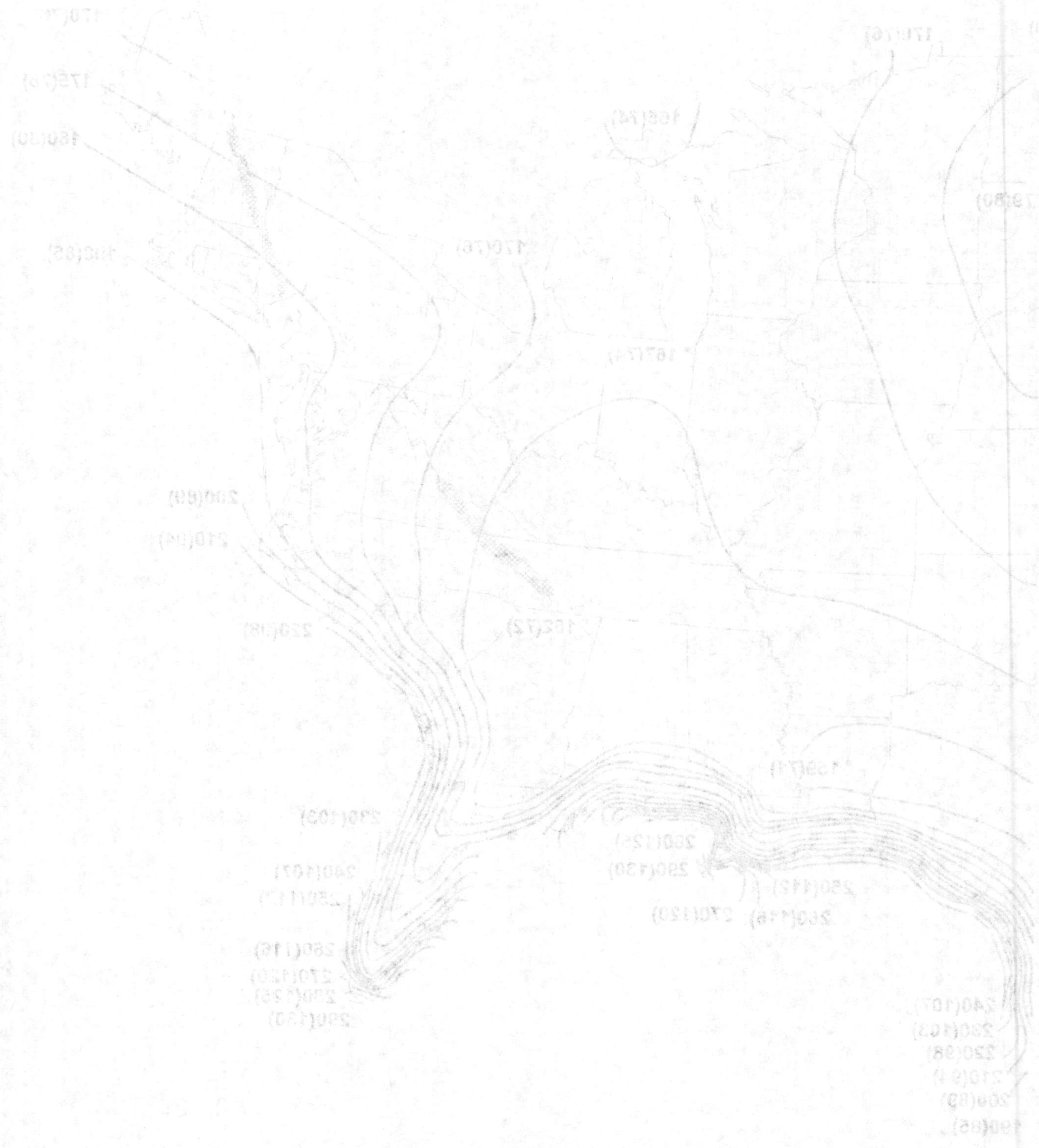

Figure F.2-3. (Continued) 1,000,000-year MRI gust wind speed, mph (m/s).

APPENDIX G
TORNADO HAZARD MAPS FOR LONG RETURN PERIODS

G.1 SCOPE

This appendix is not a mandatory part of the standard. This appendix provides tornado hazard maps for longer mean recurrence intervals (MRI) than those included in Chapter 32.

G.2 TORNADO SPEEDS

It shall be permitted to determine the tornado speed used for design tornado loads on buildings and other structures from Figures 32.A2-1 through G.2-4 as follows:

1. For 10,000-year MRI, use Figures G.2-1A through G.2-1H.
2. For 100,000-year MRI, use Figures G.2-2A through G.2-2H.
3. For 1,000,000-year MRI, use Figures G.2-3A through G.2-3H.
4. For 10,000,000-year MRI, use Figures G.2-4A through G.2-4H.

Alternatively, it shall be permitted to use the tornado speed from the ASCE Tornado Design Geodatabase. The ASCE Tornado Design Geodatabase is available at the ASCE 7 Hazard Tool (https://asce7hazardtool.online/), or approved equivalent.

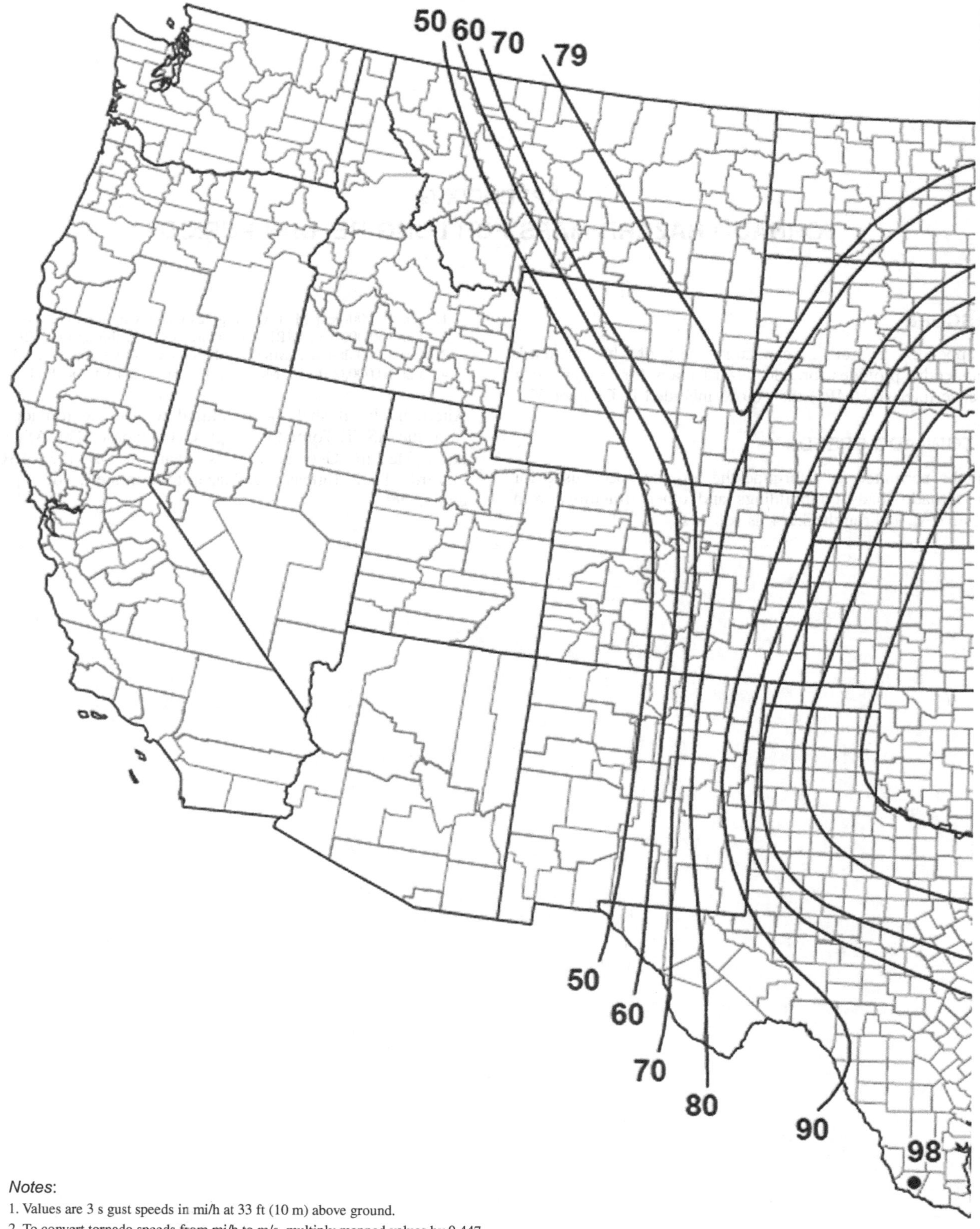

Notes:

1. Values are 3 s gust speeds in mi/h at 33 ft (10 m) above ground.
2. To convert tornado speeds from mi/h to m/s, multiply mapped values by 0.447.
3. Linear interpolation is permitted between contours. Point values (where shown) are provided to aid with interpolation.

Figure G.2-1A. Tornado speeds for 10,000-year MRI for effective plan area of 1 ft^2 (0.1 m^2).

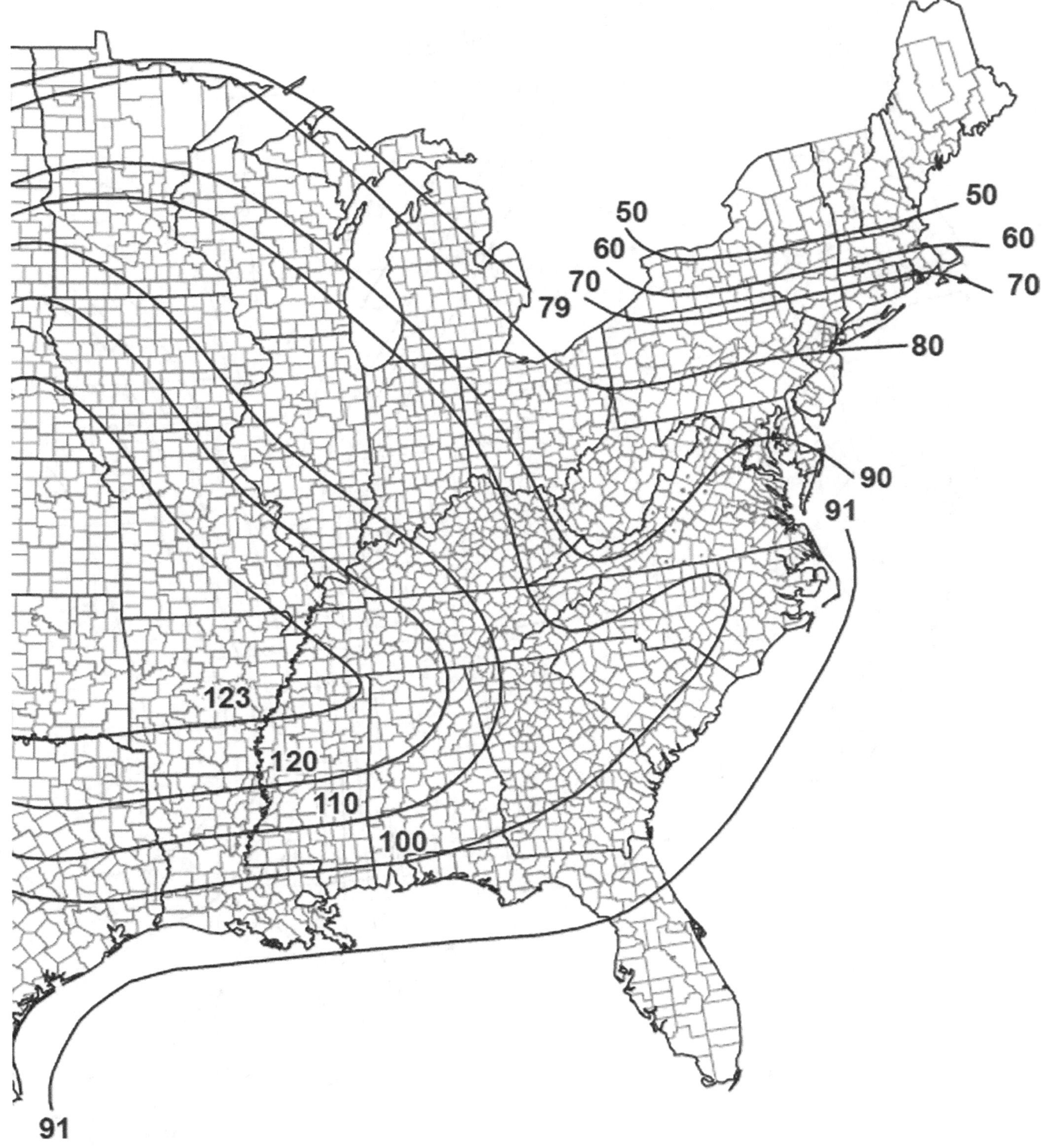

4. Islands, coastal areas, and land boundaries outside the last contour shall use the last tornado speed contour.
5. Tornado speeds correspond to approximately a 0.5% probability of exceedance in 50 years (annual exceedance probability = 0.0001, MRI = 10,000 years).
6. Location-specific tornado speed is permitted to be determined using the ASCE Tornado Design Geodatabase, available at the ASCE 7 Hazard Tool (https://asce7hazardtool.online), or approved equivalent.

Figure G.2-1A (*Continued*). Tornado speeds for 10,000-year MRI for effective plan area of 1 ft^2 (0.1 m^2).

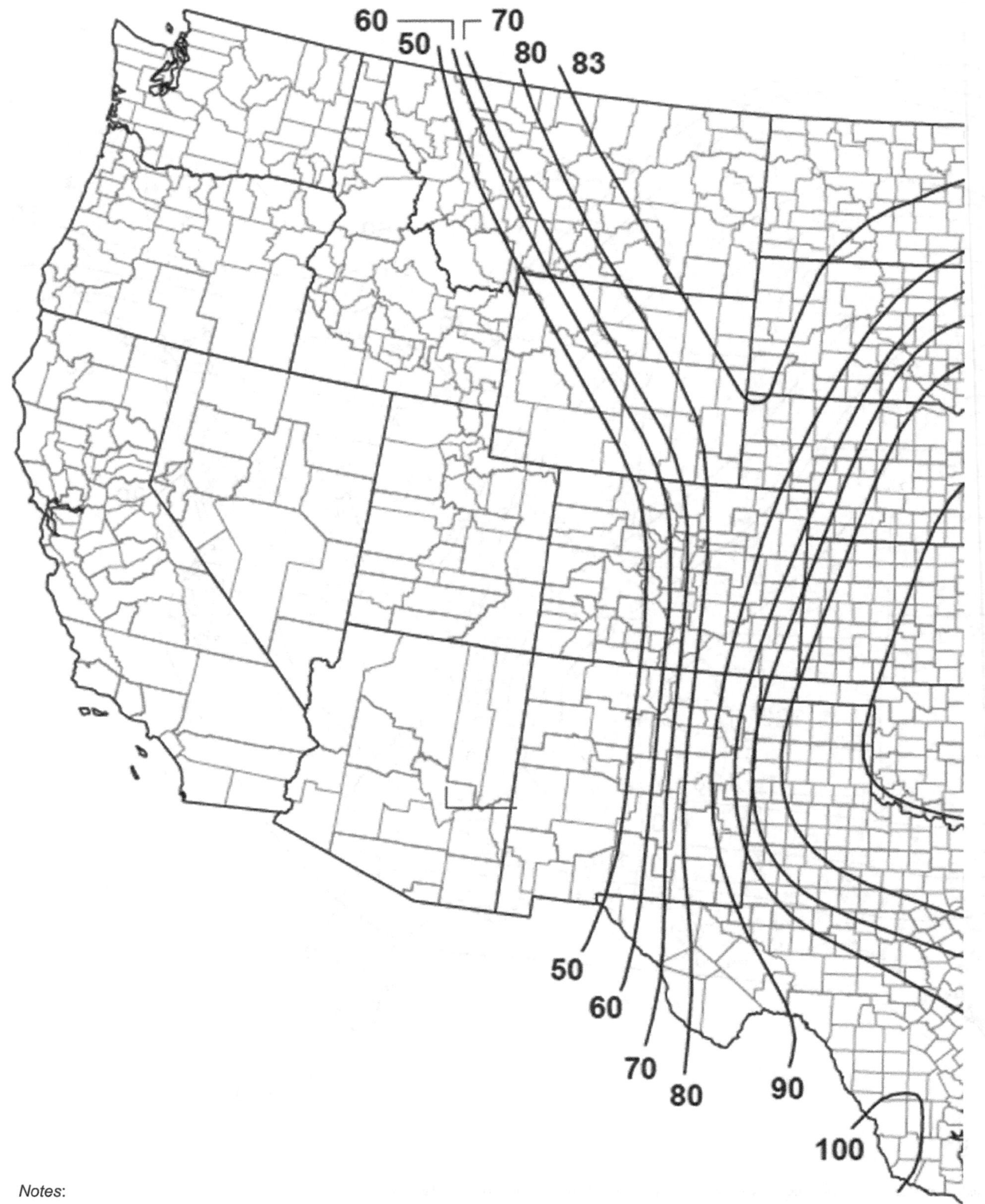

Notes:

1. Values are 3 s gust speeds in mi/h at 33 ft (10 m) above ground.
2. To convert tornado speeds from mi/h to m/s, multiply mapped values by 0.447.
3. Linear interpolation is permitted between contours. Point values (where shown) are provided to aid with interpolation.

Figure G.2-1B. Tornado speeds for 10,000-year MRI for effective plan area of 2,000 ft^2 (186 m^2).

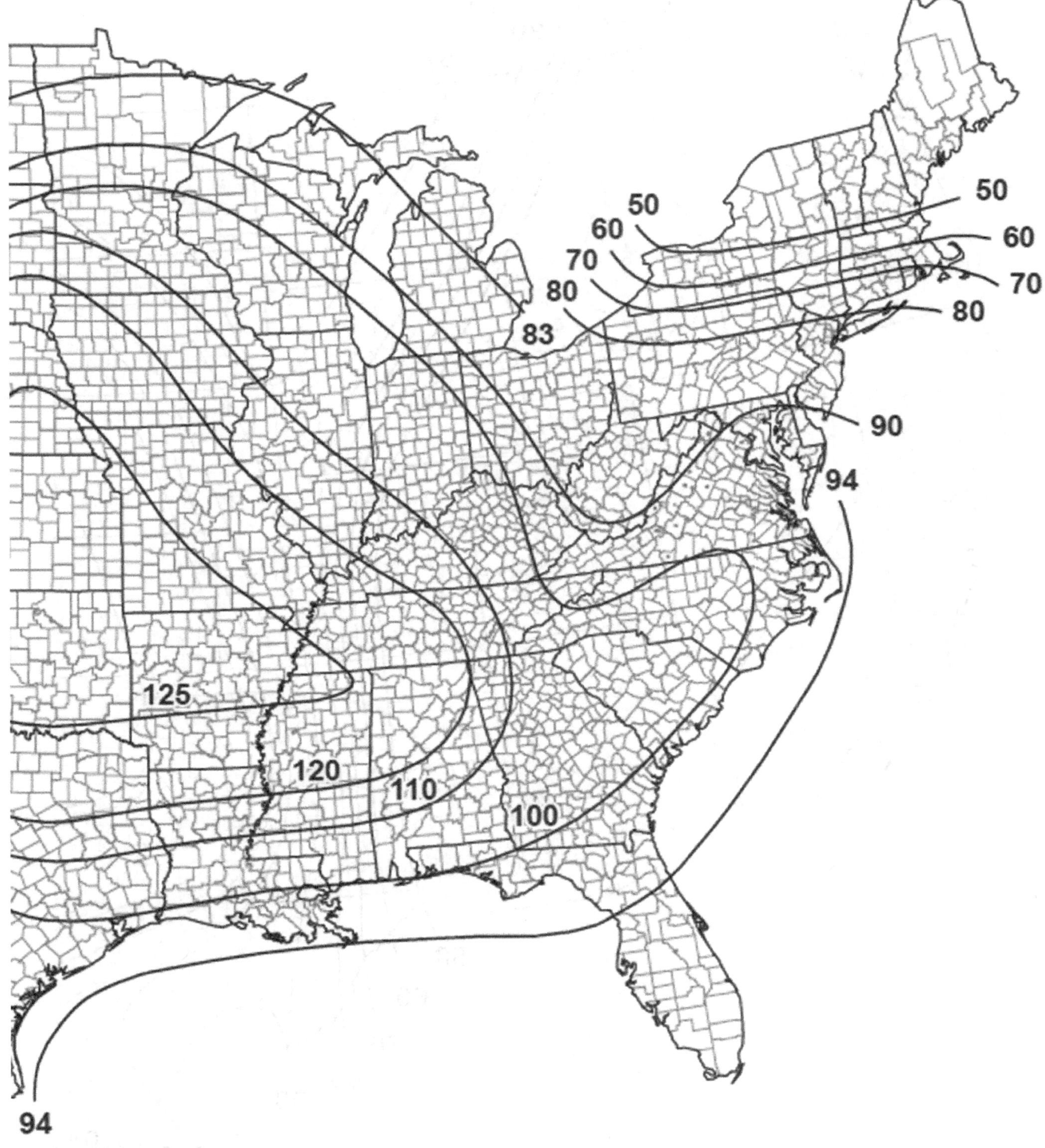

4. Islands, coastal areas, and land boundaries outside the last contour shall use the last tornado speed contour.
5. Tornado speeds correspond to approximately a 0.5% probability of exceedance in 50 years (annual exceedance probability = 0.0001, MRI = 10,000 years).
6. Location-specific tornado speed is permitted to be determined using the ASCE Tornado Design Geodatabase, available at the ASCE 7 Hazard Tool (https://asce7hazardtool.online), or approved equivalent.

Figure G.2-1B (*Continued*). Tornado speeds for 10,000-year MRI for effective plan area of 2,000 ft^2 (186 m^2).

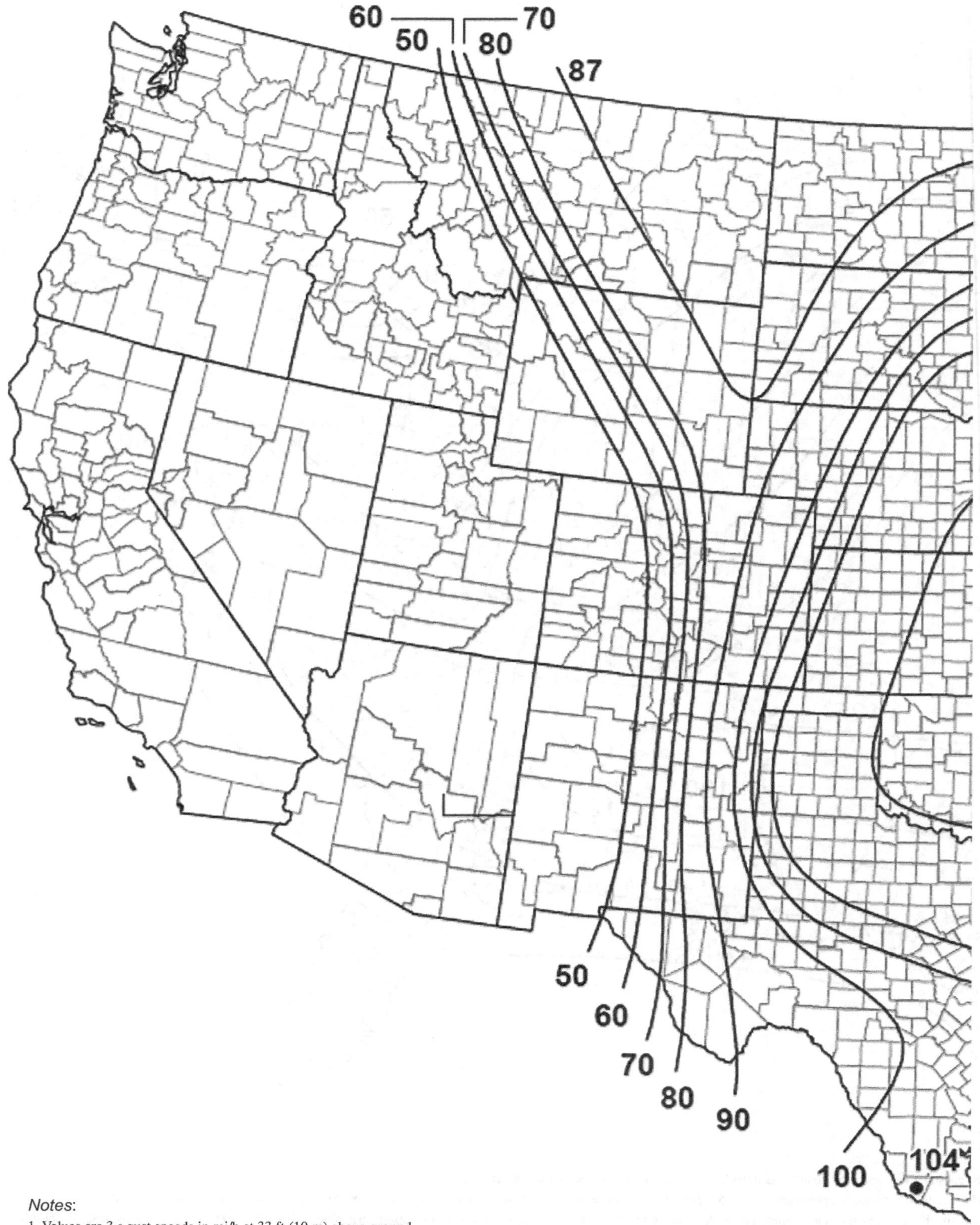

Notes:

1. Values are 3 s gust speeds in mi/h at 33 ft (10 m) above ground.
2. To convert tornado speeds from mi/h to m/s, multiply mapped values by 0.447.
3. Linear interpolation is permitted between contours. Point values (where shown) are provided to aid with interpolation.

Figure G.2-1C. Tornado speeds for 10,000-year MRI for effective plan area of 10,000 ft^2 (929 m^2).

4. Islands, coastal areas, and land boundaries outside the last contour shall use the last tornado speed contour.
5. Tornado speeds correspond to approximately a 0.5% probability of exceedance in 50 years (annual exceedance probability = 0.0001, MRI = 10,000 years).
6. Location-specific tornado speed is permitted to be determined using the ASCE Tornado Design Geodatabase, available at the ASCE 7 Hazard Tool (https://asce7hazardtool.online), or approved equivalent.

Figure G.2-1C (*Continued*). Tornado speeds for 10,000-year MRI for effective plan area of 10,000 ft^2 (929 m^2).

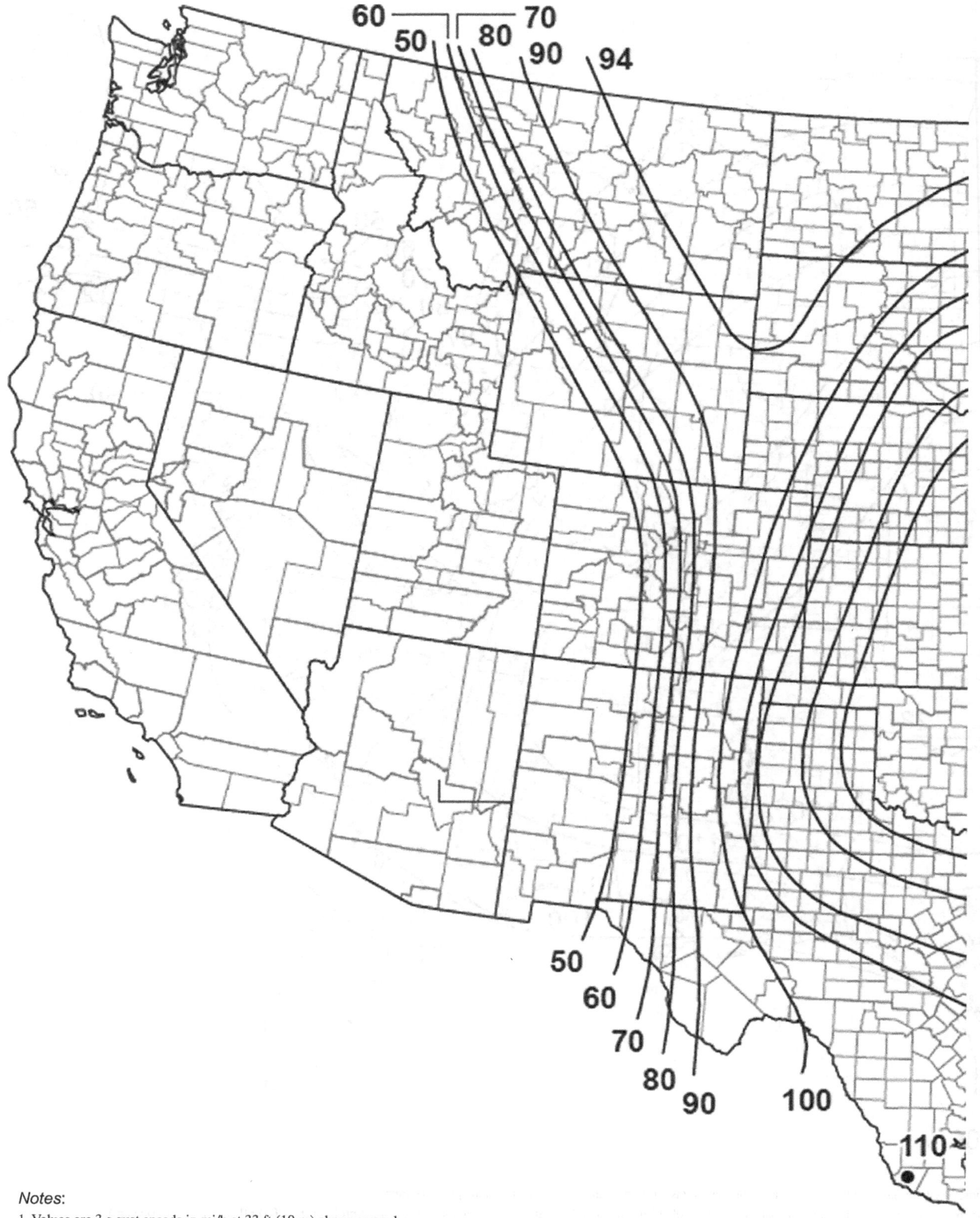

Notes:

1. Values are 3 s gust speeds in mi/h at 33 ft (10 m) above ground.
2. To convert tornado speeds from mi/h to m/s, multiply mapped values by 0.447.
3. Linear interpolation is permitted between contours. Point values (where shown) are provided to aid with interpolation.

Figure G.2-1D. Tornado speeds for 10,000-year MRI for effective plan area of 40,000 ft^2 (3,716 m^2).

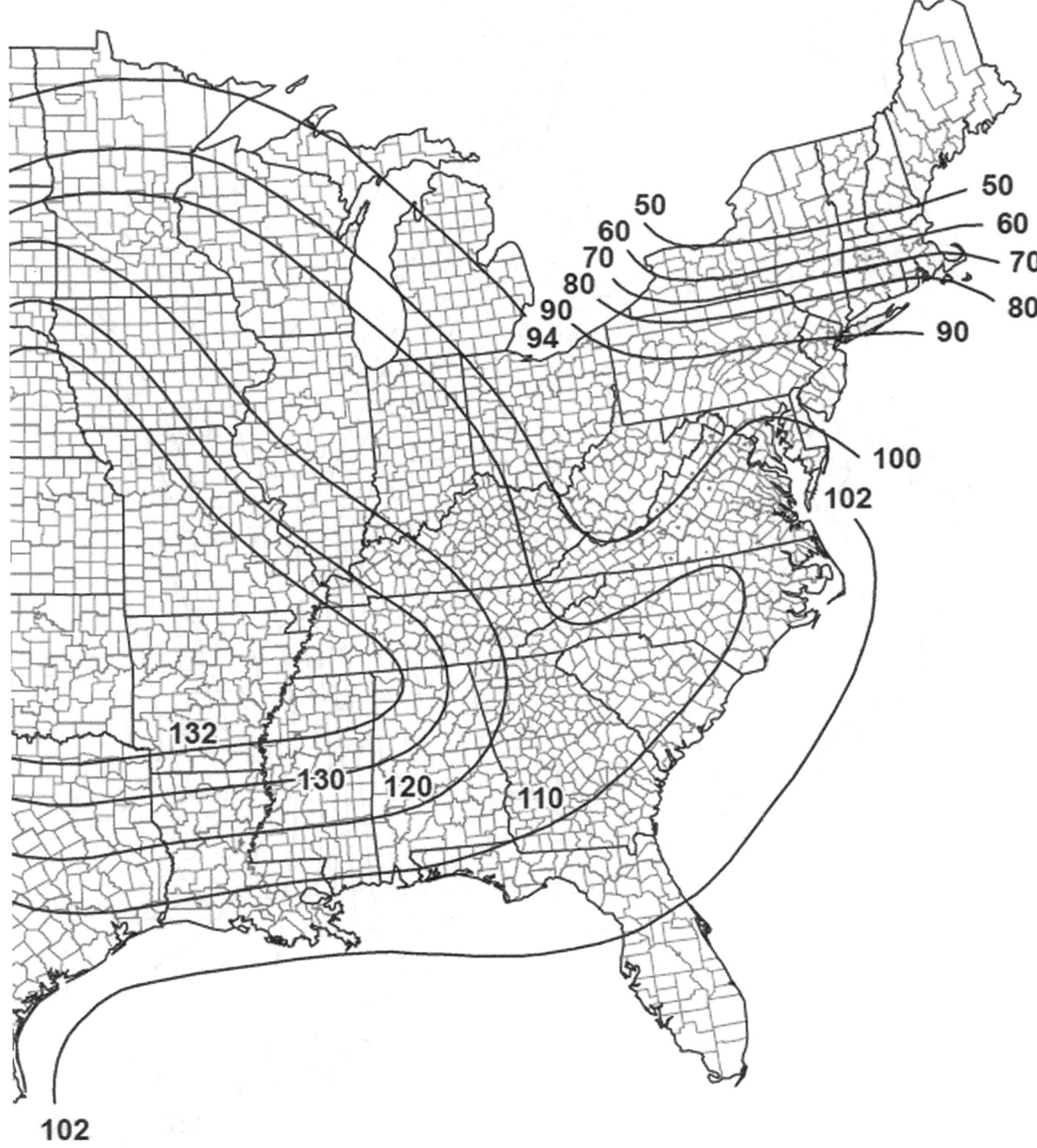

4. Islands, coastal areas, and land boundaries outside the last contour shall use the last tornado speed contour.
5. Tornado speeds correspond to approximately a 0.5% probability of exceedance in 50 years (annual exceedance probability = 0.0001, MRI = 10,000 years).
6. Location-specific tornado speed is permitted to be determined using the ASCE Tornado Design Geodatabase, available at the ASCE 7 Hazard Tool (https://asce7hazardtool.online), or approved equivalent.

Figure G.2-1D (*Continued*). Tornado speeds for 10,000-year MRI for effective plan area of 40,000 ft^2 (3,716 m^2).

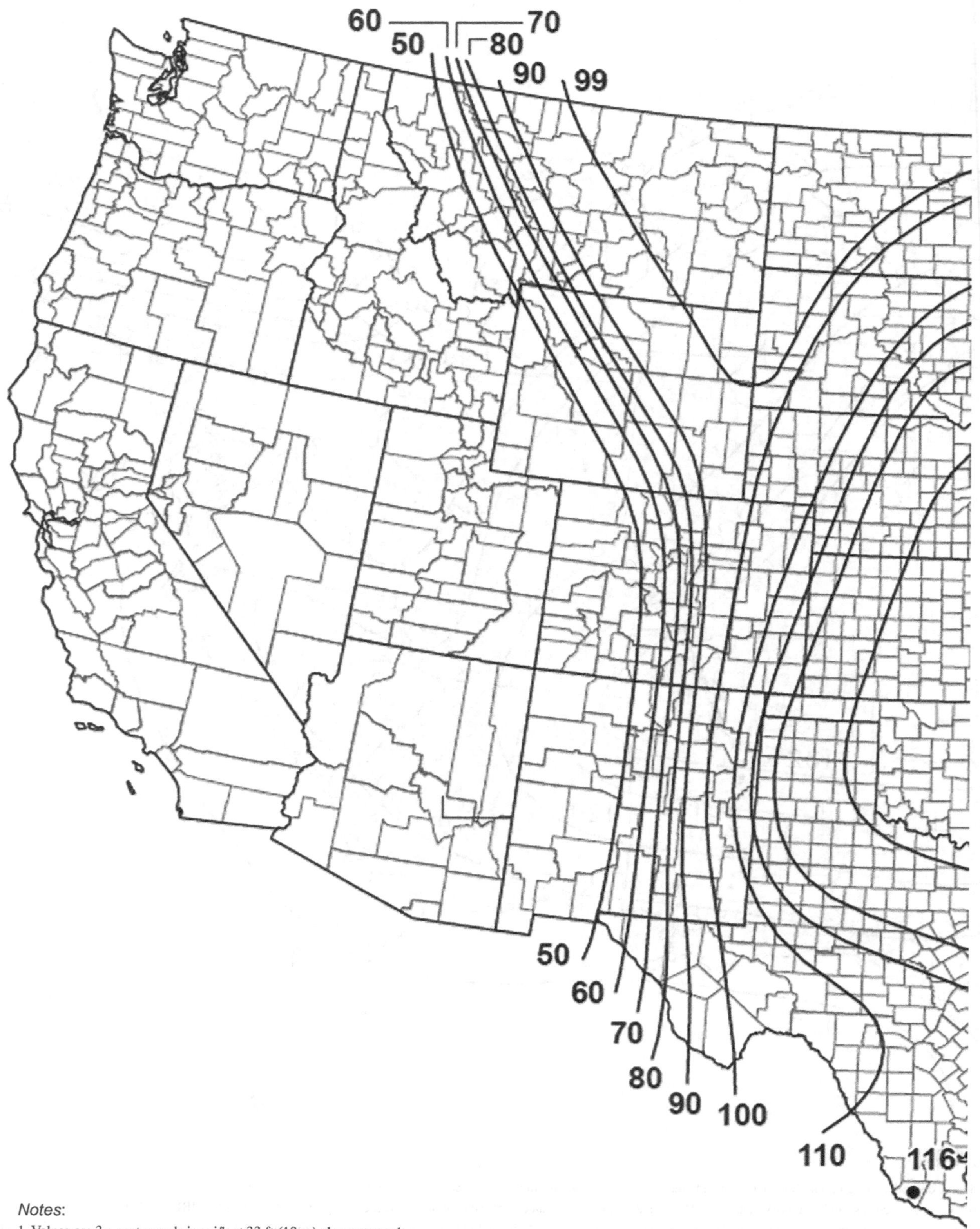

Notes:

1. Values are 3 s gust speeds in mi/h at 33 ft (10 m) above ground.
2. To convert tornado speeds from mi/h to m/s, multiply mapped values by 0.447.
3. Linear interpolation is permitted between contours. Point values (where shown) are provided to aid with interpolation.

Figure G.2-1E. Tornado speeds for 10,000-year MRI for effective plan area of 100,000 ft² (9,290 m²).

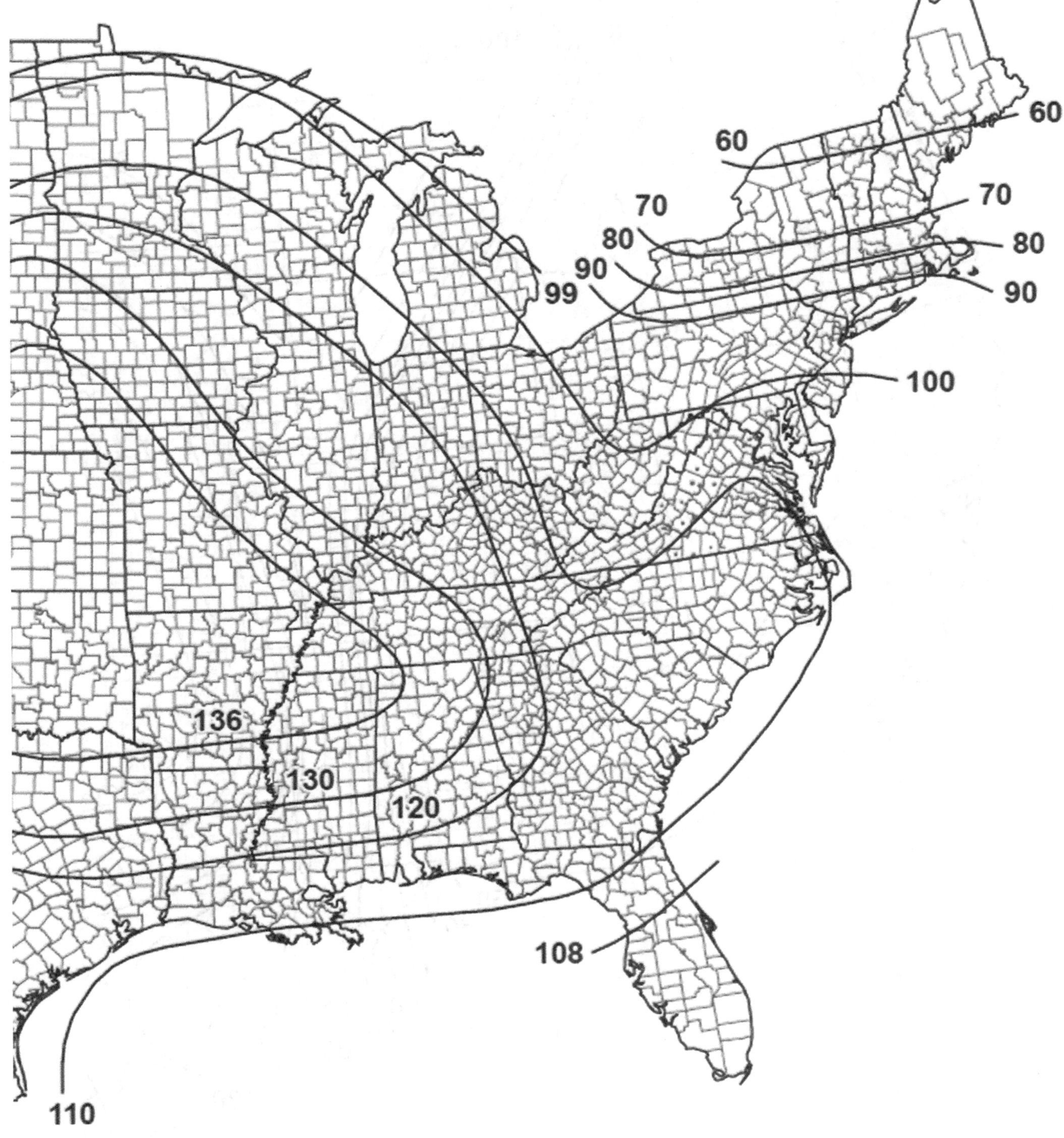

4. Islands, coastal areas, and land boundaries outside the last contour shall use the last tornado speed contour.
5. Tornado speeds correspond to approximately a 0.5% probability of exceedance in 50 years (annual exceedance probability = 0.0001, MRI = 10,000 years).
6. Location-specific tornado speed is permitted to be determined using the ASCE Tornado Design Geodatabase, available at the ASCE 7 Hazard Tool (https://asce7hazardtool.online), or approved equivalent.

Figure G.2-1E (*Continued*). Tornado speeds for 10,000-year MRI for effective plan area of 100,000 ft^2 (9,290 m^2).

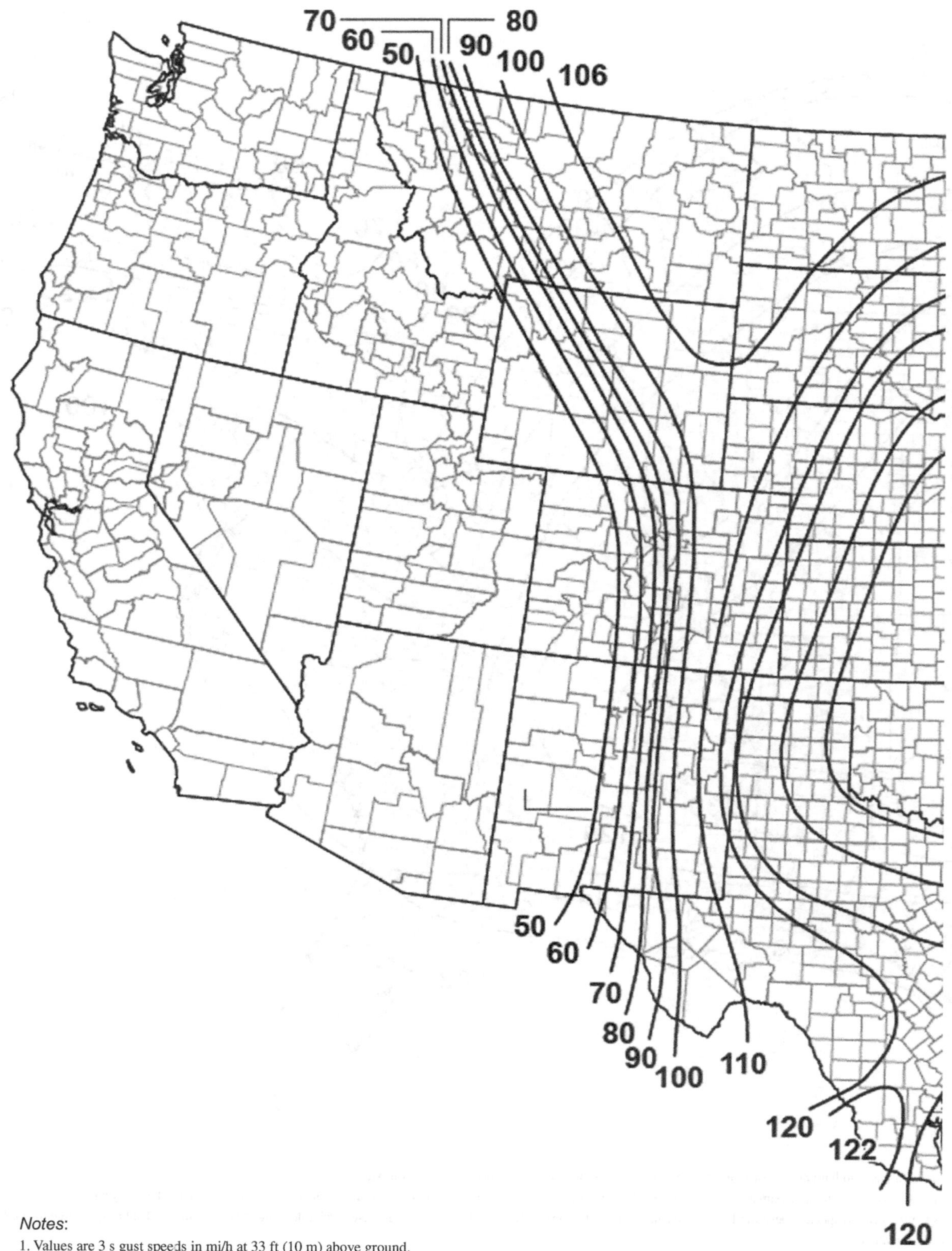

Notes:

1. Values are 3 s gust speeds in mi/h at 33 ft (10 m) above ground.
2. To convert tornado speeds from mi/h to m/s, multiply mapped values by 0.447.
3. Linear interpolation is permitted between contours. Point values (where shown) are provided to aid with interpolation.

Figure G.2-1F. Tornado speeds for 10,000-year MRI for effective plan area of 250,000 ft^2 (23,226 m^2).

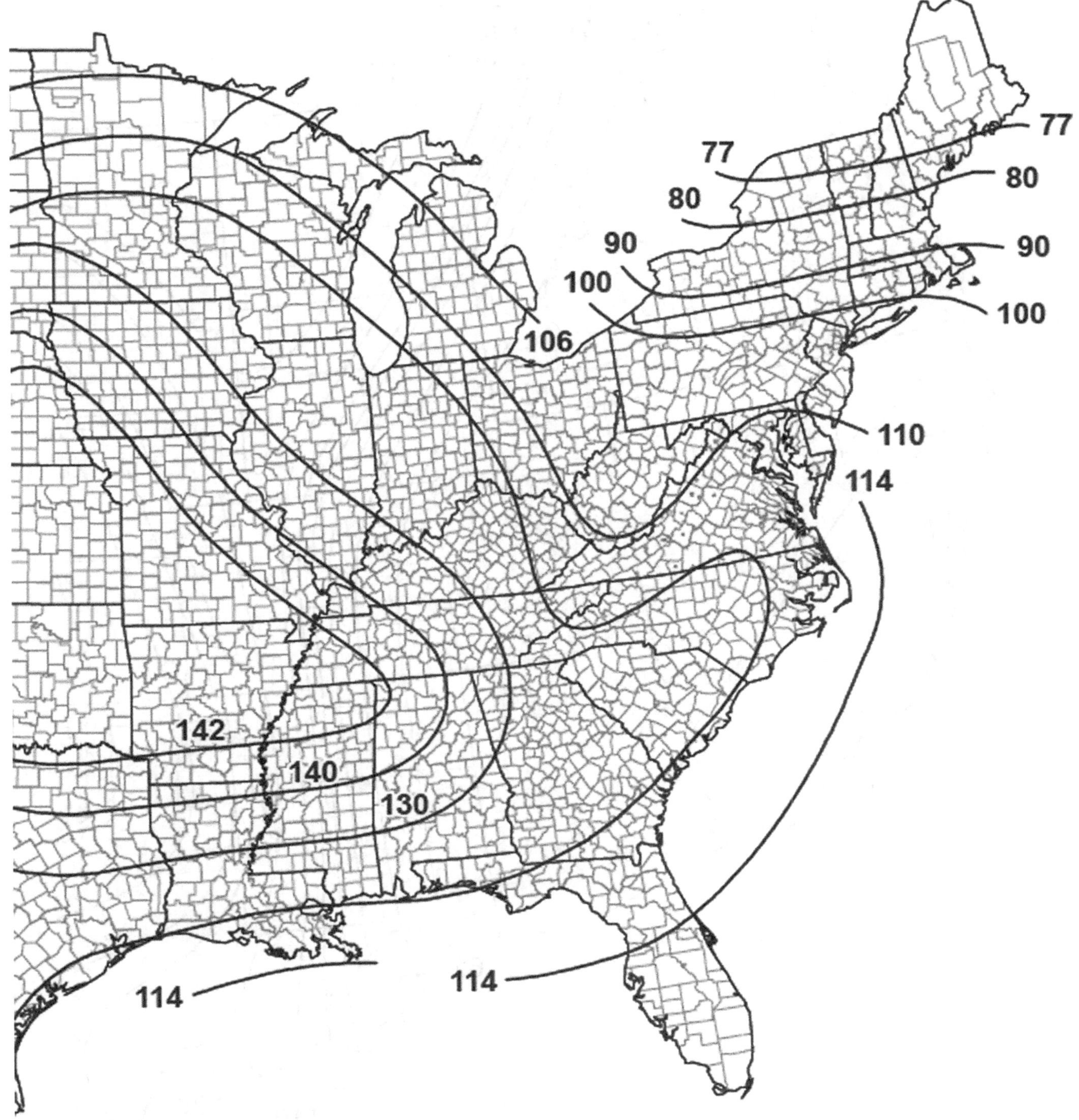

4. Islands, coastal areas, and land boundaries outside the last contour shall use the last tornado speed contour.
5. Tornado speeds correspond to approximately a 0.5% probability of exceedance in 50 years (annual exceedance probability = 0.0001, MRI = 10,000 years).
6. Location-specific tornado speed is permitted to be determined using the ASCE Tornado Design Geodatabase, available at the ASCE 7 Hazard Tool (https://asce7hazardtool.online), or approved equivalent.

Figure G.2-1F (*Continued*). Tornado speeds for 10,000-year MRI for effective plan area of 250,000 ft^2 (23,226 m^2).

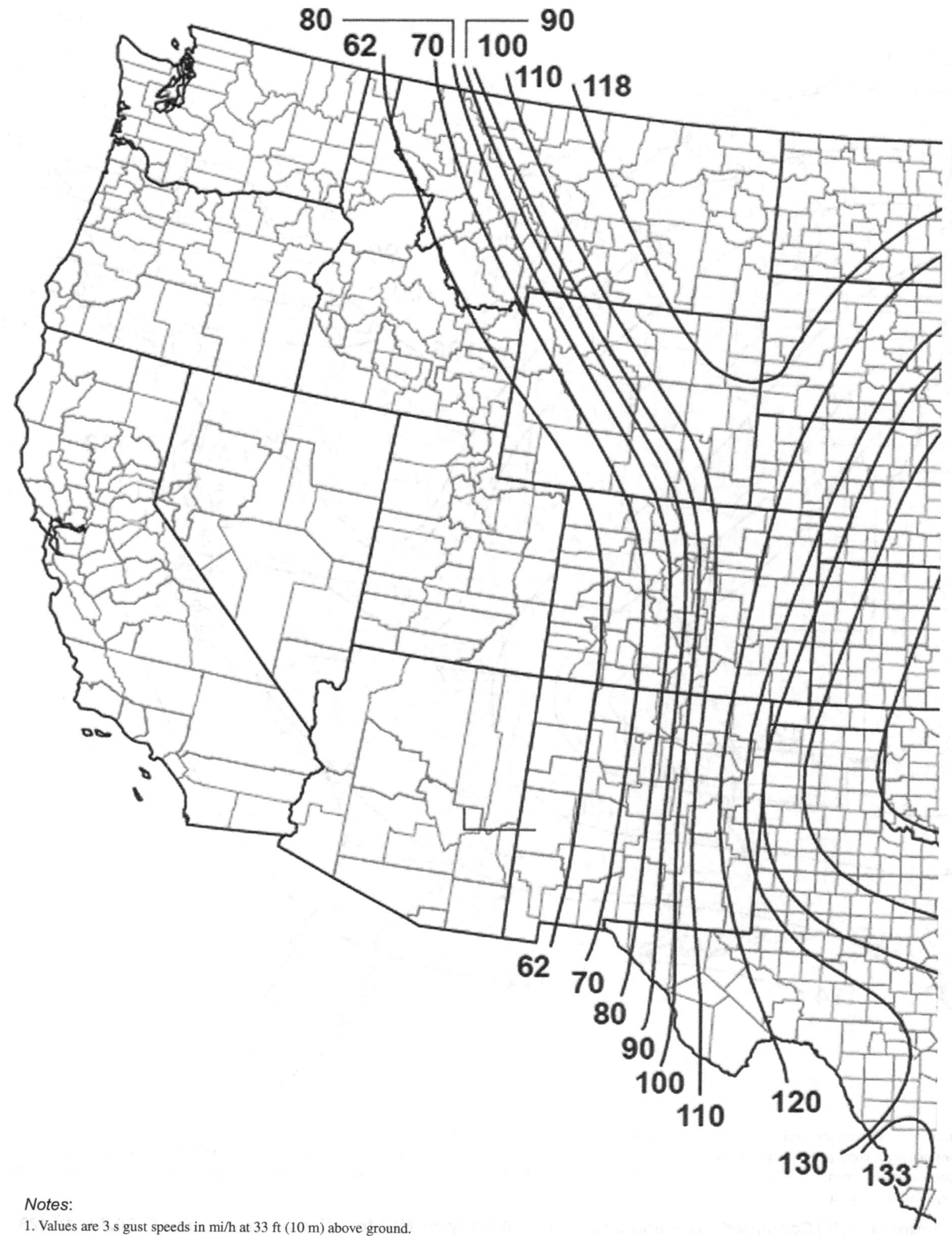

Notes:
1. Values are 3 s gust speeds in mi/h at 33 ft (10 m) above ground.
2. To convert tornado speeds from mi/h to m/s, multiply mapped values by 0.447.
3. Linear interpolation is permitted between contours. Point values (where shown) are provided to aid with interpolation.

Figure G.2-1G. Tornado speeds for 10,000-year MRI for effective plan area of 1,000,000 ft^2 (92,903 m^2).

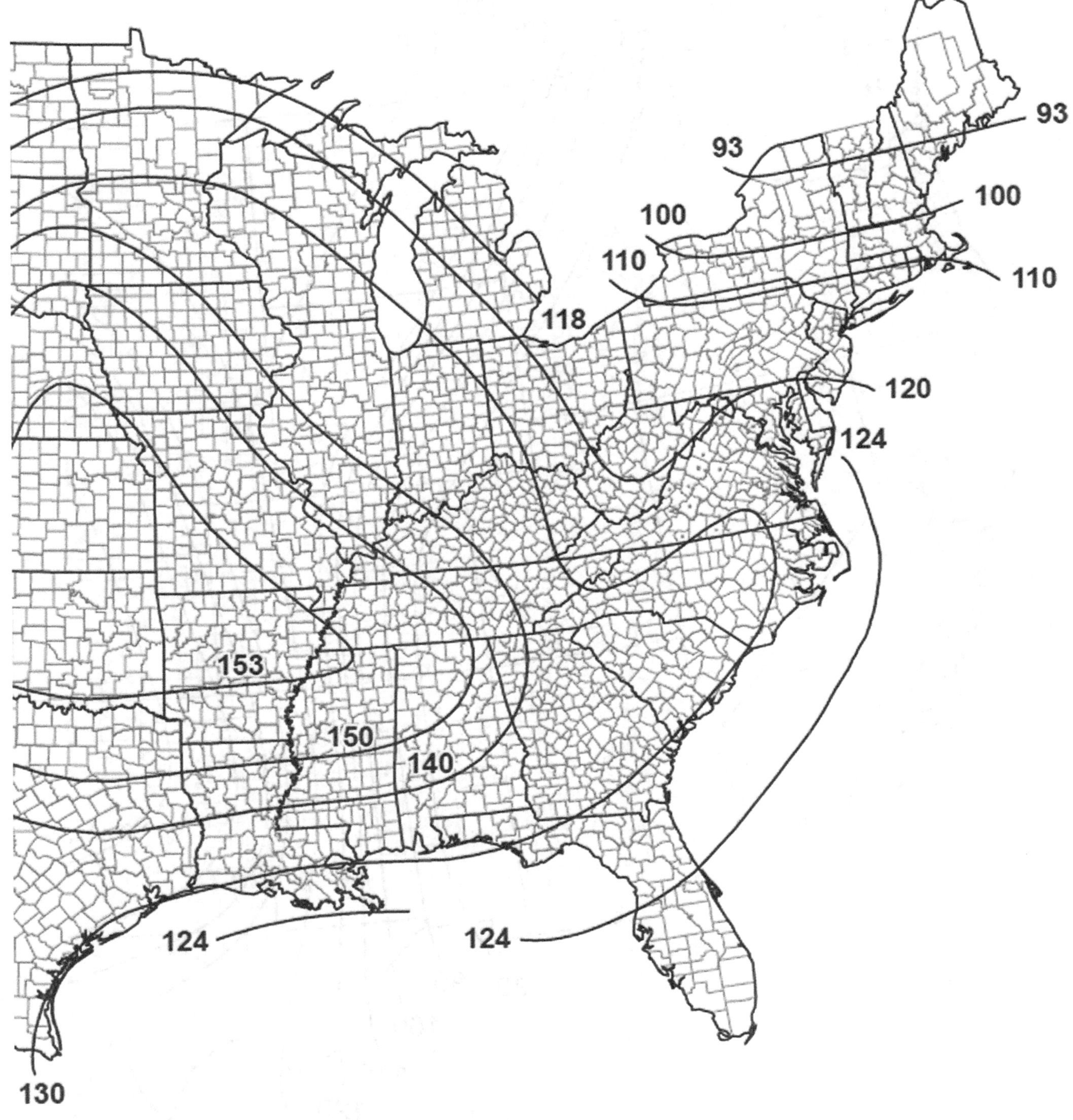

4. Islands, coastal areas, and land boundaries outside the last contour shall use the last tornado speed contour.

5. Tornado speeds correspond to approximately a 0.5% probability of exceedance in 50 years (annual exceedance probability = 0.0001, MRI = 10,000 years).

6. Location-specific tornado speed is permitted to be determined using the ASCE Tornado Design Geodatabase, available at the ASCE 7 Hazard Tool (https://asce7hazardtool.online), or approved equivalent.

Figure G.2-1G (*Continued*). Tornado speeds for 10,000-year MRI for effective plan area of 1,000,000 ft^2 (92,903 m^2).

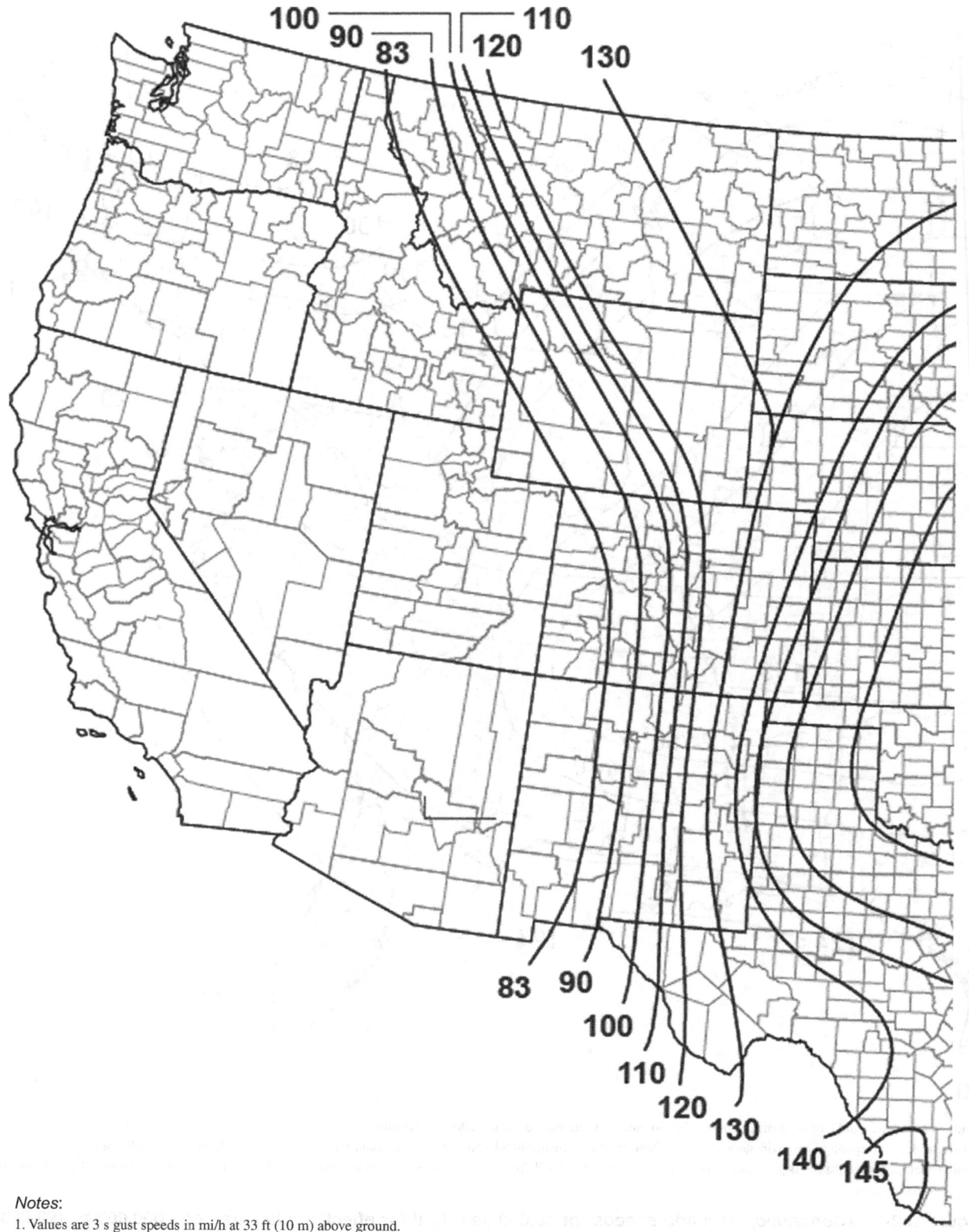

Notes:

1. Values are 3 s gust speeds in mi/h at 33 ft (10 m) above ground.
2. To convert tornado speeds from mi/h to m/s, multiply mapped values by 0.447.
3. Linear interpolation is permitted between contours. Point values (where shown) are provided to aid with interpolation.

Figure G.2-1H. Tornado speeds for 10,000-year MRI for effective plan area of 4,000,000 ft^2 (371,612 m^2).

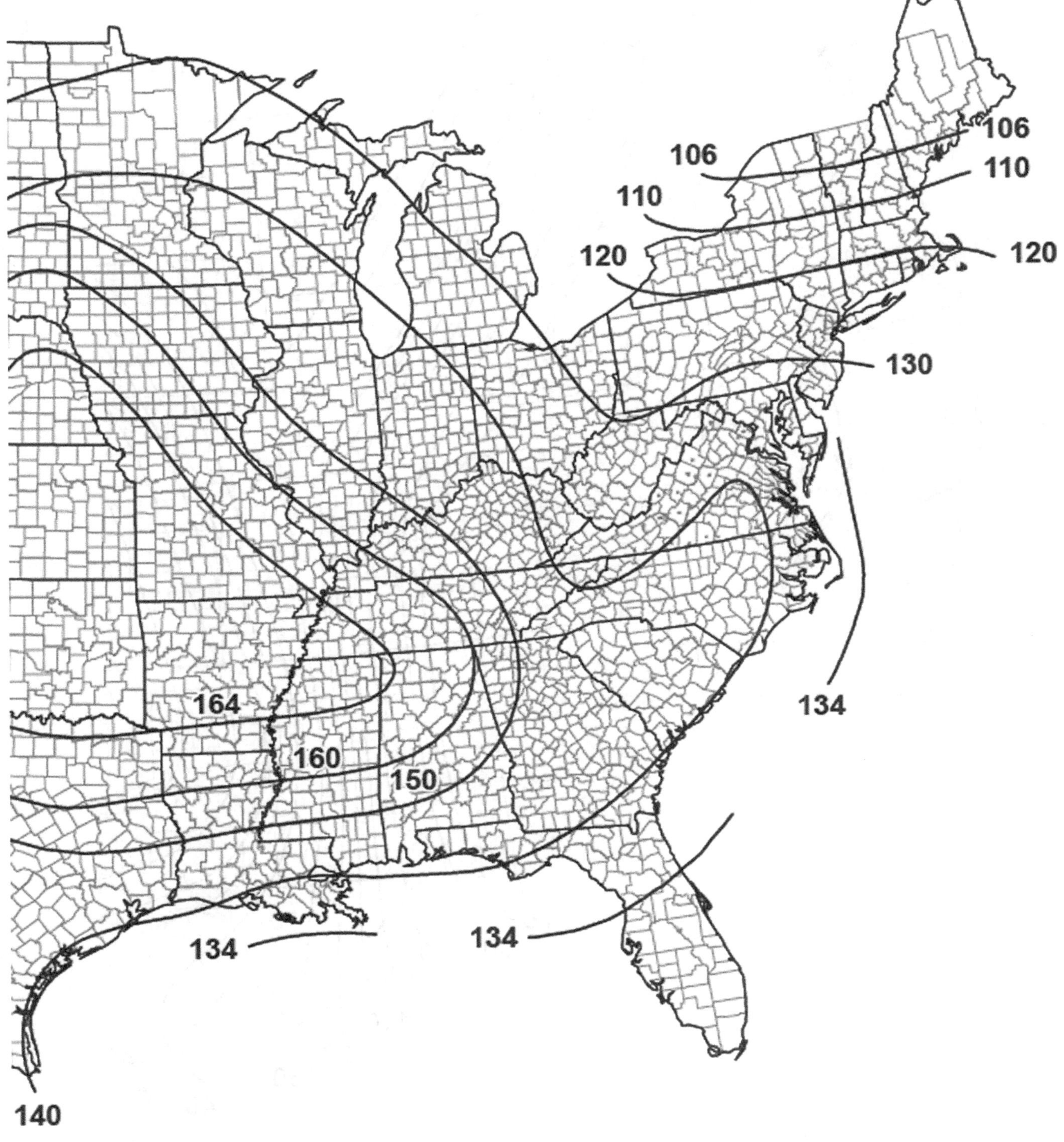

4. Islands, coastal areas, and land boundaries outside the last contour shall use the last tornado speed contour.
5. Tornado speeds correspond to approximately a 0.5% probability of exceedance in 50 years (annual exceedance probability = 0.0001, MRI = 10,000 years).
6. Location-specific tornado speed is permitted to be determined using the ASCE Tornado Design Geodatabase, available at the ASCE 7 Hazard Tool (https://asce7hazardtool.online), or approved equivalent.

Figure G.2-1H (*Continued*). Tornado speeds for 10,000-year MRI for effective plan area of 4,000,000 ft^2 (371,612 m^2).

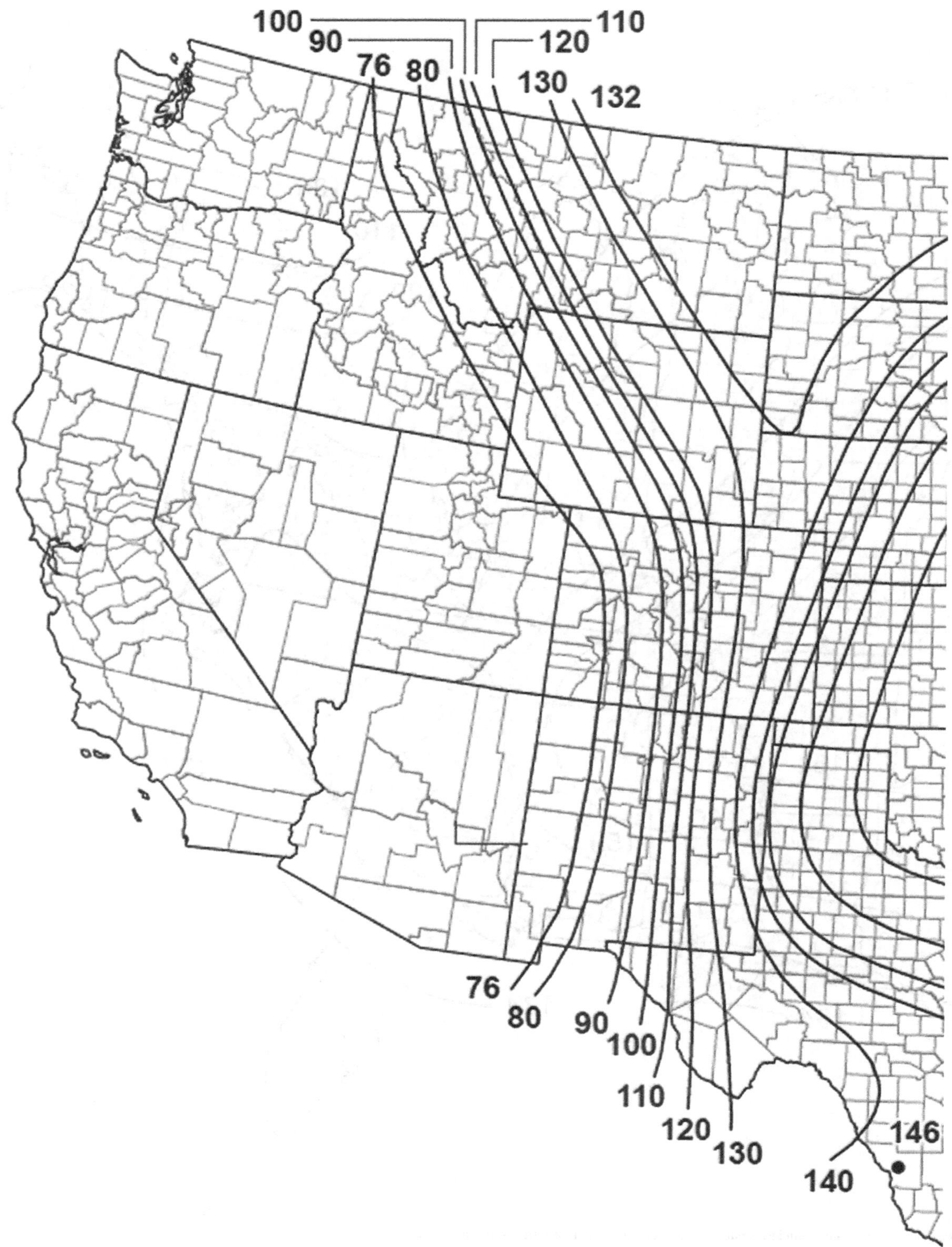

Notes:

1. Values are 3 s gust speeds in mi/h at 33 ft (10 m) above ground.
2. To convert tornado speeds from mi/h to m/s, multiply mapped values by 0.447.
3. Linear interpolation is permitted between contours. Point values (where shown) are provided to aid with interpolation.

Figure G.2-2A. Tornado speeds for 100,000-year MRI for effective plan area of 1 ft^2 (0.1 m^2).

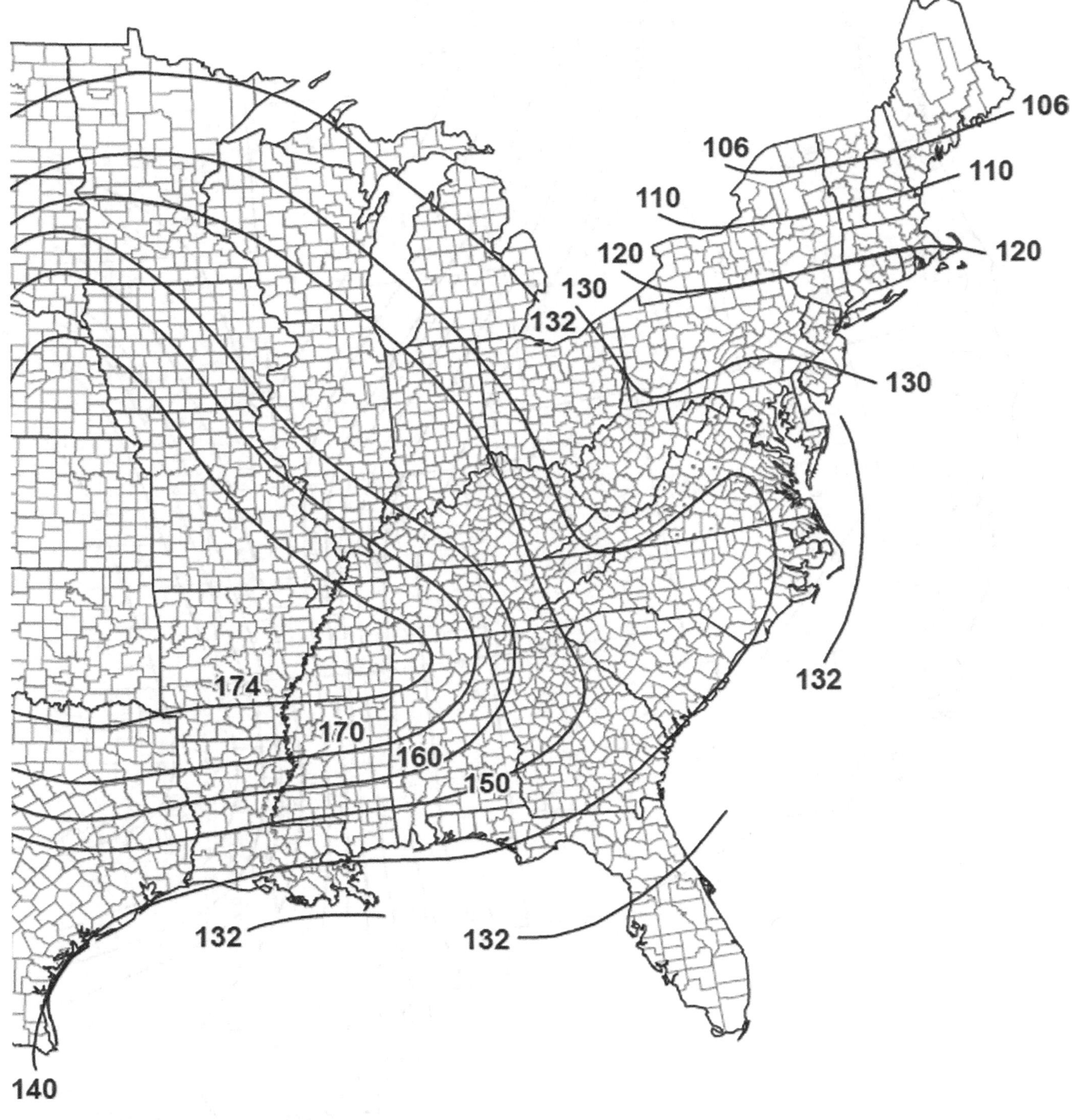

4. Islands, coastal areas, and land boundaries outside the last contour shall use the last tornado speed contour.
5. Tornado speeds correspond to approximately a 0.05% probability of exceedance in 50 years (annual exceedance probability = 0.00001, MRI = 100,000 years).
6. Location-specific tornado speed is permitted to be determined using the ASCE Tornado Design Geodatabase, available at the ASCE 7 Hazard Tool (https://asce7hazardtool.online), or approved equivalent.

Figure G.2-2A (*Continued*). Tornado speeds for 100,000-year MRI for effective plan area of 1 ft^2 (0.1 m^2).

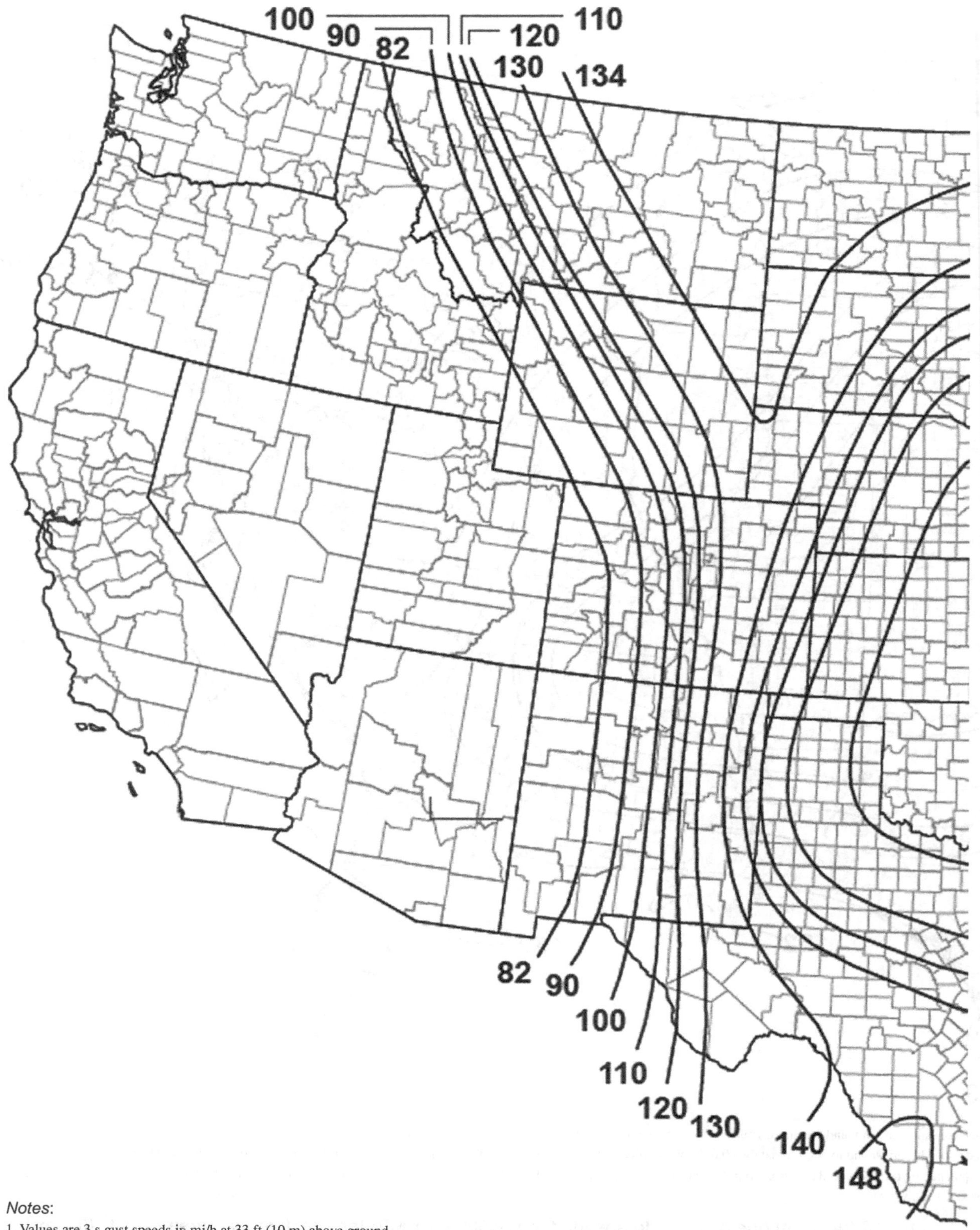

Notes:

1. Values are 3 s gust speeds in mi/h at 33 ft (10 m) above ground.
2. To convert tornado speeds from mi/h to m/s, multiply mapped values by 0.447.
3. Linear interpolation is permitted between contours. Point values (where shown) are provided to aid with interpolation.

Figure G.2-2B. Tornado speeds for 100,000-year MRI for effective plan area of 2,000 ft^2 (186 m^2).

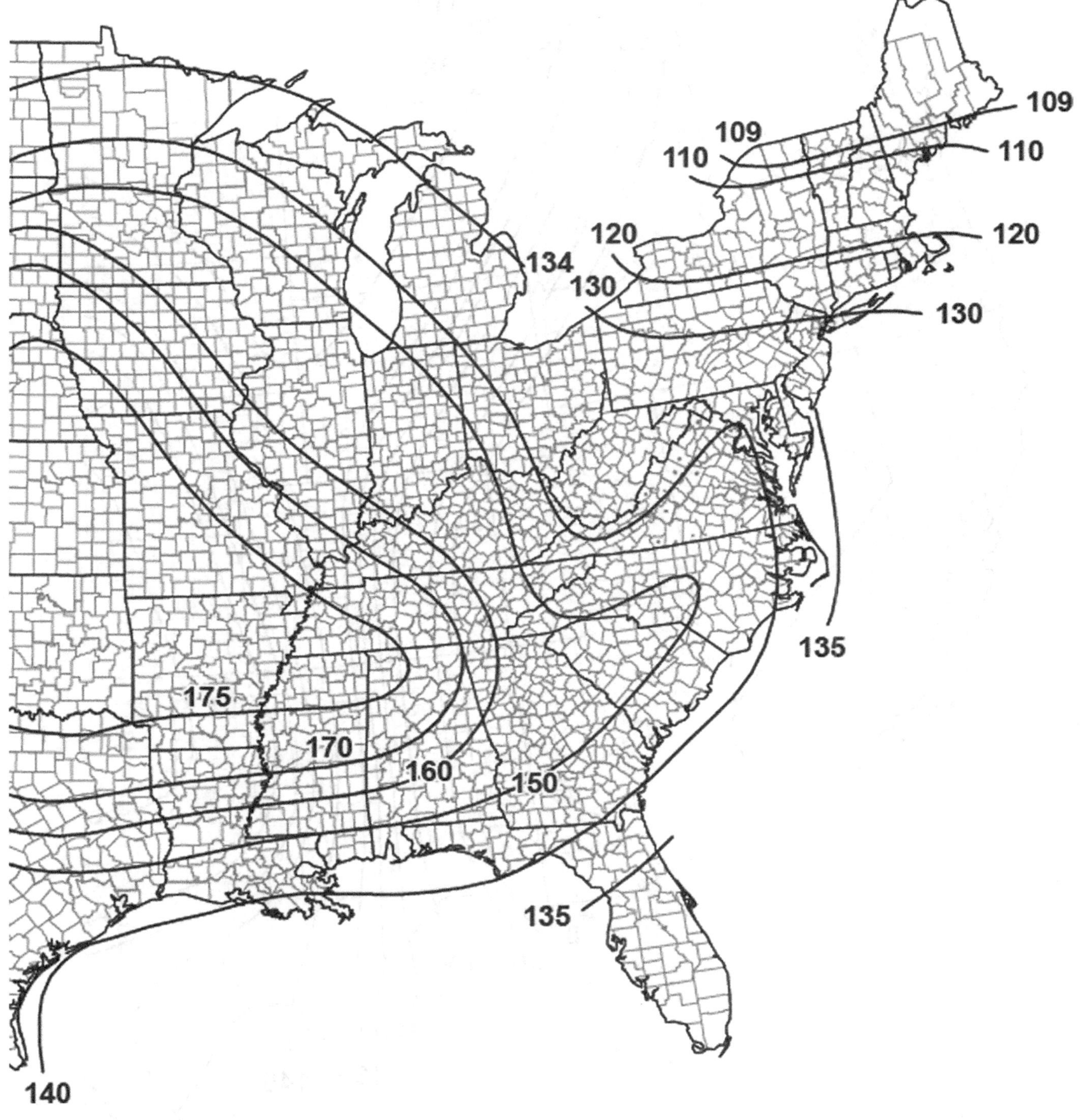

4. Islands, coastal areas, and land boundaries outside the last contour shall use the last tornado speed contour.
5. Tornado speeds correspond to approximately a 0.05% probability of exceedance in 50 years (annual exceedance probability = 0.00001, MRI = 100,000 years).
6. Location-specific tornado speed is permitted to be determined using the ASCE Tornado Design Geodatabase, available at the ASCE 7 Hazard Tool (https://asce7hazardtool.online), or approved equivalent.

Figure G.2-2B (*Continued*). Tornado speeds for 100,000-year MRI for effective plan area of 2,000 ft^2 (186 m^2).

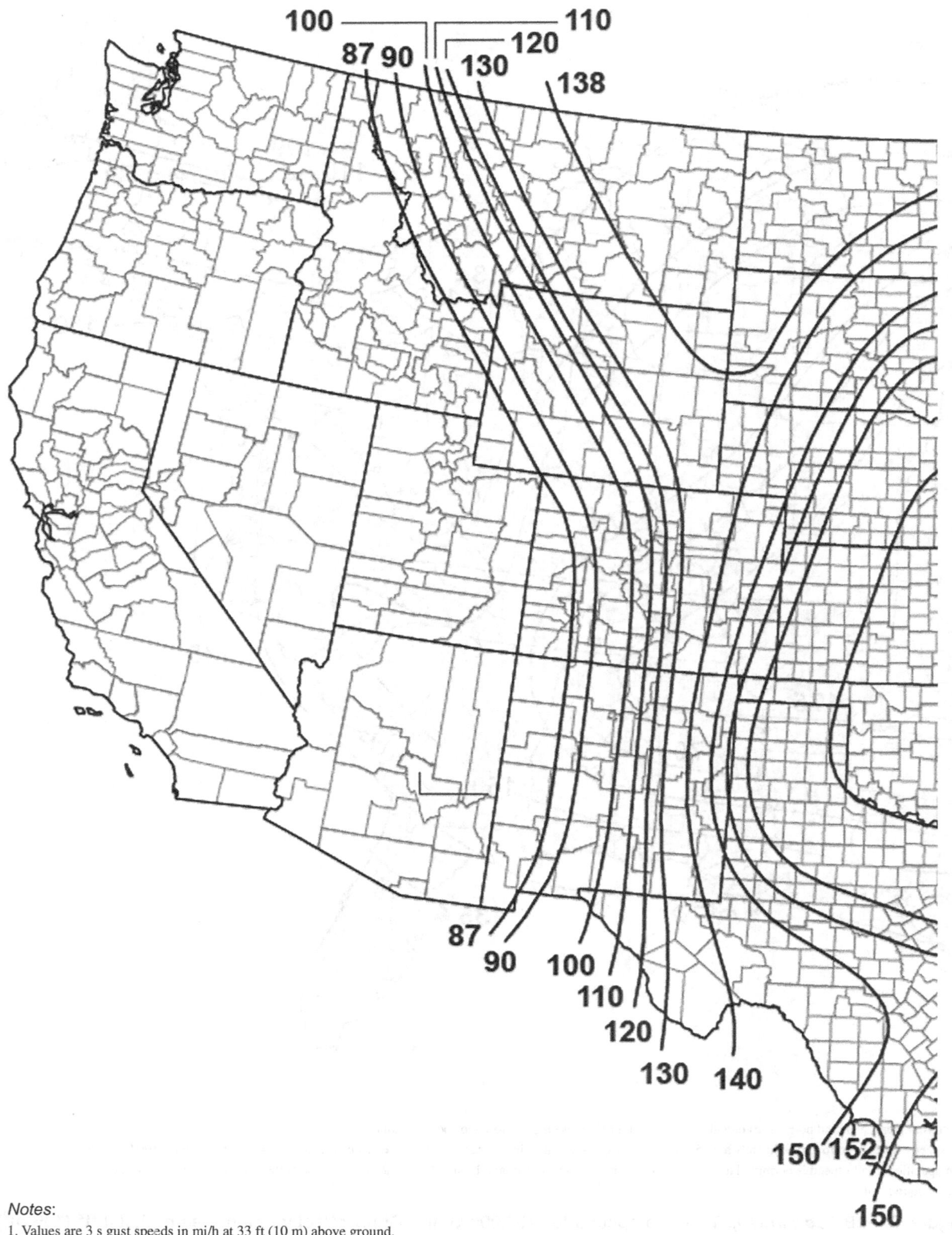

Notes:

1. Values are 3 s gust speeds in mi/h at 33 ft (10 m) above ground.
2. To convert tornado speeds from mi/h to m/s, multiply mapped values by 0.447.
3. Linear interpolation is permitted between contours. Point values (where shown) are provided to aid with interpolation.

Figure G.2-12C. Tornado speeds for 100,000-year MRI for effective plan area of 10,000 ft^2 (929 m^2).

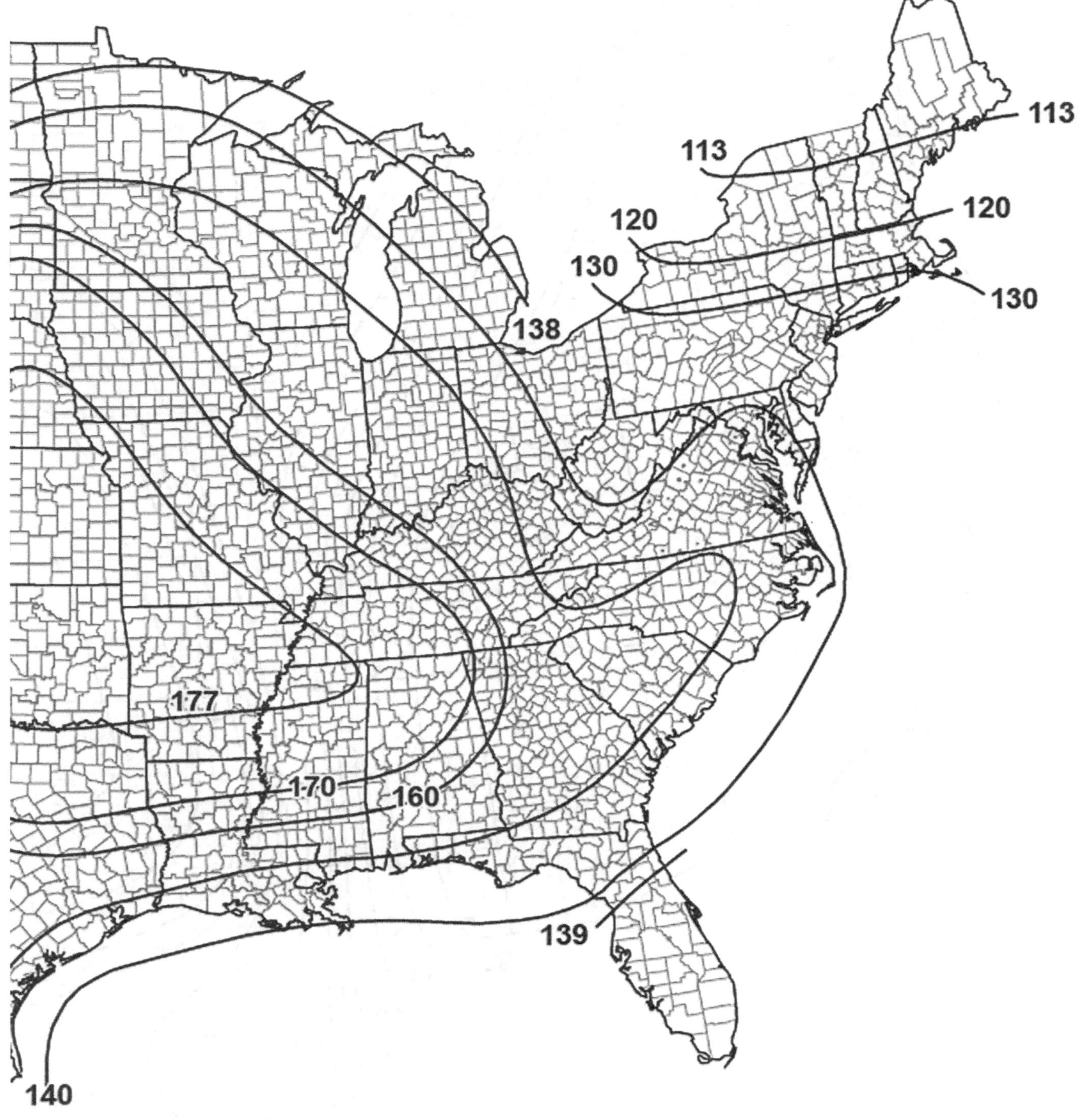

4. Islands, coastal areas, and land boundaries outside the last contour shall use the last tornado speed contour.
5. Tornado speeds correspond to approximately a 0.05% probability of exceedance in 50 years (annual exceedance probability = 0.00001, MRI = 100,000 years).
6. Location-specific tornado speed is permitted to be determined using the ASCE Tornado Design Geodatabase, available at the ASCE 7 Hazard Tool (https://asce7hazardtool.online), or approved equivalent.

Figure G.2-12C (*Continued*). Tornado speeds for 100,000-year MRI for effective plan area of 10,000 ft^2 (929 m^2).

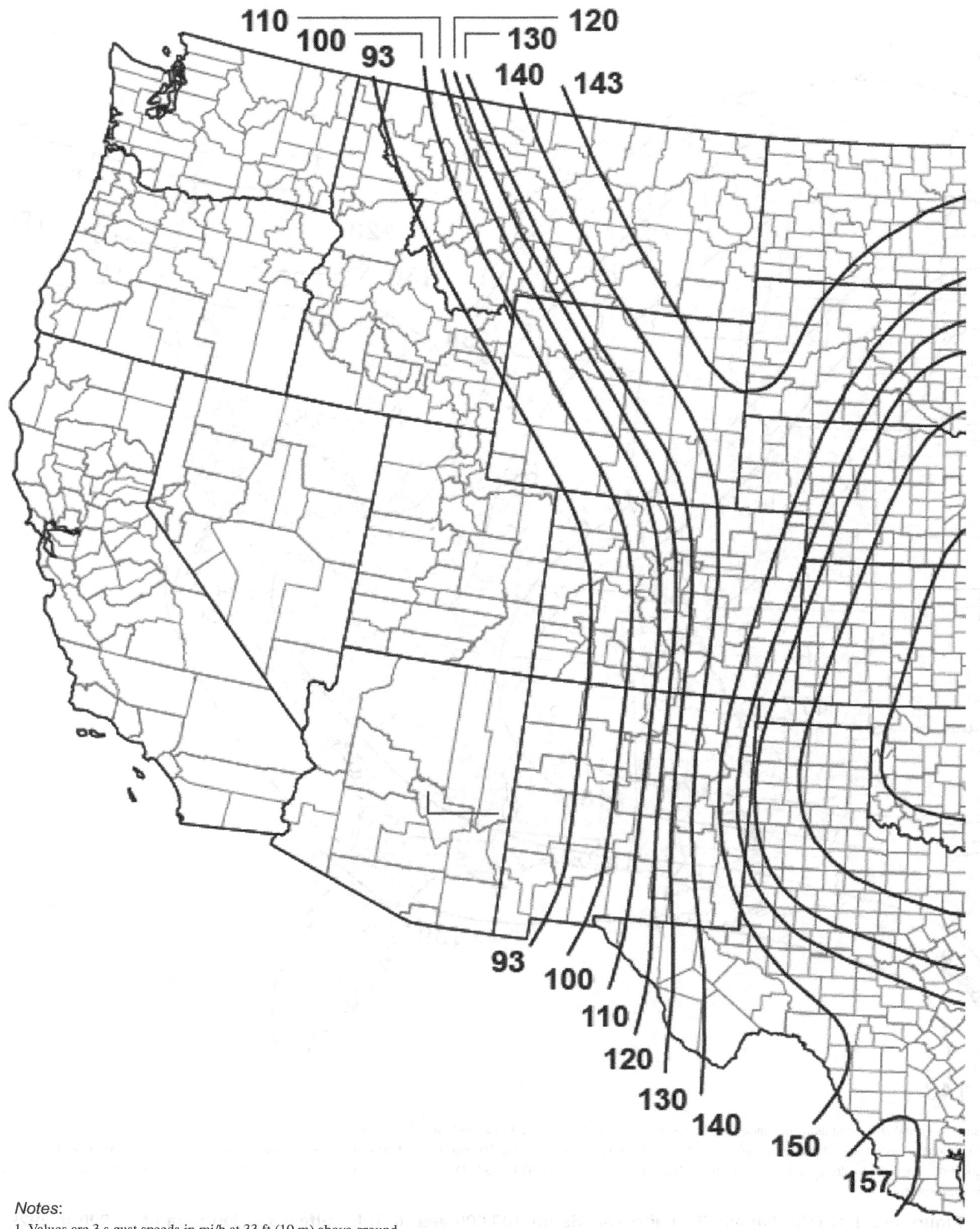

Notes:

1. Values are 3 s gust speeds in mi/h at 33 ft (10 m) above ground.
2. To convert tornado speeds from mi/h to m/s, multiply mapped values by 0.447.
3. Linear interpolation is permitted between contours. Point values (where shown) are provided to aid with interpolation.

Figure G.2-2D. Tornado speeds for 100,000-year MRI for effective plan area of 40,000 ft^2 (3,716 m^2).

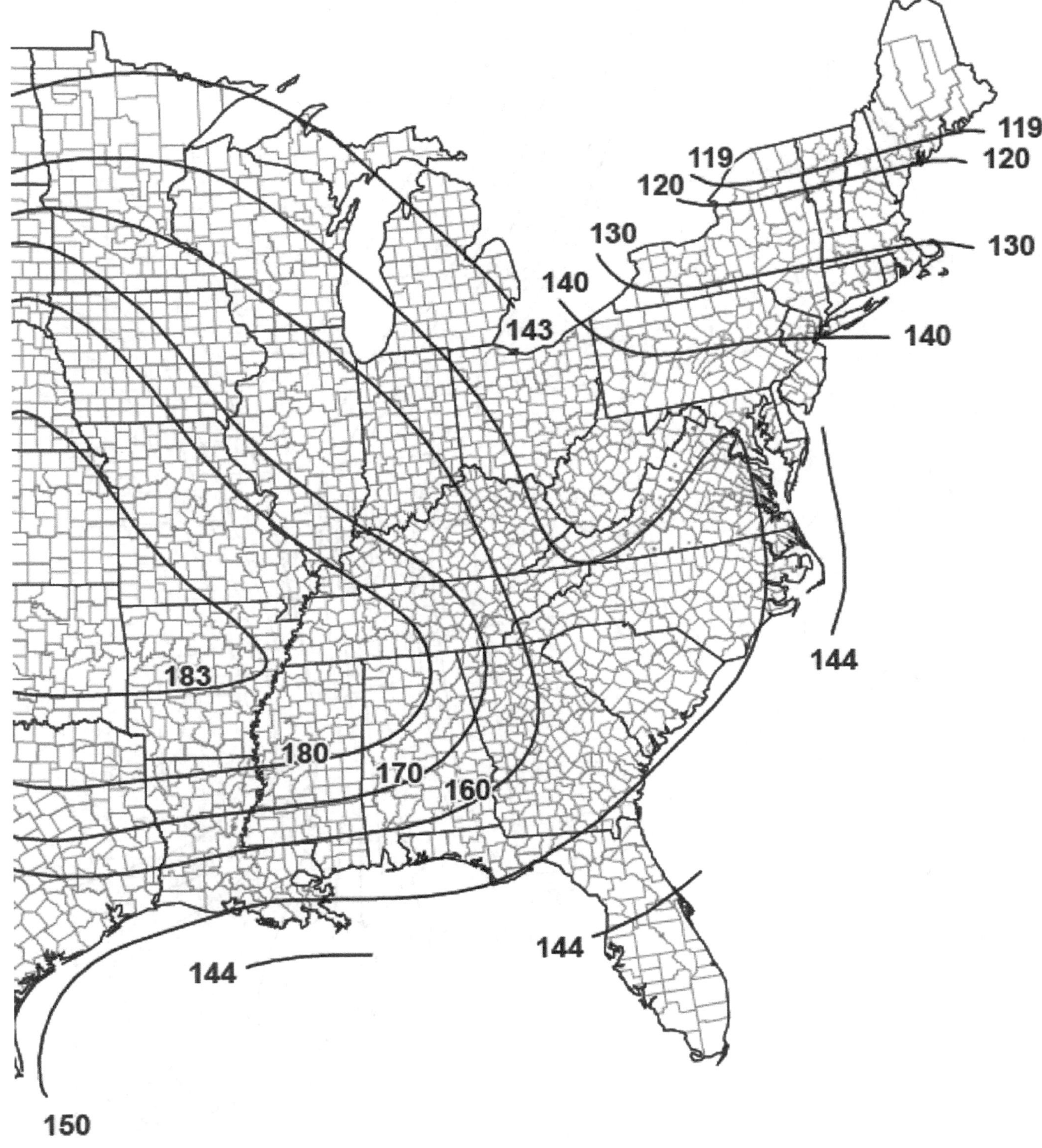

4. Islands, coastal areas, and land boundaries outside the last contour shall use the last tornado speed contour.
5. Tornado speeds correspond to approximately a 0.05% probability of exceedance in 50 years (annual exceedance probability = 0.00001, MRI = 100,000 years).
6. Location-specific tornado speed is permitted to be determined using the ASCE Tornado Design Geodatabase, available at the ASCE 7 Hazard Tool (https://asce7hazardtool.online), or approved equivalent.

Figure G.2-2D (*Continued*). Tornado speeds for 100,000-year MRI for effective plan area of 40,000 ft^2 (3,716 m^2).

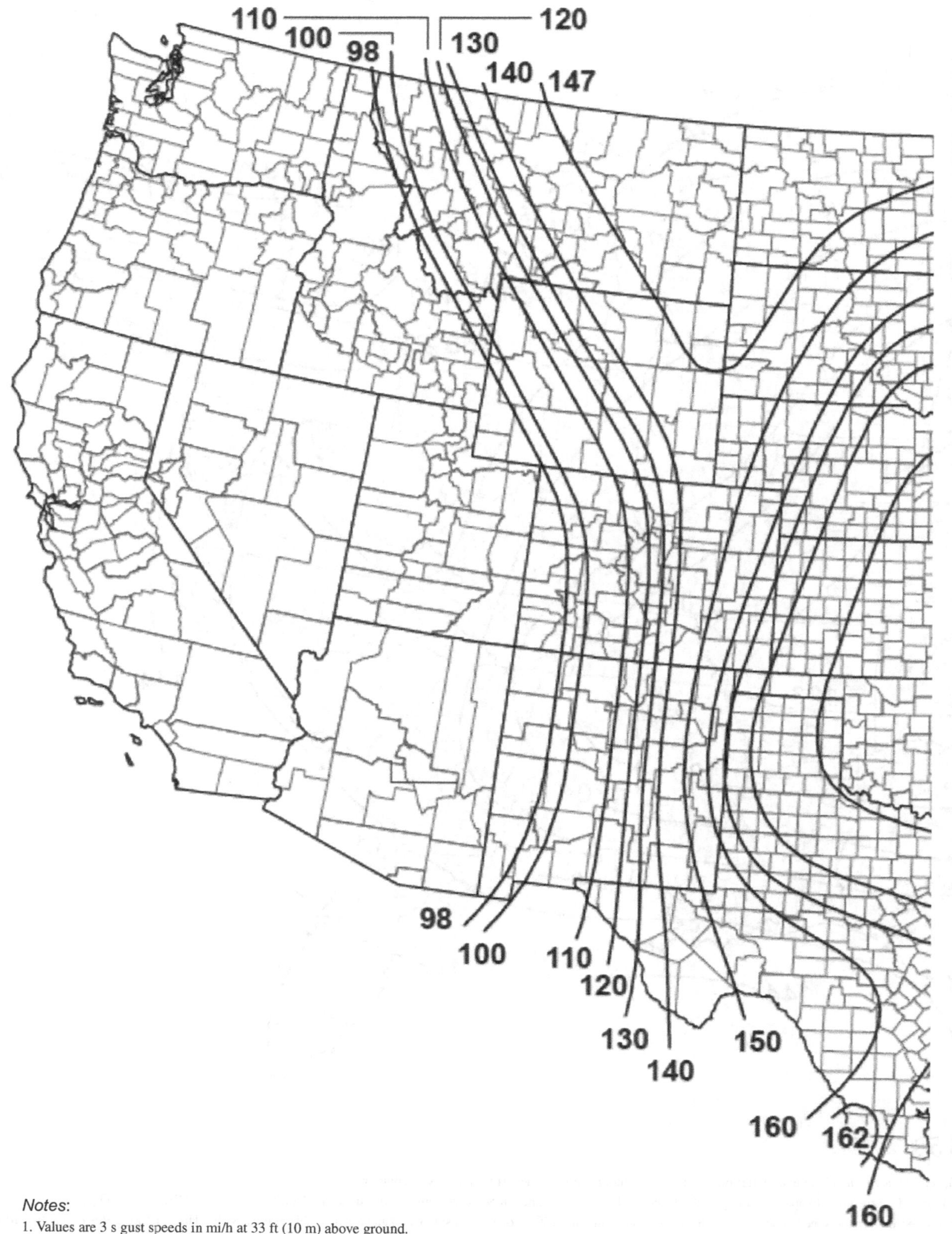

Notes:

1. Values are 3 s gust speeds in mi/h at 33 ft (10 m) above ground.
2. To convert tornado speeds from mi/h to m/s, multiply mapped values by 0.447.
3. Linear interpolation is permitted between contours. Point values (where shown) are provided to aid with interpolation.

Figure G.2-2E. Tornado speeds for 100,000-year MRI for effective plan area of 100,000 ft^2 (9,290 m^2).

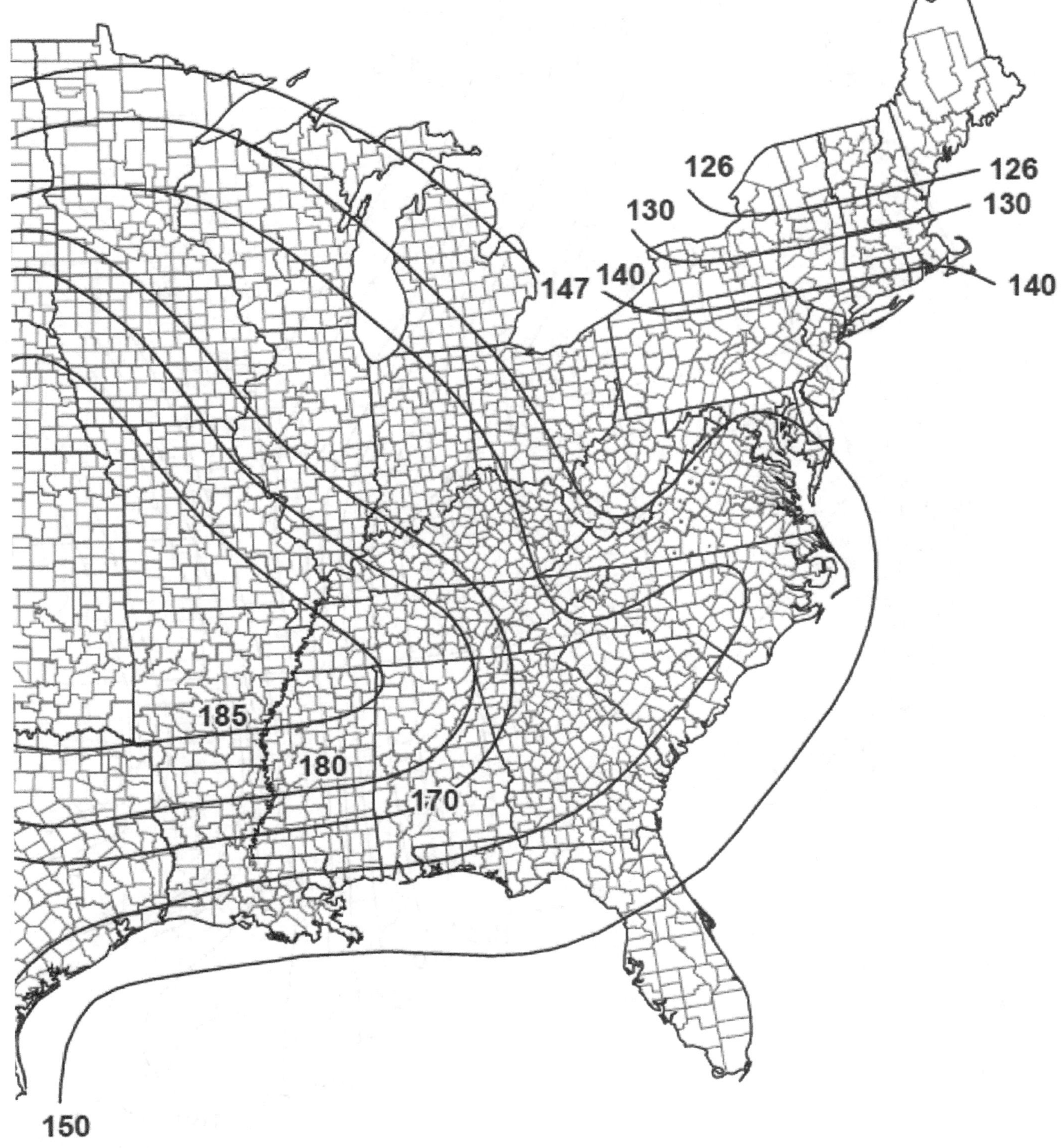

4. Islands, coastal areas, and land boundaries outside the last contour shall use the last tornado speed contour.
5. Tornado speeds correspond to approximately a 0.05% probability of exceedance in 50 years (annual exceedance probability = 0.00001, MRI = 100,000 years).
6. Location-specific tornado speed is permitted to be determined using the ASCE Tornado Design Geodatabase, available at the ASCE 7 Hazard Tool (https://asce7hazardtool.online), or approved equivalent.

Figure G.2-2E (*Continued*). Tornado speeds for 100,000-year MRI for effective plan area of 100,000 ft² (9,290 m²).

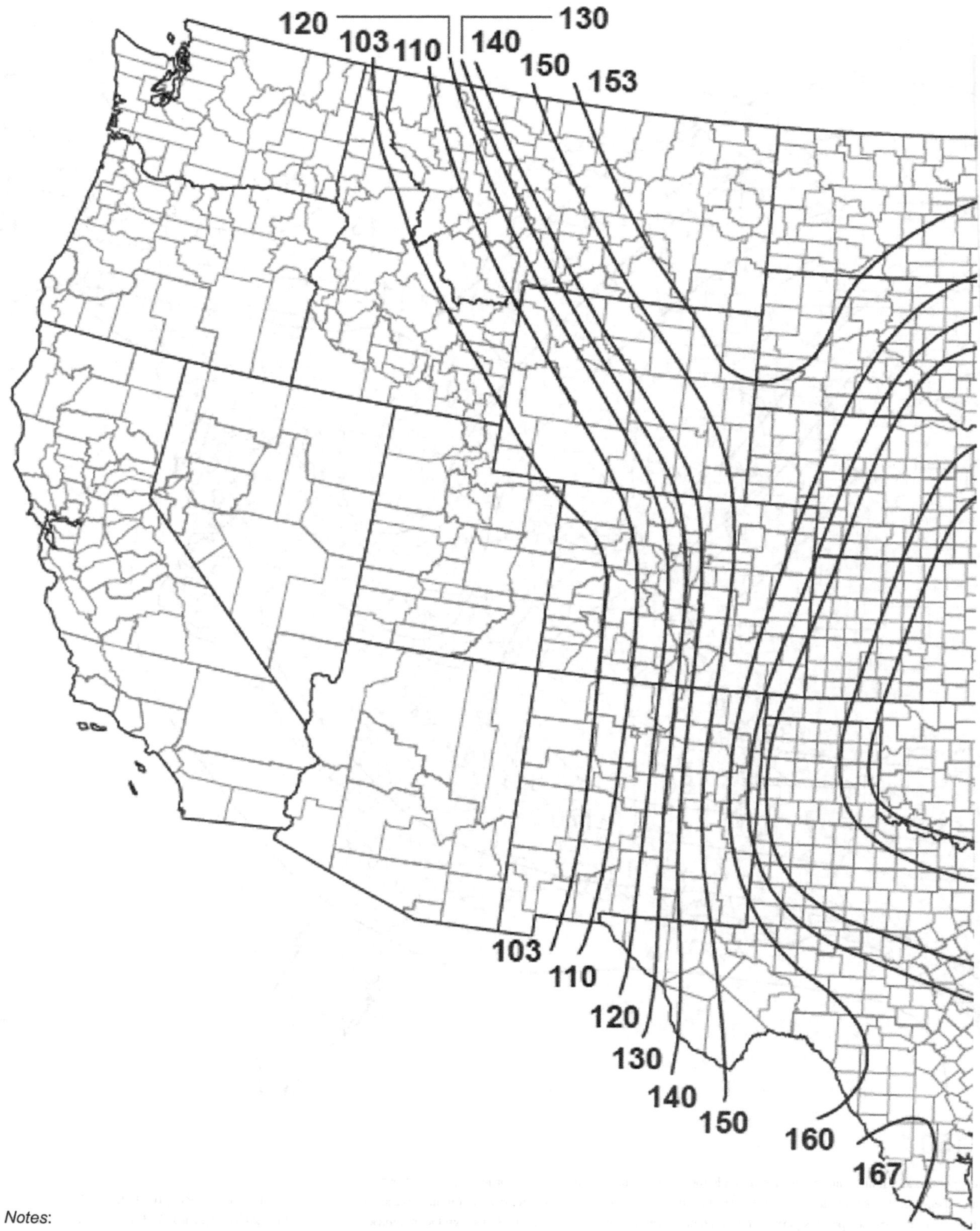

Notes:

1. Values are 3 s gust speeds in mi/h at 33 ft (10 m) above ground.
2. To convert tornado speeds from mi/h to m/s, multiply mapped values by 0.447.
3. Linear interpolation is permitted between contours. Point values (where shown) are provided to aid with interpolation.

Figure G.2-2F. Tornado speeds for 100,000-year MRI for effective plan area of 250,000 ft^2 (23,226 m^2).

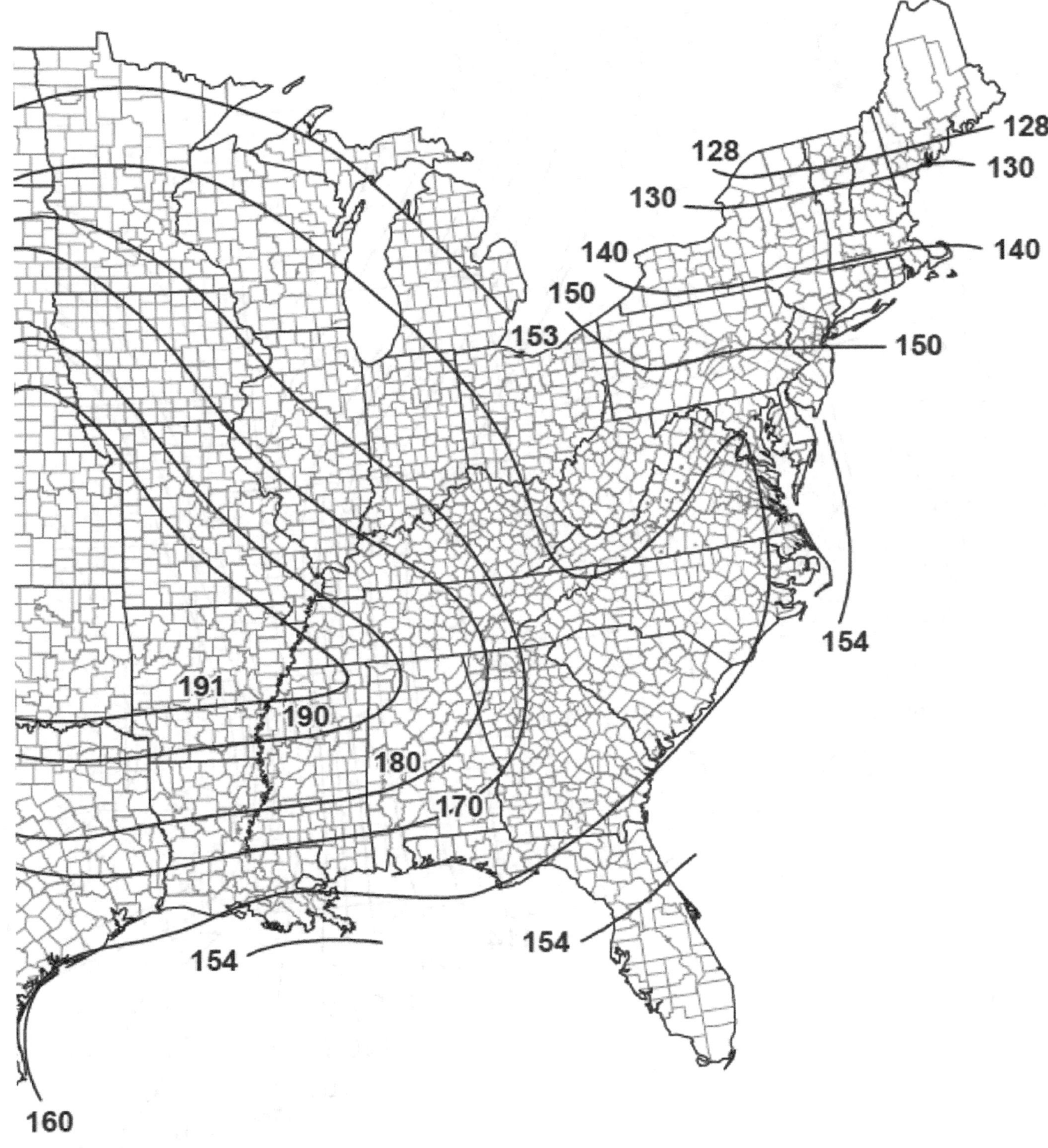

4. Islands, coastal areas, and land boundaries outside the last contour shall use the last tornado speed contour.
5. Tornado speeds correspond to approximately a 0.05% probability of exceedance in 50 years (annual exceedance probability = 0.00001, MRI = 100,000 years).
6. Location-specific tornado speed is permitted to be determined using the ASCE Tornado Design Geodatabase, available at the ASCE 7 Hazard Tool (https://asce7hazardtool.online), or approved equivalent.

Figure G.2-2F (*Continued*). Tornado speeds for 100,000-year MRI for effective plan area of 250,000 ft^2 (23,226 m^2).

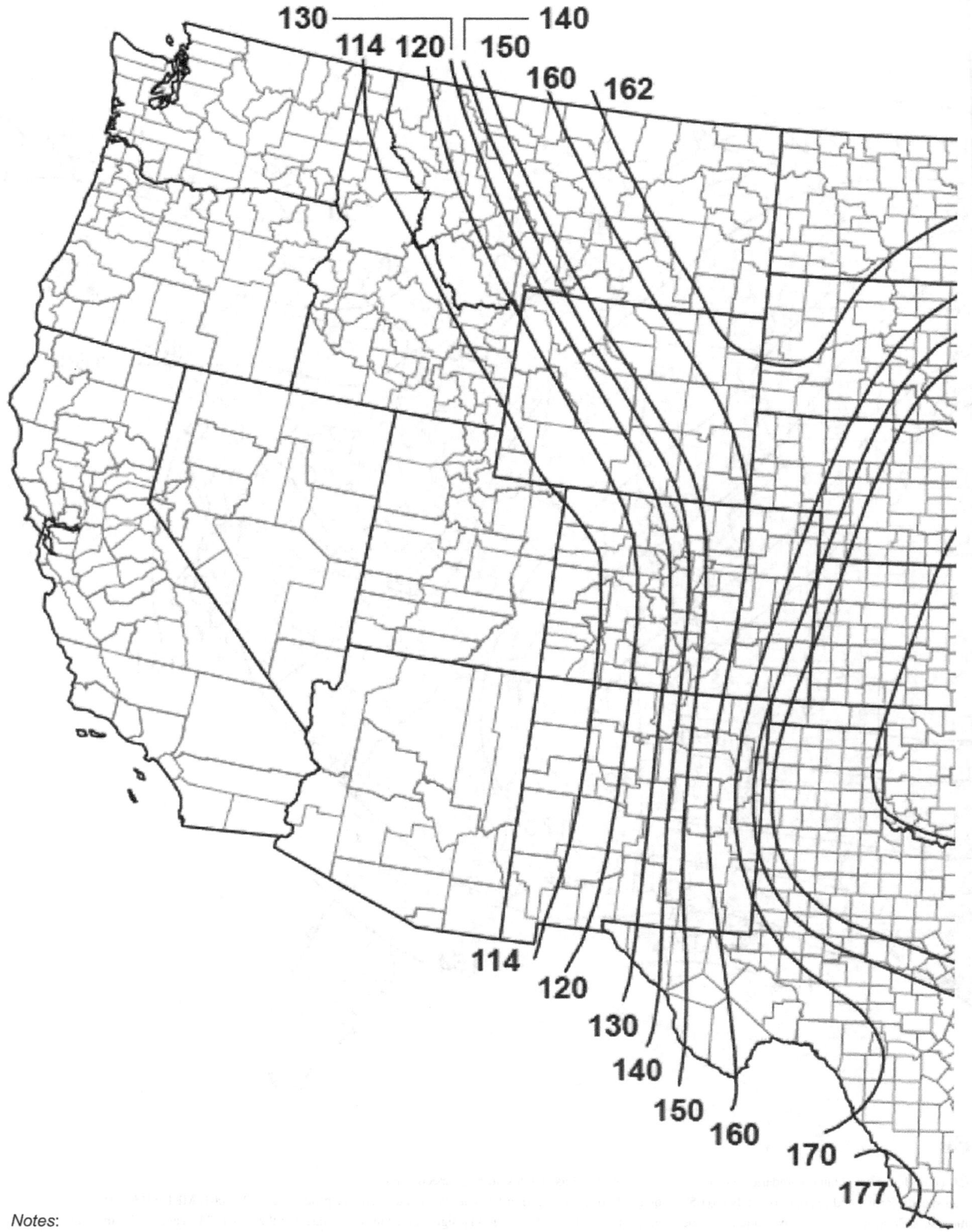

Notes:

1. Values are 3 s gust speeds in mi/h at 33 ft (10 m) above ground.
2. To convert tornado speeds from mi/h to m/s, multiply mapped values by 0.447.
3. Linear interpolation is permitted between contours. Point values (where shown) are provided to aid with interpolation.

Figure G.2-2G. Tornado speeds for 100,000-year MRI for effective plan area of 1,000,000 ft^2 (92,903 m^2).

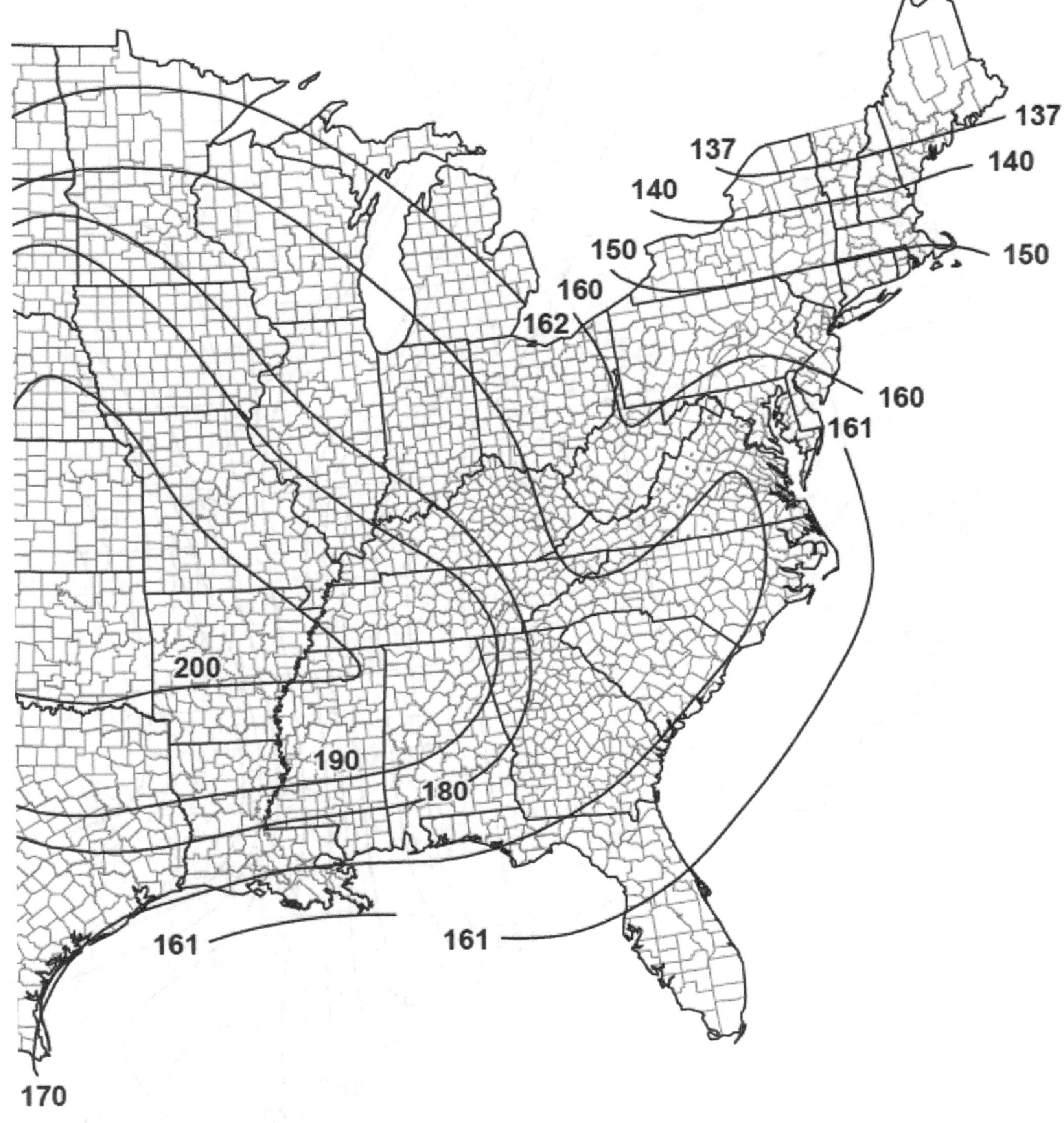

4. Islands, coastal areas, and land boundaries outside the last contour shall use the last tornado speed contour.
5. Tornado speeds correspond to approximately a 0.05% probability of exceedance in 50 years (annual exceedance probability = 0.00001, MRI = 100,000 years).
6. Location-specific tornado speed is permitted to be determined using the ASCE Tornado Design Geodatabase, available at the ASCE 7 Hazard Tool (https://asce7hazardtool.online), or approved equivalent.

Figure G.2-2G (*Continued*). Tornado speeds for 100,000-year MRI for effective plan area of 1,000,000 ft^2 (92,903 m^2).

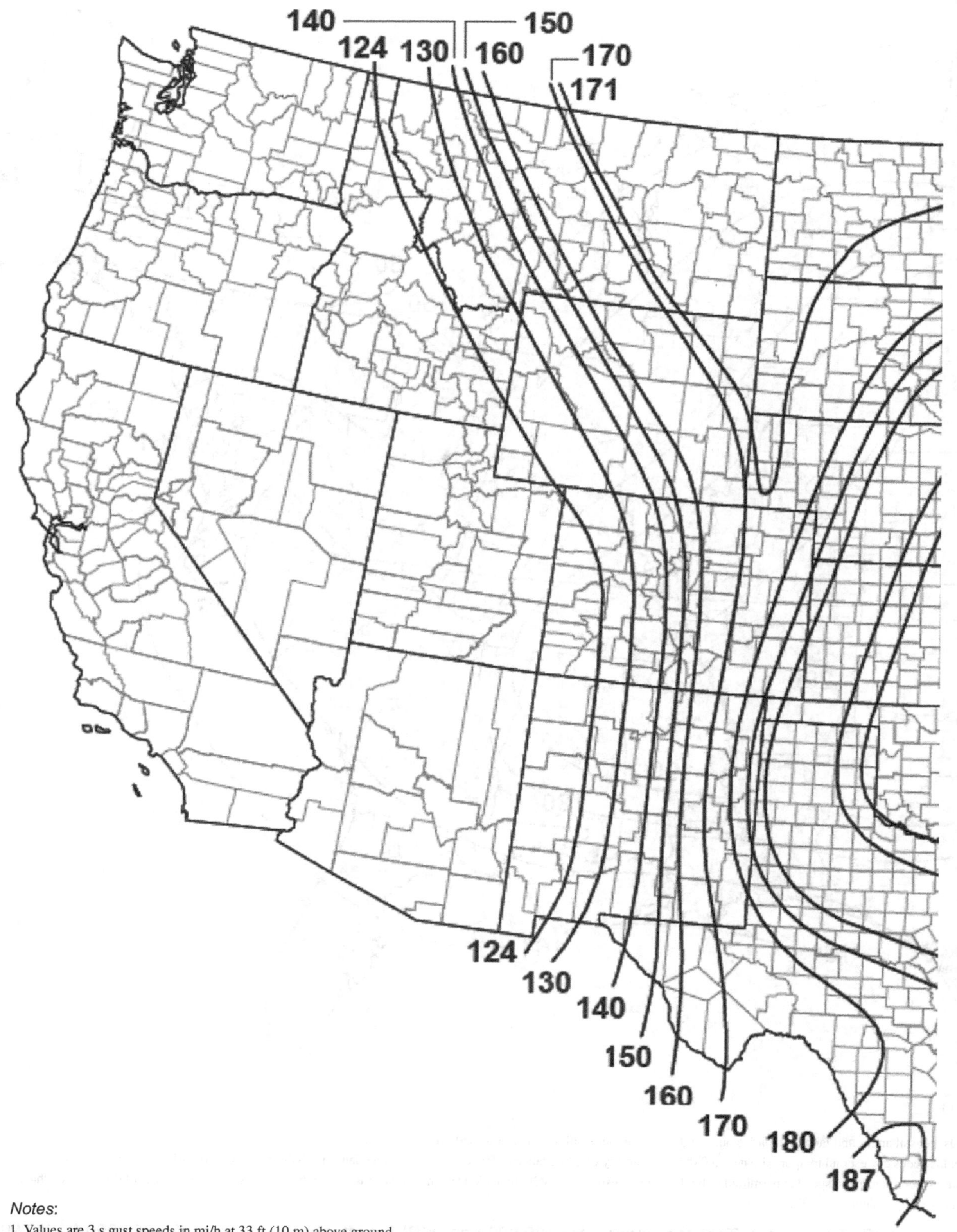

Notes:

1. Values are 3 s gust speeds in mi/h at 33 ft (10 m) above ground.
2. To convert tornado speeds from mi/h to m/s, multiply mapped values by 0.447.
3. Linear interpolation is permitted between contours. Point values (where shown) are provided to aid with interpolation.

Figure G.2-2H. Tornado speeds for 100,000-year MRI for effective plan area of 4,000,000 ft^2 (371,612 m^2).

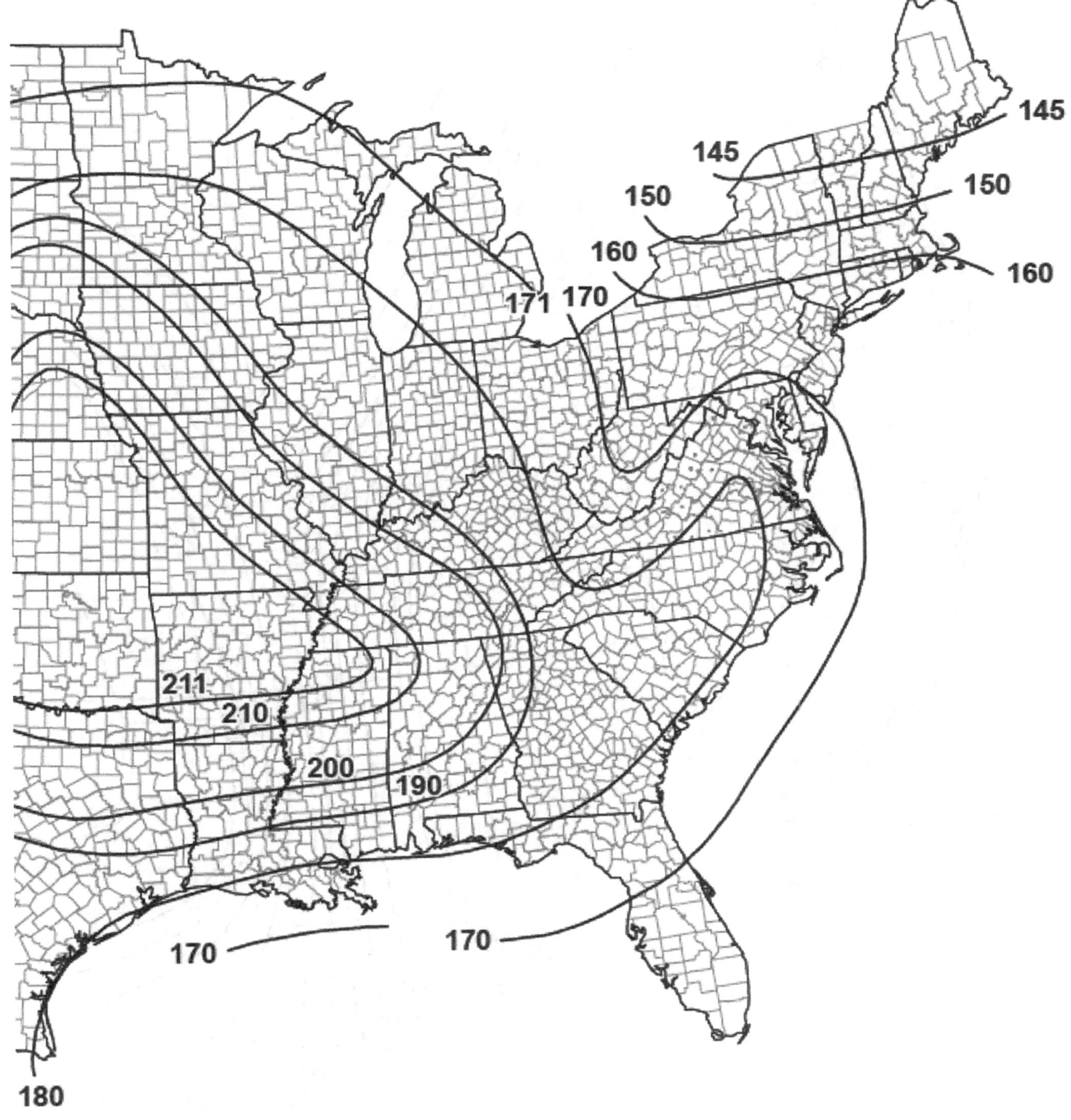

4. Islands, coastal areas, and land boundaries outside the last contour shall use the last tornado speed contour.
5. Tornado speeds correspond to approximately a 0.05% probability of exceedance in 50 years (annual exceedance probability = 0.00001, MRI = 100,000 years).
6. Location-specific tornado speed is permitted to be determined using the ASCE Tornado Design Geodatabase, available at the ASCE 7 Hazard Tool (https://asce7hazardtool.online), or approved equivalent.

Figure G.2-2H (*Continued*). Tornado speeds for 100,000-year MRI for effective plan area of 4,000,000 ft^2 (371,612 m^2).

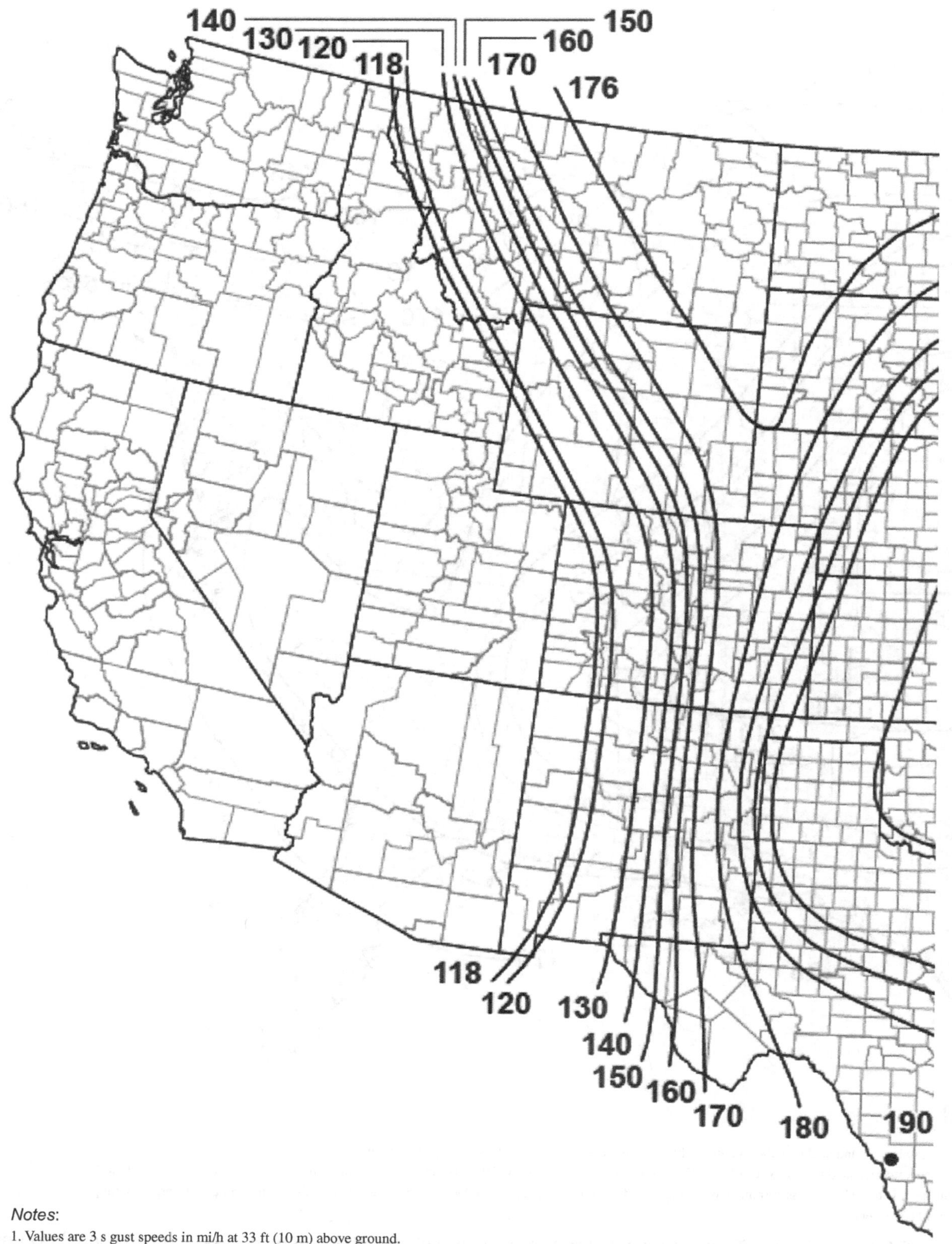

Notes:

1. Values are 3 s gust speeds in mi/h at 33 ft (10 m) above ground.
2. To convert tornado speeds from mi/h to m/s, multiply mapped values by 0.447.
3. Linear interpolation is permitted between contours. Point values (where shown) are provided to aid with interpolation.

Figure G.2-3A. Tornado speeds for 1,000,000-year MRI for effective plan area of 1 ft^2 (0.1 m^2).

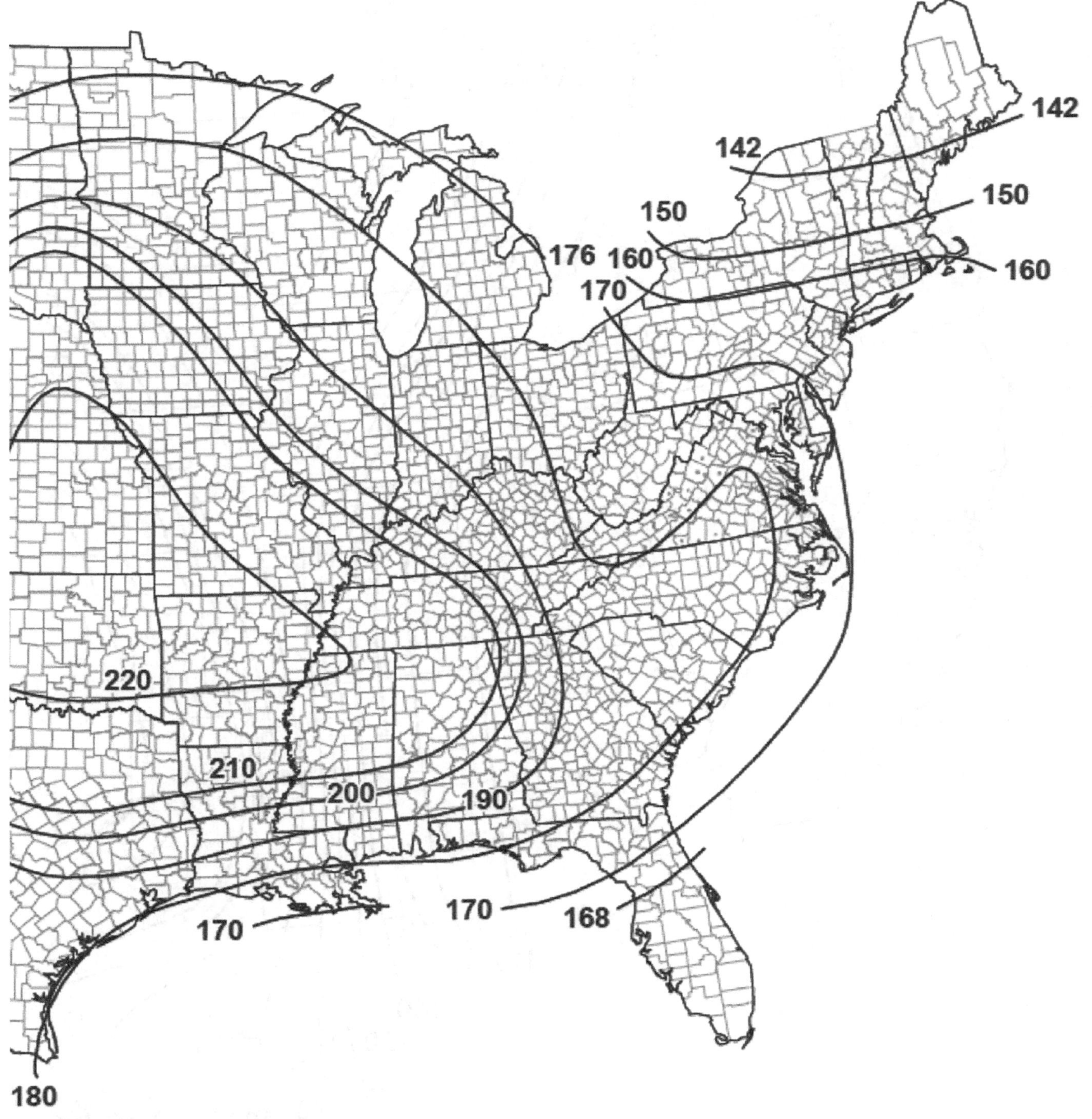

4. Islands, coastal areas, and land boundaries outside the last contour shall use the last tornado speed contour.

5. Tornado speeds correspond to approximately a 0.005% probability of exceedance in 50 years (annual exceedance probability = 0.000001, MRI = 1,000,000 years).

6. Location-specific tornado speed is permitted to be determined using the ASCE Tornado Design Geodatabase, available at the ASCE 7 Hazard Tool (https://asce7hazardtool.online), or approved equivalent.

Figure G.2-3A (*Continued*). Tornado speeds for 1,000,000-year MRI for effective plan area of 1 ft² (0.1 m²).

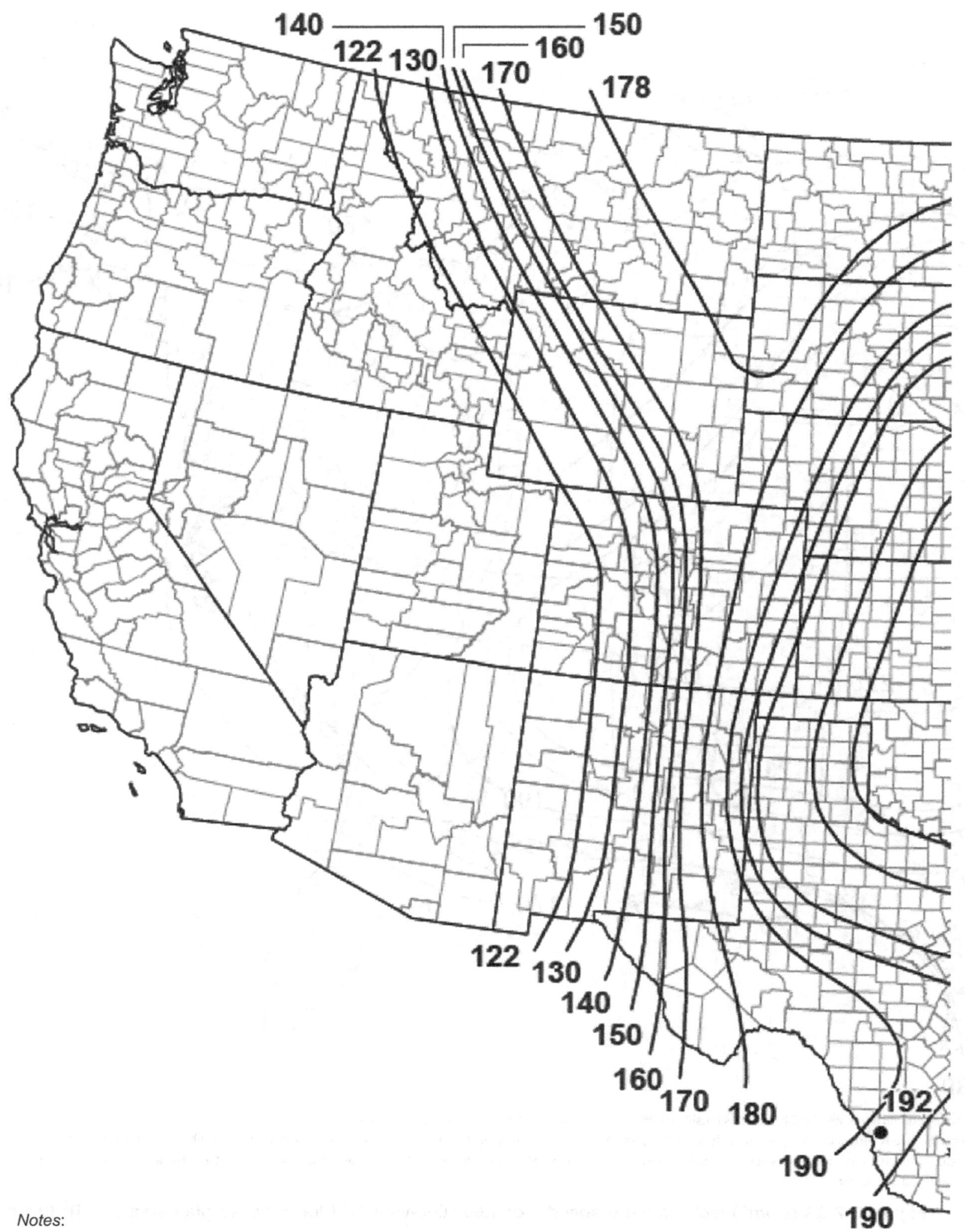

Notes:

1. Values are 3 s gust speeds in mi/h at 33 ft (10 m) above ground.
2. To convert tornado speeds from mi/h to m/s, multiply mapped values by 0.447.
3. Linear interpolation is permitted between contours. Point values (where shown) are provided to aid with interpolation.

Figure G.2-3B. Tornado speeds for 1,000,000-year MRI for effective plan area of 2,000 ft^2 (186 m^2).

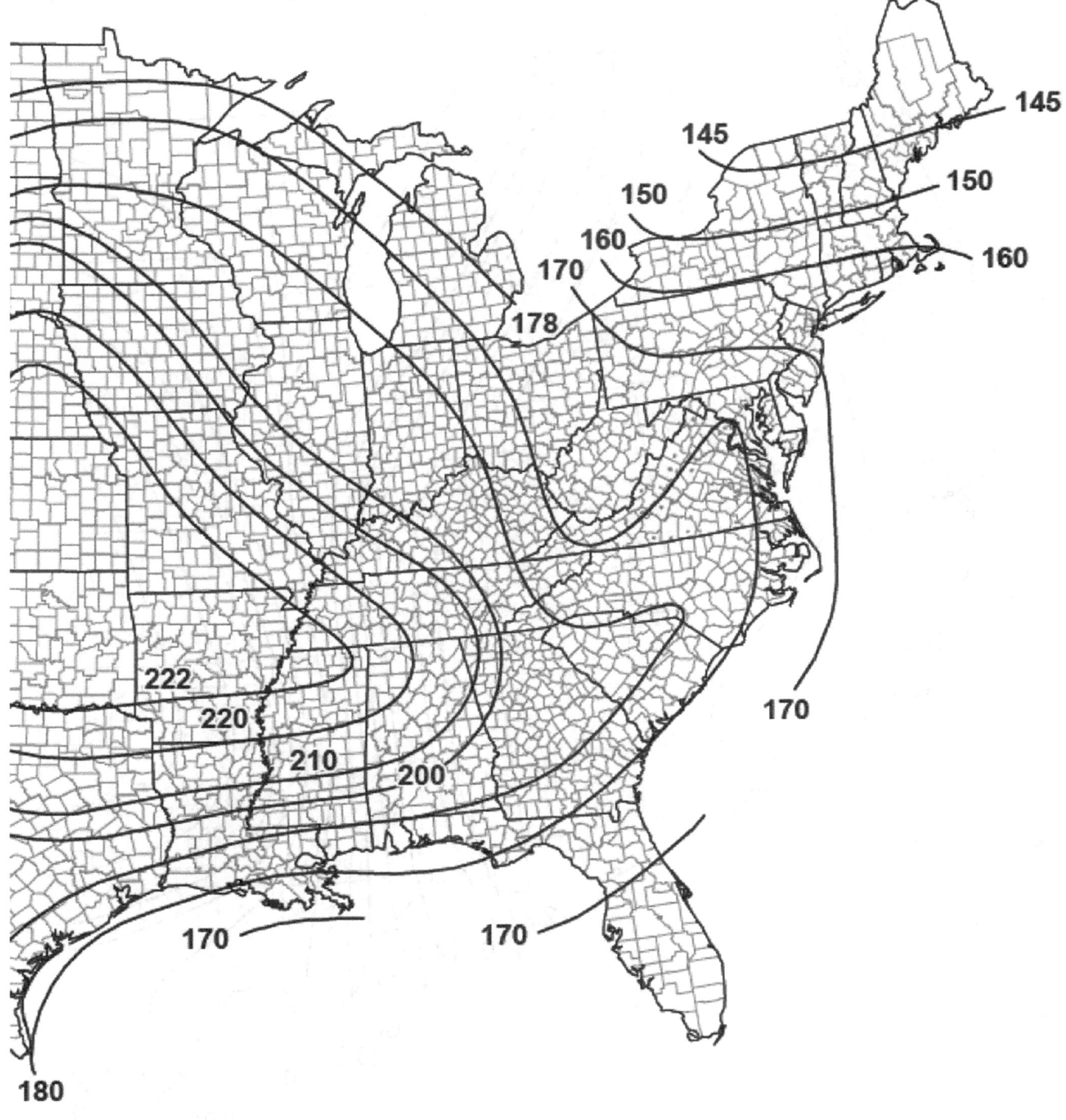

4. Islands, coastal areas, and land boundaries outside the last contour shall use the last tornado speed contour.
5. Tornado speeds correspond to approximately a 0.005% probability of exceedance in 50 years (annual exceedance probability = 0.000001, MRI = 1,000,000 years).
6. Location-specific tornado speed is permitted to be determined using the ASCE Tornado Design Geodatabase, available at the ASCE 7 Hazard Tool (https://asce7hazardtool.online), or approved equivalent.

Figure G.2-3B (*Continued*). Tornado speeds for 1,000,000-year MRI for effective plan area of 2,000 ft^2 (186 m^2).

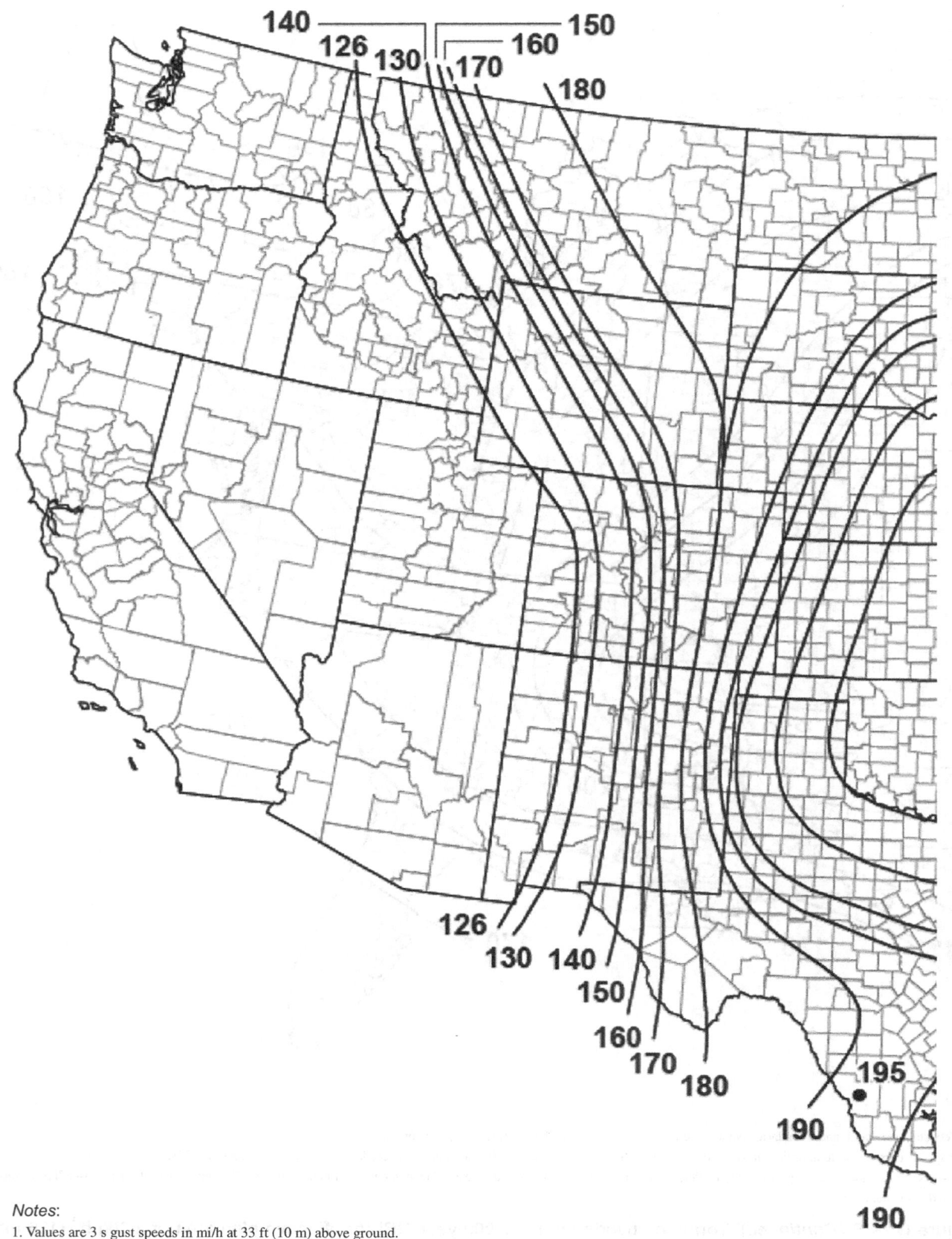

Notes:

1. Values are 3 s gust speeds in mi/h at 33 ft (10 m) above ground.
2. To convert tornado speeds from mi/h to m/s, multiply mapped values by 0.447.
3. Linear interpolation is permitted between contours. Point values (where shown) are provided to aid with interpolation.

Figure G.2-3C. Tornado speeds for 1,000,000-year MRI for effective plan area of 10,000 ft^2 (929 m^2).

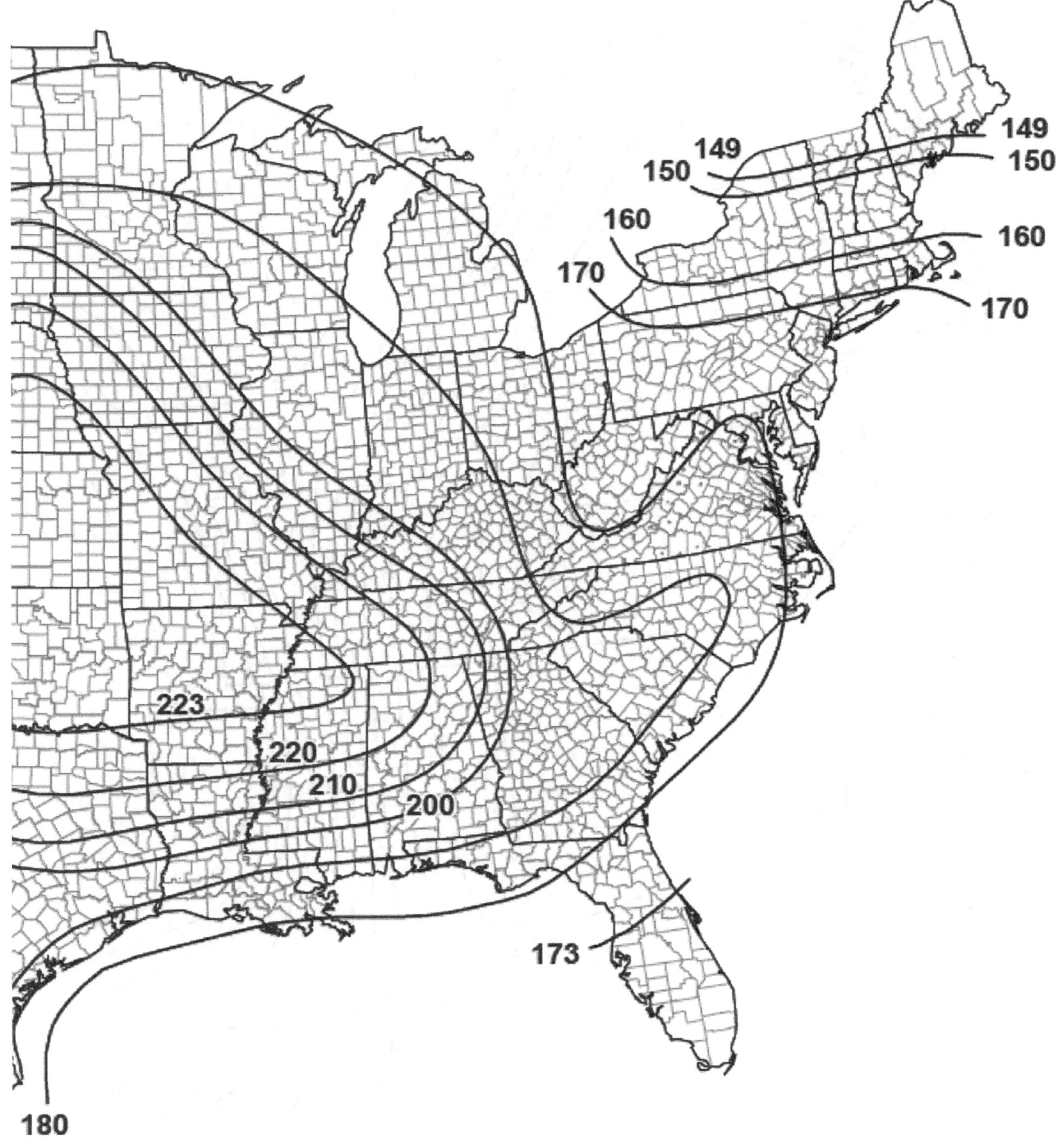

4. Islands, coastal areas, and land boundaries outside the last contour shall use the last tornado speed contour.
5. Tornado speeds correspond to approximately a 0.005% probability of exceedance in 50 years (annual exceedance probability = 0.000001, MRI = 1,000,000 years).
6. Location-specific tornado speed is permitted to be determined using the ASCE Tornado Design Geodatabase, available at the ASCE 7 Hazard Tool (https://asce7hazardtool.online), or approved equivalent.

Figure G.2-3C (*Continued*). Tornado speeds for 1,000,000-year MRI for effective plan area of 10,000 ft^2 (929 m^2).

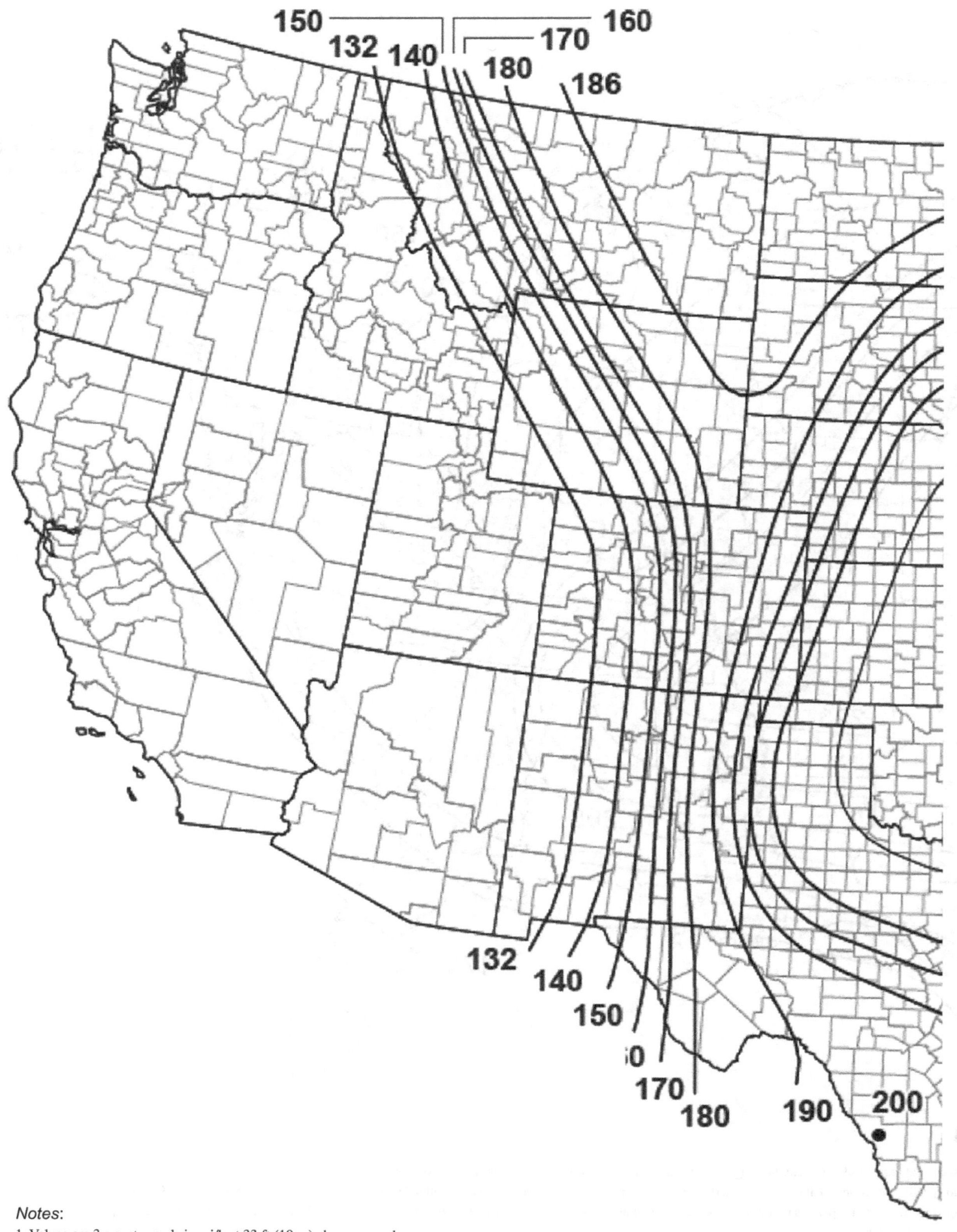

Notes:

1. Values are 3 s gust speeds in mi/h at 33 ft (10 m) above ground.
2. To convert tornado speeds from mi/h to m/s, multiply mapped values by 0.447.
3. Linear interpolation is permitted between contours. Point values (where shown) are provided to aid with interpolation.

Figure G.2-3D. Tornado speeds for 1,000,000-year MRI for effective plan area of 40,000 ft^2 (3,716 m^2).

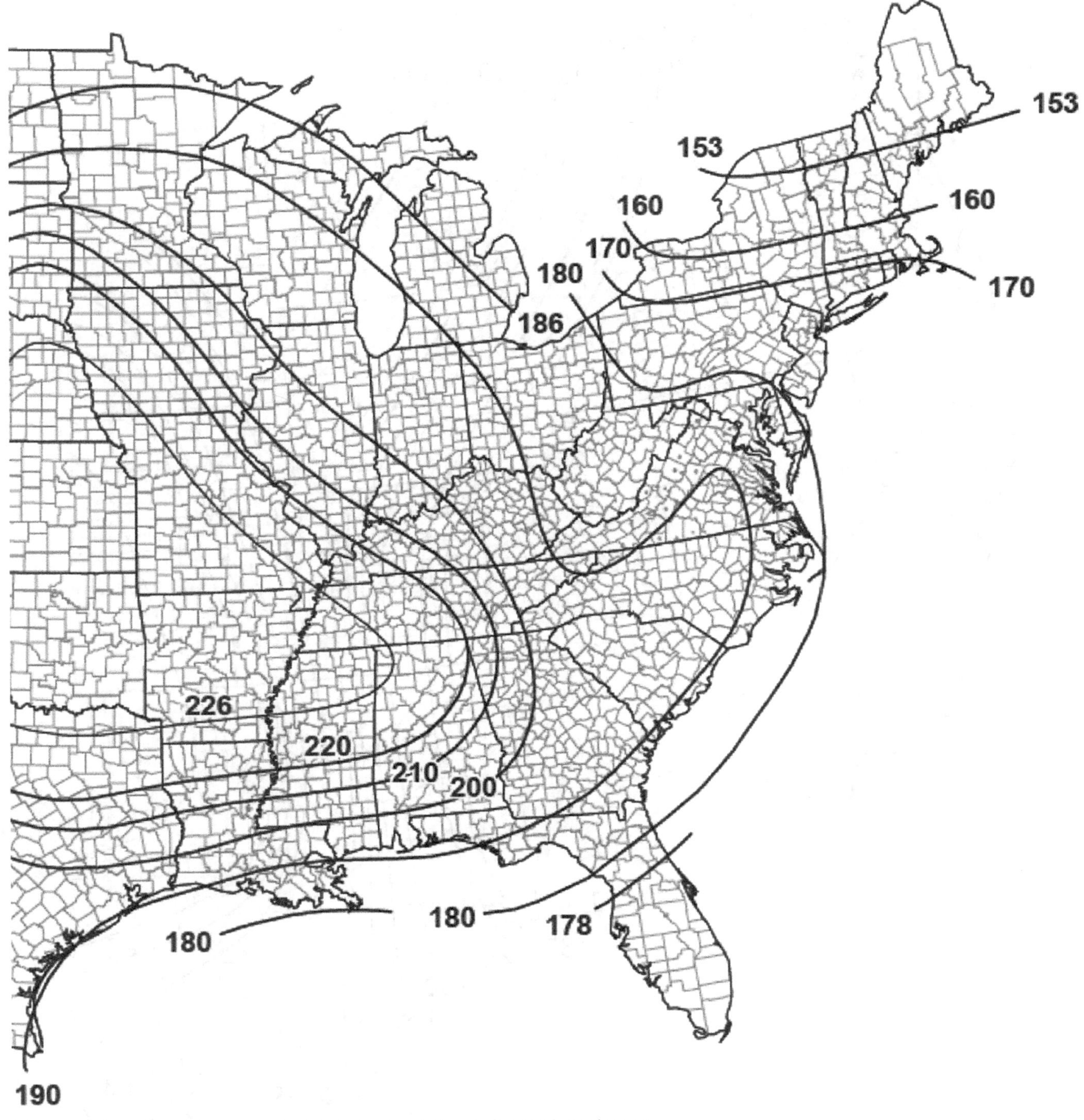

4. Islands, coastal areas, and land boundaries outside the last contour shall use the last tornado speed contour.
5. Tornado speeds correspond to approximately a 0.005% probability of exceedance in 50 years (annual exceedance probability = 0.000001, MRI = 1,000,000 years).
6. Location-specific tornado speed is permitted to be determined using the ASCE Tornado Design Geodatabase, available at the ASCE 7 Hazard Tool (https://asce7hazardtool.online), or approved equivalent.

Figure G.2-3D (*Continued*). Tornado speeds for 1,000,000-year MRI for effective plan area of 40,000 ft^2 (3,716 m^2).

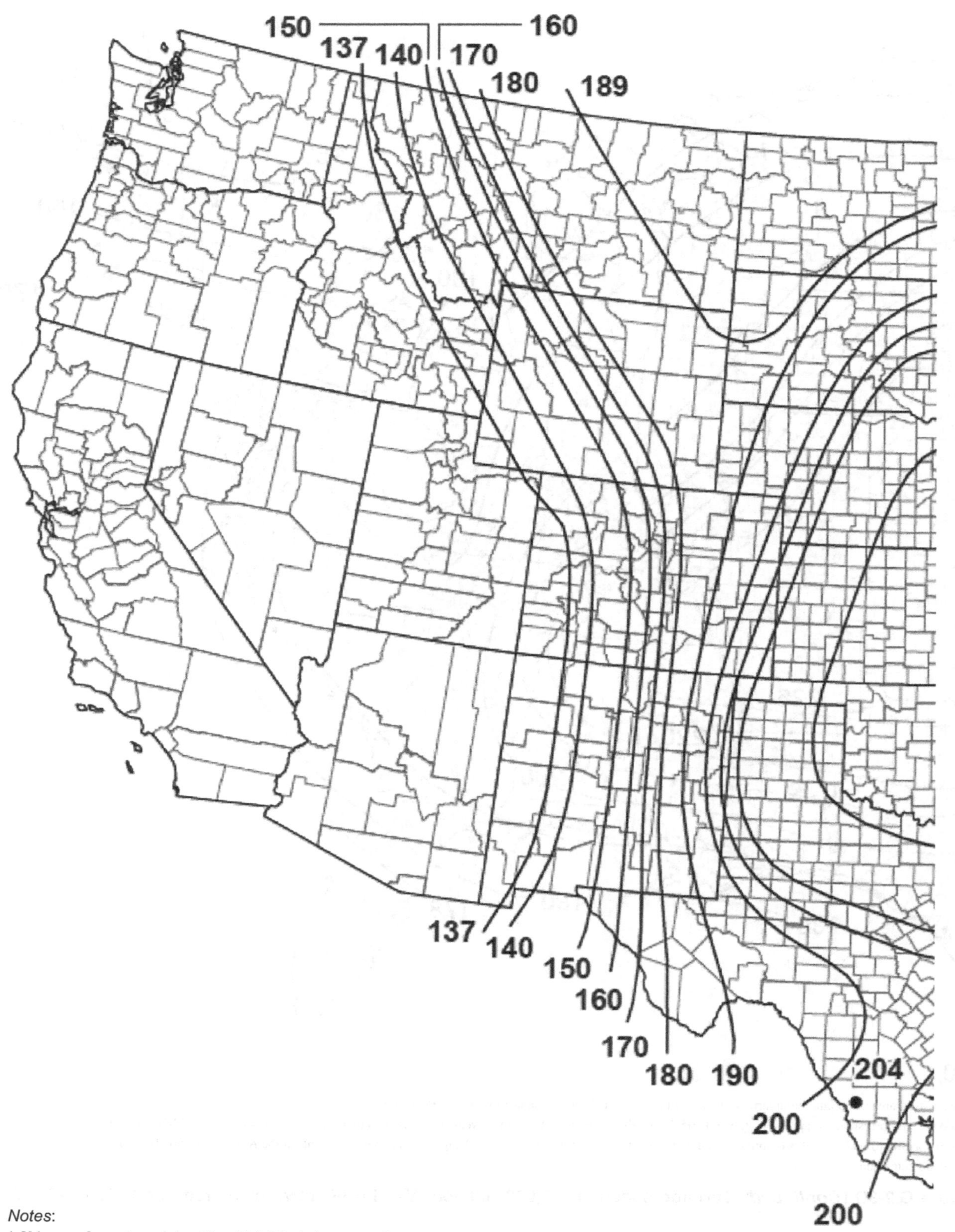

Notes:

1. Values are 3 s gust speeds in mi/h at 33 ft (10 m) above ground.
2. To convert tornado speeds from mi/h to m/s, multiply mapped values by 0.447.
3. Linear interpolation is permitted between contours. Point values (where shown) are provided to aid with interpolation.

Figure G.2-3E. Tornado speeds for 1,000,000-year MRI for effective plan area of 100,000 ft^2 (9,290 m^2).

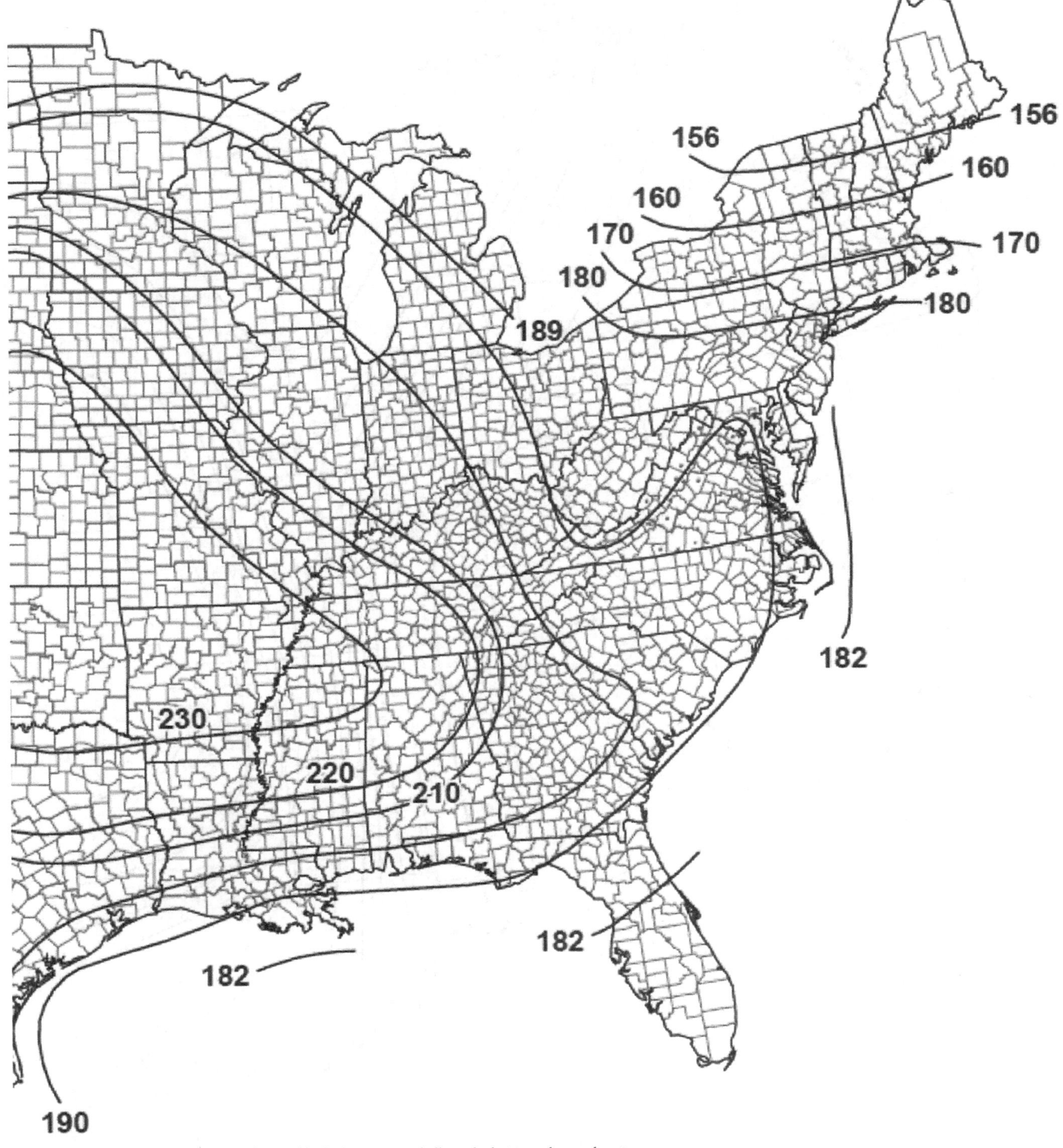

4. Islands, coastal areas, and land boundaries outside the last contour shall use the last tornado speed contour.
5. Tornado speeds correspond to approximately a 0.005% probability of exceedance in 50 years (annual exceedance probability = 0.000001, MRI = 1,000,000 years).
6. Location-specific tornado speed is permitted to be determined using the ASCE Tornado Design Geodatabase, available at the ASCE 7 Hazard Tool (https://asce7hazardtool.online), or approved equivalent.

Figure G.2-3E (*Continued*). Tornado speeds for 1,000,000-year MRI for effective plan area of 100,000 ft^2 (9,290 m^2).

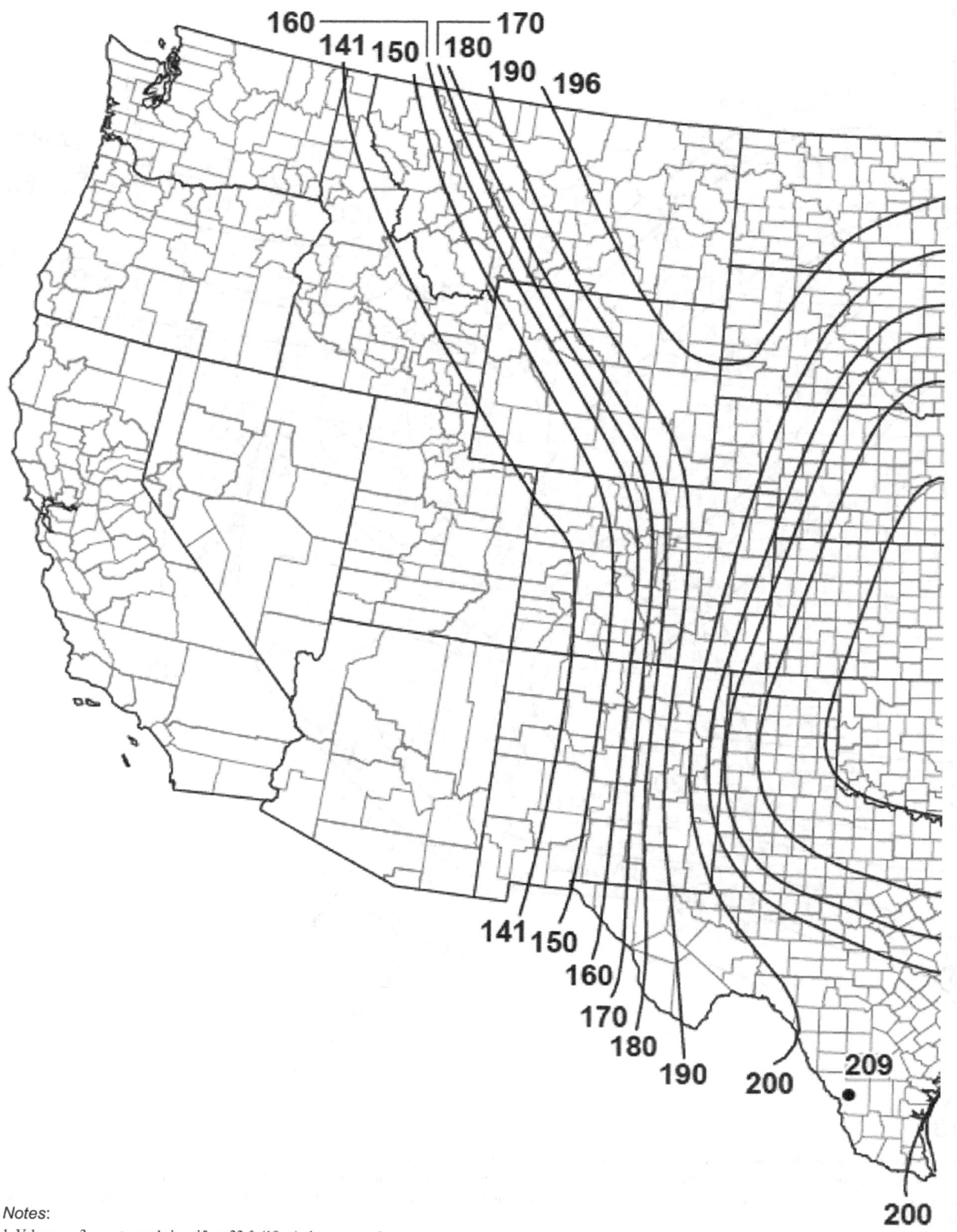

Notes:

1. Values are 3 s gust speeds in mi/h at 33 ft (10 m) above ground.
2. To convert tornado speeds from mi/h to m/s, multiply mapped values by 0.447.
3. Linear interpolation is permitted between contours. Point values (where shown) are provided to aid with interpolation.

Figure G.2-3F. Tornado speeds for 1,000,000-year MRI for effective plan area of 250,000 ft^2 (23,226 m^2).

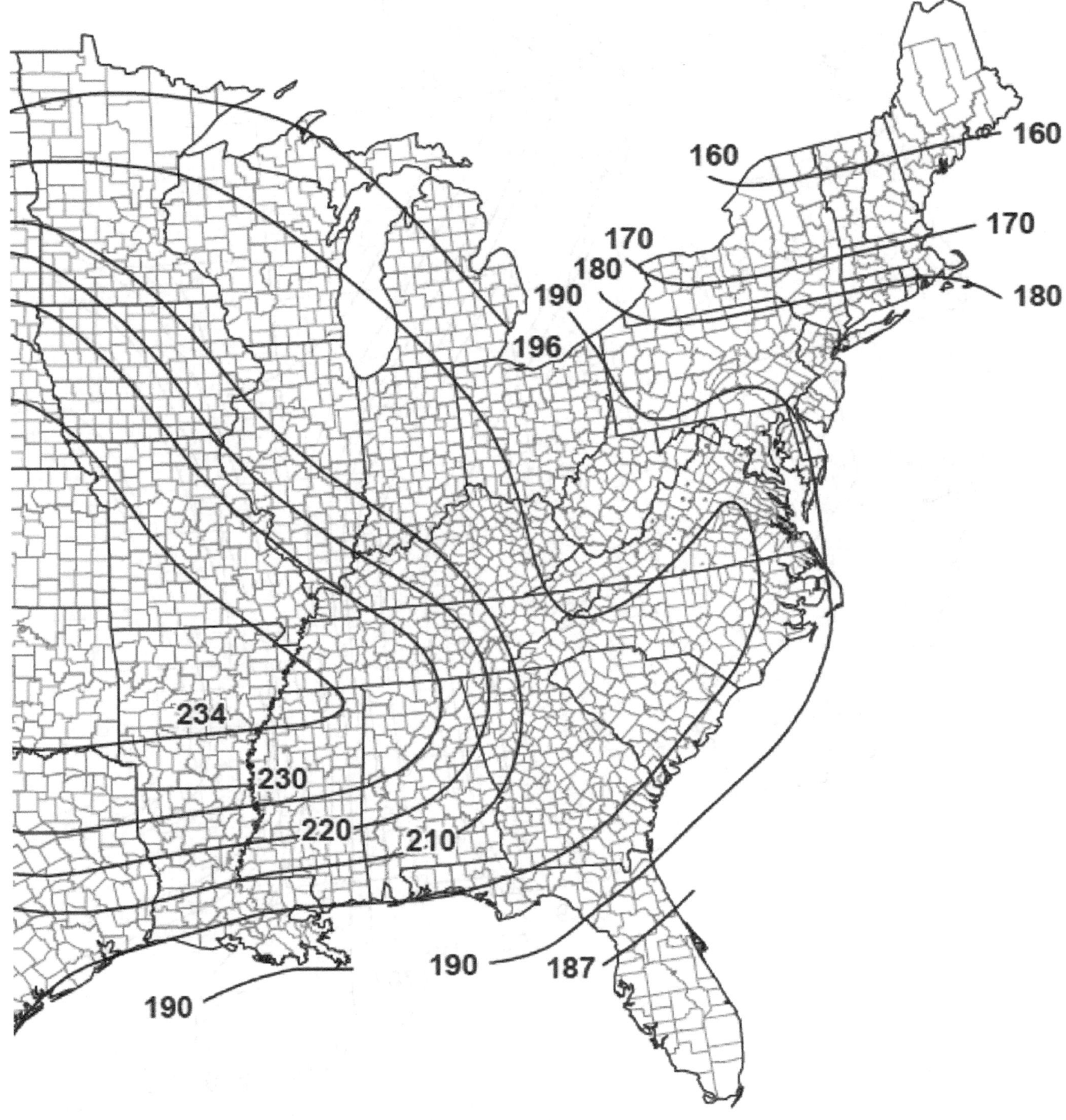

4. Islands, coastal areas, and land boundaries outside the last contour shall use the last tornado speed contour.
5. Tornado speeds correspond to approximately a 0.005% probability of exceedance in 50 years (annual exceedance probability = 0.000001, MRI = 1,000,000 years).
6. Location-specific tornado speed is permitted to be determined using the ASCE Tornado Design Geodatabase, available at the ASCE 7 Hazard Tool (https://asce7hazardtool.online), or approved equivalent.

Figure G.2-3F (*Continued*). Tornado speeds for 1,000,000-year MRI for effective plan area of 250,000 ft^2 (23,226 m^2).

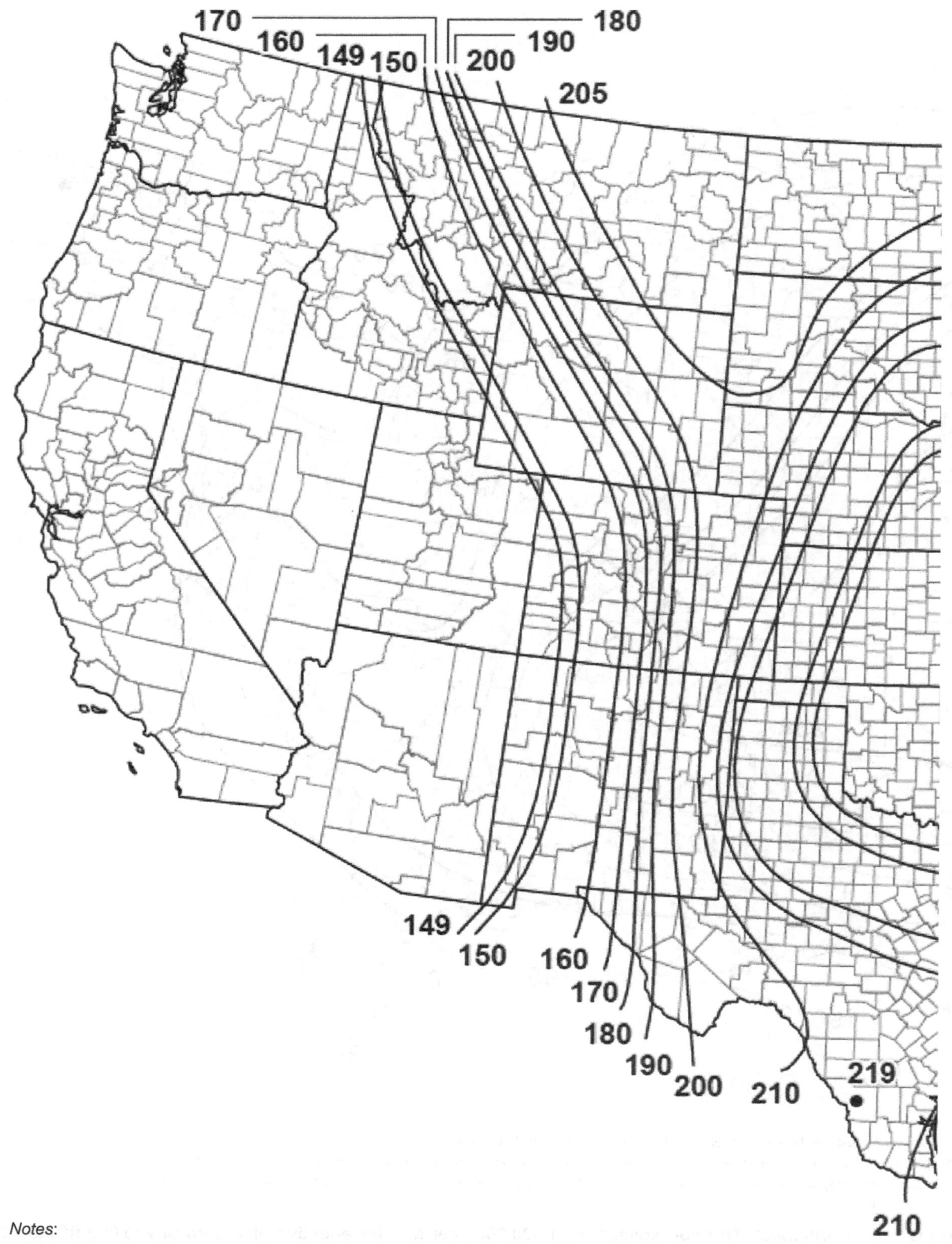

Notes:

1. Values are 3 s gust speeds in mi/h at 33 ft (10 m) above ground.
2. To convert tornado speeds from mi/h to m/s, multiply mapped values by 0.447.
3. Linear interpolation is permitted between contours. Point values (where shown) are provided to aid with interpolation.

Figure G.2-3G. Tornado speeds for 1,000,000-year MRI for effective plan area of 1,000,000 ft^2 (92,903 m^2).

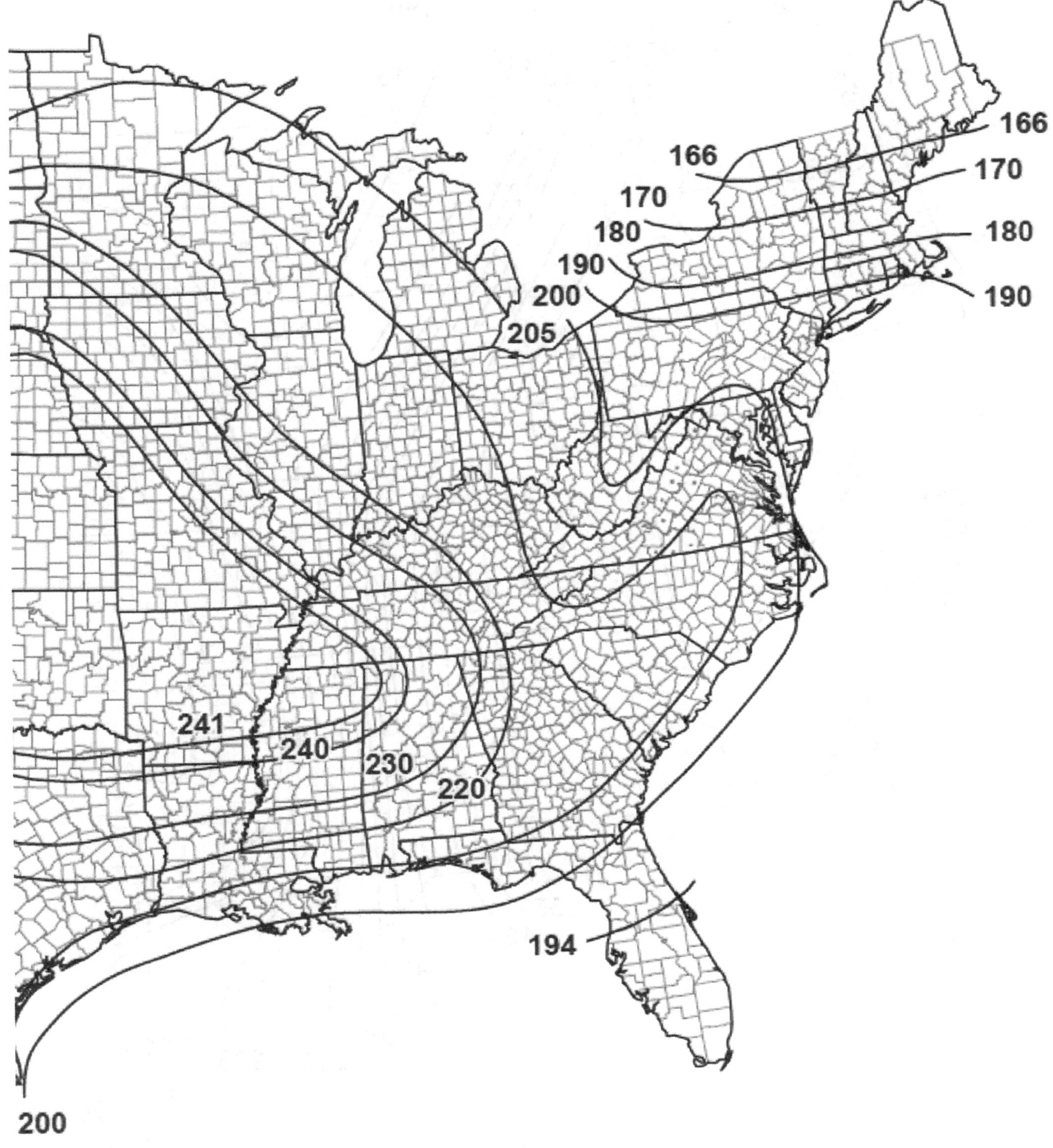

4. Islands, coastal areas, and land boundaries outside the last contour shall use the last tornado speed contour.
5. Tornado speeds correspond to approximately a 0.005% probability of exceedance in 50 years (annual exceedance probability = 0.000001, MRI = 1,000,000 years).
6. Location-specific tornado speed is permitted to be determined using the ASCE Tornado Design Geodatabase, available at the ASCE 7 Hazard Tool (https://asce7hazardtool.online), or approved equivalent.

Figure G.2-3G (*Continued*). Tornado speeds for 1,000,000-year MRI for effective plan area of 1,000,000 ft^2 (92,903 m^2).

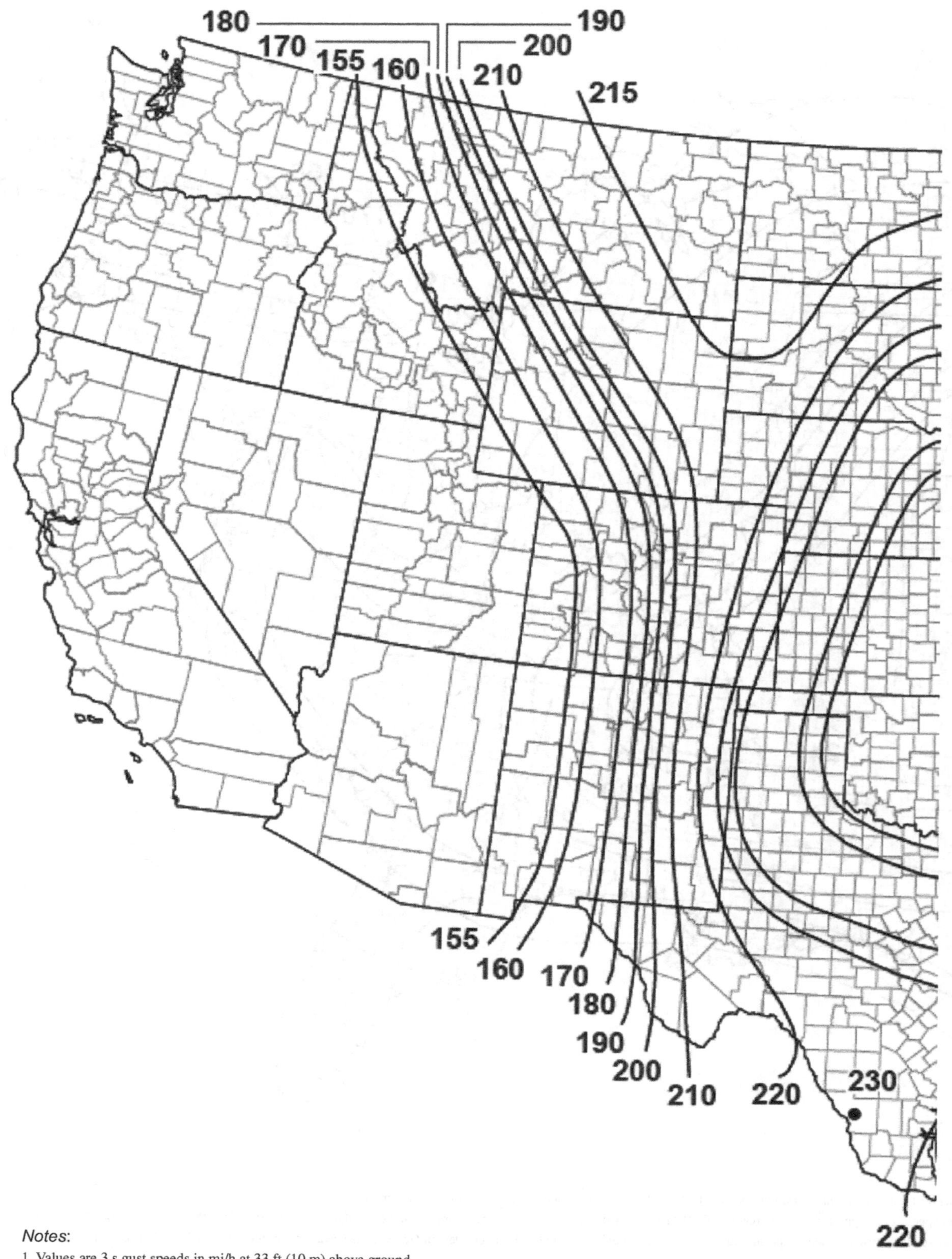

Notes:

1. Values are 3 s gust speeds in mi/h at 33 ft (10 m) above ground.
2. To convert tornado speeds from mi/h to m/s, multiply mapped values by 0.447.
3. Linear interpolation is permitted between contours. Point values (where shown) are provided to aid with interpolation.

Figure G.2-3H. Tornado speeds for 1,000,000-year MRI for effective plan area of 4,000,000 ft^2 (371,612 m^2).

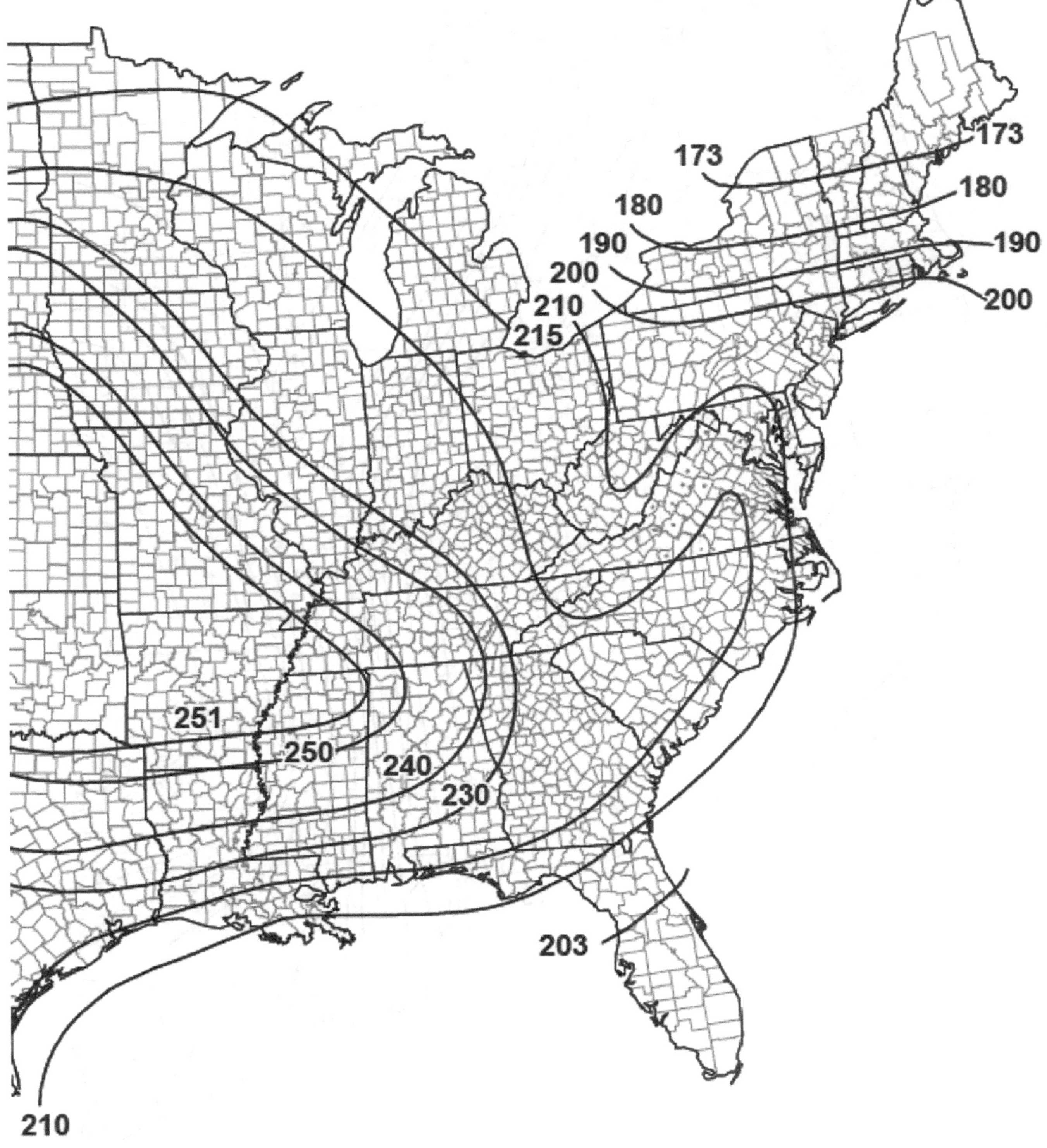

4. Islands, coastal areas, and land boundaries outside the last contour shall use the last tornado speed contour.
5. Tornado speeds correspond to approximately a 0.005% probability of exceedance in 50 years (annual exceedance probability = 0.000001, MRI = 1,000,000 years).
6. Location-specific tornado speed is permitted to be determined using the ASCE Tornado Design Geodatabase, available at the ASCE 7 Hazard Tool (https://asce7hazardtool.online), or approved equivalent.

Figure G.2-3H (*Continued*). Tornado speeds for 1,000,000-year MRI for effective plan area of 4,000,000 ft^2 (371,612 m^2).

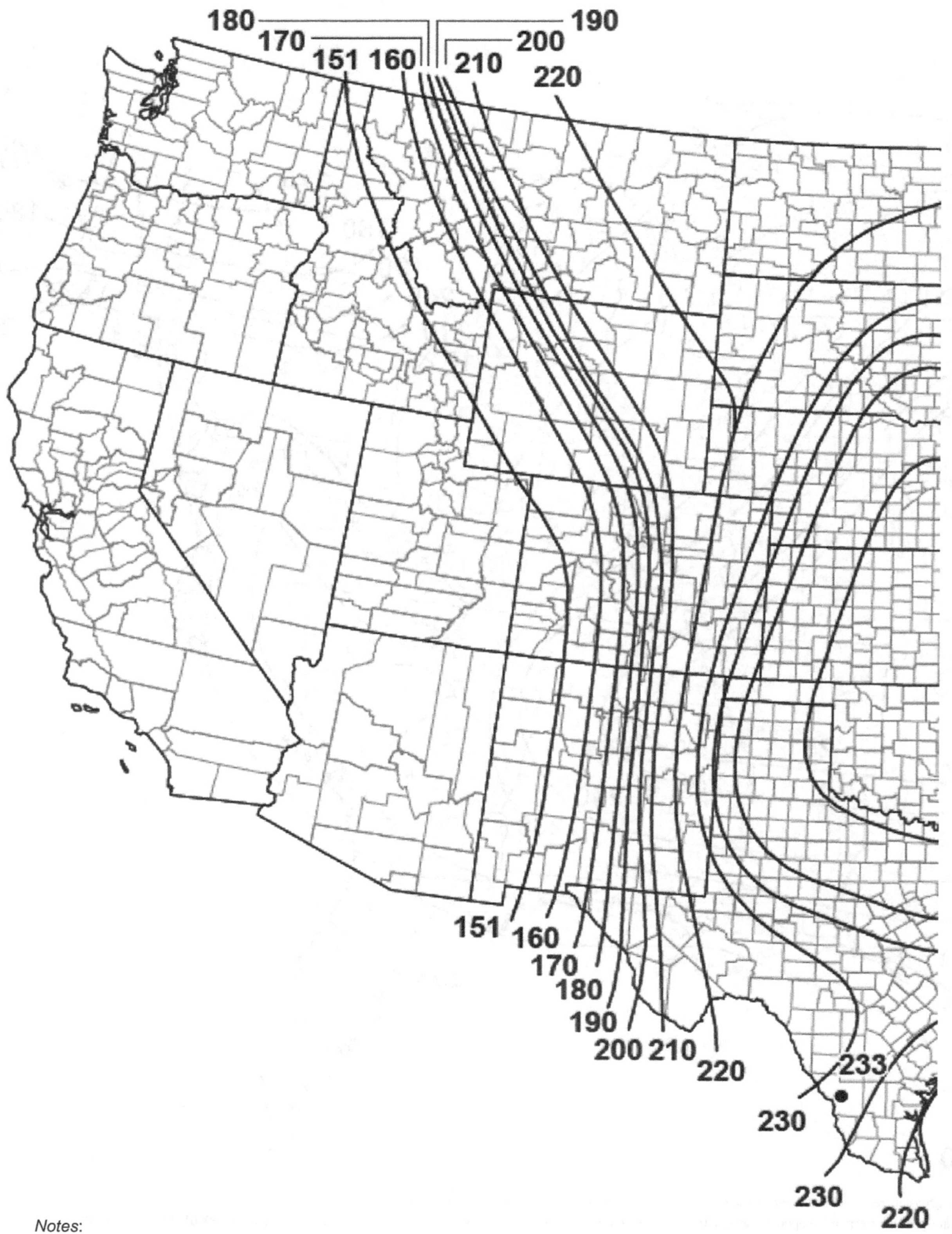

Notes:

1. Values are 3 s gust speeds in mi/h at 33 ft (10 m) above ground.
2. To convert tornado speeds from mi/h to m/s, multiply mapped values by 0.447.
3. Linear interpolation is permitted between contours. Point values (where shown) are provided to aid with interpolation.

Figure G.2-4A. Tornado speeds for 10,000,000-year MRI for effective plan area of 1 ft^2 (0.1 m^2).

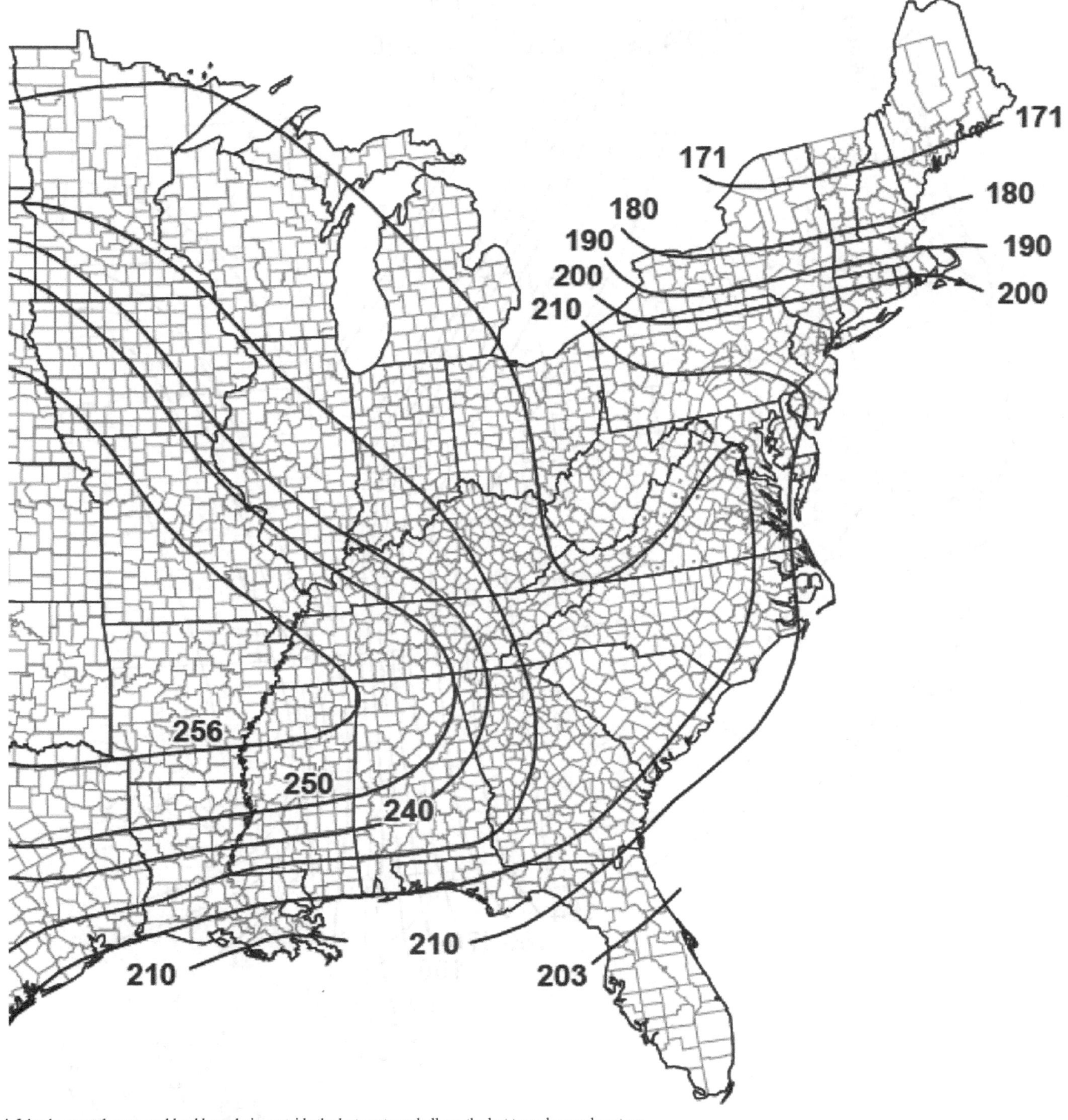

4. Islands, coastal areas, and land boundaries outside the last contour shall use the last tornado speed contour.
5. Tornado speeds correspond to approximately a 0.0005% probability of exceedance in 50 years (annual exceedance probability = 0.0000001, MRI = 10,000,000 years).
6. Location-specific tornado speed is permitted to be determined using the ASCE Tornado Design Geodatabase, available at the ASCE 7 Hazard Tool (https://asce7hazardtool.online), or approved equivalent.

Figure G.2-4A (*Continued*). Tornado speeds for 10,000,000-year MRI for effective plan area of 1 ft^2 (0.1 m^2).

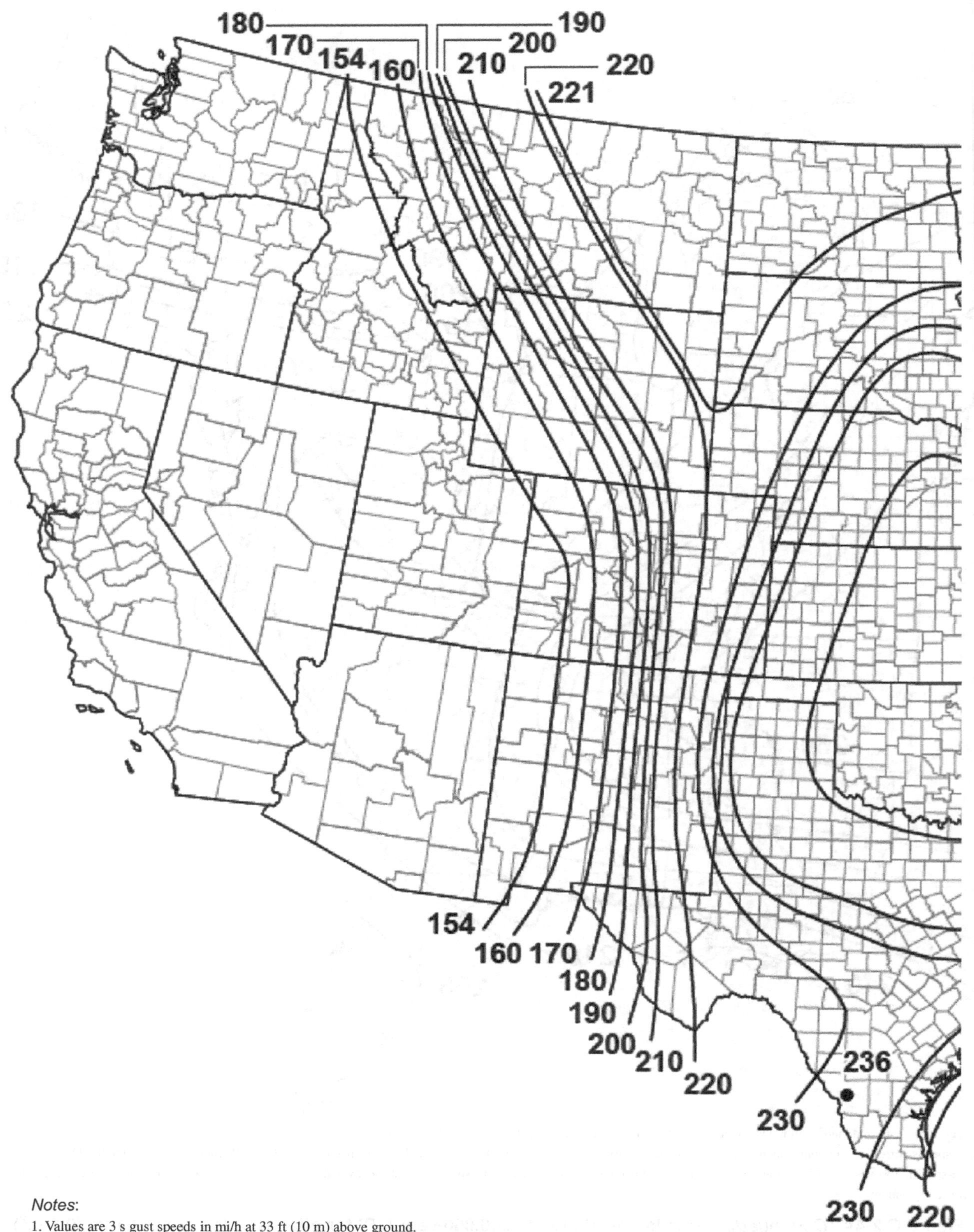

Notes:
1. Values are 3 s gust speeds in mi/h at 33 ft (10 m) above ground.
2. To convert tornado speeds from mi/h to m/s, multiply mapped values by 0.447.
3. Linear interpolation is permitted between contours. Point values (where shown) are provided to aid with interpolation.

Figure G.2-4B. Tornado speeds for 10,000,000-year MRI for effective plan area of 2,000 ft^2 (186 m^2).

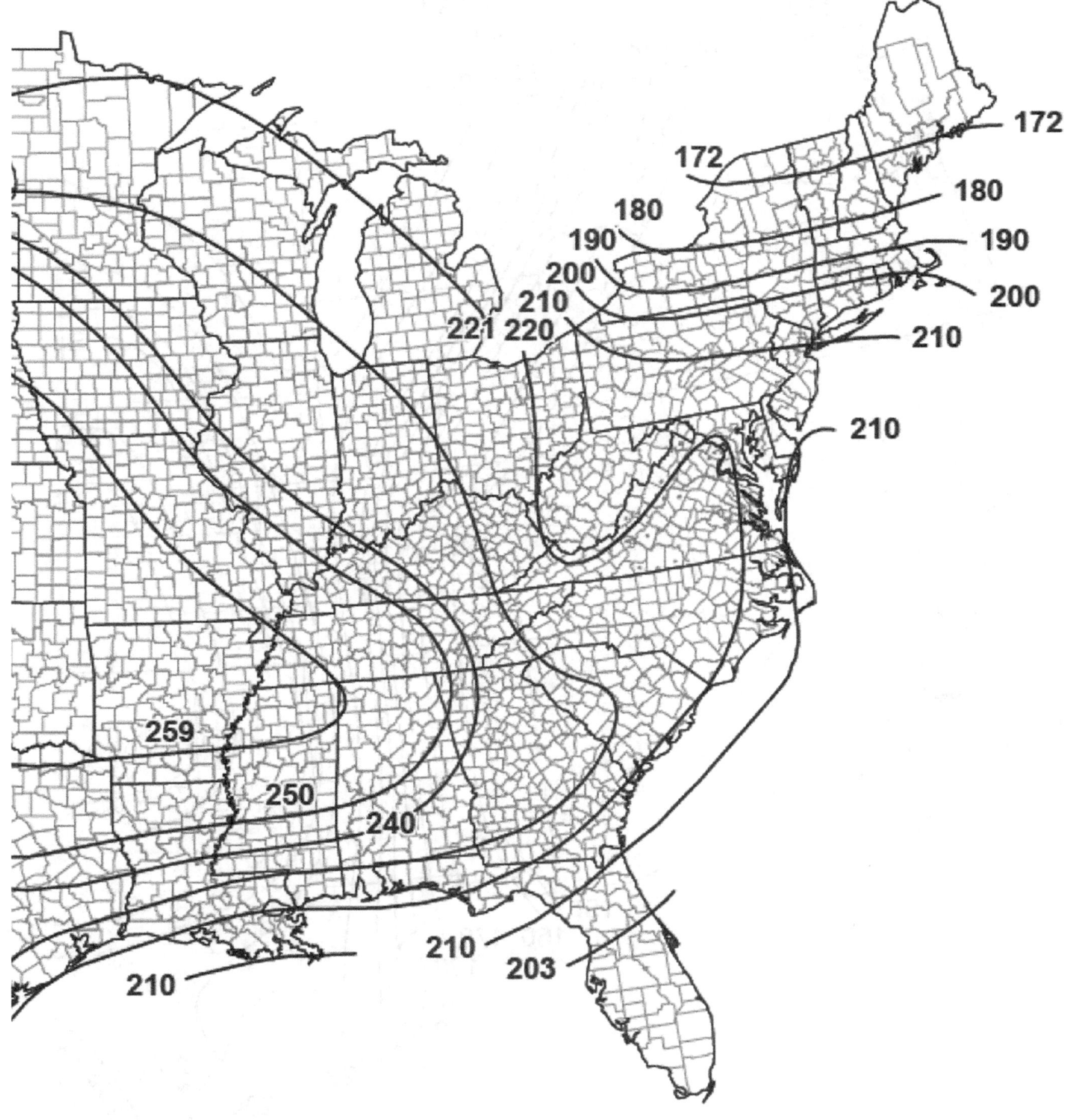

4. Islands, coastal areas, and land boundaries outside the last contour shall use the last tornado speed contour.
5. Tornado speeds correspond to approximately a 0.0005% probability of exceedance in 50 years (annual exceedance probability = 0.0000001, MRI = 10,000,000 years).
6. Location-specific tornado speed is permitted to be determined using the ASCE Tornado Design Geodatabase, available at the ASCE 7 Hazard Tool (https://asce7hazardtool.online), or approved equivalent.

Figure G.2-4B (*Continued*). Tornado speeds for 10,000,000-year MRI for effective plan area of 2,000 ft^2 (186 m^2).

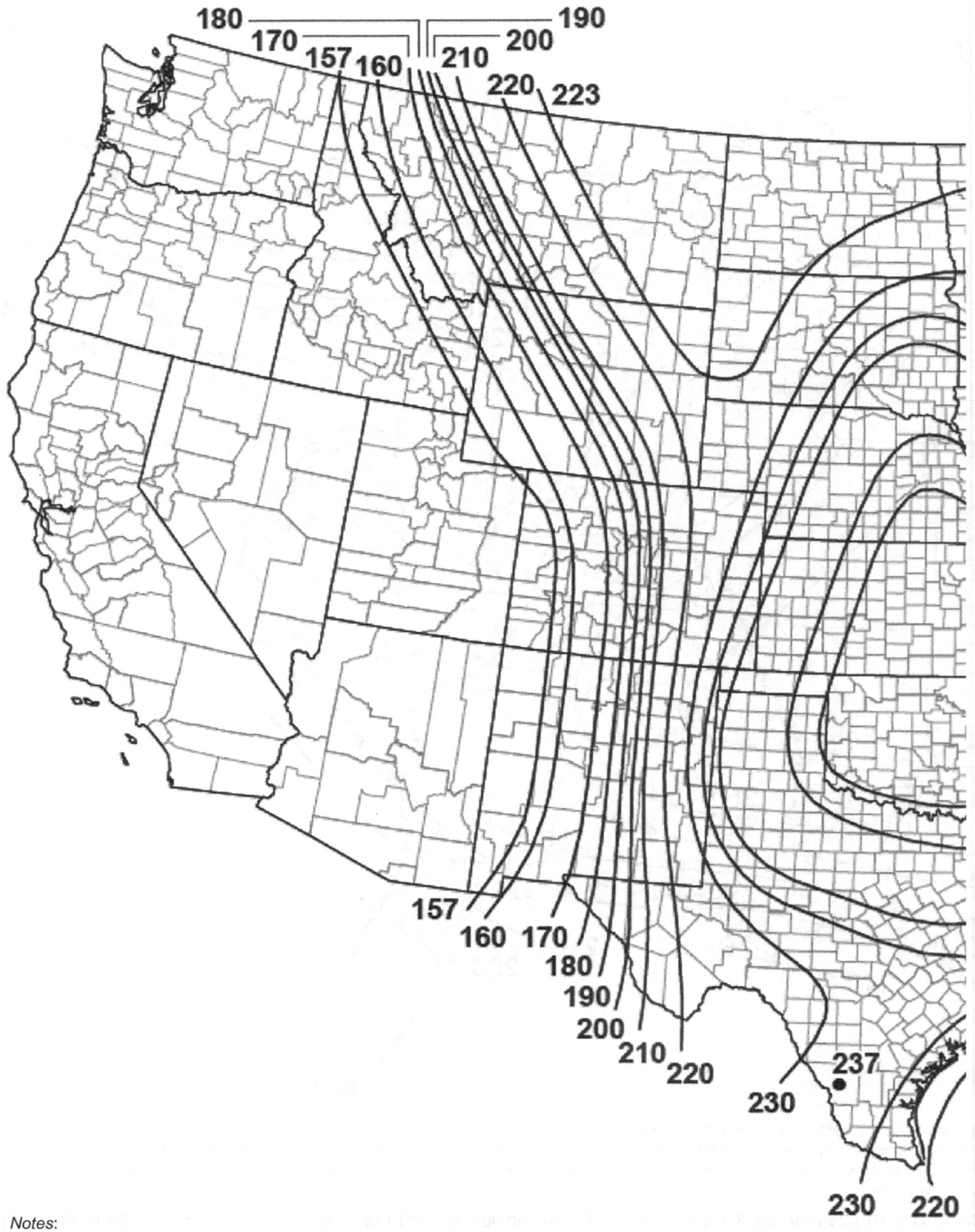

Notes:

1. Values are 3 s gust speeds in mi/h at 33 ft (10 m) above ground.
2. To convert tornado speeds from mi/h to m/s, multiply mapped values by 0.447.
3. Linear interpolation is permitted between contours. Point values (where shown) are provided to aid with interpolation.

Figure G.2-4C. Tornado speeds for 10,000,000-year MRI for effective plan area of 10,000 ft² (929 m²).

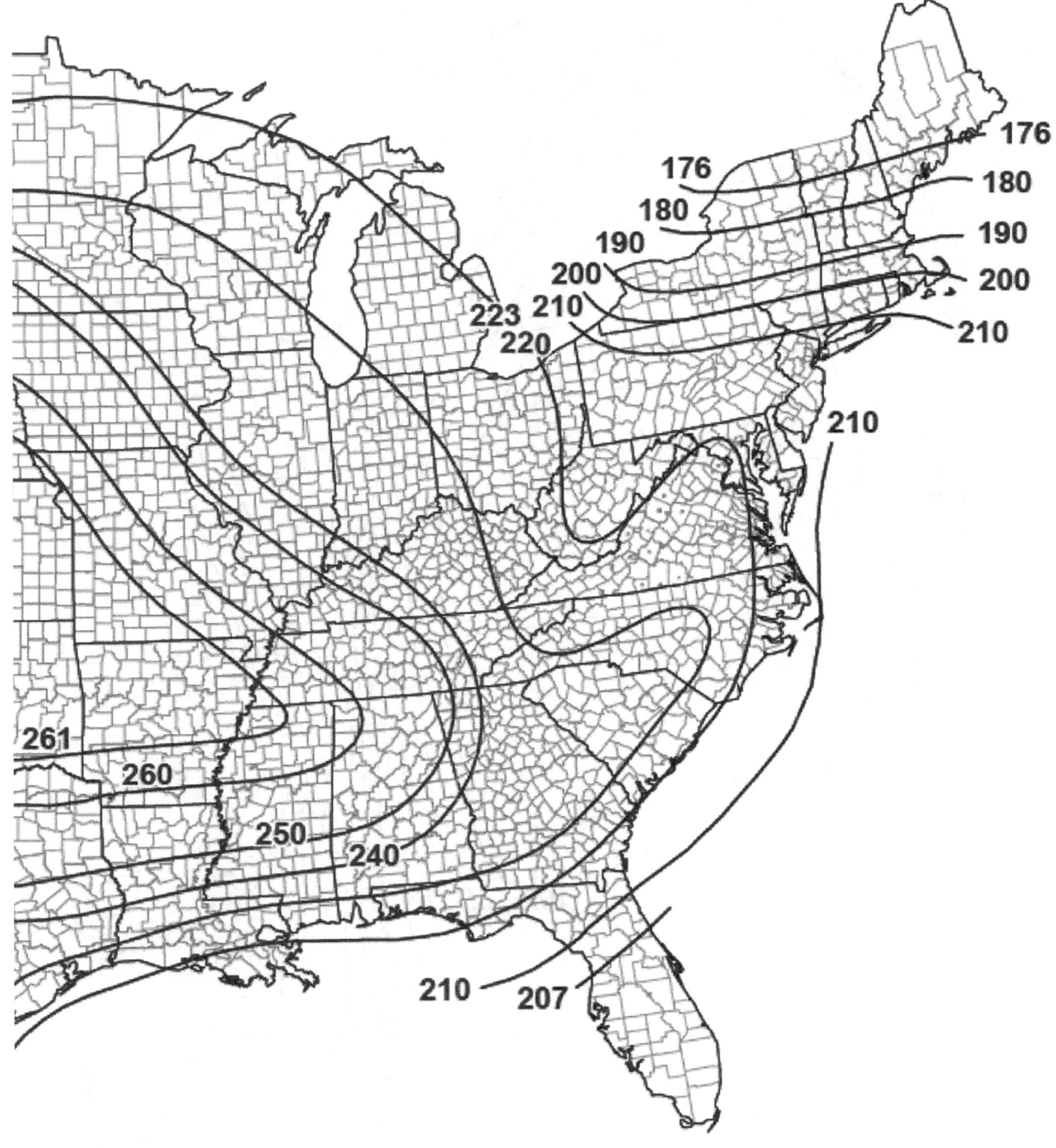

4. Islands, coastal areas, and land boundaries outside the last contour shall use the last tornado speed contour.
5. Tornado speeds correspond to approximately a 0.0005% probability of exceedance in 50 years (annual exceedance probability = 0.0000001, MRI = 10,000,000 years).
6. Location-specific tornado speed is permitted to be determined using the ASCE Tornado Design Geodatabase, available at the ASCE 7 Hazard Tool (https://asce7hazardtool.online), or approved equivalent.

Figure G.2-4C (*Continued*). Tornado speeds for 10,000,000-year MRI for effective plan area of 10,000 ft^2 (929 m^2).

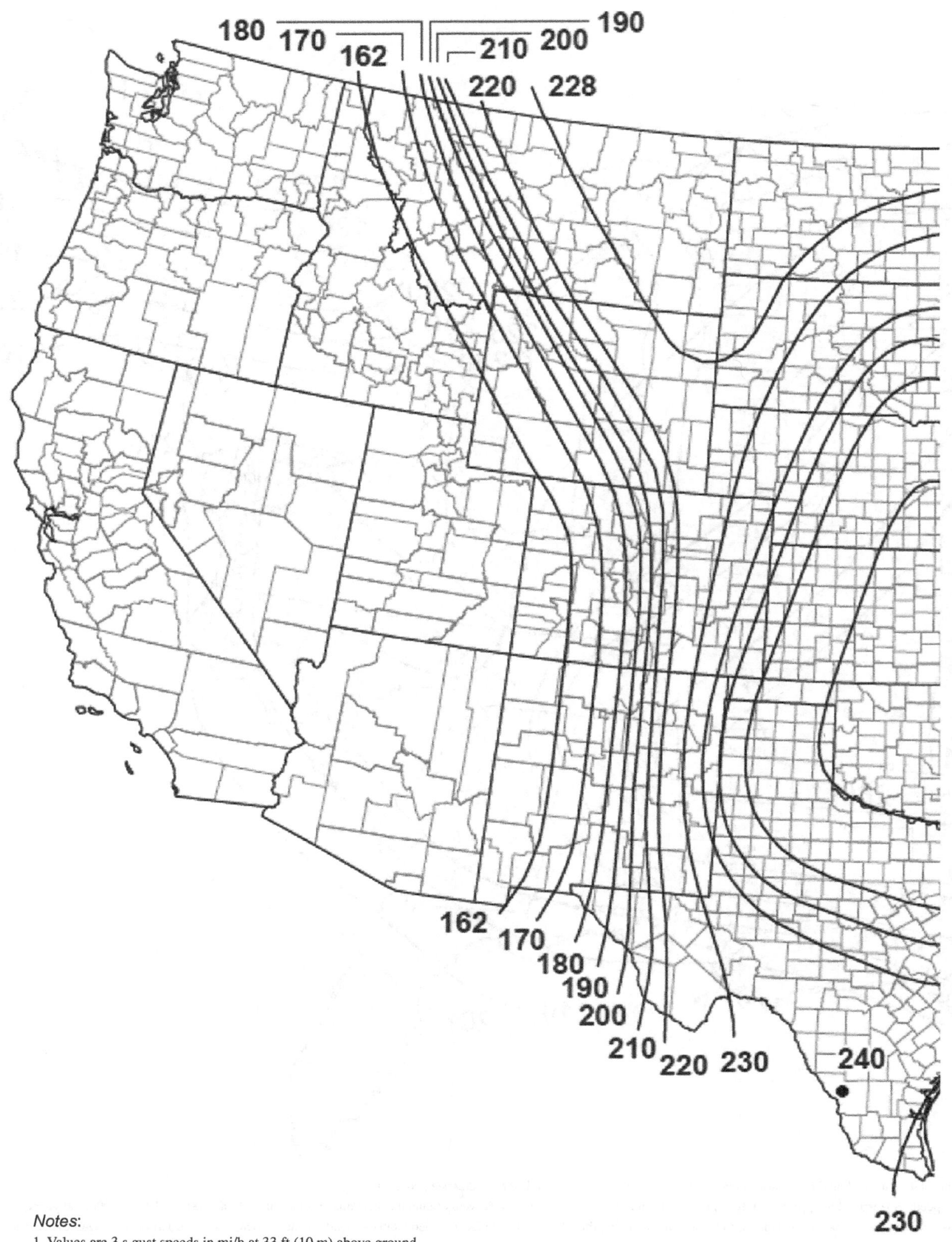

Notes:

1. Values are 3 s gust speeds in mi/h at 33 ft (10 m) above ground.
2. To convert tornado speeds from mi/h to m/s, multiply mapped values by 0.447.
3. Linear interpolation is permitted between contours. Point values (where shown) are provided to aid with interpolation.

Figure G.2-4D. Tornado speeds for 10,000,000-year MRI for effective plan area of 40,000 ft² (3,716 m²).

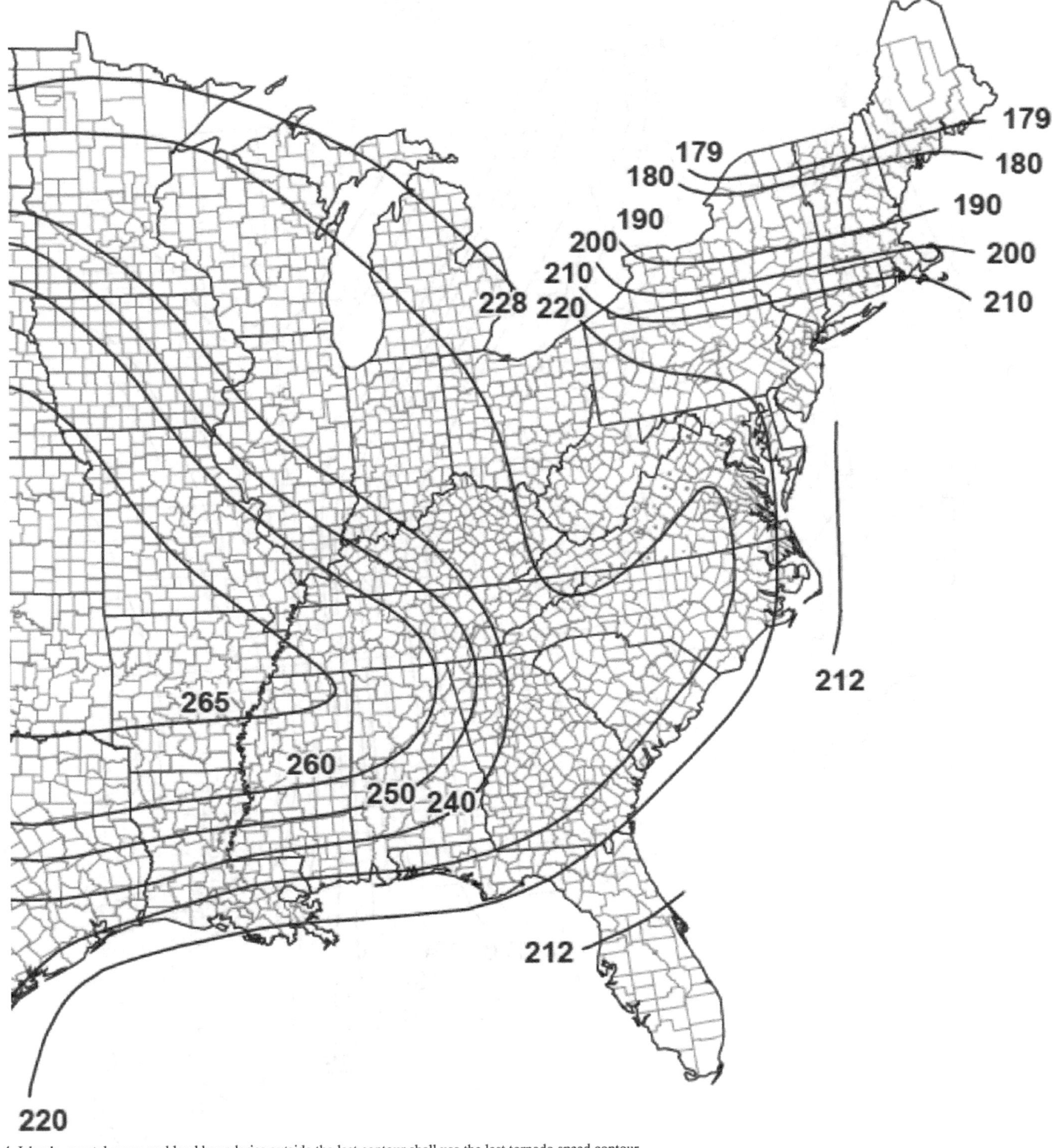

4. Islands, coastal areas, and land boundaries outside the last contour shall use the last tornado speed contour.
5. Tornado speeds correspond to approximately a 0.0005% probability of exceedance in 50 years (annual exceedance probability = 0.0000001, MRI = 10,000,000 years).
6. Location-specific tornado speed is permitted to be determined using the ASCE Tornado Design Geodatabase, available at the ASCE 7 Hazard Tool (https://asce7hazardtool.online), or approved equivalent.

Figure G.2-4D (*Continued*). Tornado speeds for 10,000,000-year MRI for effective plan area of 40,000 ft^2 (3,716 m^2).

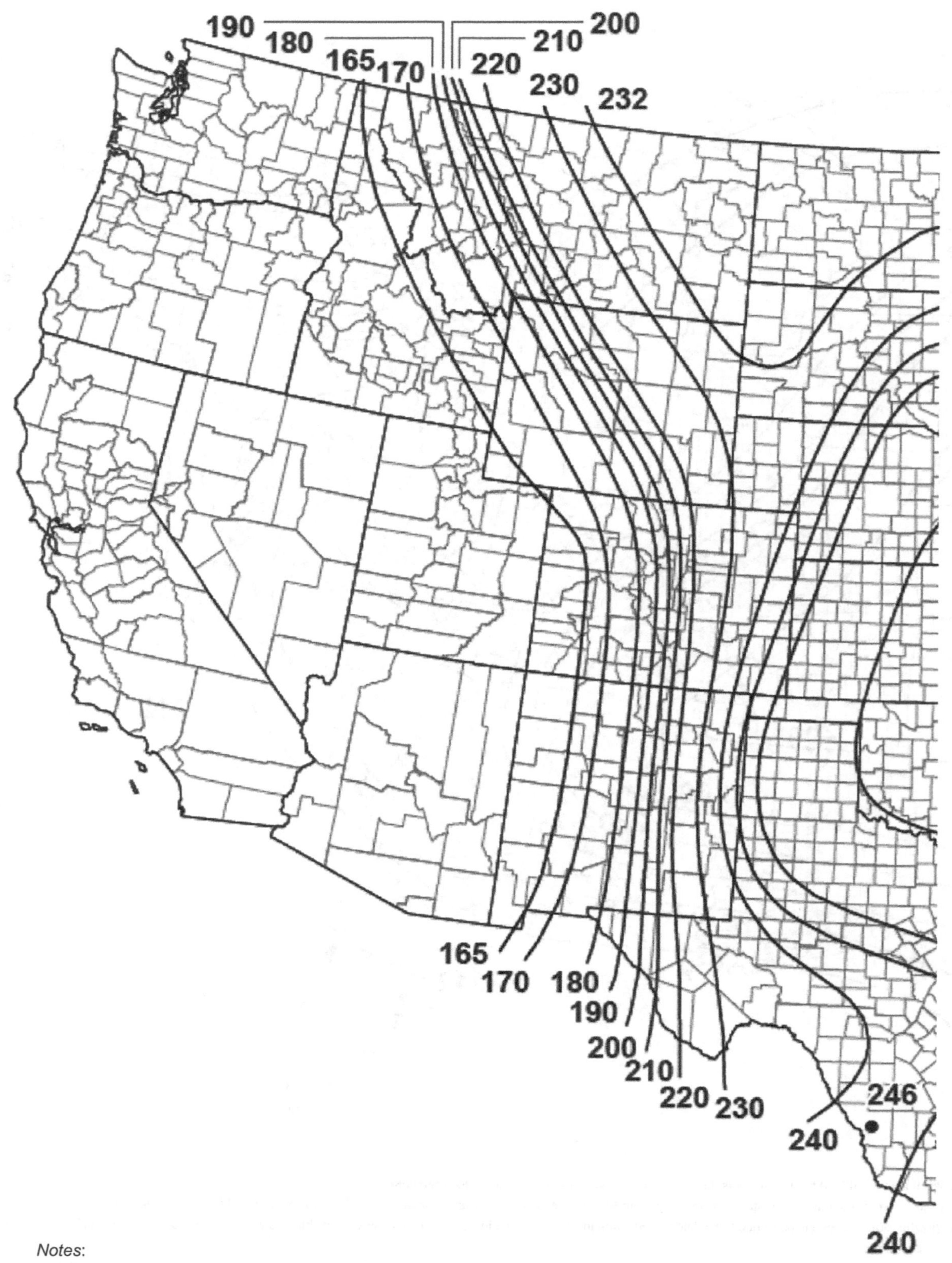

Notes:

1. Values are 3 s gust speeds in mi/h at 33 ft (10 m) above ground.
2. To convert tornado speeds from mi/h to m/s, multiply mapped values by 0.447.
3. Linear interpolation is permitted between contours. Point values (where shown) are provided to aid with interpolation.

Figure G.2-4E. Tornado speeds for 10,000,000-year MRI for effective plan area of 100,000 ft² (9,290 m²).

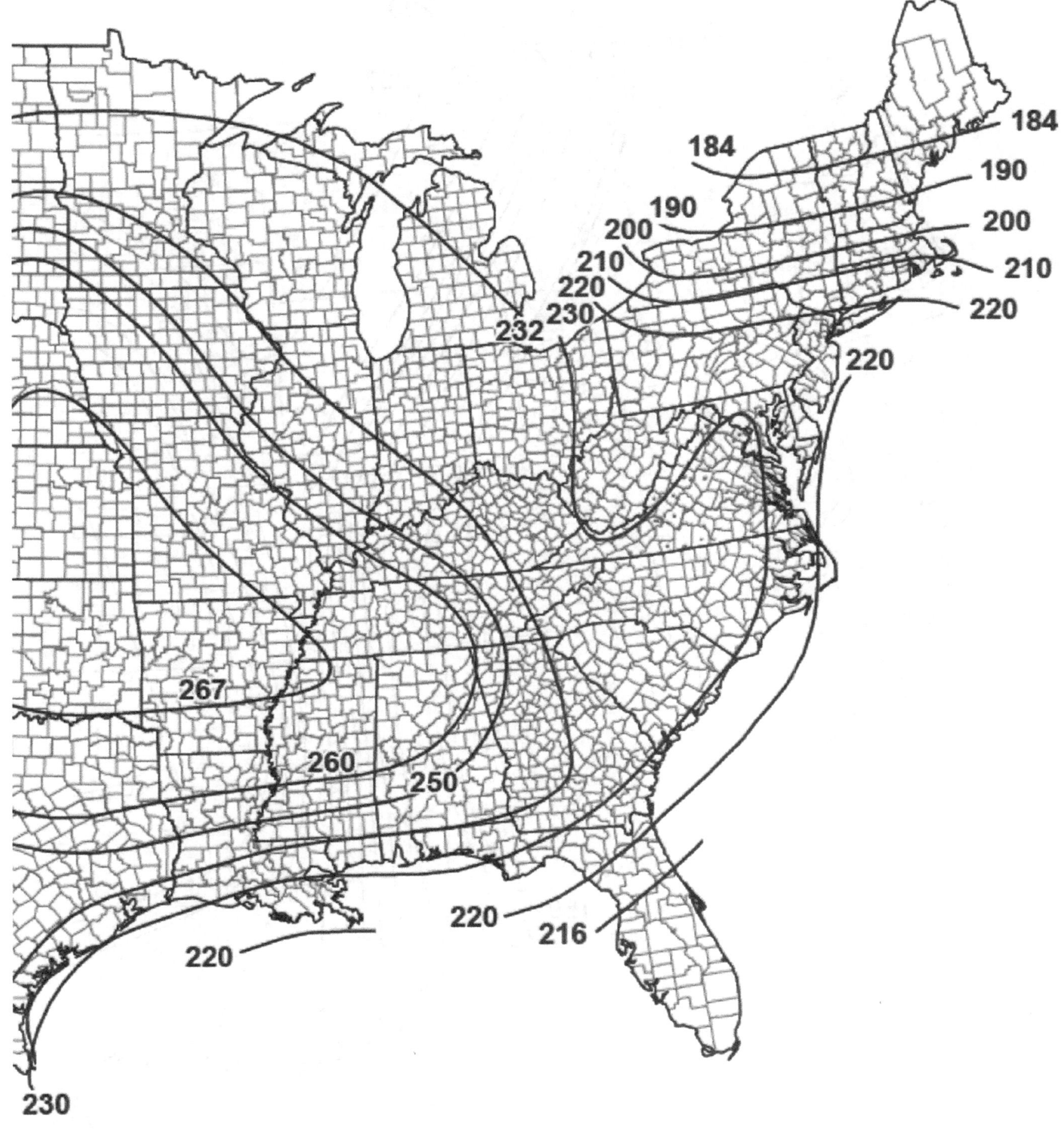

4. Islands, coastal areas, and land boundaries outside the last contour shall use the last tornado speed contour.
5. Tornado speeds correspond to approximately a 0.0005% probability of exceedance in 50 years (annual exceedance probability = 0.0000001, MRI = 10,000,000 years).
6. Location-specific tornado speed is permitted to be determined using the ASCE Tornado Design Geodatabase, available at the ASCE 7 Hazard Tool (https://asce7hazardtool.online), or approved equivalent.

Figure G.2-4E (*Continued*). Tornado speeds for 10,000,000-year MRI for effective plan area of 100,000 ft^2 (9,290 m^2).

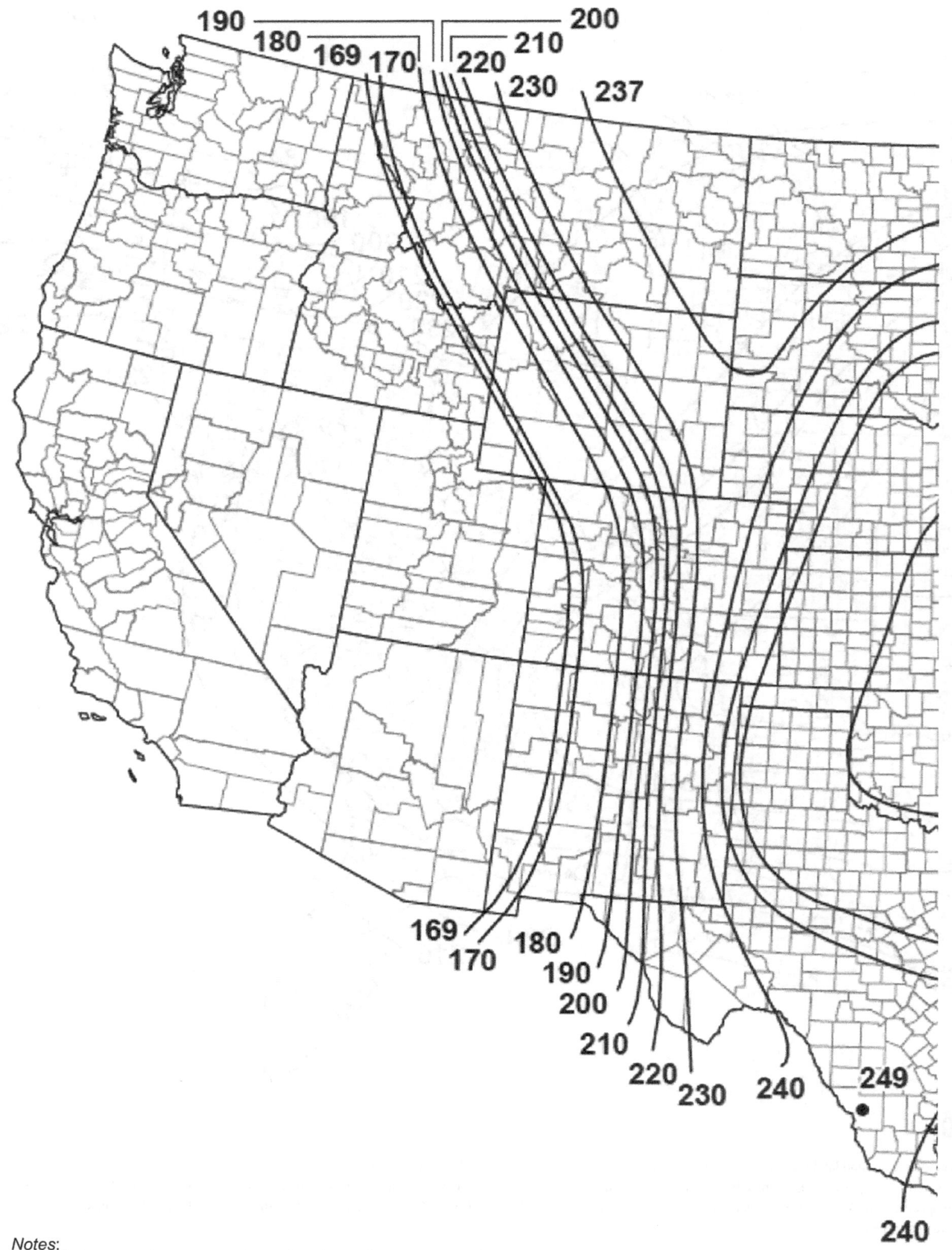

Notes:

1. Values are 3 s gust speeds in mi/h at 33 ft (10 m) above ground.
2. To convert tornado speeds from mi/h to m/s, multiply mapped values by 0.447.
3. Linear interpolation is permitted between contours. Point values (where shown) are provided to aid with interpolation.

Figure G.2-4F. Tornado speeds for 10,000,000-year MRI for effective plan area of 250,000 ft^2 (23,226 m^2).

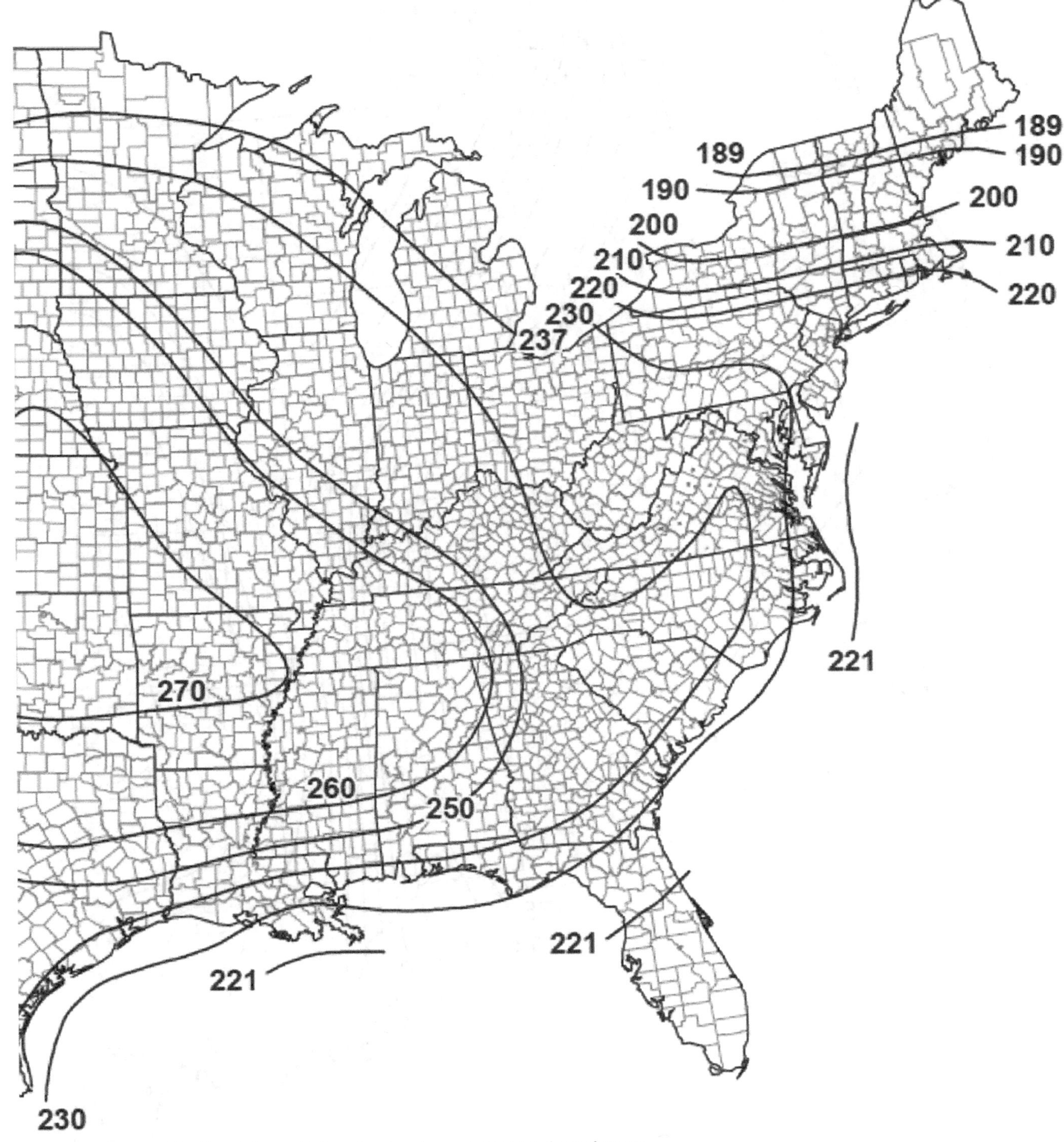

4. Islands, coastal areas, and land boundaries outside the last contour shall use the last tornado speed contour.
5. Tornado speeds correspond to approximately a 0.0005% probability of exceedance in 50 years (annual exceedance probability = 0.0000001, MRI = 10,000,000 years).
6. Location-specific tornado speed is permitted to be determined using the ASCE Tornado Design Geodatabase, available at the ASCE 7 Hazard Tool (https://asce7hazardtool.online), or approved equivalent.

Figure G.2-4F (*Continued*). Tornado speeds for 10,000,000-year MRI for effective plan area of 250,000 ft² (23,226 m²).

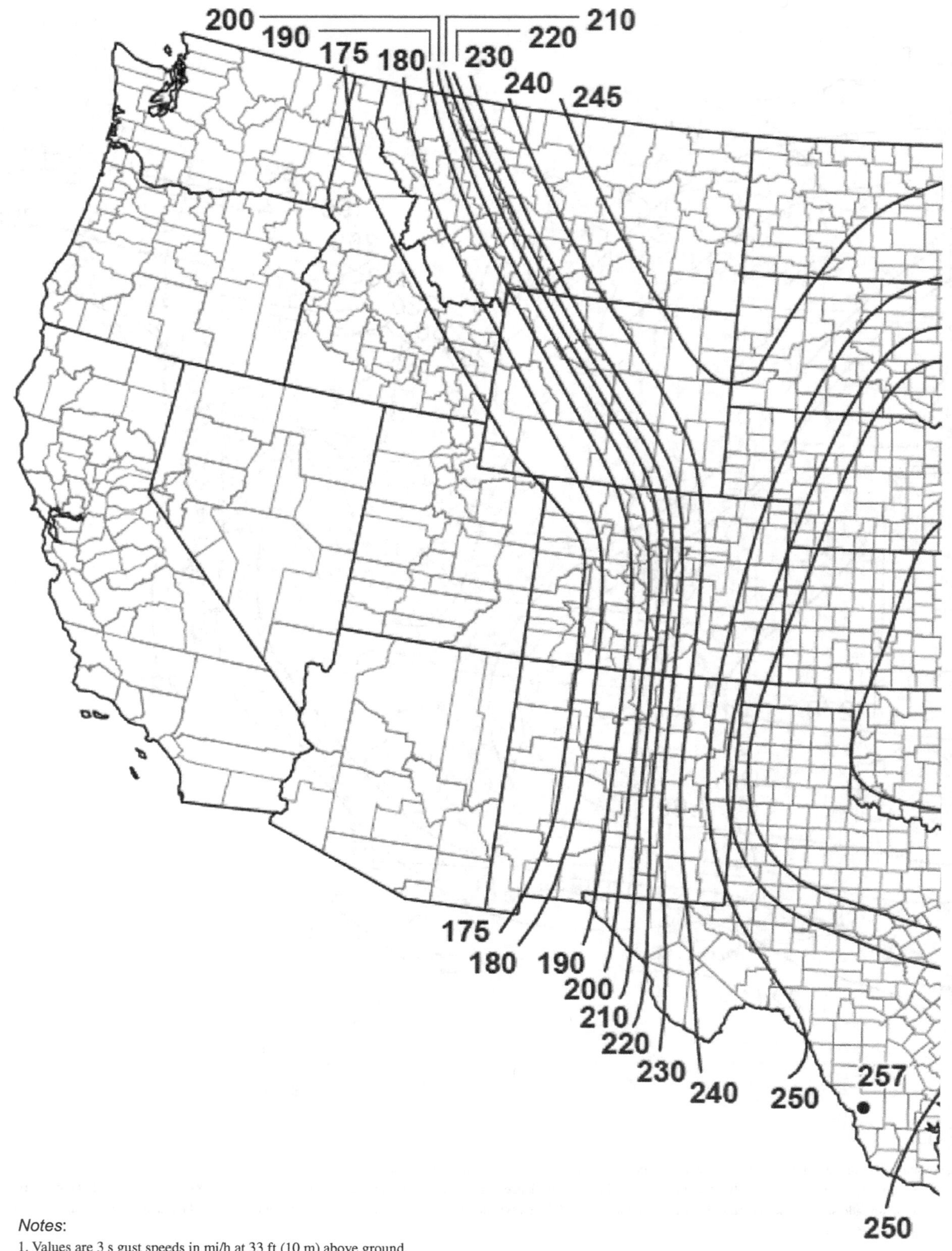

Notes:

1. Values are 3 s gust speeds in mi/h at 33 ft (10 m) above ground.
2. To convert tornado speeds from mi/h to m/s, multiply mapped values by 0.447.
3. Linear interpolation is permitted between contours. Point values (where shown) are provided to aid with interpolation.

Figure G.2-4G. Tornado speeds for 10,000,000-year MRI for effective plan area of 1,000,000 ft^2 (92,903 m^2).

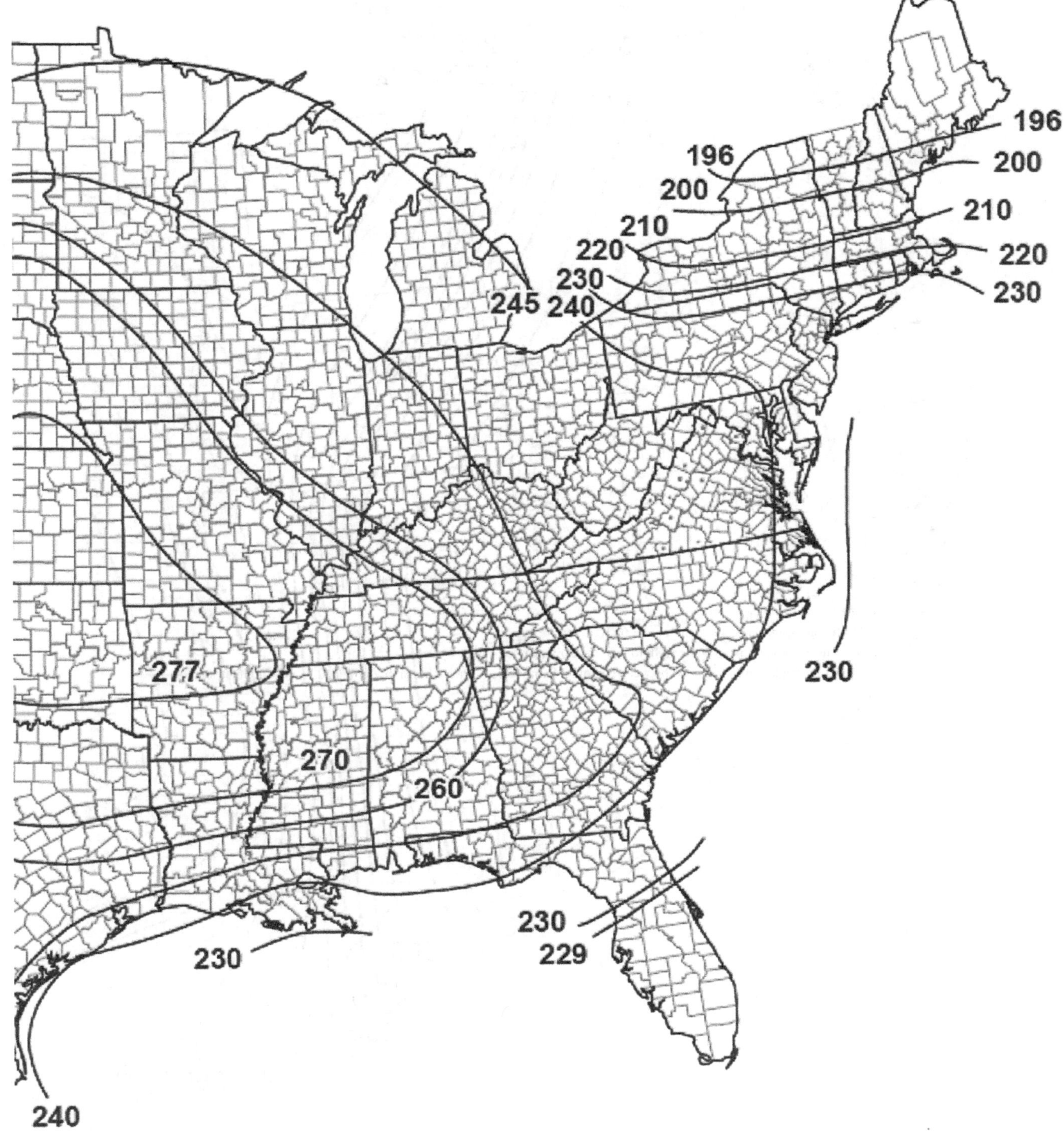

4. Islands, coastal areas, and land boundaries outside the last contour shall use the last tornado speed contour.
5. Tornado speeds correspond to approximately a 0.0005% probability of exceedance in 50 years (annual exceedance probability = 0.0000001, MRI = 10,000,000 years).
6. Location-specific tornado speed is permitted to be determined using the ASCE Tornado Design Geodatabase, available at the ASCE 7 Hazard Tool (https://asce7hazardtool.online), or approved equivalent.

Figure G.2-4G (*Continued*). Tornado speeds for 10,000,000-year MRI for effective plan area of 1,000,000 ft² (92,903 m²).

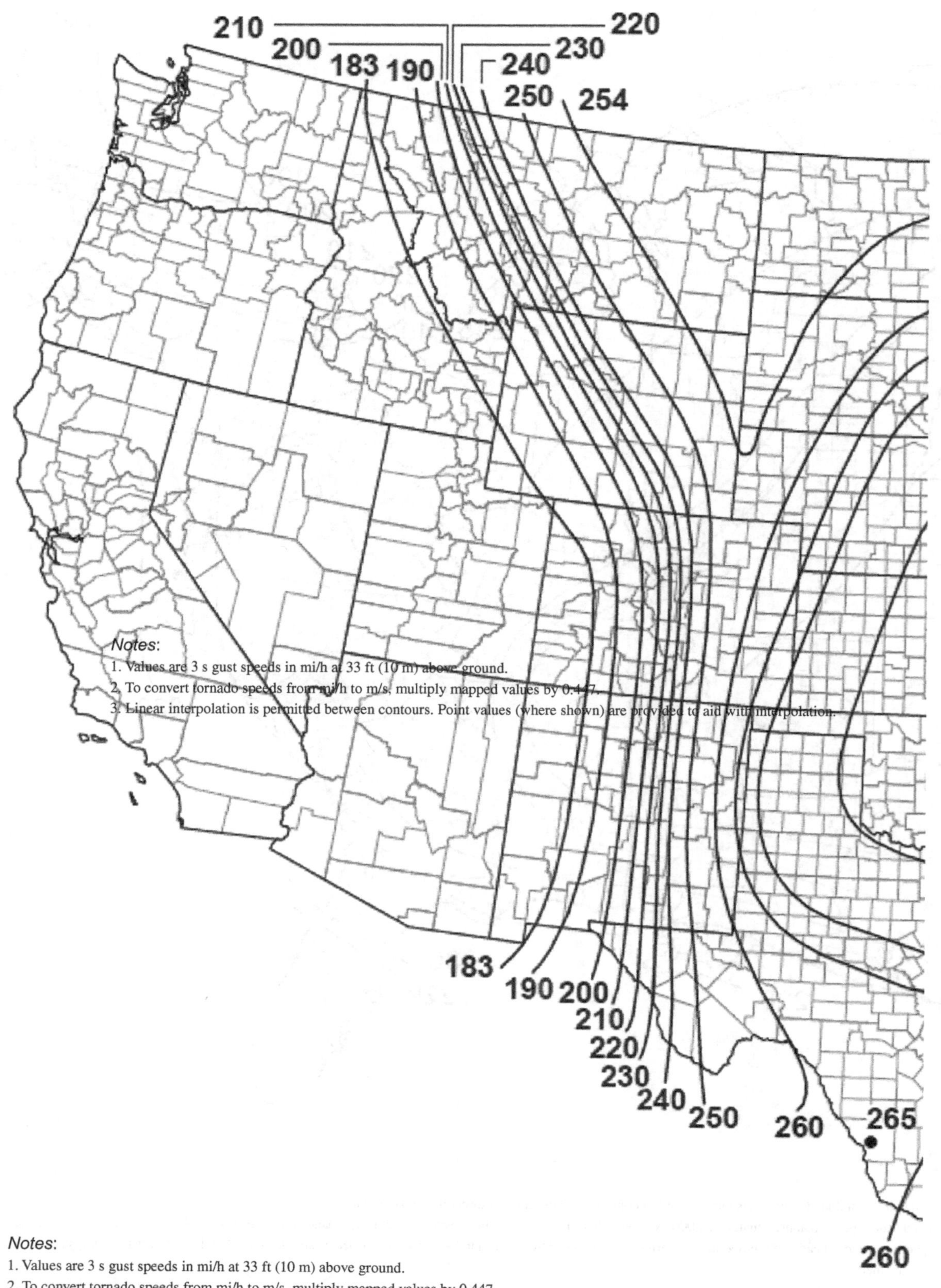

Notes:

1. Values are 3 s gust speeds in mi/h at 33 ft (10 m) above ground.
2. To convert tornado speeds from mi/h to m/s, multiply mapped values by 0.447.
3. Linear interpolation is permitted between contours. Point values (where shown) are provided to aid with interpolation.

Figure G.2-4H. Tornado speeds for 10,000,000-year MRI for effective plan area of 4,000,000 ft^2 (371,612 m^2).

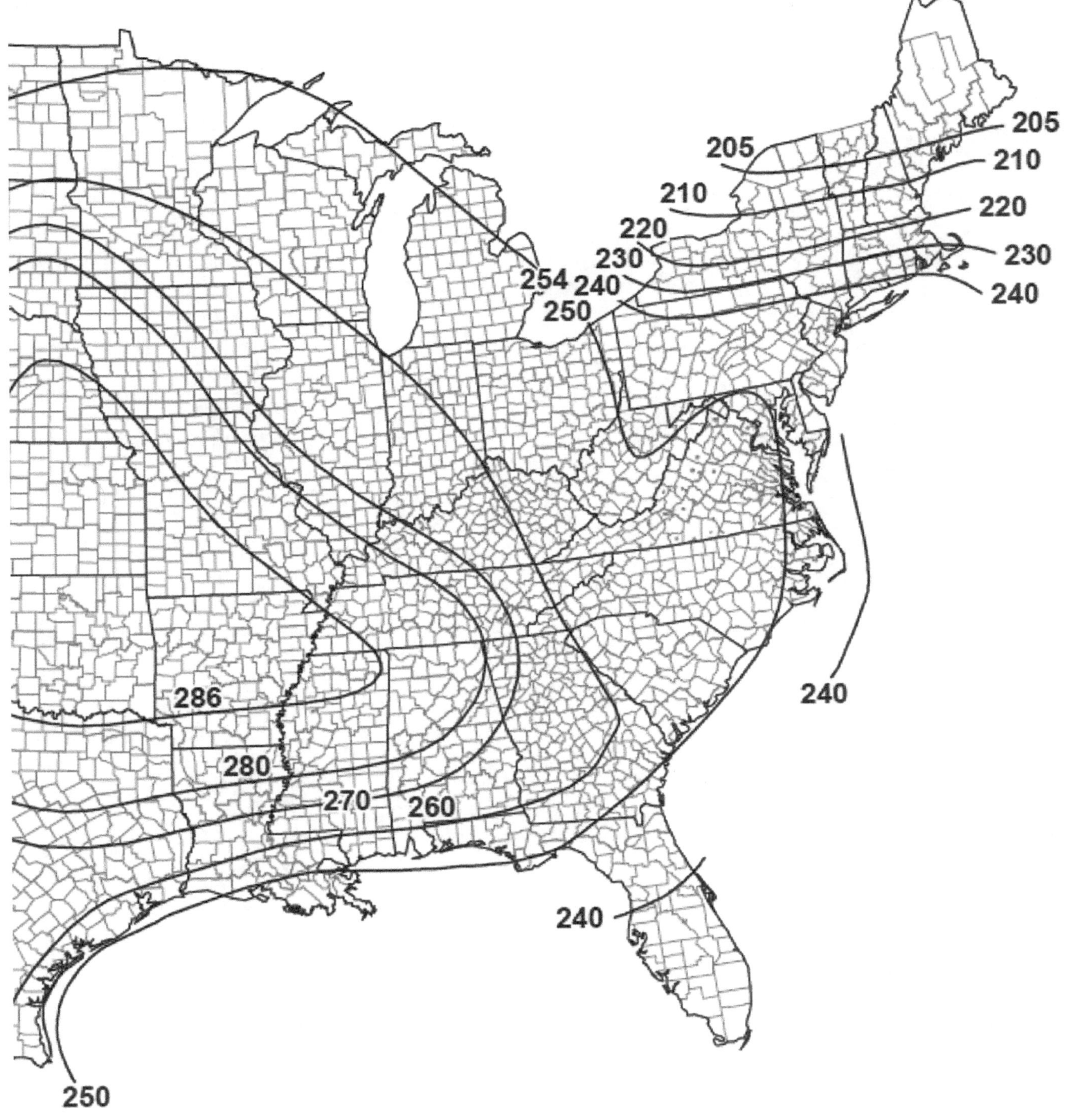

4. Islands, coastal areas, and land boundaries outside the last contour shall use the last tornado speed contour.
5. Tornado speeds correspond to approximately a 0.0005% probability of exceedance in 50 years (annual exceedance probability = 0.0000001, MRI = 10,000,000 years).
6. Location-specific tornado speed is permitted to be determined using the ASCE Tornado Design Geodatabase, available at the ASCE 7 Hazard Tool (https://asce7hazardtool.online), or approved equivalent.

Figure G.2-4H (*Continued*). Tornado speeds for 10,000,000-year MRI for effective plan area of 4,000,000 ft^2 (371,612 m^2).

ASCE STANDARD

ASCE/SEI

7-22

Minimum Design Loads and Associated Criteria for Buildings and Other Structures

COMMENTARY

PUBLISHED BY THE AMERICAN SOCIETY OF CIVIL ENGINEERS

Library of Congress Cataloging-in-Publication Data

Library of Congress Control Number: 2021951104

Published by American Society of Civil Engineers
1801 Alexander Bell Drive
Reston, Virginia, 20191-4382
www.asce.org/bookstore | ascelibrary.org

This standard was developed by a consensus standards development process which has been accredited by the American National Standards Institute (ANSI). Accreditation by ANSI, a voluntary accreditation body representing public and private sector standards development organizations in the United States and abroad, signifies that the standards development process used by ASCE has met the ANSI requirements for openness, balance, consensus, and due process.

While ASCE's process is designed to promote standards that reflect a fair and reasoned consensus among all interested participants, while preserving the public health, safety, and welfare that is paramount to its mission, it has not made an independent assessment of and does not warrant the accuracy, completeness, suitability, or utility of any information, apparatus, product, or process discussed herein. ASCE does not intend, nor should anyone interpret, ASCE's standards to replace the sound judgment of a competent professional, having knowledge and experience in the appropriate field(s) of practice, nor to substitute for the standard of care required of such professionals in interpreting and applying the contents of this standard.

ASCE has no authority to enforce compliance with its standards and does not undertake to certify products for compliance or to render any professional services to any person or entity.

ASCE disclaims any and all liability for any personal injury, property damage, financial loss, or other damages of any nature whatsoever, including without limitation any direct, indirect, special, exemplary, or consequential damages, resulting from any person's use of, or reliance on, this standard. Any individual who relies on this standard assumes full responsibility for such use.

Errata: Errata, if any, can be found at https://doi.org/10.1061/9780784415788.

ISBN 978-0-7844-1578-8 (soft cover)
ISBN 978-0-7844-8349-7 (PDF)
ASCE 7 Hazard Tool: https://asce7hazardtool.online/
Online platform: http://ASCE7.online
Manufactured in the United States of America.

27 26 25 24 23 22 1 2 3 4 5

CONTENTS

COMMENTARY TO STANDARD ASCE/SEI 7-22

Provisions contents appear in first book

COMMENTARY TO STANDARD ASCE/SEI 7-22

This commentary is not part of the ASCE standard *Minimum Design Loads and Associated Criteria for Buildings and Other Structures*. It is included for information purposes.

The commentary consists of explanatory and supplementary materials designed to assist local buildings code committees and regulatory authorities in applying the recommented requirements. In some cases, it will be necessary to adjust specific values in the standard to local condidtions. In others, a considerable amount of detailed information is needed to put the provisions into effect. This commentary provides a place for supplying material that can be used in these situations and is intended to created a better understanding of the recommended requirements through brief explanations of the reasoning employed in arriving at them.

The setions of the commentary are numbered to correspond to the sections of the provisions to which they refer. Because it is not necessary to have supplementary material for every section in the standard, there are gaps in numbering in the commentary.

CHAPTER C1
GENERAL

C1.1 SCOPE

The minimum design loads, hazard levels, associated criteria, and intended performance goals contained in this standard are derived from research and observed performance of buildings, other structures, and their nonstructural components under the effects of loads. These parameters vary depending on the relative importance of the building, other structure, or nonstructural component. The loads provided in this standard include loads from both normal operations and rare hazard events. All loads and associated criteria are prescribed to achieve an intended performance, which is defined by a reliability index or limit state exceedance probability or preservation of function during a specified hazard event.

Loads and load combinations are set forth in this document with the intent that they be used together. If one were to use loads from some other source with the load combinations set forth herein or vice versa, the reliability of the resulting design may be affected.

With the 2016 edition of the standard, the title was modified to include the words "and Associated Criteria" to acknowledge what has been in this standard for many editions. For example, earthquake loads contained herein are developed for structures that possess certain qualities of ductility and postelastic energy dissipation capability. For this reason, provisions for design, detailing, and construction are provided in Chapters 11 through 22. In some cases, these provisions modify or add to provisions contained in design specifications. However, this standard only adds associated criteria when the modification is needed to achieve the intended structural performance when subjected to the loads specified herein.

Compliance with this standard is a sufficient condition for demonstrating that changes to an existing structure would meet current standards for safety. However, existing structures were designed and built to the codes and standards in effect at the time of their construction. Those codes and standards may not be consistent with the provisions of this standard. There are evolving philosophies, methodologies, standards, and codes for making decisions concerning the proper levels for safety of new parts, modified parts, existing parts, and the existing or modified structure as a whole. Model building codes such as the *International Existing Building Code* (ICC 2021) can be used as a resource for levels of safety, evaluation, strengthening, and repair requirements for existing structures. This standard does not provide overarching direction whether existing structures undergoing alteration, additions, or repairs should comply with the loads set forth in this standard or some reduced version of those loads. There are specific provisions related to existing buildings, such as Section 7.12 on snow loads on existing roofs. There are other documents, like ASCE 41 (2017), *Seismic Evaluation and Retrofit of Existing Buildings*, or ACI 562 (2019b), *Code Requirements for Assessment, Repair, and Rehabilitation of Existing Concrete Structures and Commentary*, which may contain provisions for specific hazards or construction materials.

C1.3 BASIC REQUIREMENTS

C1.3.1 Strength and Stiffness Buildings and other structures must satisfy strength limit states in which members and components are proportioned to safely carry the design loads specified in this standard to resist buckling, yielding, fracture, and other unacceptable performance. This requirement applies not only to structural components but also to nonstructural elements, the failure of which could pose a substantial safety or other risk. Chapters 30 and 32 of this standard specify wind and tornado loads, respectively, that must be considered in the design of cladding. Chapter 13 of this standard specifies earthquake loads and deformations that must be considered in the design of nonstructural components and systems designated in that chapter.

Although strength is a primary concern of this section, strength cannot be considered independent of stiffness. In addition to considerations of serviceability, for which stiffness is a primary consideration, structures must have adequate stiffness to ensure stability. In addition, the magnitude of load imposed on a structure for some loading conditions, including earthquake, wind, and ponding, is a direct function of the structure's stiffness.

Another important consideration related to stiffness is damage to nonstructural components resulting from structural deformations. Acceptable performance of nonstructural components requires that either the structural stiffness is sufficient to prevent excessive deformations, or that the components can accommodate the anticipated deformations.

Standards produced under consensus procedures and intended for use in connection with building code requirements contain recommendations for resistance factors for use with the strength design procedures of Section 1.3.1.1 or allowable stresses (or safety factors) for the allowable stress design procedures of Section 1.3.1.2. The resistances contained in any such standards have been prepared using procedures compatible with those used to form the load combinations contained in Sections 2.3 and 2.4. When used together, these load combinations and the companion resistances are intended to provide reliabilities approximately similar to those indicated in Tables 1.3-1, 1.3-2, and 1.3-3. Some standards known to have been prepared in this manner include the following:

ACI 318. *Building Code Requirements for Structural Concrete*. American Concrete Institute.

AISC 341. *Seismic Provisions for Structural Steel Buildings*. American Institute of Steel Construction.

AISC 360. *Specification for Structural Steel Buildings*. American Institute of Steel Construction.

AISI S100. *North American Specification for the Design of Cold-Formed Steel Structural Members*. American Iron and Steel Institute.

AWC NDS. *National Design Specification for Wood Construction*. American Wood Council.

AWC SDPWS. *Special Design Provisions for Wind and Seismic*. American Wood Council.

SEI/ASCE 8. *Specification for the Design of Cold-Formed Stainless Steel Structural Members*. ASCE.

Specification for Aluminum Structures. Aluminum Association.

TMS 402. *Building Code Requirements for Masonry Structures*. The Masonry Society.

TMS 602. *Specification for Masonry Structures*. The Masonry Society.

C1.3.1.3 Performance-Based Procedures Section 1.3.1.3 introduces alternative performance-based procedures that may be used in lieu of the procedures of Sections 1.3.1.1 and 1.3.1.2 to demonstrate that a building or other structure, or parts thereof, has sufficient strength. These procedures are intended to parallel the so-called alternative means and methods procedures that have been contained in building codes for many years. Such procedures permit the use of materials, design, and construction methods that differ from the prescriptive requirements of the building code, or in this case the standard, that can be demonstrated to provide equivalent performance. Such procedures are useful in that they permit innovation and the development of new approaches before the building codes and standards have an opportunity to provide for these new approaches. In addition, these procedures permit the use of alternative methods for certain special structures, which may not be covered by code but by means of their occupancy, use, or other features, can provide acceptable performance without compliance with the prescriptive requirements.

The procedures in Section 1.3.1.3 are intended to apply to all hazards specified in the standard. Some chapters of the standard have specific provisions for performance-based design for that specific load or environmental hazard while others do not. The lack of performance-based design provisions within a chapter should not be taken as a prohibition on the use of performance-based procedures for that specific load or environmental hazard.

The reliability of a proposed design does not need to be evaluated when the standard's design procedures in Sections 1.3.1.1 and 1.3.1.2 are applied. However, when performance-based procedures are used, the reliability achieved for the proposed design should be consistent with the target reliabilities stipulated in Section 1.3.1.3.

Alternative design methods have a range of implementation levels. Such methods are addressed in standards or best practice guidelines that address performance-based design goals and methodologies that incorporate the fundamental basis of reliability analysis. A minimum level of alternative design would involve standard design procedures, with analyses based on the requirements of Section 1.3.1.3. For example, a building that exceeds the code limits for building height could be designed with applicable codes and standards, and member demand and capacities would be checked to determine their adequacy for the design loads and conditions. For seismic design, the provisions of the ASCE 41 standard and of the *Guidelines for Performance-Based Seismic Design of Tall Buildings* (PEER 2010) were either calibrated by structural performance level or were demonstrated in comparison with prescriptive design methods to provide reliabilities equal to or better than those given in Table 1.3-2.

Flood load factors in ASCE 7 may not achieve the reliability targets of Table 1.3-1. For structures in the 100-year coastal flood zone, the load factor of 2.0 was based on a beta value of 2.5 (Mehta et al. 1998) rather than 3.0. For structures not in the coastal zone, the load factor of 1.0 reflects the prescriptive minimum 100-year flood elevation for stillwater flooding; thus, this flood has a 1% annual chance of being exceeded, which is essentially a beta of 1.3. For Risk Category III and IV structures, no reliability analysis has been performed. For storm events of greater return period, the flood hazard expands both in spatial extent and in depth. For those structures and components of structures that were not subject to the prescriptive flood design requirements, any design resistance depends on that imparted from design for other hazards.

Extraordinary events arise from service or environmental conditions that are not covered in the remaining chapters of ASCE 7. Such events are characterized by a low probability of occurrence and relatively short duration. While the occurrence of any of these events may cause damage ranging from local failure to incipient collapse, their occurrence is difficult or impossible to characterize statistically from current knowledge and data. Designing for low-probability, high-consequence events requires an approach that addresses the unique risks posed by such events and recognizes that performance-based approaches often will be used for dealing with them. Accordingly, Table 1.3-5 and Section 2.5 are based on a *scenario* approach, in which the structural design criteria are based on *conditional* reliability targets, given the occurrence of extraordinary event scenarios identified by the design team and approved by the Authority Having Jurisdiction to achieve structural performance that is essential for public safety. The target reliabilities in Table 1.3-5 are consistent with the discussion of conditional reliabilities in Section C2.5 but have been expanded to cover the four risk categories in Table 1.5-1. The reader is directed to Section C2.5 for a detailed discussion of these reliability considerations.

The alternative procedures of Section 1.3.1.3 may be used to demonstrate adequacy for one or more design loads, while the standard procedures of Sections 1.3.1.1 and 1.3.1.2 are used to demonstrate adequacy for other design loads. For example, it is relatively common to use the alternative procedures to demonstrate adequate earthquake, fire, or blast resistance, while the standard prescriptive procedures of Sections 1.3.1.1 and 1.3.1.2 are used for all other loading considerations.

The alternative procedures of Section 1.3.1.3 are intended to be used in the design of individual projects, rather than as the basis for broad qualification of new structural systems, products, or components. Procedures for such qualification are beyond the scope of this standard, because the limited number of test data required in Section 1.3.1.3.2 are not appropriate for the application of new materials in a structural system or the development of prefabricated structural assemblies intended for general widespread use in structural systems. A more robust level of testing is needed for new materials or structural assemblies for general use.

Section 1.3.1.3 requires demonstration that a design has adequate strength to provide an equivalent or lower probability of failure under load than that adopted as the basis for the prescriptive requirements of this standard for buildings and structures of comparable risk category. Tables 1.3-1, 1.3-2, and 1.3-3 summarize performance goals, expressed in terms of target reliabilities, associated with protection against structural failure that approximate those notionally intended to be achieved using the design procedures of Section 2.3. The target reliability indexes are provided for a 50-year reference period, and the probabilities of failure are annual probabilities. Annualized probabilities can be applied to limit states where the loads and member resistance do not vary over the reference period. If a member is subject to degradation, such as corrosion, then degradation effects over the service period should be considered through a time-varying or stochastic process as part of the reliability analysis.

The target reliabilities have been developed and vetted by a number of consensus groups over a period of more than 30 years and have been confirmed through professional practice in AISC 360 (AISC 2022b), ACI 318 (ACI 2019a), and other standards and documents. The target reliabilities for Risk Category II in Table 1.3-1 are based on probabilistic analyses of structural member performance for strength design procedures and are documented in Ellingwood et al. (1980, 1982) and Galambos et al. (1982). The reliabilities are consistent with those adopted by the NBCC M-32A (2010), TC 250 (CEN 2002b), and ISO 2394 (1998). Structural members and connections designed using typical specifications for engineering materials (steel, reinforced concrete, masonry, and timber) were analyzed to determine their reliability for common limit states, such as yielding in tension members, formation of plastic hinges in compact laterally supported beams, or column buckling and connection fracture for a nominal service period of 50 years. The reliabilities were initially determined for load combinations involving dead, live, wind, snow, and earthquake loads. The target reliabilities listed in Table 1.3-1 are based on strength criteria for structural members. The target reliabilities listed in Table 1.3-2 for strength and deflection limit states are based on strength and deflection criteria for system response to earthquakes where inelastic behavior is assumed. The target reliabilities for Risk Categories I, III, and IV were determined by reviewing the intended performance of structural members and systems as well as target reliabilities specified by other codes and standards for similar performance criteria.

Seismic design practice has evolved in the last three decades from the original reliability basis mentioned previously. The target reliabilities in Table 1.3-2 for earthquake-resistant structural system design are defined for the response of the structural system as described in NIST (2012), which was prepared by NEHRP Consultants Joint Venture. For Risk Category I and II structures, that is, $I_e = 1.0$, acceptable Life Safety risk is defined by an "absolute" collapse probability of 1% in 50 years and a "conditional" probability of 10% given risk-targeted maximum considered earthquake (MCE_R) ground motions. The conditional probability of 10% is based on FEMA P-695 (2009) methodology. The absolute probability of 1% in 50 years and the conditional probability of 10% given MCE_R ground motions were used by the US Geological Survey to develop the probabilistic MCE_R ground motions of ASCE/SEI 7-10. The conditional probabilities of 5% (Risk Category III structures) and 2.5% (Risk Category IV structures) represent improved reliability anticipated for structures designed with an Importance Factor, I_e, greater than 1.0. Although not specifically stated by the ASCE/SEI 7-10 commentary, it may be presumed that Risk Categories III and IV structures have absolute collapse probability objectives that are less than 1% in 50 years (i.e., due to design using an Importance Factor of $I_e > 1.0$).

Engineers may need load criteria for strength design that are consistent with the requirements in this standard for situations that are not covered explicitly within this standard. They may also need to consider load criteria for special situations, as required by the client in performance-based engineering applications, in accordance with Section 1.3.1.3. In addition, groups writing standards and specifications for strength design of structural systems and elements may need to develop resistance factors that, when used with the load requirements in this standard, permit the stipulated reliability to be achieved. Such load criteria should be developed using an accepted procedure, such as that provided in Section 2.3.6, to ensure that the resulting factored design loads and load combinations are consistent with target reliabilities (or levels of performance), the common load criteria in Section 2.3.2, and existing standards and specifications governing strength design for common construction materials. Peer-reviewed statistical data for loads in this standard, adopted from Ellingwood et al. (1980, 1982) and Galambos et al. (1982), are provided in Table C1.3-1. The statistics provided are the ratio of the mean, X_m, to nominal, X_n, values of the load and the coefficient of variation (COV) of a cumulative distribution function (CDF) fitted to the 90th percentile and above of the probability distribution of the load. The parameters for S (snow) are based on data for the northeastern quadrant of the United States. The environmental hazards in this Standard do not reflect potential effects of climate variability and climate change.

Table C1.3-1. Load Distributions and Parameters.

Load	X_m/X_n	V_x	Cumulative Distribution Function (CDF)
Dead, D	1.05	0.10	Normal
Live, L	1.00	0.25	Type I
Wind, W	Site-dependent		Type I
Snow, S	Site-dependent	0.26	Generalized Extreme Value
Earthquake, E	Site-dependent		Type II

The reliability of structural members can be determined through a reliability analysis, such as a Monte Carlo analysis with random variables assigned probability distributions with mean and COV values based on statistical data. Reliability analyses can also be conducted using a nonparametric hazard curve based on data. Figure C1.3-1 provides an example of the design equation, limit state equations, and statistical variables for a compact steel flexural member designed for dead plus live load. The statistics used in this example are typical (compare Table C1.3-1). The user should determine the appropriate probabilistic models for his or her design situation.

Figure C1.3-1 presents an illustration of how the reliability index, β, is determined for the common case of a compact steel beam with full lateral support, in which the limit state is the formation of the first plastic hinge. The reliability depends on the ratio of nominal live to dead load in the limit state equation. For $L_n/D_n = 2.0$ (a typical value), the probability of failure (50-year basis) is 0.00298. The corresponding reliability index $\beta = \Phi^{-1}(1 - P_f) = 2.75$ (50-year basis), and $P_f = 6 \times 10^{-5}$ (annual basis). These reliabilities can be compared to the reliability targets in Table 1.3-1. The variation in β with L_n/D_n is very small; unlike in allowable stress design (ASD), the dead and live load factors of 1.2 and 1.6 were selected so as to properly reflect the differences in variability between dead and live load.

The system reliabilities for earthquake are different from those for other environmental hazards because the design philosophy of the standard is to prevent system collapse in the risk-targeted maximum considered earthquake (MCE_R) shaking. The R, C_d, and Ω_0 coefficients specified in Chapter 12 for seismic loading, together with the systems detailing requirements specified in the referenced standards, are intended to ensure minimum acceptable

Design Equation

$$0.9R_n = 1.2D_n + 1.6L_n$$

Limit State Equation $G(X)$

$$G(R, D, L) = R - D - L$$

$$G(X) = \left(\left[1.2 + 1.6\left(\frac{L_n}{D_n}\right)\right] / 0.9\right) X_1 - X_2 - \left(\frac{L_n}{D_n}\right) X_3$$

where

R = Strength random variable, and R_n = Nominal strength;

D = Dead load random variable, and D_n = Nominal dead load;

L = Live load random variable, and L_n = Nominal live load;

$X_1 = R/R_n$, $X_2 = D/D_n$, $X_3 = L/L_n$; and

Typical range of L/D_n is 0.5 to 4.0.

Statistics

Variable	Mean	Coefficient of Variation (COV)	Probability Density Function (PDF)
X_1	1.08	0.09	Lognormal
X_2	1.05	0.10	Normal
X_3	1.00	0.25	Type I

Figure C1.3-1. Equations and statistics of load and resistance parameters for a Monte Carlo analysis to determine achieved reliability.

probabilities of structural collapse, given the occurrence of MCE_R ground shaking. As discussed in Section C11.4, for typical structures (Risk Categories I and II), the conditional probability of collapse is assumed to be 10%, given the occurrence of the MCE_R. This assumption is based on significant research documented in FEMA P-695 (2009). The additional collapse goals of 5% for Risk Category III and 2.5% for Risk Category IV were arrived at by assuming that the seismic fragility (probabilistic model of system strength) is described by a lognormal distribution with a logarithmic standard deviation of 0.6 and adjusting the strength of the structure by the earthquake Importance Factors of 1.25 and 1.5, respectively. Since collapse is a function of loading (ground shaking) intensity, still lower but nonnegligible probabilities of collapse also exist at design shaking levels. The collapse risk for design earthquake shaking is approximately 2.5%, 1%, and 0.5% for Risk Categories II, III, and IV structures, respectively.

This standard also seeks to protect against local failure that does not result in global collapse but could result in injury risk to a few persons. Chapter 16 of the standard defines structural elements according to their criticality as critical, ordinary, and noncritical, where critical elements can lead to global collapse, ordinary elements to endangerment of a limited number of lives, and noncritical elements do not have safety consequences. For ordinary elements in Risk Category II structures, the standard accepts a 25% probability of failure given MCE_R shaking (approximately 10% probability of failure for design earthquake shaking). Failure probabilities for ordinary elements in Risk Category III and IV structures are, respectively, 15 and 9% for MCE_R shaking and 4% and 2% for DE shaking. It is anticipated that the failure probabilities for anchorage of rigid nonstructural components attached at grade to the structure may be in the same range as the probabilities for ordinary elements. However, the uncertainties associated with the reliability for anchorage of rigid nonstructural components that are elevated within a structure are much higher than for structural elements because the methods used to characterize the strength demands on nonstructural components are more approximate than those used for overall building demands, and appropriate reliability levels have not yet been established for them. Furthermore, demands on anchorage of flexible nonstructural components and distributions are significantly more complex, especially when the points of attachment of the nonstructural components are elevated within a structure and one needs to consider both inertial effects and relative displacements. Future study should seek to evaluate nonstructural reliabilities in a rigorous manner, and if consensus can be achieved regarding the appropriate reliability levels for anchorage, to adjust the design procedure for anchorage of these components to achieve these appropriate reliabilities.

The reliabilities for tsunami loads are for the performance-based design of inundated structural elements to prevent failure from hydrodynamic loading during the Maximum Considered Tsunami (MCT). The target reliabilities given in the standard have their basis in the analysis by Chock et al. (2016). Probabilistic limit-state reliabilities were computed for representative structural components carrying gravity and tsunami loads, using statistical information on the key hydrodynamic loading parameters and resistance models with specified tsunami load combination factors. The prescriptive design criteria for waterborne debris impacts and foundation scour are not probabilistic, and therefore these effects are not included in Table 1.3-4. Guidance for performance-based design for tsunami loads can be found in Chapter 6 and its associated commentary.

It is important to note that provision of adequate strength is not sufficient to ensure proper performance. Considerations of serviceability and structural integrity are also important. Use of the alternative procedures of Section 1.3.1.3 is not intended as an alternative to the requirements of Sections 1.3.2, 1.3.3, 1.3.4, 1.3.5, 1.3.6, or 1.4 of this standard.

The requirements of this standard and its companion referenced standards are intended to provide protection against structural failure. They are also intended to provide property and economic protection for small events, to the extent practical, as well as improve the probability that critical facilities will be functional after severe storms, earthquakes, and similar events. Although these goals are an important part of the requirements of this standard, at the present time there is no documentation of the reliability intended with respect to these goals. Consequently, Tables 1.3-1 and 1.3-2 address safety considerations only.

Compliance with Section 1.3.1.3 may be demonstrated by analysis, testing, or a combination of both methods. It is important to recognize that there is uncertainty as to whether the performance objectives tabulated in Tables 1.3-1 and 1.3-2 can be achieved. There is inherent uncertainty associated with prediction of the intensity of loading that a structure will experience, the actual strength of materials incorporated in construction, the quality of construction, and the condition of the structure at the time of loading. Whether testing, analysis, or a combination of these is used, provision should be made to account for these uncertainties and to ensure that the probability of poor performance is acceptably low. See Ellingwood et al. (1982) and Galambos et al. (1982) for estimates of such uncertainties.

Rigorous methods of reliability analysis can be used to demonstrate that the reliability of a design achieves the targets indicated in Tables 1.3-1 and 1.3-2; a simple illustration of such a method is provided in Figure C1.3-1. While such analyses would certainly constitute an acceptable approach to satisfy the requirements of Section 1.3.1.3, these may not be the only acceptable approaches. Consensus bodies or other standards developing organizations may develop guideline documents that provide alternative performance-based design methods or alternate prescriptive procedures that meet or exceed the reliabilities stated in this section.

Since most building officials and other Authorities Having Jurisdiction do not have the expertise necessary to judge the adequacy of designs justified using the Section 1.3.1.3 procedures, independent peer review is an essential part of this process. Peer review can help reduce the potential that the design professional of record will overlook or misinterpret one or more potential behaviors that could result in poor performance. Peer review can also help establish that an appropriate standard of care was adhered to during the design. For peer review to be effective, the reviewers must have the appropriate expertise and understanding of the types of structures, loading, analysis methods, and testing used in the procedures.

The target reliabilities listed in Table 1.3-1 for members and connections and those in Tables 1.3-2 and 1.3-3 for structural systems are included in this standard specifically for application to performance-based procedures for individual projects that are peer reviewed by experts in the field. For several reasons, these target reliabilities are not intended to be compared to reliability indexes developed by material specification groups for general structural applications. For example, reliability indexes for some materials are based on testing of small coupons of the material, supplemented by factors to account for scaling up to structural-sized elements, whereas other materials test full-sized structural members as the foundation for this analysis. Additionally, some reliability analyses use default lognormal data distribution assumptions, and others use distributional forms applicable to the material and to each load combination of interest.

C1.3.1.3.2 Project-Specific Performance Capability Testing Laboratory testing of materials and components constructed from those materials is an essential part of the process of validating the performance of structures and nonstructural components under load. Design resistances specified in the industry standards used with the strength procedures of Section 1.3.1.1 and the allowable stress procedures of Section 1.3.1.2 are based on extensive laboratory testing and many years of experience with the performance of real structures designed using these standards. Similarly, analytical modeling techniques commonly used by engineers to predict the behavior of these systems have been benchmarked and validated against laboratory testing. Similar benchmarking of resistance, component performance, and analytical models is essential when performance-based procedures are used. Where systems and components that are within the scope of the industry standards are used in a design, analytical modeling of these systems and components and their resistances should be conducted in accordance with these standards and industry practice, unless new data and testing suggest that other assumptions are more appropriate. Where new systems, components, or materials are to be used, laboratory testing must be performed to indicate appropriate modeling assumptions and resistances.

No single protocol is appropriate for use in laboratory testing of structural and nonstructural components. The appropriate number and types of tests that should be performed depend on the type of loading the component will be subjected to, the complexity of the component's behavior, the failure modes it may exhibit, the consequences of this failure, and the variability associated with the behavior. Resistances should be selected to provide an acceptably low probability of unacceptable performance. Commentary to Chapter 2 provides guidance on the calculation of load and resistance factors that may be used for this purpose, when LRFD procedures are used.

Regardless of the means used to demonstrate acceptable performance, testing should be sufficient to provide an understanding of the probable mean value and variability of resistance or component performance. For materials or components that exhibit significant variability in behavior as a result of either workmanship, material variation, or brittle modes of behavior, a very large number of tests may be required to properly characterize both the mean values and dispersion. It is seldom possible to conduct such a large number of tests as part of an individual project. Therefore, for reasons of practicality, this standard permits a small number of tests, with the number based on the observed variability. Users are cautioned to conduct tests on material that is representative of that expected to be used in the specific project and to check that all significant sources of variability are included in the test samples. When high variability is observed in these test data, the minimum requirement of six tests is not adequate to establish either the true mean or the variability with confidence, and appropriate caution should be used when developing component resistance or performance measures based on this limited testing. This is a primary reason why the procedures of this section are limited to use on the specific projects being analyzed (i.e., they are not "portable" to similar projects) and that data from these tests are not intended as a means of obtaining prequalification of new systems, materials, or components for broad application.

Some industries and industry standards have adopted standard protocols and procedures for qualification testing. For example, AISC 341 (AISC 2022a), Chapter K, specifies the required testing for qualification of connections used in certain steel seismic force-resisting systems. The wood structural panel industry has generally embraced the testing protocols developed by the Consortium of Universities for Research in Earthquake Engineering project (Krawinkler et al. 2002). When a material, component, or system is similar to those for which such an industry standard exists, the industry standard should be used, unless it can be demonstrated to the satisfaction of peer review and the Authority Having Jurisdiction that more appropriate results will be attained by using alternative procedures and protocols.

When data from Section 1.3.1.3.2 testing are used to characterize a variable in the reliability analysis, sample sizes shall be sufficient to define the mean and coefficient of variation of the test results within specific confidence bounds determined by the significance of that variable in the reliability analysis. While most testing being conducted in accordance with Section 1.3.1.3.2 will be used primarily to confirm or supplement engineering analyses, it is possible that test data will also be used to characterize one of the random variables that are part of the reliability analysis. Because each variable in the reliability analysis will influence the final computed reliability index, each test-based variable must be subjected to a reasonable amount of statistical rigor. For very small sample sizes, it is not possible to define the mean or the standard deviation precisely. However, there are well-established methods to compute confidence bounds on those parameters. For example, a conservative estimate of the mean value might be the lower 75% confidence bound.

C1.3.2 Serviceability In addition to strength limit states, buildings and other structures must also satisfy serviceability limit states that define functional performance and behavior under loads normally experienced during the lifetime of the structure or during a time defined specifically for a project or a particular limit state. Serviceability limit states include such items as deflection and vibration. In the United States, strength limit states have traditionally been specified in building codes because they control the safety of the structure. Serviceability limit states, however, are usually noncatastrophic, define a level of quality of the structure or element, and are a matter of judgment as to their application. Serviceability limit states involve the perceptions and expectations of the owner or users and are a contractual matter between the owner or users and the designer and builder. It is for these reasons, and because the benefits are often subjective and difficult to define or quantify, that serviceability limit states for the most part are not included in the model US building codes, with several notable exceptions, such as member deflection limits. In some cases, material design standards provide serviceability limit states for structural elements composed of their material.

The fact that serviceability limit states are usually not codified should not diminish their importance. Exceeding a serviceability limit state in a building or other structure usually means that its function is disrupted or impaired because of local minor damage or deterioration or because of occupant discomfort or annoyance. Therefore, this section states that serviceability limit states and the service loads associated with those limit states should be defined in the project design criteria, which would often be developed in consultation with the owner of the building or other structure. Appendix C and its commentary provide guidance to the designer on developing serviceability design criteria.

Service loads can vary significantly from the design loads specified in this standard. Often the service loads are dependent on the specific serviceability limit state being investigated. For example, beam deflection for a stiffness serviceability limit state has typically been evaluated using the live load as specified in this standard, without a load factor applied to it. While the live load used for evaluating floor vibration caused by footfall has commonly been taken as an estimated average of the actual live load present, often significantly less than the design live load is specified in this standard.

C1.3.3 Functionality Structures in Risk Category IV are intended to have some measure of protection against damage to the structure and to designated nonstructural systems that would preclude the facility from resuming its intended function following the design environmental hazard. For example, in Chapter 13 of the standard, nonstructural systems assigned to Risk Category IV must be tested or verified by analysis to be rugged enough to retain their pre-earthquake function following the design earthquake. There are additional requirements in the seismic chapters (Chapters 11 through 23) to limit structural damage and drift to preserve function of the structure. Because the provisions of this standard require that structures and nonstructural components be designed to remain essentially elastic under most other environmental hazards, preservation of function is generally provided.

When a performance-based design is elected over the prescriptive design procedures in the standard, the registered design professional should confirm that the structure has sufficient strength and stiffness to not incur damage during the design environmental hazard event that would prevent the facility from resuming its intended function, or in some cases or hazards, continue functioning during the design hazard.

Because the nature of function preservation is very broad and encompasses many different structural components and systems subjected to various hazards, there are not specific reliability targets. For that reason, the terms *reasonable probability* and *adequate structural strength* as applied to structural systems are used to indicate that the application of Section 1.1.3 is not an absolute target value. What constitutes reasonable probability depends on many factors, including the recognition that the fragilities of structural systems to ensure function are not well established and should be agreed on by the user, the client, and if applicable, the Authority Having Jurisdiction and the peer reviewers.

The designated nonstructural systems should be composed of components that have adequate strength, stiffness, and ruggedness and are adequately connected or restrained to the structure so that they do not incur damage sufficient to prohibit the function of the system within a specified period of time as the facility is brought back online. This functionality can be demonstrated through analysis or through physical testing of the nonstructural system or components.

Designated nonstructural systems may vary between structures based on the function of the facility. However, systems that are essential to Life Safety are commonly accepted as needed function preservation. Such systems may include fire detection and suppression systems, emergency exit lighting, and systems that contain explosive, toxic, or highly toxic materials. The specific systems and components that should be considered part of the designated nonstructural system should be determined by the user and may require approval by the Authority Having Jurisdiction and, if part of the design, the peer reviewers.

The requirements of this section are not to preclude damage to the structural elements or nonstructural components. In fact, there might be considerable damage to nonessential nonstructural systems and some indications of inelastic deformation of the structure. There should be no structural damage that would indicate the structure is unsafe to support the loads with reliability similar to that required to support it during normal operation. There should be no damage to the nonstructural components that prevents function, such as blocked egress routes. Designated nonstructural systems may have cosmetic damage, but the components can function as they did before the hazard.

The statement that the provisions within this standard are deemed to comply with this section means that through following the prescriptive provisions contained herein, the design professional should be able to provide the intended function preservation performance of Risk Category IV structures in a reasonable period of time consistent with the current state of the practice. In some chapters of this standard, there are other terms that refer to essentially the same components as "Designated Nonstructural System." "Designated Seismic System" in Chapter 13 and "Critical Equipment and Systems" in Chapter 6 are two examples. In those cases, the chapter-specific hazard terms are retained to allow the hazard-specific requirements to be implemented.

C1.3.4 Self-Straining Forces and Effects Indeterminate structures that experience dimensional changes develop self-straining forces and effects. Examples include moments in rigid frames that undergo differential foundation settlements, pretensioning or posttensioning forces as well as any relaxation or loss of such forces sufficient to affect structural performance, and shear forces in bearing walls that support concrete slabs that shrink. Unless provisions are made for self-straining forces and effects, stresses in structural elements, either alone or in combination with stresses from external loads, can be high enough to cause structural damage.

In many cases, the magnitude of self-straining forces can be anticipated by analyses of expected shrinkage, temperature fluctuations, foundation movement, and so forth. However, it is not always practical to calculate the magnitude of self-straining forces. Designers often specify relief joints, suitable framing systems, or other details to minimize self-straining forces and effects.

C1.3.7 Fire Resistance Where there is no applicable building code (the building code under which the structure is designed), the structural fire resistance provisions of the International Building Code (ICC 2021) should be applied. Appendix E provides performance-based design procedures for alternative means specified in Section 1.3.1.

C1.4 GENERAL STRUCTURAL INTEGRITY

Sections 1.4.1 through 1.4.4 present minimum strength criteria intended to ensure that all structures are provided with minimum interconnectivity of their elements and that a complete lateral force-resisting system is present with sufficient strength to provide stability under gravity loads and nominal lateral forces that are independent of design wind, tornado, seismic, or other anticipated loads. Conformance with these criteria provides structural integrity for normal service and minor unanticipated events that may reasonably be expected to occur throughout their lifetimes. For many structures that house large numbers of persons, or that house functions necessary to protect the public safety or occupancies that may be the subject of intentional sabotage or attack, more rigorous protection should be incorporated into designs than provided by these sections. For such structures, additional precautions can and should be taken in the design of structures to limit the effects of local collapse and to prevent or minimize progressive collapse in accordance with the procedures of Section 2.5, as charged by Section 1.4. Progressive collapse is defined as the spread of an initial local failure from element to element, resulting eventually in the collapse of an entire structure or a disproportionately large part of it.

Some authors have defined resistance to progressive collapse to be the ability of a structure to accommodate, with only local failure, the notional removal of any single structural member. Aside from the possibility of further damage that uncontrolled debris from the failed member may cause, it appears prudent to consider whether the abnormal event will fail only a single member.

Because accidents, misuse, and sabotage are normally unforeseeable events, they cannot be defined precisely. Likewise, general structural integrity is a quality that cannot be stated in simple terms. It is the purpose of Section 1.4 and this commentary to direct attention to the problem of local collapse, present guidelines for handling it that will aid the design engineer and promote consistency of treatment in all types of structures and in all construction materials. ASCE does not intend, at this time, for this standard to establish specific events to be considered during design or for this standard to provide specific design criteria to minimize the risk of progressive collapse.

Accidents, Misuse, Sabotage, and their Consequences. In addition to unintentional or willful misuse, some of the incidents that may cause local collapse (Leyendecker et al. 1976) are explosions caused by ignition of gas or industrial liquids, boiler failures, vehicle impact, impact of falling objects, effects of adjacent excavations, gross construction errors, and sabotage. In general, such abnormal events would not be a part of normal design considerations. The distinction between general collapse and limited local collapse can best be made by example, as follows.

General Collapse. The immediate deliberate demolition of an entire structure by phased explosives is an obvious instance of general collapse. Also, the failure of one column in a one-, two-, three-, or possibly even four-column structure could precipitate general collapse because the local failed column is a significant part of the total structural system at that level. Similarly, the failure of a major bearing element in the bottom story of a two- or three-story structure might cause general collapse of the whole structure. Such collapses are beyond the scope of the provisions discussed herein. There have been numerous instances of general collapse that have occurred as the result of such events as bombing, landslides, and floods.

Limited Local Collapse. An example of limited local collapse would be the containment of damage to adjacent bays and stories following the destruction of one or two neighboring columns in a multibay structure. The restriction of damage to portions of two or three stories of a higher structure following the failure of a section of bearing wall in one story is another example.

Examples of General Collapse

Ronan Point. A prominent case of local collapse that progressed to a disproportionate part of the whole building (and is thus an example of the type of failure of concern here) was the Ronan Point disaster, which brought the attention of the profession to the matter of general structural integrity in buildings. Ronan Point was a 22-story apartment building of large precast concrete load-bearing panels in Canning Town, England. In March 1968, a gas explosion in an 18th-story apartment blew out a living room wall. The loss of the wall led to the collapse of the whole corner of the building. The apartments above the 18th story, suddenly losing support from below and being insufficiently tied and reinforced, collapsed one after the other. The falling debris ruptured successive floors and walls below the 18th story, and the failure progressed to the ground. Better continuity and ductility might have reduced the amount of damage at Ronan Point.

Another example is the failure of a one-story parking garage reported in Granstrom and Carlsson (1974). Collapse of one transverse frame under a concentration of snow led to the later progressive collapse of the whole roof, which was supported by 20 transverse frames of the same type. Similar progressive collapses are mentioned in Seltz-Petrash (1979).

Alfred P. Murrah Federal Building. On April 19, 1995, a truck containing approximately 4,000 lb (1,800 kg) of fertilizer-based explosive (ammonium nitrate/fuel oil) was parked near the sidewalk next to the nine-story reinforced concrete office building (Weidlinger 1994, Engineering News-Record 1995, Longinow 1995, Glover 1996). The side facing the blast had corner columns and four other perimeter columns. The blast shock wave disintegrated one of the 20 × 36 in. (510 × 920 mm) perimeter columns and caused brittle failures of two others. The transfer girder at the third level above these columns failed, and the upper-story floors collapsed in a progressive fashion. Approximately 70% of the building experienced dramatic collapse. One hundred sixty-eight people died, many of them as a direct result of the progressive collapse. Damage might have been less had this structure not relied on transfer girders for support of upper floors, if there had been better detailing for ductility and greater redundancy, and if there had been better resistance for uplift loads on floor slabs.

There are a number of factors that contribute to the risk of damage propagation in modern structures (Breen 1976). Among them are the following:

1. There is an apparent lack of general awareness among engineers that structural integrity against collapse is important enough to be regularly considered in design.
2. To give more flexibility in floor plans and to keep costs down, interior walls and partitions are often non-load-bearing and hence may be unable to assist in containing damage.
3. In attempting to achieve economy in structure through greater speed of erection and less construction site labor, systems may be built with minimum continuity, ties between elements, and joint rigidity.
4. Unreinforced or lightly reinforced load-bearing walls in multistory structures may also have inadequate continuity, ties, and joint rigidity.
5. In roof trusses and arches, there may not be sufficient strength to carry the extra loads or sufficient diaphragm action to maintain lateral stability of the adjacent members if one collapses.
6. In eliminating excessively large safety factors, code changes over the past several decades have reduced the large margin of safety inherent in many older structures. The use of higher-strength materials permitting more slender sections compounds the problem in that modern structures may be more flexible and sensitive to load variations and, in addition, may be more sensitive to construction errors.

Experience has demonstrated that the principle of taking precautions in design to limit the effects of local collapse is realistic and can be satisfied economically. From a public safety viewpoint, it is reasonable to expect all multistory structures to possess general structural integrity comparable to that of properly designed, conventionally framed structures (Breen 1976, Burnett 1975).

Design Alternatives. There are a number of ways to obtain resistance to progressive collapse. In Ellingwood and Leyendecker (1978), a distinction is made between direct and indirect design, and the following approaches are defined:

Direct Design: Explicit consideration of resistance to progressive collapse during the design process through either

Alternate Path Method: A method that allows local failure to occur but seeks to provide alternate load paths so that the damage is absorbed and major collapse is averted.

Specific Local Resistance Method: A method that seeks to provide sufficient strength to resist failure from accidents or misuse.

Indirect Design: Implicit consideration of resistance to progressive collapse during the design process through the provision of minimum levels of strength, continuity, and ductility.

The general structural integrity of a structure may be tested by analysis to ascertain whether alternate paths around hypothetically collapsed regions exist. Alternatively, alternate-path studies may be used as guides for developing rules for the minimum levels of continuity and ductility needed to apply the indirect design approach to enhance general structural integrity. Specific local resistance may be provided in regions of high risk, because it may be necessary for some element to have sufficient strength to resist abnormal loads for the structure as a whole to develop alternate paths. Ellingwood and Leyendecker (1978) give specific suggestions for the implementation of each of the defined methods.

Guidelines for the Provision of General Structural Integrity. In general, connections between structural components should be ductile and have a capacity for relatively large deformations and energy absorption under the effect of abnormal conditions. This criterion is met in many different ways depending on the structural system used. Details that are appropriate for resistance to moderate wind loads and seismic loads often provide sufficient ductility. In 1999, ASCE issued a state-of-practice report that is a good introduction to the complex field of blast-resistant design (ASCE 1999).

Work with large precast panel structures (Schultz et al. 1977, PCI Committee on Precast Bearing Walls 1976, Fintel and Schultz 1979) provides an example of how to cope with the problem of general structural integrity in a building system that is inherently discontinuous. The provision of ties, combined with careful detailing of connections, can overcome the difficulties associated with such a system. The same kind of methodology and design philosophy can be applied to other systems (Fintel and Annamalai 1979). The ACI *Building Code Requirements for Structural Concrete* (2019a) includes such requirements in Section 4.10.

There are a number of ways of designing for the required integrity to carry loads around severely damaged walls, trusses, beams, columns, and floors. A few examples of design concepts and details are

1. Good plan layout. An important factor in achieving integrity is the proper plan layout of walls and columns. In bearing-wall structures, there should be an arrangement of interior longitudinal walls to support and reduce the span of long sections of crosswall, thus enhancing the stability of individual walls and of the structures

as a whole. In the case of local failure, this reduction will also decrease the length of wall likely to be affected.

2. Provide an integrated system of ties among the principal elements of the structural system. These ties may be designed specifically as components of secondary load-carrying systems, which often must sustain very large deformations during catastrophic events.
3. Returns on walls. Returns on interior and exterior walls make them more stable.
4. Changing directions of span of floor slab. Where a one-way floor slab is reinforced to span, with a low safety factor, in its secondary direction if a load-bearing wall is removed, the collapse of the slab will be prevented and the debris loading of other parts of the structure will be minimized. Often, shrinkage and temperature steel will be enough to enable the slab to span in a new direction.
5. Load-bearing interior partitions. The interior walls must be capable of carrying enough load to achieve the change of span direction in the floor slabs.
6. Catenary action of floor slab. Where the slab cannot change span direction, the span will increase if an intermediate supporting wall is removed. In this case, if there is enough reinforcement throughout the slab and enough continuity and restraint, the slab may be capable of carrying the loads by catenary action, though very large deflections will result.
7. Beam action of walls. Walls may be assumed to be capable of spanning an opening if sufficient tying steel at the top and bottom of the walls allows them to act as the web of a beam, with the slabs above and below acting as flanges (Schultz et al. 1977).
8. Redundant structural systems. A secondary load path allows framing to survive removal of key support elements. For example, an upper-level truss or transfer girder system may allow the lower floors of a multistory building to hang from the upper floors in an emergency.
9. Ductile detailing. Avoid low-ductility detailing in elements that might be subject to dynamic loads or very large distortions during localized failures (e.g., consider the implications of shear failures in beams or supported slabs under the influence of building weights falling from above).
10. Provide additional reinforcement to resist blast and load reversal when blast loads are considered in design (ASCE 2010).
11. Consider the use of compartmentalized construction in combination with special moment-resisting frames (as defined in FEMA 1997) in the design of new buildings when considering blast protection.

Although not directly adding structural integrity for the prevention of progressive collapse, the use of special, nonfrangible glass for fenestration can greatly reduce risk to occupants during exterior blasts (ASCE 2010). And to the extent that nonfrangible glass isolates a building's interior from blast shock waves, it can also reduce damage to interior framing elements from exterior blasts (e.g., supported floor slabs could be made to be less likely to fail due to uplift forces).

C1.5 CLASSIFICATION OF BUILDINGS AND OTHER STRUCTURES

C1.5.1 Risk Categorization In the 2010 edition of this standard, a new Table 1.5-2 was added that consolidates the various Importance Factors specified for the several types of loads throughout the standard in one location. This change was made to facilitate the process of finding values for these factors. At the same time, the Importance Factors for wind loads were deleted, because the new wind hazard maps adopted by the standard incorporated consideration of less probable design winds for structures assigned to higher risk categories, negating the need for separate Importance Factors. Similarly, no separate Importance Factors are required for tornado loads, which have been added in the 2022 edition. See Commentary Chapter 26 for further discussion.

The risk categories in Table 1.5-1 are used to relate the criteria for maximum environmental loads or distortions specified in this standard to the consequence of the loads being exceeded for the structure and its occupants. For many years, this standard used the term *occupancy category*, as have the building codes. However, the term *occupancy* as used by the building codes relates primarily to issues associated with fire and life safety protection, as opposed to the risks associated with structural failure. The term *risk category* was adopted in place of the older occupancy category term in the 2010 edition of the standard to distinguish between these two considerations. The risk category numbering is unchanged from that in the previous editions of the standard (ASCE 7-98, -02, and -05), but the criteria for selecting a category have been generalized with regard to structure and occupancy descriptions. The reason for this generalization is that the acceptable risk for a building or structure is an issue of public policy rather than purely a technical one. Model building codes such as the International Building Code (ICC 2021) and NFPA 5000 (2006) contain prescriptive lists of building types by occupancy category. Individual communities can alter these lists when they adopt local codes based on the model code, and individual owners or operators can elect to design individual buildings to higher occupancy categories based on personal risk management decisions. Classification continues to reflect a progression of the anticipated seriousness of the consequence of failure, from the lowest risk to human life (Risk Category I) to the highest (Risk Category IV). Elimination of the specific examples of buildings that fall into each category has the benefit that it eliminates the potential for conflict between the standard and locally adopted codes and also provides individual communities and development teams the flexibility to interpret acceptable risk for individual projects.

Historically, the building codes and the standard have used a variety of factors to determine the occupancy category of a building. These factors include the total number of persons who would be at risk were failure to occur, the total number of persons present in a single room or occupied area, the mobility of the occupants and their ability to cope with dangerous situations, the potential for release of toxic materials, and the loss of services vital to the welfare of the community.

Risk Category I structures generally encompass buildings and structures that normally are unoccupied and that would result in negligible risk to the public should they fail. Structures typically classified in this category have included barns, storage shelters, gatehouses, and similar small structures. Risk Category II includes the vast majority of structures, including most residential, commercial, and industrial buildings, and has historically been considered to contain all those buildings and structures not specifically classified as conforming to another category.

Risk Category III includes buildings and structures that house a large number of persons in one place, such as theaters, lecture halls, and similar assembly uses, and buildings with persons having limited mobility or ability to escape to a safe haven in the event of failure, including elementary schools, prisons, and small health-care facilities. This category has also included structures associated with utilities required to protect the health and safety of a community, including power-generating stations and water-treatment and sewage-treatment plants. It has also included structures housing hazardous substances, such as explosives or toxins, which if released in quantity could endanger the surrounding community, such as structures in petrochemical process facilities containing large quantities of hydrogen sulfide (H_2S) or ammonia.

Failures of power plants that supply electricity on the national grid can cause substantial economic losses and disruption to civilian life when their failures can trigger other plants to go offline in succession. The result can be massive and potentially extended to power outage, shortage, or both, leading to huge economic losses because of idled industries and a serious disruption of civilian life because of inoperable subways, road traffic signals, and so forth. One such event occurred in parts of Canada and the northeastern United States in August 2003.

Failures of water- and sewage-treatment facilities can cause disruption to civilian life because these failures can cause large-scale (but mostly non-life-threatening) public health risks caused by the inability to treat sewage and to provide drinking water.

Failures of major telecommunication centers can cause disruption to civilian life by depriving users of access to important emergency information (using radio, television, and phone communication) and by causing substantial economic losses associated with widespread interruption of business.

Risk Category IV has traditionally included structures the failure of which would inhibit the availability of essential community services necessary to cope with an emergency situation. Buildings and structures typically grouped in Risk Category IV include hospitals, police stations, fire stations, emergency communication centers, and buildings with similar uses.

Ancillary structures required for the operation of Risk Category IV facilities during an emergency also are included in this risk category. When deciding whether an ancillary structure or a structure that supports such functions as fire suppression is in Risk Category IV, the design professional must decide whether failure of the subject structure will adversely affect the essential function of the facility. In addition to Essential Facilities, buildings and other structures containing extremely hazardous materials have been added to Risk Category IV to recognize the potentially devastating effect a release of extremely hazardous materials may have on a population.

The criteria that have historically been used to assign individual buildings and structures to occupancy categories have not been consistent and sometimes have been based on considerations that are more appropriate to fire and life safety than to structural failure. For example, university buildings housing more than a few hundred students have been placed into a higher risk category than office buildings housing the same number of persons.

A rational basis should be used to determine the risk category for structural design, which is primarily based on the number of persons whose lives would be endangered or whose welfare would be affected in the event of failure. Figure C1.5-1 illustrates this concept.

"Lives at risk" pertains to the number of people at serious risk of life loss given a structural failure. The risk category classification is not the same as the building code occupancy capacity, which is mostly based on risk to life from fire. The lives at risk from a structural failure include persons who may be outside the structure in question who are nonetheless put at serious risk by failure of the structure. From this concept, emergency recovery facilities that serve large populations, even though the structure might shelter relatively few people, are moved into the higher risk categories.

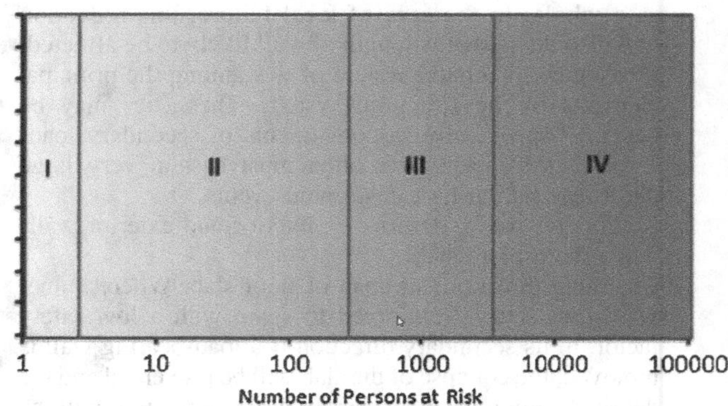

Figure C1.5-1. Approximate relationship between number of lives placed at risk by a failure and risk category.

When determining the population at risk, consideration should also be given to longer-term risks to life than those created during a structural failure. The failure of some buildings and structures, or their inability to function after a severe storm, earthquake, or other disaster, can have far-reaching impact. For example, loss of functionality in one or more fire stations could inhibit the ability of a fire department to extinguish fires, allowing fires to spread and placing many more people at risk. Similarly, the loss of function of a hospital could prevent the treatment of many patients over a period of months.

In Chapters 7, 10, and 11, Importance Factors are presented for the four risk categories identified. The specific Importance Factors differ according to the statistical characteristics of the environmental loads and the manner in which the structure responds to the loads. The principle of requiring more stringent loading criteria for situations in which the consequence of failure may be severe has been recognized in previous versions of this standard by the specification of mean recurrence interval maps for wind speed and ground snow load.

This section now recognizes that there may be situations when it is acceptable to assign multiple risk categories to a building or other structure based on its uses and the types of load conditions being evaluated. For instance, there are circumstances when a building or other structure should appropriately be designed for wind loads associated with risk categories greater than II but would be penalized unnecessarily if designed for seismic loads at that same risk category. An example would be a designated posthurricane recovery shelter in a low seismic area. The building would be classified in Risk Category IV for wind design and in Risk Category II for seismic design.

C1.5.3 Toxic, Highly Toxic, and Explosive Substances A common method of categorizing structures storing toxic, highly toxic, or explosive substances is by the use of a table of exempt amounts of these materials (EPA 1999b, ICC 2000). These references and others are sources of guidance on the identification of materials of these general classifications. A drawback to the use of tables of exempt amounts is that the method cannot handle the interaction of multiple materials. Two materials that are exempt because neither poses a risk to the public by itself may form a deadly combination if combined in a release. Therefore, an alternate and superior method of evaluating the risk to the public of a release of a material is by a hazard assessment as part of an overall risk management plan (RMP).

Buildings and other structures containing toxic, highly toxic, or explosive substances may be classified as Risk Category II structures if it can be demonstrated that the risk to the public from a release of these materials is minimal. Companies that operate industrial facilities typically perform hazard and operability (HAZOP) studies, conduct quantitative risk assessments, and develop risk management and emergency response plans. Federal regulations and local laws mandate many of these studies and plans (EPA 1999a). In addition, many industrial facilities are located in areas remote from the public and have restricted access, which further reduces the risk to the public.

The intent of Section 1.5.2 is for the RMP and the facility's design features that are critical to the effective implementation of the RMP to be maintained for the life of the facility. The RMP and its associated critical design features must be reviewed on a regular basis to ensure that the actual condition of the facility is consistent with the plan. The RMP also should be reviewed whenever consideration is given to the alteration of facility features that are critical to the effective implementation of the RMP.

The RMP generally deals with mitigating the risk to the general public. Risk to individuals outside the facility storing toxic, highly toxic, or explosive substances is emphasized because plant personnel are not placed at as high a risk as the general public because of the plant personnel's training in the handling of toxic, highly toxic, or explosive substances and because of the safety procedures implemented inside the facilities. When these elements (trained personnel and safety procedures) are not present in a facility, then the RMP must mitigate the risk to the plant personnel in the same manner as it mitigates the risk to the general public.

As a result of the prevention program portion of an RMP, buildings and other structures normally falling into Risk Category III may be classified into Risk Category II if means (e.g., secondary containment) are provided to contain the toxic, highly toxic, or explosive substances in the case of a release. To qualify, secondary containment systems must be designed, installed, and operated to prevent migration of harmful quantities of toxic, highly toxic, or explosive substances out of the system to the air, soil, groundwater, or surface water at any time during the use of the structure. This requirement is not to be construed as requiring a secondary containment system to prevent a release of any toxic, highly toxic, or explosive substance into the air. By recognizing that secondary containment shall not allow releases of "harmful" quantities of contaminants, this standard acknowledges that there are substances that might contaminate groundwater but do not produce a sufficient concentration of toxic, highly toxic, or explosive substances during a vapor release to constitute a health or safety risk to the public. Because it represents the "last line of defense," secondary containment does not qualify for the reduced classification.

If the beneficial effect of secondary containment can be negated by external forces, such as the overtopping of dike walls by floodwaters or the loss of liquid containment of an earthen dike because of excessive ground displacement during a seismic event, then the buildings or other structures in question may not be classified into Risk Category II. If the secondary containment is to contain a flammable substance, then implementation of a program of emergency response and preparedness combined with an appropriate fire suppression system would be a prudent action associated with a Risk Category II classification. In many jurisdictions, such actions are required by local fire codes.

Also, as the result of the prevention program portion of an RMP, buildings and other structures containing toxic, highly toxic, or explosive substances also could be classified as Risk Category II for hurricane wind loads when mandatory procedures are used to reduce the risk of release of toxic, highly toxic, or explosive substances during and immediately after these predictable extreme loadings. Examples of such procedures include draining hazardous fluids from a tank when a hurricane is predicted or, conversely, filling a tank with fluid to increase its buckling and overturning resistance. As appropriate to minimize the risk of damage to structures containing toxic, highly toxic, or explosive substances, mandatory procedures necessary for the Risk Category II classification should include preventive measures, such as the removal of objects that might become airborne missiles in the vicinity of the structure.

In previous editions of ASCE 7, definitions of *hazardous* and *extremely hazardous* materials were not provided. Therefore, the determination of the distinction between hazardous and extremely hazardous materials was left to the discretion of the Authority Having Jurisdiction. The change to the use of the terms *toxic* and *highly toxic* based on definitions from federal law (29 CFR 1910.1200 Appendix A with Amendments as of February 1, 2000) has corrected this problem.

Because of the highly quantitative nature of the definitions for toxic and highly toxic in 29 CFR 1910.1200 Appendix A, the General Provisions Task Committee felt that these definitions found in federal law should be directly referenced instead of repeated in the body of ASCE 7. The definitions found in 29 CFR 1910.1200 Appendix A are repeated in the following text for reference.

Highly Toxic. A chemical in any of the following categories:

1. A chemical that has a median lethal dose [LD(50)] of 50 mg or less per kilogram of body weight when administered orally to albino rats weighing between 200 and 300 g each.
2. A chemical that has a median lethal dose [LD(50)] of 200 mg or less per kilogram of body weight when administered by continuous contact for 24 h (or less if death occurs within 24 h) with the bare skin of albino rabbits weighing between 2 and 3 kg each.
3. A chemical that has a median lethal concentration [LC(50)] in air of 200 ppm by volume or less of gas or vapor, or 2 mg per liter or less of mist, fume, or dust, when administered by continuous inhalation for 1 h (or less if death occurs within 1 h) to albino rats weighing between 200 and 300 g each.

Toxic. A chemical in any of the following categories:

1. A chemical that has a median lethal dose [LD(50)] of more than 50 mg per kg, but not more than 500 mg per kg of body weight when administered orally to albino rats weighing between 200 and 300 g each.
2. A chemical that has a median lethal dose [LD(50)] of more than 200 mg per kilogram, but not more than 1,000 mg per kilogram of body weight when administered by continuous contact for 24 h (or less if death occurs within 24 h) with the bare skin of albino rabbits weighing between 2 and 3 kg each.
3. A chemical that has a median lethal concentration [LC(50)] in air of more than 200 parts per million but not more than 2,000 parts per million by volume of gas or vapor, or more than 2 mg per liter but not more than 20 mg per liter of mist, fume, or dust, when administered by continuous inhalation for 1 h (or less if death occurs within 1 h) to albino rats weighing between 200 and 300 g each.

C1.6 IN SITU LOAD TESTS

In situ load testing can be used to verify that an existing component or structure has sufficient strength to resist code-required loads with at least the minimum required factor of safety. When there is a question as to the capacity of a component or structure, an in situ load test may be required by the Authority Having Jurisdiction and is subject to review and approval by the Authority Having Jurisdiction.

Load tests involving seismic loads, tsunami loads, and extraordinary loads such as fire, blast, and vehicular impact are beyond the scope of this section.

In situ load tests are typically required to be conducted in accordance with the load test provisions in the approved material design standard (e.g., concrete components and structures should be tested in accordance with Chapter 27 of ACI 318, and steel components and structures should be tested in accordance with Section 5.4 of Appendix 5 of AISC 360). Where the approved material design standard does not provide a load test procedure and acceptance criteria, the registered design professional responsible for the test is tasked with developing and overseeing the load test. At a minimum, the load test should be equal to the factored loads found in Section 2.3.1 and should be adjusted to account for load duration effects where the material being tested is sensitive to such effects. Special care is advisable when developing tests for wood structures, because some wood strength properties are affected by duration of load (e.g., axial compression, flexural, and tensile strength) and some wood strength properties are not (e.g., compression buckling and compression perpendicular to grain). Load testing often involves engineering judgment in terms of developing an appropriate loading mechanism, including both horizontal and vertical loads, as applicable. With respect to load tests for elements that resist dynamic loads, a one-hour test duration may be excessive. For example, in the case of a fall arrest or lifeline anchorage that is used to arrest the movement of a falling person, the maximum load occurs for a fraction of a second; a properly configured load test that sustains the maximum factored design load for a few seconds is more than sufficient to demonstrate that the anchorage has the minimum required strength to serve as a fall arrest anchorage.

There may be some instances where a reduction in design load is appropriate (e.g., an older structure where design live loads are intentionally restricted, like occupancy controls or a posted load restriction on a bridge); if design loads can be reduced in this fashion, in situ tests can incorporate these reduced loads.

REFERENCES

ACI (American Concrete Institute). 2019a. *Building code requirements for structural concrete and commentary*. ACI 318. Farmington Hills, MI: ACI.

ACI. 2019b. *Code requirements for assessment, repair, and rehabilitation of existing concrete structures and commentary*. ACI 562. Farmington Hills, MI: ACI.

AISC (American Institute of Steel Construction). 2022a. *Seismic provisions for structural steel buildings*. AISC 341. Chicago: AISC.

AISC. 2022b. *Specification for structural steel buildings*. AISC 360. Chicago: AISC.

AISI (American Iron and Steel Institute). 2020. *North American specification for the design of cold-formed steel structural members*. ANSI/AISI S100. Washington, DC: AISI.

Aluminum Association. 2020. *Specification for aluminum structures*. Arlington, VA: Aluminum Association.

ASCE. 1999. *Structural design for physical security: State of the practice*. Reston, VA: ASCE.

ASCE. 2002. *Specification for the design of cold-formed stainless steel structural members*. SEI/ASCE 8-02. Reston, VA: ASCE.

ASCE. 2010. *Design of blast resistant buildings in petrochemical facilities*. Reston, VA: ASCE.

ASCE. 2017. *Seismic evaluation and retrofit of existing buildings*. ASCE 41-17. Reston, VA: ASCE.

AWC (American Wood Council). 2017. *National design specification for wood construction*. NDS-2018. Leesburg, VA: AWC.

AWC. 2020. *Special design provisions for wind and seismic*. SDPWS-2021. Leesburg, VA: AWC.

Breen, J. E., ed. 1976. *Progressive collapse of building structures*. Rep. No. PDR-182. Washington, DC: Dept. of Housing and Urban Development.

Burnett, E. F. P. 1975. *The avoidance of progressive collapse: Regulatory approaches to the problem*. NBS GCR 75-48. Washington, DC: US Dept. of Commerce.

CEN (European Committee for Standardization). 2002b. *Eurocode: Basis of structural design*. EN 1990. Brussels: CEN.

Chock, G., G. Yu, H. K. Thio, and P. Lynett. 2016. "Target structural reliability analysis for tsunami hydrodynamic loads of the ASCE 7 standard." *J. Struct. Eng*. 142 (11): 04016092. https://doi.org/10.1061/(ASCE)ST.1943-541X.0001499.

Ellingwood, B., T. V. Galambos, J. G. MacGregor, and C. A. Cornell. 1980. *Development of a probability based load criterion for American national standard A58: Special publication no. 577*. Washington, DC: National Bureau of Standards.

Ellingwood, B., and E. V. Leyendecker. 1978. "Approaches for design against progressive collapse." *J. Struct. Div*. 104 (3): 413–23. https://doi.org/10.1061/JSDEAG.0004876.

Ellingwood, B., J. G. MacGregor, T. V. Galambos, and C. A. Cornell. 1982. "Probability based load criteria: Load factors and load combinations." *J. Struct. Div*. 108 (5): 978–997. https://doi.org/10.1061/JSDEAG.0005959.

Engineering News-Record. 1995. "Moment frames avoid progressive collapse." Accessed May 1, 2013.

EPA (US Environmental Protection Agency). 1999a. *Chemical accident prevention provisions*. 40 CFR Part 68. Washington, DC: EPA.

EPA. 1999b. *Emergency planning and notification: The list of extremely hazardous substances and their threshold planning quantities*. 40 CFR Part 355. Washington, DC: EPA.

FEMA (Federal Emergency Management Agency). 1997. *NEHRP recommended provisions for seismic regulations for new buildings and other structures*. Rep. No. 302. Washington, DC: FEMA.

FEMA. 2009. *Quantification of building seismic performance factors*. FEMA P-695. Washington, DC: FEMA.

Fintel, M., and G. Annamalai. 1979. "Philosophy of structural integrity of multistory load-bearing concrete masonry structures." *Concr. Int*. 1 (5): 27–35.

Fintel, M., and D. M. Schultz. 1979. "Structural integrity of large-panel buildings." *J. Am. Concr. Inst*. 76 (5): 583–622.

Galambos, T. V., B. Ellingwood, J. G. MacGregor, and C. A. Cornell. 1982. "Probability based load criteria: Assessment of current design practice." *J. Struct. Div*. 108 (5): 959–77. https://doi.org/10.1061/JSDEAG.0005958.

Glover, N. J. 1996. *The Oklahoma City bombing: Improving building performance through multi-hazard mitigation*. FEMA 277. Washington, DC: FEMA.

Granstrom, S., and M. Carlsson. 1974. *Byggfursknіngen T3: Byggnaders beteende vid overpaverkningar" [The behavior of buildings at excessive loadings]*. Stockholm, Sweden: Swedish Institute of Building Research.

ICC (International Code Council). 2000. *International building code*. Falls Church, VA: ICC.

ICC. 2009. *International building code*. Falls Church, VA: ICC.

ICC. 2021. *International existing building code*. Falls Church, VA: ICC.

ISO (International Organization for Standardization). 1998. *General principles on reliability for structures*. ISO 2394. Geneva: ISO.

Krawinkler, H., F. Parisi, L. Ibarra, A. Ayoub, and R. Medina. 2002. *Development of a testing protocol for woodframe structures*. Richmond, CA: Consortium of Universities for Research in Earthquake Engineering.

Leyendecker, E. V., J. E. Breen, N. F. Somes, and M. Swatta. 1976. *Abnormal loading on buildings and progressive collapse: An annotated bibliography*. NBS BSS 67. Washington, DC: US Dept. of Commerce.

Longinow, A. 1995. "The threat of terrorism: Can buildings be protected?" *Build. Operating Manage*. 46–53.

Mehta, K. C., D. L. Kriebel, G. J. White, and D. A. Smith. 1998. *An investigation of load factors for flood and combined wind and flood*. Report prepared for FEMA. Washington, DC: FEMA.

NBCC (National Building Code of Canada). 2010. *National building code of Canada*. M-23A. Ottawa: NBCC.

NFPA (National Fire Protection Association). 2006. *Building construction and safety code*. NFPA 5000. Quincy, MA: NFPA.

NIST (National Institute of Standards and Technology). 2012. *Tentative framework for development of advanced seismic design criteria for new buildings*. GCR 12-917-20. Washington, DC: NIST.

PCI Committee on Precast Bearing Walls. 1976. "Considerations for the design of precast bearing-wall buildings to withstand abnormal loads." *J. Prestressed Concr. Inst.* 21 (2): 46–69.

PEER (Pacific Earthquake Engineering Research Center). 2010. *Guidelines for performance-based seismic design of tall buildings. Version 1.0*. Rep. No. 210/05. Berkeley, CA: Tall Buildings Initiative, University of California.

Schultz, D. M., E. F. P. Burnett, and M. Fintel. 1977. *A design approach to general structural integrity, design and construction of large-panel concrete structures*. Washington, DC: Dept. of Housing and Urban Development.

Seltz-Petrash, A. E. 1979. "Winter roof collapses: Bad luck or bad design." *Civ. Eng*. 49 (12): 42–45.

TMS (The Masonry Society). 2016. *Building code requirements and specification for masonry structures*. TMS 402/602-16. Longmont, CO: TMS.

Weidlinger, P. 1994. "Civilian structures: Taking the defensive." *Civ. Eng*. 64 (11): 48–50.

OTHER REFERENCES (NOT CITED)

CEN. 2002a. *Structural eurocodes*. Brussels: CEN.

FEMA. 1993. *Wet floodproofing requirements for structures located in special flood hazard areas in accordance with the National Flood Insurance Program*. Technical Bulletin 7-93. Washington, DC: FEMA.

McManamy, R. 1995. "Oklahoma blast forces unsettling design questions." *Eng. News-Rec*. 234 (17): 9.

OSHA (Occupational Safety and Health Administration). 2000. *Standards for general industry*. Washington, DC: OSHA.

Standards Australia. 2005. *General principles on reliability for structures*. AS-5104. Sydney: Standards Australia.

CHAPTER C2
COMBINATIONS OF LOADS

C2.1 GENERAL

Loads in this standard are intended for use with design specifications for conventional structural materials, including steel, concrete, masonry, and timber. Some of these specifications are based on allowable stress design, whereas others use strength (or limit states) design. In the case of allowable stress design, design specifications define allowable stresses that may not be exceeded by load effects caused by unfactored loads, that is, allowable stresses contain a factor of safety. In strength design, design specifications provide load factors and, in some instances, resistance factors. Load factors given herein were developed using a first-order probabilistic analysis and a broad survey of the reliabilities inherent in contemporary design practice (Ellingwood et al. 1982, Galambos et al. 1982). It is intended that these load factors be used by all material-based design specifications that adopt a strength design philosophy in conjunction with nominal resistances and resistance factors developed by individual materials-specification-writing groups. Ellingwood et al. (1982) also provide guidelines for materials-specification-writing groups to aid them in developing resistance factors that are compatible, in terms of inherent reliability, with load factors and statistical information specific to each structural material.

The requirement to use either allowable stress design (ASD) or load and resistance factor design (LRFD) dates back to the introduction of load combinations for strength design (LRFD) in the 1982 edition of the standard. An indiscriminate mix of the LRFD and ASD methods may lead to unpredictable structural system performance because the reliability analyses and code calibrations leading to the LRFD load combinations were based on member rather than system limit states. Registered design professionals often design (or specify) cold-formed steel and open web steel joists using ASD and, at the same time, design the structural steel in the rest of the building or other structure using LRFD. Foundations are also commonly designed using ASD, although strength design is used for the remainder of the structure. Using different design standards for these types of elements has not been shown to be a problem. This requirement is intended to permit current industry practice while, at the same time, not permitting LRFD and ASD to be mixed indiscriminately in the design of a structural frame.

C2.2 SYMBOLS

Self-straining forces and effects can be caused by the differential settlement of foundations, creep, shrinkage or expansion in concrete members after placement and similar effects that depend on the material of construction and conditions of constraint, and changes in temperature of members caused by environmental conditions or operational activities during the service life of the structure. See Section C1.3.3 for examples of when self-straining forces and effects may develop.

C2.3 LOAD COMBINATIONS FOR STRENGTH DESIGN

C2.3.1 Basic Combinations Unfactored loads, to be used with these load factors, are the nominal loads of this standard. Load factors are from Ellingwood et al. (1982), with the exception of the load factor of 1.0 for E, based on the more recent NEHRP research on seismic-resistant design (FEMA 2004), the load factor for W, based on the wind speed maps at longer return periods for each risk category, and the load factor for W_T, which was added in ASCE 7-22. The basic idea of the load combination analysis is that in addition to dead load, which is considered to be permanent, one of the principal loads (previously referred to as primary variable loads) takes on its maximum lifetime value while the other loads assume "arbitrary point-in-time" values, the latter being loads that would be measured at any instant of time (Turkstra and Madsen 1980). This is consistent with the manner in which loads actually combine in situations in which strength limit states may be approached. However, nominal loads in this standard are substantially in excess of the arbitrary point-in-time values. To avoid having to specify both a maximum and an arbitrary point-in-time value for each load type, some of the specified load factors are less than unity in combinations 2a through 5a. Load factors in Section 2.3.1, with exceptions described above for E, W, and W_T, are based on a survey of reliabilities inherent in existing design practice (Ellingwood et al. 1982, Galambos et al. 1982).

In design where first-order analysis is permitted, superposition of factored loads can be performed either before or after the analysis. However, when a second-order analysis, which considers the effects of structural deformation on member forces, is used to design members and connections, the load factors must be applied before the analysis; the second-order analysis can be accomplished using a computer program with this capability for frame effects and member effects or by amplifying the results of a first-order analysis through the use of coefficients that amplify the first-order moments for the effects of member deformations or joint displacements. Because second-order effects are nonlinear, the second-order analyses must be conducted under factored load combinations (strength design) or load combinations amplified by the factor of safety (allowable stress design). In this context, second-order effects are the effects of loads acting on the deformed configuration of a structure and include $P - \delta$ effects and $P - \Delta$ effects.

Note that each of the principal loads, including the dead load, is a random variable. The degree of variability in each load is reflected in the associated load factor for the principal load; the other load factors provide the companion values.

Table C2.3-1. Principal Loads for Strength Design Load Combinations.

	Load Combination	Principal Load
1a	$1.4D$	D
2a	$1.2D + 1.6L + (0.5L_r \text{ or } 0.3S \text{ or } 0.5R)$	L
3a	$1.2D + (1.6L_r \text{ or } 1.0S \text{ or } 1.6R) + (L \text{ or } 0.5W)$	L_r or S or R
4a	$1.2D + 1.0(W \text{ or } W_T) + L + (0.5L_r \text{ or } 0.3S \text{ or } 0.5R)$	W or W_T
5a	$0.9D + 1.0(W \text{ or } W_T)$	W or W_T
6	$1.2D + E_v + E_h + L + 0.15S$	E
7	$0.9D - E_v + E_h$	E

The principal loads in the load combinations are identified in Table C2.3-1. The load factor on wind loads, in combinations 4a and 5a of Section 2.3.1 and on earthquake loads in combinations 6 and 7 of Section 2.3.6, were changed to 1.0 in previous editions of ASCE 7 when the maps for design wind speed and design seismic acceleration were modified to support a risk-consistent design methodology, as described in Chapters 21 and 26. The tornado load, W_T, was newly added in ASCE 7-22, as described in Chapter 32.

Exception 2 permits the companion load, S, that appears in combinations 2a and 4a to be the balanced snow load defined in Sections 7.3 for flat roofs and 7.4 for sloped roofs. Drifting and unbalanced snow loads, as principal loads, are covered by combination 3a.

Exception 3 permits the removal of the companion load, S, from combination 4a where using W_T, since tornadoes are overwhelmingly a warm-weather phenomenon.

Load combinations 5a and 7 apply specifically to the case in which the structural actions, due to lateral forces and gravity loads, counteract one another.

Load combination requirements in Section 2.3 only apply to strength limit states. Serviceability limit states and associated load factors are covered in Appendix C of this standard.

Historically, this standard has provided specific procedures for determining magnitudes of dead, occupancy live, wind, snow, and earthquake loads. Other loads, not traditionally considered by this standard, may also require consideration in design. Some of these loads may be important in certain material specifications and are included in the load criteria to enable uniformity to be achieved in the load criteria for different materials. However, statistical data on these loads are limited or nonexistent, and the same procedures used to obtain load factors and load combinations in Section 2.3.1 cannot be applied at the present time. Accordingly, load factors for fluid load (F) and lateral pressure caused by soil and water in soil (H) have been chosen to yield designs that would be similar to those obtained with existing specifications, if appropriate adjustments consistent with the load combinations in Section 2.3.1 were made to the resistance factors. Further research is needed to develop more accurate load factors.

Fluid load, F, defines structural actions in structural supports, framework, or foundations of a storage tank, vessel, or similar container caused by stored liquid products. The product in a storage tank shares characteristics of both dead and live loads. It is similar to a dead load in that its weight has a maximum calculated value, and the magnitude of the actual load may have a relatively small dispersion. However, it is not permanent; emptying and filling cause fluctuating forces in the structure, the maximum load may be exceeded by overfilling, and densities of stored products in a specific tank may vary.

Uncertainties in lateral forces from bulk materials, included in H, are higher than those in fluids, particularly when dynamic effects are introduced as the bulk material is set in motion by filling or emptying operations. Accordingly, the load factor for such loads is set equal to 1.6.

Where H acts as a resistance, a factor of 0.9 is suggested if the passive resistance is computed with a conservative bias. The intent is that soil resistance be computed for a deformation limit appropriate for the structure being designed, not at the ultimate passive resistance. Thus, an at-rest lateral pressure, as defined in the technical literature, would be conservative enough. Higher resistances than at-rest lateral pressure are possible, given appropriate soil conditions. Fully passive resistance would likely never be appropriate because the deformations necessary in the soil would likely be so large that the structure would be compromised. Furthermore, there is a great uncertainty in the nominal value of passive resistance, which would also argue for a lower factor on H, should a conservative bias not be included.

When H is due to lateral earth pressure, the additional lateral earth pressure from fixed or moving surcharge loads is added to H, as required by Section 3.2.1. Uncertainties in lateral forces from earth pressure are fairly high, such that the load factor for these lateral loads is set equal to 1.6. Therefore, the load factor used for the additional lateral load from fixed or moving surcharge loads is also set equal to 1.6, even if the gravity weight of the surcharge loads is well defined.

C2.3.2 Load Combinations Including Flood Load The nominal flood load, F_a, is based on the 100-year flood (Section 5.1). The recommended flood load factor of 2.0 in V-Zones and Coastal A-Zones is based on a statistical analysis of flood loads associated with hydrostatic pressures, pressures caused by steady overland flow, and hydrodynamic pressures caused by waves, as specified in Section 5.4.

The flood load criteria were derived from an analysis of hurricane-generated storm tides produced along the United States East and Gulf coasts (Kriebel et al. 1998), where storm tide is defined as the water level above mean sea level resulting from wind-generated storm surge added to randomly phased astronomical tides. Hurricane wind speeds and storm tides were simulated at 11 coastal sites based on historical storm climatology and on accepted wind speed and storm surge models. The resulting wind speed and storm tide data were then used to define probability distributions of wind loads and flood loads using wind and flood load equations specified in Sections 5.3 and 5.4. Load factors for these loads were then obtained using established reliability methods (Ellingwood et al. 1982, Galambos et al. 1982) and achieve approximately the same level of reliability as do combinations involving wind loads acting without floods. The relatively high flood load factor stems from the high variability in floods relative to other environmental loads. The presence of $2.0F_a$ in both combinations 4b and 5b in V-Zones and Coastal A-Zones is the result of high stochastic dependence between extreme wind and flood in hurricane-prone coastal zones. The $2.0F_a$ also applies in coastal areas subject to northeasters, extratropical storms, or coastal storms other than hurricanes, where a high correlation exists between extreme wind and flood.

While flooding can occur in storms that produce tornadoes, the spatiotemporal correlation between tornadoes and flooding is much less than that between hurricane winds and flooding. Therefore, the load combinations from Section 2.3.2 maximizing flood loads do not include arbitrary point-in-time tornado loads.

Flood loads are unique in that they are initiated only after the water level exceeds the local ground elevation. As a result, the statistical characteristics of flood loads vary with ground elevation. The load factor 2.0 is based on calculations (including hydrostatic, steady flow, and wave forces) with stillwater flood depths ranging from approximately 4 to 9 ft (1.2 to 2.7 m) [average stillwater flood depth of approximately 6 ft (1.8 m)] and applies to a wide variety of flood conditions. For lesser flood depths, load factors exceed 2.0 because of the wide dispersion in flood loads relative to the nominal flood load. As an example, load factors appropriate to water depths slightly less than 4 ft (1.2 m) equal 2.8 (Mehta et al. 1998). However, in such circumstances, the flood load is generally small. Thus, the load factor 2.0 is based on the recognition that flood loads of most importance to structural design occur in situations where the depth of flooding is greatest.

The 1.6 load factor, specified for H loads in Section 2.3.1, can be applied for groundwater induced uplift pressures. Figure C2.3-1 illustrates the flood zones and load factors for F_a and H.

C2.3.3 Load Combinations Including Atmospheric Ice and Wind-on-Ice Loads Load combinations 2b and 4c in Section 2.3.3 and load combination 3b in Section 2.4.3 include the simultaneous effects of snow loads, as defined in Chapter 7 and atmospheric ice loads, as defined in Chapter 10. Load combinations 4c and 5c in Sections 2.3.3 and load combinations 3b and 7c in

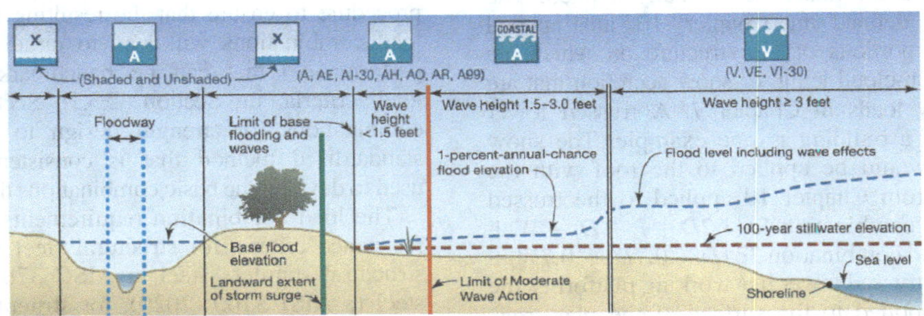

	Flood Zone			
	X Zone	A Zone	Coastal A Zone	V Zone
Strength Design Load Factor on F_a for Structural Elements	N/A	1.0	2.0	2.0
	X Zone	A Zone	Coastal A Zone	V Zone
Strength Design Load Factor on H for Structural Elements (hydrostatic uplift and lateral pressure due to groundwater)	1.6	1.6	1.6	1.6
	X Zone	A Zone	Coastal A Zone	V Zone
Allowable Stress Design Load Factor on F_a for Structural Elements	N/A	0.75	1.5	1.5
	X Zone	A Zone	Coastal A Zone	V Zone
Allowable Stress Design Load Factor on H for Structural Elements (hydrostatic uplift and lateral pressure due to groundwater)	1.0	1.0	1.0	1.0

Figure C2.3-1. Flood zones and ASCE 7 load factors for F_a and H for (from right to left in figure): coastal flood zones (V Zone, Coastal A Zone, and A Zone), areas outside the 100-year floodplain (X Zone shaded are areas inside the 500-year flood plain and X Zone unshaded are areas outside the 500-year flood plain), and riverine flood zone (A Zone). See FEMA (2005, 2011) for definitions of flood zones.

Section 2.4.3 introduce the simultaneous effect of wind on the atmospheric ice. The ice load, D_i, and the wind load on the atmospheric ice, W_i, in combination, correspond to an event with approximately a 500-year mean recurrence interval (MRI). Accordingly, the load factors on W_i and D_i are set equal to 1.0 and 0.7 in Sections 2.3.3 and 2.4.3, respectively. The 0.7 load factor on W_i and D_i in Section 2.4.3 aligns allowable stress design to have reliabilities for atmospheric ice loads consistent with the definition of wind and ice loads in Chapter 10 of this standard, which is based on strength principles. The snow loads defined in Chapter 7 are based on measurements of frozen precipitation accumulated on the ground, which includes snow, ice caused by freezing rain, and rain that falls onto snow and later freezes. Thus, the effects of freezing rain are included in the snow loads for roofs, catwalks, and other surfaces to which snow loads are normally applied. The atmospheric ice loads defined in Chapter 10 are applied simultaneously to those portions of the structure on which ice caused by freezing rain, in-cloud icing, or snow accretion that are not subject to the snow loads in Chapter 7. A trussed tower installed on the roof of a building is one example. The snow loads from Chapter 7 would be applied to the roof with the atmospheric ice loads from Chapter 10 applied to the trussed tower. Section 2.3.3 load combination 4c ($1.2D + L + D_i + W_i + 0.3S$) or Section 2.4.3 load combination 3b ($D + 0.7D_i + 0.7W_i + 0.7S$) are applicable. If a trussed tower has working platforms, the snow loads would be applied to the surface of the platforms, similar to a roof, with the atmospheric ice loads applied to the tower. Section 2.3.3 load combination 4c would reduce to $1.2D + D_i$ for cases for which the live load, wind on ice load, and snow load are zero. Section 2.4.3 load combination 3b would similarly reduce to $D + 0.7D_i$. If a sign is mounted on a roof, the snow loads would be applied to the roof and the atmospheric ice loads to the sign.

C2.3.4 Load Combinations Including Self-Straining Forces and Effects Self-straining forces and effects should be calculated based on a realistic assessment of the most probable values rather than the upper bound values of the variables. The most probable value is the value that can be expected at any arbitrary point in time.

When self-straining forces and effects are combined with dead loads as the principal action, a load factor of 1.2 may be used. However, when more than one variable load is considered and self-straining forces and effects are considered as a companion load, the load factor may be reduced if it is unlikely that the principal and companion loads will attain their maximum values at the same time. The load factor applied to T should not be taken as less than a value of 1.0.

If only limited data are available to define the magnitude and frequency distribution of the self-straining forces and effects, then its value must be estimated conservatively. Estimating the uncertainty in the self-straining forces and effects may be complicated by variation of the material stiffness of the member or structure under consideration.

When checking the capacity of a structure or structural element to withstand the effects of self-straining forces and effects, the following load combinations should be considered.

When using strength design:

$$1.2D + 1.2T + 0.5L$$

$$1.2D + 1.6L + 1.0T$$

These combinations are not all-inclusive, and judgment is necessary in some situations. For example, where roof live loads or snow loads are significant and could conceivably occur simultaneously with self-straining forces and effects, their effect should be included. The design should be based on the load combination causing the most unfavorable effect.

C2.3.5 Load Combinations for Nonspecified Loads Engineers may wish to develop load criteria for strength design that are consistent with the requirements of this standard in some situations for which the standard provides no information on loads or load combinations. They also may wish to consider loading criteria for special situations, as required by the client in performance-based engineering (PBE) applications, in accordance with Section 1.3.1.3. Groups responsible for strength design criteria for design of structural systems and elements may wish to develop resistance factors that are consistent with the standard. Such load criteria should be developed using a standardized procedure to ensure that the resulting factored design loads and load combinations will lead to target reliabilities (or levels of performance) that can be benchmarked against the common load criteria in Section 2.3.1. Section 2.3.5 permits load combinations for strength design to be developed through a standardized method that is consistent with the methodology used to develop the basic combinations that appear in Section 2.3.1.

The load combination requirements in Section 2.3.1 and the resistance criteria for structural steel in AISC 360 (2022), for structural stainless steel in AISC 370 (2021), for cold-formed steel in AISI S100 (2020), for structural concrete in ACI 318 (2019), for structural aluminum in the *Specification for Aluminum Structures* (Aluminum Association 2020), for engineered wood construction in *AWC NDS-2018 National Design Specification for Wood Construction* (2017), and for masonry in TMS 402-16, *Building Code Requirements and Specifications for Masonry Structures and Companion Commentaries* (2016), are based on modern concepts of structural reliability theory. In probability-based limit states design (PBLSD), the reliability is measured by a reliability index, β, which is related (approximately) to the limit state probability by $P_f = \varphi(-\beta)$. The approach taken in PBLSD was to

1. Determine a set of reliability objectives or benchmarks, expressed in terms of β, for a spectrum of traditional structural member designs involving steel, reinforced concrete, engineered wood, and masonry. Gravity load situations were emphasized in this calibration exercise, but wind and earthquake loads were considered as well. A group of experts from material specifications participated in assessing the results of this calibration and selecting target reliabilities. Thus, the reliability benchmarks identified are *not* the same for all limit states; if the failure mode is relatively ductile and consequences are not serious, β tends to be in the range 2.5 to 3.0, whereas if the failure mode is brittle and consequences are severe, β is 4.0 or more.
2. Determine a set of load and resistance factors that best meet the reliability objectives identified in Item 1 in an overall sense, considering the scope of structures that might be designed by this standard and the material specifications and codes that reference it.

The load combination requirements appearing in Section 2.3.1 used this approach. They are based on a "principal action–companion action" format, in which one load is taken at its maximum value while other loads are taken at their point-in-time values. Based on the comprehensive reliability analysis performed to support their development, it was found that these load factors are well approximated by

$$\gamma_Q = (\mu_Q/Q_n)(1 + \alpha_Q \beta V_Q) \qquad \text{(C2.3-1)}$$

where

μ_Q = Mean load,
Q_n = Nominal load from other chapters in this standard,
V_Q = Coefficient of variation in the load,
β = Reliability index, and
α_Q = Sensitivity coefficient that is approximately equal to 0.8 when Q is a principal action and 0.4 when Q is a companion action.

This approximation is valid for a broad range of common probability distributions used to model structural loads. The load factor is an increasing function of the bias in the estimation of the nominal load, the variability in the load, and the target reliability index, as common sense would dictate.

As an example, the load factors in combination 2a of Section 2.3.1 are based on achieving a β of approximately 3.0 for a ductile limit state with moderate consequences (e.g., formation of first plastic hinge in a steel beam). For live load acting as a principal action, $\mu_Q/Q_n = 1.0$ and $V_Q = 0.25$; for live load acting as a companion action, $\mu_Q/Q_n \approx 0.3$ and $V_Q \approx 0.6$. Substituting these statistics into Equation (C2.3-1), $\gamma_Q = 1.0[1 + 0.8(3)(0.25)] = 1.6$ (principal action) and $\gamma_Q = 0.3[1 + 0.4(3)(0.60)] = 0.52$ (companion action). ASCE Standard 7-05 (2005) stipulates 1.60 and 0.50 for these live load factors in combinations 2a and 3a. If an engineer wished to design for a limit state probability that is less than the standard case by a factor of approximately 10, β would increase to approximately 3.7, and the principal live load factor would increase to approximately 1.74.

Similarly, resistance factors that are consistent with the aforementioned load factors are well approximated for most materials by

$$\varphi = (\mu_R/R_n)\exp[-\alpha_R \beta V_R] \qquad \text{(C2.3-2)}$$

where

μ_R = Mean strength,
R_n = Code-specified strength,
V_R = Coefficient of variation in strength, and
α_R = Sensitivity coefficient equal to approximately 0.7.

For the limit state of yielding in an ASTM A992 (2011) steel tension member with specified yield strength of 50 ksi (345 MPa), $\mu_R/R_n = 1.06$ (under a static rate of load) and $V_R = 0.09$. Equation (C2.3-2) then yields $\varphi = 1.06\exp[-(0.7)(3.0)(0.09)] = 0.88$. The resistance factor for yielding in tension in Section D of the AISC Specification (2022) is 0.9. If a different performance objective were to require that the target limit state probability be decreased by a factor of 10, then φ would decrease to 0.84, a reduction of about 7%. Engineers wishing to compute alternative resistance factors for engineered wood products, and other structural components where duration-of-load effects might be significant, are advised to review the reference materials provided by their professional associations before using Equation (C2.3-2).

There are two key issues that must be addressed to use Equations (C2.3-1) and (C2.3-2): selection of reliability index, β, and determination of the load and resistance statistics.

The reliability index controls the safety level, and its selection should depend on the mode and consequences of failure. The loads and load factors in this standard do not explicitly account for higher reliability indexes normally desired for brittle failure mechanisms or more serious consequences of failure. Common standards for design of structural materials often do account for such differences in their resistance factors (for example, the design of connections under AISC or the design of columns under ACI). Tables 1.3-1 and 1.3-2 provide general guidelines for selecting target reliabilities consistent with the extensive calibration studies performed earlier to develop the load requirements in Section 2.3.1 and the resistance factors in the design standards for structural materials. The reliability indexes in those earlier studies were determined for structural members based on a service period of 50 years. System reliabilities are higher to a degree that depends on structural redundancy and ductility. The probabilities represent, *in order of magnitude*, the associated annual member failure rates for those who would find this information useful in selecting a reliability target.

The load requirements in Sections 2.3.1 through 2.3.3 are supported by extensive peer-reviewed statistical databases, and the values of mean and coefficient of variation, μ_Q/Q_n and V_Q, are well established. This support may not exist for other loads that have traditionally not been covered by this standard. Similarly, the statistics used to determine μ_R/R_n and V_R should be consistent with the underlying material specification. When statistics are based on small-batch test programs, all reasonable sources of end-use variability should be incorporated in the sampling plan. The engineer should document the basis for all statistics selected in the analysis and submit the documentation for review by the Authority Having Jurisdiction. Such documents should be made part of the permanent design record.

The engineer is cautioned that load and resistance criteria necessary to achieve a reliability-based performance objective are coupled through the common term β in Equations (C2.3-1) and (C2.3-2). Adjustments to the load factors without corresponding adjustments to the resistance factors will lead to an unpredictable change in structural performance and reliability.

C2.3.6 Basic Combinations with Seismic Load Effects The seismic load effect, E, is combined with the effects of other loads. For strength design, the load combinations with E in Section 2.3.6 include the horizontal and vertical seismic load effects of Sections 12.4.2.1 and 12.4.2.2, respectively. Similarly, the basic load combinations for allowable stress design with E in Section 2.4.5 include the same seismic load effects.

The seismic load effect including overstrength factor, E_m, is combined with other loads. The purpose for load combinations with overstrength factor is to approximate the maximum seismic load combination for the design of critical elements, including discontinuous systems, transfer beams and columns supporting discontinuous systems, and collectors. The allowable stress increase for load combinations with overstrength is to provide compatibility with past practice.

Load combination 6 addresses the combined structural actions (member forces) arising from snow and earthquake effects. The contribution of snow to effective seismic weight is defined in Section 12.7.2. The full snow load need not be considered in the effective seismic weight, as discussed in Section C12.7.2.

C2.3.7 Alternative Method for Loads from Water in Soil The effect of ground water pressure in soil, H_w, is decoupled from soil pressure and computed per Section 3.3; however, the effect of buoyancy is still included in the computation of lateral soil pressure, H_{eb}. The basic method includes H_w as part of H, which does not account for the uncertainty in the elevation of the groundwater. The 1.6 load factor in the basic method effectively adjusts the density of water.

C2.4 LOAD COMBINATIONS FOR ALLOWABLE STRESS DESIGN

C2.4.1 Basic Combinations The load combinations listed cover those loads for which specific values are given in other parts of this standard. Design should be based on the load combination causing the most unfavorable effect. In some cases, this may occur when one or more loads are not acting. No safety factors have been applied to these loads because such

Table C2.4-1. Principal Loads for Allowable Stress Design Load Combinations.

	Load Combination	Principal Load
1a	D	D
2a	$D+L$	L
3a	$D+(L_r \text{ or } 0.7S \text{ or } R)$	L_r or S or R
4a	$D+0.75L+0.75(L_r \text{ or } 0.7S \text{ or } R)$	L
5a	$D+0.6(W \text{ or } W_T)$	W or W_T
6a	$D+0.75L+0.75(0.6(W \text{ or } W_T))+0.75(L_r \text{ or } 0.7S \text{ or } R)$	W or W_T
7a	$0.6D+0.6(W \text{ or } W_T)$	W or W_T
8	$D+0.7E_v+0.7E_{mh}$	E
9	$D+0.525E_v+0.525E_{mh}+0.75L+0.1S$	E
10	$0.6D-0.7E_v+0.7E_{mh}$	E

factors depend on the design philosophy adopted by the particular material specification. The principal load, or maximum variable load, in the load combinations is identified in Table C2.4-1.

Wind, tornado, and earthquake loads need not be assumed to act simultaneously. However, the most unfavorable effects of each should be considered separately in design, where appropriate. In some instances, forces caused by wind or tornado might exceed those caused by earthquake, and ductility requirements might be determined by earthquake loads.

Load combinations 7a in Section 2.4.1 and 10 in Section 2.4.5 address the situation in which the effects of lateral or uplift forces counteract the effect of gravity loads. This action eliminates an inconsistency in the treatment of counteracting loads in allowable stress design and strength design and emphasizes the importance of checking stability. In such a situation, the reliability of structural components and systems is determined mainly by the large variability in the destabilizing load (Ellingwood and Li 2009), and the factor 0.6 on dead load is necessary for maintaining comparable reliability between strength design and allowable stress design. The earthquake load effect is multiplied by 0.7 to align allowable stress design for earthquake effects with the definition of E in Section 11.3 of this standard, which is based on strength principles.

Most loads, other than dead loads, vary significantly with time. When these variable loads are combined with dead loads, their combined effect should be sufficient to reduce the risk of unsatisfactory performance to an acceptably low level. However, when more than one variable load is considered, it is extremely unlikely that they will all attain their maximum value at the same time (Turkstra and Madsen 1980). Accordingly, some reduction in the total of the combined load effects is appropriate. This reduction is accomplished through the 0.75 load combination factor. The 0.75 factor applies only to the variable loads because the dead loads (or stresses caused by dead loads) do not vary in time.

Historically, many materials standards and building codes permitted a 1/3 increase in allowable stresses for load combinations that included either wind or seismic. Justifications for the 1/3 increase included the combined load effect as discussed, the ability of some materials to exhibit higher strength under short-term loading, and the apparent acceptability of a reduced factor of safety for these loadings. This standard addresses the low likelihood of simultaneous maximum load effect as discussed in the preceding paragraph. The acceptable probability of failure is explicitly addressed in the load factors applied in the strength design load combinations; the factors in the allowable stress design combinations are intended to provide a rough equivalence to the strength combinations. Where the strength of the material depends on rate or duration of loading, such as is the case with some soils and wood, it is permitted to adjust the allowable stresses to account for the increased strength under short-term loading, for example, the load duration factor (C_d) in AWC (2017). However, the simple 1/3 increase for all load combinations that involve wind or seismic is no longer permitted.

Exception 1 permits the companion load, S, appearing in combinations 4a and 6a to be the balanced snow load defined in Sections 7.3 for flat roofs and 7.4 for sloped roofs. Drifting and unbalanced snow loads, as principal loads, are covered by combination 3a.

When wind or tornado forces act on a structure, the structural action causing uplift at the structure–foundation interface is less than would be computed from the peak lateral force because of area averaging. Area averaging of wind forces occurs for all structures. In the method used to determine the wind forces for enclosed structures, the area-averaging effect is already taken into account in the data analysis leading to the pressure coefficients, C_p [or (GC_p)]. However, in the design of tanks and other industrial structures, the wind force coefficients, C_f, provided in the standard do not account for area averaging. For this reason, Exception 2 permits the wind or tornado loads to be reduced by 10% in the design of nonbuilding structure foundations and self-anchored ground-supported tanks. For different reasons, a similar approach is already provided for seismic actions in Section 2.4.1, Exception 2.

Exception 3 permits removal of the companion load, S, from combination 6a, where using W_T, since tornadoes are overwhelmingly a warm-weather phenomenon.

C2.4.2 Load Combinations Including Flood Load See Section C2.3.2. The multiplier on F_a aligns allowable stress design for flood load with strength design.

C2.4.3 Load Combinations Including Atmospheric Ice and Wind-on-Ice Loads See Section C2.3.3.

C2.4.4 Load Combinations Including Self-Straining Forces and Effects When using allowable stress design, determination of how self-straining forces and effects should be considered together with other loads should be based on the considerations discussed in Section C2.3.4. For typical situations, the following load combinations should be considered for evaluating the effects of self-straining forces and effects together with dead and live loads:

$$1.0D+1.0T$$

$$1.0D+0.75(L+T)$$

These combinations are not all-inclusive, and in some situations, judgment is necessary. For example, where roof live loads or snow loads are significant and could conceivably occur simultaneously with self-straining forces, their effect should be included.

The design should be based on the load combination causing the most unfavorable effect.

C2.4.5 Basic Combinations with Seismic Load Effects Load combination 9 addresses the combined structural actions (member forces) arising from snow and earthquake effects. The contribution of snow to effective seismic weight is defined in Section 12.7.2. The full snow load need not be considered in the effective seismic weight, as discussed in Section C12.7.2.

Exception 2 in Section 2.4.5, for special reinforced masonry walls, is based on the combination of three factors that yield a conservative design for overturning resistance under the seismic load combination:

1. The basic allowable stress for reinforcing steel is 40% of the specified yield.
2. The minimum reinforcement required in the vertical direction provides a protection against the circumstance where the dead and seismic loads result in a very small demand for tension reinforcement.
3. The maximum reinforcement limit prevents compression failure under overturning.

Of these, the low allowable stress in the reinforcing steel is the most significant. This exception should be deleted when and if the standard for design of masonry structures substantially increases the allowable stress in tension reinforcement.

C2.5 LOAD COMBINATIONS FOR EXTRAORDINARY EVENTS

Section 2.5 advises the structural engineer that certain circumstances might require structures to be checked for low-probability events such as fire, explosions, and vehicular impact. Since the 1995 edition of ASCE Standard 7, Commentary C2.5 has provided a set of load combinations that were derived using a probabilistic basis similar to that used to develop the load combination requirements for ordinary loads in Section 2.3. In recent years, social and political events have led to an increasing desire on the part of architects, structural engineers, project developers, and regulatory authorities to enhance design and construction practices for certain buildings to provide additional structural robustness and to lessen the likelihood of disproportionate collapse if an abnormal event were to occur. Several federal, state, and local agencies have adopted policies that require new buildings and structures to be constructed with such enhancements of structural robustness (GSA 2003, DOD 2009). Typically, robustness is assessed by notional removal of key load-bearing structural elements, followed by a structural analysis to assess the ability of the structure to bridge over the damage (often denoted alternative path analysis). Concurrently, advances in structural engineering for fire conditions (e.g., AISC 2022, Appendix 4; AISC 2021, Appendix 4 and Appendix 7) raise the prospect that new structural design requirements for fire safety will supplement the existing deemed-to-satisfy provisions in the next several years. To meet these needs, the load combinations for extraordinary events in ASCE 7-10 (ASCE 2010) were moved to Section 2.5 of this standard from Section C2.5, where they appeared in previous editions.

These provisions are not intended to supplant traditional approaches to ensure fire endurance based on standardized time–temperature curves and code-specified endurance times. Current code-specified endurance times are based on the ASTM E119 (2014) time–temperature curve under full allowable design load.

Extraordinary events arise from service or environmental conditions that are traditionally not considered explicitly in design of ordinary buildings and other structures. Such events are characterized by a low probability of occurrence and usually a short duration. Few buildings are ever exposed to such events, and statistical data to describe their magnitude and structural effects are rarely available. Included in the category of extraordinary events would be fire, explosions of volatile liquids or natural gas in building service systems, sabotage, vehicular impact, misuse by building occupants, and subsidence (not settlement) of subsoil. The occurrence of any of these events is likely to lead to structural damage or failure. If the structure is not properly designed and detailed, this local failure may initiate a chain reaction of failures that propagates throughout a major portion of the structure and leads to a potentially catastrophic partial, or total, collapse. Although all buildings are susceptible to such collapses in varying degrees, construction that lacks inherent continuity and ductility is particularly vulnerable (Taylor 1975, Breen and Siess 1979, Carper and Smilowitz 2006, Nair 2006, NIST 2007).

Good design practice requires that structures be robust and that their safety and performance not be sensitive to uncertainties in loads, environmental influences, and other situations not explicitly considered in design. The structural system should be designed in such a way that if an extraordinary event occurs, the probability of damage disproportionate to the original event is sufficiently small (Carper and Smilowitz 2006, NIST 2007). The philosophy of designing to limit the spread of damage, rather than to prevent damage entirely, is different from the traditional approach to designing to withstand dead, live, snow, and wind loads but is similar to the philosophy adopted in modern earthquake-resistant design.

In general, structural systems should be designed with sufficient continuity and ductility that alternate load paths can develop after individual member failure so that failure of the structure as a whole does not ensue. At a simple level, continuity can be achieved by requiring development of a minimum tie force, say 20 kN/m (1.37 kip/ft) between structural elements (NIST 2007). Member failures may be controlled by protective measures that ensure that no essential load-bearing member is made ineffective as a result of an accident, although this approach may be more difficult to implement. Where member failure would inevitably result in a disproportionate collapse, the member should be designed for a higher degree of reliability (NIST 2007).

Design limit states include loss of equilibrium as a rigid body, large deformations leading to significant second-order effects, yielding or rupture of members or connections, formation of a mechanism, and instability of members or the structure as a whole. These limit states are the same as those considered for other load events, but the load-resisting mechanisms in a damaged structure may be different, and sources of load-carrying capacity that normally would not be considered in ordinary ultimate limit states design, such as arch, membrane, or catenary action, may be included. The use of elastic analysis underestimates the load-carrying capacity of the structure (Marjanishvili and Agnew 2006). Materially or geometrically nonlinear or plastic analyses may be used, depending on the response of the structure to the actions.

Specific design provisions to control the effect of extraordinary loads and risk of progressive failure are developed with a probabilistic basis (Ellingwood and Leyendecker 1978, Ellingwood and Corotis 1991, Ellingwood and Dusenberry 2005). One can either reduce the likelihood of the extraordinary event or design the structure to withstand or absorb damage from the event if it occurs. Let F be the event of failure (damage or collapse) and A be the event that a structurally damaging event occurs. The probability of failure due to event A is

$$P_f = P(F|A)P(A) \quad \text{(C2.5-1)}$$

in which $P(F|A)$ is the conditional probability of failure of a damaged structure and $P(A)$ is the probability of occurrence of event A. The separation of $P(F|A)$ and $P(A)$ allows one to focus on strategies for reducing risk. $P(A)$ depends on siting, controlling the use of hazardous substances, limiting access, and other actions that are essentially independent of structural design. In contrast, $P(F|A)$ depends on structural design measures, ranging from minimum provisions for continuity to a complete post-damage structural evaluation.

The probability, $P(A)$, depends on the specific hazard. Limited data for severe fires, gas explosions, bomb explosions, and vehicular collisions indicate that the event probability depends on building size, measured in dwelling units or square footage, and ranges from about 0.2×10^{-6}/dwelling unit/year to about 8.0×10^{-6}/dwelling unit/year (NIST 2007). Thus, the probability that a building structure is affected may depend on the number of dwelling units (or square footage) in the building. If one were to set the conditional limit state probability, $P(F|A) = 0.05 - 0.10$, however, the annual probability of structural failure from Equation (C2.5-1) would be less than 10^{-6}, placing the risk in the low-magnitude background along with risks from rare accidents (Pate-Cornell 1994).

Design requirements corresponding to this desired $P(F|A)$ can be developed using first-order reliability analysis if the limit state function describing structural behavior is available (Ellingwood and Dusenberry 2005). The structural action (force or constrained deformation) resulting from extraordinary event A used in design is denoted A_k. Only limited data are available to define the frequency distribution of the load (NIST 2007; Ellingwood and Dusenberry 2005). The uncertainty in the load caused by the extraordinary event is encompassed in the selection of a conservative A_k, and thus the load factor on A_k is set equal to 1.0, as is done in the earthquake load combinations in Section 2.3. The dead load is multiplied by the factor 0.9 if it has a stabilizing effect; otherwise, the load factor is 1.2, as it is with the ordinary combinations in Sections 2.3.1 and 2.3.6. Load factors less than 1.0 on the companion actions reflect the small probability of a simultaneous occurrence of the extraordinary load and the design live, snow, or wind load. The companion actions $0.5L$ and $0.2S$ correspond, approximately, to the mean of the yearly maximum live and snow loads (Chalk and Corotis 1980, Ellingwood 1981). The companion action in Equation (2.5-1) includes only snow load because the probability of a coincidence of A_k with L_r or R, which have short durations in comparison with S, is negligible. A similar set of load combinations for extraordinary events appears in *Eurocode 1* (2006).

The term $0.2W$ that previously appeared in these combinations has been removed and has been replaced by a requirement to check lateral stability. One approach for meeting this requirement, which is based on recommendations of the Structural Stability Research Council (Galambos 1998), is to apply lateral notional forces, $N_i = 0.002\Sigma P_i$, at level i, in which ΣP_i = gravity force from Equations (2.5-1) or (2.5-2) act at level i, in combination with the loads stipulated in Equations (2.5-1) or (2.5-2). Note that Equation (1.4-1) stipulates that when checking general structural integrity, the lateral forces acting on an intact structure shall equal $0.01W_x$, where W_x is the dead load at level x.

REFERENCES

ACI (American Concrete Institute). 2019. *Building code requirements for structural concrete and commentary*. ACI 318-14/19, 318R-14/19. Farmington Hills, MI: ACI.

AISC (American Iron and Steel Institute). 2020. *North American specification for the design of cold-formed steel structural members*. ANSI/AISI S100. Chicago: AISC.

AISC. 2021. *Specification for structural stainless steel buildings*. ANSI/AISC 370-21. Chicago: AISC.

AISC. 2022. *Specification for structural steel buildings*. ANSI/AISC 360-22. Chicago: AISC.

Aluminum Association. 2020. "Specification for aluminum structures." In *Aluminum design manual*. Arlington, VA: Aluminum Association.

ASCE. 2005. *Minimum design loads for buildings and other structures*. ASCE/SEI 7-05. Reston, VA: ASCE.

ASCE. 2010. *Minimum design loads for buildings and other structures*. ASCE/SEI 7-10. Reston, VA: ASCE.

ASTM International. 2011. *Standard specification for structural steel shapes*. ASTM A992/A992M–11. West Conshohocken, PA: ASTM.

ASTM. 2014. *Standard test methods for fire tests of building construction and materials*. ASTM E119-14. West Conshohocken, PA: ASTM.

AWC (American Wood Council). 2017. *National design specification (NDS) for wood construction*, 2018 edition. Leesburg, VA: AWC.

Breen, J. E., and C. P. Siess. 1979. "Progressive collapse—Symposium summary." *ACI J.* 76 (9): 997–1004.

Carper, K., and R. Smilowitz, eds. 2006. "Mitigating the potential for progressive disproportionate collapse." *J. Perform. Constr. Facil.* 20 (4): 5–26.

CEN (European Committee for Standardization). 2006. *Eurocode 1: Actions on structures, Part 1–7: General actions–Accidental actions*. NEN-EN 1991-1-7. Brussels, Belgium: CEN.

Chalk, P. L., and R. B. Corotis. 1980. "Probability models for design live loads." *J. Struct. Div.* 106 (10): 2017–2033. https://doi.org/10.1061/JSDEAG.0005542.

DOD (US Department of Defense). 2009. *Design of buildings to resist progressive collapse*. UFC 4-023-03. Washington, DC: DOD.

Ellingwood, B. 1981. "Wind and snow load statistics for probabilistic design." *J. Struct. Div.* 107 (7): 1345–1350. https://doi.org/10.1061/JSDEAG.0006152.

Ellingwood, B., and R. B. Corotis. 1991. "Load combinations for buildings exposed to fires." *Eng. J.* 28 (1): 37–44.

Ellingwood, B. R., and D. O. Dusenberry. 2005. "Building design for abnormal loads and progressive collapse." *Comput.-Aided Civ. Inf. Eng.* 20 (5): 194–205. https://doi.org/10.1111/j.1467-8667.2005.00387.x.

Ellingwood, B. R., and E. V. Leyendecker. 1978. "Approaches for design against progressive collapse." *J. Struct. Div.* 104 (3): 413–423. https://doi.org/10.1061/JSDEAG.0004876.

Ellingwood, B. R., and Y. Li. 2009. "Counteracting structural loads: Treatment in ASCE Standard 7-05." *J. Struct. Eng.* 135 (1): 94–97. https://doi.org/10.1061/(ASCE)0733-9445(2009)135:1(94).

Ellingwood, B. R., J. G. MacGregor, T. V. Galambos, and C. A. Cornell. 1982. "Probability-based load criteria: Load factors and load combinations." *J. Struct. Div.* 108 (5): 978–997. https://doi.org/10.1061/JSDEAG.0005959.

FEMA (Federal Emergency Management Agency). 2004. *NEHRP recommended provisions for the development of seismic regulations for new buildings and other structures*. FEMA Rep. No. 450. Washington, DC: FEMA.

FEMA. 2005. *Design and construction in coastal a zones, Hurricane Katrina recovery advisory*. Washington, DC: FEMA.

FEMA. 2011. *Coastal construction manual, principles and practices of planning, siting, designing, constructing, and maintaining residential buildings in coastal areas.* FEMA P-55. Washington, DC: FEMA.

Galambos, T. V., ed. 1998. *SSRC guide to stability design criteria for metal structures*, 5th ed. New York: Wiley.

Galambos, T. V., B. Ellingwood, J. G. MacGregor, and C. A. Cornell. 1982. "Probability-based load criteria: Assessment of current design practice." *J. Struct. Div.* 108 (5): 959–977. https://doi.org/10.1061/JSDEAG.0005958.

GSA (General Services Administration). 2003. *Progressive collapse analysis and design guidelines for new federal office buildings and major modernization projects.* Washington, DC: GSA.

Kriebel, D. L., G. J. White, K. C. Mehta, and D. A. Smith. 1998. *An investigation of load factors for flood and combined wind and flood.* Washington, DC: FEMA.

Marjanishvili, S., and E. Agnew. 2006. "Comparison of various procedures for progressive collapse analysis." *J. Perform. Constr. Facil.* 20 (4): 365–374. https://doi.org/10.1061/(ASCE)0887-3828(2006)20:4(365).

Nair, R. S. 2006. "Preventing disproportionate collapse." *J. Perform. Constr. Facil.* 20 (4): 309–314. https://doi.org/10.1061/(ASCE)0887-3828(2006)20:4(309).

NIST (National Institute of Standards and Technology). 2007. *Best practices for reducing the potential for progressive collapse in buildings.* NISTIR 7396. Gaithersburg, MD: NIST.

Pate-Cornell, E. 1994. "Quantitative safety goals for risk management of industrial facilities." *Struct. Saf.* 13 (3): 145–157. https://doi.org/10.1016/0167-4730(94)90023-X.

Taylor, D. A. 1975. "Progressive collapse." *Can. J. Civ. Eng.* 2 (4): 517–529.

TMS (The Masonry Society). 2016. *Building code requirements and specifications for masonry structures and companion commentaries*, 402–406. Longmont, CO: TMS.

Turkstra, C. J., and H. O. Madsen. 1980. "Load combinations in codified structural design." *J. Struct. Div.* 106 (12): 2527–2543. https://doi.org/10.1061/JSDEAG.0005599.

OTHER REFERENCES (NOT CITED)

FEMA. 2015. *Safe rooms for tornadoes and hurricanes: Guidance for community and residential safe rooms.* FEMA P-361. Washington, DC: FEMA.

CHAPTER C3
DEAD LOADS, SOIL LOADS, AND HYDROSTATIC PRESSURE

C3.1 DEAD LOADS

C3.1.2 Weights of Materials of Construction To establish uniform practice among designers, it is desirable to present a list of materials generally used in building construction, together with their proper weights. Many building codes prescribe the minimum weights for only a few building materials, and in other instances no guide whatsoever is furnished on this subject. In some cases, the codes are drawn up so as to leave the question of what weights to use to the discretion of the building official, without providing any authoritative guide. This practice, as well as the use of incomplete lists, has been subjected to much criticism. The solution chosen has been to present, in this commentary, an extended list that will be useful to designer and official alike. However, special cases will unavoidably arise, and authority is therefore granted in the standard for the building official to deal with them.

For ease of computation, most values are given in terms of pounds per square foot, psf (kilonewtons per square meter, kN/m^2) of given thickness in Tables C3.1-1a and C3.1-1b. Pounds-per-cubic-foot, lb/ft^3 (kN/m^3), values, consistent with the pounds-per-square foot psf (kilonewtons per square meter, kN/m^2) values, are also presented in some cases in Table C3.1-2. Some constructions for which a single figure is given actually have a considerable range in weight. The average figure given is suitable for general use, but when there is reason to suspect a considerable deviation from this, the actual weight should be determined.

Engineers, architects, and building owners are advised to consider factors that result in differences between actual and calculated loads.

Engineers and architects cannot be responsible for circumstances beyond their control. Experience has shown, however, that conditions are encountered that, if not considered in design, may reduce the future utility of a building or reduce its margin of safety. Among them are

- *Dead loads*. There have been numerous instances in which the actual weights of members and construction materials have exceeded the values used in design. Care is advised in the use of tabular values. Also, allowances should be made for such factors as the influence of formwork and support deflections on the actual thickness of a concrete slab of prescribed nominal thickness.
- *Future installations*. Allowance should be made for the weight of future wearing or protective surfaces where there is a good possibility that such may be applied. Special consideration should be given to the likely types and position of partitions, as insufficient provision for partitioning may reduce the future utility of the building.

Attention is directed also to the possibility of temporary changes in the use of a building, as in the case of clearing a dormitory for a dance or other recreational purpose.

C3.1.3 Weight of Fixed Service Equipment Fixed service equipment includes but is not limited to plumbing stacks and risers; electrical feeders; heating, ventilating, and air conditioning systems; and process equipment such as vessels, tanks, piping, and cable trays. Both the empty weight of equipment and the maximum weight of contents are treated as dead load.

Section 1.3.6 indicates that when resistance to overturning, sliding, and uplift forces is provided by dead load, the dead load shall be taken as the minimum dead load likely to be in place. Therefore, liquid contents and movable trays shall not be used to resist these forces unless they are the source of the forces. For example, the liquid in a tank contributes to the seismic mass of the tank and therefore can be used to resist the seismic uplift; however, the weight of the liquid cannot be used to resist overturning, sliding, or uplift from wind loads because the liquid may not be present during the wind event.

C3.1.4 Vegetative and Landscaped Roofs Landscaping elements, such as soil, plants, and drainage layer materials, and hardscaping elements, such as walkways, fences, and walls, are intended to remain in place and are therefore considered dead loads. While the weight of hardscaping materials does not fluctuate, the weight of soil and drainage layer materials used to support vegetative growth is subject to significant variation because of its ability to absorb and retain water. Where the weight is additive to other loads, the dead load should be computed assuming full saturation of soil and drainage layer materials. Where the weight acts to counteract uplift forces, the dead load should be computed assuming dry unit weight of the soil and dry drainage layer materials.

Vegetative and landscaped roof areas may be able to retain more water than the condition where the soil and drainage layer materials are fully saturated. The water may result from precipitation or from irrigation of the vegetation. This additional amount of water should be considered rain load in accordance with Chapter 8 or snow load in accordance with Chapter 7 as applicable.

C3.1.5 Solar Panels This section clarifies that solar panel-related loads, including ballasted systems that are not permanently attached, shall be considered as dead loads for all load combinations specified in Chapter 2.

C3.2 SOIL LOADS AND HYDROSTATIC PRESSURE

C3.2.1 Lateral Pressures Table 3.2-1 includes high earth pressures, 85 pcf ($13.36\ kN/m^3$) or more, to show that certain soils are poor backfill material. In addition, when wall horizontal

Table C3.1-1a. Minimum Design Dead Loads (psf).

Component	Load (psf)
CEILINGS	
Acoustical fiberboard	1
Gypsum board (per 1/8 in. thickness)	0.55
Mechanical duct allowance	4
Plaster on tile or concrete	5
Plaster on wood lath	8
Suspended steel channel system	2
Suspended metal lath and cement plaster	15
Suspended metal lath and gypsum plaster	10
Wood furring suspension system	2.5
COVERINGS, ROOF, AND WALL	
Asbestos-cement shingles	4
Asphalt shingles	2
Cement tile	16
Clay tile (for mortar add 10 psf)	
Book tile, 2 in.	12
Book tile, 3 in.	20
Ludowici	10
Roman	12
Spanish	19
Composition:	
Three-ply ready roofing	1
Four-ply felt and gravel	5.5
Five-ply felt and gravel	6
Copper or tin	1
Corrugated asbestos-cement roofing	4
Deck, metal, 20 gauge	2.5
Deck, metal, 18 gauge	3
Decking, 2 in. wood (Douglas fir)	5
Decking, 3 in. wood (Douglas fir)	8
Fiberboard, 1/2 in.	0.75
Fiber cement siding, 5/16 in.	2.5
Gypsum sheathing, 1/2 in.	2
Insulation, wall, or roof boards (per in. thickness)	
Cellular glass	0.7
Fibrous glass	1.1
Fiberboard	1.5
Perlite	0.8
Polystyrene or polyisocyanurate foam	0.2
Urethane foam with skin	0.5
Plywood (per 1/8 in. thickness)	0.4
Rigid insulation, 1/2 in.	0.75
Skylight, metal frame, 3/8 in. wire glass	8
Slate, 3/16 in.	7
Slate, 1/4 in.	10
Vinyl siding, 0.05 in. nominal material thickness	0.6
Waterproofing membranes:	
Bituminous, gravel-covered	5.5
Bituminous, smooth surface	1.5
Liquid applied	1
Single-ply, sheet	0.7
Wood sheathing (per in. thickness)	3
Wood shakes and shingles	3
FLOOR FILL	
Cinder concrete, per in.	9
Lightweight concrete, per in.	8
Sand, per in.	8
Stone concrete, per in.	12
FLOORS AND FLOOR FINISHES	
Asphalt block (2 in.), 1/2 in. mortar	30
Cement finish (1 in.) on stone–concrete fill	32
Ceramic or quarry tile (3/4 in.) on 1/2 in. mortar bed	16

continues

Table C3.1-1a *(Continued)*. Minimum Design Dead Loads (psf).

Component					Load (psf)
Ceramic or quarry tile (3/4 in.) on 1 in. mortar bed					23
Concrete fill finish (per in. thickness)					12
Hardwood flooring, 7/8 in.					4
Linoleum or asphalt tile, 1/4 in.					1
Marble and mortar on stone–concrete fill					33
Slate (per mm thickness)					15
Solid flat tile on 1 in. mortar base					23
Subflooring, 3/4 in.					3
Terrazzo (1 – 1/2 in.) directly on slab					19
Terrazzo (1 in.) on stone–concrete fill					32
Terrazzo (1 in.), 2 in. stone concrete					32
Wood block (3 in.) on mastic, no fill					10
Wood block (3 in.) on 1/2 in. mortar base					16
FLOORS, WOOD-JOIST (NO PLASTER)					
DOUBLE WOOD FLOOR					
Joint sizes (in.)	*12 in. spacing (psf)*	*16 in. spacing (psf)*	*24 in. spacing (psf)*		
2 × 6	6	5	5		
2 × 8	6	6	5		
2 × 10	7	6	6		
2 × 12	8	7	6		
FRAME PARTITIONS					
Movable steel partitions					4
Wood or steel studs, 1/2 in. gypsum board each side					8
Wood studs, 2 × 4, unplastered					4
Wood studs, 2 × 4, plastered one side					12
Wood studs, 2 × 4, plastered two sides					20
FRAME WALLS					
Exterior wood stud walls:					
2 × 4 @ 16 in., 5/8 in. gypsum, insulated, 3/8 in. siding					11
2 × 6 @ 16 in., 5/8 in. gypsum, insulated, 3/8 in. siding					12
2 × 6 @ 16 in., 5/8 in. gypsum, insulated, stucco (lath and plaster)					18
2 × 6 @ 16 in., 5/8 in. gypsum, insulated, adhered masonry veneer					25
2 × 6 @ 16 in., 5/8 in. gypsum, insulated, anchored brick veneer					48
Windows, glass, frame, and sash					8
Clay brick wythes:					
4 in.					39
8 in.					79
12 in.					115
16 in.					155
Hollow concrete masonry unit wythes:					
Wythe thickness (in inches)	*4*	*6*	*8*	*10*	*12*
Density of unit (105 pcf) with grout spacing as follows:					
No grout	22	24	31	37	43
48 in. o.c.		29	38	47	55
40 in. o.c.		30	40	49	57
32 in. o.c.		32	42	52	61
24 in. o.c.		34	46	57	67
16 in. o.c.		40	53	66	79
Full grout		55	75	95	115
Density of unit (125 pcf) with grout spacing as follows:					
No grout	26	28	36	44	50
48 in. o.c.		33	44	54	62
40 in. o.c.		34	45	56	65
32 in. o.c.		36	47	58	68
24 in. o.c.		39	51	63	75
16 in. o.c.		44	59	73	87
Full grout		59	81	102	123
Density of unit (135 pcf) with grout spacing as follows:					
No grout	29	30	39	47	54
48 in. o.c.		36	47	57	66
40 in. o.c.		37	48	59	69

continues

Table C3.1-1a *(Continued)*. Minimum Design Dead Loads (psf).

Component					Load (psf)
32 in. o.c.		38	50	62	72
24 in. o.c.		41	54	67	78
16 in. o.c.		46	61	76	90
Full grout		62	83	105	127
Solid concrete masonry unit wythes (incl. concrete brick):					
Wythe thickness (in mm)	*4*	*6*	*8*	*10*	*12*
Density of unit (105 pcf)	32	51	69	87	105
Density of unit (125 pcf)	38	60	81	102	124
Density of unit (135 pcf)	41	64	87	110	133

Notes: Weights of masonry include mortar but not plaster. For plaster, add 5 psf for each face plastered. Values given represent averages. In some cases, there is a considerable range of weight for the same construction.

Table C3.1-1b. Minimum Design Dead Loads (kN/m²).

Component	Load (kN/m²)
CEILINGS	
Acoustical fiberboard	0.05
Gypsum board (per mm thickness)	0.008
Mechanical duct allowance	0.19
Plaster on tile or concrete	0.24
Plaster on wood lath	0.38
Suspended steel channel system	0.10
Suspended metal lath and cement plaster	0.72
Suspended metal lath and gypsum plaster	0.48
Wood furring suspension system	0.12
COVERINGS, ROOF, AND WALL	
Asbestos-cement shingles	0.19
Asphalt shingles	0.10
Cement tile	0.77
Clay tile (for mortar add 0.48kN/m²)	
Book tile, 51 mm	0.57
Book tile, 76 mm	0.96
Ludowici	0.48
Roman	0.57
Spanish	0.91
Composition:	
Three-ply ready roofing	0.05
Four-ply felt and gravel	0.26
Five-ply felt and gravel	0.29
Copper or tin	0.05
Corrugated asbestos-cement roofing	0.19
Deck, metal, 20 gauge	0.12
Deck, metal, 18 gauge	0.14
Decking, 51 mm wood (Douglas fir)	0.24
Decking, 76 mm wood (Douglas fir)	0.38
Fiberboard, 13 mm	0.04
Fiber cement siding, 8 mm	0.12
Gypsum sheathing, 13 mm	0.10
Insulation, wall, or roof boards (per mm thickness)	
Cellular glass	0.0013
Fibrous glass	0.0021
Fiberboard	0.0028
Perlite	0.0015
Polystyrene or polyisocyanurate foam	0.0004

continues

Table C3.1-1b *(Continued)*. Minimum Design Dead Loads (kN/m2).

Component				Load (kN/m²)
Urethane foam with skin				0.0009
Plywood (per mm thickness)				0.006
Rigid insulation, 13 mm				0.04
Skylight, metal frame, 10 mm wire glass				0.38
Slate, 5 mm				0.34
Slate, 6 mm				0.48
Vinyl siding, 1.3 mm nominal thickness				0.03
Waterproofing membranes:				
Bituminous, gravel-covered				0.26
Bituminous, smooth surface				0.07
Liquid applied				0.05
Single-ply, sheet				0.03
Wood sheathing (per mm thickness)				
Plywood				0.0057
Oriented strand board				0.0062
Wood shakes and shingles				0.14
FLOOR FILL				
Cinder concrete, per mm				0.017
Lightweight concrete, per mm				0.015
Sand, per mm				0.015
Stone concrete, per mm				0.023
FLOORS AND FLOOR FINISHES				
Asphalt block (51 mm), 13 mm mortar				1.44
Cement finish (25 mm) on stone–concrete fill				1.53
Ceramic or quarry tile (19 mm) on 13 mm mortar bed				0.77
Ceramic or quarry tile (19 mm) on 25 mm mortar bed				1.10
Concrete fill finish (per mm thickness)				0.023
Hardwood flooring, 22 mm				0.19
Linoleum or asphalt tile, 6 mm				0.05
Marble and mortar on stone–concrete fill				1.58
Slate (per mm thickness)				0.028
Solid flat tile on 25 mm mortar base				1.10
Subflooring, 19 mm				0.14
Terrazzo (38 mm) directly on slab				0.91
Terrazzo (25 mm) on stone–concrete fill				1.53
Terrazzo (25 mm), 51 mm stone concrete				1.53
Wood block (76 mm) on mastic, no fill				0.48
Wood block (76 mm) on 13 mm mortar base				0.77
FLOORS, WOOD-JOIST (NO PLASTER)				
DOUBLE WOOD FLOOR				
Joint sizes (mm):	*305-mm spacing* (kN/m²)	*406-mm spacing* (kN/m²)	*610-mm spacing* (kN/m²)	
51 × 152	0.29	0.24	0.24	
51 × 203	0.29	0.29	0.24	
51 × 254	0.34	0.29	0.29	
51 × 305	0.38	0.34	0.29	
FRAME PARTITIONS				
Movable steel partitions				0.19
Wood or steel studs, 13 mm gypsum board each side				0.38
Wood studs, 51 × 102, unplastered				0.19
Wood studs, 51 × 102, plastered one side				0.57
Wood studs, 51 × 102, plastered two sides				0.96
FRAME WALLS				
Exterior wood stud walls:				
51 mm × 102 mm @ 406 mm, 16 mm gypsum, insulated, 10 mm siding				0.53
51 mm × 152 mm @ 406 mm, 16 mm gypsum, insulated, 10 mm siding				0.57
51 mm × 152 mm @ 406 mm, 16 mm gypsum, insulated, stucco (lath and plaster)				0.86
51 mm × 152 mm @ 406 mm, 16 mm gypsum, insulated, adhered masonry veneer				1.20
51 mm × 152 mm @ 406 mm, 16 mm gypsum, insulated, anchored brick veneer				2.30
Windows, glass, frame, and sash				0.38
Clay brick wythes:				
102 mm				1.87

continues

Table C3.1-1b *(Continued)*. Minimum Design Dead Loads (kN/m2).

Component					Load (kN/m²)
203 mm					3.78
305 mm					5.51
406 mm					7.42
Hollow concrete masonry unit					
Wythes:					
Wythe thickness (mm)	*102*	*152*	*203*	*254*	*305*
Density of unit (16.49 kN/m³) with grout spacing as follows:					
No grout	1.05	1.29	1.68	2.01	2.35
1,219 mm		1.48	1.92	2.35	2.78
1,016 mm		1.58	2.06	2.54	3.02
813 mm		1.63	2.15	2.68	3.16
610 mm		1.77	2.35	2.92	3.45
406 mm		2.01	2.68	3.35	4.02
Full grout		2.73	3.69	4.69	5.70
Density of unit (19.64 kN/m³) with grout spacing as follows:					
No grout	1.25	1.34	1.72	2.11	2.39
1,219 mm		1.58	2.11	2.59	2.97
1,016 mm		1.63	2.15	2.68	3.11
813 mm		1.72	2.25	2.78	3.26
610 mm		1.87	2.44	3.02	3.59
406 mm		2.11	2.78	3.50	4.17
Full grout		2.82	3.88	4.88	5.89
Density of unit (21.21 kN/m³) with grout spacing as follows:					
No grout	1.39	1.68	2.15	2.59	3.02
1,219 mm		1.70	2.39	2.92	3.45
1,016 mm		1.72	2.54	3.11	3.69
813 mm		1.82	2.63	3.26	3.83
610 mm		1.96	2.82	3.50	4.12
406 mm		2.25	3.16	3.93	4.69
Full grout		3.06	4.17	5.27	6.37
Solid concrete masonry unit					
Wythe thickness (mm)	*102*	*152*	*203*	*254*	*305*
Density of unit (16.49 kN/m³)	1.53	2.35	3.21	4.02	4.88
Density of unit (19.64 kN/m³)	1.82	2.82	3.78	4.79	5.79
Density of unit (21.21 kN/m³)	1.96	3.02	4.12	5.17	6.27

Notes: Weights of masonry include mortar but not plaster. For plaster, add 0.24 kN/m² for each face plastered. Values given represent averages. In some cases, there is a considerable range of weight for the same construction.

Table C3.1-2. Minimum Densities for Design Loads from Materials.

Material	Density (lb/ft³)	Density (kN/m³)
Aluminum	170	27
Bituminous products		
Asphaltum	81	12.7
Graphite	135	21.2
Paraffin	56	8.8
Petroleum, crude	55	8.6
Petroleum, refined	50	7.9
Petroleum, benzene	46	7.2
Petroleum, gasoline	42	6.6
Pitch	69	10.8
Tar	75	11.8
Brass	526	82.6
Bronze	552	86.7
Cast-stone masonry (cement, stone, sand)	144	22.6

continues

Table C3.1-2 *(Continued)*. Minimum Densities for Design Loads from Materials.

Material	Density (lb/ft³)	Density (kN/m³)
Cement, Portland, loose	90	14.1
Ceramic tile	150	23.6
Charcoal	12	1.9
Cinder fill	57	9.0
Cinders, dry, in bulk	45	7.1
Coal		
Anthracite, piled	52	8.2
Bituminous, piled	47	7.4
Lignite, piled	47	7.4
Peat, dry, piled	23	3.6
Concrete, plain		
Cinder	108	17.0
Expanded-slag aggregate	100	15.7

continues

Table C3.1-2 *(Continued)*. Minimum Densities for Design Loads from Materials.

Material	Density (lb/ft³)	Density (kN/m³)
Haydite (burned-clay aggregate)	90	14.1
Slag	132	20.7
Stone (including gravel)	144	22.6
Vermiculite and perlite aggregate, nonload-bearing	25–50	3.9–7.9
Other light aggregate, load-bearing	70–105	11.0–16.5
Concrete, reinforced		
Cinder	111	17.4
Slag	138	21.7
Stone (including gravel)	150	23.6
Copper	556	87.3
Cork, compressed	14	2.2
Earth (not submerged)		
Clay, dry	63	9.9
Clay, damp	110	17.3
Clay and gravel, dry	100	15.7
Silt, moist, loose	78	12.3
Silt, moist, packed	96	15.1
Silt, flowing	108	17.0
Sand and gravel, dry, loose	100	15.7
Sand and gravel, dry, packed	110	17.3
Sand and gravel, wet	120	18.9
Earth (submerged)		
Clay	80	12.6
Soil	70	11.0
River mud	90	14.1
Sand or gravel	60	9.4
Sand or gravel and clay	65	10.2
Glass	160	25.1
Gravel, dry	104	16.3
Gypsum, loose	70	11.0
Gypsum, wallboard	50	7.9
Ice	57	9.0
Iron		
Cast	450	70.7
Wrought	480	75.4
Lead	710	111.5
Lime		
Hydrated, loose	32	5.0
Hydrated, compacted	45	7.1
Masonry, ashlar stone		
Granite	165	25.9
Limestone, crystalline	165	25.9
Limestone, oolitic	135	21.2
Marble	173	27.2
Sandstone	144	22.6
Masonry, brick		
Hard (low absorption)	130	20.4
Medium (medium absorption)	115	18.1
Soft (high absorption)	100	15.7
Masonry, concrete*		
Lightweight units	105	16.5
Medium weight units	125	19.6
Normal weight units	135	21.2
Masonry grout	140	22.0
Masonry, rubble stone		
Granite	153	24.0
Limestone, crystalline	147	23.1
Limestone, oolitic	138	21.7
Marble	156	24.5

continues

Table C3.1-2 *(Continued)*. Minimum Densities for Design Loads from Materials.

Material	Density (lb/ft³)	Density (kN/m³)
Sandstone	137	21.5
Mortar, cement or lime	130	20.4
Particleboard	45	7.1
Plywood	36	5.7
Riprap (not submerged)		
Limestone	83	13.0
Sandstone	90	14.1
Sand		
Clean and dry	90	14.1
River, dry	106	16.7
Slag		
Bank	70	11.0
Bank screenings	108	17.0
Machine	96	15.1
Sand	52	8.2
Slate	172	27.0
Steel, cold-drawn	492	77.3
Stone, quarried, piled		
Basalt, granite, gneiss	96	15.1
Limestone, marble, quartz	95	14.9
Sandstone	82	12.9
Shale	92	14.5
Greenstone, hornblende	107	16.8
Terra cotta, architectural		
Voids filled	120	18.9
Voids unfilled	72	11.3
Tin	459	72.1
Water		
Fresh	62	9.7
Sea	64	10.1
Wood, seasoned		
Ash, commercial white	41	6.4
Cypress, southern	34	5.3
Fir, Douglas, coast region	34	5.3
Hem fir	28	4.4
Oak, commercial reds and whites	47	7.4
Pine, southern yellow	37	5.8
Redwood	28	4.4
Spruce, red, white, and Sitka	29	4.5
Western hemlock	32	5.0
Zinc, rolled sheet	449	70.5

* Tabulated values apply to solid masonry and to the solid portion of hollow masonry.

movement is restricted at the top, the earth pressure is increased from active pressure toward earth pressure at rest, resulting in 60 pcf (9.43 kN/m³) for granular soils and 100 pcf (15.71 kN/m³) for silt and clay type soils (Terzaghi and Peck 1967). The exception permits active pressure to be used for shallow foundation walls supporting flexible floor diaphragms such as wood-framed floors and cold-formed steel joists without a cast-in-place concrete floor attached.

Expansive soils exist in many regions of the United States and may cause serious damage to foundation walls unless special design considerations are provided. Expansive soils should not be used as backfill because they can exert very high pressures against walls. Special soil testing is required to determine the magnitude of these pressures. It is preferable to excavate expansive soil and backfill with nonexpansive, freely draining sands or gravels. The excavated back slope adjacent to the wall should be no steeper than 45 degrees from the horizontal to minimize the transmission of swelling pressure from the expansive soil through the new backfill. Other special details are recommended, such as a cap of nonpervious soil on top of the backfill and provision of foundation drains. Refer to current reference books on geotechnical engineering for guidance.

C3.2.2 Uplift Loads on Floors and Foundations If expansive soils are present under floors or footings, large pressures can be exerted and must be resisted by special design. Alternatively, the expansive soil can be removed and replaced with nonexpansive material. A geotechnical engineer should make recommendations in these situations.

C3.3 ALTERNATIVE METHOD FOR LOADS FROM WATER IN SOIL

Prior to the 2022 edition, the standard did not define hazard levels for groundwater in soil. The probability levels now defined for maximum and minimum levels of groundwater correspond, in round numbers, to mean recurrence intervals of 400, 800, 1,700, and 3,300 years for the four risk categories. These are similar, but not identical to the intervals of 300, 750, 3,000, and 10,000 years specified for hydraulic structures by the US Army Corps of Engineers (USACE). The hazard levels selected for ASCE 7 are not the same as used by the USACE, because the USACE uses a different classification of risks for hydraulic structures than the risk classification used in ASCE 7. Site specific information is required to establish these groundwater levels, although conservative assumptions can be made in practice, such as setting the maximum groundwater level at the ground surface where not in a flood zone. Ground water pressure in a flood zone may or may not be correlated with flood elevation depending on many factors, including the nature of the soil and the duration of flood. Professionals responsible for establishing groundwater levels have various resources, including historical data from the USGS National Water Information System (wells monitoring known aquifers), flood zone maps, and local geological and hydrological information. Information sources will vary from region to region and site to site. The difficulty in obtaining usable data is a primary reason this method is included as an elective alternate, rather than the only approved method.

REFERENCES

Terzaghi, K., and R. B. Peck. 1967. *Soil mechanics in engineering practice*, 2nd ed. New York: Wiley.

CHAPTER C4
LIVE LOADS

C4.3 UNIFORMLY DISTRIBUTED LIVE LOADS

C4.3.1 Required Live Loads A selected list of loads for occupancies and uses more commonly encountered is given in Section 4.3.1, and the Authority Having Jurisdiction should approve on occupancies not mentioned. Tables C4.3-1 and C4.3-2 are offered as a guide in the exercise of such authority.

In selecting the occupancy and use for the design of a building or a structure, the building owner should consider the possibility of later changes of occupancy involving loads heavier than originally contemplated. The lighter loading appropriate to the first occupancy should not necessarily be selected. The building owner should ensure that a live load greater than that for which a floor or roof is approved by the Authority Having Jurisdiction is not placed, or caused or permitted to be placed, on any floor or roof of a building or other structure.

To solicit specific informed opinion regarding the design loads in Table 4.3-1, a panel of 25 distinguished structural engineers was selected. A Delphi (Corotis et al. 1981) was conducted with this panel in which design values and supporting reasons were requested for each occupancy type. The information was summarized and recirculated back to the panel members for a second round of responses. Those occupancies for which previous design loads were reaffirmed and those for which there was consensus for change were included.

It is well known that the floor loads measured in a live load survey usually are well below present design values (Peir and Cornell 1973, McGuire and Cornell 1974, Sentler 1975, Ellingwood and Culver 1977). However, buildings must be designed to resist the maximum loads they are likely to be subjected to during some reference period T, frequently taken as 50 years. Table C4.3-2 briefly summarizes how load survey data are combined with a theoretical analysis of the loading for some common occupancy types and illustrates how a design load might be selected for an occupancy not specified in Table 4.3-1 (Chalk and Corotis 1980). The floor load normally present for the intended functions of a given occupancy is referred to as the sustained load. This load is modeled as constant until a change in tenant or occupancy type occurs. A live load survey provides the statistics of the sustained load. Table C4.3-2 gives the mean, m_s, and standard deviation, σ_s, for particular reference areas. In addition to the sustained load, a building is likely to be subjected to a number of relatively short-duration, high-intensity, extraordinary, or transient loading events (caused by crowding in special or emergency circumstances, concentrations during remodeling, and the like). Limited survey information and theoretical considerations lead to the means, m_t, and standard deviations, σ_t, of single transient loads shown in Table C4.3-2.

Combination of the sustained load and transient load processes, with due regard for the probabilities of occurrence, leads to statistics of the maximum total load during a specified reference period. The statistics of the maximum total load depend on the average duration of an individual tenancy, τ, the mean rate of occurrence of the transient load, v_e, and the reference period, T. Mean values are given in Table C4.3-2. The mean of the maximum load is similar, in most cases, to Table 4.3-1 values of minimum uniformly distributed live loads and, in general, is a suitable design value.

For library stack rooms, the 150 psf (7.18 kN/m^2) uniform live load specified in Table 4.3-1 is intended to cover the range of ordinary library shelving which is described in Section 4.13. The 150 psf (7.18 kN/m^2) floor loading is also applicable to typical file cabinet installations, provided that the 36 in. (0.92 m) minimum aisle width is maintained. Five-drawer lateral or conventional file cabinets, even with two levels of bookshelves stacked above them, are unlikely to exceed the 150 psf (7.18 kN/m^2) average floor loading unless all drawers and shelves are filled to capacity with maximum density paper. Such a condition is essentially an upper bound for which the normal load factors and safety factors applied to the 150 psf (7.18 kN/m^2) criterion should still provide a safe design.

The 50 psf (2.40 kN/m^2) live load for theater follow spot, projection, and control rooms considers the use of typical theater equipment such as projectors, spotlights, and equipment to control the sound, lighting, and HVAC within the theater. When these rooms are used for storage, the live load used should account for the actual conditions.

For the 2010 version of the standard, the provision in the live load table for "Marquees" with its distributed load requirement of 75 psf (3.59 kN/m^2) was removed, along with "Roofs used for promenade purposes" and its 60 psf (2.87 kN/m^2) loading. Both *marquee* and *promenade* are considered archaic terms that are not used elsewhere in the standard or in building codes, with the exception of the listings in the live load tables. "Promenade purposes" is essentially an assembly use and is more clearly identified as such.

"Marquee" has not been defined in this standard but has been defined in building codes as a roofed structure that projects into a public right of way. However, the relationship between a structure and a right of way does not control loads that are applied to a structure. The marquee should therefore be designed with all of the loads appropriate for a roofed structure. If the arrangement of the structure is such that it invites additional occupant loading (e.g., there is window access that might invite loading for spectators of a parade), balcony loading should be considered for the design.

Balconies and decks are recognized as often having distinctly different loading patterns than most interior rooms. They are often subjected to concentrated live loads from people congregating along the edge of the structure (e.g., for viewing vantage points). This loading condition is acknowledged in Table 4.3-1 as

Table C4.3-1. Minimum Uniformly Distributed Live Loads.

Occupancy or Use	Live Load, psf (kN/m^2)	Occupancy or Use	Live Load, psf (kN/m^2)
Air conditioning (machine space)	200 (9.58)[a]	Laboratories, scientific	100 (4.79)
Amusement park structure	100 (4.79)[a]	Laundries	150 (7.18)[a]
Attic, nonresidential		Manufacturing, ice	300 (14.36)
Nonstorage	25 (1.20)	Morgue	125 (6.00)
Storage	80 (3.83)[a]	Printing plants	
Bakery	150 (7.18)	Composing rooms	100 (4.79)
Boathouse, floors	100 (4.79)[a]	Linotype rooms	100 (4.79)
Boiler room, framed	300 (14.36)[a]	Paper storage	[e]
Broadcasting studio	100 (4.79)	Press rooms	150 (7.18)[a]
Ceiling, accessible furred	10 (0.48)[b]	Railroad tracks	[f]
Cold storage		Rinks	
No overhead system	250 (11.97)[c]	Ice skating	250 (11.97)
Overhead system		Storage, hay or grain	300 (14.36)[a]
Floor	150 (7.18)	Theaters	
Roof	250 (11.97)	Dressing rooms	40 (1.92)
Computer equipment	150 (7.18)[a]	Gridiron floor or fly gallery	
Courtrooms	50–100 (2.40–4.79)	Grating	60 (2.87)
Dormitories		Well beams	250 psf (3.65 kN/m) per pair
Nonpartitioned	80 (3.83)	Header beams	1,000 psf (14.60 kN/m)
Partitioned	40 (1.92)	Pin rail	250 psf (3.65 kN/m)
Elevator machine room and control room	150 (7.18)[a]		
Fan room	150 (7.18)[a]	Transformer rooms	200 (9.58)[a]
Foundries	600 (28.73)[a]	Vaults, in offices	250 (11.97)[a]
Fuel rooms, framed	400 (19.15)		
Greenhouses	150 (7.18)		
Hangars, including seaplane ramps	150 (7.18)[d]		
Incinerator charging floor	100 (4.79)		
Kitchens, other than domestic	150 (7.18)[a]		

[a] Use weight of actual equipment or stored material when greater. Note that fixed service equipment is treated as dead load instead of live load.
[b] Accessible ceilings normally are not designed to support persons. The value in this table is intended to account for occasional light storage or suspension of items. If it may be necessary to support the weight of maintenance personnel, this shall be provided for.
[c] Plus 150 psf (7.18 kN/m^2) for trucks.
[d] Use American Association of State Highway and Transportation Officials lane loads. Also subject to not less than 100% maximum axle load.
[e] Paper storage: 50 psf (2.40 kN/m^2) of clear story height.
[f] Live load for railroad tracks and other rail transit fixed structures should be based on the requirements of the transit operating company or regulatory authority.

Table C4.3-2. Typical Live Load Statistics.

	Survey Load		Transient Load		Temporal Constants			
Occupancy or Use	m_s	σ_s[a]	m_t[a]	σ_t[a]	τ_s[b]	ν_e[c]	T[d]	Mean Maximum Load, psf (kN/m^2)[a]
	psf (kN/m^2)	psf (kN/m^2)	psf (kN/m^2)	psf (kN/m^2)	years	per year	years	
Office buildings: offices	10.9 (0.52)	5.9 (0.28)	8.0 (0.38)	8.2 (0.39)	8	1	50	55 (2.63)
Residential								
Renter occupied	6.0 (0.29)	2.6 (0.12)	6.0 (0.29)	6.6 (0.32)	2	1	50	36 (1.72)
Owner occupied	6.0 (0.29)	2.6 (0.12)	6.0 (0.29)	6.6 (0.32)	10	1	50	38 (1.82)
Hotels: guest rooms	4.5 (0.22)	1.2 (0.06)	6.0 (0.29)	5.8 (0.28)	5	20	50	46 (2.2)
Schools: classrooms	12.0 (0.57)	2.7 (0.13)	6.9 (0.33)	3.4 (0.16)	1	1	100	34 (1.63)

[a] For 200 ft^2 (18.58 m^2) area, except 1,000 ft^2 (92.9 m^2) for schools.
[b] Duration of average sustained load occupancy.
[c] Mean rate of occurrence of transient load.
[d] Reference period.

an increase of the live load for the area served, up to the point of satisfying the loading requirement for most assembly occupancies. As always, the designer should be aware of potential unusual loading patterns in the structure that are not covered by these minimum standards.

The minimum live loads applicable to roofs with vegetative and landscaped areas are dependent upon the use of the roof area. The 20 psf (0.96 kN/m^2) live load for unoccupied areas is the same load as for typical roof areas and is intended to represent the loads caused by maintenance activities and small decorative appurtenances. The 100 psf (4.79 kN/m^2) live load for roof assembly areas is the same as prescribed for interior building areas because the potential for a dense grouping of occupants is similar. Other occupancies within green roof areas should have the same minimum live load as specified in Table 4.3-1 for that occupancy. Soil and walkways, fences, walls, and other hardscaping features are considered dead loads in accordance with Section 3.1.4.

The public restroom uniform live load in Table 4.3-1 applies to restrooms for publicly accessible spaces. Public restrooms should be designed for the live load associated with the occupancy it serves, with an upper limit of 60 psf (2.87 kN/m^2). The upper limit recognizes that the fixtures within restrooms limit the space available for a dense grouping of occupants.

C4.3.2 Provision for Partitions Partition load provisions have been included in the standard since the first edition in 1988, however the 2005 edition was the first edition to provide a minimum value for the partition load. Historically, a value of 20 psf (0.96 kN/m^2) has been required by building codes. The 20 psf (0.96 kN/m^2) value, however, was typically specified as a dead load.

The live load value of 15 psf (0.72 kN/m^2) with a load factor of 1.6 gives the same structural demand as a dead load value of 20 psf (0.96 kN/m^2) with a 1.2 load factor. When partitions were considered as dead load, the partition load would counteract potential uplift conditions, such as the backspan of a cantilevered member or in the overturning of the structure due to lateral loads. As a live load, the partition load is excluded from the uplift load combinations and is also subject to the partial loading provisions in Section 4.3.

Assuming that a normal partition would be a stud wall with 0.5 in. (13 mm) gypsum board on each side, 8 psf (0.38 kN/m^2) per Table C3.1-1, and 10 ft (3.05 m) high, the wall load on the floor would be 80 psf (1.16 kN/m). If the partitions are spaced throughout the floor area creating rooms on a grid 10 ft (3.05 m) on center, which would be an extremely dense spacing over a whole bay, the average distributed load would be 16 psf (0.77 kN/m^2). A design value of 15 psf (0.72 kN/m^2) is judged to be reasonable in that the partitions are not likely to be spaced this closely over large areas. Designers should consider a larger design load for partitions if taller wall heights, heavier walls, or a higher density of partitions is anticipated.

The nature of the distribution of partitions is fundamentally different from office live loads, so the partition live load is not reducible in the design for gravity loads. Another difference with ordinary live loads is that the partition load is both semipermanent and firmly attached to the structure. Therefore, the partition weight is required to be considered as a part of the effective seismic weight for seismic design (see Chapter 12).

C4.3.3 Partial Loading Uniform live loads applied over only a portion of a structure or a member can produce greater forces than when uniform live loads are applied over the full length of the structure or member.

Partial-length loads on a simple beam or truss produce higher shear on a portion of the span than a full-length load. "Checkerboard" load patterns applied to multistory, multipanel bents, frames, and continuous members produce higher positive moments than loads applied to all spans, whereas loads on either side of a support produce greater negative moments. Loads on the half span of arches and domes or on the two central quarters can be critical.

Cantilevers must not rely on a possible live load on the back span for equilibrium.

C4.3.3.1 Partial Loading of Roofs All probable load patterns should be considered uniform for roof live loads that are reduced to less than 20 psf (0.96 kN/m^2) using Section 4.8. Where the full value of the roof live load (L_r) is used without reduction, it is considered that there is a low probability that the loads from maintenance workers, equipment, and material could occur in a patterned arrangement. Where a uniform roof live load is caused by occupancy, partial or pattern loading should be considered regardless of the magnitude of the uniform load.

C4.3.4 Interior Walls and Partitions Interior walls and partitions are subjected to lateral loads from occupants, occasional impact from moving furniture or equipment, and from pressurization from heating, ventilating and air-conditioning (HVAC) systems. The 5 psf (0.24 kN/m^2) horizontal load is judged to be reasonable to provide for out-of-plane strength and stability. Interior walls and partitions that are subjected to larger lateral loads should be designed for those loads.

C4.4 CONCENTRATED LIVE LOADS

The provision in Table 4.3-1 regarding concentrated loads supported by roof trusses or other primary roof members is intended to provide for a common situation for which specific requirements are generally lacking.

Primary roof members are main structural members such as roof trusses, girders, and frames, which are exposed to a work floor below, where the failure of such a primary member resulting from their use as attachment points for lifting or hoisting loads could lead to the collapse of the roof. Single roof purlins or rafters (where there are multiple such members placed side by side at some reasonably small center-to-center spacing, and where the failure of a single such member would not lead to the collapse of the roof) are not considered to be primary roof members.

C4.5 LOADS ON HANDRAIL, GUARD, GRAB BAR, AND VEHICLE BARRIER SYSTEMS, AND ON SHOWER SEATS AND FIXED LADDERS

C4.5.1 Handrail and Guard Systems Loads that can be expected to occur on handrail and guard systems are highly dependent on the use and occupancy of the protected area. The single concentrated load applied on the handrail or top rail represents load from one person or object, or a small number of people.

C4.5.1.1 Uniform Load. The uniform load represents distributed loads such as from a group of people. For cases in which extreme loads can be anticipated, such as long, straight runs of guard systems against which crowds can surge, appropriate increases in loading should be considered. There are three exceptions that allow the uniform load to not be applied, such as when handrail and guard systems are located in areas not typically open to the general public.

C4.5.2 Grab Bar Systems and Shower Seats The specified concentrated live load provides for the normal anticipated use of grab bars and shower seats. Shower seats include built-in seats and seats that are removable but secured to the shower or tub. The specified concentrated live load provides consistency with ICC/ANSI A117.1, *Accessible and Usable Buildings and Facilities* (2017). However, while grab bars and shower seats are commonly provided for use by persons with physical disabilities, they shall be designed to resist the concentrated load whenever they are provided.

C4.5.3 Vehicle Barrier Systems Vehicle barrier systems may be subjected to horizontal loads from moving vehicles. These horizontal loads may be applied normal to the plane of the barrier system, parallel to the plane of the barrier system, or at any intermediate angle. Loads in garages accommodating trucks and buses may be obtained from the provisions contained in AASHTO *LRFD Bridge Design Specifications* (2014/2015).

C4.5.4 Fixed Ladders This provision was introduced to the standard in 1998 and is consistent with the provisions for stairs.

Side rail extensions of fixed ladders are often flexible and weak in the lateral direction. The Occupational Safety and Health Administration (OSHA 2014a) requires side rail extensions, only with specific geometric requirements. The load provided was introduced to the standard in 1998 and has been determined on the basis of a 250 lb (1.11 kN) person standing on a rung of the ladder and accounting for reasonable angles of pull on the rail extension.

C4.6 IMPACT LOADS

Grandstands, stadiums, and similar assembly structures may be subjected to loads caused by crowds swaying in unison, jumping to their feet, or stomping. Designers are cautioned that the possibility of such loads should be considered.

Elevator loads are changed in the standard from a direct 100% impact factor to a reference to ASME A17.1 (2016). The provisions in ASME A17.1 include the 100% impact factor, along with deflection limits on the applicable elements.

C4.6.4 Elements Supporting Hoists for Façade Access and Building Maintenance Equipment The Occupational Safety and Health Administration (OSHA) requires that façade access platforms that are used for building maintenance meet the requirements of Standard 1910.66, *Powered Platforms for Building Maintenance* (OSHA 2014b). OSHA requires that building anchors and components be capable of sustaining without failure a load of at least four times the rated load of the hoist (i.e., the maximum anticipated load or total weight of the suspended platform plus occupants and equipment) applied or transmitted to the components and anchors. A design live load of 2.5 times the rated load, when combined with a live load factor of 1.6, results in a total factored load of 4.0 times the rated load, which matches OSHA's requirements. It should also be noted that when using allowable stress design (ASD), 2.5 times the rated load will result in a comparable design when a safety factor of 1.6 is used in determining the allowable stresses. This load requirement is not statistically based but is intended by OSHA to address accidental hang-up-and-fall scenarios as well as starting and stopping forces that the platforms experience on a day-to-day basis. It also provides a small margin of safety relative to situations where a suspended platform gets hung up on a façade while ascending, allowing the hoists to apply large forces on the supporting elements. OSHA permits hoists to generate in-service forces up to three times their rated loads. These loads should be applied in the same direction(s) as they are expected to occur.

OSHA (2014c) provisions (CFR 1926.451) related to "construction" activities also require supporting equipment to be able to carry at least 1.5 times the stall load of the supported hoist. Since OSHA defines "construction" rather broadly (it includes activities such as painting and hanging signs), most equipment is used for "construction" work, which means that it must have the strength required by OSHA construction provisions. The stall load times the live load factor of 1.6 slightly exceeds the OSHA 1.5 times the stall load requirement.

C4.6.5 Fall Arrest, Lifeline, and Rope Descent System Anchorages OSHA requires that anchorages that support fall arrest lines, lifelines, and rope descent systems be capable of sustaining without failure an ultimate load of 5,000 lb (22.2 kN) for each attached person. Using a design live load of 3,100 lb (13.8 kN), when combined with a live load factor of 1.6, results in a total factored load of 4,960 lb (22.1 kN), which essentially matches OSHA's requirements for lifeline anchorages. It should also be noted that when using Allowable Stress Design (ASD), a design live load of 3,100 lb (13.8 kN) results in a comparable design when a safety factor of 1.6 is used in determining the allowable stresses. This lifeline load is intended by OSHA to address the fall arrest loads that can and do reasonably occur in typical lanyards for body harnesses, which are highly variable.

When a fall arrest load is applied perpendicular to the length of a horizontal lifeline, much larger loads can develop at the end anchorages due to the geometry of the horizontal cable; it is important that these increased forces be considered in the design of the anchorages and their supports.

C4.7 REDUCTION IN UNIFORM LIVE LOADS

C4.7.1 General The concept of, and methods for, determining member live load reductions as a function of a loaded member's influence area, A_I, was first introduced into this standard in 1982 and was the first such change since the concept of live load reduction was introduced more than 40 years ago. The revised formula is a result of more extensive survey data and theoretical analysis (Harris et al. 1981). The change in format to a reduction multiplier results in a formula that is simple and more convenient to use. The use of influence area, now defined as a function of the tributary area, A_T, in a single equation has been shown to give more consistent reliability for the various structural effects. The influence area is defined as that floor area over which the influence surface for structural effects is significantly different from zero.

The factor K_{LL} is the ratio of the influence area (A_I) of a member to its tributary area (A_T), that is, $K_{LL} = A_I/A_T$, and is used to better define the influence area of a member as a function of its tributary area. Figure C4.7-1 illustrates typical influence areas and tributary areas for a structure with regular bay spacings. Table 4.7-1 has established K_{LL} values (derived from calculated K_{LL} values) to be used in Equation (4.7-1) for a variety of structural members and configurations. Calculated K_{LL} values vary for column and beam members that have adjacent cantilever construction, as is shown in Figure C4.7-1, and the Table 4.7-1 values have been set for these cases to result in live load reductions that are slightly conservative. For unusual shapes, the concept of significant influence effect should be applied.

An example of a member without provisions for continuous shear transfer normal to its span would be a precast T-beam or double-T beam that may have an expansion joint along one or

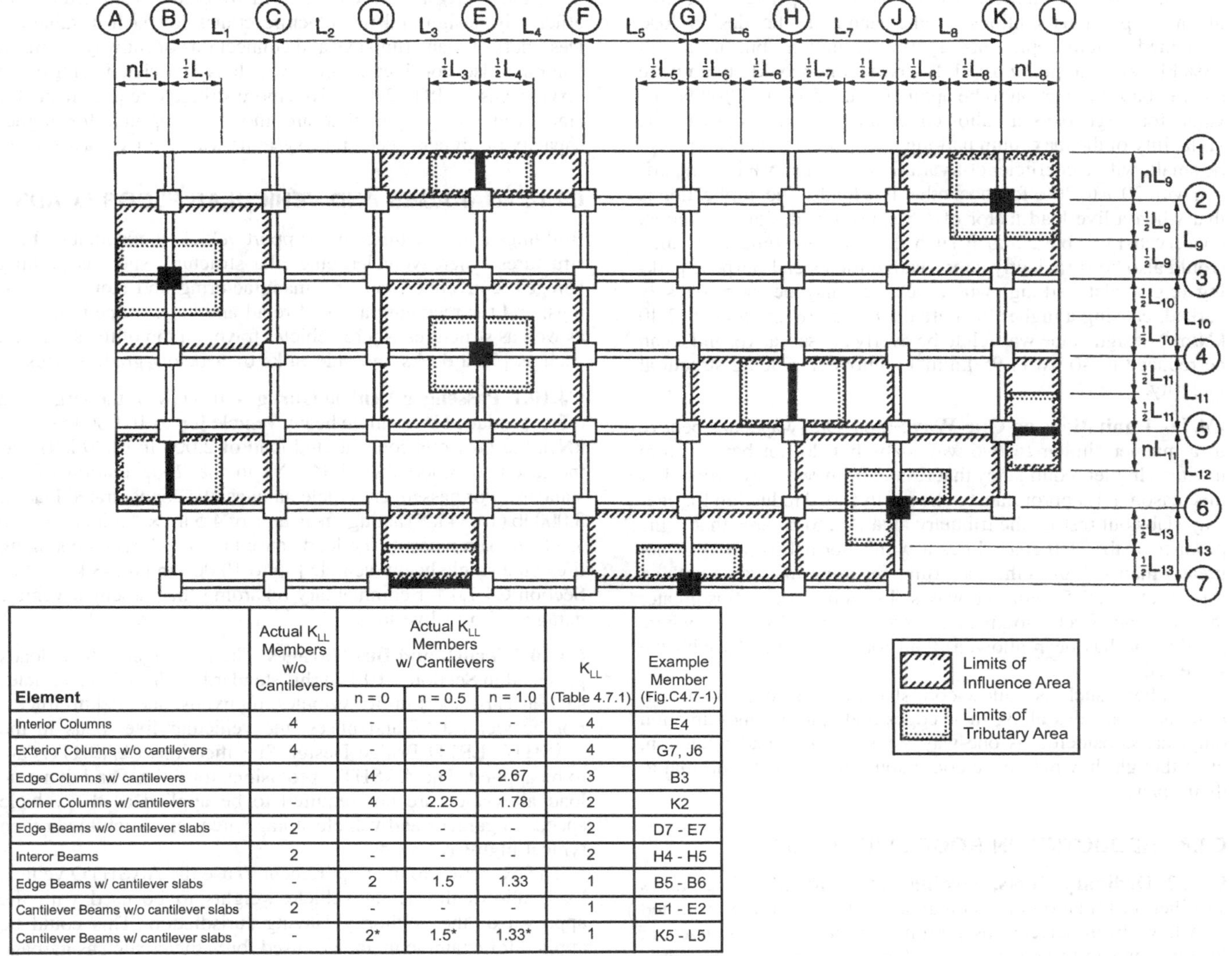

Element	Actual K_{LL} Members w/o Cantilevers	Actual K_{LL} Members w/ Cantilevers n = 0	n = 0.5	n = 1.0	K_{LL} (Table 4.7.1)	Example Member (Fig.C4.7-1)
Interior Columns	4	-	-	-	4	E4
Exterior Columns w/o cantilevers	4	-	-	-	4	G7, J6
Edge Columns w/ cantilevers	-	4	3	2.67	3	B3
Corner Columns w/ cantilevers	-	4	2.25	1.78	2	K2
Edge Beams w/o cantilever slabs	2	-	-	-	2	D7 - E7
Interor Beams	2	-	-	-	2	H4 - H5
Edge Beams w/ cantilever slabs	-	2	1.5	1.33	1	B5 - B6
Cantilever Beams w/o cantilever slabs	2	-	-	-	1	E1 - E2
Cantilever Beams w/ cantilever slabs	-	2*	1.5*	1.33*	1	K5 - L5

* The value of n for member K5-L5 is used to calculate the distance nL_{11}

Figure C4.7-1. Typical tributary and influence areas.

both flanges or that may have only intermittent weld tabs along the edges of the flanges. Such members do not have the ability to share loads located within their tributary areas with adjacent members, thus resulting in $K_{LL} = 1$ for these types of members. Reductions are permissible for two-way slabs and for beams, but care should be taken in defining the appropriate influence area. For multiple floors, areas for members supporting more than one floor are summed.

The formula provides a continuous transition from unreduced to reduced loads. The smallest allowed value of the reduction multiplier is 0.4 (providing a maximum 60% reduction), but there is a minimum of 0.5 (providing a 50% reduction) for members with a contributory load from just one floor.

C4.7.3 Heavy Live Loads In the case of occupancies involving relatively heavy basic live loads, such as storage buildings, several adjacent floor panels may be fully loaded. However, data obtained in actual buildings indicate that rarely is any story loaded with an average actual live load of more than 80% of the average rated live load. It appears that the basic live load should not be reduced for the floor-and-beam design, but that it could be reduced by up to 20% for the design of members supporting more than one floor. Accordingly, this principle has been incorporated in the recommended requirement.

C4.7.4 Passenger Vehicle Garages Unlike live loads in office and residential buildings, which are generally spatially random, parking garage loads are caused by vehicles parked in regular patterns, and the garages are often full. The rationale behind the reduction according to area for other live loads, therefore, does not apply. A load survey of vehicle weights was conducted at nine commercial parking garages in four cities of different sizes (Wen and Yeo 2001). Statistical analyses of the maximum load effects on beams and columns caused by vehicle loads over the garage's life were carried out using the survey results. Dynamic effects on the deck caused by vehicle motions, and on the ramp caused by impact, were investigated. The equivalent uniformly distributed loads (EUDL) that would produce the lifetime maximum column axial force and midspan beam bending moment are conservatively estimated at 34.8 psf (1.67 kN/m^2). The EUDL is not sensitive to bay-size variation. In view of the possible impact of very heavy vehicles in the future such as sport utility vehicles, however, a design load of 40 psf (1.95 kN/m^2) is recommended with no allowance for reduction according to bay area.

Compared with the design live load of 50 psf (2.40 kN/m^2) given in previous editions of the standard, the design load contained herein represents a 20% reduction, but it is still 33% higher than the 30 psf (1.44 kn/m^2) one would obtain were an area-based reduction to be applied to the 50 psf (2.40 kN/m^2) value for large bays as allowed in most standards. Also, the variability of the maximum parking garage load effect is found to be small, with a coefficient of variation less than 5% in comparison with 20% to 30% for most other live loads. The implication is that when a live load factor of 1.6 is used in design, additional conservatism is built into it such that the recommended value would also be sufficiently conservative for special-purpose parking (e.g., valet parking) where vehicles may be more densely parked, causing a higher load effect. Therefore, the 50 psf (2.40 kN/m^2) design value was felt to be overly conservative, and it can be reduced to 40 psf (1.95 kn/m^2) without sacrificing structural integrity.

C4.7.6 Limitations on One-Way Slabs One-way slabs behave in a manner similar to two-way slabs but do not benefit from having a higher redundancy that results from two-way action. For this reason, it is appropriate to allow a live load reduction for one-way slabs but restrict the tributary area, A_T, to an area that is the product of the slab span times a width normal to the span not greater than 1.5 times the span (thus resulting in an area with an aspect ratio of 1.5). For one-way slabs with aspect ratios greater than 1.5, the effect is to give a somewhat higher live load (where a reduction has been allowed) than for two-way slabs with the same ratio.

Members, such as hollow-core slabs, that have grouted continuous shear keys along their edges and span in one direction only, are considered as one-way slabs for live load reduction, even though they may have continuous shear transfer normal to their spans.

C4.8 REDUCTION IN ROOF LIVE LOADS

C4.8.2 Ordinary Roofs, Awnings, and Canopies The values specified in Equation (4.8-1) that act vertically on the projected area have been selected as minimum roof live loads, even in localities where little or no snowfall occurs. This is because it is considered necessary to provide for occasional loading caused by the presence of workers and materials during repair operations.

C4.8.3 Occupiable Roofs Designers should consider any additional dead loads that may be imposed by saturated landscaping materials in addition to the live load required in Table 4.3-1. Occupancy-related loads on roofs are live loads (L) normally associated with the design of floors rather than roof live loads (L_r) and may be reduced in accordance with the provisions for live loads in Section 4.7 rather than Section 4.8.

C4.9 CRANE LOADS

All support components of moving bridge cranes and monorail cranes, including runway beams, brackets, bracing, and connections, shall be designed to support the maximum wheel load of the crane and the vertical impact, lateral, and longitudinal forces induced by the moving crane. Also, the runway beams shall be designed for crane stop forces. The methods for determining these loads vary depending on the type of crane system, support, and service class. CMAA (2015a, b), MMA (2016), and MBMA (2018) describe types of bridge cranes and monorail cranes, including top running bridge cranes with top running trolley, underhung bridge cranes, and underhung monorail cranes. The bridge crane service class descriptions are based on those given in CMAA (2015a), which are used by crane manufacturers to efficiently design and manufacture cranes for each installation. Designers of crane runways and connections of runway beams to columns also consider fatigue, which is associated with crane service class. AIST (2003) gives more stringent requirements for crane runway designs that are more appropriate for higher capacity or higher speed crane systems with service class E or F.

C4.10 GARAGES AND VEHICULAR FLOOR LOADS

Buildings and structures that support vehicle loads include both structures where vehicles enter the structure, such as parking garages, convention centers, manufacturing and storage buildings, and the receiving areas of retail and large office buildings, as well as structures where vehicles travel on top of the structure, such as plaza decks and sidewalks over below-grade spaces.

C4.10.1 Passenger Vehicle Garages In view of the large load effect produced by a single heavy vehicle [up to 10,000 lb (44.48 kN)], the previous concentrated load of 2,000 lb (8.90 kN) was increased to 3,000 lb (13.34 kN) in the 2002 edition of the standard. For passenger vehicle garages, the concentrated load of 3,000 lb (13.34 kN) acting on an area of 4.5 in. × 4.5 in. (114 mm × 114 mm) represents the load caused by a jack when changing tires on a single heavy vehicle [up to 10,000 lb (44.48 kN)]. See Section C4.7.4 for commentary regarding the passenger vehicle garage uniform live load.

C4.10.2 Truck and Bus Garages The passenger vehicle loads provided in Section 4.10.1 of this standard are limited to vehicles such as cars, sport utility vehicles, minivans, and pickup trucks. For heavier trucks and buses, the vehicular live loads in the AASHTO LRFD Bridge Design Specifications (AASHTO) are to be applied. The AASHTO provisions for fatigue and dynamic load allowance are not required to be applied as the vehicle speeds in garages and vehicle storage areas are much lower than typical highway speeds.

The exception to the requirement to use the AASHTO vehicle loads allows the actual vehicle weights to be used with the approval of the Authority Having Jurisdiction. This could be applied for example in garages used for vehicles such as garbage trucks, armored trucks, and delivery trucks that are heavier than passenger vehicles but lighter than the highway semitrailer and triple-axle trucks represented by the AASHTO design truck.

C4.10.3 Sidewalks, Vehicular Driveways, and Yards Subject to Trucking The large uniform and concentrated load specified for these areas have been historically used to design city sidewalk areas over basements and vaults that can be subject to trucks driving up on the sidewalk and parking to unload.

C4.10.4 Emergency Vehicle Loads Examples of structures that are accessible to emergency vehicles include parking garages and at-grade areas such as plaza decks and parking lots over underground structures. Fire trucks and ambulances are likely to enter these areas when responding to emergencies, therefore they are required to be designed for the vehicle loads unless access is physically restricted. Moveable physical barriers are not considered to physically restrict access as they may allow emergency vehicles on the structure if the barriers are not present when the emergency occurs or if the barriers swing out of the way for the emergency vehicle.

The emergency vehicle live loads are permitted to be either the operating loads of the vehicles in the local jurisdiction or the AASHTO design truck and design tandem live loads. The operating loads are required to be obtained from the Authority Having Jurisdiction (AHJ) as they have the most knowledge of emergency vehicles in their jurisdiction and their mutual aid partner jurisdictions. The AHJ may need to consult with the vehicle manufacturer to determine the weight and outrigger loads if this information has not been previously documented. Operating loads include the vehicle weight, equipment weight, and, if present, stored water weight. The wheel and outrigger loads need to account for the distribution of the weight, the wheel and outrigger pad contact area and spacing, and the operation of integral ladders and aerial platforms when present. Outrigger reactions can be significant as industry standards permit ratings of up to 60,000 lb (266.9 kN).

C4.11 HELIPAD LOADS

C4.11.1 General Helipad provisions were added to the standard in 2010. For the standard, the term *helipads* is used to refer specifically to the structural surface. In building codes and other references, different terminology may be used when describing helipads, e.g., heliports, helistops, but the distinctions between these are not relevant to the structural loading issue addressed in ASCE 7.

Although these structures are intended to be specifically kept clear of non-helicopter occupant loads on the landing and taxi areas, the uniform load requirement is a minimum to ensure a degree of substantial construction and the potential to resist the effects of unusual events.

Additional information on helipad design can be found in Annex 14 to the *Convention on International Civil Aviation, Aerodromes, Vol. 2* (ICAO 2013).

C4.11.2 Concentrated Helicopter Loads Concentrated loads applied separately from the distributed loads are intended to cover the primary helicopter loads. The designer should always consider the geometry of the design basis helicopter for applying the design loads. A factor of 1.5 is used to address impact loads (two single concentrated loads of 0.75 times the maximum takeoff weight) to account for a hard landing with many kinds of landing gear. The designer should be aware that some helicopter configurations, particularly those with rigid landing gear, could result in substantially higher impact factors that should be considered.

The 3,000 lb (13.34 kN) concentrated load is intended to cover maintenance activities, similar to the jack load for a parking garage.

C4.13 LIBRARY STACK ROOMS

Where library shelving installation does not fall within the parameter limits that are specified in Section 4.13 and Table 4.3-1, the design should account for the actual conditions. For example, the floor loading for storage of medical X-ray film may easily exceed 200 psf (9.58 kN/m^2), mainly because of the increased depth of the shelves. Mobile library shelving that rolls on rails should also be designed to meet the actual requirements of the specific installation, which may easily exceed 300 psf (14.4 kN/m^2). The rail support locations and deflection limits should be considered in the design, and the engineer should work closely with the system manufacturer to provide a serviceable structure.

C4.14 SEATING FOR ASSEMBLY USES

The lateral loads apply to "stadiums and arenas" and to "bleachers, folding and telescopic seating, and grandstands." However, it does not apply to "gymnasiums." Consideration should be given to treating gymnasium balconies that have stepped floors for seating as arenas, and requiring the appropriate swaying forces.

C4.17 SOLAR PANEL LOADS

C4.17.1 Roof Loads at Solar Panels These provisions were added to the 2016 edition of the standard to address the installation of rooftop solar panels consistent with current practices (Blaney and LaPlante 2013). These provisions allow the offset of roof live load where the space below the solar panel is considered inaccessible. The dimension of 24 in. (610 mm) was chosen as the clear vertical distance as it is consistent with existing published requirements for solar panel systems and is also a typical minimum height permitted for access into or out of spaces.

C4.17.3 Open-Grid Roof Structures Supporting Solar Panels This section reduces the uniform roof live load for building structures such as carports and shade structures, which do not include roof deck or sheathing, to the value of the minimum uniform roof live load permitted by Section 4.8.2. The concentrated roof live load requirement in Table 4.3-1 is not modified by this section.

REFERENCES

AASHTO (American Association of State Highway and Transportation Officials). 2014–2015. *LRFD bridge design specifications*, 7th ed. Washington, DC: AASHTO.

AIST (Association of Iron and Steel Technology). 2003. *Guide for the design and construction of mill buildings*. Tech. Rep. No. 13. Warrendale, PA: AISE.

ASME (American Society of Mechanical Engineers). 2016. *Safety code for elevators and escalators. A17*. New York: ASME.

Blaney, C., and R. LaPlante. 2013. "Recommended design live loads for rooftop solar arrays." In *Proc., SEAOC Convention*, 264–278.

Chalk, P. L., and R. B. Corotis. 1980. "Probability model for design live loads." *J. Struct. Div.* 106 (10): 2017–2033. https://doi.org/10.1061/JSDEAG.0005542.

Corotis, R. B., J. C. Harris, and R. R. Fox. 1981. "Delphi methods: Theory and design load application." *J. Struct. Div.* 107 (6): 1095–1105. https://doi.org/10.1061/JSDEAG.0005722.

CMAA (Crane Manufacturers Association of America). 2015a. *Specifications for top running bridge and gantry type multiple girder electric overhead traveling cranes*. Rep. No. 70-2015. Charlotte, NC: CMAA.

CMAA. 2015b. *Specifications for top running and under running single girder electric overhead traveling cranes utilizing under running trolley hoist*. Rep. No. 74-2015. Charlotte, NC: CMAA.

Ellingwood, B. R., and C. G. Culver. 1977. "Analysis of live loads in office buildings." *J. Struct. Div.* 103 (8): 1551–1560. https://doi.org/10.1061/JSDEAG.0004693.

Harris, M. E., C. J. Bova, and R. B. Corotis. 1981. "Area-dependent processes for structural live loads." *J. Struct. Div.* 107 (5): 857–872. https://doi.org/10.1061/JSDEAG.0005709.

ICAO (International Civil Aviation Organization). 2013. Vol. 2 of *Convention on international civil aviation, aerodromes*. Montreal, QC: ICAO.

ICC/ANSI (International Code Council and American National Standards Institute). 2017. *Accessible and usable buildings and facilities. A117.1*. Washington, DC: ICC.

MBMA (Metal Building Manufacturers Association). 2018. *Metal building systems manual*. Cleveland, OH: MBMA.

McGuire, R. K., and C. A. Cornell. 1974. "Live load effects in office buildings." *J. Struct. Div.* 100 (7): 1351–1366. https://doi.org/10.1061/JSDEAG.0003816.

MMA (Monorail Manufacturers Association). 2016. *Patented track underhung cranes and monorail systems*. Rep. No. MH 27.1-2016. Charlotte, NC: MMA.

OSHA (Occupational Safety and Health Administration). 2014a. "Ladders." In *Code of federal regulations, section 1910.23*. Washington, DC: OSHA Standards.

OSHA. 2014b. "Powered platforms for building maintenance." In *Code of federal regulations, section 1910.66*. Washington, DC: OSHA Standards.

OSHA. 2014c. "Safety standards for scaffolds used in the construction industry." In *Code of federal regulations, section 1926.451*. Washington, DC: OSHA Standards.

Peir, J. C., and C. A. Cornell. 1973. "Spatial and temporal variability of live loads." *J. Struct. Div.* 99 (5): 903–922. https://doi.org/10.1061/JSDEAG.0003512.

Sentler, L. 1975. *A stochastic model for live loads on floors in buildings*. Rep. No. 60. Lund, Sweden: Lund Institute of Technology.

Wen, Y. K., and G. L. Yeo. 2001. "Design live loads for passenger cars parking garages." *J. Struct. Eng.* 127 (3): 280–289. https://doi.org/10.1061/(ASCE)0733-9445(2001)127:3(280).

CHAPTER C5
FLOOD LOADS

C5.1 GENERAL

This section presents information for the design of buildings and other structures in areas prone to flooding. Design professionals should be aware that there are important differences between flood characteristics, flood loads, and flood effects in riverine and coastal areas (e.g., the potential for wave effects is much greater in coastal areas, the depth and duration of flooding can be much greater in riverine areas, the direction of flow in riverine areas tends to be more predictable, and the nature and amount of flood-borne debris varies between riverine and coastal areas).

Much of the impetus for flood-resistant design has come about from the federal government sponsored initiatives of flood-damage mitigation and flood insurance, both through the work of the USACE and the National Flood Insurance Program (NFIP). The NFIP is based on an agreement between the federal government and participating communities that have been identified as being flood prone. The Federal Emergency Management Agency (FEMA), through the Federal Insurance and Mitigation Administration (FIMA), makes flood insurance available to the residents of communities provided that the community adopts and enforces adequate floodplain management regulations that meet the minimum FIMA requirements. Included in the NFIP requirements, found under Title 44 of the US Code of Federal Regulations (FEMA 1999b), are minimum building design and construction standards for buildings and other structures located in special flood hazard areas (SFHAs).

Special flood hazard areas are those identified by FEMA as being subject to inundation during the 100-year flood. SFHAs are shown on flood insurance rate maps (FIRMs), which are produced for flood-prone communities. SFHAs are identified on FIRMs as zones A, A1-30, AE, AR, AO, and AH, and in coastal high hazard areas as V1-30, V, and VE. The SFHA is the area in which communities must enforce NFIP-compliant, flood-damage-resistant design and construction practices.

Before designing a structure in a flood-prone area, design professionals should contact the local building official to determine if the site in question is located in an SFHA or other flood-prone area that is regulated under the community's floodplain management regulations. If the proposed structure is located within the regulatory floodplain, local building officials can explain the regulatory requirements.

Answers to specific questions on flood-resistant design and construction practices may be directed to the mitigation division of each of FEMA's regional offices, which are available to assist design professionals.

C5.2 DEFINITIONS

Three new concepts were added with ASCE 7-98. First, the concept of the design flood was introduced. The design flood will, at a minimum, be equivalent to the flood having a 1% chance of being equaled or exceeded in any given year (i.e., the base flood or 100-year flood, which served as the load basis in ASCE 7-95). In some instances, the design flood may exceed the base flood in elevation or spatial extent; this excess will occur where a community has designated a greater flood (lower frequency, higher return period) as the flood to which the community will regulate new construction.

Many communities have elected to regulate to a flood standard higher than the minimum requirements of the NFIP. Those communities may do so in a number of ways. For example, a community may require new construction to be elevated a specific vertical distance above the base flood elevation (this is referred to as "freeboard"); a community may select a lower frequency flood as its regulatory flood; or a community may conduct hydrologic and hydraulic studies, on which flood hazard maps are based, in a manner different from the Flood Insurance Study prepared by the NFIP (e.g., the community may complete flood hazard studies based on development conditions at build-out, rather than following the NFIP procedure, which uses conditions in existence at the time the studies are completed; the community may include watersheds smaller than 1 mi^2 (2.6 km^2) in size in its analysis, rather than following the NFIP procedure, which neglects watersheds smaller than 1 mi^2 (2.6 km^2).

Use of the design flood concept will ensure that the requirements of this standard are not less restrictive than a community's requirements where that community has elected to exceed minimum NFIP requirements. In instances where a community has adopted the NFIP minimum requirements, the design flood described in this standard will default to the base flood.

Second, this standard also uses the terms "flood hazard area" and "flood hazard map" to correspond to and show the areas affected by the design flood. Again, in instances where a community has adopted the minimum requirements of the NFIP, the flood hazard area defaults to the NFIP's SFHA and the flood hazard map defaults to the FIRM.

Third, the concept of a Coastal A-Zone is used to facilitate application of load combinations contained in Chapter 2 of this standard. Coastal A-Zones lie landward of V-Zones, or landward of an open coast shoreline where V-Zones have not been mapped (e.g., the shorelines of the Great Lakes). Coastal A-Zones are subject to the effects of waves, high-velocity flows, and erosion, although not to the extent that V-Zones are. Like V-Zones, flood forces in Coastal A-Zones will be highly correlated with coastal winds or coastal seismic activity.

Coastal A-Zones are not delineated on flood hazard maps prepared by FEMA, but are zones where wave forces and erosion potential should be taken into consideration by designers. The following guidance is offered to designers to determine whether or not an A-Zone in a coastal area can be considered a Coastal A-Zone.

For a Coastal A-Zone to be present, two conditions are required: (1) a still-water flood depth greater than or equal to 2.0 ft (0.61 m), and (2) breaking wave heights greater than or equal to 1.5 ft (0.46 m). Note that the still-water depth requirement is necessary, but is not sufficient by itself, to render an area a Coastal A-Zone. Many A-Zones will have still-water flood depths in excess of 2.0 ft (0.61 m) but will not experience breaking wave heights greater than or equal to 1.5 ft (0.46 m) and therefore should not be considered Coastal A-Zones. Wave heights at a given site can be determined using procedures outlined in USACE (2002) or similar references.

The 1.5 ft (0.46 m) breaking wave height criterion was developed from post-flood damage inspections, which show that wave damage and erosion often occur in mapped A-Zones in coastal areas, and from laboratory tests on breakaway walls that show that breaking waves 1.5 ft (0.46 m) in height are capable of causing structural failures in wood-frame walls (FEMA 2000).

C5.3 DESIGN REQUIREMENTS

Sections 5.3.4 (dealing with A-Zone design and construction) and 5.3.5 (dealing with V-Zone design and construction) of ASCE 7-98 were deleted in preparation of the 2002 edition of this standard. These sections summarized basic principles of flood-resistant design and construction [building elevation, anchorage, foundation, below design flood elevation (DFE) enclosures, breakaway walls, etc.]. Some of the information contained in these deleted sections was included in Section 5.3, beginning with ASCE 7-02. The design professional is also referred to ASCE/SEI Standard 24 (*Flood Resistant Design and Construction*) for specific guidance.

C5.3.1 Design Loads Wind loads and flood loads may act simultaneously at coastlines, particularly during hurricanes and coastal storms. This may also be true during severe storms at the shorelines of large lakes and during riverine flooding of long duration.

C5.3.2 Erosion and Scour The term *erosion* indicates a lowering of the ground surface in response to a flood event or in response to the gradual recession of a shoreline. The term *scour* indicates a localized lowering of the ground surface during a flood, due to the interaction of currents and/or waves with a structural element. Erosion and scour can affect the stability of foundations and can increase the local flood depth and flood loads acting on buildings and other structures. For these reasons, erosion and scour should be considered during load calculations and the design process. Design professionals often increase the depth of foundation embedment to mitigate the effects of erosion and scour and often site buildings away from receding shorelines (building setbacks).

C5.3.3 Loads on Breakaway Walls Floodplain management regulations require buildings in coastal high hazard areas to be elevated to or above the design flood elevation by a pile or column foundation. Space below the DFE must be free of obstructions to allow the free passage of waves and high-velocity waters beneath the building (FEMA 1993). Floodplain management regulations typically allow space below the DFE to be enclosed by insect screening, open lattice, or breakaway walls. Local exceptions are made in certain instances for shear walls, firewalls, elevator shafts, and stairwells. Check with the Authority Having Jurisdiction for specific requirements related to obstructions, enclosures, and breakaway walls.

Where breakaway walls are used, they must meet the prescriptive requirements of NFIP regulations or be certified by a registered professional engineer or architect as having been designed to meet the NFIP performance requirements. Meeting the NFIP performance requirements should be understood to mean that the structure to which breakaway walls are attached should withstand both of the following: (1) load combinations, including flood loads acting on the structure and the breakaway walls, up to the point of breakaway wall collapse, and (2) load combinations, including flood loads acting on the structure that remains following breakaway collapse, for flood conditions between those causing breakaway wall collapse and those associated with the design flood.

The prescriptive requirements call for breakaway wall designs that are intended to collapse at loads not less than 10 psf ($0.48\,kN/m^2$) and not more than 20 psf ($0.96\,kN/m^3$). Inasmuch as wind or earthquake loads often exceed 20 psf ($0.96\,kN/m^2$), breakaway walls may be designed for higher loads, provided the designer certifies that the walls have been designed to break away before base flood conditions are reached, without damaging the elevated building or its foundation. FEMA (1999a) provides guidance on how to meet the performance requirements for certification.

C5.4 LOADS DURING FLOODING

C5.4.1 Load Basis Water loads are the loads or pressures on surfaces of buildings and structures caused and induced by the presence of floodwaters. These loads are of two basic types: hydrostatic and hydrodynamic. Impact loads result from objects transported by floodwaters striking against buildings and structures or parts thereof. Wave loads can be considered a special type of hydrodynamic load.

C5.4.2 Hydrostatic Loads Hydrostatic loads are those caused by water either above or below the ground surface, free or confined, which is either stagnant or moves at velocities less than 5 ft/s (1.52 m/s). These loads are equal to the product of the water pressure multiplied by the surface area on which the pressure acts.

Hydrostatic pressure at any point is equal in all directions and always acts perpendicular to the surface on which it is applied. Hydrostatic loads can be subdivided into vertical downward loads, lateral loads, and vertical upward loads (uplift or buoyancy). Hydrostatic loads acting on inclined, rounded, or irregular surfaces may be resolved into vertical downward or upward loads and lateral loads based on the geometry of the surfaces and the distribution of hydrostatic pressure.

C5.4.3 Hydrodynamic Loads Hydrodynamic loads are those loads induced by the flow of water moving at moderate to high velocity above the ground level. They are usually lateral loads caused by the impact of the moving mass of water and the drag forces as the water flows around the obstruction. Hydrodynamic loads are computed by recognized engineering methods. In the coastal high hazard area, the loads from high-velocity currents due to storm surge and overtopping are of particular importance. USACE (2002) is one source of design information regarding hydrodynamic loadings.

Note that accurate estimates of flow velocities during flood conditions are very difficult to make, both in riverine and coastal flood events. Potential sources of information regarding velocities of floodwaters include local, state, and federal government agencies and consulting engineers specializing in coastal engineering, stream hydrology, or hydraulics.

As interim guidance for coastal areas, FEMA (2000) gives a likely range of flood velocities as

$$V = d_s/(1\text{ s}) \qquad \text{(C5.4-1)}$$

to

$$V = (gd_s)^{0.5} \qquad \text{(C5.4-2)}$$

where

V = Average velocity of water, ft/s (m/s);
d_s = Local still-water depth, ft (m); and
g = Acceleration due to gravity, 32.2 ft/s/s (9.81 m/s^2).

Selection of the correct value of a in Equation (5.4-1) will depend on the shape and roughness of the object exposed to flood flow, as well as the flow condition. As a general rule, the smoother and more streamlined the object, the lower the drag coefficient (shape factor). Drag coefficients for elements common in buildings and structures (round or square piles, columns, and rectangular shapes) will range from approximately 1.0 to 2.0, depending on flow conditions. However, given the uncertainty surrounding flow conditions at a particular site, ASCE 7-05 recommends a minimum value of 1.25 be used. Fluid mechanics texts should be consulted for more information on when to apply drag coefficients above 1.25.

C5.4.4 Wave Loads The magnitude of wave forces (lb/ft^2) (kN/m^2) acting against buildings or other structures can be 10 or more times higher than wind forces and other forces under design conditions. Thus, it should be readily apparent that elevating above the wave crest elevation is crucial to the survival of buildings and other structures. Even elevated structures, however, must be designed for large wave forces that can act over a relatively small surface area of the foundation and supporting structure.

Wave load calculation procedures in Section 5.4.4 are taken from USACE (2002) and Walton et al. (1989). The analytical procedures described by Equations (5.4-2) through (5.4-9) should be used to calculate wave heights and wave loads unless more advanced numerical or laboratory procedures permitted by this standard are used.

Wave load calculations using the analytical procedures described in this standard all depend on the initial computation of the wave height, which is determined using Equations (5.4-2) and (5.4-3). These equations result from the assumptions that the waves are depth limited and that waves propagating into shallow water break when the wave height equals 78% of the local still-water depth and that 70% of the wave height lies above the local still-water level. These assumptions are identical to those used by FEMA in its mapping of coastal flood hazard areas on FIRMs.

Designers should be aware that wave heights at a particular site can be less than depth-limited values in some cases (e.g., when the wind speed, wind duration, or fetch is insufficient to generate waves large enough to be limited in size by water depth, or when nearby objects dissipate wave energy and reduce wave heights). If conditions during the design flood yield wave heights at a site less than depth-limited heights, Equation (5.4-2) may overestimate the wave height and Equation (5.4-3) may underestimate the still-water depth. Also, Equations (5.4-4) through (5.4-7) may overstate wave pressures and forces when wave heights are less than depth-limited heights. More advanced numerical or laboratory procedures permitted by this section may be used in such cases, in lieu of Equations (5.4-2) through (5.4-7).

It should be pointed out that present NFIP mapping procedures distinguish between A-Zones and V-Zones by the wave heights expected in each zone. Generally speaking, A-Zones are designated where wave heights less than 3 ft (0.91 m) in height are expected. V-Zones are designated where wave heights equal to or greater than 3 ft (0.91 m) are expected. Designers should proceed cautiously, however. Large wave forces can be generated in some A-Zones, and wave force calculations should not be restricted to V-Zones. Present NFIP mapping procedures do not designate V-Zones in all areas where wave heights greater than 3 ft (0.91 m) can occur during base flood conditions. Rather than rely exclusively on flood hazard maps, designers should investigate historical flood damages near a site to determine whether or not wave forces can be significant.

C5.4.4.2 Breaking Wave Loads on Vertical Walls Equations used to calculate breaking wave loads on vertical walls contain a coefficient, C_p. Walton et al. (1989) provide recommended values of the coefficient as a function of probability of exceedance. The probabilities given by Walton et al. (1989) are not annual probabilities of exceedance, but probabilities associated with a distribution of breaking wave pressures measured during laboratory wave tank tests. Note that the distribution is independent of water depth. Thus, for any water depth, breaking wave pressures can be expected to follow the distribution described by the probabilities of exceedance in Table C5.4-2.

This standard assigns values for C_p according to building category, with the most important buildings having the largest values of C_p. Category II buildings are assigned a value of C_p corresponding to a 1% probability of exceedance, which is consistent with wave analysis procedures used by FEMA in mapping coastal flood hazard areas and in establishing minimum floor elevations. Category I buildings are assigned a value of C_p corresponding to a 50% probability of exceedance, but designers may wish to choose a higher value of C_p. Category III buildings are assigned a value of C_p corresponding to a 0.2% probability of exceedance, while Category IV buildings are assigned a value of C_p corresponding to a 0.1% probability of exceedance.

Breaking wave loads on vertical walls reach a maximum when the waves are normally incident (direction of wave approach is perpendicular to the face of the wall; wave crests are parallel to the face of the wall). As guidance for designers of coastal buildings or other structures on normally dry land (i.e., flooded only during coastal storm or flood events), it can be assumed that the direction of wave approach will be approximately perpendicular to the shoreline. Therefore, the direction of wave approach relative to a vertical wall will depend on the orientation of the wall relative to the shoreline. Section 5.4.4.4 provides a method for reducing breaking wave loads on vertical walls for waves not normally incident.

C5.4.5 Impact Loads Impact loads are those that result from logs, ice floes, and other objects striking buildings, structures, or parts thereof. USACE (1995) divides impact loads into three categories: (1) normal impact loads, which result from the isolated impacts of normally encountered objects; (2) special impact loads, which result from large objects, such as broken up ice floes and accumulations of debris, either striking or resting against a building, structure, or parts thereof; and (3) extreme impact loads, which result from very large objects, such as boats, barges, or collapsed buildings, striking the building, structure, or component under consideration. Design for extreme impact loads is not practical for most buildings and structures. However, in cases where there is a high probability that a Category III or IV structure (see Table 1.5-1) will be exposed to extreme impact loads during the design flood, and where the resulting damages will be very severe, consideration of extreme impact loads may be justified. Unlike extreme impact loads, design for special and normal impact loads is practical for most buildings and structures.

The recommended method for calculating normal impact loads has been modified beginning with ASCE 7-02. Previous editions of ASCE 7 used a procedure contained in USACE (1995) [the procedure, which had been unchanged since at least 1972, relied on an impulse-momentum approach with a 1,000 lb (4.5 kN) object striking the structure at the velocity of the floodwater and coming to rest in 1.0 s]. Work (Kriebel et al. 2000, Haehnel and Daly 2001) has been conducted to evaluate this procedure through a literature review and laboratory tests. The literature review considered riverine and coastal debris, ice floes and impacts, ship berthing and impact forces, and various methods for calculating debris loads (e.g., impulse-momentum, work-energy). The laboratory tests included log sizes ranging from 380 lb (1.7 kN) to 730 lb (3.3 kN) traveling at up to 4 ft/s (1.2 m/s).

Kriebel et al. (2000) and Haehnel and Daly (2001) conclude that (1) an impulse-momentum approach is appropriate; (2) the 1,000 lb (4.5 kN) object is reasonable, although geographic and local conditions may affect the debris object size and weight; (3) the 1.0 s impact duration is not supported by the literature or by laboratory tests—a duration of impact of 0.03 s should be used instead; (4) a half-sine curve represents the applied load and resulting displacement well; and (5) setting the debris velocity equivalent to the flood velocity is reasonable for all but the largest objects in shallow water or obstructed conditions.

Given the short-duration impulsive loads generated by flood-borne debris, a dynamic analysis of the affected building or structure may be appropriate. In some cases (e.g., when the natural period of the building is much greater than 0.03 s), design professionals may wish to treat the impact load as a static load applied to the building or structure (this approach is similar to that used by some following the procedure contained in Section C5.3.3.5 of ASCE 7-98).

In either type of analysis—dynamic or static—Equation (C5.4-3) provides a rational approach for calculating the magnitude of the impact load:

$$F = \frac{\pi W V_b C_I C_O C_D C_B R_{\max}}{2g\Delta t} \quad \text{(C5.4-3)}$$

where

F = Impact force, lb (N);
W = Debris weight, lb (N);
V_b = Velocity of object (assume equal to velocity of water, V), ft/s (m/s);
g = Acceleration due to gravity, 32.2 ft/s^2 (9.81 m/s^2);
Δt = Impact duration (time to reduce object velocity to zero), s;
C_I = Importance coefficient (see Table C5.4-1);
C_O = Orientation coefficient, 0.8;
C_D = Depth coefficient (see Table C5.4-2, Figure C5.4-1);
C_B = Blockage coefficient (see Table C5.4-3, Figure C5.4-2); and
$R_{\max}$ = Maximum response ratio for impulsive load (see Table C5.4-4).

The form of Equation (C5.4-3) and the parameters and coefficients are discussed in the following text:

Basic Equation. The equation is similar to the equation used in ASCE 7-98, except for the $\pi/2$ factor (which results from the half-sine form of the applied impulse load) and the coefficients C_I, C_O, C_D, C_B, and $R_{\max}$. With the coefficients set equal to 1.0, the equation reduces to $F = \pi W V_b / 2g\Delta t$ and calculates the maximum static load from a head-on impact of a debris object. The coefficients have been added to allow design professionals to "calibrate" the resulting force to local flood, debris, and building characteristics. The approach is similar to that used by ASCE 7 in calculating wind, seismic, and other loads. A scientifically based equation is used to match the physics, and the results are modified by coefficients to calculate realistic load magnitudes. However, unlike wind, seismic, and other loads, the body of work associated with flood-borne debris impact loads does not yet account for the probability of impact.

Table C5.4-1. Values of Importance Coefficient, C_I.

Risk Category	C_I
I	0.6
II	1.0
III	1.2
IV	1.3

Table C5.4-2. Values of Depth Coefficient, C_D.

Building Location in Flood Hazard Zone and Water Depth	C_D
Floodway or V-Zone	1.0
A-Zone, still-water depth > 5ft	1.0
A-Zone, still-water depth = 4ft	0.75
A-Zone, still-water depth = 3ft	0.5
A-Zone, still-water depth = 2ft	0.25
Any flood zone, still-water depth < 1ft	0.0

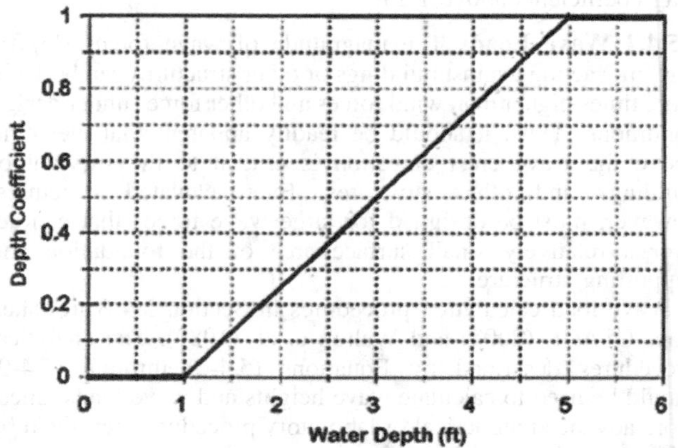

Figure C5.4-1. Depth Coefficient, C_D.

Table C5.4-3. Values of Blockage Coefficient, C_B.

Degree of Screening or Sheltering within 100 ft Upstream	C_B
No upstream screening, flow path wider than 30 ft	1.0
Limited upstream screening, flow path 20 ft wide	0.6
Moderate upstream screening, flow path 10 ft wide	0.2
Dense upstream screening, flow path less than 5 ft wide	0.0

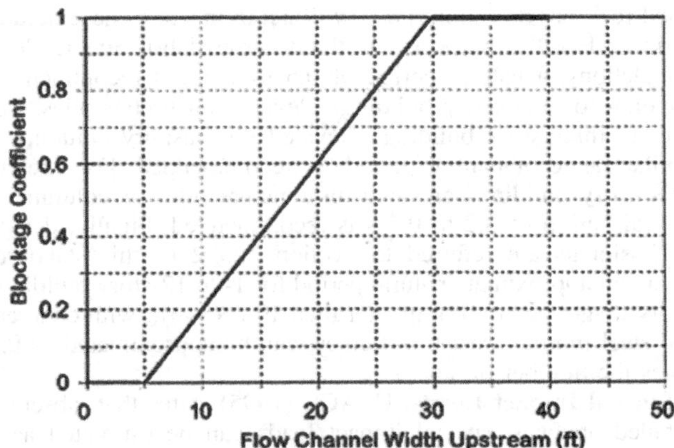

Figure C5.4-2. Blockage Coefficient, C_B.

Table C5.4-4. Values of Response Ratio for Impulsive Loads, R_{max}.

Ratio of Impact Duration to Natural Period of Structure	R_{max} (Response Ratio for Half-Sine Wave Impulsive Load)
0.00	0.0
0.10	0.4
0.20	0.8
0.30	1.1
0.40	1.4
0.50	1.5
0.60	1.7
0.70	1.8
0.80	1.8
0.90	1.8
1.00	1.7
1.10	1.7
1.20	1.6
1.30	1.6
≥1.40	1.5

Source: Adapted from Clough and Penzien (1993).

Debris Object Weight. A 1,000 lb (4.5 kN) object can be considered a reasonable average for flood-borne debris (no change from ASCE 7-98). This represents a reasonable weight for trees, logs, and other large woody debris that is the most common form of damaging debris nationwide. This weight corresponds to a log approximately 30 ft (9.1 m) long and just under 1 ft (0.3 m) in diameter. The 1,000 lb (4.5 kN) object also represents a reasonable weight for other types of debris ranging from small ice floes, to boulders, to manufactured objects.

However, design professionals may wish to consider regional or local conditions before the final debris weight is selected. The following text provides additional guidance. In riverine floodplains, large woody debris (trees and logs) predominates, with weights typically ranging from 1,000 lb (4.5 kN) to 2,000 lb (9.0 kN). In the Pacific Northwest, larger tree and log sizes suggest a typical 4,000 lb (18.0 kN) debris weight. Debris weights in riverine areas subject to floating ice typically range from 1,000 lb (4.5 kN) to 4,000 lb (18.0 kN). In arid or semiarid regions, typical woody debris may be less than 1,000 lb (4.5 kN). In alluvial fan areas, nonwoody debris (stones and boulders) may present a much greater debris hazard. Debris weights in coastal areas generally fall into three classes: in the Pacific Northwest, a 4,000 lb (18.0 kN) debris weight owing to large trees and logs can be considered typical; in other coastal areas where piers and large pilings are available locally, debris weights may range from 1,000 lb (4.5 kN) to 2,000 lb (9.0 kN); and in other coastal areas where large logs and pilings are not expected, debris will likely be derived from failed decks, steps, and building components and will likely average less than 500 lb (2.3 kN) in weight.

Debris Velocity. The velocity with which a piece of debris strikes a building or structure will depend on the nature of the debris and the velocity of the floodwaters. Small pieces of floating debris, which are unlikely to cause damage to buildings or other structures, will typically travel at the velocity of the floodwaters, in both riverine and coastal flood situations. However, large debris, such as trees, logs, pier pilings, and other large debris capable of causing damage, will likely travel at something less than the velocity of the floodwaters. This reduced velocity of large debris objects results in large part from debris dragging along the bottom and/or being slowed by prior collisions. Large riverine debris traveling along the floodway (the deepest part of the channel that conducts the majority of the flood flow) is most likely to travel at speeds approaching that of the floodwaters. Large riverine debris traveling in the floodplain (the shallower area outside the floodway) is more likely to be traveling at speeds less than that of the floodwaters for those reasons stated in the preceding text. Large coastal debris is also likely to be traveling at speeds less than that of the floodwaters. Equation (C5.4-3) should be used with the debris velocity equal to the flow velocity because the equation allows for reductions in debris velocities through application of a depth coefficient, C_D, and an upstream blockage coefficient, C_B.

Duration of Impact. A detailed review of the available literature (Kriebel et al. 2000), supplemented by laboratory testing, concluded the previously suggested 1.0 s duration of impact is much too long and is not realistic. Laboratory tests showed that measured impact durations (from initial impact to time of maximum force Δt) varied from 0.01 s to 0.05 s (Kriebel et al. 2000). Results for one test, for example, produced a maximum impact load of 8,300 lb (37,000 N) for a log weighing 730 lb (3,250 N), moving at 4 ft/s (1.2 m/s), and impacting with a duration of 0.016 s. Over all the test conditions, the impact duration averaged about 0.026 s. The recommended value for use in Equation (C5.4-3) is therefore 0.03 s.

Coefficients C_I, C_O, C_D, and C_B. The coefficients are based in part on the results of laboratory testing and in part on engineering judgment. The values of the coefficients should be considered interim, until more experience is gained with them.

The *importance coefficient*, C_I, is generally used to adjust design loads for the structure category and hazard to human life following ASCE 7-98 convention in Table 1.5-1. Recommended values given in Table C5.4-1 are based on a probability distribution of impact loads obtained from laboratory tests in Haehnel and Daly (2001).

The *orientation coefficient*, C_O, is used to reduce the load calculated by Equation (C5.4-3) for impacts that are oblique, not head on. During laboratory tests (Haehnel and Daly 2001) it was observed that although some debris impacts occurred as direct or head-on impacts that produced maximum impact loads, most impacts occurred as eccentric or oblique impacts with reduced values of the impact force. Based on these measurements, an orientation coefficient of $C_O = 0.8$ has been adopted to reflect the general load reduction observed due to oblique impacts.

The *depth coefficient*, C_D, is used to account for reduced debris velocity in shallow water due to debris dragging along the bottom. Recommended values of this coefficient are based on typical diameters of logs and trees, or on the anticipated diameter of the root mass from drifting trees that are likely to be encountered in a flood hazard zone. Kriebel et al. (2000) suggest that trees with typical root mass diameters will drag the bottom in depths of less than 5 ft (1.5 m), while most logs of concern will drag the bottom in depths of less than 1 ft (0.30 m). The recommended values for the depth coefficient are given in Table C5.4-2 and Figure C5.4-1. No test data are available to fully validate the recommended values of this coefficient. When better data are available, designers should use them in lieu of the values contained in Table C5.4-2 and Figure C5.4-1.

The *blockage coefficient*, C_B, is used to account for the reductions in debris velocities expected due to screening and sheltering provided by trees or other structures within about 10 log lengths [300 ft (91.4 m)] upstream from the building or structure of interest. Kriebel et al. (2000) quote other studies in which dense trees have been shown to act as a screen to remove debris and shelter downstream structures. The effectiveness of the screening depends primarily on the spacing of the upstream obstructions relative to the design log length of interest. For a 1,000-lb (453.6 kg) log, having a length of about 30 ft (9.1 m), it is therefore assumed that any blockage narrower than 30 ft (9.1 m) would trap some or all of the transported debris. Likewise, typical root mass diameters are on the order of 3 to 5 ft (0.91 to 1.5 m), and it is therefore assumed that blockages of this width would fully trap any trees or long logs. Recommended values for the blockage coefficient are given in Table C5.4-3 and Figure C5.4-2 based on interpolation between these limits. No test data are available to fully validate the recommended values of this coefficient.

The *maximum response ratio*, R_{max}, is used to increase or decrease the computed load, depending on the degree of compliance of the building or building component being struck by debris. Impact loads are impulsive in nature, with the force rapidly increasing from zero to the maximum value in time Δt, then decreasing to zero as debris rebounds from the structure. The actual load experienced by the structure or component will depend on the ratio of the impact duration Δt relative to the natural period of the structure or component, T_n. Stiff or rigid buildings and structures with natural periods similar to the impact duration will see an amplification of the impact load. More flexible buildings and structures with natural periods greater than approximately four times the impact duration will see a reduction of the impact load. Likewise, stiff or rigid components will see an amplification of the impact load; more flexible components will see a reduction of the impact load. Successful use of Equatoin (C5.4-3), then, depends on estimation of the natural period of the building or component being struck by flood-borne debris. Calculating the natural period can be carried out using established methods that take building mass, stiffness, and configuration into account. One useful reference is Appendix C of ACI 349 (1985). Design professionals are also referred to Chapter 9 of ASCE 7-10 for additional information.

Natural periods of buildings generally vary from approximately 0.05 s to several seconds (for high-rise, moment frame structures). For flood-borne debris impact loads with a duration of 0.03 s, the critical period (above which loads are reduced) is approximately 0.11 s (see Table C5.4-4). Buildings and structures with natural periods above approximately 0.11 s will see a reduction in the debris impact load, whereas those with natural periods below approximately 0.11 s will see an increase.

Recent shake table tests of conventional, one- to two-story wood-frame buildings have shown natural periods ranging from approximately 0.14 s (7 Hz) to 0.33 s (3 Hz), averaging approximately 0.20 s (5 Hz). Elevating these types of structures for flood-resistant design purposes will act to increase these natural periods. For the purposes of flood-borne debris impact load calculations, a natural period of 0.5 to 1.0 s is recommended for one- to three-story buildings elevated on timber piles. For one- to three-story buildings elevated on masonry columns, a similar range of natural periods is recommended. For one- to three-story buildings elevated on concrete piles or columns, a natural period of 0.2 to 0.5 s is recommended. Finally, design professionals are referred to Section 12.8.2 of this standard, where an approximate natural period for 1- to 12-story buildings [story height equal to or greater than 10 ft (3 m)], with concrete and steel moment-resisting frames, can be approximated as 0.1 times the number of stories.

Special Impact Loads. USACE (1995) states that, absent a detailed analysis, special impact loads can be estimated as a uniform load of 100 lb/ft (1.48 kN/m), acting over a 1 ft (0.31 m) high horizontal strip at the design flood elevation or lower. However, Kriebel et al. (2000) suggest that this load may be too small for some large accumulations of debris and suggest an alternative approach involving application of the standard drag force expression

$$F = (1/2)C_D \rho A V^2 \qquad \text{(C5.4-4)}$$

where

F = Drag force due to debris accumulation, lb (N);
V = Flow velocity upstream of debris accumulation, ft/s (m/s);
A = Projected area of the debris accumulation into the flow, approximated by depth of accumulation times width of accumulation perpendicular to flow, ft^2 (m^2);
ρ = Density of water, $slugs/ft^3$ (kg/m^3); and
C_D = Drag coefficient = 1.

This expression produces loads similar to the 100 lb/ft (1.48 kN/m) guidance from USACE (1995) when the debris depth is assumed to be 1 ft (0.31 m) and when the velocity of the floodwater is 10 ft/s (3m/s). Other guidance from Kriebel et al. (2000) and Haehnel and Daly (2001) suggests that the depth of debris accumulation is often much greater than 1 ft (0.31 m) and is only limited by the water depth at the structure. Observations of debris accumulations at bridge piers listed in these references show typical depths of 5 to 10 ft (1.5 to 3 m), with horizontal widths spanning between adjacent bridge piers whenever the spacing of the piers is less than the typical log length. If debris accumulation is of concern, the design professional should specify the projected area of the debris accumulation based on local observations and experience and apply Equation (C5.4-4) to predict the debris load on buildings or other structures.

REFERENCES

ACI (American Concrete Institute). 1985. *Code requirements for nuclear safety related concrete structures*. ANSI/ACI 349. Farmington Hills, MI: ACI.

Clough, R. W., and J. Penzien. 1993. *Dynamics of structures*, 2nd ed. New York: McGraw-Hill.

FEMA (Federal Emergency Management Agency). 1993. *Free-of-obstruction requirements for buildings located in coastal high hazard areas in accordance with the National Flood Insurance Program*. NFIP Technical Bulletin 5-93. Washington, DC: FEMA.

FEMA. 1999a. *Design and construction guidance for breakaway walls below elevated coastal buildings in accordance with the National Flood Insurance Program*. Technical Bulletin 9-99. Washington, DC: FEMA.

FEMA. 1999b. *National flood insurance program.* 44 CFR, Ch. 1, Parts 59 and 60. Washington, DC: FEMA.

FEMA. 2000. *Coastal construction manual*, 3rd ed. Washington, DC: FEMA.

Haehnel, R., and S. Daly. 2001. *Debris impact tests*. Reston, VA: ASCE.

Kriebel, D. L., L. Buss, and S. Rogers. 2000. *Impact loads from floodborne debris*. Reston, VA: ASCE.

USACE (US Army Corps of Engineers). 1995. *Flood proofing regulations*. EP 1165-2-314. Washington, DC: USACE.

USACE. 2002. *Coastal engineering manual*. Washington, DC: USACE.

Walton, T. L., Jr., J. P. Ahrens, C. L. Truitt, and R. G. Dean. 1989. *Criteria for evaluating coastal flood protection structures*. Technical Rep. No. CERC 89-15. Washington, DC: USACE.

CHAPTER C6
TSUNAMI LOADS AND EFFECTS

C6.1 GENERAL REQUIREMENTS

C6.1.1 Scope The 2022 edition of the ASCE 7 Tsunami Loads and Effects chapter is applicable only to the states of Alaska, Washington, Oregon, California, and Hawaii, which are Tsunami Prone Region that have quantifiable probabilistic hazards resulting from tsunamigenic earthquakes with subduction faulting (Table C6.1-1). The Tsunami Design Zone (TDZ) is the area vulnerable to being flooded or inundated by the Maximum Considered Tsunami (MCT), which is taken as having a 2% probability of being exceeded in a 50-year period, or 1:2,475 annual probability of exceedance. The MCT constitutes the design event, consisting of the inundation depths and flow velocities taken at the stages of inflow and outflow most critical to the structure (Chock 2015, 2016).

Other causes of local tsunamis include large landslides near the coast or underwater and undersea volcanic eruptions. In Alaska, there is a history of coseismic landslides generated in fjords during major earthquakes.

For other states, there is insufficient analysis at present to reliably quantify the probabilistic hazard of landslide-induced local tsunami. As of 2020, PTHA is lacking for some other regions with historic tsunamis (such as Guam, Commonwealth of the Northern Marianas, American Samoa, Puerto Rico, and US Virgin Islands), so these regions are not covered. However, for Tsunami Risk Categories III and IV buildings and other structures, it is recommended that there be consideration of performing probabilistic site-specific tsunami hazard analysis to use as a basis for implementing tsunami-resistant design with these provisions.

Table C6.1-1 provides a general assessment of the exposure of the five western states to tsunami hazard.

Because of the high hydrodynamic forces exerted by tsunamis, one- and two-family dwellings and low-rise buildings of light-frame construction do not survive significant tsunami loading, as indicated by numerous post-tsunami forensic engineering surveys of the major tsunami events of the past decade. Therefore, it is not practical to include tsunami design requirements for these types of low-rise structures. Impractical applications to low-rise light-frame construction should be avoided. As is typical in flood-prone regions, midrise light-frame construction can be built on a pedestal structure to provide sufficient elevation to enable a tsunami-resistant design.

The state or local government should determine a threshold height for where tsunami-resilient design requirements for Risk Category II buildings shall apply in accordance with the state or local statute adopting the building code with tsunami-resilient requirements. The height threshold should be chosen to be appropriate for both reasonable Life Safety and reasonable economic cost to resist tsunami loads. The threshold height would depend on their community's tsunami hazard, tsunami response procedures, and whole community disaster resilience goals. Considerations of evacuation egress time from a local community would also be a consideration. When evacuation travel times exceed the available time to tsunami arrival, there is a greater need for vertical evacuation into an ample number of nearby buildings. The inventory and design vintage of existing taller buildings that may offer some increased safety can also be a consideration for determining the shortfall of such possible refuges. In summary, multiagency stakeholders in coordination with emergency management and design and construction professionals should evaluate the tsunami risks to each local community's public safety and Critical Facilities for response and recovery from the MCT and should use that information to rationally determine the threshold height of application of the tsunami provisions to Risk Category II buildings.

The technique demonstrated by Carden et al. (2015) can be used to evaluate the systemic level of tsunami resistance provided by the ASCE 7 seismic design requirements. Past this point of structural system parity between tsunami and seismic demand, additional investment into the lateral-force-resisting system would become necessary for tsunami resistance. For any structure, it may be necessary to provide enhanced local resistance of structural components subjected to tsunami inundation, in accordance with the provisions of Chapter 6, particularly if the tsunami is close to a shipping port.

From the previous discussion, note that structural engineering and regional tsunami science expertise is necessary to evaluate several important technical factors relevant to the jurisdiction's decision to establish a threshold height of applicability for Risk Category II buildings and structures (Chock et al. 2018). The referenced paper provides initial guidance for the process of examining the entire Tsunami Design Zone for a local code jurisdiction.

Tsunami Risk Category II structures taller than the designated height threshold, all Tsunami Risk Category III, and all Tsunami Risk Category IV buildings and other structures are subject to these requirements. Risk Category II buildings shorter than the designated height threshold are not required to be designed for tsunami loads. Accordingly, it is important for the community to have Tsunami Evacuation Maps and operational response procedures, recognizing the inventory of Risk Category II buildings not designated to be designed for tsunami effects, which would therefore be vulnerable to tsunami damage.

Mitigation of tsunami risk requires a combination of emergency preparedness for evacuation in addition to providing structural resilience of Critical Facilities, infrastructure, and key resources necessary for immediate response and economic and social recovery. Risk Category I and low-rise Risk Category II buildings are not required to be designed against any tsunami

Table C6.1-1. Exposure of the Five Western States to Tsunami Hazard.

State	Population at Direct Risk[a]	Profile of Economic Assets and Critical Infrastructure
California	275,000 residents plus another 400,000 to 2 million tourists; 840 miles of coastline	Over $200 billion plus three major airports (SFO, OAK, SAN) and one military port, five very large ports, one large port, five medium ports
	Total resident population of area at immediate risk of post-tsunami impacts: 1,950,000[b]	
Oregon	25,000 residents plus another 55,000 tourists; 300 miles of coastline	$8.5 billion plus essential facilities, two medium ports, one fuel depot hub
	Total resident population of area at immediate risk of post-tsunami impacts: 100,000[b]	
Washington	45,000 residents plus another 20,000 tourists; 160 miles of coastline	$4.5 billion plus essential facilities, one military port, two very large ports, one large port, three medium ports
	Total resident population of area at immediate risk of post-tsunami impacts: 900,000[b]	
Hawaii	About 200,000 residents plus another 175,000 or more tourists and approximately 1,000 buildings directly relating to the tourism industry[c]; 750 miles of coastline	$40 billion, plus three international airports, one military port, one medium port, four other container ports, one fuel refinery intake port, three regional power plants, 100 government buildings
	Total resident population of area at immediate risk of post-tsunami impacts: 400,000[c]	
Alaska	105,000 residents, plus highly seasonal visitor count; 6,600 miles of coastline	Over $10 billion plus international airport's fuel depot, three medium ports, plus nine other container ports; 55 ports in total
	Total resident population of area at immediate risk of post-tsunami impacts: 125,000[b]	

[a] Wood (2007), Wood et al. (2007, 2013), Wood and Soulard (2008), Wood and Peters (2015). Lower-bound estimates based on present evacuation zones.
[b] National Research Council (2011). The total population at immediate risk includes those in the same census tract whose livelihood or utility and other services would be interrupted by a major tsunami with this inundation.
[c] Updated for exposure to great Aleutian tsunamis (modeling by University of Hawaii per Hawaii Emergency Management Agency).

event because they are at a higher risk of being fully inundated and collapsing during a major tsunami. Communities in the Tsunami-Prone Region should be enabled with tsunami warning systems and emergency operations plans for evacuation because Risk Category I and low-rise Risk Category II buildings should not be occupied during a tsunami. A tsunami warning and evacuation procedure consists of a plan and procedure developed and adopted by a community that has a system to act on a tsunami warning from the designated National Oceanic and Atmospheric Administration (NOAA) tsunami warning center at all hours using two independent means—a 24 h operational site to receive the warning and established methods of transmitting the warning that will be received by the affected population—and has established and designated evacuation routes for its citizens to either high ground or to a designated Tsunami Vertical Evacuation Refuge Structure. At present, the states of Alaska, Washington, Oregon, California, and Hawaii already have tsunami warning and evacuation procedures. A NOAA TsunamiReady community would include this procedure. In these states, it is recognized by federal, state, and local governments that mitigation of tsunami risk to public safety requires emergency preparedness for evacuation.

The ASCE Board authorized the SEI and COPRI funding necessary for Tsunami Design Zone mapping and Offshore Tsunami Amplitude, within the project for this standard. In 2013, the NOAA National Tsunami Hazard Mitigation Program (NTHMP) Modeling and Mapping Subcommittee convened a Probabilistic Tsunami Hazard Analysis (PTHA) working group and conducted a comparison of prototype probabilistic tsunami inundation maps produced in 2013 by two independent researchers, including a team from the University of Washington as well as the Pacific Marine Environmental Laboratory (PMEL) team performing the work for ASCE, to validate the key procedures of the PTHA methodology. The study was reviewed and advised by a broad panel of experts of the Modeling and Mapping Subcommittee of NTHMP. The USGS team involved in the scientific consensus process for seismic source characterization for the National Earthquake Hazards Reduction Program and ASCE seismic maps were also included in this panel. State geologists and tsunami modelers from all five western states also participated. The results of this 2013–2014 peer review phase (California Geological Survey 2015) were incorporated into the production process of the 2014–2015 ASCE 7 mapping project. The PTHA methodology used to develop the Tsunami Design Zone Maps is explained in Section C6.7.

The ASCE Tsunami Design Geodatabase (version 2022-1.0) includes updated Tsunami Design Zone data based on high-resolution Digital Elevation Models (DEMs), updated probabilistic tsunami hazard analysis (PTHA), and improved inundation modeling for the state of California, the Salish Sea region of the state of Washington, the City and County of Honolulu, and the County of Hawaii. The methodology for the updates complies with the requirements of Section 6.7 of ASCE 7-22. These updates included the improved Tsunami Design Zone, Runup Elevations, and Inundation Depth Points. In certain areas of California and Hawaii, zones with 3 ft (0.914 m) or less inundation depth were also identified. For Risk Categories II and III, per Section 6.1.1, tsunami structural design becomes required where the inundation depth is greater than 3 ft (0.914 m) at any location within the intended footprint of the structure. The boundaries of the California, Washington, and the County of Hawaii high-resolution Tsunami Design Zone areas were transitioned to the Runup Elevations in the surrounding lower-resolution areas of the ASCE Tsunami Design Geodatabase to provide a smooth transition between the data sets. The areas modeled in high-resolution are shown in Figure C6.1-2(a), (b), and (c). A description of the analysis and geoprocessing is provided in Thio (2019) and Wilson et al. (2020) for California, in Thio (2020) for Washington, in Robertson (2019) for Honolulu, and in Wei (2016) for the County of Hawaii.

The final step of the PTHA process involves the inundation analysis to determine the runup. Runup geodata define the ground elevation where the tsunami inundation reaches its horizontal limit. The runup data set of the ASCE Tsunami Design Geodatabase includes these geocoded points defining the locations and elevations of the runup. The inundation limit on land is the smoothed extent line formed by these discrete runup points, and the Tsunami Design Zone essentially consists of the land area between the inundation limit line and the coastline. Review of the ASCE 7-16 Tsunami Design Zone Maps by the Tsunami Loads and Effects Subcommittee also included comparison with the results of two independent modelers conducting region-specific work. *ASCE 7-22 Tsunami Design Zone Maps for Selected Locations* provides 62 Tsunami Design Zone Maps rendered in PDF format and downloadable from https://doi.org/10.1061/9780784480748. The locations of these 62 maps are indicated by circles in Figure C6.1-1(a) to (i). For the purpose of defining the Tsunami Design Zone, these PDF maps are considered equivalent to the results served from the ASCE 7 Tsunami Design Geodatabase for the corresponding areas and are produced with the runup, GIS point data, and Tsunami Design Zone data in the Geodatabase.

ASCE 7-16 Tsunami Design Zone Maps for Selected Locations (ASCE 2017a) also includes a report prepared for ASCE, "Probabilistic Tsunami Design Maps for the ASCE 7-16 Standard," that describes the development of the 2,500-year probabilistic tsunami design zone maps and is suitable as an accompanying user manual.

The key data sets of the ASCE Tsunami Design Geodatabase are the runup points. These are organized by segments of coastline for each of the five western states [the extent is shown in Figure C6.1-1(a) to (i)].

The ASCE Tsunami Design Geodatabase data is web-based (http://asce7tsunami.online) and does not require proprietary software.

The effect of seismic subsidence, typically following the recognized deformation model of Okada (1985) illustrated in Figure C6.1-3, is accounted for within the modeling used to develop the mapped inundation limits of Figure 6.1-1. Figure C6.1-3 illustrates the effects of relative sea level rise and seismic subsidence on tsunami inundation. Note that in subduction regions, the long-term trend of ground elevation may be uplift, but the engineer is not to consider extrapolating this trend to reduce the design inundation since the subduction earthquake mechanism will result in this temporal uplift being eliminated by seismic subsidence.

A general outline of the main steps in the tsunami analysis and design requirements is given in Figure C6.1-4.

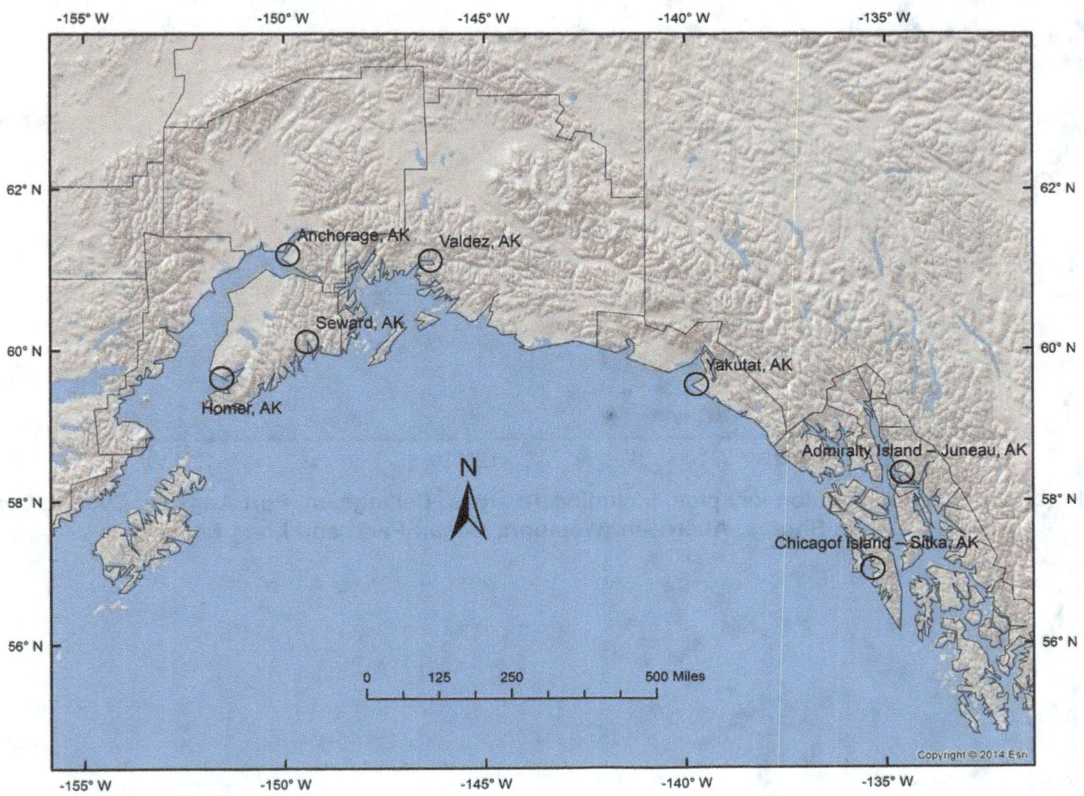

Figure C6.1-1(a). Alaska location key plan, including seven areas: Anchorage, Valdez, Seward, Yakutat, Homer, Admiralty Island–Juneau, and Chicagof Island–Sitka.

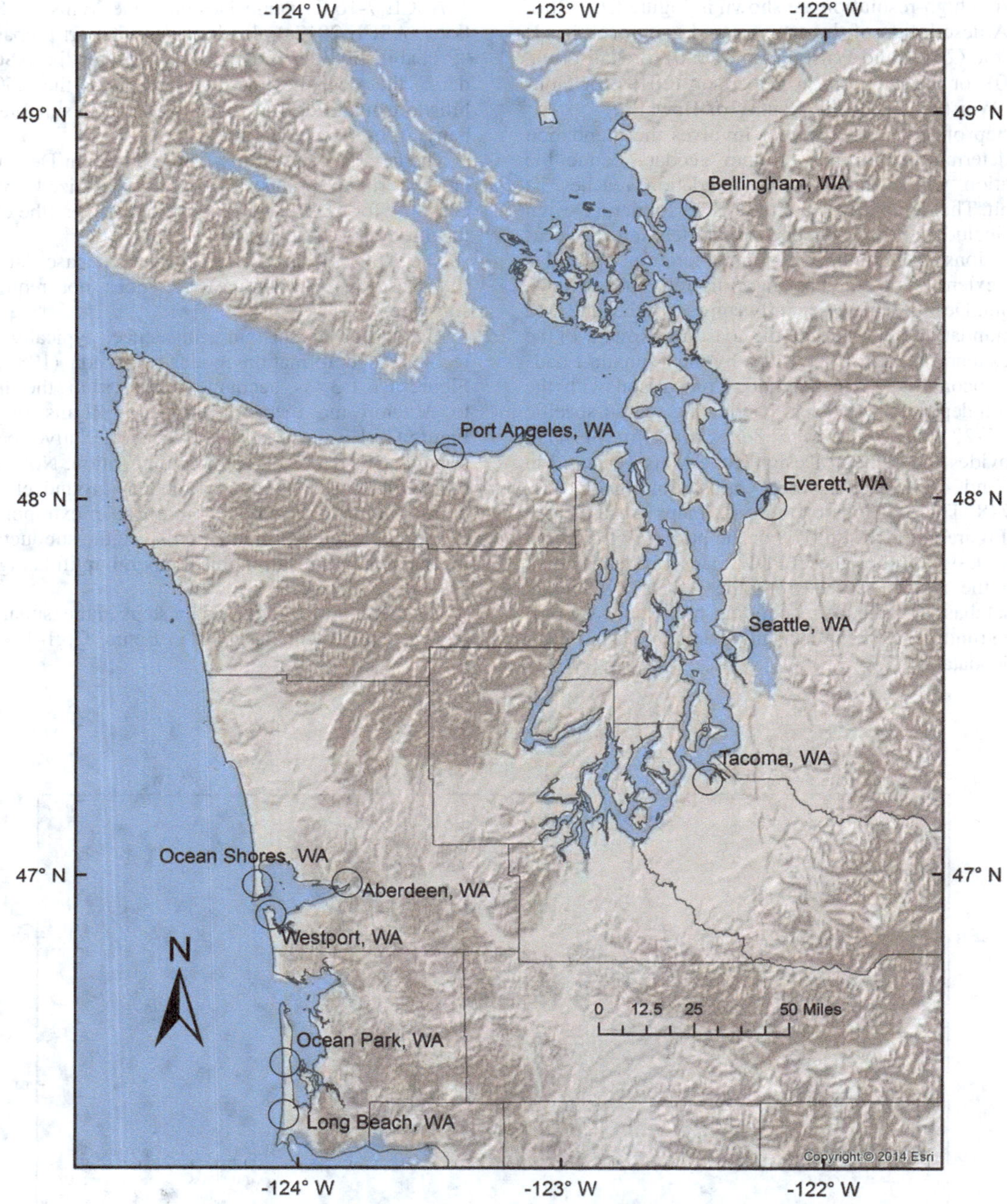

Figure C6.1-1(b). Washington location key plan, including 10 areas: Bellingham, Port Angeles, Everett, Seattle, Tacoma, Ocean Shores, Aberdeen, Westport, Ocean Park, and Long Beach.

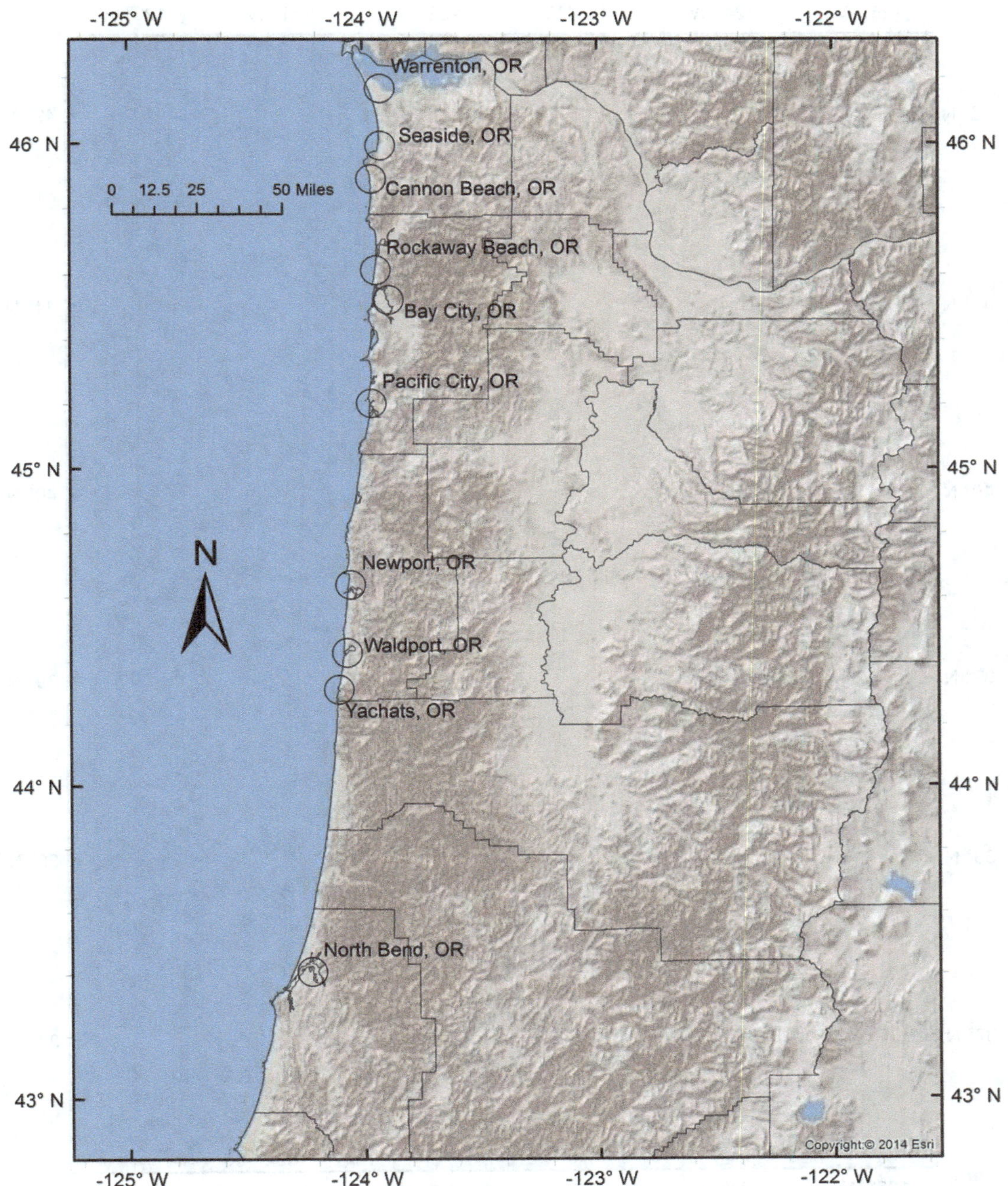

Figure C6.1-1(c). Oregon location key plan, including 10 areas: Warrenton, Seaside, Cannon Beach, Rockaway Beach, Bay City, Pacific City, Newport, Waldport, Yachats, and North Bend.

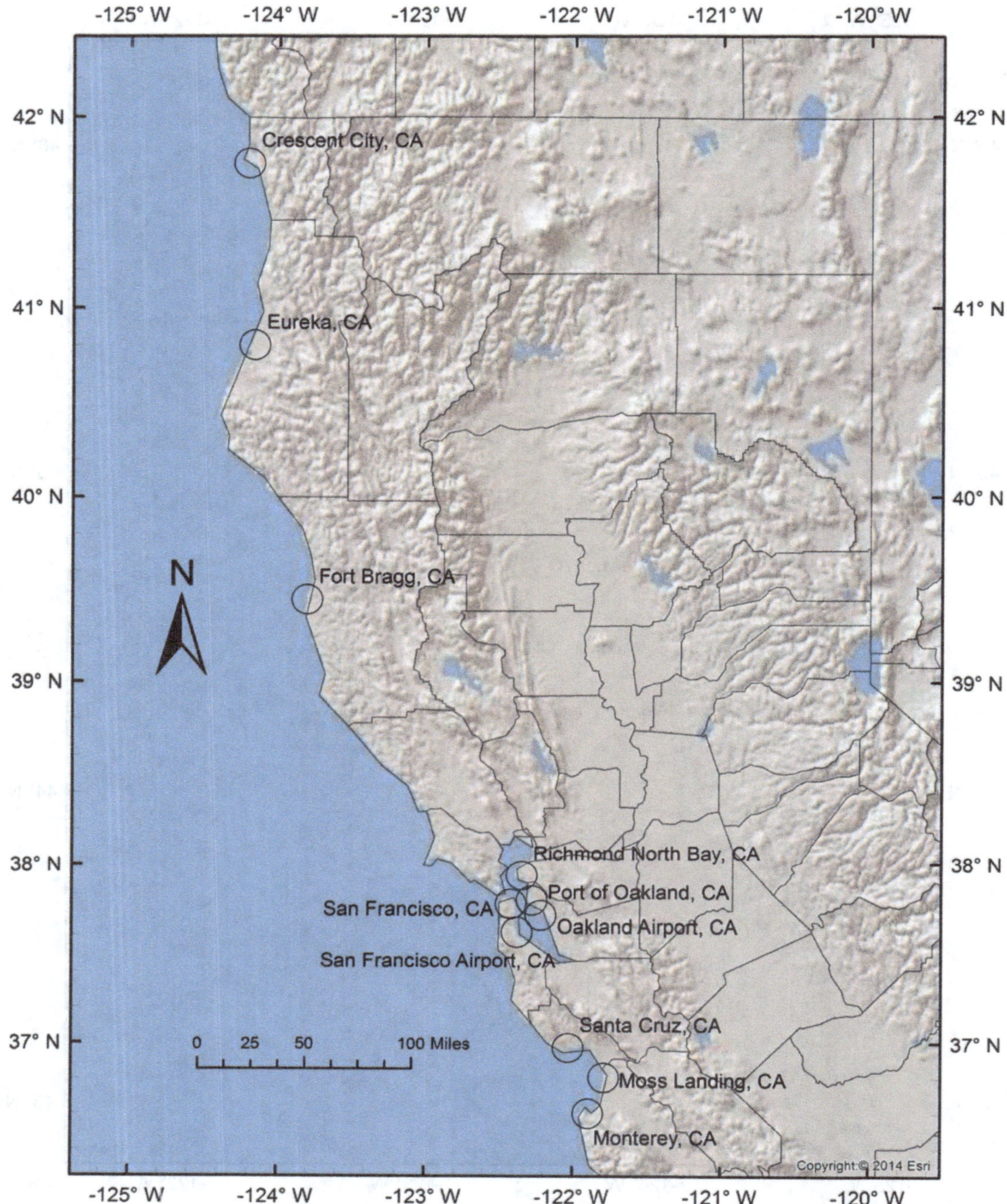

Figure C6.1-1(d). Northern California location key plan, including 11 areas: Crescent City, Eureka, Fort Bragg, Richmond North Bay, Port of Oakland, San Francisco, Oakland Airport, San Francisco Airport, Santa Cruz, Moss Landing, and Monterey.

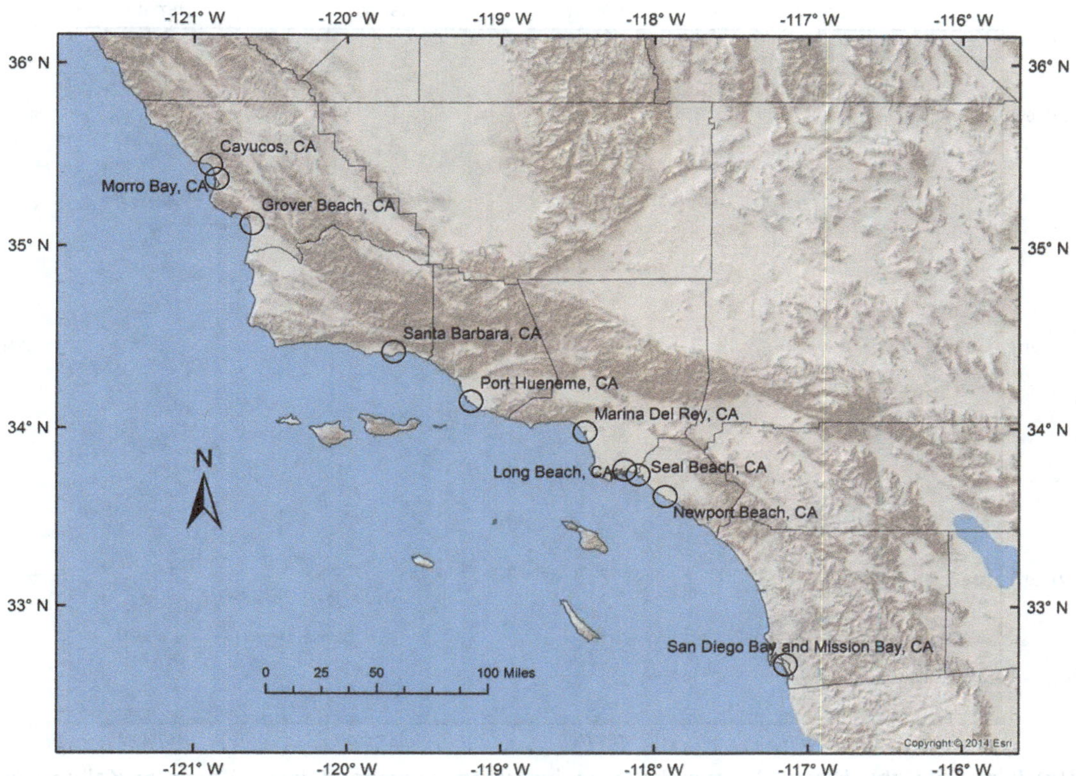

Figure C6.1-1(e). Southern California location key plan, including 10 areas: Cayucos, Morro Bay, Grover Beach, Santa Barbara, Port Hueneme, Marina del Rey, Long Beach, Seal Beach, Newport Beach, and San Diego Bay and Mission Bay.

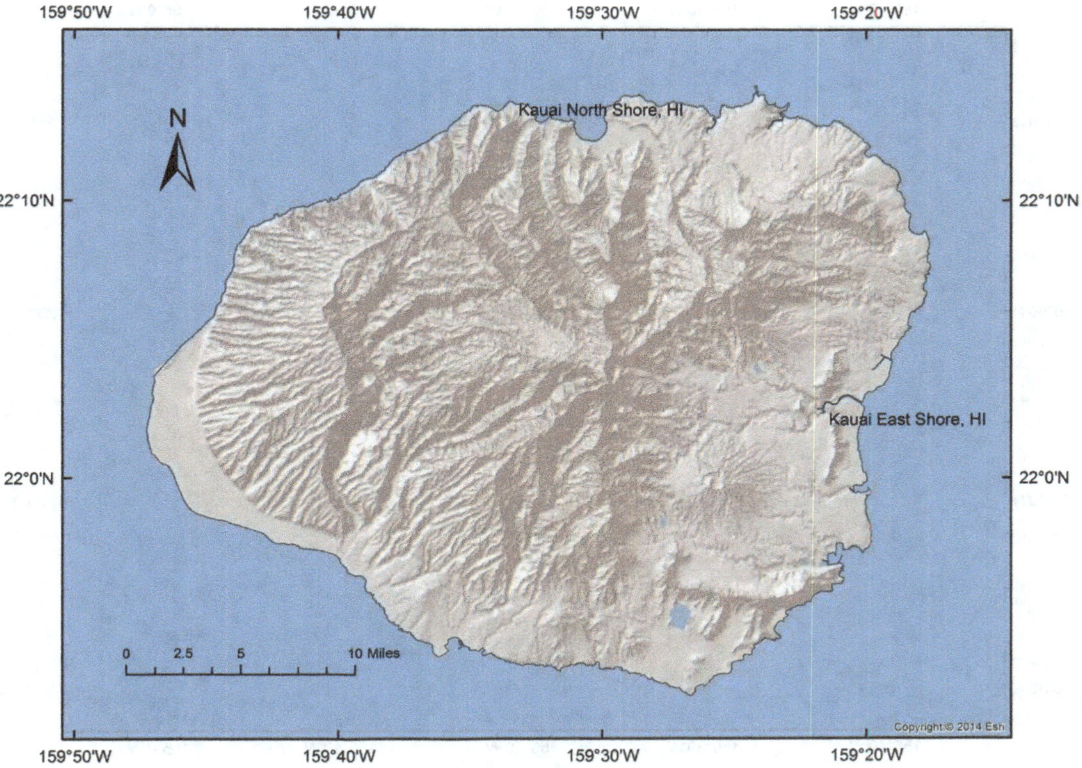

Figure C6.1-1(f). Island of Kauai, Hawaii, location key plan, including two areas: Kauai North Shore and Kauai East Shore.

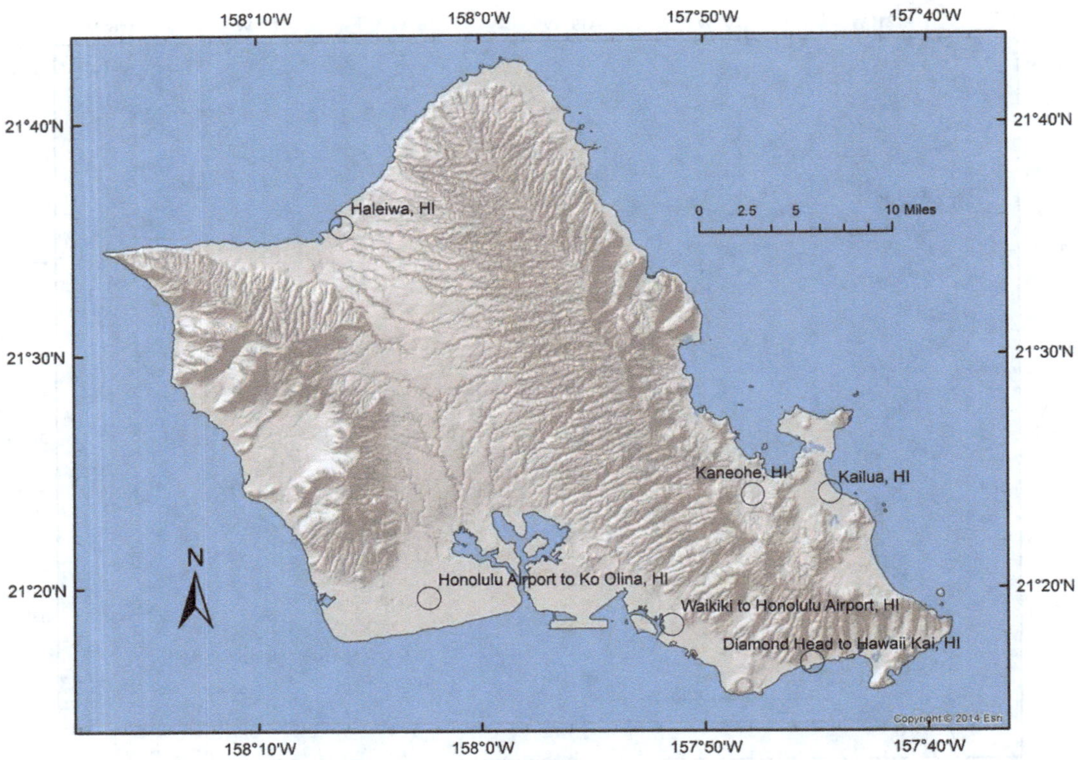

Figure C6.1-1(g). Island of Oahu, Hawaii, location key plan, including six areas: Haleiwa, Kaneohe, Kailua, Honolulu Airport to Ko Olina, Waikiki to Honolulu Airport, and Diamond Head to Hawaii Kai.

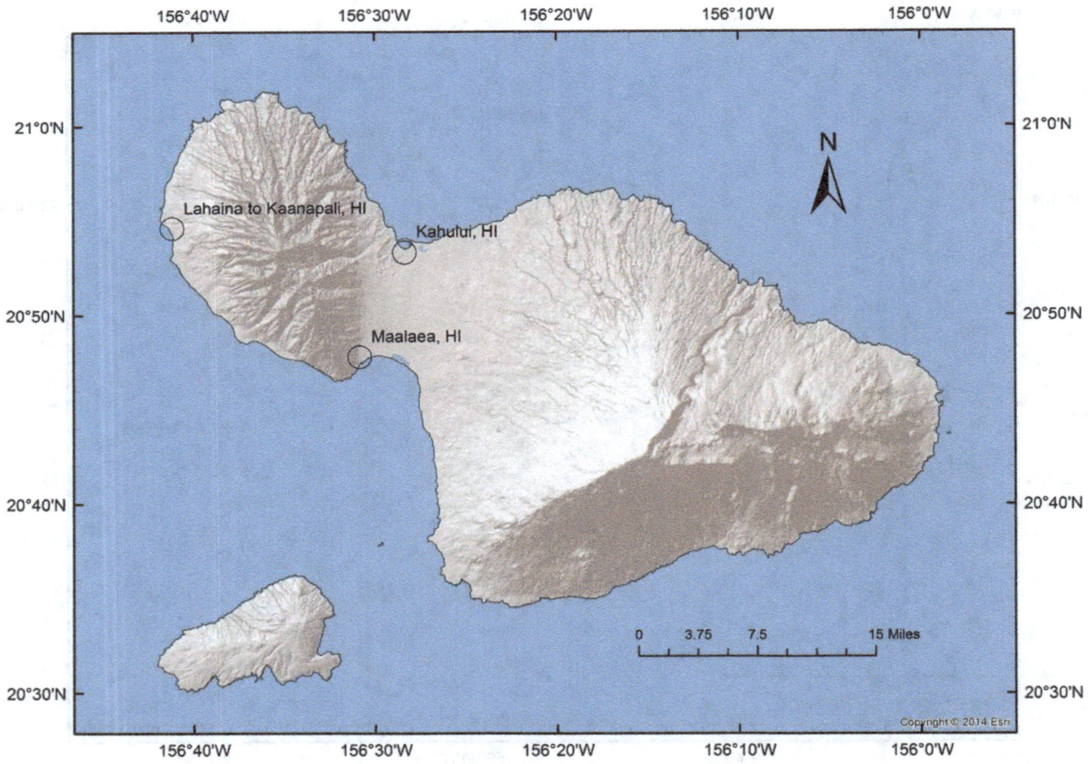

Figure C6.1-1(h). Island of Maui, Hawaii, location key plan, including three areas: Lahaina to Kaanapali, Kahului, and Maalaea.

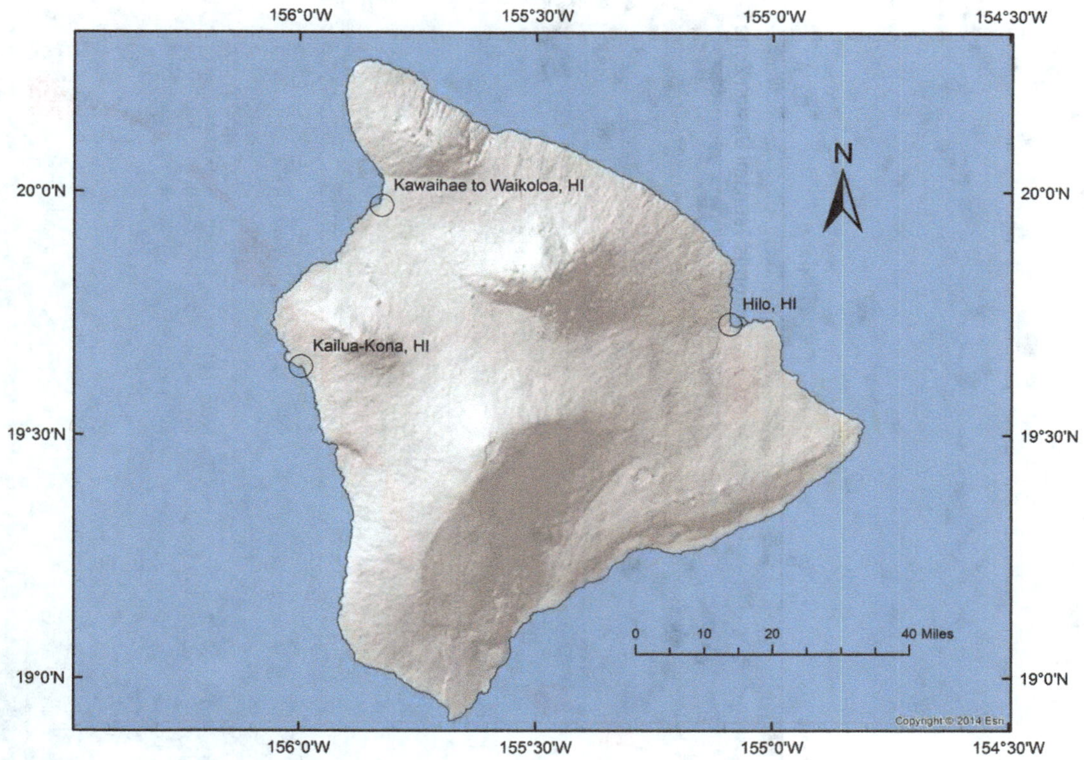

Figure C6.1-1(i). Island of Hawaii, Hawaii, location key plan, including three areas: Kawaihae to Waikoloa, Hilo, and Kailua-Kona.

C6.2 DEFINITIONS

The ASCE definitions were developed after a review of international literature before 2013 that in general did not have uniform consistency of tsunami terminology. In addition, some references may utilize waveform terminology relevant to documenting the local effects of historic tsunamis but were not directly applicable to the specification of a probabilistic tsunami for design purposes. However, the publication of a revised tsunami glossary by the Intergovernmental Oceanographic Commission (2013) should lead to greater consistency, and the ASCE terminology is compatible with that document.

In particular, as illustrated in Figure 6.2-1, key terms for tsunami definition are the Maximum Considered Tsunami, the Offshore Tsunami Amplitude, inundation depth, runup elevation, and the inundation limit.

This standard defines the size of a probabilistic Maximum Considered Tsunami at a standardized offshore depth of 328 ft (100 m), using its amplitude above or below the ambient sea level. This level is different from the peak to trough tsunami height. In site-specific tsunami hazard analysis, a Reference Sea Level of Mean High Water is primarily used in inundation modeling to account for tidal fluctuations during tsunamis.

The tsunami inundations in ASCE 7 have been computed using digital elevation models (DEMs) developed by NOAA's National Center for Environmental Information for selected US coastal regions. Bathymetric, topographic, and shoreline data used in DEM compilation are obtained from various sources, including the National Centers for Environmental Information, US National Ocean Service, US Geological Survey, US Army Corps of Engineers, Federal Emergency Management Agency (FEMA), and other federal, state, and local government agencies, academic institutions, and private companies. The DEMs are referenced to the vertical tidal datum of Mean High Water (MHW) and the horizontal datum of World Geodetic System 1984 (WGS84). MHW is the average of all the high-water heights observed over the National Tidal Datum Epoch. The present National Tidal Datum Epoch is 1983 through 2001 and is actively considered for revision every 20 to 25 years. Accordingly on land, MHW is adopted as the common reference datum for ground elevation and tsunami runup elevations resulting from the inundation analysis for the Maximum Considered Tsunami. By this datum, the "shoreline" is set at the MHW, which is thereby elevation zero. Other adopted reference data may be used in practice, including, but not limited to, the North American Vertical Datum 88, also known as NAVD 88 (Zilkoski et al. 1992), and in such cases the results of a site-specific tsunami hazard analysis should be coordinated with and converted to the appropriate local reference datum. In addition to MHW, the ASCE Tsunami Design Geodatabase also includes data referenced to NAVD88 where available. Another key definition is the Froude number, the nondimensionalized flow velocity; in this chapter, when the Froude number is specified, the engineer should recognize that this number is explicitly defining a prescribed relationship of velocity as a function of inundation depth.

As of the 2022 edition of this standard, the Tsunami-Prone Region comprises the states of Alaska, Washington, Oregon, California, and Hawaii; this region is subject to the addition of other areas once further Probabilistic Tsunami Hazard Analyses are performed that demonstrate runup of greater than 3 ft (0.914 m).

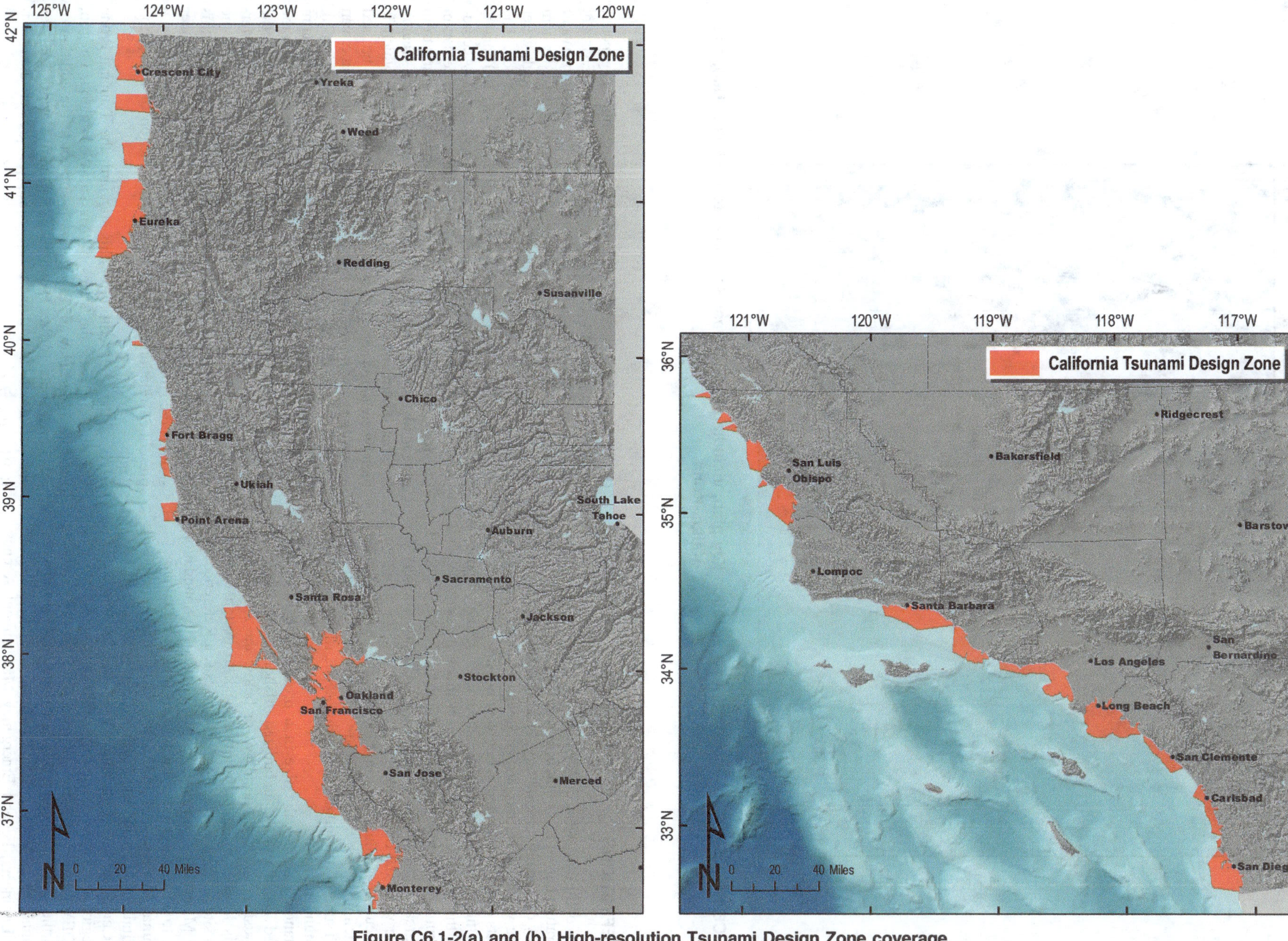

Figure C6.1-2(a) and (b). High-resolution Tsunami Design Zone coverage.
Source: CGS (2020).

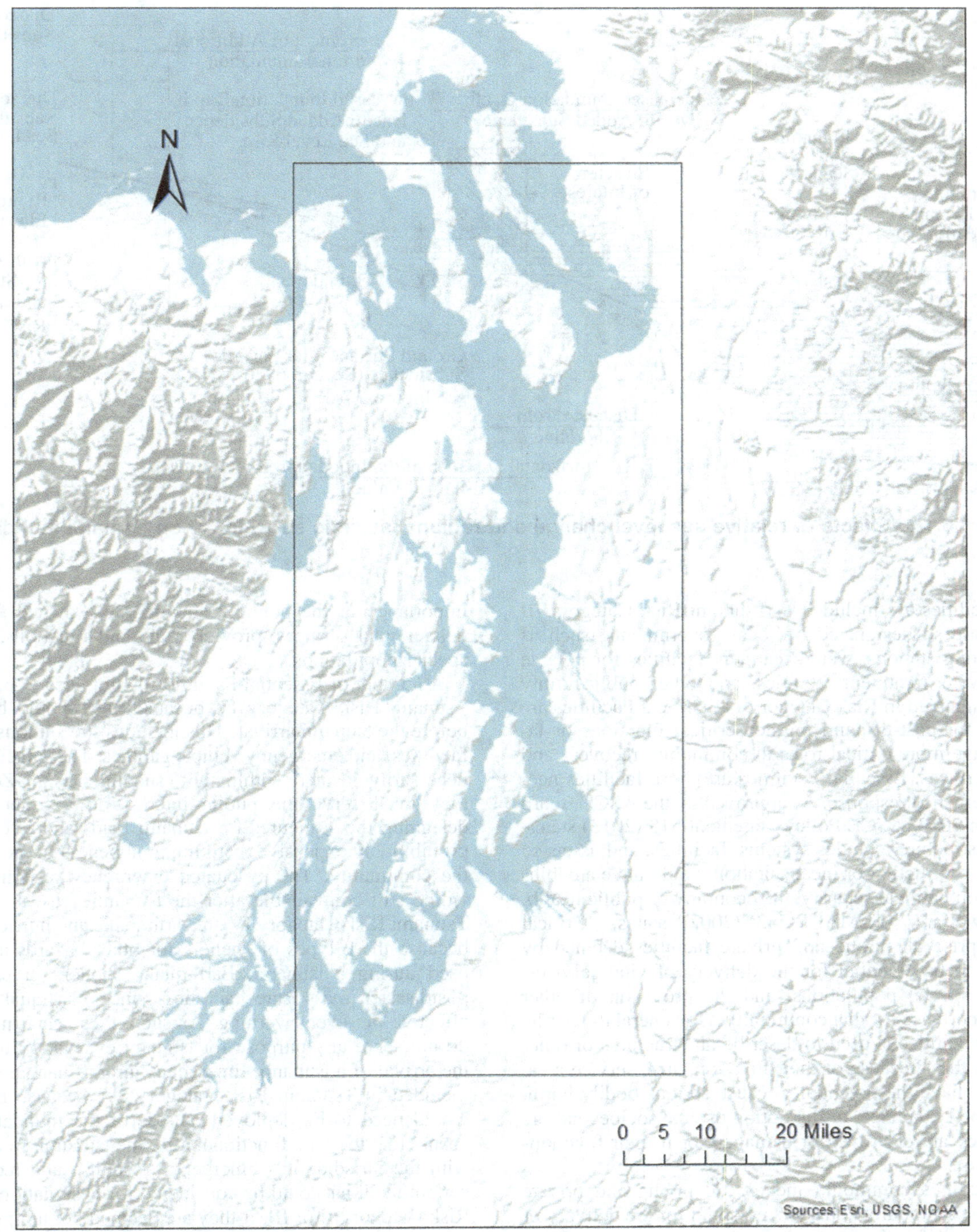

Figure C6.1-2(c). Higher-resolution Tsunami Design Zones for the Salish Sea, state of Washington.

C6.3 SYMBOLS AND NOTATION

Particular tsunami design symbology for which the user may desire clarification includes the following: V_w = General volume of displaced water that the engineer needs to calculate based on the loading condition and structural configuration. The loading equations use the symbol γ for weight density and ρ for mass density. Generally, this notation clarifies and distinguishes the fluid mechanics of hydrostatic effects from hydrodynamic effects.

C6.4 TSUNAMI RISK CATEGORIES

This standard classifies facilities in accordance with Risk Categories that recognize the importance or criticality of the facility. In the tsunami chapter, further modified definitions of the Risk Categories for Tsunami Risk Categories III and IV are included with respect to specific occupancy/functional criteria. Critical Facilities designated by local governments are included in Tsunami Risk Category III. Essential Facilities are included in Tsunami Risk Category IV.

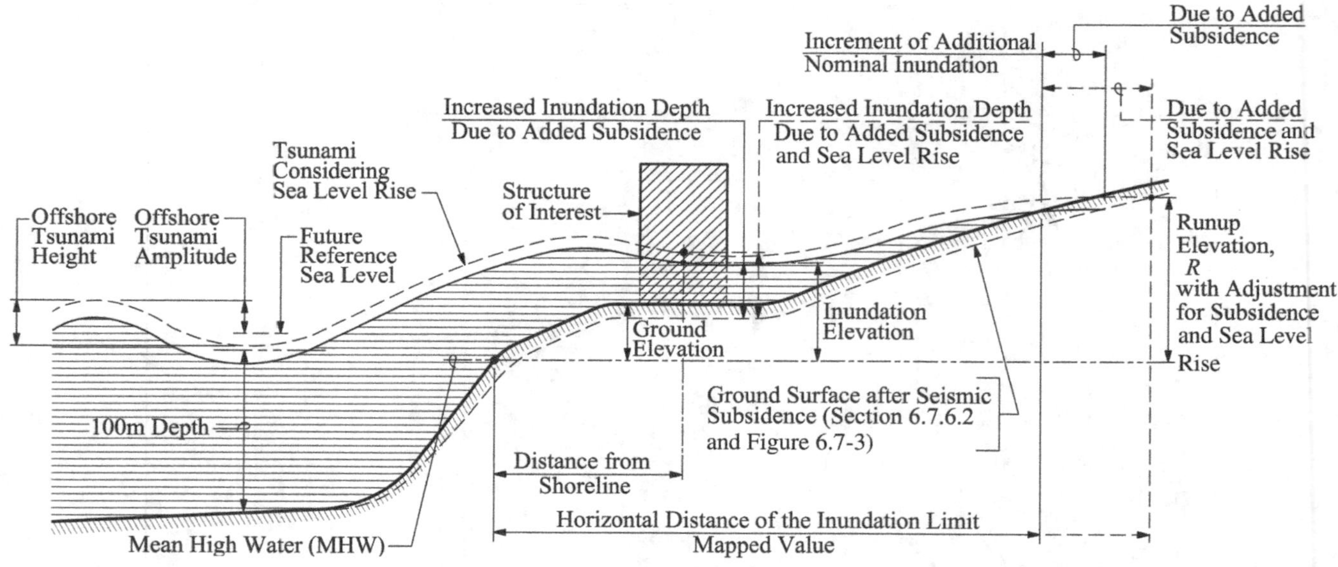

Figure C6.1-3. Effects of relative sea level change and regional seismic subsidence on tsunami inundation.

Critical Facilities are included in Tsunami Risk Category III only if they are so designated by local governments, such as power-generating stations, water-treatment facilities for potable water, wastewater-treatment facilities, and other public utility facilities not included in Risk Category IV. Critical Facilities are those needed for post-tsunami mission-critical functions or facilities that have more critical roles in community recovery and community services. Essential Facilities are those facilities necessary for emergency response. As approved by the ASCE Board of Direction in 2013, ASCE Policy Statement 518 (2013) states, "Critical infrastructure includes systems, facilities, and assets so vital that their destruction or incapacitation would have a debilitating impact on national security, the economy or public health, safety, and welfare." FEMA P-543 (2007) states, "Critical facilities comprise all public and private facilities deemed by a community to be essential for the delivery of vital services, protection of special populations, and the provision of other services of importance for that community." In general usage, the term Critical Facilities is used to describe all structures or other improvements that, because of their function, size, service area, or uniqueness, have the potential to cause serious bodily harm, extensive property damage, or disruption of vital socioeconomic activities if they are destroyed or damaged, or if their functionality is impaired.

Critical Facilities commonly include all public and private facilities that a community considers essential for the delivery of vital services and for the protection of the community. They usually include emergency response facilities (fire stations, police stations, rescue squads, and emergency operation centers, or EOCs), custodial facilities (jails and other detention centers, long-term care facilities, hospitals, and other health-care facilities), schools, emergency shelters, utilities (water supply, wastewater-treatment facilities, and power), communications facilities, and any other assets determined by the community to be of critical importance for the protection of the health and safety of the population.

The number and nature of Critical Facilities in a community can differ greatly from one jurisdiction to another, and these facilities usually include both public and private facilities. In this sense, each community needs to determine the relative importance of the publicly and privately owned facilities that deliver vital services, provide important functions, and protect special populations.

A number of Essential Facilities do not need to be included in Tsunami Risk Category IV because they should be evacuated before the tsunami arrival. This includes fire stations, ambulance facilities, and emergency vehicle garages. These facilities may be necessarily located within the Tsunami Design Zone because they must serve the public interest on a timely basis, but designing the structures for tsunami loads and effects could be prohibitively expensive, with minimal benefit to the resilience of the community. The evacuated emergency response resources would still be available after the tsunami. Also not included in Tsunami Risk Category IV are earthquake and hurricane shelters, because these types of shelters for other hazards are not to be used during a tsunami. Earthquake shelters in particular are postdisaster mass care facilities, since earthquakes have no effective pre-event warning. As such, these earthquake shelters do not serve any purpose for tsunami evacuation refuge before the arrival of a tsunami. Emergency aircraft hangars are also not included in Tsunami Risk Category IV, because these aircraft would need to be deployed outside of the inundation zone to ensure that they are functional after the tsunami. In coordination with the jurisdiction's emergency response and recovery plan, certain facilities could be considered for designation as Tsunami Risk Category II or III, if they are deemed not uniquely required for postdisaster operations or where such functionality can be sufficiently provided from a post-tsunami alternative facility.

Tsunami Vertical Evacuation Refuge Structures are included in Tsunami Risk Category IV because of their function as a safe refuge for evacuees during the tsunami. If health-care facilities with 50 or more resident patients are for some reason located in the Tsunami Design Zone, it is recommended that such facilities should also be designed in accordance with Tsunami Risk Category IV rather than Risk Category III. The height of the structure must afford a sufficient number of elevated floor levels for patient vertical evacuation, similar to a Tsunami Vertical Evacuation Refuge Structure, because of the difficulty with evacuating assisted-living patients in a timely manner before inundation.

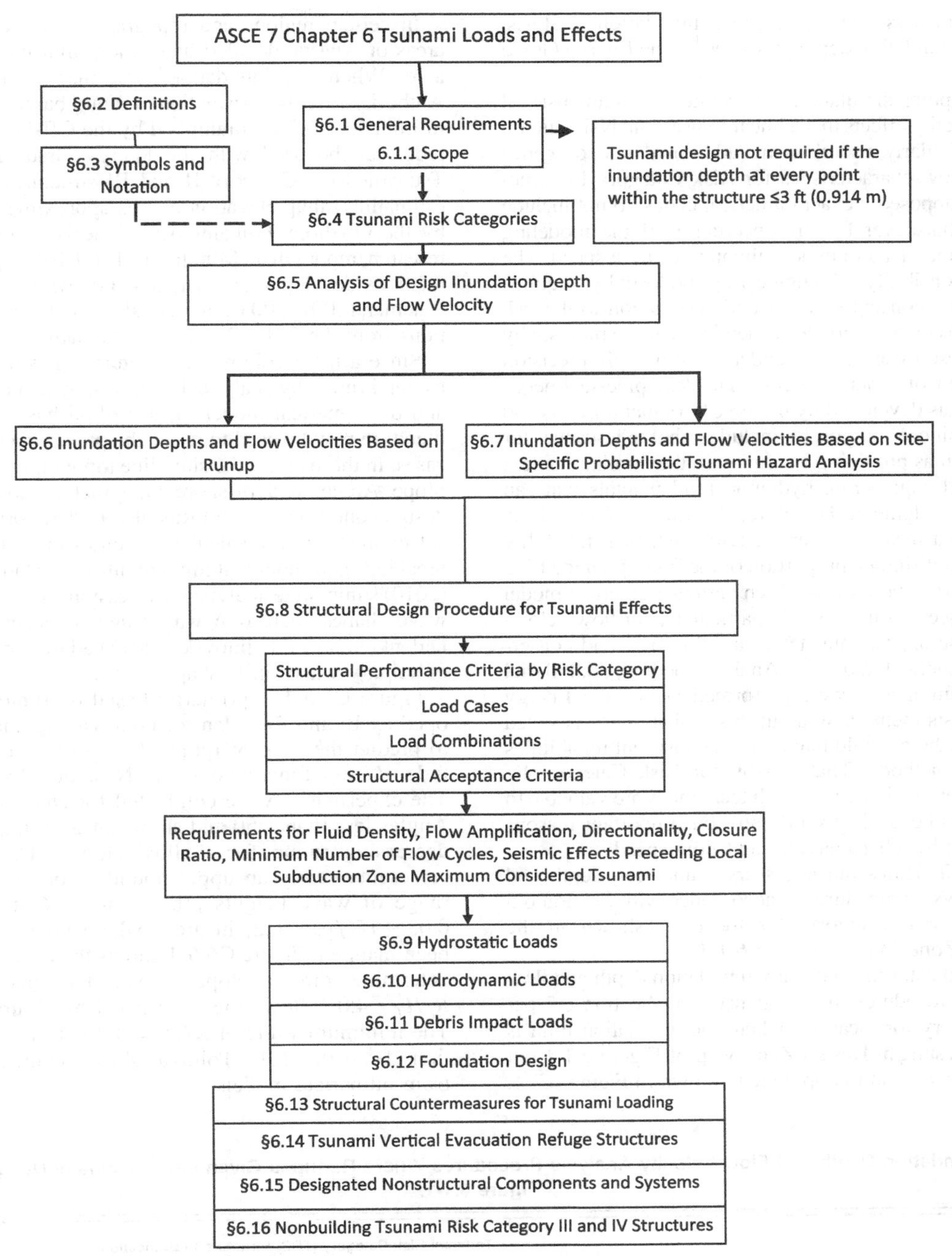

Figure C6.1-4. General organization of Chapter 6.

C6.5 ANALYSIS OF DESIGN INUNDATION DEPTH AND FLOW VELOCITY

There are two procedures for determining the inundation depth and velocities at a site: (1) Energy Grade Line Analysis, which takes the runup elevation and inundation limit indicated on the Tsunami Design Zone Map as the target solution point of a hydraulic analysis along the topographic transect from the shoreline to the point of runup; and (2) a site-specific inundation analysis of at least depth-averaged two-dimensional (2D) numerical modeling that uses the Offshore Tsunami Amplitude, the wave period that is a conserved property (during shoaling, the timescale of the overall tsunami wave hardly changes), and other waveform parameters as the input to a numerical simulation that includes a high-resolution Digital Elevation Model.

Energy Grade Line Analysis, which has been developed to produce conservative design flow parameters, is always performed for Tsunami Risk Category II, III, and IV structures. The site-specific inundation analysis procedure may or may not be required, depending on the structure's Tsunami Risk Category. Site-specific inundation analysis is not required, but may be used, for Tsunami Risk Category II and III structures. The site-specific inundation analysis procedure is performed for Tsunami Risk Category IV structures unless the Energy Grade Line Analysis shows the inundation depth to be less than 12 ft (3.66 m) at the structure. However, Tsunami Vertical Refuge

Structures shall always use site-specific inundation analysis, regardless of the inundation depth produced by the Energy Grade Line Analysis.

A precise computer simulation can capture two-dimensional flow and directionality effects that a linear transect analysis cannot, and so it is particularly useful as an additional due diligence investigation of flow characteristics for Risk Category IV structures. However, supposedly exact numerical codes do not include any allowance whatsoever for the uncertainty of the modeling technique, which may lead to underestimation of flow speed. The fundamental responsibility of engineering philosophy requires consideration of the consequences of underestimation to the risk of design failure, as opposed to the academic scientific philosophy that seeks the best mean value rendered with high precision without the burden of other concerns. The less precise Energy Grade Line Analysis developed by the ASCE Tsunami Loads and Effects Subcommitteee purposely includes that allowance for uncertainty, as well as providing a solution in the familiar context of well-established engineering hydraulic fundamentals that can aid an engineer's judgment. Therefore, the Energy Grade Line Analysis provides a measure of engineering statistical reliability that is used for establishing a proportion of the Energy Grade Line Analysis as a "floor value" below which various numerical model simulation techniques should not fall, particularly for flow velocity. To establish the appropriate statistical allowance and conservative bias in the Energy Grade Line Analysis method, 36,000 trial numerical code simulations were performed versus the Energy Grade Line Analysis method. Without this limitation, as is stated in Section 6.7.6.8, there would have been inconsistent reliabilities between the two methods. That is why for Risk Category IV structures of inherently high value, both techniques are valuable to perform, but for quite distinctly different but important reasons (two-dimensional flow characteristics and flow speed).

Table C6.5-1 indicates the necessary inundation depth and flow velocity analysis procedures in accordance with Section 6.5 for each tsunami risk category, for locations shown in the Tsunami Design Zone Map of Figure 6.1-1.

Table C6.5-2 indicates the necessary inundation depth and flow velocity analysis procedures in accordance with Section 6.5 per tsunami risk category for locations where the inundation limit is not shown in the Tsunami Design Zone Map of Figure 6.1-1 but where the Offshore Tsunami Amplitude is shown in Figure 6.7-1.

In certain regions and topographies, for example in remote areas of Alaska, detailed inundation limit mapping is not available. Where no inundation limit map is available from the Authority Having Jurisdiction, a runup based on the factor given in Equation (6.5-2), multiplied by the Offshore Tsunami Amplitude, can be used with the Energy Grade Line Analysis for Tsunami Risk Category II and III structures. The guidance of estimating runup elevation by this approximate factor multiplied by the Offshore Tsunami Amplitude is derived from Japanese research reported in Murata et al. (2010), together with runup relations developed by Synolakis (1986), Li (2000), and Li and Raichlen (2001, 2003); numerical simulations; and field observations from the 2011 Tohoku-Oki tsunami.

Since a large volume of literature exists on the prediction of tsunami runup by analytical methods, by laboratory experiments, and by numerical modeling, a method based on peer-reviewed literature may be used to refine the prediction of runup in certain cases. In these cases, the shoreline topography, Nearshore Profile Slope Angle, and offshore bathymetry should correspond to design conditions. Li and Raichlen (2001, 2003) present analytical expressions for calculating runup of solitary waves under breaking and nonbreaking conditions; Madsen and Schäffer (2010) summarize analytical expressions for runup of nonsolitary wave shapes such as *N*-waves and transient wave trains; and Didenkulova and Pelinovsky (2011) address runup amplification of solitary waves in U-shaped bays.

Figure C6.5-1 is primarily based on runup relations developed by Li and Raichlen for nonbreaking and breaking waves to predict the ratio of runup, R, to Offshore Tsunami Amplitude, H_T, as a function of Φ, the Nearshore Profile Slope Angle. The experiments were conducted for Nearshore Profile Slope Angles less than 1:50 and show a peak runup ratio near 4 and decreasing runup for shallow slopes. The envelope curve represents nearly an upper bound of predicted runup for the range of wave heights [16.4 ft to 65.6 ft (5 m to 20 m)], $0.05 \leq H_T/h_o \leq 0.2$, in areas where no inundation limit has been mapped. Figure C6.5-1 shows the runup relations used to determine the envelope curve. The maximum value of $R/H_T = 4.0$ follows the guidance from Murata et al. (2010). The minimum value of $R/H_T = 3.0$ is based on field observations from the 2011 Tohoku-Oki tsunami, as well as output from numerical models.

Table C6.5-1. Inundation Depth and Flow Velocity Analysis Procedures Where Runup is Given in the Tsunami Design Zone Maps of Figure 6.1-1.

Analysis Procedure Using the Tsunami Design Zone Map of Figure 6.1-1	Tsunami Risk Category (TRC) Structure Classification			
	TRC II	TRC III	TRC IV (excluding TVERS)	TRC IV—Tsunami Vertical Evacuation Refuge Shelter (TVERS)
Section 6.5.1.1 (R/H_T Analysis)	Not permitted	Not permitted	Not permitted	Not permitted
Section 6.6 (Energy Grade Line Analysis)	✓ If MCT inundation depth $\leq$ 3 ft (0.914 m)*, Ch. 6 does not apply	✓ If MCT inundation depth $\leq$ 3 ft (0.914 m)*, Ch. 6 does not apply	✓	✓
Section 6.7 (Site-Specific Analysis)	Permitted; If MCT inundation depth $\leq$ 3 ft (0.914 m)*, Ch. 6 does not apply	Permitted; If MCT inundation depth $\leq$ 3 ft (0.914 m)*, Ch. 6 does not apply	✓ Required if Energy Grade Line Analysis inundation depth $\geq$ 12 ft (3.66 m)*	✓

*MCT inundation depth including sea level rise component, per Section 6.5.3.

Notes: ✓ indicates a required procedure. MCT = Maximum Considered Tsunami. TVERS = Tsunami Vertical Evacuation Refuge Shelter.

Table C6.5-2. Inundation Depth and Flow Velocity Analysis Procedures Where Runup is Calculated from Figure 6.7-1.

Analysis Procedure	Tsunami Risk Category (TRC) Structure Classification			
	TRC II	TRC III	TRC IV (excluding TVERS)	TRC IV—TVERS
Section 6.5.1.1 (R/H_T) Analysis	✓	✓	Not permitted	Not permitted
Section 6.6 (Energy Grade Line Analysis)	✓	✓	✓	✓
	If MCT inundation depth ≤ 3 ft (0.914 m)*, Ch. 6 does not apply	If MCT inundation depth ≤ 3 ft (0.914 m)*, Ch. 6 does not apply		
Section 6.7 (Site-Specific Analysis)	Permitted; If MCT inundation depth ≤ 3 ft (0.914 m)*, Ch. 6 does not apply	Permitted; If MCT inundation depth ≤ 3 ft (0.914 m)*, Ch. 6 does not apply	✓	✓

*MCT inundation depth including sea level rise component, per Section 6.5.3.
Notes: ✓ indicates a required procedure. MCT = Maximum Considered Tsunami. TVERS = Tsunami Vertical Evacuation Refuge Shelter.

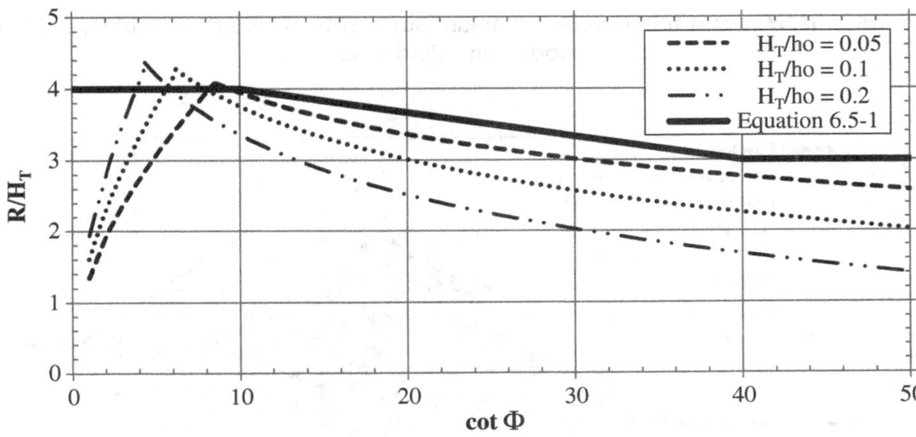

Figure C6.5-1. Runup ratio, R/H_T, as a function of the mean slope of the Nearshore Profile, cot Φ, for relative wave heights $H_T/h_o = 0.05$, 0.1, and 0.2.

Figure C6.5-2 presents the envelope equation of Figure C6.5-1 together with field observations from the 2011 Tohoku tsunami and the results of numerical and physical models of real and idealized runups, together with a large catalog of simulated runups for 36,000 possible combinations of Nearshore Profile Slope Angle and onshore topography. The field and numerical model data suggest that the relation between runup and Offshore Tsunami Amplitude is not greatly reduced as Nearshore Profile Slope Angle increases. In addition, the shallow slopes (high values of cotΦ) observed in the field fall outside of the range possible in physical experiments.

In coastal engineering practice, wave runup is commonly analyzed using a dimensionless parameter called surf similarity, or Iribarren number. The surf similarity parameter characterizes runup using a combination of wave steepness and slope that governs the way the wave evolves as it passes from offshore to onshore. Surf similarity parameter has been applied to tsunamis by Madsen and Schäffer (2010), Hughes (2004), and others. Use of the surf similarity parameter for runup is consistent with coastal engineering design worldwide.

Surf similarity parameter, ξ_{100}, for this application to tsunami engineering is defined as follows:

$$\xi_{100} = \frac{T_{TSU}}{\cot(\Phi)} \sqrt{g/2\pi H_T} \qquad \text{(C6.5-1)}$$

where Φ is the mean slope angle of the Nearshore Profile taken from the 328 ft (100 m) water depth to the Mean High Water Level elevation along the axis of the topographic transect for the site. H_T is the offshore tsunami amplitude, and T_{TSU} is the wave period of the tsunami at 328 ft (100 m) water depth. Runup ratios from field observations, as well as from physical and numerical modeling, are shown as a function of ξ_{100} in Figure C6.5-3 together with an envelope curve that encompasses most of the data. Outliers are in areas where wave focusing is expected to occur and for Monte Carlo simulations with very small crest heights relative to troughs, which alters relative runup values.

Using the surf similarity parameter allows a reduction in the relative runup height for cases where the surf similarity parameter is less than 6 or greater than 20. For the field values presented in Figure C6.5-3, 30% of the surf similarity values fall into the range that would allow for a reduction in relative runup height below the value of 4.

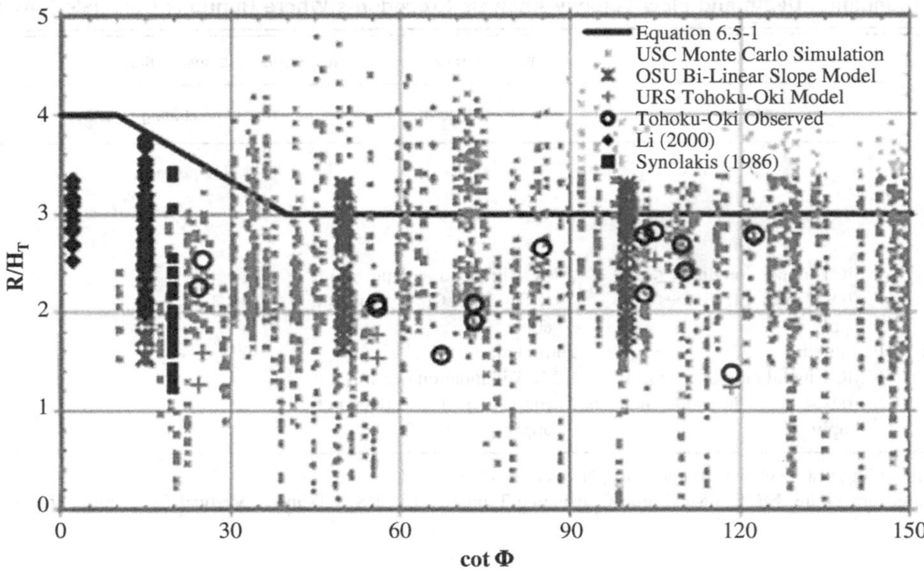

Figure C6.5-2. Runup ratio, R/H_T, as a function of the mean slope of the Nearshore Profile, cotΦ, showing laboratory, model, and field data.

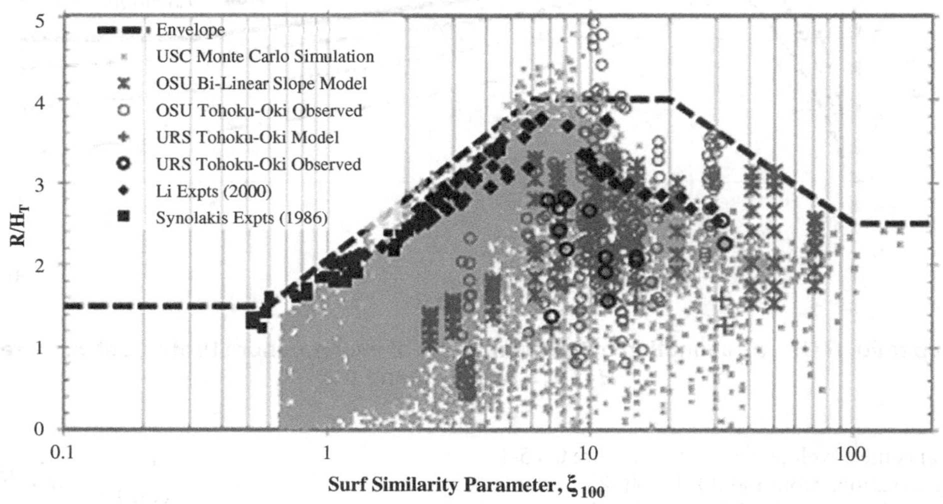

Figure C6.5-3. Runup ratio, R/H_T, as a function of the surf similarity parameter, ξ_{100}.

C6.5.3 Sea Level Change Sea level rise has not been incorporated into the Tsunami Design Zone Maps, and any additive effect on the inundation at the site should be explicitly evaluated. The approach taken for consideration of the effects of sea level rise is consistent with that given by the U.S. Army Corps of Engineers (2013). For a given Maximum Considered Tsunami, the runup elevation and inundation depth at the site are increased by at least the future projected relative sea level rise during the design life of the structure. The minimum design lifecycle of 50 years can be exceeded by the physical life span of many structures. Relative sea level change for a particular site may result from ground subsidence and long-term erosion, and change in local temperature of the ocean causing thermal expansion, as well as possible global effects by melting of ice sheets. Note that as the ASCE provisions for tsunami may later incorporate other areas with more significant relative sea level rise because of greater geologic (nonseismic) subsidence, this section on sea level rise would be more important. An example of this would be American Samoa and the other Pacific insular states and territories in the western Pacific Ocean associated with the United States, particularly where the habitable terrain elevation is low.

Historical sea level change trends over the past 150 years can be obtained from the NOAA Center for Operational Oceanographic Products and Services, which has been measuring sea level at tide stations along the US coasts. Changes in local relative mean sea level (MSL) have been computed at 128 long-term water level stations using a minimum span of 30 years of observations at each location (http://tidesandcurrents.noaa.gov/sltrends/: "Tide gauge measurements are made with respect to a local fixed reference level on land; therefore, if there is some long-term vertical land motion occurring at that location, the relative MSL trend measured there is a combination of the global sea level rate and the local vertical land motion.")

C6.6 INUNDATION DEPTHS AND FLOW VELOCITIES BASED ON RUNUP

C6.6.1 Maximum Inundation Depth and Flow Velocities Based on Runup Calculated flow velocities are subject to minimum and maximum limits based on observational data from past tsunamis of significant inundation depth. The upper limit of 50 ft/s (15.2 m/s) in Section 6.6 already includes a 1.5 factor on the velocity that is greater than any on-land flow velocity that the committee has examined from field or video evidence. The flow velocity upper limit is therefore applicable subsequent to any flow amplification factors caused by obstructions in the built environment, per Sections 6.6.5, 6.7.6.6, or 6.8.5.

In a Monte Carlo simulation of hundreds of variations of terrain profiles where the Energy Grade Line Analysis was compared to numerical models, it was found that cases of negative terrain slope gradient would cause the Energy Grade Line Analysis to fail to converge on a solution. To deal with the possibility of such cases in which local downslope acceleration of flow would occur, but to retain the simplicity of the analysis, a conservative adjustment of the runup elevation is prescribed in method 2 of Section 6.6.1, which essentially increases the energy grade line. Alternatively, a site-specific inundation analysis may be performed per method 1 of Section 6.6.1.

C6.6.2 Energy Grade Line Analysis of Maximum Inundation Depths and Flow Velocities The coastline can be approximated in behavior by the use of one-dimensional linear transects of a composite bathymetric and/or topographic profile. The tsunami inundation design parameters of inundation depth and flow velocities are determined by an energy analysis approach. In the Energy Grade Line Analysis, hydraulic analysis using Manning's coefficient for equivalent terrain macroroughness is used to account for friction, along with the profile comprised of a series of one-dimensional slopes, to determine the variation of depth and velocity-associated inundation depth across the inland profile. Velocity is assumed to be a function of inundation depth, calibrated to the prescribed Froude number, $u/\sqrt{gh}$, at the shoreline and decreasing inland. As noted in Kriebel et al. (2017), a Froude number is prescribed in the Energy Grade Line Analysis (EGLA) as a conservative means to determine the magnitudes of $u_{\max}$ and $h_{\max}$, but which in reality rarely occur simultaneously. Hence the Froude number value used within the EGLA does not represent the actual Froude number of the flow at the site, which should instead be determined using the flow depth and corresponding velocity from Figure 6.8-1 for the load case being considered. For example, the design Froude number for Load Case 2 when h is 2/3 of $h_{\max}$ will be 22% greater than the Froude number prescribed for use within the EGLA. The specified incremental horizontal distance maximum spacing of 100 ft (30.5 m) is necessary for accuracy of the hydraulic analysis.

The Energy Grade Line Analysis is based on rational hydraulic principles for a one-dimensional transect flow assumption, so it is not the same analysis technique as performed by a two-dimensional (2D) flow numerical simulation. The Energy Grade Line Analysis is simple and it is inherently less precise, but is statistically conservative for use as a practical design tool that can be performed by the engineer. One additional conservatism is to use the actual roughness of the terrain in the energy method rather than a "bare-earth" assumption. Also the parameters of "maximum" depth and flow speed are associated with the prescriptive load cases that define the combinations derived from these "maximum" flow parameters. When the actual roughness of the terrain is used in the energy method, it takes a greater initial energy at the shoreline in order to reach the runup. Therefore, conceptually, the energy at the shoreline in the prescriptive method is implicitly greater than the energy value used in the 2D inundation model run on "bare earth." The benefit of this approach is that it produces conservative design values and it implicitly accounts for flow amplification that occurs through the built environment (which the standard 2D inundation model technique does not consider). The 328 ft (100 m) depth is where the waveform based on Probabilistic Tsunami Hazard Analysis can be regionally characterized along the coastline. The Tsunami Design Zone Maps of this standard have been developed from a Probabilistic Tsunami Hazard Analysis (PTHA) using the NOAA Pacific Marine Environmental Laboratory (PMEL) Short-Term Inundation Forecasting for Tsunamis (SIFT), Method of Splitting Tsunamis (MOST), and Community Model Interface for Tsunamis (ComMIT) model for the 2D inundation analysis. The Tsunami Design Zone Maps have been produced to ensure that the inundation limit and runup elevation and the extent of the hazard zone are generally not underestimated (Montoya et al. 2017). The Energy Grade Line Analysis uses this input so that the flow characteristics at the site are not severely underestimated for design purposes. The statistical level of conservatism has been verified by doing many thousands of simulations to compare the energy method with time history outputs from the primary types of numerical inundation models.

The Energy Grade Line Analysis stepwise procedure consists of the following steps:

1. Obtain the runup and inundation limit values from the Tsunami Design Zone Map generated by the ASCE Tsunami Design Geodatabase. Determine the value of α in accordance with Sections 6.6.2 and 6.6.4.
2. Approximate the principal topographic transect by a series of x–z grid coordinates defining a series of segmented slopes, in which x is the distance inland from the shoreline to the point and z is the ground elevation of the point. The horizontal spacing of transect points should be less than 100 ft (30.5 m), and the transect elevations should be obtained from a topographic Digital Elevation Model (DEM) of at least 33 ft (10 m) resolution.
3. Compute the topographic slope, ϕ_i, of each segment as the ratio of the increments of elevation and distance from point to point in the direction of the incoming flow.
4. Obtain the Manning's coefficient, n, from Table 6.6-1 for each segment based on terrain analysis.
5. Compute the Froude number at each point on the transect using Equation (6.6-3).

   Start at the point of runup with a boundary condition of $E_R = 0$ at the point of runup.

   i. Per Section 6.6.1, where the maximum topographic elevation along the topographic transect between the shoreline and the inundation limit is greater than the runup elevation, use a runup elevation that has at least 100% of the maximum topographic elevation along the topographic transect.*
6. Select a nominally small value of inundation depth [~0.1 ft (0.03 m)] h_r at the point of runup.
7. Calculate the hydraulic friction slope, s_i, using Equation (6.6-2).
8. Compute the hydraulic head, E_i, from Equation (6.6-1) at successive points toward the shoreline.
9. Calculate the inundation depth, h_i, from the hydraulic head, E_i. Using the definition of Froude number, determine the velocity u. Repeat through the transect until the h and u are calculated at the site.

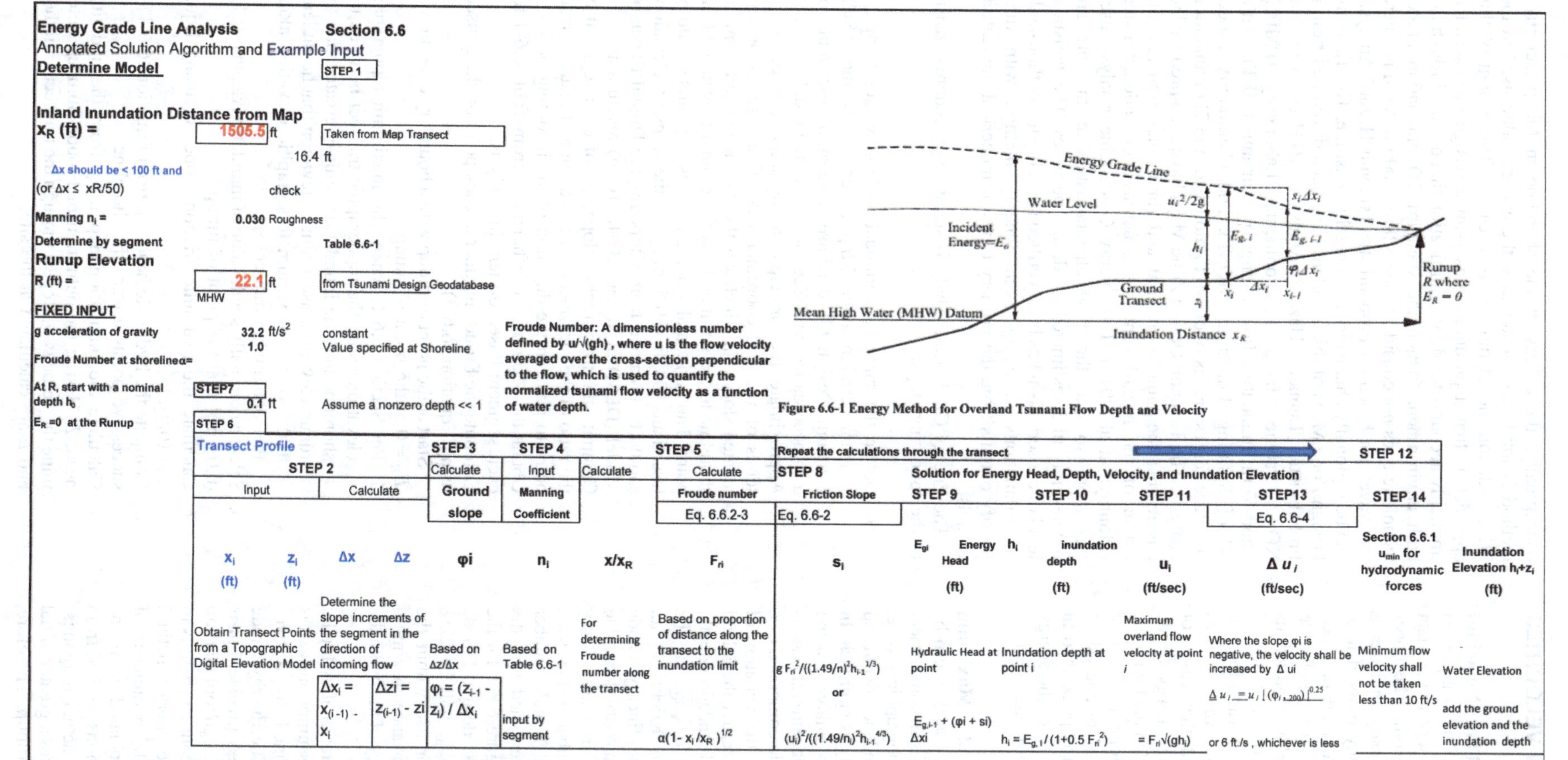

The Energy Grade Line Analysis stepwise procedure consists of the following steps:

1. Obtain the Runup and Inundation Limit values from the Tsunami Design Map
   Determine the value of α in accordance with Section 6.6.2 and 6.6.4.
2. Approximate the principal topographic transect by a series of X-Z grid coordinates defining a series of segmented slopes;
   x is the distance inland from the shoreline to the point and z is the ground elevation of the point
3. Compute the topographic slope ,φ_i, of each segment as the ratio of the increments of elevation and distance from point to point in the direction of the incoming flow.
4. Obtain the Manning's coefficient, n, from Table 6.6-1 for each segment based on terrain analysis.
5. Compute the Froude number at each point on the transect using **Equation 6.6-3**.
6. Start at the point of Runup with a boundary condition of E_R =0 at the point of runup.
7. Select a nominally small value of inundation depth (~0.1 ft.) h_R at the point of runup.
8. Calculate the hydraulic friction slope , s_i, using **Equation 6.6-2.**
9. Compute the hydraulic energy head E_i from at successive points towards the shoreline.
10. Calculate the inundation depth h_i from the hydraulic energy E_i.
11. Using the definition of the Froude number, determine the velocity u_i.
12. Repeat through the transect until the h and u are calculated at the site.
13. Then, at locations along the transect where the slope ϕi is negative, the velocity determined in accordance with Eq. (6.6-3) shall be increased to account for downslope acceleration.
14. Check that all velocity values are not less than the minimum flow velocity required by Section 6.6.1.
15. The final values are used as the maximum inundation depth, hmax, and the maximum velocity, umax, along the transect

Load Cases based on these parameters are defined in **Section 6.8.3.1.**

Figure C6.6-1 Energy Grade Line Analysis Algorithm

Figure C6.6-1. Energy Grade Line Analysis—Spreadsheet Algorithm.

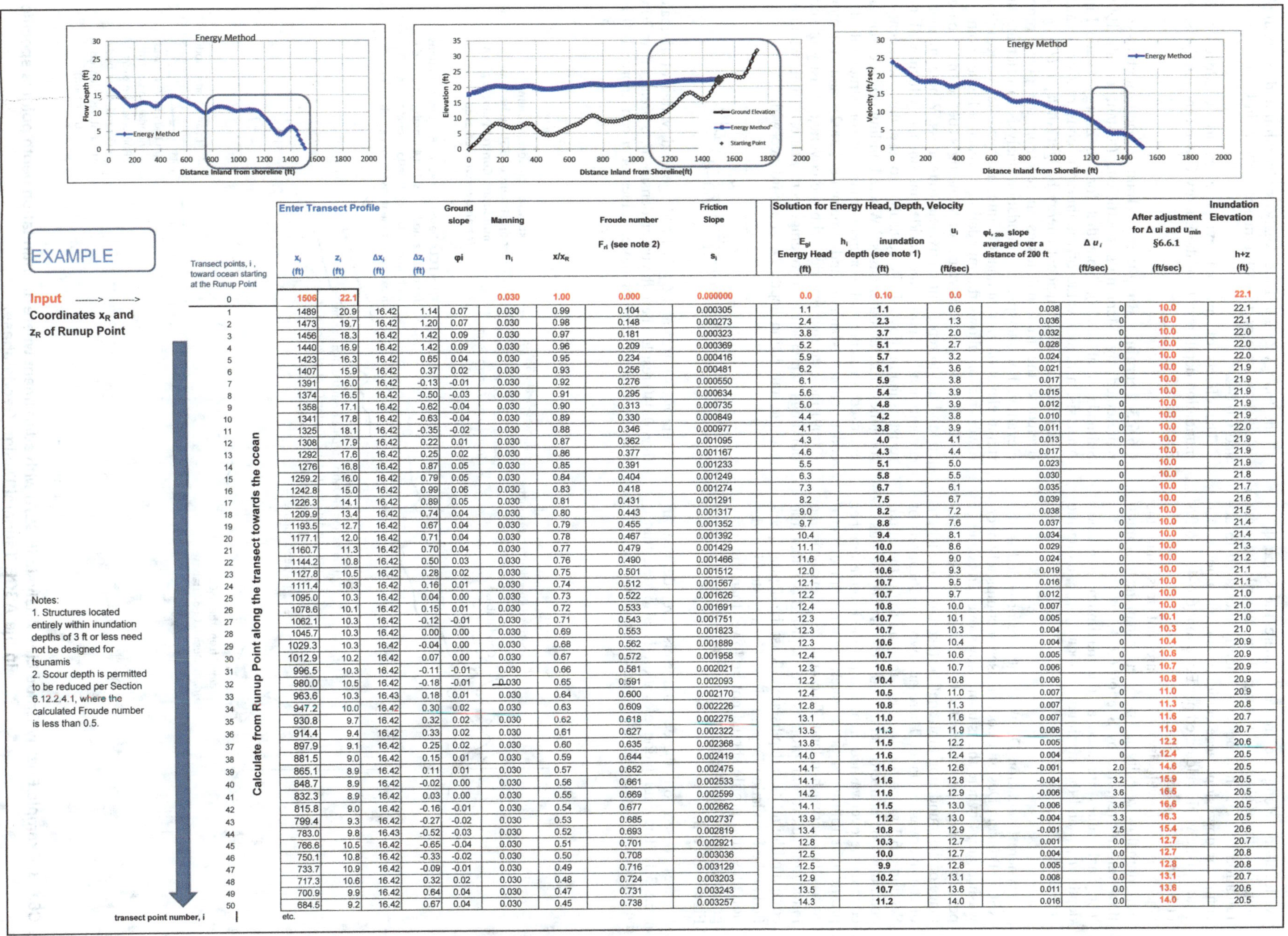

EXAMPLE

Input ------> ------>
Coordinates x_R and z_R of Runup Point

Notes:
1. Structures located entirely within inundation depths of 3 ft or less need not be designed for tsunamis
2. Scour depth is permitted to be reduced per Section 6.12.2.4.1, where the calculated Froude number is less than 0.5.

Transect points, i, toward ocean starting at the Runup Point	Enter Transect Profile x_i (ft)	z_i (ft)	Δx_i (ft)	Δz_i (ft)	Ground slope φi	Manning n_i	x/x_R	Froude number F_{ri} (see note 2)	Friction Slope s_i	Solution for Energy Head, Depth, Velocity: E_{gi} Energy Head (ft)	h_i inundation depth (see note 1) (ft)	u_i (ft/sec)	$\varphi i,_{200}$ slope averaged over a distance of 200 ft	Δu_i (ft/sec)	After adjustment for Δ ui and u_{min} §6.6.1 (ft/sec)	Inundation Elevation h+z (ft)
0	1506	22.1				0.030	1.00	0.000	0.000000	0.0	0.10	0.0				22.1
1	1489	20.9	16.42	1.14	0.07	0.030	0.99	0.104	0.000305	1.1	1.1	0.6	0.038	0	10.0	22.1
2	1473	19.7	16.42	1.20	0.07	0.030	0.98	0.148	0.000273	2.4	2.3	1.3	0.036	0	10.0	22.1
3	1456	18.3	16.42	1.42	0.09	0.030	0.97	0.181	0.000323	3.8	3.7	2.0	0.032	0	10.0	22.0
4	1440	16.9	16.42	1.42	0.09	0.030	0.96	0.209	0.000369	5.2	5.1	2.7	0.028	0	10.0	22.0
5	1423	16.3	16.42	0.65	0.04	0.030	0.95	0.234	0.000416	5.9	5.7	3.2	0.024	0	10.0	22.0
6	1407	15.9	16.42	0.37	0.02	0.030	0.93	0.256	0.000481	6.2	6.1	3.6	0.021	0	10.0	21.9
7	1391	16.0	16.42	-0.13	-0.01	0.030	0.92	0.276	0.000550	6.1	5.9	3.8	0.017	0	10.0	21.9
8	1374	16.5	16.42	-0.50	-0.03	0.030	0.91	0.295	0.000634	5.6	5.4	3.9	0.015	0	10.0	21.9
9	1358	17.1	16.42	-0.62	-0.04	0.030	0.90	0.313	0.000735	5.0	4.8	3.9	0.012	0	10.0	21.9
10	1341	17.8	16.42	-0.63	-0.04	0.030	0.89	0.330	0.000849	4.4	4.2	3.8	0.010	0	10.0	21.9
11	1325	18.1	16.42	-0.35	-0.02	0.030	0.88	0.346	0.000977	4.1	3.8	3.9	0.011	0	10.0	22.0
12	1308	17.9	16.42	0.22	0.01	0.030	0.87	0.362	0.001095	4.3	4.0	4.1	0.013	0	10.0	21.9
13	1292	17.6	16.42	0.25	0.02	0.030	0.86	0.377	0.001167	4.6	4.3	4.4	0.017	0	10.0	21.9
14	1276	16.8	16.42	0.87	0.05	0.030	0.85	0.391	0.001233	5.5	5.1	5.0	0.023	0	10.0	21.9
15	1259.2	16.0	16.42	0.79	0.05	0.030	0.84	0.404	0.001249	6.3	5.8	5.5	0.030	0	10.0	21.8
16	1242.8	15.0	16.42	0.99	0.06	0.030	0.83	0.418	0.001274	7.3	6.7	6.1	0.035	0	10.0	21.7
17	1226.3	14.1	16.42	0.89	0.05	0.030	0.81	0.431	0.001291	8.2	7.5	6.7	0.039	0	10.0	21.6
18	1209.9	13.4	16.42	0.74	0.04	0.030	0.80	0.443	0.001317	9.0	8.2	7.2	0.038	0	10.0	21.5
19	1193.5	12.7	16.42	0.67	0.04	0.030	0.79	0.455	0.001352	9.7	8.8	7.6	0.037	0	10.0	21.4
20	1177.1	12.0	16.42	0.71	0.04	0.030	0.78	0.467	0.001392	10.4	9.4	8.1	0.034	0	10.0	21.4
21	1160.7	11.3	16.42	0.70	0.04	0.030	0.77	0.479	0.001429	11.1	10.0	8.6	0.029	0	10.0	21.3
22	1144.2	10.8	16.42	0.50	0.03	0.030	0.76	0.490	0.001466	11.6	10.4	9.0	0.024	0	10.0	21.2
23	1127.8	10.5	16.42	0.28	0.02	0.030	0.75	0.501	0.001512	12.0	10.6	9.3	0.019	0	10.0	21.1
24	1111.4	10.3	16.42	0.16	0.01	0.030	0.74	0.512	0.001567	12.1	10.7	9.5	0.016	0	10.0	21.1
25	1095.0	10.3	16.42	0.04	0.00	0.030	0.73	0.522	0.001626	12.2	10.7	9.7	0.012	0	10.0	21.0
26	1078.6	10.1	16.42	0.15	0.01	0.030	0.72	0.533	0.001691	12.4	10.8	10.0	0.007	0	10.0	21.0
27	1062.1	10.3	16.42	-0.12	-0.01	0.030	0.71	0.543	0.001751	12.3	10.7	10.1	0.005	0	10.1	21.0
28	1045.7	10.3	16.42	0.00	0.00	0.030	0.69	0.553	0.001823	12.3	10.7	10.3	0.004	0	10.3	21.0
29	1029.3	10.3	16.42	-0.04	0.00	0.030	0.68	0.562	0.001889	12.3	10.6	10.4	0.004	0	10.4	20.9
30	1012.9	10.2	16.42	0.07	0.00	0.030	0.67	0.572	0.001958	12.4	10.7	10.6	0.005	0	10.6	20.9
31	996.5	10.3	16.42	-0.11	-0.01	0.030	0.66	0.581	0.002021	12.3	10.6	10.7	0.006	0	10.7	20.9
32	980.0	10.5	16.42	-0.18	-0.01	0.030	0.65	0.591	0.002093	12.2	10.4	10.8	0.006	0	10.8	20.9
33	963.6	10.3	16.43	0.18	0.01	0.030	0.64	0.600	0.002170	12.4	10.5	11.0	0.007	0	11.0	20.9
34	947.2	10.0	16.42	0.30	0.02	0.030	0.63	0.609	0.002226	12.8	10.8	11.3	0.007	0	11.3	20.8
35	930.8	9.7	16.42	0.32	0.02	0.030	0.62	0.618	0.002275	13.1	11.0	11.6	0.007	0	11.6	20.7
36	914.4	9.4	16.42	0.33	0.02	0.030	0.61	0.627	0.002322	13.5	11.3	11.9	0.006	0	11.9	20.7
37	897.9	9.1	16.42	0.25	0.02	0.030	0.60	0.635	0.002368	13.8	11.5	12.2	0.005	0	12.2	20.6
38	881.5	9.0	16.42	0.15	0.01	0.030	0.59	0.644	0.002419	14.0	11.6	12.4	0.004	0	12.4	20.5
39	865.1	8.9	16.42	0.11	0.01	0.030	0.57	0.652	0.002475	14.1	11.6	12.6	-0.001	2.0	14.6	20.5
40	848.7	8.9	16.42	-0.02	0.00	0.030	0.56	0.661	0.002533	14.1	11.6	12.8	-0.004	3.2	15.9	20.5
41	832.3	8.9	16.42	0.03	0.00	0.030	0.55	0.669	0.002599	14.2	11.6	12.9	-0.006	3.6	16.5	20.5
42	815.8	9.0	16.42	-0.16	-0.01	0.030	0.54	0.677	0.002662	14.1	11.5	13.0	-0.006	3.6	16.6	20.5
43	799.4	9.3	16.42	-0.27	-0.02	0.030	0.53	0.685	0.002737	13.9	11.2	13.0	-0.004	3.3	16.3	20.5
44	783.0	9.8	16.43	-0.52	-0.03	0.030	0.52	0.693	0.002819	13.4	10.8	12.9	-0.001	2.5	15.4	20.6
45	766.6	10.5	16.42	-0.65	-0.04	0.030	0.51	0.701	0.002921	12.8	10.3	12.7	0.001	0.0	12.7	20.7
46	750.1	10.8	16.42	-0.33	-0.02	0.030	0.50	0.708	0.003036	12.5	10.0	12.7	0.004	0.0	12.7	20.8
47	733.7	10.9	16.42	-0.09	-0.01	0.030	0.49	0.716	0.003129	12.5	9.9	12.8	0.005	0.0	12.8	20.8
48	717.3	10.6	16.42	0.32	0.02	0.030	0.48	0.724	0.003203	12.9	10.2	13.1	0.008	0.0	13.1	20.7
49	700.9	9.9	16.42	0.64	0.04	0.030	0.47	0.731	0.003243	13.5	10.7	13.6	0.011	0.0	13.6	20.6
50	684.5	9.2	16.42	0.67	0.04	0.030	0.45	0.738	0.003257	14.3	11.2	14.0	0.016	0.0	14.0	20.5

etc.

Figure C6.6-2. Example transect using the Energy Grade Line Analysis.

10. At locations along the transect where the slope ϕ_i is negative, the velocity determined in accordance with Equation (6.6-3) shall be subsequently increased to account for downslope acceleration, using an amount Δu_i equal to the value calculated per Equation (6.6-4) or 6 ft/s (1.83 m/s), whichever is less.
11. Check that all velocity values are not less than the minimum flow velocity and not greater than the maximum flow velocity required by Section 6.6.1.
12. The final values are used as the maximum inundation depth, h_{max}, and the maximum velocity, u_{max}, along the transect.

*Where Inundation Depth Points are provided in the ASCE Tsunami Design Geodatabase for completely overwashed areas, where the tsunami flows over an island or peninsula into a second water body, the horizontal distance of the inundation limit shall be taken to be the length of the overwashed land. There are two modifications of Section 6.6.1 given for this case, relating to (a) the initial conditions of nonzero depth and velocity at the inland edge of the overwashed area, and (b) the Froude number profile that follows a linear interpolation with distance across the overwashed land. To complete the analysis, the inundation elevation profiles are adjusted so that the computed inundation depths are at least the depths specified at the Inundation Depth Points. The adjusted inundation elevation profile should not be lower than the topographic elevation transect. An example is given in Figure C6.6-3.

Load cases based on these parameters are defined in Section 6.8.3.1. The inundation elevation is also determined from the inundation depth plus the ground elevation.

The Energy Grade Line Analysis algorithm and an example calculation for a portion of a transect follows in Figures C6.6-1 and C6.6-2.

The effects of any sea level rise are to be added to the inundation depth results of the Energy Grade Line Analysis, per Section 6.5.3.

C6.6.3 Terrain Roughness Terrain macroroughness is represented by Manning's coefficient (Kotani et al. 1998, Shimada et al. 2003). Note that for the Energy Grade Line Analysis, a higher value of roughness is conservative because it requires a greater incident energy budget (of $h_0 + U_0^2/2g$) to overcome the frictional losses in reaching the given runup elevation. For site-specific inundation analysis to calculate the runup elevation and inundation limit, the opposite is true, in that lower friction allows the inundation to reach a higher point.

Table 6.6-1 provides values of Manning's coefficient, n, based on land use to represent equivalent roughness. Where an area is predominantly light-frame residential houses, which will not resist tsunami flow, it is advisable to use the coefficient of roughness for "all other cases." Where building obstructions are enclosed structures of concrete, masonry, or structural steel construction, Aburaya and Imamura (2002) and Imamura (2009) indicate that n varies with the coverage density of such buildings expressed as a coverage percentage, θ, their width, w, and the inundation depth, D, in accordance with

$$n = \sqrt{n_0^2 + \frac{C_D}{2gw} \times \frac{\theta}{100-\theta} \times D^{4/3}}$$

where n_0 is 0.025 and C_D is 1.5. In general, values of n greater than 0.040 result where the urban density, θ, is greater than 20%. For densities of 80% or more, values of n can approach 0.10.

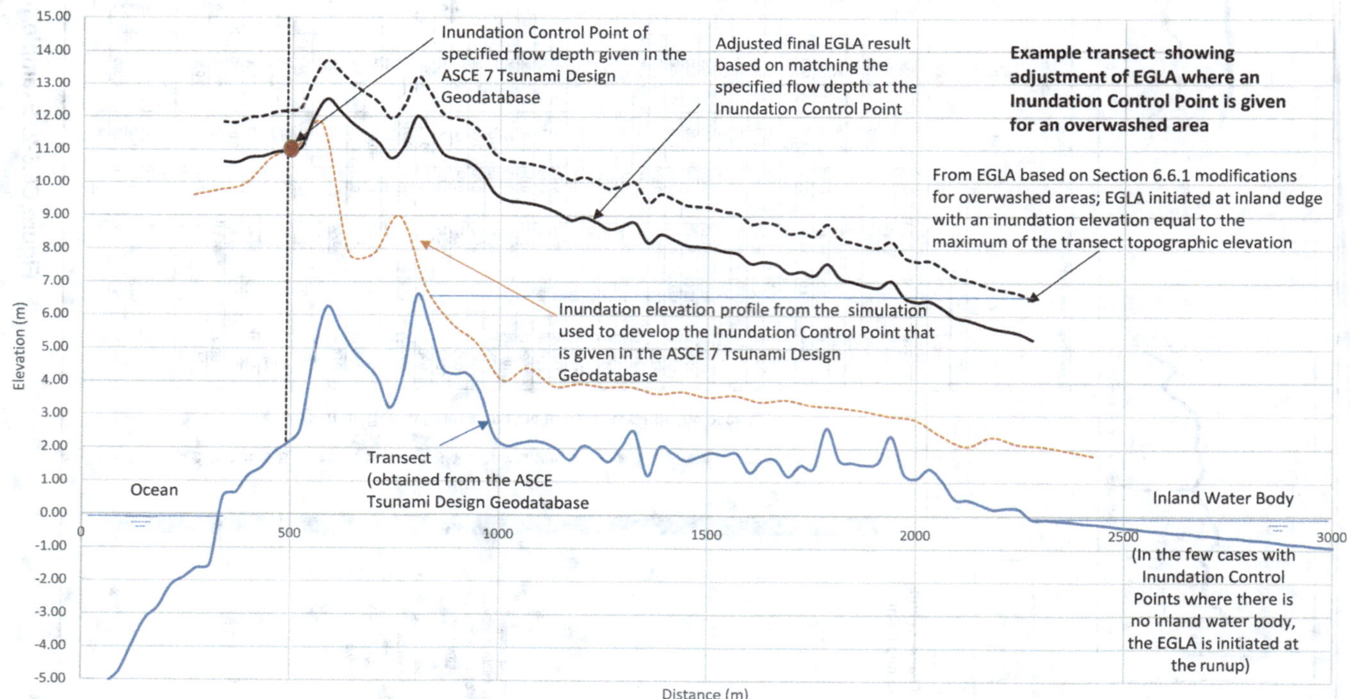

Figure C6.6-3. Example Energy Grade Line Analysis (EGLA) with adjustment where an inundation depth point is specified in the ASCE Tsunami Design Geodatabase.

Studies by the Tsunami Loads and Effects Subcommittee indicate that it is not necessary to use values of Manning's roughness greater than 0.050 in the Energy Grade Line Analysis. Values of n greater than 0.050 result in much higher statistical conservatism in calculated momentum flux by the Energy Grade Line Analysis, compared to numerical solutions.

C6.6.4 Tsunami Bores The criteria for the occurrence of bores consist of a complex interaction of the tsunami waveform and the coastal bathymetry, which can be determined through site-specific inundation numerical modeling. One such condition described in the literature arises where soliton fission waves are generated by shoaling over a long bathymetric slope of approximately 1/100 or milder (Murata et al. 2010, Madsen et al. 2008). These packets of shorter-period waves on the front face of the long-period tsunami then break into a series of bores where the soliton amplitude over depth ratio reaches 0.78 to 0.83. Another case occurs where a shoaling tsunami encounters an abruptly rising seabed floor, such as a fringing shallow reef.

C6.7 INUNDATION DEPTHS AND FLOW VELOCITIES BASED ON SITE-SPECIFIC PROBABILISTIC TSUNAMI HAZARD ANALYSIS

A method of Probabilistic Tsunami Hazard Analysis has been established in the recognized literature that is generally consistent with Probabilistic Seismic Hazard Analysis in the treatment of uncertainty.

The basics of Probabilistic Tsunami Hazard Analysis for a region are as follows:

1. Tsunamigenic subduction zones and nonsubduction seismic thrust faults are discretized into a compiled system of rectangular subfaults, each with corresponding tectonic parameters.
2. Tsunami waveform generation is modeled as a linear combination of individual tsunami waveforms, each generated from a particular subfault, with a set sequence of subfaults used to describe the earthquake rupture in location, orientation, and rupture direction and sequence.
3. A statistically weighted logic tree approach is used to account for epistemic uncertainties in the tectonic parameters. The logic tree model parameters are developed from available tectonic, geodetic, historical, and paleotsunami data and estimated plate convergence rates. A probability distribution is used to account for aleatory uncertainties (i.e., intrinsic variability in each actual earthquake event).
4. Each individual tsunami waveform is propagated in deep water to 328 ft (100 m) depth using linear long-wave equations to take into account spatial variations in seafloor depth. This set of precomputed individual tsunami waveforms is combined in a linear fashion (i.e., Green's function approach) for subsequent analysis.
5. At the 328 ft (100 m) depth, determine the highest offshore wave heights and develop wave height, period, and waveform parameters for the design level exceedance of the 2,475-year tsunami event. The still-water level may be considered to be at the Reference Sea Level for this purpose.
6. Disaggregate (i.e., separate out for use in subsequent analysis) the seismic sources and associated moment magnitudes that together contribute at least 90% to the net offshore tsunami hazard at the site under consideration for the design level mean recurrence interval.
7. Use nonlinear shallow-water wave equations to transform each disaggregated tsunami event, or the Hazard-Consistent Tsunami event(s) (see following) from 328 ft (100 m) depth toward the shore to determine the associated maximum inundation.
8. Analyze each tsunami event from the disaggregated sample to determine flow parameters for the site of interest. Manning's coefficient for equivalent terrain macroroughness is used to account for friction. Maximum runup, inundation depth, flow velocity, and/or specific momentum flux at the site of interest are permitted to be evaluated by either of the following techniques.
   (a) Determine the Hazard-Consistent Tsunami as one or more surrogate events that replicate the weighted average waveform corresponding to the offshore wave height for the return period. The inundation limit for the Hazard-Consistent Tsunami shall match the area that is inundated by tsunami waves from all the disaggregated major source zones for that particular return period. Flow parameters at the site of interest shall be determined from the Hazard-Consistent Tsunami waveform.
   (b) Develop the probabilistic distributions of flow parameters from the disaggregated sample of computed tsunamis and construct the statistical distributions of flow parameters for at least three critical load cases.
9. From the probabilistic events, capture the design flow parameters of inundation depth, flow velocity, and/or specific momentum flux at the site of interest.

Data referenced in Section 6.7 are provided by ASCE. The ASCE Tsunami Design Geodatabase is version-specific for the application of each edition of ASCE 7, and the user of the standard should take care to select the proper data set as appropriate for the building code adopted by the jurisdiction.

The ASCE Tsunami Design Geodatabase of geocoded reference points of offshore 328 ft (100 m) depth; Tsunami Amplitude, H_T; and predominant wave period, T_{TSU}, of the Maximum Considered Tsunami are available at http://asce7tsunami.online, including the following data sets:

- Offshore tsunami amplitude with predominant period, and
- Disaggregated hazard source contribution figures.

Because actual and model-simulated offshore wave amplitude time histories are not exhibited as pure signals with a singular wave period, a procedure was used to capture a practical equivalent period that would be representative of the Offshore Tsunami Amplitude. To determine half of the predominant wave period, the following steps were performed. First, the zero-crossings immediately before and after the maximum Offshore Tsunami Amplitude were found. If there was no zero-crossing before the maximum amplitude, as might be the case for a leading-elevation tsunami waveform, then the start time of the time series was considered as the first zero-crossing time. Next, the total area in the tsunami elevation time series between the two zero-crossing times was integrated. The half-period is the time between the point when 5% of the total area has passed and the point when 95% of the total area has passed.

The GIS digital map layers for probabilistic subsidence are available in the ASCE Tsunami Design Geodatabase at http://asce7tsunami.online.

An example of the ASCE Tsunami Design Geodatabase of Offshore Tsunami Amplitude, predominant wave period, and geocoded location of a database point is shown in Figure C6.7-1(a). An illustration of the derivation of the predominant wave period, T_{TSU}, is shown in Figure C6.7-1(b).

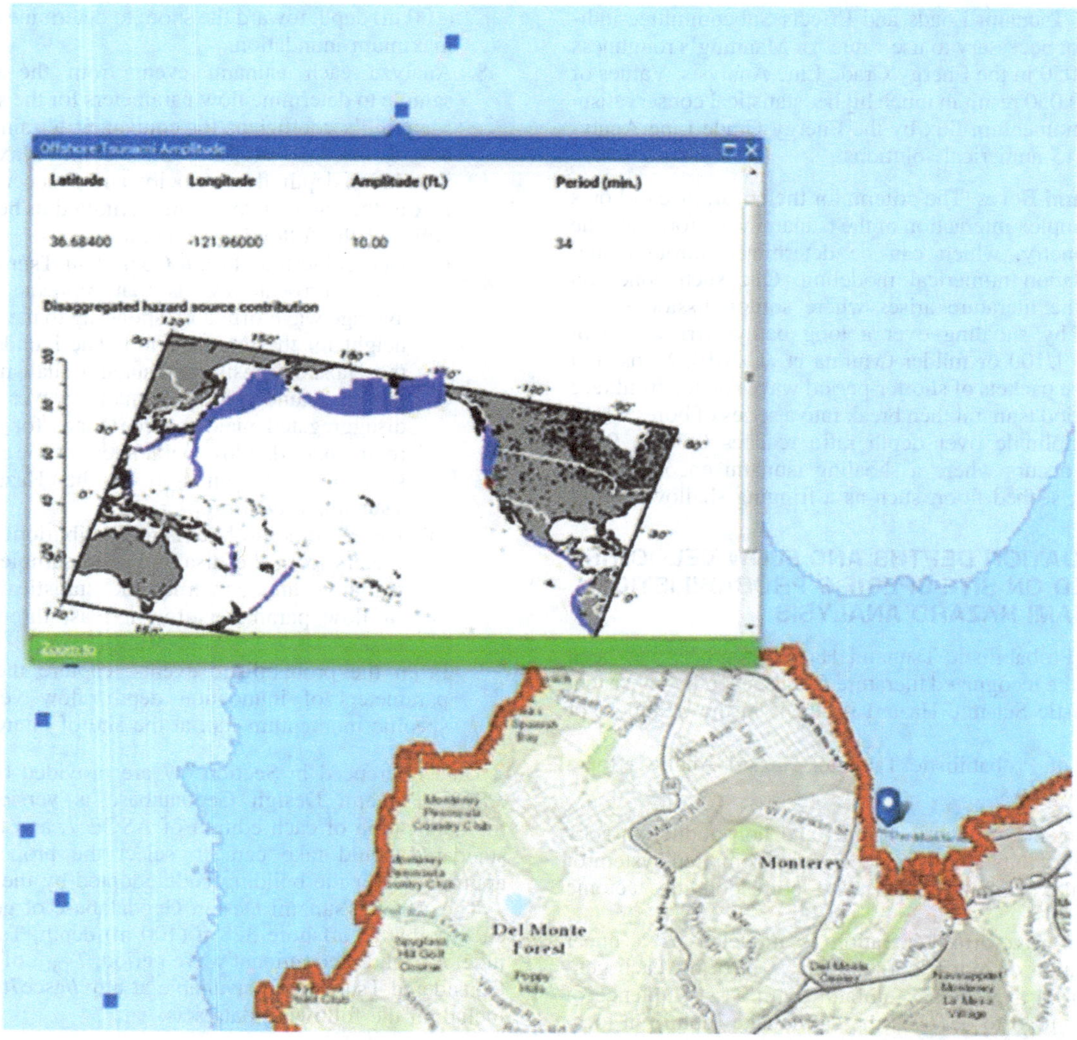

Figure C6.7-1(a). Example of the Offshore Tsunami Amplitude from the ASCE Tsunami Design Geodatabase (amplitude in feet, period in minutes).

C6.7.1 Tsunami Waveform The waveform prescribed here is a sum of a series of soliton pairs, with each pair composed of one with a positive and one with a negative amplitude. These pairs reflects (1) the case where the leading wave has a positive amplitude, and (2) the case where the largest (positive) amplitude wave follows a large trough wave. Because tsunamis consist of multiple waves interacting with the coastal bathymetry, both cases of leading or trailing depression waveforms are to be considered. The total number of soliton pairs to be included in the offshore tsunami waveform is given by the parameter N. As shown in tsunami observations and tsunami simulations, the value of N is dependent on whether the tsunami is from a local earthquake source or from a distant earthquake source, and therefore different values of N are prescribed for different regions, as given in Table 6.7-1. The range of parameter a_2 shall also be considered in developing the suite of waveforms to be analyzed by inundation modeling. The parameters a_1, a_2, and T_{TSU} can be obtained from the map (Figure 6.7-1) or from Table 6.7-1, which provides a more regionalized model based on Figure 6.7-1. The values in Table 6.7-1 were compiled from a series of scenarios modeled for each of the five western states, in which the maxima of the offshore wave amplitudes of the crest and trough were sampled at offshore locations at 328 ft (100 m) depth.

C6.7.2 Tsunamigenic Sources Tsunamigenic sources that affect the Pacific states have been identified through historical records as well as systematic modeling of all circum-Pacific subduction zones. Maximum magnitudes were derived from perceived maximum dimensions for future earthquake ruptures and scaling relations between dimensions and magnitude. In the Salish Sea of Washington State, the ASCE Tsunami Design Geodatabase has been probabilistically analyzed from a hazard-consistent source in the Cascadia Subduction Zone together with local thrust sources that include the Seattle Fault, the Tacoma Fault, and the Rosedale-Dominant Tacoma Fault, with source parameters described in Venturato et al. (2007), and other local tsunamigenic sources, such as the South Whidbey Island and Darrington-Devils Mountain Faults, that are considered potentially significant.

In Alaska, local tsunami inundations are based on suites of hypothetical earthquake scenarios and landslide scenarios modeled by the Alaska Division of Geological & Geophysical

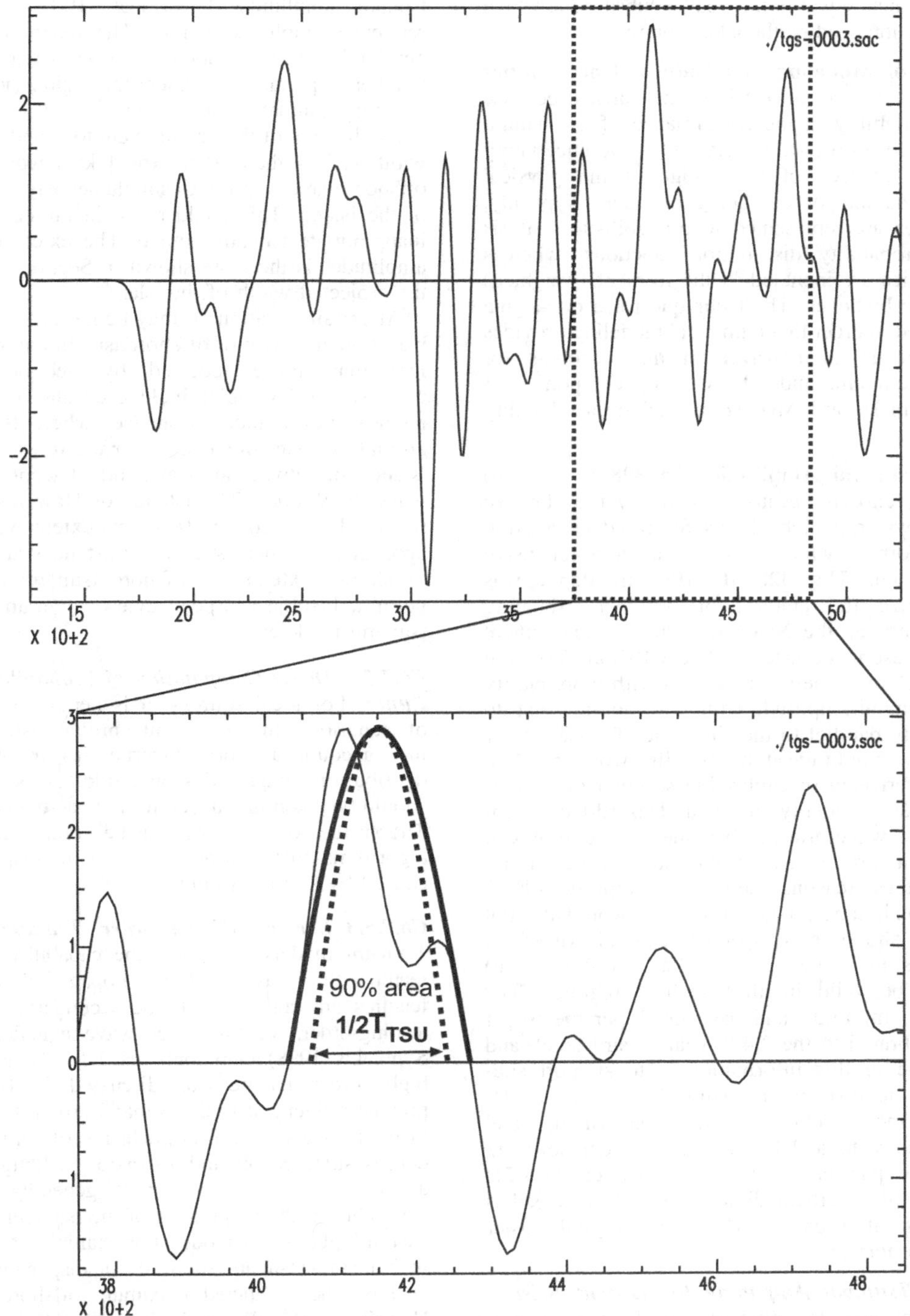

Figure C6.7-1(b). Illustration of the predominant wave period, T_{TSU}, of the offshore tsunami waveform.

Surveys for Elfin Cove, Gustavus, Haines, Homer, Hoonah, Juneau, Seldovia, Seward, Skagway, Valdez, Whittier, and Yakutat (Suleimani et al. 2010, 2015, 2016, 2018; Nicolsky et al. 2011, 2017, 2018). Because each of these maps is based on several hypothetical tsunamis of tectonic and landslide origins, none of which is the result of Probabilistic Tsunami Hazard Analysis, they do not constitute a design basis for the ASCE 7 standard and are not included in the ASCE Tsunami Design Geodatabase. The referenced reports state that "the limits of inundation shown should only be used as a guideline for emergency planning and response action," and that "these results are not intended for land-use regulation." Therefore, the ASCE Tsunami Design Geodatabase does not include any landslide-induced tsunami inundation. Nevertheless, tsunamis generated by subaerial or submarine slope failures can pose significant hazards in the fjords of coastal Alaska and other high-latitude fjord coastlines.

C6.7.3 Earthquake Rupture Unit Source Tsunami Functions for Offshore Tsunami Amplitude For the computation of offshore amplitudes [e.g., at the 328 ft (100 m) depth contour], the principle of linearity is used to compute tsunami waves as a summation of precomputed waves for smaller subsources, which in total make up the intended earthquake rupture.

C6.7.4 Treatment of Modeling and Natural Uncertainties *Aleatory* uncertainty is the uncertainty that arises because of unpredictable variability in the performance of the natural system being modeled. *Epistemic* uncertainty is the uncertainty that arises because of lack of knowledge of the physical mechanism of the natural system being modeled. Typically, aleatory uncertainties are represented in probabilistic analysis in the form of probability distribution functions, whereas epistemic uncertainties are included in the form of (weighted) alternative logic tree branches. The technique for a one-sigma allowance is specified in order that numerical modeling provides a Hazard-Consistent Tsunami waveform that accounts for uncertainty in tsunami amplitude. It is a key component of Probabilistic Tsunami Hazard Analysis to include a reliability factor.

C6.7.5 Offshore Tsunami Amplitude The 328 ft (100 m) depth contour was chosen because tsunami waves behave linearly in deeper water, which allows for rapid calculation of tsunami waveforms, which enables a comprehensive probabilistic approach. The 328 ft (100 m) depth was selected to minimize the amount of nonlinear shoaling caused by such things as the Nearshore Profile, embayment amplification, and resonance effects. The Offshore Tsunami Amplitude at this reference depth represents with more clarity the probabilistic tsunami amplitude values for an area that do not vary dramatically parallel to the shoreline. To use a fully shoaled tsunami wave amplitude at the shoreline would result in more difficulty in verifying or calibrating whether other site-specific models have actually used a Hazard-Consistent Tsunami because the waveform has become more complex at the shoreline, and in certain cases it may have broken into a series of bores; for these reasons, the tsunami amplitudes have large spatial variation in the nearshore regime. It would also not be possible at the shoreline to specify regional waveform parameters, such as in Table 6.7-1 at the 328 ft (100 m) depth, that would be valid in all nearshore regimes. The Offshore Tsunami Amplitude and associated parameters of Section 6.7 are calibrated to the 2,475-year hazard level, and the primary purpose of this information is to support site-specific inundation analysis of Tsunami Risk Category IV structures, where more detailed spatial bathymetry and topography are used with a 2D inundation model software code to develop flow parameters at the site. It is called a 2D model because the third vertical dimension of flow speed is depth-averaged, but it uses a 3D spatial digital raster representation of the terrain.

C6.7.5.1 Offshore Tsunami Amplitude for Distant Seismic Sources An efficient way to compute probabilistic tsunami exceedance amplitudes is by precomputing tsunami time series (Green's functions) for a suite of subsources with unit slip (e.g., 3.28 ft or 1 m) that together can be used to represent actual slip distribution on a larger fault. The resultant tsunami time series is then a summation of the individual Green's functions, which are each multiplied by the actual slip relative to the unit slip for each subsource. Examples of this process are given in Thio et al. (2010) and Gica et al. (2008). The Site-Specific Probabilistic Tsunami Hazard Analysis is validated by comparison to the Offshore Tsunami Amplitudes given by the ASCE Tsunami Design Geodatabase. This validation includes individual matching to at least 80% of the reference amplitude values and matching of the mean of the computed offshore tsunami amplitudes to at least 100% of the mean of the reference amplitude points. The extent of offshore tsunami amplitudes to be computed and validated requires a sufficient number of points to represent the region most influential on the resulting inundation at the site.

Application of this requirement to islands where the projected width is less than 40 mi (64.4 km) would require matching offshore wave amplitudes simultaneously on more than one side of the island. This could result in unintended influence of an inappropriate tsunami source. The extent of offshore tsunami amplitudes is therefore shown in Section 6.7.5.1.4(d) to match the projected width of the island.

At certain locations it may be necessary to generate sources based on an extent of offshore tsunami amplitude points that is less than those required by Sections 6.7.5.1.4(a) and 6.7.5.1.4(b). Examples include coastal geometries where tsunami waves induce resonance, where tsunami waves wrap around to the lee of an island, or where wave reflection between islands, or between an island and adjacent coastline, affects the wave amplitudes. The islands of Hawaii exhibit such conditions, where a 10 mi (16.1 km) extent was determined to be appropriate. Where such circumstances are demonstrated, use of shorter extents of offshore tsunami amplitude points is permitted, subject to peer review by an appropriately qualified tsunami modeler.

C6.7.5.2 Direct Computation of Probabilistic Inundation and Runup For local sources, it is necessary to compute a suite of scenarios, which represent a probabilistic distribution, taking into account aleatory uncertainties in such things as slip distribution, magnitudes, and tide levels. Ground subsidence should be taken into account in the actual tsunami calculations. The surface deformation can be computed using a variety of algorithms that compute the elastic response of a medium caused by slip on a fault.

C6.7.5.3 Use of Higher-Order Tsunami Model Features Tsunami models may have the capability to use higher-order features that may affect the propagation of a tsunami. These features are considered to be secondary enhancements to the leading-order, shallow-water wave equations (e.g., Titov and Synolakis 1998) commonly used for tsunami modeling. Such higher-order features, as discussed in Baba et al. (2017), provide refinement in the tsunami arrival time and shape of the incoming waves, with the inclusion of earth elasticity, seawater density stratification, and dispersion yielding the most significant accuracy increase. There will generally be greater spatial variability in the amplitudes of the tsunami waves generated by such a higher-order model. For example, it has been shown that with the same tsunami source event, dispersion of waves can lower or raise the computed maximum offshore amplitudes around Hawaii by 50% (Bai and Cheung 2016, Kirby et al. 2013, Bai et al. 2018), with the pattern strongly dependent on the location and size of the earthquake source.

Given that the Offshore Tsunami Amplitudes in the ASCE Tsunami Design Geodatabase were generated using a model which solves the leading-order, shallow-water wave equations without such higher-order features, independent peer review is required to ensure that the tsunami inundation resulting from the use of higher-order model features is consistent with the probabilistic hazard level of this chapter. In addition, the peer reviewer should verify that the statistical aleatory uncertainty factor in the hazard level of the model results is consistent with the reliability target parameters associated with the Offshore Tsunami Amplitude hazard, as given in Chock et. al. (2016).

C6.7.6 Procedures for Determining Tsunami Inundation and Runup

C6.7.6.1 Representative Design Inundation Parameters These parameters have been shown to be relevant for the estimation of different types of tsunami impact.

C6.7.6.2 Seismic Subsidence before Tsunami Arrival The uplift and subsidence during an earthquake that generates a tsunami also causes vertical changes to the coastal topography, and this change needs to be taken into account for modeling local tsunamis directly from the source. The effect of seismic subsidence follows the recognized deformation model of Okada (1985) illustrated in Figure C6.7-2. The subsidence can be considered instantaneous, and it occurs before the arrival of the tsunami waves. Figure 6.7-3 is the 2,475-year subsidence map, which also identifies which areas are governed by the local coseismic tsunami source. For ASCE 7-22, models with improved fault geometry were used in the probabilistic analysis for a more accurate subsidence map for Northern California, Oregon, and Washington. Because the map shows probabilistic subsidence consistent with the Probabilistic Tsunami Hazard Analysis logic tree, values cannot be directly compared with the tectonic movements resulting from specific earthquake scenarios. Section 6.7.6.2 allows the use of the subsidence shown in Figure 6.7-3 as implemented in the ASCE Tsunami Design Geodatabase, or the subsidence computed from the fault rupture mechanism associated with a Hazard-Consistent Tsunami site-specific scenario. Inundation Depth Points in the Geodatabase for overwashed Tsunami Design Zones refer to the local flow depth taking into account the modeled subsidence. Southern California and Hawaii are not adjacent to interplate subduction zones, so they are not subject to regional seismic ground subsidence before a tsunami.

C6.7.6.3 Model Macroroughness Parameter The effects of ground friction on tsunami inundation are significant and need to be considered. There are different techniques in which ground friction is incorporated in numerical schemes; the most common is the Manning's coefficient. If other approaches are used, care should be taken so that they closely follow the effect related to the appropriate Manning's coefficient.

C6.7.6.4 Nonlinear Modeling of Inundation Typically in numerical tsunami modeling, the equations solved correspond to the shallow-water approximation, with nonlinear effects such as inundation (moving boundary), bottom friction, and advection. Usually, these algorithms solve the equations of motion in two (horizontal) dimensions, with the vertical properties averaged over depth. Most of the effects mentioned here are taken into account, provided that an acceptable tsunami inundation model is used. Dispersion is taken into account in some approaches (Boussinesq-type); dispersion is required for the modeling of small dimensional sources, such as submarine landslides. Bore formation is more accurately modeled with algorithms that preserve momentum. In the case of local tsunamis, where the bore formation is particularly pronounced, these algorithms are preferred over approaches that do not explicitly conserve mass.

C6.7.6.5 Model Spatial Resolution Tsunami Digital Elevation Models for the United States are available from NOAA at http://www.ngdc.noaa.gov/mgg/inundation/. For select areas, models with resolutions of up to 32.8 ft (10 m), or 1/3 arcsec, are available as of 2015.

C6.7.6.6 Built Environment The standard practice of all tsunami inundation models is to use a low Manning's roughness parameter for the topography so that the lines for the inundation limit and runup elevation are not underestimated. The models have also been numerically benchmarked with that assumption against analytical and laboratory cases with smooth profiles. Digital Elevation Models (DEMs) for very high-resolution modeling may be available in the form of lidar (light detection and ranging) models from federal, state, county, or city agencies, or from local site surveys. In some

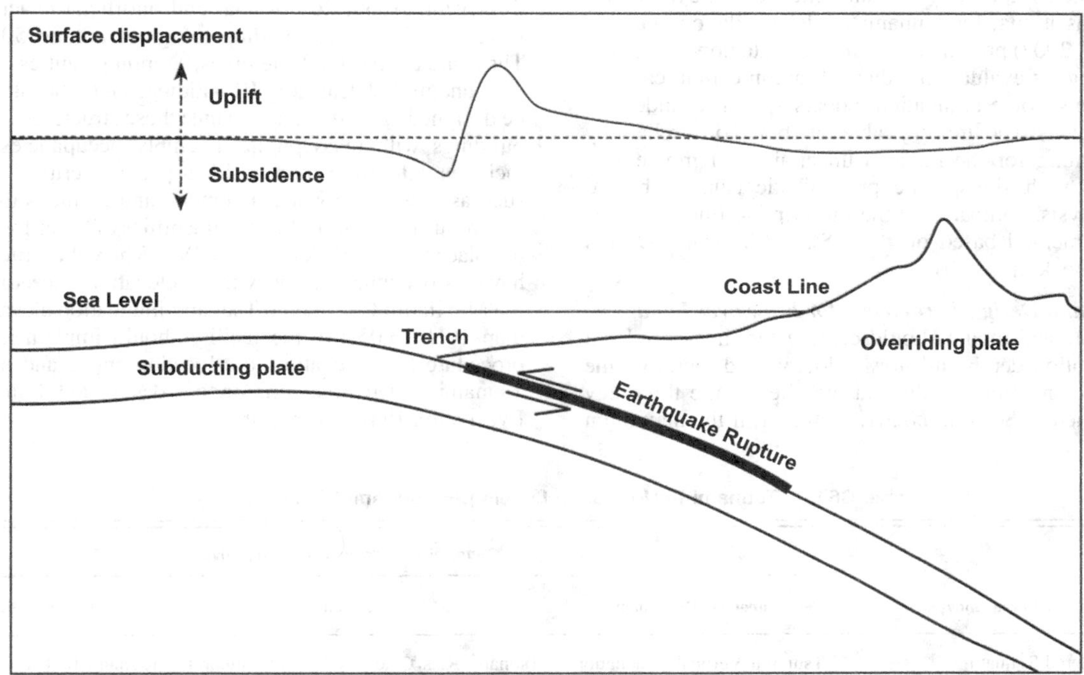

Figure C6.7-2. Subduction zone earthquake mechanism for tsunami generation.

cases, the built environment has been removed from the lidar images, which negates their usefulness for detailed inundation modeling. When the built environment is modeled as fixed bluff bodies on the terrain, it is appropriate to include only those with sufficient structural resilience to resist tsunami forces.

C6.7.6.7 Inundation Model Validation The National Tsunami Hazard Mitigation Program (NTHMP) (2012) has developed a number of benchmarking exercises that include analytical, laboratory, and real-world simulations, available at the Forecast Propagation Database (http://nctr.pmel.noaa.gov/propagation-database.html). Aida (1978) provides further information relating to the degree-of-fit criteria of K between 0.8 and 1.2 and κ less than 1.4.

C6.7.6.8 Determining Site-Specific Inundation Flow Parameters The Hazard-Consistent Tsunami is a term of craft used to refer to a means to replicate the effects of the Maximum Considered Tsunami (MCT) hazard level by one or more surrogate scenarios. Hazard-Consistent Tsunamis are devised to incorporate the net effect of an explicit analysis of uncertainty in the parameters originally used in determining the MCT through conducting the Probabilistic Tsunami Hazard Analysis. To incorporate this aleatory uncertainty in a few surrogate Hazard-Consistent Tsunamis, the slip parameters of the generating rupture of the Hazard-Consistent Tsunami are increased. This is a procedural device to produce a convenient representation of input for a more limited number of inundation analyses. As a result, the Hazard-Consistent Tsunami appears greater than a single deterministic event generated by an earthquake of that mean recurrence interval at the same source.

In urban environments, the flow velocities determined by a site-specific inundation analysis for a given structure location may not be taken as less than 90% of those determined in accordance with the Energy Grade Line Analysis method. For other terrain roughness conditions, the site-specific inundation analysis flow velocities at a given structure location shall not be taken as less than 75% of those determined in accordance with the Energy Grade Line Analysis method. These restrictions apply before any velocity adjustments caused by flow amplification. The reason for these restrictions is that the tsunami model validation standard (Synolakis et al. 2007) presently does not validate flow velocities, and the committee's evaluation indicated that in certain cases of field observations, some inundation models appear to underestimate flow velocities. The limiting values are based on a reliability analysis accounting for the reduced uncertainty of momentum flux associated with site-specific probabilistic tsunami hazard inundation analysis, compared to the prescriptive Energy Grade Line Analysis method based on the ASCE 7 Tsunami Design Zone Maps (Chock et al. 2016).

C6.7.6.9 Tsunami Design Parameters for Flow over Land The inundation numerical model should capture the time-correlated series of inundation depth and flow velocity (and therefore the representation of momentum flux) at the site for the three key load cases defined in Section 6.8.3.1. However, if the maximum momentum flux is found to occur at an inundation depth different from Load Case 2, that point in the time series should also be considered.

C6.8 STRUCTURAL DESIGN PROCEDURES FOR TSUNAMI EFFECTS

It is important to understand that building failure modes differ fundamentally between seismic loads (high-frequency dynamic effects generated on the inertial masses of a structure) and tsunami loads (externally and internally applied sustained fluid forces varying with inundation depth over long-period cycles of load reversal). Tsunami-induced failure modes of buildings have been examined in several detailed analyses of case studies taken from the Tohoku tsunami of March 11, 2011 (Chock et al. 2013a). Building components are subject simultaneously to internal forces generated by the external loading on the lateral-force-resisting system, together with high-intensity momentum pressure forces exerted on individual members. Performance objectives for structures subject to tsunami design requirements are given in Table C6.8-1.

C6.8.1 Performance of Tsunami Risk Category II and III Buildings and Other Structures Importantly, when evacuation from the inundation zone before tsunami arrival is not possible because of long evacuation distances, road congestion, or damaged infrastructure, the public will attempt to use taller buildings to escape the tsunami inundation and will inherently expect that such taller structures will not collapse during the tsunami. As just one example from the city of Ishinomaki, per a detailed investigation of the lessons for international preparedness after the 2011 Great East Japan Tsunami (Fraser et al. 2012), about 500 people sought refuge at three designated buildings in Ishinomaki. In addition to the few Tsunami Vertical Evacuation Refuge Structures available, "there was widespread use of buildings for informal (unplanned) vertical evacuation in Ishinomaki on March 11th, 2011. In addition to these three designated buildings, almost any building that is higher than a 2-story residential structure was used for vertical evacuation in this event. About 260 official and unofficial evacuation places were used in total, providing refuge to around 50,000 people. These included schools, temples, shopping centres and housing."

Tsunami Risk Category III buildings and other structures are to be designed against collapse, since these structures include school buildings with mass public assembly occupancies, health-care facilities with 50 or more resident patients, critical infrastructure such as power and water treatment, and facilities that may store hazardous materials (when the quantities of hazardous material do not place them in Risk Category IV). Where the structure does not have an occupiable floor with an elevation exceeding 1.3 times the Maximum Considered Tsunami inundation elevation by more than 10 ft (3.05 m), the facility should implement a plan and procedure for evacuation to a location above and outside of the Tsunami Evacuation Zone or to a designated Tsunami Vertical Evacuation Refuge Structure.

Table C6.8-1. Tsunami Performance Levels per Tsunami Risk Category.

Hazard Level and Tsunami Frequency	Tsunami Performance Level Objective		
	Immediate Occupancy	Damage Control	Collapse Prevention
Maximum Considered Tsunami (2,475-year mean recurrence interval)	Tsunami Vertical Evacuation Refuge Structures	Tsunami Risk Category IV and Tsunami Risk Category III Critical Facilities	Tsunami Risk Category III and Risk Category II

C6.8.2 Performance of Tsunami Risk Category III Critical Facilities and Tsunami Risk Category IV Buildings and Other Structures The state, local, or tribal government may specifically designate certain traditionally Risk Category III structures as more important Critical Facilities that have post-tsunami mission-critical functions or have more critical roles in economic recovery. Critical Facilities and Risk Category IV Essential Facilities should be located outside of the Tsunami Design Zone whenever possible. For those structures that necessarily exist to provide critical and essential services to a community within a coastal zone subject to tsunami hazard, the design provisions target better than a Life Safety performance level (that is, what is called the Damage Control performance level) for the floor levels that are not inundated. Vertical Evacuation Refuge Structures have the highest performance level objective. The Damage Control performance level is an intermediate structural performance level between Life Safety and Immediate Occupancy.

C6.8.3 Structural Performance Evaluation

C6.8.3.1 Load Cases The normalized inundation depth and depth-averaged velocity time history curves of Figure 6.8-1 (and corresponding Table C6.8-2) are based on tsunami video analysis, and they are generally consistent with numerical modeling with respect to the load cases defining critical stages of structural loading for design purposes (Ngo and Robertson 2012). Load Case 1 is for calculating the maximum buoyant force on the structure along with the associated hydrodynamic lateral force, the primary purpose of which is to check the overall stability of the structure and its foundation anchorage against net uplift. The uplift force, calculated in accordance with Section 6.9.1, depends on the differential inundation depth exterior to the structure versus the flooded depth within it. Load Case 2 is for calculating the maximum hydrodynamic forces on the structure. Load Case 3 is for calculating the hydrodynamic forces associated with the maximum inundation depth. The time history curves can be used to determine the inundation depth and velocities at other stages of inundation as a function of the maximum values determined by the Energy Grade Line Analysis. When a site-specific inundation analysis procedure is used, the local Authority Having Jurisdiction may approve a site-specific inundation and velocity time history curve, subject to the minimum values of 80% or 100% of the Energy Grade Line Analysis, as indicated.

Table C6.8-2. Values of Normalized Inundation and Flow Speed of Figure 6.8-1.

t/T_{tsu}	h/h_{max}	u/u_{max}
0.000	0.000	0.000
0.033	0.125	0.517
0.067	0.250	0.726
0.111	0.417	0.881
0.133	0.500	0.943
0.178	0.670	1.000
0.267	0.833	0.764
0.356	0.933	0.550
0.444	0.983	0.333
0.500	1.000	0.000
0.556	0.983	−0.333
0.644	0.933	−0.550
0.733	0.833	−0.764
0.822	0.670	−1.000
0.867	0.500	−0.943
0.889	0.417	−0.881
0.933	0.250	−0.726
0.967	0.125	−0.517
1.000	0.000	0.000

C6.8.3.2 Tsunami Importance Factors The values of Tsunami Importance Factors are derived from the target structural reliabilities, which were calculated using Monte Carlo simulation involving a million trial combinations of random variables independently occurring in proportion to their statistical distributions for the demand parameters of fluid density, closure ratio, Energy Grade Line Analysis momentum flux, inundation depth hazard, and the aleatory uncertainty of inundation depth. For capacity, the structural component analyzed is a beam-column member carrying gravity loads. The Importance Factors for each risk category, analyzed in combination with the other parameters as discussed, result in structural component reliabilities, given that the MCT has occurred, that are similar to seismic systemic performance given that the maximum considered earthquake (MCE) has occurred:

Tsunami Risk Category II: A tsunami Importance Factor of 1.0 results in structural component limit state exceedance probability of 7.5% $_{(MCT)}$ versus 10% $_{(MCE)}$ for collapse of the lateral-force-resisting system during an earthquake.

Tsunami Risk Category III: A tsunami Importance Factor of 1.25 results in structural component limit state exceedance probability of 4.9% $_{(MCT)}$ versus 5% $_{(MCE)}$ for collapse of the lateral-force-resisting system during an earthquake.

Tsunami Risk Category IV: A tsunami Importance Factor of 1.25 results in structural component limit state exceedance probability of 2.7% $_{(MCT)}$ versus 2.5% $_{(MCE)}$ for collapse of the lateral-force-resisting system during an earthquake.

The exceedance probability for the limit state of structural components of a Tsunami Vertical Evacuation Refuge Structure given the occurrence of the MCT would be approximately 0.8%.

Section 6.5.2, Tsunami Risk Category IV Buildings and Other Structures, requires that a site-specific Probabilistic Tsunami Hazard Analysis per Section 6.7 be performed, but this requirement has an exception where the inundation depth determined by the Energy Grade Line Analysis (EGLA) is less than 12 ft (3.66 m). The Importance Factors of Section 6.8.3.2 were determined in an analysis described in Chock et al. (2016). Site-specific inundation analysis reduces the uncertainty of the flow depth and velocities. Under the exception where site-specific inundation analysis is not performed, the Importance Factor needs to be greater, since the less accurate parametric EGLA would be used for design. Accordingly, if the engineer does not use a site-specific inundation model analysis for Category IV structures where the inundation depth is less than 12 ft (3.66 m), an Importance Factor of 1.5 is used to achieve equivalent reliability for this exception.

C6.8.3.3 Load Combinations The structural load combinations given are consistent with the Extraordinary Load Combinations of Section 2.5 of this standard by applying F_{TSU} for incoming and receding directions for A_k in the load combinations given in Equation (2.5-1), as modified to include the effect of tsunami-induced lateral earth pressure caused by water seepage through the soil, H_{TSU}, resulting in Equation (6.8-1). The inundation depths and velocities for determining F_{TSU} and H_{TSU} must be consistent at the hazard level of the Maximum Considered

Tsunami (MCT). The load factor for H_{TSU} is given as 1.0 because these load combinations occur during the defined submerged conditions of the Maximum Considered Tsunami. The Probabilistic Tsunami Hazard Analysis criteria for determining runup already include explicit mathematical consideration of uncertainty; therefore, no additional factor needs to be applied to the 2,475-year MCT flow characteristics for Risk Category II buildings and other structures.

No allowable stress design load combinations were deemed necessary. When foundation stability analysis is performed for tsunami-induced soil seepage pressures, the results are equivalent to the existing recognized US Army Corps of Engineers (USACE 2005, 2011) specifications for geotechnical limit equilibrium analysis for both overturning and uplift. For foundation design, resistance factors are given in Section 6.12.1.

C6.8.3.4 Lateral-Force-Resisting System Acceptance Criteria The use of the overstrength factor to evaluate the overall lateral-force-resisting system is permitted to be used where the Life Safety performance level is to be verified. If Life Safety criteria have been met, then the Collapse Prevention criteria have also been met as a consequence, which is the minimum requirement for Tsunami Risk Category II and Tsunami Risk Category III buildings and other structures. For Immediate Occupancy, the structural system should be explicitly analyzed.

The capacity of minimum-code seismically designed Risk Category II structures using this technique was demonstrated in the Earthquake Engineering Research Institute (EERI) special issue of March 2013, in a paper providing analysis of prototypical buildings for tsunami and seismic requirements (Chock et al. 2013a). In this paper, several prototypical buildings were investigated, with the following assumptions.

Parameters for a comparison of overall systemic resistance to tsunami loading are as follows: (1) the prototypical Risk Category II buildings of increasing height selected for illustrative purposes were 120 ft (36.5 m) long and 90 ft (27.5 m) wide and are 25% open; (2) they were located in high seismic zones (in the United States, $S_s = 1.5$ and $S_1 = 0.6$); (3) connections developed the inelastic capacities of the members; (4) tsunami flow velocity was 26.2 ft/s (8 m/s); (5) each tsunami inundation load curve represented the continuum of hydrodynamic loading cases inundation increases to the maximum depth; and (6) seismic capacity is used based on the overstrength capacity. It was found by analysis that the structural systems of larger scaled and taller buildings are inherently less susceptible to tsunamis than this prototype, provided that there is adequate foundation anchorage for resistance to scour and uplift.

C6.8.3.5 Structural Component Acceptance Criteria The design of structural components shall comply with Section 6.8.3.5.1 or in accordance with the alternative performance-based criteria of Section 6.8.3.5.2 or 6.8.3.5.3, as applicable. The alternative Sections 6.8.3.5.2 and 6.8.3.5.3 are not prescriptive methods of analysis.

C6.8.3.5.1 Acceptability Criteria by Component Design Strength The primary means of determining structural component acceptability is based on a linear elastic analysis and the evaluation of resultant actions of the load combinations compared with the design strength of the structural components and connections.

C6.8.3.5.2 Alternative Performance-Based Criteria The alternative structural component acceptability criteria use acceptance criteria of ASCE 41-17 (ASCE 2017b), *Seismic Evaluation and Retrofit of Existing Buildings*. With an adaptation of this method, strength and stability can be checked to determine that the design of the structural components is capable of withstanding the tsunami to achieve the Structural Performance level required. The tsunami adaption allows the techniques of linear static and nonlinear static analysis. For the purposes of this chapter, it may be appropriate to generically adapt modeling parameters nonspecific to earthquake ground motion from the nonlinear static procedure or simplified nonlinear static procedure of ASCE 41-17 (ASCE 2017b). For example, it should be acceptable to use the effective stiffness values of ASCE 41-17 (ASCE 2017b) for this purpose.

For the nonlinear static analysis procedure, the mathematical model of the structure is subjected to monotonically increasing loads until the required tsunami forces and applied actions are reached. The lateral-force-resisting system can be evaluated for the overall hydrodynamic drag force prescribed in Section 6.10.2.1 by simultaneously increasing both flow depth and flow velocity monotonically, in accordance with Figure 6.8-1, up to Load Case 2. When this procedure is used with a site-specific tsunami inundation analysis, the maximum load case is determined from the inundation depth and velocity time history results. These loads can be applied as concentrated loads at each floor level corresponding to the tributary height for that floor level (Figure C6.8-1a), or as distributed loads along the vertical load-bearing structural components (Figure C6.8-1b).

Individual structural components that are part of the lateral-force-resisting system must resist their portion of the overall drag force on the structure concurrently with the drag force on the individual component as specified in Sections 6.10.2.2 through 6.10.2.5. The drag force on the individual component is distributed uniformly along the length of the component and increased monotonically by simultaneously increasing both flow depth and flow velocity in accordance with Figure 6.8-1. The balance of the overall drag force on the structure is applied proportionally to the remainder of the lateral-force-resisting system. These loads can be applied as concentrated loads at each floor level corresponding to the tributary height for that floor level (Figure C6.8-2a), or as distributed loads along the vertical load-bearing structural components (Figure C6.8-2b). Application of this nonlinear static analysis procedure to a prototypical building and its structural components is demonstrated by Baiguera et al. (2020). The overall tsunami lateral drifts would generally induce some flexural and shear demands on individual components, even if they are not part of the lateral-force-resisting system.

Structural components can then be checked using the ASCE 41-17 acceptability criteria for actual tsunami loads and depths that are correct from the standpoint of fluid mechanics without load factors. Results of the analysis procedure shall not exceed the numerical acceptance criteria for linear procedures of ASCE 41-17 (Chapters 9 through 11) for the structural performance criteria required for the building or structure's tsunami risk category. For nonlinear static analysis procedures, expected deformation capacities shall be greater than or equal to the maximum deformation demands calculated at the required tsunami forces and applied actions.

C6.8.3.5.3 Alternative Acceptability by Progressive Collapse Avoidance. Where tsunami loads or effects exceed acceptability criteria for a structural element, a recognized method of checking the residual structural system gravity-load-carrying capacity is the alternate load path procedure given in *Design of Structures to Resist Progressive Collapse* (DOD 2013). An ASCE standard for progressive collapse avoidance is not yet available.

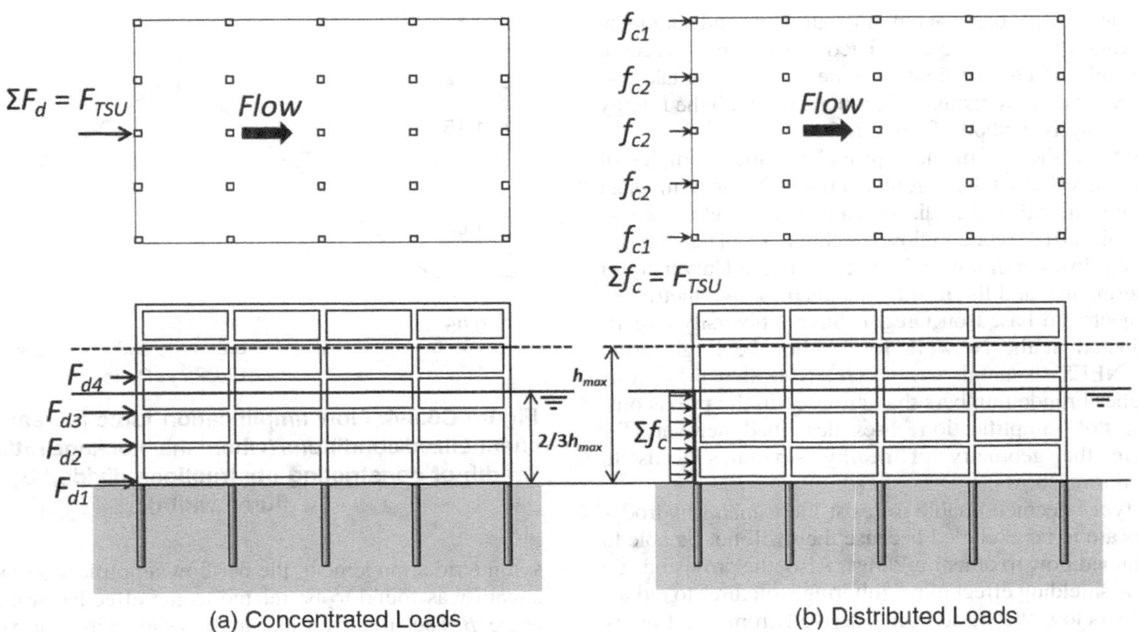

Figure C6.8-1. Load distributions for assessment of a lateral-force-resisting system.

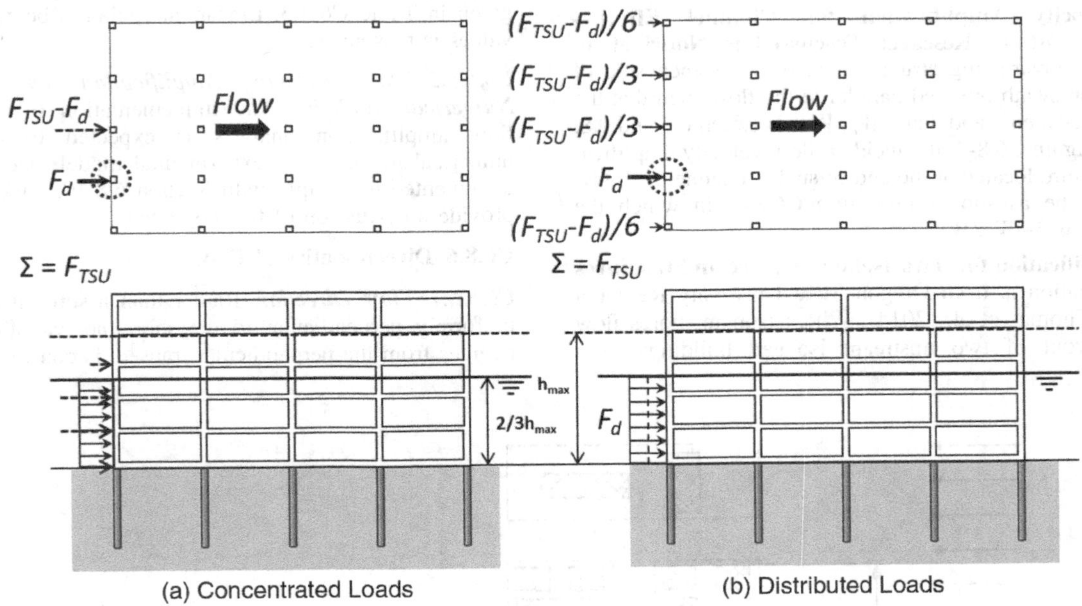

Figure C6.8-2. Load distributions for assessment of an individual column (circled).

C6.8.4 Minimum Fluid Density for Tsunami Loads The tsunami inundation conditions relevant to structural design are by no means represented by pristine seawater. Overland flow is more similar to a debris flood, with suspended soil and debris objects of various sizes and materials. The fluid density factor, k_s, is used to account for such soil and debris within equivalent fluid forces that are not accounted for in Section 6.11, which deals with debris impacts by larger objects. The value of $k_s = 1.1$ was selected to represent an equivalent sediment concentration of 7%, assuming a specific gravity of 2.5 for the suspended soil particles.

C6.8.5 Flow Velocity Amplification Tsunami Design Zone Maps are based on so-called bare-earth conditions (i.e., without discrete objects) with an equivalent Manning's roughness to represent what exists above ground. When the Energy Grade Line Analysis is used with the Manning's roughness based on built environments, more energy is required to reach the specified mapped runup. Urban environments are expected to have a higher probability of generating flow amplification effects through numerous diffraction scenarios that are not possible to specifically enumerate. Nevertheless, the Energy Grade Line Analysis implemented with an urban area's roughness produces sufficiently conservative depth and momentum flux, and additional flow amplification need not be considered. Therefore, to account for these possible effects, site flow velocities shall not be taken as less than 100% of those determined in accordance with the Energy Grade Line Analysis method of Section 6.6.

This section of the commentary provides some examples of flow diffraction effects taken from available North American research. Experimental studies discussed in Thomas et al. (2014) and Nouri et al. (2010) relate to flow amplification. The example of channelized flow conditions is inferred from the University of Ottawa experiments, and the results indicated for symmetrically arranged objects are based on Oregon State University experiments conducted in the Network for Earthquake Engineering Simulation (NEES) tsunami wave basin. Note that these tests were for higher Froude numbers than are used in the provisions. The explicit flow amplification cases described here are for places where the geometry of nearby structures leads to higher-than-normal flows.

Building types deemed unable to resist the tsunami hydrodynamic forces are to be excluded because they will not be able to redirect sustained flow to other buildings. Also, the provisions do not permit the shielding effect of neighboring structures to reduce flow velocity below the basic value derived from the Energy Grade Line Analysis method. The engineer should exercise judgment and conduct additional studies for other conditions where flow amplification would be expected based on the principles of fluid mechanics.

Flow Velocity Amplification for Channel Effect—University of Ottawa Research Discussed in Nouri et al. (2010). Where obstructing structures exist along each side of a street or open swath oriented parallel to the flow such that the ratio L/W is between 1 and 3 and W_c/W is between 0.6 and 0.8, as shown in Figure C6.8-3, the incident flow velocity amplification on a structure located at the end of such a channelized flow was found to be as shown in Figure C6.8-4, in which the blockage ratio is $1-W_c/W$.

Flow Amplification for Two Isolated Upstream Structures This information is from Oregon State University Research discussed in Thomas et al. (2014). Where tsunami bores flow through a layout of two upstream isolated buildings in a

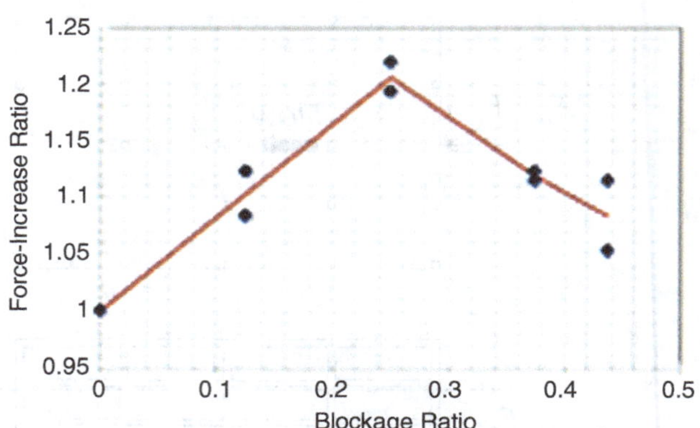

Figure C6.8-4. Flow amplification force-increase ratio for channelized conditions (where the blockage ratio is the net width of constricting obstructions divided by the total flume width).

symmetric arrangement, the net flow amplification factor downstream was found to be related to the effective wake clearance angle, β. The effective wake angle is shown in Figure C6.8-5, and a summary of results is shown in Figure C6.8-6 and Table C6.8-3. This figure also illustrates what is the center third of the width of the downstream structure.

The approximate envelope of the flow amplification factors is given in Table C6.8-3. Linear interpolation between the listed values is reasonable.

C6.8.5.2 Flow Velocity Amplification by Physical or Numerical Modeling The implementation of an alternative flow amplification analysis is expected to consist of a numerical model or an experimental model that is sufficiently documented to comply with Section 6.8.10. Park et al. (2013) provide a discussion of this approach.

C6.8.6 Directionality of Flow

C6.8.6.1 Flow Direction Rather than assume that the flow is uniformly perpendicular to the shoreline, variation by ±22.5 degrees from the perpendicular transect is considered, in which

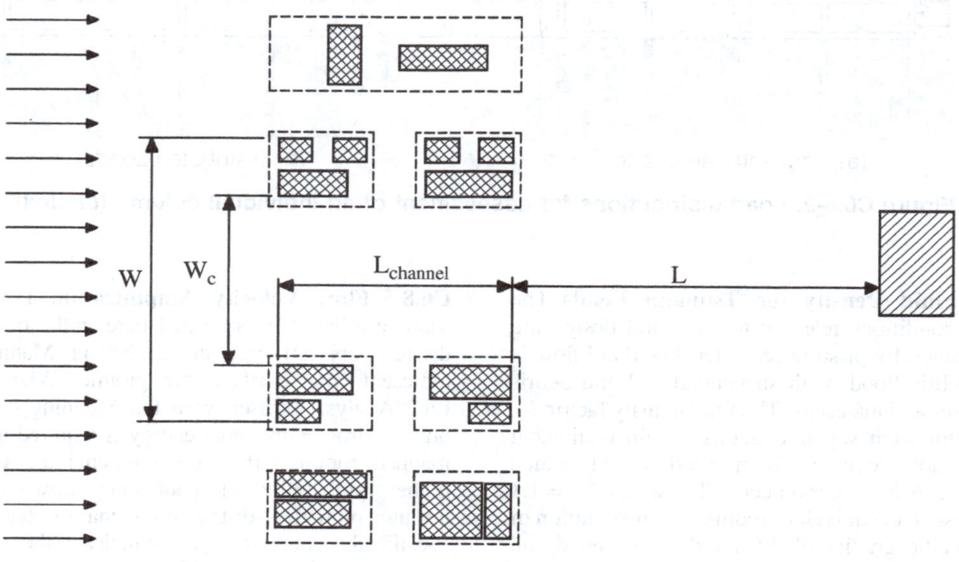

Figure C6.8-3. Flow amplification for channelized conditions.

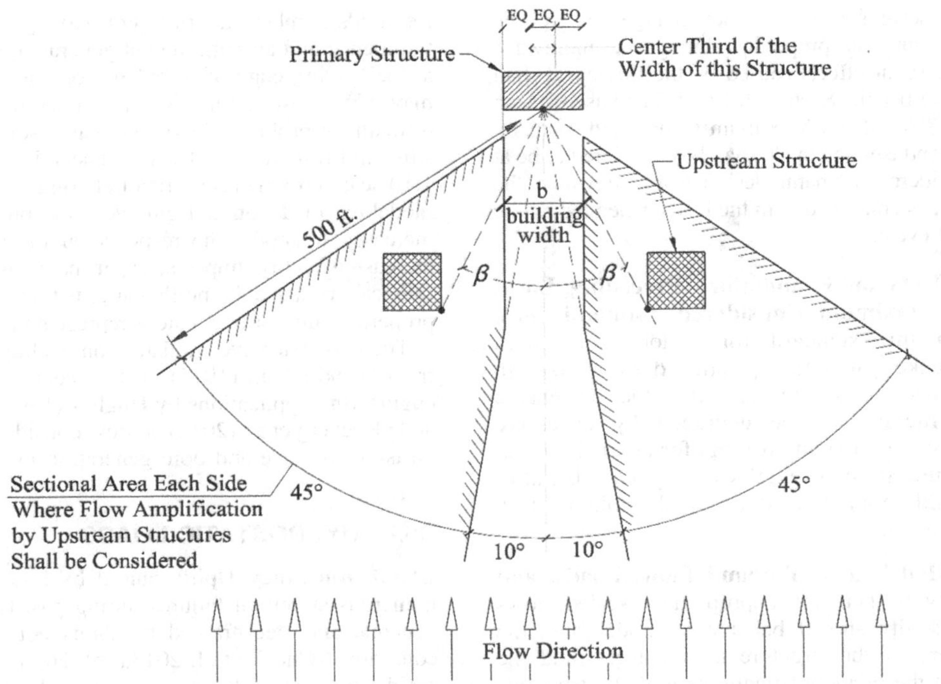

Figure C6.8-5. Effective wake clearance angle, β, for flow amplification.

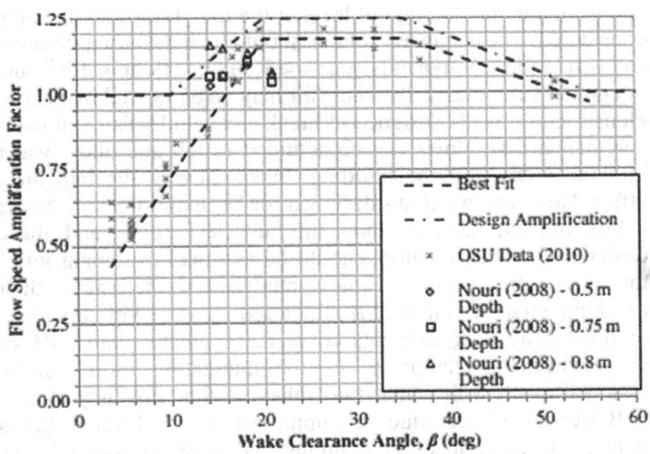

Figure C6.8-6. Amplification of flow speed vs. wake clearance angle.

Table C6.8-3. Flow Speed Amplification Factors.

Effective Wake Clearance Angle (β) in degrees	0	10	20	35	≥55
Flow speed amplification factor from two symmetrically arranged buildings located upstream	1.0	1.0	1.25	1.25	1.0

the center of rotation of the possible transects is located at the site. The first objective is to find the highest runup elevation associated with the possible design transects. The second objective is to account for variation in the load application to the structure. The governing transect that produces the maximum runup elevation may be used to perform an Energy Grade Line Analysis, and the prescribed directional variation for the resulting flow parameters may be used to compute directional loads. Alternatively, an Energy Grade Line Analysis may be computed for each possible direction.

The determination of outflow direction may also require consideration that the outflow current may be affected by existing stream and riverbeds and drainage canals, as well as additional scour and sediment transport, which may cause significant morphological changes in topography during a vigorous drawdown cycle after a large watershed is inundated.

C6.8.6.2 Site-Specific Directionality Although a site-specific inundation analysis is permitted to establish flow vectoring at the site, some uncertainty in the accuracy of this estimation is reflected in the ±10-degree variability.

C6.8.7 Minimum Closure Ratio for Load Determination Loads on buildings shall be calculated assuming a minimum closure ratio of 70% of the pressure-exposed surface area of the exterior enclosure. This assumption accounts for accumulated waterborne debris trapped against the side of the structure, as well as any internal blockage caused by building contents that cannot easily flow out of the structure. As a practical matter based on observations of buildings subjected to destructive tsunami, "breakaway" walls cannot be totally relied on to relieve structural loading, primarily because of the copious amount of external debris. Also, studs and girts may be capable of entrapping contents within a building, thus generating hydrodynamic drag forces on the internal debris, which in turn transfers those loads to the structure (Chock et al. 2013a, b).

C6.8.8 Minimum Number of Tsunami Flow Cycles Designers working on loading shall consider a minimum of two tsunami inflow-and-outflow cycles, the second of which shall be at the Maximum Considered Tsunami design level. This consideration is required because the condition of the building and its foundation may change in each tsunami inflow and outflow cycle. Therefore, building foundation designs shall consider changes in the site surface and the in situ soil properties during the multiwave tsunami event. Local scour effects caused by the first cycle shall be calculated as described in Section 6.12.2.5 but based on an inundation depth at 80% of the Maximum Considered Tsunami design level. The second tsunami cycle shall be considered to be at the Maximum Considered Tsunami design level, in which the scour of the first cycle is combined with the loads generated by the inflow of the second cycle.

C6.8.9 Seismic Effects on Foundations Preceding Local Subduction Zone Maximum Considered Tsunami Since the assumption in this standard for a local maximum considered earthquake includes plastic deformation in structural performance, it is important that the postelastic strength of the structure not be degraded by excessive ductility demand. Tsunami loading (except for debris impacts) is of a sustained nature, so the strength capacity available after the earthquake should be maintained at a predictable level by limited ductility demand.

C6.8.10 Physical Modeling of Tsunami Flow, Loads, and Effects The capacity to generate appropriately scaled flows means that a test facility should be able to model both the specific site geometry of the structure or structures, and the form and duration of the incident tsunami or multiple tsunamis with an accuracy and duration appropriate for the process being investigated. Testing of a building or multiple buildings, for example, may need to account for the relevant topography and bathymetry, as well as the scaled roughness of terrain. Testing for site-specific conditions may require separate simulations of inflow and drawdown outflow of a tsunami waveform or use of a bore generator or recirculating flume to ensure adequate duration of loading.

It is appropriate to use physical modeling together with numerical modeling to evaluate site-specific conditions. This combination includes tasks such as the calibration of numerical models with data obtained from physical models and the testing of loads on individual components in a site-specific numerical model. The spatial and temporal limitations of physical model testing may necessitate combination with numerical modeling for a site-specific analysis. For example, the required variation in wave directionality (Section 6.8.6) may not be achievable for the structure or structures under consideration, so results of physical flow tests at a single incident angle can be used to calibrate a numerical model that addresses the full variability in flow direction.

Physical models can be used to simulate the inundation depths and velocities in tsunami flows for use with the design procedures and equations in Sections 6.9, 6.10, 6.11, and 6.12. Physical modeling may also be indicated for cases where flows do not follow the assumptions used to develop the equations in Sections 6.8, 6.9, 6.10, 6.11, and 6.12 or cannot economically be addressed using a numerical model alone. This situation can be the case where the flow varies rapidly temporally or spatially and where there are dynamic pressure variations, rotational flow, or multiphase flows.

Dynamic and kinematic similarity of model and prototype should match as closely as possible. For tsunamis, the most important scaling parameter is usually taken to be the Froude number, but many other dimensionless parameters (such as the Reynolds, Euler, and Cauchy numbers) should also be constrained to obtain useful results. For example, the Reynolds number should be maintained sufficiently high to ensure fully developed turbulent flow, so that the model is insensitive to Reynolds number variation. For wave generation, the effects of turbulence and air entrainment generated in breaking waves and at the leading edge of turbulent bores are difficult to scale and may affect test results. Scaling concerns are also particularly difficult for problems involving water–soil–structure interaction, structural response, or debris impacts. For example, the limiting 1:10 scale provided for structural components is appropriate for modeling loads on a rigid element, but it may need to be increased to model the response of the structural element. In the case of debris impacts, experiments are often carried out at full scale because the nonlinear structural behavior and material properties are not adequately represented at smaller scales.

There is extensive guidance on scaling of physical models, from Buckingham (1914) up to recent commentary on coastal engineering applications by Hughes (1993). Briggs et al. (2010) and Goseberg et al. (2013) address considerations on scaling and on tsunami wave and bore generation in the laboratory.

C6.9 HYDROSTATIC LOADS

C6.9.1 Buoyancy Uplift caused by buoyancy has resulted in numerous structural failures during past tsunamis, including in concrete and steel-framed buildings not designed for tsunami conditions (Chock et al. 2013a, b). High water table levels and rapid saturation of the ground surface during tsunami inundation may enable pressure to develop below the grade level of the building. The resulting uplift force is proportional to the volume of water displaced by the structural components and any enclosed spaces below the inundation level at the time buoyancy is being considered. The displaced volume should include, as a minimum, any structural components, enclosed spaces, floor soffits, and integrated structural slabs where air may be entrapped by beams. Nonstructural walls designed to break away under tsunami loads and standard windows can be assumed to fail and allow water into the interior of the building, thus relieving the buoyancy effect. However, windows designed for large missile wind-borne debris impact, such as those in hurricane zones and those designed for blast loading, should be assumed to remain intact throughout the tsunami (unless analyzed to determine their breaking strength threshold). Enclosed spaces with openings or breakaway wall elements equal to or greater than 25% of the enclosure envelope below the inundation level can be assumed to fill with water, thus relieving the buoyancy effect.

Buoyancy effects due to submergence of elevated floors reduce the axial load on columns and walls at lower levels (Figure C6.9-1). This reduced axial load should be included in the load combinations used to evaluate the design strength of these components because of the adverse effect it may have on the component bending and shear capacities, particularly for reinforced concrete members (Del Zoppo et al. 2020).

That full hydrostatic pressure develops below the grade level of the building is a conservative assumption based on permeable foundation soils, such as silts, sands, and gravels, but may be too conservative for cohesive soils such as clay and clayey silt. Soil permeability should be evaluated in the context of the duration and pressure head of the tsunami inundation depth at the site. However, if the first-floor slab on grade has the typical isolation joints around the columns, uplift on the slab will lift the slab and yield it but will not lift the superstructure. Cohesive soils adjacent to foundation components and basement walls may provide resistance to the resulting buoyancy.

Load Case 1, defined in Section 6.8.3.1, requires that a minimum uplift condition be evaluated at an inundation depth of one story or the height of the top of the first-story windows.

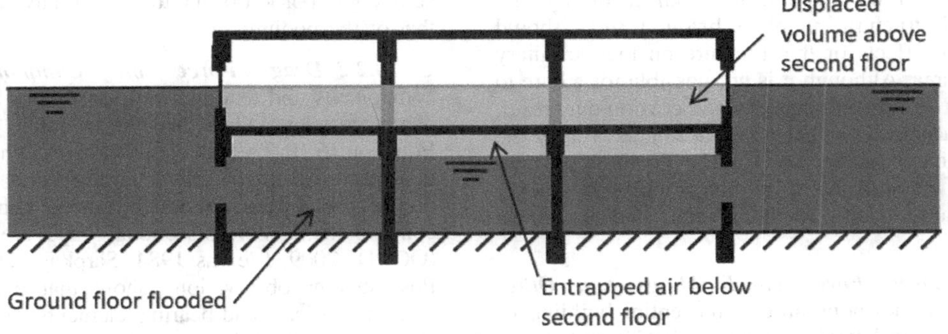

Figure C6.9-1. Buoyancy effects due to submergence of elevated floors.

The exceptions to Load Case 1 of Section 6.8.3.1 also apply. In summary, net buoyancy may be avoided by preventing the buildup of hydrostatic pressure beneath structural slabs, allowing the interior space to become flooded or designing for pressure relief or structural yielding relief of the hydrostatic head pressure, sufficient deadweight, anchorage, or a combination of the above design considerations.

C6.9.2 Unbalanced Lateral Hydrostatic Force Hydrostatic unbalanced lateral force develops on a wall element because of differences in water level on either side of the wall, irrespective of the wall orientation to the tsunami flow direction. Narrow walls or those with openings equal to or exceeding 10% of the wall area are assumed to allow water levels to equalize on opposite sides of the wall. However, for wide walls, or when perpendicular walls on either the front or back of the wall under consideration prevent water from getting to the other side of the wall, unbalanced hydrostatic loads should be considered. This condition need only be considered during inflow Load Cases 1 and 2. Equation (6.9-3) also includes a 1.2 factor to account for lack of aeration underneath the overtopping flow that induces a negative pressure, per Patil et al. (2018).

C6.9.3 Residual Water Surcharge Load on Floors and Walls During tsunami drawdown, water may not drain from elevated floor slabs that have perimeter structural components such as an upturn beam, perimeter masonry, or concrete wall or parapet. This lack of drainage results in surcharge loads on the floor slab that may exceed the slab capacity. The potential depth of water retained on the slab depends on the maximum inundation depth during the tsunami but would be limited to the height of any continuous perimeter structural components that have the capacity to survive the tsunami loads and that would retain water on an inundated floor. Nonstructural elements above this perimeter structural element are assumed to have failed during tsunami inflow so that they will not contribute to the retention of water on the slab during drawdown.

C6.9.4 Hydrostatic Surcharge Pressure on Foundation During tsunami inundation and drawdown, it is possible for different water levels to exist on opposite sides of a wall, building, or other structure under consideration. The resulting differential in hydrostatic surcharge pressure on the foundations should therefore be considered in the foundation design.

C6.10 HYDRODYNAMIC LOADS

Hydrodynamic loads develop when fluid flows around objects in the flow path. Tsunami inundation may take the form of a rapidly rising tide or surge, or a broken bore. Both of these conditions are considered here. Because tsunami waves typically break offshore, no consideration is given to the wave breaking loads typically associated with wind-driven storm waves (FEMA 2011).

C6.10.1 Simplified Equivalent Uniform Lateral Static Pressure It is anticipated that most buildings and other structures subject to the provisions of this chapter will be designed for other lateral load conditions, such as wind and seismic loads. For large or tall buildings, these other load conditions can result in greater loads on the lateral-force-resisting system than the tsunami loads, particularly in high seismic hazard regions. In such cases, it is therefore desirable to have a simplified but conservative approach to check whether or not tsunami load conditions will affect the structural system.

Equation (6.10-1) is provided as a conservative alternative to more detailed tsunami loading analysis. This equation is based on the assumption that all of the most conservative provisions presented elsewhere in this section occur simultaneously on a rectangular building with no openings. The maximum hydrodynamic loads are assumed to occur during Load Case 2 (Section 6.8.3.1), when $h=2/3h_{\max}$, assuming a conservative Froude number of $\sqrt{2}$ and a drag coefficient of $C_d=2.0$. Based on the more severe condition of bore impact per Section 6.10.2.3, the resulting lateral load per unit width of the building is given by the following:

$$f_w = 1.5 * \frac{1}{2} k_s \rho_{sw} I_{tsu} C_d C_{cx} (hu^2)_{\max}$$

$$= \frac{3}{4} * k_s \rho_{sw} I_{tsu} (2)(0.7) \left(\frac{2}{3} h_{\max}\right) \left(\sqrt{2}\sqrt{g * \frac{2}{3} h_{\max}}\right)^2$$

$$= 0.933 k_s \rho_{sw} g I_{tsu} h_{\max}^2$$

Including the worst effects of flow focusing suggested in Section C6.8.6 as a 1.25 amplification on the flow velocity (Table C6.8-3) and a 1.1 factor to allow for additional uncertainty, results in Equation (6.10-1):

$$f_w = 1.1(1.25)^2 (0.933 k_s \rho_{sw} g I_{tsu} h_{\max}^2) = 1.6 k_s \rho_{sw} g I_{tsu} h_{\max}^2$$

To account for additional buildup of water level at the forward edge of the building, this load is distributed as a rectangular pressure distribution over a height of $1.3h_{\max}$. The resulting pressure is therefore

$$P_{us} = 1.6 k_s \rho_{sw} g I_{tsu} h_{\max}^2 / 1.3 h_{\max} \approx 1.25 I_{tsu} \gamma_s h_{\max}$$

The lateral-force-resisting system should be evaluated for this pressure distribution acting over the entire width of the building perpendicular to the flow direction for both incoming and outgoing flows. All structural members below $1.3h_{max}$ should be evaluated for the effects of this pressure on their tributary width of projected area. Although it is not possible for a bore to occur on drawdown outflow, topography and erosion may result in additional flow acceleration that is not accounted for by the 1.25 amplification factor of Section C6.8.5. Therefore, it was determined that the intent of this simplified equation was better met by using the same loading for both inflow and drawdown outflow cases.

C6.10.2.1 Overall Drag Force on Buildings and Other Structures Once flow develops around the entire building or structure, the unbalanced lateral load caused by hydrodynamic effects can be estimated using Equation (6.10-2), which is based on fluid mechanics (FEMA 2011). The values of the drag coefficient, C_d, provided in Table 6.10-1 depend on the ratio between the building width perpendicular to the flow direction and the inundation depth (FEMA 2011). A wider structure results in a greater buildup of water level at the leading edge of the structure. The closure coefficient, C_{cx}, represents the vertical projected area of structural components, relative to the vertical projected area of the submerged section of the building. This ratio may not be taken as less than the value given in Section 6.8.7, so as to account for debris accumulation. Equation (6.10-2) is evaluated for all three load cases defined in Section 6.8.3.1. Because the drag coefficient, C_d, depends on the inundation depth, and the inundation depth changes for each load case, the appropriate drag coefficient is determined from Table 6.10-1 for each of the specified load cases.

Where building horizontal setbacks occur, the drag force coefficient will vary for each inundated portion of uniform width. In such cases, the B/h ratio is computed for each portion separately, in which the denominator h is taken as the combined vertical dimension of the inundated stories and any partial story thereof, for each portion of the building that is of an equivalent building width. As an example, as illustrated in Figure C6.10-1, the drag force coefficient on two inundated stories of a slender tower will be lower than the drag force coefficient on two inundated stories of a wider podium beneath it, since the submerged portion of the tower will have a lesser B/h ratio than that of the podium.

C6.10.2.2 Drag Force on Components All structural components and exterior wall assemblies below the inundation depth are subjected to the hydrodynamic drag forces given by Equation (6.10-3). This classical hydrodynamic drag expression depends on an empirically determined drag coefficient, C_d, based on the shape of the individual element. Typical values for C_d for common member cross-sectional shapes are given in Table 6.10-2 (OCADI 2009, Blevins 1984, Sarpkaya 2010, Newman 1977). Post-tsunami observations show that exterior elements, and interior vertical load-bearing elements of storage warehouses, are subject to debris accumulation that makes for an irregular shape, so for these elements, a C_d of 2.0 is used. The net force determined from Equation (6.10-3) is to be applied as a distributed pressure load on the submerged portion of the component being designed. All three load cases defined in Section 6.8.3.1 should be considered. For interior vertical load-bearing elements of storage warehouses, the specified width of 9 ft (2.74 m) represents the damming effect of contents of the warehouse. It is based on the typical width of a pallet rack commonly used in warehouses (DOD 2019). For interior vertical load-bearing elements of truck and bus garages, the specified width of 40 ft (12.19 m) represents the typical length of a standard bus, which is longer than the typical 30 ft (9.14 m) single-unit truck. However, to maintain consistency with exterior columns, the effective damming width on an interior column need not exceed the tributary width multiplied by the closure ratio value given in Section 6.8.7. Field observations during the Palu tsunami in Indonesia indicate that large portions of a steel-framed warehouse collapsed due to the increased hydrodynamic loads on interior gravity load columns due to debris damming (Stolle et al., 2019, Robertson et al., 2019). Previous observations in Japan after the Tohoku tsunami identified the tendency for materials stored in buildings to accumulate on both interior and exterior structural columns as flow passed through the building (Chock et al., 2013b). As the inundation depth increases for different load cases, the components that are inundated increase. Structural components that are a part of the lateral-force-resisting system

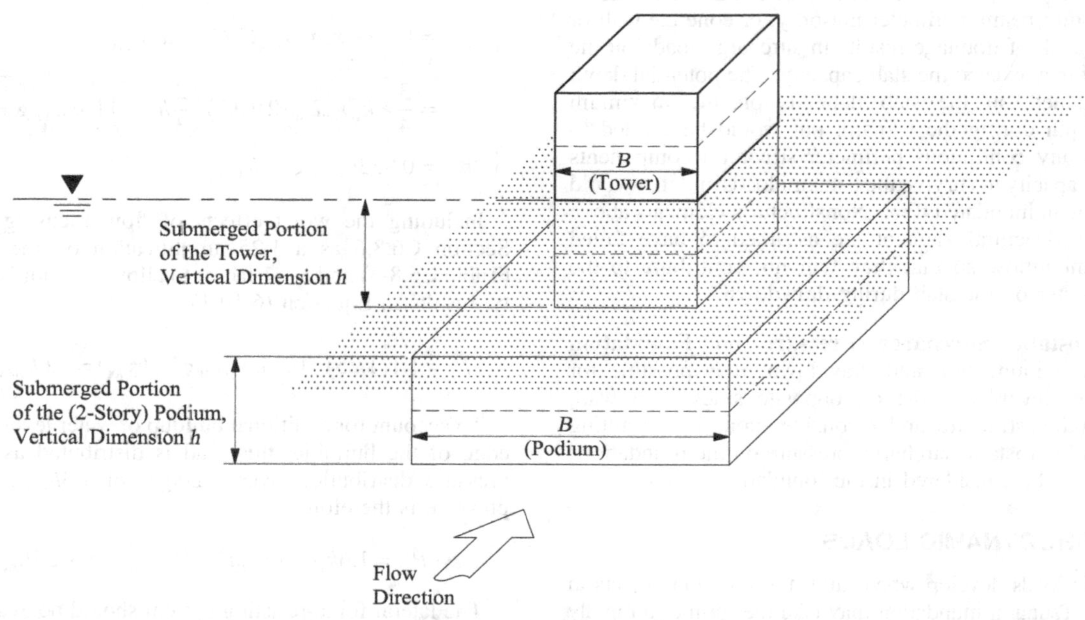

Figure C6.10-1. Illustration of dimensional parameters *B* and *h* applied to a building with a setback.

may be subject to the net resultant of their participation in resisting the overall drag force on the structure of Section 6.10.2.1 and the pressure drag caused by local flow around the component defined in this section.

C6.10.2.3 Tsunami Loads on Vertical Structural Components, F_w Laboratory experiments have shown that when the leading edge of a tsunami surge, often taking the form of a broken bore, impinges on a wide wall element, a short-duration impulsive load develops that exceeds the hydrodynamic drag force obtained from Equation (6.10-5) (Ramsden 1993, Arnason et al. 2009, Paczkowski 2011, Robertson et al. 2013). This increase is approximately 50%, resulting in the amplification factor of 1.5 applied to the steady-state drag expression in Equation (6.10-5). Where the flow initiated as a bore but the Froude number has been reduced by topography to be less than 1.0 at a site, then the bore is considered dissipated, and Equation (6.10-5b) is not invoked.

Because tsunami inflow may consist of a number of bores of varying sizes, it is necessary to check the bore loading effects for the most critical conditions. These loading conditions therefore need to be checked for the following cases: (1) when the inflow velocity is at a maximum, that is, Load Case 2 in Section 6.8.3.1, is applied to all wall elements that are wider than three times the inundation depth corresponding to Load Case 2; and (2) when the inflow inundation reaches depths equal to one-third of the structural widths of each of the wall elements, with the corresponding flow velocities. This aspect ratio is based on the bore height to specimen width ratio for which Arnason's experiments resulted in an impulsive force exceeding the subsequent steady-state drag force (Arnason et al. 2009). Because of the short duration of this impulsive load, windows and doors are assumed to be intact until the peak load is reached.

An alternative, more detailed approach to predicting the maximum impulsive force developed when a bore strikes a wall was developed by Robertson et al. (2013). Based on large-scale experiments at Oregon State University, it was determined that the lateral load per unit width of wall can be estimated as

$$F_w = k_s \rho_{sw} \left(\frac{1}{2} g h_b^2 + h_j v_j^2 + g^{\frac{1}{3}} (h_j v_j)^{\frac{4}{3}} \right)$$

where h_b is the bore height (equal to the sum of still water depth, d_s, and jump height, h_j), and v_j is the bore velocity, which can be estimated using hydraulic jump theory as

$$v_j = \sqrt{\frac{1}{2} g h_b \left[\frac{h_b}{d_s} + 1 \right]}$$

This force acts as a triangular pressure distribution over a height of h_p with a base pressure of $2F_w/h_p$, where h_p is the instantaneous ponding height at maximum load, given by

$$h_p = \left(0.25 \frac{h_j}{d_s} + 1 \right) (h_b + h_r) \leq 1.75 (h_b + h_r)$$

where $h_r = (\frac{v_j h_j}{\sqrt{g}})^{2/3}$.

Application of this expression to a large structural wall damaged during the Tohoku tsunami showed excellent agreement with the observed damage (Chock et al. 2013a). The above equations can be appropriate when detailed information on a bore strike scenario is determined from field data or estimated from a site-specific inundation model analysis.

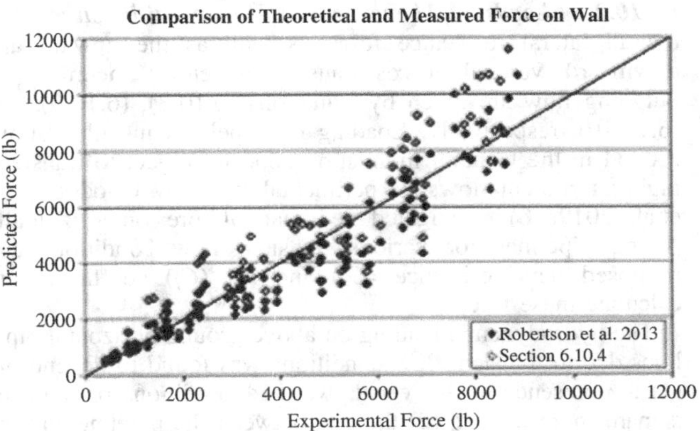

Figure C6.10-2. Predicted versus experimental bore loading on wall.

Note: 1:5 model scale.

For design purposes, comparison between the more detailed approach and that used in Section 6.10.2.3 for all of the bore-on-wall experimental cases performed at Oregon State University shows good agreement between predicted and measured results (Figure C6.10-2).

C6.10.2.4 Hydrodynamic Load on Perforated Walls, F_{pw} The impulsive force on a solid wall obtained from Section 6.10.2.3 can be reduced if there are openings in the wall through which the flow can pass. When applied to Equation (6.10-3), the 1.5 factor applied in Section 6.10.2.4 produces an effective drag coefficient, $C_d = 3.0$. Experiments performed at Oregon State University on perforated walls indicate that closure coefficients, C_{cx}, less than 20% have no effect on the force on individual wall piers (Santo and Robertson 2010). However, higher closure ratios resulted in increased loading on all wall piers. A linear transition is therefore assumed between the fully closed wall and a wall with 20% closure coefficient, as shown in Figure C6.10-3.

C6.10.2.5 Walls Angled to the Flow Equation (6.10-7) provides for a reduction in hydrodynamic loads on a wall positioned oblique to the flow. This reduction is the same as that provided for breaking waves in Chapter 5 of the 2010 edition of this standard.

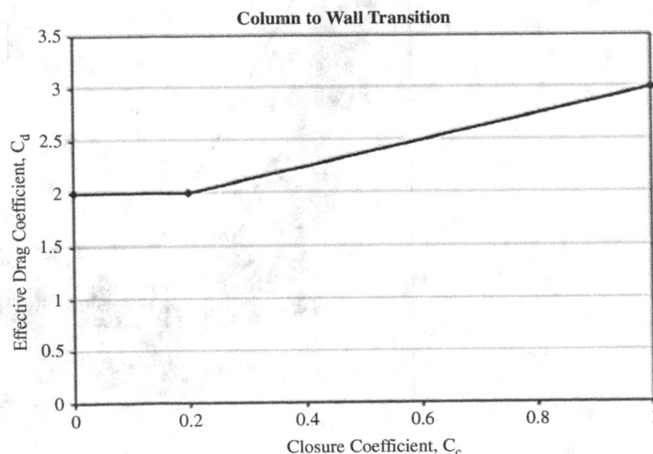

Figure C6.10-3. Hydrodynamic drag transition from individual column to solid wall.

C6.10.2.6 Loads on Above-Ground Horizontal Pipelines The overall lateral resistance force as well as the upward and downward vertical forces caused by either incoming or outgoing flow are given by Equations (6.10-8), (6.10-9), and (6.10-10), respectively. Loading on pipelines installed above ground in the tsunami inundation zone is subject to transient, highly turbulent flows. Experimental work by Ghodoosipour et al. (2019a, b) investigated the impact of bores on horizontally placed pipelines for various transient flow conditions and proposed new resistance (C_r) and lift (C_l) coefficients to calculate these forces.

The hydrodynamic loading on above-ground horizontal pipelines due to transient flow conditions was found to depend on: (1) the presence of dry versus wet bed conditions prior to the tsunami bore impact, (2) the gap between the pipeline and the soil, and (3) the degree of pipe submergence. An illustration of these parameters is provided in Figure C6.10-4. The design tables and figures of Section 6.10.2.6 for the pipe resistance coefficient and lift coefficients encompass these variations.

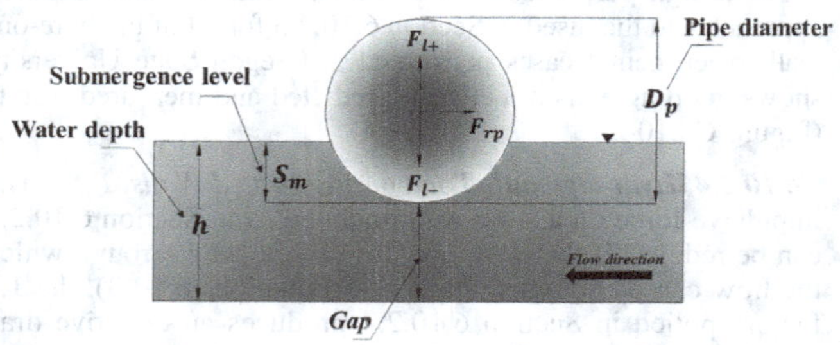

Figure C6.10-4. Definition of resistance and lift forces, pipe gap between the pipe and the soil, and degree of submergence.

Figure C6.10-5 illustrates the relationships for the calculated resistance coefficient values as a function of the Froude number in the experiments by Ghodoosipour et al. (2019b).

The values of the maximum estimated resistance coefficients were used to determine an upper envelope for the resistance coefficient, C_r. In the case of the dry bed condition, due to the higher flow velocities, the values of the resistance coefficients are found at the right side of the graph, corresponding to the larger Froude numbers which characterized the dry bed condition. Based on the results presented in Figure C6.10-5, the suggested C_r values vary between 1.0 and 3.5. Figure C6.10-5 shows that, for the case of supercritical flows ($F_r > 1$), C_r values were almost constant, while they linearly increased with a decrease in F_r number in the transitory and subcritical flow regime ($F_r < 1$). The black line in Figure C6.10-5 represents the upper envelope encompassing the entire range of experimental cases performed, from subcritical to supercritical flow conditions. As shown, it covers a variety of submergence conditions and gap space, as well as both wet and dry bed conditions.

Figures C6.10-6 illustrates relationships for the calculated lift coefficient values as a function of the Froude number for the experiments by Ghodoosipour et al. (2019b).

The maximum upward and downward lift coefficients for all experimental cases were calculated to determine two envelopes which encompassed the variation of these coefficients with the Froude number. The values of the upward lift coefficient, C_l^+, for pipelines subjected to unsteady flow conditions vary in the range of $1.0 \leq C_l^+ < 2.8$, while those of the downward lift coefficient, C_l^-, vary in the range of $-2.8 \leq C_l^- < -0.5$. Similar to the resistance coefficient behavior, the upward and downward lift coefficients remained almost constant in the supercritical flow region, while their absolute values increased as the

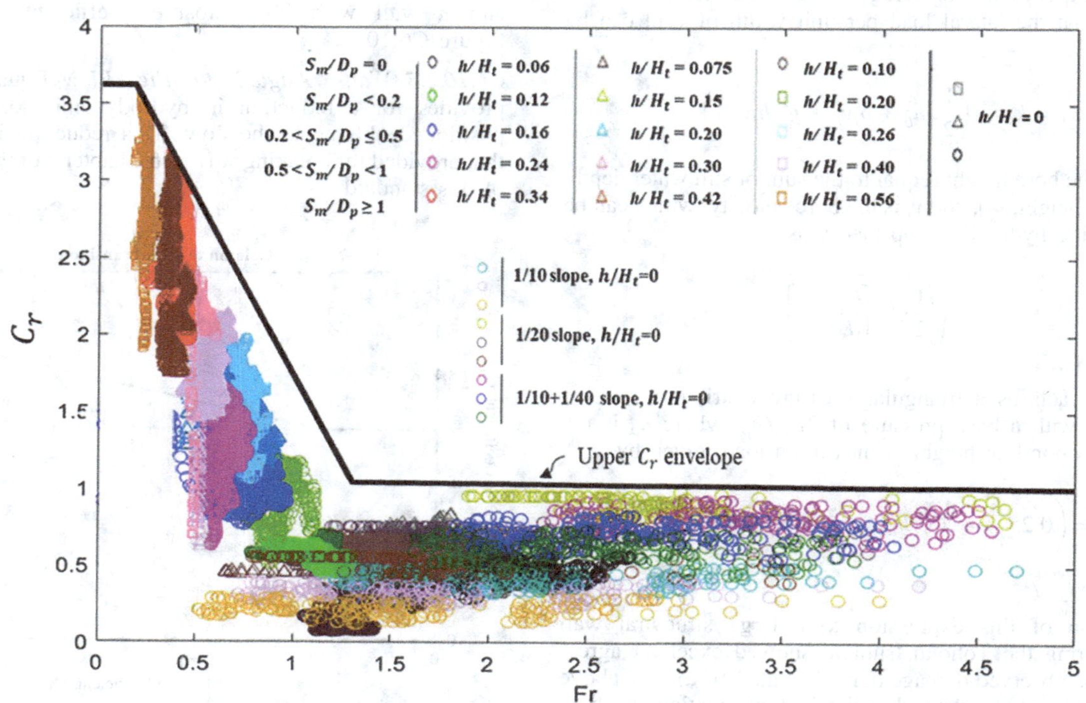

Figure C6.10-5. Resistance coefficient, C_r, versus Froude number, F_r, for different relative bore heights, gap space, and degree of submergence.

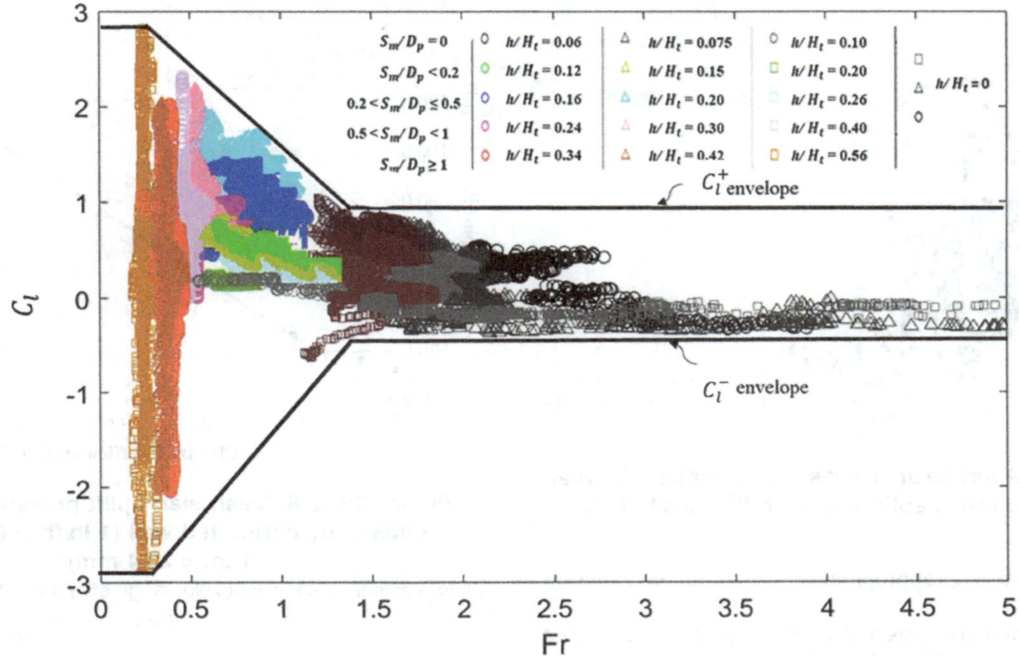

Figure C6.10-6. Upward and downward lift coefficients, C_l^+ and C_l^-, versus the Froude number, F_r, for different bore heights, gap space, and degrees of submergence.

value of F_r decreased in the transitional and subcritical flow region.

C6.10.3 Hydrodynamic Pressures Associated with Slabs This section specifies a number of conditions where hydrodynamic uplift pressures develop below elevated slabs. The resulting reduction in axial load on the structural columns and walls below these slabs should be included in the load combinations used to evaluate the design strength of these components because of the adverse effect it may have on the component bending and shear capacities, particularly for reinforced concrete members (Del Zoppo et al. 2020).

C6.10.3.1 Flow Stagnation Pressure Observations from the Tohoku tsunami indicate that structurally enclosed spaces created by structural walls on three sides and structural slabs can become pressurized by flow entering a wall-enclosed space without openings in the side or leeward walls (Chock et al. 2013b). Analysis of three of these conditions formed by reinforced concrete walls confirmed that the internal pressure reached the theoretical flow stagnation pressure given by Equation (6.10-11) (Bernoulli 1738, 342).

C6.10.3.2 Hydrodynamic Surge Uplift at Horizontal Slabs

C6.10.3.2.1 Slabs Submerged during Tsunami Inflow Experiments performed at Oregon State University on horizontal slabs with no flow obstructions above or below the slab indicate that uplift pressures of 20 lb/ft^2 (0.958 kPa) can develop on the slab soffit (Ge and Robertson 2010). Blockage in the form of structural columns or walls below the slab results in larger uplift pressures given in Section 6.10.3.3.

C6.10.3.2.2 Slabs over Sloping Grade A horizontal elevated slab located over sloping ground is subjected to upward hydrodynamic pressure if the flow reaches the slab soffit elevation. Because of the grade slope, the flow has a vertical velocity component of $u_v = u_{\max} \tan \phi$, where ϕ is the average slope of the grade below the slab being considered. Equation (6.10-12) estimates that this vertical velocity induces an upward pressure equivalent to the transient surge load on walls given by Section 6.10.2.3, assuming a C_d value of 2.0.

C6.10.3.3 Tsunami Bore Flow Entrapped in Structural Wall-Slab Recesses

C6.10.3.3.1 Pressure Load in Structural Wall-Slab Recesses Evidence from past tsunamis shows that significant uplift pressures can develop below horizontal structural slabs, such as floors and piers, if the flow below the slab is blocked by a wall or other obstruction (Saatcioglu et al. 2005, Chock et al. 2013b). Incoming flow striking the wall is diverted upward but is blocked by the slab, resulting in large pressures on both the wall and slab soffit close to the face of the wall. A series of experiments performed at small and large scales at Oregon State University and the University of Hawaii demonstrated that this effect is most severe when the inundation depth exceeds 2/3 of the slab elevation, or $h_s/h \leq 1.5$ (Ge and Robertson 2010). Figure C6.10-7 shows the results for tests at small scale (OSU tsunami wave basin and UH dam break using identical test setup at approximately 1:10 model scale) and at large scale (OSU large wave flume at approximately 1:5 model scale) with bore heights twice those in the small-scale experiments. An envelope was developed that enclosed 90% of the test data, as shown in Figure C6.10-7.

Although there is considerable variability in the uplift pressures even for identical repeat wave conditions, there appeared to be consistency between the results for both small- and large-scale tests, indicating that the laboratory results also apply at full scale. Estimation of the uplift pressures required to cause uplift failure of pier access slabs during the Tohoku tsunami indicated that the pressures ranged from 180 to 250 lb/ft^2 (8.62 to 12.0 kPa), indicated by the shaded area in Figure C6.10-7 (Chock et al. 2013b). Although the inundation depth at the time of slab failure is unknown, these uplift pressures observed at

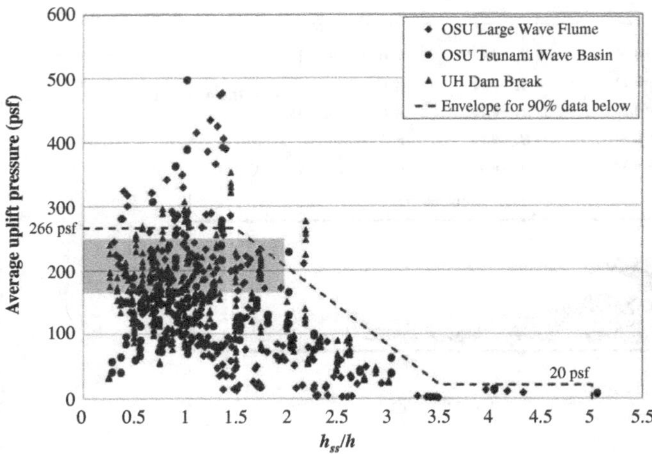

Figure C6.10-7. Average uplift pressure on slab soffit when flow is blocked by a solid wall (1 lb/ft^2 = 0.048 kPa).

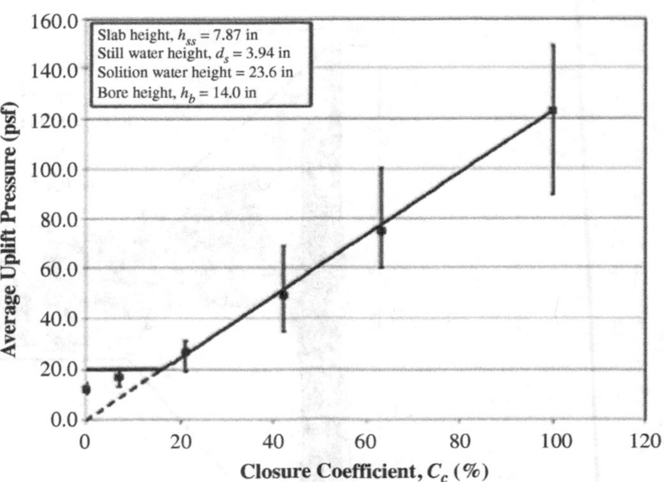

Figure C6.10-8. Mean slab uplift pressure reduction caused by perforated wall (1 lb/ft^2 = 0.048 kPa; 1 in. = 25.4 mm).

Note: Vertical bars show boxplot range of test results at each closure ratio.

full scale confirm the appropriateness of the envelope in Figure C6.10-7.

Pressure distributions measured at peak uplift indicated that the outward pressure on the wall and the uplift pressure under the slab adjacent to the wall were approximately a third greater than the average pressure, whereas the outward pressure away from the wall dropped to about half of this value. If the wall has finite width, l_w, and water is able to flow around the ends of the wall, then the uplift pressure on the slab beyond a distance $h_s + l_w$ is assumed to drop to the nominal uplift of 30 lb/ft^2 (1.436 kPa). The entire wall and the slab within h_s of the wall are therefore to be designed for an uplift pressure of 350 lb/ft^2 (16.76 kPa). Between a distance h_s and $h_s + l_w$, the slab is to be designed for an uplift pressure of 175 lb/ft^2 (8.38 kPa). Beyond $h_s + l_w$, the slab is to be designed for an uplift pressure of 30 lb/ft^2 (1.436 kPa).

C6.10.3.3.2 Reduction of Load with Inundation Depth As shown in Figure C6.10-8, when the inundation depth is less than 2/3 of the slab elevation, or $h_s/h > 1.5$, the uplift pressure envelope drops linearly. Equation (6.10-13) provides the equivalent linear decrease for the 350 lb/ft^2 (16.76 kPa) pressure prescribed in Section 6.10.3.3.1. The slab uplift pressure outside a distance of h_s from the wall also decreases proportionately. A minimum average uplift pressure of 20 lb/ft^2 (0.958 kPa) is indicated by the test data for slabs as high as five times the inundation depth. Once the slab height exceeds five times the inundation depth, the upward directed flow did not reach the slab, so no uplift pressure need be considered.

C6.10.3.3.3 Reduction of Load for Wall Openings Experiments were performed in the tsunami wave basin at Oregon State University using the same experimental setup as described in Section C6.10.3.3.1 but with a perforated wall replacing the solid wall behind the slab (Ge and Robertson 2010). These tests indicated that the slab uplift pressures decrease linearly as a function of the percentage closure provided by the perforated wall, as shown in Figure C6.10-9. A minimum uplift pressure of 20 lb/ft^2 (0.958 kPa) was observed even when the wall was removed completely, leading to the provision in Section 6.10.3.2.1.

C6.10.3.3.4 Reduction in Load for Slab Openings Experiments performed at the University of Hawaii demonstrated that the presence of an opening gap between the slab and the solid wall significantly relieves the uplift pressure on the slab.

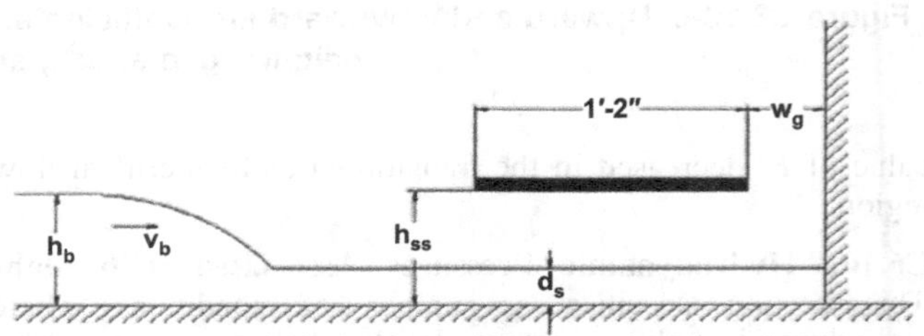

Figure C6.10-9. Test setup to study effect of opening gap or breakaway slab on uplift pressure.

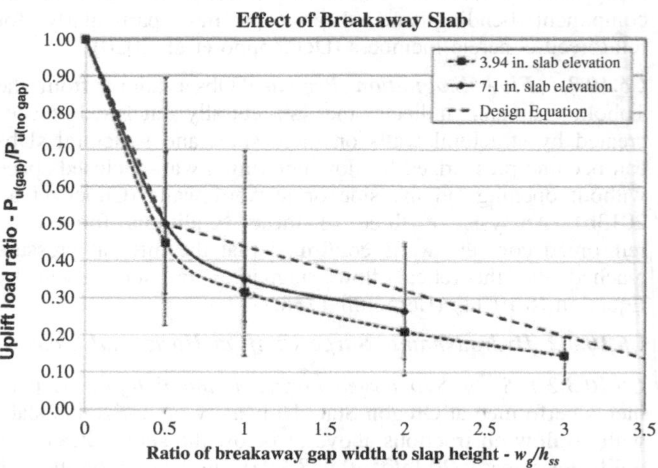

Figure C6.10-10. Reduction in slab uplift pressure caused by presence of opening gap or breakaway slab (1 in. = 25.4 mm).

Figure C6.10-9 shows the test setup used for these experiments, performed at approximately 1:12 scale. Figure C6.10-10 (from Takakura and Robertson 2010) shows the reduction in uplift pressure, P_u, for a slab with an opening gap, compared with one that has no gap, as the gap width, w_g, changes relative to the slab soffit height, h_{ss}. Equations (6.10-15), (6.10-16), and (6.10-17) are based on these data and are represented by the dashed line in Figure C6.10-10. The same effect is assumed to occur when the gap is created by means of a panel designed to break away at an uplift pressure less than 175 lb/ft^2 (8.38 kPa). Evidence from numerous piers inundated during the Tohoku tsunami indicate that pressure relief gratings and loose access panels were effective as breakaway panels for the purpose of relieving uplift pressures on the remaining slab (Chock et al. 2013b).

C6.10.3.3.5 Reduction in Load for Tsunami Breakaway Wall If the wall restricting flow below the slab is designed as a tsunami breakaway wall, then it can be assumed to fail when the pressure on the wall exceeds that required to fail the connection between the wall and the slab. This pressure will be the highest that can be experienced by the slab before failure of the wall.

C6.11 DEBRIS IMPACT LOADS

Tsunamis can transport a large volume of debris. Virtually anything in the flow path that can float given the inundation depth and that cannot withstand the water flow becomes debris. Common examples are trees, wooden utility poles, cars, and wood-frame houses and portions thereof. Some nonfloating debris, such as boulders and pieces of concrete, can also be transported if the flow is strong enough. This section covers the specification of forces and duration of the impact on structures by such debris. Debris impact forces shall be determined for the location of the structure based on the potential debris in the surrounding area that would be expected to reach the site during the tsunami. Of particular concern are the perimeter structural components oriented perpendicular to the flow direction, because they are at the greatest risk of impact and their loss may compromise the ability of the structure to support gravity loads.

In general, interior structural columns and walls are exempt from debris impact loading. However, there are certain conditions where interior vertical load-bearing elements are to be designed for debris impact.

1. Vehicles in parking garages are likely to float and impact interior structural elements. This applies to all inundated levels, including basements, where flow down parking garage ramps is likely to be sufficient to displace vehicles.
2. Portions of failed slabs-on-grade at ground level may result in tumbling debris strikes on interior structural elements (Stolle et al. 2019, Chock et al. 2013b, Robertson et al. 2010). Slabs-on-grade in basements will not experience high flow velocity conditions, so it is only slabs-on-grade at ground level that may potentially fail due to flow through the building. Damage to slabs-on-grade is more likely in Open Structures where the flow velocity through the building is high. However, if the slab-on-grade has been designed according to Section 6.12.4.2, it is deemed to no longer pose a debris impact threat.
3. To reduce the uplift forces due to buoyancy, it is permitted to design the slab-on-grade at ground level (as opposed to a basement slab-on-grade) with isolation joints so that it will fail due to uplift. However, this will generate tumbling concrete debris, so the interior vertical structural elements must be designed for the resulting impact loads.
4. Breakaway concrete slab panels provided to relieve uplift pressure on elevated structural slabs (similar to that described in Section 6.10.3.3.4) will dislodge when subjected to sufficient uplift pressure and become tumbling concrete debris. The columns and structural walls located nearest to the breakaway panels in any direction must be designed for this impact load.

The impact forces depend on the impact velocity, which for floating debris is assumed to be equal to the flow velocity for floating debris. The points of application of the impact force, which is assumed to be a concentrated force, shall be chosen to give the worst case for shear and moment for each structural member that should be considered within the inundation depth and the corresponding flow velocity. Exceptions to this are specified in subsequent sections based on specific debris characteristics.

The ubiquity of (1) logs and/or poles, (2) passenger vehicles and (3) boulders and concrete debris requires the assumption that these things will impact the structure if the inundation depth and velocity make it feasible.

Closed shipping containers float very easily, even if loaded. Therefore, for structures near a container yard, impact from floating containers should be considered. The site hazard assessment procedure in Section 6.11.5 is used to assess whether impact from shipping containers should be considered at a particular location.

Ships (including ferries) and barges are also potential debris that may impact structures. The likelihood of such "extraordinary" debris impact is most significant for structures near ports and harbors that contain these vessels. Because impact from these objects is likely to place a demand that cannot be resisted economically for many structures, only Tsunami Risk Category III Critical Facilities and Tsunami Risk Category IV buildings and structures are required to consider such impact. The site hazard assessment procedure in Section 6.11.5 is used to assess whether impact from marine vessels should be considered.

Table C6.11-1 summarizes the requirements for design, especially the threshold inundation depths at which level (or greater) it is required to consider each debris impact type.

C6.11.1 Alternative Simplified Debris Impact Static Load Designing for a conservative, prescriptive load is allowed to replace specific consideration of impact by logs, poles, vehicles, boulders, concrete debris, and shipping containers. The value of 330 kips (1,470 kN) is based on the cap of 220 kips (980 kN) in Section 6.11.6 for shipping containers, multiplied by a dynamic amplification of 1.5. Note that the maximum dynamic amplification factor from Table 6.11-1 was not used to account for the reduction of peak forces caused by inelastic response of the impacted component. If it is shown that the site is not in the

Table C6.11-1. Conditions for Which Design for Debris Impact Is Evaluated.

Debris	Buildings and Other Structures	Threshold Inundation Depth
Poles, logs, and passenger vehicles	All[a]	3 ft (0.914 m)
Boulders and concrete debris	All[a]	6 ft (1.8 m)
Shipping containers	All[a]	3 ft (0.914 m)
Ships and/or barges	Tsunami Risk Category III Critical Facilities and Category IV[b]	12 ft (3.66 m)

[a] All buildings and other structures as specified in Section 6.1.1.

[b] Tsunami Risk Category III Critical Facilities and Category IV buildings and other structures in the debris impact hazard region as determined in Section 6.11.5.

container or ship impact zone per Section 6.11.5, then the impact force is assumed to occur from a direct strike by a wood log. Lehigh University (Piran Aghl et al. 2014) tested a nominal wood log of 450 lb (204 kg) weight (see Section C6.11.2). A basic direct strike force of 165 kips (734 kN) can be used instead of the 330 kips (1,470 kN) based on the shipping container. The nominal design impact force for logs or poles is limited to the material crushing strength. This prescriptive 165 kips (734 kN) load includes a structural dynamic response factor of 1.5 and is associated with poles and logs with parallel-to-grain crushing strength of 5,000 psi (34.5 MPa) (approximately mean plus one standard deviation) for Southern Pine or Douglas Fir per ASTM D2555 (2011). The effective contact area is assumed to be 22 in.^{2} (142 cm^2), which represents about 20% of the end-on area of a 1 ft (30.5 cm) diameter pole, consistent with Piran Aghl et al. (2014). In all cases, the Orientation Factor, C_o, is applied. The net prescriptive simplified force is conservative compared with the laboratory test results, which are also considered conservative compared to likely field conditions.

C6.11.2 Wood Logs and Poles Previous provisions for debris impact forces, such as in ASCE 7-10 (2013a), Section C5.4.5, have been based principally on an impulse-momentum formulation for rigid-body impact, which requires an assumption of the duration of impact. Equation (6.11-2) is based on stress wave propagation in the debris and hence considers the flexibility of the debris and structural member. The assumptions are elastic impact and a longitudinal strike. That is, in the case of a pole or log, it hits the structure at its butt end, rather than transversely. Likewise, for a shipping container, per Section 6.11.6, the assumption is that the end of one bottom longitudinal rail of the container strikes the structural member. Full-scale testing at Lehigh University has validated the equation for a utility pole and a 20 ft (6.1 m) shipping container under these conditions (Piran Aghl et al. 2014, Riggs et al. 2014). Equation (6.11-4) for the duration is also based on elastic impact. It assumes that the impact force is constant, resulting in a rectangular force–time history. Although the duration may be underestimated somewhat, the total impulse is conservative.

The nominal 1,000 lb (454 kg) log or pole is adopted from ASCE 7-10 (2013a), Section C5.4.5, for flooding. Wood properties vary widely, but this corresponds approximately to a 30 ft (9.14 m) log with a 1 ft (30.5 cm) diameter. However, much larger trees are possible in certain geographical areas, and design professionals should consider regional and local conditions. The minimum stiffness is based on these dimensions and a modulus of elasticity of approximately 1,100 ksi (7,580 MPa). It is obtained from the well-known EA/L relation for axial stiffness.

Debris impact is clearly a dynamic event, and the structural element responds dynamically. However, an equivalent static analysis is allowed, where the static displacement is multiplied by a dynamic factor. The scaling factors as a function of the ratio of impact duration to the natural period of the structural element represent the shock spectrum. Shock spectra also depend on the shape of the force–time curve. The shock spectrum specified in Table 6.11-1 is adopted from ASCE 7-10 (2013a), Section C5.4.5, and is a modified version of the shock spectrum for a half-sine wave. The difference is that the factor in Table 6.11-1 remains constant for $t_d/T > 1.4$, whereas the half-sine wave shock spectrum decreases. However, the Lehigh experimental results show better agreement with the values in Table 6.11-1, because the force–time history is not truly a half-sine wave (Piran Aghl et al. 2014).

The value of the Orientation Factor, C_o, was derived from the data of Haehnel and Daly (2004), jointly sponsored by ASCE and FEMA. It is the mean plus one standard deviation value of the log debris impact force for trials that included glancing and direct impacts of freely floating logs.

C6.11.3 Impact by Vehicles Passenger vehicles are ubiquitous, float, and are easily transported. This standard requires the assumption that impact occurs as long as the inundation depth is sufficient to float the vehicle, which is deemed to be 3 ft (0.914 m). NCAC (2011, 2012) describes an experimental and numerical analysis of the frontal crash impact against a wall of a 2,400 lb (1,090 kg) subcompact passenger vehicle traveling at 35 mi/h (15.6 m/s). Based on the results therein, the initial stiffness of the vehicle was estimated to be 5,700 lb/in. (1 kN/mm). With an assumed velocity of 9 mi/h (4 m/s), Equation (6.11-2) results in an impact force of approximately 30 kips (133 kN) (Naito et al. 2014). Based on a more likely glancing impact with a smaller contact area, it is judged that 30 kips (133 kN) is a sufficiently conservative, prescriptive load to cover a range of possible vehicle impact scenarios. The impact can occur anywhere from 3 ft (0.914 m) up to the inundation depth.

Incipient motion, where vehicles become floating debris, may actually initiate at inundation depths less than 3 ft (0.914 m), depending on the vehicle type and flow speed, as shown in Figure C6.11-1(a). Martinez-Gomariz et al. (2016) conducted a review of the available literature of small-scale laboratory experiments measuring the threshold of motion for vehicles in steady-state flooding. Figure C6.11-1(a) shows the results, which depend on the size and clearance of the vehicle type. These studies indicate that at greater than the 3 ft (0.914 m) threshold criteria of Section 6.1.1 or for flow velocity of at least the minimum 10 ft/s (3.0 m/s) specified in Section 6.6.1, the entire range of passenger vehicle types would be expected to become floating debris.

C6.11.4 Impact by Submerged Tumbling Boulder and Concrete Debris The tumbling boulder debris impact force has been established based on a simplified static approach (Chau and Bao 2010). A "boulder" weight of 5,000 lb (2,270 kg) (either an actual boulder at the lower end of the "large boulder" classification shown in Table C6.11-2 or similar-sized debris from failed structural components) is considered within the inundation zone tumbling at a relative maximum velocity of about 13.1 ft/s (4 m/s). A dynamic amplification factor of 2 is implicitly incorporated in the force. The tumbling boulder is assumed to impact the structural element at 2 ft (0.61 m) above grade to reflect that the motion of the boulder is tumbling (rolling) along the ground surface.

C6.11.5 Site Hazard Assessment for Shipping Containers, Ships, and Barges A procedure is specified to determine if impact from these debris objects should be considered. The procedure is based on the assumption that the debris is disbursed from a linear source and then the prescriptive hazard region associated with that source is geometrically defined up to the grounding limit appropriate for the type of debris being evaluated. The angle of debris spreading with distance is based on observations by Naito et al. (2014) and experimental validation by Nistor et al. (2017). Model studies of debris spread in dam break flow support this procedure as being conservative (Stolle et al. 2018). This procedure may be used with large container yards or ports to be split into several sources, each with its own hazard region.

Transport of the debris can be limited by tsunami-resilient structures, geography, and insufficient inundation depth. For example, containers in a yard that is ringed on the leeward

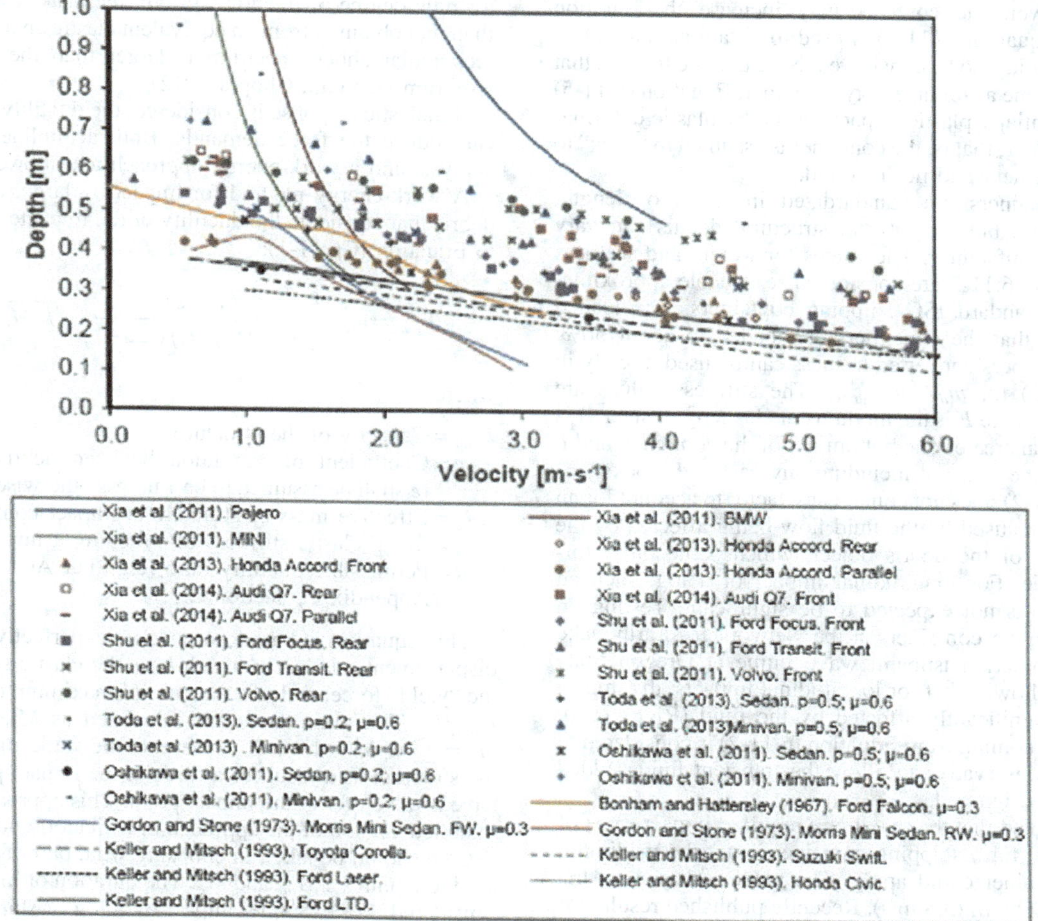

Figure C6.11-1(a). Threshold-of-motion criteria for vehicles as a function of flow depth and velocity.
Source: From Martinez-Gomariz et al. (2016).

Table C6.11-2. Boulder Class Size.

Boulder Size		
inches	**milimeters**	**Classification**
160–80	4,000–2,000	Very large boulders
80–40	2,000–1,000	Large boulders
40–20	1,000–500	Medium boulders
20–10	500–250	Small boulders

Source: Lane (1947).

side by structural steel and concrete structures will not disburse beyond the structures, as long as they cannot float over the structures. Similarly, ships with, for example, a 4 ft (1.2 m) draft will not be transported far or at significant speed in a 4 ft (1.2 m) inundation depth.

A project may utilize debris transport modeling in conjunction with a site-specific inundation analysis to determine the debris impact hazard region. While there are numerous methods to model debris motion in flows, analysis should employ massless tracers (Van Sebille et al. 2018) to capture a conservative inland extent of debris transport. Tracer trajectory should be determined using the flow patterns as provided by a site-specific inundation analysis. The furthest inland tracers provide the inland extent of the debris hazard impact region. Parameters of the debris transport model should follow practices found in the recognized literature (e.g., Lebreton et al. 2012). The debris particles would include a release and grounding threshold, which may be functions of the flow depth and velocity, consistent with parameters provided in this section. The inland limit of debris transport should be checked for numerical convergence, such that this extent is not a function of the numerical time step or the initial tracer spacing. General recommendations for these numerical parameters are to use a time step of one second or less for updating the tracer trajectory, and to use an initial tracer density of no less than one equally spaced tracer per every 110 ft^2 (10.2 m^2) of debris source area.

C6.11.6 Shipping Containers The impact force and duration equations in Section 6.11.2 are also valid for head-on (longitudinal) impact by the corner of a shipping container. See the discussion in Section C6.11.2.

The Lehigh test results (Piran Aghl et al. 2014) have shown that the mass of container contents does not significantly affect the impact force as long as the contents are not rigidly attached to the structural frame. Therefore, for shipping containers, the empty mass of the container is used in Equations (6.11-2) and

(6.11-4). However, the contents may increase the duration somewhat, so Equation (6.11-5) is used to obtain an alternative duration that should also be considered. Note that the force in that equation is the same as for an empty container. Equation (6.11-5) is based on essentially plastic impact caused by plastic deformation of the container; that is, the container is assumed to "stick" to the structural element and not rebound.

Shipping containers are standardized in terms of length, height, and width, but weight and structural details can vary somewhat by manufacturer. The values for weight and stiffness provided in Table 6.11-2 are considered reasonable approximations for most standard ISO shipping containers. The loaded weights assume that the containers are loaded to about 50%. Hence, these numbers converted to mass can be used directly in Equation (6.11-5) for $m_d + m_{\text{contents}}$. The stiffness values are based on EA/L, where E is the modulus of elasticity of steel, A is the cross-sectional area of one bottom rail of the container, and L is the length of the rail, not including any cast end blocks.

Equation (6.11-2) does not contain any factor to account for an increase in force caused by the fluid flow being affected by the sudden stoppage of the debris object, which some other formulations include. For longitudinal impact of a log, such an increase in force is not expected to be significant. Testing on scale-model shipping containers at the Network for Earthquake Engineering Simulation tsunami wave flume at Oregon State University also showed that for longitudinal impacts, the impact force was not significantly affected by the fluid (Riggs et al. 2014). The force coming from Equation (6.11-2) is considered to be sufficiently conservative to allow the transient fluid "added mass" effect to be ignored.

It should be noted that the maximum required impact force of 220 kip (980 kN) for a shipping container is not the maximum force that a container could apply. The value is based on the Lehigh tests at 8.5 mi/h (3.8 m/s). Recently published results for simulations have indicated that the maximum force may be higher (Madurapperuma and Wijeyewickrema 2013), depending on the impact scenario. The 220 kip (980 kN) force has been chosen as a reasonable value for design.

An Orientation Factor, C_o, value of 0.65 is used, assuming that there is similar randomness in the alignment of the lower corner steel chord of the shipping containers with the target structural component.

C6.11.7 Extraordinary Debris Impacts Extraordinary debris impacts, defined as impact by 88,000 lb (39,916 kg) or larger marine vessels, should be considered for Tsunami Risk Category III Critical Facilities and Tsunami Risk Category IV buildings and structures that are in the debris hazard impact region of a port or harbor, as defined in Section 6.11.5, and for which the inundation depth is 12 ft (3.66 m) or deeper. The size vessel to be used depends on the most probable size vessel typically present at the port or harbor. The harbormaster or port authority can be consulted to determine typical vessel sizes, ballasted drafts, and weight displacement under ballasted draft. Typical vessel sizes are also provided in PIANC (2014).

The nominal impact force is calculated with Equation (6.11-3), with the assumption that the vessel stiffness is larger than the structural member stiffness. Hence, the transverse structural member stiffness is to be used. The calculated force may be larger than any economical capacity of the member. Hence, the member may be assumed to have failed, in which case progressive collapse should be prevented for these important structures.

C6.11.8 Alternative Methods of Response Analysis It is also permitted to carry out a dynamic analysis. This standard specifies that a rectangular pulse be applied. For a linear elastic analysis of a single-degree-of-freedom system, the peak response is higher than that obtained from an equivalent elastic analysis because the rectangular shock spectrum is larger than the half-sine wave spectrum (see e.g., Chopra 2012).

If inelastic response is considered, the ductility of the structure can reduce the force demands. Both a nonlinear time history analysis and a work-energy approach are allowed.

A work-energy method for impact by large, essentially rigid debris that considers the ductility of the impacted structure leads to Equation (C6.11-1):

$$F_{cap} = u_{\max}\left[\frac{(1+e)}{(1+M_m/m_d)\sqrt{2\mu-1}}\right]\sqrt{k_e M_e} \qquad \text{(C6.11-1)}$$

where

F_{cap} = Capacity of the structure;
e = Coefficient of restitution between debris and structure (e shall be assumed to be 1 unless otherwise substantiated);
M_e = Effective mass of structure at impact point;
k_e = Initial elastic stiffness of structure at impact point; and
μ = Permissible ductility ratio [e.g., per ACI 349-13 (2013), Appendix C, Section C.3.7].

This equation is based on an elastic–perfectly plastic force–displacement relation in which k_e is the elastic stiffness, F_{cap} is the "yield" force, and μ is the ratio of maximum displacement to F_{cap}/k_e. The mass terms are calculated as $M_m = \sum m_i \Delta_i$ and $M_e = \sum m_i \Delta_i^2$, where Δ_i represents the static displacements of the structure as a result of a force at the impact point, scaled to have a value of 1 at the impact point. This approach is a slightly modified version of the approach in Kuilanoff and Drake (1991).

Care should be taken in choosing what part of the structure is used to obtain k_e, M_m, and M_e. The duration of impact should be considered. For example, impact by a log or shipping container occurs most likely over a very short time frame, and it is unlikely for all of a large structure to respond during such a short duration. Therefore, a section of the structure, possibly down to the structural member, should be considered. However, impact by a ship may occur over a long enough time to allow the entire structure to respond, and it may be appropriate to consider the entire structure.

For other work-energy approaches, where forces are accommodated by inelastic behavior up to some permissible ductility, the structure's initial stiffness is modified to an effective stiffness that reflects that deformation. In lieu of a nonlinear time history analysis, the secant or effective stiffness is a recognized technique of linearizing the modeled response analysis based on a performance point in the inelastic range, as shown in Figure C6.11-1(b). Sample analysis indicates that to reach

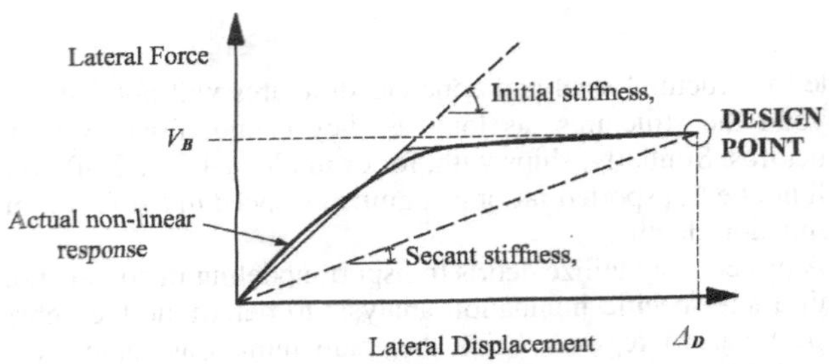

Figure C6.11-1(b). Stiffness definitions for initial elastic and effective secant stiffness.
Source: Sullivan et al. (2004).

consistent results over a broad range of inelastic behavior, the velocity applied in the work-energy method of analysis shall be u_{max} multiplied by the product of the Importance Factor, I_{tsu}, and the Orientation Factor, C_o.

C6.12 FOUNDATION DESIGN

Design of structure foundations and tsunami barriers should consider changes in the site surface and in situ soil properties during the design tsunami. In addition to the site response and geologic site hazard considerations, similar to the seismic hazard provisions of this standard, the designer should consider both topographic changes caused by scour and erosion, and the effects of surrounding natural or design elements, such as shielding and flow concentrations caused by other structures. Natural or designed countermeasures, such as barriers, berms, geotextile reinforcements, or ground improvements designed to protect the foundations and relieve them of direct loads, may be applied in the proximity of the foundation as well as in the foundation itself.

For deep foundations, the pile design procedure for tsunami loads relies on determining the maximum anticipated loss of strength and scour depth from pore pressure softening while being subjected to the assigned inundation loading. The scour depths are based on best available tsunami observation and analytical research. For comparison, the effects of general site erosion and local scour on deep foundations are also discussed for hurricane events, producing similar magnitudes of general erosion and scour, in FEMA 55 (2011), Section 10.5. Failures of deep foundations can result either from overloading the pile itself or from overloading at the pile–soil interface. Increasing a pile's embedment depth does not offset a pile with a cross section that is too small or pile material that is too weak; similarly, increasing a pile's cross section (or its material strength) does not compensate for inadequate pile embedment. The proposed approach provides a check on both cases for simple calculations. Advanced numerical modeling can assist in determining whether load combinations can be reasonably isolated and may justify more efficient designs, as in the design of piles, to resist seismic liquefaction and shaking combinations.

The basic principle for foundation design is to apply the loads under pseudostatic conditions. As with seismic loads, for Critical Facilities subject to significant tsunami loads, it may be appropriate to apply time history loading, applying various combinations of loads and effects (Section 6.12.3).

This section is organized to provide a logical progression of analytical steps to define siting effects and to apply suitable countermeasures consistent with the direct loads. For in-water or over-water structures or barriers, which are beyond the intent of this standard, it is suggested that foundation design may be approached by applying the specified tsunami loads and using appropriate offshore design methods such as USACE CEM (2011), California MOTEMS (California State Lands Commission 2005), California Building Code chapter on marine oil terminals (California State Lands Commission 2016), PIANC (2010), and API (2004).

In the tsunami load combination (as an extraordinary load in this standard, Section 2.5.2), a 1.0 factor on H_{tsu} is used rather than the 1.6 factor in Section 2.3.2, since loading and unloading of bulk material storage structures would not be done during a tsunami.

A quotation from USACE's *Stability Analysis of Concrete Structures* (2005) is relevant to the necessary coordination between geotechnical and structural engineers:

Even though stability analysis of concrete structures is a structural engineering responsibility, the analysis must be performed with input from other disciplines. It is necessary to determine hydrostatic loads consistent with water levels determined by hydraulic and hydrological engineers. Geotechnical engineers and geologists must provide information on properties of foundation materials, and must use experience and judgment to predict behavior of complex foundation conditions. To ensure that the proper information is supplied, it is important that those supplying the information understand how it will be used by the structural engineer. To ensure that the information is applied appropriately, it is important that the structural engineer understand methods and assumptions used to develop this interdisciplinary data.

Therefore, it is recommended that the report of the geotechnical investigation for a project in the Tsunami Design Zone include explanation of the derivation of design values for explicit use in strength design load combinations in accordance with Section 6.12.1.

C6.12.1 Resistance Factors for Foundation Stability Analyses Typical failure mechanisms evaluated in foundation stability analyses are the following:

- Lateral sliding due to lateral loads from tsunami forces, with the added effects of any unbalanced lateral soil pressures caused by local scour on one side and retained soils at walls;
- Uplift or flotation;
- Piping conditions caused by excess seepage stresses reducing the strength and integrity of the soil fabric;
- Slope instability caused by saturation and pore pressure softening effects of inundation and/or local scour, which affects the stability of retaining structures, berms/slopes, and slope-wall systems; and
- Bearing capacity (where soil strength properties may be affected by sustained pore pressures. However, lateral sliding failures typically occur before actual bearing stress failures, so this analysis is not expected to be governing, except where specifically cited in this chapter to be checked, for example hydrostatic fluid forces caused by differential water depth).

These failure mechanisms are dominated by tsunami effects on the foundation and soil properties. As is typical with other geotechnical problems, soil loading analysis incorporates geotechnical judgment in selecting a reduced nominal strength that is valid across the variability of the soil deposit or load-bearing strata and in recognition of the inherently nonlinear behavior of soil materials. For nonlinear materials, such as soils, a limit state is assumed to exist along some failure surface, and the resultant actions from an equilibrium analysis are compared to the reduced nominal strength for that material. Hence, this approach is commonly called limit equilibrium analysis. To ensure that the assumed failure does not occur, a resistance factor is applied to the material nominal strength. This factor does not imply an allowable stress design methodology of elastic analysis. However, the inverse of the resistance factor is often called a "factor of safety" in the recognized literature. So, equivalently,

$$\text{Applied Load or Stress Resultant} \leq \phi\text{Resistance}$$

In typical stability analyses of structures, resistance factors and factors of safety vary based on the failure mechanism, limit state, foundation type, soil type, construction methodology, load testing procedures, and other considerations. A review of multiple design standards for various structures, including bridges (AASHTO 2017), buildings (ASCE/SEI 2017, International Code Council 2018), dams and levees (USACE 1995, 2000, 2005; Bureau of Reclamation 2015), slopes (USACE 2003, WSDOT 2015), foundations (USACE 1991), retaining walls (USACE 1989, 2005), and nuclear facilities (USNRC 1978, 2003, 2013, 2014), indicates different guidance for resistance factors. (Most of the cited references prescribe target factors of safety, not resistance factors; however, for purposes of discussion the factors of safety have been converted to resistance factors by the inverse function.)

For extraordinary event conditions, higher resistance factors of 0.75 to 1.0 (lower factors of safety of 1.0 to 1.3) are common to reflect a greater risk tolerance for potential geotechnical failure.

Tsunami loading represents an extraordinary loading condition for which various design standards identify resistance factors on the order of 0.75 to 1.0 (i.e., factors of safety of 1.0 to 1.3) as being warranted for stability analysis. Tsunami Risk Category III and IV structures in particular need to be designed to remain occupiable/functional through multiple tsunami loading and unloading cycles. In recognition of the progressive consequences related to failure of such structures, a resistance factor on the lower end of the range of values recommended by other standards was deemed appropriate for the evaluation of the stability of structure foundations and slopes subject to tsunami loading. Therefore, a resistance factor of 0.8 is recommended for foundation and overall stability analyses. This resistance factor, when combined with a tsunami Importance Factor, I_{tsu}, of 1.25 for Tsunami Risk Category III and IV structures, results in an "effective factor of safety" of 1.6. For Tsunami Risk Category II structures, with an I_{tsu} of 1.0, the use of a resistance factor of 0.8 results in an effective factor of safety of 1.25. These values are similar to or higher than other values used for extreme and rare loading conditions.

The use of a resistance factor of 0.8 is further deemed appropriate, as Section 6.12.3 allows the use of performance-based design using site-specific tsunami-soil-structure numerical modelling in lieu of limit equilibrium analysis. With this more rigorous analytical approach, resistance factors are not applied, as actual deformations of structures/slopes are modeled under appropriate loading scenarios. The resistance factor of 0.8 combined with a tsunami Importance Factor of 1.25 incentivizes the more rigorous numerical modelling effort for Tsunami Vertical Evacuation Refuge Structures.

C6.12.2 Load and Effect Characterization Because of the successive cycles of tsunami inundation, foundation loads should be considered concurrent and cumulative and need to take into account landward flow of the tsunami waves, outflow of drawdown water, and the possibility that areas remain flooded between and after waves. Figure C6.12-1 presents a schematic representation of the applicable loading on a foundation element for the design condition after local scour and general erosion have occurred and pore pressure and seepage effects are present.

In addition to the shear, axial, and bending forces that are transmitted to the foundation from a structure during a tsunami, the direct hydrostatic and hydrodynamic loading of the foundation or barrier because of exposure to flow from general erosion or local scour should be considered. Loading is affected by flow blockages and flow focusing, as well as by dissipation of energy at the structure or foundation. Lateral soil pressure and pressure gradients are also affected by the tsunami inundation and may result in unbalanced loading because of flow blockage, excessive seepage, and unsteady flow effects.

C6.12.2.1 Uplift and Underseepage Forces Hydrostatic uplift and seepage forces are traditionally considered for structures such as dams and levees that are designed to control or retain water. In the case of tsunamis, however, where there is sustained overland flow as well as trapping of water in low-lying areas and behind berms, roads, and foundations, the uplift force on the base of foundation elements should be evaluated. Guidance for design in the presence of uplift and seepage is available in USACE (1989, 2005). To evaluate the potential for soil saturation, seepage, and uplift, careful site investigations should be conducted to determine soil characteristics. Soil permeability and the potential for erosion, soil blowout, and piping during tsunami inundation should be assessed. Guidance for evaluating critical gradients with potential for internal erosion and piping caused by high-velocity inundation seepage stresses is available in Zhang et al. (2010) and Jantzer and Knutsson (2013).

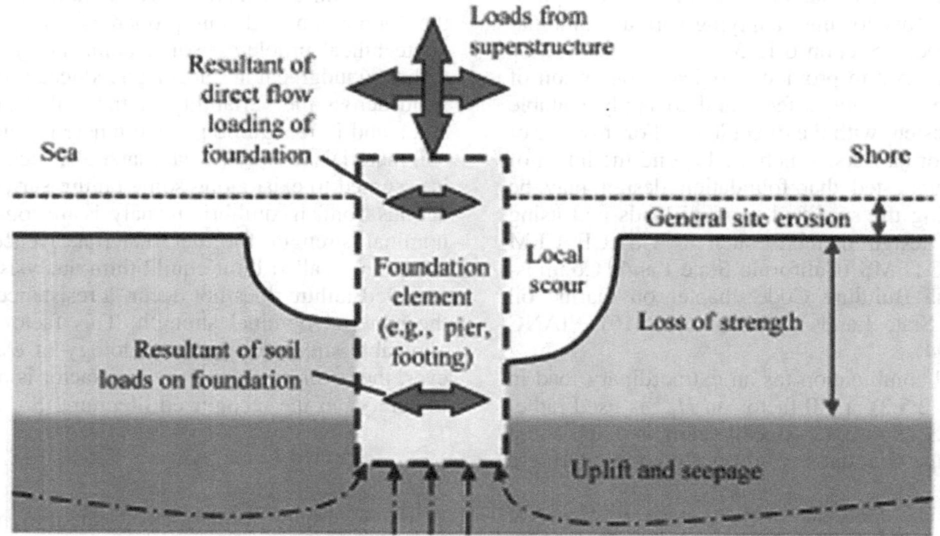

Figure C6.12-1. Schematic of tsunami loading condition for a foundation element.

Guidance for evaluating the magnitude and distribution of hydrostatic seepage uplift forces using traditional seepage flow net analysis methods is available in USACE (1993b) and Cedergren (1989). Design and analysis for the distribution of uplift pressures, including seepage time rate changes and coupled numerical modeling analysis of seepage pressures, can be performed with widely accepted methods implemented in open source software, such as SEEP2D (USACE and Tracy 1973), MODFLOW (USGS, McDonald, and Harbaugh, 1988), and a 3D platform, FEMWATER (USACE et al. 1997). These platforms are also accessible using the Department of Defense Groundwater Modeling System (USACE 2012), which provides a graphical user interface and additional post-processing features. Other suitable software programs are also available commercially. These seepage analysis methods are also used for tsunami barrier design countermeasures described in Section C6.13.

To ensure that failure does not occur, a resistance factor is applied to the resistance of anchoring elements in recognition of the uncertainty and inherent variability in the soil resisting properties in uplift and/or underseepage conditions. Current load and resistance factors are provided in accordance with the unique dynamic loading magnitude and velocity conditions associated with tsunami inundation conditions. Future research is needed to determine whether material-specific resistance factors are needed for conditions of excessive underseepage and resulting loss of strength. A methodology developing specific resistance factors for earthquake loading (Akbas and Tekin 2013) may provide an example to consider for adaptation to tsunami seepage conditions.

For foundations subject to tsunami flows and inundation, the weight of the structure and the soils that overlie the foundation act together with the foundation elements to resist uplift, as described in Equation (C6.12-1):

$$0.9D + F_{\text{tsu}} \leq \phi R \qquad \text{(C6.12-1)}$$

where

D = Counteracting downward weight of the structure, including deadweight and soil, above the bearing surface of the foundation exposed to uplift. The moist or saturated unit weight shall be used for soil above the saturation level, and the submerged unit weight shall be used for soil below the groundwater table;

F_{tsu} = Net maximum uplift caused by the distribution of hydrostatic pressure around the building as determined from the analysis of tsunami inundation. The load cases during inflow and outflow of the tsunami are defined in Section 6.8.3.1;

R = Upward design load-resisting capacity for foundation structural elements, such as piles and anchors; and

ϕ = Resistance factor, which is 0.67 for these uplift-resisting elements.

C6.12.2.2 Loss of Strength Loss of soil shear strength is a critical design consideration that may require extensive mitigative countermeasures to be incorporated in the design. Loss of shear strength can result from tsunami-induced pore pressure softening, and piping, or seismic shaking (e.g., liquefaction).

Pore pressure softening is a mechanism whereby increased pore-water pressure is generated during rapid tsunami loading and is released during drawdown. The increased pore pressure can soften the soil and decrease its effective shear strength. During tsunami inundation, the greatest loss of strength occurs near the ground surface with decreased loss of strength with increasing depth below grade (Abdollahi and Mason 2019a, b). Pore pressure softening also decreases the shear stress required to initiate sediment transport and thereby increases scour depth.

The Skempton B value is a measure of the degree of saturation, and relative compressibility of soil skeleton and pore water, that affects how the load is shared between the soil skeleton and pore water. As the Skempton B value decreases, the pore pressure softening increases, the hydraulic conductivity decreases, and entrapped air within pore fluid increases (Abdollahi and Mason 2019a, b). This mechanism is different from earthquake-induced liquefaction. Earthquake-induced pore pressure is caused by the residual buildup of excess pore pressure during cyclic shaking and initiates at some depth below the ground surface in fully saturated relatively loose, coarse-grained soils (Seed 1979, Mason and Yeh 2016). The primary differences between seismic liquefaction and tsunami-induced pore pressure softening are illustrated in Figure C6.12-2.

The methods used to evaluate loss of strength caused by pore pressure softening should utilize the fundamentals of soil mechanics, including flow through porous media. The primary interests are estimated elevated pore pressures, uplift forces on soil grains, loss of confinement (decrease in effective stress), and associated loss of shear strength. The loss of shear strength is presumed to directly follow the percent loss of confinement.

Physical or numerical modeling of soil–structure–fluid interactions may be used to determine soil shear strength loss. Alternatively, loss of strength caused by pore pressure softening may be correlated to Λ_T, the pore pressure softening parameter calculated in Equation (6.12-1), which is derived from Tonkin et al. (2003), as modified by Abdollahi and Mason (2020a). The softened soil shear strength is determined by multiplying the unsoftened soil shear strength by a factor $(1-\Lambda_T)$. The pore pressure softening parameter Λ_T approximates the fraction of the weight of soil grains supported by the excess pore water pressure. As such, it is a measure of the loss of confinement.

The loss of shear strength as it relates to both scour and structural design is presumed to directly follow theis fractional loss of confinement. This strength reduction may be applied uniformly throughout the depth of soil susceptible to strength loss or varied with depth, as determined by Equation (C6.12-2). The corresponding increased active earth pressure, decreased passive earth pressure, and reduced frictional and bearing resistances in this zone should also be evaluated.

The pore pressure softening parameter at a depth, d, below the soil surface may be estimated from Equation (C6.12-2). as

$$\Lambda_T(d) = 0.5 + \frac{(D_e - d)}{D_e}(\Lambda_T(0) - 0.5) \qquad \text{(C6.12-2)}$$

where $d \leq D_e$. D_e shall be estimated per Figure 6.12-3, and $\Lambda_T(0)$ shall be determined by Equation (6.12-1). Determination of D_e is based on the approach of Abdollahi and Mason (2020a).

The pore pressure softening during tsunami loading is also a function of the drainage factor, κ_d, which is determined per Equation (C6.12-3) and used in Equation (6.12-2).

$$\kappa_d = [B_{sk}(1 + \nu_u)]/[3(1 - \nu_u)] \qquad \text{(C6.12-3)}$$

where B_{sk} is the Skempton B parameter, and ν_u is the undrained Poisson ratio. B_{sk} is a modification of the B values for saturated soils described in Skempton (1954) for use with unsaturated soils. B_{sk} may be estimated based on Bishop (1973), Kokusho (2000), and Yang (2005). Table 6.12-1 provides representative

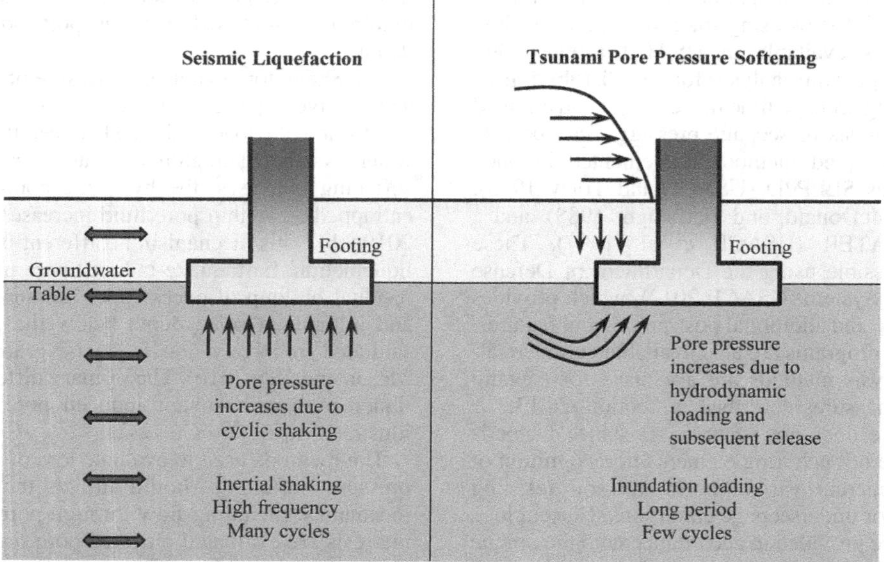

Figure C6.12-2. Schematic diagram showing differences between seismic liquefaction and tsunami-induced pore pressure softening.

values of κ_d using typical values of B_{sk} and v_u for different soil types.

Typical values of the consolidation coefficient, c_V, for sand and gravel for use in Equation (6.12-3) are as follows (Tonkin et al. 2003, Hicher 1996, Francis 2008):

- Gravel: 10 ft^2/s to 1,000 ft^2/s (approximately 1 to 100 m^2/s);
- Sand: 0.1 ft^2/s to 1 ft^2/s (approximately 0.01 to 0.1 m^2/s).

When calculating Λ_T for finer materials, c_V may be taken as 0.1 ft^2/s (approximately 0.01 m^2/s). Although much lower values are commonly used in standard geotechnical practice, such low values lead to excessive prediction of loss of strength for fine materials. Field observations provide no evidence that loss of strength or scour is substantially greater in finer soils.

The use of a site-specific inundation time history analysis (e.g., a transformed waveform) for determination of the T_{draw} and h_{draw} parameters (per Figure 6.12-1) allows for effects that modify the initial waveform, such as the influence of topography, bathymetry, wave resonance and reflection, drawdown, and secondary waves. The use of $T_{TSU}/4$ is stipulated as a default value when site-specific inundation time history analysis is not available. The steepest gradient during drawdown for the inundation waveform occurs over the time interval of $T_{TSU}/4$, which results in an appropriate design Λ_T value for induced pore pressure softening.

The potential for tsunami-induced pore pressure softening, and if needed, the susceptible depth and magnitude of soil strength loss, can be determined using the procedures outlined in this section. The following example shows the determination of the various parameters for a case where no site-specific time history is available.

C6.12.2.3 General Erosion Both general erosion and local scour can contribute to the lowering of the grade around a structure foundation. In the absence of countermeasures such as protective slabs on grade, the sum of general erosion and local scour is used in foundation design. Scour effects are discussed (nonspecific to tsunamis) in USACE (1984, 1993a), Simons and Senturk (1977), and FHWA (2012).

Evaluation of general site erosion may be based on the standard literature and models that describe flood-induced general erosion, e.g., USACE 2010. However, these approaches do not include the effects of pore pressure softening. Pore pressure softening can increase the depth of general site erosion, as described in Yeh and Li (2008) and in Xiao et al. (2010). The effect of pore pressure softening on erosion and during drawdown may be evaluated using physical scale modeling or numerical modeling of soil–structure–fluid interactions (e.g., Abdollahi and Mason 2019a, b) similar to those described in these references.

Alternatively, the increase in general site erosion during drawdown may be evaluated by multiplying the buoyant specific weight, γ_b, of the sediment or the critical shear stress by a factor of $1-\Lambda_T$. The pore pressure softening parameter Λ_T is given by Equation (C6.12-2). This approach is based on the model of Tonkin et al. (2003) and Abdollahi and Mason (2020a).

Channelized scour occurs when significant quantities of return flow collect into a channel, for example along a seawall or in a preexisting streambed. Because of this flow concentration, the scour depth in such channels can be greater than the general site erosion. If the geometry of the situation and lack of countermeasures suggest that channelized scour may be a factor, then this type of erosion should be analyzed. Pore pressure softening need not be factored into the depth of channelized scour, because pore pressure softening is associated with rapid changes in hydrodynamic loading, whereas channelized scour is associated with a longer timescale of drawdown flow collected from a wider area.

C6.12.2.4 Scour The geometry of the structure should be considered in the evaluation of local scour. Specifically, it should be determined whether tsunami flow is expected to be around the structure, causing flow acceleration around the obstruction, or whether the flow overtops the structure. Because of the very high stress gradients caused in tsunami flow, all levels of soil cohesion short of weathered rock or hard saprolite have been observed to rupture.

C6.12.2.4.1 Sustained Flow Scour Sustained flow scour is caused by tsunami flow around a structure. Numerical, physical

Table C6.12-1. Example Problem Steps (Customary Units).

Step/Parameter (units)	Value	Comment
Step 1: Determine maximum Froude number, F_r		
Maximum F_r (–)	0.9	Determined from Energy Grade Line Analysis per Section 6.6.2. Since $F_r \geq 0.5$, continue with evaluation.
Step 2: Determine pore pressure softening parameter, $\Lambda_T(0)$		
γ_s, weight density of seawater for tsunami loads (lb/ft^3)	70	Per Equation (6.8-4)
γ_b, buoyant weight density of soil (lb/ft^3)	51	Per geotechnical study, assuming moist soil weight density = 115 lb/ft^3
h_{max}, maximum inundation depth (ft)	13	Determined from Energy Grade Line Analysis per Section 6.6.2
T_{TSU}, predominant wave period(s)	2,160	From ASCE Tsunami Design Geodatabase
h_{draw}, drawdown height (ft)	13	No site-specific time history, so $h_{draw} = h_{max}$
T_{draw}, site-specific drawdown time (s)	540	No site-specific time history, so $T_{draw} = ¼\ T_{TSU}$
c_v, coefficient of consolidation (ft^2/s)	1	Per geotechnical study, assuming medium-to-coarse-grained sand
κ_d, drainage factor (–)	0.2	Per geotechnical study and Table 6.12-1, assuming medium-dense sand
F_Λ, pore pressure softening vertical flux index (ft)	14.3	Per Equation (6.12-2)
I_Λ, pore pressure softening drawdown inertia index (ft^2)	540	Per Equation (6.12-3)
Option 1: $\Lambda_T(0)$ (–)	≥0.5	Per Figure (6.12-2) with $\sqrt{I_\Lambda} = 23.2$ ft; see example figure after the table
Option 2: $\Lambda_T(0)$ (–)	0.69	Per Equation (6.12-1)
Step 3: Check susceptibility to tsunami-induced pore pressure softening		
Are F_r and $\Lambda_T(0)$ greater than or equal to 0.5?	Yes	Both values ≥0.5, therefore the site is considered susceptible
Step 4: Determine Susceptible Depth, D_e		
Option 1: D_e, susceptible depth (ft)	11.5	Per Figure (6.12-3); see example figure after the table
Option 2: D_e, susceptible depth (ft)	15.6	1.2 h_{draw}
Step 5: Determine soil strength loss		
Strength reduction factor at ground surface ($d = 0$)	0.31	Reduction factor at $d = 0$ is $1 - \Lambda_T(0)$. This reduction factor shall be applied to the soil shear strength to a depth of D_e. The reduction can be applied uniformly or varied linearly to a value of 0.5 at D_e.
Strength reduction factor at Susceptible Depth ($d = D_e$)	0.5	
Step 6: Summary of Findings		
The site has been determined to be susceptible to tsunami-induced pore pressure softening. A susceptible depth, D_e, of 11.5 ft is determined. The soil shear strength is to be multiplied by a reduction factor that varies from 31% at the ground surface to 50% at 11.5 ft ($d = D_e$). Below 11.5 ft, no reduction factor is applied.		

modeling or empirical methods may be used to evaluate sustained flow scour; the analysis method should consider the effects of pore pressure softening. The methodology used to develop Table 6.12-1 and Figure 6.12-1 is described in Tonkin et al. (2013) and is based on a comparison of post-tsunami field observations of scour around structures in Chock et al. (2013b) and Francis (2008) with the model of Tonkin et al. (2003).

The methodology used to develop Table 6.12-2 and Figure 6.12-2 for pile scour is based on experiments carried out by Razieh et al. (2017, 2021) and Lavictoire et al. (2014). Figure C6.12-4 presents experimental data from these studies regarding the vertical extent of scour to element width ratio and the Froude number.

The use of the Froude number relates the inertial energy of the flow and the accompanying flow turbulence to the associated maximum scour depth around a structural element with a given geometry. A maximum Froude number of 1.7 is specified for use in Figure 6.12-5 because this was the upper limit of the Froude number in the experimental data. As shown in the experimental data presented in Figure C6.12-4, scour around foundation elements depends on the physically based parameters of flow conditions (Froude number), sediment characteristics (sand, gravel, and soil saturation), foundation element geometries (square and circular columns were tested), and slope of the soil bed.

For deep foundation systems with pile caps, grade beams, or similar load-carrying or horizontal tie elements, the scour depth calculated using Figure 6.12-1, which includes the effects of general erosion, is used to determine whether scour extends down to the top of the pile. Subsequently, Figure 6.12-5 is used to determine the vertical extent of scour along the length of the pile itself.

The spatial extent of scour around a structure is primarily based solely on field observations in Francis (2008) and Chock et al. (2013b); a schematic representation of the extent and geometry of sustained flow scour is given in Figure C6.12-5.

C6.12.2.4.2 Plunging Scour Plunging scour is caused by tsunami flow over an overtopped structure. Numerical or physical modeling or empirical methods may be used for sustained flow scour; however, the analysis method is not required to include the effects of pore pressure softening. The methodology provided here is described in Tonkin et al. (2013) and is based on a comparison of post-tsunami field observations of scour around structures with the physical model results of Fahlbusch (1994), as described in Hoffmans and Verheij (1997).

C6.12.2.6 Displacements Calculation of displacements is performed with the same procedures as other geotechnical displacement calculations recognized in the literature for the identified cases of footing, slopes, walls, and piles. Each uses a different procedure. The calculations for tsunami cases are possible with no procedural modifications because the tsunami loads given in Section 6.12.2 are in a form consistent with other geotechnical loads.

C6.12.3 Alternative Foundation Performance-Based Design Criteria In ASCE 41-17 (2017), foundation strength and stiffness characterization parameters are generally suitable for tsunami loading, but its procedures are not suitable for direct

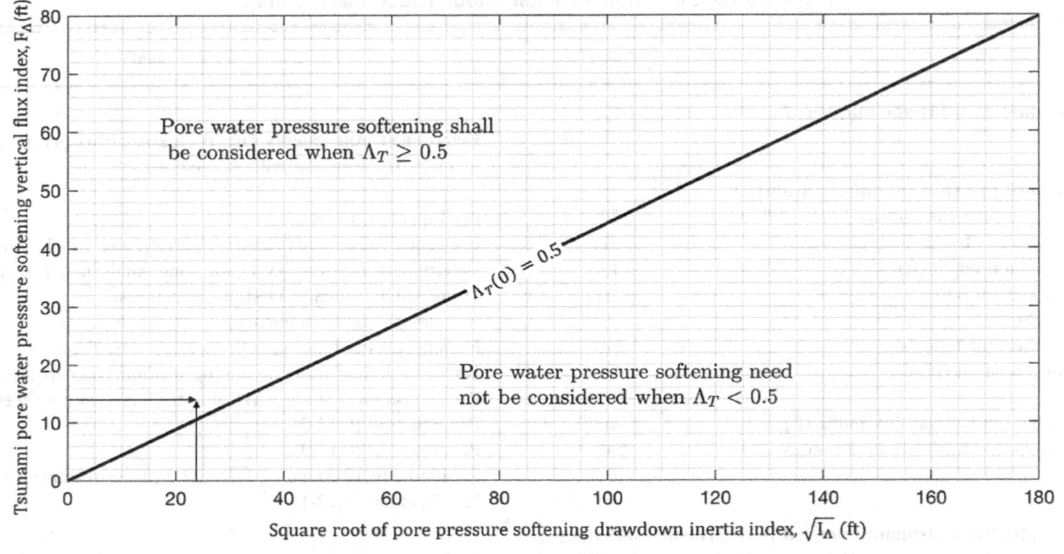

(a) $\Lambda_T(0)$ determination per Figure 6.12-2 for example problem.

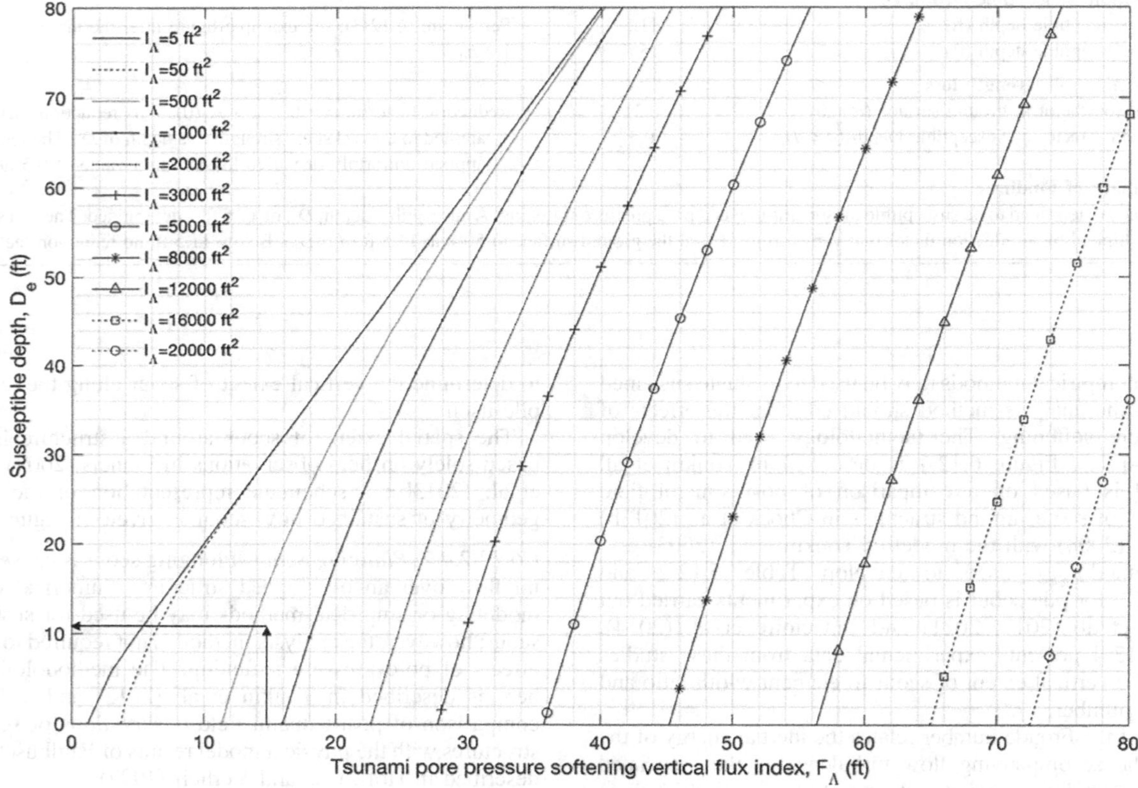

(b) D_e determination per Figure 6.12-3 for example problem.

Figure C6.12-3. Example problem solutions with Figures 6.12-2 and 6.12-3.

reference in the provisions. It may be also desirable for the foundation performance objectives for the facility to consider the role of the facility within community resilience and sustainability objectives. Tradeoffs that go beyond structural hardening include siting countermeasures, evacuation planning, and other emergency response planning provisions. In addition to Life Safety, loss of physical infrastructure and consequential damage and economic impacts to occupants and community services and commerce may be considered. These concepts are discussed in Office of the President (2013) and TISP (2012). For Tsunami Risk Category IV buildings and structures, particularly in Site Classes D, E, and F, a tsunami–soil–structure interaction modeling analysis is recommended. Such modeling analysis is typically performed by geotechnical engineers.

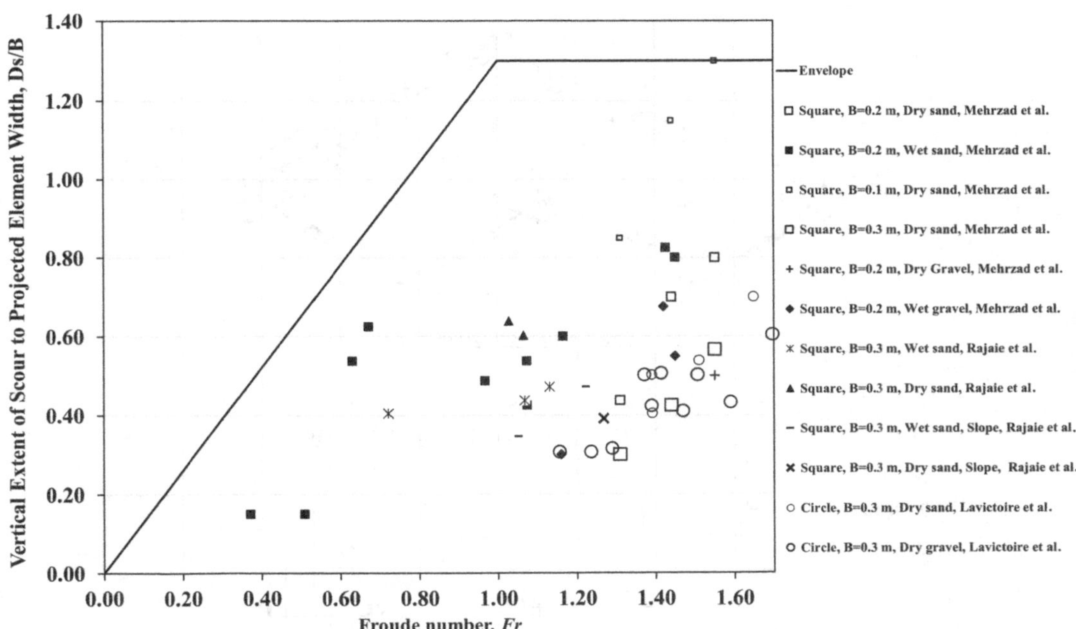

Figure C6.12-4. Vertical extent of scour to element width ratios for foundation elements with different soil conditions and Froude numbers in scaled laboratory experiments.

C6.12.4 Foundation Countermeasures

C6.12.4.1 Fill The uses of structural fill are discussed in FEMA 55 (2011), Section 10.3. Structural fill can be eroded during tsunamis, and it may not be feasible to provide adequate countermeasures in some areas without additional ground improvement or reinforcement, such as geotextiles.

C6.12.4.2 Protective Slab on Grade Exterior slab-on-grade uplift shall be assumed to occur as a preexisting condition to computation of local, sustained flow and plunging scour unless determined otherwise by a site-specific design analysis based on recognized literature. The design of stable slabs on grade under tsunami loading relies on recognizing the potential for scour at slab edges and ensuring the stability of slab sections and substrate. At slab edges, grade changes often result in rapid changes in flow speed and depth, which can carry away material and substrate, while large-scale pressure fluctuations in high-speed flows over pavers or concrete slab sections can pry sections loose and cause further damage. This type of damage, as well as failure of substrate and loss of soil strength, has been observed both during overwash by tsunamis (Yeh and Li 2008, Yeh et al. 2012) and during coastal storms (Seed et al. 2008). A schematic showing the different loading conditions and considerations is shown in Figure C6.12-6.

Seepage flow gradients and uplift pressure during tsunami drawdown under the slab may be evaluated using the analytical and numerical modeling techniques described in Section C6.12.2.1 and in Abdollahi and Mason (2020a, b).

Guidance for protective slab-on-grade design is drawn from roadway design in the coastal environment and is discussed in Douglass and Krolak (2008), with specific recommendations for best practices at slab-on-grade transitions in Clopper and Chen (1988).

C6.12.4.3 Geotextiles and Reinforced Earth Systems Use of geotextiles to provide foundation stability and erosion resistance under tsunami loading provides internal reinforcement to the soil mass through both high- and low-strength geotextiles. They are applied in various configurations, relying on composite material behavior to a predetermined geometry of improved ground bearing on strata that remain stable through the event loading. Broad use in coastal environments has proven their effectiveness, with varying levels of reinforcement used to address varying severity of water and wave loading. They can be effective for creating protective reinforcement of traditional shallow footings, slabs on grade, small retaining walls, berms, and larger structures, up to tall, mechanically stabilized earth walls as used in the transportation industry. Additional guidance for geotextile placement and design is available in FHWA NHI-10-024 (2010) and AASHTO M288-06 (2006).

C6.12.4.4 Facing Systems Facing materials in coastal structures and reinforced earth systems are critical to prevent raveling and erosion. AASHTO M288-06 provides design guidance for geotextile filter layers assuming high-energy wave conditions. Armor sizing in areas of high Froude number should take into account the high-velocity turbulent flows associated with tsunamis and the height of the incoming waves. FHWA (2009) provides methods appropriate for current flow. Esteban et al. (2014) provides an adaptation of the Hudson equation (USACE 2011) for tsunami waves. Some approaches, such as the Van der Meer equation provided in USACE 2011, recommend armor stone sizing that decreases with increasing wave periods; these approaches should not be used for design of tsunami-resistant facing systems. In areas of low Froude number, the tsunami acts more as current flow, and stone sizing may be treated accordingly using standard methods.

C6.12.4.5 Ground Improvement Soil–cement ground improvement for foundations is effective under high-velocity turbulent flows such as tsunamis because it provides both strength and erosion resistance to the improved mass. The widely used methods of deep soil mixing and jet grouting can be applied in a variety of geometries and design strengths for particular

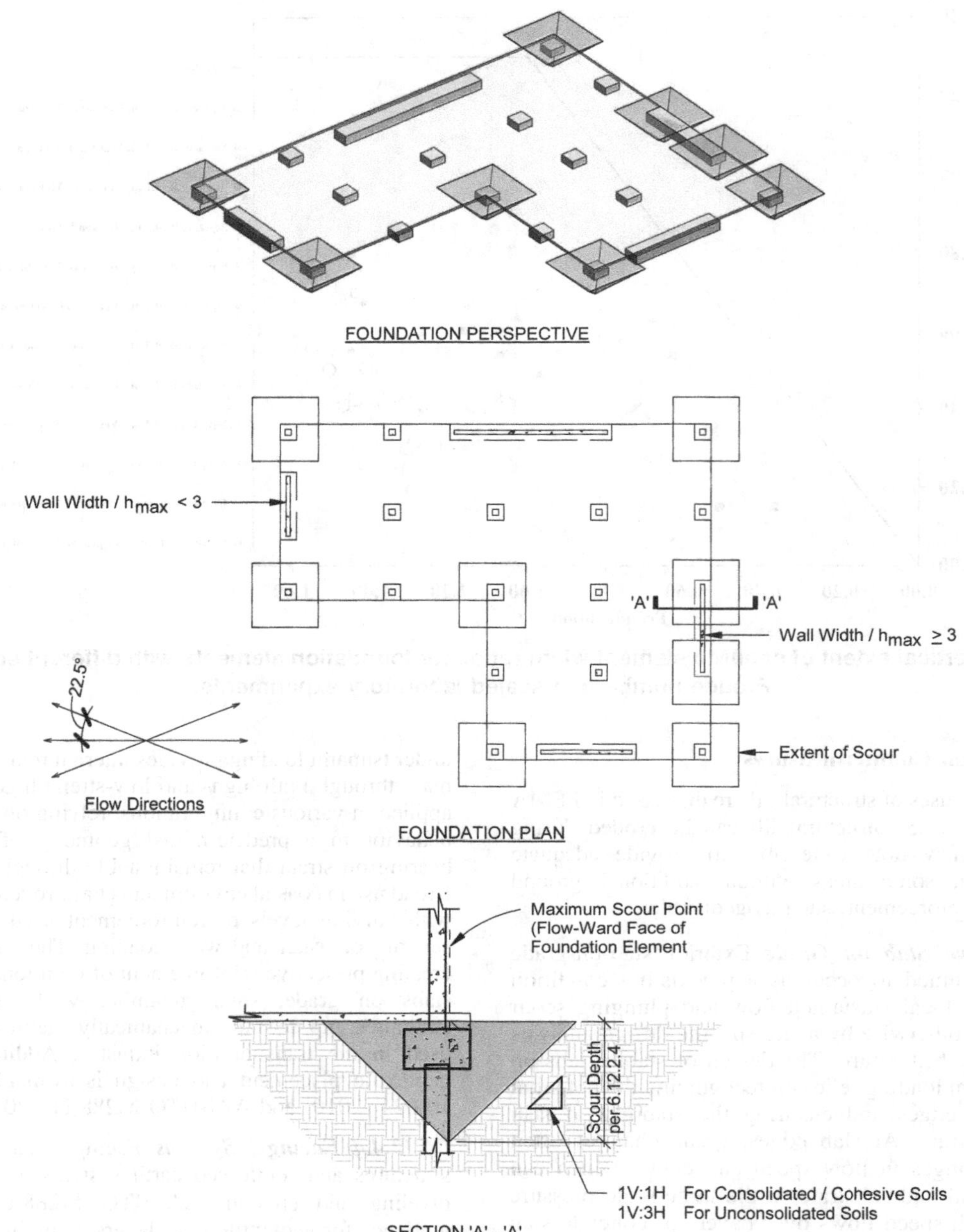

Figure C6.12-5. Schematic of sustained flow scour extents.

tsunami loading conditions. These methods, when incorporated in the modeling and analysis methods in this section, can be used to determine the optimal limits of treatment for desired performance levels. Similar applications are used for bridge scour and foundations for levees, dikes, and coastal structures. Additional guidance for soil–cement ground improvement is available in FHWA RD-99-138 (2000), USACE EM 1110-2-1913 (2000, Appendix G), and ASTM D1633-00 (2007).

C6.13 STRUCTURAL COUNTERMEASURES FOR TSUNAMI LOADING

The potential extreme magnitude or severity of tsunami loading warrants use of robust or redundant structural countermeasures, including Open Structures, retrofitting and/or alterations, and use of tsunami mitigation barriers located exterior to buildings. The type of selected countermeasures and their strength and extent of protection depend on both the performance objectives of the structure to be protected and the extent of protection achievable by countermeasures applied to the structure itself. For most sites, an alternatives evaluation of these three types of countermeasures is needed to identify the optimal protection method or blend of methods. The application of structural countermeasures should be integrated with foundation countermeasures described in Section 6.12.4.

C6.13.2 Tsunami Barriers Tsunami mitigation barriers consist of widely varying materials and designs, ranging from simple berms and engineered levees to advanced performance-based, instrumented, heavy coastal infrastructure systems of reinforced

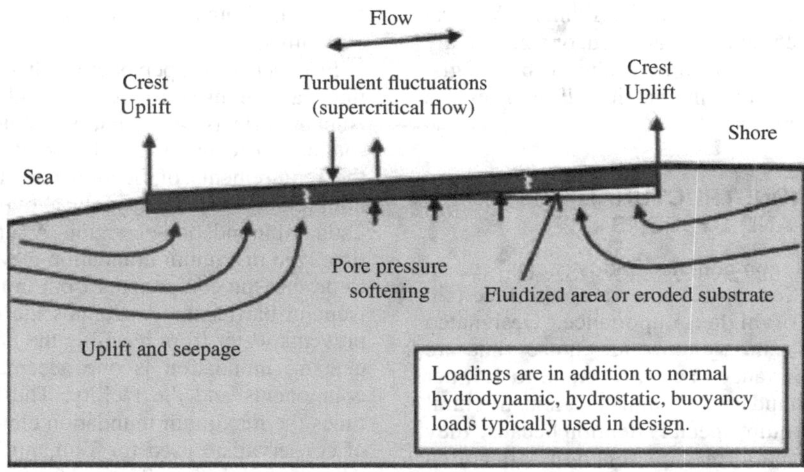

Figure C6.12-6. Schematic of tsunami-induced loading on exterior slabs on grade.

concrete barrier walls with active or passive floodgates. The design of large-scale coastal infrastructure barriers for extreme tsunami loading may involve other considerations beyond the scope of this section. Areas with existing tsunami mitigation barriers require coordination of performance objectives and interactions with new barrier designs, and design should include combined system scenario modeling.

Tsunami barriers are required in some cases to achieve reliable Life Safety performance under extreme tsunami loads, and this requirement highlights the need to consider siting design as well as traditional structure-level design. Tsunami barriers may also provide an opportunity for more cost-effective mitigation under modest-to-medium inundation loads.

Performance objectives of barriers may include the possibility of some overtopping. Overtopping may result in residual inundation of the protected structure. Design to account for the inundation caused by barrier overtopping is critical for those cases of intentional overtopping, and it is still prudent as a precaution for possible extreme events in excess of design levels.

This section focuses on adaptation of existing standards for modest-size levee design with geotextile reinforced earth and facing systems, using current best practices developed after Hurricane Katrina with consideration of the Tohoku tsunami (Kuwano et al. 2014). The standards are being evaluated and applied on a massive scale in the Gulf Coast and California (USACE 2000, California DWR 2012, CIRIA 2013). These methods incorporate specific failure mode analysis of fundamental performance requirements for stability, seepage control, and erosion and scour control, including overtopping conditions. The guidance also calls for checks of conventional levee design criteria.

C6.13.2.2 Site Layout The simplified guidance for determining the layout (location and linear extent) of tsunami mitigation barriers uses simplified shoreline setback and tsunami incident angle shielding criteria, based on general coastal engineering principles of wave inundation and barrier interaction (USACE 2011). This layout can be optimized with site-specific inundation scenario modeling. In this case, the designer should consider an analysis of various alternative barrier configurations to best address complex wave interactions during runup, drawdown, and channeling.

The requirement on radius of curvature for alignment changes is based on the avoidance of sharp corners, which may be vulnerable to scouring.

C6.14 TSUNAMI VERTICAL EVACUATION REFUGE STRUCTURES

Tsunami Vertical Evacuation Refuge Structures are a special classification of buildings and structures within the Tsunami Evacuation Zone designated as a means of alternative evacuation in communities where sufficiently high ground does not exist or where the time available after the tsunami warning is not deemed adequate for full evacuation before tsunami arrival. Such a building or structure should have the strength and resiliency needed to resist all effects of the Maximum Considered Tsunami. Despite the devastation of the March 11, 2011, Tohoku tsunami along the northeast coastline of Honshu Island of Japan, there were many tsunami evacuation buildings that provided safe refuge for thousands of survivors (Fraser et al. 2012, Chock et al. 2013b). In the United States, FEMA has published the P-646 *Guidelines for Design of Structures for Vertical Evacuation from Tsunamis* as a set of guidelines for the siting, design, and operation of vertical evacuation refuges (FEMA 2019). However, as a guideline, it is not written in the mandatory language necessary for building code and design standards. The unreduced live load of 100 lb/ft^2 (4.8 kPa) for public assembly is deemed adequate because occupancy of the designated refuge areas should not be more densely packed than exits; this unreduced live load is consistent with the basic intent of FEMA (2019).

Particularly important considerations are the elevation and height of the refuge since the refuge should provide structural Life Safety for the occupants within a portion of the refuge that is not inundated. Therefore, additional conservatism is necessary in the estimation of inundation elevation. The minimum elevation for a tsunami refuge area is, therefore, the Maximum Considered Tsunami inundation elevation at the site, multiplied by 1.3, plus 10 ft (3.05 m). Section 6.14.1 states, "This same Maximum Considered Tsunami site-specific inundation elevation, factored by 1.3, shall also be used for design of the Tsunami Vertical Evacuation Refuge Structure in accordance with Sections 6.8 to 6.12." There have been extensive comparisons of predicted versus observed heights for historic tsunamis. A plus-or-minus 30% deviation is generally described as reasonable agreement between field-observed data and model-predicted values. For this reason, the additional 30% factor is consistent with the skill level of present-day tsunami inundation models, for example as discussed in Tang et al. (2009, 2012).

In the event that it is discovered that the Tsunami Vertical Evacuation Refuge Structure is altered, damaged, or significantly deteriorated, the structure may need to be evaluated by a registered design professional to confirm that it still satisfies the requirements of this chapter.

C6.15 DESIGNATED NONSTRUCTURAL COMPONENTS AND SYSTEMS

"Designated nonstructural components and systems" is an explicitly defined term in Section 6.2; they are within certain buildings and structures of higher importance. Designated nonstructural components and systems are those that are assigned a component Importance Factor, I_p, equal to 1.5, per Section 13.1.3 of this standard. Designated nonstructural components and systems require special attention because they are needed to continue to perform their functions after both earthquake and tsunami events. For this reason, the same definition of what is considered a designated nonstructural component is used for tsunami effects as is used for earthquake effects. Nonstructural systems that are required for the continued operation of an essential building or structure in a Tsunami Design Zone need to be protected from tsunami inundation effects.

From a tsunami perspective, there are three approaches that can be used to better ensure that designated nonstructural components will perform as needed. One approach is to locate the components in the structure of concern above the Maximum Considered Tsunami inundation elevation. The second approach is to protect the components from inundation effects. Providing a tsunami barrier that surrounds the facility being protected and prevents water from reaching the component during tsunami inundation is one acceptable way of protecting the component(s) and the facility. The barrier height is set as 1.3 times the maximum inundation elevation, which is the same level of conservatism used for Tsunami Vertical Evacuation Refuge Structure design in Section 6.14. For large facilities where the tsunami barrier surrounds many structures, the height of the barrier may vary because the maximum inundation level may vary along the perimeter of the protective berm. The third approach that can be used is to allow the components to be designed directly for tsunami effects. The third approach may be suitable for pipes and vessels, which are inherently leak tight. However, it would not be suitable for mechanical or electrical equipment, where submersion in water (likely saltwater) would probably render the equipment inoperable. For the third approach, the designated nonstructural components and systems would need to be designed to resist flotation, collapse, and permanent lateral displacement caused by action of tsunami and debris loads in accordance with the earlier part of Chapter 6.

C6.16 NONBUILDING TSUNAMI RISK CATEGORY III AND IV STRUCTURES

The requirements of this section apply to nonbuilding structures that are required to be designed for tsunami effects. Risk Category II nonbuilding structures do not generally need to be designed for tsunami effects, and therefore requirements for Risk Category II nonbuilding structures are not provided. It should be noted, however, that some nonbuilding structures, such as tanks and vessels, could float if inundated, and it may be wise to tether or restrain them so that they would not cause damage to other nonbuilding critical structures in the vicinity (Naito et al. 2013). Requirements are provided for both Tsunami Risk Category III and IV nonbuilding structures.

From a tsunami perspective, there are at least four approaches that can be used to design nonbuilding structures to resist tsunami effects. The first is to design the structure and its foundation to resist the effects of tsunami forces directly, per the requirements of Section 6.8. The second is to locate the nonbuilding structure safely above the Maximum Considered Tsunami inundation elevation. A safe height is deemed to be 1.3 times the maximum inundation elevation. The third approach is to protect the components from inundation effects. Providing a tsunami barrier that surrounds the facility being protected and prevents water from reaching the nonbuilding structures during tsunami inundation is one acceptable way of protecting the components and the facility. The barrier height is set as 1.3 times the maximum inundation elevation; this is the same level of conservatism used for Tsunami Vertical Evacuation Refuge Structure design in Section 6.14. For large facilities where a tsunami barrier surrounds many structures, the height of the barrier may vary, because the maximum inundation level may vary along the perimeter of the protective berm. A fourth approach is to design a protective barrier to mitigate the inundation depth to a level sustainable by the structure (rather than keeping the structure entirely dry).

REFERENCES

AASHTO (American Association of State Highway and Transportation Officials). 2006. *Standard specification for geotextile specification for highway applications*. M288-06. Washington, DC: AASHTO.

AASHTO. 2017. *AASHTO LRFD bridge design specifications*, 8th ed. Washington, DC: AASHTO.

Abdollahi, A., and H. B. Mason. 2019a. "Coupled seepage-deformation model for predicting pore-water pressure response during tsunami loading." *J. Geotech. Geoenviron. Eng.* 145 (3): 04019002. https://doi.org/10.1061/(ASCE)GT.1943-5606.0002012.

Abdollahi, A., and H. B. Mason. 2019b. "Tsunami-induced pore water pressure response of unsaturated soil beds: numerical formulation and experiments." *Comput. Geotech.* 110 (Jun): 19–27. https://doi.org/10.1016/j.compgeo.2019.02.012.

Abdollahi, A., and H. B. Mason. 2020a. "Estimating tsunami-induced uplift pressure." *Géotech. Lett.* 10 (2): 270–276. https://doi.org/10.1680/jgele.19.00104.

Abdollahi, A., and H. B. Mason. 2020b. "Pore water pressure response during tsunami loading." *J. Geotech. Geoenviron. Eng.* 146 (3): 04020004. https://doi.org/10.1061/(ASCE)GT.1943-5606.0002205.

Aburaya, T., and F. Imamura. 2002. "The proposal of a tsunami run-up simulation using combined equivalent roughness." [In Japanese.] *Proc. Coastal Eng. Conf. (JSCE)* 49: 276–280.

ACI (American Concrete Institute). 2013. *Code requirements for nuclear safety-related concrete structures and commentary*. ACI 349-13. Farmington Hills, MI: ACI.

Aida, I. 1978. "Reliability of a tsunami source model derived from fault parameters." *J. Phys. Earth* 26 (1): 57–73. https://doi.org/10.4294/jpe1952.26.57.

Akbas, S. O., and E. Tekin. 2013. "Estimation of resistance factors for reliability-based design of shallow foundations in cohesionless soils under earthquake loading." In *Foundation engineering in the face of uncertainty: Honoring: Geotechnical special publication 229*, F. H. Kulhawy, J. L. Withiam, K.-K. Phoon, and M. Hussein, eds., 555–569. Reston, VA: ASCE.

API (American Petroleum Institute). 2004. "Recommended practice 2A-WSD." In *Section 6, foundation design*, 21st ed. Washington, DC: API.

Arnason, H., C. Petroff, and H. Yeh. 2009. "Tsunami bore impingement onto a vertical column." *J. Disaster Res.* 4 (6): 391–403.

ASCE. 2013. *Unified definitions for critical infrastructure resilience: ASCE policy statement 518*. Reston, VA: ASCE.

ASCE. 2017a. *Tsunami design zone maps for selected locations*. ASCE 7-16. Reston, VA: ASCE.

ASCE. 2017b. *Seismic evaluation and retrofit of existing buildings*. ASCE 41-17. Reston, VA: ASCE.

ASCE/SEI (ASCE Structural Engineering Institute). 2017. *Minimum design loads and associated criteria for buildings and other structures*. ASCE 7-16. Reston, VA: ASCE.

ASTM International 2007. *Standard test methods for compressive strength of molded soil–cement cylinders*. D1633-00. West Conshohocken, PA: ASTM.

ASTM. 2011. *Standard practice for establishing clear wood strength values*. D2555-06. West Conshohocken, PA: ASTM.

Baba, T., S. Allgeyer, J. Hossen, P. R. Cummins, H. Tsushima, and K. Imai, et al. 2017. "Accurate numerical simulation of the far-field tsunami caused by the 2011 Tohoku earthquake, including the effects of Boussinesq dispersion, seawater density stratification, elastic loading, and gravitational potential change." *Ocean Modell.* 111 (Mar): 46–54. https://doi.org/10.1016/j.ocemod.2017.01.002.

Bai, Y., and K. F. Cheung. 2016. "Hydrostatic versus nonhydrostatic modeling of tsunamis with implications for insular shelf and reef environments." *Coastal Eng.* 117 (Nov): 32–43. https://doi.org/10.1016/j.coastaleng.2016.07.008.

Bai, Y., Y. Yamazaki, and K. F. Cheung. 2018. "Amplification of drawdown and runup over Hawaii's insular shelves by tsunami N-waves from mega Aleutian earthquakes." *Ocean Modell.* 124 (Apr): 61–74. https://doi.org/10.1016/j.ocemod.2018.02.006.

Baiguera, M., T. Rossetto, and I. N. Robertson. 2020. "Tsunami design using nonlinear push-over analysis." In *Proc., 17th World Conf. on Earthquake Engineering*, Sendai, Japan.

Bernoulli, D. 1738. *Hydrodynamica, sive de viribus et motibus fluidorum commentarii. English version, Hydrodynamics, or commentaries on the forces and motions of fluids*. Translated by T. Carmody and H. Kobus (1968).

Bishop, A. W. 1973. "The influence of an undrained change in stress on the pore pressure in porous media of low compressibility." *Géotechnique* 23 (3): 435–442. https://doi.org/10.1680/geot.1973.23.3.435.

Blevins, R. D. 2003. *Applied fluid dynamics handbook*. Malabar, FL: Krieger.

Briggs, M. J., H. Yeh, and D. Cox. 2010. "Physical modeling of tsunami waves." In *Handbook of coastal and ocean engineering*, Y. C. Kim, ed., 1073–1106. Singapore: World Scientific.

Buckingham, E. 1914. "On physically similar systems: Illustrations of the use of dimensional analysis." *Phys. Rev.* 4 (4): 345. https://doi.org/10.1103/PhysRev.4.345.

Bureau of Reclamation and US Department of the Interior. 2015. *Reclamation: Managing water in the West. Design standards No. 13: Embankment dams*, Washington DC.

California DWR (Department of Water Resources). 2012. *Urban levee design criteria*. Sacramento, CA: California DWR.

California Geological Survey. 2015. *Evaluation and application of probabilistic tsunami hazard analysis in California*. Special Rep. No. 237. Sacramento, CA: California Geological Survey.

California Geological Survey. 2020. *Tsunami design zone maps for the California building code*. Special Rep. No. 2507. Sacramento, CA: California Geological Survey.

California State Lands Commission. 2005. *The marine oil terminal engineering and maintenance standard (MOTEMS)*. Sacramento, CA: California State Lands Commission.

California State Lands Commission. 2016. "2016 CCR, Title 24, Part Z, California Building Code, Chapter 31F." In *Marine oil terminals (effective January 1, 2017)*. Sacramento, CA: California State Lands Commission.

Carden, L., G. Chock, G. Yu, and I. Robertson. 2015. "The New ASCE tsunami design standard applied to mitigate Tohoku tsunami building structural failure mechanisms." In *Handbook of coastal disaster mitigation for engineers and planners*. Amsterdam, Netherlands: Elsevier Science and Technology.

Cedergen, H. R. 1989. *Seepage drainage and flownets*, 3rd ed. New York: Wiley.

Chau, K. T., and J. Q. Bao. 2010. "Hydrodynamic analysis of boulder transportation on Phi-Phi Island during the 2004 Indian Ocean tsunami." In *Proc., 7th Int. Conf. on Urban Earthquake Engineering (7CUEE) and 5th Int. Conf. on Earthquake Engineering (5ICEE)*, Tokyo Institute of Technology, Japan.

Chock, G. 2015. "The ASCE 7 tsunami loads and effects design standard for the United States." In *Handbook of coastal disaster mitigation for engineers and planners*. Amsterdam, Netherlands: Elsevier Science and Technology.

Chock, G. 2016. "Design for tsunami loads and effects in the ASCE 7-16 standard." *J. Struct. Eng.* 142 (11): 04016093. https://doi.org/10.1061/(ASCE)ST.1943-541X.0001565.

Chock, G., L. Carden, I. Robertson, M. J. Olsen, and G. Yu. 2013a. "Tohoku tsunami-induced building failure analysis with implications for USA tsunami and seismic design codes." Supplement, *Earthquake Spectra* 29 (S1): S99–S126. https://doi.org/10.1193/1.4000113.

Chock, G., L. Carden, I. N. Robertson, Y. Wei, R. Wilson, and J. Hooper. 2018. "Tsunami-resilient building design considerations for coastal communities of Washington, Oregon, and California." *J. Struct. Eng.* 144 (8): 04018116. https://doi.org/10.1061/(ASCE)ST.1943-541X.0002068.

Chock, G., I. Robertson, D. Kriebel, M. Francis, and I. Nistor. 2013b. *Tohoku, Japan, earthquake and tsunami of 2011: Performance of structures under tsunami loads*. Reston, VA: ASCE.

Chock, G., G. Yu, H. K. Thio, and P. Lynett. 2016. "Target structural reliability analysis for tsunami hydrodynamic loads of the ASCE 7 standard." *J. Struct. Eng.* 142 (11): 04016092. https://doi.org/10.1061/(ASCE)ST.1943-541X.0001499.

Chopra, K. A. 2012. *Dynamics of structures: Theory and applications to earthquake engineering*. 4th ed. Upper Saddle River, NJ: Prentice Hall.

CIRIA, French Ministry of Ecology, and US Army Corps Engineers. 2013. *The international levee handbook*. C731. London: CIRIA.

Clopper, P., and Y. H. Chen. 1988. *Minimizing embankment damage during overtopping flow*. FHWA-RD-88-181. Sterling, VA: Federal Highway Administration.

Del Zoppo, M., T. Rossetto, M. Di Ludovico, A. Prota, and I. Robertson. 2020. "Structural response under tsunami-induced vertical loads." In *Proc., 17th World Conf. on Earthquake Engineering*, Sendai, Japan.

DOD (US Department of Defense). 2013. *Design of structures to resist progressive collapse*, Washington, DC. UFC 4-023-03.

DOD. 2019. *Warehouses and storage facilities: Unified facilities criteria*, Washington, DC. UFC 4-440-01.

Didenkulova, I., and E. Pelinovsky. 2011. "Runup of tsunami waves in U-shaped bays." *Pure Appl. Geophys.* 168 (6): 1239–1249.

Douglass, S., and J. Krolak. 2008. *Highways in the coastal environment*. FHWA NHI-07-096. Sterling, VA: Federal Highway Administration.

Esteban, M., et al. 2014. "Stability of breakwater armor units against tsunami attacks." *J. Waterway, Port, Coastal, Ocean Eng.* 140 (2): 188–198. https://doi.org/10.1061/(ASCE)WW.1943-5460.0000227.

Fahlbusch, F. E. 1994. "Scour in rock river beds downstream of large dams." *Int. J. Hydropower Dams* 1 (4): 30–32.

FEMA (Federal Emergency Management Agency). 2007. *Design guide for improving critical facility safety from flooding and high winds*. FEMA P-543. Washington, DC: FEMA.

FEMA. 2011. *Coastal construction manual*. FEMA P-55. Washington, DC: FEMA.

FEMA. 2019. *Guidelines for design of structures for vertical evacuation from tsunamis*. P-646. Washington, DC: FEMA.

FHWA (Federal Highway Administration). 2000. *An introduction to the deep soil mixing methods as used in geotechnical applications*. FHWA-RD-99-138. Washington, DC: FHWA.

FHWA. 2009. "*Bridge scour and stream instability countermeasures*. FHWA-NHI-09-111. Washington, DC: FHWA.

FHWA. 2010. *Design and construction of mechanically stabilized earth walls and reinforced soil slopes*." FHWA-NHI-10-024. Washington, DC: FHWA.

FHWA. 2012. *Evaluating scour at bridges*. FHWA-HIF-12-003. Washington, DC: FHWA.

Francis, M. 2008. *Tsunami inundation scour of roadways, bridges and foundations: Observations and technical guidance from the Great Sumatra Andaman tsunami*. EERI/FEMA NEHRP 2006, Corvallis, Oregon: Oregon State University.

Fraser, S., G. S. Leonard, I. Matsuo, and H. Murakami. 2012. *Tsunami evacuation: Lessons from the Great East Japan earthquake and tsunami of March 11th, 2011*. GNS Science Rep. No. 2012/17. Lower Hutt, NZ: GNS Science.

Ge, M., and I. N. Robertson. 2010. *Uplift loading on elevated floor slab due to a tsunami bore*. Res. Rep. UHM/CEE/10-03. Honolulu, HI: University of Hawaii, Manoa.

Ghodoosipour, B., J. Stolle, I. Nistor, M. Mohammadian, and N. Goseberg. 2019a. "Experimental study on extreme hydrodynamic loading on pipelines. Part 1: Flow hydrodynamics." *J. Mar. Sci. Eng.* 7 (8): 251. https://doi.org/10.3390/jmse7080251.

Ghodoosipour, B., J. Stolle, I. Nistor, M. Mohammadian, and N. Goseberg. 2019b. "Experimental study on extreme hydrodynamic loading on pipelines. Part 2: Induced force analysis." *J. Mar. Sci. Eng.* 7 (8): 262. https://doi.org/10.3390/jmse7080262.

Gica, E., M. C. Spillane, V. V. Titov, C. D. Chamberlin, and J. C. Newman. 2008. *Development of the forecast propagation database for NOAA's short-term inundation forecast for tsunamis (SIFT)*. NOAA Tech. Memo OAR PMEL-139, University of Washington, Seattle, Washington.

Goseberg, N., A. Wurpts, and T. Schlurmann. 2013. "Laboratory-scale generation of tsunami and long waves." *Coastal Eng.* 79 (Sep): 57–74. https://doi.org/10.1016/j.coastaleng.2013.04.006.

Haehnel, R. B., and S. F. Daly. 2004. "Maximum impact force of woody debris on floodplain structures." *J. Hydraul. Eng.* 130 (2): 112–120. https://doi.org/10.1061/(ASCE)0733-9429(2004)130:2(112).

Hicher, P.-Y. 1996. "Elastic properties of soils." *J. Geo. Eng.* 122 (8): 641–648. https://doi.org/10.1061/(ASCE)0733-9410(1996)122:8(641).

Hoffmans, G. J. C. M., and H. J. Verheij. 1997. *Scour manual*. Boca Raton, FL: CRC Press.

Hughes, S. A. 1993. *Physical models and laboratory techniques in coastal engineering: advanced series on ocean engineering, 7*. Singapore: World Scientific.

Hughes, S. A. 2004. "Estimation of wave run-up on smooth, impermeable slopes using the wave momentum flux parameter." *Coastal Eng.* 51 (11–12): 1085–1104. https://doi.org/10.1016/j.coastaleng.2004.07.026.

Imamura, F. 2009. "Tsunami modeling: Calculating inundation and hazard maps." *The Sea*, Sendai, Japan: Tohuku University 15: 321–332.

Intergovernmental Oceanographic Commission. 2013. *Tsunami glossary, 2013: IOC technical series 85*. Paris: UNESCO.

ICC (International Code Council). 2018. *2018 international building code*. Country Club Hills, IL: ICC.

Jantzer, I., and S. Knutsson. 2013. *Critical gradients for tailings dam design*. Sweden: Luleå University of Technology.

Kirby, J. T., F. Shi, B. Tehranirad, J. C. Harris, and S. T. Grilli. 2013. "Dispersive tsunami waves in the ocean: Model equations and sensitivity to dispersion and Coriolis effects." *Ocean Modelling* 62 (Feb): 39–55. https://doi.org/10.1016/j.ocemod.2012.11.009.

Kokusho, T. 2000. "Correlation of pore pressure B-value with P-wave velocity and Poisson's ratio for imperfectly saturated sand or gravel." *Soils Found.* 40 (4): 95–102. https://doi.org/10.3208/sandf.40.4_95.

Kotani, M., F. Imamura, and N. Shuto. 1998. "Tsunami run-up calculations and damage estimation method using GIS." [In Japanese.] *Pac. Coast. Eng.* 45: 356–360.

Kriebel, D. L., P. J. Lynett, D. T. Cox, C. M. Petroff, H. R. Riggs, I. N. Robertson, and G. Y. K. Chock. 2017. "Energy method for approximating overland tsunami flows." *J. Waterway, Port, Coastal, Ocean Eng.* 143 (5): 04017014. https://doi.org/10.1061/(ASCE)WW.1943-5460.0000393.

Kuilanoff, G., and R. M. Drake. 1991. "Design of DOE facilities for wind-generated missiles." In *Proc., 3rd DOE Natural Phenomena Hazards Mitigation Conf.*, Washington, DC.

Kuwano, J., J. Koseki, and Y. Miyata. 2014. "Performance of reinforced soil walls during the 2011 Tohoku earthquake." *Geosynth. Int.* 21 (3): 179–196. https://doi.org/10.1680/gein.14.00008.

Lane, E. W. 1947. "Report of the subcommittee on sediment terminology." *Trans. Am. Geophys. Union* 28 (6): 936–938. https://doi.org/10.1029/TR028i006p00936.

Lavictoire, A., I. Nistor, and C. Rennie. 2014. "Local scour around structures due to hydraulic bores." In *Proc., Annual Conf. of the Canadian Society for Civil Engineering*, Halifax.

Lebreton, L. M., S. D. Greer, and J. C. Borrero. 2012. "Numerical modelling of floating debris in the world's oceans." *Marine Pollut. Bull.* 64 (3): 653–661. https://doi.org/10.1016/j.marpolbul.2011.10.027.

Li, Y. 2000. *Tsunamis: Non-breaking solitary wave run-up*. Rep. No. KH-R-60. Pasadena: California Institute of Technology.

Li, Y., and F. Raichlen. 2001. "Solitary wave runup on plane slopes." *J. Waterway, Port, Coastal, Ocean Eng.* 127 (1): 33–44. https://doi.org/10.1061/(ASCE)0733-950X(2001)127:1(33).

Li, Y., and F. Raichlen. 2003. "Energy balance model for breaking solitary wave runup." *J. Waterway, Port, Coastal, Ocean Eng.* 129 (2): 47–59. https://doi.org/10.1061/(ASCE)0733-950X(2003)129:2(47).

Madsen, P. A., D. R. Fuhrman, and H. A. Schaffer. 2008. "On the solitary wave paradigm for tsunamis." *J. Geo. Res.* 113 (Dec): C12012. https://doi.org/10.1029/2008JC004932.

Madsen, P. A., and H. A. Schäffer. 2010. "Analytical solutions for tsunami runup on a plane beach: Single waves, N-waves and transient waves." *J. Fluid Mech.* 645 (Feb): 27–57. https://doi.org/10.1017/S0022112009992485.

Madurapperuma, M. A. K. M., and A. C. Wijeyewickrema. 2013. "Response of reinforced concrete columns impacted by tsunami dispersed 20′ and 40′ shipping containers." *Eng. Struct.* 56 (Nov): 1631–1644. https://doi.org/10.1016/j.engstruct.2013.07.034.

Martínez-Gomariz, E., M. Gómez, B. Russo, and S. Djordjević. 2016. "Stability criteria for flooded vehicles: A state-of-the-art review." *J. Flood Risk Manage.* 11: S817–S826. https://doi.org/10.1111/jfr3.12262.

Mason, H. B., and H. Yeh. 2016. "Sediment liquefaction: A porewater pressure gradient view-point." *Bull. Seismol. Soc. Am.* 106 (4): 1908–1913. https://doi.org/10.1785/0120150296.

Montoya, L., P. J. Lynett, H. K. Thio, and W. Li. 2017. "Spatial statistics of tsunami overland flow properties." *J. Waterway, Port, Coastal, Ocean Eng.* 143 (2): 04016017. https://doi.org/10.1061/(ASCE)WW.1943-5460.0000363.

Murata, S., F. Imamura, K. Katoh, Y. Kawata, S. Takahashi, and T. Takayama. 2010. "Tsunami: To survive from tsunami." *Adv. Series Ocean Eng.* 32: 260–265.

Naito, C., C. Cercone, H. R. Riggs, and D. Cox. 2014. "Procedure for site assessment of the potential for tsunami debris impact." *J. Waterway, Port, Coastal, Ocean Eng.* 140 (2): 223–232. https://doi.org/10.1061/(ASCE)WW.1943-5460.0000222.

Naito, C., D. Cox, Q. S. Yu, and H. Brooker. 2013. "Fuel storage container performance during the 2011 Tohoku, Japan, tsunami." *J. Perform. Constr. Facil.* 27 (4): 373–380. https://doi.org/10.1061/(ASCE)CF.1943-5509.0000339.

National Research Council. 2011. *Tsunami warning and preparedness: An assessment of the U.S. tsunami program and the nation's preparedness effort.* Washington, DC: National Academies Press.

National Tsunami Hazard Mitigation Program. 2012. *Proceedings and results of the 2011 NTHMP Model benchmarking workshop.* NOAA Special Report. Boulder, CO: US Dept. of Commerc.

NCAC (National Crash Analysis Center). 2011. *Development and validation of a finite element model for the 2010 Toyota Yaris passenger sedan.* NCAC 2011-T-001. Washington, DC: George Washington University.

NCAC. 2012. *Extended validation of the finite element model for the 2010 Toyota Yaris passenger sedan.* NCAC Working Paper 2012-W-005. Washington, DC: George Washington University.

Newman, J. H. 1977. *Marine hydrodynamics.* Boston: MIT Press.

Ngo, N., and I. N. Robertson. 2012. *Video analysis of the March 2011 tsunami in Japan's coastal cities.* Research Rep. No. UHM/CEE/12-11. Honolulu, HI: University of Hawaii.

Nicolsky, D. J., E. N. Suleimani, R. A. Combellick, and R. A. Hansen. 2011. *Tsunami inundation maps of Whittier and western passage canal, Alaska.* Report of Investigation 2011-7. Alaska Division of Geological & Geophysical Surveys, Fairbanks, Alaska.

Nicolsky, D. J., E. N. Suleimani, R. D. Koehler, and J. B. Salisbury. 2017. *Tsunami inundation maps for Juneau, Alaska.* Report of Investigation 2017-9. Alaska Division of Geological & Geophysical Surveys, Fairbanks, Alaska.

Nicolsky, D. J., E. N. Suleimani, and J. B. Salisbury. 2018. *Tsunami inundation maps for Skagway and Haines, Alaska.* Report of Investigation 2018-2. Alaska Division of Geological & Geophysical Surveys, Fairbanks, Alaska.

Nistor, I., N. Goseberg, T. Mikami, T. Shibayama, J. Stolle, and R. Nakamura, et al. 2017. "Hydraulic experiments on debris dynamics over a horizontal plane." *J. Waterway, Port, Coastal, Ocean Eng.* 143 (3): 04016022.

Nouri, Y., I. Nistor, D. Palermo, and A. Cornett. 2010. "Experimental investigation of the tsunami impact on free standing structures." *Coastal Eng. J.* 52 (1): 43–70.

OCADI (Overseas Coastal Area Development Institute of Japan). 2009. *Technical standards and commentaries for port and harbour facilities in Japan.* Tokyo: Ports and Harbours Bureau.

Office of the President. 2013. *Critical infrastructure security and resilience: Presidential policy directive 21.* Washington, DC: Office of the President.

Okada, Y. 1985. "Surface deformation due to shear and tensile faults in a half-space." *Bull. Seismol. Soc. Am.* 75 (4): 1135–1154.

Paczkowski, K. 2011. "Bore impact upon vertical wall and water-driven, high-mass, low-velocity debris impact." Ph.D. dissertation, Civil and Environmental Engineering, University of Hawaii, Honolulu.

Park, H., D. T. Cox, P. J. Lynett, D. M. Wiebe, and S. Shin. 2013. "Tsunami inundation modeling in constructed environments: A physical and numerical comparison of free-surface elevation, velocity, and momentum flux." *Coastal Eng.* 79: 9–21.

Patil, A., S. D. Mudiyanselage, J. D. Bricker, W. Uijttewaal, and G. Keetels. 2018. "Effect of overflow nappe non-aeration on tsunami breakwater failure." In *Proc., Coastal Engineering 2018*, Baltimore, Maryland.

PIANC (Permanent International Association of Navigation Congresses). 2010. *Mitigation of tsunami disasters in ports: PIANC working group 53.* Brussels, Belgium: PIANC.

PIANC. 2014. *Harbour approach channels: Design guidelines.* Rep. No. 121-2014. Brussels, Belgium: PIANC.

Piran Aghl, P., C. J. Naito, and H. R. Riggs. 2014. "Full-scale experimental study of impact demands resulting from high mass, low velocity debris." *J. Struct. Eng.* 140 (5): 04014006. https://doi.org/10.1061/(ASCE)ST.1943-541X.0000948.

Ramsden, J. D. 1993. *Tsunamis: Forces on a vertical wall caused by long waves, bores, and surges on a dry bed.* Rep. No. KH-R-54. Pasadena, CA: California Institute of Technology.

Razieh, S.,, I. Nistor, and C. Rennie. 2017. "Modeling supercritical flow-induced scour around structures." In *Proc., 23rd Canadian Hydrotechnical Conf.*, Vancouver, Canada.

Razieh, S., I. Nistor, and C. Rennie. 2021. "Scour mechanics of a tsunami-like bore around a square structure." PH.D. thesis. In Localized scour around structures under transient flow conditions. University of Ottowa, Dept. of Civil Engineering. 63–100. http://hdl.handle.net/10393/41979.

Riggs, H. R., D. T. Cox, C. J. Naito, M. H. Kobayashi, P. Piran Aghl, and H. T.-S. Ko, et al. 2014. "Experimental and analytical study of water- driven debris impact forces on structures." *J. Offshore Mech. Arct. Eng.* 136 (4) 041603. https://doi.org/10.1115/1.4028338.

Robertson, I. N. 2019. *Development of high-resolution probabilistic tsunami design zone maps compatible with ASCE 7-16 for the Island of Oahu, State of Hawaii.* Oahu, HI: Hawaii State Office of Planning.

Robertson, I. N., et al. 2019. *StEER: Structural extreme event reconnaissance network Palu earthquake and tsunami, Sulawesi, indonesia field assessment team 1 (Fat-1) early access reconnaissance report (EARR).* Rep. No. PRJ-2128. Notre Dame, IN: University of Notre Dame.

Robertson, I. N., L. Carden, H. R. Riggs, S. Yim, Y. L. Young, and K. Paczkowski, et al. 2010. *Reconnaissance following the September 29, 2009 tsunami in Samoa.* Research Rep. No. UHM/CEE/10-01. Honolulu, HI: University of Hawaii at Manoa.

Robertson, I. N., K. Pacskowski, H. R. Riggs, and A. Mohamed. 2013. "Experimental investigation of tsunami bore forces on vertical walls." *J. Offshore Mech. Arct. Eng.* 135 (2): 021601-1–021601-8. https://doi.org/10.1115/1.4023149.

Saatcioglu, M., A. Ghobarah, and I. Nistor 2005. *Reconnaissance report on the 26 Dec 2004 Sumatra earthquake and tsunami*. Ottawa: Canadian Association for Earthquake Engineering.

Santo, J., and I. N. Robertson. 2010. *Lateral loading on vertical structural elements due to a tsunami bore*. Research Rep. No. UHM/CEE/10-02. Honolulu, HI: University of Hawaii at Manoa.

Sarpkaya, T. 2010. *Wave forces on offshore structures*. Cambridge, UK: Cambridge University Press.

Seed, H. B. 1979. "Soil liquefaction and cyclic mobility evaluation for level ground during earthquakes." *J. Geotech. Eng. Div.* 105 (2): 201–255.

Seed, R., et al. 2008. "New Orleans and Hurricane Katrina. II: The central region and the lower Ninth Ward." *J. Geotech. Geoenviron. Eng.* 134 (5): 718–739. https://doi.org/10.1061/(ASCE)1090-0241(2008)134:5(718).

Shimada, Y., Y. Kurachi, M. Toyota, and G. Tomidokoro. 2003. "Flood inundation simulation considering fine land categories by means of unstructured grids." In *Proc., Int. Symp. Disaster Mitigation and Basin-Wide Water Management, ISDB 2003*, Niigata, Japan, 70–77.

Simons, D. B., and F. Senturk. 1977. *Sediment transport technology*. Fort Collins, CO: Water Resources.

Skempton, A. 1954. "The pore-pressure coefficients A and B." *Géotechnique* 4 (4): 143–147.

Stolle, J., N. Goseberg, I. Nistor, and E. Petriu. 2018. "Probabilistic investigation and risk assessment of debris transport in extreme hydrodynamic conditions." *J. Waterway, Port, Coastal, Ocean Eng.* 144 (1): 04017039. https://doi.org/10.1061/(ASCE)WW.1943-5460.0000428.

Stolle, J., C. Krautwald, I. Robertson, H. Achiari, T. Mikami, and R. Nakamura, et al. 2019. "Engineering lessons from the 28 September 2018 Indonesian tsunami: Debris loading." *Can. J. Civ. Eng.* 47 (1): 1–12. https://doi.org/10.1139/cjce-2019-0049.

Suleimani, E. N., D. J. Nicolsky, and R. D. Koehler. 2015. *Tsunami inundation maps of Elfin Cove, Gustavus, and Hoonah, Alaska*. Report of Investigation 2015-1. Alaska Division of Geological & Geophysical Surveys, Fairbanks, AK.

Suleimani, E. N., D. J. Nicolsky, and R. D. Koehler. 2016. *Tsunami inundation maps for Yakutat, Alaska*. Report of Investigation 2016-2. Alaska Division of Geological & Geophysical Surveys, Fairbanks, AK.

Suleimani, E. N., D. J. Nicolsky, and J. B. Salisbury. 2018. *Updated tsunami inundation maps for Homer and Seldovia, Alaska*. Report of Investigation 2018-5. Alaska Division of Geological & Geophysical Surveys, Fairbanks, AK.

Suleimani, E. N., D. J. Nicolsky, D. A. West, R. A. Combellick, and R. A. Hansen. 2010. *Tsunami inundation maps of Seward and northern Resurrection Bay, Alaska*. Report of Investigation 2010-1. Alaska Division of Geological & Geophysical Surveys, Fairbanks, AK.

Sullivan, T. J., G. M. Calvi, and M. J. N. Priestley. 2004. "Initial stiffness versus secant stiffness in displacement based design." In *Proc., 13th World Conf. on Earthquake Engineering*, Vancouver, BC.

Synolakis, C. E. 1986. *The runup of long waves*. Rep. No. KH-R-61. Pasadena, CA: California Institute of Technology.

Synolakis, C. E., E. N. Bernard, V. V. Titov, U. Kanoglu, and F. I. Gonzalez. 2007. *Standards, criteria, and procedures for NOAA evaluation of tsunami numerical models*. NOAA Technical Memorandum OAR PMEL-135. National Tsunami Hazard Mitigation Program, Los Angeles.

Takakura, R., and I. N. Robertson. 2010. *Reducing tsunami bore uplift forces by providing a breakaway panel*. Research Report UHM/CEE/10-07. Honolulu, HI: University of Hawaii at Manoa.

Tang, L., V. V. Titov, E. Bernard, Y. Wei, C. Chamberlin, and J. C. Newman, et al. 2012. "Direct energy estimation of the 2011 Japan tsunami using deep-ocean pressure measurements." *J. Geophys. Res.* 117 (Aug): C08008. https://doi.org/10.1029/2011JC007635.

Tang, L., V. V. Titov, and C. D. Chamberlin. 2009. "Development, testing, and applications of site-specific tsunami inundation models for real-time forecasting." *J. Geophys. Res.* 114 (Dec): C12025. https://doi.org/10.1029/2009JC005476.

Thio, H. K. 2019. *Probabilistic tsunami hazard maps for the state of California (Phase 2)*. AECOM Technical Services, Los Angeles.

Thio, H. K. 2020. *Probabilistic tsunami hazard analysis for the Washington State Ferry Terminals*. AECOM Technical Services, Los Angeles.

Thio, H. K., P. G. Somerville, and L. Polet. 2010. *Probabilistic tsunami hazard in California*. Rep. No. 108, 331. Berkeley, CA: Pacific Earthquake Engineering Research Center.

Thomas, S., J. Killian, and K. Bridges. 2014. "Influence of macroroughness on tsunami loading of coastal structures." *J. Waterway, Port, Coastal, Ocean Eng.* 141 (1): 04014028. https://doi.org/10.1061/(ASCE)WW.1943-5460.0000268.

TISP (The Infrastructure Security Partnership). 2012. *A guide to regional resilience planning*, 2nd ed. Arlington, VA: Society of Military Engineers Press.

Titov, V. V., and C. E. Synolakis. 1998. "Numerical modeling of tidal wave runup." *J. Waterway, Port, Coastal, Ocean Eng.* 124 (4): 157–171. https://doi.org/10.1061/(ASCE)0733-950X(1998)124:4(157).

Tonkin, S. P., M. Francis, and J. D. Bricker. 2013. "Limits on coastal scour depths due to tsunami." In *International efforts in lifeline earthquake engineering: TCLEE monograph 38*, Davis, C., X. Du, M. Miyajima, and L Yan, eds. Reston, VA: ASCE.

Tonkin, S. P., H. Yeh, F. Kato, and S. Sato. 2003. "Tsunami scour around a cylinder." *J. Fluid Mech.* 496 (Dec): 165–192.

USACE (US Army Corps of Engineers). 1984. *Shore protection manual*. Washington, DC: Department of the Army.

USACE. 1989. *Retaining and flood walls*. Engineering and Design. Engineer Manual 1110-2-2502. Washington, DC: Department of the Army.

USACE. 1991. *Design of pile foundations*. Engineering and Design. Engineer Manual 1110-2-2906. Washington, DC: Department of the Army.

USACE. 1995. *Gravity dam design*. Engineer Manual 1110-2-2200. Washington, DC: USACE.

USACE. 2000. *Design and construction of levees*. Engineer Manual 1110-2-1913. Washington, DC: USACE.

USACE. 2003. *Slope stability*. Engineer Manual 1110-2-1902. Washington, DC: USACE.

USACE. 2005. *Stability analysis of concrete structures*. Engineer Manual 1110-2-2100. Washington, DC: USACE.

USACE. 2010. "Hydrologic engineering centers, computer programs." In *HEC-RAS, river analysis system, v. 4.1*. Washington, DC: USACE.

USACE. 2011. *Coastal engineering manual (CEM)*. Engineer Manual 1110-2-1100. Washington, DC: USACE.

USACE. 2012. *Department of defense groundwater modeling system (GMS)*. Washington, DC: USACE.

USACE. 2013. *Incorporating sea-level change considerations for civil works programs*. Engineer Regulation 1100-2-8162. Washington, DC: USACE.

USACE, H. C. Lin, D. R. Richards, G. T. Yeh, J. R. Cheng, H. P. Cheng, and N. L. Jones. 1997. *FEMWATER: A three-dimensional finite element computer model for simulating density-dependent flow and transport in variably saturated media.* Technical Rep. No. CHL-97-12. Washington, DC: USACE.

USACE and F. T. Tracy. 1973. *SEEP2D: A plane and axisymmetric finite element program for steady-state and transient seepage problems: Miscellaneous paper MP K-73-4.* Vicksburg, MS: US Army Engineer Waterways Experiment Station.

USGS (US Geological Survey), M. McDonald, and A. Harbaugh. 1988. *MODFLOW: A modular three-dimensional finite-difference ground-water flow model.* Techniques of Water-Resources Investigations 06-A1. Reston, VA: USGS.

USNRC (US Nuclear Regulatory Commission). 1978. *Regulatory guide 1.70: Standard format and content of safety analysis reports for nuclear power plants*, LWR ed. Washington, DC: USNRC.

USNRC. 2003. *Regulatory guide 1.198, procedures and criteria for assessing seismic soil liquefaction at nuclear power plant sites.* Washington, DC: USNRC.

USNRC. 2013. *Standard review plan, NUREG-0800, 3.8.5 foundations, Rev. 4.* Washington, DC: USNRC.

USNRC. 2014. *Standard review plan, NUREG-0800, 2.5.5 stability of slopes, Rev. 5.* Washington, DC: USNRC.

Van Sebille, E., S. M. Griffies, R. Abernathey, T. P. Adams, P. Berloff, and A. Biastoch, et al. 2018. "Lagrangian ocean analysis: Fundamentals and practices." *Ocean Modell.* 121 (Jan): 49–75.

Venturato, A. J., D. Arcas, V. V. Titov, H. O. Mofjeld, C. C. Chamberlin, and F. I. Gonzalez. 2007. *Tacoma, Washington, tsunami hazard mapping project: Modeling tsunami inundation from Tacoma and Seattle fault earthquakes.* NOAA Tech. Memo OAR PMEL-132. Seattle: Pacific Marine Environmental Laboratory.

Wei, Y. 2016. *Tsunami probabilistic reference maps for benchmarking Hawaii tsunami design zone maps per the ASCE 7-16 standard.* Honolulu, HI: Hawaii Emergency Management Agency.

Wilson, R. I., H. K. Thio, W. Li, J. D. J. Bott, N. A. Graehl, and T. P. McCrink, et al. 2020. *Tsunami design zone maps for the California building code.* Sacramento, CA: California Geological Survey, California Department of Conservation.

Wood, N. 2007. *Variations in city exposure and sensitivity to tsunami hazards in Oregon.* USGS Scientific Investigations Rep. No. 2007-5283. Menlo Park, CA: Western Geographic Science Center.

Wood, N., A. Church, T. Frazier, and B. Yarnal. 2007. *Variations in community exposure and sensitivity to tsunami hazards in the state of Hawai'i.* USGS Scientific Investigations Rep. No. 2007-5208. Menlo Park, CA: Western Geographic Science Center.

Wood, N., and J. Peters. 2015. "Variations in population vulnerability to tectonic and landslide-related tsunami hazards in Alaska." *Nat. Hazard.* 75 (2): 1811–1831.

Wood, N., J. Ratliff, and J. Peters. 2013. *Community exposure to tsunami hazards in California.* USGS Scientific Investigations Rep. No. 2012-5222. Menlo Park, CA: Western Geographic Science Center.

Wood, N., and C. Soulard. 2008. *Variations in community exposure and sensitivity to tsunami hazards on the open-ocean and Strait of Juan de Fuca coasts of Washington.* USGS Scientific Investigations Rep. No. 2008-5004. Menlo Park, CA: Western Geographic Science Center.

WSDOT (Washington State Department of Transportation). 2015. *Geotechnical design manual.* Publication M46-03. Seattle: WSDOT.

Xiao, H., Y. L. Young, and J. H. Prévost. 2010. "Parametric study of breaking solitary wave induced liquefaction of coastal sandy slopes." *Ocean Eng.* 37 (17): 1546–1553.

Yang, J. 2005. "Pore pressure coefficient for soil and rock and its relation to compressional wave velocity." *Géotechnique* 55 (3): 251–256. https://doi.org/10.1680/geot.2005.55.3.251.

Yeh, H., and W. Li. 2008. "Tsunami scour and sedimentation." In *Proc., 4th Int. Conf. on Scour and Erosion*, Tokyo, 95–106.

Yeh, H., S. Sato, and Y. Tajima. 2012. "The 11 March 2011 East Japan earthquake and tsunami: Tsunami effects on coastal infrastructure and buildings." *Pure Appl. Geophys.* 170 (6): 1019–1031.

Zhang, J., S. Jiang, Q. Wang, Y. Hou, and Z. Chen. 2010. "Critical hydraulic gradient of piping in sand." In *Proc., 20th Int. Soc. Offshore and Polar Engineers Conf.*, Beijing.

Zilkoski, D. B., J. H. Richards, and G. M. Young. 1992. "Results of the general adjustment of the North American vertical datum of 1988." *Surv. Land Inf. Syst.* 52 (3): 133–149.

OTHER REFERENCES (NOT CITED)

Robertson, I. N. 2020. *Tsunami loads and effects: Guide to the tsunami design provisions of ASCE 7-16.* Reston, VA: ASCE.

USACE. 1993a. *Coastal scour problems and prediction of maximum scour.* Technical Report CERC-93-8. Washington, DC: Department of the Army.

USACE. 1993b. *Seepage analysis and control of dams.* Engineer Manual 1110-2-1901. Washington, DC: Department of the Army.

USACE. 2019. *Procedures to evaluate sea level change: Impacts, responses, and adaptation.* EP 1100-2-1. Washington, DC: USACE.

CHAPTER C7
SNOW LOADS

C7.1 DEFINITIONS AND SYMBOLS

Methodology. The procedure established for determining design snow loads is as follows:

1. Determine the ground snow load for the geographic location (Sections 7.2 and C7.2).
2. Generate a flat roof snow load from the ground load, with consideration given to (1) roof exposure (Sections 7.3.1, C7.3, and C7.3.1); and (2) roof thermal condition (Sections 7.3.2, C7.3, and C7.3.2).
3. Consider roof slope (Sections 7.4 through 7.4.5 and C7.4).
4. Consider partial loading (Sections 7.5 and C7.5).
5. Consider unbalanced loads (Sections 7.6 through 7.6.4 and C7.6).
6. Consider snow drifts: (1) on lower roofs (Sections 7.7 through 7.7.2 and C7.7), and (2) from projections (Sections 7.8 and C7.8).
7. Consider sliding snow (Sections 7.9 and C7.9).
8. Consider extra loads from rain on snow (Sections 7.10 and C7.10).
9. Consider ponding loads (Section 7.11 and C7.11).
10. Consider existing roofs (Sections 7.12 and C7.12).
11. Consider other roofs and sites (Section C7.13).
12. Consider the consequences of loads in excess of the design value (see the following text).

Loads in Excess of the Design Value. The philosophy of the probabilistic approach used in this standard is to establish a design value that reduces the risk of a snow load–induced failure to an acceptably low level. Because snow loads in excess of the design value may occur, the implications of such "excess" loads should be considered. For example, if a roof is deflected at the design snow load so that slope to drain is eliminated, "excess" snow load might cause ponding (Section C7.11) and perhaps progressive failure.

The snow load/dead load ratio of a roof structure is an important consideration when assessing the implications of "excess" loads. If the design snow load is exceeded, the percentage increase in total load would be greater for a lightweight structure (i.e., one with a high snow load/dead load ratio) than for a heavy structure (i.e., one with a low snow load/dead load ratio). For example, if a 40 lb/ft^2 (1.92 kN/m^2) roof snow load is exceeded by 20 lb/ft^2 (0.96 kN/m^2) for a roof that has a 25 lb/ft^2 (1.19 kN/m^2) dead load, the total load increases by 31% from 65 to 85 lb/ft^2 (3.11 to 4.07 kN/m^2). If the roof had a 60 lb/ft^2 (2.87 kN/m^2) dead load, the total load would increase only by 20% from 100 to 120 lb/ft^2 (4.79 to 5.75 kN/m^2).

C7.2 Ground Snow Loads, p_g The ground snow load, p_g, values contained in the ASCE Design Ground Snow Load Geodatabase, mapped in Figures 7.2-1A through 7.2-1D and provided in the ASCE Hazard Tool are based on a reliability analysis consistent with the targets in Table 1.3-1 for a "failure that is not sudden and does not lead to widespread progression of damage." As with other reliability targets in this standard, it is left to the material design standards to address the higher targets for limit states associated with either sudden or widespread progression of damage through adjustments in design provisions and resistance factors. For example, in ACI 318, the resistance factors for flexural failure of a beam or slab and failure of a tied column in compression are quite different. Part of the reason for the difference is that the target reliabilities for those two limit states are set considering Table 1.3-1.

The adoption of reliability-targeted design ground snow loads represents a significant change from ASCE/SEI 7-16 and prior editions, which previously used ground snow loads with a 50-year mean recurrence interval (MRI). Reliability-targeted loads are adopted here to address the nonuniform reliability of roofs designed according to the 50-year snow load in different parts of the country, due to climatic differences. In some parts of the country, designing for the 1.6 load factor times the 50-year value does not meet the reliability targets of the standard (and, in some of these places, failures due to an underestimated ground snow load have been observed); in other places, designing for the 1.6 load factor times the 50-year value is unnecessarily conservative. With the move to reliability-targeted values, the load factor on snow loads has also been revised from 1.6 to 1.0 to appropriately represent the reliability basis of the values. Snow importance factors have also been eliminated because values now are provided for each risk category. The 0.7 factor is intended to provide roughly equivalent strength when design follows Allowable Stress Design (ASD) procedures. For some materials the ratio between design strength given by LRFD procedures and design strength given by ASD procedure is 1.5. For some other materials the ratio varies depending on the limit state being checked. The inverse of 1.5 was rounded to 0.7 for this purpose.

The new design ground snow loads are also based on nearly 30 years of additional snow load data since the previous study and updated procedures for estimating snow loads from depth-only measurements. The loads account for site-specific variability throughout the United States in both the magnitude and variation of the annual ground snow loads. Additionally, this approach incorporates advanced spatial mapping that has reduced the number and size of case study regions in mountainous areas significantly and eliminates discontinuities in design values across state boundaries.

To arrive at the reliability-targeted loads, ground snow data was collected from stations throughout the United States and in

Canada, within 60 mi of the northern border of the United States, via the global historical climatological network (GHCN). In most areas, only stations with at least 30 years of record were considered to ensure an adequate period of record for statistical analysis. However, exceptions were made for areas with few measurement stations, where the minimum years of record was reduced to 15. Outliers were screened based on quality assurance flags provided by the GHCN, as well as through manual inspections of daily records in places with anomalously high or low annual maximums.

Some of the measurements in the data set are in terms of snow water equivalent, and these measurements provide load directly. However, often, measurements only include snow depth, from which load must be estimated. There are several existing methods for conversion of snow depth-to-load, but these techniques tend to work best at the location and/or elevation from which they were derived. A new nationally applicable and verified approach was developed by Bean et al. (2021) using the random forest technique (Brieman 2001, Liaw and Wiener 2002) to represent the depth-to-load relationship across the country. This depth-to-load relationship makes use of a suite of 30-year climatic variables, such as winter precipitation, to account for climatic variation across the country.

Snow loads exhibit two typical patterns of behavior, depending on the location. In mountainous and northern locations, where snow accumulates from multiple storms throughout the snow season, the annual maximum ground snow loads are relatively symmetric about their mean, with relatively small coefficients of variation. However, in other locations, where the annual maximum snow event is defined by one (or a handful) of storms, the distribution of annual maxima tends to be right skewed with large coefficients of variation. The measured annual maximum snow loads at all stations that receive snow regularly have been fitted with a generalized extreme value (GEV) distribution that can appropriately represent both of these conditions (Jenkinson 1955, Hosking et al. 1985). To handle scarcity of data at some locations, the GEV shape parameter is also spatially smoothed across sites based on regional patterns in typical snow accumulation.

The reliability-targeted loads are determined by probabilistically estimating the likelihood that roof snow load demands exceed roof capacity, considering the underlying uncertainties in both demand and resistance through statistical simulation. Roof snow load demands are estimated by combining a statistical model for the ground-to-roof conversion factor with the ground snow load statistical GEV model. The selected ground-to-roof model was constructed using historical Canadian data for snow on roofs (e.g., Allen 1956, Kennedy and Lutes 1968). The capacity (or resistance) assumed in the reliability-targeted loads was based on steel W-shapes in flexure, fabricated from A992 steel, for the ANSI/AISC 360-16 (2016) plastic moment limit state. Statistical models for resistance were obtained from testing by Bartlett et al. (2003) on A992 steel, fabrication and professional factors obtained from Ellingwood et al. (1980). The target roof is a heated, flat roof (less than 30 degrees) of a nonslippery material in normal sheltered conditions.

For each station for which ground snow load data are available, a reliability analysis was performed for the different risk categories. The design load was incremented until the reliability analysis produced the targeted reliability index for each risk category. After reliability-targeted loads were obtained at each measurement location in the United States, regional generalized additive models with spatial smoothing were used to map values based on the available data. This modeling approach was used to estimate loads between measurement stations in each EPA-defined Ecoregion (CEC 1997) and, thus, accounts for differences in climate and mapping trends between different parts of the country. Buffer zones are used around the ecoregions to ensure smooth transitions between regions. The end result is a grid, approximately one-half-mile by one-half-mile, of snow loads for the conterminous United States.

Northern and mountainous locations tend to have reliability-targeted loads for Risk Category II less than or equal to 1.6 times the 50-year MRI load used in ASCE/SEI 7-16. However, locations with more year-to-year variability in the annual maximum ground snow loads tend to have reliability-targeted loads greater than 1.6 times the 50-year load. This is especially true of mid-latitude locations that *typically* do not receive large accumulations of snow but have the *potential* for extreme accumulations of snow (Maguire et al. 2021). This is a fundamental reason for moving to reliability-targeted loads and away from a concept that one load factor is satisfactory across the nation. The effect of the ground snow, ground-to-roof, and resistance probability distributions in the reliability calculation causes the MRI of the computed loads to vary significantly from location to location, but all satisfy the targeted reliability.

A very small fraction of the locations defined in the Geodatabase indicate that a case study must be completed to determine the ground snow load. These case-study regions are now limited and apply only to locations higher than any locally available snow measurement locations. Database ground snow load values are still provided to the user, with a warning that the estimated value lies outside the range of elevations of surrounding measurement locations. Information from local experts, from the Bean et al. (2021) report, or from Buska et al. (2020) can be used to determine values at these locations.

The contour maps in Figure 7.2-1A through Figure 7.2-1D are aggregated to an 8 mi resolution. This aggregation obfuscates small scale changes in elevation and climate, which may create discrepancies between the mapped values and the hazard tool. The data set in the hazard tool is intended to be used in locations with rapidly changing snow loads over short geographical distances. Consecutive contour increments are separated by a constant ratio of 1.18. These contour increments were chosen to increase precision in areas with small loads/load variation and reduce contour congestion in areas with high loads, to make the figure more readable.

When comparing snow loads to factored roof live loads, the engineer should also consider load duration effects that apply to some materials. Locations with less than 10 psf design loads are almost always associated with locations that have at least 80% of their winters with no snow. This is why the snow load provisions need not be considered if the ground snow load, p_g, is less than the factored roof live load. Also, additional guidance is provided concerning roofs with potential for drifting snow, which was generated by studying the total controlling gravity loads on roof secondary structural members under low values of p_g and varying values of l_u. This study did account for the reduction in the ASCE 7 roof live load based on area supported by the member.

The values in Table 7.2-1 are for specific Alaskan locations only and generally do not represent appropriate design values for other nearby locations. This variability precludes statewide mapping of ground snow loads there.

The locations and snow loads for Alaska are based on a 50-year MRI values [see Structural Engineers Association of Alaska (2019)] converted to strength-level loads based on a study of reliability-targeted load data [see Structural Engineers Association of Alaska (2020)]. These references also include recommendations for snow loads in locations other than those listed in Table 7.2-1.

C7.3 FLAT ROOF SNOW LOADS, p_f

The live load reductions in Section 4.8 should not be applied to snow loads. The minimum allowable values of p_f, presented in Section 7.3, acknowledge that in some areas, a single major storm can generate loads that exceed those developed from an analysis of weather records and snow load case studies.

The factors in this standard that account for the thermal, aerodynamic, and geometric characteristics of the structure in its particular setting were developed using the National Building Code of Canada (National Research Council of Canada 1990) as a point of reference. The case study reports in Peter et al. (1963), Schriever et al. (1967), Lorenzen (1970), Lutes and Schriever (1971), Elliott (1975), Mitchell (1978), Meehan (1979), and Taylor (1979, 1980) were examined in detail.

In addition to these published references, an extensive program of snow load case studies was conducted by eight universities in the United States, the US Army Corps of Engineers' Alaska District, and the US Army Cold Regions Research and Engineering Laboratory (CRREL) for the Corps of Engineers. The results of this program were used to modify the Canadian methodology to better fit the conditions of the United States. Measurements obtained during the severe winters of 1976–1977 and 1977–1978 are included. A statistical analysis of some of that information is presented in O'Rourke et al. (1983). The experience and perspective of many design professionals, including several with expertise in building failure analysis, have also been incorporated.

C7.3.1 Exposure Factor, C_e Except in areas of "aerodynamic shade," where loads are often increased by snow drifting, less snow is present on most roofs than on the ground. Loads in unobstructed areas of conventional flat roofs average less than 50% of ground loads in some parts of the country. The values in this standard are above-average values, chosen to reduce the risk of snow load–induced failures to an acceptably low level. Because of the variability of wind action, a conservative approach has been taken when considering load reductions by wind.

The effects of exposure are handled on two scales. First, Equation (7.3-1) contains a basic exposure factor of 0.7. Second, the type of surface roughness and the exposure of the roof are handled by exposure factor, C_e. This two-step procedure generates ground-to-roof load reductions as a function of exposure that range from 0.49 to 0.84.

Table 7.3-1 has been changed, from what appeared in a prior version of this standard, to separate regional wind issues associated with surface roughness from local wind issues associated with roof exposure. This change was made to better define categories without significantly changing the values of C_e.

Although there is a single "regional" surface roughness category for a specific site, different roofs of a structure may have different exposure factors caused by obstruction provided by higher portions of the structure or by objects on the roof. For example, in Surface Roughness Category C, an upper level roof could be fully exposed ($C_e = 0.9$) while a lower level roof would be partially exposed ($C_e = 1.0$), because of the presence of the upper level roof, as shown in Example 3 in this chapter.

The adjective "windswept" is used in the "mountainous areas" surface roughness category to preclude use of this category in those high mountain valleys that receive little wind.

The normal, combined exposure reduction in this standard is 0.70 as compared with a normal value of 0.80 for the ground-to-roof conversion factor in the 1990 National Building Code of Canada. The decrease from 0.80 to 0.70 does not represent decreased safety but arises because of increased choices of exposure and thermal classification of roofs (i.e., five surface roughness categories, three roof exposure categories, and four thermal categories in this standard vs. three exposure categories and no thermal distinctions in the Canadian code).

It is virtually impossible to establish exposure definitions that clearly encompass all possible exposures that exist across the country. Because individuals may interpret exposure categories somewhat differently, the range in exposure has been divided into several categories rather than just two or three. A difference of opinion of one category results in about a 10% "error" using these several categories and an "error" of 25% or more if only three categories are used.

C7.3.2 Thermal Factor, C_t Usually, more snow will be present on cold roofs than on warm roofs. An exception to this is discussed in the following text. The thermal condition selected from Table 7.3-2 should represent that which is likely to exist during the life of the structure. In Table 7.3-2, the term "structure kept just above freezing" represents a structure in which the internal ambient temperature is kept at approximately 40 to 50 °F (4 to 10 °C) to keep contents from freezing but is usually not intended for continuous human occupancy. Such a structure is referred to by the ASHRAE 90.1 (2019) standard as "semi-heated." Also, in Table 7.3-2, the term "cold, ventilated roof" represents a roof where an airspace is provided between the building envelope and the roof surface through which air is intended to flow in order to relieve heat buildup under the roof. The term "cold" is not intended to indicate that the location of the structure is in a cold climate region but simply references the fact that the attic airspace is located outside of the insulation envelope and will not be heated or cooled. Also, please note that if the thermal envelope of an existing structure is modified via alterations or additions, the implications on the thermal factor and the resulting design snow loads should be reviewed by the design professional responsible for those alterations or additions.

Although it is possible that a brief power interruption will cause temporary cooling of a heated structure, the joint probability of this event and a simultaneous peak snow load event is very small. Brief power interruptions and loss of heat are acknowledged in the $C_t = 1.0$ category. Although it is possible that a heated structure will subsequently be used as an unheated structure, the probability of this is rather low. Consequently, heated structures need not be designed for this unlikely event.

The values of C_t found in Table 7.3-3 are determined using the approach outlined in O'Rourke and Russell (2019), based on the study of eave ice dam formation and heat loss through an insulated roof surface found in O'Rourke et al. (2007).

C_t is larger for buildings in warmer climates with lower p_g values and smaller for buildings in colder climates with higher p_g values. In warmer climates, the 32°F (0°C) freezing point is within the roof insulation layer. Thus, there is no melting of roof snow due to heat flow up through the roof and the thermal condition is nominally the same as an unheated building with $C_t = 1.2$. Conversely, C_t is smaller for buildings with low R_{roof} values and larger for buildings with high R_{roof} values. For R_{roof} above a particular value, the 32°F (0°C) freezing point is within the roof insulation layer and again $C_t = 1.2$.

Some dwellings are not used during the winter. Although their thermal factor may increase to 1.2 at that time, they are unoccupied, risk category changes from II to I. The net effect is to require the same design load as for a heated, occupied dwelling.

Discontinuous heating of structures may cause thawing of snow on the roof and subsequent refreezing in lower areas. Drainage systems of such roofs have become clogged with ice, and extra loads associated with layers of ice several inches thick have built up in these undrained lower areas. The possibility of similar occurrences should be investigated for any intermittently heated structure.

Similar icings may build up on cold roofs subjected to meltwater from warmer roofs above. Exhaust fans and other mechanical equipment on roofs may also generate meltwater and icings.

Icicles and ice dams are a common occurrence on cold eaves of sloped roofs. They introduce problems related to leakage and to loads. Large ice dams that can prevent snow from sliding off roofs are generally produced by heat losses from within buildings. Icings associated with solar melting of snow during the day and refreezing along eaves at night are often small and transient. Although icings can occur on cold or warm roofs, roofs that are well insulated and ventilated are not commonly subjected to serious icings at their eaves. Methods of minimizing eave icings are discussed in Grange and Hendricks (1976), Klinge (1978), de Marne (1988), Mackinlay (1988), Tobiasson (1988), and Tobiasson and Buska (1993). Ventilation guidelines to prevent problematic icings at eaves have been developed for attics (Tobiasson et al. 1998) and for cathedral ceilings (Tobiasson et al. 1999).

Because ice dams can prevent load reductions by sliding on some warm ($C_t \leq 1.0$) roofs, the "unobstructed slippery surface" curve in Figure 7.4-1(a) now only applies to unventilated roofs with a thermal resistance equal to, or greater than, 30 ft^2 h °F/Btu (5.3°C m^2/W) and to ventilated roofs with a thermal resistance equal to or greater than 20 ft^2 h °F/Btu (3.5 °C m^2/W). For roofs that are well insulated and ventilated, see $C_t = 1.1$ in Table 7.3-2.

Glass, plastic, and fabric roofs of continuously heated structures are seldom subjected to much snow load because their high heat losses cause snowmelt and sliding. For such specialty roofs, knowledgeable manufacturers and designers should be consulted. The National Greenhouse Manufacturers Association (1988) recommends $C_t = 0.83$ for continuously heated greenhouses and $C_t = 1.00$ for unheated or intermittently heated greenhouses. They suggest a value of $I_s = 1.0$ for retail greenhouses and $I_s = 0.8$ for all other greenhouses. To qualify as a continuously heated greenhouse, a production or retail greenhouse must have a constantly maintained temperature of 50 °F (10 °C) or higher during winter months. In addition, it must also have a maintenance attendant on duty at all times or an adequate temperature alarm system to provide warning in the event of a heating system failure. Finally, the greenhouse roof material must have a thermal resistance, R-value, less than 2 ft^2 × h × °F/Btu (0.4° C m^2/W). In this standard, the C_t factor for such continuously heated greenhouses is set at 0.85. An unheated or intermittently heated greenhouse is any greenhouse that does not meet the requirements of a continuously heated single- or double-glazed greenhouse. Greenhouses should be designed so that the structural supporting members are stronger than the glazing. If this approach is used, any failure caused by heavy snow loads will be localized and in the glazing. This should avert progressive collapse of the structural frame. Higher design values should be used where drifting or sliding snow is expected.

Little snow accumulates on warm air-supported fabric roofs because of their geometry and slippery surface. However, the snow that does accumulate is a significant load for such structures and should be considered.

The combined consideration of exposure and thermal conditions generates ground-to-roof factors that range from a low of 0.49 to a high of 1.01. The equivalent ground-to-roof factors in the 1990 National Building Code of Canada are 0.8 for sheltered roofs, 0.6 for exposed roofs, and 0.4 for exposed roofs in exposed areas north of the tree line, all regardless of their thermal condition.

Sack (1988) and case history experience indicate that the roof snow load on open-air structures (e.g., parking structures and roofs over loading docks) and on buildings intentionally kept below freezing (e.g., freezer buildings) can be larger than the nearby ground snow load. It is thought that this effect is caused by the lack of heat flow up from the "warm" earth for these select groups of structures. Open-air structures are explicitly included with unheated structures. For freezer buildings, the thermal factor is specified to be 1.3 to account for this effect. This value is intended specifically for structures constructed to act as freezer buildings and not those that contain freezer enclosures inside.

C7.3.3 Minimum Snow Load for Low-Slope Roofs, p_m These minimums account for a number of situations that develop on low-slope roofs. They are particularly important considerations for regions where p_g is less than $p_{m,\max}$ from Table 7.3-3. In such areas, single storm events can result in loading for which the basic ground-to-roof conversion factor of 0.7, as well as the C_e and C_t factors, are not applicable.

The unbalanced load for hip and gable roofs, with an eave-to-ridge distance W of 20 ft (6.1 m) or less that have simply supported prismatic members spanning from ridge to eave, is greater than or equal to the minimum roof snow load, p_m. Hence, if such a hip and gable roof has a slope that requires unbalanced loading, the minimum snow load would not control and need not be checked for the roof.

C7.4 SLOPED ROOF SNOW LOADS, p_s

Snow loads decrease as the slopes of roofs increase. Generally, less snow accumulates on a sloped roof because of wind action. Also, such roofs may shed some of the snow that accumulates on them by sliding and improved drainage of meltwater. The ability of a sloped roof to shed snow load by sliding is related to the absence of obstructions not only on the roof but also below it, the temperature of the roof, and the slipperiness of its surface. It is difficult to define "slippery" in quantitative terms. For that reason, a list of roof surfaces that qualify as slippery and others that do not is presented in the standard. Most common roof surfaces are on that list. The slipperiness of other surfaces is best determined by comparisons with those surfaces. Some tile roofs contain built-in protrusions or have a rough surface that prevents snow from sliding. However, snow does slide off other smooth-surfaced tile roofs. When a surface may or may not be slippery, the implications of treating it either as a slippery or nonslippery surface should be determined. Because valleys obstruct sliding on slippery surfaced roofs, the dashed lines in Figures 7.4-1(a–c) should not be used in such roof areas.

Discontinuous heating of a building may reduce the ability of a sloped roof to shed snow by sliding because meltwater created during heated periods may refreeze on the roof's surface during periods when the building is not heated, thereby "locking" the snow to the roof.

All of these situations are considered in the slope reduction factors presented in Figure 7.4-1 and are supported by Taylor (1983, 1985), Sack et al. (1987), and Sack (1988). Mathematically, the information in Figure 7.4-1 can be represented as follows:

1. $C_t < 1.1$:
   (a) All other surfaces:
   0°–30° slope $C_s = 1.0$
   30°–70° slope $C_s = 1.0 - (\text{slope} - 30°)/40°$
   >70° slope $C_s = 0$

2. $1.1 \leq C_t < 1.2$
   (a) Unobstructed slippery surfaces:
      0°–10° slope $C_s = 1.0$
      10°–70° slope $C_s = 1.0 - (\text{slope} - 10°)/60$
      >70° slope $C_s = 0$
   (b) All other surfaces:
      0°–37.5° slope $C_s = 1.0$
      37.5°–70° slope $C_s = 1.0 - (\text{slope} - 37.5°)/32.5°$
      >70° slope $C_s = 0$
3. $C_t \geq 1.2$:
   (a) Unobstructed slippery surfaces:
      0°–15° slope $C_s = 1.0$
      15°–70° slope $C_s = 1.0 - (\text{slope} - 15°)/55°$
      >70° slope $C_s = 0$
   b) All other surfaces:
      0°–45° slope $C_s = 1.0$
      45°–70° slope $C_s = 1.0 - (\text{slope} - 45°)/25°$
      >70° slope $C_s = 0$

If $C_t \leq 1.1$ [Figure 7.4-1(a)], it is expected that the 32°F (0°C) freezing line will be outside of the building envelope and thus, roof snow melting will create an ice dam at the eave. The ice dam at the eave will create an obstruction that will resist snow from sliding off the roof. Therefore, Figure 7.4-1(a) does not include a line for an unobstructed, slippery surface.

If the ground (or another roof of less slope) exists near the eave of a sloped roof, snow may not be able to slide completely off the sloped roof. This may result in the elimination of snow loads on upper portions of the roof and their concentration on lower portions. Steep A-frame roofs that nearly reach the ground are subject to such conditions. Lateral and vertical loads induced by such snow should be considered for such roofs.

If the roof has snow retention devices (installed to prevent snow and ice from sliding off the roof), it should be considered an obstructed roof and the slope factor C_s should be based on the "All Other Surfaces" curves in Figure 7.4-1.

C7.4.2 Slope Factor for Curved Roofs These provisions were changed from those in the 1993 edition of this standard to cause the load to diminish along the roof as the slope increases.

C7.4.3 Slope Factor for Multiple Folded Plate, Sawtooth, and Barrel Vault Roofs Because these types of roofs collect extra snow in their valleys by wind drifting and snow creep and sliding, no reduction in snow load should be applied because of slope.

C7.4.4 Ice Dams and Icicles along Eaves The intent is to consider heavy loads from ice that form along eaves only for structures where such loads are likely to form. It is also not considered necessary to analyze the entire structure for such loads, just the eaves themselves. Eave ice dam loads with various return periods on roofs with overhangs of 4 ft (1.2 m) or less are presented in O'Rourke et al. (2007).

This provision is intended for short roof overhangs and projections, with a horizontal extent less than 5 ft (1.5 m). In instances where the horizontal extent is greater than 5 ft (1.5 m), the surcharge that accounts for eave ice damming need only extend for a maximum of 5 ft (1.5 m) from the eave of the heated structure (Figure C7.4-1).

C7.5 PARTIAL LOADING

In many situations, a reduction in snow load on a portion of a roof by wind scour, melting, or snow-removal operations simply reduces the stresses in the supporting members. However, in some cases, a reduction in snow load from an area induces higher stresses in the roof structure than occur when the entire roof is loaded. Cantilevered roof joists are a good example; removing half the snow load from the cantilevered portion increases the bending stress and deflection of the adjacent continuous span. In continuous beam roof systems, problems can occur during snow-removal operations. The nonuniform loading imposed by the removal of snow load in an indiscriminate manner can significantly alter the load distribution throughout the roof system and can result in marked increases in stresses and deflections over those experienced during uniform loading. In other situations, adverse stress reversals may result.

The simplified provisions offered for continuous beam roof systems have been adopted to mimic real loadings experienced by this common structural system. The Case 1 load scenario simulates a critical condition encountered when, for example, the process of removing a portion of the snow in each span starts at one building end and proceeds toward the other end. The Case 2 load scenario is intended to model loading nonuniformity caused by wind scour or local heat loss and snowmelt at the building edges. The Case 3 load group is intended to encompass conditions that might be encountered, for example, when removal operations must be discontinued and are restarted in a different location, or when multiple snow removal crews start at different roof locations.

The intent is not to require consideration of multiple "checkerboard" loadings.

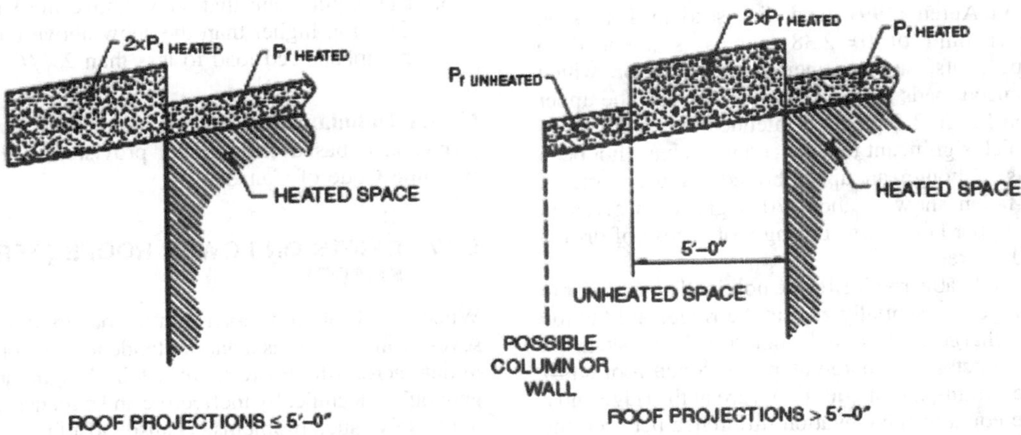

Figure C7.4-1. Eave ice dam loading.

Members that span perpendicular to the ridge in gable roofs with slopes between 2.38 degrees (½ on 12) and 30.3 degrees (7 on 12) are exempt from partial load provisions because the unbalanced load provisions of Section 7.6.1 provide for this situation.

C7.6 UNBALANCED ROOF SNOW LOADS

Unbalanced snow loads may develop on sloped roofs because of sunlight and wind. Winds tend to reduce snow loads on windward portions and increase snow loads on leeward portions. Because it is not possible to define wind direction with assurance, winds from all directions should generally be considered when establishing unbalanced roof loads.

C7.6.1 Unbalanced Snow Loads for Hip and Gable Roofs The expected shape of a gable roof drift is nominally a triangle located close to the ridgeline. Recent research suggests that the size of this nominally triangular gable roof drift is comparable to a leeward roof step drift with the same fetch. For certain simple structural systems, for example, wood or light-gauge roof rafter systems with either a ridge board or a supporting ridge beam, with small eave to ridge distances, the drift is represented by a uniform load of p_g from eave to ridge. For all other gable roofs, the drift is represented by a rectangular distribution located adjacent to the ridge. The location of the centroid for the rectangular distribution is identical to that for the expected triangular distribution. The intensity is the average of that for the expected triangular distribution.

Prior to ASCE 7-22, the drift height, h_d, was a function of the ground snow load, p_g, and the upwind fetch distance, l_u. In the 2022 version, the drift height is now also a function of a winter wind parameter, W_2. The paper by O'Rourke and Cocca (2019) demonstrates that the winter wind parameter (the percent time from October through April when the wind speed is above the 10 mph snow transport threshold) has as much influence on drift height as either the ground snow load or the upwind fetch distance. O'Rourke and Cocca (2019) present the development of Equation 7.6-1. A companion paper, O'Rourke et al. (2020), describes the development of the W_2 map in Figure 7.6-1.

Note that the winter wind parameter map was based on the W_2 parameter as evaluated at 272 airport weather stations. As such, the stations were not within special wind regions nor above the tree line. At these unusual wind locations (i.e, sites above the tree line or within special wind regions), it is suggested that the W_2 parameter be based on any available local records.

The design snow load on the windward side for the unbalanced case, $0.3p_s$, is based on case histories presented in Taylor (1979) and O'Rourke and Auren (1997) and discussed in Tobiasson (1999). The lower limit of $\theta = 2.38$ degrees is intended to exclude low-slope roofs, such as membrane roofs, on which significant unbalanced loads have not been observed. The upper bound of $\theta > 7$ on 12 (30.2 degrees) is intended to exclude high-slope roofs on which significant unbalanced loads have not been observed. That is, although an upper bound for the angle of repose for fresh-fallen snow is about 70 degrees, as given in Figure 7.4-1, the upper bound for the angle of repose of drifted snow is about 30 degrees.

As noted, observed gable roof drifts are nominally triangular in shape. The surcharge is essentially zero at the ridge, and the top surface of the surcharge is nominally horizontal. As such, an upper bound for an actual surcharge atop the sloped roof snow load, p_s, would be a triangular distribution: zero at the ridge and a height at the eave equal to the elevation difference between the eave and the ridge.

For intersecting gable roofs and similar roof geometries, some codes and standards have required a valley drift load. Such valley drift loads are not required in ASCE 7. However, valley locations are subject to unbalanced or gable roof drifts, as described in Section 7.6.1. An example of unbalanced loading on an L-shaped gable roof is presented in O'Rourke (2007).

For intersecting monoslope roofs and intersecting gable roofs with slopes greater than 7 on 12, unbalanced loads are not required in ASCE 7. However, at such valleys, snow on each side of the valley is prevented from sliding by the presence of roof snow on the other side of the valley. As such, the valley portion of the roof (drainage area upslope of the reentrant corner) is obstructed and the slope factor, C_s, should be based on the "All Other Surfaces" lines in Figure 7.4-1.

C7.6.2 Unbalanced Snow Loads for Curved Roofs The method of determining roof slope is the same as in the 1995 edition of this standard. C_s is based on the actual slope, not an equivalent slope. These provisions do not apply to roofs that are concave upward. For such roofs, see Section C7.13.

C7.6.3 Unbalanced Snow Loads for Multiple Folded Plate, Sawtooth, and Barrel Vault Roofs A minimum slope of 3/8 in./ft (1.79 degrees) has been established to preclude the need to determine unbalanced loads for most internally drained membrane roofs that slope to internal drains. Case studies indicate that significant unbalanced loads can occur when the slope of multiple gable roofs is as low as 1/2 in./ft (2.38 degrees).

The unbalanced snow load in the valley is $2p_f/C_e$, to create a total unbalanced load that does not exceed a uniformly distributed ground snow load in most situations.

Sawtooth roofs and other "up-and-down" roofs with significant slopes tend to be vulnerable in areas of heavy snowfall for the following reasons:

1. They accumulate heavy snow loads and are therefore expensive to build.
2. Windows and ventilation features on the steeply sloped faces of such roofs may become blocked with drifting snow and be rendered useless.
3. Meltwater infiltration is likely through gaps in the steeply sloped faces if they are built as walls because slush may accumulate in the valley during warm weather. This accumulation can promote progressive deterioration of the structure.
4. Lateral pressure from snow drifted against clerestory windows may break the glass.
5. The requirement that snow above the valley not be at an elevation higher than the snow above the ridge may limit the unbalanced load to less than $2p_f/C_e$.

C7.6.4 Unbalanced Snow Loads for Dome Roofs This provision is based on a similar provision in the 1990 National Building Code of Canada.

C7.7 DRIFTS ON LOWER ROOFS (AERODYNAMIC SHADE)

When a rash of snow load failures occurs during a particularly severe winter, there is a natural tendency for concerned parties to initiate across-the-board increases in design snow loads. This is generally a technically ineffective and expensive way of attempting to solve such problems because most failures associated with snow loads on roofs are caused not by moderate overloads on every square foot (square meter) of the roof, but rather by localized significant overloads caused by drifted snow.

Drifts accumulate on roofs (even on sloped roofs) in the wind shadow of higher roofs or terrain features. Parapets have the same effect. The affected roof may be influenced by a higher portion of the same structure or by another structure or terrain feature nearby if the separation is 20 ft (6.1 m) or less. When a new structure is built within 20 ft (6.1 m) of an existing structure, drifting possibilities should also be investigated for the existing structure (see Sections C7.7.2 and C7.12). The snow that forms drifts may come from the roof on which the drift forms, from higher or lower roofs, or, on occasion, from the ground.

Prior to ASCE 7-22, leeward drift load provisions were based on studies of snow drifts on roofs (Speck 1984; Taylor 1984; O'Rourke et al. 1985, 1986). Drift size was related to the amount of driftable snow, as quantified by the upwind roof length and the ground snow load. As discussed in Section C7.6.1, the drift load is now also a function of the winter wind parameter W_2. Drift loads are considered for ground snow loads as low as 5 lb/ft^2 (0.24 kN/m^2). Case studies show that, in regions with low ground snow loads, drifts 3 to 4 ft (0.9 to 1.2 m) high can be caused by a single storm accompanied by high winds. The leeward drift height limit of 60% of the lower roof length applies to canopies and other lower level roofs with small horizontal projection from the building wall. It is based on a 30 degree angle of repose for drifted snow [tan (30 degrees) = drift height/drift length = $0.577 \sim 0.6$] and is consistent with provisions in Section 7.6.1, which exclude roofs steeper than 7 on 12 from the gable roof drift (unbalanced load) provisions.

A change from a prior (1988) edition of this standard involves the width, w, when the drift height, h_d, exceeds the clear height, h_c. In this situation, the width of the drift is taken as $4h_d^2/h_c$, with a maximum value of $8h_c$. This drift width relation is based on equating the cross-sectional area of this drift (i.e., $1/2h_c \times w$) with the cross-sectional area of a triangular drift, where the drift height is not limited by h_c (i.e., $1/2h_d \times 4h_d$), as suggested by Zallen (1988). The upper limit of drift width is based on studies by Finney (1939) and Tabler (1975) that suggest that a "full" drift has a rise-to-run of about 1:6.5, and case studies (Zallen 1988) that show observed drifts with a rise-to-run greater than 1:10.

The drift height relationship in Equation (7.6-1) is based on snow blowing off a high roof upwind of a lower roof. The change in elevation where the drift forms is called a "leeward step." Drifts can also form at "windward steps." An example is the drift that forms at the downwind end of a roof that abuts a higher structure there. Figure 7.7-1 shows "windward step" and "leeward step" drifts.

As described by O'Rourke et al. (2018), leeward roof step drift formation is fairly straightforward. Wind causes upper level roof snow to be transported (i.e., blown) to the edge of the upper level roof. A percentage of that transported snow (typically taken to be 50%) remains in the region of aerodynamic shade until the wind stops, the upwind source area is depleted of snow, or the leeward drift becomes full. The situation for windward drifts is more complex. The initial trapping efficiency of a wall is nominally 100%. That is, all the transported snow initially stays upwind of the wall. If the windward drift grows large enough [i.e., wind streamlines along the snow drift surface (snow ramp) move high enough up the wall], some of the windblown snow is carried over the wall and the windward trapping efficiency drops to less than 20%. The slope of a windward drift (rise to run of 1 to 8) is different than that for a non-full leeward drift (rise-to-run of 1:4). For the same upwind fetch and ground snow loads, there are comparatively small differences between leeward and windward drift height, whereas there are comparatively large differences between leeward and windward drift width with windward drifts having the larger widths.

The ratio of windward to leeward drift height, β, is a function of whether the windward drift is in its initial phase (trapping efficiency = 100%), or in the later phase (trapping efficiency = 20%). The upwind fetch at which the transition from initial to later phase occurs, l_u^*, is given by

$$l_u^* = \left[\frac{0.162\ \gamma\ h_o^2}{P_g^{0.74} W_2^{1.7}}\right]^{1.43} \quad \text{C7.7-1}$$

For a building where the lower roof upwind fetch $l_u \leq l_u^*$ the height of the windward drift is equal to that of the leeward drift with an equal upwind fetch for the upper roof, that is $\beta = 1.0$. For $l_u > l_u^*$, the ratio β is given by

$$\beta = \left[0.2 + \frac{0.8}{\alpha^{0.7}}\right]^{0.5} \quad \text{C7.7-2}$$

where α is l_u/l_u^*. That is, for $\alpha \leq 1.0$ $\beta = 1.0$ as previously noted while for $\alpha = 4.05$, $\beta = 0.707$. For $\alpha > 4.05$, β is less than 0.707. For simplicity, the Snow and Rain Load Subcommittee decided that the windward drift height for all values of l_u should be taken as 3/4 the height of the corresponding leeward drift (i.e., $\beta = 0.75$). This maintains consistency with previous editions of ASCE 7, which used a value of 3/4 based on analysis of case histories in O'Rourke and De Angelis (2002).

Depending on wind direction, any change in elevation between roofs can be either a windward or leeward step. Thus, the height of a drift is determined for each wind direction, as shown in Example 3 in this chapter, and the larger of the two heights is used as the design drift. Prior to ASCE 7-22, the drift height relation was based on a data set of leeward roof step drifts for which the average upwind fetch distance was about 170 ft (52 m). Unfortunately, the functional form of the empirical relation provided unrealistic results for very short upwind fetch distances and very low ground snow loads. In prior versions of this load standard, this shortcoming was handled by specifying a minimum l_u of 20 ft (6.1m). Given the functional form of new Equation (7.6-1), these limits are no longer needed.

The drift load provisions cover most, but not all, situations. Finney (1939) and O'Rourke (1989) document a larger drift than would have been expected based on the length of the upper roof. The larger drift was caused when snow on a somewhat lower roof, upwind of the upper roof, formed a drift between those two roofs, allowing snow from the upwind lower roof to be carried up onto the upper roof then into the drift on its downwind side. It was suggested that the sum of the lengths of both roofs could be used to calculate the size of the leeward drift. The situation of two roof steps in series was studied by O'Rourke and Conlin (n.d.). For the roof geometry sketched in Figure C7.7-1, the effective upwind fetch for the leeward drift atop roof C caused by wind from left to right was shown to be

for $\frac{h_{AB}*}{(h_d)_A} \geq 0.707$

$$L_B + 0.5\ L_A \quad \text{C7.7-3}$$

for $\frac{h_{AB}*}{(h_d)_A} < 0.707$

$$L_B + \left[1 - \left(\frac{h_{AB}*}{(h_d)_A}\right)^2\right] L_A \quad \text{C7.7-4}$$

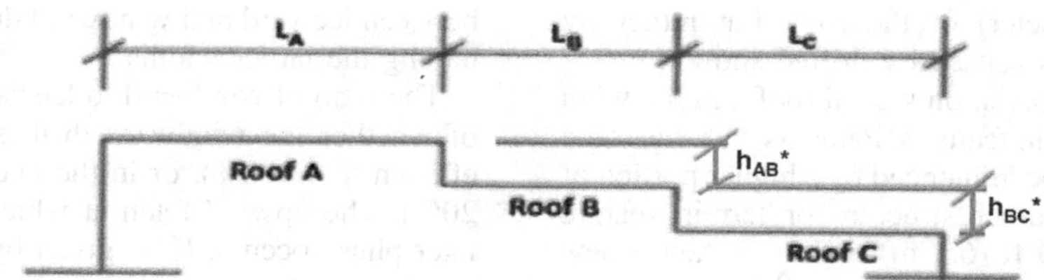

Figure C7.7-1. Roof steps in series.

where h^*_{AB} is the elevation difference between roofs A and B minus the balanced snow depth atop roof B, and $(h_d)_A$ is the expected leeward drift height for an upwind fetch of L_A.

That is, some of the snow originally on roof A ends up in the leeward drift atop roof B, thereby reducing the amount of snow available for drift formation atop roof C. For the windward snowdrift atop roof B caused by wind from right to left, the effective upwind fetch was shown to be

for $\frac{h_{BC}^*}{(h_d)_C} > 1.86$

$$L_B \qquad \text{C7.7-5}$$

for $1.86 \geq \frac{h_{BC}^*}{(h_d)_C} \geq 0.51$

$$L_B + \left[0.8 - 0.23\left(\frac{h_{BC}^*}{(h_d)_C}\right)^2\right] L_C \qquad \text{C7.7-6}$$

for $\frac{h_{BC}^*}{(h_d)_C} < 0.51$

$$L_B + \left(1 - \left(\frac{h_{BC}^*}{(h_d)_C}\right)^2\right) L_C \qquad \text{C7.7-7}$$

where h^*_{BC} is the elevation difference between roofs B and C minus the balanced snow depth atop roof C, and $(h_d)_C$ is the expected leeward drift height for an upwind fetch of L_C.

To reduce or eliminate leeward snow drifts on an adjacent or adjoining lower roof, a snow capture wall can be added to the upper roof. As developed in O'Rourke, and Conlin (2020), Equation (C7.7-8) presents the wall height, $h_{0,i}$ n feet, above the upper level roof balanced snow, as shown in Figure C7.7-2, required to fully capture the upper level roof snow. Furthermore, the expected leeward drift height h^*_d atop the lower level roof for a capture wall smaller than that in Equation (C7.7-8) is provided in Equations (C7.7-9) and (C7.7-10). Also, the addition of a parapet wall at a roof step would increase the space available for windward drift formation on the lower roof.

$$h_o > 1.86\ h_d \qquad \text{C7.7-8}$$

for $0.51 < \frac{h_o}{h_d} < 1.86$

$$h_d* = \sqrt{0.80\ h_d^2 - 0.23\ h_o^2} \qquad \text{C7.7-9}$$

for $\frac{h_o}{h_d} < 0.51$

$$h_d* = \sqrt{h_d^2 - h_o^2} \qquad \text{C7.7-10}$$

where h_d is the expected leeward drift height atop the lower level roof, using the upper level roof length in Equation (7.6-1).

In another situation (Kennedy et al. 1992), a long "spike" drift was created at the end of a long skylight with the wind about 30 degrees off the long axis of the skylight. The skylight acted as a guide or deflector that concentrated drifting snow. This action caused a large drift to accumulate in the lee of the skylight. This drift was replicated in a wind tunnel.

As shown in Figure 7.7-2, the clear height, h_c, is determined based on the assumption that the upper roof is blown clear of snow in the vicinity of the drift. This assumption is reasonable for windward drifting but does not necessarily hold for leeward

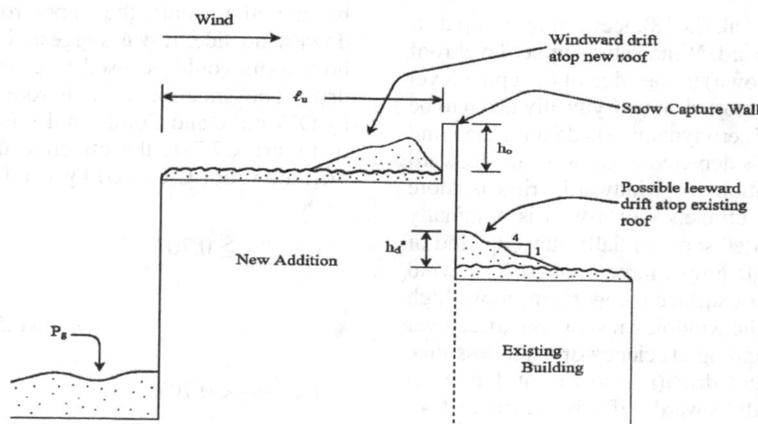

Figure C7.7-2. Capture wall to prevent or reduce leeward drift accumulation atop lower level roof.

drifting. For leeward drifting, the last portion of the upper level roof that would become blown clear of snow is the portion adjacent to the roof step. That is, there may still be snow on the upper level roof when the roof step drift has stopped growing. Nevertheless, for simplicity, the same assumption regarding clear height is used for both leeward and windward drifts.

Tests in wind tunnels (Irwin et al. 1992, Isyumov and Mikitiuk 1992) and flumes (O'Rourke and Weitman 1992) have proven quite valuable in determining patterns of snow drifting and drift loads. For roofs of unusual shape or configuration, wind tunnel or water-flume tests may be needed to help define drift loads. An ASCE standard for wind tunnel testing including procedures to assist in the determination of snow loads on roofs is currently under development.

C7.7.2 Adjacent Structures One expects a leeward drift to form on an adjacent lower roof only if the lower roof is low enough and close enough to be in the wind shadow (aerodynamic shade) region of the upper roof as sketched in Figure C7.7-3. The provisions in Section 7.7.2 are based on a wind shadow region that trails from the upper roof at a 1 downward to 6 horizontal slope.

For windward drifts, the requirements of Section 7.7.1 are to be used. However, the resulting drift may be truncated by eliminating the drift in the horizontal separation region as sketched in Figure C7.7-4.

C7.7.3 Intersecting Drifts The accumulation of drifting snow from perpendicular wind directions can occur concurrently from single or multiple wind and snow events to create an intersecting snowdrift load at reentrant corners and similar conditions. Intersecting drifts can also occur at a roof step where the lower level roof is a gable with a ridgeline that runs into the vertical sidewall of the roof step. A leeward roof step is expected for wind blowing from the direction of the higher level roof, whereas a gable roof drift is expected for wind blowing over the ridgeline of the lower gable roof. For example, if the gable roof is to the east of the upper level roof, a leeward roof step drift is expected for wind out of the west, but gable roof drift on the north side of the ridge is expected for wind out of the south. The procedures for analyzing such interesting drifts are as follows: (1) first consider each 2D drift separately, (2) then assume each 2D drift has a shape determined in accordance with this standard, and (3) within the intersecting footprints of the two 2D drifts, use the larger of the two 2D snow drift heights. Finally, the intersecting drift provisions in Section 7.7.3 do not apply to a simple gable roof subject to the gable roof drifts in Section 7.6.1. For a simple gable roof, the two unbalanced loads (actually two drift loads one each side of the ridgeline) do not have overlapping footprints.

Note that the approach for intersecting drifts in Section 7.7.3 is consistent with that for windward and leeward roof step drifts in Section 7.7.1, where the larger of these two heights shall be used in design. For the reentrant corner geometry, two resulting drifts are combined, as shown in Figure 7.7-3, where the snowdrift load at the drift intersection is based on the larger drift, not the additive effect of the two drifts (i.e., design h_d at the

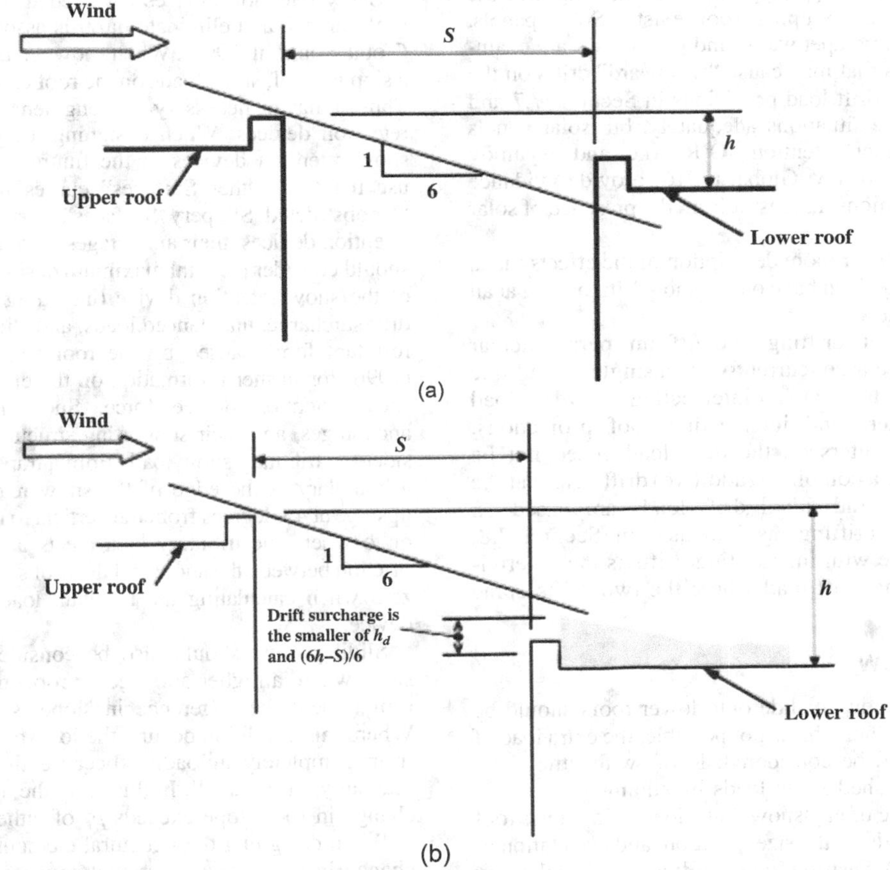

Figure C7.7-3. Leeward snow drift on adjacent roof, separation $S < 20$ ft: (a) elevation view, $S \geq 6h$; lower roof above wind shadow (aerodynamic shade) region, no leeward drift on lower roof; and (b) elevation view, $S < 6h$; lower roof within wind shadow (aerodynamic shade) region, leeward drift on lower roof; drift length is the smaller of $(6h - S)$ and $6h_D$.

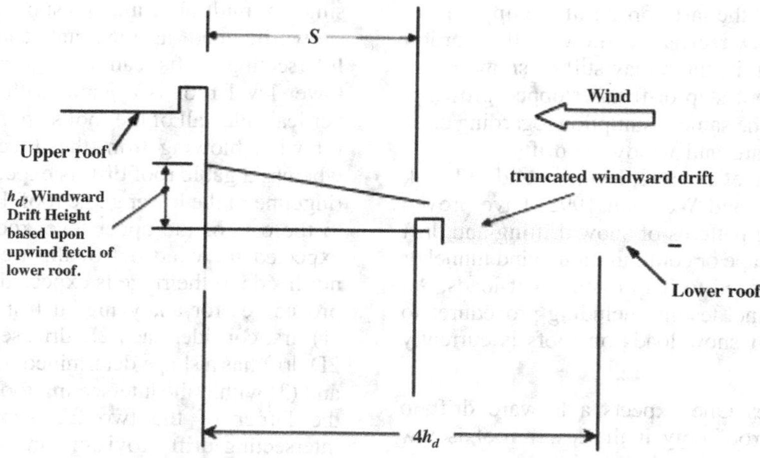

Figure C7.7-4. Windward snow drift on adjacent roof, separation $S < 20$ ft.

intersection is h_{d1} or h_{d2}, but not h_{d1} added to h_{d2}). Examples of the determination of intersecting drift loads are provided in the FEMA design guide for three-dimensional roof snow drifts (FEMA 2019).

C7.8 ROOF PROJECTIONS AND PARAPETS

Drifts around penthouses, roof obstructions, and parapet walls are also of the "windward step" type because the length of the upper roof is small or no upper roof exists. Solar panels, mechanical equipment, parapet walls, and penthouses are examples of roof projections that may cause "windward" drifts on the roof around them. The drift load provisions in Sections 7.7 and 7.8 cover most of these situations adequately, but solar panels warrant some additional attention. O'Rourke and Isyumov (2016) and Section 2.9 of FM Global (2016) provide guidance for the snow load conditions that result from the presence of solar panels on a roof.

Refer to Section C7.7 for more description of the effects that a parapet wall at a high roof can have on the snowdrift loading at an adjacent or adjoining lower roof.

The accumulation of drifting snow from perpendicular wind directions can occur concurrently from single or multiple wind and snow events to create an intersecting snowdrift load at parapet wall corners and intersecting roof projections. Where the two drifts intersect, the drift loads need not be superimposed to create a combined (additive) drift height at the drift intersection. The individual drift loads are based on windward or leeward drifting, as indicated in Section 7.8, and the only difference with intersecting drifts is the determination of the governing drift load where the two drifts come together.

C7.9 SLIDING SNOW

Situations that permit snow to slide onto lower roofs should be avoided (Paine 1988). Where this is not possible, the extra load of the sliding snow should be considered. Roofs with little slope have been observed to shed snow loads by sliding.

The final resting place of any snow that slides off a higher roof onto a lower roof depends on the size, position, and orientation of each roof (Taylor 1983). Distribution of sliding loads might vary from a uniform 5 ft (1.5 m) wide load, if a significant vertical offset exists between the two roofs, to a 20 ft (6.1 m) wide uniform load, where a low-slope upper roof slides its load onto a second roof that is only a few feet (about a meter) lower or where snowdrifts on the lower roof create a sloped surface that promotes lateral movement of the sliding snow.

In some instances, a portion of the sliding snow may be expected to slide clear of the lower roof. Nevertheless, it is prudent to design the lower roof for a substantial portion of the sliding load to account for any dynamic effects that might be associated with sliding snow.

Snow retention devices are needed on some roofs to prevent roof damage and eliminate hazards associated with sliding snow (Tobiasson et al. 1996). When snow retention devices are added to a sloping roof, snow loads on the roof can be expected to increase. Thus, it may be necessary to strengthen a roof before adding snow retention devices. When designing a roof that will likely need snow retention devices in the future, it may be appropriate to use the "All Other Surfaces" curves in Figure 7.4-1, not the "Unobstructed Slippery Surfaces" curves. The design of snow retention devices, their anchorages, and their supporting elements should consider the total maximum design roof snow load upslope of the snow retention device (including sloped roof snow load, drift surcharge, unbalanced loads, and sliding snow loads) and the resultant force caused by the roof slope. See Tobiasson et al. (1996) for further information on the effects of roof slope on the snow retention device force. Snow retention devices, their anchorages, and their supporting structural elements should consider the tributary snow loads from a trapezoidal-shaped area with a boundary at the edge of the snow retention device extending upslope at 45 degrees from the vertical to the ridge (Figure C7.9-1) or to where the tributary area meets an adjacent tributary area. Friction between the snow and the roof surface is typically taken as zero when calculating the resultant load on the snow retention device.

Sliding snow should also be considered at changes in roof slope where a higher and steeper roof meets a lower, less steep roof (where the difference in slope is greater than 2 on 12). Where this condition occurs, the low roof prevents the high roof from completely unloading because the low roof slides more gradually, or not at all. In this case, the accumulated snow at the change in roof slope exceeds p_f of either roof.

When designing the structural elements of the building at the change in roof slope, a rational approach would be to take the sliding snow ($0.4 p_f W$) and distribute it a distance of $W/4$ each side of the change in roof slope, accumulating it with the distributed design snow p_f.

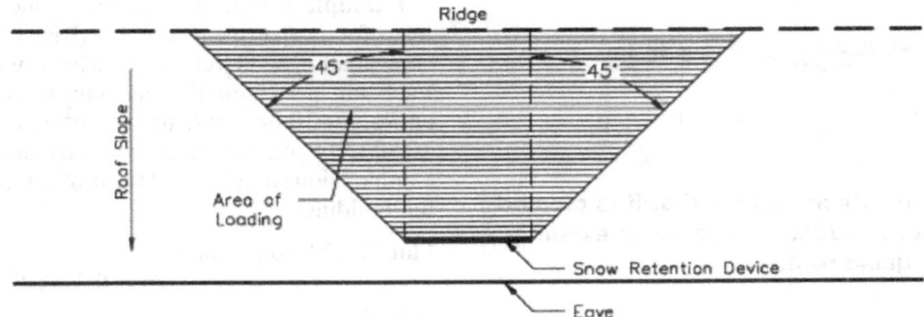

Figure C7.9-1. Plan view of tributary snow loads for snow retention device.

C7.10 RAIN-ON-SNOW SURCHARGE LOAD

The ground snow load measurements on which this standard is based contain the load effects of light rain on snow. However, because heavy rains percolate down through snow packs and may drain away, they might not be included in measured values. Where p_g is greater than the value of $p_{m,\max}$ given in Table 7.3-3, it is assumed that the full rain-on-snow effect has been measured and a separate rain-on-snow surcharge is not needed. The temporary roof load contributed by a heavy rain may be significant. Its magnitude depends on the duration and intensity of the design rainstorm, the drainage characteristics of the snow on the roof, the geometry of the roof, and the type of drainage provided. Loads associated with rain on snow are discussed in Colbeck (1977a, b) and O'Rourke and Downey (2001).

O'Rourke and Downey (2001) show that the surcharge from rain-on-snow loading is an increasing function of eave to ridge distance and a decreasing function of roof slope. That is, rain-on-snow surcharges are largest for wide, low-sloped roofs. The minimum slope reflects that functional relationship.

The following example illustrates the evaluation of the rain-on-snow surcharge. Consider a monoslope roof with slope of 1/4 on 12 and a width of 100 ft (30.5 m) with $C_e = 1.0$, $C_t = 1.1$, and $p_g = 32$ psf (1.54 kN/m^2) for a Risk Category IV structure. Because $C_s = 1.0$ for a slope of 1/4 on 12, $p_s = 0.7(1.0)(1.1)(1.0)(32) = 24.6$ psf (1.18 kN/m^2). Because the roof slope 1.19 degrees is less than $100/50 = 2.0$, the 8 psf (0.38 kN/m^2) surcharge is added to p_s, resulting in a design load of 32.6 psf (1.56 kN/m^2). Because the slope is less than 15 degrees, the minimum load, p_m, from Section 7.3.3 is the minimum of $p_m =$ 32 psf (1.54 kN/m^2) and $p_{m,\max} = 40$ psf (1.73 kN/m^2), which is 32 psf (1.54 kN/m^2). Hence, the rain-on-snow modified load controls.

C7.11 PONDING INSTABILITY

Where adequate slope to drain does not exist, or where drains are blocked by ice, snow meltwater and rain may pond in low areas. Intermittently heated structures in very cold regions are particularly susceptible to blockages of drains by ice. A roof designed without slope or one sloped with only 1/8 in./ft (0.6 degrees) to internal drains probably contains low spots away from drains by the time it is constructed. When a heavy snow load is added to such a roof, it is even more likely that undrained low spots exist. As rainwater or snow meltwater flows to such low areas, these areas tend to deflect increasingly, allowing a deeper pond to form. If the structure does not possess enough stiffness to resist this progression, failure by localized overloading can result. This mechanism has been responsible for several roof failures under combined rain and snow loads.

It is very important to consider roof deflections caused by snow loads when determining the likelihood of ponding instability from rain-on-snow or snow meltwater.

Internally drained roofs should have a slope of at least 1/4 in./ft (1.19 degrees) to provide positive drainage and to minimize the chance of ponding. Slopes of 1/4 in./ft (1.19 degrees) or more are also effective in reducing peak loads generated by heavy spring rain on snow. Further incentive to build positive drainage into roofs is provided by significant improvements in the performance of waterproofing membranes when they are sloped to drain.

Rain loads and ponding instability are discussed in detail in Chapter 8.

C7.12 EXISTING ROOFS

Numerous existing roofs have failed when additions or new buildings nearby caused snow loads to increase on the existing roof. A prior (1988) edition of this standard mentioned this issue only in its commentary where it was not a mandatory provision. The 1995 edition moved this issue to the standard.

The addition of a gable roof alongside an existing gable roof, as shown in Figure C7.12-1, most likely explains why some such metal buildings failed in the south during the winter of 1992–1993. The change from a simple gable roof to a multiple folded plate roof increased loads on the original roof, as would be expected from Section 7.6.3. Unfortunately, the original roofs were not strengthened to account for these extra loads, and they collapsed.

If the eaves of the new roof in Figure C7.12-1 had been somewhat higher than the eaves of the existing roof, the exposure factor, C_e, for the original roof may have increased, thereby increasing snow loads on it. In addition, drift loads and loads from sliding snow would also have to be considered.

C7.13 SNOW ON OPEN-FRAME EQUIPMENT STRUCTURES

Snow loads should be considered on all levels of the open-frame structures that can retain snow. The snow accumulations on the flooring, equipment, cable trays, and pipes should be considered. Snow cornicing effects impacting the accumulation of snow on surfaces such as grating, pipes, cable trays, and equipment should also be considered. Grating is considered to retain snow because of the cornicing effect of the snow between the grates.

C7.13.2 Snow at Levels below the Top Level In the absence of site-specific information, the length of the loaded zone is approximated for convenience as the vertical distance to the covering level above.

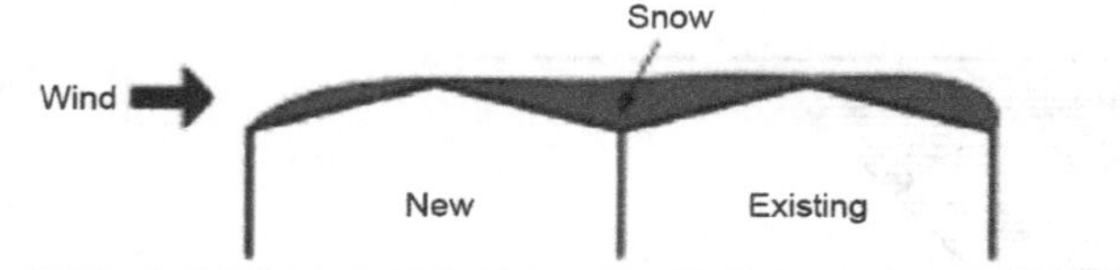

Figure C7.12-1. Valley in which snow will drift is created when a new gable roof is added alongside an existing gable roof.

The drift snow at levels below the top levels caused by wind walls may be ignored because of the limited snow accumulation on such levels.

C.7.13.3 Snow Loads on Pipes and Cable Trays The snow loading on any pipe rack or pipe bridge occurs because of the snow loading on the pipes and cable trays at each level. For pipe racks and pipe bridges wider than 12 ft (4 m), where the spaces between individual pipes are less than the pipe diameter (including insulation), Section 7.13.1 and 7.13.2 should be applied.

C7.13.4 Snow Loads on Equipment and Equipment Platforms Extended out-of-service conditions for equipment shall be considered for the loading criteria, particularly equipment that is often out of service for the whole winter.

C7.14 ALTERNATE PROCEDURE

Wind tunnel model studies, similar tests using fluids other than air, for example, water flumes, and other special experimental and computational methods, have been used with success to establish design snow loads for other roof geometries and complicated sites (Irwin et al. 1992, Isyumov and Mikitiuk 1992, O'Rourke and Weitman 1992). To be considered reliable, such methods must reproduce the mean and turbulent characteristics of the wind and the manner in which snow particles are deposited on roofs and then redistributed by wind action. Reliability should be demonstrated through comparisons with situations for which full-scale experience is available.

Examples. The following three examples illustrate the method used to establish design snow loads for some of the situations discussed in this standard. Additional examples are found in O'Rourke and Wrenn (2004).

Example 1. Determine balanced and unbalanced design snow loads for an apartment complex (Risk Category II) with a ground snow load $P_g = 30$ lb/ft^2 and a winter wind parameter, $W_2 = 0.55$. Each unit has a 6-on-12 slope unventilated gable roof insulated to an $R_{roof} = 30$ h·ft^2 °F/Btu (5.3 m^2·K/W). The building length is 100 ft (30.5 m), and the eave to ridge distance, W, is 30 ft (9.1 m). Composition shingles clad the roofs. Trees will be planted among the buildings.

Flat Roof Snow Load:

$$P_f = 0.7\ C_e\ C_t\ P_g$$

Where

$P_g = 30$ lb/ft^2 (1.44 kN/M^2) (from Figure 7.2-1B for Risk Category II),

$C_e = 1.0$ (from Table 7.3-1 for Surface Roughness Category B and a partially exposed roof), and

$C_t = 1.4$ (from Table 7.3-3 for unventilated roofs, R_{roof} = 30 h·ft^2·°F/Btu, and p_g = 30 lb/ft^2).

Thus, p_f = (0.7)(1.0)(1.14)(30) = 23.9 lb/ft^2. In SI, p_f = (0.7)(1.0)(1.14)(1.44) = 1.15 kN/m^2.

Because the roof slope is greater than 15 degrees, the minimum roof snow load, p_m, does not apply.

Sloped-Roof Snow Load:

$P_s = C_sP_f$ where $C_s = 1.0$

(using the "All Other Surfaces" [or solid] line, Figure 7.4-1b for 1.1 < = C_t <1.2)]

Thus, $p_s = 1.00(23.9) = 23.9$ lb/ft^2

($p_s = 1.0(1.15) = 1.15$ kN/m^2) (SI)

Unbalanced Snow Load: Because the roof slope is greater than 1/2 on 12 (2.38 degrees), unbalanced loads must be considered. From the problem statement $W_2 = 0.55$. Hence, for $p_g = 30$ psf (1.44 kN/m^2) $W = l_u = 30$ ft (9.14 m), and $\gamma = 17.9$ pcf (2.80 kN/m^3) from Equation (7.7-1), and $h_d = 2.46$ ft (0.75 m) from Equation (7.6-1). For a 6 on 12 roof, $S = 2.0$ and hence the intensity of the drift surcharge $h_d\gamma/\sqrt{S}$ is 31.1 psf (1.74 kN/m^2), and its horizontal extent $8\sqrt{S}h_d/3$ is 9.3 ft (2.8 m).

Rain-on-Snow Surcharge: A rain-on-snow surcharge load need not be considered because $W/50 = 0.6° < 26.6° = 6$ on 12 roof slope (see Section 7.10). See Figure C7.14-1 for both loading conditions.

Example 2. Determine the roof snow load for a vaulted theater that can seat 450 people in a location where the ground snow load for Risk Category III in Figure 7.2-1C is 27.5 psf (1.32 kN/m^2).

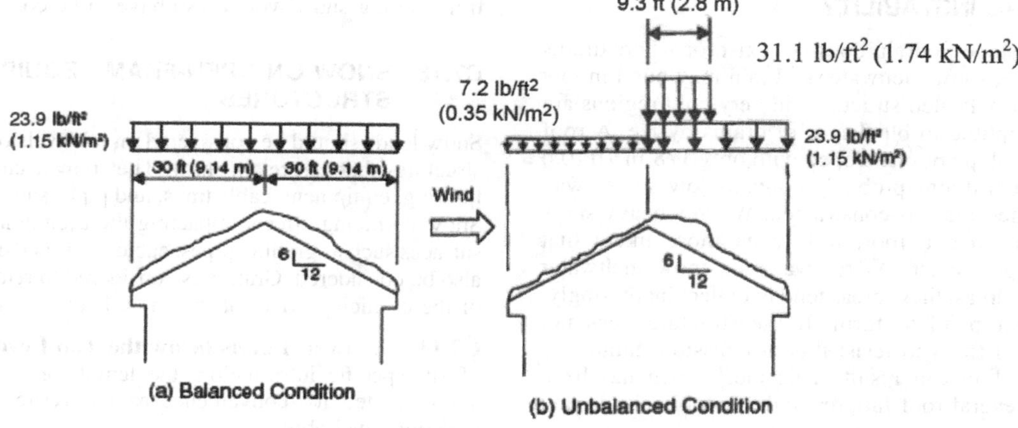

Figure C7.14-1. Design snow loads for Example 1.

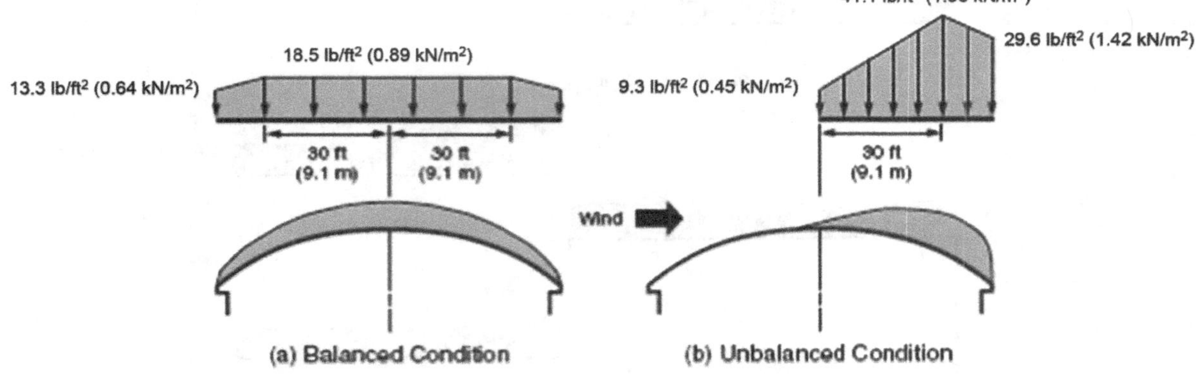

Figure C7.14-2. Design snow loads for Example 2.

The building is the tallest structure in a recreation-shopping complex surrounded by a parking lot. Two large deciduous trees are located in an area near the entrance. The building has an 80 ft (24.4 m) span and 15 ft (4.6 m) rise circular arc structural concrete roof covered with insulation and aggregate surfaced built-up roofing. The unventilated roofing system has a thermal resistance of 20 ft² hr °F/Btu (3.5 Km²/W). It is expected that the structure will be exposed to winds during its useful life.

Flat Roof Snow Load:

$$P_f = 0.7\ C_e\ C_t\ P_g$$

Where

$P_g = 27.5$ lb/ft² (1.32 kN/M²) (from problem statement),
$C_e = 0.9$ (from Table 7.3-1 for Surface Roughness Category B and a fully exposed roof), and
$C_t = 1.07$ (from Table 7.3-3), interpolating for unventilated roof, $R_{roof} = 20\ h$ ft² F/BTU and $P_g = 27.5$ lb/ft²

Thus, $P_f = 0.7\ (1.07)\ (1.08)(27.5) = 18.5$ lb/ft²
$P_f = (0.7)\ (0.9)(1.07)(1.2) = 0.81$ kN/m² (SI)

Tangent of vertical angle from eaves to crown = 15/40 = 0.375.

Angle = 21 degrees.

Because the vertical angle exceeds 10 degrees, the minimum roof snow load, p_m, does not apply per Section 7.3.4.

Sloped-Roof Snow Load:

$p_s = C_s p_f$

From Figure 7.4-1(a) (for $C_t < 1.1$), $C_s = 1.0$ until slope exceeds 30 degrees, which (by geometry) is 30 ft (9.1 m) from the centerline. In this area, $p_s = 1.0(18.5) = 18.5$ lb/ft² [in SI $p_s = 0.81(1) = 0.81$ kN/m²]. At the eaves, where the slope is (by geometry) 41 degrees, $C_s = 0.72$ and $p_s = 18.5(0.72) = 13.3$ lb/ft² [in SI $p_s = 0.81(0.72) = 0.58$ kN/m²]. Because slope at the eaves is 41 degrees, Case 2 loading applies.

Unbalanced Snow Load: Because the vertical angle from the eaves to the crown is greater than 10 degrees and less than 60 degrees, unbalanced snow loads must be considered.

Unbalanced load at crown = $0.5p_f$ = 0.5(18.5) = 9.3 lb/ft² [= 0.5(0.81) = 0.41 kN/m²] (SI).
Unbalanced load at 30 degree point = $2p_f C_s / C_e$ = 2(18.5)(1.0)/0.9 = 41.1 lb/ft² [= 2(0.81)(1.0)/0.9 = 1.8 kN/m²] (SI).
Unbalanced load at eaves = 2(18.5)(0.72)/0.9 = 29.6 lb/ft² [in SI = 2(0.81)(0.72)/0.9 = 1.30 kN/m²].

Rain-on-Snow Surcharge: A rain-on-snow surcharge load need not be considered because $W/50$ = 80/50 = 1.6 < average slope = (0° + 41°)/2 = 20.5°. See Figure C7.14-2 for both loading conditions.

Example 3. Determine design snow loads for the upper and lower flat roofs of a building located where P_g = 40 lb/ft² (1.92 kN/m²) for Risk Category II from Figure 7.2-1B and 32 lb/ft² (1.54 kN/m²) for Risk Category I from Figure 7.2-1A and $W_2 = 0.5$. The elevation difference between the roofs is 10 (3). The 100 × 100 ft (30.5 × 30.5 m) unventilated high portion is heated, and the roof is insulated to an R_{roof} = 30 h ft² F/BTU (5.3 K m²/w) and the 240 ft (73 m) wide, 100 ft (30.5 m) long low portion is an unheated storage area. The building is in an industrial park in flat open country with no trees or other structures offering shelter.

High Roof:

$$P_f = 0.7\ C_e\ C_t\ P_g$$

where

$P_g = 40$ lb/ft² (1.92 kN/m²) (given),
$C_e = 0.9$ (from Table 7.3-1), and
$C_t = 1.13$ (from Table 7.3-3 for unventilated, R_{roof} = 30 h ft² F/BTU, and P_g = 40 lb/ft²)

Thus, $P_f = 0.7\ (0.9)\ (1.13)\ (40) = 28.5$ lb/ft²
$P_f = (0.7)\ (0.9)(1.13)\ (1.92) = 1.37$ kN/m² (SI)

Because P_g = 40 lb/ft² (1.92 kN/m²) with Risk Category II, the minimum roof snow load case with uniform load = 30 lb/ft² (1.44 kN/m²) needs to be considered. However, it does not combine with drift loading and is unlikely the controlling snow load for the most highly loaded structural components.

Low Roof:

$$P_f = 0.7\ C_e\ C_t\ P_g$$

where

$P_g = 32$ lb/ft² (1.54 kN/m²) (given),
$C_e = 1.0$ (from Table 7.3-1) partially exposed because of the presence of a high roof, and
$C_t = 1.2$ (from Table 7.3-2).

Thus, $P_f = 0.7\ (1.0)\ (1.2)\ (32) = 27$ lb/ft²
$P_f = (0.7)\ (1.0)(1.2)(1.54) = 1.29$ kN/m² (SI)

Because P_g = 32 lb/ft² (1.54 kN/m²) the minimum roof snow load value $P_{m,max}$ = 25 lb/ft² (1.20 kN/m²) and hence does not control.

Drift Load Calculation:

$\gamma = 0.13(40) + 14 = 19$ lb/ft³
In SI, $\gamma = 0.426(1.92) + 2.2 = 3.02$ kN/m³.

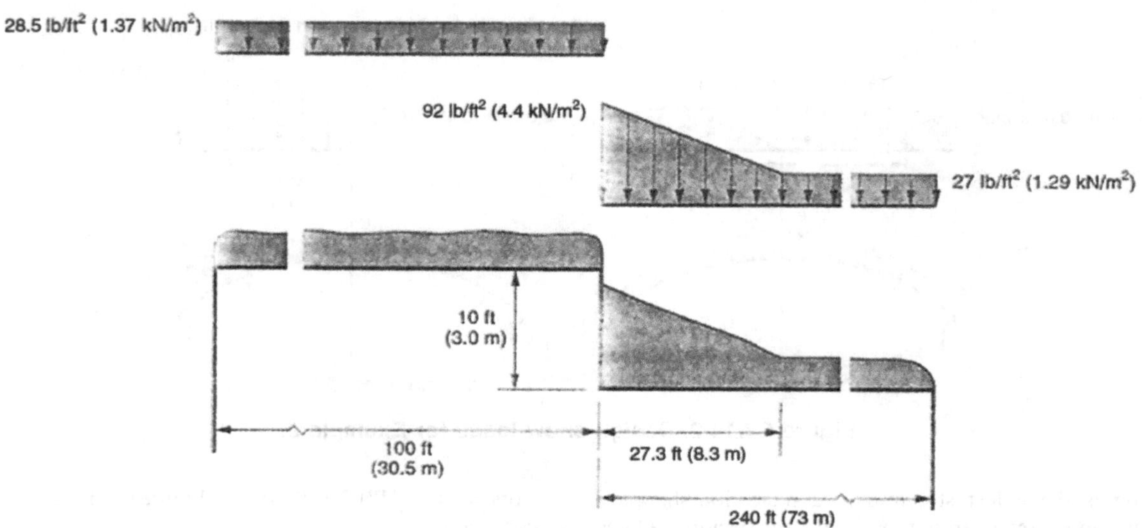

Figure C7.14-3. Design snow loads for Example 3.

$h_b = p_f/19 = 27/19 = 1.4$ ft
$h_b = 1.29/3.02 = 0.43$ m (SI).
$h_c = 10 - 1.4 = 8.6$ ft
$h_c = 3.05 - 0.43 = 2.62$ m (SI).
$h_c/h_b = 8.6/1.4 = 6.1$
$h_c/h_b = 2.62/0.43 = 6.1$) (SI).

Because $h_c/h_b \geq 0.2$, drift loads must be considered (see Section 7.7.1).

For $W_2 = 0.5$, $p_g = 32$ lb/ft² (1.54 kN/m²)(unheated storage), $l_u = 100$ ft (30.5 m), gamma = 19 pcf, and $I_s = 0.8$ (unheated storage), $h_d = 3.4$ ft (1.04 m) from Equation 7.6-1.

For the windward drift, $p_g = 32$ lb/ft² (1.54 kN/m²), l_u = length of lower roof = 240 ft (73 m), and $W_2 = 0.5$. Hence, from Equation (7.6-1), $h_d = 4.55$ ft (1.38 m). The windward drift height is then 0.75 $h_d = 3.41$ ft (1.04 m) and its drift width is 8 (3.41) = 27.3 ft (8.3 m).

The windward drift controls and the peak load at the step is the balanced load of 27 lb/ft² (1.29 kN/m²) plus the drift surcharge of 3.41(19) = 64.8 lb/ft² (3.1 kN/m²) or a total of 92 lb/ft² (4.4 kN/m²).

Rain-on-Snow Surcharge: A rain-on-snow surcharge load need not be considered because P_g for both roofs are greater than $P_{m,\max}$ in Table 7.3-3 for both roofs. See Figure C7.14-3 for snow loads on both roofs.

REFERENCES

AISC (American Institute of Steel Construction). 2016. *Specification for structural steel buildings*. ANSI/AISC 360-16. Chicago: AISC.

Allen, D. E. 1956. *Snow loads on roofs. The present requirements and a proposal for a survey of snow loads on roofs*. Ottawa, Ontario: National Research Council of Canada.

ASHRAE (American Society of Heating, Refrigerating, and Air-Conditioning Engineers) Standard 90.1. Peachtree Corners, GA: ASHRAE.

Bartlett, F. M., R. J. Dexter, M.D. Graeser, J. J. Jelinek, B. J. Schmidt, and T. V. Galambos. 2003. "Updating standard shape material properties database for design and reliability." *Eng. J.* 40(1):2.

Bean, B., M. Maguire, Y. Sun, J. Wagstaff, S. Al-Rubaye, J. Wheeler, S. Jarman, and M. Rogers. 2021. "The 2020 national snow load study." https://doi.org/10.26077/200k-pr86.

Breiman, L. 2001. "Random forests." *Mach. Learn.* 45 (1): 5–32.

Buska, J., A. Greatorex, and W. Tobiasson. 2020. *Site-specific case studies for determining ground snow loads in the United States*. ERDC/CRREL SR-20-1. Washington, DC: USACE.

CEC (Commission for Environmental Cooperation). 1997. *Ecological regions of North America: Toward a common perspective*. Montreal: CEC.

Colbeck, S. C. 1977a. "Roof loads resulting from rain-on-snow: Results of a physical model." *Can. J. Civil Eng.* 4 (December): 482–490. https://doi.org/10.1139/l77-057.

Colbeck, S. C. 1977b. *Snow loads resulting from rain-on-snow*. CRREL Rep. No. 77-12. Hanover, NH: Cold Regions Research and Engineering Laboratory.

de Marne, H. 1988. *Field experience in control and prevention of leaking from ice dams in New England*. CRREL Special Rep. No. 89-67. Hanover, NH: US Army Cold Regions Research and Engineering Laboratory.

Ellingwood, B., T. V. Galambos, J. G. MacGregor, and C. A. Cornell. 1980. "Development of a probability based load criterion for American National Standard A58." Special Publication No. 577. Washington, DC: National Bureau of Standards.

Elliott, M. 1975. "Snow load criteria for western United States, case histories and state-of-the-art." In *Proc., 1st Western States Conf. of Structural Engineer Associations*, Sun River, OR.

FEMA (Federal Emergency Management Agency). 2019. "FEMA design guide: Three-dimensional roof snowdrifts." https://www.fema.gov/media-library/assets/documents/182816. Access date July 20, 2021.

Finney, E. 1939. *Snow drift control by highway design*. Bulletin 86. Lansing, MI: Michigan State College Engineering Station.

FM Global. 2016. *Property loss prevention data sheet 1-54, roof loads for new construction*. Johnston, RI: FM Global.

Grange, H. L., and L. T. Hendricks. 1976. *Roof-snow behavior and ice-dam prevention in residential housing*. Ext. Bull. 399. St. Paul, MN: University of Minnesota.

Hosking, J. R. M., J. R. Wallis, and E. F. Wood. 1985. "Estimation of the generalized extreme-value distribution by the method of probability-weighted moments." *Technometrics* 27 (3): 251–261.

Irwin, P., C. William, S. Gamle, and R. Retziaff. 1992. *Snow prediction in Toronto and the Andes Mountains; FAE simulation capabilities*. CRREL Special Rep. No. 92-27. Hanover, NH: US Army Cold Regions Research and Engineering Laboratory.

Isyumov, N., and M. Mikitiuk. 1992. *Wind tunnel modeling of snow accumulation on large roofs*. CRREL Special Rep. No. 92-27. Hanover, NH: US Army Cold Regions Research and Engineering Laboratory.

Jenkinson, A. F. 1955. "The frequency distribution of the annual maximum (or minimum) values of meteorological elements." *Q. J. R. Meteorol. Soc.* 81 (348): 158–171. https://doi.org/10.1002/qj.49708134804.

Kennedy, D., M. Isyumov, and M. Mikitiuk. 1992. *The effectiveness of code provisions for snow accumulations on stepped roofs*. CRREL Special Rep. No. 92-27. Hanover, NH: US Army Cold Regions Research and Engineering Laboratory.

Kennedy, I., and D. Lutes. 1968. *Snow loads on roofs 1966-67: Eleventh progress report*. Quebec: National Research Council Canada, Division of Building Research.

Klinge, A. F. 1978. "Ice dams." *Popular Sci.*, 119–120, New York: Times Mirror Magazine, Inc.

Liaw, A., and M. Wiener. 2002. "Classification and regression by random forest." *R News* 2 (3): 18–22.

Lorenzen, R. T. 1970. *Observations of snow and wind loads precipitant to building failures in New York State, 1969–1970*. Paper NA 70-305. St. Joseph, MO: American Society of Agricultural Engineers.

Lutes, D. A., and W. R. Schriever. 1971. *Snow accumulation in Canada: Case histories: II. Ottawa*. DBR Technical Paper 339. Quebec: National Research Council of Canada.

Mackinlay, I. 1988. *Architectural design in regions of snow and ice*. CRREL Special Rep. No. 89-67. Hanover, NH: US Army Cold Regions Research and Engineering Laboratory.

Maguire, M., B. Bean, J. Harris, A. Liel, and S. Russell. 2021. *Ground snow loads for ASCE 7-22 – What has changed and why?* Paper 277. Logan, UT: Utah State University.

Meehan, J. F. 1979. "Snow loads and roof failures." In *Proc., Structural Engineers Association of California*, San Francisco.

Mitchell, G. R. 1978. "Snow loads on roofs—An interim report on a survey." In *Wind and snow loading*, 177–190. Lancaster, UK: Construction Press.

National Greenhouse Manufacturers Association. 1988. *Design loads in greenhouse structures*. Taylors, SC: National Greehouse Manufacturers Association.

National Research Council of Canada. 1990. *National building code of Canada 1990*. Ottawa: National Research Council of Canada.

O'Rourke, M. 1989. "Discussion of 'Roof collapse under snowdrift loading and snow drift design criteria." *J. Perform. Constr. Facil.* 3 (4): 266–268. https://doi.org/10.1061/(ASCE)0887-3828(1989)3:4(266).

O'Rourke, M. 2007. *Snow loads: A guide to the snow load provisions of ASCE 7-05*. Reston, VA: ASCE.

O'Rourke, M., and M. Auren. 1997. "Snow loads on gable roofs." *J. Struct. Eng.* 123 (12): 1645–1651.

O'Rourke, M., and J. Cocca. 2019. "Improved snow drift relations." *J. Struct. Eng.* 145 (5): 04019027. doi/abs/10.1061/%28ASCE%29ST.1943-541X.0002278.

O'Rourke, M., and J. Conlin. n.d. "Effective fetch for roof steps in series." *J. Struct. Eng.* (Submitted January 2021).

O'Rourke, M., and J. Conlin. 2020. "Snow capture at walls." *J. Struct. Eng.* 146 (9): 06020007. https://doi.org/10.1061/(ASCE)ST.1943-541X.0002763.

O'Rourke, M., and C. De Angelis. 2002. "Snow drifts at windward roof steps." *J. Struct. Eng.* 128 (10): 1330–1336. https://doi.org/10.1061/(ASCE)0733-9445(2002)128:10(1330).

O'Rourke, M., and C. Downey. 2001. "Rain-on-snow surcharge for roof design." *J. Struct. Eng.* 127 (1): 74–79. https://doi.org/10.1061/(ASCE)0733-9445(2001)127:1(74).

O'Rourke, M., M. Ganguly, and L. Thompson. 2007. *Eave ice dams*. Troy, NY: Rensselaer Polytechnic Institute.

O'Rourke, M., and N. Isyumov. 2016. *Snow loads on solar-paneled roofs*. Reston, VA: ASCE.

O'Rourke, M., P. Koch, and R. Redfield. 1983. *Analysis of roof snow load case studies: Uniform loads*. US Army, CRREL Rep. No. 83-1. Hanover, NH: Cold Regions Research and Engineering Laboratory.

O'Rourke, M., J. Potac, and T. Thiis. 2018. "Windward snow drift loads." *J. Struct. Eng.* 144 (5): 04018033. https://doi.org/10.1061/(ASCE) ST. 1943 -541X0002032.

O'Rourke, M., and S. Russell. 2019. "Snow thermal factors for structural renovations." In *Structure*. Accessed July 23, 2021. Chicago: Structure magazine. https://www.structuremag.org/?p=14566

O'Rourke M., H. Sinh, J. Cocca, and T. Williams. 2020 "Winter wind parameter for snow drifts." *J. Struct. Eng.* 146 (7): 04020124. https://doi.org/10.1061/(ASCE)ST.1943-541X.0002634.

O'Rourke, M. J., R. S., Speck Jr., and U. Stiefel. 1985. "Drift snow loads on multilevel roofs." *J. Struct. Eng.* 111 (2): 290–306. https://doi.org/10.1061/(ASCE)0733-9445(1985)111:2(290).

O'Rourke, M., W. Tobiasson, and E. Wood. 1986. "Proposed code provisions for drifted snow loads." *J. Struct. Eng.* 112 (9): 2080–2092. https://doi.org/10.1061/(ASCE)0733-9445(1986)112:9(2080).

O'Rourke, M., and N. Weitman. 1992. *Laboratory studies of snow drifts on multilevel roofs*. CRREL Special Rep. No. 92-27. Hanover, NH: US Army Cold Regions Research and Engineering Laboratory.

O'Rourke, M., and P. D. Wrenn. 2004. *Snow loads: A guide to the use and understanding of the snow load provisions of ASCE 7-02*. Reston, VA: ASCE.

Paine, J. C. 1988. *Building design for heavy snow areas*. CRREL Special Rep. No. 89-67. Hanover, NH: US Army Cold Regions Research and Engineering Laboratory.

Peter, B. G. W., W. A. Dalgliesh, and W. R. Schriever. 1963. "Variations of snow loads on roofs." *Trans. Eng. Inst. Canada* 6 (A-1): 8.

Sack, R. L. 1988. "Snow loads on sloped roofs." *J. Struct. Eng.* 114 (3): 501–517. https://doi.org/10.1061/(ASCE)0733-9445(1988)114:3(501).

Sack, R. L., D. A. Arnholtz, and J. S. Haldeman. 1987. "Sloped roof snow loads using simulation." *J. Struct. Eng.* 113 (8): 1820–1833. https://doi.org/10.1061/(ASCE)0733-9445(1987)113:8(1820).

Schriever, W. R., Y. Faucher, and D. A. Lutes. 1967. *Snow accumulation in Canada: Case histories: I. Ottawa, Ontario, Canada*. Technical Paper NRCC 9287. Ottawa: National Research Council of Canada, Division of Building Research.

Speck, R., Jr. 1984. "Analysis of snow loads due to drifting on multilevel roofs." M.S. thesis, Dept. of Civil Engineering, Rensselaer Polytechnic Institute.

Structural Engineers Association of Alaska. 2019. *Alaska snow loads for the 2022 update of ASCE 7*. Anchorage, AL: Structural Engineers Association of Alaska.

Structural Engineers Association of Alaska. 2020. *Reliability targeted Alaska ground snow loads for the 2022 edition of ASCE 7*. Anchorage, AL: Structural Engineers Association of Alaska.

Tabler, R. 1975. "Predicting profiles of snowdrifts in topographic catchments." In *Proc., 43rd Annual Western Snow Conf*, Coronado, California.

Taylor, D. 1983. "*Sliding snow on sloping roofs*. Can. Bldg. Digest 228. Ottawa: National Research Council of Canada.

Taylor, D. 1985. "Snow loads on sloping roofs: Two pilot studies in the Ottawa area." *Can. J. Civ. Eng.* 12 (2): 334–343. https://doi.org/10.1139/l85-036.

Taylor, D. A. 1979. "A survey of snow loads on roofs of arena-type buildings in Canada." *Can. J. Civ. Eng.* 6 (1): 85–96. https://doi.org/10.1139/l79-010.

Taylor, D. A. 1980. "Roof snow loads in Canada." *Can. J. Civ. Eng.* 7 (1): 1–18. https://doi.org/10.1139/l80-001.

Taylor, D. A. 1984. "Snow loads on two-level flat roofs." In *Proc. 41st Eastern Snow Conf., June 7–8*, New Carrollton, MD.

Tobiasson, W. 1988. *Roof design in cold regions*. CRREL Special Rep. No. 89-67. Hanover, NH: US Army Cold Regions Research and Engineering Laboratory.

Tobiasson, W. 1999. "Discussion of 'Snow loads on gable roofs." *J. Struct. Eng.* 125 (4): 470–471.

Tobiasson, W., and J. Buska. 1993. "Standing seam metal roofs in cold regions." In *Proc., 10th Conf. on Roofing Technology*, Buenos Aires, Argentina, 34–44.

Tobiasson, W., J. Buska, and A. Greatorex. 1996. "Snow guards for metal roofs." In *Cold regions engineering: The cold regions infrastructure—An international imperative for the 21st century*, R. F. Carlson, ed. Reston, VA: ASCE.

Tobiasson, W., J. Buska, and A. Greatorex. 1998. *Attic ventilation guidelines to minimize icings at eaves*. Raleigh, NC: Roof Consultants Institute.

Tobiasson, W., T. Tantillo, and J. Buska. 1999. "Ventilating cathedral ceilings to prevent problematic icings at their eaves." In *Proc., North American Conf. on Roofing Technology*, Hanover, New Hampshire. 34–44.

Zallen, R. M. 1988. "Roof collapse under snowdrift loading and snow drift design criteria." *J. Perform. Constr. Facil.* 2 (2): 80–98. https://doi.org/10.1061/(ASCE)0887-3828(1988)2:2(80).

OTHER REFERENCES (NOT CITED)

Al Hatailah, H., B. L. Godfrey, R. J. Nielsen, and R. L. Sack. 2015. *Ground snow loads for Idaho*. Moscow, ID: University Idaho Civil Engineering Dept. Report.

Brown, J. 1970. "An approach to snow load evaluation." In *Proc., 38th Western Snow Conf*, Victoria, British Columbia.

Curtis, J., and G. Grimes. 2004. *Wyoming climate atlas*, 81–92. Laramie, WY: University of Wyoming.

Ellingwood, B., and R. Redfield. 1983. "Ground snow loads for structural design." *J. Struct. Eng.* 109 (4): 950–964. https://doi.org/10.1061/(ASCE)0733-9445(1983)109:4(950).

Mackinlay, I., and W. E. Willis. 1965. *Snow country design*. Washington, DC: National Endowment for the Arts.

Maji, A. K. 1999. *Ground snow load database for New Mexico*. Albuquerque, NM: University of New Mexico.

Newark, M. 1984. "A new look at ground snow loads in Canada." In *Proc., 41st Eastern Snow Conf. June 7–8*, New Carrollton, MD, 37–48.

O'Rourke, M., and N. Kuskowski. 2005. "Snow drifts at roof steps in series." *J. Struct. Eng.* 131 (10): 1637–1640. https://doi.org/10.1061/(ASCE)0733-9445(2005)131:10(1637).

Placer County Building Division. 1985. "Snow load design." In *Placer County code*. Auburn, CA: Placer County Building Division.

Sack, R. L., and A. Sheikh-Taheri. 1986. *Ground and roof snow loads for Idaho*. Moscow, ID: University of Idaho.

SEAW (Structural Engineers Association of Washington). 1995. *Snow loads analysis for Washington*. Seattle: SEAW.

Structural Engineers Association of Arizona. 1981. *Snow load data for Arizona*. Tempe, AZ: Structural Engineers Association of Arizona.

Structural Engineers Association of Colorado. 2016. *Colorado design snow loads*. Denver: Structural Engineers Association of Colorado.

Structural Engineers Association of Northern California. 1981. *Snow load design data for the Lake Tahoe area*. San Francisco: Structural Engineers Association of Northern California.

Structural Engineers Association of Oregon. 2013. *Snow load analysis for Oregon*. Salem, OR: Oregon Dept. of Commerce, Building Codes Division.

Structural Engineers Association of Utah. 1992. *Utah snow load study*. Salt Lake City, UT: Structural Engineers Association of Utah.

Theisen, G. P., M. J. Keller, J. E. Stephens, F. F. Videon, and J. P. Schilke. 2004. *Snow loads for structural design in Montana*. Bozeman, MT: Montana State University.

Tobiasson, W., J. Buska, A. Greatorex, J. Tirey, J. Fisher, and S. Johnson. 2000. "Developing ground snow loads for New Hampshire." In *Snow engineering: Recent advances and developments*. Brookfield, VT: A. A. Balkema.

Tobiasson, W., J. Buska, A. Greatorex, J. Tirey, J. Fisher, and S. Johnson. 2002. *Ground snow loads for New Hampshire*. Technical Rep. No. ERDC/CRREL TR-02-6. Washington, DC: USACE.

Tobiasson, W., and A. Greatorex. 1996. "Database and methodology for conducting site specific snow load case studies for the United States." In *Proc., 3rd Int. Conf. on Snow Engineering*, Sendai, Japan, 249–256.

USDA Soil Conservation Service. 1970. *Lake Tahoe basin snow load zones*. Reno, NV: USDA, Soil Conservation Service.

CHAPTER C8
RAIN LOADS

C8.1 DEFINITIONS AND SYMBOLS

C8.1.1 Definitions

PRIMARY MEMBERS: Structural members having direct attachment to the columns, including girders, beams, and trusses.

SCUPPER: An opening in the side of a building (typically through a parapet wall) for the purpose of draining water off the roof.

C8.1.2 Symbols

A = Tributary roof area, plus one-half the wall area that diverts rainwater onto the roof, serviced by a single drain outlet in the secondary drainage system, ft^2 (m^2)
B_p = Ponding amplification factor
C_p = Flexibility coefficient related to primary members
C_s = Flexibility coefficient related to secondary members
D = Drain bowl diameter for a primary roof drain, or overflow dam or standpipe diameter for a secondary roof drain, in. (mm)
EI_p = Flexural rigidity of primary members
EI_s = Flexural rigidity of secondary members
i = Design rainfall intensity, in./h (mm/h)
L_r = Length of level roof edge that allows for free overflow drainage of rainwater when the roof edge is acting as the secondary drainage system, ft (m)
L_p = Span of primary members
L_s = Span of secondary members, in. (mm)
Q = Flow rate out of a single drainage system, gal./min(m^3/s)
S = Spacing of secondary members, in. (mm)
γ = Unit weight of water.
β = Minimum rise in inches for a run of 1 ft (i.e., a β on 12 roof slope) to achieve no local minimum

C8.2 DESIGN RAIN LOADS

Roof drainage is a structural, architectural, and mechanical (plumbing) issue. The provisions of this chapter define rain loads and do not explicitly specify any requirements for the design of drainage systems. Among the first steps in establishing the rain loads is the identification of the secondary drainage system for structural loads (SDSL). The SDSL is the roof drainage system by which water drains off the roof when the drainage systems that are required to be assumed to be blocked, listed in Section 8.2, are blocked or not working.

It is possible that a drainage system that is designated as a secondary or emergency overflow drain, by the architect or mechanical (plumbing) engineer, is required to be assumed to be blocked, per Section 8.2. Such a drainage system cannot serve as the SDSL and the next point of discharge that is not required to be assumed to be blocked will be the SDSL. As an example of an extreme case in which no overflow drainage is provided, flow of water over a parapet could serve as the SDSL. However, this likely leads to large rain loads and an inefficient structure. Cases such as this should be avoided, where possible, through design team coordination.

Rain loads are based on the condition of a blocked primary and other drainage systems (per Section 8.2) and a 15 min duration storm with return period based on the risk category of the structure. Therefore, the SDSL is of greater importance than the primary drainage system for the determination of rain loads.

If the primary drainage system is a free-draining edge, then the edge can also serve as the SDSL since it cannot become blocked like internal drains or scuppers can. Otherwise, the SDSL must be distinct from the primary drainage system. One reason for this is so that activation of the SDSL can serve as an urgent warning of blockage on the roof and the need for prompt maintenance of the primary drains. Similarly, the elevation of the SDSL must be at least 2 in. (50 mm) above that of the primary drainage system so that the SDSL is not frequently activated, which would decrease the urgency of the warning and also make the SDSL more susceptible to blockage. The *Uniform Plumbing Code* (IAPMO 2021) and FM Global (2016) have similar requirements. Greater separation may be warranted based on analysis and the desired frequency of activation of the SDSL.

In some areas of the country, ordinances are in effect that either limit the rate or delay the release of rainwater flow from roofs into storm drains. Controlled flow drains are often used on such roofs. Those roofs must be capable of supporting the stormwater temporarily stored on them similar to traditional roof drainage systems. Because controlled flow drains necessarily restrict the passage of water, they can only be primary drains and cannot be considered to act as an SDSL.

Roof drainage systems are not always designed to handle all the flow associated with intense, short-duration rainfall events. For example, the *International Plumbing Code* (ICC 2021) uses a 1 h duration event with a 100-year return period for the design of both the primary and secondary drainage systems. An adequate secondary (overflow) drainage system, which is used to limit the depth of water on the roof in the event of clogging of the primary drains, must be designed for an adequately short-duration rainfall event. Some plumbing codes use an arbitrary 1 h duration storm event for the design of roof drain systems; however, the critical duration for a roof is generally closer to 15 min (the critical duration depends on the roof geometry and drain sizes), and therefore the plumbing codes do not appropriately account for the coincidence of both blocked primary drains and short-duration rainfall events at the design mean recurrence interval (return period or frequency). Graber (2009) provides guidance for determining the critical durations for different types of roof configurations. A very severe local storm or thunderstorm in excess of the 100-year return period storm may produce a deluge of such intensity and duration that properly designed primary drainage systems are temporarily overloaded. Such temporary loads are typically covered in design when blocked drains (see Section 8.2) and a rainfall duration of 15 min is considered. The use of a 60 min duration/100-year return period rainfall event for the design of the primary drainage system and the 15 min duration/100-year return period rainfall event for the SDSL (assuming the primary drainage system is completely blocked) is consistent with the NFPA *5000 Building Construction and Safety Code* (2021). Internal gutters are typically designed for 2 to 5 min duration rainfall events because their critical duration is very short due to their limited storage volume and inability to attenuate a rainfall event. The use of longer design storm return periods for Risk Categories III and IV buildings and structures is consistent with other loads in this standard. The return periods in Table 8.2-1 are based on the collective judgment of the committee and availability of the necessary data.

The National Oceanic and Atmospheric Administration (NOAA) National Weather Service Precipitation Frequency Data Server, Hydrometeorological Design Studies Center provides rainfall intensity data in inches per hour for 15 min duration storms with various mean recurrence intervals (http://hdsc.nws.noaa.gov/hdsc/pfds/index.html). Precipitation intensity [in Equation (C8.2-1)] is in the units of inches per hour; if precipitation depth is provided, a conversion to intensity is required.

The following roof conditions adversely affect the critical duration, or increase the peak flow rate, and should be avoided or appropriately considered by the designer when determining the design rain load:

1. Roofs with internal gutters that have limited storage capacity and quickly fill with rainwater. Gutters are typically designed for 2 to 5 min duration storms, because their critical duration is much shorter than the critical duration for typical roofs with scuppers or internal drains.
2. Architecturally complex roofs with internal gutters with significant gutter slopes. Significant gutter slopes allow water to flow at high velocities, which need to be considered when designing the gutter outlets and determining rain loads.
3. Areas susceptible to a concentration of flow, for example, when an addition is added to a low-sloped gable roof (Figure C8.2-1). In this case, rainwater at the edge of the main roof cannot build up over the primary drains or the SDSL to attain the design flow rate, and the water flows into and inundates the small roof extension.
4. Small roofs adjacent to large walls, where the wall is capable of contributing substantial wind-driven rain flow (sheet flow down the wall) to the roof.
5. High roof areas that drain onto a low roof, increasing the tributary roof area and decreasing the critical rainfall duration.

Items 4 and 5 can occur when building additions occur (i.e., the new construction imposes an unfavorable condition on the existing construction conditions).

The amount of water that could accumulate on a roof from blockage of the primary drainage system is determined, and the roof is designed to withstand the load created by that water plus the uniform load caused by water that rises above the inlet of the secondary drainage systems at its design flow. If parapet walls, cant strips, expansion joints, and other features create the potential for deep water in an area, it may be advisable to install in that area secondary (overflow) drains with separate drain lines rather than overflow scuppers to reduce the magnitude of the design rain load. Where geometry permits, free discharge is the preferred form of emergency drainage.

The depth of water, d_h, above the inlet of the secondary drainage system (i.e., the hydraulic head) is a function of the rainfall intensity, i, at the site, the area of roof serviced by that drainage system, and the size of the drainage system.

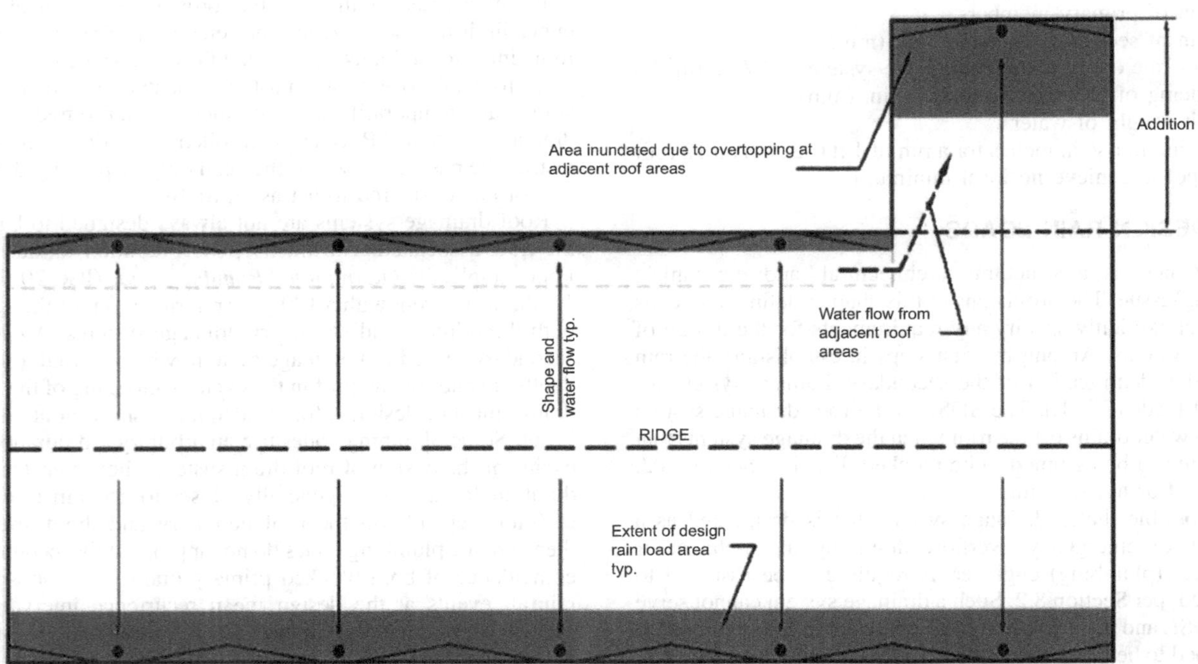

Figure C8.2-1. Low-slope gable roof with drainage condition at building extension.

The flow rate through a single drainage system is as follows:

$$Q = 0.0104\ Ai \quad \text{(C8.2-1)}$$

$$Q = 0.278 \times 10^{-6}\ Ai \quad \text{(C8.2-1.SI)}$$

The hydraulic head, d_h, is related to flow rate, Q, for various drainage systems in Table C8.2-1. This table indicates that d_h can vary considerably depending on the type and size of each drainage system and the flow rate it must handle. For this reason, the single value of 1 in. (25 mm) (i.e., 5 lb/ft^2 [0.24 kN/m^2]) used in ASCE 7-93 was eliminated.

The hydraulic head, d_h, can generally be assumed to be negligible for design purposes when the secondary drainage system is free to overflow along a roof edge where the length of the level roof edge (L_r) providing free drainage is

$$L_r \geq Ai/400 \quad \text{(C8.2-2)}$$

$$L_r \geq Ai/3{,}100 \quad \text{(C8.2-2.SI)}$$

Equation (C8.2-2) is based on the assumption that hydraulic head (d_h) of approximately 0.25 in. (6 mm) above the level roof edge, which represents a rain load of 1.3 lb/ft^2 (6.3 kg/m^2), is negligible in most circumstances.

Flow rates and corresponding hydraulic heads for roof drains are often not available in industry codes, standards, or drain manufacturers' literature for many commonly specified drain

Table C8.2-1. Flow Rate (Q) in Gallons per Minute for Secondary (Overflow) Roof Drains at Various Hydraulic Heads (d_h) above the Dam or Standpipe, in Inches.

	Hydraulic Head (in.) above Dam or Standpipe						
	Overflow Dam 8 in. Diameter			Overflow Dam 12.75 in. Diameter		Overflow Dam 17 in. Diameter	Overflow Standpipe 6 in. Diameter
	Drain Outlet Size (in.)						
	3	4	6	6	8	10	4
	Drain Bowl Depth (in.)						
Flow Rate (gal./min)	2	2	2	2	3.25	4.25	2
50	0.5	0.5	0.5	0.5	0.5	—	1.0
75	1.0	—	—	—	—	—	—
100	1.5	1.0	1.0	1.0	0.5	1.0	1.5
125	2.0	—	—	—	—	—	—
150	2.0	1.5	1.5	1.0	—	—	2.5
175	3.0	—	—	—	—	—	—
200	—	2.0	2.0	1.5	1.5	1.5	2.5
225	—	—	—	—	—	—	—
250	—	2.5	2.5	1.5	—	—	2.5
300	—	3.0	3.0	2.0	2.0	1.5	3.0
350	—	3.5	3.5	2.5	—	—	3.5
400	—	5.5	3.5	3.0	2.5	2.0	—
450	—	—	4.0	3.0	—	—	—
500	—	—	5.0	3.5	3.0	2.5	—
550	—	—	5.5	4.0	—	—	—
600	—	—	6.0	5.5	3.5	2.5	—
650	—	—	—	—	—	—	—
700	—	—	—	—	3.5	3.0	—
800	—	—	—	—	4.5	3.0	—
900	—	—	—	—	5.0	3.5	—
1,000	—	—	—	—	5.5	3.5	—
1,100	—	—	—	—	—	4.0	—
1,200	—	—	—	—	—	4.5	—

Source: Adapted from FM Global (2016).

Notes:

1. Assume that the flow regime is either weir flow or transition flow, except where the hydraulic head values are in shaded cells below the heavy line that designates orifice flow.
2. To determine total head, add the depth of water (static head, d_s) above the roof surface to the secondary drain inlet (which is the height of the dam or standpipe above the roof surface) to the hydraulic head listed in this table plus the ponding head.
3. Linear interpolation for flow rate and hydraulic head is appropriate for approximations.
4. Extrapolation is not appropriate.

types and sizes. Since the hydraulic characteristics and performance of roof drains can depend not only on the size of the drain outlet but also on the geometry of the drain body (e.g., the diameter of the drain dam and depth of the drain bowl), determining the flow rate and corresponding hydraulic head for a drain can be difficult based only on hydraulic calculations. This is particularly true when considering the difficulty in predicting the flow regime (i.e., weir flow, orifice flow, or transition between the two) and the significant effect that flow regime has on the relationship between flow rate and corresponding hydraulic head for a drain.

Based on a drain flow testing program completed by FM Global (2012), the hydraulic heads corresponding to a given range of drain flow rates are provided in Tables C8.2-1 and C8.2-2. This drain testing program included six sizes of primary roof drains and seven sizes and types of secondary (overflow) roof drains. The drains were tested with debris guards (strainers) in place and in a test basin with a relatively smooth bottom surface (waterproofing membrane) to simulate typical smooth-surface roofing material. Measurements of water depth in the test basin were made at a distance of 2 ft (0.6 m) or more from the drain, which ensured that the head measurements were not significantly affected by surface water velocity, and therefore were made where the velocity head was negligible, which was confirmed when comparing water depth based on direct depth measurements to hydraulic head based on pressure taps embedded in the bottom surface of the test basin.

Refer to Figure C8.2-2 for a schematic view of a secondary drain and the relationship between the drain, the roof surface, and the head.

The following method can be used to approximate hydraulic head for differing drain body dimensions:

(a) For weir flow and transition flow regime designations (cells that are not shaded) in Tables C8.2-1 and C8.2-2:

Table C8.2-2. SI Units, Flow Rate (Q) in cubic meters per second for Secondary (Overflow) Roof Drains at Various Hydraulic Heads (d_h) above the Dam or Standpipe, in millimeters.

	Hydraulic Head (mm) above Dam or Standpipe						
	Overflow Dam 203 mm Diameter			Overflow Dam 329 mm Diameter		Overflow Dam 432 mm Diameter	Overflow Standpipe 152 mm Diameter
	Drain Outlet Size (mm)						
	76	102	152	152	203	254	102
	Drain Bowl Depth (mm)						
Flow Rate (m³/s)	51	51	51	51	83	108	51
0.0032	13	13	13	13	13	—	25
0.0047	25	—	—	—	—	—	—
0.0063	38	25	25	25	13	25	38
0.0079	51	—	—	—	—	—	—
0.0095	51	38	38	25	—	—	64
0.0110	76	—	—	—	—	—	—
0.0126	—	51	51	38	38	38	64
0.0142	—	—	—	—	—	—	—
0.0158	—	64	64	38	—	—	64
0.0189	—	76	76	51	51	38	76
0.0221	—	89	89	64	—	—	89
0.0252	—	140	89	76	64	51	—
0.0284	—	—	102	76	—	—	—
0.0315	—	—	127	89	76	64	—
0.0347	—	—	140	102	—	—	—
0.0379	—	—	152	140	89	64	—
0.0410	—	—	—	—	—	—	—
0.0442	—	—	—	—	89	76	—
0.0505	—	—	—	—	114	76	—
0.0568	—	—	—	—	127	89	—
0.0631	—	—	—	—	140	89	—
0.0694	—	—	—	—	—	102	—
0.0757	—	—	—	—	—	114	—

Source: Adapted from FM Global (2016).

Notes:

1. Assume that the flow regime is either weir flow or transition flow, except where the hydraulic head values are in shaded cells below the heavy line that designates orifice flow.
2. To determine total head, add the depth of water (static head, d_s) above the roof surface to the secondary drain inlet (which is the height of the dam or standpipe above the roof surface) to the hydraulic head listed in this table.
3. Linear interpolation for flow rate and hydraulic head is appropriate for approximations.
4. Extrapolation is not appropriate.

Where the specified secondary (overflow) drain dam or standpipe diameter differs from what is provided in Tables C8.2-1 and C8.2-2, the hydraulic head can be adjusted based on Equation (C8.2-3), while holding flow rate constant; however, it is advisable not to use an adjusted design hydraulic head less than 80% of the hydraulic head indicated in the tables (for a given flow rate) unless flow test results are provided to justify the hydraulic head values.

$$d_{h2} = [(D_1/D_2)^{0.67}](d_{h1}) \qquad \text{(C8.2-3)}$$

where

d_{h1} = Known hydraulic head from Tables C8.2-1 and C8.2-2;
D_1 = Overflow dam or standpipe diameter for secondary (overflow) drain, corresponding to d_{h1} for a given flow rate, as shown in Tables C8.2-1 and C8.2-2;
d_{h2} = Hydraulic head to be determined for the specified secondary drain; and
D_2 = Specified overflow dam or standpipe diameter for secondary (overflow) drain corresponding to d_{h2} for a given flow rate.

Example 1. Determine the total head for an 8 in. secondary drain (8 in. outlet diameter) with a 10 in. diameter × 2 in. high overflow dam (d_s) at a flow rate (Q) of 300 gal./min.

From Table C8.2-1:

$D_1 = 12.75$ in. (dam diameter) and
$d_{h1} = 2.0$ in. for 300 gal./min, 8 in. outlet.

For the specified 10 in. diameter overflow dam on an 8 in. drain outlet:

$D_2 = 10$ in. (dam diameter).

Therefore,

$$d_{h2} = [(D_1/D_2)^{0.67}](d_{h1})$$

$d_{h2} = [(12.75\text{ in.}/10\text{ in.})^{0.67}](2.0\text{ in.}) = 2.4$ in. at $Q = 300$ gal./min
Total head $= d_{h2} + d_s = 2.4$ in. + 2 in. = 4.4 in.

(b) For orifice flow regime designations for roof drains, as shown in the shaded cells in Tables C8.2-1 and C8.2-2.

The depth of the drain bowl can affect the hydraulic head acting on the drain outlet for a given flow rate; therefore, where the depth of the specified drain bowl is less than the depth of the tested drain bowl (indicated in the tables), the difference in drain bowl depth should be added to the hydraulic head from the tables to determine the design hydraulic head and total head. Where the depth of the specified drain bowl is greater than that indicated in the tables, the difference in drain bowl depth can be subtracted from hydraulic head in the tables to determine the design hydraulic head and total head; however, it is advisable not to use an adjusted design hydraulic head less than 80% of the hydraulic head provided in the tables (for a given flow rate) unless flow test results are provided to justify the hydraulic head values.

Example 2. Determine the total head for a 4 in. secondary drain (4-in. outlet diameter) with a drain bowl depth of 1.5 in. and a 2.5 in. high overflow dam (d_s), at a flow rate (Q) of 350 gal./min.

From Table C8.2-1: When $Q = 350$ gal./min, for a 4 in. drain with an 8 in. dam, $d_h = 3.5$ in., the flow regime is orifice flow (shaded portion of the table), and the drain bowl depth is 2 in.

The specified drain bowl depth is 1.5 in., and since this is 0.5 in. less than the drain bowl depth referenced in the table, the hydraulic head from the table is increased by 0.5 in.

Therefore, for the proposed drain: $d_h = 3.5$ in. + 0.5 in. = 4.0 in.

Total head $= d_h + d_s = 4.0$ in. + 2.5 in. = 6.5 in.

Drain outlet sizes are generally standard in the industry, so it is unlikely that adjustments to hydraulic head values in Tables C8.2-1 and C8.2-2 based on differing drain outlet sizes will be needed.

Where a roof drain is installed in a sump pan located below the adjoining roof surface, reductions in hydraulic head and rain load on the adjoining roof surface should only be credited when based on hydraulic analysis from a qualified plumbing engineer.

Refer to Tables C8.2-3, C8.2-4, C8.2-5, and C8.2-6 for flow rates of rectangular and circular (pipe) roof scuppers at various hydraulic heads. Note that these tables are based on the assumption that no backwater is present (i.e., free outfall) at the discharge end of the scupper. If backwater is present, then the hydraulic head can be expected to increase for the same flow rate.

The accumulation of water on a roof can be exacerbated by the deflection of the roof, an effect known as ponding. Rain loads cause deflections, which allow the accumulation of more water,

Table C8.2-3. Flow Rate (Q) in Gallons Per Minute for Scuppers at Various Hydraulic Heads (d_h) in Inches.

	Hydraulic Head, d_h (in.)									
Drainage System	**1**	**2**	**2.5**	**3**	**3.5**	**4**	**4.5**	**5**	**7**	**8**
6 in. wide channel scupper[a]	18	50	[b]	90	[b]	140	[b]	194	321	393
24 in. wide channel scupper	72	200	[b]	360	[b]	560	[b]	776	1,284	1,572
6 in. wide, 4 in. high, closed scupper[a]	18	50	[b]	90	[b]	140	[b]	177	231	253
24 in. wide, 4-in. high, closed scupper	72	200	[b]	360	[b]	560	[b]	708	924	1,012
6 in. wide, 6 in. high, closed scupper	18	50	[b]	90	[b]	140	[b]	194	303	343
24 in. wide, 6 in. high, closed scupper	72	200	[b]	360	[b]	560	[b]	776	1,212	1,372

[a] Channel scuppers are open-topped (i.e., three-sided). Closed scuppers are four-sided.
[b] Interpolation is appropriate, including between widths of each scupper.
Source: Adapted from FM Global (2016).

Table C8.2-4. In SI Units, Flow Rate (Q) in Cubic Meters Per Second for Scuppers at Various Hydraulic Heads (d_h) in Millimeters.

	Hydraulic Head, d_h (mm)									
Drainage System	25	51	64	76	89	102	114	127	178	203
152 mm wide channel scupper[a]	0.0011	0.0032	[b]	0.0057	[b]	0.0088	[b]	0.0122	0.0202	0.0248
610 mm wide channel scupper	0.0045	0.0126	[b]	0.0227	[b]	0.0353	[b]	0.0490	0.0810	0.0992
152 mm wide, 102 mm high, closed scupper[a]	0.0011	0.0032	[b]	0.0057	[b]	0.0088	[b]	0.0112	0.0146	0.0160
610 mm wide, 102 mm high, closed scupper	0.0045	0.0126	[b]	0.0227	[b]	0.0353	[b]	0.0447	0.0583	0.0638
152 mm wide, 152 mm high, closed scupper	0.0011	0.0032	[b]	0.0057	[b]	0.0088	[b]	0.0122	0.0191	0.0216
610 mm wide, 152 mm high, closed scupper	0.0045	0.0126	[b]	0.0227	[b]	0.0353	[b]	0.0490	0.0765	0.0866

[a] Channel scuppers are open-topped (i.e., three-sided). Closed scuppers are four-sided.
[b] Interpolation is appropriate, including between widths of each scupper.
Source: Adapted from FM Global (2016).

Table C8.2-5. Flow Rate (Q) in Gallons per Minute, for Circular Scuppers at Various Hydraulic Heads (d_h) in Inches.

	Scupper Flow (gal./min.)						
	Scupper Diameter (in.)						
d_h (in.)	5	6	8	10	12	14	16
1	6	7	8	8	10	10	10
2	25	25	30	35	40	40	45
3	50	55	65	75	75	90	95
4	80	90	110	130	140	155	160
5	115	135	165	190	220	240	260
6	155	185	230	270	300	325	360
7	190	230	300	350	410	440	480
8	220	280	375	445	510	570	610

Source: Data from Carter (1957) and Bodhaine (1968).
Notes:

1. Hydraulic head (d_h) is taken above the scupper invert (design water level above base of scupper opening).
2. Linear interpolation is appropriate for approximations.
3. Extrapolation is not appropriate.

Table C8.2-6. SI Units, Flow Rate (Q) in Cubic Meters per Second for Circular Scuppers at Various Hydraulic Heads (d_h) in Millimeters.

	Scupper Flow Rate (m³/s)						
	Scupper Diameter (mm)						
d_h (mm)	127	152	203	254	305	356	406
25	0.0004	0.0004	0.0005	0.0005	0.0006	0.0006	0.0006
51	0.0016	0.0016	0.0019	0.0022	0.0025	0.0025	0.0028
76	0.0032	0.0035	0.0041	0.0047	0.0047	0.0057	0.0060
100	0.0050	0.0057	0.0069	0.0082	0.0088	0.0098	0.0101
127	0.0073	0.0085	0.0104	0.0120	0.0139	0.0151	0.0164
152	0.0098	0.0117	0.0145	0.0170	0.0189	0.0205	0.0227
178	0.0120	0.0145	0.0189	0.0221	0.0259	0.0278	0.0303
203	0.0139	0.0177	0.0237	0.0281	0.0322	0.0360	0.0385

Source: Data from Carter (1957) and Bodhaine (1968).
Notes:

1. Hydraulic head (d_h) is taken above the scupper invert (design water level above base of scupper opening).
2. Linear interpolation is appropriate for approximations.
3. Extrapolation is not appropriate.

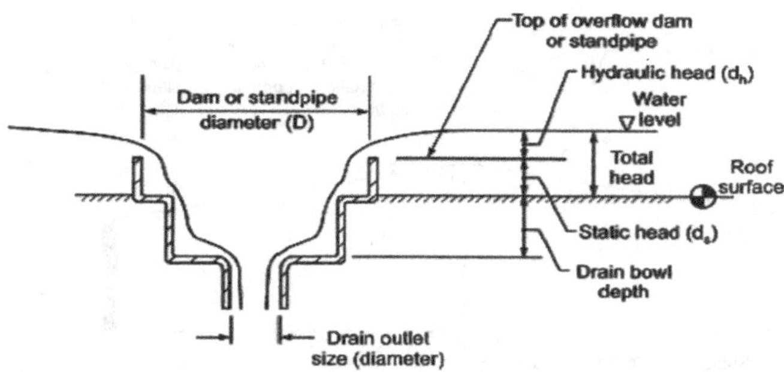

Figure C8.2-2. Schematic cross section of secondary (overflow) roof drain and total head ($d_s + d_h + d_p$). Drain debris guard (strainer) and ring clamp (gravel stop) are not shown for clarity.

which increases the rain loads, and the cycle continues until stabilization or failure occurs. Failure will result if the structure does not possess enough stiffness to limit the deflection or enough strength at the stabilized condition.

The definition of rain load was changed in the 2022 edition of the standard to include the additional load due to ponding. This eliminates the need to perform a separate ponding instability check, however, the effects of ponding need to be explicitly included in the rain load. There are several methods available to compute the additional load due to ponding, any of which are acceptable for use in design.

The additional load may be computed iteratively. The first step in such an analysis is to compute the rain load neglecting the ponding contribution (i.e., with $d_p = 0$), which serves as a first approximation of the rain load. Then, a structural analysis is performed to compute deflections under unfactored dead load and the unfactored first approximation of the rain load, excluding ponding. These deflections are used to compute d_p and update the rain load, which in turn is used to compute new deflections. Successive analyses, each with loads updated based on the deflected shape of the roof eventually converge to the correct value of d_p, or diverge, indicating ponding instability.

Other methods of computing the load include the use of springs with negative stiffness (Baber and Rigsbee 2010), closed-form solutions for simple cases (Silver 2010), or amplified first-order analysis (Denavit 2019).

For any type of analysis, all flexural, shear, and axial member deformations and all other component and connection deformations that contribute to the deflection of the roof should be considered. If an open-web steel joist or truss is modeled as a beam without shear deformations, then it is common to reduce the moment of inertia by 15% to account for shear deformations. Axial deformations of columns are typically small and can often be neglected.

For rectangular framed bays consisting of primary members supporting equally spaced secondary members, approximate results can be obtained via an amplified first-order analysis as follows. Load effects due to the total load are obtained by performing a first-order analysis with unfactored dead load and unfactored rain load excluding the ponding contribution (i.e., with $d_p = 0$) and multiplying the result by the amplification factor, B_p [Equation (C8.2-4)], which depends on flexibility coefficients, C_p [Equation (C8.2-5)], and C_s [Equation (C8.2-6)]. Using this method, rain loads are calculated by first computing the total load (dead load plus rain load) and subtracting out the dead load.

$$B_p = \frac{1}{1 - 1.15C_p - C_s} \quad \text{(C8.2-4)}$$

$$C_p = \frac{\gamma L_s L_p^4}{\pi^4 E I_p} \quad \text{(C8.2-5)}$$

$$C_s = \frac{\gamma S L_s^4}{\pi^4 E I_s} \quad \text{(C8.2-6)}$$

Example: Consider a bay on a flat roof with structural steel framing. W24x76 girders span 40 ft and support W18x35 beams that also span 40 ft and are spaced at 8 ft. The dead load is 15 psf and $d_s + d_h = 5$ in. Neglecting the effects of ponding, dead load and rain load result in a combined load of 15 psf + 5.2 psf/in. (5 in.) = 41 psf. Approximating the effects of ponding using Equation (C8.2-4), dead load and rain load result in a combined load of B_p (41 psf) = (1.44)(41 psf) = 59 psf, where B_p is computed as shown following. Thus, the rain load (sum of static head, hydraulic head, and ponding head), R, is equal to 59 psf − 15 psf = 44 psf. The total factored load for strength design in accordance with Section 2.3.1 is $1.2D + 1.6R$ = 1.2(15 psf) + 1.6(44 psf) = 88.4 psf. In this method, the ponding head, d_p, is not explicitly computed, but can be determined from the rain load.

$$C_p = \frac{\gamma L_s L_p^4}{\pi^4 E I_p} = \frac{(62.4 \frac{\text{lb}}{\text{ft}^3} \times \frac{1\ k}{1{,}000\ \text{lb.}} \times (\frac{1\ \text{ft}}{12\ \text{in.}})^3)(40\ \text{ft} \times \frac{12\ \text{in.}}{1\ \text{ft.}})(40\ \text{ft} \times \frac{12\ \text{in.}}{1\ \text{ft.}})^4}{\pi^4 (29{,}000\ \text{ksi})(2{,}100\ \text{in.}^4)} = 0.155$$

$$C_s = \frac{\gamma S L_s^4}{\pi^4 E I_s} = \frac{(62.4 \frac{\text{lb}}{\text{ft}^3} \times \frac{1\ k}{1{,}000\ \text{lb.}} \times (\frac{1\ \text{ft}}{12\ \text{in.}})^3)(8\ \text{ft} \times \frac{12\ \text{in.}}{1\ \text{ft.}})(40\ \text{ft} \times \frac{12\ \text{in.}}{1\ \text{ft.}})^4}{\pi^4 (29{,}000\ \text{ksi})(510\ \text{in.}^4)} = 0.128$$

$$B_p = \frac{1}{1 - 1.15C_p - C_s} = \frac{1}{1 - 1.15(0.155) - (0.128)} = 1.44$$

C8.3 BAYS WITH LOW SLOPE

The provisions of Section 8.3 require additional investigation of low-sloped bays which may be susceptible to ponding loads despite not being subject to rain load per Section 8.2.

An example structure is shown in Figure C8.3-1. The leftmost bay with the parapet and the center bays impound water during the design event. These bays are subject to rain load, per Section 8.2, and are susceptible to ponding instability. Ponding is addressed for these bays by the inclusion of the ponding head, d_p, in the calculation of rain load. Bays at the edge of a building with a perimeter curb would also impound water up to the height of the curb and would therefore also be subject to rain load per Section 8.2.

If a bay is above the water level or adjacent to a free-draining edge, the depth of water on the undeflected roof may be zero (i.e., $d_s = 0$), but there is still the potential for isolated ponds to form in low points in the deflected shape of the roof. However, this will only occur when the slope of the roof is low. Low points

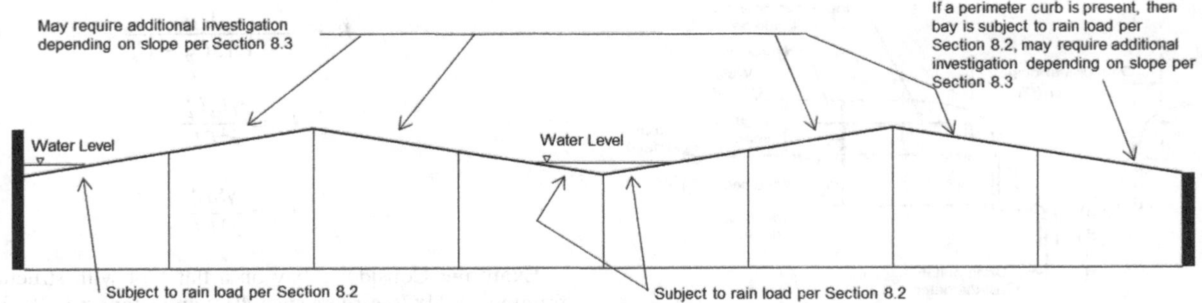

Figure C8.3-1. Identification of bays requiring additional investigation.

(i.e., local minimums) in the deflected shape can be identified by performing a structural analysis. The provisions utilize simplified limits on the roof slope to identify cases which require further investigation. These limits were developed by applying an assumed maximum deflected shape to all members. The limits were derived assuming rectangular bays with sinusoidal deflected shape with a maximum deflection of $L/240$ (where L is the span) and rigid eaves. When secondary members are perpendicular and adjacent to a free draining edge, the minimum rise (β in inches) for a run of 1 ft (i.e., a β on 12 roof slope) to achieve no local minimum is

$$\beta = \frac{1}{20}\left(\pi + \frac{L_p}{L_s}\right) \tag{C8.3-1}$$

where L_p and L_s are the spans of the primary and secondary members, respectively ($\beta = 0.23$ in. for $L_s = 40$ ft and $L_p = 60$ ft). When secondary members are parallel and adjacent

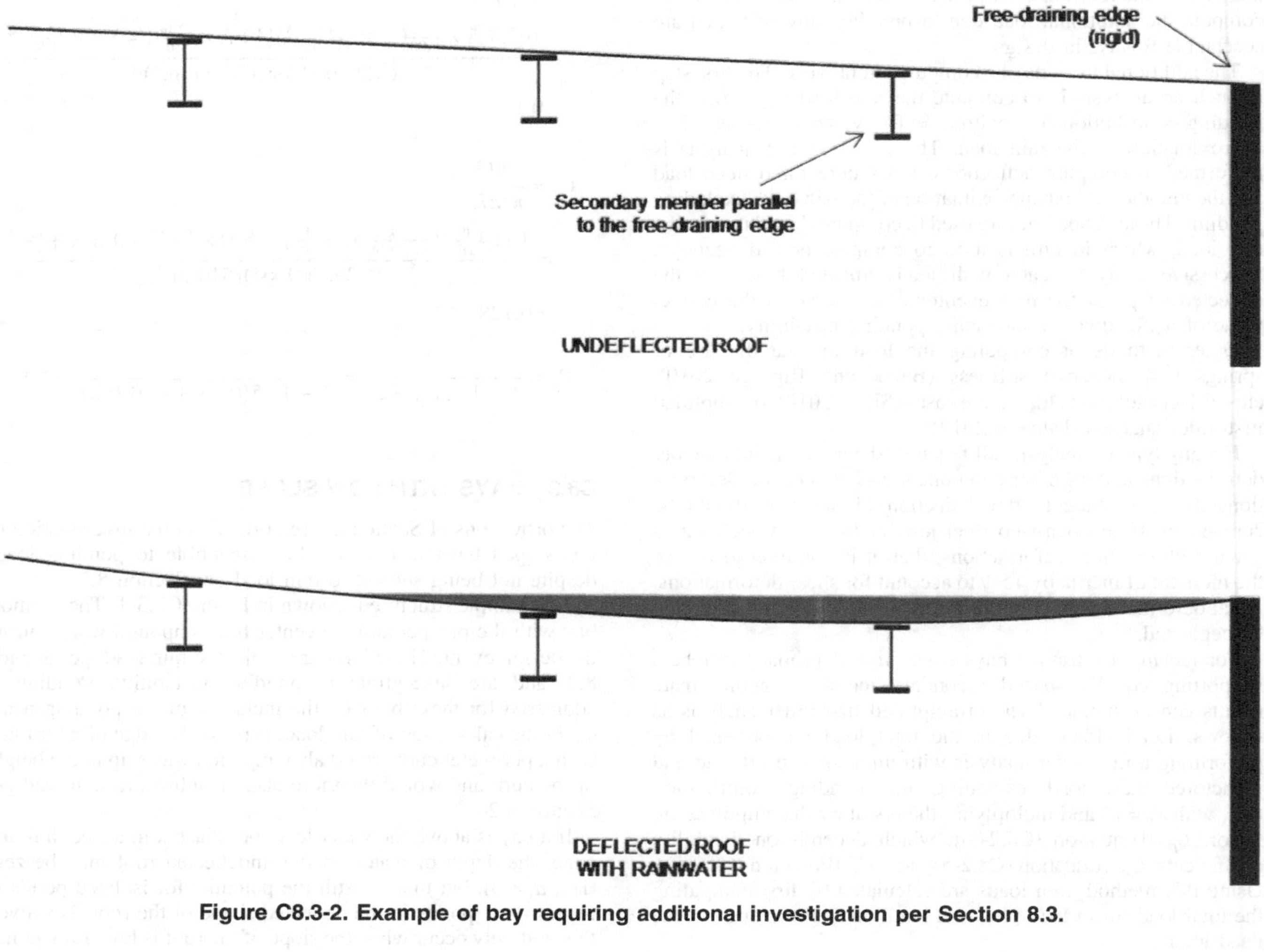

Free-draining edge (rigid)

Secondary member parallel to the free-draining edge

UNDEFLECTED ROOF

DEFLECTED ROOF WITH RAINWATER

Figure C8.3-2. Example of bay requiring additional investigation per Section 8.3.

to a free draining edge, the minimum rise (β in inches) for a run of 1 ft to achieve no local minimum is

$$\beta = \frac{1}{20}\left(\pi + \frac{L_s}{S}\right) \quad \text{(C8.3-2)}$$

where S is the spacing of the secondary members ($\beta = 0.76$ in. for $L_s = 60$ ft and $S = 5$ ft). For interior bays regardless of the orientation of the members, the minimum rise (β in inches) for a run of 1 ft to achieve no local minimum is $\beta = \pi/20$ ($\beta = 0.16$ in.) but is specified as 0.25 in. in Section 8.3.

An example of a bay that requires additional investigation per Section 8.3, is shown in Figure C8.3-2. Rainwater would simply flow off the free-draining edge if the roof was rigid. However, if the deflection of the first secondary member upslope from the free-draining edge exceeds the initial difference in elevation, then a pond can form. To determine if the bay will be subjected to rain loads, only deflection due to dead load should be considered. For an evaluation of ponding in the presence of snow per Section 7.11, deflection due to dead load plus snow load should be considered. Not all bays requiring additional investigation per Section 8.3 will have rain loads. If it is determined that positive slope to the free-draining edge is maintained even for the deflected roof, then neither rain loads nor ponding instability need to be considered in design. If the analysis determines that a pond will form, then Section 8.2 can be used to determine the rain load. In such a case, water level would be taken as the elevation of the free-draining edge plus a nominal amount of hydraulic head (e.g., $d_h = 1/4$ in.).

If a bay does not have impounded water and the slope is not low, then it is expected that any rainwater will drain off the roof and neither rain loads nor ponding instability need to be considered in design.

C8.4 DRAINAGE TO EXISTING ROOFS

In some cases, existing roofs have failed when alterations, additions, or new construction caused rain loads to increase on the existing roof. In many of these cases, the drainage systems had not been modified to account for the increased flow of water nor was the roof strengthened to account for the increased loads. Increased loads may be significant for existing roof structures designed to older codes that did not require secondary drainage systems. Installation of scuppers or other overflow drainage systems can help mitigate large increases in load.

New construction can increase rain flow to an existing roof in cases such as a new wall constructed directly adjacent to an existing roof where sheet flow down the wall diverts onto the roof.

REFERENCES

Baber, T. T., and E. D. Rigsbee. 2010. "Noniterative finite element analysis of ponding." In *Proc., Structures Congress 2010*, 1150–1159. Reston, VA: ASCE.

Bodhaine, G. L. 1968. "Measurement of peak discharge at culverts by indirect methods." In *Techniques of water-resources investigations of the United States Geological Survey: Book 3 application of hydraulics*. Washington, DC: USGS.

Carter, R. W. 1957. *Computation of peak discharge at culverts.* Geological Survey Circular 376. Washington, DC: USGS.

Denavit, M. D. 2019. "Approximate ponding analysis by amplified first-order analysis." *Eng. Struct.* 197 (Oct): 109428. https://doi.org/10.1016/j.engstruct.2019.109428.

Fisher, J. M., and M. D. Denavit. 2018. "Structural design of steel joist roofs to resist ponding loads." In *Technical digest 3.* 3rd ed. Florence, SC: Steel Joist Institute.

FM Global (Factory Mutual Global). 2012. *Loss prevention data 1–54, roof loads for new construction.* Norwood, MA: Factory Mutual Engineering Corporation.

FM Global. 2016. *Loss prevention data 1–54, roof loads for new construction*. Norwood, MA: Factory Mutual Engineering Corporation.

Graber, S. D. 2009. "Rain loads and flow attenuation on roofs." *J. Archit. Eng.* 15 (3): 91–101. https://doi.org/10.1061/(ASCE)1076-0431(2009)15:3(91).

IAPMO (International Association of Plumbing and Mechanical Officials). 2021 *Uniform plumbing code*. Ontario, CA: IAPMO.

ICC (International Code Council). 2021. *International plumbing code*. Washington, DC: ICC.

NFPA (National Fire Protection Association). 2021. *NFPA 5000 building construction and safety code*. Quincy, MA: NFPA.

Silver, E. 2010. "A strength design approach to ponding." *Eng. J.* 47 (3): 175–187.

CHAPTER C9
RESERVED FOR FUTURE COMMENTARY

CHAPTER C10
ICE LOADS - ATMOSPHERIC ICING

C10.1 GENERAL

In most of the contiguous United States, freezing rain is considered the cause of the most severe ice loads. Maps of ice thickness from freezing rain are in the standard. Values for ice thicknesses caused by in-cloud icing and snow suitable for inclusion in this standard are not currently available.

Very few sources of direct information or observations of naturally occurring ice accretions (of any type) are available. Bennett (1959) presents the geographical distribution of the occurrence of ice on utility wires from data compiled by various railroad, electric power, and telephone associations, in the nine-year period from the winter of 1928–1929 to the winter of 1936–1937. The data include measurements of all forms of ice accretion on wires, including glaze ice, rime ice, and accreted snow, but do not differentiate among them. Ice thicknesses were measured on wires of various diameters, heights above ground, and exposures. No standardized technique was used in measuring the thickness. The maximum ice thickness observed during the nine-year period in each of 975 squares, 60 mi (97 km) on a side, in a grid covering the contiguous United States is reported. In every state except Florida, thickness measurements of accretions with unknown densities of approximately one inch were reported. Information on the geographical distribution of the number of storms in this nine-year period, with ice accretions greater than specified thicknesses, is also included.

Tattelman and Gringorten (1973) reviewed ice load data, storm descriptions, and damage estimates in several meteorological publications to estimate the maximum ice thicknesses with a 50-year mean recurrence interval (MRI) in each of seven regions in the United States. *Storm Data* (NOAA 1959–Present) is a monthly publication that describes damage from storms of all sorts throughout the United States. The compilation of this qualitative information on storms causing damaging ice accretions in a particular region can be used to estimate the severity of ice and wind-on-ice loads. The Electric Power Research Institute has compiled a database of icing events from the reports in *Storm Data* (Shan and Marr 1996) and prepared damage severity maps.

C10.1.1 Site-Specific Studies

Freezing rain. In determining equivalent radial ice thicknesses from freezing rain from historical weather data, the quality, completeness, and accuracy of the data should be considered along with the robustness of the ice accretion algorithm. Weather stations may be closed by ice storms because of power outages, anemometers may be iced over, and hourly precipitation data recorded only after the storm when the ice in the rain gauge melts. These problems are likely to be more severe at automatic weather stations where observers are not available to estimate the weather parameters during power outages or correct erroneous readings. Note also that (1) air temperatures are recorded only to the nearest 1 °F (0.6 °C), at best, and may vary significantly from the recorded value in the region around the weather station; (2) the wind speed during freezing rain has a significant effect on the accreted ice load on objects oriented perpendicular to the wind direction; (3) wind speed and direction vary with terrain and exposure; (4) enhanced precipitation may occur on the windward side of mountainous terrain; and (5) ice may remain on the structure for days or weeks after freezing rain ends, subjecting the iced structure to wind speeds that may be significantly higher than those that accompanied the freezing rain. These factors should be considered both in estimating the accreted ice thickness at a weather station in past freezing rain events and in extrapolating those thicknesses to a specific site.

Bernstein and Brown (1997) and Robbins and Cortinas (1996) provide information on freezing rain climatology for the 48 contiguous states, based on recent meteorological data. Konrad (1998) discusses spatial patterns of freezing rain in the Appalachians.

In-cloud icing. In-cloud icing may cause significant loadings on ice-sensitive structures wherever supercooled clouds/fog exist, such as in mountainous regions, where supercooled clouds often occur with elevated winds, and in basins that trap freezing fog for extended periods of time under light wind conditions. In-cloud ice accretion is very sensitive to the degree of exposure to moisture-laden clouds, which is related to terrain, elevation, wind direction, and velocity. Large differences in ice load can occur over a few hundred feet and can cause severe load unbalances in overhead wire systems. Advice from a meteorologist familiar with the area is particularly valuable in these circumstances. Significant loads may also accrete on very tall structures anywhere in the country when supercooled clouds are advected past some portion of the structure.

Mulherin (1996) reports that of 120 communications tower failures in the United States caused by atmospheric icing, 38 were caused by in-cloud icing, and in-cloud icing combined with freezing rain caused an additional 26 failures. In Arizona, New Mexico, and the panhandles of Texas and Oklahoma, the United States Forest Service specifies ice loads caused by in-cloud icing for towers constructed in specific regions (US Forest Service 1994). Severe in-cloud icing has been observed in southern California (Mallory and Leavengood 1983a, b), eastern Colorado (NOAA Feb. 1978), the Pacific Northwest (Winkleman 1974; Sinclair and Thorkildson 1980, Thorkildson 2019), Alaska (Ryerson and Claffey 1991), and the Appalachians (Ryerson 1987, 1988a, b, 1990; Govoni 1990).

Snow. Snow accretions can also result in severe structural loads and may occur anywhere snow falls, even in localities that may experience only one or two snow events per year. Some examples of locations where snow accretion events resulted in significant damage to structures are Nebraska (NPPD 1976), Maryland (Mozer and West 1983), Pennsylvania (Goodwin et al. 1983), Georgia and North Carolina (Lott 1993a, b), Colorado (McCormick and Pohlman 1993), Alaska (Peabody and Wyman 2005), and the Pacific Northwest (Hall 1977).

Alaska. For Alaska, available information indicates that moderate to severe snow and in-cloud icing can be expected. The measurements made by Golden Valley Electric Association (Peabody and Wyman 2005) and Alaska Energy Authority (Peabody 1996) are consistent in magnitude with visual observations across a broad area of central Alaska (Peabody 1993). Peterka et al. (1996) used an ice accretion model to estimate ice loads for high-voltage transmission lines in Alaska.

Hawaii. In Hawaii, freezing rain (Wylie 1958), snow, and in-cloud icing are known to occur at higher elevations.

Local analyses. Local records and experience should be considered when establishing the design ice thickness or load, concurrent wind speed, concurrent temperature, and ice density. Information on ice-related damage and weather conditions in the area of interest may be available from the local electric utilities; tree companies that are called on to remove fallen ice-covered trees from roads and distribution lines; and state and county road departments. These sources would supplement damage descriptions in *Storm Data* (1959-Present) and can help establish the frequency of icing events along with any typical spatial variations in the icing severity in the area.

In using local data to determine extremes, it must also be emphasized that sampling errors can lead to large uncertainties in the specification of ice thicknesses or loads for long mean recurrence intervals (MRI). Sampling errors are those associated with the limited number of cases in the sample of extremes, related to the duration of the period of record and the frequency of icing events. When local records of limited extent are used to determine extremes, care should be exercised in their use.

Ice loads from freezing rain, in-cloud icing, or snow provided in reports by Richmond, some of which were referenced in previous editions of ASCE 7, may be excessively high because of choices made in the application of the ice accretion model that are inconsistent with the available weather data (Jones 2019).

C10.1.2 Dynamic Loads While design for dynamic loads is not specifically addressed in this edition of the standard, the effects of dynamic loads are an important consideration for some ice-sensitive structures and should be considered in the design when they are anticipated to be significant. For example, large-amplitude galloping (Rawlins 1979; Simiu and Scanlan 1996, Section 6.2) of guys and overhead cable systems occurs in many areas. The motion of the cables can cause damage because of direct impact of the cables on other cables or structures and can also cause damage because of wear and fatigue of the cables and other components of the structure (White 1999). Ice shedding from the guys on guyed masts can cause substantial dynamic loads in the mast.

C10.1.3 Exclusions Current revisions of the consensus standards and other documents referenced in this document are ASCE (2020), TIA (2019), and IEEE (2017). Additional guidance is available in Committee on Electrical Transmission Structures (1982). In Canada, the CSA (2018, 2019) standards apply.

C10.2 DEFINITIONS

FREEZING RAIN: Freezing rain occurs when warm, moist air is forced over a layer of subfreezing air at the earth's surface. The precipitation usually begins as snow that melts as it falls through the layer of warm air aloft. The drops then cool as they fall through the cold surface air layer and may then freeze on structures or the ground. Upper air data indicate that the cold surface air layer is typically between 1,000 and 3,900 ft (300 and 1,200 m) thick (Young 1978), averaging 1,600 ft (500 m) (Bocchieri 1980). The warm air layer aloft averages 5,000 ft (1,500 m) thick in freezing rain, but in freezing drizzle, the entire temperature profile may be below 32 °F (0 °C) (Bocchieri 1980).

Precipitation rates and wind speeds are typically low to moderate during freezing rain. The water impingement rate is often greater than the freezing rate. The excess water drips off and may freeze as icicles, resulting in a variety of accretion shapes that range from a smooth cylindrical sheath, through a crescent on the windward side with icicles hanging on the bottom, to large irregular protuberances; see Figure C10.2-1. The shape of an accretion depends on a combination of varying meteorological factors and the cross-sectional shape of the structural member, its spatial orientation, and flexibility.

Note that the theoretical maximum density of ice [57 lb/ft^3 (917 kg/m^3)] is unlikely to be reached in naturally formed accretions because of the presence of air bubbles.

HOARFROST: Hoarfrost, which is often confused with rime, forms by a completely different process. Hoarfrost is an accumulation of ice crystals formed by direct deposition of water vapor from the air on an exposed object. Because it forms on objects with surface temperatures that have fallen below the frost point (a dew point temperature below freezing) of the surrounding air, due to strong radiational cooling, hoarfrost is often found early in the morning after a clear, cold night. It is feathery in appearance and typically accretes up to about 1 in. (25 mm) in thickness and has very little weight. Hoarfrost does not constitute a significant loading problem; however, it is a very good collector of supercooled fog droplets. In light winds, a hoarfrost-coated wire may accrete rime faster than a bare wire (Power 1983).

ICE-SENSITIVE STRUCTURES: Ice-sensitive structures are structures for which the load effects from atmospheric icing control the design of part, or all, of the structural system. Many open structures are efficient ice collectors, so ice accretions can have a significant load effect. The sensitivity of an open structure to ice loads depends on the size and number of structural members, components, and appurtenances and also on the other loads for which the structure is designed. For example, the additional weight of ice that may accrete on a heavy wide-flange member is smaller in proportion to the dead load than the same ice thickness on a light angle member. Also, the percentage increase in projected area for wind loads is smaller for the wide-flange member than for the angle member. For some open structures, other design loads, for example, snow loads and live loads on a catwalk floor, may be larger than the design ice load.

IN-CLOUD ICING: This icing condition occurs when a cloud or fog consisting of supercooled water droplets, 100 microns or less in diameter, blown by the wind encounters a surface at, or below, freezing. It occurs in mountainous areas where adiabatic cooling causes saturation of the atmosphere to occur at temperatures below freezing, in free air in supercooled clouds, and in supercooled fogs produced by a stable air mass with a strong temperature inversion. In-cloud ice accretions can reach thicknesses of 1 ft (0.30 m) or more because the icing conditions can include high winds and that may persist or recur episodically during long periods of subfreezing temperatures. Large concentrations of supercooled droplets are not common at air temperatures below about −22 °F (−30 °C).

In-cloud ice accretions have densities ranging from that of low-density rime to glaze. When convective and evaporative cooling removes the heat of fusion as fast as it is released by the freezing droplets, the drops freeze on impact. When the cooling

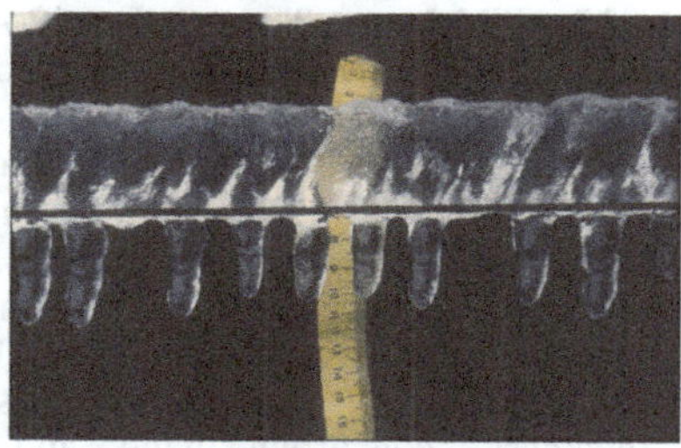

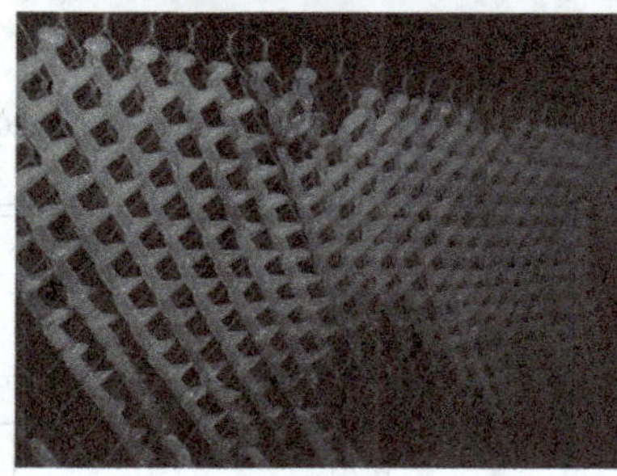

Figure C10.2-1. Glaze ice accretion caused by freezing rain.

rate is lower, the droplets do not completely freeze on impact. The unfrozen water then spreads out on the object and may flow completely around it and even drip off to form icicles. The degree to which the droplets spread as they collide with the structure and freeze governs how much air is incorporated in the accretion, and thus, its density. The density of ice accretions caused by in-cloud icing varies over a wide range from 5 to 56 lb/ft^3 (80 to 900 kg/m^3) (Macklin 1962, Jones 1990). The resulting accretion can be either white or clear, possibly with attached icicles; see Figure C10.2-2.

The amount of ice accreted during in-cloud icing depends not only on the cloud properties, but also on the size of the accreting object, the duration of the icing condition, and the wind speed. If, as often occurs, wind speed increases and air temperature decreases with height above ground, larger amounts of ice accrete on taller structures. The accretion shape depends on the flexibility of the structural member, component, or appurtenance. If it is free to rotate, such as a long guy or a long span of a single conductor or wire, the ice accretes with a roughly circular cross section. On more rigid structural members, components, and

Figure C10.2-2. Rime ice accretion caused by in-cloud icing.

Source: Courtesy of Bonneville Power Administration.

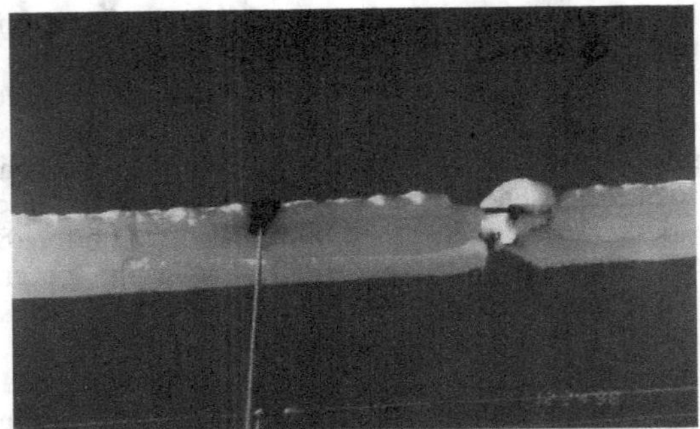

Figure C10.2-3. Snow accretion on wires.

appurtenances, the ice forms in irregular pennant shapes extending into the wind.

SNOW: Under certain conditions, snow falling on objects may adhere because of capillary forces, interparticle freezing (Colbeck and Ackley 1982), and/or sintering (Kuroiwa 1962). On objects with circular cross section, such as a wire, cable, conductor, or guy, sliding, deformation, and/or torsional rotation of the underlying cable may occur, resulting in the formation of a cylindrical sleeve, even around bundled conductors and wires; see Figure C10.2-3. When accreting snow is accompanied by high winds, the density of accretions may be much higher than the density of the same snowfall on the ground.

Damaging snow accretions have been observed at surface air temperatures ranging from about 23 to 36 °F (−5 to 2 °C). Snow with a high moisture content appears to stick more readily than drier snow. Snow falling at a surface air temperature above freezing may accrete even at wind speeds above 25 mi/h (10 m/s), producing dense 37 to 50 pcf (600 to 800 kg/m^3) accretions. Snow with a lower moisture content is not as sticky, blowing off the structure in high winds. These accreted snow densities are typically between 2.5 and 16 lb/ft^3 (40 and 250 kg/m^3) (Kuroiwa 1965).

Even apparently dry snow can accrete on structures (Gland and Admirat 1986). The cohesive strength of the dry snow is initially supplied by the interlocking of the flakes and ultimately by sintering, as molecular diffusion increases the bond area between adjacent snowflakes. These dry snow accretions appear to form only in very low winds and have densities estimated at between 5 and 10 lb/ft^3 (80 and 150 kg/m^3) (Sakamoto et al. 1990, Peabody 1993).

Recent references involving measuring and modeling snow loads on ice sensitive structures in Europe and Japan include Eliasson et al. (2013, 2015), Nygaard et al. (2013a), Lacavalla et al. (2015, 2019), Nikolov et al. (2015), Ueno et al. (2015), Faggian et al. (2019), Iversen et al. (2019), and Matsumiya et al. (2019). In Nygaard et al. (2013b), design ice loads for snow and in-cloud icing are mapped for Great Britain.

C10.4 ICE LOADS CAUSED BY FREEZING RAIN

C10.4.1 Ice Load The ice thicknesses shown in Figures 10.4-2 and 10.4-5 were determined for a horizontal cylinder oriented perpendicular to the wind. These ice thicknesses cannot be applied directly to cross sections that are not round, such as channels and angles. However, the ice area from Equation (10.4-1) is the same for all shapes for which the circumscribed circles have equal diameters (Peabody and Jones 2002, Jones and Peabody 2006). It is assumed that the maximum dimension of the cross section is perpendicular to the trajectory of the raindrops. Similarly, the ice volume in Equation (10.4-2) is for a flat plate perpendicular to the trajectory of the raindrops. The constant π in Equation (10.4-2) corrects the thickness from that on a cylinder to the thickness on a flat plate. For vertical cylinders and horizontal cylinders parallel to the wind direction, the ice area given by Equation (10.4-1) is conservative.

C10.4.2 Nominal Ice Thickness The ASCE 7-22 Atmospheric Icing Subcommittee generated the nominal ice thickness maps shown in Figures 10.4-2 (lower 48 states), 10.4-3 (Alaska), 10.4-4 (Columbia Gorge region), and 10.4-5 (Lake Superior region) for Risk Categories I, II, III, and IV, using an ice accretion model and local data. Location-specific nominal ice thicknesses may be determined using the ASCE Atmospheric Ice Geodatabase, which can be accessed at the ASCE 7 Hazard Tool (https://asce7hazardtool.online/). This geodatabase provides nominal ice thickness, concurrent wind and concurrent temperature data based on a defined location using either latitude/longitude or an address. The website results are interpolated from the paper map contours currently in the standard.

The MRI for the four risk categories come directly from matching the ASCE 7-16 Importance Factors for ice in Table 1.5-2, with the ASCE 7-16 Table C10.4-1 multipliers on 500-year ice thickness as a function of MRI, as shown in the first three columns of Table C10.4-1. As the generalized Pareto distribution is used to characterize extreme ice thicknesses from freezing rain, the tail shape parameter, k, is determined by the sample of extremes for each superstation. The tail is finite for $k > 0$, exponential for $k = 0$, and increasingly heavy for more negative k. Thus, the values in Column 2 represented the mean ice thickness multiplier for the MRI.

Historical weather data from 540 National Weather Service (NWS), military, Federal Aviation Administration (FAA), and Environment Canada weather stations were used, with the US Army's Cold Regions Research and Engineering Laboratory

Table C10.4-1. Risk Category Importance Factors and MRI from ASCE7-16 and Statistics of Ice Thickness Multipliers for the 2022 Extreme Value Analysis.

	ASCE 7-16		Statistics of ice thickness multipliers for ASCE 7-22		
Risk Category	Ice thickness multiplier (Table 1.5-2)	MRI (Table C10.4-1)	Mean multiplier	Minimum multiplier	Maximum multiplier
I	0.80	250	0.85	0.70	1.00
II	1	500	1	1	1
III	1.15	1000	1.17	1.00	1.43
IV	1.25	1400	1.26	1.00	1.70

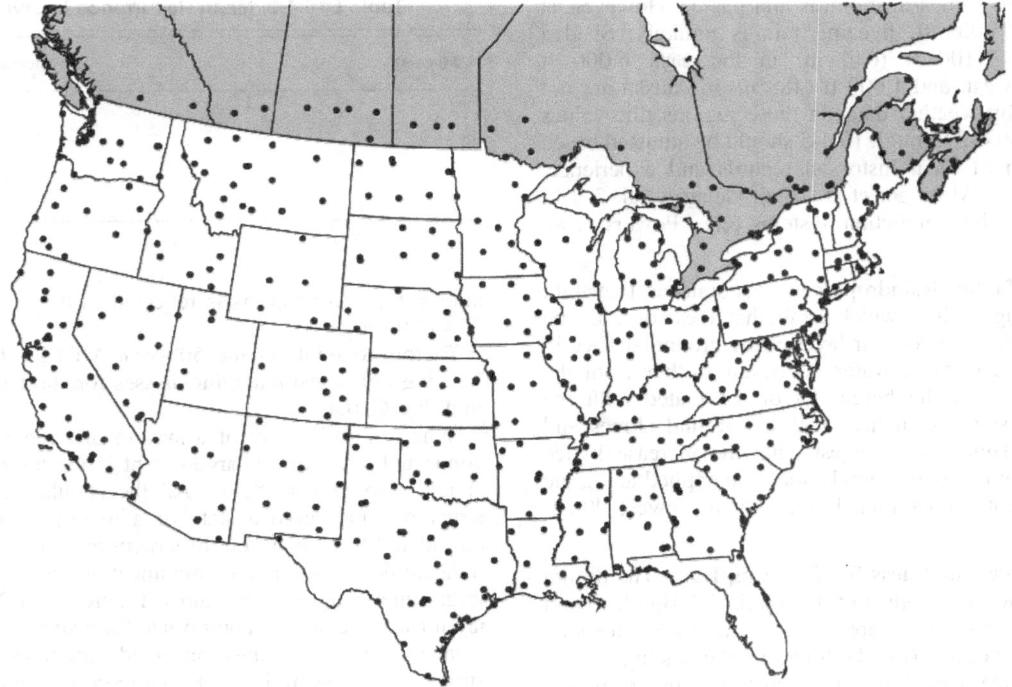

Figure C10.4-1. Locations of weather stations used in preparation of Figures 10.4-2 through 10.4-5.

(CRREL) and simple ice accretion models (Jones 1996, 1998), to estimate uniform radial glaze ice thicknesses in past freezing rain events. The models and algorithms have been applied to additional stations in Canada along the border of the lower 48 states and Alaska. The station locations are shown in Figure C10.4-1 for the 48 contiguous states and in Figure 10.4-3 for Alaska. The period of record of the meteorological data at any station is typically 20 to 50 years. The ice accretion models use weather and precipitation data to simulate the accretion of ice on cylinders 33 ft (10 m) above the ground, oriented perpendicular to the wind direction in freezing rain. Accreted ice is assumed to remain on the cylinder until after freezing rain ceases and the air temperature increases to at least 33 °F (0.6 °C). At each station, the maximum ice thickness and the maximum wind-on-ice load were determined for each event. Severe events, those with significant ice or wind-on-ice loads at one or more weather stations, were researched in *Storm Data* (NOAA 1959–Present), newspapers, and utility reports to obtain corroborating qualitative information on the extent of and damage from the storm. Yet, very little corroborating information was obtained about damaging freezing rain events in Alaska, perhaps because of the low population density and relatively sparse newspaper coverage in the state.

Extreme ice thicknesses were determined from an extreme value analysis using L-moments and the generalized Pareto distribution (Hosking and Wallis 1987, 1997; Wang 1991; Abild et al. 1992). To reduce sampling error, weather stations were grouped into superstations (Peterka 1992) based on the incidence of severe storms, the frequency of freezing rain events, latitude, proximity to large bodies of water, elevation, terrain, and similarity in the occurrence rate of ice thicknesses of at least 0.04 in. (1 mm). Concurrent wind-on-ice speeds were back-calculated from the extreme wind-on-ice load and the extreme ice thickness. The analysis of the weather data and the calculation of extreme ice thicknesses are described in more detail in Jones et al. (2002). The maps in Figures 10.4-2 through 10.4-5 represent the most consistent and best available nationwide nominal ice thicknesses. The icing model used to produce the maps has not, however, been verified with a large set of collocated measurements of meteorological data and measured equivalent radial ice thicknesses. Furthermore, the weather stations used to develop these maps are almost all at airports. Structures in more exposed locations at higher elevations or in valleys or gorges, for example, Signal and Lookout Mountains in Tennessee, the Pontotoc Ridge and the edge of the Yazoo Basin in Mississippi, the Shenandoah Valley and Poor Mountain in Virginia, Mount Washington in New Hampshire, Buffalo Ridge in Minnesota and South Dakota, Stampede Pass in Washington, Sexton Summit in Oregon, Blue Canyon in California, and Flagstaff in Arizona, may be subject to larger ice thicknesses. However, structures in more sheltered locations, for example, along the north shore of Lake Superior within 300 vertical ft (90 m) of the lake, may be subject to smaller ice thicknesses. Where complex terrain traps cold air near the surface, freezing rain may last longer than at nearby weather stations, resulting in larger local ice thicknesses than those shown on the maps. Complex terrain causing wind speed-up or precipitation enhancement during freezing rain will also cause larger ice thicknesses. These complex terrain features include, but are not limited to, hills, mountains, ridges, valleys, gorges, and basins. Loads from snow or in-cloud icing may be more severe than those from freezing rain (see Section C10.1.1).

Special Icing Regions. Special icing regions are identified by shading on the maps in Figures 10.4-2 through 10.4-5. As described previously, freezing rain occurs only under special conditions when a cold, relatively shallow layer of air at the surface is overrun by warm, moist air aloft. For this reason, severe freezing rain at high elevations in mountainous terrain typically does not occur in the same weather systems that cause severe freezing rain at the nearest airport with a weather station. In the Cascades of Oregon and Washington, ice thicknesses may exceed the mapped values in foothills and passes. However, at elevations above 5,000 ft, freezing rain is unlikely. Shaded elevations above 2,100 ft (640 m) in the east, 6,000 ft (1,829 m) in the west, and 1,600 ft (488 m) in Alaska are not well represented by weather data. In these regions, the values given in Figures 10.4-2 through 10.4-5 should be adjusted based on a combination of local historical records and experience, reanalysis data (e.g., Mesinger et al. 2006, Gelaro et al. 2017), and numerical weather prediction systems (e.g., Powers et al. 2017).

C10.4.3 Height factor Raindrops can evaporate as they fall, sometimes resulting in virga, which is rain that does not reach the ground. Thus, within the cold air layer at the ground surface in freezing rain, there is more water in the air further from the ground. This enhances the height factor associated with the increase in wind speed with height above ground (Jones and Eylander 2017). This study suggests that the increase in ice thickness with height above ground should be applied along the full structure height, rather than limited to the lowest 900 ft (275 m).

C10.4.5 Design Ice Thickness for Freezing Rain The design load on the structure is a product of the nominal design load and the load factors specified in Chapter 2. The load factors for load and resistance factor design (LRFD) for atmospheric icing are 1.0.

The studies of ice accretion, on which the maps are based, indicate that the concurrent gust speed on ice does not increase with mean recurrence interval. The lateral wind-on-ice load does, however, increase with mean recurrence interval if the ice thickness increases. The concurrent gust speed used with the nominal ice thickness is based on both the winds that occur during freezing rain and those that occur between the time the freezing rain stops and the time the temperature rises to above freezing. When the temperature rises above freezing, it is assumed that the ice melts enough to fall from the structure. In the colder northern regions, the ice generally stays on structures for a longer period of time after the end of freezing rain resulting in higher concurrent gust speeds.

The 2002 edition of ASCE 7 was the first edition to include atmospheric icing maps. At that time, 50-year MRI maps were provided in order to match the approach used for the wind and snow load maps. An ice-thickness multiplier equal to 2 was included (see Equation (10.4-5) in ASCE 7-02 to ASCE 7-10) to adjust the mapped values to a 500-year MRI for design. Thus, the adjusted 500-year MRI value would be appropriate for use with the LRFD load factor of 1.0 shown for ice loads in Chapter 2. At that time, a 500-year MRI load was consistent with those historically used for seismic and wind loads. The 2016 edition of ASCE 7 included atmospheric icing maps based on a 500-year MRI load with no ice-thickness multiplier in Equation (10.4-5). The 2016 maps were redrawn directly from the original extreme value analysis. Design load changes from the 2010 edition to the 2016 edition were caused by the maps being redrawn, not by changing the map MRI. For the 2022 edition, the weather stations are grouped into larger superstations to generate longer periods of record. Ice thickness extremes are determined for MRI from 250 to 1,400 years, using L-moments to calculate all three generalized Pareto distribution parameters, rather than setting a threshold just below the smallest value in the sample of extremes and calculating the other two parameters. The maps in ASCE 7-22 differ from those in all the previous editions in that they are contour maps rather than zone maps. Columns 4 through 6 in Table C10.4-1 show the mean, minimum and maximum values for the multiplier for the 2022 extreme value analysis. The variation in tail weight among the superstations is reflected in the contours in Figures 10.4-2 through 10.4-5.

Table C10.4-2. Mean Recurrence Interval Factors.

MRI (years)	Ice thickness multiplier
25	0.4
50	0.5
100	0.6
500	1.0

Factors to multiply the 500-year MRI ice thicknesses from freezing rain to estimate thicknesses for shorter MRI are shown in Table C10.4-2.

When the reliability of a system of structures or one interconnected structure of large extent is important, spatial effects should also be considered. All the cellular telephone antenna structures that serve a state or a metropolitan area could be considered to be a system of structures. Long overhead electric transmission lines and communication lines are examples of large, interconnected structures. Figures 10.4-2 through 10.4-5 are for ice thicknesses appropriate for a single structure of small areal extent. Large, interconnected structures and systems of structures are hit by icing storms more frequently than a single structure. The frequency of occurrence increases with the area encompassed or the linear extent. To obtain equal risks of exceeding the design load in the same icing climate, the individual structures forming the system, or the large, interconnected structure, should be designed for a larger ice load than a single structure (CEATI 2003, 2005; Chouinard and Erfani 2006; Golikova 1982; Golikova et al. 1982).

C10.5 WIND ON ICE-COVERED STRUCTURES

The ASCE 7 Atmospheric Icing Subcommittee generated the concurrent 3-s gust speed maps shown in Figures 10.5-1 (lower 48 states) and 10.5-2 (Alaska), using an ice accretion model and local data. The calculation of concurrent wind speeds is described in Sections C10.4.2 and C10.4.5.

Ice accretions on structures change the structure's wind drag coefficients. The ice accretions tend to round sharp edges, reducing the drag coefficient for such members as angles and bars. Natural ice accretions can be irregular in shape with an uneven distribution of ice around the object on which the ice has accreted. The shape varies from storm to storm and from place to place within a storm. The actual projected area of a glaze ice accretion may be larger than that obtained by assuming a uniform ice thickness.

C10.5.5 Wind on Ice-Covered Guys and Cables Wind tunnel data on force coefficients for ice covered cables are compiled and analyzed in CEATI (2009), using measurements from references dating from 1950 to 2005. Force coefficients vary with the shape of the accreted ice and the angle of attack of the wind relative to the maximum dimension of the shape. There have also been many studies of the force coefficient for cylinders without ice. The force coefficient varies with the surface roughness and the Reynolds number. At subcritical Reynolds numbers, both smooth and rough cylinders have force coefficients of approximately 1.2, as do square sections with rounded edges (Figure 4.5.5 in Simiu and Scanlan 1996). For a wide variety of stranded electrical transmission cables, the supercritical force coefficients are approximately 1.0 with subcritical values as high as 1.3 (Figure 5-2 in Shan 1997). The transition from subcritical

Table C10.5-1. Typical Reynolds Numbers for Iced Guys and Cables.

Guy or cable diameter (in.)	Design ice thickness td(in.)	Iced diameter (in.)	Concurrent gust speed (mi/h)	Reynolds number
0.3	0.25	0.8	30	22,000
2.0	0.5	3	30	84,000
4.0	0.25	4.5	40	169,000
6.0	1.5	9	40	337,000
1.0	3.5	8	40	300,000
3.0	0.5	4	50	187,000
5.0	1	7	50	328,000
4.0	0.75	5.5	60	309,000
2.0	2	6	60	337,000
1.0	0.5	2	80	150,000

Notes: To convert in. to mm, multiply by 25.4. To convert mi/h to m/s, multiply by 0.45.

to supercritical depends on the surface characteristics and takes place over a wide range of Reynolds numbers. For the stranded cables described in Shan (1997), the range is from approximately 25,000 to 150,000. For a square section with rounded edges, the transition takes place at a Reynolds number of approximately 800,000 (White 1999). The concurrent 3-s gust speed in Figure 10.5-1 for the contiguous 48 states varies from 30 to 60 mi/h (13 to 27 m/s), with speeds in Figure 10.5-2 for Alaska up to 80 mi/h (36 m/s). Table C10.5-1 shows the Reynolds numbers (using the kinematic viscosity of air at 5 °F (−15 °C) at 1 atmosphere) for a range of iced guys and cables. In practice, the Reynolds numbers range from subcritical through critical to supercritical depending on the roughness of the ice accretion. Considering that the shape of ice accretions is highly variable from relatively smooth cylindrical shapes to accretions with long icicles with projected areas greater than the equivalent radial thickness used in the maps, a single force coefficient of 1.2 has been chosen.

C10.6 DESIGN TEMPERATURES FOR FREEZING RAIN

Some ice-sensitive structures, particularly those using overhead cable systems, are also sensitive to changes in temperature. In some cases, the maximum load effect occurs around the melting point of ice (32 °F or 0 °C) and in others at the lowest temperature that occurs while the structure is loaded with ice. Figures 10.6-1 and 10.6-2 show the low temperatures to be used for design in addition to the melting temperature of ice.

The freezing rain model described in Section C10.4.2 tracked the temperature during each modeled icing event. For each event, the minimum temperature that occurred with the maximum ice thickness was recorded. The minimum temperatures for all the freezing rain events used in the extreme value analysis of ice thickness were analyzed to determine the 10th percentile temperature at each superstation (i.e., the temperature that was exceeded during 90% of the extreme icing events). These temperatures were used to make the maps shown in Figures 10.6-1 and 10.6-2.

C10.7 PARTIAL LOADING

Variations in ice thickness caused by freezing rain on objects at a given elevation are small over distances of about 1,000 ft (300 m). Therefore, partial loading of a structure from freezing rain is usually not significant (Cluts and Angelos 1977).

In-cloud icing is more strongly affected by wind speed, thus partial loading caused by differences in exposure to in-cloud icing may be significant. Differences in ice thickness over several structures or components of a single structure are associated with differences in the exposure. The exposure is a function of shielding by other parts of the structure and by the upwind terrain.

Partial loading associated with ice shedding may be significant for snow or in-cloud ice accretions and for guyed structures when ice is shed from some guys before others.

REFERENCES

Abild, J., E. Y. Andersen, and L. Rosbjerg. 1992. "The climate of extreme winds at the Great Belt, Denmark." *J. Wind Eng. Ind. Aerodyn.* 41 (1–3): 521–532.

ASCE. 2020. *Guidelines for electrical transmission line structural loading: Manual of practice 74*. 4th ed. Edited by F. Agnew. Reston, VA: ASCE.

Bennett, I. 1959. *Glaze: Its meteorology and climatology, geographical distribution and economic effects*. Environmental Protection Research Division Technical Rep. No. EP-105. Natick, MA: US Army Quartermaster, Research and Engineering Center. Brooklyn, New York.

Bernstein, B. C., and B. G. Brown. 1997. "A climatology of supercooled large drop conditions based upon surface observations and pilot reports of icing." In *Proc., 7th Conf. on Aviation, Range and Aerospace Meteorology*. Long Beach, California.

Bocchieri, J. R. 1980. "The objective use of upper air soundings to specify precipitation type." *Mon. Weather Rev.* 108 (5): 596–603. https://doi.org/10.1175/1520-0493(1980)108<0596:TOUOUA>2.0.CO;2.

CEATI (Centre for Energy Advancement through Technological Innovation). 2003. *Spatial factors for extreme ice and extreme wind: Task 1 literature review on the determination of spatial factors*. T033700-3316B. Montreal: CEATI.

CEATI. 2005. *Spatial factors for extreme ice and extreme wind: Task 2 calculation of spatial factors from ice and wind data*. T033700-3316B-2. Montreal: CEATI.

CEATI. 2009. *Technology watch for the drag coefficient of ice-covered conductors*. T083700-3360. Montreal: CEATI.

Chouinard, L., and R. Erfani. 2006. "Spatial modeling of atmospheric icing hazards." In *Proc., Int. Forum on Engineering Decision Making*. Lake Louise, Canada.

Cluts, S., and A. Angelos. 1977. "Unbalanced forces on tangent transmission structures." In *Proc., IEEE Winter Power Meeting*. Los Alamitos, CA.

Colbeck, S. C., and S. F. Ackley. 1982. "Mechanisms for ice bonding in wet snow accretions on power lines." In *Proc., 1st*

Int. Workshop on Atmospheric Icing of Structures, L. D. Minsk, ed., Hanover, NH, 25–30.

Committee on Electrical Transmission Structures. 1982. "Loadings for electrical transmission structures by the committee on electrical transmission structures." *J. Struct. Div.* 108 (5): 1088–1105. https://doi.org/10.1061/JSDEAG.0005946.

CSA (Canadian Standards Association). 2019. *Overhead transmission lines—Design criteria*. CSA C22.3 No. 60826. Toronto, ON: CSA.

CSA. 2018. *Antennas, towers and antenna-supporting structures*. CSA-S37-18. Toronto, ON: CSA.

Druez, J., S. Louchez, and P. McComber. 1995. "Ice shedding from cables." *Cold Reg. Sci. Technol.* 23 (4): 377–388. https://doi.org/10.1016/0165-232X(94)00024-R.

Elíasson, A. J., S. P. Ísaksson, H. Ágústsson, and E. Thorsteins. 2015. "Wet snow icing–Comparing simulated accretion with observational experience." In *Proc., 16th Int. Workshop on Atmospheric Icing of Structures*. Upsalla, Sweden.

Elíasson, A. J., H. Ágústsson, G. M. Hannesson, and E. Thorsteins. 2013. "Modeling wet-snow accretion; Comparison of cylindrical model to field measurements." In *Proc., 15th Int. Workshop on Atmospheric Icing of Structures*. St. Johns, Canada.

Faggian, P., R. Bonanno, and G. Pirovano. 2019. "Research activities to cope with wet snow impacts on overhead power lines in future climate over Italy." In *Proc., 18th Int. Workshop on Atmospheric Icing of Structures*. Reykjavik, Iceland.

Finstad, K. J., E. P. Lozowski, and E. M. Gates. 1988. "A computational investigation of water droplet trajectories." *J. Atmos. Oceanic Technol.* 5 (1): 160–170. https://doi.org/10.1175/1520-0426(1988)005<0160:ACIOWD>2.0.CO;2.

Gelaro, R., W. McCarty, M. J. Suárez, R. Todling, A. Molod, and L. Takacs, et al. 2017. "The modern-era retrospective analysis for research and applications, version 2 (MERRA-2)." *J. Clim.* 30 (14): 5419–5454. https://doi.org/10.1175/JCLI-D-16-0758.1.

Gland, H., and P. Admirat. 1986. "Meteorological conditions for wet snow occurrence in France—Calculated and measured results in a recent case study on 5 March 1985." In *Proc., 3rd Int. Workshop on Atmospheric Icing of Structures*, L. E. Welsh, and D. J. Armstrong, eds., Vancouver, Canada, 91–96.

Golikova, T. N. 1982. *Probability of increased ice loads on overhead lines depending on their length*. Soviet Power No. 10. Portage, WI: Ralph McElroy.

Golikova, T. N., B. F. Golikov, and D. S. Savvaitov. 1982. "Methods of calculating icing loads on overhead lines as spatial constructions." In *Proc., 1st Int. Workshop on Atmospheric Icing of Structures*, L. D. Minsk, ed. Hanover, NH, 341–345.

Goodwin, E. J., J. D. Mozer, A. M. DiGioia Jr., and B. A. Power. 1983. "Predicting ice and snow loads for transmission line design." In *Proc., 3rd Int. Workshop on Atmospheric Icing of Structures*, L. E. Welsh, and D. J. Armstrong, eds., Vancouver, Canada, 267–275.

Govoni, J. W. 1990. "A comparative study of icing rates in the White Mountains of New Hampshire, Paper A1-9." In *Proc., 5th Int. Workshop on Atmospheric Icing of Structures*. Tokyo.

Hall, E. K. 1977. "Ice and wind loading analysis of Bonneville Power Administration's transmission lines and test spans." In *Proc., IEEE Power Engineering Society Summer Meeting*. Mexico City.

Hosking, J. R. M., and J. R. Wallis. 1987. "Parameter and quantile estimation for the generalized Pareto distribution." *Technometrics* 29 (3): 339–349.

Hosking, J. R. M., and J. R. Wallis. 1997. *Regional frequency analysis: An approach based on L-moments*. New York: Cambridge University Press.

IEEE. 2017. *National electrical safety code, C2-2017*. New York: IEEE.

Iversen, E.C., G. Thompson, B. E. Nygaard, and M. Lacavalla. 2019. "Improved prediction of wet snow." In *Proc., 18th Int. Workshop on Atmospheric Icing of Structures*. Reykjavik, Iceland.

Jones, K. 1998. "A simple model for freezing rain ice loads." *Atmos. Res.* 46 (1–2): 87–97. https://doi.org/10.1016/S0169-8095(97)00053-7.

Jones, K. F. 1990. "The density of natural ice accretions related to nondimensional icing parameters." *Q. J. R. Meteorol. Soc.* 116 (492): 477–496. https://doi.org/10.1002/qj.49711649212.

Jones, K. F. 1996. *Ice accretion in freezing rain*. US Army CRREL Rep. No. 96-2. Hanover, NH: Cold Regions Research and Engineering Laboratory.

Jones, K. F., and A. B. Peabody. 2006. "The application of a uniform radial ice thickness to structural sections." *Cold Reg. Sci. Technol.* 44 (2): 145–148. https://doi.org/10.1016/j.coldregions.2005.10.002.

Jones, K. F., R. Thorkildson, and J. N. Lott. 2002. "The development of the map of extreme ice loads for ASCE Manual 74." In *Electrical transmission in a new age*, D. E. Jackman, ed., 9–31. Reston, VA: ASCE.

Jones, K. F. 2019. "How was the MRI model applied?" In *Proc., 18th Int. Workshop on Atmospheric Icing of Structures*. Reykjavik, Iceland.

Jones, K. F., and J. B. Eylander. 2017. "Vertical variation of ice loads from freezing rain." *Cold Reg. Sci. Tech.* 143 (Nov): 126–136. https://doi.org/10.1016/j.coldregions.2017.07.008.

Jones, K. F., R. Thorkildson, and J. B. Eylander. 2017. "Median volume drop diameter of in-cloud icing simulations." In *Proc., 17th Int. Workshop on Atmospheric Icing of Structures*. Chongqing, China.

Konrad, C. E. II. 1998. "An empirical approach for delineating fine scaled spatial patterns of freezing rain in the Appalachian region of the USA." *Clim. Res.* 10 (3): 217–227. https://doi.org/10.3354/cr010217.

Kuroiwa, D. 1962. *A study of ice sintering*. US Army CRREL Research Rep. No. 86. Hanover, NH: Cold Regions Research and Engineering Laboratory.

Kuroiwa, D. 1965. *Icing and snow accretion on electric wires*. US Army CRREL Research Paper No. 123. Hanover, NH: Cold Regions Research and Engineering Laboratory.

Lacavalla, M., P. Marcacci, and A. Freddo. 2015. "Wet snow research activity in Italy." In *Proc., 16th Int. Workshop on Atmospheric Icing of Structures*. Uppsala, Sweden.

Lacavalla, M., S. Sperati, R. Bonanno, and P. Marcacci. 2019. "A revision of wet snow load map for the Italian power lines with a new high resolution reanalysis dataset." In *Proc., 18th Intl. Workshop on Atmospheric Icing of Structures*. Reykjavik, Iceland.

Lott, N. 1993a. *The big one! A review of the MARCH 12-14, 1993 "Storm of the century"* NCDC Technical Rep. No. 93-01. Asheville, NC: National Climatic Data Center.

Lott, N. 1993b. *Water equivalent vs. rain gauge measurements from the March 1993 Blizzard*. NCDC Technical Rep. No. 93-03. Asheville, NC: National Climatic Data Center.

Macklin, W. C. 1962. "The density and structure of ice formed by accretion." *Q. J. R. Meteorol. Soc.* 88 (375): 30–50. https://doi.org/10.1002/qj.49708837504.

Mallory, J. H., and D. C. Leavengood. 1983a. "Extreme glaze and rime ice loads in Southern California: Part I—Rime." In *Proc., 1st Int. Workshop on Atmospheric Icing of Structures*, L. D. Minsk, ed., Hanover, NH, 299–308.

Mallory, J. H., and D. C. Leavengood. 1983b. "Extreme glaze and rime ice loads in Southern California: Part II—Glaze." In *Proc., 1st Int. Workshop on Atmospheric Icing of Structures*, L. D. Minsk, ed., Hanover, NH, 309–318.

Matsumiya, H. I., T. Aso, M. Shugo, T. Nishihara, M. Shimizu, and S. Sugimot. 2019. "Field observation of wet snow accretion and galloping on a single conductor transmission line." In *Proc., 18th Int. Workshop on Atmospheric Icing of Structures*. Reykjavik, Iceland.

McCormick, T., and J. C. Pohlman. 1993. "Study of compact 220 kV line system indicates need for micro-scale meteorological information." In *Proc., 6th Int. Workshop on Atmospheric Icing of Structures*, Budapest, Hungary, 155–159.

Mesinger, F., et al. 2006. "North American regional reanalysis." *Bull. Am. Meteorol. Soc.* 87 (3): 343–360. https://doi.org/10.1175/BAMS-87-3-343.

Mozer, J. D., and R. J. West. 1983. "Analysis of 500 kV tower failures." In *Proc., Meeting of the Pennsylvania Electric Association*. Harrisburg, PA.

Mulherin, N. D. 1996. "Atmospheric icing and tower collapse in the United States." In *Proc., 7th Int. Workshop on Atmospheric Icing of Structures*, M. Farzaneh, and J. Laflamme, eds., Chicoutimi, QC.

Nikolov, D., and L. Makkonen. 2015. "Testing six wet snow models by 30 years of observations in Bulgaria." In *Proc., 16th Int. Workshop on Atmospheric Icing of Structures*. Uppsala, Sweden.

NOAA (National Oceanic and Atmospheric Administration). 1959–Present. *Storm data*. Washington, DC: NOAA.

NPPD (Nebraska Public Power District). 1976. *The storm of March 29, 1976*. Columbus, NE: Public Relations Department.

Nygaard, B. E., H. Ágústsson, and K. Somfalvi-Tóth. 2013. "A study on the sticking efficiency of wet snow using 50 years of observations." In *Proc., 15th Int. Workshop on Atmospheric Icing of Structures*. St. Johns, Canada.

Nygaard, B. E., I. A. Seierstad, S. M. Fikke, D. Horsman, and J. B. Wareing. 2013. "The development of new maps for design ice loads for Great Britain." In *Proc., 15th Int. Workshop on Atmospheric Icing of Structures*. St. Johns, Canada.

Peabody, A. B. 1993. "Snow loads on transmission and distribution lines in Alaska." In *Proc., 6th Intl. Workshop on Atmospheric Icing of Structures*. Budapest, Hungary, 201–205.

Peabody, A. B. 1996. "Measurements of ice and snow loads on the Tyee Lake 138 kV line." In *Proc., 7th Int. Workshop on Atmospheric Icing of Structures*. Chicoutimi, Canada.

Peabody, A. B., and K. F. Jones. 2002. "Effect of wind on the variation of ice thickness from freezing rain." In *Proc., 10th Int. Workshop on Atmospheric Icing of Structures*. Brno, Czech Republic.

Peabody, A. B., and G. Wyman. 2005. "Atmospheric icing measurements in Fairbanks, Alaska." In *Proc., 11th Int. Workshop on Atmospheric Icing of Structures*, M. Farzaneh, and A. P. Goel, eds., Yokohama, Japan.

Peterka, J. A. 1992. "Improved extreme wind prediction for the United States." *J. Wind Eng. Ind. Aerodyn.* 41 (1–3): 533–541. https://doi.org/10.1002/qj.49708837504.

Peterka, J. A., K. Finstad, and A. K. Pandy. 1996. *Snow and wind loads for Tyee transmission line*. Fort Collins, CO: Cermak Peterka Petersen.

Power, B. A. 1983. *Estimation of climatic loads for transmission line design*. CEA No. ST 198. Montreal: Canadian Electric Association.

Powers, J. G., et al. 2017. "The weather research and forecasting model: Overview, system efforts, and future directions." *Bull. Am. Met. Soc.* 98 (8): 1717–1737.

Rawlins, C. B. 1979. "Galloping conductors." *Transmission line reference book, wind-induced conductor motion*, 113–168. Palo Alto, CA: Electric Power Research Institute.

Robbins, C. C., and J. V. Cortinas Jr. 1996. "A climatology of freezing rain in the contiguous United States: Preliminary results." In *Proc., 15th AMS Conf. on Weather Analysis and Forecasting*. Norfolk, VA.

Ryerson, C. 1987. *Rime meteorology in the Green Mountains*. US Army CRREL Rep. No. 87-1. Hanover, NH: Cold Regions Research and Engineering Laboratory.

Ryerson, C. 1988a. "Atmospheric icing climatologies of two New England mountains." *J. Appl. Meteorol.* 27 (11): 1261–1281. https://doi.org/10.1002/qj.49708837504.

Ryerson, C. 1988b. *New England mountain icing climatology*. CRREL Rep. No. 88-12. Hanover, NH: Cold Regions Research and Engineering Laboratory.

Ryerson, C. 1990. "Atmospheric icing rates with elevation on northern New England mountains, U.S.A." *Arctic Alpine Res.* 22 (1): 90–97.

Ryerson, C., and K. Claffey. 1991. "High latitude, West Coast mountaintop icing." In *Proc., Eastern Snow Conf.*, Guelph, ON, 221–232.

Ryerson, C., and P. Kenyon. 1996. "Ice ablation processes on Mt. Equinox, Vermont, USA." In *Proc., 7th Int. Workshop on Atmospheric Icing of Structures*, Chicoutimi, Canada, 277–281.

Sakamoto, Y., K. Mizushima, and S. Kawanishi. 1990. "Dry snow type accretion on overhead wires: Growing mechanism, meteorological conditions under which it occurs and effect on power lines." In *Proc., 5th Int. Workshop on Atmospheric Icing of Structures*. Tokyo.

Shan, L. 1997. *Wind tunnel study of drag coefficients of single and bundled conductors*. EPRI TR-108969. Palo Alto, CA: Electric Power Research Institute.

Shan, L., and L. Marr. 1996. *Ice storm data base and ice severity maps*. Palo Alto, CA: Electric Power Research Institute.

Simiu, E., and R. H. Scanlan. 1996. *Wind effects on structures: Fundamentals and applications to design*. New York: Wiley.

Sinclair, R. E., and R. M. Thorkildson. 1980. *In-cloud moisture droplet impingement angles and track clearances at the Moro UHV test site*. BPA 1200 kV Project Rep. No. ME-80-7. Portland, OR: Bonneville Power Administration.

Tattelman, P., and I. Gringorten. 1973. *Estimated glaze ice and wind loads at the earth's surface for the contiguous United States*. Rep. No. AFCRLTR-73-0646. Bedford, MA: US Air Force Cambridge Research Laboratories.

Thorkildson, R. M., K. F. Jones, and M. K. Emery. 2009. "In-cloud icing in the Columbia Basin." *Monthly Weather Review* 137 (12): 4369–4381.

Thorkildson, R. M. 2019. "Rime ice occurrences from radiation fog that impact overhead transmission lines in Central Washington State." In *Proc., 18th Int. Workshop on Atmospheric Icing of Structures*. Reykjavik, Iceland.

TIA (Telecommunication Industry Association). 2019. *structural standard for antenna supporting structures, antennas and small wind turbine support structures*. ANSI/TIA 222-H-1. Arlington VA: TIA.

Ueno, K., Y. Eguchi, T. Nishihara, S. Sugimoto, and H. Matsumiya. 2015. "Development of snow accretion simulation method for electric wires in consideration of snow melting and shedding." In *Proc., 16th Int. Workshop on Atmospheric Icing of Structures*. Uppsala, Sweden.

USFS (US Forest Service). 1994. *Forest service handbook FSH6609.14, Telecommunications handbook*. R3 Supplement 6609.14-94-2. Washington, DC: USFS.

Wang, Q. J. 1991. "The POT model described by the generalized Pareto distribution with Poisson arrival rate." *J. Hydrol.* 129 (1–4): 263–280. https://doi.org/10.1016/0022-1694(91)90054-L.

White, H. B. 1999. "Galloping of ice covered wires." In *Proc., 10th Int. Conf. on Cold Regions Engineering*, Hanover, NH, 799–804.

Winkleman, P. F. 1974. *Investigation of ice and wind loads: Galloping, vibrations and subconductor oscillations*. Portland, OR: Bonneville Power Administration.

Wylie, W. G. 1958. "Tropical ice storms—Winter invades Hawaii." *Weatherwise* 11 (3): 84–90. https://doi.org/10.1080/00431672.1958.9925023.

Young, W. R. 1978. "Freezing precipitation in the Southeastern United States." M.S. thesis, Texas A&M University, Dept. of Civil and Environmental Engineering.

CHAPTER C11
SEISMIC DESIGN CRITERIA

C11.1 GENERAL

Many of the technical changes made to the seismic provisions of the 2010 edition of this standard are primarily based on Part 1 of the 2009 edition of the *NEHRP Recommended Seismic Provisions for New Buildings and Other Structures* (FEMA 2009), which was prepared by the Building Seismic Safety Council (BSSC) under sponsorship of the Federal Emergency Management Agency (FEMA) as part of its contribution to the National Earthquake Hazards Reduction Program (NEHRP). The National Institute of Standards and Technology (NIST) is the lead agency for NEHRP, the federal government's long-term program to reduce the risks to life and property posed by earthquakes in the United States. Since 1985, the NEHRP provisions have been updated every three to five years. The efforts by the BSSC to produce the NEHRP provisions were preceded by work performed by the Applied Technology Council (ATC) under sponsorship of the National Bureau of Standards (NBS)—now NIST—which originated after the 1971 San Fernando Valley earthquake. These early efforts demonstrated the design rules of that time for seismic resistance but had some serious shortcomings. Each subsequent major earthquake has taught new lessons. The NEHRP agencies [FEMA, NIST, the National Science Foundation (NSF), and the US Geological Survey (USGS)], ATC, BSSC, ASCE, and others have endeavored to work individually and collectively to improve each succeeding document to provide state-of-the-art earthquake engineering design and construction provisions and to ensure that the provisions have nationwide applicability.

Content of Commentary. This commentary is updated from the commentary to ASCE/SEI 7-16 and includes updates identified in Part 2, Commentary, of the 2020 *NEHRP Recommended Seismic Provisions for New Buildings and Other Structures* (FEMA 2020). For additional background on the earthquake provisions contained in Chapters 11 through 23 of this standard, the reader is referred to *Recommended Lateral Force Requirements and Commentary* (SEAOC 1999).

Nature of Earthquake "Loads." Earthquakes load structures indirectly through ground motion. As the ground shakes, a structure responds. The response vibration produces structural deformations with associated strains and stresses. The computation of dynamic response to earthquake ground shaking is complex. The design forces prescribed in this standard are intended only as approximations to generate internal forces suitable for proportioning the strength and stiffness of structural elements and for estimating the deformations (when multiplied by the deflection amplification factor, C_d) that would occur in the same structure in the event of the design-level earthquake ground motion (not MCE_R).

The basic methods of analysis in the standard use the common simplification of a response spectrum. A response spectrum for a specific earthquake ground motion provides the maximum value of response for elastic single-degree-of-freedom oscillators as a function of period without the need to reflect the total response history for every period of interest. The design response spectrum specified in Section 11.4 and used in the basic methods of analysis in Chapter 12 is a smoothed and normalized approximation for many different recorded ground motions.

The design limit state for resistance to an earthquake is unlike that for any other load within the scope of ASCE 7. The earthquake limit state is based on system performance, not member performance, and considerable energy dissipation through repeated cycles of inelastic straining is assumed. The reason is the large demand exerted by the earthquake and the associated high cost of providing enough strength to maintain linear elastic response in ordinary buildings. This unusual limit state means that several conveniences of elastic behavior, such as the principle of superposition, are not applicable and make it difficult to separate design provisions for loads from those for resistance. This difficulty is the reason Chapter 14 of the standard contains so many provisions that modify customary requirements for proportioning and detailing structural members and systems. It is also the reason for the construction quality assurance requirements.

Use of Allowable Stress Design Standards. The conventional design of almost all masonry structures and many wood and steel structures has been accomplished using allowable stress design (ASD). Although the fundamental basis for the earthquake loads in Chapters 11 through 23 is a strength limit state beyond the first yield of the structure, the provisions are written such that conventional ASD methods can be used by the design engineer. Conventional ASD methods may be used in one of two ways:

1. The earthquake load as defined in Chapters 11 through 23 may be used directly in allowable stress load combinations of Section 2.4, and the resulting stresses may be compared directly with conventional allowable stresses.
2. The earthquake load may be used in strength design load combinations, and resulting stresses may be compared with amplified allowable stresses (for those materials for which the design standard gives the amplified allowable stresses, e.g., masonry).

Executive Order (EO) 13717: *Establishing a Federal Earthquake Risk Management Standard*, issued February 2016, establishes a minimum level of seismic safety compliance for new buildings that will be constructed, financed, or regulated by the federal government. NIST Technical Note 1922 (NIST 2017) provides guidance on agency compliance with the EO.

C11.1.1 Purpose The purpose of Section 11.1.1 is to clarify that the detailing requirements and limitations prescribed in this section and referenced standards are still required even when the design load combinations involving the wind forces of Chapters 26 through 29 produce greater effects than the design load combinations involving the earthquake forces of Chapters 11 through 23. This detailing is required so that the structure resists, in a ductile manner, potential seismic loads in excess of the prescribed wind loads. A proper, continuous load path is an obvious design requirement, but experience has shown that it is often overlooked and that significant damage and collapse can result. The basis for this design requirement is twofold:

1. To ensure that the design has fully identified the seismic force-resisting system and its appropriate design level, and
2. To ensure that the design basis is fully identified for the purpose of future modifications or changes in the structure.

Detailed requirements for analyzing and designing this load path are given in the appropriate design and materials chapters.

C11.1.2 Scope Certain structures are exempt for the following reasons:

Exemption 1. Detached wood-frame dwellings not exceeding two stories above grade plane constructed in accordance with the prescriptive provisions of the International Residential Code (IRC) for light-frame wood construction, including all applicable IRC seismic provisions and limitations, are deemed capable of resisting the anticipated seismic forces. Detached one- and two-story wood-frame dwellings generally have performed well even in regions of higher seismicity. Therefore, within its scope, the IRC adequately provides the level of safety required for buildings. The structures that do not meet the prescriptive limitations of the IRC are required to be designed and constructed in accordance with the International Building Code (IBC) and the ASCE 7 provisions adopted therein.

Exemption 2. Agricultural storage structures generally are exempt from most code requirements because such structures are intended only for incidental human occupancy and represent an exceptionally low risk to human life.

Exemption 3. Bridges, transmission towers, nuclear reactors, and other structures with special configurations and uses are not covered. The regulations for buildings and buildinglike structures presented in this document do not adequately address the design and performance of such special structures.

ASCE 7 is not retroactive and usually applies to existing structures only when there is an addition, change of use, or alteration. Minimum acceptable seismic resistance of existing buildings is a policy issue normally set by the Authority Having Jurisdiction. ASCE 41 (2014) provides technical guidance but does not contain policy recommendations. A chapter in the International Building Code (IBC) applies to alteration, repair, addition, and change of occupancy of existing buildings, and the International Code Council maintains the International Existing Building Code (IEBC) and associated commentary.

C11.1.3 Applicability Industrial buildings may be classified as nonbuilding structures in certain situations for the purposes of determining seismic design coefficients and factors, system limitations, height limits, and associated detailing requirements. Many industrial building structures have geometries and/or framing systems that are different from the broader class of occupied structures addressed by Chapter 12, and the limited nature of the occupancy associated with these buildings reduces the hazard associated with their performance in earthquakes. Therefore, when the occupancy is limited primarily to maintenance and monitoring operations, these structures may be designed in accordance with the provisions of Section 15.5 for nonbuilding structures similar to buildings. Examples of such structures include, but are not limited to, boiler buildings, aircraft hangars, steel mills, aluminum smelting facilities, and other automated manufacturing facilities, whereby the occupancy restrictions for such facilities should be uniquely reviewed in each case. These structures may be clad or open structures.

C11.1.4 Alternate Materials and Methods of Construction It is not possible for a design standard to provide criteria for the use of all possible materials and their combinations and methods of construction, either existing or anticipated. This section serves to emphasize that the evaluation and approval of alternate materials and methods require a recognized and accepted approval system. The requirements for materials and methods of construction contained within the document represent the judgment of the best use of the materials and methods based on well-established expertise and historical seismic performance. It is important that any replacement or substitute be evaluated with an understanding of all the ramifications of performance, strength, and durability implied by the standard.

Until needed standards and agencies are created, authorities that have jurisdiction need to operate on the basis of the best evidence available to substantiate any application for alternates. If accepted standards are lacking, applications for alternative materials or methods should be supported by test data obtained from test data requirements in Section 1.3.1. The tests should simulate expected load and deformation conditions to which the system, component, or assembly may be subjected during the service life of the structure. These conditions, when applicable, should include several cycles of full reversals of loads and deformations in the inelastic range.

C11.1.5 Quality Assurance Quality assurance (QA) requirements are essential for satisfactory performance of structures in earthquakes. QA requirements are usually incorporated in building codes as special inspections and tests or as structural observation, and they are enforced through the Authorities Having Jurisdiction. Many building code requirements parallel or reference the requirements found in standards adopted by ASCE 7. Where special inspections and testing, or structural observations are not specifically required by the building code, a level of QA is usually provided by inspectors employed by the Authority Having Jurisdiction.

Where building codes are not in force or where code requirements do not apply to or are inadequate for a unique structure or system, the registered design professional for the structure or system should develop a QA program to verify that the structure or system is constructed as designed. A QA program could be modeled on similar provisions in the building code or applicable standards.

The quality assurance plan is used to describe the QA program to the owner, the Authority Having Jurisdiction, and to all other participants in the QA program. As such, the QA plan should include definitions of the roles and responsibilities of the participants. It is anticipated that in most cases the owner of the project would be responsible for implementing the QA plan.

C11.2 DEFINITIONS

ATTACHMENTS, COMPONENTS, AND SUPPORTS: The distinction among attachments, components, and supports is

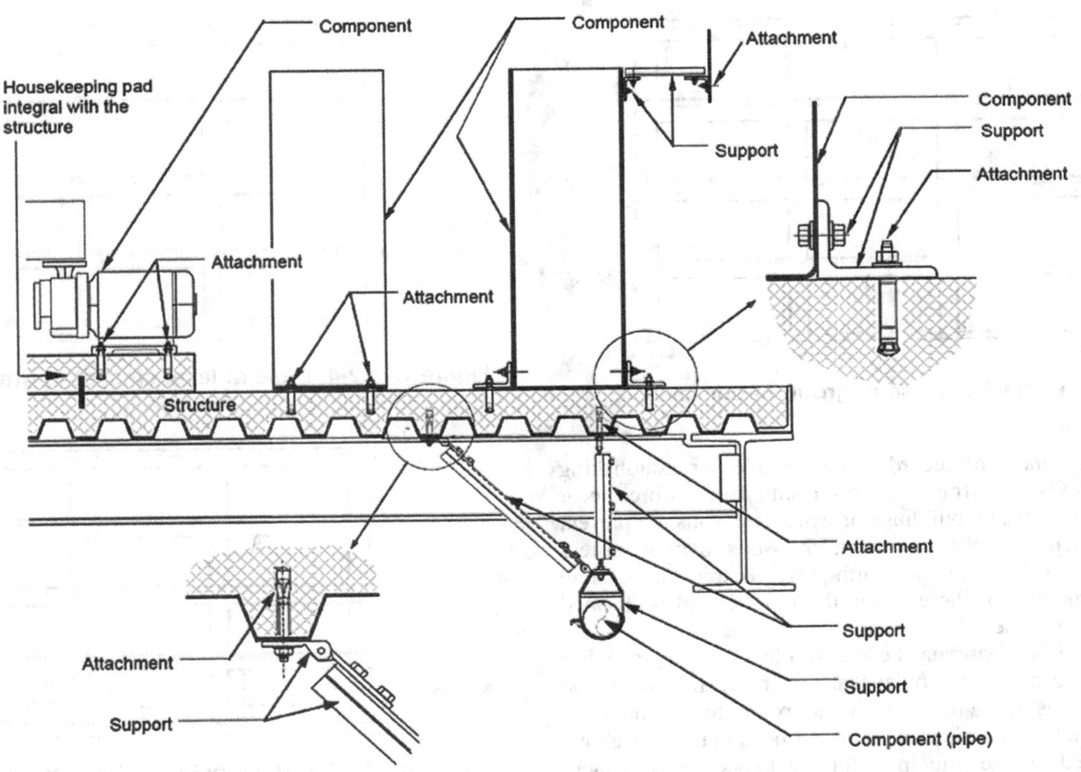

Figure C11.2-1. Examples of attachments, components, and supports.

necessary to the understanding of the requirements for nonstructural components and nonbuilding structures. Common cases associated with nonstructural elements are illustrated in Figure C11.2-1. The definitions of attachments, components, and supports are generally applicable to components with a defined envelope in the as-manufactured condition and for which additional supports and attachments are required to provide support in the as-built condition. This distinction may not always be clear, particularly when the component is equipped with prefabricated supports; therefore, judgment must be used in the assignment of forces to specific elements in accordance with the provisions of Chapter 13.

BASE: The following factors affect the location of the seismic base:

- Location of the grade relative to floor levels,
- Soil conditions adjacent to the building,
- Openings in the basement walls,
- Location and stiffness of vertical elements of the seismic force-resisting system,
- Location and extent of seismic separations,
- Depth of basement,
- Manner in which basement walls are supported,
- Proximity to adjacent buildings, and
- Slope of grade.

For typical buildings on level sites with competent soils, the base is generally close to the grade plane. For a building without a basement, the base is generally established near the ground-level slab elevation, as shown in Figure C11.2-2. Where the vertical elements of the seismic force-resisting system are supported on interior footings or pile caps, the base is the top of these elements. Where the vertical elements of the seismic force-resisting system are supported on top of perimeter foundation walls, the base is typically established at the top of the foundation walls. Often vertical elements are supported at various elevations on the top of footings, pile caps, and perimeter foundation walls. Where this occurs, the base is generally established as the lowest elevation of the tops of elements supporting the vertical elements of the seismic force-resisting system.

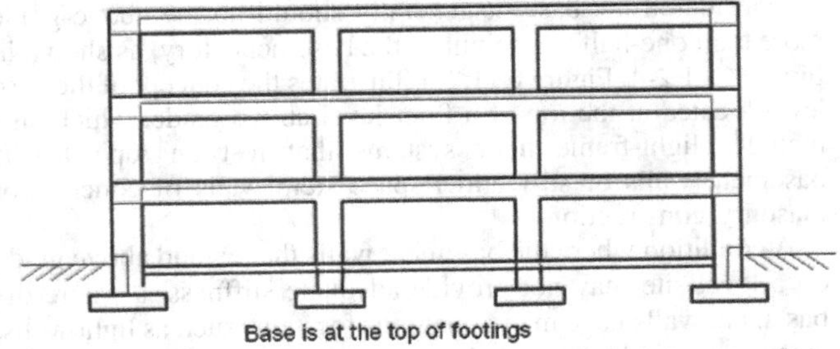

Figure C11.2-2. Base for a level site.

For a building with a basement located on a level site, it is often appropriate to locate the base at the floor closest to grade, as shown in Figure C11.2-3. If the base is to be established at the level located closest to grade, the soil profile over the depth of the basement should not be liquefiable in the MCE_G ground motion. The soil profile over the depth of the basement also should not include quick and highly sensitive clays or weakly cemented soils prone to collapse in the MCE_G ground motion. Where liquefiable soils or soils susceptible to failure or collapse in an MCE_G ground motion are located within the depth of the basement, the base may need to be located below these soils rather than close to grade. Stiff soils are required over the depth of the basement because seismic forces are transmitted to and from the building at this level and over the height of the basement

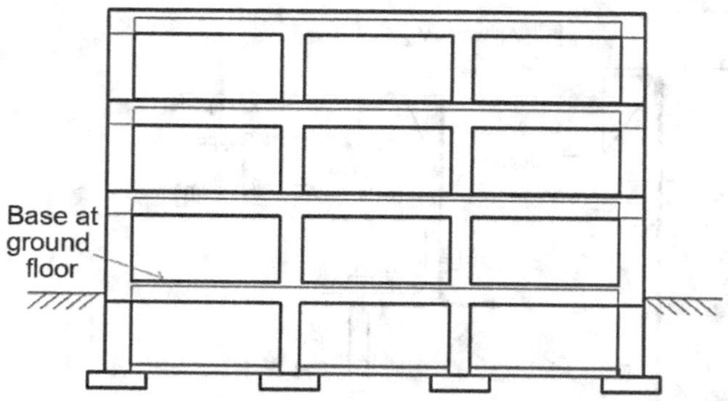

Figure C11.2-3. Base at ground floor level.

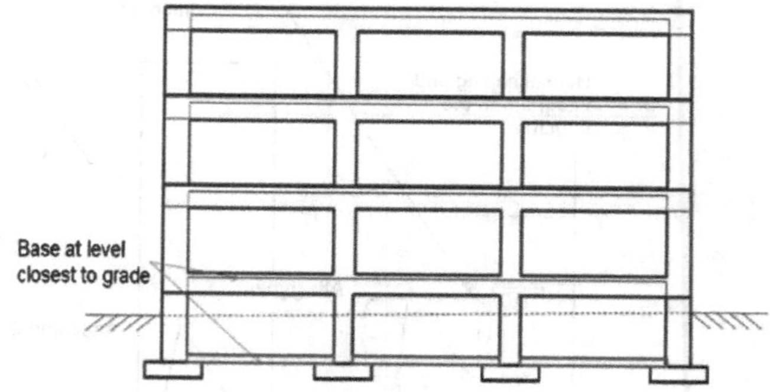

Figure C11.2-4. Base at level closest to grade elevation.

walls. The engineer of record is responsible for establishing whether the soils are stiff enough to transmit seismic forces near grade. For tall or heavy buildings or where soft soils are present within the depth of the basement, the soils may compress laterally too much during an earthquake to transmit seismic forces near grade. For these cases, the base should be located at a level below grade.

In some cases, the base may be at a floor level above grade. For the base to be located at a floor level above grade, stiff foundation walls on all sides of the building should extend to the underside of the elevated level considered the base. Locating the base above grade is based on the principles for the two-stage equivalent lateral force procedure for a flexible upper portion of a building with one-tenth the stiffness of the lower portion of the building, as permitted in Section 12.2.3.2. For a floor level above grade to be considered the base, it generally should not be above grade more than one-half the height of the basement story, as shown in Figure C11.2-4. Figure C11.2-4 illustrates the concept of the base level located at the top of a floor level above grade, which also includes light-frame floor systems that rest on top of stiff basement walls or stiff crawl space stem walls of concrete or masonry construction.

A condition where the basement walls that extend above grade on a level site may not provide adequate stiffness is where the basement walls have many openings for items such as light wells, areaways, windows, and doors, as shown in Figure C11.2-5. Where the basement wall stiffness is inadequate, the base should be taken as the level close to but below grade. If all of the vertical elements of the seismic force-resisting system are located on top of basement walls and there are many openings in the basement walls, it may be appropriate to establish the base at the bottom of the openings. Another condition where the basement walls may not be stiff enough is where the vertical elements of the seismic force-resisting system are long concrete shear walls extending over the full height and length of the building, as shown in Figure C11.2-6. For this case, the appropriate location for the base is the foundation level of the basement walls.

Where the base is established below grade, the weight of the portion of the story above the base that is partially above and below grade must be included as part of the effective seismic weight. If the equivalent lateral force procedure is used, this procedure can result in disproportionately high forces in upper levels because of a large mass at this lowest level above the base. The magnitude of these forces can often be mitigated by using the two-stage equivalent lateral force procedure where it is allowed or by using dynamic analysis to determine force distribution over the height of the building. If dynamic analysis is used, it may be necessary to include multiple modes to capture the required mass

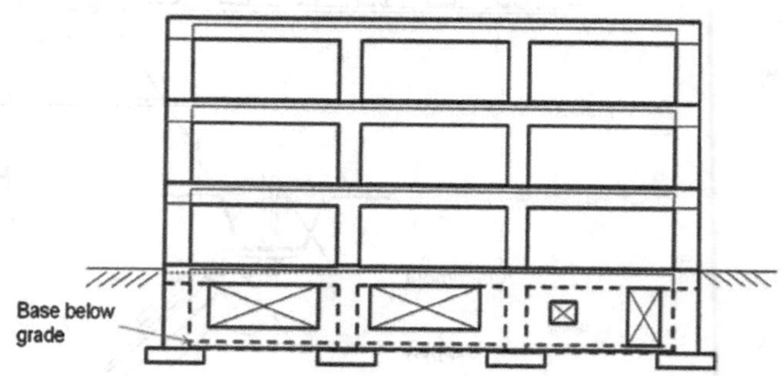

Figure C11.2-5. Base below substantial openings in basement wall.

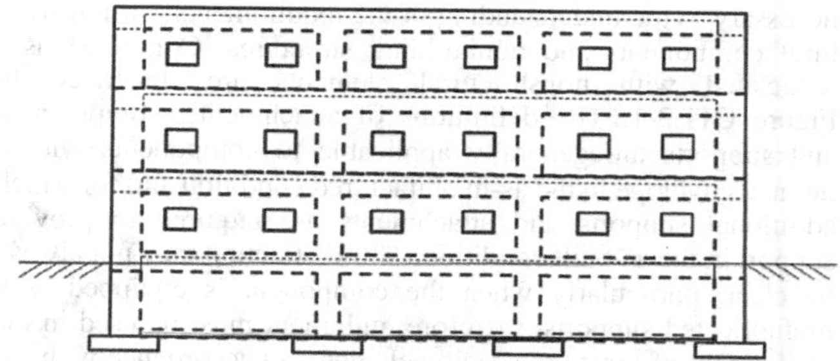
Figure C11.2-6. Base at foundation level where there are full-length exterior shear walls.

participation, unless soil springs are incorporated into the model. Incorporation of soil springs into the model generally reduces seismic forces in the upper levels. With one or more stiff stories below more flexible stories, the dynamic behavior of the structure may result in the portion of the base shear from the first mode being less than the portion of base shear from higher modes.

Other conditions may also necessitate establishing the base below grade for a building with a basement that is located on a level site. Such conditions include those where seismic separations extend through all floors, including those located close to and below grade; those where the floor diaphragms close to and below grade are not tied to the foundation wall; those where the floor diaphragms, including the diaphragm for the floor close to grade, are flexible; and those where other buildings are located nearby.

For a building with seismic separations extending through the height of the building including levels close to and below grade, the separate structures are not supported by the soil against a basement wall on all sides in all directions. If there is only one joint through the building, assigning the base to the level close to grade may still be appropriate if the soils over the depth of the basement walls are stiff and the diaphragm is rigid. Stiff soils are required so that the seismic forces can be transferred between the soils and basement walls in both bearing and side friction. If the soils are not stiff, adequate side friction may not develop for movement in the direction perpendicular to the joint.

For large footprint buildings, seismic separation joints may extend through the building in two directions and there may be multiple parallel joints in a given direction. For individual structures within these buildings, substantial differences in the location of the center of rigidity for the levels below grade relative to levels above grade can lead to torsional response. For such buildings, the base should usually be at the foundation elements below the basement or the highest basement slab level where the separations are no longer provided.

Where floor levels are not tied to foundation walls, the base may need to be located well below grade at the foundation level. An example is a building with tieback walls and posttensioned floor slabs. For such a structure, the slabs may not be tied to the wall to allow relative movement between them. In other cases, a soft joint may be provided. If shear forces cannot be transferred between the wall and a ground level or basement floor, the location of the base depends on whether forces can be transferred through bearings between the floor diaphragm and basement wall and between the basement wall and the surrounding soils. Floor diaphragms bearing against the basement walls must resist the compressive stress from earthquake forces without buckling. If a seismic or expansion joint is provided in one of these buildings, the base almost certainly needs to be located at the foundation level or a level below grade where the joint no longer exists.

If the diaphragm at grade is flexible and does not have substantial compressive strength, the base of the building may need to be located below grade. This condition is more common with existing buildings. Newer buildings with flexible diaphragms should be designed for compression to avoid the damage that can otherwise occur.

Proximity to other structures can also affect where the base should be located. If other buildings with basements are located adjacent to one or more sides of a building, it may be appropriate to locate the base at the bottom of the basement. The closer the adjacent building is to the building, the more likely it is that the base should be below grade.

For sites with sloping grade, many of the same considerations for a level site are applicable. For example, on steeply sloped sites, the earth may be retained by a tieback wall so that the building does not have to resist the lateral soil pressures. For such a case, the building is independent of the wall, so the base should be located at a level close to the elevation of grade on the side of the building where it is lowest, as shown in Figure C11.2-7. Where the building's vertical elements of the seismic force-resisting system also resist lateral soil pressures, as shown in Figure C11.2-8, the base should also be located at a level close to the elevation of grade on the side of the building where grade is low. For these buildings, the seismic force-resisting system below highest grade is often much stiffer than the system used above it, as shown in Figure C11.2-9, and the seismic weights for levels close to and below highest grade are greater than for levels above highest grade. Use of a two-stage equivalent lateral force procedure can be useful for these buildings.

Where the site is moderately sloped such that it does not vary in height by more than a story, stiff walls often extend to the underside of the level close to the elevation of high grade, and the seismic force-resisting system above grade is much more flexible above grade than it is below grade. If the stiff walls extend to the underside of the level close to high grade on all sides of the building, locating the base at the level closest to high grade may be appropriate. If the stiff lower walls do not extend to the underside of the level located closest to high grade on all sides of the building, the base should be assigned to the level closest to low grade. If there is doubt as to where to locate the base, it should conservatively be taken at the lower elevation.

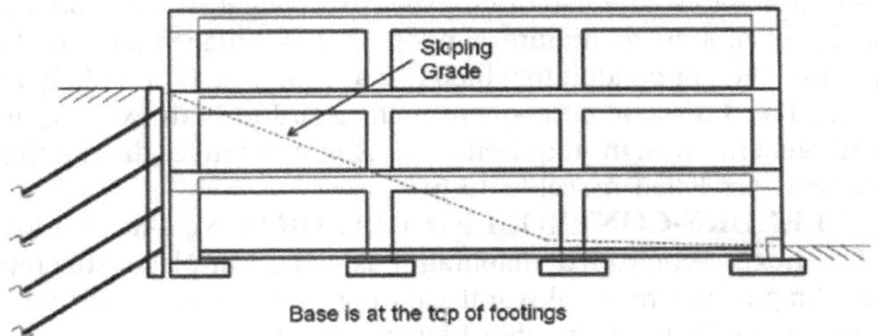

Figure C11.2-7. Building with tie-back or cantilevered retaining wall that is separate from the building.

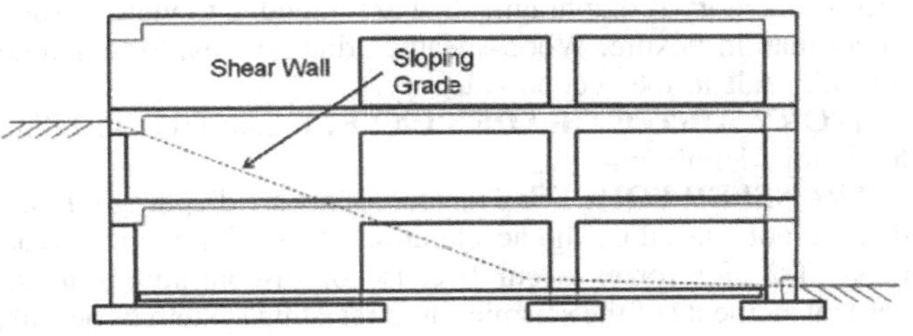

Figure C11.2-8. Building with vertical elements of the seismic force-resisting system supporting lateral earth pressures.

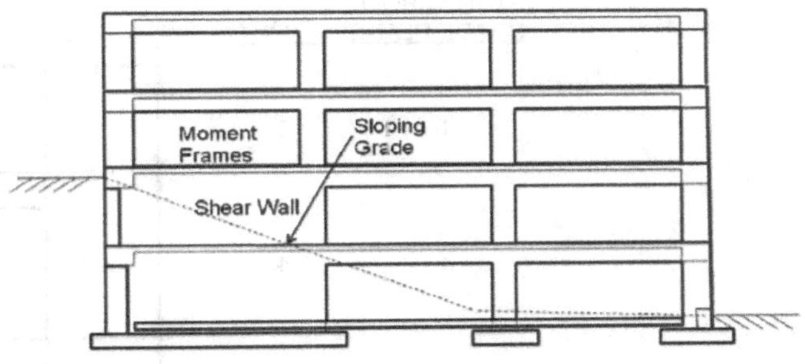

Figure C11.2-9. Building with vertical elements of the seismic force-resisting system supporting lateral earth pressures.

DISTRIBUTION SYSTEM: For the purposes of determining the anchorage of components in Chapter 13, a distribution system is characterized as a series of individual in-line mechanical or electrical components that have been physically attached together to function as an interconnected system. In general, the individual in-line components of a distribution system are comparable to those of the pipe, duct, or electrical raceway, so that the overall seismic behavior of the system is relatively uniform along its length. For example, a damper in a duct or a valve in a pipe is sufficiently similar to the weight of the duct or pipe itself, as opposed to a large fan or large heat exchanger. If a component is large enough to require support that is independent of the piping, duct, or conduit to which it is attached, it should likely be treated as a discrete component with regard to both exemptions and general design requirements. Representative distribution systems are listed in Table 13.6-1.

FLEXURE-CONTROLLED DIAPHRAGM: An example of a flexure-controlled diaphragm is a cast-in-place concrete diaphragm, where the flexural yielding mechanism would typically be yielding of the chord tension reinforcement.

SHEAR-CONTROLLED DIAPHRAGM: Shear-controlled diaphragms fall into two main categories. The first category is diaphragms that cannot develop a flexural mechanism because of aspect ratio, chord member strength, or other constraints. The second category is diaphragms that are intended to yield in shear rather than in flexure. Wood-sheathed diaphragms, for example, typically fall in the second category.

STORY ABOVE GRADE PLANE: Figure C11.2-10 illustrates this definition.

TRANSFER FORCES: Transfer forces are diaphragm forces that are not caused by the acceleration of the diaphragm inertial mass. Transfer forces occur because of discontinuities in the vertical elements of the seismic force-resisting system or because of changes in stiffness in these vertical elements from one story to the next, even if there is no discontinuity. In addition, buildings that combine frames and shear walls, which would have different deflected shapes under the same loading, also develop transfer forces in the diaphragms that constrain the frames and shear walls to deform together; this development is especially significant in dual systems.

C11.3 SYMBOLS

The provisions for precast concrete diaphragm design are intended to ensure that yielding, when it occurs, is ductile. Since yielding in shear is generally brittle at precast concrete connections, an additional overstrength factor, Ω_v, has been introduced; the required shear strength for a precast diaphragm is required to be amplified by this factor. This term is added to the symbols.

δ_{MDD}= This symbol refers to in-plane diaphragm deflection and is therefore designated with a lower-case delta. Note that the definition for δ_{MDD} refers to "lateral load" without any qualification, and the definition for Δ_{ADVE} refers to "tributary lateral load equivalent to that used in the computation of δ_{MDD}." This equivalency is an important concept that was part of the 1997 Uniform Building Code (UBC) (ICBO 1997) definition for a flexible diaphragm.

Ω_v= The provisions for precast concrete diaphragm design are intended to ensure that yielding, when it occurs, is ductile. Since yielding in shear is generally brittle at precast concrete connections, an additional overstrength factor, Ω_v, has been introduced; the required shear strength for a precast diaphragm is required to be amplified by this factor. This term is added to the symbols.

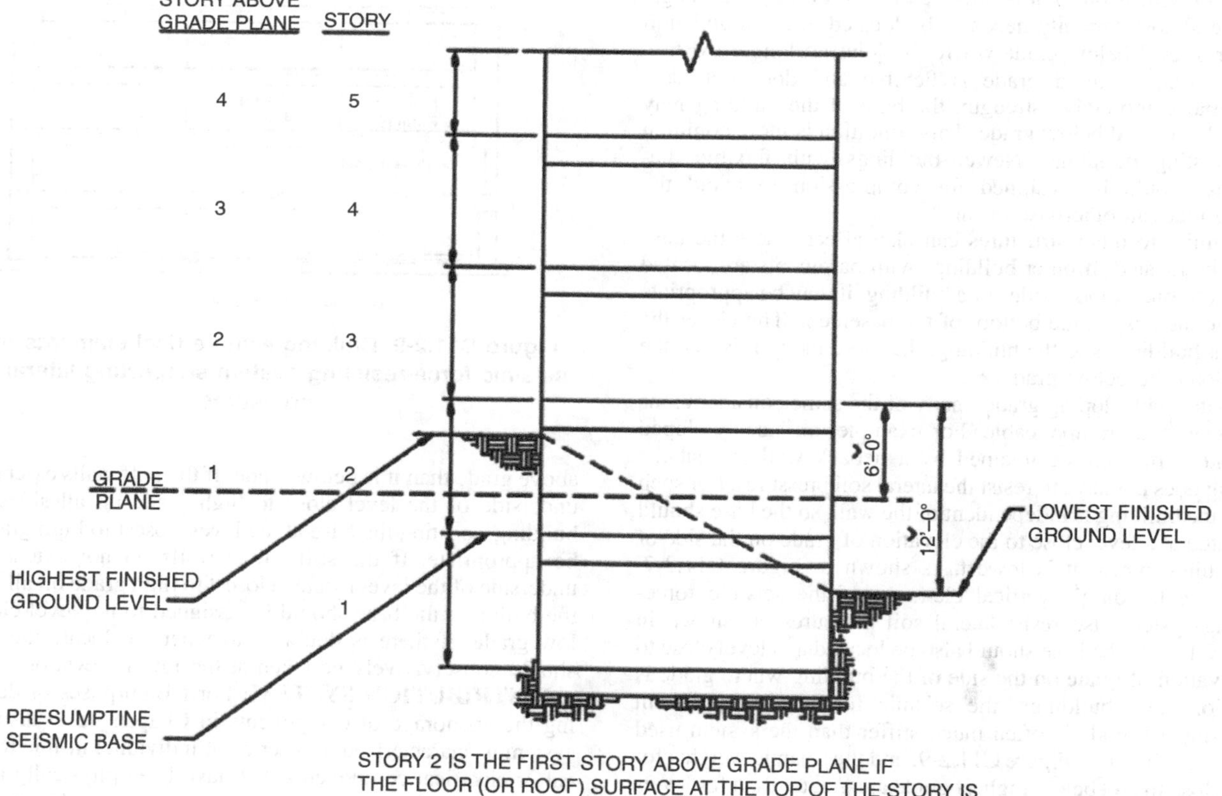

Figure C11.2-10. Illustration of definition of story above grade plane.

Note: To convert feet to millimeters, multiply by 304.8.

C11.4 SEISMIC GROUND MOTION VALUES

The theoretical basis for the mapped values of the MCE_R ground motions in the 2020 *NEHRP Recommended Provisions* (ASCE 7-22) is identical to that in the 2015 *NEHRP Recommended Provisions* (the basis of ASCE 7-16) and in the 2009 *NEHRP Recommended Provisions* (the basis of ASCE 7-10). ASCE 7-22 MCE_R ground motions (like those of ASCE 7-16 and ASCE 7-10) are significantly different from mapped values of maximum considered earthquake (MCE) ground motions in earlier editions of ASCE 7. These differences include use of (1) probabilistic ground motions that are based on uniform risk, rather than uniform hazard; (2) deterministic ground motions that are based on the 84th percentile (approximately 1.8 times median), rather than 1.5 times median response spectral acceleration for sites near active faults; and (3) ground motion intensity that is based on maximum rather than average (geometric mean) response spectral acceleration in the horizontal plane. These differences are explained in detail in the commentary of the 2009 *NEHRP Recommended Provisions*. Except for determining the MCE_G Peak Ground Acceleration (PGA) values in Chapters 11 and 21, the mapped values are given as MCE_R spectral values.

While the theoretical basis for MCE_R ground motions has not changed from ASCE 7-16 (and prior editions), ASCE 7-22 now uses a multi-period response spectra (MPRS) to improve the accuracy of the frequency content of earthquake design ground motions and to enhance the reliability of the seismic design parameters derived from these ground motions. These improvements make better use of the available earth science, which has, in general, sufficiently advanced to accurately define spectral response for different site conditions over a broad range of periods and eliminate the need for site-specific hazard analysis required by ASCE 7-16 for certain (soft soil) sites, as discussed following.

During the closing months of the 2015 cycle of the Provisions Update Committee (PUC) of the Building Seismic Safety Council, a study was undertaken of the compatibility of current Site Class coefficients F_a and F_v, with the ground motion relations used by USGS to produce the design maps (Kircher & Associates 2015). In the course of this study, it was discovered that the standard three-domain spectral shape defined by the short-period response spectral acceleration parameter, S_{DS}, the 1-second response spectral acceleration parameter, S_{D1}, and the long-period transition period, T_L, is not appropriate for soft soil sites (Site Class D or softer), in particular, where ground motion hazard is dominated by large magnitude events. Specifically, on such sites, the standard spectral shape substantially understates spectral demand for moderately long period structures. The PUC initiated a proposal to move to specification of spectral acceleration values over a range of periods, abandoning the present three-domain format, as this would provide better definition of likely ground motion demands. However, this proposal was ultimately not adopted due to both the complexity of implementing such a revision in the design procedure and time constraints. Instead, the PUC adopted a proposal prohibiting the general use of the three-parameter spectrum, and instead requiring site-specific hazard determination for longer period structures on soft soil sites.

Subsequently, Project 17 (NIBS 2019) was charged with re-evaluating the use of multiperiod response spectra as a replacement or supplement to the present three-domain spectral definition and to consider how the basic design procedures embedded in ASCE 7 should be modified for compatibility with the multi-period spectra. As a result, Project 17 developed (and unanimously approved) a comprehensive multi-period response spectra (MPRS) proposal that was subsequently adopted (with changes) in the 2020 *NEHRP Recommended Provisions* (FEMA P-2082, 2020a), which form the basis for related changes to ASCE 7-22. "Procedures for Developing Multi-Period Response Spectra of Non-Conterminous United States Sites," FEMA P-2078 (2020b), complements the changes to ASCE 7-22 by providing methods for developing MPRS of those regions (i.e., Alaska, Hawaii, Puerto Rico, Guam, and American Samoa) for which ground motion relations have not yet been used by the USGS to fully define all periods and site classes of interest.

C11.4.1 Near-Fault Sites In addition to very large accelerations, ground motions on sites located close to the zone of fault rupture of large-magnitude earthquakes can exhibit impulsive characteristics as well as unique directionality not typically recorded at sites located more distant from the zone of rupture. In past earthquakes, these characteristics have been observed to be particularly destructive. Accordingly, this standard establishes more restrictive design criteria for structures located on sites where such ground motions may occur. The standard also requires direct consideration of these unique characteristics in selection and scaling of ground motions used in nonlinear response history analysis and for the design of structures using seismic isolation or energy-dissipation devices when located on such sites.

The distance from the zone of fault rupture at which these effects can be experienced is dependent on a number of factors, including the rupture type, depth of fault, magnitude, and direction of fault rupture. Therefore, a precise definition of what constitutes a near-fault site is difficult to establish on a general basis. This standard uses two categorizations of near-fault conditions, both based on the distance of a site from a known active fault, capable of producing earthquakes of a defined magnitude or greater, and having average annual slip rates of nonnegligible amounts. These definitions were first introduced in the 1997 Uniform Building Code (ICBO 1997). Figure C11.4-1 illustrates the means of determining the distance of a site from a fault, where the fault plane dips at an angle relative to the ground surface.

C11.4.2 Site Class Site class is defined in terms of average shear wave velocity ($\overline{v}_s$) in accordance with Table 20.2-1 of Chapter 20. Table 0.2-1 includes the six site classes of ASCE 7-16 (A, B, C, D, E, and F) plus three new site classes (BC, CD, and DE) that provide better resolution of site shear wave velocity and associated site amplification for common site conditions. The new site classes allow for more accurate derivation of the amplitude and frequency content of earthquake ground motions, and their variation with shaking intensity (nonlinear effects). The additional site classes are of particular importance to the characterization of long period ground motions for softer sites.

C11.4.2.1 Default Site Conditions The "default" site condition is defined as the most critical response spectral acceleration at each period of typical soil site conditions (Site Classes C, CD, and D) to provide a conservative basis for design where site class is not known (e.g., due to insufficient geological investigation). Enveloping of Site Classes C, CD, and D is consistent with ASCE 7-16, which effectively requires the more critical of Site Class C and D to be used for design where site class is not known. Use of the default site conditions for design presumes that the site does not have soft soil site conditions (i.e., Site Class DE, E, or F site conditions) and should not be used for design where such site conditions are known to exist.

C11.4.3 Risk-Targeted Maximum Considered Earthquake (MCE_R) Spectral Response Acceleration Parameters "Mapped" values of seismic parameters S_S, S_1, S_{MS}, and

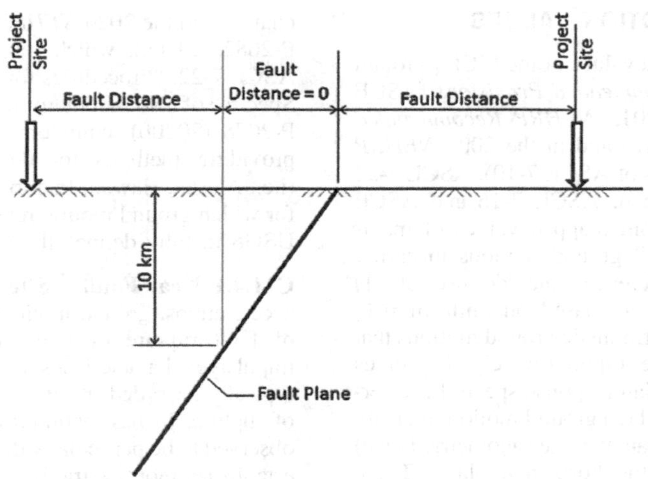

Figure C11.4-1. Fault distance for various project site locations.

Note: To convert to km to mi, multiply by 0.62.

S_{M1} are archived in the USGS Seismic Design Geodatabase at gridded locations across United States regions of interest and provided online by the USGS Seismic Design Web Service, for user-specified site location (latitude and longitude) and site class. The USGS web service spatially interpolates between the gridded values of these parameters based on site location (latitude and longitude). Chapter 22 provides print copies of seismic parameters S_{MS} and S_{M1} for default site conditions. Seismic parameters S_{MS} and S_{M1} (and S_{DS} and S_{D1}) incorporate site effects, eliminating the need for the tables of site factors F_a and F_v of ASCE 7-16.

C11.4.4 Design Spectral Acceleration Parameters Design in ASCE 7 (e.g., Chapter 12) is performed for earthquake demands that are 2/3 of MCE_R ground motions. As set forth in Section 11.4.4, two additional parameters S_{DS} and S_{D1} are used to define design spectral accelerations.

Values of the seismic parameters S_{DS} and S_{D1} ($2/3S_{MS}$ and $2/3S_{M1}$) are provided online by the USGS Seismic Design Web Service for user-specified site location (latitude and longitude) and site class. Values of the seismic parameters S_{DS} and S_{D1} provided by the USGS are based on the multiperiod design response spectrum (Section 11.4.5.1) of the site of interest and the requirements of Section 21.4 (for determining values of S_{DS} and S_{D1} from a site-specific design response spectrum).

C11.4.5 Design Response Spectrum The design response spectrum, (and MCE_R response spectrum of Section 11.4.6) are defined by either (1) a multi-period response spectrum (Section 11.4.5.1) or (2) a two-period response spectrum (Section 11.4.5.2), unless the design is based on site-specific ground motions (Section 21.3). The multi-period design response spectrum provides a more accurate representation of the frequency content of design ground motions and is the preferred characterization of spectral response. The shape of the two-period design response spectrum is the same as that of ASCE 7-16, which relies on a simpler characterization of the frequency content of design ground motions (Figure 11.4-1) based on the values of seismic parameters S_{DS} and S_{D1} (and T_L). The two-period design response spectrum is retained in ASCE 7-22 as an alternative characterization of ground motions for design where multi-period spectra are not available (e.g., from the USGS).

C11.4.5.1 Multi-Period Design Response Spectrum Sets of multi-period MCE_R response spectra (5% damping) at 22 response periods (0.0 s, 0.01 s, 0.02 s, 0.03 s, 0.05 s, 0.075 s, 0.1 s, 0.15 s, 0.2 s, 0.25 s, 0.3 s, 0.4 s, 0.5 s, 0.75 s, 1.0 s, 1.5 s, 2.0 s, 3.0 s, 4.0 s, 5.0 s, 7.5 s, and 10 s) are archived in the USGS Seismic Design Geodatabase at gridded locations across United States regions of interest. A USGS Seismic Design Web Service, for the site location and site class of interest, spatially interpolates between the gridded sets of multi-period MCE_R response spectra based on site location (latitude and longitude). Multi-period design response spectrum is constructed from two-thirds of these values by linear interpolation for response periods less than 10 s and by extrapolation for response periods greater than 10 s.

At response periods beyond 10 s, values of the multi-period design response spectrum are assumed to decrease from the value of the design response spectrum value at 10 s as the inverse of the period, T, where T is less than T_L and/or as inverse of the square of the period, T^2, where T is greater than T_L, essentially following the same approach as that used to construct the two-period spectrum (Section 11.4.5.2) at very long-periods.

C11.4.5.2 Two-Period Design Response Spectrum The two-period design response spectrum (Figure 11.4-1) consists of several segments. The constant-acceleration segment covers the period band from T_0 to T_s; response accelerations in this band are constant and equal to S_{DS}. The constant-velocity segment covers the period band from T_s to T_L, and the response accelerations in this band are proportional to $1/T$ with the response acceleration at a 1-s period equal to S_{D1}. The long-period portion of the design response spectrum is defined on the basis of the parameter T_L, the period that marks the transition from the constant-velocity segment to the constant-displacement segment of the design response spectrum. Response accelerations in the constant-displacement segment, where $T \geq T_L$, are proportional to $1/T^2$. Values of T_L are provided on maps in Figures 22-14 through 22-17.

The T_L maps were prepared following a two-step procedure. First, a correlation between earthquake magnitude and T_L was established. Then, the modal magnitude from deaggregation of the ground-motion seismic hazard at a 2 s period (a 1 s period for Hawaii) was mapped. Details of the procedure and the rationale for it are found in Crouse et al. (2006).

C11.4.7 Site-Specific Ground Motion Procedures Site-specific ground motions are permitted for design of any structure and are required for design of certain structures and certain site soil conditions. The objective of a site-specific ground motion analysis is to determine ground motions for local seismic and site conditions with higher confidence than is possible using the general procedure of Section 11.4.

As noted earlier, the site-specific procedures of Chapter 21 are the same as those used by the US Geological Survey (USGS) to develop the mapped values of MCE_R ground motions. Unless significant differences in local seismic and site conditions are determined by a site-specific analysis of earthquake hazards, site-specific ground motions would not be expected to differ significantly from those of the mapped values of MCE_R ground motions prepared by the USGS.

C11.5 IMPORTANCE FACTOR AND RISK CATEGORY

Large earthquakes are rare events that include severe ground motions. Such events are expected to result in damage to structures even if they were designed and built in accordance with the minimum requirements of the standard. The consequence of structural damage or failure is not the same for the various types of structures located within a given community. Serious damage to certain classes of structures, such as critical facilities (e.g., hospitals), disproportionately affects a community. The fundamental purpose of this section and of subsequent requirements that depend on this section is to improve the ability of a community to recover from a damaging earthquake by tailoring the seismic protection requirements to the relative importance of a structure. That purpose is achieved by requiring improved performance for structures that

1. Are necessary to response and recovery efforts immediately after an earthquake,
2. Present the potential for catastrophic loss in the event of an earthquake, or
3. House a large number of occupants or occupants less able to care for themselves than the average.

The first basis for seismic design in the standard is that structures should have a suitably low likelihood of collapse in the rare events defined as the maximum considered earthquake (MCE) ground motion. A second basis is that life-threatening damage, primarily from failure of nonstructural components in and on structures, is unlikely in a design earthquake ground motion (defined as two-thirds of the MCE). Given the occurrence of ground motion equivalent to the MCE, a population of structures built to meet these design objectives probably still experiences substantial damage in many structures, rendering these structures unfit for occupancy or use. Experience in past earthquakes around the world has demonstrated that there is an immediate need to treat injured people, to extinguish fires and prevent conflagration, to rescue people from severely damaged or collapsed structures, and to provide shelter and sustenance to a population deprived of its normal means. These needs are best met when structures essential to response and recovery activities remain functional.

This standard addresses these objectives by requiring that each structure be assigned to one of the four risk categories presented in Chapter 1 and by assigning an Importance Factor, I_e, to the structure based on that risk category. (The two lowest categories, I and II, are combined for all purposes within the seismic provisions.) The risk category is then used as one of two components in determining the seismic design category (SDC, see Section C11.6) and is a primary factor in setting drift limits for building structures under the design earthquake ground motion (see Section C12.12).

Figure C11.5-1 shows the combined intent of these requirements for design. The vertical scale is the likelihood of the ground motion; the MCE is the rarest considered. The horizontal scale is the level of performance intended for the structure and attached nonstructural components, which range from Collapse Prevention to Operational. The basic objective of Collapse Prevention at the MCE for ordinary structures (Risk Category II) is shown at the lower right by the solid triangle; protection from life-threatening damage at the design earthquake ground motion (defined by the standard as two-thirds of the MCE) is

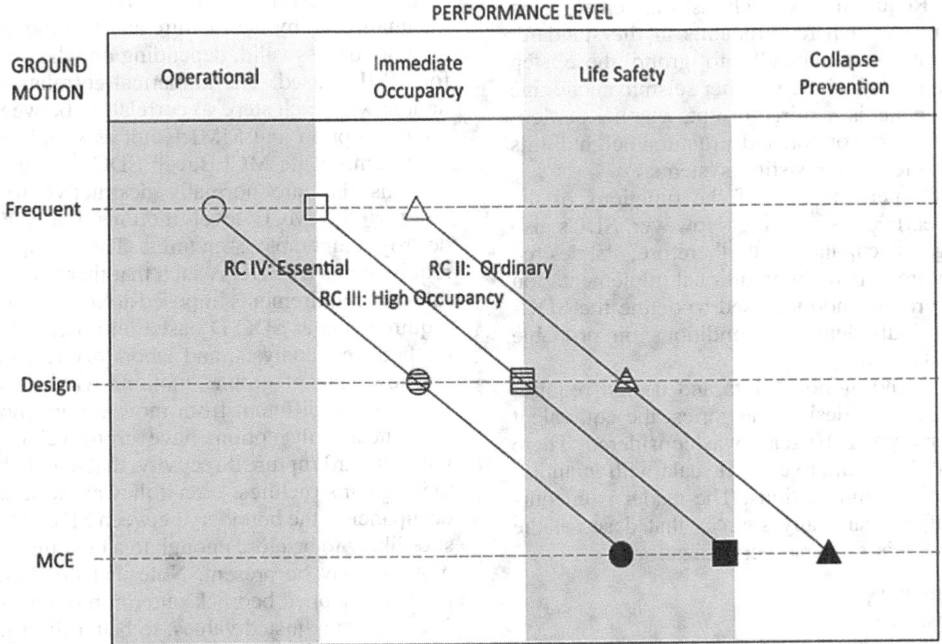

Figure C11.5-1. Expected performance as related to risk category and level of ground motion.

shown by the hatched triangle. The performance implied for the higher Risk Categories III and IV is shown by squares and circles, respectively. The performance anticipated for less severe ground motion is shown by open symbols.

C11.5.1 Importance Factor The Importance Factor, I_e, is used throughout the standard in quantitative criteria for strength. In most of those quantitative criteria, the Importance Factor is shown as a divisor on the factor R or R_p to reduce damage for important structures in addition to preventing collapse in larger ground motions. The R and R_p factors adjust the computed linear elastic response to a value appropriate for design; in many structures, the largest component of that adjustment is ductility (the ability of the structure to undergo repeated cycles of inelastic strain in opposing directions). For a given strength demand, reducing the effective R factor (by means of the Importance Factor) increases the required yield strength, thus reducing ductility demand and related damage.

C11.5.2 Protected Access for Risk Category IV Those structures considered Essential Facilities for response and recovery efforts must be accessible to carry out their purpose. For example, if the collapse of a simple canopy at a hospital could block ambulances from the emergency room admittance area, then the canopy must meet the same structural standard as the hospital. The protected access requirement must be considered in the siting of Essential Facilities in densely built urban areas.

C11.6 SEISMIC DESIGN CATEGORY

Seismic design categories (SDCs) provide a means to step progressively from simple, easily performed design and construction procedures and minimums to more sophisticated, detailed, and costly requirements as both the level of seismic hazard and the consequence of failure escalate. The SDCs are used to trigger requirements that are not scalable; such requirements are either on or off. For example, the basic amplitude of ground motion for design is scalable—the quantity simply increases in a continuous fashion as one moves from a low hazard area to a high hazard area. However, a requirement to avoid weak stories is not particularly scalable. Requirements such as this create step functions. There are many such requirements in the standard, and the SDCs are used systematically to group these step functions. (Further examples include whether seismic anchorage of nonstructural components is required or not, whether particular inspections will be required or not, and structural height limits applied to various seismic force-resisting systems.)

In this regard, SDCs perform one of the functions of the seismic zones used in earlier US building. However, SDCs also depend on a building's occupancy and, therefore, its desired performance. Furthermore, unlike the traditional implementation of seismic zones, the ground motions used to define the SDCs include the effects of individual site conditions on probable ground-shaking intensity.

In developing the ground-motion limits and design requirements for the various seismic design categories, the equivalent modified Mercalli intensity (MMI) scale was considered. There are now correlations of the qualitative MMI scale with quantitative characterizations of ground motions. The reader is encouraged to consult any of a great many sources that describe the MMIs. The following list is a coarse generalization:

MMI V: No real damage
MMI VI: Light nonstructural damage
MMI VII: Hazardous nonstructural damage
MMI VIII: Hazardous damage to susceptible structures
MMI IX: Hazardous damage to robust structures

When the current design philosophy was adopted from the 1997 NEHRP provisions and commentary (FEMA 1997a, b), the upper limit for SDC A was set at roughly one-half of the lower threshold for MMI VII, and the lower limit for SDC D was set at roughly the lower threshold for MMI VIII. However, the lower limit for SDC D was more consciously established by equating that design value (two-thirds of the MCE) to one-half of what had been the maximum design value in building codes over the period of 1975 to 1995. As more correlations between MMI and numerical representations of ground motion have been created, it is reasonable to make the following correlation between the MMI at MCE ground motion and the seismic design category (all this discussion is for ordinary occupancies):

MMI V: SDC A
MMI VI: SDC B
MMI VII: SDC C
MMI VIII: SDC D
MMI IX: SDC E

An important change was made to the determination of SDC when the current design philosophy was adopted. Earlier editions of the NEHRP provisions used the peak velocity-related acceleration, A_v, to determine a building's seismic performance category. However, this coefficient does not adequately represent the damage potential of earthquakes on sites with soil conditions other than rock. Consequently, the 1997 NEHRP provisions (FEMA 1997a) adopted the use of response spectral acceleration parameters S_{DS} and S_{D1}, which include site soil effects for this purpose.

Except for the lowest level of hazard (SDC A), the SDC also depends on the Risk Categories. For a given level of ground motion, the SDC is one category higher for Risk Category IV structures than for lower risk structures. This rating has the effect of increasing the confidence that the design and construction requirements can deliver the intended performance in the extreme event.

Note that the tables in the standard are at the design level, defined as two-thirds of the MCE level. Also recall that the MMIs are qualitative by their nature and that the above correlation will be more or less valid, depending on which numerical correlation for MMI is used. The numerical correlations for MMI roughly double with each step, so correlation between design earthquake ground motion and MMI is not as simple or convenient.

In sum, at the MCE level, SDC A structures should not see motions that are normally destructive to structural systems, whereas the MCE level motions for SDC D structures can destroy vulnerable structures. The grouping of step function requirements by SDC is such that there are a few basic structural integrity requirements imposed at SDC A, graduating to a suite of requirements at SDC D based on observed performance in past earthquakes, analysis, and laboratory research.

The nature of ground motions within a few kilometers of a fault can be different from more distant motions. For example, some near-fault motions have strong velocity pulses, associated with forward rupture directivity, that tend to be highly destructive to irregular structures, even if they are well detailed. For ordinary occupancies, the boundary between SDCs D and E is set to define sites likely to be close enough to a fault that these unusual ground motions may be present. Note that this boundary is defined in terms of mapped bedrock outcrop motions affecting response at 1 s, not site-adjusted values, to better discriminate between sites near and far from faults. Short-period response is not normally as affected as the longer period response. The additional design criteria imposed on structures in SDCs E and F specifically are intended to provide acceptable performance under these very intense near-fault ground motions.

For most buildings, the SDC is determined without consideration of the building's period. Structures are assigned to an SDC based on the more severe condition determined from 1 s acceleration and short-period acceleration. This assigning is done for several reasons. Perhaps the most important of these is that it is often difficult to estimate precisely the period of a structure using the default procedures contained in the standard. Consider, for example, the case of rigid wall–flexible diaphragm buildings, including low-rise reinforced masonry and concrete tilt-up buildings with either untopped metal deck or wood diaphragms. The formula in the standard for determining the period of vibration of such buildings is based solely on the structural height, h_n, and the length of wall present. These formulas typically indicate very short periods for such structures, often on the order of 0.2 s or less. However, the actual dynamic behavior of these buildings often is dominated by the flexibility of the diaphragm—a factor neglected by the formula for approximate fundamental period. Large buildings of this type can have actual periods on the order of 1 s or more. To avoid misclassifying a building's SDC by inaccurately estimating the fundamental period, the standard generally requires that the more severe SDC determined on the basis of short- and long-period shaking be used.

Another reason for this requirement is a desire to simplify building regulation by requiring all buildings on a given soil profile in a particular region to be assigned to the same SDC, regardless of the structural type. This assignment has the advantage of permitting uniform regulation in the selection of seismic force-resisting systems, inspection and testing requirements, seismic design requirements for nonstructural components, and similar aspects of the design process regulated on the basis of SDC, within a community.

Notwithstanding the above, it is recognized that classification of a building as SDC C instead of B or D can have a significant impact on the cost of construction. Therefore, the standard includes an exception permitting the classification of buildings that can reliably be classified as having short structural periods on the basis of short-period shaking alone.

Local or regional jurisdictions enforcing building regulations may desire to consider the effect of the maps, typical soil conditions, and seismic design categories on the practices in their jurisdictional areas. For reasons of uniformity of practice or reduction of potential errors, adopting ordinances could stipulate particular values of ground motion, particular site classes, or particular seismic design categories for all or part of the area of their jurisdiction. For example,

- An area with a historical practice of high seismic zone detailing might mandate a minimum SDC of D regardless of ground motion or site class.
- A jurisdiction with low variation in ground motion across the area might stipulate particular values of ground motion rather than requiring the use of maps.
- An area with unusual soils might require use of a particular site class unless a geotechnical investigation proves a better site class.

C11.7 DESIGN REQUIREMENTS FOR SEISMIC DESIGN CATEGORY A

The 2002 edition of the standard included a new provision of minimum lateral force for Seismic Design Category A structures. The minimum load is a structural integrity issue related to the load path. It is intended to specify design forces in excess of wind loads in heavy low-rise construction. The design calculation in Section 1.4.2 of the standard is simple and easily done to ascertain if the seismic load or the wind load governs. This provision requires a minimum lateral force of 1% of the total gravity load assigned to a story to ensure general structural integrity.

Seismic Design Category A is assigned when the MCE ground motions are below those normally associated with hazardous damage. Damaging earthquakes are not unknown or impossible in such regions, however, and ground motions close to such events may be large enough to produce serious damage. Providing a minimum level of resistance reduces both the radius over which the ground motion exceeds structural capacities and the resulting damage in such rare events. There are reasons beyond seismic risk for minimum levels of structural integrity.

The requirements for SDC A in Section 1.4 are all minimum strengths for structural elements stated as forces at the level appropriate for direct use in the strength design load combinations of Section 2.3. The two fundamental requirements are a minimum strength for a structural system to resist lateral forces (Section 1.4.2) and a minimum strength for connections of structural members (Section 1.4.3).

For many buildings, the wind force controls the strength of the lateral-force-resisting system, but for low-rise buildings of heavy construction with large plan aspect ratios, the minimum lateral force specified in Section 1.4.2 may control. Note that the requirement is for strength and not for toughness, energy-dissipation capacity, or some measure of ductility. The force level is not tied to any postulated seismic ground motion. The boundary between SDCs A and B is based on a spectral response acceleration of 25% of gravity (MCE level) for short-period structures; clearly the 1% acceleration level [from Equation (1.4-1)] is far smaller. For ground motions below the A/B boundary, the spectral displacements generally are on the order of a few inches or less depending on period. Experience has shown that even a minimal strength is beneficial in providing resistance to small ground motions, and it is an easy provision to implement in design. The low probability of motions greater than the MCE is a factor in taking the simple approach without requiring details that would produce a ductile response. Another factor is that larger design forces are specified in Section 1.4.3 for connections between main elements of the lateral force load path.

The minimum connection force is specified in three ways: a general minimum horizontal capacity for all connections; a special minimum for horizontal restraint of in-line beams and trusses, which also includes the live load on the member; and a special minimum for horizontal restraint of concrete and masonry walls perpendicular to their plane (Section 1.4.4). The 5% coefficient used for the first two is a simple and convenient value that provides some margin over the minimum strength of the system as a whole.

C11.8 GEOLOGIC HAZARDS AND GEOTECHNICAL INVESTIGATION

In addition to this commentary, Part 3 of the 2009 *NEHRP Recommended Provisions* (FEMA 2009) includes additional and more detailed discussion and guidance on evaluation of geologic hazards and determination of seismic lateral pressures.

C11.8.1 Site Limitation for Seismic Design Categories E and F Because of the difficulty of designing a structure for the direct shearing displacement of fault rupture and the relatively high seismic activity of SDCs E and F, locating a structure on an active fault that has the potential to cause rupture of the ground surface at the structure is prohibited.

C11.8.2 Geotechnical Investigation Report Requirements for Seismic Design Categories C through F Earthquake motion is only one factor in assessing potential for geologic and seismic hazards. All of the listed hazards can lead to surface ground displacements with potential adverse consequences to structures. Finally, hazard identification alone has little value unless mitigation options are also identified.

C11.8.3 Additional Geotechnical Investigation Report Requirements for Seismic Design Categories D through F Provisions for computing peak ground acceleration (PGA) for soil liquefaction and stability evaluations were introduced in this section in ASCE 7-16. Of particular note in this section is the explicitly stated requirement that liquefaction must now be evaluated for maximum considered earthquake geometric mean (MCE_G) peak ground acceleration (PGA_M), where the parameter PGA_M includes site effects. Values of the parameter PGA_M are archived in the USGS Seismic Design Geodatabase at gridded locations across United States regions of interest. Values are provided online by the USGS Seismic Design Web Service for user-specified site location (latitude and longitude) and site class, by spatially interpolating between the gridded values of PGA_M based on site location. Mapped values of PGA_M are provided in Chapter 22 for default site conditions.

PGA Provisions. Item 2 of Section 11.8.3 states that Peak Ground Acceleration (PGA) shall be determined based on either a site-specific study, taking into account soil amplification effects, or from the USGS Seismic Design Geodatabase via the USGS Seismic Design Web Service, for the site location and site class of interest. This methodology for determining Peak Ground Acceleration for liquefaction provides an alternative to conducting site response analysis using rock PGA by providing a site-adjusted ground surface acceleration (PGA_M) that can directly be applied in the widely used empirical correlations for assessing liquefaction potential. Correlations for evaluating liquefaction potential are elaborated on in Resource Paper RP 12, "Evaluation of Geologic Hazards and Determination of Seismic Lateral Earth Pressures," published in the 2009 NEHRP provisions (FEMA 2009).

There is an important difference in the derivation of the PGA maps and the maps of S_s and S_1 in ASCE 7-10. Unlike previous editions of ASCE 7, the S_s and S_1 maps in ASCE 7-10 were derived for the "maximum direction shaking" and are risk based rather than hazard based. However, the PGA maps have been derived based on the geometric mean of the two horizontal components of motion. The geometric mean was used in the PGA maps rather than the PGA for the maximum direction shaking to ensure that there is consistency between the determination of PGA and the basis of the simplified empirical field procedure for estimating liquefaction potential based on results of standard penetration tests (SPTs), cone penetrometer tests (CPTs), and other similar field investigative methods. When these correlations were originally derived, the geomean (or a similar metric) of peak ground acceleration at the ground surface was used to identify the cyclic stress ratio for sites with or without liquefaction. The resulting envelopes of data define the liquefaction cyclic resistance ratio (CRR). Rather than reevaluating these case histories for the "maximum direction shaking," it was decided to develop maps of the geomean PGA and to continue using the existing empirical methods.

Liquefaction Evaluation Requirements. Beginning with ASCE 7-02, it has been the intent that liquefaction potential be evaluated at MCE ground motion levels. There was ambiguity in the previous requirement in ASCE 7-05 as to whether liquefaction potential should be evaluated for the MCE or for the design earthquake. Paragraph 2 of Section 11.8.3 of ASCE 7-05 stated that liquefaction potential would be evaluated for the design earthquake; it also stated that in the absence of a site-specific study, peak ground acceleration shall be assumed to be equal to $S_s/2.5$ (S_s is the MCE short-period response spectral acceleration on Site Class B rock). There has also been a difference in provisions between ASCE 7-05 and the 2006 edition of the International Building Code, in which Section 1802.2.7 stated that liquefaction shall be evaluated for the design earthquake ground motions, and the default value of peak ground acceleration in the absence of a site-specific study was given as $S_{DS}/2.5$ (S_{DS} is the short-period site-adjusted design response spectral acceleration). Item 2 of Section 11.8.3, require explicitly that liquefaction potential and other effects be evaluated based on the MCE_G peak ground acceleration.

The explicit requirement in ASCE 7-10 and ASCE 7-16 to evaluate liquefaction for MCE ground motion, rather than to design earthquake ground motion, ensures that the full potential for liquefaction is addressed during the evaluation of structure stability, rather than a lesser level when the design earthquake is used. This change also ensures that, for the MCE ground motion, the performance of the structure is considered under a consistent hazard level for the effects of liquefaction, such as Collapse Prevention or Life Safety, depending on the risk category for the structure (Figure C11.5-1). By evaluating liquefaction for the MCE rather than the design earthquake peak ground acceleration, the ground motion for the liquefaction assessment increases by a factor of 1.5. This increase in peak ground acceleration to the MCE level means that sites that previously were nonliquefiable could now be liquefiable, and sites where liquefaction occurred to a limited extent under the design earthquake could undergo more liquefaction, in terms of depth and lateral extent. Some mechanisms that are directly related to the development of liquefaction, such as lateral spreading and flow or ground settlement, could also increase in severity.

This change in peak ground acceleration level for the liquefaction evaluation addressed an issue that has existed and has periodically been discussed since the design earthquake concept was first suggested in the 1990s. The design earthquake ground motion was obtained by multiplying the MCE ground motion by a factor of $2/3$ to account for a margin in capacity in most buildings. Various calibration studies at the time of code development concluded that for the design earthquake, most buildings had a reserve capacity of more than 1.5 relative to collapse. This reserve capacity allowed the spectral accelerations for the MCE to be reduced using a factor of $2/3$, while still achieving safety from collapse. However, liquefaction potential is evaluated at the selected MCE_G peak ground acceleration and is typically determined to be acceptable if the factor of safety is greater than 1.0, meaning that there is no implicit safety margin on liquefaction potential. By multiplying peak ground acceleration by a factor of $2/3$, liquefaction would be assessed at an effective return period or probability of exceedance different than that for the MCE. However, ASCE 7-10 requires that liquefaction be evaluated for the MCE.

Item 3 of Section 11.8.3 lists the various potential consequences of liquefaction, seismically induced permanent ground displacement, and soil strength loss that must be assessed; soil downdrag and loss in lateral soil reaction for pile foundations are additional consequences that have been included in this paragraph. This section of the new provisions, as in previous editions, does not present specific seismic criteria for the design of the foundation or substructure, but item 4 does state that the geotechnical report must include discussion of possible measures to mitigate these consequences.

A liquefaction resource document has been prepared in support of these revisions to Section 11.8.3. The resource document "Evaluation of Geologic Hazards and Determination of Seismic Lateral Earth Pressures" includes a summary of methods that are currently being used to evaluate liquefaction potential and the limitations of these methods. This summary appears as Resource Paper RP 12 in the 2009 NEHRP provisions (FEMA 2009). The resource document summarizes alternatives for evaluating liquefaction potential, methods for evaluating the possible consequences of liquefaction (e.g., loss of ground support and increased lateral earth pressures), and methods of mitigating the liquefaction hazard. The resource document also identifies alternate methods of evaluating liquefaction hazards, such as analytical and physical modeling. Reference is made to the use of nonlinear effective stress methods for modeling the buildup in pore water pressure during seismic events at liquefiable sites.

Evaluation of Dynamic Seismic Lateral Earth Pressures. The dynamic lateral earth pressure on basement and retaining walls during earthquake ground shaking is considered to be an earthquake load, E, for use in design load combinations. This dynamic earth pressure is superimposed on the preexisting static lateral earth pressure during ground shaking. The preexisting static lateral earth pressure is considered to be an H load.

C11.9 VERTICAL GROUND MOTIONS FOR SEISMIC DESIGN

C11.9.1 General The provisions for developing vertical spectra apply in the western US because that is the main region for which models for vertical-component response spectra are available. The boundary line of -105 degrees longitude comes from the approximate eastern limit of ground motion models for active tectonic regions, as given in Figure 1.1 of Goulet et al. (2017).

ASCE 7-16 included recommendation for developing vertical spectra in the central and eastern United States as presented by EPRI (2015, Appendix A); those recommendations apply to relatively old versions of western United States models for applications in the east. Since the application of western models in the central and eastern United States has not been demonstrated, a simple two-thirds rule is suggested in lieu of the more complex model in these provisions.

C11.9.2 MCE$_R$ Vertical Response Spectrum Recent studies of horizontal and vertical ground motions (e.g., Bozorgnia and Campbell 2004, 2016a, 2016b; Gülerce et al. 2017; Stewart et al. 2016) have shown that vertical ground motion is different from horizontal ground motion in several important respects: (1) vertical ground motion has a larger proportion of short-period (high-frequency) spectral content than horizontal ground motion, and this difference increases with decreasing soil stiffness; (2) vertical ground motion attenuates at a higher rate than horizontal ground motion, and this difference increases with decreasing distance from the earthquake; and (3) the nonlinear component of site response is stronger in the horizontal component than in the vertical component, which causes the vertical/horizontal (V/H) spectral ratio to exceed unity for soil sites at close distance to large faults where nonlinear effects are significant. The observed differences in the spectral content and attenuation rate of vertical and horizontal ground motion lead to the following observations regarding the V/H spectral ratio:

1. The V/H spectral ratio is sensitive to spectral period, distance from the earthquake, local site conditions, and earthquake magnitude and is insensitive to earthquake mechanism and sediment depth;
2. The V/H spectral ratio has a distinct peak at short periods that generally exceeds $2/3$ in the near-source region of an earthquake; and
3. The V/H spectral ratio is generally less than $2/3$ at mid-to-long periods.

The procedure for defining the MCE$_R$ vertical response is keyed to the MCE$_R$ spectral response acceleration parameter, S_{aM}. The procedure is based on the studies of horizontal and vertical ground motions by Campbell and Bozorgnia (2003) and Bozorgnia and Campbell (2004), and a series of models generated in the NGA-West2 project (Bozorgnia and Campbell 2016a, b; Gülerce et al. 2017, Stewart et al. 2016).

The specification of vertical ground motions in Section 11.9.2 is based on the product of the multi-period risk-targeted maximum considered earthquake ground motion (S_{aM}) and a simplified representation of the V/H spectral ratio that has five regions defined by the vertical period of vibration (T_v). Based on the study of Bozorgnia and Campbell (2004), the periods that define these regions are approximately constant with respect to the magnitude of the earthquake, the distance from the earthquake, and the local site conditions.

The MCE$_R$ is based on maximum direction parameters, so it must be converted back to the geometric mean component to be consistent with the studies referenced above. The NGA-West2 models used geometric mean component horizontal spectral parameters to compute the ratio of vertical to horizontal. To reduce the S_{aM} from the maximum direction to the geometric mean component, S_{aM} is divided by a factor, F_{md} [Equations (11.9-1) to (11.9-5)]. Those F_{md} factors are consistent with the commentary of Chapter 21, Resource Paper 4 in the 2015 provisions, and Shahi and Baker (2014).

The equations in Section 11.9.2 that are used to define the design vertical response spectrum are based on four considerations (adapted from Bozorgnia and Campbell 2004):

1. The short-period part of the 5% damped vertical response spectrum is controlled by the spectral acceleration at $T_v = 0.1$ s;
2. The mid-period part of the vertical response spectrum is controlled by a spectral acceleration that decays as the inverse of a power of the vertical period of vibration. This was taken as $T_v^{-0.75}$ in the 2009 NEHPR provisions and has been updated to $T_v^{-0.5}$;
3. The short-period part of the V/H spectral ratio is a function of the local site conditions (i.e., V_{S30}) and the level of seismic demand (represented in Table 11.9-1 by parameter S_s); and
4. For vertical vibration periods $T_v >$ about 0.3 to 0.5 sec, V/H spectral ratios saturate to values typically less than 0.5 that are relatively consistent with respect to period across this period range. For simplicity, V/H spectral ratios are taken as 0.5 in this range of periods.

The following description of the detailed procedure listed in Section 11.9.2 refers to the illustrated MCE$_R$ vertical response spectrum in Figure C11.9-1.

Vertical Periods Less than or Equal to 0.025 s. Equation (11.9-1) defines that part of the MCE$_R$ vertical response spectrum that is controlled by the vertical peak ground acceleration. The f_1 factor (taken as 0.65) was selected to approximately match V/H ordinates from recent NGA-West2 models for soil

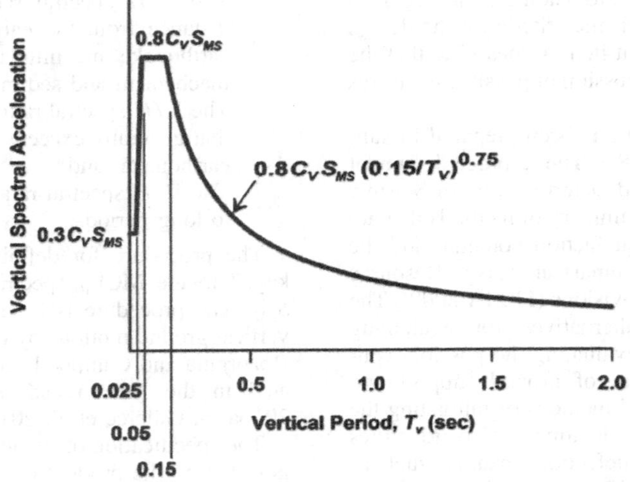

Figure C11.9-1. Illustrative example of the vertical response spectrum.

sites classes (it is somewhat unconservative in this period range for rock sites). The vertical coefficient, C_v, in Table 11.9-1 accounts for the site dependence of V/H ordinates.

Vertical Periods Greater than 0.025 s and Less than or Equal to 0.05 s. Equation (11.9-2) defines that part of the MCE_R vertical response spectrum for which the V/H ratio linearly transitions from the part of the spectrum that is controlled by the vertical peak ground acceleration to the part that is controlled by the dynamically amplified short-period spectral plateau. The factor of 16 is required to provide appropriate levels of amplification at the peak of the V/H spectral ratio plot.

Vertical Periods Greater than 0.05 s and Less than or Equal to 0.1 s. Equation (11.9-3) defines that part of the MCE_R vertical response spectrum for which the V/H spectral ratio is dynamically amplified to a short-period plateau at amplitude f_2 in Figure C11.9-2. The width of the peak from 0.05 to 0.1 s is best suited to soil sites (Classes C to DE), being conservative for rock sites (Classes BC to A).

Vertical Periods Greater than 0.1 s and Less than or Equal to 2.0 s. Equation (11.9-4) defines that part of the MCE_R vertical response spectrum for which the V/H spectral ratio decays with the inverse of the vertical period of vibration raised to the $-f_3$ power (currently taken as -0.5, formerly -0.75 in ASCE 7-16). This portion of the spectrum was constructed in a generally conservative manner, as two of the three NGA-West2 models suggest that the period range of post-peak decay is steeper than implied by the -0.5 power, with approximately flat V/H ratios at periods longer than about 0.3 to 0.5 s. The flat V/H ratios at periods beyond 0.3-0.5 s are typically less than 0.5, and a limiting value of 0.5 is suggested in the absence of site-specific analysis. This limit of 0.5 is considered a reasonable, but somewhat conservative, lower bound (Campbell and Bozorgnia 2003, Bozorgnia and Campbell 2004).

Vertical Periods Greater than 2.0 s and Less than or Equal to 10.0 s. Equation (11.9-5) defines that part of the MCE_R vertical response spectrum for which the V/H spectral ratio is roughly constant. A recommended lower limit of 0.5 is provided for this range.

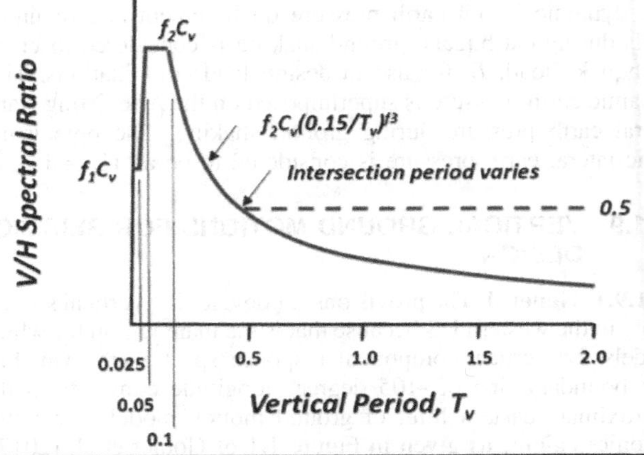

Figure C11.9-2. Illustrative example of the vertical/horizontal spectral ratio.

REFERENCES

ASCE. 2017. *Seismic evaluation and retrofit of existing buildings*. ASCE/SEI 41-17. Reston, VA: ASCE.

Bozorgnia, Y., and K. W. Campbell. 2004. "The vertical-to-horizontal response spectral ratio and tentative procedures for developing simplified V/H and vertical design spectra." *J. Earthquake Eng.* 8 (2): 175–207.

Bozorgnia, Y., and K. W. Campbell. 2016a. "Ground motion model for the vertical-to-horizontal (V/H) ratios of PGA, PGV, and response spectra." *Earthquake Spectra* 32 (2): 951–978. https://doi.org/10.1193/100614eqs151m.

Bozorgnia, Y., and K. W. Campbell. 2016b. "Vertical ground motion model for PGA, PGV, and linear response spectra using the NGA-West2 database." *Earthquake Spectra* 32 (2): 979–1004. https://doi.org/10.1193/072814eqs121m.

Campbell, K. W., and Y. Bozorgnia. 2003. "Updated near-source ground motion (attenuation) relations for the horizontal and vertical components of peak ground acceleration and acceleration response spectra." *Bull. Seismol. Soc. Am.* 93 (1): 314–331. https://doi.org/10.1785/0120020029.

Crouse, C. B., E. V. Leyendecker, P. G. Somerville, M. Power, and W. J. Silva. 2006. "Development of seismic ground-motion criteria for the ASCE/SEI 7 standard." In *Proc., 8th US Nat. Conf. on Earthquake Engineering*, San Francisco.

EPRI (Electric Power Research Institute). 2015. *High frequency program: Application guidance for functional confirmation and fragility evaluation*. Technical Rep. No. 3002004396. Washington, DC: EPRI.

Executive Order 12699. 2016. “Establishing a federal earthquake risk management standard.” http://www.whitehouse.gov/the-press-office/2016/02/02/executive-order-establishing-federal-earthquake-risk-management-standard.

FEMA (Federal Emergency Management Agency). 1997a. *NEHRP recommended provisions for seismic regulations in new buildings and other structures*. FEMA 302. Washington, DC: FEMA.

FEMA. 1997b. *NEHRP recommended provisions for seismic regulations in new buildings and other structures: Part 2, commentary*. FEMA 303. Washington, DC: FEMA.

FEMA. 2009. *NEHRP recommended seismic provisions for new buildings and other structures*. FEMA P-750. Washington, DC: FEMA.

FEMA. 2015. *NEHRP recommended seismic provisions for new buildings and other structures. Vol. II, part 3: Resource papers*. FEMA P-1050-2. Washington, DC: FEMA.

FEMA. 2020a. *NEHRP recommended seismic provisions for new buildings and other structures*. FEMA P-2082. Washington, DC: FEMA.

FEMA. 2020b. *Procedures for developing multi-period response spectra at non-conterminous United States sites*. FEMA P-2078. Washington, DC: FEMA.

Goulet, C. A., Y. Bozorgnia, N. Kuehn, L. Al Atik, R. R. Youngs, R. W. Graves, and G. M. Atkinson. 2017. *NGA-east ground-motion models for the U.S. Geological Survey national seismic hazard maps*. Rep. No. 2017/03. Berkeley, CA: Pacific Earthquake Engineering Research Center.

Gülerce, Z., R. Kamai, N. A. Abrahamson, and W. J. Silva. 2017. “Ground motion prediction equations for the vertical ground motion component based on the NGA-W2 database.” *Earthquake Spectra* 33 (2): 499–528. https://doi.org/10.1193/121814EQS213M.

ICBO (International Conference of Building Officials). 1997. *Uniform building code*. Whittier, CA: ICBO.

Kircher & Associates. 2015. *Investigation of an identified shortcoming in the seismic design procedures of ASCE 7-10 and development of recommended improvements for ASCE 7-16*. Palo Alto, CA: Kircher & Associates.

NIBS (National Institute of Buildings Sciences). 2019. *BSSC project 17 final report: Development of next generation of seismic design value maps for the 2020 NEHRP provisions*. Washington, DC: NIBS.

NIST (National Institute of Standards and Technology). 2017. *ICSSC recommended practice 9, implementation guidelines for executive order 13717: Establishing a federal earthquake risk management standard: NIST technical note 1922*. Gaithersburg, MD: NIST.

SEAOC (Structural Engineers Association of California). 1999. *Recommended lateral force requirements and commentary*. Sacramento, CA: SEAOC.

Shahi, S. K., and J. W. Baker. 2014. “NGA-West2 models for ground motion directionality.” *Earthquake Spectra* 30 (3): 1285–1300. https://doi.org/10.1193/040913EQS097M.

Silva, W. 1997. *Characteristics of vertical strong ground motions for applications to engineering design. FHWA/NCEER workshop on the national representation of seismic ground motion for new and existing highway facilities*. Technical Rep. No. NCEER-97-0010. Buffalo, NY: National Center for Earthquake Engineering Research.

United States, Executive Office of the President. 2016. “Executive order 13717: Establishing a federal earthquake risk management standard.” *Fed. Regist.* 81 (24): 6405–6410. Accessed July 16, 2021. whitehouse.gov.

OTHER REFERENCES (NOT CITED)

Abrahamson, N. A. 2000. “Effects of rupture directivity on probabilistic seismic hazard analysis.” In *Proc., 6th Int. Conf. on Seismic Zonation*, Palm Springs, CA.

ASCE. 2017. *Seismic rehabilitation of existing buildings*. ASCE/SEI 41–17. Reston, VA: ASCE.

Borcherdt, R. D. 2002. “Empirical evidence for site coefficients in building-code provisions.” *Earthquake Spectra* 18 (2): 189–217. https://doi.org/10.1193/1.1486243.

Crouse, C. B., and J. W. McGuire. 1996. “Site response studies for purposes of revising NEHRP seismic provisions.” *Earthquake Spectra* 12 (3): 129–143. https://doi.org/10.1193/1.1585891.

Dobry, R., R. Ramos, and M. S. Power. 1999. *Site factors and site categories in seismic codes*. Technical Rep. No. MCEER-99-0010. Buffalo, NY: University of Buffalo.

Field, E. H. 2000. “A modified ground motion attenuation relationship for Southern California that accounts for detailed site classification and a basin depth effect.” *Bull. Seismol. Soc. of Am.* 90: S209–S221. https://doi.org/10.1785/0120000507.

Harmsen, S. C. 1997. “Determination of site amplification in the Los Angeles urban area from inversion of strong motion records.” *Bull. Seismol. Soc. of Am.* 87 (4): 866–887.

Huang, Y.-N., A. S. Whittaker, and N. Luco. 2008. “Orientation of maximum spectral demand in the near-fault region.” *Earthqauke Spectra* 24 (3): 319–341. https://doi.org/10.1193/1.3158997

Joyner, W. B., and D. M. Boore. 2000. “Recent developments in earthquake ground motion estimation.” In *Proc., 6th Int. Conf. on Seismic Zonation*, Palm Springs, CA.

Luco, N., B. R. Ellingwood, R. O. Hamburger, J. D. Hooper, J. K. Kimball, and C. A. Kircher. 2007. “Risk-targeted vs. current seismic design maps for the conterminous United States.” In *Proc., SEAOC 76th Annual Convention*, Sacramento, CA.

Petersen, M. D., et al. 2008. *Documentation for the 2008 update of the United States national seismic hazard maps*. USGS Open File Rep. No. 2008-1128. Washington, DC: USGS.

Rodriguez-Marek, A., J. D. Bray, and N. Abrahamson. 2001. “An empirical geotechnical site response procedure.” *Earthquake Spectra* 17 (1): 65–87. https://doi.org/10.1193/1.1586167.

Seyhan, E. 2014. *Weighted average of 2014 NGA West-2 GMPEs*. Berkeley, CA: Pacific Earthquake Engineering Center.

Silva, W., R. Darragh, N. Gregor, G. Martin, N. Abrahamson, and C. Kircher. 2000. *Reassessment of site coefficients and near-fault factors for building code provisions*. Rep. No. 98-HQ-GR-1010. Washington, DC: USGS.

Somerville, P. G., N. F. Smith, R. W. Graves, and N. A. Abrahamson. 1997. “Modification of empirical strong ground motion attenuation relations to include the amplitude and duration effects of rupture directivity.” *Seismol. Res. Lett.* 68 (1): 199–222. https://doi.org/10.1785/gssrl.68.1.199.

Steidl, J. H. 2000. “Site response in Southern California for probabilistic seismic hazard analysis.” *Bull. Seismol. Soc. Am.* 90 (6B): S149–S169. https://doi.org/10.1785/0120000504.

Stewart, J. P., D. M. Boore, E. Seyhan, and G. M. Atkinson. 2016. “NGA-West2 equations for predicting vertical-component PGA, PGV, and 5%-damped PSA from shallow crustal earthquakes.” *Earthquake Spectra* 32 (2): 1005–1031. https://doi.org/10.1193/072114eqs116m.

Stewart, J. P., A. H. Liu, and Y. Choi. 2003. “Amplification factors for spectral acceleration in tectonically active regions.” *Bull. Seismol. Soc. Am.* 93 (1): 332–352. https://doi.org/10.1785/0120020049.

CHAPTER C12
SEISMIC DESIGN REQUIREMENTS FOR BUILDING STRUCTURES

C12.1 STRUCTURAL DESIGN BASIS

The performance expectations for structures designed in accordance with this standard are described in Sections C11.1 and C11.5. Structures designed in accordance with the standard are likely to have a low probability of collapse but may suffer serious structural damage if subjected to the risk-targeted maximum considered earthquake (MCE_R) or stronger ground motion.

Although the seismic requirements of the standard are stated in terms of forces and loads, there are no external forces applied to the structure during an earthquake as, for example, is the case during a windstorm. The seismic design forces are intended only as approximations to generate internal forces suitable for proportioning the strength and stiffness of structural elements and for estimating the deformations (when multiplied by the deflection amplification factor, C_d) that would occur in the same structure in the event of design earthquake (not MCE_R) ground motion.

C12.1.1 Basic Requirements Chapter 12 of the standard sets forth a set of coordinated requirements that must be used together. The basic steps in structural design of a building structure for acceptable seismic performance are as follows.

1. Select gravity- and seismic force-resisting systems appropriate to the anticipated intensity of ground shaking. Section 12.2 sets forth limitations depending on the Seismic Design Category (SDC).
2. Configure these systems to produce a continuous, regular, and redundant load path so that the structure acts as an integral unit in responding to ground shaking. Section 12.3 addresses configuration and redundancy issues.
3. Analyze a mathematical model of the structure subjected to lateral seismic motions and gravity forces. Sections 12.6 and 12.7 set forth requirements for the method of analysis and for construction of the mathematical model. Sections 12.5, 12.8, and 12.9 set forth requirements for conducting a structural analysis to obtain internal forces and displacements.
4. Proportion members and connections to have adequate lateral and vertical strength and stiffness. Section 12.4 specifies how the effects of gravity and seismic loads are to be combined to establish required strengths, and Section 12.12 specifies deformation limits for the structure.

One to three-story structures with shear wall or braced frame systems of simple configuration may be eligible for design under the simplified alternative procedure in Section 12.14. Any other deviations from the requirements of Chapter 12 are subject to approval by the Authority Having Jurisdiction and must be rigorously justified, as specified in Section 11.1.4.

The baseline seismic forces used for proportioning structural elements (individual members, connections, and supports) are static horizontal forces derived from an elastic response spectrum procedure. A basic requirement is that horizontal motion can come from any direction relative to the structure, with detailed requirements for evaluating the response of the structure provided in Section 12.5. For most structures, the effect of vertical ground motions is not analyzed explicitly; it is implicitly included by adjusting the load factors (up and down) for permanent dead loads, as specified in Section 12.4. Certain conditions requiring more detailed analysis of vertical response are defined in Chapters 13 and 15 for nonstructural components and nonbuilding structures, respectively.

The basic seismic analysis procedure uses response spectra that are representative of, but substantially reduced from, the anticipated ground motions. As a result, at the MCE_R level of ground shaking, structural elements are expected to yield, buckle, or otherwise behave inelastically. This approach has substantial historical precedent. In past earthquakes, structures with appropriately ductile, regular, and continuous systems that were designed using *reduced* design forces have performed acceptably. In the standard, such design forces are computed by dividing the forces that would be generated in a structure behaving elastically when subjected to the design earthquake ground motion by the response modification coefficient, R, and this design ground motion is taken as two-thirds of the MCE_R ground motion.

The intent of R is to reduce the demand determined, assuming that the structure remains elastic at the design earthquake, to target the development of the first significant yield. This reduction accounts for the displacement ductility demand, R_d, required by the system and the inherent overstrength, Ω, of the seismic force-resisting system (SFRS) (Figure C12.1-1). Significant yield is the point where complete plastification of a critical region of the SFRS first occurs (e.g., formation of the first plastic hinge in a moment frame), and the stiffness of the SFRS to further increase in lateral forces decreases as continued inelastic behavior spreads within the SFRS. This approach is consistent with member-level ultimate strength design practices. As such, first significant yield should not be misinterpreted as the point where first yield occurs in any member (e.g., 0.7 times the yield moment of a steel beam or either initial cracking or initiation of yielding in a reinforcing bar in a reinforced concrete beam or wall).

Figure C12.1-1 shows the relation of lateral force versus deformation for an archetypal moment frame used as an SFRS. First significant yield is shown as the lowest plastic hinge on the force–deformation diagram. Because of particular design rules and limits, including material strengths in excess of nominal or project-specific design requirements, structural elements are stronger by some degree than the strength required by analysis. The SFRS is therefore expected to reach first significant yield for forces in excess of design forces. With increased lateral loading,

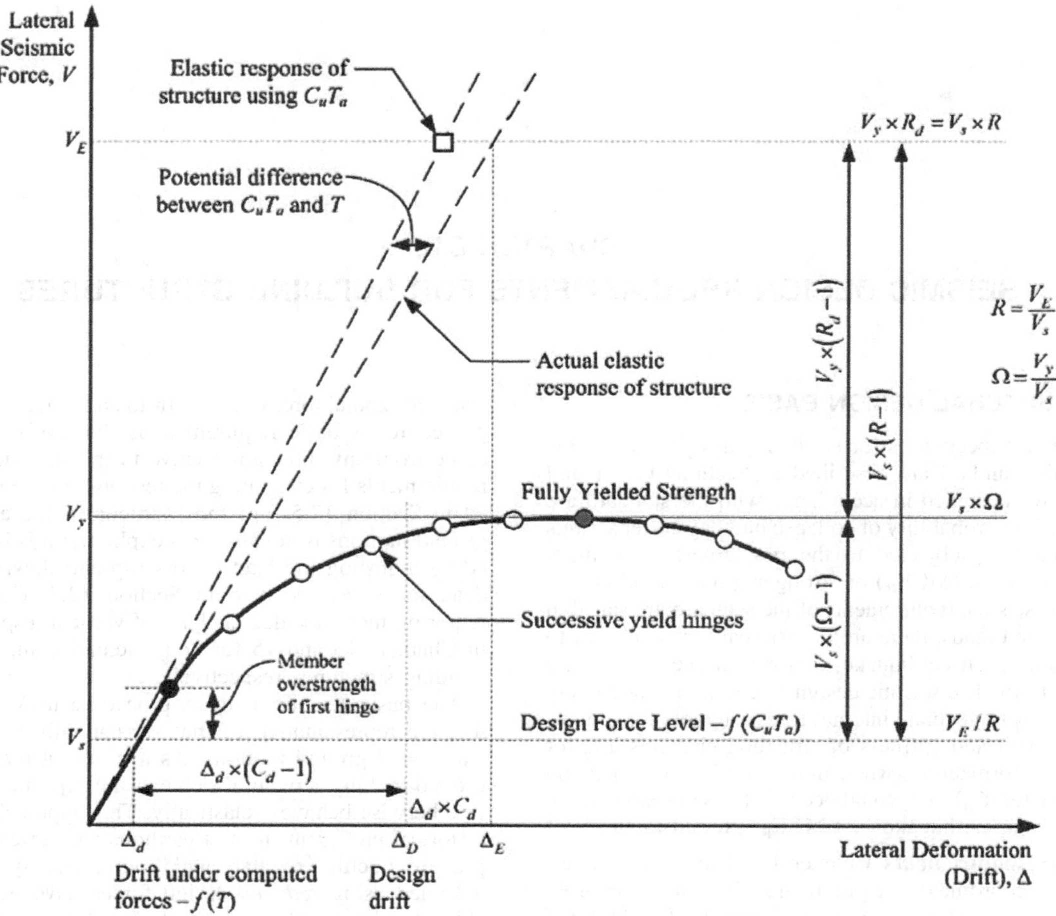

Figure C12.1-1. Inelastic force–deformation curve.

additional plastic hinges form and the resistance increases at a reduced rate (following the solid curve) until the maximum strength is reached, representing a fully yielded system. The maximum strength developed along the curve is substantially higher than that at first significant yield, and this margin is referred to as the system overstrength capacity. The ratio of these strengths is denoted as Ω. Furthermore, the figure illustrates the potential variation that can exist between the actual elastic response of a system and that considered using the limits on the fundamental period (assuming 100% mass participation in the fundamental mode—see Section C12.8.6). Although not a concern for strength design, this variation can have an effect on the expected drifts.

The system overstrength described above is the direct result of overstrength of the elements that form the SFRS and, to a lesser extent, the lateral force distribution used to evaluate the inelastic force–deformation curve. These two effects interact with applied gravity loads to produce sequential plastic hinges, as illustrated in the figure. This member overstrength is the consequence of several sources. First, material overstrength (i.e., actual material strengths higher than the nominal material strengths specified in the design) may increase the member overstrength significantly. For example, a recent survey shows that the mean yield strength of ASTM A36 steel is about 30% to 40% higher than the specified yield strength used in design calculations. Second, member design strengths usually incorporate a strength reduction or resistance factor, φ, to produce a low probability of failure under design loading. It is common to not include this factor in the member load–deformation relation when evaluating the seismic response of a structure in a nonlinear structural analysis. Third, designers can introduce additional strength by selecting sections or specifying reinforcing patterns that exceed those required by the computations. Similar situations occur where prescriptive minimums of the standard, or of the referenced design standards, control the design. Finally, the design of many flexible structural systems (e.g., moment-resisting frames) can be controlled by the drift rather than strength, with sections selected to control lateral deformations rather than to provide the specified strength.

The result is that structures typically have a much higher lateral strength than that specified as the minimum by the standard, and the first significant yielding of structures may occur at lateral load levels that are 30% to 100% higher than the prescribed design seismic forces. If provided with adequate ductile detailing, redundancy, and regularity, full yielding of structures may occur at load levels that are two to four times the prescribed design force levels.

Most structural systems have some elements whose action cannot provide reliable inelastic response or energy dissipation. Similarly, some elements are required to remain essentially elastic to maintain the structural integrity of the structure (e.g., columns supporting a discontinuous SFRS). Such elements and actions must be protected from undesirable behavior by considering that the actual forces within the structure can be significantly larger than those at first significant yield. The standard specifies an overstrength factor, Ω_0, to amplify the prescribed seismic forces for use in design of such elements and for such actions. This approach is a simplification to determining the maximum forces that could be developed in a system and the distribution of these forces within the structure. Thus, this specified overstrength factor is neither an upper nor a lower bound; it is simply an approximation specified to provide a nominal degree of protection against undesirable behavior.

The elastic deformations calculated under these reduced forces (Section C12.8.6) are multiplied by the deflection amplification factor, C_d, to estimate the deformations likely to result from the design earthquake ground motion. This factor was first introduced in the 1978 edition of ATC 3-06 (1984). For the vast majority of systems, C_d is less than R, with a few notable exceptions, where inelastic drift is strongly coupled with an increased risk of collapse (e.g., reinforced concrete bearing walls). Research over the past 30 years has illustrated that inelastic displacements may be significantly greater than Δ_E for many structures and less than Δ_E for others. Where C_d is substantially less than R, the system is considered to have damping greater than the nominal 5% of critical damping. As set forth in Section 12.12 and Chapter 13, the amplified deformations are used to assess story drifts and to determine seismic demands on elements of the structure that are not part of the seismic force-resisting system and on nonstructural components within structures.

Figure C12.1-1 illustrates the significance of seismic design parameters in the standard, including the response modification coefficient, R; the deflection amplification factor, C_d; and the overstrength factor, Ω_0. The values of these parameters, provided in Table 12.2-1, as well as the criteria for story drift and P-delta effects, have been established considering the characteristics of typical properly designed structures. The provisions of the standard anticipate an SFRS with redundant characteristics wherein significant system strength above the level of first significant yield can be obtained by plastification at other critical locations in the structure before the formation of a collapse mechanism. If excessive "optimization" of a structural design is performed with lateral resistance provided by only a few elements, the successive yield hinge behavior depicted in Figure C12.1-1 is not able to form, the actual overstrength, Ω, is small, and use of the seismic design parameters in the standard may not provide the intended seismic performance.

The response modification coefficient, R, represents the ratio of the forces that would develop under the specified ground motion if the structure had an entirely linear-elastic response to the prescribed design forces (Figure C12.1-1). The structure must be designed so that the level of significant yield exceeds the prescribed design force. The ratio R_d, expressed as $R_d = V_E/V_S$, where V_E is the elastic seismic force demand and V_S is the prescribed seismic force demand, is always larger than 1.0; thus, all structures are designed for forces smaller than those the design ground motion would produce in a structure with a completely linear-elastic response. This reduction is possible for a number of reasons. As the structure begins to yield and deform inelastically, the effective period of response of the structure lengthens, which results in a reduction in strength demand for most structures. Furthermore, the inelastic action results in a significant amount of energy dissipation (hysteretic damping) in addition to other sources of damping present below significant yield. The combined effect, which is known as the ductility reduction, explains why a properly designed structure with a fully yielded strength (V_y in Figure C12.1-1) that is significantly lower than V_E can be capable of providing satisfactory performance under the design ground motion excitations.

The energy dissipation resulting from hysteretic behavior can be measured as the area enclosed by the force–deformation curve of the structure as it experiences several cycles of excitation. Some structures have far more energy dissipation capacity than others. The extent of energy dissipation capacity available depends largely on the amount of stiffness and strength degradation the structure undergoes as it experiences repeated cycles of inelastic deformation. Figure C12.1-2 shows representative load–deformation curves for two simple substructures, such as a beam–column assembly in a frame. Hysteretic curve (a) in the figure represents the behavior of substructures that have been detailed for ductile behavior. The substructure can maintain almost all of its strength and stiffness over several large cycles of inelastic deformation. The resulting force–deformation "loops" are quite wide and open, resulting in a large amount of energy dissipation. Hysteretic curve (b) represents the behavior of a substructure that has much less energy dissipation than that for the substructure (a) but has a greater change in response period. The structural response is determined by a combination of energy dissipation and period modification.

The principles of this section outline the conceptual intent behind the seismic design parameters used by the standard. However, these parameters are based largely on engineering judgment of the various materials and performance of structural systems in past earthquakes and cannot be directly computed using the relationships presented in Figure C12.1-1. The seismic design parameters chosen for a specific project or system should be chosen with care. For example, lower values should be used

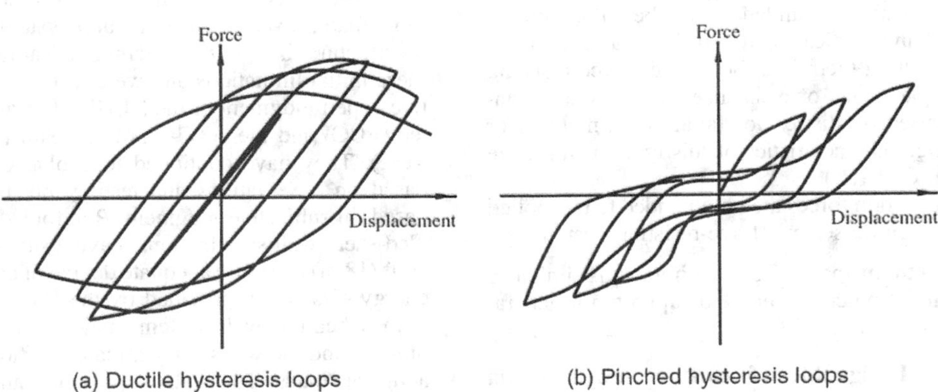

Figure C12.1-2. Typical hysteretic curves.

for structures possessing a low degree of redundancy wherein all the plastic hinges required for the formation of a mechanism may be formed essentially simultaneously and at a force level close to the specified design strength. This situation can result in considerably more detrimental P-delta effects. Because it is difficult for individual designers to judge the extent to which the value of R should be adjusted based on the inherent redundancy of their designs, Section 12.3.4 provides the redundancy factor, ρ, that is typically determined by being based on the removal of individual seismic force-resisting elements.

Higher-order seismic analyses are permitted for any structure and are required for some structures (Section 12.6); lower limits based on the equivalent lateral force (ELF) procedure may, however, still apply.

C12.1.2 Member Design, Connection Design, and Deformation Limit Given that key elements of the seismic force-resisting system are likely to yield in response to ground motions, as discussed in Section C12.1.1, it might be expected that structural connections would be required to develop the strength of connected members. Although that is a logical procedure, it is not a general requirement. The actual requirement varies by system and generally is specified in the standards for design of the various structural materials cited by reference in Chapter 14. Good seismic design requires careful consideration of this issue.

C12.1.3 Continuous Load Path and Interconnection In effect, Section 12.1.3 calls for the seismic design to be complete and in accordance with the principles of structural mechanics. The loads must be transferred rationally from their point of origin to the final point of resistance. This requirement should be obvious, but it often is overlooked by those inexperienced in earthquake engineering. Design consideration should be given to potentially adverse effects where there is a lack of redundancy. Given the many unknowns and uncertainties in the magnitude and characteristics of earthquake loading, in the materials and systems of construction for resisting earthquake loadings, and in the methods of analysis, good earthquake engineering practice has been to provide as much redundancy as possible in the seismic force-resisting system of buildings. Redundancy plays an important role in determining the ability of the building to resist earthquake forces. In a structural system without redundant elements, every element must remain operative to preserve the integrity of the building structure. However, in a highly redundant system, one or more redundant elements may fail and still leave a structural system that retains its integrity and can continue to resist lateral forces, albeit with diminished effectiveness.

Although a redundancy requirement is included in Section 12.3.4, overall system redundancy can be improved by making all joints of the vertical load-carrying frame moment resisting and incorporating them into the seismic force-resisting system. These multiple points of resistance can prevent a catastrophic collapse caused by distress or failure of a member or joint. (The overstrength characteristics of this type of frame are discussed in Section C12.1.1.)

The minimum connection forces are not intended to be applied simultaneously to the entire seismic force-resisting system.

C12.1.4 Connection to Supports The requirement is similar to that given in Section 1.4 on connections to supports for general structural integrity. See Section C1.4.

C12.1.5 Foundation Design Most foundation design criteria are still stated in terms of allowable stresses, and the forces computed in the standard are all based on the strength level of response. When developing strength-based criteria for foundations, all the factors cited in Section 12.1.5 require careful consideration. Section C12.13 provides specific guidance.

C12.1.6 Material Design and Detailing Requirements The design limit state for resistance to an earthquake is unlike that for any other load within the scope of the standard. The earthquake limit state is based on overall system performance, not member performance, where repeated cycles of inelastic straining are accepted as an energy-dissipating mechanism. Provisions that modify customary requirements for proportioning and detailing structural members and systems are provided to produce the desired performance.

C12.2 STRUCTURAL SYSTEM SELECTION

C12.2.1 Selection and Limitations For the purpose of seismic analysis and design requirements, seismic force-resisting systems are grouped into categories as shown in Table 12.2-1. These categories are subdivided further for various types of vertical elements used to resist seismic forces. In addition, the sections for detailing requirements are specified.

Specification of response modification coefficients, R, requires considerable judgment based on knowledge of actual earthquake performance and research studies. The coefficients and factors in Table 12.2-1 continue to be reviewed in light of recent research results. The values of R for the various systems were selected considering observed performance during past earthquakes, the toughness (ability to dissipate energy without serious degradation) of the system, and the amount of damping typically present in the system when it undergoes inelastic response. FEMA P-695 (2009b) has been developed with the purpose of establishing and documenting a methodology for quantifying seismic force-resisting system performance and response parameters for use in seismic design. Whereas R is a key parameter being addressed, related design parameters such as the overstrength factor, Ω_0, and the deflection amplification factor, C_d, also are addressed. Collectively, these terms are referred to as *seismic design coefficients (or factors)*. Future systems are likely to derive their seismic design coefficients (or factors) using this methodology, and existing system coefficients (or factors) also may be reviewed in light of this new procedure.

The addition of reinforced concrete ductile coupled walls is based on the development of definitions and design provisions for this system in ACI 318-19 (2019) and recent research that demonstrates adequate adjusted collapse margin ratios using the FEMA P-695 methodology (Tauberg et al. 2019).

Height limits have been specified in codes and standards for more than 50 years. The structural system limitations and limits on structural height, h_n, specified in Table 12.2-1, evolved from these initial limitations and were further modified by the collective expert judgment of the NEHRP Provisions Update Committee (PUC) and the ATC-3 project team (the forerunners of the PUC). They have continued to evolve over the past 30 years based on observations and testing, but the specific values are based on subjective judgment. Reinforced concrete ductile coupled shear wall systems must have a structural height of at least 60 ft (18 m) to ensure adequate degree of coupling and significant energy dissipation provided by the coupling beams.

In a bearing wall system, major load-carrying columns are omitted and the walls carry a major portion of the gravity (dead and live) loads. The walls supply in-plane lateral stiffness and strength to resist wind and earthquake loads and other lateral loads. In some cases, vertical trusses are used to augment lateral stiffness. In general, lack of redundancy for support of vertical and horizontal loads causes values of R to be lower for this system compared with R values of other systems.

In a building frame system, gravity loads are carried primarily by a frame supported on columns rather than by bearing walls. Some portions of the gravity load may be carried on bearing walls, but the amount carried should represent a relatively small percentage of the floor or roof area. Lateral resistance is provided by shear walls or braced frames. Light-framed walls with shear panels are intended for use only with wood and steel building frames. Although gravity load-resisting systems are not required to provide lateral resistance, most of them do. To the extent that the gravity load-resisting system provides additional lateral resistance, it enhances the building's seismic performance capability, so long as it is capable of resisting the resulting stresses and undergoing the associated deformations.

In a moment-resisting frame (MRF) system, moment-resisting connections between the columns and beams provide lateral resistance. In Table 12.2-1, such frames are classified as ordinary, intermediate, or special. In high seismic design categories, the anticipated ground motions are expected to produce large inelastic demands, so special moment frames (SMFs) designed and detailed for ductile response in accordance with Chapter 14 are required. In low Seismic Design Categories, the inherent overstrength in typical structural designs is such that the anticipated inelastic demands are somewhat reduced, and less ductile systems may be used safely. Because these less ductile ordinary framing systems do not possess as much toughness, lower values of R are specified.

The values for R, Ω_0, and C_d for the composite systems in Table 12.2-1 are similar to those for comparable systems of structural steel and reinforced concrete. Use of the tabulated values is allowed only when the design and detailing requirements in Section 14.3 are followed.

In a dual system, a three-dimensional (3D) space frame made up of columns and beams provides primary support for gravity loads. Primary lateral resistance is supplied by shear walls or braced frames, and secondary lateral resistance is provided by a moment frame complying with the requirements of Chapter 14.

Where a beam–column frame or a slab–column frame lacks special detailing, it cannot act as an effective backup to a shear wall subsystem, so there are no dual systems with ordinary moment frames (OMFs). Instead, Table 12.2-1 permits the use of a shear wall–frame interactive system with ordinary reinforced concrete moment frames and ordinary reinforced concrete shear walls. Use of this defined system, which requires compliance with Section 12.2.5.8, offers a significant advantage over a simple combination of the two constituent ordinary reinforced concrete systems. Where those systems are simply combined, Section 12.2.3.3 would require use of seismic design parameters for an ordinary reinforced concrete moment frame.

In a cantilevered column system, stability of mass at the top is provided by one or more columns with base fixity acting as a single-degree-of-freedom system.

Cantilever column systems are essentially a special class of moment-resisting frame, except that they do not possess the redundancy and overstrength that most moment-resisting frames derive from sequential formation of yield or plastic hinges. Where a typical moment-resisting frame must form multiple plastic hinges in members to develop a yield mechanism, a cantilever column system develops hinges only at the base of the columns to form a mechanism. As a result, their overstrength is limited to that provided by material overstrength and by design conservatism.

It is permitted to construct cantilever column structures using any of the systems that can be used to develop moment frames, including ordinary and special steel; ordinary, intermediate, and special concrete; and timber frames. The system limitations for cantilever column systems reflect the type of moment frame detailing provided but with a limit on structural height, h_n, of 35 ft (10.7 m).

The value of R for cantilever column systems is derived from moment-resisting frame values, where R is divided by Ω_0 but is not taken as less than 1 or greater than 2 1/2. This range accounts for the lack of sequential yielding in such systems. C_d is taken as equal to R, recognizing that damping is quite low in these systems and the inelastic displacement of these systems is not less than the elastic displacement.

Requirements for cross-laminated timber (CLT) shear walls and associated seismic design coefficients are based on research that demonstrates adequate adjusted collapse margins ratios using the FEMA P-695 methodology (van de Lindt et al. 2019). Two variants of the CLT shear wall system are addressed in Table 12.2-1: (a) CLT shear wall system with ratio of wall panel height, h, to individual wall panel length, b_s, of between 2 and 4; and (b) CLT shear wall system with shear resistance provided by high aspect ratio panels only, where high aspect ratio is defined as wall panel height to individual wall panel length ratio of 4. The system overstrength factor of $\Omega_0 = 3$ is based on nonlinear static pushover analysis results that ranged from 2.02 to 4.03 for the performance groups studied. The response modification coefficient $R = 3$ for CLT shear walls and $R = 4$ for CLT shear walls with shear resistance provided by high aspect ratio panels only recognizes improved displacement capacity associated with use of high aspect ratio CLT panels and has been validated by incremental dynamic analysis results indicating collapse probabilities of less than 10%. The deflection amplification factor of $C_d = 3$ for CLT shear walls and $C_d = 4$ for CLT shear walls of high aspect ratio only is based on $C_d = R$ in accordance with the FEMA P-695 methodology.

C12.2.1.1 Alternative Structural Systems Historically, this standard has permitted the use of alternative seismic force-resisting systems subject to satisfactory demonstration that the proposed systems' lateral force resistance and energy dissipation capacity is equivalent to structural systems listed in Table 12.2-1, for equivalent values of the response modification coefficient, R, overstrength factor, Ω_0, and deflection amplification coefficient, C_d. These design factors were established based on limited analytical and laboratory data and the engineering judgment of the developers of the standard.

Under funding from the Federal Emergency Management Agency (FEMA), the Applied Technology Council developed a rational methodology for validation of design criteria for seismic force-resisting systems under its ATC-63 project. Published as FEMA P-695 (2009b), this methodology recognizes that the fundamental goal of the seismic design rules contained in the standard is to limit collapse probability to acceptable levels. The FEMA P-695 methodology uses nonlinear response history (LRH) analysis to predict an adjusted collapse margin ratio (ACMR) for a suite of archetypical structures designed in accordance with a proposed set of system-specific design criteria and subjected to a standard series of ground motion accelerograms. The suite of archetypical structures is intended to represent the typical types and sizes of structures that are likely to incorporate the system. The ACMR relates to the conditional probability of collapse given MCE_R shaking and considers uncertainties associated with the record-to-record variability of ground motions, the quality of the design procedure, the comprehensiveness and quality of the laboratory data on which the analytical modeling is based, and uncertainties associated with the analytical modeling. Subsequent studies have been used to benchmark this methodology against selected systems in Table 12.2-1 and have demonstrated that the methodology provides rational results consistent with past engineering judgment for many systems. The FEMA P-695 methodology is therefore deemed to constitute the preferred procedure for demonstrating adequate collapse resistance for new structural systems not currently in Table 12.2-1.

Under the FEMA P-695 methodology, the archetypes used to evaluate seismic force-resisting systems are designed using the criteria for Risk Category II structures and are evaluated to demonstrate that the conditional probability of collapse of such structures conforms to the 10% probability of collapse goal stated in this section and also described in Section C1.3.1 of the commentary to this standard. It is assumed that application of the seismic Importance Factors and more restrictive drift limits associated with the design requirements for structures assigned to Risk Categories III and IV will provide such structures with the improved resistance to collapse described in Section C1.3.1 for those Risk Categories.

In addition to providing a basis for establishment of design criteria for structural systems that can be used for design of a wide range of structures, the FEMA P-695 methodology also contains a building-specific methodology intended for application to individual structures. The rigor associated with application of the FEMA P-695 methodology may not be appropriate to the design of individual structures that conform with limited and clearly defined exceptions to the criteria contained in the standard for a defined structural system. Nothing contained in this section is intended to require the use of FEMA P-695 or similar methodologies for such cases.

C12.2.1.2 Elements of Seismic Force-Resisting Systems This standard and its referenced standards specify design and detailing criteria for members and their connections (elements) of seismic force-resisting systems defined in Table 12.2-1. Substitute elements replace portions of the defined seismic force-resisting systems. Examples include proprietary products made up of special steel moment-resisting connections or proprietary shear walls for use in light-frame construction. Requirements for qualification of substitute elements of seismic force-resisting systems are intended to ensure equivalent seismic performance of the element and the system as a whole. The evaluation of suitability for substitution is based on comparison of key performance parameters of the code-defined (conforming) element and the substitute element.

FEMA P-795, *Quantification of Building Seismic Performance Factors: Component Equivalency Methodology* (2011), is an acceptable methodology to demonstrate equivalence of substitute elements and their connections and provides methods for component testing, calculation of parameter statistics from test data, and acceptance criteria for evaluating equivalency. Key performance parameters include strength ratio, stiffness ratio, deformation capacity, and cyclic strength and stiffness characteristics.

Section 12.2.1.2, item (f), requires independent design review as a condition of approval of the use of substitute elements. It is not the intent that design review be provided for every project that incorporates a substitute component, but rather that such review would be performed one time, as part of the general qualification of such substitute components. When used on individual projects, evidence of such review could include an evaluation service report or review letter indicating the conditions under which use of the substitute component is acceptable.

C12.2.2 Combinations of Framing Systems in Different Directions Different seismic force-resisting systems can be used along each of the two orthogonal axes of the structure, as long as the respective values of R, Ω_0, and C_d are used. Depending on the combination selected, it is possible that one of the two systems may limit the extent of the overall system with regard to structural system limitations or structural height, h_n; the more restrictive of these would govern.

C12.2.3 Combinations of Framing Systems in the Same Direction The intent of the provision requiring use of the most stringent seismic design parameters (R, Ω_0, and C_d) is to prevent mixed seismic force-resisting systems that could concentrate inelastic behavior in the lower stories.

C12.2.3.1 R, C_d, and Ω_0 Values for Vertical Combinations This section expands on Section 12.2.3 by specifying the requirements specific to the cases where (a) the value of R for the lower seismic force-resisting system is lower than that for the upper system, and (b) the value of R for the upper seismic force-resisting system is lower than that for the lower system.

The two cases are intended to account for all possibilities of vertical combinations of seismic force-resisting systems in the same direction. For a structure with a vertical combination of three or more seismic force-resisting systems in the same direction, Section 12.2.3.1 must be applied to the adjoining pairs of systems until the vertical combinations meet the requirements therein.

There are also exceptions to these requirements for conditions that do not affect the dynamic characteristics of the structure or that do not result in concentration of inelastic demand in critical areas.

C12.2.3.2 Two-Stage Analysis Procedure A two-stage equivalent lateral force (ELF) procedure is permitted where the lower portion of the structure has a minimum of 10 times the stiffness of the upper portion of the structure. In addition, the period of the entire structure is not permitted to be greater than 1.1 times the period of the upper portion considered as a separate structure supported at the transition from the upper to the lower portion. An example would be a concrete podium under a wood- or steel-framed upper portion of a structure. The upper portion may be analyzed for seismic forces and drifts using the values of R, Ω_0, C_d, and ρ for the upper portion as a separate structure. The seismic forces resulting from horizontal seismic loading, F_x (e.g., shear and overturning), at the base of the upper portion are applied to the top of the lower portion and scaled up by the ratio of $(R/\rho)_{upper}$ to $(R/\rho)_{lower}$. All other forces from the upper portion, including gravity forces and seismic forces resulting from vertical seismic loading, are directly applied to the lower portion without scaling. The lower portion, which now includes the seismic forces from the upper portion, may then be analyzed using the values of R, Ω_0, C_d, and ρ for the lower portion of the structure. The lower portion of a structure qualifying for the two-stage analysis procedure is not exempted from meeting the requirements of 12.2.3.1. Item (f) clarifies that the structural height limits of Table 12.2-1 are applied to the upper portion individually, and not the combination of the two portions. This is permitted because the criteria is based on stiffness and period limit dynamic interaction between the lower portion and the upper portion.

Where horizontal structural irregularity Type 4 (out-of-plane offset) or vertical structural irregularity Type 3 (in-plane discontinuity) is present at the transition from the upper portion to the lower portion, the requirement for the overstrength effects is associated with the expected peak capacity of the upper portion. The ratio of R/ρ values (that of the lower portion to that of the upper portion) is associated with the effect of the upper portion mass on the lower portion, which is a separate consideration. These two effects are required to be used in combination where applicable. Overstrength effects include but are not limited to the requirements of Section 12.3.3.4 for elements supporting discontinued walls or frames and the requirements of Sections 12.10.1 and 12.10.3.3 for diaphragm transfer forces. See also Commentary Section C12.3.3.4.

C12.2.3.3 R, C_d, and Ω_0 Values for Horizontal Combinations For almost all conditions, the least value of R of different seismic force-resisting systems in the same direction must be used in design. This requirement reflects the expectation that the entire system will undergo the same deformation with its behavior controlled by the least ductile system. However, for light-frame construction or flexible diaphragms meeting the listed conditions, the value of R for each independent line of resistance can be used. This exceptional condition is consistent with light-frame construction that uses the ground for parking with residential use above.

C12.2.3.4 Two-Stage Analysis Procedure for One-Story Structures with Flexible Diaphragms and Rigid Vertical Elements Section 12.2.3.4 incorporates a two-stage analysis for flexible diaphragms supported on rigid vertical elements that is conceptually parallel to the two-stage analysis already permitted for flexible superstructures supported on rigid base structures. This approach is based on numerical studies that are discussed in Section C12.10.4. Standard design procedures require the designer to determine seismic forces to vertical elements based on the approximate period of the vertical elements, irrespective of diaphragm response. Section 12.2.3.4, however, permits the designer to recognize the reduction in seismic design forces to the vertical elements resulting from the anticipated longer period of the diaphragm.

Item (a) requires that this provision be used in combination with the Section 12.10.4 provisions. Section 12.10.4 establishes scoping requirements that limit use to rigid vertical element–flexible diaphragm structure types, derived based on the FEMA P-1026 numerical studies. Section 12.10.4 also provides equations for determination of diaphragm design forces, F_{px}, intended to be used in the Item (b) sum of forces.

Item (b) requires that diaphragm design forces be combined with in-plane vertical element forces determined using the equivalent lateral force (ELF) procedure of Section 12.8, in order to be consistent with the FEMA P-1026 basis. If other methods are used to determine in-plane forces to the vertical elements, this could result in a duplication of the force reduction due to diaphragm flexibility.

Item (b)(1) requires that the lateral seismic forces contributed by the diaphragm be amplified by a ratio of R and ρ factors. Note that per Section 12.10.4.2, ρ for the diaphragm is to be 1.0. This amplification directly parallels the provisions of Section 12.2.3.2. Note that this amplification is applied to the seismic forces calculated in accordance with Equation (12.10-15). The shear amplification of Section 12.10.4.2.2 need not be included.

Item (b)(2) requires consideration of seismic forces applied to the vertical elements not entering through the horizontal diaphragm. Typically, these forces are from the vertical element's self-weight and other effective seismic weights in the same plane as the vertical element. This total effective seismic weight induces seismic design forces associated with the in-plane vertical element's period and response modification factor.

C12.2.4 Combination Framing Detailing Requirements This requirement is provided so that the seismic force-resisting system with the highest value of R has the necessary ductile detailing throughout. The intent is that details common to both systems be designed to remain functional throughout the response to earthquake load effects to preserve the integrity of the seismic force-resisting system.

C12.2.5 System-Specific Requirements

C12.2.5.1 Dual System The moment frame of a dual system must be capable of resisting at least 25% of the design seismic forces; this percentage is based on judgment. The purpose of the 25% frame is to provide a secondary seismic force-resisting system with higher degrees of redundancy and ductility to improve the ability of the building to support the service loads (or at least the effect of gravity loads) after strong earthquake shaking. The primary system (walls or bracing) acting together with the moment frame must be capable of resisting all of the design seismic forces. The following analyses are required for dual systems.

1. The moment frame and shear walls or braced frames must resist the design seismic forces, considering fully the force and deformation interaction of the walls or braced frames and the moment frames as a single system. This analysis must be made in accordance with the principles of structural mechanics that consider the relative rigidities of the elements and torsion in the system. Deformations imposed on members of the moment frame by their interaction with the shear walls or braced frames must be considered in this analysis.
2. The moment frame must be designed with sufficient strength to resist at least 25% of the design seismic forces.

C12.2.5.2 Cantilever Column Systems Cantilever column systems are singled out for special consideration because of their unique characteristics. These structures often have limited redundancy and, without an appropriate level of overstrength, will concentrate inelastic behavior at bases that may not have the ductility or strength to provide an adequate margin against collapse at the maximum considered earthquake (MCE). Strength and ductility are sensitive to detailing provisions associated with foundation design, anchorage design, and base connection details. As a result, they have substantially less energy dissipation capacity than other systems. A number of apartment buildings incorporating this system experienced severe damage and, in some cases, collapsed in the 1994 Northridge (California) earthquake. In general, the ductility of columns is adversely affected by high axial force. Where this effect is addressed by an axial-force limitation in the material standard for the system, the provisions defer to that limit; otherwise the provisions place a limit on individual column axial demands from the load combinations that include seismic load effects; this condition is prescribed through height limitations shown in Table 12.2-1 and requirements in Section 12.2.5.2.

Elements providing restraint at the base of cantilever columns must be designed for seismic load effects, including overstrength, so that the strength of the cantilever columns is developed. Detailing provisions, contained in materials standards, are intended to maintain structural system vertical load carrying capacity after the onset of yielding, preferably in the column element. By setting the overstrength factor equal to the response modification coefficient, behavior is consistent with this intent (e.g., intermediate reinforced concrete moment frames). For special seismic systems, seismic design factors are chosen to force yielding within the member cross section that is designed per special detailing provisions in material standards. Ideally, each cantilever column should have a robust base connection that can either develop the member's full plastic moment capacity or provide adequate resistance against collapse at the maximum considered earthquake (MCE). Some common details include embedment in deep foundations that develop full column section capacity at the transition between foundation and cantilever column.

C12.2.5.3 Inverted-Pendulum-Type Structures Inverted-pendulum-type structures do not have a unique entry in Table 12.2-1 because they can be formed from many structural systems. Inverted-pendulum-type structures have more than half of their mass concentrated near the top (producing one degree of freedom in horizontal translation) and rotational compatibility of the mass with the column (producing vertical accelerations acting in opposite directions). Dynamic response amplifies this rotation; hence, the bending moment induced at the top of the column can exceed that computed using the procedures of Section 12.8. The requirement to design for a top moment that is one-half of the base moment calculated in accordance with Section 12.8 is based on analyses of inverted pendulums covering a wide range of practical conditions.

C12.2.5.4 Increased Structural Height Limit for Steel Eccentrically Braced Frames, Steel Special Concentrically Braced Frames, Steel Buckling-Restrained Braced Frames, Steel Special Plate Shear Walls, Steel and Concrete Coupled Composite Plate Shear Walls, and Special Reinforced Concrete Shear Walls The first criterion for an increased limit on structural height, h_n, precludes a high Torsional Irregularity Ratio (TIR) because premature failure of one of the shear walls or braced frames could lead to excessive inelastic torsional response. The second criterion, which is similar to the redundancy requirements, is to limit the structural height of systems that are too strongly dependent on any single line of shear walls or braced frames. The inherent torsion resulting from the distance between the center of mass and the center of rigidity must be included, but accidental torsional effects are neglected for ease of implementation.

C12.2.5.5 Special Moment Frames in Structures Assigned to Seismic Design Categories D through F Special moment frames, either alone or as part of a dual system, are required to be used in Seismic Design Categories D through F where the structural height, h_n, exceeds 160 ft (48.8 m) or 240 ft (73.2 m) for buildings that meet the provisions of Section 12.2.5.4, as indicated in Table 12.2-1. In shorter buildings where special moment frames are not required to be used, the special moment frames may be discontinued and supported on less ductile systems as long as the requirements of Section 12.2.3 for framing system combinations are followed.

For the situation where special moment frames are required, they should be continuous to the foundation. In cases where the foundation is located below the building's base, provisions for discontinuing the moment frames can be made as long as the seismic forces are properly accounted for and transferred to the supporting structure.

C12.2.5.6 Steel Ordinary Moment Frames Steel ordinary moment frames (OMFs) are less ductile than steel special moment frames; consequently, their use is prohibited in structures assigned to Seismic Design Categories D, E, and F (Table 12.2-1). Structures with steel OMFs, however, have exhibited acceptable behavior in past earthquakes where the structures were sufficiently limited in their structural height, number of stories, and seismic mass. The provisions in the standard reflect these observations. The exception is discussed separately below. Table C12.2-1 summarizes the provisions.

C12.2.5.6.1 Seismic Design Category D or E Single-story steel OMFs are permitted, provided (a) the structural height, h_n, is a maximum of 65 ft (20 m), (b) the dead load supported by and tributary to the roof is a maximum of 20 lb/ft^2 (0.96 kN/m^2), and (c) the dead load of the exterior walls more than 35 ft (10.7 m) above the seismic base tributary to the moment frames is a maximum of 20 lb/ft^2 (0.96 kN/m^2).

In structures of light-frame construction, multistory steel OMFs are permitted, provided (a) the structural height, h_n, is

Table C12.2-1. Summary of Conditions for Ordinary Moment Frames (OMFs) and Intermediate Moment Frames (IMFs) in Structures Assigned to Seismic Design Category D, E, or F (Refer to the Standard for Additional Requirements).

Section	Frame Type	Seismic Design Category (SDC)	Max. Number Stories	Light-Frame Construction	Max. h_n (ft)	Max. roof/floor DL (lb/ft^2)	Exterior Wall DL Max. (lb/ft^2)	Exterior Wall DL Wall Height (ft)*
12.2.5.6.1(a)	OMF	D, E	1	NA	65	20	20	35
12.2.5.6.1(a)-Exc	OMF	D, E	1	NA	NL	20	20	35
12.2.5.6.1(b)	OMF	D, E	NL	Required	35	35	20	0
12.2.5.6.2	OMF	F	1	NA	65	20	20	0
12.2.5.7.1(a)	IMF	D	1	NA	65	20	20	35
12.2.5.7.1(a)-Exc	IMF	D	1	NA	NL	20	20	35
12.2.5.7.1(b)	IMF	D	NL	NA	35	NL	NL	NA
12.2.5.7.2(a)	IMF	E	1	NA	65	20	20	35
12.2.5.7.2(a)-Exc	IMF	E	1	NA	NL	20	20	35
12.2.5.7.2(b)	IMF	E	NL	NA	35	35	20	0
12.2.5.7.3(a)	IMF	F	1	NA	65	20	20	0
12.2.5.7.3(b)	IMF	F	NL	Required	35	35	20	0

*Applies to portion of wall above listed wall height.

Notes: DL = Dead Load, NL = no limit; NA = not applicable. For metric units, use 20 m for 65 ft, 10.6 m for 35 ft, 0.96 kN/m^2 for 20 lb/ft^2, and 1.68 kN/m^2 for 30 lb/ft^2.

a maximum of 35 ft (10.7 m), (b) the dead load of the roof and each floor above the seismic base supported by and tributary to the moment frames are each a maximum of 35 lb/ft^2 (1.68 kN/m^2), and (c) the dead load of the exterior walls tributary to the moment frames is a maximum of 20 lb/ft^2 (0.96 kN/m^2).

EXCEPTION: Industrial structures, such as aircraft maintenance hangars and assembly buildings, with steel OMFs have performed well in past earthquakes with strong ground motions (EQE Inc. 1983, 1985, 1986a, 1986b, 1986c, 1987); the exception permits single-story steel OMFs to be unlimited in height provided that (a) the structure is limited to the enclosure of equipment or machinery; (b) its occupants are limited to maintaining and monitoring the equipment, machinery, and their associated processes; (c) the sum of the dead load and equipment loads supported by and tributary to the roof is a maximum of 20 lb/ft^2 (0.96 kN/m^2); and (d) the dead load of the exterior wall system, including exterior columns more than 35 ft (10.7 m) above the seismic base is a maximum of 20 lb/ft^2 (0.96 kN/m^2). Though the latter two load limits (Items C and D) are similar to those described in this section, there are meaningful differences.

The exception further recognizes that these facilities often require large equipment or machinery, and associated systems, not supported by or considered tributary to the roof, that support the intended operational functions of the structure, such as top running bridge cranes, jib cranes, and liquid storage containment and distribution systems. To limit the seismic interaction between the seismic force-resisting systems and these components, the exception requires the weight of equipment or machinery that is not self-supporting (i.e., not freestanding) for all loads (e.g., dead, live, or seismic) to be included when determining compliance with the roof or exterior wall load limits. This *equivalent* equipment load shall be in addition to the loads listed above.

To determine the equivalent equipment load, the exception requires the weight to be considered fully (100%) tributary to an area not exceeding 600 ft^2 (55.8 m^2). This limiting area can be taken either to an adjacent exterior wall for cases where the weight is supported by an exterior column (which may also span to the first interior column) or to the adjacent roof for cases where the weight is supported entirely by an interior column or columns, but not both; nor can a fraction of the weight be allocated to each zone. Equipment loads within overlapping tributary areas should be combined in the same limiting area. Other provisions in the standard, as well as in past editions, require satisfying wall load limits tributary to the moment frame, but this requirement is not included in the exception in that it is based on a component-level approach that does not consider the interaction between systems in the structure. As such, the limiting area is considered to be a reasonable approximation of the tributary area of a moment frame segment for the purpose of this conversion. Although this weight allocation procedure may not represent an accurate physical distribution, it is considered to be a reasonable method for verifying compliance with the specified load limits to limit seismic interactions. The engineer must still be attentive to actual mass distributions when computing seismic loads. Further information is discussed in Section C11.1.3.

C12.2.5.6.2 Seismic Design Category F Single-story steel OMFs are permitted, provided that they meet conditions (a) and (b) in Section C12.2.5.6.1 for single-story frames, and (c) the dead load of the exterior walls tributary to the moment frames is a maximum of 20 lb/ft^2 (0.96 kN/m^2).

C12.2.5.7 Steel Intermediate Moment Frames Steel intermediate moment frames (IMFs) are more ductile than steel ordinary moment frames (OMFs) but less ductile than steel special moment frames (SMFs); consequently, restrictions are placed on their use in structures assigned to Seismic Design Category D and their use is prohibited in structures assigned to Seismic Design Categories E and F (Table 12.2-1). As with steel OMFs, steel IMFs have also exhibited acceptable behavior in past earthquakes where the structures were sufficiently limited in their structural height, number of stories, and seismic mass. The provisions in the standard reflect these observations. The exceptions are discussed separately (following). Table C12.2-1 summarizes the provisions.

C12.2.5.7.1 Seismic Design Category D Single-story steel IMFs are permitted without limitations on dead load of the roof and exterior walls, provided that the structural height, h_n, is a maximum of 35 ft (10.7 m). An increase to 65 ft (20 m) is permitted for h_n, provided that (a) the dead load supported by and tributary to the roof is a maximum of 20 lb/ft^2 (0.96 kN/m^2), and (b) the dead load of the exterior walls more than 35 ft (10.7 m) above the seismic base tributary to the moment frames is a maximum of 20 lb/ft^2 (0.96 kN/m^2).

The exception permits single-story steel IMFs to be unlimited in height, provided that they meet all of the conditions described in the exception to Section C12.2.5.6.1 for the same structures.

C12.2.5.7.2 Seismic Design Category E Single-story steel IMFs are permitted, provided that they meet all of the conditions described in Section C12.2.5.6.1 for single-story OMFs.

The exception permits single-story steel IMFs to be unlimited in height, provided that they meet all of the conditions described in Section C12.2.5.6.1 for the same structures.

Multistory steel IMFs are permitted, provided that they meet all of the conditions described in Section C12.2.5.6.1 for multistory OMFs, except that the structure is not required to be of light-frame construction.

C12.2.5.7.3 Seismic Design Category F Single-story steel IMFs are permitted, provided that (a) the structural height, h_n, is a maximum of 65 ft (20 m), (b) the dead load supported by and tributary to the roof is a maximum of 20 lb/ft^2 (0.96 kN/m^2), and (c) the dead load of the exterior walls tributary to the moment frames is a maximum of 20 lb/ft^2 (0.96 kN/m^2).

Multistory steel IMFs are permitted, provided that they meet all of the conditions described in the exception to Section C12.2.5.6.1 for multistory OMFs in structures of light-frame construction.

C12.2.5.8 Shear Wall–Frame Interactive Systems For structures assigned to Seismic Design Category A or B (where seismic hazard is low), it is usual practice to design shear walls and frames of a shear wall–frame structure to resist lateral forces in proportion to their relative rigidities, considering interaction between the two subsystems at all levels. As discussed in Section C12.2.1, this typical approach would require use of a lower response modification coefficient, R, than that defined for shear wall–frame interactive systems. Where the special requirements of this section are satisfied, more reliable performance is expected, justifying a higher value of R.

C12.3 DIAPHRAGM FLEXIBILITY, CONFIGURATION IRREGULARITIES, AND REDUNDANCY

C12.3.1 Diaphragm Flexibility Most seismic force-resisting systems have two distinct parts: the horizontal system, which distributes lateral forces to the vertical elements and the vertical system, which transmits lateral forces between the floor levels and the base of the structure.

The horizontal system may consist of diaphragms or a horizontal bracing system. For the majority of buildings, diaphragms offer the most economical and positive method of resisting and distributing seismic forces in the horizontal plane. Typically, diaphragms consist of a metal deck (with or without concrete), concrete slabs, and wood sheathing and/or decking. Although most diaphragms are flat, consisting of the floors of buildings, they also may have inclined, curved, warped, or folded configurations, and most diaphragms have openings.

The diaphragm stiffness relative to the stiffness of the supporting vertical seismic force-resisting system ranges from flexible to rigid. Provisions defining diaphragm flexibility are given in Sections 12.3.1.1 through 12.3.1.3. If a diaphragm cannot be idealized as either flexible or rigid, explicit consideration of its stiffness must be included in the analysis. However, even where the diaphragm can be idealized as either flexible or rigid in accordance with provisions of Sections 12.3.1.1 through 12.3.1.3, an analysis that explicitly includes consideration of the stiffness of the diaphragm (i.e., semirigid modeling assumption) is permissible.

The diaphragms in most buildings braced by wood light-frame shear walls can be idealized as flexible in accordance with Section 12.3.1.1. Where the design professional uses semirigid modeling, it can be accomplished by a structural model with diaphragm stiffness properties calibrated to those recognized in the American Wood Council's (2014) *Special Design Provisions for Wind and Seismic* (SDPWS), Chapter 4. An acceptable alternative to semirigid modeling is the use of an envelope analysis that evaluates force distribution with structural models using both rigid and flexible diaphragm assumptions and with each vertical resisting element taking the larger of the forces of the two analyses. It should be noted that AWC SDPWS Chapter 4 requires the use of either a semirigid diaphragm modeling assumption or rigid diaphragm modeling assumption for the purpose of complying with some cantilevered diaphragm provisions (Open Front Structures) even where the diaphragm can be idealized as flexible in accordance with Section 12.3.1.1.

C12.3.1.1 Flexible Diaphragm Condition Section 12.3.1.1 defines broad categories of diaphragms that may be idealized as flexible, regardless of whether the diaphragm meets the calculated conditions of Section 12.3.1.3. These categories include the following:

(a) Construction with relatively stiff vertical framing elements, such as steel-braced frames and concrete or masonry shear walls;
(b) One- and two-family dwellings; and
(c) Light-frame construction (e.g., construction consisting of light-frame walls and diaphragms) with or without nonstructural toppings of limited stiffness.

For Item (c) above, compliance with story drift limits along each line of shear walls is intended as an indicator that the shear walls are substantial enough to share load on a tributary area basis and not require torsional force redistribution.

C12.3.1.2 Rigid Diaphragm Condition Span-to-depth ratio limits are included in the deemed-to-comply condition as an indirect measure of the flexural contribution to diaphragm stiffness. Section 12.3.1.2 limits the conditions in which a concrete slab or slab on metal deck diaphragm can be considered rigid. Horizontal irregularities 2, 3, 4, and 5 are conditions where modeling the diaphragm as rigid may not capture the impacts of these irregularities.

C12.3.1.3 Calculated Flexible Diaphragm Condition A diaphragm is permitted to be idealized as flexible if the calculated diaphragm deflection (typically at midspan) between supports (lines of vertical elements) is greater than two times the average story drift of the vertical lateral force-resisting elements located at the supports of the diaphragm span.

Figure 12.3-1 depicts a distributed load, conveying the intent that the tributary lateral load be used to compute δ_{MDD}, consistent with the Section 11.3 symbols. A diaphragm opening is illustrated, and the shorter arrows in the portion of the diaphragm with the opening indicate lower load intensity because of lower tributary seismic mass.

C12.3.1.4 Diaphragms in Hillside Light-Frame Structures The first sentence of Section 12.3.1.4 addresses hillside light-frame structures for which design with flexible diaphragm assumptions has been found in numerical studies to provide seismic performance notably less than the performance targeted by ASCE 7 seismic provisions. For these structures the seismic forces acting in either horizontal direction will follow the stiffest load path, usually to the anchorage to the uphill foundation. Design of these structures using rigid or semirigid diaphragm assumptions is required because it properly characterizes the load path and resulting seismic response of this structure configuration. For these diaphragms, Section 12.3.1.4 overrides the Section 12.3.1.1 and Section 12.3.1.3 provisions that would otherwise allow the designer to classify these diaphragms as flexible. The structure torsional response has been identified by numerical studies to be of particular concern where downhill light-frame walls taller than 7 ft (2.1 m) clear height are paired with attachment of the floor diaphragm to the uphill foundation or to framed walls of less than 2 ft (0.61 m) (Figure C12.3-1a); response is of less concern where the downhill framed wall is 7 ft (2.1 m) or less, or where the uphill framed wall is more than 2 ft (0.61 m). See Chapter 6 of FEMA P-1100 (2018b) for discussion of the structural configuration of concern. Where the diaphragm conditions fall outside of this configuration, the last sentence of Section 12.3.1.4 allows the diaphragm to be characterized in accordance with the other provisions of Section 12.3.1.

C12.3.2 Irregular and Regular Classification The configuration of a structure can significantly affect its performance during a strong earthquake that produces the ground motion contemplated in the standard. Structural configuration can be divided into two aspects: horizontal and vertical. Most seismic design provisions were derived for buildings that have regular configurations, but earthquakes have shown repeatedly that buildings that have irregular configurations suffer greater damage. This situation prevails even with good design and construction.

There are several reasons for the poor behavior of irregular structures. In a regular structure, the inelastic response, including energy dissipation and damage, produced by strong ground shaking tends to be well distributed throughout the structure. However, in irregular structures, inelastic behavior can be concentrated by irregularities and can result in rapid failure of structural elements in these areas. In addition, some irregularities introduce unanticipated demands into the structure, which designers frequently overlook when detailing the structural system. Finally, the elastic analysis methods typically used in the design of structures often cannot predict the distribution of earthquake demands in an irregular structure very well, leading to inadequate design in the areas associated with the irregularity. For these reasons, the standard encourages regular structural configurations and prohibits gross irregularity in buildings

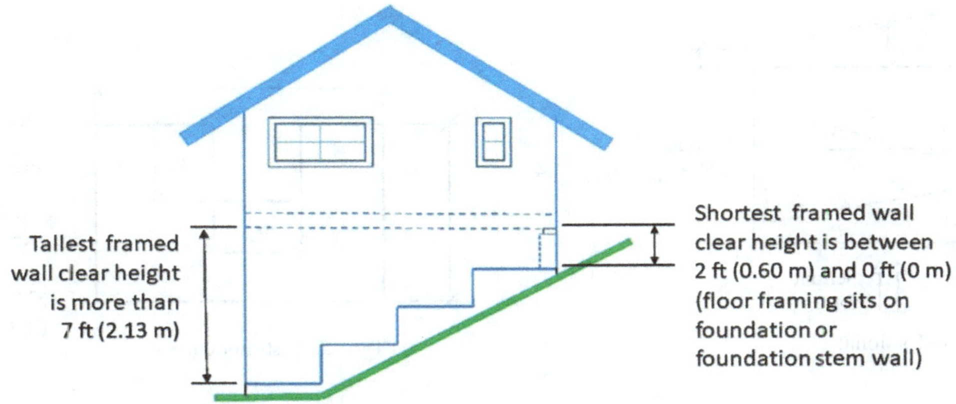

Figure C12.3-1a. Illustration of framed wall height criteria of Section 12.3.1.4, item 1.
Source: FEMA P-1100 (2018b).

located on sites close to major active faults, where very strong ground motion and extreme inelastic demands are anticipated. The termination of seismic force-resisting elements at the foundation, however, is not considered to be a discontinuity.

C12.3.2.1 Horizontal Irregularity A building may have a symmetric geometric shape without reentrant corners or wings but still be classified as irregular in plan because of its distribution of mass or vertical seismic force-resisting elements. Torsional effects in earthquakes can occur even where the centers of mass and rigidity coincide. For example, ground motion waves acting on a skew with respect to the building axis can cause torsion. Cracking or yielding in an asymmetric fashion also can cause torsion. These effects also can magnify the torsion caused by eccentricity between the centers of mass and rigidity. Torsional structural irregularities (Type 1) are defined to address this concern. The ATC123 project (FEMA P-2012 [2018a]) quantified the effect of torsional structural response on collapse. Both the Type 1 irregularity classifications in Table 12.3-1, and the design provisions triggered by an irregularity classification, are intended to produce structural designs with roughly the same collapse potential in structures with and without a torsional response.

The 75% strength requirement was added to the Type 1 irregularity based on the findings of the FEMA P-2012 project. Structures can have sufficient torsional stiffness to pass the torsional drift trigger but still lack torsional strength. Those structures rely on lines of resistance orthogonal to the ground motion direction to resist torsion, and therefore needed to be designed with the orthogonal combination rules to provide adequate collapse resistance. For structures with non-parallel seismic force resisting systems, the vector component of each seismic force resisting element in each orthogonal direction should be used to determine the 75% strength trigger.

A square or rectangular building with minor reentrant corners would still be considered regular, but large reentrant corners creating a crucifix form would produce an irregular structural configuration (Type 2). The response of the wings of this type of building generally differs from the response of the building as a whole, and this difference produces higher local forces than would be determined by application of the standard without modification. Other winged plan configurations (e.g., H-shapes) are classified as irregular even if they are symmetric because of the response of the wings.

Significant differences in stiffness between portions of a diaphragm at a level are classified as Type 3 structural irregularities because they may cause a change in the distribution of seismic forces to the vertical components and may create torsional forces not accounted for in the distribution normally considered for a regular building.

Where there are discontinuities in the path of lateral force resistance, the structure cannot be considered regular. The most critical discontinuity defined is the out-of-plane offset of vertical elements of the seismic force-resisting system (Type 4). Such offsets impose vertical and lateral load effects on horizontal elements that are difficult to provide for adequately.

Where vertical lateral force-resisting elements are not parallel to the major orthogonal axes of the seismic force-resisting system, the equivalent lateral force procedure of the standard cannot be applied appropriately, so the structure is considered to have an irregular structural configuration (Type 5).

Figure C12.3-1b illustrates horizontal structural irregularities.

C12.3.2.1.1 Torsional Irregularity Ratio The torsional irregularity ratio (TIR) is a convenient parameter for assessing the likelihood of excessive torsional response in buildings and is used to trigger design penalties or other consequences where the TIR is greater than 1.2. This parameter was used in [FEMA P-2012 (2018a)], which was the technical basis for several modifications to the Provisions.

C12.3.2.2 Vertical Irregularity Vertical irregularities in structural configuration affect the responses at various levels and induce loads at these levels that differ significantly from the distribution assumed in the equivalent lateral force (ELF) procedure given in Section 12.8.

A moment-resisting frame building might be classified as having a soft story irregularity (Type 1) if one story is much taller than the adjoining stories and the design did not compensate for the resulting decrease in stiffness that normally would occur.

FEMA P-2012 studied mass irregularities, Type 2 vertical structural irregularities as defined in ASCE 7-16 in concrete and steel moment frame buildings using the FEMA P-695 methodology. The study concluded that collapse margin ratios of structures with a mass irregularity in moment frame structures were not substantively affected by mass irregularities. Furthermore, structures proportioned using the equivalent lateral force (ELF) procedure produced very similar results as structures proportioned with the Modal Response Spectrum Analysis (MRSA) procedure. FEMA P-2012 also examined the effect of

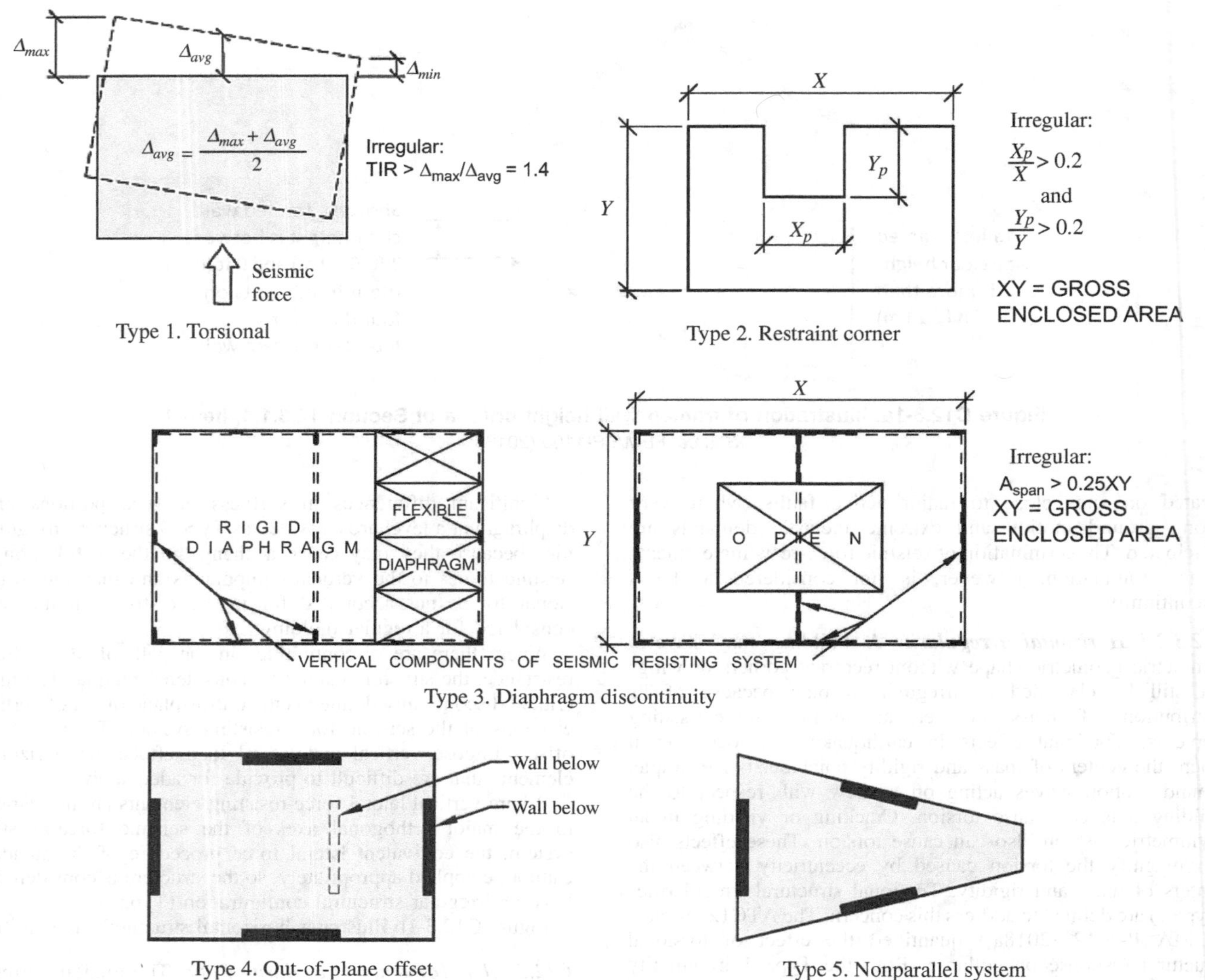

Figure C12.3-1b. Horizontal structural irregularity examples.

using ELF versus MRSA analysis on the design of non-moment frame structural types and concluded that ELF generally produces more conservative designs than MRSA.

Based on this work, the idea that structures with additional mass at a given level(s) be required to be designed using the MRSA has been rethought, and the concept of a "mass irregularity" has been removed from the standard. The user of this standard should be reminded that the spatial distribution of mass and the modeling requirements of Section 12.7.3 remain important considerations, particularly in the instance where a mass is significant and can contribute to local force and displacements related to support of a mass (e.g., hanging masses such as a large central video display in a stadium). In these instances, dynamic analyses should be used to ensure that dynamic and local effects related to seismic loads are adequately addressed in the design.

A vertical geometric irregularity (Type 2) applies regardless of whether the larger dimension is above or below the smaller one. Vertical lateral force-resisting elements at adjoining stories that are offset from each other in the vertical plane of the elements and impose overturning demands on supporting structural elements, such as beams, columns, trusses, walls, or slabs, are classified as in-plane discontinuity irregularities (Type 3).

Buildings with a weak-story irregularity (Type 4) tend to develop all of their inelastic behavior and consequent damage at the weak story, possibly leading to collapse.

Figure C12.3-2 illustrates examples of vertical structural irregularities.

C12.3.3 Limitations and Additional Requirements for Systems with Structural Irregularities

C12.3.3.1 Prohibited Vertical Irregularities for Seismic Design Category E or F The prohibitions and limits caused by structural irregularities in this section stem from poor performance in past earthquakes and the potential to concentrate large inelastic demands in certain portions of the structure. Even where such irregularities are permitted, they should be avoided whenever possible in all structures.

Prior to ASCE 7-22, extreme torsional irregularity was prohibited from Seismic Design Categories E and F, as torsional response in structural systems has been a well-documented indication of poor structural performance in past earthquakes. FEMA P-2012 rigorously studied the effect of torsional response on collapse reliability. The study found that torsionally irregular structures proportioned without the code provisions referenced in

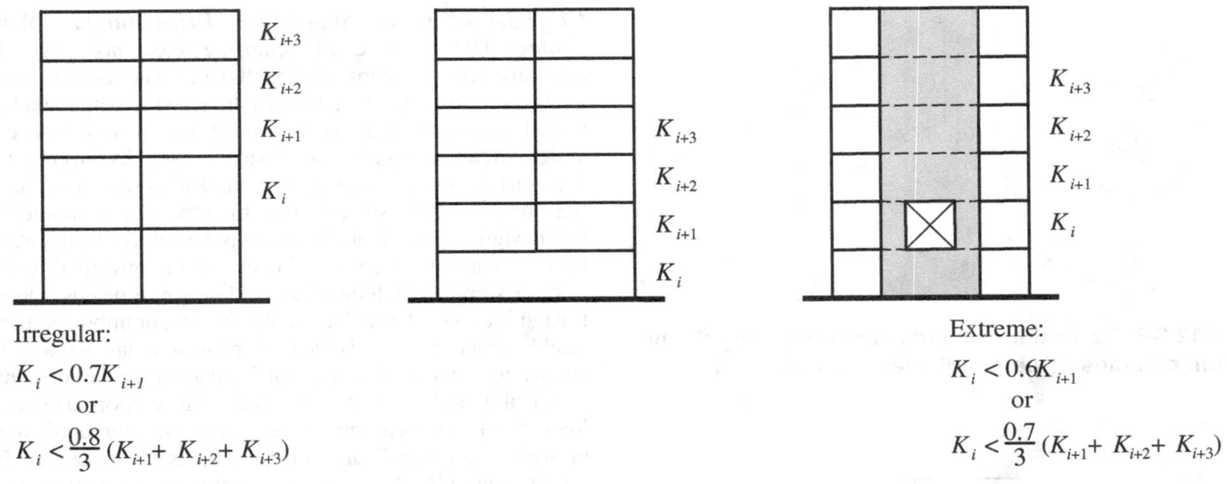

Type 1. Stiffness — Soft Story

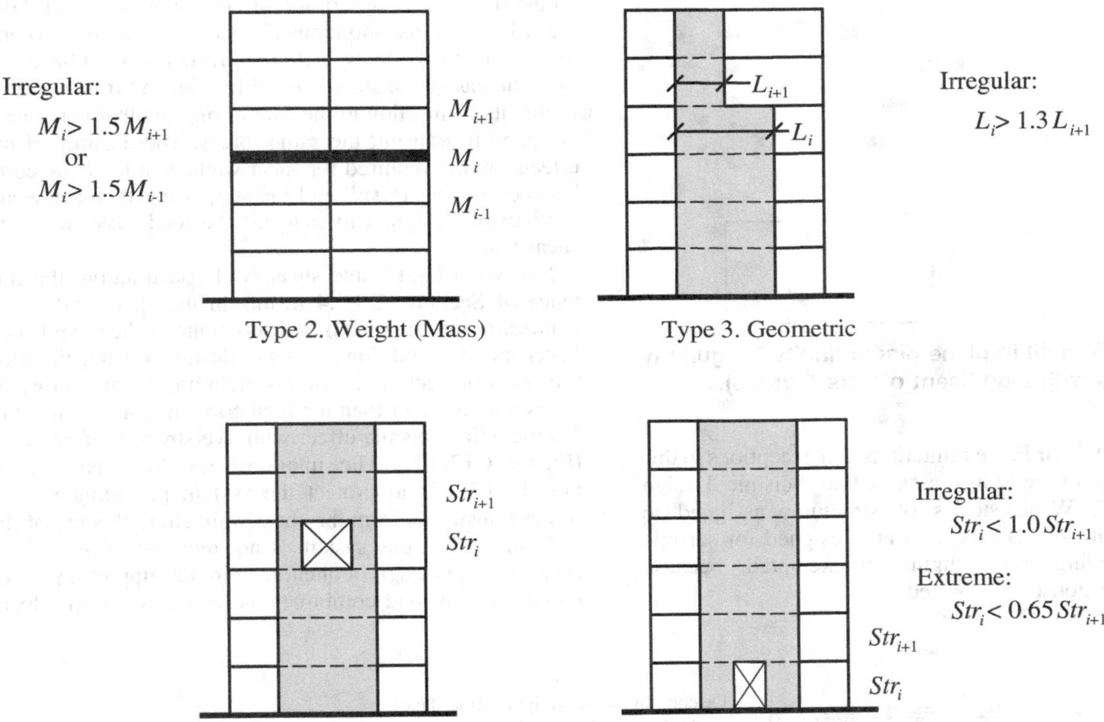

Type 4. Lateral Strength — Weak Story

Figure C12.3-2. Vertical structural irregularities.

Table 12.3-1 performed poorly, and collapsed at increasingly higher rates for increasing levels of torsional irregularity. However, the study also found that when the code provisions referenced in Table 12.3-1 were applied, the torsionally irregular structures performed at least as well as their torsionally regular counterparts. Collectively, the provisions referenced in Table 12.3-1 (including orthogonal combination, amplification of accidental torsion, redundancy provision triggers, etc.) require that torsionally irregular and extremely torsionally irregular buildings have increasingly more strength and stiffness in their lateral force resisting systems compared to their regular counterparts to meet the strength and drift requirements in ASCE 7. It is this increased strength and stiffness that protects these irregular structures from collapse. Because the code provisions referenced in Table 12.3-1 result in designs that compensate for their torsional properties with increased strength and stiffness, FEMA P-2012 found that there is no reason to prohibit these structures outright.

C12.3.3.2 Prohibited Vertical Irregularities for Seismic Design Category D See Commentary Section C12.3.3.1.

C12.3.3.3 Extreme Weak Stories for Seismic Design Category B or C Because extreme weak story irregularities are prohibited in Sections 12.3.3.1 and 12.3.3.2 for buildings located in Seismic

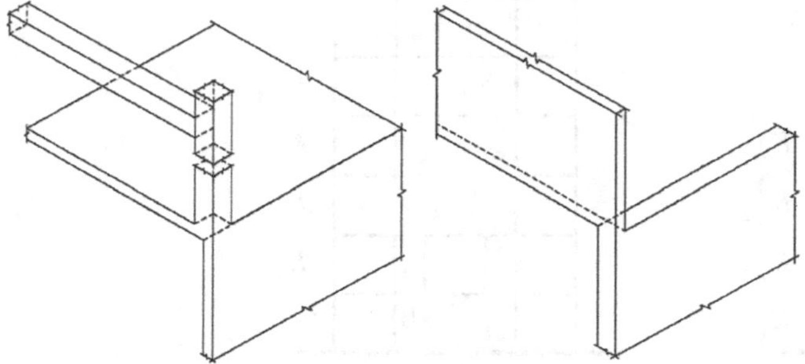

Figure C12.3-3. Vertical in-plane-discontinuity irregularity from columns or perpendicular walls (Type 3).

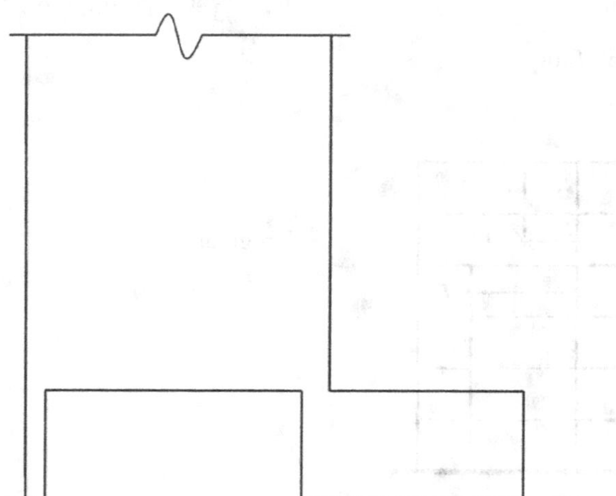

Figure C12.3-4. Vertical in-plane-discontinuity irregularity from walls with significant offsets (Type 3).

Design Categorie D, E, or F, the limitations and exceptions in this section apply only to buildings assigned to Seismic Design Category B or C. Weak stories of structures assigned to Seismic Design Category B or C that are designed for seismic load effects including overstrength are exempted, because reliable inelastic response is expected.

C12.3.3.4 Elements Supporting Discontinuous Walls or Frames The purpose of requiring elements (e.g., beams, columns, trusses, slabs, and walls) that support discontinuous walls or frames to be designed to resist seismic load effects including overstrength is to protect the gravity load-carrying system against possible overloads caused by overstrength of the seismic force-resisting system. Either columns or beams may be subject to such failure; therefore, both should include this design requirement. Beams may be subject to failure caused by overloads in either the downward or upward directions of force. Examples include reinforced concrete beams, the weaker top laminations of glued laminated beams, or unbraced flanges of steel beams or trusses. Hence, the provision has not been limited simply to downward force, but instead to the larger context of "vertical load." In addition, walls that support isolated point loads from frame columns or discontinuous perpendicular walls or walls with significant vertical offsets, as shown in Figures C12.3-3 and C12.3-4, can be subject to the same type of failure caused by overload.

The connection between the discontinuous element and the supporting member must be adequate to transmit the forces required for the design of the discontinuous element. For example, where the discontinuous element is required to comply with the seismic load effects, including overstrength in Section 12.4.3, as is the case for a steel column in a braced frame or a moment frame, its connection to the supporting member is required to be designed to transmit the same forces. These same seismic load effects are not required for shear walls, and thus, the connection between the shear wall and the supporting member would only need to be designed to transmit the loads associated with the shear wall.

For wood light-frame shear wall construction, the final sentence of Section 12.3.3.4 results in the shear and overturning connections at the base of a discontinuous shear wall (i.e., shear fasteners and hold-downs) being designed using the load combinations of Section 2.3 or 2.4 including seismic effect without overstrength rather than the load combinations of Section 2.3 or 2.4 including seismic effect with overstrength of Section 12.4.3 (Figure C12.3-5). The intent of the first sentence of Section 12.3.3.4 is to protect the system providing resistance to forces transferred from the shear wall; strengthening of the shear wall anchors to this system is not required to meet this intent. However, the design of anchorage in the supporting system may require use of load combinations including seismic load effect

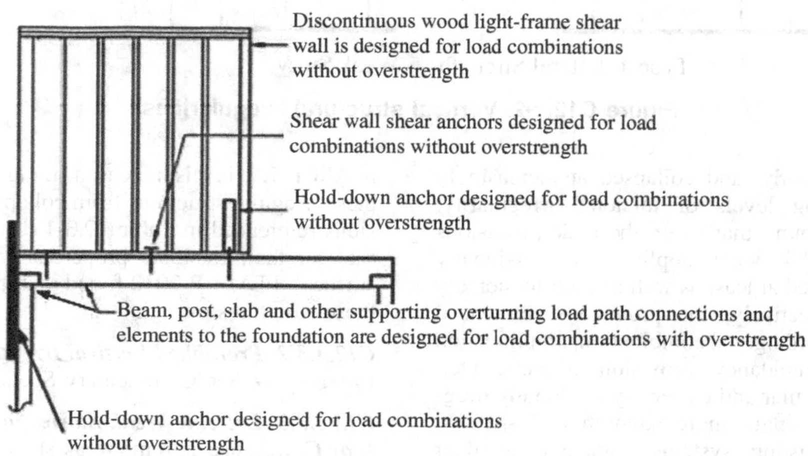

Figure C12.3-5. Elements supporting discontinuous wood light-frame shear wall.

with overstrength in accordance with the material design standard of the supporting system. For example, when anchoring to concrete, design of the concrete in accordance with ACI 318 may require use of load combinations including seismic load effect with overstrength.

C12.3.3.5 Increase in Forces Caused by Irregularities for Seismic Design Categories D through F The listed irregularities may result in loads that are distributed differently than those assumed in the equivalent lateral force procedure of Section 12.8, especially as related to the interconnection of the diaphragm with vertical elements of the seismic force-resisting system. The 25% increase in force is intended to account for this difference. The variation in diaphragm loading is only expected to significantly affect the diaphragm(s) at the level(s) of irregularities as discussed in FEMA P-2012. Consequently, the 25% increase in diaphragm forces is limited to the diaphragm level(s) where irregularities are present. Where the force is calculated using the seismic load effects including overstrength, no further increase is warranted.

C12.3.4 Redundancy The standard introduces a revised redundancy factor, ρ, for structures assigned to Seismic Design Category D, E, or F to quantify redundancy. The value of this factor is either 1.0 or 1.3. This factor has the effect of reducing the response modification coefficient, R, for less redundant structures, thereby increasing the seismic demand. The factor is specified in recognition of the need to address the issue of redundancy in the design.

The desirability of redundancy, or multiple lateral force-resisting load paths, has long been recognized. The redundancy provisions of this section reflect the belief that an excessive loss of story shear strength or development of a high TIR may lead to structural failure. The value of ρ determined for each direction may differ.

C12.3.4.1 Conditions Where Value of ρ is 1.0 This section provides a convenient list of conditions where ρ is 1.0.

C12.3.4.2 Redundancy Factor, ρ, for Seismic Design Categories D through F There are two approaches to establishing a redundancy factor, ρ, of 1.0. Where neither condition is satisfied, ρ is taken as equal to 1.3. It is permitted to take ρ equal to 1.3 without checking either condition. FEMA P-2012 showed that structures with extreme torsional irregularities in both orthogonal direction or structures with all of their lateral resistance located on one side of the center of mass, should include a ρ of 1.3 to achieve the same probability of collapse as non-irregular structures.

Redundancy is only checked for stories resisting a significant proportion of base shear. Parametric studies (conducted by Building Seismic Safety Council Technical Subcommittee 2 but unpublished) were used to select the 35% value. Those studies indicated that stories with story shears of at least 35% of the base shear include all stories of low-rise buildings (buildings up to five to six stories) and about 87% of the stories of tall buildings. The intent of this limit is to exclude the penthouses of most buildings and the uppermost stories of tall buildings from the redundancy requirements. This approach only applies where lateral-force resisting elements are located on both sides of the center of mass. The FEMA P-2012 study showed that structural configurations where all vertical resisting elements occurred on the same side of the center of mass should include a ρ of 1.3 to achieve the same probability of collapse as redundantly configured structures, and Section 12.3.4.2.a was updated in ASCE 7-22 accordingly. Note that it is possible that configurations where all lines of lateral resistance are on the same side of the center of mass could pass the requirements of Table 12.3-3 without triggering a high TIR, especially in shear wall buildings where the aspect ratio of the walls does not trigger the removal of a wall per Table 12.3-3.

The first approach requires the removal (or loss of resistance) of an individual lateral-force-resisting element or connection to determine its effect on the remaining structure. If the removal of elements, one by one, does not result in more than a 35% reduction in story strength or a high TIR, ρ may be taken as 1.0. The engineer may consider verifying that there is a viable load path to redistribute loads to the remaining elements, but this is not required by the provisions. For this evaluation, the determination of story strength may require an in-depth calculation. The intent of the check is to use a simple measure (elastic or plastic) to determine whether an individual member has a significant effect on the overall system strength. If a high TIR results, the resulting ρ is 1.3. The redundancy check for loss of story strength with the removal of an indicated element or connection is intended to be made without consideration of whether the remaining structure has sufficient capacity for a complete lateral load redistribution. The removal check is notional, meaning that this provision does not require the structure to be designed for the full loss of the element.

Table 12.3-3 describes the requirements for element or connection removal based on the lateral-force-resisting system (LFRS) type. The removal of long shear walls with height-to-length ratios less than or equal to 1.0 is not required for the evaluation of redundancy. If all lateral resisting elements in the direction considered consist of such long, stout walls, the layout is deemed redundant and ρ is 1.0. Removal of short, slender shear walls with height-to-length ratios greater than 1.0 is required to be checked, with the full length of a wall segment between openings or edges.

Lateral force resisting elements that are not noted in the table (such as concrete coupling beams) do not have requirements to be removed. However, specialty systems with elements similar to elements in Table 12.3-3 should be removed in accordance with the table. Diaphragms do not have any explicit removal requirements. However, it is noted that in certain configurations of buildings the loss of a line of lateral elements could result in a diaphragm (or its chords or collectors) that does not have adequate capacity without the removed element. While it is not required, the designer may want to consider this in the design of the building.

The second approach, Item (2), is a deemed-to-comply condition wherein the structure is regular and has a specified arrangement of seismic force-resisting elements to qualify for a ρ of 1.0. As part of the parametric study, simplified braced frame and moment frame systems were investigated to determine their sensitivity to the analytical redundancy criteria. This simple deemed-to-comply condition is consistent with the results of the study.

C12.4 SEISMIC LOAD EFFECTS AND COMBINATIONS

C12.4.1 Applicability Structural elements designated by the engineer as part of the seismic force-resisting system typically are designed directly for seismic load effects. None of the seismic forces associated with the design base shear are formally assigned to structural elements that are not designated as part of the seismic force-resisting system, but such elements must be designed using the load conditions of Section 12.4 and must accommodate the deformations resulting from application of seismic loads.

C12.4.2 Seismic Load Effect The seismic load effect includes horizontal and vertical components. The horizontal seismic load effects, E_h, are caused by the response of the structure to horizontal seismic ground motions, whereas the vertical seismic load effects are caused by the response of the structure to vertical seismic ground motions. The basic load combinations in Chapter 2 were duplicated and reformulated in Section 12.4 to clarify the intent of the provisions for the vertical seismic load effect term, E_v.

The concept of using an equivalent static load coefficient applied to the dead load to represent vertical seismic load effects was first introduced in ATC 3-06 (1978), where it was defined as simply $\pm 0.2D$. The load combinations where the vertical seismic load coefficient was to be applied assumed strength design load combinations. Neither ATC 3-06 nor the early versions of the NEHRP provisions (FEMA 2009a) clearly explained how the values of 0.2 were determined, but it is reasonable to assume that it was based on the judgment of the writers of those documents. It is accepted by the writers of this standard that vertical ground motions do occur and that the value of $\pm 0.2S_{DS}$ was determined based on consensus judgment. Many issues enter into the development of the vertical coefficient, including phasing of vertical ground motion and appropriate R factors, which make determination of a more precise value difficult. Although no specific rationale or logic is provided in editions of the NEHRP provisions (FEMA 2009a) on how the value of $0.2S_{DS}$ was determined, one possible way to rationalize the selection of the $0.2S_{DS}$ value is to recognize that it is equivalent to $(2/3)(0.3)S_{DS}$, where the 2/3 factor represents the often-assumed ratio between the vertical and horizontal components of motion, and the 0.3 factor represents the 30% in the 100%-to-30% orthogonal load combination rule used for horizontal motions.

For situations where the vertical component of ground motion is explicitly included in design analysis, the vertical ground motion spectra definition provided in Section 11.9 should be used. Following the rationale described above, the alternate vertical ground motion component determined in Section 11.9, S_{av}, is combined with the horizontal component of ground motion by using the 100%-to-30% orthogonal load combination rule used for horizontal motions, resulting in the vertical seismic load effect determined with Equation (12.4-4b), $E_v = 0.3S_{av}D$.

C12.4.2.1 Horizontal Seismic Load Effect Horizontal seismic load effects, E_h, are determined in accordance with Equation (12.4-3) as $E_h = \rho Q_E$. Q_E is the seismic load effect of horizontal seismic forces from V or F_p. The purpose of E_h is to approximate the horizontal seismic load effect from the design basis earthquake to be used in load combinations including E for the design of lateral-force-resisting elements including diaphragms, vertical elements of seismic force-resisting systems as defined in Table 12.2-1, the design and anchorage of elements such as structural walls, and the design of nonstructural components.

C12.4.2.2 Vertical Seismic Load Effect The vertical seismic load effect, E_v, is determined with Equation (12.4-4a) as $E_v = 0.2S_{DS}D$ or with Equation (12.4-4b) as $E_v = 0.3S_{av}D$. E_v is permitted to be taken as zero in Equations (12.4-1), (12.4-2), (12.4-5), and (12.4-6) for structures assigned to Seismic Design Category B and in Equation (12.4-2) for determining demands on the soil–structure interface of foundations. E_v increases the load on beams and columns supporting seismic elements and increases the axial load in the axial-moment (P–M) interaction of walls resisting seismic load effects.

C12.4.3 Seismic Load Effects Including Overstrength Some elements of properly detailed structures are not capable of safely resisting ground-shaking demands through inelastic behavior. To ensure safety, these elements must be designed with sufficient strength to remain elastic.

The horizontal load effect including overstrength may be calculated in either of two ways. The load effect may be approximated by use of an overstrength factor, Ω_0, which approximates the inherent overstrength in typical structures based on the structure's seismic force-resisting systems. This approach is addressed in Section 12.4.3.1. Alternatively, the expected system strength may be directly calculated based on actual member sizes and expected material properties, as addressed in Section 12.4.3.2.

C12.4.3.1 Horizontal Seismic Load Effect Including Overstrength Horizontal seismic load effects including overstrength, E_{mh}, are determined in accordance with Equation (12.4-7) as $E_{mh} = \Omega_0 Q_E$. Q_E is the effect of horizontal seismic forces from V, F_{px}, or F_p. The purpose of E_{mh} is to approximate the maximum seismic load for the design of critical elements, including discontinuous systems, transfer beams and columns supporting discontinuous systems, and collectors. Forces calculated using this approximate method need not be used if a more rigorous evaluation as permitted in Section 12.4.3.2 is used.

C12.4.3.2 Capacity-Limited Horizontal Seismic Load Effect The standard permits the horizontal seismic load effect including overstrength to be calculated directly using actual member sizes and expected material properties where it can be determined that yielding of other elements in the structure limits the force that can be delivered to the element in question. When calculated this way, the horizontal seismic load effect including overstrength is termed the capacity-limited seismic load effect, E_{cl}.

As an example, the axial force in a column of a moment-resisting frame results from the shear forces in the beams that connect to this column. The axial forces caused by seismic loads need never be taken as greater than the sum of the shear forces in these beams at the development of a full structural mechanism, considering the probable strength of the materials and strain-hardening effects. For frames controlled by beam hinge-type mechanisms, these shear forces would typically be calculated as $2M_{pr}/L_h$, where M_{pr} is the probable flexural strength of the beam considering expected material properties and strain hardening, and L_h is the distance between plastic hinge locations. Both ACI 318 (2019) and AISC 341 (2022) require that beams in special moment frames (SMFs) be designed for shear calculated in this manner, and both standards include many other requirements that represent the capacity-limited seismic load effect, E_{cl}, instead of the use of a factor approximating overstrength. This design approach is sometimes termed "capacity design." In this design method, the capacity (expected strength) of one or more elements is used to generate the demand (required strength) for other elements, because the yielding of the former limits the forces delivered to the latter. In this context, the capacity of the yielding element is its expected or mean anticipated strength, considering potential variation in material yield strength and strain-hardening effects. When calculating the capacity of elements for this purpose, expected member strengths should not be reduced by strength reduction or resistance factors, φ.

The capacity-limited design is not restricted to yielding limit states (axial, flexural, or shear); other examples include flexural buckling (axial compression) used in steel special concentrically braced frames, or lateral-torsional buckling in steel ordinary moment frame (OMF) beams, as confirmed by testing.

C12.4.4 Minimum Upward Force for Horizontal Cantilevers for Seismic Design Categories D through F In Seismic Design Categories D, E, and F, horizontal cantilevers are designed for an upward force that results from an effective vertical acceleration of 1.2 times gravity. This design requirement is meant to provide some minimum strength in the upward direction and to account for possible dynamic amplification of vertical ground motions resulting from the vertical flexibility of the cantilever. The requirement is not applied to downward forces on cantilevers, for which the typical load combinations are used.

C12.5 DIRECTION OF LOADING

Seismic forces are delivered to a building through ground accelerations that may approach from any direction relative to the orthogonal directions of the building; therefore, seismic effects are expected to develop in both directions simultaneously. The standard requires structures to be designed for the most critical loading effects from seismic forces applied in any direction. The procedures outlined in this section are deemed to satisfy this requirement.

For horizontal structural elements such as beams and slabs, orthogonal effects may be minimal; however, design of vertical elements of the seismic force-resisting system that participate in both orthogonal directions is likely to be governed by these effects.

C12.5.1 Direction of Loading Criteria For structures with orthogonal seismic force-resisting systems, the most critical load effects can typically be computed using a pair of orthogonal directions that coincide with the principal axes of the structure. Structures with nonparallel or nonorthogonal systems may require a set of orthogonal direction pairs to determine the most critical load effects. If a three-dimensional mathematical model is used, the analyst must be attentive to the orientation of the global axes in the model in relation to the principal axes of the structure.

C12.5.1.1 Independent Directional Procedure This procedure requires that the seismic forces be applied in two directions causing the most critical loading—typically along the principal X and Y axes of the building. There is no requirement for considering these loads simultaneously. This is allowed for structures where the vertical elements of the SFRS are unlikely to be strongly affected by ground motion from orthogonal directions and for structures in lower seismic zones.

C12.5.1.2 Orthogonal Directional Combination Procedure Section 12.5.1 requires consideration of the application of the seismic forces in the direction that causes the most critical effect for each element. While application of these forces in numerous analyses at angular increments is an acceptable means of achieving this, the orthogonal direction combination procedure of C12.5.1.2 offers more practical alternatives.

The orthogonal directional combination procedure combines the effects from 100% of the seismic load applied in one direction with 30% of the seismic load applied in the perpendicular direction. This general approximation, the "30% rule," was introduced by Rosenblueth and Contreras (1977) based on earlier work by A. S. Veletsos and also N. M. Newmark (cited by Rosenblueth and Contreras) as an alternative to other elastic-response-combination methods, such as the similar 100%+40% rule, the square root of the sum of the squares (SRSS), and the (elastic) response history analysis procedure.

Combining effects for seismic loads in each direction, and accidental torsion in accordance with Sections 12.8.4.2 and 12.8.4.3, results in the following 16 load combinations:

- $Q_E = \pm Q_{EX+AT} \pm 0.3Q_{EY}$, where Q_{EY} is the effect of Y-direction load at the center of mass (Section 12.8.4.2);
- $Q_E = \pm Q_{EX-AT} \pm 0.3Q_{EY}$, where Q_{EX} is the effect of X-direction load at the center of mass (Section 12.8.4.2);
- $Q_E = \pm Q_{EY+AT} \pm 0.3Q_{EX}$, where AT is the accidental torsion computed in accordance with Sections 12.8.4.2 and 12.8.4.3; and
- $Q_E = \pm Q_{EY-AT} \pm 0.3Q_{EX}$.

The maximum effect of seismic forces, Q_E, from orthogonal load combinations is modified by the redundancy factor, ρ, or the overstrength factor, Ω_0, where required, and the effects of vertical seismic forces, E_v, are considered in accordance with Section 12.4 to obtain the seismic load effect, E. The designer should also use the combinations above to determine the maximum story drift in each orthogonal axis as required by 12.12.1.

Though the standard permits combining effects from forces applied independently in any pair of orthogonal directions (to approximate the effects of concurrent loading), accidental torsion need not be considered in the direction that produces the lesser effect, in accordance with Section 12.8.4.2. This provision is sometimes disregarded when using a mathematical model for three-dimensional analysis that can automatically include accidental torsion, which then results in 32 load combinations.

In past standards, the square-root-of-the-sum-of-the-squares combination (SRSS) was explicitly identified as an acceptable means of satisfying the critical-direction requirement of 12.5.1. The 100%+30% rule was introduced as a modification of a 100% +40% rule based on unequal orthogonal ground motion components. (Typically, the direction of the weaker component is not defined, and so the 100%+30% rule requires 30%+100% to address all directions.) The 100%+40% rule corresponds to equal components, and was itself an approximation of the SRSS (Zimmerman et al. 2014). As such, the SRSS remains an acceptable alternative to the 100%+30% combination and is the theoretically correct method of combining the results of orthogonal Modal-Response-Spectrum Analyses (MRSAs) (Wilson 2014).

The complete quadratic combination (CQC) 3 directional combination is similar to the SRSS but allows consideration of unequal ground motion components in orthogonal directions by mathematically rotating the orientation of the strong and weak components such that every possible orientation is addressed. [Where components are equal, the CQC3 method simplifies to SRSS (Wilson 2014)]. Typically, this standard does not provide for unequal components. For near-field sites unequal components may be determined; however, the orientation of these components is known. As such, the CQC3 offers limited advantage in this standard except when used in combination with a method that may have unequal components (such as response spectra) and have significantly different response in each direction (e.g., moment frames versus braced frames).

Note that all of these methods, to varying degrees, result in force vectors larger than 100% of the required base-shear strength. The SRSS, for example, results in 141% of the required base shear, and thus may not provide suitable demands for elements that have resistance in all directions, such as piles in a mat foundation; such elements may be insensitive to the direction of loading and thus may not require any directional-combination procedure. In addition, using SRSS to determine drift values is often conservative.

Orthogonal effects can alternatively be considered by performing two- or three-dimensional (2D or 3D) response history analyses (see Section 12.9.2 and Chapter 16) with application of orthogonal ground motion pairs applied simultaneously in two orthogonal directions. If the structure is located near an active fault, the fault-normal and fault-parallel spectra may be unequal, and Chapter 16 requires that each ground motion pair be rotated to properly align with the causative fault.

These orthogonal combinations should not be confused with uniaxial modal combination rules, such as the square root of the sum of the squares (SRSS) or the complete quadratic combination (CQC) method. Modal combination rules are intended to account for the total response of a structure due to ground shaking in one horizontal direction, while directional combination rules address the total response of a structure due to ground shaking in two horizontal, orthogonal axes (Zimmerman et al. 2014).

Also note that all the preceding discussion is based on linear response, which is not accurate for response to MCE_R ground motions. The system overstrength factor, Ω_0, is required elsewhere (Section 12.4.3) to capture nonlinear effects on vulnerable elements in an approximate fashion.

C12.5.2 Seismic Design Category B Recognizing that design of structures assigned to Seismic Design Category (SDC) B is often controlled by nonseismic load effects and therefore is not sensitive to orthogonal loadings regardless of any horizontal structural irregularities, it is permitted to determine the most critical load effects by considering that the maximum response can occur in any single direction; simultaneous application of response in the orthogonal direction is not required. Typically, the two directions used for analysis coincide with the principal axes of the structure.

C12.5.3 Seismic Design Category C Design of structures assigned to Seismic Design Category C often parallels the design of structures assigned to Seismic Design Category B and therefore, as a minimum, conforms to Section 12.5.2. Although it is not likely that design of the seismic force-resisting system (SFRSs) in regular structures assigned to Seismic Design Category C would be sensitive to orthogonal loadings, special consideration must be given to structures with nonparallel or nonorthogonal systems (Type 5 horizontal structural irregularity) and structures with torsional irregularities (Type 1 horizontal structural irregularity) to avoid overstressing by different directional loadings. In these cases, the standard provides two methods to approximate simultaneous orthogonal loadings and requires a three-dimensional (3D) mathematical model of the structure for analysis in accordance with Section 12.7.3.

C12.5.4 Seismic Design Categories D through F Structures assigned to SDC D, E, or F have the highest seismic hazard. Thus, additional triggers are in place requiring the Orthogonal Directional Combination Procedure. Consistent with SDC C, structures with Type 5 horizontal irregularities (non-orthogonal systems) require the more stringent Orthogonal Directional Combination Procedure in accordance with 12.5.1.2. In addition to the trigger for a Type 5 horizontal irregularity, there are six additional triggers for requiring more rigorous analysis.

Item (b) triggers structures with horizontal Type 1 irregularities. The torsion causes significant seismic forces in the secondary direction when seismic forces are applied in the primary direction.

Item (c) relates to braced frame or moment frame systems in multiple planes that share a common column. These columns see significant increases in axial and/or flexural load when considering seismic forces in multiple directions. The trigger of 20% axial strength is for both tension and compression; this threshold is maintained from previous provisions (ASCE 7-16 and earlier). These requirements are consistent with the AISC 241 (2022) seismic provisions for steel braced frame and moment frame systems. It should be noted that AISC 341 (2022) may have more restrictive requirements for the design of intersecting frame columns (such as buckling restrained braced (BRBF) frames); these provisions do not remove the requirements set in the material standards.

Items (d), (e), and (f) relate to wall systems with elements in multiple planes that share common wall sections. These could be the intersecting elements of L, T, C, and box layouts. These configurations see significant increases in axial stress at the intersecting elements when seismic forces are applied in multiple directions. The threshold of 20% is selected to be consistent with columns [Item (c)] capturing cases where the seismic demands are significant. Therefore, wall systems with intersecting elements are required to be designed considering loading in both directions in accordance with Section 12.5.1.2.

The designer could choose to separate intersecting walls (often done in light-frame construction), each with their own tension/compression elements, to avoid the orthogonal requirement; however, it should be noted that there could be adverse effects to the performance of the building. If the end of a wall is in tension and when the adjacent end of an isolated wall is in compression, the separated (or isolated) walls impose severe local distortion on the horizontal structure (slabs, beams, joists), and design considerations similar to the deformation compatibility requirements of Section 12.12.5 are advisable, and in addition to the issue of gravity load capability, the effect on the diaphragm could be important.

Item (g) is for cantilever column systems. By their nature, cantilever columns resist loads in both directions and is important where the strength of the column is not the same in all directions. Even if columns are not designed to take loads in their weak axis they will take a portion of the load. Thus by not considering loads in both directions the column could be under designed for the flexural demands it will see.

Foundations should also be designed for orthogonal combinations if (a) the lateral-force-resisting element it supports is required to be designed for orthogonal combinations, or (b) the foundation supports multiple lateral-force-resisting elements that are not co-planar. This includes the connections to the foundations, the foundation elements, and the soil–structure interface (i.e., bearing pressure).

C12.6 ANALYSIS PROCEDURE SELECTION

ASCE 7-16 did not permit equivalent lateral force (ELF) procedure with certain height and irregularity conditions. This restriction was removed in ASCE 7-22. Chapter 12 has provisions for ELF, Modal Response Spectrum Analysis (MRSA), and Linear Response History Analysis (LRHA) procedures. The designer can select the procedure best suited to the building design needs. The only difference is that the MRSA and LRHA procedures always require 3D modeling whereas ELF permits 2D analysis in some cases.

The MRSA and LRHA procedures can provide excellent representations of the linear dynamic seismic response of buildings to a single input ground motion. Extending these linear procedures multiple ground motions by way of design spectral values, and including inelastic effects with spectral reductions for inelasticity, R, are not as reliable. A major conclusion of many studies was that the ELF procedure provided more consistent story shear, overturning moment and story drift results than the

MRSA procedure when compared to nonlinear dynamic response at design level earthquakes. However, there may be unusual situations where MRSA design values exceed those of ELF and there may be other reasons the engineer may wish to use MRSA rather than ELF.

C12.7 MODELING CRITERIA

C12.7.1 Foundation Modeling Structural systems consist of three interacting subsystems: the structural framing (girders, columns, walls, and diaphragms), the foundation (footings, piles, and caissons), and the supporting soil. The ground motion that a structure experiences, as well as the response to that ground motion, depend on the complex interaction among these subsystems.

Those aspects of ground motion that are affected by site characteristics are assumed to be independent of the structure–foundation system because these effects would occur in the free field in the absence of the structure. Hence, site effects are considered separately (Sections 11.4.3 through 11.4.5, Chapters 20 and 21).

Given a site-specific ground motion or response spectrum, the dynamic response of the structure depends on the foundation system and on the characteristics of the soil that supports the system. The dependence of the response on the structure–foundation–soil system is referred to as soil–structure interaction (SSI). Such interactions usually, but not always, result in a reduction of seismic base shear. This reduction is caused by the flexibility of the foundation–soil system and an associated lengthening of the fundamental period of vibration of the structure. In addition, the soil system may provide an additional source of damping. However, that total displacement typically increases with soil–structure interaction.

If the foundation is considered to be rigid, the computed base shears are usually conservative, and it is for this reason that rigid foundation analysis is permitted. The designer may neglect soil–structure interaction or may consider it explicitly in accordance with Section 12.13.3 or implicitly in accordance with Chapter 19.

As an example, consider a moment-frame building without a basement and with moment-frame columns supported on footings designed to support shear and axial loads (i.e., pinned column bases). If foundation flexibility is not considered, the columns should be restrained horizontally and vertically, but not rotationally. Consider a moment-frame building with a basement. For this building, horizontal restraint may be provided at the level closest to grade, as long as the diaphragm is designed to transfer the shear out of the moment frame. Because the columns extend through the basement, they may also be restrained rotationally and vertically at this level. However, it is often preferable to extend the model through the basement and provide the vertical and rotational restraints at the foundation elements, which is more consistent with the actual building geometry.

C12.7.2 Effective Seismic Weight During an earthquake, ground motion excites the structure from a static state of rest into dynamic movement, causing displacement and accelerations of the structural mass, resulting in inertial forces. Horizontal inertial forces, accumulated over the height of the structure, produce the seismic base shear.

ASCE 7 defines seismic weight, W, which has units of force. The standard defines loads as forces, and therefore it was preferred in the standard to use seismic weight (i.e., gravity load) multiplied by a unitless response coefficient, which is equivalent to multiplying a spectral acceleration by the seismic mass.

When a building vibrates during an earthquake, only that portion of the seismic weight that is strongly coupled or physically tied to the structure needs to be considered as effective. Hence, live loads (e.g., loose furniture, loose equipment, and human occupants) need not be included as seismic weight. However, certain types of live loads, such as storage loads, may develop inertial forces, particularly where they are densely packed. A minimum of 25% of the live load of stored material located on a floor above the seismic base is to be included in the effective seismic weight. Examples of occupancies or use types related to floor live loads that are commonly associated with areas used for storage include but are not limited to storage warehouses (light or heavy loading), wholesale stores (when functionally similar to warehouse use), library stack rooms, manufacturing (light or heavy loading), armories, computer rooms, office file rooms, media and archive storage rooms, cold storage, greenhouses, high-density storage systems, self-storage units, vault rooms, stockrooms, and parts storage.

Seismic weight must include the operational weight of permanent equipment. This provision affirms the inclusion of all equipment defined as dead load, such as fixed service equipment per Section 3.1.3, cranes and material handling systems per Section 3.1.1, and ballasted solar panels per Section 3.1.5. Where future installation of ballasted photovoltaic panel systems is anticipated, a collateral dead load should be considered as part of the seismic weight in the original design. In general, any fixed equipment that is well-defined and substantial enough to require specific accommodation in the gravity design or documentation on structural drawings should be included as seismic weight, whether considered as dead or live load. Where the gravity load design includes a live load allowance for fixed equipment, the weight of such equipment is still required to be included in the seismic weight, and a rational portion of the live load allowance should be included.

Fluids and bulk materials, such as water in a pool on a raised floor, or elevated areas containing granular products in a building, should be included as a rigid seismic weight unless fluid effects are explicitly considered in the analysis. This distinction is important to ensure the incorporation of seismic weight that may not be explicitly categorized as a dead or live load under the definitions in Chapters 3 and 4. Variations in fluid or bulk material loads and center of weight should be considered when significant fluctuations are expected during normal operation to determine all critical design effects in accordance with Section 1.3.6. The seismic interaction of a deformable structure with an internal fluid is known as fluid–structure interaction. Fluid–structure interaction seismic forces are categorized as sloshing (convective) and rigid (impulsive) seismic forces. Section 15.7.10.2 describes the conditions where fluid structure interaction can be applied to an elevated liquid storage tank. The seismic behavior of a liquid in an open pool located in a building is similar to the behavior of a liquid in an elevated tank. The proportion of these seismic forces depends on the geometry (height-to-diameter or height-to-width ratio) of the pool. The same conditions for elevated tanks should be met when considering the fluid–structure interaction of an open pool located in a building in determining the forces, effective period, and mass centroids of the system. These conditions are as follows:

(a) The sloshing period, T_c, is greater than $3T$, where T is the natural period of the supporting structure; and
(b) The sloshing mechanism (i.e., the percentage of convective mass and centroid) is determined for the specific configuration of the container by detailed fluid–structure interaction analysis or testing.

Fluid-interaction analysis methods can be found in many structural dynamics textbooks. See the commentary to Section 15.7 and the references cited therein.

It should be recognized that structures may be designed with special seismic systems where portions of their weight provide counteracting dynamic response, such as with the use of mass dampers. This provision is applicable for determining the total seismic weight of such a structure and its component systems, but the net effective seismic weight may appear much lower when using a dynamic analysis than for a typical structure. Note that it is required to use Chapter 18 for the design of structures that use damping systems.

Also considered as contributing to effective seismic weight are

1. All permanent equipment (e.g., air conditioners, elevator equipment, and mechanical systems);
2. Partitions to be erected or rearranged as specified in Section 4.3.2 (greater of actual partition weight and 10 lb/ft^2 (0.5 kN/m^2) of floor area);
3. 15% of significant snow load, $p_f > 45$ lb/ft^2 (2.1 kn/m^2); and
4. The weight of landscaping and similar materials.

The full snow load need not be considered because maximum snow load and maximum earthquake load are unlikely to occur simultaneously, and loose snow does not move with the roof.

C12.7.3 Structural Modeling The development of a mathematical model of a structure is always required, because the story drifts and the design forces in the structural members cannot be determined without such a model. In some cases the mathematical model can be as simple as a free-body diagram as long as the model can appropriately capture the strength and stiffness of the structure.

The most realistic analytical model is three-dimensional (3D), includes all sources of stiffness in the structure and the soil–foundation system as well as P-delta effects, and allows for nonlinear inelastic behavior in all parts of the structure–foundation–soil system. Development of such an analytical model is time-consuming, and such analysis is rarely warranted for typical building designs performed in accordance with the standard. Instead of performing a nonlinear analysis, inelastic effects are accounted for indirectly in the linear analysis methods by means of the response modification coefficient, R, and the deflection amplification factor, C_d.

Using modern software, it often is more difficult to decompose a structure into planar models than it is to develop a full three-dimensional model, so three-dimensional models are now commonplace. Increased computational efficiency also allows efficient modeling of diaphragm flexibility. Three-dimensional models are required where the structure has horizontal torsional (Type 1), out-of-plane offset (Type 4), or nonparallel system (Type 5) irregularities.

Analysis using a three-dimensional model is not required for structures with flexible diaphragms that have horizontal out-of-plane offset irregularities. It is not required because the irregularity imposes seismic load effects in a direction other than the direction under consideration (orthogonal effects) because of eccentricity in the vertical load path caused by horizontal offsets of the vertical lateral force-resisting elements from story to story. This situation is not likely to occur, however, with flexible diaphragms to an extent that warrants such modeling. The eccentricity in the vertical load path causes a redistribution of seismic design forces from the vertical elements in the story above to the vertical elements in the story below in essentially the same direction. The effect on the vertical elements in the orthogonal direction in the story below is minimal. Three-dimensional modeling may still be required for structures with flexible diaphragms caused by other types of horizontal irregularities (e.g., nonparallel system).

In general, the same three-dimensional model may be used for the equivalent lateral force (ELF), the modal response spectrum Analysis (MRSA), and the linear response history (LRH) analysis procedures. Modal response spectrum analysis (MRSA) and linear response history (LRH) analyses require a realistic modeling of structural mass; the response history method also requires an explicit representation of inherent damping. Five percent of critical damping is automatically included in the MRSA approach. Chapter 16 and the related commentary have additional information on linear and nonlinear response history analysis procedures.

It is well known that deformations in the panel zones of the beam–column joints of steel moment frames are a significant source of flexibility. Two different mechanical models for including such deformations are summarized in Charney and Marshall (2006). These methods apply to both elastic and inelastic systems. For elastic structures, centerline analysis provides reasonable, but not always conservative, estimates of frame flexibility. Fully rigid end zones should not be used, because this method always results in an overestimation of lateral stiffness in steel moment-resisting frames. Partially rigid end zones may be justified in certain cases, such as where doubler plates are used to reinforce the panel zone.

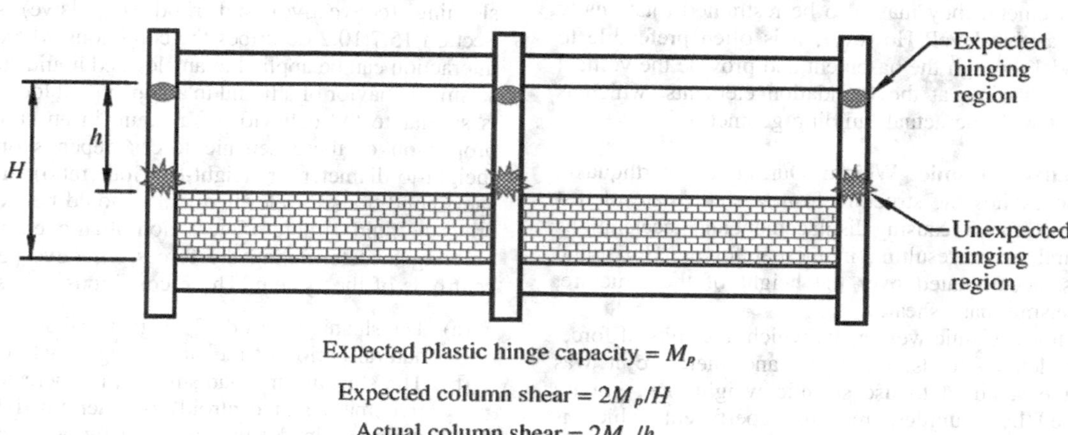

Expected plastic hinge capacity = M_p

Expected column shear = $2M_p/H$

Actual column shear = $2M_p/h$

Figure C12.7-1. Undesired interaction effects.

Including the effect of composite slabs in the stiffness of beams and girders may be warranted in some circumstances. Where composite behavior is included, due consideration should be paid to the reduction in effective composite stiffness for portions of the slab in tension (Schaffhausen and Wegmuller 1977, Liew et al. 2001).

For reinforced concrete buildings, it is important to address the effects of axial, flexural, and shear cracking in modeling the effective stiffness of the structural elements. Determining appropriate effective stiffness of the structural elements should take into consideration the anticipated demands on the elements, their geometry, and the complexity of the model. Recommendations for computing cracked section properties may be found in Paulay and Priestley (1992) and similar texts.

When dynamic analysis is performed, at least three dynamic degrees of freedom must be present at each level consistent with language in Section 16.2.2. Depending on the analysis software and modal extraction technique used, dynamic degrees of freedom and static degrees of freedom are not identical. It is possible to develop an analytical model that has many static degrees of freedom but only one or two dynamic degrees of freedom. Such a model does not capture response properly.

C12.7.4 Interaction Effects The interaction requirements are intended to prevent unexpected failures in members of moment-resisting frames. Figure C12.7-1 illustrates a typical situation where masonry infill is used and this masonry is fitted tightly against reinforced concrete columns. Because the masonry is much stiffer than the columns, hinges in a column form at the top of the column and at the top of the masonry rather than at the top and bottom of the column. If the column flexural capacity is M_p, the shear in the columns increases by the factor H/h, and this increase may cause an unexpected nonductile shear failure in the columns. Many building collapses have been attributed to this effect.

C12.8 EQUIVALENT LATERAL FORCE PROCEDURE

The equivalent lateral force (ELF) procedure provides a simple way to incorporate the effects of inelastic dynamic response into a linear static analysis.

The ELF procedure has three basic steps:

1. Determine the seismic base shear, V;
2. Distribute V vertically along the height of the structure; and
3. Distribute V horizontally across the width and breadth of the structure.

Each of these steps is based on a number of simplifying assumptions. A broader understanding of these assumptions may be obtained from any structural dynamics textbook that emphasizes seismic applications.

C12.8.1 Seismic Base Shear Treating the structure as a single-degree-of-freedom system with 100% mass participation in the fundamental mode, Equation (12.8-1) simply expresses V as the product of the effective seismic weight, W, and the seismic response coefficient, C_s, which is a period-dependent, unitless value that represents horizontal seismic spectral pseudoacceleration, divided by g. C_s is modified by the response modification coefficient, R, and the Importance Factor, I_e, as appropriate, to account for inelastic behavior and to provide improved performance for high-occupancy or essential structures.

C12.8.1.1 Calculation of Seismic Response Coefficient The multiperiod spectral method for defining ground motion permits a simplification for the equivalent lateral force (ELF) method, in which the seismic response coefficient is computed

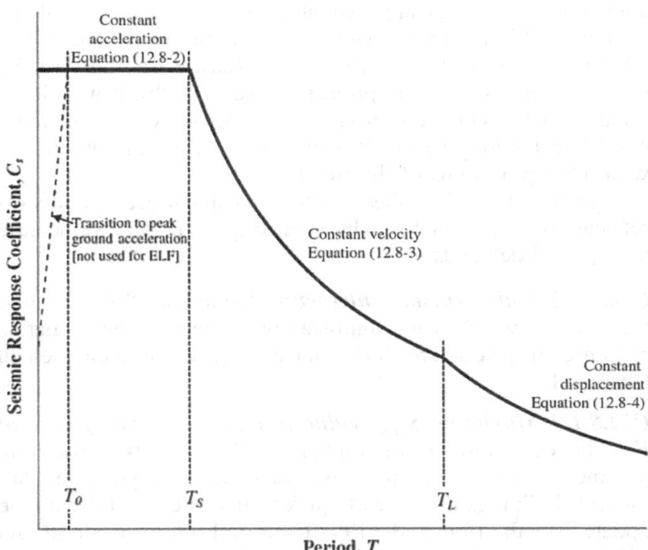

Figure C12.8-1a. Seismic response coefficient versus period.

from the design spectral acceleration S_a for the period of the structure by simply dividing by the response modification factor R, modified by the Importance Factor I_e where necessary. The upper bound on period, C_uT_a, does apply when using this procedure. It is still permissible to use the traditional design spectrum using the parameters S_{DS} and S_{D1}. The traditional method makes use of five equations for determining C_s. Equations (12.8-3), (12.8-4), and (12.8-5) are illustrated in Figure C12.8-1a.

Equation (12.8-3) controls where $0.0 < T < T_s$ and represents the constant-acceleration part of the design response spectrum (Section 11.4.5). In this region, C_s is independent of period. Although the theoretical design response spectrum shown in Figure 11.4-1 illustrates a transition in pseudoacceleration to the peak ground acceleration as the fundamental period, T, approaches zero from T_0, this transition is not used in the ELF procedure. One reason is that simple reduction of the response spectrum by $1/R$ in the short-period region would exaggerate inelastic effects.

Equation (12.8-4), representing the constant-velocity part of the spectrum, controls where $T_s < T < T_L$. In this region, the seismic response coefficient is inversely proportional to period, and the pseudovelocity (pseudoacceleration divided by circular frequency, ω, assuming steady-state response) is constant. T_L, the long-period transition period, represents the transition to constant displacement and is provided in Figures 22-12 through 22-16. T_L ranges from 4 s in the north-central conterminous states and western Hawaii to 16 s in the Pacific Northwest and western Alaska.

Equation (12.8-5), representing the constant-displacement part of the spectrum, controls where $T > T_L$. Given the current mapped values of T_L, this equation only affects long-period structures. The transition period has recently received increased attention because displacement response spectra from the 2010 magnitude 8.8 Chilean earthquake indicate that a considerably lower transition period is possible in locations controlled by subduction-zone earthquakes.

The final two equations represent minimum base shear levels for design. Equation (12.8-6) is the minimum base shear and primarily affects sites in the far field. This equation provides an allowable strength of approximately 3% of the weight of the structure. This minimum base shear was originally enacted in 1933 by the state of California (Riley Act). Based on research conducted in the ATC-63 project (FEMA 2009b), it was determined that this equation provides an adequate level of collapse resistance for long-period structures when used in conjunction with other provisions of the standard.

Equation (12.8-7) applies to sites near major active faults (as reflected by values of S_1), where pulse-type effects can increase long-period demands.

C12.8.1.2 Soil–Structure Interaction Reduction Soil–structure interaction, which can significantly influence the dynamic response of a structure during an earthquake, is addressed in Chapter 19.

C12.8.1.3 Maximum S_{DS} Value in Determination of C_s and E_v This cap on the maximum value of S_{DS} reflects engineering judgment about the performance of code-complying, regular, low-rise buildings in past earthquakes. It was created during the update from the 1994 to the 1997 Uniform Building Code (ICBO 1994, 1997) and has been carried through to this standard. At that time, near-source factors were introduced, which increased the design force for buildings in Zone 4, which is similar to Seismic Design Categories D through F in this standard. The near-source factor was based on observations of instrument recording during the 1994 Northridge earthquake and new developments in seismic hazard and ground motion science. The cap placed on S_{DS} for design reflected engineering judgment by the SEAOC Seismology Committee about the performance of code-complying low-rise structures based on anecdotal evidence from past California earthquakes, specifically the 1971 San Fernando, 1989 Loma Prieta, and 1994 Northridge earthquakes.

In the 1997 Uniform Building Code (UBC), the maximum reduction of the cap provided was 30%. Since the change from seismic zones in the 1997 UBC to probabilistic and deterministic seismic hazard in ASCE 7-02 (2003) and subsequent editions, S_{DS} values in some parts of the country can exceed $S_{DS} = 2.0$, creating reductions well beyond the original permitted reduction. That is the rationale for this provision providing a maximum reduction in design force of 30%.

The structural height, period, redundancy, presence of ductility, and regularity conditions required for use of the limit are important qualifiers. In addition, the observations of acceptable performance have been with respect to collapse and life safety, not damage control or preservation of function, so this cap on the design force is limited to Risk Category I and II structures, not Risk Category III and IV structures, where higher performance is expected. Also, because past earthquake experience has indicated that buildings on very soft soils, Site Classes E and F, have performed noticeably more poorly than buildings on more competent ground, this cap cannot be used on those sites.

C12.8.2 Period Determination The fundamental period, T, for an elastic structure is used to determine the design base shear, V, as well as the exponent, k, that establishes the distribution of V along the height of the structure (Section 12.8.3). T may be computed using a mathematical model of the structure that incorporates the requirements of Section 12.7 in a properly substantiated analysis. In general, this type of analysis is performed using a computer program that incorporates all deformational effects (e.g., flexural, shear, and axial) and accounts for the effect of gravity load on the stiffness of the structure. For many structures, however, the sizes of the primary structural members are not known at the outset of design. For preliminary design, as well as instances where a substantiated analysis is not used, the standard provides formulas to compute an approximate fundamental period, T_a (Section 12.8.2.1). These periods represent lower-bound estimates of T for different structure types. Period determination is typically computed for a mathematical model that is fixed at the base. That is, the base where seismic effects are imparted into the structure is globally restrained (e.g., horizontally, vertically, and rotationally). Column base modeling (i.e., pinned or fixed) for frame-type seismic force-resisting systems (SFRSs) is a function of frame mechanics, detailing, and foundation (soil) rigidity; attention should be given to the adopted assumption. However, this conceptual restraint is not the same for the structure as is stated above. Soil flexibility may be considered for computing T (typically assuming a rigid foundation element). The engineer should be attentive to the equivalent linear soil-spring stiffness used to represent the deformational characteristics of the soil at the base (Section 12.13.3). Similarly, pinned column bases in frame-type structures are sometimes used to conservatively account for soil flexibility under an assumed rigid foundation element. Period shifting of a fixed-base model of a structure caused by soil–structure interaction is permitted in accordance with Chapter 19.

The fundamental mode of a structure with a geometrically complex arrangement of seismic force-resisting systems (SFRSs) determined with a three-dimensional model may be associated with the torsional mode of response of the system, with mass participating in both horizontal directions (orthogonal) concurrently. The analyst must be attentive to this mass participation and recognize that the period used to compute the design base shear should be associated to the mode with the largest mass participation in the direction being considered. Often in this situation these periods are close to each other. Significant separation between the torsional mode period (when fundamental) and the shortest translational mode period may indicate an ill-conceived structural system or potential modeling error. The standard requires that the fundamental period, T, used to determine the design base shear, V, does not exceed the approximate fundamental period, T_a, times the upper limit coefficient, C_u, provided in Table 12.8-1. This period limit prevents the use of an unusually low base shear for design of a structure that is, analytically, overly flexible because of mass and stiffness inaccuracies in the analytical model.

C_u has two effects on T_a. First, recognizing that project-specific design requirements and design assumptions can influence T, C_u lessens the conservatism inherent in the empirical formulas for T_a to more closely follow the mean curve (Figure C12.8-2). Second, the values for C_u recognize that the formulas for T_a are targeted to structures in high seismic hazard locations. The stiffness of a structure is most likely to decrease in areas of lower seismicity, and this decrease is accounted for in the values of C_u. The response modification coefficient, R, typically decreases to account for reduced ductility demands, and the relative wind effects increase in lower seismic hazard locations. The design engineer must therefore be attentive to the value used for design of seismic force-resisting systems (SFRSs) in structures that are controlled by wind effects. Although the value for C_u is most likely to be independent of the governing design forces in high-wind areas, project-specific serviceability requirements may add considerable stiffness to a structure and decrease the value of C_u from considering seismic effects alone. This effect should be assessed where design forces for seismic and wind effects are almost equal. Lastly, if T from a properly substantiated analysis (Section 12.8.2) is less than $C_u T_a$, then the lower value of T and $C_u T_a$ should be used for the design of the structure.

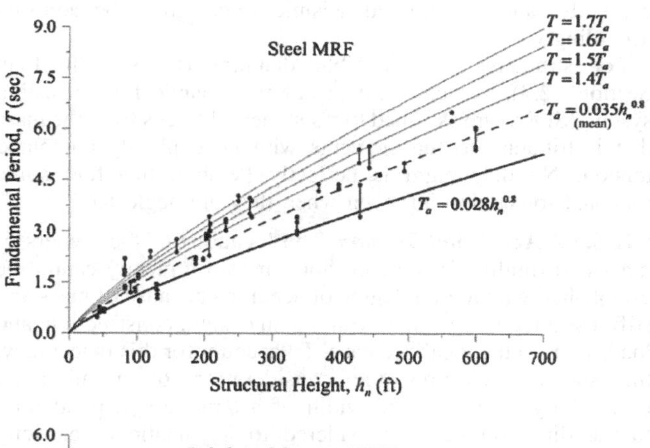

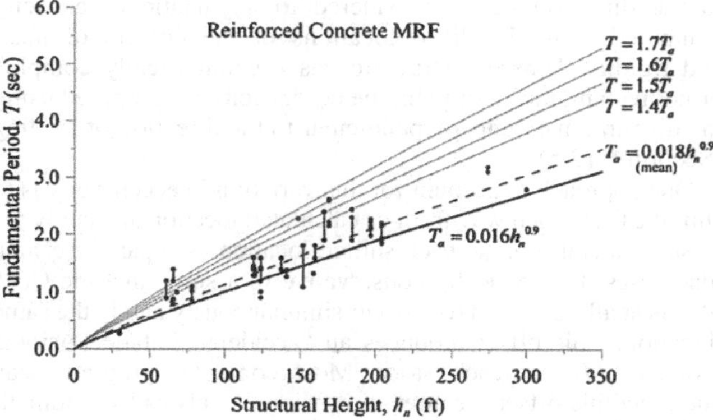

Figure C12.8-2. Variation of fundamental period with structural height.

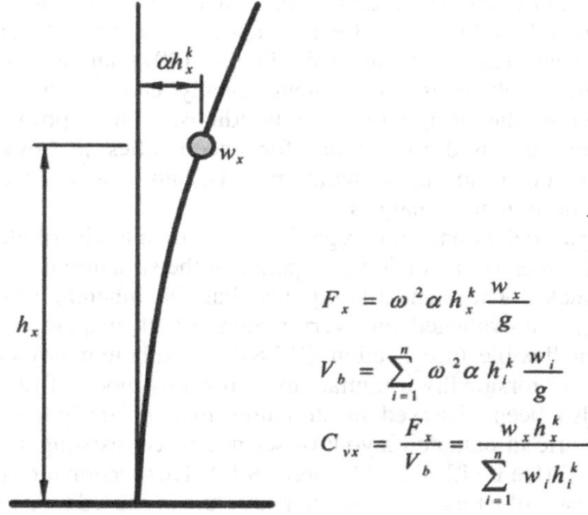

Figure C12.8-3. Basis of Equation (12.8-13).

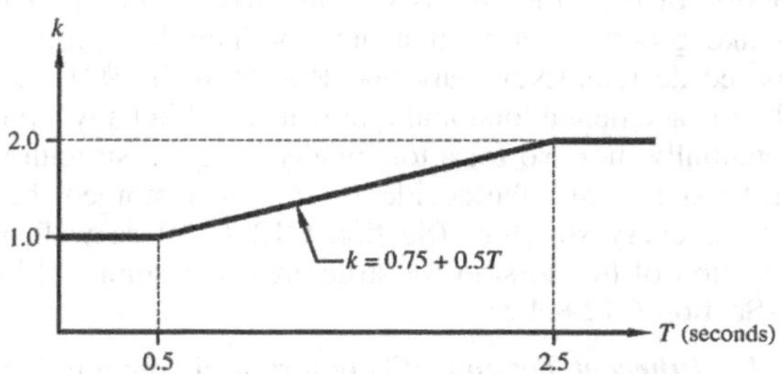

Figure C12.8-4. Variation of exponent *k* with period *T*.

C12.8.2.1 Approximate Fundamental Period Equation (12.8-8) is an empirical relationship determined through statistical analysis of the measured response of building structures in small-to-moderate-sized earthquakes, including response to wind effects (Goel and Chopra 1997, 1998). Figure C12.8-2 illustrates such data for various building structures with steel and reinforced concrete moment-resisting frames. Historically, the exponent x in Equation (12.8-8) has been taken as 0.75 and was based on the assumption of a linearly varying mode shape while using Rayleigh's method. The exponents provided in the standard, however, are based on actual response data from building structures, more accurately reflecting the influence of mode shape on the exponent. Because the empirical expression is based on the lower bound of the data, it produces a lower-bound estimate of the period for a building structure of a given height. This lower-bound period, when used in Equations (12.8-4) and (12.8-5) to compute the seismic response coefficient, C_s, provides a conservative estimate of the seismic base shear, V.

C12.8.3 Vertical Distribution of Seismic Forces Equation (12.8-13) is based on the simplified first-mode shape shown in Figure C12.8-3. In the figure, F_x is the inertial force at level x, which is simply the absolute acceleration at level x times the mass at level x. The base shear is the sum of these inertial forces, and Equation (12.8-12) simply gives the ratio of the lateral seismic force at level x, F_x, to the total design lateral force or shear at the base, V.

The deformed shape of the structure in Figure C12.8-3 is a function of the exponent k, which is related to the fundamental period of the structure, T. The variation of k with T is illustrated in Figure C12.8-4. The exponent k is intended to approximate the effect of higher modes, which are generally more dominant in structures with a longer fundamental period of vibration. Lopez and Cruz (1996) discuss the factors that influence higher modes of response. Although the actual first-mode shape for a structure is also a function of the type of seismic force-resisting system (SFRS), that effect is not reflected in these equations. Also, because T is limited to $C_u T_a$ for design, this mode shape may differ from that corresponding to the statistically based empirical formula for the approximate fundamental period, T_a. A drift analysis in accordance with Section 12.8.6 can be conducted using the actual period. Thus, k changes to account for the variation between T and the actual period.

The horizontal forces computed using Equation (12.8-12) do not reflect the actual inertial forces imparted on a structure at any particular point in time. Instead, they are intended to provide lateral seismic forces at individual levels that are consistent with enveloped results from more accurate analyses (Chopra and Newmark 1980).

C12.8.4 Horizontal Distribution of Forces Within the context of an ELF analysis, the horizontal distribution of lateral forces in a given story to various seismic force-resisting elements in that story depends on the type, geometric arrangement, and vertical extents of the structural elements and on the shape and flexibility of the floor or roof diaphragm. Because some elements of the seismic force-resisting system (SFRS) are expected to respond inelastically to the design ground motion, the distribution of forces to the various structural elements and other systems also depends on the strength of the yielding elements and their sequence of yielding (see Section C12.1.1). Such effects cannot be captured accurately by a linear-elastic static analysis (Paulay 1997), and a nonlinear dynamic analysis is too computationally cumbersome to be applied to the design of most buildings. Thus, approximate methods are used to account for uncertainties in horizontal distribution in an elastic static analysis, and to a lesser extent in elastic dynamic analysis.

Of particular concern in regard to the horizontal distribution of lateral forces is the torsional response of the structure during the earthquake. The standard requires that the inherent torsional moment be evaluated for every structure with diaphragms that are not flexible (see Section C12.8.4.1). Although primarily a factor for torsionally irregular structures, this mode of response has also been observed in structures that are designed to be symmetric in plan and layout of seismic force-resisting systems (SFRSs) (De La Llera and Chopra 1994). This torsional response in the case of a torsionally regular structure is caused by a variety of "accidental" torsional moments caused by increased eccentricities between the centers of rigidity and mass that exist because of uncertainties in quantifying the mass and stiffness distribution of the structure, as well as torsional components of earthquake ground motion that are not included explicitly in code-based designs (Newmark and Rosenblueth 1971). Consequently, the accidental torsional moment can affect any structure, and potentially more so for a torsionally irregular structure. The standard requires that the accidental torsional moment be considered for every structure (Section C12.8.4.2) as well as the amplification of this torsion for structures with torsional irregularity (Section C12.8.4.3).

C12.8.4.1 Inherent Torsion Where a rigid diaphragm is in the analytical model, the mass tributary to that floor or roof can be idealized as a lumped mass located at the resultant location on the floor or roof, termed the center of mass (CoM). This point represents the resultant of the inertial forces on the floor or roof. This diaphragm model simplifies structural analysis by reducing what would be many degrees of freedom in the two principal directions of a structure to three degrees of freedom (two horizontal and one rotational about the vertical axis). Similarly, the resultant stiffness of the structural members providing lateral stiffness to the structure tributary to a given floor or roof can be idealized as the center of rigidity (CoR).

It is difficult to accurately determine the CoR for a multistory building because the CoR for a particular story depends on the configuration of the seismic force-resisting elements above and below that story and may be load dependent (Chopra and Goel 1991). Furthermore, the location of the CoR is more sensitive to inelastic behavior than the CoM. If the CoM of a given floor or roof does not coincide with the CoR of that floor or roof, an inherent torsional moment, M_t, is created by the eccentricity between the resultant seismic force and the CoR. In addition to this *idealized* inherent torsional moment, the standard requires that an accidental torsional moment, M_{ta}, be considered (Section C12.8.4.2).

Similar principles can be applied to models of semirigid diaphragms that explicitly model the in-plane stiffness of the diaphragm, except that the deformation of the diaphragm needs to be included in computing the distribution of the resultant seismic force and inherent torsional moment to the seismic force-resisting system (SFRS).

This inherent torsion is included automatically when performing a three-dimensional analysis using either a rigid or semirigid diaphragm. If a two-dimensional planar analysis is used, where permitted, the CoR and CoM for each story must be determined explicitly, and the applied seismic forces must be adjusted accordingly.

For structures with flexible diaphragms (as defined in Section 12.3), vertical elements of the seismic force-resisting system (SFRS) are assumed to resist inertial forces from the mass that is tributary to the elements with no explicitly computed torsion. No diaphragm is perfectly flexible; therefore some torsional forces develop even when they are neglected.

C12.8.4.2 Accidental Torsion The locations of the centers of mass and rigidity for a given floor or roof typically cannot be established with a high degree of accuracy because of mass and stiffness uncertainty and deviations in design, construction, and loading from the idealized case. To account for this inaccuracy, the standard requires the consideration of a minimum eccentricity of 5% of the width of a structure perpendicular to the direction being considered to any static eccentricity computed using idealized locations of the centers of mass and rigidity. Where a structure has a geometrically complex or nonrectangular floor plan, the eccentricity is computed using the diaphragm extents perpendicular to the direction of loading (Section C12.5).

One approach to account for this variation in eccentricity is to shift the CoM each way from its calculated location and apply the seismic lateral force at each shifted location as separate seismic load cases. It is typically conservative to assume that the CoM offsets at all floors and roof occur simultaneously and in the same direction. This offset produces an "accidental" static torsional moment, M_{ta}, at each story. Most computer programs can automate this offset for three-dimensional analysis by automatically applying these static moments in the autogenerated seismic load case (along the global coordinate axes used in the computer model—see Section C12.5). Alternatively, user-defined torsional moments can be applied as separate load cases and then added to the seismic lateral force load case. For two-dimensional analysis, the accidental torsional moment is distributed to each seismic force-resisting system (SFRS) as an applied static lateral force in proportion to its relative elastic lateral stiffness and distance from the CoR.

Shifting the CoM is a static approximation and thus does not affect the dynamic characteristics of the structure, as would be the case were the CoM to be physically moved by, for example, altering the horizontal mass distribution and mass moment of inertia. Although this "dynamic" approach can be used to adjust the eccentricity, it can be too computationally cumbersome for static analysis and therefore is reserved for dynamic analysis (Section C12.9.1.5).

The previous discussion is applicable only to a rigid diaphragm model. A similar approach can be used for a semirigid diaphragm model, except that the accidental torsional moment is decoupled into nodal moments or forces that are placed throughout the diaphragm. The amount of nodal action depends on how sensitive the diaphragm is to in-plane deformation. As the in-plane stiffness of the diaphragm decreases, tending toward a flexible diaphragm, the nodal inputs decrease proportionally.

The physical significance of this mass eccentricity should not be confused with the physical meaning of the eccentricity required for representing nonuniform wind pressures acting on a structure. However, this accidental torsion also incorporates to a lesser extent the potential torsional motion input into structures with large footprints from differences in ground motion within the footprint of the structure.

Torsionally irregular structures whose fundamental mode is potentially dominated by the torsional mode of response can be more sensitive to dynamic amplification of this accidental torsional moment. Consequently, the 5% minimum can underestimate the accidental torsional moment. In these cases, the standard requires the amplification of this moment for design when using an elastic static analysis procedure, including satisfying the drift limitations (Section C12.8.4.3).

Accidental torsion results in forces that are combined with those obtained from the application of the seismic design story shears, V_x, including inherent torsional moments. All elements are designed for the maximum effects determined, considering positive accidental torsion, negative accidental torsion, and no accidental torsion (Section C12.5). Where consideration of earthquake forces applied concurrently in any two orthogonal directions is required by the standard, it is permitted to apply the 5% eccentricity of the CoM along the single orthogonal direction that produces the greater effect, but it need not be applied simultaneously in the orthogonal direction.

The exception in this section provides relief from accidental torsion requirements for buildings that are deemed to be relatively insensitive to torsion. It is supported by research (Debock et al. 2014) that compared the collapse probability (using nonlinear dynamic response history analysis) of buildings designed with and without accidental torsion requirements. The research indicated that, while accidental torsion requirements are important for most torsionally sensitive buildings (i.e., those with plan torsional irregularities arising from torsional flexibility or irregular plan layout), and especially for buildings in Seismic Design Category D, E or F, the implementation of accidental torsion provisions has little effect on collapse probability for Seismic Design Category B buildings with a torsional irregularity ratio (TIR) less than 1.4 and for Seismic Design Category D buildings without Type 1a or 1b irregularity.

C12.8.4.3 Amplification of Accidental Torsional Moment For structures with torsional irregularity (Type 1 horizontal structural irregularity) analyzed using the equivalent lateral force (ELF) procedure, the standard requires amplification of the accidental torsional moment to account for increases in the torsional moment caused by potential yielding of the perimeter seismic force-resisting systems (SFRSs) (i.e., shifting of the center of rigidity, CoR), as well as other factors potentially leading to dynamic torsional instability. For verifying torsional irregularity requirements in Table 12.3-1, story drifts resulting from the applied loads, which include both the inherent and accidental torsional moments, are used with no amplification of the accidental torsional moment ($A_x = 1$). The same process is used when computing the amplification factor, A_x, except that displacements (relative to the base) at the level being evaluated are used in lieu of story drifts. Displacements are used here to indicate that amplification of the accidental torsional moment is primarily a system-level phenomenon, proportional to the increase in acceleration at the extreme edge of the structure, and not explicitly related to an individual story and the components of the seismic force-resisting system (SFRS) contained therein.

Equation (12.8-15) was developed by the SEAOC Seismology Committee to encourage engineers to design buildings with good torsional stiffness; it was first introduced in the Uniform Building Code (ICBO 1988). Figure C12.8-5 illustrates the effect of Equation (12.8-15) for a symmetric rectangular building with various aspect ratios, L/B, where the seismic force-resisting elements are positioned at a variable distance (defined by α) from the center of mass (CoM) in each direction. Each element is assumed to have the same stiffness. The structure is loaded parallel to the short direction, with an eccentricity of $0.05L$.

For α equal to 0.5, these elements are at the perimeter of the building, and for α equal to 0.0, they are at the center (providing no torsional resistance). For a square building ($L/B = 1.00$), A_x is greater than 1.0 where α is less than 0.25 and increases to its maximum value of 3.0 where α is equal to 0.11. For a rectangular building with L/B equal to 4.00, A_x is greater than 1.0 where α is less than 0.34 and increases to its maximum value of 3.0 where α is equal to 0.15.

C12.8.5 Overturning The overturning effect on a vertical lateral-force-resisting element is computed based on the calculation of lateral seismic force, F_x, times the height from the base to the level of the horizontal lateral-force-resisting element that transfers F_x to the vertical element, summed over each story. Each vertical lateral-force-resisting element resists its

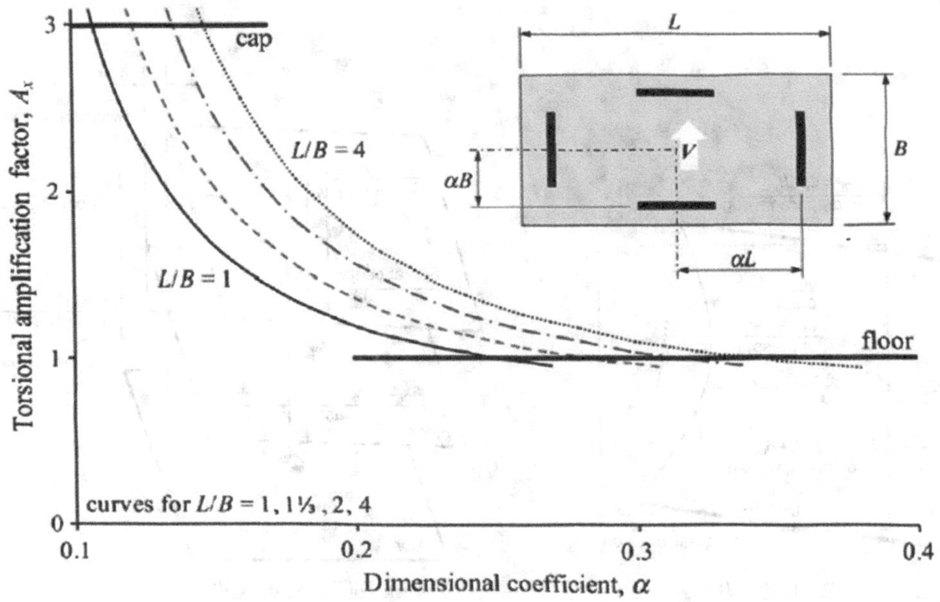

Figure C12.8-5. Torsional amplification factor for symmetric rectangular buildings.

portion of overturning based on its relative stiffness with respect to all vertical lateral-force-resisting elements in a building or structure. The seismic forces used are those from the equivalent lateral force (ELF) procedure determined in Section 12.8.3 or based on a dynamic analysis of the building or structure. The overturning forces may be resisted by dead loads and can be combined with dead and live loads or other loads, in accordance with the load combinations of Section 2.3.7.

C12.8.6 Displacement and Story Drift Determination This section defines three types of displacement or drift: the Design Earthquake Displacement (δ_{DE}); the Maximum Considered Earthquake Displacement (δ_{MCE}); and the Design Story Drift (Δ).

The Design Earthquake Displacement corresponds to the design earthquake. It is used for structural separations and deformation compatibility and is computed at the location of the element being evaluated. (Previous editions referred to this quantity as the "maximum inelastic response displacement.") There is thus a Design Earthquake Displacement at every point in the structure, although evaluations using this quantity are not required at every location. The Design Earthquake Displacement is used for structural separation (Section 12.12.2); deformation compatibility (Section 12.12.4); and nonstructural components (Section 13.3.2).

The Design Earthquake Displacement includes diaphragm deformation and rotation, as center-of-mass displacement could significantly underestimate displacement at the building corners and perimeter, and at non-rigid diaphragm locations away from the vertical elements of the seismic force-resisting system (SFRS, see Figure C12.8-1b). The diaphragm deformation corresponding to the design earthquake is required to be used. (The diaphragm deformation is represented by the term δ_{di}; diaphragm deformation is therefore not included in the elastic displacement δ_e, which, however, does include displacement and diaphragm-rotation effects of the seismic force-resisting system.) The engineer may determine that the diaphragm remains elastic under the expected demands (such as those corresponding to seismic loads including overstrength [Section 12.4.3] and considering diaphragm loading per Section 12.10.1) or may use rational methods of estimating its inelastic deformation; the engineer may determine that amplification by C_d/I_e is the appropriate method. The 2020 NEHRP Provisions provide guidance on force–displacement characteristics of diaphragms (see the white paper, "Calculation of Diaphragm Deflections under Seismic Loading," Part 3).

Maximum Considered Earthquake Displacement is used for members spanning between structures (Section 12.12.3). The Maximum Considered Earthquake Displacement corresponds to the MCE_R ground motion displacement and is similar to the Design Earthquake Displacement with two differences:

1. Displacement calculations include a factor of 1.5. This factor corrects for the 2/3 factor that is used in the calculation of seismic base shear, V, to reduce the base shear from the value based on the MCE_R ground motion (Section 11.4.4).
2. Displacements are calculated by multiplying elastic displacements by the response reduction coefficient, R, rather than the displacement amplification factor, C_d. Multiplying by R corrects for the fact that values of C_d less than R may substantially underestimate displacements for many seismic-force-resisting systems (Uang and Maarouf 1994). The degree of such underestimation and its variation among the various types of seismic-force-resisting systems is not known, and R is substituted for C_d in this provision pending more detailed information.

Design Story Drift, Δ, is a single representative value of interstory drift at each story corresponding to the design earthquake. The Design Story Drift is calculated as the difference in Design Earthquake Displacements at the center of mass (CoM) at each story (or at the diaphragm edge for torsionally irregular structures; see Figure C12.8-1b). For buildings with flexible diaphragms, the additional displacement at the reference location due to diaphragm deformation is allowed to be neglected, making the Design Story Drift calculation consistent with previous editions of the standard. In such cases, the Design Story Drift is inconsistent with the Design Earthquake Displacement, with implications addressed below. The Design Story Drift is used for comparison to allowable drift (Section 12.12.1) and for the calculation of the stability coefficient (Section 12.8.7).

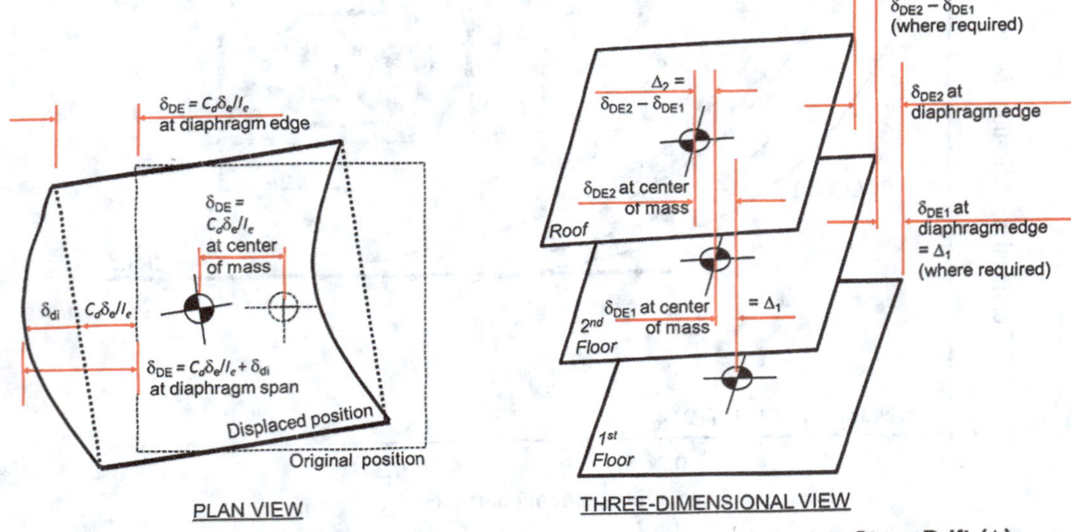

Figure C12.8-1b. Design Earthquake Displacement (δ_{DE}) and Design Story Drift (Δ).

Figure C12.8-1b shows the determination of the Design Earthquake Displacement and the Design Story Drift. In the plan view, three different Design Earthquake Displacements are shown: one at the CoM, one at the diaphragm edge (which includes the effects of diaphragm rotation), and one at the span midpoint (which includes the effects of diaphragm deformation). Each location in the structure has its own Design Earthquake Displacement, and at some locations the effects of diaphragm rotation and deformation both contribute to the total displacement. In the three-dimensional view, the determination of the Design Story Drift from either the CoM or diaphragm-edge Design Earthquake Displacement is illustrated. (The latter is only required for structures in Seismic Design Category C, D, E, or F with plan irregularity Type 1.) Diaphragm deformation is not required to be considered in the determination of the Design Story Drift and is not required for diaphragms that are permitted to be idealized as rigid per Section 12.3.1.2.

Where semirigid diaphragm modelling is performed and the engineer elects to compute the Design Story Drift at the CoM without including the diaphragm deformation, the engineer may determine the theoretical displacement at the CoM based on the displacements of the vertical elements of the seismic-force-resisting system (SFRS). In some cases, this can be approximated using a rigid-diaphragm analysis.

The drift corresponding to the Design Earthquake Displacement may exceed the Design Story Drift in certain cases, such as in buildings with highly flexible diaphragms. As the drift limits only apply to the Design Story Drift, the Design Earthquake drift (determined from the Design Earthquake Displacements) may therefore exceed the drift limit at those locations. Where this is the case the engineer should consider documenting the Design Earthquake Displacements if it is possible that the design of drift-sensitive building components, such as cladding and certain nonstructural attachments, will be performed by others under the incorrect assumption that the entire structure complies with the drift limits.

Table C12.7-1 summarizes the use of the Design Story Drift, the Design Earthquake Displacement, and the Maximum Considered Earthquake Displacement.

Where other standards refer to the Design Story Drift or design displacement, the engineer should consider whether the Design Earthquake Displacement (or the corresponding drift) is the appropriate quantity to use.

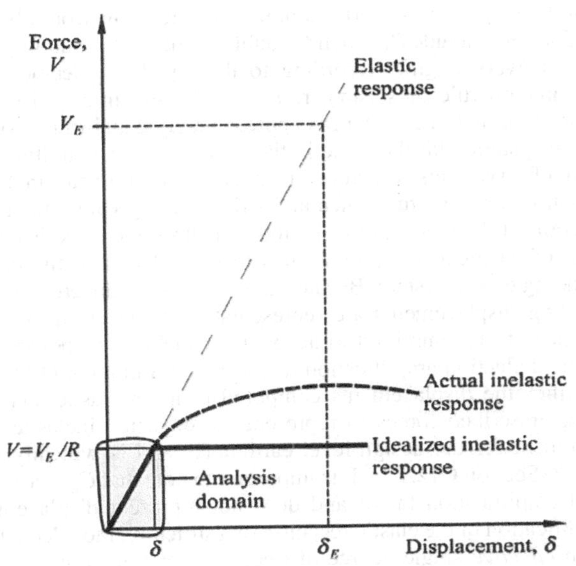

Figure C12.8-6. Displacements used to compute drift.

The Design Story Drifts must be less than the allowable story drifts, Δ_a, of Table 12.12-1. For structures without torsional irregularity, computations are performed using deflections of the centers of mass of the floors bounding the story. If the eccentricity between the centers of mass of two adjacent floors, or a floor and a roof, is more than 5% of the width of the diaphragm extents, it is permitted to compute the deflection for the bottom of the story at the point on the floor that is vertically aligned with the location of the center of mass of the top floor or roof. This situation can arise where a building has story offsets and the diaphragm extents of the top of the story are smaller than the extents of the bottom of the story. For structures assigned to Seismic Design Category C, D, E, or F that are torsionally irregular, the standard requires that deflections be computed along the edges of the diaphragm extents using two vertically aligned points.

Figure C12.8-6 illustrates the force–displacement relationships between elastic response, response to reduced design-level forces, and the expected inelastic response. If the structure remained elastic during an earthquake, the force developed

Table C12.7-1. Summary of Movement.

Movement	Δ, Design Story Drift	δDE, Design Earthquake Displacement (or Corresponding Drift)	δMCE, Maximum Considered Earthquake Displacement (or Corresponding Drift)
Location at which movement is determined	Center of mass (building edge for structures assigned to Seismic Design Categories C, D, E, or F with torsional irregularity)	Location of element	Location of element
Diaphragm deformation	Not required to be considered	Included	Included
Diaphragm rotation	Included	Included	Included
Second-order effect (12.8.7)	•		
Drift limit (12.12.1)	•		
Structural separation (12.12.2)		•	
Members spanning between structures (12.12.3)			•
Deformation compatibility (12.12.4)		•	
Nonstructural components (13.3.2)		•	

Notes: The dot (•) indicates requirement.

would be V_E, and the corresponding displacement would be δ_E. V_E does not include R, which accounts primarily for ductility and system overstrength. According to the equal-displacement-approximation rule of seismic response, the maximum displacement of an inelastic system is approximately equal to that of an elastic system with the same initial stiffness. This condition has been observed for structures idealized with bilinear inelastic response and a fundamental period, T, greater than T_s (Section 11.4.6). For shorter-period structures, the peak displacement of an inelastic system tends to exceed that of the corresponding elastic system. Because the forces are reduced by R, the resulting displacements are representative of an elastic system and need to be amplified to account for inelastic response.

The deflection amplification factor, C_d, in Equation (12.8-15) amplifies the displacements computed from an elastic analysis using prescribed forces to represent the expected inelastic displacement for the design-level earthquake and is typically less than R (Section C12.1.1). It is important to note that C_d is a story-level amplification factor and does not represent displacement amplification of the elastic response of a structure, modeled either as an *effective* single-degree-of-freedom structure (fundamental mode) or a constant amplification to represent the deflected shape of a multiple-degree-of-freedom structure, implying, in effect, that the mode shapes do not change during inelastic response. Furthermore, drift-level forces are different from the design-level forces used for strength compliance of the structural elements. Drift forces are typically lower, because the computed fundamental period can be used to compute the base shear (Section C12.8.6.2).

When conducting a drift analysis, the analyst should be attentive to the applied gravity loads used in combination with the strength-level earthquake forces so that consistency is maintained between the forces used in the drift analysis and those used for stability verification (P-delta) in Section 12.8.7, including consistency in computing the fundamental period if a second-order analysis is used. Further discussion is provided in Section C12.8.7.

The design forces used to compute the elastic deflection, δ_{xe}, include the Importance Factor, I_e, so Equation (12.8-16) includes I_e in the denominator. This inclusion is appropriate because the allowable story drifts in Table 12.12-1 (except for masonry shear wall structures) are more stringent for higher Risk Categories.

C12.8.6.1 Minimum Base Shear and Load Combination for Computing Displacement and Drift Except for period limits (as described in Section C12.8.6.2), all of the requirements of Section 12.8 must be satisfied when computing drift for an ELF analysis, except that the minimum base shear determined from applying Equation (12.8-6) does not need to be considered. This equation represents a minimum strength that needs to be provided to a system (Section C12.8.1.1). Equation (12.8-7) needs to be considered, when triggered, because it represents the increase in the response spectrum in the long-period range from near-fault effects.

C12.8.6.2 Period for Computing Displacement and Drift Where the design response spectrum of Section 11.4.6 or the corresponding equations of Section 12.8.1 are used and the fundamental period of the structure, T, is less than the long-period transition period, T_L, displacements increase with increasing period (even though forces may decrease). Section 12.8.2 applies an upper limit on T so that design forces are not underestimated, but if the lateral forces used to compute drifts are inconsistent with the forces corresponding to T, then displacements can be overestimated. To account for this variation in dynamic response, the standard allows the determination of displacements using forces that are consistent with the computed fundamental period of the structure, without the upper limit of Section 12.8.2.

The analyst must still be attentive to the period used to compute drift forces. The same analytical representation (see Section C12.7.3) of the structure used for strength design must also be used for computing displacements. Similarly, the same analysis method (Table 12.6-1) used to compute design forces must also be used to compute drift forces. It is generally appropriate to use 85% of the computed fundamental period to account for mass and stiffness inaccuracies as a precaution against overly flexible structures, but it need not be taken as less than that used for strength design. The more flexible the structure, the more likely it is that P-delta effects ultimately control the design (Section C12.8.7). Computed values of T that are significantly greater than $C_u T_a$ (perhaps more than 1.5 times in high seismic areas) may indicate a modeling error. Similar to the discussion in Section C12.8.2, the analyst should assess the value of C_u used where serviceability constraints from wind effects add significant stiffness to the structure.

C12.8.7 P-Delta Effects Figure C12.8-7 shows an idealized static force–displacement response for a simple one-story structure (e.g., idealized as an inverted-pendulum-type structure). As the top of the structure displaces laterally, the gravity load, P, supported by the structure acts through that displacement and produces an increase in overturning moment by P times the story drift, Δ, that must be resisted by the structure —the so-called P-delta (P-Δ) effect. This effect also influences the lateral displacement response of the structure from an applied lateral force, F.

The response of the structure not considering the P-delta effect is depicted by Condition 0 in the figure, with a slope of K_0 and lateral first-order yield force F_{0y}. This condition characterizes the first-order response of the structure (the response of the structure from an analysis not including P-delta effects). Where the P-delta effect is included (Condition 1 in the figure), the related quantities are K_1 and F_{1y}. This condition characterizes the second-order response of the structure (i.e., the response of the structure from an analysis including P-delta effects).

The geometric stiffness of the structure, K_G, in this example is equal to the gravity load, P, divided by the story height, h_{sx}. K_G is used to represent the change in lateral response by analytically

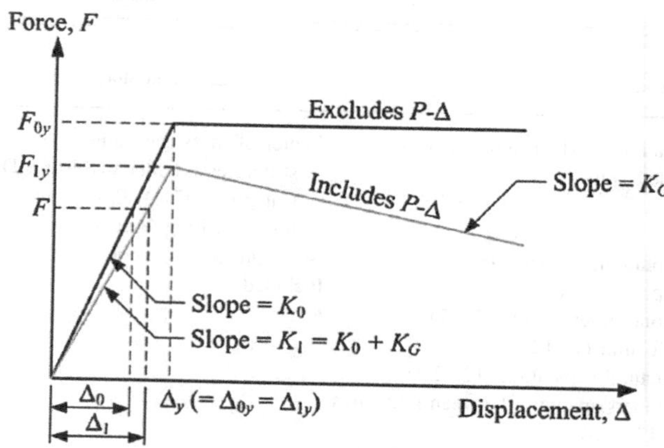

Figure C12.8-7. Idealized response of a one-story structure with and without the P-delta (P-Δ) effect.

reducing the elastic stiffness, K_0. K_G is negative where gravity loads cause compression in the structure. Because the two response conditions in the figure are for the same structure, the inherent yield displacement of the structure is the same ($\Delta_{0y} = \Delta_{1y} = \Delta_y$).

Two consequential points taken from the figure are (1) the increase in required strength and stiffness of the seismic force-resisting system (SFRS) where the P-delta effect influences the lateral response of the structure must be accounted for in design, and (2) the P-delta effect can create a negative stiffness condition during postyield response, which could initiate instability of the structure. Where the postyield stiffness of the structure may become negative, dynamic displacement demands can increase significantly (Gupta and Krawinkler 2000).

One approach that can be used to assess the influence of the P-delta effect on the lateral response of a structure is to compare the first-order response to the second-order response, which can be done using an elastic stability coefficient, θ, defined as the absolute value of K_G divided by K_0:

$$\theta = \frac{|K_G|}{K_0} = \left|\frac{P\Delta_{0y}}{F_{0y}h_{sx}}\right| \quad \text{(C12.8-1)}$$

Given this, and the geometric relationships shown in Figure C12.8-7, it can be shown that the force producing yield in Condition 1 (with P-delta effects) is

$$F_{1y} = F_{0y}(1-\theta) \quad \text{(C12.8-2)}$$

and that for a force, F, less than or equal to F_{1y},

$$\Delta_1 = \frac{\Delta_0}{1-\theta} \quad \text{(C12.8-3)}$$

Therefore, the stiffness ratio, K_0/K_1, is

$$\frac{K_0}{K_1} = \frac{1}{1-\theta} \quad \text{(C12.8-4)}$$

In the previous equations,

F_{0y} = Lateral first-order yield force;
F_{1y} = Lateral second-order yield force;
h_{sx} = Story height (or structure height in this example);
K_G = Geometric stiffness;
K_0 = Elastic first-order stiffness;
K_1 = Elastic second-order stiffness;
P = Total gravity load supported by the structure;
Δ_0 = Lateral first-order drift;
Δ_{0y} = Lateral first-order yield drift;
Δ_1 = Lateral second-order drift;
Δ_{1y} = Lateral second-order yield drift; and
θ = Elastic stability coefficient.

A physical interpretation of this effect is that to achieve the second-order response depicted in the figure, the seismic force-resisting system (SFRS) must be designed to have the increased stiffness and strength depicted by the first-order response. As θ approaches unity, Δ_1 approaches infinity and F_1 approaches zero, defining a state of static instability.

The intent of Section 12.8.7 is to determine whether P-delta effects are significant when considering the first-order response of a structure, and if so, to increase the strength and stiffness of the structure to account for the P-delta effects. Some material-specific design standards require P-delta effects to always be included in the elastic analysis of a structure and strength design of its members. The amplification of first-order member forces in accordance with Section 12.8.7 should not be misinterpreted to mean that these other requirements can be disregarded; nor should they be applied concurrently. Therefore, Section 12.8.7 is primarily used to verify compliance with the allowable drifts and check against potential postearthquake instability of the structure, while provisions in material-specific design standards are used to increase member forces for design, if provided. In so doing, the analyst should be attentive to the stiffness of each member used in the mathematical model so that synergy between standards is maintained.

Equation (12.8-18), shown as Equation (C12.8-5), is used to determine the elastic stability coefficient, θ, of each story of a structure:

$$\theta = \left|\frac{P\Delta_0}{F_0 h_{sx}}\right| = \frac{P_x/h_{sx}}{V_x/\Delta_{xe}} \quad \text{(C12.8-5)}$$

where h_{sx} and V_x are the same as defined in the standard;

F_0 = Force in a story causing $\Delta_0 = \sum F_x = V_x$;
Δ_0 = Elastic lateral first-order story drift;
V_x/Δ_{xe} = Story stiffness at level x, calculated as the ratio of the seismic shear force (V_x) divided by the corresponding elastic story drift (Δ_{xe}) calculated at the location(s) required by Section 12.8.6; and
P = Total point-in-time gravity load supported by the structure.

The provisions present the shear and displacement together as a ratio, V_x/Δ_{xe}, to make clear that the stiffness is the relevant measure to compare to the geometric, destabilizing effect of P_x/h_{sx}. It is expected that for convenience engineers will use the story shears determined using the procedures of Section 12.8 or 12.9. The engineer should use the corresponding elastic displacement, not the amplified drift defined by Equation (12.8-5).

Structures with θ less than 0.10 generally are expected to have a positive monotonic postyield stiffness. Where θ for any story exceeds 0.10, P-delta effects must be considered for the entire structure using one of the two approaches in the standard. Either first-order displacements and member forces are multiplied by $1/(1-\theta)$, or the P-delta effect is explicitly included in the structural analysis and the resulting θ is multiplied by $1/(1+\theta)$ to verify compliance with the first-order stability limit. Most commercial computer programs can perform second-order analysis. The analyst must therefore be attentive to the algorithm incorporated in the software and cognizant of any limitations, including suitability of iterative and noniterative methods, inclusion of second-order effects (P-Δ and P-δ) in automated modal analyses, and appropriateness of superposition of design forces.

Gravity load drives the increase in lateral displacements from the equivalent lateral forces. The standard requires that the total vertical design load, and the largest vertical design load for combination with earthquake loads, is given by combination 6 from Section 2.3.6, which is transformed to

$$(1.2 + 0.2S_{DS})D + 1.0L + 0.2S + 1.0E$$

where the 1.0 factor on L is actually 0.5 for many common occupancies. The provision of Section 12.8.7 allows the factor on dead load, D, to be reduced to 1.0 for the purpose of P-delta analysis under seismic loads. The vertical seismic component need not be considered for checking θ_{max}.

As explained in the commentary for Chapter 2, the 0.5 and 0.2 factors on L and S, respectively, are intended to capture the arbitrary point-in-time values of those loads. The factor 1.0 results in the dead load effect being fairly close to best estimates of the arbitrary point-in-time value for dead load. L is defined in Chapter 4 of the standard to include the reduction in live load based on floor area. Many commercially available computer programs do not include live load reduction in the basic structural analysis. In such programs, live load reduction is applied only in the checking of design criteria; this difference results in a conservative calculation with regard to the requirement of the standard.

The formulation of Equation (12.8-18) differs slightly from that in previous editions. Whereas the standard has always required the V_x and Δ used in this equation to be those occurring simultaneously, the formulation in previous editions made reference to the design story drift, requiring corrections for the displacement-amplification factor, the Importance Factor, the redundancy factor (if included in the story shear), and the potentially different base shears used to obtain story shear and story drift. The provision was revised for clarity, specifying that the shears and displacements are to be taken from the same loading. Displacements are not to be amplified by the displacement amplification factor, C_d. The magnitude of the lateral loading used for this calculation is unimportant; it is the lateral stiffness that is considered. Thus there is no reason to correct for the use of the Importance Factor or the redundancy factor in the story shear. The pattern of the lateral loading, however, may have some effect on the calculation, as the overturning from stories above contributes to the story drift. For this reason the provisions use the story shear and corresponding displacements instead of an arbitrary shear force at the story in question.

Equation (12.8-19) establishes the maximum stability coefficient, θ_{max}, permitted. The intent of this requirement is to protect structures from the possibility of instability triggered by post-earthquake residual deformation. The danger of such failures is real and may not be eliminated by apparently available overstrength. This problem is particularly true of structures designed in regions of lower seismicity.

For the idealized system shown in Figure C12.8-7, assume that the maximum displacement is $C_d\Delta_0$. Assuming that the unloading stiffness, K_u, is equal to the elastic stiffness, K_0, the residual displacement is

$$\left(C_d - \frac{1}{\beta}\right)\Delta_0 \qquad \text{(C12.8-6)}$$

In addition, assume that there is a factor of safety, FS, of 2 against instability at the maximum residual drift, $\Delta_{r,max}$. Evaluating the overturning and resisting moments ($F_0 = V_0$ in this example),

$$P\Delta_{r,max} \le \frac{V_0}{\beta FS}h \text{ where, } \beta = \frac{V_0}{V_{0y}} \le 1.0 \qquad \text{(C12.8-7)}$$

Therefore,

$$\frac{P[\Delta_0(\beta C_d - 1)]}{V_0 h} \le 0.5 \rightarrow \theta_{max}(\beta C_d - 1) = 0.5$$
$$\rightarrow \theta_{max} = \frac{0.5}{\beta C_d - 1} \qquad \text{(C12.8-8)}$$

Conservatively assume that $\beta C_d - 1 \approx \beta C_d$. Then

$$\theta_{max} = \frac{0.5}{\beta C_d} \le 0.25 \qquad \text{(C12.8-9)}$$

In the previous equations,

C_d = Displacement amplification factor,
FS = Factor of safety,
h_{sx} = Story height (or height of the structure in this example),
P = Total point-in-time gravity load supported by the structure,
V_0 = First-order story shear demand,
V_{0y} = First-order yield strength of the story,
β = Ratio of shear demand to shear capacity,
Δ_0 = Elastic lateral story drift,
$\Delta_{r,max}$ = Maximum residual drift at $V_0 = 0$, and
θ_{max} = Maximum elastic stability coefficient.

The standard requires that the computed stability coefficient, θ, not exceed 0.25 or $0.5/\beta C_d$, where βC_d is an adjusted ductility demand that takes into account the variation between the story strength demand and the story strength supplied. The story strength demand is simply V_x. The story strength supplied may be computed as the shear in the story that occurs simultaneously with the attainment of the development of first significant yield of the overall structure. To compute first significant yield, the structure should be loaded with a seismic force pattern similar to that used to compute story strength demand and iteratively increased until first yield. Alternatively, a simple and conservative procedure is to compute the ratio of demand to strength for each member of the seismic force-resisting system (SFRS) in a particular story and then use the largest such ratio as β. For an optimally designed system, β approaches φ, the resistance factor used in design. The ratio β should be calculated on the basis of specific system demands and capacities. It is incorrect to assume that $1/\beta$ is equal to the nominal overstrength value Ω_0 provided in the standard.

The principal reason for inclusion of β is to allow for a more equitable analysis of those structures in which substantial extra strength is provided, whether as a result of added stiffness for drift control, code-required wind resistance, or simply a feature of other aspects of the design. Some structures inherently possess more strength than required, but instability is not typically a concern. For many flexible structures, the proportions of the structural members are controlled by drift requirements rather than strength requirements; consequently, β is less than 1.0 because the members provided are larger and stronger than required. Equation C12.8-6 recognizes that, even when using the equal displacement theorem for yielding systems, the higher yield strength results in a smaller residual displacement, and thus β is placed as a factor on C_d.

The lower limit of $1.25/\Omega_0$ that is imposed on β is to ensure that $1/\beta$ is not greater than the nominal overstrength of the system. The factor of 1.25 is based on engineering judgement. In addition, the minimum value of Ω_0 in Table 12.2-1 is 1.25 (for Cantilever Column Systems), and the lower limit on β for these systems would be equal to the default value of 1.0.

Accurate evaluation of β would require consideration of all pertinent load combinations to find the maximum ratio of demand to capacity caused by seismic load effects in each member. A conservative simplification is to divide the total demand, with seismic load effects included, by the total capacity; this simplification covers all load combinations in which dead and live load effects add to seismic load effects. If a member is controlled by a load combination where dead load counteracts seismic load effects, to be correctly computed, β must be based only on the seismic component, not the total. The gravity load, P, in the P-delta computation would be less in such a circumstance, and therefore θ would be less. The importance of the counteracting load combination does have to be considered, but it rarely controls instability.

Although the P-delta procedure in the standard reflects a simple static idealization, as shown in Figure C12.8-7, the real issue is one of dynamic stability. To adequately evaluate second-order effects during an earthquake, a nonlinear response history analysis should be performed that reflects the variability of ground motions and system properties, including initial stiffness, strain hardening stiffness, initial strength, hysteretic behavior, and magnitude of point-in-time gravity load, P. Unfortunately, the dynamic response of structures is highly sensitive to such parameters, causing considerable dispersion to appear in the results (Vamvatsikos 2002). This dispersion, which increases dramatically with the stability coefficient, θ, is caused primarily by the incrementally increasing residual deformations (ratcheting) that occur during the response. Residual deformations may be controlled by increasing either the initial strength or the secondary stiffness. Gupta and Krawinkler (2000) give additional information.

For structural systems with a Torsional Irregularity Ratio (TIR) greater than 1.2, Section 12.7.3 of the provisions requires that the structure be modelled in 3D, and Section 12.8.6 requires that story drifts be evaluated at the edges of the structure. Where second-order effects are included in the analysis, it is possible that a significant P-theta (P-Θ) effect may occur, in which rotations about the vertical axis and edge displacements are amplified due to the presence of gravity loads (Flores et al. 2018).

The P-Θ effect can only be captured in a 3D analysis where geometric stiffness is included and where the gravity loads are accurately distributed over the surface of the diaphragms. Such an analysis also accounts for P-Delta effects. The combined second-order effect is referred to as the P-$\Delta\Theta$ effect, and analysis that includes this effect is referred to as the P-$\Delta\Theta$ analysis.

A second-order mathematical model that appropriately captures P-$\Delta\Theta$ effects in a one-story building is illustrated in Figure C12.8-8. In this structure all of the lateral-system columns and all of the gravity-only columns are explicitly modelled, and gravity forces resisted by these columns are computed on a tributary-area basis. These forces would automatically be

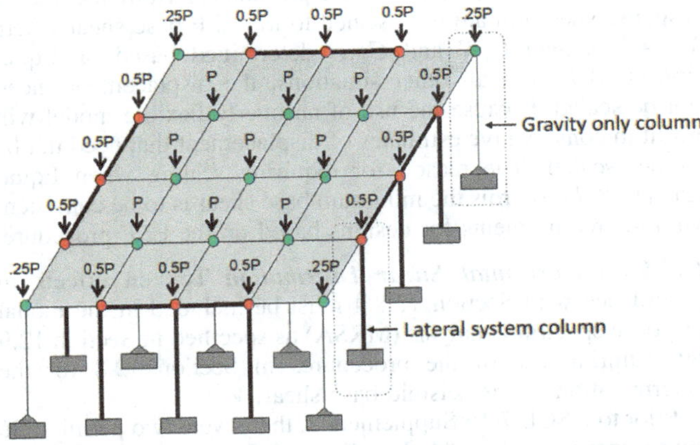

Figure C12.8-8. A mathematical model for incorporating P-Delta and P-theta effects.

determined by the computer software (or entered manually if the software does not have such capabilities). It is important to note that the gravity-only columns and walls, the gravity-only beams (where included), and the diaphragm elements should be modeled such that they do not contribute to lateral load resistance.

C12.9 LINEAR DYNAMIC ANALYSIS

C12.9.1 Modal Response Spectrum Analysis In the Modal Response Spectrum Analysis method, the structure is decomposed into a number of single-degree-of-freedom systems, each having its own mode shape and natural period of vibration. The number of modes available is equal to the number of mass degrees of freedom of the structure, so the number of modes can be reduced by eliminating mass degrees of freedom. For example, rigid diaphragm constraints may be used to reduce the number of mass degrees of freedom to one per story for planar models and to three per story (two translations and rotation about the vertical axis) for three-dimensional structures. However, where the vertical elements of the seismic force-resisting system have significant differences in lateral stiffness, rigid diaphragm models should be used with caution, because relatively small in-plane diaphragm deformations can have a significant effect on the distribution of forces.

For a given direction of loading, the displacement in each mode is determined from the corresponding spectral acceleration, modal participation, and mode shape. Because the sign (positive or negative) and the time of occurrence of the maximum acceleration are lost in creating a response spectrum, there is no way to recombine modal responses exactly. However, statistical combination of modal responses produces reasonably accurate estimates of displacements and component forces. The loss of signs for computed quantities leads to problems in interpreting force results where seismic effects are combined with gravity effects, producing forces that are not in equilibrium, and making it impossible to plot deflected shapes of the structure.

C12.9.1.1 Number of Modes The key motivation to perform modal response spectrum analysis (MRSA) is to determine how the actual distribution of mass and stiffness of a structure affects the elastic displacements and member forces. Where at least 90% of the modal mass participates in the response, the distribution of forces and displacements is sufficient for design. The scaling required by Section 12.9.1.4 controls the overall magnitude of design values so that incomplete mass participation does not produce nonconservative results.

The number of modes required to achieve 90% modal mass participation is usually a small fraction of the total number of modes. Lopez and Cruz (1996) contribute further discussion of the number of modes to use for MRSA.

In general, the provisions require modal analysis to determine all individual modes of vibration, but permit modes with periods less than or equal to 0.05 s to be collectively treated as a single, rigid mode of response with an assumed period of 0.05 s. In general, structural modes of interest to building design have periods greater than 0.05 s (frequencies greater than 20 Hz), and earthquake records tend to have little, if any, energy, at frequencies greater than 20 Hz. Thus, only "rigid" response is expected for modes with frequencies above 20 Hz. Although not responding dynamically, the "residual mass" of modes with frequencies greater than 20 Hz should be included in the analysis to avoid underestimation of earthquake design forces.

Section 4.3 of ASCE 4 (2000) provides formulas that may be used to calculate the modal properties of the residual-mass mode.

When using the formulas of ASCE 4 to calculate residual-mass mode properties, the "cutoff" frequency should be taken as 20 Hz, and the response spectral acceleration at 20 Hz (0.05 s) should be assumed to govern response of the residual-mass mode. It may be noted that the properties of the residual-mass mode are derived from the properties of modes with frequencies less than or equal to 20 Hz, such that modal analysis need only determine properties of modes of vibration with periods greater than 0.05 s (when the residual-mass mode is included in the modal analysis). The design response spectral acceleration at 0.05 s (20 Hz) should be determined using Equation (11.4-5) of this standard where the design response spectrum shown in Figure 11.4-1 is being used for the design analysis. Substituting 0.05 s for T and $0.2T_s$ for T_0 in Equation (11.4-5), one obtains the residual-mode response spectral acceleration as $S_a = S_{DS}\ (0.4 + 0.15/T_s)$. Most general-purpose linear structural analysis software has the capacity to consider residual mass modes to meet the requirements of ASCE 4.

The exception permits excluding modes of vibration when they would result in a modal mass in each orthogonal direction of at least 90% of the actual mass. This approach has been included in ASCE 7 (2003, 2010) for many years and is still considered adequate for most building structures that typically do not have significant modal mass in the very short-period range.

C12.9.1.2 Modal Response Parameters The design response spectrum (whether the general spectrum from Section 11.4.6 or a site-specific spectrum determined in accordance with Section 21.2) is representative of linear-elastic structures. Division of the spectral ordinates by the response modification coefficient, R, accounts for inelastic behavior, and multiplication of spectral ordinates by the Importance Factor, I_e, provides the additional strength needed to improve the performance of important structures. The displacements that are computed using the response spectrum that has been modified by R and I_e (for strength) must be amplified by C_d and reduced by I_e to produce the expected inelastic displacements (see Section C12.8.6.)

C12.9.1.3 Combined Response Parameters Most computer programs provide either the SRSS or the CQC method (Wilson et al. 1981) of modal combination. The two methods are identical where applied to planar structures, or where zero damping is specified for the computation of the cross-modal coefficients in the CQC method. The modal damping specified in each mode for the CQC method should be equal to the damping level that was used in the development of the design response spectrum. For the spectrum in Section 11.4.6, the damping ratio is 0.05.

The SRSS or CQC method is applied to loading in one direction at a time. Where Section 12.5 requires explicit consideration of orthogonal loading effects, the results from one direction of loading may be added to 30% of the results from loading in an orthogonal direction. Wilson (2000) suggests that a more accurate approach is to use the SRSS method to combine 100% of the results from each of two orthogonal directions where the individual directional results have been combined by SRSS or CQC, as appropriate.

The CQC4 method, as modified by ASCE 4 (2000), is specified (and is an alternative to the required use of the CQC method) where there are closely spaced modes with significant cross-correlation of translational and torsional response. The CQC4 method varies slightly from the CQC method through the use of a parameter that forces a correlation in modal responses where they are partially or completely in phase with the input motion. This difference primarily affects structures with short fundamental periods, T, that have significant components of response that are in phase with the ground motion. In these cases, using the CQC method can be nonconservative. A general overview of the various modal response combination methods can be found in the US Nuclear Regulatory Commission (2012).

The SRSS or CQC method is applied to loading in one direction at a time. Where Section 12.5 requires explicit consideration of orthogonal loading effects, the results from one direction of loading may be added to 30% of the results from loading in an orthogonal direction. Wilson (2000) suggests that a more accurate approach is to use the SRSS method to combine 100% of the results from each of two orthogonal directions where the individual directional results have been combined by SRSS or CQC, as appropriate. Menun and Der Kiureghian (1998) propose an alternate method, referred to as CQC3, which provides the critical orientation of the earthquake relative to the structure. Wilson (2000) now endorses the CQC3 method for combining the results from multiple component analyses.

C12.9.1.4 Scaling Design Values of Combined Response The modal base shear, V_t, may be less than the ELF base shear, V, because (a) the calculated fundamental period, T, may be longer than that used in computing V, (b) the response is not characterized by a single mode, or (c) the ELF base shear assumes 100% mass participation in the first mode, which is always an overestimate.

C12.9.1.4.1 Scaling of Forces The scaling required by Section 12.9.1.4.1 provides, in effect, a minimum base shear for design. This minimum base shear is provided because the computed fundamental period may be the result of an overly flexible (incorrect) analytical model. Recent studies of building collapse performance, such as those of FEMA P-695 (the ATC-63 project, 2009b), NIST GCR 10-917-8 (the ATC-76 project, NIST 2011), and NIST GCR 12-917-20 (the ATC-84 project, NIST 2012) show that designs based on the ELF procedure generally result in better collapse performance than those based on modal response spectrum analysis (MRSA) with the 15% reduction in base shear included. In addition, many of the designs using scaled MRSA did not achieve the targeted 10% probability of collapse given Maximum Considered Earthquake (MCE) ground shaking. Although scaling to 100% of the ELF base shear and to 100% of the drifts associated with Equation (12.8-7) does not necessarily achieve the intended collapse performance, it does result in performance that is closer to the stated goals of this standard.

C12.9.1.4.2 Scaling of Drifts Displacements from the modal response spectrum are only scaled to the ELF base shear where V_t is less than C_sW and C_s is determined based on Equation (12.8-7). For all other situations, the displacements need not be scaled, because the use of an overly flexible model will result in conservative estimates of displacement that need not be further scaled. The reason for requiring scaling when Equation (12.8-7) controls the minimum base shear is to be consistent with the requirements for designs based on the ELF procedure.

C12.9.1.5 Horizontal Shear Distribution Torsion effects in accordance with Section 12.8.4 must be included in the modal response spectrum analysis (MRSA) as specified in Section 12.9 by requiring use of the procedures in Section 12.8 for the determination of the seismic base shear, V.

Prior to ASCE 7-16 Supplement 2, there were two permissible approaches to account for the effects of accidental torsion for all structures. The first approach accounted for accidental torsion by applying a static accidental torsional moment to the MRSA results, and the second approach directly assessed the dynamic effects of accidental torsion by physically offsetting the center of mass (CoM) in the three-dimensional (3D) analysis model. The ATC-123 project found that the second method, physically offsetting the mass, could actually de-amplify the effects of accidental torsion for some buildings with torsional first modes, resulting in designs with significantly less collapse resistance than designs proportioned by applying a static accidental torsional moment to the MRSA results (see "Assessing Seismic Performance of Buildings with Configuration Irregularities: Calibrating Current Standards and Practices," in FEMA 2018). This occurs because the torsional response can become uncoupled from the translational response in extremely torsionally irregular structures, and Section 12.9 only requires translational response to be considered and scaled in a MRSA. This potential de-amplification effect was particularly pronounced for buildings with extreme torsional irregularities. Consequently, the second approach is now permissible only for structures with a TIR below the threshold of 1.6.

The permissible approach that can be used for all structures follows the static procedure discussed in Section C12.8.4.2, where the total seismic lateral forces obtained from MRSA—using the computed locations of the centers of mass (CoM) and rigidity—are statically applied at an artificial point offset from the CoM to compute the accidental torsional moments. Most computer programs can automate this procedure for three-dimensional analysis. Alternatively, the torsional moments can be statically applied as separate load cases and added to the results obtained from MRSA.

Because this approach is a static approximation, amplification of the accidental torsion in accordance with Section 12.8.4.3 is required. MRSA results in a single, positive response, inhibiting direct assessment of torsional response. One method to circumvent this problem is to determine the maximum and average displacements for each mode participating in the direction being considered and then apply modal combination rules (primarily the CQC method) to obtain the total displacements used to check torsional irregularity and compute the amplification factor, A_x. The analyst should be attentive about how accidental torsion is included for individual modal responses.

The alternate approach is only permissible for structures with a TIR below the threshold. This approach, which applies primarily to three-dimensional analysis, is to modify the dynamic characteristics of the structure so that dynamic amplification of the accidental torsion is directly considered. This modification can be done, for example, by either reassigning the lumped mass for each floor and roof (rigid diaphragm) to alternate points offset from the initially calculated CoM and modifying the mass moment of inertia, or physically relocating the initially calculated CoM on each floor and roof by modifying the horizontal mass distribution (typically presumed to be uniformly distributed). This approach increases the computational demand significantly, because all possible configurations have to be analyzed, primarily two additional analyses for each principal axis of the structure. The advantage of this approach is that the dynamic effects of direct loading and accidental torsion are assessed automatically. Practical disadvantages are the increased bookkeeping required to track multiple analyses and the cumbersome calculations of the mass properties.

Where this "dynamic" approach is used, in structures without an extreme torsional irregularity, amplification of the accidental torsion in accordance with Section 12.8.4.3 is not required, because repositioning the CoM increases the coupling between the torsional and lateral modal responses, directly capturing the amplification of the accidental torsion.

Most computer programs that include accidental torsion in a MRSA do so statically (permissible approach discussed above) and do not physically shift the CoM. The designer should be aware of the methodology used for consideration of accidental torsion in the selected computer program.

C12.9.1.6 P-Delta Effects The requirements of Section 12.8.7, including the stability coefficient limit, θ_{max}, apply to modal response spectrum analysis.

C12.9.1.7 Soil–Structure Interaction Reduction The standard permits including soil–structure interaction (SSI) effects in a modal response spectrum analysis (MRSA) in accordance with Chapter 19. The increased use of modal analysis for design stems from computer analysis programs automatically performing such an analysis. However, common commercial programs do not give analysts the ability to customize modal response parameters. This problem hinders the ability to include SSI effects in an automated modal analysis.

C12.9.1.8 Structural Modeling Using modern software, it often is more difficult to decompose a structure into planar models than it is to develop a full three-dimensional model. As a result, three-dimensional models are now commonplace. Increased computational efficiency also allows efficient modeling of diaphragm flexibility. Therefore, when MRSA is used, a three-dimensional model is required for all structures, including those with diaphragms that can be designated as flexible.

C12.9.2 Linear Response History Analysis

C12.9.2.1 General Requirements The linear response history (LRH) analysis method provided in this section is intended as an alternate to the modal response spectrum analysis (MRSA) method. The principal motivation for providing the LRH analysis method is that signs (positive–negative bending moments, tension–compression brace forces) are preserved, whereas they are lost in forming the SRSS and CQC combination in MRSA.

It is important to note that, like the ELF procedure and the MRSA method, the LRH analysis method is used as a basis for structural design, and not to predict how the structure will respond to a given ground motion. Thus, in the method provided in this section, spectrum-matched ground motions are used in lieu of amplitude-scaled motions. The analysis may be performed using modal superposition, or by analysis of the fully coupled equations of motion (often referred to as direct integration response history analysis).

As discussed in Section 12.9.2.3, the LRH analysis method requires the use of three sets of ground motions, with two orthogonal components in each set. These motions are then modified such that the response spectra of the modified motions closely match the shape of the target response spectrum. Thus, the maximum computed response in each mode is virtually identical to the value obtained from the target response spectrum. The only difference between the MRSA method and the LRH analysis method (as developed in this section using the spectrum-matched ground motions) is that in the MRSA method the system response is computed by statistical combination (SRSS or CQC) of the modal responses and in the LRH analysis method, the system response is obtained by direct addition of modal responses or by simultaneous solution of the full set of equations of motion.

C12.9.2.2 General Modeling Requirements Three-dimensional (3D) modeling is required for conformance with the inherent and accidental torsion requirements of Section 12.9.2.2.2.

C12.9.2.2.1 P-Delta Effects A static analysis is required to determine the stability coefficients using Equation (12.8-18). Typically, the mathematical model used to compute the quantity Δ does not directly include P-delta effects. However, Section 12.8.7 provides a methodology for checking compliance with the θ_{max} limit where P-delta effects are directly included in the model. For dynamic analysis, an ex post facto modification of results from an analysis that does not include P-delta effects to one that does (approximately) include such effects is not rational.

Given that virtually all software that performs linear response history (LRH) analysis has the capability to directly include P-delta effects, it is required that P-delta effects be included in all analyses, even when the maximum stability ratio at any level is less than 0.1. The inclusion of such effects causes a lengthening of the period of vibration of the structure, and this period should be used for establishing the range of periods for spectrum matching (Section 12.9.2.3.1) and for selecting the number of modes to include in the response (Section 12.9.2.2.4).

While the P-delta effect is essentially a nonlinear phenomenon (stiffness depends on displacements and displacements depend on stiffness), such effects are often "linearized" by forming a constant geometric stiffness matrix that is created from member forces generated from an initial gravity load analysis (Wilson and Habibullah 1987, Wilson 2004). This approach works for both the modal superposition method and the direct analysis method. It is noted, however, that there are some approximations in this method, principally the way the global torsional component of P-delta effects is handled. The method is of sufficient accuracy in analysis for which materials remain elastic. Where direct integration is used, a more accurate response can be computed by iteratively updating the geometric stiffness at each time step or by iteratively satisfying equilibrium about the deformed configuration. In either case, the analysis is in fact nonlinear, but it is considered as a linear analysis in Section 12.9.2 because the material properties remain linear.

For 3D models, it is important to use a realistic spatial distribution of gravity loads, because such a distribution is necessary to capture torsional P-delta effects.

C12.9.2.2.2 Accidental Torsion The required 5% offset of the center of mass (CoM) need not be applied in both orthogonal directions at the same time. Direct modeling of accidental torsion by offsetting the CoM is required to retain the signs (positive–negative bending moments, tension–compression forces in braces). In addition to the four mathematical models with mass offsets, a fifth model without accidental torsion (including only inherent torsion) must also be prepared. The model without accidental torsion is needed as the basis for scaling results as required in Section 12.9.2.5. Though not a requirement of the LRH analysis method, the analyst may also compare the modal characteristics (periods, mode shapes) to the systems with and without accidental mass eccentricity to gauge the sensitivity of the structure to accidental torsional response.

The mass offset approach for implementing accidental torsion in linear dynamic analysis should be used with caution for systems that have high TIR. The ATC 123 project (FEMA P-2012) recommended that the mass offset approach be prohibited for both modal MRSA and linear response history (LRH) analysis. Following this recommendation, the use of a mass offset was prohibited in MRS analysis for systems with Type 1 horizontal irregularity with Torsional Irregularity Ratio (TIR) greater than 1.6, in Supplement 2 to ASCE 7-16 and it is similarly prohibited in ASCE 7-22. In this case, accidental torsion must be implemented by equivalent lateral force (ELF) analysis with lateral forces applied at an eccentricity of $0.05A_x$ times the dimension of the building perpendicular to the direction of loading (referred to as static accidental torsion). Such a prohibition was not implemented for LRH analysis even though the underlying causes for the prohibition are the same for MRS and LRH analysis.

Where MRSA is used there is no significant disadvantage to using a static accidental torsion. In linear response history analysis, however, the use of a static accidental torsion eliminates several advantages, such as maintaining the signs of deformations and forces, evaluating force-moment interactions with concurrent rather than peak actions, and assessing total accelerations for use in developing floor spectra. Given these issues, implementing the static accidental torsion procedure in LRH analysis is not rational, and is not recommended. Thus, the only logical choice would be to prohibit linear response history analysis where the system has an extreme torsional irregularity.

Since such a prohibition has not been implemented in ASCE 7-22, the analyst should be aware of the possible implications of using a mass offset in LRH analysis. As shown in FEMA P-2012, systems with extreme torsional irregularities that are designed on the basis of MRSA with a mass offset were found to be considerably weaker, and in many cases unsafe, relative to designs using the ELF procedure with accidental torsion implemented by use of a force offset of $0.05A_x$ times the dimension of the building perpendicular to the direction of loading.

Since LRH analysis carried out in compliance with Section 12.9.2 is expected to provide displacements and forces similar to MRS analysis, there is reason to believe that the consequences of using a mass offset in LRH analysis would be similar. However, this was not specifically explored in the ATC-123 project. Until this issue can be resolved, it is recommended that LRH analysis incorporating mass offsets be used with caution in the design of building systems with Torsional Irregularity Ratio (TIR) greater than 1.6.

C12.9.2.2.3 Foundation Modeling Foundation flexibility may be included in the analysis. Where such modeling is used, the requirements of Section 12.13.3 should be satisfied. Additional guidance on modeling foundation effects may be found in *Nonlinear Structural Analysis for Seismic Design: A Guide for Practicing Engineers* (NIST 2010).

C12.9.2.2.4 Number of Modes to Include in Modal Response History Analysis Where modal response history analysis is used, it is common to analyze only a subset of the modes. In the past, the number of modes to analyze has been determined such that a minimum of 90% of the effective mass in each direction is captured in the response. An alternate procedure that produces participation of 100% of the effective mass is to represent all modes with periods less than 0.05 s in a single rigid body mode having a period of 0.05 s. In direct analysis, the question of the number of modes to include does not arise, because the system response is computed without modal decomposition.

An example of a situation where it would be difficult to obtain 90% of the mass in a reasonable number of modes is reported in Chapter 4 of FEMA P-751 (2013), which presents the dynamic analysis of a 12-story building over a one-story basement. When the basement walls and grade-level diaphragm were excluded from the model, 12 modes were sufficient to capture 90% of the effective mass. When the basement was modeled as a stiff first story, it took more than 120 modes to capture 90% of the total mass (including the basement and the ground-level diaphragm). Also, when the full structure was modeled and only 12 modes were used, the member forces and system deformations obtained were virtually identical to those obtained when 12 modes were used for the fixed-base system (modeled without the podium).

If modal response history analysis is used and it is desired to use a mathematical model that includes a stiff podium, it might be beneficial to use Ritz vectors in lieu of eigenvectors (Wilson 2004). Another approach is the "static correction method," in which the responses of the higher modes are determined by a static analysis instead of a dynamic analysis (Chopra 2007). The requirement in Section 12.9.2.2.4 of including all modes with periods of less than 0.05 s as a rigid body mode is in fact an implementation of the static correction method.

C12.9.2.2.5 Damping Where modal superposition analysis is used, 5% damping should be specified for each mode, because it is equal to the damping used in the development of the response spectrum specified in Sections 11.4.6 and 21.1.3. Where direct analysis is used, it is possible but not common to form a damping matrix that provides uniform damping across all modes (Wilson and Penzien 1972). It is more common to use a mass and stiffness proportional damping matrix (i.e., Rayleigh damping), but when this is done, the damping ratio may be specified at only two periods. Damping ratios at other periods depend on the mass and stiffness proportionality constants. At periods associated with higher modes, the damping ratios may become excessive, effectively damping out important modes of response. To control this effect, Section 12.9.2.2.5 requires the damping in all included modes (with periods as low as T_{lower}) be less than or equal to 5% critical.

C12.9.2.3 Ground Motion Selection and Modification Response spectrum matching (also called spectral matching) is the nonuniform scaling of an actual or artificial ground motion such that its pseudoacceleration response spectrum closely matches a target spectrum. In most cases, the target spectrum is the same spectrum used for scaling actual recorded ground motions (i.e., the ASCE 7 design spectrum). Spectral matching can be contrasted with amplitude scaling, in which a uniform scale factor is applied to the ground motion. The principal advantage of spectral matching is that fewer ground motions, compared to amplitude scaling, can be used to arrive at an acceptable estimate of the mean response, as recommended in NIST GCR 11-918-15 (2011). Figure C12.9-1(a) shows the response spectra of two ground motions that have been spectral matched, and Figure C12.9-1(b) shows the response spectra of the original ground motions. In both cases, the ground motions are normalized to match the target response spectrum at a period of 1.10 s. Clearly, the two amplitude-scaled records will result in significantly different responses, whereas analysis using the spectrum-matched records will be similar. As described later, however, there is enough variation in the response using spectrum-matched records to require the use of more than one record in the response history analysis.

A variety of methods are available for spectrum matching (see Hancock et al. 2006 for details). Additional information on use of spectrum-matched ground motions in response history analysis is provided by Grant and Diaferia (2012).

C12.9.2.3.1 Procedure for Spectrum Matching Experience with spectrum matching has indicated that it is easier to get a good match when the matching period extends beyond the period range of interest. It is for this reason that spectrum matching is required over the range $0.8T_{\mathrm{lower}}$ to $1.2T_{\mathrm{upper}}$. For the purposes of this section, a good match is defined when the ordinates of the average (arithmetic mean) of the computed acceleration spectrum from the matched records in each direction does not fall above or below the target spectrum by more than 10% over the period range of interest.

C12.9.2.4 Application of Ground Acceleration Histories One of the advantages of linear response history analysis is that analyses for gravity loads and for ground shaking may be computed separately and then combined in accordance with Section 12.4.2. Where linear response history analysis is performed in accordance with Section 12.9.2, it is required that each direction of response for each ground motion be computed independently. This requirement is based on the need to apply different scaling factors in the two orthogonal directions. Analyses with and without accidental torsion are required to be run for each ground motion. Thus, the total number of response histories that need to be computed is 18. (For each ground motion, one analysis is needed in each direction without mass eccentricity, and two analyses are needed in each

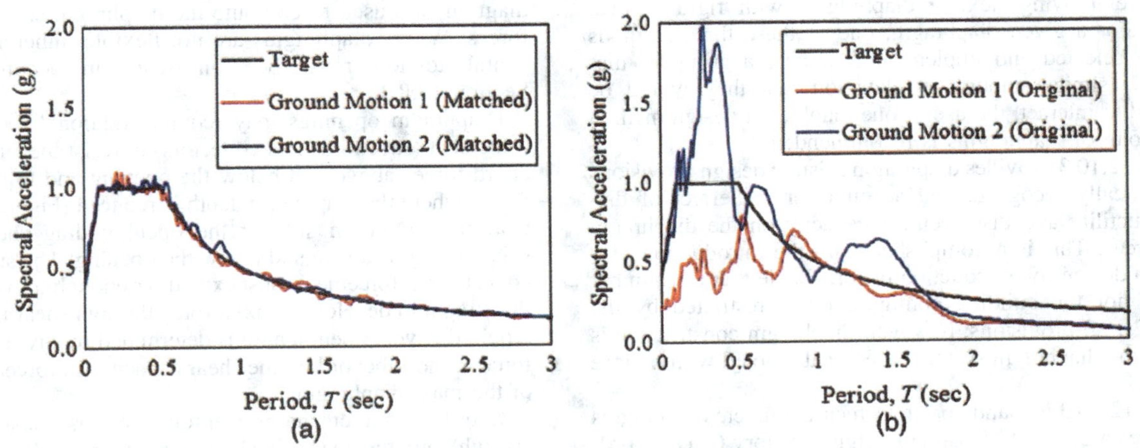

Figure C12.9-1. Spectral matching versus amplitude-scaled response spectra.

direction to account for accidental torsion. These six cases times three ground motions give 18 required analyses.)

C12.9.2.5 Modification of Response for Design The dynamic responses computed using spectrum-matched motions are elastic responses and must be modified for inelastic behavior.

For force-based quantities, the design base shear computed from the dynamic analysis must not be less than the base shear computed using the equivalent lateral force (ELF) procedure. The factors η_X and η_Y, computed in Section 12.9.2.5.2, serve that purpose. Next, the force responses must be multiplied by I_e and divided by R. This modification, together with the application of the ELF scale factors, is accomplished in Section 12.9.2.5.3.

For displacement base quantities, it is not required to normalize to ELF, and computed response history quantities need be multiplied only by the appropriate C_d/R in the direction of interest. This step is accomplished in Section 12.9.2.5.4.

Although accidental torsion is not required for determining the maximum elastic base shear, which is used only for determining the required base shear scaling, it is required for all analyses that are used to determine design displacements and member forces.

C12.9.2.6 Enveloping of Force Response Quantities Forces used in design are the envelope of forces computed from all analyses. Thus, for a brace, the maximum tension and the maximum compression forces are obtained. For a beam-column, envelope values of axial force and envelope values of bending moment are obtained, but these actions do not likely occur at the same time, and using these values in checking member capacity is not rational. The preferred approach is to record the histories of axial forces and bending moments, and to plot their traces together with the interaction diagram of the member. If all points of the force trace fall inside the interaction diagram, for all ground motions analyzed, the design is sufficient. An alternative is to record member demand-to-capacity ratio histories (also called usage ratio histories) and to base the design check on the envelope of these values.

C12.10 DIAPHRAGMS, CHORDS, AND COLLECTORS

This section permits a choice of diaphragm design in accordance with either Sections 12.10.1 and 12.10.2, Section 12.10.3, or the new provisions of Section 12.10.4. The diaphragm seismic design provisions in Sections 12.10.1 and 12.10.2 are the basic design method. Section 12.10.3 is an alternative method, first included in the 2016 edition of this standard. In Section 12.10.4, another alternative design method is provided. The Section 12.10.4 alternative method is permitted only for one-story structures employing flexible diaphragms with rigid vertical elements. For a given diaphragm, one of these three methods should be selected and implemented. Where a group of diaphragms is similar enough in elevation that they would be anticipated to interact, the use of one diaphragm design method for the group of diaphragms is recommended.

Section 12.10.3 provides diaphragm seismic design provisions that specifically recognize and account for the effect of diaphragm ductility and displacement capacity on the diaphragm design forces. This is accomplished with the introduction of a diaphragm design force reduction factor, R_s. Neither the number of stories nor the building configuration is restricted by the Section 12.10.3 provisions; however, diaphragm construction is limited to the diaphragm systems specifically noted within those provisions.

Section 12.10.3 is mandatory for precast concrete diaphragms in structures assigned to Seismic Design Category C, D, E, or F and is optional for precast concrete diaphragms in Seismic Design Category B, and cast-in-place concrete diaphragms, cast-in-place concrete equivalent precast diaphragms, wood, and bare steel deck diaphragms in structures assigned to all Seismic Design Categories. Precast concrete diaphragms and cast-in-place concrete equivalent precast diaphragms are defined in Section 14.2. The required mandatory use of Section 12.10.3 for precast diaphragm systems in Seismic Design Categories C through F buildings is based on recent research that indicates that improved earthquake performance can thus be attained. Many conventional diaphragm systems designed in accordance with Sections 12.10.1 and 12.10.2 have performed adequately. Continued use of Sections 12.10.1 and 12.10.2 is considered reasonable for diaphragm systems other than those for which Section 12.10.3 is mandated.

Section 12.10.4 introduces diaphragm seismic design provisions that are permitted to be used for one-story structures combining flexible diaphragms with rigid vertical elements. The seismic design methodology specifically recognizes the dynamic response of these structures as being dominated by dynamic response of and inelastic behavior in the diaphragm. While the most common occurrences of this structure type are the concrete tilt-up and masonry "big box" structures, the "rigid vertical element" terminology of this section recognizes a wider range of vertical elements for which this methodology is permitted to be used. The primary use of Section 12.10.4 provisions is intended to be for structures where one or more spans of the diaphragm exceed 100 ft (30.5 m); however, use for structures in which all diaphragm spans are less than 100 ft (30.5 m) is not precluded.

C12.10.1 Diaphragm Design Diaphragms are generally treated as horizontal deep beams or trusses that distribute lateral forces to the vertical elements of the seismic force-resisting system. As deep beams, diaphragms must be designed to resist the resultant shear and bending stresses. Diaphragms are commonly compared to girders, with the roof or floor deck analogous to the girder web in resisting shear, and the boundary elements (chords) analogous to the flanges of the girder in resisting flexural tension and compression. As in girder design, the chord members (flanges) must be sufficiently connected to the body of the diaphragm (web) to prevent separation and to force the diaphragm to work as a single unit.

Diaphragms may be considered flexible, semirigid, or rigid. The flexibility or rigidity of the diaphragm determines how lateral forces are distributed to the vertical elements of the seismic force-resisting system (see Section C12.3.1). Once the distribution of lateral forces is determined, shear and moment diagrams are used to compute the diaphragm shear and chord forces. Where diaphragms are not flexible, inherent and accidental torsion must be considered in accordance with Section 12.8.4.

Diaphragm openings may require additional localized reinforcement (subchords and collectors) to resist the subdiaphragm chord forces above and below the opening and to collect shear forces where the diaphragm depth is reduced (Figure C12.10-1). Collectors on each side of the opening drag shear into the subdiaphragms above and below the opening. The subchord and collector reinforcement must extend far enough into the adjacent diaphragm to develop the axial force through shear transfer. The required development length is determined by dividing the axial force in the subchord by the shear capacity (in force/unit length) of the main diaphragm.

Chord reinforcement at reentrant corners must extend far enough into the main diaphragm to develop the chord force through shear transfer (Figure C12.10-2). Continuity of the chord

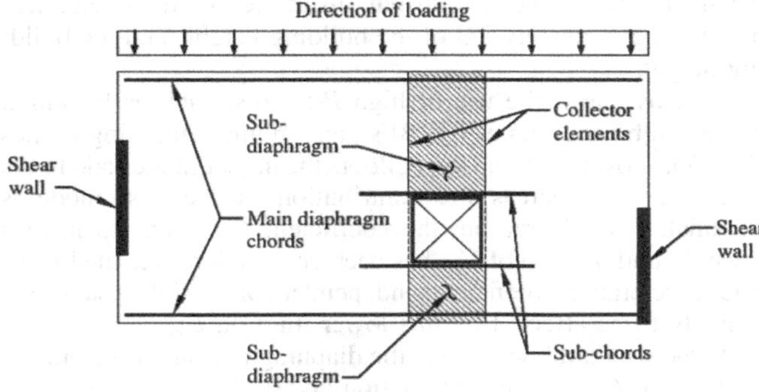

Figure C12.10-1. Diaphragm with an opening.

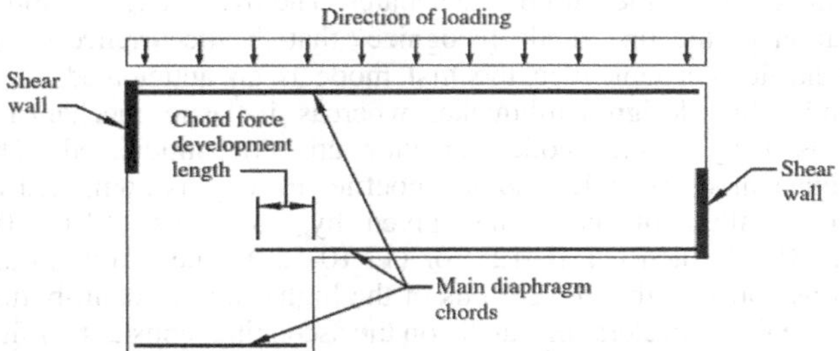

Figure C12.10-2. Diaphragm with a reentrant corner.

members also must be considered where the depth of the diaphragm is not constant.

In wood and metal deck diaphragm design, framing members are often used as continuity elements, serving as subchords and collector elements at discontinuities. These continuity members also are often used to transfer wall out-of-plane forces to the main diaphragm, where the diaphragm itself does not have the capacity to resist the anchorage force directly. For additional discussion, see Sections C12.11.2.2.3 and C12.11.2.2.4.

C12.10.1.1 Diaphragm Design Forces Diaphragms must be designed to resist inertial forces, and in addition to transfer forces acting through the diaphragm between vertical seismic force-resisting elements. Inertial forces are those seismic forces that originate at the diaphragm level, whereas transfer forces originate from locations other than the diaphragm level under consideration. Transfer forces between vertical seismic force-resisting elements may occur due to conditions such as, but not limited to, geometric offsets that cause the diaphragm to serve as an intermediary load path, differential stiffness between elements attached to a shared diaphragm through which deformation compatibility forces are induced, and interaction with external restraints. Examples of these conditions include horizontal out-of-plane offset irregularities in the vertical lateral-force-resisting system, shear-wall or brace-frame interaction effects with moment frames on separate lines in dual systems, and backstay effects. The redundancy factor, ρ (or the overstrength factor, Ω_0, where required), that is used for design of the seismic force-resisting elements also applies to diaphragm transfer forces, thus completing the load path.

The weight tributary to a diaphragm, w_{px}, may conservatively be taken as the weight of the level, however only lateral masses that directly induce diaphragm forces need to be considered. Lateral mass whose inertial forces do not pass through the diaphragm (such as the self-weight of walls parallel to the considered direction that resist their own in-plane forces) need not be included. The weight tributary to a diaphragm may be different along each orthogonal direction.

The seismic forces given in Equations (12.10-1) through (12.10-3) are usually larger than those obtained for the design of the vertical elements of the seismic force resisting system. In particular, the ELF procedure is designed to envelope the story shear demands, and may not envelope peak story accelerations. This procedure reflects the expected dynamic horizontal amplification of floor and roof motion relative to the vertical system that has been observed in earthquakes. Transfer forces are generally independent of the acceleration of diaphragm mass; thus their magnitude is not affected by Equations (12.10-1) through (12.10-3). Note that while Equation (12.10-3) limits the maximum result of Equation (12.10-1), it does not limit the design forces determined from the structural analysis, should they be larger.

C12.10.2.1 Collector Elements Requiring Load Combinations Including Overstrength for Seismic Design Categories C through F The overstrength requirement of this section is intended to keep inelastic behavior in the ductile elements of the seismic force-resisting system (consistent with the response modification coefficient, R) rather than in collector elements.

C12.10.3 Alternative Design Provisions for Diaphragms, Including Chords and Collectors The provisions of Section 12.10.3 are being mandated for precast concrete diaphragms in buildings assigned to Seismic Design Category C, D, E, or F and are offered as an alternative to those of Sections 12.10.1 and 12.10.2 for other precast concrete diaphragms, cast-in-place concrete diaphragms, and wood-sheathed diaphragms supported by wood framing. Diaphragms designed by Sections 12.10.1 and 12.10.2 have generally performed adequately in past earthquakes. The level of diaphragm design force from Sections 12.10.1 and 12.10.2 may not ensure, however, that diaphragms have sufficient strength and ductility to mobilize the inelastic behavior of vertical elements of the seismic force-resisting system. Analytical and experimental results show that actual diaphragm forces over much of the height of a structure during the design-level earthquake may be significantly greater than those from Sections 12.10.1 and 12.10.2, particularly when diaphragm response is near-elastic. There are material-specific factors that are related to overstrength and deformation capacity that may account for the adequate diaphragm performance in past earthquakes. The provisions of Section 12.10.3 consider both the significantly greater forces observed in near-elastic diaphragms and the anticipated overstrength and deformation capacity of diaphragms, improving the distribution of diaphragm strength over the height of buildings and among buildings with different types of seismic force-resisting systems.

Based on experimental and analytical data and observations of building performance in past earthquakes, changes are warranted to the procedures of Sections 12.10.1 and 12.10.2 for some types of diaphragms and for some locations within structures. Examples include the large diaphragms in some parking garages.

Section 12.10.3, Item 1, footnote b to Table 12.2-1, permits reducing Ω_0 for structures with flexible diaphragms. The lower Ω_0 results in lower diaphragm forces, which is not consistent with experimental and analytical observations. Justification for footnote b is not apparent; therefore, to avoid the inconsistency, the reduction is eliminated when using the Section 12.10.3 design provisions.

Section 12.10.3, Item 2. Section 12.3.3.5 provision requiring a 25% increase in design forces for certain diaphragm elements in buildings with several listed irregularities is eliminated when using the Section 12.10.3 design provisions, because the diaphragm design force level in this section is based on realistic assessment of anticipated diaphragm behavior. Under the Sections 12.10.1 and 12.10.2 design provisions, the 25% increase is invariably superseded by the requirement to amplify seismic design forces for certain diaphragm elements by Ω_0; the only exception is wood diaphragms, which are exempt from the Ω_0 multiplier.

Section 12.10.3, Items 3 and 4. Section 12.10.3.2 provides realistic seismic design forces for diaphragms. Section 12.10.3.4 requires that diaphragm collectors be designed for 1.5 times the force level used for diaphragm in-plane shear and flexure. Based on these forces, the use of a ρ factor greater than 1 for collector design is not necessary and would overly penalize designs. The unit value of the redundancy factor is retained for diaphragms designed by the force level given in Sections 12.10.1 and 12.10.2. This value is reflected in the deletion of Item 7 and the addition of diaphragms to Item 5. For transfer diaphragms, see Section 12.10.3.3.

C12.10.3.1 Design This provision is a rewrite of ASCE 7-10, Sections 12.10.1 and 12.10.2. The phrase "diaphragms including chords, collectors, and their connections to the vertical elements" is used consistently throughout the added or modified provisions, to emphasize that its provisions apply to all portions of a diaphragm. It is also emphasized that the diaphragm is to be designed for motions in two orthogonal directions.

C12.10.3.2 Seismic Design Forces for Diaphragms, Including Chords and Collectors Equation (12.10-4) makes the diaphragm seismic design force equal to the weight tributary to the diaphragm, w_{px}, times a diaphragm design acceleration coefficient, C_{px}, divided by a diaphragm design force reduction factor, R_s, which is material-dependent and whose background is given in Section C12.10.3.5. The background to the diaphragm design acceleration coefficient, C_{px}, is given below.

The diaphragm design acceleration coefficient at any height of the building can be determined from linear interpolation, as indicated in Figure 12.10-2. The diaphragm design acceleration coefficient at the building base, C_{p0}, equals the peak ground acceleration consistent with the design response spectrum in ASCE 7-10, Section 11.4.5, times the Importance Factor, I_e. Note that the term $0.4S_{DS}$ can be calculated from Equation (11.4-5) by making $T = 0$.

The diaphragm design acceleration coefficient at 80% of the structural height, C_{pi}, given by Equations (12.10-8) and (12.10-9), reflects the observation that at about this height, floor accelerations are largely, but not solely, contributed by the first mode of response. In an attempt to provide a simple design Equation, C_{pi} was formulated as a function of the design base shear coefficient, C_s, of ASCE 7-10, which may be determined from equivalent static analysis or modal response spectrum analysis (MRSA) of the structure. Note that C_s includes a reduction by the response modification factor, R, of the seismic force-resisting system. It is magnified back up by the overstrength factor, Ω_0, of the seismic force-resisting system because overstrength will generate higher first-mode forces in the diaphragm. In many lateral systems, at 80% of the building height, the contribution of the second mode is negligible during linear response, and during nonlinear response it is typically small, though nonnegligible. In recognition of this observation, the diaphragm seismic design coefficient at this height has been made a function of the first mode of response only, and the contribution of this mode has been factored by $0.9\Gamma_{m1}$ as a weighed value between contributions at the first-mode effective height (approximately 2/3 of the building height) and the building height.

Systems that make use of high R-factors, such as buckling, restrained braced frames (BRBFs), and moment-resisting frames (MRFs), show that in the lower floors the higher modes add to the accelerations, whereas the contribution of the first mode is minimal. For this reason, the coefficient C_{pi} needs to have a lower bound. A limit of C_{p0} has been chosen; it makes the lower-floor acceleration coefficients independent of R. Wall systems are unlikely to be affected by this lower limit on C_{pi}.

At the structural height, h_n, the diaphragm design acceleration coefficient, C_{pn}, given by Equation (12.10-7), reflects the influence of the first mode, amplified by system overstrength, and of the higher modes without amplification, on the floor acceleration at this height. The individual terms are combined using the square root of the sum of the squares. The overstrength amplification of the first mode recognizes that the occurrence of an inelastic mechanism in the first mode is an anticipated event under the design earthquake, whereas inelastic mechanisms caused by higher-mode behavior are not anticipated. The higher-mode seismic response coefficient, C_{s2}, is computed as the smallest of the values given by Equations (12.10-10), (12.10-11), and (12.10-12a) or (12.10-12b). These four Equations consider that the periods of the higher modes contributing to the floor acceleration can lie on the ascending, constant, or first descending branch of the design response spectrum shown in ASCE 7-10, Figure 11.4-1. Users are warned against extracting higher modes from their modal analysis of buildings and using them in lieu of the procedure presented in Section 12.10.3.2.1, because the higher-mode contribution to floor accelerations can come from a number of modes, particularly when there is lateral-torsional coupling of the modes.

Note that Equation (12.10-7) makes use of the modal contribution factor, defined here as the mode shape ordinate at the building height times the modal participation factor and is uniquely defined for each mode of response (Chopra 1995). A building database was compiled to obtain approximate Equations for the first-mode and higher-mode contribution factors. The first and second translational modes, as understood in the context of two-dimensional modal analysis, were extracted from the mode shapes obtained from three-dimensional modal analysis by considering modal ordinates at the center of mass. These buildings had diverse lateral systems, and the number of stories ranged from 3 to 23. Equations (12.10-13) and (12.10-14) were empirically calibrated from simple two-dimensional models of realistic frame-type and wall-type buildings and then compared with data extracted from the database (Figure C12.10-3). In Equation (12.10-7), C_{pn} is required to be no less than C_{pi}, based on judgment, to eliminate instances where the design acceleration at roof level might be lower than that at $0.8h_n$. This cap will particularly affect low-z_s systems such as BRBFs.

To validate Equation (12.10-4), C_{px} coefficients were calculated for various buildings tested on a shake table. Figures C12.10-4 and C12.10-5 plot the floor acceleration envelopes and the floor accelerations predicted from Equation (12.10-4), with $R_s = 1$, for two buildings built at full scale and tested on a shake table (Panagiotou et al. 2011, Chen et al. 2015), with C_{p0} defined as the diaphragm design acceleration coefficient at the structure base and C_{px} defined as the diaphragm design acceleration coefficient at level x. Measured floor accelerations are reasonably predicted by Equation (12.10-4). Research work by Choi et al. (2008) concluded that BRBFs are very effective in limiting floor accelerations in buildings arising from higher-mode effects. This finding is reflected in this proposal, where

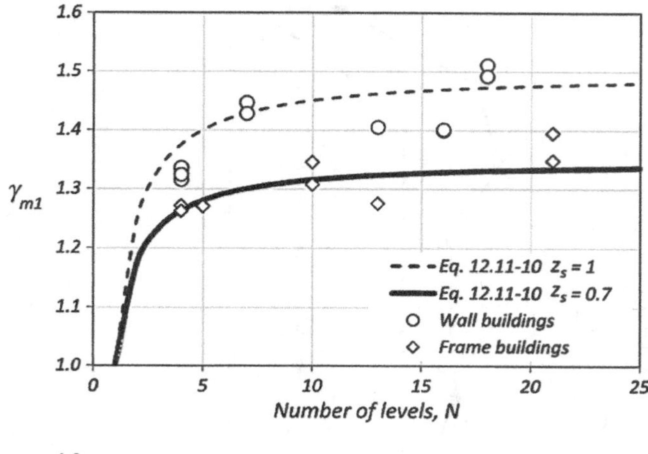

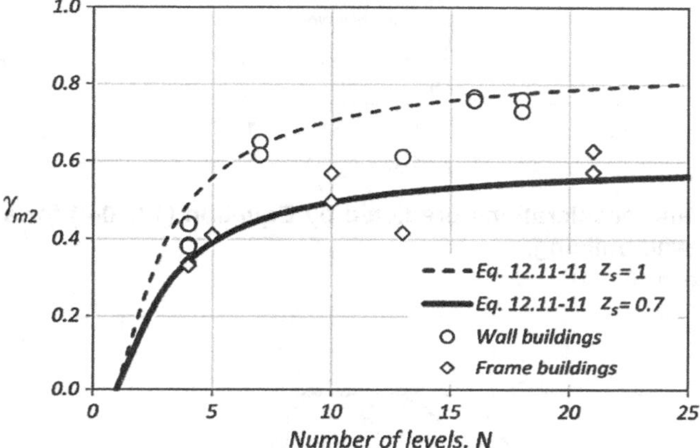

Figure C12.10-3. Comparison of factors Γ_{m1} and Γ_{m2} obtained from analytical models and actual structures with those predicted by Equations (12.10-13) and (12.10-14).

the mode shape factor, z_s, has been made the smallest for BRBF systems. Figure C12.10-6 compares average floor accelerations obtained from the nonlinear time history analyses of four buildings (two steel BRBF systems and two steel special moment frame systems) when subjected to an ensemble of spectrum-compatible earthquakes with floor accelerations computed from Equations (12.10-4) and (12.10-5). The proposed design Equations predict the accelerations in the uppermost part of the building and in the lowest levels reasonably well.

The significant difference between a low-z_s system such as the BRBF and a high-z_s system such as a bearing-wall system is that inelastic deformations are distributed throughout the height of the structure in a low-z_s system, whereas they are concentrated at the base of the structure in a high-z_s system. If rational analysis can be performed to demonstrate that inelastic deformations are in fact distributed along the height of the structure, as is often the case with eccentrically braced frame or coupled shear wall systems, then the use of a low z_s value, as has been assigned to the BRBF for such a system, would be justified.

During the calibration of the design procedure leading to Equation (12.10-4), it was found that at intermediate levels in lateral systems designed using large response modification coefficients, diaphragm design forces given by this Equation could be rather low. There was consensus within the Building Seismic Safety Council Provisions Update Committee (BSSC PUC) Issue Team that developed Section 12.10.3 that diaphragm design forces should not be taken as less than the minimum force currently prescribed by ASCE 7-10, and hence they developed Equation (12.10-5).

The procedure presented in Section 12.10.3 is based on consideration of buildings and structures whose mass distribution is reasonably uniform along the building height. Buildings or structures with tapered mass distribution along their height or with setbacks in their upper levels may experience diaphragm forces in the upper levels that are greater than those derived from Equation (12.10-4). In such buildings and structures, it is preferable to define an effective building height, h_{ne}, and a corresponding level, n_e, the level to which the structural effective height is measured. The effective number of levels in a building, n_e, is defined as level x, where the ratio $\sum_{i=1}^{x} w_i / \sum_{i=1}^{n} w_i$ first exceeds 0.95. Level 1 is defined as the first level above the base. The effective structural height, h_{ne}, is the height of the building measured from the base to level n_e. In buildings with tapered mass distribution or setbacks, the diaphragm design acceleration coefficient, C_{pn}, is calculated by interpolation and extrapolation, as shown in Figure C12.10-7, with n replaced by n_e in Equations (12.10-10) through (12.10-14).

C12.10.3.3 Transfer Forces in Diaphragms All diaphragms are subject to inertial forces caused by the weight tributary to the diaphragm. Where the relative lateral stiffnesses of vertical seismic force-resisting elements vary from story to story, or the vertical seismic force-resisting elements have out-of-plane offsets, lateral forces in the vertical elements need to be transferred through the diaphragms as part of the load path between vertical elements above and below the diaphragm. These transfer forces are in addition to the inertial forces and can at times be quite large.

For structures that have a horizontal structural irregularity of Type 4 in Table 12.3-1, the magnitude of the transfer forces is largely dependent on the overstrength in the offset vertical elements of the seismic force-resisting system. Therefore, the transfer force caused by the offset is required to be amplified by the overstrength factor, Ω_0, of the seismic force-resisting system. The amplified transfer force is to be added to the inertial force for the design of this portion of the diaphragm.

Transfer forces can develop in many other diaphragms, even within regular buildings; the design of diaphragms with such transfer forces can be for the sum of the transfer forces, unamplified, and the inertial forces.

C12.10.3.4 Collectors—Seismic Design Categories C through F For structures in Seismic Design Categories C through F, ASCE 7-10, Section 12.10.2.1, specifies the use of forces including the overstrength factor, Ω_0, for design of diaphragm collectors and their connections to vertical elements of the seismic force-resisting system. The intent of this requirement is to increase collector forces to help ensure that collectors will not be the weak links in the seismic force-resisting system.

In this section the collector force is instead differentiated by using a multiplier of 1.5. This is a smaller multiplier than has been used in the past, but it is justified because the diaphragm forces are more accurately determined by Equation (12.10-4). For collector elements of diaphragms that carry transfer forces caused by out-of-plane offsets of the vertical elements of the seismic force-resisting system, only the inertia force is amplified by 1.5; the transfer forces, already amplified by Ω_0, are not further amplified by 1.5.

Some seismic force-resisting systems, such as braced frames and moment frames, have a fairly well-defined lateral strength corresponding to a well-defined yield mechanism. When collectors deliver seismic forces to such systems, it is not sensible to

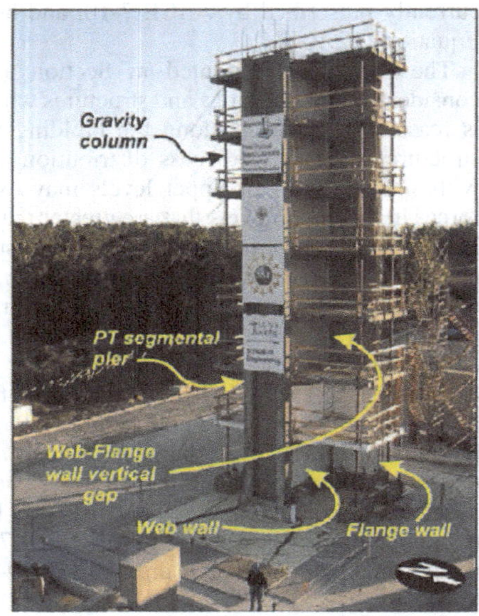

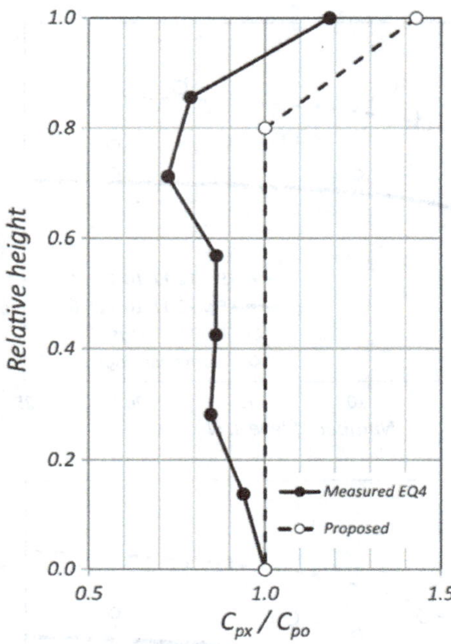

Figure C12.10-4. Comparison of measured floor accelerations and accelerations predicted by Equation (12.10-4) for a seven-story bearing-wall building.

Source: Panagiotou et al. (2011).

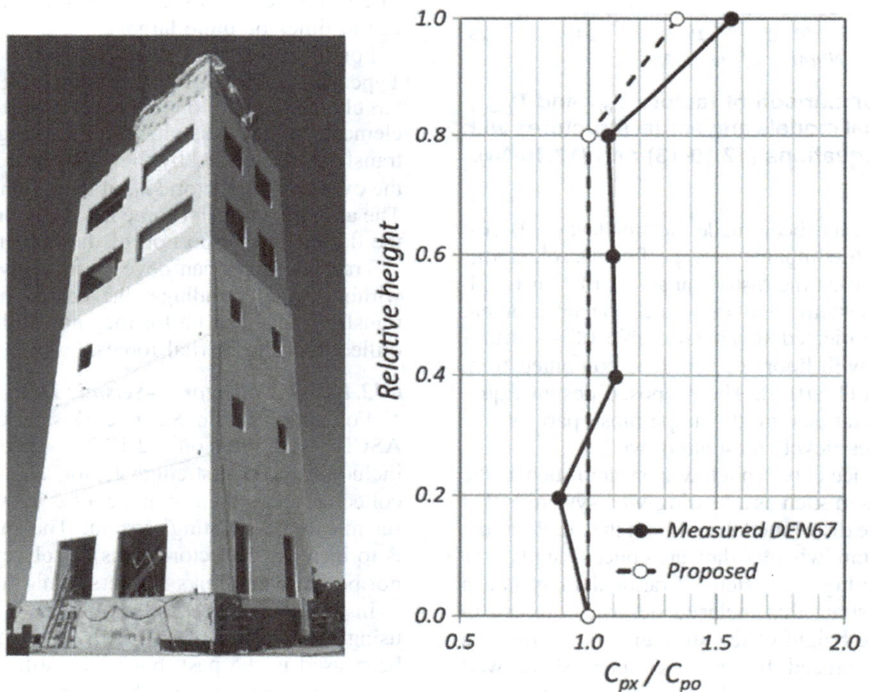

Figure C12.10-5. Comparison of measured floor accelerations and accelerations predicted by Equation (12.10-4) for a five-story special moment-resisting frame building.

Source: (left) Courtesy of Michelle Chen; (right) adapted from Chen et al. (2015).

have to design the collectors for forces higher than those corresponding to the lateral strength of the supporting elements in the story below. This is why the cap on collector design forces is included. The lateral strength of a braced frame or moment frame may be calculated using the same methods as are used for determining whether a weak-story irregularity is present (Table 12.3-2). It should be noted that only the moment frames or braced frames below the collector are to be considered in

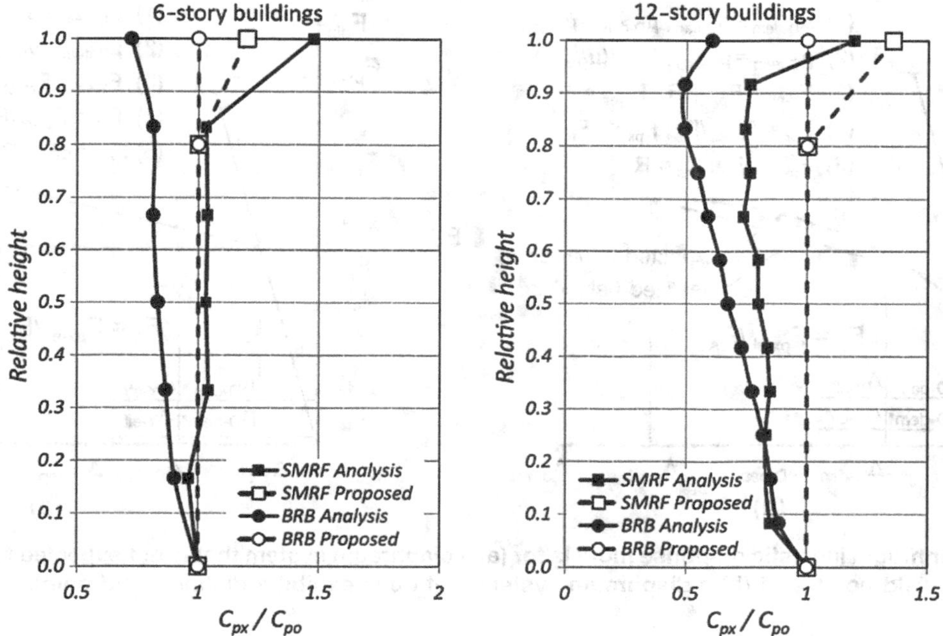

Figure C12.10-6. Comparison of measured floor accelerations with proposed Equations (12.10-4) and (12.10-5) for steel buckling-restrained braced frame and special moment-resisting frame buildings.
Source: Adapted from Choi et al. (2008).

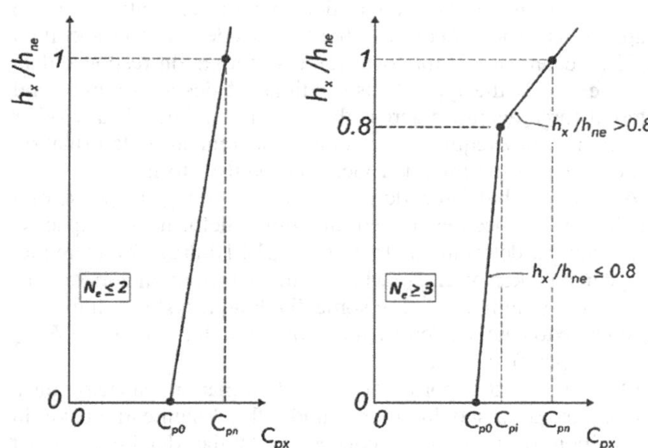

Figure C12.10-7. Diaphragm design acceleration coefficient, C_{px}, for buildings with nonuniform mass distribution.

calculating the upper-bound collector design force. The shear strength of the gravity columns and the lateral strength of the frames above are not included.

The design forces in diaphragms that deliver forces to collectors can also be limited by the maximum forces that can be generated in those collectors by the moment frames or braced frames below.

C12.10.3.5 Diaphragm Design Force Reduction Factor Although analytical and shake table studies indicate higher diaphragm accelerations than are currently used in diaphragm design, many commonly used diaphragm systems, including diaphragms designed under a number of US building codes and editions, have a history of excellent earthquake performance. With limited exceptions, diaphragms have not been reported to have performed below the life-safety intent of building code seismic design provisions in earthquakes. Based on this history, it is felt that, for many diaphragm systems, no broad revision is required to the balance between demand and capacity used for design of diaphragms under current ASCE 7 provisions. In view of this observation, it was recognized that the analytical studies and diaphragm testing from which the higher accelerations and design forces were being estimated used diaphragms that were elastic or near-elastic in their response. Commonly used diaphragm systems are recognized to have a wide range of overstrength and inelastic displacement capacity (ductility). It was recognized that the effect of the varying diaphragm systems on seismic demand required evaluation and incorporation into the proposed diaphragm design forces. Equation (12.10-4) incorporates the diaphragm overstrength and inelastic displacement capacity through the use of the diaphragm force reduction factor, R_s. This factor is most directly based on the global ductility capacity of the diaphragm system; however, the derivation of the global ductility capacity inherently also captures the effect of diaphragm overstrength.

For diaphragm systems with inelastic deformation capacity sufficient to permit inelastic response under the design earthquake, the diaphragm design force reduction factor, R_s, is typically greater than 1.0, so that the design force demand, F_{px}, is reduced relative to the force demand for a diaphragm that remains linear-elastic under the design earthquake. For diaphragm systems that do not have sufficient inelastic deformation capacity, R_s should be less than 1.0, or even 0.7, so that linear-elastic force–deformation response can be expected under the risk-targeted maximum considered earthquake (MCE_R).

Diaphragms with R_s greater than 1.0 shall have the following characteristics: (1) a well-defined, specified yield mechanism; (2) global ductility capacity for the specified yield mechanism, which exceeds anticipated ductility demand for the risk-targeted Maximum Considered Earthquake (MCE_R); and (3) sufficient

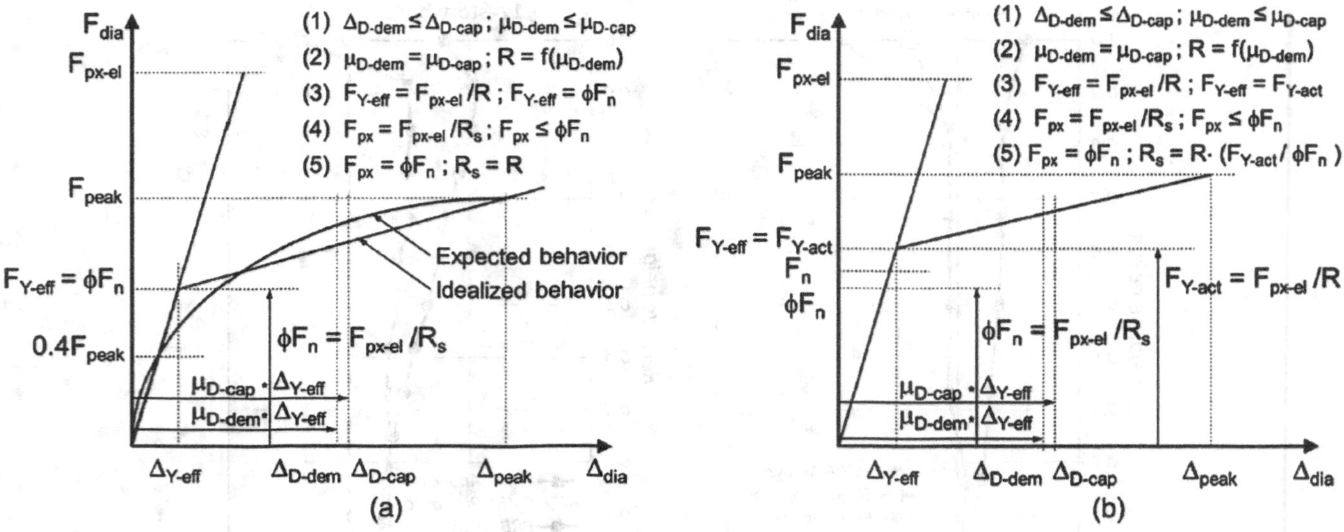

Figure C12.10-8. Diaphragm inelastic response models for (a) a diaphragm system that is not expected to exhibit a distinct yield point, and (b) a diaphragm system that does exhibit a distinct yield point.

local ductility capacity to provide the intended global ductility capacity, considering that the specified yield mechanism may require concentrated local inelastic deformations to occur. The following discussion addresses these characteristics and the development of R_s-factors in detail.

A diaphragm system with R_s greater than 1.0 should have a specified, well-defined yield mechanism, for which both the global strength and the global deformation capacity can be estimated. For some diaphragm systems, a shear-yield mechanism may be appropriate, whereas for other diaphragm systems, a flexural-yield mechanism may be appropriate.

Figure C12.10-8(a) shows schematically the force-deformation ($F_{\text{dia}} - \Delta_{\text{dia}}$) response of a diaphragm with significant inelastic deformation capacity. The figure illustrates the response of a diaphragm system, such as a wood diaphragm or a steel deck diaphragm, which is not expected to exhibit a distinct yield point, so that an effective yield point ($F_{Y\text{-eff}}$ and $\Delta_{Y\text{-eff}}$) needs to be defined. For wood diaphragms and steel deck diaphragms, the figure illustrates one way to define the effective yield point. The stiffness of a test specimen is defined by the secant stiffness through a point corresponding to 40% of the peak strength, F_{peak}. The effective yield point ($F_{Y\text{-eff}}$ and $\Delta_{Y\text{-eff}}$) for a diaphragm is defined by the secant stiffness through $0.4F_{\text{peak}}$ and the nominal diaphragm strength reduced by a strength reduction factor, φF_n, as shown in the figure. The $F_{\text{dia}} - \Delta_{\text{dia}}$ response is then idealized with a bilinear model, using the effective yield point ($F_{Y\text{-eff}}$ and $\Delta_{Y\text{-eff}}$) and F_{peak} and the corresponding deformation, Δ_{peak}, as shown in the figure.

Figure C12.10-8(b) shows schematically the force–deformation ($F_{\text{dia}} - \Delta_{\text{dia}}$) response of a diaphragm with significant inelastic deformation capacity, which is expected to have nearly linear $F_{\text{dia}} - \Delta_{\text{dia}}$ response up to a distinct yield point, such as a cast-in-place reinforced concrete diaphragm. For this type of diaphragm system, the effective yield point can be taken as the actual yield point ($F_{Y\text{-actual}}$ and $\Delta_{Y\text{-actual}}$) of the diaphragm (accounting for diaphragm material overstrength and not including a strength reduction factor, φ).

The global (or system) deformation capacity of a diaphragm system, Δ_{cap}, should be estimated from analyses of test data. The force–deformation ($F_{\text{dia}} - \Delta_{\text{dia}}$) response shown schematically in Figure C12.10-8(a, b) is the global force–deformation behavior.

In some cases, tests directly provide the global deformation capacity, but more often, tests provide only the local response, including the strength and deformation capacity, of diaphragm components and connections. When tests provide only the local deformation capacity, analyses of typical diaphragms should be made to estimate the global deformation capacity of these diaphragms. These analyses should consider (1) the specified yield mechanism, (2) the local force–deformation response data from tests, (3) the typical distributions of design strength and internal force demands across the diaphragm, and (4) any other factors that may require concentrated local inelastic deformation to occur when the intended yield mechanism forms.

After the global force–deformation ($F_{\text{dia}} - \Delta_{\text{dia}}$) response of a diaphragm has been estimated, the global deformation capacity, Δ_{cap}, can be determined. In Figure C12.10-8(a), for example, Δ_{cap} can be taken as Δ_{peak}, which is the deformation corresponding to the strength, F_{peak}. For some diaphragm systems, it may be acceptable to take the deformation corresponding to 80% of F_{peak} (i.e., postpeak) as Δ_{cap}.

Only a selected portion of the deformation capacity of a diaphragm, Δ_{cap}, should be used under the design earthquake, in recognition of two major concerns: (1) the diaphragm must perform adequately under the MCE$_R$, which has a design spectrum 50% more intense than the design earthquake design spectrum; and (2) significant inelastic deformation under the design earthquake may result in undesirable damage to the diaphragm. As a rough estimate, the diaphragm deformation capacity under the design earthquake, $\Delta_{D\text{-cap}}$, should be limited to approximately one-half to two-thirds of the deformation capacity, Δ_{cap}.

To develop the diaphragm force reduction factor, R_s, the diaphragm global deformation capacity should be expressed as a global ductility capacity, μ_{cap}, which equals the deformation capacity, Δ_{cap}, divided by the effective yield deformation, $\Delta_{Y\text{-eff}}$. The corresponding diaphragm design ductility capacity, $\mu_{D\text{-cap}}$, equals $\Delta_{D\text{-cap}}/\Delta_{Y\text{-eff}}$.

From the diaphragm global deformation capacity and corresponding ductility capacity, an appropriate R_s factor can be estimated. Use of the estimated R_s factor in design should result in diaphragm ductility demands that do not exceed the ductility capacity that was used to estimate R_s. The force reduction factor is ideally derived from system-specific studies. Where such studies are unavailable, however, some guidance on the conversion from global ductility to force reduction is available from past studies.

Expressions that provide the force reduction factor, R, for the seismic force-resisting system of a building corresponding to an expected ductility demand, μ_{dem}, have been proposed by numerous research teams. Numerous factors, including vibration period, inherent damping, deformation hardening (stiffness after the effective yield point), and hysteretic energy dissipation under cyclic loading, have been considered in developing these expressions. Two such expressions, which are based on elastoplastic force–deformation response under cyclic loading (Newmark and Hall 1982), are $R = (2\mu_{dem} - 1)^{0.5}$, applicable to short-period systems, and $R = \mu_{dem}$, applicable to systems with longer periods. The first function, known as the equal energy rule, gives a smaller value of R for a given value of μ_{dem}; the second function, known as the equal displacement rule, is also widely used.

Figure C12.10-8(a, b) summarize an approach to estimating R_s as follows:

1. For the selected value of R_s, the diaphragm deformation demand under the design earthquake, $\Delta_{D\text{-dem}}$, should not exceed the diaphragm design deformation capacity, $\Delta_{D\text{-cap}}$. This design constraint, expressed in terms of diaphragm ductility, requires that the diaphragm ductility demand under the design earthquake, $\mu_{D\text{-dem}}$, not exceed the diaphragm design ductility capacity, $\mu_{D\text{-cap}}$.
2. The largest value of R that can be justified for a given diaphragm design deformation capacity is obtained by setting the ductility demand, $\mu_{D\text{-dem}}$, equal to the design ductility capacity, $\mu_{D\text{-cap}}$, and determining R from a function that provides R for a given μ_{dem}. For example, if $\mu_{D\text{-cap}} = 2.5$, then $\mu_{D\text{-dem}}$ is set equal to 2.5, and the corresponding $R = 2$ from the equal energy rule or $R = 2.5$ from the equal displacement rule.
3. R from step (2) is the ratio of the force demand for a linear-elastic diaphragm, $F_{px\text{-el}}$, to the effective yield strength of the diaphragm, $F_{Y\text{-eff}}$. For a diaphragm system that is not expected to exhibit a distinct yield point (Figure C12.10-8a), $F_{Y\text{-eff}}$ equals the factored nominal diaphragm strength, φF_n. For a diaphragm system that is expected to exhibit a distinct yield point (Figure C12.10-8b), $F_{Y\text{-eff}}$ equals the actual yield strength, $F_{Y\text{-actual}}$, accounting for diaphragm material overstrength and not including the strength reduction factor, φ.
4. R_s, however, is defined as the ratio of the force demand for a linear-elastic diaphragm, $F_{px\text{-el}}$, to the design force demand, F_{px}. The diaphragm must be designed such that F_{px} is less than or equal to the factored nominal diaphragm strength, φF_n.
5. For a diaphragm system without a distinct yield point (Figure C12.10-8(a)) that has the minimum strength $F_{px} = \varphi F_n$, R_s equals R from step (2). For a diaphragm system with a distinct yield point (Figure C12.10-8(b)), which has the minimum strength $F_{px} = \varphi F_n$, R_s equals R from step 2 multiplied by the ratio $F_{Y\text{-eff}}/\varphi F_n$.

Diaphragm force reduction factors, R_s, have been developed for some commonly used diaphragm systems. The derivation of factors for each of these systems is explained in detail in the following commentary sections. For each, the specific design standard considered in the development of the R_s factor is specified. The resulting R_s factors are specifically tied to the design and detailing requirements of the noted standard, because these play a significant role in setting the ductility and overstrength of the diaphragm system. For this reason, the applicability of the R_s factor to diaphragms designed using other standards must be specifically considered and justified.

Cast-in-Place Concrete Diaphragms. The R_s values in Table 12.10-1 address cast-in-place concrete diaphragms designed in accordance with ACI 318.

Intended Mechanism. Flexural yielding is the intended yield mechanism for a reinforced concrete diaphragm. Where this can be achieved, designation as a flexure-controlled diaphragm and use of the corresponding R_s factor in Table 12.10-1 is appropriate. There are many circumstances, however, where the development of a well-defined yielding mechanism is not possible because of diaphragm geometry (aspect ratio or complex diaphragm configuration), in which case, designation as a shear-controlled diaphragm and use of the lower R_s factor is required.

Derivation of Diaphragm Force Reduction Factor. Test results for reinforced concrete diaphragms are not available in the literature. Test results for shear walls under cyclic lateral loading were considered. The critical regions of shear wall test specimens usually have high levels of shear force, moment, and flexural deformation demands; high levels of shear force are known to degrade the flexural ductility capacity. The flexural ductility capacity of shear wall test specimens under cyclic lateral loading was used to estimate the flexural ductility capacity of reinforced concrete diaphragms, using the previously described method based on Newmark and Hall (1982).

Based on shear wall test results, the estimated global flexural ductility capacity of a reinforced concrete diaphragm is 3, based on the actual yield displacement, $\Delta_{Y\text{-actual}}$, of the test specimens. The design ductility capacity, $\mu_{D\text{-cap}}$, is taken as 2/3 of the ductility capacity; the design ductility capacity ($\mu_{D\text{-cap}}$) is 2.

Setting the ductility demand, μ_{dem}, equal to the design ductility capacity, $\mu_{D\text{-cap}}$, and using the equal energy rule, the force reduction factor $R = (2\mu_{dem} - 1)^{0.5} = 1.73$.

R_s equals R multiplied by the ratio $F_{Y\text{-eff}}/\varphi F_n$. $F_{Y\text{-eff}}$ is taken equal to $F_{Y\text{-actual}}$, which is assumed to be $1.1F_n$, and φ equals 0.9. Therefore, $R_s = 2.11$, which is rounded to 2.

Because of the geometric characteristics of a building or other factors, such as minimum reinforcement requirements, it is not possible to design some reinforced concrete diaphragms to yield in flexure. Such diaphragms are termed "shear controlled" to indicate that they are expected to yield in shear. Shear-controlled reinforced concrete diaphragms should be designed to remain essentially elastic under the design earthquake, with their available global ductility held in reserve for safety under the MCE_R.

Based on the following considerations, R_s is specified as 1.5 for shear-controlled reinforced concrete diaphragms. Reinforced concrete diaphragms have performed well in past earthquakes. ACI-318 specifies φ of 0.75 or 0.6 for diaphragm shear strength and limits the concrete contribution to the shear strength to only $2(f'_c)^{0.5}$. In addition, reinforced concrete floor slabs often have gravity load reinforcement that is not considered in determining the diaphragm shear strength. Therefore, shear-controlled reinforced concrete diaphragms are expected to have significant overstrength. The ratio $F_{Y\text{-eff}}/\varphi F_n$ for a reinforced concrete diaphragm, where $F_{Y\text{-eff}}$ is taken equal to $F_{Y\text{-actual}}$, is expected to exceed 1.5, which is the rationale for $R_s = 1.5$, even though μ_{dem} is assumed to be 1 for the design earthquake.

Precast Concrete Diaphragms. The R_s values in Table 12.10-1 address precast concrete diaphragms designed in accordance with ACI 318.

Derivation of Diaphragm Force Reduction Factors. The diaphragm force reduction factors, R_s, in Table 12.10-1 for precast concrete diaphragms were established based on the results of analytical earthquake simulation studies conducted within a multiple-university project: Diaphragm Seismic

Table C12.10-1. Diaphragm Design Performance Targets.

	Flexure		Shear
Options	DE	MCER	DE and MCER
Elastic design objective	Elastic	Elastic	Elastic
Basic design objective	Elastic	Inelastic	Elastic
Reduced design objective	Inelastic	Inelastic	Elastic

Notes: DE = design earthquake; MCE_R = risk-targeted maximum considered earthquake.

Design Methodology (DSDM) for Precast Concrete Diaphragms (Fleischman et al. 2013). In this research effort, diaphragm design force levels have been aligned with the diaphragm deformation capacities specifically for precast concrete diaphragms. Three different design options were proposed according to different design performance targets (Table C12.10-1). The relationships between diaphragm design force levels and diaphragm local/global ductility demands have been established in the DSDM research project. These relationships have been used to derive the R_s for precast concrete diaphragms in Table 12.10-1.

Diaphragm R_{dia}–μ_{global}–μ_{local} Relationships. Extensive analytical studies have been performed (Fleischman et al. 2013) to develop the relationship of R_{dia}-μ_{global}-μ_{local}. R_{dia} is the diaphragm force reduction factor (similar to the R_s in Table 12.10-1) measured from the required elastic diaphragm design force at MCE_R level. μ_{global} is the diaphragm global ductility demand, and μ_{local} is the diaphragm local connector ductility demand measured at Maximum Considered Earthquake (MCE) level. Figure C12.10-9 shows the μ_{global}-μ_{local} and R_{dia}-μ_{global} analytical results for different diaphragm aspect ratios (AR) and proposed linear Equations derived from the data.

Diaphragm Force Reduction Factor, R_s. Using the Equations in Figure C12.10-9, R_s can be calculated for different diaphragm design options, provided that the diaphragm local reinforcement ductility capacity is known. In the DSDM research, precast diaphragm connectors have been extensively tested (Fleischman et al. 2013) and have been qualified into three categories: high-deformability elements, moderate-deformability elements, and low-deformability elements, which are required as a minimum for designs using the reduced design objective, the basic design objective, and the elastic design objective, respectively. The local deformation and ductility capacities for diaphragm connector categories are shown in Table C12.10-2. Considering that the proposed diaphragm design force level targets elastic diaphragm response at the design earthquake, which is equivalent to design using the basic design objective, where $\mu_{local} = 3.5$ at MCE_R (see Table C12.10-2), the available diaphragm global ductility capacity has to be reduced from Figure C12.10-9(a), acknowledging more severe demands in the MCE_R:

$$\mu_{global,red} = 0.17(\mu_{local} - 3.5) + 1 \quad \text{(C12.10-1)}$$

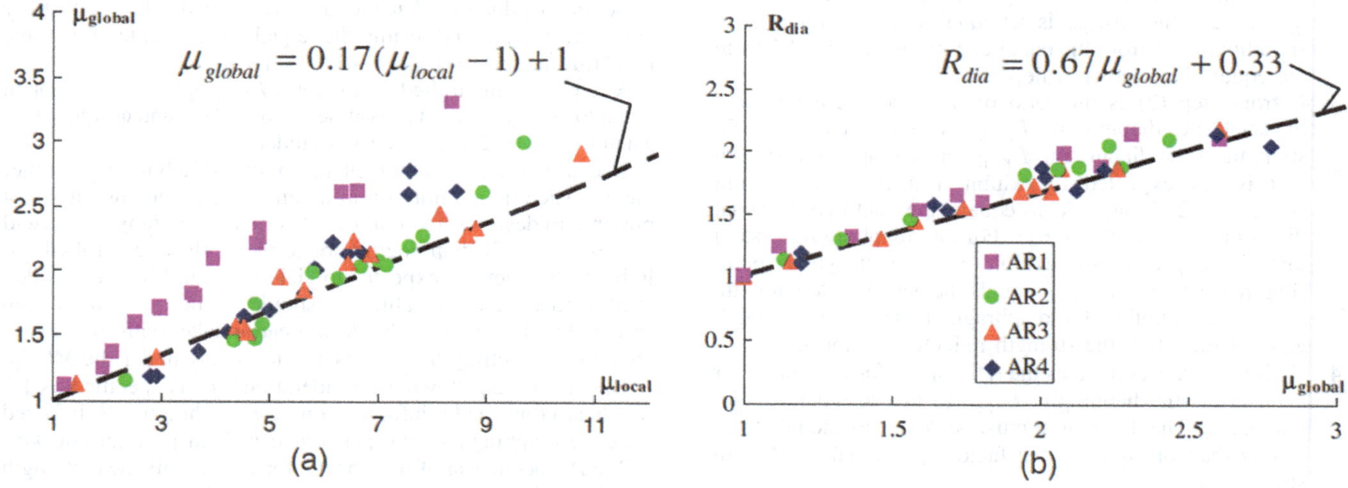

Figure C12.10-9. Relationships: (a) μ_{global}–μ_{local}; (b) R_{dia}–μ_{global}.
Note: AR = aspect ratio.

Table C12.10-2. Diaphragm Force Reduction Factors, R_s.

Options	Diaphragm Connector Category	δ_{local} (in.)	μ_{local}	μ_{global}	$\mu_{global,red}$	R_s
Elastic design objective	Low-deformability elements	0.06	1.0	1.0	0.58	0.7
Basic design objective	Moderate-deformability elements	0.2	3.5	1.4	1.0	1.0
Reduced design objective	High-deformability elements	0.4	7.0	2.0	1.6	1.4

Accordingly, the R_s-factor can be modified from Figure C12.10-9(b) (see Table C12.10-2):

$$R_s = 0.67\mu_{global,red} + 0.33 \quad (C12.10\text{-}2)$$

Wood-Sheathed Diaphragms. The R_s values given in Table 12.10-1 are for wood-sheathed diaphragms designed in accordance with AWC (2014) *Special Design Provisions for Wind and Seismic*.

Intended Mechanism. Wood-sheathed diaphragms are shear-controlled, with design strength determined in accordance with AWC (2014) and the shear behavior based on the sheathing-to-framing connections. Wood diaphragm chord members are unlikely to form flexural mechanisms (ductile or otherwise) because of the overstrength inherent in axially loaded members designed in accordance with applicable standards.

Derivation of Diaphragm Design Force Reduction Factor. An R_s factor of 3 is assigned in Table 12.10-1, based on diaphragm test data (APA 1966, 2000; DFPA 1954, 1963) and analytical studies. The available testing includes diaphragm spans (loaded as simple-span beams) ranging from 24 to 48 ft (7.3 to 14.6 m), aspect ratios ranging from 1 to 3.3, and diaphragm construction covering a range of construction types, including blocked and unblocked construction and regular and high-load diaphragms. The loading was applied with a series of point loads at varying spacing; however, the loading was reasonably close to uniform. Whereas available diaphragm testing was monotonic, based on shear wall loading protocol studies (Gatto and Uang 2002), it is believed that the monotonic load–deflection behavior is reasonably representative of the cyclic load–deflection envelope, suggesting that it is appropriate to use monotonic load–deflection behavior in the estimation of overstrength, ductility, and displacement capacity.

Analytical studies using nonlinear response history analysis evaluated the relationship between global ductility and diaphragm force reduction factor for a model wood building. The analysis identified the resulting diaphragm force reduction factor as ranging from just below 3 to significantly in excess of 5. A force reduction factor of 3 was selected so that diaphragm design force levels would generally not be less than those determined in accordance with provisions of Section 12.10.

The calibration approach for selection of R_s of 3 was considered appropriate to limit conditions where diaphragm force levels would drop below those determined in accordance with Section 12.10. This was due in part to historical experience of good diaphragm performance across a range of wood diaphragm types, even though test data showed varying levels of ductility and deformation capacity. Tests of nailed wood diaphragms showed significant but varying levels of overstrength. It is recognized that even further variation of overstrength will result from

- presence of floor coverings or toppings and their attachment or bond to diaphragm sheathing;
- presence of wall to floor framing nailing through diaphragm sheathing; and
- presence of adhesives in combination with required sheathing nailing (commonly used to mitigate floor vibration, increase floor stiffness for gravity loading, and reduce the potential for squeaking).

These sources of overstrength are not considered to be detrimental to overall diaphragm performance.

Bare Steel Deck Diaphragms. The R_s values in Table 12.10-1 address bare (untopped) steel deck diaphragms designed in accordance with AISI S400 (2020b). These include diaphragms meeting requirements for special seismic detailing, and all other bare steel deck diaphragms.

Intended Mechanism. The ductility of bare steel deck diaphragms is largely driven by the performance of the deck profile and its interaction with sidelap and structural connections. It has been found for a specific class of WR roof deck that, if the sidelap and structural connections have adequate ductility and deformation capacity, the full bare steel deck diaphragm can similarly develop productive levels of ductility with sufficient system-level deformation capacity (O'Brien et al. 2017, Schafer 2019). These findings formed the basis for prescriptive special seismic detailing requirements that are found in AISI S400 (2020b) as referenced from Section 14.1.5. AISI S400 (2020b) also provides performance-based criteria to establish that selected detailing associated with a particular bare steel deck diaphragm (e.g., new profile or new connector) meets the same performance objectives as the prescriptive system and thus is deemed to provide an intended ductile mechanism. Other bare steel deck diaphragms have fasteners and system behavior that are less ductile. As a result, the R_s factor is smaller, resulting in design for near-elastic level forces.

Derivation of Diaphragm Force Reduction Factor. The derivation of the bare steel deck diaphragm force reduction factor, R_s, is summarized in Appendix 1 of Schafer (2019). The ductility and deformation capacity of sidelap and structural connections employed in bare steel deck diaphragms are established by evaluation of cyclic shear testing (NBM 2017, 2018; Schafer 2019). The ductility of bare steel deck diaphragms is preliminarily established by assembly and evaluation of cyclic cantilever diaphragm tests (O'Brien et al. 2017). The impact of the connector and cantilever diaphragm tests on full building performance is assessed in a 3D building model, as detailed in Schafer (2019). The model shows that only bare steel deck diaphragms with connections that have sufficient ductility and deformation capacity provide adequate inelastic diaphragm performance, leading to special seismic detailing requirements. For the subset of cyclically tested diaphragms that meet the special seismic detailing requirements, the tested subsystem ductility and system overstrength are established (Schafer 2019). To establish the diaphragm system ductility, an additional correction is provided for the reduction in ductility of a roof that experiences varying shear across its width, compared with a cantilever diaphragm test which is under constant shear (O'Brien et al. 2017, Schafer 2019). From the system ductility and overstrength, the diaphragm force reduction factor R_s was developed based on μ-R relations using the method documented in ATC-19 (1995).

Concrete-Filled Steel Deck Diaphragms. The R_s values in Table 12.10-1 address concrete-filled steel deck diaphragms designed in accordance with AISC 341 (2022).

Intended Mechanism. Concrete-filled steel deck diaphragms exhibit two types of failure limit states: diagonal tension cracking in the field of the diaphragm, and failure of the shear transfer at the fasteners to the collectors. Both failure modes are considered shear-controlled limit states, and thus concrete-filled steel deck diaphragms only have a value listed for the shear-controlled category in Table 12.10-1. Previous testing by Porter and Easterling (1994) and Avellaneda et al. (2019) has demonstrated ductility and overstrength for concrete-filled steel deck diaphragms experiencing both limit states. These tests also covered a range of variables, including concrete specific weight (normal weight and lightweight), reinforcing steel (unreinforced and with reinforcing steel), concrete strength [2,500 to 6,000 psi (17,000 to 41,000 kPa)], deck height [1.5 to 3.0 in. (38 to 76 mm)], total thickness [4 to 7.5 in. (100 to 190 mm)], and perimeter fastener type (headed shear studs and arc spot welds). The most common limit state in testing was diagonal tension cracking of the concrete, but specimens experiencing other limit states, such as failure of perimeter fasteners or debonding between the deck and concrete, had similar or better ductility. Because this set of tests considered a wide range of variables, the resulting R_s factor is associated with concrete-filled steel deck diaphragms designed according to AISC 341 (2022) without any special detailing. It is anticipated that ongoing and future work may lead to a new category of concrete-filled steel deck with special detailing and larger R_s value.

Derivation of Diaphragm Force Reduction Factor. The derivation of R_s for concrete-filled steel deck diaphragms is described in Eatherton et al. (2020), O'Brien et al. (2017), and Wei et al. (2020). A two-stage procedure was used to define the R_s factor: (1) using measured ductility and overstrength obtained from cantilever diaphragm tests to calculate a value of R_s, and (2) validating the R_s factor using a computational study.

The beginning of this section's commentary describes a method for calculating the diaphragm design force reduction factor, R_s, based on measured ductility and overstrength obtained from test data. That method is essentially the same as that described in ATC-19 (1995) and the approach used to derive R_s for cast-in-place concrete diaphragms. In this method, the R_s factor is the sum of two components; the first is associated with the expected overstrength of the diaphragm, and the second is a function of the ductility. Eatherton et al. (2020) and O'Brien et al. (2017) evaluate the ductility and overstrength of 13 cantilever tests on concrete-filled steel deck diaphragm specimens conducted by Easterling and Porter (1994). The average ductility of these specimens was 3.63, using a definition of ductility as the shear angle associated with 20% loss of strength compared to the peak divided by the shear angle associated with yield. Eatherton et al. (2020) also propose a method for converting ductility from a cantilever diaphragm test into expected system-level ductility for a full-diaphragm span. Using this conversion, the average system-level ductility was found to be 2.05. Eatherton et al. (2020) calculated values of the diaphragm force reduction factor, R_s, of between 2.00 and 3.35, depending on the building period and whether the resistance factor was included in the calculation of overstrength. Based on this study, a value of R_s = 2.0 was proposed. It is noted that more recent cantilever diaphragm tests by Avellaneda et al. (2019) on concrete-filled steel deck diaphragm specimens that are more representative of contemporary construction resulted in ductility and overstrength consistent with the tests by Easterling and Porter (1994) as analyzed in FEMA (2020).

To validate this proposed value of R_s = 2.0 for concrete-filled steel deck diaphragms, a computational study was conducted to evaluate the associated effect on collapse probability (Wei et al. 2020). Eight buildings with buckling restrained braced frames as the lateral-force-resisting system were analyzed. The height of the building was varied with 1-, 4-, 8-, and 12-story buildings, and two variations were designed for each height: floor diaphragms designed using R_s = 1.0 and R_s = 2.0. A FEMA P-695 (2009b) type of study was conducted. Using the criteria in FEMA P-695, the collapse probability was found to be acceptable for the 1-, 4-, and 8-story buildings designed using R_s = 2.0 for the concrete-filled steel deck floor diaphragms. The 12-story building exceeded the limit for collapse probability by 16%, but changing the diaphragm design from R_s = 2.0 to R_s = 1.0 only reduced the collapse probability by 4%, and it had a negligible effect on median peak story drifts for all buildings. This study concluded that diaphragm design using R_s = 2.0 resulted in similar collapse performance and peak drift performance as R_s = 1.0 buildings and thus was acceptable. The study further noted that the vast majority of past FEMA P-695 studies have been conducted on 2D frames and that more research is required to define appropriate collapse performance criteria for 3D building models that consider diaphragm inelasticity.

C12.10.4 Alternative Diaphragm Design Provisions for One-Story Structures with Flexible Wood Structural Panel Diaphragms and Rigid Vertical Elements Section 12.10.4 introduces diaphragm seismic design provisions for one-story structures combining flexible diaphragms with rigid vertical elements. This approach is based on numerical studies conducted as part of the development of the 2015 guideline document *Seismic Design of Rigid Wall-Flexible Diaphragm Buildings: An Alternate Procedure* (FEMA P-1026), supplemented by additional recent steel deck diaphragm research. The numerical studies and the resulting seismic design methodology specifically recognized the dynamic response of rigid wall–flexible diaphragm structures as being dominated by dynamic response of and inelastic behavior in the diaphragm. These studies indicate that improved seismic performance can be obtained for this group of structures through use of this design methodology. While the most common occurrences of this structure type are the concrete tilt-up and masonry "big box" structures, the "rigid vertical element" terminology of this section recognizes a wider range of vertical elements with which this methodology is permitted to be used.

C12.10.4.1 Limitations This section imposes a series of limitations intended to restrict use of the methodology to flexible diaphragm–rigid vertical element structures consistent with the FEMA P-1026 basis.

Item 1 requires that when these alternative provisions are used they be used in both orthogonal directions. Performance of diaphragms designed using a mix of different design provisions for each orthogonal direction has not been studied, so the resulting seismic performance is not known.

Item 2 limits use to wood structural panel or bare steel deck diaphragms, consistent with the recommendations of the FEMA P-1026 document.

Item 2 limits wood structural panel diaphragms to those designed in accordance with the AWC (2014) SDPWS standard. The Item 2 reference to SDPWS tables further limits use to wood structural panel sheathing fastened to wood framing members or fastened to wood nailers (e.g., wood nailers attached to steel open-web joists) with nailing as specified in the SDPWS diaphragm tables. This is intended to limit fastening to nailed diaphragms and the nail types and nail patterns specified in the table. The FEMA P-1026 wood numerical studies were conducted using hysteretic descriptions derived from testing of diaphragms using full-length common nails based on the SDPWS tables. The performance of wood diaphragms using alternative fasteners and boundary members (chords, collectors, ledgers) has not been studied.

Dynamic testing of SDPWS diaphragms using short nails [nails shorter than specified by ASTM F1667 (2017), but still meeting the minimum penetration requirements of the SDPWS tables] is not readily available. APA (1966, 2007) testing of diaphragms with short nails showed little change in strength or displacement ductility. Thus the use of short nails in diaphragm construction is not prohibited. The provisions of Section 12.10.4 are, however, very narrowly limited to the nails specified in the SDPWS table. Use of shorter nail shank lengths can be justified by the alternate methods of construction provisions of the

International Building Code, Section 104.11 (ICC 2012). The provisions of Section 12.10.4 are not intended to carry over to other nail types or other fastener types (staples, screws, etc.).

Direct attachment of wood structural panel diaphragm sheathing to perimeter structural steel ledgers using power-actuated fasteners (PAFs) is believed to be very common in concrete and masonry wall buildings. Although approval of this type of fastener falls under the "alternative methods of construction" provisions of the International Building Code, it is appropriate to note that the use of PAFs at the perimeter is of lesser concern when using the Section 12.10.4 diaphragm design method because amplified shear forces in the boundary zone reduce the inelastic deformation demands at the diaphragm perimeter. Therefore, wood diaphragm use of PAFs in combination with steel members at the diaphragm boundary is not prohibited. This is, however, very narrowly limited to PAFs at the diaphragm perimeter in the amplified shear zone and where the PAFs can be justified by the "alternative methods of construction" provisions of the International Building Code, Section 104.11.

For bare steel deck diaphragms, Item 2 references the recently developed steel deck design and detailing provisions of AISI S400 (2020b) and provisions of AISI S310 (2020a).

Item 3 prohibits use of this methodology where materials installed over the diaphragm would add significant diaphragm stiffness. This limitation is included because where such materials are used, the diaphragm period and therefore seismic forces would not be appropriately estimated by this section, and the diaphragm may no longer qualify as flexible. Prohibited materials include concrete and similar materials (e.g., vermiculite concrete, cellular concrete, gypsum concrete, etc.) that provide significant stiffness and impede diaphragm deflection. Other toppings of potential concern include rigid insulation board (if bonded to the sheathing) and spray foam, as the bonding caused by these types of materials can impede the fastener behavior that is fundamental to deflection of diaphragms. Available testing of polyisocyanurate board attached with screws indicates an initial increase in stiffness only at very low load levels, beyond which the board does not affect the strength or stiffness of the diaphragm; thus it is considered acceptable to use this system. The wording used for Item 3 is taken from Section 12.3.1.1 because of the very similar intent of both sections. Where isolated pads of concrete occur on the bare steel deck (such as for rooftop equipment), the engineer will need to judge whether the concrete pads reduce the flexibility of the diaphragm such that the diaphragm falls outside the intent of Section 12.10.4.

Item 4 prohibits use with horizontal irregularities including torsional and diaphragm discontinuity irregularities. These configurations generate forces in diaphragms outside those studied in the FEMA P-1026 numerical studies, and the force levels and amplified shear boundary zones may not be appropriate with these irregularities.

Re-entrant corners are specifically permitted in structures designed using these seismic design forces. Re-entrant corners are believed to be prevalent in the existing building stock, and so anticipated to be common in buildings to be constructed under these provisions in the future. Because these diaphragms are dominated by shear deformation, it is judged that reentrant corners will not influence building seismic performance in a manner that would make use of these provisions inappropriate. Vertical irregularities are not addressed, because use of the section is limited to one-story buildings.

Horizontal irregularity Table 12.3-1 triggers two design requirements for structures in Seismic Design Categories D, E, and F that have re-entrant corners. The first is the Section 12.3.3.5 requirement for increasing diaphragm collector design forces determined per Section 12.10.1; new Section 12.10.4.2.6 identifies this requirement as not being applicable because design per Section 12.10.4 requires use of an overstrength factor, and therefore the design forces fall under the exception to Section 12.3.3.5. The second references Table 12.6-1 limitations on analytical procedures; this requirement will not affect the one-story structures intended to fall under Section 12.10.4 provisions, because they will always be significantly below the 160 ft (49 m) height limit of Table 12.6-1.

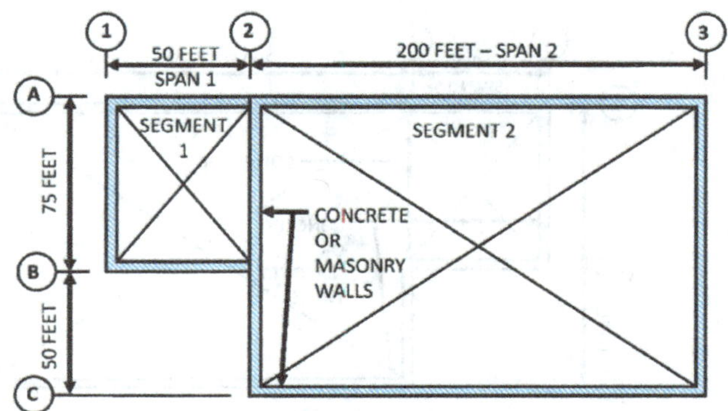

Figure C12.10.4-1a. Structure in plan view, divided into rectangular diaphragm segments for purposes of transverse seismic design. Note that identification of different segments will be required for longitudinal diaphragm forces.

Note: 1 ft = 305 mm.

Item 5 requires that rectangular buildings with interior vertical elements or buildings with nonrectangular plan configurations be broken down into a series of rectangular sections for purposes of diaphragm design. Buildings to which this methodology might be applied are often not completely rectangular in plan, and often combine both long and short diaphragm spans. This provision requires that each section of diaphragm be defined as spanning between boundaries consisting of either vertical elements or collectors, with each span referred to as a "diaphragm segment." Figure C12.10.4-1(a) illustrates diaphragm segments for transverse seismic forces, with each segment supported on all sides by concrete or masonry walls. Figure C12.10.4-1(b) illustrates the same concept, with two building plans in which, for transverse design, the diaphragm segments span to a combination of walls and collectors. The Item 5 limitation would also prohibit application to nonrectangular diaphragm segments such as triangular, trapezoidal, or curved configurations; this is because of the more complex response of these configurations, and the difficulty in defining the effective diaphragm span and resulting diaphragm period (Figure C12.10.4-2).

Item 6 limits the vertical elements to systems that are inherently rigid for in-plane forces. This list functions as a definition of rigid vertical elements for purposes of Section 12.10.4. The list is in lieu of the FEMA P-1026 recommended minimum ratio of 3 between the fundamental period of the diaphragm and vertical elements. Limitation by system was identified to be the easiest and most direct way to address this criterion; see FEMA P-1026 for further discussion. The modifiers "ordinary," "intermediate," and "special" are not included for the vertical elements, with the intent that all types are sufficiently rigid and therefore permitted.

For Item 6, questions have arisen as to why steel and composite concentric braced frames are included in the scope and why buckling restrained braced frames (BRBFs) are not. The criteria established and studied during development of FEMA

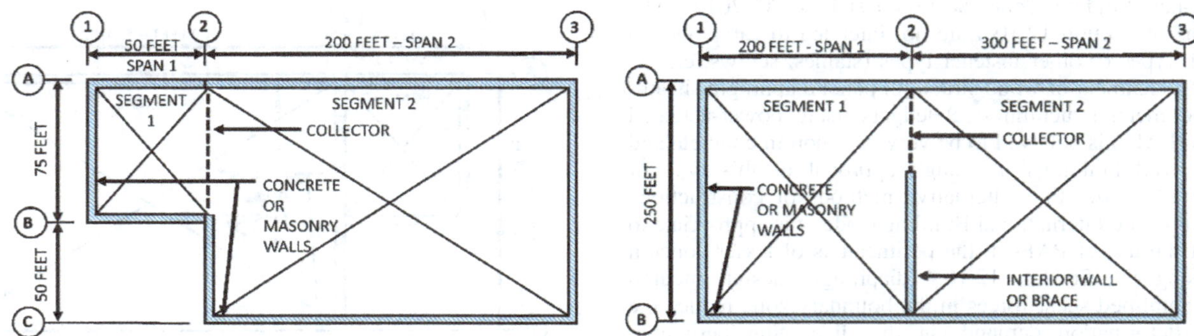

Figure C12.10.4-1b. Structures in plan view, divided into rectangular diaphragm segments for purposes of transverse seismic design. The Line 2 collector serves as a boundary between diaphragm segments for transverse direction loading. Note that identification of different segments will be required for longitudinal diaphragm forces in the building on the left.

Note: 1 ft = 305 mm.

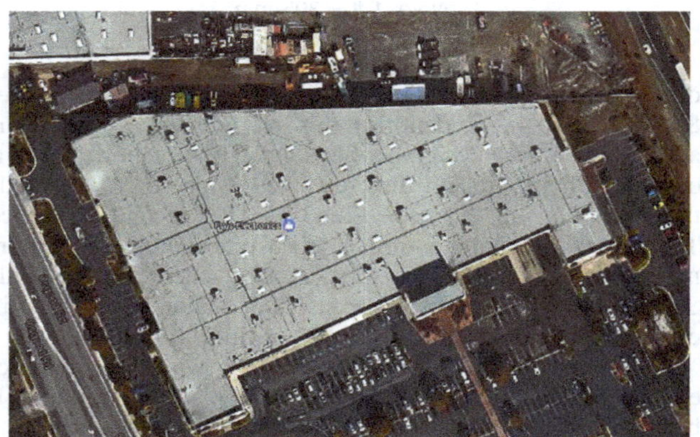

Figure C12.10.4-2. Example of a structure with nonrectangular diaphragm that is beyond the scope of the Section 12.10.4 provisions.

Source: Google Maps

P-1026 required that the diaphragm calculated period be at least three times the period of the vertical elements. When considering wood structural panel and bare steel deck diaphragms, one-story concentric braced frames designed in accordance with the requirements of ASCE 7 and adopted material standards will very clearly categorically meet these criteria, and are therefore included in the scope of these provision. Because BRBFs designed in accordance with adopted standards tend to have longer periods (more in line with steel moment frames), it is believed that they will generally not meet the FEMA P-1026 criteria, and are therefore excluded from the scope of these provisions. The drafters of the Section 12.10.4 provisions believed it important to use item 6 to limit use to vertical elements to those where use of the provisions is perceived to be of benefit.

Item 7 serves as a reminder that the provisions of Section 12.10.4 address only design of the diaphragm, and it is not intended or permitted that these force levels also be used for the vertical elements of the seismic force-resisting system. Vertical elements are to be designed using the equivalent lateral force (ELF) procedures of Section 12.8, except when designed in accordance with the two-stage analysis procedure of Section 12.2.3.2.2.

12.10.4.2 Design Section 12.10.4.2.1 provides new Equations (12.10-15), (12.10-16a), and (12.10-16b), to be used in place of Equation (12.10-1) for calculation of diaphragm design forces. These Equations incorporate the diaphragm approximate period, T_{diaph}. Where T_{diaph} is greater than T_s, this will permit C_s to be defined by the descending velocity-controlled portion of the response acceleration spectrum, thereby reducing diaphragm design forces. This is a distinct deviation from past design practices where seismic forces for diaphragm design were determined exclusively based on the approximate period and response modification factor of the vertical elements of the seismic force-resisting system. It is not necessary to set a lower bound on C_s, because the numerical studies for FEMA P-1026 investigated C_s values below the lower bound of Equation (12.10-2) and found acceptable performance. Equation (12.10-2) is provided primarily to guard against the higher diaphragm forces associated with higher-mode shapes of tall multistory buildings; however, the numerical studies for FEMA P-1026's single-story-building methodology saw limited participation of these higher modes and no reason to maintain this lower bound with these provisions.

The seismic response modification coefficient, R_{diaph}, is provided for both wood and bare steel deck roof diaphragms. The selected values for wood diaphragms and bare steel deck diaphragms with mechanical fasteners are based on studies reported in FEMA P-1026 (2015) and Koliou et al. (2015a, b). Based on the work of Schafer (2019), bare steel deck diaphragms were separated into two classes: diaphragms meeting special seismic detailing requirements, where ductile diaphragm performance can be reliably provided, are given an R_{diaph} of 4.5; and other diaphragms, where ductility is not required due to design forces representing near-elastic response, which are given an R_{diaph} of 1.5. The special seismic detailing requirements are found in AISI S400 (2020b).

Section 12.10.4.2.2 modifies the diaphragm shear forces determined per Section 12.10.4.2.1 for all diaphragms in an effort to manage the diaphragm's inelastic behavior. In larger diaphragm segments with spans of 100 ft (30.5 m) or more, the provisions create an amplified shear boundary zone at the supported ends of the diaphragm segment span (Figure 12.10.4-3, Segment 2). This boundary zone is strengthened to reduce the inelastic demand within this zone, and push inelastic behavior to the interior of the diaphragm segment. FEMA P-1026 studies found that the strengthening of the diaphragm segment's ends resulted in broadly distributed inelastic behavior toward the diaphragm segment interior and significantly improved diaphragm performance. This also served to move inelastic demand away from the diaphragm-to-vertical-element interface,

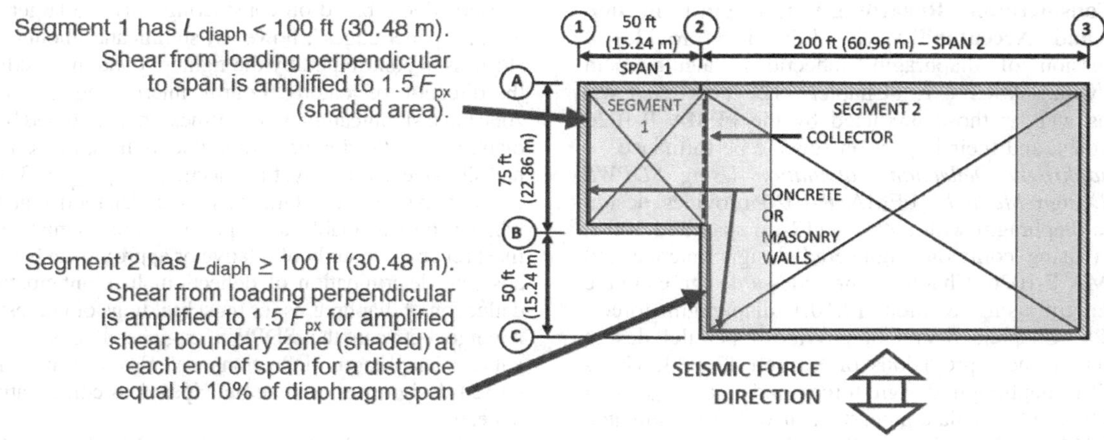

Figure C12.10.4-3. Illustration of Section 12.10.4.2.2 diaphragm shear amplification.
Note: 1 ft = 305 mm.

where it can be most damaging and most greatly affect structural performance.

Small diaphragm segments with spans less than 100 ft (30.5 m) have limited width available to distribute inelastic behavior; thus the amplification of diaphragm shears over the full diaphragm segment serves to limit inelastic behavior overall (Figure 12.10.4-3, Segment 1). The FEMA P-1026 numerical studies did not investigate diaphragm segment spans of less than 100 ft (30.5 m). The studies did, however, find a trend of the margin against collapse reducing with reduced diaphragm segment span, and the collapse margins at 100 ft (30.5 m) approaching the minimum acceptance criteria. As a result the amplified shear force level is required throughout the diaphragm segment for spans less than 100 ft (30.5 m). It is recognized that the different treatment of diaphragms based on being above or below the 100 ft (30.5 m) span introduces a step function into the design process; while ideally this step function would not exist, it is necessary based on the information currently available, and it is likely to have limited impact on design because the primary use of the methodology is intended to be diaphragms with spans greater than 100 ft (30.5 m).

When using this design procedure, it is important that strengthening of the diaphragm using amplified shear zones be limited to the zones prescribed by Section 12.10.4. Similarly, it is important that the diaphragm interior fastening zones be designed and constructed to minimize excess fastening and therefore excess shear capacity. There is a widespread belief that putting in more fasteners is always better, but in this case putting in more fasteners could result in reduced performance. This caution applies equally to wood structural panel and bare steel deck diaphragms.

Section 12.10.4.2.3 makes clear that chords are to be designed for the diaphragm design forces of Section 12.10.4.2.1, rather than the amplified forces of Section 12.10.4.2.2.

Section 12.10.4.2.4 makes clear that collectors and their connections to vertical elements are to be designed for the diaphragm design forces of Section 12.10.4.2.1, and in Seismic Design Categories D through F, amplified by an overstrength factor, $\Omega_{0\text{-diaph}}$, of 2, which was determined as part the FEMA P-1026 numerical studies.

Section 12.10.4.2.5 provides a diaphragm deflection amplification factor, $C_{d\text{-diaph}}$, intended to be used where the seismic design provisions currently require calculation of deflection. The calculation of diaphragm deflections is specified to use the seismic design forces of Section 12.10.4.2.1; it is not intended that the forces used for calculation of diaphragm deflections include the shear amplifications of Section 12.10.4.2.2. No new uses or checks of deflection are intended to be imposed by Section 12.10.4 provisions. The $C_{d\text{-diaph}}$ factor has been derived from the FEMA P-1026 nonlinear response history analysis (NLRHA) studies, in conformance with FEMA P-695 procedures (developed for vertical elements of the seismic force-resisting system), with some modifications. According to FEMA P-695, vertical systems with typical damping are assigned a deflection amplification factor, C_d, equal to the response modification factor, R. For the case where bare steel deck diaphragms with special seismic detailing and wood diaphragms are designed according to Section 12.10.4, a deflection amplification factor, $C_{d\text{-diaph}}$, of 3.0 (less than the R_{diaph} of 4.5) is deemed appropriate for several reasons: (1) the boundary zones are designed for 1.5 times the design shear value, equating to R_{diaph} of 3.0 at the boundaries; (2) inelastic deformations in diaphragms concentrate at the diaphragm perimeter and at transitions in strength and stiffness (e.g., transitions in nailing patterns), leaving large portions of the diaphragm elastic; and (3) numerical studies underlying FEMA P-1026 found that the ratio of median design basis earthquake drift to yield drift (approximation of predicted drift) was between 1.4 and 2.9, with an average of 2.1, suggesting that $C_{d\text{-diaph}}$ of 3.0 is conservative. See additional discussion at the end of Commentary Section C12.10.4 regarding calculation of diaphragm deflections and evaluation of deflection implications. For bare steel deck diaphragms not meeting the special seismic detailing requirements, the diaphragm deflection amplification factor is set to 1.5 in recognition of the near-elastic seismic demands being used for design.

Section 12.10.4.2.6 identifies two items affecting typical design that are not intended to be applicable when using Section 12.10.4. The Table 12.2-1 adjustment of vertical system Ω_0 for purposes of diaphragm design does not apply, as it is not consistent with the Section 12.10.4 formulation of diaphragm design forces. For item 2, Section 12.3.3.5 could potentially be triggered when using Section 12.10.4 for a re-entrant corner irregularity. It is clarified that modification of diaphragm design forces per Section 12.3.3.5 is not required. This is because per Section 12.10.4.2.4, the collectors and their connections to vertical elements are required to be designed for seismic loads including overstrength, meaning that Section 12.3.3.5 would never apply.

Additional Considerations Regarding Diaphragm Deflection Calculation and Acceptability The following provides a general discussion of diaphragm deflections calculated in accordance with ASCE 7, Chapter 12 and SDPWS procedures, as well as those predicted by the FEMA P-1026 numerical studies, and their impact on seismic performance.

Design Diaphragm Deflection Calculation Using SDPWS Engineered Design Methods. FEMA P-1026 provides design examples of a diaphragm with a 400 ft (12 m) span and 200 ft (60 m) width using common engineered design practice and SDPWS. FEMA P-1026, Chapter 3, provides a design example of the diaphragm using Section 12.10.1 diaphragm forces. FEMA P-1026, Chapters 5 and 6, provide a parallel design example using the new provisions of Section 12.10.4. Using Section 12.10.1 diaphragm design forces and a $C_{d\text{-diaph}}$ of 4 consistent with an intermediate precast shear wall, the estimated maximum diaphragm deflection is 29 in. (736 mm) [using the three-term Equation of SDPWS Section 4.2.2. Using Section 12.10.4 diaphragm design forces, the prescribed $C_{d\text{-diaph}}$ of 3.0, and the design assumptions used in the FEMA P-1026 examples, the estimated maximum diaphragm deflection is 19 in.(482 mm)]. Depending on calculation assumptions and calculation methods, it is anticipated that design engineers might calculate maximum diaphragm deflection as being anywhere between 10 and 19 in. (254 and 482 mm). This is a relative estimated displacement between the foundation and roof diaphragm at diaphragm midspan, which will be a maximum imposed drift on the vertical elements of the gravity system. The primary contributions to roof diaphragm deflection come from the shear deformation of the wood structural panel diaphragm (combined nail slip and panel shear deformation) and flexural deformation from tension and compression of the chord member.

Numerical Studies Using NLRHA. Numerical studies used as the basis for FEMA P-1026 provide data on analytical predictions of average peak diaphragm deflections. Diaphragm drift ratios published in Koliou et al. (2015a, b) are average peak ratios for the FEMA P-695 ground motion suite, scaled to $S_{DS} = 1.0$. The published diaphragm drift ratios correspond to an average peak roof deflection of 7 in. (178 mm) for the Chapter 3 example of the 400 ft (12 m) span and 200 ft (60 m) wide diaphragm designed for Section 12.10.1 forces. The published diaphragm drift ratios correspond to an average peak roof deflection of 10 in. (254 mm) for a structure designed using a method close to but not exactly matching Section 12.10.4 (the design of this similar building model used a period that combined diaphragm and shear wall period, modestly increasing the period, lowering the design forces, and lowering diaphragm stiffness).

The user will notice that the SDPWS engineered design estimate of peak diaphragm deflection of 19 in. (482 mm) [or the range of 10 to 19 in. (254 and 482 mm)] is generally larger than the NLRHA analytically predicted deflections of 7 and 10 in. (178 and 254 mm). A few reasons potentially contribute to this disparity. First, the FEMA P-1026 calculation conservatively computed the diaphragm's flexural deflection based on a single steel angle chord; however, numerous other building elements will engage as inadvertent chord elements (including concrete and masonry walls, wall reinforcing, and roof structure continuous ties), significantly reducing the flexural contribution to the deflection. Second, the three-term deflection Equation of SDPWS, Section 4.2.2, overestimates the diaphragm deflection compared to the more accurate four-term Equation in the SDPWS Commentary when design shears are below strength level, and when nail spacing varies between different sides of the sheathing panels. Third, the nail slip contribution of the SDPWS diaphragm deflection Equation is conservatively based on considering only the larger "nail spacing at other panel edges"; however, significant amounts of additional stiffness is contributed by the tighter "continuous edge nailing" in the direction of loading. Fourth, interior regions of each nailing zone have significantly more stiffness than assumed by the SDPWS diaphragm deflection procedure due to the stiffness nonlinearity of nail slippage. And lastly, the selection of $C_{d\text{-diaph}} = 3.0$ is potentially conservative as well. Finally, it is understood that the NLRHA, while a best available tool, provides approximate results and is most reliable for study of relative or approximate behavior, and not absolute determination of deflection. It is anticipated that actual deflection of diaphragms for most buildings of interest would fall in a range between the SDPWS engineered design and NLRHA values. Diaphragm deflections calculated using SDPWS engineered design methods are anticipated to conservatively estimate deflections.

Deflection of diaphragms is limited by Section 12.2.4, which requires that deflection be limited such that attached elements retain structural integrity. There are two primary aspects of structural integrity that should be checked. The first is the ability of the concrete or masonry walls (or other vertical elements) to maintain support of the prescribed loads through their out-of-plane rotation. Where gravity columns have rotational fixity at their top or bottom, the ability of the columns to support gravity loads in the displaced configuration should also be evaluated. Diaphragm deflection causes second-order moments in these elements which should be considered in conjunction with axial forces. The second is the ability of the connections within the gravity system to maintain strength as the vertical elements rock and rotate relative to the horizontal diaphragm; detailed discussion follows. In addition, interior full-height partitions or demising walls and other nonstructural components may suffer from racking or connection failure.

Consideration of typical roof system connections to the vertical elements can provide insight into the ability of gravity-load-carrying systems to withstand estimated roof diaphragm deflections. This discussion is affected, however, by whether the NLRHA analytically predicted diaphragm deflections or the SDPWS estimated deflections are used. Using the higher predicted mid-diaphragm deflection of 10 in. (254 mm) from the FEMA P-1026 NLRHA numerical studies, and story heights of 20 and 30 ft (6 and 9 m), this would create a gap of between 1/3 in. and 1/2 in. (8 and 13 mm) between an exterior wall and a 12 in. (305 mm) deep ledger and joist, as seen in Figure 12.10.4-4(a) for a wood-framed roof system. This amount of deformation can reasonably be taken up at several different interfaces within this connection without connection failure being likely. Similarly for wood system girder supports [Figure 12.10.4-4(b)] and interior columns [Figure 12.10.4-4(c)], the connections should be able to withstand this level of deflection. As the diaphragm deflection is increased to approximately 19 in. (482 mm) based on SDPWS calculations, the gaps increase to between 2/3 in. and 1 in. (17 to 25 mm) for the joists, which is approaching but likely not reaching gap levels that could potentially unseat rafters from hangers and cause damage to ledgers that are susceptible to cross-grain tension failure. Greater wall deflections or shorter wall heights would create gaps that could potentially push these connections to failure, and so deserve detailed consideration during design.

The larger mid-diaphragm deflection of 10 in. (254 mm) from the NLRHA numerical studies and roof diaphragm heights of 20 and 30 ft (6 and 9 m) would likely be acceptable for an open-web steel joist connection to an exterior wall, as seen in Figure 12.10.4-5(a). The behavior of this connection is considered reasonably close to pinned. The same would be true of truss

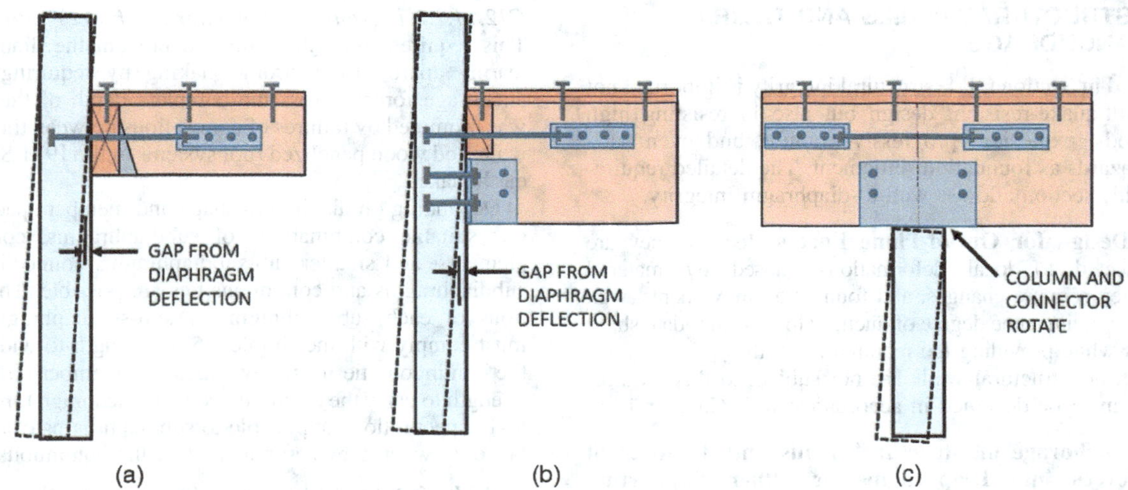

Figure C12.10.4-4. Wood-framed, wood-sheathed diaphragm connections: (a) joist to perimeter concrete or masonry wall; (b) girder connection to perimeter concrete or masonry wall; (c) girder to interior column.

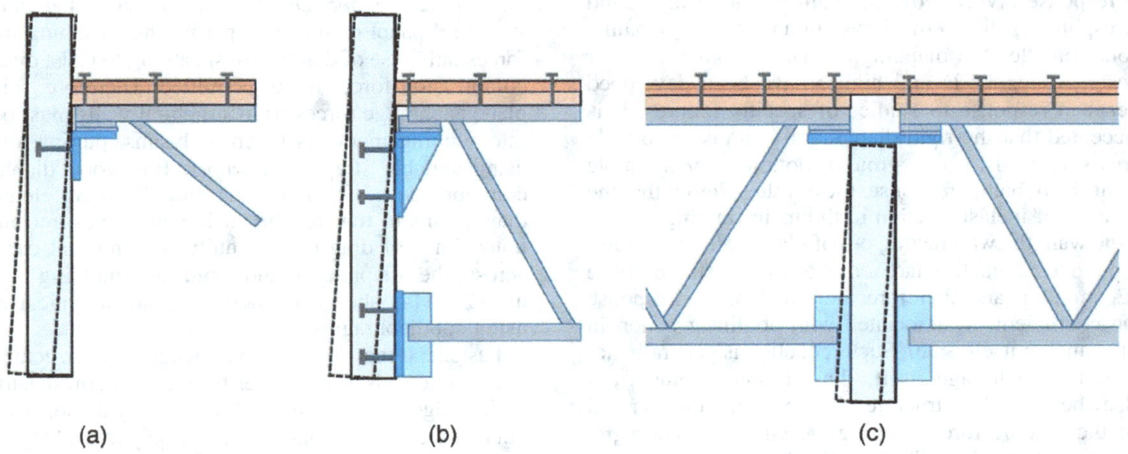

Figure C12.10.4-5. Steel open-web joist framed, wood-sheathed diaphragm connections: (a) joist to perimeter concrete or masonry wall; (b) joist girder to perimeter concrete or masonry wall; (c) joist girder to interior column.

girder connections to the exterior walls and interior columns, provided the girder truss connections are close to a pinned condition. Of concern in the steel open-web joist system is when a joist girder bottom chord has insufficient clearance or is axially connected to the wall or column [Figure 12.10.4-5(b, c)]. For a 3 ft (914 mm) girder depth, the required gap would be on the order of 1 to 1.5 in. (25 to 38 mm) at a diaphragm drift of 10 in. (254 mm). If the diaphragm drift were to be 19 in. (482 mm), the needed gap would increase to 2 to 3 in. (50 to 75 mm). If not detailed for this gap, the roof diaphragm deflection could be very damaging to the girder truss and connections. These illustrations serve as a reminder to the designer that this provision must be checked, detailing choices may be limited.

In addition to structural integrity considerations, global structural stability is a separate consideration where the diaphragm deflection's contribution could lead to P-delta instability of the system as a whole. As the roof mass horizontally translates and the gravity system rotates, secondary forces and moments develop, potentially leading to instability. Section 12.8.7 provides a methodology using a stability coefficient, θ, to determine whether the secondary effects are significant enough to require consideration; however, this section was developed expressly for buildings where the deformation is associated primarily with the vertical system, not the horizontal diaphragm. Nevertheless, the provisions can be adapted by considering P_x as the building weight tributary to the diaphragm (diaphragm weight plus half the rotating wall weight) and Δ as the weighted average diaphragm deflection. This approach is illustrated in FEMA P-1026.

The designer is reminded that diaphragms with openings are permitted to be designed per Section 12.10.4, provided the openings do not trigger a horizontal irregularity. Design for forces around openings is required per Section 12.10 and the material design standard.

The designer is reminded that diaphragm deflection causes leaning of structural wall elements, including concrete and masonry walls, creating a P-delta effect and out-of-plane wall bending moment. This effect should be considered in the wall design.

The designer is also reminded that seismic forces for the design of subdiaphragms serving as anchorage of concrete or masonry walls remains as per Section 12.11.

C12.11 STRUCTURAL WALLS AND THEIR ANCHORAGE

As discussed in Section C1.4, structural integrity is important not only in earthquake-resistant design but also in resisting high winds, floods, explosion, progressive failure, and even such ordinary hazards as foundation settlement. The detailed requirements of this section address wall-to-diaphragm integrity.

C12.11.1 Design for Out-of-Plane Forces Because they are often subjected to local deformations caused by material shrinkage, temperature changes, and foundation movements, wall connections require some degree of ductility to accommodate slight movements while providing the required strength.

Although nonstructural walls are not subject to this requirement, they must be designed in accordance with Chapter 13.

C12.11.2 Anchorage of Structural Walls and Transfer of Design Forces into Diaphragms or Other Supporting Structural Elements There are numerous instances in US earthquakes of tall, single-story, and heavy walls becoming detached from supporting roofs, resulting in collapse of walls and supported bays of roof framing (Hamburger and McCormick 2004). The response involves dynamic amplification of ground motion by response of the vertical system and further dynamic amplification from flexible diaphragms. The design forces for Seismic Design Category D and higher have been developed over the years in response to studies of specific failures. It is generally accepted that the rigid diaphragm value is reasonable for structures subjected to high ground motions. For a simple idealization of the dynamic response, these values imply that the combined effects of inelastic action in the main framing system supporting the wall, the wall (acting out of plane), and the anchor itself correspond to a reduction factor of 4.5 from elastic response to an MCE_R motion, and therefore the value of the response modification coefficient, R, associated with nonlinear action in the wall or the anchor itself is 3.0. Such reduction is generally not achievable in the anchorage itself; thus, it must come from yielding elsewhere in the structure, for example the vertical elements of the seismic force-resisting system, the diaphragm, or walls acting out of plane. The minimum forces are based on the concept that less yielding occurs with smaller ground motions, and less yielding is achievable for systems with smaller values of R, which are permitted in structures assigned to Seismic Design Categories B and C. The minimum value of R in structures assigned to Seismic Design Category D, except cantilever column systems and light-frame walls sheathed with materials other than wood structural panels, is 3.25, whereas the minimum values of R for Categories B and C are 1.5 and 2.0, respectively.

Where the roof framing is not perpendicular to anchored walls, provision needs to be made to transfer both the tension and sliding components of the anchorage force into the roof diaphragm. Where a wall cantilevers above its highest attachment to, or near, a higher level of the structure, the reduction factor based on the height within the structure, $(1 + 2z/h)/3$, may result in a lower anchorage force than appropriate. In such an instance, using a value of 1.0 for the reduction factor may be more appropriate.

C12.11.2.1 Wall Anchorage Forces Diaphragm flexibility can amplify out-of-plane accelerations so that the wall anchorage forces in this condition are twice those defined in Section 12.11.1.

C12.11.2.2 Additional Requirements for Anchorage of Concrete or Masonry Structural Walls to Diaphragms in Structures Assigned to Seismic Design Categories C through F

C12.11.2.2.1 Transfer of Anchorage Forces into Diaphragm This requirement, which aims to prevent the diaphragm from tearing apart during strong shaking by requiring transfer of anchorage forces across the complete depth of the diaphragm, was prompted by failures of connections between tilt-up concrete walls and wood panelized roof systems in the 1971 San Fernando earthquake.

Depending on diaphragm shape and member spacing, numerous suitable combinations of subdiaphragms, continuous tie elements, and smaller sub-subdiaphragms connecting to larger subdiaphragms and continuous ties are possible. The configurations of each subdiaphragm (or sub-subdiaphragm) provided must comply with the simple 2.5-to-1 length-to-width ratio, and the continuous tie must have adequate member and connection strength to carry the accumulated wall anchorage forces. The 2.5-to-1 aspect ratio is applicable to subdiaphragms of all materials, but only when they serve as part of the continuous tie system.

C12.11.2.2.2 Steel Elements of Structural Wall Anchorage System A multiplier of 1.4 has been specified for strength design of steel elements to obtain a fracture strength of almost 2 times the specified design force (where φ_t is 0.75 for tensile rupture).

C12.11.2.2.3 Wood Diaphragms Material standards for wood structural panel diaphragms permit the sheathing to resist shear forces only; use of diaphragm sheathing to resist direct tension or compression forces is not permitted. Therefore, seismic out-of-plane anchorage forces from structural walls must be transferred into framing members (such as beams, purlins, or subpurlins) using suitable straps or anchors. For wood diaphragms, it is common to use local framing and sheathing elements as subdiaphragms to transfer the anchorage forces into more concentrated lines of drag or continuity framing that carry the forces across the diaphragm and hold the building together. Figure C12.11-1 shows a schematic plan of typical roof framing using subdiaphragms.

Fasteners that attach wood ledgers to structural walls are intended to resist shear forces from diaphragm sheathing attached to the ledger that act longitudinally along the length of the ledger but not shear forces that act transversely to the ledger, which tend to induce splitting in the ledger caused by cross-grain bending. Separate straps or anchors generally are provided to transfer out-of-plane wall forces into perpendicular framing members.

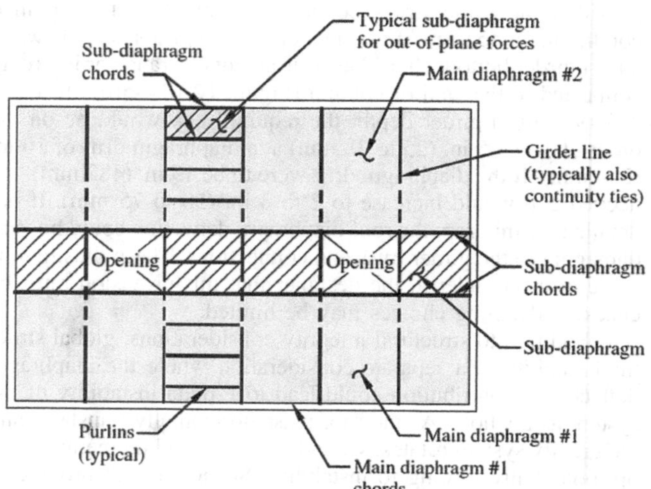

Figure C12.11-1. Typical subdiaphragm framing.

The requirements of Section 12.11.2.2.3 are consistent with the requirements of AWC SDPWS-15 (2014), Section 4.1.5.1, but also apply to wood use in diaphragms that may fall outside the scope of AWC SDPWS. Examples include use of wood structural panels attached to steel bar joists or metal deck attached to wood nailers.

C12.11.2.2.4 Metal Deck Diaphragms In addition to transferring shear forces, metal deck diaphragms often can resist direct axial forces in at least one direction. However, corrugated metal decks cannot transfer axial forces in the direction perpendicular to the corrugations and are prone to buckling if the unbraced length of the deck as a compression element is large. To manage diaphragm forces perpendicular to the deck corrugations, it is common for metal decks to be supported at 8 to 10 ft (2.4 to 3.0 m) intervals by joists that are connected to walls in a manner suitable to resist the full wall anchorage design force and to carry that force across the diaphragm. In the direction parallel to the deck corrugations, subdiaphragm systems are considered near the walls; if the compression forces in the deck become large relative to the joist spacing, small compression reinforcing elements are provided to transfer the forces into the subdiaphragms.

C12.11.2.2.5 Embedded Straps Steel straps may be used in systems where heavy structural walls are connected to wood or steel diaphragms as the wall-to-diaphragm connection system. In systems where steel straps are embedded in concrete or masonry walls, the straps are required to be bent around reinforcing bars in the walls, which improve their ductile performance in resisting earthquake load effects (i.e., the straps pull the bars out of the wall before the straps fail by pulling out without also pulling the reinforcing bars out). Consideration should be given to the probability that light steel straps have been used in past earthquakes and have developed cracks or fractures at the wall-to-diaphragm framing interface because of gaps in the framing adjacent to the walls.

C12.11.2.2.6 Eccentrically Loaded Anchorage System Wall anchors often are loaded eccentrically, either because the anchorage mechanism allows eccentricity or because of anchor bolt or strap misalignment. This eccentricity reduces the anchorage connection capacity and hence must be considered explicitly in design of the anchorage. Figure C12.11-2 shows a one-sided roof-to-wall anchor that is subjected to severe eccentricity because of a misplaced anchor rod. If the detail were designed as a concentric two-sided connection, this condition would be easier to correct.

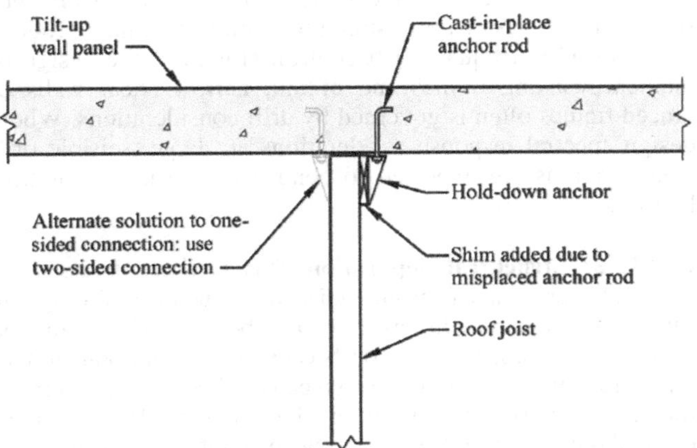

Figure C12.11-2. Plan view of wall anchor with misplaced anchor rod.

C12.11.2.2.7 Walls with Pilasters The anchorage force at pilasters must be calculated considering two-way bending in wall panels. It is customary to anchor the walls to the diaphragms assuming one-way bending and simple supports at the top and bottom of the wall. However, where pilasters are present in the walls, their stiffening effect must be taken into account. The panels between pilasters are typically supported along all panel edges. Where this support occurs, the reaction at the top of the pilaster is the result of two-way action of the panel and is applied directly to the anchorage supporting the top of the pilaster. The anchor load at the pilaster generally is larger than the typical uniformly distributed anchor load between pilasters. Figure C12.11-3 shows the tributary area typically used to determine the anchorage force for a pilaster.

Anchor points adjacent to the pilaster must be designed for the full tributary loading, conservatively ignoring the effect of the adjacent pilaster.

C12.12 DRIFT AND DEFORMATION

As used in the standard, *deflection* is the absolute lateral displacement of any point in a structure relative to its base, and *design story drift*, Δ, is the difference in deflection along the height of a story (i.e., the deflection of a floor relative to that of the floor below). The drift, Δ, is calculated according to the procedures of Section 12.8.6. (Sections 12.9.2 and 16.1 give procedures for calculating displacements for modal response spectrum and linear response history analysis procedures, respectively; the definition of Δ in Section 11.3 should be used.)

Calculated story drifts generally include torsional contributions to deflection (i.e., additional deflection at locations of the center of rigidity at other than the center of mass caused by diaphragm rotation in the horizontal plane). The provisions allow these contributions to be neglected where they are not significant, such as in cases where the calculated drifts are much less than the allowable story drifts, Δ_a, no torsional irregularities exist, and more precise calculations are not required for structural separations (see Sections C12.12.2 and C12.12.3).

The deflections and design story drifts are calculated using the design earthquake ground motion, which is two-thirds of the risk-targeted maximum considered earthquake (MCE_R) ground motion. The resulting drifts are therefore likely to be underestimated.

The design base shear, V, used to calculate Δ is reduced by the response modification coefficient, R. Multiplying displacements by the deflection amplification factor, C_d, is intended to correct for this reduction and approximate inelastic drifts corresponding to the design response spectrum unreduced by R. However, it is recognized that use of values of C_d less than R underestimates deflections (Uang and Maarouf 1994). Also, Sections C12.8.6.2 and C12.9.1.4 deal with the appropriate base shear for computing displacements.

For these reasons, the displacements calculated may not correspond well to MCE_R ground motions. However, they are appropriate for use in evaluating the structure's compliance with the story drift limits put forth in Table 12.12-1 of the standard.

There are many reasons to limit drift; the most significant are to address the structural performance of member inelastic strain and system stability and to limit damage to nonstructural components, which can be life-threatening. Drifts provide a direct but imprecise measure of member strain and structural stability. Under small lateral deformations, secondary stresses caused by the P-delta effect are normally within tolerable limits (see Section C12.8.7). The drift limits provide indirect control of structural performance.

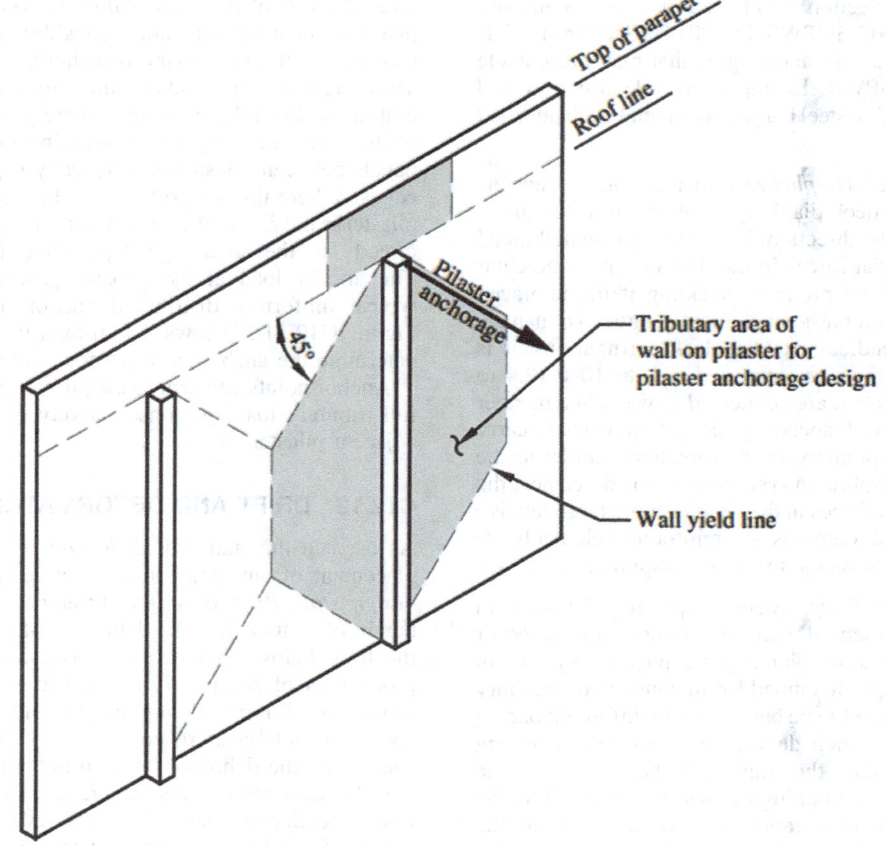

Figure C12.11-3. Tributary area used to determine anchorage force at pilaster.

Buildings subjected to earthquakes need drift control to limit damage to partitions, shaft and stair enclosures, glass, and other fragile nonstructural components. The drift limits have been established without regard to economic considerations such as a comparison of present worth of future repairs with additional structural costs to limit drift. These are matters for building owners and designers to address.

The allowable story drifts, Δ_a, of Table 12.12-1 reflect the consensus opinion of the ASCE 7 Committee, taking into account the life safety and damage control objectives described in the aforementioned commentary. Because the displacements induced in a structure include inelastic effects, structural damage as the result of a design-level earthquake is likely. This may be seen from the values of Δ_a in Table 12.12-1. For other structures assigned to Risk Category I or II, the value of Δ_a is $0.02h_{sx}$, which is about 10 times the drift ordinarily allowed under wind loads. If deformations well in excess of Δ_a were to occur repeatedly, structural elements of the seismic force-resisting system could lose so much stiffness or strength that they would compromise the safety and stability of the structure.

To provide better performance for structures assigned to Risk Category III or IV, their allowable story drifts, Δ_a, generally are more stringent than for those assigned to Risk Category I or II. However, those limits are still greater than the damage thresholds for most nonstructural components. Therefore, though the performance of structures assigned to Risk Category III or IV should be improved, there may be considerable damage from a design-level earthquake.

The allowable story drifts, Δ_a, for structures a maximum of four stories above the base are relaxed somewhat, provided that the interior walls, partitions, and ceilings have been designed to accommodate story drifts. The type of structure envisioned by footnote *c* in Table 12.12-1 would be similar to a prefabricated steel structure with metal skin.

As noted in the preceding paragraph, the more relaxed drift limits require that interior walls, partitions, and ceilings be designed to accommodate story drifts to limit damage for these components, which are not otherwise required to accommodate the story drift per Chapter 13. Prior provisions that required exterior wall systems to accommodate the more relaxed story drift limits have been removed, since this is required by Section 13.5.3 regardless of the applicable drift limit.

The values of Δ_a in Table 12.12-1 apply to each story. For some structures, satisfying strength requirements may produce a system with adequate drift control. However, the design of moment-resisting frames and of tall, narrow shear walls or braced frames often is governed by drift considerations. Where design spectral response accelerations are large, seismic drift considerations are expected to control the design of midrise buildings.

C12.12.2 Structural Separation This section addresses the potential for impact from adjacent structures during an earthquake. Such conditions may arise because of construction on or near a property line or because of the introduction of separations within a structure (typically called "seismic joints") for the purpose of permitting their independent response to earthquake ground motion. Such joints may effectively eliminate irregularities and large force transfers between portions of the building with different dynamic properties.

The standard requires the distance to be "sufficient to avoid damaging contact under total deflection." It is recommended that the distance be computed using the square root of the sum of the squares (SRSS) of the lateral deflections. Such a combination method treats the deformations as linearly independent variables. The deflections used are the expected displacements (e.g., the anticipated maximum inelastic deflections including the effects of torsion and diaphragm deformation). Just as these displacements increase with height, so does the required separation. If the effects of impact can be shown not to be detrimental, the required separation distances can be reduced.

For rigid shear wall structures with rigid diaphragms whose lateral deflections cannot be reasonably estimated, the NEHRP provisions (FEMA 2009a) suggest that older code requirements for structural separations of at least 1 in. (25 mm) plus 1/2 in. (13 mm) for each 10 ft (3 m) of height above 20 ft (6 m) be followed.

C12.12.3 Members Spanning between Structures Where a portion of the structure is seismically separated from its support, the design of the support requires attention to ensure that support is maintained as the two portions move independently during earthquake ground motions. To prevent local collapse due to loss of gravity support for members that bridge between the two portions, the relative displacement must not be underestimated. Design Earthquake Displacements may be insufficient for this purpose. The provision thus requires that maximum considered earthquake displacements be used and that the absolute sum of displacements of the two portions be used instead of a modal combination, such as with Equation (12.12-2), which would represent a probable value.

It is recognized that displacements so calculated are likely to be conservative. However, the consequences of loss of gravity support are likely to be severe, and some conservatism is deemed appropriate.

C12.12.4 Deformation Compatibility for Seismic Design Categories D through F In regions of high seismicity, many designers apply ductile detailing requirements to elements that are intended to resist seismic forces but neglect such practices for nonstructural components, or for structural components that are designed to resist only gravity forces but must undergo the same lateral deformations as the designated seismic force-resisting system. Even where elements of the structure are not intended to resist seismic forces and are not detailed for such resistance, they can participate in the response and may suffer severe damage as a result. This provision requires the designer to provide a level of ductile detailing or proportioning to all elements of the structure appropriate to the calculated deformation demands at the Design Earthquake Displacement, δ_{DE}, and the associated story drift. This provision may be accomplished by applying details in gravity members similar to those used in members of the seismic force-resisting system or by providing sufficient strength in those members, or by providing sufficient stiffness in the overall structure to preclude ductility demands in those members.

In the 1994 Northridge (California) earthquake, such participation was a cause of several failures. A preliminary reconnaissance report of that earthquake (EERI 1994, pp. 56–57) states the following:

> *Of much significance is the observation that six of the seven partial collapses (in modern precast concrete parking structures) seem to have been precipitated by damage to the gravity load system. Possibly, the combination of large lateral deformation and vertical load caused crushing in poorly confined columns that were not detailed to be part of the lateral load resisting system. Punching shear failures were observed in some structures at slab-to-column connections, such as at the Four Seasons building in Sherman Oaks. The primary lateral load resisting system was a perimeter ductile frame that performed quite well. However, the interior slab–column system was incapable of undergoing the same lateral deflections and experienced punching failures.*

This section addresses such concerns. Rather than relying on designers to assume appropriate levels of stiffness, this section explicitly requires that the stiffening effects of adjoining rigid structural and nonstructural elements be considered and that a rational value of member and restraint stiffness be used for the design of structural components that are not part of the seismic force-resisting system.

This section also includes a requirement to address shears that can be induced in structural components that are not part of the seismic force-resisting system because sudden shear failures have been catastrophic in past earthquakes.

The exception is intended to encourage the use of intermediate or special detailing in beams and columns that are not part of the seismic force-resisting system. In return for better detailing, such beams and columns are permitted to be designed to resist moments and shears from unamplified deflections. This design approach reflects observations and experimental evidence that well-detailed structural components can accommodate large drifts by responding inelastically without losing significant vertical load-carrying capacity.

C12.13 FOUNDATION DESIGN

C12.13.1 Design Basis In traditional geotechnical engineering practice, foundation design is based on allowable stresses, with allowable foundation load capacities, Q_{as}, for dead plus live loads based on limiting static settlements, which provides a large factor of safety against exceeding ultimate capacities. In this practice, allowable soil stresses for dead plus live loads often are increased arbitrarily by one-third for load combinations that include wind or seismic forces. That approach is overly conservative and not entirely consistent with the design basis prescribed in Section 12.1.5, since it is not based on explicit consideration of the expected strength and dynamic properties of the site soils. Strength design of foundations in accordance with Section 12.13.5 facilitates more direct satisfaction of the design basis.

Section 12.13.1.1 provides horizontal load effect, E_h, values that are used in Section 12.4.2 to determine foundation load combinations that include seismic effects. Vertical seismic load effects are still determined in accordance with Section 12.4.2.2.

Foundation horizontal seismic load effect values specified in Section 12.13.1.1 are intended to be used with horizontal seismic forces, Q_E, defined in Section 12.4.2.1.

C12.13.3 Foundation Load–Deformation Characteristics For linear static and dynamic analysis methods, where foundation flexibility is included in the analysis, the load–deformation behavior of the supporting soil should be represented by an equivalent elastic stiffness using soil properties that are compatible with the soil strain levels associated with the design earthquake motion. The strain-compatible shear modulus, G, and the associated strain-compatible shear wave velocity, v_s, needed for the evaluation of equivalent elastic stiffness are specified in Chapter 19 of the standard or can be based on a site-specific study. Although inclusion of soil flexibility tends to lengthen the fundamental period of the structure, it should not change the maximum-period limitations applied when calculating the required base shear of a structure.

A mathematical model incorporating a combined superstructure and foundation system is necessary to assess the effect of foundation and soil deformations on the superstructure elements. Typically, frequency-independent linear springs are included in the mathematical model to represent the load–deformation characteristics of the soil, and the foundation components are either explicitly modeled (e.g., mat foundation supporting a configuration of structural walls) or are assumed to be rigid (e.g., spread footing supporting a column). In specific cases, a spring may be used to model both the soil and the foundation component (e.g., grade beams or individual piles).

For dynamic analysis, the standard requires a parametric evaluation with upper- and lower-bound soil parameters to account for the uncertainty in as-modeled soil stiffness and in situ soil variability and to evaluate the sensitivity of these variations on the superstructure. Varying the strain-compatible shear modulus with an upper bound equal to $G(1 + C_{vs})$ and a lower bound equal to $G/(1 + C_{vs})$ produces a range that is consistent with a lognormal distribution, and where the upper- and lower-bound properties correspond to the best-estimate strain-compatible shear modulus plus or minus one standard deviation in the logarithmic domain. This approach is consistent with ASCE 4-16, Section 5.1.7, and ASCE 41-17, Section 8.4.2.3. Sources of uncertainty include variability in the rate of loading, including the cyclic nature of building response, level of strain associated with loading at the design earthquake (or stronger), idealization of potentially nonlinear soil properties as elastic, and variability in the estimated soil properties. To a lesser extent, this variation accounts for variability in the performance of the foundation components, primarily when a rigid foundation is assumed or distribution of cracking of concrete elements is not explicitly modeled.

Commonly used analysis procedures tend to segregate the "structural" components of the foundation (e.g., footing, grade beam, pile, and pile cap) from the supporting (e.g., soil) components. The "structural" components are typically analyzed using standard strength design load combinations and methodologies, whereas the adjacent soil components are analyzed using allowable stress design (ASD) practices, in which earthquake forces (which have been reduced by R) are considered using ASD load combinations, to make comparisons of design forces versus allowable capacities. These "allowable" soil capacities are typically based on expected strength divided by a factor of safety, for a given level of potential deformations.

When design of the superstructure and foundation components is performed using strength-level load combinations, this traditional practice of using allowable stress design to verify soil compliance can become problematic for assessing the behavior of foundation components. The 2009 NEHRP provisions (FEMA 2009a) contain two resource papers (RP 4 and RP 8) that provide guidance on the application of ultimate strength design procedures in the geotechnical design of foundations and the development of foundation load–deformation characterizations for both linear and nonlinear analysis methods. Additional guidance on these topics is given in ASCE 41 (2014b).

C12.13.4 Reduction of Foundation Overturning Since the vertical distribution of horizontal seismic forces prescribed for use with the equivalent lateral force (ELF) procedure is intended to envelope story shears, the resulting base overturning forces can be exaggerated in some cases (Section C12.13.3). Such overturning will be overestimated where multiple vibration modes are excited, so a 25% reduction in overturning effects is permitted for verification of soil stability. This reduction is not permitted for inverted-pendulum or cantilevered-column-type structures, which typically have a single mode of response.

Since the modal response spectrum analysis (MRSA) procedure more accurately reflects the actual distribution of base shear and overturning moment, the permitted reduction is reduced to 10%.

C12.13.5 Strength Design for Foundation Geotechnical Capacity This section provides guidance for determination of nominal strengths, resistance factors, and acceptance criteria when the strength design load combinations of Section 12.4.2 are used, instead of allowable stress load combinations, to check stresses at the soil–foundation interface.

C12.13.5.1 Soil Strength Parameters If soils are saturated or anticipated to become so, undrained soil properties might be used for transient seismic loading, even though drained strengths may have been used for static or more sustained loading. For competent soils that are not expected to degrade in strength during seismic loading (e.g., due to partial or total liquefaction of cohesionless soils or strength reduction of sensitive clays), use of static soil strengths is recommended for determining the nominal foundation geotechnical capacity, Q_{ns}, of foundations. Use of static strengths is somewhat conservative for such soils because rate-of-loading effects tend to increase soil strengths for transient loading. Such rate effects are neglected because they may not result in significant strength increase for some soil types and are difficult to estimate confidently without special dynamic testing programs. The assessment of the potential for soil liquefaction or other mechanisms that reduce soil strengths is critical, because these effects may reduce soil strengths greatly below static strengths for susceptible soils.

The best estimated nominal strength of footings, Q_{ns}, should be determined using accepted foundation engineering practice. In the absence of moment loading, the ultimate vertical load capacity of a rectangular footing of width B and length L may be written as $Q_{ns} = q_c(BL)$, where q_c is the ultimate soil bearing pressure.

For rigid footings subject to moment and vertical load, contact stresses become concentrated at footing edges, particularly at the edge opposite to where footing uplift occurs. Although the nonlinear behavior of soils causes the actual soil pressure beneath a footing to become nonlinear, resulting in an ultimate foundation strength that is slightly greater than the strength that is determined by assuming a simplified trapezoidal or triangular soil pressure distribution with a maximum soil pressure equal to the ultimate soil pressure, q_c, the difference between the nominal ultimate foundation strength and the effective ultimate strength calculated using these simplified assumptions is not significant.

Lateral resistance may be determined from test data, or by a combination of lateral bearing, lateral friction, and cohesion values. The lateral bearing values may represent values determined from the passive strength values of soil or rock, or they may represent a reduced "allowable" value determined to meet a defined deformation limit. Lateral friction values may represent side-friction values caused by uplift or movement of a foundation against soils, such as for pile uplift or a side friction caused by lateral foundation movement, or they may represent the lateral friction resistance that may be present beneath a foundation caused by the gravity weight of loads that is bearing on the supporting material.

The lateral foundation geotechnical capacity of a footing may be assumed to be equal to the sum of the best estimated soil passive resistance against the vertical face of the footing plus the best estimated soil friction force on the footing base. The determination of passive resistance should consider the potential contribution of friction on the vertical face.

For piles, the best estimated vertical strength (for both axial compression and axial tensile loading) should be determined using accepted foundation engineering practice. The moment capacity of a pile group should be determined assuming a rigid pile cap, leading to an initial triangular distribution of axial pile loading from applied overturning moments. However, the full expected axial capacity of piles may be mobilized when computing moment capacity, in a manner analogous to that described for a footing. The strength provided in pile caps and intermediate connections should be capable of transmitting the best estimated pile forces to the supported structure. When evaluating axial tensile strength, consideration should be given to the capability of pile cap and splice connections to resist the factored tensile loads.

The lateral foundation geotechnical capacity of a pile group may be assumed to be equal to the best estimated passive resistance acting against the face of the pile cap plus the additional resistance provided by piles.

When the nominal foundation geotechnical capacity, Q_{ns}, is determined by in situ testing of prototype foundations, the test program, including the appropriate number and location of test specimens, should be provided to the Authority Having Jurisdiction by a registered design professional, based on the scope and variability of geotechnical conditions present at the site.

C12.13.5.2 Resistance Factors Resistance factors, φ, are provided to reduce nominal foundation geotechnical capacities, Q_{ns}, to design foundation geotechnical capacities, φQ_{ns}, to verify foundation acceptance criteria. The values of φ recommended here have been based on the values presented in the AASHTO *LRFD Bridge Design Specifications* (2010). The AASHTO values have been further simplified by using the lesser values when multiple values are presented. These resistance factors account not only for unavoidable variations in design, fabrication, and erection, but also for the variability that often is found in site conditions and test methods (AASHTO 2010).

C12.13.5.3 Acceptance Criteria The design foundation geotechnical capacity, φQ_{ns}, is used to assess acceptability for the linear analysis procedures. The mobilization of ultimate capacity in nonlinear analysis procedures does not necessarily lead to unacceptable performance, because structural deformations caused by foundation displacements may be tolerable. For the nonlinear analysis procedures, Section 12.13.3 also requires evaluation of structural behavior using parametric variation of foundation strength to identify potential changes in structural ductility demands.

C12.13.6 Allowable Stress Design for Foundation Geotechnical Capacity In traditional geotechnical engineering practice, foundation design is based on allowable stresses, with allowable foundation load capacities, Q_{as}, for dead plus live loads based on limiting static settlements, which provides a large factor of safety against exceeding ultimate capacities. In this practice, allowable soil stresses for dead plus live loads often are increased arbitrarily by one-third for load combinations that include wind or seismic forces. That approach may be both more conservative and less consistent than the strength design basis prescribed in Section 12.1.5, since it is not based on explicit consideration of the expected strength and dynamic properties of the site soils.

C12.13.7 Requirements for Structures Assigned to Seismic Design Category C

C12.13.7.1 Pole-Type Structures The high contact pressures that develop between an embedded pole and soil as a result of lateral loads make pole-type structures sensitive to earthquake motions. Pole-bending strength and stiffness, the soil lateral bearing capacity, and the permissible deformation at grade level are key considerations in the design. For further discussion of pole–soil interaction, see Section C12.13.8.7.

C12.13.7.2 Foundation Ties One important aspect of adequate seismic performance is that the foundation system should act as an integral unit, not permitting one column or wall to move appreciably relative to another. To attain this performance, the standard requires that pile caps be tied together. This requirement is especially important where the use of deep foundations is driven by the existence of soft surface soils. Pilecaps should be tied together in a way that precludes significant differential lateral movement so that the foundation essentially moves as a unit. This reduces unequal deformation demands on vertically oriented lateral-force-resisting elements due to relative foundation movement and reduces the likelihood that the structural system will behave in unanticipated and detrimental ways if the lateral-force-resisting elements are not acting in concert.

This requirement has typically been applied by tying pilecaps together with an orthogonal system of grade beams, although structural slabs and non-orthogonal systems of grade beams capable of resisting forces in two orthogonal directions may also be used. ACI 318 (2019), Section 18.13.4, requires such foundation ties (or equivalent restraint) under certain conditions; that provision specifies the requirement in orthogonal directions. Neither the ASCE 7 nor the ACI 318 provisions explicitly requires a foundation system that constrains pilecaps to move as a unit except on the tie-beam lines, which are typically in an orthogonal grid. Where there is no foundation-level diaphragm or slab capable of providing such constraint, there may be significant relative movement between the foundations not on the same tie-beam line. Such relative foundation displacement effectively adds to the displacement of the seismic-force resisting system above the foundation and should be considered for deformation compatibility, deformation of nonstructural elements, and drift where applicable.

Multistory buildings often have major columns that run the full height of the building adjacent to smaller columns that support only one level; the calculated tie force should be based on the heavier column load.

The standard permits alternate methods of tying foundations together when appropriate. Relying on lateral soil pressure on pile caps to provide the required restraint is not a recommended method because ground motions are highly dynamic and may occasionally vary between structure support points during a design-level seismic event.

C12.13.7.3 Pile Anchorage Requirements The pile anchorage requirements are intended to prevent brittle failures of the connection to the pile cap under moderate ground motions. Moderate ground motions can result in pile tension forces or bending moments that could compromise shallow anchorage embedment. Loss of pile anchorage could result in increased structural displacements from rocking, overturning instability, and loss of shearing resistance at the ground surface. A concrete bond to a bare steel pile section usually is unreliable, but connection

The 15-story Banco Central Building was not damaged due to its strong foundations when subjected to a 7 in. (0.18 m) surface fault rupture in the 1972 Nicaragua earthquake. However, in the 1999 Chi-Chi earthquake, 3 to 12 ft (1 to 4 m) of reverse fault rupture resulted in significant damage to a number of structures, including the collapse of a building with a weak foundation. Structures with stronger foundations tied together suffered less damage and did not collapse. However, the Attaturk Basketball Court in Turkey suffered significant damage from 5 ft (1.5 m) of normal fault movement in the 1999 Kocaeli earthquake because its pile foundation could not resist the forces generated by the piles tying the structure to both sides of the fault.

In the 1994 Northridge earthquake, several hundred structures built on fill suffered settlement and lateral extension of up to 8 in. (0.2 m) primarily resulting from seismically induced compression of the unsaturated fill. The structural damage varied from negligible to significant, depending on the amount of differential ground movement.

C12.13.9.1 Foundation Design Foundations are not allowed to lose their strength capacity to support vertical reactions. This requirement is intended to prevent bearing capacity failure of shallow foundations and axial load failure of deep foundations. Settlement in the event of such failures cannot be accurately estimated and has potentially catastrophic consequences. Such failures can be prevented by using ground improvement or adequately designed deep foundations.

Seismically induced differential settlement can result from variations in the thickness, relative density, or fines content of potentially liquefiable layers across the footprint of the structure. When planning a field exploration program for a site that may be subject to seismically induced differential ground displacement, the geotechnical engineer must have information on the proposed layout of the building(s) on the site. This information is essential to properly locating and spacing exploratory holes to obtain an appropriate estimate of anticipated differential settlement. One acceptable method for dealing with unacceptable liquefaction-induced settlements is by performing ground improvement. There are many acceptable methods for ground improvement.

C12.13.9.2 Shallow Foundations Shallow foundations are permitted where individual footings are tied together, so that they have the same horizontal displacements, and differential settlements are limited, or where the expected differential settlements can be accommodated by the structure and the foundation system. The lateral spreading limits provided in Table 12.13-2 are based on engineering judgment and are the judged upper limits of lateral spreading displacements that can be tolerated while still achieving the desired performance for each Risk Category, presuming that the foundation is well tied together. Differential settlement is defined as δ_v/L, where δ_v and L are illustrated for an example structure in Figure C12.13-1. The differential settlement limits specified in Table 12.13-3 are intended to provide collapse resistance for Risk Category II and III structures.

The limit for one-story Risk Category II structures with concrete or masonry structural walls is consistent with the drift limit in ASCE 41 (2014b) for concrete shear walls to achieve collapse prevention. The limit for taller structures is more restrictive because of the effects the tilt would have on the floors of upper levels. This more restrictive limit is consistent with "moderate to severe damage" for multistory masonry structures, as indicated in Boscardin and Cording (1989).

The limits for structures without concrete or masonry structural walls are less restrictive and are consistent with the drift limits in ASCE 41 (2014b) for high-ductility concrete frames to maintain collapse prevention. Frames of lower ductility are not permitted in Seismic Design Categories C and above, which are the only categories where liquefaction hazards need to be assessed.

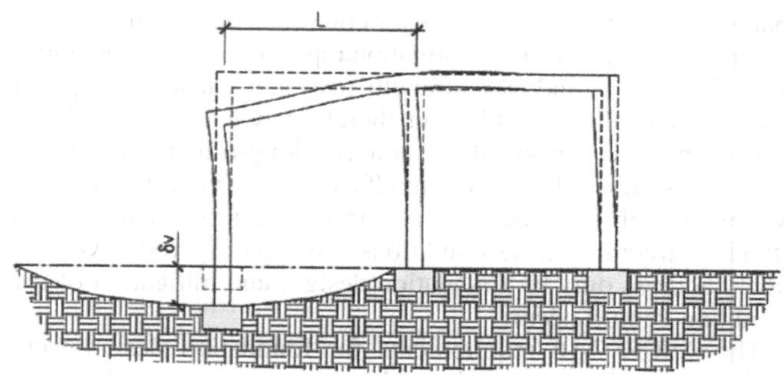

Figure C12.13-1. Example showing differential settlement terms δ_v and L.

The limits for Risk Category III structures are two-thirds of those specified for Risk Category II.

The limits for Risk Category IV are intended to maintain differential settlements less than the distortion that will cause doors to jam in the design earthquake. The numerical value is based on the median value of drift (0.0023) at the onset of the damage state for jammed doors developed for the ATC-58 project (ATC 2012), multiplied by 1.5 to account for the dispersion and scaled to account for the higher level of shaking in the maximum considered earthquake (MCE) relative to the design earthquake (DE).

Shallow foundations are required to be interconnected by ties, regardless of the effects of differential permanent seismically induced ground displacement. The additional detailing requirements in this section are intended to provide moderate ductility in the behavior of the ties, because the adjacent foundations may settle differentially. The tie force required to accommodate lateral ground displacement is intended to be a conservative assessment to overcome the maximum frictional resistance that could occur between footings along each column or wall line. The tie force assumes that the lateral spreading displacement occurs abruptly, midway along the column or wall line. The coefficient of friction between the footings and underlying soils may be taken conservatively as 0.50. This requirement is intended to maintain continuity throughout the substructure in the event of lateral ground displacement affecting a portion of the structure. The required tie force should be added to the force determined from the lateral loads for the design earthquake in accordance with Sections 12.8, 12.9, 12.14, or Chapter 16.

C12.13.9.3 Deep Foundations Pile foundations are intended to remain elastic under axial loadings, including those from gravity, seismic, and downdrag loads. Since geotechnical design is most frequently performed using allowable stress design methods, and downdrag from seismically induced ground displacement is assessed at an ultimate level, the requirements state that the downdrag is considered as a reduction in the ultimate capacity. Since structural design is most frequently performed using load and resistance factor design methods, and the downdrag is considered as a load for the pile structure to resist, the requirements clarify that the downdrag is considered as a seismic axial load, to which a factor of 1.0 would be applied for design.

Although downdrag load is to be factored as a seismic load, it is not intended to be considered concurrently with seismic loads

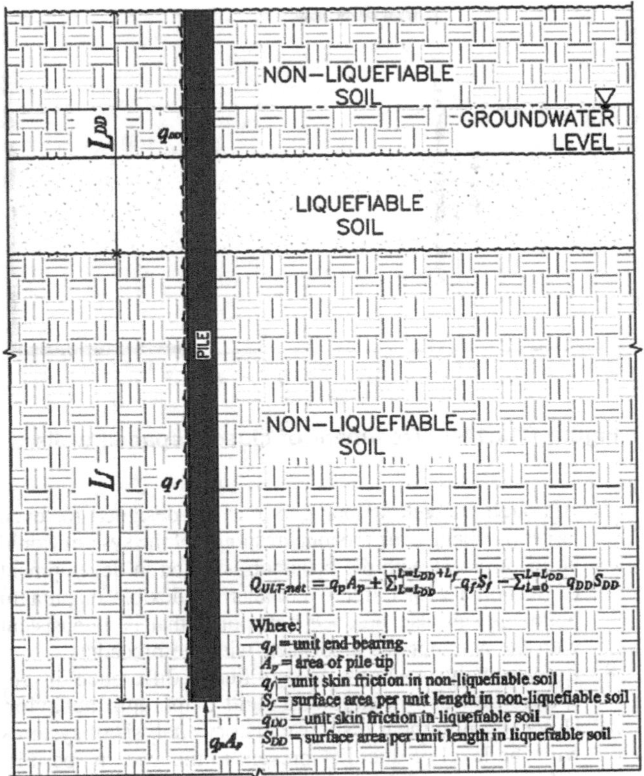

Figure C12.13-2. Determination of ultimate pile capacity in liquefiable soils.

due to inertial response of the structure. Significant excess pore pressure dissipation and settlement occur after the cessation of shaking. This effect has been borne out in the laboratory, as documented by Wilson et al. (1997).

The ultimate geotechnical capacity of the pile should be determined using only the contribution from the soil below the liquefiable layer. The net ultimate capacity is the ultimate capacity reduced by the downdrag load (Figure C12.13-2).

The above provisions, depicted in Figure C12.13-2, were introduced in ASCE 7-16 and treat downdrag as a pile capacity issue rather than as a settlement performance issue. Treating downdrag as a pile capacity issue per Figure C12.13-2 involves subtracting the pile skin friction above the lowest liquefiable layer from the pile capacity below the lowest liquefiable layer to determine the net ultimate geotechnical capacity of the pile. For some projects this method can be very conservative and could potentially result in unnecessary foundation costs, such as at a site where a significant nonliquefiable layer is present on top of a liquefiable layer with relatively minor settlement potential. Therefore, ASCE 7-22 introduced an exception to the treatment of downdrag as a pile capacity issue if a rational analysis is used, such as a neutral plane analysis.

In the case of a neutral plane analysis (Fang 1991, 511–536; Fellenius and Siegel 2008), pile capacity is still checked, but the net ultimate geotechnical capacity of the pile may consider the contribution of all soil layers, including those above liquefiable layers, without reduction of the downdrag load and without consideration of the downdrag load as an applied geotechnical load. Any pile capacity derived from liquefiable layers should consider potential reductions in strength as a result of liquefaction. The foundation will therefore need to be designed to accommodate the expected liquefaction-induced foundation settlement. A deep foundation will generally reduce the expected liquefaction-induced differential foundation settlement relative to a shallow foundation. However, at the same level of differential foundation settlement, damage to the structure is not expected to be significantly different between shallow and deep foundations. Therefore, the differential settlement requirements introduced in ASCE 7-16 for shallow foundations on liquefiable soils were adopted for use in ASCE 7-22 for deep foundations, except that a cap of $0.0075L$ was added to preclude larger settlements. Nothing in these provisions should be construed as preventing more stringent settlement limits from being used to meet project performance objectives, where using superstructure systems that may be more sensitive to differential foundation movement, or for other reasons. To meet these requirements, individual piles or individual pile groups should be treated as individual footings, and a pile foundation connected with a continuous pile cap should be treated as a mat foundation. For the structural design of the piles, the maximum downdrag load in a pile is determined at the neutral plane and considered along with other loads from the structure by treating the downdrag load as a seismic load in applicable load combinations.

The reader is referred to publications such as Fang (1991) and Fellenius and Siegel (2008) for estimating liquefaction-induced foundation settlement and for use of the neutral plane method. Other publications may be referenced, and other rational analysis (e.g., finite element analysis) may be used, for estimating net geotechnical ultimate capacity, downdrag load, and liquefaction-induced foundation settlement.

Lateral resistance of the foundation system includes resistance of the piles as well as passive pressure acting on walls, pile caps, and grade beams. Analysis of the lateral resistance provided by these disparate elements is usually accomplished separately. For these analyses to be applicable, the displacements used must be compatible. Lateral pile analyses commonly use nonlinear soil properties. Geotechnical recommendations for passive pressure should include the displacement at which the pressure is applicable, or they should provide a nonlinear mobilization curve. Liquefaction and other reductions in soil strength occurring in near-surface layers may substantially reduce the ability to transfer lateral inertial forces from foundations to the subgrade, potentially resulting in damaging lateral deformations to piles. Ground improvement of surface soils may be considered for pile-supported structures to provide additional passive resistance to be mobilized on the sides of embedded pile caps and grade beams, as well as to increase the lateral resistance of piles. Otherwise, the check for transfer of lateral inertial forces is the same as for structures on nonliquefiable sites.

The International Building Code (ICC 2012), Section 1810.2.1, requires that deep foundation elements in fluid (potentially liquefiable) soil be considered unsupported for lateral resistance until a point 5 ft (1.5 m) into stiff soil or 10 ft (3.1 m) into soft soil, unless otherwise approved by the Authority Having Jurisdiction on the basis of a geotechnical investigation by a registered design professional. Where liquefaction is predicted to occur, the geotechnical engineer should provide the dimensions (depth and length) of the unsupported length of the pile or should indicate if the liquefied soil will provide adequate resistance such that the length is considered laterally supported in this soil. The geotechnical engineer should develop these dimensions by performing an analysis of the nonlinear resistance of the soil to lateral displacement of the pile ($p - y$ analysis).

Concrete pile detailing includes transverse reinforcing requirements for columns in ACI 318-19 (2019). This is intended to provide ductility within the pile similar to that required for columns.

Where permanent ground displacement is indicated, piles are not required to remain elastic when subjected to this displacement. The provisions are intended to provide ductility and maintain vertical capacity, including flexure-critical behavior of concrete piles.

The required tie force specified in Section 12.13.9.3.5 should be added to the force determined from the lateral loads for the design earthquake in accordance with Sections 12.8, 12.9, 12.14, or Chapter 16.

C12.14 SIMPLIFIED ALTERNATIVE STRUCTURAL DESIGN CRITERIA FOR SIMPLE BEARING WALL OR BUILDING FRAME SYSTEMS

C12.14.1 General In recent years, engineers and building officials have become concerned that the seismic design requirements in codes and standards, though intended to make structures perform more reliably, have become so complex and difficult to understand and implement that they may be counterproductive. Because the response of buildings to earthquake ground shaking is complex (especially for irregular structural systems), realistically accounting for these effects can lead to complex requirements. There is a concern that the typical designers of small, simple buildings, which may represent more than 90% of construction in the United States, have difficulty understanding and applying the general seismic requirements of the standard.

The simplified procedure presented in this section of the standard applies to low-rise, stiff buildings. The procedure, which was refined and tested over a five-year period, was developed to be used for a defined set of buildings deemed to be sufficiently regular in structural configuration to allow a reduction of prescriptive requirements. For some design elements, such as foundations and anchorage of nonstructural components, other sections of the standard must be followed, as referenced within Section 12.14.

C12.14.1.1 Simplified Design Procedure The reasons for the limitations of the simplified design procedure of Section 12.14 are as follows.

1. The procedure was developed to address adequate seismic performance for standard occupancies. Because it was not developed for the higher levels of performance associated with structures assigned to Risk Categories III and IV, no Importance Factor (I_e) is used.
2. Site Class E and F soils require specialized procedures that are beyond the scope of the procedure.
3. The procedure was developed for stiff, low-rise buildings, where higher-mode effects are negligible.
4. Only stiff systems where drift is not a controlling design criterion may use the procedure. Because of this limitation, drifts are not computed. The response modification coefficient, R, and the associated system limitations are consistent with those found in the general Chapter 12 requirements.
5. To achieve a balanced design and a reasonable level of redundancy, two lines of resistance are required in each of the two major axis directions. Because of this stipulation, no redundancy factor (ρ) is applied.
6. When combined with the requirements in items 7 and 8, this requirement reduces the potential for dominant torsional response.

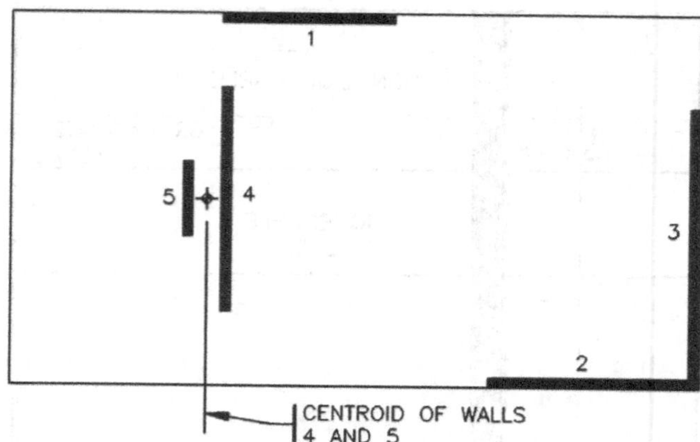

Figure C12.14-1. Treatment of closely spaced walls.

7. Although concrete diaphragms may be designed for even larger overhangs, the torsional response of the system would be inconsistent with the behavior assumed in development of Section 12.14. Large overhangs for flexible diaphragm buildings can also produce a response that is inconsistent with the assumptions associated with the procedure.
8. Linear analysis shows a significant difference in response between flexible and rigid diaphragm behavior. However, nonlinear response history analysis of systems with the level of ductility present in the systems permitted in Table 12.14-1 for the higher Seismic Design Categories has shown that a system that satisfies these layout and proportioning requirements provides essentially the same probability of collapse as a system with the same layout but proportioned based on rigid diaphragm behavior (BSSC 2015). This procedure avoids the need to check for torsional irregularity, and calculation of accidental torsional moments is not required. Figure C12.14-1 shows a plan with closely spaced walls in which the method permitted in subparagraph (c) should be implemented. In that circumstance, the flexible diaphragm analysis would first be performed as if there were one wall at the location of the centroid of walls 4 and 5, then the force computed for that group would be distributed to walls 4 and 5 based on an assessment of their relative stiffnesses.
9. An essentially orthogonal orientation of lines of resistance effectively uncouples response along the two major axis directions, so orthogonal effects may be neglected.
10. Where the simplified design procedure is chosen, it must be used for the entire design in both major axis directions.
11. Because in-plane and out-of-plane offsets generally create large demands on diaphragms, collectors, and discontinuous elements, which are not addressed by the procedure, these irregularities are prohibited.
12. Buildings that exhibit weak-story behavior violate the assumptions used to develop the procedure.

C12.14.3 Seismic Load Effects and Combinations The Equations for seismic load effects in the simplified design procedure are consistent with those for the general procedure, with one notable exception: the overstrength factor (corresponding to Ω_0 in the general procedure) is set at 2.5 for all systems, as indicated in Section 12.14.3.2.1. Given the limited set of systems that can use the simplified design procedure, specifying unique overstrength factors was deemed unnecessary.

C12.14.7 Design and Detailing Requirements The design and detailing requirements outlined in this section are similar to those for the general procedure. The few differences include the following:

1. Forces used to connect smaller portions of a structure to the remainder of the structure are taken as 0.20 times the short-period design spectral response acceleration, S_{DS}, rather than the general procedure value of 0.133 (Section 12.14.7.1).
2. Anchorage forces for concrete or masonry structural walls for structures with diaphragms that are not flexible are computed using the requirements for nonstructural walls (Section 12.14.7.5).

C12.14.8 Simplified Lateral Force Analysis Procedure

C12.14.8.1 Seismic Base Shear The seismic base shear in the simplified design procedure, as given by Equation (12.14-11), is a function of the short-period design spectral response acceleration, S_{DS}. The value for F in the base shear Equation addresses changes in dynamic response for buildings that are two or three stories above grade plane (see Section 11.2 for definitions of "grade plane" and "story above grade plane"). As in the general procedure (Section 12.8.1.3), S_{DS} may be computed for short, regular structures with S_S taken as no greater than 1.5.

C12.14.8.2 Vertical Distribution The seismic forces for multistory buildings are distributed vertically in proportion to the weight of the respective floor. Given the slightly amplified base shear for multistory buildings, this assumption, along with the limit of three stories above grade plane for use of the procedure, produces results consistent with the more traditional triangular distribution without introducing that more sophisticated approach.

C12.14.8.5 Drift Limits and Building Separation For the simplified design procedure, which is restricted to stiff shear wall and braced frame buildings, drift need not be calculated. Where drifts are required (such as for structural separations and cladding design), a conservative drift value of 1% is specified.

REFERENCES

AASHTO (American Association of State Highway and Transportation Officials). 2010. *LRFD bridge design specifications.* Washington, DC: AASHTO.

AISC (American Institute of Steel Construction). 2022. *Seismic provisions for structural steel buildings.* AISC 341-10. Chicago: AISC.

AISI (American Iron and Steel Institute). 2020a. *North American standard for the design of profiled steel diaphragm panels.* ANSI/AISI S310-20. Washington, DC: AISI/ANSI.

ANSI/AISI. 2020b. *North American standard for seismic design of cold-formed steel structural systems.* ANSI/AISI S400-20. Washington, DC: AISI.

APA (American Plywood Association). 1966. *1966 horizontal plywood diaphragm tests.* Laboratory Rep. No. 106. Tacoma, WA: APA.

ATC (Applied Technology Council). 2012. *Seismic performance assessment of buildings.* ATC 58. Redwood City, CA: ATC.

Avellaneda, R. E., W. S. Easterling, B. W. Schafer, J. F. Hajjar, and M. R. Eatherton. 2019. "Cyclic testing of composite concrete on metal deck diaphragms undergoing diagonal tension cracking." In *Proc., 12th Canadian Conf. on Earthquake Engineering*, Quebec City.

AWC (American Wood Council). 2014. *Special design provisions for wind and seismic.* AWC SDPWS-15. Leesburg, VA: AWC.

Boscardin, M. D., and E. J. Cording. 1989. "Building response to excavation-induced settlement." *J. Geotech. Eng.* 115 (1): 1–21. https://doi.org/10.1061/(ASCE)0733-9410(1989)115:1(1).

BSSC (Building Seismic Safety Council). 2015. *Development of simplified seismic design provisions. Working group 1 report: Simplifying section 12.14.* Washington, DC: National Institute of Building Sciences.

Charney, F. A., and J. Marshall. 2006. "A comparison of the Krawinkler and scissors models for including beam-column joint deformations in the analysis of moment-resisting frames." *AISC Eng. J.* 43 (1): 31–48.

Chen, M., et al. 2015. "Full-scale structural and nonstructural building system performance during earthquakes, Part I: Specimen description, test protocol and structural response." *Earthquake Spectra* 32 (2): 737–770. https://doi.org/10.1193/012414eqs016m.

Choi, H., J. Erochko, C. Christopoulos, and R. Tremblay. 2008. *Comparison of the seismic response of steel buildings incorporating self-centering energy-dissipative braces, buckling restrained braces and moment-resisting frames.* Research Rep. 05-2008. Toronto: University of Toronto.

Chopra, A. K. 1995. *Structural dynamics.* New York: Prentice Hall.

Chopra, A. K. 2007. *Dynamics of structures.* 4th ed. Upper Saddle River, NJ: Prentice Hall.

Chopra, A. K., and R. K. Goel. 1991. "Evaluation of torsional provisions in seismic codes." *J. Struct. Eng.* 117 (12): 3762–3782. https://doi.org/10.1061/(ASCE)0733-9445(1991)117:12(3762).

Chopra, A. K., and N. M. Newmark. 1980. "Analysis." In *Design of earthquake resistant structures*, E. Rosenblueth, ed. New York: Wiley.

Czerniak, E. 1957. "Resistance to overturning of single, short piles." *J. Struct. Div.* 83 (2): 1–25. https://doi.org/10.1061/JSDEAG.0000096.

De La Llera, J. C., and A. K. Chopra. 1994. "Evaluation of code accidental-torsion provisions from building records." *J. Struct. Eng.* 120 (2): 597–616. https://doi.org/10.1061/(ASCE)0733-9445(1994)120:2(597).

Debock, D. J., A. B. Liel, C. B. Haselton, J. D. Hopper, and R. Henige. 2014. "Importance of seismic design accidental torsion requirements for building collapse." *Earthquake Eng. Struct. Dyn.* 43 (6): 831–850. https://doi.org/10.1002/eqe.2375.

DFPA (Douglas Fir Plywood Association). 1954. *Horizontal plywood diaphragm tests.* Laboratory Rep. No. 63a. Tacoma, WA: DFPA.

DFPA. 1963. *Lateral tests on plywood sheathed diaphragms.* Laboratory Rep. No. 55. Tacoma, WA: DFPA.

Easterling, W. S., and M. Porter. 1994. "Steel-deck-reinforced concrete diaphragms. I." *J. Struct. Eng.* 120 (2): 560–576. https://doi.org/10.1061/(ASCE)0733-9445(1994)120:2(577).

Eatherton, M. R., P. E. O'Brien, and W. S. Easterling. 2020. "An examination of ductility and seismic diaphragm design force reduction factors for steel deck and composite diaphragms." *J. Struct. Eng.* 146 (11): 04020231. https://doi.org/10.1061/(ASCE)ST.1943-541X.0002797.

EERI (Earthquake Engineering Research Institute). 1994. *Northridge earthquake, January 17, 1994: Preliminary reconnaissance report*, J. F. Hall, ed. Oakland, CA: EERI.

Ensoft, Inc. 2004a. *GROUP, version 6.0: A program for the analysis of a group of piles subjected to axial and lateral loading. User's manual and technical manual.* Austin, TX: Ensoft.

Ensoft, Inc. 2004b. *LPILE plus, version 5.0: A program for the analysis of piles and drilled shafts under lateral loads. User's manual and technical manual.* Austin, TX: Ensoft.

EQE Engineering. 1991. *Structural concepts and details for seismic design.* UCRL-CR-106554. Washington, DC: DOE.

EQE. 1983. *The effects of the May 2, 1983, Coalinga, California, earthquake on industrial facilities.* Newport Beach, CA: EQE.

EQE. 1985. *Summary of the September 19, 1985, Mexico earthquake.* Newport Beach, CA: EQE.

EQE. 1986a. *Summary of the March 3, 1985, Chile earthquake.* Newport Beach, CA: EQE.

EQE. 1986b. *The effects of the March 3, 1985, Chile earthquake on power and industrial facilities.* Newport Beach, CA: EQE.

EQE. 1986c. *Power and industrial facilities in the epicentral area of the 1985 Mexico earthquake.* Newport Beach, CA: EQE.

EQE. 1987. *Summary of the 1987 Bay of Plenty, New Zealand, earthquake.* Newport Beach, CA: EQE.

Fang, H. Y. 1991. *Foundation engineering handbook.* Berlin: Springer, 511–536.

Fellenius, B. H., and T. C. Siegel. 2008. "Pile drag load and downdrag in a liquefaction event." *J. Geotech. Geoenviron.* 134 (9): 1412–1416. https://doi.org/10.1061/(ASCE)1090-0241(2008)134:9(1412).

FEMA (Federal Emergency Management Agency). 2009a. *NEHRP recommended seismic provisions for new buildings and other structures.* FEMA P-750. Washington, DC: FEMA.

FEMA. 2009b. *Quantification of building seismic performance factors.* FEMA P-695. Washington, DC: FEMA.

FEMA. 2011. *Quantification of building seismic performance factors: Component equivalency methodology.* FEMA P-795. Washington, DC: FEMA.

FEMA. 2013. *2009 NEHRP recommended seismic provisions: Design examples.* FEMA P-751CD. Washington, DC: FEMA.

FEMA. 2015. *Seismic design of rigid wall-flexible diaphragm buildings: An alternate procedure.* FEMA P-1026. Washington, DC: FEMA.

FEMA. 2018a. *Assessing seismic performance of buildings with configuration irregularities: Calibrating current standards and practices.* FEMA P-2012. Washington, DC: FEMA.

FEMA. 2018b. *Vulnerability-based seismic assessment and retrofit of one- and two-family dwellings.* FEMA P-1100. Washington, DC: FEMA.

FEMA. 2020. *NEHRP recommended seismic provisions for new buildings and other structures. Vol. II, Part 3: Resource papers.* FEMA P-2082. Washington, DC: FEMA.

Fleischman, R. B., J. I. Restrepo, C. J. Naito, R. Sause, D. Zhang, and M. Schoettler. 2013. "Integrated analytical and experimental research to develop a new seismic design methodology for precast concrete diaphragms." *J. Struct. Eng.* 139 (7): 1192–1204. https://doi.org/10.1061/(ASCE)ST.1943-541X.0000734.

Flores, F., F. A. Charney, and D. Lopez-Garcia. 2018. "The influence of accidental torsion on the inelastic dynamic response of buildings during earthquakes." *Earthquake Spectra* 34 (1): 21–53. https://doi.org/10.1193/100516EQS169M.

Gatto, K., and C.-M. Uang. 2002. *Cyclic response of woodframe shearwalls: Loading protocol and rate of loading effects.* Richmond, CA: Consortium of Universities for Research in Earthquake Engineering.

Gerwick, B. Jr., and G. Fotinos. 1992. "Drilled piers and driven piles for foundations in areas of high seismicity." In *Proc., SEAONC Fall Seminar, October 29*, San Francisco.

Goel, R. K., and A. K. Chopra. 1997. "Period formulas for moment-resisting frame buildings." *J. Struct. Eng.* 123 (11): 1454–1461. https://doi.org/10.1061/(ASCE)0733-9445(1997)123:11(1454)

Goel, R. K., and A. K. Chopra. 1998. "Period formulas for concrete shear wall buildings." *J. Struct. Eng.* 124 (4): 426–433. https://doi.org/10.1061/(ASCE)0733-9445(1998)124:4(426).

Grant, D., and R. Diaferia. 2012. "Assessing adequacy of spectrum matched ground motions for response history analysis." *Earthquake Eng. Struct. Dyn.* 42 (9): 1265–1280. https://doi.org/10.1002/eqe.2270.

Gupta, A., and H. Krawinkler. 2000. "Dynamic P-delta effects for flexible inelastic steel structures." *J. Struct. Eng.* 126 (1): 145–154.

Hamburger, R. O., and D. L. McCormick. 2004. "Implications of the January 17, 1994, Northridge earthquake on tilt-up and masonry buildings with wood roofs." In *Proc., 63rd Annual Convention Structural Engineers Association of California*, Lake Tahoe, CA, 243–255.

Hancock, J., J. Watson-Lamprey, N. A. Abrahamson, J. J. Bommer, A. Markatis, E. McCoy, and R. Mendis. 2006. "An improved method of matching response spectra of recorded earthquake ground motion using wavelets." *J. Earthquake Eng.* 10 (1): 67–89. https://doi.org/10.1080/13632460609350629.

ICBO (International Conference of Building Officials). 1988. *Uniform building code.* Whittier, CA: ICBO.

ICBO. 1994. *Uniform building code.* Whittier, CA: ICBO.

ICBO. 1997. *Uniform building code.* Whittier, CA: ICBO.

ICC (International Code Council). 2009. *International building code.* Country Club Hills, IL: ICC.

ICC. 2012. *International building code.* Country Club Hills, IL: ICC.

ICC. 2015. *International building code.* Country Club Hills, IL: ICC.

Koliou, M., A. Filiatrault, D. Kelly, and J. Lawson. 2015a. "Buildings with rigid walls and flexible diaphragms I: Evaluation of current US seismic provisions." *J. Struct. Eng.* 142 (3): 04015166. https://doi.org/10.1061/(ASCE)ST.1943-541X.0001438.

Koliou, M., A. Filiatrault, D. Kelly, and J. Lawson. 2015b. "Buildings with rigid walls and flexible diaphragms II: Evaluation of a new seismic design approach based on distributed diaphragm yielding." *J. Struct. Eng.* 142 (3): 04015167. https://doi.org/10.1061/(ASCE)ST.1943-541X.0001439.

Lam, I., and V. Bertero. 1990. "A seismic design of pile foundations for port facilities." In *Proc., POLA Seismic Workshop on Seismic Engineering*, San Pedro, CA.

Liew, J. Y. R., H. Chen, and N. E. Shanmugam. 2001. "Inelastic analysis of steel frames with composite beams." *J. Struct. Eng.* 127 (2): 194–202. https://doi.org/10.1061/(ASCE)0733-9445(2001)127:2(194).

Lopez, O. A., and M. Cruz. 1996. "Number of modes for the seismic design of buildings." *Earthquake Eng. Struct. Dyn.* 25 (8): 837–856. https://doi.org/10.1002/(SICI)1096-9845(199608)25:8<837::AID-EQE585>3.0.CO;2-L.

Margason, E., and M. Holloway. 1977. "Pile bending during earthquakes." In *Proc., 6th World Conf. on Earthquake Engineering*, New Delhi, India.

Menun, C., and A. Der Kiureghian. 1998. "A replacement for the 30%, 40%, and SRSS rules for multicomponent seismic analysis." *Earthquake Spectra* 14 (1): 153–163. https://doi.org/10.1193/1.1585993.

Mylonakis, G. 2001. "Seismic pile bending at soil-layer interfaces." *Soils Found.* 41 (4): 47–58.

Newmark, N. M., and W. J. Hall. 1982. *Earthquake spectra and design: EERI monograph series*. Oakland, CA: Earthquake Engineering Research Institute.

Newmark, N. M., and E. Rosenblueth. 1971. *Fundamentals of earthquake engineering*. Englewood Cliffs, NJ: Prentice Hall.

NIST (National Institute of Standards and Technology). 2010. *Nonlinear structural analysis for seismic design: A guide for practicing engineers*. NIST GCR 10-917-5. Gaithersburg, MD: NIST.

NIST. 2010. *Evaluation of the FEMA P-695 methodology for quantification of building seismic performance factors*. NIST GCR 10-917-8. Gaithersburg, MD: NIST.

NIST. 2011. *Selecting and scaling earthquake ground motions for performing response-history analyses*. NIST GCR 11-918-15. Gaithersburg, MD: NIST.

NIST. 2012. *Tentative framework for development of advanced seismic design criteria for new buildings*. NIST GCR 10-917-8. Gaithersburg, MD: NIST.

Norris, G. M. 1994. "Seismic bridge pile foundation behavior." In Vol. 1 of *Proc., Int. Conf. on Design and Construction of Deep Foundations*. Orlando, Florida.

O'Brien, P., M. R. Eatherton, and W. S. Easterling. 2017. *Characterizing the load-deformation behavior of steel deck diaphragms using past test data*. CFSRC Rep. 2017-02. Baltimore, MD: Cold-Formed Steel Research Consortium.

Oettle, N. K., and J. D. Bray. 2013. "Geotechnical mitigation strategies for earthquake surface fault rupture." *J. Geotech. Geoenviron.* 139 (11): 1864–1874.

Panagiotou, M., J. I. Restrepo, and J. P. Conte. 2011. "Shake-table test of a full-scale 7-story building slice. Phase I: Rectangular wall." *J. Struct. Eng.* 137 (6): 691–704. https://doi.org/10.1061/(ASCE)ST.1943-541X.0000332.

Paulay, T. 1997. "Are existing seismic torsion provisions achieving design aims?" *Earthquake Spectra* 13 (2): 259–280. https://doi.org/10.1193/1.1585945.

Paulay, T., and M. J. N. Priestley. 1992. *Seismic design of reinforced concrete and masonry structures*. New York: Wiley.

Rollins, K. M., K. T. Peterson, T. J. Weaver, and A. E. Sparks. 1999. *Static and dynamic lateral load behavior on a full-scale pile group in clay*. Salt Lake City, UT: Brigham Young University.

Rosenblueth, E., and H. Contreras. 1977. "Approximate design for multicomponent earthquakes." *J. Eng. Mech. Div.* 103 (5): 881–893. https://doi.org/10.1061/JMCEA3.0002280.

Schafer, B. 2019. "Research on the seismic performance of rigid wall flexible diaphragm buildings with bare steel deck diaphragms." Accessed April 13, 2021. https://jscholarship.library.jhu.edu/handle/1774.2/60360.

Schaffhausen, R., and A. Wegmuller. 1977. "Multistory rigid frames with composite girders under gravity and lateral forces." *AISC Eng. J.* 14 (2): 68–77.

Sheppard, D. A. 1983. "Seismic design of prestressed concrete piling." *PCI J.* 28 (2): 20–49. https://doi.org/10.15554/pcij.03011983.20.49.

Song, S. T., Y. H. Chai, and T. H. Hale. 2005. "Analytical model for ductility assessment of fixed-head concrete piles." *J. Struct. Eng.* 131 (7): 1051–1059. https://doi.org/10.1061/(ASCE)0733-9445(2005)131:7(1051).

Tauberg, N. A., K. Kolozvari, and J. W. Wallace. 2019. *Ductile reinforced concrete coupled walls: P695 study*. Rep. No. SEERL 2019/01. Los Angeles: University of California, Los Angeles.

Uang, C.-M., and A. Maarouf. 1994. "Deflection amplification factor for seismic design provisions." *J. Struct. Eng.* 120 (8): 2423–2436. https://doi.org/10.1061/(ASCE)0733-9445(1994)120:8(2423).

US Nuclear Regulatory Commission. 2012. *Combining modal responses and spatial components in seismic response analysis*. Regulatory Guide 1.92. Upton, NY: Brookhaven National Laboratory.

Vamvatsikos, D. 2002. "Seismic performance, capacity and reliability of structures as seen through incremental dynamic analysis." Ph.D. dissertation, Dept. of Civil and Environmental Engineering, Stanford University.

van de Lindt, J., D. Rammer, P. Line, M. Amini, S. Pei, and M. Popovski. 2019. *Determination of seismic performance factors for cross-laminated timber shear walls based on the FEMA P695 methodology*. WCTE 2016 Vienna, Austria.

Wei, G., M. R. Eatherton, H. Foroughi, S. Torabian, and B. W. Schafer. 2020. *Seismic behavior of steel BRBF buildings including consideration of diaphragm inelasticity*. CFSRC Rep. No. 2020-04. Baltimore, MD: Cold-Formed Steel Research Consortium.

Wilson, D. W., R. W. Boulanger, B. L. Kutter, and A. Abghari. 1997. *Aspects of dynamic centrifuge testing of soil-pile-superstructure interaction: ASCE Geotechnical Special Publication 64*, 47–63. New York: ASCE.

Wilson, E. L. 2000. *Three-dimensional static and dynamic analysis of structures*. Berkeley, CA: Computers and Structures.

Wilson, E. L. 2004. *Static and dynamic analysis of structures*. Berkeley, CA: Computers and Structures.

Wilson, E. L., A. Der Kiureghian, and E. P. Bayo. 1981. "A replacement for the SRSS method in seismic analysis." *Earthquake Eng. Struct. Dyn.* 9 (2): 187–194. https://doi.org/10.1002/eqe.4290090207.

Wilson, E. L., and A. Habibullah. 1987. "Static and dynamic analysis of multi-story buildings, including P-delta effects." *Earthquake Spectra* 3 (2): 289–298.

Wilson, E. L., and J. Penzien. 1972. "Evaluation of orthogonal damping matrices." *Int. J. Numer. Methods* 4 (1): 5–10. https://doi.org/10.1002/nme.1620040103.

OTHER REFERENCES (NOT CITED)

ACI (American Concrete Institute). 2019. *Building code requirements for structural concrete and commentary*. ACI 318-19. Farmington Hills, MI: ACI.

APA (American Plywood Association). 2000. *Plywood diaphragms*. Research Rep. No. 138. Tacoma, WA: APA.

APA. 2007. *Diaphragms and shear walls: Design/construction guide*. Form L350A. Tacoma, WA: APA.

ASCE. 2000. *Seismic analysis of safety-related nuclear structures and commentary*. ASCE 4-98. Reston, VA: ASCE.

ASCE. 2003. *Minimum design loads for buildings and other structures*. ASCE 7-02. Reston, VA: ASCE.

ASCE. 2007. *Seismic rehabilitation of existing buildings*. ASCE/SEI 41-06. Reston, VA: ASCE.

ASCE. 2010. *Minimum design loads for buildings and other structures*. ASCE 7-10. Reston, VA: ASCE.

ASCE. 2014b. *Seismic evaluation and retrofit of existing buildings*. ASCE 41-13. Reston, VA: ASCE.

ASCE. 2017. "Seismic rehabilitation of existing buildings". ASCE/SEI 41-17. Reston, VA.

ASTM International. 2017. *Standard specification for driven fasteners: Nails, spikes, and staples*. ASTM F1667. West Conshohocken, PA: ASTM.

ATC. 1984. *Tentative provisions for the development of seismic regulations for buildings*. ATC 3-06. Redwood City, CA: ATC.

ATC. 1995. *Structural response modification factors*. ATC-19. Redwood City, CA: ATC.
AWC. 2008. *Special design provisions for wind and seismic*. AWC SDPWS-2008. Leesburg, VA: AWC.
Bernal, D. 1987. "Amplification factors for inelastic dynamic P-delta effects in earthquake analysis." *Earthquake Eng. Struct. Dyn.* 18 (5): 635–681. https://doi.org/10.1002/eqe.4290150508.
California Geological Survey. 2008. *Guidelines for evaluation and mitigation of seismic hazards in California*. SP 117A. Sacramento, CA: California Geological Survey.
Charney, F. A. 1990. "Wind drift serviceability limit state design of multistory buildings." *J. Wind Eng. Ind. Aerodyn.* 36 (Jan): 203–212. https://doi.org/10.1016/0167-6105(90)90305-V.
City of Newport Beach. 2012. *Minimum liquefaction mitigation measures*.
Degenkolb, H. J. 1987. "A study of the P-delta effect." *Earthquake Spectra* 3 (1).
FEMA. 2012. *Seismic performance assessment of buildings*. FEMA P-58. Washington, DC: FEMA.
Grant, D., and R. Diaferia. 2012. "Assessing adequacy of spectrum matched ground motions for response history analysis." *Earthquake Eng. Struct. Dyn.* 42 (9): 1265–1280. https://doi.org/10.1002/eqe.2270.
Greater Vancouver Liquefaction Task Force. 2007. *Geotechnical guidelines for buildings on liquefiable sites in accordance with the NBC 2005 for Greater Vancouver*. Vancouver, BC: University of British Columbia.
Griffis, L. 1993. "Serviceability limit states under wind load." *Eng. J. Am. Inst. Steel Constr.* 30 (1): 1–16.
ICC. 2000. *International building code*. Country Club Hills, IL: ICC.
NBM International. 2017. *Cyclic performance and characterization of steel deck connections*. NBM Technologies Rep. No. AISI/SDI/SJI Blacksburg, VA: NBM Technologies.
NBM Technologies. 2018. *Button punch sidelaps (cyclic testing program)*. Project No. 103-042-18.
Newmark, N. M., and W. J. Hall. 1978. *Development of criteria for seismic review of selected nuclear power plants*. NUREG/CR-0098. Washington DC: US Nuclear Regulatory Commission.
Newmark, N. M., and W. J. Hall. 1982. *Earthquake spectra and design: EERI monograph series*. Oakland, CA: Earthquake Engineering Research Institute.
Pantoli, E., et al. 2013. *Full-scale structural and nonstructural building system performance during earthquakes and post-earthquake fire–test results*. BNCS Rep. No. #2. San Diego: University of California, San Diego.
Post Tensioning Institute. 2012. *Standard requirements of the design and analysis of post-tensioned concrete foundations on expansive soils*. PTI DC10.5. Phoenix, AZ: Post Tensioning Institute.
US Nuclear Regulatory Commission. 1999. *Reevaluation of regulatory guidance on modal response combination methods for seismic response spectrum analysis*. NUREG/CR-6645. Upton, NY: Brookhaven National Laboratory.
Wilson, E. L., I. Suhawardy, and A. Habibullah. 1995. "A clarification of the orthogonal effects in a three-dimensional seismic analysis." *Earthquake Spectra* 11 (4): 659–666. https://doi.org/10.1193/1.1585831.

CHAPTER C13
SEISMIC DESIGN REQUIREMENTS FOR NONSTRUCTURAL COMPONENTS

C13.1 GENERAL

Chapter 13 defines minimum design criteria for architectural, mechanical, electrical, and other nonstructural systems and components, recognizing structure use, occupant load, the need for operational continuity, and the interrelation of structural, architectural, mechanical, electrical, and other nonstructural components. Nonstructural components are designed for design earthquake ground motions, as defined in Section 11.2 and determined in Section 11.4.5 of the standard. In contrast to structures, which are implicitly designed for a low probability of collapse when subjected to risk-targeted maximum considered earthquake (MCE_R) ground motions, there are no implicit performance goals associated with the MCE_R for nonstructural components. Performance goals associated with the design earthquake are discussed in Section C13.1.3.

Suspended or attached nonstructural components that could detach either in full or in part from the structure during an earthquake are referred to as falling hazards and may represent a serious threat to property and life safety. Critical attributes that influence the hazards posed by these components include their weight, their attachment to the structure, their failure or breakage characteristics (e.g., nonshatterproof glass), and their location relative to occupied areas (e.g., over an entry or exit, a public walkway, an atrium, or a lower adjacent structure). Architectural components that pose potential falling hazards include parapets, cornices, canopies, marquees, glass, large ornamental elements (e.g., chandeliers), and building cladding. In addition, suspended mechanical and electrical components (e.g., mixing boxes, piping, and ductwork) may represent serious falling hazards. Figures C13.1-1 through C13.1-4 show damage to nonstructural components in past earthquakes.

Components whose collapse during an earthquake could result in blockage of the means of egress deserve special consideration. The term *means of egress* is used commonly in building codes with respect to fire hazard. Egress paths may include intervening aisles, doors, doorways, gates, corridors, exterior exit balconies, ramps, stairways, pressurized enclosures, horizontal exits, exit passageways, exit courts, and yards. Items whose failure could jeopardize the means of egress include walls around stairs and corridors, veneers, cornices, canopies, heavy partition systems, ceilings, architectural soffits, light fixtures, and other ornaments above building exits or near fire escapes. Examples of components that generally do not pose a significant falling hazard include fabric awnings and canopies. Architectural, mechanical, and electrical components that, if separated from the structure, fall in areas that are not accessible to the public (e.g., into a mechanical shaft or light well) also pose little risk to egress routes.

For some architectural components, such as exterior cladding elements, wind design forces may exceed the calculated seismic design forces. Nevertheless, seismic detailing requirements may still govern the overall structural cladding design. Where this is a possibility, it must be investigated early in the structural design process.

The seismic design of nonstructural components may involve consideration of nonseismic requirements that are affected by seismic bracing. For example, accommodation of thermal expansion in pressure piping systems often is a critical design consideration, and seismic bracing for these systems must be arranged in a manner that accommodates thermal movements. Particularly in the case of mechanical and electrical systems, the design for seismic loads should not compromise the functionality, durability, or safety of the overall design; this method requires collaboration among the various disciplines of the design and construction team.

For various reasons (e.g., business continuity), it may be desirable to consider higher performance than that required by the building code. For example, to achieve continued operability of a piping system, it is necessary to prevent unintended operation of valves or other in-line components, in addition to preventing collapse and providing leak tightness. Higher performance also is required for components containing substantial quantities of hazardous contents (as defined in Section 11.2). These components must be designed to prevent uncontrolled release of those materials.

The requirements of Chapter 13 are intended to apply to nonstructural components in new construction and tenant improvements installed at any time during the life of the structure, provided that they are listed in Table 13.5-1 or 13.6-1. Furthermore, they are intended to reduce (not eliminate) the risk to occupants and to improve the likelihood that Essential Facilities remain functional. Although property protection (in the sense of investment preservation) is a possible consequence of implementation of the standard, it is not currently a stated or implied goal; a higher level of protection may be advisable if such protection is desired or required.

C13.1.1 Scope The requirements for seismic design of nonstructural components apply to the nonstructural component and to its supports and attachments, regardless of whether it is within or supported by a building or nonbuilding structure, or if it is outside of a structure. Figure 13.1-5 illustrates possible locations of nonstructural components. Components outside the structure must be connected to it by mechanical or electrical systems to be considered part of the structure. Components without such connections should be considered nonbuilding structures. The phrase "connection by a mechanical or electrical system" is not intended to require analysis or design of the components or systems of public utilities serving the structure. However, the physical connections themselves of utilities to the structure are

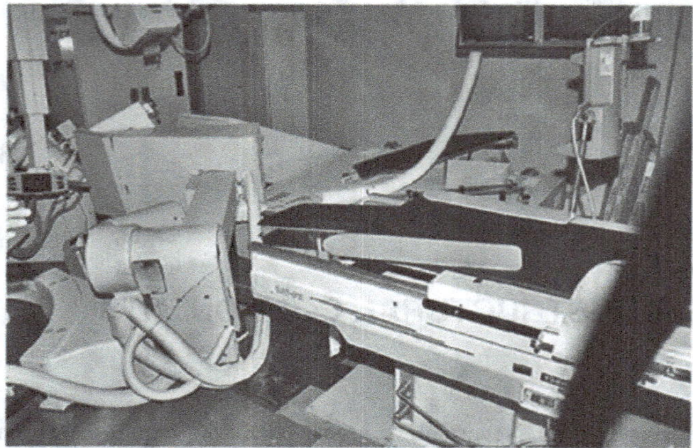

Figure C13.1-1. Hospital imaging equipment that fell from overhead mounts.

Figure C13.1-2. Damaged ceiling system.

Figure C13.1-3. Collapsed light fixtures.

Figure C13.1-4. Toppled storage cabinets.

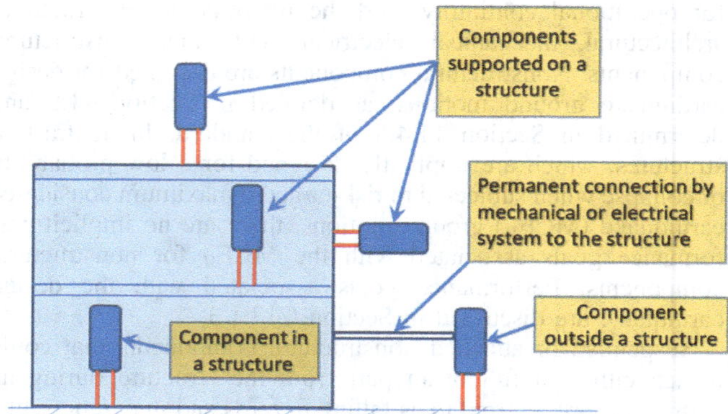

Figure 13.1-5. Possible locations of nonstructural components.

considered in Section 13.6.9, where utilities pass between adjacent structures, or when structures are located on sites where seismically induced differential settlement between the structure and the utilities may occur. Architectural components that are part of the egress system may also be outside the structure and require seismic design, even if structurally separate from the structure they serve. In some cases, as defined in Section 13.2, it is necessary to consider explicitly the performance characteristics of the component. The requirements are intended to apply only to permanently attached components, not to furniture, temporary items, or mobile units. Furniture, such as tables, chairs, and desks, may shift during strong ground shaking but generally poses minimal hazards, provided they do not obstruct emergency egress routes. Storage cabinets, tall bookshelves, and other items of significant mass do not fall into this category and should be anchored or braced in accordance with this chapter.

Temporary items are those that remain in place for short periods of time (months, not years). Components that are expected to remain in place for periods of six months or longer, even if they are designed to be movable, should be considered permanent for the purposes of this section. Modular office systems are considered permanent because they ordinarily remain in place for long periods. In addition, they often include storage units that have significant capacity and may topple in an earthquake. They are subject to the provisions of Section 13.5.8 for partitions if they exceed 6 ft (1.8 m) high. Mobile units include components that are moved from one point in the structure to another during ordinary use. Examples include desktop computers, office equipment, and other components that are not permanently attached to the building utility systems (Figure C13.1-6). Components that are mounted on wheels to

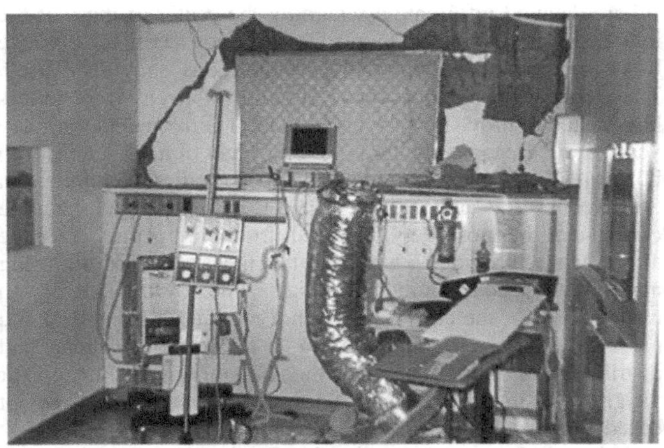

Figure C13.1-6. Collapsed duct and HVAC diffuser.

facilitate periodic maintenance or cleaning but that otherwise remain in the same location (e.g., server racks) are not considered mobile for the purposes of anchorage and bracing. Likewise, skid-mounted components (as shown in Figure C13.1-7), as well as the skids themselves, are considered permanent equipment.

With the exception of solar panels satisfying the provisions of Section 13.6.12, equipment must be anchored if it is permanently attached to utility services (electricity, gas, and water). For the purposes of this requirement, "permanently attached" should be understood to include all electrical connections except NEMA 5-15 and 5-20 straight-blade connectors (duplex receptacles).

C13.1.2 Seismic Design Category The requirements for nonstructural components are based in part on the seismic design category to which they are assigned. As the seismic design category is established considering factors not unique to specific nonstructural components, all nonstructural components are assigned to the same seismic design category as the structure they are located in or supported by, or to the structure to which they are permanently connected by mechanical or electrical systems. Components located in a structure with a lower seismic design category than the structure that the components serve should be assigned the higher seismic design category.

Figure C13.1-7 Skid-mounted components.

C13.1.3 Component Importance Factor Performance expectations for nonstructural components often are defined in terms of the functional requirements of the structure to which the components are attached. Although specific performance goals for nonstructural components have yet to be defined in building codes, the component Importance Factor, I_p, implies performance levels for specific cases. For noncritical nonstructural components (those with a component I_p of 1.0), the following behaviors are anticipated for shaking of different levels of intensity.

1. Minor earthquake ground motions: minimal damage; not likely to affect functionality.
2. Moderate earthquake ground motions: some damage that may affect functionality.
3. Design earthquake ground motions: major damage but significant falling hazards are avoided; likely loss of functionality.

Components with Importance Factors greater than 1.0 are expected to remain in place, sustain limited damage, and when necessary, function after an earthquake (see Section C.13.2.3). These components can be located in structures that are not assigned to Risk Category IV. For example, fire sprinkler piping systems have an Importance Factor, I_p, of 1.5 in all structures because these essential systems should function after an earthquake. Egress stairways are assigned an I_p of 1.5 as well, although in many cases the design of these stairways is dictated by differential displacements, not inertial force demands.

The component Importance Factor is intended to represent the greater of the life-safety importance of the component and the hazard-exposure importance of the structure. It indirectly influences the survivability of the component via required design forces and displacement levels, as well as component attachments and detailing. Although this approach provides some degree of confidence in the seismic performance of a component, it may not be sufficient in all cases. For example, individual ceiling tiles may fall from a ceiling grid that has been designed for larger forces. This problem may not represent a serious falling hazard if the ceiling tiles are made of lightweight materials, but it may lead to blockage of critical egress paths or disruption of the facility function. When higher levels of confidence in performance are required, the component is classified as a designated seismic system (Section 11.2), and in certain cases, seismic qualification of the component or system is necessary. Seismic qualification approaches are provided in Sections 13.2.6 and 13.2.7. In addition, seismic qualification approaches presently in use by the Department of Energy (DOE) can be applied.

Most Risk Category IV structures are Essential Facilities and are intended to be functional after a design earthquake; critical nonstructural components and equipment in such structures are designed with I_p equal to 1.5. This requirement applies to most components and equipment because damage to vulnerable unbraced systems or equipment may disrupt operations after an earthquake even if they are not directly classified as essential to life safety. The nonessential and nonhazardous components are themselves not affected by this requirement. Instead, requirements focus on the supports and attachments. UFC 3-310-04 (US Department of Defense 2007) has additional guidance for improved performance.

Some Risk-Category IV structures are not Essential Facilities, and their continued operation is not needed for public safety. These nonessential facilities may be rated as Risk-Category IV structures due to their hazardous materials (for example, many computer-chip fabrication facilities). It is not in the public's interest for safety that such a facility continues its manufacturing operations immediately following an earthquake, but it is in the public's interest for safety that the hazardous materials within these facilities be contained.

C13.1.4 Exemptions Several classes of nonstructural components are exempted from the requirements of Chapter 13. The exemptions are made on the assumption that, either because of their inherent strength and stability or the lower level of earthquake demand (accelerations and relative displacements), or both, these nonstructural components and systems can achieve the performance goals described earlier in this commentary without explicitly satisfying the requirements of this chapter.

The requirements are intended to apply only to permanent components, not furniture and temporary or mobile equipment. Components that are mounted on wheels to facilitate periodic maintenance or cleaning but that otherwise remain in the same location (e.g., server racks) are not considered mobile for the purposes of anchorage and bracing. Furniture (with the exception of more massive elements like storage cabinets) may shift during strong ground shaking but pose minimal hazards. With the exception of solar panels satisfying the provisions of Section 13.6.12, equipment must be anchored if it is permanently attached to the structure utility services, such as electricity, gas, or water. For the purposes of this requirement, "permanently attached" includes all electrical connections except plugs for common electrical outlets, such as receptacles with a maximum capacity of 50 A, which is common for small appliances and equipment.

Temporary items are those that remain in place for 180 days or less. Modular office systems are considered permanent, since they ordinarily remain in place for long periods. In addition, they often include storage units of significant capacity, which may topple in earthquakes. Mobile units include components that are moved from one point in the structure to another during ordinary use. Examples include desktop computers, office equipment, and other components that are not permanently attached to the building utility systems. Components mounted on wheels to facilitate periodic maintenance or cleaning but that otherwise remain in the same location are not considered mobile for the purposes of anchorage and bracing.

Furniture resting on floors, such as tables, chairs, and desks, may shift during strong ground shaking, but they generally pose minimal hazards, provided they do not obstruct emergency egress routes. Examples also include desktop computers, office equipment, and other components that are not permanently attached to the building utility systems.

With the exception of parapets supported by bearing walls or shear walls, all components in Seismic Design Categories A and B are exempt because of the low levels of ground shaking expected. Parapets are not exempt because experience has shown that these items can fail and pose a significant falling hazard, even at low-level shaking levels.

Discrete components are generally understood to be standalone items such as cabinets, pumps, electrical boxes, lighting, and signage. Discrete architectural or mechanical, weighing 20 lb (89 N) or less generally do not pose a risk and are exempted, provided they are positively attached to the structure, regardless of whether they carry an Importance Factor, I_p, of 1.5 or not. Larger items up to 400 lb (1,780 N) in weight with $I_p = 1.0$ have historically been exempted, provided they are positively attached and have flexible connections. The exemption for mechanical and electrical components in Seismic Design Category D, E, or F, based on weight and location of the center of mass, is particularly applicable to vertical equipment racks and similar components. Where detailed information regarding the center of mass of the intended installation is unavailable, a conservative estimate based on potential equipment configurations should be used. The exemption for components weighing 400 lb (1,780 N) or less has existed in provisions for nonstructural components for many years and corresponds roughly to the weight of a 40 gal. (150 L) hot water tank. Coupled with this and the other exemptions in Seismic Design Categories D, E, and F is a requirement that the component be positively attached to the structure. Positive attachment is provided when the attachment is carried out using appropriate structural-grade materials, whereby explicit design calculations for the anchorage are not required. Mechanical and electrical components (including both discrete components and distribution systems) meeting the Importance Factor and/or weight restrictions in Seismic Design Category C are exempt based on the definition of a component in Chapter 11.

Although the exemptions listed in Section 13.1.4 are intended to waive bracing requirements for nonstructural components that are judged to pose negligible life-safety hazard, in some cases it may nevertheless be advisable to consider (in consultation with the owner) bracing for exempted components to minimize repair costs and/or disproportionate loss (e.g., artworks of great value).

The bracing exemptions for short hangers have been moved to the respective sections in which they apply. These exemptions are based on the assumption that the hangers have sufficient ductility to undergo plastic deformations without failure while at the same time providing sufficient stiffness to limit lateral displacement to a reasonable level. This assumption extends to the anchors, so the design and detailing of the connections to the structure should take this into account. Raceways, ducts, and piping systems must be able to accommodate the relative displacement demands calculated in Section 13.3.2, since these displacements can be substantially greater than those that occur at connections to equipment. At seismic separation joints between structures, large displacements may occur over a short distance.

Short hangers fabricated from threaded rods resist lateral force primarily through bending and are prone to failure through cyclic fatigue. Tests by Soulages and Weir (2011) suggest that low-cycle fatigue is not an issue when the ductility ratios for the rods are less than about 4. The testing also indicated that swivel connections are not required, provided the load and rod length limitations are observed. The limits on unbraced trapezes and hangers are based on limiting the ductility ratios to reasonable levels, when subject to the maximum force demands in the highest seismic risk regions. It should be noted that in areas of lower seismic risk, less restrictive criteria could be used.

The exemption for short hangers is limited to the case where every hanger in the raceway run is less than 12 in. (305 mm) because of the need to carefully consider the seismic loads and compatible displacement limits for the portions of raceways with longer hanger supports.

The historical exemption for trapeze-supported conduit less than 2.5 in. (64 mm) trade size has been removed, since its application to specific cases, such as a trapeze supporting multiple conduit runs, was unclear.

The exemption for trapezes with short rod hangers applies only to trapezes configured with the rod hangers attached directly to the trapeze and the structural framing. Where one or more rod hangers for a trapeze are supported from another trapeze, the bracing exemption does not apply.

C13.1.5 Premanufactured Modular Mechanical and Electrical Systems Large premanufactured modular mechanical and electrical systems (as shown in Figure C13.1-8) should be

Figure C13.1-8. Premanufactured modular mechanical systems.
Source: Courtesy of Matthew Tobolski.

considered nonbuilding structures for the purposes of the enveloping structural system design, unless the module has been prequalified in accordance with Section 13.2.3. However, where the premanufactured module has not been prequalified, the nonstructural components contained in the module should be addressed using the requirements of Chapter 13. Note that this provision is not intended to address skid-mounted equipment assemblies not equipped with an enclosure, nor does it address single large components, such as air handlers, cooling towers, chillers, and boilers.

C13.1.6 Application of Nonstructural Component Requirements to Nonbuilding Structures At times, a nonstructural component should be treated as a nonbuilding structure. When the physical characteristics associated with a given class of nonstructural components vary widely, judgment is needed to select the appropriate design procedure and coefficients. For example, cooling towers vary from small packaged units with an operating weight of 2,000 lb (8.9 kN) or less to structures the size of buildings. Consequently, design coefficients for the design of "cooling towers" are found in both Tables 13.6-1 and 15.4-2. Small cooling towers are best designed as nonstructural components using the provisions of Chapter 13, whereas large ones are clearly nonbuilding structures that are more appropriately designed using the provisions of Chapter 15. Similar issues arise for other classes of nonstructural components (e.g., boilers and bins). Guidance on determining whether an item should be treated as a nonbuilding structure or a nonstructural component for the purpose of seismic design is provided in Bachman and Dowty (2008).

The specified weight limit for nonstructural components (25% relative to the combined weight of the structure and component) relates to the condition at which dynamic interaction between the component and the supporting structural system is potentially significant. Section 15.3.2 contains requirements for addressing this interaction in design.

C13.1.7 Reference Documents Professional and trade organizations have developed nationally recognized codes and standards for the design and construction of specific mechanical and electrical components. These documents provide design guidance for normal and upset (abnormal) operating conditions and for various environmental conditions. Some of these documents include earthquake design requirements in the context of the overall mechanical or electrical design. It is the intent of the standard that seismic requirements in referenced documents be used. The developers of these documents are familiar with the expected performance and failure modes of the components; however, the documents may be based on design considerations not immediately obvious to a structural design professional. For example, in the design of industrial piping, stresses caused by seismic inertia forces typically are not added to those caused by thermal expansion.

Where reference documents have been adopted specifically by this standard as meeting the force and displacement requirements of this chapter with or without modification, they are considered a part of the standard.

There is a potential for misunderstanding and misapplication of reference documents for the design of mechanical and electrical systems. A registered design professional familiar with both the standard and the reference documents used should be involved in the review and acceptance of the seismic design.

Even when reference documents for nonstructural components lack specific earthquake design requirements, mechanical and electrical equipment constructed in accordance with industry-standard reference documents have performed well historically when properly anchored. Nevertheless, manufacturers of mechanical and electrical equipment are expected to consider seismic loads in the design of the equipment itself, even when such consideration is not explicitly required by this chapter.

Although some reference documents provide requirements for seismic capacity appropriate to the component being designed, the seismic demands used in design may not be less than those specified in the standard.

Specific guidance for selected mechanical and electrical components and conditions is provided in Section 13.6.

Unless exempted in Section 13.1.4, components should be anchored to the structure and, to promote coordination, required supports and attachments should be detailed in the construction documents. Reference documents may contain explicit instruction for anchorage of nonstructural components. The anchorage requirements of Section 13.4 must be satisfied in all cases, however, to ensure a consistent level of robustness in the attachments to the structure.

C13.1.8 Reference Documents Using Allowable Stress Design Many nonstructural components are designed using specifically developed reference documents that are based on allowable stress loads and load combinations and generally permit increases in allowable stresses for seismic loading. Although Section 2.4.1 of the standard does not permit increases in allowable stresses, Section 13.1.8 explicitly defines the conditions for stress increases in the design of nonstructural components where reference documents provide a basis for earthquake-resistant design.

C13.2 GENERAL DESIGN REQUIREMENTS

C13.2.1 Applicable Requirements for Architectural, Mechanical, and Electrical Components, Supports, and Attachments Compliance with the requirements of Chapter 13 may be accomplished by project-specific design or by a manufacturer's certification of seismic qualification of a system or component. When compliance is by manufacturer's certification, the items must be installed in accordance with the manufacturer's requirements. Evidence of compliance may be provided in the form of a signed statement from a representative of the manufacturer or from the registered design professional indicating that the component or system is seismically qualified. One or more of the following options for evidence of compliance may be applicable:

1. An analysis (e.g., of a distributed system such as piping) that includes derivation of the forces used for the design of the system, the derivation of displacements and reactions, and the design of the supports and anchorages;
2. A test report, including the testing configuration and boundary conditions used (where testing is intended to address a class of components, the range of items covered by the testing performed should also include the justification of similarities of the items that make this certification valid); and/or
3. An experience data report.

Components addressed by the standard include individual simple units and assemblies of simple units for which reference documents establish seismic analysis or qualification requirements. Also addressed by the standard are complex architectural, mechanical, and electrical systems for which reference documents either do not exist or exist for only elements of the system. In the design and analysis of both simple components and complex systems, the concepts of flexibility and ruggedness often can assist the designer in determining the necessity for analysis and, when analysis is necessary, the extent and methods by which seismic adequacy may be determined. These concepts are discussed in Section C13.6.1.

C13.2.3 Special Certification Requirements for Designated Seismic Systems This section addresses the qualification of active designated seismic equipment, its supports, and attachments with the goals of improving survivability and achieving a high level of confidence that a facility will be functional after a design earthquake. Where components are interconnected, the qualification should provide the permissible forces (e.g., nozzle loads) and, as applicable, anticipated displacements of the component at the connection points to facilitate assessment for consequential damage, in accordance with Section 13.2.4. Active equipment has parts that rotate, move mechanically, or are energized during operation. Active designated seismic equipment constitutes a limited subset of designated seismic systems. Failure of active designated seismic equipment may itself pose a significant hazard. For active designated seismic equipment, failure of structural integrity and loss of function are to be avoided.

Examples of active designated seismic equipment include mechanical (components of HVACR systems and piping systems) or electrical (power supply distribution) equipment, medical equipment, fire pump equipment, and uninterruptible power supplies for hospitals. It is generally understood that fire protection sprinkler piping systems designed and installed per NFPA 13 (2007) are deemed to comply with the special certification requirements of Section 13.2.3; see also Section 13.6.7.2.

There are practical limits on the size of a component that can be qualified via shake table testing. Components too large to be qualified by shake table testing need to be qualified by a combination of structural analysis and qualification testing or empirical evaluation through a subsystem approach. Subsystems of large, complex components (e.g., large chillers, skid-mounted equipment assemblies, and boilers) can be qualified individually, and the overall structural frame of the component can be evaluated by structural analysis.

Evaluating postearthquake operational performance for active equipment by analysis generally involves sophisticated modeling with experimental validation and may not be reliable. Therefore, the use of analysis alone for active or energized components is not permitted unless a comparison can be made to components that have otherwise been deemed as rugged. As an example, a transformer is energized but contains components that can be shown to remain linearly elastic and are inherently rugged. However, switching equipment that contains fragile components is similarly energized but not inherently rugged, and it therefore cannot be certified solely by analysis. For complex components, testing or experience may therefore be the only practical way to ensure that the equipment will be operable after a design earthquake. Past earthquake experience has shown that much active equipment is inherently rugged. Therefore, evaluation of experience data, together with analysis of anchorage, is adequate to demonstrate compliance of active equipment such as pumps, compressors, and electric motors. In other cases, such as for motor control centers and switching equipment, shake table testing may be required.

With some exceptions (e.g., elevator motors), experience indicates that active mechanical and electrical components that contain electric motors of more than 10 hp (7.4 kW) or that have a thermal exchange capacity greater than 200 MBH are unlikely to merit the exemption from shake table testing on the basis of inherent ruggedness. Components with less motor horsepower and thermal exchange capacity are generally considered small active components and are deemed rugged. Exceptions to this rule may be appropriate for specific cases, such as elevator motors that have higher horsepower but have been shown by experience to be rugged. Analysis is still required to ensure the structural integrity of the nonactive components. For example, a 15-ton condenser would require analysis of the load path between the condenser fan and the coil to the building structure attachment.

Where certification is accomplished by analysis, the type and sophistication of the required analysis varies by specific equipment type and construction. Static analysis using the total force specified in Section 13.3 considering applicable load combinations may be appropriate for single components where the structural frame is the only item to be certified and where internal dynamic effects are shown to be negligible. For single components where dynamic effects may be significant, or for assemblages of components, dynamic analysis is strongly suggested. Either modal analysis or response history procedures may be used, but care should be exercised when using modal analysis to ensure that the significant interactions between individual components are properly captured. In all analyses, it is essential that the stiffness, mass, and applied load be distributed in accordance with the component properties, and in sufficient detail (number of degrees of freedom) to allow the desired forces, deformations, and accelerations to be accurately determined. Input motions for dynamic procedures should reflect the expected motion at the attachment points of the component. Nonlinear behavior of the component is typically not advisable in the certification analysis in the absence of well-documented test results for specific components. In general, the input motion is (a) a generic floor response spectrum such as that provided in ICC-ES AC156 (2010), (b) location- and structure-specific floor spectra generated using the procedures of Section 13.3.1, or (c) acceleration time histories developed using dynamic analysis procedures similar to those specified in Chapter 16 or Section 12.9. Horizontal and vertical inputs are usually applied simultaneously when performing these types of dynamic analyses. As with all structural analysis, judgment is required to ensure that the results are applicable and representative of the behavior anticipated for the input motions.

C13.2.4 Consequential Damage Although the components identified in Tables 13.5-1 and 13.6-1 are listed separately, significant interrelationships exist and must be considered. Consequential damage occurs because of interaction between components and systems. Even "braced" components displace, and the displacement between lateral supports can be significant in distributed systems such as piping systems, cable and conduit systems, and other linear systems. It is the intent of the standard that the seismic displacements considered include both relative displacement between multiple points of support (addressed in Section 13.3.2) and, for mechanical and electrical components, displacement within the component assemblies. Impact of components must be avoided, unless the components are fabricated of ductile materials that have been shown to be capable of accommodating the expected impact loads. With protective coverings, ductile mechanical and electrical components and many more fragile components are expected to survive all but the most severe impact loads. Flexibility and ductility of the connections between distribution systems and the equipment to which they attach are essential to the seismic performance of the system.

The determination of the displacements that generate these interactions is not addressed explicitly in Section 13.3.2.1. That section concerns relative displacement of support points. Consequential damage may occur because of displacement of components and systems between support points. For example, in older suspended-ceiling installations, excessive lateral displacement of a ceiling system may fracture sprinkler heads that project through the ceiling. A similar situation may arise if sprinkler heads projecting from a small-diameter branch line pass through a rigid ceiling system. Although the branch line may be properly restrained, it may still displace sufficiently between lateral support points to affect other components or systems. Similar interactions occur where a relatively flexible distributed system connects to a braced or rigid component.

The potential for impact between components that are in contact with or close to other structural or nonstructural components must be considered. However, when considering these potential interactions, the designer must determine whether the potential interaction is both credible and significant. For example, the fall of a ceiling panel located above a motor control center is a credible interaction because the falling panel in older suspended ceiling installations can reach and impact the motor control center. An interaction is significant if it can result in damage to the target. Impact of a ceiling panel on a motor control center may not be significant because of the light weight of the ceiling panel. Special design consideration is appropriate where the failure of a nonstructural element could adversely influence the performance of an adjacent critical nonstructural component, such as an emergency generator.

C13.2.5 Flexibility In many cases, flexibility is more important than strength in the performance of distributed systems, such as piping and ductwork. A good understanding of the displacement demand on the system, as well as its displacement capacity, is required. Components or their supports and attachments must be flexible enough to accommodate the full range of expected differential movements; some localized inelasticity is permitted in accommodating the movements. Relative movements in all directions must be considered. For example, even a braced branch line of a piping system may displace, so it needs to be connected to other braced or rigid components in a manner that accommodates the displacements without failure

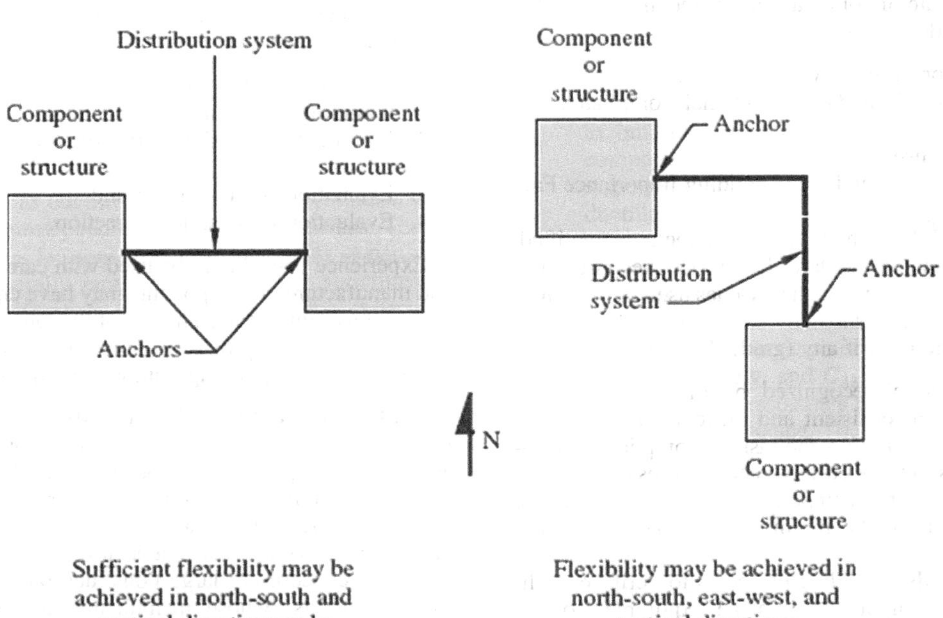

Figure C13.2-1. Schematic plans illustrating branch line flexibility.

results were not performed, the number of studies performed was substantial, and engineering judgment was used in parameter studies and final equation setting to target a general design level such that the proposed equations are approximately mean plus one standard deviation above archetype results. Due to the number of parameters involved and their potential for variation, in some cases, there will be a higher level of reliability; in other cases, there will be a lower level of reliability.

- *Component Strength.* Like building structural systems, the component and its attachments to the structure typically have some inherent overstrength. This is captured by the variable R_{po}. It serves to reduce the design force needed.
- *Ground versus Superstructure.* The amplification of PCA/PFA as the ratio of component period to building period approaches unity comes from narrow-band filtering of response by the dynamic properties of the building. Components that are ground supported can see dynamic amplification due to component flexibility, based on structural dynamics, but this amplification is typically less than what occurs in the building. Given that there are both theoretical and numerical differences between the ground and superstructure cases, it was decided to distinguish the two.

Revisions Made from NIST (2018)
Building Ductility

Determination of the Structure Ductility Reduction Factor, R_u, relies on the R and Ω_o values in Tables 12.2-1, 15.4-1, and 15.4-2, and the Importance Factor, I_e, as prescribed in Section 11.5.1. While these variables were not originally intended to be used in the determination of lateral forces on nonstructural components, they provide a reasonable basis for estimating the ductility and strength of building lateral force-resisting systems commonly used in Seismic Design Categories D and higher. The Importance Factor, I_e, accounts for the higher lateral strength required for Risk Category III and IV structures, which results in lower ductility demands for the Design Earthquake. For nonbuilding structures, the tabulated values of R and Ω_0 were assigned both for technical considerations and to facilitate inclusion of low-ductility systems into the building codes. In regions of high seismicity the low values of R that are used, especially for nonbuilding structures not similar to buildings, do not reflect behaviors such as sliding and rocking that reduce floor accelerations in these structures. To reflect this, a lower limit of 1.3 is placed on the value of R_u.

When determining the value of R_u, several alternative situations can arise, including the following.

- The Importance Factor, I_e, can be determined based on the occupancy of the Risk Category of the structure using Tables 1.5-1 and 1.5-2.
- If a seismic force-resisting system (SFRS) is not listed in these tables or the SFRS does not conform to the associated requirements for the system, then use $R_u = 1.3$. This situation can apply to existing buildings with detailing provisions that do not meet current requirements and have less available ductility.
- If the SFRS in which the component will be placed is not known, but it is known that it will be a system that complies with Tables 12.2-1, 15.4-1, or 15.4-2, then use the lowest value of R_u in the applicable table for the applicable seismic design category. This situation can arise when the component anchorage and bracing are designed to be able to be installed in a range of potential code-conforming buildings. It can also apply when the engineer responsible for anchorage and bracing of the component is told it is for a new building, but information about the building's SFRS is not provided.
- If an alternative system has been developed with associated R and Ω_o values and approved by the Authority Having Jurisdiction, then use those values to determine R_u.
- If the SFRS category is known, but the details are unknown, for example it is known that the system for an office building is a braced frame, but not which type of braced frame, then use the braced frame system with the lowest value of R_u, which is the steel ordinary concentrically braced frame with $R_u = [1.1R/(I_e\Omega_o)]^{1/2} = [(1.1)(3.25)/(1.0)(2.0)]^{1/2} = 1.3$.
- If nonlinear response history analysis is performed, then the procedure in Section 13.1.3.1.5 can be used, and R_u need not be calculated.

Increasing building ductility generally reduces nonstructural component seismic demands, and reduced ductility generally increases component demands. Buildings with higher design forces (such as those using a seismic Importance Factor, I_e, greater than 1.0) and/or higher levels of overstrength will have less ductility demand for the same level of seismic shaking than those designed to code minimums and with limited overstrength. The ATC-120 project analyses included building archetypes with both limited overstrength and substantial overstrength. Calibration studies were done to reasonably bound results from the archetypes with substantial overstrength.

An alternative procedure for diaphragm design was introduced in ASCE 7-16. The procedure, presented in Section 12.10.3, considers the strength and ductility of the diaphragm and provides procedures to obtain more realistic diaphragm design forces. The development of that procedure is described in Commentary Section C12.10.3. Both the diaphragm procedures and nonstructural component force equations can produce floor acceleration profiles over the height of the structure. These profiles will differ, due to differences in formulation of the acceleration profiles. The less complex procedures in Chapter 13 produce more conservative estimates of acceleration. The acceleration profile in Equation (13.3-1) is described by $\frac{H_f}{R_\mu}$, where H_f is a factor for force amplification determined using the approximate fundamental period, T_a, and R_μ is a structure ductility reduction factor determined using R and Ω_0. The diaphragm design procedure in Chapter 12 is a substantially more complex process and produces lower floor acceleration estimates due to the use of the period T, which is generally larger than the approximate period T_a, the use of S_{D1} for longer-period structures, and due to a more refined method for considering modal effects. That procedure has not been adopted for the determination of floor accelerations used to design nonstructural component and system restraints, since the determination of F_p is correlated to the use of the approximate fundamental period, T_a (NIST 2018).

Determination of Likelihood of Being in Resonance

Nonstructural components have been categorized as likely or unlikely to be in resonance by engineering judgment. An underlying concept is that if the component period, T_p, is less than 0.06 s, then resonance is unlikely regardless of the building period, since the building period will typically be well above that level. In the 2016 and earlier editions of ASCE 7, components with $T_p \leq 0.06$ s were termed "rigid" and did not receive any amplification of PFA (while those with $T_p > 0.06$ were termed "flexible" and received an increase of 2.5 times PFA). A second underlying concept is that, when the ratio of component period to building period is relatively low or relatively large, then

Table C13.3-1. Component Ductility Categories.

Ductility Category	Assumed Component Ductility, μ_{comp}	PCA/PFA (C_{AR}) Supported At or Below Grade Plane	PCA/PFA (C_{AR}) Supported Above Grade Plane by a Structure
Elastic	1	2.5	4.0
Low	1.25	2.0	2.8
Moderate	1.5	1.8	2.2
High	2.0	1.4	1.4

resonance is also unlikely. A criterion of $T_p/T_a < 0.5$ or $T_p/T_a > 1.5$ can be used, as suggested by NIST (2018) as well as by extrapolation of results from Hadjian and Ellison (1986). Distribution systems may experience resonance, but its effect is judged to be minimized due to reduced mass participation caused by multiple points of support.

Component Ductility Categories

Nonstructural components have been assigned to one of three categories of component ductility. In the ATC-120 studies, the underlying relationships used are shown in Table C13.3-1.

As discussed in NIST (2018), the "elastic" category is used for reference only. It is assumed that typical nonstructural components and their attachments to the structure systems used in practice have at least "low" component ductility.

Alternative C_{AR} values can be developed for components that are not in Table C13.3-1 or for which substantiating data is available. Such data can be derived from instrumented shake table component tests that compare the acceleration experienced by the component (PCA) at its effective center of mass versus the acceleration of the shake table (PFA). Such tests need to include realistic attachments between the component and the anchors, and realistic anchorage.

Equipment Support Conditions

Design coefficients are assigned to mechanical and electrical equipment based on the properties of the equipment item, but may be supported on a support structure or platform that has structural properties that are substantially different from the component itself. In some cases this can be beneficial, such as when a moderate- or low-ductility component is mounted on a support structure or platform with high ductility. In this case, the support structure or platform will limit the shaking demands on supported components, by providing a structure with ductile behavior in the load path.

Determination of Ground versus Superstructure Category

Nonstructural components supported by slabs or foundation elements at grade that are not part of a building use the ground category. Similarly, nonstructural components supported by slabs or foundations, or other elements of the superstructure located at or below grade, use the ground category. For the definition of grade, including sloping sites, see the definition of "Grade Plane" in Section 11.2.

Design Forces for Elements in the Load Path with Limited Ductility

Anchors in concrete or masonry that cannot develop a ductile yield mechanism are required to use design forces increased by the Ω_{op} factor. Designers should consider amplifying design forces by an overstrength factor for elements in the load path between the component and the anchor that have limited ductility.

Upper- and Lower-Bound Seismic Design Forces

Equation (13.3-3) establishes a minimum seismic design force, F_p, which is consistent with current practice. Equation (13.3-2) provides a simple maximum value of F_p that prevents multiplication of the individual factors from producing a design force that would be unreasonably high, considering the expected nonlinear response of support and component. Figure C13.3-1 illustrates the distribution of the specified lateral design forces. Note that depending on the nature of the supporting structure, the value of F_p calculated using Equation (13.3-1) may not vary linearly with height.

For components with points of attachment at more than one height, it is recommended that design be based on the average of values of F_p determined individually at each point of attachment (but with the entire component weight, W_p) using Equations (13.3-1) through (13.3-3).

Alternatively, for each point of attachment, a force F_p may be determined using Equations (13.3-1) through (13.3-3), with the portion of the component weight, W_p, tributary to the point of attachment. For design of the component, the attachment force F_p must be distributed relative to the component's mass

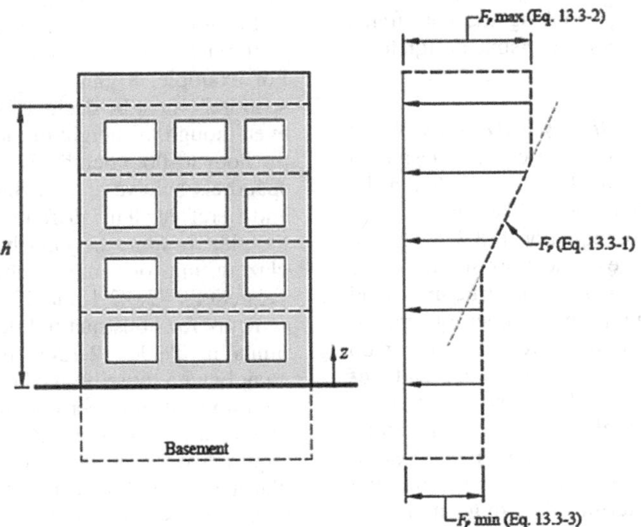

Figure C13.3-1. Lateral seismic design force magnitude over height.

distribution over the area used to establish the tributary weight. To illustrate these options, consider a solid exterior nonstructural wall panel, supported top and bottom, for a one-story building with a rigid diaphragm. The values of F_p computed for the top and bottom attachments using Equations (13.3-1) through (13.3-3), respectively, are $0.48S_{DS}I_pW_p$ and $0.30S_{DS}I_pW_p$. In the recommended method, a uniform load is applied to the entire panel based on $0.39S_{DS}I_pW_p$. In the alternative method, a trapezoidal load is applied, varying from $0.48S_{DS}I_pW_p$ at the top to $0.30S_{DS}I_pW_p$ at the bottom. Each anchorage force is then determined considering the static equilibrium of the complete component subject to all the distributed loads.

Cantilever parapets that are part of a continuous element should be checked separately for parapet forces. The seismic design force on any component must be applied at the center of gravity of the component and must be assumed to act in any horizontal direction. Vertical forces on nonstructural components equal to $\pm0.2S_{DS}W_p$ are specified in Section 13.3.1 and are intended to be applied to all nonstructural components and not just cantilevered elements. Nonstructural concrete or masonry walls laterally supported by flexible diaphragms must be anchored out of plane in accordance with Section 12.11.2.

It is expected that the nonstructural component be evaluated for the maximum load effect from F_p applied in any horizontal direction. The most critical load effects can typically be computed using a pair of orthogonal directions that coincide with the principal axes of the component. Components with nonparallel or nonorthogonal geometry may require a set of orthogonal direction pairs to determine the most critical load effects. See Sections C12.5.1 and C12.5.3 for additional background.

C13.3.1.1 Amplification with Height, H_f The FEMA P-58 / BD-3.7.17 report observed that while the approximate fundamental period T_a may underestimate the period of a structure, the fundamental period of the structure T determined in Section 12.8.2 tends to overestimate it. Therefore, the use of T in lieu of T_a is not permitted for components supported by buildings. In buildings, the gravity system, partitions, and cladding all act to reduce the fundamental period, which will increase the seismic design force on nonstructural components. T_a was recommended by the ATC-120 team to provide a reasonable estimate of the building fundamental period for the purposes of computing forces on nonstructural components. Nonbuilding structures generally lack partitions and cladding, and the gravity system is less extensive than that found in buildings, making the bare-frame period of the structure suitable for component seismic design force calculations.

C13.3.1.5 Nonlinear Response History Analysis When nonlinear response history analysis is used to design a structure, there are several options available for calculating the seismic demands on nonstructural components. The forces can be determined using the basic equation, Equation (13.3-1), or the designer may choose to take advantage of the more sophisticated analysis procedures that were used to design the structure found in Chapter 16, 17, or 18. The minimum number of ground motions used for design of the structure differs depending on which chapter is used. The intent is that the entire suite of motions used to design the structure shall also be used to determine the forces on nonstructural components. In some cases, a structure may be analyzed using the procedures of Chapter 12, but it is desired that the forces on the nonstructural components be determined using nonlinear response history analysis. In such a case, a minimum of seven motions shall be used. Structures with significant horizontal irregularities may experience large torsional response. During the ATC-120 project, the influence of torsional response of the structure on floor accelerations experienced by the component was investigated, and it was concluded that due to the complexity of the problem and the limited information available, additional study is needed before it is included in the design equations.

C13.3.2 Seismic Relative Displacements The equations in this section are for use in design of cladding, stairways, windows, piping systems, sprinkler components, and other components connected to one structure at multiple levels or to multiple structures. Two equations are given for each situation. Equations (13.3-9) and (13.3-11) produce structural displacements as determined by elastic analysis, unreduced by the structural response modification factor, R. Because the actual displacements may not be known when a component is designed or procured, Equations (13.3-10) and (13.3-12) provide upper-bound displacements based on structural drift limits. Use of upper-bound equations may facilitate timely design and procurement of components, but may also result in costly added conservatism. Designers should be aware that some buildings are designed without drift limits; this is permitted in note *a* of Table 12.12-1. Similarly, for buildings designed with drift limits, those limits apply only at the center of mass of the structure (or, in certain cases, at the building perimeter). The drift limit does not apply to the Design Earthquake Displacement, and thus the displacement demand on a component may exceed the drift limit, unless the drift limit has been applied to the entire structure.

The value of seismic relative displacements is taken as the calculated displacement, D_p, times the Importance Factor, I_e, because the elastic displacement calculated in accordance with Equation (12.8-16) to establish δ_x (and thus D_p) is adjusted for I_e in keeping with the philosophy of displacement demand for the structure. For component design, the unreduced elastic displacement is appropriate.

The standard does not provide explicit acceptance criteria for the effects of seismic relative displacements, except for glazing. Damage to nonstructural components caused by relative displacement is acceptable, provided the performance goals defined elsewhere in the chapter are achieved.

The design of some nonstructural components that span vertically in the structure can be complicated when supports for the element do not occur at horizontal diaphragms. The language in Section 13.3.2 was previously amended to clarify that story drift must be accommodated in the elements that actually distort. For example, a glazing system supported by precast concrete spandrels must be designed to accommodate the full-story drift, even though the height of the glazing system is only a fraction of the floor-to-floor height. This condition arises because the precast spandrels behave as rigid bodies relative to the glazing system, and therefore all the drift must be accommodated by anchorage of the glazing unit, the joint between the precast spandrel and the glazing unit, or some combination of the two.

Sections 13.3.2.1 and 13.3.2.2 permit seismic relative displacements for nonstructural components to be capped at the drift limits in Table 12.12-1, because the structure must already comply with those limits; however, one-story structures designed with no drift limit per note *a* of Table 12.12-1 are not permitted to use this approach. For example, a piping system that is required to be designed to accommodate seismic relative displacements and that is attached to tanks and equipment at grade and supported by the roof of a long-span single-story structure designed without drift limits in accordance with note *a* should consider the actual diaphragm displacements. These

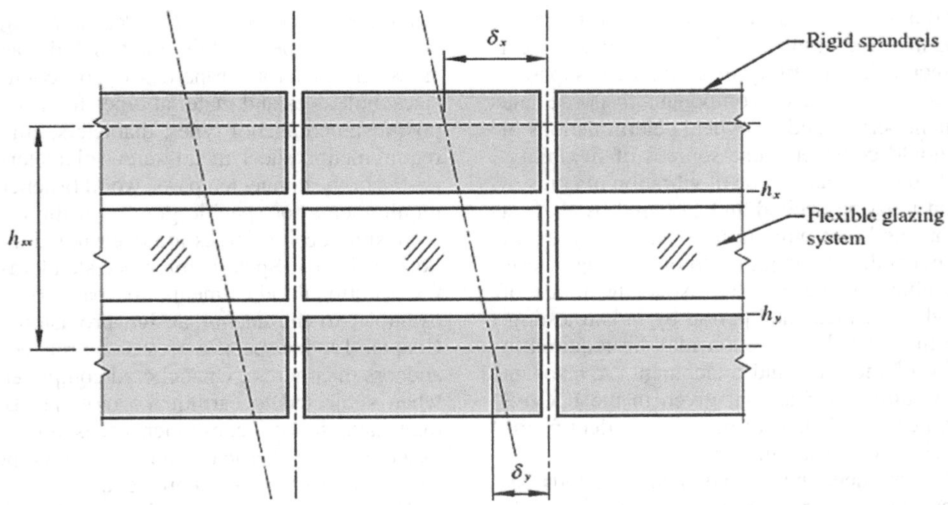

Figure C13.3-4. Displacements over less than story height.

displacements may exceed the drift limits in Table 12.12-1, resulting in unanticipated damage to the piping system if the piping system is unable to accommodate that drift.

C13.3.2.1 Displacements within Structures Seismic relative displacements can subject components or systems to unacceptable stresses. The potential for interaction resulting from component displacements (in particular for distributed systems) and the resulting impact effects should also be considered (see Section 13.2.4).

These interrelationships may govern the clearance requirements between components or between components and the surrounding structure. Where sufficient clearance cannot be provided, consideration should be given to the ductility and strength of the components and associated supports and attachments to accommodate the potential impact.

Where nonstructural components are supported between, rather than at, structural levels, as frequently occurs for glazing systems, partitions, stairs, veneers, and mechanical and electrical distributed systems, the height over which the displacement demand, D_p, must be accommodated may be less than the story height, h_{sx}, and should be considered carefully. For example, consider the glazing system supported by rigid precast concrete spandrels shown in Figure C13.3-4. The glazing system may be subjected to full-story drift, D_p, although its height $(h_x - h_y)$ is only a fraction of the story height. The design drift must be accommodated by anchorage of the glazing unit, or the joint between the precast spandrel and the glazing unit, or some combination of the two. Similar displacement demands arise where pipes, ducts, or conduits that are braced to the floor or roof above are connected to the top of a tall, rigid, floor-mounted component.

For ductile components, such as steel piping fabricated with welded connections, the relative seismic displacements between support points can be more significant than inertial forces. Ductile piping can accommodate relative displacements by local yielding, with strain accumulations well below failure levels. However, for components fabricated using less ductile materials, where local yielding must be avoided to prevent unacceptable failure consequences, relative displacements must be accommodated by flexible connections.

C13.3.2.2 Displacements between Structures A component or system connected to two structures must accommodate horizontal movements in any direction, as illustrated in Figure C13.3-5.

C13.3.3 Component Period Component period is used to classify components as rigid ($T \leq 0.06$ s) or flexible ($T > 0.06$ s). Determination of the fundamental period of an architectural, mechanical, or electrical component using

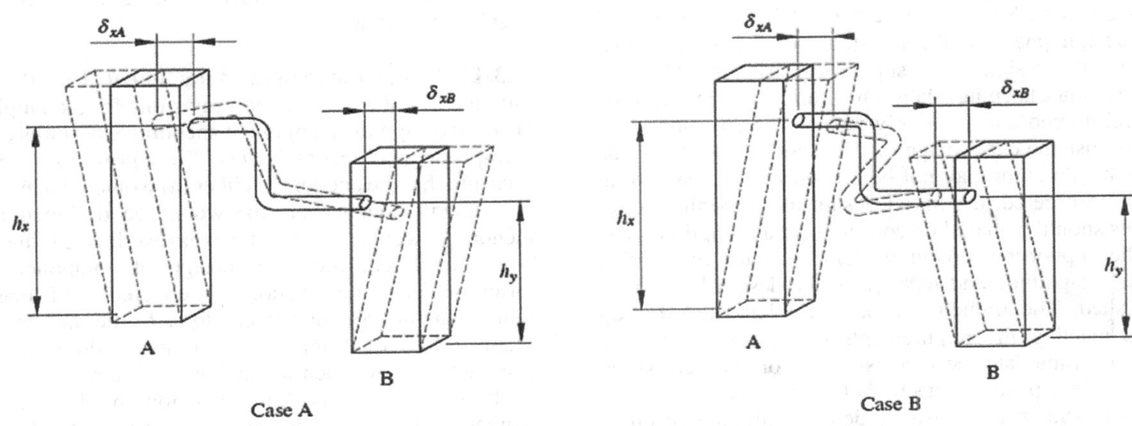

Figure C13.3-5. Displacements between structures.

analytical or test methods is often difficult and, if not properly performed, may yield incorrect results. In the case of mechanical and electrical equipment, the flexibility of component supports and attachments typically dominates component response and fundamental component period, and analytical determinations of component period should consider those sources of flexibility. Where testing is used, the dominant mode of vibration of concern for seismic evaluation must be excited and captured by the test setup. The dominant mode of vibration for these types of components cannot generally be acquired through in situ tests that measure only ambient vibrations. To excite the mode of vibration with the highest fundamental period by in situ testing, relatively significant input levels of motion may be required to activate the flexibility of the base and attachment. A resonant frequency search procedure, such as that given in the ICC-ES AC156 acceptance criteria (2010), may be used to identify the dominant modes of vibration of a component.

Many mechanical components have fundamental periods below 0.06 s and may be considered rigid. Examples include horizontal pumps, engine generators, motor generators, air compressors, and motor-driven centrifugal blowers. Other types of mechanical equipment, while relatively stiff, have fundamental periods (up to about 0.13 s) that do not permit automatic classification as rigid. Examples include belt-driven and vane axial fans, heaters, air handlers, chillers, boilers, heat exchangers, filters, and evaporators. Where such equipment is mounted on vibration isolators, the fundamental period is substantially increased.

Electrical equipment cabinets can have fundamental periods ranging from 0.06 to 0.3 s, depending on the supported weight and its distribution, the stiffness of the enclosure assembly, the flexibility of the enclosure base, and the load path through to the attachment points. Tall, narrow motor control centers and switchboards lie at the upper end of this period range. Low- and medium-voltage switchgear, transformers, battery chargers, inverters, instrumentation cabinets, and instrumentation racks usually have fundamental periods of 0.1 to 0.2 s. Braced battery racks, stiffened vertical control panels, bench boards, electrical cabinets with top bracing, and wall-mounted panelboards generally have fundamental periods ranging from 0.06 to 0.1 s.

C13.4 NONSTRUCTURAL COMPONENT ANCHORAGE

Unless exempted in Section 13.1.4 or 13.6.9, components must be anchored to the structure, and all required supports and attachments must be detailed in the construction documents. To satisfy the load path requirement of this section, the detailed information given in Section C13.2.8 must be communicated during the design phase to the registered design professional responsible for the design of the supporting structure. The load path includes housekeeping slabs and curbs, which must be adequately reinforced and positively fastened to the supporting structure. Because the exact magnitude and location of the loads imposed on the structure may not be known until nonstructural components are ordered, the initial design of supporting structural elements should be based on conservative assumptions. The design of the supporting structural elements must be verified once the final magnitude and location of the design loads have been established. The limited exception for ballasted rooftop solar panels meeting the requirements of Section 13.6.12 is intended to accommodate the increasing use of such arrays on roof systems where positive attachment is difficult.

Design documents should provide details with enough information that compliance with these provisions can be verified. Parameters such as C_{AR}, R_{po}, R_{μ}, I_p, I_p, S_{DS}, and W_p should be noted. Attachment details may include, as appropriate, dimensions and material properties of the connecting material, weld sizes, bolt sizes and material types for steel-to-steel connections, post-installed anchor types, diameters, embedments, installation requirements, sheet metal screw diameters and material thicknesses of the connected parts, wood fastener types, and minimum requirements for specific gravity of the base materials.

Seismic design forces are determined using the provisions of Section 13.3.1. Specific reference standards should be consulted for additional adjustments to loads or strengths. Refer, for example, to the anchor design provisions of ACI 318 (2014), Chapter 17, for specific provisions related to seismic design of anchors in concrete. Unanchored components often rock or slide when subjected to earthquake motions. Because this behavior may have serious consequences, is difficult to predict, and is exacerbated by vertical ground motions, positive restraint must be provided for each component.

The effective seismic weight used in design of the seismic force-resisting system must include the weight of supported components. To satisfy the load path requirements of this section, localized component demand must also be considered. Compliance may be accomplished by checking the capacity of the first structural element in the load path (for example, a floor beam directly under a component) for combined dead, live, operating, and seismic loads, using the horizontal and vertical loads from Section 13.3.1 for the seismic demand, and repeating this procedure for each structural element or connection in the load path until the load case, including horizontal and vertical loads from Section 13.3.1, no longer governs design of the element. The load path includes housekeeping slabs and curbs, which must be adequately reinforced and positively fastened to the supporting structure.

Because the exact magnitude and location of loads imposed on the structure may not be known until nonstructural components are ordered, the initial design of supporting structural elements should be based on conservative assumptions. The design of the supporting structural elements may need to be verified once the final magnitude and location of the design loads have been established.

Tests have shown that there are consistent shear ductility variations between bolts installed in drilled or punched plates with nuts and connections using welded shear studs. The need for reductions in allowable loads for particular anchor types to account for loss of stiffness and strength may be determined through appropriate dynamic testing. Although comprehensive design recommendations are not available at present, this issue should be considered for critical connections subject to dynamic or seismic loading.

C13.4.1 Design Force in the Attachment The 2016 and earlier editions of ASCE 7 included provisions for the amplification of forces to design the component anchorage, or limits of the values of response modification factors. These provisions were intended to ensure that the anchorage either (a) would respond to overload in a ductile manner or, (b) would be designed so that the anchorage would not be the weakest link in the load path. While amplified forces for design of component anchorage currently focus on anchors to concrete and masonry, any component anchorage subject to a brittle failure mechanism where a loss of component stability could result should be designed to avoid such an anchorage failure.

The revisions to the force equations produce more accurate estimates of seismic demands on nonstructural components, and the component resonance ductility factors (C_{AR}) for high-ductility components are all the same, eliminating the need for a cap on some components.

C13.4.2 Anchors in Concrete or Masonry Design capacity for anchors in concrete must be determined in accordance with ACI 318, Chapter 17. Design capacity for anchors in masonry is determined in accordance with TMS 402. Anchors must be designed to have ductile behavior or to provide a specified amount of excess strength. Depending on the specifics of the design condition, ductile design of anchors in concrete may satisfy one or more of the following objectives:

1. Adequate load redistribution between anchors in a group,
2. Allowance for anchor overload without brittle failure, or
3. Energy dissipation.

Achieving deformable, energy-absorbing behavior in the anchor itself is often difficult. Unless the design specifically addresses the conditions influencing desirable hysteretic response (e.g., adequate gauge length, anchor spacing, edge distance, and steel properties), anchors cannot be relied on for energy dissipation. Simple geometric rules, such as restrictions on the ratio of anchor embedment length to depth, may not be adequate to produce reliable ductile behavior. For example, a single anchor with sufficient embedment to force ductile tension failure in the steel body of the anchor bolt may still experience concrete fracture (a nonductile failure mode) if the edge distance is small, the anchor is placed in a group of tension-loaded anchors with reduced spacing, or the anchor is loaded in shear instead of tension. In the common case where anchors are subject primarily to shear, response governed by the steel element may be nonductile if the deformation of the anchor is constrained by rigid elements on either side of the joint. Designing the attachment so that its response is governed by a deformable link in the load path to the anchor is encouraged. This approach provides ductility and overstrength in the connection while protecting the anchor from overload. Ductile bolts should be relied on as the primary ductile mechanism of a system only if the bolts are designed with enough gauge length (using the unbonded strained length of the bolt) to accommodate the anticipated nonlinear displacements of the system at the design earthquake. Guidance for determining the gauge length can be found in Part 3 of the NEHRP (2009) provisions.

The revised force equations allow correlation between the component resonance ductility factor, C_{AR}, and the anchorage overstrength factor, Ω_{0p}. In general, components unlikely to be in resonance, and high-ductility components likely to be in resonance, are assigned the highest anchorage overstrength factor. These components are designed for lower lateral forces, and an extra margin of strength in anchorage to concrete and masonry is warranted in the event that some resonance does occur or the component ductility is lower than anticipated. Low-ductility components that are likely to be in resonance are designed for high seismic design force levels. Since minimal reductions in response due to ductile behavior are expected, the seismic design force is less likely to be exceeded in a design earthquake, warranting a lower anchorage overstrength factor.

Anchors used to support towers, masts, and equipment are often provided with double nuts for leveling during installation. Where base-plate grout is specified at anchors with double nuts, it should not be relied on to carry loads, because it can shrink and crack or be omitted altogether. The design should include the corresponding tension, compression, shear, and flexure loads.

Post-installed anchors in concrete and masonry should be qualified for seismic loading through appropriate testing. The requisite tests for expansion and undercut anchors in concrete are given in ACI 355.2, *Qualification of Post-Installed Mechanical Anchors in Concrete and Commentary* (2007). Testing and assessment procedures based on the ACI standard that address expansion, undercut, screw, and adhesive anchors are incorporated in ICC-ES acceptance criteria. ICC-ES AC193, *Acceptance Criteria for Mechanical Anchors in Concrete Elements* (2012c), and AC308, *Acceptance Criteria for Post-Installed Adhesive Anchors in Concrete Elements* (2012d), refer to ACI 355.4-11, *Qualification of Post-Installed Adhesive Anchors in Concrete and Commentary* (2011). These criteria, which include specific provisions for screw anchors and adhesive anchors, also reference ACI qualification standards for anchors. For postinstalled anchors in masonry, seismic prequalification procedures are given in ICC-ES AC01, *Acceptance Criteria for Expansion Anchors in Masonry Elements* (2012b), AC58, *Acceptance Criteria for Adhesive Anchors in Masonry Elements* (2012a), and AC106, *Acceptance Criteria for Predrilled Fasteners (Screw Anchors) in Masonry* (2012e).

Other references to adhesives (such as in Section 13.5.7.2) apply not to adhesive anchors but to steel plates and other structural elements bonded or glued to the surface of another structural component with adhesive; such connections are generally nonductile.

C13.4.3 Installation Conditions Prying forces on anchors, which result from a lack of rotational stiffness in the connected part, can be critical for anchor design and must be considered explicitly.

For anchorage configurations that do not provide a direct mechanism to transfer compression loads (for example, a base plate that does not bear directly on a slab or deck but is supported on a threaded rod), the design for overturning must reflect the actual stiffness of base plates, equipment, housing, and other elements in the load path when computing the location of the compression centroid and the distribution of uplift loads to the anchors.

C13.4.4 Multiple Attachments Although the standard does not prohibit the use of single anchor connections, it is good practice to use at least two anchors in any load-carrying connection whose failure might lead to collapse, partial collapse, or disruption of a critical load path.

C13.4.5 Power-Actuated Fasteners Restrictions on the use of power-actuated fasteners are based on observations of failures of sprinkler pipe runs in the 1994 Northridge (California) earthquake. Although it is unclear from the record to what degree the failures occurred because of poor installation, product deficiency, overload, or consequential damage, the capacity of power-actuated fasteners in concrete often varies more than that of drilled postinstalled anchors. The shallow embedment, small diameter, and friction mechanism of these fasteners make them particularly susceptible to the effects of concrete cracking. The suitability of power-actuated fasteners to resist tension in concrete should be demonstrated by simulated seismic testing in cracked concrete.

Where properly installed in steel, power-actuated fasteners typically exhibit reliable cyclic performance. Nevertheless, they should not be used singly to support suspended elements. Where used to attach cladding and metal decking, subassembly testing may be used to establish design capacities, because the interaction among the decking, the subframe, and the fastener can only be estimated crudely by currently available analysis methods.

The exception permits the use of power-actuated fasteners for specific light-duty applications, with upper limits on the load that can be resisted in these cases. All fasteners must have adequate capacity for the calculated loads, including prying forces.

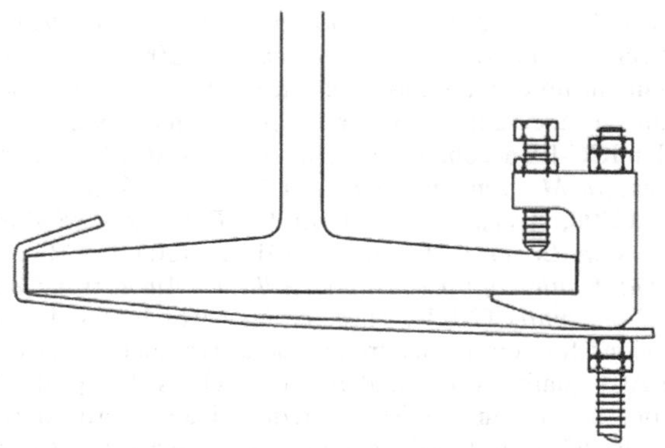

Figure C13.4-1. C-type beam clamp equipped with a restraining strap.

The exception allows the continued use of power-actuated fasteners in concrete for the vertical support of suspended acoustical tile or lay-in panel ceilings and for other light distributed systems, such as small-diameter conduit held to the concrete surface with C-clips. Experience indicates that these applications have performed satisfactorily because of the high degree of redundancy and light loading. Other than ceilings, hung systems should not be included in this exception because of the potential for bending in the fasteners.

The exception for power-actuated fasteners in steel provides a conservative limit on loading. Currently, no accepted procedure exists for the qualification of power-actuated fasteners to resist earthquake loads.

C13.4.6 Friction Clips The term "friction clip" is defined in Section 11.2 in a general way to encompass C-type beam clamps, as well as cold-formed metal channel (strut) connections. Friction clips are suitable to resist seismic forces provided they are properly designed and installed, but under no circumstances should they be relied on to resist sustained gravity loads. C-type clamps must be provided with restraining straps, as shown in Figure C13.4-1.

C13.5 ARCHITECTURAL COMPONENTS

For structures in Risk Categories I through III, the requirements of Section 13.5 are intended to reduce property damage and life-safety hazards posed by architectural components and caused by loss of stability or integrity. When subjected to seismic motion, components may pose a direct falling hazard to building occupants or to people outside the building (as in the case of parapets, exterior cladding, and glazing). Failure or displacement of interior components (such as partitions and ceiling systems in exits and stairwells) may block egress.

For structures in Risk Category IV, the potential disruption of essential function caused by component failure must also be considered.

Architectural component failures in earthquakes can be caused by deficient design or construction of the component, interrelationship with another component that fails, interaction with the structure, or inadequate attachment or anchorage. For architectural components, attachment and anchorage are typically the most critical concerns related to their seismic performance. Concerns regarding loss of function are most often associated with mechanical and electrical components. Architectural damage, unless severe, can be accommodated temporarily. Severe architectural damage is often accompanied by significant structural damage.

C13.5.1 General Suspended architectural components are not required to satisfy the force and displacement requirements of Chapter 13, where prescriptive requirements are met. The requirements were relaxed in the 2005 edition of the standard to better reflect the consequences of the expected behavior. For example, impact of a suspended architectural ornament with a sheet metal duct may only dent the duct, without causing a credible danger (assuming that the ornament remains intact). The reference to Section 13.2.4 allows the designer to consider such consequences in establishing the design approach.

Nonstructural components supported by chains or otherwise suspended from the structure are exempt from lateral bracing requirements, provided they are designed not to inflict damage to themselves or any other component when subject to seismic motion. However, for the 2005 edition, it was determined that clarifications were needed on the type of nonstructural components allowed by these exceptions and the acceptable consequences of interaction between components. In ASCE 7-02, certain nonstructural components that could represent a fire hazard after an earthquake were exempted from the Section 9.6.1 requirements. For example, gas-fired space heaters clearly pose a fire hazard after an earthquake, but were permitted to be exempted from the ASCE 7-02 Section 9.6.1 requirements. The fire hazard after the seismic event must be given the same level of consideration as the structural failure hazard when considering components to be covered by this exception. In addition, the ASCE 7-02 language was sometimes overly restrictive because it did not distinguish between credible seismic interactions and incidental interactions. In ASCE 7-02, if a suspended lighting fixture could hit and dent a sheet metal duct, it would have to be braced, although no credible danger is created by the impact. The new reference in Section 13.2.3 of ASCE 7-05 allowed the designer to consider whether the failures of the component and/or the adjacent components are likely to occur if contact is made. These provisions were carried into ASCE 7-10.

C13.5.2 Forces and Displacements Partitions and interior and exterior glazing must accommodate story drift without failure that will cause a life-safety hazard. Design judgment must be used to assess potential life-safety hazards and the likelihood of life-threatening damage. Special detailing to accommodate drift for typical gypsum board or demountable partitions is unlikely to be cost-effective, and damage to these components poses a low hazard to life safety. Damage in these partitions occurs at low drift levels but is inexpensive to repair.

If they must remain intact after strong ground motion, nonstructural fire-resistant enclosures and fire-rated partitions require special detailing that provides isolation from the adjacent or enclosing structure for deformation equivalent to the calculated drift (relative displacement). In-plane differential movement between structure and wall is permitted. Provision must be made for out-of-plane restraint. These requirements are particularly important in steel or concrete moment-frame structures, which experience larger drifts. The problem is less likely to be encountered in stiff structures, such as those with shear walls.

Differential vertical movement between horizontal cantilevers in adjacent stories (such as cantilevered floor slabs) has occurred in past earthquakes. The possibility of such effects should be considered in the design of exterior walls.

C13.5.3 Exterior Nonstructural Wall Elements and Connections Nonbearing wall panels that are attached to and enclose the structure must be designed to resist seismic (inertial) forces, wind forces, and gravity forces and to accommodate movements of the structure resulting from lateral forces and temperature change. The connections must allow wall panel movements caused by thermal and moisture changes and must be designed to prevent the loss of load-carrying capacity in the event of significant yielding. Where wind loads govern, common practice is to design connectors and panels to allow for not less than two times the story drift caused by wind loads, determined using a return period appropriate to the site location.

Design to accommodate seismic relative displacements often presents a greater challenge than design for strength. Story drifts can amount to 2 in. (51 mm) or more. Separations between adjacent panels are intended to limit contact and resulting panel misalignment or damage under all but extreme building response. Section 13.5.3, Item 1, calls for a minimum separation of 1.5 in. (13 mm). For practical joint detailing and acceptable appearance, separations typically are limited to about 0.75 in. (19 mm). Manufacturing and construction tolerances for both wall elements and the supporting structure must be considered in establishing design joint dimensions and connection details.

Cladding elements, which are often stiff in-plane, must be isolated so that they do not restrain and are not loaded by drift of the supporting structure. Slotted connections can provide isolation, but connections with long rods that flex achieve the desired behavior without requiring precise installation. Such rods must be designed to resist tension and compression in addition to induced flexural stresses and brittle, low-cycle fatigue failure.

Full-story wall panels are usually rigidly attached to and move with the floor structure nearest the panel bottom and isolated at the upper attachments. Panels also can be vertically supported at the top connections, with isolation connections at the bottom. An advantage of this configuration is that failure of an isolation connection is less likely to result in complete detachment of the panel, because it tends to rotate into the structure rather than away from it.

To minimize the effects of thermal movements and shrinkage on architectural cladding panels, connection systems are generally detailed to be statically determinate. Because the resulting support systems often lack redundancy, exacerbating the consequences of a single connection failure, fasteners must be designed for amplified forces, and connecting members must be ductile. The intent is to keep inelastic behavior in the connecting members while the more brittle fasteners remain essentially elastic. To achieve this intent, the tabulated design coefficients produce fastener design forces that are about 2.5 times those for the connecting members.

Limited-deformability curtain walls, such as aluminum systems, are generally light and can undergo large deformations without separating from the structure. However, care must be taken in design of these elements so that low-deformability components (as defined in Section 11.2) that may be part of the system, such as glazing panels, are detailed to accommodate the expected deformations without failure.

In Table 13.5-1, veneers are classified as either limited- or low-deformability elements. Veneers with limited deformability, such as vinyl siding, pose little risk. Veneers with low deformability, such as brick and ceramic tile, are highly sensitive to the performance of the supporting substrate. Significant distortion of the substrate results in veneer damage, possibly including separation from the structure. The resulting risk depends on the size and weight of fragments likely to be dislodged and on the height from which the fragments would fall. Detachment of large portions of veneer can pose a significant risk to life. Such damage can be reduced by isolating the veneer from displacements of the supporting structure. For structures with flexible lateral force-resisting systems, such as moment frames and buckling-restrained braced frames, approaches used to design nonbearing wall panels to accommodate story drift should be applied to veneers.

The limits on length-to-diameter ratios are needed to ensure proper connection performance. Full-scale building shake table tests conducted at the University of California, San Diego, demonstrated that sliding connections perform well when the rod is short. Longer rods in sliding connections bind if there is significant bending and rotation in the rod, which may lead to a brittle failure. For rods that accommodate drift by flexure, longer rods reduce inelastic bending demands and provide better performance. Since anchor rods used in sliding and bending may undergo inelastic action, the use of mild steel improves ductility.

Threaded rods subjected to bending have natural notches (the threads) and are therefore a concern for fatigue. In high-seismic applications, the response may induce a high bending demand and low-cycle fatigue. Cold-worked threaded rod has significantly less ductility unless annealed. Rods meeting the requirements of ASTM F1554, Grade 36, in their as-fabricated condition (i.e., after threading) provide the desired level of performance. ASTM F1554 rods that fulfill the requirements of Supplement 1 for Grade 55 bars and anchor bolts are also acceptable. Other grades that may also be acceptable include ASTM A36, A307, A572, Grade 50, and A588. Other connection configurations and materials may be used, provided they are approved in accordance with ASCE 7-16 Section 1.3.1.3 and are designed to accommodate the story drift without brittle failure.

The reference to D_p has been changed to D_{pI} to reflect consideration of the earthquake Importance Factor on drift demands. Connections should include a means for accommodating erection tolerance so that the required connection capacity is maintained.

C13.5.4 Glass Glass is commonly secured to the window system framing by a glazing pocket built into the framing. This is commonly referred to as a *mechanically captured* or *dry-glazed* window system. Glass can also be secured to the window system framing with a structural silicone sealant; this is commonly referred to as a *wet-glazed* window system. Imposed loads are transferred from the glass to the window system framing through the adhesive bond of the structural silicone sealant. ASTM C1401 *Standard Guide for Structural Sealant Glazing* (2014b) provides guidance and reference standards for the manufacture, testing, design, and installation of structural silicone sealant. This standard addresses glazing sloped to a maximum of 15 degrees from vertical. For glazing slopes exceeding 15 degrees, additional general building code requirements pertaining to sloped glazing and skylights apply.

C13.5.5 Out-of-Plane Bending The effects of out-of-plane application of seismic forces (defined in Section 13.3.1) on nonstructural walls, including the resulting deformations, must be considered. Where weak or brittle materials are used, conventional deflection limits are expressed as a proportion of the span. The intent is to preclude out-of-plane failure of heavy materials (such as brick or block) or applied finishes (such as stone or tile).

C13.5.6 Suspended Ceilings Suspended-ceiling systems are fabricated from a wide range of building materials with differing characteristics. Some systems (such as gypsum board, screwed or nailed to suspended members) are fairly homogeneous and should be designed as light-frame diaphragm assemblies, using the forces of Section 13.3 and the applicable material-specific design provisions of Chapter 14.

Others are composed of discrete elements laid into a suspension system and are the subject of this section.

The seismic performance of ceiling systems with lay-in or acoustical panels depends on support of the grid and individual panels at walls and expansion joints, integrity of the grid and panel assembly, interaction with other systems (such as fire sprinklers), and support for other nonstructural components (such as light fixtures and HVACR systems). Observed performance problems include dislodgement of tiles because of impact with walls and water damage (sometimes leading to loss of occupancy) because of interaction with fire sprinklers.

Suspended lath-and-plaster ceilings are not exempted from the requirements of this section because of their more significant mass and the greater potential for harm associated with their failure. However, the prescriptive seismic provisions of Section 13.5.6.2 and ASTM E580 for acoustical tile and lay-in panel ceilings, including the use of compression posts, are not directly applicable to these systems, primarily because of their behavior as a continuous diaphragm and their greater mass. As such, they require more attention to design and detailing, in particular for the attachment of the hanger wires to the structure and main carriers, the attachment of the cross-furring channels to main carriers, and the attachment of lath to cross-furring channels. Attention should also be given to the attachment of light fixtures and diffusers to the ceiling structure. Bracing should consider both horizontal and vertical movement of the ceiling, as well as discontinuities and offsets. The seismic design and detailing of lath-and-plaster ceilings should use rational engineering methods to transfer seismic design ceiling forces to the building structural elements.

The performance of ceiling systems is affected by the placement of seismic bracing and the layout of light fixtures and other supported loads. Dynamic testing has demonstrated that splayed wires, even with vertical compression struts, may not adequately limit lateral motion of the ceiling system caused by straightening of the end loops. Construction problems include slack installation or omission of bracing wires caused by obstructions. Other testing has shown that unbraced systems may perform well where the system can accommodate the expected displacements, by providing both sufficient clearance at penetrations and wide closure members, which are now required by the standard.

With reference to the exceptions in Section 13.5.6,

- The first exemption is based on the presumption that lateral support is accomplished by the surrounding walls for areas equal to or less than 144 ft^2 (13.4 m^2) (e.g., a 12 ft × 12 ft (3.7 m × 3.7 m) room). The 144 ft^2 (13.4 m^2) limit corresponds historically to an assumed connection strength of 180 lb (4.5 N) and forces associated with requirements for suspended ceilings that first appeared in the 1976 Uniform Building Code (ICC 1976).
- The second exemption assumes that planar, horizontal drywall ceilings behave as diaphragms (i.e., develop in-plane strength). This assumption is supported by the performance of drywall ceilings in past earthquakes.

C13.5.6.1 Seismic Forces Where the weight of the ceiling system is distributed nonuniformly, that condition should be considered in the design, because the typical T-bar ceiling grid has limited ability to redistribute lateral loads.

C13.5.6.2 Industry Standard Construction for Acoustical Tile or Lay-In Panel Ceilings The key to good seismic performance is sufficiently wide closure angles at the perimeter to accommodate relative ceiling motion, and adequate clearance at penetrating components (such as columns and piping) to avoid concentrating restraining loads on the ceiling system.

Table C13.5-1 provides an overview of the combined requirements of ASCE 7 and ASTM E580 (2014a). Careful review of both documents is required to determine the actual requirements.

C13.5.6.2.1 Seismic Design Category C The prescribed method for Seismic Design Category C is a floating ceiling. The design assumes a small displacement of the building structure caused by the earthquake at the ceiling and isolates the ceiling from the perimeter. The vertical hanger wires are not capable of transmitting significant movement or horizontal force into the ceiling system, and therefore the ceiling does not experience significant force or displacement as long as the perimeter gap is not exceeded. All penetrations and services must be isolated from the building structure for this construction method to be effective. If this isolation is impractical or undesirable, the prescribed construction for Seismic Design Categories D, E, and F may be used.

C13.5.6.2.2 Seismic Design Categories D through F The industry-standard construction addressed in this section relies on ceiling contact with the perimeter wall for restraint.

Typical splay wire lateral bracing allows some movement before it effectively restrains the ceiling. The intent of the 2 in. (51 mm) perimeter closure wall angle is to permit back-and-forth motion of the ceiling during an earthquake without loss of support, thus the width of the closure angle is important to good performance. This standard has been experimentally verified by large-scale testing conducted by ANCO Engineers in 1983.

Extensive shake table testing by major manufacturers of suspended ceilings using the protocol in ICC-ES AC156 has been used to justify the use of perimeter clips designed to accommodate the same degree of movement as the closure angle while supporting the tee ends. These tests are conducted on 16 ft × 16 ft (4.9 m × 4.9 m) ceiling installations. Testing on larger ceiling systems reported by Rahmanishamsi et al. (2014) and Soroushian et al. (2012, 2014) indicates that the use of approved perimeter clips may lead to damage to the grid members and seismic clips, crushing of wall angles, and deformation of grid latches at moderate ground motion levels if the grid member loses contact with the horizontal leg of the closure angle or channel. A requirement has been added to screw the clips to the closure angle or channel to prevent this type of damage.

The requirement for a 1 in. (25 mm) clearance around sprinkler drops in Section 13.5.6.2.2 (e) of ASCE 7-05 is maintained and is contained in ASTM E580.

This seismic separation joint is intended to break the ceiling into isolated areas, preventing large-scale force transfer across the ceiling. The new requirement to accommodate 0.75 in. (19 mm) axial movement specifies the performance requirement for the separation joint.

The requirement for seismic separation joints to limit ceiling areas to 2,500 ft^2 (230 m^2) is intended to prevent overload of the connections to the perimeter angle. Limiting the ratio of long to short dimensions to 4:1 prevents dividing the ceiling into long and narrow sections, which could defeat the purpose of the separation.

C13.5.6.3 Integral Construction Ceiling systems that use integral construction are constructed of modular pre-engineered components that integrate lights, ventilation components, fire sprinklers, and seismic bracing into a complete system. They may include aluminum, steel, and PVC components and may be designed using integral construction of ceiling and wall. They often use rigid grid and

Table C13.5-1. Summary of Requirements for Acoustical Tile or Lay-In Panel Ceilings.

Item	Seismic Design Category C	Seismic Design Categories D, E, and F
Area less than or equal to 144 ft^2 (13 m^2)		
N/A	No requirements (§1.4)	No requirements (§1.4)
Area greater than 144 ft^2 (13 m^2) but less than or equal to 1,000 ft^2 (93 m^2)		
Duty rating	Only intermediate or heavy-duty load rated grid as defined by ASTM C635 (2004) may be used for commercial ceilings (ASTM C635 sections 4.1.3.1, 4.1.3.2, 4.1.3.3)	Heavy-duty load rating as defined in ASTM C635 is required (§5.1.1)
Grid connections	Minimum main tee connection and cross tee intersection strength of 60 lb (27 kg) (§4.1.2)	Minimum main tee connection and cross tee intersection strength of 180 lb (4.5 N) (§5.1.2)
Vertical suspension wires	Vertical hanger wires must be a minimum of 12 gauge (§4.3.1)	Vertical hanger wire must be a minimum of 12 gauge (§5.2.7.1)
	Vertical hanger wires maximum 4 ft (1.2 m) on center (§4.3.1)	Vertical hanger wires maximum 4 ft (1.2 m) on center (§5.2.7.1)
	Vertical hanger wires must be sharply bent and wrapped with three turns in 3 in. (76 mm) or less (§4.3.2)	Vertical hanger wires must be sharply bent and wrapped with three turns in 3 in. (76 mm) or less (§5.2.7.2)
	All vertical hanger wires may not be more than 1 in 6 out of plumb without having additional wires counter-splayed (§4.3.3)	All vertical hanger wires may not be more than 1 in 6 out of plumb without having additional wires counter-splayed (§5.2.7.3)
	Any connection device from the vertical hanger wire to the structure above must sustain a minimum load of 90 lb (41 kg) (§4.3.2)	Any connection device from the vertical hanger wire to the structure above must sustain a minimum load of 90 lb (41 kg) (§5.2.7.2)
	Wires may not attach to or bend around interfering equipment without the use of trapezes (§4.3.4)	Wires may not attach to or bend around interfering equipment without the use of trapezes (§5.2.7.4)
Lateral bracing	Lateral bracing is not permitted; ceiling is intended to "float" relative to balance of structure; tee connections may be insufficient to maintain integrity if braces were included, see section C13.5.6.2.1	Not required under 1,000 ft^2 (93 m^2), where perimeter and tee connections are presumed to be strong enough to maintain integrity whether bracing is installed or not (§5.2.8.1)
Perimeter	Perimeter closure (molding) width must be a minimum of 7/8 in. (22 mm) (§4.2.2)	Perimeter closure (molding) width must be a minimum of 2 in. (51 mm) (§5.2.2)
	Perimeter closures with a support ledge of less than 7/8 in. (22 mm) shall be supported by perimeter vertical hanger wires not more than 8 in. (200 mm) from the wall (§4.2.3)	Two adjacent sides must be connected to the wall or perimeter closure (§5.2.3)
	A minimum clearance of 3/8 in. (9.5 mm) must be maintained on all four sides (§4.2.4)	A minimum clearance of 3/4 in. (19 mm) must be maintained on the other two adjacent sides (§5.2.3)
	Permanent attachment of grid ends is not permitted (§4.2.6)	Perimeter tees must be supported by vertical hanger wires not more than 8 in. (200 mm) from the wall (§5.2.6)
	Perimeter tee ends must be prevented from spreading (§4.2.5)	Perimeter tee ends must be prevented from spreading (§5.2.4)
Light fixtures	Light fixtures must be positively attached to the grid by at least two connections, each capable of supporting the weight of the lighting fixture (§4.4.1 and NEC)	Light fixtures must be positively attached to the grid by at least two connections, each capable of supporting the weight of the lighting fixture (§5.3.1) (NEC)
	Surface-mounted light fixtures shall be positively clamped to the grid (§4.4.2)	Surface-mounted light fixtures shall be positively clamped to the grid (§5.3.2)
	Clamping devices for surface-mounted light fixtures shall have safety wires to the ceiling hanger or to the structure above (§4.4.2)	Clamping devices for surface-mounted light fixtures shall have safety wires to the ceiling wire or to the structure above (§5.3.2)
	Light fixtures and attachments weighing up to 10 lb (4.5 kg) require one 12-gauge (minimum) hanger wire connected to the housing (e.g., canister light fixture); this wire may be slack (§4.4.3)	When cross tees with a load-carrying capacity of less than 16 lb/ft (24 kg/m) are used, supplementary hanger wires are required (§5.3.3)
	Light fixtures that weigh more than 10 lb (4.5 kg) and up to 56 lb (25 kg) require two 12-gauge (minimum) hanger wires connected to the housing; these wires may be slack (§4.4.4)	Light fixtures and attachments weighing up to 10 lb (4.5 kg) require one 12-gauge minimum hanger wire connected to the housing and connected to the structure above; this wire may be slack (§5.3.4)
	Light fixtures that weigh more than 56 lb (25 kg) require independent support from the structure (§4.4.5)	Light fixtures that weigh more than 10 lb (4.5 kg) and up to 56 lb (25 kg) require two 12-gauge minimum hanger wires attached to the fixture housing and connected to the structure above; these wires may be slack (§5.3.5)
	Pendant-hung light fixtures shall be supported by a minimum 9-gauge wire or other approved alternative (§4.4.6)	Light fixtures that weigh more than 56 lb (25 kg) require independent support from the structure by approved hangers (§5.3.6)

continues

Table 13.5-1 *(Continued).*

Item	Seismic Design Category C	Seismic Design Categories D, E, and F
	Rigid conduit is not permitted for the attachment of fixtures (§4.4.7)	Pendant-hung light fixtures shall be supported by a minimum 9-gauge wire or other approved support (§5.3.7) Rigid conduit is not permitted for the attachment of fixtures (§5.3.8)
Mechanical services	Flexibly mounted mechanical services weighing up to 20 lb (89 N) must be positively attached to main runners or cross runners with the same load-carrying capacity as the main runners (§4.5.1) Flexibly mounted mechanical services weighing more than 20 lb (89 N) and up to 56 lb (25 kg) must be positively attached to main runners or cross runners with the same load-carrying capacity as the main runners and require two 12-gauge (minimum) hanger wires; these wires may be slack (§4.5.2) Flexibly mounted mechanical services heavier than 56 lb (25 kg) require direct support from the structure (§4.5.3)	Flexibly mounted mechanical services weighing up to 20 lb (89 N) must be positively attached to main runners or cross runners with the same load-carrying capacity as the main runners (§5.4.1) Flexibly mounted mechanical services weighing more than 20 lb (89 N) and up to 56 lb (25 kg) must be positively attached to main runners or cross runners with the same load-carrying capacity as the main runners and require two 12-gauge (minimum) hanger wires; these wires may be slack (§5.4.2) Flexibly mounted mechanical services heavier than 56 lb (25 kg) require direct support from the structure (§5.4.3)
Supplemental requirements	All ceiling penetrations must have a minimum of 3/8 in. (9.5 mm) clearance on all sides (§4.2.4)	Direct concealed systems must have stabilizer bars or mechanically connected cross tees a maximum of 60 in. (1.5 m) on center with stabilization within 24 in. (61 cm) of the perimeter (§5.2.5) Bracing is required for ceiling plane elevation changes (§5.2.8.6) Cable trays and electrical conduits shall be supported independently of the ceiling (§5.2.8.7) All ceiling penetrations and independently supported fixtures or services must have closures that allow for a 1 in. (25 mm) movement (§5.2.8.5) An integral ceiling sprinkler system may be designed by the licensed design professional to eliminate the required spacing of penetrations (§5.2.8.8) A licensed design professional must review the interaction of nonessential ceiling components with essential ceiling components to prevent the failure of the essential components (§5.7.1)
Partitions	The ceiling may not provide lateral support to partitions (§4.6.1) Partitions attached to the ceiling must use flexible connections to avoid transferring force to the ceiling (§4.6.1)	Partition attached to the ceiling and all partitions greater than feet in height shall be laterally braced to the building structure; this bracing must be independent of the ceiling (§5.5.1)
Exceptions	The ceiling must weigh less than 2.5 lb/ft^2 (12 kg/m^2); otherwise the prescribed construction for Seismic Design Categories D, E, and F must be used (§4.1.1)	None
Area greater than 1,000 ft^2 (93 m^2) but less than or equal to 2,500 ft^2 (230 m^2)		
Lateral bracing	No additional requirements	Lateral force bracing (four 12-gauge splay wires) is required within 2 in. (51 mm) of main tee/cross tee intersection and splayed 90 degrees apart in the plan view, at a maximum 45-degree angle from the horizontal and located 12 ft (3.7 m) on center in both directions, starting 6 ft (1.8 m) from walls (§5.2.8.1 & §5.2.8.2) Lateral force bracing must be spaced a minimum of 6 in. (15 cm) from unbraced horizontal piping or ductwork (§5.2.8.3) Lateral force bracing connection strength must be a minimum of 250 lb (113 kg) (§5.2.8.3) Rigid bracing designed to limit deflection at the point of attachment to less than 0.25 in. (6.4 mm) may be used in place of splay wires.

continues

Table 13.5-1 *(Continued)*.

Item	Seismic Design Category C	Seismic Design Categories D, E, and F
		Unless rigid bracing is used or calculations have shown that lateral deflection is less than 0.25 in. (6.4 mm), sprinkler heads and other penetrations shall have a minimum of 1 in. (25 mm) clear space in all directions (§5.2.8.5)
Area greater than 2,500 ft^2 (230 m^2)		
Special considerations	No additional requirements	Seismic separation joints with a minimum or 3/4 in. (19 mm) axial movement, bulkhead, or full-height partitions with the usual 2 in. (51 mm) closure and other requirements (§5.2.9.1) Areas defined by seismic separation joints, bulkheads, or full-height partitions must have a ratio of long to short dimensions of less than or equal to 4 (§5.2.9.1)

Notes: There are no requirements for suspended ceilings located in structures assigned to Seismic Design Categories A and B. Unless otherwise noted, all section references (§) refer to ASTM E580 (2014).

bracing systems, which provide lateral support for all the ceiling components, including sprinkler drops. This bracing reduces the potential for adverse interactions among components and eliminates the need to provide clearances for differential movement.

C13.5.7 Access Floors

C13.5.7.1 General In past earthquakes and in cyclic load tests, some typical raised access floor systems behaved in a brittle manner and exhibited little reserve capacity beyond initial yielding or failure of critical connections. Testing shows that unrestrained individual floor panels may pop out of the supporting grid unless they are mechanically fastened to supporting pedestals or stringers. This condition may be a concern, particularly in egress pathways.

For systems with floor stringers, it is accepted practice to calculate the seismic force, F_p, for the entire access floor system within a partitioned space and then distribute the total force to the individual braces or pedestals. For stringerless systems, the seismic load path should be established explicitly.

Overturning effects subject individual pedestals to vertical loads well in excess of the weight, W_p, used in determining the seismic force, F_p. It is unconservative to use the design vertical load simultaneously with the design seismic force for design of anchor bolts, pedestal bending, and pedestal welds to base plates. "Slip-on" heads that are not mechanically fastened to the pedestal shaft and thus cannot transfer tension are likely unable to transfer to the pedestal the overturning moments generated by equipment attached to adjacent floor panels.

To preclude brittle failure, each element in the seismic load path must have energy-absorbing capacity. Buckling failure modes should be prevented. Lower seismic force demands are allowed for special access floors that are designed to preclude brittle and buckling failure modes.

C13.5.7.2 Special Access Floors An access floor can be a "special access floor" if the registered design professional opts to comply with the requirements of Section 13.5.7.2. Special access floors include construction features that improve the performance and reliability of the floor system under seismic loading. The provisions focus on providing an engineered load path for seismic shear and overturning forces. Special access floors are designed for smaller lateral forces, and their use is encouraged at facilities with higher nonstructural performance objectives.

C13.5.8 Partitions Partitions subject to these requirements must have independent lateral support bracing from the top of the partition to the building structure or to a substructure attached to the building structure. Some partitions are designed to span vertically from the floor to a suspended ceiling system. The ceiling system must be designed to provide lateral support for the top of the partition. An exception to this condition is provided to exempt bracing of light (gypsum board) partitions where the load does not exceed the minimum partition lateral load. Experience has shown that partitions subjected to the minimum load can be braced to the ceiling without failure.

C13.5.9 Glass in Glazed Curtain Walls, Glazed Storefronts, and Glazed Partitions The performance of glass in earthquakes falls into one of four categories:

1. The glass remains unbroken in its frame or anchorage.
2. The glass cracks but remains in its frame or anchorage while continuing to provide a weather barrier and to be otherwise serviceable.
3. The glass shatters but remains in its frame or anchorage in a precarious condition, likely to fall out at any time.
4. The glass falls out of its frame or anchorage, either in shards or as whole panels.

Categories 1 and 2 satisfy both Immediate Occupancy and Life Safety Performance Objectives. Although the glass is cracked in Category 2, immediate replacement is not required. Categories 3 and 4 cannot provide for immediate occupancy, and their provision of life safety depends on the post-breakage characteristics of the glass and the height from which it can fall. Tempered glass shatters into multiple, pebble-size fragments that fall from the frame or anchorage in clusters. These broken glass clusters are relatively harmless to humans when they fall from limited heights, but they could be harmful when they fall from greater heights.

C13.5.9.1 General Equation (13.5-2) is derived from Sheet Glass Association of Japan (1982) and is similar to an equation in Bouwkamp and Meehan (1960) that permits calculation of the story drift required to cause glass-to-frame contact in a given rectangular window frame. Both calculations are based on the principle that a rectangular window frame (specifically, one that is anchored mechanically to adjacent stories of a structure) becomes a parallelogram as a result of story drift, and that glass-to-frame contact occurs when the length of the shorter diagonal of the parallelogram is equal to the diagonal of the glass panel itself. The variable $\Delta_{fallout}$ represents the displacement capacity of the system, and D_p represents the displacement demand.

The 1.25 factor in these requirements described above reflects uncertainties associated with calculated inelastic seismic displacements of building structures. Wright (1989) states that post-elastic deformations calculated using the structural analysis process may well underestimate the actual building deformation by up to 30%. It would therefore be reasonable to require the curtain wall glazing system to withstand 1.25 times the computed maximum interstory displacement to verify adequate performance.

The reason for the second exception to Equation (13.5-2) is that the tempered glass, if shattered, would not produce an overhead falling hazard to adjacent pedestrians, although some pieces of glass may fall out of the frame.

C13.5.9.2 Seismic Drift Limits for Glass Components As an alternative to the prescriptive approach of Section 13.5.9.1, the deformation capacity of glazed curtain wall systems may be established by test.

C13.5.10 Egress Stairs and Ramps In the Christchurch earthquake of February 22, 2011, several buildings using precast concrete stairs with a sliding joint at one end experienced stair collapse (Canterbury Earthquakes Royal Commission 2012). In one notable case, the 18-story Forsyth Barr office building, the structure was otherwise largely undamaged. In all cases, the primary cause of the stair collapse was loss of vertical bearing at the end connection due to building drift that exceeded the support detail capacity. In general, these stairs were intended to serve as egress routes, and occupants were trapped in some of these buildings following the earthquake. In US practice, precast stairs (Figure C13.5-1) are less common than steel-framed stairs (Figure C13.5-2), which are generally considered to be inherently flexible. But in shake table tests conducted at the University of California, San Diego, as part of the Network for Earthquake Engineering Simulation project, "Full-Scale Structural and Nonstructural Building System Performance during Earthquakes and Post-Earthquake Fire" (2011), connections of the commercial metal stair included in the test structure were shown to be brittle and susceptible to damage. Considering the critical nature of egress for life safety, specific attention to the ability of egress stairs to accept building drift demands is warranted. Effective sliding joints in typical steel stairs are complex to design and construct. Ductile connections, capable of accepting the drift without loss of vertical load-carrying capacity, are often preferred. In such cases, sufficient ductility must be provided in these connections to accommodate multiple cycles at anticipated maximum drift levels. If drift is to be accommodated with full sliding connections lacking a fail-safe stop, additional length of bearing is required to prevent collapse where displacements exceed design levels. Where stair systems are rigidly attached to the structure, they must be included in the structure model, and the resultant forces must be accommodated, with overstrength, in the stair design.

Figure C13.5-1. Precast stair.
Source: Courtesy of Tindall Corp.

Figure C13.5-2. Steel-framed exit stair.
Source: Courtesy of Tara Hutchinson.

These requirements do not apply to egress stair systems and ramps that are integral with the building structure, since it is assumed that the seismic resistance of these systems is addressed in the overall building design. Examples include stairs and ramps comprising monolithic concrete construction, light-frame wood and cold-formed metal stair systems in multiunit residential construction, and integrally constructed masonry stairs.

C13.5.11 Penthouses and Rooftop Structures Penthouses and rooftop structures can vary from small enclosures at the top of stairs and elevators to large structures covering 30% of the roof area. In the 2016 and earlier editions of ASCE 7, penthouses were designed to the requirements of Chapter 13 as a nonstructural component, without any restrictions on the design of their lateral force-resisting systems. For ASCE 7-22, the requirements for rooftop structures have been revised. The seismic design force for rooftop structures is now dependent on the lateral force-resisting system selected for the rooftop structure. Force-resisting systems may be selected from Chapter 12 or 15, and are subject to the system limitations and detailing requirements for the system selected. The seismic design force is determined in Section 13.3, using design coefficients obtained from Table 13.6-1. The values of these design coefficients depend on the value of R listed in Chapter 12 or 15 for the force-resisting system chosen. Chapter 15 permits the use of lower-ductility force-resisting systems in regions of high seismicity, but their use will result in higher lateral forces.

An option for the use of an undefined force-resisting system is provided, with a height limit of 28 ft (8.5 m) above the roof deck. The height limit was applied to penthouses designed using an undefined lateral system, since such a penthouse located in regions of high seismicity could potentially have low ductility. Height limits are applied to low-ductility lateral systems permitted in Chapters 12 and 15, and such systems are in some cases limited to a single story. A height limit of 28 ft (8.5 m) was selected to permit a penthouse with an undefined lateral system on buildings other than Type 1 construction, which could be used to enclose tanks or elevators that travel to the roof level. This height limit is consistent with the requirements of Section 1510.2 of the 2018 International Building Code. The exception for penthouses and rooftop structures framed by an extension of the building frame and designed in accordance with Chapter 12 was retained.

C13.6 MECHANICAL AND ELECTRICAL COMPONENTS

These requirements, focused on design of supports and attachments, are intended to reduce the hazard to life posed by loss of component structural stability or integrity. The requirements increase the reliability of component operation but do not directly address functionality. For critical components where operability is vital, Section 13.2.3 provides methods for seismically qualifying the component.

Traditionally, mechanical components (such as tanks and heat exchangers) without rotating or reciprocating components are directly anchored to the structure. Mechanical and electrical equipment components with rotating or reciprocating elements are often isolated from the structure by vibration isolators (using rubber acting in shear, springs, or air cushions). Heavy mechanical equipment (such as large boilers) may not be restrained at all, and electrical equipment other than generators, which are normally isolated to dampen vibrations, usually is rigidly anchored (for example, switch gear and motor control centers).

Two distinct levels of earthquake safety are considered in the design of mechanical and electrical components. At the usual safety level, failure of the mechanical or electrical component itself because of seismic effects poses no significant hazard. In this case, design of the supports and attachments to the structure is required to avoid a life-safety hazard. At the higher safety level, the component must continue to function acceptably after the design earthquake. Such components are defined as designated seismic systems in Section 11.2 and may be required to meet the special certification requirements of Section 13.2.3.

Not all equipment or parts of equipment need to be designed for seismic forces. Where I_p is specified to be 1.0, damage to, or even failure of, a piece or part of a component does not violate these requirements as long as a life-safety hazard is not created. The restraint or containment of a falling, breaking, or toppling component (or its parts) by means of bumpers, braces, guys, wedges, shims, tethers, or gapped restraints to satisfy these requirements often is acceptable, although the component itself may suffer damage.

Judgment is required to fulfill the intent of these requirements; the key consideration is the threat to life safety. For example, a nonessential air handler package unit that is less than 4 ft (1.2 m) tall and bolted to a mechanical room floor is not a threat to life as long as it is prevented from significant displacement by having adequate anchorage. In this case, seismic design of the air handler itself is unnecessary. However, a 10 ft (3.0 m) tall tank on 6 ft (1.8 m) long angles used as legs, mounted on a roof near a building exit, does pose a hazard. The intent of these requirements is that the supports and attachments (tank legs, connections between the roof and the legs, and connections between the legs and the tank), and possibly even the tank itself, be designed to resist seismic forces. Alternatively, restraint of the tank by guys or bracing could be acceptable.

It is not the intent of the standard to require the seismic design of shafts, buckets, cranks, pistons, plungers, impellers, rotors, stators, bearings, switches, gears, non-pressure-retaining casings and castings, or similar items. Where the potential for a hazard to life exists, the design effort should focus on equipment supports, including base plates, anchorages, support lugs, legs, feet, saddles, skirts, hangers, braces, and ties.

Many mechanical and electrical components consist of complex assemblies of parts that are manufactured in an industrial process that produces similar or identical items. Such equipment may include manufacturers' catalog items and often is designed by empirical (trial-and-error) means for functional and transportation loadings. A characteristic of such equipment is that it may be inherently rugged. The term "rugged" refers to an ampleness of construction that provides such equipment with the ability to survive strong motions without significant loss of function. By examining such equipment, an experienced design professional usually should be able to confirm such ruggedness. The results of an assessment of equipment ruggedness may be used in determining an appropriate method and extent of seismic design or qualification effort.

The revisions to Table 13.6-1 in ASCE 7-10 were the result of work done in recent years to better understand the performance of mechanical and electrical components and their attachment to the structure. The primary concepts of flexible and rigid equipment and ductile and rugged behavior are drawn from SEAOC (1999), Commentary Section C107.1.7. Material on HVACR is based on ASHRAE (2000). Other material on industrial piping, boilers, and pressure vessels is based on the American Society of Mechanical Engineers' codes and standards publications (ASME 2007, 2010a, 2010b).

C13.6.1 General The seismic demand requirements are based on component structural attributes of flexibility (or rigidity) and ruggedness. Table 13.6-1 provides seismic coefficients based on judgments of component ductility, reserve strength, and likelihood of being in resonance. The properties of the attachment system that provides seismic restraint may also influence seismic demand.

Entries for components and systems in Table 13.6-1 are grouped and described to improve clarity of application. Components are divided into three broad groups, within which they are further classified depending on the type of construction or expected seismic behavior. For example, mechanical components include "air-side" components (such as fans and air handlers) that experience dynamic amplification but are light and deformable; "wet-side" components, which generally contain liquids (such as boilers and chillers) and are more rigid and somewhat ductile; and rugged components (such as engines, turbines, and pumps) that are of massive construction because of demanding operating loads and that generally perform well in earthquakes, if adequately anchored.

C13.6.2 Mechanical Components and C13.6.3 Electrical Components Most mechanical and electrical equipment is inherently rugged and, where properly attached to the structure, has performed well in past earthquakes. Because the operational and transportation loads for which the equipment is

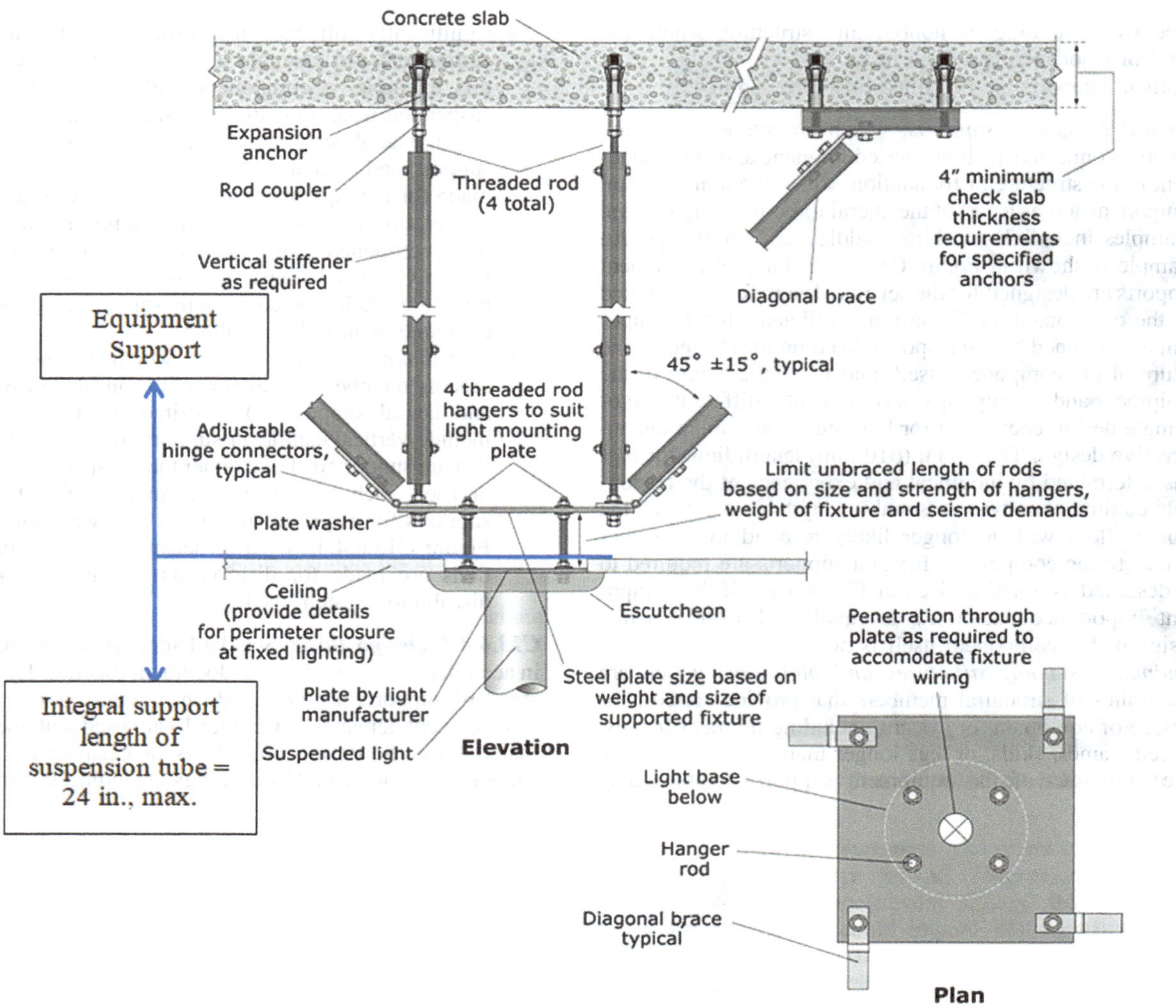

Figure C13.6.4.2. Example of an equipment support structure for a heavy light fixture.

Source: FEMA E-74 (2015).

Note: 1 in. = 25 mm.

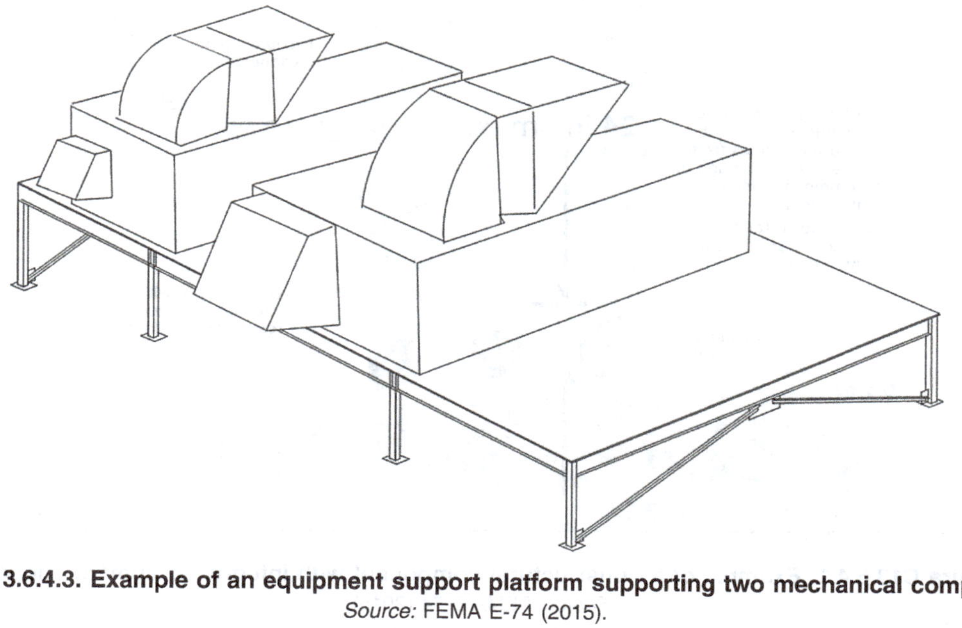

Figure C13.6.4.3. Example of an equipment support platform supporting two mechanical components.

Source: FEMA E-74 (2015).

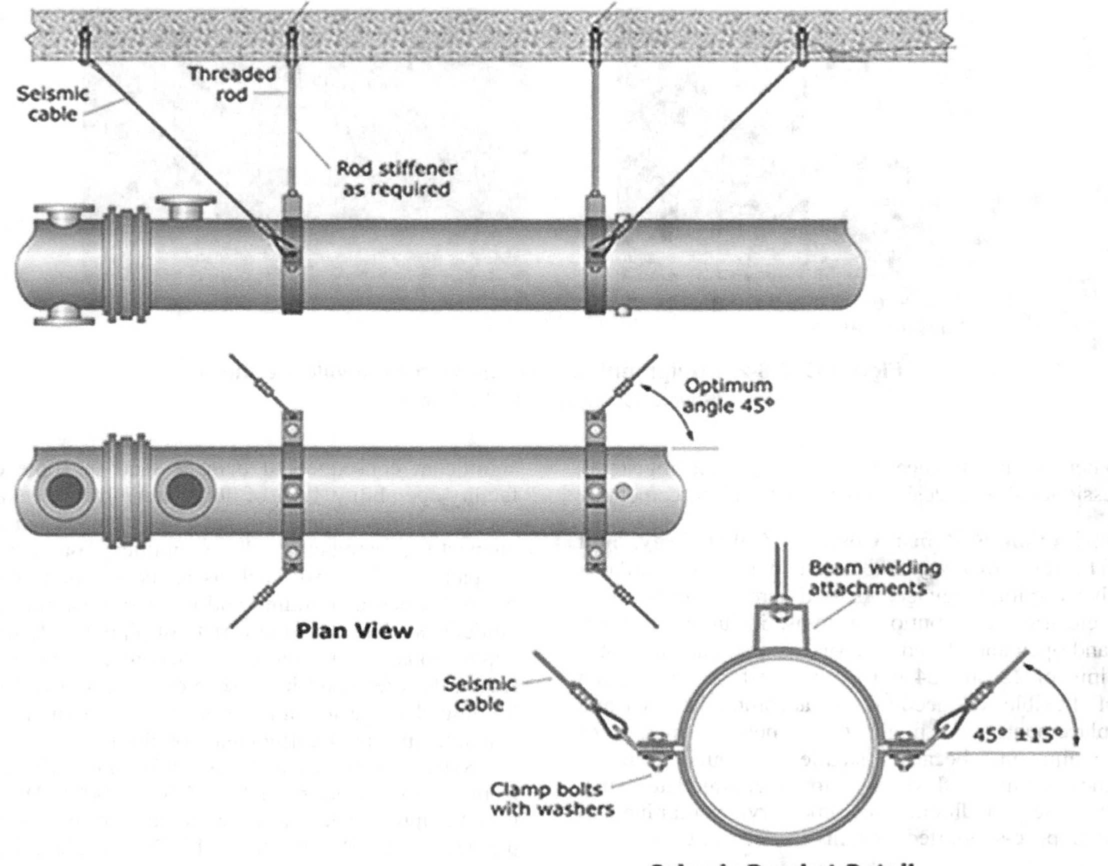

Figure C13.6.4.4. Example of a distribution system support for piping.
Source: FEMA E-74 (2015).

such as snubbers, isolators, and sway bracing, are typically based on testing in accordance with nationally recognized standards such as ANSI/FM 1950 (2016), ANSI/ASHRAE 171 (2017), or calculations based on material-specific design standards. Alternative test methods or design procedures may be acceptable to the Authority Having Jurisdiction.

C13.6.4.2 Design for Relative Displacement For some items, such as piping, seismic relative displacements between support points are of more significance than inertial forces. Components made of high-deformability materials such as steel or copper can accommodate relative displacements inelastically, provided the connections also provide high deformability. Threaded and soldered connections exhibit poor ductility under inelastic displacements, even for ductile materials. Components made of less ductile materials can accommodate relative displacement effects only if appropriate flexibility or flexible connections are provided.

Detailing distribution systems that connect separate structures with bends and elbows makes them less prone to damage and less likely to fracture and fall, provided the supports can accommodate the imposed loads.

C13.6.4.3 Support Attachment to Component As used in this section, "integral" relates to the manufacturing process, not the location of installation. For example, both the legs of a cooling tower and the attachment of the legs to the body of the cooling tower must be designed, even if the legs are provided by the manufacturer and installed at the plant. Also, if the cooling tower has an I_p of 1.5, the design must address not only the attachments (e.g., welds and bolts) of the legs to the component but also local stresses imposed on the body of the cooling tower by the support attachments.

C13.6.4.5 Additional Requirements As reflected in this section of the standard and in note *b* to Table 13.6-1, vibration-isolated equipment with snubbers is subject to amplified loads as a result of dynamic impact.

Most sheet metal connection points for seismic anchorage do not exhibit the same mechanical properties as bolted connections with structural elements. The use of Belleville washers improves the seismic performance of connections to equipment enclosures fabricated from sheet metal 7 gauge [0.18 in. (5 mm)] or thinner by distributing the stress over a larger surface area of the sheet metal connection interface, allowing bolted connections to be torqued to recommended values for proper preload while reducing the tendency for local sheet metal tearing or bending failures or loosening of the bolted connection (Figure C13.6-2). The intrinsic spring loading capacity of the Belleville washer assists with long-term preload retention to maintain integrity of the seismic anchorage.

Manufacturers test or design their equipment to handle seismic loads at the equipment "hard points" or anchor locations. The results of this design qualification effort are typically reflected in installation instructions provided by the manufacturer. It is imperative that the manufacturer's installation instructions be

Failure of sheet metal base anchored with standard washer

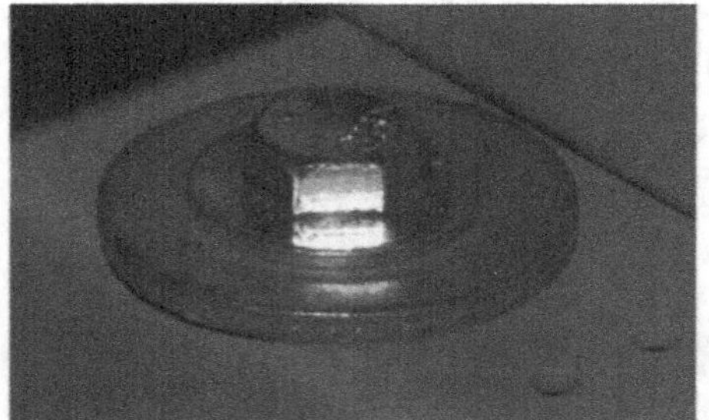

Anchorage equipped with Belleville washer

Figure C13.6-2. Equipment anchorage with Belleville washers.
Source: Courtesy of Philip Caldwell.

followed. Where such guidance does not exist, the registered design professional should design appropriate reinforcement.

C13.6.5 Distribution Systems: Conduit, Cable Tray, and Raceways The term *raceway* is defined in several standards with somewhat varying language. As used here, it is intended to describe all electrical distribution systems, including conduit, cable trays, and open and closed raceways. Experience indicates that a size limit of 2.5 in. (64 mm) can be established for the provision of flexible connections to accommodate seismic relative displacements that might occur between pieces of connected equipment, because smaller conduit normally possesses the required flexibility to accommodate such displacements. See additional commentary pertaining to exemption of trapeze-supported systems in Section C13.1.4.

C13.6.6 Distribution Systems: Duct Systems Experience in past earthquakes has shown that HVACR duct systems are rugged and perform well in strong ground shaking. Bracing in accordance with ANSI/SMACNA 001 (2000) has been effective in limiting damage to duct systems. Typical failures have affected only system function, and major damage or collapse has been uncommon. Therefore, industry-standard practices should be adequate for most installations. Expected earthquake damage is limited to opening of duct joints and tears in ducts. Connection details that are prone to brittle failures, especially hanger rods subject to large amplitude cycles of bending stress, should be avoided. See additional commentary in Section C13.1.4.

Duct systems that carry hazardous materials or must remain operational during and after an earthquake are assigned a value of $I_p = 1.5$, and they require a detailed engineering analysis addressing leak tightness.

Lighter in-line components may be designed to resist the forces from Section 13.3 as part of the overall duct system design, whereby the duct attached to the in-line component is explicitly designed for the forces generated by the component. Where in-line components are more massive, the component must be supported and braced independently of the ductwork to avoid failure of the connections.

The requirement for flexible connections of unbraced piping to in-line components, such as reheat coils, applies regardless of the component weight.

C13.6.7 Distribution Systems: Piping and Tubing Systems Because of the typical redundancy of piping system supports, documented cases of total collapse of piping systems in earthquakes are rare; however, pipe leakage resulting from excessive displacement or overstress often results in significant consequential damage, and in some cases loss of facility operability. Loss of fluid containment (leakage) normally occurs at discontinuities such as threads, grooves, bolted connectors, geometric discontinuities, or locations where incipient cracks exist, such as at the toe or root of a weld or braze. Numerous building and industrial national standards and guidelines address a wide variety of piping system materials and applications. Construction in accordance with the national standards referenced in these provisions is usually effective in limiting damage to piping systems and avoiding loss of fluid containment under earthquake conditions.

ASHRAE (2000) and MSS (2001) are derived in large part from the predecessors of SMACNA (2008). These documents may be appropriate references for use in the seismic design of piping systems. Because the SMACNA standard does not refer to pipe stresses in the determination of hanger and brace spacing, however, a supplementary check of pipe stresses may be necessary when this document is used. ASME piping rules as given in ASME (2010a) and ASME B31, Parts 1, 3, 5, 9, and 12, are normally used for high-pressure, high-temperature piping but can also conservatively be applied to lower-pressure, lower-temperature piping systems. Code-compliant seismic design manuals prepared specifically for proprietary systems may also be appropriate references.

Although seismic design in accordance with Section 13.6.8 generally ensures that effective seismic forces do not fail piping, seismic displacements may be underestimated such that impact with nearby structural, mechanical, or electrical components could occur. In marginal cases, it may be advisable to protect the pipe with wrapper plates where impacts could occur, including at gapped supports. Insulation may in some cases also serve to protect the pipe from impact damage. Piping systems are typically designed for pressure containment, and piping designed with a factor of safety of 3 or more against pressure failure (rupture) may be inherently robust enough to survive impact with nearby structures, equipment, and other piping, particularly if the piping is insulated. Piping that has less than standard-weight wall thickness may require the evaluation of the effects of impact locally on the pipe wall and may necessitate means to protect the pipe wall.

It is usually preferable for piping to be detailed to accommodate seismic relative displacements between the first seismic support upstream or downstream from connections and other seismically supported components or headers. This accommodation is preferably achieved by means of pipe flexibility or, where pipe flexibility is not possible, flexible supports. Piping not otherwise detailed to accommodate such seismic relative displacements must be provided with connections that have sufficient flexibility in the connecting element or in the component or header to avoid failure of the piping. The option to use a flexible connecting element may be less desirable because of the need for greater maintenance efforts to ensure continued proper function of the flexible element.

Grooved couplings, ball joints, resilient gasket compression fittings, other articulating-type connections, bellows expansion joints, and flexible metal hose are used in many piping systems and can serve to increase the rotational and lateral deflection design capacity of the piping connections.

Grooved couplings are classified as either rigid or flexible. Flexible grooved couplings demonstrate limited free rotational capacity. The free rotational capacity is the maximum articulating angle where the connection behaves essentially as a pinned joint, with limited or negligible stiffness. The remaining rotational capacity of the connection is associated with conventional joint behavior, and design force demands in the connection are determined by traditional means.

Rigid couplings are typically used for high-pressure applications and usually are assumed to be stiffer than the pipe. Alternatively, rigid coupling may exhibit bilinear rotational stiffness, with the initial rotational stiffness affected by installation.

Coupling flexibilities vary significantly between manufacturers, particularly for rigid couplings. Manufacturer's data may be available. Industrywide procedures for the determination of coupling flexibility are not currently available; however, some guidance for couplings may be found in the provisions for fire sprinkler piping, where grooved couplings are classified as either rigid or flexible on the basis of specific requirements on angular movement. In Section 3.5.4 of NFPA (2007), flexible couplings are defined as follows:

> *A listed coupling or fitting that allows axial displacement, rotation, and at least 1 degree of angular movement of the pipe without inducing harm on the pipe. For pipe diameters of 8 in. (203.2 mm) and larger, the angular movement shall be permitted to be less than 1 degree but not less than 0.5 degrees.*

Couplings determined to be flexible on this basis are listed with either FM Global (2007) or UL (2004).

Piping component testing suggests that the ductility capacity of carbon steel threaded and flexible grooved piping component joints ranges between 1.4 and 3.0, implying an effective stress intensification of approximately 2.5. These types of connections have been classified as having limited deformability.

The allowable stresses for piping constructed with ductile materials (assumed to be materials with high deformability), and not designed in accordance with an applicable standard or recognized design basis, are based on values consistent with industrial piping and structural steel standards for comparable piping materials.

The allowable stresses for piping constructed with low-deformability materials, and not designed in accordance with an applicable standard or recognized design basis, are derived from values consistent with ASME standards for comparable piping materials.

For typical piping materials, pipe stresses may not be the governing parameter in determining the hanger and other support spacing. Other considerations—such as the capacity of the hanger and other support connections to the structure, limits on the lateral displacements between hangers and other supports to avoid impacts, the need to limit pipe sag between hangers to avoid the pooling of condensing gases, and the loads on connected equipment—may govern the design. Nevertheless, seismic span tables, based on limiting stresses and displacements in the pipe, can be a useful adjunct for establishing seismic support locations.

Piping systems' service loads of pressure and temperature also need to be considered in conjunction with seismic inertia loads. The potential for low ambient and lower-than-ambient operating temperatures should be considered in the designation of the piping system materials as having high or low deformability. High deformability may often be assumed for steels (particularly ASME listed materials operating at high temperatures), copper and copper alloys, and aluminum. Low deformability should be assumed for any piping material that exhibits brittle behavior, such as glass, ceramics, and many plastics.

Piping should be designed to accommodate relative displacements between the first rigid piping support and connections to equipment or piping headers often assumed to be anchors. Barring such design, the equipment or header connection could be designed to have sufficient flexibility to avoid failure. The specification of such flexible connections should consider the necessity of connection maintenance.

Where appropriate, a walkdown of the finally installed piping system by an experienced design professional familiar with seismic design is recommended, particularly for piping greater than 6 in. (15 cm) nominal pipe size, high-pressure piping, piping operating at higher-than-ambient temperatures, and piping containing hazardous materials. The need for a walkdown may also be related to the scope, function, and complexity of the piping system, as well as the expected performance of the facility. In addition to providing a review of seismic restraint location, orientation, and attachment to the structure, the walkdown verifies that the required separation exists between the piping and nearby structures, equipment, and other piping in the as-built condition.

C13.6.7.1 ASME Pressure Piping Systems In Table 13.6-1, the increased R_{po} values listed for ASME B31-compliant piping systems are intended to reflect the more rigorous design, construction, and quality control requirements, as well as the intensified stresses associated with ASME B31 designs.

Materials meeting ASME toughness requirements may be considered high-deformability materials.

C13.6.7.2 Fire Protection Sprinkler Piping Systems The lateral design procedures of NFPA 13 (2007) have been revised for consistency with the ASCE 7 design approach while retaining traditional sprinkler system design concepts. Using conservative upper-bound values of the various design parameters, a single lateral force coefficient, C_p, was developed. It is a function of the mapped short-period response parameter, S_s. Stresses in the pipe and connections are controlled by limiting the maximum reaction at bracing points as a function of pipe diameter.

Other components of fire protection systems, for example pumps and control panels, are subject to the general requirements of ASCE 7.

Experience has shown that interaction of other nonstructural components and sprinkler drops and sprigs is a significant source of damage and can result in serious consequential damage, as well as compromising the performance of the fire protection system. Clearance for sprinkler drops and sprigs and other nonstructural components needs to be addressed beyond NFPA 13. The minimum clearance value provided is based on judgment observations in past earthquakes. It is not the intent of this committee to require that sprinkler systems be field modified to accommodate these installed clearances if supports or equipment are installed after the sprinkler system is installed (i.e., the burden should not necessarily be on the sprinkler contractor to make the field modifications). It is the intent of this committee that the installation of permanently attached equipment, distribution systems, supports, and fire sprinkler systems be coordinated such that the minimum clearance is maintained after their installation. As building information systems become more widely used and nonstructural components and systems are detailed in the design phase of the project, maintaining these clearances should become easier to ensure by design.

C13.6.7.3 Exceptions The conditions under which the force requirements of Section 13.3 may be waived are based on observed performance in past earthquakes. The limits on the maximum hanger or trapeze drop (hanger rod length) must be met by all the hangers or trapezes supporting the piping system. See additional commentary in Section C13.1.4.

C13.6.9 Utility and Service Lines For essential facilities (Risk Category IV), auxiliary on-site mechanical and electrical utility sources are recommended.

Where utility lines pass through the interface of adjacent, independent structures, they must be detailed to accommodate differential displacement computed in accordance with Section 13.3.2 and including the C_d factor of Section 12.2.1.

As specified in Section 13.1.3, nonessential piping whose failure could damage essential utilities in the event of pipe rupture may also be considered designated seismic systems.

C13.6.10 Boilers and Pressure Vessels Experience in past earthquakes has shown that boilers and pressure vessels are rugged and perform well in strong ground motion. Construction in accordance with current requirements of the ASME *Boiler and Pressure Vessel Code* (2010a) has been shown to be effective in limiting damage and avoiding loss of fluid containment in boilers and pressure vessels under earthquake conditions. It is therefore the intent of the standard that nationally recognized codes be used to design boilers and pressure vessels, provided the seismic force and displacement demands are equal to or exceed those outlined in Section 13.3. Where nationally recognized codes do not yet incorporate force and displacement requirements comparable to the requirements of Section 13.3, it is nonetheless the intent to use the design acceptance criteria and construction practices of those codes.

C13.6.11 Elevator and Escalator Design Requirements The ASME *Safety Code for Elevators and Escalators* (A17.1, 2007) has adopted many requirements to improve the seismic response of elevators; however, they do not apply to some regions covered by this chapter. These changes are to extend force requirements for elevators to be consistent with the standard.

C13.6.11.3 Seismic Controls for Elevators ASME A17.1, Section 8.4.10.1.2, specifies the requirements for the location and sensitivity of seismic switches to achieve the following goals: (a) safe shutdown in the event of an earthquake severe enough to impair elevator operations, (b) rapid and safe reactivation of elevators after an earthquake, and (c) avoidance of unintended elevator shutdowns. This level of safety is achieved by requiring the switches to be in or near the elevator equipment room, by using switches located on or near building columns that respond to vertical accelerations that would result from P and S waves, and by setting the sensitivity of the switches at a level that avoids false shutdowns because of nonseismic sources of vibration. The trigger levels for switches with horizontal sensitivity (for cases where the switch cannot be located near a column) are based on the experience with California hospitals in the Northridge earthquake of 1994. Elevators in which the seismic switch and counterweight derail device have triggered should not be put back into service without a complete inspection. However, in the case where the loss of use of the elevator creates a life-safety hazard, an attempt to put the elevator back into service may be attempted. Operating the elevator before inspection may cause severe damage to the elevator or its components.

The building owner should have detailed written procedures in place defining for the elevator operator and/or maintenance personnel which elevators in the facility are necessary from a post-earthquake, life-safety perspective. It is highly recommended that these procedures be in place, with appropriate personnel training, before an event occurs that is strong enough to trip the seismic switch.

C13.6.11.4 Retainer Plates The use of retainer plates is a low-cost provision to improve the seismic response of elevators.

C13.6.12 Rooftop Solar Panels Rooftop solar panels without positive attachment to the roof structure are limited to low-profile panels with a low height-to-depth ratio that respond by sliding on the roof surface without overturning. The amount of roof slope is limited because studies show that panels on sloped surfaces tend to displace in the downslope direction when subjected to seismic shaking, and the displacement increases with greater roof slope.

Displacement-based design of panels includes verifying that the panel remains safe if displaced. It needs to be verified that there is roof capacity to support the weight of the displaced panel and that wiring to the panel can accommodate the design panel displacement without damage.

Equation (13.6-1) conservatively assumes a minimum coefficient of friction between the solar panel and the roof of 0.4. In cold-weather regions, the effects on the friction coefficient should be considered for Seismic Design Categories D, E, and F.

Structural interconnection between portions of a solar array must be of adequate design strength, in tension or compression, and stiffness in order to account for the potential that frictional resistance to sliding will be different under some portions of the panel as a result of varying normal force and actual instantaneous values of friction coefficient for a given roof surface material. The interconnection force is typically resisted by continuous rails or members to transfer these forces and tie the array together so it slides on the roof surface as one unit. Solar panels are typically composed of glass and solar cells bound by a light frame. Historically, they have been tested to national standards for out-of-plane wind loading and other loads normal to the panel surface but not typically tested for in-plane axial, shear, or point loads that are associated with the interconnection loads specified herein. Therefore, the solar panels are not permitted to transfer this interconnection load unless they are specifically evaluated or tested for such loading.

The requirement for unattached panels to be bounded by a curb or parapet is usually satisfied by a curb at the roof edge. In lieu of being bounded by curbs or parapets at roof edges and offsets, the panel may be set back farther from the edge.

Analytical and experimental studies of the seismic response of unattached solar panels are reported by Schellenberg et al. (2012) and Maffei et al. (2013).

Shake table testing and nonlinear time history analysis may also be used to predict panel displacements; however, for unattached panels, it is necessary to use input motions appropriate for predicting sliding displacement, which can be affected by content in the low-frequency range. See SEAOC (2012) for guidance on the performance of such testing and analysis.

C13.6.13 Other Mechanical and Electrical Components The material properties set forth in Item 2 of this section are similar to those allowed in the ASME *Boiler and Pressure Vessel Code* (2010a) and reflect the high factors of safety necessary for seismic, service, and environmental loads.

REFERENCES

ACI (American Concrete Institute). 2007. *Qualification of post-installed mechanical anchors in concrete and commentary.* ACI 355.2. Farmington Hills, MI: ACI.

ACI. 2011. *Qualification of post-installed adhesive anchors in concrete and commentary.* ACI 355.4. Farmington Hills, MI: ACI.

ACI. 2014. *Building code requirements for structural concrete and commentary.* ACI 318. Farmington Hills, MI: ACI.

ANCO Engineers. 1983. *Seismic hazard assessment of non-structural components: Phase I.* Culver City, CA: National Science Foundation from ANCO Engineers.

ANSI/ASHRAE (American National Standards Institute and American Society of Heating, Refrigerating and Air-Conditioning Engineers). 2017. *Method of testing for rating seismic and wind restraints.* Standard 171-2017. Atlanta, GA: ASHRAE.

ANSI/FM (American National Standards Institute and FM Approvals). 1950–2016. *American national standard for seismic sway braces for pipe, tubing and conduit.* Norwood, MA: FM Approvals.

ANSI/SMACNA (American National Standards Institute and Sheet Metal and Air-Conditioning Contractors National Association). 2000. *Seismic restraint manual: Guidelines for mechanical systems.* ANSI/SMACNA 001. Chantilly, VA: SMACNA.

ASCE. 2011. *Guidelines for seismic evaluation and design of petrochemical facilities*, 2nd ed. Reston, VA: ASCE.

ASHRAE (American Society of Heating, Refrigerating, and Air-Conditioning Engineers). 2000. *Practical guide to seismic restraint.* RP-812. Atlanta, GA: ASHRAE.

ASME (American Society of Mechanical Engineers). 2007. *Safety code for elevators and escalators.* A17.1. New York: ASME.

ASME. 2010a. *Boiler and pressure vessel code.* New York: ASME.

ASME. 2010b. *Standard for the seismic design and retrofit of above-ground piping systems.* New York: ASME.

ASTM International. 2004. *Standard specification for the manufacture, performance, and testing of metal suspension systems for accoustical tile and lay-in panel ceiling.* ASTM C635. West Conshohocken, PA: ASTM.

ASTM 2014a. *Standard guide for structural sealant glazing.* ASTM C1401. West Conshohocken, PA: ASTM.

ASTM. 2014b. *Standard practice for installation of ceiling suspension systems for acoustical tile and lay-in panels for areas subject to earthquake ground motion.* ASTM E580/E580M-14. West Conshohocken, PA: ASTM.

Bachman, R. E., and S. M. Dowty. 2008. "Nonstructural component or nonbuilding structure?" *Build. Saf. J* (April/May).

Bouwkamp, J. G., and J. F. Meehan. 1960. "Drift limitations imposed by glass." In *Proc., 2nd World Conf. on Earthquake Engineering*, Tokyo, 1763–1778.

Canterbury Earthquakes Royal Commission. 2012. *The performance of Christchurch CBD buildings.* Final Rep. No. 2. Wellington, NZ: Canterbury.

Dowell, R. K., and T. P. Johnson. 2013. *Evaluation of seismic overstrength factors for anchorage into concrete via dynamic shaking table tests.* Rep. No. SERP–13/09. San Diego: San Diego State University.

FEMA (Federal Emergency Management Agency). 2015. *Reducing the risks of nonstructural earthquake damage: A practical guide.* FEMA E-74. Washington, DC: FEMA.

FM Global. 2007. "Approved standard for pipe couplings and fittings for aboveground fire protection systems." Accessed April 15, 2021. http://www.fmglobal.com.

Hadjian, A., and E. Ellison. 1986. "Decoupling of secondary systems for seismic analysis." *J. of Press. Vessel Technol.* 108 (1): 78–85. https://doi.org/10.1115/1.3264755.

ICC (International Code Council). 1976. *Uniform building code.* Whittier, CA: ICC.

ICC-ES (International Code Council Evaluation Service). 2010. *Seismic qualification by shake-table testing of nonstructural components and systems.* ICC-ES AC156. Whittier, CA: ICC-ES.

ICC-ES. 2012a. *Acceptance criteria for adhesive anchors in masonry elements.* ICC-ES AC58. Whittier, CA: ICC-ES.

ICC-ES. 2012b. *Acceptance criteria for expansion anchors in masonry elements.* ICC-ES AC01. Whittier, CA: ICC-ES.

ICC-ES. 2012c. *Acceptance criteria for mechanical anchors in concrete elements.* ICC-ES AC193. Whittier, CA: ICC-ES.

ICC-ES. 2012d. *Acceptance criteria for post-installed adhesive anchors in concrete elements.* ICC-ES AC308. Whittier, CA: ICC-ES.

ICC-ES. 2012e. *Acceptance criteria for predrilled fasteners (screw anchors) in masonry.* ICC-ES AC106. Whittier, CA: ICC-ES.

IEEE. 2005. *IEEE recommended practices for seismic design of substations.* IEEE 693-2005. New York: IEEE.

Maffei, J., S. Fathali, K. Telleen, R. Ward, and A. Schellenberg. 2013. "Seismic design of ballasted solar arrays on low-slope roofs." *J. Struct. Eng.* 140 (1): 04013020. https://doi.org/10.1061/(ASCE)ST.1943-541X.0000865.

MSS (Manufacturers Standardization Society of the Valve and Fitting Industry). 2001. *Bracing for piping systems: Seismic–wind–dynamic design, selection, application.* MSS SP-127. Vienna, VA: MSS.

NFPA. 2011. *National electric code.* NFPA 70. Quincy, MA: NFPA.

NEES (Network for Earthquake Engineering Simulation). 2011. *Full-scale structural and nonstructural building system performance during earthquake and post-earthquake fire.* San Diego: University of California, San Diego.

NEHRP (National Earthquake Hazards Reduction Program). 2009. *NEHRP recommended provisions for seismic regulations for new buildings and other structures.* Washington, DC: NEHRP.

NFPA (National Fire Protection Association). 2007. *Standard for the installation of sprinkler systems.* NFPA 13. Quincy, MA: NFPA.

NIST (National Institute of Standards and Technology). 2018. *Recommendations for improved seismic performance of non-structural components.* GCR 18-917-43. Gaithursburg, MD: NIST.

Rahmanishamsi, E., S. Soroushian, and E. Maragakis. 2014. "Seismic response of ceiling/piping/partition systems in NEESR-GC system-level experiments." In *Proc., ASCE Structures Congress*, Boston.

Schellenberg, A., J. Maffei, K. Miller, M. Williams, R. Ward, and M. Dent. 2012. "Shake-table testing of unattached rooftop solar arrays: Interim report, subtask 4.1." Accessed April 15, 2021. https://www.cpuc.ca.gov/General.aspx?id=6043.

SEAOC (Structural Engineers Association of California). 1999. *Recommended lateral force requirements and commentary.* Sacramento, CA: SEAOC.

SEAOC. 2012. *Structural seismic requirements and commentary for rooftop solar photovoltaic arrays*. Rep. No. SEAOC-PV1-2012. Sacramento, CA: SEAOC.

SGAJ (Sheet Glass Association of Japan). 1982. *Earthquake safety design of windows*. Tokyo: SGAJ.

SMACNA (Sheet Metal and Air-Conditioning Contractors National Association). 2008. *Seismic restraint manual: Guidelines for mechanical systems,* 3rd ed. Chantilly, VA: SMACNA.

Soroushian, S., E. Rahmanishamsi, K. P. Ryu, E. M. Maragakis, and A. M. Reinhorn. 2014. "A comparative study of subsystem and system level experiments of suspension ceiling systems." In *Proc., 10th US National Conf. on Earthquake Engineering*, Anchorage, AK.

Soroushian, S., A. Reinhorn, E. Rahmanishamsi, K. Ryu, and M. Maragakis. 2012. "Seismic response of ceiling/sprinkler piping nonstructural systems in NEES TIPS/NEES nonstructural/NIED collaborative tests on a full scale 5-story building." In *Proc., ASCE Structures Congress*, Chicago.

Soulages, J. R., and R. Weir. 2011. "Cyclic testing of pipe trapezes with rigid hanger assemblies." In *Proc., 80th Annual Convention, Structural Engineers Association of California*, Las Vegas, NV.

UL (Underwriters Laboratories). 2004. *Rubber gasketed fittings for fire-protection service*. UL 213. Northbrook, IL: UL.

US Department of Defense. 2007. *Seismic design for buildings*. Unified Facilities Criteria (UFC) 3-310-04. Washington, DC: US Department of Defense.

Wright, P. D. 1989. *The development of a procedure and rig for testing the racking resistance of curtain wall glazing*. Study Rep. No. 17. Porirua, NZ: Building Research Association of New Zealand.

OTHER REFERENCES (NOT CITED)

ACI. 2011a. *Building code requirements and specification for masonry structures and related commentaries*. ACI 530/530.1. Farmington Hills, MI: ACI.

ACI. 2011b. *Building code requirements for structural concrete and commentary*. ACI 318. Farmington Hills, MI: ACI.

ASTM. 2007. *Standard specification for the manufacture, performance, and testing of metal suspension systems for acoustical tile and lay-in panel ceilings*. ASTM C635/C635M. West Conshohocken, PA: ASTM.

ASTM. 2013a. *Standard practice for installation of metal ceiling suspension systems for acoustical tile and lay-in panels*. ASTM C636/C636M-13. West Conshohocken, PA: ASTM.

ASTM. 2013b. *Standard specification for the manufacture, performance, and testing of metal suspension systems for acoustical tile and lay-in panel ceilings*. ASTM C635/C635M-13a. West Conshohocken, PA: ASTM.

Bachman, R. E., and R. M. Drake. 1996. *A study to empirically validate the component response modification factors in the 1994 NEHRP Provisions design force equations for architectural, mechanical, and electrical components*. Taipei: National Center for Earthquake Engineering Research.

Bachman, R. E., R. M. Drake, and P. J. Richter. 1993. *1994 update to 1991 NEHRP Provisions for architectural, mechanical, and electrical components and systems*. Taipei: National Center for Earthquake Engineering Research.

Behr, R. A., and A. Belarbi. 1996. "Seismic test methods for architectural glazing systems." *Earthquake Spectra* 12 (1): 129–143. https://doi.org/10.1193/1.1585871.

Behr, R. A., A. Belarbi, and A. T. Brown. 1995. "Seismic performance of architectural glass in a storefront wall system." *Earthquake Spectra* 11 (3): 367–391. https://doi.org/10.1193/1.1585819.

Drake, R. M., and R. E. Bachman. 1994. "1994 NEHRP provisions for architectural, mechanical, and electrical components." In *Proc., 5th United States National Conf. on Earthquake Engineering*, Chicago.

Drake, R. M., and R. E. Bachman. 1995. "Interpretation of instrumented building seismic data and implications for building codes." In *Proc., 1995 SEAOC Annual Convention*, Squaw Creek, CA.

Drake, R. M., and R. E. Bachman. 1996. "NEHRP provisions for 1994 for nonstructural components." *J. Arch. Eng.* 2 (1): 26–31. https://doi.org/10.1061/(ASCE)1076-0431(1996)2:1(26).

Fleischmann, R. B., J. I. Restrepo, and S. Pampanin. 2014. "Damage evaluations of precast concrete structures in the 2010–2011 Canterbury earthquake sequence." *EERI Earthquake Spectra* 30 (1): 277–306. https://doi.org/10.1061/(ASCE)1076-0431(1996)2:1(26).

Gates, W. E., and G. McGavin. 1998. *Lessons learned from the 1994 Northridge earthquake on the vulnerability of nonstructural systems*. ATC-29-1. Redwood City, CA: Applied Technology Council.

Haroun, M. A., and G. W. Housner. 1981. "Seismic design of liquid storage tanks." *J. Tech. Councils ASCE* 107 (1): 191–207. https://doi.org/10.1061/JTCAD9.0000080.

Higgins, C. 2009. "Prefabricated steel stair performance under combined seismic and gravity loads." *J. Struct. Eng.* 135 (2): 122–129. https://doi.org/10.1061/(ASCE)0733-9445(2009)135:2(122).

Kehoe, B., and M. Hachem. 2003. "Procedures for estimating floor accelerations." In *Proc., Seminar on Seismic Design, Performance, and Retrofit of Nonstructural Components in Critical Facilities*, Newport Beach, CA, 361–374.

NFPA. 2011. *National electric code*. NFPA 70. Quincy, MA: NFPA.

Pantelides, C. P., K. Z. Truman, R. A. Behr, and A. Belarbi. 1996. "Development of a loading history for seismic testing of architectural glass in a shop-front wall system." *Eng. Struct.* 18 (12): 917–935. https://doi.org/10.1016/0141-0296(95)00224-3.

Pantoli, E., M. Chen, T. Hutchinson, G. Underwood, and M. Hildebrand. 2013. "Shake table testing of a full-scale five-story building: Seismic performance of precast concrete cladding panels." In *Proc., 4th ECCOMAS Thematic Conf. on Computational Methods in Structural Dynamics and Earthquake Engineering (COMPDYN 2013)*, Kos Island, Greece.

Trautner, C., T. Hutchinson, and P. Grosser. 2014. "Cyclic behavior of structural base plate connections with ductile fastening failure: Component test results." In *Proc., 10th US National Conf. on Earthquake Engineering*, Anchorage, AK.

US Army. 1986. *Seismic design guidelines for essential buildings*. TM 5-809-1. Washington, DC: Joint Dept. of the Army, Navy, and Air Force.

CHAPTER C14
MATERIAL-SPECIFIC SEISMIC DESIGN AND DETAILING REQUIREMENTS

Because seismic loading is expected to cause nonlinear behavior in structures, seismic design criteria require not only provisions to govern loading but also provisions to define the required configurations, connections, and detailing to produce material and system behavior consistent with the design assumptions. Thus, although ASCE 7 is primarily a loading standard, compliance with Chapter 14, which covers material-specific seismic design and detailing, is required. In general, Chapter 14 adopts material design and detailing standards developed by material standards organizations. These material standards organizations maintain complete commentaries covering their standards, and such material is not duplicated here.

C14.0 SCOPE

The scoping statement in this section clarifies that foundation elements are subject to all of the structural design requirements of the standard.

C14.1 STEEL

C14.1.1 Reference Documents This section lists a series of structural standards published by the American Institute of Steel Construction (AISC), the American Iron and Steel Institute (AISI), ASCE, the Steel Deck Institute (SDI), and the Steel Joist Institute (SJI), which are to be applied in the seismic design of steel members and connections in conjunction with the requirements of ASCE 7. The AISC references are available free of charge in electronic format at www.aisc.org; the AISI references are available at www.aisistandards.org; the SDI references are available as a free download at www.sdi.org; and the SJI references are available as a free download at www.steeljoist.org.

C14.1.2 Structural Steel

C14.1.2.1 General This section adopts AISC 360 (2022b) by direct reference. The specification applies to the design of the structural steel system or systems with structural steel acting compositely with reinforced concrete. In particular, the document sets forth criteria for the design, fabrication, and erection of structural steel buildings and other structures, where "other structures" are defined as structures designed, fabricated, and erected in a manner similar to buildings, with building-like vertical and lateral load-resisting elements. The document includes extensive commentary.

C14.1.2.2 Seismic Requirements for Structural Steel Structures

C14.1.2.2.1 Seismic Design Categories B and C For the lower Seismic Design Categories (SDCs) B and C, a range of options are available in the design of a structural steel lateral force-resisting system. The first option is to design the structure to meet the design and detailing requirements in AISC 341 (2022a) for structures assigned to higher SDCs, with the corresponding seismic design parameters (R, Ω_0, and C_d). The second option, presented in the exception, is to use an R factor of 3 (resulting in an increased base shear), an Ω_0 of 3, and a C_d value of 3 but without the specific seismic design and detailing required in AISC 341. The basic concept underlying this option is that design for a higher base shear force results in essentially elastic response that compensates for the limited ductility of the members and connections. The resulting performance is considered comparable to that of more ductile systems.

C14.1.2.2.2 Seismic Design Categories D through F For the higher SDCs, the engineer must follow the seismic design provisions of AISC 341 using the seismic design parameters specified for the chosen structural system, except as permitted in Table 15.4-1. For systems other than those identified in Table 15.4-1, it is not considered appropriate to design structures without specific design and detailing for seismic response in these high SDCs.

C14.1.3 Cold-Formed Steel

C14.1.3.1 General This section adopts two standards by direct reference: AISI S100, *North American Specification for the Design of Cold-Formed Steel Structural Members* (2020c), and ASCE 8, *Specification for the Design of Cold-Formed Stainless Steel Structural Members* (2002).

Both of the adopted reference documents have specific limits of applicability. AISI S100, Section A1.1, applies to the design of structural members that are cold-formed to shape from carbon or low-alloy steel sheet, strip, plate, or bar not more than 1 in. (25 mm) thick. ASCE 8, Section 1.1.1, governs the design of structural members that are cold-formed to shape from annealed and cold-rolled sheet, strip, plate, or flat bar stainless steels. Both documents focus on load-carrying members in buildings; however, allowances are made for applications in nonbuilding structures, if dynamic effects are considered appropriately.

Within each document, there are requirements related to general provisions for the applicable types of steel; design of elements, members, structural assemblies, connections, and joints; and mandatory testing. In addition, AISI S100 contains a chapter on the design of cold-formed steel structural members and connections undergoing cyclic loading. Both standards contain extensive commentaries.

C14.1.3.2 Seismic Requirements for Cold-Formed Steel Structures This section adopts three standards by direct reference: AISI S100 (2020c), ASCE 8 (2002), and AISI S400 (2020b). Cold-formed steel and stainless steel members that are part of a seismic force-resisting system listed in Table 12.2-1 must be detailed in accordance with the appropriate base standard: AISI S100 or ASCE 8.

AISI S400 includes additional design provisions for a specific cold-formed steel seismic force-resisting system, the "cold-formed steel—special bolted moment frame" or CFS-SBMF. Sato and Uang (2007) have shown that this system experiences inelastic deformation at the bolted connections because of slip and bearing during significant seismic events. To develop the designated mechanism, requirements based on capacity design principles are provided for the design of the beams, columns, and associated connections. The document has specific requirements for the application of quality assurance and quality control procedures.

C14.1.4 Cold-Formed Steel Light-Frame Construction

C14.1.4.1 General This subsection of cold-formed steel relates to light-frame construction, which is defined as a method of construction where the structural assemblies are formed primarily by a system of repetitive wood or cold-formed steel framing members, or subassemblies of these members (Section 11.2 of this standard). It adopts AISI S240 (2020a), *North American Standard for Cold-Formed Steel Structural Framing,* by reference, which includes a commentary to aid users in the correct application of the requirements.

C14.1.4.2 Seismic Requirements for Cold-Formed Steel Light-Frame Construction Cold-formed steel structural members and connections in seismic force-resisting systems and diaphragms must be designed in accordance with the additional provisions of AISI S400 (2020b) in seismic design categories (SDC) D, E, or F, or wherever the seismic response modification coefficient, R, used to determine the seismic design forces is taken as other than 3. In particular, this requirement includes all entries from Table 12.2-1 of this standard for "light-frame (cold-formed steel) walls sheathed with wood structural panels rated for shear resistance or steel sheets," "light-frame walls with shear panels of all other materials" (e.g., gypsum board and fiberboard panels), and "light-frame (cold-formed steel) wall systems using flat strap bracing."

C14.1.4.3 Prescriptive Cold-Formed Steel Light-Frame Construction This section adopts AISI S230 (2019), *Standard for Cold-Formed Steel Framing: Prescriptive Method for One- and Two-Family Dwellings*, which applies to the construction of detached one- and two-family dwellings, townhouses, and other attached single-family dwellings not more than three stories in height using repetitive in-line framing practices (Section A1). This document includes a commentary to aid the user in the correct application of its requirements.

C14.1.5 Cold-Formed Steel Deck Diaphragms This section adopts the applicable standards for the general design of cold-formed steel deck diaphragms and steel roof, noncomposite floor, and composite floor deck. The SDI standards also reference AISI S100 (2020c) for materials and determination of cold-formed steel cross-section strength and specify additional requirements specific to steel deck design and installation.

In addition, design of cold-formed steel deck diaphragms is to be based on AISI S310 (2020d). All fastener design values (welds, screws, power-actuated fasteners, and button punches) for attaching deck sheet to deck sheet or for attaching the deck to the building framing members must be per AISI S310 with specific testing procedures as prescribed in AISI S310. All cold-formed steel deck diaphragm and fastener design properties not specifically included in AISI S310 must be approved for use by the Authority Having Jurisdiction in whose jurisdiction the construction project occurs. Deck diaphragm in-plane design forces (seismic, wind, or gravity) must be determined per ASCE 7, Section 12.10.1. Cold-formed steel deck manufacturer test reports prepared in accordance with this provision can be used where adopted and approved by the Authority Having Jurisdiction for the building project. The *Diaphragm Design Manual* produced by the Steel Deck Institute (2015) is also a reference for design values.

Cold-formed steel deck has a corrugated profile consisting of alternating up and down flutes that are manufactured in various widths and heights. Use of flat sheet metal as the overall floor or roof diaphragm is permissible when designed by engineering principles, but it is beyond the scope of this section. Flat or bent sheet metal may be used as closure pieces for small gaps or penetrations or for shear transfer over short distances in the deck diaphragm where diaphragm design forces are considered.

Cold-formed steel deck diaphragm analysis must include design of chord members at the perimeter of the diaphragm and around interior openings in the diaphragm. Chord members may be steel beams attached to the underside of the steel deck designed for a combination of axial loads and bending moments caused by acting gravity and lateral loads.

Where diaphragm design loads exceed the bare steel deck diaphragm design capacity, then either horizontal steel trusses or a structurally designed concrete topping slab placed over the deck must be provided to distribute lateral forces. Where horizontal steel trusses are used, the cold-formed steel deck must be designed to transfer diaphragm forces to the steel trusses. Where a deck topped with structural concrete is used as the diaphragm, the diaphragm chord members at the perimeter of the diaphragm and edges of interior openings must be either (a) designed with flexural reinforcing steel placed in the structural concrete topping, or (b) steel beams located under the deck, with connectors (that provide a positive connection) as required to transfer design shear forces between the concrete topping and steel beams.

The stiffness and available strength (factored resistance) of steel deck diaphragms are provided in AISI S310. However, AISI S310 does not cover seismic design considerations. AISI S400 (2020b) recognizes that in some situations, the applicable building code may require that the diaphragm provide energy dissipation for desired structural performance. For example, in rigid wall–flexible diaphragm (RWFD) structures, research has shown the benefits of and demands for energy dissipation in the roof diaphragm (FEMA 2015; Koliou et al. 2016a, b). ASCE 7 provides an alternative design method for RWFD structures in Section 12.10.4, where forces in the diaphragm may be reduced if special seismic detailing is provided for bare steel deck diaphragms. Further, for all other structures, the alternative diaphragm design provisions of Section 12.10.3 also provide a means to reduce diaphragm forces when special seismic detailing is provided. The provisions of AISI S400, Section F3.5, are specifically intended to meet these special seismic detailing requirements.

Traditional equivalent lateral force (ELF)–based seismic design of bare steel deck diaphragms per Section 12.10.1 allows diaphragm forces to be reduced based on the response modification factor, R, for the particular vertical seismic force-resisting system, subject to minimum diaphragm force levels as defined in this standard. The reduction in the diaphragm force levels is independent of the ductility or deformation capacity of the diaphragm. Analysis of a large-scale RWFD archetype building under high demand with precast tilt-up walls and bare steel deck diaphragm roofs that either meet or violate the special seismic detailing requirements was completed by Schafer (2019). He found that a mechanically fastened roof that met the special seismic detailing requirements of AISI S400, Section F3.5, had approximately one-half the roof shear angle demands and one-half the anchorage demands of an equivalent welded bare steel deck diaphragm roof that did not meet the special seismic detailing requirements. If the designer desires (for force reduction) or expects (due to the nature of the structure) inelastic demands in a bare steel deck diaphragm, the special seismic detailing requirements provide a means to ensure ductility and deformation capacity in the diaphragm.

In addition to special seismic detailing, standard installation and construction procedures are necessary for successful performance. SDI (2022) provides QC/QA criteria for steel deck installation, and SDI (2016) provides additional construction guidance. The QC/QA provisions include required special inspection for steel deck installation, both with and without special seismic detailing.

C14.1.6 Concrete-Filled Steel Deck Diaphragms Testing by Porter and Easterling (1988) and Avellaneda et al. (2019) has demonstrated sufficient ductility and overstrength to permit the tabulated R_s factor. The specimens tested included a range of variations, such as concrete specific weight (normal weight and lightweight), reinforcing steel (unreinforced and with reinforcing steel), concrete strength [2,500 to 6,000 psi (17 to 41 Gpa)], deck height [1.5 to 3.0 in. (38 to 76 mm)], total thickness [4 to 7.5 in. (100 to 190 mm)], and perimeter fastener type (headed shear studs and arc spot welds). Because this set of test specimens, which was used to determine the diaphragm design force reduction factor, R_s, covered a wide range of configurations, there are few limitations associated with detailing. This section references AISC 341 (2022a) for the design of the diaphragm and design of the shear transfer to the supports.

C14.1.7 Steel Cables These provisions reference ASCE 19, *Structural Applications of Steel Cables for Buildings* (2010), for the determination of the design strength of steel cables.

C14.1.8 Additional Detailing Requirements for Steel Piles in Seismic Design Categories D through F Steel piles used in higher SDCs are expected to yield just under the pile cap or foundation because of combined bending and axial load. Design and detailing requirements of AISC 341 (2022a) for H-piles are intended to produce stable plastic hinge formation in the piles. Because piles can be subjected to tension caused by overturning moment, mechanical means to transfer such tension must be designed for the required tension force, but not less than 10% of the pile compression capacity.

C14.2 CONCRETE

The section adopts by reference ACI 318 (2019) for structural concrete design and construction. In addition, modifications to ACI 318-19 are made that are needed to coordinate the provisions of that material design standard with the provisions of ASCE 7. Work is ongoing to better coordinate the provisions of the two documents (ACI 318 and ASCE 7) such that the provisions in Section 14.2 will be progressively reduced in future editions of ASCE 7.

C14.2.2.1 Definitions Two definitions included here describe wall types for which definitions currently do not exist in ACI 318. These definitions are essential to the proper interpretation of the R and C_d factors for each wall type specified in Table 12.2-1.

The addition of precast concrete diaphragm and cast-in-place concrete equivalent precast diaphragm definitions is meant to clarify that Section 12.10.3 applies to precast concrete diaphragms as defined in this section and not to cast-in-place concrete equivalent precast diaphragms.

C14.2.2.2 ACI 318, Section 10.7.6 ACI 318-19, Section 10.7.6.1.5, prescribes details of transverse reinforcement around anchor bolts in the top of a column or pedestal. This modification prescribes additional details for transverse reinforcement around such anchor bolts in structures assigned to Seismic Design Categories C through F.

C14.2.2.3 Scope This provision describes how the ACI 318-19 provisions should be interpreted for consistency with the ASCE 7 provisions.

C14.2.2.4 Intermediate Precast Structural Walls Section 18.5 of ACI 318-19 imposes requirements on precast walls for moderate seismic risk applications. Ductile behavior is to be ensured by yielding of the steel elements or reinforcement between panels or between panels and foundations. This provision requires the designer to determine the deformation in the connection corresponding to the Design Earthquake Displacement and then to check from experimental data that the connection type used can accommodate that deformation without significant strength degradation.

Several steel element connections have been tested under simulated seismic loading, and the adequacy of their load–deformation characteristics and strain capacity have been demonstrated (Schultz and Magana 1996). One such connection was used in the five-story building test that was part of the Precast Seismic Structural Systems (PRESSS) Phase 3 research. The connection was used to provide damping and energy dissipation, and it demonstrated a very large strain capacity (Nakaki et al. 2001). Since then, several other steel element connections have been developed that can achieve similar results (Banks and Stanton 2005, Nakaki et al. 2005). In view of these results, it is appropriate to allow yielding in steel elements that have been shown experimentally to have adequate strain capacity to maintain at least 80% of their yield force through the full design displacement of the structure.

C14.2.2.6 Foundations The intention is that there should be no conflicts between the provisions of ACI 318-19, Section 18.13, and ASCE 7, Sections 12.1.5 and 12.13.

C14.2.2.7 Detailed Plain Concrete Shear Walls Design requirements for plain masonry walls have existed for many years, and the corresponding type of concrete construction is the plain concrete wall. To allow the use of such walls as the lateral force-resisting system in Seismic Design Categories A and B, this provision requires such walls to contain at least the minimal reinforcement specified in ACI 318-19, Section 14.6.2.2.

C14.3 COMPOSITE STEEL AND CONCRETE STRUCTURES

This section provides guidance on the design of composite and hybrid steel–concrete structures. Composite structures are defined as those incorporating structural elements made of steel and concrete connected integrally throughout the structural element by mechanical connectors, bonds, or both. Hybrid structures are defined as consisting of steel and concrete structural elements connected together at discrete points. Composite and hybrid structural systems mimic many of the existing steel (moment and braced frame) and reinforced concrete (moment frame and wall) configurations but are given their own design coefficients and factors in Table 12.2-1. Their design is based on ductility and energy dissipation concepts comparable to those used in conventional steel and reinforced concrete structures, but the design requires special attention to the interaction of the two materials, because it affects the stiffness, strength, and inelastic behavior of the members, connections, and systems.

C14.3.1 Reference Documents Seismic design for composite structures is governed by AISC 341 (2022a), which in turn refers to AISC 360 and other related specifications as needed. Both AISC 360 (2022b) and AISC 341 (2022a) refer to ACI 318 (2019) for much of the design and detailing of reinforced concrete members in composite structures and detailing of composite members. AISC 341 (2022a) primarily points to AISC 360 (2022b) for the design of composite structures assigned to Seismic Design Category A, B, or C. Seismic design for composite structures assigned to Seismic Design Category D, E, or F is governed primarily by AISC 341 (2022a).

C14.3.4 Metal-Cased Concrete Piles Design of metal-cased concrete piles, which are analogous to circular concrete filled tubes, is governed by Section 18.13.5.8 of ACI 318 (2019). The intent of these provisions is to require metal-cased concrete piles to have confinement and protection against long-term deterioration comparable to that for uncased concrete piles.

C14.4 MASONRY

This section adopts by reference and then makes modifications to TMS 402 (2016a) and TMS 602 (2016b). In past editions of this standard, modifications to the TMS referenced standards were also made. During the development of the 2016 edition of the TMS standards, each of these modifications was considered by the TMS 402/602 committee. Some were incorporated directly into the TMS standards. Those modifications have accordingly been removed from this standard. Work is ongoing to better coordinate the provisions of the two documents so that the provisions in Section 14.4 are significantly reduced or eliminated in future editions.

C14.5 WOOD

C14.5.1 Reference Documents Two national consensus standards are adopted for seismic design of engineered wood structures: *National Design Specification* (AWC NDS-18, 2018), and *Special Design Provisions for Wind and Seismic* (AWC SDPWS-21, 2020). Both of these standards are presented in dual formats, for allowable stress design (ASD) and load and resistance factor design (LRFD). Both standards reference secondary standards for related items such as wood materials and fasteners. AWC NDS addresses requirements for member and connection design, and AWC SDPWS addresses requirements for shear wall and diaphragm design.

REFERENCES

ACI (American Concrete Institute). 2019. *Building code requirements for structural concrete and commentary.* ACI 318. Farmington Hills, MI: ACI.

AISC (American Institute of Steel Construction). 2022a. *Seismic provisions for structural steel buildings.* ANSI/AISC 341. Chicago: AISC.

AISC. 2022b. *Specification for structural steel buildings.* ANSI/AISC 360. Chicago: AISC.

AISI (American Iron and Steel Institute). 2019. *Standard for cold-formed steel framing: Prescriptive method for one- and two-family dwellings.* ANSI/AISI S230-19. Washington, DC: AISI.

AISI. 2020a. *North American standard for cold-formed steel structural framing.* ANSI/AISI S240-20. Washington, DC: AISI.

AISI. 2020b. *North American standard for seismic design of cold-formed steel structural systems.* ANSI/AISI S400-20. Washington, DC: AISI.

AISI. 2020c. *North American specification for the design of cold-formed steel structural members.* 2016 ed., reaffirmed 2020 with Supplement 2, 2020 edition. ANSI/AISI S100-16 (2020) w/S2-20. Washington, DC: AISI.

AISI. 2020d. *North American standard for the design of profiled steel diaphragm panels.* ANSI/AISI S310-20. Washington, DC: AISI.

ASCE. 2002. *Specification for the design of cold-formed stainless steel structural members.* ASCE/SEI 8-02. Reston, VA: ASCE.

ASCE. 2010. *Structural applications of steel cables for buildings.* ASCE 19-10. Reston, VA: ASCE.

Avellaneda, R. E., W. S. Easterling, B. W. Schafer, J. F. Hajjar, and M. R. Eatherton. 2019. "Cyclic testing of composite concrete on metal deck diaphragms undergoing diagonal tension cracking." In *Proc., 12th Canadian Conf. on Earthquake Engineering*, Quebec.

AWC (American Wood Council). 2017. *National design specification (NDS) for wood construction with commentary.* ANSI/AWC NDS-2018. Leesburg, VA: AWC.

AWC. 2020. *Special design provisions for wind and seismic.* ANSI/AWC SDPWS-2021. Leesburg, VA: AWC.

Banks, G., and J. Stanton. 2005. "Panel-to-panel connections for hollow-core shear walls subjected to seismic loading." In *Proc., 2005 PCI Convention*, Chicago.

FEMA (Federal Emergency Management Agency). 2015. *Seismic design of rigid wall–flexible diaphragm buildings: An alternate procedure.* FEMA P-1026. Washington, DC: FEMA.

Koliou, M., A. Filiatrault, D. J. Kelly, and J. Lawson. 2016a. "Buildings with rigid walls and flexible roof diaphragms. I: Evaluation of current U.S. seismic provisions." *J. Struct. Eng.* 142 (3): 04015166. https://doi.org/10.1061/(ASCE)ST.1943-541X.0001438.

Koliou, M., A. Filiatrault, D. J. Kelly, and J. Lawson. 2016b. "Buildings with rigid walls and flexible roof diaphragms. II: Evaluation of a new seismic design approach based on distributed diaphragm yielding." *J. Struct. Eng.* 142 (3): 04015167. https://doi.org/10.1061/(ASCE)ST.1943-541X.0001439.

Nakaki, S., R. Becker, M. G. Oliva, and D. Paxson. 2005. "New connections for precast wall systems in high seismic regions." In *Proc., 2005 PCI Convention*, Chicago.

Nakaki, S., J. F. Stanton, and S. Sritharan. 2001. "The PRESSS five-story precast concrete test building, University of California, San Diego, La Jolla, California." *PCI J.* 46 (5): 20–26.

Sato, A., and C.-M. Uang. 2007. *Development of a seismic design procedure for cold-formed steel bolted frames.* Rep. No. SSRP-07/16. San Diego: University of California, San Diego.

Schafer, B. W. 2019. *Research on the seismic performance of rigid wall flexible diaphragm buildings with bare steel deck diaphragms.* CFSRC Rep. No. R-2019-02. Baltimore: Cold-Formed Steel Research Consortium, Johns Hopkins University.

Schultz, A. E., and R. A. Magana. 1996. *Seismic behavior of connections in precast concrete walls.* SP-162. Farmington Hills, MI: American Concrete Institute.

SDI (Steel Deck Institute). 2015. *Diaphragm design manual*, 4th ed. Glenview, PA: SDI.

SDI. 2016. *Manual of construction with steel deck*, 3rd ed. Glenview, PA: SDI.

SDI. 2022. *Standard for quality control and quality assurance for installation of steel deck.* ANSI/SDI QA/QC-2022. Glenview, PA: SDI.

TMS (The Masonry Society). 2016a. *Building code requirements and specification for masonry structures.* TMS 402-16. Longmont, CO: TMS.

TMS. 2016b. *Specification for masonry structures.* TMS 602-16. Longmont, CO: TMS.

OTHER REFERENCES (NOT CITED)

American Institute of Timber Construction. 2005. *Timber construction manual*, 5th ed. New York: Wiley.

APA (The Engineered Wood Association). 1994. *Northridge California earthquake.* T94-5. Tacoma, WA: APA.

APA. 2004. *Diaphragms and shear walls design/construction guide.* L350. Tacoma, WA: APA.

ATC (Applied Technology Council). 1981. *Guidelines for the design of horizontal wood diaphragms.* ATC-7. Redwood City, CA: ATC.

Bora, C., M. G. Oliva, S. D. Nakaki, and R. Becker. 2007. "Development of a precast concrete shear-wall system requiring special code acceptance." *PCI J.* 52 (1): 122–135.

Breyer, D., K. Fridley, Jr., D. Pollack, and K. Cobeen. 2006. *Design of wood structures ASD/LRFD*, 6th ed. New York: McGraw-Hill.

Charles Pankow Foundation. 2014. *Seismic design methodology document for precast concrete diaphragms.* Vancouver, WA: CPF.

Cobeen, K. 2004. "Recent developments in the seismic design and construction of woodframe buildings." In *Earthquake engineering: From engineering seismology to performance-based engineering*, edited by Y. Bozorgia and V. Bertero. Boca Raton, FL: CRC Press.

CUREE (Consortium of Universities for Research in Earthquake Engineering). 2004. *Recommendations for earthquake resistance in the design and construction of woodframe buildings.* CUREE W-30. Richmond, CA: CUREE.

CWC (Canadian Wood Council). 1995. *Wood reference handbook.* Ottawa: CWC.

CWC. 2005. *Wood design manual.* Ottawa: CWC.

Dolan, J. D. 2003. "Wood structures." In *Earthquake engineering handbook*, edited by W.-F. Chen and C. Scawthorn. Boca Raton, FL: CRC Press.

Easterling, W. S., and M. Porter. 1994. "Steel-deck-reinforced concrete diaphragms. I." *J. Struct. Eng.* 120 (2): 560–576. https://doi.org/10.1061/(ASCE)0733-9445(1994)120:2(577).

EERI (Earthquake Engineering Research Institute). 1996. "Northridge earthquake reconnaissance report." In Vol. 11 of *Earthquake spectra.* Oakland, CA: EERI.

Faherty, K. F., and T. G. Williamson. 1989. *Wood engineering and construction handbook.* New York: McGraw-Hill.

FEMA. 2003. *NEHRP recommended provisions for seismic regulations for new buildings and other structures.* FEMA 450. Washington, DC: FEMA.

FEMA. 2005. *Coastal construction manual.* FEMA 55. Washington, DC: FEMA.

Forest Products Laboratory. 1986. *Wood: Engineering design concepts.* University Park, PA: Pennsylvania State University.

Goetz, K. H., D. Hoor, K. Moehler, and J. Natterer. 1989. *Timber design and construction source book: A comprehensive guide to methods and practice.* New York: McGraw-Hill.

Hoyle, R. J., and F. E. Woeste. 1989. *Wood technology in the design of structures.* Ames, IA: Iowa State University Press.

ICC (International Code Council). 2006. *ICC standard on the design and construction of log structures, 3rd draft.* Country Club Hills, IL: ICC.

Ishizuka, T., and N. M. Hawkins. 1987. *Effect of bond deterioration on the seismic response of reinforced and partially prestressed concrete ductile moment resistant frames.* Rep. No. SM 87-2. Seattle: University of Washington.

Karacabeyli, E., and M. Popovsky. 2003. "Design for earthquake resistance." In *Timber engineering*, edited by H. Larsen and S. Thelandersson. New York: Wiley.

Keenan, F. J. 1986. *Limit states design of wood structures.* North York, ON: Morrison Hershfield.

Lee, N. H., K. S. Kim, C. J. Bang, and K. R. Park. 2007. "Tensile-headed anchors with large diameter and deep embedment in concrete." *ACI Struct. J.* 104 (4): 479–486.

Lee, N. H., K. R. Park, and Y. P. Suh. 2010. "Shear behavior of headed anchors with large diameters and deep embedments." *ACI Struct. J.* 107 (2): 146–156.

Nakaki, S. D., J. F. Stanton, and S. Sritharan. 1999. "An overview of the PRESSS five-story precast test building." *PCI J.* 44 (2): 26–39.

NOAA (National Oceanic and Atmospheric Administration). 1971. *San Fernando, California, earthquake of February 9, 1971.* Washington, DC: NOAA.

Park, R., and K. J. Thompson. 1977. "Cyclic load tests on prestressed and partially prestressed beam-column joints." *PCI J.* 22 (5): 84–110. https://doi.org/10.15554/pcij.09011977.84.110.

PCI (Precast/Prestressed Concrete Institute). 2004. *Precast/prestressed concrete piles.* Bridge design manual BM-20-04. Chicago: PCI.

Ren, R., and C. J. Naito. 2013. "Precast concrete diaphragm connector performance database." *J. Struct. Eng.* 139 (1): 15–27. https://doi.org/10.1061/(ASCE)ST.1943-541X.0000598.

SEAOC (Structural Engineers Association of California). 1999. *Recommended lateral force requirements and commentary.* Sacramento, CA: SEAOC.

SEAONC (Structural Engineers Association of Northern California). 2005. *Guidelines for seismic evaluation and rehabilitation of tilt-up buildings and other rigid wall/flexible diaphragm structures.* Sacramento, CA: SEAONC.

Sherwood, G. E., and R. C. Stroh. 1989. *Wood-frame house construction: Agricultural handbook 73.* Washington, DC: US Government Printing Office.

Somayaji, S. 1992. *Structural wood design.* St. Paul, MN: West.

Stalnaker, J. J., and E. C. Harris. 1996. *Structural design in wood*, 2nd ed. New York: McGraw-Hill.

US Departments of the Army, Navy, and Air Force. 1992. *Seismic design for buildings.* Tri-Services Manual TM5-809-10. Washington, DC: US Government Printing Office.

CHAPTER C15
SEISMIC DESIGN REQUIREMENTS FOR NONBUILDING STRUCTURES

C15.1 GENERAL

C15.1.1 Nonbuilding Structures Building codes traditionally have been perceived as minimum standards for the design of nonbuilding structures, and building code compliance of these structures is required by building officials in many jurisdictions. However, requirements in the industry reference documents are often at odds with building code requirements. In some cases, the industry documents need to be altered, whereas in other cases, the building codes need to be modified. Registered design professionals are not always aware of the numerous accepted documents within an industry and may not know whether the accepted documents are adequate. One of the intents of Chapter 15 of the standard is to bridge the gap between building codes and existing industry reference documents.

Differences between the ASCE 7 design approaches for buildings and industry document requirements for steel multi-legged water towers (Figure C15.1-1) are representative of this inconsistency. Historically, such towers have performed well when properly designed in accordance with American Water Works Association (AWWA) standards and industry practices. Those standards and practices differ from the ASCE 7 treatment of buildings in that tension-only rods are allowed, upset rods are preloaded at the time of installation, and connection forces are not amplified.

Chapter 15 also provides an appropriate link so that the industry reference documents can be used with the seismic ground motions established in the standard. Some nonbuilding structures are similar to buildings and can be designed using sections of the standard directly, whereas other nonbuilding structures require special analysis unique to the particular type of nonbuilding structure.

Building structures, vehicular bridges, electrical transmission towers, hydraulic structures (e.g., dams), buried utility lines and their appurtenances, and nuclear reactors are excluded from the scope of the nonbuilding structure requirements, although industrial buildings are permitted per Chapter 11 to use the provisions in Chapter 15 for nonbuilding structures with structural systems similar to buildings, provided specific conditions, identified in Section 11.1.3, are met. The excluded structures are covered by other well-established design criteria (e.g., electrical transmission towers and vehicular bridges), are not under the jurisdiction of local building officials (e.g., nuclear reactors and dams), or require technical considerations beyond the scope of the standard (e.g., buried utility lines and their appurtenances).

C15.1.2 Design Nonbuilding structures and building structures have much in common with respect to design intent and expected performance, but there are also important differences. Chapter 15 relies on other portions of the standard where possible and provides special notes where necessary.

There are two types of nonbuilding structures: those with structural systems similar to buildings and those with structural systems not similar to buildings. Specific requirements for these two cases appear in Sections 15.5 and 15.6, respectively.

C15.1.3 Structural Analysis Procedure Selection Nonbuilding structures that are similar to buildings are subject to the same analysis procedure limitations as building structures. Nonbuilding structures that are not similar to buildings are subject to those limitations and are subject to procedure limitations prescribed in applicable specific reference documents.

For many nonbuilding structures supporting flexible system components, such as pipe racks (Figure C15.1-2), the supported piping and platforms generally are not regarded as rigid enough to redistribute seismic forces to the supporting frames.

For nonbuilding structures supporting very stiff (i.e., rigid) system components, such as steam turbine generators (STGs) and heat recovery steam generators (HRSGs) (Figure C15.1-3), the supported equipment, ductwork, and other components (depending on how they are attached to the structure) may be rigid enough to redistribute seismic forces to the supporting frames. Torsional effects may need to be considered in such situations.

Section 12.6 presents seismic analysis procedures for building structures based on the Seismic Design Category (SDC); the fundamental period, T; and the presence of certain horizontal or vertical irregularities in the structural system. Where the fundamental period is greater than or equal to $3.5T_s$ (where $T_s = S_{D1}/S_{DS}$), the use of the equivalent lateral force (ELF) procedure is not permitted in Seismic Design Categories D, E, and F. This requirement is based on the fact that, unlike the dominance of the first-mode response in case of buildings with lower first mode period, higher vibration modes do contribute more significantly in situations when the first mode period is larger than $3.5T_s$. For buildings that exhibit classic flexural deformation patterns (such as slender shear-wall or braced-frame systems), the second mode frequency is at least 3.5 times the first mode frequency, so where the fundamental period exceeds $3.5T_s$, the higher modes have larger contributions to the total response because they occur near the peak of the design response spectrum.

It follows that dynamic analysis (modal response spectrum analysis or response history analysis) may be necessary to properly evaluate building-like nonbuilding structures if the first mode period is larger than $3.5T_s$ and the ELF analysis is sufficient for nonbuilding structures that respond as single-degree-of-freedom systems.

The recommendations for nonbuilding structures provided in the following are intended to supplement the designer's judgment and experience. The designer is given considerable latitude in selecting a suitable analysis method for nonbuilding structures.

Figure C15.1-1. Steel multilegged water tower.
Source: Courtesy of CB&I LLC; reproduced with permission.

Figure C15.1-2. Steel pipe rack.
Source: Courtesy of CB&I LLC; reproduced with permission.

Building-Like Nonbuilding Structures. Table 12.6-1 is used in selecting analysis methods for building-like nonbuilding structures, but, as illustrated in the following three conditions, the relevance of key behavior must be considered carefully.

1. **Irregularities.** Table 12.6-1 requires dynamic analysis for Seismic Design Category D, E, and F structures that have certain horizontal or vertical irregularities. Some of these building irregularities (defined in Section 12.3.2) are relevant to nonbuilding structures. The weak- and soft-story vertical irregularities (Types 1a, 1b, 4a, and 4b of Table 12.3-2) are pertinent to the behavior of building-like nonbuilding structures. Other vertical and horizontal irregularities may or may not be relevant:
   (a) *Horizontal irregularities* of Types 1a and 1b affect the choice of analysis method, but these irregularities apply only where diaphragms are rigid or semirigid, and some building-like nonbuilding structures have either no diaphragms or flexible diaphragms.
   (b) *Vertical irregularity* Type 2 addresses large differences in the horizontal dimension of the seismic force-resisting system in adjacent stories because the resulting stiffness distribution can produce a fundamental mode shape unlike that assumed in the development of the equivalent lateral force procedure. Because the concern relates to stiffness distribution, the horizontal dimension of the seismic force-resisting system, not of the overall structure, is important.
2. **Arrangement of supported masses.** Even where a nonbuilding structure has a building-like appearance, it may not behave like a building, depending on how masses are attached. For example, the response of nonbuilding structures with suspended vessels and boilers cannot be determined reliably using the equivalent lateral force procedure because of the pendulum modes associated with the significant mass of the suspended components. The resulting pendulum modes, while potentially reducing story shears and base shear, may require large clearances to allow pendulum motion of the supported components and may produce excessive demands on attached piping. Dynamic analysis is highly recommended in such cases, with consideration for appropriate impact forces in the absence of adequate clearances.
3. **Relative rigidity of beams.** Even where a classic building model may seem appropriate, the equivalent lateral force

Figure C15.1-3. Heat recovery steam generators.
Source: Courtesy of CB&I LLC; reproduced with permission.

procedure may underpredict the total response if the beams are flexible relative to the columns (of moment frames) or the braces (of braced frames). This underprediction occurs because higher modes associated with beam flexure may contribute more significantly to the total response (even if the first mode response is at a period less than $3.5T_s$). This situation of flexible beams can be especially pronounced for nonbuilding structures because the "normal" floors common to buildings may be absent. Therefore, the dynamic analysis procedures are suggested for building-like nonbuilding structures with flexible beams.

Nonbuilding Structures Not Similar to Buildings. The (static) equivalent lateral force procedure is based on classic building dynamic behavior, which differs from the behavior of many nonbuilding structures not similar to buildings. As discussed next, several issues should be considered for selecting either an appropriate method of dynamic analysis or a suitable distribution of lateral forces for static analysis.

1. *Structural geometry*. The dynamic response of nonbuilding structures with a fixed base and a relatively uniform distribution of mass and stiffness, such as bottom-supported vertical vessels, stacks, and chimneys, can be represented adequately by a cantilever (shear building) model. For these structures, the equivalent lateral force procedure provided in the standard is suitable. This procedure treats the dynamic response as being dominated by the first mode. In such cases, it is necessary to identify the first mode shape (using, for instance, the Rayleigh–Ritz method or other classical methods from the literature) for distribution of the dynamic forces. For some structures, such as tanks with low height-to-diameter ratios storing granular solids, it is conservative to assume a uniform distribution of forces. Dynamic analysis is recommended for structures that have neither a uniform distribution of mass and stiffness nor an easily determined first mode shape.
2. *Number of lateral supports*. Cantilever models are obviously unsuitable for structures with multiple supports. Figure C15.1-4(a) shows a nonbuilding braced frame structure that provides nonuniform horizontal support to a piece of equipment. In such cases, the analysis should include coupled model effects. For such structures, an application of the equivalent lateral force method could be used, depending on the number and locations of the supports. For example, most beam-type configurations lend themselves to application of the equivalent lateral force method. For adjacent nonbuilding structures connected by nonstructural components [Figure 15.1-4(b)], a combined dynamic analysis may be required, as indicated in Section 15.2.1.
3. *Method of supporting dead weight*. Certain nonbuilding structures (such as power boilers) are supported from the top. They may be idealized as pendulums with uniform mass distribution. In contrast, a suspended platform may be idealized as a classic pendulum with concentrated mass. In either case, these types of nonbuilding structures can be analyzed adequately using the equivalent lateral force method by calculating the appropriate frequency and mode shape. Figure C15.1-5 shows a nonbuilding structure containing lug-supported equipment with W_P greater than $0.20(W_S + W_P)$. In such cases, the analysis should include a coupled system, with the mass of the equipment and the local flexibility of the supports considered in the model. Where the support is located near the nonbuilding

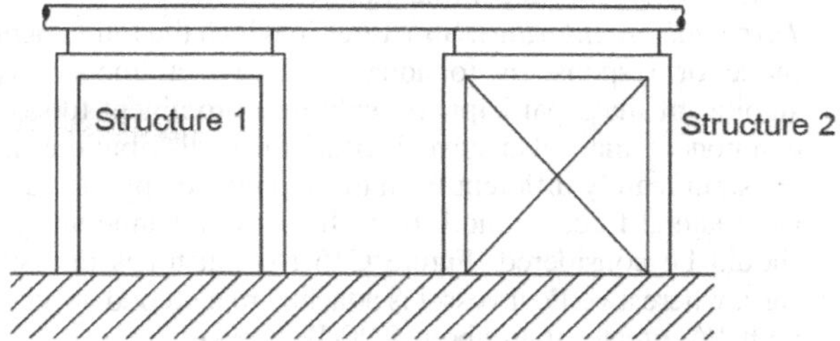

Figure C15.1-4(b). Adjacent nonbuilding structures connected by nonstructural components.

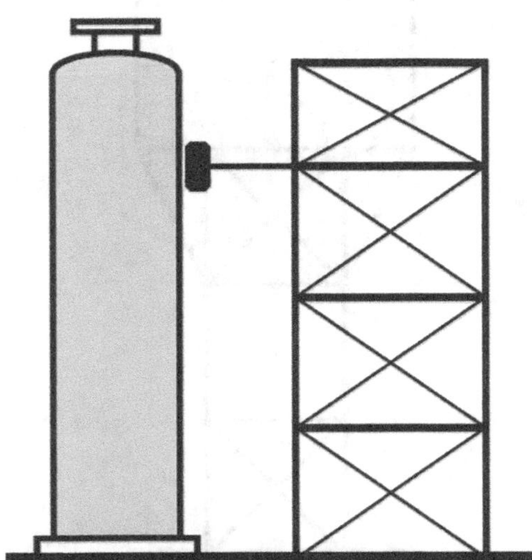

Figure C15.1-4(a). Multiple lateral supports.

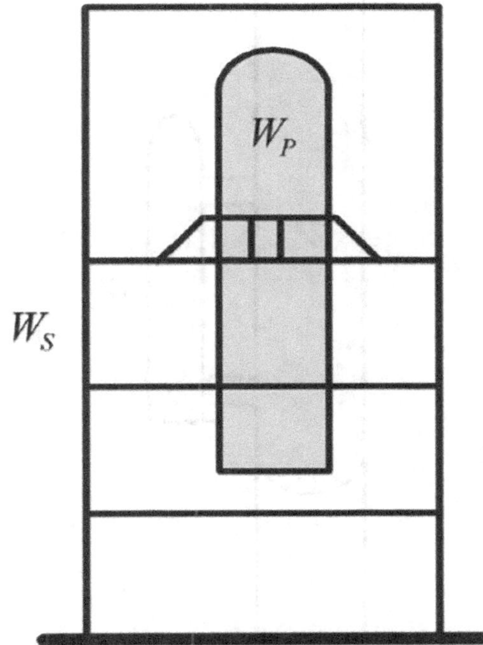

Figure C15.1-5. Unusual support of dead weight.

structure's vertical location of the center of mass, a dynamic analysis is recommended.

4. *Torsional irregularities*. Structures in which the fundamental mode of response is torsional or in which modes with significant mass participation exhibit a prominent torsional component may also have inertial force distributions that are significantly different from those predicted by the equivalent lateral force method. In such cases, dynamic analyses should be considered. Figure C15.1-6 illustrates one such case, where a vertical vessel is attached to a secondary vessel with W_2 greater than about $0.20(W_1 + W_2)$.
5. *Stiffness and strength irregularities*. Just as for building-like nonbuilding structures, abrupt changes in the distribution of stiffness or strength in a nonbuilding structure not similar to buildings can result in substantially different inertial forces from those indicated by the equivalent lateral force method. Figure C15.1-7 represents one such case. For structures that have such configurations, consideration should be given to the use of dynamic analysis procedures. Even where dynamic analysis is required, the standard does not define in any detail the degree of modeling; an adequate model may have a few dynamic degrees of freedom or tens of thousands of dynamic degrees of freedom. The important point is that the model captures the significant dynamic response features so that the resulting lateral force distribution is valid for design. The designer is responsible for determining whether dynamic analysis is warranted and, if so, the degree of detail required to adequately address the seismic performance.
6. *Coupled response*. Where the weight of the supported structure is large compared with the weight of the supporting structure, the combined response can be affected significantly by the flexibility of the supported nonbuilding structure. In that case, dynamic analysis of the coupled system is recommended. Examples of such structures are shown in Figure C15.1-8(a and b). Figure C15.1-8(a) shows a flexible nonbuilding structure with W_p greater than $0.20(W_S + W_P)$, supported by a relatively flexible structure; the flexibility of the supports and attachments should be considered. Figure C15.1-8(b) shows flexible equipment connected by a large-diameter, thick-walled pipe and supported by a flexible structure; the structures should be modeled as a coupled system including the pipe.

Exceptions previously allowed in ASCE 7-16 for the scaling of base shear and drifts determined using the modal analysis procedure of Section 12.9.1 for distributed-mass cantilever structures have been removed in ASCE 7-22. The exceptions were removed because the modal response spectrum analysis procedure (MRSA) of Section 12.9.1 does not do a good job of predicting the story shear, overturning moment, and story drift when compared to nonlinear response history analysis and ELF analysis. Scaling of base shears and drifts determined using the modal analysis procedure of Section 12.9.1 for distributed-mass cantilever structures is intended to follow the rules of Section 12.9.1.

C15.1.4 Nonbuilding Structures Sensitive to Vertical Ground Motions Traditionally, ASCE 7 did not provide guidance to address designing for a separate vertical ground motion. Historically, this omission has not been a problem for buildings because there is inherent strength in the vertical direction because of the margin that is developed when the dead load and live load are applied. However, this is not necessarily the case for nonbuilding structures. Many nonbuilding structures are

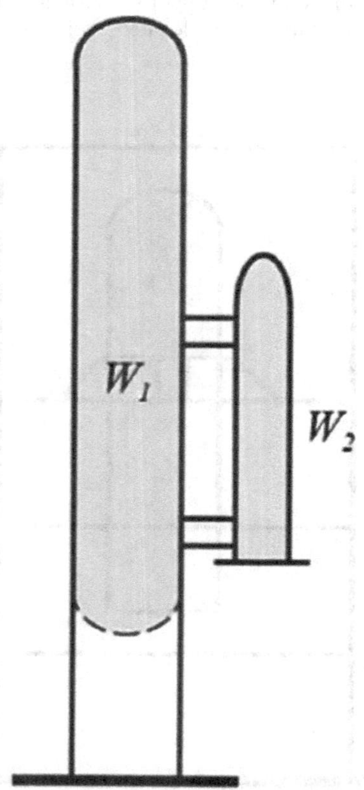

Figure C15.1-6. Torsional irregularity.

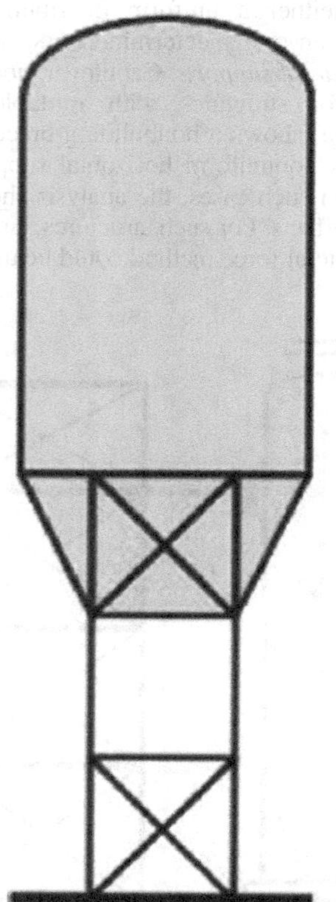

Figure C15.1-7. Soft-story irregularity.

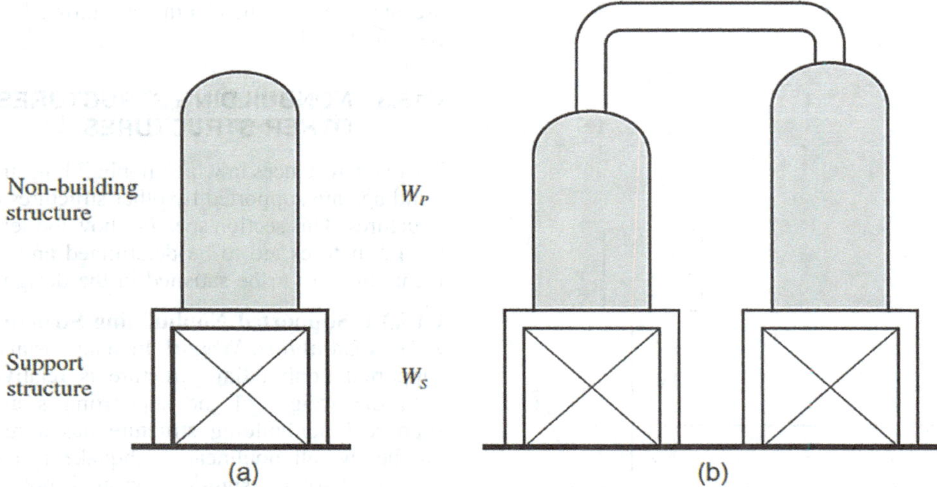

Figure C15.1-8. Coupled system.

sensitive to vertical motions and do not have the benefit of the inherent strength that exists in buildings. Examples of some structures are liquid and granular storage tanks or vessels, suspended structures (such as boilers), and nonbuilding structures incorporating horizontal cantilevers. Such structures are required to incorporate Section 11.9 into the design of the structure in lieu of applying the traditional vertical ground motion of $0.2S_{DS}$.

C15.2 NONBUILDING STRUCTURES CONNECTED BY NONSTRUCTURAL COMPONENTS TO OTHER ADJACENT STRUCTURES

C15.2.1 General Requirements The ASCE 7-22 edition of this standard added coupled analysis requirements for adjacent structures that are connected by nonstructural components to other adjacent structures. This scenario commonly occurs in the case of industrial structures, such as pipe racks and equipment structures. Before this edition of the standard, little guidance was given on the design of adjacent structures connected by nonstructural components. Work by Wey and Mejia (2019) proposed decoupling triggers based on the fundamental periods of the connected structures and the stiffness of the interconnecting elements. The use of fundamental periods as decoupling triggers follows similar reasoning as the work of Hadjian and Ellison (1986).

Nonstructural components spanning between structures that have a small stiffness in the direction of seismic loading compared to the stiffnesses of both the nonbuilding and adjacent structures do not significantly alter the fundamental periods of the connected structures. Therefore, when the stiffness of the interconnecting element in the direction of motion is small compared to the nonbuilding and adjacent structures, the adjacent structures can be designed independently, without needing to consider the structural characteristics of the nonstructural components. Electrical cable trays, small-bore piping, and light platforming are examples of nonstructural components that can be considered flexible with respect to the supporting structures. However, these components should be designed to accommodate the relative displacement of the nonbuilding structure and the adjacent structure. Typical platform support members with significant axial stiffness in the direction of motion have either slotted or sliding connections at one end to minimize the transfer of seismic loading. An additional intent of this provision is to encourage the use of detailing and devices, such as slide plates and expansion joints, that work to reduce the stiffness of the connecting element in the direction of motion.

As the connecting nonstructural component stiffness increases, coupling effects become more pronounced, and the fundamental period of the combined structure tends to dominate the seismic response. When structures with similar dynamic characteristics in the direction of motion are connected by stiff elements, the resulting fundamental period of the combined system will fall between the fundamental periods of the independently analyzed structures. As a result, interconnected structures with similar dynamic characteristics in the direction of motion will tend to behave similarly when connected by stiff components, provided ground motions are sufficiently uniform. Large-bore pipe, thick-walled ductwork, and large mechanical equipment are examples of nonstructural components that can be considered rigid with respect to the supporting structures.

Given the complex nature of structures with horizontal and vertical irregularities, such as those shown in Section C15.1.3, the irregularities of one or both connected structures can lead to inaccuracies in the magnitude and distribution of the resulting lateral forces. The designer should acknowledge that, in such scenarios, there is limited applicability to the coupling exceptions provided in this section, and the designer is responsible for accounting for the combined behavior of the connected structures.

The exceptions only apply to structures with vertical mass and stiffness regularity with similar fundamental periods. Figure C15.2-1 (a stack connected to a supported tank by large ductwork) provides an example for which the exceptions do not apply. The tank support structure has both mass and soft-story irregularities (Figure C15.1-7). Connection of the ductwork to the tank support introduces an additional stiffness irregularity to the structure. Design of the structures shown in Figure C15.2-1 requires a coupled analysis.

C15.2.2 Nonstructural Components Spanning between Nonbuilding Structures Depending on how they are attached to the supporting structures, nonstructural components spanning between structures may be rigid enough to redistribute seismic forces between them. The rigidity of the interconnecting elements and any possible interaction caused by out-of-phase motion between structures may result in significant induced

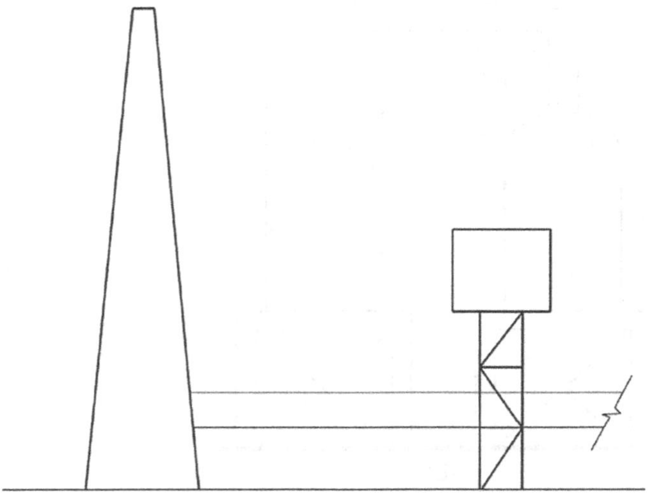

Figure C15.2-1. Stack connected to tower by large duct.

forces and displacements. Design of nonstructural components for these seismic demands requires careful consideration of the relative forces and displacements acting on the nonstructural component in any direction, including out-of-phase motion, as illustrated in Figure C13.3-5, and should be compatible with the assumptions used to define the effective component stiffness in the direction of motion. The design of such components and their anchorage should incorporate the requirements set forth in Section 13.3.2.2, regardless of whether the connected structures are analyzed independently, as allowed by the exceptions in Section 15.2.1.

C15.3 NONBUILDING STRUCTURES SUPPORTED BY OTHER STRUCTURES

There are instances in which nonbuilding structures not similar to buildings are supported by other structures or other nonbuilding structures. This section specifies how the seismic design loads for such structures are to be determined and the detailing requirements that are to be satisfied in the design.

C15.3.1 Supported Nonbuilding Structures with Less Than 20% of Combined Weight In many instances, the weight of the supported nonbuilding structure is relatively small compared with the weight of the supporting structure, such that the supported nonbuilding structure has a relatively small effect on the overall nonlinear earthquake response of the primary structure during design-level ground motions. It is permitted to treat such structures as nonstructural components and to use the requirements of Chapter 13 for their design. The ratio of secondary component weight to total weight of 20% at which this treatment is permitted was not originally documented in any commentary but appears to be based on judgment and the work of Hadjian and Ellison (1986). Figure C15.3-1 shows the decoupling criteria originally proposed by Hadjian and Ellison (1986). Combinations of frequency ratio and mass ratio falling to the left of the curve would be exempt from a coupled analysis. The original triggering ratio of 25%, now revised to 20%, was introduced by the SEAOC Seismology Committee into provisions in the 1988 *Uniform Building Code* (ICBO 1988). The original proposal was based on consideration of the ratio of

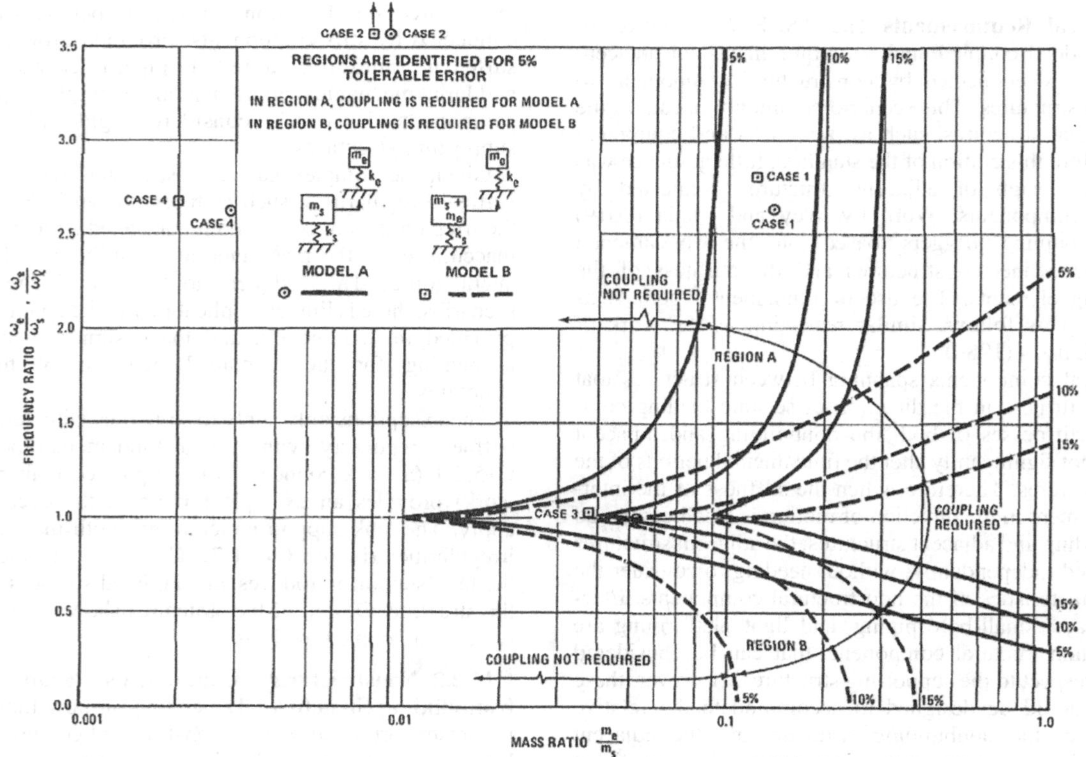

Fig. 1 System frequency errors due to decoupling of subsystems

Figure C15.3-1. Decoupling criteria.

Source: From Hadjian and Ellison (1986).

secondary component weight to the weight of the supporting structure and a single point of attachment, whereby multiple closely spaced attachments can be represented as a single point of attachment. In ASCE 7-02, the weight, W, was amended to reflect the total weight of component and supporting structure, but the 25% limit remained unchanged. ASCE 7-22 revised the triggering ratio downward to 20% to better align with the original research of Hadjian and Ellison (1986). Analytical studies, typically based on linear-elastic primary and secondary structures, support a lower triggering ratio, but the SEAOC Seismology Committee judged that the 20% ratio is appropriate where primary and secondary structures exhibit nonlinear behavior because this will lessen the effects of resonance and interaction. Therefore, a vertical line, originally set at a mass ratio of 25% but now revised to 20%, was used instead of the curve shown in Figure C15.3-1 to determine when a coupled analysis was required. The 20% ratio also reflects a 15% tolerance error on frequency. In cases in which a nonbuilding structure (or nonstructural component) is supported by another structure, it may be appropriate to analyze in a single model. In such cases, it is intended that seismic design loads and detailing requirements be determined following the procedures of Section 15.3.2. Where there are multiple large nonbuilding structures, such as vessels supported on a primary nonbuilding structure, and the weight of an individual supported nonbuilding structure does not exceed the 20% limit but the combined weight of the supported nonbuilding structures does, it is recommended that the combined analysis and design approach of Section 15.3.2 be used. It is also suggested that dynamic analysis be performed in such cases, because the equivalent lateral force procedure may not capture some important response effects in some members of the supporting structure.

Where the weight of the supported nonbuilding structure does not exceed the 20% limit and a combined analysis is performed, the following procedure should be used to determine the F_p force of the supported nonbuilding structure based on Equation (13.3-7):

1. A modal analysis should be performed in accordance with Section 12.9 as modified by section 15.1.3.
2. For a component supported at level i, the acceleration at that level should be taken as a_i, the total shear just below level i divided by the seismic weight at and above level i.
3. Next, the elastic value of the component shear force coefficient should be determined, as the shear force from the modal analysis at the point of attachment of the component to the structure, divided by the weight of the component. This value is taken as $a_i[\frac{C_{AR}}{R_{po}}]$
4. The resulting value of $a_i[\frac{C_{AR}}{R_{po}}]$ should be used in Equation (13.3-7); the resulting value of F_p is subject to the maximum and minimum values of Equations (13.3-2) and (13.3-3), respectively.

C15.3.2 Supported Nonbuilding Structures with Greater Than or Equal to 20% of Combined Weight Where the weight of the supported structure is relatively large compared with the weight of the supporting structure, the overall response can be affected significantly. Previously, the standard set forth two analysis approaches, depending on the rigidity of the nonbuilding structure. The determination of what was deemed rigid or flexible was based on the same criteria used for nonstructural components. Few if any nonbuilding structures with weights greater than or equal to 20% of the total weight are rigid. Therefore, the case in which the supported nonbuilding structure is rigid has been removed from the standard. Instead, exceptions have been added introducing new decoupling criteria for nonbuilding structures that are sufficiently detuned (that is, where the fundamental period of the supporting structure, including the lumped weight of the nonbuilding structure, is sufficiently small or large compared to that of the nonbuilding structure). These criteria follow work initially performed by Hadjian and Ellison (1986), with specific detuning thresholds proposed by Wey and Mejia (2019). The criteria are in line with current practice in seismic design of industrial structures, such as steel vertical vessels supported on concrete tabletops, where the concrete structure's fundamental period is calculated including the lumped weight of the steel vessel, and the structure is designed without knowledge of the structural characteristics of the supported vessel. When the fundamental period of the nonbuilding structure is more than twice that of the supporting structure, this standard permits the nonbuilding structure to be modeled as attached to a rigid base.

A combined model of the supporting structure and the supported nonbuilding structure is now used in all cases, with exceptions given in Section 15.3.2. The design loads and detailing are determined based on the lower R value of the supported nonbuilding structure or the supporting structure. The use of the lower R value of the supported nonbuilding structure or supporting structure is based on the rules for vertical combinations found in Section 12.2.3.1, Item 2.

Although not specifically mentioned in Section 15.3.2, another approach is permitted. A nonlinear response history analysis of the combined system can be performed in accordance with Chapter 16, and the results can be used for the design of both the supported and supporting nonbuilding structures. This option should be considered where standard static and dynamic elastic analysis approaches may be inadequate to evaluate the earthquake response (such as for suspended boilers). This option should be used with extreme caution, because modeling and interpretation of results require considerable judgment. Because of this sensitivity, Chapter 16 requires independent design review.

C15.4 STRUCTURAL DESIGN REQUIREMENTS

This section specifies the basic coefficients and minimum design forces to be used to determine seismic design loads for nonbuilding structures. It also specifies height limits and restrictions. As with building structures, it presumes that the first step in establishing the design forces is to determine the design base shear for the structure.

There are two types of nonbuilding structures: those with structural systems similar to buildings and those with structural systems not similar to buildings. Specific requirements for these two cases appear in Sections 15.5 and 15.6, respectively.

Table 15.4-1 gives the response modification coefficient, R, for nonbuilding structures similar to buildings. Table 15.4-2 gives the response modification coefficient for nonbuilding structures not similar to buildings. Every response modification coefficient has associated design and detailing requirements to ensure the required ductility associated with that response modification coefficient value (e.g., AISC 341). Some structures, such as pipe racks, do not resemble a traditional building in that they do not house people or have such things as walls and bathrooms. These structures have lateral force-resisting systems composed of braced frames and moment frames, similar to a traditional building. Therefore, pipe racks are considered nonbuilding structures similar to buildings. The response modification coefficient for a pipe rack should be taken from Table 15.4-1 for the appropriate lateral force-resisting system used, and the braced frames and/or moment frames used must meet all of the design and detailing requirements associated with the R value selected (see Section 15.5.2, "Pipe Racks").

Most major power distribution facility (power island) structures, such as heat recovery steam generator (HRSG) support structures, steam turbine pedestals, coal boiler support structures, pipe racks, air inlet structures, and duct support structures, also resist lateral forces predominantly by use of building-like framing systems such as moment frames, braced frames, or cantilever column systems. Therefore, their response modification coefficient should be selected from Table 15.4-1, and they must meet all the design and detailing requirements associated with the response modification coefficient selected.

Many nonbuilding structures, such as flat-bottom tanks, silos, and stacks, do not use braced frames or moment frames similar to those found in buildings to resist seismic loads. Therefore, they have their own unique response modification coefficient, which can be found in Table 15.4-2.

For nonbuilding structures with lateral systems composed predominantly of building-like framing systems, such as moment frames, braced frames, or cantilever column systems, it would be inappropriate to extrapolate the descriptions in Table 15.4-2, which would give inappropriately high response modification coefficients and the elimination of detailing requirements.

Once a response modification coefficient is selected from the tables, Section 15.4.1 provides additional guidance.

C15.4.1 Design Basis Separate tables provided in this section identify the basic coefficients, associated detailing requirements, and height limits and restrictions for the two types of nonbuilding structures.

For nonbuilding structures similar to buildings, the design seismic loads are determined using the same procedures used for buildings as specified in Chapter 12, with two exceptions: fundamental periods are determined in accordance with Section 15.4.4, and Table 15.4-1 provides additional options for structural systems. Although only Section 12.8 (the equivalent lateral force procedure) is specifically mentioned in Section 15.4.1, Section 15.1.3 provides the analysis procedures that are permitted for nonbuilding structures.

In Table 15.4-1, seismic coefficients, system restrictions, and height limits are specified for a few nonbuilding structures similar to buildings. The values of R, Ω_0, and C_d, the detailing requirement references, and the structural system height limits are the same as those in Table 12.2-1 for the same systems, except for ordinary moment frames. In Chapter 12, increased height limits for ordinary moment frame structural systems apply to metal building systems, whereas in Chapter 15 they apply to pipe racks with end plate bolted moment connections. The seismic performance of pipe racks was judged to be similar to that of metal building structures with end plate bolted moment connections, so the height limits were made the same as those specified in previous editions.

Table 15.4-1 also provides lower R values with less restrictive height limits in Seismic Design Categories D, E, and F based on good performance in past earthquakes. For some options, no seismic detailing is required if very low values of R (and corresponding high seismic design forces) are used. The concept of extending this approach to other structural systems is the subject of future research using the methodology developed in FEMA P695 (2009).

For nonbuilding structures not similar to buildings, the seismic design loads are determined as in Chapter 12, with three exceptions: the fundamental periods are determined in accordance with Section 15.4.4, the minima are those specified in Section 15.4.1, Item 2, and the seismic coefficients are those specified in Table 15.4-2.

Some entries in Table 15.4-2 may seem to be conflicting or confusing. For example, the first major entry is for elevated tanks, vessels, bins, or hoppers. A subset of this entry is for tanks on braced or unbraced legs. This subentry is intended for structures with supporting columns that are integral with the shell (such as an elevated water tank). Tension-only bracing is allowed for such a structure. If the tank or vessel is supported by building-like frames, the frames are to be designed in accordance with all of the restrictions normally applied to building frames. Section 15.3 provides provisions for nonbuilding structures supported by building-like frames. Beginning with the 2005 edition of ASCE 7, Table 15.4-2 contained an entry for "Tanks or vessels supported on structural towers similar to buildings." Under certain circumstances, text provided with this table entry conflicted with the requirements of Section 15.3. If the weight of the nonbuilding structure is relatively small compared to the weight of the structure (less than 25% of the weight of the structure) or the nonbuilding structure is rigid, the supported nonbuilding structure can be treated as a nonstructural component, and the values of the supporting structure seismic coefficients can be taken from Table 15.4-1. Under these circumstances, the deleted entry was correct. However, if the weight of the supported nonbuilding structure is not small and the nonbuilding structure is flexible (which is generally the case, especially when you consider the vertical and rocking flexibility of supporting floor beams), the seismic coefficients are determined as the most conservative.

The accidental torsion requirements of Section 12.8.4.2 were formulated primarily for use in building structures. The primary factors that contribute to accidental torsion are lateral force-resisting systems that are located primarily near the center of the structure rather than the perimeter, disproportionate concentration of inelastic demands in system components, the effects of nonstructural elements, uncertainties in defining the structure's stiffness characteristics, and spatial variation (and rotational components of ground motions) of horizontal input motions applied to long structures. Inherently torsionally resistant systems, as defined in Section 15.4.1, Item 5, with R values less than or equal to 3.5 are not expected to have inelastic demands of a level that would require additional consideration of accidental torsion. In addition, nonbuilding structures rarely contain significant nonstructural elements that are not accounted for explicitly in the design of these structures, and they typically have very well-known mass and stiffness characteristics. Nonbuilding structures also rarely, if ever, have their lateral force-resisting systems located at the center of the structure in plan rather than at the perimeter. The requirement that the calculated center of rigidity at each diaphragm is greater than 5% of the plan dimension of the diaphragm in each direction from the calculated center of mass of the diaphragm prevents configurations of lateral force-resisting elements that are inherently susceptible to the effects of torsion from being exempted from the effects of accidental torsion. Spatial variations of ground motions should be considered in the design of structures of considerable length. If there are significant variations between the full and empty weights of the structure, the inherent torsion caused by these variations should be considered in the design of the structure. If there is a nonuniform distribution of mass in silos or bins storing bulk materials because of multiple filling or discharge points, multiple hoppers, nonuniform funnel flow, bulk material behavior, or other operational considerations, the inherent torsion caused by these conditions should be considered in the design of the silo or bin.

For nonbuilding structures not similar to a building, Table 15.4-2 provides an appropriate entry for other self-supporting structures, tanks, or vessels not covered in Table 15.4-2 or a reference standard. Nonbuilding structures outside the scope of Chapter 15 that may be covered by other regulations are exempted in Section 11.1.2.

C15.4.1.1 Importance Factor The Importance Factor for a nonbuilding structure is based on the risk category defined in Chapter 1 of this standard or the building code being used in conjunction with the standard. In some cases, reference standards provide a higher Importance Factor, in which case the higher Importance Factor is used.

If the Importance Factor is taken as 1.0 based on a hazard and operability (HAZOP) analysis performed in accordance with Chapter 1, the third paragraph of Section 1.5.3 requires careful consideration; worst-case scenarios (instantaneous release of a vessel or piping system contents) must be considered. HAZOP risk analysis consultants often do not make such assumptions, so the design professional should review the HAZOP analysis with the consultant to confirm that such assumptions have been made to validate adjustment of the Importance Factor. Clients may not be aware that HAZOP consultants do not normally consider the worst-case scenario of instantaneous release but tend to focus on other, more hypothetical, limited-release scenarios, such as those associated with a 2 in.^2 (1,290 mm^2) hole in a tank or vessel.

C15.4.2 Rigid Nonbuilding Structures The definition of rigid (having a natural period of less than 0.06 s) was selected judgmentally. Below that period, the energy content of seismic ground motion is generally believed to be very low, and therefore the building response is not likely to be excessively amplified. Also, it is unlikely that any building will have a first mode period as low as 0.06 s, and it is even more unusual for a second mode period to be that low. Thus, the likelihood of either resonant behavior or excessive amplification becomes quite small for equipment that has periods below 0.06 s.

The analysis to determine the period of the nonbuilding structure should include the flexibility of the soil subgrade.

C15.4.3 Loads As is done for buildings, the seismic weight must include the range of design operating weight of permanent equipment.

C15.4.4 Fundamental Period A significant difference between building structures and nonbuilding structures is that the approximate period formulas and limits of Section 12.8.2.1 may not be used for nonbuilding structures. In lieu of calculating a specific period for a nonbuilding structure for determining seismic lateral forces, it is of course conservative to assume a period of $T = T_s$, which results in the largest lateral design forces. Computing the fundamental period is not considered a significant burden, because most commonly used computer analysis programs can perform the required calculations.

C15.4.5 Drift Limit Unlike buildings, nonbuilding structures do not have explicit drift limits. Drift in nonbuilding structures must simply be limited to a level that does not adversely affect structural stability or attached or interconnected components and elements such as walkways and piping.

C15.4.6 P-Delta P-delta effects are evaluated differently for nonbuilding structures similar to buildings and nonbuilding structures not similar to buildings. For nonbuilding structures similar to buildings, the same requirements used for buildings from Section 12.8.7 are applied. For nonbuilding structures not similar to buildings, a much more conservative approach to P-delta is taken. Many nonbuilding structures not similar to buildings are distributed-mass cantilever structures or inverted-pendulum structures for which plate buckling is the primary failure mode. Therefore, the effects of P-delta are based on displacements determined by an elastic analysis multiplied by C_d/I_e, the expected inelastic displacement. An example of a nonbuilding structure not similar to buildings where P-delta is critical to the function and stability of the structure is a single-pedestal elevated water tank. AWWA D100 (2006a) uses this approach for the seismic design of single pedestal-type steel elevated water tanks. An example of this type of structure is shown in Figure C15.7-2. Examples of nonbuilding structures not similar to buildings for which P-delta is not critical to the function or stability of the structure are flat-bottom ground-supported tanks and chimneys and stacks (with no equipment weight added to the top of the structure, such as a scrubber). An exhaustive list of nonbuilding structures not similar to buildings for which P-delta is not critical to the function or stability of the structure cannot be compiled because many of these structures in their standard configuration can be modified (usually by attaching large masses near the top of the structure), thus requiring that a P-delta analysis be performed.

C15.4.7 Drift, Deflection, and Structure Separation Nonbuilding structure drifts, deflection, and structure separation are calculated using strength design factored load combinations to be compatible with the seismic load definition and the definition of the C_d factors. This philosophy is consistent with that of drift, deflections, and structure separation for buildings defined in Chapter 12.

C15.4.8 Site-Specific Response Spectra Where site-specific response spectra are required, they should be developed in accordance with Chapter 21 of the standard. If determined for other recurrence intervals, Section 21.1 applies, but Sections 21.2 through 21.4 apply only to risk-targeted maximum considered earthquake (MCE_R) determinations. Where other recurrence intervals are used, it should be demonstrated that the requirements of Chapter 15 also are satisfied.

C15.4.9 Anchors in Concrete or Masonry Many nonbuilding structures rely on the ductile behavior of anchor bolts to justify the response modification factor, *R*, assigned to the structure. Nonbuilding structures typically rely more heavily on anchorage to provide system ductility. The additional requirements of Section 15.4.9 provide additional anchorage strength and ductility to support the response modification factors assigned to these systems. The addition of Section 15.4.9 provides a consistent treatment of anchorage for nonbuilding structures.

C15.4.9.4 ASTM F1554 Anchors ASTM F1554 (2020) contains a requirement that is not consistent with the anchor requirements in Chapter 15. Section 6.4 of ASTM F1554 allows the anchor supplier to substitute weldable Grade 55 anchors for Grade 36 anchors without the approval of the registered design professional. Because many nonbuilding structures rely on the ability of the anchors to stretch to justify the response modification factor, *R*, assigned to the structure, a higher-yield anchor cannot be allowed to be substituted for a lower-yield anchor without the approval of the registered design professional. Except where anchors are specified and are designed as ductile steel anchors in accordance with ACI 318 (2019), Section 17.10.5.3(a), or where the design must meet the requirements of Section 15.7.5 or Section 15.7.11.7(b), this provision does not prohibit ductility from being provided by another element of the structure. In that case, the ASTM F1554 anchors would be designed for the corresponding forces.

C15.4.10 Requirements for Nonbuilding Structure Foundations on Liquefiable Sites Section 12.13.9 allows shallow foundations to be built on liquefiable soils, with a number of restrictions. Many nonbuilding structures are sensitive to large foundation settlements. This sensitivity is caused by restraint imposed by interconnecting piping and equipment and the buckling sensitivity of shell structures. Therefore, to build these structures on shallow foundations on liquefiable soils, it must be demonstrated that the foundation, the nonbuilding structure not similar to buildings, and the connecting systems, can be designed for the soil strength loss, the anticipated settlements from lateral spreading, and the total and differential settlements induced by MCE_G ground motions.

C15.5 NONBUILDING STRUCTURES SIMILAR TO BUILDINGS

C15.5.1 General Although certain nonbuilding structures exhibit behavior similar to that of building structures, their functions and occupancies are different. Section 15.5 of the standard addresses the differences.

C15.5.2 Pipe Racks Freestanding pipe racks supported at or below grade with framing systems that are similar to building systems are designed in accordance with Section 12.8 or 12.9 and Section 15.4. Single-column pipe racks that resist lateral loads should be designed as inverted pendulums.

Based on good performance in past earthquakes, Table 15.4-1 sets forth the option of lower R values and less restrictive height limits for structural systems commonly used in pipe racks. The R-value-versus-height-limit tradeoff recognizes that the size of some nonbuilding structures is determined by factors other than traditional loadings and results in structures that are much stronger than required for seismic loadings. Therefore, the ductility demand is generally much lower than that for a corresponding building. The intent is to obtain the same structural performance at the increased heights. This option is economical in most situations because of the relative cost of materials and construction labor. The lower R values and increased height limits of Table 15.4-1 apply to nonbuilding structures similar to buildings; they cannot be applied to building structures. Table C15.5-1 illustrates the R values and height limits for a 70 ft (21.3 m) high steel ordinary moment frame (OMF) pipe rack.

C15.5.3.1 Steel Storage Racks The two approaches to the design of steel storage racks set forth by the standard are intended to produce comparable results. The specific revisions to the Rack Manufacturers Institute (RMI) specification cited in earlier editions of this standard and the detailed requirements of the ANSI/RMI MH 16.1 (RMI 2012) specification reflect the recommendations of FEMA 460 (2005).

Although the ANSI/RMI MH 16.1 specification reflects the recommendations of FEMA 460, the anchorage provisions of the ANSI/RMI MH 16.1 specification are not in conformance with ASCE 7. Therefore, specific anchorage requirements were added in Sections 15.5.3.1.1 and 15.5.3.1.2.

These recommendations address the concern that storage racks in warehouse-type retail stores may pose a greater seismic risk to the general public than exists in low-occupancy warehouses or more conventional retail environments. Under normal conditions, retail stores have a far higher occupant load than an ordinary warehouse of a comparable size. Failure of a storage rack system in a retail environment is much more likely to cause personal injury than would a similar failure in a storage warehouse. To provide an appropriate level of additional safety in areas open to the public, an Importance Factor of 1.50 is specified. Storage rack contents, although beyond the scope of the standard, may pose a serious threat to life should they fall from the shelves in an earthquake. It is recommended that restraints be provided to prevent the contents of rack shelving open to the general public from falling during strong ground shaking (Figure C15.5-1).

C15.5.3.2 Steel Cantilevered Storage Racks The two approaches to the design of steel cantilevered storage racks set forth by the standard are intended to produce comparable results. The specific development of a new RMI standard to include the detailed requirements of the new ANSI/RMI MH 16.3 specification (RMI 2016), reflects the unique characteristics of this structural storage system, along with the recommendations of FEMA 460.

The values of R, C_d, and Ω_0 added to Table 15.4-1 for Steel Cantilever Storage Racks were taken directly from Table 2.7.2.2.3(1) of ANSI/RMI MH 16.3.

The anchorage provisions of the ANSI/RMI MH 16.3 specification are not in conformance with ASCE 7. Therefore, specific anchorage requirements were added in Section 15.5.3.2.1.

These recommendations address the concern that steel cantilevered storage racks in warehouse-type retail stores may pose a greater seismic risk to the general public than exists in low-occupancy warehouses or more conventional retail environments. Under normal conditions, retail stores have a far higher occupant load than an ordinary warehouse of a comparable size. Failure of a steel cantilevered storage rack system in a retail environment is much more likely to cause personal injury than would a similar failure in a storage warehouse. To provide an appropriate level of additional safety in areas open to the public, an Importance Factor of 1.50 is specified. Steel cantilevered storage rack contents, although beyond the scope of the standard,

Table C15.5-1. *R* Value Selection Example for Steel Ordinary Moment Frame Pipe Racks.

Seismic Design Category	R	Table in ASCE 7-22	System	Seismic Detailing Requirements
C	3.5	12.2-1 or 15.4-1	Steel OMF	AISC 341
C	3	12.2-1	Structural steel systems not specifically detailed for seismic resistance	None
D or E	2.5	15.4-1	Steel OMF with permitted height increase	AISC 341
D, E, or F	1	15.4-1	Steel OMF with unlimited height	AISC 341

Note: OMF = ordinary moment frame.

Figure C15.5-1. Merchandise restrained by netting.
Source: FEMA 460 (2005).

may pose a serious threat to life should they fall from the shelves in an earthquake. It is recommended that restraints be provided to prevent the contents of rack shelving open to the general public from falling during strong ground shaking (Figure C15.5-2).

All systems in ANSI/RMI MH 16.3, Table 2.7.2.2.3(1), are ordinary systems. For all systems in Seismic Design Categories B and C, the values in ANSI/RMI MH 16.3 (RMI 2016), Table 2.7.2.2.3(1) for R, Ω_0, and C_d correspond to the values shown in Table 12.2-1 for steel systems not specifically detailed for seismic resistance, excluding cantilever column systems. No seismic detailing is required. For hot-rolled steel systems in Seismic Design Categories D, E, and F, the values in ANSI/RMI MH 16.3, Table 2.7.2.2.3(1) for R, Ω_0, and C_d correspond to the values in Table 15.4-1 for ordinary systems with permitted height increase, except that no height limits apply. The hot-rolled steel systems are detailed to AISC 341. For cold-formed steel systems in Seismic Design Categories D, E, and F, the values in ANSI/RMI MH 16.3 (RMI 2016), Table 2.7.2.2.3(1) for R, Ω_0, and C_d correspond to the values shown in Table 15.4-1 for ordinary systems with unlimited height. Seismic detailing is not required for the cold-formed steel systems.

C15.5.4 Electrical Power-Generating Facilities Electrical power plants closely resemble building structures, and their performance in seismic events has been good. For reasons of mechanical performance, lateral drift of the structure must be limited. The lateral bracing system of choice has been the concentrically braced frame. In the past, the height limits on braced frames in particular have been an encumbrance to the design of large power-generating facilities. Based on acceptable past performance, Table 15.4-1 permits the use of ordinary concentrically braced frames with both lower R values and less restrictive height limits. This option is particularly effective for boiler buildings, which generally are 300 ft (91.4 m) or more high. A peculiarity of large boiler buildings is the general practice of suspending the boiler from the roof structures; this practice results in an unusual mass distribution, as discussed in Section C15.1.3.

C15.5.5 Structural Towers for Tanks and Vessels The requirements of this section apply to structural towers that are not integral with the supported tank. Elevated water tanks designed in accordance with AWWA D100 are not subject to Section 15.5.5. A structural tower supporting a tank or vessel is considered integral with the supported tank or vessel where the tank or vessel shell acts as a part of the seismic force-resisting system of the supporting tower.

Examples of structural towers that are not integral with the supported tank are shown in Figure C15.5-2. Examples of structural towers that are integral with the supported tank are shown in Figure C15.5-3. Examples of structural towers that are integral with the supported tank include column-supported elevated water tanks designed to AWWA D100 and column-supported liquid and gas spheres designed to the ASME *Boiler and Pressure Vessel Code* (2007), Section VIII.

C15.5.6 Piers and Wharves Current industry practice recognizes the distinct differences between the two categories

Figure C15.5-2. Examples of structural towers that are not integral with the supported tank.
Source: Left, Courtesy of Chevron, reproduced with permission; *right,* Courtesy of CB&I LLC, reproduced with permission.

Figure C15.5-3. Examples of structural towers that are integral with the supported tank.
Source: Courtesy of CB&I LLC; reproduced with permission.

of piers and wharves described in the standard. Piers and wharves with public occupancy, described in Section 15.5.6.2, are commonly treated as the "foundation" for buildings or building-like structures; design is performed using the standard, likely under the jurisdiction of the local building official. Piers and wharves without occupancy by the general public are often treated differently and are outside the scope of the standard; in many cases, these structures do not fall under the jurisdiction of building officials, and design is performed using other industry-accepted approaches.

Design decisions associated with these structures often reflect economic considerations by both owners and local, regional, or state jurisdictional entities with interest in commercial development. Where building officials have jurisdiction but lack experience analyzing pier and wharf structures, reliance on other industry-accepted design approaches is common.

Where occupancy by the general public is not a consideration, seismic design of structures at major ports and marine terminals often uses a performance-based approach, with criteria and methods that are very different from those used for buildings, as provided in the standard. Design approaches most commonly used are generally consistent with the practices and criteria described in *Seismic Design Guidelines for Port Structures* (Maritime Navigation Commission 2001), Ferritto et al. (1999), Priestley et al. (1996), Werner (1998), and *Marine Oil Terminal Engineering and Maintenance Standards* (California State Lands Commission 2005).

These alternative approaches have been developed over many years by working groups within the industry, and they reflect the historical experience and performance characteristics of these structures, which are very different from those of building structures.

The main emphasis of the performance-based design approach is to provide criteria and methods that depend on the economic importance of a facility. Adherence to the performance criteria in the documents listed do not seek to provide uniform margins of collapse for all structures; their application is expected to provide at least as much inherent life safety as for buildings designed using the standard. The reasons for the higher inherent level of life safety for these structures include the following.

1. These structures have relatively infrequent occupancy, with few working personnel and very low density of personnel. Most of these structures consist primarily of open area, with no enclosed structures that can collapse onto personnel. Small control buildings on marine oil terminals or similar secondary structures are commonly designed in accordance with the local building code.
2. These pier or wharf structures typically are constructed of reinforced concrete, prestressed concrete, or steel and are highly redundant because of the large number of piles supporting a single wharf deck unit. Tests done at the University of California, San Diego, for the Port of Los Angeles have shown that high ductilities (10 or more) can be achieved in the design of these structures using practices currently used in California ports.
3. Container cranes, loading arms, and other major structures or equipment on piers or wharves are specifically designed not to collapse in an earthquake. Typically, additional piles and structural members are incorporated into the wharf or pier specifically to support such items.
4. Experience has shown that seismic "failure" of wharf structures in zones of strong seismicity is indicated not by collapse but by economically irreparable deformations of the piles. The wharf deck generally remains level or slightly tilting but shifts out of position. Earthquake loading on properly maintained marine structures has never induced complete failure that could endanger life safety.
5. The performance-based criteria of the listed documents address reparability of the structure. These criteria are much more stringent than collapse prevention criteria and create a greater margin for life safety.

Lateral load design of these structures in low, or even moderate, seismic regions often is governed by other marine conditions.

C15.6 GENERAL REQUIREMENTS FOR NONBUILDING STRUCTURES NOT SIMILAR TO BUILDINGS

Nonbuilding structures not similar to buildings exhibit behavior markedly different from that of building structures. Most of these types of structures have reference documents that address their unique structural performance and behavior. The ground motion in the standard requires appropriate translation to allow use with industry standards.

C15.6.1 Earth-Retaining Structures Section C11.8.3 presents commonly used approaches for the design of nonyielding walls and yielding walls for bending, overturning, and sliding, taking into account the varying soil types, importance, and site seismicity.

C15.6.2 Trussed Towers, Chimneys, and Stacks

C15.6.2.1 General In Table 15.4-2 the entry for trussed towers, guyed stacks, and guyed chimneys gives a seismic response modification coefficient, R, of 3, and other parameters and detailing requirements. These are intended for lattice-type trussed structures (not similar to buildings) that are relatively lightweight and support relatively lightweight components, such as stacks, chimneys, or electrical components, similar to a telecommunication tower (Figure C15.6-1).

For nonbuilding structures similar to buildings, the seismic response modification coefficient, R, and other parameters, including detailing, should be per Table 15.4-1. An example is a tall, open, stair tower with vertical columns, horizontal beams, and diagonal vertical bracing, a structural system comparable to that found in building structures (Figure C15.6-2). A stair tower like this should use Table 15.4-1, not Table 15.4-2.

The design of stacks and chimneys to resist natural hazards generally is governed by wind design considerations. The exceptions to this general rule involve locations with high seismicity, stacks and chimneys with large elevated masses, and stacks and chimneys with unusual geometries. It is prudent to evaluate the effect of seismic loads in all areas but those with the lowest seismicity. Although not specifically required, it is recommended that the special seismic details required elsewhere in the standard be considered for application to stacks and chimneys.

C15.6.2.2 Concrete Chimneys and Stacks Concrete chimneys typically possess low ductility, and their performance is especially critical in the regions around large (breach) openings because of reductions in strength and loss of confinement for vertical reinforcement in the jamb regions around the openings. Earthquake-induced chimney failures have occurred in recent history (in Turkey in 1999) and have been attributed to strength and detailing problems (Kilic and Sozen 2003). Therefore, the R value of 3 traditionally used in ASCE 7-05 for concrete stacks and chimneys was reduced to 2, and detailing requirements for breach openings were added in the 2010 edition of this standard.

Figure C15.6-1. Typical trussed tower supporting a stack.
Source: Courtesy of Thomas Heausler.

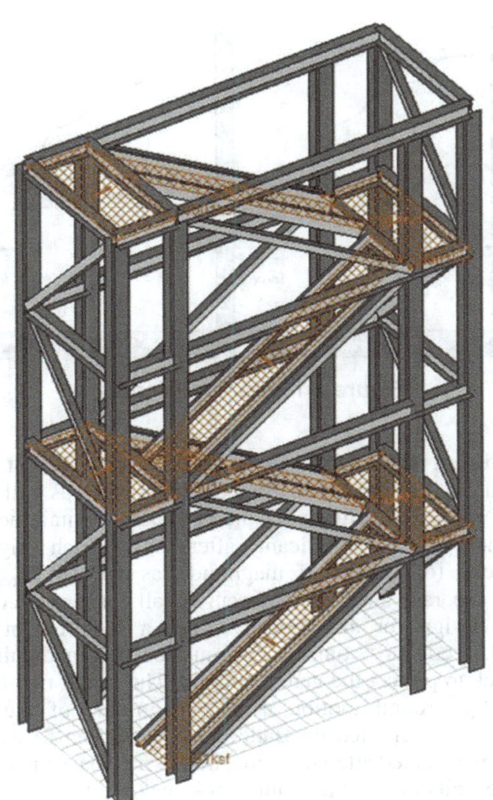

Figure C15.6-2. Typical exterior access stair tower with a structural system similar to a building.
Source: Courtesy of Thomas Heausler.

C15.6.2.3 Steel Chimneys and Stacks Guyed steel stacks and chimneys generally are lightweight. As a result, the design loads caused by natural hazards generally are governed by wind. In some cases, large flares or other elevated masses located near the top may require in-depth seismic analysis. Although it does not specifically address seismic loading, Chapter 6 of Troitsky (1990) provides a methodology appropriate for resolution of the seismic forces defined in the standard, in addition to the requirements found in ASME STS-1 (2011).

C15.6.4 Special Hydraulic Structures The most common special hydraulic structures are the baffle walls and weirs used in water treatment and wastewater treatment plants. Because there are openings in the walls, during normal operations the fluid levels are equal on each side of the wall, exerting no net horizontal force. But sloshing during a seismic event can exert large forces on the wall as illustrated in Figure C15.6-3. The walls can fail unless they are designed properly to resist the dynamic fluid forces.

C15.6.5 Secondary Containment Systems This section reflects the judgment that designing all impoundment dikes for the MCE_R ground motion when full and sizing all

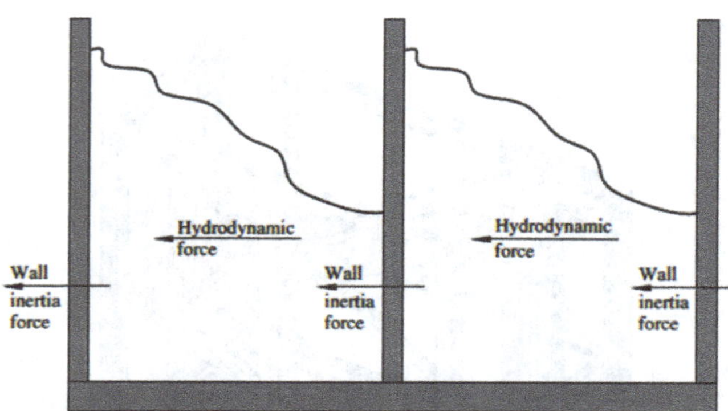

Figure C15.6-3. Wall forces.

impoundment dikes for the sloshing liquid height is too conservative. Designing an impoundment dike as full for the MCE_R assumes failure of the primary containment and occurrence of a significant aftershock. Such significant aftershocks (of the same magnitude as the MCE_R ground motion) are rare and do not occur in all locations. Although explicit design for aftershocks is not a requirement of the standard, secondary containment must be designed full for an aftershock to protect the general public. The use of two-thirds of the MCE_R ground motion as the magnitude of the design aftershock is supported by Bath's law, according to which the maximum expected aftershock magnitude may be estimated to be 1.2 scale units below the main shock magnitude.

The risk assessment and risk management plan described in Section 1.5.2 are used to determine where the secondary containment must be designed full for the MCE_R. The decision to design secondary containment for this more severe condition should be based on the likelihood of a significant aftershock occurring at the particular site, considering the risk posed to the general public from the release of hazardous material from the secondary containment.

Secondary containment systems must be designed to contain the sloshing liquid height where the release of liquid would place the general public at risk by exposing them to hazardous materials, by scouring of foundations of adjacent structures, or by causing other damage to adjacent structures.

C15.6.5.1 Freeboard Equation (15.6-1) was revised in ASCE 7-10 (2010) to return to the more exact theoretical formulation for sloshing liquid height instead of the rounded value introduced in ASCE 7-05. The rounded value in part accounted for maximum direction of response effects. Because the ground motion definition in ASCE 7-10 was changed and the maximum direction of response is now directly accounted for, it is no longer necessary to account for these effects by rounding up the theoretical sloshing liquid height factor in Equation (15.6-1).

C15.6.6 Telecommunication Towers Telecommunication towers support small masses, and their design generally is governed by wind forces. Although telecommunication towers have a history of experiencing seismic events without failure or significant damage, seismic design in accordance with ANSI/TIA-222-H, as modified by ASCE 7, is required.

C15.6.7 Steel Tubular Support Structures for Onshore Wind Turbine Generator Systems The most common support structures for large onshore wind turbine generator systems are steel tubular towers. Recommendations for the design of these structures can be found in ASCE/AWEA RP2011 (2011). ASCE/AWEA (2011) applies to wind turbines that have a rotor-swept area greater than 2,153 ft^2 (200 m^2). These recommendations are to be used in conjunction with seismic lateral forces determined in accordance with Section 15.4. A typical steel tubular support structures for an onshore wind turbine generator system is shown in Figure C15.6-4.

Figure C15.6-4. Typical steel tubular support structure for onshore wind turbine generator systems.

Source: Courtesy of GE Power; reproduced with permission.

C15.6.8 Ground-Supported Cantilever Walls or Fences Ground-supported cantilever walls and fences constructed from masonry, concrete, timber, or a combination of materials, including steel, are common. Such walls are often used as sound barrier walls or to limit access to residential subdivisions. Ground-supported cantilever walls and fences include walls supported by a footing and pier and panel/pilaster and panel wall systems (Figure C15.6-5) as long as these systems are not supported laterally in the out-of-plane

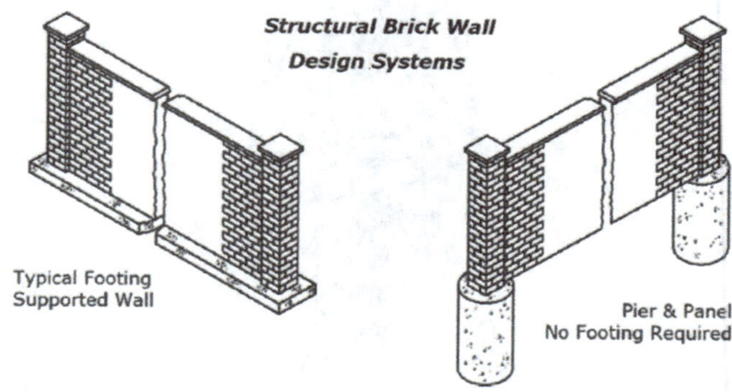

Figure C15.6-5. Typical cantilever wall systems falling under the requirements of Section 15.6.8.

Source: Courtesy of J. G. Soules; reproduced with permission.

Figure C15.6-6. Typical masonry ground-supported cantilever wall.

direction above grade. An example of a masonry ground-supported cantilever wall is shown in Figure C15.6-6. Many improperly designed ground-supported cantilever walls and fences constructed from masonry or concrete have experienced problems and have failed during seismic events, as evidenced in Section 6.3.9.1 of FEMA E-74 (2012).

The provisions for ground-supported cantilever walls and fences more than 6 ft (1.83 m) high were contained in prior editions of the Uniform Building Code, including the 1997 edition (ICBO 1997). When the International Building Code was developed, the provisions were inadvertently dropped and were not incorporated in ASCE 7. Walls of all heights should be properly designed. The 6 ft (1.83 m) height has been retained from the 1997 Uniform Building Code as the minimum height at which these provisions apply, because walls less than 6 ft (1.83 m) high are not deemed to present as significant a risk to life safety.

The seismic design parameters chosen for this system are based on those given in Table 15.4-2 for "all other self-supporting structures, tanks, or vessels not covered above or by reference standards that are not similar to buildings," except that all height limits were changed to no limit (NL), considering that the structure is a cantilever wall. Cantilever walls covered by these provisions can be of any material or combination of materials; therefore, a relatively low value of R was chosen to account for these material combinations. In addition, the pilasters incorporated in many of these wall systems are essentially ordinary cantilever columns. Ordinary cantilever columns in ASCE 7 tend to have low R values irrespective of the material used.

A decision was made by the ASCE 7 Seismic Subcommittee that a ground-supported freestanding wall or fence was a nonbuilding structure not similar to a building and should fall under the provisions of Chapter 15 instead of Chapter 13.

C15.6.9 Reinforced Concrete Tabletop Structure for Rotating Equipment and Process Vessels or Drums A typical tabletop structure is nominally 25 to 50 ft (7.6 to 15.2 m) above its mat foundation. It is typically used to support elevated rotating equipment in heavy industrial facilities, such as engines, turbines, generators, compressors, pumps, or fans, and process vessels or drums in petrochemical facilities, such as coke drums. The geometric layout of a tabletop structure is usually dictated by functional requirements of the heavy equipment it supports. Rotating equipment can load the supporting structure with reciprocating or impulsive forces. Similarly, some elevated process vessels or drums, such as coke drums in petrochemical facilities, are subject to random vibration and impact forces. For this reason, concrete tabletop structures are usually designed to have ample mass and high stiffness to control the dynamic response generated by the supported equipment.

Reinforced concrete tabletop structures are in essence moment-frame structures in which the columns deform in double curvature under lateral loads, similar to a bridge bent.

The R factor for a structure not including any seismic detailing ($R = 1.5$) is based on near-linear response at the design ground motion, caused by overstrength. At the MCE_R ground motion, the degree of yielding and strain hardening will be mild. Where there are enough columns to generate significant redistribution, the factor is increased. For tabletops detailed to meet the requirements of ACI 318-19, Chapter 18, for ordinary or intermediate moment frames (Section 18.3 or 18.4), the R factor is reduced somewhat to allow for extending the applicability of those detailing rules to designs for higher ground motions than the fundamental basis for those systems.

In practice it is extremely difficult, if not impossible, to detail reinforced concrete tabletop structures conforming to special moment frames, in particular the "strong column/weak beam" provisions (ACI 318-19, Sections 18.6, 18.7, and 18.8). Given that tabletop structures are designed with high stiffness and ample mass for smooth operation of rotating equipment and process vessels, they are usually much larger than the size needed to resist earthquake loading.

C15.6.10 Steel Lighting System Support Pole Structures Lighting system support pole structures respond dynamically to wind and earthquake loads. Wind loads can also cause fatigue problems, a concern that is addressed in ASCE 72.

C15.7 TANKS AND VESSELS

C15.7.1 General Methods for seismic design of tanks, currently adopted by a number of reference documents, have evolved from earlier analytical work by Jacobsen (1949), Housner (1963), Veletsos (1974), Haroun and Housner (1981), and others. The procedures used to design flat-bottom storage tanks and liquid containers are based on the work of Housner (US Department of Energy 1963) and Wozniak and Mitchell (1978). The reference documents for tanks and vessels have specific requirements to safeguard against catastrophic failure of the primary structure based on observed behavior in seismic events since the 1930s. Other methods of analysis, using flexible-shell models, have been proposed but at present are beyond the scope of this standard.

The industry-accepted design methods use three basic steps:

1. Dynamic modeling of the structure and its contents. When a liquid-filled tank is subjected to ground acceleration, the lower portion of the contained liquid, identified as the impulsive component of mass, W_i, acts as if it were a solid mass rigidly attached to the tank wall. As this mass accelerates, it exerts a horizontal force, P_i, on the wall; this force is directly proportional to the maximum acceleration of the tank base. This force is superimposed on the inertia force of the accelerating wall itself, P_s. Under the influence of the same ground acceleration, the upper portion of the contained liquid responds as if it were a solid mass flexibly attached to the tank wall. This portion, which oscillates at its own natural frequency, is identified

as the convective component, W_c, and exerts a horizontal force, P_c, on the wall. The convective component oscillations are characterized by sloshing, whereby the liquid surface rises above the static level on one side of the tank and drops below that level on the other side.
2. Determination of the period of vibration, T_i, of the tank structure and the impulsive component and determination of the natural period of oscillation (sloshing), T_c, of the convective component.
3. Selection of the design response spectrum. The response spectrum may be site-specific, or it may be constructed on the basis of seismic coefficients given in national codes and standards. Once the design response spectrum is constructed, the spectral accelerations corresponding to T_i and T_c are obtained and are used to calculate the dynamic forces P_i, P_s, and P_c.

Detailed guidelines for the seismic design of circular tanks, incorporating these concepts to varying degrees, have been the province of at least four industry reference documents: AWWA D100 for welded steel tanks (since 1964); API 650 (2014c) for petroleum storage tanks; AWWA D110 for prestressed, wire-wrapped tanks (since 1986); and AWWA D115 (2006b) for prestressed concrete tanks stressed with tendons (since 1995). In addition, API 650 and API 620 (2014a) contain provisions for petroleum, petrochemical, and cryogenic storage tanks. The detail and rigor of analysis prescribed in these documents have evolved from a semistatic approach in the early editions to a more rigorous approach at present, reflecting the need to include the dynamic properties of these structures.

The requirements in Section 15.7 are intended to link the latest procedures for determining design-level seismic loads with the allowable stress design procedures based on the methods in the standard. These requirements, which in many cases identify specific substitutions to be made in the design equations of the reference documents, will assist users of the standard in making consistent interpretations.

The American Concrete Institute has published ACI 350.3-06 (2006), *Seismic Design of Liquid-Containing Concrete Structures*. This document, which addresses all types of concrete tanks (prestressed and non-prestressed, circular and rectilinear), has provisions that unfortunately are not consistent with the seismic criteria of ASCE 7. However, when combined with the modifications required in Section 15.7.7.3, it serves as both a practical "how-to" loading reference and a guide to supplement application of ACI 318, Chapter 18.

C15.7.2 Design Basis In the seismic design of nonbuilding structures, standardization requires adjustments to industry reference documents to minimize the inconsistencies among them, while recognizing that structures designed and built over the years in accordance with these documents have performed well in earthquakes of varying severity. Of the inconsistencies among reference documents, the most important to seismic design relate to the base shear equation. The traditional base shear takes the following form:

$$V = \frac{ZIS}{R_w} CW \tag{C15.7-1}$$

An examination of those terms as used in the different references reveals the following.

1. The seismic zone coefficient, Z, has been rather consistent among all the documents because it usually has been obtained from the seismic zone designations and maps in the model building codes. However, the soil profile coefficient, S, does vary from one document to another. In some documents, these two terms are combined.
2. The Importance Factor, I, has varied from one document to another, but this variation is unavoidable and understandable because of the multitude of uses and degrees of importance of tanks and vessels.
3. The coefficient C represents the dynamic amplification factor that defines the shape of the design response spectrum for any given ground acceleration. Because C is primarily a function of the frequency of vibration, inconsistencies in its derivation from one document to another stem from at least two sources: differences in the equations for the determination of the natural frequency of vibration, and differences in the equation for the coefficient itself. [For example, for the shell/impulsive liquid component of lateral force, the steel tank documents use a constant design spectral acceleration (constant$C$) that is independent of the "impulsive" period, T.] In addition, the value of C varies depending on the damping ratio assumed for the vibrating structure (usually between 2% and 7% of critical).
4. Where a site-specific response spectrum is available, calculation of the coefficient C is not necessary, except in the case of the convective component (C_c), which is assumed to oscillate with 0.5% of critical damping and whose period of oscillation is usually long (greater than 2.5 s). Because site-specific spectra are usually constructed for high damping values (3% to 7% of critical) and because the site-specific spectral profile may not be well defined in the long-period range, an equation for C_c applicable to a 0.5% damping ratio is necessary to calculate the convective component of the seismic force.
5. The response modification factor, R_w, is perhaps the most difficult to quantify, for a number of reasons. Although R_w is a compound coefficient that is supposed to reflect the ductility, energy-dissipating capacity, and redundancy of the structure, it is also influenced by serviceability considerations, particularly in the case of liquid-containing structures.

In the standard, the base shear equation for most structures has been reduced to $V = C_s W$, where the seismic response coefficient, C_s, replaces the product ZSC/R_w. C_s is determined from the design spectral response acceleration parameters S_{DS} and S_{D1} (for short periods and a period of 1 s, respectively), which in turn are obtained from the mapped MCE_R spectral accelerations S_s and S_1. As in the prevailing industry reference documents, where a site-specific response spectrum is available, C_s is replaced by the actual values of that spectrum.

The standard contains several bridging equations, each designed to allow proper application of the design criteria of a particular reference document in the context of the standard. The bridging equations associated with particular types of liquid-containing structures, and the corresponding reference documents, are discussed below. Calculation of the periods of vibration of the impulsive and convective components is in accordance with the reference documents, and the detailed resistance and allowable stresses for structural elements of each industry structure are unchanged, except where new information has led to additional requirements.

It is expected that the bridging equations of Section 15.7.7.3 will be eliminated as the relevant reference documents are updated to conform to this standard. The bridging equations previously provided for AWWA D100 and API 650 have already been eliminated following updates of these documents.

Tanks and vessels are sensitive to vertical ground motions. Traditionally, the approach has been to apply a vertical seismic coefficient equal to $0.2S_{DS}$ to the design. This design approach came from the process used to design buildings and may underestimate the vertical response of the tank and its contents. For noncylindrical tanks, the increase in the hydrostatic pressure caused by vertical excitation has taken the form of $0.4S_{av}$, where S_{av} is determined in accordance with Sections 15.7.2 and 11.9. This pressure is combined directly with the hydrodynamic loads induced from lateral ground motions. The result is equal to 100% horizontal plus 40% vertical. The response of cylindrical tanks to vertical motions is well known and documented in various papers. Unless otherwise specified in a reference document, the vertical period, T_v, may be determined by

$$T_v = 2\pi\sqrt{\frac{\gamma_L R H_L^2}{gtE}} \tag{C15.7-2}$$

where

γ_L = Unit weight of stored liquid,
R = Tank radius to the inside of the wall,
H_L = Liquid height inside the tank,
g = Acceleration caused by gravity in consistent units,
t = Average shell thickness, and
E = Modulus of elasticity of the shell.

Equation (C15.7-2) comes from ACI 350.3 (2006) and is based on a rigid response of the liquid to vertical ground motions. Additional documents, such as Section 7.7.1 of ASCE (1984), provide solutions to determine the response of a flexible tank to vertical ground motions. The response of the structure itself is set equal to 0.4 times the peak of the vertical response spectra to recognize the vertical stiffness of the tank walls. This load is combined directly with loads produced from lateral ground motions. The result is equal to 100% horizontal plus 40% vertical. It should also be noted that R has been added to Equation (15.7-1). R is included in the calculation of hoop stress because the response of the tank shell caused by the added hoop tension from vertical ground motions is no different than the response of the tank shell caused by the added hoop tension from horizontal ground motions. ACI 350.3 has used this philosophy for many years.

C15.7.3 Strength and Ductility For tanks and vessels as for building structures, ductility and redundancy in lateral support systems are desirable and necessary for good seismic performance. Tanks and vessels are not highly redundant structural systems, and therefore ductile materials and well-designed connection details are needed to increase the capacity of the vessel to absorb more energy without failure. The critical performance of many tanks and vessels is governed by shell stability requirements rather than by yielding of the structural elements. For example, contrary to building structures, ductile stretching of anchor bolts is a desirable energy absorption component where tanks and vessels are anchored. The performance of cross-braced towers is highly dependent on the ability of the horizontal compression struts and connection details to develop fully the tension yielding in the rods. In such cases, it is also important to preclude both premature failure in the threaded portion of the connection and failure of the connection of the rod to the column before yielding of the rod.

The changes made to Section 15.7.3, Item (a), are intended to ensure that anchors and anchor attachments are designed such that the anchor yields (stretches) before the anchor attachment to the structure fails. The changes also clarify that the anchor rod embedment requirements are to be based on the requirements of Section 15.7.5, not Section 15.7.3, Item (a).

C15.7.4 Flexibility of Piping Attachments Poor performance of piping connections (tank leakage and damage) caused by seismic deformations is a primary weakness observed in seismic events. Although commonly used piping connections can impart mechanical loads to the tank shell, proper design in seismic areas results in only negligible mechanical loads on tank connections subject to the displacements shown in Table 15.7-1. API 650 treats the values shown in Table 15.7-1 as allowable-stress-based values and therefore requires that these values be multiplied by 1.4 where strength-based capacity values are required for design.

The displacements shown in Table 15.7-1 are based on movements observed during past seismic events. The vertical tank movements listed are caused by stretch of the mechanical anchors or steel tendons (in the case of a concrete tank) for mechanically anchored tanks, or the deflection caused by bending of the bottom of self-anchored tanks. The horizontal movements listed are caused by the deformation of the tank at the base.

In addition, interconnected equipment, walkways, and bridging between multiple tanks must be designed to resist the loads and accommodate the displacements imposed by seismic forces. Unless connected tanks and vessels are founded on a common rigid foundation, the calculated differential movements must be assumed to be out of phase.

C15.7.5 Anchorage Many steel tanks can be designed without anchors by using annular plate detailing in accordance with reference documents. Where tanks must be anchored because of overturning potential, proper anchorage design provides both a shell attachment and an embedment detail that allows the bolt to yield without tearing the shell or pulling the bolt out of the foundation. Properly designed anchored tanks have greater reserve strength to resist seismic overload than do unanchored tanks.

To ensure that the bolt yields (stretches) before failure of the anchor embedment, the anchor embedment must be designed in accordance with ACI 318, Equation (17.6.1.2), and must be provided with a minimum gauge length of eight bolt diameters. Gauge length is the length of the bolt that is allowed to stretch; it may include part of the embedment length into the concrete that is not bonded to the bolt (Figure C15.7-1). Two options are allowed for designing the anchor embedment. The option of providing anchor reinforcement according to ACI 318, Section 17.5.2.1, simply reflects the outcome of developing the steel strength of the anchor per ACI 318, Equation (17.6.1.2), for larger bolt sizes. The option of using the provisions of ACI 318, Section 17.10.5.3(a), acknowledges that the embedment design for small anchor bolt sizes may not require anchor reinforcement, and it also provides an option for post-installed anchors where it is not practical to provide anchor reinforcement.

To ensure that the bolt stretches, it is also important that the bolt not be significantly oversized. The prohibition on using the load combinations with overstrength of Section 12.4.3 is intended to accomplish this goal. For small tanks and vessels [less than or equal to 5 ft (1.5 m) in diameter or width and less than 10 ft (3.0 m) in height], the practical bolt sizes used to anchor these small tanks and vessels are too large to yield under the seismic loads they will experience. Therefore, for these small tanks and vessels, the other anchorage options allowed in ACI 318, Chapter 17, may be used. The definition of a small tank and vessel was derived by determining the diameter that would limit the convective mass to 10% of the total liquid mass while holding the height to approximately one story.

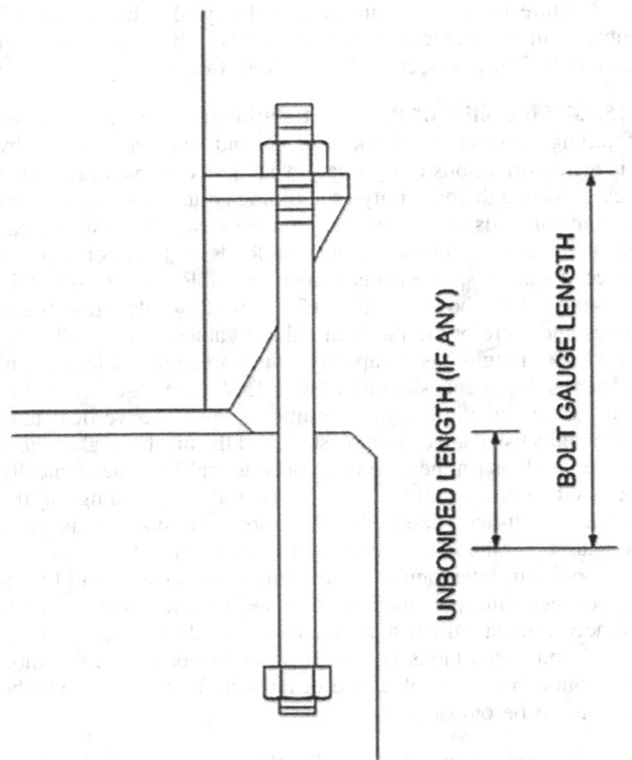

Figure C15.7-1. Bolt gauge length.

Where anchor bolts and attachments are misaligned such that the anchor nut or washer does not bear evenly on the attachment, additional bending stresses in threaded areas may cause premature failure before anchor yielding.

C15.7.6 Ground-Supported Storage Tanks for Liquids

C15.7.6.1 General For the two tank design options listed in this section, the difference in behavior between small diameter or width and large diameter or width tanks is recognized.

Tanks and vessels with small diameter or width are more susceptible to overturning and vertical buckling. As a general rule, a greater height-to-diameter (*H*/*D*) ratio produces lower resistance to vertical buckling. Where *H*/*D* is greater than 2, overturning approaches "impulsive mass" behavior: the sloshing (convective) mass is small. The definition of a small tank and vessel was derived by determining the diameter that would limit the convective mass to 10% of the total liquid mass while holding the height to approximately one story. Large-diameter tanks may be governed by additional hydrodynamic hoop stresses in the middle regions of the shell. Option (b) is best suited for larger-diameter tanks.

Option (c) has been deleted from this section in ASCE 7-22 because the force and displacement requirements of Section 15.4 were not specific enough to apply to tank design.

C15.7.6.2 Design Basis The response of ground storage tanks to earthquakes is well documented by Housner (1963), Wozniak and Mitchell (1978), Veletsos (1974), and others. Unlike building structures, the structural response of these tanks is influenced strongly by the fluid–structure interaction. Fluid–structure interaction forces are categorized as sloshing (convective) and rigid (impulsive) forces. The proportion of these forces depends on the geometry (*H*/*D* ratio) of the tank. API 650, API 620, AWWA D100-11, AWWA D110, AWWA D115, and ACI 350.3 provide the data necessary to determine the relative masses and moments for each of these contributions.

The standard requires that these structures be designed in accordance with the prevailing reference documents, except that the height of the sloshing wave, δ_s, must be calculated using Equation (15.7-13). API 650 and AWWA D100-11 include this requirement in their latest editions.

Equations (15.7-10) and (15.7-11) provide the spectral acceleration of the sloshing liquid for the constant-velocity and constant-displacement regions of the response spectrum, respectively. The 1.5 factor in these equations is an adjustment for 0.5% damping. An exception in the use of Equation (15.7-11) was added for the 2010 edition of this standard. The mapped values of T_L were judged by the ASCE 7 Seismic Subcommittee to be unnecessarily conservative in light of actual site-specific studies carried out since the introduction of the T_L requirements of ASCE 7-05. These studies suggest that the mapped values of T_L are very conservative based on observations during recent large earthquakes, especially the 2010 M_w 8.8 Chilean earthquake, where the large amplifications at very long periods (6 to 10 s) were not evident either in the ground motion records or in the behavior of long-period structures (particularly sloshing in tanks). Because a revision of the T_L maps is a time-consuming task that was not possible during the 2010 update cycle, Exception 3 was added to allow the use of site-specific values that are less than the mapped values, with a floor of 4 s or one-half the mapped value of T_L. The exception was added under Section 15.7.6 because, for nonbuilding structures, the overly conservative values for T_L are primarily an issue for tanks and vessels. Discussion of the site-specific procedures can be found in Chapter C22.

Exceptions 1 and 2 allow the determination of impulsive and convective spectral accelerations from the multiperiod spectral method for defining ground motions of Sections 11.4.5.1 or Chapter 21. It is still permissible to use the traditional design spectrum using the parameters S_{DS} and S_{D1}. Exception 1 makes use of the impulsive period, T_i. The impulsive period (the natural period of the tank components and the impulsive component of the liquid) is typically in the 0.25 to 0.6 s range. Many methods are available for calculating the impulsive period. The Veletsos flexible-shell method is used by many tank designers; see for example Veletsos (1974) and Malhotra et al. (2000).

C15.7.6.2.1 Distribution of Hydrodynamic and Inertial Forces Most of the reference documents for tanks define reaction loads at the base of the shell–foundation interface, without indicating the distribution of loads on the shell as a function of height. ACI 350.3 specifies the vertical and horizontal distribution of such loads.

The overturning moment at the base of the shell in the industry reference documents is only the portion of the moment that is transferred to the shell. The total overturning moment also includes the variation in bottom pressure, which is an important consideration for design of pile caps, slabs, or other support elements that must resist the total overturning moment. Wozniak and Mitchell (1978) and US Department of Energy TID-7024 (1963) provide additional information.

C15.7.6.2.2 Sloshing In past earthquakes, sloshing contents in ground storage tanks have caused both leakage and noncatastrophic damage to the roof and internal components. Even this limited damage and the associated costs and inconvenience can be significantly mitigated where the following items are considered:

1. Effective masses and hydrodynamic forces in the container;
2. Impulsive and pressure loads at
   (a) the sloshing zone (that is, the upper shell and edge of the roof system),
   (b) the internal supports (such as roof support columns and tray supports), and
   (c) the internal equipment (such as distribution rings, access tubes, pump wells, and risers); and
3. Freeboard (which depends on the sloshing wave height).

When no freeboard is required, a minimum freeboard of $0.7\delta_s$ is recommended for economic considerations. Freeboard is always required for tanks assigned to Risk Category IV.

Tanks and vessels storing biologically or environmentally benign materials typically do not require freeboard to protect the public health and safety. However, providing freeboard in areas of frequent seismic occurrence for vessels normally operated at or near top capacity may lessen damage (and the cost of subsequent repairs) to the roof and upper container. The exception to the minimum required freeboard per Table 15.7-3 for open-top tanks was added because it is rare for damage to occur that would impair the functionality of the facility when water or municipal wastewater overtops an open-top tank, provided measures have been taken to intercept and properly handle the resulting overflow.

The sloshing liquid height specified in Section 15.7.6.2.2 is based on the design earthquake defined in the standard. For economic reasons, freeboard for tanks assigned to Risk Category I, II, or III may be calculated using a fixed value of T_L equal to 4 s [as indicated in Section 15.7.6.2.2, Item (c)] but using the appropriate Importance Factor taken from Table 1.5-2. Because of life-safety and operational functionality concerns, freeboard for tanks assigned to Risk Category IV must be based on the mapped value of T_L. Because use of the mapped value of T_L results in the theoretical maximum value of freeboard, the calculation of freeboard in the case of Risk Category IV tanks is based on an importance factor equal to 1.0 [as indicated in Section 15.7.6.2.2, Item (b)].

If the freeboard provided is less than the computed sloshing height, δ_s, the sloshing liquid impinges on the roof in the vicinity of the roof-to-wall joint, subjecting it to a hydrodynamic force. This force may be approximated by considering the sloshing wave as a hypothetical static liquid column of height δ_s. The pressure exerted at any point along the roof at a distance y_s above the at-rest surface of the stored liquid may be assumed to be equal to the hydrostatic pressure exerted by the hypothetical liquid column at a distance $\delta_s - y_s$ from the top of that column. A better approximation of the pressure exerted on the roof is found in Malhotra (2005, 2006).

Another effect of a less-than-full freeboard is that the restricted convective (sloshing) mass "converts" into an impulsive mass, increasing the impulsive forces. This effect should be taken into account in the tank design. A method for converting the restricted convective mass into an impulsive mass is found in Malhotra (2005, 2006). It is recommended that sufficient freeboard to accommodate the full sloshing height be provided wherever possible.

Equation (15.7-13) was revised to use the theoretical formulation for sloshing wave height instead of the rounded value introduced in ASCE 7-05. The rounded value of Equation (15.6-1) increased the required freeboard by approximately 19%, significantly increasing the cost of both secondary containment and large-diameter, ground-supported storage tanks. See Section C15.6.5.1 for additional commentary on freeboard.

C15.7.6.2.4 Internal Elements Wozniak and Mitchell (1978) provide a recognized analysis method for determining the lateral loads on internal components caused by sloshing liquid.

C15.7.6.2.5 Sliding Resistance Historically, steel ground-supported tanks full of product have not slid off their foundations. A few unanchored, empty tanks or bulk storage tanks without steel bottoms have moved laterally during earthquake ground shaking. In most cases, these tanks could be returned to their proper locations. Resistance to sliding is obtained from the frictional resistance between the steel bottom and the sand cushion on which bottoms are placed. Because tank bottoms usually are crowned upward toward the tank center and are constructed of overlapping, fillet-welded, individual steel plates (resulting in a rough bottom), it is reasonably conservative to take the ultimate coefficient of friction on concrete as 0.70 (AISC 1986), and therefore a value of tan 30 degrees (about 0.577) for sand is used in design. The value of 30 degrees represents the internal angle of friction of sand and is conservatively used in design. The vertical weight of the tank and contents, as reduced by the component of vertical acceleration, provides the net vertical load. An orthogonal combination of vertical and horizontal seismic forces, following the procedure in Section 12.5.3, may be used. In recent years, the prevention of subsurface pollution caused by tank-bottom corrosion and leakage has been a significant issue. To prevent this problem, liners are often used with the tank foundation. When some of these liners are used, sliding of the tank and/or foundation caused by the seismic base shear may be an issue. If the liner is completely contained within a concrete ringwall foundation, the liner's surface is not the critical plane to check for sliding. If the liner is placed within an earthen foundation or is placed above or completely below a concrete foundation, it is imperative that sliding be evaluated. It is recommended that the sliding resistance factor of safety be at least 1.5.

C15.7.6.2.6 Local Shear Transfer The transfer of seismic shear from the roof to the shell and from the shell to the base is accomplished by a combination of membrane shear and radial shear in the wall of the tank. For steel tanks, the radial (out-of-plane) seismic shear is very small and usually is neglected; thus, the shear is assumed to be resisted totally by membrane (in-plane) shear. For concrete walls and shells, which have a greater radial shear stiffness, the shear transfer may be shared. The ACI 350.3-06 (2006) commentary provides further discussion.

C15.7.6.2.7 Pressure Stability Internal pressure may increase the critical buckling capacity of a shell. Provision to include pressure stability in determining the buckling resistance of the shell for overturning loads is included in AWWA D100-11 (2011). Recent testing on conical and cylindrical shells with internal pressure yielded a design methodology for resisting permanent loads in addition to temporary wind and seismic loads (Miller et al. 1997).

C15.7.6.2.8 Shell Support Anchored steel tanks should be shimmed and grouted to provide proper support for the shell and to reduce impact on the anchor bolts under reversible loads. The high bearing pressures on the toe of the tank shell may cause inelastic deformations in compressible material (such as fiberboard), creating a gap between the anchor and the attachment. As the load reverses, the bolt is no longer snug, and an impact of the attachment on the anchor can occur. Grout is a structural element and should be installed and inspected as an important part of the vertical and lateral force-resisting system.

C15.7.6.2.9 Repair, Alteration, or Reconstruction During their service life, storage tanks are frequently repaired, modified, or relocated. Repairs often are related to corrosion, improper operation, or overload from wind or seismic events. Modifications are made for changes in service, updates to safety equipment for changing regulations, or installation of additional process piping connections. It is imperative that these repairs and modifications be designed and implemented properly to maintain the structural integrity of the tank or vessel for seismic loads and the design operating loads.

The petroleum steel tank industry has developed specific guidelines (API 2014b) that are statutory requirements in some states. It is recommended that the provisions of API 653 (2014b) also be applied to other liquid storage tanks (e.g., water, wastewater, and chemical) as they relate to repairs, modifications, or relocation that affect the pressure boundary or lateral force-resisting system of the tank or vessel.

C15.7.7 Water Storage and Water Treatment Tanks and Vessels The AWWA design requirements for ground-supported steel water storage structures use allowable stress design procedures that conform to the requirements of the standard.

C15.7.7.1 Welded Steel AWWA D100 refers to ASCE 7-05 and repeats the ASCE 7-05 seismic design ground motion maps within the body of the document. A requirement is added in this section to point the user to the ground motions in the current version of ASCE 7. The clause “unless otherwise specified” in AWWA D100, Section 13.5.4.4, in the context of the determination of seismic freeboard, can result in seismic freeboard below that required by ASCE 7 and is therefore disallowed.

C15.7.7.2 Bolted Steel A clarification on the ground motions to use in design is added, and restrictions are added on the use of Type 6 tanks, in AWWA D103 (2009). AWWA D103 refers to ASCE 7-05 and repeats the ASCE 7-05 ground motion maps within the body of the document. Therefore, a clarifying statement is added to point the user to the seismic design ground motions in the current version of ASCE 7. A Type 6 tank is a concrete-bottom bolted steel shell tank with an embedded steel base setting ring. Type 6 tanks are considered to be mechanically anchored. There are no requirements for the anchorage design or bottom design (other than ACI 318) in AWWA D103. For the tank to be considered mechanically anchored, the tank bottom cannot uplift. In this case, the tank bottom is the foundation. If the bottom/foundation uplifts, the tank is now a self-anchored tank, and the additional shell compression that develops must be taken into account in the design. That is why J in Equation (14–32) of AWWA D103 (2009) is limited to 0.785.

C15.7.7.3 Reinforced and Prestressed Concrete A review of ACI 350.3, *Seismic Design of Liquid-Containing Concrete Structures and Commentary* (2006), revealed that this document is not in general agreement with the seismic provisions of ASCE 7-10.

This section was clarified to note that the Importance Factor, I, and the response modification factor, R, are to be specified by ASCE 7 and not the reference document. The descriptions used in ACI 350.3 to determine the applicable values of the Importance Factor and response modification factor do not match those used in ASCE 7.

It was noted that the ground motions for determining the convective (sloshing) seismic forces specified in ACI 350.3 were not the same and are actually lower than those specified by ASCE 7. ACI 350.3 essentially redefines the long-period transition period, T_L. This alternative transition period allows large-diameter tanks to have significantly lower convective forces and lower seismic freeboard than those permitted by the provisions of ASCE 7. Therefore, Section 15.7.7.3 was revised to require that the convective acceleration be determined according to the procedure in Section 15.7.6.1.

C15.7.7.4 Corrugated Steel Corrugated steel tanks, once used only for bulk product storage, are increasingly used for water storage. These tanks are often located in public areas, such as schools or grocery stores, or used for fire protection. Currently, there is no reference document that addresses the seismic design of corrugated steel liquid storage tanks. Because some corrugated steel water tanks are small and used in agriculture, only corrugated steel tanks assigned to Risk Category II, III, or IV are included in these provisions.

Provisions have been added to provide a level of safety equivalent to that provided for other types of tanks covered by this standard. Component allowable stresses and materials are limited to those found in AWWA D103 and AISI 100, as appropriate to provide this equivalent level of safety. Vertical elements are required to be added to the tank shell to carry seismic compressive forces, because corrugated steel shells do not have significant vertical load capacity. These vertical elements are commonly found on corrugated steel grain tanks. The assumption of impulsive mass is commonly used for small tanks and provides some additional conservatism. The additional conservatism is necessary given the lack of an accepted industry reference standard for the design of corrugated steel tanks. Many corrugated steel tanks do not have steel bottoms. The liquid is contained in a liner. Therefore, the base of the tank needs to be confined by a curb or by embedment in the foundation to prevent ovalling of the shell, and shear anchorage is required because there is no tank bottom to develop sliding resistance. Ovalling is the tank shell going out-of-round. Mechanical anchorage of the tank to the foundation is required to resist seismic uplift forces and to provide a yielding mechanism required to justify the R value assigned in Table 15.4-2.

C15.7.8 Petrochemical and Industrial Tanks and Vessels Storing Liquids

C15.7.8.1 Welded Steel The American Petroleum Institute (API) uses an allowable stress design procedure that conforms to the requirements of the standard.

The most common damage to tanks observed during past earthquakes includes the following:

1. Buckling of the tank shell near the base because of excessive axial membrane forces. This buckling damage is usually evident as “elephant foot” buckles a short distance above the base or as diamond-shaped buckles in the lower ring. Buckling of the upper ring also has been observed.
2. Damage to the roof caused by impingement on the underside of the roof of sloshing liquid with insufficient freeboard.
3. Failure of piping or other attachments that are overly restrained.
4. Foundation failures.

Other than the aforementioned damage, the seismic performance of floating roofs during earthquakes has generally been good, with damage usually confined to the rim seals, gauge poles, and ladders. However, floating roofs have sunk in some earthquakes because of lack of adequate freeboard or the proper buoyancy and strength required by API 650. Similarly, the performance of open-top tanks with top wind girder stiffeners designed per API 650 has been generally good.

C15.7.8.2 ***Bolted Steel*** Bolted steel tanks are often used for temporary functions. Where use is temporary, it may be acceptable to the jurisdictional authority to design bolted steel tanks for no seismic loads or for reduced seismic loads based on a reduced return period. For such reduced loads based on reduced exposure time, the owner should include a signed removal contract with the fixed removal date as part of the submittal to the Authority Having Jurisdiction.

C15.7.8.4 ***Corrugated Steel*** See C15.7.7.4.

C15.7.8.5 ***Reinforced Thermoset Plastic and Fiber-Reinforced Plastic*** Reinforced thermoset plastic and fiber-reinforced plastic tanks are used to store corrosive and otherwise hazardous materials. For the purpose of this section, the terms "reinforced thermoset plastic" and "fiber-reinforced plastic" may be used interchangeably. Until ASCE 7-22, Table 15.4-2 had always contained an entry for fiber-reinforced tanks, but Chapter 15 contained no specific design requirements for these types of tanks. ASME RTP-1 (2019) is the industry document most used for these types of tanks, but it does not contain mandatory seismic requirements. Therefore, API 650, Annex E, is used to determine the seismic base shear and overturning moment.

C15.7.9 Ground-Supported Storage Tanks for Granular Materials

C15.7.9.1 ***General*** The response of a ground-supported storage tank storing granular materials to a seismic event is highly dependent on its *H/D* ratio and the characteristics of the stored product. The effects of intergranular friction are described in more detail in Sections C15.7.9.3.1 (Increased Lateral Pressure), C15.7.9.3.2 (Effective Mass), and C15.7.9.3.3 (Effective Density).

Long-term increases in shell hoop tension because of temperature changes after the product has been compacted also must be included in the analysis of the shell; Anderson (1966) provides a suitable method.

C15.7.9.2 ***Lateral Force Determination*** Seismic forces acting on ground-supported liquid storage tanks are divided between impulsive and convective (sloshing) components. However, in a ground-supported storage tank for granular materials, all seismic forces are of the impulsive type and relate to the period of the storage tank itself. Because of the relatively short period of a tank shell, the response is normally in the constant-acceleration region of the response spectrum, which relates to S_{DS}. Therefore, the seismic base shear is calculated as follows:

$$V = \frac{S_{DS}}{\left(\frac{R}{I}\right)} W_{\text{effective}} \tag{C15.7-3}$$

where V, S_{DS}, I, and R have been previously defined, and $W_{\text{effective}}$ is the gross weight of the stored product multiplied by an effective mass factor and an effective density factor, as described in Sections C15.7.9.3.2 and C15.7.9.3.3, plus the dead weight of the tank. Unless substantiated by testing, it is recommended that the product of the effective mass factor and the effective density factor be taken as no less than 0.5, given the limited test data and the highly variable properties of the stored product.

C15.7.9.3 Force Distribution to Shell and Foundation

C15.7.9.3.1 Increased Lateral Pressure In a ground-supported tank storing granular materials, increased lateral pressures develop as a result of rigid-body forces that are proportional to ground acceleration. Information concerning design for such pressure is scarce. Trahair et al. (1983) describe both a simple, conservative method and a difficult, analytical method using failure wedges based on the Mononobe–Okabe modifications of the classical Coulomb method.

C15.7.9.3.2 Effective Mass For ground-supported tanks storing granular materials, much of the lateral seismic load can be transferred directly into the foundation, via intergranular shear, before it can reach the tank shell. The effective mass that loads the tank shell is highly dependent on the *H/D* ratio of the tank and the characteristics of the stored product. Quantitative information concerning this effect is scarce, but Trahair et al. (1983) describe a simple, conservative method to determine the effective mass. That method presents reductions in effective mass, which may be significant, for *H/D* ratios less than 2. This effect is absent for elevated tanks.

C15.7.9.3.3 Effective Density Granular material stored in tanks (both ground-supported and elevated) does not behave as a solid mass. Energy loss through intergranular movement and grain-to-grain friction in the stored material effectively reduces the mass subject to horizontal acceleration. This effect may be quantified by an effective density factor less than 1.0.

Based on Chandrasekaran and Jain (1968) and on shake table tests reported in Chandrasekaran et al. (1968), ACI 313 (1997) recommends an effective density factor of not less than 0.8 for most granular materials. According to Chandrasekaran and Jain (1968), an effective density factor of 0.9 is more appropriate for materials with high moduli of elasticity, such as aggregates and metal ores. These effective density factors are appropriate for the design of silos storing bulk materials, except for materials stored in concrete silos and stacking tubes governed by ACI 313 (2016). Section 6.3.8 of ACI 313 (2016) effectively sets the effective density factor to 1.0 by requiring that the full density of the stored material be used in the computation of seismic forces.

C15.7.9.3.4 Lateral Sliding Most ground-supported steel storage tanks for granular materials rest on a base ring and do not have a steel bottom. To resist seismic base shear, a partial bottom or annular plate is used in combination with anchor bolts or a curb angle. An annular plate can be used alone to resist the seismic base shear through friction between the plate and the foundation, in which case the friction limits of Section 15.7.6.2.5 apply. The curb angle detail serves to keep the base of the shell round while allowing it to move and flex under seismic load. Various base details are shown in Figure 13 of Kaups and Lieb (1985).

C15.7.9.3.5 Combined Anchorage Systems This section is intended to apply to combined anchorage systems that share loads based on their relative stiffnesses, and not to systems where sliding is resisted completely by one system (such as a steel annular plate) and overturning is resisted completely by another system (such as anchor bolts).

C15.7.10 Elevated Tanks and Vessels for Liquids and Granular Materials

C15.7.10.1 ***General*** The three basic lateral load-resisting systems for elevated water tanks are defined by their support structure:

1. Multilegged braced steel tanks (trussed towers, Figure C15.1-1);
2. Small-diameter, single-pedestal steel tanks (cantilever column, Figure C15.7-2); and
3. Large-diameter, single-pedestal tanks of steel or concrete construction (load-bearing shear walls, Figure C15.7-3).

Figure C15.7-2. Small-diameter, single-pedestal steel tank.
Source: Courtesy of CB&I LLC; reproduced with permission.

(a) Steel (b) Concrete

Figure C15.7-3. Large-diameter, single-pedestal tank.
Source: Courtesy of CB&I LLC; reproduced with permission.

Unbraced multilegged tanks are uncommon. These types of tanks differ in their behavior, redundancy, and resistance to overload. Multilegged and small-diameter pedestal tanks have longer fundamental periods (typically greater than 2 s) than the shear wall type tanks (typically less than 2 s). The lateral load failure mechanisms usually are brace failure for multilegged tanks, compression buckling for small-diameter steel tanks, compression or shear buckling for large-diameter steel tanks, and shear failure for large-diameter concrete tanks. Connection, welding, and reinforcement details require careful attention to mobilize the full strength of these structures. To provide a greater margin of safety, R factors used with elevated tanks typically are smaller than those for other comparable lateral load-resisting systems.

C15.7.10.4 Transfer of Lateral Forces into Support Tower The vertical loads and shears transferred at the base of a tank or vessel supported by grillage or beams typically vary around the base because of the relative stiffness of the supports, settlements, and variations in construction. Such variations must be considered in the design for vertical and horizontal loads.

C15.7.10.5 Evaluation of Structures Sensitive to Buckling Failure Nonbuilding structures that are designed with limited structural redundancy for lateral loads may be susceptible to total failure when loaded beyond the design loads. This phenomenon is particularly true for shell-type structures that exhibit unstable postbuckling behavior, such as tanks and vessels supported on shell skirts or pedestals. Evaluation for this critical condition ensures stability of the nonbuilding structure for governing design loads.

The design spectral response acceleration, S_a, used in this evaluation includes site factors. The I/R coefficient is taken as 1.0 for this critical check. The structural capacity of the shell is taken as the critical buckling strength (that is, the factor of safety is 1.0). Vertical and orthogonal combinations need not be considered for this evaluation, because the probability of peak values occurring simultaneously is very low.

The intent of Section 15.7.10.5 and Table 15.4-2 is that skirt-supported vessels must be checked for seismic loads based on $I_e/R = 1.0$ if the structure falls in Risk Category IV or if an R value of 3.0 is used in the design of the vessel. For the purposes of this section, a skirt is a thin-walled steel cylinder or cone used to support the vessel in compression. Skirt-supported vessels fail in buckling, which is not a ductile failure mode. Therefore, a more conservative design approach is required. The $I_e/R = 1.0$ check typically governs the design of the skirt over using loads determined with an R factor of 3 in an area of moderate-to-high seismic activity. The only benefit of using an R factor of 3 in this case is in the design of the foundation. The foundation is not required to be designed for the $I_e/R = 1.0$ load. Section 15.7.10.5, Item (b), states that resistance of the structure shall be defined as the critical buckling resistance of the element for the $I_e/R = 1.0$ load. This stipulation means that the support skirt can be designed based on critical buckling (factor of safety of 1.0). The critical buckling strength of a skirt can be determined using a number of published sources. The two most common methods for determining the critical buckling strength of a skirt are given in the ASME *Boiler and Pressure Vessel Code* (2007), Section VIII, Division 2, 2008 Addenda, Paragraph 4.4, using a factor of safety of 1.0; and in AWWA D100-05 (2006b), Section 13.4.3.4. To use these methods, the radius, length, and thickness of the skirt, modulus of elasticity of the steel, and yield strength of the steel are required. These methods take into account both local buckling and slenderness effects of the skirt. Under no circumstance should the theoretical buckling strength of a cylinder, found in many engineering mechanics texts, be used to determine the critical buckling strength of the skirt. The theoretical value, based on a perfect cylinder, does not take into account the imperfections built into real skirts. The theoretical buckling value is several times greater than the actual value measured in tests. The buckling values found in the aforementioned references are based on actual tests.

Examples of applying the ASME BPVC (2007), Section VIII, Division 2, 2008 Addenda, Paragraph 4.4, and AWWA D100-05 (2006a), Section 13.4.3.4, buckling rules are shown in Figures C15.7-4 and C15.7-5, respectively.

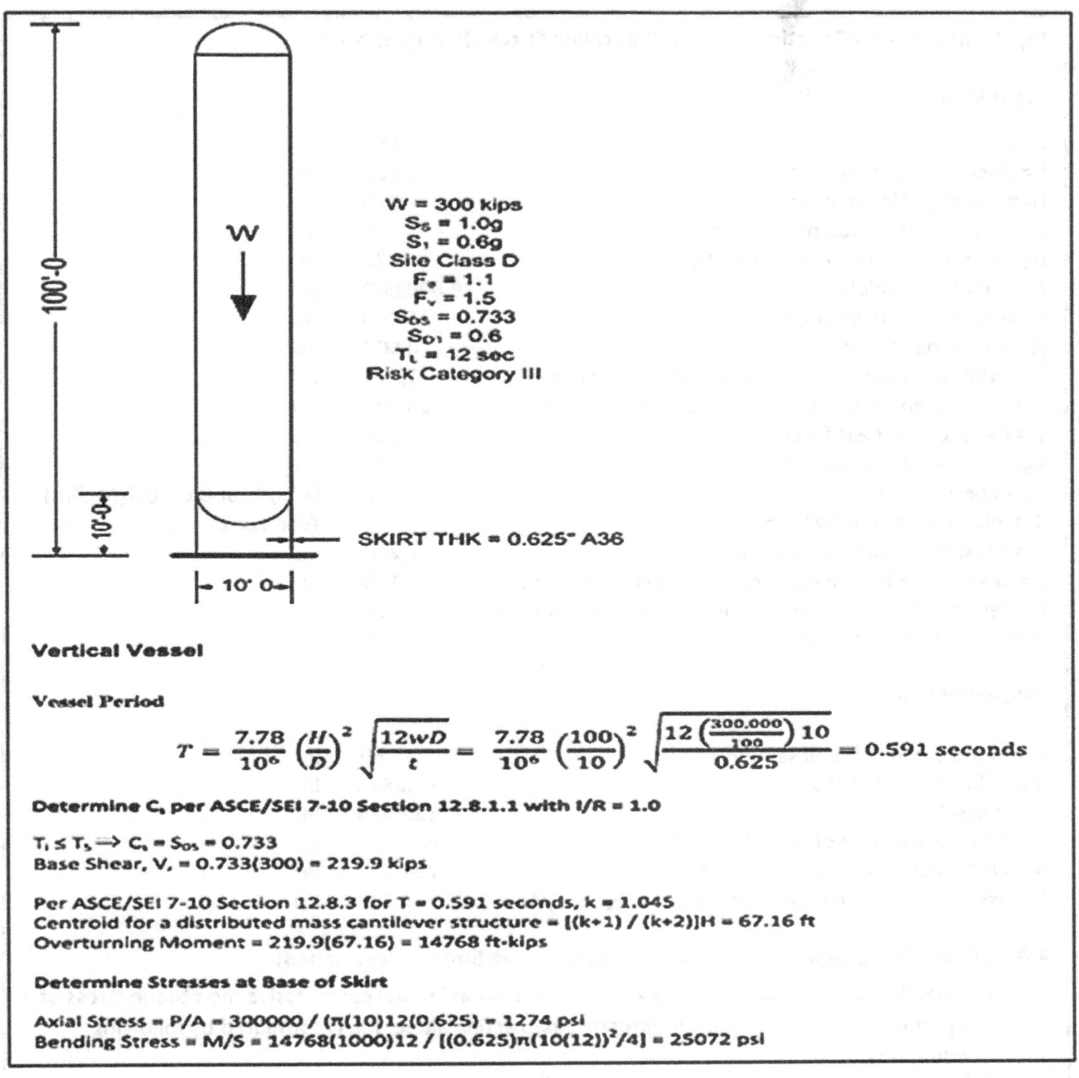

Figure C15.7-4. Example problem using the ASME *Boiler and Pressure Vessel Code* (2007), Section VIII, Division 2, 2008 Addenda, Paragraph 4.4.

C15.7.10.7 Concrete Pedestal (Composite) Tanks A composite elevated water storage tank is composed of a welded steel tank for watertight containment, a single-pedestal concrete support structure, a foundation, and accessories. The lateral load-resisting system is a load-bearing concrete shear wall. ACI 371R (1998), referenced in editions of ASCE 7 before 2016, was replaced with AWWA D107 (2010) in ASCE 7-16. Because AWWA D107-10 was based on the seismic design ground motions from ASCE 7-05, a provision was added in Section 15.7.10.7 requiring the use of the seismic design ground motions from Section 11.4.

C15.7.11 Boilers and Pressure Vessels The support system for boilers and pressure vessels must be designed for the seismic forces and displacements presented in the standard. Such design must include consideration of the support, the attachment of the support to the vessel (even if "integral"), and the body of the vessel itself, which is subject to local stresses imposed by the support connection.

C15.7.12 Liquid and Gas Spheres The commentary in Section C15.7.11 also applies to liquid and gas spheres.

C15.7.13 Refrigerated Gas Liquid Storage Tanks and Vessels Even though some refrigerated storage tanks and vessels, such as those storing liquefied natural gas, are required to be designed for ground motions and performance goals in excess of those found in the standard, all such structures must also meet the requirements of this standard as a minimum. All welded steel refrigerated storage tanks and vessels must be designed in accordance with the requirements of the standard and of API 620.

C15.7.14 Horizontal, Saddle-Supported Vessels for Liquid or Vapor Storage Past practice has been to assume that a horizontal, saddle-supported vessel (including its contents) behaves as a rigid structure (with natural period, T, less than 0.06 s). For this situation, seismic forces would be determined using the requirements of Section 15.4.2. For large horizontal, saddle-supported vessels (length-to-diameter ratio of 6 or more), this assumption can be unconservative, so Section 15.7.14.3 requires that the natural period be determined assuming the vessel to be a simply supported beam.

Input Data for ASME Section VIII, Div. 2 Buckling Checks (Paragraph 4.4)

Input Values

Course =	Skirt	
t = thickness of vessel section =	0.625	in.
H_T = top elevation of course =	120	in.
H_B = bottom elevation of course =	0	in.
D_O = outer diameter of vessel section =	120	in.
E_y = material modulus of elasticity =	29,000,000	psi
S_y = material yield strength =	36,000	psi
P_{ext} = external pressure =	0.000	psi
f_a = axial comp membrane stress from axial load =	1,274	psi
f_b = axial comp membrane stress from bending =	25,072	psi
V = net section shear force =	219,900	lbs
V_{phi} = applied shear force angle =	90	deg.
C_m = coefficient =	1	0.85, 1, or $0.6 - 0.4(M_1/M_2)$
K_u = effective length factor =	2.1	Free-Fixed
L_u = maximum unbraced length =	1,200	in.
L = design length vessel section for external pressure =	120	in.
L = design length vessel section for axial compression =	120	in.
FS = Input Factor of Safety =	1.0	

Calculated Values

R_o = outer radius of shell section =	60	in.
R = radius to center of shell =	59.6875	in.
R_m = vessel mean radius =	59.6875	in.
r = radius of gyration of cyl = $(1/4)(D_o^2 + D_i^2)^{0.5}$ =	42.2	in.
A = Cross sectional area of cylinder =	234.4	in^2
f_q = axial compressive membrane stress = $P_{ext}\pi D_i^2/4A$ =	0.0	psi

4.4.12 Combined Loadings and Allowable Compressive Stresses (continued)

b) Axial Compressive Stress Acting Alone - The allowable axial compressive membrane stress of a cylinder subject to an axial compressive load acting alone, F_{xa}, is computed by following equations.

1) For lambda$_c \le 0.15$ (Local Buckling)

$F_{xa} = \min\,[F_{xa1}, F_{xa2}]$		(4.4.61)
$F_{xa1} = S_y / FS$	for $D_o/t \le 135$	(4.4.62)
$F_{xa1} = 466\,S_y / [(331 + D_o/t)\,FS]$	for $135 < D_o/t < 600$	(4.4.63)
$F_{xa1} = 0.5\,S_y / FS$	for $135 < D_o/t < 600$	(4.4.64)
$F_{xa2} = F_{xe} / FS$		(4.4.65)
where: $F_{xe} = C_x E_y t/D_o$		(4.4.66)
$C_x = \min[409\,c/(389 + D_o/t), 0.9]$	for $D_o/t < 1247$	(4.4.67)
$C_x = 0.25\,c$	for $1247 \le D_o/t \le 2000$	(4.4.68)
$c = 2.64$	for $M_x \le 1.5$	(4.4.69)
$c = 3.13 / M_x^{0.42}$	for $1.5 < M_x < 15$	(4.4.70)
$c = 1.0$	for $M_x \ge 15$	(4.4.71)

Figure C15.7-4 *(Continued).* Example problem using the ASME *Boiler and Pressure Vessel Code* (2007), Section VIII, Division 2, 2008 Addenda, Paragraph 4.4.

4.4.12 Combined Loadings and Allowable Compressive Stresses (continued)

$M_x = L/(R_o t)^{1/2}$ (4.4.124)

where L is the design length of a vessel section between lines of support

$D_o/t = 192.00$ $135 < D_o/t < 600$

$M_x = L/(R_o t)^{1/2} = 19.60 > 15$ $c = 1.0000$

$D_o/t < 1247$ $C_x = \min\,[409\,c\,/\,[389 + D_o/t], 0.9] = 0.7039587$ (4.4.67)

$F_{xe} = C_x E_y t/D_o = 106{,}327$ psi (4.4.66)

Calculate F_{xa1}

$F_{ic} = 466{*}S_y/(331 + D_o/t) = 32{,}076$ psi

Use Input FS = 1.0

$F_{xa1} = 466{*}S_y/[(331 + D_o/t)FS] = 32{,}076$ psi

Calculate F_{xa2}

$F_{ic} = 106{,}327$ psi

Use Input FS = 1.0

$F_{xa2} = F_{xe}/FS = 106{,}327$ psi (4.4.65)

Calculate F_{xa}

$F_{xa} = \min\,[F_{xa1}, F_{xa2}] = 32{,}076$ psi (4.4.61)

$\text{lambda}_c = (K)(L_u)\,/\,[(\pi)(r_g)]\,[(F_{xa} I\,(FS)\,/\,E]^{0.5} = 0.6321$ $0.15 < \text{lambda}_c < 1.147$

2) For $\text{lambda}_c > 0.15$ and $K_u L_u/r_g < 200$ (Column Buckling)

$F_{ca} = F_{xa}\,[1 - 0.74\,(\text{lambda}_c - 0.15)]^{0.3}$ $0.15 < \text{lambda}_c < 1.147$ (4.4.72)

$F_{ca} = 0.88\,F_{xa}\,/\,(\text{lambda}_c)^2$ $\text{lambda}_c \geq 1.147$ (4.4.73)

$KL_u/r_g = 59.7 < 200$

$\text{Lambda}_c > 0.15$ and $KL_u/r_g < 200$ therefore:

$F_{ca} = F_{xa}\,[1 - 0.74\,(\text{lambda}_c - 0.15)]^{0.3} = 28{,}100$ psi

c) Compressive Bending Stress - The allowable axial compressive membrane stress of a cylindrical shell subject to a bending moment acting across the full circular cross section F_{ba}, is computed using the following equations.

$F_{ba} = F_{xa}$ for $135 \leq D_o/t \leq 2000$ (4.4.74)

$F_{ba} = 466\,S_y\,/\,[(331 + D_o/t)\,FS]$ for $100 < D_o/t < 135$ (4.4.75)

$F_{ba} = 1.081\,S_y\,/\,FS$ for $D_o/t < 100$ and $\gamma > 0.11$ (4.4.76)

$F_{ba} = (1.4 - 2.9\,\gamma)\,S_y\,/\,FS$ for $D_o/t < 100$ and $\gamma < 0.11$ (4.4.77)

where: $\gamma = S_y D_o\,/\,E_y t$ (4.4.78)

$D_o/t = 192$

$\gamma = S_y D_o\,/\,E_y t = 0.2383$

$F_{ic} = F_{xa}$

$F_{ic} = 32{,}076$ psi

$D_o/t = 192 > 135$ (see Sect. 3.1.1)

Use Input FS = 1.0

$F_{ba} = F_{xa} = 32{,}076$ psi

Figure C15.7-4 ***(Continued).*** **Example problem using the ASME** ***Boiler and Pressure Vessel Code*** **(2007), Section VIII, Division 2, 2008 Addenda, Paragraph 4.4.**

d) **Shear Stress - The allowable shear stress of a cylindrical shell, F_{va}, is computed using the following equations.**

$F_{va} = n_v F_{ve} / FS$ (4.4.79)

where: $F_{ve} = a_v C_v E t / D_o$ (4.4.80)

$C_v = 4.454$	for $Mx < 1.5$	(4.4.81)
$C_v = (9.64 / M_x^2)(1 + 0.0239 M_x^3)^{1/2}$	for $1.5 < M_x < 26$	(4.4.82)
$C_v = 1.492 / (M_x)^{1/2}$	for $26 < M_x < 4.347 D_o / t$	(4.4.83)
$C_v = 0.716 (t / D_o)^{1/2}$	for $M_x > 4.347 D_o / t$	(4.4.84)
$a_v = 0.8$	for $D_o / t < 500$	(4.4.85)
$a_v = 1.389 - 0.218 \log_{10} (D_o / t)$	for $D_o / t > 500$	(4.4.86)
$n_v = 1.0$	for $F_{ve} / S_y < 0.48$	(4.4.87)
$n_v = 0.43 S_y / F_{ve} + 0.1$	for $0.48 < F_{ve} / S_y < 1.7$	(4.4.88)
$n_v = 0.6 S_y / F_{ve}$	for $F_{ve} / S_y > 1.7$	(4.4.89)

$D_o / t = 192$

$M_x = L / (R_o t)^{0.5} = 19.596$

$1.5 < M_x < 26$ $\quad C_v = (9.64/M_x^2)(1+0.0239M_x^3)^{0.5} = 0.3376$

$D_o / t < 500$ $\quad a_v = 0.8000$

$F_{ve} = a_v C_v E t / D_o = 40{,}793$ psi

$F_{ve} / S_y = 1.13$ $\quad 0.48 < F_{ve} / S_y < 1.7$

$n_v = 0.43 S_y / F_{ve} + 0.1 = 0.4795$

$F_{ic} = 19{,}559$ psi

Use Input FS = 1.000

$F_{va} = n_v F_{ve} / FS = 19{,}559$ psi

4.4.12 Combined Loadings and Allowable Compressive Stresses (continued)

Axial Compressive Stress, Compressive Bending Stress, and Shear - the allowable compressive stress for the combination of uniform axial compression, axial compression due to bending, and shear in the absence of hoop compression.

Let $K_s = 1 - (f_v / F_{va})^2$ (4.4.105)

For $0.15 < \text{lambda}_c < 1.2$

$\text{lambda}_c = 0.63$ (Section 3.2) $\quad 0.15 < \text{lambda}_c < 1.2$ OK

$f_a /(K_s F_{ca}) + (8 / 9)(\text{delta}) f_b /(K_s F_{ba}) < 1.0$ $\quad f_a /(K_s F_{ca}) > 0.2$ (4.4.112)

$f_a /(2 K_s F_{ca}) + (\text{delta}) f_b /(K_s F_{ba}) < 1.0$ $\quad f_a /(K_s F_{ca}) < 0.2$ (4.4.113)

$K_s = 1 - (f_v / F_{va})^2 = 0.9977$

$F_e = (\pi)^2 E / (K_s L_u / r)^2 = 80{,}287$ psi

$\text{delta} = C_m / [1 - f_a FS / F_e] = 1.0161$

$f_a /(K_s F_{ca}) = 0.045442959 < 0.2$

$f_a /(2 K_s F_{ca}) + (\text{delta}) f_b /(K_s F_{ba}) = 0.82 < 1.0$ OK!

Figure C15.7-4 ***(Continued).*** **Example problem using the ASME** ***Boiler and Pressure Vessel Code*** **(2007), Section VIII, Division 2, 2008 Addenda, Paragraph 4.4.**

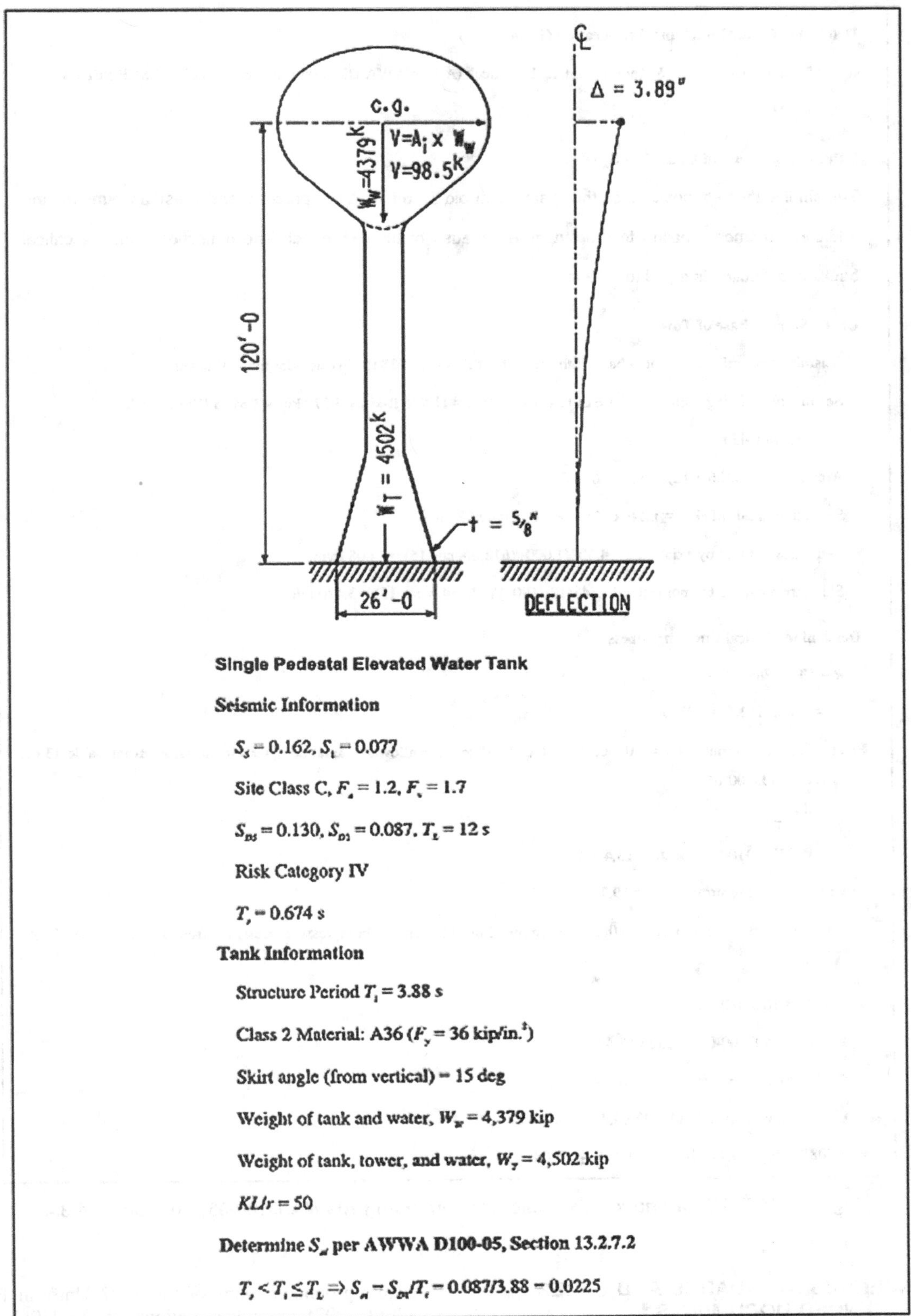

Figure C15.7-5. Example problem using AWWA D100-05, Section 13.4.3.4.

Determine Critical Buckling Acceleration ($I/1.4R_i = 1$)

Per Section 13.4.3.4, $A_i = S_{ai}$ for critical buckling check (A_i in AWWA D100-05 is the same as C_s in ASCE/SEI 7-10)

$A_i = 0.0225$

Lateral Displacement Caused by S_{ai} (P-Δ)

The final deflected position of the water centroid is an iterative process and must account for the additional moment applied to the structure because of the P-Δ effect. The deflection from the critical buckling deflection is equal to 3.89 in.

Check Skirt at Base of Tower

Seismic overturning moment at base of tower without P-Δ = 11,928 ft-kip (includes mass of tower).

Seismic overturning moment at base of tower with P-Δ = 11,928 ft-kip + 4,379 kip × 3.89 in./12 in. per ft

= 13,348 ft-kip

Area of skirt = $\pi(26 \times 12)(0.625) = 612.6\ \text{in.}^2$

Section modulus of skirt = $\pi(26 \times 12)^2/4 \times 0.625 = 47,784\ \text{in.}^3$

Skirt stress caused by axial load = 4,502(1,000)/(612.6 × cos 15) = 7,608 lb/in.^{2}

Skirt stress caused by moment = 13,348(12)(1000)/(47,784 × cos 15) = 3,470 lb/in.^{2}

Determine Critical Buckling Stress

$R = 13 \times 12/\cos 15 = 161.5$ in.

$t/R = 0.625/161.5 = 0.0039$

For Class 2 material, $KL/r = 50$, and $t/R = 0.0039$, determine allowable axial compressive stress, F_a, from Table 13 of AWWA D100-05.

$F_a = 9,882\ \text{lb/in.}^2$

Per AWWA D100-05, Section 13.4.3.4,

Critical buckling stress = $2F_a$ = 19,764 lb/in.^{2}

For Class 2 material and $t/R = 0.0039$, determine allowable bending compressive stress, F_b, from Table 11 of AWWA D100-05.

$F_b = F_L = 10,380\ \text{lb/in.}^2$

Per AWWA D100-05, Section 13.4.3.4,

Critical bending stress = $2F_b$ = 20,760 lb/in.^{2}

Check unity per AWWA D100-05, Section 3.3.1:

$7,608/19,764 + 3,470/20,760 = 0.552 \leq 1.0$ OK

Figure C15.7-5 *(Continued)*. Example problem using AWWA D100-05, Section 13.4.3.4.

C15.8 CONSENSUS STANDARDS AND OTHER REFERENCED DOCUMENTS

Chapter 15 of this standard makes extensive use of reference documents in the design of nonbuilding structures for seismic forces; see Chapter 23. The documents referenced in Chapter 15 are industry documents commonly used to design specific types of nonbuilding structures. The vast majority of these reference documents contain seismic provisions that are based on the seismic ground motions of the 1997 Uniform Building Code (ICBO 1997) or earlier editions of the UBC. To use these reference documents, Chapter 15 modifies the seismic force provisions of these reference documents through the use of "bridging equations." The standard only modifies industry documents that specify seismic demand and capacity. The bridging equations are intended to be used directly with the other provisions of the specific reference documents. Unlike the other provisions of the standard, if the reference documents are written

Table C15.8-1. Usage of Reference Documents in Conjunction with Section 15.4.1.

Subject	Requirement
R, Ω_0, and C_d values, detailing requirements, and height limits	Use values and limits in Tables 12.2-1, 15.4-1, or 15.4-2 as appropriate. Values from the reference document are not to be used.
Minimum base shear	Use the appropriate value from Equation (15.4-1) or (15.4-2) for nonbuilding structures not similar to buildings. For structures containing liquids, gases, and granular solids supported at the base, the minimum seismic force cannot be less than that required by the reference document.
Importance factor	Use the value from Section 15.4.1.1 based on risk category. Importance factors from the reference document are not to be used unless they are greater than those provided in the standard.
Vertical distribution of lateral load	Use requirements of Section 12.8.3 or Section 12.9 or the applicable reference document.
Seismic provisions of reference documents	The seismic force provisions of reference documents may be used only if they have the same basis as Section 11.4 and the resulting values for total lateral force and total overturning moment are no less than 80% of the values obtained from the standard.
Load combinations	Load combinations specified in Section 2.3 [load and resistance factor design (LRFD)] or Chapter 15 [includes the allowable stress design (ASD), load combinations of Section 2.4] must be used.

in terms of allowable stress design, then the bridging equations are shown in allowable stress design format. In addition, the detailing requirements referenced in Tables 15.4-1 and Table 15.4-2 are expected to be followed, as well as the general requirements found in Section 15.4.1. The usage of reference documents in conjunction with the requirements of Section 15.4.1 is summarized in Table C15.8-1.

Currently, only four reference documents have been revised to meet the seismic requirements of the standard. AWWA D100, API 620, API 650, and ANSI/RMI MH 16.1 have been adopted by reference in the standard without modification, except that height limits are imposed on "elevated tanks on symmetrically braced legs (not similar to buildings)" in AWWA D100, and the anchorage requirements of Section 15.4.9 are imposed on steel storage racks in ANSI/RMI MH 16.1. Three of these reference documents apply to welded steel liquid storage tanks.

REFERENCES

ACI (American Concrete Institute). 1997. *Standard practice for the design and construction of concrete silos and stacking tubes for storing granular materials*. ACI 313. Farmington Hills, MI: ACI.

ACI. 1998. *Guide for the analysis, design, and construction of concrete-pedestal water towers*. ACI 371R. Farmington Hills, MI: ACI.

ACI. 2006. *Seismic design of liquid-containing concrete structures*. ACI 350.3-06. Farmington Hills, MI: ACI.

ACI. 2016. *Design specification for concrete silos and stacking tubes for storing granular materials and commentary*. ACI 313. Farmington Hills, MI: ACI.

ACI. 2019. *Building code requirements for structural concrete and commentary*. ACI 318. Farmington Hills, MI: ACI.

AISC. 1986. *Load and resistance factor design specification for structural steel buildings*. Chicago: AISC.

AISC. 2016. *Seismic provisions for structural steel buildings*. AISC 341. Chicago: AISC.

AISI. 2020. *ANSI/AISI S100, North American specification for the design of cold-formed steel structural members*, 2016 Edition (Reaffirmed 2020), with Supplement 2, 2020 Edition, AISI 100-16 (2020) with Supplement 2, 2020, Washington, DC: AISI.

Anderson, P. F. 1966. "Temperature stresses in steel grain storage tanks." *Civ. Eng*. 36 (1): 74.

API (American Petroleum Institute). 2014a. *Design and construction of large, welded, low pressure storage tanks*. 12th ed., Addendum 1. API 620. Washington, DC: API.

API. 2014b. *Tank inspection, repair, alteration, and reconstruction*, 5th ed. API 653. Washington, DC: API.

API. 2014c. *Welded steel tanks for oil storage* 12th ed., Addendum 1. API 650. Washington, DC: API.

ASCE. 1984. *Guidelines for the seismic design of oil and gas pipeline systems*. Reston, VA: ASCE.

ASCE. 2005. *Minimum design loads for buildings and other structures*. ASCE/SEI 7-05 including Supplement 2. Reston, VA: ASCE.

ASCE. 2010. *Minimum design loads for buildings and other structures*. ASCE/SEI 7-10 including Supplement 2. Reston, VA: ASCE.

ASCE/AWEA (ASCE/American Wind Energy Association). 2011. *Recommended practice for large land-based wind turbine support structures*. RP2011. Reston, VA: ASCE.

ASME (American Society of Mechanical Engineers). 2007. *Boiler and pressure vessel code*. New York: ASME.

ASME. 2011. *Steel stacks*. ASME STS-1. New York: ASME.

ASME. 2019. *Reinforced thermoset plastic corrosion-resistant equipment*. RTP-1. New York: ASME.

ASTM. 2020. *Standard specification for anchor bolts, steel, 36, 55, and 105-ksi yield strength*. ASTM F1554. West Conshohocken, PA: ASTM.

AWWA (American Water Works Association). 2006a. *Welded steel tanks for water storage*. AWWA D100-05. Denver: AWWA.

AWWA. 2009. *Factory-coated bolted steel tanks for water storage*. AWWA D103. Denver: AWWA.

AWWA. 2010. *Composite elevated tanks for water storage*. AWWA D107. Denver: AWWA.

AWWA. 2011. *Welded steel tanks for water storage*. AWWA D100. Denver: AWWA.

AWWA. 2013. *Wire- and strand-wound, circular, prestressed concrete water tanks*. AWWA D110. Denver: AWWA.

California State Lands Commission. 2005. *Marine oil terminal engineering and maintenance standards*. California Building Code. Sacramento, CA: California State Lands Commission.

Chandrasekaran, A. R., and P. C. Jain. 1968. "Effective live load of storage materials under dynamic conditions." *Indian Concr. J.* 42 (9): 364–365.

Chandrasekaran, A. R., S. S. Saini, and I. C. Jhamb. 1968. "Live load effects on dynamic behavior of structures." *J. Inst. Eng.* 48: 850–859.

FEMA (Federal Emergency Management Agency). 2005. *Seismic considerations for steel storage racks located in areas accessible to the public*. FEMA 460. Washington, DC: Building Seismic Safety Council, National Institute of Building Sciences.

FEMA. 2009. *Quantification of building seismic performance factors*. FEMA P695. Redwood City, CA: Applied Technology Council.

FEMA. 2012. *Reducing the risks of nonstructural earthquake damage: A practical guide*. FEMA E-74. Redwood City, CA: Applied Technology Council.

Ferritto, J., S. Dickenson, N. Priestley, S. Werner, C. Taylor, D. Burke, W. Seeling, and S. Kelly. 1999. *Seismic criteria for California marine oil terminals.* Technical Rep. No. TR-2103-SHR. Port Hueneme, CA: Naval Facilities Engineering Service Center.

Hadjian, A. H., and B. Ellison. 1986. "Decoupling of secondary systems for seismic analysis." *J. Pressure Vessel Technol.* 108 (1): 78–85. https://doi.org/10.1115/1.3264755.

Haroun, M. A., and G. W. Housner. 1981. "Seismic design of liquid storage tanks." *J. Tech. Counc.* 107 (1): 191–207.

Housner, G. W.. 1963. "The dynamic behavior of water tanks." *Bull. Seismol. Soc. Am.* 53(2): 381–387.

ICBO (International Conference of Building Officials). 1988. *Uniform building code*. Whittier, CA: ICBO.

ICBO. 1997. *Uniform building code*. Whittier, CA: ICBO.

Jacobsen, L. S. 1949. "Impulsive hydrodynamics of fluid inside a cylindrical tank and of fluid surrounding a cylindrical pier." *Bull. Seismol. Soc. Am.* 39(3): 189–203.

Kaups, T., and J. M. Lieb. 1985. *A practical guide for the design of quality bulk storage bins and silos*. Plainfield, IL: Chicago Bridge & Iron.

Kilic, S., and M. Sozen. 2003. "Evaluation of effect of August 17, 1999, Marmara earthquake on two tall reinforced concrete chimneys." *ACI Struct. J.* 100 (3): 357–364. https://doi.org/10.14359/12611.

Malhotra, P. K. 2005. "Sloshing loads in liquid-storage tanks with insufficient freeboard." *Earthquake Spectra* 21 (4): 1185–1192. https://doi.org/10.1193/1.2085188.

Malhotra, P. K. 2006. "Earthquake induced sloshing in cone and dome roof tanks with insufficient freeboard." *J. Struct. Eng. Int.* 16 (3): 222–225. https://doi.org/10.2749/101686606778026466.

Malhotra, P. K., T. Wenk, and M. Wieland. 2000. "Simple procedure for seismic analysis of liquid-storage tanks." *J. Struct. Eng. Int.* 10 (3): 197–201. https://doi.org/10.2749/101686600780481509.

Maritime Navigation Commission. 2001. *Seismic design guidelines for port structures*. Working group No. 34. Brussels, Belgium: PIANC General Secretariat.

Miller, C. D., S. W. Meier, and W. J. Czaska. 1997. "Effects of internal pressure on axial compressive strength of cylinders and cones." In *Proc., Structural Stability Research Council Annual Technical Meeting*. Reston, VA: ASCE.

Priestley, M. J. N., F. Siebel, and G. M. Calvi. 1996. *Seismic design and retrofit of bridges*. New York: Wiley.

RMI (Rack Manufacturers Institute). 2012. *Specification for the design, testing, and utilization of industrial steel storage racks*. ANSI/RMI MH 16.1. Charlotte, NC: RMI.

RMI. 2016. *Specification for the design, testing, and utilization of industrial steel cantilevered storage racks*. ANSI/RMI MH 16.3. Charlotte, NC: RMI.

TIA (Telecommunications Industry Association). 2019. *Structural standard for antenna supporting structures, antennas and small wind turbine support structures*, ANSI/TIA-222-H, Addendum 1. Arlington, VA: TIA.

Trahair, M. S., A. Abel, P. Ansourian, H. M. Irvine, and J. M. Rotter. 1983. *Structural design of steel bins for bulk solids.* Sydney: Australian Institute of Steel Construction.

Troitsky, M. S. 1990. *Tubular steel structures: Theory and design.* Mentor, OH: James F. Lincoln Arc Welding Foundation.

US Department of Energy. 1963. *Nuclear reactors and earthquakes*. TID-7024. Washington, DC: US Atomic Energy Commission.

Veletsos, A. S. 1974. "Seismic effects in flexible liquid storage tanks." In *Proc., 5th World Conf. on Earthquake Engineering*, Rome, 630–639.

Werner, S. D., ed. 1998. *Seismic guidelines for ports: Monograph no. 12*. Reston, VA: ASCE.

Wey, E. H., and F. Mejia. 2019. "Factors triggering combined analysis of coupled industrial structures subjected to seismic loading." In *Proc., ASCE/SEI Structures Congress 2019*, Orlando, FL.

Working Group No. 34, Maritime Navigation Commission. 2001. *Seismic design guidelines for port structures*. Lisse, Belgium: Maritime Navigation Commission, PIANC.

Wozniak, R. S., and W. W. Mitchell. 1978. "Basis of seismic design provisions for welded steel oil storage tanks." In *Proc., Session on Advances in Storage Tank Design, American Petroleum Institute, Refining, 43rd Midyear Meeting*, Toronto.

OTHER REFERENCES (NOT CITED)

ACI. 2008. *Code requirements for reinforced concrete chimneys and commentary*. ACI 307. Farmington Hills, MI: ACI.

API. 2014. *Specification for bolted tanks for storage of production liquids*. API 12B. Washington, DC: API.

ASCE. 1997. *Design of secondary containment in petrochemical facilities*. Reston, VA: ASCE.

ASCE. 2011. *Guidelines for seismic evaluation and design of petrochemical facilities,* 2nd ed. Reston, VA: ASCE.

AWWA. 2006b. *Tendon-prestressed concrete water tanks*. AWWA D115. Denver: AWWA.

Drake, R. M., and R. J. Walter. 2010. "Design of structural steel pipe racks." *AISC Eng. J.* 47 (4): 241–251.

NFPA (National Fire Protection Association). 2013. *Standard for the production, storage, and handling of liquefied natural gas (LNG)*. NFPA 59A. Quincy, MA: NFPA.

Soules, J. G. 2006. "The seismic provisions of the 2006 IBC: Nonbuilding structure criteria." In *Proc., 8th National Conf. on Earthquake Engineering*, San Francisco.

CHAPTER C16
NONLINEAR RESPONSE HISTORY ANALYSIS

C16.1 GENERAL REQUIREMENTS

C16.1.1 Scope Response history analysis is a form of dynamic analysis in which response of the structure to a suite of ground motions is evaluated through numerical integration of the equations of motions. In nonlinear response history analysis, the structure's stiffness matrix is modified throughout the analysis to account for the changes in element stiffness associated with hysteretic behavior and P-delta effects. When nonlinear response history analysis is performed, the R, C_d, and Ω_0 coefficients considered in linear procedures are not applied because the nonlinear analysis directly accounts for the effects represented by these coefficients.

Nonlinear response history analysis is permitted to be performed as part of the design of any structure and is specifically required to be performed for the design of certain structures incorporating seismic isolation or energy dissipation systems. Nonlinear response history analysis is also frequently used for the design of structures that use alternative structural systems or do not fully comply with the prescriptive requirements of the standard in one or more ways. Before this edition, ASCE 7 specified that nonlinear response history analyses had to be performed using ground motions scaled to the design earthquake level and that design acceptance checks be performed to ensure that mean element actions do not exceed two-thirds of the deformations at which loss of gravity-load-carrying capacity would occur. In this edition of ASCE 7, a complete reformulation of these requirements was undertaken to require analysis at the Risk-Targeted Maximum Considered Earthquake (MCE_R) level and also to be more consistent with the target reliabilities indicated in Section 1.3.1.3.

The target collapse reliabilities given in Table 1.3-2 are defined such that when a building is subjected to MCE_R ground motion, not greater than a 10% probability of collapse exists for Risk Category I and II structures. For Risk Category III and IV structures, these maximum collapse probabilities are reduced to 5% and 2.5%, respectively.

There are additional performance expectations for Risk Category III and IV structures that go beyond the collapse safety performance goals (e.g., limited damage and postearthquake functionality for lower ground motion levels). These enhanced performance goals are addressed in this chapter by enforcing an $I_e > 1.0$ in the linear design step (which is consistent with the approach taken in the other design methods of Chapter 12) and also by considering I_e in acceptance checks specified in Section 16.4.

It is conceptually desirable to create a Chapter 16 response history analysis (RHA) design process that explicitly evaluates the collapse probability and ensures that the performance goal is fulfilled. However, explicit evaluation of collapse safety is a difficult task requiring (a) a structural model that is able to directly simulate the collapse behavior, (b) use of numerous nonlinear response history analyses, and (c) proper treatment of many types of uncertainties. This process is excessively complex and lengthy for practical use in design. Therefore, Chapter 16 maintains the simpler approach of *implicitly* demonstrating adequate performance through a prescribed set of analysis rules and acceptance criteria. Even so, this implicit approach does not preclude the use of more advanced procedures that explicitly demonstrate that a design fulfills the collapse safety goals. Such more advanced procedures are permitted by Section 1.3.1.3 of this standard. An example of an advanced explicit procedure is the building-specific collapse assessment methodology in Appendix F of FEMA P-695 (FEMA 2009b).

C16.1.2 Linear Analysis As a precondition to performing nonlinear response history analysis, a linear analysis, in accordance with the requirements of Chapter 12, is required. Any of the linear procedures allowed in Chapter 12 may be used. The purpose of this requirement is to ensure that structures designed using nonlinear response history analyses meet the minimum strength and other criteria of Chapter 12, with a few exceptions. In particular, when performing the Chapter 12 evaluations, it is permitted to take the value of Ω_0 as 1.0 because it is felt that values of demand obtained from the nonlinear procedure are a more accurate representation of the maximum forces that will be delivered to critical elements, considering structural overstrength, than does the application of the judgmentally derived factors specified in Chapter 12. Similarly, it is permitted to use a value of 1.0 for the redundancy factor, ρ, because it is felt that the inherent nonlinear evaluation of response to MCE_R shaking required by this chapter provides improved reliability relative to the linear procedures of Chapter 12. For Risk Category I, II, or III structures, it is permitted to neglect the evaluation of story drift when using the linear procedure because it is felt that the drift evaluation performed using the nonlinear procedure provides a more accurate assessment of the structure's tolerance to earthquake-induced drift. However, linear drift evaluation is required for Risk Category IV structures because it is felt that this level of drift control is important to attaining the enhanced performance desired for such structures.

As with other simplifications permitted in the linear analysis required under this section, it is also permitted to use a value of 1.0 for the torsional amplification, A_x, when performing a nonlinear analysis if accidental torsion is explicitly modeled in the nonlinear analysis. Although this does simplify the linear analysis somewhat, designers should be aware that the resulting structure may be more susceptible to torsional instability when performing the nonlinear analysis. Therefore, some designers may find it expedient to use a value of A_x consistent with the linear procedures as a means of providing a higher likelihood that the nonlinear analysis will result in acceptable outcomes.

C16.1.3 Vertical Response Analysis Most structures are not sensitive to the effects of response to vertical ground shaking, and there is little evidence of the failure of structures in earthquakes resulting from vertical response. However, some nonbuilding structures and building structures with long spans, cantilevers, prestressed construction, or vertical discontinuities in their gravity-load-resisting systems can experience significant vertical earthquake response that can cause failures. The linear procedures of Chapter 12 account for these effects in an approximate manner, through use of the $0.2S_{DS}D$ term in the load combinations. When nonlinear response history analysis is performed for structures with sensitivity to vertical response, direct simulation of this response is more appropriate than the use of the approximate linear procedures. However, in order to properly capture vertical response to earthquake shaking, it is necessary to accurately model the stiffness and distribution of mass in the vertical load system, including the flexibility of columns and horizontal framing. This effort can considerably increase the complexity of analytical models. Rather than requiring this extra effort in all cases where vertical response can be significant, this chapter continues to rely on the approximate approach embedded in Chapter 12 for most cases. However, where the vertical load path is discontinuous and where vertical response analysis is required by Chapter 15, Chapter 16 does require explicit modeling and analysis of vertical response. Since in many cases the elements sensitive to vertical earthquake response are not part of the seismic force-resisting system, it is often possible to decouple the vertical and lateral response analyses, using separate models for each.

Appropriate accounting for the effects of vertical response to ground shaking requires that horizontal framing systems, including floor and roof systems, be modeled with distributed masses and sufficient vertical degrees of freedom to capture their out-of-plane dynamic characteristics. This increased fidelity in modeling of the structure's vertical response characteristics will significantly increase the size and complexity of models. As a result, the chapter requires direct simulation of vertical response only for certain structures sensitive to those effects and relies on the procedures of Chapter 12 to safeguard the vertical response of other structures.

C16.1.4 Documentation By its nature, most calculations performed using nonlinear response history analysis are contained within the input and output of computer software used to perform the analysis. This section requires documentation, beyond the computer input and output, of the basic assumptions, approaches, and conclusions so that thoughtful review may be performed by others including peer reviewers and the Authority Having Jurisdiction. This section requires submittal and review of some of these data before the analyses are performed, in order to ensure that the engineer performing the analysis/design and the reviewers are in agreement before substantive work is performed.

C16.2 GROUND MOTIONS

C16.2.1 Target Response Spectrum The target response spectrum used for nonlinear dynamic analysis is the maximum direction MCE_R spectrum determined in accordance with Chapter 11 or Chapter 21. Typical spectra, determined in accordance with those procedures, are derived from uniform hazard spectra (UHSs) and modified to provide a uniform risk spectrum (URS), or alternatively, a deterministic MCE spectrum. Since the 1980s, UHSs have been used as the target spectra in design practice. The UHS is created for a given hazard level by enveloping the results of seismic hazard analysis for each period (for a given probability of exceedance). Accordingly, it is generally a conservative target spectrum if used for ground motion selection and scaling, especially for large and rare ground motions, unless the structure exhibits only elastic first-mode response. This inherent conservatism comes from the fact that the spectral values at each period are not likely to all occur in a single ground motion. This limitation of the UHS has been noted for many years (e.g., Bommer et al. 2000, Naeim and Lew 1995, Reiter 1990). The same conservatism exists for the URS and deterministic MCE spectra that serve as the basis for Method 1.

Method 2 uses the conditional mean spectrum (CMS), an alternative to the URS that can be used as a target for ground motion selection in nonlinear response history analysis (e.g., Baker and Cornell 2006, Baker 2011, Al Atik and Abrahamson 2010).

To address the conservatism inherent in analyses using URSs as a target for ground motion selection and scaling, the CMS instead conditions the spectrum calculation on a spectral acceleration at a single period and then computes the mean (or distribution of) spectral acceleration values at other periods. This conditional calculation ensures that the resulting spectrum is reasonably likely to occur and that ground motions selected to match the spectrum have an appropriate spectral shape consistent with naturally occurring ground motions at the site of interest. The calculation is no more difficult than the calculation of a URS and is arguably more appropriate for use as a ground motion selection target in risk assessment applications. The spectrum calculation requires disaggregation information, making it a site-specific calculation that cannot be generalized to other sites. It is also period-specific, in that the conditional response spectrum is based on a spectral acceleration value at a specified period. The shape of the conditional spectrum also changes as the spectral amplitude changes (even when the site and period are fixed). Figure C16.2-1 provides examples of CMSs for an example site in Palo Alto, California, anchored at four different candidate periods. The UHS for this example site is also provided for comparison.

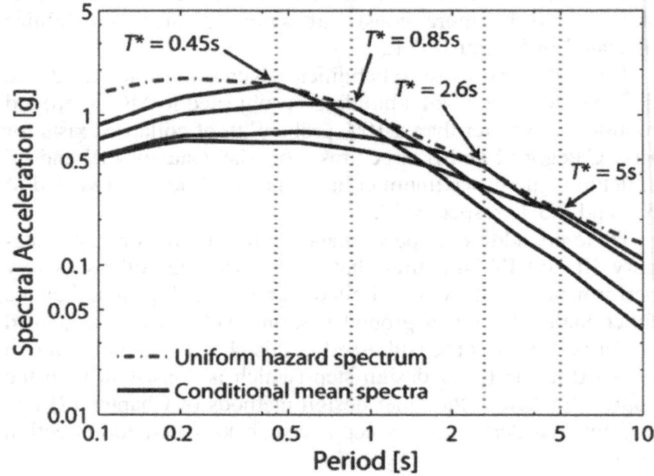

Figure C16.2-1. Example conditional mean spectra for a Palo Alto site, anchored for 2% in 50-year motion at $T = 0.45s$, 0.85s, 2.6s, and 5s.

Source: NIST (2011).

As previously discussed, the URS is a conservative target spectrum for ground motion selection, and the use of CMS target spectra is more appropriate for representing anticipated MCE_R ground motions at a specified period. A basic CMS-type approach was used in the analytical procedures of the FEMA P-695 (FEMA 2009b) project, the results of which provided the initial basis for establishing the 10% probability of collapse goal shown in Table 1.3-2. Therefore, the use of CMS target spectra in the Chapter 16 RHA design procedure is also internally consistent with how the collapse probability goals of Table 1.3-2 were developed.

The URS (or deterministic MCE) target spectrum is retained in Section 16.2.1.1 (as a simpler and more conservative option) as the specified target spectrum, and the CMS is permitted as an alternate in Section 16.2.1.2. Whereas CMS appropriately captures the earthquake energy and structural response at a particular period resulting from a particular scenario earthquake, it is not capable of capturing the MCE_R level response associated with other scenarios that are relevant to the MCE_R spectrum. Therefore, when using CMS, it may be necessary to use several conditioning periods and associated targets to develop conditional mean spectra, in order to fully capture the structure's response to different earthquake scenarios. The recommended procedure includes the following steps for creating the site-specific scenario response spectra.

1. Select those periods that correspond to periods of vibration that significantly contribute to the building's inelastic dynamic response. This selection includes a period near the fundamental period of the building, or perhaps a slightly longer period to account for inelastic period lengthening (e.g., $1.5T_1$). In buildings where the fundamental response periods in each of two orthogonal axes is significantly different, a conditioning period associated with each direction is needed. It also likely requires periods near the translational second-mode periods. When selecting these significant periods of response, the elastic periods of response should be considered (according to the level of mass participation for each of these periods), and the amount of first-mode period increase caused by inelastic response effects should also be considered.
2. For each period selected above, create a scenario spectrum that matches or exceeds the MCE_R value at that period. When developing the scenario spectrum, (a) perform site-specific disaggregation to identify earthquake events likely to result in MCE_R ground shaking, and then (b) develop the scenario spectrum to capture one or more spectral shapes for dominant magnitude and distance combinations revealed by the disaggregation.
3. Enforce that the envelope of the scenario spectra not be less than 75% of the MCE_R spectrum (from Method I) for any period within the period range of interest (as defined in Section 16.2.3.1).

After the target spectra are created, each target response spectrum is then used in the remainder of the response history analyses process and the building must be shown to meet the acceptance criteria for each of the scenarios.

The primary purpose of the 75% limit value is to provide a basis for determining how many target spectra are needed for analysis. For small period ranges, fewer targets are needed, and more target spectra are needed for buildings where a wider range of periods are important to the structural response (e.g., taller buildings). When creating the target spectra, some spectral values can also be artificially increased to meet the requirements of this 75% limit. A secondary reason for the 75% limit is to enforce a reasonable lower bound. The specific 75% threshold value was determined using several examples. The intention is that this 75% floor requirement will be fulfilled through the use of two target spectra, in most cases. From the perspective of collapse risk, the requirement of being within 75% of the MCE_R at all periods may introduce some conservatism, but the requirement adds robustness to the procedure by ensuring that the structure is subjected to ground motions with near-MCE_R-level intensities at all potentially relevant periods. In addition, this requirement ensures that demands unrelated to collapse safety, such as higher mode-sensitive force demands, can be reasonably determined from the procedure.

C16.2.2 Ground Motion Selection Before ASCE 7-16, Chapter 16 required a minimum of three ground motions for nonlinear response history analysis. If three ground motions were used, the procedures required evaluation of structural adequacy using the maximum results obtained from any of the ground motions. If seven or more motions were used, mean results could be used for evaluation. Neither three nor seven motions are sufficient to accurately characterize either mean response or the record-to-record variability in response. In the 2016 edition of the standard, the minimum number of motions was increased to 11. The requirement for this larger number of motions was not based on detailed statistical analyses, but rather was judgmentally selected to balance the competing objectives of more reliable estimates of mean structural responses (through use of more motions) against computational effort (reduced by using fewer motions). An advantage of using this larger number of motions is that if unacceptable response is found for more than one of the 11 motions, this indicates a significant probability that the structure will fail to meet the 10% target collapse reliability for Risk Categories I and II structures of Section 1.3.1.3. This advantage is considered in the development of acceptance criteria discussed in Section C16.4.

All real ground motions include three orthogonal components. For most structures, it is only necessary to consider response to horizontal components of ground shaking. However, consideration of vertical components is necessary for structures defined as sensitive to vertical earthquake effects.

Section 11.4.1 defines near-fault sites as sites located within 9.3 mi (15 km) of the surface projection of faults capable of producing earthquakes of magnitude 7.0 or greater and within 6.2 mi (10 km) of the surface projection of faults capable of producing earthquakes of magnitude 6.0 or greater, where the faults must meet minimum annual slip rate criteria. Such near-fault sites have a reasonable probability of experiencing ground motions strongly influenced by rupture directivity effects. These effects can include pulse-type ground motions (e.g., Shahi et al. 2011) observable in velocity histories and polarization of ground motions, such that the maximum direction of response tends to be in the direction normal to the fault strike. The issue of pulse-type ground motions affects the manner by which individual ground motions should be selected for the site and applied to the structure.

Selection of Ground Motions for Sites That Are Not Near-Fault. The traditional approach has been to select (and/or simulate) ground motions that have magnitudes, fault distances, source mechanisms, and site soil conditions that are roughly similar to those likely to cause the ground motion intensity level of interest (e.g., Stewart et al. 2002) and not to consider the spectral shape in the ground motion selection. In many cases, the response spectrum is the property of a ground motion most correlated with the structural response (Bozorgnia et al. 2009) and should be considered when selecting ground motions.

When spectral shape is considered in the ground motion selection, the allowable range of magnitudes, distances, and site conditions can be relaxed so that a sufficient number of ground motions with appropriate spectral shapes are available.

The selection of recorded motions typically occurs in two steps, as explained in the following illustration. Step 1 involves preselecting the ground motion records in the database (e.g., Anchenta et al. 2015) that have reasonable source mechanisms, magnitude, site soil conditions, a range of usable frequencies, and site-to-source distance. In completing this preselection, it is permissible to use relatively liberal ranges because Step 2 can involve selecting motions that provide good matches to a target spectrum of interest (and matching to a target spectrum tends to implicitly account for many of the above issues). Step 2 in the selection process is to select the final set of motions from those preselected in Step 1.

In the first step, the following criteria should be used to filter out ground motions that should not be considered as candidates for the final selection process:

- **Source Mechanism:** Ground motions from differing tectonic regimes (e.g., subduction versus active crustal regions) often have substantially differing spectral shapes and durations, so recordings from appropriate tectonic regimes should be used whenever possible.
- **Magnitude:** Earthquake magnitude is related to the duration of ground shaking, so using ground motions from earthquakes with appropriate magnitudes should already have approximately the appropriate durations. Earthquake magnitude is also related to the shape of the resulting ground motion's response spectrum, though spectral shape is considered explicitly in Step 2 of the process, so this is not a critical factor when identifying ground motions from appropriate magnitude earthquakes.
- **Site Soil Conditions:** Site soil conditions (Site Class) exert a large influence on ground motions but are already reflected in the spectral shape used in Step 2. For Step 1, reasonable limits on site soil conditions should be imposed but should not be too restrictive as to unnecessarily limit the number of candidate motions.
- **Usable Frequency of the Ground Motion:** Only processed ground motion records should be considered for RHA. Processed motions have a usable frequency range; in active regions, the most critical parameter is the lowest usable frequency. It is important to verify that the usable frequencies of the record (after filtering) accommodate the range of frequencies important to the building response; this frequency (or period) range is discussed in the next section on scaling.
- **Period/Frequency Sampling:** Ground motion recordings are discretized representations of continuous functions. The sampling rate for the recorded data can vary from as little as 0.001 s to as much as 0.02 s, depending on the recording instrument and processing. If the sampling rate is too coarse, important characteristics of the motion, particularly in the high-frequency range, can be lost. On the other hand, the finer the sampling rate, the longer the analysis will take. Particularly for structures with significant response at periods less than 0.1 s, caution should be used to ensure that the sampling rate is sufficiently fine to capture the motion's important characteristics. As a general guideline, discretization should include at least 100 points per decade of significant response. Thus, for a structure with significant response at a period of 0.1 s, time steps should not be greater than 0.001 s.
- **Site-to-Source Distance:** The distance is a lower priority parameter to consider when selecting ground motions. Studies investigating this property have all found that response history analyses performed using ground motions from different site-to-source distances, but otherwise equivalent properties, produce practically equivalent demands on structures.

Once the preselection process has been completed, Step 2 is undertaken to select the final set of ground motions according to the following criteria:

- **Spectral Shape:** The shape of the response spectrum is a primary consideration when selecting ground motions.
- **Scale Factor:** It is also traditional to select motions such that the necessary scale factor is limited; an allowable scale factor limit of approximately 0.25 to 4 is not uncommon.
- **Maximum Motions from a Single Event:** Many also think it important to limit the number of motions from a single seismic event, such that the ground motion set is not unduly influenced by the single event. This criterion is deemed less important than limiting the scale factor, but imposing a limit of only three or four motions from a single event would not be unreasonable for most cases.

Further discussion of ground motion selection is available in NIST GCR 11-917-15 (NIST 2011), *Selecting and Scaling Earthquake Ground Motions for Performing Response-History Analyses*.

Near-fault sites have a probability of experiencing pulse-type ground motions. This probability is not unity, so only a certain fraction of selected ground motions should exhibit pulselike characteristics, while the remainder can be nonpulse records, selected according to the standard process defined previously. The probability of experiencing pulselike characteristics is dependent principally on (1) distance of site from fault, (2) fault type (e.g., strike slip or reverse), and (3) location of hypocenter relative to site, such that rupture occurs toward or away from the site.

Criteria (1) and (2) are available from conventional disaggregation of probabilistic seismic hazard analysis. Criterion (3) can be computed as well in principle, but is generally not provided in a conventional hazard analysis. However, for the long ground motion return periods associated with MCE_R spectra, it is conservative and reasonable to assume that the fault rupture is toward the site for the purposes of evaluating pulse probabilities. Empirical relations for evaluating pulse probabilities in consideration of these criteria are given in NIST GCR 11-917-15 (2011) and in Shahi et al. (2011).

Once the pulse probability is identified, the proper percentage of pulselike records should be enforced in the ground motion selection. For example, if the pulse probability is 30% and 11 records are to be used, then 3 or 4 records in the set should exhibit pulselike characteristics in at least one of the horizontal components. The PEER Ground Motion Database can be used to identify records with pulse-type characteristics. The other criteria described in the previous section should also be considered to identify pulselike records that are appropriate for a given target spectrum and set of disaggregation results.

C16.2.3 Ground Motion Modification Two procedures for modifying ground motions for compatibility with the target spectrum are available: amplitude scaling and spectral matching. Amplitude scaling consists of applying a single scaling factor to the entire ground motion record, such that the variation of earthquake energy with structural period found in the original record is preserved. Amplitude scaling also preserves record-to-record variability; however, individual ground motions that are amplitude scaled can significantly exceed the response input of the target spectrum at some periods, which can tend to overstate the importance of higher mode response in some structures. In spectral matching techniques, shaking amplitudes are modified by differing amounts at differing periods, and in some cases additional wavelets of energy are added to or subtracted from the motions, such that the response spectrum of the modified motion closely resembles the target spectrum. Some spectral matching techniques are incapable of preserving important characteristics of velocity pulses in motions and should not be used for near-fault sites where these effects are important. Spectral matching does not generally preserve the record-to-record response variability observed when evaluating a structure for unmodified motions, but it can capture the mean response well, particularly if the nonlinear response is moderate.

Vertical response spectra of earthquake records are typically significantly different from the horizontal spectra. Therefore, regardless of whether amplitude scaling or spectral matching is used, separate scaling of horizontal and vertical effects is required.

C16.2.3.1 Period Range for Scaling or Matching The period range for scaling of ground motions is selected such that the ground motions accurately represent the MCE_R hazard at the structure's fundamental response periods, periods somewhat longer than this to account for period lengthening effects associated with nonlinear response and shorter periods associated with a higher mode response. Before the 2016 edition of the standard, ground motions were required to be scaled between periods of $0.2T$ and $1.5T$. The lower bound was selected to capture higher mode response, and the upper bound, period elongation effects. In the 2016 edition, nonlinear response history analyses were performed at the MCE_R ground motion level. Greater inelastic response is anticipated at this level compared to the design spectrum, so the upper bound period has accordingly been raised from $1.5T$ to $0.2T$, where T is redefined as the *maximum* fundamental period of the building (i.e., the maximum of the fundamental periods in both translational directions and the fundamental torsional period). This increase in the upper bound period is also based on recent research, which has shown that the $1.5T$ limit is too low for assessing ductile frame buildings subjected to MCE_R motions (Haselton and Baker 2006).

For the lower bound period, the $0.2T$ requirement is now supplemented with an additional requirement that the lower bound should also capture the periods needed for 90% mass participation in both directions of the building. This change is made to ensure that when used for tall buildings and other long-period structures, the ground motions are appropriate to capture response in higher modes that have significant response.

In many cases, the substructure is included in the structural model, and this inclusion substantially affects the mass participation characteristics of the system. Unless the foundation system is being explicitly designed using the results of the response history analyses, the above 90% modal mass requirement pertains only to the superstructure behavior; the period range does not need to include the very short periods associated with the subgrade behavior.

C16.2.3.2 Amplitude Scaling This procedure is similar to those found in earlier editions of the standard, but with the following changes:

1. Scaling is based directly on the maximum direction spectrum, rather than the square root of the sum of the squares spectrum. This change was made for consistency with the MCE_R ground motion now being explicitly defined as a maximum direction motion.
2. The approach of enforcing that the average spectrum "does not fall below" the target spectrum is replaced with requirements that (a) the average spectrum "matches the target spectrum" and (b) the average spectrum does not fall below 90% of the target spectrum for any period within the period range of interest. This change was made to remove the conservatism associated with the average spectrum being required to *exceed* the target spectrum at *every* period within the period range.

The scaling procedure requires that a maximum direction response spectrum be constructed for each ground motion. For some ground motion databases, this response spectrum definition is already precomputed and publicly available (e.g., Ancheta 2012). The procedure basically entails computing the maximum acceleration response to each ground motion pair for a series of simple structures that have a single mass. This procedure is repeated for structures of different periods, allowing construction of the spectrum. A number of software tools can automatically compute this spectrum for a given time–history pair.

Figure C16.2-2 shows an example of the scaling process for an example site and structure. This figure shows how the average of the maximum direction spectra meets the target spectrum (a) and shows more detail for a single Loma Prieta motion in the scaled ground motion set (b).

C16.2.3.3 Spectral Matching Spectral matching of ground motions is defined as the modification of a real recorded earthquake ground motion in some manner, such that its response spectrum matches a desired target spectrum across a period range of interest. There are several spectral matching procedures in use, as described in the NIST GCR 11-917-15 report (NIST 2011). The recommendations in this report should be followed regarding appropriate spectral matching techniques to be applied.

This section requires that when spectral matching is applied, the average of the maximum direction spectra of the matched motions must exceed the target spectrum over the period range of interest; this is intentionally a more stringent requirement compared to the requirement for scaled unmatched motions, because the spectral matching removes variability in the ground motion spectra and also has the potential to predict lower mean response (e.g., Luco and Bazzurro 2007, Grant and Diaferia 2012).

The specific technique used to perform spectral matching is not prescribed. It is possible to match both components of motion to a single target spectrum or to match the individual components to different spectra, as long as the average maximum direction spectra for the matched records meets the specified criteria.

Spectral matching is not allowed for near-fault sites, unless the pulse characteristics of the ground motions are retained after the matching process has been completed. This is based on the concern that when common spectral matching methods are used, the pulse characteristics of the motions may not be appropriately retained.

C16.2.4 Application of Ground Motions to the Structural Model This section explains the guidelines for ground motion application for both non-near-fault and near-fault sites.

Sites That Are Not Near-Fault. In this standard, the maximum direction spectral acceleration is used to describe the ground motion intensity. This spectral acceleration definition causes a

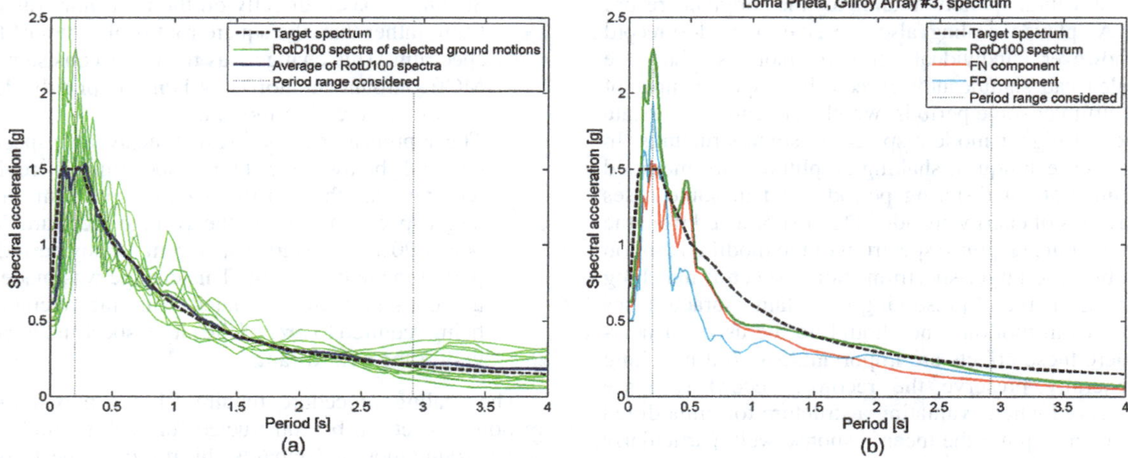

Figure C16.2-2. Ground motion scaling for an example site and structure, showing (a) the ground motion spectra for all 11 motions and (b) an example for the Loma Prieta, Gilroy array No. 3 motion.

perceived directional dependence to the ground motion. However, the direction in which the maximum spectral acceleration occurs is random at distances beyond 5 km (3.1 mi) from the fault (Huang et al. 2008), does not necessarily align with a principal direction of the building, and is variable from period to period. Accordingly, for the analysis to result in an unbiased prediction of structural response, the ground motions should be applied to the structure in a random orientation to avoid causing a biased prediction of structural response. True random orientation is difficult to achieve. Instead, the standard specifies that the average of the spectra applied in each direction should be similar to each other, such that unintentional bias in the application of motion, with one building axis experiencing greater demand than the other, is avoided.

Near-Fault Sites. Some recorded ground motions obtained from instruments located near zones of fault rupture have exhibited motions of significantly different character in one direction compared to the other. When this effect, known as directionality, occurs, it is common for the component of motion perpendicular to the fault to be stronger than that parallel to the fault and also for the fault-normal component to exhibit large velocity pulses. Sites located close to faults that can experience motion having these characteristics are termed near-fault in this standard. For such sites, the fault-normal and fault-parallel components of recorded ground motions should be maintained and applied to the corresponding orientations of the structure.

It is important to note that not all near-fault records exhibit these characteristics and also that when records do have these characteristics, the direction of maximum motion is not always aligned perpendicular to the fault strike. If an appropriate selection of records is performed, some of the records used in the analysis should have these characteristics and some should not. For those records that exhibit directionality, the direction of strong shaking is generally aligned at varying azimuths, as occurred in the original recordings. It is also important to note that because ground motions have considerable variability in their characteristics, it is specifically not intended that buildings be designed weaker in the fault-parallel direction than in the fault-normal direction.

ASCE 7-16 required all ground motions to be rotated to the fault-normal and fault-parallel directions for near-fault sites regardless of whether the selected motions were representative of a nearby fault source or a distant fault source. This provision only applied to sites where the hazard was solely from nearby faults; however, it was subsequently recognized that in some areas, the hazard is controlled by both near-fault and non near-fault seismic sources. For these areas, the suite of selected ground motions will generally consist of a combination of records from near-fault and non near-fault seismic sources. Therefore, the current version of the standard requires that for ground motions that have been selected to be representative of a nearby fault source, the fault-normal and fault-parallel components of recorded ground motions should be maintained and applied to the corresponding orientations of the structure. For ground motions selected to be representative of a distant fault source, the standard specifies that the average of the spectra applied in each direction should be similar to each other, such that unintentional bias in the application of motion, with one building axis experiencing greater demand than the other, is avoided.

C16.3 MODELING AND ANALYSIS

C16.3.1 Modeling Nonlinear response history analysis offers several advantages over linear response history analysis, including the ability to model a wide variety of nonlinear material behaviors, geometric nonlinearities (including P-delta and large displacement effects), gap opening and contact behavior, and nonlinear viscous damping, and to identify the likely spatial and temporal distributions of inelasticity. Nonlinear response history analysis has several disadvantages, however including an increased effort to develop the analytical model, increased time to perform the analysis (which is often complicated by difficulties in obtaining converged solutions), sensitivity of computed response to system parameters, large amounts of analysis results to evaluate, and the inapplicability of superposition to combine live, dead, and seismic load effects.

While computation of collapse probability is not necessary, it is important to note that mathematical models used in the analysis should have the capability to determine if collapse occurs when the structure is subjected to MCE_R level ground motions. The ability to predict collapse is important because the global acceptance criteria in Section 16.4.1.1 allow collapse (or unacceptable response) to occur for only one of the 11 ground motions for Risk Category I and II buildings and allows no such responses for Risk Categories III and IV buildings. Development of models with the ability to predict collapse requires attributes such as cyclic loss of strength and stiffness, low cycle fatigue failure, and geometric nonlinearity.

Although analytical models used to perform linear analysis, in accordance with Chapter 12, typically do not include representation of elements other than those that compose the intended lateral-force-resisting system, the gravity-load-carrying system and some nonstructural components can add significant stiffness and strength. Because the goal of nonlinear response history analysis is to accurately predict the building's probable performance, it is important to include such elements in the analytical model and also to verify that the behavior of these elements will be acceptable. This inclusion may mean that the contribution of stiffness and strength from elements considered as nonparticipating elements in other portions of this standard should be included in the response history analysis model. Since structures designed using nonlinear response history analysis must also be evaluated using linear analyses, this analysis ensures that the strength of the intended seismic force-resisting system is not reduced relative to that of structures designed using only the linear procedures.

Expected material properties are used in the analysis model, attempting to characterize the expected performance as closely as possible. It is suggested that expected properties be selected considering actual test data for the proposed elements. Where test data are not readily available, the designer may consider estimates as found in ASCE 41 and the PEER TBI Guidelines (Bozorgnia et al. 2009). Guidance on important considerations in modeling may also be found in *Nonlinear Structural Analysis for Seismic Design,* NIST GCR 10-917-5 (NIST 2010).

Two-dimensional structural models may be useful for initial studies and for checking some specific issues in a structure; however, the final structural model used to confirm the structural performance should be three-dimensional.

For certain structures, the response under both horizontal and vertical ground motions should be considered. NIST GCR 11-917-15 (NIST 2011) provides some guidance to designers considering the application of vertical ground motions. To properly capture the nonlinear dynamic response of structures where vertical dynamic response may have a significant influence on structural performance, it is necessary to include vertical mass in the mathematical model. Typically, the vertical mass must be distributed across the floor and roof plates to properly capture vertical response modes. Additional degrees of freedom (e.g., nodes at quarter points along the span of a beam) need to be added to capture this effect, or horizontal elements need to be modeled with consistent mass. Numerical convergence problems caused by large oscillatory vertical accelerations have been noted (NIST 2012), where base rotations caused by wall cracking in fiber wall models are the primary source of vertical excitation. See also the Commentary on Chapter 22.

Consideration of the additional vertical load of $(0.2S_{DS}) * D$, as per Section 12.4.2, is inappropriate for response history analysis. Response history analyses are desired to reflect actual building response to the largest extent possible. Applying an artificial vertical load to the analysis model before application of a ground motion results in an offset in the yield point of elements carrying gravity load because of the initial artificial stress. Similarly, applying an artificial vertical load to the model at the conclusion of a response history analysis is not indicative of actual building response. If vertical ground motions are expected to significantly affect response, application of vertical shaking to the analysis model is recommended. It should be noted that vertical response often occurs at higher frequencies than lateral response, and hence, a finer analysis time-step might be required when vertical motions are included.

For structures composed of planar seismic force-resisting elements connected by floor and roof diaphragms, the diaphragms should be modeled as semirigid in plane, particularly where the vertical elements of the seismic force-resisting system are of different types (such as moment frames and walls). Biaxial bending and axial force interaction should be considered for corner columns, nonrectangular walls, and other similar elements.

Nonlinear response history analysis is load path-dependent, with the results depending on combined gravity and lateral load effects. The MCE shaking and design gravity load combinations required in ASCE 7 have a low probability of occurring simultaneously. Therefore, the gravity load should instead be a realistic estimate of the expected loading on a typical day in the life of the structure. In this chapter, two gravity load cases are used. One includes an expected live loading characterizing probable live loading at the time of the MCE shaking, and the other, no live load. The case without live load is required to be considered only for those structures where live load constitutes an appreciable amount of the total gravity loading. In those cases, structural response modes can be significantly different, depending on whether the live load is present. The dead load used in this analysis should be determined in a manner consistent with the determination of seismic mass. When used, the live load is reduced from the nominal design live load to reflect both the low probability of the full design live load occurring simultaneously throughout the building and the low probability that the design live load and MCE shaking will occur simultaneously.

The reduced live load values of $0.8L_0$ for live loads that exceed $100\,\text{lb/ft}^2$ ($4.79\,\text{kN/m}^2$) and $0.4L_0$ for all other live loads, were simply taken as the maximum reduction allowable in Sections 4.7.2 and 4.7.3.

Gravity loads are to be applied to the nonlinear model first and then ground shaking simulations are applied. The initial application of gravity load is critical to the analysis, so member stresses and displacements caused by ground shaking are appropriately added to the initially stressed and displaced structure.

C16.3.3 P-delta Effects P-delta effects should be realistically included, regardless of the value of the elastic story stability coefficient $\theta = P\Delta I_e/(Vh)$. The elastic story stability coefficient is not a reliable indicator of the importance of P-delta during large inelastic deformations. This problem is especially important for dynamic analyses with large inelastic deformations because significant ratcheting can occur. During these types of analyses, when the global stiffness starts to deteriorate and the tangent stiffness of story shear to story drift approaches zero or becomes negative, P-delta effects can cause significant ratcheting (which is a precursor to dynamic instability) of the displacement response in one direction. The full reversal of drifts is no longer observed, and the structural integrity is compromised. To ascertain the full effect of P-delta effects for a given system, a comparison of static pushover curves from a P-delta model and non-P-delta model can be compared.

When including P-delta effects, it is important to capture not only the second-order behavior associated with lateral displacements but also with global torsion about the vertical axis of the system. In addition, the gravity load used in modeling P-delta effects must include 100% of the gravity load in the structure. For these reasons, the use of a single "leaning column," where much of a structure's vertical weight is lumped at a single vertical coordinate, is discouraged, and instead, the structure's vertical load should be distributed throughout the structure in a realistic manner, either through direct modeling of the gravity system or by appropriately distributed "leaning columns."

In some structures, in addition to considering P-delta effects associated with global structural deformation, it is also important to consider local P-delta effects associated with the local deformation of members. This is particularly important for slender elements that are subject to buckling.

C16.3.4 Torsion Inherent torsion is actual torsion caused by differences in the location of the center of mass and center of rigidity throughout the height of the structure. Accidental torsion effects, as per Section 12.8.4.2, are artificial effects that attempt to account for actual variations in load and material strengths during building operation that differ from modeling assumptions. Some examples of this difference would be nonuniformity of the actual mass in the building, openings in the diaphragm that are unaccounted for, torsional foundation input motion caused by the ground motion being out of phase at various points along the base, the lateral stiffness of the gravity framing, variation in material strength and stiffness caused by typical construction tolerances, and incidental stiffness contribution by the nonstructural elements.

When the provision for accidental torsion was first introduced, it was to address buildings that have no inherent torsion but are sensitive to torsional excitation. Common examples of this type of configuration are cruciform core or I-shaped core buildings. In reality, many things can cause such a building to exhibit some torsional response. None of the aforementioned items are typically included in the analysis model; therefore, the accidental torsion approach was introduced to ensure that the structure has some minimum level of resistance to incidental twisting under seismic excitation.

The accidental torsion also serves as an additional check to provide more confidence in the torsional stability of the structure. During the initial proportioning of the structure using linear analysis (as per Section 16.1.1), accidental torsion is required to be enforced in accordance with Section 12.8.4.2. When there is no inherent torsion in the building, accidental torsion is a crucial step in the design process because this artificial offset in the center of mass is a simple way to force a minimum level of twisting to occur in the building. The accidental torsion step (i.e., the required 5% force offsets) is also important when checking for plan irregularities in symmetric and possibly, torsionally flexible buildings. Where there is already inherent torsion in the building, additional accidental torsion is not generally a crucial requirement (though still required, in accordance with Section 12.8.4.2) because the building model will naturally twist during analysis, and no additional artificial torsion is required for this twisting to occur. However, for buildings exhibiting either torsional or extreme torsional irregularities, inclusion of accidental torsion in the nonlinear analysis is required by this standard to assist in identification of potential nonlinear torsional instability.

C16.3.5 Damping Viscous damping can be represented by combined mass and stiffness (Rayleigh) damping. To ensure that the viscous damping does not exceed the target level in the primary response modes, the damping is typically set at the target level for two periods, one above the fundamental period and one below the highest mode frequency of significance. For very tall buildings, the second and even third modes can have significant contributions to response; in this case, the lower multiple on T_1 may need to be reduced to avoid excessive damping in these modes.

Viscous damping may alternatively be represented by modal damping, which allows for the explicit specification of the target damping in each mode.

Various studies have shown that the system damping may vary with time as the structure yields, and in some cases, damping well above the target levels can temporarily exist. Zareian and Medina (2010) provide recommendations for the implementation of damping in such a way that the level of viscous damping remains relatively constant throughout the response.

The level of structural damping caused by component-level hysteresis can vary significantly based on the degree of inelastic action. Typically, hysteretic damping provides a contribution less than or equal to 2.5% of critical damping.

Damping and/or energy dissipation caused by supplemental damping and energy dissipation elements should be explicitly accounted for with component-level models and not included in the overall viscous damping term.

C16.3.6 Explicit Foundation Modeling The PEER TBI guidelines (Bozorgnia et al. 2009) and NIST GCR 12-917-21 (NIST 2012) both recommend the inclusion of subterranean building levels in the mathematical model of the structure. The modeling of the surrounding soil has several possible levels of sophistication, two of which are depicted below in (b) and (c) of Figure C16.3-1, which are considered most practical for current practice. For a MCE_R-level assessment, which is the basis for the Chapter 16 RHA procedure, the rigid bathtub model is preferred by PEER TBI (Bozorgnia et al. 2009) and NIST (2012) (Figure C16.3-1c). This model includes soil springs and dashpots, and identical horizontal ground motions are input at each level of the basement. Such a modeling approach, where the soil is modeled in the form of springs and/or dashpots (or similar methods) placed around the

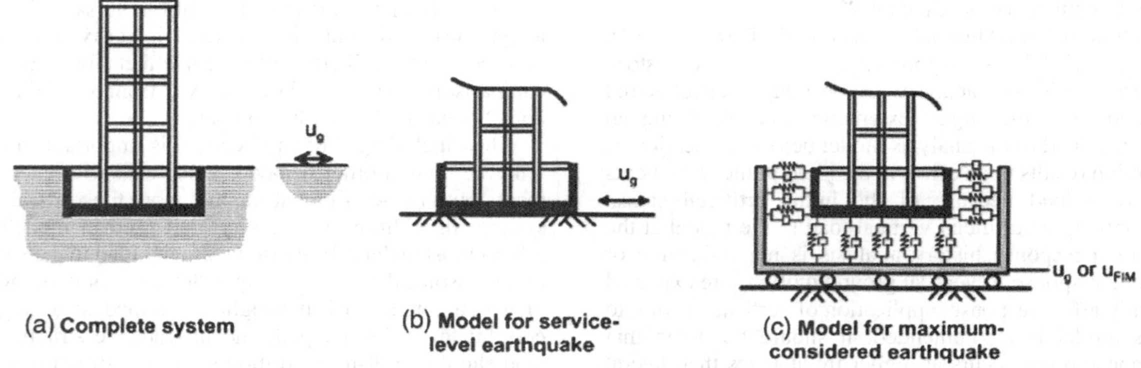

(a) **Complete system** (b) **Model for service-level earthquake** (c) **Model for maximum-considered earthquake**

Figure C16.3-1. Illustration of the method of inputting ground motions into the base of the structural model.
Source: NIST (2011).

foundation, is encouraged but is not required. When spring and dashpot elements are included in the structural model, horizontal input ground motions are applied to the ends of the horizontal soil elements rather than being applied to the foundation directly. A simpler but less accurate model is to exclude the soil springs and dashpots from the numerical model and apply the horizontal ground motions at the bottom level of the basement (Figure C16.3-1b), which is fixed at the base. Either the fixed-base (Figure C16.3-1b) or bathtub (Figure C16.3-1c) approach is allowed, but the bathtub approach is encouraged because it is more accurate.

For the input motions, the PEER TBI (Bozorgnia et al. 2009) guidelines allow the use of either the free-field motion, which is the motion defined in Section 16.2.2, or a foundation input motion modified for kinematic interaction effects. Guidelines for modeling kinematic interaction are contained in NIST (2012).

More sophisticated procedures for soil–structure interaction modeling, including the effects of multisupport excitation, can also be applied in RHA. Such analyses should follow the guidelines presented in NIST (2012).

Approximate procedures for the evaluation of foundation springs are provided in Chapter 19 of this standard.

C16.4.1 Global Acceptance Criteria

C16.4.1.1 Unacceptable Response This section summarizes the criteria for determining unacceptable response and how the criteria were developed. It must be made clear that these unacceptable response acceptance criteria are not the primary acceptance criteria that ensure adequate collapse safety of the building but rather the story drift criteria and the element-level criteria discussed later in Section C16.4. The unacceptable response acceptance criteria were developed to be a secondary protection to supplement the primary criteria. Unacceptable responses result in instabilities and loss of gravity load support. Consequently, if it can be shown that after a deformation-controlled element reaches its (collapse prevention) limit, the model is able to redistribute demands to other elements, this would not constitute an unacceptable response. The acceptance criteria were intentionally structured in this manner because there is high variability in unacceptable response (as described in this section) and the other primary acceptance criteria are much more stable and reliable (because they are based on mean values of 11 motions rather than the extreme response of 11 motions).

When performing nonlinear analysis for a limited suite of ground motions, the observance of a single unacceptable response (or, conversely, the observance of no unacceptable responses) is statistically insignificant. That is, it is reasonably probable that no collapses will be observed in a small suite of analyses, even if the structure has a greater than 10% chance of collapse at MCE_R shaking levels. It is also possible that a structure with less than a 10% chance of collapse at MCE_R shaking levels will still produce an unacceptable response for one ground motion in a small suite. In order for statistics on the number of unacceptable responses in a suite of analyses to produce a meaningful indication of collapse probability, a very large suite of analyses must be performed. Furthermore, the observance or nonobservance of an unacceptable response depends heavily on how the ground motions were selected and scaled (or spectrally matched) to meet the target spectrum.

Since the observance or nonobservance of an unacceptable response is not statistically meaningful, the standard does not rely heavily on the prohibition of unacceptable responses in the attempt to "prove" adequate collapse safety. The many other acceptance criteria of Section 16.4 are relied upon to implicitly ensure adequate collapse safety of the building. If one desired to expand the unacceptable response acceptance criteria to provide true meaningful collapse safety information about the building, a more complex statistical inference approach would need to be used. This is discussed further below.

The statistical insignificance of unacceptable response in a small suite of analyses leaves a large open question about how to interpret the meaning of such responses when they occur. Even though occurrence of a single unacceptable response is statistically meaningless, the occurrence of many unacceptable responses (e.g., 5 of 11) indicates that the collapse probability is significantly in excess of 10%. In addition, a conscientious structural designer is concerned about such occurrences, and the occurrences of unacceptable responses may provide the designer some insight into possible vulnerabilities in the structural design.

Some engineers presume that the acceptance criteria related to *average* response effectively disallow any unacceptable responses (because you cannot average in an infinite response scenario), while others presume that *average* can also be interpreted as *median,* which could allow almost half of the ground motions to cause an unacceptable response.

The statistics presented below are provided to help better interpret the meaning of observance of a collapse of the model analyzed or other type of unacceptable response in a suite of analyses. These simple statistics are based on predicting the occurrence of collapse (or other unacceptable response) using a binomial distribution, based on the following assumptions:

- The building's collapse probability is exactly 10% at the MCE_R level.
- Collapse probability is lognormally distributed and has a dispersion (lognormal standard deviation) of 0.6. This value includes all sources of uncertainty and variability (e.g., record-to-record variability, modeling uncertainty). The value of 0.6 is the same value used in creating the risk-consistent hazard maps for ASCE 7-10 (FEMA 2009a) and is consistent with the values used in FEMA P-695 (FEMA 2009b).
- The record-to-record variability ranges from 0.25 to 0.40. This is the variability in the collapse capacity that would be expected from the analytical model. This value is highly dependent on the details of the ground motion selection and scaling; values of 0.35 to 0.45 are expected for motions that do not fit tightly to the target spectrum, and values of 0.2 to 0.3 are expected for spectrally matched motions (FEMA 2009b).

Figure C16.4-1 shows collapse fragility curves for a hypothetical building that has a 10% collapse probability conditioned on MCE_R motion ($P[C|MCE_R] = 10\%$), with an assumed record-to-record collapse uncertainty of 0.40 and a total collapse uncertainty of 0.60. The figure shows that the median collapse capacity must be a factor of 2.16 above the MCE_R ground motion level, that the probability of collapse is 10% at the MCE_R when the full variability is included (as required), but that the probability of collapse is only 2.7% at the MCE_R when only the record-to-record variability is included. This 2.7% collapse probability is what would be expected from the structural model that is used in the RHA assessment procedure.

Table C16.4-1 shows the probability of observing *n* collapses in a suite of 11 ground motions for a structure that has different values of $P[C|MCE_R]$.

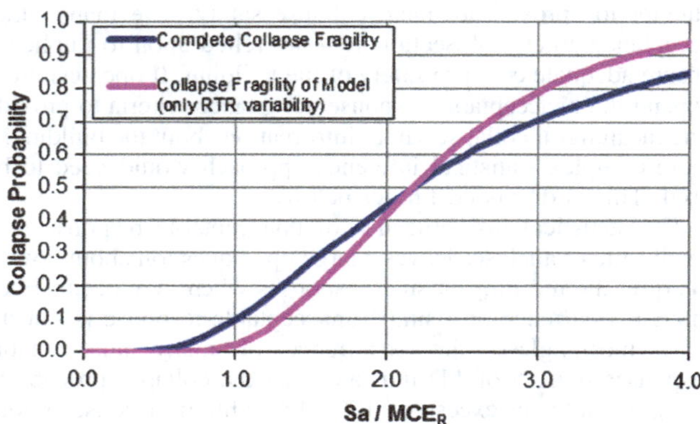

Figure C16.4-1. Collapse fragilities for a building with $P[C|MCE_R] = 10\%$ and $\beta_{COL,RTR} = 0.40$.

Table C16.4-1 shows that for a building meeting the $P[C|MCE_R] = 10\%$ performance goal, there is a 74% chance of observing no collapses, a 23% chance of observing one collapse, a 3% chance of observing two collapses, and virtually no chance of observing more than two collapses. In comparison, for a building with $P[C|MCE_R] = 20\%$, there is a 30% chance of observing no collapses, a 38% chance of observing one collapse, a 22% chance of observing two collapses, and a 10% chance of observing more than two collapses.

This table illustrates that

- Even if no collapses are observed in a set of 11 records, this does not, in any way, prove that the $P[C|MCE_R] = 10\%$ performance goal has been met. For example, even for a building with $P[C|MCE_R] = 20\%$, there is still a 30% chance that no collapses will be observed in the analysis. Therefore, the other noncollapse acceptance criteria (e.g., criteria for drifts and element demands) must be relied on to enforce the 10% collapse probability goal.
- If the $P[C|MCE_R] = 10\%$ performance goal is met, it is highly unlikely (only a 3% chance) that two collapses will be observed in the set of 11 records. Therefore, an acceptance criterion that prohibits two collapses is reasonable.

The collapse likelihoods shown in Table C16.4-1 are based on a relatively large record-to-record variability value of 0.40. Table C16.4-2 illustrates similar statistics for the case when the record-to-record variability is suppressed in ground motion selection and scaling, such as occurs with spectral matching. This table shows that, for a building meeting the $P[C|MCE_R] = 10\%$ performance goal and with record-to-record variability taken as 0.25, the likelihood of observing a collapse response is very low. This is why no unacceptable responses are permitted in the suite of analyses when spectral matching is used.

For Risk Categories I and II structures, if more than 11 ground motions are used for analysis, then additional unacceptable responses may be permissible. Two unacceptable responses would be permissible if 20 or more motions are used, and three unacceptable responses would be permissible if 30 or more motions are used. For Risk Categories III and IV structures, the collapse probability goals are 6% and 3%, respectively, at the MCE_R level. When the above computations are redone using these lower collapse probability targets, this shows that the acceptance criteria should require that no motions of the 11 produce an unacceptable response for these categories.

Typically, mean building response values (story drifts, element deformations, and forces) are used in acceptance evaluations, where the "mean" is the simple statistical average for the response parameter of interest. When an unacceptable response occurs, it is not possible to compute a mean value of the building response values because one of the 11 response quantities is

Table C16.4-1. Likelihood of Observing Collapses in 11 Analyses, Given Various MCE_R Collapse Probabilities and a Record-to-Record Uncertainty of 0.4.

Number of Collapses	Likelihood for Various $P[C\|MCE_R]$ Values				
	0.05	0.10	0.15	0.20	0.30
0 of 11	0.93	**0.74**	0.51	0.30	0.07
1 of 11	0.07	**0.23**	0.36	0.38	0.21
2 of 11	0	**0.03**	0.11	0.22	0.29
3 of 11	0	**0**	0.02	0.08	0.24
4 of 11	0	**0**	0	0.02	0.13
5 of 11	0	**0**	0	0	0.05

Table C16.4-2. Likelihood of Observing Collapses in 11 Analyses, Given Various MCE_R Collapse Probabilities and a Record-to-Record Uncertainty of 0.25.

Number of Collapses	Likelihood for Various $P[C\|MCE_R]$ Values				
	0.05	0.10	0.15	0.20	0.30
0 of 11	1.00	**0.99**	0.93	0.79	0.30
1 of 11	0	**0.01**	0.07	0.19	0.38
2 of 11	0	**0**	0	0.02	0.22
3 of 11	0	**0**	0	0	0.08
4 of 11	0	**0**	0	0	0.02
5 of 11	0	**0**	0	0	0

undefined. In this case, rather than the mean, the standard requires use of the counted median response multiplied by 1.2 but not less than the mean response from the remaining motions.

To compute the median value, the unacceptable response is assumed as larger than the other responses and then, assuming that 11 analyses were performed, the counted median value is taken to be the 6th largest value from the set of 11 responses. The 1.2 factor is based on a reasonable ratio of mean to median values for a lognormal distribution ($\beta = 0.4$ results in mean/median = 1.08, $\beta = 0.5$ results in mean/median = 1.13, $\beta = 0.6$ results in mean/median = 1.20, and $\beta = 0.7$ results in mean/median = 1.28).

The requirement to also check the mean of the remaining 10 response results is simply an added safeguard to ensure that the 1.2 × median value does not underpredict the mean response values that should be used when checking the acceptance criteria.

Although currently the purpose of this acceptance criterion is not meant to quantify the structure's collapse probability under MCE_R ground motions, the acceptance criterion can be recast to do so in future provisions. The collapse probability can be inferred from analysis results and compared to the target value (e.g., 10% for structures in Risk Category I or II). In this alternate light, existing statistical inference theories can be used to determine the number of acceptable responses, and the number of ground motions required to conclude that the proposed design may have an acceptable collapse probability.

As discussed in the previous section, analysis results can be thought of as following a binomial distribution. Based on this distribution, one could use the observed counts of collapsed and noncollapsed responses (indicated by unacceptable and acceptable responses) to estimate the collapse probability of the proposed design in a manner that accounts for the uncertainty in the estimated collapse probability. This uncertainty depends on the total number of ground motions. If few ground motions are used, there is a large uncertainty in the collapse probability. If many ground motions are used, there is a small uncertainty. For example, compare a set of 11 ground motions with one unacceptable response to a set of 110 ground motions with ten unacceptable responses. Both sets most likely have a unacceptable response probability of 9.1%. The design with one unacceptable and ten acceptable responses has only a 34% chance that its unacceptable response probability is 10% or less. The design with 10 unacceptable and 100 acceptable responses has a 56% chance that its unacceptable response probability is 10% or less.

In the current acceptance criterion, the choice to require 11 ground motions follows from the need to have confidence in the average values of the resulting element-level and story-level responses (Section C16.2.3.1). These element-level and story-level responses are then used to *implicitly* demonstrate adequate collapse safety. If future provisions seek to *explicitly* ensure that the proposed design has an acceptable collapse probability, then this unacceptable response acceptance criterion should be revised using statistical inference theory to establish the number of required ground motions and the maximum number of unacceptable responses, as well as the element- and story-level response limits.

C16.4.1.2 Transient Story Drift The limit on mean story drift was developed to be consistent with the linear design procedures of this standard. To this end, the basic Table 12.12-1 story drift limits are adopted with the following adjustments:

- Increased by a factor of 1.5, to reflect the analysis being completed at the MCE_R ground motion level rather than at 2/3 of the MCE_R level, and
- Increased by another factor of 1.25, to reflect an average ratio of R/C_d.

These two increases are the basis for the requirement that the mean story drift be limited to 1.9 (which was rounded to 2.0) of the standard Table 12.12-1 limits.

The masonry-specific drift limits of Table 12.12-1 are not enforced in this section because the component-level acceptance criteria of Section 16.4.2 are expected to result in equivalent performance (i.e., a masonry building designed in accordance with Chapter 16 is expected to have similar performance to a masonry building designed using linear analysis methods and the more stringent drift limits of Table 12.12-1).

For tall structures, more restrictive drift limits are adopted than permitted under Table 12.2.1. When performing nonlinear response history analysis, the model is required to include all elements that significantly affect the stiffness and strength of the structure to resist lateral forces. This is in stark contrast to the intent of Chapter 12 analysis procedures, which consider only the seismic force-resisting system in the analysis. As a result, nonlinear response history analyses will include elements normally considered to carry only gravity loads, but which, particularly in tall buildings, add substantial stiffness to the analytical model. Analysis of these models will result in lower predictions of drift than the typical models used for linear analysis. If the same drift limits were applied, the effect would be to permit more flexible structures than permitted by Chapter 12. The drift limit applied for buildings over 240 ft is the same as that in the 2017 PEER Tall Buildings Initiative, Guidelines for Performance-Based Seismic Design of Tall Buildings, which is commonly used for the design of tall structures in the US. For buildings with heights between 100 ft (30 m) and 240 ft (73 m), Equation (16.4-1) applies and allows for a smooth transition between the requirements for tall buildings and the doubled limits of Table 12.12-1 requirement, as applied to Risk Category I and II mid-rise structures, other than masonry. For building heights above 240 ft (73 m), the story drift limit of 0.03 hsx controls over the value calculated per Equation (16.4-1).

C16.4.1.3 Residual Story Drift The standard does not require checks on residual drift for structures less than 240 ft (73 m) in height. Residual drift is the unrecoverable portion of transient drift that commonly occurs when a structure undergoes inelastic response. It is computed as the absolute value of the largest difference of the deflections of vertically aligned points at the top and bottom of the story under consideration along any edge of the structure, following completion of the structure's response to earthquake motion. Residual drifts are an indicator of potential incipient dynamic instability, and the prudent engineer checks for this instability. Limiting residual drifts is an important consideration for post-earthquake operability and for limiting financial losses, but such performance goals are not included in the scope of the ASCE 7 standard. For Risk Category I and II buildings, the ASCE 7 standard is primarily meant to ensure the protection of life safety.

Design practice for tall buildings, however, has routinely included limits on residual drift. One important reason for this is that many tall buildings are located in dense urban areas. Residual drift is commonly used to judge the post-earthquake safety of damaged buildings. If tall buildings exhibit large residual drifts after an earthquake and consequently appear to be unstable, this could prompt building officials to cordon off large areas, imposing hardship on many people and businesses. The tall building community therefore designs to reduce the probability of such occurrence. ASCE 7 adopted the acceptance directly from the recent practice for tall buildings exceeding 240 ft (73 m) in the 2022 edition to be compatible with common design practice for these structures.

C16.4.2 Element-Level Acceptance Criteria The element-level acceptance criteria requires classification of each element action as either force-controlled or deformation-controlled, similar to the procedures of ASCE 41. Note that this is done for each *element action*, rather than for each *element*. For example, for a single column element, the flexural behavior may be classified as a deformation-controlled action, whereas the axial behavior may be classified as a force-controlled action.

Deformation-controlled actions are those that have reliable inelastic deformation capacity. Force-controlled actions pertain to brittle modes where inelastic deformation capacity cannot be ensured. Based on how the acceptance criteria are structured, any element action that is modeled elastically must be classified as being force-controlled.

Some examples of force-controlled actions are

- Shear in reinforced concrete (other than diagonally reinforced coupling beams);
- Axial compression in columns;
- Punching shear in slab–column joints without shear reinforcing;
- Connections that are not explicitly designed for the strength of the connected component, such as some braces in braced frames;
- Displacement of elements resting on a supporting element without rigid connection (such as slide bearings); and
- Axial forces in diaphragm collectors.

Some examples of deformation-controlled actions are

- Shear in diagonally reinforced coupling beams;
- Flexure in reinforced concrete columns and walls;
- Axial yielding in buckling restrained braces; and
- Flexure in special moment frames.

Section 16.4.2 further requires categorization of component actions as critical, ordinary, or noncritical based on the consequence of their exceeding strength or deformation limits. Because of the differences in consequence, the acceptance criteria are developed differently for each of the above classifications of component actions. An element's criticality is judged based on the extent of collapse that may occur, given the element's failure, and is also judged on whether the effect of the element's failure on seismic resistance is substantive. An element's failure could be judged to have substantial effect on the structure's seismic resistance if analysis of a model of the building without the element present predicts unacceptable performance, while analysis with the element present does not.

Limits placed on response quantities are correlated to building performance and structural reliability. In order for compliance with these limits to meaningfully characterize overall performance and reliability, grouping of certain component actions for design purposes may be appropriate. For example, while symmetrical design forces may be obtained for symmetrical structures using equivalent lateral force and modal response spectrum analysis procedures, there is no guarantee that component actions in response history analysis of symmetrical models will be the same—or even similar—for identical components arranged symmetrically. Engineering judgment should be applied to the design to maintain symmetry by using the greater demands (that is, the demands on the more heavily loaded component determined using the appropriate factor on its mean demand) for the design of both components. For this purpose, using the mean demands of the pair of components would not be appropriate because this method would reduce the demand used for design of the more heavily loaded component.

Though this point is perhaps trivial in the case of true symmetry, it is also a concern in nonsymmetrical structures. For these buildings, it may be appropriate to group together structural components that are highly similar either in geometric placement or purpose. The demands determined using the suite mean (the mean response over all ground motions within a suite) may be very different for individual components within this grouping. This is a result both of the averaging process and the limited explicit consideration of ground motion to structure orientation in the provisions. Although the analysis may indicate that only a portion of the grouped components do not meet the provisions, the engineer ought to consider whether such nonconformance should also suggest redesign in other similar elements. Thus, response history analysis places a higher burden on the judgment of the engineer to determine the appropriate methods for extracting and interpreting meaningful response quantities for design purposes.

C16.4.2.1 Force-Controlled Actions The application of the load combinations and resistance factors in Equations (16.4-2), (16-4-2), (16.4-3), and (16.4-4) are limited to load effects determined using NLRHA for structures subject to an MCE_R level event. These load and resistance factors should not be used for structural design with the design earthquake effects, E, determined in accordance with Chapter 12 of the standard. When evaluating earthquake effects in accordance with Chapter 12 of the standard, the load combinations of Chapter 2 apply.

The acceptance criteria for force-controlled actions Equations 16.4-1 and 16.4-2 follow the same framework that underlies the load combinations contained in Chapter 2 of the standard, except that both the load factors and the capacity (resistance) factors have been adjusted for consistency with the NLRHA approach and to maintain compatibility with the target reliability in Table 1.3-2. The load factor of 1.3Ie on seismic demands was determined assuming a demand dispersion associated with record-to-record variability of 0.3, based on values observed for several real buildings and additional modeling uncertainty of 0.2, which was selected based on engineering judgment consistent with approaches used in FEMA P695. Based on data presented in National Bureau of Standard Special Publication 577, the typical bias in resistance, (i.e. the ratio of the expected strength of an element to the nominal strength), was taken as 1.1, while a 0.15 coefficient of variation on resistance was assumed. The materials standards have resistance factors that generally vary from values of approximately 0.7 to 0.9. An average value of 0.85 was used in computing the 1.3 load factor on seismic demand.

Exception 2 to the section permits the use of an alternative set of load combinations (16.4-3 and 16.4-4) to evaluate force-controlled actions when demands are limited by the development of a plastic mechanism. For example, shear in a column in a moment frame cannot exceed the sum of the plastic moments at the ends of the column, divided by the free height of the column. A load factor of unity is used on the capacity-limited earthquake demand in this case, recognizing the very low probability that demand can exceed the computed value and also for consistency with similar criteria in ACI 318 and AISC 341.

To determine appropriate values of λ, we begin with the collapse probability goals of Table 1.3-2 (for Risk Categories I and II) for MCE_R motions. These collapse probability goals include a 10% chance of a total or partial structural collapse and a 25% chance of a failure that could result in endangerment of individual lives. For the assessment of collapse, we then make the somewhat conservative assumption that the failure of a single

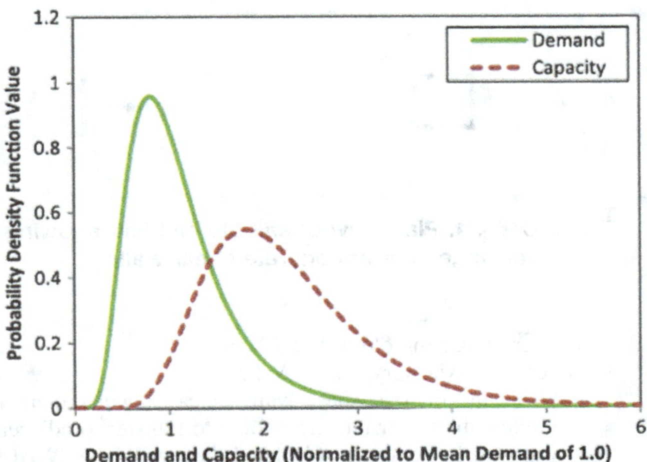

Figure C16.4-2. Illustration of component capacity and demand lognormal distributions (normalized to a mean capacity of 1.0); the mean component capacity is calibrated to achieve P[C|MCE$_R$] = 10%.

critical force-controlled component would result in a total or partial structural collapse of the building.

Focusing first on the goal of a 10% chance of a total or partial structural collapse, we assume that the component force demand and component capacity both follow a lognormal distribution and that the estimate of $F_{n,e}$ represents the true expected strength of the component. We then calibrate the λ value required to achieve the 10% collapse probability goal. This value is depicted in Figure C16.4-2, which shows the lognormal distributions of component capacity and component demand.

The calibration process is highly dependent on the uncertainties in component demand and capacity. Table C16.4-3a shows typical uncertainties in force *demand* for analyses at the MCE$_R$ ground motion level, for both the general case and the case where the response parameter is limited by a well-defined yield mechanism. Table C16.4-3b shows typical uncertainty values for the component *capacity*. The values are based on reference materials, as well as the collective experience and professional judgment of the development team.

In the calibration process, the λ and φ values both directly affect the required component strength. Therefore, the calibration is completed to determine the required value of λ/φ needed to fulfill the 10% collapse safety objective. This calibration is done by assuming a value of λ/φ, convolving the lognormal distributions of demand and capacity and iteratively determining the capacity required to meet the 10% collapse safety objective by adjusting λ/φ. Table C16.4-4 reports the final λ/φ values that come from such integration.

It should be clearly stated that this approach of calibrating the λ/φ ratio means that the final acceptance criterion is independent of the φ value specified by a material standard. If it is desirable for the acceptance criteria to be partially dependent on the value of φ, then the uncertainty factors of Table C16.4-3b would need to be made dependent on the φ value in some manner.

Since the values in Table C16.4-4 are similar, for simplicity, the acceptance criterion is based on $\lambda/\varphi = 2.0$ for all cases, and a separate criterion for the existence of a well-defined mechanism is not included. In addition, the strength term is defined slightly differently. For Risk Categories III and IV, this full calculation was redone using the lower collapse probability goals of 6% and 3%, respectively, and it was found that scaling the force demands by I_e sufficiently achieves these lower collapse probability goals.

This statistical calculation was then repeated for the goal of 25% chance of a failure that could result in endangerment of individual lives. This resulted in a required ratio of 1.5 for such force-controlled failure modes; deemed as "ordinary."

Force-controlled actions are deemed noncritical if the failure does not result in structural collapse or any meaningful endangerment to individual lives. This occurs in situations where gravity forces can reliably redistribute to an alternate load path and no failure will ensue. For noncritical force-controlled components, the acceptance criteria allow the use of $\lambda = 1.0$.

Where an industry standard does not define expected strength, expected (or mean) strength, F_e, is computed as follows. First, a standard strength-prediction equation is used from a material standard, using a strength reduction factor, φ, of 1.0; the expected material properties are also used in place of nominal material properties. In some cases, this estimate of strength ($F_{n,e}$) may still be conservative in comparison with the mean expected strength shown by experimental tests (F_e) caused by inherent conservatism in the strength equations adopted by the materials standards.

Table C16.4-3a Assumed Variability and Uncertainty Values for Component Force Demand.

Demand Dispersion (β_D)		
General	Well-Defined Mechanism	Variabilities and Uncertainties in the Force Demand
0.40	0.20	Record-to-record variability (for MCE$_R$ ground motions)
0.20	0.20	Uncertainty from estimating force demands using structural model
0.13	0.06	Variability from estimating force demands from mean of only 11 ground motions
0.46	**0.29**	$\beta_{D\text{-Total}}$

Table C16.4-3b. Assumed Variability and Uncertainty Values for Component Force Capacity.

Capacity Dispersion (β_C)		
General	Well-Defined Mechanism	Variabilities and Uncertainties in the Final As-Built Capacity of the Component
0.30	0.30	Typical variability in strength equation for $F_{n,e}$ (from available data)
0.10	0.10	Typical uncertainty in strength equation for $F_{n,e}$ (extrapolation beyond available data)
0.20	0.20	Uncertainty in as-built strength because of construction quality and possible errors
0.37	**0.37**	$\beta_{C\text{-Total}}$

Table C16.4-4. Required Ratios of λ/φ to Achieve the 10% Collapse Probability Objective.

Dispersion	Required Ratios of λ/φ
General	2.1
Well-Defined Mechanism	1.9

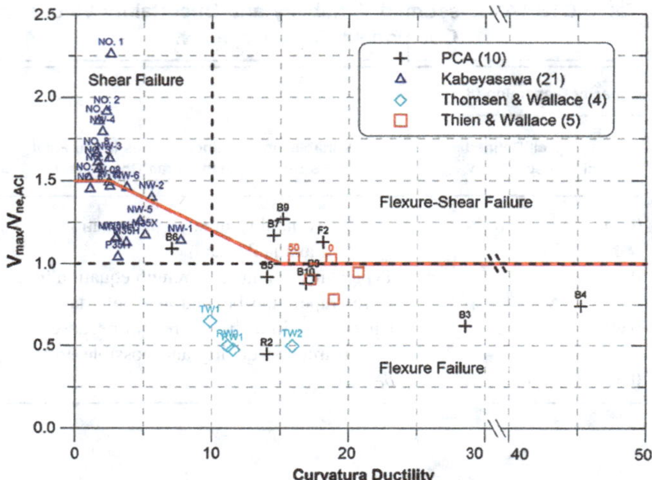

Figure C16.4-3. Expected shear strengths (in terms of $F_e/F_{n,e}$) for reinforced concrete shear walls when subjected to various levels of flexural ductility.
Source: Courtesy of John Wallace.

If such conservatism exists, the $F_{n,e}$ value may be multiplied by a "component reserve strength factor" greater than 1.0 to produce the estimate of the mean expected strength (F_e). This process is illustrated in Figure C16.4-3, which shows the $F_e/F_{n,e}$ ratios for the shear strengths from test data of reinforced concrete shear walls (Wallace et al. 2013). This figure shows that the ratio of $F_e/F_{n,e}$ depends on the flexural ductility of the shear wall, demonstrating that $F_e = 1.0\ F_{n,e}$ is appropriate for the shear strength in the zone of high flexural damage and $F_e = 1.5\ F_{n,e}$ may be appropriate in zones with no flexural damage.

For purposes of comparison, Equation (C16.4-1) is comparable to the PEER TBI acceptance criteria (Bozorgnia et al. 2009) for the case that $\varphi = 0.75$ and $F_e = 1.0F_{n,e}$.

The exception allows for the use of the capacity design philosophy for force-controlled components that are "protected" by inelastic fuses, such that the force delivered to the force-controlled component is limited by the strength of the inelastic fuse.

The following are some examples of force-controlled actions that are deemed to be critical actions:

- Steel Moment Frames (SMF):
  - Axial compression forces in columns caused by combined gravity and overturning forces
  - Combined axial force, bending moments, and shear in column splices
  - Tension in column base connections (unless modeled inelastically, in which case it would be a deformation-controlled component);
- Steel Braced Frames (BRBF - Buckling Restrained Braced Frame, SCBF - Special Concentrically Braced Frames):
  - Axial compression forces in columns caused by combined gravity and overturning forces
  - Combined axial force, bending moments, and shear in column splices
  - Tension in brace and beam connections;
  - Column base connections (unless modeled inelastically)
- Concrete Moment Frames:
  - Axial compression forces in columns caused by combined gravity and overturning forces

Figure C16.4-4. Plan view of sample building showing arrangement of concrete shear walls.

  - Shear force in columns and beams;
- Concrete or Masonry Shear Walls:
  - Shear in concrete shear wall, in cases when there is limited ability for the shear force to transfer to adjacent wall panels. For cases of isolated shear walls (i.e., WALL #1 in Figure C16.4-4), the shear force in this isolated wall is deemed as a critical action. In contrast, the shear force in a one-wall pier that is in a group of wall piers (e.g., panel #2 of Figure C16.4-5) need not be deemed a critical action (especially when determining whether an analysis is deemed to represent an unacceptable response). For this case of a group of wall piers, it may be appropriate to consider the sum of the wall shears to be the critical action (e.g., the sum of wall shears in PANELS #1, #2, and #3 of Figure C16.4-5).
  - Axial (plus flexural) compression in concrete shear wall (for most cases)
  - Axial compression in outrigger columns
  - Axial (plus flexural) tension in outrigger column splices;
- Other Types of Components:
  - Shear forces in piles and pile cap connections (unless modeled inelastically)
  - Shear forces in shallow foundations (unless modeled inelastically)
  - Punching shear in slabs without shear reinforcing (unless modeled inelastically)
  - Diaphragms that transfer a substantial amount of force (from more than one story)
  - Elements supporting discontinuous frames and walls;

The following are some examples of force-controlled actions that are deemed to be ordinary actions:

- Steel Moment Frames (SMF):
  - Shear force in beams and columns
  - Column base connections (unless modeled inelastically)

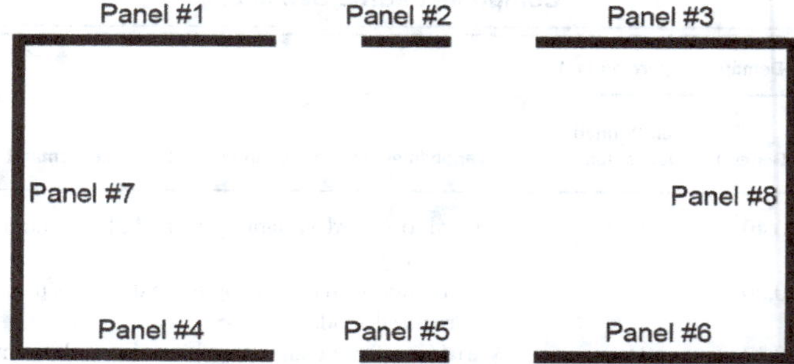

Figure C16.4-5. Plan view of sample building showing components of a reinforced concrete core shear wall.

- Welded or bolted joints (as distinct from the inelastic action of the overall connection) between moment frame beams and columns;
- Steel Braced Frames (BRBF, SCBF):
  - Axial tension forces in columns caused by overturning forces (unless modeled inelastically);
- Concrete Moment Frames:
  - Splices in longitudinal beam and column reinforcement;
- Concrete or Masonry Shear Walls:
  - An ordinary classification would only apply in special cases where failure would not cause widespread collapse and would cause minimal reduction in the building seismic resistance.
- Other Types of Components:
  - Axial forces in diaphragm collectors (unless modeled inelastically)
  - Shear and chord forces in diaphragms (unless modeled inelastically)
  - Pile axial forces;

The following are some examples of force-controlled actions that could be deemed noncritical actions:

- Any component where the failure would not result in either collapse or substantive loss of the seismic resistance of the structure.

C16.4.2.2 Deformation-Controlled Actions ASCE 7-16 included requirements to evaluate the adequacy of deformation-controlled actions against limiting deformation levels at which laboratory testing suggested either that element behavior was unpredictable or that loss of vertical load-carrying ability would occur. While substantive data exist to indicate the capacity of force-controlled actions, there are relatively few laboratory data to indicate the deformation at which a deformation-controlled element action reaches a level where loss of vertical load-carrying capacity occurs. There are a number of reasons for this, including the following: (1) the deformation at which such loss occurs can be very large and beyond the practical testing capability of typical laboratory equipment; (2) many researchers have tested such components with the aim of quantifying useful capacity for elements of a seismic force-resisting system and have terminated testing after substantial degradation in strength has occurred, even though actual failure has not yet been experienced; and (3) testing of gravity-load-bearing elements to failure can be dangerous and destructive of test equipment. In ASCE 7-22, this requirement was dropped in favor of an approach in which the demand imposed on deformation-controlled actions is evaluated against the valid range of modeling for that element in each analysis. The valid range of modeling is defined as that limiting deformation at which the model used in the analysis can be benchmarked against suitable laboratory test data. This approach has been successfully used in the design of tall buildings, using the procedures specified in the Pacific Earthquake Engineering Research Center, Guidelines for Performance-based Seismic Design of Tall Buildings, and was found to produce acceptable designs compared with prior practice. In addition, this approach will be facilitated with the adoption of data indicating the valid range of nonlinear modeling for deformation-controlled elements by both ACI and AISC in their standards.

C16.4.2.3 Elements of the Gravity Force-Resisting System The basic deformation-compatibility requirement of ASCE 7-16, Section 12.12.4, is imposed for gravity-system components, which are not part of the established seismic force-resisting system, using the deformation demands predicted from response history analysis under MCE_R-level ground motions, as opposed to evaluation under linear analysis.

If an analyst wanted to further investigate the performance of the gravity system (which is not required), the most direct and complete approach (but also the most time-consuming) would be to directly model the gravity system components as part of the structural model and then impose the same acceptance criteria used for the components of the seismic force-resisting system. An alternative approach (which is more common) would be to model the gravity system in a simplified manner and verify that the earthquake-imposed force demands do not control over the other load combinations and/or to verify that the mean gravity system deformations do not exceed the deformation limits for deformation-controlled components.

REFERENCES

Al Atik, L., and N. Abrahamson. 2010. "An improved method for nonstationary spectral matching." *Earthquake Spectra* 26 (3): 601–617. https://doi.org/10.1193/1.3459159.

Ancheta, T. D., et al. 2012. *PEER NGA-West 2 database*. Berkeley, CA: Pacific Earthquake Engineering Research Center.

Baker, J. W. 2011. "Conditional mean spectrum: Tool for ground motion selection." *J. Struct. Eng.* 137 (3): 322–331. https://doi.org/10.1061/(ASCE)ST.1943-541X.0000215.

Baker, J. W., and C. A. Cornell. 2006. "Correlation of response spectral values for multi-component ground motions." *Bull. Seismol. Soc. Am.* 96 (1): 215–227. https://doi.org/10.1785/0120050060.

Bommer, J. J., S. G. Scott, and S. K. Sarma. 2000. "Hazard-consistent earthquake scenarios." *Soil Dyn. Earthquake Eng.* 19 (4): 219–231. https://doi.org/10.1016/S0267-7261(00)00012-9.

Bozorgnia, Y., C. B. Crouse, R. O. Hamburger, R. Klemencic, H. Krawinkler, J. O. Malley, J. P. Moehle, F. Naeim, and J. P. Stewart. 2009. *Guidelines for performance-based seismic design of tall buildings*. Berkeley, CA: Pacific Earthquake Engineering Research Center.

FEMA (Federal Emergency Management Agency). 2009a. *NEHRP recommended provisions for seismic regulation for buildings and other structures. FEMA P-753*. Washington, DC: FEMA.

FEMA. 2009b. *Quantification of building seismic performance factors. FEMA P-695*. Washington, DC: FEMA.

Grant, D. N., and R. Diaferia. 2012. "Assessing adequacy of spectrum-matched ground motions for response history analysis." *Earthquake Eng. Struct. Dyn.* 42 (9): 1265–1280. https://doi.org/10.1002/eqe.2270.

Haroun, M. A., and G. W. Housner. 1981. "Seismic design of liquid storage tanks." *J. Tech. Councils ASCE* 107 (1): 191–207.

Housner, G. W. 1963. "The dynamic behavior of water tanks." *Bull. Seismol. Soc. Am.* 53 (2): 381–387.

Huang, Y. N., A. S. Whittaker, and N. Luco. 2008. "Maximum spectral demands in the near fault region." *Earthquake Spectra* 24 (1): 319–341. https://doi.org/10.1193/1.2830435.

Jacobsen, L. S. 1949. "Impulsive hydrodynamics of fluid inside a cylindrical tank and of fluid surrounding a cylindrical pier." *Bull. Seismol. Soc. Am.* 39 (3): 189–203. https://doi.org/10.1785/BSSA0390030189.

Luco, N., and P. Bazzurro. 2007. "Does amplitude scaling of ground motion records result in biased nonlinear structural drift responses?" *Earthquake Eng. Struct. Dyn.* 36 (13): 1813–1835. https://doi.org/10.1002/eqe.695.

Naeim, F., and M. Lew. 1995. "On the use of design spectrum compatible time histories." *Earthquake Spectra* 11 (1): 111–127. https://doi.org/10.1193/1.1585805.

NIST (National Institute of Standards and Technology). 2010. *Nonlinear structural analysis for seismic design. NIST GCR 10-917-5*. Gaithersburg, MD: NIST.

NIST. 2011. *Selecting and scaling earthquake ground motions for performing response-history analyses. NIST GCR 11-917-15*. Gaithersburg, MD: NIST.

NIST. 2012. *Soil-structure interaction for building structures. GCR 12-917-21*. Gaithersburg, MD: NIST.

Reiter, L. 1990. *Earthquake hazard analysis: Issues and insights*. New York: Columbia University Press.

Schnabel, P. B., J. Lysmer, and H. B. Seed. 1972. *SHAKE—A computer program for earthquake response analysis of horizontally layered sites*. Rep. No. EERC 72-12. Berkeley, CA: Earthquake Engineering Research Center, University of California.

Shahi, S. K., T. Ling, J. W. Baker, and N. Jayaram. 2011. *New ground motion selection procedures and selected motions for the PEER transportation research program*. Berkeley, CA: Pacific Earthquake Research Center.

Stewart, J. P., S. J. Chiou, J. D. Bray, R. W. Graves, P. G. Sommerville, and N. A. Abrahamson. 2002. "Ground motion evaluation procedures for performance-based design." *J. Soil Dyn. Earthquake Eng.* 22 (9–22): 9–12. https://doi.org/10.1016/S0267-7261(02)00097-0.

Wallace, J. W., C. Segura, and T. Tran. 2013. "Shear design of structural walls." In *Proc., 10th Int. Conf. on Urban Earthquake Engineering*, Tokyo.

Zareian, F., and R. Medina. 2010. "A practical method for proper modeling of structural damping in inelastic plane structural systems." *J. Comput. Struct.* 88 (1–2): 45–53. https://doi.org/10.1016/j.compstruc.2009.08.001.

OTHER REFERENCES (NOT CITED)

ASCE. 2017. *Seismic evaluation and retrofit of existing buildings*. ASCE/SEI Standard 41-13. Reston, VA: ASCE.

Haselton, C. B., and G. G. Deierlein. 2007. *Assessing seismic collapse safety of modern reinforced concrete frame buildings*. PEER Rep. No. 2007/08. Berkeley, CA: Pacific Earthquake Engineering Research Center, University of California.

CHAPTER C17
SEISMIC DESIGN REQUIREMENTS FOR SEISMICALLY ISOLATED STRUCTURES

C17.1 GENERAL

Seismic isolation, also referred to as base isolation because of its common use at the base of building structures, is a design method used to substantially decouple the response of a structure from potentially damaging horizontal components of earthquake motions. This decoupling can result in a response that is significantly reduced from that of a conventional, fixed-base building.

Over the last three decades, the significant damage to buildings and infrastructure following large earthquakes has led to the rapid growth of seismic isolation technology and the development of specific guidelines for the design and construction of seismically isolated buildings and bridges in the United States, as well as standardized testing procedures of isolation devices.

Design requirements for seismically isolated building structures were first codified in the United States, as an appendix to the 1991 Uniform Building Code (ICBO 1991), based on "General Requirements for the Design and Construction of Seismic-Isolated Structures" developed by the State Seismology Committee of the Structural Engineers Association of California. In the intervening years, those provisions have developed, along two parallel tracks, into the design requirements in Chapter 17 of the ASCE 7 standard and the rehabilitation requirements in Section 9.2 of ASCE 41 (2007), *Seismic Rehabilitation of Existing Buildings*. The design and analysis methods of both standards are similar, but ASCE 41 allows more relaxed design requirements for the superstructure of rehabilitated buildings. The basic concepts and design principles of seismic isolation of highway bridge structures were developed in parallel and were first codified in the United States in the 1990 AASHTO provisions *Guide Specifications for Seismic Isolation Design*. The subsequent version of this code (AASHTO 1999) provides a systematic approach to determining bounding limits for analysis and design of isolator mechanical properties.

The present edition of the ASCE 7, Chapter 17 provisions contains significant modifications with respect to superseded versions, intended to facilitate the design and implementation process of seismic isolation, thus promoting the expanded use of the technology. Rather than addressing a specific method of seismic isolation, the standard provides general design requirements applicable to a wide range of seismic isolation systems. Because the design requirements are general, testing of isolation-system hardware is required to confirm the engineering parameters used in the design and to verify the overall adequacy of the isolation system. Use of isolation systems whose adequacy is not proved by testing is prohibited. In general, acceptable systems (1) maintain horizontal and vertical stability when subjected to design displacements, (2) have an inherent restoring force defined as increasing resistance with increasing displacement, (3) do not degrade significantly under repeated cyclic load, and (4) have quantifiable engineering parameters (such as force-deflection characteristics and damping).

The lateral force-displacement behavior of isolation systems can be classified into four categories, as shown in Figure C17.1-1, in which each idealized curve has the same design displacement, D_D.

A linear isolation system (Curve A) has an effective period that is constant and independent of the displacement demand, where the force generated in the superstructure is directly proportional to the displacement of the isolation system.

A hardening isolation system (Curve B) has a low initial lateral stiffness (or equivalently, a long effective period) followed by a relatively high second stiffness (or a shorter effective period) at higher displacement demands. Where displacements exceed the design displacement, the superstructure is subjected to increased force demands, while the isolation system is subject to reduced displacements, compared to an equivalent linear system with equal design displacement, as shown in Figure C17.1-1.

A softening isolation system (Curve C) has a relatively high initial stiffness (short effective period), followed by a relatively low second stiffness (longer effective period) at higher displacements. Where displacements exceed the design displacement, the superstructure is subjected to reduced force demands, while the isolation system is subject to increased displacement demand than for a comparable linear system.

The response of a purely sliding isolation system without lateral restoring force capabilities (Curve D) is governed by friction forces developed at the sliding interface. With increasing displacements, the effective period lengthens while loads on the superstructure remain constant. For such systems, the total displacement caused by repeated earthquake cycles is highly dependent on the characteristics of the ground motion and may exceed the design displacement, D_D. Because these systems do not have increasing resistance with increasing displacement, which helps to recenter the structure and prevent collapse, the procedures of the standard cannot be applied, and use of the system is prohibited.

Chapter 17 establishes isolator design displacements, shear forces for structural design, and other specific requirements for seismically isolated structures, based on MCE_R only. All other design requirements, including loads (other than seismic), load combinations, allowable forces and stresses, and horizontal shear distribution, are the same as those for conventional, fixed-base structures. The main changes incorporated in this edition of the provisions include the following:

- Modified calculation procedure for the elastic design base shear forces from the design earthquake (DE) event to the MCE_R event, using a consistent set of upper and lower bound stiffness properties and displacements. This modification simplifies the design and analysis process by focusing only on the MCE_R event;

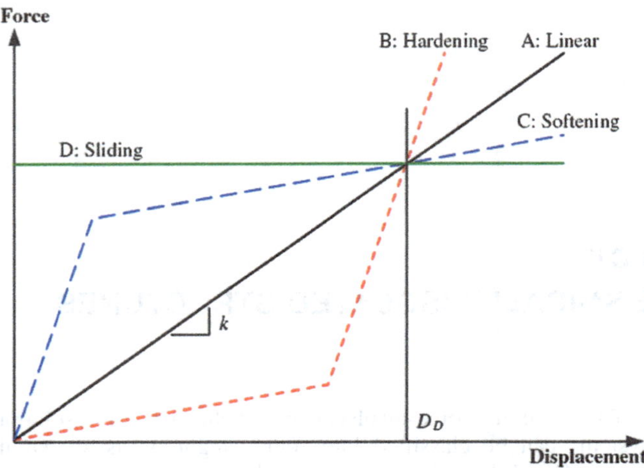

Figure C17.1-1. Idealized force-deflection relationships for isolation systems (stiffness effects of sacrificial wind-restraint systems not shown for clarity).

- Relaxed permissible limits and criteria for the use of the equivalent lateral force (ELF) procedure. This modification minimizes the need to perform complex and computationally expensive nonlinear time history analyses to design the superstructure and isolation system on many base-isolated structures;
- Enhanced definitions of design properties of the isolation system;
- Use of nominal properties in the design process of typical isolation bearings specified by the manufacturers based on prior prototype testing;
- These nominal properties are adjusted using the newly incorporated AASHTO (1999) lambda factor concept to account for response uncertainties and obtain upper and lower bound properties of the isolation system for the design process;
- New method for the vertical distribution of lateral forces associated with the ELF method of design;
- Simplified approach for incorporating a 5% accidental mass eccentricity in nonlinear time history analyses;
- Reduction in the required number of peer reviewers on a seismic isolation project from the current three to five, to a minimum of one peer reviewer. Also, peer reviewers are not required to attend the prototype tests; and
- Calculation procedure to estimate permanent residual displacements that may occur in seismic isolation applications with relatively long period high yield/friction levels, and small yield displacements under a wide range of earthquake intensity.

C17.2 GENERAL DESIGN REQUIREMENTS

In an ideal seismic isolation application, the lateral displacement of the structure is primarily accommodated through large lateral displacement or deformation of the isolation system rather than internal deformation of the superstructure above. Accordingly, the lateral force-resisting system of the superstructure above the isolation system is designed to have sufficient stiffness and strength to prevent large, inelastic displacements. Therefore, the standard contains criteria that limit the inelastic response of the superstructure. Although damage control is not an explicit objective of the standard, design to limit inelastic response of the structural system directly reduces the level of damage that would otherwise occur during an earthquake. In general, isolated structures designed in accordance with the standard are expected to

1. Resist minor and moderate levels of earthquake ground motion without damage to structural elements, nonstructural components, or building contents; and
2. Resist major levels of earthquake ground motion without failure of the isolation system, significant damage to structural elements, extensive damage to nonstructural components, or major disruption to facility function.

Table C17.2-1. Performance Expected for Minor, Moderate, and Major Earthquakes.

Performance Measure	Earthquake Ground Motion Level* Minor	Moderate	Major
Life safety: Loss of life or serious injury is not expected	F, I	F, I	F, I
Structural damage: Significant structural damage is not expected	F, I	F, I	I
Nonstructural damage: Significant nonstructural or content damage is not expected	F, I	I	I

*F indicates fixed-base; I indicates isolated.

Isolated structures are expected to perform considerably better than fixed-based structures during moderate and major earthquakes. Table C17.2-1 compares the expected performance of isolated and fixed-based structures designed in accordance with the standard. Actual performance of an isolated structure should be determined by performing nonlinear time history analyses and computing interstory drifts and floor acceleration demands for an array of ground motions. Those results can be used to compute postearthquake repair costs of the structure using the FEMA P-58 performance-based earthquake engineering (PBEE) methodology (FEMA 2012) and/or large-scale simulations of direct and indirect costs using HAZUS software (FEMA 1999). Evaluation of seismic performance enhancement using seismic isolation should include its impact on floor accelerations, as well as interstory drifts, because these elements are key engineering demand parameters affecting damage in mechanical, electrical, and plumbing (MEP) equipment, ceilings and partitions, and building contents.

Loss of function or discontinued building operation is not included in Table C17.2-1. For certain fixed-based facilities, loss of function would not be expected unless there is significant structural and nonstructural damage that causes closure or restricted access to the building. In other cases, a facility with only limited or no structural damage would not be functional as a result of damage to vital nonstructural components or contents. Seismic isolation, designed in accordance with these provisions, would be expected to mitigate structural and nonstructural damage and to protect the facility against loss of function. The postearthquake repair time required to rehabilitate the structure can also be determined through a FEMA P-58 PBEE evaluation.

Observed structural or nonstructural damage in fixed-based buildings caused by moderate and large earthquakes around the world have typically been associated with high-intensity lateral ground motion excitation rather than vertical acceleration. Gravity design procedures for typical structures result in structural sections and dimensions with relatively high safety factors for seismic resistance. Therefore, current code provisions for fixed-based (or isolated) buildings only require use of a vertical earthquake component, E_v, obtained from static analysis procedures per Sections 12.2.4.6 and 12.2.7.1, defined as $0.2S_{DS}D$ under the design earthquake, where D is the tributary dead load rather than explicit incorporation of vertical ground motions in the design analysis process. For seismic isolation, it should be noted that the term $0.2S_{DS}$ is replaced with $0.2S_{MS}$.

However, similar to fixed-based buildings, consideration of horizontal ground motion excitation alone may underestimate the acceleration response of floors and other building components. Portions of fixed-based and isolated structures may be especially sensitive to adverse structural response amplification induced by vertical ground motions including long spans, vertical discontinuities, or large cantilever elements. Certain nonstructural components, such as acoustic tile suspended ceiling systems, are also particularly vulnerable to the combination of vertical and horizontal ground motion effects. These building subassemblies or components may warrant additional vertical considerations. In addition, isolators with relatively low tributary gravity load, and isolators located below columns that form part of the lateral force-resisting system, can potentially have net uplift or tensile displacements caused by combined large vertical ground motion accelerations and global overturning. This uplift or bearing tension may induce high impact forces on the substructure, jeopardize the stability of the bearings, or result in bearing.

Base-isolated structures located near certain faults with characteristics that produce large vertical accelerations (e.g., hanging wall in reverse and reverse/oblique faults) are also more vulnerable and therefore may also require consideration of vertical ground motion excitation.

Vertical ground acceleration may affect the behavior of axial load-dependent isolation systems in the horizontal direction caused by potential coupling between horizontal and vertical response of the building structure.

Building response parameters that are expected to be affected by vertical excitation are vertical floor spectra and axial load demand on isolation bearings and columns, as discussed in Section C17.2.4.6. Isolated buildings with significant horizontal–vertical coupling are also expected to impart additional horizontal accelerations to the building at the frequencies of coupled modes that match the vertical motions.

If it is elected to investigate the effect of vertical ground motion acceleration on building response, one of the following analysis methods is suggested:

- Response spectrum analysis using horizontal and vertical spectrum (upward and downward).
- Response spectrum analysis using a vertical spectrum, combined with horizontal response spectrum analysis results, using orthogonal combinations corresponding to the 100%–30%–30% rule.
- Three-dimensional response history analysis following the recommendations of Section C17.3.3, with explicit inclusion of vertical ground motion acceleration records.
- Horizontal response history analysis following the provisions of Section 17.3.3, considering the two limiting initial gravity load conditions, as defined per Section 17.2.7.1. Note that this analysis affects the effective characteristics of axial load-dependent isolators with resulting changes in base shear and displacement demands.

The structural model in these analyses should be capable of capturing the effects of vertical response and vertical mass participation and should include the modeling recommendations in Section C17.6.2.

C17.2.4 Isolation System

C17.2.4.1 Environmental Conditions Environmental conditions that may adversely affect isolation system performance must be investigated thoroughly. Specific requirements for environmental considerations on isolators are included in the new Section 17.2.8. Unlike conventional materials, whose properties do not vary substantially with time, the materials used in seismic isolators are typically subject to significant aging effects over the life span of a building structure. Because the testing protocol of Section 17.8 does not account for the effects of aging, contamination, scragging (temporary degradation of mechanical properties with repeated cycling), temperature, velocity effects, and wear, the designer must account for these effects by explicit analysis. The approach to accommodate these effects, introduced in the AASHTO specifications (AASHTO 1999), is to use property modification factors as specified in Section 17.2.8.4.

C17.2.4.2 Wind Forces Lateral displacement over the over the height of the isolation system resulting from wind loads must be limited
to a value similar to that required for other stories of the superstructure.

C17.2.4.3 Fire Resistance Where fire may adversely affect the lateral performance of the isolation system, the system must be protected to maintain the gravity-load resistance and stability required for the other elements of the superstructure supported by the isolation system.

C17.2.4.4 Lateral Restoring Force The restoring force requirement is intended to limit residual displacements in the isolation system resulting from any earthquake event so that the isolated structure will adequately withstand aftershocks and future earthquakes. The potential for residual displacements is addressed in Section C17.2.6.

C17.2.4.5 Displacement Restraint The use of a displacement restraint to limit displacements beyond the design displacement is discouraged. Where a displacement restraint system is used, explicit nonlinear response history analysis of the isolated structure for the MCE_R level is required using the provisions of Chapter 16 to account for the effects of engaging the displacement restraint.

C17.2.4.6 Vertical-Load Stability The vertical loads used to assess the stability of a given isolator should be calculated using bounding values of dead load, live load, and the peak earthquake demand at the MCE_R level. Because earthquake loads are reversible in nature, peak earthquake load should be combined with bounding values of dead and live load, in a manner that produces both the maximum downward force and the maximum upward force on any isolator. Stability of each isolator should be verified for these two extreme values of vertical load at peak MCE_R displacement of the isolation system. In addition, all elements of the isolation system require testing or equivalent measures that demonstrate their stability for the MCE_R ground motion levels. This stability can be demonstrated by performing a nonlinear static analysis for an MCE_R response displacement of the entire structural system, including the isolation system, and showing that lateral and vertical stability are maintained. Alternatively, this stability can be demonstrated by performing a nonlinear dynamic analysis for the MCE_R motions using the same inelastic reductions as for the Design Earthquake (DE) and acceptable capacities, except that member and connection strengths can be taken as their nominal strengths with resistance factors, ϕ, taken as 1.0.

Vertical ground motion excitation affects bounding axial loads on isolation bearings and vertical stability design checks. The E component of load combination 5 of Section 2.3.2 should consider the maximum of E_v per this standard or the dynamic amplification from analysis when significant vertical acceleration is anticipated, per Section C17.2.

C17.2.4.7 Overturning The intent of this requirement is to prevent both global structural overturning and overstress of elements caused by localized uplift. Isolator uplift is acceptable as long as the isolation system does not disengage from its horizontal-resisting connection details. The connection details used in certain isolation systems do not develop tension resistance, a condition which should be accounted for in the analysis and design. Where the tension capacity of an isolator is used to resist uplift forces, design and testing, in accordance with Sections 17.2.4.6 and 17.8.2.5, must be performed to demonstrate the adequacy of the system to resist tension forces at the total maximum displacement.

C17.2.4.8 Inspection and Replacement Although most isolation systems do not require replacement following an earthquake event, access for inspection, repair, and replacement must be provided. In some cases (Section 17.2.6), recentering may be required. The isolation system should be inspected periodically as well as following significant earthquake events, and any damaged elements should be repaired or replaced.

C17.2.4.9 Quality Control A testing and inspection program is necessary for both fabrication and installation of the isolator units. Because of the rapidly evolving technological advances of seismic isolation, reference to specific standards for testing and inspection is difficult for some systems, while reference for some systems is possible, for example, elastomeric bearings should follow ASTM D4014 requirements (ASTM 2012). Similar standards are yet to be developed for other isolation systems. Special inspection procedures and load testing to verify manufacturing quality should therefore be developed for each project. The requirements may vary depending on the type of isolation system used. Specific requirements for quality control testing are now given in Section 17.8.5.

C17.2.5 Structural System

C17.2.5.2 Minimum Building Separations A minimum separation between the isolated structure and other structures or rigid obstructions is required to allow unrestricted horizontal translation of the superstructure in all directions during an earthquake event. The separation dimension should be determined based on the total design displacement of the isolation system, the maximum lateral displacement of the superstructure above the isolation, and the lateral deformation of the adjacent structures.

C17.2.5.4 Steel Ordinary Concentrically Braced Frames Section 17.5.4.2 of this standard implies that only seismic force-resisting systems permitted for fixed-based building applications are permitted to be used in seismic isolation applications. Table 12.2-1 limits the height of steel ordinary concentrically braced frames (OCBFs) in fixed-based multistory buildings assigned to Seismic Design Categories D and E to 35 ft (10.7 m) and does not permit them in buildings assigned to Seismic Design Category F. Section 17.2.5.4 permits them to be used for seismic isolation applications to heights of 160 ft (48.8 m) in buildings assigned to Seismic Design Categories D, E, and F, provided that certain additional requirements are satisfied. The additional design requirements that must be satisfied include that the building must remain elastic at the design earthquake level (i.e., $R_I = 1.0$), that the moat clearance displacement, D_{TM}, be increased by 20%, and that the braced frame be designed to satisfy Section F1.7 of AISC 341. It should be noted that currently permitted OCBFs in seismically isolated buildings assigned to Seismic Design Categories D and E also need to satisfy Section F1.7 of AISC 341.

Seismic isolation has the benefit of absorbing most of the displacement of earthquake ground motions, allowing the seismic force-resisting system to remain essentially elastic. Restrictions in Chapter 17 on the seismic force-resisting system limit the inelastic reduction factor to a value of 2 or less to ensure essentially elastic behavior. A steel OCBF provides the benefit of providing a stiff superstructure with reduced drift demands on drift-sensitive nonstructural components while providing significant cost savings compared to special systems. Steel OCBFs have been used in the United States for numerous (perhaps most) new seismically isolated essential facility buildings since seismic isolation was first introduced in the 1980s. Some of these buildings had heights of as much as 130 ft (39.6 m). The 160 ft (48.8 m) height limit was permitted for seismic isolation with OCBFs in high seismic zones when seismic isolation was first introduced in the building code as an appendix to the Uniform Building Code in 1991. When height limits were restricted for fixed-based OCBFs in the 2000 NEHRP Recommended Provisions (FEMA 2003), the effect the restriction could have on the design of seismically isolated buildings was not recognized. The Section 17.2.5.4 change rectifies that oversight. It is the judgment of this committee that height limits should be increased to the 160 ft (48.8 m) level, provided that the additional conditions are met.

The AISC Seismic Committee (Task Committee-9) studied the concept of steel OCBFs in building applications to heights of 160 ft (48.8 m) in high seismic areas. They decided that additional detailing requirements are necessary, which are found in Section F1.7 of AISC 341.

There has been some concern that steel ordinary concentrically braced frames may have an unacceptable collapse hazard if ground motions greater than MCE_R cause the isolation system to impact the surrounding moat wall. While there has not been a full FEMA P-695 (FEMA 2009a) study of ordinary steel concentrically braced frame systems, a recent conservative study of one structure using OCBFs with $R_I = 1$ on isolation systems, performed by Armin Masroor at SUNY Buffalo (Masroor and Mosqueda 2015), indicates that an acceptable risk of collapse (10% risk of collapse given MCE ground motions) is achieved if a 15% to 20% larger isolator displacement is provided. The study does not include the backup capacity of gravity connections or the influence of concrete-filled metal deck floor systems on the collapse capacity. Even though there is no requirement to consider ground motions beyond the risk-targeted maximum considered earthquake ground motion in design, it was the judgment of this committee to provide additional conservatism by requiring 20% in moat clearance. It is possible that further P-695 studies will demonstrate that the additional 1.2 factor of displacement capacity may not be needed.

C17.2.5.5 Isolation System Connections This section addresses the connections of the structural elements that join isolators together. The isolators, joining elements, and connections compose the isolation system. The joining elements are typically located immediately above the isolators; however, there are many ways to provide this framing, and this section is not meant to exclude other types of systems. It is important to note that the elements and the connections of the isolation system are designed for V_b level forces, while elements immediately above the isolation system are designed for V_s level forces.

Although ductility detailing for the connections in the isolation system is not required, and these elements are designed to remain elastic with V_b level forces using $R = 1.0$. In some cases, it may still be prudent to incorporate ductility detailing in these connections (where possible) to protect against unforeseen loading. This incorporation has been accomplished in the past by providing connection details similar to those used for a seismic force-resisting system of Table 12.2-1, with connection moment and shear strengths beyond the code minimum requirements. Ways of accomplishing this include factoring up the design forces for these connections or providing connections with moment and shear strengths capable of developing the expected plastic moment strength of the beam, similar to AISC 341 or ACI 318 requirements for ordinary moment frames (OMFs).

C17.2.6 Elements of Structures and Nonstructural Components To accommodate the differential horizontal and vertical movement between the isolated building and the ground, flexible utility connections are required. In addition, stiff elements crossing the isolation interface (such as stairs, elevator shafts, and walls) must be detailed to accommodate the total maximum displacement without compromising life safety provisions.

The effectiveness and performance of different isolation devices in building structures under a wide range of ground motion excitations have been assessed through numerous experimental and analytical studies (Kelly et al. 1980, Kelly and Hodder 1981, Kelly and Chaloub 1990, Zayas et al. 1987, Constantinou et al. 1999, Warn and Whittaker 2006, Buckle et al. 2002, Kelly and Konstantinidis 2011). The experimental programs included in these studies have typically consisted of reduced-scale test specimens constructed with relatively high precision under laboratory conditions. These studies initially focused on elastomeric bearing devices, although in recent years the attention has shifted to the single- and multi-concave friction pendulum bearings. The latter system provides the option for longer isolated periods.

Recent full-scale shake table tests (Ryan et al. 2012) and analytical studies (Katsaras 2008) have shown that the isolation systems included in these studies with a combination of longer periods, relatively high yield/friction levels, and small yield displacements will result in postearthquake residual displacements. In these studies, residual displacements ranging from 2 to 6 in. (50 to 150 mm) were measured and computed for isolated building structures with a period of 4 s or greater and a yield level in the range of 8% to 15% of the structure's weight. This permanent offset may affect the serviceability of the structure and possibly jeopardize the functionality of elements crossing the isolation plane (such as fire protection and weatherproofing elements, egress/entrance details, elevators, and joints of primary piping systems). Because it may not be possible to recenter some isolation systems, isolated structures with such characteristics should be detailed to accommodate these permanent offsets.

The Katsaras report (2008) provides recommendations for estimating the permanent residual displacement in any isolation system based on an extensive analytical and parametric study. The residual displacements measured in full-scale tests (Ryan et al. 2012) are reasonably predicted by this procedure, which uses an idealized bilinear isolation system, shown in Figure C17.2-1. The three variables that affect the residual

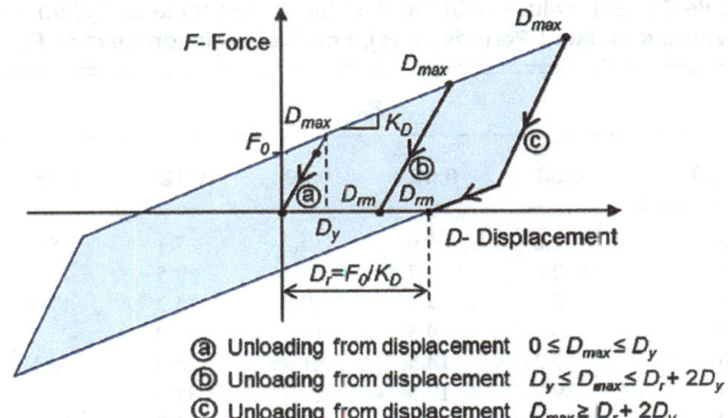

Figure C17.2-1. Definitions of static residual displacement, D_{rm}, for a bilinear hysteretic system.

Table C17.2-2. Values of Static Residual Displacement, D_{rm}.

Range of Maximum Displacement, D_{max}	Static Residual Displacement, D_{rm}
$0 \leq D_{\max} \leq D_y$	0
$D_y \leq D_{\max} < D_r + 2D_y$	$D_r(D_{\max} - D_y)/(D_r + D_y)$
$D_r + 2D_y \leq D_{\max}$	D_r

displacement are the isolated period (based on the second slope stiffness K_D), the yield/friction level (F_0), and the yield displacement (D_y).

The procedure for estimating the permanent residual displacement, D_{rd} [see Equation (C17.2-1)], is a function of the system yield displacement D_y, the static residual displacement, $D_r = F_0/K_p$, and D_{rm}, which is a function of D_m, the maximum earthquake displacement shown in Table C17.2-2. For most applications, D_{rm} is typically equal to D_r

$$D_{rd} = \frac{0.87 D_{rm}}{\left(1 + 4.3 \frac{D_{rm}}{D_r}\right)\left(1 + 31.7 \frac{D_y}{D_r}\right)} \quad \text{(C17.2-1)}$$

Thus, there is a simple two-step process to estimate the permanent residual displacement, D_{rd}:

1. Calculate the static residual displacement, D_r, based on the isolated period (using the second slope stiffness, K_D) and the yield/friction levels. Table C17.2-3 provides values of D_r for a range of periods from 2.5 to 20 s and a range of yield/friction levels from $0.03W$ to $0.15W$.
2. Using the value of D_r calculated for the isolation system and the yield displacement, D_y, of the system, the permanent residual displacement, D_{rd}, can be calculated from Equation (C17.2-1), and Tables C17.2-4 and C17.2-5 provide the residual displacements for earthquake displacements (D_m) of 10 and 20 in. (250 and 508 mm), respectively.

The cells with bold type in Tables C17.2-4 and C17.2-5 correspond to permanent residual displacements exceeding 2.0 in. (50 mm). Note that for yield displacements of

Table C17.2-3. Values of Static Residual Displacement, D_r (in.), for Various Isolated Periods, T (s), and Yield/Friction Levels, F_0.

	F_0				
T (s)	0.03	0.06	0.09	0.12	0.15
2.5	1.8	3.6	5.3	7.1	8.9
2.8	2.4	4.7	7.1	9.5	11.9
3.5	3.6	7.1	10.7	14.2	17.8
4.0	4.7	9.5	14.2	19.0	23.7
5.0	7.2	14.5	21.7	28.9	36.1
5.6	9.2	18.5	27.7	37.0	46.2
6.0	10.7	21.3	32.0	42.7	53.3
7.0	14.2	28.4	42.7	56.9	71.1
8.0	18.7	37.4	56.2	74.9	93.6
9.0	23.7	47.4	71.1	94.8	118.5
20.1	118.5	237.0	355.5	474.0	592.5

Notes: 1 in. = 25 mm.

Table C17.2-4. Permanent Residual Displacement, D_{rd}, for a Maximum Earthquake Displacement, D_m, of 10 in. (250 mm).

	D_y (in.)							
D_r (in.)	0.005	0.01	0.02	0.20	0.39	0.59	0.98	1.97
4.0	0.63	0.60	0.56	0.25	0.16	0.11	0.07	0.04
7.9	1.28	1.25	1.21	0.73	0.50	0.39	0.26	0.14
11.9	1.86	1.84	1.79	1.22	0.90	0.71	0.50	0.27
15.8	**2.32**	**2.30**	**2.25**	1.67	1.29	1.04	0.75	0.43
19.8	**2.72**	**2.70**	**2.66**	**2.07**	1.65	1.37	1.01	0.59
23.7	**3.08**	**3.06**	**3.02**	**2.43**	1.99	1.68	1.27	0.76
27.7	**3.39**	**3.37**	**3.34**	**2.75**	**2.30**	1.97	1.51	0.92
31.6	**3.68**	**3.66**	**3.62**	**3.05**	**2.59**	**2.24**	1.75	1.09
35.6	**3.93**	**3.91**	**3.87**	**3.32**	**2.85**	**2.49**	1.97	1.25
39.5	**4.16**	**4.14**	**4.11**	**3.56**	**3.09**	**2.73**	**2.19**	1.41

Notes: 1 in. = 25 mm.
Bold values designate D_{rd} values of 2 in. or more.

Table C17.2-5. Permanent Residual Displacements, D_{rd}, for a Maximum Earthquake Displacement, D_m, of 20 in. (508 mm).

	D_y (in.)							
D_r (in.)	0.005	0.01	0.02	0.20	0.39	0.59	0.98	1.97
4.0	0.63	0.60	0.56	0.25	0.16	0.11	0.07	0.04
7.9	1.28	1.25	1.21	0.73	0.50	0.39	0.26	0.15
11.9	1.93	1.90	1.85	1.28	0.95	0.76	0.54	0.31
15.8	**2.58**	**2.55**	**2.50**	1.86	1.45	1.19	0.87	0.52
19.8	**3.23**	**3.20**	**3.15**	**2.47**	1.98	1.65	1.24	0.75
23.7	**3.75**	**3.72**	**3.67**	**2.97**	**2.45**	**2.08**	1.59	0.99
27.7	**4.22**	**4.20**	**4.15**	**3.45**	**2.90**	**2.50**	1.95	1.24
31.6	**4.67**	**4.64**	**4.60**	**3.90**	**3.33**	**2.90**	**2.30**	1.50
35.6	**5.08**	**5.06**	**5.02**	**4.32**	**3.74**	**3.30**	**2.65**	1.76
39.5	**5.47**	**5.45**	**5.41**	**4.72**	**4.13**	**3.67**	**2.99**	**2.02**

Notes: 1 in. = 25 mm.
Bold values designate D_{rd} values of 2 in. or more.

approximately 2.0 in. (50 mm), residual displacements will not occur for most isolation systems.

C17.2.8 Isolation System Properties This section defines and combines sources of variability in isolation system mechanical properties measured by prototype testing, permitted by manufacturing specification tolerances, and occurring over the life span of the structure because of aging and environmental effects. Upper bound and lower bound values of isolation system component behavior (e.g., for use in response history analysis procedures) and maximum and minimum values of isolation system effective stiffness and damping based on these bounding properties (e.g., for use in equivalent lateral force procedures) are established in this section. Values of property modification factors vary by product and cannot be specified generically in the provisions. Typical "default" values for the more commonly used systems are provided below. The designer and peer reviewer are responsible for determining appropriate values of these factors on a project-specific and product-specific basis.

This section also refines the concept of bounding (upper bound and lower bound) values of isolation system component behavior by

1. Explicitly including variability caused by manufacturing tolerances, aging, and environmental effects; ASCE 7-10 only addressed variability associated with prototype testing; and
2. Simplifying design by basing bounding measures of amplitude-dependent behavior on only MCE_R ground motions; ASCE 7-10 used both DE and MCE_R ground motions.

The new section also refines the concept of maximum and minimum effective stiffness and damping of the isolation system by use of revised formulas that

1. Define effective properties of the isolation system on bounding values of component behavior (i.e., same two refinements, described previously), and
2. Eliminates the intentional conservatism of ASCE 7-10 that defines minimum effective damping in terms of maximum effective stiffness.

C17.2.8.2 Isolator Unit Nominal Properties Isolator manufacturers typically supply nominal design properties that are reasonably accurate and can be confirmed by prototype tests in the design and construction phases. These nominal properties should be based on past prototype tests, as defined in Section 17.8.2; see Figure C17.2-2.

C17.2.8.3 Bounding Properties of Isolation System Components The methodology for establishing lower and upper bound values for isolator basic mechanical properties based on property modification factors was first presented in Constantinou et al. (1999). Since then, it has been revised in Constantinou et al. (2007) based on the latest knowledge of lifetime behavior of isolators. The methodology presented uses property modification factors to adjust isolator nominal properties based on considerations of natural variability in properties, effects of heating during cyclic motion, and the effects of aging, contamination, ambient temperature and duration of exposure to that temperature, and history of loading. The nominal mechanical properties should be based on prototype (or representative) testing on isolators not previously tested, at normal temperature and under dynamic loading.

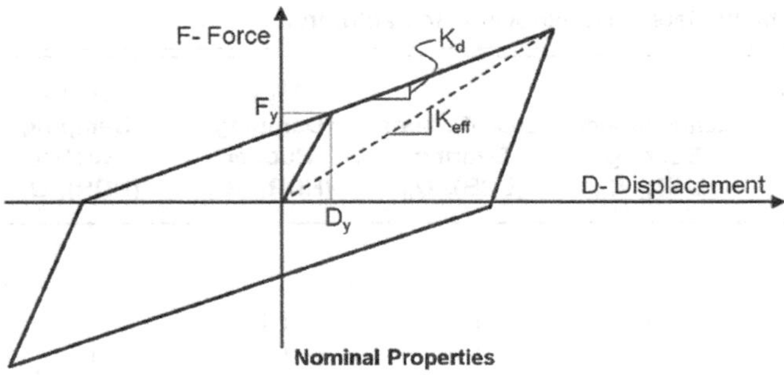

Figure C17.2-2. Example of the nominal properties of a bilinear force deflection system.

The methodology also modifies the property modification factors to account for the unlikely situation of having several events of low probability of occurrence occur at the same time (i.e., maximum earthquake, aging, and low temperature) by use of property adjustment factors that are dependent on the significance of the structure analyzed (values range from 0.66 for a typical structure to 1.0 for a critical structure). This standard presumes that the property adjustment factor is 0.75. However, the registered design professional may opt to use the value of 1.0 based on the significance of the structure (e.g., health-care facilities or emergency operation centers) or based on the number of extreme events considered in the establishment of the property modification factor. For example, if only aging is considered, then a property adjustment factor of unity is appropriate.

Examples of application in the analysis and design of bridges may be found in Constantinou et al. (2011). These examples may serve as guidance in the application of the methodology in this standard. Constantinou et al. (2011) also presents procedures for estimating the nominal properties of lead-rubber and friction pendulum isolators, again based on the assumption that prototype test data are not available. Data used in the estimation of the range of properties were based on available test data, all of which were selected to heighten heating effects. Such data would be appropriate for cases of high-velocity motion and large lead core size or high friction values.

Recommended values for the specification tolerance on the average properties of all isolators of a given size isolator are typically in the $\pm 10\%$ to $\pm 15\%$ range. For a $\pm 10\%$ specification tolerance, the corresponding lambda factors would be $\lambda_{(\text{spec,max})} = 1.10$ and $\lambda_{(\text{spec,min})} = 0.90$. Variations in individual isolator properties are typically greater than the tolerance on the average properties of all isolators of a given size as presented in Section 17.2.8.4. It is recommended that the isolator manufacturer be consulted when establishing these tolerance values.

Section 17.2.8.4 requires the isolation system to be designed with consideration given to environmental conditions, including aging effects, creep, fatigue, and operating temperatures. The individual aging and environmental factors are multiplied together and then the portion of the lambda factor differing from unity is reduced by 0.75, based on the assumption that not all of the maximum values will occur simultaneously. As part of the design process, it is important to recognize that there will be additional variations in the nominal properties because of manufacturing. The next section specifies the property modification factors corresponding to the manufacturing process or default values if manufacturer-specific data are not available. These factors are combined with the property modification factors (Section 17.2.8.4) to determine the maximum and minimum properties of the isolators (Section 17.2.8.5) for use in the design and analysis process.

The lambda-test values, $\lambda_{\text{test,max}}$ and $\lambda_{\text{test,min}}$, are determined from prototype testing and shall bound the variability and degradation in properties caused by speed of motion, heating effects, and scragging from Item 2 of Section 17.8.2.2. The registered design professional (RDP) shall specify whether this testing is performed quasi-statically, as in Item 2(a), or dynamically, as in Item 2(b). When testing is performed quasi-statically, the dynamic effects shall be accounted for in analysis and design using appropriate adjustment of the lambda-test values.

Item 3 of the testing requirements of Section 17.8.2.2 is important for property determination because it is common to Item 2. Using this testing, the lambda-test values, $\lambda_{\text{test,max}}$ and $\lambda_{\text{test,min}}$, may be determined by three fully reversed cycles of dynamic (at the effective period T_M) loading at the maximum displacement $1.0D_M$ on full-scale specimens. This test regime incorporates the effects of high-speed motion. The upper and lower bound values of K_d shall also envelop the $0.67D_M$ and $1.0D_M$ tests of Item 2 of Section 17.8.2.2. Therefore, the lambda-test values bound the effects of heating and scragging. As defined by Section 17.2.8.2, the nominal property of interest is defined as the average among the three cycles of loading. $\lambda_{\text{test,max}}$ shall be determined as the ratio of the first cycle property to the nominal property value. $\lambda_{\text{test,min}}$ shall be determined as the ratio of the property value at a representative cycle, determined by the RDP, to the nominal property value. The number of cycles shall be representative of the accepted performance of the isolation system for the local seismic hazard conditions, with the default cycle being the third cycle. A critique and guidance are provided in McVitty and Constantinou (2015).

C17.2.8.4 Property Modification Factors The lambda factors are used to establish maximum and minimum mathematical models for analysis, the simplest form of which is the linear static procedure used to assess the minimum required design base shear and system displacements. More complex mathematical models account for various property variation effects explicitly (e.g., velocity, axial load, bilateral displacement, and instantaneous temperature). In this case, the cumulative effect of the lambda factors reduces (the combined lambda factor is closer to 1.0). However, some effects, such as specification tolerance and aging, are likely to always remain because they cannot be accounted for in mathematical models. Default lambda factors are provided in Table C17.2-6 as isolators from unknown manufacturers that do not have qualification test data. Default lambda factors are provided in Table C17.2-7 for most common types of isolators fabricated by quality manufacturers. Note that this table does not have any values of property modification factors for the actual stiffness (K_d) of sliding isolators. It is presumed that sliding isolators, whether flat or spherical, are produced with sufficiently high accuracy that their actual stiffness characteristics are known. The RDP may assign values of property modification factors different than unity for the actual stiffness of sliding bearings on the basis of data obtained in the prototype testing or on the basis of lack of experience with unknown manufacturers. Also note that this table provides values of property modification factors to approximately account for uncertainties in the materials and manufacturing methods used. These values presume lack of test data or incomplete test data and unknown manufacturers. For example, the values in Table C17.2-6 for sliding bearings presume unknown materials for the sliding interfaces so that there is considerable uncertainty in the friction coefficient values. Also, the data presume that elastomers used in elastomeric

Table C17.2-6. Default Upper and Lower Bound Multipliers for Unknown Manufacturers.

Variable	Unlubricated Interfaces, μ or Q_d	Lubricated (Liquid) Interfaces, μ or Q_d	Plain Low-Damping Elastomeric, K	Lead-Rubber Bearing (LRB), K_d	Lead-Rubber Bearing (LRB), Q_d	High-Damping Rubber (HDR), K_d	High-Damping Rubber (HDR), Q_d
Example: Aging and Environmental Factors							
Aging, λ_a	1.3	1.8	1.3	1.3	1	1.4	1.3
Contamination, λ_c	1.2	1.4	1	1	1	1	1
Example Upper Bound, $\lambda_{(ae,max)}$	1.56	2.52	1.3	1.3	1	1.4	1.3
Example Lower Bound, $\lambda_{(ae,min)}$	1	1	1	1	1	1	1
Example: Testing Factors							
All Cyclic Effects, Upper	1.3	1.3	1.3	1.3	1.6	1.5	1.3
All Cyclic Effects, Lower	0.7	0.7	0.9	0.9	0.9	0.9	0.9
Example Upper Bound, $\lambda_{(test,max)}$	1.3	1.3	1.3	1.3	1.6	1.5	1.3
Example Lower Bound, $\lambda_{(test,min)}$	0.7	0.7	0.9	0.9	0.9	0.9	0.9
$\lambda_{(PM,max)} = (1 + (0.75 * (\lambda_{(ae,max)} - 1))) * \lambda_{(test,max)}$	1.85	2.78	1.59	1.59	1.6	1.95	1.59
$\lambda_{(PM,min)} = (1 - (0.75 * (1 - \lambda_{(ae,min)})) * \lambda_{(test,min)}$	0.7	0.7	0.9	0.9	0.9	0.9	0.9
Lambda Factor for Spec. Tolerance, $\lambda_{(spec,max)}$	1.15	1.15	1.15	1.15	1.15	1.15	1.15
Lambda Factor for Spec. Tolerance, $\lambda_{(spec,min)}$	0.85	0.85	0.85	0.85	0.85	0.85	0.85
Upper Bound Design Property Multiplier	2.12	3.2	1.83	1.83	1.84	2.24	1.83
Lower Bound Design Property Multiplier	0.6	0.6	0.77	0.77	0.77	0.77	0.77
Default Upper Bound Design Property Multiplier	2.1	3.2	1.8	1.8	1.8	2.2	1.8
Default Lower Bound Design Property Multiplier	0.6	0.6	0.8	0.8	0.8	0.8	0.8

Notes: λ_{PM} is the lambda value for testing and environmental effects.

bearings have significant scragging and aging. Moreover, for lead-rubber bearings, the data in the table presume that there is considerable uncertainty in the starting value (before any hysteretic heating effects) of the effective yield strength of lead.

Accordingly, there is a considerable range in the upper and lower values of the property modification factors. Yet, these values should be used with caution because low-quality fabricators could use materials and vulcanization and manufacturing processes that result in even greater property variations. The preferred approach for establishing property modification factors is through rigorous qualification testing of materials and manufacturing methods by a quality manufacturer, and dynamic prototype testing of full-size specimens, and by quality control testing at project-specific loads and displacements. These test data on similar-sized isolators take precedence over the default values.

For elastomeric isolators, lambda factors and prototype tests may need to address axial–shear interaction, bilateral deformation, load history including first cycle effects and the effects of scragging of virgin elastomeric isolators, ambient temperature, other environmental loads, and aging effects over the design life of the isolator.

For sliding isolators, lambda factors and prototype tests may need to address contact pressure, rate of loading or sliding velocity, bilateral deformation, ambient temperature, contamination, other environmental loads, and aging effects over the design life of the isolator.

Rate of loading or velocity effects are best accounted for by dynamic prototype testing of full-scale isolators. Property modification factors for accounting for these effects may be used in lieu of dynamic testing.

In general, ambient temperature effects can be ignored for most isolation systems if they are in conditioned space where the expected temperature varies between 30°F and 100°F.

The following comments are provided in the approach to be followed for the determination of the bounding values of mechanical properties of isolators:

1. Heating effects (hysteretic or frictional) may be accounted for on the basis of a rational theory (e.g., Kalpakidis and Constantinou 2008, 2009; Kalpakidis et al. 2010) so that only the effects of uncertainty in the nominal values of the properties, aging, scragging, and contamination need to be considered. This is true for lead-rubber bearings for which lead of high purity and of known thermomechanical properties is used. For sliding bearings, the composition of the sliding interface affects the relation of friction to temperature and therefore cannot be predicted by theory alone. Moreover, heating generated during high-speed motion may affect the bond strength of liners. Given that there are numerous sliding interfaces (and that they are typically proprietary) that heating effects in sliding bearings are directly dependent on pressure and velocity, and that size is important in the heating effects (Constantinou et al. 2007), full-scale dynamic prototype and production testing are very important for sliding bearings.
2. Heating effects are important for sliding bearings and the lead core in lead-rubber bearings. They are not important

Table C17.2-7. Default Upper and Lower Bound Multipliers for Quality Manufacturers.

Variable	Unlubricated PTFE, μ	Lubricated PTFE, μ	Rolling/ Sliding, $K2$	Plain Elastomerics, K	Lead-Rubber bearing (LRB), $K2$	Lead-Rubber bearing (LRB), Q_d	High-Damping Rubber (HDR), Q_d	High-Damping Rubber (HDR), K_d
Example: Aging and Environmental Factors								
Aging, λ_a	1.10	1.50	1.00	1.10	1.10	1.00	1.20	1.20
Contamination, $\lambda+$	1.10	1.10	1.00	1.00	1.00	1.00	1.00	1.00
Example Upper Bound, $\lambda_{(ae,max)}$	1.21	1.65	1.00	1.10	1.10	1.00	1.20	1.20
Example Lower Bound, $\lambda_{(ae,min)}$	1.00	1.00	1.00	1.00	1.00	1.00	1.00	1.00
Example: Testing Factors								
All Cyclic Effects, Upper	1.20	1.30	1.00	1.03	1.03	1.30	1.50	1.30
All Cyclic Effects, Lower	0.95	0.95	1.00	0.98	0.98	0.95	0.95	0.95
Example Upper Bound, $\lambda_{(test,max)}$	1.20	1.30	1.00	1.03	1.03	1.30	1.50	1.30
Example Lower Bound, $\lambda_{(test,min)}$	0.95	0.95	1.00	0.98	0.98	0.95	0.95	0.95
$\lambda_{(PM,max)} = (1 + (0.75 * (\lambda_{(ae,max)} - 1))) * \lambda_{(test,max)}$	1.39	1.93	1.00	1.11	1.11	1.30	1.73	1.50
$\lambda_{(PM,min)} = (1 - (0.75 * (1 - \lambda_{(ae,min)})) * \lambda_{(test,min)}$	0.95	0.95	1.00	0.98	0.98	0.95	0.95	0.95
Lambda Factor for Spec. Tolerance, $\lambda_{(spec,max)}$	1.15	1.15	1.00	1.15	1.15	1.15	1.15	1.15
Lambda Factor for Spec. Tolerance, $\lambda_{(spec,min)}$	0.85	0.85	1.00	0.85	0.85	0.85	0.85	0.85
Upper Bound Design Property Multiplier	1.60	2.22	1.00	1.27	1.27	1.50	1.98	1.72
Lower Bound Design Property Multiplier	0.81	0.81	1.00	0.83	0.83	0.81	0.81	0.81
Default Upper Bound Design Property Multiplier	1.6	2.25	1	1.3	1.3	1.5	2	1.7
Default Lower Bound Design Property Multiplier	0.8	0.8	1	0.8	0.8	0.8	0.8	0.8

Notes: λ_{PM} is the lambda value for testing and environmental effects.

and need not be considered for elastomeric bearings of either low or high damping. The reason for this is described in Constantinou et al. (2007), in which it has been shown, based on theory and experimental evidence, that the rise in temperature of elastomeric bearings during cyclic motion (about 1°C per cycle) is too small to significantly affect their mechanical properties. Prototype and production testing of full-size specimens at the expected loads and displacements should be sufficient to detect poor material quality and poor material bonding in plain elastomeric bearings, even if done quasi-statically.

3. Scragging and recovery to the virgin rubber properties (see Constantinou et al. 2007 for details) are dependent on the rubber compound, size of the isolator, the vulcanization process, and the experience of the manufacturer. Also, it has been observed that scragging effects are more pronounced for rubber of low shear modulus and that the damping capacity of the rubber has a small effect. It has also been observed that some manufacturers are capable of producing low-modulus rubber without significant scragging effects, whereas others cannot. It is therefore recommended that the manufacturer should present data on the behavior of the rubber under virgin conditions (not previously tested and immediately after vulcanization) so that scragging property modification factors can be determined. This factor is defined as the ratio of the effective stiffness in the first cycle to the effectiveness stiffness in the third cycle, typically obtained at a representative rubber shear strain (e.g., 100%). It has been observed that this factor can be as high as, or can exceed, a value of 2.0 for shear-modulus rubber less than or equal to 65 psi (0.45 MPa). Also, it has been observed that some manufacturers can produce rubber with a shear modulus of 65 psi (0.45 MPa) and a scragging factor of approximately 1.2 or less. Accordingly, it is preferred to establish this factor by testing for each project or to use materials qualified in past projects.
4. Aging in elastomeric bearings has in general small effects (typically increases in stiffness and strength of the order of 10% to 30% over the lifetime of the structure), provided that scragging is also minor. It is believed that scragging is mostly the result of incomplete vulcanization, which is thus associated with aging as chemical processes in the rubber continue over time. Inexperienced manufacturers may produce low shear modulus elastomers by incomplete vulcanization, which should result in significant aging.
5. Aging in sliding bearings depends on the composition of the sliding interface. There are important concerns with bimetallic interfaces (Constantinou et al. 2007), even in the absence of corrosion, so that they should be penalized by large aging property modification factors or simply not used. Also, lubricated interfaces warrant higher aging and contamination property modification factors. The designer can refer to Constantinou et al. (2007) for detailed values of the factor depending on the conditions of operation and the environment of exposure. Note that lubrication is meant to

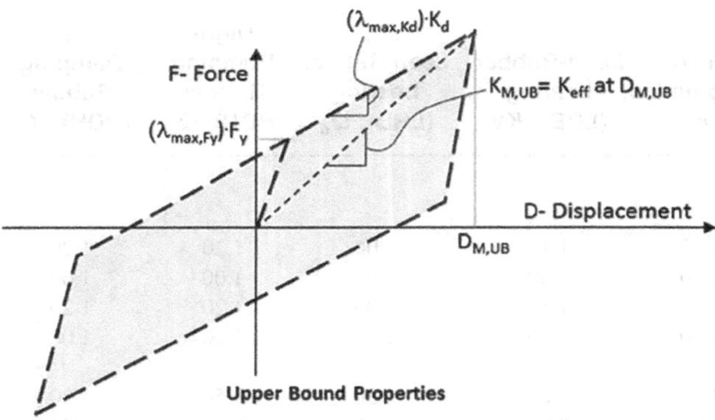

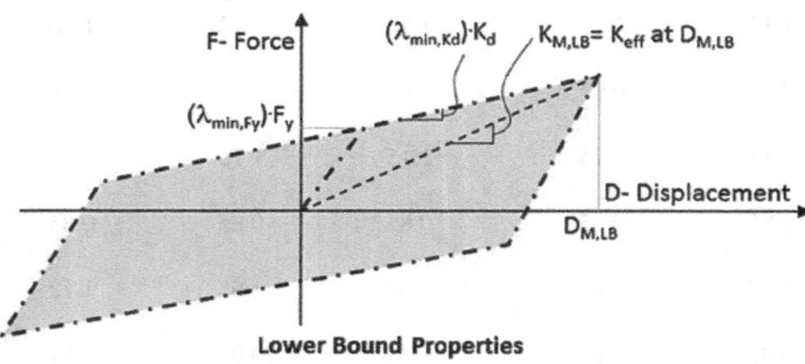

Figure C17.2-3. Example of the upper and lower bound properties of a bilinear force deflection system.

be *liquid* lubrication, typically applied either directly at the interface or within dimples. Solid lubrication in the form of graphite, or similar materials that are integrated in the fabric of liners and used in contact with stainless steel for the sliding interface, does not have the problems experienced by liquid lubrication.

C17.2.8.5 Upper Bound and Lower Bound Force-Deflection Behavior of Isolation System Components An upper and lower bound representation of each type of isolation system component shall be developed using the lambda factors developed in Section 17.2.8.4. An example of a bilinear force deflection loop is shown in Figure C17.2-2. In Figure C17.2-3, the upper and lower bound lambda factors are applied to the nominal properties of the yield/friction level and the second or bilinear slope of the lateral force-displacement curve to determine the upper and lower bound representation of an isolation system component. The nomenclature shown in Figure C17.2-3 is important to note. The effective stiffness and effective damping are calculated for both the upper and lower bound properties at the corresponding D_M. The maximum and minimum effective stiffness and effective damping are then developed from these upper and lower bound lateral force-displacement relationships in Section 17.2.8.6.

C17.3 SEISMIC HAZARD

C17.3.1 Spectral Response Acceleration Parameters and Response Spectrum The MCE_R spectral response acceleration parameters for short periods (S_{MS}) and at 1 s (S_{M1}) are used in the equivalent lateral force procedure either directly for design or to establish minimum lateral displacements and forces for dynamic procedures. The MCE_R spectrum is used for response spectrum analysis.

C17.3.2 Ground Motions for Response History Analysis Development of target spectra and ground motions for seismically isolated structures designed using response history analysis generally follows the requirements for the design of fixed-base structures using response history analysis. However, the period range of interest for seismically isolated versus fixed-base structures differs. The lower bound period range of interest defined in these provisions, while suitable for estimating isolation system displacement and building base shear, may not be sufficiently low to capture higher modes of response, which have a strong effect on floor spectra. Where calculation of floor spectra from the nonlinear response history analysis is an objective (e.g., for determining nonstructural component demands in Chapter 13), it may be necessary to reduce the lower bound period range of interest to include these higher modes. The other sections of Chapter 16 that are not specifically referenced do not apply to seismically isolated structures.

C17.4 ANALYSIS PROCEDURE SELECTION

Three different analysis procedures are available for determining design-level seismic loads: the equivalent lateral force procedure, the response spectrum procedure, and the response history procedure. For the ELF procedure, simple equations computing the lateral force demand at each level of the building structure (similar to those for conventional, fixed-base structures) are used to determine peak lateral displacement and design forces as a function of spectral acceleration and isolated-structure period and damping. The provisions of this section permit the increased use of the ELF procedure, recognizing that the ELF procedure is adequate for isolated structures whose response is dominated by a single translational mode of vibration and whose superstructure is designed to remain essentially elastic (limited ductility demand and inelastic deformations), even for MCE_R level ground motions. The ELF procedure is now permitted for the design of isolated structures at all sites (except Site Class F), as long as the superstructure is regular (as defined in new Section 17.2.2), has a fixed-base period (T) that is well separated from the isolated period (T_{min}), and the isolation system meets certain "response predictability" criteria with which typical and commonly used isolation systems comply.

The design requirements for the structural system are based on the forces and drifts obtained from the MCE_R earthquake, using a consistent set of upper and lower bound isolation system properties, as discussed in Section C17.5. The isolation system—including all connections, supporting structural elements, and the "gap"—is required to be designed (and tested) for 100% of MCE_R demand. Structural elements above the isolation system are now designed to remain essentially elastic for the MCE_R earthquake. A similar fixed-base structure would be designed for design earthquake loads ($2/3MCE_R$) reduced by a factor of 6 to 8 rather than the MCE_R demand reduced by a factor of up to 2 for a base-isolated structure.

C17.5 EQUIVALENT LATERAL FORCE PROCEDURE

The lateral displacements given in this section approximate peak earthquake displacements of a single degree of freedom, linear-elastic system of period, T, and effective damping, β. Equations (17.5-1) and (17.5-3) of ASCE 7-10 provided the peak displacement in the isolation system at the center of mass for both the DE and MCE_R earthquakes, respectively. In these prior equations, as well as the current equation, the spectral acceleration terms at the isolated period are based on the premise that the longer period portion of the response spectra decayed as $1/T$. This is a conservative assumption and is the same as that required for design of a conventional, fixed-base structure of period T_M. A damping factor, B, is used to decrease (or increase) the computed displacement demand where the effective damping coefficient of the isolation system is greater (or smaller) than 5% of critical damping. A comparison of values obtained from Equation (17.5-1) and those obtained from nonlinear time history analyses are given in Kircher et al. (1988) and Constantinou et al. (1993).

The ELF formulas in this new edition compute minimum lateral displacements and forces required for isolation system design based only on MCE_R level demands, rather than on a combination of design earthquake and MCE_R levels, as in earlier editions of the provisions.

The calculations are performed separately for upper bound and lower bound isolation system properties, and the governing case shall be considered for design. Upper bound properties typically, but not always, result in a lower maximum displacement (D_M), higher damping (β_M), and higher lateral forces (V_b, V_{st}, V_s, and k).

Section 17.2.8 relates bounding values of effective period, stiffness, and damping of the isolation system to upper bound and lower bound lateral force-displacement behavior of the isolators.

C17.5.3 Minimum Lateral Displacements Required for Design

C17.5.3.1 Maximum Displacement The provisions of this section reflect the MCE_R-only basis for design and define maximum MCE_R displacement in terms of MCE_R response spectral acceleration, S_{M1}, at the appropriate T.

In addition, and of equal significance, the maximum displacement (D_M) and the damping modification factor (B_M) are determined separately for upper bound and lower bound isolation system properties. In previous provisions, the maximum displacement (D_M) was defined only in terms of the damping associated with lower bound displacement, and this damping was combined with the upper bound stiffness to determine the design forces. This change is theoretically more correct, but it removes a significant conservatism in the ELF design of the superstructure. This reduction in superstructure design conservatism is offset by the change from design earthquake to MCE_R ground motions as the basis for superstructure design forces.

C17.5.3.2 Effective Period at the Maximum Displacement The provisions of this section are revised to reflect the MCE_R-only basis for design and associated changes in terminology (although maintaining the concept of effective period). The effective period T_M is also determined separately for the upper and lower bound isolation properties.

C17.5.3.3 Total Maximum Displacement The provisions of this section are revised to reflect the MCE_R-only basis for design and associated changes in terminology. In addition, the formula for calculating total (translational and torsional) maximum MCE_R displacement has been revised to include a term and corresponding equations that reward isolation systems configured to resist torsion.

The isolation system for a seismically isolated structure should be configured to minimize eccentricity between the center of mass of the superstructure and the center of rigidity of the isolation system, thus reducing the effects of torsion on the displacement of isolation elements. For conventional structures,

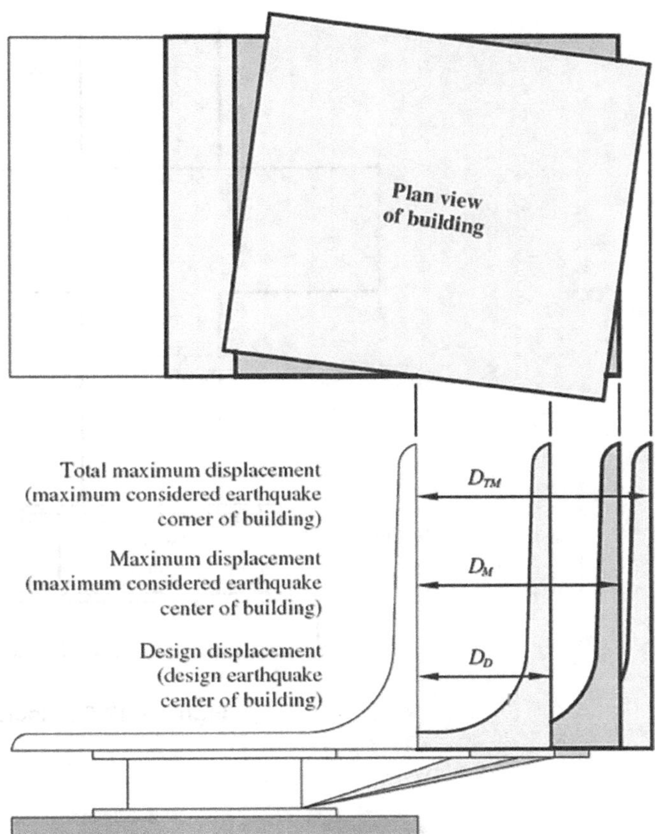

Figure C17.5-1. Displacement terminology.

an allowance must be made for accidental eccentricity in both horizontal directions. Figure C17.5-1 illustrates the terminology used in the standard. Equation (17.5-3) provides a simplified formula for estimating the response caused by torsion in lieu of a more refined analysis. The additional component of displacement caused by torsion increases the design displacement at the corner of a structure by about 15% (for one perfectly square in plan) to about 30% (for one long and rectangular in plan) if the eccentricity is 5% of the maximum plan dimension. These calculated torsional displacements correspond to structures with an isolation system whose stiffness is uniformly distributed in plan. Isolation systems that have stiffness concentrated toward the perimeter of the structure, or certain sliding systems that minimize the effects of mass eccentricity, result in smaller torsional displacements. The standard permits values of D_{TM} as small as $1.15D_M$, with proper justification.

C17.5.4 Minimum Lateral Forces Required for Design

Figure C17.5-2 illustrates the terminology for elements at, below, and above the isolation system. Equation (17.5-5) specifies the peak elastic seismic shear for the design of all structural elements at or below the isolation system (without reduction for ductile response). Equation (17.5-7) specifies the peak elastic seismic shear for the design of structural elements above the isolation system. For structures that have appreciable inelastic-deformation capability, this equation includes an effective reduction factor ($R_I = 3\,R/8$, not exceeding 2). This factor ensures essentially elastic behavior of the superstructure above the isolators.

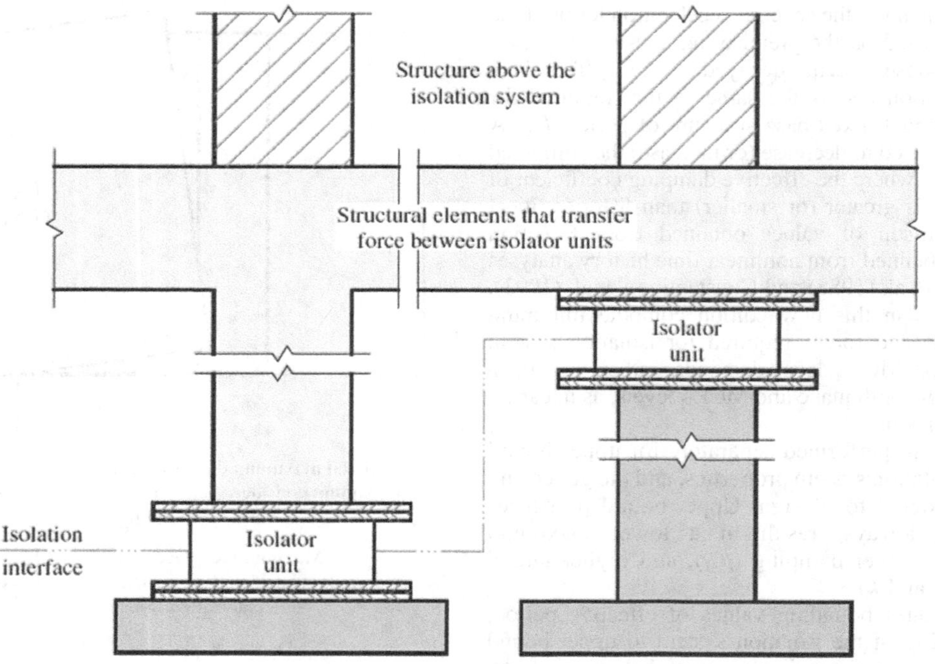

Figure C17.5-2. Isolation system terminology.

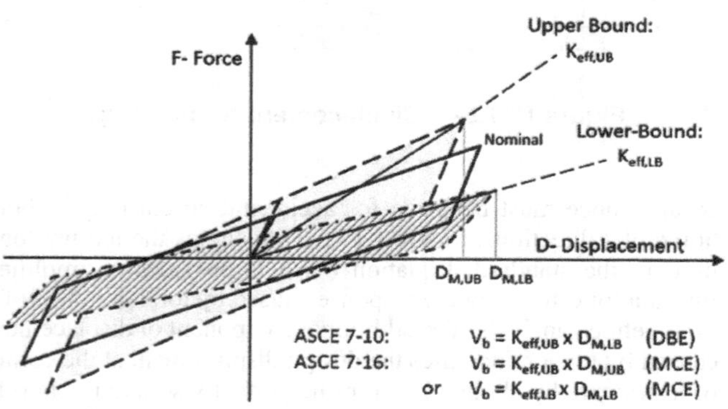

Figure C17.5-3. Nominal, upper bound, and lower bound bilinear hysteretic properties of typical isolator bearing.

These provisions include two significant philosophic changes in the method of calculating the elastic base shear for the structure. In ASCE 7-10 and earlier versions of the provisions, the elastic design base shear forces were determined from the design earthquake (DE), using a mixture of the upper bound effective stiffness and the maximum displacement obtained using the lower bound properties of the isolation system, as shown schematically in Figure C17.5-3. This was known to be conservative. The elastic design base shear is now calculated from the MCE_R event with a consistent set of upper and lower bound stiffness properties, as shown in Equation (17.5-5) and Figure C17.5-3.

A comparison of the old elastic design base shears for a range of isolation system design parameters and lambda factors using the ASCE 7-10 provisions, and those using these new provisions, is shown in Table C17.5-1. This comparison assumes that the DE is $2/3$ the MCE_R and the longer period portion of both spectra decay as S_1/T. Table C17.5-1 shows a comparison between elastic design base shear calculated using the ASCE 7-10 and 7-16 editions for a range of yield levels, second slopes, and bounding property multipliers.

The dark gray cells in Table C17.5-1 indicate that the new elastic design base shears are more than 10% higher than the old provisions; the light gray cells indicate that the new elastic base shears are 0 to 10% higher than the old provisions; and the Empty cells indicate that the new elastic base shears are less than the old provisions.

C17.5.4.1 Isolation System and Structural Elements below the Base Level The provisions of this section are revised to reflect the MCE_R-only basis for design and associated changes in terminology. A new paragraph was added to this section to clarify that unreduced lateral loads should be used to determine overturning forces on the isolation system.

C17.5.4.2 Structural Elements above the Base Level The provisions of this section are revised to reflect the MCE_R-only basis for design and associated changes in terminology, including the new concept of the "base level" as the first floor immediately above the isolation system.

An exception has been added to allow values of R_I to exceed the current limit of 2.0, provided that the pushover strength of the superstructure at the MCE_R drift or $0.015h_{sx}$ story drift exceeds (by 10%) the maximum MCE_R force at the isolation interface (V_b). This exception directly addresses required strength and associated limits on inelastic displacement for MCE_R demands. The pushover method is addressed in ASCE 41 (2007).

A new formula [Equation (17.5-7)] now defines lateral force on elements above the base level in terms of reduced seismic weight (seismic weight excluding the base level), and the effective damping of the isolation system (York and Ryan 2008). In this formulation, it is assumed that the base level is located

Table C17.5-1. Comparison of Elastic Design Base Shears between ASCE 7-10 and 7-16.

		Upper Bound Multipliers		K_d	Yield Level		Lower Bound Multipliers		K_d	Yield Level
MCE$_R$ S_1 = 1.5				**1.15**	**1.6**				**0.85**	**0.85**
$T2$ (s)	2.00	2.00	3.00	3.00	4.00	4.00	5.00	5	6	6
Yield Level	0.05	0.10	0.05	0.10	0.05	0.10	0.05	0.1	0.05	0.1
New, V_b/W	0.80	0.66	0.47	0.42	0.33	0.33	0.26	0.28	0.21	0.26
ASCE 7-16/ASCE 7-10	1.14	1.02	1.08	0.91	1.02	0.84	0.96	0.83	0.91	0.82
		1.0	**1.6**		**1.0**	**0.85**				
New, V_b/W	0.77	0.71	0.52	0.42	0.35	0.31	0.26	0.27	0.21	0.25
ASCE 7-16/ASCE 7-10	1.32	1.25	1.39	1.01	1.25	0.88	1.24	1.02	1.16	1.12
MCE$_R$ S_1 = 1.0		**1.15**	**1.6**		**0.85**	**0.85**				
$T2$ (s)	2.00	2.00	3.00	3.00	4.00	4.00	5.00	5	6	6
Yield Level	0.05	0.10	0.05	0.10	0.05	0.10	0.05	0.1	0.05	0.1
New, V_b/W	0.47	0.43	0.29	0.30	0.21	0.23	0.17	0.23	0.15	0.21
ASCE 7-16/ASCE 7-10	1.08	0.91	0.99	0.83	0.91	0.65	0.84	0.76	0.84	0.71
		1.35	**1.5**		**0.85**	**0.85**				
New, V_b/W	0.54	0.47	0.33	0.32	0.24	0.29	0.19	0.22	0.16	0.20
ASCE 7-16/ASCE 7-10	1.12	0.99	1.05	0.90	0.99	0.92	0.94	0.82	0.90	0.81
		1.3	**1.3**		**0.85**	**0.85**				
New, V_b/W	0.55	0.47	0.33	0.31	0.24	0.24	0.18	0.20	0.15	0.18
ASCE 7-16/ASCE 7-10	1.22	1.10	1.16	1.01	1.10	0.94	1.05	0.91	1.01	0.89

Notes: Dark gray cells indicate that the new elastic design base shears are more than 10% higher than the old provisions; light gray cells indicate 0% to 10% higher than old provisions.

immediately [within 3.0 ft (0.9 m) of top of isolator] above the isolation interface. When the base level is not located immediately above the isolation interface (e.g., there is no floor slab just above the isolators), the full (unreduced) seismic weight of the structure above the isolation interface is used in Equation (17.5-7) to conservatively define lateral forces on elements above the base level.

C17.5.4.3 Limits on V_s The provisions of this section are revised to reflect the MCE$_R$-only basis for design and associated changes in terminology.

In Section 17.5.4.3, the limits given on V_s are revised to clarify that the force required to fully activate the isolation system should be based on either the upper bound force-deflection properties of the isolation system or 1.5 times nominal properties, whichever is greater. Other limits include (1) the yield/friction level to fully activate the isolation system, and (2) the ultimate capacity of a sacrificial wind-restraint system that is intended to fail and release the superstructure during significant lateral load.

These limits are needed so that the superstructure does not yield prematurely before the isolation system has been activated and significantly displaced.

C17.5.5 Vertical Distribution of Force The provisions of this section are revised to incorporate a more accurate distribution of shear over height considering the period of the superstructure and the effective damping of the isolation system. The specified method for vertical distribution of forces calculates the force at the base level immediately above the base isolation plane, then distributes the remainder of the base shear among the levels above (i.e., the mass of the "base slab" above the isolators is not included in the vertical distribution of forces).

The proposed revision to the vertical force distribution is based on analytical studies (York and Ryan 2008 in collaboration with the Structural Engineers Association of Northern California's Protective Systems Subcommittee PSSC). Linear theory of base isolation predicts that base shear is uniformly distributed over the height of the building, while the equivalent lateral force procedure of ASCE 7-10 prescribes a distribution of lateral forces that increase linearly with increasing height. The uniform distribution is consistent with the first mode shape of an isolated building, and the linear distribution is consistent with the first mode shape of a fixed-base building. However, a linear distribution may be overly conservative for an isolated building structure, especially for one- or two-story buildings with heavy base mass relative to the roof.

The principle established in the York and Ryan (2008) study was to develop two independent equations: one to predict the superstructure base shear V_{st} relative to the base shear across the isolators V_b, and a second to distribute V_{st} over the height of the building. Considering a reduction in V_{st} relative to V_b allowed for the often significant inertial forces at the base level, which can be amplified because of disproportionate mass at the base level, to be accounted for in design. The study also assumed that the superstructure base shear was distributed over the height using a k distribution (i.e., lateral force $\propto w_x h_x^k$ where w_x is the weight and h_x the height to level x), where $k = 0$ is a uniform distribution and $k = 1$ is a linear distribution. In the study, representative base-isolated multistory single-bay frame models were developed, and response history analysis was performed with a suite of 20 motions scaled to a target spectrum corresponding to the effective isolation system parameters. Regression analysis was performed to develop a best fit (relative to median results from response history analysis) of the superstructure to base shear ratio and k factor as a function of system parameters. The equations recommended in York and Ryan (2008) provided the best "goodness of fit" among several considered, with R^2 values exceeding 0.95. Note that Equations (17.5-8) and (17.5-11) in the code change are the same as Equations (15) and (17) in York and Ryan (2008), with one modification: the coefficient for k in Equation (17.5-11) has been modified to reflect that the reference plane for determining height should be taken as the plane of isolation, which is below the isolated base slab.

Table C17.5-2. Comparison of "Strongly Bilinear" and "Weakly Bilinear" Isolation Systems.

System Type and Equation Term[a]	Pre- to Postyield Transition Characteristics	Cyclic Behavior Below Bilinear Yield/Slip Deformation	Example of Hysteresis Loop Shape	Example Systems[b]
"Strongly bilinear" $(1–3.5\beta_M)$	Abrupt transition from preyield or preslip to postyield or postslip	Essentially linear elastic, with little energy dissipation	Figure C17.5-4(a)	Flat sliding isolators with rigid backing Single-concave FPS Double-concave FPS with same friction coefficients top and bottom
"Weakly bilinear" $(1–2.5\beta_M)$	Smooth or multistage transition from preyield or preslip to postyield or postslip	Exhibits energy dissipation caused by yielding or initial low-level friction stage slip	Figure C17.5-4(b)	Elastomeric and viscous dampers Triple-concave FPS High-damping rubber Lead-rubber Elastomeric-backed sliders

[a] Equation term refers to the exponent in Equation (17.5-11).
[b] FPS = friction pendulum system.

It is difficult to confirm in advance whether the upper bound or lower bound isolation system response will govern the design of the isolation system and structure. It is possible, and even likely, that the distribution corresponding to upper bound isolation system properties will govern the design of one portion of the structure, and the lower bound distribution will govern another. For example, lower bound isolation system response may produce a higher displacement, D_M, a lower damping, β_M, but also a higher base shear, V_b. This difference could result in a vertical force distribution that governs for the lower stories of the building. The corresponding upper bound case, with lower displacement, D_M, but higher damping, β_M, might govern design of the upper part of the structure, even though the base shear, V_b, is lower.

The proposal to adopt the approach in York and Ryan (2008) is part of an overall revamp that will permit the equivalent static force method to be extended to a wider class of buildings. In York and Ryan (2008), the current method was shown to be quite conservative for systems with low to medium levels of damping combined with stiff superstructures but unconservative for highly damped systems or systems with relatively flexible superstructures.

The proposal has undergone a high level of scrutiny by the code committee. First, regression analysis was performed using the original York and Ryan (2008) response history data set to fit several alternative distributions suggested by code committee members that were intuitively more appealing. In all cases, the equations recommended in York and Ryan (2008) were shown to best fit the data. Second, a few code committee members appropriately attempted to validate the equations using independently generated response history analysis data sets. Much discussion ensued following the discovery that the equations were unconservative for a class of one- and two-story buildings with long isolation periods and high levels of effective damping in the isolation system. This was most noticeable for one- and two-story buildings (i.e., with relatively low W_{st}/W ratios), predominantly single-mode fixed-base response, and where T_{fb} aligned with the period based on the initial stiffness of the isolation system, T_{k1}. The York and Ryan (2008) data set was confirmed to contain cases similar to those generated independently, and the unconservatism was rationalized as a natural outcome of the regression approach. In an attempt to remove the unconservatism, equations were fit to the 84th percentile (median $+1\sigma$) vertical force distributions based on the original York and Ryan (2008) data set. However, the resulting distributions were unacceptably conservative and thus rejected.

The York and Ryan (2008) data set was subsequently expanded to broaden the range of fixed-base periods for low-rise structures and to provide additional confirmation of the independent data set. In addition, isolation system hysteresis loop shape was identified as the most significant factor in the degree of higher mode participation, resulting in increased V_{st}/V_b ratio and k factor. The provisions now identify this variable as needing a more conservative k factor.

When computing the vertical force distribution using the equivalent linear force procedure, the provisions now divide isolation systems into two broad categories according to the shape of the hysteresis loop. Systems that have an abrupt transition between preyield and postyield response (or preslip and postslip for friction systems) are described as "strongly bilinear" and have been found to typically have higher superstructure accelerations and forces. Systems with a gradual or multistage transition between preyield and postyield response are described as "weakly bilinear" and were observed to have relatively lower superstructure accelerations and forces, at least for systems that fall within the historically adopted range of system strength/friction values (nominal isolation system force at zero displacement, $F_o = 0.03 \times W$ to $0.07 \times W$).

This limitation is acceptable because isolation systems with strength levels that fall significantly outside the upper end of this range are likely to have upper bound properties that do not meet the limitations of Section 17.4.1, unless the postyield stiffness or hazard level is high. Care should also be taken when using the equations to assess the performance of isolation systems at lower hazard levels because the equivalent damping can increase beyond the range of applicability of the original work.

An additional description of the two hysteresis loop types is provided in Table C17.5-2. An example of a theoretical loop for each system type is shown in Figure C17.5-4.

Capturing this acceleration and force increase in the equivalent linear force procedure requires an increase in the V_{st}/V_b ratio [Equation (17.5-7)] and the vertical force distribution k factor [Equation (17.5-11)]. Consequently, the provisions require a different exponent to be used in Equation (17.5-7) for a system that exhibits "strongly bilinear" behavior. Similar differences were observed in the k factor [Equation (17.5-11)], but these findings were judged to be insufficiently well developed to include in the provisions at this time, and the more conservative value for "strongly bilinear" systems was adopted for both system types.

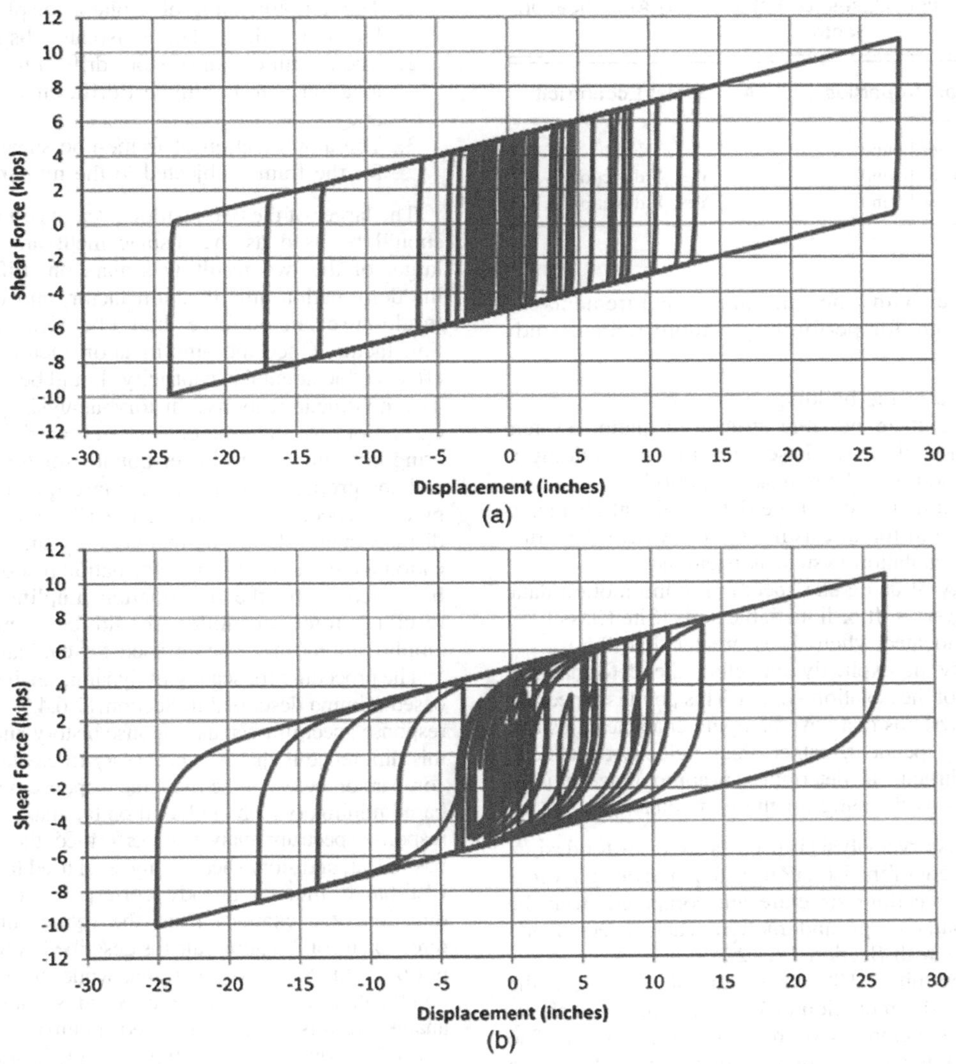

Figure C17.5-4. Example isolation system example loops.

The exception in Section 17.5.5 is a tool to address the issue identified in the one- and two-story buildings on a project-specific basis and to simplify the design of seismically isolated structures by eliminating the need to perform time-consuming and complex response history analysis of complete 3D building models each time the design is changed. At the beginning of the project, a response history analysis of a simplified building model (e.g., a stick model on isolators) is used to establish a custom inertia force distribution for the project. The analysis of the 3D building model can then be accomplished using simple static analysis techniques.

The limitations on use of the equivalent linear force procedure (Section 17.4.1) and on the response spectrum analysis procedure (Section 17.4.2.1) provide some additional limits. Item 7a in Section 17.4.1 requires a minimum restoring force, which effectively limits postyield stiffness to $K_d > F_o/D_M$ and also limits effective damping to 32% for a bilinear system.

Items 2 and 3 in Section 17.4.1 limit the effective period, $T_M \leq 4.5s$ and effective damping, $\beta_M \leq 30\%$ explicitly.

C17.5.6 Drift Limits Drift limits are divided by C_d/R for fixed-base structures since displacements calculated for lateral loads reduced by R are multiplied by C_d before checking drift. The C_d term is used throughout the standard for fixed-base structures to approximate the ratio of actual earthquake response to response calculated for reduced forces. Generally, C_d is $1/2$ to $4/5$ the value of R. For isolated structures, the R_I factor is used both to reduce lateral loads and to increase displacements (calculated for reduced lateral loads) before checking drift. Equivalency would be obtained if the drift limits for both fixed-base and isolated structures were based on their respective R factors. It may be noted that the drift limits for isolated structures are generally more conservative than those for conventional, fixed-base structures, even where fixed-base structures are assigned to Risk Category IV. The maximum story drift permitted for design of isolated structures is constant for all risk categories.

C17.6 DYNAMIC ANALYSIS PROCEDURES

This section specifies the requirements and limits for dynamic procedures.

A more detailed or refined study can be performed in accordance with the analysis procedures described in this

Table C17.6-1. Analysis Cases for Establishing Amplification Factors.

Case	Isolator Properties	Accidental Eccentricity
I	Lower bound	No
IIa	Lower bound	Yes, X direction
IIb	Lower bound	Yes, Y direction

section, compatible with the minimum requirements of Section 17.5. Reasons for performing a more refined study include

1. The importance of the building.
2. The need to analyze possible structure-isolation system interaction where the fixed-base period of the building is greater than one-third of the isolated period.
3. The need to explicitly model the deformational characteristics of the lateral force-resisting system where the structure above the isolation system is irregular.
4. The desirability of using site-specific ground motion data, especially for very soft or liquefiable soils (Site Class F) or for structures located where S_1 is greater than 0.60.
5. The desirability of explicitly modeling the deformational characteristics of the isolation system. This point is especially important for systems that have damping characteristics that are amplitude-dependent, rather than velocity-dependent, because it is difficult to determine an appropriate value of equivalent viscous damping for these systems.

Where response history analysis is used as the basis for design, the design displacement of the isolation system and design forces in elements of the preceding structure are computed from the average of seven pairs of ground motion, each selected and scaled in accordance with Section 17.3.2.

The provisions permit a 10% reduction of V_b below the isolation system and 20% reduction of V_b for the structure above the isolators if the structure is of regular configuration. The displacement reduction should not be greater than 20% if a dynamic analysis is performed.

In order to avoid the need to perform a large number of nonlinear response history analyses that include the suites of ground motions, the upper and lower bound isolator properties, and five or more locations of the center of mass, this provision allows the center-of-mass analysis results to be scaled and used to account for the effects of mass eccentricity in different building quadrants.

The following is a recommended method of developing appropriate amplification factors for deformations and forces for use with center-of-mass nonlinear response history analyses (NRHAs), which account for the effects of accidental torsion. The use of other rationally developed amplification factors is permitted.

The most critical directions for shifting the calculated center of mass are such that the accidental eccentricity adds to the inherent eccentricity in each orthogonal direction at each level. For each of these two eccentric mass positions, and with lower bound isolator properties, the suite of NRHAs should be run and the results processed in accordance with Section 17.6.3.4. The analysis cases are defined in Table C17.6-1.

The results from Cases IIa and IIb are then compared in turn to those from Case I. The following amplification factors (ratio of Case IIa or IIb response to Case I response) are computed:

1. The amplification of isolator displacement at the plan location with the largest isolator displacement;
2. The amplification of story drift in the structure at the plan location with the highest drift, enveloped over all stories; and
3. The amplification of frame-line shear forces at each story for the frame subjected to the maximum drift.

The larger of the two resulting scalars on isolator displacement should be used as the displacement amplification factor, the larger of the two resulting scalars on drift should be used as the deformation amplification factor, and the larger of the two resulting scalars on force should be used as the force amplification factor. Once the amplification factors are established, the effects of accidental eccentricity should be considered as follows: The nonlinear response history analysis procedure should be carried out for the inherent mass eccentricity case only, considering both upper and lower bound isolator properties. For each isolator property variation, response quantities should be computed in accordance with Section 17.6.3.4. All resulting isolator displacements should be increased by the displacement amplification factor, all resulting deformation response quantities should be increased by the deformation amplification factor, and all resulting force quantities should be increased by the force amplification before being used for evaluation or design.

The procedure for scaling of dynamic analysis results to the ELF-based minima described in Section 17.6.4.3 is slightly different for response spectrum versus response history analysis. The reason for this difference is that it is necessary to create a consistent basis of comparison between the dynamic response quantities and the ELF-based minima (which are based on the maximum direction). When response spectrum analysis is performed, the isolator displacement, base shear, and story shear at any level used for comparison with the ELF-based minima already correspond to a single, maximum direction of excitation. Thus, the vector sum of the 100%/30% directional combination rule (as described in Section 17.6.3.3) need not be used. Note, however, that while the 100%/30% directional combination rule is not required in scaling response spectrum analysis results to the ELF-based minima of Section 17.6.4.3, the 100%/30% directional combination rule is still required for design of the superstructure by response spectrum analysis, per Section 17.6.3.3. When nonlinear response history analysis is performed, the isolator displacement and base shear for each ground motion is calculated as the maximum of the vector sum of the two orthogonal components (of displacement or base shear) at each time step. The average of the maxima over all ground motions of these displacement and base shear vector-sum values is then used for comparison with the ELF-based minimum displacement and base shear, per Section 17.6.4.3.

C17.6.2 Modeling Capturing the vertical response of a building structure with a high degree of confidence may be a challenging task. Nonetheless, when the effects of vertical shaking are to be included in the analysis and/or design process of an isolated building structure, the following modeling recommendations are provided:

1. Vertical mass: All beams, columns, shear walls, and slabs should be included in the model, and the vertical mass should be distributed appropriately across the footprint of each floor.
2. Foundation properties: A range of soil properties and foundation damping should be considered in the analysis procedure since horizontal and vertical ground motion excitation can significantly affect building response.

3. Soil–foundation–structure interaction effects: Foundation damping, embedment, and base slab averaging may alter the vertical motions imparted on the structure compared to the free-field motions.
4. Degrees of freedom: Additional degrees of freedom (e.g., nodes along the span of a beam or slab) will need to be added to the model to capture vertical effects.
5. Reduced time step: Because vertical ground motion excitation and building response often occur at higher frequencies than lateral excitation and response, a finer analysis time step might be required when vertical motions are included.

C17.7 DESIGN REVIEW

The provisions allow for a single peer reviewer to evaluate the isolation system design. The reviewer should be a registered design professional (RDP), and if the Engineer of Record (EOR) is required to be a structural engineer (SE), the owner may consider ensuring that there is one SE on the peer review team. On more significant structures, it is likely that the design review panel may include two or three individuals, but for many isolated structures, a single, well-qualified peer reviewer is sufficient. If a manufacturer with unknown experience in the United States is selected as the supplier, the building owner may require the design reviewer to attend prototype tests.

The standard requires peer review to be performed by RDPs who are independent of the design team and other project contractors. The reviewer or review panel should include individuals with special expertise in one or more aspects of the design, analysis, and implementation of seismic isolation systems.

The peer reviewer or review panel should be identified before the development of design criteria (including site-specific ground-shaking criteria) and isolation system design options. Furthermore, the review panel should have full access to all pertinent information and the cooperation of the general design team and regulatory agencies involved in the project.

C17.8 TESTING

The design displacements and forces determined using the standard assume that the deformational characteristics of the isolation system have been defined previously by comprehensive testing. If comprehensive test data are not available for a system, major design alterations in the structure may be necessary after the tests are complete. This change would result from variations in the isolation system properties assumed for design and those obtained by test. Therefore, it is advisable that prototype tests of systems be conducted during the early phases of design if sufficient prototype test data are not available from a given manufacturer.

The design displacements and forces determined using the standard are based on the assumption that the deformational characteristics of the isolation system have been defined previously by comprehensive qualification and prototype testing. Variations in isolator properties are addressed by the use of property variation factors that account for expected variation in isolator and isolation system properties from the assumed nominal values. In practice, past prototype test data are very likely to have been used to develop the estimated nominal values and associated lambda factors used in the design process, as described in Section 17.2.8.4.

When prototype testing is performed in accordance with Section 17.8.2, it serves to validate and check the assumed nominal properties and property variation factors used in the design. Where project-specific prototype testing is not performed, it is possible to perform a subset of the checks described subsequently on the isolator unit and isolation system test properties using data from the quality control test program, described in Section 17.8.5.

C17.8.2.2 Sequence and Cycles Section 17.2.8.4 describes the method by which minimum and maximum isolator properties for design and analysis are established using property variation or lambda (λ) factors to account for effects such as specification tolerance, cyclic degradation, and aging. The structural analysis is therefore performed twice, and the resulting demands are enveloped for design. For force-based design parameters and procedures, this requirement is relatively straightforward, as typically one case the upper bound controls. However, for components dependent on both force and deformation (e.g., the isolators), two sets of axial load and displacement values exist for each required test. Lower bound properties typically result in larger displacements and smaller axial loads, whereas upper bound properties typically result in smaller displacements and larger axial loads. To avoid requiring that a complete set of duplicate tests be performed for the lower and upper bound conditions, Section 17.8.2.2 requires the results to be enveloped, combining the larger axial demands from one case with the larger displacements from the other. Strictly, these demands and displacement do not occur simultaneously, but the enveloping process is conservative.

The enveloping process typically results in test axial loads that correspond to the maximum properties and displacements that correspond to minimum properties. Hence, the test results determined using the enveloped demands may not directly relate to the design properties or analysis results determined for maximum and minimum properties separately. However, because the test demands envelop the performance range for the project, the registered design professional is able to use them to determine appropriate properties for both linear and nonlinear analysis using the same philosophy as provided here.

Two alternate testing protocols are included in Section 17.8.2.2. The traditional three-cycle tests are preserved in Item 2(a) for consistency with past provisions. These tests can be performed dynamically but have often been performed at slow speed consistent with the capability of manufacturers' testing equipment. The alternate test sequence provided in Item 2(b) is more suited to full-scale dynamic cyclic testing.

The Item 3, test displacement has been changed from D_D to D_M, reflecting the focus of the provisions on only the MCE_R event. Because this test is common to both test sequences 2(a) and 2(b), it becomes important for property determination. This is the only test required to be repeated at different axial loads when isolators are also axial load-carrying elements, which is typically the case. This change was made to counter the criticism that the total test sequence of past provisions represented the equivalent energy input of many MCE_R events back to back and that prototype test programs could not be completed in a reasonable time if any provision for isolator cooling and recovery was included.

The current test program is therefore more reflective of code-minimum required testing. The RDP and/or the isolator manufacturer may wish to perform additional testing to more accurately characterize the isolator for a wider range of axial loads and displacements than is provided here. For example, this might include performing the Item 2(b) dynamic test at additional axial loads once the code-required sequence is complete.

Heat effects for some systems may become significant, and misleading, if insufficient cooling time is included between tests. As a consequence, in Test Sequence 4, only five cycles of continuous dynamic testing are required as this is a limit of most test equipment. The first-cycle or scragging effects observed in some isolators may recover with time, so back-to-back testing may result in an underestimation of these effects. Refer to Constantinou et al. (2007) and Kalpakidis and Constantinou (2008) for additional information. The impact of this behavior may be mitigated by basing cyclic lambda factors on tests performed relatively early in the sequence before these effects become significant.

C17.8.2.3 Dynamic Testing Section 17.8.2.3 clarifies when dynamic testing is required. Many isolator exhibit velocity dependence. However, this testing can be expensive and can only be performed by a limited number of test facilities. The intent is not that dynamic testing of isolators be performed for every project. Sufficient dynamic test data must be available to characterize the cyclic performance of the isolator, in particular, the change in isolator properties during the test (i.e., with respect to the test average value). Dynamic testing must therefore be used to establish the $\lambda_{(\text{test,min})}$ and $\lambda_{(\text{test,max})}$ values used in Section 17.2.8.4, because these values are typically underestimated from slow-speed test data. If project prototype or production testing is to be performed at slow speeds, this testing would also be used to establish factors that account for the effect of velocity and heating on the test average values of k_{eff}, k_d, and E_{loop}. These factors can either be thought of as a separate set of velocity-correction factors to be applied on test average values, or they can be incorporated into the $\lambda_{(\text{test,min})}$ and $\lambda_{(\text{test,max})}$ values themselves.

It may also be possible to modify the isolator mathematical model, for example, to capture some or all of the isolator velocity dependence, for example, the change in yield level of the lead core in a lead-rubber bearing (LRB).

If project-specific prototype testing is undertaken, it may be necessary to adjust the test sequence in recognition of the capacity limitations of the test equipment, and this notion is now explicitly recognized in Section 17.8.2.2. For example, tests that simultaneously combine maximum velocity and maximum displacement may exceed the capacity of the test equipment and may also not be reflective of earthquake shaking characteristics. A more detailed examination of analysis results may be required to determine the maximum expected velocity corresponding to the various test deformation levels and to establish appropriate values for tests. Refer to Constantinou et al. (2007) for additional information.

C17.8.2.4 Units Dependent on Bilateral Load All types of isolators have bilateral load dependence to some degree. The mathematical models used in the structural analysis may include some or all of the bilateral load characteristics for the particular isolator type under consideration. If not, it may be necessary to examine prototype test data to establish the impact on the isolator force-deformation response as a result of the expected bilateral loading demands. A bounding approach using lambda (λ) factors is one method of addressing bilateral load effects that cannot be readily incorporated in the isolator mathematical model.

Bilateral isolator testing is complex, and only a few test facilities are capable of performing these tests. Project-specific bilateral load testing has not typically been performed for isolation projects completed to date. In lieu of performing project-specific testing, less restrictive similarity requirements may be considered by the registered design professional compared to those required for test data submitted to satisfy similarity for Sections 17.8.2.2 and 17.8.2.5. Refer to Constantinou et al. (2007) for additional information.

C17.8.2.5 Maximum and Minimum Vertical Load The exception to Section 17.8.2.5 permits that the tests may be performed twice, once with demands resulting from upper bound properties and once with lower bound properties. This option may be preferable for these isolator tests performed at D_{TM} since the isolator will be closer to its ultimate capacity.

C17.8.2.7 Testing Similar Units Section 17.8.2.7 now provides specific limits related to the acceptability of data from testing of similar isolators. A wider range of acceptability is permitted for dynamic test data.

1. The submitted test data should demonstrate the manufacturers' ability to successfully produce isolators that are comparable in size to the project prototypes, for the relevant dimensional parameters, and to test them under force and displacement demands equal to or comparable to those required for the project.
2. It is preferred that the submitted test data necessary to satisfy the registered design professional and design review be for as few different isolator types and test programs as possible. Nonetheless, it may be necessary to consider data for Isolator A to satisfy one aspect of the required project prototype test program, and data from Isolator B for another.
3. For more complex types of testing, it may be necessary to accept a wider variation of isolator dimension or test demands than for tests that more fundamentally establish the isolator nominal operating characteristics [e.g., the testing required to characterize the isolator for loading rate dependence (Section 17.8.2.3) and bilateral load dependence (Section 17.8.2.4)].
4. The registered design professional is not expected to examine quality control procedures in detail to determine whether the proposed isolators were manufactured using sufficiently similar methods and materials. Rather, it is the responsibility of the manufacturer to document the specific differences, if any, preferably via traceable quality control documentation and to substantiate that any variations are not significant.
5. In some cases, the manufacturer may not wish to divulge proprietary information regarding methods of isolator fabrication, materials, or quality control procedures. These concerns may or may not be alleviated by confidentiality agreements or other means to limit the distribution and publication of sensitive material. Regardless, the final acceptability of the test information of similar units is at the sole discretion of the registered design professional and the design review, and not the manufacturer.
6. Similarity can be especially problematic in a competitive bid situation, when successful selection may hinge on the success of one supplier in eliminating the need to fabricate and test project-specific prototype isolators. This requirement can be addressed by determining acceptability of similarity data before the bid or by including more detailed similarity acceptance provisions in the bid documentation than have been provided herein.

Refer to Constantinou et al. (2007) and Shenton (1996) for additional information.

C17.8.3 Determination of Force-Deflection Characteristics The method of determining the isolator effective stiffness and effective damping ratio is specified in Equations (17.8-1) and (17.8-2). Explicit direction is provided for the establishment of effective stiffness and effective damping ratio for each cycle of test. A procedure is also provided for fitting a bilinear loop to a given test cycle, or to an average test loop to determine the postyield stiffness, k_d. This process can be performed in several different ways; however, the fitted bilinear loop should also match the effective stiffness and energy dissipated per cycle from the test. Once k_d is established, the other properties of the bilinear loop (e.g., f_y, f_o) all follow from the bilinear model.

Depending on the isolator type and the degree of sophistication of the isolator hysteresis loop adopted in the analysis, additional parameters may also be calculated, such as different friction coefficients, tangent stiffness values, or trilinear loop properties.

These parameters are used to develop a mathematical model of the isolator test hysteresis that replicates, as near as possible, the observed test response for a given test cycle. The model should result in a very close match to the effective stiffness and effective damping ratio and should result in a good visual fit to the hysteresis loop with respect to the additional parameters. The mathematic loop model must, at a minimum, match the effective stiffness and loop area from the test within the degree of variation adopted within the $\lambda_{(spec,min)}$ to $\lambda_{(spec,max)}$ range.

Data from the first cycle (or half cycle) of testing are not usually representative of full-cycle behavior and is typically discarded by manufacturers during data processing. An additional cycle (or half cycle) is added at the end to provide the required number of test cycles from which data can be extracted. However, the first cycle of a test is often important when establishing upper bound isolator properties and should be included when determining the $\lambda_{(test,min)}$ and $\lambda_{(test,max)}$ factors. The form of the test loop, however, is different to that of a full-scale loop, particularly for multistage isolator systems such as the double- or triple-concave friction pendulum system. This form may require different hysteresis parameters to be considered than the ones described by the bilinear model in Figure 17.8-1. The provisions permit the use of different methods for fitting the loop, such as a straight-line fit of k_d directly to the hysteresis curve extending to D_M and then determining k_1 to match E_{loop}, or an alternate is defining D_y and F_y by visual fit and then determining k_d to match E_{loop}.

The effective stiffness and effective damping ratio are required in linear static and linear response spectrum analysis. However, even if a nonlinear response history analysis is performed, these parameters are still required to check the required minimum lateral displacements and lateral forces of Sections 17.5.3 and 17.5.4, respectively.

C17.8.4 Test Specimen Adequacy For each isolator type, the effective stiffness and effective damping ratio for a given test axial load, test displacement, and cycle of test are determined in accordance with Section 17.8.3. For the dynamic test sequence in Item 2(a) of Section 17.8.2.2, there are two cycles at each increment of test displacement; for the traditional slow-speed sequence, there are three.

However, as part of a seismic isolation system, the axial load on a given isolator varies during a single complete cycle of loading. The required range of variation is assumed to be defined by the test load combinations required in Section 17.2.4.6, and the appropriate properties for analysis are assumed to be the average of the properties at the three axial loads. The test performed for Item 3 in Section 17.8.2.2 is critical to this evaluation since it is the three-cycle test performed at all three axial loads common to both the dynamic and slow-speed sequence.

In addition, since all isolators must undergo the same total horizontal cyclic loading as part of the same system, it is therefore assumed to be appropriate to assemble the total seismic isolation system properties using the following sequence:

1. Average the test results for a given isolator and cycle of loading across the three test axial loads. Also, compute corresponding test lambda factors for each isolator type.
2. Sum up the total isolation system properties for each cycle of loading according to the number of isolators of each type.
3. Determine the maximum and minimum values of total system effective stiffness over the required three cycles of testing and the corresponding values of the effective damping ratio. Also, compute the test lambda factors for the overall isolation system.

Two sets of test lambda factors emerge from this process, those applicable to individual isolators determined in 1 and those applicable to the overall isolation system properties determined in 3. In general, the test lambda factors for individual isolator tests are similar to those for each isolator type, which are similar to that for the overall isolation system. If this is the case, it may be more convenient to simplify the lambda factors assumed during design to reflect reasonable envelope values to be applied to all isolator types.

However, if the test lambda factors that emerge from project-specific prototype testing differ significantly from those assumed during design, it may be helpful to build up the previously described system properties, since the unexpectedly high test lambda factors for one isolator type may be offset by test lambda factors for another isolator type that were lower than the assumed values. In this circumstance, the prototype test results may be considered acceptable, provided that the torsional behavior of the system is not significantly affected and that the isolator connection and adjacent members can accommodate any resulting increase in local force demands.

Also, note that a subset of the isolation system properties can be determined from quality assessment and quality control (production) testing. This testing is typically performed at an axial load corresponding to the average $D + 0.5L$ axial load for the isolator type and to a displacement equal to $2/3(D_M)$. Keep in mind that isolator properties with target nominal three-cycle values estimated to match the average test value across three axial loads may not exactly match the values from production testing at the average dead load.

This result is most commonly observed with effective stiffness and effective damping ratio values for friction-based isolators because the average of the three test axial loads required in Section 17.8.2.2 does not exactly match that present in the isolator during the lateral analysis (the seismic weight, typically $1.0 \times$ dead load). In this case, some additional adjustment of properties may be required. Once the test effective stiffness and effective damping ratio of the isolation system have been established, these are compared to the values assumed for design in Section 17.2.8.4, defined by the nominal values and the values of $\lambda_{(test,max)}$ and $\lambda_{(test,min)}$.

In practice, instead of performing prototype tests for direct use in analysis, it may be simpler to use prototype test data or data from acceptable past testing of similar units (see Section 17.8.2.7) to establish isolator property dependence relationships for such things as axial load or velocity. If relationships are established for applicable hysteresis-loop parameters, such as yield force, friction ratio, initial stiffness, and postyield stiffness, these can be used to generate the required isolator unit and isolation system effective stiffness and effective damping ratios for the project over the required operating range.

C17.8.5 Production Tests All production isolation units have to be tested in combined compression and shear. Both quasi-static and dynamic tests are acceptable for all types of isolators. If a quasi-static test is used, it must have been performed as a part of the prototype tests. The RDP is responsible for defining in the project specifications the scope of the manufacturing quality control test program. The RDP decides on the acceptable range of variations in the measured properties of the production isolation units. All (100%) of the isolators of a given type and size are tested in combined compression and shear, and the allowable variation of the mean should be within the specified tolerance of Section 17.2.8.4 (typically ±10% or ±15%). Individual isolators may be permitted a wider variation (±15% or ±20%) from the nominal design properties. For example, the mean of the characteristic strength, Q, for all tested isolators might be permitted to vary no more than ±10% from the specified value of Q, but the characteristic strength for any individual isolation unit might be permitted to vary no more than ±15% from the specified value of Q. Another commonly specified allowable range of deviation from specified properties is ±15% for the mean value of all tested isolation units and ±20% for any single isolation unit.

The combined compression and shear testing of the isolators reveals the most relevant characteristics of the completed isolation unit and permits the RDP to verify that the production isolation units provide load-deflection behavior that is consistent with the structural design assumptions. Although vertical load-deflection tests have sometimes been specified in quality control testing programs, these test data are typically of little value. Consideration should be given to the overall cost and schedule effects of performing multiple types of quality control tests, and only those tests that are directly relevant to verifying the design properties of the isolation units should be specified.

Where project-specific prototype testing in accordance with Section 17.8.2 is not performed, the production test program should evaluate the performance of each isolator unit type for the property variation effects from Section 17.2.8.4.

REFERENCES

AASHTO (American Association of State Highway and Transportation Officials). 1990. *Guide specifications for seismic isolation design*. Washington, DC: AASHTO.

AASHTO. 1999. *Guide specifications for seismic isolation design*. Washington, DC: AASHTO.

AISC (American Institute of Steel Construction). 2022. *Seismic provisions for structural steel buildings*. AISC 341. Chicago: AISC.

ASCE. 2007. *Seismic rehabilitation of existing buildings*. ASCE 41-06. Reston, VA: ASCE.

ASTM International. 2012. *Standard specification for plain and steel-laminated elastomeric bearings for bridges*. ASTM D4014. West Conshohocken, PA: ASTM.

Buckle, I. G., S. Nagarajaiah, and K. Ferrel. 2002. "Stability of elastomeric isolation bearings: Experimental study." *J. Struct. Eng.* 128 (1): 3–11. https://doi.org/10.1061/(ASCE)0733-9445(2002)128:1(3).

Constantinou, M. C., I. Kalpakidis, A. Filiatrault, and R. A. Ecker Lay. 2011. *LRFD-based analysis and design procedures for bridge bearings and seismic isolators*. Rep. No. MCEER-11-0004. Buffalo, NY: Multidisciplinary Center for Earthquake Engineering Research (MCEER).

Constantinou, M. C., P. Tsopelas, A. Kasalanati, and E. D. Wolff. 1999. *Property modification factors for seismic isolation bearings*. MCEER-99-0012. Buffalo, NY: MCEER.

Constantinou, M. C., A. S. Whittaker, Y. Kalpakidis, D. M. Fenz, and G. P. Warn. 2007. *Performance of seismic isolation hardware under service and seismic loading*. MCEER-07-0012. Buffalo, NY: MCEER.

Constantinou, M. C., C. W. Winters, and D. Theodossiou. 1993. "Evaluation of SEAOC and UBC analysis procedures. Part 2: Flexible superstructure." In *Proc., Seminar on Seismic Isolation, Passive Energy Dissipation and Active Control, ATC Report 17-1*. Redwood City, CA: *Applied Technology Council*.

FEMA (Federal Emergency Management Agency). 1999. *HAZUS software*. Washington, DC: FEMA.

FEMA. 2003. *NEHRP recommended seismic provisions for new buildings and other structures*. Washington, DC: FEMA.

FEMA. 2012. *Seismic performance assessment of buildings*. P-58. Washington, DC: FEMA.

ICBO (International Council of Building Officials). 1991. *Uniform building code*. Whither, CA: ICBO.

Kalpakidis, I. V., and M. C. Constantinou. 2008. *Effects of heating and load history on the behavior of lead-rubber bearings*. MCEER-08-0027. Buffalo, NY: MCEER.

Kalpakidis, I. V., and M. C. Constantinou. 2009. "Effects of heating on the behavior of lead-rubber bearings. I: Theory." *J. Struct. Eng.* 135 (12): 1440–1449. https://doi.org/10.1061/(ASCE)ST.1943-541X.0000072.

Kalpakidis, I. V., M. C. Constantinou, and A. S. Whittaker. 2010. "Modeling strength degradation in lead-rubber bearings under earthquake shaking." *Earthquake Eng. Struct. Dyn.* 39 (13): 1533–1549. https://doi.org/10.1002/eqe.1039.

Katsaras, A. 2008. *Evaluation of current code requirements for displacement restoring capability of seismic isolation systems and proposals for revisions*. Project No. GOCE-CT-2003-505488. European Commission with 6th Framework.

Kelly, J. M., and M. S. Chaloub. 1990. *Earthquake simulator testing of a combined sliding bearing and rubber bearing isolation system*. Rep. No. UCB/EERC-87/04. Berkeley, CA: University of California.

Kelly, J. M., and S. B. Hodder. 1981. *Experimental study of lead and elastomeric dampers for base isolation systems*. Rep. No. UCB/EERC- 81/16. Berkeley, CA: University of California.

Kelly, J. M., and D. A. Konstantinidis. 2011. *History of multilayered rubber bearings*. New York: Wiley.

Kelly, J. M., M. S. Skinner, and K. E. Beucke. 1980. *Experimental testing of an energy absorbing seismic isolation system*. Rep. No. UCB/EERC-80/35. Berkeley, CA: University of California.

Kircher, C. A., B. Lashkari, R. L. Mayes, and T. E. Kelly. 1988. "Evaluation of nonlinear response in seismically isolated buildings." In *Proc., Symp. on Seismic, Shock and Vibration Isolation, ASME Pressure Vessels and Piping Conf.*, New York.

Masroor, A., and G. Mosqueda. 2015. "Assessing the collapse probability of base-isolated buildings considering pounding to moat walls using the FEMA P695 methodology." *Earthquake Spectra* 31 (4): 2069–2086. https://doi.org/10.1193/092113EQS256M.

McVitty, W., and M. C. Constantinou. 2015. *Property modification factors for seismic isolators: Design guidance for buildings*. MCEER Rep. No. 000-2015.

Ryan, K. L., C. B. Coria, and N. D. Dao. 2012. *Large scale earthquake simulation for hybrid lead rubber isolation system designed with consideration for nuclear seismicity*. US Nuclear Regulatory Commission CCEER 13-09, Washington, DC: Nuclear Regulatory Commission.

Shenton, H. W., III. 1996. *Guidelines for pre-qualification, prototype, and quality control testing of seismic isolation systems*. NISTIR 5800. Gaithersburg, MD: NIST.

York, K., and K. Ryan. 2008. "Distribution of lateral forces in base-isolated buildings considering isolation system nonlinearity." *J. Earthquake Eng.* 12 (7): 1185–1204. https://doi.org/10.1080/13632460802003751.

Zayas, V., S. Low, and S. Mahin. 1987. *The FPS earthquake resisting system*. Rep. No. UCB/EERC-87-01. Berkeley, CA: University of California.

OTHER REFERENCES (NOT CITED)

ATC (Applied Technology Council). 1982. *An investigation of the correlation between earthquake ground motion and building performance*. ATC Rep. No. 10. Redwood City, CA: ATC.

FEMA. 2009a. *Quantification of building seismic performance factors*. P-695. Washington, DC: FEMA.

FEMA. 2009b. *NEHRP recommended seismic provisions for new buildings and other structures*. Washington, DC: FEMA.

Lashkari, B., and C. A. Kircher. 1993. "Evaluation of SEAOC and UBC analysis procedures. Part 1: Stiff superstructure." In *Proc., Seminar on Seismic Isolation, Passive Energy Dissipation and Active Control.*

NIST (National Institute of Standards and Technology). 2011. *Selecting and scaling earthquake ground motions for performing response-history analyses*. GCR 11-917-15. Gaithersburg, MD: NIST.

Warn, G. P., and A. S. Whittaker. 2004. "Performance estimates in seismically isolated bridge structures." *Eng. Struct.* 26 (9): 1261–1278. https://doi.org/10.1016/j.engstruct.2004.04.006.

CHAPTER C18
SEISMIC DESIGN REQUIREMENTS FOR STRUCTURES WITH DAMPING SYSTEMS

C18.1 GENERAL

The requirements of this chapter apply to all types of damping systems, including both displacement-dependent damping devices of hysteretic or friction systems and velocity-dependent damping devices of viscous or viscoelastic systems (Soong and Dargush 1997, Constantinou et al. 1998, Hanson and Soong 2001). Compliance with these requirements is intended to produce performance comparable to that for a structure with a conventional seismic force-resisting system, but the same methods can be used to achieve higher performance.

The damping system (DS) is defined separately from the seismic force-resisting system (SFRS), although the two systems may have common elements. As illustrated in Figure C18.1-1, the DS may be external or internal to the structure and may have no shared elements, some shared elements, or all elements in common with the SFRS. Elements common to the DS and the SFRS must be designed for a combination of the loads of the two systems. When the DS and SFRS have no common elements, the damper forces must be collected and transferred to members of the SFRS.

C18.2 GENERAL DESIGN REQUIREMENTS

C18.2.1 System Requirements Structures with a DS must have an SFRS that provides a complete load path. The SFRS must comply with all the height, Seismic Design Category, and redundancy limitations and with the detailing requirements specified in this standard for the specific SFRS. The SFRS without the damping system (as if damping devices were disconnected) must be designed to have not less than 75% of the strength required for structures without a DS that have that type of SFRS (and not less than 100% if the structure is horizontally or vertically irregular). The damping systems, however, may be used to meet the drift limits (whether the structure is regular or irregular). Having the SFRS designed for a minimum of 75% of the strength required for structures without a DS provides safety in the event of damping system malfunction and produces a composite system with sufficient stiffness and strength to have a controlled lateral displacement response.

The analysis and design of the SFRS under the base shear, $V_{\min}$, from Equations (18.2-1) or (18.2-2) or, if the exception applies, under the unreduced base shear, V, should be based on a model of the SFRS that excludes the damping system.

C18.2.1.2 Damping System The DS must be designed for the actual (unreduced) MCE_R forces (such as peak force occurring in damping devices) and deflections. For certain elements of the DS (such as the connections or the members into which the damping devices frame), other than damping devices, limited yielding is permitted provided that such behavior does not affect damping system function or exceed the amount permitted for elements of conventional structures by the standard.

Furthermore, force-controlled actions in elements of the DS must consider seismic forces that are 1.2 times the computed average MCE_R response. Note that this increase is applied for each *element action*, rather than for each *element*. Force-controlled actions are associated with brittle failure modes where inelastic deformation capacity cannot be ensured. The 20% increase in seismic force for these actions is required to safeguard against undesirable behavior.

C18.2.2.1 Spectral Response Acceleration Parameters and Response Spectrum The design earthquake and MCE_R spectral response acceleration parameters (S_{DS}, S_{D1}, S_{MS}, and S_{M1}) are used in the equivalent lateral force procedure. The design earthquake and MCE_R spectra are used for response spectrum analysis.

C18.2.2.2 Ground Motions for Response History Analysis Development of target spectra and ground motions for structures with damping systems analyzed using response history methods generally follows the corresponding requirements for non-damped structures. The maximum and minimum damper properties, required elsewhere in this chapter, need not be considered to determine the period range of interest. Although amplitude scaling is used to adjust the ground motions from MCE_R down to design-earthquake, either amplitude scaling or spectral matching are available methods in Chapter 16 for developing the MCE_R motions. Other sections of Chapter 16 that are not specifically referenced do not apply to structures with damping systems.

C18.2.3 Procedure Selection The nonlinear response history procedure for structures incorporating supplemental damping systems is the preferred procedure, and Chapter 18 is structured accordingly. This method, consistent with the majority of current practice, provides the most realistic predictions of the seismic response of the combined seismic force-resisting system and damping system.

However, via the exception, response spectrum (RS) and equivalent lateral force (ELF) analysis methods can be used for design of structures with damping systems that meet certain configuration and other limiting criteria (e.g., at least two damping devices at each story configured to resist torsion). The analysis methods of damped structures are based on nonlinear static "pushover" characterization of the structure and calculation of peak response, using effective (secant) stiffness and effective damping properties of the first (pushover) mode in the direction of interest. These concepts are used in Chapter 17 to characterize the force-deflection properties of isolation systems, modified to explicitly incorporate the effects of ductility demand (postyield

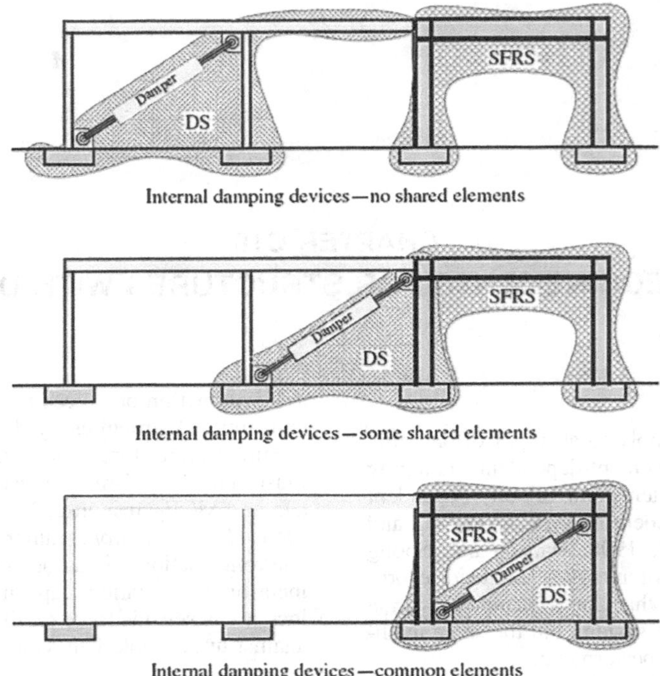

Figure C18.1-1. Damping system and seismic force-resisting system configurations.

response) and higher mode response of structures with dampers. Similar to conventional structures, damped structures generally yield during strong ground shaking, and their performance can be influenced strongly by the response of higher modes.

The RS and ELF procedures presented in Chapter 18 have several simplifications and limits, outlined as follows:

1. A multiple degree of freedom (MDOF) structure with a damping system can be transformed into equivalent single degree of freedom (SDOF) systems using modal decomposition procedures. This procedure assumes that the collapse mechanism for the structure is an SDOF mechanism so that the drift distribution over height can be estimated reasonably, using either the first mode shape or another profile, such as an inverted triangle. Such procedures do not strictly apply to either yielding buildings or buildings that are nonproportionally damped.
2. The response of an inelastic SDOF system can be estimated using equivalent linear properties and a 5% damped response spectrum. Spectra for damping greater than 5% can be established using damping coefficients, and velocity-dependent forces can be established either by using the pseudovelocity and modal information or by applying correction factors to the pseudovelocity.
3. The nonlinear response of the structure can be represented by a bilinear hysteretic relationship with zero postelastic stiffness (elastoplastic behavior).
4. The yield strength of the structure can be estimated either by performing simple plastic analysis or by using the specified minimum seismic base shear and values of R, Ω_0, and C_d.
5. Higher modes need to be considered in the equivalent lateral force procedure to capture their effects on velocity-dependent forces. This requirement is reflected in the residual mode procedure.

FEMA 440 (2005) presents a review of simplified procedures for the analysis of yielding structures. The combined effects of the simplifications mentioned previously are reported by Ramirez et al. (2001) and Pavlou and Constantinou (2004), based on studies of three-story and six-story buildings with damping systems designed by the procedures of the standard. The RS and ELF procedures of the standard are found to provide conservative predictions of drift and predictions of damper forces and member actions that are of acceptable accuracy when compared to results of nonlinear dynamic response history analysis. When designed in accordance with the standard, structures with damping systems are expected to have structural performance at least as good as that of structures without damping systems. Pavlou and Constantinou (2006) report that structures with damping systems, designed in accordance with the standard, provide the benefit of reduced secondary system response, although this benefit is restricted to systems with added viscous damping.

C18.2.4.1 Device Design Damping devices may operate on a variety of principles and may use materials that affect their short-term and long-term performance. This commentary provides guidance on the behavior of some of these devices, in order to justify the language in the standard and to assist the engineer in deciding on the upper and lower bound values of mechanical properties of the devices for use in analysis and design.

Damping devices that have found applications or have potential for application may be classified as follows:

1. Fluid viscous dampers (or oil dampers) that operate on the principle of orificing of fluid, typically some form of oil (Constantinou et al. 2007). These devices are typically highly engineered and precision made so that their properties are known within a narrow range. That is, when the devices are tested, their properties show small variability. One issue is heating that may have significant effects (Makris et al. 1998), which can be alleviated or eliminated by using accumulators or by using materials with varying thermal expansion properties so that the orifice size is automatically adjusted with varying temperature.

However, their long-term behavior may be affected by a variety of potential problems:

(a) Devices using accumulators include valves that may fail over time, depending on the quality of construction and history of operation. It is not possible to know if and when a valve may fail.

(b) Fluid is maintained in the device by seals between the body and the moving piston of the device, which may leak either as a result of wear caused by excessive cumulative travel or poor construction. For buildings, excessive cumulative travel is rarely an issue. When seals leak, the output of the device reduces, depending on the reduction of internal pressure of the device. It is recommended that potential leakage of oil not be considered in establishing lower bound values of property modification factors (as it is not possible to know) but rather a periodic inspection and maintenance program recommended by the manufacturer be used to detect problems and make corrections.

(c) Orifices may be very small in diameter and therefore may clog when impure oil is used or the if oil is contaminated by particles of rubber used in the sealing of fluid in poorly constructed devices or by metal particles resulting from internal corrosion or because of oil cavitation when poor-quality materials are used. Typically, rubber should not be used in sealing and parts should be threaded rather than welded or connected by posttensioning. Larger diameter orifices are preferred.

2. Viscoelastic fluid or solid devices. These devices operate on the principle of shearing of highly viscous fluids or viscoelastic solids. These viscous fluids and viscoelastic solids have a strong dependence on properties such as frequency and temperature. These effects should be assessed by qualification testing. Their long-term behavior is determined by the behavior of the fluid or solid used, both of which are expected to harden with time. The engineer should ask the supplier for data on the aging of the material based on observations in real time. Information based on accelerated aging is not useful and should not be used (Constantinou et al. 2007).
3. Metallic yielding devices. Yielding steel devices are typically manufactured of steel with yield properties that are known within a narrow range. Nevertheless, the range of values of the yield strength can be determined with simple material tests. Also, testing some of the devices should be used to verify the information obtained in coupon testing. Aging is of least concern because corrosion may only slightly reduce the section geometrical properties. An inspection and maintenance program should eliminate the concern for aging.
4. Friction devices. Friction devices operate on the principle of preloaded sliding interfaces. There are two issues with such devices:

   (a) The preload may reduce over time because of creep in sliding interface materials or the preloading arrangement, or wear in the sliding interface when there is substantial service-load related motion or after high-speed seismic motion. It is not possible to know what the preload may be within the lifetime of the structure, but the loss may be minimized when high-strength bolts are used and high-strength/low-wear materials are used for the sliding interface.

   (b) The friction coefficient at the sliding interface may substantially change over time. The engineer is directed to Constantinou et al. (2007) for a presentation on the nature of friction and the short-term and long-term behavior of some sliding interfaces. In general, reliable and predictable results in the long-term friction may be obtained when the sliding interface consists of a highly polished metal (typically stainless steel) in contact with a nonmetallic softer material that is loaded to high pressure under confined conditions, so that creep is completed in a short time. However, such interfaces also result in low friction (and thus are typically used in sliding isolation bearings). The engineer is referred to Chapter 17 and the related commentary for such cases. Desirable high friction (from a performance standpoint) may be obtained by use of metal to metal sliding interfaces. However, some of these interfaces are absolutely unreliable because they promote severe additional corrosion and they should never be used (BSI 1983). Other bimetallic interfaces have the tendency to form solid solutions or intermetallic compounds with one another when in contact without motion. This tendency leads to cold welding (very high adhesion or very high friction). Such materials are identified by compatibility charts (Rabinowicz 1995). The original Rabinowicz charts categorized pairs of metals as incompatible (low adhesion) to compatible and identical (high adhesion). Based on this characterization, identical metals and most bimetallic interfaces should be excluded from consideration in sliding interfaces. Excluding interfaces that include lead (too soft), molybdenum, silver, and gold (too expensive), only interfaces of tin–chromium, cadmium–aluminum, and copper–chromium are likely to have low adhesion. Of these, the tin–chromium interface has problems of additional corrosion (BSI 1983) and should not be used. Accordingly, only bimetallic interfaces of cadmium–aluminum and copper–chromium may be useful. The materials in these interfaces have similar hardness so that creep-related effects are expected to be important, leading to increased true area of contact and increased friction force over time (Constantinou et al. 2007). This increase leads to the conclusion that all bimetallic interfaces result in significant changes in friction force over time that are not possible to predict, and therefore, these types of interfaces should not be used.
5. Lead extrusion devices. These devices operate on the principle of extruding lead through an orifice. The behavior of the device is dependent on the rate of loading and temperature, and its force output reduces with increasing cycling because of heating effects. These effects can be quantified by testing so that the nominal properties and property modification factors can be established. Leakage of lead during the lifetime of the device is possible during operation and provided that the seals fail, although the effects cannot be expected to be significant. Leakage is preventable by the use of proven sealing technologies and by qualification testing to verify (Skinner et al. 1993).

The registered design professional (RDP) must define the ambient temperature and the design temperature range. The ambient temperature is defined as the normal in-service temperature of the damping device. For devices installed in interior spaces, this temperature may be taken as 70°F, and the design temperature range could come from the project mechanical engineer. For devices installed exposed to exterior temperature variation, the ambient temperature may be taken as the annual average temperature at the site, and the design temperature range may be taken as the annual minimum and maximum temperatures. Because the design temperature range is implicitly tied to MCE_R analysis through λ factors for temperature, the use of maximum and minimum temperatures over the design life of the structure are considered too severe.

C18.2.4.4 Nominal Design Properties Device manufacturers typically supply nominal design properties that are reasonably accurate based on previous prototype test programs. The nominal properties can be confirmed by project-specific prototype tests during either the design or construction phases of the project.

C18.2.4.5 Maximum and Minimum Damper Properties

Specification Tolerance on Nominal Design Properties. As part of the design process, it is important to recognize that there are variations in the production damper properties from the nominal properties. This difference is caused by manufacturing variation. Recommended values for the specification tolerance on the average properties of all devices of a given type and size are typically in the ±10% to ±15% range. For a ±10% specification tolerance, the corresponding λ factors would be $\lambda_{(spec,max)} = 1.1$ and $\lambda_{(spec,min)} = 0.9$. Variations for individual device properties may be greater than the tolerance on the average properties of all devices of a given type and size. It is recommended that the device manufacturer be consulted when establishing these tolerance values.

Property Variation (λ) Factors and Maximum and Minimum Damper Properties. Section 18.2.4.5 requires the devices to be analyzed and designed with consideration given to environmental conditions, including the effects of aging, creep, fatigue, and operating temperatures. The individual aging and environmental factors are multiplied together, and then the portion of the resulting λ factor (λ_{ae}) differing from unity is reduced by 0.75 based on the assumption that not all the maximum/minimum aging and environmental values occur simultaneously.

Results of prototype tests may also indicate the need to address device behavior whereby tested properties differ from the nominal design properties because of test-related effects. Such behavior may include velocity effects, first cycle effects, and any other testing effects that cause behavior different from the nominal design properties. This behavior is addressed through a testing λ factor (λ_{test}), which is a multiple of all the individual testing effects.

The specification (λ_{spec}), environmental (λ_{ae}) and testing (λ_{test}) factors are used to establish maximum (λ_{max}) and minimum (λ_{min}) damper properties for each device type and size for use in mathematical models of the damped structure, in accordance with Equations (18.2-3a) and (18.2-3b). These factors are typically applied to whatever parameters govern the mathematical representation of the device.

It should be noted that more sophisticated mathematical models account for various property variation effects directly (e.g., velocity or temperature). When such models are used, the cumulative effect of the λ factors reduce (become closer to 1.0), because some of the typical behaviors contributing to λ_{max} and λ_{min} are already included explicitly in the model. Some effects, such as specification tolerance and aging, will likely always remain because they cannot be accounted for in mathematical models.

Example

Data from prototype testing, as defined in Section 18.6.2, are used to illustrate the λ factors and the maximum and minimum values to be used in analysis and design. The fluid viscous damper under consideration has the following nominal force-velocity constitutive relationship, with kips and inch units:

$$F = C\,\mathrm{sgn}(V)|V|^{\alpha} = 128\,\mathrm{sgn}(v)|V|^{0.38}$$

The solid line in Figure C18.2-1 depicts the nominal force-displacement relationship.

Prototype tests of damper corresponding to the following conditions were conducted:

- Force-velocity characteristic tests, all conducted at ambient temperature of 70°F (21.1°C).
  - 10 full cycles performed at various amplitudes.
- Temperature tests, three fully reversible cycles conducted at various velocities at the following temperatures:
  - 40 °F (4.4 °C)
  - 70 °F (21.1 °C)
  - 100 °F (37.8 °C)

The data from prototype tests for each cycle (maximum and negative) are shown as data points in Figure C18.2-1.

Also shown in Figure C18.2-1 are the variations from nominal in the force-velocity relationships for this damper. The relationships are obtained by changing the damper constant (C) value. No variation is considered for the velocity exponent, α. The following diagrams are shown:

- A pair of lines corresponding to damper nominal constitutive relationship computed with the C value increased or decreased by 10%. These lines account for the λ_{test} factors, as defined in Section 18.2.4.5: $\lambda_{(test,min)} = 0.9$.

  For these particular devices, the variation in properties caused by aging and environmental factors is taken as ±5% ($\lambda_{(ae,max)} = 1.05, \lambda_{(ae,min)} = 0.95$), and the specification tolerance is set at ±5% ($\lambda_{(spec,max)} = 1.05$, $\lambda_{(spec,min)} = 0.95$). These values should be developed in conjunction with the device manufacturer based on their history of production damper test data and experience with aging and other environmental effects. Using these values in Equations (18.2-3a) and (18.2-3b) results in $\lambda_{max} = 1.20$ and $\lambda_{min} = 0.82$. These values satisfy the minimum

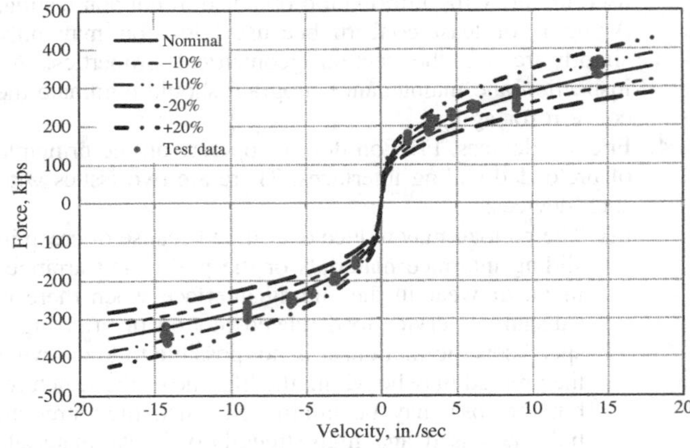

Figure C18.2-1. Force-velocity relationship for a nonlinear viscous damper.

variation requirements of Section 18.2.4.5. They are rounded to $\lambda_{max} = 1.2$ and $\lambda_{min} = 0.8$.
- A pair of lines corresponding to the cumulative maximum and minimum λ values (accounting for testing, specification tolerance, and other factors listed in Section 18.2.4.5) computed with the nominal C value increased or decreased by 20%.

For this example, analysis with minimum and maximum damper properties should be conducted by using 80% and 120% of the nominal value for C, respectively. The analysis with maximum damper properties typically produces larger damper forces for use in the design of members and connections, whereas the analysis with minimum damper properties typically produces less total energy dissipation and hence larger drifts.

C18.2.4.6 Damping System Redundancy This provision is intended to discourage the use of damping systems with low redundancy in any story. At least four damping devices should be provided in each principal direction, with at least two devices in each direction on each side of the center of stiffness to control torsional response. In cases where there is low damping system redundancy by this definition, all damping devices in all stories must be capable of sustaining increased displacements (with associated forces) and increased velocities (with associated displacements and forces) relative to a system with adequate redundancy. The penalty is 130%.

C18.3 NONLINEAR RESPONSE HISTORY PROCEDURE

Those elements of the SFRS and the DS that respond essentially elastically at MCE_R (based on a limit of 1.5 times the expected strength calculated using $\phi = 1$) are permitted to be modeled elastically. Modeling parameters and acceptance criteria provided in ASCE 41, with a performance objective defined in Table 2.2, as modified in this chapter, are deemed satisfactory to meet the requirements of this section.

The hardware of all damping devices (e.g., the cylinder of a piston-type device) and the connections between the damping devices and the remainder of the structure must remain elastic at MCE_R (see Section 18.2.1.2). The nonlinear behavior of all other elements of both the SFRS and the DS must be modeled based on test data, which must not be extrapolated beyond the tested deformations. Strength and stiffness degradation must be included if such behavior is indicated. However, the damping system must not become nonlinear to such an extent that its function is impaired.

Nonlinear response history analysis (NRHA) is performed at both the design earthquake (DE) and the MCE_R levels. Accidental eccentricity is included at MCE_R but need not be included at the DE level, because the SFRS design checks from Section 18.2.1.1 include accidental eccentricity. However, the results from the NRHA at DE, using a model of the combined SFRS and DS, must be used to recheck all elements of the SFRS, because the checks of Section 18.2.1.1 are conducted using a representation of the structure excluding the damping system. This requirement is defined in Section 18.4.1. The damping system is designed and evaluated based on the results of the MCE_R analyses, as defined in Section 18.4.2.

C18.3.2 Accidental Mass Eccentricity In order to avoid the need to perform a large number of nonlinear response history analyses that include the suites of ground motions, the upper and lower bound damper properties, and five or more locations of the center of mass, the exception in this provision allows the center-of-mass analysis results to be scaled and used to account for the effects of mass eccentricity in different building quadrants.

The following is one suggested method of developing appropriate amplification factors for deformations and forces for use with center-of-mass NRHAs to account for the effects of accidental eccentricity. The use of other rationally developed amplification factors is permitted and encouraged, given that the artificial shift of the center of mass changes the dynamic characteristics of the analyzed structure and may lead to the paradox of reduced torsional response with increasing accidental eccentricity (Basu et al. 2014).

The most critical directions for moving the calculated center of mass are such that the accidental eccentricity adds to the inherent eccentricity in each orthogonal direction at each level. For each of these two eccentric mass positions, and with minimum damper properties, the suite of NRHAs should be run and the results processed in accordance with Section 18.3.3. The analysis cases are defined in Table C18.3-1.

The results from Cases IIa and IIb are then compared, in turn, to those from Case I. The following amplification factors (ratio of Case IIa or IIb response to Case I response) are computed:

(a) The amplification for story drift in the structure at the plan location with the highest drift, enveloped over all stories; and
(b) The amplification for frame-line shear forces at each story for the frame subjected to the maximum drift.

The larger of the two resulting scalars on drift should be used as the deformation amplifier, and the larger of the two resulting scalars on force should be used as the force amplifier. Once the amplification factors are established, the effects of accidental eccentricity should be considered as follows.

The NRHA procedure should be run for the inherent mass eccentricity case only, considering both maximum and minimum damper properties. For each damper property variation, response quantities should be computed in accordance with Section 18.3.3. All resulting deformation response quantities should be increased by the deformation amplifier, and all resulting force quantities should be increased by the force amplifier before being used for evaluation or design.

C18.4 SEISMIC LOAD CONDITIONS AND ACCEPTANCE CRITERIA FOR NONLINEAR RESPONSE HISTORY PROCEDURE

C18.4.1 Seismic Force-Resisting System All elements of the SFRS are checked under two conditions. First, the SFRS (excluding the damping system) is checked under the minimum base shear requirements of Section 18.2.1.1. Second, the demands from the NRHA at DE (with a model of the combined SFRS and DS) must be used to recheck all elements of the SFRS.

There are three limiting values for the analytically computed drift ratios at the MCE_R. Table 12.12-1 lists the allowable drifts for structures. These limiting drift ratios are checked against drift

Table C18.3-1. Analysis Cases for Establishing Amplification Factors.

Case	Damper Properties	Accidental Eccentricity
I	Minimum	No
IIa	Minimum	Yes, X-direction
IIb	Minimum	Yes, Y-direction

ratio demands computed from the code procedure. Because the code design is an implied DE intensity, the drift ratios in the table are also intended to be used at analysis conducted at this level.

1. 3% limit: For most common structures, the DE allowable drift ratio (Δ_a/h) is 2%. Because for most cases, the ratio of MCE_R to DE intensity is 1.5, then the allowable drift ratio at MCE_R becomes 3% ($1.5 \times 2\%$).
2. 1.9 factor: When NRHA is used, the commentary to Chapter 16 assumes a factor of 1.25 (to reflect an average ratio of R/Cd) can be applied to the DE drift ratio limits of Table 12.12-1. Therefore, the MCE_R drift ratios are limited to 1.9 (approximately equal to 1.5×1.25) of limits of Table 12.12-1.
3. $1.5\,R/C_d$ factor: The deflections δ_x of Equation (12.8-15) are computed by amplifying the deflections computed from analysis by the deflection amplification factor (C_d). The elastic deflections used in Chapter 12 are computed at DE intensity using elastic analysis with forces that are reduced by the response modification factor, R. Thus, for the purpose of comparing drift ratios computed from NRHA with Table 12.12-1, the entries of the table need to be modified by the R/C_d factor for comparison at DE level. Therefore, the allowable drift ratios at MCE_R correspond to $1.5R/C_d$ of entries of the table.

Example: Five-Story Steel Special Moment Frames in Risk Category I or II

- Allowable drift ratio from Table 12.12-1 $= 2\%$.
- Allowable drift ratio for structures with dampers using NRHA then would be the smallest of
  - 3%,
  - $1.9 \times 2\% = 3.8\%$, and
  - $1.5 \times (8/5.5) \times 2\% = 4.4\%$.
- 3% controls. Thus, all computed drift ratios from NRHA should be 3% or less at MCE_R.

C18.5 DESIGN REVIEW

The independent design review of many structures incorporating supplemental damping may be performed adequately by one registered and appropriately experienced design professional. However, for projects involving significant or critical structures, it is recommended that a design review panel consisting of two or three registered and appropriately experienced design professionals be used.

C18.6 TESTING

C18.6.1.1 Qualification Tests Property modification factors are used to bound the mechanical properties of damping devices. These factors are device- and manufacturer-dependent and are derived from testing of devices of the physical size and internal construction (e.g., orifices and seals for fluid viscous dampers), and with the force-velocity-displacement response proposed for construction. If damping devices smaller than full-scale are used for qualification testing, principles of scaling and similarity in the peer-reviewed literature for seismic protective devices may be used to interpret test data.

C18.6.2.2 Sequence and Cycles of Testing The use of $1/(1.5T_1)$ as the testing frequency is based on a softening of the combined SFRS and DS associated with a system ductility of approximately 2. Test 2(d) in Section 18.6.2.2 ensures that the prototype damper is tested at the maximum force from analysis.

It should be noted that velocity-dependent devices (e.g., those devices characterized by $F = C_v^\alpha$) are not intended to be characterized as frequency-dependent under item 4 of this section.

C18.6.2.3 Testing Similar Devices For existing prototype test data to be used to satisfy the requirement of Section 18.6.2, the conditions of this provision must be satisfied. It is imperative that identical manufacturing and quality control procedures be used for the preexisting prototype and the project-specific production damping
devices. The precise interpretations of "similar dimensional characteristics, internal construction, and static and dynamic internal pressures" and "similar maximum strokes and forces" are left to the RDP and the design review team. However, variations in these characteristics of the preexisting prototype device beyond approximately $\pm 20\%$ from the corresponding project-specific values should be cause for concern.

C18.6.2.4 Determination of Force-Velocity-Displacement Characteristics When determining nominal properties (Item 2) for damping devices whose first-cycle test properties differ significantly from the average properties of the first three cycles, an extra cycle may be added to the test, and the nominal properties may be determined from the average value using data from the second through fourth cycles. In this case, the effect of first-cycle properties must be addressed explicitly and included in the λ_{max} factor. It should be noted that if the property variation methodology of Sections 18.2.4.4 and 18.2.4.5 is applied consistently, the maximum and minimum design properties [Equations (18.2-4a) and (18.2-4b)] will be identical, regardless of whether the nominal properties are taken from the average of Cycles 1 through 3 or Cycles 2 through 4.

C18.6.3 Production Tests The registered design professional is responsible for defining, in the project specifications, the scope of the production damper test program, including the allowable variation in the average measured properties of the production damping devices. The registered design professional must decide on the acceptable variation of damper properties on a project-by-project basis. This range must agree with the specification tolerance from Section 18.2.4.5. The standard requires that all production devices of a given type and size be tested.

Individual devices may be permitted a wider variation (typically $\pm 15\%$ or $\pm 20\%$) from the nominal design properties. For example, in a device characterized by $F = C_v^\alpha$, the mean of the force at a specified velocity for all tested devices might be permitted to vary no more than $\pm 10\%$ from the specified value of force, but the force at a specified velocity for any individual device might be permitted to vary no more than $\pm 15\%$ from the specified force.

The production dynamic cyclic test is identical (except for three versus five cycles) to one of the prototype tests of Section 18.6.2.2, so that direct comparison of production and prototype damper properties is possible.

The exception is intended to cover those devices that would undergo yielding or be otherwise damaged under the production test regime. The intent is that piston-type devices be 100% production tested, because their properties cannot be shown to meet the requirements of the project specifications without testing. For other types of damping devices, whose properties can be demonstrated to be in compliance with the project specifications by other means (e.g., via material testing and a manufacturing quality control program), the dynamic cyclic testing of 100% of the devices is not required. However, in this case, the RDP must establish an alternative production test

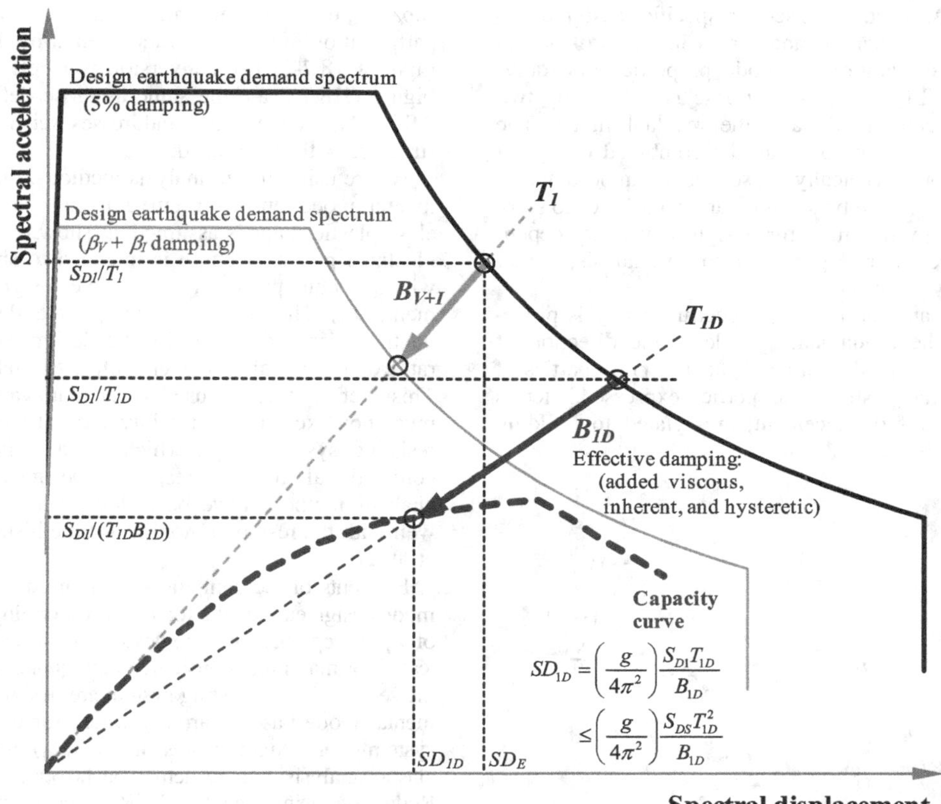

Figure C18.7-1. Effective damping reduction of design demand.

program to ensure the quality of the production devices. Such a program would typically focus on such things as manufacturing quality control procedures (identical between prototype and production devices), material testing of samples from a production run, welding procedures, and dimensional control. At least one production device must be tested at 0.67 times the $\mathrm{MCE_R}$ stroke at a frequency equal to $1/(1.5T_1)$, unless the complete project-specific prototype test program has been performed on an identical device. If such a test results in inelastic behavior in the device, or the device is otherwise damaged, that device cannot be used for construction.

C18.7 ALTERNATE PROCEDURES AND CORRESPONDING ACCEPTANCE CRITERIA

This section applies only to those cases where either the RS or the ELF procedure is adopted.

C18.7.1 Response-Spectrum Procedure and C18.7.2 Equivalent Lateral Force Procedure

Effective Damping In the standard, the reduced response of a structure with a damping system is characterized by the damping coefficient, B, based on the effective damping, β, of the mode of interest. This approach is the same as that used for isolated structures. Like isolation, effective damping of the fundamental mode of a damped structure is based on the nonlinear force-deflection properties of the structure. For use with linear analysis methods, nonlinear properties of the structure are inferred from the overstrength factor, Ω_0, and other terms.

Figure C18.7-1 illustrates a reduction in design earthquake response of the fundamental mode caused by increased effective damping (represented by coefficient, B_{1D}). The capacity curve is a plot of the nonlinear behavior of the fundamental mode in spectral acceleration-displacement coordinates. The reduction caused by damping is applied at the effective period of the fundamental mode of vibration (based on the secant stiffness).

In general, effective damping is a combination of three components:

1. Inherent Damping (β_I)—Inherent damping of the structure at or just below yield, excluding added viscous damping (typically assumed to be 2% to 5% of critical for structural systems without dampers).
2. Hysteretic Damping (β_H)—Postyield hysteretic damping of the seismic force-resisting system and elements of the damping system at the amplitude of interest (taken as 0% of critical at or below yield).
3. Added Viscous Damping (β_V)—The viscous component of the damping system (taken as 0% for hysteretic or friction-based damping systems).

Both hysteretic damping and added viscous damping are amplitude-dependent, and the relative contributions to total effective damping change with the amount of postyield response of the structure. For example, adding dampers to a structure decreases postyield displacement of the structure and hence decreases the amount of hysteretic damping provided by the seismic force-resisting system. If the displacements are reduced to the point of yield, the hysteretic component of effective damping is zero and the effective damping is equal to inherent damping plus added viscous damping. If there is no damping system (as in a conventional structure), effective damping simply equals inherent damping.

Linear Analysis Methods. The section specifies design earthquake displacements, velocities, and forces in terms of design earthquake spectral acceleration and modal properties. For equivalent lateral force (ELF) analysis, response is defined by two modes: the fundamental mode and the residual mode. The residual mode is used to approximate the combined effects of higher modes. Although typically of secondary importance to story drift, higher modes can be a significant contributor to story velocity and hence are important for design of velocity-dependent damping devices. For response spectrum analysis, higher modes are explicitly evaluated.

For both the ELF and the response spectrum analysis procedures, response in the fundamental mode in the direction of interest is based on assumed nonlinear (pushover) properties of the structure. Nonlinear (pushover) properties, expressed in terms of base shear and roof displacement, are related to building capacity, expressed in terms of spectral coordinates, using mass participation and other fundamental-mode factors shown in Figure C18.7-2. The conversion concepts and factors shown in Figure C18.7-2 are the same as those defined in Chapter 9 of ASCE 41 (2014), which addresses seismic rehabilitation of a structure with damping devices.

Where using linear analysis methods, the shape of the fundamental mode pushover curve is not known, so an idealized elastoplastic shape is assumed, as shown in Figure C18.7-3. The idealized pushover curve is intended to share a common point with the actual pushover curve at the design earthquake displacement, D_{1D}. The idealized curve permits definition of the global ductility demand caused by the design earthquake, μ_D, as the ratio of design displacement, D_{1D}, to yield displacement, D_Y. This ductility factor is used to calculate various design factors; it must not exceed the ductility capacity of the seismic force-resisting system, μ_{max}, which is calculated using factors for conventional structural response. Design examples using linear analysis methods have been developed and found to compare well with the results of nonlinear time history analysis (Ramirez et al. 2001).

Elements of the damping system are designed for fundamental mode design earthquake forces corresponding to a base shear value of V_Y (except that damping devices are designed and prototypes are tested for maximum considered earthquake response). Elements of the seismic force-resisting system are designed for reduced fundamental mode base shear, V_1, where force reduction is based on system overstrength (represented by Ω_0), multiplied by C_d/R for elastic analysis (where actual pushover strength is not known). Reduction, using the ratio C_d/R, is necessary because the standard provides values of C_d that are less than those for R. Where the two parameters have equal value and the structure is 5% damped under elastic conditions, no adjustment is necessary. Because the analysis methodology is based on calculating the actual story drifts and damping device displacements (rather than the displacements calculated for elastic conditions at the reduced base shear and then multiplied by C_d), an adjustment is needed. Because actual story drifts are calculated, the allowable story drift limits of Table 12.12-1 are multiplied by R/C_d before they are used.

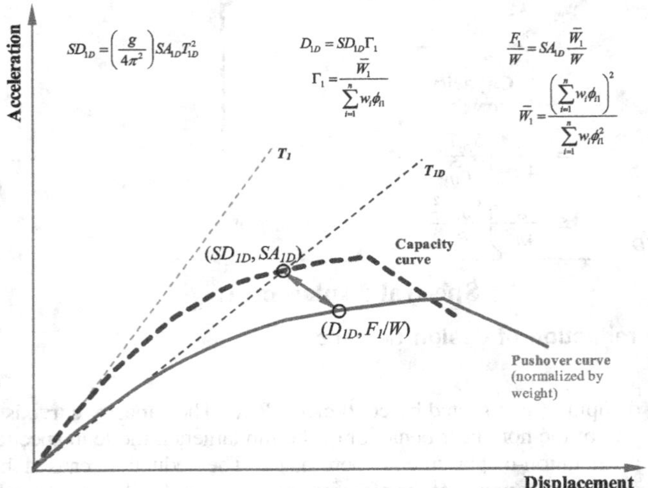

Figure C18.7-2. Pushover and capacity curves.

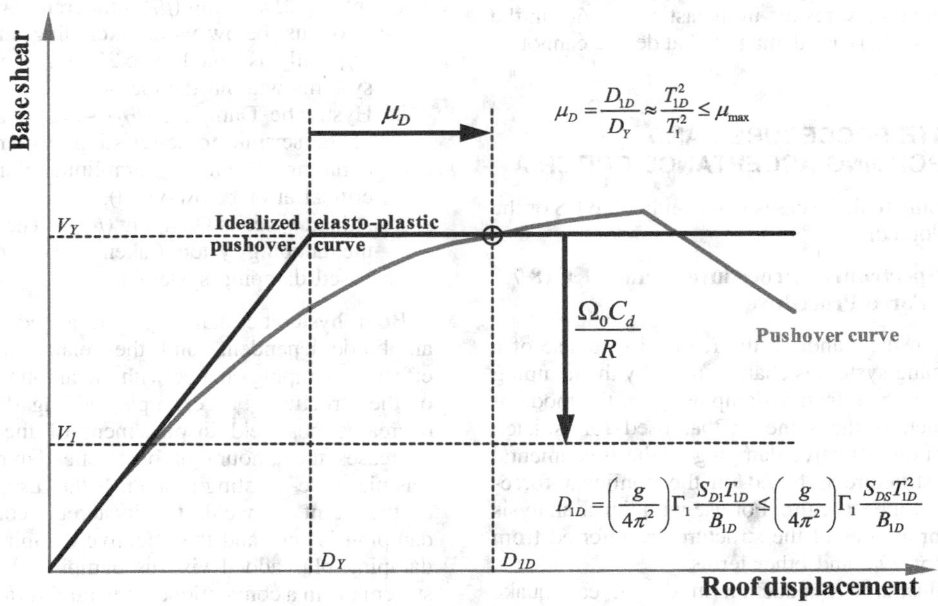

Figure C18.7-3. Pushover and capacity curves.

Table C18.7-1. Values of Damping Coefficient, *B*.

Effective Damping, β (%)	Table 17.5-1 of ASCE/SEI 7 (2010), AASHTO (2010), CBSC (2013) (seismically isolated structures)	Table 18.6-1 of ASCE/SEI 7 (2010) (structures with damping systems)	FEMA 440 (2005)	Eurocode 8 (2005)
2	0.8	0.8	0.8	0.8
5	1.0	1.0	1.0	1.0
10	1.2	1.2	1.2	1.2
20	1.5	1.5	1.5	1.6
30	1.7	1.8	1.8	1.9
40	1.9	2.1	2.1	2.1
50	2.0	2.4	2.4	2.3

C18.7.3 Damped Response Modification

C18.7.3.1 Damping Coefficient Values of the damping coefficient, *B*, in Table 18.7-1 for the design of damped structures are the same as those in Table 17.5-1 for isolated structures at damping levels up to 20% but extend to higher damping levels based on results presented in Ramirez et al. (2001). Table C18.7-1 compares values of the damping coefficient as found in the standard and various resource documents and codes. FEMA 440 (2005) and Eurocode 8 (2005) present equations for the damping coefficient, *B*, whereas the other documents present values of *B* in tabular format.

The equation in FEMA 440 is $B = \frac{4}{5.6 - \ln(100\beta)}$

The equation in Eurocode 8 (2005) is $B = \sqrt{\frac{0.05+\beta}{0.10}}$

C18.7.3.2 Effective Damping The effective damping is calculated assuming that the structural system exhibits perfectly bilinear hysteretic behavior characterized by the effective ductility demand, μ, as described in Ramirez et al. (2001). Effective damping is adjusted using the hysteresis loop adjustment factor, q_H, which is the actual area of the hysteresis loop divided by the area of the assumed perfectly bilinear hysteretic loop. In general, values of this factor are less than unity. In Ramirez et al. (2001), expressions for this factor (which is called Quality Factor) are too complex to serve as a simple rule. Equation (18.7-49) provides a simple estimate of this factor. The equation predicts correctly the trend in the constant acceleration domain of the response spectrum, and it is believed to be conservative for flexible structures.

C18.7.4 Seismic Load Conditions and Acceptance Criteria for RSA and ELF Procedures

C18.7.4.5 Seismic Load Conditions and Combination of Modal Responses Seismic design forces in elements of the damping system are calculated at three distinct stages: maximum displacement, maximum velocity, and maximum acceleration. All three stages need to be checked for structures with velocity-dependent damping systems. For displacement-dependent damping systems, the first and third stages are identical, whereas the second stage is inconsequential.

Force coefficients C_{mFD} and C_{mFV} are used to combine the effects of forces calculated at the stages of maximum displacement and maximum velocity to obtain the forces at maximum acceleration. The coefficients are presented in tabular form based on analytic expressions presented in Ramirez et al. (2001) and account for nonlinear viscous behavior and inelastic structural system behavior.

REFERENCES

AASHTO (American Association of State Highway and Transportation Officials). 2010. *Guide specifications for seismic isolation design*. Washington, DC: AASHTO.

ASCE. 2010. *Minimum design loads for buildings and other structures*. ASCE 7-10. Reston, VA: ASCE.

ASCE. 2014. *Seismic evaluation and retrofit of existing buildings*. ASCE/SEI 41-13. Reston, VA: ASCE.

Basu, D., M. C. Constantinou, and A. S. Whittaker. 2014. "An equivalent accidental eccentricity to account for the effects of torsional ground motion on structures." *Eng. Struct.* 69: 1–11. https://doi.org/10.1016/j.engstruct.2014.02.038

BSI (British Standards Institution). 1983. *Commentary on corrosion at bimetallic contacts and its alleviation*. PD6484:1979. London: BSI.

CBSC (California Building Standards Commission). 2013. *California building code*. Sacramento, CA: CBSC.

CEN (European Committee for Standardization). 2005. *Design of structures for earthquake resistance. Part 2: Bridges*. Eurocode 8. Brussels, Belgium: CEN.

Constantinou, M. C., T. T. Soong, and G. F. Dargush. 1998. *Passive energy dissipation systems for structural design and retrofit, monograph 1*. Buffalo, NY: University of Buffalo.

Constantinou, M. C., A. S. Whittaker, Y. Kalpakidis, D. M. Fenz, and G. P. Warn. 2007. *Performance of seismic isolation hardware under service and seismic loading*. Rep. No. MCEER-07-0012. Buffalo, NY: Multidisciplinary Center for Earthquake Engineering Research.

FEMA (Federal Emergency Management Agency). 2005. *Improvement of nonlinear static seismic analysis procedures*. FEMA 440. Washington, DC: FEMA.

Hanson, R. D., and T. T. Soong. 2001. *Seismic design with supplemental energy dissipation devices*. MNO-8. Oakland, CA: Earthquake Engineering Research Institute.

Makris, N., Y. Roussos, A. S. Whittaker, and J. M. Kelly. 1998. "Viscous heating of fluid dampers. I: Large-amplitude motions." *J. Eng. Mech.*

Pavlou, E., and M. C. Constantinou. 2004. "Response of elastic and inelastic structures with damping systems to near-field and soft-soil ground motions." *Eng. Struct.* 26 (9): 1217–1230. https://doi.org/10.1016/j.engstruct.2004.04.001.

Pavlou, E., and M. C. Constantinou. 2006. "Response of nonstructural components in structures with damping systems." *J. Struct. Eng.* 132 (7): 1108–1117. https://doi.org/10.1061/(ASCE)0733-9445(2006)132:7(1108).

Rabinowicz, E. 1995. *Friction and wear of materials*. New York: Wiley.

Ramirez, O. M., M. C. Constantinou, C. A. Kircher, A. Whittaker, M. Johnson, J. D. Gomez, and C. Z. Chrysostomou.

2001. *Development and evaluation of simplified procedures of analysis and design for structures with passive energy dissipation systems*. Technical Rep. MCEER-00-0010. Buffalo, NY: University of Buffalo.

Skinner, R. I., W. H. Robinson, and G. H. McVerry. 1993. *An introduction to seismic isolation*. Chichester, UK: Wiley.

Soong, T. T., and G. F. Dargush. 1997. *Passive energy dissipation systems in structural engineering*. London: Wiley.

OTHER REFERENCES (NOT CITED)

Miyamoto, H. K., A. S. J. Gilani, A. Wada, and C. Ariyaratana. 2011. "Identifying the collapse hazard of steel special moment-frame buildings with viscous dampers using the FEMA P-695 methodology." *Earthquake Spectra* 27 (4): 1147–1168. https://doi.org/10.1193/1.3651357.

Newmark, N. M., and W. J. Hall. 1969. "Seismic design criteria for nuclear reactor facilities." In *Proc., 4th World Conf. in Earthquake Engineering*, Santiago, Chile.

Ramirez, O. M., M. C. Constantinou, J. Gomez, A. S. Whittaker, and C. Z. Chrysostomou. 2002a. "Evaluation of simplified methods of analysis of yielding structures with damping systems." *Earthquake Spectra* 18 (3): 501–530. https://doi.org/10.1193/1.1509763.

Ramirez, O. M., M. C. Constantinou, A. S. Whittaker, C. A. Kircher, and C. Z. Chrysostomou. 2002b. "Elastic and inelastic seismic response of buildings with damping systems." *Earthquake Spectra* 18 (3): 531–547. https://doi.org/10.1193/1.1509762.

Ramirez, O. M., M. C. Constantinou, A. S. Whittaker, C. A. Kircher, M. W. Johnson, and C. Z. Chrysostomou. 2003. "Validation of 2000 NEHRP provisions equivalent lateral force and modal analysis procedures for buildings with damping systems." *Earthquake Spectra* 19 (4): 981–999. https://doi.org/10.1193/1.1622392.

SEAOC (Structural Engineers Association of California). 2013. *2012 IBC SEAOC structural/seismic design manual Volume 5: Examples for seismically isolated buildings and buildings with supplemental damping*. Sacramento, CA: SEAOC.

Whittaker, A. S., M. C. Constantinou, O. M. Ramirez, M. W. Johnson, and C. Z. Chrysostomou. 2003. "Equivalent lateral force and modal analysis procedures of the 2000 NEHRP provisions for buildings with damping systems." *Earthquake Spectra* 19 (4): 959–980. https://doi.org/10.1193/1.1622391.

CHAPTER C19
SOIL–STRUCTURE INTERACTION FOR SEISMIC DESIGN

C19.1 GENERAL

In an earthquake, the shaking is transmitted up through the structure from the geologic media underlying and surrounding the foundation. The response of a structure to earthquake shaking is affected by interactions among three linked systems: the structure, the foundation, and the geologic media underlying and surrounding the foundation. The analysis procedures in Chapters 12 and 15 idealize the response of the structure by applying forces to the structure, which is typically assumed to have a fixed base at the foundation–soil interface. In some cases, the flexibility of the foundation elements and underlying soils is included in the analysis model. The forces that are applied to the structure are devised based on parameters representing free-field ground motions. The term *free-field* refers to motions not affected by structural vibrations or the foundation characteristics of the specific structure and represents the condition for which the design spectrum is derived using the procedures given in Chapters 11 and 21. In most cases, however, the motions at the foundation that are imparted to the structure are different from the free-field motions. This difference is caused by the effects of the interaction of the structure and the geologic media. A seismic soil–structure interaction (SSI) analysis evaluates the collective response of these systems to a specified free-field ground motion.

SSI effects are absent for the theoretical condition of rigid geologic media, which is typical of analytical models of structures. Accordingly, SSI effects reflect the differences between the actual response of the structure and the response for the theoretical, rigid base condition. Visualized within this context, the three following SSI effects can significantly affect the response of structures:

1. **Foundation Deformations**. Flexural, axial, and shear deformations of foundation elements occur as a result of loads applied by the superstructure and the supporting geologic media. In addition, the underlying geologic media deforms because of loads from the foundations. Such deformations represent the seismic demand for which foundation components should be designed. These deformations can also significantly affect the overall system behavior, especially with respect to damping.
2. **Inertial SSI Effects**. Inertia developed in a vibrating structure gives rise to base shear, moment, and torsional excitation, and, in turn, these loads cause displacements and rotations of the foundation relative to the free-field displacement. These relative displacements and rotations are only possible because of flexibility in the soil–foundation system, which can significantly contribute to the overall structural flexibility in some cases. Moreover, the relative foundation free-field motions give rise to energy dissipation via radiation damping (i.e., damping associated with wave propagation into the ground away from the foundation, which acts as the wave source) and hysteretic soil damping, and this energy dissipation can significantly affect the overall damping of the soil–foundation–structure system. Because these effects are rooted in the structural inertia, they are referred to as inertial interaction effects.
3. **Kinematic SSI Effects**. Kinematic SSI results from the presence of foundation elements on or in soil that are much stiffer than the surrounding soil. This difference in stiffness causes foundation motions to deviate from free-field motion as a result of base slab averaging and embedment effects.

Chapter 19 addresses both types of SSI effects. Procedures for calculating kinematic and inertial SSI effects were taken from recommendations in NIST GCR 12-917-21 (NIST 2012). Further discussion of SSI effects can be found in this NIST document and some of the references cited therein.

Substantial revisions have been made to Chapter 19 in this edition of ASCE 7. They include

1. The introduction of formulas for the stiffness and damping of rectangular foundations;
2. Revisions to the formulas for the reduction of base shear caused by SSI;
3. Reformulation of the effective damping ratio of the SSI system;
4. Introduction of an effective period lengthening ratio, which appears in the formula for the effective damping ratio of the SSI system, and which depends on the expected structural ductility demand; and
5. The introduction of kinematic SSI provisions.

Most of these revisions come from the NIST GCR 12-917-21 (NIST 2012) report on SSI. However, the basic model of the inertial SSI system has remained the same since SSI provisions were first introduced in the ATC 3-06 report (ATC 1978).

The first effect, foundation deformation, is addressed by explicitly requiring the design professional to incorporate the deformation characteristics of the foundation into their analysis model. Including foundation deformations is essential for the understanding of SSI. Therefore, the flexibility of the foundation must be modeled to capture the translational and rotational movement of the structure at the soil–foundation interface.

For the linear procedures, this requirement to model the flexibility of the foundation and soil means that springs should be placed in the model to approximate the effective linear stiffness of the deformations of the underlying geologic media and the foundation elements. This could be done by placing isolated spring elements under the columns and walls, by explicitly modeling the foundation elements and geologic media in the mathematical model, or some combination of the two. For the response history procedure, this would mean that, in addition to the stiffness of the subsurface media and foundation elements, the nonlinear parameters of those materials would be incorporated into the analytical model. Because of the uncertainty in estimating the stiffness and deformation capacity of geologic media, upper and lower bound estimates of the properties should be used and the condition that produces the more conservative change in response parameters from a fixed-base structure must be used.

Inertial interaction effects are addressed through the consideration of foundation damping. Inertial interaction in structures tends to be important for stiff structural systems such as shear walls and braced frames, particularly where the foundation soil is relatively soft. The provisions provide a method for estimating radiation damping and soil hysteretic damping.

The two main kinematic interaction effects are included in these provisions: base slab averaging and embedment effects. The kinematic interaction effects cause the motion input into the structure to be different from the free-field motions. The provisions provide a means by which a free-field, site-specific response spectrum can be modified to account for these kinematic interaction effects to produce a foundation-input spectrum.

Site Classes A and B are excluded from Chapter 19 because the dynamic interaction between structures and rock is minimal and based on theory. Furthermore, there are no empirical data to indicate otherwise.

Section 19.1.1 prohibits using the cap of S_s included in Section 12.8.1.3 because of the belief that structures meeting the requirements of that section have performed satisfactorily in past earthquakes, partially because of SSI effects. Taking advantage of that predetermined cap on S_s, and then subsequently reducing the base shear caused by SSI effects, may therefore amount to double-counting the SSI effects.

Section 19.1.1 prohibits the reduction in foundation overturning in Section 12.13.4 with SSI effects because there is not adequate research to demonstrate that the response parameter reductions are additive. The reduction in overturning in Section 12.13.4 is based on the equivalent lateral force procedure, and to a lesser extent, the modal response spectrum procedure, potentially overpredicting the overturning moments in multi-degree-of-freedom systems. The equations that predict the radiation damping reductions are based on single-degree-of-freedom models on elastic media, creating a very specific two-degree-of-freedom system. In these models, the multi-degree-of-freedom building is idealized as an equivalent single-degree-of-freedom model. The idealization does not take into account the overturning reduction, so there is no basis for further response parameter reductions. Therefore, the committee chose to prohibit the direct combination of the two reductions until further research is conducted.

C19.2 SSI ADJUSTED STRUCTURAL DEMANDS

When the equivalent lateral force procedure is used, the equivalent lateral force is computed using the period of the flexible base structure and is modified for the SSI system damping. For the modal analysis procedure, a response spectrum, which has been modified for the SSI system damping and then divided by (R/I_e), is input into the mathematical model. The lower bound limit on the design base shear based on the equivalent lateral force procedures per Section 12.9.1.4 still applies, but the equivalent lateral force base shear modified to account for SSI effects replaces the base shear for the fixed-base case.

For both the equivalent lateral force and response spectrum procedures, the total reduction caused by SSI effects is limited to a percentage of the base shear, determined in accordance with Section 12.8.1, which varies based on the R factor. This limitation on potential reductions caused by SSI reflects the limited understanding of how the effects of SSI interact with the R factor. All of the SSI effects presented herein are based on theoretical linear elastic models of the structure and geologic media. That is why reductions of 30% are permitted for $R = 3$ or less. It is believed that those systems exhibit limited inelastic response and therefore a larger reduction in the design force caused by SSI should be permitted. For higher R factor systems, where significant damping caused by structural yielding is expected, the contribution of foundation damping is assumed to have little effect on the reduction of the response. Some reduction is permitted because of (1) an assumed period lengthening resulting from the incorporation of base flexibility, (2) potential reduction in mass participation in the fundamental mode because two additional degrees of freedom are present caused by translation and rotation of the base, and (3) limited foundation damping interacting with the structural damping.

Reductions to the response spectrum caused by the SSI system damping and kinematic SSI effects are for the elastic 5% damped response spectrum typically provided to characterize free-field motion. In addition, studies have indicated that there is a fair amount of uncertainty in the amount of kinematic SSI when measured reductions between the free-field motion and the foundation input motion are compared with the theoretical models (Stewart 2000).

Reductions for kinematic SSI effects are not permitted for the equivalent lateral force and modal response spectrum procedures. The equations for predicting the kinematic SSI effects are based on modifications to the linear elastic response spectrum. Studies have not been performed to verify if they are similarly valid for inelastic response spectra, on which the R factor procedures are based. In addition, the amount of the reduction for kinematic SSI effects is dependent on the period of the structure, with the greatest modifications occurring in the short period range. Because the fundamental periods of most structures lengthen as they yield, what would potentially be a significant reduction at the initial elastic period may become a smaller reduction as the structure yields. Without an understanding of how the period may lengthen in the equivalent lateral force or modal response spectrum procedures, there is a potential for a user to overestimate the reduction in the response parameters caused by kinematic SSI effects. Thus, their use is not permitted.

All types of SSI effects are permitted to be considered in a response history analysis per Chapter 16. If SSI effects are considered, the site-specific response spectrum should be used as the target to which the acceleration histories are scaled. The requirement to use a site-specific response spectrum was placed in the provisions because of the belief that it provided a more realistic definition of the earthquake shaking than is provided by the design response spectrum and MCE_R response spectrum in accordance with Sections 11.4.6 and 11.4.7. A more realistic spectrum was required for proper consideration of SSI effects, particularly kinematic SSI effects. The design response spectrum and MCE_R response spectrum, in accordance with Sections 11.4.6 and 11.4.7, use predetermined factors to modify the probabilistic or deterministic response spectrum for the soil conditions. These factors are sufficient for most design situations. However, if SSI effects are to be considered and the response spectrum modified accordingly, then more accurate representations of how the underlying geologic media alter the spectral ordinates should be included before the spectrum is modified because of the SSI effects.

A site-specific response spectrum that includes the effects of SSI can be developed with explicit consideration of SSI effects by modifying the spectrum developed for free-field motions through the use of the provisions in Sections 19.3 and 19.4. If the foundation damping is not specifically modeled in the analytical model of the structure, the input response spectrum can include the effects of foundation damping. Typically, the base slab averaging effect is not explicitly modeled in the development of a site-specific response spectrum and the provisions in Section 19.4.1 are used to modify the free-field, site-specific response spectrum to obtain the foundation input spectrum. Embedment effects can be modeled directly by developing the site-specific spectrum at the foundation base level, as opposed to the ground surface. Alternatively, the site-specific spectrum for the free field can be developed at the ground level and the provisions of Section 19.4.2 can be used to adjust it to the depth corresponding to the base of foundation.

The limitations on the reductions from the site-specific, free-field spectrum to the foundation input spectrum are based on several factors. The first is the scatter between measured ratios of foundation input motion to free-field motion versus the ratios from theoretical models (Stewart 2000). The second is the inherent variability of the properties of the underlying geologic media over the footprint of the structure. Whereas there is a requirement to bound the flexibility of the soil and foundation springs, there are no corresponding bounding requirements applied to the geologic media parameters used to compute the foundation damping and kinematic SSI. The last factor is the aforementioned lack of research into the interaction between SSI effects and yielding structures. Some studies have shown that there are reductions for most cases of SSI when coupled with an *R* factor-based approach (Jarernprasert et al. 2013).

A limitation was placed on the maximum reduction for an SSI modified site-specific response spectrum with respect to the response spectrum developed based on the USGS ground-motion parameters and the site coefficients. This limitation is caused by similar concerns expressed in Section C21.3 regarding the site-specific hazard studies generating unreasonably low response spectra. There is a similar concern that combining SSI effects with site-specific ground motions could significantly reduce the seismic demand from that based on the USGS ground-motion parameters and the site coefficients. However, it was recognized that these modifications are real and the limit could be relaxed, but not eliminated, if there were (1) adequate peer review of the site-specific seismic hazard analysis and the methods used to determine the reductions attributable to SSI effects and (2) approval of the Authority Having Jurisdiction.

The peer review should include, but not be limited to, the following:

1. Confirmation of the process for the site-specific response spectrum used to scale the ground motions;
2. Confirmation that the determination of foundation stiffness and damping, including the properties of the underlying subsurface media used in the determination is appropriate; and
3. Confirmation that the base slab and first slab above the base are sufficiently rigid to allow base slab averaging to occur, including verification that the base slab is detailed to act as a diaphragm; and
4. Validating the assumptions used in the development of the soil and radiation damping ratios.

The SSI effects can be used in a response history analysis, per Chapter 16. Two options for the modeling of the SSI are as follows:

1. Create a nonlinear finite-element (FE) model of the structure, foundation, and geologic media. The mesh for the geologic media should extend to an appropriate depth and horizontal distance away from the foundation with transmitting boundaries along the sides to absorb outgoing seismic waves generated by the foundation. The motion should be input at the base of the FE model and should propagate upward as shear waves. The free-field response spectrum can be reduced for kinematic SSI only per the provisions in Section 19.4, but embedment effects would not be allowed in the reduction because the waves propagating up from the depth of the foundation to the surface would automatically include kinematic effects of embedment.
2. Create a nonlinear finite element model of the structure and foundation, with springs and dashpots attached to the perimeter walls and base of the foundation to account for the soil–foundation interaction. Guidance on the development of dashpots can be found in NIST GCR 12-917-21 (NIST 2012). The free-field response spectrum can be reduced for kinematic SSI per Section 19.4, but embedment effects may or may not be allowed in the reduction depending on whether or not (1) the motion is allowed to vary with depth along the embedded portion of the foundation, and (2) the free-field motion used as input motion is defined at the ground surface or at the bottom of the basement. The dashpots would account for the radiation and hysteretic damping of the geologic media, either per Section 19.3 or more detailed formulations.

C19.3 FOUNDATION DAMPING EFFECTS

The procedures in Section 19.3 are used to estimate an SSI system damping ratio, β_0, based on the underlying geologic media and interaction of the structure and its foundation with this geologic media.

Tables 19.3-1, 19.3-2, and 19.3-3 provide values for three parameters that are used in the evaluation of damping in an SSI system: (1) effective shear wave velocity ratios (Table 19.3-1), (2) effective shear modulus ratios (Table 19.3-2), and (3) soil hysteretic damping ratios (Table 19.3-3). These parameters represent different effects of soil nonlinearity, which has a fundamental dependence on shear strain. Strain levels are indirectly represented in the tables by different site classes and different ranges of effective peak acceleration. All other factors being equal, strains (and nonlinear effects on the respective parameters) increase as site conditions soften and effective peak accelerations increase. For each of the three identified tables, new values of the associated ratios were added to account for the new site classes added in Chapter 20.

There are two main components that contribute to foundation damping: soil hysteretic damping and radiation damping. The provisions in this section provide simplified ways to approximate these effects. However, they are complex phenomena and there are considerably more detailed methods to predict their effects on structures. The majority of the provisions in this section are based on material in NIST GCR 12-917-21 (NIST 2012). Detailed explanations of the background of these provisions, supplemental references, and more sophisticated methods for predicting radiation damping can be found in that report. However, those references do not provide the derivation of the effective period lengthening ratio, $(\tilde{T}/T)_{\mathrm{eff}}$ given by Equation (19.3-2). This ratio appears in the equation for β_0 [Equation (19.3-1)], and it is derived from the total displacement of the mass of the SSI oscillator model, resulting from a horizontal force applied to the mass. A component of this displacement is the displacement of the mass relative to its base, and it is equal to the ductility demand, μ, times the elastic displacement of the mass relative to the base. The other components of the total displacement arise from displacement of the translational foundation spring (K_y or K_r) and the translation resulting from the rotational foundation spring (K_{xx} or K_{rr}). The period lengthening ratio, $(\tilde{T}/T)$ appearing in Equation (19.3-2), is derived in the same manner assuming that $\mu = 1$.

Radiation damping refers to energy dissipation from wave propagation away from the vibrating foundation. As the ground shaking is transmitted into the structure's foundation, the structure itself begins to translate and rock. The motion of the foundation relative to the free-field motion creates waves in the geologic media, which can act to counter the waves being transmitted through the geologic media caused by the earthquake shaking. The interference is dependent on the stiffness of the geologic media and the structure, the size of the foundation, type of underlying geologic media, and period of the structure. The equations for radiation damping in Section 19.3.3 were taken from NIST GCR 12-917-21 (NIST 2012); details of the derivation are found in Givens (2013).

In Section 19.3.3, the equations for K_y and K_{xx}, for rectangular foundations, and the associated damping ratios, β_y and β_{xx}, come from Pais and Kausel (1988) and are listed in Table 2-2a and Table 2-3a in NIST (2012). The corresponding static stiffness equations for circular foundations in Section 19.3.4 were taken from Veletsos and Verbic (1973); the other equations appearing in Section 19.3.4 were adapted from equations in NIST (2012). The foundation stiffness and damping equations in these two sections apply to surface foundations. The reasons for excluding embedment effects are explained in the third paragraph from the end of this subsection.

Soil hysteretic damping occurs because of shearing within the soil and at the soil–foundation interface. Values of the equivalent viscous damping ratio, β_s, to model the hysteretic damping can be obtained from site response analysis or Table 19.3-3.

Foundation damping effects, modeled by β_f, tend to be important for stiff structural systems such as shear walls and braced frames, particularly where they are supported on relatively soft soil sites, such as Site Classes D, DE, and E. This effect is determined by taking the ratio of the fundamental period of the structure, including the flexibility of the foundation and underlying subsurface media (flexible-base model) and the fundamental period of the structure assuming infinite rigidity of the foundation and underlying subsurface media (fixed-base model). Analytically, this ratio can be determined by computing the period of the structure with the foundation/soil springs in the model and then replacing those springs with rigid support.

Figure C19.3-1 illustrates the effect of the period ratio, $\tilde{T}/T$, on the radiation damping, β_r, which typically accounts for most of the foundation damping. $\tilde{T}/T$ is the ratio of the fundamental period of the SSI system to the period of the fixed-base structure. Figure C19.3-1 shows that for structures with larger height, h, to foundation half-width, B, aspect ratios, the effects of foundation damping become less. In this figure, the aspect ratio of the foundation is assumed to be square.

These inertial interaction effects are influenced considerably by the shear modulus of the underlying subgrade, specifically the modulus that coincides with the seismic shaking being considered. As noted in the standard, shear modulus G can be evaluated from small-strain shear wave velocity as $G = (G/G_o) G_o = (G/G_o)\gamma v_{so}^2/g$ (all terms defined in the standard). Shear wave velocity, v_{so}, should be evaluated as the average small-strain shear wave velocity within the effective depth of influence below the foundation. The effective depth should be taken as half the lesser dimension of the foundation, which, in the provisions, is defined as B. Methods for measuring v_{so} (preferred) or

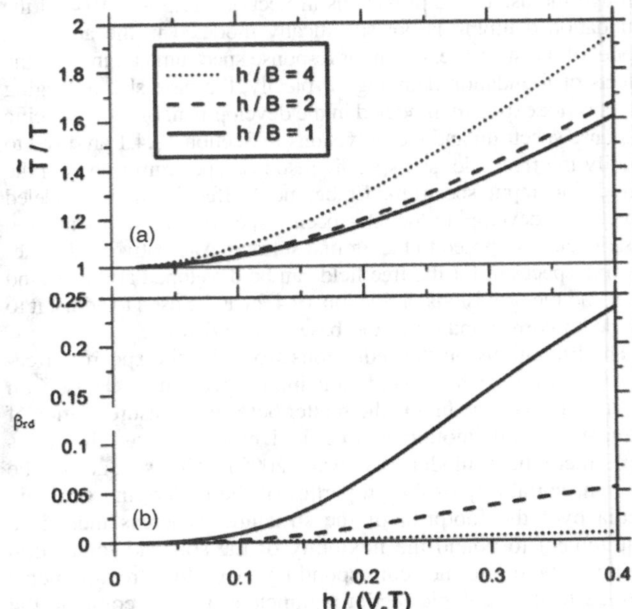

Figure C19.3-1. Example of radiation damping.
Source: NIST (2012).

estimating it from other soil properties are summarized elsewhere (e.g., Kramer 1996).

The radiation damping procedure is conservative and underestimates the foundation damping for shaking in the long direction where the foundation aspect ratios exceed 2:1 but could potentially be unconservative where wall and frame elements are close enough so that waves emanating from distinct foundation components destructively interfere with each other across the period range of interest. That is why the limit of spacing of the vertical lateral force-resisting elements is imposed on the use of these provisions.

For structures supported on footings, the formulas for radiation damping can generally be used with B and L calculated using the footprint dimensions of the entire structure, provided that the footings are interconnected with grade beams and/or a sufficiently rigid slab on grade. An exception can occur for structures with both shear walls and frames, for which the rotation of the foundation beneath the wall may be independent of that for the foundation beneath the column (this type is referred to as weak rotational coupling). In such cases, B and L are often best calculated using the dimensions of the wall footing. Very stiff foundations like structural mats, which provide strong rotational coupling, are best described using B and L values that reflect the full foundation dimension. Regardless of the degree of rotational coupling, B and L should be calculated using the full foundation dimension if foundation elements are interconnected or continuous. Further discussion can be found in FEMA 440 (FEMA 2005) and NIST GCR 12-917-21 (NIST 2012).

The radiation damping provisions conservatively exclude the effects of embedment. Embedment typically increases the amount of radiation damping if the basement or below-grade foundation stays in contact with the soil on all sides. Because there is typically some gapping between the soil and the sides of the basement or foundation, these embedment effects may be less than the models predict. There are some additional issues with the procedures for embedded foundations. In the case where the embedment is significant but the soils along the sides are much more flexible than the bearing soils, a high impedance contrast between the first two layers is recognized as a potential problem regardless of the embedment. The NIST GCR 12-917-21 (NIST 2012) report therefore recommends ignoring the additional contributions caused by embedment but still using the soil properties derived below the embedded base.

The equations in Sections 19.3.3 and 19.3.4 are for shallow foundations. This is not to say that radiation damping does not occur with deep (pile or caisson) foundation systems, but the phenomenon is more complex. Soil layering and group effects are important, and there are the issues of the possible contributions of the bottom structural slab and pile caps. Because the provisions are based on the impedance produced by a rigid plate in soil, these items cannot be easily taken into account. Therefore, more detailed modeling of the soil and the embedded foundations is required to determine the foundation impedances. The provisions permit such modeling but do not provide specific guidance for it. Guidance can be found, for example, in NIST GCR 12-917-21 (NIST 2012) and its references.

Soil hysteretic damping occurs as seismic waves propagate through the subsurface media and reach the base of the structure, and it can have an effect on the overall system damping when the soil strains are high. Table 19.3-3 was derived based on relationships found in EPRI (1993) and Vucetic and Dobry (1991) that relate the ratio between G/G_0 to cyclic shear strain in the soil and then to soil damping. The values in the table are based on conservative assumptions about overburden pressures on granular soils and plasticity index of clayey soils. This simplified approach does not preclude the geotechnical engineer from providing more detailed estimates of soil damping. However, the cap on reductions in the seismic demand are typically reached at around an additional 5% hysteretic damping ratio (10% total damping ratio), and further reductions would require peer review.

C19.4 BASE SLAB AVERAGING AND EMBEDMENT (KINEMATIC) SSI EFFECTS

Kinematic SSI effects are broadly defined as the difference between the ground motion measured in a free-field condition and the motion which would be measured at the structure's foundation, assuming that it and the structure were massless (i.e., inertial SSI was absent). The differences between free-field and foundation input motions are caused by the characteristics of the structure foundation, exclusive of the soil and radiation damping effects in the preceding section. There are two main types of kinematic interaction effects: base slab averaging and embedment. The provisions provide simplified methods for capturing these effects. The basis for the provisions and additional background material can be found in FEMA 440 (FEMA 2005) and NIST GCR 12-917-21 (NIST 2012).

FEMA 440 (FEMA 2005) specifically recommends against applying these provisions to very soft soil sites such as E and F. These provisions allow kinematic SSI for Site Class E but retain the prohibition for Site Class F. That is not to say that kinematic interaction effects are not present at Site Class F sites, but that these specific provisions should not be used; rather, more detailed site-specific assessments are permitted to be used to determine the possible modifications at those sites.

In addition to the prescriptive methods contained in the standard, there are also provisions that allow for direct computation of the transfer function of the free-field motion to a foundation input motion caused by base slab averaging or embedment. Guidance on how to develop these transfer functions can be found in NIST GCR 12-917-21 (NIST 2012) and the references contained therein.

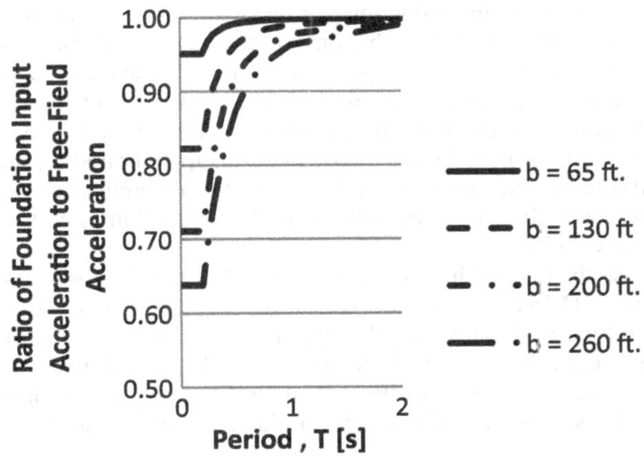

Figure C19.4-1. Example of base slab averaging response spectra ratios.

C19.4.1 Base Slab Averaging Base slab averaging refers to the filtering of high-frequency portions of the ground shaking caused by the incongruence of motion over the base. For this filtering to occur, the base of the structure must be rigid or semirigid with respect to the vertical lateral force-resisting elements and the underlying soil. If the motions are out of phase from one end of the foundation to the other and the foundation is sufficiently rigid, then the motion on the foundation would be different from the ground motion at either end. The ground motions at any point under the structure are not in phase with ground motions at other points along the base of the structure. That incongruence leads to interference over the base of the structure, which translates into the motions imparted to the foundation, which are different from the ground motions. Typically, this phenomenon results in a filtering out of short-period motions, which is why the reduction effect is much more pronounced in structures with short fundamental periods, as shown in Figure C19.4-1.

Figure C19.4-1 illustrates the increase in reduction as the base area parameter, b_e, increases. This parameter is computed as the square root of the foundation area. Therefore, for larger foundations, base slab averaging effects are more significant.

For base slab averaging effects to occur, foundation components must be interconnected with grade beams or a concrete slab that is sufficiently stiff to permit the base to move as a unit and allow this filtering effect to occur. That is why requirements are placed on the rigidity of the foundation diaphragm relative to the vertical lateral force-resisting elements at the first story. In addition, requirements are placed on the floor diaphragm or roof diaphragm, in the case of a one-story structure needing to be stiff in order for this filtering of ground motion to occur. FEMA 440 (FEMA 2005) indicates that there is a lack of data regarding this effect when either the base slab is not interconnected or the floor diaphragms are flexible. It is postulated that reductions between the ground motion and the foundation input motion may still occur. Because cases like this have not been studied in FEMA 440 (FEMA 2005) and NIST GCR 12-917-21 (NIST 2012) explicitly, the requirements for foundation connectivity and stiff or rigid diaphragms above the foundation have been incorporated into the provisions.

The underlying models have only been studied up to an effective base size of 260 ft (79.2 m), which is why that limitation has been placed on Equation (19.4-4). FEMA 440 (FEMA 2005) postulates that this effect is likely to still occur for larger base areas, but there has not been sufficient study to compare the underlying equations to data at larger effective base sizes.

Also, because the reduction can become quite significant and because studies of these phenomena have indicated variability between the theoretically predicted modifications and actual measured modifications (Stewart et al. 1999, Stewart 2000), a 0.75 factor is applied to the equations that are found in NIST GCR 12-917-21 (NIST 2012) to provide an upper bound estimate of the reduction factors with respect to the theoretical models. This is why the equations differ from those found in FEMA 440 (FEMA 2005).

Last, the method has not been rigorously studied for structures on piles (NIST 2012); however, it is considered reasonable to extend the application to pile-supported structures in which the pile caps are in contact with the soil and are laterally connected to one another. Another justification is that some of the empirical data for kinematic SSI come from pile-supported structures.

C19.4.2 Embedment The kinematic interaction effects caused by embedment occur because the seismic motions vary with depth below the ground surface. It is common for these effects to be directly considered in a site-specific response spectrum by generating response spectra and acceleration histories at the embedded base of the structure instead of the ground surface. If that is not done, then these effects can be accounted for using the provisions in this section. However, these provisions should not be used if the response spectrum has already been developed at the embedded base of the structure. The embedment effect model was largely based on studies of structures with basements. The provisions can also be applied to structures with embedded foundations without basements where the foundation is laterally connected at the plane taken as the embedment depth. However, the provisions are not applicable to embedded individual spread footings.

As with base slab averaging, the reduction can become quite significant, and studies of these phenomena have indicated variability between the theoretically predicted modifications and actual measured modifications (Stewart et al. 1999). Again, a 0.75 factor is applied to the equations found in NIST GCR 12-917-21 (NIST 2012) to provide a slightly conservative estimate of the reductions with respect to the theoretical models. This is why the equations differ from those found in FEMA 440 (FEMA 2005) and NIST GCR 12-917-21 (NIST 2012). In addition, the underlying models on which the provisions are based have only been validated in NIST GCR 12-917-21 (NIST 2012) up to an effective embedment depth of approximately 20 ft (6.096 m), which is why a depth limitation has been placed on Equation (19.2-3).

REFERENCES

ATC (Applied Technology Council). 1978. *Tentative provisions for the development of seismic regulations for buildings*. ATC-3-06. Redwood City, CA: ATC.

EPRI (Electrical Power Research Institute). 1993. *Guidelines for determining design basis ground motions*. EPRI TR-102293. Palo Alto, CA: EPRI.

FEMA (Federal Emergency Management Agency). 2005. *Improvement of nonlinear static seismic analysis procedures*. FEMA 440. Washington, DC: FEMA.

Givens, M. J. 2013. "Dynamic soil-structure interaction of instrumented buildings and test structures." Ph.D. thesis, Dept. of Civil and Environmental Engineering, University of California, Los Angeles.

Jarernprasert, S., E. Bazan-Zurita, and J. Bielak. 2013. "Seismic soilstructure interaction response of inelastic structures." *Soil Dyn. Earthquake Eng.* 47 (Apr): 132–143. https://doi.org/10.1016/j.soildyn.2012.08.008.

Kramer, S. L. 1996. *Geotechnical earthquake engineering*. Upper Saddle River, NJ: Prentice Hall.

NIST (National Institute of Standards and Technology). 2012. *Soil-structure interaction for building structures*. NIST GCR 12-917-21. Gaithersburg, MD: NIST.

Pais, A., and E. Kausel. 1988. "Approximate formulas for the dynamic stiffness of rigid foundations." *Soil Dyn. Earthquake Eng.* 7 (4): 213–227. https://doi.org/10.1016/S0267-7261(88)80005-8.

Stewart, J. P. 2000. "Variations between foundation-level and free-field earthquake ground motions." *Earthquake Spectra* 16 (2): 511–532. https://doi.org/10.1193/1.1586124.

Stewart, J. P., R. B. Seed, and G. L. Fenves. 1999. "Seismic soil-structure interaction in buildings. II: Empirical findings." *J. Geotech. Geoenviron. Eng.* 125 (1): 38–48. https://doi.org/10.1061/(ASCE)1090-0241(1999)125:1(38).

Veletsos, A. S., and B. Verbic. 1973. "Vibration of viscoelastic foundations." *Earthquake Eng. Struct. Dyn.* 2 (1): 87–105. https://doi.org/10.1002/eqe.4290020108.

Vucetic, M., and R. Dobry. 1991. "Effect of soil plasticity on cyclic response." *J. Geotech. Eng.* 117 (1): 89–107. https://doi.org/10.1061/(ASCE)0733-9410(1991)117:1(89).

CHAPTER C20
SITE CLASSIFICATION PROCEDURE FOR SEISMIC DESIGN

C20.1 SITE CLASSIFICATION

Site classification procedures are given in Chapter 20 for the purpose of classifying the site, which is required for the development of site-compatible, risk-targeted, maximum considered earthquake ground motions, in accordance with Section 11.4.3. Site classification procedures are also used to define the site conditions for which site-specific site response analyses are required to obtain site ground motions, in accordance with Section 11.4.7 and Chapter 21.

Site class is defined fundamentally in terms of ranges of site shear wave velocity ($\overline{v}_s$). Table 20.2-1 includes the six site classes of ASCE 7-16 (A, B, C, D, E, and F) plus three new site classes (BC, CD, and DE) that provide better resolution of site shear wave velocity and associated site amplification for common site conditions. The site-specific data required to classify a site can be developed from a geotechnical investigation, which may include seismic velocity testing and/or the development of other data on the site profile that can be used to estimate shear wave velocity as a function of depth. In cases where there is inadequate information on which to base a site classification, a "default" condition is defined. Section 11.4.2.1 defines the default site class as the most critical spectral response acceleration response for site conditions of Site Classes C, CD, and D, and the MCE_R response spectrum for default site conditions would be determined as the envelope of Site Class C, CD, and D MCE_R response spectra.

C20.2 SITE CLASS DEFINITIONS

C20.2.1 Site Class F The ground motion models used to develop MCE_R spectral ordinates in Section 11.4.3 are derived in part from ground motion recordings, which span a range of conditions of $\overline{v}_s$ between about 150 and 1,500 m/s. Site Class F conditions are largely not represented in the databases on which these ground motions are based; hence, the models are generally thought to be ineffective for such conditions. For this reason, site-specific site response analyses are required for Site Class F soils.

This section defines the types of site conditions for which Site Class F is assigned. For three of the categories of Site Class F soils—Category 1: liquefiable soils, Category 3: very high plasticity clays, and Category 4: very thick soft/medium stiff clays—exceptions to the requirement to conduct site response analyses are given, provided that certain conditions and requirements are satisfied. These exceptions are discussed as follows.

Category 1. For liquefiable soils in Category 1, an exception is made for short-period structures, defined for purposes of the exception as having fundamental periods of vibration equal to or less than 0.5 s. For such structures, it is permissible to develop ground motion under the assumption that liquefaction does not occur. This exception is based on the observation that ground motion data obtained in liquefied soil areas during earthquakes indicate that short-period ground motions are generally reduced in amplitude because of liquefaction, whereas long-period ground motions may be amplified by liquefaction (e.g., Youd and Carter 2005). Since 2005, other work has confirmed the amplification effects at long periods but has not always found reductions of short period ground motion (e.g., Gingery et al. 2015). Note, however, that this exception does not affect the requirement in Section 11.8 to assess liquefaction potential as a geologic hazard and develop hazard mitigation measures, if required.

Categories 3 and 4. For very high plasticity clays in Category 3 and very thick soft/medium stiff clays in Category 4, site-specific response analyses are required because of the potential for large site-specific amplification effects that are concentrated at one or more site periods. Such effects are not captured well by the site amplification effects incorporated into ground motion models. Procedures for conducting such analyses are described in Chapter 21 and in literature on non-ergodic (i.e., site-specific) site response (e.g., Stewart et al. 2014, 2017; Rodriguez-Marek et al. 2014; NCHRP 2012). Exceptions for Categories 3 and 4 were limited to sites of expected low amplitude ground motions (i.e., Seismic Design Categories A and B, as defined in Tables 11.6-1 and 11.6-2).

Sections C20.2.2 through C20.2.5. These sections and Table 20.2-1 provide definitions for Site Classes A through E. Except for the additional definitions for Site Class E in Section 20.2.2, the site classes are defined fundamentally in terms of the average small-strain shear wave velocity measured from the ground surface to a depth of 100 ft (30 m) of the soil or rock profile.

In general, the soil profile should not be taken from the foundation-level elevation downward when that elevation is below the natural ground surface, which is sometimes done with the intent of accounting for foundation embedment effects on ground motions. Ground motions at the foundation level of embedded structures are lower than those at the ground surface. Such reductions can be especially pronounced for sites with a soft soil layer overlying a much stiffer material, and for which the planned structure is embedded and bearing on the stiffer layer. These effects can be accounted for in an approximate manner using models for kinematic soil-structure interaction in Chapter 19 and NIST (2012) or with site-specific ground response analyses. In the case of site-specific ground response analyses, ground motions should be computed at the foundation-level elevations and at the ground surface, and the ratio of the response spectral ordinates computed. This ratio can be applied to the ground surface spectrum to estimate the foundation-level spectrum. The use of either approximate or site-specific methods will reduce short period spectral ordinates (i.e., periods shorter than the fundamental period of the portion of the profile within the embedment depth of the structure).

If shear wave velocities are available for the site, they should be used to evaluate $\overline{v}_s$ for site classification per Section 20.4.

When measured shear wave velocities in soil materials are not available for the site, shear wave velocity can be estimated based on appropriate correlations (Section 20.3). If these correlation relationships are used, mean shear wave velocity should be estimated as a function of depth, and $\overline{v}_s$ should be computed from the mean profile, using procedures in Section 20.4.

When measured shear wave velocities in competent rock with moderate weathering and/or fracturing are not available for the site, Site Class BC should be assigned (Section 20.2.4). For softer rock with a higher degree of fracturing and/or weathering, if $\overline{v}_s$ is not based on measurement, Site Class C should be assigned. Site Classes A and B can only be assigned on the basis of seismic velocity measurements (Section 20.2.5).

C20.3 ESTIMATION OF SHEAR WAVE VELOCITY PROFILES

When measured shear wave velocities are not available for a site, shear wave velocity can be estimated from geotechnical data using appropriate correlations. Correlation relationships predict the mean shear wave velocity and variability (typically expressed as a standard deviation or coefficient of variation) given a series of independent variables. Some correlations are based on large data sets encompassing multiple regions ("global" models), while other correlations are derived from local data specific to the geology of a particular region. Where local models are available, and the models are of good quality with appropriate documentation, their use is preferred to global models.

When $\overline{v}_s$ is estimated in this manner, the associated uncertainty has been found to be approximately 0.22 to 0.26 in natural logarithmic units when local shear wave velocity correlation models are used, that is, for California (Brandenberg et al. 2010) and Japan (Kwak et al. 2015). This indicates that there is approximately a 68% chance that the actual value of $\overline{v}_s$ is between $\overline{v}_s/1.3$ and 1.3 $\overline{v}_s$. This will result in two or three site classes being assigned to the site. When a global correlation model is used to estimate shear wave velocity, or a combination of local models is used outside of their calibration region, the associated uncertainty in $\overline{v}_s$ is unknown. However, to avoid undue complexity in the implementation of estimated $\overline{v}_s$ values, the same factor of 1.3 was retained for this case.

The independent variables used in correlation models take two major forms: (1) parameters derived from laboratory tests (shear strength, void ratio, water content), and (2) penetration resistance in combination with either depth or effective stress along with information on soil type. Examples of local and geologic unit-specific models of the first type are provided for the Los Angeles area by Fumal and Tinsley (1985) and various clay units in the San Francisco Bay Region by Dickenson (1994), as shown in Figures C20.3-1 through C20.3-3.

Correlation models that use penetration resistance consider three main types of penetration tests: standard penetration testing (SPT), cone penetration testing (CPT), and Becker penetration testing (BPT). SPT is most effective when applied to sandy soils with little or no gravel content. CPT is suitable for fine-grained soils (clays, silts) and sands, but not coarser-grained materials such as gravels, due to difficulties with penetration. The BPT is preferred for gravels.

A general form for V_S prediction models based on penetration resistance is

$$lnV_s = c_0 + c_1 ln(PR) + c_2 ln f_z + \varepsilon\sigma_{lnVs} \qquad \text{(C20.3-1)}$$

where

PR = Measure of penetration resistance;

f_z = A parameter related to depth, either taken as depth directly or as an effective stress parameter; and

ε = The standard normal variate (taken as 0 for the mean and +1 for one standard deviation above the mean), and σ_{lnVs} is the total standard deviation of lnV_s.

One such local model derived using data in California (Brandenberg et al. 2010) takes PR as the SPT blow count at 60% energy efficiency (N_{60}) and f_z as the vertical effective stress, σ'_v, in units of kPa. V_s is provided in units of m/s. The coefficients are soil type-dependent, as given in Table C20.2-1. A variation on this approach for Japanese data has coefficients conditioned on surface geology (Kwak et al. 2015), and future versions of the

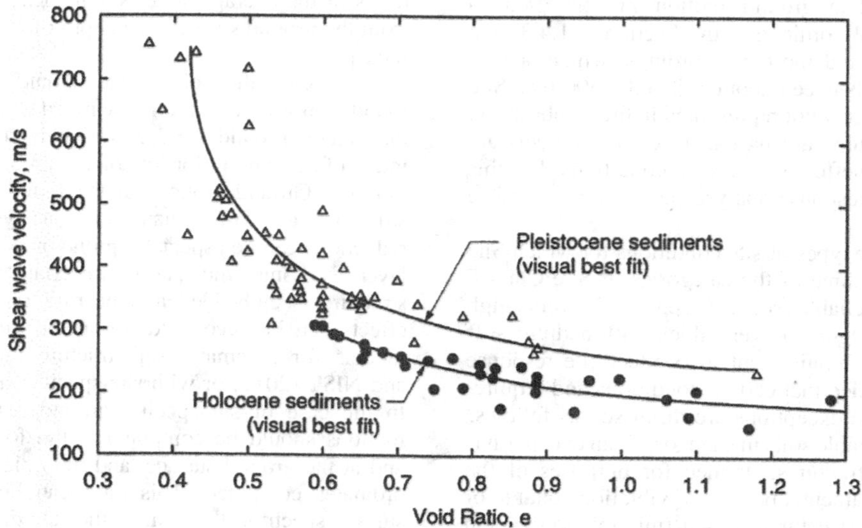

Figure C20.3-1. Variation of shear wave velocity with void ratio for Los Angeles area sediments.
Source: Modified from Fumal and Tinsley (1985).

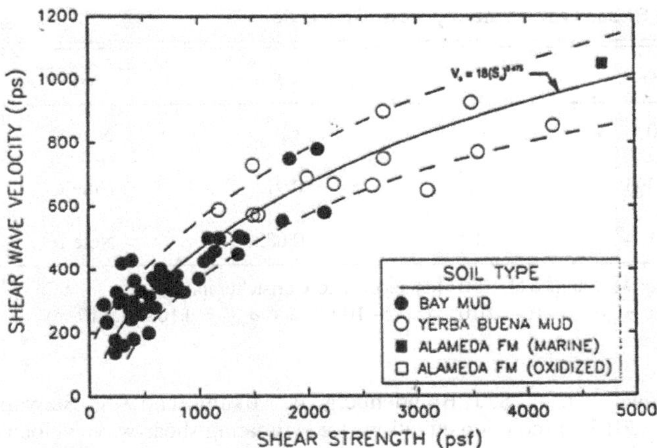

Figure C20.3-2. Variation of shear wave velocity with undrained shear strength (from monotonic triaxial compression testing) for San Francisco Bay Area sediments.
Source: Dickenson (1994).

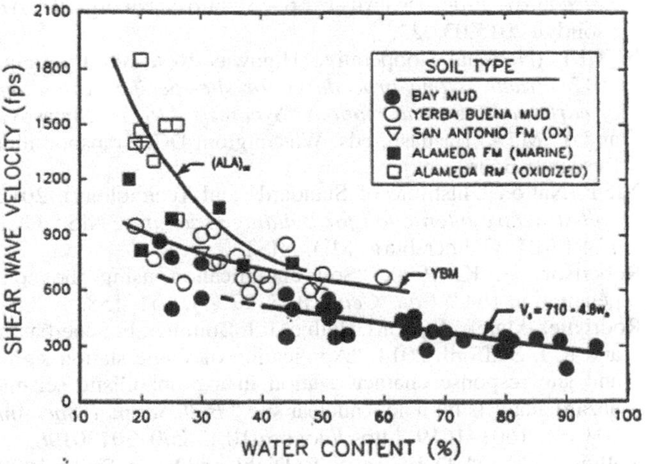

Figure C20.3-3. Variation of shear wave velocity with water content for San Francisco Bay Area sediments.
Source: Dickenson (1994).

California model will likely include this feature. The standard deviation term representing data variability relative to the model is $\sigma_{lnVs} = \sqrt{\tau^2 + \sigma^2}$, where τ is the inter-boring standard deviation and σ is the intra-boring standard deviation. Coefficients for both standard deviations are given in Table C20.3-1.

Table C20.3-1. Coefficients for Estimation of Shear Wave Velocity in Local Model Applicable to California.

					σ	
Soil type	c_0	c_1	c_2	τ	$\sigma'_v \le 200$ kPa	$\sigma'_v > 200$ kPa
Sand	4.045	0.096	0.236	0.217	$0.57 - 0.07 \ln \sigma'_v$	0.20
Silt	3.783	0.178	0.231	0.227	$0.31 - 0.03 \ln \sigma'_v$	0.15
Clay	3.996	0.230	0.164	0.227	$0.21 - 0.01 \ln \sigma'_v$	0.16

Source: After Brandenberg et al. (2010).

A variation on Equation C.20.3-1 is typically applied for CPT data:

$$\ln V_s = c_0 + c_1 \ln(q_t/p_a) + c_2 \ln f_z + c_3 \ln f_m \qquad \text{(C20.3-2)}$$

where

q_t = The CPT tip resistance,
p_a = Atmospheric pressure in the same units as q_t, and
f_m = A parameter used to assess material type (typically friction ratio, R_f, or soil behavior type index, I_c) (Robertson 1990).

In CPT models of this type, coefficients c_0 to c_3 are generally fixed (they are not soil type dependent). An exception is Hegazy and Mayne (2006), in which coefficient c_2 depends on I_c. That model, which is based on data from the United States, Italy, and Japan (and hence can be considered to be global), takes $f_z = p_a/\sigma'_v$, and $= exp(1.786 I_c)$. Andrus et al. (2007) present a model that is derived primarily from data from South Carolina and California, in which f_z is depth in meters and f_m is taken as I_c. McGann et al. (2015) presents a local model for Christchurch, New Zealand, in which f_z is depth in meters and $f_m = f_s/p_a$. Coefficients for these models are given in Table C20.3-2, with the resulting V_S in units of m/s.

Correlations based on BPT are limited in the literature, but one such model is provided by Rollins et al. (1998).

The correlation models described here are not intended to encompass all the models that might be used for prediction. Rather, the intent is to describe the general form of contemporary models to help engineers and regulatory officials judge the efficacy of a candidate model and to provide a set of models that may be applied in the absence of preferred alternate models for a particular region.

Although local models are preferred to global models, there are many regions for which local models either may not exist or may have unknown efficacy. In such cases, global models, such as Hegazy and Mayne (2006), could be used, or a series of local models for other regions could be applied. In the latter case, for each depth in the profile, a geometric mean velocity should be taken, and the resulting profile used to compute $\overline{v}_s$.

Equation (20.4-1) requires a profile that extends to a depth of 100 ft (30 m) or greater. If the available geophysical or geotechnical information for the site ends at shallower depths [50 ft or greater (15 m or greater)], it is possible to estimate $\overline{v}_s$ based on the information over the available depth range if the geological conditions at the site are suitable. Such suitability is judged based on the potential for velocity inversions (i.e., a marked decrease in velocity as depth increases); where such inversions could reasonably be anticipated based on the site geology, the use of shallow site characteristics to assess $\overline{v}_s$ is not recommended. In such cases, geotechnical characterization should extend to at least 100 ft (30 m). For sites where velocity inversions are not expected, a reasonable estimate of $\overline{v}_s$ can generally be obtained by extending the velocity of the last layer in the profile to 100 ft (30 m) for use in Equation (20.4-1). This approach provides an estimate of $\overline{v}_s$ that is usually slightly lower than alternative methods that implicitly account for the effects of velocity gradient. A summary of such methods, and current best practices, is provided in Kwak et al. (2017).

C20.4 DEFINITIONS OF SITE CLASS PARAMETERS

Section 20.4 provides formulas for computing $\overline{v}_s$ from a shear wave velocity profile, which in turn is used to define site classes in accordance with definitions in Section 20.2 and Table 20.2-1. Equation (20.4-1) is for determining the effective average

Table C20.3-2. Coefficients for Estimation of Shear Wave Velocity from CPT Data.

Study	c_0	c_1	c_2	c_3	σ_{lnVs}
Hegazy and Mayne (2006); granular soils ($I_c < 2.6$)	−2.488	1.0	0.25	1.0	Note a
Hegazy and Mayne (2006); cohesive soils ($I_c < 2.6$)			0.50		
Andrus et al. (2007); Holocene materials	2.699	0.395	0.124	0.912	Note a
Andrus et al. (2007); Pleistocene materials	2.896				
McGann et al. (2015); Christchurch, NZ	3.959	0.144	0.278	0.0832	Note b

Source: Hegazy and Mayne (2006) and Andrus et al. (2007) are not region-specific; McGann et al. (2015) is specific to Christchurch, NZ.
Notes: (a) Standard deviation model not provided; (b) $\sigma_{lnVs} = 0.162$ (depth $z < 5$ m), $\sigma_{lnVs} = 0.216–0.0108z$ (z = 5–10 m), and $\sigma_{lnVs} = 0.108$ ($z > 10$ m)

small-strain shear wave velocity, $\bar{v}_s$, to a depth of 100 ft (30 m) at a site. This equation defines $\bar{v}_s$ as 100 ft (30 m) divided by the sum of the times for a shear wave to travel through each layer within the upper 100 ft (30 m), where travel time for each layer is calculated as the layer thickness divided by the small-strain shear wave velocity for the layer.

REFERENCES

Andrus, R. D., N. P. Mohanan, P. Piratheepan, B. S. Ellis, and T. L. Holzer. 2007. "Predicting shear-wave velocity from cone penetration resistance." In *Proc., 4th Int. Conf. Earthquake Geotechnical Engineering*, Thessaloniki, Greece.

Brandenberg, S. J., N. Bellana, and T. Shantz. 2010. "Shear wave velocity as function of standard penetration test resistance and vertical effective stress at California bridge sites." *Soil Dyn. Earthquake Eng.* 30 (10): 1026–1035. https://doi.org/10.1016/j.soildyn.2010.04.014.

Dickenson, S. A. 1994. "The dynamic response of soft and deep cohesive soils during the Loma Prieta earthquake of October 17, 1989." Ph.D. dissertation, Dept. of Civil Engineering, University of California.

Fumal, T. E., and J. C. Tinsley. 1985. "Mapping shear wave velocities of near surface geologic materials." In *Proc., Evaluating Earthquake Hazards in the Los Angeles Region —, a an Earth-Science Perspective, Paper 1360*, Reston VA: USGS.

Gingery, G. R., A. Elgamal, and J. D. Bray. 2015. "Response spectra at liquefaction sites during shallow crustal earthquakes." *Earthquake Spectra* 31 (4): 2325–2349. https://doi.org/10.1193/101813EQS272M.

Hegazy, Y. A., and P. W. Mayne. 2006. "A global statistical correlation between shear wave velocity and cone penetration data." In *Proc., Site and Geomaterial Characterization*, 243–248. https://doi.org/10.1061/40861(193)31

Kwak, D. Y., T. D. Ancheta, D. Mitra, S. K. Ahdi, P. Zimmaro, G. A. Parker, et al. 2017. "Performance evaluation of VSZ-to-VS30 correlation methods using global VS profile database." In *Proc., 3rd Int. Conf. on Performance-Based Design in Earthquake Geotechnical Engineering*, Vancouver, British Colombia. M. Taiebat, D. Wijewickreme, A. Athanasopoulos-Zekkos, and R. W. Boulanger, eds. Los Angeles: UCLA.

Kwak, D. Y., S. J. Brandenberg, A. Mikami, and J. P. Stewart. 2015. "Prediction equations for estimating shear-wave velocity from combined geotechnical and geomorphic indexes based on Japanese data set." *Bull. Seismol. Soc. Am.* 105 (4): 1919–1930. https://doi.org/10.1785/0120140326.

McGann, C. R., B. A. Bradley, M. L. Taylor, L. M. Wotherspoon, and M. Cubrinovski. 2015. "Development of an empirical correlation for predicting shear wave velocity of Christchurch soils from cone penetration test data." *Soil Dyn. Earthquake Eng.* 75 (Aug): 66–75. https://doi.org/10.1016/j.soildyn.2015.03.023.

NCHRP (National Cooperative Highway Research Program). 2012. *Practices and procedures for site-specific evaluations of earthquake ground motions: Synthesis 428*, N. Matasovic and Y. M. A. Hashash, eds. Washington, DC: Transportation Research Board.

NIST (National Institute of Standards and Technology). 2012. *Soil-structure interaction for building structures*. NIST GCR 12-917-21. Gaithersburg, MD: NIST.

Robertson, P. K. 1990. "Soil classification using the cone penetration test." *Can. Geotech. J.* 27 (1): 151–158.

Rodriguez-Marek, A., E. M. Rathje, J. J. Bommer, F. Scherbaum, and P. J. Stafford. 2014. "Application of single-station sigma and site response characterization in a probabilistic seismic hazard analysis for a new nuclear site." *Bull. Seismol. Soc. Am.* 104 (4): 1601–1619. https://doi.org/10.1785/0120130196.

Rollins, K. M., M. D. Evans, N. B. Diehl, and W. D. Daily. 1998. "Shear modulus and damping relationships for gravels." *J. Geotech. Geoenviron. Eng.* 124 (5): 396–405. https://doi.org/10.1061/(ASCE)1090-0241(1998)124:5(396).

Stewart, J. P., K. Afshari, and C. A. Goulet. 2017. "Non-ergodic site response in seismic hazard analysis." *Earthquake Spectra* 33 (4): 1385–1414. https://doi.org/10.1193/081716eqs135m.

Stewart, J. P., K. Afshari, and Y. M. A. Hashash. 2014. *Guidelines for performing hazard-consistent one-dimensional ground response analysis for ground motion prediction*. PEER Rep. No. 2014/16. Berkeley, CA: Pacific Earthquake Engineering Research Center.

Youd, T. L., and B. L. Carter. 2005. "Influence of soil softening and liquefaction on spectral acceleration." *J. Geotech. Geoenviron. Eng.* 131 (7): 811–825. https://doi.org/10.1061/(ASCE)1090-0241(2005)131:7(811).

CHAPTER C21
SITE-SPECIFIC GROUND MOTION PROCEDURES FOR SEISMIC DESIGN

Site-specific procedures for computing earthquake ground motions include dynamic site response analyses and probabilistic and deterministic seismic hazard analyses (PSHA and DSHA), which may include dynamic site response analysis as part of the calculation. Use of site-specific procedures may be required in lieu of the general procedure in Sections 11.4.2 through 11.4.7; Section C11.4.8 in ASCE 7-16 explains the conditions under which the use of these procedures is required. Such studies must be comprehensive and must incorporate current scientific interpretations. Because there is typically more than one scientifically credible alternative for models and parameter values used to characterize seismic sources and ground motions, it is important to formally incorporate these uncertainties in a site-specific analysis. For example, uncertainties may exist in seismic source location, extent, and geometry; maximum earthquake magnitude; earthquake recurrence rate; ground motion attenuation; local site conditions, including soil layering and dynamic soil properties; and possible two- or three-dimensional wave-propagation effects. The use of peer review for a site-specific ground motion analysis is encouraged.

Site-specific ground motion analysis can consist of one of the following approaches: (1) PSHA and possibly DSHA if the site is near an active fault, (2) PSHA/DSHA followed by dynamic site response analysis, and (3) dynamic site response analysis only. The first approach is used to compute ground motions for bedrock or stiff soil conditions (not softer than Site Class D). In this approach, if the site consists of stiff soil overlying bedrock, for example, the analyst has the option of either computing the bedrock motion from the PSHA/DSHA and then using the site coefficient (F_a and F_v) tables in Section 11.4.3 to adjust for the stiff soil overburden or computing the response spectrum at the ground surface directly from the PSHA/DSHA. The latter requires the use of attenuation equations for computing stiff soil-site response spectra (instead of bedrock response spectra).

The second approach is used where softer soils overlie the bedrock or stiff soils. The third approach assumes that a site-specific PSHA/DSHA is not necessary but that a dynamic site response analysis should, or must, be performed. This analysis requires the definition of an outcrop ground motion, which can be based on the 5% damped response spectrum computed from the PSHA/DSHA or obtained from the general procedure in Section 11.4. A representative set of acceleration time histories is selected and scaled to be compatible with this outcrop spectrum. Dynamic site response analyses, using these acceleration histories as input, are used to compute motions at the ground surface. The response spectra of these surface motions are used to define a maximum considered earthquake (MCE) ground motion response spectrum.

The approaches described in the aforementioned have advantages and disadvantages. In many cases, user preference governs the selection; geotechnical conditions at the site may dictate the use of one approach over the other. If bedrock is at a depth much greater than the extent of the site geotechnical investigations, the direct approach of computing the ground surface motion in the PSHA/DSHA may be more reasonable. On the other hand, if bedrock is shallow and a large impedance contrast exists between it and the overlying soil (i.e., density times shear wave velocity of bedrock is much greater than that of the soil), the two-step approach might be more appropriate.

Use of peak ground acceleration as the anchor for a generalized site-dependent response spectrum is discouraged because sufficiently robust ground motion attenuation relations are available for computing response spectra in western and eastern United States tectonic environments.

C21.1 SITE RESPONSE ANALYSIS

C21.1.1 Base Ground Motions Ground motion acceleration histories that are representative of horizontal rock motions at the site are required as input to the soil model. Where a site-specific ground motion hazard analysis is not performed, the MCE_R base response spectrum is developed assuming a site condition representative of base-condition average shear wave velocity, $\bar{v}_s$. USGS national seismic hazard mapping project website (www.earthquake.usgs.gov/hazards/products/conterminous) includes hazard deaggregation options that can be used to evaluate the predominant types of earthquake sources, magnitudes, and distances contributing to the probabilistic ground motion hazard. Sources of recorded acceleration time histories include the databases of the Consortium of Organizations for Strong Motion Observation Systems (COSMOS) Virtual Data Center (www.cosmos-eq.org), the Pacific Earthquake Engineering Research (PEER) Center Strong Motion Database (https://peer.berkeley.edu/products/strong_ground_motion_db.html), and the US National Center for Engineering Strong Motion Data (NCESMD) (https://www.strongmotioncenter.org). Ground motion acceleration histories at these sites generally were recorded at the ground surface and hence apply for an outcropping condition and should be specified as such in the input to the site response analysis code (for additional details, see Kwok et al. 2007).

C21.1.2 Site Condition Modeling Modeling criteria are established by site-specific geotechnical investigations that should include (1) borings with sampling; (2) standard penetration tests (SPTs), cone penetrometer tests (CPTs), and/or other subsurface investigative techniques; and (3) laboratory testing to establish the soil types, properties, and layering. The depth to rock or to stiff soil material should be established from these investigations. Investigation should extend to bedrock or, for very deep soil profiles, to material in which the model is terminated. Although it is preferable to measure shear wave velocities in all soil layers, it is also possible to estimate shear wave velocities based on measurements available for similar soils in the local area or through correlations with soil types and properties. A number of such correlations are summarized by Kramer (1996).

Typically, a one-dimensional soil column extending from the ground surface to bedrock is adequate to capture first-order site response characteristics. For very deep soils, the model of the soil columns may extend to very stiff or very dense soils at depth in the column. Two- or three-dimensional models should be considered for critical projects when two- or three-dimensional wave propagation effects may be significant (e.g., sloping ground sites). The soil layers in a one-dimensional model are characterized by their total unit weights and shear wave velocities from which low-strain (maximum) shear moduli may be obtained and by relationships defining the nonlinear shear stress–strain behavior of the soils. The required relationships for analysis are often in the form of curves that describe the variation of soil shear modulus with shear strain (modulus reduction curves) and by curves that describe the variation of soil damping with shear strain (damping curves). In a two- or three-dimensional model, compression wave velocities or moduli or Poisson ratios are also required. In an analysis to estimate the effects of liquefaction on soil site response, the nonlinear soil model must also incorporate the buildup of soil pore water pressures and the consequent reductions of soil stiffness and strength. Typically, modulus reduction curves and damping curves are selected on the basis of published relationships for similar soils (e.g., Vucetic and Dobry 1991, EPRI 1993, Darendeli 2001, Menq 2003, Zhang et al. 2005). Site-specific laboratory dynamic tests on soil samples to establish nonlinear soil characteristics can be considered where published relationships are judged to be inadequate for the types of soils present at the site. Shear and compression wave velocities and associated maximum moduli should be selected based on field tests to determine these parameters or, if such tests are not possible, on published relationships and experience for similar soils in the local area. The uncertainty in the selected maximum shear moduli, modulus reduction and damping curves, and other soil properties should be estimated (Darendeli 2001, Zhang et al. 2008). Consideration of the ranges of stiffness prescribed in Section 12.13.3 is recommended.

C21.1.3 Site Response Analysis and Computed Results Analytical methods may be equivalently linear or nonlinear. Frequently used computer programs for one-dimensional analysis include the equivalent linear program SHAKE (Schnabel et al. 1972, Idriss and Sun 1992) and the nonlinear programs FLAC (Itasca 2005), DESRA-2 (Lee and Finn 1978), MARDES (Chang et al. 1991), SUMDES (Li et al. 1992), D-MOD_2 (Matasovic 2006), DEEPSOIL (Hashash and Park 2001), TESS (Pyke 2000), and OpenSees (Ragheb 1994, Parra 1996, Yang 2000). If the soil response induces large strains in the soil (such as for high acceleration levels and soft soils), nonlinear programs may be preferable to equivalent linear programs. For analysis of liquefaction effects on site response, computer programs that incorporate pore water pressure development (effective stress analyses) should be used (e.g., FLAC, DESRA-2, SUMDES, D-MOD_2, TESS, DEEPSOIL, OpenSees). Response spectra of output motions at the ground surface are calculated as the ratios of response spectra of ground surface motions to input outcropping rock motions. Typically, an average of the response spectral ratio curves is obtained and multiplied by the input MCE response spectrum to obtain the MCE ground surface response spectrum. Alternatively, the results of site response analyses can be used as part of the PSHA using procedures described by Goulet et al. (2007) and programmed for use in OpenSHA (https://www.opensha.org; Field et al. 2005). Sensitivity analyses to evaluate effects of soil-property uncertainties should be conducted and considered in developing the final MCE response spectrum.

C21.2 RISK-TARGETED MAXIMUM CONSIDERED EARTHQUAKE (MCE_R) GROUND MOTION HAZARD ANALYSIS

Site-specific risk-targeted maximum considered earthquake (MCE_R) ground motions are based on separate calculations of site-specific probabilistic and site-specific deterministic ground motions.

Both the probabilistic and deterministic ground motions are defined in terms of 5% damped spectral response in the maximum direction of horizontal response. The maximum direction in the horizontal plane is considered the appropriate ground motion intensity parameter for seismic design, using the equivalent lateral force (ELF) procedure of Section 12.8, whose primary intent is avoiding collapse of the structural system.

Most ground motion relations are defined in terms of average (geometric mean) horizontal response. Maximum response in the horizontal plane is greater than average response by an amount that varies with period. Maximum response may be reasonably estimated by factoring average response by period-dependent factors of 1.2 at periods less than or equal to 0.2 s, by 1.3 for a period of 1.0 s, and by 1.5 for periods greater than or equal to 10 s (Resource Paper 4, NEHRP Provisions 2015, which is based on Shahi and Baker 2013). The maximum direction was adopted as the ground motion intensity parameter for use in seismic design, in lieu of explicit consideration of directional effects.

When performing a ground motion hazard analysis, it is important to consider the regional tectonic setting, geology, and seismology. An important geologic consideration is topography, which has been shown to contribute significantly to the effects local geology has on ground motions by increasing the amplitude of ground motions relative to flat ground conditions and causing damage to structures near the crests of slopes (e.g., Assimaki et al. 2005). Because most existing ground motion models do not account for the effects of site topography (Rai et al. 2017), the ground motion hazard analysis should consider topographic effects where appropriate. Topographic amplification is typically considered significant where all of the following are present: (1) slope angles greater than about 17 degrees (Bouckovalas and Papadimitriou 2005), (2) the horizontal or vertical dimensions of the topographic feature are comparable to the seismic wavelengths of interest (e.g., the shear wave velocity of the feature divided by the frequencies that impact structural response), and (3) the location of interest is within a horizontal distance approximately equal to one wavelength from the slope crest. Topographic amplification can be estimated by semi-analytical solutions (e.g., Bouckovalas and Papadimitriou 2005), numerical modeling (e.g., Jeong et al. 2019), laboratory experiments (e.g., Jeong et al. 2019), field experiments (e.g., Wood and Cox 2016), and ground motion models (e.g., Rai et al. 2017). Engineers should be cognizant of the large degree of spatial variability in the effects of surface topography on ground motions (Wood and Cox 2016) and the limitations, applicability, and uncertainty of each method of analysis.

C21.2.1 Probabilistic (MCE_R) Ground Motions Probabilistic seismic hazard analysis (PSHA) methods and subsequent computations of risk-targeted probabilistic ground motions, based on the output of PSHA, are sufficient to define MCE_R ground motion at all locations except those near highly active faults. Descriptions of current PSHA methods can be found in McGuire (2004). The primary output of PSHA methods is a so-called hazard curve, which provides mean annual frequencies of exceeding various user-specified ground motion amplitudes. Risk-targeted probabilistic ground motions are derived from hazard curves as described in Luco et al. (2007). In summary, the hazard curve is combined with a collapse fragility (or probability distribution of the ground motion amplitude that causes collapse) that depends on the risk-targeted probabilistic ground motion itself. The combination quantifies the risk of collapse. Iteratively, the risk-targeted probabilistic ground motion is modified until combination of the corresponding collapse fragility with the hazard curve results in a risk of collapse of 1% in 50 years. This target is based on the average collapse risk across the western United States that is expected to result from design for the probabilistic MCE ground motions in ASCE 7.

C21.2.2 Deterministic (MCE_R) Ground Motions ASCE 7-16 and prior editions stated that "the largest ... acceleration for the *characteristic* earthquakes on all known *active* faults ... shall be used." The concept of "characteristic earthquakes" is not included in the version of the Uniform California Earthquake Rupture Forecast that is used for the MCE_R and MCE_G ground motion maps of this standard (i.e., "UCERF3" by the Working Group on California Earthquake Probabilities 2013). Characteristic earthquakes are no longer specified because they are considered to be inconsistent with recent earthquakes such as the 2010 Baja California event. Accordingly, in the definition of deterministic ground motions in this standard, the requirement for characteristic earthquakes has been replaced by "scenario earthquakes" that are determined from hazard deaggregations for the probabilistic ground motions at the site. At each spectral response period, the deaggregation provides a mean earthquake magnitude for each fault, which is termed a scenario earthquake. These scenario earthquakes average over all of the magnitudes that could occur on each fault. This is in contrast to the single magnitude that needed to be chosen for each of the characteristic earthquakes called for in ASCE 7-16. Examples of scenario earthquakes from hazard deaggregations are given in Table C21.2.2-1.

Whereas ASCE 7-16 stipulated "active" faults, albeit without a definition, this standard instead specifies that scenario earthquakes contributing less than 10% of the largest contributor at each period shall be ignored. The contribution of each scenario earthquake is an output of the same hazard deaggregations used to determine the earthquake magnitudes. In effect, this specification defines what constitutes an active fault, in a way that ensures that deterministic ground motions are only calculated for faults contributing significantly for the probabilistic ground motions at the site. For example, at the San Jose site considered in Table C21.2.2-1, deterministic ground motions (at spectral response periods from 0.20 to 5.0 seconds) would be calculated for the Hayward, Calaveras, and San Andreas faults but not for the Silver Creek fault.

For consistency, the same attenuation equations and ground motion variability used in the PSHA should be used in the deterministic seismic hazard analysis (DSHA). Adjustments for directivity and/or directional effects should also be made, when appropriate. In some cases, ground motion simulation methods may be appropriate for the estimation of long-period motions at sites in deep sedimentary basins or from great ($M \geq 8$) or giant ($M \geq 9$) earthquakes, for which recorded ground motion data are lacking.

Deterministic (MCE_R) ground motions are taken as the greater of those calculated for the site of interest and the deterministic lower limit MCE_R ground motions of Table 21.2-1 (i.e., a "lower limit" below which deterministic ground motions are not required for design). Table 21.2-1 defines lower limit MCE_R deterministic ground motions in terms of multi-period spectra for the site-specific value of Site Class. These response spectra represent 84th percentile ground motions of a magnitude M8.0 shallow crustal earthquake at a distance of about 12.5 km from fault rupture (FEMA, 2020). These fault properties were selected such that short-period (0.2-second) and 1.0-second response spectral accelerations for Site Class BC, 1.5 g and 0.6 g, would be the same as the values of the spectral parameters, $1.5F_a$ (g) and $0.6F_v/T$ (g), of the two-period lower limit deterministic MCE_R response spectrum of Figure 12.2-1 of ASCE 7-16.

Where the probabilistic MCE_R ground motions (Section 21.2.1) are less than the deterministic lower limit MCE_R ground motions of Table 21.2-1 for the site class of interest, at all

Table C21.2.2-1. Examples of Scenario Earthquake from Hazard Deaggregations at a Site in SanJose, California.

	Scenario Earthquake							
Period T (s)	Hayward M	Calaveras Contribution	M	San Andreas Contribution	M	Silver Creek Contribution	M	Contribution
0.20	7.0	53%	7.2	16%	7.9	11%	6.9	3%
0.25	7.0	52%	7.2	16%	7.9	12%	6.9	3%
0.30	7.0	52%	7.2	16%	7.9	13%	6.9	3%
0.40	7.0	52%	7.2	16%	7.9	15%	7.0	3%
0.50	7.0	51%	7.3	16%	7.9	16%	7.0	3%
0.75	7.1	49%	7.3	16%	7.9	19%	7.0	3%
1.0	7.1	48%	7.3	16%	7.9	20%	7.1	2%
1.5	7.1	43%	7.3	16%	8.0	24%	7.1	1%
2.0	7.2	39%	7.3	16%	8.0	27%	7.2	1%
3.0	7.2	34%	7.4	15%	8.0	34%	7.2	1%
4.0	7.3	29%	7.4	14%	8.0	40%	NA	<1%
5.0	7.4	24%	7.4	13%	8.0	44%	NA	<1%

response periods, probabilistic MCE_R ground motions govern at the site of interest and deterministic MCE_R ground motions need not be calculated. Such is typically the case for sites in the Central and Eastern U.S.

C21.2.3 Site-Specific MCE_R Response Spectrum Because of the deterministic lower limit on the MCE_R spectrum (Table 21.2-1), the site-specific MCE_R ground motion is equal to the corresponding risk-targeted probabilistic ground motion wherever it is less than the deterministic limit. Where the probabilistic ground motions are greater than the lower limits, the deterministic ground motions sometimes govern, but only if they are less than their probabilistic counterparts. The site-specific MCE_R response spectrum may be taken as equal to the MCE_R response spectrum obtained from the USGS web service without performing site-specific calculations and cannot be taken as less than 80% of the MCE_R response spectrum obtained from the USGS web service. These requirements recognize the considerable earthquake hazard and ground motion expertise of the USGS and rely on the MCE_R ground motions available from the USGS (1) to provide an acceptable alternative to calculating site-specific ground motions (which can be a burden for smaller projects), and (2) to establish a "safety net" on site-specific ground motions.

Eighty percent of the MCE_R response spectrum obtained from the USGS web service was established as the lower limit to prevent the possibility of site-specific studies generating unreasonably low ground motions from potential misapplication of site-specific procedures or misinterpretation or mistakes in the quantification of the basic inputs to these procedures. Even if site-specific studies were correctly performed and resulted in ground motion response spectra less than the 80% lower limit, the uncertainty in the seismic potential and ground motion attenuation across the United States was recognized in setting this limit. Under these circumstances, the allowance of up to a 20% reduction in the design response spectrum based on site-specific studies was considered reasonable.

Although the 80% lower limit is reasonable for sites not classified as Site Class F, an exception has been introduced at the end of this section to permit a site class other than E to be used in establishing this limit when a site is classified as F. This revision eliminates the possibility of an overly conservative design spectrum on sites that would normally be classified as Site Class C or D.

C21.3 DESIGN RESPONSE SPECTRUM

The site-specific design response spectrum is defined as two-thirds of the site-specific MCE_R response spectrum consistent with the requirements of Section 11.5.4.1.

C21.4 DESIGN ACCELERATION PARAMETERS

The S_{DS} criteria of Section 21.4 are based on the premise that the value of the parameter S_{DS} should be taken as 90% of peak value of site-specific response spectral acceleration, regardless of the period (greater than or equal to 0.2 s) at which the peak value of response spectral acceleration occurs. Consideration of periods beyond 0.2 s recognizes that site-specific studies (e.g., softer site conditions) can produce response spectra with ordinates at periods greater than 0.2 s that are significantly greater than those at 0.2 s. Periods less than 0.2 s are excluded for consistency with the 0.2 s period definition of the short-period ground motion parameter, S_S, and recognizing that certain sites, such as Central and Eastern United States (CEUS) sites, could have peak response at very short periods that would be inappropriate for defining the value of the parameter S_{DS}. The upper bound limit of 5 s precludes unnecessary checking of response at periods that cannot govern the peak value of site-specific response spectral acceleration. Ninety percent (rather than 100%) of the peak value of site-specific response spectral acceleration is considered appropriate for defining the parameter S_{DS} (and the domain of constant acceleration) because most short-period structures have a design period that is not at, or near, the period of peak response spectral acceleration. Away from the period of peak response, response spectral accelerations are less, and the domain of constant acceleration is adequately described by 90% of the peak value. For those short-period structures with a design period at, or near, the period of peak response spectral acceleration, anticipated yielding of the structure during MCE_R ground motions effectively lengthens the period and shifts dynamic response to longer periods at which spectral demand is always less than that at the peak of the spectrum.

The S_{D1} criteria of Section 21.4 are based on the premise that the value of the parameter S_{D1} should be taken as the larger of (1) 100% of the response at 1 s and (2) 90% of maximum value of the product TS_a for a period range, $1\text{ s} \le T \le 2\text{ s}$, for stiffer sites $\overline{v}_s > 1{,}450$ ft/s ($\overline{v}_s > 442$ m/s) for a period range, $1\text{ s} \le T \le 5\text{ s}$, for softer sites $\overline{v}_s \le 1{,}450$ ft/s ($\overline{v}_s \le 442$ m/s), which are expected to have peak values of response spectral velocity at periods greater than 2 s. The criteria use the maximum value of the product, TS_a, over the period range of interest to effectively identify the period at which the peak value of response spectral velocity occurs. Consideration of periods beyond 1 s accounts for the possibility that the assumed $1/T$ proportionality for the constant velocity portion of the design response spectrum begins at periods greater than 1 s or is actually $1/T_n$ (where $n < 1$). Periods less than 1 s are excluded for consistency with the definition of the 1 s ground motion parameter, S_{M1}. Peak velocity response is expected to occur at periods less than or equal to 5 s, and periods beyond 5 s are excluded by the criteria to avoid potential misuse of very long period ground motions that may not be reliable. Ninety percent of the peak value of site-specific response spectral acceleration at the period of peak velocity response is considered appropriate for defining the value of the parameter S_{D1} for sites where ground motions have peak velocity response significantly beyond 1 s (e.g., sites whose hazard is governed by larger magnitude earthquakes). For these sites, away from the period of peak velocity response, response spectral acceleration is less than that of the assumed $1/T$ shape and the domain of constant velocity is adequately described by 90% of the response at the period of peak velocity response. One hundred percent of 1-s response is considered appropriate for defining the value of the parameter S_{DI} for sites where ground motions have peak velocity response at or near 1 s (e.g., sites whose hazard if governed by smaller magnitude earthquakes). For these sites, at periods beyond 1 s response spectral acceleration is typically similar to that of the assumed $1/T$ shape and the domain of constant velocity would not be adequately described by less than 100% of 1-s response.

C21.5 MAXIMUM CONSIDERED EARTHQUAKE GEOMETRIC MEAN (MCE_G) PEAK GROUND ACCELERATION

Site-specific requirements for determination of peak ground acceleration (PGA) are provided in a new Section 21.5 that is parallel to the procedures for developing site-specific response spectra in Section 21.2. The site-specific MCE peak ground acceleration, PGA_M, is taken as the lesser of the probabilistic geometric mean peak ground acceleration of Section 21.5.1 and the deterministic geometric mean peak ground acceleration of Section 21.5.2. Similar to the provisions for site-specific spectra, a deterministic lower limit is prescribed for PGA_M, with the intent to limit application of deterministic ground motions to the site regions containing active faults where probabilistic ground motions are relatively high. However, consistent with the lower limit of ASCE 7-16 (e.g., 0.5 F_{PGA}), the deterministic lower limit for PGA_M (in g) is set at a lower value, (e.g., $PGA_G = 0.5$ for Site Class BC, Table 21.2-1), than the value set for the zero-period response spectral acceleration, (e.g., 0.66 for Site Class BC, Table 21.2-1). The rationale for the value of the lower deterministic limit for spectra is based on the desire to limit minimum spectral values, for structural design purposes, to the values given by the 1997 Uniform Building Code (UBC) for Zone 4 (multiplied by a factor of 1.5 to adjust to the MCE level). This rationale is not applicable to PGA_M for geotechnical applications, and therefore a lower value was selected. Section 21.5.3 of ASCE 7-22 states that the site-specific MCE peak ground acceleration cannot be less than 80% of the value of PGA_M obtained from the USGS web service. The 80% limit is a long-standing base for site-specific analyses in recognition of the uncertainties and limitations associated with the various components of a site-specific evaluation.

REFERENCES

Assimaki, D., G. Gazetas, and E. Kausel. 2005. "Effects of local soil conditions on the topographic aggravation of seismic motion: Parametric investigation and recorded field evidence from the 1999 Athens earthquake." *Bull. Seismol. Soc. Am.* 95 (3): 1059–1089. https://doi.org/10.1785/0120040055.

Bouckovalas, G. D., and A. G. Papadimitriou. 2005. "Numerical evaluation of slope topography effects on seismic ground motion." *Soil Dyn. Earthquake Eng.* 25 (7–10): 547–558. https://doi.org/10.1016/j.soildyn.2004.11.008.

Chang, C.-Y., C. M. Mok, M. S. Power, and Y. K. Tang. 1991. *Analysis of ground response at Lotung large-scale soil-structure interaction experiment site*. Rep. No. NP-7306-SL. Palo Alto, CA: EPRI.

Darendeli, M. 2001. "Development of a new family of normalized modulus reduction and material damping curves." Ph.D. dissertation, Dept. of Civil Engineering, University of Texas.

EPRI (Electric Power Research Institute). 1993. *Guidelines for determining design basis ground motions*. Rep. No. EPRI TR-102293. Palo Alto, CA: EPRI.

FEMA (Federal Emergency Management Agency). 2020. *Procedures for developing multi-period response spectra at non-conterminous United States sites*. FEMA P-2078, Washington, DC: FEMA.

Field, E. H., N. Gupta, V. Gupta, M. Blanpied, P. Maechling, and T. H. Jordan. 2005. "Hazard calculations for the WGCEP-2002 forecast using OpenSHA and distributed object technologies." *Seismol. Res. Lett.* 76 (2): 161–167. https://doi.org/10.1785/gssrl.76.2.161.

Goulet, C. A., J. P. Stewart, P. Bazzurro, and E. H. Field. 2007. "Integration of site-specific ground response analysis results into probabilistic seismic hazard analyses." In *Proc., 4th Int. Conf. on Earthquake Geotechnical Engineering*, Thessaloniki, Greece.

Hashash, Y. M. A., and D. Park. 2001. "Non-linear one-dimensional seismic ground motion propagation in the Mississippi embayment." *Eng. Geol.* 62 (1–3): 185–206. https://doi.org/10.1016/S0013-7952(01)00061-8.

Huang, Y.-N., A. S. Whittaker, and N. Luco. 2008. "Maximum spectral demands in the near-fault region." *Earthquake Spectra* 24 (1): 319–341. https://doi.org/10.1193/1.2830435.

Idriss, I. M., and J. I. Sun. 1992. *User's manual for SHAKE91*. Davis, CA: University of California.

Itasca Consulting Group. 2005. *FLAC, fast Lagrangian analysis of continua, version 5.0*. Minneapolis: Itasca Consulting Group.

Jeong, S., D. Asimaki, J. Dafni, and J. Wartman. 2019. "How topography-dependent are topographic effects?" Complementary numerical modeling of centrifuge experiments." *Soil Dyn. Earthquake Eng.* 116 (Jan): 654–667. https://doi.org/10.1016/j.soildyn.2018.10.028.

Kramer, S. L. 1996. *Geotechnical earthquake engineering*. Englewood Cliffs, NJ: Prentice-Hall.

Kwok, A. O. L., J. P. Stewart, Y. M. A. Hashash, N. Matasovic, R. Pyke, Z. Wang, and Z. Yang. 2007. "Use of exact solutions of wave propagation problems to guide implementation of nonlinear seismic ground response analysis procedures." *J. Geotech. Geoenviron. Eng.* 133 (11): 1385–1398. https://doi.org/10.1061/(ASCE)1090-0241(2007)133:11(1385).

Lee, M. K. W., and W. D. L. Finn. 1978. *DESRA-2, Dynamic effective stress response analysis of soil deposits with energy transmitting boundary including assessment of liquefaction potential: Soil mechanics series 36*. Vancouver, BC: University of British Columbia.

Li, X. S., Z. L. Wang, and C. K. Shen. 1992. *SUMDES, a nonlinear procedure for response analysis of horizontally-layered sites subjected to multi-directional earthquake loading*. Davis, CA: University of California.

Luco, N., B. R. Ellingwood, R. O. Hamburger, J. D. Hooper, J. K. Kimball, and C. A. Kircher. 2007. "Risk-targeted versus current seismic design maps for the conterminous United States." In *Proc., SEAOC 76th Annual Convention*, Sacramento, CA.

Matasovic, N. 2006. "D-MOD_2—A computer program for seismic response analysis of horizontally layered soil deposits, earthfill dams, and solid waste landfills." In *User's manual,* 20. Lacey, WA: GeoMotions, LLC.

McGuire, R. K. 2004. *Seismic hazard and risk analysis*. Monograph MNO-10. Oakland, CA: Earthquake Engineering Research Institute.

Menq, F. 2003. "Dynamic properties of sandy and gravely soils." Ph.D. dissertation, Dept. of Civil Engineering, University of Texas.

Parra, E. 1996. "Numerical modeling of liquefaction and lateral ground deformation including cyclic mobility and dilation response in soil systems." Ph.D. dissertation, Dept. of Civil Engineering, Rensselaer Polytechnic Institute.

Pyke, R. M. 2000. *TESS: A computer program for nonlinear ground response analyses*. Lafayette, CA: TAGA Engineering Systems & Software.

Ragheb, A. M. 1994. "Numerical analysis of seismically induced deformations in saturated granular soil strata." Ph.D. dissertation, Dept. of Civil Engineering, Rensselaer Polytechnic Institute.

Rai, M., A. Rodriguez-Marek, and B. S. Chiou. 2017. "Empirical terrain-based topographic modification factors for use in ground motion prediction." *Earthquake Spectra* 33 (1): 157–177. https://doi.org/10.1193/071015eqs111m.

Schnabel, P. B., J. Lysmer, and H. B. Seed. 1972. *SHAKE: A computer program for earthquake response analysis of horizontally layered sites*. Rep. No. EERC 72-12. Berkeley, CA: University of California.

Shahi, S. K., and J. W. Baker. 2013. *NGA-West2 models for ground-motion directionality*, PEER Report 2013/10. Berkeley, CA: Pacific Earthquake Engineering Research Center, University of California.

Vucetic, M., and R. Dobry. 1991. "Effect of soil plasticity on cyclic response." *J. Geotech. Eng.* 117 (1): 89–107. https://doi.org/10.1061/(ASCE)0733-9410(1991)117:1(89).

Wood, C. M., and B. R. Cox. 2016. "Comparison of field data processing methods for the evaluation of topographic effects." *Earthquake Spectra* 32 (4): 2127–2147. https://doi.org/10.1193/111515EQS170M.

Working Group on California Earthquake Probabilities (WGCEP). (2013). "Uniform California earthquake rupture forecast, v. 3 (UCERF3) -The time-independent model." USGS Open File Report 2013-1165.

Yang, Z. 2000. "Numerical modeling of earthquake site response including dilation and liquefaction." Ph.D. dissertation, Dept. of Civil Engineering and Engineering Mechanics, Columbia University.

Zhang, J., R. D. Andrus, and C. H. Juang. 2005. "Normalized shear modulus and material damping ratio relationships." *J. Geotech. Geoenviron. Eng.* 131 (4): 453–464. https://doi.org/10.1061/(ASCE)1090-0241(2005)131:4(453).

Zhang, J., R. D. Andrus, and C. H. Juang. 2008. "Model uncertainty in normalized shear modulus and damping relationships." *J. Geotech. Geoenviron. Eng.* 134 (1): 24–36. https://doi.org/10.1061/(ASCE)1090-0241(2008)134:1(24).

OTHER REFERENCES (NOT CITED)

Abrahamson, N. A. 2000. "Effects of rupture directivity on probabilistic seismic hazard analysis." In *Proc., 6th Int. Conf. on Seismic Zonation*, Oakland, CA.

Huang, Y.-N., A. S. Whittaker, and N. Luco. (2008). "Maximum spectral demands in the near-fault region." *Earthq. Spectra*, 24 (1): 319–341.

Kircher, C. A. (2015). Investigation of an identified short-coming in the seismic design procedures of ASCE 7-16 and development of recommended improvements for ASCE 7-16, prepared for Building Seismic Safety Council, National Institute of Building Sciences, Washington, DC, prepared by Kircher & Associates, Consulting Engineers, Palo Alto, CA, March 15, 2015. c.ymcdn.com/sites/www.nibs.org/resource/resmgr/BSSC2/Seismic_Factor_Study.pdf

Somerville, P. G., N. F. Smith, R. W. Graves, and N. A. Abrahamson. 1997. "Modification of empirical strong ground motion attenuation relations to include the amplitude and duration effects of rupture directivity." *Seismol. Res. Lett.* 68 (1): 199–222. https://doi.org/10.1785/gssrl.68.1.199.

CHAPTER C22
SEISMIC GROUND MOTION LONG-PERIOD TRANSITION MAPS

This standard contains risk-targeted maximum considered earthquake (MCE_R) spectral response acceleration maps (Figures 22-1 through 22-8) and maximum considered earthquake geometric mean (MCE_G) peak ground acceleration maps (Figures 22-9 through 22-13), both introduced in the 2009 NEHRP provisions and ASCE 7-10. Long-period transition period (T_L) maps are also provided (Figures 22-14 through 22-17), as introduced in ASCE 7-05. The MCE_R and MCE_G maps, but not T_L, maps, are updated with respect to those in ASCE 7-16, based on recommendations of an effort referred to as "Project 17" (BSSC 2019) and on the 2018 US Geological Survey (USGS 2018) National Seismic Hazard Model (NSHM). Furthermore, the risk coefficient maps of ASCE 7-16 have been removed for consistency with the revised site-specific ground motion procedures of Section 21.2.1 of this standard. The updates to the MCE_R and MCE_G maps are summarized next, followed by a summary of map development details and a description of USGS resources for retrieving Chapter 22 values.

Modifications to MCE_R and MCE_G ground motions from Project '17 recommendations. Project '17 was a collaboration between the Building Seismic Safety Council (BSSC), with funding from FEMA, and the USGS. The project focused on improvements to the seismic maps of ASCE 7-16 for the 2020 NEHRP provisions (FEMA 2020a) and this standard and resulted in recommendations documented in BSSC (2019). The recommendations led to the four modifications to the MCE_R and/or MCE_G ground motions summarized here. Examples of the resulting changes to the MCE_R and MCE_G values are given in subsequent subsections.

1. Whereas ASCE 7-16 (and earlier editions back to ASCE 7-98) calculated MCE_R and MCE_G ground motions for each site class (S_{MS}, S_{M1}, and PGA_M) by adjusting mapped ground motions (S_S, S_1, and PGA) with site coefficients (F_a, F_v, and F_{PGA}), this standard provides S_{MS}, S_{M1}, and PGA_M values that come directly from the USGS NSHM (or an interim approximation outside of the conterminous United States). In effect, this changes the site class effects. The changes are most significant in the eastern United States, because the prior ASCE 7-16 site coefficients were based on western United States data and simulations (Seyhan and Stewart 2014, Kircher & Associates 2015). Note that rather than providing maps of S_{MS}, S_{M1}, and PGA_M for all of the site classes, this standard provides maps for default site conditions and an online USGS Seismic Design Geodatabase archive for specific site classes. Although S_S, S_1, and PGA are no longer needed to calculate S_{MS}, S_{M1}, and PGA_M, values of S_S and S_1 are also contained in the USGS geodatabase for use elsewhere in this standard (e.g., for vertical ground motions and seismic design category).
2. Whereas the S_{MS} and S_{M1} spectral response accelerations of ASCE 7-16 (and back to ASCE 7-98) were at the same periods as the previously mapped MCE_R ground motions (S_S and S_1), namely, 0.2 and 1.0 s, this standard adheres to the definitions of S_{MS} and S_{M1} in Chapter 21 (Section 21.4), which consider ranges of periods. Like the ground motions for each site class (discussed previously), the spectral response accelerations at the range of periods come directly from the USGS NSHM (or an interim approximation outside of the conterminous United States). In general, this modification to the periods considered increases S_{M1} values at locations where relatively frequent, large-magnitude crustal earthquakes dominate the seismic hazard. The impact on S_{MS} values is generally smaller.
3. Whereas the deterministic caps considered for the MCE_R and MCE_G ground motions of ASCE 7-16 (and back to ASCE 7-98) were from "characteristic earthquakes on all known active faults," this standard calls for scenario earthquakes from hazard deaggregation, as explained in the commentary of Chapter 21 (Section C21.2.2). For each earthquake, the way in which the 84th percentile ground motion is calculated has also been modified. These deterministic ground motions are now approximated via so-called epsilons (corresponding to ground motion percentiles) from the same hazard deaggregations that determine the scenario earthquakes. Furthermore, this standard has modified the lower limits imposed on the deterministic ground motions, although they are anchored to the limits used for the ASCE 7-16 MCE_R and MCE_G maps; see Section C21.2.2. Together, the three modifications to deterministic capping change the MCE_R and MCE_G values in the conterminous United States. Pending updates to the USGS NSHM for the other states and territories, only the modification to the deterministic lower limits has been applied there.
4. The maximum-response scale factors used for the MCE_R maps of ASCE 7-16 (and ASCE 7-10) have been updated for this standard, adopting the proposal of Resource Paper 4 of NEHRP (2015), which is based on Shahi and Baker (2013). Over the range of spectral response periods considered for the MCE_R ground motions of this standard, the change in the scale factors ranges from a 9% increase to a 15% decrease.

Table C22-1 charts which of these four modifications affect each of the MCE_R and MCE_R ground motion parameters (S_S, S_1, PGA_M, S_{MS}, and S_{M1}). As mentioned previously the MCE_R and MCE_G ground motions outside of the conterminous United States

Table C22-1. Modifications to MCE_R and MCE_G Ground Motions from ASCE 7-16 to ASCE 7-22 Based on Project '17 Recommendations.

	Conterminous United States			Alaska, Hawaii, and Territories		
Modification	S_S and S_1	PGA_M	S_{MS} and S_{M1}	S_S and S_1	PGA_M	S_{MS} and S_{M1}
1. Site class effects		X*	X*		X*	X*
2. Spectral periods			X			X
3a. Deterministic earthquakes	X	X	X			
3b. 84th percentile calculation	X	X	X			
3c. Deterministic lower limits		X*	X*		X*	X*
4. Max-direction factors	X		X	X		X

* For Site Class BC, the site class effects and deterministic lower limits have not changed from ASCE 7-16 to this standard.

in Alaska, Hawaii, and the territories do not yet incorporate the modifications to deterministic earthquakes and 84th percentile ground motions, pending forthcoming USGS updates for those states and territories. Moreover, the modifications to site class effects and spectral periods are currently approximated outside of the conterminous United States by means of a procedure documented in FEMA (2020b). For the MCE_R ground motions in Alaska, Hawaii, Puerto Rico, and the US Virgin Islands, the collapse-fragility logarithmic standard deviation (or "beta value") has also been updated from 0.8 to 0.6 for consistency with the other states and territories.

Note: For international locations where a multi-period and multi-site-class hazard model is not available, and where the aforementioned FEMA (2020b) approximation is not applicable, a local or regional study could develop ground motions that are consistent with this standard.

Modifications to MCE_R and MCE_G ground motions from 2018 USGS NSHM update. Whereas the MCE_R and MCE_G maps of ASCE 7-16 were derived from the 2014 USGS National Seismic Hazard Model (Petersen et al. 2014), the MCE_R and MCE_G ground motions of this standard are derived from the 2018 USGS NSHM. The four modifications in the 2018 NSHM, all for the conterminous United States, are summarized here. Details are presented in the documentation of the 2018 update (Petersen et al. 2019). Examples of the resulting changes to the values of MCE_R and MCE_G ground motions are given in the ensuing subsections.

1. For the central and eastern United States (CEUS), the ground motion models incorporated into the 2018 NSHM are from the NGA-East project of the Pacific Earthquake Engineering Research Center (PEER). This was a multi-year project funded by the US Nuclear Regulatory Commission that gathered data applicable to the CEUS and developed models of median ground motions (PEER 2015), site effects (Stewart et al. 2017), the random or aleatory variability in ground motions (e.g., Al Atik 2015), and the epistemic uncertainty in median ground motions (Goulet et al. 2017). The NGA-East ground motion models have made it possible to produced NSHM spectral response accelerations in the CEUS for periods from 0 (for peak ground acceleration) to 10 s and for Site Classes A through E. The CEUS ground motion models used for the 2014 NSHM only included a relatively narrow range of periods and site classes. For these few periods and site classes, the ground motion values of the 2018 NSHM are larger at most, but not all, CEUS locations.
2. In the Los Angeles, Seattle, San Francisco, and Salt Lake City regions—where there are published models that are deemed applicable (from Lee et al. 2014, Stephenson 2007, Aagaard et al. 2008, Magistrale et al. 2008, respectively)—basin depths have been incorporated into the 2018 NSHM. The depths are input to the ground motion models used for the western United States, although in the Seattle region, this entails modification of the models for subduction zone earthquakes. At sites where the basin depths are relatively large (see Petersen et al. 2019 for details), there is scientific consensus that longer-period ground motions are amplified based on observations and simulations. Thus, at these deeper-basin sites, the 2018 NSHM ground motions are amplified. The amount of amplification increases with basin depth and spectral response period, and it decreases with site class (from A to E). At sites where the depths are relatively small, currently there is a lack of scientific consensus on basin effects. Thus, at these shallower basin sites and outside of the four regions, a default basin depth that is estimated from site class (by each of the NGA-West2 ground motion models) is assumed in the 2018 NSHM, and ground motions are unaffected. In the 2014 NSHM, used for ASCE 7-16, the effects of deep basins were not included; only a relatively shallow default basin depth, corresponding to the BC site class of the NSHM, was considered.
3. Outside of California (because the Uniform California Earthquake Rupture Forecast has not been modified), the catalog of past earthquakes has been updated for the 2018 NSHM. Seismicity catalogs are used to calculate spatially smoothed rates of occurrence of future earthquakes on unmodeled (or unknown) faults. In addition to appending earthquakes that occurred in 2013 through 2017, other updates have been made to the catalog and the smoothed earthquake rates (see Petersen et al. 2019). The amount of change to the NSHM caused by these updates is generally smaller than the updates as described previously.
4. For the western United States, two of the ground motion models that were included in the 2014 NSHM have been excluded for the 2018 update. One of the two models

cannot be used for softer site classes (Idriss 2014), and the other cannot be used for spectral response periods longer that 3.0 s (Atkinson and Boore 2003, 2008). For the harder site classes and shorter periods covered by these ground motion models, excluding them only slightly changes the NSHM by an amount that depends on location, spectral response period, and site class.

For Alaska, Hawaii, and territories outside of the conterminous United States, the USGS NSHM has not been updated with respect to what was used for ASCE 7-16. There, changes to the MCE_R and MCE_G ground motions from ASCE 7-16 to this standard are a result of the Project 17 modifications, summarized in the preceding subsection.

Examples of changes in MCE_R and MCE_G values. The combined impacts of the Project '17 and USGS NSHM modifications summarized as on MCE_R and MCE_G values are demonstrated in Tables C22-3 through C22-5 and Figures C22-1 through C22-3, for the 34 example locations listed in Table C22-2. These locations are the same as those considered in ASCE 7-16, which were first introduced in the 2009 NEHRP Provisions (NEHRP 2009). In the MCE_R and MCE_G tables and figures, ground motions of this standard are compared with those from ASCE 7-16 and ASCE 7-10. The S_{MS}, S_{M1}, and PGA_M ground motions are for default site conditions, so that the comparisons capture the most common site class effects. Corresponding seismic design categories

Table C22-2. Latitudes and Longitudes at Which MCE_R and MCE_G Ground Motions and Corresponding Seismic Design Categories from This Standard Are Compared.

	City and Location of Site			County or Metropolitan Statistical Area	
Region	Name	Latitude	Longitude	Name	Population
Southern California	Los Angeles	34.05	−118.25	Los Angeles	9,948,081
	Century City	34.05	−118.40		
	Northridge	34.20	−118.55		
	Long Beach	33.80	−118.20		
	Irvine	33.65	−117.80	Orange	3,002,048
	Riverside	33.95	−117.40	Riverside	2,026,603
	San Bernardino	34.10	−117.30	San Bernardino	1,999,332
	San Luis Obispo	35.30	−120.65	San Luis Obispo	257,005
	San Diego	32.70	−117.15	San Diego	2,941,454
	Santa Barbara	34.45	−119.70	Santa Barbara	400,335
	Ventura	34.30	−119.30	Ventura	799,720
	Total Population—S. California		22,349,098	Population—8 Counties	21,374,788
Northern California	Oakland	37.80	−122.25	Alameda	1,502,759
	Concord	37.95	−122.00	Contra Costa	955,810
	Monterey	36.60	−121.90	Monterey	421,333
	Sacramento	38.60	−121.50	Sacramento	1,233,449
	San Francisco	37.75	−122.40	San Francisco	776,733
	San Mateo	37.55	−122.30	San Mateo	741,444
	San Jose	37.35	−121.90	Santa Clara	1,802,328
	Santa Cruz	36.95	−122.05	Santa Cruz	275,359
	Vallejo	38.10	−122.25	Solano	423,473
	Santa Rosa	38.45	−122.70	Sonoma	489,290
	Total Population—N. California		14,108,451	Population—10 Counties	8,621,978
Pacific Northwest	Seattle	47.60	−122.30	King, WA	1,826,732
	Tacoma	47.25	−122.45	Pierce, WA	766,878
	Everett	48.00	−122.20	Snohomish, WA	669.887
	Portland	45.50	−122.65	Portland Metro, OR (3)	1,523,690
	Total Population—OR and WA		10,096,556	Population—6 Counties	4,787,187
Other WUS	Salt Lake City	40.75	−111.90	Salt Lake, UT	978,701
	Boise	43.60	−116.20	Ada/Canyon, ID (2)	532,337
	Reno	39.55	−119.80	Washoe, NV	396,428
	Las Vegas	36.20	−115.15	Clarke, NV	1,777,539
	Total Population—ID/UT/NV		6,512,057	Population—5 Counties	3,685,005
CEUS	St. Louis	38.60	−90.20	St. Louis MSA (16)	2,786,728
	Memphis	35.15	−90.05	Memphis MSA (8)	1,269,108
	Charleston	32.80	−79.95	Charleston MSA (3)	603,178
	Chicago	41.85	−87.65	Chicago MSA (7)	9,505,748
	New York	40.75	−74.00	New York MSA (23)	18,747,320
	Total Population—MO/TN/SC/IL/NY		48,340,918	Population—57 Counties	32,912,082

Notes: ASCE 7-16 and ASCE 7-10 are compared in Tables C22-3 through C22-6.

Table C22-3. Comparison of Short-Period MCE_R Spectral Response Acceleration Values from This Standard, ASCE 7-16, and ASCE 7-10.

	ASCE/SEI 7-10		ASCE/SEI 7-16		ASCE/SEI 7-22	
Location Name	S_S (g)	S_{MS} (g)	S_S (g)	S_{MS} (g)	S_S (g)	S_{MS} (g)
Los Angeles, CA	2.40	2.40	1.97	2.36	2.25	2.37
Century City, CA	2.17	2.17	2.11	2.53	2.37	2.49
Northridge, CA	1.69	1.69	1.74	2.08	2.09	2.26
Long Beach, CA	1.64	1.64	1.68	2.02	1.90	2.03
Irvine, CA	1.55	1.55	1.25	1.50	1.43	1.68
Riverside, CA	1.50	1.50	1.50	1.80	1.50	1.67
San Bernardino, CA	2.37	2.37	2.33	2.79	2.78	2.97
San Luis Obispo, CA	1.12	1.18	1.09	1.31	1.23	1.45
San Diego, CA	1.25	1.25	1.58	1.89	1.74	1.80
Santa Barbara, CA	2.83	2.83	2.12	2.54	2.37	2.44
Ventura, CA	2.38	2.38	2.02	2.42	2.25	2.38
Oakland, CA	1.86	1.86	1.88	2.26	2.07	2.21
Concord, CA	2.08	2.08	2.22	2.67	2.76	2.84
Monterey, CA	1.53	1.53	1.33	1.59	1.50	1.75
Sacramento, CA	0.67	0.85	0.57	0.76	0.66	0.97
San Francisco, CA	1.50	1.50	1.50	1.80	1.57	1.81
San Mateo, CA	1.85	1.85	1.80	2.16	2.10	2.30
San Jose, CA	1.50	1.50	1.50	1.80	1.50	1.65
Santa Cruz, CA	1.52	1.52	1.59	1.91	1.65	1.86
Vallejo, CA	1.50	1.50	1.50	1.80	2.33	2.42
Santa Rosa, CA	2.51	2.51	2.41	2.90	2.81	2.93
Seattle, WA	1.37	1.37	1.40	1.68	1.56	1.76
Tacoma, WA	1.30	1.30	1.36	1.63	1.50	1.62
Everett, WA	1.27	1.27	1.20	1.44	1.36	1.55
Portland, OR	0.98	1.09	0.89	1.07	0.94	1.15
Salt Lake City, UT	1.54	1.54	1.55	1.85	1.76	1.81
Boise, ID	0.31	0.48	0.31	0.48	0.33	0.50
Reno, NV	1.50	1.50	1.47	1.76	1.54	1.74
Las Vegas, NV	0.50	0.70	0.65	0.83	0.70	0.82
St Louis, MO	0.44	0.64	0.46	0.66	0.63	0.68
Memphis, TN	1.01	1.11	1.03	1.23	1.32	1.25
Charleston, SC	1.15	1.20	1.42	1.70	1.66	1.52
Chicago, IL	0.14	0.22	0.12	0.19	0.15	0.20
New York City, NY	0.28	0.44	0.29	0.46	0.28	0.31

Notes: The S_{MS} values are for default site conditions.

(SDCs) are compared in Table C22-6. For the locations where the SDCs have changed from ASCE 7-16 to this standard, and where the MCE_R and/or MCE_G ground motions for default site conditions have changed by greater than 15%, an explanation of which Project '17 and/or USGS modifications are the predominant causes is as follows: Changes less than 15%, while potentially impactful, tend to be combinations of several of the Project '17 and/or USGS modifications, and hence do not lend themselves to clear explanations. For changes to example locations outside of the conterminous United States, see FEMA (2020).

These 34 locations are the same as those considered in ASCE 7-16 and ASCE 7-10, which were first introduced in NEHRP (2009). It is important to note that these locations are each just one of many in the named cities, and their ground motions may be significantly different than those at other locations in the cities.

As seen in Table C22-3 and Figure C22-1, which are for the short-period MCE_R spectral response accelerations, the S_{MS} values change from ASCE 7-16 to this standard by less than 15% at all but 3 of the 34 locations, which is explained as follows. From ASCE 7-10 to ASCE 7-16, the S_{MS} values changed by more than 15% at 20 of the 34 locations.

- The increase at the Sacramento location (from 0.76g to 0.97g) is mostly due to the Project '17 modification to site class effects. At this location, the ratio from this standard of the 0.2 s spectral response acceleration for default site conditions divided by that for Site Class BC is approximately 20% greater than the corresponding ASCE 7-16 site coefficient.
- The increase at the Vallejo location (from 1.80g to 2.42g) is mostly due to the Project '17 modification to deterministic capping. Whereas ASCE 7-16 excluded the nearby Franklin fault from deterministic capping, based on its geologic rate of slip alone, this standard includes it, based on the hazard deaggregations now used for the caps. In addition to geologic slip rates, the hazard deaggregations account for geodetic data [from the Global Positioning System (GPS)] that are used by the 2014 and 2018 USGS NSHMs. The deaggregations for the Vallejo location reveal that the Franklin fault is a primary contributor to the hazard.

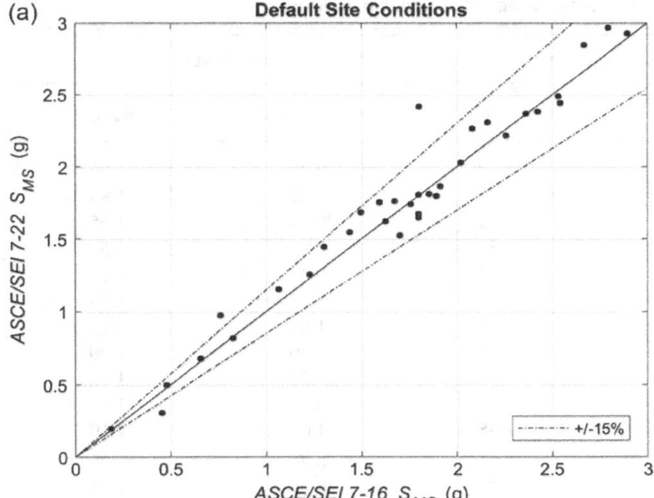

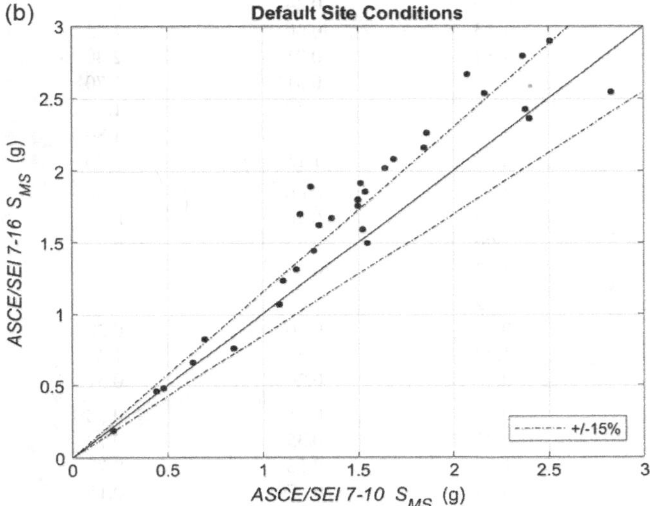

Figure C22-1. Comparison of short-period MCE_R spectral response acceleration (S_{MS}) values from this standard versus (a) ASCE 7-16 and (b) ASCE 7-16 versus ASCE/SEI 7-10 for default site conditions and 34 example locations.

- The decrease at the New York location (from 0.46*g* to 0.31*g*) is mostly due to a combination of the USGS NSHM update and the Project '17 modification to site class effects. At this location, the ratio from this standard of the 0.2 s spectral response acceleration for default site conditions divided by that for Site Class BC is approximately 20% less than the corresponding ASCE 7-16 site coefficient. Even for Site Class BC, the NGA-East ground motion models decrease the 0.2 s spectral response acceleration at this location.

As seen in Table C22-4 and Figure C22-2, which are for the 1.0 s MCE_R spectral response accelerations, the S_{M1} values changed from ASCE 7-16 to this standard by more than 15% at 11 of the 34 locations: and at 8 of these 11 locations, the S_{M1} values decreases. Note that these decreases are relative to ASCE 7-16 values that include the 1.5 multiplier of the applicable Section 11.4.8 exception. With this multiplier, from ASCE 7-10 to ASCE 7-16, the S_{M1} values changed by more than 15% at all but 3 of the 34 locations.

- The increases at the locations in San Bernardino (from 2.37*g* to 2.83*g*) and San Mateo (from 1.88*g* to 2.30*g*) are mostly due to the Project '17 modification to the spectral periods considered for S_{M1}. At a period of 1.0 s, without the consideration of longer periods, the spectral response accelerations of this standard (namely, 2.39*g* and 1.84*g* at the San Bernardino and San Mateo locations, respectively) are within a few percentage points of those from ASCE 7-16.
- The decreases at the locations in Tacoma (from 1.29*g* to 0.96*g*), Everett (from 1.20*g* to 0.97*g*), and Portland (from 1.13*g* to 0.78*g*) are mostly due to the Project '17 modification to site class effects. At these Pacific Northwest locations (and similarly at the Seattle location), the ratios from this standard of the 1.0 s spectral response acceleration for default site conditions divided by that for Site Class BC are approximately 25% less than the corresponding ASCE 7-16 site coefficients. Whereas the latter were based on data and simulations for shallow crustal earthquakes, the former also take into account site class effects for the subduction zone earthquakes of the Pacific Northwest.
- The decrease at the Las Vegas location (from 0.68*g* to 0.51*g*) is mostly due to the Project '17 modification to site class effects. Whereas the ASCE 7-16 site coefficient in this case is approximately 3.3 (including the 1.5 multiplier of Section 11.4.8), the corresponding ratio from this standard of the 1.0 s spectral response acceleration for default site conditions divided by that for Site Class BC is approximately 2.6. Note that the S_1 value of ASCE 7-16 is 0.21*g*, slightly larger than the 0.2*g* threshold, above which the 1.5 multiplier of Section 11.4.8 is applied; if this S_1 value were instead slightly less than 0.2*g*, the ASCE 7-16 site coefficients would result in an S_{M1} value within 10% of the corresponding value of this standard.
- The decreases at the locations in Memphis (from 1.02*g* to 0.72*g*) and Charleston (from 1.17*g* to 0.77*g*) are again mostly due to the Project '17 modification to site class effects. At these high-hazard CEUS locations, the ratios of the 1.0 s spectral response acceleration for default site conditions divided by that for Site Class BC are 30% less than the corresponding ASCE 7-16 site coefficients. Whereas the latter were based on western United States data and simulations, the former come from the NGA-East ground motion models for the CEUS.
- The changes at the locations in Sacramento (from 0.79*g* to 0.63*g*), Vallejo (from 1.53*g* to 1.80*g*), and New York City (from 0.14*g* to 0.11*g*) locations are mostly due to the same reasons explained previously for S_{MS}, with the following exceptions: Whereas the S_{MS} value increases at the Sacramento location, the S_{M1} value decreases, but still mostly due to the Project '17 modification to site class effects. At the Vallejo location, in addition to the Project '17 modification to deterministic capping, the S_{M1} value also increases as a result of the USGS NSHM update and its incorporation of basin depths.

As seen in Table C22-5 and Figure C22-3, which are for the MCE_G peak ground accelerations, the PGA_M values changed from ASCE 7-16 to this standard by more than 15% at 7 of the 34 locations. All but one of these changes was a decrease. From ASCE 7-10 to ASCE 7-16, the PGA_M values changed by more than 15% at 27 of the 34 locations, all but one of which was an increase.

Table C22-4. Comparison of 1.0 s MCE_R Spectral Response Acceleration Values.

Location Name	ASCE/SEI 7-10		ASCE/SEI 7-16		ASCE/SEI 7-22	
	S_1 (g)	S_{M1} (g)	S_1 (g)	S_{M1} (g)	S_1 (g)	S_{M1} (g)
Los Angeles, CA	0.84	1.26	0.70	1.79	0.72	1.66
Century City, CA	0.80	1.21	0.75	1.92	0.92	1.84
Northridge, CA	0.60	0.90	0.60	1.53	0.70	1.65
Long Beach, CA	0.62	0.93	0.61	1.55	0.75	1.58
Irvine, CA	0.57	0.86	0.45	1.24	0.45	1.12
Riverside, CA	0.60	0.90	0.58	1.50	0.58	1.45
San Bernardino, CA	1.08	1.63	0.93	2.37	1.07	2.83
San Luis Obispo, CA	0.43	0.67	0.40	1.14	0.40	1.00
San Diego, CA	0.48	0.73	0.53	1.40	0.53	1.23
Santa Barbara, CA	0.99	1.49	0.78	1.98	0.87	1.89
Ventura, CA	0.90	1.35	0.76	1.95	1.01	2.10
Oakland, CA	0.75	1.12	0.72	1.83	0.72	1.70
Concord, CA	0.74	1.10	0.67	1.72	0.91	1.91
Monterey, CA	0.56	0.84	0.50	1.34	0.49	1.21
Sacramento, CA	0.29	0.53	0.25	0.79	0.24	0.63
San Francisco, CA	0.64	0.96	0.60	1.53	0.60	1.75
San Mateo, CA	0.86	1.29	0.74	1.88	0.81	2.30
San Jose, CA	0.60	0.90	0.60	1.53	0.61	1.70
Santa Cruz, CA	0.60	0.90	0.60	1.54	0.59	1.47
Vallejo, CA	0.60	0.90	0.60	1.53	0.89	1.80
Santa Rosa, CA	1.04	1.55	0.94	2.39	1.15	2.40
Seattle, WA	0.53	0.79	0.49	1.32	0.66	1.38
Tacoma, WA	0.51	0.76	0.47	1.29	0.46	0.96
Everett, WA	0.48	0.73	0.43	1.20	0.47	0.97
Portland, OR	0.42	0.66	0.39	1.13	0.36	0.78
Salt Lake City, UT	0.56	0.84	0.56	1.45	0.62	1.28
Boise, ID	0.11	0.25	0.11	0.26	0.10	0.28
Reno, NV	0.52	0.78	0.52	1.38	0.54	1.32
Las Vegas, NV	0.17	0.36	0.21	0.68	0.20	0.51
St Louis, MO	0.17	0.36	0.16	0.39	0.18	0.38
Memphis, TN	0.35	0.60	0.35	1.02	0.35	0.72
Charleston, SC	0.37	0.61	0.41	1.17	0.38	0.77
Chicago, IL	0.062	0.15	0.063	0.15	0.070	0.15
New York City, NY	0.072	0.17	0.060	0.14	0.051	0.11

Source: From ASCE 7-16 and ASCE 7-10.
Notes: The S_{M1} values are for default site conditions. The ASCE 7-16 S_{M1} values include the 1.5 multiplier of the applicable section 11.4.8 exception.

- The decreases at the locations in Oakland (from 0.95*g* to 0.78*g*), San Jose (from 0.69*g* to 0.56*g*), Santa Cruz (from 0.81*g* to 0.65*g*), Santa Rosa (1.22*g* to 1.04*g*), and Reno (from 0.74*g* to 0.59*g*) were due to combinations of less than 15% decreases caused by the Project '17 modifications to deterministic capping and less than 10% decreases caused by the Project '17 modification to site effects. Both decreases can be related to the relatively large epsilons of peak ground accelerations compared to longer-period spectral response accelerations.
- The changes at the locations in Vallejo (from 0.74*g* to 0.88*g*) and New York City (from 0.26*g* to 0.18*g*) were due to the same reasons explained previously for S_{MS}.

As seen from Table C22-6, four of the SDCs changed from SDC D in ASCE 7-16 to SDC E in this standard, as a result of increases in S_1 from below 0.75*g* to at or above it. At the Vallejo location, the increase can be attributed to the Project '17 deterministic capping change (as elaborated upon previously) and the incorporation of basin depths in the USGS NSHM. At the Concord location, the increase was mostly a result of the modification to deterministic capping. At the Long Beach location, the increase was mostly because of the incorporation of basin depths. At the San Mateo location, the S_1 increase was relatively small, approximately 10%, but it can be attributed to a relatively small change in deterministic capping. Note that one of the SDCs assigned using the short-period Table 11.6-1 alone changed from ASCE 7-16 to this standard. At the Boise location, the short-period SDC increased from B to C, owing to a 4% increase in the S_{MS} value. By itself, recall that the Project '17 modification to the maximum-response scale factors increased short-period spectral response accelerations by as much as 9%.

In summary, the S_{MS} values changed by more than 15% at only three of the locations, each due to different reasons. The S_{M1} values changed (by more than 15%) at 11 of the locations, eight of which were decreases, all owing to the Project '17 modification to site class effects. Last, the PGA_M values changed (by more than 15%) at seven of the 34 locations, five of which were decreases resulting from combinations of smaller changes caused

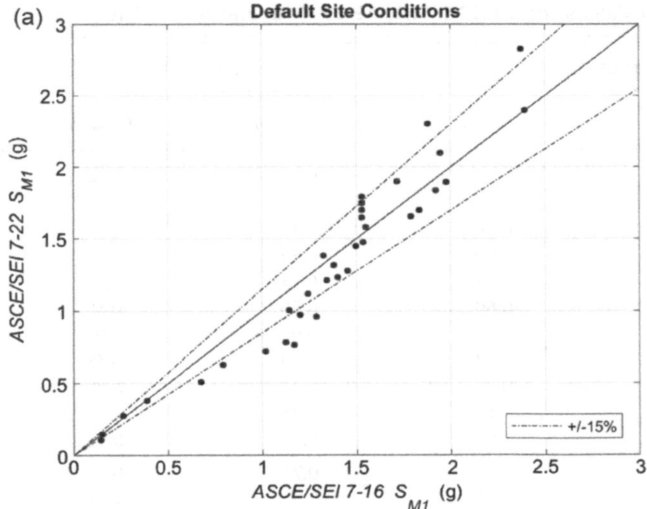

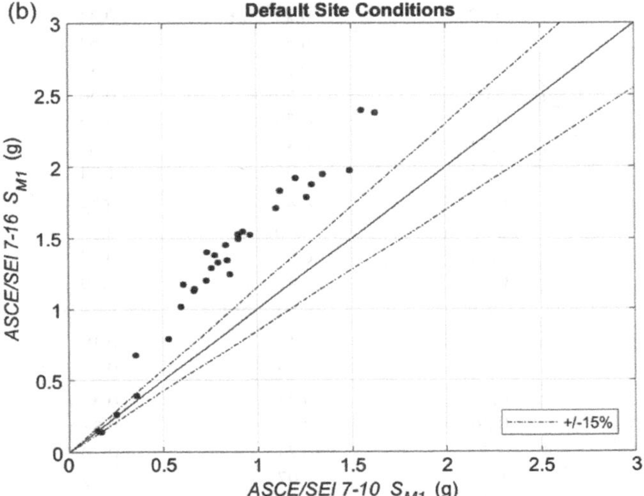

Figure C22-2. Comparison of 1.0 s MCE_R spectral response acceleration (S_{M1}) values from this standard versus (a) ASCE 7-16 and (b) ASCE 7-16 versus ASCE 7-10 for default site conditions and 34 example locations. The ASCE 7-16 values include the 1.5 multiplier of the applicable Section 11.4.8 exception.

by the Project '17 modifications to deterministic capping and site class effect. Only a handful of the SDCs at the 34 example locations changed from ASCE 7-16 to this standard, mostly from SDC D to E.

RISK-TARGETED MAXIMUM CONSIDERED EARTHQUAKE (MCE_R) SPECTRAL RESPONSE ACCELERATION MAPS

Using the most recent USGS National Seismic Hazard Models (NSHMs) for the conterminous United States and the other US states and territories, MCE_R spectral response accelerations of this standard have been prepared in accordance with the site-specific procedures of Sections 21.2 through 21.4, with approximations outside of the conterminous United States. To prepare the MCE_R maps in this chapter, which are for default site conditions, the site-specific procedures are followed for Site Classes C, CD, and D. In accordance with Section 11.4.2.1, the resulting multi-period response spectra are enveloped. It is from the envelope spectrum that S_{MS} for default site conditions is computed in accordance with Section 21.4. To derive values of the parameters S_S and S_1, the site-specific procedures are followed for spectral periods of 0.2 and 1.0 s and Site Class BC. As specified in Section 21.2.3, the MCE_R spectral response acceleration for each site class and period represents the lesser of a probabilistic ground motion defined in Section 21.2.1 and a deterministic ground motion defined in Section 21.2.2. The preparation of the probabilistic and deterministic ground motions are described as follows.

For the conterminous United States, the probabilistic ground motions for each spectral response period (see Section 11.4.5.1 and Site Classes A through E) have been calculated using corresponding USGS hazard curves for a grid of locations. The USGS hazard curves are first converted from geometric mean ground motions (output by the ground motion models available to USGS) to ground motions in the direction of maximum horizontal spectral response acceleration. The conversions were done by applying the factors specified in the site-specific procedures commentary (Section C21.2) (e.g., 1.2 at 0.2 s and 1.25 at 1.0 s). See Section C21.2.1 and Luco et al. (2007) for information on the development of risk-targeted probabilistic ground motions from hazard curves.

For the states and territories outside of the conterminous United States, the probabilistic ground motions for spectral response periods of 0.2 and 1.0 s and the BC site class have been calculated as described in the preceding paragraph, with one exception. The USGS hazard curves for Hawaii, without conversion, are deemed to represent the maximum response ground motions because of the ground motion models applied there. The probabilistic ground motions for the other spectral response periods and site classes are estimated from T_L and the probabilistic ground motions for spectral response periods of 0.2 and 1.0 s and the BC site class by means of a procedure documented in FEMA (2020a). This approximation is temporary, awaiting future USGS updates for the states and territories outside of the conterminous United States.

The deterministic ground motions for the conterminous United States, for each spectral response period and site class, have been calculated via USGS hazard deaggregations for the corresponding probabilistic ground motions. As explained in Section C21.2.2, the hazard deaggregation at each site (and period and site class) provides a mean earthquake magnitude for each fault in the region, which is termed a scenario earthquake. The requisite 84th percentile spectral response acceleration for each scenario earthquake can be calculated using the same ground motion models used for the probabilistic spectral response acceleration and the deaggregation. Alternatively, the mean epsilon for each scenario that is output from the deaggregation can be used to estimate the 84th percentile ground motion. The epsilon corresponds to the percentile of the probabilistic ground motion, per the normal cumulative distribution function used in ground motion models. For example, epsilons of 0, 1, and 2 correspond to the 50th, 84th, and 98th percentiles, respectively. With a logarithmic standard deviation of ground motion for a given scenario earthquake, σ, the 84th percentile ground motion can be calculated from the probabilistic ground motion and its epsilon, ε. This is done by dividing the probabilistic ground motion: $\exp(\varepsilon\cdot\sigma) \div \exp(1\cdot\sigma)$. As the MCE_R (and MCE_G) maps of ASCE 7-16 (and back to the NEHRP 2009) used a σ value of 0.6 to estimate 84th percentile ground motions from medians (50th percentiles), via a multiplier of $\exp(0.6) = 1.8$, the same σ of 0.6 is used for the maps of this standard. Examples of the calculation of 84th percentile ground motions are given in

Table C22-5. Comparison of MCE_G Peak Ground Acceleration Values from ASCE 7-22, ASCE 7-16, and ASCE 7-10.

	ASCE/SEI 7-10		ASCE/SEI 7-16		ASCE/SEI 7-22	
Location Name	PGA (*g*)	PGA_M (*g*)	PGA (*g*)	PGA_M (*g*)	"*PGA*" (*g*)	PGA_M (*g*)
Los Angeles, CA	0.91	0.91	0.84	1.01	0.85	0.94
Century City, CA	0.81	0.81	0.91	1.09	0.91	0.99
Northridge, CA	0.62	0.62	0.71	0.86	0.72	0.80
Long Beach, CA	0.64	0.64	0.74	0.89	0.74	0.82
Irvine, CA	0.60	0.60	0.53	0.63	0.54	0.64
Riverside, CA	0.50	0.50	0.50	0.60	0.50	0.58
San Bernardino, CA	0.91	0.91	0.98	1.18	0.95	1.04
San Luis Obispo, CA	0.44	0.47	0.48	0.58	0.49	0.58
San Diego, CA	0.57	0.57	0.72	0.86	0.71	0.78
Santa Barbara, CA	1.09	1.09	0.93	1.12	0.93	1.01
Ventura, CA	0.91	0.91	0.89	1.06	0.87	0.95
Oakland, CA	0.72	0.72	0.79	0.95	0.71	0.78
Concord, CA	0.79	0.79	0.90	1.07	0.95	1.01
Monterey, CA	0.59	0.59	0.58	0.69	0.58	0.68
Sacramento, CA	0.23	0.31	0.24	0.32	0.25	0.34
San Francisco, CA	0.57	0.57	0.58	0.70	0.54	0.63
San Mateo, CA	0.73	0.73	0.78	0.93	0.72	0.81
San Jose, CA	0.50	0.50	0.58	0.69	0.50	0.56
Santa Cruz, CA	0.59	0.59	0.67	0.81	0.58	0.65
Vallejo, CA	0.51	0.51	0.62	0.74	0.81	0.88
Santa Rosa, CA	0.97	0.97	1.02	1.22	0.95	1.04
Seattle, WA	0.56	0.56	0.60	0.72	0.61	0.73
Tacoma, WA	0.50	0.50	0.50	0.60	0.50	0.58
Everett, WA	0.52	0.52	0.51	0.62	0.54	0.64
Portland, OR	0.43	0.46	0.40	0.48	0.40	0.50
Salt Lake City, UT	0.67	0.67	0.70	0.84	0.72	0.77
Boise, ID	0.12	0.19	0.14	0.21	0.13	0.19
Reno, NV	0.50	0.50	0.62	0.74	0.54	0.59
Las Vegas, NV	0.20	0.28	0.28	0.37	0.27	0.34
St Louis, MO	0.23	0.31	0.27	0.36	0.33	0.33
Memphis, TN	0.50	0.50	0.61	0.73	0.74	0.70
Charleston, SC	0.75	0.75	0.93	1.12	1.07	0.97
Chicago, IL	0.068	0.11	0.059	0.094	0.072	0.081
New York City, NY	0.17	0.25	0.18	0.26	0.17	0.18

Notes: The PGA_M values are for default site conditions. This standard has replaced the notation PGA of ASCE 7-10 and ASCE 7-16 with "PGA_M for Site Class BC," Abbreviated here as "PGA."

Table C22-7. Starting with the same grid of locations considered for the probabilistic ground motions, such calculations have been done for all of the locations and site classes where the probabilistic ground motions exceed, at one or more spectral response periods, the deterministic lower limits of Table 21.2-1. For more information on the development of the deterministic ground motions for the MCE_R maps of this standard, see Section C21.2.2 and Luco et al. (2020).

The deterministic ground motions outside of the conterminous United States have been calculated in accordance with the site-specific procedures of ASCE 7-16, for spectral response periods of 0.2 and 1.0 s and the BC site class (denoted here as $S_{0.2D}$ and $S_{1.0D}$). More specifically, they have been calculated from "characteristic earthquake on all known active faults." The largest characteristic magnitude considered by USGS on each fault, excluding any lower weighted magnitudes from the USGS logic tree for epistemic uncertainty, is used for the deterministic ground motions. For each characteristic earthquake, USGS has computed median (50th percentile), geometric mean ground motions. To convert to maximum response ground motions, the same scale factors as mentioned for probabilistic ground motions are applied. To approximately convert to 84th percentile ground motions, the maximum response ground motions are multiplied by 1.8. The probabilistic ground motions for the other spectral response periods and site classes are estimated from $S_{0.2D}$ and $S_{1.0D}$ using FEMA (2020).

MAXIMUM CONSIDERED EARTHQUAKE GEOMETRIC MEAN (MCE_G) PEAK GROUND ACCELERATIONS

Using the most recent USGS NSHMs for the conterminous United States and the other US states and territories, the MCE_G peak ground accelerations (i.e., the PGA_M values for each site class) of this standard have been prepared in accordance with the site-specific procedures of Section 21.5, with two approximations outside of the conterminous United States. There, the underlying probabilistic and deterministic peak ground accelerations are estimated from those for Site Class BC using FEMA

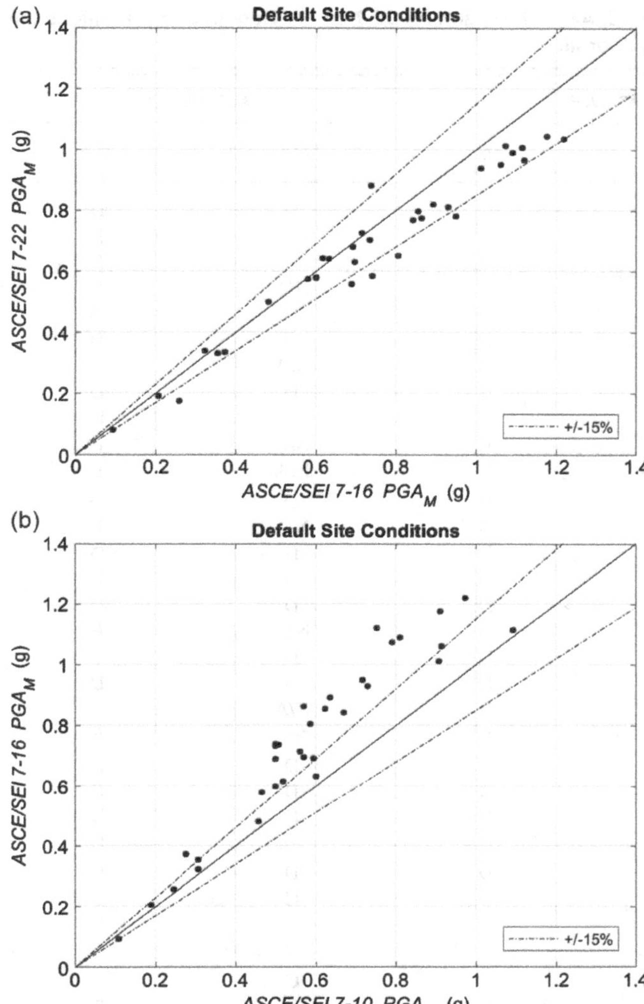

Figure C22-3. Comparison of MCE_G peak ground acceleration (PGA_M) values from this standard versus (a) ASCE 7-16 and (b) ASCE 7-16 versus ASCE 7-10 for default site conditions and 34 example locations.

(2020). Even for Site Class BC, the deterministic ground motions are not yet calculated from scenario earthquakes from hazard deaggregations but rather from the characteristic earthquakes called for in previous editions of ASCE 7. These two approximations outside of the conterminous United States are temporary, awaiting future USGS updates. For the conterminous United States, the PGA_M values are in full accordance with the site-specific procedures. Because the MCE_G maps (like the MCE_G maps) are for default site conditions, the site-specific procedures (or approximations outside of the conterminous United States) have been followed for Site Classes C, CD, and D, and the largest of the three PGA_M values was mapped.

LONG-PERIOD TRANSITION MAPS

The maps of the long-period transition period, T_L (Figures 22-14 through 22-17), were introduced in ASCE 7-05. They were prepared by the USGS in response to Building Seismic Safety Council recommendations and were subsequently included in NEHRP (2003). See Section C11.4.6 for a discussion of the technical basis of these maps. The value of T_L obtained from these maps is used in Equation (11.4-7) to determine values of S_a for periods greater than T_L.

The exception in Section 15.7.6.1, regarding the calculation of S_{ac}, the convective response spectral acceleration for tank response, is intended to provide the user the option of computing this acceleration with three different types of site-specific procedures: (1) the procedures in Chapter 21, provided that they cover the natural period band containing T_c, the fundamental convective period of the tank-fluid system; (2) ground motion simulation methods using seismological models; and (3) analysis of representative accelerogram data. Elaboration of these procedures is provided as follows.

With regard to the first procedure, attenuation equations have been developed for the western United States (Next Generation Attenuation) (e.g., Power et al. 2008) and for the central and eastern United States (e.g., Somerville et al. 2001) that cover the period band, 0 to 10 s. Thus, for $T_c \leq 10$ s, the fundamental convective period range for nearly all storage tanks, these attenuation equations can be used in the same probabilistic seismic hazard analysis (PSHA) and deterministic seismic hazard analysis (DSHA) procedures described in Chapter 21 to compute $S_a(T_c)$. The 1.5 factor in Equation (15.7-11), which converts a 5% damped spectral acceleration to a 0.5% damped value, could then be applied to obtain S_{ac}. Alternatively, this factor could be established by statistical analysis of 0.5% damped and 5% damped response spectra of accelerograms representative of the ground motion expected at the site.

In some regions of the United States, such as the Pacific Northwest and southern Alaska, where subduction zone earthquakes dominate the ground motion hazard, attenuation equations for these events only extend to periods between 3 and 5 s, depending on the equation. Thus, for tanks with T_c greater than these periods, other site-specific methods are required.

The second site-specific method to obtain S_a at long periods is simulation through the use of seismological models of fault rupture and wave propagation (e.g., Graves and Pitarka 2004, Hartzell and Heaton 1983, Hartzell et al. 1999, Liu et al. 2006, Zeng et al. 1994). These models could range from simple seismic source theory and wave propagation models, which currently form the basis for many of the attenuation equations used in the central and eastern United States, for example, to more complex numerical models that incorporate finite fault rupture for scenario earthquakes and seismic wave propagation through 2D or 3D models of the regional geology, which may include basins. These models are particularly attractive for computing long-period ground motions from great earthquakes ($M_w \geq \sim 8$) because ground motion data are limited for these events. Furthermore, the models are more accurate for predicting longer period ground motions because (1) seismographic recordings may be used to calibrate these models, and (2) the general nature of the 2D or 3D regional geology is typically well resolved at these periods and can be much simpler than would be required for accurate prediction of shorter period motions.

A third site-specific method is the analysis of the response spectra of representative accelerograms that have accurately recorded long-period motions to periods greater than T_c. As T_c increases, the number of qualified records decreases. However, as digital accelerographs continue to replace analog accelerographs, more recordings with accurate long-period motions are becoming available. Nevertheless, a number of analog and digital recordings of large and great earthquakes are available that have accurate long-period motions to 8 s and beyond. Subsets of these records,

Table C22-6. Comparison of Seismic Design Categories from this Standard, ASCE 7-16, and ASCE 7-10, for Default Site Conditions and Risk Category I, II, or III.

	ASCE/SEI 7-10		ASCE/SEI 7-16		ASCE/SEI 7-22	
Location Name	"SDC_S"	SDC	"SDC_S"	SDC	"SDC_S"	SDC
Los Angeles, CA	N/A	E	D	D	D	D
Century City, CA	N/A	E	N/A	E	N/A	E
Northridge, CA	D	D	D	D	D	D
Long Beach, CA	D	D	D	D	N/A	E
Irvine, CA	D	D	D	D	D	D
Riverside, CA	D	D	D	D	D	D
San Bernardino, CA	N/A	E	N/A	E	N/A	E
San Luis Obispo, CA	D	D	D	D	D	D
San Diego, CA	D	D	D	D	D	D
Santa Barbara, CA	N/A	E	N/A	E	N/A	E
Ventura, CA	N/A	E	N/A	E	N/A	E
Oakland, CA	D	D	D	D	D	D
Concord, CA	D	D	D	D	N/A	E
Monterey, CA	D	D	D	D	D	D
Sacramento, CA	D	D	D	D	D	D
San Francisco, CA	D	D	D	D	D	D
San Mateo, CA	N/A	E	D	D	N/A	E
San Jose, CA	D	D	D	D	D	D
Santa Cruz, CA	D	D	D	D	D	D
Vallejo, CA	D	D	D	D	N/A	E
Santa Rosa, CA	N/A	E	N/A	E	N/A	E
Seattle, WA	D	D	D	D	D	D
Tacoma, WA	D	D	D	D	D	D
Everett, WA	D	D	D	D	D	D
Portland, OR	D	D	D	D	D	D
Salt Lake City, UT	D	D	D	D	D	D
Boise, ID	B	C	B	C	C	C
Reno, NV	D	D	D	D	D	D
Las Vegas, NV	C	D	D	D	D	D
St Louis, MO	C	D	C	D	C	D
Memphis, TN	D	D	D	D	D	D
Charleston, SC	D	D	D	D	D	D
Chicago, IL	A	B	A	B	A	B
New York City, NY	B	B	B	B	B	B

Note: The SDC_s are determined from Table 11.6-1 (Seismic Design Category based on short-period response acceleration parameter) alone but only where $S_1 \leq 0.75g$

representative of the earthquake(s) controlling the ground motion hazard at a site, can be selected. The 0.5% damped response spectra of the records can be scaled using seismic source theory to adjust them to the magnitude and distance of the controlling earthquake. The levels of the scaled response spectra at periods around T_c can be used to determine S_{ac}. If the subset of representative records is limited, then this method should be used in conjunction with the aforementioned simulation methods.

USGS SEISMIC DESIGN GEODATABASE AND WEB SERVICE

Because maps of MCE_R and MCE_G ground motions for all the site classes (and all the MCE_R spectral periods) are too numerous to include in this standard, gridded values of the ground motions are contained in the online USGS Seismic Design Geodatabase archive, defined in Section 11.2. To spatially interpolate between these gridded ground motions for a user-specified latitude and longitude (and site class), the USGS has developed a web service. The web service returns S_S, S_1, S_{MS}, S_{M1}, PGA_M, and T_L, values, as well as MCE_R multi-period response spectra. For visualization, the USGS has also prepared online maps for all the site classes. As summarized previously, all these online resources have been developed following the site-specific procedures of Chapter 21. They are accessible via https://doi.org/10.5066/F7NK3C76.

REFERENCES

Aagaard, B. T., T. M. Brocher, D. Dolenc, D. Dreger, R. W. Graves, S. Harmsen, et al. 2008. "Ground-motion modeling of the 1906 San Francisco earthquake, part II: Ground motion estimates for the 1906 earthquake and scenario events." *Bull. Seismol. Soc. Am.* 98 (2): 1012–1046. https://doi.org/10.1785/0120060410.

Al Atik, L. 2015. *NGA-east: Ground-motion standard deviation models for central and Eastern North America*. PEER Rep. No. 2015/07. Berkeley, CA: Pacific Earthquake Engineering Research Center (PEER).

Table C22-7. Examples of 84th Percentile Ground Motions

		Scenario Earthquakes					
		Hayward		Calaveras		San Andreas	
Period T (s)	Prob. GM (g)	ε	84th GM (g)	ε	84th GM (g)	ε	84th GM (g)
0.20	2.36	1.87	**1.40**	1.88	1.39	2.07	1.24
0.25	2.19	1.84	1.32	1.84	**1.32**	1.99	1.21
0.30	2.03	1.83	1.23	1.81	**1.25**	1.95	1.15
0.40	1.73	1.81	1.06	1.80	**1.07**	1.91	1.00
0.50	1.53	1.80	0.95	1.78	**0.96**	1.85	0.92
0.75	1.15	1.77	0.72	1.75	**0.73**	1.76	0.73
1.0	0.92	1.77	0.58	1.75	0.59	1.74	**0.59**
1.5	0.62	1.72	0.40	1.73	0.40	1.67	**0.41**
2.0	0.47	1.69	0.31	1.71	0.31	1.60	**0.33**
3.0	0.32	1.60	0.22	1.63	0.22	1.39	**0.25**
4.0	0.25	1.49	0.19	1.50	0.19	1.15	**0.23**
5.0	0.20	1.36	0.16	1.41	0.16	0.97	**0.20**

Notes: GM = ground motions. Examples are calculated using epsilons from hazard deaggregations (ε) and the corresponding probabilistic ground motions, at a Site Class BC location in San Jose, California.

ASCE. 2016. *Minimum design loads for buildings and other structures*. ASCE/SEI 7-16. Reston, VA: ASCE.

Atkinson, G. M., and D. M. Boore. 2003. "Empirical ground-motion relations for subduction-zone earthquakes and their application to Cascadia and other regions." *Bull. Seismol. Soc. Am.* 93 (4): 1703–1729. https://doi.org/10.1785/0120020156.

Atkinson, G. M., and D. M. Boore. 2008. "Erratum to empirical ground-motion relations for subduction zone earthquakes and their application to Cascadia and other regions." *Bull. Seismol. Soc. Am.* 98 (5): 2567–2569. https://doi.org/10.1785/0120080108.

BSSC (Building Seismic Safety Council). 2019. *BSSC Project 17 final report: Development of next generation of seismic design value maps for the 2020 NEHRP provisions*. Washington, DC: National Institute of Building Sciences.

BSSC. 2020. *NEHRP recommended seismic provisions for new buildings and other structures*. FEMA P-750/2020. Washington, DC: FEMA.

FEMA (Federal Emergency Management Agency). 2020a. *NEHRP recommended seismic provisions for new buildings and other structures*. FEMA P-2082. Washington, DC: FEMA.

FEMA. 2020b. *Procedures for developing multi-period response spectra at non-conterminous United States sites*. FEMA P-2078. Washington, DC: FEMA.

Goulet, C. A., Y. Bozorgnia, N. Kuehn, L. Al Atik, R. R. Youngs, R. W. Graves, and G. M. Atkinson. 2017. *NGA-east ground-motion models for the U.S. Geological Survey national seismic hazard maps*. PEER Rep. No. 2017/03. Berkeley, CA: Pacific Earthquake Engineering Research Center.

Graves, R. W., and A. Pitarka. 2004. "Broadband time history simulation using a hybrid approach." In *Proc., 13th World Conf. on Earthquake Engineering*, Vancouver, Canada.

Hartzell, S., S. Harmsen, A. Frankel, and S. Larsen. 1999. "Calculation of broadband time histories of ground motion: Comparison of methods and validation using strong ground motion from the 1994 Northridge earthquake." *Bull. Seismol. Soc. Am.* 89 (6): 1484–1504.

Hartzell, S., and T. Heaton. 1983. "Inversion of strong ground motion and teleseismic waveform data for the fault rupture history of the 1979 Imperial Valley, California, earthquake." *Bull. Seismol. Soc. Am.* 73 (6A): 1553–1583.

Huang, Y.-N., A. S. Whittaker, and N. Luco. 2008. "Maximum spectral demands in the near-fault region." *Earthquake Spectra* 24 (1): 319–341.

Idriss, I. M. 2014. "An NGA-West2 empirical model for estimating the horizontal spectral values generated by shallow crustal earthquakes." *Earthquake Spectra* 30 (3): 1155–1177. https://doi.org/10.1193/070613EQS195M.

Kircher & Associates. 2015. *Investigation of an identified shortcoming in the seismic design procedures of ASCE 7-10 and development of recommended improvements for ASCE 7-16*. Washington, DC: National Institute of Building Sciences.

Klein, F., A. D. Frankel, C. S. Mueller, R. L.Wesson, and P. Okubo. 2001. "Seismic hazard in Hawaii: High rate of large earthquakes and probabilistic ground-motion maps." *Bull. Seismol. Soc. Am.* 91: 479–498.

Lee, E.-J., P. Chen, T. H. Jordan, P. J. Maechling, M. Denolle, and G. C. Beroza. 2014. "Full-3-D tomography for crustal structure in southern California based on the scattering-integral and the adjoint-wavefield methods." *J. Geophys. Res.* 119 (8): 6421–6451. https://doi.org/10.1002/2014JB011346.

Liu, P., R. J. Archuleta, and S. H. Hartzell. 2006. "Prediction of broadband ground-motion time histories: Hybrid low/high-frequency method with correlated random source parameters." *Bull. Seismol. Soc. Am.* 96 (6): 2118–2130. https://doi.org/10.1785/0120060036.

Luco, N., et al. 2020. "Using probabilistic hazard disaggregation for the deterministic ground motion capping in U.S. building codes." *Earthquake Spectra*.

Luco, N., B. R. Ellingwood, R. O. Hamburger, J. D. Hooper, J. K. Kimball, and C. A. Kircher. 2007. "Risk-targeted versus current seismic design maps for the conterminous United States." In *Proc., Structural Engineers Association of California 76th Annual Convention*, Sacramento, CA.

Magistrale, H., K. B. Olsen, and J. C. Pechmann. 2008. *Construction and verification of a Wasatch Front community velocity model*. USGS Technical Rep. No. 06HQGR0012.

Mueller, C. S., A. D. Frankel, M. D. Petersen, and E. V. Leyendecker. 2003. *Documentation for the 2003 USGS seismic hazard maps for Puerto Rico and the U.S. Virgin Islands*. U.S. Geological Survey Open-File Rep. No. 3-379. Washington, DC: USGS.

Mueller, C. S., K. M. Haller, N. Luco, M. D. Petersen, and A. D. Frankel. 2012. *Seismic hazard assessment for Guam and the Northern Mariana Islands*. U.S. Geological Survey Open-File Rep. No. 2012-1015. Washington, DC: USGS.

NEHRP (National Earthquake Hazard Reduction Program). 2003. *Recommended provisions for seismic regulations for new buildings and other structures*. FEMA 450. Washington, DC: National Institute of Building Sciences.

NEHRP. 2009. *Recommended provisions for seismic regulations for new buildings and other structures*. FEMA 750. Washington, DC: National Institute of Building Sciences.

NEHRP. 2015. *Recommended provisions for seismic regulations for new buildings and other structures*. FEMA 1050. Washington, DC: National Institute of Building Sciences.

PEER (Pacific Earthquake Engineering Research Center). 2015. *NGA-East: Adjustments to median ground-motion models for central and eastern North America*. PEER Rep. No. 2015/08. Berkeley, CA: PEER.

Petersen, M. D., S. C. Harmsen, K. S. Rukstales, C. S. Mueller, D. E. McNamara, N. Luco, et al. 2012. *Seismic hazard of American Samoa and neighboring South Pacific Islands: Data, methods, parameters, and results*. U.S. Geological Survey Open-File Rep. No. 2012-1087. Washington, DC: USGS.

Petersen, M. D., et al. 2014. *Documentation for the 2014 update of the United States national seismic hazard maps*. USGS Open File Rep. No. 2014-1091. Washington, DC: USGS.

Petersen, M. D., A. M. Shumway, P. M. Powers, C. S. Mueller, M. P. Moschett, and A. D. Frankel, et al. 2019. "2018 update of the U.S. National seismic hazard model: Overview of model and implications." *Earthquake Spectra* 36 (1): 5–41. https://doi.org/10.1177/8755293019878199.

Power, M., B. Chiou, N. Abrahamson, Y. Bozorgnia, T. Shantz, and C. Roblee. 2008. "An overview of the NGA project." *Earthquake spectra special issue on the next generation of ground motion attenuation (NGA) project*. Oakland, CA: Earthquake Engineering Research Institute.

Seyhan, E., and J. P. Stewart. 2014. "Semi-empirical nonlinear site amplification from NGAWest2 data and simulations." *Earthquake Spectra* 30 (3): 1241–1256. https://doi.org/10.1193/063013EQS181M.

Shahi, S. K., and J. W. Baker. 2013. "NGA-West2 models for ground motion directionality." *Earthquake Spectra* 30 (3). https://doi.org/10.1193/040913EQS097M.

Somerville, P. G., N. Collins, N. Abrahamson, R. Graves, and C. Saikia. 2001. *Earthquake source scaling and ground motion attenuation relations for the central and eastern United States*. Final Report to the USGS under Contract 99HQGR0098. Washington, DC: USGS.

Stephenson, W. J. 2007. *Velocity and density models incorporating the Cascadia subduction zone for 3D earthquake ground motion simulations*. USGS Open-File Rep. No. 2007-1348. Washington, DC: USGS.

Stewart, J. P., G. A. Parker, J. A. Harmon, G. M. Atkinson, D. M. Boore, R. B. Darragh, W. J. Silva, and Y. M. Hashash. 2017. *Expert panel recommendations for ergodic site amplification in central and eastern North America*. PEER Rep. No. 2017/04. Berkeley, CA: Pacific Earthquake Engineering Research Center.

USGS (US Geological Survey). 2018. *National seismic hazard model*. Reston, VA: USGS.

Wesson, R. L., O. S. Boyd, C. S. Mueller, C. G. Bufe, A. D. Frankel, and M. D. Petersen. 2007. *Revision of time-independent probabilistic seismic hazard maps for Alaska*. U.S. Geological Survey Open-File Rep. No. 2007-1043. Washington, DC: USGS.

Zeng, Y., J. G. Anderson, and G. Yu. 1994. "A composite source model for computing synthetic strong ground motions." *Geophys. Res. Lett.* 21 (8): 725–728. https://doi.org/10.1029/94GL00367.

CHAPTER C23
SEISMIC DESIGN REFERENCE DOCUMENTS

There is no commentary for Chapter 23.

CHAPTER C24
RESERVED FOR FUTURE COMMENTARY

CHAPTER C25
RESERVED FOR FUTURE COMMENTARY

CHAPTER C26
WIND LOADS: GENERAL REQUIREMENTS

C26.1 PROCEDURES

Chapter 26 is the first of six chapters devoted to the wind load provisions. It provides the basic wind design parameters that are applicable to the various wind load determination methodologies contained in Chapters 27 through 31. Specific items covered in Chapter 26 include definitions, basic wind speed, exposure categories, internal pressures, elevation effects, enclosure classification, gust effects, and topographic factors.

C26.1.1 Scope The procedures specified in this standard provide wind pressures and forces for the design of the main wind force resisting system (MWFRS) and of components and cladding (C&C) of buildings and other structures. The procedures involve the determination of wind directionality and velocity pressure, the selection or determination of an appropriate gust-effect factor, and the selection of appropriate pressure or force coefficients. The procedures account for the level of structural reliability required, the effects of differing wind exposures, the speed-up effects of certain topographic features such as hills and escarpments, and the size and geometry of the building or other structure under consideration. The procedures differentiate between rigid and flexible buildings and other structures, and the results generally envelop the most critical load conditions for the design of the MWFRS, as well as C&C.

The pressure and force coefficients provided in Chapters 27, 28, 29, and 30 have been assembled from the latest boundary-layer wind tunnel and full-scale tests and from previously available literature. Because the boundary-layer wind tunnel results were obtained for specific types of building, such as low-rise or high-rise buildings and buildings that have specific types of structural framing systems, the designer is cautioned against indiscriminate interchange of values among the figures and tables.

Tornado loads are treated separately from wind loads, as described in Section C32.1.

Storm Shelters. Storm shelters and safe rooms are designed with the intent of providing an enhanced level of life-safety protection for a design storm event that is greater than typical for conventional buildings and structures, including those designed for the wind loads in this chapter. Since 2009, the model building and residential codes have required that structures designated as storm shelters be designed according to ICC 500 (ICC 2020), the *ICC/NSSA Standard for the Design and Construction of Storm Shelters* (the lastest version of which is the 2020 edition), which applies to hurricane, tornado, and combined hurricane and tornado shelters. The load provisions of ICC 500 directly reference ASCE 7, with some modifications including greater hazard levels.

FEMA also publishes guidance on the design and construction of structures for storm protection in FEMA P-361, *Safe Rooms for Tornadoes and Hurricanes: Guidance for Community and Residential Safe Rooms* (2021a), and FEMA P-320, *Taking Shelter from the Storm: Building or Installing a Safe Room for Your Home* (2021b). FEMA P-361 provides comprehensive guidance for the design of community and residential safe rooms, as well as operations and maintenance guidance. Although the criteria in FEMA P-361 and ICC 500 are quite similar, "safe room" is FEMA terminology for storm shelters that also meet the criteria in FEMA P-361. All safe room criteria in FEMA P-361 meet the storm shelter requirements in ICC 500, but a few design and performance criteria in FEMA P-361 are more restrictive than those in ICC 500 and are required for safe room projects to be eligible for federal grant funding. FEMA P-320 provides prescriptive solutions for residential safe rooms.

Storm shelters and safe rooms are designed for more extreme hazard levels than conventional buildings. For example, where Chapter 26 uses wind speeds with return periods of 700, 1,700, and 3,000 years to determine the design wind loads for Risk Category II, III, and IV buildings, respectively, the wind loads for hurricane shelters and safe rooms are based on hurricane wind speed maps having a 10,000-year mean recurrence interval (MRI). "The design wind speeds chosen by FEMA for safe room guidance were determined with the intent of specifying near-absolute protection with an emphasis on life safety" (FEMA 2021a). Rain loads and flood loads are also based on hazard levels with greater return periods than used for Risk Category II through IV buildings. In addition, hurricane shelters and safe rooms have much more stringent wind-borne debris requirements than ASCE 7. The cumulative effect of these and other differences between ICC 500 and ASCE 7 results in storm shelters providing greater life-safety protection from hurricanes than buildings designed to ASCE 7 Risk Category IV. Although ICC 500 does not explicitly state a performance objective for shelters designed to that standard, safe rooms constructed to the almost-identical FEMA criteria are intended to "provide near-absolute protection from wind and wind-borne debris for occupants." Since 1998, when FEMA first published guidance for residential and community safe rooms, "thousands of safe rooms have been built, and a growing number of these safe rooms have already saved lives in actual events. There has not been a single reported failure of a safe room constructed to FEMA criteria" (FEMA 2021a). Similarly, failures of hurricane shelters constructed to ICC 500 criteria (first published in 2008) have not been reported.

C26.1.2 Permitted Procedures The wind load provisions provide several procedures (as illustrated in Figure 26.1-1) from which the designer can choose.

For MWFRS:

1. Directional Procedure for buildings of all heights (Chapter 27).
2. Envelope Procedure for low-rise buildings (Chapter 28).
3. Directional Procedure for building appurtenances and other structures (Chapter 29).
4. Wind Tunnel Procedure for all buildings and other structures (Chapter 31).

For C&C:

1. Analytical Procedure for buildings and building appurtenances (Chapter 30).
2. Wind Tunnel Procedure for all buildings and other structures (Chapter 31).

The "simplified methods" for the determination of the MWFRS design pressures that were contained in previous editions of this standard have been removed from this edition. The Wind Load Subcommittee evaluated the need for the tabular methods and concluded that ASCE 7 should generally not tabulate calculated design pressures for specific buildings and that the tabular methods belong in a guide or other resource.

Limitations. The provisions given under Section 26.1.2 apply to the majority of site locations and to buildings and other structures, but for some projects, these provisions may be inadequate. Examples of site locations and buildings and other structures (or portions thereof) that may require other approved standards, special studies using applicable recognized literature pertaining to wind effects, or using the Wind Tunnel Procedure of Chapter 31, include

1. Site locations that have channeling effects or wakes from upwind obstructions. Channeling effects can be caused by topographic features (e.g., a mountain gorge) or buildings (e.g., a neighboring tall building or a cluster of tall buildings). Wakes can be caused by hills, buildings, or other structures. These effects can increase the local wind speed and alter the action of wind on a subject building or structure. For flexible or dynamically sensitive buildings, such interference effects can increase the resonant dynamic wind loads and responses. Situations where interference effects may be significant are listed in Item 4.
2. Buildings with unusual or irregular geometric shape, including barrel vaults, arched roofs, and others whose shape (in plan or vertical cross section) differs significantly from the shapes in Figures 27.3-1, 27.3-2, 27.3-3, 27.3-7, 28.3-1, and 30.3-1 through 30.3-7. Unusual or irregular geometric shapes include buildings with multiple setbacks, curved façades, or irregular plans resulting from significant indentations or projections, openings through the building, or multitower buildings connected by bridges.
3. Buildings or other structures with response characteristics that result in substantial vortex-induced and/or torsional dynamic effects, or dynamic effects resulting from aeroelastic instabilities such as flutter or galloping. Such dynamic effects are difficult to anticipate, being dependent on many factors, but should be considered when any one or more of the following apply:
   - Height of the building or other structure is more than 400 ft (122 m).
   - Height of the building or other structure is greater than four times its minimum effective width, $B_{\min}$, as defined subsequently.
   - Lowest natural frequency of the building or other structure is less than $n_1 = 0.25\ Hz$.
   - Reduced velocity $\overline{V}_{\bar{z}}/(n_1 B_{\min}) > 5$, where $\bar{z}$ is $0.6h$ and $\overline{V}_{\bar{z}}$ is the mean hourly velocity in ft/s (m/s) at height $\bar{z}$.

   The minimum effective width $B_{\min}$ is defined as the minimum value of $\sum h_i B_i / \sum h_i$ considering all wind directions. The summations are performed over the height of the building or other structure for each wind direction, where h_i is the height above grade of level i, and B_i is the width at level i normal to the wind direction.
4. Flexible or dynamically sensitive buildings with one or more upwind building(s) for which interference effects, such as channeling effects or wakes, may be significant. Interference effects can increase the resonant dynamic loads and response of a dynamically sensitive building, and can be significant when all of the following characteristics apply to the building under study:
   - The height of the subject building is greater than 2 times the average height of surrounding buildings within a distance of $10B_{\min}$, where $B_{\min}$ is defined in Item 3 above.
   - The presence of one or more upwind building(s) with a height comparable to or greater than the height of the subject building, and located within a distance of less than $10B_{\min}$ of the subject building.
   - The height of the building or other structure is greater than 4 times its minimum effective width, and reduced velocity is greater than 5 ft/s (m/s), and the lowest natural frequency (n_1) is less than 0.25 Hz (i.e., $n_1 < 0.25$ Hz).

   Interference effects are generally more significant for site locations characterized by wind with lower turbulence (e.g., relatively smooth upstream fetches with few obstructions.
5. Bridges, cranes, electrical transmission lines, guyed masts, highway signs and lighting structures, telecommunication towers, and flagpoles.

When undertaking detailed studies of the dynamic response to wind forces, the fundamental frequencies of the building or other structure in each direction under consideration should be established using the structural properties and deformational characteristics of the resisting elements in a properly substantiated analysis, and not using approximate equations based on height.

Shielding. Because of the lack of reliable analytical procedures for predicting the effects of shielding provided by buildings and other structures or by topographic features, reductions in velocity pressure caused by shielding are not permitted under the provisions of this chapter. However, this does not preclude the determination of shielding effects and the corresponding reductions in velocity pressure by means of the Wind Tunnel Procedure in Chapter 31.

C26.1.3 Performance-Based Procedures Performance-based procedures for wind design can quantify service interruption or loss due to wind, improve wind resistance, or perform an enhanced evaluation of the MWFRS and/or envelope. Performance-based procedures identify the source of loss or service interruption and seek to mitigate undesirable outcomes due to design decisions informed by design objectives. Selection of design objectives should consider the ramifications of structural and nonstructural response for the building occupants, the ability of the building or structure to provide services following and/or during a wind event, and maintenance of structural safety expressed by Table 1.3-1. The evaluation should consider strength-level wind effects, service-level wind effects, and occupant-comfort wind effects, which all influence building or facility response and the resulting ability to maintain functionality following and/or during a wind event (also termed *resilience*).

Performance design objectives should consider that, historically, a large proportion of damage from high winds has occurred through envelope component damage, including water penetration through leakage and breaches in the building envelope. In addition, direct loss or interruption of service can occur due to structural motions, causing occupant discomfort and/or nonstructural damage. Using approved methods, the designer is permitted to evaluate the performance of a building, facility, or structural system using performance-based wind engineering procedures documented in the recognized literature, such as the ASCE/SEI *Prestandard for Performance-Based Wind Design* (2019). The performance results should demonstrate consistency with approved performance criteria. An independent peer review should be conducted as part of a review of the performance-based design where the design takes exception to provisions of Chapters 26 through 31.

Major advancements introduced by ASCE/SEI (2019) include limited inelasticity in the MWFRS elements, nonlinear dynamic analysis for wind design, system-based performance criteria, and enhanced design criteria for the building envelope.

The maps provided in Appendix F can be referenced where a performance-based wind structural reliability assessment requires wind hazards with long return periods.

C26.2 DEFINITIONS

Several important definitions given in the standard are discussed in the following text. These terms are used throughout the standard and are provided to clarify application of the standard provisions.

BUILDING, ENCLOSED; BUILDING, OPEN; BUILDING, PARTIALLY ENCLOSED; BUILDING, PARTIALLY OPEN: These definitions relate to the proper selection of internal pressure coefficients, (GC_{pi}). "Enclosed," "open," and "partially enclosed" buildings are specifically defined. All other buildings are considered to be "partially open" by definition, although there may be large openings in two or more walls. An example of this would be a parking garage through which the wind can easily pass but which does not meet the definition for either an open or a partially enclosed building. The internal pressure coefficient for such a building would be ± 0.18, and the internal pressures would act on the solid areas of the walls and roof. The standard also specifies that a building that meets both the "open" and "partially enclosed" definitions should be considered "partially open."

BUILDING OR OTHER STRUCTURE, FLEXIBLE: A building or other structure is considered "flexible" if it contains a significant dynamic resonant response. Resonant response depends on the gust structure contained in the approaching wind, on wind loading pressures generated by the wind flow about the building, and on the dynamic properties of the building or structure. Gust energy in the wind is smaller at frequencies above about 1 Hz. Therefore, the resonant response of most buildings and structures with their lowest natural frequency above 1 Hz are sufficiently small that resonant response can often be ignored. The natural frequency of buildings or other structures greater than 60 ft (18.3 m) in height is determined in accordance with Sections 26.11.1 and 26.11.2. When buildings or other structures have a height exceeding 4 times the least horizontal dimension or when there is reason to believe that the natural frequency is less than 1 Hz (natural period greater than 1 s), the natural frequency of the structure should be investigated. Approximate equations for natural frequency or period for various building and structure types in addition to those given in Section 26.11.2 for buildings are contained in Commentary Section C26.11.

BUILDING OR OTHER STRUCTURE, REGULAR-SHAPED: Defining the limits of applicability of the various procedures in the standard requires a balance between the practical need to use the provisions past the range for which data have been obtained and restricting use of the provisions past the range of realistic application. Wind load provisions are based primarily on wind tunnel tests on shapes shown in Figures 27.3-1, 27.3-2, 27.3-3, 27.3-7, 28.3-1, and 30.3-1 through 30.3-7. Extensive wind tunnel tests on actual structures under design show that relatively large changes from these shapes can, in many cases, have minor changes in wind load, while in other cases seemingly small changes can have relatively large effects, particularly on cladding pressures. Wind loads on complicated shapes are frequently smaller than those on the simpler shapes of Figures 27.3-1, 27.3-2, 27.3-7, 28.3-1, and 30.3-1 through 30.3-7, and so wind loads determined from these provisions are expected to envelop most structure shapes. Buildings or other structures that are clearly unusual should be designed using the Wind Tunnel Procedure of Chapter 31.

BUILDING OR OTHER STRUCTURE, RIGID: The defining criterion for "rigid," in comparison to "flexible," is that the natural frequency is greater than or equal to 1 Hz. A general guidance is that most rigid buildings and structures have height-to-minimum-width less than 4. The provisions of Sections 26.11.1 and 26.11.2 provide methods for calculating natural frequency (period = 1/natural frequency), and Commentary Section C26.11 provides additional guidance.

COMPONENTS AND CLADDING (C&C): Components receive wind loads directly or from cladding and transfer the load to the MWFRS. Cladding receives wind loads directly. Examples of components include, but are not limited to, fasteners, purlins, girts, studs, sheathing, roof decking, certain trusses, and elements of trusses receiving wind loads from cladding. Examples of cladding include, but are not limited to, wall coverings, curtain walls, roof coverings, sheathing, roof decking, exterior windows, and doors. Components can be part of the MWFRS when they act as elements in shear walls or roof diaphragms, but they may also be loaded directly by wind as individual elements. The designer should use appropriate loads for design of components, which may require certain components to be designed for more than one type of wind loading; for example, long-span roof trusses should be designed for loads associated with MWFRS, and individual members of trusses should also be designed for C&C loads (Mehta and Marshall 1998).

DIAPHRAGM: This definition for diaphragm in wind load applications, for the case of untopped steel decks, differs somewhat from the definition used in Section 12.3 because diaphragms under wind loads are expected to remain essentially elastic.

EFFECTIVE WIND AREA, *A*: Effective wind area is the area of the building surface used to determine (GC_p). This area does not necessarily correspond to the area of the building surface contributing to the force being considered. Two cases arise. In the usual case, the effective wind area does correspond to the area tributary to the force component being considered. For example, for a cladding panel, the effective wind area may be equal to the total area of the panel. For a cladding fastener, the effective wind area is the area of cladding secured by a single fastener. A mullion may receive wind from several cladding panels. In this case, the effective wind area is the area associated with the wind load that is transferred to the mullion.

The second case arises where components such as roofing panels, wall studs, or roof trusses are closely spaced. The area served by the component may become long and narrow. To better approximate the actual load distribution in such cases, the width of the effective wind area used to evaluate (GC_p) need not be taken as less than one-third the length of the area. This increase in effective wind area has the effect of reducing the average wind pressure acting on the component. Note, however, that this effective wind area should only be used in determining (GC_p) in Figures 30.3-1 through 30.3-6. The induced wind load should be applied over the actual area tributary to the component being considered.

For membrane roof systems, the effective wind area is the area of an insulation board (or deck panel if insulation is not used) if the boards are fully adhered (or the membrane is adhered directly to the deck). If the insulation boards or membrane are mechanically attached or partially adhered, the effective wind area is the area of the board or membrane secured by a single fastener or individual spot or row of adhesive.

For windows, doors, and other fenestration assemblies, the effective wind area for typical single-unit assemblies can be taken as the overall area of the assembly. For assemblies comprised of more than one unit mulled together or for more complex fenestration systems, it is recommended that the fenestration product manufacturer be consulted for guidance on the appropriate effective wind area to use when calculating the design wind pressure for product specification purposes.

The definition of effective wind area for rooftop solar panels and arrays is similar to that for C&C. As with C&C, the width of the effective wind area need not be less than one-third its length (which is typically equal to the span of the framing element being considered). The induced wind pressure is calculated per Figure 29.4-4 using this effective wind area, and the wind pressure is then applied over the actual area tributary to the element.

Effective wind area is equal to the tributary area, except in cases where the exception is invoked that the width of the effective wind area need not be less than one-third its length. In such cases, the effective wind area can be taken as larger than the tributary area.

Tributary area for a spanning structural member of a solar array depends on the span length of that member times the perpendicular distances to adjacent parallel members. For a support point or fastener, tributary area depends on the span of members framing into that support point.

Tributary area (and effective wind area) can depend on the characteristics of the solar array support system and the load path. For a roof-bearing system that has different load paths for upward, downward, and lateral forces, the appropriate effective wind area for each direction of forces is used.

If the support system for the solar array has adequate strength, stiffness, and interconnectedness to span across a support or ballast point that is subject to yielding or uplift, the effective wind area can be correspondingly increased, provided strengths are not governed by brittle failure and the deformation of the array is evaluated and does not result in adverse performance. It should be noted that effective wind areas for uplift are usually much smaller than for lateral (drag) forces of ballasted arrays.

MAIN WIND FORCE RESISTING SYSTEM (MWFRS): The MWFRS can consist of a structural frame or an assemblage of structural elements that work together to transfer wind loads acting on the entire building or structure to the ground. Structural elements such as cross-bracing, shear walls, roof trusses, and roof diaphragms are part of the MWFRS when they assist in transferring overall loads (Mehta and Marshall 1998).

WIND-BORNE DEBRIS REGIONS: Wind-borne debris regions are defined to alert the designer to areas requiring consideration of missile impact design. These areas are located within hurricane-prone regions where there is a high risk of glazing failure caused by the impact of wind-borne debris.

C26.3 SYMBOLS

The following additional symbols and notation are used herein:

A_{ob} = Average area of open ground surrounding each obstruction

n = Reference period, years

P_a = Annual probability of wind speed exceeding a given magnitude [Equation (C26.5-3)]

P_n = Probability of exceeding design wind speed during n years [Equation (C26.5-3)]

S_{ob} = Average frontal area presented to the wind by each obstruction

V_t = Wind speed averaged over t seconds (see Figure C26.5-1), mi/h (m/s)

$V_{3,600}$ = Mean wind speed averaged over 1 h (see Figure C26.5-1), mi/h (m/s)

β = Damping ratio (percentage of critical damping)

C26.4 GENERAL

C26.4.3 Wind Pressures Acting on Opposite Faces of Each Building Surface Section 26.4.3 is included in the standard to ensure that internal and external pressures acting on a building surface are taken into account by determining a net pressure from the algebraic sum of those pressures. For additional information on the application of the net C&C wind pressure acting across a multilayered building envelope system, including air-permeable cladding, refer to Section C30.1.1.

C26.5 WIND HAZARD MAP

C26.5.1 Basic Wind Speed All the wind speed maps in ASCE 7-16 have been updated, based on (1) a new analysis of nonhurricane wind data available through 2010, and (2) improvements to the hurricane simulation model, which better account for the translation speed effects of fast-moving storms and the transition from hurricanes to extratropical storms in the northern latitudes (i.e., transition from warm-core to cold-core low-pressure systems). Separate wind speed maps are now provided for Risk Category III and IV buildings and structures,

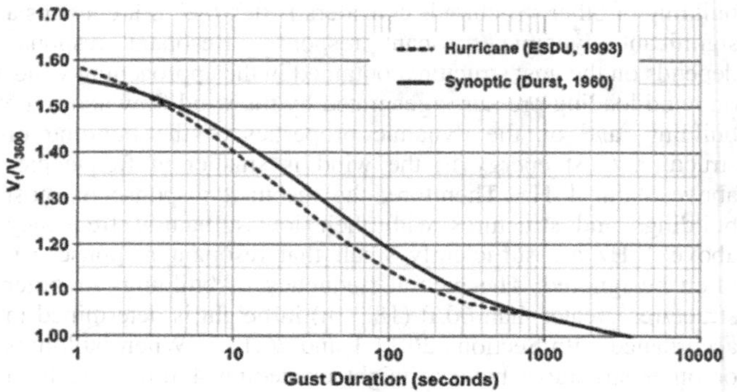

Figure C26.5-1. Maximum speed averaged over *t* (s) to hourly mean speed.

recognizing the higher reliabilities required for essential facilities and facilities whose failure could pose a substantial hazard to the community. Location-specific basic wind speeds may be determined from the ASCE Wind Design Geodatabase using the ASCE 7 Hazard Tool (https://asce7hazardtool.online/). This website provides wind speeds to the nearest mile per hour based on a location defined using either latitude/longitude or an address. The website results use the same data used to develop the paper maps currently in the standard, as well as wind speeds for Hawaii, Puerto Rico, and the US Virgin Islands. Wind speeds are provided to the user for each of the risk categories and each of the serviceability periods.

In the 2016 edition, microzoned "effective" wind speed maps for Hawaii were added in accordance with the strength design return periods, including the effect of topography. The Hawaii effective wind speeds are algebraically formulated to include the macroscale and mesoscale terrain-normalized values of K_{zt} and K_d (Chock et al. 2005); that is, $V_{\text{effective}}$ is the basic wind speed V multiplied by $\sqrt{(K_{zt} \times K_d/0.85)}$, so that the engineer is permitted to more conveniently use the standard values of K_{zt} of 1.0 and K_d as given in Table 26.6-1. Note that local site conditions of finer toposcale, such as ocean promontories and local escarpments, should still be examined. Spatial resolution scales for digital modeling, including terrain effects, are conventionally described in the recognized literature as follows:

Scale	Spatial Resolution
Toposcale	32–656 ft (10–200 m)
Mesoscale	656 ft–3.1 mi (200 m–5 km)
Macroscale	3.1–311 mi (5–500 km)

In the 2022 edition, microzoned "effective" wind speed maps for US Virgin Islands, Puerto Rico, and the municipal islands of Culebra and Vieques were added in accordance with the strength design return periods. The effective wind speeds include the topographic effects which were quantified by the factor of K_{zt}. Thus, the design professional is permitted to evaluate effective wind speeds from these microzoned wind speed maps as a simplified method to determine wind loads and pressures (using $K_{zt} = 1.0$ during calculation) on a building or structure.

To develop the microzoned wind maps, a wind speed-up model has been developed to evaluate wind speed-ups in 16 directional sectors for 36,973 locations in the US Virgin Islands and 828,924 locations distributed among Puerto Rico and the municipal islands of Culebra and Vieques (ARA 2019). These locations comprise a grid with a resolution of 0.00095 degrees (about 100 m or 328 ft). A 1/3-arcsecond digital elevation model was used to create elevation profiles for use in the wind speed-up calculations. The directional wind speed-up data were combined with the directional hurricane winds from a hurricane simulation model to develop effective wind speeds which include the effect of topographically induced wind speed-ups.

The decision in ASCE 7-10 to move to separate strength design wind speed maps for different risk categories in conjunction with a wind load factor of 1.0, instead of using a single map in conjunction with an Importance Factor and a load factor of 1.6, relied on several considerations:

1. A strength-level design wind speed map brings the wind loading approach in line with that used for seismic loads, in that they both are aimed at achieving uniform risk rather than uniform hazard, thus eliminating the use of a load factor for strength design.
2. Having separate maps removes inconsistencies that occurred with the use of Importance Factors, which varied with location, and allows the geographical description of zones affected by nonhurricane winds only and by both hurricane and nonhurricane winds as a function of MRI.
3. Each map has the same MRI for design wind speeds in hurricane and nonhurricane zones.
4. By providing the design wind speed directly, the maps more clearly inform owners and their consultants about the storm intensities for which the buildings and other structures are designed.

Selection of Return Periods. The methodology for selection of the return periods used in ASCE 7-10 (Vickery et al. 2010) has been modified for ASCE 7-16. To determine a return period for each risk category consistent with the target reliabilities in Table C1.3-1, the ASCE 7 Load Combinations Subcommittee conducted a reliability analysis that incorporated new data on the directionality factor. The nominal design value $K_d = 0.85$ was based on a relatively simple directional analysis conducted as part of the original ANSI A58/ASCE 7 load factor development. One of the underlying assumptions of the original analysis was that the wind directionality factor, $K_d = 0.85$, was unbiased because only limited data on the effects of wind directionality were available at the time. More recent research by Isyumov et al. (2013), simulating three building geometries at two different locations, indicates that the ASCE 7 nominal values of K_d are affected by a bias (defined as the ratio of the mean value, μK_d, to the nominal value, K_{dn}). The nominal value of K_d is conservative for both tropical and extratropical winds because the mean value is less than the nominal value. Additional reliability analyses were performed to examine the effect of K_d on the return period and associated reliability. The subcommittee found that the following return periods for each risk category are consistent with the target reliabilities in the first row of Table 1.3-1: Risk Category I, 300 years; Risk Category II, 700 years; Risk Category III, 1,700 years; Risk Category IV, 3,000 years.

Wind Speed. The wind speed maps of Figures 26.5-1A through 26.5-1D present basic wind speeds for the contiguous United States, Alaska, Hawaii, and other selected locations. In ASCE 7-22, wind speeds are included for the first time for the US territory of the Northern Mariana Islands, which match those of nearby Guam. The wind speeds correspond to 3 s gust speeds at 33 ft (10 m) above ground for Exposure Category C. Because the wind speeds of Figures 26.5-1A through 26.5-1D reflect conditions at airports and similar open-country exposures, they do not account for the effects of significant topographic features such as those described in Section 26.8. In ASCE 7-16, wind speeds in nonhurricane-prone areas of the contiguous United States are mapped using contours to better reflect regional variations in the extreme wind climate. Point values are provided to aid interpolation, in a style similar to that used in the ASCE 7 seismic hazard maps. Summaries of the data and methods used to estimate both the nonhurricane and hurricane wind speeds are given next along with a description of how these wind speeds are combined to make the final maps. Detailed descriptions are provided by Pintar et al. (2015) and Lombardo et al. (2016).

Nonhurricane Wind Speeds. The nonhurricane wind speeds for the contiguous United States were estimated from peak gust speed data collected at 575 meteorological stations. The data at each station were extracted from the meteorological records and classified by storm type, thunderstorm or non-thunderstorm after removal of gusts associated with tropical cyclones (i.e., hurricanes and tropical storms). Recorded peak gusts from each station were corrected as needed to standardize the observations to equivalent 3 s peak gusts at 33 ft (10 m) height over open (Exposure C) terrain. At each station, there were at least 15 years of data, and there were sufficient numbers of both thunderstorm and nonthunderstorm observations to account for their potential differences when estimating wind speeds with specified MRIs. The estimation was performed in two stages. In the first stage, a peaks-over-thresholds (POT) model was fitted to the data from each station. The POT model used was the Poisson process model first described by Pickands (1971) and extended by Smith (1989) to allow the parameters of the Poisson process to be time-dependent. This model allowed differentiation between thunderstorm and nonthunderstorm winds. The Poisson process has a tail length parameter that may be set to zero, leading to Gumbel-like tails for the distribution of wind speeds. Such distributional tails were used in this work, consistent with past practice in wind engineering. The fitted POT models allowed the estimation of wind speeds for any required MRI at all stations. In the second stage, local regression (Cleveland and Devlin 1988) was used to interpolate wind speeds at all points of a fine regular grid covering the contiguous United States for all required MRIs. This had the effect of spatially smoothing the noisy station estimates. The smoothed wind speed estimates provided the basis for creating the isotach maps.

Limited data were available on the Washington and Oregon coast. In this region, a special wind region was defined to permit local jurisdictions to select speeds based on local knowledge and analysis. Speeds in the Aleutian Islands and in the interior of Alaska were established from gust data. Insufficient data were available for a detailed coverage of the mountainous regions, so gust data in Alaska were not corrected for potential terrain influence. It is possible that wind speeds in parts of Alaska would be lower if the topographic wind speed-up effect on recorded wind speeds were taken into account. In Alaska, the maps for each return period were determined by multiplying the 50-year MRI contours given in ASCE 7-10, Appendix Figure CC-3 by a factor, F_{RA}

$$F_{RA} = 0.45 + 0.085 \ln(12T) \qquad \text{(C26.5-1)}$$

where T is the return period in years (Peterka and Shahid 1998). The resulting contours were interpolated to the nearest 10 mi/h, except for the innermost and outermost contours, which were rounded to the nearest 5 mi/h.

Hurricane Wind Speeds. The hurricane wind speeds are based on the results of a Monte Carlo simulation model generally described by ARA (2001), Vickery and Wadhera (2008a, b), and Vickery et al. (2009a, b, 2010). The hurricane simulation model used to develop the wind speeds in ASCE 7-16 includes two updates to the model used for ASCE 7-10. A reduced translation speed effect for fast-moving storms (USNRC 2011) was incorporated, and a simple extratropical transition model was also implemented, where the surface winds are reduced linearly by up to 10% over the latitude range 37 N to 45 N. This reduction approximates transitioning from a hurricane boundary layer to an extratropical storm boundary layer. The effects of the model revisions are to slightly reduce hurricane speeds in the northeast, extending from Maine to Virginia.

In addition to the aforementioned model changes as noted, since the 2010 edition of ASCE 7 a number of minor improvements have been made to the hurricane hazard and wind field models.

The first change includes a reduction in the translation speed effect for storms translating at speeds of 22 mi/h (10 m/s) or more (formerly the reduction in the translation speed effect was used for storms translating at 33.6 mi/h (15 m/s) or more, which reduced wind speeds in the New York / New England area). This change affects wind speeds associated with storms in the Gulf of Mexico that have recurved, or are recurving, to have a northerly through northeasterly heading, typically affecting a portion of storms making landfall along the middle and eastern portions of the Gulf of Mexico coastline. The translation speed change also affects storms moving in a north to northeasterly direction along the Atlantic coast, particularly for storms making landfall along Long Island, New York, and points further north.

The hurricane contours along the US coastline for locations south of the North Carolina–Virginia border in the 2016 edition of ASCE 7 are the same as those given in the 2010 edition of ASCE 7 (except for the 3,000-year contours) and were developed using a hurricane climatology based on the period 1900 through 2006. The hurricane contours were blended with the nonhurricane wind speeds adjacent to the hurricane coastline, which, at the time of the 2010 edition of ASCE 7, were a uniform 115 mi/h (51 m/s) (700-year). At the time of the 2016 edition of ASCE 7 the nonhurricane wind speeds adjacent to the hurricane coastline varied with location, but were approximately 105 mi/h (47 m/s) (700-year). In the 2016 edition of ASCE 7, the hurricane contours south of the NC–VA border were left unchanged, and the transition between the new, lower nonhurricane wind speeds was performed using judgment, likely resulting in an overestimate of wind speeds in parts of the transition region. For example, along the south Texas coast the ASCE 7-22 contours increased by approximately 10 mi/h (4.5 m/s), from 150 mi/h (67 m/s) in ASCE 7-10 and ASCE 7-16 to 160 mi/h (72 m/s) in ASCE 7-22. However, the 130 mi/h (58 m/s) contour did not move, and locations inland of the 130 mi/h (58 m/s) contour in ASCE 7-22 are lower than those given in ASCE 7-16.

The wind field model used in the hurricane simulation model was modified where the inflow angle was increased by 0.15 radian (8.6 degrees) to be consistent with the inflow angle used for most of the hurricane validation studies that have been performed. The change in the inflow angle changes how the rotational winds interact with the hurricane translation speed, resulting in a small reduction of wind speeds on the right side of the hurricane and a small increase on the left side of the hurricane. The net effect of the inflow angle change is about a 2% to 3% reduction in wind speed for most return periods.

The updated model uses historical track and landfall data for the period 1900 to 2018, a 13-year increase of hurricane landfall data, resulting in an increase in the hurricane hazard (defined by landfalling hurricanes ranked by central pressure) along the Texas coast and the Florida Panhandle, and a decrease along the mid-Gulf coastline.

The net effect of all the hurricane model updates yields increases in wind speeds in parts of Texas and the Florida Panhandle, and decreases in wind speeds in parts of Louisiana, along the Mississippi and Alabama coastlines, and in parts of the Northeast.

Wind speeds on the Florida peninsula were left unchanged from those given in the 2010 and 2016 editions of ASCE 7, making the ASCE 7-22 maps consistent with those used in the Florida Building Code.

Combination of Nonhurricane and Hurricane Wind Speed Data. Nonhurricane wind speeds and hurricane wind speeds were estimated for return periods ranging from 10 years to 100,000 years. The nonhurricane and hurricane winds were then combined as statistically independent events using Equation (C26.5-2), the same general approach that has been used in previous editions of ASCE 7:

$$P_a(v > V) = 1 - P_{NH}(v < V)P_H(v < V) \quad \text{(C26.5-2)}$$

where

$P_a(v > V)$ = Annual exceedance probability for the combined wind hazards,

$P_{NH}(v < V)$ = Annual nonexceedance probability for nonhurricane winds, and

$P_H(v < V)$ = Annual nonexceedance probability for hurricane winds.

The combined winds were interpolated to yield the combined wind hazard curves for MRIs associated with each of the wind speed maps. In cases where the hurricane contours are unchanged from ASCE 7-10, the shape files from these previous maps were used to ensure continuity between the maps.

Correlation of Basic Wind Speed Map with the Saffir–Simpson Hurricane Wind Scale. Hurricane intensities are reported by the National Hurricane Center (NHC 2015) according to the Saffir–Simpson Hurricane Wind Scale, shown in Table C26.5-1. This scale has found broad use by hurricane forecasters and local and federal agencies responsible for short-range evacuation of residents during hurricane alerts, as well as long-range disaster planners and the news media. The scale contains five categories of hurricanes and distinguishes them based on wind speed intensity.

The wind speeds used in the Saffir–Simpson Hurricane Wind Scale are defined in terms of a sustained wind speed with a 1 min averaging time at 33 ft (10 m) over open water. The ASCE 7 standard by comparison uses a 3 s gust speed at 33 ft (10 m) above ground in Exposure C (defined as the basic wind speed, and shown in the wind speed map, Figure 26.5-1). The sustained wind speed over water in Table C26.5-2 cannot be converted to a peak gust wind speed using the Durst curve of Figure C26.5-1, which is only valid for wind blowing over open terrain (Exposure C). An approximate relationship between the wind speeds in ASCE 7 and the Saffir–Simpson scale, based on recent data which indicate that the sea surface roughness remains approximately constant for mean hourly speeds in excess of 67 mi/h (30 m/s), is shown in Table C26.5-2. The table provides the sustained wind speeds of the Saffir–Simpson Hurricane Wind Scale over water, equivalent-intensity gust wind speeds over water, and equivalent-intensity gust wind speeds over land. For a storm of a given intensity, Table C26.5-2 takes into consideration both the reduction in wind speed as the storm moves from over water to over land because of changes in surface roughness, and the change in the gust factor as the storm moves from over water to over land (Vickery et al. 2009a, Simiu et al. 2007).

Table C26.5-3 shows the design wind speed from the ASCE 7 basic wind speed maps (Figure 26.5-1) for various locations along the hurricane coastline from Maine to Texas, and for Hawaii, Puerto Rico, and US Virgin Islands. Tables C26.5-4 through C26.5-6 show the basic wind speeds for Risk Category II, III, and IV buildings and other structures in terms of the hurricane category equivalents on the Saffir–Simpson Hurricane Wind Scale. These wind speeds represent an approximate limit state; structures designed to withstand the wind loads specified in this standard, which are also appropriately constructed and maintained, should have a high probability of surviving hurricanes of the intensities shown in Tables C26.5-4 through C26.5-6 without serious structural damage from wind pressure alone.

Tables C26.5-2 through C26.5-6 are intended to help users better understand design wind speeds as used in this standard in relation to wind speeds reported by weather forecasters and the news media, who commonly use the Saffir–Simpson Hurricane Wind Scale. The Saffir–Simpson hurricane category equivalent

Table C26.5-1. Saffir–Simpson Hurricane Wind Scale.

Hurricane Category	Sustained Wind Speed, mi/h (m/s)*	Types of Damage Due to Hurricane Winds
1	74–95 (33–42)	Very dangerous winds will produce some damage
2	96–110 (43–49)	Extremely dangerous winds will cause extensive damage
3	111–129 (50–57)	Devastating damage will occur
4	130–156 (58–69)	Catastrophic damage will occur
5	≥157 (70)	Highly catastrophic damage will occur

*1 min average wind speed at 33 ft (10 m) above open water.

Table C26.5-2. Approximate Relationship between Wind Speeds in ASCE 7 and Saffir–Simpson Hurricane Wind Scale.

Saffir–Simpson Hurricane Category	Sustained Wind Speed over Water[a]		Gust Wind Speed over Water[b]		Gust Wind Speed over Land[c]	
	mi/h	m/s	mi/h	m/s	mi/h	m/s
1	74–95	33–42	90–116	40–51	81–105	36–47
2	96–110	43–49	117–134	52–59	106–121	48–54
3	111–129	50–57	135–157	60–70	122–142	55–63
4	130–156	58–69	158–190	71–84	143–172	64–76
5	>157	>70	>191	>85	>173	>77

[a] 1 min average wind speed at 33 ft (10 m) above open water.
[b] 3 s gust wind speed at 33 ft (10 m) above open water.
[c] 3 s gust wind speed at 33 ft (10 m) above open ground in Exposure Category C. This column has the same basis (averaging time, height, and exposure) as the basic wind speed from Figure 26.5-1.

Table C26.5-3. Basic Wind Speeds at Selected Coastal Locations in Hurricane-Prone Areas.

Location	Coordinates (decimal degrees)		Basic Wind Speeds (mi/h)		
	Latitude	Longitude	Risk Cat. II (700-year)	Risk Cat. III (1,700-year)	Risk Cat. IV (3,000-year)
Bar Harbor, Maine	44.3813	−68.1968	109	119	121
Hampton Beach, New Hampshire	42.9107	−70.8102	113	124	125
Boston, Massachusetts	42.3578	−71.0012	116	125	129
Hyannis, Massachusetts	41.6359	−70.2901	123	139	141
Newport, Rhode Island	41.453	−71.3058	124	139	139
New Haven, Connecticut	41.2803	−72.9327	120	129	133
Southhampton, New York	40.871	−72.3844	129	138	140
Manhattan, New York	40.7005	−74.0135	116	127	130
Atlantic City, New Jersey	39.3536	−74.4336	126	135	138
Rehoboth Beach, Delaware	38.7167	−75.0752	122	131	136
Ocean City, Maryland	38.3314	−75.0835	128	136	139
Virginia Beach, Virginia	36.8306	−75.9691	125	132	138
Wrightsville Beach, North Carolina	34.1973	−77.8014	146	156	160
Folly Beach, South Carolina	32.6496	−79.9512	149	158	165
Sea Island, Georgia	31.179	−81.3472	131	145	153
Jacksonville Beach, Florida	30.2836	−81.387	129	140	149
Melbourne Beach, Florida	28.0684	−80.5564	152	162	172
Miami Beach, Florida	25.7643	−80.1309	171	183	191
Key West, Florida	24.5477	−81.7843	176	200	200
Clearwater, Florida	27.9658	−82.8042	146	154	160
Panama City Beach, Florida	30.1558	−85.7744	141	146	162
Gulf Shores, Alabama	30.2486	−87.6808	159	172	181
Biloxi, Mississippi	30.3924	−88.8887	157	176	177
Slidell, Louisiana	30.2174	−89.824	138	152	155
Cameron, Louisiana	29.7761	−93.2921	141	154	157
Galveston, Texas	29.2663	−94.826	151	159	166
Port Aransas, Texas	27.8346	−97.0446	159	157	174

Notes:

1. All wind speeds in Table C26.5-3 are 3 s gust wind speeds at 33 ft (10 m) above ground for Exposure Category C.
2. Basic wind speed in hurricane-prone regions can vary significantly over a city or county; the values shown are for randomly selected points along the coast for communities listed in the tables. Wind speeds at other locations in those communities may be greater or less than the values shown in the table.
3. Conversion of mi/h to m/s: mi/h × 0.44704 = m/s.

Exposure C gust wind speed values given in Tables C26.5-2 through C26.5-6, which are associated with a given sustained wind speed, should be used as a guide only. The gust wind speeds associated with a given sustained wind speed may vary with storm size and intensity, as suggested by Vickery et al. (2009a).

Wind Speeds for Long Return Periods. Wind hazard maps for longer return periods than the 3,000-year MRI map provided in Figure 26.5-1D are provided in Appendix F, which may be needed for some performance-based wind designs and other applications.

Wind Speeds for Serviceability Design. For serviceability applications such as drift and habitability, the Appendix C commentary presents maps of peak gust wind speeds at 33 ft (10 m) above ground in Exposure C conditions for return periods of 10, 25, 50, and 100 years (Figures CC.2-1 through CC.2-4).

The probability, P_n, that the wind speed associated with a certain annual probability, P_a, will be equaled or exceeded at least once during an exposure period of n years is

$$P_n = 1-(1-P_a)^n \quad \text{(C26.5-3)}$$

where

$$P_a = 1-e^{(-1/\mathrm{MRI})} \quad \text{(C26.5-4)}$$

For MRIs of about 10 years or longer, P_a is very closely approximated by the reciprocal of the MRI: $P_a \approx 1/\mathrm{MRI}$.

As an example, if a wind speed is based on $P_a = 0.02$/year (50-year MRI), the probability that this speed will be equaled or exceeded (at least once) during a 25-year period is 0.40 (i.e., 40%), and the probability of being equaled or exceeded in a 50-year period is 64%. Similarly, if a wind speed is based on $P_a = 0.00143$ (700-year MRI), the probability that this speed will be equaled or exceeded during a 25-year period is 3.5%, and the probability of being equaled or exceeded in a 50-year period is 6.9%.

Some products have been evaluated, and test methods have been developed, based on design wind speeds that are consistent with the unfactored load effects typically used in allowable stress

Table C26.5-4. Basic Wind Speed for Risk Category II Buildings and Other Structures at Selected Locations in Hurricane-Prone Areas.

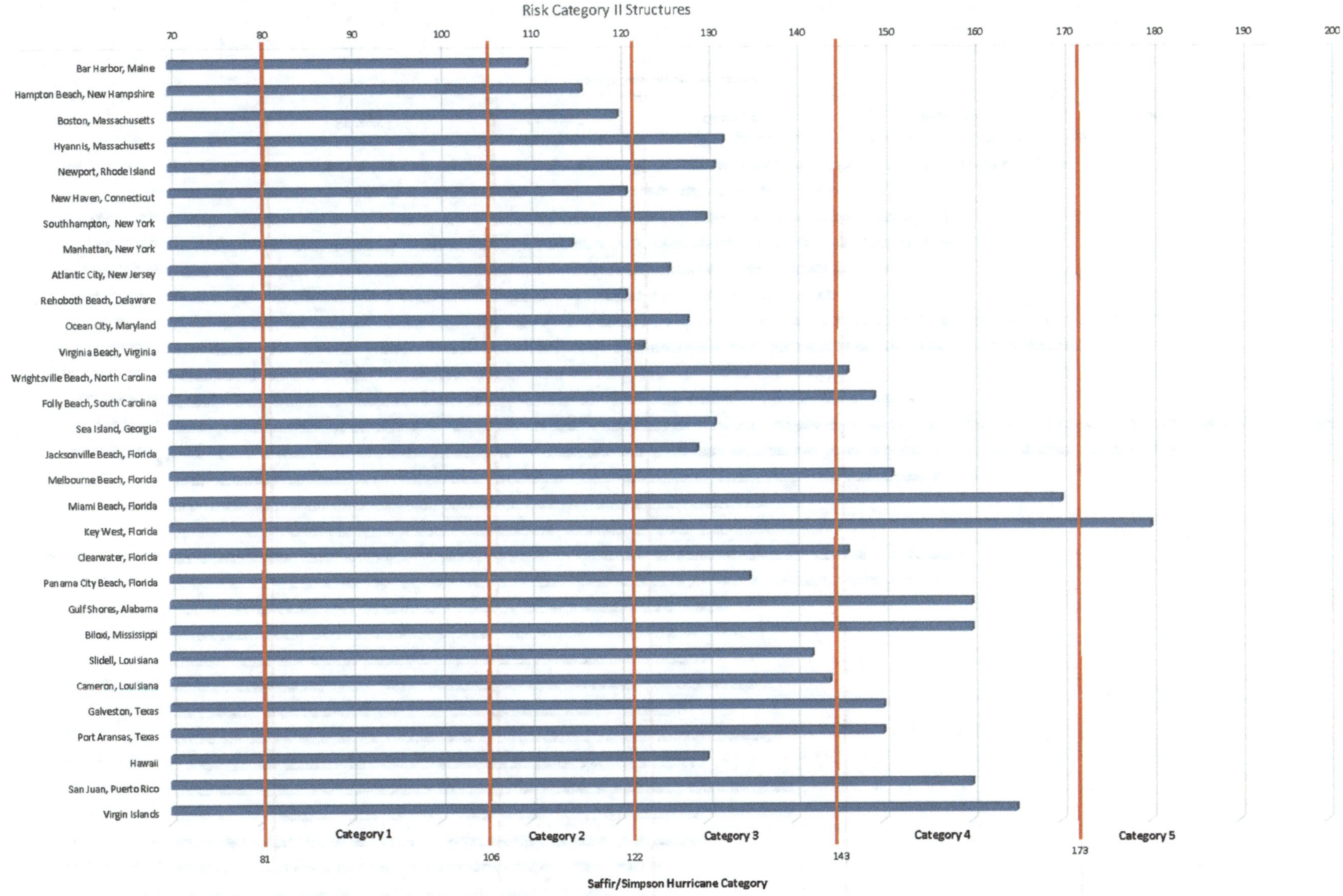

Risk Category II Structures

70 80 90 100 110 120 130 140 150 160 170 180 190 200

Bar Harbor, Maine
Hampton Beach, New Hampshire
Boston, Massachusetts
Hyannis, Massachusetts
Newport, Rhode Island
New Haven, Connecticut
Southhampton, New York
Manhattan, New York
Atlantic City, New Jersey
Rehoboth Beach, Delaware
Ocean City, Maryland
Virginia Beach, Virginia
Wrightsville Beach, North Carolina
Folly Beach, South Carolina
Sea Island, Georgia
Jacksonville Beach, Florida
Melbourne Beach, Florida
Miami Beach, Florida
Key West, Florida
Clearwater, Florida
Panama City Beach, Florida
Gulf Shores, Alabama
Biloxi, Mississippi
Slidell, Louisiana
Cameron, Louisiana
Galveston, Texas
Port Aransas, Texas
Hawaii
San Juan, Puerto Rico
Virgin Islands

Category 1 Category 2 Category 3 Category 4 Category 5

81 106 122 143 173

Saffir/Simpson Hurricane Category

Table C26.5-5. Basic Wind Speed for Risk Category III Buildings and Other Structures at Selected Locations in Hurricane-Prone Areas.

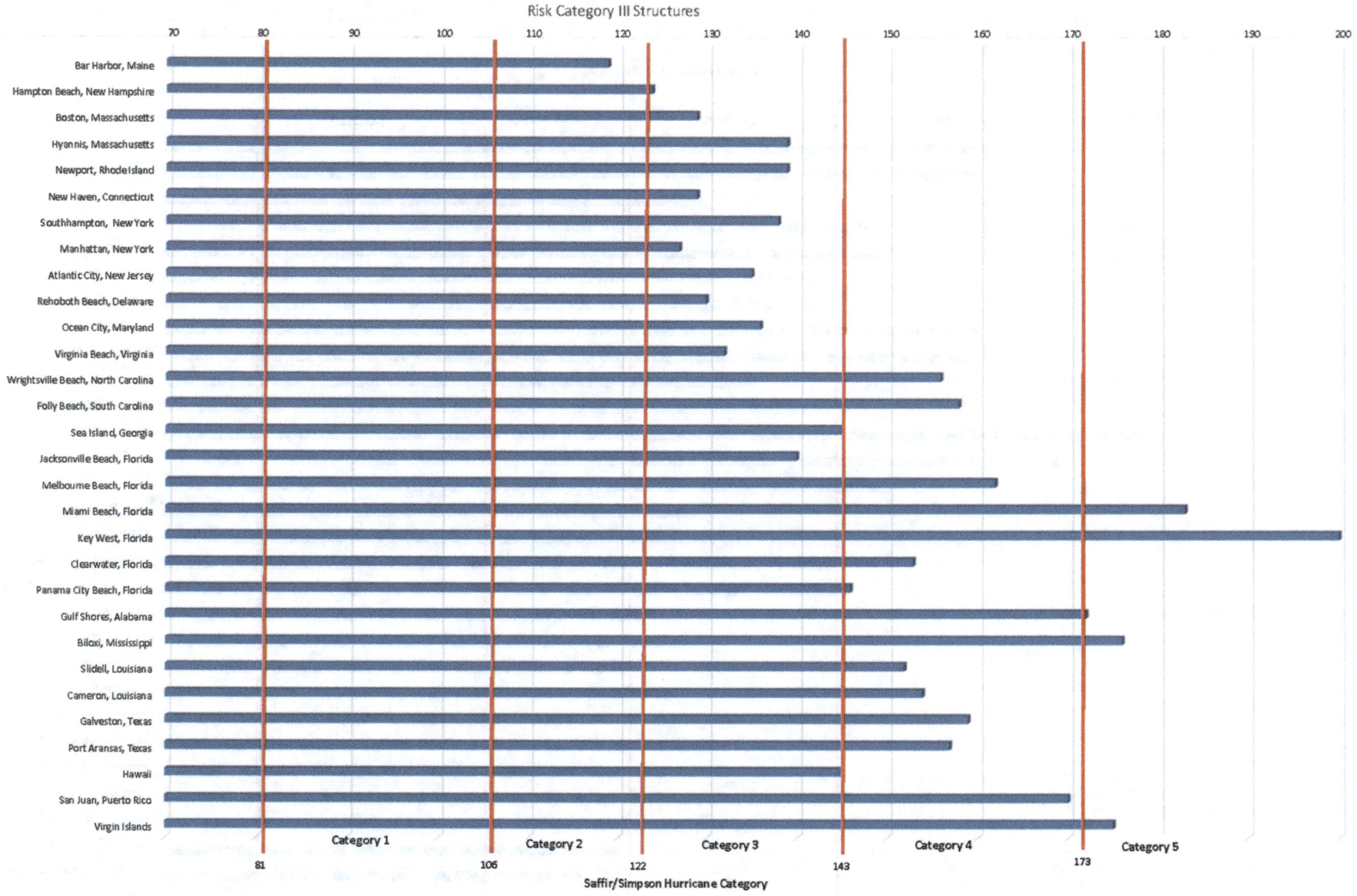

Table C26.5-6. Basic Wind Speed for Risk Category IV Buildings and Other Structures at Selected Locations in Hurricane-Prone Areas.

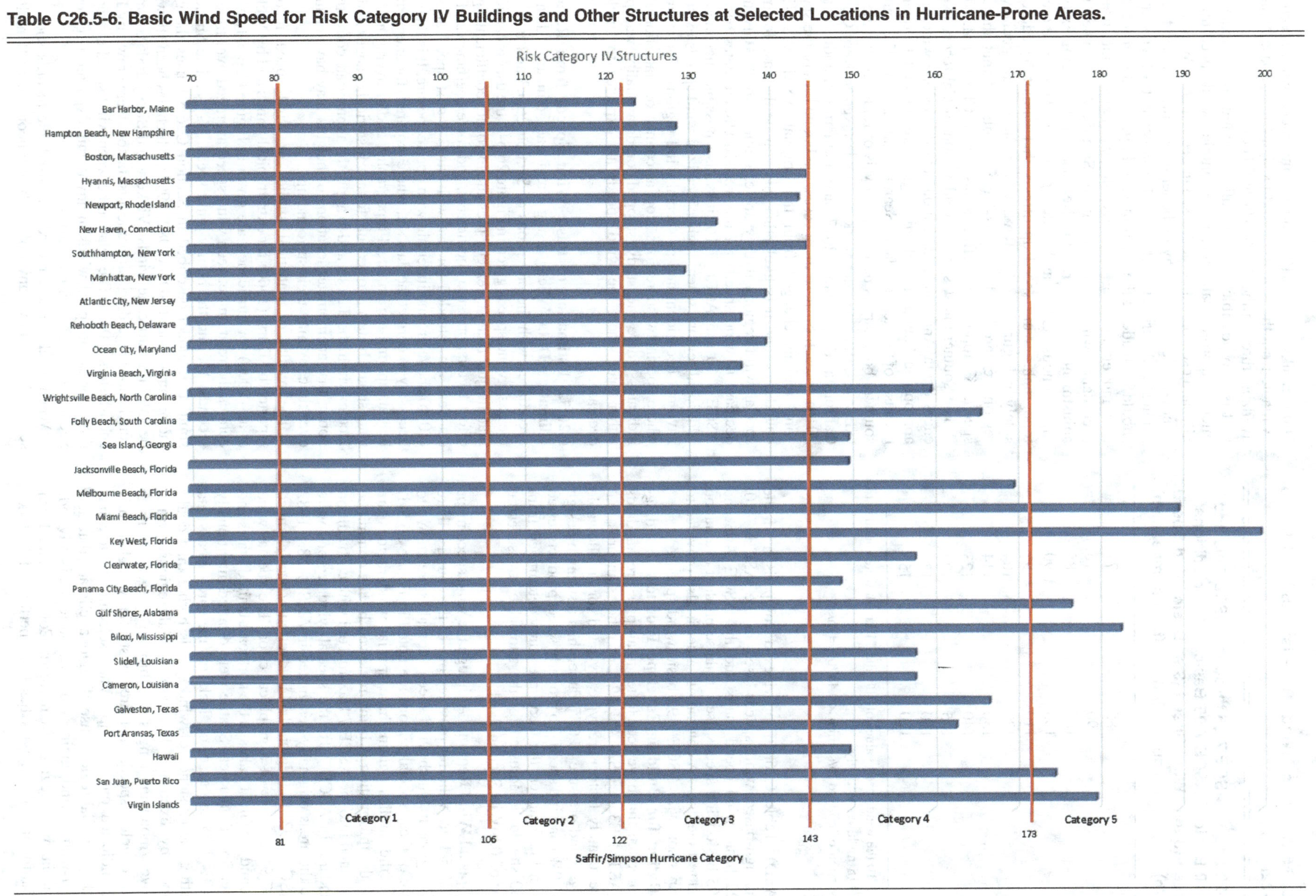

Table C26.5-7. Basic Wind Speeds: ASCE 7-93 to ASCE 7-22.

ASCE 7-22 and ASCE 7-10 Basic Wind Speed (3 s gust, mi/h)	ASCE 7-05 through ASCE 7-95 Basic Wind Speed (3 s gust, mi/h)	ASCE 7-93 and Prior Editions Basic Wind Speed (fastest mile, mi/h)
108*	85	71
114*	90	76
126	100	85
133	105	90
139	110	95
152	120	104
164	130	114
177	140	123
183	145	128
190	150	133
215	170	152

* In ASCE 7-10 the wind speed values of 108 and 114 mi/h were rounded to 110 and 115 mi/h, respectively.

Note: Conversion of mi/h to m/s: mi/h × 0.44704 = m/s.

design (ASD). Table C26.5-7 provides a comparison of the strength design-based wind speeds used in the ASCE 7-10, 7-10, 7-22 basic wind speed maps and the ASCE 7-05 and ASCE 7-95 basic wind speeds used in these product evaluation reports and test methods. A column of values is also provided to allow comparison with ASCE 7-93 basic wind speeds.

C&C Rating for Building-Envelope Products. Building-envelope products that have been tested to air pressure standards (such as ASTM E330, CSA A123.21, or other standards that incorporate a safety factor) are typically rated for an allowable stress design wind pressure ($0.6W$) rather than a strength design pressure ($1.0W$) or wind speed. To properly select products tested and rated in this manner, the C&C pressures determined from Chapter 30 should be adjusted for the allowable stress design load factor of $0.6W$ in Section 2.4.1.

C26.5.2 Special Wind Regions Although the wind speed maps of Figure 26.5-1 are valid for most regions of the country, there are special regions in which wind speed anomalies are known to exist. Some of these special regions are noted in Figure 26.5-1. As site-specific studies within these special wind regions are performed, the information is being reviewed by the committee and incorporated into the ASCE 7 Hazard Tool with each edition of the standard. In ASCE 7-16, the special wind regions were restored to the areas originally designated in ANSI A58.1-1982 through ASCE 7-93. The regions had inadvertently been changed in ASCE 7-95 because of a graphical error. The special wind regions around the Great Lakes in the northeast United States and in the Puget Sound area near Seattle were omitted intentionally from ASCE 7-95 because of lack of meteorological data demonstrating the existence of wind speeds higher than could be explained by exposure. In these special regions, winds blowing over mountain ranges or through gorges or river valleys can develop speeds that are substantially higher than the values indicated on the map. When selecting basic wind speeds in these special regions, use of regional climatic data and consultation with a wind engineer or meteorologist is advised. For the special wind regions in northern Colorado and Kern County, California, this analysis has been performed (Peterka 2006, Banks et al. 2019), and recommended wind speeds are available from the ASCE 7 Wind Design Geodatabase, which is available through the ASCE Hazard Tool.

It is also possible that anomalies in wind speeds exist on a micrometeorological scale. For example, wind speed-up over hills and escarpments is addressed in Section 26.8. Wind speeds over complex terrain may be better determined by wind tunnel studies, as described in Chapter 31. Adjustments of wind speeds should be made at the micrometeorological scale on the basis of wind engineering or meteorological advice and used in accordance with the provisions of Section 26.5.3 when such adjustments are warranted. Because of the complexity of mountainous terrain and valley gorges in Hawaii, there are topographic wind speed-up effects that cannot be addressed solely by Figure 26.8-1 (ARA 2001). In the Hawaii special wind region, research and analysis have established that there are special K_{zt} topographic-effect adjustments (Chock et al. 2005).

The southernmost special wind region in California experiences Santa Ana winds (dry, mountain downslope winds). The appropriate boundaries of this region are difficult to quantify because of a lack of data (Searer et al. 2010). The current southern boundary is based on analysis of regional climatic data (SEAOC 2017). Similar work was undertaken in Kern County, California (Kern County 2017) and northern Colorado (Peterka 2010). In each of these three cases, the boundaries of the special wind regions changed from the original ANSI A58.1-1982 boundaries, and it is likely that this will be the case for many special wind regions if they are examined closely. Given the decrease in wind speeds for the western United States between ASCE 7-10 and ASCE 7-16, it is also possible that there are additional regions where higher localized wind speeds exist that have not been identified as special wind regions. For example, where special wind regions stop at county or state lines, this indicates where the study stopped, or data quality was poor, rather than where the wind speeds are expected to suddenly decrease.

C26.5.3 Estimation of Basic Wind Speeds from Regional Climatic Data When using regional climatic data in accordance with the provisions of Section 26.5.3 and in lieu of the basic wind speeds given in Figure 26.5-1, the user is cautioned that the gust factors, velocity pressure exposure coefficients, gust-effect factors, pressure coefficients, and force coefficients of this standard are intended for use with the 3 s gust speed at 33 ft (10 m) above ground in terrain with open exposure. It is therefore necessary to correct the data set, where required, for anemometer height, upwind terrain, local topography, and/or averaging time. Care should be taken to account for any historical changes in these factors within the data set and to apply the appropriate corrections, noting that anemometer locations or heights may have changed over time, as may their surroundings.

The results of statistical studies of wind speed records, reported by Durst (1960) for extratropical winds and by Vickery et al. (2000) for hurricanes, are given in Figure C26.5-1, which defines the relation between wind speed averaged over time t in seconds, V_t, and the hourly wind speed, $V_{3,600}$. The hurricane simulation model described in Section C26.5.1 uses the gust-factor curve from ESDU (1982, 1993), which has been shown to be valid for hurricane winds (Vickery and Skerlj 2005). Similar conclusions regarding hurricane gust factors were drawn by Jung and Masters (2013). The relation between wind speed averaging times for extratropical winds in any terrain exposure is given in Section 11.2.4.2 of Simiu (2011).

When local climatic records are used, the records should be screened to remove any obviously erroneous data. If multiple locations are available, then the data from each should be compared for consistency. In areas with mixed wind climates, where multiple storm types contribute to the extreme wind climate, storm separation should be conducted on the data sets and individual extreme value fits applied to each storm type before combining to determine the overall risk.

In using local data, it should be emphasized that sampling errors can lead to large uncertainties in specification of wind speed. Sampling errors are the errors associated with the limited size of the climatological data samples (e.g., years of record of extreme speeds or number of storms). When short local records are used to estimate extreme wind speeds, care and conservatism should be exercised. Different methods, such as periodic independent maxima or the Method of Independent Storms, may be used to increase the effective size of the data set in comparison with just the use of annual maxima, as may the use of super-stations where multiple data sets are combined.

An examination of statistical sampling error can be used to ensure the reliability of the statistical fits. It is recommended that design wind speeds for use in strength design be based on the best-estimate wind speed plus a minimum of two to three times the standard deviation of the sampling error (Simiu and Scanlan 1996), equating to 95% and 99% confidence intervals, respectively. This is particularly important if justifying a wind speed lower than the basic wind speed from Figure 26.5-1. The preceding analyses should only be conducted by professionals experienced in extreme wind climate analysis for wind loading applications. Where values significantly below those from Figure 26.5-1 are being proposed, an independent peer review by a suitably experienced professional may be advisable.

C26.6 WIND DIRECTIONALITY FACTOR

The wind load factor 1.3 in ASCE 7-95 included a "wind directionality factor" with a nominal value of 0.85 (Ellingwood 1981. Ellingwood et al. 1982). This factor accounts for two effects: (1) The reduced probability of maximum winds coming from any given direction, and (2) the reduced probability of the maximum pressure coefficient occurring for any given wind direction. The nominal wind directionality factor (denoted by K_d in the standard) is tabulated in Table 26.6-1 for different structure types. As new research becomes available, this factor can be directly modified. Nominal values for the factor were established from references in the literature and collective committee judgment. A value of 0.85 might be more appropriate if a triangular trussed frame is shrouded in a round cover. A value of 0.95 might be more appropriate for a round structure that has a nonaxisymmetrical lateral load resistance system.

C26.7 EXPOSURE

The descriptions of the surface roughness categories and exposure categories in Section 26.7 have been expressed as far as possible in easily understood descriptive terms that are sufficiently precise for most practical applications. Upwind surface roughness conditions required for Exposures B and D are shown schematically in Figures C26.7-1 and C26.7-2, respectively. Aerial photographs showing examples of Exposures B, C, and D are shown in Figures C26.7-5 through C26.7-7. For cases where the designer wishes to make a more detailed assessment of the surface roughness category and exposure category, the following more mathematical description is offered for guidance (Irwin 2006). The ground surface roughness is best measured in terms of a roughness length parameter called z_0. Each of the surface roughness categories B through D corresponds to a range of values of this parameter, as does the even rougher category A used in older versions of the standard in heavily built-up urban areas but removed in the recent editions. The range of z_0 for each terrain category is given in Table C26.7-1. Exposure A has been included in the table as a reference that may be useful when using the Wind Tunnel Procedure. Further information on values of z_0 in different types of terrain can be found in Simiu and Yeo (2019) and in Table C26.7-2, based on Davenport et al. (2000) and Wieringa et al. (2001).

The roughness classifications in Table C26.7-2 are not intended to replace the use of exposure categories as required in the standard for structural design purposes. However, the terrain roughness classifications in Table C26.7-2 may be related to exposure categories by comparing z_0 values between Tables C26.7-1 and C26.7-2. For example, the z_0 values for Classes 3 and 4 in Table C26.7-2 fall within the range of z_0 values for Exposure C in Table C26.7-1. Similarly, the z_0 values for Classes 5 and 6 in Table C26.7-2 fall within the range of z_0 values for Exposure B in Table C26.7-1.

Research described by Powell et al. (2003), Donelan et al. (2004), and Vickery et al. (2009a) showed that the drag coefficient over the ocean in high winds in hurricanes does not continue to increase with increasing wind speed, as was previously believed (e.g., Powell 1980). These studies showed that the sea surface drag coefficient, and hence the aerodynamic roughness of the ocean, reached a maximum at mean wind speeds of about 67 mi/h (30 m/s). There is some evidence that the drag coefficient actually decreases (i.e., the sea surface becomes aerodynamically smoother) as the wind speed increases further (Powell et al. 2003) or as the hurricane radius decreases (Vickery et al. 2009a). The consequences of these studies are that the surface roughness over the ocean in a hurricane is consistent with that of Exposure D rather than Exposure C. Consequently, the use of Exposure D along the hurricane coastline is required.

For Exposure B, previous editions of ASCE 7, beginning with 7-02, used tabulated values of K_z corresponding to $z_0 =$ 0.66 ft (0.2 m), which is below the typical value of 1 ft (0.3 m), whereas for Exposures C and D they have always corresponded to the typical values of z_0. The reason for the difference in Exposure B is that this category of terrain, which is applicable to suburban areas,

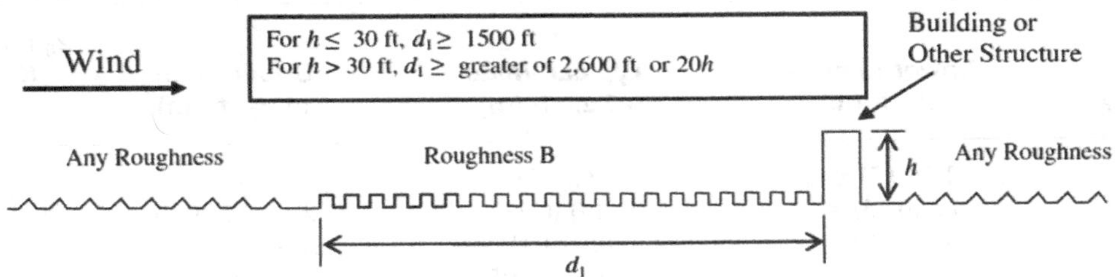

Figure C26.7-1. Upwind surface roughness conditions required for Exposure B.

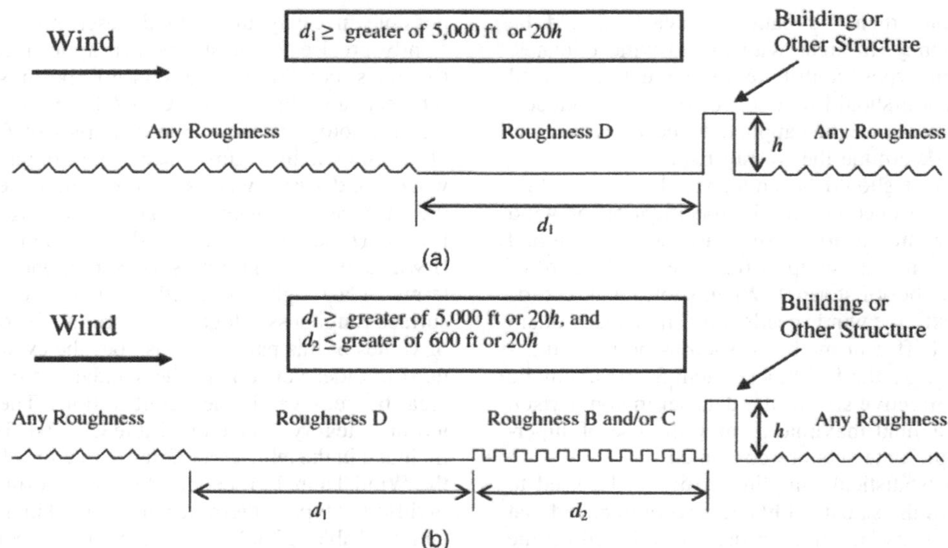

Figure C26.7-2. Upwind surface roughness conditions required for Exposure D, for the cases with (a) Surface Roughness D immediately upwind of the building, and (b) Surface Roughness B and/or C immediately upwind of the building.

often contains open patches, such as highways, parking lots, and playing fields. These open patches cause local increases in the wind speeds at their edges. By using an exposure coefficient corresponding to a lower-than-typical value of z_0, some allowance is made for this. The alternative would be to introduce a number of exceptions to the use of Exposure B in suburban areas, which would add an undesirable level of complexity. ASCE 7-22 uses z_0 = 1.0 ft (0.3 m) for Exposure B, corresponding to the typical value of z_0. Although resulting K_z values for Exposure B are generally reduced by using the typical value of z_0, the updated ASCE 7-22 power law equation results in K_z values comparable to those determined in earlier editions of ASCE 7 up to 60 ft (18.3 m). Thus, the design intent of considering open patches applicable to suburban areas in Exposure B has been maintained for the majority of low-rise building heights where these terrain features will have the largest impact on wind speed. For taller buildings and other structures taller than 60 ft (18.3 m), the influence of suburban open patches on wind speed will decrease with height, which is reflected in corresponding values of K_z based on the typical value of z_0 and the updated ASCE 7-22 power law.

For Exposure C, ASCE 7-22 updated the typical value of z_0 from 0.066 ft (0.02 m) to 1 ft (0.3 m), to maintain consistency with the value of z_0 used in the hazard maps for both hurricane and nonhurricane winds. The tabulated values of K_z also correspond to the updated typical value of z_0.

The value of z_0 for a particular terrain can be estimated from the typical dimensions of surface roughness elements and their spacing on the ground area using an empirical relationship found by Lettau (1969):

$$z_0 = 0.5 H_{ob} \frac{S_{ob}}{A_{ob}} \quad \text{(C26.7-1)}$$

where

H_{ob} = Average height of the roughness in the upwind terrain;
S_{ob} = Average vertical frontal area per obstruction presented to the wind; and
A_{ob} = Average area of ground occupied by each obstruction, including the open area surrounding it.

Vertical frontal area is defined as the area of the projection of the obstruction onto a vertical plane normal to the wind direction. The area S_{ob} may be estimated by summing the approximate vertical frontal areas of all obstructions within a selected area of upwind fetch and dividing the sum by the number of obstructions in the area. The average height, H_{ob}, may be estimated in a similar way by averaging the individual heights rather than using the frontal areas. Likewise, A_{ob} may be estimated by dividing the size of the selected area of upwind fetch by the number of obstructions in it.

As an example, if the upwind fetch consists primarily of single-family homes with typical height H_{ob} = 20 ft (6 m), vertical frontal area (including some trees on each lot) of 1,000 ft² (100 m²), and ground area per home of 10,000 ft² (1,000 m²),

Table C26.7-1. Range of z_0 by Exposure Category.

Exposure Category	Lower Limit of z_0, ft (m)	Typical Value of z_0, ft (m)	Upper Limit of z_0, ft (m)	z_0 Inherent in Tabulated K_z Values in Table 26.10-2, ft (m)
A	2.3 (0.7) ≤ z_0	6.6 (2)	—	
B	0.5 (0.15) ≤ z_0	1.0 (0.3)	z_0 < 2.3 (0.7)	1.0 (0.3)
C	0.033 (0.01) ≤ z_0	0.1 (0.03)	z_0 < 0.5(0.15)	0.1 (0.03)
D	—	0.016 (0.005)	z_0 < 0.033(0.01)	0.016 (0.005)

Table C26.7-2. Davenport Classification of Effective Terrain Roughness.

Class	z_0, ft (m)[a]	α[b]	z_g', ft (m)[b]	z_d (ft or m)[c]	Wind Flow and Landscape Description[d]
1	0.0007 (0.0002)	12.9	509 (155)	$z_d = 0$	*Sea:* Open sea or lake (irrespective of wave size), tidal flat, snow-covered flat plain, featureless desert, tarmac, and concrete, with a free fetch of several kilometers.
2	0.016 (0.005)	11.4	760 (232)	$z_d = 0$	*Smooth:* Featureless land surface without any noticeable obstacles and with negligible vegetation (e.g., beaches, pack ice without large ridges, marsh, and snow-covered or fallow open country).
3	0.1 (0.03)	9.0	952 (290)	$z_d = 0$	*Open:* Level countryside with low vegetation (e.g., grass) and isolated obstacles with separations of at least 50 obstacle heights (e.g., grazing land without windbreaks, heather, moor, and tundra, runway area of airports). Ice with ridges across-wind.
4	0.33 (0.10)	7.7	1,107 (337)	$z_d = 0$	*Roughly open:* Cultivated or natural area with low crops or plant covers, or moderately open country with occasional obstacles (e.g., low hedges, isolated low buildings, or trees) at relative horizontal distances of at least 20 obstacle heights.
5	0.82 (0.25)	6.8	1,241 (378)	$z_d = 0.2z_H$	*Rough:* Cultivated or natural area with high crops or crops of varying height and scattered obstacles at relative distances of 12 to 15 obstacle heights for porous objects (e.g., shelterbelts) or 8 to 12 obstacle heights for low solid objects (e.g., buildings).
6	1.64 (0.5)	6.2	1,354 (413)	$z_d = 0.5z_H$	*Very rough:* Intensely cultivated landscape with many rather large obstacle groups (large farms, clumps of forest) separated by open spaces of about 8 obstacle heights. Low, densely planted major vegetation like bushland, orchards, young forest. Also, area moderately covered by low buildings with interspaces of 3 to 7 building heights and no high trees.
7	3.3 (1.0)	5.7	1,476 (450)	$z_d = 0.7z_H$	*Skimming:* Landscape regularly covered with similar-size large obstacles, with open spaces of the same order of magnitude as obstacle heights (e.g., mature regular forests, densely built-up area without much building height variation).
8	≥*matu* (≥*mat*)	5.2	1,610 (490)	Analysis by wind tunnel advised	*Chaotic:* City centers with mixture of low-rise and high-rise buildings or large forests of irregular height with many clearings. (Analysis by wind tunnel advised.)

[a] The surface roughness length, z_0, represents the physical effect that roughness objects (obstacles to wind flow) on the Earth's surface have on the shape of the atmospheric boundary-layer wind velocity profile as determined by the logarithmic law and used in the ESDU model.

[b] The power law uses the exponent α and is valid up to the height at which geostrophic wind flow begins to occur (see Section C26.10.1). The values provided in this table are based on z_0 values originally proposed by Davenport (1960), and the equations ($\alpha = c_1 z_0^{-0.133}$ and $z_g' = c_2 z_0^{0.125}$) can be used to calculate α and historial gradient height z_g', respectively, where $c_1 = 6.62$ and $c_2 = 1,273$ when the units of z_0 and z_g' are ft, and $c_1 = 5.65$ and $c_2 = 450$ when the units of z_0 and z_g' are m, as described in the commentary of ASCE 7-16.

[c] The zero plane displacement height, z_d, is the elevation above ground to which the base of the logarithmic law (and power law) wind profile must be elevated to accurately depict the boundary-layer wind flow. Below z_d and less than some fraction of the typical height, z_H, of obstacles causing roughness, the near-ground wind flow is characterized as a turbulent exchange with the boundary-layer wind flow above, resulting in significant shielding effects under uniform to moderately uniform roughness conditions (e.g., Classes 5 through 7 in this table). In this condition, the effective mean roof height, h_{eff}, may then be determined as h–z_d [but not less than 15 ft (4.6 m)] for the purpose of determining MWFRS wind loads acting on a building structure located within such a roughness class. Appropriate values of z_d for a given site may vary widely, and those shown in this table should be used with professional judgment. Because of the presence of highly turbulent flow at elevations near or below z_d (except perhaps structures embedded in uniform Class 7 roughness), use of an effective mean roof height should not be applied for the determination of C&C wind loads. In Class 8 roughness, where wind flow disruptions can be highly nonuniform, channeling effects and otherwise "chaotic" wind flow patterns can develop between and below the height of obstacles to wind flow. For this reason, a wind tunnel study is generally advised.

[d] Use of these wind flow and landscape descriptions should result in no greater than one roughness class error, corresponding to a maximum +/− of these winds q_h.

then z_0 is calculated to be $0.5 \times 20 \times 1{,}000/10{,}000$ ft = 1 ft (0.3 m), which falls into Exposure Category B, according to Table C26.7-1.

Trees and bushes are porous and are deformed by strong winds, which reduce their effective frontal areas (ESDU 1993). For conifers and other evergreens, no more than 50%, and for deciduous trees and bushes, no more than 15%, of their gross frontal area can be taken to be effective in obstructing the wind.

Gross frontal area is defined in this context as the projection onto a vertical plane (normal to the wind) of the area enclosed by the envelope of the tree or bush.

Ho (1992) estimated that the majority of buildings (perhaps as much as 60% to 80%) have an exposure category corresponding to Exposure B. While the relatively simple definition in the standard normally suffices for most practical applications, the designer is often in need of additional information, particularly with regard to the effect of large openings or clearings (e.g., large parking lots, freeways, or tree clearings) in the otherwise "normal" ground Surface Roughness B. The following is offered as guidance for these situations:

1. The simple definition of Exposure B given in the body of the standard, using the surface roughness category definition, is shown pictorially in Figure C26.7-1. This definition applies for the Surface Roughness B condition prevailing 2,630 ft (800 m) upwind with insufficient "open patches," as defined in the following procedure to disqualify the use of Exposure B. This procedure on the net effect of these open patches applies where the prevailing exposure beyond 2,600 ft (792 m) is Exposure B.
2. An open area in the Surface Roughness B large enough to have a significant effect on the exposure category determination is defined as an "open patch." To be considered an open patch, an open area meets the following:
   (a) Open areas should be greater than the minimum areas given by Figure C26.7-4. Interpolation shall be used between the reference distances of 500, 1,500, and 2,600 ft (152, 457, and 790 m) to determine the intermediate minimum open patch area criteria.
   (b) The open area shall have minimum dimensions given by conditions i, ii, or iii and have length-to-width ratios between 0.5 and 2.0.
      (i) Within 500 ft (152 m) of the building or structure, an open area greater than or equal to approximately 164 ft (50 m) in length or width.
      (ii) At 1,500 ft (457 m) upwind from the building or structure, an open area greater than or equal to approximately 328 ft (100 m) in length or width.
      (iii) At 2,600 ft (790 m) upwind from the building or structure, an open area greater than or equal to approximately 500 ft (152 m) in length or width.
3. Open patches separated by less than the along-wind dimension of the larger patch shall be treated as equivalent to a single open patch with length equal to the sum of the individual patch along-wind dimensions and width determined to provide an area equal to the sum of the individual open patch areas.
4. A *circular sector* is an area defined by a limiting radius from the center and an arc, in this case, 45 degrees, per Section 26.7.4. If the proportion of open patch within any of the sectors defined by the preceding three radii is less than 25% of the sector area, the sector is considered to meet the requirements for Exposure Category B. Where the proportion of open patches within any 45-degree sector within any of the three radii of 500 ft (152 m), 1,500 ft (457 m), or the greater of 2,600 ft (790 m) or 20 times the height of the building or structure exceeds 25% of the sector area but is not greater than 50%, the values of K_z are taken as the average of the Exposure B and C values within 100 ft (31 m) height above grade. Above 100 ft (31 m), Exposure B values of K_z shall still apply. Where the proportion of open patches within any of the sectors defined by the three radii of the building or structure exceeds 50%, the values of K_z shall be based on Exposure C.
5. The procedure for evaluation of the net effect of open patches of Surface Roughness C or D on the use of Exposure Category B is shown pictorially in Figures C26.7-3 and C26.7-4. Note that the plan location of any open patch may have a different effect for different wind directions.

This procedure is a simplification derived from a boundary-layer model, and therefore more exact results for the velocity profile may be achieved through direct use of an accepted boundary-layer model that is capable of addressing the effects of open areas within a regime defined by the surface roughness parameters given in Table C26.7-2.

Aerial photographs, representative of each exposure type, are included in Figures C26.7-5 through C26.7-7 to aid the user in establishing the proper exposure for a given site. Obviously, the proper assessment of the exposure is a matter of good engineering judgment. This fact is particularly true in light of the possibility that the exposure could change in one or more wind directions due to future demolition and/or development.

C26.7.4 Exposure Requirements Section 26.5.1 of the standard requires that a structure be designed for winds from all directions. A rational procedure to determine directional wind loads is described here. Wind load for buildings using Section 27.3.1 and Figures 27.3-1, 27.3-2, or 27.3-3 are determined for eight wind directions at 45-degree intervals, with four falling along primary building axes, as shown in Figure C26.7-8. For each of the eight directions, upwind exposure is determined for each of two 45-degree sectors, one on each side of the wind direction axis. The sector with the exposure giving the highest loads is used to define wind loads for that direction. For example, for winds from the north, the exposure from Sector 1 or 8, whichever gives the higher load, is used. For wind from the east, the exposure from Sector 2 or 3, whichever gives the highest load, is used. For wind coming from the northeast, the more exposed of Sectors 1 or 2 is used to determine full *x* and *y* loading individually, and then 75% of these loads is to be applied in each direction at the same time, according to the requirements of Section 27.3.5 and Figure 27.3-8. The procedure defined in this section for determining wind loads in each design direction is not to be confused with the determination of the wind directionality factor, K_d. The K_d factor determined from Section 26.6 and Table 26.6-1 applies for all design wind directions. See Section C26.6.

C&C loads for all buildings and MWFRS loads for low-rise buildings are determined using the upwind exposure for the single surface roughness in one of the eight sectors of Figure C26.7-8 that gives the highest wind loads.

C26.8 TOPOGRAPHIC EFFECTS

This section specifies when topographic effects need to be applied to a particular structure (Means et al. 1996). In some wind loading guidelines, hills of sufficient size have been assumed to eliminate topographic effects for up to 2 mi downwind. Research by Almeida et al. (1993) indicates that very little sheltering is expected from nearby (3*H* upwind) hills of comparable size. Some sheltering is expected downwind of significantly larger hills, but evidence from studies or literature is needed to quantify this.

Condition 3 specifies a lower height, *H*, for consideration of topographic effects in Exposures C and D than for Exposure B (Means et al. 1996), and observation of actual wind damage has shown that the affected height, *H*, is less in Exposures C and D.

Buildings sited on the upper half of an isolated hill or escarpment may experience significantly higher wind speeds than buildings on level ground. The topographic feature (2D ridge or escarpment, or 3D axisymmetrical hill) is described by two parameters, *H* and L_h. *H* is the height of the hill or difference in elevation between the crest and that of the

Diagrams

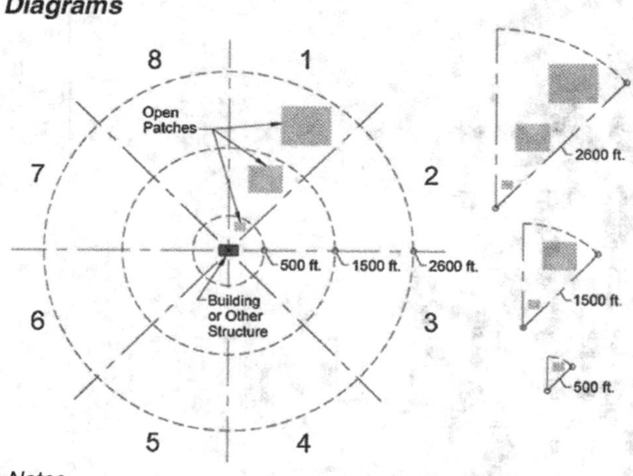

Notes:

1. For each selected wind direction at which the wind loads are to be determined, the exposure of the building or structure shall be determined for the two upwind sectors extending 45 degrees to either side of the selected wind direction.

2. Consider open patches of sizes equal to or greater than the areas given in Figure C26.7-4 per Commentary Section C26.7.

3. Determine the proportion of open patches in any 45-degree sector within radii of 500 ft (152 m), 1,500 ft (457 m), or the greater of 2,600 ft (790 m) or 20 times the height of the structure.

4. If the proportion of open patch within any of the three radii is less than 25% of the sector area, the sector is considered to meet the requirements for Exposure B. Where the proportion within any of the three radii exceeds 25% of the sector area but is not greater than 50%, the values of Kz are taken as the average of the Exposure B and C values within 100 ft (31 m) height above grade. Above 100 ft (31 m), Exposure B values shall still apply. Where the proportion of open patches within any of the three radii of the structure exceeds 50%, the values of Kz shall be based on Exposure C.

5. Apply the exposure requirements of Section 26.7.4 once the directional exposures are determined for each sector. See Commentary Section C26.7.4.

Figure C26.7-3. Sector analysis for Exposure B with upwind open patches.

upwind terrain. L_h is the distance upwind of the crest to where the ground elevation is equal to half the height of the hill. K_{zt} is determined from three multipliers, K_1, K_2, and K_3, which are obtained from Figure 26.8-1. K_1 is related to the shape of the topographic feature; the maximum speed-up near the crest, K_2, accounts for the reduction in speed-up with distance upwind or downwind of the crest; and K_3 accounts for the reduction in speed-up with height above the local ground surface.

The multipliers listed in Figure 26.8-1 are based on the assumption that the wind approaches the hill along the direction of maximum slope, causing the greatest speed-up near the crest. The average maximum upwind slope of the hill is approximately $H/2L_h$, and measurements have shown that hills with slopes of less than about 0.10 ($H/L_h < 0.20$) are unlikely to produce significant speed-up of the wind. For values of $H/L_h > 0.5$, the speed-up effect is assumed to be independent of slope. The speed-up principally affects the mean wind speed rather than the amplitude of the turbulent fluctuations, and this fact has been accounted for in the values

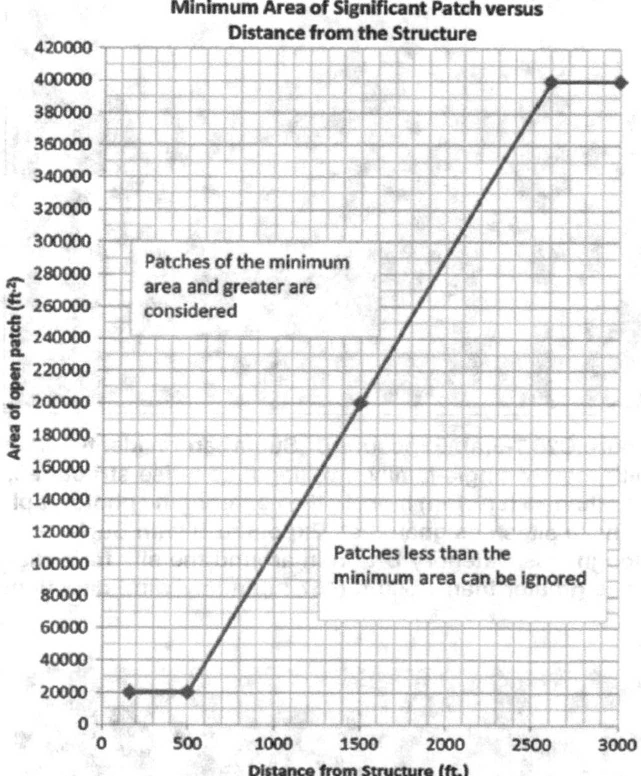

Figure C26.7-4. Minimum area of individual open patches affecting qualification of Exposure B.

of K_1, K_2, and K_3 given in Figure 26.8-1. Therefore, values of K_{zt} obtained from Figure 26.8-1 are intended for use with velocity pressure exposure coefficients, K_h and K_z, which are based on gust speeds.

It is not the intent of Section 26.8 to address the general case of wind flow over hilly or complex terrain, for which engineering judgment, expert advice, or the Wind Tunnel Procedure as described in Chapter 31 may be required. Background material on topographic speed-up effects may be found in the literature (Jackson and Hunt 1975, Lemelin et al. 1988, Walmsley et al. 1986).

The designer is cautioned that, at present, the standard contains no provision for vertical wind speed-up because of a topographic effect, even though this phenomenon is known to exist and can cause additional uplift on roofs. Additional research is required to quantify this effect before it can be incorporated into the standard.

As illustrated in the example (in US customary units) that follows, there is a minimal difference in the calculated K_{zt} for Exposure B, C, or D. For this reason, the table of topographic multipliers shall be used for any exposure unless there is a need for increased accuracy in the topographic multiplier.

An example using the topographic provisions follows for Exposures B, C, and D. Using the graphic of a hill, escarpment, or 3D escarpment shown in Figure 26.8-1, the topographic parameters are

$H = 500$ ft,
$L_h = 1{,}500$ ft,
$x = 300$ ft, and
$z = 50$ ft.

And then evaluation downwind of the crest results in the following:

Figure C26.7-5(a). Exposure B: Suburban residential area with mostly single-family dwellings. Low-rise structures, less than 30 ft (9.1 m) high, in the center of the photograph have sites designated as Exposure B with Surface Roughness Category B terrain around the site for a distance greater than 1,500 ft (457 m) in any wind direction.

Figure C26.7-5(b). Exposure B: Urban area with numerous closely spaced obstructions having the size of single-family dwellings or larger. For all structures shown, terrain representative of Surface Roughness Category B extends more than 20 times the height of the structure or 2,600 ft (792 m), whichever is greater, in the upwind direction.

C26.9 GROUND ELEVATION FACTOR

The ratio of air pressure and density at elevation z relative to the standard values at $z = 0$, with constant temperature, is given by the barometric formula

$$p_z/p_0 = \rho_z/\rho_0 = e^{-gz/RT}$$

where

g = Acceleration of gravity [32.174 ft/s^2 (9.807 m/s^2)],
R = Gas constant of air [1,718 lb-ft/slug/°R (287 N-m/kg/K)], and
T = Absolute temperature [518° R (288° K)].

Figure C26.7-5(c). Exposure B: Buildings and other structures in the foreground are located in Exposure B. Buildings and other structures in the center top of the photograph adjacent to the clearing to the left, which meets the "open patch" criteria defined in Section C26.7, are located in Exposure C when wind comes from the left over the clearing.

With these values and elevation z (elevation of ground above sea level), the ratio is determined from the formulas given in Table 26.9-1, where $K_e = \rho_z/\rho_0$. For reference, a more complete version of Table 26.9-1, including air density values, is provided in Table C26.9-1.

$K_e = 1.0$ is permitted in all cases. While this is somewhat unconservative for elevations below sea level, the committee believes it is reasonable to permit this since the effect is very small for all areas below sea level in the United States [0 to −300 ft (−91 m) in Death Valley, for a maximum of 1% increase in air density], and is likely to be reduced even more because of higher average temperatures.

C26.10 VELOCITY PRESSURE

C26.10.1 Velocity Pressure Exposure Coefficient The velocity pressure exposure coefficient K_z can be obtained using these equations:

$$K_z = 2.41\left(\frac{15}{z_g}\right)^{2/\alpha} \quad \text{for } z < 15\,\text{ft} \qquad \text{(C26.10-1)}$$

$$K_z = 2.41\left(\frac{4.6}{z_g}\right)^{2/\alpha} \quad \text{for } z < 4.6\,\text{m} \qquad \text{(C26.10-1.SI)}$$

$$K_z = 2.41\left(\frac{z}{z_g}\right)^{2/\alpha} \quad \text{for } 15\,\text{ft}\ (4.6\,\text{m}) \le z \le z_g \qquad \text{(C26.10-2)}$$

$$K_z = 2.41 \quad \text{for } z_g < z \le 3,280\,\text{ft}\ (1,000\,\text{m}) \qquad \text{(C26.10-3)}$$

in which values of α and z_g are given in Table 26.11-1. These formulas are also given in Table 26.10-1 to aid the user.

Figure C26.7-6(a). Exposure C: Flat open grassland with scattered obstructions having heights generally less than 30 ft (9.1 m).

Figure C26.7-6(b). Exposure C: Open terrain with scattered obstructions having heights generally less than 30 ft (9.1 m). For most wind directions, all structures with a mean roof height less than 30 ft (9.1 m) in the photograph are less than 1,500 ft (457 m) from an open field, which prevents the use of Exposure B.

Figure C26.7-7. Exposure D: A building at the shoreline with wind flowing over open water for a distance of at least 5,000 ft (1,524 m) or 20 times the building or structure height, whichever is greater.

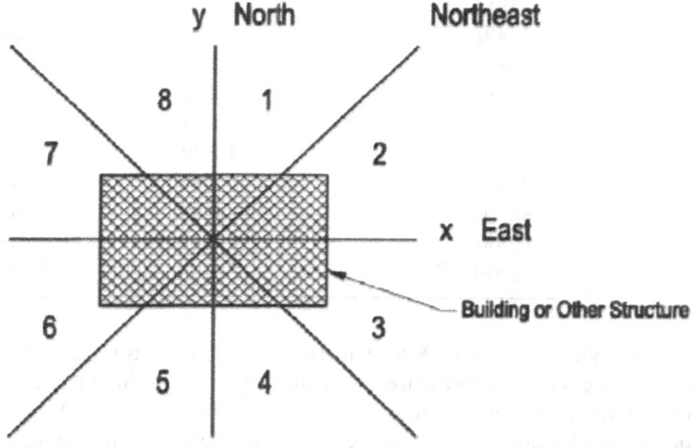

Figure C26.7-8. Determination of wind loads from different directions.

Previous editions of the standard, from ASCE 7-95 to ASCE 7-16, effectively established a maximum value for K_z of 2.01 by limiting the height z to the nominal height of the atmospheric boundary layer, also known as gradient height z_g. The boundary-layer height was based on research of the 1950s and 1960s (e.g., Davenport 1960). Since the 1970s, studies by Harris and Deaves (1981) and Zilitinkevich and Esau (2002) have shown that atmospheric boundary-layer heights in large synoptic storms are much greater [e.g., 6,700 to 13,000 ft (2 to 4 km)] than the z_g values from previous editions of ASCE 7, and wind speeds in such storms are known to increase above the previous values of z_g (Irwin 2006, Simiu et al. 2016). Recent studies of tropical cyclones (Tse et al. 2013) have also shown boundary-layer heights greater than the z_g values in ASCE 7-16 and previous editions. As the power-law equation in use from ASCE 7-95 to 7-16 deviates significantly at high elevations from wind speed profiles based on atmospheric boundary-layer theory and measurements (Panofsky and Dutton 1984), and as wind speed profiles at elevations of up to 3,280 ft (1,000 m) are known to be reasonably accurate (Irwin 2006), the value of z_g in ASCE 7-22 was increased to 3,280 ft (1,000 m), 2,460 ft (750 m), and 1,935 ft (590 m) for Exposures B, C, and D, respectively. The power law exponents were recalibrated accordingly using more recent wind speed profile models (Harris and Deaves 1981, Kelly et al. 2019). At height z_g the value of K_z is limited to a maximum of 2.41 for the exposure being considered. Above this height, and up to 3,280 ft (1,000 m), K_z is assumed to remain constant; however, for buildings and other structures with height greater than z_g, or for types of wind that differ significantly from straight (synoptic) winds, special studies are warranted. As the profiles for Exposures B, C, and D are simplified to work within the constraints of the standard, it is acceptable to use site-specific profiles determined by a rational analysis method defined in the recognized literature, such as the ESDU model, in their place.

Research by Vickery et al. (2009a) suggests that in hurricanes, specifically near the eyewall, the boundary layer height is lower

Feature	Exp.	H (ft)	L_h (ft)	H/L_h	K_1	x (ft)	μ (dn)	K_2	γ	z (ft)	K_3	K_{zt}
2D escarp.	B	500	1,500	0.33	0.25	300	4	0.95	2.5	50	0.92	1.48
2D escarp.	C	500	1,500	0.33	0.28	300	4	0.95	2.5	50	0.92	1.56
2D escarp.	D	500	1,500	0.33	0.32	300	4	0.95	2.5	50	0.92	1.63
3D axi.	B	500	1,500	0.33	0.32	300	1.5	0.87	4	50	0.88	1.54
3D axi.	C	500	1,500	0.33	0.35	300	1.5	0.87	4	50	0.88	1.60
3D axi.	D	500	1,500	0.33	0.38	300	1.5	0.87	4	50	0.88	1.67
2D ridge	B	500	1,500	0.33	0.43	300	1.5	0.87	3	50	0.90	1.80
2D ridge	C	500	1,500	0.33	0.48	300	1.5	0.87	3	50	0.90	1.90
2D ridge	D	500	1,500	0.33	0.52	300	1.5	0.87	3	50	0.90	1.97

Table C26.9-1. Gound Elevation Factor including Air Density.

Ground Elevation, z_e		Air Density, ρ		Ratio, K_e
(ft)	(m)	(Slug/ft^3)	(kg/m^3)	(ρ/ρ_0)
–10,000	–305	0.000247	1.269	1.04
0	0	0.000238	1.224	1.00
1,000	305	0.000229	1.180	0.96
2,000	610	0.000221	1.138	0.93
3,000	914	0.000213	1.098	0.90
4,000	1,219	0.000206	1.059	0.86
5,000	1,524	0.000198	1.021	0.83
6,000	1,829	0.000191	0.985	0.80
7,000	2,134	0.000185	0.950	0.78
8,000	2,438	0.000178	0.916	0.75
9,000	2,743	0.000172	0.883	0.72
10,000	3,048	0.000166	0.852	0.70

than for synoptic winds. Where hurricane wind speeds govern the design, the velocity pressure exposure coefficients in this standard, which are based on synoptic winds, may conservatively model wind loads at elevations exceeding a few hundred meters (about 600 to 700 ft) above the ground, especially in Exposure D. With proper documentation and approval by the Authority Having Jurisdiction, the use of alternative exposure coefficients that appropriately reflect this lower boundary layer in hurricane-prone regions is acceptable.

Table 26.10-1 was adjusted in ASCE 7-22 according to the updated K_z formulas presented earlier; however, where updated values of K_z were within 0.01 of the tabular values presented in previous editions of ASCE 7, the previous values were retained. Variances in K_z of 0.01 or less occurred at heights of up to 30 ft (9.1 m) for Exposure B, up to 120 ft (36.6 m) for Exposure C, and up to the table limit of 500 ft (152.4 m) for Exposure D. The intent was to dismiss insignificant changes to K_z values in Table 26.10-1 at heights covering the majority of low-rise buildings and other structures for which tabular values of K_z are most often used.

The values of α given in Table 26.11-1 (7.5, 9.8, and 11.5 for Exposures B, C, and D, respectively) define gust profiles used to calculate K_z and were updated in ASCE 7-22 for the first time since 1995. The updated $\bar{\alpha}$ values of mean hourly profiles in Table 26.11-1 based on 4.5, 6.4, and 8.0 are used only to calculate the gust factor, G_f, for flexible structures in Section 26.11.5.

Changes were implemented in ASCE 7-98, including truncation of K_z values for Exposures A and B below heights of 100 ft (30.5 m) and 30 ft (9.1 m), respectively, applicable to C&C and the Envelope Procedure. Exposure A was eliminated in the 2002 edition.

In the ASCE 7-05 standard, the K_z expressions were unchanged from ASCE 7-98. However, the possibility of interpolating between the standard exposures using a rational method was added in the ASCE 7-05 edition. One rational method is provided in the following text.

To a reasonable approximation, the empirical exponent α and gradient height z_g in the preceding expressions [Equations (C26.10-1) through (C26.10-3)] for exposure coefficient K_z may be related to the roughness length z_0 (defined in Section C26.7) by the relations

$$\propto = c_1 z_0^{-0.104} \quad \text{(C26.10-4)}$$

and

$$z_g = c_2 z_0^{0.127} \quad \text{(C26.10-5)}$$

where c_1 is 6.69 and c_2 is 1,166 when the units of z_0 and z_g are meters, and c_1 is 7.59 and c_2 is 3,284 when the units of z_0 and z_g are feet.

The preceding relationships are based on matching the empirical boundary-layer models proposed by Harris and Deaves (1981) and Kelly et al. (2019) with the power law relationship in Equations (C26.10-1) through (C26.10-3). The models correspond to a gradient wind of 168 mi/h (75 m/s) at latitude 40 degrees. If z_0 has been determined for a particular upwind fetch, Equations (C26.10-1) through (C26.10-5) can be used to evaluate K_z. The correspondence between z_0 and the parameters α and z_g implied by these relationships does not align exactly with that described in Table 26.11-1; however, the differences are relatively small and not of practical consequence. The following simplified method (Irwin 2006) allows evaluation of K_z following a transition from one surface roughness to another.

In uniform terrain, the wind travels a sufficient distance over the terrain for the planetary boundary layer to reach an equilibrium state. The exposure coefficient values in Table 26.10-1 are intended for this condition. Suppose that the site is x miles downwind of a change in terrain. The equilibrium value of the exposure coefficient at height z for the terrain roughness downwind of the change will be denoted by K_{zd}, and the equilibrium value for the terrain roughness upwind of the change will be denoted by K_{zu}. The effect of the change in terrain roughness on the exposure coefficient at the site can be represented by adjusting K_{zd} by an increment ΔK, thus arriving at a corrected value K_z for the site:

$$K_z = K_{zd} + \Delta K \quad \text{(C26.10-6)}$$

In this expression, ΔK is calculated using

$$\Delta K = (K_{33,u} - K_{33,d}) \frac{K_{zd}}{K_{33,d}} F_{\Delta K}(x) \qquad |\Delta K| \le |K_{zu} - K_{zd}| \quad \text{(C26.10-7)}$$

where $K_{33,d}$ and $K_{33,u}$ are the downwind and upwind equilibrium values, respectively, of the exposure coefficient at 33 ft (10 m) height, and the function $F_{\Delta K}(x)$ is given by

$$F_{\Delta K}(x) = \frac{\log_{10}\left(\frac{x_1}{x}\right)}{\log_{10}\left(\frac{x_1}{x_0}\right)} \quad \text{(C26.10-8)}$$

For $x_0 < x < x_1$

$$F_{\Delta k}(x) = 1 \quad \text{for } x < x_0 \qquad F_{\Delta k}(x) = 0 \quad \text{for } x > x_1$$

In the preceding relationships

$$x_0 = c_3 \times 10^{-(K_{33,d} - K_{33,u})^2 - 2.3} \quad \text{(C26.10-9)}$$

The constant c_3 equals 0.621 mi (1.0 km). The length x_1 is 6.21 mi (10 km) for $K_{33,d} < K_{33,u}$ (wind going from smoother terrain upwind to rougher terrain downwind) or x_1 is 62.1 mi (100 km) for $K_{33,d} > K_{33,u}$ (wind going from rougher terrain upwind to smoother terrain downwind).

The preceding description is in terms of a single roughness change. The method can be extended to multiple roughness changes. The extension of the method is best described by an example. Figure C26.10-1 shows wind with an initial profile characteristic of Exposure D encountering an expanse of B roughness, followed by a further expanse of D roughness, and then some more B roughness again, before it arrives at the building site. This situation is representative of wind from the sea flowing over an outer strip of land, then a coastal waterway, and then some suburban roughness before arriving at the building site.

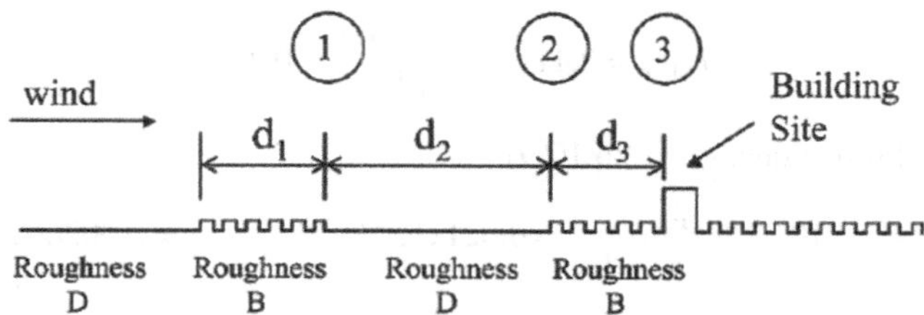

Figure C26.10-1. Multiple roughness changes caused by coastal waterway.

The aforementioned method for a single roughness change is first used to compute the profile of K_z at Station 1 in Figure C26.10-1. Call this profile $K_z^{(1)}$. The value of ΔK for the transition between Stations 1 and 2 is then determined using the equilibrium value of $K_{33,u}$ for the roughness immediately upwind of Station 1, that is, as though the roughness upwind of Station 1 extended to infinity. This value of ΔK is then added to the equilibrium value $K_{zd}^{(2)}$ of the exposure coefficient for the roughness between Stations 1 and 2 to obtain the profile of K_z at Station 2, which we will call $K_z^{(2)}$. Note, however, that the value of $K_z^{(2)}$ obtained in this way cannot be any lower than $K_z^{(1)}$. The process is then repeated for the transition between Stations 2 and 3. Thus, ΔK for the transition from Station 2 to Station 3 is calculated using the value of $K_{33,u}$ for the equilibrium profile of the roughness immediately upwind of Station 2 and the value of $K_{33,d}$ for the equilibrium profile of the roughness downwind of Station 2. This value of ΔK is then added to $K_{zd}^{(2)}$ to obtain the profile $K_z^{(3)}$ at Station 3, with the limitation that the value of $K_z^{(3)}$ cannot be any higher than $K_z^{(2)}$.

Example 1 (US Customary Units): Single Roughness Change. Suppose that the building is 66 ft high, and its local surroundings are suburban, with a roughness length z_0 = 1 ft. However, the site is 0.37 mi downwind of the edge of the suburbs, beyond which the terrain is characteristic of open country, with z_0 = 0.1 ft. From Equations (C26.10-4) and (C26.10-5), for the open terrain,

$$\alpha = c_1 z_0^{-0.104} = 7.59 \times 0.1^{-0.104} = 9.6$$

$$z_g = c_2 z_0^{0.127} = 3.284 \times 0.1^{0.127} = 2,451 \text{ ft}$$

Therefore, applying Equation (C26.10-2) at 66 ft (20 m) and 33 ft (10 m) heights

$$K_{zu} = 2.41 \left(\frac{66}{2,451}\right)^{2/9.6} = 1.13$$

and

$$K_{33,u} = 2.41 \left(\frac{33}{2,451}\right)^{2/9.6} = 0.98$$

Similarly, for the suburban terrain

$$\alpha = c_1 z_0^{-0.104} = 7.59 \times 0.1^{-0.104} = 7.6$$

$$z_g = c_2 z_0^{0.127} = 3.284 \times 0.1^{0.127} = 3,284 \text{ ft}$$

Therefore,

$$K_{zd} = 2.41 \left(\frac{66}{3,284}\right)^{2/7.6} = 0.86$$

and

$$K_{33,d} = 2.41\left(\frac{33}{3,284}\right)^{2/7.6} = 0.72$$

From Equation (C26.10-9)

$$x_0 = c_2 x 10^{-(K_{33,d}-K_{33,u})^2-2.3} = 0.621x10^{-(0.72-0.98)^2-2.3} = 0.00266 \text{ mi}$$

From Equation (C26.10-8)

$$F_{\Delta K}(x) = \frac{\log_{10}(6.21/0.37)}{\log_{10}(6.21/0.00266)} = 0.36$$

Therefore, from Equation (C26.10-7)

$$\Delta K = (0.98-0.72)\frac{0.86}{0.72}0.36 = 0.11$$

Note that because $|\Delta K|$ is 0.11, which is less than the 0.26 value of $\Delta|K_{33,u}-K_{33,d}|$, 0.11 is retained. Finally, from Equation (C26.10-6), the value of K_z is

$$K_z = K_{zd} + \Delta K = 0.86 + 0.11 = 0.97$$

Because the value 0.97 for K_z lies between the values 0.85 and 1.15, which would be derived from Table 26.10-1 for Exposures B and C, respectively, it is an acceptable interpolation. If it falls below the Exposure B value, then the Exposure B value of K_z is to be used. The value $K_z = 0.97$ may be compared with the value 1.15 that would be required by the simple 2,600 ft fetch length requirement of Section 26.7.3.

The most common case of a single roughness change where an interpolated value of K_z is needed is for the transition from Exposure C to Exposure B, as in the example just described. For this particular transition, using the typical values of z_0 of 0.1 ft and 1.0 ft, the preceding formulas can be simplified to

$$K_z = K_{zd}\left[1 + 0.107\log_{10}\left(\frac{6.21}{x}\right)\right] \qquad \text{(C26.10-10)}$$

$$K_{zB} \leq K_z \leq K_{zC}$$

where x is in miles, and K_{zd} is computed using $\alpha = 7.6$.

K_{zB} and K_{zC} are the exposure coefficients in the standard Exposures B and C, respectively. Figure C26.10-2 illustrates the transition from Terrain Roughness C to Terrain Roughness B from this expression. Note that it is acceptable to use the typical z_0 rather than the lower limit for Exposure B in deriving this formula because the rate of transition of the wind profiles depends on average roughness over significant distances, not local roughness anomalies. The potential effects of local roughness anomalies, such as parking lots and playing fields, are covered by using the standard Exposure B value of exposure coefficient, K_{zB}, as a lower limit to the calculated value of K_z.

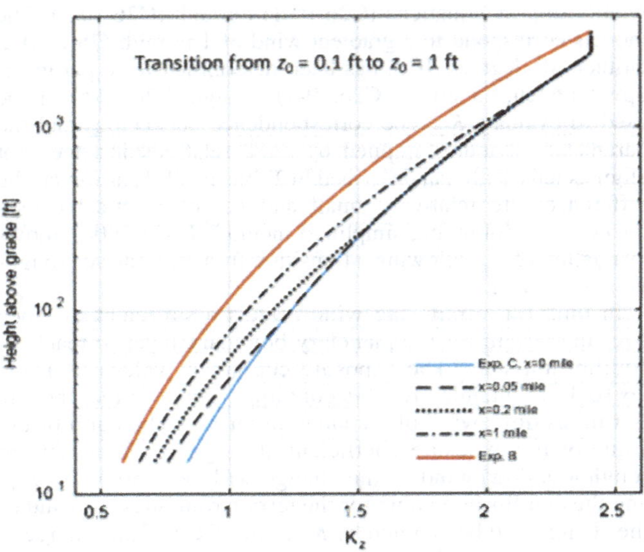

Figure C26.10-2. Transition from Terrain Roughness C to Terrain Roughness B, Equation (C26.10-10).

Example 2: Multiple Roughness Change. Suppose we have a coastal waterway situation, as illustrated in Figure C26.10-1, where the wind comes from open sea with roughness type D, for which we assume $z_0 = 0.01$ ft, and passes over a strip of land 1 mi wide, which is covered in buildings that produce typical B-type roughness, that is, $z_0 = 1$ ft. It then passes over a 2 mi wide strip of coastal waterway, where the roughness is again characterized by the open-water value, $z_0 = 0.01$ ft. It then travels over 0.1 mi of roughness type B ($z_0 = 1$ ft) before arriving at the building site, Station 3 in Figure C26.10-1, where the exposure coefficient is required at the 50 ft height. The exposure coefficient at Station 3 at a 50 ft height is calculated as shown in Table C26.10-1.

The value of the exposure coefficient at 50 ft at Station 3 is seen from the table to be 1.029. This is above the value in Table 26.10-1 for Exposure B, which would be 0.79, but well below that for Exposure D, which would be 1.27, and somewhat below that for Exposure C, which would be 1.09.

Table C26.10-1. Tabulated Exposure Coefficients.

Transition from sea to Station 1	$K_{33,u}$ 1.251	$K_{33,d}$ 0.717	$K_{50,d}$ 0.800	$F_{\Delta K}$ 0.221	ΔK_{50} 0.132	$K_{50}^{(1)}$ 0.932
Transition from Station 1 to Station 2	$K_{33,u}$ 0.835	$K_{33,d}$ 1.251	$K_{50,d}$ 1.339	$F_{\Delta K}$ 0.334	ΔK_{50} –0.149	$K_{50}^{(2)}$ 1.191
Transition from Station 2 to Station 3	$K_{33,u}$ 1.113	$K_{33,d}$ 0.717	$K_{50,d}$ 0.800	$F_{\Delta K}$ 0.519	ΔK_{50} 0.229	$K_{50}^{(3)}$ 1.029

Notes: The equilibrium values of the exposure coefficients, $K_{33,u}$, $K_{33,d}$ and $K_{50,d}$ (downwind value of K_z at 50 ft), were calculated from Equation (C26.10-2) using α and z_g values obtained from Equations (C26.10-4) and (C26.10-5) with the roughness values given. Then $F_{\Delta K}$ is calculated using Equations (C26.10-8) and (C26.10-9), and then the value of ΔK at 50 ft height, ΔK_{50}, is calculated from Equation (C26.10-7). $K_{33,u}$, $K_{33,d}$, and $K_{50,d}$ from Station 1 to Station 3 include the computed Exposure K_z determined in the calculation rows upwind. Finally, the exposure coefficient at 50 ft at Station i, $K_{50}^{(i)}$, is obtained at each transition as well as at the site from Equation (C26.10-6).

C26.10.2 Velocity Pressure The basic wind speed is converted to velocity pressure q_z in lb/ft^2 (N/m^2) at height z by the use of Equation (26.10-1).

The constant in this equation reflects the mass density of air for the standard atmosphere, that is, temperature of 59°F (15°C) and sea level pressure of 29.92 in. (101.325 kPa) of mercury, equal to 0.0765 lbm/ft^3 or 0.002378 slug/ft^3 or 0.002378 lb s^2/ft^4 (1.225 kg/m^3), and dimensions associated with wind speed in mi/h (m/s). The constant is obtained as follows.

Dynamic pressure from Bernoulli's law:

$$p = \frac{1}{2}\rho V^2$$

with V in mi/h

$$p = \tfrac{1}{2}(0.002378\ \text{lb s}^2/\text{ft}^4)[V\ \text{mi/h}\,(88\ \text{ft/s}\,/\,60\ \text{mi/h})^2] = 0.00256V^2\ \text{lb/ft}^2$$

with V in m/s

$$p = \tfrac{1}{2}(1.225\ \text{kg/m}^3)(1\ \text{N}/1\ \text{kg m/s}^2)(V\ \text{m/s})^2 = 0.613V^2\text{N/m}^2 \quad \text{(SI)}$$

Values of air density other than the preceding standard-atmosphere values may be adjusted using the factor K_e as described in Section C26.9.

In ASCE 7-22 the wind directionality factor, K_d, was removed from the velocity pressure equation and placed in the individual wind pressure equations and force equations. Prior to ASCE 7-98, the effect of K_d had been incorporated into the wind load factor. When K_d was removed from the load factor it was explicitly introduced into the standard in ASCE 7-98 through the calculation of the velocity pressure. K_d is a function of (1) the directionality of the extreme wind climate, and (2) the shape of the building or other structure and the directional characteristics of the structural system, so it is more appropriate to incorporate K_d into the individual design wind pressure and force equations. The effect of building or other structure shape on K_d is evident in Table 26.6-1. In the velocity pressure equation used before ASCE 7-22, any building or other structure that comprises multiple shapes will have velocity design pressures which is not consistent with the physics.

C26.11 GUST EFFECTS

This standard specifies a single, conservative, gust-effect factor of 0.85 for rigid buildings. As an option, the designer can incorporate specific features of the wind environment and building size to more accurately calculate an alternate but more accurate gust-effect factor that accounts for the decorrelation of wind gusts over the size of the structure. One such procedure is in the body of the standard (Solari and Kareem 1998). Neither of these factors accounts for dynamic amplification caused by vibration of the structure, but they are considered acceptable for rigid structures as defined in the standard. The alternate calculated gust factor is 5% to 10% lower than the value of 0.85 permitted in the standard without calculation.

A third gust-effect factor, G_f, is provided for flexible buildings and structures that do not meet the requirements that the fundamental natural frequency, n_1, is greater than or equal to 1 Hz. This factor accounts for the building size and gust size in the same manner as the alternate calculated factor for rigid buildings, but it also accounts for dynamic amplification caused by the design wind speed, the fundamental natural frequency of vibration, and the damping ratio. The natural frequency should account for the minimum expected structural stiffness under the design wind event V, be it the ultimate wind speed or any desired serviceability speed, and the mean building mass expected under the design wind event, including self-weight, dead load, and sustained live load. The damping ratio should account for inherent structural damping under the design wind event (see Structural Damping subsection). It may also include damping added by supplementary devices if expected to be operational under the design wind event. The in-service reliability of such devices should be ensured through an established and documented maintenance program.

Example: Calculation of the gust-effect factors for a subject building is demonstrated in Table C26.11-1. The frequency-dependent relationship among all factors is illustrated in the graph at the end of this table. The flexible factor, G_f, may be used for all cases but is required when $n_1 < 1$. This factor gradually approaches the alternate calculated factor for rigid cases, G, as the natural frequency exceeds 1, especially for higher levels of damping, but it always exceeds G. The difference is deemed negligible for $n_1 > 1$, so G, which is considerably simpler to calculate, is offered as an acceptable alternative. The default value of $G = 0.85$, which requires no calculation, is offered as an even more convenient alternative when $n_1 > 1$, if the greater conservatism is acceptable to the designer. In addition, the default value results in a large abrupt change in the gust-effect value for cases that have a natural frequency close to 1, which may be awkward for a designer to reconcile. A designer is free to use any other rational procedure in the approved literature, as stated in Section 26.11.5.

The gust-effect factors account for loading effects in the along-wind direction caused by wind turbulence–structure interaction. They do not include allowances for across-wind loading effects, vortex shedding, instability caused by galloping or flutter, or amplification of aerodynamic torsion caused by building vibration in a pure torsional mode. For structures susceptible to loading effects that are not accounted for in the gust-effect factor, information should be obtained from recognized literature (Kareem 1992, 1985; Gurley and Kareem 1993; Solari 1993a, b; Zhou et al. 2002; Chen and Kareem 2004; Bernardini et al. 2013a) or from wind tunnel tests.

Along-Wind Response. The maximum along-wind displacement response can be approximated by a static analysis of the structure under the action of loads multiplied by the appropriate gust-effect factor, as defined in the standard. Such displacements are based on the static elastic curve of the structure and are reasonably accurate when the resonant response is small compared to the mean and background responses. For highly flexible structures, where the response is dominated by resonance, more accurate values, including variation with height and dynamic responses such as acceleration, can be calculated as described in the following sections. These response components are needed for strength and serviceability limit states.

The maximum along-wind displacement as a function of height above the ground surface is given by

$$X_{\max}(z) = \frac{\phi(z)\rho BhC_{fx}\hat{V}_{\bar{z}}^2}{2m_1(2\pi n_1)^2}KG_f \quad \text{(C26.11-1)}$$

where

$\phi(z)$ = Fundamental model shape, $\phi(z) = (z/h)^{\xi}$;
ξ = Mode shape power-law exponent;
ρ = Air density;
C_{fx} = Mean along-wind force coefficient;
m_1 = Modal mass = $\int_0^h \mu(z)\phi^2(z)dz$;

Table C26.11-1. Gust-Effect Factors: Example Calculation.

Item	Value	Source
DEFAULT FACTOR FOR RIGID BUILDING (requires $n_1 \geq 1$)		
G Gust-effect factor	0.85	Equation (26.11.1)
ALTERNATE CALCULATED FACTOR FOR RIGID BUILDING (requires $n_1 \geq 1$)		
h Mean roof height	600 ft (183 m)	User spec
B Width normal to wind	100 ft (30 m)	User spec
D Depth parallel to wind	100 ft (30 m)	User spec
$\bar{z}$ Effective structure height	360 ft (110 m)	0.6h (26.11.4)
Exposure category	B	
c Turbulence intensity factor at 10 m	0.3	Table 26.11-1
$I_{\bar{z}}$ Turbulence intensity at effective height	0.201	Equation (26.11-7)
ℓ Turbulence length scale at 33 ft (10 m)	320 ft (98 m)	Table 26.11-1
$\bar{\varepsilon}$ Power law exponent of turbulent length scale profile	1/3	Table 26.11-1
$L_{\bar{z}}$ Turbulence length scale at effective height	710 ft (216 m)	Equation (26.11-9)
Q^2 Background response (squared)	0.616	Equation (26.11-8)
g_Q Background load peak factor	3.4	Equation (26.11.4)
g_v Velocity peak factor	3.4	Equation (26.11.4)
G Calculated gust-effect factor	0.860	Equation (26.11-6)
ADDITIONAL CALCULATIONS FOR FLEXIBLE BUILDING (all n_1)		
V Basic wind speed	115 mi/h (51 m/s)	
n_1 Fundamental natural frequency in direction of wind	0.2 Hz	Analysis or rational approximation
β Damping ratio	0.01	Rational assignment
$\bar{\alpha}$ Power law exponent of mean wind speed profile	1/4.5	Table 26.11-1
$\bar{b}$ Gust factor $1/F$ at 33 ft (10 m)	0.47	Table 26.11-1
$\bar{V}_{\bar{z}}$ Mean wind speed at effective height	135 ft/s (41.1 m/s)	Equation (26.11-16)
N_1 Reduced natural frequency	1.053	Equation (26.11-14)
R_n Resonance response factor for n	0.128	Equation (26.11-13)
η_h Vertical decay parameter	4.094	26.11.5: $4.6n_1h/\bar{V}_{\bar{z}}$
η_B Cross-wind decay parameter	0.682	26.11.5: $4.6n_1B/\bar{V}_{\bar{z}}$
η_L Along-wind decay parameter	2.285	26.11.5: $15.4n_1L/\bar{V}_{\bar{z}}$
R_h Resonant factor for h	0.214	Equation (26.11-15a)
R_B Resonant factor for B	0.666	Equation (26.11-15a)
R_L Resonant factor for L	0.343	Equation (26.11-15a)
R^2 Resonant response factor (squared)	1.261	Equation (26.11-12)
g_R Resonant peak factor	3.787	Equation (26.11-11)
G_f Gust-effect factor	1.162	Equation (26.11-10)

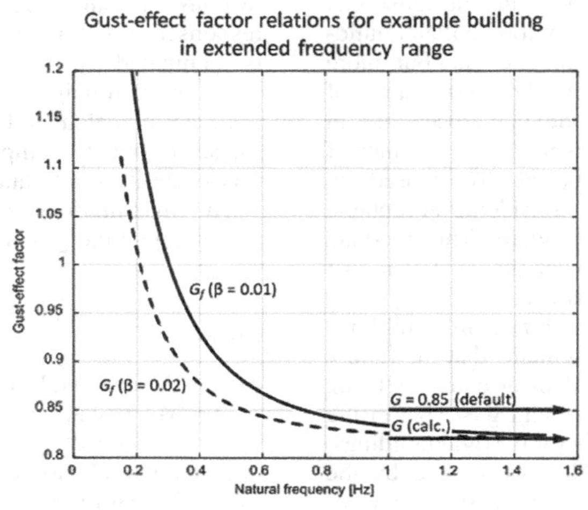

n_1 = Fundamental natural frequency;
$\mu(z)$ = Mass per unit height: $K = (1.65)^{\hat{\alpha}}/(\hat{\alpha} + \xi + 1)$; and
$\hat{V}_{\bar{z}}$ = 3 s gust speed at height $\bar{z}$.

$$\hat{V}_{\bar{z}} = \hat{b}(z/33)^{\hat{\alpha}} V\left(\frac{88}{60}\right)$$

where V is the 3 s gust speed in Exposure C (mi/h) at the reference height (obtained from Figure 26.5-1); and $\hat{b}$ and $\hat{\alpha}$ are given in Table 26.11-1.

The root-mean-square (RMS) along-wind acceleration $\sigma_{\ddot{x}}(z)$ as a function of height above the ground surface is given by

$$\sigma_{\ddot{x}}(z) = \frac{0.85\varphi(z)\rho BhC_{fx}\overline{V}_{\bar{z}}^2}{m_1} I_{\bar{z}}KR \qquad \text{(C26.11-2)}$$

where $\overline{V}_{\bar{z}}$ is the mean hourly wind speed at height $\bar{z}$ (ft/s).

$$\overline{V}_{\bar{z}} = \bar{b}\left(\frac{\bar{z}}{33}\right)^{\bar{\alpha}} V\left(\frac{88}{60}\right) \qquad \text{(C26.11-3)}$$

where $\bar{b}$ and $\bar{\alpha}$ are defined in Table 26.11-1.

The maximum along-wind acceleration as a function of height above the ground surface is given by

$$\ddot{X}_{\max}(z) = g_{\ddot{x}}\sigma_{\ddot{x}}(z) \qquad \text{(C26.11-4)}$$

where

$$g_{\ddot{x}} = \sqrt{2\ln(n_1 T)} + \frac{0.5772}{\sqrt{2\ln(n_1 T)}} \qquad \text{(C26.11-5)}$$

where T is the length of time over which the minimum acceleration is computed, usually taken to be 3,600 s to represent 1 h.

Example calculations of maximum along-wind displacement, RMS along-wind acceleration, and maximum along-wind acceleration are given in Table C26.11-2.

Approximate Fundamental Frequency. To estimate the dynamic response of structures, knowledge of the fundamental frequency (lowest natural frequency) of the structure is essential. The frequency indicates whether the dynamic response estimates are necessary. Most software used in the analysis of structures are able to provide estimates of the natural frequency of the structure being analyzed. For the preliminary design stages, some empirical relationships for building period T_a ($T_a = 1/n_1$) are available in the earthquake-related chapters of this standard. However, these expressions are based on recommendations for earthquake design with inherent bias toward higher estimates of fundamental frequencies (Goel and Chopra 1997, 1998). For wind design applications, these values may be unconservative because an estimated

Table C26.11-2. Along Wind Response, Example.

Item	Value	Source
V Basic wind speed	115 mi/h	Figure 26.5-1A
ρ Air density	0.0024 slug/ft^3	Site elevation near sea level
C_{pw} External pressure coefficient, windward wall	0.8	Figure 27.4-1
C_{pw} External pressure coefficient, leeward wall	−0.8	Figure 27.4-1
C_{fx} Along-wind force coefficient	1.3	
ξ Mode shape power law exponent	1	Analysis of structure
$\hat{\alpha}$ Peak velocity power law exponent	1/7.5	Table 26.11-1
$\hat{b}$ Velocity profile parameter	0.84	Table 26.11-1
$\hat{V}_{\bar{z}}$ 3 s gust velocity at height $\bar{z}$	195 ft/s	Equation (C26.11-1a)
K Modal load parameter	0.501	
ρ_b Building density	12 lbm/ft^3, 0.3727 slug/ft^3	Building design
μ Building mass per unit height	3,727 slug/ft	
m_1 Modal mass	745,400 slug	
MAXIMUM ALONG-WIND DISPLACEMENT AT BUILDING TOP		
$\phi(h)$ Mode shape at $z = h$	1.0	
$X_{\max}(h)$	1.76 ft	Equation (C26.11-1)
RMS ALONG-WIND ACCELERATION AT BUILDING TOP		
V Basic wind speed, 10-year MRI	76 mi/h	Figure CC-1
$\overline{V}_{\bar{z}}$ Mean velocity at height $\bar{z}$	89.1 ft/s	Equation (C26.11-3)
R Resonant response factor	1.123	Equation (26.11-12)
$\sigma_{\ddot{x}}(h)$	0.176 ft/s^2	Equation (C26.11-2)
MAXIMUM ALONG-WIND ACCELERATION AT BUILDING TOP		
T Time period for maximum	3,600 s	Traditional
$g_{\ddot{x}}$ Peak factor	3.79	Equation (C26.11-5)
$\ddot{X}_{\max}(h)$	0.668 ft/s^2	Equation (C26.11-4)

Note: See Table C26.11-1 for additional items not shown.

frequency higher than the actual frequency would yield lower values of the gust-effect factor and thus a lower design wind pressure. However, Goel and Chopra also cite lower-bound estimates of frequency that are more suited for use in wind applications and are now given in Section 26.11.2; graphs of these expressions are shown in Figure C26.11-1.

Because these expressions are based on regular buildings, limitations based on height and slenderness are required. The effective length, L_{eff}, uses a height-weighted average of the along-wind length of the building for slenderness evaluation. The top portion of the building is most important; hence, the height-weighted average is appropriate. This method is an appropriate first-order equation for addressing buildings with setbacks. Explicit calculation of the gust-effect factor by the other methods given in Section 26.11 can still be performed.

Observations from wind tunnel testing of buildings where frequency is calculated using analysis software show that the following expression for frequency can be used for steel and concrete buildings less than about 400 ft (122 m) in height:

For the average value of the frequency

$$n_1 = 100/h \quad \text{(C26.11-6)}$$

$$n_1 = 30.48/h \quad \text{(C26.11-6.SI)}$$

For the lower bound value of the frequency

$$n_1 = 75/h \quad \text{(C26.11-7)}$$

$$n_1 = 22.86/h \quad \text{(C26.11-7.SI)}$$

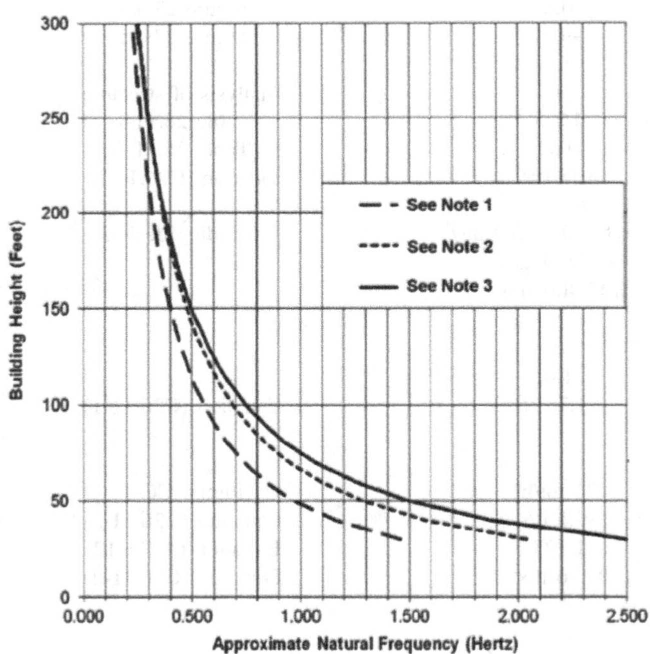

Notes:

1. Equation (26.11-2): $22.2/h^{0.8}$
2. Equation (26.11-3): $43.5/h^{0.9}$
3. Equation (26.11-4): $75/h$

Figure C26.11-1. Equations for approximate lower-bound natural frequency n_a versus building height.

Equations (C26.11-7) and (C26.11-7.SI) for the lower-bound value are provided in Section 26.11.3.

Based on full-scale measurements of buildings under the action of wind, the following expression has been proposed for wind applications (Zhou and Kareem 2001a, Zhou et al. 2002):

$$n_1 = 150/h \quad \text{(C26.11-8)}$$

$$n_1 = 45.72/h \quad \text{(C26.11-8.SI)}$$

This frequency expression is based on older buildings and overestimates the frequency common in US construction for buildings less than 400 ft (122 m) in height, but it becomes more accurate for tall buildings greater than 400 ft (122 m) in height.

Studies in Japan involving a suite of buildings under low-amplitude excitations have led to the following expressions for natural frequencies of buildings (Sataka et al. 2003):

For concrete buildings

$$n_1 = 220/h \quad \text{(C26.11-9)}$$

$$n_1 = 67/h \quad \text{(C26.11-9.SI)}$$

For steel buildings

$$n_1 = 164/h \quad \text{(C26.11-10)}$$

$$n_1 = 50/h \quad \text{(C26.11-10.SI)}$$

These expressions result in higher frequency estimates than those obtained from the general expression given in Equations (C26.11-6) through (C26.11-8), particularly since the Japanese data set has limited observations for the more flexible buildings sensitive to wind effects, and Japanese construction tends to be stiffer.

For cantilevered masts or poles of uniform cross section (in which bending action dominates)

$$n_1 = (0.56/h^2)\sqrt{(EI/m)}$$

where EI is the bending stiffness of the section, and m is the mass per unit height. This formula may be used for masts with a slight taper, using average value of EI and m (ECCS 1978).

An approximate formula for cantilevered, tapered, circular poles (ECCS 1978) is

$$n_1 \approx [\lambda/(2\pi h^2)]\sqrt{(EI/m)} \quad \text{(C26.11-12)}$$

where h is the height of the pole; and E, I, and m are calculated for the cross section at the base. λ depends on the wall thicknesses at the tip and base (e_t and e_b) and external diameter at the tip and base (d_t and d_b), according to the formula

$$\lambda = \left[1.9\exp\left(\frac{-4d_t}{d_b}\right)\right] + \left[\frac{6.65}{0.9 + (e_t/e_b)^{0.666}}\right] \quad \text{(C26.11-13)}$$

Equation (C26.11-12) reduces to Equation (C26.11-11) for uniform masts. For freestanding lattice towers (without added ancillaries, such as antennas or lighting frames)

$$n_1 \approx 4{,}921 w_a / h^2 \quad \text{(C26.11-14)}$$

$$n_1 \approx 1{,}500 w_a / h^2 \quad \text{(C26.11-14.SI)}$$

where w_a is the average width of the structure, and h is the tower height (Standards Australia 1994). An alternative formula for lattice towers (with added ancillaries) is

$$n_1 = \left(\frac{L_N}{h}\right)^{2/3} \left(\frac{w_b}{h}\right)^{1/2} \quad \text{(C26.11-15)}$$

where w_b is the tower base width and L_N = 885.8 ft (270 m) for square base towers, or 7,546 ft (230 m) for triangular base towers (Wyatt 1984).

Structural Damping. Structural damping is a measure of energy dissipation in a vibrating structure that results in bringing the structure to a quiescent state. It is defined as the ratio of the energy dissipated in one oscillation cycle to the maximum amount of energy in the structure in that cycle. There are as many structural damping mechanisms as there are modes of converting mechanical energy into heat. The most important mechanisms are material damping and interfacial damping.

In engineering practice, the damping mechanism is often approximated as viscous damping because it leads to a linear equation of motion. This damping measure, in terms of the damping ratio, is usually assigned based on the construction material, for example, steel or concrete. The calculation of dynamic load effects requires damping ratio as an input. In wind applications, damping ratios of 1% and 2% are typically used in the United States for steel and concrete buildings, respectively, at serviceability levels, while ISO 4354 (1997) suggests 1% and 1.5% for steel and concrete, respectively. Damping ratios for buildings under ultimate strength design conditions may be significantly higher, and 2.5% to 3% is commonly assumed. Damping values for steel support structures for signs, chimneys, and towers may be much lower than for buildings and may fall in the range of 0.15% to 0.5%. Damping values of special structures like steel stacks can be as low as 0.2% to 0.6% and 0.3% to 1.0% for unlined and lined steel chimneys, respectively (ASME 1992, CICIND 1999). These values may provide some guidance for design. Damping levels used in wind load applications are smaller than the 5% damping ratios common in seismic applications because buildings subjected to wind loads respond essentially elastically, whereas buildings subjected to design-level earthquakes respond inelastically at higher damping levels.

Because the level of structural response in the strength and serviceability limit states is different, the damping values associated with these states may differ. Furthermore, because of the number of mechanisms responsible for damping, the limited full-scale data manifest a dependence on factors such as material, height, and type of structural system and foundation (Kijewski-Correa et al. 2013). The Committee on Damping of the Architectural Institute of Japan suggests different damping values for these states based on a large damping database described in Sataka et al. (2003).

The NatHaz Group developed an interactive database of full-scale experimentally determined modal damping ratios based on the database (Kareem et al. 2012). The database is publicly available at https://vortex-winds.org. It has a query-based web interface for the rapid identification of modal damping ratios that satisfy specific requirements, such as geometric form, structural system, construction material, foundation type, and building use. A recent publication offers a data-driven model of damping that has been validated with several full-scale studies (Spence and Kareem 2014).

In addition to structural damping, aerodynamic damping may be experienced by a structure oscillating in air. In general, the aerodynamic damping contribution is quite small compared to the structural damping, and it is positive in low-to-moderate wind speeds. Depending on the structural shape, at some wind velocities the aerodynamic damping may become negative, which can lead to unstable oscillations. In these cases, reference should be made to recognized literature or a wind tunnel study.

Alternate Procedure to Calculate Wind Loads. The concept of the gust-effect factor implies that the effect of gusts can be adequately accounted for by multiplying the mean wind load distribution with height by a single factor. This is an approximation. If a more accurate representation of gust effects is required, the alternative procedure in this section can be used. It takes account of the fact that the inertial forces created by the building's mass, as it moves under wind action, have a different distribution with height than the mean wind loads or the loads caused by the direct actions of gusts (Zhou and Kareem 2001a, Chen and Kareem 2004). The alternate formulation of the equivalent static load distribution uses the peak base bending moment and expresses it in terms of inertial forces at different building levels. A base bending moment, instead of the base shear as in earthquake engineering, is used for the wind loads because it is less sensitive to deviations from a linear mode shape. For a more detailed discussion on this wind-loading procedure, see Zhou and Kareem (2001a, b) and Chen and Kareem (2004).

Alternate Procedure: Along-Wind Equivalent Static Wind Loading. The equivalent static wind loading for the mean, background, and resonant components is obtained using the procedure outlined in this section.

Mean wind load component, $\overline{P}_j$, is given by

$$\overline{P}_j = q_j \times C_p \times A_j \times \overline{G} \quad \text{(C26.11-16)}$$

where

j = Floor level,
q_j = Velocity pressure at height z_j,
z_j = Height of the jth floor above the ground level,
C_p = External pressure coefficient, and
$\overline{G} = 0.925 \times (1 + 1.7 g_v I_{\bar{z}})^{-1}$ = gust velocity factor.

Peak background wind load component, $\hat{P}_{Bj}$, at the jth floor level is given similarly by

$$\hat{P}_{Bj} = \overline{P}_j \times G_B / \overline{G} \quad \text{(C26.11-17)}$$

where $G_B = 0.925 \times (1.7 I_{\bar{z}} \times g_Q Q)/(1 + 1.7 g_v I_{\bar{z}})$ is the background component of the gust-effect factor.

Peak resonant wind load component, $\hat{P}_{Rj}$, at the jth floor level is obtained by distributing the resonant base bending moment response to each level:

$$\hat{P}_{Rj} = C_{Mj} \hat{M}_R \quad \text{(C26.11-18)}$$

$$C_{Mj} = \frac{w_j \varphi_j}{\sum w_j \varphi_j z_j} \quad \text{(C26.11-19)}$$

Table C26.11-3. Along-Wind, Across-Wind, Torsional Moments, and Acceleration Response.

	Survivability Design				Serviceability Design				
	Aerodynamic Load Coefficient		Base Moments (10^6 kips-ft)		Aerodynamic Load Coefficient		Accelerations (milli-g or rad/s^2)		
								Corner	
Load components	σ_{CM}	f_1	$C_M(f_1)$	M	f_1	$C_M(f_1)$	σ_a	X	Y
ASCE 7-10	–	–	–	1.73	–	–	1.95	2.77	3.24
Along-wind	0.109	0.193	0.046	1.72	0.292	0.022	2.03		
Across-wind	0.133	0.193	0.093	1.82	0.292	0.024	2.64		
Torsional	0.044	0.337	0.040	0.086	0.512	0.043	0.0001		

Note: As this database is experimental in nature, thus has limitation in scope, it can be conveniently expanded using additional data as it becomes available.

$$\hat{M}_R = \overline{M} \bullet G_R/\overline{G} \quad \text{(C26.11-20)}$$

$$\overline{M} = \sum_{j=1,n} \overline{P}_j \bullet z_j \quad \text{(C26.11-21)}$$

where

C_{Mj} = Vertical load distribution factor,
$\hat{M}_R$ = Peak resonant component of the base bending moment response,
w_j = Portion of the total gravity load of the building located or assigned to level j,
φ_j = First structural mode shape value at level j,
$\overline{M}$ = Mean base bending produced by mean wind load,
$G_R = 0.925 \times [(1.7 I_{\bar{z}} \times g_R R)/(1 + 1.7 g_v I_{\bar{z}})]$ = Resonant component of the gust-effect factor, and
n = Total stories in the building.

Alternate Procedure: Along-Wind Response. Through a simple static analysis, the peak building response in the along-wind direction can be obtained by

$$\hat{r} = \overline{r} + \sqrt{\hat{r}_B^2 + \hat{r}_R^2} \quad \text{(C26.11-22)}$$

where $\overline{r}$, $\hat{r}_B$, and $\hat{r}_R$ are the mean, peak background, and resonant response components of interest, for example, shear forces, moment, or displacement. Once the equivalent static wind load distribution is obtained, any response component, including acceleration, can be obtained using a simple static analysis. Caution is suggested when combining the loads instead of response according to the preceding expression, as for example in

$$\hat{P}_j = \overline{P}_j + \sqrt{\hat{P}_{Bj}^2 + \hat{P}_{Rj}^2} \quad \text{(C26.11-23)}$$

because the background and the resonant load components have normally different distributions along the building height. Additional background can be found in Zhou and Kareem (2001b); Zhou et al. (2002), and Chen and Kareem (2004).

Aerodynamic Loads on Tall Buildings: An Interactive Database. Under the action of wind, tall buildings oscillate simultaneously in the along-wind, across-wind, and torsional directions. While the along-wind loads have been successfully treated in terms of gust loading factors based on quasi-steady and strip theories, the across-wind and torsional loads cannot be treated in this manner because these loads cannot be related in a straightforward manner to fluctuations in the approach flow. As a result, most current codes and standards provide little guidance for the across-wind and torsional response (Zhou et al. 2002, Chen and Kareem 2004, Kwon and Kareem 2013, Bernardini et al. 2013a).

To provide some guidance at the preliminary design stages of buildings, an interactive aerodynamic loads database for assessing dynamic wind-induced loads on a suite of generic isolated buildings is introduced (Zhou et al. 2002, Kwon et al. 2008). Although the analysis based on this experimental database is not intended to replace wind tunnel testing in the final design stages, it provides users a methodology to approximate the previously untreated across-wind and torsional responses in the early design stages. The database consists of high-frequency base balance measurements involving seven rectangular building models, with side ratio (D/B, where D is the depth of the building section along the oncoming wind direction, and B is the width of the building) from 1/3 to 3, and three aspect ratios for each building model in two approach flows, namely, BL_1 ($\overline{\alpha} = 0.16$) and BL_2 ($\overline{\alpha} = 0.35$), corresponding to an open and an urban environment, respectively. The data are accessible with a user-friendly, Java-based internet applet, the NatHaz Aerodynamic Loads Database, version 2.0 (NALD 2012). Through the use of this interactive portal, users can select the geometry and dimensions of a model building from the available choices and specify an urban or suburban condition. The aerodynamic load spectra for the along-wind, across-wind, or torsional directions are then displayed with a Java interface permitting users to specify a reduced frequency [(building frequency × building dimension)/wind velocity] of interest and automatically obtain the corresponding spectral value. When coupled with the supporting web documentation, examples, and concise analysis procedure, the database provides a comprehensive tool for computation of wind-induced response of tall buildings, suitable as a design guide in the preliminary stages.

Example: An example tall building is used to demonstrate the analysis using the database. The building is a square steel tall building with size $H \times W_1 \times W_2 = 656 \times 131 \times 131$ ft ($200 \times 40 \times 40$ m) and an average radius of gyration of 59 ft (18 m).

The three fundamental mode frequencies, f_1, are 0.2, 0.2, and 0.35 Hz in the X-, Y-, and Z-directions, respectively; the mode shapes are all linear, or $\beta = 1.0$, and there is no modal coupling. The building density is 0.485 slugs/ft^3 (250 kg/m^3). This building is located in Exposure A or close to the BL_2 test condition of the internet-based database (Zhou et al. 2002, Kwon et al. 2008). In this location, the 3 s design gust speed at a 700-year recurrence interval is 115 mi/h (51 m/s) in ASCE 7-16. For serviceability requirements, 3 s design gust speed with 10-year MRI is 76 mi/h (34 m/s) in ASCE 7-16. For the sake of illustration only, the first-mode critical structural damping ratio, ζ_1, is to be 0.01 for both survivability and serviceability design.

Using these aerodynamic data and the procedures provided by Zhou et al. (2002) and Kwon et al. (2008), the wind load effects are evaluated and the results are presented in Table C26.11-3. This table includes base moments and acceleration response in the along-wind direction obtained by the procedure in ASCE 7-16. It should be pointed out that the building experiences higher across-wind load effects when compared to the along-wind response for this example, which reiterates the significance of wind loads and their effects in the across-wind direction and the need for such database-enabled design approaches.

C26.12 ENCLOSURE CLASSIFICATION

The magnitude and sense of internal pressure depend on the magnitude and location of openings around the building envelope with respect to a given wind direction. Accordingly, the standard requires that a determination be made of the amount of openings in the envelope to assess enclosure classification (enclosed, partially enclosed, partially open, or open). *Openings* are defined as holes that allow air to flow through the building envelope during a design wind event. Holes in the building envelope, including air intake and exhaust vents for ventilation systems, should always be considered as openings. Whether doors, operable windows and skylights, and flexible and operable louvers are considered as openings depends on their intended use during a storm event. For example, a door of a fire station or hospital ambulance bay should be considered an opening if the intended function for such essential facilities requires the door to remain in an open position during a design wind event. However, if doors, windows, and skylights are likely be in a closed position during a design wind event, they do not need to be considered as openings. The porosity of an "enclosed building" is such that there are not sufficient openings in the exterior building envelope to allow significant airflow into the building. The porosity of a "partially enclosed" building is such that there are sufficient openings in the building envelope windward wall to allow wind to enter the building; however, there are not sufficient openings in the remaining portions of the building envelope to allow airflow out of the building without a buildup of internal pressure. The porosity of a "partially open" building is such that there exist sufficient openings in the building envelope windward wall to allow airflow into the building, and sufficient openings exist in the remaining portions of the building envelope to allow some airflow out of the building, but with some buildup of internal pressure. The porosity for the "open building" is such that air flow can enter and exit the building without a significant buildup of internal pressure. The classification of a "partially open" building has been added to the standard to help the user in the understanding that a building with openings and significant porosity (such as an open parking garage, for example) that does not meet the requirements of the "partially enclosed" classification does not automatically classify the building as "open" or "enclosed." Once the enclosure classification is known, the designer enters Table 26.13-1 to select the appropriate internal pressure coefficient.

Wind-Borne Debris. This version of the standard has five terms applicable to enclosure: wind-borne debris regions, glazing, impact-resistant glazing, impact-protective system, and unprotected glazing. *Wind-borne debris regions* are specified to alert the designer to areas requiring consideration of test-missile-impact design and potential openings in the building envelope. *Glazing* is defined as any glass or transparent or translucent plastic sheet used in windows, doors, skylights, or curtain walls. *Impact-resistant glazing* is specifically defined as glazing that has been shown by testing to withstand the impact of test missiles. *Impact-protective systems* over glazing can be shutters or screens designed to withstand wind-borne debris impact. *Unprotected glazing* is glazing that is not impact-resistant and does not have an impact-protective system. Impact resistance of glazing and protective systems can be tested using the test method specified in ASTM E1886, with test missiles, impact speeds, and pass/fail criteria specified in ASTM E1996. Glazing in sectional doors, rolling doors, and flexible doors can be tested for impact resistance with test missiles, impact speeds, and pass-fail criteria specified in ANSI/DASMA 115. Sectional doors, rolling doors, and flexible doors are commonly used for vehicle access and material transportation in warehouses, retail buildings, and other structures requiring garage-type entry. A sectional door, sometimes known as a garage door, is a door made of two or more horizontal sections, spanning between supports at the sides of the opening, hinged together so as to provide a door capable of closing the entire opening, and which is raised and lowered by means of the track rollers, in tracks which are in turn supported by the structure. A rolling door is a vertically operating, coiling door typically used in commercial or industrial applications. A flexible door is a door in which a flexible fabric or other flexible sheet material forms the panel portion, even though it may have a rigid frame, rigid reinforcements, rigid support means for one or more edges thereof, or combinations of these features. Other methods involving opening protection of building envelope systems are acceptable when approved by the Authority Having Jurisdiction. The origins of the test-missile-impact provisions contained in these standards are summarized by Minor (1994) and Twisdale et al. (1996).

Section 26.12.3 requires glazing in Risk Category II, III, and IV buildings in wind-borne debris regions to be protected with an impact-protective system or to be made of impact-resistant glazing to reduce the amount of interior wind and water damage to buildings during design windstorm events (Surry et al. 1977, Reinhold 1982, Stubbs and Perry 1993). An exception in Section 26.12.3.1 permits glazing to be unprotected when it is located more than 60 ft (18.3 m) above grade and more than 30 ft (9.1 m) above aggregate-surfaced roofs located within 1,500 ft (457 m) of the building. Unprotected glazing permitted by the Section 26.12.3.1 exception is not considered an opening. Unprotected glazing higher than 60 ft (18.3 m) above grade may be broken by wind-borne debris when a debris source is present. At these higher elevations, loose roof aggregate has been the predominant source of debris damage in previous wind events. This includes gravel or stone used as ballast that is not protected by a sufficiently high parapet. Accordingly, if an existing aggregate-surfaced roof is within 1,500 ft (457 m) of a new building, glazing in the new building, from 30 ft (9.1 m) above the source building to grade, needs to be protected with an impact-protective system or be made of impact-resistant glazing. If loose roof aggregate is proposed for the new building, it too should be considered as a debris source, because aggregate can be blown off the roof and be propelled into glazing on the leeward side of the building. The requirement for protection 30 ft (9.1 m) above the roof aggregate debris source is to account for debris that can be lifted during flight. The following references provide further information regarding debris damage to glazing: Beason et al. (1984), Minor (1985, 1994), Kareem (1986), and Behr and Minor (1994), and Minor and Behr (1993). See C26.13 for discussion regarding breakage of unprotected glazing higher than 60 ft (18.3 m) above grade by wind-borne debris other than roof aggregate.

Although wind-borne debris can occur in just about any condition, the level of risk in comparison to the postulated debris regions and test-missile-impact criteria may also be lower than that determined for the purpose of standardization. For example, individual buildings may be sited away from likely debris sources that would generate significant risk of impacts similar in magnitude to pea gravel (i.e., small roof aggregate as simulated by 2 g steel balls in impact tests) or butt-on 2×4 impacts, as required in impact-testing criteria. This situation describes a condition of low vulnerability only as a result of limited debris sources in the vicinity of the building. In other cases, potential sources of debris may be present, but extenuating conditions can lower the risk. These extenuating conditions include the type of materials and surrounding construction, the level of protection offered by surrounding exposure conditions, and the basic wind speed. Therefore, the risk of impact may differ from those postulated as a result of the conditions specifically enumerated in the standard and the referenced impact standards. There are vastly differing opinions regarding the significance of these parameters that are not fully considered in developing standardized debris regions or referenced test-missile-impact criteria.

The definition of the wind-borne debris regions for Risk Category II buildings and structures was chosen such that the coastal areas included in the wind-borne debris regions are approximately consistent with those given in ASCE 7-05 and prior editions. Thus, the wind speed contours that define the wind-borne debris regions in Section 26.12.3.1 are not direct conversions of the wind speed contours that are defined in ASCE 7-05, as shown in Table C26.5-7.

While the coastal areas included in the wind-borne debris regions for Risk Category II are approximately consistent with those given in ASCE 7-05, significant reductions in the area of wind-borne debris regions for this Risk Category occur around Jacksonville, Fla., in the eastern part of the Florida Panhandle, and inland from the coast of North Carolina.

The introduction in ASCE 7-10 of separate maps for different Risk Categories provided a means for achieving a more risk-consistent approach for defining wind-borne debris regions. The approach selected was to link the geographical definition of the wind-borne debris regions to the wind speed contours in the maps that correspond to the particular Risk Category, resulting in expansion of the wind-borne debris region for some Risk Category III and all Risk Category IV buildings and structures. A review of the types of buildings and structures currently included in Risk Category III suggests that in the expanded wind-borne debris region, life-safety issues would be most important for health care facilities. Consequently, the committee chose to apply the expanded wind-borne debris protection requirement to these types of Risk Category III facilities and not to all Risk Category III buildings and structures.

C26.13 INTERNAL PRESSURE COEFFICIENTS

The internal pressure coefficient values in Table 26.13-1 were obtained from wind tunnel tests (Stathopoulos et al. 1979) and full-scale data (Yeatts and Mehta 1993). Even though the wind tunnel tests were conducted primarily for low-rise buildings, the internal pressure coefficient values are assumed to be valid for buildings of any height. The values $(GC_{pi}) = +0.18$ and -0.18 are for enclosed buildings. It is assumed that the building has no dominant opening or openings and that the small leakage paths that do exist are essentially uniformly distributed over the building's envelope. The internal pressure coefficient values for partially enclosed buildings assume that the building has a dominant opening or openings. For such a building, the internal pressure is dictated by the exterior pressure at the opening and is typically increased substantially as a result. Net loads (i.e., the combination of the internal and exterior pressures) are therefore also significantly increased on the building surfaces that do not contain the opening. Therefore, higher (GC_{pi}) values of $+0.55$ and -0.55 are applicable to this case. These values include a reduction factor to account for the lack of perfect correlation between the internal pressure and the external pressures on the building surfaces not containing the opening (Irwin 1987, Beste and Cermak 1996). Taken in isolation, the internal pressure coefficients can reach values of ± 0.8 (or possibly even higher on the negative side).

For partially enclosed buildings containing a large unpartitioned space, the response time of the internal pressure is increased, and this increase reduces the ability of the internal pressure to respond to rapid changes in pressure at an opening. The gust factor applicable to the internal pressure is therefore reduced. Equation (26.13-1), which is based on Vickery and Bloxham (1992) and Irwin and Dunn (1994), is provided as a means of adjusting the gust factor for this effect on buildings or other structures with large internal spaces, such as stadiums and arenas.

Breached Glazing. Unprotected glazing may be breached by wind-borne debris during any type of windstorm. However, the vast majority of damage has occurred during hurricanes and tornadoes. See C26.12 for discussion of the requirements in Section 26.12.3 for protection of glazing openings. The following discussion provides guidance on determining the internal pressure coefficient with respect to glazing breaches caused by wind-borne debris.

Outside, and within, wind-borne debris regions, it is common to design new buildings for enclosed conditions. Where unprotected glazing is permitted in wind-borne debris regions [i.e., greater than 60 ft (18.3 m) above grade and not susceptible to roof aggregate per Section 26.12.3.1], the glazing can be considered as not being an opening. Numerous wind-damage investigations have shown that extensive breakage of unprotected glazing higher than 60 ft (18.3 m) above grade by wind-borne debris other than roof aggregate is very unlikely. It is possible for a few, or several, panes of permitted unprotected glazing to be breached by debris. The breaching may result in development of increased internal pressure in the area (compartment) behind the breached glazing. The increased pressure may overload non-structural building envelope elements, as illustrated in Figures C26.13-1 and C26.13-2.

At the building shown in Figure C26.13-1, some glazing was protected by shutters (blue dashed line), but most of the glazing was unprotected. Some of the glazing was breached by debris (which was not roof aggregate), and some glazing assemblies collapsed (red dashed line) because they were inadequately attached to the wall framing. See Figure C26.13-2 for a view of the sidewall in the area shown by the red and blue lines (the exterior of the sidewall experienced suction pressures).

The wall shown by the red line is at the floor with the collapsed glazing assembly. The wall was composed of stucco/gypsum board, metal framing, and gypsum board. In some areas, the entire wall blew away. At the area shown by the yellow line, the

Figure C26.13-1. Glazing breaches and non-load-bearing wall failures (Hurricane Ivan).
Source: FEMA (2004).

Figure C26.13-2. View of sidewall damage.
Source: FEMA.

stud track detached from the floor slab and deformed outward. The wall shown by the blue line is at the floor with the protected glazing. At this wall area, the metal framing and most of the interior and exterior gypsum board was still in place. The difference between the amount of damage between the two wall areas was judged to be due to development of increased internal pressure caused by collapse of the glazing assembly. The wall was also judged to have inadequate resistance to wind loads derived from ASCE 7 for an enclosed building.

Although the probability is low, buildings that are designed in accordance with ASCE 7, using the internal pressure coefficients for an enclosed building, could experience breaching of permitted unprotected glazing that results in development of increased internal pressure. However, if properly constructed and maintained, damage to other portions of the building envelope or the MWFRS would not be expected during a design event because of the additional strength provided by safety and load factors.

Regarding distribution of breached glazing, the dominant opening(s) will initially be on a windward façade. However, during hurricanes, the angle of attack may change and result in breached glazing in more than one façade.

The ASCE/SEI *Prestandard for Performance-Based Wind Design* (2019) provides glazing guidance for designers retained by building owners that desire to achieve higher wind performance.

Compartmentalization. The influence of compartmentalization on the distribution of increased internal pressure has not been researched. If the space behind large breaches of the building envelope (such as broken glazing or collapsed doors or walls) is separated from the remainder of the building by a sufficiently strong and reasonably airtight compartment, the increased internal pressure would likely be confined to that compartment. However, if the compartment is breached (e.g., by an open corridor door or by collapse of the compartment wall), the increased internal pressure will spread beyond the initial compartment quite rapidly. The next compartment may contain the higher pressure, or it too could be breached, allowing the high internal pressure to continue to propagate. For buildings with more than one floor, increased internal pressure may spread to other floors.

Aircraft Hangers. Because of the great amount of air leakage that often occurs at large hangar doors, designers of hangars should consider using the internal pressure coefficients in Table 26.13-1 for partially enclosed buildings.

Lobby Entry Vestibules. Vestibules can be exposed to increased internal pressure due to operation or breach of exterior entry doors under high winds, subjecting interior vestibule doors, walls, and ceilings to external pressures. For those Risk Category III and IV buildings where operation of exterior entry doors either during or after a high wind event is critical to building function, the interior vestibule envelope could be designed for the same pressures as the exterior wall and doors. For other buildings, the interior vestibule envelope could be reasonably designed to withstand external pressures at wind speeds expected during normal operation of exterior entry doors, which would otherwise be closed in high wind events.

Remedial Work on Existing Buildings Located in a Wind-Borne Debris Region. If existing glazing does not comply with the requirements of Section 26.12.3, then the existing glazing should be replaced with new impact-resistant glazing or retrofitted with impact-protective systems in accordance with Section 26.12.3. If the existing noncompliant glazing is to remain, then it should be assumed that it can be breached by wind-borne debris, and load calculations for new building-envelope elements should include a confirmation of building enclosure (i.e., enclosed versus partially enclosed) to determine the correct internal pressure coefficient. If replacing a roof system, uplift resistance of the roof deck and deck support structure should be considered.

REFERENCES

Almeida, G. P., D. F. G. Durao, and M. V. Heitor. 1993. "Wake flows behind two-dimensional model hills." *Exp. Therm. Fluid Sci.* 7(1): 87–101.

ANSI/DASMA (Door and Access Systems Manufacturers Association). 2017. *Standard method for testing sectional garage doors: determination of structural performance under missile impact and cyclic wind pressure*. ANSI/DASMA 115-17. Cleveland, OH: DASMA.

ARA (Applied Research Associates). 2001. *Hazard mitigation study for the Hawaii Hurricane Relief Fund*. Rep. No. 0476. Raleigh, NC: ARA.

ARA. 2019. *Development of wind speed-ups and hurricane hazard maps for Puerto Rico.* Final Draft Rep. Raleigh, NC: ARA.

ASCE/SEI. 2019. *Prestandard for performance-based wind design*. Reston, VA: ASCE.

ASME. 1992. *Steel stacks*. STS-1. New York: ASME.

ASTM International. 2014. *Standard test method for structural performance of exterior windows, doors, skylights and curtain walls by uniform static air pressure difference*. ASTM E330/E330M-14. West Conshohocken, PA: ASTM

ASTM. 2017. *Standard specification for performance of exterior windows, curtain walls, doors, and impact protective systems impacted by windborne debris in hurricanes*. ASTM E1996-17. West Conshohocken, PA: ASTM.

ASTM. 2019. *Standard test method for performance of exterior windows, curtain walls, doors, and impact protective systems impacted by missile(s) and exposed to cyclic pressure differentials*. ASTM E1886-19. West Conshohocken, PA: ASTM.

Banks, D., N. Ellison, and W. Esterday. 2019. *Kern County special wind region study*. Final Rep. CPP Project 9413 Rev02.

Beason, W. L., G. E. Meyers, and R. W. James. 1984. "Hurricane related window glass damage in Houston." *J. Struct. Eng*. 110 (12): 2843–2857. https://doi.org/10.1061/(ASCE)0733-9445(1984)110:12(2843).

Behr, R. A., and J. E. Minor. 1994. "A survey of glazing system behavior in multi-story buildings during Hurricane Andrew." *Struct. Des. Tall Build.* 3 (3): 143–161. https://doi.org/10.1002/tal.4320030302.

Bernardini, E., S. M. J. Spence, and A. Kareem. 2013a. "A probabilistic approach for the full response estimation of t all buildings with 3D modes using the HFFB." *Struct. Saf.* 44 (Sep): 91–101. https://doi.org/10.1016/j.strusafe.2013.06.002.

Bernardini, E., S. M. J. Spence, and A. Kareem. 2013b. "An efficient performance-based design approach to the high frequency force balance." In *Proc., 11th Int. Conf. on Structural Safety and Reliability*, ICOSSAR 2013, edited by G. Deodatis, B. Ellingwood, and D. Frangopol, 1777–1784. London: Taylor & Francis.

Beste, F., and J. E. Cermak. 1996. "Correlation of internal and area-averaged wind pressures on low-rise buildings." In *Proc., 3rd Int. Colloquium on Bluff Body Aerodynamics and Applications*. Blacksburg, VA: Virginia Polytechnic Institute.

Chen, X., and A. Kareem. 2004. "Equivalent static wind loads on buildings: New model." *J. Struct. Eng*. 130 (10): 1425–1435. https://doi.org/10.1061/(ASCE)0733-9445(2004)130:10(1425).

Chock, G., L. Peterka, and G. Yu. 2005. "Topographic wind speed-up and directionality factors for use in the city and county of Honolulu building code." In *Proc., 10th Americas Conf. on Wind Engineering*. Baton Rouge, LA: Louisiana State University.

CICIND (Comité International des Cheminées Industrielles). 1999. *Model code for steel chimneys, Revision 1-1999.* Zurich, Switzerland: CICIND.

Cleveland, W. S., and S. J. Devlin. 1988. "Locally weighted regression: An approach to regression analysis by local fitting." *J. Am. Stat. Assoc.* 83 (403): 596–610. https://doi.org/10.2307/2289282.

CSA (Canadian Standard Association). 2014. *Standard test method for the dynamic wind uplift resistance of membrane-roofing systems*. CSA A123.21-14. CSA.

Davenport, A. G. 1960. "Rationale for determining design wind velocities." *J. Struct. Div.* 86 (5): 39–68.

Davenport, A. G., C. S. B. Grimmond, T. R. Oke, and J. Wieringa. 2000. "Estimating the roughness of cities and sheltered country." In *Proc., 12th AMS Conf. Applied Climatology*, 96–99. Boston: American Meteorological Society.

Donelan, M. A., B. K. Haus, N. Reul, W. J. Plant, M. Stiassnie, H. C. Graber, O. B. Brown, and E. S. Saltzman. 2004. "On the limiting aerodynamic roughness of the ocean in very strong winds." *Geophys. Res. Lett.* 31 (18): 1–5. https://doi.org/10.1029/2004GL019460.

Durst, C. S. 1960. "Wind speeds over short periods of time." *Meteorol. Mag.* 89: 181–187.

ECCS (European Convention for Structural Steelwork). 1978. *Recommendations for the calculation of wind effects on buildings and structures.* Technical Committee T12. Brussels, Belgium: ECCS.

Ellingwood, B. 1981. "Wind and snow load statistics for probabilistic design." *J. Struct. Div.* 107 (7): 1345–1350. https://doi.org/10.1061/JSDEAG.0006152.

Ellingwood, B., J. G. MacGregor, T. V. Galambos, and C. A. Cornell. 1982. "Probability based load criteria: Load factors and load combinations." *J. Struct. Div.* 108 (5): 978–997. https://doi.org/10.1061/JSDEAG.0005959.

ESDU (Engineering Sciences Data Unit). 1982. *Strong winds in the atmospheric boundary layer, Part 1: Mean hourly wind speed.* ESDU 82026. London: ESDU.

ESDU. 1993. *Strong winds in the atmospheric boundary layer, Part 2: Discrete gust speeds.* ESDU 83045. London: ESDU.

FEMA (Federal Emergency Management Agency). 2021a. *Safe rooms for tornadoes and hurricanes: Guidance for community and residential safe rooms.* P-361, 4th ed. Washington, DC: FEMA. https://www.fema.gov/emergency-managers/risk-management/safe-rooms.

FEMA. 2021b. *Taking shelter from the storm: Building or installing a safe room for your home*. P-320, 5th ed. Washington, DC: FEMA. https://www.fema.gov/emergency-managers/risk-management/safe-rooms.

Goel, R. K., and A. K. Chopra. 1997. "Period formulas for moment-resisting frame buildings." *J. Struct. Eng.* 123 (11): 1454–1461. https://doi.org/10.1061/(ASCE)0733-9445(1997)123:11(1454).

Goel, R. K., and A. K. Chopra. 1998. "Period formulas for concrete shear wall buildings." *J. Struct. Eng.* 124 (4): 426–433. https://doi.org/10.1061/(ASCE)0733-9445(1998)124:4(426).

Gurley, K., and A. Kareem. 1993. "Gust loading factors for tension leg platforms." *Appl. Ocean Res.* 15 (3): 137–154. https://doi.org/10.1016/0141-1187(93)90037-X.

Haan, F., P. Sarkar, and W. Gallus. 2008. Design, construction and performance of a large tornado simulator for wind engineering applications. *Engineering Structures*. 30. 1146–1159. https://doi.org/10.1016/j.engstruct.2007.07.010.

Harris, R. I., and D. M. Deaves. 1981. "The structure of strong winds." In *Proc., CIRIA Conf. Wind Engineering in the*

Eighties. London: Construction Industry Research and Information Association.
Ho, E. 1992. "Variability of low building wind lands." Ph.D. thesis, Dept. of Civil Engineering, University of Western Ontario.
ICC (International Code Council). 2020. *ICC/NSSA standard for the design and construction of storm shelters*. ICC 500-2020. Washington, DC: ICC.
Irwin, P. A. 2006. "Exposure categories and transitions for design wind loads." *J. Struct. Eng*. 132 (11): 1755–1763. https://doi.org/10.1061/(ASCE)0733-9445(2006)132:11(1755).
Irwin, P. A., and G. E. Dunn. 1994. *Review of internal pressures on low-rise buildings*. RWDI Rep. No. 93-270. Canadian Sheet Building Institute.
Irwin, P. A. 1987. "Pressure model techniques for cladding loads." *J. Wind Eng. Ind. Aerodyn*. 29 (1–3): 69–78. https://doi.org/10.1016/0167-6105(88)90146-8.
ISO. 1997. *Wind actions on structures*. ISO 4354. Geneva: ISO.
Isyumov, N., E. Ho, and P. Case. 2013. "Influence of wind directionality on wind loads and responses." In *Proc., 12th Americas Conf. Wind Engineering*, edited by D. Reed and A. Jain.
Jackson, P. S., and J. C. R. Hunt. 1975. "Turbulent wind flow over a low hill." *Q. J. R. Meteorol. Soc*. 101 (430): 929–955. https://doi.org/10.1002/qj.49710143015.
Jung, S., and F. J. Masters. 2013. "Characterization of open and suburban boundary layer wind turbulence in 2008 Hurricane Ike." *Wind Struct.* 17 (430): 135–162. https://doi.org/10.1002/qj.49710143015.
Kareem, A. 1985. "Lateral-torsional motion of tall buildings." *J. Struct. Eng*. 111 (11): 2479–2496. https://doi.org/10.1061/(ASCE)0733-9445(1985)111:11(2479).
Kareem, A. 1986. "Performance of cladding in Hurricane Alicia." *J. Struct. Eng*. 112 (12): 2679–2693. https://doi.org/10.1061/(ASCE)0733-9445(1986)112:12(2679).
Kareem, A. 1992. "Dynamic response of high-rise buildings to stochastic wind loads." *J. Wind Eng. Indust. Aerodyn*. 42 (1–3): 1101–1112. https://doi.org/10.1016/0167-6105(92)90117-S.
Kareem, A., D. K. Kwon, and Y. Tamura. 2012. "Cyberbased analysis, modeling and simulation of wind load effects in VORTEX-winds." In *Proc., 3rd American Association for Wind Engineering Workshop*. Hyannis, MA.
Kelly, M., R. A. Cersosimo, and J. Berg. 2019. "A universal wind profile for the inversion-capped neutral atmospheric boundary layer." *Q. J. R. Meteorol. Soc*. 145 (720): 982–992. https://doi.org/10.1002/qj.3472.
Kern County. 2017. "Title 17 buildings and construction, Chapter 17.08 Building Code.17.08.420, Section 1609.3 amended, ultimate design wind speed." https://library.municode.com/ca/kern_county/codes/code_of_ordinances?nodeId=TIT17BUCO_CH17.08BUCO.
Kijewski-Correa, T., A. Kareem, Y. L. Guo, R. Bashor, and T. Weigand. 2013. "Performance of tall buildings in urban zones: Lessons learned from a decade of full-scale monitoring." *Int. J. High-Rise Build.* 2 (3): 179–192. https://doi.org/10.21022/IJHRB.2013.2.3.179.
Kwon, D.-K., and A. Kareem. 2013. "Comparative study of major international wind codes and standards for wind effects on tall buildings." *Eng. Struct.* 51 (Jun): 23–25. https://doi.org/10.1016/j.engstruct.2013.01.008.
Kwon, D.-K., T. Kijewski-Correa, and A. Kareem. 2008. "e-Analysis of high-rise buildings subjected to wind loads." *J. Struct. Eng*. 133 (7): 1139–1153. https://doi.org/10.1061/(ASCE)0733-9445(2008)134:7(1139).
Lemelin, D. R., D. Surry, and A. G. Davenport. 1988. "Simple approximations for wind speed-up over hills." *J. Wind Eng. Ind. Aerodyn*. 28 (1–3): 117–127. https://doi.org/10.1016/0167-6105(88)90108-0.
Lettau, H. 1969. "Note on aerodynamic roughness element description." *J. Appl. Meteorol*. 8: 828–832.
Lombardo, F., A. Pintar, P. J. Vickery, E. Simiu, and M. Levitan. 2016. *Development of new wind speed maps for ASCE 7-16*. NIST Special Publication.
Means, B., T. A. Reinhold, and D. C. Perry. 1996. "Wind loads for low-rise buildings on escarpments." In *Building an international community of structural engineers*, edited by S. K. Ghosh and J. Mohammadi, 1045–1052. Reston, VA: ASCE.
Mehta, K. C., and R. D. Marshall. 1998. *Guide to the use of the wind load provisions of ASCE 7-95*. Reston, VA: ASCE.
Minor, J. E. 1985. "Window glass performance and hurricane effects." In *Hurricane Alicia: One year later*, edited by A. Kareem, 151–167. Reston, VA: ASCE.
Minor, J. E. 1994. "Windborne debris and the building envelope." *J. Wind Eng. Ind. Aerodyn*. 53 (1–2): 207–227. https://doi.org/10.1016/0167-6105(94)90027-2.
Minor, J. E., and R. A. Behr. 1993. "Improving the performance of architectural glazing in hurricanes." In *Hurricanes of 1992: Lessons learned and implications for the future*, 476–485. Reston, VA: ASCE.
NALD (NatHaz Aerodynamic Loads Database). 2012. "NatHaz aerodynamic loads database, version 2." http://aerodata.ce.nd.edu/.
NHC (National Hurricane Center). 2015. "Saffir-Simpson hurricane wind scale." http://www.nhc.noaa.gov/aboutsshws.php.
Palutikof, J. P., B. B. Brabson, D. H. Lister, and S. T. Adcock. 1999. "A review of methods to calculate extreme wind speeds." *Meteorol. Appl*. 6 (2): 119–132. https://doi.org/10.1017/S1350482799001103.
Panofsky, H. A., and J.A. Dutton. 1984. *Atmospheric turbulence: models and methods for engineering applications*. 1st ed. New York: Wiley-Interscience.
Peterka, J. A. 2006. "Colorado front range gust map: CPP report for use by Colorado Structural Engineers Association and Colorado Front Range communities." https://bit.ly/2006ColoradoFrontRangeGustMap.
Peterka, J. A. 2010. "Colorado front range gust map – ASCE 7-10 compatible." https://seacolorado.org/docs/FINAL-COLORADO-FRONT-RANGE-GUST-MAP-2013.pdf.
Peterka, J. A., and S. Shahid. 1998. "Design gust wind speeds in the United States." *J. Struct. Eng*. 124 (2): 207–214. https://doi.org/10.1061/(ASCE)0733-9445(1998)124:2(207).
Pickands, J., III. 1971. "The two-dimensional poisson process and extremal processes." *J. Appl. Prob*. 8 (4): 745–56. https://doi.org/10.2307/3212238.
Pintar, A., E. Simiu, F. Lombardo, and M. Levitan. 2015. *Maps of nonhurricane non-tornadic wind speeds with specified mean recurrence intervals for the contiguous united states using a two-dimensional poisson process extreme value model and local regression*. Special Publication 500-301. Gaithersburg, MD: National Institute of Standards and Technology.
Powell, M. D. 1980. "Evaluations of diagnostic marine boundary-layer models applied to hurricanes." *Monthly Weather Rev*. 108 (6): 757–766. https://doi.org/10.1175/1520-0493(1980)108<0757:EODMBL>2.0.CO;2.
Powell, M. D., P. J. Vickery, and T. A. Reinhold. 2003. "Reduced drag coefficients for high wind speeds in tropical

cyclones." *Nature* 422 (6929): 279–283. https://doi.org/10.1038/nature01481.

Reinhold, T. A., ed. 1982. "Wind tunnel modeling for civil engineering applications." In *Proc., Int. Workshop on Wind Tunnel Modeling Criteria and Techniques in Civil Engineering Applications*. Gaithersburg, MD: Cambridge University Press.

Sataka, N., K. Suda, T. Arakawa, A. Sasaki, and Y. Tamura. 2003. "Damping evaluation using full-scale data of buildings in Japan." *J. Struct. Eng.* 129 (4): 470–477. https://doi.org/10.1061/(ASCE)0733-9445(2003)129:4(470).

SEAOC (Structural Engineers Association of California). 2017. "Recommendations for action to address design wind speeds in California." https://www.seaoc.org/store/ListProducts.aspx?catid=597243.

Searer, et al. 2010. *SEAOSC summary report: Study of historical and design wind speeds in the Los Angeles area*. Los Angeles: Structural Engineers Association of Southern California.

Simiu, E. 2011. *Design of buildings for wind: A guide for ASCE 7-10 Standard users and designers of special structures.* 2nd ed. New York: Wiley.

Simiu, E., and R. H. Scanlan. 1996. *Wind effects on structures*. 3rd ed. New York: Wiley.

Simiu, E., L. Shi, and D. Yeo. 2016. "Planetary boundary-layer modelling and tall building design." *Boundary-Layer Meteorol.* 159 (1): 173–181. https://doi.org/10.1007/s10546-015-0106-9.

Simiu, E., P. Vickery, and A. Kareem. 2007. "Relation between Saffir–Simpson hurricane scale wind speeds and peak 3-s gust speeds over open terrain." *J. Struct. Eng.* 133 (7): 1043–1045. https://doi.org/10.1061/(ASCE)0733-9445(2007)133:7(1043).

Simiu, E., and D. Yeo. 2019. *Wind effects on structures: Modern structural design for wind*. 4th ed. Oxford, UK: Wiley-Blackwell.

Smith, R. L. 1989. "Extreme value analysis of environmental time series: An application to trend detection in ground-level ozone." *Stat. Sci.* 4 (4): 367–377.

Solari, G. 1993a. "Gust buffeting. I: Peak wind velocity and equivalent pressure." *J. Struct. Eng.* 119 (2): 365–382. https://doi.org/10.1061/(ASCE)0733-9445(1993)119:2(365).

Solari, G. 1993b. "Gust buffeting. II: Dynamic alongwind response." *J. Struct. Eng.* 119 (2): 383–398. https://doi.org/10.1061/(ASCE)0733-9445(1993)119:2(383).

Solari, G., and A. Kareem. 1998. "On the formulation of ASCE 7-95 gust effect factor." *J. Wind Eng. Ind. Aerodyn.* 77–78 (Sep): 673–684. https://doi.org/10.1016/S0167-6105(98)00182-2.

Spence, S., and A. Kareem. 2014. "Tall buildings and damping: A concept-based data driven model." *J. Struct. Eng.* 140 (5): 04014005. https://doi.org/10.1061/(ASCE)ST.1943-541X.0000890.

Standards Australia. 1994. *Design of steel lattice towers and masts*. AS3995-1994. North Sydney: Standards Australia.

Stathopoulos, T., D. Surry, and A. G. Davenport. 1979. "Wind-induced internal pressures in low buildings." In *Proc., 5th Int. Conf. on Wind Engineering*, edited by J. E. Cermak. Fort Collins, CO: Colorado State University.

Stubbs, N., and D. C. Perry. 1993. "Engineering of the building envelope: To do or not to do." In *Hurricanes of 1992: Lessons learned and implications for the future*, edited by R. A. Cook and M. Sotani, 10–30. Reston, VA: ASCE.

Surry, D., R. B. Kitchen, and A. G. Davenport. 1977. "Design effectiveness of wind tunnel studies for buildings of intermediate height." *Can. J. Civ. Eng.* 4 (1): 96–116. https://doi.org/10.1139/l77-010.

Tse, K. T., S. W. Li, P. W. Chan, H. Y. Mok, and A. U. Weerasuriya. 2013. "Wind profile observations in tropical cyclone events using wind-profilers and doppler SODARs." *J. Wind Eng. Ind. Aerodyn.* 115 (Apr): 93–103. https://doi.org/10.1016/j.jweia.2013.01.003.

Twisdale, L. A., P. J. Vickery, and A. C. Steckley. 1996. *Analysis of hurricane windborne debris impact risk for residential structures*. Bloomington, IL: State Farm Mutual Automobile Insurance Companies.

USNRC (US Nuclear Regulatory Commission). 2011. *Technical basis for regulatory guidance on design-basis hurricane wind speeds for nuclear power plants.* NUREG/CR-7005. Washington, DC: US Nuclear Regulatory Commission.

Vickery, B. J., and C. Bloxham. 1992. "Internal pressure dynamics with a dominant opening." *J. Wind Eng. Ind. Aerodyn.* 41–44 (1–3): 193–204. https://doi.org/10.1016/0167-6105(92)90409-4.

Vickery, P. J., and P. F. Skerlj. 2005. "Hurricane gust factors revisited." *J. Struct. Eng.* 131 (5): 825–832. https://doi.org/10.1061/(ASCE)0733-9445(2005)131:5(825).

Vickery, P. J., P. F. Skerlj, A. C. Steckley, and L. A. Twisdale. 2000. "Hurricane wind field model for use in hurricane simulations." *J. Struct. Eng.* 126 (10): 1203–1221. https://doi.org/10.1061/(ASCE)0733-9445(2000)126:10(1203).

Vickery, P. J., and D. Wadhera. 2008a. *Development of design wind speed maps for the Caribbean for application with the wind load provisions of ASCE 7*. ARA Rep. No. 18108-1 Washington, DC: Pan American Health Organization.

Vickery, P. J., and D. Wadhera. 2008b. "Statistical models of the Holland pressure profile parameter and radius to maximum winds of hurricanes from flight level pressure and H* wind data." *J. Appl. Meteorol.* 47 (10): 2497–2517. https://doi.org/10.1175/2008JAMC1837.1.

Vickery, P. J., D. Wadhera, J. Galsworthy, J. A. Peterka, P. A. Irwin, and L. A. Griffis. 2010. "Ultimate wind load design gust wind speeds in the United States for use in ASCE-7." *J. Struct. Eng.* 136 (5): 613–625. https://doi.org/10.1061/(ASCE)ST.1943-541X.0000145.

Vickery, P. J., D. Wadhera, M. D. Powell, and Y. Chen. 2009a. "A hurricane boundary layer and wind field model for use in engineering applications." *J. Appl. Meteorol.* 48 (2): 381–405.

Vickery, P. J., D. Wadhera, L. A. Twisdale, Jr., and F. M. Lavelle. 2009b. "U.S. hurricane wind speed risk and uncertainty." *J. Struct. Eng.* 135 (3): 301–320. https://doi.org/10.1061/(ASCE)0733-9445(2009)135:3(301).

Walmsley, J. L., P. A. Taylor, and T. Keith. 1986. "A simple model of neutrally stratified boundary-layer flow over complex terrain with surface roughness modulations." *Boundary-Layer Meteorol.* 36 (1): 157–186. https://doi.org/10.1007/BF00117466.

Wieringa, J., A. G. Davenport, C. S. B. Grimmond, and T. R. Oke. 2001. "New revision of Davenport roughness classification." In *Proc.*, 3EACWE, 285–292. Eindhoven, Netherlands.

Wyatt, T. A. 1984. "Sensitivity of lattice towers to fatigue induced by wind gusts." *Eng. Struct.* 6: 262–267.

Yeatts, B. B., and K. C. Mehta. 1993. "Field study of internal pressures." In Vol. 2 of *Proc., 7th U.S. Natural Conf. on Wind Engineering*, 889–897.

Zhou, Y., and A. Kareem. 2001a. "Gust loading factor: New model." *J. Struct. Eng.* 127 (2): 168–175. https://doi.org/10.1061/(ASCE)0733-9445(2001)127:2(168).

Zhou, Y., and A. Kareem. 2001b. "Equivalent static lateral forces on buildings under seismic and wind effects." *J. Wind Eng.* 89 (Oct): 605–608.

Zhou, Y., T. Kijewski, and A. Kareem. 2002. "Along-wind load effects on tall buildings: Comparative study of major international codes and standards." *J. Struct. Eng.* 128 (6): 788–796. https://doi.org/10.1061/(ASCE)0733-9445(2002)128:6(788).

Zilitinkevich, S. S., and I. N. Esau. 2002. "On integral measures of the neutral barotropic planetary boundary layer." *Boundary-Layer Meteorol.* 104 (3): 371–379. https://doi.org/10.1023/a:1016540808958.

OTHER REFERENCES (NOT CITED)

ASCE. 1987. "Wind tunnel model studies of buildings and structures." MOP 67. Reston, VA: ASCE.

ASTM. 2006. *Standard specification for rigid poly(vinyl chloride) (PVC) siding*. ASTM D3679-06a. West Conshohocken, PA: ASTM.

ASTM. 2007. *Standard test method for wind resistance of sealed asphalt shingles (uplift force/uplift resistance method)*. ASTM D7158-07. West Conshohocken, PA: ASTM.

Cook, N. 1985. *The designer's guide to wind loading of building structures, Part I: Background, damage survey, wind data and structural classification*. London: Building Research Establishment and Butterworths.

ESDU. 1990. *Characteristics of atmospheric turbulence near the ground. Part II: Single point data for strong winds (neutral atmosphere)*. ESDU 85020. London: ESDU.

Georgiou, P. N. 1985. "Design wind speeds in tropical cyclone regions." Ph.D. dissertation, Dept. of Civil Engineering, University of Western Ontario.

Georgiou, P. N., A. G. Davenport, and B. J. Vickery. 1983. "Design wind speeds in regions dominated by tropical cyclones." *J. Wind Eng. Ind. Aerodyn.* 13 (1–3): 139–152. https://doi.org/10.1016/0167-6105(83)90136-8.

Holmes, J. D. 2001. *Wind loads on structures*. New York: SPON Press/Taylor & Francis.

Isyumov, N. 1982. "The aeroelastic modeling of tall buildings." In *Proc., Int. Workshop on Wind Tunnel Modeling Criteria and Techniques in Civil Engineering Applications*, edited by T. Reinhold, 373–407. New York: Cambridge University Press.

Jeary, A. P., and B. R. Ellis. 1983. "On predicting the response of tall buildings to wind excitation." *J. Wind Eng. Ind. Aerodyn.* 13 (1–3): 173–182. https://doi.org/10.1016/0167-6105(83)90139-3.

Kala, S., T. Stathopoulos, and K. Kumar. 2008. "Wind loads on rainscreen walls: Boundary-layer wind tunnel experiments." *J. Wind Eng. Ind. Aerodyn.* 96 (6–7): 1058–1073. https://doi.org/10.1016/j.jweia.2007.06.028.

Kareem, A., and C. W. Smith. 1994. "Performance of off-shore platforms in Hurricane Andrew." In *Hurricanes of 1992: Lessons learned and implications for the future*, edited by R. A. Cook and M. Soltani, 577–586. Reston, VA: ASCE.

Kijewski, T., and Kareem, A. 1998. "Dynamic wind effects: A comparative study of provisions in codes and standards with wind tunnel data." *J. Wind Struct.* 1 (1): 77–109.

Krayer, W. R., and R. D. Marshall. 1992. "Gust factors applied to hurricane winds." *Bull. Am. Meteorol. Soc.* 73 (5): 613–617. https://doi.org/10.1175/1520-0477(1992)073<0613:GFATHW>2.0.CO;2.

Kwon, D.-K., and A. Kareem. 2013. "A multiple database-enabled design module with embedded features of international codes and standards." *Int. J. High-Rise Build.* 2 (3): 257–269. https://doi.org/10.21022/IJHRB.2013.2.3.257.

Kwon, K. K., S. M. J. Spence, and A. Kareem. 2014. "A cyberbased data-enabled design framework for high-rise buildings driven by synchronously measured surface pressures." *Adv. Eng. Software* 77 (Nov): 13–27. https://doi.org/10.1016/j.advengsoft.2014.07.001.

Liu, H. 1999. *Wind engineering: A handbook for structural engineers*. New York: Prentice Hall.

MacDonald, P. A., K. C. S. Kwok, and J. H. Holmes. 1986. *Wind loads on isolated circular storage bins, silos and tanks: Point pressure measurements*. Research Rep. No. R529. Sydney, Australia: University of Sydney.

Myer, M. F., and G. F. White. 2003. "Communicating damage potentials and minimizing hurricane damage." In *Hurricane! Coping with disaster: Progress and challenges since Galveston, 1900*, edited by R. Simpson. Washington, DC: American Geophysical Union.

Perry, D. C., N. Stubbs, and C. W. Graham. 1993. "Responsibility of architectural and engineering communities in reducing risks to life, property and economic loss from hurricanes." In *Hurricanes of 1992: Lessons learned and implications for the future*. Reston, VA: ASCE.

Standards Australia. 1989. *Australian standard SAA loading code, Part 2: Wind loads*. North Sydney, Australia: Standards House.

Stathopoulos, T., D. Surry, and A. G. Davenport. 1980. "A simplified model of wind pressure coefficients for low-rise buildings." In *Proc., 4th Colloquium on Industrial Aerodynamics*, 17–31.

Stubbs, N., and A. C. Boissonnade. 1993. "A damage simulation model for building contents in a hurricane environment." In Vol. 2 of *Proc., 7th US National Conf. on Wind Engineering*, 759–771.

Vickery, B. J., A. G. Davenport, and D. Surry. 1984. "Internal pressures on low-rise buildings." In *Proc., 4th Canadian Workshop on Wind Engineering*. Ottawa: Canadian Wind Engineering Association.

Vickery, P. J., and P. F. Skerlj. 1998. "On the elimination of exposure D along the hurricane coastline in ASCE-7." In *Report for Andersen corporation by applied research associates, project 4667*. Raleigh, NC: Applied Research Associates.

Vickery, P. J., P. F. Skerlj, and L. A. Twisdale. 2000. "Simulation of hurricane risk in the U.S. using empirical track model." *J. Struct. Eng.* 126 (10): 1222–1237. https://doi.org/10.1061/(ASCE)0733-9445(2000)126:10(1222).

Vickery, P. J., and L. A. Twisdale. 1995a. "Prediction of hurricane wind speeds in the United States." *J. Struct. Eng.* 121 (11): 1691–1699. https://doi.org/10.1061/(ASCE)0733-9445(1995)121:11(1691).

Vickery, P. J., and L. A. Twisdale. 1995b. "Wind-field and filling models for hurricane wind-speed predictions." *J. Struct. Eng.* 121 (11): 1700–1709. https://doi.org/10.1061/(ASCE)0733-9445(1995)121:11(1700).

Womble, J. A., B. B. Yeatts, and K. C. Mehta. 1995. "Internal wind pressures in a full and small scale building." In *Proc., 9th Int. Conf. on Wind Engineering*. New Delhi, India: Wiley Eastern.

Zhou, Y., and A. Kareem. 2003. "Aeroelastic balance." *J. Eng. Mech.* 129 (3): 283–292. https://doi.org/10.1061/(ASCE)0733-9399(2003)129:3(283).

Zhou, Y., A. Kareem, and M. Gu. 2000. "Equivalent static buffeting loads on structures." *J. Struct. Eng.* 126 (8): 989–992. https://doi.org/10.1061/(ASCE)0733-9445(2000)126:8(989).

Zhou, Y., T. Kijewski, and A. Kareem. 2003. "Aerodynamic loads on tall buildings: Interactive database." *J. Struct. Eng.* 129 (3): 394–404. https://doi.org/10.1061/(ASCE)0733-9445(2003)129:3(394).

CHAPTER C27

WIND LOADS ON BUILDINGS: MAIN WIND FORCE RESISTING SYSTEM (DIRECTIONAL PROCEDURE)

The Directional Procedure is the former "buildings of all heights" provision in Method 2 of ASCE 7-05, for the main wind force resisting system (MWFRS). The Directional Procedure is considered the traditional approach in that the pressure coefficients reflect the actual loading on each surface of the building as a function of wind direction, namely, winds perpendicular or parallel to the ridge line.

C27.1 SCOPE

C27.1.5 Minimum Design Wind Loads This section specifies a minimum wind load to be applied horizontally on the entire vertical projection of the building, as shown in Figure C27.1-1. This load case is to be applied as a separate load case in addition to the normal load cases specified in other portions of this chapter.

ENCLOSED, PARTIALLY ENCLOSED, PARTIALLY OPEN, AND OPEN BUILDINGS OF ALL HEIGHTS

C27.2 GENERAL REQUIREMENTS

The Directional Procedure of Chapter 27 applies to "regular-shaped buildings," as defined in Chapter 26 and its commentary. For low-rise buildings with non rectangular plan shapes, such as L, T, and U, with single- or multilevel roofs, and/or multi-ridge/hip/gable roof systems, it is reasonable for the designer to use Chapter 27 provisions and pressure coefficients, in conjunction with engineering judgment, unless the Wind Tunnel Procedure of Chapter 31 is utilized. When utilizing engineering judgment to apply the ASCE 7 external pressure coefficients to a low-rise nonrectangular building, the designer should consider the advice of an engineer knowledgeable of wind pressure distribution for the location and application of the external pressure coefficients, as well as the guidance available in literature recognized by the applicable building code.

Most building codes adopted in the United States specifically recognize high-wind design standards, guides, and manuals that include guidance for low-rise buildings with nonrectangular plan shapes, specifically, utilizing the so-called Inscribed Structure Method and Separate Structures Method. Examples of such reference documents include ICC 600-2014 *Standard for Residential Construction in High-Wind Regions* (ICC 2014), AISI S230-19 *Standard for Cold-Formed Steel Framing* (AISI 2019), and the *Wood Frame Construction Manual* (AWC 2018).

C27.3 WIND LOADS: MAIN WIND FORCE RESISTING SYSTEM

C27.3.1 Enclosed, Partially Enclosed, and Partially Open Rigid and Flexible Buildings In Equations (27.3-1) and (27.3-2), a velocity pressure term, q_i, appears that is defined as the "velocity pressure for internal pressure determination." The positive internal pressure is dictated by the positive exterior pressure on the windward face at the point where there is an opening. The positive exterior pressure at the opening is governed by the value of q at the level of the opening, not q_h. For positive internal pressure evaluation, q_i may conservatively be evaluated at height $h(q_i = q_h)$. For low buildings, this evaluation does not make much difference, but for the example of a 300 ft (91.4 m) tall building in Exposure B with a highest opening at 60 ft (18.2 m), the difference between q_{300} and q_{60} represents a 59% increase in internal pressure. This difference is unrealistic and represents an unnecessary degree of conservatism. Accordingly, $q_i = q_z$ for positive internal pressure evaluation in partially enclosed buildings, where height z is defined as the level of the highest opening in the building that could affect the positive internal pressure.

Figure 27.3-1. The pressure coefficients for MWFRSs are separated into two categories:

1. Directional Procedure for buildings of all heights (Figure 27.3-1), as specified in Chapter 27 for buildings that meet the requirements specified therein.
2. Envelope Procedure for low-rise buildings that have a height less than, or equal to, 60 ft (18.3 m) (Figure 28.3-1), as specified in Chapter 28 for buildings that meet the requirements specified therein.

In generating these coefficients, two distinctly different approaches were used. For the pressure coefficients given in Figure 27.3-1, the more traditional approach was followed, and the pressure coefficients reflect the actual loading on each surface of the building as a function of wind direction, namely, winds perpendicular or parallel to the ridge line.

Observations in wind tunnel tests show that areas of very low negative pressure, and even slightly positive pressure, can occur in all roof structures, particularly as the distance from the windward edge increases and the wind streams reattach to the surface. These pressures can even occur for relatively flat or low-slope roof structures. Experience and judgment from wind tunnel studies have been used to specify either zero or slightly negative pressures (−0.18), depending on the negative pressure coefficient. These values require the designer to consider a zero or slightly positive net wind pressure in the load combinations of Chapter 2.

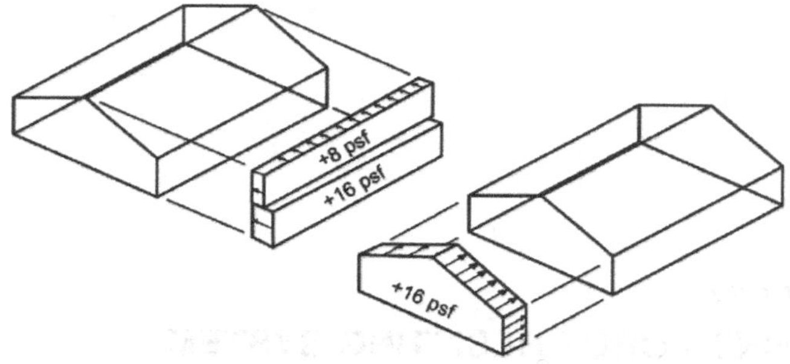

Figure C27.1-1. Application of minimum wind load.
Note: Shown in customary and SI units: 1.0 lb/ft² (0.0479 kN/m²).

Figure 27.3-2. Frame loads on dome roofs are adapted from the Eurocode (1995). The loads are based on data obtained in a modeled atmospheric boundary-layer flow that does not fully comply with requirements for wind tunnel testing specified in this standard (Blessman 1971). Loads for three domes ($h_D/D = 0.5$, $f/D = 0.5$), ($h_D/D = 0$, $f/D = 0.5$), and ($h_D/D = 0$, $f/D = 0.33$) are roughly consistent with the data of Taylor (1991), who used an atmospheric boundary layer as required in this standard. Two load cases are defined, one of which has a linear variation of pressure from A to B as in the Eurocode (1995), and one in which the pressure at A is held constant from 0 to 25 degrees; these two cases are based on a comparison of the Eurocode provisions with Taylor (1991). In many cases, Case A (the Eurocode calculation) is necessary to define maximum uplift. Case B is necessary to properly define positive pressures in some cases, which cannot be isolated with current information and which result in maximum base shear. For domes larger than 200 ft (61 m) in diameter, the designer should consider the use of wind tunnel testing. Resonant response is not considered in these provisions. Wind tunnel testing should be used to consider resonant response. Local bending moments in the dome shell may be larger than predicted by this method because of the difference between instantaneous local pressure distributions and those predicted by Figure 27.3-2. If the dome is supported on vertical walls directly below, it is appropriate to consider the walls as a "chimney," using Figure 29.4-1.

Figure 27.3-3. The pressure and force coefficient values in these tables were taken from ANSI A58.1-1972 (1972). Additional information was added in ANSI A58.1-1982 (1982). The coefficients specified in these tables are based on wind tunnel tests conducted under conditions of uniform flow and low turbulence, and their validity in turbulent boundary-layer flows has yet to be completely established. Additional pressure coefficients for conditions not specified herein may be found in SIA (1956) and ASCE (1961).

C27.3.1.1 Elevated Buildings This section provides guidance for designers of buildings that are elevated on structural elements, which allow wind to pass beneath the building. It was developed for typical elevated beach houses as well as taller elevated buildings. A basic assumption is that the elevated building is supported by a number of distributed structural columns and that it may include one or more relatively small enclosed, partially enclosed, or partially open spaces that accommodate stairs, elevators, and reception areas. It draws on judgment and experience, as well as research conducted by Marshall (1977) on manufactured homes with and without skirting. Limited testing of elevated residential model buildings, using a five-component high-frequency force balance, was performed as part of an independent study by students at Clemson University, where lateral loads, uplift loads, and overturning moments on the elevated buildings were found to be consistent with loads calculated using Section 27.3.1. More recent tests at Florida International University reported by Kim et al. (2020) showed bottom surface C&C loads on elevated low-rise buildings to be of similar magnitude to the roof C&C loads. Judgment has been used to establish practical limits (Limitations 1 and 2 in Section 27.3.1.1), where it is prudent to simply treat the entire building as an enclosed, partially enclosed, or partially open building when determining the overall MWFRS loads. Figures C27.3-1 and C27.3-2 illustrate the determination of area ratios used to determine whether the elevated building meets the two limitations.

Use of a net force coefficient of 1.3 includes windward and leeward pressures on the elements and is conservative, up to L/B ratios of about 2.5, relative to the net overall drag force coefficients determined for relatively porous (solidity ratio less than 0.5) petrochemical and industrial buildings, as shown in Figure 5B.3 in ASCE (2011). Equations developed in that reference indicate that the relative porosity must be increased for larger L/B ratios in order for the net force coefficient of 1.3 to be appropriate. The reductions in solidity ratio needed when a net force coefficient of 1.3 is used are provided in the table associated with Limitation 1. For L/B ratios greater than 5, engineering judgment and analysis are required. Some guidance to assist in applying engineering judgment and analysis can be found in ASCE (2011).

Tall buildings generally have a stagnation point somewhere close to 60% of their height. Above that elevation, wind tends to be directed over and around the building, whereas below that level, there is a tendency for wind to flow around the sides of the building and down the face of the building. Acceleration of winds beneath the enclosed, partially enclosed, or partially open building are accounted for by specifying use of q_z, where z is set equal to the height above grade of the bottom horizontal surface of the elevated building plus 25% of the distance from the bottom of the elevated portion of the building to the mean roof height. Constriction of the wind passing below the enclosed, partially enclosed, or partially open building will potentially reduce the size of the flow separation bubble that would form due to flow over the roof where no restriction exists. In addition to flow separation at the edges of the bottom surface, accelerated flow will result in a decrease in static pressure, leading to generally negative pressures on the bottom surface unless blocked by relatively large elements that can produce locally positive pressures or by enough blockage to create a partially enclosed case where positive pressures typical of wall pressures would be exerted upward on the bottom surface. The limits provided in Section 27.3.1.1 will eliminate cases where large MWFRS uplift loads could be applied to the bottom surface of the structure. The use of h, equal to the height of the elevating members in the calculation of zone limits for coefficients obtained from Figure 27.3-1, is intended to reduce the extent of higher edge loads on the bottom surface when the height of the elevation is relatively short.

C27.3.2 Open Buildings with Monoslope, Pitched, or Troughed Free Roofs Figures 27.3-4 through 27.3-6 and Figures 30.7-1 through 30.7-3 are presented for wind loads on MWFRSs and C&C of open buildings with roofs as shown, respectively. This work is based on the Australian standard AS1170.2-2002, Part 2: Wind Actions, with modifications to the MWFRS pressure coefficients based on studies current at the time of publication (Altman 2001, Uematsu and Stathopoulos 2003).

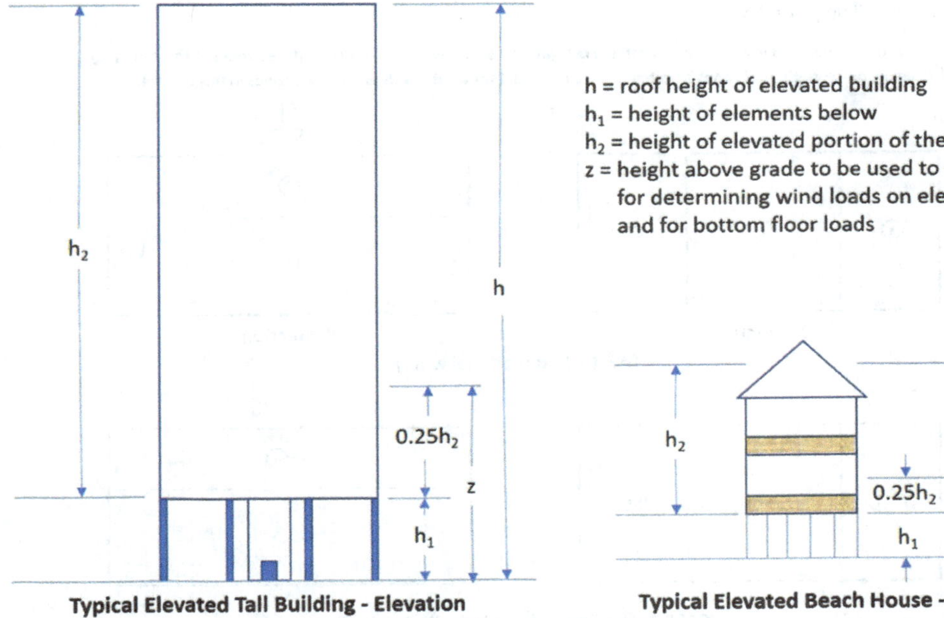

Figure C27.3-1. High-rise and low-rise examples of dimensions and heights used in Section 27.3.1.1.

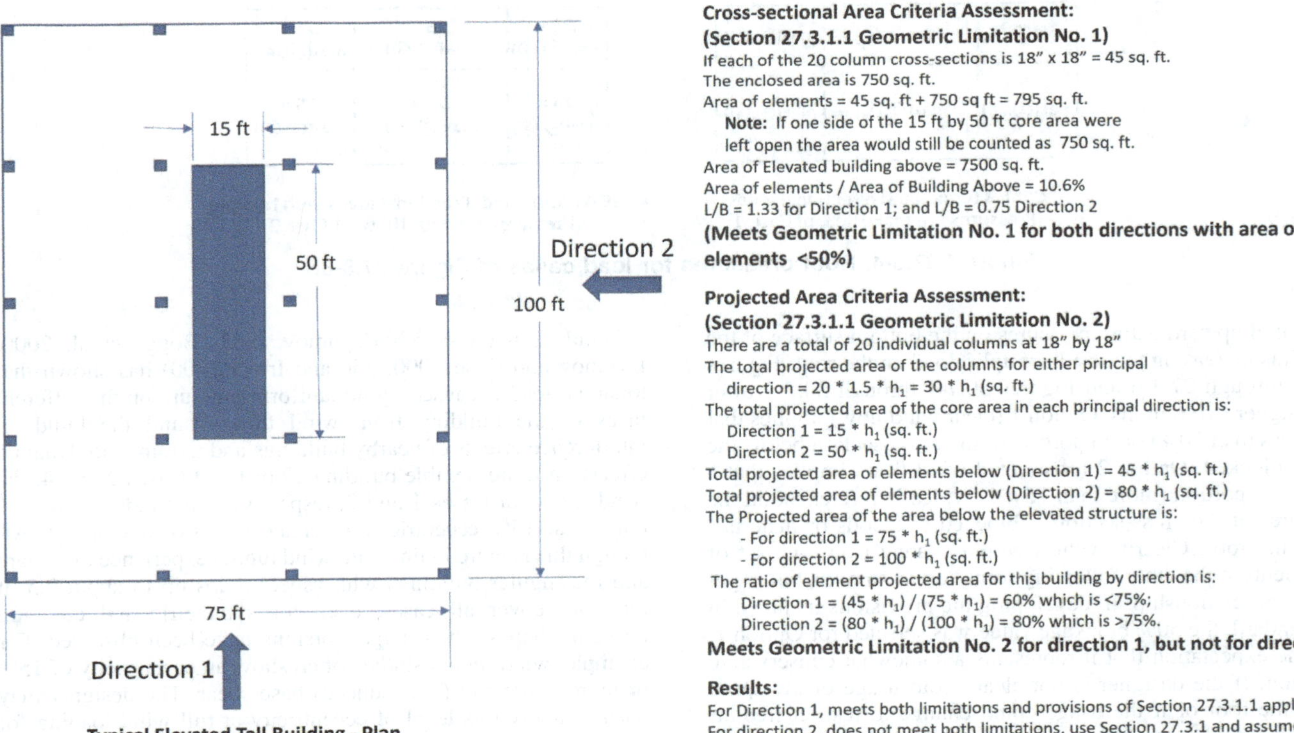

Figure C27.3-2. Example determination of plan cross-sectional areas and vertical projected areas for evaluation of Section 27.3.1.1 limitations on elevated building criteria.

Two load cases, A and B, are given in Figures 27.3-4 through 27.3-6. These pressure distributions provide loads that envelop the results from detailed wind tunnel measurements of simultaneous normal forces and moments. Application of both load cases is required to envelop the combinations of maximum normal forces and moments that are appropriate for the particular roof shape and blockage configuration.

The roof wind loading on open building roofs is highly dependent on whether goods or materials are stored under the roof and restrict the wind flow. Restricting the flow can introduce

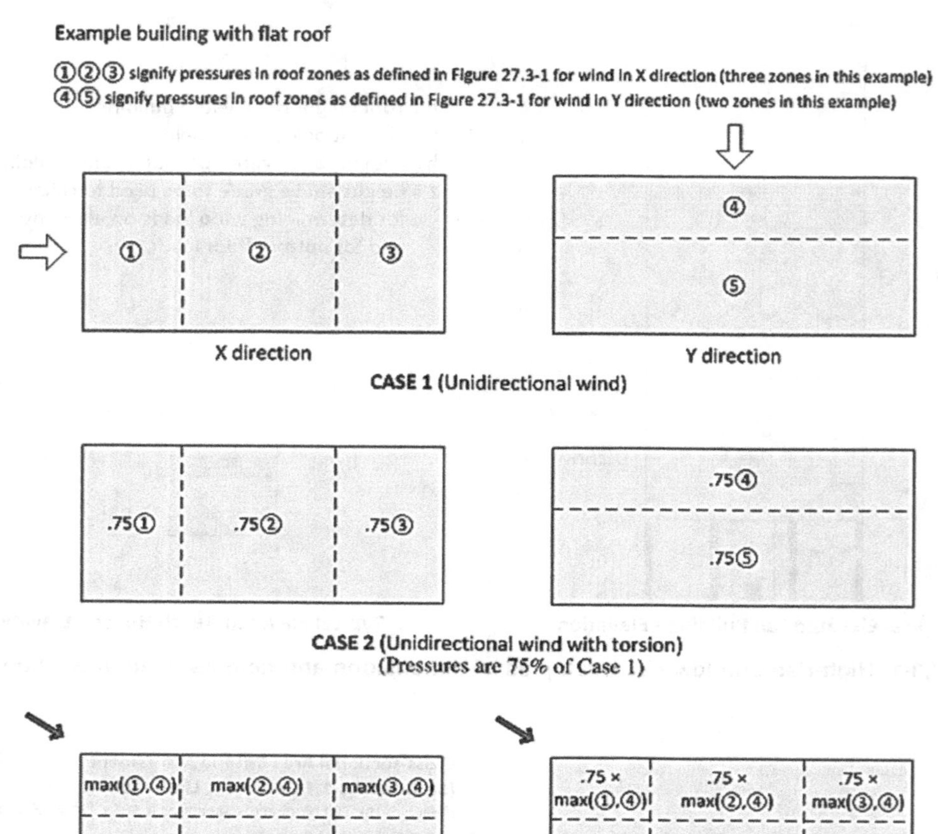

Figure C27.3-4. Roof pressures for load cases of Figure 27.3-8.

substantial upward-acting pressures on the bottom surface of the roof, thus increasing the resultant uplift load on the roof. Figures 27.3-4 through 27.3-6 and Figures 30.7-1 through 30.7-3 offer the designer two options. Option 1 (clear wind flow) implies that little (less than 50%) or no portion of the cross-section below the roof is blocked. Option 2 (obstructed wind flow) implies that a significant portion (more than 75% is typically referenced in the literature) of the cross-section is blocked by goods or materials below the roof. Clearly, values would change from one set of coefficients to the other following some sort of smooth, but as yet unknown, relationship. In developing the provisions included in this standard, the 50% blockage value was selected for Option 1, with the expectation that it represents a somewhat conservative transition. If the designer is not clear about usage of the space below the roof or if the usage could change to restrict free air flow, then design loads for both options should be used.

See Section 28.3.5 for an explanation of the horizontal wind loads on open buildings with transverse frames and pitched roofs that act in combination with the loads calculated in Section 27.3.3.

C27.3.5 Design Wind Load Cases Figure 27.3-8 has been modified in this 2022 version to clarify the proper application of both wall and roof pressures for the four load cases shown in the figure. Roof pressures are now explicitly included as a requirement to be applied simultaneously with wall pressures for all four load cases.

Wind tunnel research (Isyumov 1983, Boggs et al. 2000, Isyumov and Case 2000, Xie and Irwin 2000) has shown that torsional load is caused by nonuniform pressure on the different faces of the building from wind flow around the building, interference effects of nearby buildings and terrain, and dynamic effects on more flexible buildings. For Load Cases 2 and 4, the wind shears of Cases 1 and 2, respectively, are reduced to 75% but act at 15% eccentricity to create a torsional moment. Although this is more in line with wind tunnel experience on square and rectangular buildings with aspect ratios up to about 2.5, it may not cover all cases, even for symmetric and common building shapes where larger torsions have been observed. For example, wind tunnel studies often show an eccentricity of 15% or more under full (not reduced) base shear. The designer may wish to apply this level of eccentricity at full wind loading for certain more critical buildings, even though it is not required by the standard. The present more moderate torsional load requirements can, in part, be justified by the fact that the design wind forces tend to be upper bound for most common building shapes.

In buildings with some structural systems, more severe loading can occur when the resultant wind load acts diagonally to the building. This can be the result of a quartering-oriented wind, or more generally, the joint action of multiple random variables (fluctuating wind actions) that are not fully correlated. To account for this effect and the fact that many buildings exhibit maximum response in the across-wind direction (the standard currently has no analytical procedure for this case), a structural system should be capable of resisting 75% of the design wind load applied simultaneously along each principal axis, as required by Case 3 in Figure 27.3-8. Because this combination is often the result of a quartering wind direction that can produce rooftop vortexes emanating from the upwind corner, and recent research made available to the committee, which is in the process of publication, shows that the largest overall roof uplift loads are often caused by quartering winds, the roof pressures in this case are not subject to a 75% reduction factor. Cases 3 and 4 now specify how the diagonal wind roof pressures are to be applied simultaneously with the wall pressures. For the diagonal wind pressure application in Case 3, roof area zones specified in Case 1 are superimposed to create sub zones that include both principal axis roof zones. For each of these subzones, the larger wind pressure coefficient from either of the principal axis cases is applied to each zone. Case 4 is derived from Case 2 in the same manner (Figure C27.3-4).

For flexible buildings, wind dynamic effects can increase torsional loading. Additional torsional loading can occur because of eccentricity between the elastic shear center and the center of mass at each level of the building. Equation (27.3-4) accounts for this effect.

It is important to note that significant torsion can also occur on low-rise buildings (Isyumov and Case 2000), and therefore, the wind loading requirements of Section 27.3.5 are now applicable to buildings of all heights.

As discussed in Chapter 31, the wind tunnel procedure should always be considered for buildings with unusual shapes, rectangular buildings with larger aspect ratios, and dynamically sensitive buildings, as well as any building size or shape where the wind tunnel procedure can be economically justified. The effects of torsion and diagonal wind pressures acting on both wall and roof pressures simultaneously can more accurately be determined for these cases, and for the more normal building shapes using the wind tunnel procedure.

REFERENCES

AISI (American Iron and Steel Institute). 2019. *Standard for cold-formed steel framing prescriptive method for one- and two- family dwellings*. AISI S230-19. Washington, DC: AISI.

Altman, D. R. 2001. "Wind uplift forces on roof canopies." Master's thesis, Clemson Univ., Dept. of Civil Engineering.

ANSI (American National Standards Institute). 1972. *Minimum design loads for buildings and other structures*. ANSI A58.1-1972. Washington, DC: ANSI.

ANSI. 1982. *Minimum design loads for buildings and other structures*. ANSI A58.1-1982. Washington, DC: ANSI.

ASCE. 1961. "Wind forces on structures." *Trans. ASCE* 126 (2): 1124–1198.

ASCE. 1998. *Minimum design loads for buildings and other structures*. ASCE 7-95. Reston, VA: ASCE.

ASCE. 2011. "Wind loads for petrochemical and other industrial facilities." In *Task committee on wind-induced forces*. Reston, VA: ASCE.

AWC (American Wood Council). 2018. *Wood frame construction manual for one- and two-family dwellings 2018 edition*. Leesburg, VA: AWS.

Blessman, J. 1971. "Pressures on domes with several wind profiles." In *Proc., 3rd Int. Conf. on Wind Effects on Buildings and Structures*, 317–326. Tokyo: Japanese Organizing Committee.

Boggs, D. W., N. Hosoya, and L. Cochran. 2000. "Sources of torsional wind loading on tall buildings: Lessons from the wind tunnel." In *Proc., Structures Congress 2000, Advanced Technology in Structural Engineering*, edited by P. E. Mohamad Elgaaly. Reston, VA: ASCE.

CEN (European Committee for Standardization). 1995. *Eurocode 1: Basis of design and actions on structures, Part 2-4: Actions on structures–wind actions*. ENV 1991-2-4. Brussels, Belgium: CEN.

ICC (International Code Council). 2014. *Standard for residential construction in high-wind regions*. ICC 600-2014. Washington, DC: ICC.

Isyumov, N. 1983. "Wind induced torque on square and rectangular building shapes." *J. Wind Eng. Ind. Aerodyn.* 13 (1–3): 183–196. https://doi.org/10.1016/0167-6105(83)90140-X.

Isyumov, N., and P. C. Case. 2000. "Wind-induced torsional loads and responses of buildings." In *Proc., Structures Congress 2000: Advanced Technology in Structural Engineering*, edited by P. E. Mohamad Elgaaly. Reston, VA: ASCE.

Kim, J. H., M. Moravej, E. J. Sutley, A. Chowdhury, and T. N. Dao. 2020. "Observations and analysis of wind pressures on the floor underside of elevated buildings." *Eng. Struct.* 221 (2020): 111101. https://doi.org/10.1016/j.engstruct.2020.111101.

Marshall, R. D. 1977. *The measurement of wind loads on a full-scale mobile home*. NBSIR 77-1289. Gaithersburg, MD: National Institute of Standards and Technology.

SIA (Swiss Society of Engineers and Architects). 1956. *Normen fur die Belastungsannahmen, die Inbetriebnahme und die Uberwachung der Bauten*. SIA Technische Normen No. 160. Zurich, Switzerland: SIA.

Standards Australia. 2002. *Structural design actions, Part 2: Wind actions*. AS/NZS 1170.2:2002. Sydney, NSW: Standards Australia.

Taylor, T. J. 1991. "Wind pressures on a hemispherical dome." *J. Wind Eng. Ind. Aerodyn.* 40 (2): 199–213. https://doi.org/10.1016/0167-6105(92)90365-H.

Uematsu, Y., and T. Stathopoulos. 2003. "Wind loads on free-standing canopy roofs: A review." *J. Wind Eng. Jpn. Assoc. Wind Eng.* 28 (2): 95_245–95_256.

Xie, J., and P. A. Irwin. 2000. "Key factors for torsional wind response of tall buildings." In *Proc., Structures Congress 2000: Advanced Technology in Structural Engineering*, edited by P. E. Mohamad Elgaaly. Reston, VA: ASCE.

OTHER REFERENCES (NOT CITED)

Cook, N. J. 1990a. *The designer's guide to wind loading of building structures, Part 1*. London: Butterworths.

Cook, N. J. 1990b. *The designer's guide to wind loading of building structures, Part 2 static structures*. London: Butterworths.

Twisdale, L. A., P. J. Vickery, and A. C. Steckley. 1996. *Analysis of hurricane windborne debris impact risk for residential structures*. Bloomington, IL: State Farm Mutual Automobile Insurance Companies.

Vickery, B. J., and C. Bloxham. 1992. "Internal pressure dynamics with a dominant opening." *J. Wind Eng. Ind. Aerodyn.* 41–44 (1–3): 193–204. https://doi.org/10.1016/0167-6105(92)90409-4.

Yeatts, B. B., and K. C. Mehta. 1993. "Field study of internal pressures." In Vol. 2 of *Proc., 7th US National Conf. on Wind Engineering*, edited by G. Hart, 889–897.

CHAPTER C28
WIND LOADS ON BUILDINGS: MAIN WIND FORCE RESISTING SYSTEM (ENVELOPE PROCEDURE)

The Envelope Procedure is the former "Low-Rise Buildings" provision in Method 2 of ASCE 7-05 for the Main Wind Force Resisting System (MWFRS). The provisions are not intended for buildings with arched, barrel, or unusually shaped roofs.

ENCLOSED, PARTIALLY ENCLOSED, AND PARTIALLY OPEN LOW-RISE BUILDINGS

C28.2 GENERAL REQUIREMENTS

The Envelope Procedure shown in Chapter 28 applies to "regular-shaped buildings," as defined in Chapter 26 and its commentary. For low-rise buildings with non rectangular plan shapes, such as L, T, and U, with single-level or multilevel roofs, and/or multi-ridge/hip/gable roof systems, it is reasonable for the designer to use Chapter 28 provisions and pressure coefficients in conjunction with engineering judgment, unless the Wind Tunnel Procedure of Chapter 31 is used. When using engineering judgment to apply the ASCE 7 external pressure coefficients to a low-rise non rectangular building, the designer should consider the advice of an engineer knowledgeable of wind pressure distribution for location and application of the external pressure coefficients, as well as guidance available in literature recognized by the applicable building code.

Most building codes adopted in the United States specifically recognize high-wind design standards, guides, and manuals that include guidance for low-rise buildings with non rectangular plan shapes, specifically, using the "Inscribed Structure Method" and "Separate Structures Method." Examples of such reference documents include ICC 600-2014, *Standard for Residential Construction in High-Wind Regions* (ICC 2014); AISI S230-19, *Standard for Cold-Formed Steel Framing* (AISI 2019); and the *Wood Frame Construction Manual* (AWC 2018).

C28.3 WIND LOADS: MAIN WIND FORCE RESISTING SYSTEM

C28.3.1 Design Wind Pressure for Low-Rise Buildings See Section C26.10.1 for information related to determining appropriate velocity pressures.

Loads on Main Wind Force Resisting Systems. The pressure coefficients for the MWFRS are basically separated into two categories:

1. Directional Procedure for buildings of all heights (Figure 27.3-1), as specified in Chapter 27 for buildings meeting the requirements specified therein; and
2. Envelope Procedure for low-rise buildings (Figure 28.3-1), as specified in Chapter 28 for buildings meeting the requirements specified therein.

In generating these coefficients, two distinctly different approaches were used. For the pressure coefficients for buildings of all heights given in Figure 27.3-1, the more traditional approach was followed, and the pressure coefficients reflect the actual loading on each surface of the building as a function of wind direction, namely, winds perpendicular or parallel to the ridge line.

For the pressure coefficients for low-rise buildings, however, the values of (GC_{pf}) represent "pseudo" loading conditions that, when applied to the building, envelop the desired structural actions (bending moment, shear, thrust) independent of wind direction. To capture all appropriate structural actions, the building must be designed for all wind directions by considering in turn each corner of the building as the windward or reference corner shown in the sketches of Figure 28.3-1. At each corner, two load patterns are applied, one for each wind direction range. In each case, the roof end zone will remain at the end of the building that is perpendicular to the ridge line. The end zone creates the required structural actions in the end frame or bracing. There are two load cases with four basic scenarios for each case. Note also that for all roof slopes, all eight load scenarios must be considered individually to determine the critical loading for a given structural assemblage or component thereof (Figure C28.3-1).

To develop the appropriate pseudovalues of (GC_{pf}), investigators at the University of Western Ontario (Davenport et al. 1978) used an approach that consisted essentially of permitting the building model to rotate in the wind tunnel through a full 360 degrees while simultaneously monitoring the loading conditions on each of the surfaces (Figure C28.3-2). Exposures B and C were both considered. Using influence coefficients for rigid frames, it was possible to spatially average and time-average the surface pressures to ascertain the maximum induced external force components to be resisted. More specifically, the following structural actions were evaluated:

1. Total uplift,
2. Total horizontal shear,
3. Bending moment at knees (two-hinged frame),
4. Bending moment at knees (three-hinged frame), and
5. Bending moment at ridge (two-hinged frame).

The next step involved developing sets of pseudo-pressure coefficients to generate loading conditions that would envelop the maximum induced force components to be resisted for all possible wind directions and exposures. Note, for example, that the wind azimuth producing the maximum bending moment at the knee would not necessarily produce the maximum total uplift. The maximum induced external force components determined for each of the preceding five actions were used to develop the coefficients. The end result was a set of coefficients that represent

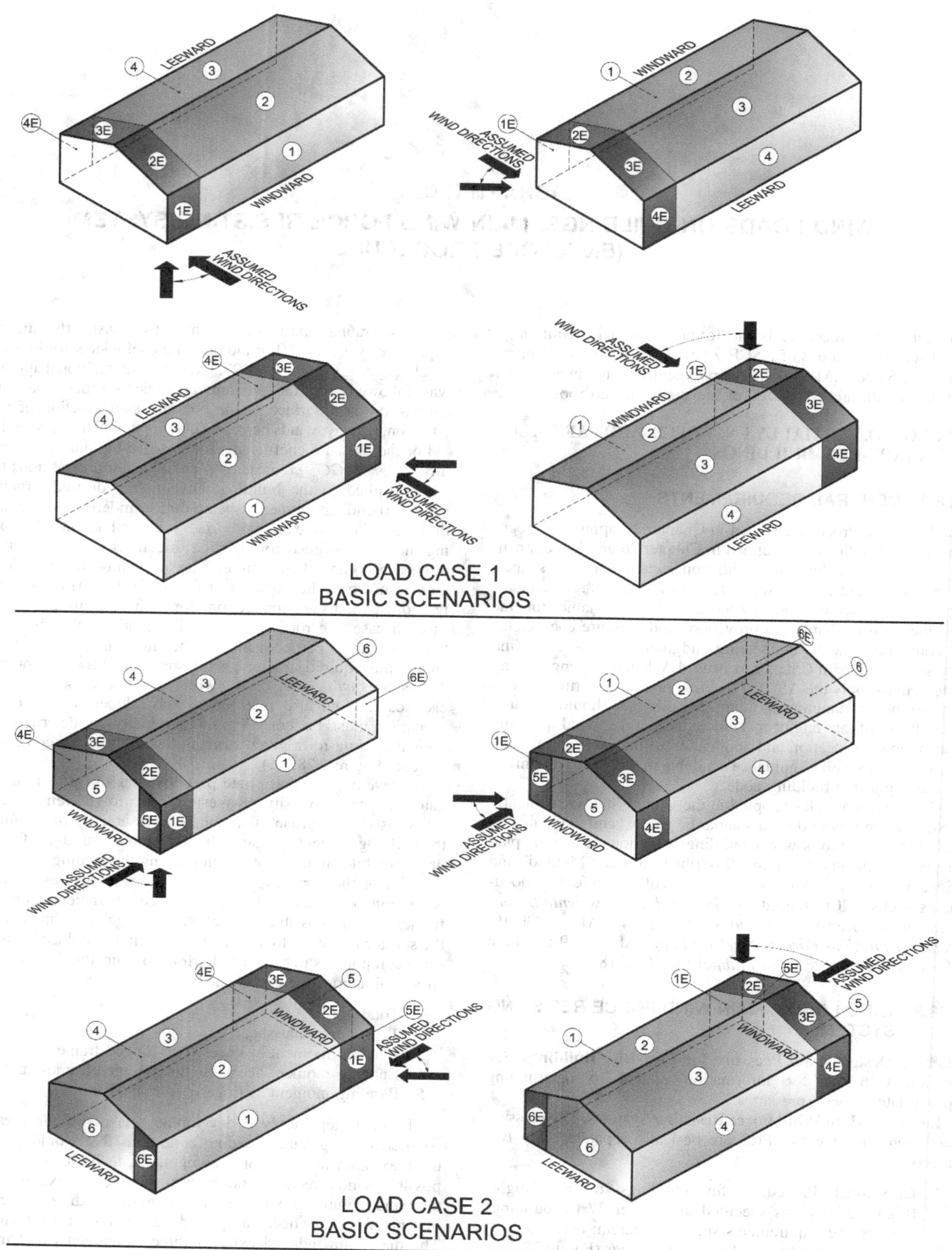

Figure C28.3-1. Illustration of load application in Figure 28.3-1.

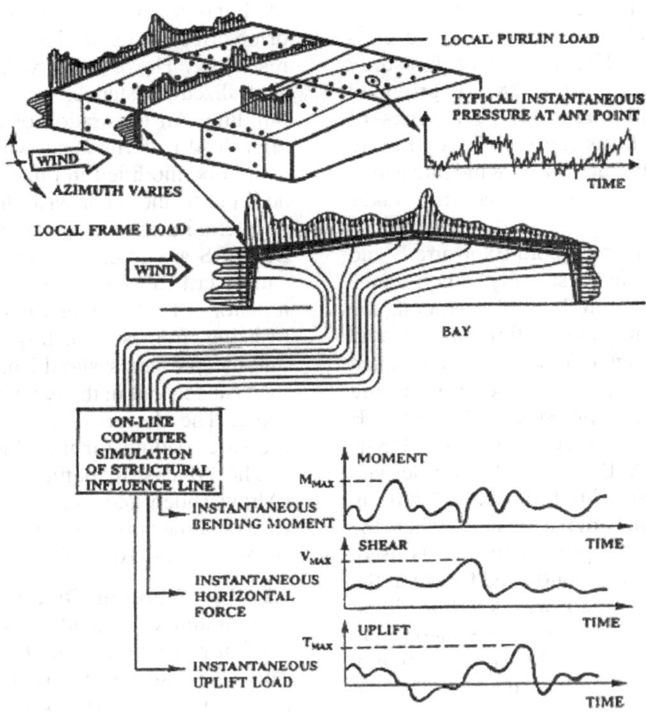

Figure C28.3-2. Unsteady wind loads on low buildings for a given wind direction.

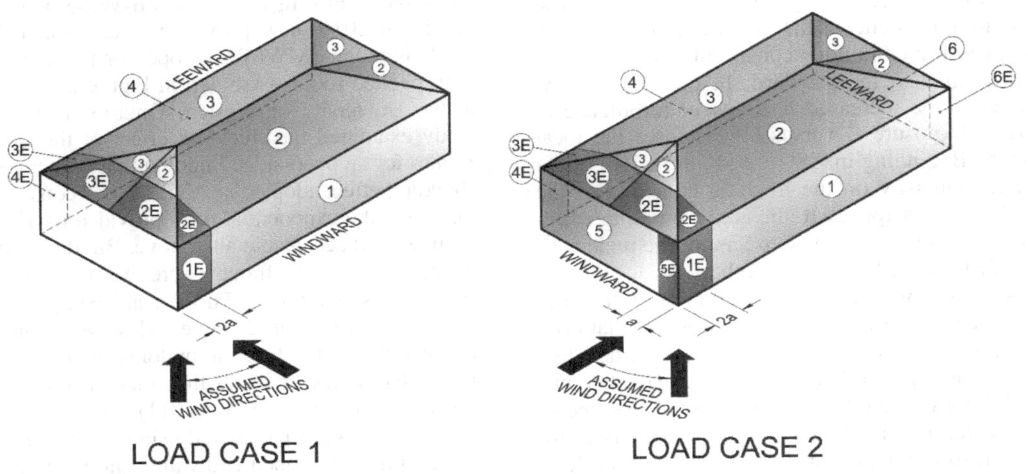

Notes:

1. Adapt the loadings shown in Figure 28.3-1 for hip-roofed buildings as shown. For a given hip roof pitch, use the roof coefficients from the Case 1 table for both Load Case 1 and Load Case 2.
2. The total horizontal shear shall not be less than that determined by neglecting the wind forces on roof surfaces.

Figure C28.3-3. Loads application for hip-roofed low-rise buildings.

fictitious loading conditions but that conservatively envelop the maximum induced force components (bending moment, shear, and thrust) to be resisted, independent of wind direction.

The original set of coefficients was generated for the framing of conventional pre-engineered buildings, that is, single-story, moment-resisting frames in one of the principal directions and bracing in the other principal direction. The approach was later extended to single-story, moment-resisting frames with interior columns (Kavanagh et al. 1983).

Subsequent wind tunnel studies (Isyumov and Case 1995) have shown that the (GC_{pf}) values of Figure 28.3-1 are also applicable to low-rise buildings with structural systems other than moment-resisting frames. That work examined the instantaneous wind pressures on a low-rise building with a 4:12 pitched gable roof and the resulting wind-induced forces on its MWFRS. Two different MWFRSs were evaluated. One consisted of shear walls and roof trusses at different spacings. The other had moment-resisting frames in one direction, positioned at the same spacings as the roof trusses, and diagonal wind bracing in the other direction. Wind tunnel tests were conducted for both Exposures B and C. The findings of this study showed that the (GC_{pf}) values of Figure 28.3-1 provided satisfactory estimates of the wind forces for both types of structural systems. This work confirms the validity of Figure 28.3-1, which reflects the combined action of wind pressures on different external surfaces of a building and thus takes advantage of spatial averaging.

In the original wind tunnel experiments, both Exposure B and Exposure C terrains were checked. In these early experiments, Exposure B did not include nearby buildings. In general, the force components, bending moments, and so forth were found comparable in both exposures, although (GC_{pf}) values associated with Exposure B terrain would be higher than those for Exposure C terrain because of reduced velocity pressure in Exposure B terrain. The (GC_{pf}) values given in Figures 28.3-1, 30.3-1, 30.3-2A–C, 30.3-3, 30.3-4, 30.3-5A–B, and 30.3-6 are derived from wind tunnel studies modeled with Exposure C terrain. However, they may also be used in other exposures when the velocity pressure representing the appropriate exposure is used.

In comprehensive wind tunnel studies conducted by Ho at the University of Western Ontario (Ho 1992), it was determined that when low buildings [$h < 60$ ft ($h < 18.3$ m)] are embedded in suburban terrain (Exposure B, which included nearby buildings), the pressures in most cases are lower than those currently used in existing standards and codes, although the values show a very large scatter because of high turbulence and many other variables. The results seem to indicate that some reduction in pressures for buildings located in Exposure B is justified. The Wind Loads Subcommittee believes that it is desirable to design buildings for the exposure conditions consistent with the exposure designations defined in the standard. In the case of low buildings, the effect of the increased intensity of turbulence in rougher terrain (i.e., Exposure B versus C) increases the local pressure coefficients. Beginning in ASCE 7-98, the effect of the increased turbulence intensity on the loads is treated with the truncated profile. Using this approach, the actual building exposure is used, and the profile truncation corrects for the underestimate in the loads that would be obtained otherwise.

Figure 28.3-1 is most appropriate for low buildings with width greater than twice their height and a mean roof height that does not exceed 33 ft (10 m). The original database included low buildings with widths no greater than 5 times their eave heights, and eave height did not exceed 33 ft (10 m). In the absence of more appropriate data, Figure 28.3-1 may also be used for buildings with mean roof height that does not exceed the least horizontal dimension and is less than or equal to 60 ft (18.3 m). Beyond these extended limits, Figure 27.3-1 should be used.

All the research was carried out to develop and refine the low-rise building method for MWFRS loads was done on gable-roofed buildings. In the absence of research on hip-roofed buildings, the committee developed a rational method of applying Figure 28.3-1 to hip roofs based on its collective experience, intuition, and judgment. This method is presented in Figure C28.3-3.

Research indicated that the low-rise method alone underestimates the torsion caused by wind loads (Isyumov 1983, Isyumov and Case 2000). In ASCE 7-02, torsional requirements were added to Figure 28.3-1 to account for this torsional effect. In ASCE 7-22, the torsional requirements were separated into their own section. The reduction in loading on only 50% of the building results in a torsional load case without an increase in the predicted base shear for the building. This reduction in loading results in equivalent torsion that agrees well with the wind tunnel measurements carried out by Elsharawy et al. (2012, 2015) and Stathopoulos et al. (2013). In general, the provision will have little or no effect on the design of MWFRSs that have well-distributed resistance. However, it will affect the design of systems with centralized resistance, such as a single core in the center of the building. An illustration of the intent of the note on two of the eight load patterns is shown in Figure 28.3-2. All eight patterns should be modified in this way as a separate set of load cases in addition to the eight basic load cases.

Internal pressure coefficients, (GC_{pi}), to be used for loads on MWFRS are given in Table 26.13-1. The internal pressure load can be critical in one-story, moment-resisting frames and in the top story of a building where the MWFRS consists of moment-resisting frames. Loading cases with positive and negative internal pressures should be considered. The internal pressure load cancels out in the determination of total lateral load and base shear. The designer can use judgment in the use of internal pressure loading for the MWFRS of high-rise buildings.

The edge strip definition was modified following research (Alrawashdeh and Stathopoulos 2015) showing that the definition of dimension "a" in ASCE 7-10 led to unduly large edge strips and end zones for very large buildings.

C28.3.6 Minimum Design Wind Loads This section specifies a minimum wind load to be applied horizontally on the entire vertical projection of the building, as shown in Figure C27.1-1. This load case is to be applied as a separate load case in addition to the normal load cases specified in other portions of this chapter.

C28.3.7 Horizontal Wind Loads on Open or Partially Enclosed Buildings with Transverse Frames and Pitched Roofs In 2016, new provisions were added for wind loads on the longitudinal MWFRS of open or partially enclosed buildings with pitched roofs, as shown in Figure 28.3-3, based on research at the University of Western Ontario (Kopp et al. 2010). This study examined the shielding effect of these multiple transverse frames for an open-sided building that was covered by a roof. The shielding effect adopted in ASCE 7 was conservatively simplified; therefore, the exponential form provided by Kopp et al. is a more accurate and acceptable alternative. Building models consisting of three, six, and nine frames were evaluated. A building with two frames was not tested. Further, an extrapolation using $n = 2$ is not necessarily conservative. However, this method can be conservatively used for a building with two frames by using $n = 3$. Examples of evaluating these additional wind forces are given by Shoemaker et al. (2011).

The wind loads calculated using Section 28.4.5 are applicable to buildings with open end walls, end walls with the gable filled with cladding, and with additional end wall cladding; however, the area used is always the total end wall area, A_f. The effective solid area of a frame, A_s, is the projected area of any portion of the end wall that would be exposed to the wind.

The measured peak base shear coefficients were used as the basis for the design drag loads in the direction parallel to the ridge (i.e., wind directions in the range 0 to 45 degrees). These loads include the effects of both friction drag and pressure drag. However, to put this in a format consistent with ASCE 7-10, it was reasonable to use the enclosed pseudo load coefficients, (GCw), and then apply factors to account for the parameters that affect the load coefficients on open buildings (i.e., building size, solidity ratio, and number of frames). This method yielded conservative results for all experimental wind tunnel data points. The force from Equation (28.3-3), calibrated to the measured base shear, does not reflect a direct load path from the calculated end wall pressure but is to be used to calculate the longitudinal bracing requirement. For the building configurations evaluated, the University of Western Ontario study showed that the force measured in the bracing was equal to 70% of the total base shear. The remaining base shear was transferred directly at the column bases.

The wind tunnel studies used to develop the provisions of Section 28.3.7 did not evaluate the effect of obstructed flow due to materials or objects sheltered by the building. Barring an unusual arrangement of materials that could produce a Venturi effect, it is judged that obstructed flow would decrease the wind loads on the longitudinal MWFRS. However, as noted in previous studies (Altman 2001, Uematsu and Stathopolous 2003), the roof wind loads are more sensitive to the effect of obstructed flow.

The wind load in the transverse direction (perpendicular to the ridge) for this type of open building is a separate loading case and is due to the horizontal pressure from the roof load calculated using Section 27.3.2, with C_N from Figure 27.3-5, and additional pressures acting on the projected areas of any surfaces exposed to the transverse wind.

REFERENCES

AISI (American Iron and Steel Institute). 2019. *Standard for cold-formed steel framing: prescriptive method for one and two family dwellings. AISI S230-19*. Washington, DC: AISI.

Alrawashdeh, H., and T. Stathopoulos. 2015. "Wind pressures on large roofs of low buildings and wind codes and standards." *J. Wind Eng. Ind. Aerodyn.* 147 (Dec): 212–225.

Altman, D. R. 2001. "Wind uplift forces on roof canopies." Master's thesis, Dept. of Civil Engineering, Clemson University.

AWC (American Wood Council). 2018. *Wood frame construction manual for one- and two-family dwellings*. Leesburg, VA: AWC.

Davenport, A. G., D. Surry, and T. Stathopoulos. 1978. *Wind loads on low-rise buildings*. Final report on Phase III, BLWT-SS4. London, ON, Canada: University of Western Ontario.

Elsharawy, M., K. Galal, and T. Stathopoulos. 2015. "Torsional and shear wind loads on flat-roofed buildings." *Eng. Struct.* 84 (2): 313–324. https://doi.org/10.1016/j.engstruct.2014.11.028.

Elsharawy, M., T. Stathopoulos, and K. Galal. 2012. "Wind-induced torsional loads on low buildings." *J. Wind Eng. Ind. Aerodyn.* 104–106: 40–48. https://doi.org/10.1016/j.jweia.2012.03.011.

Ho, E. 1992. "Variability of low building wind lands." Ph.D. thesis, Dept. of Civil and Environmental Engineering, Univ. of Western Ontario.

ICC (International Code Council). 2014. *Standard for residential construction in high-wind regions*. ICC 600-2014. Washington, DC: ICC.

Isyumov, N. 1983. "Wind induced torque on square and rectangular building shapes." *J. Wind Eng. Ind. Aerodyn.* 13 (1–3): 183–186.

Isyumov, N., and P. Case. 1995. *Evaluation of structural wind loads for low-rise buildings contained in ASCE Standard 7-95*. BLWT-SS17-1995. London, ON, Canada: University of Western Ontario.

Isyumov, N., and P. C. Case. 2000. "Wind-induced torsional loads and responses of buildings." In *Advanced technology in structural engineering*, edited by P. E. Mohamad Elgaaly. Reston, VA: ASCE.

Kavanagh, K. T., D. Surry, T. Stathopoulos, and A. G. Davenport. 1983. *Wind loads on low-rise buildings*. Phase IV, BLWT-SS14. London, ON: University of Western Ontario.

Kopp, G. A., J. Galsworthy, and J. H. Oh. 2010. "Horizontal wind loads on open-frame low-rise buildings." *J. Struct. Div.* 136 (1): 98–105. https://doi.org/10.1061/(ASCE)ST.1943-541X.0000082.

Shoemaker, W. L., G. A. Kopp, and J. Galsworthy. 2011. "Design of braced frames in open buildings for wind loading." *AISC Eng. J.* 48 (3): 225–233.

Stathopoulos, T., M. Elsharawy, and K. Galal. 2013. "Wind load combinations including torsion for rectangular medium-rise building." *Int. J. High-Rise Build.* 2 (3): 1–11.

Uematsu, Y., and T. Stathopoulos. 2003. "Wind loads on free-standing canopy roofs: A review." *J. Wind Eng.* 28 (2): 95_245–95_256. https://doi.org/10.5359/jwe.28.95_245.

OTHER REFERENCES (NOT CITED)

Cook, N. J. 1990a. *The designer's guide to wind loading of building structures, part 1*. London: Butterworths.

Cook, N. J. 1990b. *The designer's guide to wind loading of building structures, part 2: Static structures*. London: Butterworths.

Krayer, W. R., and R. D. Marshall. 1992. "Gust factors applied to hurricane winds." *Bull. Am. Meteorol. Soc.* 73 (5): 613–617. https://doi.org/10.1175/1520-0477(1992)073<0613:GFATHW>2.0.CO;2.

CHAPTER C29

WIND LOADS ON BUILDING APPURTENANCES AND OTHER STRUCTURES: MAIN WIND FORCE RESISTING SYSTEM (DIRECTIONAL PROCEDURE)

C29.3 DESIGN WIND LOADS: SOLID FREESTANDING WALLS AND SOLID SIGNS

C29.3.1 Solid Freestanding Walls and Solid Freestanding Signs (see Section C26.10.1) The risk category for rooftop equipment or appurtenances is required to be not less than that for the building on which the equipment is located, nor that for any other facility to which the equipment provides a necessary service. For example, if a solar array is located on the roof of a hospital, the design wind load for the solar array is based on the risk category for the hospital, even if the solar array is not needed for the functioning of the hospital. In another example, if an antenna provides critical communication service for a hospital and the antenna is located on top of a parking structure separate from the hospital building, the design wind load for the antenna is based on the risk category for the hospital, which is greater than the risk category for the parking structure.

Figure 29.3-1. The provisions in Figure 29.3-1 are based on the results of boundary-layer wind tunnel studies (Letchford 1985, 2001; Holmes 1986; Letchford and Holmes 1994; Ginger et al. 1998a, b; Letchford and Robertson 1999; Mehta et al. 2012).

A surface curve fit to Letchford's (2001) and Holmes's (1986) area averaged mean net pressure coefficient data (equivalent to mean force coefficients in this case) is given by the following equation (Fox and Levitan 2005):

$$C_f = \{1.563 + 0.008542\ln(x) - 0.06148y + 0.009011[\ln(x)]^2 \\ -0.2603y^2 - 0.08393y[\ln(x)]\}/0.85$$

where x is B/s, and y is s/h.

The 0.85 term in the denominator modifies the wind tunnel-derived force coefficients into a format where the gust-effect factor, as defined in Section 26.11, can be used.

Force coefficients for Cases A and B were generated from the preceding equation, then rounded off to the nearest 0.05. That equation is only valid within the range of the B/s and s/h ratios given in the figure for Cases A and B.

Of all the pertinent studies on single-faced signs, only Letchford (2001) specifically addressed eccentricity (i.e., Case B). Letchford reported that the data provided a reasonable match to Cook's (1990) recommendation for using an eccentricity of 0.25 times the average width of the sign. However, the data were too limited in scope to justify changing the existing eccentricity value of 0.2 times the average width of the sign, which was also used in the 2011 Australian/New Zealand Standard (Standards Australia 2011).

Mehta et al. (2012) tested a variety of aspect ratios B/s and clearance ratios s/h for double-faced signs with all sides enclosed to address current industry practice. The study included both wind tunnel testing and a full-scale field test to calibrate the wind tunnel models (Zuo et al. 2014, Smith et al. 2014). These sign configurations exhibited an average reduction of 16% in mean force coefficients, with a range of 9% to 22% compared to single-faced sign force coefficients given by the preceding equation. These tests also showed that the eccentricity of 0.2 times the width of the structure is overly conservative. Eccentricities reported in the study ranged from 0.039 to 0.105 times the width of the structure, with an average of 0.061. Testing by Giannoulis et al. (2012) and Meyer et al. (2017) supports the findings in Mehta et al. (2012).

ASCE 7-22 clarifies that it is not necessary to apply Case B to the design of freestanding walls or signs with aspect ratio $B/s \geq 2$, as the required Case C already considers a torsional effect or an eccentricity consistent with the aforementioned research findings (Mehta et al. 2012, Giannoulis et al. 2012, Zuo et al. 2014, Smith et al. 2014, Meyer et al. 2017).

Case C was added to account for the higher pressures observed in both wind tunnel studies (Letchford 1985, 2001; Holmes 1986; Letchford and Holmes 1994; Ginger et al. 1998a, b; Letchford and Robertson 1999) and full-scale studies (Robertson et al. 1997) near the windward edge of a freestanding wall or sign for oblique wind directions. Linear regression equations were fit to the local mean net pressure coefficient data (for wind direction 45 degrees) from the referenced wind tunnel studies to generate force coefficients for square regions starting at the windward edge. Pressures near this edge increase significantly as the length of the structure increases. No data were available on the spatial distribution of pressures for structures with low aspect ratios ($B/s < 2$).

The sample illustration for Case C at the top of Figure 29.3-1 is for a sign with an aspect ratio $B/s = 4$. For signs of differing B/s ratios, the number of regions is equal to the number of force coefficient entries located below each B/s column heading.

For oblique wind directions (Case C), increased force coefficients have been observed on above ground signs compared to the same aspect ratio walls on ground (Letchford 1985, 2001; Ginger et al. 1998a). The ratio of force coefficients between above ground and on-ground signs (i.e., $s/h = 0.8$ and 1.0, respectively) is 1.25, which is the same ratio used in the Australian/New Zealand Standard (Standards Australia 2002). Note 5 of Figure 29.3-1 provides for linear interpolation between these two cases.

For walls and signs on the ground ($s/h = 1$), the mean vertical center of pressure ranged from $0.5h$ to $0.6h$ (Holmes 1986; Letchford 1989; Letchford and Holmes 1994; Robertson et al. 1995, 1996; Ginger et al. 1998a); $0.55h$ was the average value.

$(GC_{rn})_{\text{nom}}$ values are for both positive and negative values. Wind tunnel test data show similar positive and negative pressures for solar panels (which are very different to typical roof member design wind loads).

Parapets typically worsen the wind loads on solar panels, particularly on wider buildings. The parapets lift the vortices higher above the roof surface and push them closer together, inward from the edges. It is not entirely clear why the vortex effects are more severe in this situation, but tests show that this can result in wind loads that are significantly greater than in the absence of a parapet, particularly for unshrouded tilted panels. The parapet height factor, γ_p, accounts for this effect.

Solar panels are typically installed in large arrays with closely spaced rows, and the end rows and panels experience larger wind pressures than interior panels, which are sheltered by adjacent panels. To account for the higher loading at the end rows and panels, an array edge increase factor is applied, taken from SEAOC (2012). However, single rows of solar panels can be determined using this section, taking into account that all solar panels are defined as being exposed.

Rooftop equipment and structures, such as HVAC units, screens, or penthouses, can provide some sheltering benefits to solar arrays located directly downwind of the object; conversely, however, the regions around the edges of such structures can have accelerated wind flow under varied wind directions. Accordingly, the edge increase factor ignores such structures and is calculated based on the distance to the building edge or adjacent array, neglecting any intervening rooftop structures. This results in the panels adjacent to rooftop objects being designed for higher wind loads to account for the accelerated wind flow.

The requirements can be used for arrays in any plan orientation relative to building axes or edges; the dimensions, d_1 and d_2 are measured parallel to the principal axes of the array being considered. The requirement, in Figure 29.4-7, for array panels to be set back from the roof edge is meant to ensure that the panels are out of the high-speed wind in the separated shear layers at the edge. If the array is made up of a single row of solar panels, or a single panel, then d_2 is undefined and $\gamma_E = 1.5$.

Wind tunnel studies have shown that the wind loads on rooftop solar panels need not be applied simultaneously to the roof C&C wind loads for portions of the roof that are covered by the panel. Where a portion of the span of a roof member is covered by a solar array and the remainder is not covered, then the roof member should be designed with the solar array wind load on the covered portion, with simultaneous application of roof C&C load on the uncovered portion. In a separate load case, the member should also be checked for C&C wind loads, assuming that the photovoltaic panels are not present. For installations of new panels on existing buildings, this separate load case, to check the capacity of the existing roof structure to resist the roof C&C wind loads applied over the entire roof area (i.e., assuming that the solar panels are not present), is not required.

The wind loads here were obtained for solar arrays without aerodynamic treatments, such as shrouds or deflectors. Uplifting wind load for arrays with shrouding may be lower, but because of the range of possible results and sensitivity to design details, such arrays would need to be wind tunnel tested in order to use reduced loads from those specified here. It should also be noted that horizontal (drag) loads could increase with the use of shrouds or deflectors.

Procedure for Using Figure 29.4-7. To simplify the use of the figure, the following is a step-by-step procedure:

Step 1: Confirm applicability of the figure to the solar installation and building.

Step 2: For panels with $\omega \leq 2°$ and $h_2 \leq 10$ in. (254 mm), the procedure using d_2, Section, 29.4.4, may be used.

Step 3: Confirm that layout provides minimum distance from roof edge.

Step 4: Determine roof zones.

Step 5: Determine effective wind area and normalized wind area for each element being evaluated.

Step 6: Compute $(GC_{rn})_{\text{nom}}$ from applicable chart, using linear interpolation for values of ω between 5 and 15 degrees.

Step 7: Apply chord length adjustment factor, γ_c.

Step 8: Apply the edge factor d_2, γ_E, if necessary.

Step 9: Apply parapet height factor, γ_p.

Step 10: Calculate (GC_{rn}).

Step 11: Calculate pressure, p, using Equation (29.4-5).

C29.4.4 Rooftop Solar Panels Parallel to the Roof Surface on Buildings of All Heights and Roof Slopes Wind loads of roof-mounted, planar solar panels that are close to and parallel to the roof surface tend to be lower than the loads on a bare roof because of pressure equalization (Kopp et al. 2012, Kopp 2013), except on the perimeter of the array. The solar array pressure equalization factor, γ_a, accounts for this reduction, based, in particular, on data from Stenabaugh et al. (2015). For pressure equalization to occur, the panels cannot be too large; there needs to be a minimum gap between the panels, and the height above the roof surface cannot be too large. The requirements in ASCE 7-16 were based on panel sizes up to 6.7 ft (2.0 m) long for maximum heights above the roof surface that are less than 10 in. (254 mm) and a minimum gap around the panels of 0.25 in. (6.35 mm). Larger gaps and lower heights above the roof surface could further decrease the wind loads and this is quantified by the changes in Figure 29.4-8. For metal roof panels, the maximum distance above the roof surface is measured from the flat portion of the panels, rather than from the top of the roof panel ribs. The value of γ_a, between 10 ft^2 and 100 ft^2 for the solid line is 1.2-(0.4*log10A). The value of γ_a, between the effective wind area of 10 ft^2 and 100 ft^2 for the dashed line is 0.8-(0.2*log10A).

Panels around the edge of the array may experience higher wind loads.

Bilinear interpolation of the air equalization factor may be done as follows:

1. Interpolate based on the gap width between panels.
2. Interpolate based on h_2.
3. Average both values.

More substantiation is contained in Sections 5.2.3, 5.3.3, and 5.3.5 of the report, "Wind Design for Solar Arrays" (SEAOC 2017) 2017, and the report, "Wind Loads on Photovoltaic Arrays Mounted Parallel to Sloped Roofs on Low-Rise Buildings" (Stenabaugh et al. 2015) provides further commentary on why these changes are appropriate for this proposal.

C29.4.5 Ground-Mounted Fixed-Tilt Solar Panel Systems The requirements in Section 29.4.5 are primarily based on wind tunnel data presented in Browne et al. (2020) and augmented by other proprietary wind tunnel tests. The coefficients enveloped the available data for a wide range of geometric parameters (e.g., tilt angle, row spacing, chord length, and ground clearance) and, thus are typically expected to be conservative. All wind tunnel tests were based on solid flat plates; thus, gaps between modules must be limited to ensure the surface is nominally solid.

The design forces and moments derived using the combined static and dynamic wind load coefficients in this section are appropriate for fixed-tilt ground-mounted systems meeting the geometric requirements of Section 29.4.5.1, assuming the wind-induced deflections are small, and thus, the system is not flexible. Flexible structures, such as unrestrained single-axis trackers, are prone to aeroelastic effects and torsional instabilities, as discussed in Taylor and Browne (2020), which must be considered in the design of such systems.

In addition to the provisions in Section 29.4.5, the Japanese Industrial Standard C 8955 (JIS 2017) provides approximate formulas to obtain uniform positive and negative pressures on ground-mounted PV arrays for tilt angles ranging from 5 to 60 degrees, which may be multiplied by 0.6 for the central parts of arrays. Currently, there are no other codes/standards or other publications in the public domain that include provisions for ground-mounted solar arrays. For variables or configurations that are outside the scope of these requirements, testing in accordance with Chapter 31 and ASCE 49 may be considered.

A building or other structure is defined as "rigid" in Chapter 26 when the fundamental frequency is greater than or equal to 1 Hz. This definition applies to structures with large characteristic dimensions, which is not appropriate for small structures, such as ground-mounted solar photovoltaic (PV) systems. For these types of structures, the inertial wind force associated with modes of vibration with frequencies much larger than 1 Hz can be substantial. The study by Strobel and Banks (2014) on the wind loading due to potential resonance with vortex shedding from the panels highlighted the need to consider wind loading on arrays beyond the typical limit of 1 Hz used for structures exposed to wind. Moreover, this study clearly implies that the gust effect factor (G) in Section 26.11 or equivalent factors in any other current code/standard does not address the potential wake effect resonance caused by upwind rows within a multi-row solar array close to the ground. This sentiment is echoed in Section 8.2 of SEAOC PV2 (2017). For this reason, dynamic net pressure and moment coefficients have been provided in Figure 29.4-11 for a wide range of reduced frequencies. For stiffer structures with reduced frequencies greater than 0.8, the dynamic coefficients from a reduced frequency of 0.8 may be conservatively used.

The dynamic coefficients represent an allowance for the resonant loading, while the static coefficients include both the mean and background loading components. The mean loading is the time-averaged component, the background loading is caused by fluctuating wind gusts or turbulence acting on the structure, whereas the resonant or inertial component is caused by the structure's inertia as it moves in its fundamental modes of vibration. The background and resonant parts together make up the dynamic response of a structure to wind. It should be noted that the maximum total design forces and moments are well predicted using the combined static and dynamic wind load coefficients in this section. The dynamic component has been determined by subtracting the sum of the mean and background components from the total component, which has been obtained by summing the mean and the square root of the sum of squares of the background and resonant components, as shown in Browne et al. (2020) and is similar to Equation (C26.11-22). Consequently, the dynamic component is a mixture of the effects of resonance and background turbulence and is numerically smaller than the actual resonant dynamic component. Upward corrections to the dynamic load are necessary in order to determine the resonant component alone. This underestimate becomes significant when the resonant and background dynamic components are of comparable magnitude. In such cases, a correction factor up to 1.5 may be applied to the provided dynamic coefficients, should estimates of the full resonant component be required. The dynamic coefficients are also based on the conservative assumption that the mode shapes equal unity at all locations, which does not consider structural details, such as constraints at the support system. A more comprehensive dynamic analysis for a specific structure will likely result in lower dynamic wind loads compared to those obtained from the provisions in Section 29.4.5.

The dynamic coefficients in Figure 29.4-11 have been provided for a damping ratio of $\beta = 0.01$ and an approximate adjustment for higher values of damping (up to a maximum value of 0.05) is provided in Note 1 of the same figure. This adjustment is physically correct for the resonant component alone, but is an approximation when applied to the provided dynamic coefficients, which also include the effect of the background turbulence. The error in the peak total wind load, which can result from the use of this extrapolation, is expected to be in the range of ±5% for typical values of background and resonant loading.

To select the appropriate dynamic wind load coefficients from Figure 29.4-11, knowledge of the lowest natural frequency and structural damping in the primary vibration modes of the system is essential. A practical approach to obtain these dynamic properties for ground-mounted fixed-tilt solar panel systems is to conduct field measurements on representative full-scale systems. As discussed in Section C26.11, most computer software used in the analysis of structures would provide estimates of the natural frequencies of the structure being analyzed. General guidance on damping ratios for various types of structures is also provided in Section C26.11. The damping ratio used in Figure 29.4-11 should be the structural damping, rather than the total damping including aerodynamic damping. Aerodynamic damping may be experienced by a structure oscillating in air. The aerodynamic damping can be positive, adding to the structural damping, or negative, which can lead to unstable oscillations.

The wind forces and moments, derived using the combined static and dynamic wind load coefficients in this section, are appropriate for fixed-tilt ground-mounted systems meeting the requirements in Section 29.4.5 and assuming the wind-induced deflections are small, and thus, the system is not flexible. Flexible structures, such as unrestrained single-axis trackers, are prone to aeroelastic effects and torsional instabilities as discussed in Taylor and Browne (2020), which must be considered in the design of such systems.

C29.4.5.1 Scope The simulated scaled PV models used in the wind tunnel research were installed on a flat surface in the wind tunnel. For locations such as Hawaii, where the topographic effects are considered in the basic wind speed map and K_{zt} is assumed to be 1.0, the K_{zt} should be calculated in accordance with Equation (26.8-1). Where K_{zt} exceeds 1.0 based on Equation (26.8-1), Zone 1 should be designed as Zone 2.

C29.5 PARAPETS

Before the 2002 edition of ASCE 7, no provisions for the design of parapets were included because of the lack of direct research. In the 2002 edition of ASCE 7, a rational method was added based on the committee's collective experience, intuition, and judgment. In the 2005 edition, the parapet provisions were updated as a result of research performed at the University of Western Ontario (Mans et al. 2000, 2001) and at Concordia University (Stathopoulos et al. 2002a, b).

Wind pressures on a parapet are a combination of wall and roof pressures, depending on the location of the parapet and the direction of the wind (Figure C29.5-1). A windward parapet experiences the positive wall pressure on its front surface

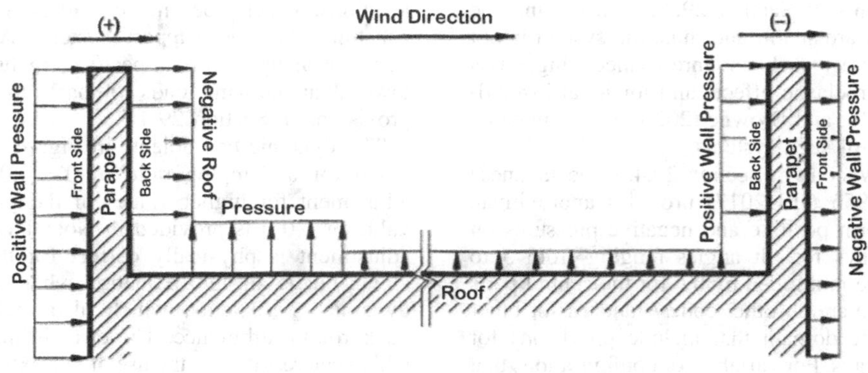

Methodology used to Develop External Parapet Pressures
(Main Wind Force Resisting Systems and Components and Cladding)

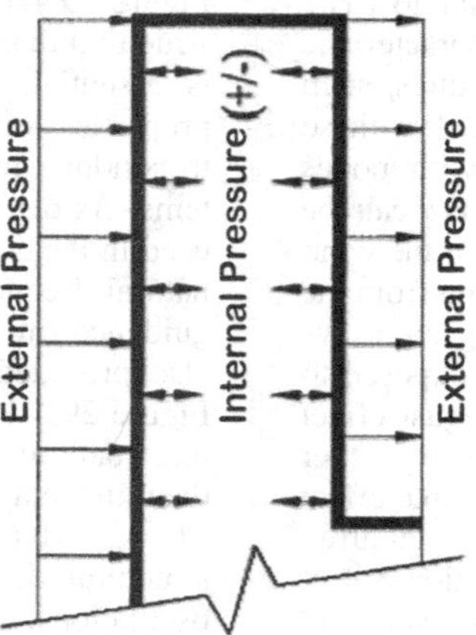

External and Internal Parapet Pressures
(Component and Cladding Only)

Figure C29.5-1. Design wind pressures on parapets.

(exterior side of the building) and the negative roof edge zone pressure on its back surface (roof side). This behavior is based on the concept that the zone of suction caused by the wind stream separation at the roof eave moves up to the top of the parapet when one is present. Thus, the same suction that acts on the roof edge also acts on the back of the parapet.

The leeward parapet experiences a positive wall pressure on its back surface (roof side) and a negative wall pressure on its front surface (exterior side of the building). There should be no reduction in the positive wall pressure to the leeward parapet caused by shielding by the windward parapet because, typically, they are too far apart to experience this effect. Because all parapets would be designed for all wind directions, each parapet would in turn be the windward and leeward parapet and, therefore, must be designed for both sets of pressures.

For the design of the main wind force resisting system (MWFRS), the pressures used describe the contribution of the parapet to the overall wind loads on that system. For simplicity, the front and back pressures on the parapet have been combined into one coefficient for MWFRS design. Typically, the designer should not need the separate front and back pressures for MWFRS design. The internal pressures inside the parapet cancel out in the determination of the combined coefficient. The summation of these external and internal, front and back pressure coefficients is a new term, (GC_{pn}), the combined net pressure coefficient for a parapet.

For the design of the C&C, a similar approach was used. However, it is not possible to simplify the coefficients because of the increased complexity of the C&C pressure coefficients. In addition, the front and back pressures are not combined because the designer may be designing separate elements on each face of the parapet. The internal pressure is required to determine the net pressures on the windward and leeward surfaces of the parapet. The provisions guide the designer to the correct (GC_p) and velocity pressure to use for each surface, as shown in Figure C29.5-1.

Interior walls that protrude through the roof, such as party walls and fire walls, should be designed as windward parapets for both MWFRS and C&C.

The internal pressure that may be present inside a parapet is highly dependent on the porosity of the parapet envelope. In other words, it depends on the likelihood of the wall surface materials to leak air pressure into the internal cavities of the parapet. For solid parapets, such as concrete or masonry, the internal pressure is zero because there is no internal cavity. Certain wall materials may be impervious to air leakage, and, as such, have little or no internal pressure or suction, so using the value of (GC_{pi}) for an enclosed building may be appropriate. However, certain materials and systems used to construct parapets containing cavities are more porous, thus justifying the use of the (GC_{pi}) values for partially enclosed buildings or higher. Another factor in the internal pressure determination is whether the parapet cavity connects to the internal space of the building, allowing the building's internal pressure to propagate into the parapet. Selection of the appropriate internal pressure coefficient is left to the judgment of the design professional.

C29.7 MINIMUM DESIGN WIND LOADING

This section specifies a minimum wind load to be applied horizontally on the entire vertical projection of the other structures, as shown in Figure C27.1-1. This load case is to be applied as a separate load case in addition to the normal load cases specified in other portions of this chapter.

REFERENCES

AASHTO (American Association of State Highway and Transportation Officials). 2013. *Standard specifications for structural supports for highway signs, luminaires, and traffic signals*. AASHTO LTS-6. Washington, DC: AASHTO.

ANSI (American National Standards Institute). 1972. *Minimum design loads for buildings and other structures*. ANSI A58.1-1972. Washington, DC: ANSI.

ANSI. 1982. *Minimum design loads for buildings and other structures*. ANSI A58.1-1982. Washington, DC: ANSI.

ASCE. 1961. "Wind forces on structures." *Trans. ASCE* 126 (2): 1124–1198.

ASCE. 1994. *Minimum design loads for buildings and other structures*. New York: ASCE.

ASCE Task Committee on Wind-Induced Forces. 2011. *Wind loads for petrochemical and other industrial facilities*. Reston, VA: ASCE.

Banks, D. 2012. "Wind loads on tilted flat panels on commercial roofs: The effects of corner vortices." In *Advances in hurricane engineering*, edited by C. P. Jones, and L. G. Griffis. Reston, VA: ASCE.

Browne, M. T. L., Z. J. Taylor, S. H. Li, and S. Gamble. 2020. "A wind load design method for ground-mounted multi-row solar arrays based on a compilation of wind tunnel experiments." *J. Wind Eng. Ind. Aerodyn.* 205 (Oct): 104294. https://doi.org/10.1016/j.jweia.2020.104294.

Cook, N. J. 1990. *The designer's guide to wind loading of building structures, Part II*. London: Butterworths.

Erwin, J. W., A. G. Chowdhury, and G. Bitsuamlak. 2011. "Wind loads on rooftop equipment mounted on a flat roof." *J. Wind Eng.* 8 (1): 23–42.

Fox, T., and M. Levitan. 2005. "A comprehensive look at wind loading on freestanding walls and signs. In *Proc., 10th Americas Conf. on Wind Engineering*. Baton Rouge, LA.

Giannoulis, A., T. Stathopoulos, D. Briassoulis, and A. Mistriotis. 2012. "Wind loading on vertical panels with different permeabilities." *J. Wind Eng. Ind. Aerodyn.* 107 (Aug): 1–16. https://doi.org/10.1016/j.jweia.2012.02.014.

Ginger, J. D., G. F. Reardon, and B. L. Langtree. 1998a. "Wind loads on fences and hoardings." In *Proc., Australasian Structure and Engineering Conf.*, Townsville, Australia: James Cook University Cyclone Structural Testing Station. 983–990.

Ginger, J. D., G. F. Reardon, and B. L. Langtree. 1998b. *Wind loads on fences and hoardings*. Townsville, QLD: James Cook University.

Holmes, J. D. 1986. *Wind tunnel tests on free-standing walls at CSIRO*. Internal Rep. No. 86/47. Clayton, South, VIC: CSIRO Division of Building Research.

Hosoya, N., J. E. Cermak, and C. Steele. 2001. "A wind-tunnel study of a cubic rooftop AC unit on a low building." In *Proc., Americas Conf. on Wind Engineering*. Fort Collins, CO: American Association for Wind Engineering.

IASS (International Association for Shell and Spatial Structures). 1981. *Recommendations for guyed masts*. Madrid, Spain: IASS.

JIS (Japanese Standards Association). 2017. *Load design guide on structures for photovoltaic array*. C 8955. Tokyo: JIS.

Kopp, G. A. 2013. "Wind loads on low profile, tilted, solar arrays placed on large, flat, low-rise building roofs." *J. Struct. Eng.* 140 (2): 04013057. https://doi.org/10.1061/(ASCE)ST.1943-541X.0000821.

Kopp, G. A., S. Farquhar, and M. J. Morrison. 2012. "Aerodynamic mechanisms for wind loads on tilted, roof-mounted, solar arrays." *J. Wind Eng. Ind. Aerodyn.* 111 (Dec): 40–52. https://doi.org/10.1016/j.jweia.2012.08.004.

Kopp, G. A., and G. Traczuk. 2007. *Wind loads on a roof-mounted cube*. BLWT-SS47_2007. London: Boundary Layer Wind Tunnel Laboratory.

Letchford, C. W. 1985. *Wind loads on free-standing walls*. Rep. No. OUEL 1599/85. Oxford, UK: University of Oxford.

Letchford, C. W. 1989. "Wind loads and overturning moments on free standing walls." In *Proc., 2nd Asia Pacific Symp. on Wind Engineering*. Kanagawa, Japan.

Letchford, C. W. 2001. "Wind loads on rectangular signboards and hoardings." *J. Wind Eng. Ind. Aerodyn.* 89 (Feb): 135–151. https://doi.org/10.1016/S0167-6105(00)00068-4.

Letchford, C. W., and J. D. Holmes. 1994. "Wind loads on free-standing walls in turbulent boundary layers." *J. Wind Eng. Ind. Aerodyn.* 51 (1): 1–27. https://doi.org/10.1016/0167-6105(94)90074-4.

Letchford, C. W., and A. P. Robertson. 1999. "Mean wind loading at the leading ends of free-standing walls." *J. Wind Eng. Ind. Aerodyn.* 79 (1): 123–134. https://doi.org/10.1016/S0167-6105(97)00292-4.

Macdonald, P. A., J. D. Holmes, and K. C. S. Kwok. 1990. "Wind loads on circular storage bins, silos and tanks. II. Effect of grouping." *J. Wind Eng. Ind. Aerodyn.* 34 (1): 77–95. https://doi.org/10.1016/0167-6105(90)90150-B.

Macdonald, P. A., K. C. S. Kwok, and J. D. Holmes. 1988. "Wind loads on circular storage bins, silos and tanks: 1. Point pressure measurements on isolated structures." *J. Wind Eng. Ind. Aerodyn.* 31 (2–3): 165–187. https://doi.org/10.1016/0167-6105(88)90003-7.

Mans, C., G. Kopp, and D. Surry. 2000. *Wind loads on parapets, Part 1*. BLWTL-SS23-2000. London, ONT: University of Western Ontario.

Mans, C., G. Kopp, and D. Surry. 2001. *Wind loads on parapets, Parts 2 and 3*. BLWT-SS37-2001 and BLWT-SS38-2001. London, ON: University of Western Ontario.

Mehta, K. C., D. A. Smith, and D. Zuo. 2012. *Field and wind tunnel testing of signs, final report, test procedures and outcomes*. Lubbock, TX: Texas Tech University.

Meyer, D., I. Zisis, B. Hajra, A. G. Chowdhury, and P. Irwin. 2017. "An experimental study in the wind-induced response of variable message signs." *Front. Built Environ.* 3 (Nov): 1–10. https://doi.org/10.3389/fbuil.2017.00066.

NAAMM (National Association of Architectural Metal Manufacturers). 2007. *Guide specifications for design of metal flagpoles*. ANSI/NAAMM FP 1001-13. Glen Ellyn, IL: NAAMM.

Robertson, A. P., R. P. Hoxey, J. L. Short, and W. A. Ferguson. 1997. "Full scale measurements and computational predictions of wind loads on free standing walls." *J. Wind Eng. Ind.*

Aerodyn. 67–68 (Apr): 639–646. https://doi.org/10.1016/S0167-6105(96)00106-7.

Robertson, A. P., R. P. Hoxey, J. L. Short, W. A. Ferguson, and S. Osmond. 1995. "Wind loads on free-standing walls: A full-scale study." In *Proc., 9th Int. Conf. on Wind Engineering*, 457–468.

Robertson, A. P., R. P. Hoxey, J. L. Short, W. A. Ferguson, and S. Osmond. 1996. "Full-scale testing to determine the wind loads on free-standing walls." *J. Wind Eng. Ind. Aerodyn.* 60 (1): 123–137. https://doi.org/10.1016/0167-6105(96)00028-1.

Sabransky, I. J., and W. H. Melbourne. 1987. "Design pressure distribution on circular silos with conical roofs." *J. Wind Eng. Ind. Aerodyn.* 26 (1): 65–84. https://doi.org/10.1016/0167-6105(87)90036-5.

SEAOC (Structural Engineers Association of California). 2012. *Wind loads on low profile solar photovoltaic system on flat roofs.* Rep. No. SEAOC-PV2-2012. Sacramento, CA: SEAOC.

SEAOC. 2017. *Wind design for solar arrays.* PV2-2017. San Diego: SEAOC.

SIA (Swiss Society of Engineers and Architects). 1956. *Normen fur die Belastungsannahmen, die Inbetriebnahme und die Uberwachung der Bauten.* SIA Technische Normen No. 160. Zurich, Switzerland: SIA.

Smith, D. A., D. Zuo, and K. C. Mehta. 2014. "Characteristics of wind induced net force and torque on a rectangular sign measured in the field." *J. Wind Eng. Ind. Aerodyn.* 133 (Oct): 80–91. https://doi.org/10.1016/j.jweia.2014.07.010.

Standards Australia. 2002. *Structural design actions, Part 2: Wind actions.* AS/NZS 1170.2:2002. Sydney, NSW: Standards Australia.

Standards Australia. 2011. *Structural design actions—Wind actions.* AS/NZS 1170.2:2011. Sydney, NSW: Standards Australia.

Stathopoulos, T., P. Saathoff, and R. Bedair. 2002a. "Wind pressures on parapets of flat roofs." *J. Archit. Eng.* 8 (2): 49–54. https://doi.org/10.1061/(ASCE)1076-0431(2002)8:2(49).

Stathopoulos, T., P. Saathoff, and X. Du. 2002b. "Wind loads on parapets." *J. Wind Eng. Ind. Aerodyn.* 90 (4–5): 503–514. https://doi.org/10.1016/S0167-6105(01)00206-9.

Stenabaugh, S. E., Y. Iida, G. A. Kopp, and P. Karava. 2015. "Wind loads on photovoltaic arrays mounted on sloped roofs of low-rise building, parallel to the roof surface." *J. Wind Eng. Ind. Aerodyn.* 139 (4): 16–26. https://doi.org/10.1016/j.jweia.2015.01.007.

Strobel, K., and D. Banks. 2014. "Effects of vortex shedding in arrays of long inclined flat plates and ramifications for ground-mounted photovoltaic arrays." *J. Wind Eng. Ind. Aerodyn.* 133 (Oct): 146–149. https://doi.org/10.1016/j.jweia.2014.06.013.

Taylor, Z. J., and M. T. L. Browne. 2020. "Hybrid pressure integration and buffeting analysis for multi-row wind loading in an array of single-axis trackers." *J. Wind Eng. Ind. Aerodyn.* 197 (Feb): 104056. https://doi.org/10.1016/j.jweia.2019.104056.

TIA (Telecommunications Industry Association). 1991. *Structural standards for steel antenna towers and antenna supporting structures.* ANSI/EIA/TIA 222-E. Arlington, VA: TIA.

Zuo, D., C. W. Letchford, and S. Wayne. 2011. "Wind tunnel study of wind loading on rectangular louvered panels." *Wind Struct.* 14 (5): 449–463.

Zuo, D., D. A. Smith, and K. C. Mehta. 2014. "Experimental study of wind loading of rectangular sign structures." *J. Wind Eng. Ind. Aerodyn.* 130 (0): 62–74. https://doi.org/10.1016/j.jweia.2014.04.005.

CHAPTER C30
WIND LOADS: COMPONENTS AND CLADDING

In developing the set of pressure coefficients applicable for the design of components and cladding (C&C) as given in Figures 30.3-1, 30.3-2A–G, 30.3-3, 30.3-4, 30.3-5A–B, and 30.3-6, an envelope approach was followed but using different methods than for the main wind force resisting system (MWFRS) of Figure 28.3-1. Because of the small effective area that may be involved in the design of a particular component (consider, for example, the effective area associated with the design of a fastener), the pointwise pressure fluctuations may be highly correlated over the effective area of interest.

Consider the local purlin loads shown in Figure C28.3-1. The approach involved spatial averaging and time averaging of the point pressures over the effective area transmitting loads to the purlin while the building model was permitted to rotate in the wind tunnel through 360 degrees. As the induced localized pressures may also vary widely as a function of the specific location on the building, height above ground level, exposure, and, more important, local geometric discontinuities and the location of the element relative to the boundaries in the building surfaces (e.g., walls, roof lines), these factors were also enveloped in the wind tunnel tests. Thus, for the pressure coefficients given in Figures 30.3-1, 30.3-2A–G, 30.3-3, 30.3-4, 30.3-5A–B, and 30.3-6, the directionality of the wind and influence of exposure have been removed and the surfaces of the building have been "zoned" to reflect an envelope of the peak pressures possible for a given design application.

For ASCE 7-16, the roof zones and pressure coefficients for Figure 30.3-2A were modified based on the analysis by Kopp and Morrison (2014), which made use of the extensive wind tunnel database developed by Ho et al. (2005). St. Pierre et al. (2005) provided an evaluation of this database compared to earlier data by Davenport et al. (1977, 1978) and ASCE 7 (2002), while Ho et al. (2005) compared the data to full-scale field data from Texas Tech University (Mehta and Levitan 1998). All source data used in the study are publicly accessible through the NIST website (see e.g. Main and Fritz 2006). Compared to previous versions of ASCE 7, the pressure coefficients have been increased and are now more consistent with coefficients for buildings higher than 60 ft (18.3 m). Roof zone sizes are also modified from those of earlier versions to minimize the increase of pressure coefficients in Zones 1 and 2. The data indicate that for these low-rise buildings, the size of the roof zones depends primarily on the building height, h. Zone 1 now occurs for large buildings, which accounts for the lower wind loads in the middle of the roof. Zone 3 (roof corner) is an L shape, consistent with the shape of Zone 3 for buildings higher than 60 ft (18.3 m) and consistent with the wind loading data. Four potential zone configurations based on the ratios of the smallest and largest building plan dimensions are illustrated in Figure C30-1. In addition, when the greatest horizontal dimension is less than $0.4h$ (the building does not correspond to a typical low-rise building shape), there is a single roof zone (Zone 3). Detailed explanations can be found in Kopp and Morrison (2014).

As indicated in the discussion for Figure 28.3-1, the wind tunnel experiments checked both Exposure B and C terrains. Basically, (GC_p) values associated with Exposure B terrain would be higher than those for Exposure C terrain because of reduced velocity pressure in Exposure B terrain. The (GC_p) values given in Figures 30.3-1, 30.3-2A–G, 30.3-3, 30.3-4, 30.3-5A–B, and 30.3-6 are associated with Exposure C terrain as obtained in the wind tunnel. However, they may also be used for any exposure when the correct velocity pressure representing the appropriate exposure is used. The (GC_p) values given in Figure 30.3-2A–C are associated with wind tunnel tests performed in both Exposures B and C.

For Figure 30.3-2A, the coefficients apply equally to Exposures B and C, based on wind tunnel data that show insignificant differences in (GC_p) for Exposures B and C. Consequently, the truncation for K_z in Table 30.3-1 of ASCE 7-10 is not required for buildings below 30 ft (9.1 m), and the lower K_z values may be used.

The pressure coefficients given in Figure 30.4-1 for buildings with mean height greater than 60 ft (18.3 m) were developed following a similar approach, but the influence of exposure was not enveloped (Stathopoulos and Dumitrescu-Brulotte 1989). Therefore, Exposure Category B, C, or D may be used with the values of (GC_p) in Figure 30.4-1, as appropriate.

C30.1 SCOPE

C30.1.1 Building Types Guidance for determining C_f and A_f for C&C of structures found in petrochemical and other industrial facilities that are not otherwise addressed in ASCE 7 can be found in *Wind Loads for Petrochemical and Other Industrial Facilities* (ASCE Task Committee on Wind-Induced Forces 2011). The 2011 edition references ASCE 7-05, and the user needs to make appropriate adjustments where compliance with the ASCE 7-10 standard is required.

C30.1.5 Air-Permeable Cladding Air-permeable roof or wall claddings allow partial air pressure equalization between their exterior and interior surfaces. Examples include siding, pressure-equalized rain screen walls, shingles, tiles (including modular vegetative roof assemblies), concrete roof pavers, and aggregate roof surfacing.

The peak pressure acting across an air-permeable cladding material is dependent on the characteristics of other components or layers of a building envelope assembly. At any given instant, the total net pressure across a building envelope assembly is equal to the sum of the partial pressures across the individual layers, as shown in Figure C30.1-1. However, the proportion of

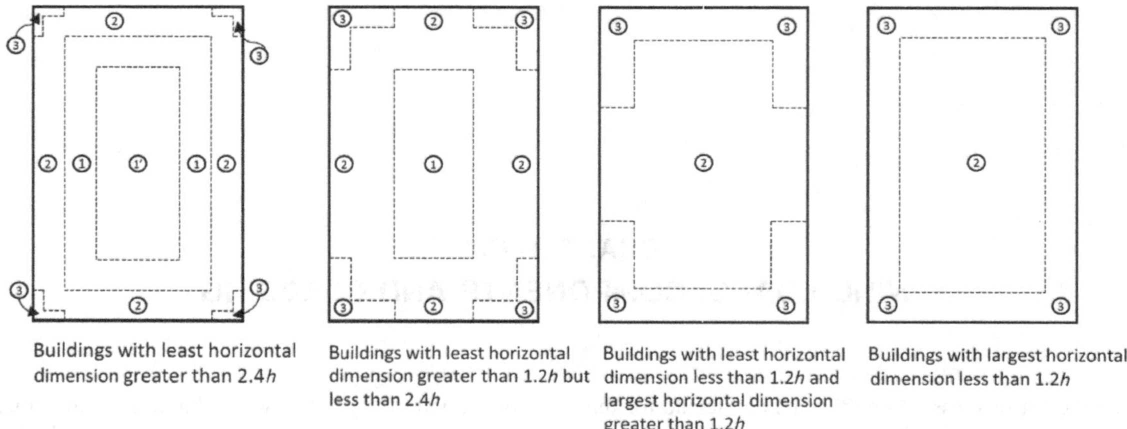

Figure C30-1. Four possible scenarios for roof zones, which depend on the ratios of the least and largest horizontal plan dimensions to the mean roof height, *h*.

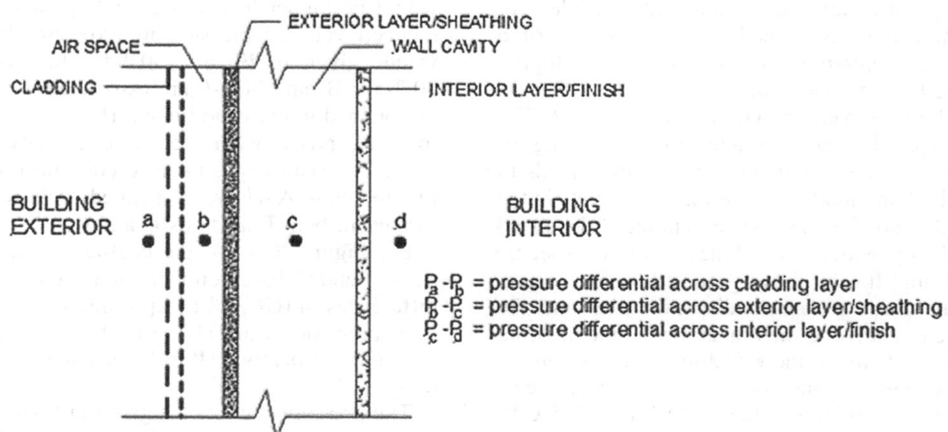

Figure C30.1-1. Distribution of net components and cladding pressure acting on a building surface (building envelope) composed of three components (layers).

the total net pressure borne by each layer varies from instant to instant because of fluctuations in the external and internal pressures and depends on the porosity and stiffness of each layer, as well as the volumes of the air spaces between the layers. As a result, although there is load sharing among the various layers, the sum of the peak pressures across the individual layers typically exceeds the peak pressure across the entire system. In the absence of detailed information on the division of loads, a simple, conservative approach is to assign the entire differential pressure to each layer designed to carry load.

To maximize pressure equalization (reduction) across any cladding system (irrespective of the permeability of the cladding itself), the layer or layers behind the cladding should be

- Relatively stiff in comparison to the cladding material; and
- Relatively air-impermeable in comparison to the cladding material.

Furthermore, the air space between the cladding and the next adjacent building envelope surface behind the cladding (e.g., the exterior sheathing) should be as small as practicable and compartmentalized to avoid communication or venting between different pressure zones of a building's surfaces.

The design wind pressures derived from Chapter 30 represent the pressure differential between the exterior and interior surfaces of the exterior envelope (wall or roof system). Because of the partial air-pressure equalization provided by air-permeable claddings, the C&C pressures derived from Chapter 30 can overestimate the load on air-permeable cladding elements. The designer may elect either to use the loads derived from Chapter 30 or to use loads derived by an approved alternative method. If the designer desires to determine the pressure differential across a specific cladding element in combination with other elements comprising a specific building envelope assembly, appropriate pressure measurements should be made on the applicable building envelope assembly, or reference should be made to recognized literature (Cheung and Melbourne 1986; Haig 1990; Baskaran 1992; SBCCI 1994; Peterka et al. 1997; ASTM 2006, 2007; Kala et al. 2008; Baskaran et al. 2012; Kopp and Gavanski 2012; Cope et al. 2012) for documentation pertaining to wind loads. Such alternative methods may vary according to a given cladding product or class of cladding products or assemblies, because each has unique features that affect pressure equalization. It is important to consider the methodology used to determine wind pressure distribution through a multilayered

Table C30.3-1. Walls for Buildings with $h \leq 60$ ft (18.3 m) (Figure 30.3-1).

Positive:	$GC_p = 1.0$	$A = 10\,\text{ft}^2$
Zones 4 and 5	$GC_p = 1.1766 - 0.1766 \log A$	$10 < A \leq 500\,\text{ft}^2$
	$GC_p = 0.7$	$A > 500\,\text{ft}^2$
Negative:	$GC_p = -1.1$	$A = 10\,\text{ft}^2$
Zone 4	$GC_p = -1.2766 + 0.1766 \log A$	$10 < A \leq 500\,\text{ft}^2$
	$GC_p = -0.8$	$A > 500\,\text{ft}^2$
Negative:	$GC_p = -1.4$	$A = 10\,\text{ft}^2$
Zone 5	$GC_p = -1.7532 + 0.3532 \log A$	$10 < A \leq 500\,\text{ft}^2$
	$GC_p = -0.8$	$A > 500\,\text{ft}^2$

Table C30.3-2. Gable Roof, $\theta \leq 7°$ (Figure 30.3-2A).

	Positive with and without overhang	
All zones	$GC_p = 0.3$	$A \leq 10\,\text{ft}^2$
	$GC_p = 0.4000 - 0.1000 \log A$	$10 \leq A \leq 100\,\text{ft}^2$
	$GC_p = 0.2$	$A \geq 100\,\text{ft}^2$
	Negative without overhang	
Zone 1′	$GC_p = -0.9$	$A \leq 100\,\text{ft}^2$
	$GC_p = -1.9000 + 0.5000 \log A$	$100 \leq A \leq 1{,}000\,\text{ft}^2$
	$GC_p = 1.1766 - 0.1766 \log A$	$A \geq 1{,}000\,\text{ft}^2$
Zone 1	$GC_p = -1.7$	$A \leq 10\,\text{ft}^2$
	$GC_p = -2.1120 + 0.4120 \log A$	$10 \leq A \leq 500\,\text{ft}^2$
	$GC_p = -1.0$	$A \geq 500\,\text{ft}^2$
Zone 2	$GC_p = -2.3$	$A \leq 10\,\text{ft}^2$
	$GC_p = -2.8297 + 0.5297 \log A$	$10 \leq A \leq 500\,\text{ft}^2$
	$GC_p = -1.4$	$A \geq 500\,\text{ft}^2$
Zone 3	$GC_p = -3.2$	$A \leq 10\,\text{ft}^2$
	$GC_p = -4.2595 + 1.0595 \log A$	$10 \leq A \leq 500\,\text{ft}^2$
	$GC_p = -1.4$	$A \geq 500\,\text{ft}^2$
	Negative with overhang	
Zones 1 and 1′	$GC_p = -1.7$	$A \leq 10\,\text{ft}^2$
	$GC_p = -1.8000 + 0.1000 \log A$	$10 \leq A \leq 100\,\text{ft}^2$
	$GC_p = -3.3168 + 0.8584 \log A$	$100 \leq A \leq 500\,\text{ft}^2$
	$GC_p = -1.0$	$A \geq 500\,\text{ft}^2$
Zone 2	$GC_p = -2.3$	$A \leq 10\,\text{ft}^2$
	$GC_p = -3.0063 + 0.7063 \log A$	$10 \leq A \leq 500\,\text{ft}^2$
	$GC_p = -1.1$	$A \geq 500\,\text{ft}^2$
Zone 3	$GC_p = -3.2$	$A \leq 10\,\text{ft}^2$
	$GC_p = -4.4360 + 1.2360 \log A$	$10 \leq A \leq 500\,\text{ft}^2$
	$GC_p = -1.1$	$A \geq 500\,\text{ft}^2$

Table C30.3-3. Gable Roof, $7° < \theta \leq 20°$ (Figure 30.3-2B).

	Positive	
All zones	$GC_p = 0.6$	$A \leq 10\,\text{ft}^2$
	$GC_p = 0.8306 - 0.2306 \log A$	$10 \leq A \leq 200\,\text{ft}^2$
	$GC_p = 0.3$	$A \geq 200\,\text{ft}^2$
	Negative	
Zone 1	$GC_p = -2.0$	$A \leq 10\,\text{ft}^2$
	$GC_p = -3.0155 + 1.0155 \log A$	$10 \leq A \leq 300\,\text{ft}^2$
	$GC_p = -0.5$	$A \geq 300\,\text{ft}^2$
Zone 2	$GC_p = -2.7$	$A \leq 10\,\text{ft}^2$
	$GC_p = -4.0067 + 1.3066 \log A$	$10 \leq A \leq 200\,\text{ft}^2$
	$GC_p = -1.0$	$A \geq 200\,\text{ft}^2$
Zone 3	$GC_p = -3.6$	$A \leq 10\,\text{ft}^2$
	$GC_p = -5.400 + 1.800 \log A$	$10 \leq A \leq 100\,\text{ft}^2$
	$GC_p = -1.8$	$A \geq 100\,\text{ft}^2$

assembly including an air-permeable cladding layer. Recent full-scale wind tunnel tests have shown that an accurate distribution of the wind pressure in a multilayered exterior wall assembly must account for the spatial and temporal (dynamic) fluctuations in wind pressure representative of actual wind flow conditions (Cope et al. 2012). Other factors to consider include the influence of airflow pathways through the assembly (e.g., openings or penetrations through any given layer) and appropriate methods of enveloping peak pressure coefficients for each layer of a multi-layered assembly (e.g., Cope et al. 2012) to ensure system reliability and consistency with the characterization of peak pressure coefficients in this standard.

Modular Vegetative Roof Assemblies consist of vegetation and other components integrated as a tray. These trays have vertical air gaps [a minimum of 0.25 in. (6.25 mm)] between the module and roofing system and horizontal air gaps between them. These air gaps allow partial air pressure equalization.

C30.3 BUILDING TYPES

C30.3.1 Conditions For velocity pressure, see commentary, Section C26.10.1.

C30.3.2 Design Wind Pressures For velocity pressure, see commentary, Section C26.10.1.

Figures 30.3-1 and 30.3-2A–G. The pressure coefficient values provided in these figures are to be used for buildings with a mean roof height of 60 ft (18.3 m) or less. The values were obtained from wind tunnel tests conducted at the University of Western Ontario (Davenport et al. 1977, 1978; Ho et al. 2005; St. Pierre et al. 2005; Kopp and Morrison 2014; Vickery et al. 2011; Gavanski et al. 2013). The negative roof GC_p values given in these figures are greater (in magnitude) than those given in previous versions (2010 and earlier) but are consistent with those given by Ho et al. (2005). The GC_p values given in the figures are given in equation form in Tables C30.3-1 to C30.3-10. Note that the GC_p values given in Figure 30.3-2A–G are a function of the roof slope. There has been an effort made for ASCE 7-22 to make the graphs and roof zones simpler than in ASCE 7-16. Most of the highest and lowest (GC_p) values have not changed except where zones have been merged or where the zones were modified to better fit the actual wind tunnel data. The smallest effective wind areas (EWAs) have been truncated at 10 ft^2 (0.93 m^2). There was not a large amount of wind tunnel data for EWAs smaller than 10 ft^2 (0.93 m^2) available for the graphs used in ASCE 7-16, and thus the GC_p values for some shapes and slopes have been reduced, since the GC_p values do not exceed that value established at the 10 ft^2 (0.93 m^2) cutoff. Some changes have resulted in greater pressures in ASCE 7-22 than in ASCE 7-16; primarily these increases are evident for gable roofs with a slope of 7 to 20 degrees, and Zones 2 and 3 on hip roofs with a slope of 27 to 45 degrees. To interpolate GC_p values for hip roofs with slopes between 27 and 45 degrees, the effective wind areas for both slopes must be the same. It is the judgement of the Wind Load Subcommittee that with small EWAs, there is significant load that occurs between cladding elements, thus distributing the effects of the high localized

Table C30.3-4. Gable Roofs, $20° < \theta \le 27°$ (Figure 30.3-2C).

	Positive	
All zones	$GC_p = 0.6$	$A \le 10\ \text{ft}^2$
	$GC_p = 0.8306 - 0.2306 \log A$	$10 \le A \le 200\ \text{ft}^2$
	$GC_p = 0.3$	$A \ge 200\ \text{ft}^2$
	Negative	
Zone 1	$GC_p = -1.5$	$A \le 10\ \text{ft}^2$
	$GC_p = -2.0380 + 0.5380 \log A$	$10 \le A \le 200\ \text{ft}^2$
	$GC_p = -0.8$	$A \ge 200\ \text{ft}^2$
Zone 2	$GC_p = -2.5$	$A \le 10\ \text{ft}^2$
	$GC_p = 3.800 + 1.300 \log A$	$10 \le A \le 100\ \text{ft}^2$
	$GC_p = -1.2$	$A \ge 100\ \text{ft}^2$
Zone 3	$GC_p = -3.0$	$A \le 10\ \text{ft}^2$
	$GC_p = 4.600 + 1.600 \log A$	$10 \le A \le 100\ \text{ft}^2$
	$GC_p = -1.4$	$A \ge 100\ \text{ft}^2$

Table C30.3-5. Gable Roofs, $27° < \theta \le 45°$ (Figure 30.3-2D).

	Positive	
All zones	$GC_p = 0.9$	$A \le 10\ \text{ft}^2$
	$GC_p = 1.2074 - 0.3074 \log A$	$10 \le A \le 200\ \text{ft}^2$
	$GC_p = 0.5$	$A \ge 200\ \text{ft}^2$
	Negative	
Zone 1	$GC_p = -1.8$	$A \le 10\ \text{ft}^2$
	$GC_p = -2.800 + 1.000 \log A$	$10 \le A \le 100\ \text{ft}^2$
	$GC_p = -0.8$	$A \ge 100\ \text{ft}^2$
Zone 2	$GC_p = -2.0$	$A \le 10\ \text{ft}^2$
	$GC_p = -2.7686 + 0.76862 \log A$	$10 \le A \le 200\ \text{ft}^2$
	$GC_p = -1.0$	$A \ge 200\ \text{ft}^2$
Zone 3	$GC_p = -2.5$	$A \le 10\ \text{ft}^2$
	$GC_p = -3.6529 + 1.1529 \log A$	$10 \le A \le 200\ \text{ft}^2$
	$GC_p = -1.0$	$A \ge 200\ \text{ft}^2$

Table C30.3-6. Hip Roofs, $7° < \theta \le 20°$ (Figure 30.3-2E).

	Positive	
All zones	$GC_p = 0.7$	$A \le 10\ \text{ft}^2$
	$GC_p = 1.100 - 0.400 \log A$	$10 \le A \le 100\ \text{ft}^2$
	$GC_p = 0.3$	$A \ge 100\ \text{ft}^2$
	Negative	
Zone 1	$GC_p = -1.8$	$A \le 10\ \text{ft}^2$
	$GC_p = -2.5686 + 0.7686 \log A$	$10 \le A \le 200\ \text{ft}^2$
	$GC_p = -0.8$	$A \ge 200\ \text{ft}^2$
Zone 2	$GC_p = -2.4$	$A \le 10\ \text{ft}^2$
	$GC_p = -3.2455 + 0.8455 \log A$	$10 \le A \le 200\ \text{ft}^2$
	$GC_p = -1.3$	$A \ge 200\ \text{ft}^2$
Zone 3	$GC_p = -2.6$	$A \le 10\ \text{ft}^2$
	$GC_p = -3.5224 + 0.9223 \log A$	$10 \le A \le 200\ \text{ft}^2$
	$GC_p = -1.4$	$A \ge 200\ \text{ft}^2$

pressures presented in ASCE 7-16. Some of the characteristics of the values in the figure are as follows:

1. Values are combined values of (GC_p). Gust-effect factors from these values should not be separated.

Table C30.3-8. Hip Roofs, $20° < \theta \le 27°$ (Figure 30.3-2F).

	Positive	
All zones	$GC_p = 0.7$	$A \le 10\ \text{ft}^2$
	$GC_p = 1.100 - 0.400 \log A$	$10 \le A \le 100\ \text{ft}^2$
	$GC_p = 0.3$	$A \ge 100\ \text{ft}^2$
	Negative	
Zone 1	$GC_p = -1.4$	$A \le 10\ \text{ft}^2$
	$GC_p = -2.000 + 0.600 \log A$	$10 \le A \le 100\ \text{ft}^2$
	$GC_p = -0.8$	$A \ge 100\ \text{ft}^2$
Zones 2	$GC_p = -2.0$	$A \le 10\ \text{ft}^2$
and 3	$GC_p = -3.000 + 1.000 \log A$	$10 \le A \le 100\ \text{ft}^2$
	$GC_p = -1.0$	$A \ge 100\ \text{ft}^2$

Table C30.3-9. Hip Roofs, $\theta = 45°$ (Figure 30.3-2G).

	Positive	
Zone 1	$GC_p = 0.7$	$A \le 10\ \text{ft}^2$
	$GC_p = 1.100 - 0.400 \log A$	$10 \le A \le 100\ \text{ft}^2$
	$GC_p = 0.3$	$A \ge 100\ \text{ft}^2$
	Negative	
Zone 1	$GC_p = -1.5$	$A \le 10\ \text{ft}^2$
	$GC_p = -2.300 + 0.800 \log A$	$10 \le A \le 100\ \text{ft}^2$
	$GC_p = -0.7$	$A \ge 100\ \text{ft}^2$
Zone 2	$GC_p = -1.8$	$A \le 10\ \text{ft}^2$
	$GC_p = -2.800 + 1.000 \log A$	$10 \le A \le 100\ \text{ft}^2$
	$GC_p = -0.8$	$A \ge 100\ \text{ft}^2$
Zone 3	$GC_p = -2.4$	$A \le 10\ \text{ft}^2$
	$GC_p = -3.800 + 1.400 \log A$	$10 \le A \le 100\ \text{ft}^2$
	$GC_p = -1.0$	$A \ge 100\ \text{ft}^2$

2. Velocity pressure, q_h, evaluated at mean roof height should be used with all values of (GC_p).
3. Values provided in the figure represent the upper bounds of the most severe values for any wind direction. The reduced probability that the design wind speed may not occur in the particular direction for which the worst pressure coefficient is recorded has not been included in the values shown in the figure.
4. Wind tunnel values, as measured, were based on the mean hourly wind speed. Values provided in the figures are the measured values divided by the 3 s dynamic gust pressure at mean roof height to adjust for the reduced pressure coefficient values associated with a 3 s gust speed.

Each C&C element should be designed for the maximum positive and negative pressures (including applicable internal pressures) acting on it. The pressure coefficient values should be determined for each C&C element on the basis of its location on the building and the effective area for the element. Research indicates that the pressure coefficients provided generally apply to façades with architectural features, such as balconies, ribs, and various façade textures (Stathopoulos and Zhu 1988, 1990).

Overhang pressures are determined by adding the appropriate roof coefficient from the graphs shown in Figures 30.3-2A–G to the adjacent positive wall coefficient from the graph shown in Figure 30.3-1. For example, for the total overhang uplift on an edge zone (Zone 3) of a hip roof with a slope of 27 degrees [$GC_p = -2.0$ for an EWA of $10\ \text{ft}^2$ ($0.93\ \text{m}^2$)], add the adjacent

Diagram

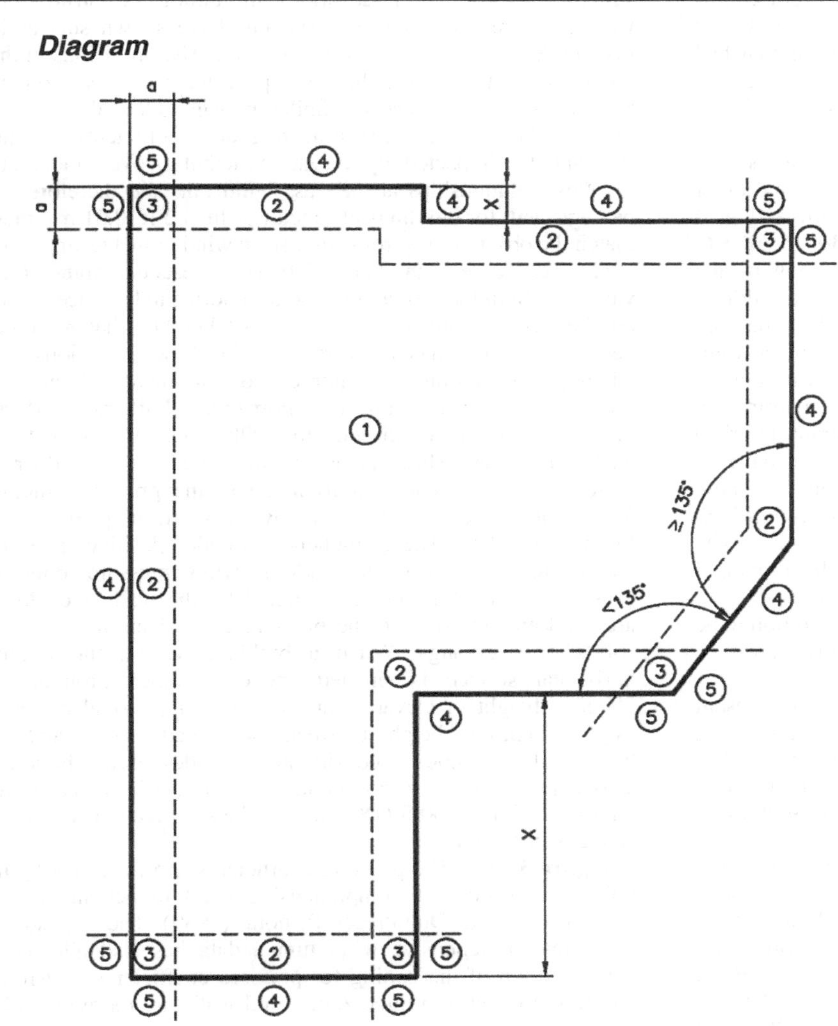

Notation

a = 10% of least horizontal dimension or $0.4h$, whichever is smaller, but not less than either 4% of least horizontal dimension or 3 ft (0.9 m).

X = Offset dimension, ft (m).

Notes

1. Labels marked on the roof plan indicate roof zones; labels marked outside the roof plan indicate wall zones.
2. If $X \leq a$, then Zone 3 and Zone 5 need not be applied at that corner.
3. If the interior angle is ≥135°, then Zone 3 and Zone 5 need not be applied at that corner.

Figure C30.3-2. Plan view of roof and wall zones for component and cladding loads on buildings with nonrectangular plans.

positive wall coefficient to determine the total uplift. The GC_p for Zone 4 on the wall is +1.0 for an EWA of 10 ft^2 (0.93 m^2), so the total uplift has a GC_p coefficient of −3.0. The positive wall pressure creates a positive or windward pressure on the underside of the overhang. That positive upward pressure is added to the uplift pressure on the upper side of the roof surface.

Although the smallest EWA shown on any graph is 10 ft^2 (0.93 m^2), the practitioner may still need to determine what the approximate GC_p should be for roof systems (e.g., tile roof) that may have EWAs of less than 10 ft^2 (0.93 m^2). This may be done by extending the sloped portion of the graphed coefficient line up to the required EWA and read the Y-axis for GC_p for the reduced EWA, while considering any load-sharing reduction factors from relevant studies or testing.

The coefficients for hip roof slopes of 20 to 27 degrees and 45 degrees have been modified so that there are the same number of zones with the same EWA limits; therefore, the roof coefficients can now be determined by interpolation for roof slopes between 27 and 45 degrees, eliminating the equations that had been needed to determine the slope before determining the roof GC_p coefficients. Zones 2 and 3 on a 45-degree slope should both be interpolated to Zone 3 on a 27-degree slope.

The following guidance is based on the collective judgment of the Wind Load Committee. For L-shaped, T-shaped, and other "irregular" shapes, Figure C30.3-2 depicts the roof and wall zones for use with Figures 30.3-1, 30.3-2, 30.3-4, 30.3-5, 30.3-6, and 30.4-1 for wind loads on C&C of buildings, showing the applicability to buildings that are rectangular in plan. To address buildings with nonrectangular plans, Figure C30.3-2 can be used for guidance in applying the requirements. When an outward corner protrudes less than the distance a from the wall, neither Zone 3 nor 5 are required; however, when the outward protrusion is greater than a, Zones 3 and 5 are required. Reentrant (interior) corners do not require Zones 3 or 5. For corners that have an included interior angle greater than 135 degrees, neither Zone 3 nor 5 is required. To determine the length of a, a rectangle which encloses the building is drawn over the building plan. The dimensions of this rectangle are used to determine the horizontal dimensions for the calculation of a.

Figure 30.3-3. This has been updated in the 2022 edition to be consistent with the changes to the flat roof C&C external pressure zones that appeared in the 2016 edition.

Figures 30.3-4, 30.3-5A, and 30.3-5B. These figures present values of (GC_p) for the design of roof C&C for buildings with multispan gable roofs and buildings with monoslope roofs. The coefficients are based on wind tunnel studies (Stathopoulos and Mohammadian 1986, Surry and Stathopoulos 1988, Stathopoulos and Saathoff 1991).

Figure 30.3-6. The values of (GC_p) in this figure are for the design of roof C&C for buildings with sawtooth roofs and mean roof height, h, less than or equal to 60 ft (18.3 m). Note that the coefficients for corner zones on segment A differ from those coefficients for corner zones on the segments designated as B, C, and D. Also, when the roof angle is less than or equal to 10 degrees, values of (GC_p) for regular gable roofs (Figure 30.3-2A) are to be used. The coefficients included in Figure 30.3-6 are based on wind tunnel studies reported by Saathoff and Stathopoulos (1992).

Figure 30.3-7. This figure for cladding pressures on dome roofs is based on Taylor (1991). Negative pressures are to be applied to the entire surface because they apply along the full arc that is perpendicular to the wind direction and that passes through the top of the dome. Users are cautioned that only three shapes were available to define values in this figure: $h_D/D=0.5$, $f/D=0.5$; $h_D/D=0.0, f/D=0.5$; and $h_D/D=0.0, f/D=0.33$.

Figure 30.3-8. The pressure and force coefficient values in these tables were based on the pressure and force coefficient values from ANSI A58.1-1982 (1982) and multiplied by 1.2. That multiplier was changed from 1.2 to 0.87 in ASCE 7-95 (1998), but no substantiation was provided for the change. The multiplier was changed back to 1.2 in ASCE 7-16 (2017) and incorporated into the values in this figure. The coefficients specified in these tables are based on wind tunnel tests conducted under conditions of uniform flow and low turbulence, and their validity in turbulent boundary-layer flows has yet to be completely established.

C30.3.2.1 Bottom Horizontal Surface of Elevated Buildings This section addresses the design wind pressures for the underside of the bottom flat horizontal surface of elevated buildings. The elevation of buildings on piers or other supporting structures exposes the underside of the building to airflow and wind pressures, and field observations have shown substantial floor underside cladding loss due to wind. The provisions in this section indicate that (1) the wind pressure coefficients on the bottom horizontal surface are similar in magnitude to those on the roof of the building, (2) roof pressure coefficients are not substantially impacted by elevation height above grade, and (3) these relationships are not as significant, and therefore can be neglected, for buildings elevated less than 2 ft (0.61 m); these match the observations from large-scale wind tunnel testing (Kim et al. 2020, Abdelfatah et al. 2020). Those tests determined GC_p values for both the roof and the bottom horizontal surfaces based on the velocity pressure at mean roof height. That reference velocity pressure has been retained in these provisions. An additional provision was introduced for increased positive pressure coefficients on the horizontal building surfaces directly above and adjacent to walls, and within partially enclosed spaces. These areas are illustrated by shaded areas around the small, enclosed room under the primary structure and the area labeled *partially enclosed area* in the plan view in Figure 30.3-1A. These higher-magnitude positive pressure coefficients are meant to address wind pressure build-up caused by wind flow being restricted by the wall or enclosed area and are set equal to the pressure coefficients used for the soffits of overhangs. When a building with a flat bottom horizontal surface is situated above a sloped ground, the effective height of elevation above grade, h_B, should be taken as the maximum height between the sloped ground and the bottom of the building considering all sides of the building. This approach leads to conservative wind coefficient zones for buildings with $h \leq 60$ ft (18.3 m), but does not lead to significant changes in loading.

Figure 30.4-1. The pressure coefficients shown in this figure reflect the results of comprehensive wind tunnel studies by Stathopoulos and Dumitrescu-Brulotte (1989). The availability of more-comprehensive wind tunnel data has also allowed a simplification of the zoning for pressure coefficients: flat roofs are now divided into three zones, and walls are represented by two zones.

The external pressure coefficients and zones given in Figure 30.4-1 were established by wind tunnel tests on isolated "boxlike" buildings (Akins and Cermak 1975, Peterka and Cermak 1975). Boundary-layer wind tunnel tests on high-rise buildings (mostly in downtown city centers) have identified variations in pressure coefficients and the distribution of pressure on the different building façades (Templin and Cermak 1978). These variations are caused by building geometry, low attached buildings, nonrectangular cross sections, setbacks, and sloping surfaces. Surrounding buildings also contribute to the variations in pressure. Wind tunnel tests indicate that pressure coefficients are not distributed symmetrically and can give rise to torsional wind loading on the building.

Boundary-layer wind tunnel tests that include modeling of surrounding buildings permit the establishment of more exact magnitudes and distributions of (GC_p) for buildings that are not isolated or "boxlike" in shape.

PART 2: BUILDINGS WITH $h > 60$ ft (18.3 m)

In Equation (30.4-1) a velocity pressure term, q_i, appears that is defined as the "velocity pressure for internal pressure determination." The positive internal pressure is dictated by the positive exterior pressure on the windward face at the point where there is an opening. The positive exterior pressure at the opening is governed by the value of q at the level of the opening, not q_h.

For positive internal pressure evaluation, q_i may conservatively be evaluated at height $h(q_i = g_h)$. For low buildings, this height does not make much difference, but for the example of a 300 ft (91.4 m) building in Exposure B with the highest opening at 60 ft (18.3 m), the difference between q_{300} and q_{60} represents a 59% increase in internal pressure. This increase is unrealistic and represents an unnecessary degree of conservatism. Accordingly, $q_i = q_z$ for positive internal pressure evaluation in partially enclosed buildings, where height z is defined as the level of the highest opening in the building that could affect the positive internal pressure.

C30.4.2.1 Bottom Horizontal Surface of Elevated Buildings This section provides guidance for determining design pressures for the bottom flat horizontal surface of elevated buildings with $h > 60$ ft (18.3 m). It follows the logic used to develop main wind force resisting system design wind loads for the bottom surface of elevated buildings (see C27.3.1.1), except that the GC_p values follow the rules for C&C pressure coefficients and are obtained from Figure 30.5-1. In contrast to C&C loads for the bottom horizontal surface of low-rise buildings, which use the velocity pressure at mean roof height, these provisions use a velocity pressure at a height equal to the elevation of the bottom surface of the elevated building plus 25% of the height of the building above, which is consistent with the provisions of Section 27.3.1.1 and as explained and illustrated in Section C27.3.1.1. The provision, which calls for increased positive pressure coefficients on the horizontal building surfaces directly above and adjacent to walls, and within partially enclosed spaces, used for low-rise buildings has also been applied to buildings with $h > 60$ ft (18.3 m), as explained in Section C30.3.2.1. These areas are illustrated by shaded areas around the small, enclosed room under the primary structure and the area labeled *partially enclosed area* in the plan view in Figure 30.5-1A. When a building with a flat bottom horizontal surface is situated above a sloped ground, the effective height of elevation above grade, h_B, should be taken as the maximum height between the sloped ground and the bottom of the building considering all sides of the building. This approach is used to ensure conservative velocity pressures for such buildings with $h > 60$ ft (18.3 m). For buildings elevated above a parking garage, the height of the elevation of the building is h_B, the distance between the bottom of the building above and the top of the parking deck surface below. However, the height used to calculate the reference wind pressure used to determine the magnitude of the wind loads should still be based on the height of the bottom surface of the elevated building above grade plus 25% of the height of the building above.

PART 3: OPEN BUILDINGS

C30.5 BUILDING TYPES

In determining loads on C&C elements for open building roofs using Figures 30.5-1, 30.5-2, and 30.5-3, it is important for the designer to note that the net pressure coefficient, C_N, is based on contributions from the top and bottom surfaces of the roof—that is, the element receives load from both surfaces. Such would not be the case if the surface below the roof were separated structurally from the top roof surface. In this case, the pressure coefficient should be separated for the effect of top and bottom pressures, or conservatively, each surface could be designed using the C_N value from Figures 30.5-1, 30.5-2, and 30.5-3.

PART 4: BUILDING APPURTENANCES, ROOFTOP STRUCTURES AND EQUIPMENT

C30.9 ATTACHED CANOPIES ON BUILDINGS

The provisions result from wind tunnel test results on pressures applied on horizontal canopies described by Zisis and Stathopoulos (2010), Zisis et al. (2011), Candelario et al. (2014), Zisis et al. (2017), Sakib et al. (n.d.), and Naeiji et al. (2020). Restrictions to canopies that are essentially flat (maximum slope: 2%) are based on a lack of test data. Figures 30.9-1A and 30.9-1B are to be used for buildings under 60 ft (18.3 m) high, and Figures 30.9-2A and 30.9-2B are to be used for buildings over 60 ft (18.3 m) high. Canopies are different from roof overhangs, which are simply extensions of the roof surfaces at the same slope with the roof.

In a canopy with two physical surfaces, both Figures 30.9-1A and 30.9-1B [buildings with height <60 ft (<18.3 m)] and both Figures 30.9-2A and 30.9-2B [buildings with height >60 ft (>18.3 m)] would be needed.

Figures 30.9-1A and 30.9-2A, which provide the coefficients on separate surfaces, would be used to design the fasteners of the top and soffit elements. Figures 30.9-1B and 30.9-2B would be used to design the structure of the canopy (e.g., joists, posts, and building fasteners). In a canopy with one physical surface, only Figures 30.9-1B and 30.9-2B are needed.

The (GC_p) values given in the figures are given in equation form in Tables C30.9-1 to C30.9-4.

Table C30.9-1. Pressure Coefficients on Separate Surfaces of Attached Canopies on Buildings with $h \le 60$ ft (18.3 m) (Figure 30.9-1A).

Negative:	$GC_p = -1.15$	$A = 10\text{ ft}^2$
Upper surface	$GC_p = -1.55 + 0.1737\log(A)$	$10 < A \le 100\text{ ft}^2$
	$GC_p = -0.75$	$A > 100\text{ ft}^2$
Negative:	$GC_p = -0.8$	$A = 10\text{ ft}^2$
Lower surface	$GC_p = -0.95 + 0.0651\log(A)$	$10 < A \le 100\text{ ft}^2$
	$GC_p = -0.65$	$A > 100\text{ ft}^2$
Positive:	$GC_p = 0.8$	$A = 10\text{ ft}^2$
Upper and	$GC_p = 1.0 - 0.087\log(A)$	$10 < A \le 100\text{ ft}^2$
lower surfaces	$GC_p = 0.6$	$A > 100\text{ ft}^2$

Table C30.9-2. Net Pressure Coefficients Canopies Considering Simultaneous Contributions from Upper and Lower Surfaces on Attached Canopies on Buildings with $h \le 60$ ft (18.3 m) (Figure 30.9-1B).

Negative:	$GC_p = -1.4$	$A = 10\text{ ft}^2$
$0.9 \le h_c/$	$GC_p = -1.7 + 0.1303\log(A)$	$10 < A \le 100\text{ ft}^2$
$h_e \le 1$	$GC_p = -1.1$	$A > 100\text{ ft}^2$
Negative:	$GC_p = -0.9$	$A = 10\text{ ft}^2$
$0.5 < h_c/$	$GC_p = -1.15 + 0.1086\log(A)$	$10 < A \le 100\text{ ft}^2$
$h_e < 0.9$	$GC_p = -0.65$	$A > 100\text{ ft}^2$
Negative:	$GC_p = -0.6$	$A = 10\text{ ft}^2$
$h_c/h_e \le 0.5$	$GC_p = -0.7 + 0.0434\log(A)$	$10 < A \le 100\text{ ft}^2$
	$GC_p = -0.5$	$A > 100\text{ ft}^2$
Positive:	$GC_p = 0.9$	$A = 10\text{ ft}^2$
All h_c/h_e	$GC_p = 1.15 - 0.109\log(A)$	$10 < A \le 100\text{ ft}^2$
	$GC_p = 0.65$	$A > 100\text{ ft}^2$

Table C30.9-3. Pressure Coefficients on Separate Surfaces of Attached Canopies on Buildings with $h > 60$ ft (18.3 m) (Figure 30.9-2A).

Negative:	$GCp = -1.9$	$A = 10\ \text{ft}^2$
Upper	$GCp = -2.1 + 0.0869 \log(A)$	$10 < A \leq 100\ \text{ft}^2$
surface	$GCp = -3.1 + 0.304 \log(A)$	$100\ \text{ft}^2 < A \leq 1000\ \text{ft}^2$
Negative:	$GCp = -1.0$	$A = 10\ \text{ft}^2$
Lower	$GCp = -1.2 + 0.0869 \log(A)$	$10 < A \leq 100\ \text{ft}^2$
surface	$GCp = -1.4 + 0.1303 \log(A)$	$100\ \text{ft}^2 < A \leq 1000\ \text{ft}^2$
Positive:	$GCp = 0.8$	$A = 10\ \text{ft}^2$
Upper and	$GCp = 1.0 - 0.087 \log(A)$	$10 < A \leq 100\ \text{ft}^2$
lower	$GCp = 0.6$	$A > 100\ \text{ft}^2$
surfaces		

Table C30.9-4. Net Pressure Coefficients Considering Simultaneous Contributions from Upper and Lower Surfaces on Attached Canopies on Buildings with $h > 60$ ft (18.3 m) (Figure 30.9-2B).

Negative:	$GC_p = -2.3$	$A = 10\ \text{ft}^2$
$0.9 \leq h_c/$	$GC_p = -2.5 + 0.0869 \log(A)$	$10 < A \leq 100\ \text{ft}^2$
$h_e \leq 1$	$GC_p = -3.9 + 0.3909 \log(A)$	$100\ \text{ft}^2 < A \leq 1000\ \text{ft}^2$
Negative:	$GC_p = -1.3$	$A = 10\ \text{ft}^2$
$0.1 < h_c/$	$GC_p = -1.85 + 0.2389 \log(A)$	$10 < A \leq 100\ \text{ft}^2$
$h_e < 0.9$	$GC_p = -0.75$	$A > 100\ \text{ft}^2$
Positive:	$GC_p = 0.9$	$A = 10\ \text{ft}^2$
All h_c/h_e	$GC_p = 1.15 - 0.109 \log(A)$	$10 < A \leq 100\ \text{ft}^2$
	$GC_p = 0.65$	$A > 100\ \text{ft}^2$

PART 5: NONBUILDING STRUCTURES

C30.10 CIRCULAR BINS, SILOS, AND TANKS WITH $H \leq 120$ FT (36.5 M)

Section 30.10 contains the provisions for determining wind pressures on silo and tank walls and roofs. The results for isolated and grouped silos are largely based on Australian standards (Standards Australia 2011) and the wind tunnel tests by Sabransky and Melbourne (1987) and Macdonald et al. (1988, 1990). Significant increases in the mean pressures of grouped silos were found in the wind tunnel tests, so the provisions of grouped tanks and silos are specified in this section.

C30.10.2 External Walls of Isolated Circular Bins, Silos, and Tanks This section specifies the external pressure coefficients, $GC_{p(\alpha)}$, for the walls of circular bins, silos, and tanks. The pressure coefficients for isolated silos are adopted from Australian standards (Standards Australia 2011).

C30.10.3 Internal Surface of Exterior Walls of Isolated Open-Topped Circular Bins, Silos, and Tanks This section specifies the internal pressure coefficients, (GC_{pi}), for the walls of circular bins, silos, and tanks. The internal pressure coefficients (GC_{pi}) are adopted from Standards Australia (2011). Based on the wind tunnel test results, mean pressures on walls for open-topped bins, silos, and tanks are different from the values of circular bins, silos, and tanks with flat or conical roofs. Table C30.10-1 lists the mean pressure coefficients $(GC_p - GC_{pi})$ for open-topped circular bins, silos, and tanks, based on Equations (30.10-2) and (30.10-5).

Table C30.10-1. Mean Pressure Coefficients $(GC_p - GC_{pi})$ for Open-Topped Tanks.

	Aspect Ratio H/D					
Angle α	0.25	0.50	1	2	3	4
0°	1.69	1.80	1.9	2	2.07	2.11
15°	1.39	1.50	1.6	1.7	1.77	1.81
30°	0.99	1.10	1.2	1.3	1.37	1.41
45°	0.39	0.50	0.6	0.7	0.77	0.81
60°	−0.01	−0.01	−0.01	−0.1	−0.13	−0.09
75°	−0.11	−0.31	−0.5	−0.7	−0.83	−0.89
90°	−0.11	−0.31	−0.5	−0.7	−0.83	−0.89
105°	−0.01	−0.11	−0.2	−0.3	−0.33	−0.29
120°	0.09	0.10	0.2	0.2	0.27	0.21
135°	0.29	0.30	0.4	0.5	0.47	0.51
150°	0.29	0.40	0.5	0.5	0.57	0.61
165°	0.29	0.40	0.5	0.5	0.57	0.61
180°	0.29	0.40	0.5	0.5	0.57	0.61

The distribution of the external pressure around the perimeter of the wall is shown in Figure C30.10-1.

C30.10.4 Roofs of Isolated Circular Bins, Silos, and Tanks This section specifies the external pressure coefficients, (GC_p) for the roofs of circular bins, silos, and tanks. Two conditions are covered, as shown in Figure 30.10-2: Class 1 roofs have the roof angle $\theta < 10°$, and Class 2 roofs have $10° \leq \theta < 30°$. Zone 1 pressures are defined differently, increasing either with the increment of the silo height for Class 1 roofs, or with the silo or tank diameter for Class 2 roofs. For cladding design, Zone 3 pressures are specified for the local pressures near the windward edges applicable to all classes, and Zone 4 is specified for the region near the cone apex used for Class 2b roofs only. Figure C30.10-2 is the graphic presentation of the elevation views for the external pressure coefficients, (GC_p). For Class 1 roofs, the external pressure coefficients are based on comparisons of domed roofs and flat roofs from Chapter 27 of ASCE 7-10 for maximum uplift conditions. The results of Class 2 roofs are consistent with the data of Sabransky and Melbourne (1987) and Macdonald et al. (1988).

C30.10.6 Roofs and Walls of Grouped Circular Bins, Silos, and Tanks For grouped silos, (GC_p) values for roofs and walls are largely based on AS/NZS 1170.2 (Standards Australia 2011) and wind tunnel tests by Sabransky and Melbourne (1987) and Macdonald et al. (1990). Test results of an in-line group of three silos with a clear spacing of $0.25D$ between nearest adjacent walls ($1.25D$ center-to-center) by Sabransky and Melbourne (1987) indicated that the mean pressure coefficient between the gaps increased by 70% compared to the one for the isolated silo. A similar result was observed for the roof near the wall of the silo. It was concluded that a clear spacing of $0.25D$ produced the maximum interference between two finite cylinders.

Test results of an in-line group of five silos with various center-to-center spacings by Macdonald et al. (1990) indicated that the region of positive pressure on the windward side spans a larger angular sector of the circumference than that for an isolated silo, and high negative mean pressures occur near the point of shortest distance between the adjacent silos and at the outside corners of the groups.

Diagrams

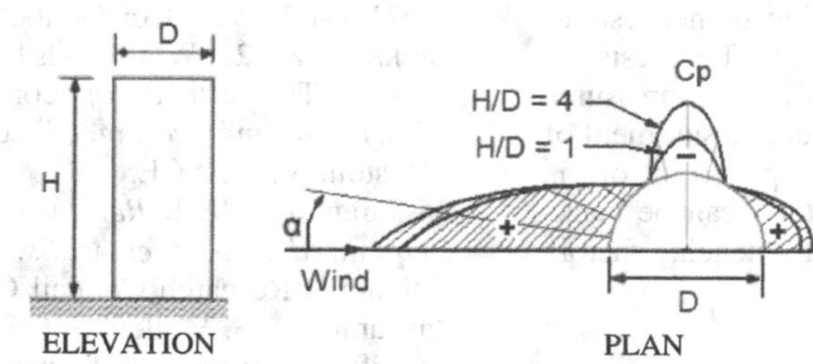

Notation

C_p = External pressure coefficient to be used in determination of wind loads for buildings.
D = Diameter of a circular structure, ft (m).
H = Height of solid cylinder, ft (m).
α = Angle from the wind direction to a point on the wall of a circular bin, silo, or tank, degrees (see Section 30.12.2).

Figure C30.10-1. Mean pressure coefficients (GC_p) – (GC_{pi}) for open-topped tanks.

Note: D = diameter of circular structure, ft (m); H = height ft (m); α = angle from the wind direction to a point on the wall of a circular bin, silo, or tank, in degrees.

Diagrams

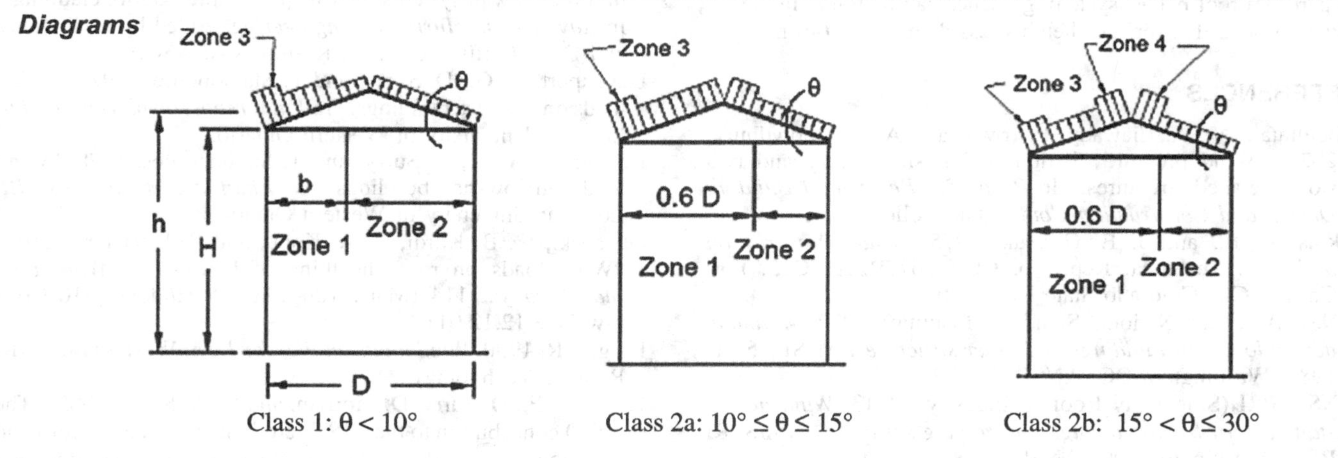

Notation

b = Horizontal dimension specified for Zone 1 of a conical roof, ft (m). For roof angles less than 10 degrees, b is calculated from the table of external pressure coefficients in Fig. 30.12-2 (e.g., $b = 0.5D$ for $H/D = 0.5$.) For roof angles equal to or larger than 10 degrees, $b = 0.6D$. (So for Class 2a and 2b, $b = 0.6D$).
D = Diameter of a circular structure, ft (m).
h = Mean roof height, ft (m).
H = Height of the solid cylinder, ft (m).
θ = Angle of plane of roof from horizontal, degrees.

Figure C30.10-2. External pressure coefficients, (GC_p), for roofs.

C30.12 ROOF PAVERS FOR BUILDINGS OF ALL HEIGHTS WITH ROOF SLOPES LESS THAN OR EQUAL TO 7 DEGREES

Loose-laid roof pavers are often placed on the roof with gaps in between them and with spacing underneath the pavers above the roof membrane using pedestals, tabs, or integrated legs. The net uplift pressure on pavers is substantially affected by pressure equalization between the top and bottom surfaces of the pavers, via the gaps between the solid pavers and by other openings that may be present on pavers that are not solid (Irwin et al. 2012; Asghari Mooneghi et al. 2014, 2016, 2017). The equalization effect is similar to that observed on solar panels mounted parallel to a roof surface (Kopp 2013, Stenabaugh et al. 2015, Banks 2012). The highest uplifts generally occur near roof edges, particularly near roof corners, where strong vortices cause very localized high suctions. Pressure equalization, which depends on the size of the gaps between pavers and on the height of pedestals on which the pavers are frequently mounted, helps reduce the net uplift compared with the external pressure calculated for a roof. Thus, using roof external pressure coefficients for designing pavers, option (a) in Section 30.12 generally results in a conservative design.

Alternative (b) in Section 30.12 permits wind tunnel tests to determine pressure equalization effects on pavers. These tests are of two types: measurement of $C_{L_{net}}$ by integration of net pressures between the top and bottom surfaces; and direct measurement of the wind speeds at roof level at which the paver lift-off is initiated, from which the effective value of $C_{L_{net}}$ can be back-calculated. In wind tunnel tests the following influencing factors need to be considered:

- Building and roof geometry;
- Ratio of the size of the gaps between the pavers to the height of pedestals on which the pavers are mounted (d_g/h_g);
- Interconnection of the pavers by straps or other means, which increases the effective area over which wind uplift is spread and the weight that must be lifted;
- Dynamic effects that prevent the pavers from responding instantaneously to short-duration uplift forces; and
- Effect of parapet height.

Alternative (c) in Section 30.12 permits use of methods in the recognized literature. The methods should take account of the five factors listed above. For pavers laid directly on the roof membrane as part of a roof ballast system, guidance can be found in ANSI/SPRI RP-4 (2013) for roof heights less than 150 ft (46 m).

REFERENCES

Abdelfatah, N., A. Elawady, P. Irwin, and A. G. Chowdhury. 2020. "Wind pressure distribution on single-story and two story elevated structures." In *Proc., 5th Residential Building Design and Construction Conf.*, State College, PA.

Akins, R. E., and J. E. Cermak. 1975. *Wind pressures on buildings*. Technical Rep. No. CER 7677REAJEC15. Fort Collins, CO: Colorado State University.

ANSI (American National Standards Institute). 1982. *Minimum design loads for buildings and other structures*. ANSI A58.1-1982. Washington, DC: ANSI.

ANSI/SPRI (Single Ply Roofing Industry). 2013. *Wind design standard for ballasted single ply roofing systems*. ANSI/SPRI RP-4. Washington, DC: ANSI.

ASCE. 1998. *Minimum design loads for buildings and other structures*. ASCE 7-95. Reston, VA: ASCE.

ASCE. 2002. *Minimum design loads for buildings and other structures*. ASCE/SEI 7-02. Reston, VA: ASCE.

ASCE. 2011. *Wind loads for petrochemical and other industrial facilities*. Reston, VA: ASCE.

ASCE. 2017. *Minimum design loads and associated criteria for buildings and other structures*. ASCE/SEI 7-16. Reston, VA: ASCE.

Asghari Mooneghi, M., P. Irwin, and A. G. Chowdhury. 2014. "Large-scale testing on wind uplift of roof pavers." *J. Wind Eng. Ind. Aerod.* 128 (May): 22–36. https://doi.org/10.1016/j.jweia.2014.03.001.

Asghari Mooneghi, M., P. Irwin, and A. G. Chowdhury. 2016. "Towards guidelines for design of loose-laid roof pavers for wind uplift." *Wind Struct. Int. J.* 22 (2): 133–160.

Asghari Mooneghi, M., T. Smith, P. Irwin, and A. G. Chowdhury. 2017. "Concrete roof pavers: Wind uplift aerodynamic mechanisms and design guidelines. A proposed addition to ANSI/SPRI RP-4." In *Proc., 32nd Int. Convention and Trade Show*. Anaheim, CA.

ASTM International. 2006. *Standard specification for rigid poly(vinyl chloride) (PVC) siding*. ASTM D3679-06a. West Conshohocken, PA: ASTM.

ASTM. 2007. *Standard test method for wind resistance of sealed asphalt shingles (uplift force/uplift resistance method)*. ASTM D7158-07. West Conshohocken, PA: ASTM.

Banks, D. 2012. "Wind loads on tilted flat panels on commercial roofs: The effects of corner vortices." In *Advances in Hurricane engineering*, edited by C. P. Jones and L. G. Griffis. Reston, VA: ASCE.

Baskaran, A. 1992. *Review of design guidelines for pressure equalized rainscreen walls*. Internal Rep. No. 629. Ottawa: National Research Council Canada.

Baskaran, A., S. Molleti, S. Ko, and L. Shoemaker. 2012. "Wind uplift performance of composite metal roof assemblies." *J. Archit. Eng.* 18 (1): 2–15. https://doi.org/10.1061/(ASCE)AE.1943-5568.0000042.

Candelario, J. D., T. Stathopoulos, and I. Zisis. 2014. "Wind loading on attached canopies: Codification study." *Struct. Eng.* 140 (5): 04014007. https://doi.org/10.1061/(ASCE)ST.1943-541X.0001007.

Cheung, J. C. J., and W. H. Melbourne. 1986. "Wind loadings on porous cladding." In *Proc., 9th Australian Conf. on Fluid Mechanics,* 308.

Cope, A., L. Crandell, D. Johnston, V. Kochkin, Z. Liu, L. Stevig, and T. Reinhold. 2012. "Wind loads on components of multi-layer wall systems with air-permeable exterior cladding." In *Advances in Hurricane engineering,* edited by C. P. Jones and L. G. Griffis, 238–257. Reston, VA: ASCE.

Davenport, A. G., D. Surry, and T. Stathopoulos. 1977. "Wind loads on low-rise buildings." In *Final report on phases I and II*. London: University of Western Ontario.

Davenport, A. G., D. Surry, and T. Stathopoulos. 1978. "Wind loads on low-rise buildings." In *Final report on phase III*. London: University of Western Ontario.

Gavanski, E., B. Kordi, G. A. Kopp, and P. J. Vickery. 2013. "Wind loads on roof sheathing of houses." *J. Wind Eng. Ind. Aerodyn.* 114 (Mar): 106–121. https://doi.org/10.1016/j.jweia.2012.12.011.

Haig, J. R. 1990. *Wind loads on tiles for USA*. West Sussex, UK: Redland Technology.

Ho, T. C. E., D. Surry, D. Morrish, and G. A. Kopp. 2005. "The UWO contribution to the NIST aerodynamic database for wind loads on low buildings. Part 1. Basic aerodynamic data and archiving." *J. Wind Eng. Ind. Aerodyn.* 93 (1): 1–30. https://doi.org/10.1016/j.jweia.2004.07.006.

Irwin, P., C. Dragoiescu, M. Cicci, and G. Thompson. 2012. "Wind tunnel model studies of aerodynamic lifting of roof pavers." In *Proc., ATC & SEI Conf. on Advances in Hurricane Engineering*. Miami, FL.

Kala, S., T. Stathopoulos, and K. Kumar. 2008. "Wind loads on rainscreen walls: Boundary-layer wind tunnel experiments." *J. Wind Eng. Ind. Aerodyn.* 96 (6–7): 1058–1073. https://doi.org/10.1016/j.jweia.2007.06.028.

Kim, J. H., M. Moravej, E. J. Sutley, A. Chowdhury, and T. N. Dao. 2020. "Observations and analysis of wind pressures on the floor underside of elevated buildings." *Eng. Struct.* 221 (Oct): 111101. https://doi.org/10.1016/j.engstruct.2020.111101.

Kopp, G. A. 2013. "Wind loads on low profile, tilted, solar arrays placed on large, flat, low-rise building roofs." *J. Struct. Eng.* 140 (2): 04013057. https://doi.org/10.1061/(ASCE)ST.1943-541X.0000821.

Kopp, G., and E. Gavanski. 2012. "Effects of pressure equalization on the performance of residential wall systems under extreme wind loads." *J. Struct. Eng.* 138 (4): 526–538. https://doi.org/10.1061/(ASCE)ST.1943-541X.0000476.

Kopp, G. A., and M. J. Morrison. 2014. "Component and cladding pressures and zones for the roofs of low-rise buildings." In *Boundary layer wind tunnel report*. London: University of Western Ontario.

Macdonald, P. A., J. D. Holmes, and K. C. S. Kwok. 1990. "Wind loads on circular storage bins, silos and tanks. II. Effect of grouping." *J. Wind Eng. Ind. Aerodyn.* 34 (1): 77–95. https://doi.org/10.1016/0167-6105(90)90150-B.

Macdonald, P. A., K. C. S. Kwok, and J. D. Holmes. 1988. "Wind loads on circular storage bins, silos and tanks: 1. Point pressure measurements on isolated structures." *J. Wind Eng. Ind. Aerodyn.* 31 (2–3): 165–187. https://doi.org/10.1016/0167-6105(88)90003-7.

Main, J. A., and W. P. Fritz. 2006. *Database-assisted design for wind: Concepts, software, and examples for rigid and flexible buildings: Building science series 180*. National Institute of Standards and Technology.

Mehta, K. C., and M. L. Levitan. 1998. *Field experiments for wind pressures*. Texas Tech University.

Naeiji, A., M. Moravej, M. Matus, and I. Zisis. 2020. *Codification study of wind-induced loads on canopies attached to mid-rise buildings*.

Peterka, J. A., and J. E. Cermak. 1975. "Wind pressures on buildings: Probability densities." *J. Struct. Div.* 101 (6): 1255–1267. https://doi.org/10.1061/JSDEAG.0004076.

Peterka, J. A., J. E. Cermak, L. S. Cochran, B. C. Cochran, N. Hosoya, R. G. Derickson, C. Harper, J. Jones, and B. Metz. 1997. "Wind uplift model for asphalt shingles." *J. Arch. Eng.* 3 (4): 147–155. https://doi.org/10.1061/(ASCE)1076-0431(1997)3:4(147).

Saathoff, P. J., and T. Stathopoulos. 1992. "Wind loads on buildings with Sawtooth roofs." *J. Struct. Eng.* 118 (2): 429–446. https://doi.org/10.1061/(ASCE)0733-9445(1992)118:2(429).

Sabransky, I. J., and W. H. Melbourne. 1987. "Design pressure distribution on circular silos with conical roofs." *J. Wind Eng. Ind. Aerodyn.* 26 (1): 65–84.

Sakib, F. A., T. Stathopoulos, and A. Bhowmick. n.d. "Wind-induced loads on canopies attached on the walls of tall buildings." *Draft in preparation*.

SBCCI (Southern Building Code Congress International). 1994. *Standard building code*. Janesville, WI: SBCCI.

Standards Australia. 2011. *Structural design actions: Wind actions*. AS/NZS 1170.2:2011. North Sydney, Australia: Standards Australia.

Stathopoulos, T., and M. Dumitrescu-Brulotte. 1989. "Design recommendations for wind loading on buildings of intermediate height." *Can. J. Civ. Eng.* 16 (6): 910–916. https://doi.org/10.1139/l89-134.

Stathopoulos, T., and A. R. Mohammadian. 1986. "Wind loads on low buildings with mono-sloped roofs." *J. Wind Eng. Ind. Aerodyn.* 23: 81–97. https://doi.org/10.1016/0167-6105(86)90034-6.

Stathopoulos, T., and P. Saathoff. 1991. "Wind pressures on roofs of various geometries." *J. Wind Eng. Ind. Aerodyn.* 38 (2–3): 273–284. https://doi.org/10.1016/0167-6105(91)90047-z.

Stathopoulos, T., and X. Zhu. 1988. "Wind pressures on buildings with appurtenances." *J. Wind Eng. Ind. Aerodyn.* 31: 265–281.

Stathopoulos, T., and X. Zhu. 1990. "Wind pressures on buildings with mullions." *J. Struct. Eng.* 116 (8): 2272–2291. https://doi.org/10.1061/(ASCE)0733-9445(1990)116:8(2272).

Stenabaugh, S. E., Y. Iida, G. A. Kopp, and P. Karava. 2015. "Wind loads on photovoltaic arrays mounted on sloped roofs of low-rise building, parallel to the roof surface." *J. Wind Eng. Ind. Aerodyn.* 139 (4): 16–26. https://doi.org/10.1016/j.jweia.2015.01.007.

St. Pierre, L. M., G. A. Kopp, D. Surry, and T. C. E. Ho. 2005. "The UWO contribution to the NIST aerodynamic database for wind loads on low buildings: Part 2. Comparison of data with wind load provisions." *J. Wind Eng. Ind. Aerodyn.* 93 (1): 31–59. https://doi.org/10.1016/j.jweia.2004.07.007

Surry, D., and T. Stathopoulos. 1988. *The wind loading of buildings with monosloped roofs*. Final Rep. No. BLWT-SS38. London: University of Western Ontario.

Taylor, T. J. 1991. "Wind pressures on a hemispherical dome." *J. Wind Eng. Ind. Aerodyn.* 40 (2): 199–213. https://doi.org/10.1016/0167-6105(92)90365-H.

Templin, J. T., and J. E. Cermak. 1978. *Wind pressures on buildings: Effect of mullions*. Technical Rep. No. CER76-77JTT-JEC24. Fort Collins, CO: Colorado State University.

Vickery, P. J., G. A. Kopp, and L. A. Twisdale Jr. 2011. "Component and cladding wind pressures on hip and gable roofs: Comparisons to the U.S. wind loading provisions." In *Proc., 13th Int. Conf. on Wind Engineering*. Amsterdam, Netherlands.

Zisis, I., F. Raji, and J. D. Candelario. 2017. "Large-scale wind tunnel tests of canopies attached to low-rise buildings." *ASCE J. Archit. Eng.* 23 (1): B4016005. https://doi.org/10.1061/(ASCE)AE.1943-5568.0000235.

Zisis, I., and T. Stathopoulos. 2010. "Wind-induced pressures on patio covers." *Struct. Eng.* 136 (9): 1172-1181. https://doi.org/10.1061/(ASCE)ST.1943-541X.0000210.

Zisis, I., T. Stathopoulos, and J. D. Candelario. 2011. "Codification of wind loads on a patio cover based on a parametric wind tunnel study." In *Proc., 13th Int. Conf. on Wind Engineering*. Amsterdam, Netherlands.

OTHER REFERENCES (NOT CITED)

Batts, M. E., M. R. Cordes, L. R. Russell, J. R. Shaver, and E. Simiu. 1980. *Hurricane wind speeds in the United States: NBS building science series 124*. Washington, DC: National Bureau of Standards.

Best, R. J., and J. D. Holmes. 1978. *Model study of wind pressures on an isolated single-story house*. Wind Engineering Rep. No. 3/78. North Queensland, Australia: James Cook University.

Beste, F., and J. E. Cermak. 1996. "Correlation of internal and area-averaged wind pressures on low-rise buildings." In *Proc., 3rd Int. Colloquium on Bluff Body Aerodynamics and Applications*.

Chock, G., J. Peterka, and G. Yu. 2005. "Topographic wind speed-up and directionality factors for use in the city and county of Honolulu building code." In *Proc., 10th Americas Conf. on Wind Engineering*. Baton Rouge, LA.

CSA Group. 2015. *Standard test method for the dynamic wind uplift resistance of vegetated roof assemblies*. CSA A123.24-15. Toronto: CSA Group.

Davenport, A. G., C. S. B. Grimmond, T. R. Oke, and J. Wieringa. 2000. "Estimating the roughness of cities and sheltered country." In *Proc., 12th AMS Conf. on Applied Climatology*, 96–99.

Eaton, K. J., and J. R. Mayne. 1975. "The measurement of wind pressures on two-story houses at Aylesbury." *J. Ind. Aerodyn.* 1 (1): 67–109. https://doi.org/10.1016/0167-6105(75)90007-0.

Ellingwood, B. 1981. "Wind and snow load statistics for probabilistic design." *J. Struct. Div.* 107 (7): 1345–1350. https://doi.org/10.1061/JSDEAG.0006152.

ESDU (Engineering Sciences Data Unit). 1990. *Strong winds in the atmospheric boundary layer. Part 1: Mean hourly wind speeds*. Item no. 82026. London: ESDU.

Ho, E. 1992. "Variability of low building wind lands." Doctoral dissertation, University of Western Ontario.

Marshall, R. D. 1977. *The measurement of wind loads on a full-scale mobile home*. NBSIR 77-1289. Washington, DC: National Bureau of Standards.

McDonald, J. R. 1983. *A methodology for tornado hazard probability assessment*. NUREG/CR3058. Washington, DC: US Nuclear Regulatory Commission.

Peterka, J. A., and S. Shahid. 1993. "Extreme gust wind speeds in the U.S." In Vol. 2 of *Proc., 7th US National Conf. on Wind Engineering,* edited by G. Hart, 503–512.

Powell, M. D. 1980. "Evaluations of diagnostic marine boundary-layer models applied to hurricanes." *Mon. Weather Rev.* 108 (6): 757–766. https://doi.org/10.1175/1520-0493(1980)108<0757:EODMBL>2.0.CO;2.

Sataka, N., K. Suda, T. Arakawa, A. Sasaki, and Y. Tamura. 2003. "Damping evaluation using full-scale data of buildings in Japan." *J. Struct. Eng.* 129 (4): 470–477. https://doi.org/10.1061/(ASCE)0733-9445(2003)129:4(470).

SPRI (Single Ply Roofing Industry). 2013. *Wind design standard for vegetative roofing systems*. ANSI/SPRI RP-14. Waltham, MA: SPRI.

Stathopoulos, T. 1981. "Wind loads on eaves of low buildings." *J. Struct. Div.* 107 (10): 1921–1934. https://doi.org/10.1061/JSDEAG.0005793.

Stathopoulos, T., and H. Luchian. 1990. "Wind pressures on building configurations with stepped roofs." *Can. J. Civ. Eng.* 17 (4): 569–577. https://doi.org/10.1139/l90-065.

Stathopoulos, T., and H. Luchian. 1992. "Wind-induced forces on eaves of low buildings." In *Proc., Wind Engineering Society Inaugural Conf.* Cambridge, UK.

Stathopoulos, T., D. Surry, and A. G. Davenport. 1979. "Wind-induced internal pressures in low buildings." In *Proc., 5th Int. Conf. on Wind Engineering,* edited by J. E. Cermak.

Stathopoulos, T., K. Wang, and H. Wu. 1999. "Wind standard provisions for low building gable roofs revisited." In *Proc., 10th Int. Conf. on Wind Engineering*, edited by J. E. Cermak.

Stathopoulos, T., K. Wang, and H. Wu. 2000. "Proposed new Canadian wind provisions for the design of gable roofs." *Can. J. Civ. Eng.* 27 (5): 1059–1072. https://doi.org/10.1139/l00-023.

Stathopoulos, T., K. Wang, and H. Wu. 2001. "Wind pressure provisions for gable roofs of intermediate roof slope." *Wind Struct.* 4 (2).

Stubbs, N., and D. C. Perry. 1993. "Engineering of the building envelope: To do or not to do." In *Hurricanes of 1992: Lessons learned and implications for the future,* edited by R. A. Cook and M. Sotani, 10–30. Reston, VA: ASCE Press.

Surry, D., R. B. Kitchen, and A. G. Davenport. 1977. "Design effectiveness of wind tunnel studies for buildings of intermediate height." *Can. J. Civ. Eng.* 4 (1): 96–116. https://doi.org/10.1139/l77-010.

Twisdale, L. A., P. J. Vickery, and A. C. Steckley. 1996. *Analysis of hurricane windborne debris impact risk for residential structures.* Bloomington, IL: State Farm Mutual Automobile Insurance Companies.

CHAPTER C31
WIND TUNNEL PROCEDURE

Wind tunnel testing is specified when a building or other structure contains any of the characteristics defined in Sections 27.1.3, 28.1.3, 29.1.3, or 30.1.3, or when the designer wishes to more accurately determine the design wind loads. For some building or structure shapes, wind tunnel testing can reduce the conservatism caused by the enveloping of wind loads inherent in the Directional Procedure, Envelope Procedure, or Analytical Procedure for Components and Cladding (C&C). Also, wind tunnel testing can account for shielding or channeling and can more accurately determine wind loads for a complex building or structure shape than the Directional Procedure, Envelope Procedure, or Analytical Procedure for C&C. It is the intent of the standard that any building or other structure can be allowed to use the wind tunnel testing method to determine wind loads.

Requirements for proper wind tunnel testing are given in ASCE 49. Such a standard is a prerequisite for determining wind loads, either via numerical or physical wind tunnel testing. While Computational Fluid Dynamics (CFD) simulations are increasingly being used in Wind Engineering applications, ASCE 49 does not explicitly identify all procedures necessary for CFD. While awaiting a similar standard documenting the procedures needed to obtain reliable and accurate wind loads using CFD tools, any use of CFD to determine design main wind force resisting system (MWFRS), C&C or other structures' wind loads, requires peer review and a verification and validation (V&V) study (Yeo 2020). In the absence of a standard, this is necessary to address quality assurance and quality control of this method.

When using CFD as a numerical wind tunnel, many of the requirements described in ASCE 49 for physical wind tunnel testing are applicable. For example, a suitable approach flow, accurate geometry, the inclusion of significant nearby structures, and consideration of the potential for modal excitation and aeroelastic effects are all also needed in the numerical model. Once validated against a base-case physical model, the CFD simulation can help resolve details that cannot be measured in the physical model, and/or allow for sensitivity analysis for parametric changes. In the absence of such validation, the numerical wind tunnel simulations can only be considered as qualitative information.

It is common practice to resort to wind tunnel tests when design data are required for the following wind-induced loads:

1. Curtain wall pressures resulting from irregular geometry;
2. Across-wind and/or torsional loads;
3. Periodic loads caused by vortex shedding; and
4. Loads resulting from instabilities, such as flutter or galloping.

Boundary-layer wind tunnels, capable of developing flows that meet the conditions stipulated in Section 31.2, typically have test-section dimensions in the following ranges: width of 6 to 12 ft (2 to 4 m), height of 6 to 10 ft (2 to 3 m), and length of 50 to 100 ft (15 to 30 m). Maximum wind speeds are ordinarily in the range of 25 to 100 mi/h (10 to 45 m/s) The wind tunnel may be either an open-circuit or closed-circuit type.

Three basic types of wind tunnel test models are commonly used. These are designated as follows: (1) rigid pressure model (PM), (2) rigid high-frequency base balance model (H-FBBM), and (3) aeroelastic model (AM). One or more of the models may be used to obtain design loads for a particular building or structure. The PM provides local peak pressures for design of elements, such as cladding and mean pressures, for the determination of overall mean loads. The H-FBBM measures overall fluctuating loads (aerodynamic admittance), for the determination of dynamic responses. When motion of a building or structure influences the wind loading, the AM is used for direct measurement of overall loads, deflections, and accelerations. Each of these models, together with a model of the surroundings (proximity model), can provide information other than wind loads, such as snow loads on complex roofs, wind data to evaluate environmental impact on pedestrians, and concentrations of air pollutant emissions for environmental impact determinations. Several references provide detailed information and guidance for the determination of wind loads and other types of design data by wind tunnel tests (Cermak 1977, Reinhold 1982, ASCE 1999, Boggs and Peterka 1989).

Wind tunnel tests frequently measure wind loads that are significantly lower than those required by Chapters 26, 27, 28, 29, and 30 because of the shape of the building or other structure; the likelihood that the highest wind speeds occur at directions where the building or structure's shape or pressure coefficients are less than their maximum values; specific buildings or structures included in a detailed proximity model that may provide shielding in excess of that implied by exposure categories; and necessary conservatism in enveloping load coefficients in Chapters 28 and 30. In some cases, adjacent buildings or structures may shield the subject building or structure sufficiently so that removal of one or two of the adjacent buildings or structures could significantly increase wind loads. Additional wind tunnel testing without specific nearby buildings or structures (or with additional buildings or structures if they might cause increased loads through channeling or buffeting) is an effective method for determining the influence of adjacent buildings or structures.

For this reason, the standard limits the reduction that can be accepted from wind tunnel tests to 80% of the result obtained from Chapter 27, Chapter 28, or Chapter 30, if the wind tunnel proximity model included any specific influential buildings or other objects that, in the judgment of an experienced wind engineer, are likely to have substantially influenced the results beyond those characteristic of the general surroundings. If there are any such buildings or objects, supplemental testing can be performed to quantify their effect on the original results and possibly justify a limit lower than 80%, by removing them from the detailed proximity model and replacing them with characteristic ground roughness consistent with the adjacent roughness. A specific influential building or object is one within the detailed proximity model that protrudes well above its surroundings, is unusually close to the subject building, or may otherwise cause substantial sheltering effect or magnification of the wind loads. When these supplemental test results are included with the original results, the acceptable results are then considered to be the higher of both conditions.

However, the absolute minimum reduction permitted is 65% of the baseline result for C&C and 50% for the MWFRS. A higher reduction is permitted for MWFRS because C&C loads are more subject to changes caused by local channeling effects when surroundings change, and they can easily be dramatically increased when a new adjacent building is constructed. It is also recognized that cladding failures are much more common than failures of the MWFRS. In addition, in the case of MWFRS, it is easily demonstrated that the overall drag coefficient for certain common building shapes, such as circular cylinders (especially with rounded or domed tops), is one-half or less of the drag coefficient for the rectangular prisms that form the basis of Chapters 27, 28, and 30.

For C&C, the 80% limit is defined by the interior Zones 1 and 4 in Figures 30.3-1, 30.3-2A–C, 30.3-3, 30.3-4, 30.3-5A–B, 30.3-6, 30.3-7, and 30.4-1. This limitation recognizes that pressures in the edge zones are the ones most likely to be reduced by the specific geometry of real buildings, compared with the rectangular prismatic buildings assumed in Chapter 30. Therefore, pressures in edge and corner zones are permitted to be as low as 80% of the interior pressures in Chapter 30 without the supplemental tests. The 80% limit based on Zone 1 is directly applicable to all roof areas, and the 80% limit based on Zone 4 is directly applicable to all wall areas.

The limitation on MWFRS loads is more complex because at any point, the load effects (e.g., member stresses or forces, deflections) are the combined effect of a vector of applied loads instead of a simple scalar value. In general, the ratio of forces, moments, or torques (force eccentricity), at various floors throughout the building using a wind tunnel study will not be the same as those ratios determined from Chapter 27 and 28, and therefore comparison between the two methods is not well defined. Requiring each load effect from a wind tunnel test to be no less than 80% of the same effect resulting from Chapters 27 and 28 is impractical and unnecessarily complex and detailed, given the approximate nature of the 80% value. Instead, the intent of the limitation is effectively implemented by applying it only to a simple index that characterizes the overall loading. For flexible (tall) buildings, the most descriptive index of overall loading is the base overturning moment. For other buildings, the overturning moment can be a poor characterization of the overall loading, and the base shear is recommended instead.

C31.4 SITE SPECIFIC LOAD EFFECTS FOR BUILDINGS, OTHER STRUCTURES, AND COMPONENTS

C31.4.1 Mean Recurrence Intervals of Load Effects Examples of analysis methods for combining directional wind tunnel data with directional meteorological data, or probabilistic models based thereon, are described in Lepage and Irwin (1985), Rigato et al. (2001), Isyumov et al. (2013), Irwin et al. (2005), Simiu and Filliben (2005), and Simiu and Miyata (2006).

C31.4.2 Limitations on Wind Speeds Section 31.4.2 specifies that the statistical methods used to analyze historical wind speed and direction data for wind tunnel studies shall be subject to the same limitations specified in Section 31.4.2 that apply to the Analytical Method.

Database-Assisted Design Wind tunnel aerodynamics databases that contain records of pressures measured synchronously at large numbers of locations on the exterior surface of building models have been developed by wind researchers, such as Simiu et al. (2003) and Main and Fritz (2006). Such databases include data that permit a designer to determine, without specific wind tunnel tests, wind-induced forces and moments in MWFRSs and C&C of selected shapes and sizes of buildings. A public domain set of such databases, recorded in tests conducted at the University of Western Ontario (Ho et al. 2005, St. Pierre et al. 2005) for buildings with gable roofs is available on the National Institute of Standards and Technology website, www.nist.gov/wind (NIST 2012). Interpolation software for buildings with similar shape, and dimensions close to and intermediate between those included in the set of databases, is also available on that website. Because the database results are for generic surroundings, as permitted in ASCE 49, interpolation or extrapolation from these databases should only be used if Condition 2 of Section 27.1.2 is true. Extrapolations from available building shapes and sizes are not permitted, and in some instances, interpolations may not be advisable. For these reasons, the guidance of an engineer who is experienced in wind loads on buildings and is familiar with the usage of these databases is recommended.

All databases must have been obtained using testing methodology that meets the requirements for wind tunnel testing, as specified in Chapter 31.

C31.4.3 Wind Directionality The variability of wind speed determined for particular azimuth intervals is greater than that of the wind speed determined regardless of wind direction (Isyumov et al. 2013). Consequently, wind loads and wind-induced effects, determined by allowing for wind directionality, are inherently less certain. Several methods for combining data from wind tunnel model studies with information on wind speed and direction at the project site are currently in use (Isyumov et al. 2013, Yeo and Simiu 2011, Simiu 2011). Whichever method is used shall be clearly described to allow scrutiny by the designer and the Authority Having Jurisdiction. A common approach for allowing for uncertainties in the wind direction is to rotate the project wind climate relative to the orientation of the building or structure. This rotation of the wind climate at the building location is intended to ensure that the wind loads determined for design are not unconservative and shall be considered, regardless of the method used for arriving at the design wind speeds. The appropriate magnitude of wind climate rotation varies depending on the quality and resolution of the directional wind climate data at the project site.

C31.4.5 Limitations on Wind Loads for Ground-Mounted Fixed-Tilt Solar Panel Systems The minimum C&C and MWFRS wind loads, as indicated in ASCE 7, are primarily applicable to buildings and are not applicable to ground-mounted fixed-tilt solar panel systems. The limitations contained herein are to establish the lower bound wind loads for wind tunnel studies of conditions similar to those addressed in Section 29.4.5. The limits on wind tunnel results shown in Figure 29.4-10 represent an envelope of the static wind pressures measured in the wind tunnel. The limits on wind tunnel results shown in Figure 29.4-11 represent an envelope of the dynamic wind loads derived from the wind tunnel pressure data, which necessarily included simplifying (conservative) assumptions regarding the supporting structure and dynamic properties of the system. Specific installations or system geometries/supporting structures may give significantly lower loads than those in Figures 29.4-10 and 29.4-11; limits are imposed to prevent too much deviation from the enveloped results. Ground-mounted fixed-tilt solar PV systems can have wind-tunnel-based wind loads less than the lower bound thresholds indicated in Sections 31.4.5. In order to use these lower values, a peer review of the test and report is required.

C31.5 LOAD EFFECTS FOR BUILDINGS, OTHER STRUCTURES, AND COMPONENTS USED AT MULTIPLE SITES

C31.5.1 Wind Loads For ASCE 7-22, the wind tunnel test requirements were redefined to account for wind tunnel tests on generic buildings, other structures, and components that are used in multiple locations or on multiple buildings. Roof-mounted solar panels are but one such example. Other examples include building-mounted components, such as sunshades, HVAC units, screen walls, or could be free-standing, such as ground-mounted solar trackers, gazebos, and fences.

In determining wind loads on generic buildings, other structures, and components, the approach needs to be similar to that used to develop the (GC_p) figures in ASCE 7 by modeling the generic buildings with various features to capture a wide range of effects. The objective of such testing is to evaluate aerodynamic effects accounted for by pressure coefficients (in contrast to site-specific wind tunnel testing, which also evaluates the effect of surrounding structures and terrain). Nearby buildings should not be included, unless they are to be a part of every design application for such buildings, other structures, or components.

Wind tunnel testing must include a sufficiently large test matrix to address an appropriate range of the relevant variables that affect wind loads, as listed in the provisions. Guidance for testing is provided in ASCE 49. Wind loads are expressed as coefficients usable in Chapters 27, 29, and 30 to produce loads in engineering units. Alternately, a different formulation of nondimensional load coefficients may be used, provided that the analysis procedure is clearly defined in the test report.

C31.5.2 Limitations on Wind Loads for Rooftop Solar Collectors In ASCE 7-22, the scope of wind tunnel testing for buildings, other structures, and components that are used at multiple sites was added. A peer review is required for the use of this approach, except for items that are already covered in the standard, such as roof-mounted solar panels. A peer review is also required when the loads fall below the minimum threshold.

Regarding rooftop solar panels, the minimum components and cladding wind load pressures indicated in ASCE 7 are primarily applicable to the building envelope and are not applicable to rooftop solar collectors. The limitations contained herein are meant to establish the lower bound wind pressures for wind tunnel studies of conditions similar to those addressed in Figure 29. 4-7. The limits on wind tunnel results shown in Figure 29.4-7 represent an envelope of wind loads measured in the wind tunnel without deflectors or shrouds that are commonly used to lower wind loads. Specific installations or collector geometries may give significantly lower loads than Figure 29.4-7; limits are imposed to prevent too much deviation from the enveloped results. Solar panel systems that have aerodynamic devices or more efficient profiles can have wind-tunnel-based wind loads less than the lower bound thresholds indicated in Sections 31.5.2 and 31.5.3. In order to use these lower values, a peer review of the test and report is required.

C31.5.3 Peer Review Requirements for Wind Tunnel Tests of Buildings, Other Structures, and Components at Multiple Sites This section provides the requirements for peer reviews of wind tunnel studies. The peer reviewer's qualifications and requirements are included to promote consistencies among the various jurisdictions so that a peer review could be accepted by multiple enforcement agencies. The peer reviewer's qualifications are intended to be those of a wind tunnel expert who is familiar with wind tunnel testing of buildings and the applicability of the ASCE 7 provisions to determine generalized wind design coefficients for roof-mounted solar panels. One source for peer reviewers is the American Association for Wind Engineering's (AAWE) boundary layer wind tunnels list (http://www.aawe.org/info/wind_tunnels.php).

REFERENCES

ASCE. 1999. *Wind tunnel model studies of buildings and structures: ASCE manuals and reports of engineering practice no. 67*. Reston, VA: ASCE.

ASCE. 2012. *Wind tunnel testing for buildings and other structures*. ASCE/SEI 49-12. Reston, VA: ASCE.

Boggs, D. W., and J. A. Peterka. 1989. "Aerodynamic model tests of tall buildings." *J. Eng. Mech.* 115 (3): 618–635. https://doi.org/10.1061/(ASCE)0733-9399(1989)115:3(618).

Cermak, J. E. 1977. "Wind-tunnel testing of structures." *J. Eng. Mech. Div.* 103 (6): 1125–1140. https://doi.org/10.1061/JMCEA3.0002301.

Ho, T. C. E., D. Surry, D. Morrish, and G. A. Kopp. 2005. "The UWO contribution to the NIST aerodynamic database for wind loads on low buildings: Part 1. Archiving format and basic aerodynamic data." *J. Wind Eng. Ind. Aerodyn.* 93 (1): 1–30. https://doi.org/10.1016/j.jweia.2004.07.006.

Irwin, P., J. Garber, and E. Ho. 2005. "Integration of wind tunnel data with full scale wind climate." In *Proc., 10th Americas Conf. on Wind Engineering*. Baton Rouge, LA.

Isyumov, N., E. Ho, and P. Case. 2013. "Influence of wind directionality on wind loads and responses." In Vol. 141 of *Proc., 12th Americas Conf. on Wind Engineering*.

Isyumov, N., M. Mikitiuk, P. Case, G. Lythe, and A. Welburn. 2013. "Predictions of wind loads and responses from simulated tropical storm passages." In *Proc., 11th Int. Conf. on Wind Engineering*, edited by D. A. Smith and C. W. Letchford.

Kopp, G., and D. Banks. 2013. "Use of the wind tunnel test method for obtaining design wind loads on roof-mounted solar arrays." *J. Struct. Eng.* 139 (2): 284–287 https://doi.org/10.1061/(ASCE)ST.1943-541X.0000654.

Kopp, G., J. Maffei, and C. Tilley. 2011. *Rooftop solar arrays and wind loading: A primer on using wind tunnel testing as a basis for code compliant design per ASCE 7*. London: University of Western Ontario.

Kopp, G. A., S. Farquhar, and M. J. Morrison. 2012. "Aerodynamic mechanisms for wind loads on tilted, roof-mounted, solar arrays." *J. Wind Eng. Ind. Aerodyn.* 111 (Dec): 40–52. https://doi.org/10.1016/j.jweia.2012.08.004.

Lepage, M. F., and P. A. Irwin. 1985. "A technique for combining historical wind data with wind tunnel tests to predict extreme wind loads." In *Proc., 5th US National Conf. on Wind Engineering*, edited by M. Mehta.

Main, J. A., and W. P. Fritz. 2006. *Database-assisted design for wind: Concepts, software, and examples for rigid and flexible buildings: NIST building science series 180*. Washington, DC: National Institute of Standards and Technology.

NIST (National Institute of Standards and Technology). 2012. "Extreme winds and wind effects on structures." Accessed March 5, 2012. www.nist.gov/wind.

Reinhold, T. A., ed. 1982. "Wind tunnel modeling for civil engineering applications." In *Proc., Int. Workshop on Wind Tunnel Modeling Criteria and Techniques in Civil Engineering Applications.*

Rigato, A., P. Chang, and E. Simiu. 2001. "Database-assisted design, standardization, and wind direction effects." *J. Struct. Eng.* 127 (8): 855–860. https://doi.org/10.1061/(ASCE)0733-9445(2001)127:8(855).

Simiu, E. 2011. *Design of building for wind.* Hoboken, NJ: Wiley.

Simiu, E., and J. J. Filliben. 2005. "Wind tunnel testing and the sector-by-sector approach to wind directionality effects." *J. Struct. Eng.* 131 (7): 1143–1145. https://doi.org/10.1061/(ASCE)0733-9445(2005)131:7(1143).

Simiu, E., and T. Miyata. 2006. *Design of buildings and bridges for wind: A practical guide for ASCE Standard 7 users and designers of special structures.* Hoboken, NJ: Wiley.

Simiu, E., F. Sadek, T. A. Whalen, S. Jang, L.-W. Lu, S. M. C. Diniz, A. Grazini, and M. A. Riley. 2003. "Achieving safer and more economical buildings through database-assisted, reliability-based design for wind." *J. Wind Eng. Ind. Aerodyn.* 91 (12–15): 1587–1611. https://doi.org/10.1016/j.jweia.2003.09.017.

St. Pierre, L. M., G. A. Kopp, D. Surry, and T. C. E. Ho. 2005. "The UWO contribution to the NIST aerodynamic database for wind loads on low buildings: Part 2. Comparison of data with wind load provisions." *J. Wind Eng. Ind. Aerodyn.* 93 (1): 31–59. https://doi.org/10.1016/j.jweia.2004.07.007.

Yeo, D. 2020. *A summary of industrial verification, validation, and uncertainty quantification procedures in computational fluid dynamics.* NIST Internal Rep. No. 8298. Gaithersburg, MD: National Institute of Standards and Technology.

Yeo, D., and Simiu, E. 2011. "High-rise reinforced concrete structures: Database-assisted design for wind," *J. Struct. Eng.* 127 (11): 1340–1349. https://doi.org/10.1061/(ASCE)ST.1943-541X.0000394.

CHAPTER C32
TORNADO LOADS

C32.1 PROCEDURES

C32.1.1 Scope Historically, tornadoes kill more people per year in the United States than hurricanes and earthquakes combined (NIST 2014), and the average annual insured catastrophe losses for events involving tornadoes exceed those for hurricanes and tropical storms combined (Insurance Information Institute 2020). Despite these statistics, building codes and standards have not previously included tornado hazards in their scope except for specialized applications, such as for storm shelters and nuclear facilities. Recent research on tornado climatology has shown that tornadoes occur with much greater frequency and intensity than had previously been quantified (Twisdale et al. 2021). The requirements introduced in Chapter 32, along with corresponding changes in Chapters 1, 2, and 26, provide for determination of tornado loads on the main wind force resisting system (MWFRS) and components and cladding (C&C) and incorporation of those loads into the design process.

While tornadoes are a type of windstorm, and many of the tornado load procedures are similar to those in Chapters 26 through 31, there are significantly different characteristics between tornadoes and other windstorms. For instance, tornadic winds have significant vertical components; rapid atmospheric pressure changes that can induce loads; and load combinations including tornadoes are not always the same as those including wind. Hence, tornado loads are treated separately from wind loads, not as a subset of wind loads. This is analogous in some ways to the separate treatment of flood loads and tsunami loads.

Determining When Design for Tornado Loads Is Required. Under this standard, most buildings and other structures are not required to be designed for tornado loads. A flowchart describing the four-step process to determine whether tornado loads need to be checked is provided in Figure 32.1-2. Risk Category I and II facilities are exempt from tornado load requirements (Step 1), as are Risk Category III and IV buildings and other structures located outside of the tornado-prone region shown in Figure 32.1-1 (Step 2). For buildings and structures not already eliminated, the tornado speed, V_T, is next determined in accordance with Section 32.5.1. If $V_T < 60$ mi/h (26.8 m/s), design for tornado loads is not required per Section 32.5.2 (Step 3). Lastly, if V_T is less than a certain percentage of V (as a function of exposure), where V is the basic wind speed from Chapter 26, design for tornado loads is also not required (Step 4). The rationales underlying these four criteria are described elsewhere in the commentary: risk category limitations are discussed in Section C32.5.1; the definition of the tornado-prone region is covered later in this section; and limitations on V_T in Steps 3 and 4 are addressed in Section C35.5.2. While tornado loads must be checked for all buildings and other structures not meeting any of these four exclusionary conditions, tornado loads will still not control in many cases. The criteria used in Steps 3 and 4 represent the approximate threshold speeds at which tornado loads can begin to control some aspect of the wind load design, based on the worst-case combinations of geographic location, exposure, effective plan area, mean roof height, enclosure classification, building shape, and other parameters.

Tornado-Prone Region. The tornado-prone region (Figure 32.1-1) is approximately the same as the portion of the conterminous United States that lies east of the Continental Divide. While tornadoes can and do occur to the west of the tornado-prone region, tornado speeds associated with Risk Category III and IV return periods in those locations are such that tornado loads would not control over wind loads computed using Chapters 26 through 31. However, at longer return periods, tornado speeds in the western United States do increase significantly (see Appendix G). Tornadoes are rare in Alaska, Hawaii, and US territories, and when they do occur, are typically weak. Therefore, those locations are not included in the tornado-prone region.

The boundary of the tornado-prone region in Figure 32.1-1 corresponds to the 60 mi/h (26.8 m/s) contour in Figure 32.5-2H, the longest return period (Risk Category IV, 3,000 years) map for the largest effective plan area, 4 million ft^2 (371,612 m^2). Because this figure shows the greatest extent of tornado speeds over the largest area of the country it represents the worst case scenario that defines the tornado-prone region. The 60 mi/h (26.8 m/s) contour was chosen because it represents the approximate threshold speed at which tornado loads can begin to control some aspect of the wind load design.

Storm Shelters. Storm shelters and safe rooms are designed with the intent of providing an enhanced level of life safety protection for a design storm event that is greater than typical for conventional buildings and structures, including those designed to the tornado loads in this chapter. A report (CDC 2012) on fatalities during the April 25–28, 2011 tornado outbreak illustrates the importance of providing tornado protection for building occupants. There were 351 recorded tornadoes during the outbreak, where 338 fatalities were caused by 27 of the tornadoes. According to the report, 90% of the people who died were in buildings at the time the tornadoes struck. Similarly, during the May 22, 2011 tornado in Joplin Missouri, 135 (84%) of the 161 fatalities occurred inside buildings, with nearly all deaths caused by impact-related factors, such as multiple blunt force trauma to the body (NIST 2014).

The design of storm shelters, including community and residential storm shelters for both tornadoes and hurricanes, has been regulated by model building and residential codes since 2009. These codes all directly reference ICC 500, the *ICC/NSSA Standard for the Design and Construction of Storm Shelters* (the latest version of which is the 2020 edition (ICC 2020)). The load provisions of ICC 500 directly reference ASCE 7, with some modifications including greater hazard levels. Beginning with the 2015 International Building Code (ICC 2014) and continuing is subsequent editions, certain types of critical facilities located in the 250 mi/h (400 km/h) zone in Figure C32.1-1 have been required to include tornado shelters compliant with ICC 500. This includes many Group E occupancies (i.e., K–12 schools), emergency operations centers, and fire, rescue, ambulance, police, and 911 call stations. The 2018 International Existing Building Code (ICC 2017) extended these requirements to include additions to certain existing Group E occupancy buildings.

FEMA also publishes guidance on the design and construction of structures for storm protection in FEMA P-361, *Safe Rooms for Tornadoes and Hurricanes: Guidance for Community and Residential Safe Rooms* (2021a), and FEMA P-320, *Taking Shelter from the Storm: Building or Installing a Safe Room for Your Home* (2021b). FEMA P-361 provides comprehensive guidance for the design of community and residential safe rooms, as well as operations and maintenance guidance. While the criteria in FEMA P-361 and ICC 500 are quite similar, "safe room" is FEMA terminology for storm shelters that also meet the criteria in FEMA P-361. All safe room criteria in FEMA P-361 meet the storm shelter requirements in ICC 500, but a few design and performance criteria in FEMA P-361 are more restrictive than those in ICC 500 and are required for safe room projects to be eligible for federal grant funding. FEMA P-320 provides prescriptive solutions for residential safe rooms.

Storm shelters and safe rooms are designed for more extreme hazard levels than conventional buildings, including wind-borne debris. All elements of the tornado shelter envelope must resist the impact of a 15 lb (6.8 kg) sawn lumber 2×4 missile traveling at speeds of 53 to 100 mi/h (23.7 to 44.7 m/s), depending on the shelter design wind speed (Figure C32.1-1) and the orientation of the shelter envelope surface (vertical or horizontal). In terms of tornado speed, both ICC 500 and FEMA P-361 use the same map to determine design speeds for tornado shelters and safe rooms. The map in Figure C32.1-1 was developed using a spatial analysis of raw tornado climate data from 1950 to 2006. The 250 mi/h (400 km/h) zone represents the region of the country that most frequently experiences EF4 and EF5 tornadoes. The 200, 160, and 130 mi/h zones correspond to areas where the most intense tornadoes are usually EF3, EF2, and EF1, respectively. The ICC 500 / FEMA P-361 tornado speeds and resultant tornado design pressures and loads represent something approaching a worst-case scenario: "The design wind speeds chosen by FEMA for safe room guidance were determined with the intent of specifying near-absolute protection with an emphasis on life safety" (FEMA 2021a). Given the deterministic nature of the tornado shelter design wind speed map used in ICC 500 and FEMA P-361, the annual probability of exceedance (and its approximate reciprocal, the MRI) of these tornado speeds are not uniform across the country and are therefore undefined. In contrast, the tornado speed maps in Chapter 32 are probabilistic in nature (see Section C32.5.1). Tornado speeds for Risk Category III and IV buildings are based on return periods of 1,700 and 3,000 years, respectively, two to three orders of magnitude smaller than the return periods estimated for the tornado shelter design speeds.

The tornado loads specified in ICC 500 and FEMA P-361 are greater than ASCE 7 requirements in Chapter 32, consistent with the differences in intended levels of life-safety protection and building performance. While ICC 500 does not explicitly state a performance objective for shelters designed to that standard, safe

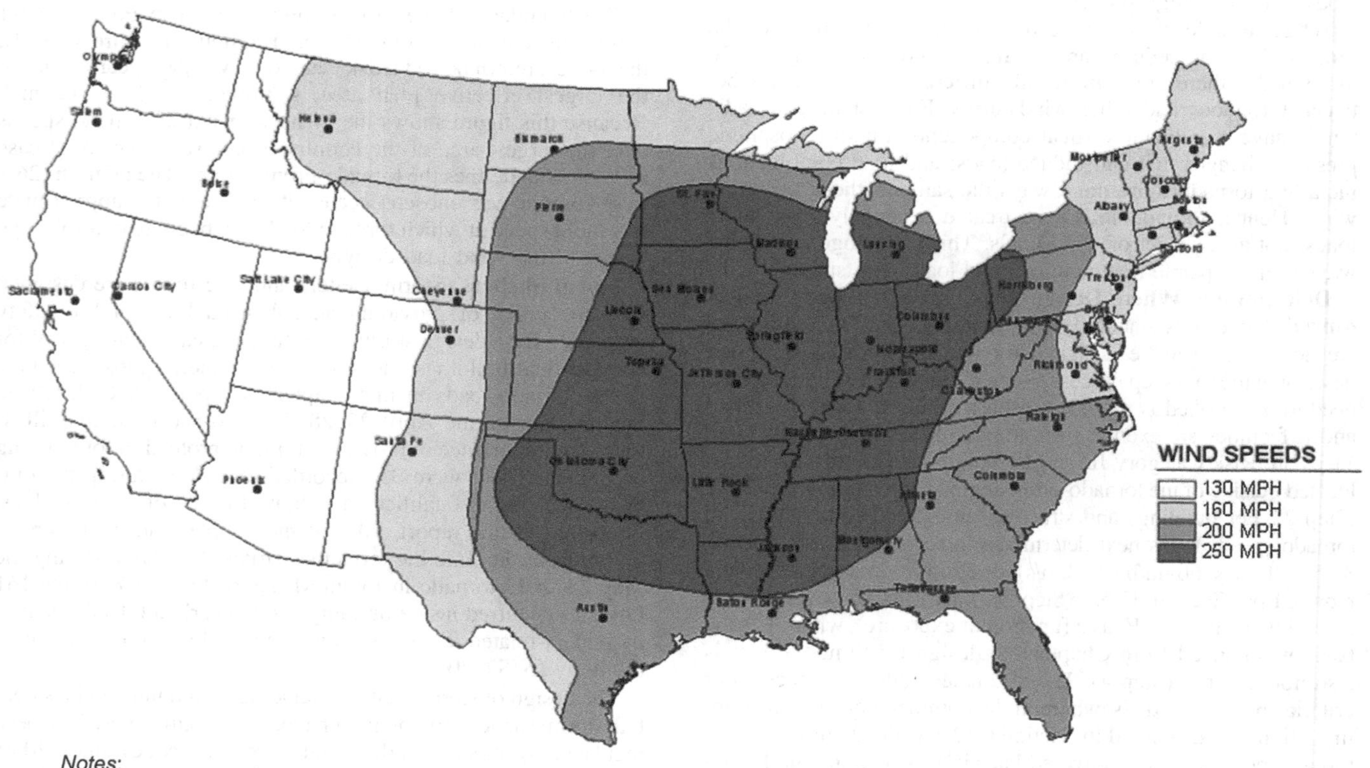

Notes:

1. Values are nominal 3 s gust speeds in miles per hour at 33 feet above ground for Exposure Category C.
2. Multiply miles per hour by 0.477 to obtain meters per second.
3. Location-specific storm shelter design wind speeds shall be permitted to be determined using the ATC Hazards by Location website, https://hazards.atcouncil.org/.

Figure C32.1-1. Tornado wind speeds for design of tornado shelters and tornado safe rooms (ICC 2020).

rooms constructed to the almost identical FEMA criteria are intended to "provide near-absolute protection from wind and wind-borne debris for occupants" (FEMA 2021a). Since 1998, when FEMA first published guidance for residential and community safe rooms, "thousands of safe rooms have been built, and a growing number of these safe rooms have already saved lives in actual events. There has not been a single reported failure of a safe room constructed to FEMA criteria" (FEMA 2021a). Similarly, failures of tornado shelters constructed to ICC 500 criteria (first published in 2008) have not been reported.

C32.1.2 Permitted Procedures The basic framework for determination of tornado loads is adapted from the wind load procedures specified in Chapters 26 through 31, given the general similarities between tornadic winds and other windstorm types and how they both interact with buildings and other structures. There are a number of differences in the details of the procedures between tornado loads and wind loads that are incorporated in this chapter. Most of the load determination procedures from Chapters 27 through 31 have been adapted to work for tornadoes, except for the Chapter 28 Envelope Procedure, which is not compatible with the tornado load methodology. The MWFRS wind load coefficients given in Chapter 28 were developed by combining influence coefficients for structural actions in a building by pneumatically averaging and weighting the wind pressures measured on model buildings in a boundary layer wind tunnel. The integration of the loads incorporates the effects of the correlation of the gusts over various portions of the building associated with the size of the gusts in an atmospheric boundary layer. The size of the gusts in a tornado are likely much smaller than those in a standard atmospheric boundary layer, suggesting that the procedures in Chapter 28 are not appropriate for adaptation to tornado loads. No changes introduced in Chapter 32 restrict the applicability of Chapter 28 for use in determining nontornadic wind loads.

The organization of Chapter 32 follows that of the wind load chapters on which so many of the tornado load provisions are based. Sections 32.1 through 32.13 provide tornado-specific versions of those same sections in Chapter 26, as applicable. This structure provides parallel construction with the familiar layout of the wind loading provisions. Section 32.14 introduces a new factor to modify external pressure coefficients developed for boundary layer winds to work with tornadic winds. Sections 32.15 through 32.18 provide the tornado loading procedures adapted from Chapters 27, 29, 30, and 31, respectively. Lastly, Appendix G provides additional tornado hazard maps for longer return periods than are used in Chapter 32, which can be of use for performance-based tornado designs and other applications.

C32.1.3 Performance-Based Procedures Performance-Based Procedures can quantify service interruption or loss due to tornadoes, improve tornado resistance, or perform an enhanced evaluation of the MWFRS and/or envelope. See Section C26.1.3 for discussion of performance-based wind design, which could be adapted for use in tornado design. For nontornadic winds, the *Prestandard for Performance-Based Wind Design* (ASCE 2019) provides guidance on the process to perform the design and provides examples of acceptance criteria. For a tornado performance-based design, the acceptance criteria of the Prestandard might need modification based on the facilities' use and owner's requirements. The determinations of the wind hazard characteristics for tornado loading will require consultation with a wind engineering expert. For applications where tornado hazard maps at longer return periods than those provided in Chapter 32 are needed, see Appendix G.

C32.3 SYMBOLS AND NOTATION

The following notations apply to the Commentary for Chapter 32:

ABL = Atmospheric boundary layer
APC = Atmospheric pressure change
DI = Damage indicator
DOD = Degree of damage
EF Scale = Enhanced Fujita Scale
FEMA = Federal Emergency Management Agency
FR12 = One-and-two-family residence DI in the EF Scale
ICC = International Code Council
MRI = Mean recurrence interval
NWS = National Weather Service
PV = Photovoltaic

C32.5 TORNADO HAZARD MAPS

C32.5.1 Tornado Speed The tornado hazard maps provided in Section 32.5.1 and Appendix G are based on the first-ever engineering-derived tornado wind speed maps produced for the contiguous United States, as reported by Twisdale et al. (2021). Tornado databases (for the years of 1950 through 2016) and physiographic data were analyzed to identify large-scale spatial patterns with similar tornado characteristics, using multivariate statistical analysis of 11 tornado and physiographic variables. From these patterns, nine broad US regions with distinct tornado climatologies were developed. Regional and national data were analyzed to produce probabilistic models for population bias; Enhanced Fujita Scale (EF-Scale) distribution; tornado path length, width, direction, and translational speed; radius of maximum winds; tornado path length intensity variation; variable path widths within a tornado; mean to maximum path width ratios; and maximum damage widths relative to local path width.

Since tornado wind speeds that result in an assigned EF rating for each tornado are estimated based on observed damage in the field, development of probability distributions that relate tornado wind speed to observed damage was an important element of the work. A probabilistic load and resistance modeling framework was used to develop engineering-derived wind speeds from the EF-Scale tornado intensity rating system (Texas Tech University 2006) by analyzing the most common Damage Indicator (DI) used to rate EF3 to EF5 tornadoes, namely that of one- and two-family residences (FR12). Engineering models were developed to enable tornado wind speed estimation for the EF-Scale Degrees of Damage (DOD) for FR12. This work used load path quality considerations that the National Weather Service (NWS) employs in its damage assessment process to estimate wind speeds and assign EF ratings to tornadoes. Field investigations of tornadoes were conducted to gain a better understanding of the field process used by the NWS, to investigate failure modes of various DIs, and to obtain data for validation of the models used to estimate tornado wind speed from observed damage. The developed wind speed distributions, based on 44 3D models of FR12 structures, are broad and encompass the original judgment-based EF-Scale wind speeds. These data were used to support development of a probabilistic tornado wind field model. Monte Carlo methods were then used to simulate tornadoes, produce damage swaths, score wind speed exceedances numerically over a wide range of wind speeds, and develop regional tornado hazard curves. Since tornadoes often have modest path widths relative to the size of the structure, tornado wind speed risk depends on structure size (see Section C32.5.4). Therefore, the tornado hazard curves were developed as a function of structure size.

The final step was the development of the tornado hazard maps, which reflect the spatial variation in risk across the contiguous United States and within the context of the tornado climatology regions. A one-degree grid was used to map the regional tornado wind speeds for a given return period. Gaussian smoothing was applied to the grid to reflect epistemic uncertainties in the location of the region boundaries. ArcGIS kriging routines were next used to produce the tornado speed contours from the smoothed grid. These contours were further smoothed using the Polynomial Approximation with Exponential Kernel (PAEK) method routine in ArcGIS.

Tornado hazard maps for return periods of 1,700, 3,000, 10,000, 100,000, 1,000,000, and 10,000,000 years are included in ASCE 7. These six return periods span the range of design needs for conventional buildings and structures, Critical and Essential Facilities, and nuclear power plants. The 1,700-year and 3,000-year maps are provided in Section 32.5 for use in determining tornado loads on Risk Category III and IV buildings, other structures, and facilities. These are the same mean recurrence intervals (MRIs) used for Risk Category III and IV basic wind speeds in Chapter 26 (see Return Periods later in this commentary section for information on the reliability analysis underpinning selection of these MRIs). Tornado hazard maps for the 10,000-year and longer return periods are provided in Appendix G, which may be needed for performance-based tornado designs and other applications.

Maps were produced for eight structure sizes at each return period, ranging from a point target of 1 ft^2 (0.9 m^2) to 4 million ft^2 (371,612 m^2). An example of a very small target would be a freestanding communications tower. The maps for this small structure size are also helpful for interpolation between this and the next largest available mapped size of 2,000 ft^2 (186 m^2) (see Section 32.5.4 for determination of structure size, referred to as Effective Plan Area).

Tornado speeds are contoured at 10 mi/h (4.5 m/s) intervals except for the innermost (greatest speed) and outermost (least speed) contours, which are shown to 1 mi/h (0.45 m/s) to aid with interpolation. In South Texas there is often a narrow ridge of slightly higher tornado speeds that occurs parallel to the coast, a few counties inland. This ridge is represented by either a short contour or a point value, for situations where the ridge maximum value (near the border with Mexico) is either a small plateau or a local peak, respectively.

The tornado speeds shown on the maps represent the maximum 3 s gust horizontal wind speed at a height of 33 ft (10 m) anywhere within the effective plan area of the target building or other structure produced by the translating tornado (Twisdale et al. 2021). The tornado speed is defined independently of terrain.

The tornado hazard maps were developed through probabilistic models that are "best estimates" rather than "conservatively based." This approach follows the intent of ASCE 7 wind speed maps for other wind hazards, including hurricanes (Vickery et al. 2009) and straight winds (Pintar et al. 2015), as provided in Section 26.5. Epistemic (modeling) uncertainties were considered in the tornado hazard map development process. As a result, the maps are intended to provide results that are also applicable to the nuclear power industry in the United States, where both aleatory (randomness) and epistemic uncertainties are required in the risk analysis of nuclear power plants.

Return Periods. An ad hoc working group composed of members from both the ASCE 7-22 Load Combinations and Wind Load Subcommittees conducted a study to identify appropriate return periods for the tornado hazard maps in Chapter 32 (Levitan et al. 2021). The methodology used to determine wind speed map return periods for each Risk Category in ASCE 7-16 [see Section C26.5.1 and McAllister et al. (2018)] was adapted for use with tornado loads. A reliability analysis was conducted for roof uplift, the primary mode of structural failure in tornadic events, where roof framing members and/or connections fail due to significant wind pressure resulting from the combination of internal and external pressures. The reliability analysis procedures used Monte Carlo analysis, in which significant uncertainties for system demands and capacity were identified and quantified in the form of random variables with defined probability distributions. Investigation over a range of return periods found that the 1,700 and 3,000-year MRIs used for Risk Category III and IV wind hazard maps in Chapter 26 were also suitable for the tornado hazard maps. The results of a series of risk-informed analyses show that the tornadic load criteria of Chapter 32 provide reasonable consistency with the reliability delivered by the existing criteria in Chapters 26 and 27 for MWFRSs (Levitan et al. 2021). Tornado speeds at return periods of 300 and 700 years are generally so low that tornado loads would not control over Chapter 26 wind loads, so design for tornadoes is not required for Risk Category I and II buildings and other structures. It should be recognized that the tornado hazard maps and tornado load design methods are new to ASCE 7-22, that the profession needs experience with their application, and that changes may be needed in the future.

The tornado speeds required by Chapter 32 depend on the geographic location, Risk Category, and Effective Plan Area of the building or other structure. For many Risk Category III buildings and other structures located in the area having the greatest tornado hazard (roughly the area between northern Texas, South Dakota/Minnesota, and northern Alabama, and extending father east for large target sizes), the tornado speed will be in the EF1 intensity range (86 to 110 mi/h [140 to 180 km/h]). For Risk Category IV, the tornado speeds will typically be of EF1 or EF2 intensity (86 to 135 mi/h [140 to 220 km/h]) in the same geographic area. While this clearly does not include the most intense tornadoes, it does include the majority of tornadoes that occur in the United States. From 1995 to 2016, 61.3% of the over 1,200 recorded tornadoes per year were rated EF0, 27.8% were EF1, 8.0% EF2, 2.3% EF3, 0.52% EF4, and 0.05% EF5 (Twisdale et al. 2021). In addition, most of the area impacted by a tornado does not experience the maximum winds on which the tornado is rated. For example, in the 2011 EF5 Joplin, Missouri, tornado, an estimated 72% of the area swept by the tornado experienced EF0 to EF2 winds, while just 28% experienced EF3 to EF5 winds (NIST 2014). While the aggregate number of fatalities caused by lower-intensity tornadoes is clearly smaller than for the more intense tornadoes, property damage follows a different trend. Losses per individual tornado increase dramatically with increasing EF rating; however, since there are so many more lower-intensity tornadoes, the aggregate losses caused by all EF1 tornadoes are very similar in magnitude to aggregate losses for all EF2s, all EF3s, all EF4s, and all EF5s (NIST 2014, Section 2.4.2.2).

Tornado Speed Determination. Tornado speeds for Risk Category III and IV buildings and other structures are shown in Figures 32.5-1 and 32.5-2, respectively. Maps for eight effective plan area sizes, ranging from a very small or point target of 1 ft^2 (0.9 m^2) to a very large target of 4 million ft^2 (371,612 m^2), are provided for each Risk Category. Interpolation is permitted for effective plan area sizes between mapped sizes. Linear interpolation of tornado speed should be calculated using the logarithm of the effective plan area, rather than the plan area itself.

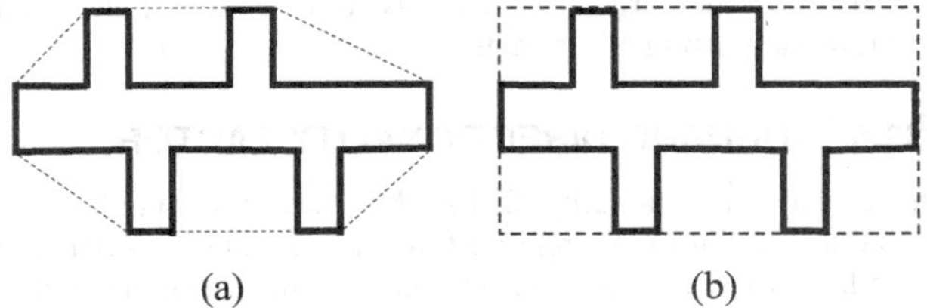

Figure C32.5-1. Effective plan areas for a building determined using (a) the smallest convex polygon enclosing the building plan, and (b) a rectangle enclosing the same building plan.

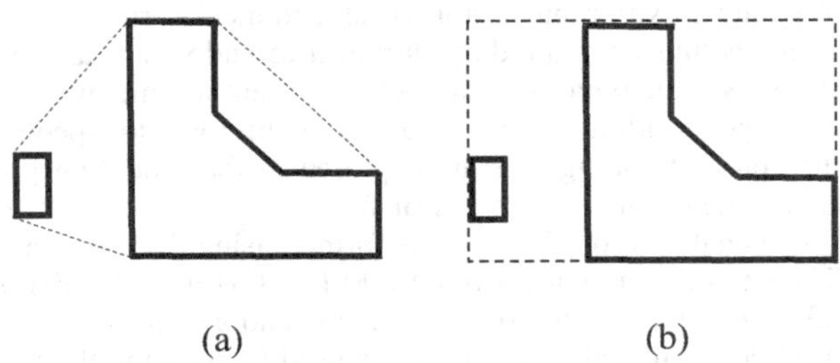

Figure C32.5-2. Effective plan area for a hospital and its central utility plant determined using (a) the smallest convex polygon enclosing the facility, and (b) a rectangle enclosing the facility.

Alternatively, location-specific tornado speeds may be determined using the ASCE Tornado Design Geodatabase, available through the ASCE 7 Hazard Tool (https://asce7hazardtool.online) or approved equivalent. The ASCE Hazard Tool website provides tornado speeds to the nearest mile per hour based on a defined location using either latitude/longitude or an address, risk category, and effective plan area.

Example (in customary units). Determine V_T for a Risk Category IV hospital building located in East Dubuque, Illinois (at the far northwest corner of the state), having an effective plan area, $A_e = 200{,}000\ \text{ft}^2$.

Single-map solution:

(a) Select which map to use. The smallest effective plan area that is larger than that of the hospital would be 250,000 ft^2, so use Figure 32.5-2F.

(b) Determine V_T using the selected map. East Dubuque is located at the extreme northwest corner of the state, between the 90 and 100 mi/h contours. By linear interpolation between the contours, $V_T = 99$ mi/h.

Alternate solution, interpolating between two maps:

(a) Determine the tornado speed for the smallest A_e that is larger than the hospital, which is 99 mi/h per the single map solution.

(b) Determine the tornado speed using the largest A_e that is smaller than the hospital, which would be 100,000 ft^2. From Figure 32.5-2E, the tornado speed is 94 mi/h by linear interpolation between the adjacent contours.

(c) Use linear interpolation on the logarithm of A_e from results of Steps (a) and (b) to determine V_T.

$$V_T = 94 + (\log 200{,}000 - \log 100{,}000) \times \frac{(99 - 94)}{(\log 250{,}000 - \log 100{,}000)} = 98\ \text{mi/h}.$$

C32.5.2 Design for Tornado Loads Not Required For Risk Category III and IV buildings, other structures, and facilities having a tornado speed $V_T < 60$ mi/h (26.8 m/s), the tornado loads will be of similar or lesser magnitude compared to the wind loads specified in Chapter 26. Therefore, calculation of tornado loads is not necessary. In addition, when V_T is less than a certain percentage of the basic wind speed, V, at the same geographic location, the determination of tornado loads is not necessary. These threshold speeds depend on V and exposure, both determined per Chapter 26. These criteria in this section provide approximate limits on V_T, above which tornado loads begin to control some aspect of the wind load design. These threshold speeds were established through analysis of worst-case combinations of geographic location, exposure, effective plan area, mean roof height, enclosure classification, building shape, and other parameters.

C32.5.3 Direction of Tornadic Wind While tornado paths do have a strong directional preference, the tornadic wind at any point on or near the building or other structure can come from any direction.

C32.5.4 Effective Plan Area In addition to risk category (with its associated return period) and geographic location, tornado speed is also a function of the size and shape of the footprint of the building, other structure, or facility. This is due to the comparatively small width of a tornado relative to the plan size of the structure; the width of tornado can be less than the size of the structure in some cases, particularly for large structures. At any given return period, tornado strike probabilities and associated maximum tornado speeds impacting the structure increase with increasing plan (i.e., footprint) area. The relevant size of the structure does not always correspond to the area of the footprint. For example, the tornado strike probability for a building that has a U-shaped plan is essentially the same as the strike probability for a building that has a rectangular plan that encloses the U shape, but those two buildings could have very different actual plan areas. Therefore, an effective plan area is used, which accounts for both plan size and shape. The effective plan area, A_e, of a building, other structure or facility is equal to the area of the smallest convex polygon enclosing the plan or footprint, meaning that A_e will always be as large or larger than the actual plan area of the building, other structure, or facility. A polygon is convex if all interior angles are less than 180 degrees. For rectangular plan buildings, a rectangle is the smallest convex polygon. For hexagonal and similar multi sided plan-shaped buildings, the smallest convex polygon is the area of the respective hexagon or similar shape. For circular or oval plan-shaped buildings, the areas of the smallest convex polygons enclosing these shapes are approximately equal to respective areas of the circle or oval. For a building with a more complex plan configuration, Figure C32.5-1(a) shows an example of the smallest convex polygon (dashed line) enclosing the maximum plan area (i.e., building footprint). Alternatively, A_e can simply and conservatively be calculated as the area of the smallest rectangle that encloses the maximum plan area, as shown in Figure C32.5-1(b).

For facilities comprised of multiple adjacent buildings and/or other structures, which all need to survive the tornado strike in order for the facility to remain operational, the convex polygon should be constructed to enclose the group of buildings/

structures. An example of this situation would be a hospital building and its nearby central utility plant (see Figure C32.5-2). This will yield a larger effective plan area, that would then be used to determine a single, larger tornado speed, V_T, for all of the critical buildings and other structures within the bounds of the polygon. An emergency operations center having an adjacent tower for emergency communications and parking structure for emergency response vehicles needed following a disaster would be another example.

C32.5.4.1 Essential Facilities For facilities comprising multiple adjacent buildings and/or other structures, which need to survive the tornado strike in order for the facility to remain operational, the designer should consider the convex polygon that encloses the Essential Facility and all of the buildings and other structures required to maintain the functionality of the Essential Facility, to capture the facility as a single site. This will yield a larger effective plan area used to determine a single, larger tornado speed, V_T, for all of the relevant buildings and other structures within the bounds of the polygon. An example of this situation would be a hospital building and its nearby central utility plant (see Figure C32.5-2). A second example would be an emergency operations center having an adjacent tower for emergency communications and parking structure for emergency response vehicles needed following a disaster.

C32.5.4.2 Other than Essential Facilities Where a building (or other structure) consists of multiple, structurally independent sections (e.g., buildings with expansion or seismic joints), a reduction in the effective plan area is permitted. The effective plan area for the entire building can be reduced to equal the effective plan area of the largest structurally independent section of the building. The tornado speed determined using this reduced A_e would be applied to the design of the entire building.

C32.5.4.3 Ground-Mounted Photovoltaic Panel Systems Most ground-mounted photovoltaic (PV) panel systems are designed, permitted, and inspected as Risk Category I. This includes large-scale systems (also referred to as "utility scale") PV facilities. Where the nonbuilding structures supporting PV panels are designed as Risk Category III or IV structures, tornado provisions apply. As PV facilities are an intermittent power source, generating power only when the sun is shining, they are not suitable for emergency backup power for Risk Category IV buildings unless paired with an energy storage system.

Large-scale PV facilities can cover hundreds of acres of land, yet they are composed of hundreds or thousands of small, structurally independent "tables" of PV panels, each with its own independent foundation system. The PV panels on these independent nonbuilding structures are linked by electrical conductors to central inverters that convert DC power to AC power. Large-scale PV facilities can have dozens to hundreds of independent central inverters. If an electrical fault is detected, only the inverter associated with that fault is shut down, and the remainder of the facility remains operational. The entire PV facility will shut down only if the electrical substation is shut down, or if the system otherwise detects a loss of the AC signal from the grid. Substations and grids are outside the scope of ASCE 7.

While there is little data on tornado strikes on large-scale PV facilities, in two known cases the tornado strike caused only isolated, localized damage. These facilities typically remain operational with localized damage. For ground-mounted PV installations, the effective plan area, A_e, should be the effective plan area of the largest structurally independent nonbuilding structure supporting PV panels.

C32.6 TORNADO DIRECTIONALITY FACTOR

The tornado directionality factor, K_{dT}, was computed using a procedure similar to that reported by Vickery and Liu (2018) that was adapted for tornado loads (Vickery et al. 2021). For both the main wind force resisting system (MWFRS) and component and cladding (C&C) load cases, the direct wind-induced pressures were computed assuming that the pressure coefficients derived from boundary layer wind tunnel tests are also valid for tornadic winds (e.g., Roueche et al. 2020, Kopp and Wu 2020). The effects of the vertical component of the tornadic winds on the pressure coefficients were considered separately, as described in Section C32.14. Unlike straight-line winds (including those from hurricanes, extratropical storms, and thunderstorms), the wind speeds acting on a building during a tornado can vary significantly over the building at any given instant in time, particularly for large buildings. These variations in tornado speed as a function of building size are captured in the modeling process used to determine the values of K_{dT}.

The analyses for K_{dT} were performed using three buildings of different sizes, ranging from 1,800 ft^2 (170 m^2) to 250,000 ft^2 (23,000 m^2), including both low-slope and gabled roofs. A total of 5,000 simulated tornadoes were used to compute the tornado loads over a range of tornado speeds and sizes. C&C loads are based on uplift loads computed for two different deck element sizes, considering elements located in roof Zones 1′, 1, 2, and 3 for the low-slope roof buildings per Figure 30.3-2A and Zones 1, 2, and 3 for the pitched-roof building per Figure 30.3-2B. The MWFRS loads are based on uplift loads computed for the whole roof (for the small building) and roof trusses or large roof panels (for the bigger buildings).

In the computation of K_{dT} for C&C loads, K_{dT} is taken as the ratio of the maximum wind induced pressure at a point produced by the tornado, divided by the maximum (GC_p) (over all wind directions) multiplied by the maximum value of the velocity pressure at the same point produced by the passage of the tornado. The wind speed associated with the maximum velocity pressure at the point at which K_{dT} is being computed is always equal to or less than the maximum tornadic wind experienced somewhere on the building. The tornado directionality factor for the MWFRS is computed in a similar manner, except the normalizing wind load is the maximum MWFRS C_p coupled with the maximum dynamic pressures associated with the maximum wind speed experienced somewhere on the structural element (e.g., roof truss, frame section, etc.). Since the entire structural element does not necessarily experience the maximum wind speed (unlike components, which are small), the effective value of K_{dT} generally decreases as the structure or structural-element tributary area increases, although it is a weak trend. The effective value of K_{dT} for C&C behaves in an opposite manner, slightly increasing as the size of the structure increases, as a greater portion of the building is associated with Zone 1 (GC_p) values, which are less directionally sensitive.

Results of the simulations for C&C loads showed that for all buildings and all roof zones except 1, the mean K_{dT} values typically ranged between about 0.65 to 0.75, with a few outliers above and below. In the case of Zone 1, K_{dT} ranged from slightly over 0.8 for the intermediate-size building to 0.97 for the large building. In addition to the building size effect, the higher K_{dT} for Zone 1 is also a result of the fact that there is little variation of (GC_p) with wind direction for locations well removed from the edge of the building, away from any flow separation zones. The effect of tornado size may also play a role. While no studies were performed examining K_{dT} for components on surfaces other than roofs, the K_{dT} value of 0.75 for Zones 1, 2, and 3 was also recommended for all other zones, including walls (Vickery et al. 2021). This is believed to be conservative, since, as shown in Vickery and Liu (2018), K_d values for C&C on walls are lower than those on roofs. For the MWFRS, the value of $K_{dT} = 0.8$ was adopted based on the results of the analysis (Vickery et al. 2021). For arched roofs, circular domes, and all other structures, K_{dT} is determined using Table 26.6-1, given the lack of tornado-specific data or simulations.

Buildings and other structures intended to remain operational in the event of extreme environmental loading from a tornado include not only Essential Facilities, as defined in Section 1.2.1, but also the buildings and structures required to maintain the functionality of those Essential Facilities (which comprise a subset of Risk Category IV buildings and other structures, per Table 1.5-1). For both, it is critical to avoid breaches in the building envelope. Loss of *any* wall cladding, roof decking, or other element of the envelope can permit intrusion of wind, windborne debris, wind-driven rain, and/or falling rain into the building. Even minor damage to these and other elements, such as roof coverings, can allow significant amounts of rainwater to enter the building. This often results in serious damage to building interiors, contents, and mechanical and electrical systems, rendering the facility or parts of it nonoperational. This means that all, or a portion of, the building envelope should be considered a Designated Nonstructural System, as defined in Section 1.2.1. The directionality factor accounts for "the reduced probability of maximum winds coming from any given direction, and the maximum pressure coefficient for a specific load effect occurring for any given wind direction," per Section C26.6. It is likely that for a design tornado event that one or more portions of the building envelope are loaded with the maximum pressure, so if the "load effect" is "pressure anywhere on the building envelope," then there would be no reduction in probability. When all or most of the C&C are part of this Designated Nonstructural System, a K_{dT} value of 1.0 is more appropriate, as some portion of the envelope is nearly certain to have its maximum pressure coefficient at the direction of the maximum winds.

C32.7 TORNADO EXPOSURE

Effects of exposure on tornado characteristics are difficult to resolve near the surface with observational techniques such as radar. Therefore, research has turned to wind tunnel and numerical (e.g., computational fluid dynamics) studies. Most available research suggests that tornado characteristics are modified by exposure, and these modifications are greatest near the ground. However, great variability exists on exactly what those modifications are (e.g., Refan and Hangan 2018) due to both the complexity of tornadoes and the challenge of creating the variety of realistic terrain environments in the experiments. Based on this variability, the standard does not define tornado exposure. This results in use of a single tornado velocity pressure profile per Section 32.10.1.

Several wind tunnel and numerical studies (e.g., Zhang and Sarkar 2012, Wang et al. 2017) suggest a reduction in tangential wind speed, an increase in radial wind speed, and a reduction of the core radius size for rougher exposures. However, the magnitude of both mean and maximum horizontal wind speeds is unaffected in some simulations (e.g., Nolan et al. 2017). Other studies suggest an increase in wind speed above the surface due to rougher exposure (Davies-Jones et al. 2001). Intensity and path changes have also been noted in some numerical simulations, but the direction of the changes depends on the setup (Lewellen 2014). The peak horizontal wind speed shifts upward (i.e., higher above ground) in simulations with rougher exposure (e.g., Nolan et al. 2017), which is inconsistent with other simulations (e.g., Matsui and Tamura 2009).

C32.8 TOPOGRAPHIC FACTOR

Similar to exposure, the effects of topography on tornado wind speeds are difficult to resolve near the surface with observational techniques such as radar. Observational data, where available, defy generalization when considering topography (Houser et al. 2020). Changes in tornado intensity, size, and path were noted in both numerical (Lewellen 2012) and wind tunnel (Razavi and Sarkar 2018) experiments. The sign of the changes in these studies varied, but the most significant changes occurred very close to the ground. A few references suggest an increase in wind speeds in valleys due to channeling, from both observational studies (Karstens et al. 2013) and numerical studies (Satrio et al. 2020). However, similar to studies of tornado exposure (Section C32.7), the challenge of creating the variety of realistic topographic cases in the simulations makes generalization through numerical simulation difficult. Simulations using various topographic setups show great variability in topographic effects on horizontal tornado winds (Satrio et al. 2020). Based on this variability, the standard does not define a topographic factor for tornadoes at this time.

C32.9 GROUND ELEVATION FACTOR

The effect of ground elevation on air density is not dependent on the type of windstorm; therefore, the K_e provisions of Section 26.9 are appropriate, and there is no need to modify these provisions.

C32.10 TORNADO VELOCITY PRESSURE

C32.10.1 Tornado Velocity Pressure Exposure Coefficient For determination of a wind loading profile for tornadoes (represented by K_{zTor}), data from a number of sources were reviewed and analyzed, including field (Refan et al. 2017, Kosiba and Wurman 2013), wind tunnel (Refan and Hangan 2018, Tang et al. 2018), and numerical studies (Dahl and Nolan 2017, Reinhart et al. 2018). Tornadoes from field, wind tunnel, and numerical studies used in preliminary analysis were generally "strong" tornadoes (wind speeds associated with an EF2 tornado or higher). The profiles included tangential wind speed only (perpendicular to pressure gradient) and horizontal (i.e., combination of tangential and radial) wind speed. Since these studies were interested in wind speed, initial analysis focused on the behavior of the wind speed profile with height (i.e., $\sqrt{K_{zTor}}$ which is a normalized wind speed profile) and not the wind loading profile (i.e., K_{zTor}). Although significant variability in the tornado wind speed profiles was noted at all heights, all studies showed a peak wind speed occurring relatively close to surface near the horizontal (radial) location of strongest winds (i.e., radius of maximum winds) from the tornado "center." Tornado wind speed profiles at the radius of maximum winds routinely (but not always) displayed a nose-like profile, where the wind speed is lower both above and below the nose. This behavior is different from the atmospheric boundary layer (ABL) flows, in which the wind speed increases monotonically with height until a maximum wind speed is reached at the gradient height.

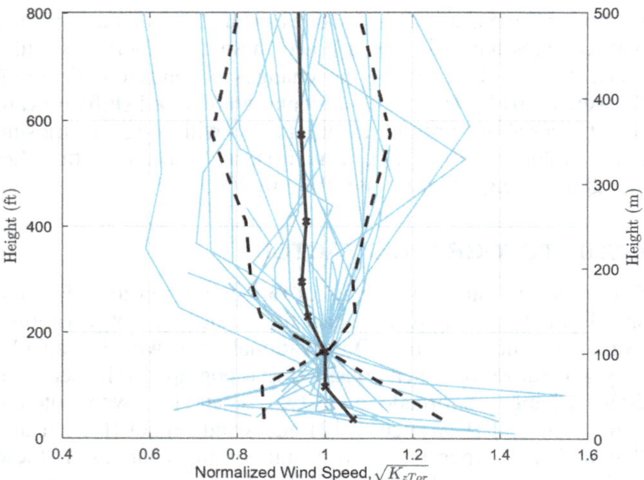

Figure C32.10-1. Vertical profiles of tornado winds plotted as the square root of velocity pressure exposure coefficient (Normalized wind speed, $\sqrt{K_{zTor}}$) for tornadoes, showing all the radar profiles (gray), the median profile (black solid line), and plus and minus one standard deviation from the median (black dashed line).

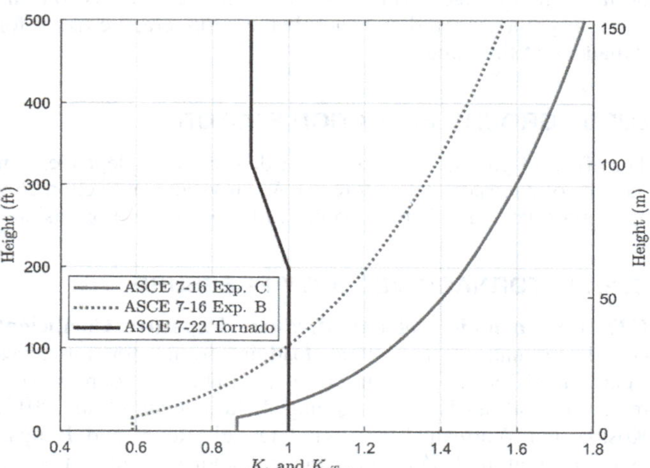

Figure C32.10-2. Vertical profiles of tornado velocity pressure (K_{zTor}) versus that of Exposure B and Exposure C for nontornadic winds (K_z) in Chapter 26 for the lowest 500 ft (152.4 m).

After preliminary analysis of the tornado wind speed profiles it was determined that only radar profiles would be used to create the ASCE 7 profile, as this was the largest and most consistent data set. These 36 profiles consisted of the maximum radar-estimated wind speed over the time of the tornado. The sampling of the maximum wind speed was over a short time period and therefore did not employ any averaging schemes. No modifications for exposure were made. Radar measurements were typically collected in open, flat terrain. Tornadoes from all studies had different swirl ratios (i.e., vortex structure). However, the averaging techniques employed, the difficulty of assessing the parameter in the field (Kosiba and Wurman 2013), and varying definitions depending on experimental technique precluded any detailed profile analysis conditional on swirl ratio values.

The next step was to determine the approximate height where the nose occurs and then to taper the profile away from this height. To better reflect behavior of tornado wind speed profiles close to the ground, profiles that had a minimum observation height below 200 ft (61 m) were kept for analysis. Given this condition, the wind speeds at any height for all studies were normalized by the wind speed at the closest observation height to 164 ft (50 m). It should be noted that the normalized profile behavior is sensitive to the minimum height of data collection. For purposes of statistical analysis, the radar measurements were also placed in data bins in height ranges of 66 ft (20 m) for elevations below 328 ft (100 m), and 164 ft (50 m) for higher elevations. Where less than 10 observations were found in a bin, the bin was extended in height until the condition was met, to minimize bin-to-bin variation and the effect of outliers. The results of the normalization and binning processes yield tornado speed profiles ($\sqrt{K_{zTor}}$) for 36 tornadoes as shown by blue lines in Figure C32.10-1. The median values (solid black line) of $\sqrt{K_{zTor}}$, as well as median plus and minus one standard deviation (dashed lines), are plotted at the mean height of each bin.

When considering the wind loading profile, the committee chose to approximately follow the median radar profile from Figure C32.10-1, resulting in the K_{zTor} profile shown in Figure C32.10-2. Below 200 ft (61 m), K_{zTor} is uniform, with a value of 1.0, effectively "anchored" by the $K_z = 1$ value for Exposure C at 33 ft (10 m), where the tornado and Exposure C profiles cross. From the 200 ft (61 m) to 328 ft (100 m) levels, a linear function was used to represent the decreasing tapered behavior which is apparent in the median profiles. Above 328 ft (100 m), K_{zTor} was kept at 0.9, as the median profiles tended to stabilize around that value as height increased, and the amount of available data also decreased above this height.

C32.10.2 Tornado Velocity Pressure The mass density of air in the tornado velocity pressure relation, Equation (32.10-1), which is a component of the constant at the beginning of that equation, is based on standard atmospheric conditions. The wind velocity pressure relation, Equation (26.1-1), uses this same approach (Section C26.10.2). The mass density of air is a function of elevation (Section C26.9), atmospheric pressure, temperature, and humidity. Although atmospheric pressure drops in a tornado, this effect is not currently incorporated in the formulation of tornado velocity pressure, similar to how the low pressure in hurricanes is not incorporated in the wind velocity pressure equation when used in hurricane-prone regions.

C32.11 TORNADO GUST EFFECTS

C32.11.1 Tornado Gust-Effect Factor The gust-effect factor for rigid structures takes into account the lack of correlation of the gusts in a turbulent wind field, which results in gust-effect factors lower than 1.0. In the case of tornadic winds, while the gustiness due to turbulence may be less than in atmospheric boundary layer (ABL) winds, the tornado winds vary in both direction and speed over the building or structure, resulting in lower peak loads compared to ABL winds, whose mean wind speed and direction are constant over the building or structure. The duration of a tornado is sufficiently short such that the gust factor provisions of Section 26.11.5 for flexible or dynamically sensitive buildings and other structures do not apply. Consequently, the rigid building/structure gust factor was selected for use with the tornado MWFRS loads.

C32.12 TORNADO ENCLOSURE CLASSIFICATION

C32.12.1 General Sealed structures are particularly susceptible to loads induced by atmospheric pressure change associated with the passage of a tornado. A new enclosure classification of "sealed" was introduced to enable capturing of these effects, which are applicable to other structures, including tanks and vessels that have controlled ventilation such that tornado-induced atmospheric pressure changes will not be transmitted to the inside of the structure. Additional information on atmospheric pressure change and its effects is provided in Section C32.13.

The wind-borne debris protection requirements of this section cover glazed openings, not other elements of the building envelope, because wind-borne debris impacts through unprotected glazing are likely to shatter the glazing and create large openings, whereas other building envelope elements are likely to only have penetrations the size of the wind-borne debris, which would not change the enclosure classification. See Section C32.17 for more information.

C32.12.2 Openings Wind-borne debris hazards are greater for tornadoes than for hurricanes. The updrafts in a tornado can loft debris higher in the air, creating the opportunity for more, larger, and faster traveling debris compared to hurricanes, where debris is mainly transported horizontally. Comparisons of observed window breakage rates between several hurricanes and from the 2011 Joplin, Missouri, tornado demonstrated greater breakage rates for tornadoes over hurricanes in areas that experienced approximately the same maximum wind speeds. This finding is consistent with observations from many field investigations that showed failure of glazing is common in tornadoes (e.g., Roueche and Prevatt 2013). Since unprotected windows are prone to failure during tornadoes, glazed openings that do not meet the protection requirements of Section 32.12.3.1 are considered openings for the purposes of determining enclosure classification, as provided in Section 32.12.2. This section requires that for buildings and other structures that would otherwise be classified as enclosed, the criteria for partial enclosure classification are reevaluated for the case that unprotected glazing on each assumed windward wall fails and is considered to be open.

C32.12.3 Protection of Glazed Openings Failure of unprotected glazing is common in tornadoes. The resultant intrusion of wind, wind-borne debris, and wind-driven rain can result in hazards to life safety and significant interior damage, including collapse of interior ceilings and walls (see also Section C32.6). For buildings and other structures intended to remain operational in the event of a tornado (including Essential Facilities) and where a building owner's tornado performance objective is to reduce occupancy disruption and interior damage during a tornado, it is important to specify glazing assemblies that have been tested for wind-borne debris impact, in addition to meeting static pressure requirements.

C32.12.3.1 Protection Requirements for Glazed Openings Assemblies that resist penetration by the ASTM E1996 test missile D are likely to resist penetration of most wind-borne debris generated by design tornado events with speeds per Section 32.5. Greater protection is provided by assemblies that resist penetration by test missile E. For significantly enhanced protection to resist breaching of exterior glazing by more intense tornadoes, glazing assemblies can be specified that have been tested in accordance with AAMA 512 (2011), using tornado test missiles given in ICC 500 (2020) or FEMA P-361 (2021a).

Impact resistant glazing and permanently anchored impact-protective systems do not require human intervention in order to achieve wind-borne debris protection. Impact-protective systems that require power for deployment (e.g., motorized shutters) or human intervention for deployment should only be specified when the building owner has procedures that will ensure rapid deployment prior to a tornado. Given the limited warning time for tornadoes (typically ranging from no warning to tens of minutes), many operable impact protections systems used in hurricane-prone regions would not be applicable for tornado protection given the time it takes to deploy them or the need to deploy them from outside the building. Use of operable systems should be limited to facilities that are staffed 24 hours a day, seven days a week, with staff that maintain situational awareness (such as a fire station or emergency operations center) and are trained to deploy the systems. As remote or automatic deployment of systems becomes available with "smart building" designs, these could be used in place of the 24 hours a day, 7 days a week, staffing requirement.

C32.13 TORNADO INTERNAL PRESSURE COEFFICIENTS

For nontornadic winds, internal pressures are caused solely by external wind-induced pressures affecting the interior through openings in the building envelope. In tornadoes, there is a second mechanism that also affects internal pressures, called atmospheric pressure change (APC). The atmospheric (or static) pressure at the center of the tornado is much lower than the ambient values, similar to a hurricane but on a much smaller spatial scale. As the core of the tornado moves near or over the building, the atmospheric pressure outside of the building drops rapidly. If the air permeability of the building is low, such that this drop in static pressure cannot be swiftly transmitted into the building, this creates a differential static pressure between the exterior and interior of the building, which effectively behaves as an increased positive internal pressure.

To determine the tornado internal pressure coefficient, (GC_{piT}), tornado wind-induced internal pressures, APC-induced internal pressures, and their combined effects were computed for sealed, enclosed (assuming uniform distribution of leakage), and partially enclosed (dominant opening) buildings, as described by Vickery et al. (2021). A total of 5,000 simulated tornadoes were used to generate wind loading data for models of each of the three enclosure classifications. For the enclosed and partially enclosed cases, internal pressures were computed with and without consideration of APC. Each simulated tornado produced a time series of tornado-induced pressures on the roof and wall panels of the model building. Using these time series of wind pressures, the internal pressure and the APC associated with the time at which the tornado-induced net pressure on an element reached its maximum (or minimum) value was retained and used in subsequent analyses. In the computation of the internal pressures due to tornadic winds only, it was assumed that there was no net flow into or out of the building, and that the balance of the net flow occurs instantaneously. This "conventional" internal pressure has a value near the weighted average of the external pressures. The effective (or net) internal pressure is a combination of (1) the conventional *internal* pressure due to tornadic wind-induced pressure and external pressure change, and (2) the differential between the *external* static pressure due to the APC and the internal static pressure. The effective internal pressure can also vary significantly over the exterior of the building, particularly if the tornado producing the APC is small compared to the building. Partially open and open buildings and other structures are not subject to APC-induced pressures.

The tornado internal pressure coefficient, (GC_{piT}), given in Table 32.13-1 accounts for the effective internal pressure induced by the tornado, i.e., the combined external and internal pressures due to the APC as well as internal pressures due to building porosity and the direct action of the tornadic wind-induced exterior pressures. For enclosed buildings, $(GC_{piT}) = +0.55$ and –0.18, where the increased positive internal pressure compared to the $(GC_{pi}) = +0.18$ for wind loads (per Table 26.13-1) is driven by the contribution of APC. Although analyses were also conducted for a (theoretical) completely sealed building, in actuality, no buildings are completely sealed. Therefore, the sealed case $(GC_{piT}) = +1.0$ in Table 32.13-1 is only required for sealed other structures, such as tanks and vessels having controlled ventilation such that tornado-induced atmospheric pressure changes will not be transmitted to the inside of the structure. It is possible that for buildings (or portions thereof) having tightly controlled ventilation, such as certain facilities housing hazardous substances that pose a threat to the public if released, use of a positive (GC_{piT}) between the values for the enclosed and sealed conditions may be appropriate.

C32.14 TORNADO EXTERNAL PRESSURE COEFFICIENTS

There are strong updrafts that occur near the radius of maximum winds or corner flow region in tornadoes, so the mean winds are no longer always horizontal. This vertical component of the wind changes the aerodynamics of wind flow around the building and, consequently, the pressures on the building, particularly over the roof where it can result in increased uplift. The tornado pressure coefficient adjustment factor for vertical winds, K_{vT}, is used to modify roof uplift pressure coefficients that were previously developed for boundary layer winds to account for these effects (see Section C32.6 for more information on the applicability of pressure coefficients developed for boundary layer winds to tornadic winds). The parameter K_{vT} takes into account the effect of the vertical component of the wind speed within the core of a tornado. Equations for MWFRS and C&C tornado loads in Sections 32.15 through 32.17 that are applicable to uplift on roofs and some rooftop appurtenances include K_{vT} as a modifier on the external pressure coefficient term.

The development of K_{vT} was documented by Vickery et al. (2021) and summarized here. The vertical component of the tornadic wind changes the vertical angle of attack of the wind with respect to the surface of the roof. The pressures resulting from the 5,000 tornado simulations were calculated with and without the effect of vertical winds. During each simulated tornado, horizontal and vertical wind speeds, atmospheric pressure, and wind-induced pressures are computed at every time step as the modeled tornado passes by a modeled building. The difference between the two sets of load data is the change in the pressure due to the effect of vertical winds. Each of the 5,000 tornado simulations yields a single value of K_{vT} for each roof deck element, at the element centroid. These K_{vT} values are computed for each roof deck element as the maximum value of the tornado-induced external wind pressure including the effects of vertical winds, divided by the maximum pressure without the effects of vertical winds.

For sloped-roof buildings, the effects of the change in the angle of attack are handled by altering the effective slope of the roof. The rate of change of (GC_p) with roof slope for sloped-roof buildings was developed using (GC_p) data for roof slopes of 4:12 and 7:12. These data use directional values of (GC_p), which are available for eight different zones (all of which can be mapped to one of the three main roof C&C zones used in Figure 30.3-2B). (GC_p) data are modeled using the same methodology used in Hazus (Vickery et al. 2006). For low-slope roofs, experiments were performed at the boundary layer wind tunnel at the University of Western Ontario. The leeward side of the model was tilted down into the floor of the tunnel, such that the horizontal wind flow over the windward roof of the building had a negative roof slope (Vickery et al. 2020). These experiments were repeated for a range of negative roof slopes in order to develop a relationship between the roof slope and (GCp) and then determine the K_{vT} values. For both the low-slope and steep-slope cases, K_{vT} values for C&C were determined for Roof Zones 1, 2, and 3 (using zone definitions from Figures 30.3-2A through 30.3-2D). For the MWFRS, K_{vT} was computed using the same buildings and tornado simulations described in Section C32.6.

For all tornado simulations used to develop these initial K_{vT} values, at least some portion of the building was within the core of the tornado. However, in reality, not all buildings will be impacted by the core of a tornado when affected by a design tornado event. The likelihood of being affected by the core increases with increasing design tornado speed. For a small building with a design tornado speed of 100 mi/h (44.7 m/s), the building is not in the core in 33% of the tornadoes, so K_{vT} would be 1.0 for these cases. For the same small building having a design tornado speed of 180 mi/h (80.5 m/s) (which occurs at around 100,000-year return periods, as shown in Appendix G), the building is not in the core in just 4% of the tornadoes. The final K_{vT} values shown in Table 32.14-1 for both C&C and MWFRS were adjusted downward to account for the fraction of the design event tornadoes where the building is within the core, considering the mapped tornado speeds for Risk Category III and IV return periods. These adjustments were typically on the order of 5% decreases. When designing to the higher tornado speeds associated with longer return periods as provided in Appendix G, see Vickery et al. (2021) for recommendations on K_{vT} values.

The effects of tornado updrafts on C&C pressure coefficients are modest at the edges and corners of low-slope roofs (C&C Zones 2 and 3 in Figure 30.2-3A), but increase in these zones for steeper roof slopes (see Table 32.14-1). Roof slope does not have much effect on Zone 1 pressure coefficients. For the MWFRS, $K_{vT} = 1.1$ for roof uplift. For downward-acting loads on roofs and all surfaces other than roofs, the values of K_{vT} are 1.0 for both MWFRS and C&C. The K_{vT} values for building roofs were adapted for use on the roofs of bins, silos, and tanks. The C&C zone shapes and locations for these structures are not the same as for buildings, but Section 32.17.5 provides a mapping to the coefficients for building roof zones provided in Table 32.14-1. One phenomenon not currently accounted for is the spatial gradient of the static pressure and its potential net effects on drag and base-shear loads, a good topic for future research.

C32.15 TORNADO LOADS ON BUILDINGS: MAIN WIND FORCE RESISTING SYSTEM

C32.15.1 Enclosed, Partially Enclosed, and Partially Open Buildings The equation for determination of tornado loads on the MWFRS is similar to that used in Chapter 27, with tornado versions of the parameters substituted for the original terms. The external pressure coefficients, C_p, are multiplied by K_{vT} to account for the effects of the vertical component of tornadic winds (see Section C32.14). The velocity pressure, q_i, for use with the tornado internal pressure coefficient (GC_{piT}) is not multiplied by K_{dT}, since APC contributes the most to (GC_{piT}) and since APC is not a function of the direction of the tornadic winds. Internal pressure evaluation of the roof and walls in partially enclosed buildings uses $q_i = q_{zop}$, where the height

z_{op} is defined as the level of the lowest opening in the building that could affect the positive internal pressure. This height is used in contrast to q_z at the level of the highest opening in Equation (27.3-1) for wind loads, since the tornado velocity pressure decreases with height, compared to the wind velocity pressure, which increases with height.

C32.15.5 Design Load Cases In cases where a tornado is smaller in plan area than the building or other structure it strikes, the load cases from Figure 27.3-8 are likely to be conservative, since the maximum tornado speed does not extend across the full plan area of the building. The asymmetry of the tornado wind field with respect to the building would tend to increase torsional loads compared to other types of windstorms, hence the removal of the exception in Section 27.3.5 that eliminates the torsional load cases in Figure 27.3-8.

C32.16 TORNADO LOADS ON BUILDING APPURTENANCES AND OTHER STRUCTURES: MAIN WIND FORCE RESISTING SYSTEM

C32.16.1 General Requirements The tornado speed for the building appurtenance should equal that used for design of the building. Building appurtenances such as rooftop equipment and rooftop solar (i.e., PV) panels are typically much smaller than the building itself. By using the same tornado speed as the building, the potential for failure of these elements and any associated risk of their being dislodged from the building is reduced to a level consistent with the reliability goals of the standard. Where building appurtenances or other structures support the continued operation of an Essential Facility, the tornado speed should equal that used in design of the Essential Facility. See also Section C32.5.4.1.

The effective plan area of freestanding walls and signs is based on the data used in the development of the tornado hazard maps (Section C32.5). A sensitivity analysis was performed to determine the aspect ratio at which a building's aspect ratio would begin to influence the tornado speed at a given return period. This aspect ratio was found to be approximately 20:1. Beyond this ratio, the structure's azimuth begins to influence the strike probability, due to the directional nature of tornadoes (predominantly travelling from southwest to northeast). For this reason, long linear structures such as transmission lines are outside the scope of this document.

C32.16.2 Solid Freestanding Walls and Solid Signs For taller walls and signs, the bullnose profile of the velocity pressure exposure coefficient for tornadoes can lead to higher velocity pressures at midheight of the sign as compared to the top of the sign or wall.

C32.16.3 Other Structures

C32.16.3.1 Trussed Towers Damage investigations have shown that wind-borne debris can cling to trussed communications towers (FEMA 2012). Clinging debris such as metal roof panels and chain-link fence with privacy slats can increase A_f, the projected surface subject to wind loading, thus increasing the wind load on the structure, with tower collapse as a possible result. The values for clinging debris in Section 32.16.4.1 were taken from ASCE 7-16, Commentary Section C26.14.6, which were based on tornado damage investigations reported by FEMA (2012). To minimize collapse potential, towers and other lattice-type open structures should be designed for the additional tornado load caused by clinging debris.

C32.16.3.3 Roofs of Isolated Circular Bins, Silos, and Tanks Unlike buildings, the tank envelope is not often punctured to alleviate the pressure differential due to atmospheric pressure change. This can result in amplified roof uplift pressures on bins, silos, and tanks. Compounding the effect on the roof is the vertical component of the tornadic winds (Section C32.14).

C32.16.3.4 Rooftop Solar Panels for Buildings of All Heights with Flat Roofs or Gable or Hip Roofs with Slopes Less than 7 Degrees The provisions in Chapter 29 for wind loads on rooftop PV panels not parallel to the roof are based on wind tunnel test data of straight-line winds (Section C29.4.3). Additional testing would be required to evaluate the nominal net pressure coefficient (GC_{rn}) as it relates to tornado loads. However, committee judgement suggests that the vertical component of the tornado winds would increase the net uplift pressure on the rooftop PV panel. The K_{vT} factor could be applied to Equation (32.16-6) to account for this increase.

C32.16.3.5 Rooftop Solar Panels Parallel to the Roof Surface on Buildings of All Heights and Roof Slopes The provisions in Chapter 29 for rooftop PV panels parallel to the roof surface are calculated based on the external pressure coefficient for C&C for the roof zone below the panels. The additional vertical component of the tornadic winds amplifies this net uplift pressure (Section C32.14).

C32.17 TORNADO LOADS: COMPONENTS AND CLADDING

The procedures for determining tornado design pressures on C&C are based on the methods of Chapter 30, with several modifications. Tornado velocity pressures replace the velocity pressures specified in Chapter 30. For enclosed, partially enclosed, and partially open buildings, the velocity pressure used for determination of internal pressures is not multiplied by the tornado directionality factor, since APC contributions to internal pressure are not directional as described in Section C32.15.1. Internal pressure coefficients, (GC_{pi}), for these buildings are replaced with (GC_{piT}), and external pressure coefficients for equations that include roof uplift are multiplied by K_{vT} to account for the increased uplift due to vertical updrafts in the core of the tornado.

Compared to MWFRS, less is known about the C&C loads in tornadoes. C&C loads are more sensitive to other aspects of the wind field, especially turbulence, which is not well understood in tornadoes but is a critical aspect of wind loading and bluff-body aerodynamics.

The development of the K_{vT} factors by Vickery et al. (2021) consisted of research on enclosed and partially enclosed buildings (Section C32.14). Therefore, K_{vT} is excluded from the equations for open buildings and attached canopies until additional data become available.

Wind-borne Debris: The only requirement for protection from wind-borne debris pertains to exterior glazing, per Section 32.12.3. However, during a design tornado event, wind-borne debris can damage other portions of the building envelope. Some debris may penetrate the roof and/or exterior walls. Although such penetrations typically do not result in a change of enclosure classification for enclosed buildings to partially enclosed conditions, the penetration can create a pathway for rain to enter the building. For design guidance to minimize debris and/or rain penetration through roof or wall assemblies, see FEMA (2011, 2012).

C32.17.5.1 Isolated Circular Bins, Silos, and Tanks The K_{vT} values for design of C&C for roofs of bins, silos, and tanks are based on committee judgement. Zones 3 and 4 in Figure 30.12-2 are edge zones similar to Zone 2 for a building roof, while Zones 1 and 2 in Figure 30.12-2 are analogous to the interior Zone 1 region of a building roof.

C32.18 TORNADO LOADS: WIND TUNNEL PROCEDURE

The use of the Wind Tunnel Procedure is limited to determination of pressure and force coefficients for use with MWFRS tornado load provisions for buildings and other structures in Sections 32.15 and 32.16, respectively, and with the C&C load provisions in Section 32.17. Sections C32.6 and C32.14 include information on the applicability of pressure coefficients developed for boundary-layer winds to tornadic winds.

Testing of isolated models in Exposure Category C is specified to be consistent with typical test conditions used for development of pressure and force coefficients in Chapters 27, 29, and 30. Traditional wind tunnel tests would commonly include placing the test building in a proximity model to capture interference and shielding effects from nearby buildings and structures. However, such effects would not be the same between boundary layer winds and tornadic winds, due to the rapid wind direction changes in tornadoes; hence the requirement for testing isolated models.

While wind tunnels have been used for many years to estimate ABL wind loads on buildings and structures, procedures are currently being developed for determination of aerodynamic loads experienced by buildings and structures in tornadoes using simulators. Several tornado simulators have been designed and constructed over the past years, including the WindEEE Dome at the University of Western Ontario (Refan and Hangan 2018), VorTECH at Texas Tech University (Tang et al. 2018), and the Iowa State University simulator (Haan et al. 2008). Increasingly available field data from tornadoes, along with proposed scaling methods (Refan et al. 2014), have contributed to reproducing the velocity field of tornado vortices in simulators. Despite this progress, many challenges remain. Little is known about turbulence levels and its effects on wind loads in tornadoes. Although these laboratory simulations have proven to be successful in replicating the velocity field and atmospheric pressure change in tornadoes, there is little consensus over test conditions, test methodology, and data analysis (e.g., definition of the pressure coefficient for tornadic winds and how to compare these coefficients with the ones previously defined for ABL winds; see discussion by Kopp and Wu 2020). Therefore, at this time, the use of tornado simulators exclusively is not considered suitable for determination of pressure and force coefficients for buildings and other structures.

REFERENCES

AAMA (American Architectural Manufacturers Association). 2011. *Voluntary specifications for tornado hazard mitigating fenestration products*. AAMA 512-11. Schaumburg, IL: AAMA.

ASCE. 2019. *Prestandard for performance-based wind design*. Reston, VA: ASCE.

CDC (Centers for Disease Control and Prevention). 2012. "Tornado-related fatalities: Five states, southeastern United States, April 25-28, 2011." *Morbidity Mon. Weekly Rep.* 61 (28): 529–533.

Dahl, N. A., D. S. Nolan, G. H. Bryan, and R. Rotunno. 2017. "Using high-resolution simulations to quantify underestimates of tornado intensity from in situ observations." *Mon. Weather Rev.* 145 (5): 1963–1982. https://doi.org/10.1175/MWR-D-16-0346.1.

Davies-Jones, R., R. J. Trapp, and H. B. Bluestein. 2001. "Tornadoes and tornadic storms." In *Severe convective storms*, C. A. Doswell III, ed., 167–221. Boston: American Meteorological Society.

FEMA (Federal Emergency Management Agency). 2011. *Critical facilities located in tornado-prone regions: Recommendations for architects and engineers*. Washington, DC: FEMA.

FEMA. 2012. *Mitigation assessment team report: Spring 2011 tornadoes: April 25–28 and May 22. Building performance observations, recommendations and technical guidance*. Washington, DC: FEMA.

FEMA. 2021a. *Safe rooms for tornadoes and hurricanes: Guidance for community and residential safe rooms*. 4th ed. Washington, DC: FEMA.

FEMA. 2021b. *Taking shelter from the storm: Building or installing a safe room for your home*. 5th ed. Washington, DC: FEMA.

Haan, F., P. Sarkar, and W. Gallus. 2008. "Design, construction and performance of a large tornado simulator for wind engineering applications." *Eng. Struct.* 30 (4): 1146–1159. https://doi.org/10.1016/j.engstruct.2007.07.010.

Houser, J. B., N. McGinnis, K. M. Butler, H. B. Bluestein, J. C. Snyder, and M. M. French. 2020. "Statistical and empirical relationships between tornado intensity and both topography and land cover using rapid-scan radar observations and GIS." *Mon. Weather Rev.* 48 (10): 4313–4338. https://doi.org/10.1175/MWR-D-19-0407.1.

ICC (International Code Council). 2014. *2015 international building code*. Washington, DC: ICC.

ICC. 2017. *2018 international existing building code*. Washington, DC: ICC.

ICC. 2020. *ICC/NSSA Standard for the design and construction of storm shelters*. ICC 500-2020. Washington, DC: ICC.

Insurance Information Institute. 2020. "Spotlight on: Catastrophes: Insurance issues." Accessed April 28. https://www.iii.org/article/spotlight-on-catastrophes-insurance-issues.

Karstens, C. D., W. A. Gallus Jr., B. D. Lee, and C. A. Finley. 2013. "Analysis of tornado-induced tree fall using aerial photography from the Joplin, Missouri, and Tuscaloosa–Birmingham, Alabama, tornadoes of 2011." *J. Appl. Meteorol. Climatol.* 52 (5): 1049–1068. https://doi.org/10.1175/JAMC-D-12-0206.1.

Kopp, G. A., and C.-H. Wu. 2020. "A framework to compare wind loads on low-rise buildings in tornadoes and atmospheric boundary layers." *J. Wind Eng. Ind. Aerodyn.* 204 (Sep): 104269. https://doi.org/10.1016/j.jweia.2020.104269.

Kosiba, K. A., and J. Wurman. 2013. "The three-dimensional structure and evolution of a tornado boundary layer." *Weather Forecasting* 28 (6): 1552–1561. https://doi.org/10.1175/WAF-D-13-00070.1.

Levitan, M. L., B. Ellingwood, P. J. Vickery, Y. Li, T. McAllister, and J. R. Harris et al. 2021. *Tornado load criteria for ASCE Standard 7-22: A probabilistic approach*. National Institute of Standards and Technology. Gaithersburg, MD.

Lewellen, D. C. 2012. "Effects of topography on tornado dynamics: A simulation study." In *Proc., 26th Conf. on Severe Local Storms*. Nashville, TN.

Lewellen, D. C. 2014. "Local roughness effects on tornado dynamics." In *Proc., 27th Conf. on Severe Local Storms*. Morgantown, WV.

Matsui, M., and Y. Tamura. 2009. "Influence of swirl ratio and incident flow conditions on generation of tornado-like vortex." In *Proc., 5th European and African Conference on Wind Engineering*, Florence, Italy.

McAllister, T. P., N. Wang, and B. R. Ellingwood. 2018. "Risk-informed mean recurrence intervals for updated wind maps in ASCE 7-16." *J. Struct. Eng.* 144 (5): 06018001. https://doi.org/10.1061/(ASCE)ST.1943-541X.0002011.

NIST (National Institute of Standards and Technology). 2014. "Final report: NIST technical investigation of the May 22, 2011, tornado in Joplin Missouri." http://www.nist.gov/manuscript-publication-search.cfm?pub_id=915628.

Nolan, D. S., N. A. Dahl, G. H. Bryan, and R. Rotunno. 2017. "Tornado vortex structure, intensity, and surface wind gusts in large-eddy simulations with fully developed turbulence." *J. Atmos. Sci.* 74 (5): 1573–1597. https://doi.org/doi/10.1175/JAS-D-16-0258.1.

Pintar, A., E. Simiu, F. Lombardo, and M. Levitan. 2015. *Maps of non-hurricane non-tornadic wind speeds with specified mean recurrence intervals for the contiguous United States using a two-dimensional Poisson process extreme value model and local regression.* Special Publication 500-301. National Institute of Standards and Technology. Gaithersburg, MD.

Refan, M., and H. Hangan. 2018. "Near surface experimental exploration of tornado vortices." *J. Wind Eng. Ind. Aerodyn.* 175 (Apr): 120–135. https://doi.org/10.1016/j.jweia.2018.01.042.

Refan, M., H. Hangan, and J. Wurman. 2014. "Reproducing tornadoes in laboratory using proper scaling." *J. Wind Eng. Ind. Aerodyn.* 135 (Dec): 136–148. https://doi.org/10.1016/j.jweia.2014.10.008.

Refan, M., H. Hangan, J. Wurman, and K. Kosiba. 2017. "Doppler radar-derived wind field of five tornado events with application to engineering simulations." *Eng. Struct.* 148 (Oct): 509–521. https://doi.org/10.1016/j.engstruct.2017.06.068.

Razavi, A., and P. P. Sarkar. 2018. "Laboratory study of topographic effects on the near-surface tornado flow field." *Boundary-Layer Meteorol.* Stowe, Vermont, 168 (2): 189–212.

Reinhart, A. E., D. J. Bodine, and F. T. Lombardo. 2018. "The impact of terrain on supercells using idealized numerical simulations." In *Proc., 29th Conf. on Severe Local Storms.* Ponte Vedra Beach, Florida.

Roueche, D. B., and D. O. Prevatt. 2013. "Residential damage patterns following the 2011 Tuscaloosa, AL and Joplin, MO tornadoes." *J. Disaster Res.* 8 (6): 1061–1067.

Roueche, D. B., D. O. Prevatt, and F. Haan. 2020. "Tornado-induced and straight-line wind loads on a low-rise building with consideration of internal pressure." *Front. Built Environ.* 6 (Feb): 18. https://doi.org/10.3389/fbuil.2020.00018.

Satrio, M. A., D. J. Bodine, A. E. Reinhart, T. Maruyama, and F. T. Lombardo. 2020. "Understanding how complex terrain impacts tornado dynamics using a suite of high-resolution numerical simulations." *J. Atmos. Sci.* 77 (10): 3277–3300. https://doi.org/10.1175/JAS-D-19-0321.1.

Tang, Z., C. Feng, L. Wu, D. Zuo, and D. L. James. 2018. "Characteristics of tornado-like vortices simulated in a large-scale ward-type simulator." *Boundary-Layer Meteorol.* 166 (2): 327–350.

Texas Tech University. 2006. *A recommendation for an enhanced Fujita scale.* Lubbock, TX: Wind Science and Engineering Center.

Twisdale, L. A., S. Banik, L. Mudd, S. Quayyum, F. Liu, M. Faletra, and M. Hardy, et al. 2021. *Tornado risk maps for building design: Research and development of tornado hazard risk assessment methodology.* National Institute of Standards and Technology. Gaithersburg, MD.

Vickery, P. J., S. Banik, M. L. Levitan, and L. A. Twisdale. 2021. *Development of new K-factors and internal pressure coefficients for use in the tornado provisions of ASCE 7-22.* National Institute of Standards and Technology. Gaithersburg, MD

Vickery, P. J., S. Banik, and F. Liu. 2020. *Tornado induced loads on lowrise flat roofed buildings with comparisons to wind loads derived in atmospheric boundary layer wind tunnels.* Final report to the National Institute of Standards and Technology on grant number 70NANB17H253. Applied Research Associates. Arlington, VA.

Vickery, P. J., J. Lin, P. Skerlj, L.A. Twisdale, and K. Huang. 2006. "HAZUS-MH hurricane model methodology. I: Hurricane hazard, terrain, and wind load modeling." *Nat. Hazards Rev.* 7 (2): 82–93.

Vickery, P. J., and F. Liu. 2018. *Development of wind directionality factors for component and cladding and main wind force resisting loads in hurricane and non-hurricane prone regions of the United States.* Raleigh, NC: Applied Research Associates.

Vickery, P. J., D. Wadhera, L. A. Twisdale Jr., and F. M. Lavelle. 2009. "U.S. hurricane wind speed risk and uncertainty." *J. Struct. Eng.* 135 (3): 301–320.

Wang, J., S. Cao, W. Pang, and J. Cao. 2017. "Experimental study on effects of ground roughness on flow characteristics of tornado-like vortices." *Boundary-Layer Meteorol.* 162 (2): 319–339.

Zhang, W., and P. P. Sarkar. 2012. "Near-ground tornado-like vortex structure resolved by particle image velocimetry (PIV)." *Exp. Fluids* 52 (2): 479–493.

APPENDIX CA
RESERVED FOR FUTURE COMMENTARY

APPENDIX CB
RESERVED FOR FUTURE COMMENTARY

APPENDIX CC
SERVICEABILITY CONSIDERATIONS

CC.1 SERVICEABILITY CONSIDERATIONS

Serviceability limit states are conditions in which the functions of a building or other structure are impaired because of local damage, deterioration, or deformation of building components, or because of occupant discomfort. Although in general safety is not an issue with serviceability limit states (one exception would be for cladding that falls off a building because of excessive story drift under wind load), they nonetheless may have severe economic consequences. The increasing use of the computer as a design tool, the use of stronger (but not stiffer) construction materials, the use of lighter architectural elements, and the uncoupling of the nonstructural elements from the structural frame may result in building systems that are relatively flexible and lightly damped. Limit state design emphasizes the fact that serviceability criteria are essential (as they always have been) to ensure functional performance and economy of design for such building structural systems (Ad Hoc Committee on Serviceability Research 1986, NBCC 1990, West and Fisher 2003).

In general, serviceability is diminished by

1. Excessive deflections or rotation that may affect the appearance, functional use, or drainage of the structure or may cause damaging transfer of load to non-load-supporting elements and attachments;
2. Excessive vibrations produced by the activities of building occupants, mechanical equipment, or the wind, which may cause occupant discomfort or malfunction of building service equipment; and
3. Deterioration, including weathering, corrosion, rotting, and discoloration.

In checking serviceability, the designer is advised to consider appropriate service loads, the response of the structure, and the reaction of the building occupants.

Service loads that may require consideration include static loads from the occupants and their possessions, snow or rain on roofs, temperature fluctuations, and dynamic loads from human activities, wind-induced effects, or the operation of building service equipment. The service loads are those loads that act on the structure at an arbitrary point in time. (In contrast, the nominal loads have a small probability of being exceeded in any year; factored loads have a small probability of being exceeded in 50 years.) Appropriate service loads for checking serviceability limit states may be only a fraction of the nominal loads.

The response of the structure to service loads can usually be analyzed assuming linear elastic behavior. However, members that accumulate residual deformations under service loads may require examination with respect to this long-term behavior. The service loads used in analyzing creep or other long-term effects may not be the same as those used to analyze elastic deflections or other short-term or reversible structural behavior.

Serviceability limits depend on the function of the building and on the perceptions of its occupants. In contrast to the ultimate limit states, it is difficult to specify general serviceability limits that are applicable to all building structures. The serviceability limits presented in Sections CC.2.1, CC.2.2, and CC.2.3 provide general guidance and have usually led to acceptable performance in the past. However, serviceability limits for a specific building should be determined only after a careful analysis by the engineer and architect, in conjunction with the building owner, of all functional and economic requirements and constraints. It should be recognized that building occupants are able to perceive structural deflections, motion, cracking, and other signs of possible distress at levels that are much lower than those that would indicate that structural failure was impending. Such signs of distress may be taken incorrectly as indications that the building is unsafe and may diminish its commercial value.

CC.2 DEFLECTION, VIBRATION, AND DRIFT

CC.2.1 Vertical Deflections Excessive vertical deflections and misalignment arise primarily from three sources: (1) gravity loads, such as dead, live, and snow loads; (2) effects of temperature, creep, and differential settlement; and (3) construction tolerances and errors. Such deformations may be visually objectionable; may cause separation, cracking, or leakage of exterior cladding, doors, windows, and seals; and may cause damage to interior components and finishes. Appropriate limiting values of deformations depend on the type of structure, detailing, and intended use (Galambos and Ellingwood 1986). Historically, common deflection limits for horizontal members have been 1/360 of the span for floors subjected to full nominal live load and 1/240 of the span for roof members. Deflections of about 1/300 of the span (for cantilevers, 1/150 of the length) are visible and may lead to general architectural damage or cladding leakage. Deflections greater than 1/200 of the span may impair operation of movable components such as doors, windows, and sliding partitions.

In certain long-span floor systems, it may be necessary to place a limit (independent of span) on the maximum deflection to minimize the possibility of damage of adjacent nonstructural elements (ISO 1977). For example, damage to non-load-bearing partitions may occur if vertical deflections exceed more than about 3/8 in. (10 mm) unless special provision is made for differential movement (Cooney and King 1988); however, many components can and do accept larger deformations.

Load combinations for checking static deflections can be developed using first-order reliability analysis (Galambos and Ellingwood 1986). Current static deflection guidelines for floor and roof systems are adequate for limiting surficial damage in most buildings. A combined load with an annual probability of 0.05 of being exceeded would be appropriate in most instances.

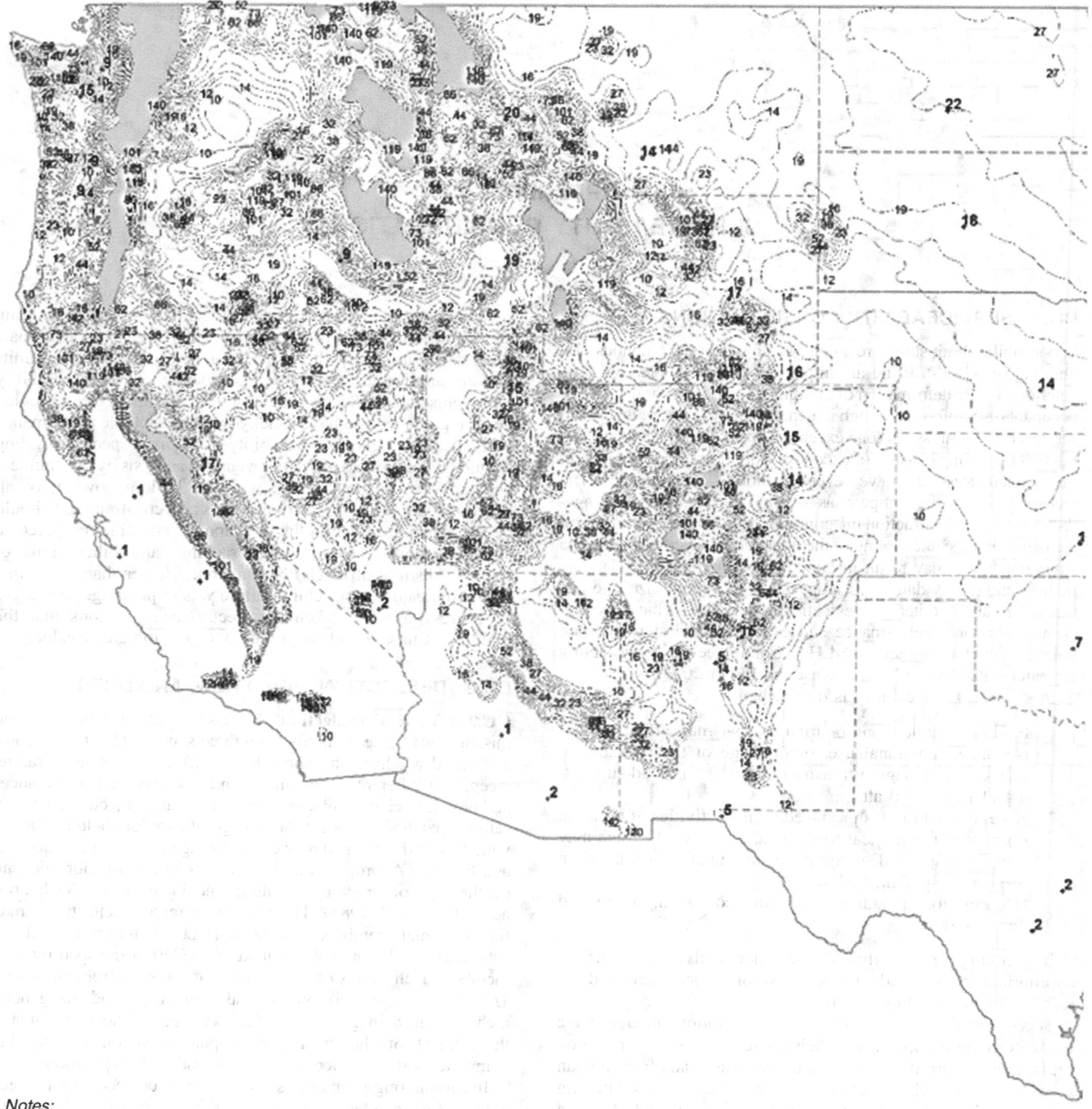

Notes:

1. This figure is a representation of data from the Ground Snow Loads Geodatabase of geocoded design found snow load values, available at https://asce7hazardtool.online.
2. Values for specific locations can most accurately be determined by accessing Ground Snow Loads Geodatabase.
3. Lines shown on the figure are contours separated by a constant ration of 1.18 with values of 10.12, 14, 16, 19, 23, 27, 32, 38, 44, 52, 62, 73, 86, 101, 119, and 140 psf.

Figure CC.2-1. 20-year MRI ground snow loads, p_g, for the conterminous US (lb/ft^2).

4. Values denoted with a "+" symbol indicate design ground snow loads at state capitals or other cities with large populations.
5. Areas show in gray represent areas with ground snow loads exceeding 140 psf. Ground snow loads values for these locations shall be determined using the Ground Snow Loads Geodatabase.

Figure CC.2-1 (*Continued*). 20-year MRI ground snow loads, p_g, for the conterminous US (lb/ft^2).

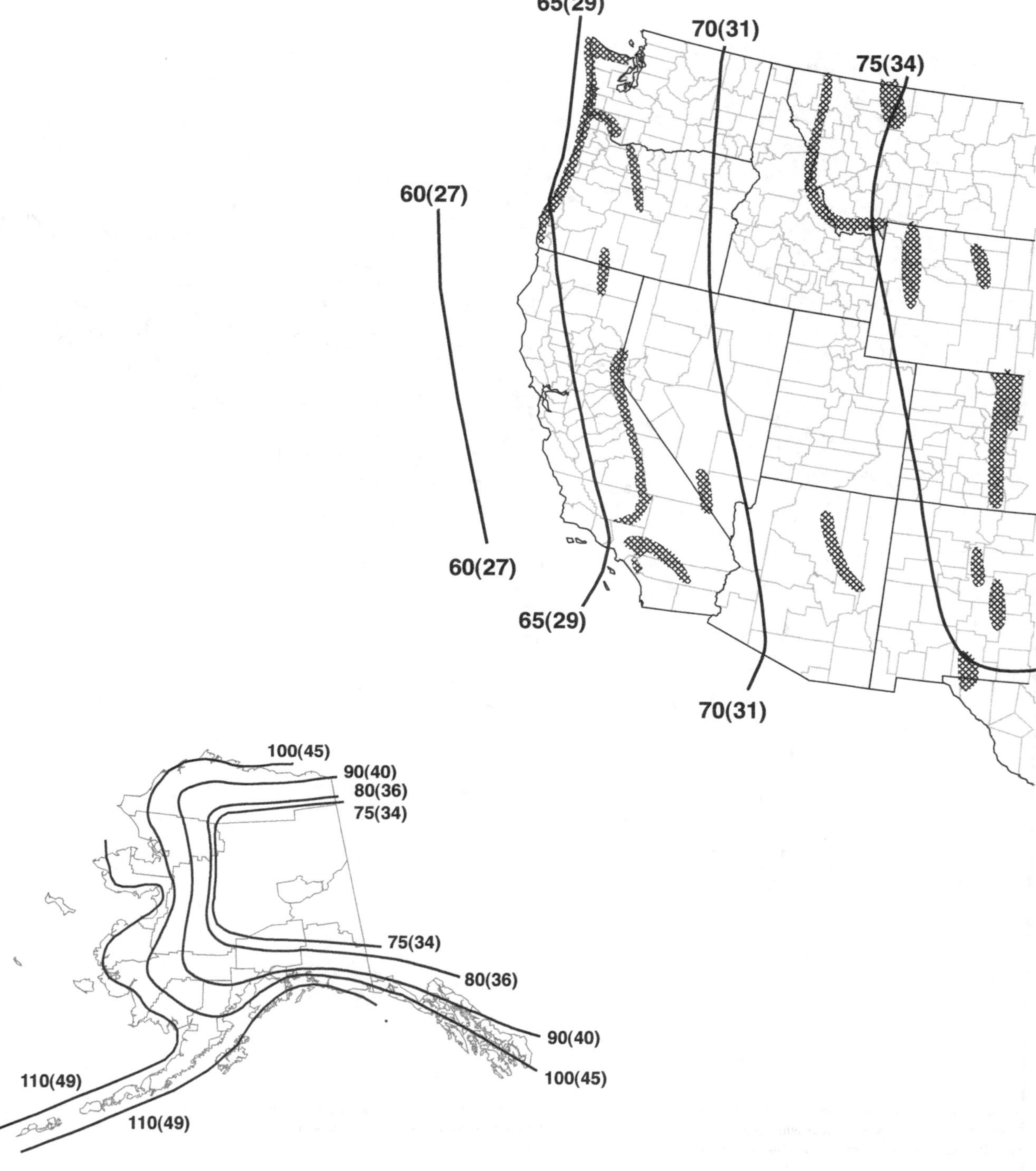

Notes:

1. Values are 3 s gust wind speeds in mi/h (m/s) at 33 ft (10 m) above ground for Exposure Category C.
2. Linear interpolation between contours is permitted. Point values are provided to aid with interpolation.
3. Islands, coast areas, and land boundaries outside the last contour shall use the last wind speed contour.
4. Mountainous terrain, gorges, ocean promontories, and special wind regions shall be examined for unusual wind conditions.
5. Wind speeds correspond to approximately a 99% probability of exceedance in 50 years (Annual Exceedance Probability = 0.1, MRI = 10 years).
6. Location-specific basic wind speeds shall be permitted to be determined using the ASCE Wind Design Geodatabase.
7. The ASCE Wind Design Geodatabase can be accessed at the ASCE 7 Hazard Tool (https://asce7hazardtool.online) or approved equivalent.

Figure CC.2-2. 10-year MRI 3 s gust wind speed in mi/h (m/s) at 33 ft (10 m) above ground in Exposure C.

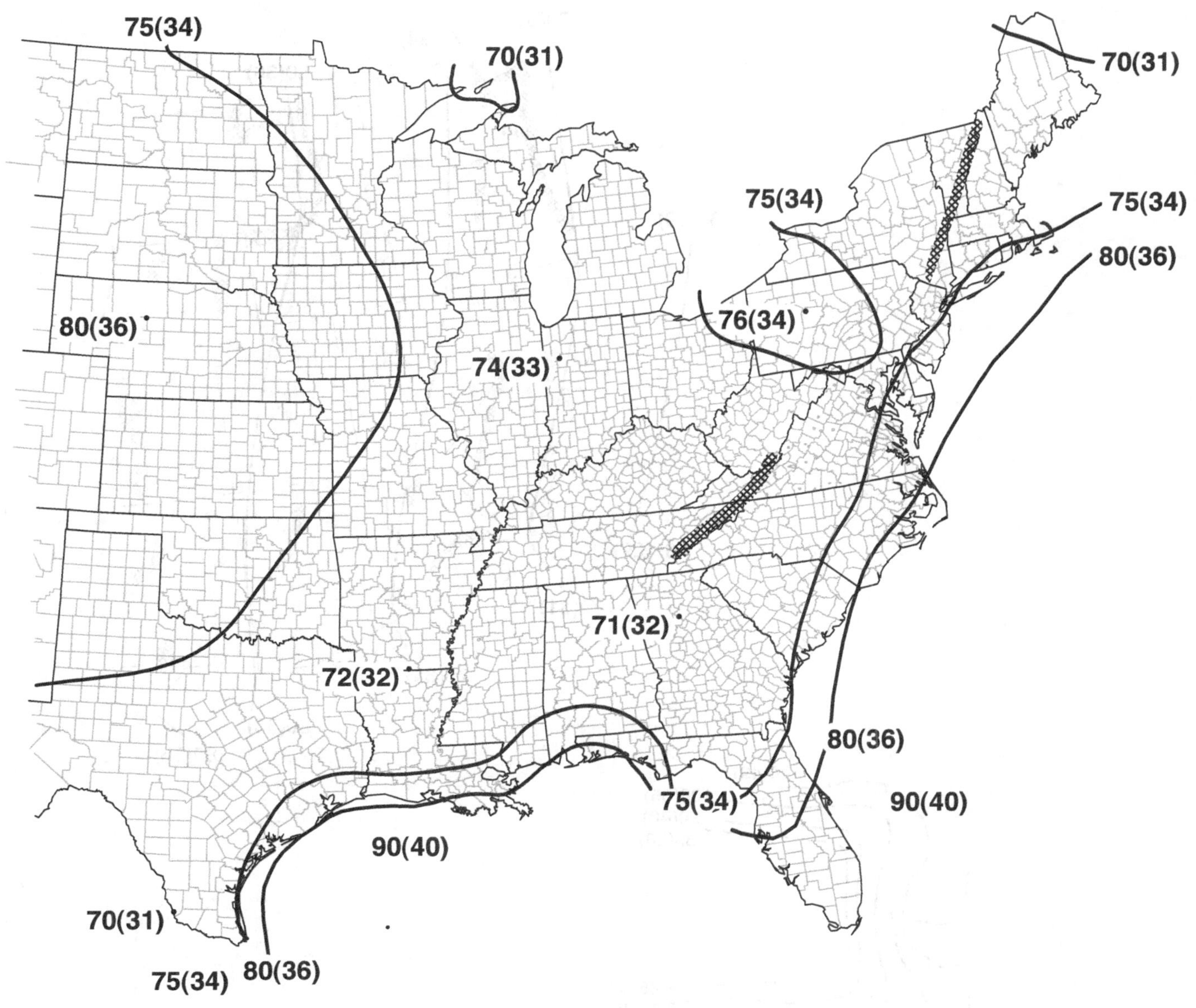

Location	Vmph	(m/s)
American Samoa	105	(47)
Guam	110	(49)
Hawaii	ASCE Wind Design Geodatabase	
Puerto Rico	ASCE Wind Design Geodatabase	
U.S. Virgin Islands	ASCE Wind Design Geodatabase	

Figure CC.2-2 (*Continued*). 10-year MRI 3 s gust wind speed in mi/h (m/s) at 33 ft (10 m) above ground in Exposure C.

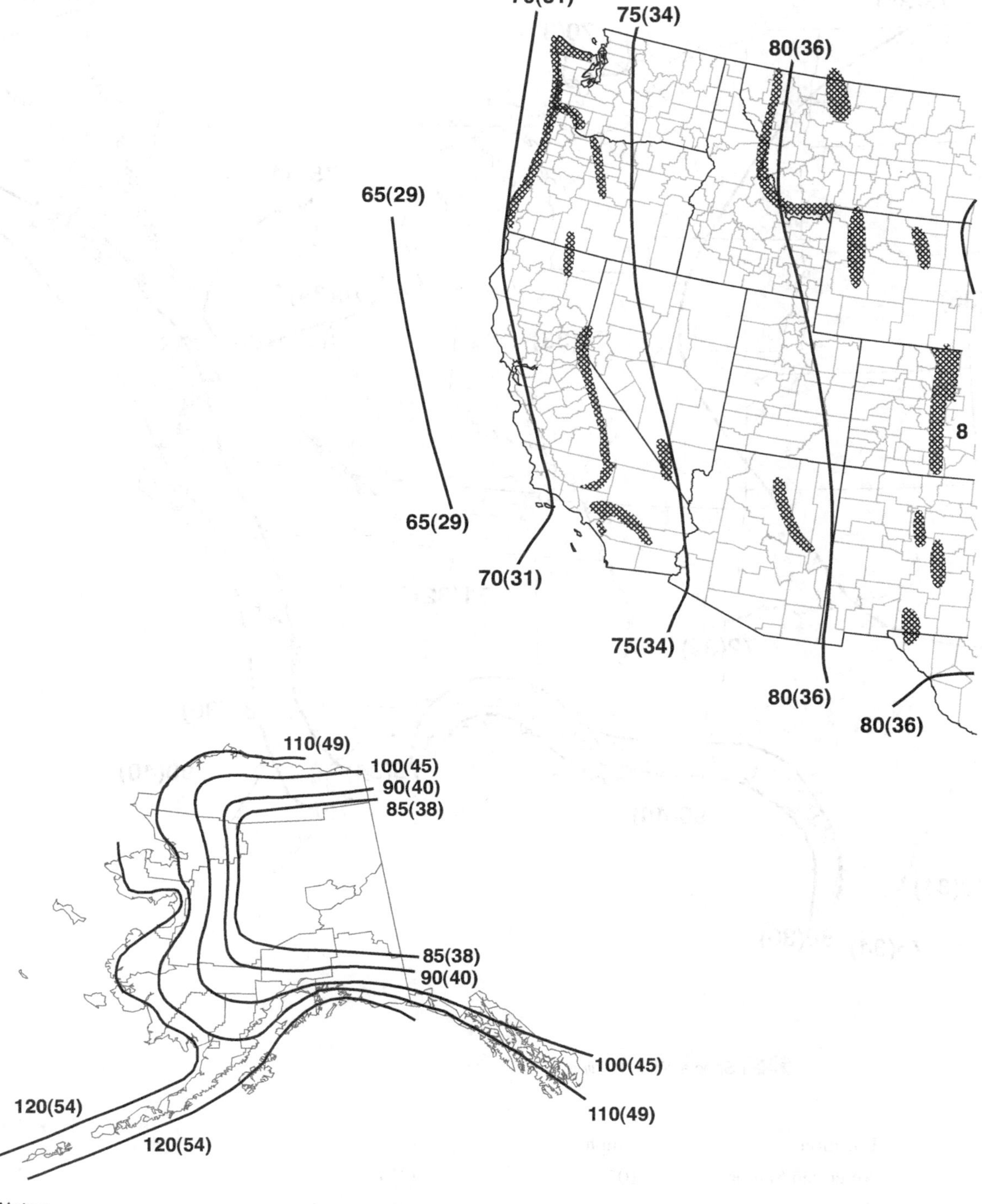

Notes:

1. Values are 3 s gust wind speeds in mi/h (m/s) at 33 ft (10 m) above ground for Exposure Category C.
2. Linear interpolation between contours is permitted. Point values are provided to aid with interpolation.
3. Islands, coast areas, and land boundaries outside the last contour shall use the last wind speed contour.
4. Mountainous terrain, gorges, ocean promontories, and special wind regions shall be examined for unusual wind conditions.
5. Wind speeds correspond to approximately a 87% probability of exceedance in 50 years (Annual Exceedance Probability = 0.04, MRI = 25 years).
6. Location-specific basic wind speeds shall be permitted to be determined using the ASCE Wind Design Geodatabase.
7. The ASCE Wind Design Geodatabase can be accessed at the ASCE 7 Hazard Tool (https://asce7hazardtool.online) or approved equivalent.

Figure CC.2-3. 25-year MRI 3 s gust wind speed in mi/h (m/s) at 33 ft (10 m) above ground in Exposure C.

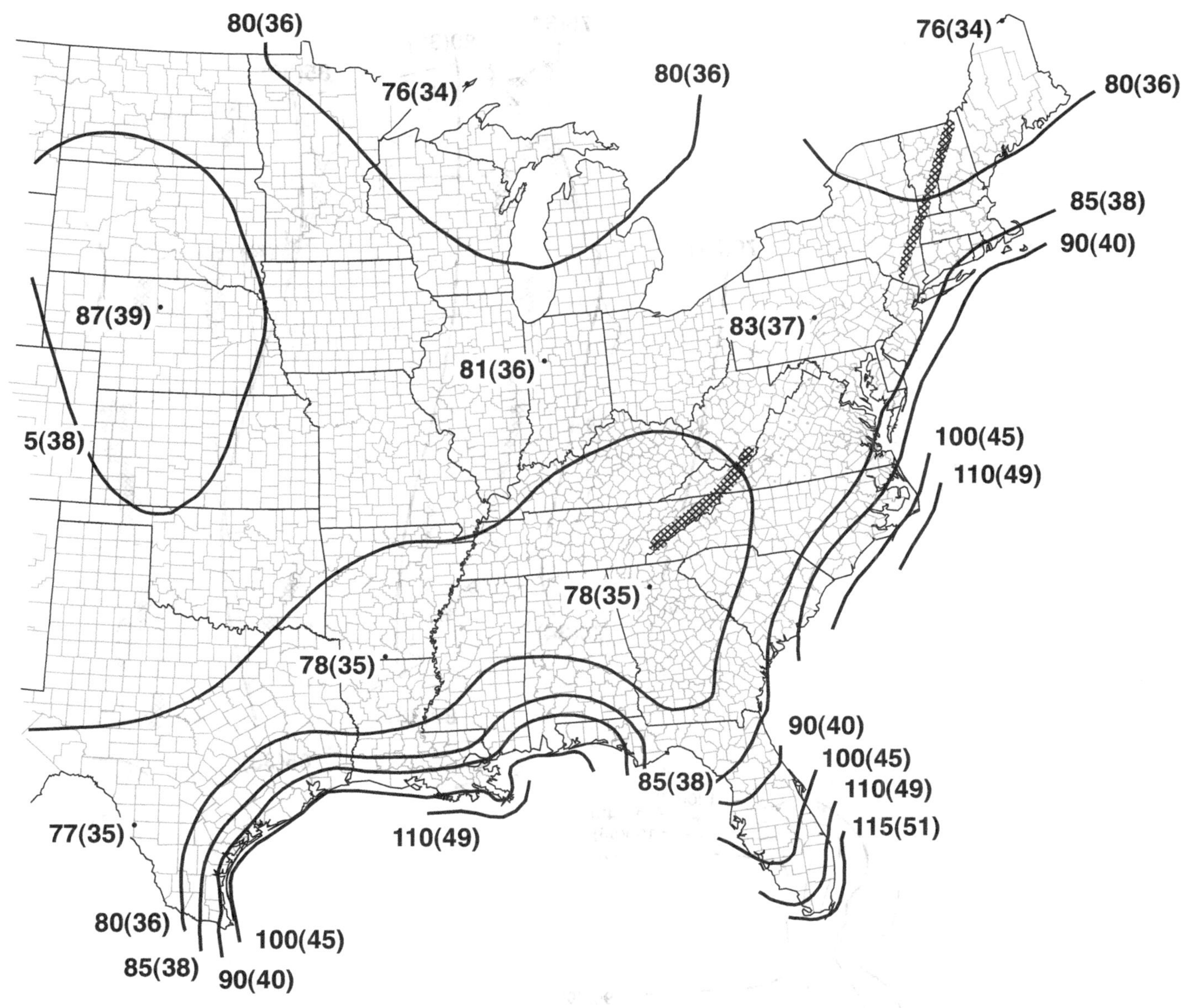

Location	Vmph	(m/s)
American Samoa	116	(52)
Guam	132	(59)
Hawaii	ASCE Wind Design Geodatabase	
Puerto Rico	ASCE Wind Design Geodatabase	
U.S. Virgin Islands	ASCE Wind Design Geodatabase	

Figure CC.2-3 (*Continued*). 25-year MRI 3 s gust wind speed in mi/h (m/s) at 33 ft (10 m) above ground in Exposure C.

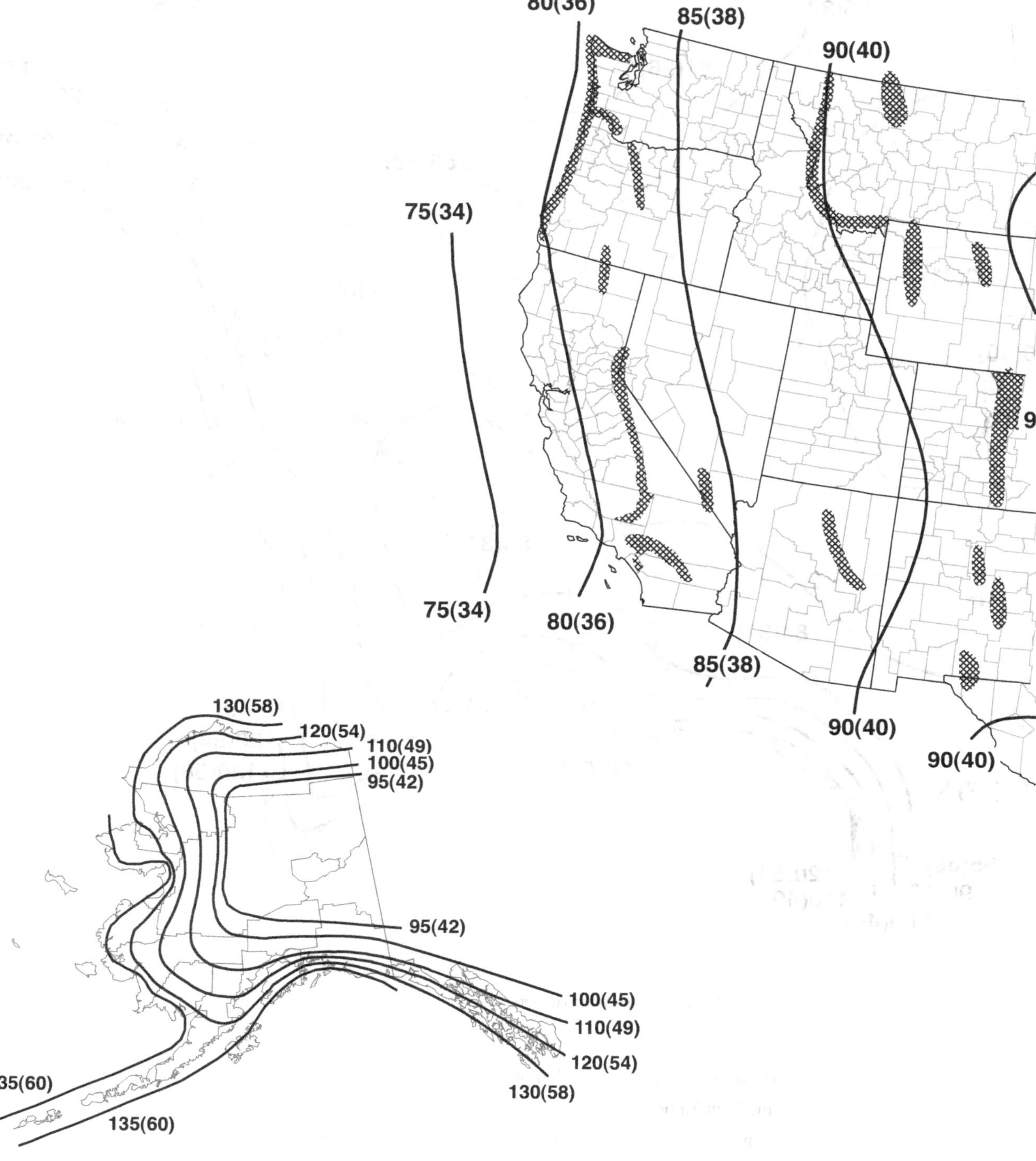

Notes:

1. Values are 3 s gust wind speeds in mi/h (m/s) at 33 ft (10 m) above ground for Exposure Category C.
2. Linear interpolation between contours is permitted. Point values are provided to aid with interpolation.
3. Islands, coast areas, and land boundaries outside the last contour shall use the last wind speed contour.
4. Mountainous terrain, gorges, ocean promontories, and special wind regions shall be examined for unusual wind conditions.
5. Wind speeds correspond to approximately a 40% probability of exceedance in 50 years (Annual Exceedance Probability = 0.01, MRI = 100 years).
6. Location-specific basic wind speeds shall be permitted to be determined using the ASCE Wind Design Geodatabase.
7. The ASCE Wind Design Geodatabase can be accessed at the ASCE 7 Hazard Tool (https://asce7hazardtool.online) or approved equivalent.

Figure CC.2-5. 100-year MRI 3 s gust wind speed in mi/h (m/s) at 33 ft (10 m) above ground in Exposure C.

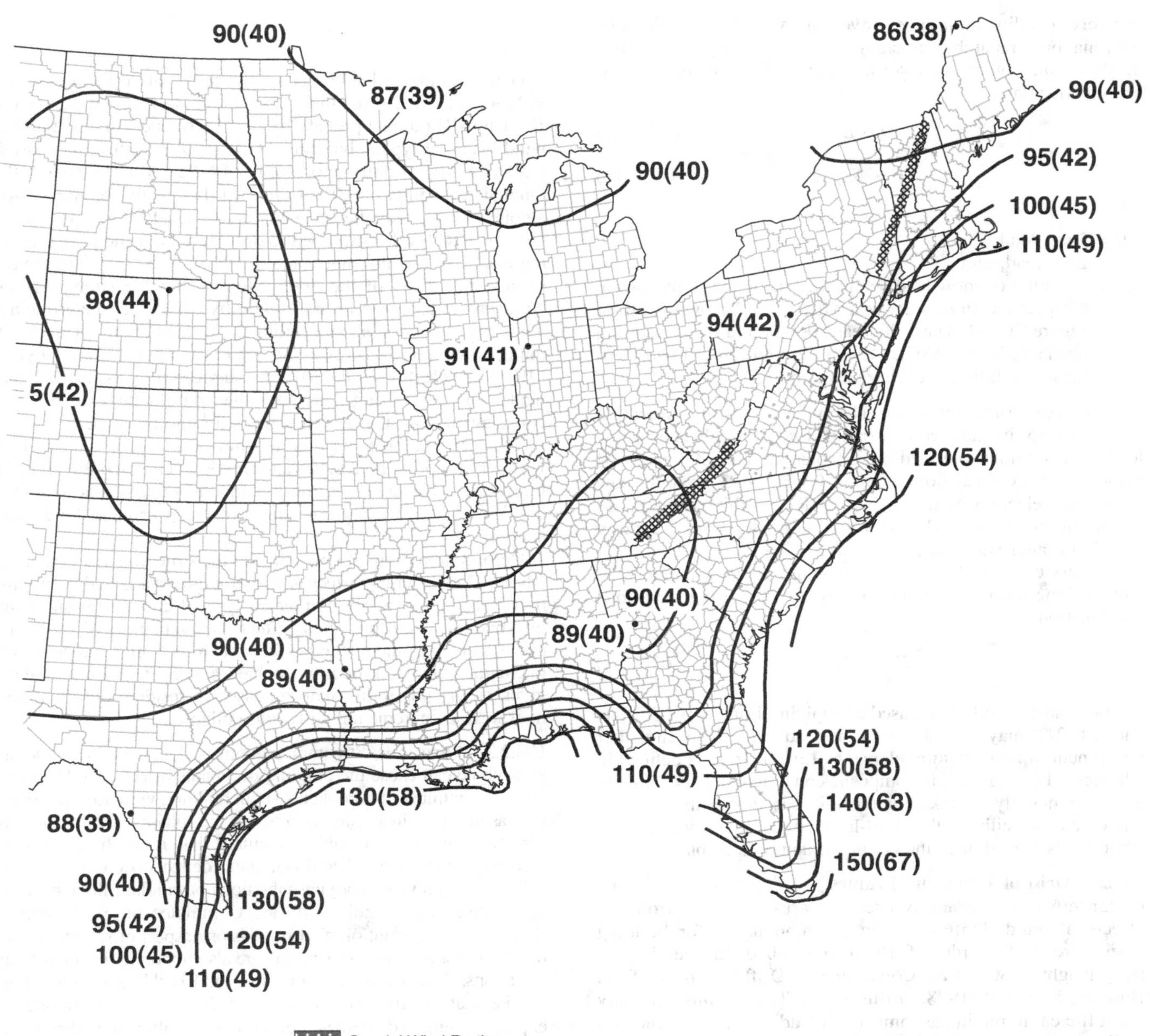

Location	Vmph	(m/s)
American Samoa	134	(60)
Guam	163	(73)
Hawaii	ASCE Wind Design Geodatabase	
Puerto Rico	ASCE Wind Design Geodatabase	
U.S. Virgin Islands	ASCE Wind Design Geodatabase	

Figure CC.2-5 (*Continued*). 100-year MRI 3 s gust wind speed in mi/h (m/s) at 33 ft (10 m) above ground in Exposure C.

For serviceability limit states involving visually objectionable deformations, reparable cracking or other damage to interior finishes, and other short-term effects, the suggested load combinations are

$$D + L \qquad \text{(CC.2-1a)}$$

$$D + S_{ser} \qquad \text{(CC.2-1b)}$$

where

D = Dead load;
L = Live load; and
S_{ser} = Design roof snow load determined using the provisions in Chapter 7 with p_g = 20-year MRI ground snow load from Figure CC.2-1. These 20-year MRI ground snow loads are available in the ASCE Design Ground Snow Load Geodatabase (https://asce7hazardtool.online/).

It is appropriate for serviceability checks using Equation CC.2-1(b) to include consideration for all applicable snow loading conditions outlined in Chapter 7. The 20-year MRI value is, on average, 80% of the 50-year MRI ground snow load. The selection of the 20-year MRI for this purpose was based on judgment and review of snow load serviceability criteria in other standards.

For serviceability limit states involving creep, settlement, or similar long-term or permanent effects, the suggested load combination is

$$D + 0.5L \qquad \text{(CC.2-2)}$$

The dead-load effect, D, used in applying Equations (CC.2-1) and (CC.2-2) may be that portion of dead load that occurs after attachment of nonstructural elements. Live load, L, is defined in Chapter 4. For example, in composite construction, the dead-load effects frequently are taken as those imposed after the concrete has cured; in ceilings, the dead-load effects may include only those loads placed after the ceiling structure is in place.

CC.2.2 Drift of Walls and Frames Drifts (lateral deflections) of concern in serviceability checking arise primarily from the effects of wind. Drift limits in common usage for building design are on the order of 1/600 to 1/400 of the building or story height (ASCE Task Committee on Drift Control of Steel Building Structures 1988, Griffis 1993). These limits generally are sufficient to minimize damage to cladding and nonstructural walls and partitions. Smaller drift limits may be appropriate if the cladding is brittle. Larger drift limits may be acceptable when the connections between the structural and non-structural elements are detailed to accommodate the relative movement without damage. West and Fisher (2003) contains recommendations for higher drift limits that have successfully been used in low-rise buildings with various cladding types. It also contains recommendations for buildings containing cranes. An absolute limit on story drift may also need to be imposed in light of evidence that damage to nonstructural partitions, cladding, and glazing may occur if the story drift exceeds about 10 mm (3/8 in.) unless special detailing practices are made to tolerate movement (Freeman 1977, Cooney and King 1988). Many components can accept deformations that are significantly larger.

It is excessively conservative to use the basic wind speed for the project's risk category to check serviceability because the corresponding MRI for design winds is very large. The following load combination, derived similarly to Equations (CC.2-1a) and (CC.2-1b), can be used to check short-term effects:

$$D + 0.5 + W_a \qquad \text{(CC.2-3)}$$

Here W_a is wind load based on the serviceability wind speeds in Figures CC.2-2 through CC.2-5. Some designers have used a 10-year MRI (annual probability of 0.1) for checking drift under wind loads for typical buildings (Griffis 1993); others have used a 50-year MRI (annual probability of 0.02) or a 100-year MRI (annual probability of 0.01) for more drift-sensitive buildings. As a point of reference, the 2005 edition of this standard specified a 50-year MRI for determining the strength-level wind pressures in regions that are not hurricane-prone. It was then suggested [Equation (CC-3) of the 2005 edition] that the strength-level wind pressures could be multiplied by 0.70 when determining wind pressures for serviceability drift calculations. This 0.70 multiplication of the pressure resulted in an approximate 10-year MRI. The selection of the MRI for serviceability evaluation is a matter of engineering judgment that should be exercised in consultation with the design team and building owner.

The maps included in this appendix are appropriate for use with serviceability limit states and should not be used for strength limit states. Because of its transient nature, wind load need not be considered in analyzing the effects of creep or other long-term actions.

Deformation limits should apply to the structural assembly as a whole. The stiffening effect of nonstructural walls and partitions may be taken into account in the analysis of drift if substantiating information regarding their effect is available. Where load cycling occurs, under wind effects such as vortex shedding or ratcheting from inelastic frame behavior, consideration should be given to the possibility that increases in residual deformations may lead to incremental structural collapse.

CC.2.3 Vibrations Structural motions of floors or of the building as a whole can cause the building occupants discomfort. In recent years, the number of complaints about building vibrations has been increasing. This increasing number of complaints is associated in part with the more flexible structures that result from modern construction practice. Traditional static deflection checks are not sufficient to prevent annoying vibrations of building floor systems or buildings as a whole (Ad Hoc Committee on Serviceability Research 1986). Control of stiffness is one aspect of serviceability, but mass distribution and damping are also important in controlling vibrations. The use of new materials and building systems may require that the dynamic response of the system be considered explicitly. Simple dynamic models often are sufficient to determine whether there is a potential problem and to suggest possible remedial measurements (Bachmann and Ammann 1987, Ellingwood 1989).

Excessive structural motion is mitigated by measures that limit building or floor accelerations to levels that are not disturbing to the occupants or do not damage service equipment. The perception and tolerance of individuals to vibration depend on their expectation of building performance (related to building occupancy) and to their level of activity at the time the vibration occurs (ANSI 1983). Individuals find continuous vibrations more objectionable than transient vibrations. Continuous vibrations (over a period of minutes) with acceleration on the order of 0.005 g to 0.01 g are annoying to most people engaged in quiet activities, whereas those engaged in physical activities or spectator events may tolerate steady-state accelerations on the order of 0.02 g to 0.05 g. Thresholds of annoyance for transient vibrations (lasting only a few seconds) are considerably higher and depend on the amount of structural damping present (Murray 1991). For a finished floor with (typically) 5% damping or more, peak transient accelerations of 0.05 g to 0.1 g may be tolerated.

Many common human activities impart dynamic forces to a floor at frequencies (or harmonics) in the range of 2 Hz to 6 Hz (Allen and Rainer 1976; Allen et al. 1985; Allen 1990a, b). If the fundamental frequency of vibration of the floor system is in this range and if the activity is rhythmic in nature (e.g., dancing, aerobic exercise, or cheering at spectator events), resonant amplification may occur. To prevent resonance from rhythmic activities, the floor system should be tuned so that its natural frequency is well removed from the harmonics of the excitation frequency. As a general rule, the natural frequency of structural elements and assemblies should be greater than 2.0 times the frequency of any steady-state excitation to which they are exposed unless vibration isolation is provided. Damping is also an effective way of controlling annoying vibration from transient events because studies have shown that individuals are more tolerant of vibrations that damp out quickly than those that persist (Murray 1991).

Several studies have shown that a simple and relatively effective way to minimize objectionable vibrations from walking and other common human activities is to control the floor stiffness, as measured by the maximum deflection independent of span. Justification for limiting the deflection to an absolute value rather than to some fraction of span can be obtained by considering the dynamic characteristics of a floor system modeled as a uniformly loaded simple span. The fundamental frequency of vibration, f_o, of this system is given by

$$f_o = \frac{\pi}{2l^2}\sqrt{\frac{EI}{\rho}} \qquad \text{(CC.2-4)}$$

where

EI = Flexural rigidity of the floor;
l = Span;
ρ = Mass per unit length = w/g,
where g = Acceleration due to gravity = (32.17 ft/s^2) (9.81 m/s^2), and
w = Dead load plus participating live load.

The maximum deflection caused by w is

$$\delta = (5/384)(wl^4/EI) \qquad \text{(CC.2-5)}$$

Substituting EI from this equation into Equation (CC.2-3), we obtain

$$f_o \approx 18/\sqrt{\delta} \quad (\delta \text{ in mm}) \qquad \text{(CC.2-6)}$$

This frequency can be compared to minimum natural frequencies for mitigating walking vibrations in various occupancies (Allen and Murray 1993). For example, Equation (CC.2-6) indicates that the static deflection caused by uniform load, w, must be limited to about 0.2 in. (5 mm), independent of span, if the fundamental frequency of vibration of the floor system is to be kept above about 8 Hz. Many floors that do not meet this guideline are perfectly serviceable; however, this guideline provides a simple means for identifying potentially troublesome situations where additional consideration in design may be warranted.

CC.3 DESIGN FOR LONG-TERM DEFLECTION

Under sustained loading, structural members may exhibit additional time-dependent deformations caused by creep, which usually occur at a slow but persistent rate over long periods of time. In certain applications, it may be necessary to limit deflection under long-term loading to specified levels. This limitation can be done by multiplying the immediate deflection by a creep factor, as provided in material standards, that ranges from about 1.5 to 2.0. This limit state should be checked using the load combination in Equation (CC.2-2).

CC.4 CAMBER

Where required, camber should be built into horizontal structural members to give proper appearance and drainage and to counteract anticipated deflection from loading and potential ponding.

CC.5 EXPANSION AND CONTRACTION

Provisions should be made in design so that if significant dimensional changes occur, the structure will move as a whole and differential movement of similar parts and members that meet at joints will be at a minimum. Design of expansion joints to allow for dimensional changes in portions of a structure separated by such joints should take both reversible and irreversible movements into account. Structural distress in the form of wide cracks has been caused by restraint of thermal, shrinkage, and prestressing deformations. Designers are advised to provide for such effects through relief joints or by controlling crack widths.

CC.6 DURABILITY

Buildings and other structures may deteriorate in certain service environments. This deterioration may be visible on inspection (e.g., as weathering, corrosion, and staining) or may result in undetected changes in the material. The designer should either provide a specific amount of damage tolerance in the design or specify adequate protection systems and/or planned maintenance to minimize the likelihood that such problems will occur. Water infiltration through poorly constructed or maintained wall or roof cladding is considered beyond the realm of designing for damage tolerance. Waterproofing design is beyond the scope of this standard. For portions of buildings and other structures exposed to weather, the design should eliminate pockets in which moisture can accumulate.

REFERENCES

Ad Hoc Committee on Serviceability Research. 1986. “Structural serviceability: A critical appraisal and research needs.” *J. Struct. Eng.* 112 (12): 2646–2664.

Allen, D. E. 1990a. “Building vibrations from human activities.” *Concr. Int.* 12 (6): 66–73.

Allen, D. E. 1990b. “Floor vibrations from aerobics.” *Can. J. Civ. Eng.* 19 (4): 771–779. https://doi.org/10.1139/l90-090.

Allen, D. E., and T. M. Murray. 1993. “Design criterion for vibrations due to walking.” *Eng. J.* 30 (4): 117–129.

Allen, D. E., and J. H. Rainer. 1976. “Vibration criteria for long-span floors.” *Can. J. Civ. Eng.* 3 (2): 165–173. https://doi.org/10.1139/l76-017.

Allen, D. E., J. H. Rainer, and G. Pernica. 1985. “Vibration criteria for assembly occupancies.” *Can. J. Civ. Eng.* 12 (3): 617–623. https://doi.org/10.1139/l85-069.

ANSI (American National Standards Institute). 1983. *Guide to the evaluation of human exposure to vibration in buildings. ANSI S3.29-1983*. New York: ANSI.

ASCE Task Committee on Drift Control of Steel Building Structures. 1988. “Wind drift design of steel-framed buildings: State-of-the-art report.” *J. Struct. Eng.* 114 (9): 2085–2108. https://doi.org/10.1061/(ASCE)0733-9445(1988)114:9(2085).

Bachmann, H., and W. Ammann. 1987. *Vibrations in structures*. 3rd ed. Zurich, Switzerland: International Association for Bridge and Structural Engineering.
Cooney, R. C., and A. B. King. 1988. *Serviceability criteria for buildings*. BRANZ Rep. No. SR14. Porirua, NZ: Building Research Association of New Zealand.
Ellingwood, B. 1989. "Serviceability guidelines for steel structures." *Eng. J.* 26 (1): 1–8. http://worldcat.org/issn/00138029.
Freeman, S. A. 1977. "Racking tests of high-rise building partitions." *J. Struct. Div.* 103 (8): 1673–1685. https://doi.org/10.1061/JSDEAG.0004702.
Galambos, T. U., and B. Ellingwood. 1986. "Serviceability limit states: Deflection." *J. Struct. Eng.* 112 (1): 67–84. https://doi.org/10.1061/(ASCE)0733-9445(1986)112:1(67).
Griffis, L. G. 1993. "Serviceability limit states under wind load." *Eng. J.* 30 (1): 1–16.
ISO (International Organization for Standardization). 1977. *Bases for the design of structures: Deformations of buildings at the serviceability limit states. ISO 4356*. Geneva: ISO.
Murray, T. 1991. "Building floor vibrations." *Eng. J.* 28 (3): 102–109.
Murray, T.M., D.E. Allen and E.E. Ungar. 2016. *Vibrations of steel-framed structural systems due to human activity, design guide 11*. 2nd ed. Chicago: AISC.
NBCC (National Building Code of Canada). 1990. *Commentary A: Serviceability criteria for deflections and vibrations*. Ottawa: National Research Council.
West, M., and J. Fisher. 2003. *Serviceability design considerations for steel buildings: Steel design guide no. 3*. 2nd ed. Chicago: American Institute of Steel Construction.

OTHER REFERENCES (NOT CITED)

Ellingwood, B., and A. Tallin. 1984. "Structural serviceability: Floor vibrations." *J. Struct. Eng.* 110 (2): 401–418. https://doi.org/10.1061/(ASCE)0733-9445(1984)110:2(401).
Ohlsson, S. 1988. "Ten years of floor vibration research: A review of aspects and some results." In *Proc., Symp. Serviceability of Buildings*, 435–450.
Tallin, A. G., and B. Ellingwood. 1984. "Serviceability limit states: Wind induced vibrations." *J. Struct. Eng.* 110 (10): 2424–2437. https://doi.org/10.1061/(ASCE)0733-9445(1984)110:10(2424).

APPENDIX CD
BUILDINGS EXEMPTED FROM TORSIONAL WIND LOAD CASES

As discussed in Section C27.3.6, a building will experience torsional loads caused by nonuniform pressures on different faces of the building. Because of these torsional loads, the four load cases, as defined in Figure 27.3-8, must be investigated except for buildings with flexible diaphragms and for buildings with diaphragms that are not flexible meeting the requirements for spatial distribution and stiffness of the main wind force resisting system (MWFRS).

The requirements for spatial distribution and stiffness of the MWFRS for the simple cases shown are necessary to ensure that wind torsion does not control the design. Appendix D presents the different requirements which, if met by a building's MWFRS, torsional wind load cases need not be investigated. Many other configurations are also possible, but it becomes too complex to describe their limitations in a simple way.

In general, the designer should place and proportion the vertical elements of the MWFRS in each direction so that the center of pressure from wind forces at each story is located near the center of rigidity of the MWFRS, thereby minimizing the inherent torsion from wind on the building. In buildings with rigid diaphragms, a torsional eccentricity larger than about 5% of the building width should be avoided to prevent large shear forces from wind torsion effects and to avoid torsional story drift that can damage interior walls and cladding.

The following information is provided to aid designers in determining whether the torsional wind load cases (Figure 27.3-8, Load Cases 2 and 4) control the design. Reference is made to Figure CD-1. The equations shown in the figure for the general case of a square or rectangular building having inherent eccentricity e_1 or e_2 about principal axis 1 and 2, respectively, can be used to determine the required stiffness and location of the MWFRS in each principal axis direction.

Using the equations contained in Figure CD-1, it can be shown that regular buildings (as defined in Chapter 12, Section 12.3.2), which at each story meet the requirements specified for the eccentricity between the center of mass (or alternatively, center of rigidity) and the geometric center with the specified ratio of seismic to wind design story shears can safely be exempted from the wind torsion load cases of Figure 27.3-6. It is conservative to measure the eccentricity from the center of mass to the geometric center rather than from the center of rigidity to the geometric center. Buildings that have an inherent eccentricity between the center of mass and center of rigidity and are designed for code seismic forces have a higher torsional resistance than if the center of mass and rigidity are coincident.

Using the equations contained in Figure CD-1 and a building drift analysis to determine the maximum displacement at any story, it can be shown that buildings with diaphragms that are not flexible, and that are defined as torsionally regular under wind load, need not be designed for the torsional load cases of Figure 27.3-6. Furthermore, it is permissible to increase the basic wind load case proportionally so that the maximum displacement at any story is not less than the maximum displacement under the torsional load case. The building can then be designed for the increased basic loading case without the need for considering the torsional load cases.

Diagram

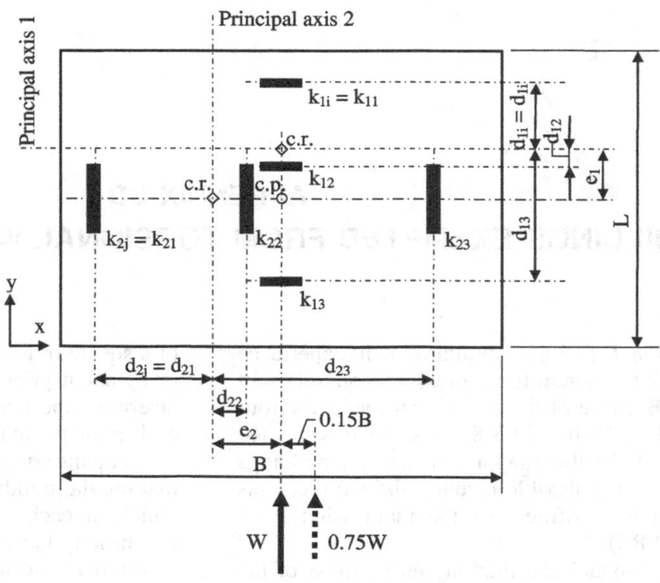

Notation

B = Horizontal plan dimension of the building normal to the wind.
L = Horizontal plan dimension of the building parallel to the wind.
c.r. = Center of rigidity.
c.p. = Center of wind pressure.
k_{1i} = Stiffness of frame i parallel to major axis 1.
k_{2j} = Stiffness of frame j parallel to major axis 2.
d_{1i} = Distance of frame i to c.r. perpendicular to major axis 1.
d_{2j} = Distance of frame j to c.r. perpendicular to major axis 2.
e_1 = Distance from c.p. to c.r. perpendicular to major axis 1.
e_2 = Distance from c.p. to c.r. perpendicular to major axis 2.
J = Polar moment of inertia of all MWFRS wind frames in the building.
W = Wind load as required by standard.
V_{1i} = Wind force in frame i parallel to major axis 1.
V_{2j} = Wind force in frame j parallel to major axis 2.
x_0, y_0 = Coordinates for center of rigidity from the origin of any convenient x, y axes.

Equations

$$x_0 = \frac{\sum_{i=1}^{n} x_{1i} k_{1i}}{\sum_{i=1}^{n} k_{1i}} \qquad y_0 = \frac{\sum_{i=1}^{n} y_{1i} k_{1i}}{\sum_{i=1}^{n} k_{1i}}$$

$$J = \sum_{i=1}^{n} k_{1i} d_{1i}^{\,2} + \sum_{j=1}^{m} k_{2j} d_{2j}^{\,2}$$

$$V_{1i} = \frac{(0.75W)\,k_{1i}}{\sum_{i=1}^{n} k_{1i}} + \frac{(0.75W)(e_1 + 0.15B)\,k_{1i}\,d_{1i}}{J}$$

$$V_{2j} = \frac{(0.75W)\,k_{2j}}{\sum_{j=1}^{m} k_{2j}} + \frac{(0.75W)(e_2 + 0.15B)\,k_{2j}\,d_{2j}}{J}$$

Figure CD-1. Exemption from torsional load cases.

APPENDIX CE
PERFORMANCE-BASED DESIGN PROCEDURES FOR FIRE EFFECTS ON STRUCTURES

CE.1 SCOPE

Design approaches that consider fire effects on structures are generally categorized as either (1) standard fire resistance design (also referred to as the prescriptive approach), or (2) performance-based design (PBD). Although this appendix does not pertain to and should not be used for standard fire resistance design, it is discussed relative to PBD in this commentary.

Designers may elect to use PBD procedures for fire effects on a structure to meet stakeholder design objectives as permitted by Section 1.3.6 and alternative materials, design, and methods of construction and equipment provision in the building codes. When PBD procedures are used, the structure is typically designed for primary gravity and environmental loads and is then evaluated for fire exposure. An alternative material, design, or method of construction typically requires approval where the Authority Having Jurisdiction finds that the proposed design is satisfactory and complies with the intent of the building code.

Structural fire resistance is the structure's ability to carry loads during exposure to fire conditions as well as provide a barrier to fire spread. Structural performance during fire exposure is often simply expressed as Fire Resistance > Fire Effects (Buchanan 2002). Three design philosophies are used for comparing fire resistance and fire effects, which are based on measures of time, temperature, and strength.

Time is used for standard fire resistance ratings in the building codes where a particular configuration is shown, by testing or equivalent analysis, to provide adequate resistance to a standard fire exposure under test conditions for a period of time. *Temperature* is used in situations where it is postulated that a particular temperature will cause failure in a component or subsystem. A maximum allowable temperature is specified, and thermal protection is provided for defined fire exposures to ensure that the limiting temperature is not reached within the specified fire rating or period of time. *Strength* compares applied gravity loads and fire effects (e.g., thermally induced forces and deformations in structural members) to the fire resistance (e.g., temperature-dependent stiffness and strength) of the heated structural members and connections.

CE.2 DEFINITIONS

STANDARD FIRE RESISTANCE DESIGN: Standard fire resistance design methods are based on either *time* or *temperature*. Fire resistance is most often defined as an hourly rating (e.g., 2 h fire rated assembly) based on either results of a standard fire test or equivalent analysis methods.

Standard fire resistance testing provides a method of rank ordering through comparative testing of different structural assemblies under controlled laboratory conditions. Each test uses the same standard time–temperature curve, which continually rises in temperature, to heat the structural members and assemblies for an established set of failure criteria. Standard time–temperature curves include ASTM E119 (2012) and ANSI/UL 263 (UL 2011b); international standards include ISO 834 (1999), CAN/ULC S101 (ULC 2007), and BS 476-20 (British Standards Institution 1987). There are also standard fire curves for hydrocarbon pool fires, such as ASTM E1529 (2010b) and UL 1709 (UL 2011a).

The standard furnace test has a standard time–temperature curve that provides severe heating conditions for test assemblies representative of field construction. Because of size limitations of furnaces, members and assembly sizes are limited. For instance, floor assemblies are typically tested at spans no greater than 17 ft (5.2 m), whereas an actual floor span may be much greater.

The fire rating of a member or assembly is based on the first failure criterion reached. For thermal response, there are temperature failure criteria measured by thermocouples applied to structural members. For flame and heat passage, the tested assembly cannot allow for ignition of cotton waste on the unexposed surface. For structural response, member deflections cannot become excessive (though excessive deflection is usually not explicitly defined in standard fire test methods). Fire resistance directories (e.g., the *UL Fire Resistance Directory* 2013) provide a list of rated assemblies based on standard testing but do not provide information about the failure criterion on which the listing is based.

When members or assemblies and their passive fire protection are similar to those already tested, methods to calculate equivalent performance for a standard fire test can be used to determine the fire resistance rating [see, e.g., ASCE/SEI/SFPE 29 (ASCE 2007)]. Analysis methods are available for structural members and fire barrier assemblies made from structural steel, plain concrete, reinforced concrete, timber and wood, concrete masonry, and clay masonry.

Standard fire resistance testing and equivalence computations do not include member connections, structural system response, or natural fire exposure. Standard fire resistance testing and equivalent analyses of hourly ratings do not provide the information needed to predict the actual performance of a structural system during structural design fires.

PERFORMANCE-BASED STRUCTURAL FIRE DESIGN: Performance-based structural fire design is useful for cases where standard fire resistance design would not address the design objectives of stakeholders. For example, a performance-based approach is appropriate for cases where performance of a structure during structural design fires needs to be quantified to properly assess risks to life safety and/or property protection. Building code variances for structural fire protection may require a performance-based approach to demonstrate the adequacy of an alternative design. Performance-based analyses also provide opportunities to develop alternative designs that are optimized for aesthetics, functionality, and/or costs.

Acceptance of performance-based designs is subject to approval by the Authority Having Jurisdiction. The designer must demonstrate that the design provides a level of safety that is equivalent or superior to that which would be attained by a design that conforms to the code's prescriptive criteria. Performance-based structural fire design provides a level of safety that is based on evaluation of structural system demand and capacity under fire conditions. Since the prescriptive approach is based solely on standard testing and does not consider structural system performance, the level of safety provided cannot be quantified. Consequently, there exists no practical method to quantitatively compare the level of safety provided by a PBD to that provided by the prescriptive approach. Therefore, it is the responsibility of the designer to properly demonstrate to the Authority Having Jurisdiction that the PBD satisfies the required performance objectives and complies with the general intent of the building code.

The Authority Having Jurisdiction should be contacted before initiating a PBD process to determine if their office is capable of and willing to support such alternative means, or if they have any concerns or requirements that need to be addressed. A peer review by an independent qualified party may be required by the Authority Having Jurisdiction as part of the PBD process. Section 1.3.1.3 addresses PBD requirements for all types of load and performance requirements.

CE.3 GENERAL REQUIREMENTS

The frequency of major building fires is relatively low because of the small probability of ignition reaching flashover conditions (see Section CE.5.2 for a discussion of flashover). Occupant or fire department intervention and/or fire suppression system extinguishment of fire typically limit fire development before it becomes uncontrolled (Hall 2013). However, certain events and/or circumstances may result in uncontrolled fires that affect the structural system. In such cases, structural integrity should be maintained to ensure occupant life safety (see Section E.4.1). This assumes that the structural system is not significantly damaged by another hazard event, such as an earthquake or an explosion.

The term *fire effects* includes thermal response and corresponding structural deformations and loads induced by heating and cooling of structural systems during fire exposure, as well as temperature-dependent changes in structural stiffness and strength, nonlinear geometric and material responses, and restrained thermal expansion or contraction. All structural materials need to be evaluated for heating effects, but some materials, such as wood, may not experience additional strains or stresses from cooling effects. However, steel connections and fasteners used in wood construction may need to be evaluated for cooling. Fire effects may result in significant forces, rotations, deflections, and deformations of members and connections. Section loss resulting from fire exposure (e.g., because of spalling or charring) may also contribute to these effects.

PBD includes development of quantifiable performance objectives that are evaluated with appropriate analysis methods. The portion of the structure affected by fire, which includes members and connections as well as cooler surrounding sections that may provide restraint against thermal expansion, should be considered to determine the structural system performance and failure modes.

Analysis techniques used to evaluate fire effects on structures range in complexity from single element analyses to finite element models that represent structural systems. Single members (columns or beams) and their connections can be analyzed in isolation for a structural design fire if reasonable assumptions can be made about restraint beyond the member and its connections, such as whether the rest of the structure provides rigid restraint to a heated member or whether the restraint of shear studs in a composite floor should be included or ignored.

Temperature-dependent strength and stiffness properties of materials may be based on peak temperatures for a structural design fire only if thermal expansion, restraint, and cooling effects can be conservatively neglected. Such considerations should include whether inelastic deformations, such as local buckling that is induced by restraint of thermal expansion, affect the member and connection behavior during cooling. For instance, bolt tear-out may occur during cooling of steel framing, which may result in member or subsystem failure. For composite floors, the thermal expansion of steel beams will be restrained as long as their shear stud connections remain intact.

Unlike single element analyses, finite element models of structural systems are able to capture the effects of thermal expansion, alternative load paths, secondary load-carrying mechanisms (e.g., catenary action), nonlinear material response, large displacement response, and connection performance. It is essential that the scope and complexity of the analysis techniques used to analyze structural response to structural design fires address the performance objectives in Section E.4 and are acceptable to the Authority Having Jurisdiction.

The achievement of adequate performance and sufficient continuity and ductility for alternate load paths following member failure due to fire effects caused by a structural design fire requires consideration of connection capacity between structural members and application of the load combination in Equation (2.5-1) with the appropriate resistance factors and member capacities for the given construction materials, as discussed in Section C2.5.

Since design and evaluation of structures for fire conditions is inherently multidisciplinary, multiple design professionals may be required. Design professionals may include structural engineers, fire protection engineers, architects, and others. When design professionals are be involved, the role and responsibility of each design professional should be clearly stated in contract documents.

For a standard fire resistance design, the architect usually serves as the responsible party for satisfying code requirements for structural fire protection. As such, the architect typically selects qualified fire resistance assemblies from available listings, perhaps with the consultation of fire protection engineers. For performance-based structural fire design, a team consisting of architects, fire protection engineers, and structural engineers is typically required. The fire protection engineer, or a design professional with similar qualifications, quantifies the fuel load, evaluates structural design fires, and estimates the temperature histories of structural systems. The structural engineer's primary responsibility is to evaluate the response of the structural system to fire effects based on the provided temperature histories. The structural engineer may also assist the fire protection engineer in determining which structural systems should be evaluated for the structural design fires and in computing deformations of structural elements that may adversely affect the integrity of fire resistance rated assemblies, such as fire barriers.

CE.4 PERFORMANCE OBJECTIVES

Performance objectives primarily address structural stability and load path continuity, and requirements related to occupant egress.

Project-specific performance objectives may also need to be considered.

CE.4.1 Structural Integrity Structural integrity supports life safety during fire in buildings and other structures. Accordingly, structural systems that support evacuation routes (e.g., corridors and exit stairs) and refuge areas should be evaluated for stability and load path continuity during structural design fires. Evaluation of stability and load path continuity should consider all supporting structural members and connections. For instance, a column under fire exposure may become unstable if lateral support elements lose their stiffness or load path continuity.

Stairways, horizontal exits, or even entire building floors may be designated as areas of refuge so that occupants can remain safely within the building during a fire. For instance, mobility-impaired occupants may need to remain within an area of refuge during a fire while awaiting rescue or evacuation assistance from emergency responders. Since areas of refuge are meant to serve as a place of safety according to the building codes, it is essential that structural systems that support these areas maintain stability, provide a continuous load path, and limit deformations throughout the heating and cooling of the structure under fire exposure.

Building codes limit egress travel distances to exits (e.g., stairways) but generally do not limit the total evacuation time. As the vertical remoteness of occupants from the point of discharge to a public way (e.g., a public street) is increased, the time required to evacuate the building increases. Hence, special consideration should be given to cases in which phased evacuation procedures are expected and longer occupant egress times are anticipated. For instance, in very tall buildings, occupants may be expected to remain on upper floors for hours, and even if those occupants are directed to use the stairways, the total evacuation time may exceed 1 hour (SFPE 2013). In these cases, the structural performance of vertical exit stairways may represent a paramount concern to designers.

Building codes typically do not mandate how a building is to be evacuated, but there is often a requirement for certain buildings, such as tall buildings, to develop evacuation plans (SFPE 2013). Determination of the time frame necessary for occupant egress and the intended function of refuge areas generally requires the expertise of a fire protection engineer or a design professional with similar qualifications.

CE.4.2 Project-Specific Performance Objectives In addition to the minimum requirements for structural integrity, project-specific performance objectives may be required. Project-specific performance objectives may address issues such as resilience aspects that consider recovery, property protection, business continuity, environmental protection, adequate structural support of fire resistance rated assemblies to limit fire and smoke spread, and/or structural support of ingress routes for first responders.

A greater level of structural performance may be required than that specified by Section E.4.1. For instance, the following example performance objectives may be applied to limit structural damage based on risk category:

- For buildings and other structures that meet Risk Category I criteria for low risk to human life in the event of failure, it may be necessary to avoid structural collapse from fire effects if the collapse is likely to damage valuable property within in the building or surrounding properties, including other buildings and infrastructure systems.
- For buildings and other structures that meet Risk Category II or III criteria, or for any risk category that would likely damage surrounding properties if structural collapse occurred, it may be necessary for the structure to endure structural design fires such that the primary structural system (e.g., columns, structural members having direct connections to columns, and lateral bracing members) remains stable with a continuous load path to supporting members during the heating and cooling of the structure. Damage to structural members or assemblies that do not compromise the stability of the primary structural system or continuity of the load path could be allowed.
- For buildings and other structures that meet Risk Category IV criteria, it may be necessary for the structure to endure structural design fires such that the entire structural system remains stable with a continuous load path to supporting members during the heating and cooling of the structure.

When designing to limit structural damage, buildings and other structures that meet Risk Category IV criteria may require that structural integrity be maintained for the entire structural system for structural design fires. By maintaining load path continuity and structural stability, enhanced property protection of adjacent areas of the building may be achieved, allowing for rapid reoccupation of areas not directly affected by fire exposure. For instance, if a critical facility experiences a severe fire in a given area, if structural integrity is maintained during and after the fire, such that there is no localized collapse and smoke damage and flame spread are contained, repair and recovery efforts will be limited primarily to the fire-affected areas.

Environmental protection objectives may include limiting the release and spread of hazardous or toxic chemicals to the air, ground and surfaces, or waterways because of loss of structural integrity.

It is desirable for fire-rated resistance assemblies to remain functional (resist fire spread and maintain adequate strength and stiffness for structural integrity) during structural design fires. Three limit states would need to be considered to evaluate fire resistance rated assemblies (e.g., fire barriers): (1) heat transmission leading to unacceptable rise of temperature on the unexposed surface, (2) breach of the barrier caused by loss of support, cracking, or loss of integrity, and (3) loss of load-bearing capacity. All three contribute to performance of fire resistance rated assemblies.

It is desirable for load-carrying elements (e.g., fire-rated floors and walls) that also serve as fire barriers to not have their fire resistance impaired because of deformations or other fire effects. When a fire resistance rated assembly is not load-bearing, the deformation of structural members supporting the assembly should not compromise its performance. However, criteria for limiting deformation of structural members supporting nonstructural fire resistance rated assemblies are not readily available. Fire resistance rated systems are qualified based on results of standard testing in which the supporting boundaries of the furnace (e.g., concrete floor) do not undergo deformation during heating. Addressing such performance objectives may require the designer to develop specific performance criteria per the discretion of the Authority Having Jurisdiction.

CE.5 THERMAL ANALYSIS OF FIRE EFFECTS

Section E.5 provides requirements for determining structural design fires and the thermal response of structural members and connections. Structural design fires have the potential to affect the integrity and stability of a structure. Development of structural design fires involves consideration of compartment layout, boundary materials, ventilation, and fuel load that combine to create conditions that potentially threaten the structural system. The designer should consider a sufficient number of structural design fires to properly address the risks with consideration of uncertainty associated with heating parameters. Based on the time-dependent thermal boundary conditions from fire exposure, the thermal response of structural members and connections can be determined based on principles of heat transfer.

CE.5.1 Fuel Load Structural design fires depend on the fuel load and its distribution. Fuel load is commonly expressed as a fuel load density, or fuel load per unit floor area. NFPA 557 (2012b) establishes a basis for selecting fuel load density and distribution. Other methods may be used if acceptable to the Authority Having Jurisdiction. Although the fuel load density concept implies a uniform distribution of combustibles in compartments, the actual distribution of combustibles may need to be addressed for analyses of structural design fires where localized heating effects may be significant.

The fuel load based on the contents of a building, space, or area typically varies because combustible materials release different levels of thermal energy when burning. For instance, plastics generally release more energy per unit mass than wood products. The conversion of building contents into an equivalent mass based on their potential energy provides a consistent basis for determining the total energy of the fuel load. The fuel load in equivalent mass measured in pound mass (lbm) [kilograms (kg)] can be readily converted into total energy measured in British thermal units (Btu) (kilojoules (kJ)) for use in characterization of structural design fires.

CE.5.2 Structural Design Fires Structural design fires are structurally significant fires based on the physical parameters, such as building layout, compartment boundary materials (e.g., walls), ventilation openings (e.g., doors and windows), and fuel load that are specific to a particular building space or spaces. Structurally significant fires include those that are not controlled by active fire protection systems, such as automatic fire sprinklers or firefighting efforts. Other design fires considered for fire detection, evacuation, or other fire-related issues may not be structurally significant fires.

Provided that there is sufficient oxygen available to support combustion, the duration of a structural design fire depends on the heat release rate history of the fire and the total energy of the fuel load. A structural design fire reaches burnout when either the available fuel load is fully consumed or there is insufficient oxygen to support combustion. The materials involved in a fire significantly affect the heat release rate history. Hence, the heat release rate history of a fire is not necessarily correlated with the total energy of the fuel load.

Structural design fires should be evaluated using methods acceptable to the Authority Having Jurisdiction. In certain cases, it may be necessary to perform fire modeling to evaluate structural design fires. Fire modeling generally requires the expertise of a fire protection engineer or a design professional with similar qualifications. Most fire models simulate the effects of fire (e.g., hot air and smoke flows) and not the phenomena of combustion and flame spread. Where fire modeling is required to determine time-dependent thermal boundary conditions on the structural system, the designer should substantiate the model according to SFPE G.06 (2011a).

Based on the fuel load, ignition(s), and arrangement of compartments and ventilation openings, a structural design fire may be broadly defined as one of the following types: enclosure fire, localized fire, exterior fire, or traveling fire. SFPE S.01 (2011b) provides methods to determine time-dependent thermal boundary conditions on a structural system caused by a structural design fire for either enclosure or localized fires. NFPA 80A (2012a) provides similar methods for exterior fires.

Events such as an earthquake or flood may result in enclosure, localized, and/or traveling fires within a building. Postevent fires may have compounding factors such as dispersed flammable contents, electrical malfunctions, interrupted power and/or water supplies, damage to fire sprinkler systems, or overextended emergency responders. Also, structural damage caused by a severe event may be exacerbated by a fire (e.g., reinforcing steel exposed by concrete spalling at a connection or loss of member insulating materials).

Enclosure Fire. An enclosure fire is affected by the compartment(s) in which it is contained. As a hot upper gas layer forms with the progression of fire, it reradiates heat back to the fire and fuel packages. The compartment boundaries may radiate inward as well. These conditions may eventually lead to flashover, at which point the fire is considered to be a fully developed fire. Flashover occurs when there is a rapid transition from localized burning to simultaneous burning of all combustible materials within the enclosure. Flashover can only occur in an enclosed compartment with sufficient fuel and ventilation, where the ceiling can trap hot gases that lead to radiant heating of all fuels to the point of combustion.

In most fires, there is variability in the fire size, depending on which items are ignited first and how the fire grows and spreads. Focusing on fully developed fires eliminates much of this variability, since a fully developed (postflashover) fire is less sensitive to which items are ignited first and how the fire grows. For PBD analyses, neglecting the heating of the structure during the growth stage of a fire is usually a reasonable assumption since the heating of the structure during the fully developed stage is much greater than heating during the growth stage.

Localized Fire. A localized fire burns combustibles at a given location and does not reach flashover. Localized burning occurs in open exposures, large spaces, areas with high ceilings, or other locations that are not conducive to flashover. This typically occurs in relatively large compartments or spaces, when the fuel is concentrated within a region. Fires that do not reach flashover may produce localized heating on the structure.

Exterior Fire. Exterior fires may lead to ignition and subsequent fire exposure within a building, possibly on multiple floors. For example, buildings in close proximity may mutually increase the risk of a large fire exposure through heat exchange between buildings. Flame impingement and convective heat transfer from exterior fires, sometimes from fires extending out windows from lower floors, may also create a fire hazard.

Traveling Fire. A traveling fire is characterized by the spread of fire from combustible to combustible across an open plan that does not burn simultaneously throughout the entire compartment. These fires move across areas as flames spread, burning over a limited area at any one time. Traveling fires are characterized by areas with combustibles that are not yet burning, a fire front with generally intense heating, and a trailing region where fuels have been largely consumed.

CE.5.3 Heat Transfer Analysis The thermal response of the structural system depends on the structural design fire and the three modes of heat transfer: conduction, convection, and radiation. All three modes of heat transfer typically occur when a structure is heated by fire.

Heat transfer analysis methods are specific to the material's physical and chemical response to heat and are used for homogeneous and nonhomogeneous materials. Material responses may include charring, intumescing, dehydration, phase changes, and chemical reactions. These material responses and associated properties may significantly affect how heat transfer analyses are conducted.

Relevant material thermal properties for heat transfer analyses include density, thermal conductivity, emissivity for exposed surfaces, and specific heat (which may include heat effects caused by phase change, if any). Many of these material properties have strong temperature dependence. Sources with temperature-dependent thermal properties for steel, concrete, masonry, and timber are listed here.

- American Concrete Institute. 2007. *Code Requirements for Determining Fire Resistance of Concrete and Masonry Construction Assemblies.* ACI 216.1-07/TMS-216-07.
- American Wood Council. 2015. *National Design Specification (NDS) for Wood Construction.* AWC NDS-2015.
- European Committee for Standardisation. 2003. *Eurocode 6: Design of Masonry Structures. Part 1-2: General Rules—Structural Fire Design.* EN 1996-1-2.
- European Committee for Standardisation. 2004a. *Eurocode 2: Design of Concrete Structures. Part 1–2: General Rules—Structural Fire Design.* EN 1992-1-2.
- European Committee for Standardisation. 2004b. *Eurocode 5: Design of Timber Structures. Part 1–2: General—Structural Fire Design.* EN 1995-1-1.
- European Committee for Standardisation. 2005. *Eurocode 3: Design of Steel Structures. Part 1–2: General Rules—Structural Fire Design.* EN 1993-1-2.
- Forest Products Laboratory. 2010. *Wood Handbook: Wood as an Engineering Material.* General Technical Report FPL-GTR-190.
- V. Kodur and T. Harmathy. 2008. "Properties of Building Materials." In *Handbook of Fire Protection Engineering,* 277–324. Society of Fire Protection Engineers.

For design purposes, constant values of thermal properties can be used if they yield conservative results. Depending on the heating or cooling conditions at exposed surfaces, applicable values for convection heat transfer coefficient should be used. Thermal insulation should be analyzed using the specified minimum thickness.

Heat transfer analyses inherently assume that the materials remain in place during the fire exposure. If insulating materials are expected to fail during a structural design fire, either the heat transfer analyses should account for the resulting increased heating of the structure or the insulation design should be modified. The deformations of structural members during structural design fires may need to be considered as part of the evaluation of the mechanical integrity of fire resistance rated assemblies.

The temperature histories of structural members and connections comprising the structural system should be determined using heat transfer analyses as permitted in SFPE S.02 (2014). Other approved methods may be used if acceptable to the Authority Having Jurisdiction.

CE.6 STRUCTURAL ANALYSIS OF FIRE EFFECTS

Structural analysis of fire effects requires consideration of the heated members and connections and the entire structural system. A single-member analysis may be justified where only a single member is affected by a fire without consequential effects from or to surrounding members. A systems approach requires consideration of thermal expansion of heated sections and restraint by cooler adjacent framing, thermally induced forces and displacements on connections, the response of floor systems, and thermally induced failure modes across the structure (McAllister et al. 2013). Structural elements may have large deflections that are an order of magnitude greater than deflection limits normally anticipated for structures (McAllister et al. 2012). Large deflections may induce forces in adjacent structural assemblies (e.g., members and connections).

Thermal Expansion and Restraint. Floor systems may experience thermal restraint during heating from columns and cooler adjacent floor members; interior bays typically experience more thermal restraint than exterior bays. Columns, on the other hand, typically do not experience significant thermal restraint from floor systems. However, if there is lateral bracing in place, it may impose some thermal restraint on the braced column section, depending on the framing geometry and member temperatures, if the bracing members and adjacent column are at a significantly lower temperature.

A temperature gradient through the depth or thickness of a structural element causes differential thermal expansion between the hotter and cooler external surfaces. Differential expansion results in curvature for simply supported members. For members with partial or full rotational end restraint, a temperature gradient results in a strain gradient through the depth of the member because of thermal restraint.

The effect of thermal expansion and contraction needs to be carefully considered. Thermal expansion and contraction of construction materials may generate forces sufficient to cause yielding or fracture, depending on the temperature reached and the degree of restraint provided by the surrounding structural system to the thermally induced actions (Gillie et al. 2002). In fact, thermal restraint may dominate the behavior of framing systems, particularly floor systems, with degradation of stiffness and strength a secondary factor (Bailey et al. 1999). Fire-exposed elements that have experienced plastic deformations caused by weakening and thermal restraint may experience tensile strain as the structure cools and may induce forces in adjacent structural assemblies (e.g., connections), depending on the level of thermal restraint.

Columns. Fire effects on steel columns include loss of strength and stiffness, thermal expansion, and P-delta effects under thermal gradients, which may affect global and local column buckling strength. Design procedures for fire effects on steel compression members are provided in Appendix 4 of AISC (2010). The equations are based on analyses conducted by Takagi and Deierlein (2007). The effects of thermal gradients on the axial-moment capacity of steel wide-flange sections are discussed by Garlock and Quiel (2008). Analyses by Seif and McAllister (2013) indicate when elevated temperatures may result in local and global buckling modes of steel wide-flange sections.

Fire effects on concrete columns include loss of strength and stiffness in both the concrete and the reinforcement. The primary causes of fire damage to concrete are deterioration in mechanical properties of the cement paste and aggregate, cracking, and spalling (Khoury 2001). Spalling occurs in normal-weight concrete as well as high-strength concrete (Hertz 2003). Concrete cover serves as insulation for reinforcement, so concrete cracking or spalling allows direct heating of reinforcement. Lie and Irwin (1993) provide models based on experimental data for predicting the performance of reinforced concrete columns with rectangular sections, based on axial deformations and temperature through the concrete section. Kodur and McGrath (2003) present experimental results for high-strength concrete columns that include the effects of concrete materials, loading, and spacing of ties. Kodur and Phan (2007) describe the factors that influence spalling in high-strength concrete members.

Floor Systems. A study of steel beams and composite floor systems exposed to a range of heating scenarios (Moss et al. 2004) found that the system behavior at elevated temperatures caused interrelated changes in the deflected shape, axial force, bending moments, and internal stresses that varied with the type of support condition and thermal restraint.

Composite floor systems tested at Cardington (British Steel 1999) and for the FRACOF (Zanon et al. 2011) and COSSFIRE (Zhao and Roosefid 2011) programs found that composite floors with beam lengths less than 30 ft (9 m) did not experience failures during the heating phase, but connection failures did occur during the cooling phase if significant deformation occurred in the floor beams during heating. Wang et al. (2011) found that connection types and axial restraint for floor beams influenced the response of floor systems, primarily during the cooling phase. In contrast, the WTC 7 numerical analysis of composite floors predicted that shear stud and beam connections failed during the heating phase of long-span composite floor systems with one-sided beam-to-girder floor beams of 50 to 56 ft (15 to 17 m) length (McAllister et al. 2012). This numerical prediction was not verified through observation or physical fire testing because of the large size of the floor bays and inability to scale structural responses to fire. Shear stud failure has not been observed in any fire event or in any structural fire tests. The lack of observations may be caused by the limitations of full-scale structural fire tests to date. In structural fire tests, floor beam length is typically less than 20 to 30 ft (7 to 10 m), columns and connections are often protected from heating, and/or substantial restraint of the steel beam and concrete slab prevents thermal expansion of the composite section. Failure modes for many structural fire tests with shorter, restrained floor beams tend to occur during the cooling phase at the beam-to-column connection (Bisby et al. 2013). Bailey et al. (1999) and Elghazouli and Izzuddin (2001) assessed the performance of the Cardington floor system and identified response mechanisms of the floor system to heating and cooling effects. Bailey (2004) presents a performance-based design approach that considers membrane action in the composite floor system.

Reinforced concrete floors, including cast-in-place and precast/prestressed construction, are typically designed by providing a specified cover thickness over reinforcement. In general, there is little guidance for the PBD of concrete floor systems for structural design fires. However, PCI (2011) provides design guidance for precast and prestressed concrete floor systems. Whereas heating of reinforcement is to be avoided, the heating of prestressed strands in floor systems should be of particular concern, as loss of prestress may significantly degrade the floor system performance.

Floor Connections. The performance of connections needs to be considered in the structural analysis of fire effects, particularly connections in floor systems. For instance, steel shear connections may experience bolt shear, local buckling or tear-out of connection plates, or local buckling of the beam flange near the connection. Parametric studies of single plate shear (fin) connections (Yu et al. 2009, Selamet and Garlock 2010, Hu and Engelhardt 2011) and double angle connections (Pakala et al. 2012) identified critical dimensions and component interactions that control connection behavior at elevated temperatures. Huang et al. (1999) evaluated the role of shear stud connections by comparing models with varying levels of composite action against test data from the Cardington fire tests of composite beams.

Heating of concrete and reinforcement in concrete framing (cast-in-place or precast) may result in concrete spalling that accelerates reinforcement heating and loss of strength in the reinforcement.

Failure Modes. Fire-induced failure modes include large deflections, member buckling (local, global, or lateral torsional), connection failures (bearing, bolt tear-out, bolt shear, or weld failure), reinforcement and anchorage failures, and section loss (concrete spalling, cracking, or crushing or charred timber sections).

As temperatures increase sufficiently to reduce the strength and stiffness of the cross section, yielding or buckling modes may occur at arbitrary point-in-time (service) load levels. If such weakening occurs for members with temperature gradients, the resulting gradient in member stiffness and strength can alter the combined axial load and moment resistance of the member (Garlock and Quiel 2008). The strength and stiffness gradient may also cause the member centroid (i.e., center of strength and stiffness) to shift toward the cooler (i.e., stronger) side of the cross section. This centroidal shift induces moments in axially loaded elements.

CE.6.1 Temperature History for Structural Members and Connections Temperature histories of structural members and connections depend on the thermal response for structural design fires. Temperature histories may include thermal gradients across a section or along a member length.

Thermal finite element analyses typically use 2D or 3D models with a fine mesh of solid elements (element size on the order of inches or centimeters), whereas structural analyses typically use a coarser mesh of shell and/or beam elements (element sizes on the order of feet or meters). Careful consideration should be given to the tradeoff between optimal model features for each analysis versus mapping of temperatures between two sets of nodal data. For example, a simplistic heat transfer model, such as the lumped mass method, provides uniform temperatures; uniform temperatures may be inappropriate for a structural system that is likely to experience significant temperature gradients. There may be situations where it is appropriate to use the same mesh discretization in the thermal and structural models, so that transfer of nodal temperature data is seamless.

In most cases, results from heat transfer models are mapped to significantly fewer nodes of beam and shell elements in structural models. The average rate of temperature changes in structural members and connections typically occurs on the order of minutes, and temperature data sets for the structural system can be input at set intervals to reflect the progress of heating and cooling. Temperatures may be interpolated linearly between the data sets during the structural analysis.

CE.6.2 Temperature-Dependent Properties At elevated temperatures, the strength and stiffness of the material(s) comprising a structural assembly change. Sources for strength and stiffness degradation for steel (including prestressing steel), concrete, masonry, and timber are listed here.

- American Concrete Institute. 2001. *Guide for Determining the Fire Endurance of Concrete Elements*. ACI 216R-89 (Reapproved 2001).
- American Concrete Institute. 2007. *Code Requirements for Determining Fire Resistance of Concrete and Masonry Construction Assemblies*. ACI 216.1-07/TMS-216-07.
- American Institute of Steel Construction. 2010. "Structural Design for Fire Conditions." Appendix A4 in *Specification for Structural Steel Buildings*, 14th ed.
- American Wood Council. 2003. *Calculating the Fire Resistance of Exposed Wood Members*. AWC TR10.

- American Wood Council. 2015. *National Design Specification (NDS) for Wood Construction*. ANSI/AWC.
- European Committee for Standardisation. 2004a. *Eurocode 2: Design of Concrete Structures. Part 1–2: General Rules—Structural Fire Design*. EN 1992-1-2.
- European Committee for Standardisation. 2005. *Eurocode 3: Design of Steel Structures. Part 1–2: General Rules—Structural Fire Design*. EN 1993-1-2.
- National Institute of Standards and Technology. 2010. *Best Practices Guidelines for Structural Fire Resistant Design of Steel and Concrete Buildings*.
- Precast/Prestressed Concrete Institute. 2011. *Design for Fire Resistance of Precast/Prestressed Concrete*, 3rd ed. MNL-124-11.
- Stephen S. Szoke. 2006. "Resistance to fire and high temperature." Chapter 27 in *ASTM STP 169D, Significance of Tests and Properties of Concrete and Concrete-Making Materials*. West Conshohocken, PA: ASTM International.

The Eurocode defines separate curves for material properties at elevated temperatures for the proportional limit and the yield stress. However, the yield stress at elevated temperatures is often defined at a 0.02 strain in the Eurocode. For ambient temperatures, the yield stress is defined at a 0.002 offset strain (other methods are also permitted as defined in ASTM A6/A6M, 2010a), and the proportional limit is not defined by an ASTM test method. However, values for the proportional limit are given in AISC (2010), Appendix 4, in terms of the ratio to the yield strength. The analyst should be careful to note these distinctions in material models.

Where possible, temperature-dependent material properties should be obtained from consensus standards. Alternative sources of data, such as research studies and independent tests, may provide useful data. However, reliance on a single test or source of data should be avoided. At a minimum, test data from several sources should be collected and used to develop a representative set of temperature-dependent material properties.

CE.6.3 Load Combinations For extraordinary events, such as structurally significant fires, load combinations 2.5-1 and 2.5-2 are used to evaluate the structural system performance. This load combination was developed for extraordinary events that may lead to ultimate limit states such as gross inelastic deformation or partial collapse.

Load combination 2.5-1 is used to perform a safety check on a structure designed for the basic load combinations at room-temperature conditions and to evaluate the effect of elevated temperatures. The force in structural members caused by fire effects, A_k, has a load factor of 1.0 (Ellingwood and Corotis 1991, Ellingwood 2005). The live load factor of 0.5 is intended for typical occupancies and arbitrary point-in-time live loads that likely exist during a significant fire. The 0.5 live load factor is also used in other load combinations in Section 2.3 when live load is a companion load and not the principal load. Note that the live load in this load combination differs from the approach used in standard fire test methods, where the assembly is loaded to its design limit for member stress during the standard fire exposure, which represents the application of the full dead and live load.

Whereas gravity loads for the structure remain constant during most fires (assuming that most of the building contents are not burning), time-dependent temperature histories may result in time-varying member strength and thermally induced forces, depending on the temperatures reached by structural members.

REFERENCES

AISC (American Institute of Steel Construction). 2010. "Structural design for fire conditions." In *Specification for structural steel buildings*. 14th ed. Chicago: AISC.

ASCE. 2007. *Standard calculation methods for structural fire protection*. ASCE/SFPE 29. Reston, VA: ASCE.

ASTM International. 2010a. *Standard specification for general requirements for rolled structural steel bars, plates, shapes, and sheet piling*. ASTM A6/A6M. West Conshohocken, PA: ASTM International.

ASTM. 2010b. *Standard test methods for determining effects of large hydrocarbon pool fires on structural members and assemblies*. ASTM E1529-10. West Conshohocken, PA: ASTM International.

ASTM. 2012. *Standard test methods for fire tests of building construction and materials*. ASTM E119-12a. West Conshohocken, PA: ASTM International.

Bailey, C. G. 2004. "Membrane action of slab/beam composite floor systems in fire." *Eng. Struct.* 26 (12): 1691–1703. https://doi.org/10.1016/j.engstruct.2004.06.006.

Bailey, C. G., T. Lennon, and D. B. Moore. 1999. "The behaviour of full-scale steel-framed buildings subjected to compartment fires." *Struct. Eng.* 77 (8): 1182–1192.

Bisby, L., J. Gales, and C. Maluk. 2013. "A contemporary review of large-scale non-standard structural fire testing." *Fire Sci. Rev.* 2 (1): 1–27. https://doi.org/10.1186/2193-0414-2-1.

British Standards Institution. 1987. *Fire tests on building materials and structures: Method for determination of the fire resistance of elements of construction (general principles)*. BS 476-20. London: BSI.

British Steel. 1999. *The behaviour of multi-storey steel framed buildings in fire*. South Yorkshire: British Steel.

Ellingwood, B. R. 2005. "Load combination requirements for fire resistant structural design." *J. Fire Protect. Eng.* 15 (2): 43–61. https://doi.org/10.1177/1042391505045582.

Ellingwood, B. R., and R. B. Corotis. 1991. "Load combinations for buildings exposed to fires." *Eng. J.* 28 (1): 37–44.

Garlock, M. E. M., and S. E. Quiel. 2008. "Plastic axial load and moment interaction curves for fire-exposed steel sections with thermal gradients." *J. Struct. Eng.* 134 (6): 874. https://doi.org/10.1061/(ASCE)0733-9445(2008)134:6(874).

Hertz, K. D. 2003. "Limits of spalling of fire-exposed concrete." *Fire Saf. J.* 38 (2): 103–116. https://doi.org/10.1016/S0143-974X(01)00066-9.

Hu, G., and M. D. Engelhardt. 2011. "Investigations on the behavior of steel single plate beam end framing connections in fire." *J. Struct. Fire Eng.* 2 (3): 195–204.

Huang, Z., I. W. Burgess, and R. J. Plank. 1999. "The influence of shear connections on the behavior of composite steel-framed buildings in fire." *J. Constr. Steel Res.* 51 (3): 219–237. https://doi.org/10.1016/S0143-974X(99)00028-0.

Khoury, G. A. 2001. "Effect of fire on concrete and concrete structures." *Prog. Struct. Eng. Mater.* 2 (4): 429–447. https://doi.org/10.1002/pse.51.

Kodur, V., and R. McGrath. 2003. "Fire endurance of high strength concrete columns." *Fire Tech.* 39 (Jan): 73–87.

Kodur, V. K. R., and L. Phan. 2007. "Critical factors governing the fire performance of high strength concrete systems." *Fire Saf. J.* 42 (6–7): 482–488. https://doi.org/10.1016/j.firesaf.2006.10.006.

Lie, T. T., and R. J. Irwin. 1993. "Method to calculate the fire resistance of reinforced concrete columns with rectangular cross section." *ACI Struct. J.* 90 (1): 52–60.

McAllister, T. P., J. L. Gross, F. Sadek, S. Kirkpatrick, R. A. MacNeill, M. Zarghamee, et al. 2013. "Structural response of

World Trade Center buildings 1, 2, and 7 to impact and fire damage." *Fire Tech.* 49 (3): 709–739.

McAllister, T. P., R. MacNeill, O. O. Erbay, A. T. Sarawit, M. S. Zarghamee, S. Kirkpatrick, et al. 2012. "Analysis of structural response of WTC 7 to fire and sequential failures leading to collapse." *J. Struct. Eng.* 138 (1): 109–117. https://doi.org/10.1061/(ASCE)ST.1943-541X.0000398.

NFPA (National Fire Protection Association). 2012a. *Recommended practice for protection of buildings from exterior fire exposures.* NFPA 80A. Quincy, MA: NFPA.

NFPA. 2012b. *Standard for determination of fire loads for use in structural fire protection design.* NFPA 557. Quincy, MA: NFPA.

Pakala, P., V. Kodur, and M. Dwaikat. 2012. "Critical factors influencing the fire performance of bolted double angle connections." *Eng. Struct.* 42 (Sep): 106–114. https://doi.org/10.1016/j.engstruct.2012.04.011.

Seif, M. S., and T. P. McAllister. 2013. "Stability of wide flange structural steel columns at elevated temperatures." *J. Constr. Steel Res.* 84 (5): 17–26. https://doi.org/10.1016/j.jcsr.2013.02.002.

Selamet, S., and M. E. Garlock. 2010. "Robust fire design of single plate shear connections." *Eng. Struct.* 32 (8): 2367–2378. https://doi.org/10.1016/j.engstruct.2010.04.011.

SFPE (Society of Fire Protection Engineers). 2011a. *Engineering guidelines for substantiating a fire model for a given application.* SFPE G.06. Gaithersburg, MD: SFPE.

SFPE. 2011b. *Engineering standard on calculating fire exposures to structures.* SFPE S.01. Gaithersburg, MD: SFPE.

SFPE. 2013. *Engineering guide: Fire safety for very tall buildings.* Gaithersburg, MD: SFPE.

SFPE. 2014. *Engineering standard on the development and use of methodologies to predict the thermal performance of structural and fire resistive assemblies.* SFPE S.02. Gaithersburg, MD: SFPE.

Takagi, J., and G. G. Deierlein. 2007. "Strength design criteria for steel members at elevated temperatures." *J. Constr. Steel Res.* 63 (8): 1036–1050. https://doi.org/10.1016/j.jcsr.2006.10.005.

UL (Underwriters Laboratories). 2011a. *Standard for rapid rise fire tests of protection materials for structural steel.* ANSI/UL 1709. Northbrook, IL: UL.

UL. 2011b. *UL standard for safety for fire tests of building construction and materials.* ANSI/UL 263. Northbrook, IL: UL.

UL. 2013. *UL fire resistance directory.* Northbrook, IL: UL.

ULC (Underwriters Laboratories of Canada). 2007. *Standard methods of fire endurance tests of building construction and materials.* CAN/ULC-S101. Ottawa: ULC.

Wang, Y. C., X. H. Dai, and C. G. Bailey. 2011. "An experimental study of relative structural fire behaviour and robustness of different types of steel joint in restrained steel frames." *J. Constr. Steel Res.* 67(7): 1149–1163. https://doi.org/10.1016/j.jcsr.2011.02.008.

Yu, H., I. W. Burgess, J. B. Davison, and R. J. Plank. 2009. "Experimental investigation of the behaviour of fin plate connections in fire." *J. Constr. Steel Res.* 65 (3): 723–736.

Zanon, R., M. Sommavilla, O. Vassart, B. Zhao, and J. M. Franssen. 2011. "FRACOF: Fire resistance assessment of partially protected steel-concrete composite floors." In *Proc., XXIII Giornate italiane della costruzione in acciaio,* Napoli, Italy, 527–536.

Zhao, B., and M. Roosefid. 2011. "Experimental and numerical investigations of steel and concrete floors subjected to ISO fire condition." *J. Struct. Fire Eng.* 2 (4): 301–310.

OTHER REFERENCES (NOT CITED)

American Concrete Institute. 2001. *Guide for determining the fire endurance of concrete elements.* ACI 216R-89. Farmington Hills, MI: ACI.

American Concrete Institute. 2007. *Code requirements for determining fire resistance of concrete and masonry construction assemblies.* ACI 216.1-07/TMS-216-07. Farmington Hills, MI: ACI.

American Wood Council. 2003. *Calculating the fire resistance of exposed wood members.* TR10-2003. Leesburg, VA: AWC.

American Wood Council. 2015. *National design specification for wood construction: ASD/LRFD manual.* Leesburg, VA: AWC.

Buchanan, A. H. 2002. *Structural design for fire safety.* New York: Wiley.

CEN (European Committee for Standardisation). 2003. *Eurocode 6: Design of masonry structures. Part 1–2: General rules—Structural fire design.* EN 1996-1-2. Brussels, Belgium: CEN.

CEN. 2004a. *Eurocode 2: Design of concrete structures. Part 1–2: General rules—Structural fire design.* EN 1992-1-2. Brussels, Belgium: CEN.

CEN. 2004b. *Eurocode 5: Design of timber structures. Part 1–2: General—Structural fire design.* EN 1995-1-1. Brussels, Belgium: CEN.

CEN. 2005. *Eurocode 3: Design of steel structures. Part 1–2: General rules—Structural fire design.* EN 1993-1-2. Brussels, Belgium: CEN.

Elghazouli, A. Y., and B. A. Izzuddin. 2001. "Analytical assessment of the structural performance of composite floors subject to compartment fires." *Fire Saf. J.* 3 (8): 769–793. https://doi.org/10.1016/S0379-7112(01)00039-X.

Forest Products Laboratory. 2010. *Wood handbook: Wood as an engineering material.* FPLG Technical Rep. No. FPL-GTR-190. Madison, WI: Forest Service, USDA.

Gillie, M., A. S. Usmani, and J. M. Rotter. 2002. "A structural analysis of the Cardington British steel corner test." *J. Constr. Steel Res.* 58 (4): 427–442. https://doi.org/10.1016/S0143-974X(01)00066-9.

Hall, J. 2013. *The total cost of fire in the United States.* Quincy, MA: National Fire Protection Association.

ISO (International Organization for Standardization). 1999. *Fire resistance tests: Elements of building construction. Part 1: General requirements.* ISO 834-1:1999. Geneva: ISO.

Kodur, V., and T. Harmathy. 2008. "Properties of building materials." In *SFPE handbook of fire protection engineering,* P. J. DiNenno, ed., 277–324. Quincy, MA: National Fire Protection Association.

Moss, J. M., A. H. Buchanan, J. Septro, C. Wastney, and R. Welsh. 2004. "Effect of support conditions on the fire behaviour of steel and composite beams." *Fire Mater.* 28 (Apr): 159–175. https://doi.org/10.1002/fam.855.

NIST (National Institute of Standards and Technology). 2010. *Best practices guidelines for structural fire resistant design of steel and concrete buildings.* Tech. Note 1681. Gaithersburg, MD: NIST.

PCI (Precast/Prestressed Concrete Institute). 2011. *Design for fire resistance of precast/prestressed concrete.* MNL-124-11. Chicago: PCI.

Szoke, S. S. 2006. "Resistance to fire and high temperature." Chap. 27 in *Significance of tests and properties of concrete and concrete-making materials.* West Conshohocken, PA: ASTM International.

APPENDIX CF
WIND HAZARD MAPS FOR LONG RETURN PERIODS

CF.1 SCOPE

This appendix provides wind hazard maps for longer mean recurrence intervals (MRIs) than those included in Chapter 26 in support of performance-based wind design applications where use of hazard levels above the minimum specified in the standard may be appropriate or design scenarios are required in which extreme performance is explicitly assessed (e.g., incipient collapse of the main wind force resisting system). The explicit consideration of such design scenarios requires estimates of the wind hazard with MRIs well beyond those associated with the standard risk categories, for which current practice essentially requires elastic component performance. These maps will provide input for developing acceptance criteria associated with such design scenarios. In addition, design for longer MRIs may be needed for wind load designs required by other codes and standards. The maps presented in this appendix are not tied to a specific risk category like the maps in Figures 26.5-1 A–D, which represent preselected MRIs of 300, 700, 1,700, and 3,000 years for Risk Categories I through IV, respectively. For the longer-MRI maps in this appendix, it is expected that the owner, perhaps in consultation with the designer, will select the appropriate map for their particular application. In some unique situations, the Authority Having Jurisdiction may require a higher MRI. For example, certain liquefied natural gas facilities require design for winds "having a probability of exceedance in a 50-year period of 0.5 percent or less" (i.e., 10,000-year MRI winds; PHMSA 2019).

Where the longer return period wind hazard maps of this appendix are being used, the designer should also refer to Appendix G to determine a tornado speed at the same return period for use with the tornado load provisions of Chapter 32, where applicable.

CF.2 WIND SPEEDS

The mapped wind speeds for the contiguous United States shown in Figures F.2-1 through F.2-3 were developed using the same methodology as the basic wind speed maps in Figures 26.5-1A–D, described in commentary Sections C26.5.1 and C26.5.2. The special wind regions shown on Appendix F wind hazard maps are unchanged from those on Chapter 26 maps.

Uncertainties in the estimated wind speeds increase with increasing wind speeds and increasing return periods. For the non hurricane component of the extreme wind climate, standard errors of the estimated speeds range from approximately 10 to 13 mi/h (4.5 to 5.8 m/s) for 10,000-year MRI and from 13 to 16 mi/h (5.8 to 7.2 m/s) for 100,000-year MRI. This compares to standard errors of approximately 7 to 9 and 9 to 11 mi/h (3.1 to 4.0 and 4.0 to 4.9 m/s) for 700 and 3,000-year MRI, respectively [see Appendix C.1 in Pintar et al. 2015) for maps of standard errors at these and other MRI]. In the case of hurricane winds, no formal uncertainty studies were undertaken; however, as discussed in Vickery et al. (2009), the uncertainties associated with the 50-year and 100-year MRI wind speed vary with location along the coastline, exceeding 10% for locations north of New York. It is expected that for return periods longer than about 1,000 years, wind speed uncertainties along the hurricane coastline would be comparable to those in the non-hurricane-prone regions. The hurricane winds include the effect of wind field modeling uncertainties. The wind speed uncertainty term is the same as that used in the development of the ASCE 7-10 and ASCE 7-16 wind speed maps. This error term increases the 100-year return period wind speed by a few percent, and the longer MRIs by approximately 10% (Vickery et al. 2011).

A procedure for estimating wind speeds in Alaska as a function of return period is described by Peterka and Shahid (1998).

The 10,000-year MRI wind speeds associated with hurricanes in Hawaii and US territories (known as typhoons in Guam and the Northern Mariana Islands and cyclones in American Samoa) are provided in Table CF.2-1. Only single wind speed values are available for Hawaii, Puerto Rico, and US Virgin Islands at this return period. All wind speed values in Table CF.2-1 are for 3 s gusts at 33 ft (10 m) above ground for Exposure Category C. These values were determined using the hurricane modeling procedures described in Section C26.5.1. Winds associated with events other than tropical cyclones have not been explicitly considered but are assumed to have negligible contribution to the extreme wind climate at the 10,000-year MRI level.

Table CF.2-1. 10,000-Year MRI Gust Hurricane Wind Speeds for Hawaii and US Territories.

	Wind Speed*	
Location	**mi/h**	**m/s**
American Samoa	190	85
Guam	235	105
Hawaii	165	74
Northern Mariana Islands	235	105
Puerto Rico	200	89
US Virgin Islands	190	85

*Wind speeds are 3 s gust speeds at 33 ft (10 m) above ground for Exposure Category C.

REFERENCES

Peterka, J. A., and S. Shahid. 1998. "Design gust wind speeds in the United States." *J. Struct. Eng.* 124 (2): 207–214. https://doi.org/10.1061/(ASCE)0733-9445(1998)124:2(207).

PHMSA (Pipeline and Hazardous Materials Safety Administration). 2019. "Transportation, liquefied natural gas facilities: Federal Safety Standards, Subpart B – Siting requirements, wind forces." Accessed October 1, 2019. https://www.govinfo.gov/content/pkg/CFR-2019-title49-vol3/xml/CFR-2019-title49-vol3-part193.xml#seqnum193.2067.

Pintar, A., E. Simiu, F. Lombardo, and M. Levitan. 2015. *Maps of non-hurricane non-tornadic wind speeds with specified mean recurrence intervals for the contiguous United States using a two-dimensional Poisson process extreme value model and local regression.* NIST Special Publication 500-301. Gaithersburg, MD: National Institute of Standards and Technology.

Vickery, P. J., D. Wadhera, and L. A. Twisdale Jr. 2011. *Technical basis for regulatory guidance on design-basis hurricane wind speeds for nuclear power plants.* NUREG/CR-7005. Rockville, MD: US Nuclear Regulatory Commission.

Vickery, P. J., D. Wadhera, L. A. Twisdale Jr., and F. M. Lavelle. 2009. "U.S. hurricane wind speed risk and uncertainty." *J. Struct. Eng.* 135 (3): 301–320. https://doi.org/10.1061/(ASCE)0733-9445(2009)135:3(301).

APPENDIX CG
TORNADO HAZARD MAPS FOR LONG RETURN PERIODS

CG.1 SCOPE

This appendix provides tornado hazard maps for longer mean recurrence intervals (MRIs) than those included in Chapter 32, in support of performance-based wind design applications where use of hazard levels above the minimum specified in the standard may be appropriate. Design for longer MRIs may also be needed for tornado load designs required by other codes and standards. The maps presented in this appendix are not tied to a specific Risk Category like the maps in Figures 32.5-1 and 32.5-2, which represent preselected MRIs of 1,700 and 3,000 years for Risk Categories III and IV, respectively. For the longer MRI maps in this appendix, it is expected that the owner, perhaps in consultation with the designer, will select the appropriate map for their particular application. In some situations, the Authority Having Jurisdiction may require a longer MRI. For example, certain liquefied natural gas facilities require design for 10,000-year-MRI winds, "if adequate wind data are available and the probabilistic methodology is reliable" (PHMSA 2019). As another example, nuclear power plants and certain other energy facilities require design for tornadic wind speeds with MRI up to 10 million years (NRC 2007, DOE 2016, ANS 2016).

Where the longer-return-period tornado hazard maps of this appendix are being used, the designer should also refer to Appendix F to determine a basic wind speed at the same return period (where available) for use with the wind load provisions of Chapters 26 through 31.

CG.2 TORNADO SPEEDS

The tornado maps in this appendix were produced using the same methodology as described in Section C32.5.1. As with nontornadic winds, the uncertainties in tornado speeds increase with increasing return periods. "The uncertainties are judged to be modest at the 3,000-year return period, say ± 10 mph, but are likely on the order of ± 30 mph for 10 million-year return periods" (Twisdale et al. 2021). Tornado speeds are contoured at 10 mi/h (4.5 m/s) intervals except for the innermost (greatest speed) and outermost (least speed) contours, which are shown to 1 mi/h (0.45 m/s) to aid with interpolation.

REFERENCES

ANS (American Nuclear Society). 2016. *Estimating tornado, hurricane, and extreme straight-line wind characteristics at nuclear facility sites*. ANSI/ANS 2.3-2011 (R2016). La Grange Park, IL: ANS.

DOE (US Department of Energy). 2016. *Natural phenomena hazards analysis and design criteria for DOE facilities*. DOE-STD-1020-2016. Washington, DC: DOE.

NRC (Nuclear Regulatory Commission). 2007. *Design-basis tornado and tornado missiles for nuclear power plants*. Regulatory Guide 1.76. Rockville, MD: NRC.

PHMSA (Pipeline and Hazardous Materials Safety Administration). 2019. "Transportation, liquefied natural gas facilities: Federal safety standards, Subpart B – Siting requirements, wind forces." Accessed October 1, 2019. https://www.govinfo.gov/content/pkg/CFR-2019-title49-vol3/xml/CFR-2019-title49-vol3-part193.xml.

Twisdale, L. A., S. Banik, L. Mudd, S. Quayyum, F. Liu, M. Faletra, M. Hardy, P. Vickery, M. Levitan, and L. Phan. 2021. *Tornado risk maps for building design: Research and development of tornado hazard risk assessment methodology*. Gaithersburg, MD: NIST.

INDEX

Provisions appear on pages 1–479

Commentary appears on pages 483–975

Page numbers followed by *e*, *f*, or *t* indicate equations, figures, or tables.